Greek Alphabet

Alpha	A	α	Iota	I	ι	Rho	P	ρ
Beta	B	β	Kappa	K	κ	Sigma	Σ	σ
Gamma	Γ	γ	Lambda	Λ	λ	Tau	T	τ
Delta	Δ	δ	Mu	M	μ	Upsilon	Υ	υ
Epsilon	E	ε	Nu	N	ν	Phi	Φ	ϕ
Zeta	Z	ζ	Xi	Ξ	ξ	Chi	X	χ
Eta	H	η	Omicron	O	o	Psi	Ψ	ψ
Theta	Θ	θ	Pi	Π	π	Omega	Ω	ω

Conversion Table for Units

Length

meter (SI unit)	m	
centimeter	cm	$= 10^{-2}$ m
ångström	Å	$= 10^{-10}$ m
micron	μ	$= 10^{-6}$ m

Volume

cubic meter (SI unit)	m³	
liter	L	$= \text{dm}^3 = 10^{-3}$ m³

Mass

kilogram (SI unit)	kg	
gram	g	$= 10^{-3}$ kg
metric ton	t	$= 1000$ kg

Energy

joule (SI unit)	J	
erg	erg	$= 10^{-7}$ J
rydberg	Ry	$= 2.179\ 87 \times 10^{-18}$ J
electron volt	eV	$= 1.602\ 176\ 565 \times 10^{-19}$ J
inverse centimeter	cm⁻¹	$= 1.986\ 455\ 684 \times 10^{-23}$ J
calorie (thermochemical)	Cal	$= 4.184$ J
liter atmosphere	l atm	$= 101.325$ J

Pressure

pascal (SI unit)	Pa	
atmosphere	atm	$= 101325$ Pa
bar	bar	$= 10^5$ Pa
torr	Torr	$= 133.322$ Pa
pounds per square inch	psi	$= 6.894\ 757 \times 10^3$ Pa

Power

watt (SI unit)	W	
horsepower	hp	$= 745.7$ W

Angle

radian (SI unit)	rad	
degree	°	$= \dfrac{2\pi}{360}\ \text{rad} = \left(\dfrac{1}{57.295\ 78}\right)\text{rad}$

Electrical dipole moment

C m (SI unit)		
debye	D	$= 3.335\ 64 \times 10^{-30}$ C m

Physical Chemistry

THIRD EDITION

Thomas Engel
University of Washington

Philip Reid
University of Washington

Chapter 26, "Computational Chemistry,"
was contributed by

Warren Hehre
CEO, Wavefunction, Inc.

PEARSON

Boston   Columbus   Indianapolis
New York   San Francisco   Upper Saddle River
Amsterdam   Cape Town   Dubai   London
Madrid   Milan   Munich   Paris   Montréal   Toronto
Delhi   Mexico City   São Paulo   Sydney
Hong Kong   Seoul   Singapore   Taipei   Tokyo

Editor in Chief: Adam Jaworski
Executive Editor: Jeanne Zalesky
Senior Marketing Manager: Jonathan Cottrell
Associate Editor: Jessica Neumann
VP/Executive Director, Development: Carol Trueheart
Development Editor: Michael Sypes
Editorial Assistant: Lisa Tarabokjia
Marketing Assistant: Nicola Houston
Managing Editor, Chemistry and Geosciences: Gina M. Cheselka
Senior Project Manager, Production: Beth Sweeten
Associate Media Producer: Ashley Eklund
Full Service/Compositor: GEX Publishing Services
Senior Technical Art Specialist: Connie Long
Illustrator: Imagineering Media Services, Inc.
Design Manager: Mark Ong
Interior Designer: Integra Software Services, Inc.
Cover Designer: Jodi Notowitz
Photo Manager and Researcher: Maya Melenchuk
Text Permissions Manager: Beth Wollar
Text Permissions Researcher: Beth Keister
Operations Specialist: Jeff Sargent
Cover Photo Credit: PhotoDisc

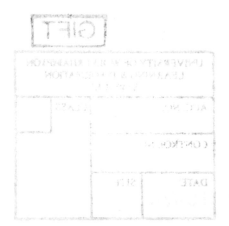

1 2 3 4 5 6 7 8 9 10— CRK—15 14 13 12 11

ISBN-10: 0-321-81719-2; ISBN-13: 978-0-321-81719-8

To Walter and Juliane,
my first teachers,
and to Gloria,
Alex,
and Gabrielle.

Thomas Engel

To my family.

Philip Reid

Brief Contents

Contents

35 Elementary Chemical Kinetics 909

36 Complex Reaction Mechanisms 955

About the Authors

Thomas Engel has taught chemistry at the University of Washington for more than 20 years, where he is currently professor emeritus of chemistry. Professor Engel received his bachelor's and master's degrees in chemistry from the Johns Hopkins University, and his Ph.D. in chemistry from the University of Chicago. He then spent 11 years as a researcher in Germany and Switzerland, in which time he received the Dr. rer. nat. habil. degree from the Ludwig Maximilians University in Munich. In 1980, he left the IBM research laboratory in Zurich to become a faculty member at the University of Washington.

Professor Engel's research interests are in the area of surface chemistry, and he has published more than 80 articles and book chapters in this field. He has received the Surface Chemistry or Colloids Award from the American Chemical Society and a Senior Humboldt Research Award from the Alexander von Humboldt Foundation.

Philip Reid has taught chemistry at the University of Washington since 1995. Professor Reid received his bachelor's degree from the University of Puget Sound in 1986, and his Ph.D. from the University of California, Berkeley in 1992. He performed postdoctoral research at the University of Minnesota, Twin Cities before moving to Washington.

Professor Reid's research interests are in the areas of atmospheric chemistry, ultrafast condensed-phase reaction dynamics, and organic electronics. He has published more than 100 articles in these fields. Professor Reid is the recipient of a CAREER Award from the National Science Foundation, is a Cottrell Scholar of the Research Corporation, and is a Sloan Fellow. He received the University of Washington Distinguished Teaching Award in 2005.

The third edition of this book builds on user and reviewer comments on the previous editions. Our goal remains to provide students with an accessible overview of the whole field of physical chemistry while focusing on basic principles that unite the subdisciplines of the field. We continue to present new research developments in the field to emphasize the vibrancy of physical chemistry. Many chapters have been extensively revised as described below. We include additional end-of-chapter concept problems and most of the numerical problems have been revised. The target audience remains undergraduate students majoring in chemistry, biochemistry, and chemical engineering, as well as many students majoring in the atmospheric sciences and the biological sciences. The following objectives, illustrated with brief examples, outline our approach to teaching physical chemistry.

- **Focus on teaching core concepts.** The central principles of physical chemistry are explored by focusing on core ideas, and then extending these ideas to a variety of problems. The goal is to build a solid foundation of student understanding rather than cover a wide variety of topics in modest detail.

- **Illustrate the relevance of physical chemistry to the world around us.** Many students struggle to connect physical chemistry concepts to the world around them. To address this issue, example problems and specific topics are tied together to help the student develop this connection. Fuel cells, refrigerators, heat pumps, and real engines are discussed in connection with the second law of thermodynamics. The particle in the box model is used to explain why metals conduct electricity and why valence electrons rather than core electrons are important in chemical bond formation. Examples are used to show the applications of chemical spectroscopies. Every attempt is made to connect fundamental ideas to applications that are familiar to the

U.S. 2002 Carbon Dioxide Emissions from Energy
Consumption – 5,682* Million Metric Tons of CO_2**

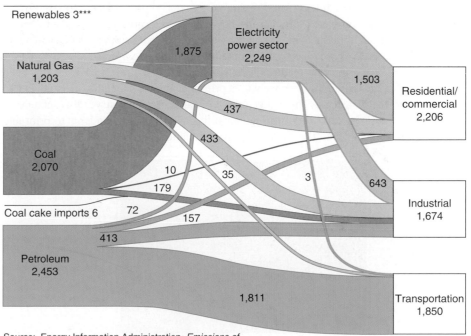

Source: Energy Information Administration. *Emissions of Greenhouse Gases in the United States 2002*. Tables 4–10.
*Includes adjustments of 42.9 million metric tons of carbon dioxide from U.S. territories, less 90.2 $MtCO_2$ from international and military bunker fuels.
**Previous versions of this chart showed emissions in metric tons of carbon, not of CO_2.
***Municipal solid waste and geothermal energy.
Note: Numbers may not equal sum of components because of independent rounding.

student. Art is used to convey complex information in an accessible manner as in the images here of U.S. carbon dioxide emissions.

- **Present exciting new science in the field of physical chemistry.** Physical chemistry lies at the forefront of many emerging areas of modern chemical research. Recent applications of quantum behavior include band-gap engineering, quantum dots, quantum wells, teleportation, and quantum computing. Single-molecule spectroscopy has led to a deeper understanding of chemical kinetics, and heterogeneous catalysis has benefited greatly from mechanistic studies carried out using the techniques of modern surface science. Atomic scale electrochemistry has become possible through scanning tunneling microscopy. The role of physical chemistry in these and other emerging areas is highlighted throughout the text. The following figure shows direct imaging of the arrangement of the atoms in pentacene as well as imaging of a delocalized molecular orbital using scanning tunneling and atomic force miscroscopies.

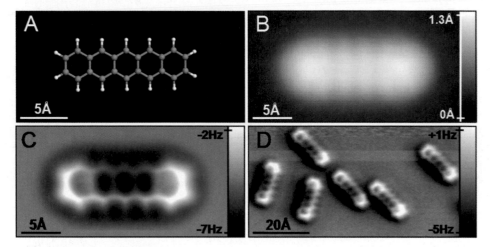

- **Web-based simulations illustrate the concepts being explored and avoid math overload.** Mathematics is central to physical chemistry; however, the mathematics can distract the student from "seeing" the underlying concepts. To circumvent this problem, web-based simulations have been incorporated as end-of-chapter problems throughout the book so that the student can focus on the science and avoid a math overload. These web-based simulations can also be used by instructors during lecture. An important feature of the simulations is that each problem has been designed as an assignable exercise with a printable answer sheet that the student can submit to the instructor. The Study Area in MasteringChemistry® also includes a graphing routine with a curve-fitting capability, which allows students to print and submit graphical data. The 50 web-based simulations listed in the end-of-chapter

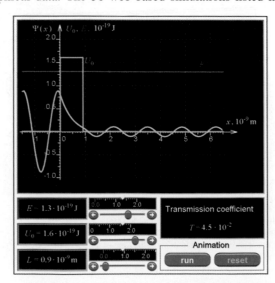

problems are available in the Study Area of MasteringChemistry® for Physical Chemistry. MasteringChemistry® also includes a broad selection of end-of-chapter problems with answer-specific feedback.

* **Show that learning problem-solving skills is an essential part of physical chemistry.**   Many example problems are worked through in each chapter. They introduce the student to a useful method to solve physical chemistry problems.

EXAMPLE PROBLEM 2.5

A system containing 2.50 mol of an ideal gas for which $C_{V,m} = 20.79$ J mol^{-1} K^{-1} is taken through the cycle in the following diagram in the direction indicated by the arrows. The curved path corresponds to $PV = nRT$, where $T = T_1 = T_3$.

a. Calculate q, w, ΔU, and ΔH for each segment and for the cycle assuming that the heat capacity is independent of temperature.

b. Calculate q, w, ΔU, and ΔH for each segment and for the cycle in which the direction of each process is reversed.

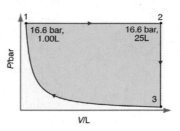

* **The End-of-Chapter Problems cover a range of difficulties suitable for students at all levels.**

P8.6   A P–T phase diagram for potassium is shown next.

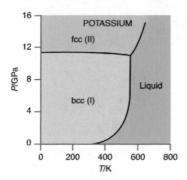

Source: Phase Diagrams of the Elements by David A. Young. © 1991 Regents of the University of California. Published by the University of California Press.

a. Which phase has the higher density, the fcc or the bcc phase? Explain your answer.

b. Indicate the range of P and T in the phase diagram for which fcc and liquid potassium are in equilibrium. Does fcc potassium float on or sink in liquid potassium? Explain your answer.

c. Redraw this diagram for a different pressure range and indicate where you expect to find the vapor phase. Explain how you chose the slope of your liquid–vapor coexistence line.

* **Conceptual questions at the end of each chapter ensure that students learn to express their ideas in the language of science.**

Conceptual Problems

Q21.1   Why does the effective nuclear charge for the $1s$ orbital increase by 0.99 in going from oxygen to fluorine but only increases by 0.65 for the $2p$ orbital?

Q21.2   There are more electrons in the $n = 4$ shell than for the $n = 3$ shell in krypton. However, the peak in the radial distribution in Figure 21.6 is smaller for the $n = 4$ shell than for the $n = 3$ shell. Explain this fact.

Q21.3   How is the effective nuclear charge related to the size of the basis set in a Hartree–Fock calculation?

Q21.4   The angular functions, $\Theta(\theta)\Phi(\phi)$, for the one-electron Hartree–Fock orbitals are the same as for the hydrogen atom, and the radial functions and radial probability functions are similar to those for the hydrogen atom. The contour coloring is explained in the caption to figure 20.7. The following figure shows (a) a contour plot in the xy plane with the y axis being the vertical axis, (b) the radial function, and (c) the radial probability distribution for a one-electron orbital. Identify the orbital ($2s$, $4d_{xz}$, and so on).

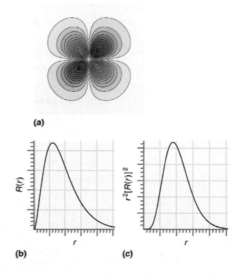

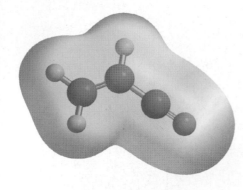

- **Integrate computational chemistry into the standard curriculum.** The teaching of quantum mechanics has not taken advantage of the widespread availability of Ab Initio Software. Many chapters include computational problems for which detailed instructions for the student are available in the Study Area in MasteringChemistry®. It is our experience that students welcome this material, (see L. Johnson and T. Engel, *Journal of Chemical Education* 2011, 88 [569-573]) which transforms the teaching of chemical bonding and molecular structure from being qualitative to quantitative. For example, an electrostatic potential map of acetonitrile built in Spartan Student is shown here.

- **Key equations.** Physical chemistry is a chemistry subdiscipline that is mathematics intensive in nature. Key equations that summarize fundamental relationships between variables are colored in red for emphasis.

- **Green boxes.** Fundamental principles such as the laws of thermodynamics and the quantum mechanical postulates are displayed in green boxes.

- **Updated graph design.** Color is used in graphs to clearly display different relationships in a single figure as shown in the heat capacity for oxygen as a function of temperature and important transitions in the electron spectroscopy of molecules.

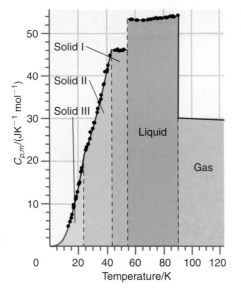

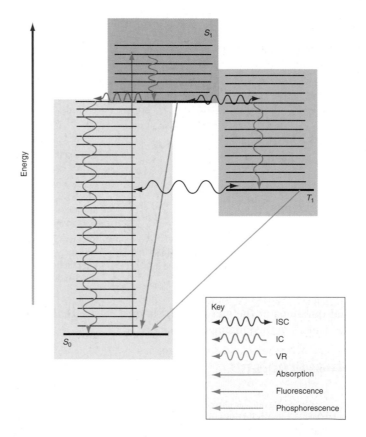

    This text contains more material than can be covered in an academic year, and this is entirely intentional. Effective use of the text does not require a class to proceed sequentially through the chapters, or to include all sections. Some topics are discussed in supplemental sections that can be omitted if they are not viewed as essential to the course. Also, many sections are self contained so that they can be readily omitted if they do not serve the needs of the instructor. This text is constructed to be flexible to your needs, not the other way around. We welcome the comments of both students and instructors how the material was used and how the presentation can be improved.

Thomas Engel
University of Washington

Philip Reid
University of Washington

New to This Edition

The third edition of *Physical Chemistry* includes changes at several levels. The most far-reaching change is the introduction of MasteringChemistry® for Physical Chemistry. Over 460 tutorials will augment the example problems in the book and enhance active learning and problem solving. Selected end of chapter problems are now assignable within MasteringChemistry® and numerical, equation and symbolic answer types are automatically graded.

The art program has been updated and expanded, and several levels of accuracy checking have been incorporated to increase accuracy throughout the text. Many new conceptual problems have been added to the book and most of the numerical problems have been revised. Significant content updates include moving part of the kinetic gas theory to Chapter 1 to allow a molecular level discussion of P and T. The heat capacity discussion previously in sections 2.5 and 3.2 have been consolidated in Chapter 2, and a new section on doing work and changing the system energy from a molecular level perspective has been added. The discussion of differential scanning calorimetry in Chapter 4 has been expanded and a molecular level discussion of entropy has been added to Chapter 5. The discussion of batteries and fuel cells in Chapter 11 has been revised and updated. Problems have been added to the end of Chapter 14 and a new section entitled on superposition wave functions has been added. A new section on traveling waves and potential energy barriers has been added to Chapter 16. The discussion of the classical harmonic oscillator and rigid rotor has been better integrated by placing these sections before the corresponding quantum models in Chapter 18. Chapter 23 has been revised to better introduce molecular orbital theory. A new section on computational results and a set of new problems working with molecular orbitals has been added to Chapter 24. The number and breadth of the numerical problems has been increased substantially in Chapter 25. The content on transition state theory in Chapter 32 has been updated. A discussion of oscillating reactions has been added to Chapter 36 and the material on electron transfer has been expanded.

Acknowledgments

Many individuals have helped us to bring the text into its current form. Students have provided us with feedback directly and through the questions they have asked, which has helped us to understand how they learn. Many of our colleagues including Peter Armentrout, Doug Doren, Gary Drobny, Graeme Henkelman, Lewis Johnson, Tom Pratum, Bill Reinhardt, Peter Rosky, George Schatz, Michael Schick, Gabrielle Varani, and especially Wes Borden and Bruce Robinson have been invaluable in advising us. Paul Siders generously provided problems for Chapter 24. We are also fortunate to have access to some end-of-chapter problems that were originally presented in *Physical Chemistry,* 3rd edition, by Joseph H. Noggle and in *Physical Chemistry,* 3rd edition, by Gilbert W. Castellan. The reviewers, who are listed separately, have made many suggestions for improvement, for which we are very grateful. All those involved in the production process have helped to make this book a reality through their efforts. Special thanks are due to Jim Smith, who helped initiate this project, to our editors Jeanne Zalesky and Jessica Neumann, and to the staff at Pearson, who have guided the production process.

3RD EDITION

MANUSCRIPT REVIEWERS

Nathan Hammer,
  The University of Mississippi
Geoffrey Hutchinson,
  University of Pittsburgh
George Kaminski,
  Central Michigan University

Herve Marand,
  *Virginia Polytechnic Institute and
  State University*
Paul Siders,
  University of Minnesota–Duluth

ACCURACY REVIEWERS

Alexander Angerhofer,
  University of Florida
Clayton Baum,
  Florida Institute of Technology

Jennifer Mihalik,
  University of Wisconsin–Oshkosh
David Zax,
  Cornell University

PRESCRIPTIVE REVIEWERS

Geoffrey Hutchinson,
  University of Pittsburgh
William Lester,
  University of California–Berkeley
Herve Marand,
  *Virginia Polytechnic Institute and
  State University*

Thomas Mason,
  University of California–Los Angeles
Paul Siders,
  University of Minnesota–Duluth

2ND EDITION

PRESCRIPTIVE REVIEWERS

David L. Cedeño,
  Illinois State University
Rosemarie Chinni,
  Alvernia College
Allen Clabo,
  Francis Marion University
Lorrie Comeford,
  Salem State College
John M. Jean,
  Regis University
Martina Kaledin,
  Kennesaw State University
Daniel Lawson,
  University of Michigan–Dearborn
Dmitrii E. Makarov,
  University of Texas at Austin

Enrique Peacock-López,
  Williams College
Anthony K. Rappe,
  Colorado State University
Markku Räsänen,
  University of Helsinki
Richard W. Schwenz,
  University of Northern Colorado
Jie Song,
  University of Michigan–Flint
Michael E. Starzak,
  Binghamton University
Liliya Vugmeyster,
  University of Alaska–Anchorage
James E. Whitten,
  University of Massachusetts–Lowell

ART REVIEWER

Lorrie Comeford,
Salem State College

MATH REVIEWER

Leon Gerber,
St. John's University

MANUSCRIPT REVIEWERS

Alexander Angerhofer,
University of Florida

Martha Bruch,
State University of New York at Oswego

Stephen Cooke,
University of North Texas

Douglas English,
University of Maryland–College Park

Sophya Garashchuk,
University of South Carolina

Cynthia Hartzell,
Northern Arizona University

George Kaminski,
Central Michigan University

Herve Marand,
Virginia Polytechnic Institute and State University

Thomas Pentecost,
University of Colorado

Rajeev Prabhakar,
University of Miami

Sanford Safron,
Florida State University

Ali Sezer,
California University of Pennsylvania

Andrew Teplyakov,
University of Delaware

Daniel Zeroka,
Lehigh University

MasteringChemistry®

www.masteringchemistry.com

MasteringChemistry® is designed with a single purpose: to help students reach the moment of understanding. The Mastering online homework and tutoring system delivers self-paced tutorials that provide students with individualized coaching set to your course objectives. MasteringChemistry® helps students arrive better prepared for lecture and lab.

Engaging Experiences

MasteringChemistry® promotes interactivity in Physical Chemistry. Research shows that Mastering's immediate feedback and tutorial assistance helps students understand and master concepts and skills in Chemistry—allowing them to retain more knowledge and perform better in this course and beyond.

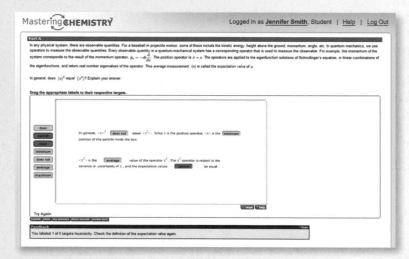

STUDENT TUTORIALS

MasteringChemistry® is the only system to provide instantaneous feedback specific to individual student entries. Students can submit an answer and receive immediate, error-specific feedback. Simpler sub-problems—hints—help students think through the problem. Over 460 tutorials will be available with MasteringChemistry® for Physical Chemistry including new ones on The Cyclic Rule, Particle in a Box, and Components of U.

END-OF-CHAPTER CONTENT AVAILABLE IN MASTERINGCHEMISTRY®:

Selected end-of-chapter problems are assignable within MasteringChemistry®, including:

- **Numerical answers** with hints and feedback

- **Equation and Symbolic answer types** so that the results of a self-derivation can be entered to check for correctness, feedback, and assistance

- A **Solution View** that allows students to see intermediate steps involved in calculations of the final numerical result

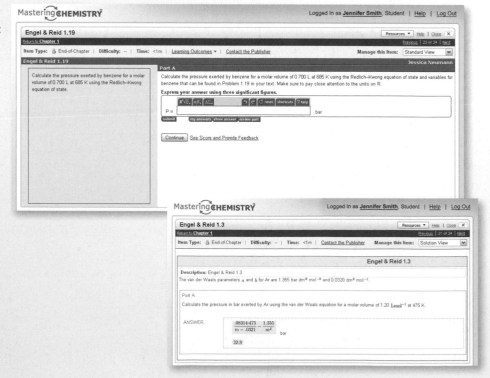

Trusted Partner

The Mastering platform was developed by scientists for science students and instructors, and has a proven history with over 10 years of student use. Mastering currently has more than 1.5 million active registrations with active users in all 50 states and in 41 countries. The Mastering platform has 99.8% server reliability.

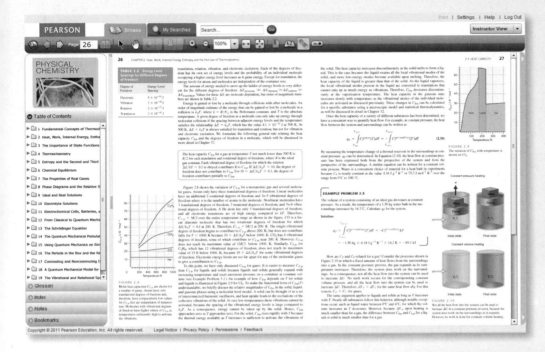

PEARSON ETEXT

Pearson eText provides access to the text when and wherever students have access to the Internet. eText pages look exactly like the printed text, offering powerful new functionality. Users can create notes, highlight the text in different colors, create bookmarks, zoom, click hyperlinked words and phrases to view definitions, view as single or two-pages. Pearson eText also links students to associated media files, enabling them to view an animation as they read the text, and offers a full text search and the ability to save and export notes.

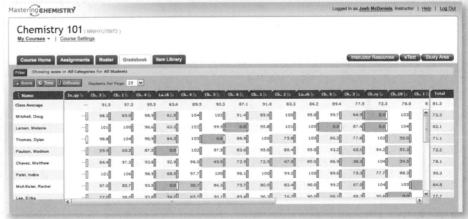

GRADEBOOK

Every assignment is automatically graded. Shades of red highlight vulnerable students and challenging assignments.

GRADEBOOK DIAGNOSTICS

This screen provides you with your favorite diagnostics. With a single click, charts summarize the most difficult problems, vulnerable students, grade distribution, and even score improvement over the course.

Fundamental Concepts of Thermodynamics

Thermodynamics provides a description of matter on a macroscopic scale using bulk properties such as pressure, density, volume, and temperature. This chapter introduces the basic concepts employed in thermodynamics including system, surroundings, intensive and extensive variables, adiabatic and diathermal walls, equilibrium, temperature, and thermometry. The macroscopic variables pressure and temperature are also discussed in terms of a molecular level model. The usefulness of equations of state, which relate the state variables of pressure, volume, and temperature, is also discussed for real and ideal gases.

1.1 What Is Thermodynamics and Why Is It Useful?

Thermodynamics is the branch of science that describes the behavior of matter and the transformation between different forms of energy on a **macroscopic scale,** or the human scale and larger. Thermodynamics describes a system of interest in terms of its bulk properties. Only a few such variables are needed to describe the system, and the variables are generally directly accessible through measurements. A thermodynamic description of matter does not make reference to its structure and behavior at the microscopic level. For example, 1 mol of gaseous water at a sufficiently low density is completely described by two of the three **macroscopic variables** of pressure, volume, and temperature. By contrast, the microscopic scale refers to dimensions on the order of the size of molecules. At the microscopic level, water would be described as a dipolar triatomic molecule, H_2O, with a bond angle of $104.5°$ that forms a network of hydrogen bonds.

In this book, we first discuss thermodynamics and then statistical thermodynamics. Statistical thermodynamics (Chapters 31 and 32) uses atomic and molecular properties to calculate the macroscopic properties of matter. For example, statistical thermodynamics can show that liquid water is the stable form of aggregation at a pressure of 1 bar and a temperature of $90°C$, whereas gaseous water is the stable form at 1 bar and $110°C$. Using statistical thermodynamics, the macroscopic properties of matter are calculated from underlying molecular properties.

Given that the microscopic nature of matter is becoming increasingly well understood using theories such as quantum mechanics, why is a macroscopic science like thermodynamics relevant today? The usefulness of thermodynamics can be illustrated by describing four applications of thermodynamics which you will have mastered after working through this book:

- You have built a plant to synthesize NH_3 gas from N_2 and H_2. You find that the yield is insufficient to make the process profitable and decide to try to improve the NH_3 output by changing the temperature and/or the pressure. However, you do not know whether to increase or decrease the values of these variables. As will be shown in Chapter 6, the ammonia yield will be higher at equilibrium if the temperature is decreased and the pressure is increased.

- You wish to use methanol to power a car. One engineer provides a design for an internal combustion engine that will burn methanol efficiently according to the reaction $CH_3OH(l) + 3/2 O_2(g) \rightarrow CO_2(g) + 2H_2O(l)$. A second engineer designs an electrochemical fuel cell that carries out the same reaction. He claims that the vehicle will travel much farther if powered by the fuel cell than by the internal combustion engine. As will be shown in Chapter 5, this assertion is correct, and an estimate of the relative efficiencies of the two propulsion systems can be made.

- You are asked to design a new battery that will be used to power a hybrid car. Because the voltage required by the driving motors is much higher than can be generated in a single electrochemical cell, many cells must be connected in series. Because the space for the battery is limited, as few cells as possible should be used. You are given a list of possible cell reactions and told to determine the number of cells needed to generate the required voltage. As you will learn in Chapter 11, this problem can be solved using tabulated values of thermodynamic functions.

- Your attempts to synthesize a new and potentially very marketable compound have consistently led to yields that make it unprofitable to begin production. A supervisor suggests a major effort to make the compound by first synthesizing a catalyst that promotes the reaction. How can you decide if this effort is worth the required investment? As will be shown in Chapter 6, the maximum yield expected under equilibrium conditions should be calculated first. If this yield is insufficient, a catalyst is useless.

1.2 The Macroscopic Variables Volume, Pressure, and Temperature

We begin our discussion of thermodynamics by considering a bottle of a gas such as He or CH_4. At a macroscopic level, the sample of known chemical composition is completely described by the measurable quantities volume, pressure, and temperature for which we use the symbols V, P, and T. The volume V is just that of the bottle. What physical association do we have with P and T?

Pressure is the force exerted by the gas per unit area of the container. It is most easily understood by considering a microscopic model of the gas known as the kinetic theory of gases. The gas is described by two assumptions: the atoms or molecules of an ideal gas do not interact with one another, and the atoms or molecules can be treated as point masses. The pressure exerted by a gas on the container confining the gas arises from collisions of randomly moving gas molecules with the container walls. Because the number of molecules in a small volume of the gas is on the order of Avogadro's number N_A, the number of collisions between molecules is also large. To describe pressure, a molecule is envisioned as traveling through space with a velocity vector $\mathbf{v}$ that can be decomposed into three Cartesian components: $\mathbf{v}_x$, $\mathbf{v}_y$, and $\mathbf{v}_z$ as illustrated in Figure 1.1.

The square of the magnitude of the velocity v^2 in terms of the three velocity components is

$$v^2 = \mathbf{v} \cdot \mathbf{v} = v_x^2 + v_y^2 + v_z^2 \qquad (1.1)$$

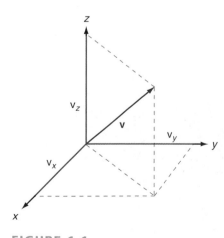

FIGURE 1.1
Cartesian components of velocity. The particle velocity $\mathbf{v}$ can be decomposed into three velocity components: $\mathbf{v}_x$, $\mathbf{v}_y$, and $\mathbf{v}_z$.

The particle kinetic energy is $1/2\ mv^2$ such that

$$\varepsilon_{Tr} = \frac{1}{2}mv^2 = \frac{1}{2}mv_x^2 + \frac{1}{2}mv_y^2 + \frac{1}{2}mv_z^2 = \varepsilon_{Tr_x} + \varepsilon_{Tr_y} + \varepsilon_{Tr_z} \quad \textbf{(1.2)}$$

where the subscript Tr indicates that the energy corresponds to translational motion of the particle. Furthermore, this equation states that the total translational energy is the sum of translational energy along each Cartesian dimension.

Pressure arises from the collisions of gas particles with the walls of the container; therefore, to describe pressure we must consider what occurs when a gas particle collides with the wall. First, we assume that the collisions with the wall are **elastic collisions,** meaning that translational energy of the particle is conserved. Although the collision is elastic, this does not mean that nothing happens. As a result of the collision, linear momentum is imparted to the wall, which results in pressure. The definition of pressure is force per unit area and, by Newton's second law, force is equal to the product of mass and acceleration. Using these two definitions, the pressure arising from the collision of a single molecule with the wall is expressed as

$$P = \frac{F}{A} = \frac{ma_i}{A} = \frac{m}{A}\left(\frac{dv_i}{dt}\right) = \frac{1}{A}\left(\frac{dmv_i}{dt}\right) = \frac{1}{A}\left(\frac{dp_i}{dt}\right) \quad \textbf{(1.3)}$$

In Equation (1.3), F is the force of the collision, A is the area of the wall with which the particle has collided, m is the mass of the particle, v_i is the velocity component along the i direction ($i = x, y,$ or z), and p_i is the particle linear momentum in the i direction. Equation (1.3) illustrates that pressure is related to the change in linear momentum with respect to time that occurs during a collision. Due to conservation of momentum, any change in particle linear momentum must result in an equal and opposite change in momentum of the container wall. A single collision is depicted in Figure 1.2. This figure illustrates that the particle linear momentum change in the x direction is $-2mv_x$ (note there is no change in momentum in the y or z direction). Given this, a corresponding momentum change of $2mv_x$ must occur for the wall.

The pressure measured at the container wall corresponds to the sum of collisions involving a large number of particles that occur per unit time. Therefore, the total momentum change that gives rise to the pressure is equal to the product of the momentum change from a single particle collision and the total number of particles that collide with the wall:

$$\Delta p_{Total} = \frac{\Delta p}{\text{molecule}} \times (\text{number of molecules}) \quad \textbf{(1.4)}$$

How many molecules strike the side of the container in a given period of time? To answer this question, the time over which collisions are counted must be considered. Consider a volume element defined by the area of the wall A times length Δx as illustrated in Figure 1.3. The collisional volume element depicted in Figure 1.3 is given by

$$V = A\Delta x \quad \textbf{(1.5)}$$

The length of the box Δx is related to the time period over which collisions will be counted Δt and the component of particle velocity parallel to the side of the box (taken to be the x direction):

$$\Delta x = v_x \Delta t \quad \textbf{(1.6)}$$

In this expression, v_x is for a single particle; however, an average of this quantity will be used when describing the collisions from a collection of particles. Finally, the number of particles that will collide with the container wall N_{coll} in the time interval Δt is equal to the number density $\tilde{N}$. This quantity is equal to the number of particles in the container N divided by the container volume V and multiplied by the collisional volume element depicted in Figure 1.3:

$$N_{coll} = \tilde{N} \times (Av_x\Delta t)\left(\frac{1}{2}\right) = \frac{nN_A}{V}(Av_x\Delta t)\left(\frac{1}{2}\right) \quad \textbf{(1.7)}$$

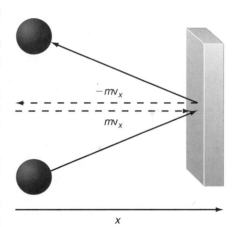

FIGURE 1.2
Collision between a gas particle and a wall. Before the collision, the particle has a momentum of mv_x in the x direction, and after the collision the momentum is $-mv_x$. Therefore, the change in particle momentum resulting from the collision is $-2mv_x$. By conservation of momentum, the change in momentum of the wall must be $2mv_x$. The incoming and outgoing trajectories are offset to show the individual momentum components.

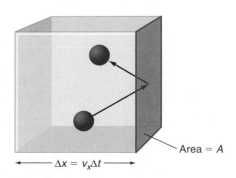

FIGURE 1.3
Volume element used to determine the number of collisions with the wall per unit time.

We have used the equality $N = n N_A$ where N_A is Avogadro's number and n is the number of moles of gas in the second part of Equation (1.7). Because particles travel in either the $+x$ or $-x$ direction with equal probability, only those molecules traveling in the $+x$ direction will strike the area of interest. Therefore, the total number of collisions is divided by two to take the direction of particle motion into account. Employing Equation (1.7), the total change in linear momentum of the container wall imparted by particle collisions is given by

$$\Delta p_{Total} = (2m\text{v}_x)(N_{coll})$$

$$= (2m\text{v}_x)\left(\frac{nN_A}{V} \frac{A\text{v}_x\Delta t}{2}\right)$$

$$= \frac{nN_A}{V} A\Delta t\, m\langle\text{v}_x^2\rangle \tag{1.8}$$

In Equation (1.8), angle brackets appear around v_x^2 to indicate that this quantity represents an average value since the particles will demonstrate a distribution of velocities. This distribution is considered in detail later in Chapter 30. With the total change in linear momentum provided in Equation (1.8), the force and corresponding pressure exerted by the gas on the container wall [Equation (1.3)] are as follows:

$$F = \frac{\Delta p_{Total}}{\Delta t} = \frac{nN_A}{V} Am\langle\text{v}_x^2\rangle$$

$$P = \frac{F}{A} = \frac{nN_A}{V} m\langle\text{v}_x^2\rangle \tag{1.9}$$

Equation (1.9) can be converted into a more familiar expression once $1/2\, m\langle\text{v}_x^2\rangle$ is recognized as the translational energy in the x direction. In Chapter 31, it will be shown that the average translational energy for an individual particle in one dimension is

$$\frac{m\langle\text{v}_x^2\rangle}{2} = \frac{k_B T}{2} \tag{1.10}$$

where T is the gas temperature.

Substituting this result into Equation (1.9) results in the following expression for pressure:

$$P = \frac{nN_A}{V} m\langle\text{v}_x^2\rangle = \frac{nN_A}{V} kT = \frac{nRT}{V} \tag{1.11}$$

We have used the equality $N_A k_B = R$ where k_B is the **Boltzmann constant** and R is the **ideal gas constant** in the last part of Equation (1.11). Equation (1.11) is the **ideal gas law.** Although this relationship is familiar, we have derived it by employing a classical description of a single molecular collision with the container wall and then scaling this result up to macroscopic proportions. We see that the origin of the pressure exerted by a gas on its container is the momentum exchange of the randomly moving gas molecules with the container walls.

What physical association can we make with the temperature T? At the microscopic level, temperature is related to the mean kinetic energy of molecules as shown by Equation (1.10). We defer the discussion of temperature at the microscopic level until Chapter 30 and focus on a macroscopic level discussion here. Although each of us has a sense of a "temperature scale" based on the qualitative descriptors *hot* and *cold*, a more quantitative and transferable measure of temperature that is not grounded in individual experience is needed. The quantitative measurement of temperature is accomplished using a **thermometer.** For any useful thermometer, the measured temperature, T, must be a single-valued, continuous, and monotonic function of some thermometric system property such as the volume of mercury confined to a narrow capillary, the electromotive force generated at the junction of two dissimilar metals, or the electrical resistance of a platinum wire.

The simplest case that one can imagine is when T is linearly related to the value of the thermometric property x:

$$T(x) = a + bx \qquad (1.12)$$

Equation (1.12) defines a **temperature scale** in terms of a specific thermometric property, once the constants a and b are determined. The constant a determines the zero of the temperature scale because $T(0) = a$ and the constant b determines the size of a unit of temperature, called a degree.

One of the first practical thermometers was the mercury-in-glass thermometer. This thermometer utilizes the thermometric property that the volume of mercury increases monotonically over the temperature range $-38.8°C$ to $356.7°C$ in which Hg is a liquid. In 1745, Carolus Linnaeus gave this thermometer a standardized scale by arbitrarily assigning the values 0 and 100 to the freezing and boiling points of water, respectively. Because there are 100 degrees between the two calibration points, this scale is known as the **centigrade scale.**

The centigrade scale has been superseded by the **Celsius scale.** The Celsius scale (denoted in units of °C) is similar to the centigrade scale. However, rather than being determined by two fixed points, the Celsius scale is determined by one fixed reference point at which ice, liquid water, and gaseous water are in equilibrium. This point is called the triple point (see Section 8.2) and is assigned the value $0.01°C$. On the Celsius scale, the boiling point of water at a pressure of 1 atmosphere is $99.975°C$. The size of the degree is chosen to be the same as on the centigrade scale.

Although the Celsius scale is used widely throughout the world today, the numerical values for this temperature scale are completely arbitrary, because a liquid other than water could have been chosen as a reference. It would be preferable to have a temperature scale derived directly from physical principles. There is such a scale, called the **thermodynamic temperature scale** or **absolute temperature scale.** For such a scale, the temperature is independent of the substance used in the thermometer, and the constant a in Equation (1.12) is zero. The **gas thermometer** is a practical thermometer with which the absolute temperature can be measured. A gas thermometer contains a dilute gas under conditions in which the ideal gas law of Equation (1.11) describes the relationship among P, T, and the molar density $\rho_m = n/V$ with sufficient accuracy:

$$P = \rho_m RT \qquad (1.13)$$

Equation (1.13) can be rewritten as

$$T = \frac{P}{\rho_m R} \qquad (1.14)$$

showing that for a gas thermometer, the thermometric property is the temperature dependence of P for a dilute gas at constant V. The gas thermometer provides the international standard for thermometry at very low temperatures. At intermediate temperatures, the electrical resistance of platinum wire is the standard, and at higher temperatures the radiated energy emitted from glowing silver is the standard. The absolute temperature is shown in Figure 1.4 on a logarithmic scale together with associated physical phenomena.

Equation (1.14) implies that as $T \rightarrow 0$, $P \rightarrow 0$. Measurements carried out by Guillaume Amontons in the 17th century demonstrated that the pressure exerted by a fixed amount of gas at constant V varies linearly with temperature as shown in Figure 1.5. At the time of these experiments, temperatures below $-30°C$ were not attainable in the laboratory. However, the P versus T_C data can be extrapolated to the limiting T_C value at which $P \rightarrow 0$. It is found that these straight lines obtained for different values of V intersect at a common point on the T_C axis that lies near $-273°C$.

The data in Figure 1.5 show that at constant V, the thermometric property P varies with temperature as

$$P = a + bT_C \qquad (1.15)$$

where T_C is the temperature on the Celsius scale, and a and b are experimentally obtained proportionality constants. Figure 1.5 shows that all lines intersect at a single

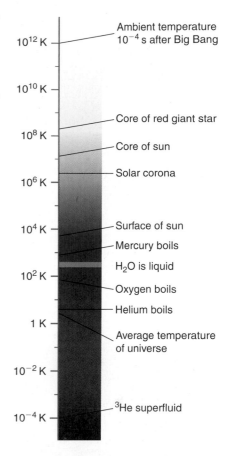

FIGURE 1.4

The absolute temperature is shown on a logarithmic scale together with the temperature of a number of physical phenomena.

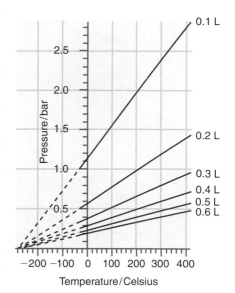

FIGURE 1.5

The pressure exerted by 5.00×10^{-3} mol of a dilute gas is shown as a function of the temperature measured on the Celsius scale for different fixed volumes. The dashed portion indicates that the data are extrapolated to lower temperatures than could be achieved experimentally by early investigators.

point, even for different gases. This suggests a unique reference point for temperature, rather than the two reference points used in constructing the centigrade scale. The value zero is given to the temperature at which $P \rightarrow 0$, so that $a = 0$. However, this choice is not sufficient to define the temperature scale, because the size of the degree is undefined. By convention, the size of the degree on the absolute temperature scale is set equal to the size of the degree on the Celsius scale. With these two choices, the absolute and Celsius temperature scales are related by Equation (1.16). The scale measured by the ideal gas thermometer is the absolute temperature scale used in thermodynamics. The unit of temperature on this scale is called the kelvin, abbreviated K (without a degree sign):

$$T/\text{K} = T_C/°\text{C} + 273.15 \qquad (1.16)$$

1.3 Basic Definitions Needed to Describe Thermodynamic Systems

Having discussed the macroscopic variables pressure, volume, and temperature, we introduce some important concepts used in thermodynamics. A thermodynamic **system** consists of all the materials involved in the process under study. This material could be the contents of an open beaker containing reagents, the electrolyte solution within an electrochemical cell, or the contents of a cylinder and movable piston assembly in an engine. In thermodynamics, the rest of the universe is referred to as the **surroundings.** If a system can exchange matter with the surroundings, it is called an **open system;** if not, it is a **closed system.** Living cells are open systems (see Figure 1.6). Both open and closed systems can exchange energy with the surroundings. Systems that can exchange neither matter nor energy with the surroundings are called **isolated systems.**

The interface between the system and its surroundings is called the **boundary.** Boundaries determine if energy and mass can be transferred between the system and the surroundings and lead to the distinction between open, closed, and isolated systems. Consider Earth's oceans as a system, with the rest of the universe being the surroundings. The system–surroundings boundary consists of the solid–liquid interface between the continents and the ocean floor and the water–air interface at the ocean surface. For an open beaker in which the system is the contents, the boundary surface is just inside the inner wall of the beaker, and it passes across the open top of the beaker. In this case, energy can be exchanged freely between the system and surroundings through the side and bottom walls, and both matter and energy can be exchanged between the system and surroundings through the open top boundary. The portion of the boundary formed by the beaker in the previous example is called a **wall.** Walls can be rigid or movable and permeable or nonpermeable. An example of a movable wall is the surface of a balloon. An example of a selectively permeable wall is the fabric used in raingear, which is permeable to water vapor, but not liquid water.

The exchange of energy and matter across the boundary between system and surroundings is central to the important concept of **equilibrium.** The system and surroundings can be in equilibrium with respect to one or more of several different **system variables** such as pressure (P), temperature (T), and concentration. **Thermodynamic equilibrium** refers to a condition in which equilibrium exists with respect to P, T, and concentration. What conditions are necessary for a system to come to equilibrium with its surroundings? Equilibrium is established with respect to a given variable only if that variable does not change with time, and if it has the same value in all parts of the system and surroundings. For example, the interior of a soap bubble[1] (the system) and the surroundings (the room) are in equilibrium with respect to P because the movable wall (the bubble) can reach a position where P on both sides of the wall is the same, and because P has the same value throughout the system and surroundings. Equilibrium with respect to concentration exists only if transport of all species across the boundary in both directions is possible. If the boundary is a movable wall that is not permeable to

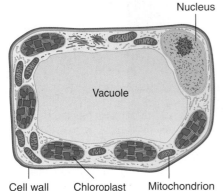

Nucleus

Vacuole

Cell wall   Chloroplast   Mitochondrion

Plant cell

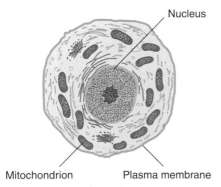

Nucleus

Mitochondrion   Plasma membrane

Animal cell

FIGURE 1.6

Animal and plant cells are open systems. The contents of the animal cell include the cytosol fluid and the numerous organelles (e.g., nucleus, mitochondria, etc.) that are separated from the surroundings by a lipid-rich plasma membrane. The plasma membrane acts as a boundary layer that can transmit energy and is selectively permeable to ions and various metabolites. A plant cell is surrounded by a cell wall that similarly encases the cytosol and organelles, including chloroplasts, that are the sites of photosynthesis.

---

[1]For this example, the surface tension of the bubble is assumed to be so small that it can be set equal to zero. This is in keeping with the thermodynamic tradition of weightless pistons and frictionless pulleys.

all species, equilibrium can exist with respect to P, but not with respect to concentration. Because N_2 and O_2 cannot diffuse through the (idealized) bubble, the system and surroundings are in equilibrium with respect to P, but not to concentration. Equilibrium with respect to temperature is a special case that is discussed next.

Two systems that have the same temperature are in **thermal equilibrium.** We use the concepts of temperature and thermal equilibrium to characterize the walls between a system and its surroundings. Consider the two systems with rigid walls shown in Figure 1.7a. Each system has the same molar density and is equipped with a pressure gauge. If we bring the two systems into direct contact, two limiting behaviors are observed. If neither pressure gauge changes, as in Figure 1.7b, we refer to the walls as being **adiabatic.** Because $P_1 \neq P_2$, the systems are not in thermal equilibrium and, therefore, have different temperatures. An example of a system surrounded by adiabatic walls is coffee in a Styrofoam cup with a Styrofoam lid.[2] Experience shows that it is not possible to bring two systems enclosed by adiabatic walls into thermal equilibrium by bringing them into contact, because adiabatic walls insulate against the transfer of "heat." If we push a Styrofoam cup containing hot coffee against one containing ice water, they will not reach the same temperature. Rely on experience at this point regarding the meaning of heat; a thermodynamic definition will be given in Chapter 2.

The second limiting case is shown in Figure 1.7c. In bringing the systems into intimate contact, both pressures change and reach the same value after some time. We conclude that the systems have the same temperature, $T_1 = T_2$, and say that they are in thermal equilibrium. We refer to the walls as being **diathermal.** Two systems in contact separated by diathermal walls reach thermal equilibrium because diathermal walls conduct heat. Hot coffee stored in a copper cup is an example of a system surrounded by diathermal walls. Because the walls are diathermal, the coffee will quickly reach room temperature.

The **zeroth law of thermodynamics** generalizes the experiment illustrated in Figure 1.7 and asserts the existence of an objective temperature that can be used to define the condition of thermal equilibrium. The formal statement of this law is as follows:

> Two systems that are separately in thermal equilibrium with a third system are also in thermal equilibrium with one another.

The unfortunate name assigned to the "zeroth" law is due to the fact that it was formulated after the first law of thermodynamics, but logically precedes it. The zeroth law tells us that we can determine if two systems are in thermal equilibrium without bringing them into contact. Imagine the third system to be a thermometer, which is defined more precisely in the next section. The third system can be used to compare the temperatures of the other two systems; if they have the same temperature, they will be in thermal equilibrium if placed in contact.

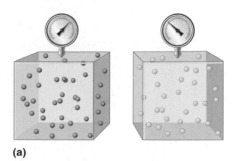

(a)

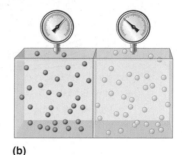

(b)

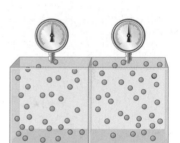

(c)

FIGURE 1.7
(a) Two separated systems with rigid walls and the same molar density have different temperatures. (b) Two systems are brought together so that their adiabatic walls are in intimate contact. The pressure in each system will not change unless heat transfer is possible. (c) As in part (b), two systems are brought together so that their diathermal walls are in intimate contact. The pressures become equal.

1.4 Equations of State and the Ideal Gas Law

Macroscopic models in which the system is described by a set of variables are based on experience. It is particularly useful to formulate an **equation of state,** which relates the state variables. A dilute gas can be modeled as consisting of point masses that do not interact with one another; we call this an **ideal gas.** The equation of state for an ideal gas was first determined from experiments by the English chemist Robert Boyle. If the pressure of He is measured as a function of the volume for different values of temperature, the set of nonintersecting hyperbolas as shown in Figure 1.8 is obtained. The curves in this figure can be quantitatively fit by the functional form

$$PV = \alpha T \qquad\qquad (1.17)$$

---

[2]In this discussion, Styrofoam is assumed to be a perfect insulator.

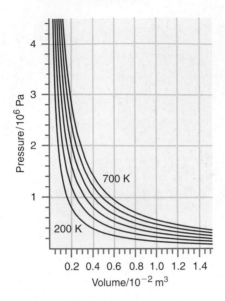

FIGURE 1.8

Illustration of the relationship between pressure and volume of 0.010 mol of He for fixed values of temperature, which differ by 100 K.

where T is the absolute temperature as defined by Equation (1.16), allowing α to be determined. The constant α is found to be directly proportional to the mass of gas used. It is useful to separate this dependence by writing $\alpha = nR$, where n is the number of moles of the gas, and R is a constant that is independent of the size of the system. The result is the ideal gas equation of state

$$PV = Nk_BT = nRT \tag{1.18}$$

as derived in Equation (1.11). The equation of state given in Equation (1.18) is familiar as the ideal gas law. Because the four variables P, V, T, and n are related through the equation of state, any three of these variables is sufficient to completely describe the ideal gas.

Of these four variables, P and T are independent of the amount of gas, whereas V and n are proportional to the amount of gas. A variable that is independent of the size of the system (for example, P and T) is referred to as an **intensive variable,** and one that is proportional to the size of the system (for example, V) is referred to as an **extensive variable.** Equation (1.18) can be written in terms of intensive variables exclusively:

$$P = \rho_m RT \tag{1.13}$$

For a fixed number of moles, the ideal gas equation of state has only two independent intensive variables: any two of P, T, and ρ_m.

For an ideal gas mixture

$$PV = \sum_i n_i RT \tag{1.19}$$

because the gas molecules do not interact with one another. Equation (1.19) can be rewritten in the form

$$P = \sum_i \frac{n_i RT}{V} = \sum_i P_i = P_1 + P_2 + P_3 + \dots \tag{1.20}$$

In Equation (1.20), P_i is the **partial pressure** of each gas. This equation states that each ideal gas exerts a pressure that is independent of the other gases in the mixture. We also have

$$\frac{P_i}{P} = \frac{\dfrac{n_i RT}{V}}{\sum_i \dfrac{n_i RT}{V}} = \frac{\dfrac{n_i RT}{V}}{\dfrac{nRT}{V}} = \frac{n_i}{n} = x_i \tag{1.21}$$

which relates the partial pressure of a component in the mixture P_i with its **mole fraction,** $x_i = n_i/n$, and the total pressure P.

In the SI system of units, pressure is measured in Pascal (Pa) units, where $1\,\text{Pa} = 1\,\text{N/m}^2$. The volume is measured in cubic meters, and the temperature is measured in kelvin. However, other units of pressure are frequently used, and these units are related to the Pascal as indicated in Table 1.1. In this table, numbers that are not exact have been given to five significant figures. The other commonly used unit of volume is the liter (L), where $1\,\text{m}^3 = 10^3\,\text{L}$ and $1\,\text{L} = 1\,\text{dm}^3 = 10^{-3}\,\text{m}^3$.

TABLE 1.1   Units of Pressure and Conversion Factors

Unit of Pressure	Symbol	Numerical Value
Pascal	Pa	$1\,\text{N m}^{-2} = 1\,\text{kg m}^{-1}\,\text{s}^{-2}$
Atmosphere	atm	$1\,\text{atm} = 101{,}325\,\text{Pa (exactly)}$
Bar	bar	$1\,\text{bar} = 10^5\,\text{Pa}$
Torr or millimeters of Hg	Torr	$1\,\text{Torr} = 101{,}325/760 = 133.32\,\text{Pa}$
Pounds per square inch	psi	$1\,\text{psi} = 6{,}894.8\,\text{Pa}$

EXAMPLE PROBLEM 1.1

Starting out on a trip into the mountains, you inflate the tires on your automobile to a recommended pressure of 3.21×10^5 Pa on a day when the temperature is –5.00°C. You drive to the beach, where the temperature is 28.0°C. (a) What is the final pressure in the tires, assuming constant volume? (b) Derive a formula for the final pressure, assuming more realistically that the volume of the tires increases with increasing pressure as $V_f = V_i\left(1 + \gamma\left[P_f - P_i\right]\right)$ where γ is an experimentally determined constant.

Solution

a. Because the number of moles is constant,

$$\frac{P_iV_i}{T_i} = \frac{P_fV_f}{T_f}; \quad P_f = \frac{P_iV_iT_f}{V_fT_i};$$

$$P_f = \frac{P_iV_iT_f}{V_fT_i} = 3.21 \times 10^5 \text{Pa} \times \frac{V_i}{V_i} \times \frac{(273.15 + 28.0)}{(273.15 - 5.00)} = 3.61 \times 10^5 \text{Pa}$$

b.
$$\frac{P_iV_i}{T_i} = \frac{P_fV_i\left(1 + \gamma\left[P_f - P_i\right]\right)}{T_f};$$

$$P_iT_f = P_fT_i\left(1 + \gamma\left[P_f - P_i\right]\right)$$

$$P_f^2T_i\gamma + P_fT_i(1 - P_i\gamma) - P_iT_f = 0$$

$$P_f = \frac{-T_i(1 - P_i\gamma) \pm \sqrt{T_i^2(1 - P_i\gamma)^2 + 4T_iT_f\gamma P_i}}{2T_i\gamma}$$

We leave it to the end-of-chapter problems to show that this expression for P_f has the correct limit as $\gamma \rightarrow 0$.

In the SI system, the constant R that appears in the ideal gas law has the value 8.314 J K^{-1} mol^{-1}, where the joule (J) is the unit of energy in the SI system. To simplify calculations for other units of pressure and volume, values of the constant R with different combinations of units are given in Table 1.2.

EXAMPLE PROBLEM 1.2

Consider the composite system, which is held at 298 K, shown in the following figure. Assuming ideal gas behavior, calculate the total pressure and the partial pressure of each component if the barriers separating the compartments are removed. Assume that the volume of the barriers is negligible.

TABLE 1.2 The Ideal Gas Constant, R, in Various Units
$R = 8.314$ J K^{-1} mol^{-1}
$R = 8.314$ Pa m^3 K^{-1} mol^{-1}
$R = 8.314 \times 10^{-2}$ L bar K^{-1} mol^{-1}
$R = 8.206 \times 10^{-2}$ L atm K^{-1} mol^{-1}
$R = 62.36$ L Torr K^{-1} mol^{-1}

He
2.00 L
1.50 bar

Ne
3.00 L
2.50 bar

Xe
1.00 L
1.00 bar

Solution

The number of moles of He, Ne, and Xe is given by

$$n_{He} = \frac{PV}{RT} = \frac{1.50\,bar \times 2.00\,L}{8.314 \times 10^{-2}\,L\,bar\,K^{-1}\,mol^{-1} \times 298\,K} = 0.121\,mol$$

$$n_{Ne} = \frac{PV}{RT} = \frac{2.50\,bar \times 3.00\,L}{8.314 \times 10^{-2}\,L\,bar\,K^{-1}\,mol^{-1} \times 298\,K} = 0.303\,mol$$

$$n_{Xe} = \frac{PV}{RT} = \frac{1.00\,bar \times 1.00\,L}{8.314 \times 10^{-2}\,L\,bar\,K^{-1}\,mol^{-1} \times 298\,K} = 0.0403\,mol$$

$$n = n_{He} + n_{Ne} + n_{Xe} = 0.464$$

The mole fractions are

$$x_{He} = \frac{n_{He}}{n} = \frac{0.121}{0.464} = 0.261$$

$$x_{Ne} = \frac{n_{Ne}}{n} = \frac{0.303}{0.464} = 0.653$$

$$x_{Xe} = \frac{n_{Xe}}{n} = \frac{0.0403}{0.464} = 0.0860$$

The total pressure is given by

$$P = \frac{(n_{He} + n_{Ne} + n_{Xe})RT}{V}$$

$$= \frac{0.464\,mol \times 8.314 \times 10^{-2}\,L\,bar\,K^{-1}\,mol^{-1} \times 298\,K}{6.00\,L}$$

$$= 1.92\,bar$$

The partial pressures are given by

$$P_{He} = x_{He}P = 0.261 \times 1.92\,bar = 0.501\,bar$$

$$P_{Ne} = x_{Ne}P = 0.653 \times 1.92\,bar = 1.25\,bar$$

$$P_{Xe} = x_{Xe}P = 0.0860 \times 1.92\,bar = 0.165\,bar$$

1.5 A Brief Introduction to Real Gases

The ideal gas law provides a first look at the usefulness of describing a system in terms of macroscopic parameters. However, we should also emphasize the downside of not taking the microscopic nature of the system into account. For example, the ideal gas law only holds for gases at low densities. In practice, deviations from the ideal gas law that occur for real gases must be taken into account in such applications as a gas thermometer. If data were obtained from a gas thermometer using He, Ar, and N_2 for a temperature very near the temperature at which the gas condenses to form a liquid, they would exhibit the behavior shown in Figure 1.9. We see that the temperature only becomes independent of P and of the gas used in the thermometer if the data are extrapolated to zero pressure. It is in this limit that the gas thermometer provides a measure of the thermodynamic temperature. In practice, gas-independent T values are only obtained below $P \sim 0.01$ bar.

For most applications, calculations based on the ideal gas law are valid to much higher pressures. Real gases will be discussed in detail in Chapter 7. However, because we need to take nonideal gas behavior into account in Chapters 2 through 6, we introduce an equation of state that is valid to higher densities in this section.

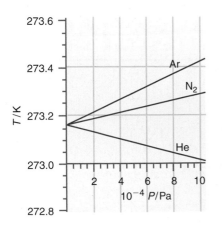

FIGURE 1.9
The temperature measured in a gas thermometer is independent of the gas used only in the limit that $P \rightarrow 0$.

The ideal gas assumptions that the atoms or molecules of a gas do not interact with one another and can be treated as point masses have a limited range of validity, which can be discussed using the potential energy function typical for a real gas, as shown in Figure 1.10. This figure shows the potential energy of interaction of two gas molecules as a function of the distance between them. The intermolecular potential can be divided into regions in which the potential energy is essentially zero ($r > r_{transition}$), negative (attractive interaction) ($r_{transition} > r > r_{V=0}$), and positive (repulsive interaction) ($r < r_{V=0}$). The distance $r_{transition}$ is not uniquely defined and depends on the energy of the molecule. It is on the order of the molecular size.

As the density is increased from very low values, molecules approach one another to within a few molecular diameters and experience a long-range attractive van der Waals force due to time-fluctuating dipole moments in each molecule. This strength of the attractive interaction is proportional to the polarizability of the electron charge in a molecule and is, therefore, substance dependent. In the attractive region, P is lower than that calculated using the ideal gas law. This is the case because the attractive interaction brings the atoms or molecules closer than they would be if they did not interact. At sufficiently high densities, the atoms or molecules experience a short-range repulsive interaction due to the overlap of the electron charge distributions. Because of this interaction, P is higher than that calculated using the ideal gas law. We see that for a real gas, P can be either greater or less than the ideal gas value. Note that the potential becomes repulsive for a value of r greater than zero. As a consequence, the volume of a gas even well above its boiling temperature approaches a finite limiting value as P increases. By contrast, the ideal gas law predicts that $V \rightarrow 0$ as $P \rightarrow \infty$.

Given the potential energy function depicted in Figure 1.10, under what conditions is the ideal gas equation of state valid? A real gas behaves ideally only at low densities for which $r > r_{transition}$, and the value of $r_{transition}$ is substance dependent. The **van der Waals equation of state** takes both the finite size of molecules and the attractive potential into account. It has the form

$$P = \frac{nRT}{V - nb} - \frac{n^2 a}{V^2} \tag{1.22}$$

This equation of state has two parameters that are substance dependent and must be experimentally determined. The parameters b and a take the finite size of the molecules and the strength of the attractive interaction into account, respectively. (Values of a and b for selected gases are listed in Table 7.4.) The van der Waals equation of state is more accurate in calculating the relationship between P, V, and T for gases than the ideal gas law because a and b have been optimized using experimental results. However, there are other more accurate equations of state that are valid over a wider range than the van der Waals equation, as will be discussed in Chapter 7.

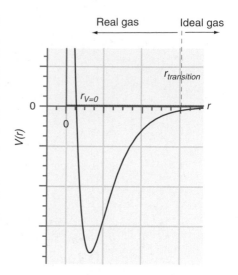

FIGURE 1.10
The potential energy of interaction of two molecules or atoms is shown as a function of their separation r. The red curve shows the potential energy function for an ideal gas. The dashed blue line indicates an approximate r value below which a more nearly exact equation of state than the ideal gas law should be used. $V(r) = 0$ at $r = r_{V=0}$ and as $r \rightarrow \infty$.

EXAMPLE PROBLEM 1.3

Van der Waals parameters are generally tabulated with either of two sets of units:
a. $Pa \, m^6 \, mol^{-2}$ or $bar \, dm^6 \, mol^{-2}$
b. $m^3 \, mol^{-1}$ or $dm^3 \, mol^{-1}$
Determine the conversion factor to convert one system of units to the other. Note that $1 \, dm^3 = 10^{-3} \, m^3 = 1 \, L$.

Solution

$$Pa \, m^6 \, mol^{-2} \times \frac{bar}{10^5 \, Pa} \times \frac{10^6 \, dm^6}{m^6} = 10 \, bar \, dm^6 \, mol^{-2}$$

$$m^3 \, mol^{-1} \times \frac{10^3 \, dm^3}{m^3} = 10^3 \, dm^3 \, mol^{-1}$$

In Example Problem 1.4, a comparison is made of the molar volume for N_2 calculated at low and high pressures, using the ideal gas and van der Waals equations of state.

EXAMPLE PROBLEM 1.4

a. Calculate the pressure exerted by N_2 at 300. K for molar volumes of 250. L mol^{-1} and 0.100 L mol^{-1} using the ideal gas and the van der Waals equations of state. The values of parameters a and b for N_2 are 1.370 bar dm^6 mol^{-2} and 0.0387 dm^3 mol^{-1}, respectively.

b. Compare the results of your calculations at the two pressures. If P calculated using the van der Waals equation of state is greater than those calculated with the ideal gas law, we can conclude that the repulsive interaction of the N_2 molecules outweighs the attractive interaction for the calculated value of the density. A similar statement can be made regarding the attractive interaction. Is the attractive or repulsive interaction greater for N_2 at 300. K and $V_m = 0.100$ L?

Solution

a. The pressures calculated from the ideal gas equation of state are

$$P = \frac{nRT}{V} = \frac{1\,\text{mol} \times 8.314 \times 10^{-2}\,\text{L bar mol}^{-1}\text{K}^{-1} \times 300.\,\text{K}}{250.\,\text{L}} = 9.98 \times 10^{-2}\,\text{bar}$$

$$P = \frac{nRT}{V} = \frac{1\,\text{mol} \times 8.314 \times 10^{-2}\,\text{L bar mol}^{-1}\text{K}^{-1} \times 300.\,\text{K}}{0.100\,\text{L}} = 249\,\text{bar}$$

The pressures calculated from the van der Waals equation of state are

$$P = \frac{nRT}{V - nb} - \frac{n^2 a}{V^2}$$

$$= \frac{1\,\text{mol} \times 8.314 \times 10^{-2}\,\text{L bar mol}^{-1}\text{K}^{-1} \times 300.\,\text{K}}{250.\,\text{L} - 1\,\text{mol} \times 0.0387\,\text{dm}^3\,\text{mol}^{-1}}$$

$$- \frac{(1\,\text{mol})^2 \times 1.370\,\text{bar dm}^6\,\text{mol}^{-2}}{(250.\,\text{L})^2}$$

$$= 9.98 \times 10^{-2}\,\text{bar}$$

$$P = \frac{1\,\text{mol} \times 8.314 \times 10^{-2}\,\text{L bar mol}^{-1}\text{K}^{-1} \times 300\,\text{K}}{0.100\,\text{L} - 1\,\text{mol} \times 0.0387\,\text{dm}^3\,\text{mol}^{-1}}$$

$$- \frac{(1\,\text{mol})^2 \times 1.370\,\text{bar dm}^6\,\text{mol}^{-2}}{(0.100\,\text{L})^2}$$

$$= 270.\,\text{bar}$$

b. Note that the result is identical with that for the ideal gas law for $V_m = 250.$ L, and that the result calculated for $V_m = 0.100$ L deviates from the ideal gas law result. Because $P^{real} > P^{ideal}$, we conclude that the repulsive interaction is more important than the attractive interaction for this specific value of molar volume and temperature.

Vocabulary

absolute temperature scale	Celsius scale	elastic collision
adiabatic	centigrade scale	equation of state
Boltzmann constant	closed system	equilibrium
boundary	diathermal	extensive variable

gas thermometer	macroscopic variables	temperature scale
ideal gas	mole fraction	thermal equilibrium
ideal gas constant	open system	thermodynamic equilibrium
ideal gas law	partial pressure	thermodynamic temperature scale
intensive variable	surroundings	thermometer
isolated system	system	van der Waals equation of state
kelvin	system variables	wall
macroscopic scale	temperature	zeroth law of thermodynamics

Conceptual Problems

Q1.1 Real walls are never totally adiabatic. Use your experience to order the following walls in increasing order with respect to their being diathermal: 1-cm-thick concrete, 1-cm-thick vacuum, 1-cm-thick copper, 1-cm-thick cork.

Q1.2 The parameter a in the van der Waals equation is greater for H_2O than for He. What does this say about the difference in the form of the potential function in Figure 1.10 for the two gases?

Q1.3 Give an example based on molecule–molecule interactions excluding chemical reactions, illustrating how the total pressure upon mixing two real gases could be different from the sum of the partial pressures.

Q1.4 Can temperature be measured directly? Explain your answer.

Q1.5 Explain how the ideal gas law can be deduced for the measurements shown in Figures 1.5 and 1.8.

Q1.6 The location of the boundary between the system and the surroundings is a choice that must be made by the thermodynamicist. Consider a beaker of boiling water in an airtight room. Is the system open or closed if you place the boundary just outside the liquid water? Is the system open or closed if you place the boundary just inside the walls of the room?

Q1.7 Give an example of two systems that are in equilibrium with respect to only one of two state variables.

Q1.8 At sufficiently high temperatures, the van der Waals equation has the form $P \approx RT/(V_m - b)$. Note that the attractive part of the potential has no influence in this expression. Justify this behavior using the potential energy diagram of Figure 1.10.

Q1.9 Give an example of two systems separated by a wall that are in thermal but not chemical equilibrium.

Q1.10 Which of the following systems are open? (a) a dog, (b) an incandescent light bulb, (c) a tomato plant, (d) a can of tomatoes. Explain your answers.

Q1.11 Which of the following systems are isolated? (a) a bottle of wine, (b) a tightly sealed, perfectly insulated thermos bottle, (c) a tube of toothpaste, (d) our solar system. Explain your answers.

Q1.12 Why do the z and y components of the velocity not change in the collision depicted in Figure 1.2?

Q1.13 If the wall depicted in Figure 1.2 were a movable piston, under what conditions would it move as a result of the molecular collisions?

Q1.14 The mass of a He atom is less than that of an Ar atom. Does that mean that because of its larger mass, Argon exerts a higher pressure on the container walls than He at the same molar density, volume, and temperature? Explain your answer.

Q1.15 Explain why attractive interactions between molecules in gas make the pressure less than that predicted by the ideal gas equation of state.

Numerical Problems

Problem numbers in **red** indicate that the solution to the problem is given in the *Student's Solutions Manual*.

P1.1 Approximately how many oxygen molecules arrive each second at the mitochondrion of an active person with a mass of 84 kg? The following data are available: Oxygen consumption is about 40. mL of O_2 per minute per kilogram of body weight, measured at $T = 300.$ K and $P = 1.00$ atm. In an adult there are about 1.6×10^{10} cells per kg body mass. Each cell contains about 800. mitochondria.

P1.2 A compressed cylinder of gas contains 2.74×10^3 g of N_2 gas at a pressure of 3.75×10^7 Pa and a temperature of 18.7°C. What volume of gas has been released into the atmosphere if the final pressure in the cylinder is 1.80×10^5 Pa? Assume ideal behavior and that the gas temperature is unchanged.

P1.3 Calculate the pressure exerted by Ar for a molar volume of 1.31 L mol^{-1} at 426 K using the van der Waals equation of state. The van der Waals parameters a and b for Ar are 1.355 bar dm^6 mol^{-2} and 0.0320 dm^3 mol^{-1}, respectively. Is the attractive or repulsive portion of the potential dominant under these conditions?

P1.4 A sample of propane (C_3H_8) is placed in a closed vessel together with an amount of O_2 that is 2.15 times the amount needed to completely oxidize the propane to CO_2 and

H_2O at constant temperature. Calculate the mole fraction of each component in the resulting mixture after oxidation, assuming that the H_2O is present as a gas.

P1.5   A gas sample is known to be a mixture of ethane and butane. A bulb having a 230.0 cm³ capacity is filled with the gas to a pressure of 97.5×10^3 Pa at 23.1°C. If the mass of the gas in the bulb is 0.3554 g, what is the mole percent of butane in the mixture?

P1.6   One liter of fully oxygenated blood can carry 0.18 liters of O_2 measured at $T = 298$ K and $P = 1.00$ atm. Calculate the number of moles of O_2 carried per liter of blood. Hemoglobin, the oxygen transport protein in blood has four oxygen binding sites. How many hemoglobin molecules are required to transport the O_2 in 1.0 L of fully oxygenated blood?

P1.7   Yeast and other organisms can convert glucose $(C_6H_{12}O_6)$ to ethanol (CH_3CH_2OH) by a process called alchoholic fermentation. The net reaction is

$$C_6H_{12}O_6(s) \rightarrow 2C_2H_5OH(l) + 2CO_2(g)$$

Calculate the mass of glucose required to produce 2.25 L of CO_2 measured at $P = 1.00$ atm and $T = 295$ K.

P1.8   A vessel contains 1.15 g liq H_2O in equilibrium with water vapor at 30.°C. At this temperature, the vapor pressure of H_2O is 31.82 torr. What volume increase is necessary for all the water to evaporate?

P1.9   Consider a 31.0 L sample of moist air at 60.°C and one atm in which the partial pressure of water vapor is 0.131 atm. Assume that dry air has the composition 78.0 mole percent N_2, 21.0 mole percent O_2, and 1.00 mole percent Ar.

a. What are the mole percentages of each of the gases in the sample?

b. The percent relative humidity is defined as %RH = $P_{H_2O}/P^*_{H_2O}$ where P_{H_2O} is the partial pressure of water in the sample and $P^*_{H_2O} = 0.197$ atm is the equilibrium vapor pressure of water at 60.°C. The gas is compressed at 60.°C until the relative humidity is 100.%. What volume does the mixture contain now?

c. What fraction of the water will be condensed if the total pressure of the mixture is isothermally increased to 81.0 atm?

P1.10   A typical diver inhales 0.450 liters of air per breath and carries a 25 L breathing tank containing air at a pressure of 300. bar. As she dives deeper, the pressure increases by 1 bar for every 10.08 m. How many breaths can the diver take from this tank at a depth of 35 m? Assume that the temperature remains constant.

P1.11   Use the ideal gas and van der Waals equations to calculate the pressure when 2.25 mol H_2 are confined to a volume of 1.65 L at 298 K. Is the gas in the repulsive or attractive region of the molecule–molecule potential?

P1.12   A rigid vessel of volume 0.400 m³ containing H_2 at 21.25°C and a pressure of 715×10^3 Pa is connected to a second rigid vessel of volume 0.750 m³ containing Ar at 30.15°C at a pressure of 203×10^3 Pa. A valve separating the two

vessels is opened and both are cooled to a temperature of 12.2°C. What is the final pressure in the vessels?

P1.13   A mixture of oxygen and hydrogen is analyzed by passing it over hot copper oxide and through a drying tube. Hydrogen reduces the CuO according to the reaction $CuO(s) + H_2(g) \rightarrow Cu(s) + H_2O(l)$, and oxygen reoxidizes the copper formed according to $Cu(s) + 1/2 \, O_2(g) \rightarrow CuO(s)$. At 25°C and 750. Torr, 172.0 cm³ of the mixture yields 77.5 cm³ of dry oxygen measured at 25°C and 750. Torr after passage over CuO and the drying agent. What is the original composition of the mixture?

P1.14   An athlete at high performance inhales ∼3.75 L of air at 1.00 atm and 298 K. The inhaled and exhaled air contain 0.50 and 6.2% by volume of water, respectively. For a respiration rate of 32 breaths per minute, how many moles of water per minute are expelled from the body through the lungs?

P1.15   Devise a temperature scale, abbreviated G, for which the magnitude of the ideal gas constant is 5.52 J G^{-1} mol^{-1}.

P1.16   Aerobic cells metabolize glucose in the respiratory system. This reaction proceeds according to the overall reaction

$$6O_2(g) + C_6H_{12}O_6(s) \rightarrow 6CO_2(g) + 6H_2O(l)$$

Calculate the volume of oxygen required at STP to metabolize 0.025 kg of glucose $(C_6H_{12}O_6)$. STP refers to standard temperature and pressure, that is, $T = 273$ K and $P = 1.00$ atm. Assume oxygen behaves ideally at STP.

P1.17   An athlete at high performance inhales ∼3.75 L of air at 1.0 atm and 298 K at a respiration rate of 32 breaths per minute. If the exhaled and inhaled air contain 15.3 and 20.9% by volume of oxygen respectively, how many moles of oxygen per minute are absorbed by the athlete's body?

P1.18   A mixture of 2.10×10^{-3} g of O_2, 3.88×10^{-3} mol of N_2, and 5.25×10^{20} molecules of CO are placed into a vessel of volume 5.25 L at 12.5°C.

a. Calculate the total pressure in the vessel.

b. Calculate the mole fractions and partial pressures of each gas.

P1.19   Calculate the pressure exerted by benzene for a molar volume of 2.00 L at 595 K using the Redlich-Kwong equation of state:

$$P = \frac{RT}{V_m - b} - \frac{a}{\sqrt{T}} \frac{1}{V_m(V_m + b)}$$

$$= \frac{nRT}{V - nb} - \frac{n^2a}{\sqrt{T}} \frac{1}{V(V + nb)}$$

The Redlich-Kwong parameters a and b for benzene are 452.0 bar dm⁶ mol^{-2} $K^{1/2}$ and 0.08271 dm³ mol^{-1}, respectively. Is the attractive or repulsive portion of the potential dominant under these conditions?

P1.20   In the absence of turbulent mixing, the partial pressure of each constituent of air would fall off with height above sea level in Earth's atmosphere as $P_i = P_i^0 e^{-M_i gz/RT}$ where P_i is the partial pressure at the height z, P_i^0 is the partial pressure of component i at sea level, g is the acceleration of gravity, R is the gas constant, T is the absolute temperature, and M_i is the molecular mass of the gas. As a result of turbulent mixing, the composition of Earth's atmosphere is constant below an altitude of 100 km, but the total pressure decreases with altitude as $P = P^0 e^{-M_{ave}gz/RT}$ where M_{ave} is the mean molecular weight of air. At sea level, $x_{N_2} = 0.78084$ and $x_{He} = 0.00000524$ and $T = 300.$ K.

a. Calculate the total pressure at 8.5 km, assuming a mean molecular mass of 28.9 g mol^{-1} and that $T = 300.$ K throughout this altitude range.

b. Calculate the value that x_{N_2}/x_{He} would have at 8.5 km in the absence of turbulent mixing. Compare your answer with the correct value.

P1.21 An initial step in the biosynthesis of glucose $C_6H_{12}O_6$ is the carboxylation of pyruvic acid $CH_3COCOOH$ to form oxaloacetic acid $HOOCCOCH_2COOH$

$$CH_3COCOOH(s) + CO_2(g) \rightarrow HOOCCOCH_2COOH(s)$$

If you knew nothing else about the intervening reactions involved in glucose biosynthesis other than no further carboxylations occur, what volume of CO_2 is required to produce 1.10 g of glucose? Assume $P = 1$ atm and $T = 298$ K.

P1.22 Consider the oxidation of the amino acid glycine NH_2CH_2COOH to produce water, carbon dioxide, and urea NH_2CONH_2:

$$NH_2CH_2COOH(s) + 3O_2(g) \rightarrow$$
$$NH_2CONH_2(s) + 3CO_2(g) + 3H_2O(l)$$

Calculate the volume of carbon dioxide evolved at $P = 1.00$ atm and $T = 305$ K from the oxidation of 0.022 g of glycine.

P1.23 Assume that air has a mean molar mass of 28.9 g mol^{-1} and that the atmosphere has a uniform temperature of 25.0°C. Calculate the barometric pressure in Pa in Santa Fe, for which $z = 7000.$ ft. Use the information contained in Problem P1.20.

P1.24 When Julius Caesar expired, his last exhalation had a volume of 450. cm^3 and contained 1.00 mole percent argon. Assume that $T = 300.$ K and $P = 1.00$ atm at the location of his demise. Assume further that T has the same value throughout Earth's atmosphere. If all of his exhaled Ar atoms are now uniformly distributed throughout the atmosphere, how many inhalations of 450. cm^3 must we make to inhale one of the Ar atoms exhaled in Caesar's last breath? Assume the radius of Earth to be 6.37×10^6 m. [*Hint:* Calculate the number of Ar atoms in the atmosphere in the simplified geometry of a plane of area equal to that of Earth's surface. See Problem P1.20 for the dependence of the barometric pressure and the composition of air on the height above Earth's surface.

P1.25 Calculate the number of molecules per m^3 in an ideal gas at the standard temperature and pressure conditions of 0.00°C and 1.00 atm.

P1.26 Consider a gas mixture in a 1.50 dm^3 flask at 22.0°C. For each of the following mixtures, calculate the partial pressure of each gas, the total pressure, and the composition of the mixture in mole percent:

a. 3.06 g H_2 and 2.98 g O_2

b. 2.30 g N_2 and 1.61 g O_2

c. 2.02 g CH_4 and 1.70 g NH_3

P1.27 A mixture of H_2 and NH_3 has a volume of 139.0 cm^3 at 0.00°C and 1 atm. The mixture is cooled to the temperature of liquid nitrogen at which ammonia freezes out and the remaining gas is removed from the vessel. Upon warming the vessel to 0.00°C and 1 atm, the volume is 77.4 cm^3. Calculate the mole fraction of NH_3 in the original mixture.

P1.28 A sealed flask with a capacity of 1.22 dm^3 contains 4.50 g of carbon dioxide. The flask is so weak that it will burst if the pressure exceeds 9.500×10^5 Pa. At what temperature will the pressure of the gas exceed the bursting pressure?

P1.29 A balloon filled with 11.50 L of Ar at 18.7°C and 1 atm rises to a height in the atmosphere where the pressure is 207 Torr and the temperature is –32.4°C. What is the final volume of the balloon? Assume that the pressure inside and outside the balloon have the same value.

P1.30 Carbon monoxide competes with oxygen for binding sites on the transport protein hemoglobin. CO can be poisonous if inhaled in large quantities. A safe level of CO in air is 50. parts per million (ppm). When the CO level increases to 800. ppm, dizziness, nausea, and unconsciousness occur, followed by death. Assuming the partial pressure of oxygen in air at sea level is 0.20 atm, what proportion of CO to O_2 is fatal?

P1.31 The total pressure of a mixture of oxygen and hydrogen is 1.65 atm. The mixture is ignited and the water is removed. The remaining gas is pure hydrogen and exerts a pressure of 0.190 atm when measured at the same values of T and V as the original mixture. What was the composition of the original mixture in mole percent?

P1.32 Suppose that you measured the product PV of 1 mol of a dilute gas and found that $PV = 24.35$ L atm at 0.00°C and 33.54 L atm at 100.°C. Assume that the ideal gas law is valid, with $T = t(°C) + a$, and that the values of R and a are not known. Determine R and a from the measurements provided.

P1.33 Liquid N_2 has a density of 875.4 kg m^{-3} at its normal boiling point. What volume does a balloon occupy at 298 K and a pressure of 1.00 atm if 3.10×10^{-3} L of liquid N_2 is injected into it? Assume that there is no pressure difference between the inside and outside of the balloon.

P1.34 Calculate the volume of all gases evolved by the complete oxidation of 0.375 g of the amino acid alanine CH3CH(NH$_2$)COOH if the products are liquid water, nitrogen gas, and carbon dioxide gas; the total pressure is 1.00 atm; and $T = 298$ K.

P1.35 As a result of photosynthesis, an acre of forest (1 acre = 4047 square meters) can take up 1000. kg of CO_2. Assuming air is 0.0314% CO_2 by volume, what volume of air is required to provide 350. kg of CO_2? Assume $T = 310$ K and $P = 1.00$ atm.

P1.36 A glass bulb of volume 0.198 L contains 0.457 g of gas at 759.0 Torr and 134.0°C. What is the molar mass of the gas?

P1.37 Use L'Hôpital's rule, $\lim\left[f(x)/g(x)\right]_{x\to 0} = \lim\left[\dfrac{df(x)/dx}{dg(x)/dx}\right]_{x\to 0}$ to show that the expression derived for P_f in part (b) of Example Problem 1.1 has the correct limit as $\gamma \to 0$.

P1.38 A 455 cm^3 vessel contains a mixture of Ar and Xe. If the mass of the gas mixture is 2.245 g at 25.0°C and the pressure is 760. Torr, calculate the mole fraction of Xe in the mixture.

P1.39 Many processes such as the fabrication of integrated circuits are carried out in a vacuum chamber to avoid reaction of the material with oxygen in the atmosphere. It is difficult to routinely lower the pressure in a vacuum chamber below 1.0×10^{-10} Torr. Calculate the molar density at this pressure at 300. K. What fraction of the gas phase molecules initially present for 1.0 atm in the chamber are present at 1.0×10^{-10} Torr?

P1.40 Rewrite the van der Waals equation using the molar volume rather than V and n.

Heat, Work, Internal Energy, Enthalpy, and the First Law of Thermodynamics

In this chapter, the internal energy U is introduced. The first law of thermodynamics relates ΔU to the heat (q) and work (w) that flows across the boundary between the system and the surroundings. Other important concepts introduced include heat capacity, the difference between state and path functions, and reversible versus irreversible processes. The enthalpy H is introduced as a form of energy that can be directly measured by the heat flow in a constant pressure process. We show how ΔU, ΔH, q, and w can be calculated for processes involving ideal gases.

2.1 The Internal Energy and the First Law of Thermodynamics

This section focuses on the change in energy of the system and surroundings during a thermodynamic process such as an expansion or compression of a gas. In thermodynamics, we are interested in the internal energy of the system, as opposed to the energy associated with the system relative to a particular frame of reference. For example, a container of gas in an airplane has a kinetic energy relative to an observer on the ground. However, the internal energy of the gas is defined relative to a coordinate system fixed on the container. Viewed at a molecular level, the internal energy can take on a number of forms such as

*   the translational energy of the molecules.
*   the potential energy of the constituents of the system; for example, a crystal consisting of polarizable molecules will experience a change in its potential energy as an electric field is applied to the system.
*   the internal energy stored in the form of molecular vibrations and rotations.
*   the internal energy stored in the form of chemical bonds that can be released through a chemical reaction.
*   the potential energy of interaction between molecules.

The total of all these forms of energy for the system of interest is given the symbol U and is called the **internal energy.**[1]

The **first law of thermodynamics** is based on our experience that energy can be neither created nor destroyed, if the energies of both the system and the surroundings are taken into account. This law can be formulated in a number of equivalent forms. Our initial formulation of this law is as follows:

> The internal energy U of an isolated system is constant.

This form of the first law looks uninteresting because it suggests that nothing happens in an isolated system when viewed from outside the system. How can the first law tell us anything about thermodynamic processes such as chemical reactions? Consider separating an isolated system into two subsystems, the system and the surroundings. When changes in U occur in a system in contact with its surroundings, ΔU_{total} is given by

$$\Delta U_{total} = \Delta U_{system} + \Delta U_{surroundings} = 0 \qquad (2.1)$$

Therefore, the first law becomes

$$\Delta U_{system} = -\Delta U_{surroundings} \qquad (2.2)$$

For any decrease of U_{system}, $U_{surroundings}$ must increase by exactly the same amount. For example, if a gas (the system) is cooled, the temperature of the surroundings must increase.

How can the energy of a system be changed? There are many ways to alter U, several of which are discussed in this chapter. Experience has shown that all changes in a closed system in which no chemical reactions or phase changes occur can be classified only as heat, work, or a combination of both. Therefore, the internal energy of such a system can only be changed by the flow of heat or work across the boundary between the system and surroundings. For example, U for a gas can be increased by putting it in an oven or by compressing it. In both cases, the temperature of the system increases. This important recognition leads to a second and more useful formulation of the first law:

$$\Delta U = q + w \qquad (2.3)$$

where q and w designate heat and work, respectively. We use ΔU without a subscript to indicate the change in internal energy of the system. What do we mean by heat and work? In the following two sections, we define these important concepts and discuss how they differ.

The symbol Δ is used to indicate a change that occurs as a result of an arbitrary process. The simplest processes are those in which one of P, V, or T remains constant. A constant temperature process is referred to as **isothermal,** and the corresponding terms for constants P and V are **isobaric** and **isochoric,** respectively.

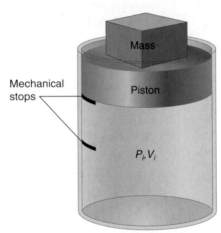

Mass

Mechanical stops

Piston

P_i, V_i

Initial state

Mass

Piston

P_f, V_f

Final state

FIGURE 2.1
A system is shown in which compression work is being done on a gas. The walls are adiabatic.

2.2 Work

In this and the next section, we discuss the two ways in which the internal energy of a system can be changed. **Work** in thermodynamics is defined as any quantity of energy that "flows" across the boundary between the system and surroundings as a result of a force acting through a distance. Examples are moving an ion in a solution from one region of electrical potential to another, inflating a balloon, or climbing stairs. In each of these examples, there is a force along the direction of motion. Consider another example, a gas in a piston and cylinder assembly as shown in Figure 2.1. In this example, the system is defined as the gas alone. Everything else shown in the figure is in the surroundings. As the gas is compressed, the height of the mass in the surroundings is lowered and the initial and final volumes are defined by the mechanical stops indicated in the figure.

---

[1]We could include other terms such as the binding energy of the atomic nuclei but choose to include only the forms of energy that are likely to change in simple chemical reactions.

Work has several important characteristics:

- Work is transitory in that it only appears during a change in state of the system and surroundings. Only energy, and not work, is associated with the initial and final states of the systems.

- The net effect of work is to change U of the system and surroundings in accordance with the first law. If the only change in the system results from a force acting through a distance (as for example the movement of the mass in Figure 2.1), work has flowed between the system and the surroundings. Work can usually be represented by a mass in the surroundings that has been raised or lowered in Earth's gravitational field.

- The quantity of work can be calculated using the definition

$$w = \int_{x_i}^{x_f} \mathbf{F} \cdot d\mathbf{x} \qquad (2.4)$$

Note that because of the scalar product in the integral, the work will be zero unless the force has a component along the displacement direction.

- The sign of the work follows from evaluating the preceding integral. If $w > 0$, $\Delta U > 0$ for an adiabatic process. It is common usage to say that if w is positive, work is done on the system by the surroundings. If w is negative, work is done by the system on the surroundings. The quantity of work can also be calculated from the change in potential energy of the mass in the surroundings, $\Delta E_{potential} = mg\Delta h = -w$, where g is the gravitational acceleration and Δh is the change in the height of the mass m.

Using the definition of pressure as the force per unit area (A), the work done in moving the mass in Figure 2.1 is given by

$$w = \int_{x_i}^{x_f} \mathbf{F} \cdot d\mathbf{x} = -\int_{x_i}^{x_f} P_{external} A dx = -\int_{V_i}^{V_f} P_{external}\, dV \qquad (2.5)$$

The minus sign appears because $\mathbf{F}$ and $d\mathbf{x}$ are vectors that point in opposite directions. Note that the pressure that appears in this expression is the external pressure $P_{external}$ which need not equal the system pressure P.

An example of another important kind of work, electrical work, is shown in Figure 2.2, in which the content of the cylinder is the system. Electrical current flows through a conductive aqueous solution and water undergoes electrolysis to produce H_2 and O_2 gas. The current is produced by a generator, like that used to power a light on a bicycle through the mechanical work of pedaling. As current flows, the mass that drives the generator is lowered. In this case, the surroundings do the electrical work on the system. As a result, some of the liquid water is transformed to H_2 and O_2. From electrostatics, the work done in transporting a charge Q through an electrical potential difference ϕ is

$$w_{electrical} = Q\phi \qquad (2.6)$$

For a constant current I that flows for a time t, $Q = It$. Therefore,

$$w_{electrical} = I\phi t \qquad (2.7)$$

The system also does work on the surroundings through the increase in the volume of the gas phase at the constant external pressure P_i, as shown by the raised mass on the piston. The total work done is

$$w = w_{P-V} + w_{electrical} = I\phi t - \int_{V_i}^{V_f} P_{external}\, dV = I\phi t - P_{external}(V_f - V_i) \qquad (2.8)$$

Other forms of work include the work of expanding a surface, such as a soap bubble, against the surface tension. Table 2.1 shows the expressions for work for four different cases. Each of these different types of work poses a requirement on the walls separating the system and surroundings. To be able to carry out the first three types of work, the walls must be movable, whereas for electrical work, they must be conductive. Several examples of work calculations are given in Example Problem 2.1.

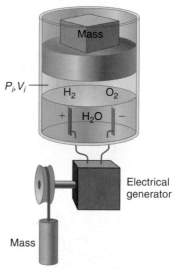

Initial state

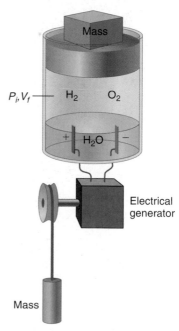

Final state

FIGURE 2.2
Current produced by a generator is used to electrolyze water and thereby do work on the system as shown by the lowered mass linked to the generator.

TABLE 2.1   Types of Work

Types of Work	Variables	Equation for Work	Conventional Units
Volume expansion	Pressure (P), volume (V)	$w = -\int_{V_i}^{V_f} P_{external}\, dV$	Pa m^3 = J
Stretching	Force (F), length (l)	$w = \int_{x_i}^{x_f} \mathbf{F} \cdot d\mathbf{l}$	N m = J
Surface expansion	Surface tension (γ), area (σ)	$w = \int_{\sigma_i}^{\sigma_f} \gamma \cdot d\sigma$	(N m^{-1})(m^2) = J
Electrical	Electrical potential (ϕ), electrical charge (Q)	$w = \int_0^Q \phi\, dQ'$	V C = J

EXAMPLE PROBLEM 2.1

a. Calculate the work involved in expanding 20.0 L of an ideal gas to a final volume of 85.0 L against a constant external pressure of 2.50 bar.

b. A bubble is expanded from a radius of 1.00 cm to a radius of 3.25 cm. The surface tension of water is 71.99 N m^{-1}. How much work is done in increasing the area of the bubble?

c. A current of 3.20 A is passed through a heating coil for 30.0 s. The electrical potential across the resistor is 14.5 V. Calculate the work done on the coil.

d. If the force to stretch a fiber a distance x is given by $F = -kx$ with $k = 100.$ N cm^{-1}, how much work is done to stretch the fiber 0.15 cm?

Solution

a. $w = -\int_{V_i}^{V_f} P_{external}\, dV = -P_{external}(V_f - V_i)$

$= -2.50\text{ bar} \times \dfrac{10^5\text{ Pa}}{\text{bar}} \times (85.0\text{ L} - 20.0\text{ L}) \times \dfrac{10^{-3}\text{ m}^3}{\text{L}} = -16.3\text{ kJ}$

b. A factor of 2 is included in the following calculation because a bubble has an inner and an outer surface. We consider the bubble and its contents to be the system. The vectors γ and σ point in opposite directions, giving rise to the negative sign in the second integral.

$w = \int_{\sigma_i}^{\sigma_f} \gamma \cdot d\sigma = -\int_{\sigma_i}^{\sigma_f} \gamma d\sigma = 2\gamma 4\pi (r_f^2 - r_i^2)$

$= -8\pi \times 71.99\text{ N m}^{-1}(3.25^2\text{ cm}^2 - 1.00^2\text{ cm}^2) \times \dfrac{10^{-4}\text{ m}^2}{\text{cm}^2}$

$= -1.73\text{ J}$

c. $w = \int_0^Q \phi\, dQ' = \phi Q = I\phi t = 14.5\text{ V} \times 3.20\text{ A} \times 30.0\text{ s} = 1.39\text{ kJ}$

d. We must distinguish between $\mathbf{F}$, the restoring force on the fiber, and $\mathbf{F}'$, the force exerted by the person stretching the fiber. They are related by $\mathbf{F} = -\mathbf{F}'$. If we calculate the work done on the fiber, $\mathbf{F}' \cdot d\mathbf{l} = F'dl$ because the vectors $\mathbf{F}'$ and dl point in the same direction

$w = \int_{x_0}^{x_f} \mathbf{F}' \cdot d\mathbf{l} = \int_{x_0}^{x_f} kx\, dx = \left[\dfrac{kx^2}{2}\right]_{x_0}^{x_f} = \left[\dfrac{100.\text{ N m}^{-1} \times x^2}{2}\right]_0^{0.15} = 1.1\text{ J}$

If we calculate the work done by the fiber, the sign of w is reversed because $\mathbf{F}$ and $d\mathbf{l}$ point in opposite directions.

2.3 Heat

Heat[2] is defined in thermodynamics as the quantity of energy that flows across the boundary between the system and surroundings because of a temperature difference between the system and the surroundings. Heat always flows spontaneously from regions of high temperature to regions of low temperature. Just as for work, several important characteristics of heat are of importance:

- Heat is transitory, in that it only appears during a change in state of the system and surroundings. Heat is not associated with the initial and final states of the system and the surroundings.

- The net effect of heat is to change the internal energy of the system and surroundings in accordance with the first law. If the only change in the surroundings is a change in temperature of a reservoir, heat has flowed between the system and the surroundings. The quantity of heat that has flowed is directly proportional to the change in temperature of the reservoir.

- The sign convention for heat is as follows. If the temperature of the system is raised, q is positive; if it is lowered, q is negative. It is common usage to say that if q is positive, heat is withdrawn from the surroundings and deposited in the system. If q is negative, heat is withdrawn from the system and deposited in the surroundings.

Defining the surroundings as the rest of the universe is impractical because it is not realistic to search through the whole universe to see if a mass has been raised or lowered and if the temperature of a reservoir has changed. Experience shows that in general only those parts of the universe close to the system interact with the system. Experiments can be constructed to ensure that this is the case, as shown in Figure 2.3. Imagine that we are interested in an exothermic chemical reaction that is carried out in a rigid sealed container with diathermal walls. We define the system as consisting solely of the reactant and product mixture. The vessel containing the system is immersed in an inner water bath separated from an outer water bath by a container with rigid diathermal walls. During the reaction, heat flows out of the system ($q < 0$), and the temperature of the inner water bath increases to T_f. Using an electrical heater, the temperature of the outer water bath is increased so that at all times, $T_{outer} = T_{inner}$. Because of this condition, no heat flows across the boundary between the two water baths, and because the container enclosing the inner water bath is rigid, no work flows across this boundary. Therefore, $\Delta U = q + w = 0 + 0 = 0$ *for the composite system* made up of the inner water bath and everything within it. Therefore, this composite system is an isolated system that does not interact with the rest of the universe. To determine q and w for the reactant and product mixture, we need to examine only the composite system and can disregard the rest of the universe.

To emphasize the distinction between q and w and the relationship between q, w, and ΔU, we discuss the two processes shown in Figure 2.4. They are each carried out in an isolated system, divided into two subsystems, I and II. In both cases, system I consists solely of the liquid in the beaker, and everything else including the rigid adiabatic walls is in system II. We refer to system I as the system and system II as the surroundings in the following discussion. We assume that the temperature of the liquid is well below its boiling point so that its vapor pressure is negligibly small. This ensures that no liquid is vaporized in the process, and the system is closed. We also assume that the change in temperature of the system is very small. System II can be viewed as the surroundings for system I and vice versa.

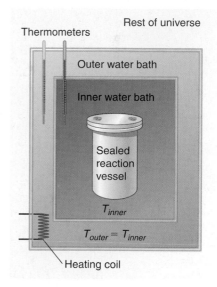

FIGURE 2.3

An isolated composite system is created in which the surroundings to the system of interest are limited in extent. The walls surrounding the inner water bath are rigid.

[2]Heat is perhaps the most misused term in thermodynamics as discussed by Robert Romer ["Heat is not a Noun." *American Journal of Physics*, 69 (2001), 107–109]. In common usage, it is incorrectly referred to as a substance as in the phrase "Close the door; you're letting the heat out!" An equally inappropriate term is heat capacity (discussed in Section 2.5), because it implies that materials have the capacity to hold heat, rather than the capacity to store energy. We use the terms *heat flow* or *heat transfer* to emphasize the transitory nature of heat. However, you should not think of heat as a fluid or a substance.

FIGURE 2.4
Two subsystems, I and II, are enclosed in a rigid adiabatic enclosure. System I consists solely of the liquid in the beaker for each case. System II consists of everything else in the enclosure, and is the surroundings for system I. **(a)** The liquid is heated using a flame. **(b)** The liquid is heated using a resistive coil through which an electrical current flows.

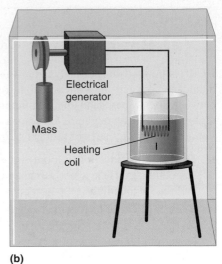

(a)   (b)

In Figure 2.4a, a Bunsen burner fueled by a propane canister is used to heat the liquid (system). The boundary between system and surroundings is the surface that encloses the liquid, and heat can flow all across this boundary. It is observed that the temperature of the liquid increases in the process. The temperature of the surroundings also increases because the system and surroundings are in thermal equilibrium. From Section 1.2, we know that the internal energy of a monatomic gas increases linearly with T. This result can be generalized to state that U is a monotonically increasing function of T for a uniform single-phase system of constant composition. Therefore, because $\Delta T > 0$, $\Delta U > 0$.

We next consider the changes in the surroundings. From the first law, $\Delta U_{surroundings} = -\Delta U < 0$. No forces oppose changes in the system. We conclude that $w = 0$. Therefore, if $\Delta U > 0$, $q > 0$ and $q_{surroundings} < 0$.

We now consider Figure 2.4b. In this case, the boundary between system and surroundings lies just inside the inner wall of the beaker, across the open top of the beaker, and just outside of the surface of the heating coil. Note that the heating coil is entirely in the surroundings. Heat can flow across the boundary surface. Upon letting the mass in the surroundings fall, electricity flows through the heating coil. It is our experience that the temperature of the liquid (system) will increase. We again conclude that $\Delta U > 0$. What values do q and w have for the process?

To answer this question, consider the changes in the surroundings. From the first law, $\Delta U_{surroundings} = -\Delta U < 0$. We see that a mass has been lowered in the surroundings. Can we conclude that $w > 0$? *No, because work is being done only on the heating coil, which is in the surroundings. The current flow never crosses the boundary between system and surroundings.* Therefore, $w = 0$ because no work is done on the system. However, $\Delta U > 0$, so if $w = 0$ we conclude that $q > 0$. The increase in U is due to heat flow from the surroundings to the system caused by the difference between the temperature of the heating filament and the liquid and not by the electrical work done on the filament. Note that because $\Delta U + \Delta U_{surroundings} = 0$, the heat flow from the surroundings to the system can be calculated from the electrical work done entirely within the surroundings, $q = -w_{surroundings} = I\phi t$.

These examples show that the distinction between heat and work must be made carefully with a clear knowledge of the position and nature of the boundary between the system and the surroundings.

EXAMPLE PROBLEM 2.2

A heating coil is immersed in a 100. g sample of H_2O liquid which boils at 99.61°C in an open insulated beaker on a laboratory bench at 1 bar pressure. In this process, 10.0% of the liquid is converted to the gaseous form at a pressure of 1 bar. A current of 2.00 A flows through the heater from a 12.0 V battery for 1.00×10^3 s to effect the transformation. The densities of liquid and gaseous water under these conditions are 997 and 0.590 kg m^{-3}, respectively.

a. It is often useful to replace a real process by a model that exhibits the important features of the process. Design a model system and surroundings, like those shown in Figures 2.1 and 2.2, that would allow you to measure the heat and work associated with this transformation. For the model system, define the system and surroundings as well as the boundary between them.

b. How can you define the system for the open insulated beaker on the laboratory bench such that the work is properly described?

c. Calculate q and w for the process.

Solution

a. The model system is shown in the following figure. The cylinder walls and the piston form adiabatic walls. The external pressure is held constant by a suitable weight.

b. Define the system as the liquid in the beaker and the volume containing only molecules of H_2O in the gas phase. This volume will consist of disconnected volume elements dispersed in the air above the laboratory bench.

c. In the system shown, the heat input to the liquid water can be equated with the work done on the heating coil. Therefore,

$$q = I\phi t = 2.00\ \text{A} \times 12.0\ \text{V} \times 1.00 \times 10^3\ \text{s} = 24.0 \text{kJ}$$

As the liquid is vaporized, the volume of the system increases at a constant external pressure. Therefore, the work done by the system on the surroundings is

$$w = -P_{external}(V_f - V_i) = -10^5\ \text{Pa}$$

$$\times \left(\frac{10.0 \times 10^{-3}\ \text{kg}}{0.590\,\text{kg\,m}^{-3}} + \frac{90.0 \times 10^{-3}\text{kg}}{997\,\text{kg\,m}^{-3}} - \frac{100.0 \times 10^{-3}\text{kg}}{997\,\text{kg\,m}^{-3}} \right)$$

$$= -1.70 \text{kJ}$$

Note that the electrical work done on the heating coil is much larger than the P–V work done in the expansion.

$P_{external} = 1.00$ atm

Mass

Heating coil

12 V

(A)

$H_2O\ (g)$

$H_2O\ (l)$

2.4 Doing Work on the System and Changing the System Energy from a Molecular Level Perspective

Our discussion so far has involved changes in energy for macroscopic systems, but what happens at the molecular level if energy is added to the system? In shifting to a molecular perspective, we move from a classical to a quantum mechanical description of matter. For this discussion, we need two results that will be discussed elsewhere in this textbook. First, as will be discussed in Chapter 15, in general the energy levels of quantum mechanical systems are discrete and molecules can only possess amounts of energy that correspond to these values. By contrast, in classical mechanics the energy is a continuous variable. Second, we use a result from statistical mechanics that the relative probability of a molecule being in a state corresponding to the allowed energy values ε_1 and ε_2 at temperature T is given by $e^{-([\varepsilon_2 - \varepsilon_1]/k_B T)}$. This result will be discussed in Chapter 30.

To keep the mathematics simple, consider a very basic model system: a gas consisting of a single He atom confined in a one-dimensional container with a length of 10. nm. Quantum mechanics tells us that the translational energy of the He atom confined in this box can only have the discrete values shown in Figure 2.5a. We lower the temperature of this one-dimensional He gas to 0.20 K. The calculated probability of the atom being in a given energy level is shown in Figure 2.5a. If we now do work on the system by compressing the box to half its original length at constant temperature, the energy levels will change to those shown in Figure 2.5b. Note

FIGURE 2.5
Energy levels are
shown for the box of
length **(a)** 10. nm, and
(b) 5.0 nm. The circles
indicate the probability
that the He atom has
an energy correspond-
ing to each of the
energy levels at 0.2 K.
Each circle indicates a
probability of 0.010.
For example, the prob-
ability that the energy
of the He atom corre-
sponds to the lowest
energy level in
Figure 2.5a is 0.22.
Note the different
scales for energy in
each graph.

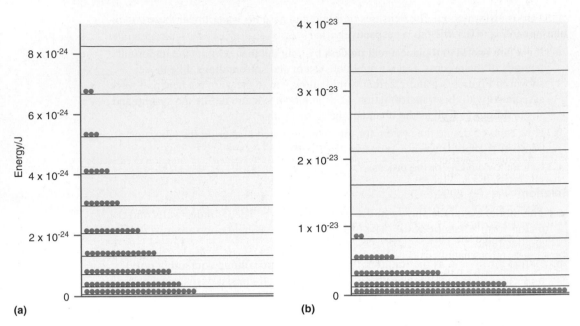

that all the energy levels are shifted to higher values as the container is made smaller
(an effect that will be fully explored in Chapter 15). If we keep the temperature constant at 0.20 K during this compression, the distribution of the atoms among the
energy levels changes to that shown in Figure 2.5b. This redistribution occurs
because the total translational energy of the He atom remains constant in the compression if the temperature is kept constant, assuming ideal gas behavior.

What happens if we keep the container at the smaller length and raise the system
energy by increasing T to 0.40 K? In this case, the energy levels are unchanged because
they depend on the container length, but not on the temperature of the gas. However,
the energy of the system increases and as shown in Figure 2.6, this occurs by a redistribution of the probability of finding the He atom in the energy levels. The increase in
system energy comes from an increase in the probability of the He atom populating
higher-energy levels and a corresponding decrease in the probability of populating
lower-energy levels.

Just as for a calculation of the energy of a gas using classical mechanics, the quantum mechanical system energy increases through an increase in the translational energy

FIGURE 2.6
Energy levels are
shown for the 5.0 nm
box. The circles indi-
cate the probability
that a He atom has an
energy corresponding
to each of the energy
levels at **(a)** 0.20 K
and **(b)** 0.40 K.

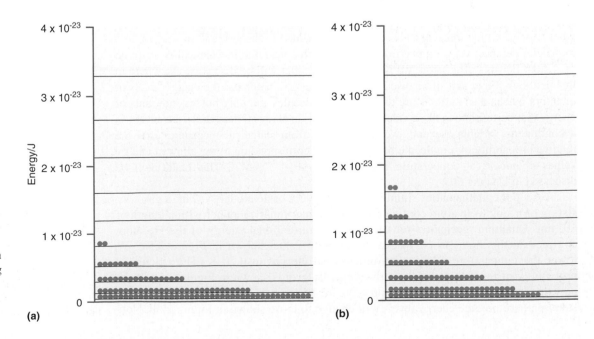

of the atom. However, the discrete energy level structure influences how the system can take up energy. This will become clearer in the next section when we consider how molecules can take up energy through rotation and vibration.

2.5 Heat Capacity

The process shown in Figure 2.4b provides a way to quantify heat flow in terms of the easily measured electrical work done on the heating coil, $w = I\phi t$. The response of a single-phase system of constant composition to heat input is an increase in T as long as the system does not undergo a phase change such as the vaporization of a liquid.

The thermal response of the system to heat flow is described by a very important thermodynamic property called the **heat capacity,** which is a measure of energy needed to change the temperature of a substance by a given amount. The name heat capacity is unfortunate because it implies that a substance has the capacity to take up heat. A much better name would be energy capacity.

Heat capacity is a material-dependent property, as will be discussed later. Mathematically, heat capacity is defined by the relation

$$C = \lim_{\Delta T \to 0} \frac{q}{T_f - T_i} = \frac{đq}{dT} \tag{2.9}$$

where C is in the SI unit of J K^{-1}. It is an extensive quantity that doubles as the mass of the system is doubled. Often, the molar heat capacity C_m is used in calculations. It is an intensive quantity with the units of J K^{-1} mol^{-1}. Experimentally, the heat capacity of fluids is measured by immersing a heating coil in the gas or liquid and equating the electrical work done on the coil with the heat flow into the sample. For solids, the heating coil is wrapped around the solid. The significance of the notation $đq$ for an incremental amount of heat is explained in the next section.

The value of the heat capacity depends on the experimental conditions under which it is determined. The most common conditions are constant V or P, for which the heat capacity is denoted C_V and C_P, respectively. Values of $C_{P,m}$ at 298.15 K for pure substances are tabulated in Tables 2.2 and 2.3 (see Appendix B, Data Tables), and formulas for calculating $C_{P,m}$ at other temperatures for gases and solids are listed in Tables 2.4 and 2.5, respectively.

We next discuss heat capacities using a molecular level model, beginning with gases. Figure 2.7 illustrates the energy level structure for a molecular gas. Molecules can take up energy by moving faster, by rotating in three-dimensional space, by periodic oscillations (known as vibrations) of the atoms around their equilibrium structure, and by electronic excitations. These energetic **degrees of freedom** are referred to as

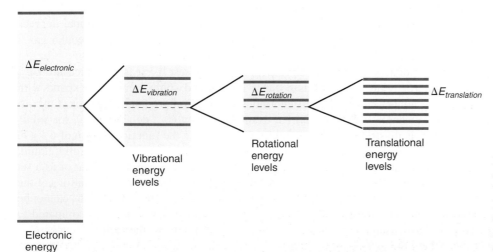

FIGURE 2.7
Energy levels are shown schematically for each degree of freedom. The gray area between electronic energy levels on the left indicates what appear to be a continuous range of allowed energies. However, as the energy scale is magnified stepwise, discrete energy levels for vibration, rotation, and translation can be resolved.

⊥E 2.2 Energy Level pacings for Different Degrees of Freedom

Degree of Freedom	Energy Level Spacing
Electronic	5×10^{-19} J
Vibration	2×10^{-20} J
Rotation	2×10^{-23} J
Translation	2×10^{-41} J

translation, rotation, vibration, and electronic excitation. Each of the degrees of freedom has its own set of energy levels and the probability of an individual molecule occupying a higher energy level increases as it gains energy. Except for translation, the energy levels for atoms and molecules are independent of the container size.

The amount of energy needed to move up the ladder of energy levels is very different for the different degrees of freedom: $\Delta E_{electronic} \gg \Delta E_{vibration} \gg \Delta E_{rotation} \gg \Delta E_{translation}$. Values for these ΔE are molecule dependent, but order of magnitude numbers are shown in Table 2.2.

Energy is gained or lost by a molecule through collisions with other molecules. An order of magnitude estimate of the energy that can be gained or lost by a molecule in a collision is $k_B T$, where $k = R/N_A$ is the Boltzmann constant, and T is the absolute temperature. A given degree of freedom in a molecule can only take up energy through molecular collisions if the spacing between adjacent energy levels and the temperature satisfies the relationship $\Delta E \approx k_B T$, which has the value 4.1×10^{-21} J at 300 K. At 300 K, $\Delta E \approx k_B T$ is always satisfied for translation and rotation, but not for vibration and electronic excitation. We formulate the following general rule relating the heat capacity $C_{V,m}$ and the degrees of freedom in a molecule, which will be discussed in more detail in Chapter 32.

> The heat capacity $C_{V,m}$ for a gas at temperature T not much lower than 300 K is $R/2$ for each translation and rotational degree of freedom, where R is the ideal gas constant. Each vibrational degree of freedom for which the relation $\Delta E/kT < 0.1$ is obeyed contributes R to $C_{V,m}$. If $\Delta E/k_B T > 10$, the degree of freedom does not contribute to $C_{V,m}$. For $10 > \Delta E/k_B T > 0.1$, the degree of freedom contributes partially to $C_{V,m}$.

Figure 2.8 shows the variation of $C_{V,m}$ for a monatomic gas and several molecular gases. Atoms only have three translational degrees of freedom. Linear molecules have an additional 2 rotational degrees of freedom and $3n$-5 vibrational degrees of freedom where n is the number of atoms in the molecule. Nonlinear molecules have 3 translational degrees of freedom, 3 rotational degrees of freedom, and $3n$-6 vibrational degrees of freedom. A He atom has only 3 translational degrees of freedom, and all electronic transitions are of high energy compared to kT. Therefore, $C_{V,m} = 3R/2$ over the entire temperature range as shown in the figure. CO is a linear diatomic molecule that has two rotational degrees of freedom for which $\Delta E/k_B T < 0.1$ at 200. K. Therefore, $C_{V,m} = 5R/2$ at 200. K. The single vibrational degree of freedom begins to contribute to $C_{V,m}$ above 200. K, but does not contribute fully for $T < 1000$. K because $10 > \Delta E/k_B T$ below 1000. K. CO_2 has 4 vibrational degrees of freedom, some of which contribute to $C_{V,m}$ near 200. K. However, $C_{V,m}$ does not reach its maximum value of $13R/2$ below 1000. K. Similarly, $C_{V,m}$ for C_2H_4, which has 12 vibrational degrees of freedom, does not reach its maximum value of 15 R below 1000. K, because $10 > \Delta E/k_B T$ for some vibrational degrees of freedom. Electronic energy levels are too far apart for any of the molecular gases to give a contribution to $C_{V,m}$.

To this point, we have only discussed $C_{V,m}$ for gases. It is easier to measure $C_{P,m}$ than $C_{V,m}$ for liquids and solids because liquids and solids generally expand with increasing temperature and exert enormous pressure on a container at constant volume (see Example Problem 3.2.) An example of how $C_{P,m}$ depends on T for solids and liquids is illustrated in Figure 2.9 for Cl_2. To make the functional form of $C_{P,m}(T)$ understandable, we briefly discuss the relative magnitudes of $C_{P,m}$ in the solid, liquid, and gaseous phases using a molecular level model. A solid can be thought of as a set of interconnected harmonic oscillators, and heat uptake leads to the excitations of the collective vibrations of the solid. At very low temperatures these vibrations cannot be activated, because the spacing of the vibrational energy levels is large compared to $k_B T$. As a consequence, energy cannot be taken up by the solid. Hence, $C_{P,m}$ approaches zero as T approaches zero. For the solid, $C_{P,m}$ rises rapidly with T because the thermal energy available as T increases is sufficient to activate the vibrations of

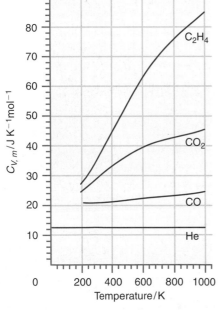

FIGURE 2.8
Molar heat capacities $C_{V,m}$ are shown for a number of gases. Atoms have only translational degrees of freedom and, therefore, have comparatively low values for $C_{V,m}$ that are independent of temperature. Molecules with vibrational degrees of freedom have higher values of $C_{V,m}$ at temperatures sufficiently high to activate the vibrations.

the solid. The heat capacity increases discontinuously as the solid melts to form a liquid. This is the case because the liquid retains all the local vibrational modes of the solid, and more low-energy modes become available upon melting. Therefore, the heat capacity of the liquid is greater than that of the solid. As the liquid vaporizes, the local vibrational modes present in the liquid are converted to translations that cannot take up as much energy as vibrations. Therefore, $C_{P,m}$ decreases discontinuously at the vaporization temperature. The heat capacity in the gaseous state increases slowly with temperature as the vibrational modes of the individual molecules are activated as discussed previously. These changes in $C_{P,m}$ can be calculated for a specific substance using a microscopic model and statistical thermodynamics, as will be discussed in detail in Chapter 32.

Once the heat capacity of a variety of different substances has been determined, we have a convenient way to quantify heat flow. For example, at constant pressure, the heat flow between the system and surroundings can be written as

$$q_P = \int_{T_{sts,i}}^{T_{sys,f}} C_P^{system}(T)dT = -\int_{T_{surr,i}}^{T_{surr,f}} C_P^{surroundings}(T)dT \qquad (2.10)$$

By measuring the temperature change of a thermal reservoir in the surroundings at constant pressure, q_P can be determined. In Equation (2.10), the heat flow at constant pressure has been expressed both from the perspective of the system and from the perspective of the surroundings. A similar equation can be written for a constant volume process. Water is a convenient choice of material for a heat bath in experiments because C_P is nearly constant at the value 4.18 J g^{-1} K^{-1} or 75.3 J mol^{-1} K^{-1} over the range from 0°C to 100.°C.

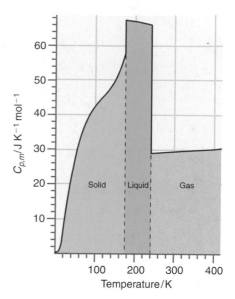

FIGURE 2.9
The variation of $C_{P,m}$ with temperature is shown for Cl_2.

EXAMPLE PROBLEM 2.3

The volume of a system consisting of an ideal gas decreases at constant pressure. As a result, the temperature of a 1.50 kg water bath in the surroundings increases by 14.2°C. Calculate q_P for the system.

Solution

$$q_P = \int_{T_{surr,i}}^{T_{surr,f}} C_P^{surroundings}(T)dT = -C_P^{surroundings}\Delta T$$

$$= -1.50 \text{ kg} \times 4.18 \text{ J g}^{-1}\text{ K}^{-1} \times 14.2 \text{ K} = -89.1 \text{ kJ}$$

How are C_P and C_V related for a gas? Consider the processes shown in Figure 2.10 in which a fixed amount of heat flows from the surroundings into a gas. In the constant pressure process, the gas expands as its temperature increases. Therefore, the system does work on the surroundings. As a consequence, not all the heat flow into the system can be used to increase ΔU. No such work occurs for the corresponding constant volume process, and all the heat flow into the system can be used to increase ΔU. Therefore, $dT_P < dT_V$ for the same heat flow $đq$. For this reason, $C_P > C_V$ for gases.

The same argument applies to liquids and solids as long as V increases with T. Nearly all substances follow this behavior, although notable exceptions occur, such as liquid water between 0°C and 4°C, for which the volume increases as T decreases. However, because ΔV_m upon heating is much smaller than for a gas, the difference between $C_{P,m}$ and $C_{V,m}$ for a liquid or solid is much smaller than for a gas.

Constant pressure heating

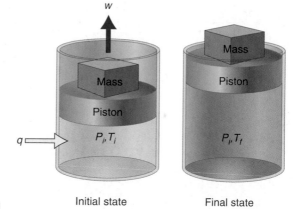

Initial state   Final state

Constant volume heating

Initial state   Final state

FIGURE 2.10
Not all the heat flow into the system can be used to increase ΔU in a constant pressure process, because the system does work on the surroundings as it expands. However, no work is done for constant volume heating.

The preceding remarks about the difference between C_P and C_V have been qualitative in nature. However, the following quantitative relationship, which will be proved in Chapter 3, holds for an ideal gas:

$$C_P - C_V = nR \quad \text{or} \quad C_{P,m} - C_{V,m} = R \tag{2.11}$$

2.6 State Functions and Path Functions

An alternate statement of the first law is that ΔU is independent of the path connecting the initial and final states, and depends only on the initial and final states. We make this statement plausible for the kinetic energy, and the argument can be extended to the other forms of energy listed in Section 2.1. Consider a single molecule in the system. Imagine that the molecule of mass m initially has the speed v_1. We now change its speed incrementally following the sequence $v_1 \rightarrow v_2 \rightarrow v_3 \rightarrow v_4$. The change in the kinetic energy along this sequence is given by

$$\begin{aligned}
\Delta E_{kinetic} &= \left(\frac{1}{2} m(v_2)^2 - \frac{1}{2} m(v_1)^2 \right) + \left(\frac{1}{2} m(v_3)^2 - \frac{1}{2} m(v_2)^2 \right) \\
&\quad + \left(\frac{1}{2} m(v_4)^2 - \frac{1}{2} m(v_3)^2 \right) \\
&= \left(\frac{1}{2} m(v_4)^2 - \frac{1}{2} m(v_1)^2 \right)
\end{aligned} \tag{2.12}$$

Even though v_2 and v_3 can take on any arbitrary values, they still do not influence the result. We conclude that the change in the kinetic energy depends only on the initial and final speed and that it is independent of the path between these values. Our conclusion remains the same if we increase the number of speed increments in the interval between v_1 and v_2 to an arbitrarily large number. Because this conclusion holds for all molecules in the system, it also holds for ΔU.

This example supports the assertion that ΔU depends only on the final and initial states and not on the path connecting these states. Any function that satisfies this condition is called a **state function,** because it depends only on the state of the system and not the path taken to reach the state. This property can be expressed in a mathematical form. Any state function, for example U, must satisfy the equation

$$\Delta U = \int_i^f dU = U_f - U_i \tag{2.13}$$

where i and f denote the initial and final states. This equation states that in order for ΔU to depend only on the initial and final states characterized here by i and f, the value of the integral must be independent of the path. If this is the case, U can be expressed as an infinitesimal quantity, dU, that when integrated, depends only on the initial and final states. The quantity dU is called an **exact differential.** We defer a discussion of exact differentials to Chapter 3.

It is useful to define a cyclic integral, denoted by the symbol $\oint$, as applying to a **cyclic path** such that the initial and final states are identical. For U or any other state function,

$$\oint dU = U_f - U_f = 0 \tag{2.14}$$

because the initial and final states are the same in a cyclic process.

We next show that q and w are not state functions. The state of a single-phase system at fixed composition is characterized by any two of the three variables P, T, and V. The same is true of U. Therefore, for a system of fixed mass, U can be written in any of the three forms $U(V,T)$, $U(P,T)$, or $U(P,V)$. Imagine that a gas characterized by V_1 and T_1 is confined in a piston and cylinder system that is isolated from the surroundings. There is a thermal reservoir in the surroundings at a temperature $T_3 < T_1$. We do compression work on the system starting from an initial state in which the volume is V_1 to an intermediate state in which the volume is V_2 using a constant external pressure $P_{external}$ where $V_2 < V_1$. The work is given by

$$w = -\int_{V_i}^{V_f} P_{external}\, dV = -P_{external}\int_{V_i}^{V_f} dV = -P_{external}(V_f - V_i) = -P_{external}\Delta V \quad (2.15)$$

Because work is done on the system in the compression (see Figure 2.11), w is positive and U increases. Because the system consists of a uniform single phase, U is a monotonic function of T, and T also increases. The change in volume ΔV has been chosen such that the temperature of the system T_2 in the intermediate state after the compression satisfies the inequality $T_1 < T_2 < T_3$.

We next lock the piston in place and let an amount of heat q flow between the system and surroundings at constant V by bringing the system into contact with the reservoir at temperature T_3. The final state values of T and V after these two steps are T_3 and V_2.

This two-step process is repeated for different values of the external pressure by changing the mass resting on the piston. In each case the system is in the same final state characterized by the variables V_2 and T_3. The sequence of steps that takes the system from the initial state V_1, T_1 to the final state V_2, T_3 is referred to as a **path**. By changing the mass, a set of different paths is generated, all of which originate from the state V_1, T_1, and end in the state V_2, T_3. According to the first law, ΔU for this two-step process is

$$\Delta U = U(T_3, V_2) - U(T_1, V_1) = q + w \quad (2.16)$$

Because ΔU is a state function, its value for the two-step process just described is the same for each of the different values of the mass.

Are q and w also state functions? For this two step process,

$$w = -P_{external}\Delta V \quad (2.17)$$

and $P_{external}$ is different for each value of the mass or for each path, whereas ΔV is constant. Therefore, w is also different for each path; we can choose one path from V_1, T_1 to V_2, T_3 and a different path from V_2, T_3 back to V_1, T_1. Because the work is different along these paths, the cyclic integral of work is not equal to zero. Therefore, w is not a state function.

Using the first law to calculate q for each of the paths, we obtain the result

$$q = \Delta U - w = \Delta U + P_{external}\Delta V \quad (2.18)$$

Because ΔU is the same for each path, and w is different for each path, we conclude that q is also different for each path. Just as for work, the cyclic integral of heat is not equal to zero. Therefore, neither q nor w are state functions, and they are called **path functions**.

Because both q and w are path functions, there are no exact differentials for work and heat unless the path is specified. Incremental amounts of these quantities are denoted by $đq$ and $đw$, rather than dq and dw, to emphasize the fact that incremental amounts of work and heat are not exact differentials. Because $đq$ and $đw$ are not exact differentials, there are no such quantities as Δq, q_f, q_i and Δw, w_f, w_i. One cannot refer to the work or heat possessed by a system or to the change in work or heat associated with a process. After a transfer of heat and/or work between the system and surroundings is completed, the system and surroundings possess internal energy, but not heat or work.

The preceding discussion emphasizes that it is important to use the terms *work* and *heat* in a way that reflects the fact that they are not state functions. Examples of systems of interest to us are batteries, fuel cells, refrigerators, and internal combustion engines. In each case, the utility of these systems is that work and/or heat flows between the system and surroundings. For example, in a refrigerator, electrical energy is used to extract heat from the inside of the device and to release it in the surroundings. One can speak of the refrigerator as having the capacity to extract heat, but it would be wrong to speak of it as having heat. In the internal combustion engine, chemical energy contained in the bonds of the fuel molecules and in O_2 is released in forming CO_2 and H_2O. This change in internal energy can be used to rotate the wheels of the vehicle, thereby inducing a flow of work between the vehicle and the surroundings. One can refer to the capability of the engine to do work, but it would be incorrect to refer to the vehicle or the engine as containing or having work.

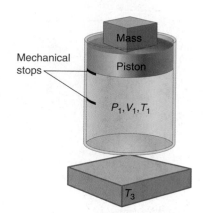

Mechanical stops

Initial state

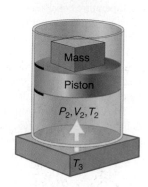

Intermediate state

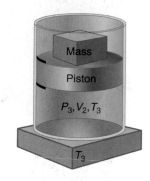

Final state

FIGURE 2.11

A system consisting of an ideal gas is contained in a piston and cylinder assembly. An external pressure is generated by the mass resting on the piston. The gas in the initial state V_1, T_1 is compressed to an intermediate state, whereby the temperature increases to the value T_2. It is then brought into contact with a thermal reservoir at T_3, leading to a further rise in temperature. The final state is V_2, T_3. The mechanical stops allow the system volume to be only V_1 or V_2.

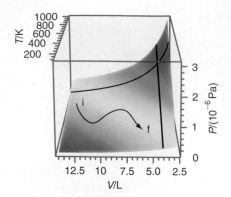

FIGURE 2.12

All combinations of pressure, volume, and temperature consistent with 1 mol of an ideal gas lie on the colored surface. All combinations of pressure and volume consistent with $T = 800$ K and all combinations of pressure and temperature consistent with a volume of 4.0 L are shown as black curves that lie in the P–V–T surface. The third curve corresponds to a path between an initial state i and a final state f that is neither a constant temperature nor a constant volume path.

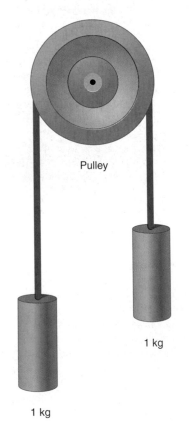

FIGURE 2.13

Two masses of exactly 1 kg each are connected by a wire of zero mass running over a frictionless pulley. The system is in mechanical equilibrium and the masses are stationary.

2.7 Equilibrium, Change, and Reversibility

Thermodynamics can only be applied to systems in internal equilibrium, and a requirement for equilibrium is that the overall rate of change of all processes such as diffusion or chemical reaction be zero. How do we reconcile these statements with our calculations of q, w, and ΔU associated with processes in which there is a macroscopic change in the system? To answer this question, it is important to distinguish between the system and surroundings each being in internal equilibrium, and the system and surroundings being in equilibrium with one another.

We first discuss the issue of internal equilibrium. Consider a system made up of an ideal gas, which satisfies the equation of state, $P = nRT/V$. All combinations of P, V, and T consistent with this equation of state form a surface in P–V–T space as shown in Figure 2.12. All points on the surface correspond to states of internal equilibrium, meaning that the system is uniform on all length scales and is characterized by single values of T, P, and concentration. Points that are not on the surface do not correspond to any equilibrium state of the system because the equation of state is not satisfied. Nonequilibrium situations cannot be represented on such a plot, because T, P, and concentration do not have unique values for a system that is not in equilibrium. An example of a system that is not in internal equilibrium is a gas that is expanding so rapidly that different regions of the gas have different values for the density, pressure, and temperature.

Next, consider a process in which the system changes from an initial state characterized by P_i, V_i, and T_i to a final state characterized by P_f, V_f, and T_f as shown in Figure 2.12. If the rate of change of the macroscopic variables is negligibly small, the system passes through a succession of states of internal equilibrium as it goes from the initial to the final state. Such a process is called a **quasi-static process,** in which internal equilibrium is maintained in the system. If the rate of change is large, the rates of diffusion and intermolecular collisions may not be high enough to maintain the system in a state of internal equilibrium. Thermodynamic calculations for such a process are valid only if it is meaningful to assign a single value of the macroscopic variables P, V, T, and concentration to the system undergoing change. The same considerations hold for the surroundings. We only consider quasi-static processes in this text.

We now visualize a process in which the system undergoes a major change in terms of a directed path consisting of a sequence of quasi-static processes, and distinguish between two very important classes of quasi-static processes, namely reversible and irreversible processes. It is useful to consider the mechanical system shown in Figure 2.13 when discussing reversible and irreversible processes. Because the two masses have the same value, the net force acting on each end of the wire is zero, and the masses will not move. If an additional mass is placed on either of the two masses, the system is no longer in mechanical equilibrium, and the masses will move. In the limit in which the incremental mass approaches zero, the velocity at which the initial masses move approaches zero. In this case, one refers to the process as being **reversible,** meaning that the direction of the process can be reversed by placing the infinitesimal mass on the other side of the pulley.

Reversibility in a chemical system can be illustrated by a system consisting of liquid water in equilibrium with gaseous water that is surrounded by a thermal reservoir. The system and surroundings are both at temperature T. An infinitesimally small increase in T results in a small increase of the amount of water in the gaseous phase, and a small decrease in the liquid phase. An equally small decrease in the temperature has the opposite effect. Therefore, fluctuations in T give rise to corresponding fluctuations in the composition of the system. If an infinitesimal opposing change in the variable that drives the process (temperature in this case) causes a reversal in the direction of the process, the process is reversible.

If an infinitesimal change in the driving variable does not change the direction of the process, one says that the process is **irreversible.** For example, if a large stepwise temperature increase is induced in the system using a heat pulse, the amount of water in the gas phase increases abruptly. In this case, the composition of the system cannot be returned to its initial value by an infinitesimal temperature decrease. This relationship is characteristic of an irreversible process. Although any process that takes place at a rapid rate in the real world is irreversible, real processes can approach reversibility in the appropriate limit. For example, a slow increase in the electrical potential in an electrochemical cell can convert reactants to products in a nearly reversible process.

2.8 Comparing Work for Reversible and Irreversible Processes

We concluded in Section 2.6 that w is not a state function and that the work associated with a process is path dependent. This statement can be put on a quantitative footing by comparing the work associated with the reversible and irreversible expansion and the compression of an ideal gas. This process is discussed next and illustrated in Figure 2.14.

Consider the following irreversible process, meaning that the internal and external pressures are not equal. A quantity of an ideal gas is confined in a cylinder with a weightless movable piston. The walls of the system are diathermal, allowing heat to flow between the system and surroundings. Therefore, the process is isothermal at the temperature of the surroundings, T. The system is initially defined by the variables T, P_1, and V_1. The position of the piston is determined by $P_{external} = P_1$, which can be changed by adding or removing weights from the piston. Because the weights are moved horizontally, no work is done in adding or removing them. The gas is first expanded at constant temperature by decreasing $P_{external}$ abruptly to the value P_2 (weights are removed), where $P_2 < P_1$. A sufficient amount of heat flows into the system through the diathermal walls to keep the temperature at the constant value T. The system is now in the state defined by T, P_2, and V_2, where $V_2 > V_1$. The system is then returned to its original state in an **isothermal process** by increasing $P_{external}$ abruptly to its original value P_1 (weights are added). Heat flows out of the system into the surroundings in this step. The system has been restored to its original state and, because this is a cyclic process, $\Delta U = 0$. Are q_{total} and w_{total} also zero for the cyclic process? The total work associated with this cyclic process is given by the sum of the work for each individual step:

$$w_{total} = \sum_i - P_{external,i}\Delta V_i = w_{expansion} + w_{compression}$$

$$= -P_2(V_2 - V_1) - P_1(V_1 - V_2)$$

$$= -(P_2 - P_1) \times (V_2 - V_1) > 0 \quad \text{because } P_2 < P_1 \quad \text{and } V_2 > V_1 \quad \textbf{(2.19)}$$

The relationship between P and V for the process under consideration is shown graphically in Figure 2.14, in what is called an **indicator diagram.** An indicator diagram is useful because the work done in the expansion and contraction steps can be evaluated from the appropriate area in the figure, which is equivalent to evaluating the integral $w = -\int P_{external}\,dV$. Note that the work done in the expansion is negative because $\Delta V > 0$, and that done in the compression is positive because $\Delta V < 0$. Because $P_2 < P_1$, the magnitude of the work done in the compression process is more than that done in the expansion process and $w_{total} > 0$. What can one say about q_{total}? The first law states that because $\Delta U = q_{total} + w_{total} = 0$, $q_{total} < 0$.

The same cyclical process is carried out in a reversible cycle. A necessary condition for reversibility is that $P = P_{external}$ at every step of the cycle. This means that P changes during the expansion and compression steps. The work associated with the expansion is

$$w_{expansion} = -\int P_{external}\,dV = -\int P\,dV = -nRT\int \frac{dV}{V} = -nRT\ln\frac{V_2}{V_1} \quad \textbf{(2.20)}$$

This work is shown schematically as the red area in the indicator diagram of Figure 2.15.

If this process is reversed and the compression work is calculated, the following result is obtained:

$$w_{compression} = -nRT\ln\frac{V_1}{V_2} \quad \textbf{(2.21)}$$

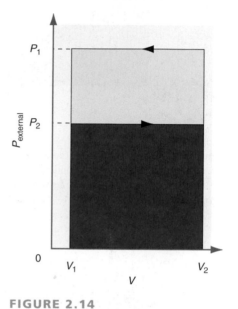

FIGURE 2.14
The work for each step and the total work can be obtained from an indicator diagram. For the compression step, w is given by the total area in red and yellow; for the expansion step, w is given by the red area. The arrows indicate the direction of change in V in the two steps. The sign of w is opposite for these two processes. The total work in the cycle is the yellow area.

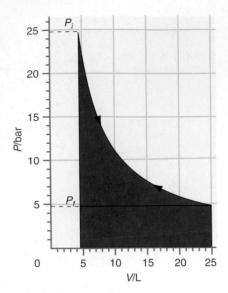

FIGURE 2.15
Indicator diagram for a reversible process. Unlike Figure 2.14, the areas under the P–V curves are the same in the forward and reverse directions.

The magnitudes of the work in the forward and reverse processes are equal. The total work done in this cyclical process is given by

$$w = w_{expansion} + w_{compression} = -nRT \ln \frac{V_2}{V_1} - nRT \ln \frac{V_1}{V_2}$$

$$= -nRT \ln \frac{V_2}{V_1} + nRT \ln \frac{V_2}{V_1} = 0 \qquad (2.22)$$

Therefore, the work done in a reversible isothermal cycle is zero. Because $\Delta U = q + w$ is a state function, $q = -w = 0$ for this reversible isothermal process. Looking at the heights of the weights in the surroundings at the end of the process, we find that they are the same as at the beginning of the process. To compare the work for reversible and irreversible processes, the state variables need to be given specific values as is done in Example Problem 2.4.

EXAMPLE PROBLEM 2.4

In this example, 2.00 mol of an ideal gas undergoes isothermal expansion along three different paths: (1) reversible expansion from $P_i = 25.0$ bar and $V_i = 4.50$ L to $P_f = 4.50$ bar, (2) a single-step irreversible expansion against a constant external pressure of 4.50 bar, and (3) a two-step irreversible expansion consisting initially of an expansion against a constant external pressure of 11.0 bar until $P = P_{external}$, followed by an expansion against a constant external pressure of 4.50 bar until $P = P_{external}$.

Calculate the work for each of these processes. For which of the irreversible processes is the magnitude of the work greater?

Solution

The processes are depicted in the following indicator diagram:

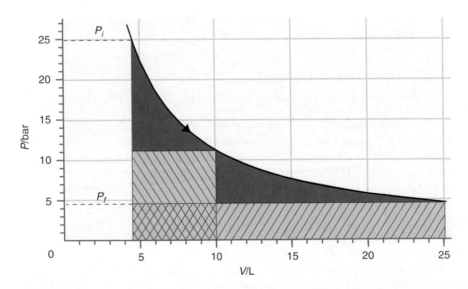

We first calculate the constant temperature at which the process is carried out, the final volume, and the intermediate volume in the two-step expansion:

$$T = \frac{P_i V_i}{nR} = \frac{25.0 \text{ bar} \times 4.50 \text{ L}}{8.314 \times 10^{-2} \text{ L bar mol}^{-1} \text{ K}^{-1} \times 2.00 \text{ mol}} = 677 \text{ K}$$

$$V_f = \frac{nRT}{P_f} = \frac{8.314 \times 10^{-2} \text{ L bar mol}^{-1} \text{ K}^{-1} \times 2.00 \text{ mol} \times 677 \text{ K}}{4.50 \text{ bar}} = 25.0 \text{ L}$$

$$V_{int} = \frac{nRT}{P_{int}} = \frac{8.314 \times 10^{-2} \text{ L bar mol}^{-1} \text{ K}^{-1} \times 2.00 \text{ mol} \times 677 \text{ K}}{11.0 \text{ bar}} = 10.2 \text{ L}$$

The work of the reversible process is given by

$$w = -nRT_1 \ln \frac{V_f}{V_i}$$

$$= -2.00 \text{ mol} \times 8.314 \text{ J mol}^{-1} \text{ K}^{-1} \times 677 \text{ K} \times \ln \frac{25.0 \text{ L}}{4.50 \text{ L}} = -19.3 \times 10^3 \text{ J}$$

We next calculate the work of the single-step and two-step irreversible processes:

$$w_{single} = -P_{external}\Delta V = -4.50 \text{ bar} \times \frac{10^5 \text{ Pa}}{\text{bar}} \times (25.0 \text{ L} - 4.50 \text{ L}) \times \frac{10^{-3} \text{ m}^3}{\text{L}}$$

$$= -9.23 \times 10^3 \text{ J}$$

$$w_{two\text{-}step} = -P_{external}\Delta V = -11.0 \text{ bar} \times \frac{10^5 \text{ Pa}}{\text{bar}} \times (10.2 \text{ L} - 4.50 \text{ L}) \times \frac{10^{-3} \text{ m}^3}{\text{L}}$$

$$-4.50 \text{ bar} \times \frac{10^5 \text{ Pa}}{\text{bar}} \times (25.0 \text{ L} - 10.2 \text{ L}) \times \frac{10^{-3} \text{ m}^3}{\text{L}}$$

$$= -12.9 \times 10^3 \text{ J}$$

The magnitude of the work is greater for the two-step process than for the single-step process, but less than that for the reversible process.

Example Problem 2.4 shows that the magnitude of w for the irreversible expansion is less than that for the reversible expansion, but also suggests that the magnitude of w for a multistep expansion process increases with the number of steps. This is indeed the case, as shown in Figure 2.16. Imagine that the number of steps n increases indefinitely.

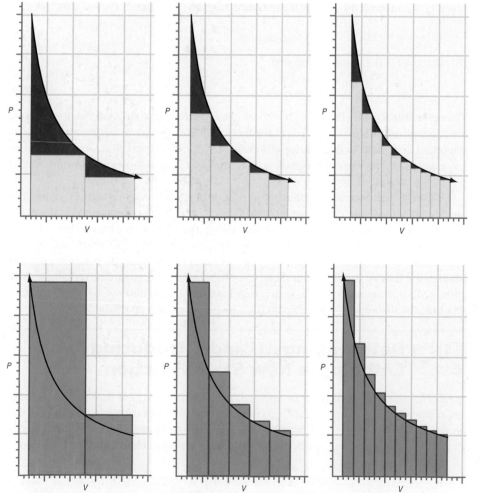

FIGURE 2.16
The work done in an expansion (red plus yellow areas) is compared with the work done in a multistep series of irreversible expansion processes at constant pressure (yellow area) in the top panel. The bottom panel shows analogous results for the compression, where the area under the black curve is the reversible compression work. Note that the total work done in the irreversible expansion and compression processes approaches that of the reversible process as the number of steps becomes large.

As n increases, the pressure difference $P_{external} - P$ for each individual step decreases. In the limit that $n \rightarrow \infty$, the pressure difference $P_{external} - P \rightarrow 0$, and the total area of the rectangles in the indicator diagram approaches the area under the reversible curve. In this limit, the irreversible process becomes reversible and the value of the work equals that of the reversible process.

By contrast, the magnitude of the irreversible compression work exceeds that of the reversible process for finite values of n and becomes equal to that of the reversible process as $n \rightarrow \infty$. The difference between the expansion and compression processes results from the requirement that $P_{external} < P$ at the beginning of each expansion step, whereas $P_{external} > P$ at the beginning of each compression step.

On the basis of these calculations for the reversible and irreversible cycles, we introduce another criterion to distinguish between reversible and irreversible processes. Suppose that a system undergoes a change through one or more individual steps, and that the system is restored to its initial state by following the same steps in reverse order. The system is restored to its initial state because the process is cyclical. If the surroundings are also returned to their original state (all masses at the same height and all reservoirs at their original temperatures), the process is reversible. If the surroundings are not restored to their original state, the process is irreversible.

We are often interested in extracting work from a system. For example, it is the expansion of the fuel–air mixture in an automobile engine upon ignition that provides the torque that eventually drives the wheels. Is the capacity to do work similar for reversible and irreversible processes? This question is answered using the indicator diagrams of Figures 2.14 and 2.15 for the specific case of isothermal expansion work, noting that the work can be calculated from the area under the P–V curve. We compare the work for expansion from V_1 to V_2 in a single stage at constant pressure to that for the reversible case. For the single-stage expansion, the constant external pressure is given by $P_{external} = nRT/V_2$. However, if the expansion is carried out reversibly, the system pressure is always greater than this value. By comparing the areas in the indicator diagrams of Figure 2.16, it is seen that

$$|w_{reversible}| \geq |w_{irreversible}| \tag{2.23}$$

By contrast, for the compression step,

$$|w_{reversible}| \leq |w_{irreversible}| \tag{2.24}$$

The reversible work is the lower bound for the compression work and the upper bound for the expansion work. This result for the expansion work can be generalized to an important statement that holds for all forms of work: *The maximum work that can be extracted from a process between the same initial and final states is that obtained under reversible conditions.*

Although the preceding statement is true, it suggests that it would be optimal to operate an automobile engine under conditions in which the pressure inside the cylinders differs only infinitesimally from the external atmospheric pressure. This is clearly not possible. A practical engine must generate torque on the drive shaft, and this can only occur if the cylinder pressure is appreciably greater than the external pressure. Similarly, a battery is only useful if one can extract a sizable rather than an infinitesimal current. To operate such devices under useful irreversible conditions, the work output is less than the theoretically possible limit set by the reversible process.[3]

2.9 Determining ΔU and Introducing Enthalpy, a New State Function

Measuring the energy taken up or released in a chemical reaction is of particular interest to chemists. How can the ΔU for a thermodynamic process be measured? This will be the topic of Chapter 4. However, this topic is briefly discussed here in order to enable you to carry out calculations on ideal gas systems in the end-of-chapter

---

[3]For a more detailed discussion of irreversible work, see D. Kivelson and I. Oppenheim, "Work in Irreversible Expansions," *Journal of Chemical Education* 43 (1966): 233.

problems. The first law states that $\Delta U = q + w$. Imagine that the process is carried out under constant volume conditions and that nonexpansion work is not possible. Because under these conditions $w = -\int P_{external}dV = 0$,

$$\Delta U = q_V \tag{2.25}$$

Equation (2.25) states that ΔU can be experimentally determined by measuring the heat flow between the system and surroundings in a constant volume process.

Chemical reactions are generally carried out under constant pressure rather than constant volume conditions. It would be useful to have an energy state function that has a relationship analogous to Equation (2.25), but at constant pressure conditions. Under constant pressure conditions, we can write

$$dU = d\hspace{-0.3em}\bar{}\,q_P - P_{external}dV = d\hspace{-0.3em}\bar{}\,q_P - P\,dV \tag{2.26}$$

Integrating this expression between the initial and final states:

$$\int_i^f dU = U_f - U_i = \int d\hspace{-0.3em}\bar{}\,q_P - \int P\,dV = q_P - P(V_f - V_i)$$

$$= q_P - \left(P_f V_f - P_i V_i\right) \tag{2.27}$$

Note that in order to evaluate the integral involving P, we must know the functional relationship $P(V)$, which in this case is $P_i = P_f = P$ where P is constant. Rearranging the last equation, we obtain

$$(U_f + P_f V_f) - (U_i + P_i V_i) = q_P \tag{2.28}$$

Because P, V, and U are all state functions, $U + PV$ is a state function. This new state function is called **enthalpy** and is given the symbol H.

$$H \equiv U + PV \tag{2.29}$$

As is the case for U, H has the units of energy, and it is an extensive property. As shown in Equation (2.30), ΔH for a process involving only P–V work can be determined by measuring the heat flow between the system and surroundings at constant pressure:

$$\Delta H = q_P \tag{2.30}$$

This equation is the constant pressure analogue of Equation (2.25). Because chemical reactions are much more frequently carried out at constant P than constant V, the energy change measured experimentally by monitoring the heat flow is ΔH rather than ΔU. When we classify a reaction as being exothermic or endothermic, we are talking about ΔH, not ΔU.

2.10 Calculating q, w, ΔU, and ΔH for Processes Involving Ideal Gases

In this section we discuss how ΔU and ΔH, as well as q and w, can be calculated from the initial and final state variables if the path between the initial and final state is known. The problems at the end of this chapter ask you to calculate q, w, ΔU, and ΔH for simple and multistep processes. Because an equation of state is often needed to carry out such calculations, the system will generally be an ideal gas. Using an ideal gas as a surrogate for more complex systems has the significant advantage that the mathematics is simplified, allowing one to concentrate on the process rather than the manipulation of equations and the evaluation of integrals.

What does one need to know to calculate ΔU? The following discussion is restricted to processes that do not involve chemical reactions or changes in phase. Because U is a state function, ΔU is independent of the path between the initial and final states. To describe a fixed amount of an ideal gas (i.e., n is constant), the values of two of the variables P, V, and T must be known. Is this also true for ΔU for processes involving ideal gases? To answer this question, consider the expansion of an ideal gas from an initial state V_1, T_1 to a final state V_2, T_2. We first assume that U is a function of both V and T. Is this assumption correct? Because ideal gas atoms or molecules do not interact with one another, U will not depend on the distance between the atoms or molecules. Therefore, U is not a function of V, and we conclude that ΔU must be a function of T only for an ideal gas, $\Delta U = \Delta U(T)$.

We also know that for a temperature range over which C_V is constant,

$$\Delta U = q_V = C_V(T_f - T_i) \qquad (2.31)$$

Is this equation only valid for constant V? Because U is a function of only T for an ideal gas, Equation (2.31) is also valid for processes involving ideal gases in which V is not constant. Therefore, if one knows C_V, T_1, and T_2, ΔU can be calculated, regardless of the path between the initial and final states.

How many variables are required to define ΔH for an ideal gas? We write

$$\Delta H = \Delta U(T) + \Delta(PV) = \Delta U(T) + \Delta(nRT) = \Delta H(T) \qquad (2.32)$$

We see that ΔH is also a function of only T for an ideal gas. In analogy to Equation (2.31),

$$\Delta H = q_P = C_P(T_f - T_i) \qquad (2.33)$$

Because ΔH is a function of only T for an ideal gas, Equation (2.33) holds for all processes involving ideal gases, whether P is constant or not, as long as it is reasonable to assume that C_P is constant. Therefore, if the initial and final temperatures are known or can be calculated, and if C_V and C_P are known, ΔU and ΔH can be calculated *regardless of the path* for processes involving ideal gases using Equations (2.31) and (2.33), as long as no chemical reactions or phase changes occur. Because U and H are state functions, the previous statement is true for both reversible and irreversible processes. Recall that for an ideal gas $C_P - C_V = nR$, so that if one of C_V and C_P is known, the other can be readily determined.

We next note that the first law links q, w, and ΔU. If any two of these quantities are known, the first law can be used to calculate the third. In calculating work, often only expansion work takes place. In this case one always proceeds from the equation

$$w = -\int P_{external} \, dV \qquad (2.34)$$

This integral can only be evaluated if the functional relationship between $P_{external}$ and V is known. A frequently encountered case is $P_{external} = $ constant, such that

$$w = -P_{external}(V_f - V_i) \qquad (2.35)$$

Because $P_{external} \neq P$, the work considered in Equation (2.35) is for an irreversible process.

A second frequently encountered case is that the system and external pressure differ only by an infinitesimal amount. In this case, it is sufficiently accurate to write $P_{external} = P$, and the process is reversible:

$$w = -\int \frac{nRT}{V} \, dV \qquad (2.36)$$

This integral can only be evaluated if T is known as a function of V. The most commonly encountered case is an isothermal process, in which T is constant. As was seen in Section 2.2, for this case

$$w = -nRT \int \frac{dV}{V} = -nRT \ln \frac{V_f}{V_i} \qquad (2.37)$$

In solving thermodynamic problems, it is very helpful to understand the process thoroughly before starting the calculation, because it is often possible to obtain the value of one or more of q, w, ΔU, and ΔH without a calculation. For example, $\Delta U = \Delta H = 0$ for an isothermal process because ΔU and ΔH depend only on T. For an adiabatic process, $q = 0$ by definition. If only expansion work is possible, $w = 0$ for a constant volume process. These guidelines are illustrated in the following two example problems.

EXAMPLE PROBLEM 2.5

A system containing 2.50 mol of an ideal gas for which $C_{V,m} = 20.79$ J mol^{-1} K^{-1} is taken through the cycle in the following diagram in the direction indicated by the arrows. The curved path corresponds to $PV = nRT$, where $T = T_1 = T_3$.

a. Calculate q, w, ΔU, and ΔH for each segment and for the cycle assuming that the heat capacity is independent of temperature.

b. Calculate q, w, ΔU, and ΔH for each segment and for the cycle in which the direction of each process is reversed.

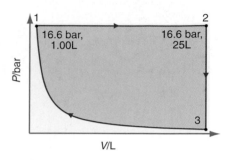

Solution

We begin by asking whether we can evaluate q, w, ΔU, or ΔH for any of the segments without any calculations. Because the path between states 1 and 3 is isothermal, ΔU and ΔH are zero for this segment. Therefore, from the first law, $q_{3\rightarrow1} = -w_{3\rightarrow1}$. For this reason, we only need to calculate one of these two quantities. Because $\Delta V = 0$ along the path between states 2 and 3, $w_{2\rightarrow3} = 0$. Therefore, $\Delta U_{2\rightarrow3} = q_{2\rightarrow3}$. Again, we only need to calculate one of these two quantities. Because the total process is cyclic, the change in any state function is zero. Therefore, $\Delta U = \Delta H = 0$ for the cycle, no matter which direction is chosen. We now deal with each segment individually.

Segment $1 \rightarrow 2$

The values of n, P_1 and V_1, and P_2 and V_2 are known. Therefore, T_1 and T_2 can be calculated using the ideal gas law. We use these temperatures to calculate ΔU as follows:

$$\Delta U_{1\rightarrow2} = nC_{V,m}(T_2 - T_1) = \frac{nC_{V,m}}{nR}(P_2V_2 - P_1V_1)$$

$$= \frac{20.79 \text{ J mol}^{-1}\text{K}^{-1}}{0.08314 \text{ L bar K}^{-1}\text{mol}^{-1}}$$

$$\times (16.6 \text{ bar} \times 25.0 \text{ L} - 16.6 \text{ bar} \times 1.00 \text{ L})$$

$$= 99.6 \text{ kJ}$$

The process takes place at constant pressure, so

$$w = -P_{external}(V_2 - V_1) = -16.6 \text{ bar} \times \frac{10^5 \text{ N m}^{-2}}{\text{bar}}$$

$$\times (25.0 \times 10^{-3} \text{ m}^3 - 1.00 \times 10^{-3} \text{ m}^3)$$

$$= -39.8 \text{ kJ}$$

Using the first law,

$$q = \Delta U - w = 99.6 \text{ kJ} + 39.8 \text{ kJ} = 139.4 \text{ kJ}$$

We next calculate T_2:

$$T_2 = \frac{P_2V_2}{nR} = \frac{16.6 \text{ bar} \times 25.0 \text{L}}{2.50 \text{ mol} \times 0.08314 \text{ L bar K}^{-1} \text{mol}^{-1}} = 2.00 \times 10^3 \text{ K}$$

We next calculate $T_3 = T_1$ and then $\Delta H_{1\rightarrow2}$:

$$T_1 = \frac{P_1V_1}{nR} = \frac{16.6 \text{ bar} \times 1.00 \text{L}}{2.50 \text{ mol} \times 0.08314 \text{ L bar K}^{-1} \text{mol}^{-1}} = 79.9 \text{ K}$$

$$\Delta H_{1\rightarrow2} = \Delta U_{1\rightarrow2} + \Delta(PV) = \Delta U_{1\rightarrow2} + nR(T_2 - T_1)$$

$$= 99.6 \times 10^3 \text{ J} + 2.5 \text{ mol} \times 8.314 \text{ J mol}^{-1}\text{K}^{-1}$$

$$\times (2000 \text{ K} - 79.9 \text{ K}) = 139.4 \text{ kJ}$$

Segment 2 → 3

As previously noted, $w = 0$, and

$$\Delta U_{2\to3} = q_{2\to3} = C_V(T_3 - T_2)$$
$$= 2.50\,\text{mol} \times 20.79\,\text{J mol}^{-1}\text{K}^{-1}(79.9\,\text{K} - 2000\,\text{K})$$
$$= -99.6\,\text{kJ}$$

The numerical result is equal in magnitude, but opposite in sign to $\Delta U_{1\to2}$ because $T_3 = T_1$. For the same reason, $\Delta H_{2\to3} = -\Delta H_{1\to2}$.

Segment 3 → 1

For this segment, $\Delta U_{3\to1} = 0$ and $\Delta H_{3\to1} = 0$ as noted earlier, and $w_{3\to1} = -q_{3\to1}$. Because this is a reversible isothermal compression,

$$w_{3\to1} = -nRT\ln\frac{V_1}{V_3} = -2.50\,\text{mol} \times 8.314\,\text{J mol}^{-1}\text{K}^{-1} \times 79.9\,\text{K}$$

$$\times \ln\frac{1.00 \times 10^{-3}\,\text{m}^3}{25.0 \times 10^{-3}\,\text{m}^3}$$

$$= 5.35\,\text{kJ}$$

The results for the individual segments and for the cycle in the indicated direction are given in the following table. If the cycle is traversed in the reverse fashion, the magnitudes of all quantities in the table remain the same, but all signs change.

Path	q (kJ)	w (kJ)	ΔU (kJ)	ΔH (kJ)
1 → 2	139.4	−39.8	99.6	139.4
2 → 3	−99.6	0	−99.6	−139.4
3 → 1	−5.35	5.35	0	0
Cycle	34.5	−34.5	0	0

EXAMPLE PROBLEM 2.6

In this example, 2.50 mol of an ideal gas with $C_{V,m} = 12.47$ J mol^{-1} K^{-1} is expanded adiabatically against a constant external pressure of 1.00 bar. The initial temperature and pressure of the gas are 325 K and 2.50 bar, respectively. The final pressure is 1.25 bar. Calculate the final temperature, q, w, ΔU, and ΔH.

Solution

Because the process is adiabatic, $q = 0$, and $\Delta U = w$. Therefore,

$$\Delta U = nC_{v,m}(T_f - T_i) = -P_{external}(V_f - V_i)$$

Using the ideal gas law,

$$nC_{v,m}(T_f - T_i) = -nRP_{external}\left(\frac{T_f}{P_f} - \frac{T_i}{P_i}\right)$$

$$T_f\left(nC_{v,m} + \frac{nRP_{external}}{P_f}\right) = T_i\left(nC_{v,m} + \frac{nRP_{external}}{P_i}\right)$$

$$T_f = T_i\frac{\left(C_{v,m} + \dfrac{RP_{external}}{P_i}\right)}{\left(C_{v,m} + \dfrac{RP_{external}}{P_f}\right)}$$

$$= 325\,\text{K} \times \left(\frac{12.47\,\text{J mol}^{-1}\,\text{K}^{-1} + \dfrac{8.314\,\text{J mol}^{-1}\,\text{K}^{-1} \times 1.00\,\text{bar}}{2.50\,\text{bar}}}{12.47\,\text{J mol}^{-1}\,\text{K}^{-1} + \dfrac{8.314\,\text{J mol}^{-1}\,\text{K}^{-1} \times 1.00\,\text{bar}}{1.25\,\text{bar}}}\right) = 268\,\text{K}$$

We calculate $\Delta U = w$ from

$$\Delta U = nC_{V,m}(T_f - T_i) = 2.5 \text{ mol} \times 12.47 \text{ J mol}^{-1}\text{K}^{-1} \times (268 \text{ K} - 325 \text{ K})$$

$$= -1.78 \text{ kJ}$$

Because the temperature falls in the expansion, the internal energy and enthalpy decreases:

$$\Delta H = \Delta U + \Delta(PV) = \Delta U + nR(T_2 - T_1)$$

$$= -1.78 \times 10^3 \text{ J} + 2.5 \text{ mol} \times 8.314 \text{ J mol}^{-1}\text{K}^{-1}$$

$$\times (268 \text{ K} - 325 \text{ K}) = -2.96 \text{ kJ}$$

2.11 The Reversible Adiabatic Expansion and Compression of an Ideal Gas

The adiabatic expansion and compression of gases is an important meteorological process. For example, the cooling of a cloud as it moves upward in the atmosphere can be modeled as an adiabatic process because the heat transfer between the cloud and the rest of the atmosphere is slow on the timescale of its upward motion.

Consider the adiabatic expansion of an ideal gas. Because $q = 0$, the first law takes the form

$$\Delta U = w \quad \text{or} \quad C_V dT = -P_{external}dV \tag{2.38}$$

For a reversible adiabatic process, $P = P_{external}$, and

$$C_V dT = -nRT\frac{dV}{V} \quad \text{or, equivalently,} \quad C_V\frac{dT}{T} = -nR\frac{dV}{V} \tag{2.39}$$

Integrating both sides of this equation between the initial and final states,

$$\int_{T_i}^{T_f} C_V \frac{dT}{T} = -nR \int_{V_i}^{V_f} \frac{dV}{V} \tag{2.40}$$

If C_V is constant over the temperature interval $T_f - T_i$, then

$$C_V \ln \frac{T_f}{T_i} = -nR \ln \frac{V_f}{V_i} \tag{2.41}$$

Because $C_P - C_V = nR$ for an ideal gas, Equation (2.41) can be written in the form

$$\ln\left(\frac{T_f}{T_i}\right) = -(\gamma - 1)\ln\left(\frac{V_f}{V_i}\right) \quad \text{or, equivalently,} \quad \frac{T_f}{T_i} = \left(\frac{V_f}{V_i}\right)^{1-\gamma} \tag{2.42}$$

where $\gamma = C_{P,m}/C_{V,m}$. Substituting $T_f/T_i = P_f V_f/P_i V_i$ in the previous equation, we obtain

$$P_i V_i^\gamma = P_f V_f^\gamma \tag{2.43}$$

for the adiabatic reversible expansion or compression of an ideal gas. Note that our derivation is only applicable to a reversible process, because we have assumed that $P = P_{external}$.

Reversible adiabatic compression of a gas leads to heating, and reversible adiabatic expansion leads to cooling. Adiabatic and isothermal expansion and compression are compared in Figure 2.17, in which two systems containing 1 mol of an ideal gas have the same volume at $P = 1$ atm. One system undergoes adiabatic compression or expansion, and the other undergoes isothermal compression or expansion. Under isothermal conditions, heat flows out of the system as it is compressed to $P > 1$ atm, and heat flows into the system as it is expanded to $P < 1$ atm to keep T constant. Because no heat flows into or out of the system under adiabatic conditions, its temperature increases in compression and decreases in expansion. Because $T > T_{isothermal}$ for a compression starting at 1 atm, $P_{adiabatic} > P_{isothermal}$ for a given volume of

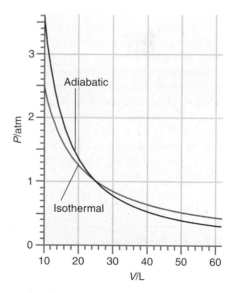

FIGURE 2.17

Two systems containing 1 mol of N_2 have the same P and V values at 1 atm. The red curve corresponds to reversible expansion and compression under adiabatic conditions. The blue curve corresponds to reversible expansion and compression under isothermal conditions.

the gas. Similarly, in a reversible adiabatic expansion originating at 1 atm, $P_{adiabatic} < P_{isothermal}$ for a given volume of the gas.

EXAMPLE PROBLEM 2.7

A cloud mass moving across the ocean at an altitude of 2000. m encounters a coastal mountain range. As it rises to a height of 3500. m to pass over the mountains, it undergoes an adiabatic expansion. The pressure at 2000. m and 3500. m is 0.802 and 0.602 atm, respectively. If the initial temperature of the cloud mass is 288 K, what is the cloud temperature as it passes over the mountains? Assume that $C_{P,m}$ for air is 28.86 J K^{-1} mol^{-1} and that air obeys the ideal gas law. If you are on the mountain, should you expect rain or snow?

Solution

$$\ln\left(\frac{T_f}{T_i}\right) = -(\gamma - 1)\ln\left(\frac{V_f}{V_i}\right)$$

$$= -(\gamma - 1)\ln\left(\frac{T_f}{T_i}\frac{P_i}{P_f}\right) = -(\gamma - 1)\ln\left(\frac{T_f}{T_i}\right) - (\gamma - 1)\ln\left(\frac{P_i}{P_f}\right)$$

$$= -\frac{(\gamma - 1)}{\gamma}\ln\left(\frac{P_i}{P_f}\right) = -\frac{\left(\dfrac{C_{P,m}}{C_{P,m} - R} - 1\right)}{\dfrac{C_{P,m}}{C_{P,m} - R}}\ln\left(\frac{P_i}{P_f}\right)$$

$$= -\frac{\left(\dfrac{28.86 \text{ J K}^{-1}\text{ mol}^{-1}}{28.86 \text{ J K}^{-1}\text{ mol}^{-1} - 8.314 \text{ J K}^{-1}\text{ mol}^{-1}} - 1\right)}{\dfrac{28.86 \text{ J K}^{-1}\text{ mol}^{-1}}{28.86 \text{ J K}^{-1}\text{ mol}^{-1} - 8.314 \text{ J K}^{-1}\text{ mol}^{-1}}} \times \ln\left(\frac{0.802 \text{ atm}}{0.602 \text{ atm}}\right)$$

$$= -0.0826$$

$$T_f = 0.9207 \, T_i = 265 \text{ K}$$

You can expect snow.

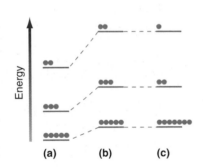

(a) ... **(b)** ... **(c)**

FIGURE 2.18
(a) A system has the translational energy levels shown before undergoing an adiabatic compression. **(b)** After the compression, the energy levels are shifted upward but the occupation probability of the levels shown on the horizontal axis is unchanged. **(c)** Subsequent cooling to the original temperature at constant V restores the energy to its original value by decreasing the probability of occupying higher energy states. Each circle corresponds to a probability of 0.10.

It is instructive to consider an adiabatic compression or expansion from a microscopic point of view. In an adiabatic compression, the energy levels are all raised, but the probability that a given level is accessed is unchanged. This behavior is observed in contrast to that shown in Figure 2.5 because in an adiabatic compression, T and therefore U increases. For more details, see R. E. Dickerson, *Molecular Thermodynamics*, W.A. Benjamin, Menlo Park, 1969. If the gas is subsequently cooled at constant V, the energy levels remain unchanged, but the probability that higher energy states are populated decreases as shown in Figure 2.18c.

Vocabulary

cyclic path

degrees of freedom

enthalpy

exact differential

first law of thermodynamics

heat

heat capacity

indicator diagram

internal energy

irreversible

isobaric

isochoric

isothermal

path

path function

quasi-static process

reversible

state function

work

Conceptual Problems

Q2.1 Electrical current is passed through a resistor immersed in a liquid in an adiabatic container. The temperature of the liquid is varied by 1°C. The system consists solely of the liquid. Does heat or work flow across the boundary between the system and surroundings? Justify your answer.

Q2.2 Two ideal gas systems undergo reversible expansion under different conditions starting from the same P and V. At the end of the expansion, the two systems have the same volume. The pressure in the system that has undergone adiabatic expansion is lower than in the system that has undergone isothermal expansion. Explain this result without using equations.

Q2.3 You have a liquid and its gaseous form in equilibrium in a piston and cylinder assembly in a constant temperature bath. Give an example of a reversible process.

Q2.4 Describe how reversible and irreversible expansions differ by discussing the degree to which equilibrium is maintained between the system and the surroundings.

Q2.5 For a constant pressure process, $\Delta H = q_P$. Does it follow that q_P is a state function? Explain.

Q2.6 A cup of water at 278 K (the system) is placed in a microwave oven and the oven is turned on for 1 minute during which the water begins to boil. State whether each of q, w, and ΔU is positive, negative, or zero.

Q2.7 In the experiments shown in Figure 2.4a and 2.4b, $\Delta U_{surroundings} < 0$, but $\Delta T_{surroundings} > 0$. Explain how this is possible.

Q2.8 What is wrong with the following statement? *Burns caused by steam at 100°C can be more severe than those caused by water at 100°C because steam contains more heat than water.* Rewrite the sentence to convey the same information in a correct way.

Q2.9 Why is it incorrect to speak of the heat or work associated with a system?

Q2.10 You have a liquid and its gaseous form in equilibrium in a piston and cylinder assembly in a constant temperature bath. Give an example of an irreversible process.

Q2.11 What is wrong with the following statement? *Because the well-insulated house stored a lot of heat, the temperature didn't fall much when the furnace failed.* Rewrite the sentence to convey the same information in a correct way.

Q2.12 Explain how a mass of water in the surroundings can be used to determine q for a process. Calculate q if the temperature of a 1.00 kg water bath in the surroundings increases by 1.25°C. Assume that the surroundings are at a constant pressure.

Q2.13 A chemical reaction occurs in a constant volume enclosure separated from the surroundings by diathermal walls. Can you say whether the temperature of the surroundings increases, decreases, or remains the same in this process? Explain.

Q2.14 Explain the relationship between the terms *exact differential* and *state function.*

Q2.15 In the experiment shown in Figure 2.4b, the weight drops in a very short time. How will the temperature of the water change with time?

Q2.16 Discuss the following statement: If the temperature of the system increased, heat must have been added to it.

Q2.17 Discuss the following statement: Heating an object causes its temperature to increase.

Q2.18 An ideal gas is expanded reversibly and adiabatically. Decide which of q, w, ΔU, and ΔH are positive, negative, or zero.

Q2.19 An ideal gas is expanded reversibly and isothermally. Decide which of q, w, ΔU, and ΔH are positive, negative, or zero.

Q2.20 An ideal gas is expanded adiabatically into a vacuum. Decide which of q, w, ΔU, and ΔH are positive, negative, or zero.

Q2.21 A bowling ball (a) rolls across a table, and (b) falls on the floor. Is the work associated with each part of this process positive, negative, or zero?

Q2.22 A perfectly insulating box is partially filled with water in which an electrical resistor is immersed. An external electrical generator converts the change in potential energy of a mass m that falls by a vertical distance h into electrical energy that is dissipated in the resistor. What value do q, w, and ΔU have if the system is defined as the resistor and the water? Everything else is in the surroundings.

Q2.23 A student gets up from her chair and pushes a stack of books across the table. They fall to the floor. Is the work associated with each part of this process positive, negative, or zero?

Q2.24 Explain why ethene has a higher value for $C_{V,m}$ at 800 K than CO.

Q2.25 Explain why $C_{P,m}$ is a function of temperature for ethane, but not for argon in a temperature range in which electronic excitations do not occur.

Q2.26 What is the difference between a quasi-static process and a reversible process?

Numerical Problems

Problem numbers in **red** indicate that the solution to the problem is given in the *Student's Solutions Manual.*

P2.1 A 3.75 mole sample of an ideal gas with $C_{V,m} = 3R/2$ initially at a temperature $T_i = 298$ K and $P_i = 1.00$ bar is enclosed in an adiabatic piston and cylinder assembly. The gas is compressed by placing a 725 kg mass on the piston of diameter 25.4 cm. Calculate the work done in this process and the distance that the piston travels. Assume that the mass of the piston is negligible.

P2.2 The temperature of 1.75 moles of an ideal gas increases from 10.2°C to 48.6°C as the gas is compressed adiabatically. Calculate q, w, ΔU, and ΔH for this process, assuming that $C_{V,m} = 3R/2$.

P2.3 A 2.50 mole sample of an ideal gas, for which $C_{V,m} = 3R/2$, is subjected to two successive changes in state: (1) From 25.0°C and 125×10^3 Pa, the gas is expanded isothermally against a constant pressure of 15.2×10^3 Pa to twice the initial volume. (2) At the end of the previous process, the gas is cooled at constant volume from 25.0°C to –29.0°C. Calculate q, w, ΔU, and ΔH for each of the stages. Also calculate q, w, ΔU, and ΔH for the complete process.

P2.4 A hiker caught in a thunderstorm loses heat when her clothing becomes wet. She is packing emergency rations that if completely metabolized will release 35 kJ of heat per gram of rations consumed. How much rations must the hiker consume to avoid a reduction in body temperature of 2.5 K as a result of heat loss? Assume the heat capacity of the body equals that of water and that the hiker weighs 51 kg. State any additional assumptions.

P2.5 Count Rumford observed that using cannon boring machinery a single horse could heat 11.6 kg of ice water $(T = 273\text{ K})$ to $T = 355$ K in 2.5 hours. Assuming the same rate of work, how high could a horse raise a 225 kg weight in 2.5 minutes? Assume the heat capacity of water is 4.18 J K^{-1} g^{-1}.

P2.6 A 1.50 mole sample of an ideal gas at 28.5°C expands isothermally from an initial volume of 22.5 dm^3 to a final volume of 75.5 dm^3. Calculate w for this process (a) for expansion against a constant external pressure of 1.00×10^5 Pa, and (b) for a reversible expansion.

P2.7 Calculate q, w, ΔU, and ΔH if 2.25 mol of an ideal gas with $C_{V,m} = 3R/2$ undergoes a reversible adiabatic expansion from an initial volume $V_i = 5.50$ m^3 to a final volume $V_f = 25.0$ m^3. The initial temperature is 275 K.

P2.8 Calculate w for the adiabatic expansion of 2.50 mol of an ideal gas at an initial pressure of 2.25 bar from an initial temperature of 450. K to a final temperature of 300. K. Write an expression for the work done in the isothermal reversible expansion of the gas at 300. K from an initial pressure of 2.25 bar. What value of the final pressure would give the same value of w as the first part of this problem? Assume that $C_{P,m} = 5R/2$.

P2.9 At 298 K and 1 bar pressure, the density of water is 0.9970 g cm^{-3}, and $C_{P,m} = 75.3$ J K^{-1} mol^{-1}. The change in volume with temperature is given by $\Delta V = V_{initial} \beta \Delta T$ where β, the coefficient of thermal expansion, is 2.07×10^{-4} K^{-1}. If the temperature of 325 g of water is increased by 25.5 K, calculate w, q, ΔH, and ΔU.

P2.10 A muscle fiber contracts by 3.5 cm and in doing so lifts a weight. Calculate the work performed by the fiber. Assume the muscle fiber obeys Hooke's law $F = -k\,x$ with a force constant k of 750. N m^{-1}.

P2.11 A cylindrical vessel with rigid adiabatic walls is separated into two parts by a frictionless adiabatic piston. Each part contains 45.0 L of an ideal monatomic gas with $C_{V,m} = 3R/2$. Initially, $T_i = 300.$ K and $P_i = 1.75 \times 10^5$ Pa in each part. Heat is slowly introduced into the left part using an electrical heater until the piston has moved sufficiently to

the right to result in a final pressure $P_f = 4.00$ bar in the right part. Consider the compression of the gas in the right part to be a reversible process.

a. Calculate the work done on the right part in this process and the final temperature in the right part.

b. Calculate the final temperature in the left part and the amount of heat that flowed into this part.

P2.12 In the reversible adiabatic expansion of 1.75 mol of an ideal gas from an initial temperature of 27.0°C, the work done on the surroundings is 1300. J. If $C_{V,m} = 3R/2$, calculate q, w, ΔU, and ΔH.

P2.13 A system consisting of 82.5 g of liquid water at 300. K is heated using an immersion heater at a constant pressure of 1.00 bar. If a current of 1.75 A passes through the 25.0 ohm resistor for 100. s, what is the final temperature of the water?

P2.14 A 1.25 mole sample of an ideal gas is expanded from 320. K and an initial pressure of 3.10 bar to a final pressure of 1.00 bar, and $C_{P,m} = 5R/2$. Calculate w for the following two cases:

a. The expansion is isothermal and reversible.

b. The expansion is adiabatic and reversible.

Without resorting to equations, explain why the result to part (b) is greater than or less than the result to part (a).

P2.15 A bottle at 325 K contains an ideal gas at a pressure of 162.5×10^3 Pa. The rubber stopper closing the bottle is removed. The gas expands adiabatically against $P_{external} = 120.0 \times 10^3$ Pa, and some gas is expelled from the bottle in the process. When $P = P_{external}$, the stopper is quickly replaced. The gas remaining in the bottle slowly warms up to 325 K. What is the final pressure in the bottle for a monatomic gas, for which $C_{V,m} = 3R/2$, and a diatomic gas, for which $C_{V,m} = 5R/2$?

P2.16 A 2.25 mole sample of an ideal gas with $C_{V,m} = 3R/2$ initially at 310. K and 1.25×10^5 Pa undergoes a reversible adiabatic compression. At the end of the process, the pressure is 3.10×10^6 Pa. Calculate the final temperature of the gas. Calculate q, w, ΔU, and ΔH for this process.

P2.17 A vessel containing 1.50 mol of an ideal gas with $P_i = 1.00$ bar and $C_{P,m} = 5R/2$ is in thermal contact with a water bath. Treat the vessel, gas, and water bath as being in thermal equilibrium, initially at 298 K, and as separated by adiabatic walls from the rest of the universe. The vessel, gas, and water bath have an average heat capacity of $C_P = 2450.$ J K^{-1}. The gas is compressed reversibly to $P_f = 20.5$ bar. What is the temperature of the system after thermal equilibrium has been established?

P2.18 An ideal gas undergoes an expansion from the initial state described by P_i, V_i, T to a final state described by P_f, V_f, T in (a) a process at the constant external pressure P_f, and (b) in a reversible process. Derive expressions for the largest mass that can be lifted through a height h in the surroundings in these processes.

P2.19 An ideal gas described by $T_i = 275$ K, $P_i = 1.10$ bar, and $V_i = 10.0$ L is heated at constant volume until $P = 10.0$ bar. It then undergoes a reversible isothermal expansion until $P = 1.10$ bar. It is then restored to its original state by the extraction of heat at constant pressure. Depict this closed-cycle process in a P–V diagram. Calculate w for each step and for the total process. What values for w would you calculate if the cycle were traversed in the opposite direction?

P2.20 In an adiabatic compression of one mole of an ideal gas with $C_{V,m} = 5R/2$, the temperature rises from 278 K to 450. K. Calculate q, w, ΔH, and ΔU.

P2.21 The heat capacity of solid lead oxide is given by

$$C_{P,m} = 44.35 + 1.47 \times 10^{-3} \frac{T}{K} \text{ in units of J K}^{-1} \text{ mol}^{-1}$$

Calculate the change in enthalpy of 1.75 mol of PbO(s) if it is cooled from 825 K to 375 K at constant pressure.

P2.22 A 2.25 mole sample of carbon dioxide, for which $C_{P,m} = 37.1$ J K^{-1} mol^{-1} at 298 K, is expanded reversibly and adiabatically from a volume of 4.50 L and a temperature of 298 K to a final volume of 32.5 L. Calculate the final temperature, q, w, ΔH, and ΔU. Assume that $C_{P,m}$ is constant over the temperature interval.

P2.23 A 1.75 mole sample of an ideal gas for which $P = 2.50$ bar and $T = 335$ K is expanded adiabatically against an external pressure of 0.225 bar until the final pressure is 0.225 bar. Calculate the final temperature, q, w, ΔH, and ΔU for (a) $C_{V,m} = 3R/2$, and (b) $C_{V,m} = 5R/2$.

P2.24 A 3.50 mole sample of N_2 in a state defined by $T_i = 250.$ K and $V_i = 3.25$ L undergoes an isothermal reversible expansion until $V_f = 35.5$ L Calculate w, assuming (a) that the gas is described by the ideal gas law, and (b) that the gas is described by the van der Waals equation of state. What is the percent error in using the ideal gas law instead of the van der Waals equation? The van der Waals parameters for N_2 are listed in Table 7.4.

P2.25 A major league pitcher throws a baseball with a speed of 162 kilometers per hour. If the baseball weighs 235 grams and its heat capacity is 1.7 J g^{-1} K^{-1}, calculate the temperature rise of the ball when it is stopped by the catcher's mitt. Assume no heat is transferred to the catcher's mitt and that the catcher's arm does not recoil when he or she catches the ball. Also assume that the kinetic energy of the ball is completely converted into thermal energy.

P2.26 A 2.50 mol sample of an ideal gas for which $C_{V,m} = 3R/2$ undergoes the following two-step process: (1) From an initial state of the gas described by $T = 13.1°C$ and $P = 1.75 \times 10^5$ Pa, the gas undergoes an isothermal expansion against a constant external pressure of 3.75×10^4 Pa until the volume has doubled. (2) Subsequently, the gas is cooled at constant volume. The temperature falls to $-23.6°C$. Calculate q, w, ΔU, and ΔH for each step and for the overall process.

P2.27 A 2.35 mole sample of an ideal gas, for which $C_{V,m} = 3R/2$, initially at 27.0°C and 1.75×10^6 Pa, undergoes a two-stage transformation. For each of the stages described in the following list, calculate the final pressure, as well as q, w, ΔU, and ΔH. Also calculate q, w, ΔU, and ΔH for the complete process.

a. The gas is expanded isothermally and reversibly until the volume triples.

b. Beginning at the end of the first stage, the temperature is raised to 105°C at constant volume.

P2.28 A 3.50 mole sample of an ideal gas with $C_{V,m} = 3R/2$ is expanded adiabatically against a constant external pressure of 1.45 bar. The initial temperature and pressure are $T_i = 310.$ K and $P_i = 15.2$ bar. The final pressure is $P_f = 1.45$ bar. Calculate q, w, ΔU, and ΔH for the process.

P2.29 A nearly flat bicycle tire becomes noticeably warmer after it has been pumped up. Approximate this process as a reversible adiabatic compression. Assume the initial pressure and temperature of the air before it is put in the tire to be $P_i = 1.00$ bar and $T_i = 280.$ K The final pressure in the tire is $P_f = 3.75$ bar. Calculate the final temperature of the air in the tire. Assume that $C_{V,m} = 5R/2$.

P2.30 For 1.25 mol of an ideal gas, $P_{external} = P = 350. \times 10^3$ Pa. The temperature is changed from 135°C to 21.2°C, and $C_{V,m} = 3R/2$. Calculate q, w, ΔU, and ΔH.

P2.31 Suppose an adult is encased in a thermally insulating barrier so that all the heat evolved by metabolism of foodstuffs is retained by the body. What is her temperature increase after 2.5 hours? Assume the heat capacity of the body is 4.18 J g^{-1}K^{-1} and that the heat produced by metabolism is 9.4 kJ kg^{-1}hr^{-1}.

P2.32 Consider the isothermal expansion of 2.35 mol of an ideal gas at 415 K from an initial pressure of 18.0 bar to a final pressure of 1.75 bar. Describe the process that will result in the greatest amount of work being done by the system with $P_{external} \geq 1.75$ bar, and calculate w. Describe the process that will result in the least amount of work being done by the system with $P_{external} \geq 1.75$ bar, and calculate w. What is the least amount of work done without restrictions on the external pressure?

P2.33 An automobile tire contains air at 225×10^3 Pa at 25.0°C. The stem valve is removed and the air is allowed to expand adiabatically against the constant external pressure of one bar until $P = P_{external}$. For air, $C_{V,m} = 5R/2$. Calculate the final temperature. Assume ideal gas behavior.

P2.34 One mole of an ideal gas is subjected to the following changes. Calculate the change in temperature for each case if $C_{V,m} = 3R/2$.

a. $q = -425$ J, $w = 185$ J

b. $q = 315.$ J, $w = -315$ J

c. $q = 0$, $w = 225$ J

P2.35 Consider the adiabatic expansion of 0.500 mol of an ideal monatomic gas with $C_{V,m} = 3R/2$. The initial state is described by $P = 6.25$ bar and $T = 300.$ K.

a. Calculate the final temperature if the gas undergoes a reversible adiabatic expansion to a final pressure of $P = 1.25$ bar.

b. Calculate the final temperature if the same gas undergoes an adiabatic expansion against an external pressure of $P = 1.25$ bar to a final pressure $P = 1.25$ bar.

Explain the difference in your results for parts (a) and (b).

P2.36 A pellet of Zn of mass 31.2 g is dropped into a flask containing dilute H_2SO_4 at a pressure of $P = 1.00$ bar and a temperature of $T = 300.$ K. What is the reaction that occurs? Calculate w for the process.

P2.37   Calculate ΔH and ΔU for the transformation of 2.50 mol of an ideal gas from 19.0°C and 1.00 atm to 550.°C and 19.5 atm if $C_{P,m} = 20.9 + 0.042 \dfrac{T}{K}$ in units of J K^{-1} mol^{-1}.

P2.38   A 1.75 mole sample of an ideal gas for which $C_{V,m} = 20.8$ J K^{-1} mol^{-1} is heated from an initial temperature of 21.2°C to a final temperature of 380.°C at constant volume. Calculate q, w, ΔU, and ΔH for this process.

P2.39   An ideal gas undergoes a single-stage expansion against a constant external pressure $P_{external} = P_f$ at constant temperature from T, P_i, V_i, to T, P_f, V_f.

a. What is the largest mass m that can be lifted through the height h in this expansion?

b. The system is restored to its initial state in a single-state compression. What is the smallest mass m' that must fall through the height h to restore the system to its initial state?

c. If $h = 15.5$ cm, $P_i = 1.75 \times 10^6$ Pa, $P_f = 1.25 \times 10^6$ Pa, $T = 280.$ K, and $n = 2.25$ mol, calculate the values of the masses in parts (a) and (b).

P2.40   The formalism of the Young's modulus is sometimes used to calculate the reversible work involved in extending or compressing an elastic material. Assume a force F is applied to an elastic rod of cross-sectional area A_0 and length L_0. As a result of this force the rod changes in length by ΔL. The Young's modulus E is defined as

$$E = \frac{\text{tensile stress}}{\text{tensile strain}} = \frac{F/A_0}{\Delta L/L_0} = \frac{FL_0}{A_0\Delta L}$$

a. Relate k in Hooke's Law to the Young's modulus expression just given.

b. Using your result in part (a) show that the magnitude of the reversible work involved in changing the length L_0 of an elastic cylinder of cross-sectional area A_0 by ΔL is

$$w = \frac{1}{2}\left(\frac{\Delta L}{L_0}\right)^2 EA_0L_0.$$

P2.41   The Young's modulus (see Problem P2.40) of muscle fiber is approximately 2.80×10^7 Pa. A muscle fiber 3.25 cm in length and 0.125 cm in diameter is suspended with a mass M hanging at its end. Calculate the mass required to extend the length of the fiber by 10%.

P2.42   DNA can be modeled as an elastic rod that can be twisted or bent. Suppose a DNA molecule of length L is bent such that it lies on the arc of a circle of radius R_c. The reversible work involved in bending DNA without twisting is $w_{bend} = \dfrac{BL}{2R_c^2}$ where B is the bending force constant. The DNA in a nucleosome particle is about 680. Å in length. Nucleosomal DNA is bent around a protein complex called the histone octamer into a circle of radius 55 Å. Calculate the reversible work involved in bending the DNA around the histone octamer if the force constant $B = 2.00 \times 10^{-28}$ J m.

P2.43   A 1.75 mole sample of an ideal gas is compressed isothermally from 62.0 L to 19.0 L using a constant external pressure of 2.80 atm. Calculate q, w, ΔU, and ΔH.

P2.44   Assume the following simplified dependence of the pressure in a ventricle of the human heart as a function of volume of blood pumped.

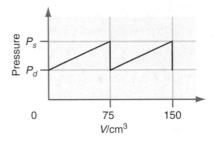

P_s, the systolic pressure, is 120. mm Hg, corresponding to 0.158 atm. P_d, the diastolic pressure, is 80.0 mm Hg, corresponding to 0.105 atm. If the volume of blood pumped in one heartbeat is 75.0 cm^3, calculate the work done in a heartbeat.

Web-Based Simulations, Animations, and Problems

W2.1   A simulation is carried out in which an ideal gas is heated under constant pressure or constant volume conditions. The quantities ΔV (or ΔP), w, ΔU, and ΔT are determined as a function of the heat input. The heat taken up by the gas under constant P or V is calculated and compared with ΔU and ΔH.

W2.2   The reversible isothermal compression and expansion of an ideal gas is simulated for different values of T. The work w is calculated from the T and V values obtained in the simulation. The heat q and the number of moles of gas in the system are calculated from the results.

W2.3   The reversible isobaric compression and expansion of an ideal gas is simulated for different values of pressure gas as heat flows to/from the surroundings. The quantities q, w, and ΔU are calculated from the ΔT and ΔV values obtained in the simulation.

W2.4   The isochoric heating and cooling of an ideal gas is simulated for different values of volume. The number of moles of gas and ΔU are calculated from the constant V value and from the T and P values obtained in the simulation.

W2.5   Reversible cyclic processes are simulated in which the cycle is either rectangular or triangular on a P–V plot. For each segment and for the cycle, ΔU, q, and w are determined. For a given cycle type, the ratio of work done on the surroundings to the heat absorbed from the surroundings is determined for different P and V values.

W2.6   The reversible adiabatic heating and cooling of an ideal gas is simulated for different values of the initial temperature. The quantity $\gamma = C_{P,m}/C_{V,m}$ as well as $C_{P,m}$ and $C_{V,m}$ are determined from the P, V values of the simulation; ΔU and ΔU are calculated from the V, T, and P values obtained in the simulation.

The Importance of State Functions: Internal Energy and Enthalpy

CHAPTER 3

The mathematical properties of state functions are utilized to express the infinitesimal quantities dU and dH as exact differentials. By doing so, expressions can be derived that relate the change of U with T and V and the change in H with T and P to experimentally accessible quantities such as the heat capacity and the coefficient of thermal expansion. Although both U and H are functions of any two of the variables P, V, and T, the dependence of U and H on temperature is generally far greater than the dependence on P or V. As a result, for most processes involving gases, liquids, and solids, U and H can be regarded as functions of T only. An exception to this statement is the cooling on the isenthalpic expansion of real gases, which is commercially used in the liquefaction of N_2, O_2, He, and Ar.

3.1 The Mathematical Properties of State Functions

In Chapter 2 we demonstrated that U and H are state functions and that w and q are path functions. We also discussed how to calculate changes in these quantities for an ideal gas. In this chapter, the path independence of state functions is exploited to derive relationships with which ΔU and ΔH can be calculated as functions of P, V, and T for real gases, liquids, and solids. In doing so, we develop the formal aspects of thermodynamics. We will show that the formal structure of thermodynamics provides a powerful aid in linking theory and experiment. However, before these topics are discussed, the mathematical properties of state functions need to be outlined.

The thermodynamic state functions of interest here are defined by two variables from the set P, V, and T. In formulating changes in state functions, we will make extensive use of partial derivatives, which are reviewed in the Math Supplement (Appendix A). The following discussion does not apply to path functions such as w and q because a functional

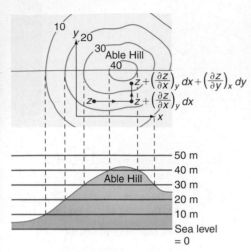

FIGURE 3.1

Starting at the point labeled z on the hill, a person first moves in the positive x direction and then along the y direction. If dx and dy are sufficiently small, the change in height dz is given by

$$dz = \left(\frac{\partial z}{\partial x}\right)_y dx + \left(\frac{\partial z}{\partial y}\right)_x dy.$$

relationship such as Equation (3.1) does not exist for path-dependent functions. Consider 1 mole of an ideal gas for which

$$P = f(V,T) = \frac{RT}{V} \tag{3.1}$$

Note that P can be written as a function of the two variables V and T. The change in P resulting from a change in V or T is proportional to the following **partial derivatives:**

$$\left(\frac{\partial P}{\partial V}\right)_T = \lim_{\Delta V \to 0} \frac{P(V + \Delta V, T) - P(V,T)}{\Delta V} = -\frac{RT}{V^2}$$

$$\left(\frac{\partial P}{\partial T}\right)_V = \lim_{\Delta T \to 0} \frac{P(V, T + \Delta T) - P(V,T)}{\Delta T} = \frac{R}{V} \tag{3.2}$$

The subscript T in $(\partial P/\partial V)_T$ indicates that T is being held constant in the differentiation with respect to V. The partial derivatives in Equation (3.2) allow one to determine how a function changes when the variables change. For example, what is the change in P if the values of T and V both change? In this case, P changes to $P + dP$ where

$$dP = \left(\frac{\partial P}{\partial T}\right)_V dT + \left(\frac{\partial P}{\partial V}\right)_T dV \tag{3.3}$$

Consider the following practical illustration of Equation (3.3). A person is on a hill and has determined his or her altitude above sea level. How much will the altitude (denoted by z) change if the person moves a small distance east (denoted by x) and north (denoted by y)? The change in z as the person moves east is the slope of the hill in that direction, $(\partial z/\partial x)_y$, multiplied by the distance dx that he or she moves. A similar expression can be written for the change in altitude as the person moves north. Therefore, the total change in altitude is the sum of these two changes or

$$dz = \left(\frac{\partial z}{\partial x}\right)_y dx + \left(\frac{\partial z}{\partial y}\right)_x dy$$

These changes in the height z as the person moves first along the x direction and then along the y direction are illustrated in Figure 3.1. Because the slope of the hill is a nonlinear function of x and y, this expression for dz is only valid for small changes dx and dy. Otherwise, higher order derivatives need to be considered.

Second or higher derivatives with respect to either variable can also be taken. The mixed second partial derivatives are of particular interest. Consider the mixed partial derivatives of P:

$$\left(\frac{\partial}{\partial T}\left(\frac{\partial P}{\partial V}\right)_T\right)_V = \frac{\partial^2 P}{\partial T \partial V} = \left(\frac{\partial\left(\frac{\partial\left[\frac{RT}{V}\right]}{\partial V}\right)_T}{\partial T}\right)_V = \left(\frac{\partial\left[-\frac{RT}{V^2}\right]}{\partial T}\right)_V = -\frac{R}{V^2}$$

$$\left(\frac{\partial}{\partial V}\left(\frac{\partial P}{\partial T}\right)_V\right)_T = \frac{\partial^2 P}{\partial V \partial T} = \left(\frac{\partial\left(\frac{\partial\left[\frac{RT}{V}\right]}{\partial T}\right)_V}{\partial V}\right)_T = \left(\frac{\partial\left[\frac{R}{V}\right]}{\partial V}\right)_T = -\frac{R}{V^2} \tag{3.4}$$

For all state functions f and for our specific case of P, the order in which the function is differentiated does not affect the outcome. For this reason,

$$\left(\frac{\partial}{\partial T}\left(\frac{\partial f(V,T)}{\partial V}\right)_T\right)_V = \left(\frac{\partial}{\partial V}\left(\frac{\partial f(V,T)}{\partial T}\right)_V\right)_T \tag{3.5}$$

Because Equation (3.5) is only satisfied by state functions f, it can be used to determine if a function f is a state function. If f is a state function, one can write $\Delta f = \int_i^f df = f_{final} - f_{initial}$. This equation states that f can be expressed as an infinitesimal quantity df that when integrated depends only on the initial and final states; df is called an **exact differential.** An example of a state function and its exact differential is U and $dU = đq - P_{external} dV$.

EXAMPLE PROBLEM 3.1

a. Calculate

$$\left(\frac{\partial f}{\partial x}\right)_y, \left(\frac{\partial f}{\partial y}\right)_x, \left(\frac{\partial^2 f}{\partial x^2}\right)_y, \left(\frac{\partial^2 f}{\partial y^2}\right)_x, \left(\frac{\partial \left(\frac{\partial f}{\partial x}\right)_y}{\partial y}\right)_x, \text{ and } \left(\frac{\partial \left(\frac{\partial f}{\partial y}\right)_x}{\partial x}\right)_y$$

for the function $f(x, y) = ye^x + xy + x\ln y$.

b. Determine if $f(x, y)$ is a state function of the variables x and y.

c. If $f(x, y)$ is a state function of the variables x and y, what is the total differential df?

Solution

a. $\left(\frac{\partial f}{\partial x}\right)_y = ye^x + y + \ln y, \qquad \left(\frac{\partial f}{\partial y}\right)_x = e^x + x + \frac{x}{y}$

$\left(\frac{\partial^2 f}{\partial x^2}\right)_y = ye^x, \qquad \left(\frac{\partial^2 f}{\partial y^2}\right)_x = -\frac{x}{y^2}$

$\left(\frac{\partial \left(\frac{\partial f}{\partial x}\right)_y}{\partial y}\right)_x = e^x + 1 + \frac{1}{y}, \qquad \left(\frac{\partial \left(\frac{\partial f}{\partial y}\right)_x}{\partial x}\right)_y = e^x + 1 + \frac{1}{y}$

b. Because we have shown that

$$\left(\frac{\partial \left(\frac{\partial f}{\partial x}\right)_y}{\partial y}\right)_x = \left(\frac{\partial \left(\frac{\partial f}{\partial y}\right)_x}{\partial x}\right)_y$$

$f(x, y)$ is a state function of the variables x and y. Generalizing this result, any well-behaved function that can be expressed in analytical form is a state function.

c. The total differential is given by

$$df = \left(\frac{\partial f}{\partial x}\right)_y dx + \left(\frac{\partial f}{\partial y}\right)_x dy$$

$$= (ye^x + y + \ln y)dx + \left(e^x + x + \frac{x}{y}\right)dy$$

Two other important results from differential calculus will be used frequently. Consider a function $z = f(x, y)$ that can be rearranged to $x = g(y, z)$ or $y = h(x, z)$. For example, if $P = nRT/V$, then $V = nRT/P$ and $T = PV/nR$. In this case

$$\left(\frac{\partial x}{\partial y}\right)_z = \frac{1}{\left(\frac{\partial y}{\partial x}\right)_z} \tag{3.6}$$

The **cyclic rule** will also be used:

$$\left(\frac{\partial x}{\partial y}\right)_z \left(\frac{\partial y}{\partial z}\right)_x \left(\frac{\partial z}{\partial x}\right)_y = -1 \tag{3.7}$$

It is called the cyclic rule because x, y, and z in the three terms follow the orders x, y, z; y, z, x; and z, x, y. Equations (3.6) and (3.7) can be used to reformulate Equation (3.3) shown next below:

$$dP = \left(\frac{\partial P}{\partial T}\right)_V dT + \left(\frac{\partial P}{\partial V}\right)_T dV$$

Suppose this expression needs to be evaluated for a specific substance, such as N_2 gas. What quantities must be measured in the laboratory in order to obtain numerical values for $(\partial P/\partial T)_V$ and $(\partial P/\partial V)_T$? Using Equations (3.6) and (3.7),

$$\left(\frac{\partial P}{\partial V}\right)_T \left(\frac{\partial V}{\partial T}\right)_P \left(\frac{\partial T}{\partial P}\right)_V = -1$$

$$\left(\frac{\partial P}{\partial T}\right)_V = -\left(\frac{\partial P}{\partial V}\right)_T \left(\frac{\partial V}{\partial T}\right)_P = -\frac{\left(\dfrac{\partial V}{\partial T}\right)_P}{\left(\dfrac{\partial V}{\partial P}\right)_T} = \frac{\beta}{\kappa} \quad \text{and}$$

$$\left(\frac{\partial P}{\partial V}\right)_T = -\frac{1}{\kappa V} \tag{3.8}$$

where β and κ are the readily measured **isobaric volumetric thermal expansion coefficient** and the **isothermal compressibility,** respectively, defined by

$$\beta = \frac{1}{V}\left(\frac{\partial V}{\partial T}\right)_P \quad \text{and} \quad \kappa = -\frac{1}{V}\left(\frac{\partial V}{\partial P}\right)_T \tag{3.9}$$

Both $(\partial V/\partial T)_P$ and $(\partial V/\partial P)_T$ can be measured by determining the change in volume of the system when the pressure or temperature is varied, while keeping the second variable constant.

The minus sign in the equation for κ is chosen so that values of the isothermal compressibility are positive. For small changes in T and P, Equations (3.9) can be written in the more compact form: $V(T_2) = V(T_1)(1 + \beta[T_2 - T_1])$ and $V(P_2) = V(P_1)(1 - \kappa[P_2 - P_1])$. Values for β and κ for selected solids and liquids are shown in Tables 3.1 and 3.2, respectively.

Equation (3.8) is an example of how seemingly abstract partial derivatives can be directly linked to experimentally determined quantities using the mathematical properties of state functions. Using the definitions of β and κ, Equation (3.3) can be written in the form

$$dP = \frac{\beta}{\kappa} dT - \frac{1}{\kappa V} dV \tag{3.10}$$

which can be integrated to give

$$\Delta P = \int_{T_i}^{T_f} \frac{\beta}{\kappa} dT - \int_{V_i}^{V_f} \frac{1}{\kappa V} dV \approx \frac{\beta}{\kappa}(T_f - T_i) - \frac{1}{\kappa} \ln\frac{V_f}{V_i} \tag{3.11}$$

TABLE 3.1 Volumetric Thermal Expansion Coefficient for Solids and Liquids at 298 K

Element	$10^6\, \beta/(\text{K}^{-1})$	Element or Compound	$10^4\, \beta/(\text{K}^{-1})$
Ag(s)	57.6	Hg(l)	1.81
Al(s)	69.3	$CCl_4(l)$	11.4
Au(s)	42.6	$CH_3COCH_3(l)$	14.6
Cu(s)	49.5	$CH_3OH(l)$	14.9
Fe(s)	36.9	$C_2H_5OH(l)$	11.2
Mg(s)	78.3	$C_6H_5CH_3(l)$	10.5
Si(s)	7.5	$C_6H_6(l)$	11.4
W(s)	13.8	$H_2O(l)$	2.04
Zn(s)	90.6	$H_2O(s)$	1.66

Sources: Benenson, W., Harris, J. W., Stocker, H., and Lutz, H. *Handbook of Physics.* New York: Springer, 2002; Lide, D. R., ed. *Handbook of Chemistry and Physics.* 83rd ed. Boca Raton, FL: CRC Press, 2002; Blachnik, R., ed. *D'Ans Lax Taschenbuch für Chemiker und Physiker.* 4th ed. Berlin: Springer, 1998.

TABLE 3.2	Isothermal Compressibility at 298 K		
Substance	$10^6 \, \kappa/\mathrm{bar}^{-1}$	Substance	$10^6 \, \kappa/\mathrm{bar}^{-1}$
Al(s)	1.33	$Br_2(l)$	64
$SiO_2(s)$	2.57	$C_2H_5OH(l)$	110
Ni(s)	0.513	$C_6H_5OH(l)$	61
$TiO_2(s)$	0.56	$C_6H_6(l)$	94
Na(s)	13.4	$CCl_4(l)$	103
Cu(s)	0.702	$CH_3COCH_3(l)$	125
C(graphite)	0.156	$CH_3OH(l)$	120
Mn(s)	0.716	$CS_2(l)$	92.7
Co(s)	0.525	$H_2O(l)$	45.9
Au(s)	0.563	Hg(l)	3.91
Pb(s)	2.37	$SiCl_4(l)$	165
Fe(s)	0.56	$TiCl_4(l)$	89
Ge(s)	1.38		

Sources: Benenson, W., Harris, J. W., Stocker, H., and Lutz, H. *Handbook of Physics.* New York: Springer, 2002; Lide, D. R., ed. *Handbook of Chemistry and Physics.* 83rd ed. Boca Raton FL: CRC Press, 2002; Blachnik, R., ed. *D'Ans Lax Taschenbuch für Chemiker und Physiker.* 4th ed. Berlin: Springer, 1998.

The second expression in Equation (3.11) holds if ΔT and ΔV are small enough that β and κ are constant over the range of integration. Example Problem 3.2 shows a useful application of this equation.

EXAMPLE PROBLEM 3.2

You have accidentally arrived at the end of the range of an ethanol-in-glass thermometer so that the entire volume of the glass capillary is filled. By how much will the pressure in the capillary increase if the temperature is increased by another $10.0°C$? $\beta_{glass} = 2.00 \times 10^{-5}(°C)^{-1}$, $\beta_{ethanol} = 11.2 \times 10^{-4}(°C)^{-1}$, and $\kappa_{ethanol} = 11.0 \times 10^{-5}(\mathrm{bar})^{-1}$. Will the thermometer survive your experiment?

Solution

Using Equation (3.11),

$$\Delta P = \int \frac{\beta_{ethanol}}{\kappa} dT - \int \frac{1}{\kappa V} dV \approx \frac{\beta_{ethanol}}{\kappa} \Delta T - \frac{1}{\kappa} \ln \frac{V_f}{V_i}$$

$$= \frac{\beta_{ethanol}}{\kappa} \Delta T - \frac{1}{\kappa} \ln \frac{V_i(1 + \beta_{glass}\Delta T)}{V_i} \approx \frac{\beta_{ethanol}}{\kappa} \Delta T - \frac{1}{\kappa} \frac{V_i \beta_{glass}\Delta T}{V_i}$$

$$= \frac{(\beta_{ethanol} - \beta_{glass})}{\kappa} \Delta T$$

$$= \frac{(11.2 - 0.200) \times 10^{-4}(°C)^{-1}}{11.0 \times 10^{-5}(\mathrm{bar})^{-1}} \times 10.0°C = 100. \, \mathrm{bar}$$

In this calculation, we have used the relations $V(T_2) = V(T_1)(1 + \beta[T_2 - T_1])$ and $\ln(1 + x) \approx x$ if $x \ll 1$.

The glass is unlikely to withstand such a large increase in pressure.

3.2 The Dependence of *U* on *V* and *T*

In this section, the fact that dU is an exact differential is used to establish how U varies with T and V. For a given amount of a pure substance or a mixture of fixed composition, U is determined by any two of the three variables P, V, and T. One could choose other combinations of variables to discuss changes in U. However, the following discussion will demonstrate that it is particularly convenient to choose the variables T and V. Because U is a state function, an infinitesimal change in U can be written as

$$dU = \left(\frac{\partial U}{\partial T}\right)_V dT + \left(\frac{\partial U}{\partial V}\right)_T dV \qquad (3.12)$$

This expression says that if the state variables change from T, V to $T + dT$, $V + dV$, the change in U, dU, can be determined in the following way. We determine the slopes of $U(T,V)$ with respect to T and V and evaluate them at T, V. Next, these slopes are multiplied by the increments dT and dV, respectively, and the two terms are added. As long as dT and dV are infinitesimal quantities, higher order derivatives can be neglected.

How can numerical values for $(\partial U/\partial T)_V$ and $(\partial U/\partial V)_T$ be obtained? In the following, we only consider P–V work. Combining Equation (3.12) and the differential expression of the first law,

$$đq - P_{external}dV = \left(\frac{\partial U}{\partial T}\right)_V dT + \left(\frac{\partial U}{\partial V}\right)_T dV \qquad (3.13)$$

The symbol $đq$ is used for an infinitesimal amount of heat as a reminder that heat is not a state function. We first consider processes at constant volume for which $dV = 0$, so that Equation (3.13) becomes

$$đq_V = \left(\frac{\partial U}{\partial T}\right)_V dT \qquad (3.14)$$

Note that in the previous equation, $đq_V$ is the product of a state function and an exact differential. Therefore, $đq_V$ behaves like a state function, but only because the path (constant V) is specified. The quantity $đq$ is *not* a state function.

Although the quantity $(\partial U/\partial T)_V$ looks very abstract, it can be readily measured. For example, imagine immersing a container with rigid diathermal walls in a water bath, where the contents of the container are the system. A process such as a chemical reaction is carried out in the container and the heat flow to the surroundings is measured. If heat flow $đq_V$ occurs, a temperature increase or decrease dT is observed in the system and the water bath surroundings. Both of these quantities can be measured. Their ratio, $đq_V/dT$, is a special form of the heat capacity discussed in Section 2.5:

$$\frac{đq_V}{dT} = \left(\frac{\partial U}{\partial T}\right)_V = C_V \qquad (3.15)$$

where $đq_V/dT$ corresponds to a constant volume path and is called the **heat capacity at constant volume.**

The quantity C_V is extensive and depends on the size of the system, whereas $C_{V,m}$ is an intensive quantity. As discussed in Section 2.5, $C_{V,m}$ is different for different substances under the same conditions. Observations show that $C_{V,m}$ is always positive for a single-phase, pure substance or for a mixture of fixed composition, as long as no chemical reactions or phase changes take place in the system. For processes subject to these constraints, U increases monotonically with T.

With the definition of C_V, we now have a way to experimentally determine changes in U with T at constant V for systems of pure substances or for mixtures of constant composition in the absence of chemical reactions or phase changes. After C_V has been determined as a function of T as discussed in Section 2.5, the following integral is numerically evaluated:

$$\Delta U_V = \int_{T_1}^{T_2} C_V dT = n \int_{T_1}^{T_2} C_{V,m}\, dT \tag{3.16}$$

Over a limited temperature range, $C_{V,m}$ can often be regarded as a constant. If this is the case, Equation (3.16) simplifies to

$$\Delta U_V = \int_{T_1}^{T_2} C_V dT = C_V \Delta T = nC_{V,m}\,\Delta T \tag{3.17}$$

which can be written in a different form to explicitly relate q_V and ΔU:

$$\int_i^f đq_V = \int_i^f \left(\frac{\partial U}{\partial T}\right)_V dT \quad \text{or} \quad q_V = \Delta U \tag{3.18}$$

Although $đq$ is not an exact differential, the integral has a unique value if the path is defined, as it is in this case (constant volume). Equation (3.18) shows that ΔU for an arbitrary process in a closed system in which only P–V work occurs can be determined by measuring q under constant volume conditions. As discussed in Chapter 4, the technique of bomb calorimetry uses this approach to determine ΔU for chemical reactions.

Next consider the dependence of U on V at constant T, or $(\partial U/\partial V)_T$. This quantity has the units of $\mathrm{J/m^3} = (\mathrm{J/m})/\mathrm{m^2} = \mathrm{kg\,ms^{-2}}/\mathrm{m^2} = \text{force/area} = \text{pressure}$ and is called the **internal pressure.** To explicitly evaluate the internal pressure for different substances, a result will be used that is derived in the discussion of the second law of thermodynamics in Section 5.12:

$$\left(\frac{\partial U}{\partial V}\right)_T = T\left(\frac{\partial P}{\partial T}\right)_V - P \tag{3.19}$$

Using this equation, the total differential of the internal energy can be written as

$$dU = dU_V + dU_T = C_V\, dT + \left[T\left(\frac{\partial P}{\partial T}\right)_V - P\right]dV \tag{3.20}$$

In this equation, the symbols dU_V and dU_T have been used, where the subscript indicates which variable is constant. Equation (3.20) is an important result that applies to systems containing gases, liquids, or solids in a single phase (or mixed phases at a constant composition) if no chemical reactions or phase changes occur. The advantage of writing dU in the form given by Equation (3.20) over that in Equation (3.12) is that $(\partial U/\partial V)_T$ can be evaluated in terms of the system variables P, V, and T and their derivatives, all of which are experimentally accessible.

Once $(\partial U/\partial V)_T$ and $(\partial U/\partial T)_V$ are known, these quantities can be used to determine dU. Because U is a state function, the path taken between the initial and final states is unimportant. Three different paths are shown in Figure 3.2, and dU is the same for these and any other paths connecting V_i, T_i and V_f, T_f. To simplify the calculation, the path chosen consists of two segments, in which only one of the variables changes in a given path segment. An example of such a path is $V_i, T_i \rightarrow V_f, T_i \rightarrow V_f, T_f$. Because T is constant in the first segment,

$$dU = dU_T = \left[T\left(\frac{\partial P}{\partial T}\right)_V - P\right]dV$$

Because V is constant in the second segment, $dU = dU_V = C_V dT$. Finally, the total change in U is the sum of the changes in the two segments, $dU_{total} = dU_V + dU_T$.

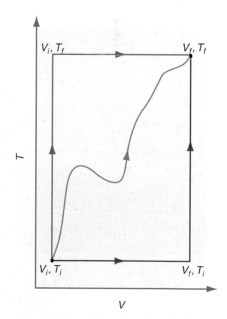

FIGURE 3.2
Because U is a state function, all paths connecting V_i, T_i and V_f, T_f are equally valid in calculating ΔU. Therefore, a specification of the path is irrelevant.

3.3 Does the Internal Energy Depend More Strongly on V or T?

Chapter 2 demonstrated that U is a function of T alone for an ideal gas. However, this statement is not true for real gases, liquids, and solids for which the change in U with V must be considered. In this section, we ask if the temperature or the volume dependence of U is most important in determining ΔU for a process of interest. To answer this question, systems consisting of an ideal gas, a real gas, a liquid, and a solid are considered separately. Example Problem 3.3 shows that Equation (3.19) leads to a simple result for a system consisting of an ideal gas.

EXAMPLE PROBLEM 3.3

Evaluate $(\partial U/\partial V)_T$ for an ideal gas and modify Equation (3.20) accordingly for the specific case of an ideal gas.

Solution

$$\left(\frac{\partial U}{\partial V}\right)_T = T\left(\frac{\partial P}{\partial T}\right)_V - P = T\left(\frac{\partial[nRT/V]}{\partial T}\right)_V - P = \frac{nRT}{V} - P = 0$$

Therefore, $dU = C_V dT$, showing that for an ideal gas, U is a function of T only.

Example Problem 3.3 shows that U is only a function of T for an ideal gas. Specifically, U is not a function of V. This result is understandable in terms of the potential function of Figure 1.10. Because ideal gas molecules do not attract or repel one another, no energy is required to change their average distance of separation (increase or decrease V):

$$\Delta U = \int_{T_i}^{T_f} C_V(T)\, dT \tag{3.21}$$

Recall that because U is only a function of T, Equation (3.21) holds for an ideal gas even if V is not constant.

Next consider the variation of U with T and V for a real gas. The experimental determination of $(\partial U/\partial V)_T$ was carried out by James Joule using an apparatus consisting of two glass flasks separated by a stopcock, all of which were immersed in a water bath. An idealized view of the experiment is shown in Figure 3.3. As a valve between the volumes is opened, a gas initially in volume A expands to completely fill the volume A + B. In interpreting the results of this experiment, it is important to understand where the boundary between the system and surroundings lies. Here, the decision was made to place the system boundary so that it includes all the gas. Initially, the boundary lies totally within V_A, but it moves during the expansion so that it continues to include all gas molecules. With this choice, the volume of the system changes from V_A before the expansion to $V_A + V_B$ after the expansion has taken place.

The first law of thermodynamics [Equation (3.13)] states that

$$đq - P_{external}\, dV = \left(\frac{\partial U}{\partial T}\right)_V dT + \left(\frac{\partial U}{\partial V}\right)_T dV$$

However, all the gas is contained in the system; therefore, $P_{external} = 0$ because a vacuum cannot exert a pressure. Therefore Equation (3.13) becomes

$$đq = \left(\frac{\partial U}{\partial T}\right)_V dT + \left(\frac{\partial U}{\partial V}\right)_T dV \tag{3.22}$$

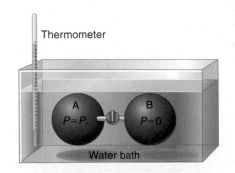

FIGURE 3.3
Schematic depiction of the Joule experiment to determine $(\partial U/\partial V)_T$. Two spherical vessels, A and B, are separated by a valve. Both vessels are immersed in a water bath, the temperature of which is monitored. The initial pressure in each vessel is indicated.

To within experimental accuracy, Joule found that $dT_{surroundings} = 0$. Because the water bath and the system are in thermal equilibrium, $dT = dT_{surroundings} = 0$. With this observation, Joule concluded that $đq = 0$. Therefore, Equation (3.22) becomes

$$\left(\frac{\partial U}{\partial V}\right)_T dV = 0 \tag{3.23}$$

Because $dV \neq 0$, Joule concluded that $(\partial U/\partial V)_T = 0$. Joule's experiment was not definitive because the experimental sensitivity was limited, as shown in Example Problem 3.4.

EXAMPLE PROBLEM 3.4

In Joule's experiment to determine $(\partial U/\partial V)_T$, the heat capacities of the gas and the water bath surroundings were related by $C_{surroundings}/C_{system} \approx 1000$. If the precision with which the temperature of the surroundings could be measured is $\pm 0.006°C$, what is the minimum detectable change in the temperature of the gas?

Solution

View the experimental apparatus as two interacting systems in a rigid adiabatic enclosure. The first is the volume within vessels A and B, and the second is the water bath and the vessels. Because the two interacting systems are isolated from the rest of the universe,

$$q = C_{water\ bath}\Delta T_{water\ bath} + C_{gas}\Delta T_{gas} = 0$$

$$\Delta T_{gas} = -\frac{C_{water\ bath}}{C_{gas}}\Delta T_{water\ bath} = -1000 \times (\pm 0.006°C) = \mp 6°C$$

In this calculation, ΔT_{gas} is the temperature change that the expanded gas undergoes to reach thermal equilibrium with the water bath, which is the negative of the temperature change during the expansion.

Because the minimum detectable value of ΔT_{gas} is rather large, this apparatus is clearly not suited for measuring small changes in the temperature of the gas in an expansion.

More sensitive experiments were carried out by Joule in collaboration with William Thomson (Lord Kelvin). These experiments, which are discussed in Section 3.8, demonstrate that $(\partial U/\partial V)_T$ is small, but nonzero for real gases.

Example Problem 3.3 has shown that $(\partial U/\partial V)_T = 0$ for an ideal gas. We next calculate $(\partial U/\partial V)_T$ and $\Delta U_T = \int_{V_{m,i}}^{V_{m,f}}(\partial U/\partial V)_T dV_m$ for a real gas, in which the van der Waals equation of state is used to describe the gas, as illustrated in Example Problem 3.5.

EXAMPLE PROBLEM 3.5

For a gas described by the van der Waals equation of state,
$P = nRT/(V - nb) - an^2/V^2$. Use this equation to complete these tasks:

a. Calculate $(\partial U/\partial V)_T$ using $(\partial U/\partial V)_T = T(\partial P/\partial T)_V - P$.

b. Derive an expression for the change in internal energy, $\Delta U_T = \int_{V_i}^{V_f}(\partial U/\partial V)_T dV$, in compressing a van der Waals gas from an initial molar volume V_i to a final molar volume V_f at constant temperature.

Solution

a. $T\left(\frac{\partial P}{\partial T}\right)_V - P = T\left(\frac{\partial\left[\dfrac{nRT}{V - nb} - \dfrac{n^2a}{V^2}\right]}{\partial T}\right)_V - P = \frac{nRT}{V - nb} - P$

$= \frac{nRT}{V - nb} - \frac{nRT}{V - nb} + \frac{n^2a}{V^2} = \frac{n^2a}{V^2}$

b. $\Delta U_T = \int_{V_i}^{V_f} \left(\frac{\partial U}{\partial V}\right)_T dV = \int_{V_i}^{V_f} \frac{n^2 a}{V^2} dV = n^2 a\left(\frac{1}{V_i} - \frac{1}{V_f}\right)$

Note that ΔU_T is zero if the attractive part of the intermolecular potential is zero.

Example Problem 3.5 demonstrates that in general $(\partial U/\partial V)_T \neq 0$, and that ΔU_T can be calculated if the equation of state of the real gas is known. This allows the relative importance of $\Delta U_T = \int_{V_i}^{V_f} (\partial U/\partial V)_T dV$ and $\Delta U_V = \int_{T_i}^{T_f} C_V dT$ to be determined in a process in which both T and V change, as shown in Example Problem 3.6.

EXAMPLE PROBLEM 3.6

One mole of N_2 gas undergoes a change from an initial state described by $T = 200.$ K and $P_i = 5.00$ bar to a final state described by $T = 400.$ K and $P_f = 20.0$ bar. Treat N_2 as a van der Waals gas with the parameters $a = 0.137$ Pa m^6 mol^{-2} and $b = 3.87 \times 10^{-5}$ m^3 mol^{-1}. We use the path N_2 $(g, T = 200.$ K, $P = 5.00$ bar) $\rightarrow$ $N_2(g, T = 200.$ K, $P = 20.0$ bar) $\rightarrow N_2(g, T = 400.$ K, $P = 20.0$ bar), keeping in mind that all paths will give the same answer for ΔU of the overall process.

a. Calculate $\Delta U_T = \int_{V_i}^{V_f} (\partial U/\partial V)_T dV$ using the result of Example Problem 3.5. Note that $V_i = 3.28 \times 10^{-3}$ m^3 and $V_f = 7.88 \times 10^{-4}$ m^3 at 200. K, as calculated using the van der Waals equation of state.

b. Calculate $\Delta U_V = n\int_{T_i}^{T_f} C_{V,m} dT$ using the following relationship for $C_{V,m}$ in this temperature range:

$$\frac{C_{V,m}}{J\,K^{-1}\,mol^{-1}} = 22.50 - 1.187 \times 10^{-2}\frac{T}{K} + 2.3968 \times 10^{-5}\frac{T^2}{K^2} - 1.0176 \times 10^{-8}\frac{T^3}{K^3}$$

The ratios T^n/K^n ensure that $C_{V,m}$ has the correct units.

c. Compare the two contributions to ΔU. Can ΔU_T be neglected relative to ΔU_V?

Solution

a. Using the result of Example Problem 3.5,

$$\Delta U_T = n^2 a\left(\frac{1}{V_{m,i}} - \frac{1}{V_{m,f}}\right) = 0.137\text{ Pa m}^6$$

$$\times\left(\frac{1}{3.28 \times 10^{-3}\text{ m}^3} - \frac{1}{7.88 \times 10^{-4}\text{ m}^3}\right) = -132\text{ J}$$

b. $\Delta U_V = n\int_{T_i}^{T_f} C_{V,m}\,dT$

$$= \int_{200.}^{400.}\left(22.50 - 1.187 \times 10^{-2}\frac{T}{K} + 2.3968 \times 10^{-5}\frac{T^2}{K^2} -1.0176 \times 10^{-8}\frac{T^3}{K^3}\right)d\left(\frac{T}{K}\right)\text{ J}$$

$$= (4.50 - 0.712 + 0.447 - 0.0610)\text{kJ} = 4.17\text{ kJ}$$

c. ΔU_T is 3.2% of ΔU_V for this case. In this example, and for most processes, ΔU_T can be neglected relative to ΔU_V for real gases.

The calculations in Example Problems 3.5 and 3.6 show that to a good approximation $\Delta U_T = \int_{V_i}^{V_f} (\partial U/\partial V)_T dV \approx 0$ for real gases under most conditions. Therefore, it is sufficiently accurate to consider U as a function of T only $[U = U(T)]$ for real gases in processes that do not involve unusually high gas densities.

Having discussed ideal and real gases, what can be said about the relative magnitude of $\Delta U_T = \int_{V_i}^{V_f} (\partial U / \partial V)_T dV$ and $\Delta U_V = \int_{T_i}^{T_f} C_V dT$ for processes involving liquids and solids? From experiments, it is known that the density of liquids and solids varies only slightly with the external pressure over the range in which these two forms of matter are stable. This conclusion is not valid for extremely high pressure conditions such as those in the interior of planets and stars. However, it is safe to say that dV for a solid or liquid is very small in most processes. Therefore,

$$\Delta U_T^{solid,\, liq} = \int_{V_1}^{V_2} \left(\frac{\partial U}{\partial V}\right)_T dV \approx \left(\frac{\partial U}{\partial V}\right)_T \Delta V \approx 0 \qquad (3.24)$$

because $\Delta V \approx 0$. This result is valid even if $(\partial U / \partial V)_T$ is large.

The conclusion that can be drawn from this section is as follows. Under most conditions encountered by chemists in the laboratory, U can be regarded as a function of T alone for all substances. The following equations give a good approximation even if V is not constant in the process under consideration:

$$U(T_f, V_f) - U(T_i, V_i) = \Delta U = \int_{T_i}^{T_f} C_V dT = n \int_{T_i}^{T_f} C_{V,m} dT \qquad (3.25)$$

Note that Equation (3.25) is only applicable to a process in which there is no change in the phase of the system, such as vaporization or fusion, and in which there are no chemical reactions. Changes in U that arise from these processes will be discussed in Chapters 4 and 8.

3.4 The Variation of Enthalpy with Temperature at Constant Pressure

As for U, H can be defined as a function of any two of the three variables P, V, and T. It was convenient to choose U to be a function of T and V because this choice led to the identity $\Delta U = q_V$. Using a similar reasoning, we choose H to be a function of T and P. How does H vary with P and T? The variation of H with T at constant P is discussed next, and a discussion of the variation of H with P at constant T is deferred to Section 3.6.

Consider the constant pressure process shown schematically in Figure 3.4. For this process defined by $P = P_{external}$,

$$dU = đq_P - P\, dV \qquad (3.26)$$

Although the integral of $đq$ is in general path dependent, it has a unique value in this case because the path is specified, namely, $P = P_{external} = $ constant. Integrating both sides of Equation (3.26),

$$\int_i^f dU = \int_i^f đq_P - \int_i^f P\, dV \qquad \text{or} \qquad U_f - U_i = q_P - P(V_f - V_i) \quad (3.27)$$

Because $P = P_f = P_i$, this equation can be rewritten as

$$(U_f + P_f V_f) - (U_i + P_i V_i) = q_P \quad \text{or} \quad \Delta H = q_P \qquad (3.28)$$

The preceding equation shows that the value of ΔH can be determined for an arbitrary process at constant P in a closed system in which only P–V work occurs by simply measuring q_P, the heat transferred between the system and surroundings in a constant pressure

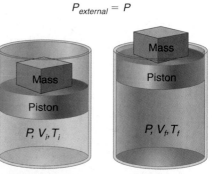

$P_{external} = P$

Mass / Piston / P, V_i, T_i

Mass / Piston / P, V_f, T_f

Initial state   Final state

FIGURE 3.4

The initial and final states are shown for an undefined process that takes place at constant pressure.

process. Note the similarity between Equations (3.28) and (3.18). For an arbitrary process in a closed system in which there is no work other than P–V work, $\Delta U = q_V$ if the process takes place at constant V, and $\Delta H = q_P$ if the process takes place at constant P. These two equations are the basis for the fundamental experimental techniques of bomb calorimetry and constant pressure calorimetry discussed in Chapter 4.

A useful application of Equation (3.28) is in experimentally determining ΔH and ΔU of fusion and vaporization for a given substance. Fusion (solid $\rightarrow$ liquid) and vaporization (liquid $\rightarrow$ gas) occur at a constant temperature if the system is held at a constant pressure and heat flows across the system–surroundings boundary. In both of these phase transitions, attractive interactions between the molecules of the system must be overcome. Therefore, $q > 0$ in both cases and $C_P \rightarrow \infty$. Because $\Delta H = q_P$, ΔH_{fusion} and $\Delta H_{vaporization}$ can be determined by measuring the heat needed to effect the transition at constant pressure. Because $\Delta H = \Delta U + \Delta(PV)$, at constant P,

$$\Delta H_{vaporization} - \Delta U_{vaporization} = P\Delta V_{vaporization} > 0 \qquad (3.29)$$

The change in volume upon vaporization is $\Delta V_{vaporization} = V_{gas} - V_{liquid} \gg 0$; therefore, $\Delta U_{vaporization} < \Delta H_{vaporization}$. An analogous expression to Equation (3.29) can be written relating ΔU_{fusion} and ΔH_{fusion}. Note that ΔV_{fusion} is much smaller than $\Delta V_{vaporization}$ and can be either positive or negative. Therefore, $\Delta U_{fusion} \approx \Delta H_{fusion}$. The thermodynamics of fusion and vaporization will be discussed in more detail in Chapter 8.

Because H is a state function, dH is an exact differential, allowing us to link $(\partial H / \partial T)_P$ to a measurable quantity. In analogy to the preceding discussion for dU, dH is written in the form

$$dH = \left(\frac{\partial H}{\partial T}\right)_P dT + \left(\frac{\partial H}{\partial P}\right)_T dP \qquad (3.30)$$

Because $dP = 0$ at constant P, and $dH = \mathit{d}q_P$ from Equation (3.28), Equation (3.30) becomes

$$\mathit{d}q_P = \left(\frac{\partial H}{\partial T}\right)_P dT \qquad (3.31)$$

Equation (3.31) allows the **heat capacity at constant pressure** C_P to be defined in a fashion analogous to C_V in Equation (3.15):

$$C_P = \frac{\mathit{d}q_P}{dT} = \left(\frac{\partial H}{\partial T}\right)_P \qquad (3.32)$$

Although this equation looks abstract, C_P is a readily measurable quantity. To measure it, one need only measure the heat flow to or from the surroundings for a constant pressure process together with the resulting temperature change in the limit in which dT and $\mathit{d}q$ approach zero and form the ratio $\lim_{dT \to 0} (\mathit{d}q/dT)_P$.

As was the case for C_V, C_P is an extensive property of the system and varies from substance to substance. The temperature dependence of C_P must be known in order to calculate the change in H with T. For a constant pressure process in which there is no change in the phase of the system and no chemical reactions,

$$\Delta H_P = \int_{T_i}^{T_f} C_P(T)dT = n\int_{T_i}^{T_f} C_{P,m}(T)dT \qquad (3.33)$$

If the temperature interval is small enough, it can usually be assumed that C_P is constant. In that case,

$$\Delta H_P = C_P \Delta T = nC_{P,m}\Delta T \qquad (3.34)$$

The calculation of ΔH for chemical reactions and changes in phase will be discussed in Chapters 4 and 8.

EXAMPLE PROBLEM 3.7

A 143.0 g sample of C(s) in the form of graphite is heated from 300. to 600. K at a constant pressure. Over this temperature range, $C_{P,m}$ has been determined to be

$$\frac{C_{P,m}}{J\,K^{-1}mol^{-1}} = -12.19 + 0.1126\frac{T}{K} - 1.947 \times 10^{-4}\frac{T^2}{K^2} + 1.919 \times 10^{-7}\frac{T^3}{K^3}$$

$$-7.800 \times 10^{-11}\frac{T^4}{K^4}$$

Calculate ΔH and q_P. How large is the relative error in ΔH if we neglect the temperature-dependent terms in $C_{P,m}$ and assume that $C_{P,m}$ maintains its value at 300. K throughout the temperature interval?

Solution

$$\Delta H = \frac{m}{M}\int_{T_i}^{T_f} C_{P,m}(T)dT$$

$$= \frac{143.0\text{ g}}{12.00\text{ g mol}^{-1}}\frac{J}{\text{mol}}\int_{300.}^{600.}\left(\begin{array}{c}-12.19 + 0.1126\dfrac{T}{K} - 1.947 \times 10^{-4}\dfrac{T^2}{K^2} + 1.919 \\[2mm] \times 10^{-7}\dfrac{T^3}{K^3} - 7.800 \times 10^{-11}\dfrac{T^4}{K^4}\end{array}\right)d\frac{T}{K}$$

$$= \frac{143.0}{12.00} \times \left[\begin{array}{c}-12.19\dfrac{T}{K} + 0.0563\dfrac{T^2}{K^2} - 6.49 \times 10^{-5}\dfrac{T^3}{K^3} + 4.798 \\[2mm] \times 10^{-8}\dfrac{T^4}{K^4} - 1.56 \times 10^{-11}\dfrac{T^5}{K^5}\end{array}\right]_{300.}^{600.} J = 46.9\text{ kJ}$$

From Equation (3.28), $\Delta H = q_P$.

If we had assumed $C_{P,m} = 8.617$ J mol^{-1} K^{-1}, which is the calculated value at 300. K, $\Delta H = 143.0$ g/12.00 g mol^{-1} × 8.617 J K^{-1} mol^{-1} × [600. K − 300. K] = 30.8 kJ. The relative error is 100 × (30.8 kJ − 46.9 kJ)/46.9 kJ = −34.3%. In this case, it is not reasonable to assume that $C_{P,m}$ is independent of temperature.

3.5 How Are C_P and C_V Related?

To this point, two separate heat capacities, C_P and C_V, have been defined. How are these quantities related? To answer this question, the differential form of the first law is written as

$$đq = C_V\,dT + \left(\frac{\partial U}{\partial V}\right)_T dV + P_{external}\,dV \tag{3.35}$$

Consider a process that proceeds at constant pressure for which $P = P_{external}$. In this case, Equation (3.35) becomes

$$đq_P = C_V\,dT + \left(\frac{\partial U}{\partial V}\right)_T dV + P\,dV \tag{3.36}$$

Because $đq_P = C_P dT$,

$$C_P = C_V + \left(\frac{\partial U}{\partial V}\right)_T\left(\frac{\partial V}{\partial T}\right)_P + P\left(\frac{\partial V}{\partial T}\right)_P = C_V + \left[\left(\frac{\partial U}{\partial V}\right)_T + P\right]\left(\frac{\partial V}{\partial T}\right)_P$$

$$= C_V + T\left(\frac{\partial P}{\partial T}\right)_V\left(\frac{\partial V}{\partial T}\right)_P \tag{3.37}$$

To obtain Equation (3.37), both sides of Equation (3.36) have been divided by dT, and the ratio dV/dT has been converted to a partial derivative at constant P. Equation (3.19) has been used in the last step. Using Equation (3.9) and the cyclic rule, one can simplify Equation (3.37) to

$$C_P = C_V + T\left(\frac{\partial P}{\partial T}\right)_V \left(\frac{\partial V}{\partial T}\right)_P = C_V - T\frac{\left(\frac{\partial V}{\partial T}\right)_P^2}{\left(\frac{\partial V}{\partial P}\right)_T}$$

$$C_P = C_V + TV\frac{\beta^2}{\kappa} \quad \text{or} \quad C_{P,m} = C_{V,m} + TV_m\frac{\beta^2}{\kappa} \tag{3.38}$$

Equation (3.38) provides another example of the usefulness of the formal theory of thermodynamics in linking seemingly abstract partial derivatives with experimentally available data. The difference between $C_{P,m}$ and $C_{V,m}$ can be determined at a given temperature knowing only the molar volume, the isobaric volumetric thermal expansion coefficient, and the isothermal compressibility.

Equation (3.38) is next applied to ideal and real gases, as well as liquids and solids, in the absence of phase changes and chemical reactions. Because β and κ are always positive for real and ideal gases, $C_P - C_V > 0$ for these substances. First, $C_P - C_V$ is calculated for an ideal gas, and then it is calculated for liquids and solids. For an ideal gas, $(\partial U/\partial V)_T = 0$ as shown in Example Problem 3.3, and

$$T\left(\frac{\partial P}{\partial T}\right)_V \left(\frac{\partial V}{\partial T}\right)_P = T\left(\frac{nR}{V}\right)\left(\frac{nR}{P}\right) = nR \text{ so that Equation (3.37) becomes}$$

$$C_P - C_V = nR \tag{3.39}$$

This result was stated without derivation in Section 2.4. The partial derivative $(\partial V/\partial T)_P = V\beta$ is much smaller for liquids and solids than for gases. Therefore, generally

$$C_V \gg \left[\left(\frac{\partial U}{\partial V}\right)_T + P\right]\left(\frac{\partial V}{\partial T}\right)_P \tag{3.40}$$

so that $C_P \approx C_V$ for a liquid or a solid. As shown earlier in Example Problem 3.2, it is not feasible to carry out heating experiments for liquids and solids at constant volume because of the large pressure increase that occurs. Therefore, tabulated heat capacities for liquids and solids list $C_{P,m}$ rather than $C_{V,m}$.

3.6 The Variation of Enthalpy with Pressure at Constant Temperature

In the previous section, we learned how H changes with T at constant P. To calculate how H changes as both P and T change, $(\partial H/\partial P)_T$ must be calculated. The partial derivative $(\partial H/\partial P)_T$ is less straightforward to determine in an experiment than $(\partial H/\partial T)_P$. As will be seen, for many processes involving changes in both P and T, $(\partial H/\partial T)_P \, dT \gg (\partial H/\partial P)_T \, dP$ and the pressure dependence of H can be neglected relative to its temperature dependence. However, the knowledge that $(\partial H/\partial P)_T$ is not zero is essential for understanding the operation of a refrigerator and the liquefaction of gases. The following discussion is applicable to gases, liquids, and solids.

Given the definition $H = U + PV$, we begin by writing dH as

$$dH = dU + P\,dV + V\,dP \tag{3.41}$$

Substituting the differential forms of dU and dH,

$$C_P dT + \left(\frac{\partial H}{\partial P}\right)_T dP = C_V \, dT + \left(\frac{\partial U}{\partial V}\right)_T dV + P\,dV + V\,dP$$

$$= C_V \, dT + \left[\left(\frac{\partial U}{\partial V}\right)_T + P\right]dV + V\,dP \tag{3.42}$$

For isothermal processes, $dT = 0$, and Equation (3.42) can be rearranged to

$$\left(\frac{\partial H}{\partial P}\right)_T = \left[\left(\frac{\partial U}{\partial V}\right)_T + P\right]\left(\frac{\partial V}{\partial P}\right)_T + V \tag{3.43}$$

Using Equation (3.19) for $(\partial U/\partial V)_T$,

$$\left(\frac{\partial H}{\partial P}\right)_T = T\left(\frac{\partial P}{\partial T}\right)_V\left(\frac{\partial V}{\partial P}\right)_T + V$$

$$= V - T\left(\frac{\partial V}{\partial T}\right)_P = V(1 - T\beta) \tag{3.44}$$

The second formulation of Equation (3.44) is obtained through application of the cyclic rule [Equation (3.7)]. This equation is applicable to all systems containing pure substances or mixtures at a fixed composition, provided that no phase changes or chemical reactions take place. The quantity $(\partial H/\partial P)_T$ is evaluated for an ideal gas in Example Problem 3.8.

EXAMPLE PROBLEM 3.8

Evaluate $(\partial H/\partial P)_T$ for an ideal gas.

Solution

$(\partial P/\partial T)_V = (\partial[nRT/V]/\partial T)_V = nR/V$ and $(\partial V/\partial P)_T = (d[nRT/P]/dP)_T = -nRT/P^2$ for an ideal gas. Therefore,

$$\left(\frac{\partial H}{\partial P}\right)_T = T\left(\frac{\partial P}{\partial T}\right)_V\left(\frac{\partial V}{\partial P}\right)_T + V = T\frac{nR}{V}\left(-\frac{nRT}{P^2}\right) + V = -\frac{nRT}{P}\frac{nRT}{nRT} + V = 0$$

This result could have been derived directly from the definition $H = U + PV$. For an ideal gas, $U = U(T)$ only and $PV = nRT$. Therefore, $H = H(T)$ for an ideal gas and $(\partial H/\partial P)_T = 0$.

Because Example Problem 3.8 shows that H is only a function of T for an ideal gas,

$$\Delta H = \int_{T_i}^{T_f} C_P(T)\,dT = n\int_{T_i}^{T_f} C_{P,m}(T)\,dT \tag{3.45}$$

for an ideal gas. Because H is only a function of T, Equation (3.45) holds for an ideal gas even if P is not constant. This result is also understandable in terms of the potential function of Figure 1.10. Because ideal gas molecules do not attract or repel one another, no energy is required to change their average distance of separation (increase or decrease P).

Equation (3.44) is next applied to several types of systems. We have seen that $(\partial H/\partial P)_T = 0$ for an ideal gas. For liquids and solids, $1 \gg T\beta$ for $T < 1000$ K as can be seen from the data in Table 3.1. Therefore, for liquids and solids, $(\partial H/\partial P)_T \approx V$ to a good approximation, and dH can be written as

$$dH \approx C_P\,dT + V\,dP \tag{3.46}$$

for systems that consist only of liquids or solids.

EXAMPLE PROBLEM 3.9

Calculate the change in enthalpy when 124 g of liquid methanol initially at 1.00 bar and 298 K undergoes a change of state to 2.50 bar and 425 K. The density of liquid methanol under these conditions is 0.791 g cm^{-3}, and $C_{P,m}$ for liquid methanol is 81.1 J K^{-1} mol^{-1}.

Solution

Because H is a state function, any path between the initial and final states will give the same ΔH. We choose the path methanol (l, 1.00 bar, 298 K) $\rightarrow$ methanol (l, 1.00 bar, 425 K) $\rightarrow$ methanol (l, 2.50 bar, 425 K). The first step is isothermal, and the second step is isobaric. The total change in H is

$$\Delta H = n \int_{T_i}^{T_f} C_{P,m} dT + \int_{P_i}^{P_f} V dP \approx nC_{P,m}(T_f - T_i) + V(P_f - P_i)$$

$$= 81.1 \text{ J K}^{-1} \text{ mol}^{-1} \times \frac{124 \text{ g}}{32.04 \text{ g mol}^{-1}} \times (425 \text{ K} - 298 \text{ K})$$

$$+ \frac{124 \text{ g}}{0.791 \text{ g cm}^{-3}} \times \frac{10^{-6} \text{ m}^3}{\text{cm}^3} \times (2.50 \text{ bar} - 1.00 \text{ bar}) \times \frac{10^5 \text{ Pa}}{\text{bar}}$$

$$= 39.9 \times 10^3 \text{ J} + 23.5 \text{ J} \approx 39.9 \text{ kJ}$$

Note that the contribution to ΔH from the change in T is far greater than that from the change in P.

Example Problem 3.9 shows that because molar volumes of liquids and solids are small, H changes much more rapidly with T than with P. Under most conditions, H can be assumed to be a function of T only for solids and liquids. Exceptions to this rule are encountered in geophysical or astrophysical applications, for which extremely large pressure changes can occur.

The following conclusion can be drawn from this section: under most conditions encountered by chemists in the laboratory, H can be regarded as a function of T alone for liquids and solids. It is a good approximation to write

$$H(T_f, P_f) - H(T_i, P_i) = \Delta H = \int_{T_1}^{T_2} C_P dT = n \int_{T_1}^{T_2} C_{P,m} dT \qquad \textbf{(3.47)}$$

even if P is not constant in the process under consideration. The dependence of H on P for real gases is discussed in Section 3.8 and Section 3.9 in the context of the Joule-Thomson experiment.

Note that Equation (3.47) is only applicable to a process in which there is no change in the phase of the system, such as vaporization or fusion, and in which there are no chemical reactions. Changes in H that arise from chemical reactions and changes in phase will be discussed in Chapters 4 and 8.

Having dealt with solids, liquids, and ideal gases, we are left with real gases. For real gases, $(\partial H/\partial P)_T$ and $(\partial U/\partial V)_T$ are small, but still have a considerable effect on the properties of the gases upon expansion or compression. Conventional technology for the liquefaction of gases and for the operation of refrigerators is based on the fact that $(\partial H/\partial P)_T$ and $(\partial U/\partial V)_T$ are not zero for real gases. To derive a useful formula for calculating $(\partial H/\partial P)_T$ for a real gas, the Joule-Thomson experiment is discussed first in the next section.

3.7 The Joule-Thomson Experiment

If the valve on a cylinder of compressed N_2 at 298 K is opened fully, it will become covered with frost, demonstrating that the temperature of the valve is lowered below the freezing point of H_2O. A similar experiment with a cylinder of H_2 leads to a considerable increase in temperature and, potentially, an explosion. How can these effects be understood? To explain them, we discuss the **Joule-Thomson experiment.**

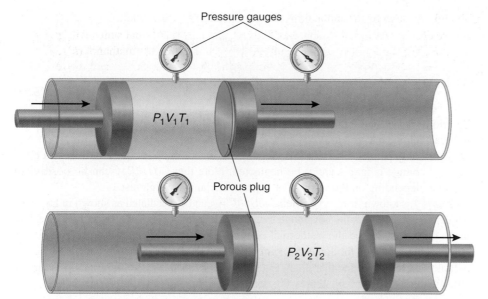

Pressure gauges

$P_1V_1T_1$

Porous plug

$P_2V_2T_2$

FIGURE 3.5
In the Joule-Thomson experiment, a gas is
forced through a porous plug using a piston and cylinder mechanism. The pistons
move to maintain a constant pressure in
each region. There is an appreciable pressure drop across the plug, and the temperature change of the gas is measured. The
upper and lower figures show the initial
and final states, respectively. As shown in
the text, if the piston and cylinder assembly forms an adiabatic wall between the
system (the gases on both sides of the
plug) and the surroundings, the expansion
is isenthalpic.

The Joule-Thomson experiment shown in Figure 3.5 can be viewed as an improved
version of the Joule experiment because it allows $(\partial U/\partial V)_T$ to be measured with a
much higher sensitivity than in the Joule experiment. In this experiment, gas flows
from the high-pressure cylinder on the left to the low-pressure cylinder on the right
through a porous plug in an insulated pipe. The pistons move to keep the pressure
unchanged in each region until all the gas has been transferred to the region to the right
of the porous plug. If N_2 is used in the expansion process $(P_1 > P_2)$, it is found that
$T_2 < T_1$; in other words, the gas is cooled as it expands. What is the origin of this
effect? Consider an amount of gas equal to the initial volume V_1 as it passes through the
apparatus from left to right. The total work in this expansion process is the sum of the
work performed on each side of the plug separately by the moving pistons:

$$w = w_{left} + w_{right} = -\int_{V_1}^{0} P_1\,dV - \int_{0}^{V_2} P_2\,dV = P_1V_1 - P_2V_2 \qquad \textbf{(3.48)}$$

Because the pipe is insulated, $q = 0$, and

$$\Delta U = U_2 - U_1 = w = P_1V_1 - P_2V_2 \qquad \textbf{(3.49)}$$

This equation can be rearranged to

$$U_2 + P_2V_2 = U_1 + P_1V_1 \quad \text{or} \quad H_2 = H_1 \qquad \textbf{(3.50)}$$

Note that the enthalpy is constant in the expansion; the expansion is **isenthalpic.** For
the conditions of the experiment using N_2, both dT and dP are negative, so
$(\partial T/\partial P)_H > 0$. The experimentally determined limiting ratio of ΔT to ΔP at constant
enthalpy is known as the **Joule-Thomson coefficient:**

$$\mu_{J-T} = \lim_{\Delta P \to 0}\left(\frac{\Delta T}{\Delta P}\right)_H = \left(\frac{\partial T}{\partial P}\right)_H \qquad \textbf{(3.51)}$$

If μ_{J-T} is positive, the conditions are such that the attractive part of the potential dominates, and if μ_{J-T} is negative, the repulsive part of the potential dominates. Using
experimentally determined values of μ_{J-T}, $(\partial H/\partial P)_T$ can be calculated. For an isenthalpic process,

$$dH = C_P\,dT + \left(\frac{\partial H}{\partial P}\right)_T dP = 0 \qquad \textbf{(3.52)}$$

TABLE 3.3 Joule-Thomson Coefficients for Selected Substances at 273 K and 1 atm

Gas	μ_{J-T} (K/MPa)
Ar	3.66
C_6H_{14}	−0.39
CH_4	4.38
CO_2	10.9
H_2	−0.34
He	−0.62
N_2	2.15
Ne	−0.30
NH_3	28.2
O_2	2.69

Source: Linstrom, P. J., and Mallard, W. G., eds. *NIST Chemistry Webbook: NIST Standard Reference Database Number 69.* Gaithersburg, MD: National Institute of Standards and Technology. Retrieved from *http://webbook.nist.gov.*

Dividing through by dP and making the condition $dH = 0$ explicit,

$$C_P \left(\frac{\partial T}{\partial P}\right)_H + \left(\frac{\partial H}{\partial P}\right)_T = 0$$

giving

$$\left(\frac{\partial H}{\partial P}\right)_T = -C_P \mu_{J-T} \tag{3.53}$$

Equation (3.53) states that $(\partial H/\partial P)_T$ can be calculated using the measurement of material-dependent properties C_P and μ_{J-T}. Because μ_{J-T} is not zero for a real gas, the pressure dependence of H for an expansion or compression process for which the pressure change is large cannot be neglected. Note that $(\partial H/\partial P)_T$ can be positive or negative, depending on the value of μ_{J-T} at the P and T of interest.

If μ_{J-T} is known from experiment, $(\partial U/\partial V)_T$ can be calculated as shown in Example Problem 3.10. This has the advantage that a calculation of $(\partial U/\partial V)_T$ based on measurements of C_P, μ_{J-T} and the isothermal compressibility κ is much more accurate than a measurement based on the Joule experiment. Values of μ_{J-T} are shown for selected gases in Table 3.3. Keep in mind that μ_{J-T} is a function of P and ΔP, so the values listed in the table are only valid for a small pressure decrease originating at 1 atm pressure.

EXAMPLE PROBLEM 3.10

Using Equation (3.43), $(\partial H/\partial P)_T = [(\partial U/\partial V)_T + P](\partial V/\partial P)_T + V$, derive an expression giving $(\partial U/\partial V)_T$ entirely in terms of measurable quantities for a gas.

Solution

$$\left(\frac{\partial H}{\partial P}\right)_T = \left[\left(\frac{\partial U}{\partial V}\right)_T + P\right]\left(\frac{\partial V}{\partial P}\right)_T + V$$

$$\left(\frac{\partial U}{\partial V}\right)_T = \frac{\left(\frac{\partial H}{\partial P}\right)_T - V}{\left(\frac{\partial V}{\partial P}\right)_T} - P$$

$$= \frac{C_P \mu_{J-T} + V}{\kappa V} - P$$

In this equation, κ is the isothermal compressibility defined in Equation (3.9).

EXAMPLE PROBLEM 3.11

Using Equation (3.43),

$$\left(\frac{\partial H}{\partial P}\right)_T = \left[\left(\frac{\partial U}{\partial V}\right)_T + P\right]\left(\frac{\partial V}{\partial P}\right)_T + V$$

show that $\mu_{J-T} = 0$ for an ideal gas.

Solution

$$\mu_{J-T} = -\frac{1}{C_P}\left(\frac{\partial H}{\partial P}\right)_T = -\frac{1}{C_P}\left[\left(\frac{\partial U}{\partial V}\right)_T\left(\frac{\partial V}{\partial P}\right)_T + P\left(\frac{\partial V}{\partial P}\right)_T + V\right]$$

$$= -\frac{1}{C_P}\left[0 + P\left(\frac{\partial V}{\partial P}\right)_T + V\right]$$

$$= -\frac{1}{C_P}\left[P\left(\frac{\partial[nRT/P]}{\partial P}\right)_T + V\right] = -\frac{1}{C_P}\left[-\frac{nRT}{P} + V\right] = 0$$

In this calculation, we have used the result that $(\partial U/\partial V)_T = 0$ for an ideal gas.

Example Problem 3.11 shows that for an ideal gas, μ_{J-T} is zero. It can be shown that for a van der Waals gas in the limit of zero pressure

$$\mu_{J-T} = \frac{1}{C_{P,m}}\left(\frac{2a}{RT} - b\right) \qquad (3.54)$$

3.8 Liquefying Gases Using an Isenthalpic Expansion

For real gases, the Joule-Thomson coefficient μ_{J-T} can take on either negative or positive values in different regions of P–T space. If μ_{J-T} is positive, a decrease in pressure leads to a cooling of the gas; if it is negative, the expansion of the gas leads to a heating. Figure 3.6 shows the variation of μ_{J-T} with T and P for N_2 and H_2. All along the solid curve, $\mu_{J-T} = 0$. To the left of each curve, μ_{J-T} is positive, and to the right, it is negative. The temperature for which $\mu_{J-T} = 0$ is referred to as the inversion temperature. If the expansion conditions are kept in the region in which μ_{J-T} is positive, ΔT can be made sufficiently large as ΔP decreases in the expansion to liquefy the gas. Note that Equation (3.54) predicts that the inversion temperature for a van der Waals gas is independent of P, which is not in agreement with experiment.

The results in Figure 3.6 are in accord with the observation that a high-pressure ($100 < P < 500$ atm) expansion of N_2 at 300 K leads to cooling and that similar conditions for H_2 lead to heating. To cool H_2 in an expansion, it must first be precooled below 200 K, and the pressure must be less than 160 atm. He and H_2 are heated in an isenthalpic expansion at 300 K for $P < 200$ atm.

The Joule-Thomson effect can be used to liquefy gases such as N_2, as shown in Figure 3.7. The gas at atmospheric pressure is first compressed to a value of 50 atm to 200 atm, which leads to a substantial increase in its temperature. It is cooled and subsequently passed through a heat exchanger in which the gas temperature decreases to a value within ~50 K of the boiling point. At the exit of the heat exchanger, the gas expands through a nozzle to a final pressure of 1 atm in an isenthalpic expansion. The cooling that occurs because $\mu_{J-T} > 0$ results in liquefaction. The gas that boils away passes back through the heat exchanger in the opposite direction than the gas to be liquefied is passing. The two gas streams are separated, but in good thermal contact. In this process, the gas to be liquefied is effectively precooled, enabling a single-stage expansion to achieve liquefaction.

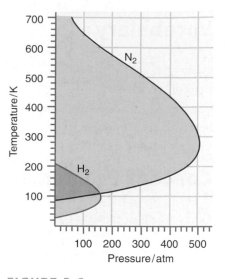

FIGURE 3.6
All along the curves in the figure, $\mu_{J-T} = 0$, and μ_{J-T} is positive to the left of the curves and negative to the right. To experience cooling upon expansion at 100. atm, T must lie between 50. K and 150. K for H_2. The corresponding temperatures for N_2 are 100. K and 650. K.

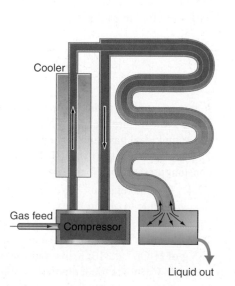

FIGURE 3.7
Schematic depiction of the liquefaction of a gas using an isenthalpic Joule-Thomson expansion. Heat is extracted from the gas exiting from the compressor. It is further cooled in the countercurrent heat exchanger before expanding through a nozzle. Because its temperature is sufficiently low at the exit to the countercurrent heat exchanger, liquefaction occurs.

Vocabulary

cyclic rule	internal pressure	isothermal compressibility
exact differential	isenthalpic	Joule-Thomson coefficient
heat capacity at constant pressure	isobaric volumetric thermal expansion	Joule-Thomson experiment
heat capacity at constant volume	coefficient	partial derivatives

Concept Problems

Q3.1   The heat capacity $C_{P,m}$ is less than $C_{V,m}$ for $H_2O(l)$ near 4°C. Explain this result.

Q3.2   What is the physical basis for the experimental result that U is a function of V at constant T for a real gas? Under what conditions will U decrease as V increases?

Q3.3   Why didn't Joule change his experiment to make $C_{surroundings}/C_{system} \approx 0.001$ to increase the sensitivity of the apparatus?

Q3.4   Why does the relation $C_P > C_V$ always hold for a gas? Can $C_P < C_V$ be valid for a liquid?

Q3.5   Why can q_V be equated with a state function if q is not a state function?

Q3.6   Explain without using equations why $(\partial H/\partial P)_T$ is generally small for a real gas.

Q3.7   Why is it reasonable to write $dH \approx C_P dT + V dP$ for a liquid or solid sample?

Q3.8   Refer to Figure 1.10 and explain why $(\partial U/\partial V)_T$ is generally small for a real gas.

Q3.9   Can a gas be liquefied through an isenthalpic expansion if $\mu_{J-T} = 0$?

Q3.10   Why is $q_V = \Delta U$ only for a constant volume process? Is this formula valid if work other than $P-V$ work is possible?

Q3.11   Classify the following variables and functions as intensive or extensive: T, P, V, q, w, U, H.

Q3.12   Why are q and w not state functions?

Q3.13   Why is the equation $\Delta H = \int_{T_i}^{T_f} C_P(T) \, dT = n \int_{T_i}^{T_f} C_{P,m}(T) \, dT$ valid for an ideal gas even if P is not constant in the process? Is this equation also valid for a real gas? Why or why not?

Q3.14   What is the relationship between a state function and an exact differential?

Q3.15   Is the following statement always, never, or sometimes valid? Explain your reasoning: ΔH is only defined for a constant pressure process.

Q3.16   Is the following statement always, never, or sometimes valid? Explain your reasoning: a thermodynamic process is completely defined by the initial and final states of the system.

Q3.17   Is the following statement always, never, or sometimes valid? Explain your reasoning: $q = 0$ for a cyclic process.

Q3.18   The molar volume of $H_2O(l)$ decreases with increasing temperature near 4°C. Can you explain this behavior using a molecular level model?

Q3.19   Why was the following qualification made in Section 3.7? Note that Equation (3.47) is only applicable to a process in which there is no change in the phase of the system, such as vaporization or fusion, and in which there are no chemical reactions.

Q3.20   Is the expression $\Delta U_V = \int_{T_1}^{T_2} C_V dT = n \int_{T_1}^{T_2} C_{V,m} dT$ only valid for an ideal gas if V is constant?

Numerical Problems

Problem numbers in **red** indicate that the solution to the problem is given in the *Student's Solutions Manual*.

P3.1   Obtain an expression for the isothermal compressibility $\kappa = -1/V \, (\partial V/\partial P)_T$ for a van der Waals gas.

P3.2   Use the result of Problem P3.26 to show that $(\partial C_V/\partial V)_T$ for the van der Waals gas is zero.

P3.3   The molar heat capacity $C_{P,m}$ of $SO_2(g)$ is described by the following equation over the range $300 \text{ K} < T < 1700 \text{ K}$:

$$\frac{C_{P,m}}{R} = 3.093 + 6.967 \times 10^{-3} \frac{T}{K} - 45.81 \times 10^{-7} \frac{T^2}{K^2}$$
$$+ 1.035 \times 10^{-9} \frac{T^3}{K^3}$$

In this equation, T is the absolute temperature in kelvin. The ratios T^n/K^n ensure that $C_{P,m}$ has the correct dimension. Assuming ideal gas behavior, calculate q, w, ΔU, and ΔH if 1.50 moles of $SO_2(g)$ is heated from 22.5°C to 1140.°C at a constant pressure of 1 bar. Explain the sign of w.

P3.4   Use the relation $(\partial U/\partial V)_T = T(\partial P/\partial T)_V - P$ and the cyclic rule to obtain an expression for the internal pressure, $(\partial U/\partial V)_T$, in terms of P, β, T, and κ.

P3.5   A mass of 34.05 g of $H_2O(s)$ at 273 K is dropped into 185 g of $H_2O(l)$ at 310. K in an insulated container at 1 bar of pressure. Calculate the temperature of the system once equilibrium has been reached. Assume that $C_{P,m}$ for H_2O is constant at its values for 298 K throughout the temperature range of interest.

P3.6 A vessel is filled completely with liquid water and sealed at 13.56°C and a pressure of 1.00 bar. What is the pressure if the temperature of the system is raised to 82.0°C? Under these conditions, $\beta_{water} = 2.04 \times 10^{-4}\,K^{-1}$, $\beta_{vessel} = 1.42 \times 10^{-4}\,K^{-1}$, and $\kappa_{water} = 4.59 \times 10^{-5}\,bar^{-1}$.

P3.7 Integrate the expression $\beta = 1/V\,(\partial V/\partial T)_P$ assuming that β is independent of temperature. By doing so, obtain an expression for V as a function of T and β at constant P.

P3.8 A mass of 32.0 g of $H_2O(g)$ at 373 K is flowed into 295 g of $H_2O(l)$ at 310. K and 1 atm. Calculate the final temperature of the system once equilibrium has been reached. Assume that $C_{P,m}$ for H_2O is constant at its values for 298 K throughout the temperature range of interest. Describe the state of the system.

P3.9 Because $(\partial H/\partial P)_T = -C_P\mu_{J-T}$, the change in enthalpy of a gas expanded at constant temperature can be calculated. To do so, the functional dependence of μ_{J-T} on P must be known. Treating Ar as a van der Waals gas, calculate ΔH when 1 mole of Ar is expanded from 325 bar to 1.75 bar at 375 K. Assume that μ_{J-T} is independent of pressure and is given by $\mu_{J-T} = [(2a/RT) - b]/C_{P,m}$, and $C_{P,m} = 5R/2$ for Ar. What value would ΔH have if the gas exhibited ideal gas behavior?

P3.10 Derive the following expression for calculating the isothermal change in the constant volume heat capacity: $(\partial C_V/\partial V)_T = T(\partial^2 P/\partial T^2)_V$.

P3.11 A 75.0 g piece of gold at 650. K is dropped into 180. g of $H_2O(l)$ at 310. K in an insulated container at 1 bar pressure. Calculate the temperature of the system once equilibrium has been reached. Assume that $C_{P,m}$ for Au and H_2O is constant at their values for 298 K throughout the temperature range of interest.

P3.12 Calculate w, q, ΔH, and ΔU for the process in which 1.75 moles of water undergoes the transition $H_2O(l, 373\,K) \rightarrow H_2O(g, 610.\,K)$ at 1 bar of pressure. The volume of liquid water at 373 K is $1.89 \times 10^{-5}\,m^3\,mol^{-1}$ and the molar volume of steam at 373 K and 610. K is 3.03 and $5.06 \times 10^{-2}\,m^3\,mol^{-1}$, respectively. For steam, $C_{P,m}$ can be considered constant over the temperature interval of interest at $33.58\,J\,mol^{-1}\,K^{-1}$.

P3.13 Equation (3.38), $C_P = C_V + TV(\beta^2/\kappa)$, links C_P and C_V with β and κ. Use this equation to evaluate $C_P - C_V$ for an ideal gas.

P3.14 Use the result of Problem P3.26 to derive a formula for $(\partial C_V/\partial V)_T$ for a gas that obeys the Redlich-Kwong equation of state,

$$P = \frac{RT}{V_m - b} - \frac{a}{\sqrt{T}}\frac{1}{V_m(V_m + b)}$$

P3.15 The function $f(x, y)$ is given by $f(x, y) = xy \sin 5x + x^2\sqrt{y}\ln y + 3e^{-2x^2}\cos y$. Determine

$$\left(\frac{\partial f}{\partial x}\right)_y, \left(\frac{\partial f}{\partial y}\right)_x, \left(\frac{\partial^2 f}{\partial x^2}\right)_y, \left(\frac{\partial^2 f}{\partial y^2}\right)_x, \left(\frac{\partial}{\partial y}\left(\frac{\partial f}{\partial x}\right)_y\right)_x,$$

and $\left(\dfrac{\partial}{\partial x}\left(\dfrac{\partial f}{\partial y}\right)_x\right)_y$. Is $\left(\dfrac{\partial}{\partial y}\left(\dfrac{\partial f}{\partial x}\right)_y\right)_x = \left(\dfrac{\partial}{\partial x}\left(\dfrac{\partial f}{\partial y}\right)_x\right)_y$?

Obtain an expression for the total differential df.

P3.16 The Joule coefficient is defined by $(\partial T/\partial V)_U = (1/C_V)[P - T(\partial P/\partial T)_V]$. Calculate the Joule coefficient for an ideal gas and for a van der Waals gas.

P3.17 Using the result of Equation (3.8), $(\partial P/\partial T)_V = \beta/\kappa$, express β as a function of κ and V_m for an ideal gas, and β as a function of b, κ, and V_m for a van der Waals gas.

P3.18 Show that the expression $(\partial U/\partial V)_T = T(\partial P/\partial T)_V - P$ can be written in the form

$$\left(\frac{\partial U}{\partial V}\right)_T = T^2\left(\partial\left[\frac{P}{T}\right]\Big/\partial T\right)_V = -\left(\partial\left[\frac{P}{T}\right]\Big/\partial\left[\frac{1}{T}\right]\right)_V$$

P3.19 Derive an expression for the internal pressure of a gas that obeys the Bethelot equation of state,

$$P = \frac{RT}{V_m - b} - \frac{a}{TV_m^2}$$

P3.20 Because U is a state function, $(\partial/\partial V\,(\partial U/\partial T)_V)_T = (\partial/\partial T\,(\partial U/\partial V)_T)_V$. Using this relationship, show that $(\partial C_V/\partial V)_T = 0$ for an ideal gas.

P3.21 Starting with the van der Waals equation of state, find an expression for the total differential dP in terms of dV and dT. By calculating the mixed partial derivatives $(\partial(\partial P/\partial V)_T/\partial T)_V$ and $(\partial(\partial P/\partial T)_V/\partial V)_T$, determine if dP is an exact differential.

P3.22 Use $(\partial U/\partial V)_T = (\beta T - \kappa P)/\kappa$ to calculate $(\partial U/\partial V)_T$ for an ideal gas.

P3.23 Derive the following relation,

$$\left(\frac{\partial U}{\partial V_m}\right)_T = \frac{3a}{2\sqrt{T}V_m(V_m + b)}$$

for the internal pressure of a gas that obeys the Redlich-Kwong equation of state,

$$P = \frac{RT}{V_m - b} - \frac{a}{\sqrt{T}}\frac{1}{V_m(V_m + b)}$$

P3.24 A differential $dz = f(x, y)dx + g(x, y)dy$ is exact if the integral $\int f(x, y)dx + \int g(x, y)dy$ is independent of the path. Demonstrate that the differential $dz = 2xydx + x^2dy$ is exact by integrating dz along the paths $(1,1) \rightarrow (1,8) \rightarrow (6,8)$ and $(1,1) \rightarrow (1,3) \rightarrow (4,3) \rightarrow (4,8) \rightarrow (6,8)$. The first number in each set of parentheses is the x coordinate, and the second number is the y coordinate.

P3.25 Show that $d\rho/\rho = -\beta dT + \kappa dP$ where ρ is the density $\rho = m/V$. Assume that the mass m is constant.

P3.26 For a gas that obeys the equation of state

$$V_m = \frac{RT}{P} + B(T)$$

derive the result

$$\left(\frac{\partial H}{\partial P}\right)_T = B(T) - T\frac{dB(T)}{dT}$$

P3.27 Because V is a state function, $(\partial(\partial V/\partial T)_P/\partial P)_T = (\partial(\partial V/\partial P)_T/\partial T)_P$. Using this relationship, show that the isothermal compressibility and isobaric expansion coefficient are related by $(\partial\beta/\partial P)_T = -(\partial\kappa/\partial T)_P$.

P3.28 Use the relation

$$C_{P,m} - C_{V,m} = T\left(\frac{\partial V_m}{\partial T}\right)_P \left(\frac{\partial P}{\partial T}\right)_V$$

the cyclic rule, and the van der Waals equation of state to derive an equation for $C_{P,m} - C_{V,m}$ in terms of V_m, T, and the gas constants R, a, and b.

P3.29 For the equation of state $V_m = RT/P + B(T)$, show that

$$\left(\frac{\partial C_{P,m}}{\partial P}\right)_T = -T\frac{d^2B(T)}{dT^2}$$

[*Hint:* Use Equation (3.44) and the property of state functions with respect to the order of differentiation in mixed second derivatives.]

P3.30 Starting with $\beta = (1/V)(\partial V/\partial T)_P$, show that $\rho = -(1/\rho)(\partial \rho/\partial T)_P$, where ρ is the density.

P3.31 This problem will give you practice in using the cyclic rule. Use the ideal gas law to obtain the three functions $P = f(V, T)$, $V = g(P, T)$, and $T = h(P, V)$. Show that the cyclic rule $(\partial P/\partial V)_T (\partial V/\partial T)_P (\partial T/\partial P)_V = -1$ is obeyed.

P3.32 Regard the enthalpy as a function of T and P. Use the cyclic rule to obtain the expression

$$C_P = -\left(\frac{\partial H}{\partial P}\right)_T \Big/ \left(\frac{\partial T}{\partial P}\right)_H$$

P3.33 Using the chain rule for differentiation, show that the isobaric expansion coefficient expressed in terms of density is given by $\beta = -1/\rho(\partial\rho/\partial T)_P$.

P3.34 Derive the equation $(\partial P/\partial V)_T = -1/(\kappa V)$ from basic equations and definitions.

P3.35 Derive the equation $(\partial H/\partial T)_V = C_V + V\beta/\kappa$ from basic equations and definitions.

P3.36 For an ideal gas, $\left(\dfrac{\partial U}{\partial V}\right)_T$ and $\left(\dfrac{\partial H}{\partial P}\right)_T = 0$. Prove that C_V and C_P are independent of volume and pressure.

P3.37 Prove that $C_V = -\left(\dfrac{\partial U}{\partial V}\right)_T \left(\dfrac{\partial V}{\partial T}\right)_U$

P3.38 Show that $\left(\dfrac{\partial C_V}{\partial V}\right)_T = T\left(\dfrac{\partial^2 P}{\partial T^2}\right)_V$

P3.39 Show that $\left(\dfrac{\partial C_V}{\partial V}\right)_T = 0$ for an ideal and for a van der Waals gas.

Thermochemistry

Thermochemistry is the branch of thermodynamics that investigates the heat flow into or out of a reaction system and deduces the energy stored in chemical bonds. As reactants are converted into products, energy can either be taken up by the system or released to the surroundings. For a reaction that takes place at constant volume, the heat that flows to or out of the system is equal to ΔU for the reaction. For a reaction that takes place at constant pressure, the heat that flows to or out of the system is equal to ΔH for the reaction. The enthalpy of formation is defined as the heat flow into or out of the system in a reaction between pure elements that leads to the formation of 1 mol of product. Because H is a state function, the reaction enthalpy can be written as the enthalpies of formation of the products minus those of the reactants. This property allows ΔH and ΔU for a reaction to be calculated for many reactions without carrying out an experiment.

4.1 Energy Stored in Chemical Bonds Is Released or Taken Up in Chemical Reactions

A significant amount of the internal energy or enthalpy of a molecule is stored in the form of chemical bonds. As reactants are transformed to products in a chemical reaction, energy can be released or taken up as bonds are made or broken, respectively. For example, consider a reaction in which $N_2(g)$ and $H_2(g)$ dissociate into atoms, and the atoms recombine to form $NH_3(g)$. The enthalpy changes associated with individual steps and with the overall reaction $1/2\ N_2(g) + 3/2\ H_2(g) \longrightarrow NH_3(g)$ are shown in Figure 4.1. Note that large enthalpy changes are associated with the individual steps but the enthalpy change in the overall reaction is much smaller.

The change in enthalpy or internal energy resulting from chemical reactions appears in the surroundings in the form of a temperature increase or decrease resulting from heat flow and/or in the form of expansion or nonexpansion work. For example, the combustion of gasoline in an automobile engine can be used to do expansion work on the surroundings. Nonexpansion electrical work is possible if the chemical

FIGURE 4.1
Enthalpy changes are shown for individual steps in the overall reaction $1/2\ N_2(g) + 3/2\ H_2(g) \longrightarrow NH_3(g)$.

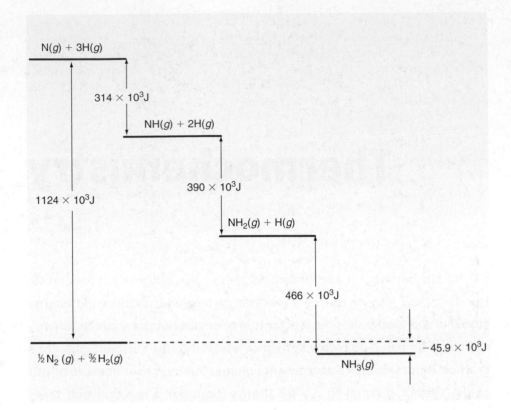

reaction is carried out in an electrochemical cell. In Chapters 6 and 11, the extraction of nonexpansion work from chemical reactions will be discussed. In this chapter, the focus is on using measurements of heat flow to determine changes in U and H due to chemical reactions.

4.2 Internal Energy and Enthalpy Changes Associated with Chemical Reactions

In the previous chapters, we discussed how ΔU and ΔH are calculated from work and heat flow between the system and the surroundings for processes that do not involve phase changes or chemical reactions. In this section, this discussion is extended to reaction systems.

Imagine that a reaction involving a stoichiometric mixture of reactants (the system) is carried out in a constant pressure reaction vessel with diathermal walls immersed in a water bath (the surroundings). If the temperature of the water bath increases, heat has flowed from the system (the contents of the reaction vessel) to the surroundings (the water bath and the vessel). In this case, we say that the reaction is **exothermic.** If the temperature of the water bath decreases, the heat has flowed from the surroundings to the system, and we say that the reaction is **endothermic.**

Consider the reaction in Equation (4.1):

$$Fe_3O_4(s) + 4\ H_2(g) \longrightarrow 3\ Fe(s) + 4\ H_2O(l) \qquad \textbf{(4.1)}$$

Note that the phase (solid, liquid, or gas) for each reactant and product has been specified because U and H are different for each phase. This reaction will only proceed at a measurable rate at elevated temperatures. However, as we show later, it is useful to tabulate values for ΔH for reactions at a pressure of 1 bar and a specified temperature, generally 298.15 K. The pressure value of 1 bar defines a **standard state,** and changes in H and U at the standard pressure of 1 bar are indicated by a superscript ° as in $\Delta H°$ and $\Delta U°$. The standard state for gases is a hypothetical state in which the gas behaves ideally at a pressure of 1 bar. For most gases, deviations from ideal behavior are very small. The **enthalpy of reaction,** ΔH_R, at specific values of T and P is defined as the heat exchanged between the system and the surroundings as the reactants are transformed into products at conditions of constant T and P. By convention, heat flowing into the system is given a positive sign. ΔH_R is, therefore, a negative quantity for an exothermic reaction and a positive quantity for an endothermic reaction. The **standard enthalpy of reaction,** ΔH_R°, refers to one mole of the specified reaction at a pressure of 1 bar, and unless indicated otherwise, to $T = 298.15$ K.

How can the reaction enthalpy and internal energy be determined? We proceed in the following way. The reaction is carried out at 1 bar pressure, and the temperature change ΔT that occurs in a finite size water bath, initially at 298.15 K, is measured. The water bath is large enough that ΔT is small. If ΔT is negative as a result of the reaction, the bath is heated to return it, the reaction vessel, and the system to 298.15 K using an electrical heater. By doing so, we ensure that the initial and final states are the same and therefore the measured ΔH is equal to ΔH_R°. The electrical work done on the heater that restores the temperature of the water bath and the system to 298.15 K is equal to ΔH_R°. If the temperature of the water bath increases as a result of the reaction, the electrical work done on a heater in the water bath at 298.15 K that increases its temperature and that of the system by ΔT in a separate experiment is measured. In this case, ΔH_R° is equal to the negative of the electrical work done on the heater.

Although an experimental method for determining ΔH_R° has been described, to tabulate the reaction enthalpies for all possible chemical reactions would be a monumental undertaking. Fortunately, ΔH_R° can be calculated from tabulated enthalpy values for individual reactants and products. This is advantageous because there are far fewer reactants and products than there are reactions among them. Consider ΔH_R° for the reaction of Equation (4.1) at $T = 298.15$ K and $P = 1$ bar. These values for P and T are chosen because thermodynamic values are tabulated for these values. However, ΔH_R at other values of P and T can be calculated as discussed in Chapters 2 and 3. In principle, we could express ΔH_R° in terms of the individual enthalpies of reactants and products:

$$\Delta H_R^\circ = H_{products}^\circ - H_{reactants}^\circ$$
$$= 3H_m^\circ(\text{Fe},s) + 4H_m^\circ(\text{H}_2\text{O},l) - H_m^\circ(\text{Fe}_3\text{O}_4,s) - 4H_m^\circ(\text{H}_2,g) \qquad \textbf{(4.2)}$$

The m subscripts refer to molar quantities. Although Equation (4.2) is correct, it does not provide a useful way to calculate ΔH_R°. There is no experimental way to determine the absolute enthalpy for any element or compound because there is no unique reference zero against which individual enthalpies can be measured. Only ΔH and ΔU, as opposed to H and U, can be determined in an experiment.

Equation (4.2) can be transformed into a more useful form by introducing the enthalpy of formation. The **standard enthalpy of formation,** ΔH_f°, is defined as the enthalpy change of the reaction in which the only reaction product is 1 mol of the species of interest, and only pure elements in their most stable state of aggregation under the standard state conditions appear as reactants. We refer to these species as being in their **standard reference state.** For example, the standard reference state of water and carbon at 298.15 K are $\text{H}_2\text{O}(l)$ and solid carbon in the form of graphite. Note that with this definition, $\Delta H_f^\circ = 0$ for an element in its standard reference state because the reactants and products are identical.

We next illustrate how reaction enthalpies can be expressed in terms of formation enthalpies. The only compounds that are produced or consumed in the reaction $\text{Fe}_3\text{O}_4(s) + 4\,\text{H}_2(g) \longrightarrow 3\,\text{Fe}(s) + 4\,\text{H}_2\text{O}(l)$ are $\text{Fe}_3\text{O}_4(s)$ and $\text{H}_2\text{O}(l)$. All elements that appear in the reaction are in their standard reference states. The formation reactions for the compounds at 298.15 K and 1 bar are

$$\text{H}_2(g) + \frac{1}{2}\,\text{O}_2(g) \longrightarrow \text{H}_2\text{O}(l)$$

$$\Delta H_R^\circ = \Delta H_f^\circ\,(\text{H}_2\text{O},\,l) = H_m^\circ(\text{H}_2\text{O},\,l) - H_m^\circ(\text{H}_2,\,g) - \frac{1}{2}\,H_m^\circ(\text{O}_2,\,g) \qquad \textbf{(4.3)}$$

$$3\,\mathrm{Fe}(s) + 2\,\mathrm{O}_2(g) \longrightarrow \mathrm{Fe}_3\mathrm{O}_4(s)$$

$$\Delta H_R^\circ = \Delta H_f^\circ(\mathrm{Fe}_3\mathrm{O}_4, s) = H_m^\circ(\mathrm{Fe}_3\mathrm{O}_4, s) - 3H_m^\circ(\mathrm{Fe}, s) - 2H_m^\circ(\mathrm{O}_2, g) \quad \textbf{(4.4)}$$

If Equation (4.2) is rewritten in terms of the enthalpies of formation, a simple equation for the reaction enthalpy is obtained:

$$\Delta H_R^\circ = 4\Delta H_f^\circ(\mathrm{H}_2\mathrm{O}, l) - \Delta H_f^\circ(\mathrm{Fe}_3\mathrm{O}_4, s) \quad \textbf{(4.5)}$$

Note that elements in their standard reference state do not appear in this equation because $\Delta H_f^\circ = 0$ for these species. This result can be generalized to any chemical transformation

$$\nu_A A + \nu_B B + \ldots \longrightarrow \nu_X X + \nu_Y Y + \ldots \quad \textbf{(4.6)}$$

which we write in the form

$$0 = \sum_i \nu_i X_i \quad \textbf{(4.7)}$$

The X_i refer to all species that appear in the overall equation. The unitless stoichiometric coefficients ν_i are positive for products and negative for reactants. The enthalpy change associated with this reaction is

$$\Delta H_R^\circ = \sum_i \nu_i \Delta H_{f,i}^\circ \quad \textbf{(4.8)}$$

The rationale behind Equation (4.8) can also be depicted as shown in Figure 4.2. Two paths are considered between the reactants A and B and the products C and D in the reaction $\nu_A A + \nu_B B \longrightarrow \nu_C C + \nu_D D$. The first of these is a direct path for which $\Delta H^\circ = \Delta H_R^\circ$. In the second path, A and B are first broken down into their elements, each in its standard reference state. Subsequently, the elements are combined to form C and D. The enthalpy change for the second route is $\Delta H_R^\circ = \Sigma_i \nu_i \Delta H_{f,products}^\circ - \Sigma_i |\nu_i| \Delta H_{f,reactants}^\circ = \Sigma_i \nu_i \Delta H_{f,i}^\circ$. Because H is a state function, the enthalpy change is the same for both paths. This is stated in mathematical form in Equation (4.8).

Writing ΔH_R° in terms of formation enthalpies is a great simplification over compiling measured values of reaction enthalpies. Standard formation enthalpies for atoms and inorganic compounds at 298.15 K are listed in Table 4.1, and standard formation enthalpies for organic compounds are listed in Table 4.2 (Appendix B, Data Tables).

Another thermochemical convention is introduced at this point in order to calculate enthalpy changes involving electrolyte solutions. The solution reaction that occurs when a salt such as NaCl is dissolved in water is

$$\mathrm{NaCl}(s) \longrightarrow \mathrm{Na}^+(aq) + \mathrm{Cl}^-(aq)$$

Because it is not possible to form only positive or negative ions in solution, the measured enthalpy of solution of an electrolyte is the sum of the enthalpies of all anions and cations formed. To be able to tabulate values for enthalpies of formation of individual ions, the enthalpy for the following reaction is set equal to zero at $P = 1$ bar for all temperatures:

$$1/2\,\mathrm{H}_2(g) \longrightarrow \mathrm{H}^+(aq) + \mathrm{e}^-(\text{metal electrode})$$

In other words, solution enthalpies of formation of ions are measured relative to that for $\mathrm{H}^+(aq)$. The thermodynamics of electrolyte solutions will be discussed in detail in Chapter 10.

As the previous discussion shows, only the ΔH_f° of each reactant and product is needed to calculate ΔH_R°. Each ΔH_f° is a *difference* in enthalpy between the compound and its constituent elements, rather than an absolute enthalpy. However, there is a convention that allows absolute enthalpies to be specified using the experimentally determined values of the ΔH_f° of compounds. In this convention, the absolute enthalpy of each pure element in its standard reference state is set equal to zero. With this convention, the absolute molar enthalpy of any chemical species in its standard reference state H_m° is equal to ΔH_f° for that species. To demonstrate this convention, the reaction in Equation (4.4) is considered:

$$\Delta H_R^\circ = \Delta H_f^\circ(\mathrm{Fe}_3\mathrm{O}_4, s) = H_m^\circ(\mathrm{Fe}_3\mathrm{O}_4, s) - 3H_m^\circ(\mathrm{Fe}, s) - 2H_m^\circ(\mathrm{O}_2, g) \quad \textbf{(4.4)}$$

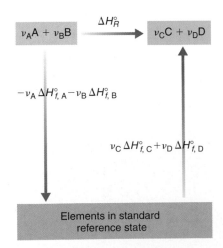

FIGURE 4.2

Equation (4.8) follows from the fact that ΔH for both paths is the same because they connect the same initial and final states.

Setting $H_m^\circ = 0$ for each element in its standard reference state,

$$\Delta H_f^\circ(Fe_3O_4, s) = H_m^\circ(Fe_3O_4, s) - 3 \times 0 - 2 \times 0 = H_m^\circ(Fe_3O_4, s) \qquad \textbf{(4.9)}$$

The value of ΔH_R° for any reaction involving compounds and elements is unchanged by this convention. In fact, one could choose a different number for the absolute enthalpy of each pure element in its standard reference state, and it would still not change the value of ΔH_R°. However, it is much more convenient (and easier to remember) if one sets $H_m^\circ = 0$ for all elements in their standard reference state. This convention will be used again in Chapter 6 when the chemical potential is discussed.

4.3 Hess's Law Is Based on Enthalpy Being a State Function

As discussed in the previous section, it is extremely useful to have tabulated values of ΔH_f° for chemical compounds at one fixed combination of P and T. Tables 4.1 and 4.2 list this data for 1 bar and 298.15 K. With access to these values of ΔH_f°, ΔH_R° can be calculated for all reactions among these elements and compounds at 1 bar and 298.15 K.

But how is ΔH_f° determined? Consider the formation reaction for $C_2H_6(g)$:

$$2\,C(graphite) + 3\,H_2(g) \longrightarrow C_2H_6(g) \qquad \textbf{(4.10)}$$

Graphite is the standard reference state for carbon at 298.15 K and 1 bar because it is slightly more stable than diamond under these conditions. However, it is unlikely that one would obtain only ethane if the reaction were carried out as written. Given this experimental hindrance, how can ΔH_f° for ethane be determined? To determine ΔH_f° for ethane, we take advantage of the fact that ΔH is path independent. In this context, path independence means that the enthalpy change for any sequence of reactions that sum to the same overall reaction is identical. This statement is known as **Hess's law.** Therefore, one is free to choose any sequence of reactions that leads to the desired outcome. Combustion reactions are well suited for these purposes because in general they proceed rapidly, go to completion, and produce only a few products. To determine ΔH_f° for ethane, one can carry out the following combustion reactions:

$$C_2H_6(g) + 7/2\,O_2(g) \longrightarrow 2\,CO_2(g) + 3\,H_2O(l) \quad \Delta H_I^\circ \qquad \textbf{(4.11)}$$

$$C(graphite) + O_2(g) \longrightarrow CO_2(g) \qquad\qquad\quad \Delta H_{II}^\circ \qquad \textbf{(4.12)}$$

$$H_2(g) + 1/2\,O_2(g) \longrightarrow H_2O(l) \qquad\qquad\qquad \Delta H_{III}^\circ \qquad \textbf{(4.13)}$$

These reactions are combined in the following way to obtain the desired reaction:

$$2 \times [C(graphite) + O_2(g) \longrightarrow CO_2(g)] \qquad\qquad 2\Delta H_{II}^\circ \qquad \textbf{(4.14)}$$

$$2\,CO_2(g) + 3\,H_2O(l) \longrightarrow C_2H_6(g) + 7/2\,O_2(g) \qquad -\Delta H_I^\circ \qquad \textbf{(4.15)}$$

$$3 \times [H_2(g) + 1/2\,O_2(g) \longrightarrow H_2O(l)] \qquad\qquad\qquad 3\Delta H_{III}^\circ \qquad \textbf{(4.16)}$$

$$2\,C(graphite) + 3\,H_2(g) \longrightarrow C_2H_6(g) \qquad\qquad 2\Delta H_{II}^\circ - \Delta H_I^\circ + 3\Delta H_{III}^\circ$$

We emphasize again that it is not necessary for these reactions to be carried out at 298.15 K. The reaction vessel is immersed in a water bath at 298.15 K and the combustion reaction is initiated. If the temperature in the vessel rises during the course of the reaction, the heat flow that restores the system and surroundings to 298.15 K after completion of the reaction is measured, allowing ΔH_R° to be determined at 298.15 K.

Several points should be made about enthalpy changes in relation to balanced overall equations describing chemical reactions. First, because H is an extensive function, multiplying all stoichiometric coefficients with any number changes ΔH_R° by the same factor. Therefore, it is important to know which set of stoichiometric coefficients has been assumed if a numerical value of ΔH_R° is given. Second, because the units of ΔH_f° for all compounds in the reaction are kJ mol^{-1}, the units of the reaction enthalpy ΔH_R° are also kJ mol^{-1}. One might pose the question "per mole of what?" given that all the stoichiometric coefficients may differ from each other and from one. The answer to this question is *per mole of the reaction as written*. Doubling all the stoichiometric coefficients doubles ΔH_R°.

EXAMPLE PROBLEM 4.1

The average **bond enthalpy** of the O—H bond in water is defined as one-half of the enthalpy change for the reaction $H_2O(g) \longrightarrow 2\,H(g) + O(g)$. The formation enthalpies, ΔH_f°, for H(g) and O(g) are 218.0 and 249.2 kJ mol^{-1}, respectively, at 298.15 K, and ΔH_f° for $H_2O(g)$ is –241.8 kJ mol^{-1} at the same temperature.

  a. Use this information to determine the average bond enthalpy of the O—H bond in water at 298.15 K.

  b. Determine the average **bond energy** ΔU of the O—H bond in water at 298.15 K. Assume ideal gas behavior.

Solution

  a. We consider the sequence

  $H_2O(g) \longrightarrow H_2(g) + 1/2\,O_2(g)$ $\Delta H^\circ = 241.8$ kJ mol^{-1}

  $H_2(g) \longrightarrow 2\,H(g)$ $\Delta H^\circ = 2 \times 218.0$ kJ mol^{-1}

  $1/2\,O_2(g) \longrightarrow O(g)$ $\Delta H^\circ = 249.2$ kJ mol^{-1}

  ─────────────────────────────

  $H_2O(g) \longrightarrow 2\,H(g) + O(g)$ $\Delta H^\circ = 927.0$ kJ mol^{-1}

  This is the enthalpy change associated with breaking both O—H bonds under standard conditions. We conclude that the average bond enthalpy of the O—H bond in water is $\frac{1}{2} \times 927.0\,\text{kJ mol}^{-1} = 463.5\,\text{kJ mol}^{-1}$. We emphasize that this is the average value because the values of ΔH for the transformations $H_2O(g) \longrightarrow H(g) + OH(g)$ and $OH(g) \longrightarrow O(g) + H(g)$ differ.

  b. $\Delta U^\circ = \Delta H^\circ - \Delta(PV) = \Delta H^\circ - \Delta nRT$

  $\qquad = 927.0$ kJ mol$^{-1} - 2 \times 8.314$ J mol^{-1}K$^{-1} \times 298.15$ K

  $\qquad = 922.0$ kJ mol^{-1}

  The average value for ΔU° for the O—H bond in water is $\frac{1}{2} \times 922.0$ kJ mol^{-1} $= 461.0$ kJ mol^{-1}. The bond energy and the bond enthalpy are nearly identical.

Example Problem 4.1 shows how bond energies can be calculated from reaction enthalpies. The value of a bond energy is of particular importance for chemists in estimating the thermal stability of a compound as well as its stability with respect to reactions with other molecules. Values of bond energies tabulated in the format of the periodic table together with the electronegativities are shown in Table 4.3 [Kildahl, N. K. "Bond Energy Data Summarized." *Journal of Chemical Education*. 72 (1995): 423]. The value of the single bond energy, ΔU_{A-B}, for a combination A–B not listed in the table can be estimated using the empirical relationship due to Linus Pauling in Equation (4.17):

$$\Delta U_{A-B} = \sqrt{\Delta U_{A-A} \times \Delta U_{B-B}} + 96.48(\chi_A - \chi_B)^2 \qquad (4.17)$$

where χ_A and χ_B are the electronegativities of atoms A and B.

TABLE 4.3 Mean Bond Energies

1	2	Selected Bond Energies (kJ/mol)	13	14	15	16	17	18
H,2.20 432 --- 432 459 565								**He**
Li,0.98 105 --- 243 --- 573	**Be**,1.57 208 --- --- ---,444 632		**B**,2.04 293 --- 389 536,636 613	**C**,2.55 346 602,835 411 358,799 485	**N**,3.04 167 418,942 386 201,607 283	**O**,3.44 142 494 459 142,494 190	**F**,3.98 155 --- 565 --- 155	**Ne**
Na,0.93 72 --- 197 --- 477	**Mg**,1.31 129 --- --- ---,377 513		**Al**,1.61 -- --- 272 --- 583	**Si**,1.90 222 318 318 452,640 565	**P**,2.19 ≈220 ---,481 322 335,544 490	**S**,2.58 240 425 363 ---,523 284	**Cl**,3.16 240 --- 428 218 249	**Ar**
K,0.82 49 --- 180 --- 490	**Sr**,1.00 105 --- --- ---,460 550		**Ga**,1.81 113 --- --- --- ≈469	**Ge**,2.01 188 272 --- --- ≈470	**As**,2.18 146 ---,380 247 301,389 ≈440	**Se**,2.55 172 272 276 --- ≈351	**Br**,2.96 190 --- 362 201 250	**Kr** 50
Rb,0.82 45 --- 163 --- 490	**Sr**,0.95 84 --- --- ---,347 553		**In**,1.78 100 --- --- --- ≈523	**Sn**,1.80 146 --- --- --- ≈450	**Sb**,2.05 121 ---,295 --- --- ≈420	**Te**,2.10 126 218 238 --- ≈393	**I**,2.66 149 -- 295 201 278	**Xe** 84 ≈131
Cs,0.79 44 --- 176 --- 502	**Ba**,0.89 44 --- --- 467,561 578		**Tl**,2.04 --- --- --- --- 439	**Pb**,2.33 --- --- --- --- ≈360	**Bi**,2.02 --- ---,192 --- ≈350	**Po**,2.00	**At**,2.20 116	**Rn**

KEY

Element symbol	**C**,2.55	Electronegativity
C—C	346	Single bond with self
C=C, C≡C	602,835	Double, triple bond with self
H—C	411	Bond with H
O—C, O=C	358,799	Single, double bond with O
C—F	485	Bond with F

4.4 The Temperature Dependence of Reaction Enthalpies

Suppose that we plan to carry out a reaction that is mildly exothermic at 298.15 K at another temperature. Is the reaction endothermic or exothermic at the second temperature? To answer this question, it is necessary to determine ΔH_R° at the second temperature. We assume that no phase changes occur in the temperature interval of interest. The enthalpy for each reactant and product at temperature T is related to the value at 298.15 K by Equation (4.18), which accounts for the energy supplied in order to heat the substance to the new temperature at constant pressure:

$$H_T^\circ = H_{298.15\,K}^\circ + \int_{298.15\,K}^{T} C_P(T')dT' \tag{4.18}$$

The prime in the integral indicates a "dummy variable" that is otherwise identical to the temperature. This notation is needed because T appears in the upper limit of the integral. In Equation (4.18), $H_{298.15\,K}^{\circ}$ is the absolute enthalpy at 1 bar and 298.15 K. However, because there are no unique values for absolute enthalpies, it is useful to combine similar equations for all reactants and products with the appropriate stoichiometric coefficients to obtain the following equation for the reaction enthalpy at temperature T:

$$\Delta H_{R,T}^{\circ} = \Delta H_{R,298.15\,K}^{\circ} + \int_{298.15\,K}^{T} \Delta C_P(T')dT' \tag{4.19}$$

where

$$\Delta C_P(T') = \sum_i \nu_i C_{P,i}(T') \tag{4.20}$$

Recall that in our notation, ΔH_R° or ΔH_f° without an explicit temperature value implies that $T = 298.15$ K. In Equation (4.20), the sum is over all reactants and products, *including both elements and compounds*. A calculation of ΔH_R° at an elevated temperature is shown in Example Problem 4.2.

EXAMPLE PROBLEM 4.2

Calculate $\Delta H_{R,\,1450\,K}^{\circ}$ for the reaction $1/2\ H_2(g) + 1/2\ Cl_2(g) \longrightarrow HCl(g)$ and 1 bar pressure given that $\Delta H_f^{\circ}(HCl,g) = -92.3$ kJ mol^{-1} at 298.15 K and that

$$C_{P,m}(H_2,g) = \left(29.064 - 0.8363 \times 10^{-3}\frac{T}{K} + 20.111 \times 10^{-7}\frac{T^2}{K^2}\right) J\ K^{-1} mol^{-1}$$

$$C_{P,m}(Cl_2,g) = \left(31.695 + 10.143 \times 10^{-3}\frac{T}{K} - 40.373 \times 10^{-7}\frac{T^2}{K^2}\right) J\ K^{-1} mol^{-1}$$

$$C_{P,m}(HCl,g) = \left(28.165 + 1.809 \times 10^{-3}\frac{T}{K} + 15.464 \times 10^{-7}\frac{T^2}{K^2}\right) J\ K^{-1} mol^{-1}$$

over this temperature range. The ratios T/K and T^2/K^2 appear in these equations in order to have the right units for the heat capacity.

Solution

$$\Delta H_{R,1450K}^{\circ} = \Delta H_{R,298.15K}^{\circ} + \int_{298.15}^{1450} \Delta C_P(T)\,dT$$

$$\Delta C_P(T) = \left[28.165 + 1.809 \times 10^{-3}\frac{T}{K} + 15.464 \times 10^{-7}\frac{T^2}{K^2} \right.$$

$$- \frac{1}{2}\left(29.064 - 0.8363 \times 10^{-3}\frac{T}{K} + 20.111 \times 10^{-7}\frac{T^2}{K^2}\right)$$

$$\left. - \frac{1}{2}\left(31.695 + 10.143 \times 10^{-3}\frac{T}{K} - 40.373 \times 10^{-7}\frac{T^2}{K^2}\right) \right] J\ K^{-1} mol^{-1}$$

$$= \left(-2.215 - 2.844 \times 10^{-3}\frac{T}{K} + 25.595 \times 10^{-7}\frac{T^2}{K^2}\right) J\ K^{-1} mol^{-1}$$

$$\Delta H_{R,1450K}^{\circ} = -92.3\,kJ\,mol^{-1}$$

$$+ \int_{298.15}^{1450} \left(-2.215 - 2.844 \times 10^{-3}\frac{T}{K} + 25.595 \times 10^{-7}\frac{T^2}{K^2}\right) \times d\frac{T}{K} J\,mol^{-1}$$

$$= -92.3\,kJ\,mol^{-1} - 2.836\,kJ\,mol^{-1} = -95.1\,kJ\,mol^{-1}$$

In this particular case, the change in the reaction enthalpy with T is not large. This is the case because $\Delta C_P(T)$ is small and not because an individual $C_{P,i}(T)$ is small.

4.5 The Experimental Determination of ΔU and ΔH for Chemical Reactions

For chemical reactions, ΔU and ΔH are generally determined through experiment. In this section, we discuss how these experiments are carried out. If some or all of the reactants or products are volatile, it is necessary to contain the reaction mixture for which ΔU and ΔH are being measured. Such an experiment can be carried out in a **bomb calorimeter,** shown schematically in Figure 4.3. In a bomb calorimeter, the reaction is carried out at constant volume. The motivation for doing so is that if $dV = 0$, $\Delta U = q_v$. Therefore, a measurement of the heat flow normalized to 1 mole of the specified reaction provides a direct measurement of ΔU_R. Bomb calorimetry is restricted to reaction mixtures containing gases because it is impractical to carry out chemical reactions at constant volume for systems consisting solely of liquids and solids, as shown in Example Problem 3.2. In the following, we describe how ΔU_R and ΔH_R are determined for an experiment in which a single liquid or solid reactant undergoes combustion in an excess of $O_2(g)$.

The bomb calorimeter is a good illustration of how one can define the system and surroundings to simplify the analysis of an experiment. The system is defined as the contents of a stainless steel thick-walled pressure vessel, the pressure vessel itself, and the inner water bath. Given this definition of the system, the surroundings consist of the container holding the inner water bath, the outer water bath, and the rest of the universe. The outer water bath encloses the inner bath and, through a heating coil, its temperature is always held at the temperature of the inner bath. Therefore, no heat flow will occur between the system and surroundings, and $q = 0$. Because the combustion experiment takes place at constant volume, $w = 0$. Therefore, $\Delta U = 0$. These conditions describe an isolated system of finite size that is not coupled to the rest of the universe. We are only interested in one part of this system, namely, the reaction mixture.

What are the individual components that make up ΔU? Consider the system as consisting of three subsystems: the reactants in the calorimeter, the calorimeter vessel, and the inner water bath. These three subsystems are separated by rigid diathermal walls and are in thermal equilibrium. Energy is redistributed among the subsystems as reactants are converted to products, the temperature of the inner water bath changes, and the temperature of the calorimeter changes.

$$\Delta U = \frac{m_s}{M_s}\Delta U_{combustion} + \frac{m_{H_2O}}{M_{H_2O}} \times C_{P,m}(H_2O) \times \Delta T + C_{calorimeter} \times \Delta T = 0 \,(\textbf{4.21})$$

In Equation (4.21), ΔT is the change in the temperature of the three subsystems. The mass of water in the inner bath, m_{H_2O}; its molecular weight, M_{H_2O}; its heat capacity, $C_{P,m}(H_2O)$; the mass of the sample, m_s; and its molecular weight, M_s, are known. $\Delta U_{combustion}$ is defined per mole of the combustion reaction, but because the reaction includes exactly 1 mole of reactant, the factor m_s/M_s in Equation (4.21) is appropriate. We wish to measure $\Delta U_{combustion}$. However, to determine $\Delta U_{combustion}$, the heat capacity of the calorimeter, $C_{calorimeter}$, must first be determined by carrying out a reaction for which ΔU_R is already known, as illustrated in Example Problem 4.3. To be more specific, we consider a combustion reaction between a compound and an excess of O_2.

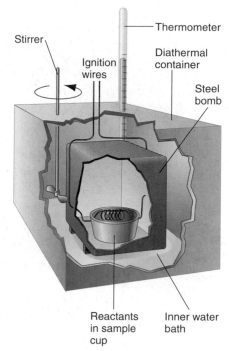

FIGURE 4.3
Schematic diagram of a bomb calorimeter. The liquid or solid reactant is placed in a cup suspended in the thick-walled steel bomb, which is filled with O_2 gas. The vessel is immersed in an inner water bath, and its temperature is monitored. The diathermal container is immersed in an outer water bath (not shown) whose temperature is maintained at the same value as the inner bath through a heating coil. By doing so, there is no heat exchange between the inner water bath and the rest of the universe.

EXAMPLE PROBLEM 4.3

When 0.972 g of cyclohexane undergoes complete combustion in a bomb calorimeter, ΔT of the inner water bath is 2.98°C. For cyclohexane, $\Delta U_{combustion}$ is -3913 kJ mol^{-1}. Given this result, what is the value for $\Delta U_{combustion}$ for the combustion of benzene if ΔT is 2.36°C when 0.857 g of benzene undergoes complete combustion in the same calorimeter? The mass of the water in the inner bath is 1.812×10^3 g, and the $C_{P,m}$ of water is 75.3 J K^{-1} mol^{-1}.

Solution

To calculate the calorimeter constant through the combustion of cyclohexane, we write Equation (4.21) in the following form:

$$C_{calorimeter} = \frac{-\dfrac{m_s}{M_s}\Delta U_{combustion} - \dfrac{m_{H_2O}}{M_{H_2O}}C_{P,m}(H_2O)\Delta T}{\Delta T}$$

$$= \frac{\dfrac{0.972\,g}{84.16\,g\,mol^{-1}}\times 3913\times 10^3\,J\,mol^{-1} - \dfrac{1.812\times 10^3\,g}{18.02\,g\,mol^{-1}}\times 75.3\,J\,mol^{-1}\,K^{-1}\times 2.98°C}{2.98°C}$$

$$= 7.59\times 10^3\,J\,(°C)^{-1}$$

In calculating $\Delta U_{combustion}$ for benzene, we use the value for $C_{calorimeter}$:

$$\Delta U_{combustion} = -\frac{M_s}{m_s}\left(\frac{m_{H_2O}}{M_{H_2O}}C_{P,m}(H_2O)\Delta T + C_{calorimeter}\Delta T\right)$$

$$= -\frac{78.12\,g\,mol^{-1}}{0.857\,g}\times\left(\begin{array}{l}\dfrac{1.812\times 10^3\,g}{18.02\,g\,mol^{-1}}\times 75.3\,J\,mol^{-1}\,K^{-1}\times 2.36°C\\[2mm] + 7.59\times 10^3\,J(°C)^{-1}\times 2.36°C\end{array}\right)$$

$$= -3.26\times 10^6\,J\,mol^{-1}$$

Once $\Delta U_{combustion}$ has been determined, $\Delta H_{combustion}$ can be determined using the following equation:

$$\Delta H_{combustion} = \Delta U_{combustion} + \Delta(PV) \tag{4.22}$$

For reactions involving only solids and liquids, $\Delta U \gg \Delta(PV)$ and $\Delta H \approx \Delta U$. If some of the reactants or products are gases, the small change in the temperature that is measured in a calorimetric experiment can generally be ignored and $\Delta(PV) = \Delta(nRT) = \Delta nRT$

$$\Delta H_{combustion} = \Delta U_{combustion} + \Delta nRT \tag{4.23}$$

where Δn is the change in the number of moles of gas in the overall reaction. For the first reaction of Example Problem 4.3,

$$C_6H_{12}(l) + 9\,O_2(g) \longrightarrow 6\,CO_2(g) + 6\,H_2O(l) \tag{4.24}$$

and $\Delta n = -3$. Note that at $T = 298.15$ K, the most stable form of cyclohexane and water is a liquid.

$$\Delta H_{combustion} = \Delta U_{combustion} - 3RT = -3913\times 10^3\,kJ\,mol^{-1}$$
$$-3\times 8.314\,J\,K^{-1}mol^{-1}\times 298.15\,K$$
$$= -3920\times 10^3\,J\,mol^{-1} \tag{4.25}$$

For this reaction, $\Delta H_{combustion}$ and $\Delta U_{combustion}$ differ by only 0.2%. Note that because the contents of the bomb calorimeter are not at 1 bar pressure, $\Delta U_{combustion}$ rather than $\Delta U°_{combustion}$ is measured. The difference is small, but it can be calculated.

If the reaction under study does not involve gases or highly volatile liquids, there is no need to operate under constant volume conditions. It is preferable to carry out the reaction at constant P using a **constant pressure calorimeter**. ΔH is directly determined because $\Delta H = q_P$. A vacuum-insulated vessel with a loosely fitting stopper as shown in Figure 4.4 is adequate for many purposes and can be treated as an isolated composite system. Equation (4.21) takes the following form for constant pressure calorimetry involving the solution of a salt in water:

$$\Delta H° = \frac{m_{salt}}{M_{salt}}\Delta H°_{solution} + \frac{m_{H_2O}}{M_{H_2O}}C_{P,m}(H_2O)\Delta T + C_{calorimeter}\Delta T = 0 \tag{4.26}$$

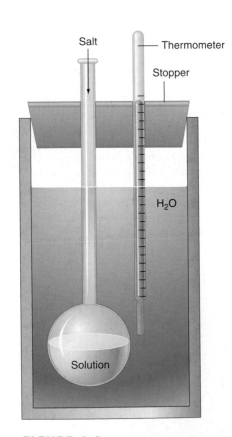

Salt — Thermometer
Stopper
H₂O
Solution

FIGURE 4.4

Schematic diagram of a constant pressure calorimeter suitable for measuring the enthalpy of solution of a salt in water.

$\Delta H^\circ_{solution}$ is defined per mole of the solution reaction, but because the reaction includes exactly 1 mole of reactant, the factor m_{salt}/M_{salt} in Equation (4.21) is appropriate. Because $\Delta(PV)$ is negligibly small for the solution of a salt in a solvent, $\Delta U^\circ_{solution} = \Delta H^\circ_{solution}$. The solution must be stirred to ensure that equilibrium is attained before ΔT is measured.

EXAMPLE PROBLEM 4.4

The enthalpy of solution for the reaction

$$Na_2SO_4(s) \xrightarrow{H_2O(l)} 2\,Na^+(aq) + SO_4^{2-}(aq)$$

is determined in a constant pressure calorimeter. The calorimeter constant was determined to be 342.5 J K^{-1}. When 1.423 g of Na_2SO_4 is dissolved in 100.34 g of $H_2O(l)$, $\Delta T = 0.031$ K. Calculate $\Delta H^\circ_{solution}$ for Na_2SO_4 from these data. Compare your result with that calculated using the standard enthalpies of formation in Table 4.1 (Appendix B, Data Tables) and in Chapter 10 in Table 10.1.

Solution

$$\Delta H^\circ_{solution} = -\frac{M_{salt}}{m_{salt}}\left(\frac{m_{H_2O}}{M_{H_2O}}C_{P,m}(H_2O)\Delta T + C_{calorimeter}\Delta T\right)$$

$$= -\frac{142.04 \text{ g mol}^{-1}}{1.423 \text{ g}} \times \left(\begin{array}{c}\dfrac{100.34 \text{ g}}{18.02 \text{ g mol}^{-1}} \times 75.3 \text{ J K}^{-1}\text{ mol}^{-1} \times 0.031 \text{ K} \\ + 342.5 \text{ J K}^{-1} \times 0.031 \text{ K}\end{array}\right)$$

$$= -2.4 \times 10^3 \text{ J mol}^{-1}$$

We next calculate $\Delta H^\circ_{solution}$ using the data tables.

$$\Delta H^\circ_{solution} = 2\Delta H^\circ_f(Na^+,aq) + \Delta H^\circ_f(SO_4^{2-},aq) - \Delta H^\circ_f(Na_2SO_4,s)$$

$$= 2 \times (-240.1 \text{ kJ mol}^{-1}) - 909.3 \text{ kJ mol}^{-1} + 1387.1 \text{ kJ mol}^{-1}$$

$$= -2.4 \text{ kJ mol}^{-1}$$

The agreement between the calculated and experimental results is good.

SUPPLEMENTAL

4.6 Differential Scanning Calorimetry

Differential scanning calorimetry (DSC) is a form of constant pressure calorimetry that is well suited to routine laboratory tests in pharmaceutical and material sciences. It is also used to study chemical changes such as polymer cross-linking, melting, and unfolding of protein molecules in which heat is absorbed or released in the transition. The experimental apparatus for such an experiment is shown schematically in Figure 4.5. The word *differential* appears in the name of the technique because the uptake of heat is measured relative to that for a reference material, and *scanning* refers to the fact that the temperature of the sample is varied, usually linearly with time.

The temperature of the enclosure T_E is increased linearly with time using a power supply. Heat flows from the enclosure through the disk to the sample because of the temperature gradient generated by the heater. Because the sample and reference are equidistant from the enclosure, the heat flow to each sample is the same. The reference material is chosen such that its melting point is not in the range of that of the samples.

We consider a simplified one-dimensional model of the heat flow in the DSC in Figure 4.6 following the treatment of Höhne, Hemminger, and Flammersheim in *Differential Scanning Calorimetry*, 2nd Edition, Berlin: Springer, 2003. The electrical current through the resistive heater increases the temperature of the calorimeter enclosure T_E to a value greater than that of the sample and reference, T_S, and T_R. The

FIGURE 4.5
A heat flux differential scanning
calorimeter consists of an insulated
massive enclosure and lid that are heated
to the temperature T_E using a resistive
heater. A support disk in good thermal
contact with the enclosure supports the
sample and reference materials. The
temperatures of the sample and reference
are measured with a thermocouple. In
practice, the reference is usually an empty
sample pan.

FIGURE 4.5
A heat flux differential scanning
calorimeter consists of an insulated
massive enclosure and lid that are heated
to the temperature T_E using a resistive
heater. A support disk in good thermal
contact with the enclosure supports the
sample and reference materials. The
temperatures of the sample and reference
are measured with a thermocouple. In
practice, the reference is usually an empty
sample pan.

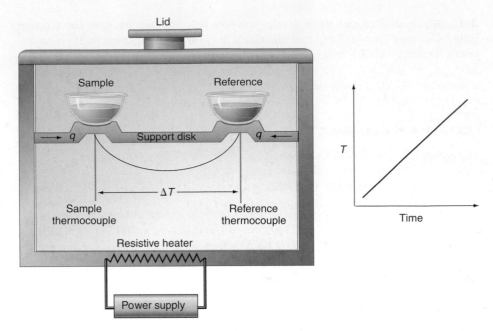

heat flow per unit time from the enclosure to the sample and reference are designated by Φ_{ES} and Φ_{ER}, respectively, which typically have the units of J g^{-1} s^{-1}.

Assume that a process such as melting occurs in the sample and not in the reference. The heat flow per unit time associated with the process is given by $\Phi(t)$. It is time dependent because, using melting as an example, heat flow associated with the phase change begins at the onset of melting and ceases when the sample is completely in the liquid state. This additional heat flow changes the sample temperature by dT_S and consequently both $T_E - T_S$ and the heat flow rate to the sample Φ_{ES} change. In the experiment $\Delta T(t) = T_S(t) - T_R(t)$ is measured. The change in the heat flow to the sample resulting from the process is

$$C_S \frac{dT_S(t)}{dt} = \Phi_{ES}(t) - \Phi(t) \tag{4.27}$$

where C_S is the constant pressure heat capacity of the sample. Equation (4.27) relates the sample temperature to the heat flow generated by the process of interest. If the process is exothermic, $\Phi(t) < 0$ and if it is endothermic, $\Phi(t) > 0$. We rewrite Equation (4.27) to explicitly contain the experimentally accessible function $\Delta T(t)$.

$$C_S \frac{dT_R(t)}{dt} + C_S \frac{d\Delta T(t)}{dt} = \Phi_{ES}(t) - \Phi(t) \tag{4.28}$$

FIGURE 4.6
ΔT is shown as a function of time for an
exothermic process occuring in the sample.
For this example, it is assumed that the
heat capacities of the sample and reference
are constant over the temperature range
shown. The heat associated with the
process on interest is proportional to the
blue area. The green area arises from
the difference in heat capacities of sample
and reference. The zero line is obtained
without material in the crucibles.

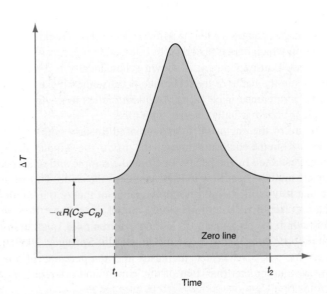

The corresponding equation for the reference in which only heating occurs is

$$C_R \frac{dT_R(t)}{dt} = \Phi_{ER}(t) \tag{4.29}$$

The essence of DSC is that a difference between the heat flux to the sample and reference is measured. We therefore subtract Equation (4.29) from Equation (4.28) and obtain

$$\Phi_{ES}(t) - \Phi_{ER}(t) = (C_S - C_R)\frac{dT_R(t)}{dt} + C_S\frac{d\Delta T(t)}{dt} + \Phi(t) \tag{4.30}$$

The quantities Φ_{ES} and Φ_{ER} are directly proportional to the temperature differences $T_E - T_S$ and $T_E - T_R$ and are inversely proportional to the thermal resistance $R_{thermal}$ between the heated enclosure and the sample and reference. The latter are equal because of the symmetry of the calorimeter. The thermal resistance $R_{thermal}$ is an intrinsic property of the calorimeter and can be obtained through calibration of the instrument.

$$\Phi_{ES} = \frac{T_E - T_S}{R_{thermal}} \text{ and } \Phi_{ER} = \frac{T_E - T_R}{R_{thermal}} \tag{4.31}$$

Therefore,

$$\Phi(t) = -\frac{\Delta T(t)}{R_{thermal}} - (C_S - C_R)\frac{dT_R(t)}{dt} - C_S\frac{d\Delta T(t)}{dt} \tag{4.32}$$

Equation (4.32) links the heat flow per unit mass generated by the process of interest, $\Phi(t)$, to the measured quantity $\Delta T(t)$. Typically T_R is increased linearly with time,

$$T_R(t) = T_0 + \alpha t$$

and Equation (4.32) simplifies to

$$\Phi(t) = -\frac{\Delta T(t)}{R_{thermal}} - \alpha(C_S - C_R) - C_S\frac{d\Delta T(t)}{dt} \tag{4.33}$$

We see that $\Delta T(t)$ is not simply related to $\Phi(t)$, which is to be determined in the experiment. The second term in Equation (4.33) arises because the heat capacities of the sample and reference are not equal. Additionally, the heat capacity of the sample changes as it undergoes a phase change.

The third term in Equation (4.33) is proportional to $d\Delta T(t)/dt$ and has the effect of broadening $\Delta T(t)$ relative to $\Phi(t)$ and shifting it to longer times. A schematic picture of a DSC scan in this model is shown in Figure 4.6.

Because $C_R\alpha = \Phi_{ER}(t)$ from Equation (4.29) and ΔT are proportional as shown in Equation (4.31), a graph of C_R versus time has the same shape as shown in Figure 4.6. An analogous equation can be written for the sample, so that

$$\Phi_{ES} - \Phi_{ER} = \alpha(C_S - C_R)$$

Therefore, a graph of $(C_S - C_R)$ versus time also has the same shape as Figure 4.6.

The goal of the experiment is to determine the heat absorbed or evolved in the process per unit mass, which is given by

$$q_P = \Delta H = \int_{t_1}^{t_2} \Phi(t)dt$$

$$= -\frac{1}{R_{thermal}}\int_{t_1}^{t_2}\Delta T(t)dt - \frac{1}{R_{thermal}}\int_{t_1}^{t_2}\left(-\alpha\left(C_S - C_R\right)\right)dt \tag{4.34}$$

In Equation (4.34), the baseline contribution to the integral has been subtracted as it has no relevance for the process of interest.

Interpreting DSC curves in terms of heat capacities must be done with caution as illustrated for a melting transition in Figure 4.7. At the melting temperature, the enthalpy rises abruptly as discussed in Chapter 8 and as shown in Figure 4.7a. The heat capacity is the derivative of the enthalpy with respect to temperature and has the form shown in Figure 4.7b. As discussed in Section 2.5, the heat capacity of the liquid is higher than that of the solid. However, a measured DSC curve has the shape shown in Figure 4.7d rather than in Figure 4.7b.

There are two reasons for this discrepancy. Heat is taken up by the sample during the melting transition, but the sample temperature does not change. However, the

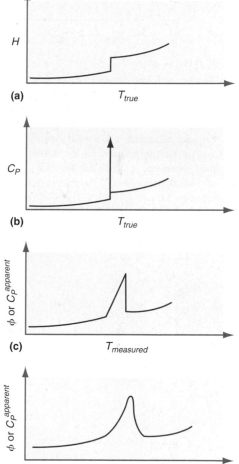

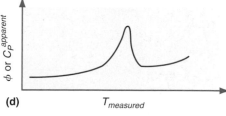

FIGURE 4.7

(a) ΔH is shown as a function of the true temperature for a melting process. (b) The heat capacity becomes infinite at the transition temperature because the temperature remains constant as heat flows into the sample. (c) The heat capacity curve in (b) is distorted because the measured temperature increases while the sample temperature remains constant during the melting transition. (d) Further distortions to the hypothetical scan (c) and a shift of the peak to higher temperature occur because of the thermal inertia of the calorimeter.

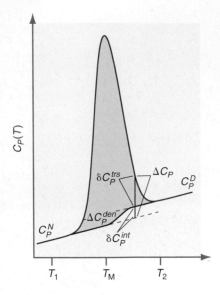

FIGURE 4.8
The apparent heat capacity of a protein undergoing a reversible, thermal denaturation within the temperature interval T_1–T_2 is shown. The temperature at the heat capacity maximum is to a good approximation the melting temperature T_m, defined as the temperature at which half the total protein has been denatured. The excess heat capacity $\Delta C_P = C_P(T) - C_P^N(T)$ is composed of two parts; the intrinsic excess heat capacity δC_P^{int} and the transition excess heat capacity δC_P^{trs}. See the text.

sample crucible temperature continues to increase as heat flows to the sample. Therefore the apparent heat capacity would have the shape shown in Figure 4.7c if there were no resistances to heat flow in the calorimeter. The thermal resistances present in the calorimeter smear out the curve shown in Figure 4.7c and shift it to higher temperatures as shown in Figure 4.7d. Clearly, the temperature dependence of the apparent heat capacity obtained directly from a DSC scan is significantly different from the temperature dependence of the true heat capacity. Deconvolution procedures must be undertaken to obtain accurate heat capacities. Because the heat absorbed or evolved in the process per unit mass is proportional to the area under the DSC scan, it is less affected by the instrument distortions than the heat capacity.

DSC is the most direct method for determining the energetics of biological macromolecules undergoing conformational transitions, which is important in understanding their biological activity. In particular, DSC has been used to determine the temperature range over which proteins undergo the conformational changes associated with reversible **denaturation,** a process in which a protein unfolds. The considerations for a melting process discussed earlier apply as well as for denaturation, although melting enthalpies are generally much larger than denaturation enthalpies. From the preceding discussion, when a protein solution is heated at a constant rate and at constant pressure, DSC reports the apparent heat capacity of the protein solution. Protein structures are stabilized by the cooperation of numerous weak forces. Assume that a protein solution is heated from a temperature T_1 to a temperature T_2. If the protein denatures within this temperature interval reversibly and cooperatively, the $C_P(T)$ versus T curve has the form shown in Figure 4.8.

The heat capacity of denaturation is defined as

$$\Delta C_P^{den} = C_P^D(T) - C_P^N(T) \tag{4.35}$$

which is the difference between the heat capacity of the denatured protein, $C_P^D(T)$, and the heat capacity of the native (i.e., structured) protein, $C_P^N(T)$. The value of ΔC_P^{den} can be determined as shown in Figure 4.8. The value of ΔC_P^{den} is a positive quantity and for many globular proteins varies from 0.3 to 0.7 J K^{-1} g^{-1}. The higher heat capacity of the protein in the denatured state can be understood in the following way. In the structured state, internal motions of the protein are characterized by coupled bending and rotations of several bonds that occur at frequencies on the order of $k_B T/h$, where k_B is Boltzmann's constant and h is Planck's constant. When the protein denatures, these "soft" vibrations are shifted to lower frequencies, higher frequency bond vibrations are excited, and the heat capacity increases. In addition to increased contributions from vibrational motions, when a protein structure is disrupted, nonpolar amino acid residues formerly isolated from solvent within the interior of the protein are exposed to water. Water molecules now order about these nonpolar amino acids, further increasing the heat capacity.

The **excess heat capacity** is defined as

$$\Delta C_P = C_P(T) - C_P^N(T) \tag{4.36}$$

where ΔC_P is obtained at a given temperature from the difference between the point on the heat capacity curve $C_P(T)$ and the linearly extrapolated value for the heat capacity of the native state $C_P^N(T)$; see Figure 4.8. The excess heat capacity is in turn composed of two parts:

$$\Delta C_P(T) = \delta C_P^{int} + \delta C_P^{trs} \tag{4.37}$$

Within the interval T_1–T_2 the protein structure does not unfold suddenly. There occurs instead a gradual unfolding of the protein such that at any given temperature a fraction of the structured protein remains, and an ensemble of unfolded species is produced. The heat capacity and enthalpy change observed arise from this ensemble of physical forms of the partially unfolded protein, and as such these properties are averages over possible configurations of the protein. The component of the heat capacity that accounts for the sum of molecular species produced in the course of making a transition from the folded to the unfolded state is the **intrinsic excess heat capacity** or δC_P^{int}. The second component of the heat capacity is called the **transition excess heat capacity** δC_P^{trs}. The transition excess heat capacity results from fluctuations of the system as the protein changes from different states in the course of denaturation.

The intrinsic and transition excess heat capacities are determined by first extrapolating the functions $C_P^N(T)$ and $C_P^D(T)$ into the transition zone between T_1 and T_2. This is indicated in Figure 4.8 by the solid line connecting the heat capacity baselines above and below T_m. Once this baseline extrapolation is accomplished, δC_P^{int} is the difference at a given temperature between the extrapolated baseline curve and $C_P^N(T)$. The value of δC_P^{trs} is obtained from the difference between $C_P(T)$ and the extrapolated baseline curve. The excess enthalpy of thermal denaturation is finally given by $\Delta H_{den} = \int_{T_1}^{T_2} \delta C_P^{trs} dT$. In other words the excess enthalpy of thermal denaturation is the area under the peak in Figure 4.8 above the extrapolated baseline curve.

Vocabulary

bomb calorimeter

bond energy

bond enthalpy

constant pressure calorimeter

denaturation

differential scanning calorimetry

endothermic

enthalpy of fusion

enthalpy of reaction

excess heat capacity

exothermic

Hess's law

intrinsic excess heat capacity

standard enthalpy of formation

standard enthalpy of reaction

standard reference state

standard state

transition excess heat capacity

Conceptual Problems

Q4.1 In calculating ΔH_R° at 285.15 K, only the ΔH_f° of the compounds that take part in the reactions listed in Tables 4.1 and 4.2 (Appendix B, Data Tables) are needed. Is this statement also true if you want to calculate ΔH_R° at 500. K?

Q4.2 What is the point of having an outer water bath in a bomb calorimeter (see Figure 4.3), especially if its temperature is always equal to that of the inner water bath?

Q4.3 Is the following statement correct? If not rewrite it so that it is correct. The standard state of water is $H_2O(g)$.

Q4.4 Does the enthalpy of formation of $H_2O(l)$ change if the absolute enthalpies of $H_2(g)$ and $O_2(g)$ are set equal to 100. kJ mol^{-1} rather than to zero? Answer the same question for $CO_2(g)$. Will ΔH_R° for the reaction $H_2O(l) + CO_2(g) \longrightarrow H_2CO_3(l)$ change as a result of this change in the enthalpy of formation of the elements?

Q4.5 Why are elements included in the sum in Equation (4.14) when they are not included in calculating ΔH_R° at 298 K?

Q4.6 Why are heat capacities of reactants and products required for calculations of ΔH_R° at elevated temperatures?

Q4.7 Is the following statement correct? If not rewrite it so that it is correct. The superscript zero in ΔH_f° means that the reactions conditions are 298.15 K.

Q4.8 Why is it valid to add the enthalpies of any sequence of reactions to obtain the enthalpy of the reaction that is the sum of the individual reactions?

Q4.9 In a calorimetric study, the temperature of the system rises to 325 K before returning to its initial temperature of 298 K. Why does this temperature rise not affect your measurement of ΔH_R° at 298 K?

Q4.10 Is the following statement correct? If not rewrite it so that it is correct. Because the reaction $H_2(g) + O_2(g) \longrightarrow H_2O(l)$ is exothermic, the products are at a higher temperature than the reactants.

Q4.11 The reactants in the reaction $2NO(g) + O_2(g) \longrightarrow 2NO_2(g)$ are initially at 298 K. Why is the reaction enthalpy the same if (a) the reaction is constantly kept at 298 K or (b) if the reaction temperature is not controlled and the heat flow to the surroundings is measured only after the temperature of the products is returned to 298 K?

Q4.12 What is the advantage of a differential scanning calorimeter over a bomb calorimeter in determining the enthalpy of fusion of a series of samples?

Q4.13 You wish to measure the heat of solution of NaCl in water. Would the calorimetric technique of choice be at constant pressure or constant volume? Why?

Q4.14 Is the following statement correct? If not rewrite it so that it is correct. Because the enthalpy of formation of elements is zero, $\Delta H_f^\circ (O(g)) = 0$.

Q4.15 If the ΔH_f° for the chemical compounds involved in a reaction is available at a given temperature, how can ΔH_R° be calculated at another temperature?

Q4.16 Is the following statement correct? If not rewrite it so that it is correct. If ΔH_R° for a chemical reaction does not change appreciably with temperature, the heat capacities for reactants and products must be small.

Q4.17 Under what conditions are ΔH and ΔU for a reaction involving gases and/or liquids or solids identical?

Q4.18 Dogs cool off in hot weather by panting. Write a chemical equation to describe this process and calculate ΔH_R°.

Q4.19 Is ΔH for breaking the first C—H bond in methane equal to the average C—H bond enthalpy in this molecule? Explain your answer.

Q4.20 Humans cool off through perspiration. How does the effectiveness of this process depend on the relative humidity?

Numerical Problems

Problem numbers in **red** indicate that the solution to the problem is given in the *Student's Solutions Manual*.

P4.1 Given the data in Table 4.1 (Appendix B, Data Tables) and the following information, calculate the single bond enthalpies and energies for Si–F, Si–Cl, C–F, N–F, O–F, H–F:

Substance	$SiF_4(g)$	$SiCl_4(g)$	$CF_4(g)$	$NF_3(g)$	$OF_2(g)$	$HF(g)$
ΔH_f° (kJ mol^{-1})	−1614.9	−657.0	−925	−125	−22	−271

P4.2 At 1000. K, $\Delta H_R^\circ = -123.77$ kJ mol^{-1} for the reaction $N_2(g) + 3 H_2(g) \longrightarrow 2 NH_3(g)$, with $C_{P,m} = 3.502R$, $3.466R$, and $4.217R$ for $N_2(g)$, $H_2(g)$, and $NH_3(g)$, respectively. Calculate ΔH_f° of $NH_3(g)$ at 450. K from this information. Assume that the heat capacities are independent of temperature.

P4.3 A sample of $K(s)$ of mass 2.740 g undergoes combustion in a constant volume calorimeter at 298.15 K. The calorimeter constant is 1849 J K^{-1}, and the measured temperature rise in the inner water bath containing 1450. g of water is 1.60 K. Calculate ΔU_f° and ΔH_f° for K_2O.

P4.4 Calculate ΔH_f° for $NO(g)$ at 975 K, assuming that the heat capacities of reactants and products are constant over the temperature interval at their values at 298.15 K.

P4.5 The total surface area of the earth covered by ocean is 3.35×10^8 km^2. Carbon is fixed in the oceans via photosynthesis performed by marine plants according to the reaction $6 CO_2(g) + 6 H_2O(l) \longrightarrow C_6H_{12}O_6(s) + 6 O_2(g)$. A lower range estimate of the mass of carbon fixed in the oceans is 44.5 metric tons/km^2. Calculate the annual enthalpy change resulting from photosynthetic carbon fixation in the ocean given earlier. Assume $P = 1$ bar and $T = 298$ K.

P4.6 Derive a formula for $\Delta H_R^\circ(T)$ for the reaction $CO(g) + 1/2 O_2(g) \longrightarrow CO_2(g)$ assuming that the heat capacities of reactants and products do not change with temperature.

P4.7 Given the data in Table 4.3 and the data tables, calculate the bond enthalpy and energy of the following:

a. The C—H bond in CH_4

b. The C—C single bond in C_2H_6

c. The C=C double bond in C_2H_4

Use your result from part (a) to solve parts (b) and (c).

P4.8 Use the following data at 298.15 K to complete this problem:

	ΔH_R° (kJ mol^{-1})
$1/2 H_2(g) + 1/2 O_2(g) \longrightarrow OH(g)$	38.95
$H_2(g) + 1/2 O_2(g) \longrightarrow H_2O(g)$	−241.814
$H_2(g) \longrightarrow 2 H(g)$	435.994
$O_2(g) \longrightarrow 2 O(g)$	498.34

Calculate ΔH_R° for

a. $OH(g) \longrightarrow H(g) + O(g)$

b. $H_2O(g) \longrightarrow 2 H(g) + O(g)$

c. $H_2O(g) \longrightarrow H(g) + OH(g)$

Assuming ideal gas behavior, calculate ΔH_R° and ΔU_R° for all three reactions.

P4.9 Calculate the standard enthalpy of formation of $FeS_2(s)$ at 600.°C from the following data at 298.15 K. Assume that the heat capacities are independent of temperature.

Substance	Fe(s)	$FeS_2(s)$	$Fe_2O_3(s)$	S(rhombic)	$SO_2(g)$
ΔH_f°(kJ mol^{-1})		−824.2			−296.81
$C_{P,m}/R$	3.02	7.48		2.72	

You are also given that for the reaction $2 FeS_2(s) + 11/2 O_2(g) \longrightarrow Fe_2O_3(s) + 4 SO_2(g)$, $\Delta H_R^\circ = -1655$ kJ mol^{-1}.

P4.10 The following data are a DSC scan of a solution of a T4 lysozyme mutant. From the data determine T_m. Determine also the excess heat capacity ΔC_P at $T = 308$ K. Determine also the intrinsic δC_P^{int} and transition δC_P^{trs} excess heat capacities at $T = 308$ K. In your calculations use the extrapolated curves, shown as dotted lines in the DSC scan.

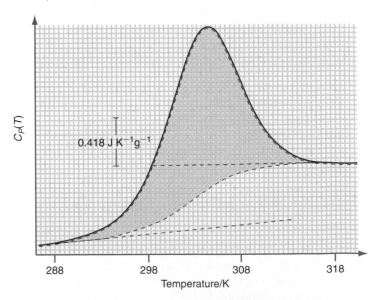

P4.11 At 298 K, $\Delta H_R^\circ = 131.28$ kJ mol^{-1} for the reaction $C(graphite) + H_2O(g) \longrightarrow CO(g) + H_2(g)$, with $C_{P,m} = 8.53$, 33.58, 29.12, and 28.82 J K^{-1} mol^{-1} for graphite, $H_2O(g)$, $CO(g)$, and $H_2(g)$, respectively. Calculate ΔH_R° at 240°C from this information. Assume that the heat capacities are independent of temperature.

P4.12 Consider the reaction $TiO_2(s) + 2\ C(graphite) + 2\ Cl_2(g) \longrightarrow 2\ CO(g) + TiCl_4(l)$ for which $\Delta H_{R,298K}^\circ = -80.$ kJ mol^{-1}. Given the following data at 25°C, (a) calculate ΔH_R° at 135.8°C, the boiling point of TiCl$_4$, and (b) calculate ΔH_f° for TiCl$_4(l)$ at 25°C:

Substance	$TiO_2(s)$	$Cl_2(g)$	$C(graphite)$	$CO(g)$	$TiCl_4(l)$
ΔH_f° (kJ mol^{-1})	−945			−110.5	
$C_{P,m}$ (J K^{-1} mol^{-1})	55.06	33.91	8.53	29.12	145.2

Assume that the heat capacities are independent of temperature.

P4.13 Calculate ΔH_R° and ΔU_R° for the oxidation of benzene. Also calculate

$$\frac{\Delta H_R^\circ - \Delta U_R^\circ}{\Delta H_R^\circ}$$

P4.14 Several reactions and their standard reaction enthalpies at 298.15 K are given here:

	ΔH_R° (kJ mol^{-1})
$CaC_2(s) + 2\ H_2O(l) \longrightarrow Ca(OH)_2(s) + C_2H_2(g)$	−127.9
$Ca(s) + 1/2\ O_2(g) \longrightarrow CaO(s)$	−635.1
$CaO(s) + H_2O(l) \longrightarrow Ca(OH)_2(s)$	−65.2

The standard enthalpies of combustion of graphite and $C_2H_2(g)$ are −393.51 and −1299.58 kJ mol^{-1}, respectively. Calculate the standard enthalpy of formation of $CaC_2(s)$ at 25°C.

P4.15 Benzoic acid, 1.35 g, is reacted with oxygen in a constant volume calorimeter to form $H_2O(l)$ and $CO_2(g)$ at 298 K. The mass of the water in the inner bath is 1.55×10^3 g. The temperature of the calorimeter and its contents rises 2.76 K as a result of this reaction. Calculate the calorimeter constant.

P4.16 The total surface area of Asia consisting of forest, cultivated land, grass land, and desert is 4.46×10^7 km^2. Every year, the mass of carbon fixed by photosynthesis by vegetation covering this land surface according to the reaction $6\ CO_2(g) + 6\ H_2O(l) \longrightarrow C_6H_{12}O_6(s) + 6\ O_2(g)$ is about $455. \times 10^3$ kg km^{-2}. Calculate the annual enthalpy change resulting from photosynthetic carbon fixation over the land surface given earlier. Assume $P = 1$ bar and $T = 298$ K.

P4.17 Calculate ΔH_R° and ΔU_R° at 298.15 K for the following reactions:

a. $4\ NH_3(g) + 6\ NO(g) \longrightarrow 5\ N_2(g) + 6\ H_2O(g)$

b. $2\ NO(g) + O_2(g) \longrightarrow 2\ NO_2(g)$

c. $TiCl_4(l) + 2\ H_2O(l) \longrightarrow TiO_2(s) + 4\ HCl(g)$

d. $2\ NaOH(aq) + H_2SO_4(aq) \longrightarrow Na_2SO_4(aq) + 2\ H_2O(l)$
   Assume complete dissociation of NaOH, H$_2$SO$_4$, and Na$_2$SO$_4$

e. $CH_4(g) + H_2O(g) \longrightarrow CO(g) + 3\ H_2(g)$

f. $CH_3OH(g) + CO(g) \longrightarrow CH_3COOH(l)$

P4.18 A sample of Na$_2$SO$_4(s)$ is dissolved in 225 g of water at 298 K such that the solution is 0.325 molar in Na$_2$SO$_4$. A temperature rise of 0.146°C is observed. The calorimeter

constant is 330. J K^{-1}. Calculate the enthalpy of solution of Na$_2$SO$_4$ in water at this concentration. Compare your result with that calculated using the data in Table 4.1 (Appendix B, Data Tables).

P4.19 Nitrogen is a vital component of proteins and nucleic acids, and thus is necessary for life. The atmosphere is composed of roughly 80% N$_2$, but most organisms cannot directly utilize N$_2$ for biosynthesis. Bacteria capable of "fixing" nitrogen (i.e., converting N$_2$ to a chemical form, such as NH$_3$, which can be utilized in the biosynthesis of proteins and nucleic acids) are called diazotrophs. The ability of some plants like legumes to fix nitrogen is due to a symbiotic relationship between the plant and nitrogen-fixing bacteria that live in the plant's roots. Assume that the hypothetical reaction for fixing nitrogen biologically is

$$N_2(g) + 3\ H_2O(l) \longrightarrow 2\ NH_3(aq) + \tfrac{3}{2}O_2(g)$$

a. Calculate the standard enthalpy change for the biosynthetic fixation of nitrogen at $T = 298$ K. For NH$_3(aq)$, ammonia dissolved in aqueous solution, $\Delta H_f^\circ = -80.3$ kJ mol^{-1}.

b. In some bacteria, glycine is produced from ammonia by the reaction

$$NH_3(g) + 2\ CH_4(g) + \tfrac{5}{2}O_2(g)$$
$$\longrightarrow NH_2CH_2COOH(s) + H_2O(l)$$

Calculate the standard enthalpy change for the synthesis of glycine from ammonia. For glycine, $\Delta H_f^\circ = -537.2$ kJ mol^{-1}. Assume $T = 298$ K.

c. Calculate the standard enthalpy change for the synthesis of glycine from nitrogen, water, oxygen, and methane.

P4.20 If 3.365 g of ethanol $C_2H_5OH(l)$ is burned completely in a bomb calorimeter at 298.15 K, the heat produced is 99.472 kJ.

a. Calculate $\Delta H_{combustion}^\circ$ for ethanol at 298.15 K.

b. Calculate ΔH_f° of ethanol at 298.15 K.

P4.21 From the following data, calculate $\Delta H_{R,391.4\ K}^\circ$ for the reaction $CH_3COOH(g) + 2\ O_2(g) \longrightarrow 2\ H_2O(g) + 2\ CO_2(g)$:

	ΔH_R° (kJ mol^{-1})
$CH_3COOH(l) + 2\ O_2(g) \longrightarrow 2\ H_2O(l) + 2\ CO_2(g)$	−871.5
$H_2O(l) \longrightarrow H_2O(g)$	40.656
$CH_3COOH(l) \longrightarrow CH_3COOH(g)$	24.4

Values for ΔH_R° for the first two reactions are at 298.15 K, and for the third reaction at 391.4 K.

Substance	$CH_3COOH(l)$	$O_2(g)$	$CO_2(g)$	$H_2O(l)$	$H_2O(g)$
$C_{P,m}/R$	14.9	3.53	4.46	9.055	4.038

P4.22 A 0.1429 g sample of sucrose $C_{12}H_{22}O_{11}$ is burned in a bomb calorimeter. In order to produce the same temperature rise in the calorimeter as the reaction, 2353 J must be expended.

a. Calculate ΔU and ΔH for the combustion of 1 mole of sucrose.

b. Using the data tables and your answer to (a), calculate ΔH_f° for sucrose.

c. The rise in temperature of the calorimeter and its contents as a result of the reaction is 1.743 K. Calculate the heat capacity of the calorimeter and its contents.

P4.23 Calculate ΔH_R° at 675 K for the reaction $4\,NH_3(g) + 6\,NO(g) \longrightarrow 5\,N_2(g) + 6\,H_2O(g)$ using the temperature dependence of the heat capacities from the data tables. Compare your result with ΔH_R° at 298.15. Is the difference large or small? Why?

P4.24 From the following data at 298.15 K as well as data in Table 4.1 (Appendix B, Data Tables), calculate the standard enthalpy of formation of $H_2S(g)$ and of $FeS_2(s)$:

	ΔH_R° (kJ mol^{-1})
$Fe(s) + 2\,H_2S(g) \longrightarrow FeS_2(s) + 2\,H_2(g)$	-137.0
$H_2S(g) + 3/2\,O_2(g) \longrightarrow H_2O(l) + SO_2(g)$	-562.0

P4.25 Using the protein DSC data in Problem P4.10, calculate the enthalpy change between $T = 288$ K and $T = 318$ K. Give your answer in units of kJ per mole. Assume the molecular weight of the protein is 14,000. grams. [*Hint: You can perform the integration of the heat capacity by estimating the area under the DSC curve and above the dotted baseline in Problem P4.10. This can be done by dividing the area up into small rectangles and summing the areas of the rectangles. Comment on the accuracy of this method.*]

P4.26 Given the following heat capacity data at 298 K, calculate ΔH_f° of $CO_2(g)$ at 525 K. Assume that the heat capacities are independent of temperature.

Substance	C(*graphite*)	$O_2(g)$	$CO_2(g)$
$C_{P,m}$/J mol^{-1}K^{-1}	8.52	28.8	37.1

P4.27 Calculate ΔH for the process in which $Cl_2(g)$ initially at 298.15 K at 1 bar is heated to 690. K at 1 bar. Use the temperature-dependent heat capacities in the data tables. How large is the relative error if the molar heat capacity is assumed to be constant at its value of 298.15 K over the temperature interval?

P4.28 From the following data at 298.15 K C, calculate the standard enthalpy of formation of $FeO(s)$ and of $Fe_2O_3(s)$:

	ΔH_R° (kJ mol^{-1})
$Fe_2O_3(s) + 3\,C(graphite) \longrightarrow 2\,Fe(s) + 3\,CO(g)$	492.6
$FeO(s) + C(graphite) \longrightarrow Fe(s) + CO(g)$	155.8
$C(graphite) + O_2(g) \longrightarrow CO_2(g)$	-393.51
$CO(g) + 1/2\,O_2(g) \longrightarrow CO_2(g)$	-282.98

P4.29 Calculate the average C—H bond enthalpy in methane using the data tables. Calculate the percent error in equating the average C—H bond energy in Table 4.3 with the bond enthalpy.

P4.30 Use the average bond energies in Table 4.3 to estimate ΔU for the reaction $C_2H_4(g) + H_2(g) \longrightarrow C_2H_6(g)$. Also calculate ΔU_R° from the tabulated values of ΔH_f° for reactant and products (Appendix B, Data Tables). Calculate the percent error in estimating ΔU_R° from the average bond energies for this reaction.

P4.31 Use the tabulated values of the enthalpy of combustion of benzene and the enthalpies of formation of $CO_2(g)$ and $H_2O(l)$ to determine ΔH_f° for benzene.

P4.32 Compare the heat evolved at constant pressure per mole of oxygen in the combustion of sucrose ($C_{12}H_{22}O_{11}$) and palmitic acid ($C_{16}H_{32}O_2$) with the combustion of a typical protein, for which the empirical formula is $C_{4.3}H_{6.6}NO$. Assume for the protein that the combustion yields $N_2(g)$, $CO_2(g)$, and $H_2O(l)$. Assume that the enthalpies for combustion of sucrose, palmitic acid, and a typical protein are 5647 kJ mol^{-1}, 10,035 kJ mol^{-1}, and 22.0 kJ g^{-1}, respectively. Based on these calculations, determine the average heat evolved per mole of oxygen consumed, assuming combustion of equal moles of sucrose, palmitic acid, and protein.

P4.33 A camper stranded in snowy weather loses heat by wind convection. The camper is packing emergency rations consisting of 58% sucrose, 31% fat, and 11% protein by weight. Using the data provided in Problem P4.32 and assuming the fat content of the rations can be treated with palmitic acid data and the protein content similarly by the protein data in Problem P4.32, how much emergency rations must the camper consume in order to compensate for a reduction in body temperature of 3.5 K? Assume the heat capacity of the body equals that of water. Assume the camper weighs 67 kg. State any additional assumptions.

P4.34 In order to get in shape for mountain climbing, an avid hiker with a mass of 60. kg ascends the stairs in the world's tallest structure, the 828 m tall Burj Khalifa in Dubai, United Arab Emirates. Assume that she eats energy bars on the way up and that her body is 25% efficient in converting the energy content of the bars into the work of climbing. How many energy bars does she have to eat if a single bar produces 1.08×10^3 kJ of energy upon metabolizing?

P4.35 We return to the 60. kg hiker of P4.34, who is climbing the 828 m tall Burj Khalifa in Dubai. If the efficiency of converting the energy content of the bars into the work of climbing is 25%, the remaining 75% of the energy released through metabolism is heat released to her body. She eats two energy bars and a single bar produces 1.08×10^3 kJ of energy upon metabolizing. Assume that the heat capacity of her body is equal to that for water. Calculate the increase in her temperature at the top of the structure. Is your result reasonable? Can you think of a mechanism by which her body might release energy to avoid a temperature increase?

Entropy and the Second and Third Laws of Thermodynamics

Real-world processes have a natural direction of change. Heat flows from hotter bodies to colder bodies, and gases mix rather than separate. Entropy, designated by S, is the state function that predicts the direction of natural, or spontaneous, change and entropy increases for a spontaneous change in an isolated system. For a spontaneous change in a system interacting with its environment, the sum of the entropy of the system and that of the surroundings increases. In this chapter, we introduce entropy, derive the conditions for spontaneity, and show how S varies with the macroscopic variables P, V, and T.

5.1 The Universe Has a Natural Direction of Change

To this point, we have discussed q and w, as well as U and H. The first law of thermodynamics states that in any process, the total energy of the universe remains constant. However, it does not predict which of several possible energy conserving processes will occur. Consider the following two examples. A metal rod initially at a uniform temperature could, in principle, undergo a spontaneous transformation in which one end becomes hot and the other end becomes cold without being in conflict with the first law, as long as the total energy of the rod remains constant. However, experience demonstrates that this does not occur. Similarly, an ideal gas that is uniformly distributed in a rigid adiabatic container could undergo a spontaneous transformation such that all of the gas moves to one-half of the container, leaving a vacuum in the other half. For an ideal gas, $(\partial U / \partial V)_T = 0$, therefore the energy of the initial and final states is the same. Neither of these transformations violates the first law of thermodynamics—and yet neither occurs.

Experience tells us that there is a natural direction of change in these two processes. A metal rod with a temperature gradient reaches a uniform temperature at some time after it has been isolated from a heat source. A gas confined to one-half of a container with a vacuum in the other half distributes itself uniformly throughout the container if a valve separating the two parts is opened. The transformations described in the previous paragraph are **unnatural transformations.** The word *unnatural* is used to indicate that such an energy-conserving process can occur but is extremely unlikely. By contrast, the reverse processes, in which the temperature gradient along the rod disappears and the gas becomes distributed uniformly throughout the container, are **natural transformations,** also called **spontaneous processes,** which are extremely likely. Spontaneous does not mean that the process occurs immediately, but rather that it will occur with high probability if any barrier to the change is overcome. For example, the transformation of a piece of wood to CO_2 and H_2O in the presence of oxygen is spontaneous, but it only occurs at elevated temperatures because an activation energy barrier must be overcome for the reaction to proceed.

Our experience is sufficient to predict the direction of spontaneous change for the two examples cited, but can the direction of spontaneous change be predicted in less obvious cases? In this chapter, we show that there is a thermodynamic function called entropy that allows us to predict the direction of spontaneous change for a system in a given initial state. For example, assume that a reaction vessel contains a given number of moles of N_2, H_2, and NH_3 at 600 K and at a total pressure of 280 bar. An iron catalyst is introduced that allows the mixture of gases to equilibrate according to $1/2 N_2 + 3/2 H_2 \rightleftharpoons NH_3$. What is the direction of spontaneous change, and what are the partial pressures of the three gases at equilibrium? The answer to this question is obtained by calculating the entropy change in the system and the surroundings.

Most students are initially uncomfortable when working with entropy because entropy is further removed from direct experience than energy, work, or heat. Historically, entropy was introduced by Clausius in 1850, several decades before entropy was understood at a microscopic level by Boltzmann. Boltzmann's explanation of entropy will be presented in Chapter 32, and we briefly state his conclusions here. At the microscopic level, matter consists of atoms or molecules that have energetic degrees of freedom (i.e., translational, rotational, vibrational, and electronic), each of which is associated with discrete energy levels that can be calculated using quantum mechanics. Quantum mechanics also characterizes a molecule by a state associated with a set of quantum numbers and a molecular energy. Entropy is a measure of the number of quantum states accessible to a macroscopic system at a given energy. Quantitatively, $S = k \ln W$, where W is the number of states accessible to the system, and $k = R/N_A$. As demonstrated later in this chapter, the entropy of an isolated system is maximized at equilibrium. Therefore, the approach to equilibrium can be envisioned as a process in which the system achieves the distribution of energy among molecules that corresponds to a maximum value of W and, correspondingly, to a maximum in S.

5.2 Heat Engines and the Second Law of Thermodynamics

The development of entropy presented here follows the historical route by which this state function was first introduced. The concept of entropy arose as 19th-century scientists attempted to maximize the work output of engines. An automobile engine operates in a cyclical process of fuel intake, compression, ignition and expansion, and exhaust, which occurs several thousand times per minute and is used to perform work on the surroundings. Because the work produced by such an engine is a result of the heat released in a combustion process, it is referred to as a **heat engine.** An idealized version of a heat engine is depicted in Figure 5.1. The system consists of a working substance (in this case an ideal gas) confined in a piston and cylinder assembly with diathermal walls. This assembly can be brought into contact with a hot reservoir at T_{hot} or a cold reservoir at T_{cold}. The expansion or contraction of the gas caused by changes in its temperature drives the piston in or out of the cylinder. This linear motion is converted to circular motion as shown in Figure 5.1, and the rotary motion is used to do work in the surroundings.

The efficiency of a heat engine is of particular interest in practical applications. Experience shows that work can be converted to heat with 100% efficiency. Consider an example from calorimetry discussed in Chapter 4, in which electrical work is done on a resistive heater immersed in a water bath. We observe that all of the electrical work done on the heater has been converted to heat, resulting in an increase in the temperature

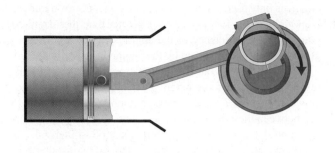

FIGURE 5.1
A schematic depiction of a heat engine is shown. Changes in temperature of the working substance brought about by contacting the cylinder with hot or cold reservoirs generate a linear motion that is mechanically converted to a rotary motion, which is used to do work.

of the water and the heater. What is the maximum theoretical efficiency of the reverse process, the conversion of heat to work? As shown later, it is less than 100%. *There is a natural asymmetry in the efficiency of converting work to heat and converting heat to work. Thermodynamics provides an explanation for this asymmetry.*

As discussed in Section 2.7, the maximum work output in an isothermal expansion occurs in a reversible process. For this reason, we next calculate the efficiency of a reversible heat engine, because the efficiency of a reversible engine is an upper bound to the efficiency of a real engine. This reversible engine converts heat into work by exploiting the spontaneous tendency of heat to flow from a hot reservoir to a cold reservoir. It does work on the surroundings by operating in a cycle of reversible expansions and compressions of an ideal gas in a piston and cylinder assembly. We discuss automotive engines in Section 5.11.

The cycle for a reversible heat engine is shown in Figure 5.2 in a P–V diagram. The expansion and compression steps are designed so that the engine returns to its initial state after four steps. Recall from Section 2.7 that the area within the cycle equals the work done by the engine. As discussed later, four separate isothermal and adiabatic steps are needed to make the enclosed area in the cycle greater than zero. Beginning at point a, the first segment is a reversible isothermal expansion in which the gas absorbs heat from the reservoir at T_{hot}, and does work on the surroundings. In the second segment, the gas expands further, this time adiabatically. Work is also done on the surroundings in this step. At the end of the second segment, the gas has cooled to the temperature T_{cold}. The third segment is an isothermal compression in which the surroundings do work on the system and heat is absorbed by the cold reservoir. In the final segment, the gas is compressed to its initial volume, this time adiabatically. Work is done on the system in this segment, and the temperature returns to its initial value, T_{hot}. In summary, heat is taken up by the engine in the first segment at T_{hot}, and released to the surroundings in the third segment at T_{cold}. Work is done on the

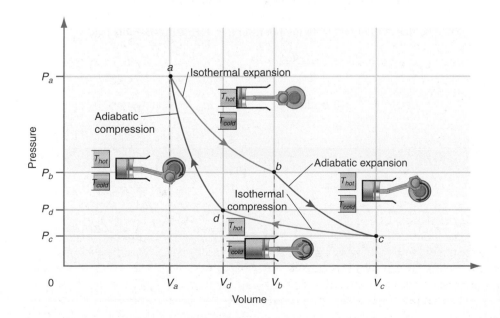

FIGURE 5.2
A reversible Carnot cycle for a sample of an ideal gas working substance is shown on an indicator diagram. The cycle consists of two adiabatic and two isothermal segments. The arrows indicate the direction in which the cycle is traversed. The insets show the volume of gas and the coupling to the reservoirs at the beginning of each successive segment of the cycle. The coloring of the contents of the cylinder indicates the presence of the gas and not its temperature. The volume of the cylinder shown is that at the beginning of the appropriate segment.

surroundings in the first two segments and on the system in the last two segments. An engine is only useful if net work is done on the surroundings, that is, if the magnitude of the work done in the first two steps is greater than the magnitude of the work done in the last two steps. The efficiency of the engine can be calculated by comparing the net work per cycle with the heat taken up by the engine from the hot reservoir.

Before carrying out this calculation, we discuss the rationale for the design of this reversible cycle in more detail. To avoid losing heat to the surroundings at temperatures between T_{hot} and T_{cold}, adiabatic segments 2 ($b \rightarrow c$) and 4 ($d \rightarrow a$) are used to move the gas between these temperatures. To absorb heat only at T_{hot} and release heat only at T_{cold}, segments 1 ($a \rightarrow b$) and 3 ($c \rightarrow d$) must be isothermal. The reason for using alternating isothermal and adiabatic segments is that no two isotherms at different temperatures intersect, and no two adiabats starting from two different temperatures intersect. Therefore, it is impossible to create a closed cycle of nonzero area in an indicator diagram out of isothermal or adiabatic segments alone. However, net work can be done using alternating adiabatic and isothermal segments. The reversible cycle depicted in Figure 5.2 is called a **Carnot cycle,** after the French engineer who first studied such cycles.

The efficiency of the Carnot cycle can be determined by calculating q, w, and ΔU for each segment of the cycle, assuming that the working substance is an ideal gas. The results are shown first in a qualitative fashion in Table 5.1. The appropriate signs for q and w are indicated. If $\Delta V > 0$, $w < 0$ for the segment, and the corresponding entry for work has a negative sign. For the isothermal segments, q and w have opposite signs because $\Delta U = q + w = 0$. From Table 5.1, it is seen that

$$w_{cycle} = w_{cd} + w_{da} + w_{ab} + w_{bc} \quad \text{and} \quad q_{cycle} = q_{ab} + q_{cd} \qquad \textbf{(5.1)}$$

Because $\Delta U_{cycle} = 0$,

$$w_{cycle} = -(q_{cd} + q_{ab}) \qquad \textbf{(5.2)}$$

By comparing the areas under the two expansion segments with those under the two compression segments in the indicator diagram in Figure 5.2, we can see that the total work as seen from the system is negative, meaning that work is done on the surroundings in each cycle. Using this result, we arrive at an important conclusion that relates the heat flow in the two isothermal segments

$$w_{cycle} < 0, \text{ therefore } |q_{ab}| > |q_{cd}| \qquad \textbf{(5.3)}$$

More heat is withdrawn from the hot reservoir than is deposited in the cold reservoir as is also seen by rearranging Equation (5.2) to the form $q_{ab} + w_{cycle} = -q_{cd}$. It is useful to make a model of this heat engine that indicates the relative magnitude and direction of the heat and work flow, as shown in Figure 5.3a. The figure makes it clear that not all of the heat withdrawn from the higher temperature reservoir is converted to work done by the system (the engine) on the surroundings.

The efficiency, ε, of the reversible Carnot engine is defined as the ratio of the work output to the heat withdrawn from the hot reservoir. Referring to Table 5.1,

$$\varepsilon = -\frac{w_{cycle}}{q_{ab}} = \frac{q_{ab} + q_{cd}}{q_{ab}}$$

$$= 1 - \frac{|q_{cd}|}{|q_{ab}|} < 1 \text{ because } |q_{ab}| > |q_{cd}|, q_{ab} > 0, \text{ and } q_{cd} < 0 \qquad \textbf{(5.4)}$$

TABLE 5.1	Heat, Work, and ΔU for the Reversible Carnot Cycle				
Segment	Initial State	Final State	q	w	ΔU
$a \rightarrow b$	P_a, V_a, T_{hot}	P_b, V_b, T_{hot}	q_{ab} (+)	w_{ab} (−)	$\Delta U_{ab} = 0$
$b \rightarrow c$	P_b, V_b, T_{hot}	P_c, V_c, T_{cold}	0	w_{bc} (−)	$\Delta U_{bc} = w_{bc}$(−)
$c \rightarrow d$	P_c, V_c, T_{cold}	P_d, V_d, T_{cold}	q_{cd} (−)	w_{cd} (+)	$\Delta U_{cd} = 0$
$d \rightarrow a$	P_d, V_d, T_{cold}	P_a, V_a, T_{hot}	0	w_{da} (+)	$\Delta U_{da} = w_{da}$(+)
Cycle	P_a, V_a, T_{hot}	P_a, V_a, T_{hot}	$q_{ab} + q_{cd}$ (+)	$w_{ab} + w_{bc} + w_{cd} + w_{da}$ (−)	$\Delta U_{cycle} = 0$

Both q_{ab} and q_{cd} are nonzero because the corresponding processes are isothermal. Equation (5.4) shows that the efficiency of a heat engine operating in a reversible Carnot cycle is always less than one. Equivalently, not all of the heat withdrawn from the hot reservoir can be converted to work. This conclusion is valid for all engines and illustrates the asymmetry in converting heat to work and work to heat.

These considerations on the efficiency of reversible heat engines led to the Kelvin–Planck formulation of the **second law of thermodynamics:**

> It is impossible for a system to undergo a cyclic process whose sole effects are the flow of heat into the system from a heat reservoir and the performance of an equal amount of work by the system on the surroundings.

The second law asserts that the heat engine depicted in Figure 5.3b cannot be constructed. Any heat engine must eject heat into the cold reservoir as shown in Figure 5.3a. The second law has been put to the test many times by inventors who have claimed that they have invented an engine that has an efficiency of 100%. No such claim has ever been validated. To test the assertion made in this statement of the second law, imagine that such an engine has been invented. We mount it on a boat in Seattle and set off on a journey across the Pacific Ocean. Heat is extracted from the ocean, which is the single heat reservoir, and is converted entirely to work in the form of a rapidly rotating propeller. Because the ocean is huge, the decrease in its temperature as a result of withdrawing heat is negligible. By the time we arrive in Japan, not a gram of diesel fuel has been used, because all the heat needed to power the boat has been extracted from the ocean. The money that was saved on fuel is used to set up an office and begin marketing this wonder engine. Does this scenario sound too good to be true? It is. Such an impossible engine is called a **perpetual motion machine of the second kind** because it violates the second law of thermodynamics. A **perpetual motion machine of the first kind** violates the first law.

The first statement of the second law can be understood using an indicator diagram. For an engine to produce work, the area of the cycle in a P–V diagram must be greater than zero. However, this is impossible in a simple cycle using a single heat reservoir. If $T_{hot} = T_{cold}$ in Figure 5.2, the cycle $a \rightarrow b \rightarrow c \rightarrow d \rightarrow a$ collapses to a line, and the area of the cycle is zero. An arbitrary reversible cycle can be constructed that does not consist of individual adiabatic and isothermal segments. However, as shown in Figure 5.4, any reversible cycle can be approximated by a succession of adiabatic and isothermal segments, an approximation that becomes exact as the length of each segment approaches zero. It can be shown that the efficiency of such a cycle is also given by Equation (5.9) so that the efficiency of all heat engines operating in any reversible cycle between the same two temperatures, T_{hot} and T_{cold}, is identical.

A more useful form than Equation (5.4) for the efficiency of a reversible heat engine can be derived by assuming that the working substance in the engine is an ideal gas. Calculating the work flow in each of the four segments of the Carnot cycle using the results of Sections 2.7 and 2.9,

$$w_{ab} = -nRT_{hot} \ln \frac{V_b}{V_a} \qquad w_{ab} < 0 \text{ because } V_b > V_a$$

$$w_{bc} = nC_{V,m}(T_{cold} - T_{hot}) \quad w_{bc} < 0 \text{ because } T_{cold} < T_{hot}$$

$$w_{cd} = -nRT_{cold} \ln \frac{V_d}{V_c} \qquad w_{cd} > 0 \text{ because } V_d < V_c \tag{5.5}$$

$$w_{da} = nC_{V,m}(T_{hot} - T_{cold}) \quad w_{da} > 0 \text{ because } T_{hot} > T_{cold}$$

As derived in Section 2.10, the volume and temperature in the reversible adiabatic segments are related by

$$T_{hot}V_b^{\gamma-1} = T_{cold}V_c^{\gamma-1} \quad \text{and} \quad T_{cold}V_d^{\gamma-1} = T_{hot}V_a^{\gamma-1} \tag{5.6}$$

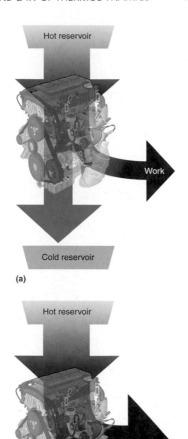

(a)

(b)

FIGURE 5.3
(a) A schematic model of the heat engine operating in a reversible Carnot cycle. The relative widths of the two paths leaving the hot reservoir show the partitioning between work and heat injected into the cold reservoir. (b) The second law of thermodynamics asserts that it is impossible to construct a heat engine that operates using a single heat reservoir and converts the heat withdrawn from the reservoir into work with 100% efficiency as shown.

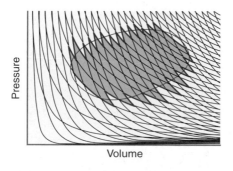

FIGURE 5.4
An arbitrary reversible cycle, indicated by the ellipse, can be approximated to any desired accuracy by a sequence of alternating adiabatic and isothermal segments.

You will show in the end-of-chapter problems that V_c and V_d can be eliminated from the set of Equations (5.5) to yield

$$w_{cycle} = -nR(T_{hot} - T_{cold}) \ln \frac{V_b}{V_a} < 0 \tag{5.7}$$

Because $\Delta U_{a \to b} = 0$, the heat withdrawn from the hot reservoir is

$$q_{ab} = -w_{ab} = nRT_{hot} \ln \frac{V_b}{V_a} \tag{5.8}$$

and the efficiency of the reversible Carnot heat engine with an ideal gas as the working substance can be expressed solely in terms of the reservoir temperatures.

$$\varepsilon = \frac{|w_{cycle}|}{q_{ab}} = \frac{T_{hot} - T_{cold}}{T_{hot}} = 1 - \frac{T_{cold}}{T_{hot}} < 1 \tag{5.9}$$

The efficiency of this reversible heat engine can approach one only as $T_{hot} \to \infty$ or $T_{cold} \to 0$, neither of which can be accomplished in practice. Therefore, heat can never be totally converted to work in a reversible cyclic process. Because w_{cycle} for an engine operating in an irreversible cycle is less than the work attainable in a reversible cycle, $\varepsilon_{irreversible} < \varepsilon_{reversible} < 1$.

EXAMPLE PROBLEM 5.1

Calculate the maximum work that can be done by a reversible heat engine operating between 500. and 200. K if 1000. J is absorbed at 500. K.

Solution

The fraction of the heat that can be converted to work is the same as the fractional fall in the absolute temperature. This is a convenient way to link the efficiency of an engine with the properties of the absolute temperature.

$$\varepsilon = 1 - \frac{T_{cold}}{T_{hot}} = 1 - \frac{200.\,K}{500.\,K} = 0.600$$

$$w = \varepsilon q_{ab} = 0.600 \times 1000.\,J = 600.\,J$$

In this section, only the most important features of heat engines have been discussed. It can also be shown that the efficiency of a reversible heat engine is independent of the working substance. For a more in-depth discussion of heat engines, the interested reader is referred to *Heat and Thermodynamics,* seventh edition, by M. W. Zemansky and R. H. Dittman (McGraw-Hill, 1997). We will return to a discussion of the efficiency of engines when we discuss refrigerators, heat pumps, and real engines in Section 5.11.

5.3 Introducing Entropy

Equating the two formulas for the efficiency of the reversible heat engine given in Equations (5.4) and (5.9),

$$\frac{T_{hot} - T_{cold}}{T_{hot}} = \frac{q_{ab} + q_{cd}}{q_{ab}} \quad \text{or} \quad \frac{q_{ab}}{T_{hot}} + \frac{q_{cd}}{T_{cold}} = 0 \tag{5.10}$$

The last expression in Equation (5.10) is the sum of the quantity $q_{reversible}/T$ around the Carnot cycle. This result can be generalized to any reversible cycle made up of any number of segments to give the important result stated in Equation (5.11):

$$\oint \frac{dq_{reversible}}{T} = 0 \tag{5.11}$$

This equation can be regarded as the mathematical statement of the second law. What conclusions can be drawn from Equation (5.11)? Because the cyclic integral of $đq_{reversible}/T$ is zero, this quantity must be the exact differential of a state function. This state function is called the **entropy,** and given the symbol S

$$dS \equiv \frac{đq_{reversible}}{T} \tag{5.12}$$

For a macroscopic change,

$$\Delta S = \int \frac{đq_{reversible}}{T} \tag{5.13}$$

Note that whereas $đq_{reversible}$ is not an exact differential, multiplying this quantity by $1/T$ makes the differential exact.

EXAMPLE PROBLEM 5.2

a. Show that the following differential expression is not an exact differential:

$$\frac{RT}{P}dP + RdT$$

b. Show that $RT\,dP + PR\,dT$, obtained by multiplying the function in part (a) by P, is an exact differential.

Solution

a. For the expression $f(P,T)dP + g(P,T)dT$ to be an exact differential, the condition $(\partial f(P,T)/\partial T)_P = (\partial g(P,T)/\partial P)_T$ must be satisfied as discussed in Section 3.1. Because

$$\left(\frac{\partial\left(\frac{RT}{P}\right)}{\partial T}\right)_P = \frac{R}{P} \quad \text{and} \quad \frac{\partial R}{\partial P} = 0$$

the condition is not fulfilled.

b. Because $(\partial(RT)/\partial T)_P = R$ and $(\partial(RP)/\partial P)_T = R$, $RT\,dP + RP\,dT$ is an exact differential.

Keep in mind that it has only been shown that S is a state function. It has not yet been demonstrated that S is a suitable function for measuring the natural direction of change in a process that the system may undergo. We will do so in Section 5.5.

5.4 Calculating Changes in Entropy

The most important thing to remember in doing entropy calculations using Equation (5.13) is that ΔS *must be calculated along a reversible path*. In considering an **irreversible process,** ΔS must be calculated for a reversible process that proceeds between the same initial and final states corresponding to the irreversible process. Because S is a state function, ΔS is necessarily path independent, *provided that the transformation is between the same initial and final states in both processes.*

We first consider two cases that require no calculation. For any reversible adiabatic process, $q_{reversible} = 0$, so that $\Delta S = \int (đq_{reversible}/T) = 0$. For any cyclic process, $\Delta S = \oint (đq_{reversible}/T) = 0$, because the change in any state function for a cyclic process is zero.

Next consider ΔS for the reversible isothermal expansion or compression of an ideal gas, described by $V_i, T_i \rightarrow V_f, T_i$. Because $\Delta U = 0$ for this case,

$$q_{reversible} = -w_{reversible} = nRT \ln \frac{V_f}{V_i} \text{ and} \tag{5.14}$$

$$\Delta S = \int \frac{đq_{reversible}}{T} = \frac{1}{T} \times q_{reversible} = nR \ln \frac{V_f}{V_i} \tag{5.15}$$

Note that $\Delta S > 0$ for an expansion ($V_f > V_i$) and $\Delta S < 0$ for a compression ($V_f < V_i$). Although the preceding calculation is for a reversible process, ΔS has exactly the same value for any reversible or irreversible isothermal path that goes between the same initial and final volumes and satisfies the condition $T_f = T_i$. This is the case because S is a state function.

Why does the entropy increase with increasing V at constant T if the system is viewed at a microscopic level? As discussed in Section 15.2, the translational energy levels for atoms and molecules are all shifted to lower energies as the volume of the system increases. Therefore, more states of the system can be accessed at constant T as V increases. This is a qualitative argument that does not give the functional form shown in Equation (5.15). The logarithmic dependence arises because S is proportional to the logarithm of the number of states accessible to the system rather than to the number of states.

Consider next ΔS for an ideal gas that undergoes a reversible change in T at constant V or P. For a reversible process described by $V_i, T_i \rightarrow V_i, T_f, đq_{reversible} = C_V dT$, and

$$\Delta S = \int \frac{đq_{reversible}}{T} = \int \frac{nC_{V,m} dT}{T} \approx nC_{V,m} \ln \frac{T_f}{T_i} \tag{5.16}$$

For a constant pressure process described by $P_i, T_i \rightarrow P_i, T_f, đq_{reversible} = C_P dT$, and

$$\Delta S = \int \frac{đq_{reversible}}{T} = \int \frac{nC_{P,m} dT}{T} \approx nC_{P,m} \ln \frac{T_f}{T_i} \tag{5.17}$$

The last expressions in Equations (5.16) and (5.17) are valid if the temperature interval is small enough that the temperature dependence of $C_{V,m}$ and $C_{P,m}$ can be neglected. Again, although ΔS has been calculated for a reversible process, Equations (5.16) and (5.17) hold for any reversible or irreversible process between the same initial and final states for an ideal gas.

We again ask what a microscopic model would predict for the dependence of S on T. As discussed in Chapter 30, the probability of a molecule accessing a state with energy E_i is proportional to $\exp(-E_i/k_B T)$. This quantity increases exponentially as T increases, so that more states become accessible to the system as T increases. Because S is a measure of the number of states the system can access, it increases with increasing T. Again, the logarithmic dependence arises because S is proportional to the logarithm of the number of states accessible to the system rather than to the number of states.

The results of the last two calculations can be combined in the following way. Because the macroscopic variables V, T or P, T completely define the state of an ideal gas, any change $V_i, T_i \rightarrow V_f, T_f$ can be separated into two segments, $V_i, T_i \rightarrow V_f, T_i$ and $V_f, T_i \rightarrow V_f, T_f$. A similar statement can be made about P and T. Because S is a state function, ΔS is independent of the path. Therefore, any reversible or irreversible process for an ideal gas described by $V_i, T_i \rightarrow V_f, T_f$ can be treated as consisting of two segments, one of which occurs at constant volume and the other of which occurs at constant temperature. For this two-step process, ΔS is given by

$$\Delta S = nR \ln \frac{V_f}{V_i} + nC_{V,m} \ln \frac{T_f}{T_i} \tag{5.18}$$

Similarly, for any reversible or irreversible process for an ideal gas described by $P_i, T_i \rightarrow P_f, T_f$

$$\Delta S = -nR \ln \frac{P_f}{P_i} + nC_{P,m} \ln \frac{T_f}{T_i} \tag{5.19}$$

In writing Equations (5.18) and (5.19), it has been assumed that the temperature dependence of $C_{V,m}$ and $C_{P,m}$ can be neglected over the temperature range of interest.

EXAMPLE PROBLEM 5.3

Using the equation of state and the relationship between $C_{P,m}$ and $C_{V,m}$ for an ideal gas, show that Equation (5.18) can be transformed into Equation (5.19).

Solution

$$\Delta S = nR \ln \frac{V_f}{V_i} + nC_{V,m} \ln \frac{T_f}{T_i} = nR \ln \frac{T_f P_i}{T_i P_f} + nC_{V,m} \ln \frac{T_f}{T_i}$$

$$= -nR \ln \frac{P_f}{P_i} + n(C_{V,m} + R) \ln \frac{T_f}{T_i} = -nR \ln \frac{P_f}{P_i} + nC_{P,m} \ln \frac{T_f}{T_i}$$

Next consider ΔS for phase changes. Experience shows that a liquid is converted to a gas at a constant boiling temperature through heat input if the process is carried out at constant pressure. Because $q_P = \Delta H$, ΔS for the reversible process is given by

$$\Delta S_{vaporization} = \int \frac{dq_{reversible}}{T} = \frac{q_{reversible}}{T_{vaporization}} = \frac{\Delta H_{vaporization}}{T_{vaporization}} \qquad (5.20)$$

Similarly, for the phase change solid $\rightarrow$ liquid,

$$\Delta S_{fusion} = \int \frac{dq_{reversible}}{T} = \frac{q_{reversible}}{T_{fusion}} = \frac{\Delta H_{fusion}}{T_{fusion}} \qquad (5.21)$$

Finally, consider ΔS for an arbitrary process involving real gases, solids, and liquids for which the isobaric volumetric thermal expansion coefficient β and the isothermal compressibility κ, but not the equation of state, are known. The calculation of ΔS for such processes is described in Supplemental Sections 5.12 and 5.13, in which the properties of S as a state function are fully exploited. The results are stated here. For the system undergoing the change $V_i, T_i \rightarrow V_f, T_f$,

$$\Delta S = \int_{T_i}^{T_f} \frac{C_V}{T} dT + \int_{V_i}^{V_f} \frac{\beta}{\kappa} dV = C_V \ln \frac{T_f}{T_i} + \frac{\beta}{\kappa}(V_f - V_i) \qquad (5.22)$$

In deriving the last result, it has been assumed that κ and β are constant over the temperature and volume intervals of interest. For the system undergoing a change $P_i, T_i \rightarrow P_f, T_f$,

$$\Delta S = \int_{T_i}^{T_f} \frac{C_P}{T} dT - \int_{P_i}^{P_f} V\beta \, dP \qquad (5.23)$$

For a solid or liquid, the last equation can be simplified to

$$\Delta S = C_P \ln \frac{T_f}{T_i} - V\beta(P_f - P_i) \qquad (5.24)$$

if C_P, V, and β are assumed constant over the temperature and pressure intervals of interest. The integral forms of Equations (5.22) and (5.23) are valid for ideal and real gases, liquids, and solids. Examples of calculations using these equations are given in Example Problems 5.4 through 5.6.

EXAMPLE PROBLEM 5.4

One mole of CO gas is transformed from an initial state characterized by $T_i = 320.$ K and $V_i = 80.0$ L to a final state characterized by $T_f = 650.$ K and $V_f = 120.0$ L. Using Equation (5.22), calculate ΔS for this process. Use the ideal gas values for β and κ. For CO,

$$\frac{C_{V,m}}{\text{J mol}^{-1}\text{K}^{-1}} = 31.08 - 0.01452 \frac{T}{K} + 3.1415 \times 10^{-5} \frac{T^2}{K^2} - 1.4973 \times 10^{-8} \frac{T^3}{K^3}$$

Solution

For an ideal gas,

$$\beta = \frac{1}{V}\left(\frac{\partial V}{\partial T}\right)_P = \frac{1}{V}\left(\frac{\partial[nRT/P]}{\partial T}\right)_P = \frac{1}{T} \quad \text{and}$$

$$\kappa = -\frac{1}{V}\left(\frac{\partial V}{\partial P}\right)_T = -\frac{1}{V}\left(\frac{\partial[nRT/P]}{\partial P}\right)_T = \frac{1}{P}$$

Consider the following reversible process in order to calculate ΔS. The gas is first heated reversibly from 320. to 650. K at a constant volume of 80.0 L. Subsequently, the gas is reversibly expanded at a constant temperature of 650. K from a volume of 80.0 L to a volume of 120.0 L. The entropy change for this process is obtained using the integral form of Equation (5.22) with the values of β and κ cited earlier. The result is

$$\Delta S = \int_{T_i}^{T_f} \frac{C_V}{T} dT + nR \ln \frac{V_f}{V_i}$$

$$\Delta S =$$

$$1 \text{ mol} \times \int_{320.}^{650.} \frac{\left(31.08 - 0.01452 \frac{T}{K} + 3.1415 \times 10^{-5} \frac{T^2}{K^2} - 1.4973 \times 10^{-8} \frac{T^3}{K^3}\right)}{\frac{T}{K}} d\frac{T}{K}$$

$$+ 1 \text{ mol} \times 8.314 \text{ J K}^{-1} \text{mol}^{-1} \times \ln \frac{120.0 \text{ L}}{80.0 \text{ L}}$$

$$= 22.025 \text{ J K}^{-1} - 4.792 \text{ J K}^{-1} + 5.028 \text{ J K}^{-1}$$

$$- 1.207 \text{ J K}^{-1} + 3.371 \text{ J K}^{-1}$$

$$= 24.4 \text{ J K}^{-1}$$

EXAMPLE PROBLEM 5.5

In this problem, 2.50 mol of CO_2 gas is transformed from an initial state characterized by $T_i = 450.$ K and $P_i = 1.35$ bar to a final state characterized by $T_f = 800.$ K and $P_f = 3.45$ bar. Using Equation (5.23), calculate ΔS for this process. Assume ideal gas behavior and use the ideal gas value for β. For CO_2,

$$\frac{C_{P,m}}{\text{J mol}^{-1}\text{K}^{-1}} = 18.86 + 7.937 \times 10^{-2} \frac{T}{K} - 6.7834 \times 10^{-5} \frac{T^2}{K^2} + 2.4426 \times 10^{-8} \frac{T^3}{K^3}$$

Solution

Consider the following reversible process in order to calculate ΔS. The gas is first heated reversibly from 450. to 800. K at a constant pressure of 1.35 bar. Subsequently, the gas is reversibly compressed at a constant temperature of 800. K from a pressure of

1.35 bar to a pressure of 3.45 bar. The entropy change for this process is obtained using Equation (5.23) with the value of $\beta = 1/T$ from Example Problem 5.4.

$$\Delta S = \int_{T_i}^{T_f} \frac{C_P}{T} dT - \int_{P_i}^{P_f} V\beta \, dP = \int_{T_i}^{T_f} \frac{C_P}{T} dT - nR \int_{P_i}^{P_f} \frac{dP}{P} = \int_{T_i}^{T_f} \frac{C_P}{T} dT - nR \ln \frac{P_f}{P_i}$$

$$= 2.50 \times \int_{450.}^{800.} \frac{\left(18.86 + 7.937 \times 10^{-2} \frac{T}{K} - 6.7834 \times 10^{-5} \frac{T^2}{K^2} + 2.4426 \times 10^{-8} \frac{T^3}{K^3}\right)}{\frac{T}{K}} d\frac{T}{K}$$

$$-2.50 \, \text{mol} \times 8.314 \text{ J K}^{-1}\text{mol}^{-1} \times \ln\frac{3.45 \text{ bar}}{1.35 \text{ bar}}$$

$$= 27.13 \text{ J K}^{-1} + 69.45 \text{ J K}^{-1} - 37.10 \text{ J K}^{-1} + 8.57 \text{ J K}^{-1}$$

$$-19.50 \text{ J K}^{-1}$$

$$= 48.6 \text{ J K}^{-1}$$

EXAMPLE PROBLEM 5.6

In this problem, 3.00 mol of liquid mercury is transformed from an initial state characterized by $T_i = 300.$ K and $P_i = 1.00$ bar to a final state characterized by $T_f = 600.$ K and $P_f = 3.00$ bar.

a. Calculate ΔS for this process; $\beta = 1.81 \times 10^{-4}$ K^{-1}, $\rho = 13.54$ g cm^{-3}, and $C_{P,m}$ for Hg(l) = 27.98 J mol^{-1}K^{-1}.

b. What is the ratio of the pressure-dependent term to the temperature-dependent term in ΔS? Explain your result.

Solution

a. Because the volume changes only slightly with temperature and pressure over the range indicated,

$$\Delta S = \int_{T_i}^{T_f} \frac{C_P}{T} dT - \int_{P_i}^{P_f} V\beta \, dP \approx nC_{P,m} \ln\frac{T_f}{T_i} - nV_{m,i}\beta(P_f - P_i)$$

$$= 3.00 \, \text{mol} \times 27.98 \text{ J mol}^{-1}\text{K}^{-1} \times \ln\frac{600. \text{ K}}{300. \text{ K}}$$

$$-3.00 \, \text{mol} \times \frac{200.59 \text{ g mol}^{-1}}{13.54 \text{ g cm}^{-3} \times \frac{10^6 \text{ cm}^3}{\text{m}^3}} \times 1.81 \times 10^{-4}\text{K}^{-1} \times 2.00 \text{ bar}$$

$$\times 10^5 \text{ Pa bar}^{-1}$$

$$= 58.2 \text{ J K}^{-1} - 1.61 \times 10^{-3} \text{ J K}^{-1} = 58.2 \text{ J K}^{-1}$$

b. The ratio of the pressure-dependent to the temperature-dependent term is -3×10^{-5}. Because the volume change with pressure is very small, the contribution of the pressure-dependent term is negligible in comparison with the temperature-dependent term.

As Example Problem 5.6 shows, ΔS for a liquid or solid as both P and T change is dominated by the temperature dependence of S. *Unless the change in pressure is very large, ΔS for liquids and solids can be considered to be a function of temperature only.*

FIGURE 5.5

Two systems at constant P, each consisting of a metal rod, are placed in thermal contact. The temperatures of the two rods differ by ΔT. The composite system is contained in a rigid adiabatic enclosure (not shown) and is, therefore, an isolated system.

Initial state   Final state
Irreversible process

(a)

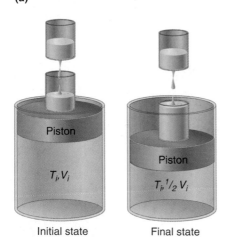

Initial state   Final state
Reversible process

(b)

FIGURE 5.6

(a) An irreversible process is shown in which an ideal gas confined in a container with rigid adiabatic walls is spontaneously reduced to half its initial volume. **(b)** A reversible isothermal compression is shown between the same initial and final states as for the irreversible process. Reversibility is achieved by adjusting the rate at which the beaker on top of the piston is filled with water relative to the evaporation rate.

5.5 Using Entropy to Calculate the Natural Direction of a Process in an Isolated System

To show that ΔS is useful in predicting the direction of spontaneous change, we now return to the two processes introduced in Section 5.1. The first process concerns the natural direction of change in a metal rod with a temperature gradient. Will the gradient become larger or smaller as the system approaches its equilibrium state? To model this process, consider the *isolated* composite system shown in Figure 5.5. Two systems, in the form of metal rods with uniform, but different, temperatures $T_1 > T_2$, are brought into thermal contact.

In the following discussion, heat is withdrawn from the left rod. (The same reasoning would hold if the direction of heat flow were reversed.) To calculate ΔS for this irreversible process using the heat flow, one must imagine a reversible process in which the initial and final states are the same as for the irreversible process. In the imaginary reversible process, the rod is coupled to a reservoir whose temperature is lowered very slowly. The temperatures of the rod and the reservoir differ only infinitesimally throughout the process in which an amount of heat, q_P, is withdrawn from the rod. The total change in temperature of the rod, ΔT, is related to q_P by

$$ đq_P = C_P dT \quad \text{or} \quad \Delta T = \frac{1}{C_P} \int đq_P = \frac{q_P}{C_P} \tag{5.25} $$

It has been assumed that $\Delta T = T_2 - T_1$ is small enough that C_P is constant over the interval.

Because the path is defined (constant pressure), $\int đq_P$ is independent of how rapidly the heat is withdrawn (the path); it depends only on C_P and ΔT. More formally, because $q_P = \Delta H$ and because H is a state function, q_P is independent of the path between the initial and final states. Therefore, $q_P = q_{reversible}$ if the temperature increment ΔT is identical for the reversible and irreversible processes.

Using this result, the entropy change for this irreversible process in which heat flows from one rod to the other is calculated. Because the composite system is isolated, $q_1 + q_2 = 0$, and $q_1 = -q_2 = q_P$. The entropy change of the composite system is the sum of the entropy changes in each rod

$$ \Delta S = \frac{q_{reversible,1}}{T_1} + \frac{q_{reversible,2}}{T_2} = \frac{q_1}{T_1} + \frac{q_2}{T_2} = q_P \left(\frac{1}{T_1} - \frac{1}{T_2} \right) \tag{5.26} $$

Because $T_1 > T_2$, the quantity in parentheses is negative. This process has two possible directions:

- If heat flows from the hotter to the colder rod, the temperature gradient will become smaller. In this case, $q_P < 0$ and $\Delta S > 0$.

- If heat flows from the colder to the hotter rod, the temperature gradient will become larger. In this case, $q_P > 0$ and $\Delta S < 0$.

Note that ΔS has the same magnitude, but a different sign, for the two directions of change. ΔS appears to be a useful function for measuring the direction of natural change in an isolated system. Experience tells us that the temperature gradient will become less with time. *It can be concluded that the process in which S increases is the direction of natural change in an isolated system.*

Next, consider the second process introduced in Section 5.1 in which an ideal gas spontaneously collapses to half its initial volume without a force acting on it. This process and its reversible analog are shown in Figure 5.6. Recall that U is independent of V for an ideal gas. Because U does not change as V increases, and U is a function of T only for an ideal gas, the temperature remains constant in the irreversible process. Therefore, the spontaneous irreversible process shown in Figure 5.6a is both adiabatic and isothermal and is described by $V_i, T_i \to 1/2 V_i, T_i$. The imaginary reversible process that we use to carry out the calculation of ΔS is shown in

Figure 5.6b. In this process, which must have the same initial and final states as the irreversible process, water is slowly and continuously added to the beaker on the piston to ensure that $P = P_{external}$. The ideal gas undergoes a reversible isothermal transformation described by $V_i, T_i \rightarrow 1/2V_i, T_i$. Because $\Delta U = 0$, $q = -w$. We calculate ΔS for this process:

$$\Delta S = \int \frac{đq_{reversible}}{T} = \frac{q_{reversible}}{T_i} = -\frac{w_{reversible}}{T_i} = nR \ln \frac{\frac{1}{2}V_i}{V_i} = -nR \ln 2 < 0 \text{ (5.27)}$$

For the opposite process, in which the gas spontaneously expands so that it occupies twice the volume, the reversible model process is an isothermal expansion for which

$$\Delta S = nR \ln \frac{2V_i}{V_i} = nR \ln 2 > 0 \tag{5.28}$$

Again, the process with $\Delta S > 0$ is the direction of natural change in this isolated system. The reverse process for which $\Delta S < 0$ is the unnatural direction of change.
The results obtained for isolated systems are generalized in the following statement:

> For any irreversible process in an isolated system, there is a unique direction of spontaneous change: $\Delta S > 0$ for the spontaneous process, $\Delta S < 0$ for the opposite or nonspontaneous direction of change, and $\Delta S = 0$ only for a reversible process. In a quasi-static reversible process, there is no direction of spontaneous change because the system is proceeding along a path, each step of which corresponds to an equilibrium state.

We cannot emphasize too strongly that $\Delta S > 0$ is a criterion for spontaneous change *only* if the system does not exchange energy in the form of heat or work with its surroundings. Note that if any process occurs in the isolated system, it is by definition spontaneous and the entropy increases. Whereas U can neither be created nor destroyed, S for an isolated system can be created $(\Delta S > 0)$, but not destroyed $(\Delta S < 0)$.

5.6 The Clausius Inequality

In the previous section, it was shown using two examples that $\Delta S > 0$ provides a criterion to predict the natural direction of change in an isolated system. This result can also be obtained without considering a specific process. Consider the differential form of the first law for a process in which only $P–V$ work is possible

$$dU = đq - P_{external} dV \tag{5.29}$$

Equation (5.29) is valid for both reversible and irreversible processes. If the process is reversible, we can write Equation (5.29) in the following form:

$$dU = đq_{reversible} - P dV = T dS - P dV \tag{5.30}$$

Because U is a state function, dU is independent of the path, and Equation (5.30) holds for both reversible and irreversible processes, as long as there are no phase transitions or chemical reactions, and only $P–V$ work occurs.
To derive the Clausius inequality, we equate the expressions for dU in Equations (5.29) and (5.30):

$$đq_{reversible} - đq = (P - P_{external}) dV \tag{5.31}$$

If $P - P_{external} > 0$, the system will spontaneously expand, and $dV > 0$. If $P - P_{external} < 0$, the system will spontaneously contract, and $dV < 0$. In both possible cases, $(P - P_{external}) dV > 0$. Therefore, we conclude that

$$đq_{reversible} - đq = T dS - đq \geq 0 \text{ or } T dS \geq đq \tag{5.32}$$

The equality holds only for a reversible process. We rewrite the **Clausius inequality** in Equation (5.32) for an irreversible process in the form

$$dS > \frac{đq}{T} \tag{5.33}$$

For an irreversible process in an isolated system, $đq = 0$. *Therefore, we have again proved that for any irreversible process in an isolated system, $\Delta S > 0$.*

How can the result from Equations (5.29) and (5.30) that $dU = đq - P_{external}dV = TdS - PdV$ be reconciled with the fact that work and heat are path functions? The answer is that $đw \geq -PdV$ and $đq \leq TdS$, where the equalities hold only for a reversible process. The result $đq + đw = TdS - PdV$ states that the amount by which the work is greater than $-PdV$ and the amount by which the heat is less than TdS in an irreversible process involving only PV work are exactly equal. *Therefore, the differential expression for dU in Equation (5.30) is obeyed for both reversible and irreversible processes.* In Chapter 6, the Clausius inequality is used to generate two new state functions, the Gibbs energy and the Helmholtz energy. These functions allow predictions to be made about the direction of change in processes for which the system interacts with its environment.

The Clausius inequality is next used to evaluate the cyclic integral $\oint đq/T$ for an arbitrary process. Because $dS = đq_{reversible}/T$, the value of the cyclic integral is zero for a reversible process. Consider a process in which the transformation from state 1 to state 2 is reversible, but the transition from state 2 back to state 1 is irreversible:

$$\oint \frac{đq}{T} = \int_1^2 \frac{đq_{reversible}}{T} + \int_2^1 \frac{đq_{irreversible}}{T} \tag{5.34}$$

The limits of integration on the first integral can be interchanged to obtain

$$\oint \frac{đq}{T} = -\int_2^1 \frac{đq_{reversible}}{T} + \int_2^1 \frac{đq_{irreversible}}{T} \tag{5.35}$$

Exchanging the limits as written is only valid for a state function. Because $đq_{reversible} > đq_{irreversible}$

$$\oint \frac{đq}{T} \leq 0 \tag{5.36}$$

where the equality only holds for a reversible process. Note that the cyclic integral of an exact differential is always zero, but the integrand in Equation (5.36) is only an exact differential for a reversible cycle.

5.7 The Change of Entropy in the Surroundings and $\Delta S_{total} = \Delta S + \Delta S_{surroundings}$

As shown in Section 5.6, the entropy of an isolated system increases in a spontaneous process. Is it always true that a process is spontaneous if ΔS for the system is positive? As shown later, this statement is only true for an isolated system. In this section, a criterion for spontaneity is developed that takes into account the entropy change in both the system and the surroundings.

In general, a system interacts only with the part of the universe that is very close. Therefore, one can think of the system and the interacting part of the surroundings as forming an interacting composite system that is isolated from the rest of the universe. The part of the surroundings that is relevant for entropy calculations is a thermal reservoir at a fixed temperature, T. The mass of the reservoir is sufficiently large that its temperature is only changed by an infinitesimal amount dT when heat is transferred between the system and the surroundings. Therefore, the surroundings always remain in internal equilibrium during heat transfer.

Next consider the entropy change of the surroundings, whereby the surroundings are at either constant V or constant P. We assume that the system and surroundings are at the same temperature. If this were not the case, heat would flow across the boundary until T is the same for system and surroundings, unless the system is surrounded by adiabatic walls, in which case $q_{surroundings} = 0$. The amount of heat absorbed by the surroundings, $q_{surroundings}$, depends on the process occurring in the system. If the surroundings are at constant V, $q_{surroundings} = \Delta U_{surroundings}$, and if the surroundings are at constant P, $q_{surroundings} = \Delta H_{surroundings}$. Because H and U are state functions, the amount of heat entering the surroundings is independent of the path; q is the same whether the transfer occurs reversibly or irreversibly. Therefore,

$$dS_{surroundings} = \frac{đq_{surroundings}}{T} \text{ or for a macroscopic change,}$$

$$\Delta S_{surroundings} = \frac{q_{surroundings}}{T} \tag{5.37}$$

Note that the heat that appears in Equation (5.37) is the *actual* heat transferred because the heat transferred to the surroundings is independent of the path as discussed earlier. By contrast, in calculating ΔS for the system, $đq_{reversible}$ for a *reversible* process that connects the initial and final states of the system must be used, *not* the actual $đq$ for the process. *It is essential to understand this reasoning in order to carry out calculations for ΔS and $\Delta S_{surroundings}$.*

This important difference is discussed in calculating the entropy change of the system as opposed to the surroundings with the aid of Figure 5.7. A gas (the system) is enclosed in a piston and cylinder assembly with diathermal walls. The gas is reversibly compressed by the external pressure generated by droplets of water slowly filling the beaker on top of the piston. The piston and cylinder assembly is in contact with a water bath thermal reservoir that keeps the temperature of the gas fixed at the value T. In Example Problem 5.7, ΔS and $\Delta S_{surroundings}$ are calculated for this reversible compression.

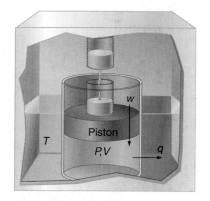

FIGURE 5.7

A sample of an ideal gas (the system) is confined in a piston and cylinder assembly with diathermal walls. The assembly is in contact with a thermal reservoir that holds the temperature at a value of 300 K. Water dripping into the beaker on the piston increases the external pressure slowly enough to ensure a reversible compression. The value of the pressure is determined by the relative rates of water filling and evaporation from the beaker. The directions of work and heat flow are indicated.

EXAMPLE PROBLEM 5.7

One mole of an ideal gas at 300. K is reversibly and isothermally compressed from a volume of 25.0 L to a volume of 10.0 L. Because the water bath thermal reservoir in the surroundings is very large, T remains essentially constant at 300. K during the process. Calculate ΔS, $\Delta S_{surroundings}$, and ΔS_{total}.

Solution

Because this is an isothermal process, $\Delta U = 0$, and $q_{reversible} = -w$. From Section 2.7,

$$q_{reversible} = -w = nRT \int_{V_i}^{V_f} \frac{dV}{V} = nRT \ln \frac{V_f}{V_i}$$

$$= 1.00 \text{ mol} \times 8.314 \text{ J mol}^{-1}\text{K}^{-1} \times 300. \text{ K} \times \ln \frac{10.0 \text{ L}}{25.0 \text{ L}} = -2.285 \times 10^3 \text{ J}$$

The entropy change of the system is given by

$$\Delta S = \int \frac{đq_{reversible}}{T} = \frac{q_{reversible}}{T} = \frac{-2.285 \times 10^3 \text{ J}}{300. \text{ K}} = -7.62 \text{ J K}^{-1}$$

The entropy change of the surroundings is given by

$$\Delta S_{surroundings} = \frac{q_{surroundings}}{T} = -\frac{q_{system}}{T} = \frac{2.285 \times 10^3 \text{ J}}{300. \text{ K}} = 7.62 \text{ J K}^{-1}$$

The total change in the entropy is given by

$$\Delta S_{total} = \Delta S + \Delta S_{surroundings} = -7.62 \text{ J K}^{-1} + 7.62 \text{ J K}^{-1} = 0$$

Because the process in Example Problem 5.7 is reversible, there is no direction of spontaneous change and, therefore, $\Delta S_{total} = 0$. In Example Problem 5.8, this calculation is repeated for an irreversible process that goes between the same initial and final states of the system.

EXAMPLE PROBLEM 5.8

One mole of an ideal gas at 300. K is isothermally compressed by a constant external pressure equal to the final pressure in Example Problem 5.7. At the end of the process, $P = P_{external}$. Because $P \neq P_{external}$ at all but the final state, this process is irreversible. The initial volume is 25.0 L and the final volume is 10.0 L. The temperature of the surroundings is 300. K. Calculate ΔS, $\Delta S_{surroundings}$, and ΔS_{total}.

Solution

We first calculate the external pressure and the initial pressure in the system

$$P_{external} = \frac{nRT}{V} = \frac{1 \text{ mol} \times 8.314 \text{ J mol}^{-1}\text{K}^{-1} \times 300. \text{ K}}{10.0 \text{ L} \times \dfrac{1 \text{ m}^3}{10^3 \text{ L}}} = 2.494 \times 10^5 \text{ Pa}$$

$$P_i = \frac{nRT}{V} = \frac{1 \text{ mol} \times 8.314 \text{ J mol}^{-1}\text{K}^{-1} \times 300. \text{ K}}{25.0 \text{ L} \times \dfrac{1 \text{ m}^3}{10^3 \text{ L}}} = 9.977 \times 10^4 \text{ Pa}$$

Because $P_{external} > P_i$, we expect that the direction of spontaneous change will be the compression of the gas to a smaller volume. Because $\Delta U = 0$,

$$q = -w = P_{external}(V_f - V_i) = 2.494 \times 10^5 \text{ Pa}$$
$$\times (10.0 \times 10^{-3} \text{ m}^3 - 25.0 \times 10^{-3} \text{ m}^3) = -3.741 \times 10^3 \text{ J}$$

The entropy change of the surroundings is given by:

$$\Delta S_{surroundings} = \frac{q_{surroundings}}{T} = -\frac{q}{T} = \frac{3.741 \times 10^3 \text{ J}}{300. \text{ K}} = 12.47 \text{ J K}^{-1}$$

The entropy change of the system must be calculated on a reversible path and has the value obtained in Example Problem 5.7

$$\Delta S = \int \frac{dq_{reversible}}{T} = \frac{q_{reversible}}{T} = \frac{-2.285 \times 10^3 \text{ J}}{300. \text{ K}} = -7.62 \text{ J K}^{-1}$$

It is seen that $\Delta S < 0$, and $\Delta S_{surroundings} > 0$. The total change in the entropy is given by

$$\Delta S_{total} = \Delta S + \Delta S_{surroundings} = -7.62 \text{ J K}^{-1} + 12.47 \text{ J K}^{-1} = 4.85 \text{ J K}^{-1}$$

The previous calculations lead to the following conclusion: *if the system and the part of the surroundings with which it interacts are viewed as an isolated composite system, the criterion for spontaneous change is $\Delta S_{total} = \Delta S + \Delta S_{surroundings} > 0$.* A decrease in the entropy of the universe will never be observed, because $\Delta S_{total} \geq 0$ for any change that actually occurs. *Any process* that occurs in the universe is by definition spontaneous and leads to an increase of S_{total}. Therefore, $\Delta S_{total} > 0$ as time increases, which defines a unique direction of time. Consider the following example to illustrate the connection between entropy and time: we view a movie in which two ideal gases are mixed, and then run the movie backward, separation of the gases occurs. We cannot decide which direction corresponds to real time (the spontaneous process) on the basis of the first law. However, using the criterion $\Delta S_{total} \geq 0$, the direction of real time can be established. The English astrophysicist Eddington coined the phrase "entropy is time's arrow" to emphasize this relationship between entropy and time.

Note that a spontaneous process in a system that interacts with its surroundings is not characterized by $\Delta S > 0$, but by $\Delta S_{total} > 0$. The entropy of the system can decrease in a spontaneous process, as long as the entropy of the surroundings increases by a greater amount. In Chapter 6, the spontaneity criterion $\Delta S_{total} = \Delta S + \Delta S_{surroundings} > 0$ will be used to generate two state functions, the Gibbs energy and the Helmholtz energy. These functions allow one to predict the direction of change in systems that interact with their environment using *only* the changes in system state functions.

5.8 Absolute Entropies and the Third Law of Thermodynamics

All elements and many compounds exist in three different states of aggregation. One or more solid phases are the most stable forms at low temperature, and when the temperature is increased to the melting point, a constant temperature transition to the liquid phase is observed. After the temperature is increased further, a constant temperature phase transition to a gas is observed at the boiling point. At temperatures higher than the boiling point, the gas is the stable form.

The entropy of an element or a compound is experimentally determined using heat capacity data through the relationship $d q_{reversible,P} = C_P dT$. Just as for the thermochemical data discussed in Chapter 4, entropy values are generally tabulated for a standard temperature of 298.15 K and a standard pressure of 1 bar. We describe such a determination for the entropy of O_2 at 298.15 K, first in a qualitative fashion, and then quantitatively in Example Problem 5.9.

The experimentally determined heat capacity of O_2 is shown in Figure 5.8 as a function of temperature for a pressure of 1 bar. O_2 has three solid phases, and transitions between them are observed at 23.66 and 43.76 K. The solid form that is stable above 43.76 K melts to form a liquid at 54.39 K. The liquid vaporizes to form a gas at 90.20 K. These phase transitions are indicated in Figure 5.8. Experimental measurements of $C_{P,m}$ are available above 12.97 K. Below this temperature, the data are extrapolated to zero kelvin by assuming that in this very low temperature range $C_{P,m}$ varies with temperature as T^3. This extrapolation is based on a model of the vibrational spectrum of a crystalline solid that will be discussed in Chapter 32. The explanation for the dependence of $C_{P,m}$ on T is the same as that presented for Cl_2 in Section 2.4.

Under constant pressure conditions, the molar entropy of the gas can be expressed in terms of the molar heat capacities of the solid, liquid, and gaseous forms and the enthalpies of fusion and vaporization as

$$S_m(T) = S_m(0\,\text{K}) + \int_0^{T_f} \frac{C_{P,m}^{solid}\, dT'}{T'} + \frac{\Delta H_{fusion}}{T_f} + \int_{T_f}^{T_b} \frac{C_{P,m}^{liquid}\, dT'}{T'}$$

$$+ \frac{\Delta H_{vaporization}}{T_b} + \int_{T_b}^{T} \frac{C_{P,m}^{gas}\, dT'}{T'} \qquad (5.38)$$

If the substance has more than one solid phase, each will give rise to a separate integral. Note that the entropy change associated with the phase transitions solid → liquid and liquid → gas discussed in Section 5.4 must be included in the calculation. To obtain a numerical value for $S_m(T)$, the heat capacity must be known down to zero kelvin, and $S_m(0\,\text{K})$ must also be known.

We first address the issue of the entropy of a solid at zero kelvin. The **third law of thermodynamics** can be stated in the following form, due to Max Planck:

> The entropy of a pure, perfectly crystalline substance (element or compound) is zero at zero kelvin.

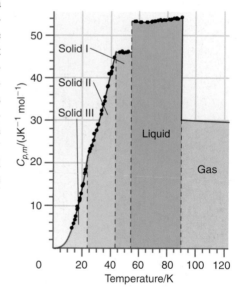

FIGURE 5.8
The experimentally determined heat capacity for O_2 is shown as a function of temperature below 125 K. The dots are data from Giauque and Johnston [*J. American Chemical Society* 51 (1929), 2300]. The red solid lines below 90 K are polynomial fits to these data. The red line above 90 K is a fit to data from the *NIST Chemistry Webbook*. The blue line is an extrapolation from 12.97 to 0 K as described in the text. The vertical dashed lines indicate constant temperature-phase transitions, and the most stable phase at a given temperature is indicated in the figure.

FIGURE 5.9

C_P/T as a function of temperature for O_2. The vertical dashed lines indicate constant temperature-phase transitions, and the most stable phase at a given temperature is indicated in the figure.

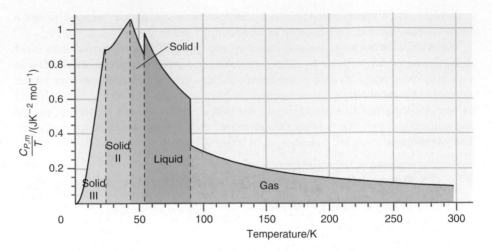

A more detailed discussion of the third law using a microscopic model will be presented in Chapter 32. Recall that in a perfectly crystalline atomic (or molecular) solid, the position of each atom is known. Because the individual atoms are indistinguishable, exchanging the positions of two atoms does not lead to a new state. Therefore, a perfect crystalline solid has only one state at zero kelvin and $S = k \ln W = k \ln 1 = 0$. The importance of the third law is that it allows calculations of the *absolute* entropies of elements and compounds to be carried out for any value of T. To calculate S at a temperature T using Equation (5.38), the $C_{P,m}$ data of Figure 5.8 are graphed in the form $C_{P,m}/T$ as shown in Figure 5.9.

The entropy can be obtained as a function of temperature by numerically integrating the area under the curve in Figure 5.9 and adding the entropy changes associated with phase changes at the transition temperatures. Calculations of S_m° for O_2 at 298.15 K are carried out in Example Problem 5.9 and $S_m^{\circ}(T)$ is shown in Figure 5.10.

One can also make the following general remarks about the relative magnitudes of the entropy of different substances and different phases of the same substance. These remarks will be justified on the basis of a microscopic model in Chapter 32.

- Because C_P/T in a single phase region and ΔS for melting and vaporization are always positive, S_m for a given substance is greatest for the gas-phase species. The molar entropies follow the order $S_m^{gas} >> S_m^{liquid} > S_m^{solid}$.

- The molar entropy increases with the size of a molecule because the number of degrees of freedom increases with the number of atoms. A non-linear gas-phase molecule has three translational degrees of freedom, three rotational degrees of freedom, and $3n - 6$ vibrational degrees of freedom. A linear molecule has three translational, two rotational, and $3n - 5$ vibrational degrees of freedom. For a molecule in a liquid, the three translational degrees of freedom are converted to local vibrational modes because of the attractive interaction between neighboring molecules.

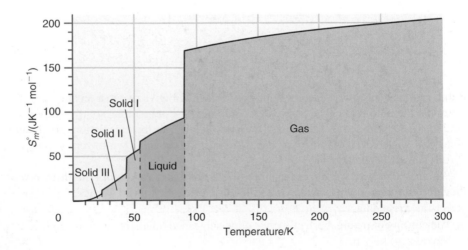

FIGURE 5.10

The molar entropy for O_2 is shown as a function of temperature. The vertical dashed lines indicate constant temperature-phase transitions, and the most stable phase at a given temperature is indicated in the figure.

- A solid has only vibrational modes. It can be modeled as a three-dimensional array of coupled harmonic oscillators as shown in Figure 5.11. This solid has a wide spectrum of vibrational frequencies, and solids with a large binding energy have higher frequencies than more weakly bound solids. Because modes with high frequencies are not activated at low temperatures, S_m^{solid} is larger for weakly bound solids than for strongly bound solids at low and moderate temperatures.

- The entropy of all substances is a monotonically increasing function of temperature.

EXAMPLE PROBLEM 5.9

The heat capacity of O_2 has been measured at 1 atm pressure over the interval 12.97 K $< T < 298.15$ K. The data have been fit to the following polynomial series in T/K, in order to have a unitless variable:

0 K $< T <$ 12.97 K:

$$\frac{C_{P,m}(T)}{\text{J mol}^{-1}\text{K}^{-1}} = 2.11 \times 10^{-3}\frac{T^3}{K^3}$$

12.97 K $< T <$ 23.66 K:

$$\frac{C_{P,m}(T)}{\text{J mol}^{-1}\text{K}^{-1}} = -5.666 + 0.6927\frac{T}{K} - 5.191 \times 10^{-3}\frac{T^2}{K^2} + 9.943 \times 10^{-4}\frac{T^3}{K^3}$$

23.66 K $< T <$ 43.76 K:

$$\frac{C_{P,m}(T)}{\text{J mol}^{-1}\text{K}^{-1}} = 31.70 - 2.038\frac{T}{K} + 0.08384\frac{T^2}{K^2} - 6.685 \times 10^{-4}\frac{T^3}{K^3}$$

43.76 K $< T <$ 54.39 K:

$$\frac{C_{P,m}(T)}{\text{J mol}^{-1}\text{K}^{-1}} = 46.094$$

54.39 K $< T <$ 90.20 K:

$$\frac{C_{P,m}(T)}{\text{J mol}^{-1}\text{K}^{-1}} = 81.268 - 1.1467\frac{T}{K} + 0.01516\frac{T^2}{K^2} - 6.407 \times 10^{-5}\frac{T^3}{K^3}$$

90.20 K $< T <$ 298.15 K:

$$\frac{C_{P,m}(T)}{\text{J mol}^{-1}\text{K}^{-1}} = 32.71 - 0.04093\frac{T}{K} + 1.545 \times 10^{-4}\frac{T^2}{K^2} - 1.819 \times 10^{-7}\frac{T^3}{K^3}$$

The transition temperatures and the enthalpies for the transitions indicated in Figure 5.8 are as follows:

Solid III → solid II	23.66 K	93.8 J mol^{-1}
Solid II → solid I	43.76 K	743 J mol^{-1}
Solid I → liquid	54.39 K	445.0 J mol^{-1}
Liquid → gas	90.20 K	6815 J mol^{-1}

a. Using these data, calculate S_m° for O_2 at 298.15 K.

b. What are the three largest contributions to S_m°?

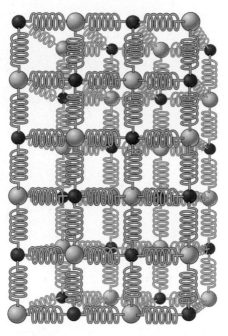

FIGURE 5.11
A useful model of a solid is a three-dimensional array of coupled harmonic oscillators. In solids with a high binding energy, the atoms are coupled by stiff springs.

Solution

a. $$S^\circ_m(298.15\,\text{K}) = \int_0^{23.66} \frac{C_{P,m}^{solid,III}\,dT}{T} + \frac{93.80\,\text{J}}{23.66\,\text{K}} + \int_{23.66}^{43.76} \frac{C_{P,m}^{solid,II}\,dT}{T} + \frac{743\,\text{J}}{43.76\,\text{K}}$$

$$+ \int_{43.76}^{54.39} \frac{C_{P,m}^{solid,I}\,dT}{T} + \frac{445.0\,\text{J}}{54.39\,\text{K}} + \int_{54.39}^{90.20} \frac{C_{P,m}^{liquid}\,dT}{T} + \frac{6815\,\text{J}}{90.20\,\text{K}}$$

$$+ \int_{90.20}^{298.15} \frac{C_{P,m}^{gas}\,dT}{T}$$

$$= 8.182\,\text{J}\,\text{K}^{-1} + 3.964\,\text{J}\,\text{K}^{-1} + 19.61\,\text{J}\,\text{K}^{-1} + 16.98\,\text{J}\,\text{K}^{-1}$$

$$+ 10.13\,\text{J}\,\text{K}^{-1} + 8.181\,\text{J}\,\text{K}^{-1} + 27.06\,\text{J}\,\text{K}^{-1} + 75.59\,\text{J}\,\text{K}^{-1}$$

$$+ 35.27\,\text{J}\,\text{K}^{-1}$$

$$= 204.9\,\text{J}\,\text{mol}^{-1}\text{K}^{-1}$$

There is an additional small correction for nonideality of the gas at 1 bar. The currently accepted value is $S^\circ_m(298.15\,\text{K}) = 205.152\,\text{J}\,\text{mol}^{-1}\text{K}^{-1}$ (Linstrom, P. J., and Mallard, W. G., eds. *NIST Chemistry Webbook: NIST Standard Reference Database Number 69*. Gaithersburg, MD: National Institute of Standards and Technology. Retrieved from http://webbook.nist.gov.)

b. The three largest contributions to S°_m are ΔS for the vaporization transition, ΔS for the heating of the gas from the boiling temperature to 298.15 K, and ΔS for heating of the liquid from the melting temperature to the boiling point.

The preceding discussion and Example Problem 5.9 show how numerical values of the entropy for a specific substance can be determined at a standard pressure of 1 bar for different values of the temperature. These numerical values can then be used to calculate entropy changes in chemical reactions, as will be shown in Section 5.10.

5.9 Standard States in Entropy Calculations

As discussed in Chapter 4, changes in U and H are calculated using the result that ΔH_f values for pure elements in their standard state at a pressure of 1 bar and a temperature of 298.15 K are zero. For S, the third law provides a natural definition of zero, namely, the crystalline state at zero kelvin. Therefore, the absolute entropy of a compound as a given temperature can be experimentally determined from heat capacity measurements as described in the previous section. The entropy is a also a function of pressure, and tabulated values of entropies refer to a standard pressure of 1 bar. The value of S varies most strongly with P for a gas. From Equation (5.19), for an ideal gas at constant T,

$$\Delta S_m = R\,\ln\frac{V_f}{V_i} = -R\,\ln\frac{P_f}{P_i} \tag{5.39}$$

Choosing $P_i = P^\circ = 1$ bar,

$$S_m(P) = S^\circ_m - R\,\ln\frac{P(\text{bar})}{P^\circ} \tag{5.40}$$

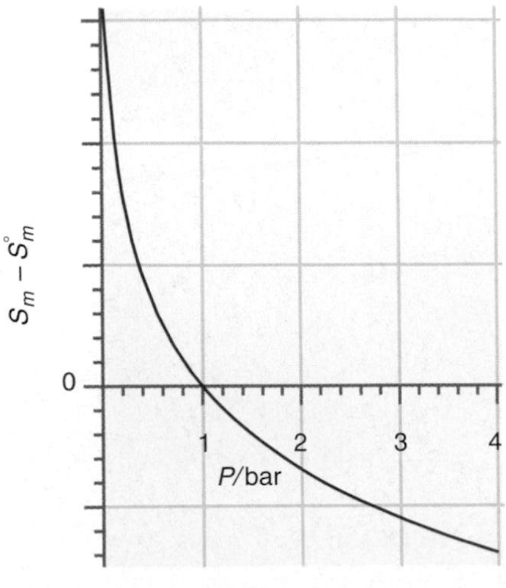

Figure 5.12 shows a plot of the molar entropy of an ideal gas as a function of pressure. It is seen that as $P \to 0$, $S_m \to \infty$. This is a consequence of the fact that as $P \to 0$, $V \to \infty$. As Equation (5.18) shows, the entropy becomes infinite in this limit.

FIGURE 5.12
The molar entropy of an ideal gas is shown as a function of the gas pressure. By definition, at 1 bar, $S_m = S^\circ_m$, the standard state molar entropy.

Equation (5.40) provides a way to calculate the entropy of a gas at any pressure. For solids and liquids, S varies so slowly with P, as shown in Section 5.4 and Example Problem 5.6, that the pressure dependence of S can usually be neglected.

5.10 Entropy Changes in Chemical Reactions

The entropy change in a chemical reaction is a major factor in determining the equilibrium concentration in a reaction mixture. In an analogous fashion to calculating ΔH_R° and ΔU_R° for chemical reactions, ΔS_R° is equal to the difference in the entropies of products and reactants, which can be written as

$$\Delta S_R^\circ = \sum_i v_i S_i^\circ \tag{5.41}$$

In Equation (5.41), the stoichiometric coefficients v_i are positive for products and negative for reactants. For example, in the reaction

$$Fe_3O_4(s) + 4\,H_2(g) \rightarrow 3\,Fe(s) + 4\,H_2O(l) \tag{5.42}$$

the entropy change under standard state conditions of 1 bar and 298.15 K is given by

$$\Delta S_{R,298.15}^\circ = 3 S_{298.15}^\circ(Fe, s) + 4 S_{298.15\,K}^\circ(H_2O, l) - S_{298.15}^\circ(Fe_3O_4, s) - 4 S_{298.15}^\circ(H_2, g)$$

$$= 3 \times 27.28\ \text{J K}^{-1}\text{mol}^{-1} + 4 \times 69.61\ \text{J K}^{-1}\text{mol}^{-1} - 146.4\ \text{J K}^{-1}\text{mol}^{-1}$$

$$- 4 \times 130.684\ \text{J K}^{-1}\text{mol}^{-1}$$

$$= -308.9\ \text{J K}^{-1}\text{mol}^{-1}$$

For this reaction, ΔS_R° is large and negative, primarily because gaseous species are consumed in the reaction, and none are generated. ΔS_R° is generally positive for $\Delta n > 0$, and negative for $\Delta n < 0$ where Δn is the change in the number of moles of gas in the overall reaction.

Tabulated values of S_m° are generally available at the standard temperature of 298.15 K, and values for selected elements and compounds are listed in Tables 4.1 and 4.2 (see Appendix B, Data Tables). However, it is often necessary to calculate ΔS° at other temperatures. Such calculations are carried out using the temperature dependence of S discussed in Section 5.4:

$$\Delta S_{R,T}^\circ = \Delta S_{R,298.15}^\circ + \int_{298.15}^{T} \frac{\Delta C_P^\circ}{T'}\,dT' \tag{5.43}$$

This equation is only valid if no phase changes occur in the temperature interval between 298.15 K and T. If phase changes occur, the associated entropy changes must be included as they were in Equation (5.38).

EXAMPLE PROBLEM 5.10

The standard entropies of CO, CO_2, and O_2 at 298.15 K are

$$S_{298.15}^\circ(CO, g) = 197.67\ \text{J K}^{-1}\text{mol}^{-1}$$

$$S_{298.15}^\circ(CO_2, g) = 213.74\ \text{J K}^{-1}\text{mol}^{-1}$$

$$S_{298.15}^\circ(O_2, g) = 205.138\ \text{J K}^{-1}\text{mol}^{-1}$$

The temperature dependence of constant pressure heat capacity for CO, CO_2, and O_2 is given by

$$\frac{C_{P,m}(CO, g)}{\text{J K}^{-1}\text{mol}^{-1}} = 31.08 - 1.452 \times 10^{-2}\frac{T}{K} + 3.1415 \times 10^{-5}\frac{T^2}{K^2} - 1.4973 \times 10^{-8}\frac{T^3}{K^3}$$

$$\frac{C_{P,m}(CO_2, g)}{J\,K^{-1}\,mol^{-1}} = 18.86 + 7.937 \times 10^{-2}\frac{T}{K} - 6.7834 \times 10^{-5}\frac{T^2}{K^2} + 2.4426 \times 10^{-8}\frac{T^3}{K^3}$$

$$\frac{C_{P,m}(O_2, g)}{J\,K^{-1}\,mol^{-1}} = 30.81 - 1.187 \times 10^{-2}\frac{T}{K} + 2.3968 \times 10^{-5}\frac{T^2}{K^2}$$

Calculate ΔS_R° for the reaction $CO(g) + 1/2\,O_2(g) \rightarrow CO_2(g)$ at 475.0 K.

Solution

$$\frac{\Delta C_{P,m}}{J\,K^{-1}\,mol^{-1}} = \left(18.86 - 31.08 - \frac{1}{2} \times 30.81\right)$$

$$+ \left(7.937 + 1.452 + \frac{1}{2} \times 1.187\right) \times 10^{-2}\frac{T}{K}$$

$$- \left(6.7834 + 3.1415 + \frac{1}{2} \times 2.3968\right) \times 10^{-5}\frac{T^2}{K^2}$$

$$+ (2.4426 + 1.4973) \times 10^{-8}\frac{T^3}{K^3}$$

$$= -27.625 + 9.9825 \times 10^{-2}\frac{T}{K} - 1.1123 \times 10^{-4}\frac{T^2}{K^2} + 3.9399 \times 10^{-8}\frac{T^3}{K^3}$$

$$\Delta S_R^\circ = S_{R\,298.15}^\circ(CO_2, g) - S_{298.15}^\circ(CO, g) - \frac{1}{2} \times S_{298.15}^\circ(O_2, g)$$

$$= 213.74\,J\,K^{-1}\,mol^{-1} - 197.67\,J\,K^{-1}\,mol^{-1} - \frac{1}{2} \times 205.138\,J\,K^{-1}\,mol^{-1}$$

$$= -86.50\,J\,K^{-1}\,mol^{-1}$$

$$\Delta S_{R,T}^\circ = \Delta S_{R\,298.15}^\circ + \int_{298.15}^{T} \frac{\Delta C_p}{T'}\,dT'$$

$$= -86.50\,J\,K^{-1}\,mol^{-1}$$

$$+ \int_{298.15}^{475} \frac{\left(-27.63 + 9.983 \times 10^{-2}\frac{T}{K} - 1.112 \times 10^{-4}\frac{T^2}{K^2} + 3.940 \times 10^{-8}\frac{T^3}{K^3}\right)}{\frac{T}{K}}\,d\frac{T}{K}\,J\,K^{-1}\,mol^{-1}$$

$$= -86.50\,J\,K^{-1}\,mol^{-1} + (-12.866 + 17.654 - 7.604 + 1.0594)\,J\,K^{-1}\,mol^{-1}$$

$$= -86.50\,J\,K^{-1}\,mol^{-1} - 1.757\,J\,K^{-1}\,mol^{-1} = -88.26\,J\,K^{-1}\,mol^{-1}$$

The value of ΔS_R° is negative at both temperatures because the number of moles of gaseous species is reduced in the reaction.

SUPPLEMENTAL

5.11 Energy Efficiency: Heat Pumps, Refrigerators, and Real Engines

Thermodynamics provides the tools to study the release of energy through chemical reactions or physical processes such as light harvesting to do work or generate heat. As Earth's population continues to increase in the coming decades and per capita energy consumption also rises, the need for energy in various forms will rise rapidly. Electricity is a major component of this energy demand and fossil fuels are expected to be the major source for electricity production for the foreseeable future. The increased

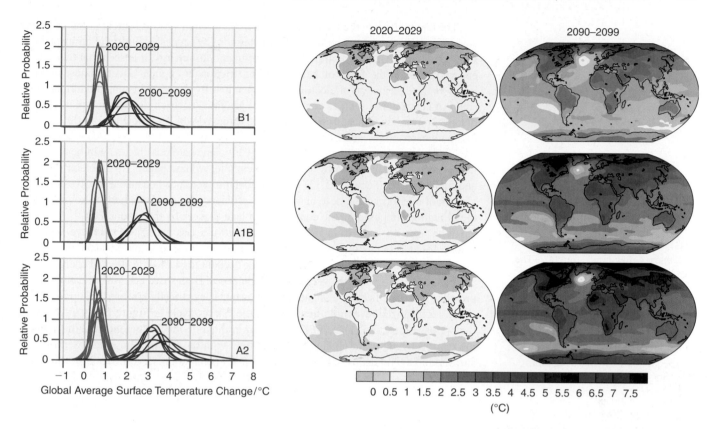

FIGURE 5.13

Projected surface temperature increase (right panel) for different scenarios of greenhouse emissions (left panel). Each curve in the left panel represents a different scenario. The uncertainty in the predicted temperature increase is represented by the width of a curve. Note that the most probable temperature increase is similar for nearly all scenarios.

Source: Climate Change 2007: The Physical Science Basis. Working Group I Contribution to the Fourth Assessment Report of the Intergovernmental Panel on Climate Change, Figure SPM.6. Cambridge University Press.

combustion of fossil fuels will continue the rapid increase in the CO_2 concentration in the atmosphere that began with the Industrial Revolution.

The increase in the atmospheric CO_2 concentration is expected to lead to a significant increase in the global average surface temperature as shown for different scenarios of the rate of atmospheric CO_2 increase in Figure 5.13. For details on the scenarios and the possible consequences of this temperature increase, which include flooding of land areas near sea level, acidification of the oceans leading to coral reef disappearance, and more severe droughts and storms, see the Intergovernmental Panel on Climate Change website at www.ipcc.ch/.

In order to slow or to reverse the buildup of greenhouse gases in the atmosphere, we must move to energy sources that do not generate greenhouse gases, and develop ways to capture greenhouses gases produced in the combustion of fossil fuels. We must also find ways to do more with less energy input. Table 5.2 shows that the per capita energy use in the United States is substantially higher than in other regions and nations with similar climates and living standards.

TABLE 5.2 Per Capita Energy Use 2005		
Nation or region	Per capita energy use/(mega BTU) 1 J = 1054 BTU	Ratio to U.S. per capita use
United States	340	1
Eurasia	160	0.429
Europe	146	0.471
Africa	16.1	0.0473
World	71.8	0.211
France	181	0.532
Germany	176	0.518

Source: U.S. Energy Information Administration

The energy flow in the United States for 2002 through various sectors of the economy is shown in Figure 5.14. The parallel pathway for CO_2 emissions into the atmosphere is shown in Figure 5.15. By comparing the useful and lost energy, it is seen that 62% of the energy is lost in the form of heat including friction in engines and turbines, resistive losses in the distribution of electricity, and heat loss from poorly insulated buildings. Our study of heat engines such as coal or natural gas fired electricity generation plants shows that the conversion of heat to work cannot be achieved with an efficiency of 100% even if dissipative processes such as friction are neglected. Therefore, significant losses are inevitable, but substantial increases in energy efficiency are possible, as will be discussed later. Electricity generation by wind turbines, photovoltaic panels, hydroelectric plants, and fuel cells (See Section 11.13) is of particular importance because it involves the conversion of one form of work to another. The efficiency for these methods is not subject to the limitations imposed on the conversion of heat to work by the second law.

What can we learn from thermodynamics to use less energy and achieve the same goals? To address this question, consider how energy is used in the residential sector as shown in Figure 5.16. Space and water heating account for 47% of residential energy usage, and electrical energy is often used for these purposes. Both space and water heating can be provided with significantly less energy input using an electrically powered heat pump.

In an idealized reversible **heat pump,** the Carnot cycle in Figure 5.2 is traversed in the opposite direction. The signs of w and q in the individual segments and the signs of the overall w and q are changed. Heat is now withdrawn from the cold reservoir (the surroundings) and deposited in the home, which is the hot reservoir. Because this is not

FIGURE 5.14

U.S. energy flow trends for 2002. The energy contributions from different sources to each economic sector such as electrical power are shown together with net useful and lost energy for each sector. Note that 62% of the total consumed energy in all sectors is lost.

Source: Lawrence Livermore National Laboratory, U.S. Department of Energy

U.S. Energy Flow Trends – 2002
Net Primary Resource Consumption ~103 Exajoules

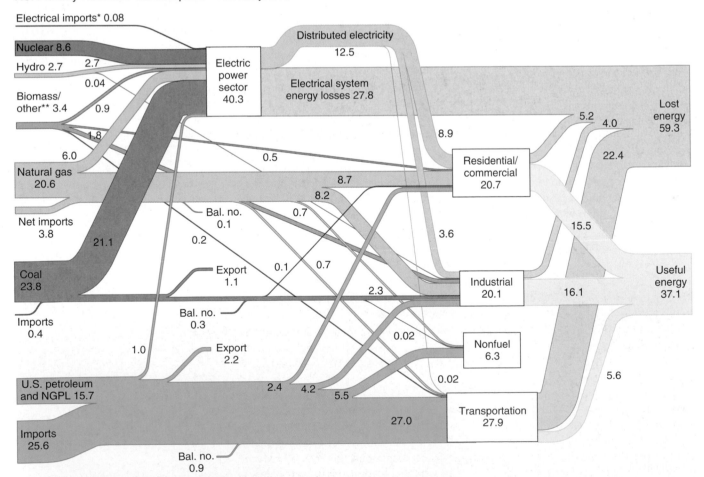

Source: Production and end-use data from Energy Information Administration, *Annual Energy Review 2002.*
*Net fossil-fuel electrical imports.
**Biomass/other includes wood, waste, alcohol, geothermal, solar, and wind.

U.S. 2002 Carbon Dioxide Emissions from Energy
Consumption – 5,682* Million Metric Tons of CO_2**

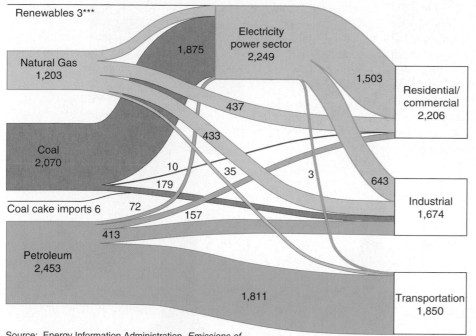

Source: Energy Information Administration. *Emissions of
Greenhouse Gases in the United States 2002*. Tables 4–10.
*Includes adjustments of 42.9 million metric tons of carbon dioxide
from U.S. territories, less 90.2 $MtCO_2$ from international and military bunker fuels.
**Previous versions of this chart showed emissions in metric tons of carbon, not of CO_2.
***Municipal solid waste and geothermal energy.
Note: Numbers may not equal sum of components because of independent rounding.

FIGURE 5.15

U.S. carbon dioxide emissions from
energy consumption for 2002. The pathways of CO_2 generation in three economic sectors are shown together with a
breakdown in the three fossil fuel sources
of natural gas, coal, and petroleum.
Source: Lawrence Livermore National
Laboratory, U.S. Department of Energy

a spontaneous process, work must be done on the system to effect this direction of heat
flow. The heat and work flow for a heat pump is shown in Figure 5.17b.

A heat pump is used to heat a building by extracting heat from a colder thermal
reservoir such as a lake, the ground, or the ambient air. The **coefficient of performance**
of a heat pump, η_{hp}, is defined as the ratio of the heat pumped into the hot reservoir to
the work input to the heat pump:

$$\eta_{hp} = \frac{q_{hot}}{w} = \frac{q_{hot}}{q_{hot} + q_{cold}} = \frac{T_{hot}}{T_{hot} - T_{cold}} \tag{5.44}$$

Assume that T_{hot} = 294 K and T_{cold} = 278 K, typical for a mild winter day. The maximum η_{hp} value is calculated to be 18. Such high values cannot be attained for real heat
pumps operating in an irreversible cycle with dissipative losses. Typical values for
commercially available heat pumps lie in the range of 3 to 4. This means that for every
joule of electrical work supplied to the heat pump, 3 to 4 J of heat are made available
for space heating.

Heat pumps become less effective as T_{cold} decreases, as shown by Equation (5.44).
Therefore, geothermal heat pumps, which use ~55°F soil 6–10 feet below Earth's
surface rather than ambient air as the cold reservoir, are much more efficient in cold
climates than air source heat pumps. Note that a coefficient of performance of 3.5
for a heat pump means that a house can be heated using 29% of the electrical power
consumption that would be required to heat the same house using electrical baseboard heaters. This is a significant argument for using heat pumps for residential
heating. Heat pumps can also be used to heat water. Assuming a heat pump coefficient of performance of 3.5, the energy usage of a typical household can be reduced
by 35% simply by replacing electrical water heaters and electrical baseboard heaters
by heat pumps, which pay for the increased initial cost by lower monthly electricity
bills within a few years. The flow of CO_2 into the atmosphere is reduced by the same

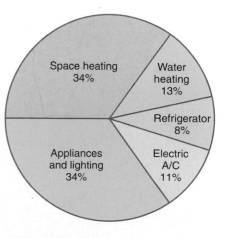

FIGURE 5.16

Distribution of U.S. residential energy use
among household needs.
Source: U.S. Energy Information Agency

FIGURE 5.17
Reverse heat engines can be used to induce heat flow from a cold reservoir to a hot reservoir with the input of work. **(a)** Refrigerator: the cold reservoir is the interior of a refrigerator, and the hot reservoir is the room in which the refrigerator is located. **(b)** Heat pump: the cold reservoir is water-filled pipes buried in the ground, and the hot reservoir is the interior of the house to be heated. The relative widths of the two paths entering the hot reservoir show that a small amount of work input can move a larger amount of heat from the cold to the hot reservoir. In both cases, the engine is a compressor.

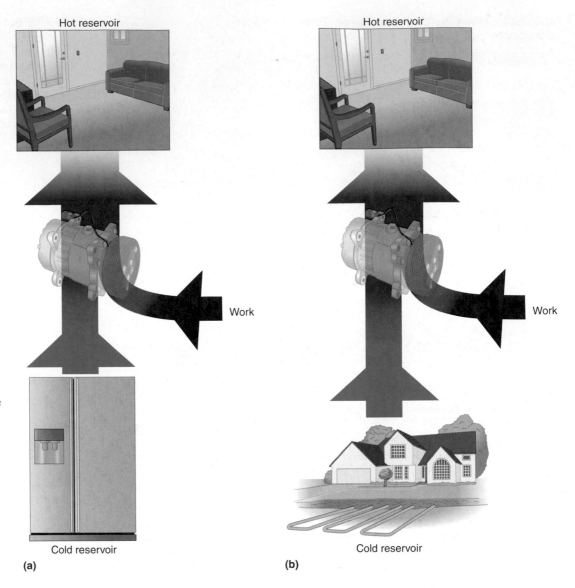

amount. Houses can be designed to significantly reduce total energy use. The "Passive House" uses a small fraction of the energy needed to operate a typical house. More than 15,000 of these houses have been built in Europe and as of 2010, only 13 have been built in the United States. For more information on the Passive House concept, see http://www.passivehouse.us/passiveHouse/PHIUSHome.html.

A **refrigerator** is also a heat engine operated in reverse. (Figure 5.17a). The interior of the refrigerator is the system, and the room in which the refrigerator is situated is the hot reservoir. Because more heat is deposited in the room than is withdrawn from the interior of the refrigerator, the overall effect of such a device is to increase the temperature of the room. However, the usefulness of the device is that it provides a cold volume for food storage. The coefficient of performance, η_r, of a reversible Carnot refrigerator is defined as the ratio of the heat withdrawn from the cold reservoir to the work supplied to the device:

$$\eta_r = \frac{q_{cold}}{w} = \frac{q_{cold}}{q_{hot} + q_{cold}} = \frac{T_{cold}}{T_{hot} - T_{cold}} \tag{5.45}$$

This formula shows that as T_{cold} decreases from $0.9\ T_{hot}$ to $0.1\ T_{hot}$, η_r decreases from 9 to 0.1. Equation (5.44) states that if the refrigerator is required to provide a lower temperature, more work is required to extract a given amount of heat.

A household refrigerator typically operates at 255 K in the freezing compartment, and 277 K in the refrigerator section. Using the lower of these temperatures for the cold reservoir and 294 K as the temperature of the hot reservoir (the room), the maximum η_r value is 6.5. This means that for every joule of work done on the system, 6.5 J of heat can be extracted from the contents of the refrigerator. This is the maximum coefficient of performance, and it is only applicable to a refrigerator operating in a reversible Carnot cycle with no dissipative losses. Taking losses into account, it is difficult to achieve η_r values greater than ~1.5 in household refrigerators. This shows the significant loss of efficiency in an irreversible dissipative cycle.

It is also instructive to consider refrigerators and heat pumps from an entropic point of view. Transferring an amount of heat, q, from a cold reservoir to a hot reservoir is not a spontaneous process in an isolated system, because

$$\Delta S = q\left(\frac{1}{T_{hot}} - \frac{1}{T_{cold}}\right) < 0 \qquad (5.46)$$

However, work can be converted to heat with 100% efficiency. Therefore, the coefficient of performance, η_r, can be calculated by determining the minimum amount of work input required to make ΔS for the withdrawal of q from the cold reservoir, together with the deposition of $q + w$ in the hot reservoir, a spontaneous process.

We have seen that a heat engine cannot convert heat into electricity with 100% efficiency because some of the heat is injected into the cold reservoir. The cold reservoir in electrical generation plants is generally the atmosphere. Generally, heat and electricity are produced separately as shown in Figure 5.18a. However, it is possible to increase the efficiency of the overall process by collecting the waste heat from electricity production and using it to heat buildings as shown in Figure 5.18b in a process called cogeneration. An example of cogeneration is to burn a fuel such as coal to generate steam that is used to drive a turbine that generates electricity. The steam exiting the turbine is still at a high temperature and can be used to heat buildings. In New York City, many buildings are heated using steam generated to produce electricity. Because of losses in piping heat over long distances, cogeneration is most suitable for urban areas. The U.S. Department of Energy has set a goal of having 20% of the electricity produced by cogeneration by the year 2030.

Further possible increases in energy efficiency in the household sector include cooking and lighting. Traditional electric cooktops use either natural gas or electricity to create heat, which is transferred to the cooking pot and its contents by conduction and convection. About 45% of the energy produced by the combustion of natural gas

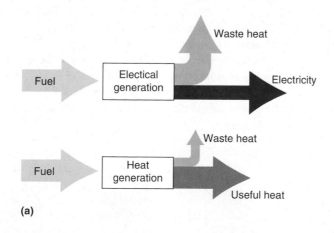

(a)

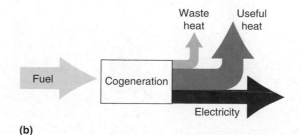

(b)

FIGURE 5.18
(a) Conventionally, electricity and heat for buildings are produced separately. Waste heat is produced in each process. (b) Using cogeneration, most of the waste heat generated in electricity production is used to heat buildings.

and 35% of the electrical energy is lost because the air rather than the pot is heated. The most efficient method for cooking is induction cooking, in which an induction coil generates a magnetic field that induces heating in metal cookware placed on top of it. Although the efficiency of this method is 90%, the initial investment is significantly greater than for a gas or electric range. Traditional lighting technology is very inefficient, with incandescent lights converting only 2% of the electrical work required to heat the tungsten filament to visible light. The remaining 98% of the radiated energy appears as heat. Fluorescent lights, which are much more efficient, contain a small amount of Hg vapor that emits UV light from an excited state created by an electrical discharge. The UV light is absorbed by a fluorescent coating on the surface of the bulb, which radiates the light in the visible spectrum. The fraction of the emitted light in the visible spectrum is much larger than in incandescent lighting. For this reason, fluorescent lighting is a factor of 5 to 6 more efficient. Australia has plans to ban the use of incandescent lighting in favor of fluorescent lighting because of its low efficiency. LED lighting (light emitting diodes) is a rapidly growing sector of lighting. LED lights have an efficiency similar to fluorescent lighting, but are more compact and can be highly directed or diffuse sources of light.

Because the transportation sector is a major user of energy, we next discuss real engines, using the Otto engine, typically used in automobiles, and the diesel engine as examples. The **Otto engine** is the most widely used engine in automobiles. The engine cycle consists of four strokes as shown in Figure 5.19. The intake valve opens as the piston is moving downward, drawing a fuel–air mixture into the cylinder. The intake valve is closed, and the mixture is compressed as the piston moves upward. Just after the piston has reached its highest point, the fuel–air mixture is ignited by a spark plug, and the rapid heating resulting from the combustion process causes the gas to expand and the pressure to increase. This drives the piston down in a power stroke. Finally, the combustion products are forced out of the cylinder by the upward-moving piston as the exhaust valve is opened. To arrive at a maximum theoretical efficiency for the Otto engine, the reversible Otto cycle shown in Figure 5.20a is analyzed, assuming reversibility.

The reversible Otto cycle begins with the intake stroke $e \rightarrow c$, which is assumed to take place at constant pressure. At this point, the intake valve is closed, and the piston compresses the fuel–air mixture along the adiabatic path $c \rightarrow d$ in the second step. This path can be assumed to be adiabatic because the compression occurs too rapidly to allow much heat to be transferred out of the cylinder. Ignition of the fuel–air mixture takes place at d. The rapid increase in pressure takes place at constant volume. In this reversible cycle, the combustion is modeled as a quasi-static heat transfer from a series of reservoirs at temperatures ranging from T_d to T_a. The power stroke is modeled as the adiabatic expansion $a \rightarrow b$. At this point, the exhaust valve opens and the gas is expelled. This step is modeled as the constant volume pressure decrease $b \rightarrow c$. The upward movement of the piston expels the remainder of the gas along the line $c \rightarrow e$, after which the cycle begins again.

FIGURE 5.19

Illustration of the four-stroke cycle of an Otto engine, as explained in the text. The left valve is the intake valve, and the right valve is the exhaust valve.

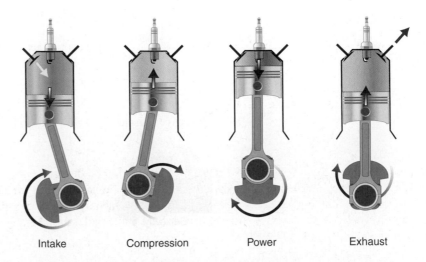

Intake   Compression   Power   Exhaust

The efficiency of this reversible cyclic engine can be calculated as follows. Assuming C_V to be constant along the segments $d \rightarrow a$ and $b \rightarrow c$, we write

$$q_{hot} = C_V(T_a - T_d) \quad \text{and} \quad q_{cold} = C_V(T_b - T_c) \tag{5.47}$$

The efficiency is given by

$$\varepsilon = \frac{q_{hot} + q_{cold}}{q_{hot}} = 1 - \frac{q_{cold}}{q_{hot}} = 1 - \left(\frac{T_b - T_c}{T_a - T_d}\right) \tag{5.48}$$

The temperatures and volumes along the reversible adiabatic segments are related by

$$T_c V_c^{\gamma-1} = T_d V_d^{\gamma-1} \quad \text{and} \quad T_b V_c^{\gamma-1} = T_a V_d^{\gamma-1} \tag{5.49}$$

because $V_b = V_c$ and $V_d = V_e$. Recall that $\gamma = C_P/C_V$. T_a and T_b can be eliminated from Equation (5.48) to give

$$\varepsilon = 1 - \frac{T_c}{T_d} \tag{5.50}$$

where T_c and T_d are the temperatures at the beginning and end of the compression stroke $c \rightarrow d$. Temperature T_c is fixed at ~300 K, and the efficiency can be increased only by increasing T_d. This is done by increasing the compression ratio, V_c/V_d. However, if the compression ratio is too high, T_d will be sufficiently high that the fuel ignites before the end of the compression stroke. A reasonable upper limit for T_d is 600 K, and for $T_c = 300$ K, $\varepsilon = 0.50$. This value is an upper limit, because a real engine does not operate in a reversible cycle, and because heat is lost along the $c \rightarrow d$ segment. Achievable efficiencies in Otto engines used in passenger cars lie in the range of 0.20 to 0.30. However, additional losses occur in the drive chain, in tire deformation and in displacing air as the vehicle moves. As shown in Figure 5.14, the overall efficiency of the transportation sector is only 20%.

In the **diesel engine** depicted in Figure 5.20b, higher compression ratios are possible because only air is let into the cylinder in the intake stroke. The fuel is injected into the cylinder at the end of the compression stroke, thus avoiding spontaneous ignition of the fuel–air mixture during the compression stroke $d \rightarrow a$. Because T~950 K for the compressed air, combustion occurs spontaneously without a spark plug along the constant pressure segment $a \rightarrow b$ after fuel injection. Because the fuel is injected over a time period in which the piston is moving out of the cylinder, this step can be modeled as a constant-pressure heat intake. In this segment, it is assumed that heat is absorbed from a series of reservoirs at temperatures between T_a and T_b in a quasi-static process. In the other segments, the same processes occur as described for the reversible Otto cycle.

The heat intake along the segment $a \rightarrow b$ is given by

$$q_{hot} = C_p(T_b - T_a) \tag{5.51}$$

and q_{cold} is given by Equation (5.47). Therefore, the efficiency is given by

$$\varepsilon = 1 - \frac{1}{\gamma}\left(\frac{T_c - T_d}{T_b - T_a}\right) \tag{5.52}$$

Similarly to the treatment of the Otto engine, T_b and T_c can be eliminated from Equation (5.52), and the expression

$$\varepsilon = 1 - \frac{1}{\gamma}\frac{\left(\frac{V_b}{V_d}\right)^{\gamma} - \left(\frac{V_a}{V_d}\right)^{\gamma}}{\left(\frac{V_b}{V_d}\right) - \left(\frac{V_a}{V_d}\right)} \tag{5.53}$$

can be derived. For typical values $V_b/V_d = 0.2, V_a/V_d = 1/15$, $\gamma = 1.5$, and $\varepsilon = 0.64$. The higher efficiency achievable with the diesel cycle in comparison with the Otto cycle is a result of the higher temperature attained in the compression cycle. Real diesel engines used in trucks and passenger cars have efficiencies in the range of 0.30 to 0.35.

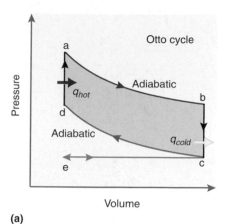

(a)

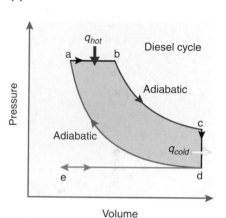

(b)

FIGURE 5.20
Idealized reversible cycles of the **(a)** Otto and **(b)** diesel engines.

Just as for space heating, the energy usage in the U.S. transportation sector can be drastically reduced. The U.S. personal vehicle fleet currently averages 27.5 miles per gallon of gasoline. The corresponding figure for the European Union, where gasoline prices are more than twice as high as in the United States, is 43 miles per gallon, corresponding to 130 g CO_2 emission per kilometer. Hybrid gasoline–electric vehicles (Figure 5.21b) that use electrical motors to power the vehicle at low speeds, switch automatically to an Otto engine at higher speeds, and use regenerative braking to capture the energy used to slow the vehicle are currently the most efficient vehicles, averaging more than 50 miles per gallon. Purely electric vehicles (Figure 5.21a) and plug-in hybrids, which will allow the batteries that power the electrical operation mode to be recharged at the owner's residence are expected to be available in 2011. Advances in battery technology are likely to make the use of electrical vehicles more widespread in the near future.

Automobiles that use fuel cells (See Section 11.13) as a power source are of particular interest for future development because the work arising from a chemical reaction can be converted to work with 100% efficiency (ignoring dissipative losses), rather than first capturing the heat from the reaction and subsequently (partially) converting the heat to work.

Reversing global climate change requires that energy be generated without also producing CO_2. Hydrogen meets this need and can be used in internal combustion engines without substantial modifications. Currently, 96% of the H_2 consumed worldwide is produced using fossil fuels, primarily through steam reforming of methane. Reversing global climate change would require that H_2 be produced using water electrolysis and electricity from renewable resources such as wind, solar, hydroelectric, or nuclear power. At present, such processes are not cost competitive, but rising oil prices and the falling cost of renewable energy may change this balance. A further unsolved problem for vehicular use is H_2 storage. The volumetric energy density of liquid H_2 (boiling temperature 20 K) is a factor of 4 below that of gasoline, and solid hydrides from which H_2 can be released at ambient temperature have a much lower energy density.

A possible alternative to the current fossil fuel economy is a methanol economy (see G. A. Olah et al., *Beyond Oil and Gas: The Methanol Economy*, Wiley-VCH, 2006). Methanol is a suitable fuel for internal combustion engines and has half the volumetric energy density of gasoline. It is also a good source for producing synthetic hydrocarbons, which could replace petroleum feedstocks widely used in the chemical industry. As shown in Figure 5.22, methanol could be synthesized by recycling CO_2 if hydrogen generated using renewable energy were available in sufficient quantities. An initial source of CO_2 could be coal-fired electrical plants, and it is conceivable that in the future CO_2 could be extracted from the atmosphere (See the previous reference for details.). These pathways would slow or reverse global climate change.

FIGURE 5.21

The Nissan Leaf **(a)** is a purely electric vehicle with an estimated range between 47 and 138 miles between recharging cycles, depending on driving conditions. The Chevrolet Volt **(b)** is an electric vehicle that has a small gasoline motor that can recharge the batteries used to power the vehicle. The U.S. Environmental Protection Agency has given the Leaf and Volt energy consumption ratings of 99 and 60 miles per gallon, respectively, using 33.7 kW-hrs as being equivalent in energy to one gallon of gasoline.

(a) Nissan Leaf

(b) Chevrolet Volt

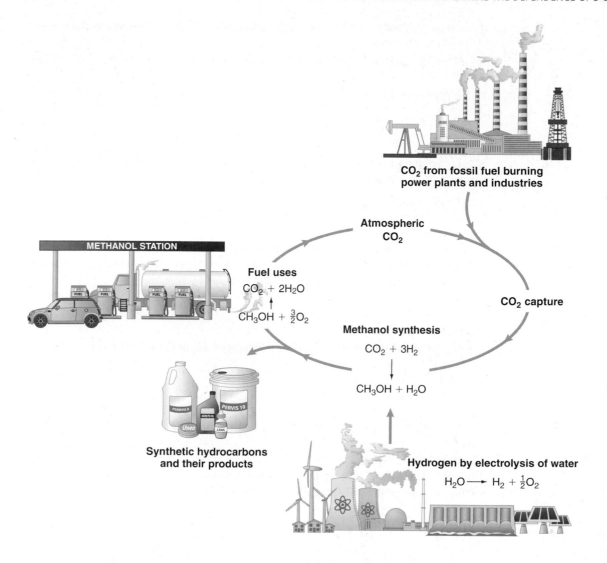

FIGURE 5.22
A cyclic model of a methanol economy includes reaction of recaptured CO_2 with H_2 generated using renewable electricity to produce methanol as a primary step. The methanol can be converted into synthetic hydrocarbons using known industrial processes.

Source: G. A. Olah et al., *Beyond Oil and Gas: The Methanol Economy*, Wiley-VCH, 2006.

SUPPLEMENTAL

5.12 Using the Fact that S Is a State Function to Determine the Dependence of S on V and T

Section 5.4 showed how the entropy varies with P, V, and T for an ideal gas. In this section, we derive general equations for the dependence of S on V and T that can be applied to solids, liquids, and real gases. We do so by using the property that dS is an exact differential. A similar analysis of S as a function of P and T is carried out in Section 5.13. Consider Equation (5.30), rewritten in the form

$$dS = \frac{1}{T}dU + \frac{P}{T}dV \qquad (5.54)$$

Because $1/T$ and P/T are greater than zero, the entropy of a system increases with the internal energy at constant volume, and increases with the volume at constant internal energy. However, because internal energy is not generally a variable under experimental control, it is more useful to obtain equations for the dependence of dS on V and T.

We first write the total differential dS in terms of the partial derivatives with respect to V and T:

$$dS = \left(\frac{\partial S}{\partial T}\right)_V dT + \left(\frac{\partial S}{\partial V}\right)_T dV \qquad (5.55)$$

To evaluate $(\partial S/\partial T)_V$ and $(\partial S/\partial V)_T$, Equation (5.54) for dS is rewritten in the form

$$dS = \frac{1}{T}\left[C_V dT + \left(\frac{\partial U}{\partial V}\right)_T dV\right] + \frac{P}{T}dV = \frac{C_V}{T}dT + \frac{1}{T}\left[P + \left(\frac{\partial U}{\partial V}\right)_T\right]dV \quad \textbf{(5.56)}$$

Equating the coefficients of dT and dV in Equations (5.55) and (5.56),

$$\left(\frac{\partial S}{\partial T}\right)_V = \frac{C_V}{T} \text{ and } \left(\frac{\partial S}{\partial V}\right)_T = \frac{1}{T}\left[P + \left(\frac{\partial U}{\partial V}\right)_T\right] \quad \textbf{(5.57)}$$

The temperature dependence of entropy at constant volume can be calculated straightforwardly using the first equality in Equation (5.57):

$$dS = \frac{C_V}{T}dT, \text{ constant } V \quad \textbf{(5.58)}$$

The expression for $(\partial S/\partial V)_T$ in Equation (5.57) is not in a form that allows for a direct comparison with experiment to be made. A more useful relation follows from the fact that dS is an exact differential (see Section 3.1):

$$\left(\frac{\partial}{\partial T}\left(\frac{\partial S}{\partial V}\right)_T\right)_V = \left(\frac{\partial}{\partial V}\left(\frac{\partial S}{\partial T}\right)_V\right)_T \quad \textbf{(5.59)}$$

Taking the mixed second derivatives of the expressions in Equation (5.57),

$$\left(\frac{\partial}{\partial V}\left(\frac{\partial S}{\partial T}\right)_V\right)_T = \frac{1}{T}\left(\frac{\partial}{\partial V}\left(\frac{\partial U}{\partial T}\right)_V\right)_T$$

$$\left(\frac{\partial}{\partial T}\left(\frac{\partial S}{\partial V}\right)_T\right)_V = \frac{1}{T}\left[\left(\frac{\partial P}{\partial T}\right)_V + \left(\frac{\partial}{\partial T}\left(\frac{\partial U}{\partial V}\right)_T\right)_V\right] - \frac{1}{T^2}\left[P + \left(\frac{\partial U}{\partial V}\right)_T\right] \quad \textbf{(5.60)}$$

Substituting the expressions for the mixed second derivatives in Equation (5.60) into Equation (5.59), canceling the double mixed derivative of U that appears on both sides of the equation, and simplifying the result, the following equation is obtained:

$$P + \left(\frac{\partial U}{\partial V}\right)_T = T\left(\frac{\partial P}{\partial T}\right)_V \quad \textbf{(5.61)}$$

This equation provides the expression for $(\partial U/\partial V)_T$ that was used without a derivation in Section 3.2. It provides a way to calculate the internal pressure of the system if the equation of state for the substance is known.

Comparing the result in Equation (5.61) with the second equality in Equation (5.57), a practical equation is obtained for the dependence of entropy on volume under constant temperature conditions:

$$\left(\frac{\partial S}{\partial V}\right)_T = \left(\frac{\partial P}{\partial T}\right)_V = -\frac{(\partial V/\partial T)_P}{(\partial V/\partial P)_T} = \frac{\beta}{\kappa} \quad \textbf{(5.62)}$$

where β is the coefficient for thermal expansion at constant pressure, and κ is the isothermal compressibility coefficient. Both of these quantities are readily obtained from experiments. In simplifying this expression, the cyclic rule for partial derivatives, Equation (3.7) has been used.

The result of these considerations is that dS can be expressed in terms of dT and dV as

$$dS = \frac{C_V}{T}dT + \frac{\beta}{\kappa}dV \quad \textbf{(5.63)}$$

Integrating both sides of this equation along a reversible path yields

$$\Delta S = \int_{T_i}^{T_f}\frac{C_V}{T}dT + \int_{V_i}^{V_f}\frac{\beta}{\kappa}dV \quad \textbf{(5.64)}$$

This result applies to a single-phase system of a liquid, solid, or gas that undergoes a transformation from the initial result T_i, V_i to T_f, V_f, provided that no phase changes or chemical reactions occur in the system.

SUPPLEMENTAL

5.13 The Dependence of S on T and P

Because chemical transformations are normally carried out at constant pressure rather than constant volume, we need to know how S varies with T and P. The total differential dS is written in the form

$$dS = \left(\frac{\partial S}{\partial T}\right)_P dT + \left(\frac{\partial S}{\partial P}\right)_T dP \tag{5.65}$$

Starting from the relation $U = H - PV$, we write the total differential dU as

$$dU = T\,dS - P\,dV = dH - P\,dV - V\,dP \tag{5.66}$$

This equation can be rearranged to give an expression for dS:

$$dS = \frac{1}{T}dH - \frac{V}{T}dP \tag{5.67}$$

The previous equation is analogous to Equation (5.54), but contains the variable P rather than V.

$$dH = \left(\frac{\partial H}{\partial T}\right)_P dT + \left(\frac{\partial H}{\partial P}\right)_T dP = C_P dT + \left(\frac{\partial H}{\partial P}\right)_T dP \tag{5.68}$$

Substituting this expression for dH into Equation (5.67),

$$dS = \frac{C_P}{T}dT + \frac{1}{T}\left[\left(\frac{\partial H}{\partial P}\right)_T - V\right]dP = \left(\frac{\partial S}{\partial T}\right)_P dT + \left(\frac{\partial S}{\partial P}\right)_T dP \tag{5.69}$$

Because the coefficients of dT and dP must be the same on both sides of Equation (5.69),

$$\left(\frac{\partial S}{\partial T}\right)_P = \frac{C_P}{T} \quad \text{and} \quad \left(\frac{\partial S}{\partial P}\right)_T = \frac{1}{T}\left[\left(\frac{\partial H}{\partial P}\right)_T - V\right] \tag{5.70}$$

The ratio C_P/T is positive for all substances, allowing us to conclude that S is a monotonically increasing function of the temperature.

Just as for $(\partial S/\partial V)_T$ in Section 5.12, the expression for $(\partial S/\partial P)_T$ is not in a form that allows a direct comparison with experimental measurements to be made. Just as in our evaluation of $(\partial S/\partial V)_T$, we equate the mixed second partial derivatives of $(\partial S/\partial T)_P$ and $(\partial S/\partial P)_T$:

$$\left(\frac{\partial}{\partial T}\left(\frac{\partial S}{\partial P}\right)_T\right)_P = \left(\frac{\partial}{\partial P}\left(\frac{\partial S}{\partial T}\right)_P\right)_T \tag{5.71}$$

These mixed partial derivatives can be evaluated using Equation (5.70):

$$\left(\frac{\partial}{\partial P}\left(\frac{\partial S}{\partial T}\right)_P\right)_T = \frac{1}{T}\left(\frac{\partial C_P}{\partial P}\right)_T = \frac{1}{T}\left(\frac{\partial}{\partial P}\left(\frac{\partial H}{\partial T}\right)_P\right)_T \tag{5.72}$$

$$\left(\frac{\partial}{\partial T}\left(\frac{\partial S}{\partial P}\right)_T\right)_P = \frac{1}{T}\left[\left(\frac{\partial}{\partial T}\left(\frac{\partial H}{\partial P}\right)_T\right)_P - \left(\frac{\partial V}{\partial T}\right)_P\right] - \frac{1}{T^2}\left[\left(\frac{\partial H}{\partial P}\right)_T - V\right] \tag{5.73}$$

Equating Equations (5.72) and (5.73) yields

$$\frac{1}{T}\left(\frac{\partial}{\partial T}\left(\frac{\partial H}{\partial P}\right)_T\right)_P = \frac{1}{T}\left[\left(\frac{\partial}{\partial T}\left(\frac{\partial H}{\partial P}\right)_T\right)_P - \left(\frac{\partial V}{\partial T}\right)_P\right] - \frac{1}{T^2}\left[\left(\frac{\partial H}{\partial P}\right)_T - V\right] \tag{5.74}$$

Simplifying this equation results in

$$\left(\frac{\partial H}{\partial P}\right)_T - V = -T\left(\frac{\partial V}{\partial T}\right)_P \tag{5.75}$$

Using this result and Equation (5.70), the pressure dependence of the entropy at constant temperature can be written in a form that easily allows an experimental determination of this quantity to be made:

$$\left(\frac{\partial S}{\partial P}\right)_T = -\left(\frac{\partial V}{\partial T}\right)_P = -V\beta \tag{5.76}$$

Using these results, the total differential dS can be written in terms of experimentally accessible parameters as

$$dS = \frac{C_P}{T}dT - V\beta\,dP \tag{5.77}$$

Integrating both sides of this equation along a reversible path yields

$$\Delta S = \int_{T_i}^{T_f}\frac{C_P}{T}dT - \int_{P_i}^{P_f}V\beta\,dP \tag{5.78}$$

This result applies to a single-phase system of a pure liquid, solid, or gas that undergoes a transformation from the initial result T_i, P_i to T_f, P_f, provided that no phase changes or chemical reactions occur in the system.

S U P P L E M E N T A L

5.14 The Thermodynamic Temperature Scale

The reversible Carnot cycle provides a basis for the **thermodynamic temperature scale,** a scale that is independent of the choice of a particular thermometric substance. This is the case because all reversible Carnot engines have the same efficiency, regardless of the working substance. The basis for the thermodynamic temperature scale is the fact that the heat withdrawn from a reservoir is a thermometric property. Both on experimental and theoretical grounds, it can be shown that

$$q = a\theta \tag{5.79}$$

where θ is the thermodynamic temperature, and a is an arbitrary scale constant that sets numerical values for the thermodynamic temperature. Using Equations (5.9) and (5.10) for the efficiency of the reversible Carnot engine,

$$\varepsilon = \frac{q_{hot} + q_{cold}}{q_{hot}} = \frac{\theta_{hot} + \theta_{cold}}{\theta_{hot}} \tag{5.80}$$

This equation is the fundamental equation establishing an absolute temperature scale. To this point, we have no numerical values for this scale. Note, however, that $q \rightarrow 0$ as $\theta \rightarrow 0$, so that there is a natural zero for this temperature scale. Additionally, if we choose one value of θ to be positive, all other values of θ must be greater than zero. Otherwise, we could find conditions under which the heats q_{hot} and q_{cold} have the same sign. This would lead to a perpetual motion machine. Both of these characteristics fit the requirements of an absolute temperature scale.

A numerical scale for the thermodynamic temperature scale can be obtained by assigning the value 273.16 to the θ value corresponding to the triple point of water, and by making the size of a degree equal to the size of a degree on the Celsius scale. With this choice, the thermodynamic temperature scale becomes numerically equal to the absolute temperature scale based on the ideal gas law. However, the thermodynamic temperature scale is the primary scale because it is independent of the nature of the working substance.

Vocabulary

Carnot cycle

Clausius inequality

coefficient of performance

diesel engine

entropy

heat engine

heat pump

irreversible process

natural transformations

Otto engine

perpetual motion machine of the first kind

perpetual motion machine of the second kind

refrigerator

second law of thermodynamics

spontaneous process

thermodynamic temperature scale

third law of thermodynamics

unnatural transformations

Conceptual Problems

Q5.1 Under what conditions is $\Delta S < 0$ for a spontaneous process?

Q5.2 Why are ΔS_{fusion} and $\Delta S_{vaporization}$ always positive?

Q5.3 An ideal gas in thermal contact with the surroundings is cooled in an irreversible process at constant pressure. Are ΔS, $\Delta S_{surroundings}$, and ΔS_{total} positive, negative, or zero? Explain your reasoning.

Q5.4 The amplitude of a pendulum consisting of a mass on a long wire is initially adjusted to have a very small value. The amplitude is found to decrease slowly with time. Is this process reversible? Would the process be reversible if the amplitude did not decrease with time?

Q5.5 A process involving an ideal gas and in which the temperature changes at constant volume is carried out. For a fixed value of ΔT, the mass of the gas is doubled. The process is repeated with the same initial mass and ΔT is doubled. For which of these processes is ΔS greater? Why?

Q5.6 You are told that $\Delta S = 0$ for a process in which the system is coupled to its surroundings. Can you conclude that the process is reversible? Justify your answer.

Q5.7 Under what conditions does the equality $\Delta S = \Delta H/T$ hold?

Q5.8 Is the following statement true or false? If it is false, rephrase it so that it is true. The entropy of a system cannot increase in an adiabatic process.

Q5.9 Which of the following processes is spontaneous?

a. The reversible isothermal expansion of an ideal gas.

b. The vaporization of superheated water at 102°C and 1 bar.

c. The constant pressure melting of ice at its normal freezing point by the addition of an infinitesimal quantity of heat.

d. The adiabatic expansion of a gas into a vacuum.

Q5.10 One Joule of work is done on a system, raising its temperature by one degree centigrade. Can this increase in temperature be harnessed to do one Joule of work? Explain.

Q5.11 Your roommate decides to cool the kitchen by opening the refrigerator. Will this strategy work? Explain your reasoning.

Q5.12 An ideal gas undergoes an adiabatic expansion into a vacuum. Are ΔS, $\Delta S_{surroundings}$, and ΔS_{total} positive, negative, or zero? Explain your reasoning.

Q5.13 When a saturated solution of a salt is cooled, a precipitate crystallizes out. Is the entropy of the crystalline precipitate greater or less than the dissolved solute? Explain why this process is spontaneous.

Q5.14 A system undergoes a change from one state to another along two different pathways, one of which is reversible and the other of which is irreversible. What can you say about the relative magnitudes of $q_{reversible}$ and $q_{irreversible}$?

Q5.15 An ideal gas in a piston and cylinder assembly with adiabatic walls undergoes an expansion against a constant external pressure. Are ΔS, $\Delta S_{surroundings}$, and ΔS_{total} positive, negative, or zero? Explain your reasoning.

Q5.16 Is the equation

$$\Delta S = \int_{T_i}^{T_f} \frac{C_V}{T} dT + \int_{V_i}^{V_f} \frac{\beta}{\kappa} dV = C_V \ln \frac{T_f}{T_i} + \frac{\beta}{\kappa}(V_f - V_i)$$

valid for an ideal gas?

Q5.17 Why is the efficiency of a Carnot heat engine the upper bound to the efficiency of an internal combustion engine?

Q5.18 Two vessels of equal volume, pressure and temperature both containing Ar are connected by a valve. What is the change in entropy when the valve is opened, allowing mixing of the two volumes? Is ΔS the same if one of the volumes contained Ar, and the other contained Ne?

Q5.19 Without using equations, explain why ΔS for a liquid or solid is dominated by the temperature dependence of S as both P and T change.

Q5.20 Solid methanol in thermal contact with the surroundings is reversibly melted at the normal melting point at a pressure of 1 atm. Are ΔS, $\Delta S_{surroundings}$, and ΔS_{total} positive, negative, or zero? Explain your reasoning.

Q5.21 Can incandescent lighting be regarded as an example of cogeneration during the heating season? In a season where air conditioning is required?

Numerical Problems

Problem numbers in **red** indicate that the solution to the problem is given in the *Student's Solutions Manual*.

P5.1 Consider the formation of glucose from carbon dioxide and water (i.e., the reaction of the photosynthetic process): $6CO_2(g) + 6H_2O(l) \rightarrow C_6H_{12}O_6(s) + 6O_2(g)$.

The following table of information will be useful in working this problem:

$T = 298K$	$CO_2\ (g)$	$H_2O\ (l)$	$C_6H_{12}O_6(s)$	$O_2(g)$
ΔH_f° kJ mol^{-1}	−393.5	−285.8	−1273.1	0.0
S° J mol^{-1}K^{-1}	213.8	70.0	209.2	205.2
$C_{P,m}^\circ$ J mol^{-1}K^{-1}	37.1	75.3	219.2	29.4

Calculate the entropy and enthalpy changes for this chemical system at $T = 298$ K and $T = 310$. K. Calculate also the entropy change of the surroundings and the universe at both temperatures, assuming that the system and surroundings are at the same temperature.

P5.2 The Chalk Point, Maryland, generating station supplies electrical power to the Washington, D.C., area. Units 1 and 2 have a gross generating capacity of 710. MW (megawatt). The steam pressure is 25×10^6 Pa, and the superheater outlet temperature (T_h) is 540.°C. The condensate temperature (T_c) is 30.0°C.

a. What is the efficiency of a reversible Carnot engine operating under these conditions?

b. If the efficiency of the boiler is 91.2%, the overall efficiency of the turbine, which includes the Carnot efficiency and its mechanical efficiency, is 46.7%, and the efficiency of the generator is 98.4%, what is the efficiency of the total generating unit? (Another 5.0% needs to be subtracted for other plant losses.)

c. One of the coal-burning units produces 355 MW. How many metric tons (1 metric ton = 1×10^6 g) of coal per hour are required to operate this unit at its peak output if the enthalpy of combustion of coal is 29.0×10^3 kJ kg^{-1}?

P5.3 An electrical motor is used to operate a Carnot refrigerator with an interior temperature of 0.00°C. Liquid water at 0.00°C is placed into the refrigerator and transformed to ice at 0.00°C. If the room temperature is 300. K, what mass of ice can be produced in one day by a 0.50-hp motor that is running continuously? Assume that the refrigerator is perfectly insulated and operates at the maximum theoretical efficiency.

P5.4 An air conditioner is a refrigerator with the inside of the house acting as the cold reservoir and the outside atmosphere acting as the hot reservoir. Assume that an air conditioner consumes 1.70×10^3 W of electrical power, and that it can be idealized as a reversible Carnot refrigerator. If the coefficient of performance of this device is 3.30, how much heat can be extracted from the house in a day?

P5.5 One mole of $H_2O(l)$ is compressed from a state described by $P = 1.00$ bar and $T = 350$. K to a state described by $P = 590$. bar and $T = 750$. K. In addition, $\beta = 2.07 \times 10^{-4}$ K^{-1} and the density can be assumed to be constant at the value 997 kg m^{-3}. Calculate ΔS for this transformation, assuming that $\kappa = 0$.

P5.6 2.25 moles of an ideal gas with $C_{V,m} = 3R/2$ undergoes the transformations described in the following list from an initial state described by $T = 310$. K and $P = 1.00$ bar. Calculate q, w, ΔU, ΔH, and ΔS for each process.

a. The gas is heated to 675 K at a constant external pressure of 1.00 bar.

b. The gas is heated to 675 K at a constant volume corresponding to the initial volume.

c. The gas undergoes a reversible isothermal expansion at 310. K until the pressure is one third of its initial value.

P5.7 Consider the reversible Carnot cycle shown in Figure 5.2 with 1.25 mol of an ideal gas with $C_V = 5R/2$ as the working substance. The initial isothermal expansion occurs at the hot reservoir temperature of $T_{hot} = 740$. K from an initial volume of 3.75 L (V_a) to a volume of 12.8 L (V_b). The system then undergoes an adiabatic expansion until the temperature falls to $T_{cold} = 310$ K. The system then undergoes an isothermal compression and a subsequent adiabatic compression until the initial state described by $T_a = 740$. K and $V_a = 3.75$ L is reached.

a. Calculate V_c and V_d.

b. Calculate w for each step in the cycle and for the total cycle.

c. Calculate ε and the amount of heat that is extracted from the hot reservoir to do 1.00 kJ of work in the surroundings.

P5.8 The average heat evolved by the oxidation of foodstuffs in an average adult per hour per kilogram of body weight is 7.20 kJ kg^{-1} hr^{-1}. Assume the weight of an average adult is 62.0 kg. Suppose the total heat evolved by this oxidation is transferred into the surroundings over a period lasting one week. Calculate the entropy change of the surroundings associated with this heat transfer. Assume the surroundings are at $T = 293$ K.

P5.9 Calculate ΔS, ΔS_{total}, and $\Delta S_{surroundings}$ when the volume of 150. g of CO initially at 273 K and 1.00 bar increases by a factor of two in (a) an adiabatic reversible expansion, (b) an expansion against $P_{external} = 0$, and (c) an isothermal reversible expansion. Take $C_{P,m}$ to be constant at the value 29.14 J mol^{-1}K^{-1} and assume ideal gas behavior. State whether each process is spontaneous. The temperature of the surroundings is 273 K.

P5.10 The maximum theoretical efficiency of an internal combustion engine is achieved in a reversible Carnot cycle. Assume that the engine is operating in the Otto cycle and that $C_{V,m} = 5R/2$ for the fuel–air mixture initially at 273 K (the temperature of the cold reservoir). The mixture is compressed by a factor of 6.9 in the adiabatic compression step. What is the maximum theoretical efficiency of this engine? How much would the efficiency increase if the compression ratio could be increased to 15? Do you see a problem in doing so?

P5.11 2.25 moles of an ideal gas with $C_{V,m} = 5R/2$ are transformed from an initial state $T = 680.$ K and $P = 1.15$ bar to a final state $T = 298$ K and $P = 4.75$ bar. Calculate ΔU, ΔH, and ΔS for this process.

P5.12 1.10 moles of N_2 at 20.5°C and 6.20 bar undergoes a transformation to the state described by 215°C and 1.75 bar. Calculate ΔS if

$$\frac{C_{P,m}}{J\,mol^{-1}K^{-1}} = 30.81 - 11.87 \times 10^{-3}\frac{T}{K} + 2.3968 \times 10^{-5}\frac{T^2}{K^2}$$
$$-1.0176 \times 10^{-8}\frac{T^3}{K^3}$$

P5.13 Calculate ΔS for the isothermal compression of 1.75 mole of Cu(s) from 2.15 bar to 1250. bar at 298 K. $\beta = 0.492 \times 10^{-4}$ K^{-1}, $\kappa = 0.78 \times 10^{-6}$ bar^{-1}, and the density is 8.92g cm^{-3}. Repeat the calculation assuming that $\kappa = 0$.

P5.14 Calculate $\Delta S°$ for the reaction $3H_2(g) + N_2(g) \rightarrow 2NH_3(g)$ at 725 K. Omit terms in the temperature-dependent heat capacities higher than T^2/K^2.

P5.15 Using the expression $dS = \dfrac{C_P}{T}dT - V\beta dP$, calculate the decrease in temperature that occurs if 2.25 moles of water at 310. K and 1650. bar is brought to a final pressure of 1.30 bar in a reversible adiabatic process. Assume that $\kappa = 0$.

P5.16 3.75 moles of an ideal gas with $C_{V,m} = 3/2R$ undergoes the transformations described in the following list from an initial state described by $T = 298$ K and $P = 4.50$ bar. Calculate q, w, ΔU, ΔH, and ΔS for each process.

a. The gas undergoes a reversible adiabatic expansion until the final pressure is one third its initial value.

b. The gas undergoes an adiabatic expansion against a constant external pressure of 1.50 bar until the final pressure is one third its initial value.

c. The gas undergoes an expansion against a constant external pressure of zero bar until the final pressure is equal to one third of its initial value.

P5.17 The interior of a refrigerator is typically held at 36°F and the interior of a freezer is typically held at 0.00°F. If the room temperature is 65°F, by what factor is it more expensive to extract the same amount of heat from the freezer than from the refrigerator? Assume that the theoretical limit for the performance of a reversible refrigerator is valid in this case.

P5.18 Using your results from Problem P5.7, calculate q, ΔU, and ΔH for each step in the cycle and for the total cycle described in Figure 5.2.

P5.19 At the transition temperature of 95.4°C, the enthalpy of transition from rhombic to monoclinic sulfur is 0.38 kJ mol^{-1}.

a. Calculate the entropy of transition under these conditions.

b. At its melting point, 119°C, the enthalpy of fusion of monoclinic sulfur is 1.23 kJ mol^{-1}. Calculate the entropy of fusion.

c. The values given in parts (a) and (b) are for 1 mol of sulfur; however, in crystalline and liquid sulfur, the molecule is present as S_8. Convert the values of the enthalpy and entropy of fusion in parts (a) and (b) to those appropriate for S_8.

P5.20 One mole of a van der Waals gas at 25.0°C is expanded isothermally and reversibly from an initial volume of 0.010 m^3 to a final volume of 0.095 m^3. For the van der Waals gas, $(\partial U/\partial V)_T = a/V_m^2$. Assume that $a = 0.556$ Pa m^6 mol^{-2}, and that $b = 64.0 \times 10^{-6}$ m^3 mol^{-1}. Calculate q, w, ΔU, ΔH, and ΔS for the process.

P5.21 From the following data, derive the absolute entropy of crystalline glycine at $T = 300.$ K.

You can perform the integration numerically using either a spreadsheet program or a curve-fitting routine and a graphing calculator (see Example Problem 5.9).

T (K)	$C_{P,m}$ (J K^{-1} mol^{-1})
10.	0.30
20.	2.4
30.	7.0
40.	13.0
60.	25.1
80.	35.2
100.	43.2
120.	50.0
140.	56.0
160.	61.6
180.	67.0
200.	72.2
220.	77.4
240.	82.8
260.	88.4
280.	94.0
300.	99.7

P5.22 Calculate ΔH and ΔS if the temperature of 1.75 moles of Hg(l) is increased from 0.00°C to 75.0°C at 1 bar. Over this temperature range, $C_{P,m}$(J K^{-1} mol^{-1}) $= 30.093 - 4.944 \times 10^{-3}T/K$.

P5.23 Calculate ΔS if the temperature of 2.50 mol of an ideal gas with $C_V = 5/2R$ is increased from 160. to 675 K under conditions of (a) constant pressure and (b) constant volume.

P5.24 Beginning with Equation (5.5), use Equation (5.6) to eliminate V_c and V_d to arrive at the result $w_{cycle} = nR(T_{hot} - T_{cold}) \ln V_b/V_a$.

P5.25 Calculate $\Delta S_R°$ for the reaction $H_2(g) + Cl_2(g) \rightarrow 2HCl(g)$ at 870. K. Omit terms in the temperature-dependent heat capacities higher than T^2/K^2.

P5.26 A 22.0 g mass of ice at 273 K is added to 136 g of $H_2O(l)$ at 310. K at constant pressure. Is the final state of the system ice or liquid water? Calculate ΔS for the process. Is the process spontaneous?

P5.27 Under anaerobic conditions, glucose is broken down in muscle tissue to form lactic acid according to the reaction $C_6H_{12}O_6 \rightarrow 2CH_3CHOHCOOH$. Thermodynamic data at $T = 298$ K for glucose and lactic acid are given in the following table:

	$\Delta H_f^\circ(\text{kJ mol}^{-1})$	$C_{P,m}(\text{J K}^{-1}\text{mol}^{-1})$	$S_m^\circ(\text{J K}^{-1}\text{mol}^{-1})$
Glucose	−1273.1	219.2	209.2
Lactic Acid	−673.6	127.6	192.1

Calculate, ΔS for the system, the surroundings, and the universe at $T = 325$ K. Assume the heat capacities are constant between $T = 298$ K and $T = 330$. K.

P5.28 The amino acid glycine dimerizes to form the dipeptide glycylglycine according to the reaction

$$2\text{Glycine}(s) \rightarrow \text{Glycylglycine}(s) + H_2O(l)$$

Calculate ΔS, ΔS_{surr}, and $\Delta S_{universe}$ at $T = 298$ K. Useful thermodynamic data follow:

	Glycine	Glycylglycine	Water
$\Delta H_f^\circ(\text{kJ mol}^{-1})$	−537.2	−746.0	−285.8
$S_m^\circ(\text{J K}^{-1}\text{mol}^{-1})$	103.5	190.0	70.0

P5.29 One mole of $H_2O(l)$ is supercooled to −3.75°C at 1 bar pressure. The freezing temperature of water at this pressure is 0.00°C. The transformation $H_2O(l) \rightarrow H_2O(s)$ is suddenly observed to occur. By calculating ΔS, $\Delta S_{surroundings}$, and ΔS_{total}, verify that this transformation is spontaneous at −3.75°C. The heat capacities are given by $C_{P,m}(H_2O(l)) = 75.3$ J K^{-1} mol^{-1} and $C_{P,m}(H_2O(s)) = 37.7$ J K^{-1} mol^{-1}, and $\Delta H_{fusion} = 6.008$ kJ mol^{-1} at 0.00°C. Assume that the surroundings are at −3.75°C. [*Hint:* Consider the two pathways at 1 bar: (a) $H_2O(l, -3.75°C) \rightarrow H_2O(s, -3.75°C)$ and (b) $H_2O(l, -3.75°C) \rightarrow H_2O(l, 0.00°C) \rightarrow H_2O(s, 0.00°C) \rightarrow H_2O(s, -3.75°C)$. Because S is a state function, ΔS must be the same for both pathways.]

P5.30 Calculate ΔS, ΔS_{surr}, and $\Delta S_{universe}$ per second for the air-conditioned room described in Problem 5.4. Assume that the interior temperature is 65°F and the exterior temperature is 99°F.

P5.31 The following heat capacity data have been reported for L-alanine:

T(K)	10.	20.	40.	60.	80.	100.	140.	180.	220.	260.	300.
$C_{P,m}$(J K^{-1} mol^{-1})	0.49	3.85	17.45	30.99	42.59	52.50	68.93	83.14	96.14	109.6	122.7

By a graphical treatment, obtain the molar entropy of L-alanine at $T = 300$. K. You can perform the integration numerically using either a spreadsheet program or a curve-fitting routine and a graphing calculator (see Example Problem 5.9).

P5.32 Calculate $\Delta S_{surroundings}$ and ΔS_{total} for the processes described in parts (a) and (b) of Problem P5.16. Which of the processes is a spontaneous process? The state of the surroundings for each part is 298 K, 1.50 bar.

P5.33 A refrigerator is operated by a 0.25-hp (1 hp = 746 watts) motor. If the interior is to be maintained at 4.50°C and the room temperature on a hot day is 38°C, what is the maximum heat leak (in watts) that can be tolerated? Assume that the coefficient of performance is 50% of the maximum theoretical value. What happens if the leak is greater than your calculated maximum value?

P5.34 Using your results from Problems P5.18 and P5.7, calculate ΔS, $\Delta S_{surroundings}$, and ΔS_{total} for each step in the cycle and for the total Carnot cycle described in Figure 5.2.

P5.35 Between 0°C and 100°C, the heat capacity of Hg(l) is given by

$$\frac{C_{P,m}(\text{Hg},l)}{\text{J K}^{-1}\text{mol}^{-1}} = 30.093 - 4.944 \times 10^{-3}\frac{T}{\text{K}}$$

Calculate ΔH and ΔS if 2.25 moles of Hg(l) are raised in temperature from 0.00° to 88.0°C at constant P.

P5.36 Calculate $\Delta S_{surroundings}$ and ΔS_{total} for part (c) of Problem P5.6. Is the process spontaneous? The state of the surroundings is $T = 310$. K, $P = 0.500$ bar.

P5.37 Calculate the entropy of one mole of water vapor at 175°C and 0.625 bar using the information in the data tables.

P5.38 The heat capacity of α-quartz is given by

$$\frac{C_{P,m}(\alpha\text{-quartz},s)}{\text{J K}^{-1}\text{mol}^{-1}} = 46.94 + 34.31 \times 10^{-3}\frac{T}{\text{K}} - 11.30 \times 10^{-5}\frac{T^2}{\text{K}^2}$$

The coefficient of thermal expansion is given by $\beta = 0.3530 \times 10^{-4}$ K^{-1} and $V_m = 22.6$ cm^3 mol^{-1}. Calculate ΔS_m for the transformation α-quartz (15.0°C, 1 atm) $\rightarrow \alpha$-quartz (420. °C, 925 atm).

P5.39

a. Calculate ΔS if 1.00 mol of liquid water is heated from 0.00° to 10.0°C under constant pressure and if $C_{P,m} = 75.3$ J K^{-1} mol^{-1}.

b. The melting point of water at the pressure of interest is 0.00°C and the enthalpy of fusion is 6.010 kJ mol^{-1}. The boiling point is 100.°C and the enthalpy of vaporization is 40.65 kJ mol^{-1}. Calculate ΔS for the transformation $H_2O(s, 0°C) \rightarrow H_2O(g, 100.°C)$.

P5.40 21.05 g of steam at 373 K is added to 415 g of $H_2O(l)$ at 298 K at a constant pressure of 1 bar. Is the final state of the system steam or liquid water? Calculate ΔS for the process.

P5.41 Using your result from fitting the data in Problem 5.31, extrapolate the absolute entropy of L-alanine to physiological conditions, $T = 310.$ K.

P5.42 The mean solar flux at Earth's surface is ~2.00 J cm^{-2} min^{-1}. In a nonfocusing solar collector, the temperature reaches a value of 79.5°C. A heat engine is operated using the collector as the hot reservoir and a cold reservoir at 298 K. Calculate the area of the collector needed to produce 1000. W. Assume that the engine operates at the maximum Carnot efficiency.

P5.43 An ideal gas sample containing 1.75 moles for which $C_{V,m} = 5R/2$ undergoes the following reversible cyclical process from an initial state characterized by $T = 275$ K and $P = 1.00$ bar:

a. It is expanded reversibly and adiabatically until the volume triples.

b. It is reversibly heated at constant volume until T increases to 275 K.

c. The pressure is increased in an isothermal reversible compression until $P = 1.00$ bar.

Calculate q, w, ΔU, ΔH, and ΔS for each step in the cycle, and for the total cycle.

P5.44 For protein denaturation, the excess entropy of denaturation is defined as $\Delta S_{den} = \displaystyle\int_{T_1}^{T_2} \dfrac{\delta C_P^{trs}}{T} dT$, where δC_P^{trs} is the transition excess heat capacity. The way in which δC_P^{trs} can be extracted from differential scanning calorimetry (DSC) data is discussed in Section 4.6 and shown in Figure 4.7. The following DSC data are for a protein mutant that denatures between $T_1 = 288$ K and $T_2 = 318$ K. Using the equation for ΔS_{den} given previously, calculate the excess entropy of denaturation. In your calculations, use the dashed curve below the yellow area as the heat capacity base line that defines δC_P^{trs} as shown in Figure 4.8. Assume the molecular weight of the protein is 14,000. grams.

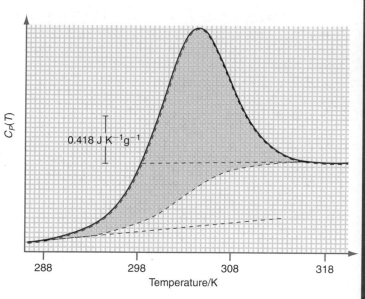

You can perform the integration numerically by counting squares.

P5.45 The standard entropy of Pb(s) at 298.15 K is 64.80 J K^{-1} mol^{-1}. Assume that the heat capacity of Pb(s) is given by

$$\frac{C_{P,m}(\text{Pb},s)}{\text{J mol}^{-1}\text{K}^{-1}} = 22.13 + 0.01172\frac{T}{K} + 1.00 \times 10^{-5}\frac{T^2}{K^2}$$

The melting point is 327.4°C and the heat of fusion under these conditions is 4770. J mol^{-1}. Assume that the heat capacity of Pb(l) is given by

$$\frac{C_{P,m}(\text{Pb},l)}{\text{J K}^{-1}\text{mol}^{-1}} = 32.51 - 0.00301\frac{T}{K}$$

a. Calculate the standard entropy of Pb(l) at 725°C.

b. Calculate ΔH for the transformation Pb(s, 25.0°C) → Pb(l, 725°C).

Web-Based Simulations, Animations, and Problems

W5.1 The reversible Carnot cycle is simulated with adjustable values of T_{hot} and T_{cold}, and ΔU, q, and w are determined for each segment and for the cycle. The efficiency is also determined for the cycle.

Chemical Equilibrium

In the previous chapter, criteria for the spontaneity of arbitrary processes were developed. In this chapter, spontaneity is discussed in the context of the approach to equilibrium of a reactive mixture of gases. Two new state functions are introduced that express spontaneity in terms of the properties of the system only, so that the surroundings do not have to be considered explicitly. The Helmholtz energy provides a criterion for determining if a reaction mixture will evolve toward reactants or products if the system is at constant V and T. The Gibbs energy is the criterion for determining if a reaction mixture will evolve toward reactants or products if the system is at constant P and T. Using the Gibbs energy, a thermodynamic equilibrium constant K_P is derived that predicts the equilibrium concentrations of reactants and products in a mixture of reactive ideal gases.

6.1 The Gibbs Energy and the Helmholtz Energy

In Chapter 5, it was shown that the direction of spontaneous change for an arbitrary process is predicted by $\Delta S + \Delta S_{surroundings} > 0$. In this section, this spontaneity criterion is used to derive two new state functions, the Gibbs and Helmholtz energies. These new state functions provide the basis for all further discussions of spontaneity. Formulating spontaneity in terms of the Gibbs and Helmholtz energies rather than $\Delta S + \Delta S_{surroundings} > 0$ has the important advantage that spontaneity and equilibrium can be defined using only properties of the system rather than of the system and surroundings. We will also show that the Gibbs and Helmholtz energies allow us to calculate the maximum work that can be extracted from a chemical reaction.

The fundamental expression governing spontaneity is the Clausius inequality [Equation (5.33)], written in the form

$$TdS \geq \mathit{d}q \tag{6.1}$$

The equality is satisfied only for a reversible process. Because $\mathit{d}q = dU - \mathit{d}w$,

$$TdS \geq dU - \mathit{d}w \text{ or, equivalently, } -dU + \mathit{d}w + TdS \geq 0 \tag{6.2}$$

As discussed in Section 2.2, a system can do different types of work on the surroundings. It is particularly useful to distinguish between expansion work, in which the work arises from a volume change in the system, and nonexpansion work (for example, electrical work). We rewrite Equation (6.2) in the form

$$-dU - P_{external} \, dV + dw_{nonexpansion} + T dS \geq 0 \qquad \textbf{(6.3)}$$

Equation 6.3 expresses the condition of spontaneity for an arbitrary process in terms of the changes in the state functions U, V, S, and T as well as the path-dependent functions $P_{external} \, dV$ and $dw_{nonexpansion}$.

To make a connection with the discussion of spontaneity in Chapter 5, consider a special case of Equation (6.3). For an isolated system, $w = 0$ and $dU = 0$. Therefore, Equation (6.3) reduces to the familiar result derived in Section 5.5:

$$dS \geq 0 \qquad \textbf{(6.4)}$$

Because chemists are interested in systems that interact with their environment, we define equilibrium and spontaneity for such systems. As was done in Chapters 1 through 5, it is useful to consider transformations at constant temperature and either constant volume or constant pressure. *Note that constant T and P (or V) does not imply that these variables are constant throughout the process, but rather that they are the same for the initial and final states of the process.*

For isothermal processes, $T dS = d(TS)$, and Equation (6.3) can be written in the following form:

$$-dU + T dS \geq -dw_{expansion} - dw_{nonexpansion} \text{ or, equivalently,}$$

$$d(U - TS) \leq dw_{expansion} + dw_{nonexpansion} \qquad \textbf{(6.5)}$$

The combination of state functions $U - TS$, which has the units of energy, defines a new state function that we call the **Helmholtz energy,** abbreviated A. Using this definition, the general condition of spontaneity for isothermal processes becomes

$$dA - dw_{expansion} - dw_{nonexpansion} \leq 0 \qquad \textbf{(6.6)}$$

Because the equality applies for a reversible transformation, Equation (6.6) provides a way to calculate the maximum work that a system can do on the surroundings in an isothermal process.

$$dw_{total} = dw_{expansion} + dw_{nonexpansion} \geq dA \qquad \textbf{(6.7)}$$

The equality holds for a reversible process. Example Problem 6.1 illustrates the usefulness of A for calculating the maximum work available through carrying out a chemical reaction.

EXAMPLE PROBLEM 6.1

You wish to construct a fuel cell based on the oxidation of a hydrocarbon fuel. The two choices for a fuel are methane and octane. Calculate the maximum work available through the combustion of these two hydrocarbons, on a per mole and a per gram basis at 298.15 K and 1 bar pressure. The standard enthalpies of combustion are $\Delta H^\circ_{combustion}(CH_4, g) = -891$ kJ mol^{-1}, $\Delta H^\circ_{combustion}(C_8H_{18}, l) = -5471$ kJ mol^{-1}, and $S^\circ_m(C_8H_{18}, l) = 361.1$ J mol^{-1} k^{-1}. Use tabulated values of S°_m in Appendix B for these calculations. Are there any other factors that should be taken into account in making a decision between these two fuels?

Solution

At constant T, ΔA and ΔH are related by $\Delta A = \Delta U - T\Delta S = \Delta H - \Delta(PV) - T\Delta S = \Delta H - \Delta nRT - T\Delta S$

The combustion reactions are

$$\text{Methane: } CH_4(g) + 2O_2(g) \rightarrow CO_2(g) + 2H_2O(l)$$

$$\text{Octane: } C_8H_{18}(l) + 25/2 \, O_2(g) \rightarrow 8CO_2(g) + 9H_2O(l)$$

$$\Delta A^{\circ}_{combustion}(CH_4, g) = \Delta U^{\circ}_{combustion}(CH_4, g) - T\left(\begin{matrix} S^{\circ}_m(CO_2, g) + 2S^{\circ}_m(H_2O, l) \\ -S^{\circ}_m(CH_4, g) - 2S^{\circ}_m(O_2, g) \end{matrix} \right)$$

$$= \Delta H^{\circ}_{combustion}(CH_4, g) - \Delta n RT$$

$$-T\,(S^{\circ}_m(CO_2, g) + 2S^{\circ}_m(H_2O, l) - S^{\circ}_m(CH_4, g)$$

$$- 2S^{\circ}_m(O_2, g))$$

$$= -891 \times 10^3\,J\,mol^{-1} + 2 \times 8.314\,J\,mol^{-1}\,K^{-1} \times 298.15\,K$$

$$-298.15\,K \times \left(\begin{matrix} 213.8\,J\,mol^{-1}\,K^{-1} + 2 \times 70.0\,J\,mol^{-1}\,K^{-1} \\ -186.3\,J\,mol^{-1}\,K^{-1} - 2 \times 205.2\,J\,mol^{-1}\,K^{-1} \end{matrix} \right)$$

$$\Delta A^{\circ}_{combustion}(CH_4, g) = -814\,kJ\,mol^{-1}$$

$$\Delta A^{\circ}_{combustion}(C_8H_{18}, l) = \Delta U^{\circ}_{combustion}(C_8H_{18}, l)$$

$$-T\left(8S^{\circ}_m(CO_2, g) + 9S^{\circ}_m(H_2O, l) - S^{\circ}_m(C_8H_{18}, l) \right.$$

$$\left. - \frac{25}{2} S^{\circ}_m(O_2, g) \right)$$

$$= \Delta H^{\circ}_{combustion}(C_8H_{18}, l) - \Delta n RT$$

$$-T\left(8S^{\circ}_m(CO_2, g) + 9S^{\circ}_m(H_2O, l) - S^{\circ}_m(C_8H_{18}, l) \right.$$

$$\left. - \frac{25}{2} S^{\circ}_m(O_2, g) \right)$$

$$= -5471 \times 10^3\,J\,mol^{-1} + \frac{9}{2} \times 8.314\,J\,mol^{-1}\,K^{-1} \times 298.15\,K$$

$$-298.15\,K \times \left(\begin{matrix} 8 \times 213.8\,J\,mol^{-1}\,K^{-1} + 9 \times 70.0\,J\,mol^{-1}\,K^{-1} \\ - 361.1\,J\,mol^{-1}\,K^{-1} - \frac{25}{2} \times 205.2\,J\,mol^{-1}\,K^{-1} \end{matrix} \right)$$

$$\Delta A^{\circ}_{combustion}(C_8H_{18}, l) = -5285\,kJ\,mol^{-1}$$

On a per mol basis, octane (molecular weight 114.25 g mol^{-1}) is capable of producing a factor of 6.5 more work than methane (molecular weight 16.04 g mol^{-1}). However, on a per gram basis, methane and octane are nearly equal in their ability to produce work (–50.6 kJ g^{-1} versus –46.3 kJ g^{-1}). You might want to choose octane because it can be stored as a liquid at atmospheric pressure. By contrast, a pressurized tank is needed to store methane as a liquid at 298.15 K.

In discussing the Helmholtz energy, $dT = 0$ was the only constraint applied. We now apply the additional constraint for a constant volume process, $dV = 0$. This condition implies $đw_{expansion} = 0$, because $dV = 0$. If nonexpansion work also is not possible in the transformation $đw_{nonexpansion} = đw_{expansion} = 0$, the condition that defines spontaneity and equilibrium becomes

$$dA \le 0 \tag{6.8}$$

Chemical reactions are more commonly studied under constant pressure than constant volume conditions. Therefore, the condition for spontaneity is considered next for an isothermal constant pressure process. At constant P and T, $PdV = d(PV)$ and $TdS = d(TS)$. In this case, using the relation $H = U + PV$, Equation (6.3) can be written in the form

$$d(U + PV - TS) = d(H - TS) \le đw_{nonexpansion} \tag{6.9}$$

The combination of state functions $H - TS$, which has the units of energy, defines a new state function called the **Gibbs energy,** abbreviated G. Using the Gibbs energy, the condition for spontaneity and equilibrium for an isothermal process at constant pressure becomes

$$dG - đw_{nonexpansion} \leq 0 \tag{6.10}$$

For a reversible process, the equality holds, and the change in the Gibbs energy is a measure of the maximum nonexpansion work that can be produced in the transformation.

We next consider a transformation at constant P and T for which nonexpansion work is not possible, for example, the burning of fuel in an internal combustion engine. In this case, Equation 6.10 becomes

$$dG \leq 0 \tag{6.11}$$

What is the advantage of using the state functions G and A as criteria for spontaneity rather than entropy? We answer this question by considering the Clausius inequality, which can be written in the form

$$dS - \frac{đq}{T} \geq 0 \tag{6.12}$$

As was shown in Section 5.8, $dS_{surroundings} = -đq/T$. Therefore, the Clausius inequality is equivalent to the spontaneity condition:

$$dS + dS_{surroundings} \geq 0 \tag{6.13}$$

By introducing G and A, the fundamental conditions for spontaneity have not been changed. However, G and A are expressed only in terms of the macroscopic state variables of the system. By introducing G and A, it is no longer necessary to consider the surroundings explicitly. Knowledge of ΔG and ΔA for the system alone is sufficient to predict the direction of natural change.

Apart from defining the condition of spontaneity, Equation (6.10) is very useful because it allows one to calculate the **maximum nonexpansion work** that can be produced by a chemical transformation. A particularly important application of this equation is to calculate the electrical work produced by a reaction in an electrochemical cell or fuel cell as shown in Example Problem 6.2. This topic will be discussed in detail in Chapter 11. The redox current that flows between the two half cells is used to do work. By contrast, only expansion work is possible in a conventional combustion process such as that in an automobile engine.

EXAMPLE PROBLEM 6.2

Calculate the maximum nonexpansion work that can be produced by the fuel cell oxidation reactions in Example Problem 6.1.

Solution

$$\Delta G^{\circ}_{combustion}(CH_4, g) = \Delta H^{\circ}_{combustion}(CH_4, g)$$

$$- T\left(\begin{array}{c} S^{\circ}_m(CO_2, g) + 2S^{\circ}_m(H_2O, l) \\ -S^{\circ}_m(CH_4, g) - 2S^{\circ}_m(O_2, g) \end{array} \right)$$

$$= -890. \times 10^3 \text{ J mol}^{-1}$$

$$-298.15K \times \left(\begin{array}{c} 213.8 \text{ J mol}^{-1}\text{K}^{-1} + 2 \times 70.0 \text{ J mol}^{-1}\text{K}^{-1} \\ -186.3 \text{ J mol}^{-1}\text{K}^{-1} - 2 \times 205.2 \text{ J mol}^{-1}\text{K}^{-1} \end{array} \right)$$

$$\Delta G^{\circ}_{combustion}(CH_4, g) = -818 \text{ kJ mol}^{-1}$$

$$\Delta G^{\circ}_{combustion}(C_8H_{18}, l) = \Delta H^{\circ}_{combustion}(C_8H_{18}, l)$$

$$-T\left(8S^{\circ}_m(CO_2, g) + 9S^{\circ}_m(H_2O, l) - S^{\circ}_m(C_8H_{18}, l)\right.$$

$$\left. - \frac{25}{2}S^{\circ}_m(O_2, g)\right)$$

$$= -5471 \times 10^3 \, J \, mol^{-1}$$

$$-298.15K \times \left(\begin{array}{c} 8 \times 213.8 \, J \, mol^{-1} \, K^{-1} + 9 \times 70.0 \, J \, mol^{-1}K^{-1} \\ - 361.1 \, J \, mol^{-1}K^{-1} - \frac{25}{2} \times 205.2 \, J \, mol^{-1} \, K^{-1} \end{array} \right)$$

$$\Delta G^{\circ}_{combustion}(C_8H_{18}, l) = -5296 \, kJ \, mol^{-1}$$

Compare this result with that of Example Problem 6.1. What can you conclude about the relative amounts of expansion and nonexpansion work available in these reactions?

It is useful to compare the available work that can be done by a reversible heat engine with the electrical work done by an electrochemical fuel cell using the same chemical reaction. In the reversible heat engine, the maximum available work is the product of the heat withdrawn from the hot reservoir and the efficiency of the heat engine. Consider a heat engine with $T_{hot} = 600. \, K$ and $T_{cold} = 300. \, K$, which has an efficiency of 0.50, and set $q_{hot} = \Delta H^{\circ}_{combustion}$. For these values, the maximum work available from the heat engine is 54% of that available in the electrochemical fuel cell for the combustion of methane. The corresponding value for octane is 52%. Why are these values less than 100%? If the oxidation reactions of Example Problem 6.1 can be carried out as redox reactions using two physically separated half cells, electrical work can be harnessed directly and converted to mechanical work. All forms of work can theoretically (but not practically) be converted to other forms of work with 100% efficiency. However, if chemical energy is converted to heat by burning the hydrocarbon fuel, and the heat is subsequently converted to work using a heat engine, the theoretical efficiency is less than 100% as discussed in Section 5.2. It is clear from this comparison why considerable research and development effort is currently being spent on fuel cells.

After this discussion of the usefulness of G in calculating the maximum nonexpansion work available through a chemical reaction, our focus turns to the use of G to determine the direction of spontaneous change in a reaction mixture. For macroscopic changes at constant P and T in which no nonexpansion work is possible, the condition for spontaneity is $\Delta G_R < 0$ where

$$\Delta G_R = \Delta H_R - T\Delta S_R \qquad (6.14)$$

The subscript R is a reminder that the process of interest is a chemical reaction. Note that there are two contributions to ΔG_R that determine if an isothermal chemical transformation is spontaneous. They are the energetic contribution ΔH_R and the entropic contribution $T\Delta S_R$.

The following conclusions can be drawn based on Equation (6.14):

- The entropic contribution to ΔG_R is greater for higher temperatures.
- A chemical transformation is always spontaneous if $\Delta H_R < 0$ (an exothermic reaction) and $\Delta S_R > 0$.
- A chemical transformation is never spontaneous if $\Delta H_R > 0$ (an endothermic reaction) and $\Delta S_R < 0$.
- For all other cases, the relative magnitudes of ΔH_R and $T\Delta S_R$ determine if the chemical transformation is spontaneous.
- If the chemical reaction, for example $CH_4(g) + 2O_2(g) \rightarrow CO_2(g) + 2H_2O(l)$ is not spontaneous, then the reverse process $CO_2(g) + 2H_2O(l) \rightarrow CH_4(g) + 2O_2(g)$ is spontaneous.
- If $\Delta G_R = 0$, the reaction mixture is at equilibrium, and neither direction of change is spontaneous.

ΔG_R depends on the concentrations of reactants and products in the reaction vessel, and a reaction mixture will evolve until $\Delta G_R = 0$ when equilibrium is achieved unless the reaction rate is zero. At this point, we have not developed a framework to calculate the concentrations of a reaction mixture at equilibrium. As we will see in Section 6.8, the function $exp\left(-\Delta G_R^\circ/RT\right)$ will allow us to do so.

For macroscopic changes at constant V and T in which no nonexpansion work is possible, the condition for spontaneity is $\Delta A_R < 0$, where

$$\Delta A_R = \Delta U_R - T\Delta S_R \tag{6.15}$$

Again, two contributions determine if an isothermal chemical transformation is spontaneous: ΔU_R is an energetic contribution, and $T\Delta S_R$ is an entropic contribution to ΔA_R. The same conclusions can be drawn from this equation as for those listed for ΔG_R, with U substituted for H.

6.2 The Differential Forms of *U*, *H*, *A*, and *G*

To this point, the state functions U, H, A, and G, all of which have the units of energy have been defined. The functions U and H are used to calculate changes in energy for processes, and A and G are used to calculate the direction in which processes evolve and the maximum work the reactions can produce. In this section, we discuss how these state functions depend on the macroscopic system variables. To do so, the differential forms dU, dH, dA, and dG are developed. As we will see, these differential forms are essential in calculating how U, H, A, and G vary with state variables such as P and T. Starting from the definitions

$$H = U + PV$$

$$A = U - TS$$

$$G = H - TS = U + PV - TS \tag{6.16}$$

the following total differentials can be formed:

$$dU = TdS - PdV \tag{6.17}$$

$$dH = TdS - PdV + PdV + VdP = TdS + VdP \tag{6.18}$$

$$dA = TdS - PdV - TdS - SdT = -SdT - PdV \tag{6.19}$$

$$dG = TdS + VdP - TdS - SdT = -SdT + VdP \tag{6.20}$$

These differential forms express the internal energy as $U(S,V)$, the enthalpy as $H(S,P)$, the Helmholtz energy as $A(T,V)$, and the Gibbs energy as $G(T,P)$. Although other combinations of variables can be used, these **natural variables** are used because the differential expressions are compact.

What information can be obtained from the differential expressions in Equations (6.17) through (6.20)? Because U, H, A, and G are state functions, two different equivalent expressions such as those written for dU here can be formulated:

$$dU = TdS - PdV = \left(\frac{\partial U}{\partial S}\right)_V dS + \left(\frac{\partial U}{\partial V}\right)_S dV \tag{6.21}$$

For Equation (6.21) to be valid, the coefficients of dS and dV on both sides of the equation must be equal. Applying this reasoning to Equations (6.17) through (6.20), the following expressions are obtained:

$$\left(\frac{\partial U}{\partial S}\right)_V = T \text{ and } \left(\frac{\partial U}{\partial V}\right)_S = -P \tag{6.22}$$

$$\left(\frac{\partial H}{\partial S}\right)_P = T \text{ and } \left(\frac{\partial H}{\partial P}\right)_S = V \tag{6.23}$$

$$\left(\frac{\partial A}{\partial T}\right)_V = -S \text{ and } \left(\frac{\partial A}{\partial V}\right)_T = -P \qquad \textbf{(6.24)}$$

$$\left(\frac{\partial G}{\partial T}\right)_P = -S \text{ and } \left(\frac{\partial G}{\partial P}\right)_T = V \qquad \textbf{(6.25)}$$

These expressions state how U, H, A, and G vary with their natural variables. For example, because T and V always have positive values, Equation (6.23) states that H increases if either the entropy or the pressure of the system increases. We discuss how to use these relations for macroscopic changes in the system variables in Section 6.4.

There is also a second way in which the differential expressions in Equations (6.17) through (6.20) can be used. From Section 3.1, we know that because dU is an exact differential:

$$\left(\frac{\partial}{\partial V}\left(\frac{\partial U\,(S,V)}{\partial S}\right)_V\right)_S = \left(\frac{\partial}{\partial S}\left(\frac{\partial U\,(S,V)}{\partial V}\right)_S\right)_V$$

Equating the mixed second partial derivative derived from Equations (6.17) through (6.20), we obtain the following four **Maxwell relations:**

$$\left(\frac{\partial T}{\partial V}\right)_S = -\left(\frac{\partial P}{\partial S}\right)_V \qquad \textbf{(6.26)}$$

$$\left(\frac{\partial T}{\partial P}\right)_S = \left(\frac{\partial V}{\partial S}\right)_P \qquad \textbf{(6.27)}$$

$$\left(\frac{\partial S}{\partial V}\right)_T = \left(\frac{\partial P}{\partial T}\right)_V = \frac{\beta}{\kappa} \qquad \textbf{(6.28)}$$

$$-\left(\frac{\partial S}{\partial P}\right)_T = \left(\frac{\partial V}{\partial T}\right)_P = V\beta \qquad \textbf{(6.29)}$$

Equations (6.26) and (6.27) refer to a partial derivative at constant S. What conditions must a transformation at constant entropy satisfy? Because $dS = \mathit{d}q_{reversible}/T$, a transformation at constant entropy refers to a reversible adiabatic process.

EXAMPLE PROBLEM 6.3

Show that $\left(\dfrac{\partial T}{\partial V}\right)_S = -\left(\dfrac{\partial P}{\partial S}\right)_V$

Solution

Because U is a state function, $\left(\dfrac{\partial}{\partial V}\left(\dfrac{\partial U\,(S,V)}{\partial S}\right)_V\right)_S = \left(\dfrac{\partial}{\partial S}\left(\dfrac{\partial U\,(S,V)}{\partial V}\right)_S\right)_V .$

Substituting $dU = TdS - PdV$ in the previous expression,

$$\left(\frac{\partial}{\partial V}\left(\frac{\partial[TdS - PdV]}{\partial S}\right)_V\right)_S = \left(\frac{\partial}{\partial S}\left(\frac{\partial[TdS - PdV]}{\partial V}\right)_S\right)_V$$

$$\left(\frac{\partial T}{\partial V}\right)_S = -\left(\frac{\partial P}{\partial S}\right)_V$$

The Maxwell relations have been derived using only the property that U, H, A, and G are state functions. These four relations are extremely useful in transforming seemingly obscure partial derivatives in other partial derivatives that can be directly measured. For example, these relations will be used to express U, H, and heat capacities solely in terms of measurable quantities such as κ, β and the state variables P, V, and T in Supplemental Section 6.15.

6.3 The Dependence of the Gibbs and Helmholtz Energies on *P*, *V*, and *T*

The state functions A and G are particularly important for chemists because of their roles in determining the direction of spontaneous change in a reaction mixture. For this reason, we need to know how A changes with T and V, and how G changes with T and P.

We begin by asking how A changes with T and V. From Section 6.2,

$$\left(\frac{\partial A}{\partial T}\right)_V = -S \text{ and } \left(\frac{\partial A}{\partial V}\right)_T = -P \tag{6.30}$$

where S and P always take on positive values. Therefore, the general statement can be made that the Helmholtz energy of a pure substance decreases as either the temperature or the volume increases.

Because most reactions of interest to chemists are carried out under constant pressure rather than constant volume conditions, we will devote more attention to the properties of G than to those of A. From Section 6.2,

$$\left(\frac{\partial G}{\partial T}\right)_P = -S \text{ and } \left(\frac{\partial G}{\partial P}\right)_T = V \tag{6.25}$$

Whereas the Gibbs energy decreases with increasing temperature, it increases with increasing pressure.

How can Equation (6.25) be used to calculate changes in G with macroscopic changes in the variables T and P? In doing so, each of the variables is considered separately. The total change in G as both T and P are varied is the sum of the separate contributions, because G is a state function. We first discuss the change in G with P.

For a macroscopic change in P at constant T, the second expression in Equation (6.25) is integrated at constant T:

$$\int_{P^\circ}^{P} dG = G(T, P) - G^\circ(T, P^\circ) = \int_{P^\circ}^{P} V dP' \tag{6.31}$$

where we have chosen the initial pressure to be the standard state pressure $P^\circ = 1$ bar. This equation takes on different forms for liquids and solids and for gases. For liquids and solids, the volume is, to a good approximation, independent of P over a limited range in P and

$$G(T, P) = G^\circ(T, P^\circ) + \int_{P^\circ}^{P} V dP' \approx G^\circ(T, P^\circ) + V(P - P^\circ) \tag{6.32}$$

By contrast, the volume of a gaseous system changes appreciably with pressure. In calculating the change of G_m with P at constant T, any path connecting the same initial and final states gives the same result. Choosing the reversible path and assuming ideal gas behavior,

$$G(T, P) = G^\circ(T) + \int_{P^\circ}^{P} V dP' = G^\circ(T) + \int_{P^\circ}^{P} \frac{nRT}{P'} dP' = G^\circ(T) + nRT \ln \frac{P}{P^\circ} \tag{6.33}$$

The functional dependence of the molar Gibbs energy G_m on P for an ideal gas is shown in Figure 6.1, where G_m approaches minus infinity as the pressure approaches zero. This is a result of the volume dependence of S that was discussed in Section 5.5, $\Delta S = nR \ln(V_f/V_i)$ at constant T. As $P \rightarrow 0$, $V \rightarrow \infty$. Because the volume available to a gas molecule is maximized as $V \rightarrow \infty$, $S \rightarrow \infty$ as $P \rightarrow 0$. Therefore, $G = H - TS \rightarrow -\infty$ in this limit.

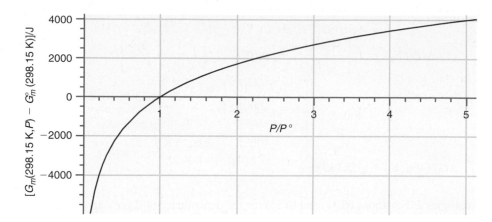

FIGURE 6.1
The molar Gibbs energy of an ideal gas relative to its standard state value is shown as a function of the pressure at 298.15 K.

We next investigate the dependence of G on T. As we see in the next section, the thermodynamic equilibrium constant K is related to G/T. Therefore, it is more useful to obtain an expression for the temperature dependence of G/T than for the temperature dependence of G. Using the chain rule, this dependence is given by

$$\left(\frac{\partial[G/T]}{\partial T}\right)_P = \frac{1}{T}\left(\frac{\partial G}{\partial T}\right)_P + G\frac{d[1/T]}{dT}$$

$$= \frac{1}{T}\left(\frac{\partial G}{\partial T}\right)_P - \frac{G}{T^2} = -\frac{S}{T} - \frac{G}{T^2} = -\frac{G + TS}{T^2} = -\frac{H}{T^2} \quad \text{(6.34)}$$

In the second line of Equation (6.34), we have used Equation (6.25), $(\partial G/\partial T)_P = -S$, and the definition $G = H - TS$. Equation (6.34) is known as the **Gibbs–Helmholtz equation.** Because

$$\frac{d(1/T)}{dT} = -\frac{1}{T^2}$$

the Gibbs–Helmholtz equation can also be written in the form

$$\left(\frac{\partial[G/T]}{\partial[1/T]}\right)_P = \left(\frac{\partial[G/T]}{\partial T}\right)_P\left(\frac{dT}{d[1/T]}\right) = -\frac{H}{T^2}(-T^2) = H \quad \text{(6.35)}$$

The preceding equation also applies to the change in G and H associated with a process such as a chemical reaction. Replacing G by ΔG and integrating Equation (6.35) at constant P,

$$\int_{T_1}^{T_2} d\left(\frac{\Delta G}{T}\right) = \int_{T_1}^{T_2} \Delta H d\left(\frac{1}{T}\right)$$

$$\frac{\Delta G(T_2)}{T_2} = \frac{\Delta G(T_1)}{T_1} + \Delta H(T_1)\left(\frac{1}{T_2} - \frac{1}{T_1}\right) \quad \text{(6.36)}$$

It has been assumed in the second equation that ΔH is independent of T over the temperature interval of interest. If this is not the case, the integral must be evaluated numerically, using tabulated values of ΔH_f° and temperature-dependent expressions of $C_{P,m}$ for reactants and products.

EXAMPLE PROBLEM 6.4

The value of ΔG_f° for Fe(g) is 370.7 kJ mol^{-1} at 298.15 K, and ΔH_f° for Fe(g) is 416.3 kJ mol^{-1} at the same temperature. Assuming that ΔH_f° is constant in the interval 250–400 K, calculate ΔG_f° for Fe(g) at 400. K.

Solution

$$\Delta G_f^\circ (T_2) = T_2 \left[\frac{\Delta G_f^\circ (T_1)}{T_1} + \Delta H_f^\circ (T_1) \times \left(\frac{1}{T_2} - \frac{1}{T_1} \right) \right]$$

$$= 400.\ \text{K} \times \left[\begin{array}{c} \dfrac{370.7 \times 10^3\ \text{J mol}^{-1}}{298.15\ \text{K}} + 416.3 \times 10^3\ \text{J mol}^{-1} \\[2mm] \times \left(\dfrac{1}{400.\ \text{K}} - \dfrac{1}{298.15\ \text{K}} \right) \end{array} \right]$$

$$\Delta G_f^\circ (400.\ \text{K}) = 355.1\ \text{kJ mol}^{-1}$$

Analogies to Equations (6.31) and (6.36) for the dependence of A on V and T are left to the end-of-chapter problems.

6.4 The Gibbs Energy of a Reaction Mixture

To this point, the discussion has been limited to systems at a fixed composition. Our focus in the rest of the chapter is in using the Gibbs energy to understand equilibrium in a reaction mixture under constant pressure conditions that correspond to typical laboratory experiments. Because reactants are consumed and products are generated in chemical reactions, the expressions derived for state functions such as U, H, S, A, and G must be revised to include changes in composition. We focus on G in the following discussion.

For a reaction mixture containing species 1, 2, 3, . . . , G is no longer a function of the variables T and P only. Because it depends on the number of moles of each species, G is written in the form $G = G(T, P, n_1, n_2, n_3, \ldots)$. The total differential dG is

$$dG = \left(\frac{\partial G}{\partial T} \right)_{P,n_1,n_2\ldots} dT + \left(\frac{\partial G}{\partial P} \right)_{T,n_1,n_2\ldots} dP + \left(\frac{\partial G}{\partial n_1} \right)_{T,P,n_2\ldots} dn_1$$

$$+ \left(\frac{\partial G}{\partial n_2} \right)_{T,P,n_1\ldots} dn_2 + \ldots. \tag{6.37}$$

Note that if the concentrations do not change, all of the $dn_i = 0$, and Equation (6.37) reduces to $dG = \left(\dfrac{\partial G}{\partial T} \right)_P dT + \left(\dfrac{\partial G}{\partial P} \right)_T dP$.

Equation (6.37) can be simplified by defining the **chemical potential** μ_i as

$$\mu_i = \left(\frac{\partial G}{\partial n_i} \right)_{P,T,n_j \neq n_i} \tag{6.38}$$

It is important to realize that although μ_i is defined mathematically in terms of an infinitesimal change in the amount dn_i of species i, the chemical potential μ_i is the change in the Gibbs energy per mole of substance i added *at constant concentration*. These two requirements are not contradictory. To keep the concentration constant, one adds a mole of substance i to a huge vat containing many moles of the various species. In this case, the slope of a plot of G versus n_i is the same if the differential $(\partial G / \partial n_i)_{P,T,n_j \neq n_i}$ is formed, where $dn_i \to 0$, or the ratio $(\Delta G / \Delta n_i)_{P,T,n_j \neq n_i}$ is formed, where Δn_i is 1 mol. Using the notation of Equation (6.38), Equation (6.37) can be written as follows:

$$dG = \left(\frac{\partial G}{\partial T} \right)_{P,n_1,n_2\ldots} dT + \left(\frac{\partial G}{\partial P} \right)_{T,n_1 n_2\ldots} dP + \sum_i \mu_i dn_i \tag{6.39}$$

Now imagine integrating Equation (6.39) at constant composition and at constant T and P from an infinitesimal size of the system where $n_i \to 0$ and therefore $G \to 0$ to a macroscopic size where the Gibbs energy has the value G. Because T and P are constant, the first two terms in Equation (6.39) do not contribute to the integral. Because the composition is constant, μ_i is constant:

$$\int_0^G dG' = \sum_i \mu_i \int_0^{n_i} dn_i'$$

$$G = \sum_i n_i \mu_i \qquad (6.40)$$

Note that because μ_i depends on the number of moles of each species present, it is a function of concentration. If the system consists of a single pure substance A, $G = n_A G_{m,A}$ because G is an extensive quantity. Applying Equation (6.38),

$$\mu_A = \left(\frac{\partial G}{\partial n_A}\right)_{P,T} = \left(\frac{\partial [n_A G_{m,A}]}{\partial n_A}\right)_{P,T} = G_{m,A}$$

showing that μ_A is an intensive quantity equal to the molar Gibbs energy of A *for a pure substance*. As shown later, this statement is not true for mixtures.

Why is μ_i called the chemical potential of species i? This can be understood by assuming that the chemical potential for species i has the values μ_i^I in region I, and μ_i^{II} in region II of a given mixture with $\mu_i^I > \mu_i^{II}$. If dn_i moles of species i are transported from region I to region II at constant T and P, the change in G is given by

$$dG = -\mu_i^I dn_i + \mu_i^{II} dn_i = (\mu_i^{II} - \mu_i^I)dn_i < 0 \qquad (6.41)$$

Because $dG < 0$, this process is spontaneous. *For a given species, transport will occur spontaneously from a region of high chemical potential to one of low chemical potential. The flow of material will continue until the chemical potential has the same value in all regions of the mixture.* Note the analogy between this process and the flow of mass in a gravitational potential or the flow of charge in an electrostatic potential. Therefore, the term *chemical potential* is appropriate. In this discussion, we have defined a new criterion for equilibrium in a multicomponent mixture: *at equilibrium, the chemical potential of each species is the same throughout a mixture.*

6.5 The Gibbs Energy of a Gas in a Mixture

In the next three sections, the conditions for equilibrium in a reactive mixture of ideal gases are derived in terms of the μ_i of the chemical constituents. Imagine that we have a reaction vessel in which all reactants are initially confined to separate volumes through barriers that are suddenly removed to let the reaction proceed. Two important processes occur, namely (i) the mixing of reactants and (ii) the conversion of reactants to products. In both processes, the concentration of individual species and their chemical potentials change. In this section, we describe how the chemical potential of a reactant or product species changes as its concentration in the reaction mixture changes. In the next section, we describe how the chemical potential of a reactant or product species changes through mixing with other species. Using these results, we show in Section 6.7 that the partial pressures of all constituents of a gaseous reaction mixture are related by the thermodynamic equilibrium constant K_P.

Consider first the simple system consisting of two volumes separated by a semipermeable membrane, as shown in Figure 6.2. On the left side, the gas consists solely of pure H_2. On the right side, H_2 is present as one constituent of a mixture. The membrane allows only H_2 to pass in both directions.

Once equilibrium has been reached with respect to the concentration of H_2 throughout the system, the hydrogen pressure (but not the total pressure) is the same on both sides of the membrane and therefore

$$\mu_{H_2}^{pure} = \mu_{H_2}^{mixture} \qquad (6.42)$$

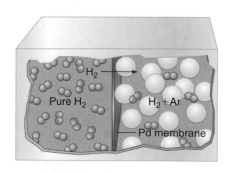

FIGURE 6.2
An isolated system consists of two subsystems. Pure H_2 gas is present on the left of a palladium membrane that is permeable to H_2, but not to argon. The H_2 is contained in a mixture with Ar in the subsystem to the right of the membrane.

Recall from Section 6.3 that the molar Gibbs energy of a pure ideal gas depends on its pressure as $G(T, P) = G°(T) + nRT \ln(P/P°)$. Therefore, Equation (6.42) can be written in the form

$$\mu_{H_2}^{pure}(T, P_{H_2}) = \mu_{H_2}^{mixture}(T, P_{H_2}) = \mu_{H_2}^{\circ}(T) + RT \ln\frac{P_{H_2}}{P°} \quad \textbf{(6.43)}$$

The chemical potential of a gas in a mixture depends logarithmically on its partial pressure. Equation (6.43) applies to any mixture, not just to those for which an appropriate semipermeable membrane exists. We therefore generalize the discussion by referring to a component of the mixture as A. The partial pressure of species A in the gas mixture P_A can be expressed in terms of x_A, its mole fraction in the mixture, and the total pressure P:

$$P_A = x_A P \quad \textbf{(6.44)}$$

Using this relationship, Equation (6.43) becomes

$$\mu_A^{mixture}(T, P) = \mu_A^{\circ}(T) + RT \ln\frac{P}{P°} + RT \ln x_A$$

$$= \left(\mu_A^{\circ}(T) + RT \ln\frac{P}{P°}\right) + RT \ln x_A \text{ or}$$

$$\mu_A^{mixture}(T, P) = \mu_A^{pure}(T, P) + RT \ln x_A \quad \textbf{(6.45)}$$

Because $x_A < 0$, we find that the chemical potential of a gas in a mixture is less than that of the pure gas if the total pressure P is the same for the pure sample and the mixture. Because $\mu_A^{mixture}(T, P) < \mu_A^{pure}(T, P)$, diffusion of H_2 from the left side to the right side of the system in Figure 6.2 will continue until the partial pressures of H_2 on both sides of the membrane are equal. There is a further important conclusion that can be drawn from Equation (6.45): mixing of the two subsystems in Figure 6.2 would be spontaneous if they were not separated by the membrane.

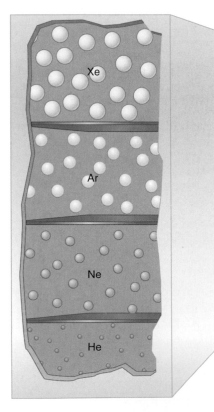

FIGURE 6.3
An isolated system consists of four separate subsystems containing He, Ne, Ar, and Xe, each at a pressure of 1 bar. The barriers separating these subsystems can be removed, leading to mixing.

6.6 Calculating the Gibbs Energy of Mixing for Ideal Gases

In the previous section, we showed how the chemical potential of a species in a mixture is related to the chemical potential of the pure species and the mole fraction of the species in the mixture. We also demonstrated that the mixing of gases is a spontaneous process, so that ΔG_{mixing} must be negative. We now obtain a quantitative relationship between ΔG_{mixing} and the mole fractions of the individual constituents of the mixture. Consider the system shown in Figure 6.3. The four compartments contain the gases He, Ne, Ar, and Xe at the same temperature and pressure. The volumes of the four compartments differ. To calculate ΔG_{mixing}, we must compare G for the initial state shown in Figure 6.3 and the final state in which all gases are uniformly distributed in the container. For the initial state, in which we have four pure separated substances,

$$G_i = G_{He} + G_{Ne} + G_{Ar} + G_{Xe} = n_{He}G_{m,He} + n_{Ne}G_{m,Ne} + n_{Ar}G_{m,Ar} + n_{Xe}G_{m,Xe} \quad \textbf{(6.46)}$$

For the final state in which all four components are dispersed in the mixture, from Equation (6.45),

$$G_f = n_{He}(G_{m,He} + RT \ln x_{He}) + n_{Ne}(G_{m,Ne} + RT \ln x_{Ne})$$
$$+ n_{Ar}(G_{m,Ar} + RT \ln x_{Ar}) + n_{Xe}(G_{m,Xe} + RT \ln x_{Xe}) \quad \textbf{(6.47)}$$

The Gibbs energy of mixing is $G_f - G_i$ or

$$\Delta G_{mixing} = RTn_{He} \ln x_{He} + RTn_{Ne} \ln x_{Ne} + RTn_{Ar} \ln x_{Ar} + RTn_{Xe} \ln x_{Xe}$$

$$= RT \sum_i n_i \ln x_i = nRT \sum_i x_i \ln x_i \quad \textbf{(6.48)}$$

Note that because all the $x_i < 1$, each term in the last expression of Equation (6.48) is negative, so that $\Delta G_{mixing} < 0$, showing that mixing is a spontaneous process.

Equation (6.48) allows us to calculate ΔG_{mixing} for any given set of the mole fractions x_i. It is easiest to graphically visualize the results for a binary mixture of species A and B. To simplify the notation, we set $x_A = x$, so that $x_B = 1 - x$. It follows that

$$\Delta G_{mixing} = nRT\left[x \ln x + (1 - x)\ln(1 - x)\right] \tag{6.49}$$

A plot of ΔG_{mixing} versus x is shown in Figure 6.4 for a binary mixture. Note that ΔG_{mixing} is zero for $x_A = 0$ and $x_A = 1$ because only pure substances are present in these limits. Also, ΔG_{mixing} has a minimum for $x_A = 0.5$, because the largest decrease in G arises from mixing when A and B are present in equal amounts.

The entropy of mixing can be calculated from Equation (6.48):

$$\Delta S_{mixing} = -\left(\frac{\partial \Delta G_{mixing}}{\partial T}\right)_P = -nR\sum_i x_i \ln x_i \tag{6.50}$$

As shown in Figure 6.5, the entropy of mixing increases in the range $x_A = 0$ to $x_A = 0.5$ and is greatest for $x_A = 0.5$. What is the cause of the initial increase in entropy? Each of the components of the mixture expands from its initial volume to a larger final volume. Therefore, ΔS_{mixing} arises from the dependence of S on V at constant T. We verify this assertion using the expression derived in Section 5.4 for the change of S with V, and obtain Equation (6.50).

$$\Delta S = R\left(n_A \ln\frac{V_f}{V_{iA}} + n_B \ln\frac{V_f}{V_{iB}}\right) = R\left(nx_A \ln\frac{1}{x_A} + nx_B \ln\frac{1}{x_B}\right)$$

$$= -nR(x_A \ln x_A + x_B \ln x_B)$$

if both components and the mixture are at the same pressure and temperature. The maximum at $x_A = 0.5$ occurs because at that value, each component contributes equally to ΔS.

EXAMPLE PROBLEM 6.5

Consider the system shown in Figure 6.3. Assume that the separate compartments contain 1.0 mol of He, 3.0 mol of Ne, 2.0 mol of Ar, and 2.5 mol of Xe at 298.15 K. The pressure in each compartment is 1 bar.

  a. Calculate ΔG_{mixing}.

  b. Calculate ΔS_{mixing}.

Solution

  a. $\Delta G_{mixing} = RT n_{He} \ln x_{He} + RT n_{Ne} \ln x_{Ne} + RT n_{Ar} \ln x_{Ar} + RT n_{Xe} \ln x_{Xe}$

$$= RT\sum_i n_i \ln x_i = nRT\sum_i x_i \ln x_i$$

$$= 8.5 \text{ mol} \times 8.314 \text{ J K}^{-1}\text{ mol}^{-1} \times 298.15 \text{ K}$$

$$\times \left(\frac{1.0}{8.5}\ln\frac{1.0}{8.5} + \frac{3.0}{8.5}\ln\frac{3.0}{8.5} + \frac{2.0}{8.5}\ln\frac{2.0}{8.5} + \frac{2.5}{8.5}\ln\frac{2.5}{8.5}\right)$$

$$= -2.8 \times 10^4 \text{ J}$$

  b. $\Delta S_{mixing} = -nR\sum_i x_i \ln x_i$

$$= -8.5 \text{ mol} \times 8.314 \text{ J K}^{-1}\text{ mol}^{-1}$$

$$\times \left(\frac{1.0}{8.5}\ln\frac{1.0}{8.5} + \frac{3.0}{8.5}\ln\frac{3.0}{8.5} + \frac{2.0}{8.5}\ln\frac{2.0}{8.5} + \frac{2.5}{8.5}\ln\frac{2.5}{8.5}\right)$$

$$= 93 \text{ J K}^{-1}$$

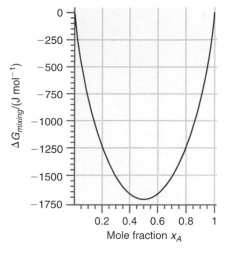

FIGURE 6.4
The Gibbs energy of mixing of the ideal gases A and B as a function of x_A, with $n_A + n_B = 1$ mole and $T = 298.15$ K.

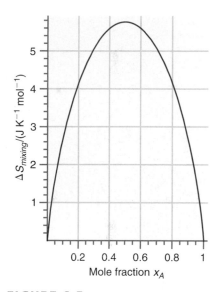

FIGURE 6.5
The entropy of mixing of the ideal gases A and B is shown as a function of the mole fraction of component A at 298.15 K, with $n_A + n_B = 1$.

What is the driving force for the mixing of gases? We saw in Section 6.1 that there are two contributions to ΔG, an enthalpic contribution ΔH and an entropic contribution $T\Delta S$. By calculating ΔH_{mixing} from $\Delta H_{mixing} = \Delta G_{mixing} + T\Delta S_{mixing}$ using Equations (6.49) and (6.50), you will see that for the mixing of ideal gases,

$\Delta H_{mixing} = 0$. Because the molecules in an ideal gas do not interact, there is no enthalpy change associated with mixing. We conclude that the mixing of ideal gases is driven entirely by ΔS_{mixing} as shown earlier.

Although the mixing of gases is always spontaneous, the same is not true of liquids. Liquids can be either miscible or immiscible. How can this fact be explained? For gases or liquids, $\Delta S_{mixing} > 0$. Therefore, if two liquids are immiscible, $\Delta G_{mixing} > 0$ and $\Delta H_{mixing} = \Delta G_{mixing} + T\Delta S_{mixing} > 0$, showing that there are repulsive interactions between the components to be mixed. If two liquids mix, $\Delta G_{mixing} < 0$ so that $\Delta H_{mixing} < T\Delta S_{mixing}$. In this case, there can be a weak repulsive interaction between the components if it is outweighed by the entropic contribution to ΔG_{mixing}.

6.7 Calculating ΔG_R° for a Chemical Reaction

Given a set of initial number of moles of all pure separated gaseous components in a reaction mixture, there is only one set of partial pressures P_i that corresponds to equilibrium. In this section, we combine the results of the two previous sections to identify the equilibrium position in a reaction mixture.

Consider the balanced chemical reaction

$$\alpha A + \beta B + \chi C + ... \rightarrow \delta M + \varepsilon N + \gamma O + ... \qquad (6.51)$$

in which Greek letters represent stoichiometric coefficients and uppercase Roman letters represent reactants and products. We can write an abbreviated expression for this reaction in the form

$$0 = \sum_i v_i X_i \qquad (6.52)$$

where v_i represents the stoichiometric coefficients and X_i represents the reactants and products. In Equation (6.52), the v_i of the products are positive, and those of the reactants are negative.

What determines the equilibrium partial pressures of the reactants and products? Imagine that the reaction proceeds in the direction indicated in Equation (6.51) by an infinitesimal amount. The change in the Gibbs energy is given by

$$dG = \sum_i \mu_i dn_i \qquad (6.53)$$

In this equation, the individual dn_i are not independent, because they are linked by the stoichiometric equation.

The direction of spontaneous change is determined by ΔG_R. How can ΔG_R be calculated for a reaction mixture? We distinguish between calculating ΔG_R° for the standard state in which $P_i = P^\circ = 1$ bar for all species and calculating ΔG_R for the set of P_i corresponding to equilibrium. Tabulated values of ΔG_f° for compounds are used to calculate ΔG_R° at $P^\circ = 1$ bar and $T = 298.15$ K as shown in Example Problem 6.6. Just as for enthalpy, ΔG_f° for a pure element in its standard reference state is equal to zero because the reactants and products in the formation reaction are identical.

$$\Delta G_R^\circ = \sum_i v_i \Delta G_{f,i}^\circ \qquad (6.54)$$

EXAMPLE PROBLEM 6.6

Calculate ΔG_R° for the reaction $2NO_2(g) \rightarrow N_2O_4(g)$ at 298.15 K. Is the reaction spontaneous if the pressure of $NO(g)$ is 1 bar at this temperature?

Solution

$$\Delta G_R^\circ = \Delta G_f^\circ(N_2O_4, g) - 2\Delta G_f^\circ(NO_2, g)$$

$$= 99.8 \text{ kJ mol}^{-1} - 2 \times 51.3 \text{ kJ mol}^{-1}$$

$$= -2.80 \text{ kJ mol}^{-1}$$

Because $\Delta G_R^\circ < 0$, the reaction is spontaneous at 298.15 K.

At other temperatures, ΔG_R° is calculated as shown in Example Problem 6.7.

EXAMPLE PROBLEM 6.7

Calculate ΔG_R° for the reaction in Example Problem 6.6 at 325 K. Is the reaction spontaneous at this temperature?

Solution

To calculate ΔG_R° at the elevated temperature, we use Equation (6.36) assuming that ΔH_R° is independent of T:

$$\Delta G_R^\circ \left(T_2 \right) = T_2 \left[\frac{\Delta G_R^\circ \left(298.15\ \text{K} \right)}{298.15\ \text{K}} + \Delta H_R^\circ (298.15\ \text{K}) \left(\frac{1}{T_2} - \frac{1}{298.15\ \text{K}} \right) \right]$$

$$\Delta H_R^\circ \left(298.15\ \text{K} \right) = \Delta H_f^\circ \left(N_2O_4, g \right) - 2\Delta H_f^\circ \left(NO_2, g \right)$$

$$= 11.1\ \text{kJ mol}^{-1} - 2 \times 33.2\ \text{kJ mol}^{-1} = -55.3\ \text{kJ mol}^{-1}$$

$$\Delta G_R^\circ \left(325\ \text{K} \right) = 325\ \text{K} \times \left[\begin{array}{c} \dfrac{-2.80 \times 10^3 \text{J mol}^{-1}}{298.15\ \text{K}} \\[2mm] -55.3 \times 10^3 \text{J mol}^{-1} \left(\dfrac{1}{325\ \text{K}} - \dfrac{1}{298.15\ \text{K}} \right) \end{array} \right]$$

$$\Delta G_R^\circ \left(325\ \text{K} \right) = 1.92\ \text{kJ mol}^{-1}$$

Because $\Delta G_R^\circ > 0$, the reaction is not spontaneous at 325 K.

6.8 Introducing the Equilibrium Constant for a Mixture of Ideal Gases

Spontaneity for a reaction does not imply that the reaction goes to completion, but rather that the equilibrium reaction mixture contains more products than reactants. In this and the next section, we establish a framework for calculating the partial pressures of reactants and products in a chemical reaction at equilibrium.

In a reaction mixture with a total pressure of 1 bar, the partial pressures of each species P_i is less than the standard state pressure of 1 bar. In this section, we introduce the pressure dependence of μ_i in order to calculate ΔG_R for a typical reaction mixture. Consider the following reaction that takes place between ideal gas species A, B, C, and D:

$$\alpha A(g) + \beta B(g) \rightleftharpoons \gamma C(g) + \delta D(g) \tag{6.55}$$

Because all four species are present in the reaction mixture, the reaction Gibbs energy for the arbitrary set of partial pressures P_A, P_B, P_C, and P_D, is given by

$$\Delta G_R = \sum_i \nu_i \Delta G_{f,i} = \gamma \mu_C^\circ + \gamma RT \ln \frac{P_C}{P^\circ} + \delta \mu_D^\circ + \delta RT \ln \frac{P_D}{P^\circ}$$

$$- \alpha \mu_A^\circ - \alpha RT \ln \frac{P_A}{P^\circ} - \beta \mu_B^\circ - \beta RT \ln \frac{P_B}{P^\circ} \tag{6.56}$$

The terms in the previous equation can be separated into those at the standard condition of $P^\circ = 1$ bar and the remaining terms:

$$\Delta G_R = \Delta G_R^\circ + \gamma RT \ln \frac{P_C}{P^\circ} + \delta RT \ln \frac{P_D}{P^\circ} - \alpha RT \ln \frac{P_A}{P^\circ} - \beta RT \ln \frac{P_B}{P^\circ}$$

$$= \Delta G_R^\circ + RT \ln \frac{\left(\dfrac{P_C}{P^\circ} \right)^\gamma \left(\dfrac{P_D}{P^\circ} \right)^\delta}{\left(\dfrac{P_A}{P^\circ} \right)^\alpha \left(\dfrac{P_B}{P^\circ} \right)^\beta} \tag{6.57}$$

where

$$\Delta G_R^\circ = \gamma \mu_C^\circ \left(T \right) + \delta \mu_D^\circ \left(T \right) - \alpha \mu_A^\circ \left(T \right) - \beta \mu_B^\circ(T) = \sum_i \nu_i \Delta G_{f,i}^\circ \tag{6.58}$$

Recall from Example Problem 6.6 that standard Gibbs energies of formation rather than chemical potentials are used to calculate ΔG_R°.

The combination of the partial pressures of reactants and products in Equation (6.57) is called the **reaction quotient of pressures,** which is abbreviated Q_P and defined as follows:

$$Q_P = \frac{\left(\dfrac{P_C}{P^\circ}\right)^\gamma \left(\dfrac{P_D}{P^\circ}\right)^\delta}{\left(\dfrac{P_A}{P^\circ}\right)^\alpha \left(\dfrac{P_B}{P^\circ}\right)^\beta} \tag{6.59}$$

With these definitions of ΔG_R° and Q_P, Equation (6.56) becomes

$$\Delta G_R = \Delta G_R^\circ + RT \ln Q_P \tag{6.60}$$

Note this important result that ΔG_R can be separated into two terms, only one of which depends on the partial pressures of reactants and products.

We next show how Equation (6.60) can be used to predict the direction of spontaneous change for a given set of partial pressures of the reactants and products. If the partial pressures of the reactants A and B are large, and those of the products C and D are small compared to their values at equilibrium, Q_P will be small. As a result, $RT \ln Q_P$ will be large and negative. In this case, $\Delta G_R = \Delta G_R^\circ + RT \ln Q_P < 0$ and the reaction will be spontaneous as written from left to right, as some of the reactants combine to form products.

Next, consider the opposite extreme. If the partial pressures of the products C and D are large, and those of the reactants A and B are small compared to their values at equilibrium, Q_P will be large. As a result, $RT \ln Q_P$ will be large and positive, and $\Delta G_R = \Delta G_R^\circ + RT \ln Q_P > 0$. In this case, the reaction as written in Equation (6.55) from left to right is not spontaneous, but the reverse of the reaction is spontaneous.

EXAMPLE PROBLEM 6.8

$NO_2(g)$ and $N_2O_4(g)$ are in a reaction vessel with partial pressures of 0.350 and 0.650 bar, respectively, at 298 K. Is this system at equilibrium? If not, will the system move toward reactants or products to reach equilibrium?

Solution

The reaction of interest is $2\,NO_2(g) \rightleftharpoons N_2O_4(g)$. We calculate ΔG_R using $\Delta G_R = \Delta G_R^\circ + RT \ln Q_P$

$$\Delta G_R = 99.8\ \text{kJ mol}^{-1} - 2 \times 51.3\ \text{kJ mol}^{-1} + 8.314\ \text{J mol}^{-1}\ \text{K}^{-1}$$

$$\times\ 298\ \text{K} \ln\frac{0.650\ \text{bar}/1\ \text{bar}}{(0.350\ \text{bar}/1\ \text{bar})^2}$$

$$= -2.80 \times 10^3\ \text{J mol}^{-1} + 4.13 \times 10^3\ \text{J mol}^{-1}$$

$$= 1.33 \times 10^3\ \text{J mol}^{-1}$$

Since $\Delta G_R \neq 0$, the system is not at equilibrium, and because $\Delta G_R > 0$, the system moves toward reactants.

Whereas the two cases that we have considered lead to a change in the partial pressures of the reactants and products, the most interesting case is equilibrium, for which $\Delta G_R = 0$. At equilibrium, $\Delta G_R^\circ = -RT \ln Q_P$. We denote this special system configuration by adding a superscript eq to each partial pressure and renaming Q_P as K_P. The quantity K_P is called the **thermodynamic equilibrium constant:**

$$0 = \Delta G_R^\circ + RT \ln \frac{\left(\dfrac{P_C^{eq}}{P^\circ}\right)^\gamma \left(\dfrac{P_D^{eq}}{P^\circ}\right)^\delta}{\left(\dfrac{P_A^{eq}}{P^\circ}\right)^\alpha \left(\dfrac{P_B^{eq}}{P^\circ}\right)^\beta} \quad \text{or, equivalently,}\ \Delta G_R^\circ = -RT \ln K_P$$

or

$$\ln K_P = -\frac{\Delta G_R^\circ}{RT} \tag{6.61}$$

Because ΔG_R° is a function of T only, K_P is also a function of T only. *The thermodynamic equilibrium constant K_P does not depend on the total pressure or the partial pressure of reactants or products.* Note also that K_P is a dimensionless number because each partial pressure is divided by P°.

It is important to understand that the equilibrium condition of Equation (6.61) links the partial pressures of all reactants and products. Imagine that a quantity of species D is added to the system initially at equilibrium. The system will reach a new equilibrium state in which P_A^{eq}, P_B^{eq}, P_C^{eq} and P_D^{eq} all have new values because the value of

$$K_P = \left(\frac{P_C^{eq}}{P^\circ}\right)^\gamma \left(\frac{P_D^{eq}}{P^\circ}\right)^\delta \Big/ \left(\frac{P_A^{eq}}{P^\circ}\right)^\alpha \left(\frac{P_B^{eq}}{P^\circ}\right)^\beta \text{ has not changed.}$$

EXAMPLE PROBLEM 6.9

a. Using data from Table 4.1 (see Appendix B, Data Tables), calculate K_P at 298.15 K for the reaction $CO(g) + H_2O(l) \rightarrow CO_2(g) + H_2(g)$.

b. Based on the value that you obtained for part (a), do you expect the mixture to consist mainly of $CO_2(g)$ and $H_2(g)$ or mainly of $CO(g) + H_2O(l)$ at equilibrium?

Solution

a. $\ln K_P = -\dfrac{1}{RT} \Delta G_R^\circ$

$$= -\frac{1}{RT}\left(\begin{array}{l} \Delta G_f^\circ(CO_2, g) + \Delta G_f^\circ(H_2, g) - \Delta G_f^\circ(H_2O, l) \\ -\Delta G_f^\circ(CO, g) \end{array}\right)$$

$$= -\frac{1}{8.314 \text{ J mol}^{-1}\text{ K}^{-1} \times 298.15 \text{ K}}$$

$$\times \left(\begin{array}{l} -394.4 \times 10^3 \text{ J mol}^{-1} + 0 + 237.1 \\ \times 10^3 \text{ J mol}^{-1} + 137.2 \times 10^3 \text{ J mol}^{-1} \end{array}\right)$$

$$= 8.1087$$

$$K_P = 3.32 \times 10^3$$

b. Because $K_P \gg 1$, the mixture will consist mainly of the products $CO_2(g) + H_2(g)$ at equilibrium.

6.9 Calculating the Equilibrium Partial Pressures in a Mixture of Ideal Gases

As shown in the previous section, the partial pressures of the reactants and products in a mixture of gases at equilibrium cannot take on arbitrary values, because they are related through K_P. In this section, we show how the equilibrium partial pressures can be calculated for ideal gases. Similar calculations for real gases will be discussed in the next chapter. In Example Problem 6.10, we consider the dissociation of chlorine:

$$Cl_2(g) \rightleftharpoons 2Cl(g) \tag{6.62}$$

It is useful to set up the calculation in a tabular form as shown in the following example problem.

EXAMPLE PROBLEM 6.10

In this example, n_0 moles of chlorine gas are placed in a reaction vessel whose temperature can be varied over a wide range, so that molecular chlorine can partially dissociate to atomic chlorine.

a. Define the degree of dissociation as $\alpha = \delta_{eq}/n_0$, where $2\delta_{eq}$ is the number of moles of $Cl(g)$ present at equilibrium, and n_0 represents the number of moles of $Cl_2(g)$ that would be present in the system if no dissociation occurred. Derive an expression for K_P in terms of n_0, δ_{eq}, and P.

b. Derive an expression for α as a function of K_P and P.

Solution

a. We set up the following table:

	$Cl_2(g)$ $\rightleftharpoons$	$2Cl(g)$
Initial number of moles	n_0	0
Moles present at equilibrium	$n_0 - \delta_{eq}$	$2\delta_{eq}$
Mole fraction present at equilibrium, x_i	$\dfrac{n_0 - \delta_{eq}}{n_0 + \delta_{eq}}$	$\dfrac{2\delta_{eq}}{n_0 + \delta_{eq}}$
Partial pressure at equilibrium, $P_i = x_i P$	$\left(\dfrac{n_0 - \delta_{eq}}{n_0 + \delta_{eq}}\right)P$	$\left(\dfrac{2\delta_{eq}}{n_0 + \delta_{eq}}\right)P$

We next express K_P in terms of n_0, δ_{eq}, and P:

$$K_P(T) = \frac{\left(\dfrac{P_{Cl}^{eq}}{P^\circ}\right)^2}{\left(\dfrac{P_{Cl_2}^{eq}}{P^\circ}\right)} = \frac{\left[\left(\dfrac{2\delta_{eq}}{n_0 + \delta_{eq}}\right)\dfrac{P}{P^\circ}\right]^2}{\left(\dfrac{n_0 - \delta_{eq}}{n_0 + \delta_{eq}}\right)\dfrac{P}{P^\circ}}$$

$$= \frac{4\delta_{eq}^2}{\left(n_0 + \delta_{eq}\right)\left(n_0 - \delta eq\right)}\frac{P}{P^\circ} = \frac{4\delta_{eq}^2}{(n_0)^2 - \delta_{eq}^2}\frac{P}{P^\circ}$$

This expression is converted into one in terms of α:

$$K_P(T) = \frac{4\delta_{eq}^2}{(n_0)^2 - \delta_{eq}^2}\frac{P}{P^\circ} = \frac{4\alpha^2}{1 - \alpha^2}\frac{P}{P^\circ}$$

b. $\left(K_P(T) + 4\dfrac{P}{P^\circ}\right)\alpha^2 = K_P(T)$

$$\alpha = \sqrt{\frac{K_P(T)}{K_P(T) + 4\dfrac{P}{P^\circ}}}$$

Because $K_P(T)$ depends strongly on temperature, α will also be a strong function of temperature. Note that α also depends on both K_P and P for this reaction.

Whereas $K_P(T)$ is independent of P, α as calculated in Example Problem 6.10 does depend on P. In the particular case considered, α decreases as P increases for constant T. As will be shown in Section 6.12, α depends on P whenever $\Delta\nu \neq 0$ for a reaction.

6.10 The Variation of K_P with Temperature

In Section 6.8, we showed that K_P and ΔG_R° are related by $\ln K_P = -\Delta G_R^\circ/RT = -\Delta H_R^\circ/RT + \Delta S_R^\circ$. We illustrate how these individual functions vary with temperature for two reactions of industrial importance, beginning with

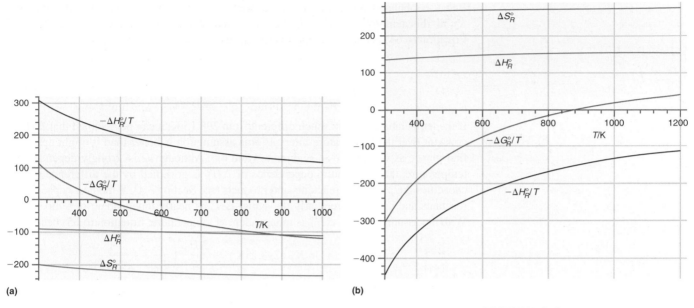

(a)

(b)

FIGURE 6.6
$-\Delta H_R^\circ/T$, ΔS_R°, ΔH_R°, and $-\Delta G_R^\circ/T$ are shown as a function of temperature. The units of ΔH_R° and ΔS_R° are kJ mol^{-1} and J mol^{-1} K^{-1}, respectively. The units of $-\Delta H_R^\circ/T$ and $-\Delta G_R^\circ/T$ are kJ mol^{-1} K^{-1}. **(a)** Ammonia synthesis. **(b)** Coal gasification.

ammonia synthesis. This reaction is used to produce almost all of the fertilizer used in growing foodcrops and will be discussed in detail in Section 6.14. The overall reaction is

$$3\,H_2(g) + N_2(g) \rightarrow 2NH_3(g)$$

Figure 6.6a shows that ΔH_R° is large, negative, and weakly dependent on temperature. Because the number of moles of gaseous species decreases as the reaction proceeds, ΔS_R° is negative and is weakly dependent on temperature. The temperature dependence of $\ln K_P = -\Delta G_R^\circ/RT$ is determined by $-\Delta H_R^\circ/RT$. Because ΔH_R° is negative, K_P decreases as T increases. We see that $K_P > 1$ for $T < 450$ K, and it decreases with increasing T throughout the temperature range shown.

Figure 6.6b shows analogous results for the coal gasification reaction

$$H_2O(g) + C(s) \rightarrow CO(g) + H_2(g)$$

This reaction produces a mixture of $CO(g)$ and $H_2(g)$, which is known as syngas. It can be used to produce diesel engine fuel in the form of alkanes through the Fischer-Tropsch process in the reaction $(2n + 1)H_2(g) + nCO(g) \rightarrow C_nH_{(2n+2)}(l) + nH_2O(l)$. For the coal gasification reaction, the number of moles of gaseous species increases as the reaction proceeds so that ΔS_R° is positive. ΔH_R° is large, positive, and not strongly dependent on temperature. Again, the temperature dependence of K_P is determined by $-\Delta H_R^\circ/RT$. Because ΔH_R° is positive, K_P increases as T increases.

The following conclusions can be drawn from Figure 6.6:

- The temperature dependence of K_P is determined by ΔH_R°.
- If $\Delta H_R^\circ > 0$, K_P increases with increasing T; if $\Delta H_R^\circ < 0$, K_P decreases with increasing T.
- ΔS_R° has essentially no effect on the dependence of K_P on T.
- If $\Delta S_R^\circ > 0$, $\ln K_P$ is increased by the same amount relative to $-\Delta H_R^\circ/RT$ at all temperatures. If $\Delta S_R^\circ < 0$, $\ln K_P$ is decreased by the same amount relative to $-\Delta H_R^\circ/RT$ at all temperatures.

We can also derive an expression for the dependence of $\ln K_P$ on T. Starting with Equation (6.61), we write

$$\frac{d\ln K_P}{dT} = -\frac{d\left(\Delta G_R^\circ/RT\right)}{dT} = -\frac{1}{R}\frac{d\left(\Delta G_R^\circ/T\right)}{dT} \qquad \textbf{(6.63)}$$

Using the Gibbs–Helmholtz equation [Equation (6.34)], the preceding equation reduces to

$$\frac{d\ln K_P}{dT} = -\frac{1}{R}\frac{d(\Delta G_R^\circ/T)}{dT} = \frac{\Delta H_R^\circ}{RT^2} \qquad \textbf{(6.64)}$$

Because tabulated values of ΔG_f° are available at 298.15 K, we can calculate ΔG_R° and K_P at this temperature; K_P can be calculated at the temperature T_f by integrating Equation (6.64) between the appropriate limits:

$$\int_{K_P(298.15\,\text{K})}^{K_P(T_f)} d\ln K_P = \frac{1}{R} \int_{298.15\,\text{K}}^{T_f} \frac{\Delta H_R^\circ}{T^2} dT \tag{6.65}$$

If the temperature T_f is not much different from 298.15 K, it can be assumed that ΔH_R° is constant over the temperature interval. This assumption is better than it might appear at first glance as was shown in Figures 6.5 and 6.6. Although H is strongly dependent on temperature, the temperature dependence of ΔH_R° is governed by the difference in heat capacities ΔC_P between reactants and products (see Section 3.4). If the heat capacities of reactants and products are nearly the same, ΔH_R° is nearly independent of temperature. With the assumption that ΔH_R° is independent of temperature, Equation (6.65) becomes

$$\ln K_P(T_f) = \ln K_P(298.15\,\text{K}) - \frac{\Delta H_R^\circ}{R}\left(\frac{1}{T_f} - \frac{1}{298.15\,\text{K}}\right) \tag{6.66}$$

EXAMPLE PROBLEM 6.11

Using the result of Example Problem 6.10 and the data tables, consider the dissociation equilibrium $Cl_2(g) \rightleftharpoons 2Cl(g)$.

a. Calculate K_P at 800. K, 1500. K, and 2000. K for $P = 0.010$ bar.

b. Calculate the degree of dissociation α at 300. K, 1500. K, and 2000. K.

Solution

a. $\Delta G_R^\circ = 2\Delta G_f^\circ(\text{Cl}, g) - \Delta G_f^\circ(\text{Cl}_2, g) = 2 \times 105.7 \times 10^3\,\text{J mol}^{-1} - 0$

$\qquad = 211.4\,\text{kJ mol}^{-1}$

$\Delta H_R^\circ = 2\Delta H_f^\circ(\text{Cl}, g) - \Delta H_f^\circ(\text{Cl}_2, g) = 2 \times 121.3 \times 10^3\,\text{J mol}^{-1} - 0$

$\qquad = 242.6\,\text{kJ mol}^{-1}$

$$\ln K_P(T_f) = -\frac{\Delta G_R^\circ}{RT} - \frac{\Delta H_R^\circ}{R}\left(\frac{1}{T_f} - \frac{1}{298.15\,\text{K}}\right)$$

$$= -\frac{211.4 \times 10^3\,\text{J mol}^{-1}}{8.314\,\text{J K}^{-1}\,\text{mol}^{-1} \times 298.15\,\text{K}} - \frac{242.6 \times 10^3\,\text{J mol}^{-1}}{8.314\,\text{J K}^{-1}\,\text{mol}^{-1}}$$

$$\times \left(\frac{1}{T_f} - \frac{1}{298.15\,\text{K}}\right)$$

$\ln K_P(800.\,\text{K})$

$$= -\frac{211.4 \times 10^3\,\text{J mol}^{-1}}{8.314\,\text{J K}^{-1}\,\text{mol}^{-1} \times 298.15\,\text{K}} - \frac{242.6 \times 10^3\,\text{J mol}^{-1}}{8.314\,\text{J K}^{-1}\,\text{mol}^{-1}}$$

$$\times \left(\frac{1}{800.\,\text{K}} - \frac{1}{298.15\,\text{K}}\right) = -23.888$$

$K_P(800.\,\text{K}) = 4.22 \times 10^{-11}$

The values for K_P at 1500. and 2000. K are 1.03×10^{-3} and 0.134, respectively.

b. The value of α at 2000. K is given by

$$\alpha = \sqrt{\frac{K_P(T)}{K_P(T) + 4\dfrac{P}{P^\circ}}} = \sqrt{\frac{0.134}{0.134 + 4 \times 0.01}} = 0.878$$

The values of α at 1500. and 800. K are 0.159 and 3.23×10^{-5}, respectively.

The degree of dissociation of Cl_2 increases with temperature as shown in Example Problem 6.11. This is always the case for an endothermic reaction.

6.11 Equilibria Involving Ideal Gases and Solid or Liquid Phases

In the preceding sections, we discussed chemical equilibrium in a homogeneous system of ideal gases. However, many chemical reactions involve a gas phase in equilibrium with a solid or liquid phase. An example is the thermal decomposition of $CaCO_3(s)$:

$$CaCO_3(s) \rightleftharpoons CaO(s) + CO_2(g) \tag{6.67}$$

In this case, a pure gas is in equilibrium with two solid phases at the pressure P. At equilibrium,

$$\Delta G_R = \sum_i n_i \mu_i = 0$$

$$0 = \mu_{eq}(CaO, s, P) + \mu_{eq}(CO_2, g, P) - \mu_{eq}(CaCO_3, s, P) \tag{6.68}$$

Because the equilibrium pressure is $P \neq P^\circ$, the pressure dependence of μ for each species must be taken into account. From Section 6.3, we know that the pressure dependence of G for a solid or liquid is very small:

$$\mu_{eq}(CaO, s, P) \approx \mu^\circ(CaO, s) \text{ and } \mu_{eq}(CaCO_3, s, P) \approx \mu^\circ(CaCO_3, s) \tag{6.69}$$

You will verify the validity of Equation (6.69) in the end-of-chapter problems. Using the dependence of μ on P for an ideal gas, Equation (6.68) becomes

$$0 = \mu^\circ(CaO, s) + \mu^\circ(CO_2, g) - \mu^\circ(CaCO_3, s) + RT \ln \frac{P_{CO_2}}{P^\circ} \text{ or}$$

$$\Delta G_R^\circ = \mu^\circ(CaO, s) + \mu^\circ(CO_2, g) - \mu^\circ(CaCO_3, s) = -RT \ln \frac{P_{CO_2}}{P^\circ} \tag{6.70}$$

Rewriting this equation in terms of K_P, we obtain

$$\ln K_P = \ln \frac{P_{CO_2}}{P^\circ} = -\frac{\Delta G_R^\circ}{RT} \tag{6.71}$$

We generalize this result as follows. *K_P for an equilibrium involving gases with liquids and/or solids contains only contributions from the gaseous species. However, ΔG_R° is calculated taking all species into account.*

EXAMPLE PROBLEM 6.12

Using the preceding discussion and the tabulated values of ΔG_f° and ΔH_f° in Appendix B, calculate the $CO_2(g)$ pressure in equilibrium with a mixture of $CaCO_3(s)$ and $CaO(s)$ at 1000., 1100., and 1200. K.

Solution

For the reaction $CaCO_3(s) \rightleftharpoons CaO(s) + O_2(g)$, $\Delta G_R^\circ = 131.1 \text{ kJ mol}^{-1}$ and $\Delta H_R^\circ = 178.5 \text{ kJ mol}^{-1}$ at 298.15 K. We use Equation (6.71) to calculate $K_P(298.15 \text{ K})$ and Equation (6.66) to calculate K_P at elevated temperatures.

$$\ln \frac{P_{CO_2}}{P^\circ}(1000. \text{ K}) = \ln K_P(1000. \text{ K})$$

$$= \ln K_P(298.15 \text{ K}) - \frac{\Delta H_{R,298.15 \text{ K}}^\circ}{R}\left(\frac{1}{1000. \text{ K}} - \frac{1}{298.15 \text{ K}}\right)$$

$$= \frac{-131.1 \times 10^3 \, \text{J mol}^{-1}}{8.314 \, \text{J K}^{-1} \, \text{mol}^{-1} \times 298.15 \, \text{K}} - \frac{178.5 \times 10^3 \, \text{J mol}^{-1}}{8.314 \, \text{J K}^{-1} \, \text{mol}^{-1}}$$

$$\times \left(\frac{1}{1000. \, \text{K}} - \frac{1}{298.15 \, \text{K}} \right)$$

$$= -2.348; \quad P_{CO_2}(1000. \, \text{K}) = 0.0956 \, \text{bar}$$

The values for P_{CO_2} at 1100. K and 1200. K are 0.673 bar and 1.23 bar.

If the reaction involves only liquids or solids, the pressure dependence of the chemical potential is generally small and can be neglected. However, it cannot be neglected if $P \gg 1$ bar as shown in Example Problem 6.13.

EXAMPLE PROBLEM 6.13

At 298.15 K, $\Delta G_f^\circ(\text{C}, graphite) = 0$, and $\Delta G_f^\circ(\text{C}, diamond) = 2.90 \, \text{kJ mol}^{-1}$. Therefore, graphite is the more stable solid phase at this temperature at $P = P^\circ = 1$ bar. Given that the densities of graphite and diamond are 2.25 and 3.52 kg/L, respectively, at what pressure will graphite and diamond be in equilibrium at 298.15 K?

Solution

At equilibrium $\Delta G = G(\text{C}, graphite) - G(\text{C}, diamond) = 0$. Using the pressure dependence of G, $(\partial G_m/\partial P)_T = V_m$, we establish the condition for equilibrium:

$$\Delta G = \Delta G_f^\circ(\text{C}, graphite) - \Delta G_f^\circ(\text{C}, diamond)$$

$$+ (V_m^{graphite} - V_m^{diamond})(\Delta P) = 0$$

$$0 = 0 - 2.90 \times 10^3 + (V_m^{graphite} - V_m^{diamond})(P - 1 \, \text{bar})$$

$$P = 1 \, \text{bar} + \frac{2.90 \times 10^3}{M_C \left(\dfrac{1}{\rho_{graphite}} - \dfrac{1}{\rho_{diamond}} \right)}$$

$$= 1 \, \text{bar} + \frac{2.90 \times 10^3}{12.00 \times 10^{-3} \, \text{kg mol}^{-1} \times \left(\dfrac{1}{2.25 \times 10^3 \, \text{kg m}^{-3}} - \dfrac{1}{3.52 \times 10^3 \, \text{kg m}^{-3}} \right)}$$

$$= 10^5 \, \text{Pa} + 1.51 \times 10^9 \, \text{Pa} = 1.51 \times 10^4 \, \text{bar}$$

Fortunately for all those with diamond rings, although the conversion of diamond to graphite at 1 bar and 298 K is spontaneous, the rate of conversion is vanishingly small.

6.12 Expressing the Equilibrium Constant in Terms of Mole Fraction or Molarity

Chemists often find it useful to express the concentrations of reactants and products in units other than partial pressures. Two examples of other units that we will consider in this section are mole fraction and molarity. Note, however, that this discussion is still limited to a mixture of ideal gases. The extension of chemical equilibrium to include neutral and ionic species in aqueous solutions will be made in Chapter 10, after the concept of activity has been introduced.

We first express the equilibrium constant in terms of mole fractions. The mole fraction x_i and the partial pressure P_i are related by $P_i = x_i P$. Therefore,

$$K_P = \frac{\left(\dfrac{P_C^{eq}}{P^\circ}\right)^\gamma \left(\dfrac{P_D^{eq}}{P^\circ}\right)^\delta}{\left(\dfrac{P_A^{eq}}{P^\circ}\right)^\alpha \left(\dfrac{P_B^{eq}}{P^\circ}\right)^\beta} = \frac{\left(\dfrac{x_C^{eq}P}{P^\circ}\right)^\gamma \left(\dfrac{x_D^{eq}P}{P^\circ}\right)^\delta}{\left(\dfrac{x_A^{eq}P}{P^\circ}\right)^\alpha \left(\dfrac{x_B^{eq}P}{P^\circ}\right)^\beta} = \frac{(x_C^{eq})^\gamma (x_D^{eq})^\delta}{(x_A^{eq})^\alpha (x_B^{eq})^\beta} \left(\frac{P}{P^\circ}\right)^{\gamma+\delta-\alpha-\beta}$$

$$= K_x \left(\frac{P}{P^\circ}\right)^{\Delta\nu}$$

$$K_x = K_P \left(\frac{P}{P^\circ}\right)^{-\Delta\nu} \tag{6.72}$$

Recall that $\Delta\nu$ is the difference in the stoichiometric coefficients of products and reactants. Note that just as for K_P, K_x is a dimensionless number.

Because the molarity c_i is defined as $c_i = n_i/V = P_i/RT$, we can write $P_i/P^\circ = (RT/P^\circ)c_i$. To work with dimensionless quantities, we introduce the ratio c_i/c°, which is related to P_i/P° by

$$\frac{P_i}{P^\circ} = \frac{c^\circ RT}{P^\circ} \frac{c_i}{c^\circ} \tag{6.73}$$

Using this notation, we can express K_c in terms of K_P:

$$K_P = \frac{\left(\dfrac{P_C^{eq}}{P^\circ}\right)^\gamma \left(\dfrac{P_D^{eq}}{P^\circ}\right)^\delta}{\left(\dfrac{P_A^{eq}}{P^\circ}\right)^\alpha \left(\dfrac{P_B^{eq}}{P^\circ}\right)^\beta} = \frac{\left(\dfrac{c_C^{eq}}{c^\circ}\right)^\gamma \left(\dfrac{c_D^{eq}}{c^\circ}\right)^\delta}{\left(\dfrac{c_A^{eq}}{c^\circ}\right)^\alpha \left(\dfrac{c_B^{eq}}{c^\circ}\right)^\beta} \left(\frac{c^\circ RT}{P^\circ}\right)^{\gamma+\delta-\alpha-\beta} = K_c \left(\frac{c^\circ RT}{P^\circ}\right)^{\Delta\nu}$$

$$K_c = K_P \left(\frac{c^\circ RT}{P^\circ}\right)^{-\Delta\nu} \tag{6.74}$$

Equations (6.72) and (6.74) show that K_x and K_c are in general different from K_P. They are only equal in the special case that $\Delta\nu = 0$.

6.13 The Dependence of the Extent of Reaction on *T* and *P*

In previous sections, we have discussed how K_P varies with T and how K_x varies with P. In this section, we consider how the extent of reaction varies with T and P in a more general framework that encompasses the previous discussions. It is useful at this point to introduce a parameter, ξ, called the **extent of reaction** in discussing the shift in the position of equilibrium of a chemical reaction if T and/or P change. If the reaction advances by ξ moles, the number of moles of each species i changes according to

$$n_i = n_i^{initial} + \nu_i\xi \tag{6.75}$$

Differentiating this equation leads to $dn_i = \nu_i d\xi$. By inserting this result in Equation (6.53), we can write dG in terms of ξ. At constant T and P,

$$dG = \left(\sum_i \nu_i \mu_i\right)d\xi = \Delta G_R d\xi \tag{6.76}$$

An advancing reaction can be described in terms of the partial derivative of G with ξ:

$$\left(\frac{\partial G}{\partial \xi}\right)_{T,P} = \sum_i \nu_i \mu_i = \Delta G_R \tag{6.77}$$

The direction of spontaneous change is that in which ΔG_R is negative. This direction corresponds to $(\partial G/\partial \xi)_{T,P} < 0$. At a given composition of the reaction mixture, the partial pressures of the reactants and products can be determined, and the μ_i can be calculated. Because $\mu_i = \mu_i(T, P, n_A, n_B,...)$, $\sum_i \nu_i \mu_i = \Delta G_R$ must be evaluated at specific values of $T, P, n_A, n_B,...$. Based on the value of ΔG_R obtained, the following conclusions can be drawn:

- If $\Delta G_R = (\partial G/\partial \xi)_{T,P} < 0$, the reaction proceeds spontaneously as written.
- If $\Delta G_R = (\partial G/\partial \xi)_{T,P} > 0$, the reaction proceeds spontaneously in the opposite direction.
- If $\Delta G_R = (\partial G/\partial \xi)_{T,P} = 0$, the reaction system is at equilibrium, and there is no direction of spontaneous change.

Suppose that we have a mixture of reactive gases at equilibrium. Does the equilibrium shift toward reactants or products as T or P is changed? Because ξ_{eq} increases if K_P or K_x increases, and decreases if K_P or K_x decreases, we need only consider the dependence of K_P on T or K_x on T and P. We first consider the dependence of K_P on T. We know from Equation (6.64) that

$$\frac{d \ln K_P}{dT} = \frac{\Delta H_R^\circ}{RT^2} \tag{6.78}$$

so that K_P will change differently with temperature for an exothermic reaction than for an endothermic reaction. For an exothermic reaction, $d \ln K_P/dT < 0$, and ξ_{eq} will shift toward the reactants as T increases. For an endothermic reaction, $d \ln K_P/dT > 0$, and ξ_{eq} will shift toward the products as T increases.

The dependence of ξ_{eq} on pressure can be determined from the relationship between K_x and P:

$$K_x = K_P \left(\frac{P}{P^\circ}\right)^{-\Delta \nu} \tag{6.79}$$

Because K_P is independent of pressure, the pressure dependence of K_x arises solely from $(P/P^\circ)^{-\Delta \nu}$. If $\Delta \nu > 0$, the number of moles of gaseous products increases as the reaction proceeds. In this case, K_x decreases as P increases, and ξ_{eq} shifts back toward the reactants. If the number of moles of gaseous products decreases as the reaction proceeds, K_x increases as P increases, and ξ_{eq} shifts forward toward the products. If $\Delta \nu = 0$, ξ_{eq} is independent of pressure.

A combined change in T and P leads to a superposition of the effects just discussed. According to the French chemist Le Chatelier, reaction systems at chemical equilibrium respond to an outside stress, such as a change in T or P, by countering the stress. Consider the $Cl_2(g) \rightleftharpoons 2Cl(g)$ reaction discussed in Example Problems 6.9 and 6.10 for which $\Delta H_R > 0$. There, ξ_{eq} responds to an increase in T in such a way that heat is taken up by the system; in other words, more $Cl_2(g)$ dissociates. This counters the stress imposed on the system by an increase in T. Similarly, ξ_{eq} responds to an increase in P in such a way that the volume of the system decreases. Specifically, ξ_{eq} changes in the direction that $\Delta \nu < 0$, and for the reaction under consideration, $Cl(g)$ is converted to $Cl_2(g)$. This shift in the position of equilibrium counters the stress brought about by an increase in P.

SUPPLEMENTAL

6.14 A Case Study: The Synthesis of Ammonia

In this section, we discuss a specific reaction and show both the power and the limitations of thermodynamics in developing a strategy to maximize the rate of a desired reaction. Ammonia synthesis is the primary route to the manufacture of fertilizers in agriculture. The commercial process used today is essentially the same as that invented by the German chemists Robert Bosch and Fritz Haber in 1908. In terms of its impact on human life, the Haber-Bosch synthesis may be the most important chemical process ever invented, because the increased food production possible with fertilizers based on NH_3 has allowed the large increase in Earth's population in the 20th century to occur. In 2000, more than 2 million tons of ammonia were produced per week using the Haber-Bosch process, and more than 98% of the inorganic nitrogen input to soils used in agriculture worldwide as fertilizer is generated using the Haber-Bosch process.

What useful information can be obtained about the ammonia synthesis reaction from thermodynamics? The overall reaction is

$$1/2 \, N_2(g) + 3/2 \, H_2(g) \rightleftharpoons NH_3(g) \tag{6.80}$$

Because the reactants are pure elements in their standard states at 298.15 K,

$$\Delta H_R^\circ = \Delta H_f^\circ(NH_3, g) = -45.9 \times 10^3 \, \text{J mol}^{-1} \text{ at } 298.15 \text{ K} \tag{6.81}$$

$$\Delta G_R^\circ = \Delta G_f^\circ(NH_3, g) = -16.5 \times 10^3 \, \text{J mol}^{-1} \text{ at } 298.15 \text{ K} \tag{6.82}$$

Assuming ideal gas behavior, the equilibrium constant K_p is given by

$$K_P = \frac{\left(\dfrac{P_{NH_3}}{P^\circ}\right)}{\left(\dfrac{P_{N_2}}{P^\circ}\right)^{\frac{1}{2}}\left(\dfrac{P_{H_2}}{P^\circ}\right)^{\frac{3}{2}}} = \frac{(x_{NH_3})}{(x_{N_2})^{\frac{1}{2}}(x_{H_2})^{\frac{3}{2}}}\left(\frac{P}{P^\circ}\right)^{-1} \tag{6.83}$$

We will revisit this equilibrium and include deviations from ideal behavior in Chapter 7. Because the goal is to maximize x_{NH_3}, Equation (6.83) is written in the form

$$x_{NH_3} = (x_{N_2})^{\frac{1}{2}} (x_{H_2})^{\frac{3}{2}} \left(\frac{P}{P^\circ}\right) K_P \tag{6.84}$$

What conditions of temperature and pressure are most appropriate to maximizing the yield of ammonia? Because

$$\frac{d \ln K_P}{dT} = \frac{\Delta H_R^\circ}{RT^2} < 0 \tag{6.85}$$

K_P and, therefore, x_{NH_3} decrease as T increases. Because $\Delta \nu = -1$, x_{NH_3} increases with P. We conclude that the best conditions to carry out the reaction are high P and low T. In fact, industrial production is carried out for $P > 100$ atm and $T \sim 700$ K.

Our discussion has focused on reaching equilibrium because this is the topic of inquiry in thermodynamics. However, note that many synthetic processes are deliberately carried out far from equilibrium in order to selectively generate a desired product. For example, ethylene oxide, which is produced by the partial oxidation of ethylene, is an industrially important chemical used to manufacture antifreeze, surfactants, and fungicides. Although the complete oxidation of ethylene to CO_2 and H_2O has a more negative value of ΔG_R° than the partial oxidation, it is undesirable because it does not produce useful products.

These predictions provide information about the equilibrium state of the system but lack one important component. How long will it take to reach equilibrium? Unfortunately, thermodynamics gives us no information about the rate at which equilibrium is attained. Consider the following example. Using values of ΔG_f° from the data tables, convince yourself that the oxidation of potassium and coal (essentially carbon) are both spontaneous processes. For this reason, potassium has to be stored in an environment free of oxygen. However, coal is stable in air indefinitely. Thermodynamics tells us that both processes are spontaneous at 298.15 K, but experience shows us that only one proceeds at a measurable rate. However, by heating coal to an elevated temperature, the oxidation reaction becomes rapid. This result implies that there is an energetic **activation barrier** to the reaction that can be overcome with the aid of thermal energy. As we will see, the same is true of the ammonia synthesis reaction. Although Equations (6.84) and (6.85) predict that x_{NH_3} is maximized as T approaches 0 K, the reaction rate is vanishingly small unless $T > 500$ K.

To this point, we have only considered the reaction system using state functions. However, a chemical reaction consists of a number of individual steps, called a **reaction mechanism.** How might the ammonia synthesis reaction proceed? One possibility is the gas-phase reaction sequence:

$$N_2(g) \rightarrow 2N(g) \tag{6.86}$$

$$H_2(g) \rightarrow 2H(g) \tag{6.87}$$

$$N(g) + H(g) \rightarrow NH(g) \tag{6.88}$$

$$NH(g) + H(g) \rightarrow NH_2(g) \tag{6.89}$$

$$NH_2(g) + H(g) \rightarrow NH_3(g) \tag{6.90}$$

In each of the recombination reactions, other species that are not involved in the reaction can facilitate energy transfer. The enthalpy changes associated with each of the individual steps are known and are indicated in Figure 6.7. This diagram is useful because it gives much more information than the single-value ΔH_R°. Although the overall process is slightly exothermic, the initial step, which is the dissociation of $N_2(g)$ and $H_2(g)$, is highly endothermic. Heating a mixture of $N_2(g)$ and $H_2(g)$ to high enough temperatures to dissociate a sizable fraction of the $N_2(g)$ would require such a high temperature that all the $NH_3(g)$ formed would dissociate into reactants. This can be shown by carrying out a calculation such as that in Example Problem 6.11. The conclusion is that the gas phase reaction between $N_2(g)$ and $H_2(g)$ is not a practical route to ammonia synthesis.

The way around this difficulty is to find another reaction sequence for which the enthalpy changes of the individual steps are not prohibitively large. For the ammonia synthesis reaction, such a route is a heterogeneous catalytic reaction, using iron as a catalyst. The mechanism for this path between reactants and products is

$$N_2(g) + \square \rightarrow N_2(a) \tag{6.91}$$

$$N_2(a) + \square \rightarrow 2N(a) \tag{6.92}$$

$$H_2(g) + 2\square \rightarrow 2H(a) \tag{6.93}$$

$$N(a) + H(a) \rightarrow NH(a) + \square \tag{6.94}$$

$$NH(a) + H(a) \rightarrow NH_2(a) + \square \tag{6.95}$$

$$NH_2(a) + H(a) \rightarrow NH_3(a) + \square \tag{6.96}$$

$$NH_3(a) \rightarrow NH_3(g) + \square \tag{6.97}$$

The symbol $\square$ denotes an ensemble of neighboring Fe atoms, also called surface sites, which are capable of forming a chemical bond with the indicated entities. The designation (a) indicates that the chemical species is adsorbed (chemically bonded) to a surface site.

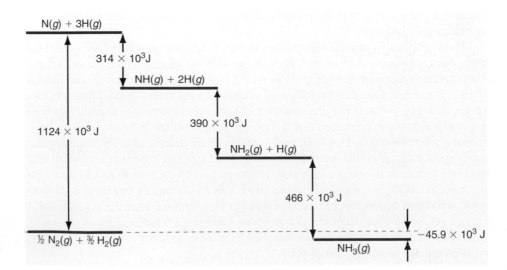

FIGURE 6.7

An enthalpy diagram is shown for the reaction mechanism in Equations (6.86) through (6.90). The successive steps in the reaction proceed from left to right in the diagram.

The enthalpy change for the overall reaction $N_2(g) + 3/2\, H_2(g) \rightarrow NH_3(g)$ is the same for the mechanisms in Equations (6.86) through (6.90) and (6.91) through (6.97) because H is a state function. This is a characteristic of a catalytic reaction. A catalyst can affect the *rate* of the forward and backward reaction but not the *position of equilibrium* in a reaction system. The enthalpy diagram in Figure 6.8 shows that the enthalpies of the individual steps in the gas phase and surface catalyzed reactions are very different. These enthalpy changes have been determined in experiments but are not generally listed in thermodynamic tables.

In the catalytic mechanism, the enthalpy for the dissociation of the gas-phase N_2 and H_2 molecules is greatly reduced by the stabilization of the atoms through their bond to the surface sites. However, a small activation barrier to the dissociation of adsorbed N_2 remains. This barrier is the rate-limiting step in the overall reaction. As a result of this barrier, only one in a million gas-phase N_2 molecules incident on the catalyst surface per unit time dissociates and is converted into ammonia. Such a small reaction probability is not uncommon for industrial processes based on heterogeneous catalytic reactions. In industrial reactors, on the order of $\sim 10^7$ collisions of the reactants with the catalyst occur in the residence time of the reactants in the reactor. Therefore, even reactions with a reaction probability per collision of 10^{-6} can be carried out with a high yield.

How was Fe chosen as the catalyst for ammonia synthesis? Even today, catalysts are usually optimized using a trial-and-error approach. Haber and Bosch tested hundreds of catalyst materials, and more than 2500 substances have been tried since. On the basis of this extensive screening, it was concluded that a successful catalyst must fulfill two different requirements: it must bind N_2 strongly enough that the molecule can dissociate into N atoms. However, it must not bind N atoms too strongly. If that were the case, the N atoms could not react with H atoms to produce gas-phase NH_3. As was known by Haber and Bosch, both osmium and ruthenium are better catalysts than iron for NH_3 synthesis, but these elements are far too expensive for commercial production. Why are these catalysts better than Fe? Quantum mechanical calculations by Jacobsen *et al.* [*J. American Chemical Society* 123(2001), 8404–8405] were able to explain the "volcano plot" of the activity versus the N_2 binding energy shown in Figure 6.9.

Jacobsen et al. concluded that molybdenum is a poor catalyst because N is bound too strongly, whereas N_2 is bound too weakly. Nickel is a poor catalyst for the opposite reason: N is bound too weakly and N_2 is bound too strongly. Osmium and ruthenium are the best elemental catalysts because they fulfill both requirements well. Jacobsen et al. suggested that a catalyst that combined the favorable properties of Mo and Co to create Co-Mo surface sites would be a good catalyst, and they concluded that the ternary compound Co_3Mo_3N best satisfied these requirements. On the basis of

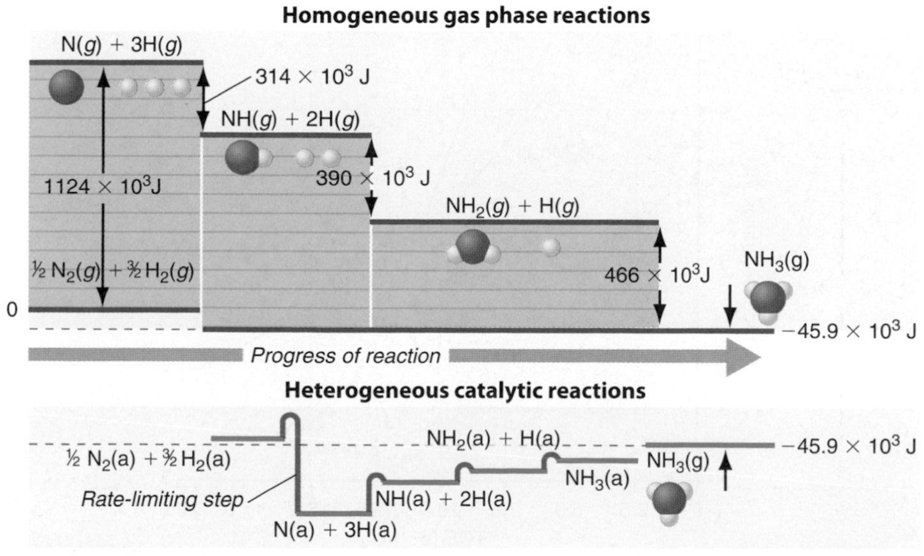

FIGURE 6.8
Enthalpy diagram that compares the homogeneous gas phase and heterogeneous catalytic reactions for the ammonia synthesis reaction. The activation barriers for the individual steps in the surface reaction are shown. The successive steps in the reaction proceed from left to right in the diagram.

Source: Ertl, G. "Surface Science and Catalysis—Studies on the Mechanism of Ammonia *Synthesis*: The P.H. Emmett Award Address." *Catalysis Reviews—Science and Engineering* 21, no.2 (1980): 201–223.

FIGURE 6.9

The calculated turnover frequency (TOF) for NH_3 production (molecules of NH_3 per surface site per second) as a function of the calculated binding energy of N_2 on the surface, relative to that on ruthenium. The reaction conditions are 400°C, 50 bar total pressure, and gas composition $H_2:N_2 = 3:1$ containing 5% NH_3.

Source: Jacobsen, C. J. H., Dahl, S., Clausen, B. S., Bahn, S., Logadottir, A., and Norskov, J. K. "Catalyst Design by Interpolation in the Periodic Table: Bimetallic Ammonia Synthesis Catalysts." *Journal of the American Chemical Society* 123, no. 34 (2001): 8404–8405.

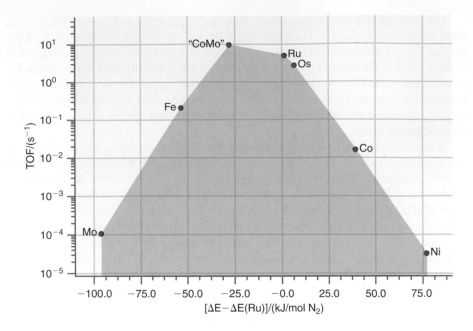

these calculations, the catalyst was synthesized with the results shown in Figure 6.10. The experiments show that Co_3Mo_3N is far more active than Fe, and more active than Ru over most of the concentration range. These experiments also show that the turnover frequency for Co_3Mo_3N lies at the point shown on the volcano plot of Figure 6.8. This catalyst is one of the few that have been developed on the basis of theoretical calculations as opposed to the trial and error method.

An unresolved issue is the nature of the surface sites indicated by □ in Equations (6.91) through (6.97). Are all adsorption sites on an NH_3 synthesis catalyst surface equally reactive? To answer this question, it is important to know that metal surfaces at equilibrium consist of large flat terraces separated by monatomic steps. It was long suspected that the steps rather than the terraces play an important role in ammonia synthesis. This was convincingly shown by Dahl et al. [*Physical Review Letters* 83(1999), 1814–1817] in an elegant experiment. They evaporated a small amount of gold on a Ru crystal. As had been demonstrated by Hwang et al. [*Physical Review Letters* 67(1991), 3279], the inert Au atoms migrate to the step sites on Ru and thereby block these sites, rendering them inactive in NH_3 synthesis.

FIGURE 6.10

Experimental data for the TOF for NH_3 production as a function of the NH_3 concentration in the mixture. The Co_3Mo_3N catalyst is superior to both Fe and Ru as predicted by the calculations. The reaction conditions are those for Figure 6.9. The number of active surface sites is assumed to be 1% of the total number of surface sites.

Source: Jacobsen, C. J. H., Dahl, S., Clausen, B. S., Bahn, S., Logadottir, A., and Norskov, J. K. "Catalyst Design by Interpolation in the Periodic Table: Bimetallic Ammonia Synthesis Catalysts." *Journal of the American Chemical Society* 123, no. 34 (2001): 8404–8405.

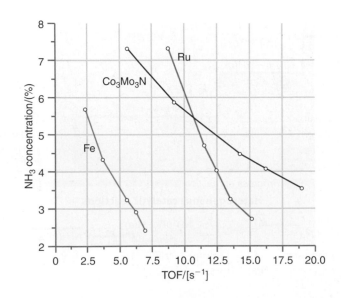

This site blocking is demonstrated using scanning tunneling microscopy as shown in Figure 6.11.

Dahl et al. showed that the rate of ammonia synthesis was lowered by a factor of 10^9 after the step sites, which make up only 1% to 2% of all surface sites, were blocked by Au atoms! This result clearly shows that in this reaction only step sites are active. They are more active because they have fewer nearest neighbors than terrace sites. This enables the step sites to bind N_2 more strongly, promoting dissociation into N atoms that react rapidly with H atoms to form NH_3.

This discussion has shown the predictive power of thermodynamics in analyzing chemical reactions. Calculating ΔG_R° allowed us to predict that the reaction of N_2 and H_2 to form NH_3 is a spontaneous process at 298.15 K. Calculating ΔH_R° allowed us to predict that the yield of ammonia increases with increasing pressure and decreases with increasing temperature. However, thermodynamics cannot be used to determine how rapidly equilibrium is reached in a reaction mixture. If thermodynamic tables of species such as $N_2(a)$, $N(a)$, $NH(a)$, $NH_2(a)$, and $NH_3(a)$ were available, it would be easier for chemists to develop synthetic strategies that are likely to proceed at an appreciable rate. However, whether an activation barrier exists in the conversion of one species to another—and how large the barrier is—are not predictable within thermodynamics. Therefore, synthetic strategies must rely on chemical intuition, experiments, and quantum mechanical calculations to develop strategies that reach thermodynamic equilibrium at a reasonable rate.

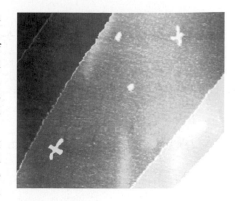

FIGURE 6.11
This scanning tunneling microscope image of a Ru surface shows wide, flat terraces separated by monatomic steps. The bright "ribbon" at the step edge is due to Au atoms, which are also seen in the form of four islands on the middle terrace.
Source: T. Diemant, T. Hager, H. E. Hoster, H. Rauscher, and R. J. Behm, *Surf. Sci* 141(2003) 137, fig. 1a.

SUPPLEMENTAL

6.15 Expressing *U* and *H* and Heat Capacities Solely in Terms of Measurable Quantities

In this section, we use the Maxwell relations derived in Section 6.2 to express *U*, *H*, and heat capacities solely in terms of measurable quantities such as κ, β and the state variables *P*, *V*, and *T*. Often, it is not possible to determine an equation of state for a substance over a wide range of the macroscopic variables. This is the case for most solids and liquids. However, material constants such as heat capacities, the thermal expansion coefficient, and the isothermal compressibility may be known over a limited range of their variables. If this is the case, the enthalpy and internal energy can be expressed in terms of the material constants and the variables *V* and *T* or *P* and *T*, as shown later. The Maxwell relations play a central role in deriving these formulas.

We begin with the following differential expression for *dU* that holds for solids, liquids, and gases provided that no nonexpansion work, phase transitions, or chemical reactions occur in the system:

$$dU = TdS - PdV \tag{6.98}$$

Invoking a transformation at constant temperature and dividing both sides of the equation by *dV*,

$$\left(\frac{\partial U}{\partial V}\right)_T = T\left(\frac{\partial S}{\partial V}\right)_T - P = T\left(\frac{\partial P}{\partial T}\right)_V - P \tag{6.99}$$

To arrive at the second expression in Equation (6.99), we have used the third Maxwell relation Equation (6.28). We next express $(\partial P/\partial T)_V$ in terms of β and κ, where β is the isobaric volumetric thermal expansion coefficient at constant pressure, and κ is the isothermal compressibility:

$$\left(\frac{\partial P}{\partial T}\right)_V = -\frac{(\partial V/\partial T)_P}{(\partial V/\partial P)_T} = \frac{\beta}{\kappa} \tag{6.100}$$

Equation (6.100) allows us to write Equation (6.99) in the form

$$\left(\frac{\partial U}{\partial V}\right)_T = \frac{\beta T - \kappa P}{\kappa} \tag{6.101}$$

Through this equation, we have linked $(\partial U/\partial V)_T$, which is called the internal pressure, to easily measured material properties. This allows us to calculate the internal pressure of a liquid or a gas as shown in Example Problem 6.14.

EXAMPLE PROBLEM 6.14

At 298 K, the thermal expansion coefficient and the isothermal compressibility of liquid water are $\beta = 2.04 \times 10^{-4}\,\mathrm{K}^{-1}$ and $\kappa = 45.9 \times 10^{-6}\,\mathrm{bar}^{-1}$.

a. Calculate $(\partial U/\partial V)_T$ for water at 320. K and $P = 1.00$ bar.

b. If an external pressure equal to $(\partial U/\partial V)_T$ were applied to 1.00 m³ of liquid water at 320. K, how much would its volume change? What is the relative change in volume?

c. As shown in Example Problem 3.5, $(\partial U/\partial V_m)_T = a/V_m^2$ for a van der Waals gas. Calculate $(\partial U/\partial V_m)_T$ for $N_2(g)$ at 320 K and $P = 1.00$ atm given that $a = 1.35\,\mathrm{atm\,L^2\,mol^{-2}}$. Compare the value that you obtain with that of part (a) and discuss the difference.

Solution

a. $\left(\dfrac{\partial U}{\partial V}\right)_T = \dfrac{\beta T - \kappa P}{\kappa}$

$$= \frac{2.04 \times 10^{-4}\,\mathrm{K}^{-1} \times 320.\,\mathrm{K} - 45.9 \times 10^{-6}\,\mathrm{bar}^{-1} \times 1.00\,\mathrm{bar}}{45.9 \times 10^{-6}\,\mathrm{bar}^{-1}}$$

$$= 1.4 \times 10^3\,\mathrm{bar}$$

b. $\kappa = -\dfrac{1}{V}\left(\dfrac{\partial V}{\partial P}\right)_T$

$$\Delta V_T = -\int_{P_i}^{P_f} V(T)\kappa\,dP \approx -V(T)\kappa(P_f - P_i)$$

$$= -1.00\,\mathrm{m}^3 \times 45.9 \times 10^{-6}\,\mathrm{bar} \times (1.4 \times 10^3\,\mathrm{bar} - 1\,\mathrm{bar})$$

$$= -6.4 \times 10^{-2}\,\mathrm{m}^3$$

$$\frac{\Delta V_T}{V} = -\frac{6.4 \times 10^{-2}\,\mathrm{m}^3}{1\,\mathrm{m}^3} = -6.4\%$$

c. At atmospheric pressure, we can calculate V_m using the ideal gas law and

$$\left(\frac{\partial U}{\partial V_m}\right)_T = \frac{a}{V_m^2} = \frac{aP^2}{R^2 T^2} = \frac{1.35\,\mathrm{atm\,L^2\,mol^{-2}} \times (1.00\,\mathrm{atm})^2}{(8.206 \times 10^{-2}\,\mathrm{atm\,L\,mol^{-1}\,K^{-1}} \times 320.\,\mathrm{K})^2}$$

$$= 1.96 \times 10^{-3}\,\mathrm{atm}$$

The internal pressure is much smaller for N_2 gas than for H_2O liquid. This is the case because H_2O molecules in a liquid are relatively strongly attracted to one another, whereas the interaction between gas phase N_2 molecules is very weak. Gases, which have a small internal pressure, are more easily compressed by an external pressure than liquids or solids, which have a high internal pressure.

Using the result in Equation (6.101) for the internal pressure, we can write dU in the form

$$dU = \left(\frac{\partial U}{\partial T}\right)_V dT + \left(\frac{\partial U}{\partial V}\right)_T dV = C_V\,dT + \frac{\beta T - \kappa P}{\kappa}\,dV \qquad \textbf{(6.102)}$$

The usefulness of Equation (6.102) arises from the fact that it contains only the material constants C_V, β, κ, and the macroscopic variables T and V. In Example Problems 6.14

and 6.15 we show that ΔU, ΔH and $C_P - C_V$ can be calculated even if the equation of state for the material of interest is not known, as long as the thermal expansion coefficient and the isothermal compressibility are known.

EXAMPLE PROBLEM 6.15

A 1000. g sample of liquid Hg undergoes a reversible transformation from an initial state $P_i = 1.00$ bar, $T_i = 300.$ K to a final state $P_f = 300.$ bar, $T_f = 600.$ K. The density of Hg is 13,534 kg m^{-3}, $\beta = 1.81 \times 10^{-4}$ K^{-1}, $C_P = 27.98$ J mol^{-1} K^{-1}, and $\kappa = 3.91 \times 10^{-6}$ bar^{-1}, in the range of the variables in this problem. Recall from Section 3.5 that $C_P - C_V = (TV \beta^2/\kappa)$, and that $C_P \approx C_V$ for liquids and solids.

a. Calculate ΔU and ΔH for the transformation. Describe the path assumed for the calculation.

b. Compare the relative contributions to ΔU and ΔH from the change in pressure and the change in temperature. Can the contribution from the change in pressure be neglected?

Solution

Because U is a state function, ΔU is independent of the path. We choose the reversible path $P_i = 1.00$ bar, $T_i = 300.$ K $\rightarrow P_i = 1.00$ bar, $T_f = 600.$ K $\rightarrow P_f = 300.$ bar, $T_f = 600.$ K. Note that because we have assumed that all the material constants are independent of P and T in the range of interest, the numerical values calculated will depend slightly on the path chosen. However, if the P and T dependence of the material constant were properly accounted for, all paths would give the same value of ΔU.

$$\Delta U = \int_{T_i}^{T_f} C_P dT + \int_{V_i}^{V_f} \frac{\beta T - \kappa P}{\kappa} dV$$

Using the definition $\kappa = -1/V(\partial V/\partial P)_T$, $\ln V_f/V_i = -\kappa(P_f - P_i)$, and $V(P) = V_i e^{-\kappa(P-P_i)}$. We can also express $P(V)$ as $P(V) = P_i - (1/\kappa)\ln(V/V_i)$

$$\Delta U = \int_{T_i}^{T_f} C_P dT + \int_{V_i}^{V_f} \left[\frac{\beta T}{\kappa} - \left(P_i - \frac{1}{\kappa} \ln \frac{V}{V_i} \right) \right] dV$$

$$= \int_{T_i}^{T_f} C_P dT + \left(\frac{\beta T}{\kappa} - P_i \right) \int_{V_i}^{V_f} dV + \frac{1}{\kappa} \int_{V_i}^{V_f} \ln \frac{V}{V_i} dV$$

Using the standard integral $\int \ln(x/a)dx = -x + x \ln(x/a)$

$$\Delta U = \int_{T_i}^{T_f} C_P dT + \left(\frac{\beta T}{\kappa} - P_i \right)(V_f - V_i) + \frac{1}{\kappa} \left[-V + V \ln \frac{V}{V_i} \right]_{V_i}^{V_f}$$

$$= C_P(T_f - T_i) + \left(\frac{\beta T}{\kappa} - P_i \right)(V_f - V_i) + \frac{1}{\kappa} \left(V_i - V_f + V_f \ln \frac{V_f}{V_i} \right)$$

$$V_i = \frac{m}{\rho} = \frac{1.000 \, \text{kg}}{13534 \, \text{kg m}^{-3}} = 7.389 \times 10^{-5} \, \text{m}^3$$

$$V_f = V_i e^{-\kappa(P_f - P_i)} = 7.389 \times 10^{-5} \, \text{m}^3 \exp(-3.91 \times 10^{-6} \, \text{bar}^{-1} \times 299 \, \text{bar})$$

$$= 7.380 \times 10^{-5} \, \text{m}^3$$

$$V_f - V_i = -9 \times 10^{-8} \, \text{m}^3$$

$$\Delta U = \frac{1000.\text{ g}}{200.59 \text{ g mol}^{-1}} \times 27.98 \text{ J mol}^{-1} \text{ K}^{-1} \times (600.\text{ K} - 300.\text{ K})$$

$$-\left(\frac{1.81 \times 10^{-4} \text{ K}^{-1} \times 300.\text{ K}}{3.91 \times 10^{-6} \text{ bar}^{-1}} - 1.00 \text{ bar}\right) \times \frac{10^5 \text{ Pa}}{1 \text{ bar}} \times 9 \times 10^{-8} \text{ m}^3$$

$$+ \frac{1}{3.91 \times 10^{-6} \text{ bar}^{-1}} \times \frac{10^5 \text{ Pa}}{1 \text{ bar}} \times \left(\begin{array}{c} 9 \times 10^{-8} \text{ m}^3 + 7.380 \times 10^{-5}\text{m}^3 \\ \times \ln \dfrac{7.380 \times 10^{-5} \text{ m}^3}{7.389 \times 10^{-5} \text{ m}^3} \end{array}\right)$$

$$= 41.8 \times 10^3 \text{ J} - 100 \text{ J} + 1 \text{ J} \approx 41.7 \times 10^3 \text{ J}$$

As shown in Section 3.1,

$$V(P, 600.\text{ K}) = [1 + \beta(600.\text{ K} - 300.\text{ K})]V(P_i, 300.\text{ K})e^{-\kappa(P-Pi)} = V' e^{-\kappa(P-Pi)}$$

$$\Delta H = \int_{T_i}^{T_f} C_P dT + \int_{P_i}^{P_f} V dP = \int_{T_i}^{T_f} C_P dT + V'e^{\kappa P_i}\int_{P_i}^{P_f} e^{-\kappa P}\, dP$$

$$= nC_{P,m}(T_f - T_i) + \frac{V_i' e^{\kappa P_i}}{\kappa}\left(e^{-\kappa P_i} - e^{-\kappa P_f}\right)$$

$$= \frac{1000.\text{ g}}{200.59 \text{ g mol}^{-1}} \times 27.98 \text{ J mol}^{-1}\text{K}^{-1} \times (600.\text{ K} - 300.\text{ K})$$

$$+ 7.389 \times 10^{-5} \text{ m}^3 \times (1 + 300.\text{ K} \times 1.81 \times 10^{-4} \text{ K}^{-1})$$

$$\times \frac{\exp(3.91 \times 10^{-6}\text{bar}^{-1} \times 1 \text{ bar})}{3.91 \times 10^{-6} \text{ bar}^{-1} \times \left(1 \text{ bar}/10^5 \text{ Pa}\right)}$$

$$\times \left[\exp(-3.91 \times 10^{-6} \text{ bar}^{-1} \times 1 \text{ bar})\right.$$

$$\left. - \exp(-3.91 \times 10^{-6} \text{ bar}^{-1} \times 300.\text{ bar})\right]$$

$$\Delta H = 41.8 \times 10^3 \text{ J} + 2.29 \times 10^3 \text{ J} = 44.1 \times 10^3 \text{ J}$$

The temperature dependent contribution to ΔU is 98% of the total change in U, and the corresponding value for ΔH is $\sim$ 95%. The contribution from the change in pressure to ΔU and to ΔH is small.

EXAMPLE PROBLEM 6.16

In the previous Example Problem, it was assumed that $C_P = C_V$.

a. Use equation 3.38, $C_{P,m} = C_{V,m} + TV_m\dfrac{\beta^2}{\kappa}$, and the experimentally determined value $C_{P,m} = 27.98 \text{ J mol}^{-1} \text{ K}^{-1}$ to obtain a value for $C_{V,m}$ for Hg(l) at 300. K.

b. Did the assumption that $C_{P,m} \approx C_{V,m}$ in Example Problem 6.15 introduce an appreciable error in your calculation of ΔU and ΔH?

Solution

a. $C_{P,m} - C_{V,m} = \dfrac{TV_m\beta^2}{\kappa}$

$$= \frac{300.\text{ K} \times \dfrac{0.20059 \text{ kg mol}^{-1}}{13534 \text{ kgm}^3} \times (1.81 \times 10^{-4} \text{ K}^{-1})^2}{3.91 \times 10^{-6} \text{ bar}^{-1} \times \left(1 \text{ bar}/10^5 \text{ Pa}\right)}$$

$$= 3.73 \text{ J K}^{-1} \text{ mol}^{-1}$$

$$C_{V,m} = C_{P,m} - \frac{TV_m\beta^2}{\kappa}$$

$$= 27.98\,\text{J K}^{-1}\,\text{mol}^{-1} - 3.73\,\text{J K}^{-1}\,\text{mol}^{-1} = 24.25\,\text{J K}^{-1}\,\text{mol}^{-1}$$

b. ΔU is in error by ~ 15% and ΔH is unaffected by assuming that $C_{P,m} \approx C_{V,m}$. It is a reasonable approximation to set $C_P = C_V$ for a liquid or solid.

SUPPLEMENTAL

6.16 Measuring ΔG for the Unfolding of Single RNA Molecules

In general, ΔG, ΔH, and ΔU are measured using themochemical methods as discussed in Chapter 4. However, many biochemical systems are irreversibly changed upon heating so that calorimetric methods are less suitable for such systems. With the advent of atomic scale manipulation of molecules, which began with the invention of scanning probe microscopies (see Chapter 16), it has become possible to measure ΔG for a change using an individual molecule. An interesting illustration is the measurement of ΔG for the unfolding of single RNA molecules. RNA molecules must fold into very specific shapes in order to carry out their function as catalysts for biochemical reactions. It is not well understood how the molecule achieves its equilibrium shape, and the energetics of the folding is a key parameter in modeling the folding process.

Liphardt et al. [*Science* 292, (2001), 733] have obtained ΔG for this process by unzipping a single RNA molecule. The measurement relies on Equation (6.9), which states that Gibbs energy is a measure of the maximum nonexpansion work that can be produced in a transformation. Figure 6.12 shows how the force associated with the unfolding of RNA into two strands can be measured. Each of the two strands that make up the RNA is attached to a chemical "handle" that allows the individual strands to be firmly linked to 2-μm-diameter polystyrene beads. Using techniques based on scanning probe microscopies, the distance between the beads is increased, and the force needed to do so is measured.

The force versus distance plot for the P5abcΔA RNA is shown in Figure 6.13 for two different rates at which the force was increased with time. For a rate of 10 pN s^{-1} (right panel), the folding and unfolding curves have different forms, showing that the process is not reversible. However, for a rate of 1 pN s^{-1} (left panel), the red and black curves are superimposable, showing that the process is essentially reversible. (No process that occurs at a finite rate is truly reversible.) We next discuss the form of the force versus distance curve.

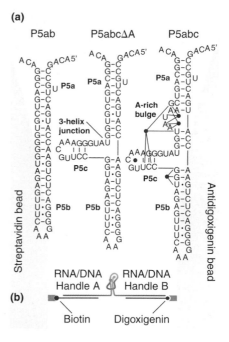

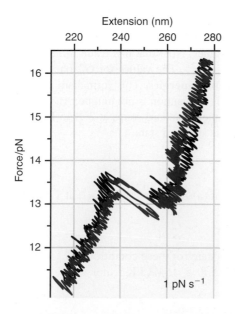

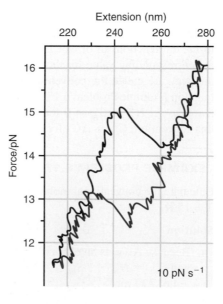

FIGURE 6.12

The three types of RNA studied by Liphardt et al. are shown. The letters G, C, A, and U refer to the bases guanine, cytosine, adenine, and uracil. The strands are held together by hydrogen bonding between the complementary base pairs A and U, and G and C. The pronounced bulge at the center of P5abcΔA is a hairpin loop. Such loops are believed to play a special role in the folding of RNA. (b) An individual RNA molecule is linked by handles attached to the end of each strand to functionalized polystyrene spheres that are moved relative to one another to induce unfolding of the RNA mechanically.

Source: Figures 1 and 2 from "Reversible Unfolding of Single RNA Molecules by Mechanical Force" by Jan Liphardt, et al., in *Science* 27, Vol. 292 No. 5517: 733-737, April 2001. Copyright © 2001, The American Association for the Advancement of Science. Reprinted with permission from AAAS.

FIGURE 6.13

The measured force versus distance curve is shown for the P5abcΔA form of RNA. The black trace represents the stretching of the molecule, and the red curve represents the refolding of the molecule as the beads are brought back to their original separation. The left panel shows data for an increase in force of 1 pN s^{-1}, and the right panel shows the corresponding results for 10 pN s^{-1}.

Source: Figures 1 and 2 from "Reversible Unfolding of Single RNA Molecules by Mechanical Force" by Jan Liphardt, et al., in *Science* 27, Vol. 292 No. 5517: 733–737, April 2001. Copyright © 2001, The American Association for the Advancement of Science. Reprinted with permission from AAAS.

The initial increase in distance between ~210 and ~225 nm nm is due to the stretching of the handles. However, a marked change in the shape of the curve is observed for $F \approx 13$ pN, for which the distance increases abruptly. This behavior is the signature of the unfolding of the two strands. The negative slope of the curve between 230 and 260 nm is determined by the experimental technique used to stretch the RNA. In a "perfect" experiment, this portion of the curve would be a horizontal line, corresponding to an increase in length at constant force. The observed length increase is identical to the length of the molecule, showing that the molecule unfolds all at once in a process in which all bonds between the complementary base pairs are broken. The analysis of the results to obtain ΔG from the force curves is complex and will not be discussed here. The experimentally obtained value for ΔG is 193 kJ/mol, showing that a change in the thermodynamic state function G can be determined by a measurement of work on a single molecule. The unfolding can also be carried out by increasing the temperature of the solution containing the RNA to ~80°C, in which case the process is referred to as melting. In the biochemical environment, the unfolding is achieved through molecular forces exerted by enzymes called DNA polymerases, which act as molecular motors.

SUPPLEMENTAL

6.17 The Role of Mixing in Determining Equilibrium in a Chemical Reaction

Three factors determine the equilibrium partial pressures of gases in a reaction system: the Gibbs energies of the reactants, the Gibbs energies of the products, and the Gibbs energy of mixing of reactants. In this section, we examine the role of each of these factors.

It is generally not necessary to assign values to the molar Gibbs energies G_m°, because only differences in the Gibbs energy rather than absolute values can be obtained from experiments. However, in order to make this discussion quantitative, we introduce a convention that allows an assignment of numerical values to G_m°. With the convention from Section 4.2 that $H_m^\circ = 0$ for an element in its standard state,

$$G_m^\circ = H_m^\circ - TS_m^\circ = -TS_m^\circ \qquad (6.103)$$

for a pure *element* in its standard state at 1 bar. To obtain an expression for G_m° for a pure *compound* in its standard state, consider ΔG° for the formation reaction. Recall that 1 mol of the product appears on the right side in the formation reaction, and only elements in their standard states appear on the left side:

$$\Delta G_f^\circ = G_{m, \, product}^\circ + \sum_i v_i G_{m, \, reactant \, i}^\circ$$

$$G_{m, \, product}^\circ = \Delta G_f^\circ + \overset{elements \, only}{\sum_i} v_i TS_{m,i}^\circ \qquad (6.104)$$

where the v_i are the stoichiometric coefficients of the elemental reactants in the balanced formation reaction, all of which have negative values. Note that this result differs from the corresponding value for the enthalpy, $H_m^\circ = \Delta H_f^\circ$. The values for G_m° obtained in this way are called the **conventional molar Gibbs energies.** This formalism is illustrated in Example Problem 6.17. We note that this convention is not unique. An alternative is to set $H_m^\circ = G_m^\circ = 0$ for all elements in their standard states at 298.15 K. With this convention, $G_m^\circ = \Delta G_f^\circ$ for a compound in its standard state.

EXAMPLE PROBLEM 6.17

Calculate the conventional molar Gibbs energy of (a) Ar(g) and (b) $H_2O(l)$ at 1 bar and 298.15 K.

Solution

a. Because Ar(g) is an element in its standard state for these conditions, $H_m^\circ = 0$ and $G_m^\circ = H_m^\circ - TS_m^\circ = -TS_m^\circ = -298.15$ K $\times$ 154.8 J K^{-1}mol^{-1} = -4.617 kJ mol^{-1}.

b. The formation reaction for $H_2O(l)$ is $H_2(g) + 1/2\, O_2(g) \rightarrow H_2O(l)$

$$G^\circ_{m,\,product} = \Delta G^\circ_f\,(product) + \overset{elements\ only}{\sum_i v_i T S^\circ_{m,i}}$$

$$G^\circ_m\,(H_2O,\,l) = \Delta G^\circ_f(H_2O,\,l) - \left[TS^\circ_m(H_2,\,g) + \frac{1}{2} TS^\circ_m(O_2,\,g) \right]$$

$$= -237.1\ \text{kJ mol}^{-1} - 298.15\ \text{K} \times \begin{bmatrix} 130.7\ \text{J K}^{-1}\ \text{mol}^{-1} + \frac{1}{2} \times \\ 205.2\ \text{J K}^{-1}\ \text{mol}^{-1} \end{bmatrix}$$

$$= -306.6\ \text{kJ mol}^{-1}$$

These calculations give G°_m at 298.15 K. To calculate the conventional molar Gibbs energy at another temperature, use the Gibbs–Helmholtz equation as discussed next.

Now that we have a way to assign numerical values to molar Gibbs energies, we consider the reaction system $2\,NO_2(g) \rightleftharpoons N_2O_4(g)$, with $(2 - 2\xi)$ moles of $NO_2(g)$ and ξ moles of $N_2O_4(g)$ present in a vessel at a constant pressure of 1 bar and 298 K. The parameter ξ could in principle take on any value between zero and one, corresponding to pure $NO_2(g)$ and pure $N_2O_4(g)$, respectively. The Gibbs energy of the pure unmixed reagents and products G_{pure} is given by

$$G_{pure} = (2 - 2\xi)G^\circ_m(NO_2,\,g) + \xi G^\circ_m(N_2O_4,\,g) \qquad \textbf{(6.105)}$$

where G°_m is the conventional molar Gibbs energy defined by Equation (6.104). Note that G_{pure} varies linearly with ξ, as shown in Figure 6.14. Because the reactants and products are mixed throughout the range accessible to ξ, G_{pure} is not equal to $G_{mixture}$, which is given by

$$G_{mixture} = G_{pure} + \Delta G_{mixing} \qquad \textbf{(6.106)}$$

Recall that $G_{mixture}$ is minimized at equilibrium. Equation (6.106) shows that if G_{pure} alone determined ξ_{eq}, either $\xi_{eq} = 0$ or $\xi_{eq} = 1$ mole, depending on whether G_{pure} were lower for reagents or products. For the reaction under consideration, $\Delta G^\circ_f\,(N_2O_4,\,g) < \Delta G^\circ_f(NO_2,\,g)$ so that $\xi_{eq} = 1$ mole if G_{pure} alone determined ξ_{eq}. More generally, if G_{pure} alone determined ξ_{eq}, every chemical reaction would go to completion, or the reverse reaction would go to completion. The value of ΔG_{mixing} can be calculated using Equation (6.48). You will show in the end-of-chapter problems that ΔG_{mixing} for this reaction system has a minimum at $\xi_{eq} = 0.55$ moles.

If G_{pure} alone determined ξ_{eq}, then $\xi_{eq} = 1$ mole, and if ΔG_{mixing} alone determined ξ_{eq}, then $\xi_{eq} = 0.55$ mole. However, the minimum in ΔG_R is determined by the minimum in $G_{pure} + \Delta G_{mixing}$ rather than in the minimum of the individual components. For our specific case, $\xi_{eq} = 0.72$ moles at 298 K. We see that ΔG_{mixing} plays a critical role in determining the position of equilibrium in this chemical reaction. How sensitive is ξ_{eq} to the values of the Gibbs energies of reactants and products? Figure 6.15 shows the same calculation carried out earlier for the reaction $2A \rightleftharpoons B$ setting $G^\circ_B = 100.$ kJ mol^{-1} for different values of G°_A.

For $2G^\circ_A \approx G^\circ_B$ as is the case for $G^\circ_A = 50.5$ kJ mol^{-1}, ξ_{eq} is largely determined by ΔG_{mixing}. However, as $2G^\circ_A$ becomes much larger than G°_B, ξ_{eq} approaches one. However, because of the contribution of ΔG_{mixing} to $G_{mixture}$, which is entirely entropic in origin, ξ_{eq} is always less than one and only approaches one in the limit that $G^\circ_{reactants} - G^\circ_{products} \rightarrow \infty$.

We summarize the roles of G_{pure} and ΔG_{mixing} in determining ξ_{eq}. For reactions in which $G^{reactants}_{pure}$ and $G^{products}_{pure}$ are very similar, ξ_{eq} will be largely determined by ΔG_{mixing}. For reactions in which $G^{reactants}_{pure}$ and $G^{products}_{pure}$ are very different, ξ_{eq} will not be greatly influenced by ΔG_{mixing}, and the equilibrium mixture approaches but never reaches one of the two extremes of pure reactants or pure products.

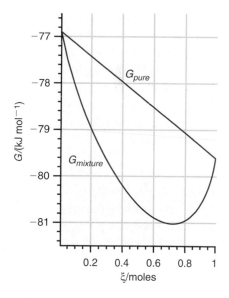

FIGURE 6.14

$G_{mixture}$ and G_{pure} are depicted for the $2\,NO_2(g) \rightleftharpoons N_2O_4(g)$ equilibrium. For this reaction system, the equilibrium position would correspond to $\xi = 1$ mole in the absence of mixing. The mixing contribution to $G_{mixture}$ shifts ξ_{eq} to values smaller than one. Note the correspondence between the three bulleted relations for $(\partial G/\partial \xi)_{T,P}$ following Equation (6.77) and the slope of $G_{mixture}$ vs ξ curve.

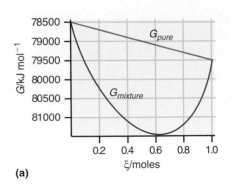

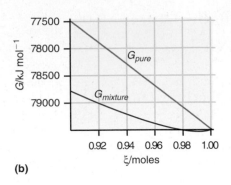

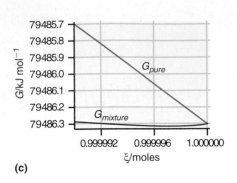

FIGURE 6.15
The corresponding graph to Figure 6.14 is shown for $G_B^\circ = 100.$ kJ mol^{-1} and G_A° values of (a) 50.5 kJ mol^{-1}, (b) 60.0 kJ mol^{-1}, and (c) 80.0 kJ mol^{-1}. Only the portion of the graph for ξ values near one is shown for (b) and (c) so that the minimum in $G_{mixture}$ can be seen. Note that ξ_{eq} approaches one as G_A° increases.

Vocabulary

activation barrier

chemical potential

conventional molar Gibbs energies

extent of reaction

Gibbs energy

Gibbs–Helmholtz equation

Helmholtz energy

maximum nonexpansion work

Maxwell relations

natural variables

reaction mechanism

reaction quotient of pressures

thermodynamic equilibrium constant

Conceptual Problems

Q6.1 K_P is independent of T for a particular chemical reaction. What does this tell you about the reaction?

Q6.2 The reaction A + B $\rightleftharpoons$ C + D is at equilibrium for $\xi = 0.1$. What does this tell you about the variation of G_{pure} with ξ?

Q6.3 Under what condition is $K_P = K_x$?

Q6.4 What is the relationship between the K_P for the two reactions $3/2\,H_2 + 1/2\,N_2 \rightleftharpoons NH_3$ and $3H_2 + N_2 \rightleftharpoons 2NH_3$?

Q6.5 Under what conditions is $dA \leq 0$ a condition that defines the spontaneity of a process?

Q6.6 By invoking the pressure dependence of the chemical potential, show that if a valve separating a vessel of pure A from a vessel containing a mixture of A and B is opened, mixing will occur. Both A and B are ideal gases, and the initial pressure in both vessels is 1 bar.

Q6.7 Under what conditions is $dG \leq 0$ a condition that defines the spontaneity of a process?

Q6.8 Can equilibrium with respect to the concentration of Ar and H_2 be attained in the system shown in Figure 6.2? If so, what can you say about the partial pressure in each part of the system?

Q6.9 Is the equation $(\partial U/\partial V)_T = (\beta T - \kappa P)/\kappa$ valid for liquids, solids, and gases?

Q6.10 Why is it reasonable to set the chemical potential of a pure liquid or solid substance equal to its standard state chemical potential at that temperature independent of the pressure in considering chemical equilibrium?

Q6.11 The reaction A + B $\rightleftharpoons$ C + D is at equilibrium for $\xi = 0.5$. What does this tell you about the variation of G_{pure} with ξ?

Q6.12 Which thermodynamic state function gives a measure of the maximum electric work that can be carried out in a fuel cell?

Questions 6.13–6.18 refer to the reaction system
CO(g) + 1/2 O$_2$(g) $\rightleftharpoons$ CO$_2$(g) at equilibrium for which
$\Delta H_R^\circ = -283.0$ **kJ mol^{-1}.**

Q6.13 Predict the change in the partial pressure of CO_2 as the temperature is increased at constant total pressure.

Q6.14 Predict the change in the partial pressure of CO_2 as the pressure is increased at constant temperature.

Q6.15 Predict the change in the partial pressure of CO_2 as Xe gas is introduced into the reaction vessel at constant pressure and temperature.

Q6.16 Predict the change in the partial pressure of CO_2 as Xe gas is introduced into the reaction vessel at constant volume and temperature.

Q6.17 Predict the change in the partial pressure of CO_2 as a platinum catalyst is introduced into the reaction vessel at constant volume and temperature.

Q6.18 Predict the change in the partial pressure of CO_2 as O_2 is removed from the reaction vessel at constant pressure and temperature.

Q6.19 Calculate the maximum expansion work that is available in carrying out the combustion reactions in

Example Problems 6.1 and 6.2. Explain both the magnitude and the sign of the work.

Q6.20 Is the equation $\Delta G = \Delta H - T\Delta S$ applicable to all processes?

Q6.21 Is the equation $\Delta A = \Delta U - T\Delta S$ applicable to all processes?

Q6.22 Under what conditions is $K_x > K_P$?

Q6.23 Under what conditions is the distribution of products in an ideal gas reactions system at equilibrium unaffected by an increase in the pressure?

Q6.24 If K_P is independent of pressure, why does the degree of dissociation in the reaction $Cl_2(g) \rightleftharpoons 2Cl(g)$ depend on presure?

Questions 6.25–6.28 refer to the reaction system $H_2(g) + Cl_2(g) \rightleftharpoons 2HCl(g)$ **at equilibrium. Assume ideal gas behavior.**

Q6.25 How does the total number of moles in the reaction system change as T increases?

Q6.26 If the reaction is carried out at constant V, how does the total pressure change if T increases?

Q6.27 Is the partial pressure of $H_2(g)$ dependent on T? If so, how will it change as T decreases?

Q6.28 If the total pressure is increased at constant T, how will the relative amounts of of $H_2(g)$ and $HCl(g)$ change?

Q6.29 If additional $Cl_2(g)$ is added to the reaction system at constant total pressure and temperature, how will the partial pressures of $H_2(g)$ and $HCl(g)$ change?

Q6.30 If additional $Cl_2(g)$ is added to the reaction system at constant V, how will the degree of dissociation of $HCl(g)$ change?

Q6.31 If T is increased at constant total pressure, how will the degree of dissociation of $HCl(g)$ change?

Numerical Problems

Problem numbers in red indicate that the solution to the problem is given in the *Student's Solutions Manual*.

P6.1 Calculate ΔA_R° and ΔG_R° for the reaction $C_6H_6(l) + 15/2\, O_2(g) \rightarrow 6CO_2(g) + 3H_2O(l)$ at 298 K from the combustion enthalpy of benzene and the entropies of the reactants and products.

P6.2 Calculate K_P at 298 and 490. K for the reaction $NO(g) + 1/2\, O_2(g) \rightarrow NO_2(g)$ assuming that ΔH_R° is constant over the interval 298–600. K. Do you expect K_P to increase or decrease as the temperature is increased to 600. K?

P6.3 A sample containing 2.75 moles of N_2 and 6.25 mol of H_2 are placed in a reaction vessel and brought to equilibrium at 52.0 bar and 690. K in the reaction $1/2\, N_2(g) + 3/2\, H_2(g) \rightleftharpoons NH_3(g)$.

a. Calculate K_P at this temperature.

b. Set up an equation relating K_P and the extent of reaction as in Example Problem 6.10.

c. Using numerical equation solving software, calculate the number of moles of each species present at equilibrium.

P6.4 Consider the equilibrium $NO_2(g) \rightleftharpoons NO(g) + 1/2\, O_2(g)$. One mole of $NO_2(g)$ is placed in a vessel and allowed to come to equilibrium at a total pressure of 1 bar. An analysis of the contents of the vessel gives the following results:

T	700. K	800. K
P_{NO}/P_{NO_2}	0.872	2.50

a. Calculate K_P at 700. and 800. K.

b. Calculate ΔG_R° and ΔH_R° for this reaction at 298.15 K, using only the data in the problem. Assume that ΔH_R° is independent of temperature.

c. Calculate ΔG_R° and ΔH_R° using the data tables and compare your answer with that obtained in part (b).

P6.5 The shells of marine organisms contain calcium carbonate, $CaCO_3$, largely in a crystalline form known as calcite.

There is a second crystalline form of calcium carbonate known as aragonite. Physical and thermodynamic properties of calcite and aragonite are given in the following table.

Properties ($T = 298$ K, P = 1 bar)	Calcite	Aragonite
ΔH_f° (kJ mol^{-1})	−1206.9	−1207.0
ΔG_f° (kJ mol^{-1})	−1128.8	−1127.7
S° (J K^{-1} mol^{-1})	92.9	88.7
$C_{P,m}^\circ$ (J K^{-1} mol^{-1})	81.9	81.3
Density (g mL^{-1})	2.710	2.930

a. Based on the thermodynamic data given, would you expect an isolated sample of calcite at $T = 298$ K and $P = 1$ bar to convert to aragonite, given sufficient time? Explain.

b. Suppose the pressure applied to an isolated sample of calcite is increased. Can the pressure be increased to the point that isolated calcite will be converted to aragonite? Explain.

c. What pressure must be achieved to induce the conversion of calcite to aragonite at $T = 298$ K. Assume both calcite and aragonite are incompressible at $T = 298$ K.

d. Can calcite be converted to aragonite at $P = 1$ bar if the temperature is increased? Explain.

P6.6 Consider the equilibrium $C_2H_6(g) \rightleftharpoons C_2H_4(g) + H_2(g)$. At 1000. K and a constant total pressure of 1.00 bar, $C_2H_6(g)$ is introduced into a reaction vessel. The total pressure is held constant at 1 bar and at equilibrium the composition of the mixture in mole percent is $H_2(g)$: 26.0%, $C_2H_4(g)$: 26.0% and $C_2H_6(g)$: 48.0%.

a. Calculate K_P at 1000. K.

b. If $\Delta H_R^\circ = 137.0$ kJ mol^{-1}, calculate the value of K_P at 298.15K.

c. Calculate ΔG_R° for this reaction at 298.15 K.

P6.7 The pressure dependence of G is quite different for gases and condensed phases. Calculate ΔG_m for the processes (C, *solid, graphite*, 1 bar, 298.15 K) → (C, *solid, graphite*, 325 bar, 298.15 K) and (He, g, 1 bar, 298.15 K) → (He, g, 325 bar, 298.15 K). By what factor is ΔG_m greater for He than for graphite?

P6.8 Many biological macromolecules undergo a transition called denaturation. Denaturation is a process whereby a structured, biological active molecule, called the native form, unfolds or becomes unstructured and biologically inactive. The equilibrium is

$$\text{native (folded)} \rightleftharpoons \text{denatured (unfolded)}$$

For a protein at pH = 2 the enthalpy change at 298 K associated with denaturation is $\Delta H° = 418.0$ kJ mol^{-1} and the entropy change at 298 K is $\Delta S° = 1.30$ kJ K^{-1} mol^{-1}.

a. Calculate the Gibbs energy change for the denaturation of the protein at pH = 2 and $T = 310.$ K. Assume the enthalpy and entropy are temperature-independent between 298 K and 303 K.

b. Calculate the equilibrium constant for the denaturation of the protein at pH 2 and $T = 310.$ K.

c. Based on your answer for parts (a) and (b), is the protein structurally stable at pH 2 and $T = 310.$ K?

P6.9 Assume that a sealed vessel at constant pressure of 1 bar initially contains 2.00 mol of $NO_2(g)$. The system is allowed to equilibrate with respect to the reaction $2 NO_2(g) \rightleftharpoons N_2O_4(g)$. The number of moles of $NO_2(g)$ and $N_2O_4(g)$ at equilibrium is $2.00 - 2\xi$ and ξ, respectively, where ξ is the extent of reaction.

a. Derive an expression for the entropy of mixing as a function of ξ.

b. Graphically determine the value of ξ for which ΔS_{mixing} has its maximum value.

c. Write an expression for G_{pure} as a function of ξ. Use Equation 6.104 to obtain values of $G_m°$ for NO_2 and N_2O_4.

d. Plot $G_{mixture} = G_{pure} + \Delta G_{mixing}$ as a function of ξ for $T = 298$ K and graphically determine the value of ξ for which $G_{mixture}$ has its minimum value. Is this value the same as for part (b)?

P6.10 Calculate K_P at 600. K for the reaction $N_2O_4(l) \rightleftharpoons 2NO_2(g)$ assuming that $\Delta H_R°$ is constant over the interval 298–725 K.

P6.11 Consider the equilibrium $CO(g) + H_2O(g) \rightleftharpoons CO_2(g) + H_2(g)$. At 1150. K, the composition of the reaction mixture is

Substance	$CO_2(g)$	$H_2(g)$	$CO(g)$	$H_2O(g)$
Mole %	20.3	20.3	29.7	29.7

a. Calculate K_P and $\Delta G_R°$ at 1150. K.

b. Given the answer to part (a), use the $\Delta H_f°$ of the reaction species to calculate $\Delta G_R°$ at 298.15 K. Assume that $\Delta H_R°$ is independent of temperature.

P6.12 For the reaction C(*graphite*) + $H_2O(g) \rightleftharpoons CO(g) + H_2(g)$, $\Delta H_R° = 131.28$ kJ mol^{-1} at 298.15 K. Use the values of $C_{P,m}°$ at 298.15 K in the data tables to calculate $\Delta H_R°$ at 125.0°C.

P6.13 $Ca(HCO_3)_2(s)$ decomposes at elevated temperatures according to the stoichiometric equation $Ca(HCO_3)_2(s) \rightleftharpoons CaCO_3(s) + H_2O(g) + CO_2(g)$.

a. If pure $Ca(HCO_3)_2(s)$ is put into a sealed vessel, the air is pumped out, and the vessel and its contents are heated, the total pressure is 0.290 bar. Determine K_P under these conditions.

b. If the vessel also contains 0.120 bar $H_2O(g)$ at the final temperature, what is the partial pressure of $CO_2(g)$ at equilibrium?

P6.14 Calculate ΔA for the isothermal compression of 2.95 mol of an ideal gas at 325 K from an initial volume of 60.0 L to a final volume of 20.5 L. Does it matter whether the path is reversible or irreversible?

P6.15 Nitrogen is a vital element for all living systems, but except for a few types of bacteria, blue-green algae, and some soil fungi, most organisms cannot utilize N_2 from the atmosphere. The formation of "fixed" nitrogen is therefore necessary to sustain life and the simplest form of fixed nitrogen is ammonia NH_3.

A possible pathway for ammonia synthesis by a living system is

$$\tfrac{1}{2} N_2(g) + \tfrac{3}{2} H_2O(l) \rightleftharpoons NH_3(aq) + \tfrac{3}{4} O_2(g)$$

where (aq) means the ammonia is dissolved in water and $\Delta G_f°(NH_3, aq) = -80.3$ kJ mol^{-1}.

a. Calculate $\Delta G°$ for the biological synthesis of ammonia at 298 K.

b. Calculate the equilibrium constant for the biological synthesis of ammonia at 298 K.

c. Based on your answer to part (b), is the pathway a spontaneous reaction?

P6.16 Collagen is the most abundant protein in the mammalian body. It is a fibrous protein that serves to strengthen and support tissues. Suppose a collagen fiber can be stretched reversibly with a force constant of $k = 10.0$ N m^{-1} and that the force **F** (see Table 2.1) is given by $\mathbf{F} = k \cdot l$. When a collagen fiber is contracted reversibly, it absorbs heat $q_{rev} = 0.050$ J. Calculate the change in the Helmholtz energy ΔA as the fiber contracts isothermally from $l = 0.20$ to 0.10 m. Calculate also the reversible work performed w_{rev}, ΔS, and ΔU. Assume that the temperature is constant at $T = 310.$ K.

P6.17 Calculate $\mu_{O_2}^{mixture}$ (298.15 K, 1 bar) for oxygen in air, assuming that the mole fraction of O_2 in air is 0.210. Use the conventional molar Gibbs energy defined in Section 6.17.

P6.18 Calculate the maximum nonexpansion work that can be gained from the combustion of benzene(l) and of $H_2(g)$ on a per gram and a per mole basis under standard conditions. Is it apparent from this calculation why fuel cells based on H_2 oxidation are under development for mobile applications?

P6.19 You wish to design an effusion source for Br atoms from $Br_2(g)$. If the source is to operate at a total pressure of

7.5 Torr, what temperature is required to produce a degree of dissociation of 0.20? What value of the pressure would increase the degree of dissociation to 0.65 at this temperature?

P6.20 Calculate ΔG for the isothermal expansion of 2.25 mol of an ideal gas at 325 K from an initial pressure of 12.0 bar to a final pressure of 2.5 bar.

P6.21 You place 3.00 mol of $NOCl(g)$ in a reaction vessel. Equilibrium is established with respect to the decomposition reaction $NOCl(g) \rightleftharpoons NO(g) + 1/2\ Cl_2(g)$.

a. Derive an expression for K_P in terms of the extent of reaction ξ.

b. Simplify your expression for part (a) in the limit that ξ is very small.

c. Calculate ξ and the degree of dissociation of NOCl in the limit that ξ is very small at 375 K and a pressure of 2.00 bar.

d. Solve the expression derived in part (a) using a numerical equation solver for the conditions stated in the previous part. What is the relative error in ξ made using the approximation of part (b)?

P6.22 A sample containing 2.50 moles of He (1 bar, 350. K) is mixed with 1.75 mol of Ne (1 bar, 350. K) and 1.50 mol of Ar (1 bar, 350. K). Calculate ΔG_{mixing} and ΔS_{mixing}.

P6.23 A hard-working horse can lift a 350 lb weight 100 ft in one minute. Assuming the horse generates energy to accomplish this work by metabolizing glucose:

$$C_6H_{12}O_6(s) + 6O_2(g) \rightarrow 6CO_2(g) + 6H_2O(l)$$

Calculate how much glucose a horse must metabolize to sustain this rate of work for one hour at 298 K.

P6.24 Consider the reaction $FeO(s) + CO(g) \rightleftharpoons Fe(s) + CO_2(g)$ for which K_P is found to have the following values:

T	700.°C	1200°C
K_P	0.688	0.310

a. Using this data, calculate ΔG_R°, ΔS_R°, and ΔH_R° for this reaction at 700.°C. Assume that ΔH_R° is independent of temperature.

b. Calculate the mole fraction of $CO_2(g)$ present in the gas phase at 700.°C.

P6.25 Derive an expression for $A(V,T)$ analagous to that for $G(T,P)$ in Equation (6.33).

P6.26 Show that

$$\left[\frac{\partial(A/T)}{\partial(1/T)}\right]_V = U$$

Write an expression analogous to Equation (6.36) that would allow you to relate ΔA at two temperatures.

P6.27 A gas mixture with 4.50 mol of Ar, x moles of Ne, and y moles of Xe is prepared at a pressure of 1 bar and a temperature of 298 K. The total number of moles in the mixture is five times that of Ar. Write an expression for ΔG_{mixing} in terms of x. At what value of x does the magnitude of ΔG_{mixing} have its minimum value? Answer this part graphically or by using an equation solver. Calculate ΔG_{mixing} for this value of x.

P6.28 In Example Problem 6.9, K_P for the reaction $CO(g) + H_2O(l) \rightleftharpoons CO_2(g) + H_2(g)$ was calculated to be 3.32×10^3 at 298.15 K. At what temperature is $K_P = 5.50 \times 10^3$? What is the highest value that K_P can have by changing the temperature? Assume that ΔH_R° is independent of temperature.

P6.29 Assuming that ΔH_f° is constant in the interval 275 K – 600. K, calculate ΔG° for the process $(H_2O, g, 298 K) \rightarrow (H_2O, g, 600. K)$. Calculate the relative change in the Gibbs energy.

P6.30 Calculate the degree of dissociation of N_2O_4 in the reaction $N_2O_4(g) \rightleftharpoons 2NO_2(g)$ at 300. K and a total pressure of 1.50 bar. Do you expect the degree of dissociation to increase or decrease as the temperature is increased to 550. K? Assume that ΔH_R° is independent of temperature.

P6.31 Oxygen reacts with solid glycylglycine $C_4H_8N_2O_3$ to form urea CH_4N_2O, carbon dioxide, and water:

$$3O_2(g) + C_4H_8N_2O_3(s) \rightleftharpoons$$
$$CH_4N_2O(s) + 3CO_2(g) + 2H_2O(l)$$

At $T = 298$ K and 1.00 atm solid glycylglycine has the following thermodynamic properties:

$$\Delta G_f^\circ = -491.5\ \text{kJ mol}^{-1}, \Delta H_f^\circ = -746.0\ \text{kJ mol}^{-1},$$
$$S^\circ = 190.0\ \text{J K}^{-1}\ \text{mol}^{-1}$$

Calculate ΔG_R° at $T = 298.15$ K and at $T = 310.0$ K. State any assumptions that you make.

P6.32 Calculate ΔG_R° for the reaction $CO(g) + 1/2\ O_2(g) \rightarrow CO_2(g)$ at 298.15 K. Calculate ΔG_R° at 600. K assuming that ΔH_R° is constant in the temperature interval of interest.

P6.33 A sample containing 2.50 mol of an ideal gas at 325 K is expanded from an initial volume of 10.5 L to a final volume of 60.0 L. Calculate the final pressure. Calculate ΔG and ΔA for this process for (a) an isothermal reversible path and (b) an isothermal expansion against a constant external pressure equal to the final pressure. Explain why ΔG and ΔA do or do not differ from one another.

P6.34 You have containers of pure O_2 and N_2 at 298 K and 1 atm pressure. Calculate ΔG_{mixing} relative to the unmixed gases of

a. a mixture of 10. mol of O_2 and 10. mol of N_2

b. a mixture of 10. mol of O_2 and 20. mol of N_2

c. Calculate ΔG_{mixing} if 10. mol of pure N_2 is added to the mixture of 10. mol of O_2 and 10. mol of N_2.

P6.35 In this problem, you calculate the error in assuming that ΔH_R° is independent of T for the reaction $2CuO(s) \rightleftharpoons 2Cu(s) + O_2(g)$.

The following data are given at 25° C:

Compound	CuO(s)	Cu(s)	$O_2(g)$
ΔH_f° (kJ mol^{-1})	–157		
ΔG_f° (kJ mol^{-1})	–130.		
$C_{P,m}$ (J K^{-1} mol^{-1})	42.3	24.4	29.4

a. From Equation (6.65),

$$\int_{K_P(T_0)}^{K_P(T_f)} d\ln K_P = \frac{1}{R}\int_{T_0}^{T_f}\frac{\Delta H_R^\circ}{T^2}dT$$

To a good approximation, we can assume that the heat capacities are independent of temperature over a limited range in temperature, giving $\Delta H_R^\circ(T) = \Delta H_R^\circ(T_0) + \Delta C_P(T - T_0)$ where $\Delta C_P = \Sigma_i \nu_i C_{P,m}(i)$. By integrating Equation (6.65), show that

$$\ln K_P(T) = \ln K_P(T_0) - \frac{\Delta H_R^\circ(T_0)}{R}\left(\frac{1}{T} - \frac{1}{T_0}\right)$$
$$+ \frac{T_0 \times \Delta C_P}{R}\left(\frac{1}{T} - \frac{1}{T_0}\right)$$
$$+ \frac{\Delta C_P}{R}\ln\frac{T}{T_0}$$

b. Using the result from part (a), calculate the equilibrium pressure of oxygen over copper and $CuO(s)$ at 1275 K. How is this value related to K_P for the reaction $2CuO(s) \rightleftharpoons 2Cu(s) + O_2(g)$?

c. What value of the equilibrium pressure would you obtain if you assumed that ΔH_R° were constant at its value for 298.15 K up to 1275 K?

P6.36   Consider the equilibrium in the reaction $3O_2(g) \rightleftharpoons 2O_3(g)$. Assume that ΔH_R° is independent of temperature.

a. Without doing a calculation, predict whether the equilibrium position will shift toward reactants or products as the pressure is increased.

b. Using only the data tables, predict whether the equilibrium position will shift toward reactants or products as the temperature is increased.

c. Calculate K_P at 600. and 700. K. Compare your results with your answer to part (b).

d. Calculate K_x at 600. K and pressures of 1.00 and 2.25 bar. Compare your results with your answer to part (a).

P6.37 N_2O_3 dissociates according to the equilibrium $N_2O_3(g) \rightleftharpoons NO_2(g) + NO(g)$. At 298 K and one bar pressure, the degree of dissociation defined as the ratio of moles of $NO_2(g)$ or $NO(g)$ to the moles of the reactant assuming no dissociation occurs is 3.5×10^{-3}. Calculate ΔG_R° for this reaction.

P6.38   If the reaction $Fe_2N(s) + 3/2\,H_2(g) \rightleftharpoons 2Fe(s) + NH_3(g)$ comes to equilibrium at a total pressure of 1 bar, analysis of the gas shows that at 700. and 800. K, $P_{NH_3}/P_{H_2} = 2.165$ and 1.083, respectively, if only $H_2(g)$ was initially present in the gas phase and $Fe_2N(s)$ was in excess.

a. Calculate K_P at 700. and 800. K.

b. Calculate ΔS_R° at 700. K and 800. K and ΔH_R° assuming that it is independent of temperature.

c. Calculate ΔG_R° for this reaction at 298.15 K.

P6.39   Assume the internal energy of an elastic fiber under tension (see Problem P6.16) is given by $dU = T\,dS - P\,dV - F\,d\ell$. Obtain an expression for $(\partial G/\partial\ell)_{P,T}$ and calculate the maximum nonexpansion work obtainable when a collagen fiber contracts from $\ell = 20.0$ to 10.0 cm at constant P and T. Assume other properties as described in Problem P6.16.

P6.40   Under anaerobic conditions, glucose is broken down in muscle tissue to form lactic acid according to the reaction: $C_6H_{12}O_6(s) \rightleftharpoons 2CH_3CHOHCOOH(aq)$. Thermodynamic data at $T = 298$ K for glucose and lactic acid are given in the following table.

	ΔH_f° (kJ mol^{-1})	C_{Pm} (J K^{-1} mol^{-1})	S_m° (J K^{-1} mol^{-1})
Glucose	−1273.1	219.2	209.2
Lactic Acid	−673.6	127.6	192.1

Calculate ΔG_R° at $T = 298$ K. and $T = 310$. K. In your calculation at 310. K, assume (a) that ΔH_R° and ΔS_R° are constant in this temperature interval and (b) calculate ΔH_R° and ΔS_R° at 310. K using the data in the previous table. Assume all heat capacities are constant in this temperature interval.

P6.41   Consider the equilibrium $3O_2(g) \rightleftharpoons 2O_3(g)$.

a. Using the data tables, calculate K_P at 298 K.

b. Assuming that the extent of reaction at equilibrium is much less than one, show that the degree of reaction defined as half the number of moles of $O_3(g)$ divided by the initial number moles of $O_2(g)$ present before dissociation is given by $\xi_{eq} = (1/2)\sqrt{K_P \times P/P^\circ}$.

c. Calculate the degree of reaction at 298 K and a pressure of 5.00 bar.

d. Calculate K_x at 298 K and a pressure of 5.00 bar.

P6.42   Use the equation $C_{P,m} - C_{V,m} = TV_m\beta^2/\kappa$ and the data tables to determine $C_{V,m}$ for $H_2O(l)$ at 298 K. Calculate $(C_{P,m} - C_{V,m})/C_{P,m}$.

P6.43   As shown in Example Problem 3.5, $(\partial U_m/\partial V)_T = a/V_m^2$ for a van der Waals gas. In this problem, you will compare the change in energy with temperature and volume for N_2, treating it as a van der Waals gas.

a. Calculate ΔU per mole of $N_2(g)$ at 1 bar pressure and 298 K if the volume is increased by 1.00% at constant T. Approximate the molar volume as the ideal gas value.

b. Calculate ΔU per mole of $N_2(g)$ at 1 bar pressure and 298 K if the temperature is increased by 1.00% at constant V.

c. Calculate the ratio of your results in part (a) to the result in part (b). What can you conclude about the relative importance of changes in temperature and volume on ΔU?

Web-Based Simulations, Animations, and Problems

W6.1 The equilibrium $A \rightleftharpoons B$ is simulated. The variables $\Delta G_{reactants}^\circ$ and $\Delta H_{products}^\circ$ and T can be varied independently using sliders. The position of the equilibrium as well as the time required to reach equilibrium is investigated as a function of these variables.

The Properties of Real Gases

The ideal gas law is only accurate for gases at low values of the density. To design production plants that use real gases at high pressures, equations of state valid for gases at higher densities are needed. Such equations must take the finite volume of a molecule and the intermolecular potential into account. They accurately describe the *P–V* relationship of a given gas at a fixed value of *T* within their range of validity using parameters that are specific to a given gas. An important consequence of nonideality is that the chemical potential of a real gas must be expressed in terms of its fugacity rather than its partial pressure. Fugacities rather than pressures must also be used in calculating the thermodynamic equilibrium constant K_P for a real gas.

7.1 Real Gases and Ideal Gases

To this point, the ideal gas equation of state has been assumed to be sufficiently accurate to describe the *P–V–T* relationship for a real gas. This assumption has allowed calculations of expansion work and of the equilibrium constant K_P in terms of partial pressures using the ideal gas law. In fact, the ideal gas law provides an accurate description of the *P–V–T* relationship for many gases, such as He, for a wide range of *P*, *V*, and *T* values. However, it describes the *P–V* relationship for water for a wide range of *P* and *V* values within ±10% only for *T* > 1300 K, as shown in Section 7.4. What is the explanation for this different behavior of He and H_2O? Is it possible to derive a "universal" equation of state that can be used to describe the *P–V* relationship for gases as different as He and H_2O?

In Section 1.5, the two main deficiencies in the microscopic model on which the ideal gas law is based were discussed. The first assumption is that gas molecules are point masses. However, molecules occupy a finite volume; therefore, a real gas cannot be compressed to a volume that is less than the total molecular volume. The second assumption is that the molecules in the gas do not interact, but molecules in a real gas do interact with one another through a potential as depicted in Figure 1.10. Because the potential has a short range, its effect is negligible at low densities, which correspond to large distances between molecules. Additionally, at low densities, the molecular volume is negligible compared with the volume that the gas occupies. Therefore, the *P–V–T* relationship of a real gas is the same as that for an ideal gas at sufficiently low densities and high temperatures. At higher densities and low temperatures, molecular interactions cannot be neglected. Because of these interactions, the pressure of a real gas can be higher or lower than that for an ideal gas at the same density and temperature. What determines which of these two cases applies? The questions raised in this section are the major themes of this chapter.

7.2 Equations of State for Real Gases and Their Range of Applicability

In this section, several equations of state for real gases and the range of the variables P, V, and T over which they accurately describe a real gas are discussed. Such equations of state must exhibit a limiting P–V–T behavior identical to that for an ideal gas at low density. They must also correctly model the deviations for ideal gas behavior that real gases exhibit at moderate and high densities. The first two equations of state considered here include two parameters, a and b, that must be experimentally determined for a given gas. The parameter a is a measure of the strength of the attractive part of the intermolecular potential, and b is a measure of the minimum volume that a mole of molecules can occupy. Real gas equations of state are best viewed as empirical equations whose functional form has been chosen to fit experimentally determined P–V–T data.

The most widely used is the **van der Waals equation of state**:

$$P = \frac{RT}{V_m - b} - \frac{a}{V_m^2} = \frac{nRT}{V - nb} - \frac{n^2 a}{V^2} \tag{7.1}$$

A second useful equation of state is the **Redlich-Kwong equation of state**:

$$P = \frac{RT}{V_m - b} - \frac{a}{\sqrt{T}} \frac{1}{V_m(V_m + b)} = \frac{nRT}{V - nb} - \frac{n^2 a}{\sqrt{T}} \frac{1}{V(V + nb)} \tag{7.2}$$

Although the same symbols are used for parameters a and b in both equations of state, they have different values for a given gas.

Figure 7.1 shows that the degree to which the ideal gas, van der Waals, and Redlich-Kwong equations of state correctly predict the P–V behavior of CO_2 depends on P, V, and T. At 426 K, all three equations of state reproduce the correct P–V behavior reasonably well over the range shown, with the ideal gas law having the largest error. By contrast, the three equations of state give significantly different results at 310. K. The ideal gas law gives unacceptably large errors, and the Redlich-Kwong equation of state is more accurate than is the van der Waals equation. We will have more to say about the range over which the ideal gas law is reasonably accurate when discussing the compression factor in Section 7.3.

A third widely used equation of state for real gases is the **Beattie-Bridgeman equation of state.** This equation uses five experimentally determined parameters to fit P–V–T data. Because of its complexity, it will not be discussed further.

$$P = \frac{RT}{V_m^2}\left(1 - \frac{c}{V_m T^3}\right)(V_m + B) - \frac{A}{V_m^2} \quad \text{with}$$

$$A = A_0\left(1 - \frac{a}{V_m}\right) \quad \text{and} \quad B = B_0\left(1 - \frac{b}{V_m}\right) \tag{7.3}$$

A further important equation of state for real gases has a different form than any of the previous equations. The **virial equation of state** is written in the form of a power series in $1/V_m$

$$P = RT\left[\frac{1}{V_m} + \frac{B(T)}{V_m^2} + \ldots\right] \tag{7.4}$$

The power series does not converge at high pressures where V_m becomes small. The $B(T)$, $C(T)$, and so on are called the second, third, and so on virial coefficients. This equation is more firmly grounded in theory than the previously discussed three equations because a series expansion is always valid in its convergence range. In practical use, the series is usually terminated after the second virial coefficient because values for the higher coefficients

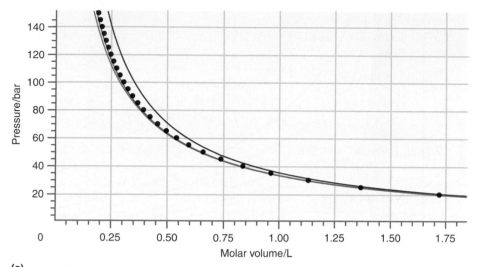

(a)

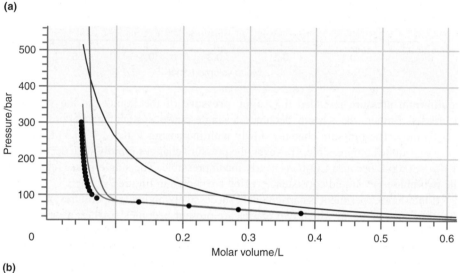

(b)

FIGURE 7.1
Isotherms for CO_2 are shown at (a) 426 K
and (b) 310 K using the van der Waals
equation of state (purple curve), the
Redlich-Kwong equation of state (blue
curve), and the ideal gas equation of state
(red curve). The black dots are accurate
values taken from the *NIST Chemistry
Webbook*.

are not easily obtained from experiments. Table 7.1 (see Appendix B, Data Tables) lists
values for the second virial coefficient for selected gases for different temperatures. If $B(T)$
is negative (positive), the attractive (repulsive) part of the potential dominates at that value
of T. Statistical thermodynamics can be used to relate the virial coefficients with the intermolecular potential function. As you will show in the end-of-chapter problems, $B(T)$ for a
van der Waals gas is given by $B(T) = b - (a/RT)$.

The principal limitation of the ideal gas law is that it does not predict that a gas can
be liquefied under appropriate conditions. Consider the approximate $P–V$ diagram for
CO_2 shown in Figure 7.2. (It is approximate because it is based on the van der Waals
equation of state, rather than experimental data.) Each of the curves is an isotherm corresponding to a fixed temperature. The behavior predicted by the ideal gas law is shown
for $T = 334$ K. Consider the isotherm for $T = 258$ K. Starting at large values of V, the
pressure rises as V decreases and then becomes constant over a range of values of V.
The value of V at which P becomes constant depends on T. As the volume of the system
is decreased further, the pressure suddenly increases rapidly as V decreases.

The reason for this unusual dependence of P on V_m becomes clear when the CO_2
compression experiment is carried out in the piston and transparent cylinder assembly
shown in Figure 7.3. The system consists of either a single phase or two phases separated by a sharp interface, depending on the values of T and V_m. For points a, b, c, and d
on the 258 K isotherm of Figure 7.2, the system has the following composition: at
point a, the system consists entirely of $CO_2(g)$. However, at points b and c, a sharp
interface separates $CO_2(g)$ and $CO_2(l)$. Along the line linking points b and c, the system contains $CO_2(g)$ and $CO_2(l)$ in equilibrium with one another. The proportion of
liquid to gas changes, but the pressure remains constant. The temperature-dependent

FIGURE 7.2
Calculated isotherms are shown for CO_2, modeled as a van der Waals gas. The gas and liquid (blue) regions and the gas–liquid (yellow) coexistence region are shown. The dashed curve was calculated using the ideal gas law. The isotherm at $T = 304.12$ K is at the critical temperature and is called the critical isotherm.

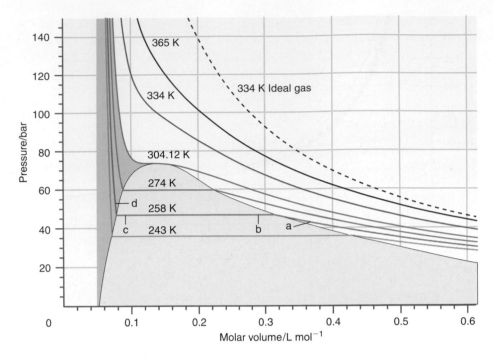

equilibrium pressure is called the **vapor pressure** of the liquid. As the volume is decreased further, the system becomes a single-phase system again, consisting of $CO_2(l)$ only. The pressure changes slowly with increasing V if $V_m > 0.33$ L because $CO_2(g)$ is quite compressible. However, the pressure changes rapidly with decreasing V if $V_m < 0.06$ L because $CO_2(l)$ is nearly incompressible. The single-phase regions and the two-phase gas–liquid coexistence region are shown in Figure 7.2.

If the same experiment is carried out at successively higher temperatures, it is found that the range of V_m in which two phases are present becomes smaller, as seen in the 243, 258, and 274 K isotherms of Figure 7.2. The temperature at which the range of V_m has shrunk to a single value is called the **critical temperature, T_c.** For CO_2, $T_c = 304.12$ K. At $T = T_c$, the isotherm exhibits an inflection point so that

$$\left(\frac{\partial P}{\partial V_m}\right)_{T=T_c} = 0 \quad \text{and} \quad \left(\frac{\partial^2 P}{\partial V_m^2}\right)_{T=T_c} = 0 \tag{7.5}$$

What is the significance of the critical temperature? Critical behavior will be discussed in Chapter 8. At this point, it is sufficient to know that as the critical point is approached, the density of $CO_2(l)$ decreases and the density of $CO_2(g)$ increases, and at $T = T_c$ the densities are equal. Above T_c, no interface is observed in the experiment depicted in Figure 7.3, and liquid and gas phases can no longer be distinguished. The term *supercritical fluid* is used instead. As will be discussed in Chapter 8, T_c and the corresponding values P_c and V_c, which together are called the **critical constants**, take on particular significance in describing the phase diagram of a pure substance. The critical constants for a number of different substances are listed in Table 7.2 (see Appendix B, Data Tables).

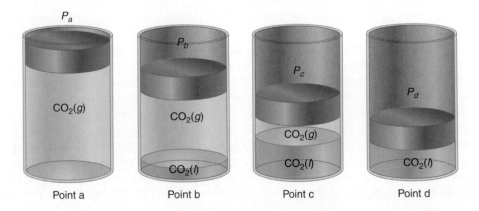

FIGURE 7.3
The volume and the composition of a system containing CO_2 at 258 K is shown at points a, b, c, and d indicated in Figure 7.2. The liquid and gas volumes are not shown to scale.

The parameters a and b for the van der Waals and Redlich-Kwong equations of state are chosen so that the equation of state best represents real gas data. This can be done by using the values of P, V, and T at the critical point T_c, P_c, and V_c as shown in Example Problem 7.1. Parameters a and b calculated from critical constants in this way are listed in Table 7.4 (see Appendix B, Data Tables).

EXAMPLE PROBLEM 7.1

At $T = T_c$, $(\partial P/\partial V_m)_{T=T_c} = 0$ and $(\partial^2 P/\partial V_m^2)_{T=T_c} = 0$. Use this information to determine a and b in the van der Waals equation of state in terms of the experimentally determined values T_c and P_c.

Solution

$$\left(\dfrac{\partial\left[\dfrac{RT}{V_m - b} - \dfrac{a}{V_m^2}\right]}{\partial V_m}\right)_{T=T_c} = -\dfrac{RT_c}{(V_{mc} - b)^2} + \dfrac{2a}{V_{mc}^3} = 0$$

$$\left(\dfrac{\partial^2\left[\dfrac{RT}{V_m - b} - \dfrac{a}{V_m^2}\right]}{\partial V_m^2}\right)_{T=T_c} = \left(\dfrac{\partial\left[-\dfrac{RT}{(V_m - b)^2} + \dfrac{2a}{V_m^3}\right]}{\partial V_m}\right)_{T=T_c}$$

$$= \dfrac{2RT_c}{(V_{mc} - b)^3} - \dfrac{6a}{V_{mc}^4} = 0$$

Equating RT_C from these two equations gives

$$RT_c = \dfrac{2a}{V_{mc}^3}(V_{mc} - b)^2 = \dfrac{3a}{V_{mc}^4}(V_{mc} - b)^3, \text{ which simplifies to}$$

$$\dfrac{3}{2V_{mc}}(V_{mc} - b) = 1$$

The solution to this equation is $V_{mc} = 3b$. Substituting this result into $(\partial P/\partial V_m)_{T=T_c} = 0$ gives

$$-\dfrac{RT_c}{(V_{mc} - b)^2} + \dfrac{2a}{V_{mc}^3} = -\dfrac{RT_c}{(2b)^2} + \dfrac{2a}{(3b)^3} = 0 \quad \text{or} \quad T_c = \dfrac{2a}{(3b)^3}\dfrac{(2b)^2}{R} = \dfrac{8a}{27Rb}$$

Substituting these results for T_c and V_{mc} in terms of a and b into the van der Waals equation gives the result $P_c = a/27b^2$. We only need two of the critical constants T_c, P_c, and V_{mc} to determine a and b. Because the measurements for P_c and T_c are more accurate than for V_{mc}, we use these constants to obtain expressions for a and b. These results for P_c and T_c can be used to express a and b in terms of P_c and T_c. The results are

$$b = \dfrac{RT_c}{8P_c} \quad \text{and} \quad a = \dfrac{27R^2T_c^2}{64P_c}$$

A similar analysis for the Redlich-Kwong equation gives

$$a = \dfrac{R^2T_c^{5/2}}{9P_c(2^{1/3} - 1)} \quad \text{and} \quad b = \dfrac{(2^{1/3} - 1)RT_c}{3P_c}$$

What is the range of validity of the van der Waals and Redlich-Kwong equations? No two-parameter equation of state is able to reproduce the isotherms shown in Figure 7.2 for $T < T_c$ because it cannot reproduce either the range in which P is constant or the discontinuity in $(\partial P/\partial V)_T$ at the ends of this range. This failure is illustrated in Figure 7.4, in which isotherms calculated using the van der Waals equation of state are plotted for some of the values of T as in Figure 7.2. Below T_c, all calculated isotherms have an oscillating region that is unphysical because V increases as P increases. In the **Maxwell construction**, the oscillating region is replaced by the horizontal line for which the areas above and below the line are equal, as indicated in Figure 7.4. The Maxwell construction is used in generating the isotherms shown in Figure 7.2.

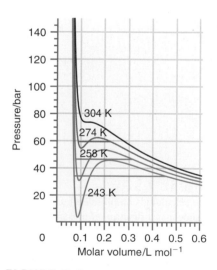

FIGURE 7.4

Van der Waals isotherms show the relationship between pressure and molar volume for CO_2 at the indicated temperatures. The Maxwell construction is shown. No oscillations in the calculated isotherms occur for $T \geq T_c$.

The Maxwell construction can be justified on theoretical grounds, but the equilibrium vapor pressure determined in this way for a given value of T is only in qualitative agreement with experiment. The van der Waals and Redlich-Kwong equations of state do a good job of reproducing P–V isotherms for real gases only in the single-phase gas region $T > T_c$ and for densities well below the critical density, $\rho_c = M/V_{mc}$, where V_{mc} is the molar volume at the critical point. The Beattie-Bridgeman equation, which has three more adjustable parameters, is accurate above T_c for higher densities.

7.3 The Compression Factor

How large is the error in P–V curves if the ideal gas law is used rather than the van der Waals or Redlich-Kwong equations of state? To address this question, it is useful to introduce the **compression factor**, z, defined by

$$z = \frac{V_m}{V_m^{ideal}} = \frac{PV_m}{RT} \tag{7.6}$$

For the ideal gas, $z = 1$ for all values of P and V_m. If $z > 1$, the real gas exerts a greater pressure than the ideal gas, and if $z < 1$, the real gas exerts a smaller pressure than the ideal gas for the same values of T and V_m.

The compression factor for a given gas is a function of temperature, as shown in Figure 7.5. In this figure, z has been calculated for N_2 at two values of T using the van der Waals and Redlich-Kwong equations of state. The dots are calculated from accurate values of V_m taken from the *NIST Chemistry Workbook*. Although the results calculated using these equations of state are not in quantitative agreement with accurate results for all P and T, the trends in the functional dependence of z on P for different T values are correct. Because it is inconvenient to rely on tabulated data for individual gases, we focus on $z(P)$ curves calculated from real gas equations of state. Both the van der Waals and Redlich-Kwong equations predict that for $T = 200$ K, z initially decreases with pressure. The compression factor only becomes greater than the ideal gas value of one for pressures in excess of 200 bar. For $T = 400.$ K, z increases linearly with T. This functional dependence is also predicted by the Redlich-Kwong equation. The van der Waals equation predicts that the initial slope is zero, and z increases slowly with P. For both temperatures, $z \rightarrow 1$ as $P \rightarrow 0$. This result shows that the ideal gas law is obeyed if P is sufficiently small.

To understand why the low pressure value of the compression factor varies with temperature for a given gas, we use the van der Waals equation of state. Consider the variation of the compression factor with P at constant T,

$$\left(\frac{\partial z}{\partial P}\right)_T = \left(\frac{\partial z}{\partial [RT/V_m]}\right)_T = \frac{1}{RT}\left(\frac{\partial z}{\partial [1/V_m]}\right)_T$$

FIGURE 7.5
The calculated compression factor for N_2 is shown as a function of pressure for $T = 200.$ and 400. K and compared with accurate values. The purple and red curves have been calculated using the van der Waals and Redlich-Kwong equations of state, respectively. The Redlich-Kwong equation does not give physically meaningful solutions above $P = 175$ bar for 200. K, where $\rho = 2.5\rho_c$. The dots are calculated from accurate values for V_m taken from the *NIST Chemistry Webbook*.

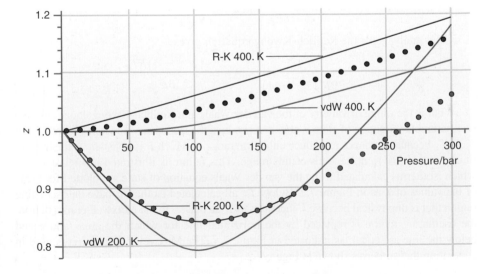

for a van der Waals gas in the ideal gas limit as $1/V_m \rightarrow 0$. As shown in Example Problem 7.2, the result $(\partial z/\partial P)_T = b/RT - a/(RT)^2$ is obtained.

EXAMPLE PROBLEM 7.2

Show that the slope of z as a function of P as $P \rightarrow 0$ is related to the van der Waals parameters by

$$\lim_{p \to 0} \left(\frac{\partial z}{\partial P} \right)_T = \frac{1}{RT} \left(b - \frac{a}{RT} \right)$$

Solution

Rather than differentiating z with respect to P, we transform the partial derivative to one involving V_m:

$$z = \frac{V_m}{V_m^{ideal}} = \frac{PV_m}{RT} = \frac{\left(\dfrac{RT}{V_m - b} - \dfrac{a}{V_m^2} \right) V_m}{RT} = \frac{V_m}{V_m - b} - \frac{a}{RTV_m}$$

$$\left(\frac{\partial z}{\partial P} \right)_T = \left(\frac{\dfrac{\partial z}{\partial V_m}}{\dfrac{\partial P}{\partial V_m}} \right)_T = \left(\frac{\dfrac{\partial}{\partial V_m}\left[\dfrac{V_m}{V_m - b} - \dfrac{a}{RTV_m} \right]}{\dfrac{\partial}{\partial V_m}\left(\dfrac{RT}{V_m - b} - \dfrac{a}{V_m^2} \right)} \right)_T$$

$$= \left(\frac{\dfrac{1}{V_m - b} - \dfrac{V_m}{(V_m - b)^2} + \dfrac{a}{RTV_m^2}}{-\dfrac{RT}{(V_m - b)^2} + 2\dfrac{a}{V_m^3}} \right)$$

$$= \left(\frac{\dfrac{V_m - b}{(V_m - b)^2} - \dfrac{V_m}{(V_m - b)^2} + \dfrac{a}{RTV_m^2}}{-\dfrac{RT}{(V_m - b)^2} + 2\dfrac{a}{V_m^3}} \right)$$

$$= \left(\frac{\dfrac{-b}{(V_m - b)^2} + \dfrac{a}{RTV_m^2}}{-\dfrac{RT}{(V_m - b)^2} + 2\dfrac{a}{V_m^3}} \right) \rightarrow \left(\frac{\dfrac{-b}{V_m^2} + \dfrac{a}{RTV_m^2}}{-\dfrac{RT}{V_m^2}} \right) \rightarrow$$

$$\rightarrow \frac{1}{RT} \left(b - \frac{a}{RT} \right) \text{ as } V_m \rightarrow \infty$$

In the second to last line, $-\dfrac{RT}{(V_m - b)^2} + 2\dfrac{a}{V_m^3} \rightarrow -\dfrac{RT}{V_m^2}$ as $V_m \rightarrow \infty$

because the second term approaches zero more rapidly than the first term and $V_m \gg b$.

From Example Problem 7.2, the van der Waals equation predicts that the initial slope of the z versus P curve is zero if $b = a/RT$. The corresponding temperature is known as the **Boyle temperature** T_B

$$T_B = \frac{a}{Rb} \tag{7.7}$$

Values for the Boyle temperature of several gases are shown in Table 7.3.

Because the parameters a and b are substance dependent, T_B is different for each gas. Using the van der Waals parameters for N_2, $T_B = 425$ K, whereas the experimentally

TABLE 7.3 Boyle Temperatures of Selected Gases

Gas	T_B (K)	Gas	T_B (K)
He	23	O_2	400.
H_2	110.	CH_4	510.
Ne	122	Kr	575
N_2	327	Ethene	735
CO	352	H_2O	1250

Source: Calculated from data in Lide, D. R., ed. *CRC Handbook of Thermophysical and Thermochemical Data.* Boca Raton, FL: CRC Press, 1994.

determined value is 327 K. The agreement is qualitative rather than quantitative. At the Boyle temperature both $z \to 1$, and $(\partial z/\partial P)_T \to 0$ as $P \to 0$, which is the behavior exhibited by an ideal gas. It is only at $T = T_B$ that a real gas exhibits ideal behavior as $P \to 0$ with respect to $\lim_{P \to 0} (\partial z/\partial P)_T$. Above the Boyle temperature, $(\partial z/\partial P)_T > 0$ as $P \to 0$, and below the Boyle temperature, $(\partial z/\partial P)_T < 0$ as $P \to 0$. These inequalities provide a criterion to predict whether z increases or decreases with pressure at low pressures for a given value of T.

Note that the initial slope of a z versus P plot is determined by the relative magnitudes of b and a/RT. Recall that the repulsive interaction is represented through b, and the attractive part of the potential is represented through a. From the previous discussion, $\lim_{P \to 0}(\partial z/\partial P)_T$ is always positive at high temperatures, because $b - (a/RT) \to b > 0$ as $T \to \infty$. This means that molecules primarily feel the repulsive part of the potential for $T \gg T_B$. Conversely, for $T \ll T_B$, $\lim_{P \to 0} (\partial z/\partial P)_T$ will always be negative because the molecules primarily feel the attractive part of the potential. For high enough values of P, $z > 1$ for all gases, showing that the repulsive part of the potential dominates at high gas densities, regardless of the value of T.

Next consider the functional dependence of z on P for different gases at a single temperature. Calculated values for oxygen, hydrogen, and ethene obtained using the van der Waals equation at $T = 400$ K are shown in Figure 7.6 together with accurate results for these gases. It is seen that $\lim_{P \to 0} (\partial z/\partial P)_T$ at 400 K is positive for H_2, negative for ethane, and approximately zero for O_2. These trends are correctly predicted by the van der Waals equation. How can this behavior be explained? It is useful to compare the shape of the curves for H_2 and ethene curves with those in Figure 7.5 for N_2 at different temperatures. This comparison suggests that at 400. K the gas temperature is well above T_B for H_2 so that the repulsive part of the potential dominates. By contrast, it suggests that 400. K is well below T_B for ethene, and the attractive part of the potential dominates. Because the initial slope is nearly zero, the data in Figure 7.6 suggests that 400. K is near T_B for oxygen. As is seen in Table 7.3, $T_B = 735$ K for ethene, 400. K for O_2, and 110. K for H_2. These values are consistent with the previous explanation.

The results shown in Figure 7.6 can be generalized as follows. If $\lim_{P \to 0} (\partial z/\partial P)_T < 0$ for a particular gas, $T < T_B$, and the attractive part of the potential dominates. If $\lim_{P \to 0} (\partial z/\partial P)_T > 0$ for a particular gas, $T > T_B$, and the repulsive part of the potential dominates.

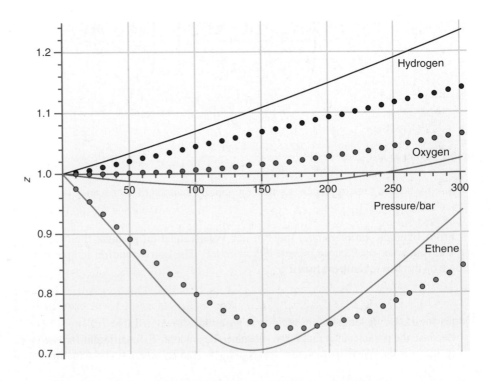

FIGURE 7.6

The compression factor is shown as a function of pressure for $T = 400$. K for three different gases. The solid lines have been calculated using the van der Waals equation of state. The dots are calculated from accurate values for V_m taken from the *NIST Chemistry Webbook*.

7.4 The Law of Corresponding States

As shown in Section 7.3, the compression factor is a convenient way to quantify deviations from the ideal gas law. In calculating z using the van der Waals and Redlich-Kwong equations of state, different parameters must be used for each gas. Is it possible to find an equation of state for real gases that does not explicitly contain material-dependent parameters? Real gases differ from one another primarily in the value of the molecular volume and in the depth of the attractive potential. Because molecules that have a stronger attractive interaction exist as liquids to higher temperatures, one might think that the critical temperature is a measure of the depth of the attractive potential. Similarly, one might think that the critical volume is a measure of the molecular volume. If this is the case, different gases should behave similarly if T, V, and P are measured relative to their critical values.

These considerations suggest the following hypothesis. Different gases have the same equation of state if each gas is described by the dimensionless reduced variables $T_r = T/T_c$, $P_r = P/P_c$, and $V_{mr} = V_m/V_{mc}$, rather than by T, P, and V_m. The preceding statement is known as the **law of corresponding states.** If two gases have the same values of T_r, P_r, and V_{mr}, they are in corresponding states. The values of P, V, and T can be very different for two gases that are in corresponding states. For example, H_2 at 12.93 bar and 32.98 K in a volume of 64.2×10^{-3} L and Br_2 at 103 bar and 588 K in a volume of 127×10^{-3} L are in the same corresponding state.

Next we justify the law of corresponding states and show that it is obeyed by many gases. The parameters a and b can be eliminated from the van der Waals equation of state by expressing the equation in terms of the reduced variables T_r, P_r, and V_{mr}. This can be seen by writing the van der Waals equation in the form

$$P_r P_c = \frac{RT_r T_c}{V_{mr} V_{mc} - b} - \frac{a}{V_{mr}^2 V_{mc}^2} \tag{7.8}$$

Next replace T_c, V_{mc}, and P_c by the relations derived in Example Problem 7.1:

$$P_c = \frac{a}{27b^2}, \quad V_{mc} = 3b, \quad \text{and} \quad T_c = \frac{8a}{27Rb} \tag{7.9}$$

Equation (7.8) becomes

$$\frac{aP_r}{27b^2} = \frac{8aT_r}{27b\,(3bV_{mr} - b)} - \frac{a}{9b^2 V_{mr}^2} \quad \text{or}$$

$$P_r = \frac{8T_r}{3V_{mr} - 1} - \frac{3}{V_{mr}^2} \tag{7.10}$$

Equation (7.10) relates T_r, P_r, and V_{mr} without reference to the parameters a and b. Therefore, it has the character of a universal equation, like the ideal gas equation of state. To a good approximation, the law of corresponding states is obeyed for a large number of different gases as shown in Figure 7.7, as long as $T_r > 1$. However, Equation (7.10) is not really universal because the material-dependent quantities enter through values of P_c, T_c, and V_c rather than through a and b.

The law of corresponding states implicitly assumes that two parameters are sufficient to describe an intermolecular potential. This assumption is best for molecules that are nearly spherical, because for such molecules the potential is independent of the molecular orientation. It is not nearly as good for dipolar molecules such as HF, for which the potential is orientation dependent.

The results shown in Figure 7.7 demonstrate the validity of the law of corresponding states. How can this law be applied to a specific gas? The goal is to calculate z for specific values of P_r and T_r, and to use these z values to estimate the error in using the ideal gas law. A convenient way to display these results is in the form of a graph. Calculated results for z using the van der Waals equation of state as a function of P_r for different values of T_r are shown in Figure 7.8. For a given gas and specific P and

FIGURE 7.7
Values for the compression factor are shown as a function of the reduced pressure P_r for seven different gases at the six values of reduced temperatures indicated in the figure. The solid curves are drawn to guide the eye.

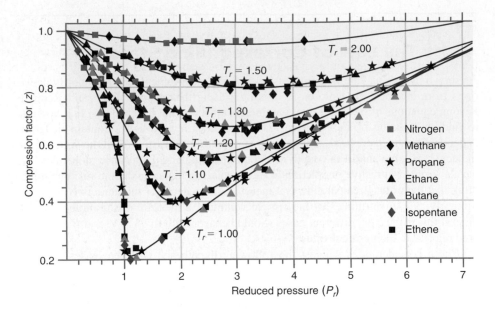

T values, P_r and T_r can be calculated. A value of z can then be read from the curves in Figure 7.8.

From Figure 7.8, we see that for $P_r < 5.5$, $z < 1$ as long as $T_r < 2$. This means that the real gas exerts a smaller pressure than an ideal gas in this range of T_r and P_r. We conclude that for these values of T_r and P_r, the molecules are more influenced by the attractive part of the potential than the repulsive part that arises from the finite molecular volume. However, $z > 1$ for $P_r > 7$ for all values of T_r as well as for all values of P_r if $T_r > 4$. Under these conditions, the real gas exerts a larger pressure than an ideal gas. The molecules are more influenced by the repulsive part of the potential than the attractive part.

Using the compression factor, the error in assuming that the pressure can be calculated using the ideal gas law can be defined by

$$\text{Error} = 100\% \frac{z-1}{z} \tag{7.11}$$

Figure 7.8 shows that the ideal gas law is in error by less than 30% in the range of T_r and P_r where the repulsive part of the potential dominates if $P_r < 8$. However, the error can be as great as –300% in the range of T_r and P_r where the attractive part of the potential dominates. The error is greatest near $T_r = 1$ because at this value of the reduced temperature, the liquid and gaseous phases can coexist. At or slightly above

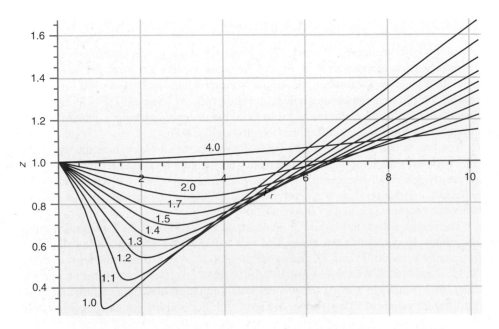

FIGURE 7.8
Compression factor z as a function of P_r for the T_r values indicated. The curves were calculated using the van der Waals equation of state.

the critical temperature, a real gas is much more compressible than an ideal gas. These curves can be used to estimate the temperature range over which the molar volume for $H_2O(g)$ predicted by the ideal gas law is within 10% of the result predicted by the van der Waals equation of state over a wide range of pressure. From Figure 7.8, this is true only if $T_r > 2.0$, or if $T > 1300$ K.

EXAMPLE PROBLEM 7.3

Using Figure 7.8 and the data in Table 7.2 (see Appendix B, Data Tables), calculate the volume occupied by 1.000 kg of CH_4 gas at $T = 230.$ K and $P = 68.0$ bar. Calculate $V - V_{ideal}$ and the relative error in V if V were calculated from the ideal gas equation of state.

Solution

From Table 7.2, T_r and P_r can be calculated:

$$T_r = \frac{230.\text{ K}}{190.56\text{ K}} = 1.21 \text{ and } P_r = \frac{68.0\text{ bar}}{45.99\text{ bar}} = 1.48$$

From Figure 7.8, $z = 0.63$.

$$V = \frac{znRT}{P} = \frac{0.63 \times \dfrac{1000.\text{ g}}{16.04\text{ g mol}^{-1}} \times 0.08314\text{ L bar K}^{-1}\text{mol}^{-1} \times 230.\text{ K}}{68.0\text{ bar}} = 11.0\text{ L}$$

$$V - V_{ideal} = 11.0\text{ L} - \frac{11.0\text{ L}}{0.63} = -6.5\text{ L}$$

$$\frac{V - V_{ideal}}{V} = -\frac{6.5\text{ L}}{11.0\text{ L}} = -58\%$$

Because the critical variables can be expressed in terms of the parameters a and b as shown in Example Problem 7.1, the compression factor at the critical point can also be calculated. For the van der Waals equation of state,

$$z_c = \frac{P_c V_c}{RT_c} = \frac{1}{R} \times \frac{a}{27b^2} \times 3b \times \frac{27Rb}{8a} = \frac{3}{8} \tag{7.12}$$

Equation (7.12) predicts that the critical compressibility is independent of the parameters a and b and should, therefore, have the same value for all gases. A comparison of this prediction with the experimentally determined value of z_c in Table 7.2 shows qualitative but not quantitative agreement. A similar analysis using the critical parameters obtained from the Redlich-Kwong equation also predicts that the critical compression factor should be independent of a and b. In this case, $z_c = 0.333$. This value is in better agreement with the values listed in Table 7.2 than that calculated using the van der Waals equation.

7.5 Fugacity and the Equilibrium Constant for Real Gases

As shown in the previous section, the pressure exerted by a real gas can be greater or less than that for an ideal gas. We next discuss how this result affects the value of the equilibrium constant for a mixture of reactive gases. For a pure ideal gas, the chemical potential as a function of the pressure has the form (see Section 6.3)

$$\mu(T, P) = \mu°(T) + RT \ln \frac{P}{P°} \tag{7.13}$$

To construct an analogous expression for a real gas, we write

$$\mu(T, P) = \mu°(T) + RT \ln \frac{f}{f°} \tag{7.14}$$

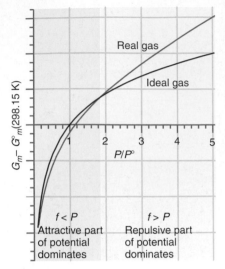

FIGURE 7.9
For densities corresponding to the attractive range of the potential, $P < P_{ideal}$. Therefore, $G_m^{real} < G_m^{ideal}$ and $f < P$. The inequalities are reversed for densities corresponding to the repulsive range of the potential.

where the quantity f is called the **fugacity** of the gas. The fugacity can be viewed as the effective pressure that a real gas exerts. For densities corresponding to the attractive range of the intermolecular potential, $G_m^{real} < G_m^{ideal}$ and $f < P$. For densities corresponding to the repulsive range of the intermolecular potential, $G_m^{real} > G_m^{ideal}$ and $f > P$. These relationships are depicted in Figure 7.9.

The fugacity has the limiting behavior that $f \rightarrow P$ as $P \rightarrow 0$. The standard state of fugacity, denoted f°, is defined as the value that the fugacity would have if the gas behaved ideally at 1 bar pressure. This is equivalent to saying that $f^\circ = P^\circ$. This standard state is a hypothetical standard state, because a real gas will not exhibit ideal behavior at a pressure of one bar. However, the standard state defined in this way makes Equation (7.14) become identical to Equation (7.13) in the ideal gas limit $f \rightarrow P$ as $P \rightarrow 0$.

The preceding discussion provides a method to calculate the fugacity for a given gas. How are the fugacity and the pressure related? For any gas, real or ideal, at constant T,

$$dG_m = V_m dP \tag{7.15}$$

Therefore,

$$d\mu_{ideal} = V_m^{ideal} dP$$
$$d\mu_{real} = V_m^{real} dP \tag{7.16}$$

Equation (7.16) shows that because $V_m^{ideal} \neq V_m^{real}$, the chemical potential of a real gas will change differently with pressure than the chemical potential of an ideal gas. We form the difference

$$d\mu_{real} - d\mu_{ideal} = (V_m^{real} - V_m^{ideal})dP \tag{7.17}$$

and integrate Equation (7.17) from an initial pressure P_i to a final pressure P:

$$\int_{P_i}^{P} (d\mu_{real} - d\mu_{ideal}) = [\mu_{real}(P) - \mu_{real}(P_i)] - [\mu_{ideal}(P) - \mu_{ideal}(P_i)]$$

$$= \int_{P_i}^{P} (V_m^{real} - V_m^{ideal})dP' \tag{7.18}$$

The previous equations allow us to calculate the difference in chemical potential between a real and an ideal gas at pressure P. Now let $P_i \rightarrow 0$. In this limit, $\mu_{real}(P_i) = \mu_{ideal}(P_i)$ because all real gases approach ideal gas behavior at a sufficiently low pressure. Equation (7.18) becomes

$$\mu_{real}(P) - \mu_{ideal}(P) = \int_{0}^{P} (V_m^{real} - V_m^{ideal})dP' \tag{7.19}$$

Equation (7.19) provides a way to calculate the fugacity of a real gas. Using Equations (7.13) and (7.14) for $\mu_{real}(P)$ and $\mu_{ideal}(P)$, P and f are related at the final pressure by

$$\ln f = \ln P + \frac{1}{RT}\int_{0}^{P} (V_m^{real} - V_m^{ideal})dP' \tag{7.20}$$

Because tabulated values of the compression factor z of real gases are widely available, it is useful to rewrite Equation (7.20) in terms of z by substituting $z = V_m^{real}/V_m^{ideal}$. The result is

$$\ln f = \ln P + \int_{0}^{P} \frac{z-1}{P'}dP' \text{ or } f = P\exp\left[\int_{0}^{P}\left(\frac{z-1}{P'}\right)dP'\right] \text{ or } f = \gamma(P,T)P \tag{7.21}$$

Equation (7.21) provides a way to calculate the fugacity if z is known as a function of pressure. It is seen that f and P are related by the proportionality factor γ, which is called the **fugacity coefficient**. However, γ is not a constant; it depends on both P and T.

EXAMPLE PROBLEM 7.4

For $T > T_B$, the equation of state $P(V_m - b) = RT$ is an improvement over the ideal gas law because it takes the finite volume of the molecules into account. Derive an expression for the fugacity coefficient for a gas that obeys this equation of state.

Solution

Because $z = PV_m/RT = 1 + Pb/RT$, the fugacity coefficient is given by

$$\ln \gamma = \ln \frac{f}{P} = \int_0^P \frac{z-1}{P'} dP' = \int_0^P \frac{b}{RT} dP' = \frac{bP}{RT}$$

Equivalently, $\gamma = e^{bP/RT}$. Note that $\gamma > 1$ for all values of P and T because the equation of state does not take the attractive part of the intermolecular potential into account.

Calculated values for $\gamma (P,T)$ for H_2, N_2, and NH_3 are shown in Figure 7.10 for $T = 700.$ K. It can be seen that $\gamma \rightarrow 1$ as $P \rightarrow 0$. This must be the case because the fugacity and pressure are equal in the ideal gas limit. The following general statements can be made about the relationship between γ and z: $\gamma (P,T) > 1$ if the integrand of Equation (7.21) satisfies the condition $(z - 1)/P > 0$ for all pressures up to P. Similarly, $\gamma (P,T) < 1$ if $(z - 1)/P < 0$ for all pressures up to P.

These predictions can be related to the Boyle temperature. If $T > T_B$, then $\gamma (P,T) > 1$ for all pressures, the fugacity is greater than the pressure, and $\gamma > 1$. However, if $T < T_B$, then $\gamma (P,T) < 1$, and the fugacity is smaller than the pressure. The last statement does not hold for very high values of P_r, because as shown in Figure 7.8, $z > 1$ for all values of T at such high relative pressures. Are these conclusions in accord with the curves in Figure 7.10? The Boyle temperatures for H_2, N_2, and NH_3 are 110, 327, and 995 K, respectively. Because at 700. K, $T > T_B$ for H_2 and N_2, but $T < T_B$ for NH_3, we conclude that $\gamma > 1$ at all pressures for H_2 and N_2, and $\gamma < 1$ at all but very high pressures for NH_3. These conclusions are consistent with the results shown in Figure 7.10.

The fugacity coefficient can also be graphed as a function of T_r and P_r. This is convenient because it allows γ for any gas to be estimated once T and P have been expressed as reduced variables. Graphs of γ as functions of P_r and T_r are shown in Figure 7.11. The curves have been calculated using Beattie-Bridgeman parameters for N_2, and are known to be accurate for N_2 over the indicated range of reduced temperature and pressure. Their applicability to other gases assumes that the law of corresponding states holds. As Figure 7.7 shows, this is generally a good assumption.

What are the consequences of the fact that except in the dilute gas limit $f \neq P$? Because the chemical potential of a gas in a reaction mixture is given by Equation (7.14), the thermodynamic equilibrium constant for a real gas K_f must be expressed in terms of the fugacities. Therefore K_f for the reaction $3/2\ H_2 + 1/2\ N_2 \rightarrow NH_3$ is given by

$$K_f = \frac{\left(\dfrac{f_{NH_3}}{f^\circ}\right)}{\left(\dfrac{f_{N_2}}{f^\circ}\right)^{1/2}\left(\dfrac{f_{H_2}}{f^\circ}\right)^{3/2}} = \frac{\left(\dfrac{\gamma_{NH_3} P_{NH_3}}{P^\circ}\right)}{\left(\dfrac{\gamma_{N_2} P_{N_2}}{P^\circ}\right)^{1/2}\left(\dfrac{\gamma_{H_2} P_{H_2}}{P^\circ}\right)^{3/2}}$$

$$= K_P \frac{\gamma_{NH_3}}{(\gamma_{N_2})^{1/2}(\gamma_{H_2})^{3/2}} \qquad (7.22)$$

We next calculate the error in using K_P rather than K_f to calculate an equilibrium constant, using the ammonia synthesis reaction at 700. K and a total pressure of 400. bar as an example. In the industrial synthesis of ammonia, the partial pressures of H_2, N_2, and NH_3 are typically 270, 90, and 40 bar, respectively. The calculated fugacity coefficients under these conditions are $\gamma_{H_2} = 1.11$, $\gamma_{N_2} = 1.04$, and $\gamma_{NH_3} = 0.968$ using the Beattie-Bridgeman equation of state. Therefore,

$$K_f = K_P \frac{(0.968)}{(1.04)^{1/2}(1.11)^{3/2}} = 0.917\ K_P \qquad (7.23)$$

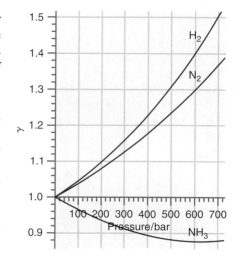

FIGURE 7.10

The fugacity coefficients for H_2, N_2, and NH_3 are plotted as a function of the partial pressure of the gases for $T = 700.$ K. The calculations were carried out using the Beattie-Bridgeman equation of state.

FIGURE 7.11
The fugacity coefficient is plotted as a function of the reduced pressure for the indicated values of the reduced temperature.

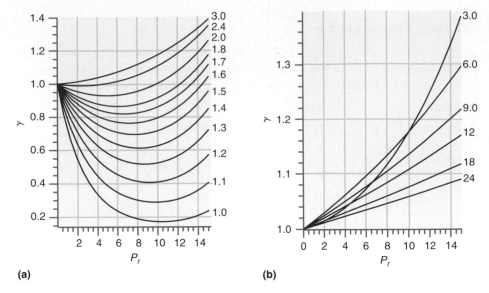

(a)

(b)

Note that K_f is smaller than K_p for the conditions of interest. The NH_3 concentration at equilibrium is proportional to the equilibrium constant, as shown in Section 6.14.

$$x_{NH_3} = (x_{N_2})^{1/2}(x_{H_2})^{3/2}\left(\frac{P}{P^\circ}\right)K_f \qquad (7.24)$$

What conclusions can be drawn from this calculation? If K_P were used rather than K_f, the calculated mole fraction of ammonia would be in error by -9%, which is a significant error in calculating the economic viability of a production plant. However, for pressures near 1 bar, activity coefficients for most gases are very close to unity as can be seen in Figure 7.11. Therefore, under typical laboratory conditions, the fugacity of a gas can be set equal to its partial pressure if P, V, and T are not close to their critical values.

Vocabulary

Beattie-Bridgeman equation of state
Boyle temperature
compression factor
critical constants
critical temperature

fugacity
fugacity coefficient
law of corresponding states
Maxwell construction

Redlich-Kwong equation of state
van der Waals equation of state
vapor pressure
virial equation of state

Conceptual Problems

Q7.1   Using the concept of the intermolecular potential, explain why two gases in corresponding states can be expected to have the same value for z.

Q7.2   Consider the comparison made between accurate results and those based on calculations using the van der Waals and Redlich-Kwong equations of state in Figures 7.1 and 7.5. Is it clear that one of these equations of state is better than the other under all conditions?

Q7.3   Why is the standard state of fugacity, f°, equal to the standard state of pressure, P°?

Q7.4   Explain the significance of the Boyle temperature.

Q7.5   A van der Waals gas undergoes an isothermal reversible expansion under conditions such that $z > 1$. Is the work done more or less than if the gas followed the ideal gas law?

Q7.6   For a given set of conditions, the fugacity of a gas is greater than the pressure. What does this tell you about the interaction between the molecules of the gas?

Q7.7   What can you conclude about the ratio of fugacity to pressure for N_2, H_2, and NH_3 at 500. bar using the data in Figure 7.10?

Q7.8   Is the ratio of fugacity to pressure greater to or less than 1 if the attractive part of the interaction potential between gas molecules dominates?

Q7.9   A gas is slightly above its Boyle temperature. Do you expect z to increase or decrease as P increases?

Q7.10   Explain why the oscillations in the two-phase coexistence region using the Redlich-Kwong and van der Waals equations of state (see Figure 7.4) do not correspond to reality.

Q7.11 A van der Waals gas undergoes an isothermal reversible expansion under conditions such that $z < 1$. Is the work done more or less than if the gas followed the ideal gas law?

Q7.12 The value of the Boyle temperature increases with the strength of the attractive interactions between molecules. Arrange the Boyle temperatures of the gases Ar, CH_4, and C_6H_6 in increasing order.

Q7.13 A system containing argon gas is at pressure P_1 and temperature T_1. How would you go about estimating the fugacity coefficient of the gas?

Q7.14 By looking at the a and b values for the van der Waals equation of state, decide whether 1 mole of O_2 or H_2O has the higher pressure at the same value of T and V.

Q7.15 Will the fugacity coefficient of a gas above the Boyle temperature be less than 1 at low pressures?

Q7.16 Show that the van der Waals and Redlich-Kwong equations of state reduce to the ideal gas law in the limit of low gas density.

Q7.17 Which of Ne or Ar has the larger van der Waals parameter a? Explain your reasoning.

Q7.18 Which of Ne or Ar has the larger van der Waals parameter b? Explain your reasoning.

Q7.19 You have calculated the pressure exerted by ethane using the ideal gas law and the Redlich-Kwong equations of state. How do you decide if the repulsive or attractive part of the molecular potential dominates under the given conditions?

Q7.20 Equation (1.19) states that the total pressure in a mixture of gases is equal to the sum of the partial pressures. Is this equation valid for real gases? If so, under what conditions?

Numerical Problems

Problem numbers in red indicate that the solution to the problem is given in the *Student's Solutions Manual*.

P7.1 A van der Waals gas has a value of $z = 1.00061$ at 410. K and 1 bar and the Boyle temperature of the gas is 195 K. Because the density is low, you can calculate V_m from the ideal gas law. Use this information and the result of Problem P7.28, $z \approx 1 + (b - a/RT)(1/V_m)$, to estimate a and b.

P7.2

a. Using the relationships derived in Example Problem 7.1 and the values of the critical constants for water from Table 7.2, calculate values for the van der Waals parameters a, b, and R from z_c, T_c, P_c, and V_c. Do your results agree with those in Tables 1.2 and 7.4?

b. Calculate the van der Waals parameters a and b using the critical constants for water and the correct value for R. Do these results agree with those in Tables 1.2 and 7.4?

P7.3 Assume that the equation of state for a gas can be written in the form $P(V_m - b(T)) = RT$. Derive an expression for $\beta = 1/V \, (\partial V/\partial T)_P$ and $\kappa = -1/V \, (\partial V/\partial P)_T$ for such a gas in terms of $b(T)$, $db(T)/dT$, P, and V_m.

P7.4 One mole of Ar initially at 310. K undergoes an adiabatic expansion against a pressure $P_{external} = 0$ from a volume of 8.5 L to a volume of 82.0 L. Calculate the final temperature using the ideal gas and van der Waals equations of state. Assume $C_{V,m} = 3R/2$.

P7.5 Calculate the P and T values for which $Br_2(g)$ is in a corresponding state to $Xe(g)$ at 330. K and 72.0 bar.

P7.6 For values of z near 1, it is a good approximation to write $z(P) = 1 + (\partial z/\partial P)_T P$. If $z = 1.00104$ at 298 K and 1 bar, and the Boyle temperature of the gas is 155 K, calculate the values of a, b, and V_m for the van der Waals gas.

P7.7 For a gas at a given temperature, the compression factor is described by the empirical equation

$$z = 1 - 8.50 \times 10^{-3} \frac{P}{P°} + 3.50 \times 10^{-5} \left(\frac{P}{P°}\right)^2$$

where $P° = 1$ bar. Calculate the fugacity coefficient for $P = 150., 250., 350., 450.,$ and 550. bar. For which of these values is the fugacity coefficient greater than 1?

P7.8 The experimentally determined density of O_2 at 140. bar and 298 K is 192 g L^{-1}. Calculate z and V_m from this information. Compare this result with what you would have estimated from Figure 7.8. What is the relative error in using Figure 7.8 for this case?

P7.9 At 725 K and 280. bar, the experimentally determined density of N_2 is 4.13 mol L^{-1}. Compare this with values calculated from the ideal and Redlich-Kwong equations of state. Use a numerical equation solver to solve the Redlich-Kwong equation for V_m or use an iterative approach starting with V_m equal to the ideal gas result. Discuss your results.

P7.10 A 1.75 mole sample of Ar undergoes an isothermal reversible expansion from an initial volume of 2.00 L to a final volume of 85.0 L at 310. K. Calculate the work done in this process using the ideal gas and van der Waals equations of state. What percentage of the work done by the van der Waals gas arises from the attractive potential?

P7.11 Show that the second virial coefficient for a van der Waals gas is given by

$$B(T) = \frac{1}{RT}\left(\frac{\partial z}{\partial \frac{1}{V_m}}\right)_T = b - \frac{a}{RT}$$

P7.12 The volume of a spherical molecule can be estimated as $V = b/(4N_A)$ where b is the van der Waals parameter and N_A is Avogadro's number. Justify this relationship by considering a spherical molecule of radius r, with volume $V = (4/3)\pi r^3$. What is the volume centered at the molecule that is excluded for the center of mass of a second molecule in terms of V? Multiply this volume by N_A and set it equal to b. Apportion this volume equally among the molecules to arrive at $V = b/(4N_A)$. Calculate the radius of a methane molecule from the value of its van der Waals parameter b.

P7.13 Show that the van der Waals and Redlich-Kwong equations of state reduce to the ideal gas equation of state in the limit of low density.

P7.14 Use the law of corresponding states and Figure 7.8 to estimate the molar volume of propane at $T = 500.$ K and $P = 75.0$ bar. The experimentally determined value is 0.438 mol L^{-1}. What is the relative error of your estimate?

P7.15 Another equation of state is the Berthelot equation, $V_m = (RT/P) + b - (a/RT^2)$. Derive expressions for $\beta = 1/V\,(\partial V/\partial T)_P$ and $\kappa = -1/V\,(\partial V/\partial P)_T$ from the Berthelot equation in terms of V, T, and P.

P7.16 Show that $P\kappa = 1 - P\left(\dfrac{\partial \ln z}{\partial P}\right)_T$ for a real gas where κ is the isothermal compressibility.

P7.17 Calculate the van der Waals parameters of carbon dioxide from the values of the critical constants and compare your results with the values for a and b in Table 7.4.

P7.18 Calculate the Redlich-Kwong parameters of fluorine from the values of the critical constants and compare your results with the values for a and b in Table 7.4.

P7.19 Calculate the critical volume for ethane using the data for T_c and P_c in Table 7.2 (see Appendix B, Data Tables) assuming (a) the ideal gas equation of state and (b) the van der Waals equation of state. Use an iterative approach to obtain V_c from the van der Waals equation, starting with the ideal gas result. How well do the calculations agree with the tabulated values for V_c?

P7.20 Show that $T\beta = 1 + T\left(\dfrac{\partial \ln z}{\partial T}\right)_P$ for a real gas where β is the volumetric thermal expansion coefficient.

P7.21 At what temperature does the slope of the z versus P curve as $P \rightarrow 0$ have its maximum value for a van der Waals gas? What is the value of the maximum slope?

P7.22 Calculate the density of $O_2(g)$ at 480. K and 280. bar using the ideal gas and the van der Waals equations of state. Use a numerical equation solver to solve the van der Waals equation for V_m or use an iterative approach starting with V_m equal to the ideal gas result. Based on your result, does the attractive or repulsive contribution to the interaction potential dominate under these conditions? The experimentally determined result is 208 g L^{-1}. What is the relative error of each of your two calculations?

P7.23 Show that $T\beta = 1 + T(\partial \ln z/\partial T)_P$ and that $P\kappa = 1 - P(\partial \ln z/\partial P)_T$.

P7.24 A sample containing 42.1 g of Ar is enclosed in a container of volume 0.0885 L at 375 K. Calculate P using the ideal gas, van der Waals, and Redlich-Kwong equations of state. Based on your results, does the attractive or repulsive contribution to the interaction potential dominate under these conditions?

P7.25 The experimental critical constants of CH_4 are found in Table 7.2. Use the values of P_c and T_c to calculate V_c. Assume that CH_4 behaves as (a) an ideal gas, (b) a van der Waals gas, and (c) a Redlich-Kwong gas at the critical point. For parts (b) and (c), use the formulas for the critical compression factor. Compare your answers with the experimental value. Are any of your calculated results close to the experimental value in Table 7.2?

P7.26 The observed Boyle temperatures of H_2, N_2, and CH_4 are 110., 327, and 510. K, respectively. Compare these values with those calculated for a van der Waals gas with the appropriate parameters.

P7.27 For the Berthelot equation, $V_m = (RT/P) + b - (a/RT^2)$, find an expression for the Boyle temperature in terms of a, b, and R.

P7.28 For a van der Waals gas, $z = V_m/(V_m - b) - a/RTV_m$. Expand the first term of this expression in a Taylor series in the limit $V_m \gg b$ to obtain $z \approx 1 + (b - a/RT)(1/V_m)$.

Web-Based Simulations, Animations, and Problems

W7.1 In this problem, the student gains facility in using the van der Waals equation of state. A set of isotherms will be generated by varying the temperature and initial volume using sliders for a given substance. Buttons allow a choice among more than 20 gases.

a. The student generates a number of P–V curves at and above the critical temperature for the particular gas, and explains trends in the ratio P_{vdW}/P_{ideal} as a function of V for a given T, and as a function of T for a given V.

b. The compression factor z is calculated for two gases at the same value of T_r and is graphed versus P and P_r. The degree to which the law of corresponding states is valid is assessed.

W7.2 A quantitative comparison is made between the ideal gas law and the van der Waals equation of state for 1 of more than 20 different gases. The temperature is varied using sliders, and P_{ideal}, P_{vdW}, the relative error $P_{vdW}/P_{vdW} - P_{ideal}$, and the density of gas relative to that at the critical point are calculated. The student is asked to determine the range of pressures and temperatures in which the ideal gas law gives reasonably accurate results.

W7.3 The compression factor and molar volume are calculated for an ideal and a van der Waals gas as a function of pressure and temperature. These variables can be varied using sliders. Buttons allow a choice among more than 20 gases. The relative error $V_{vdW}/V_{vdW} - V_{ideal}$ and the density of gas relative to that at the critical point are calculated. The student is asked to determine the range of pressures and temperatures in which the ideal gas law gives reasonably accurate results for the molar volume.

W7.4 The fugacity and fugacity coefficient are determined as a function of pressure and temperature for a model gas. These variables can be varied using sliders. The student is asked to determine the Boyle temperature and also the pressure range in which the fugacity is either more or less than the ideal gas pressure for the temperature selected.

Phase Diagrams and the Relative Stability of Solids, Liquids, and Gases

I t is our experience that the solid form of matter is most stable at low temperatures and that most substances can exist in liquid and gaseous phases at higher temperatures. In this chapter, criteria are developed that allow one to determine which of these phases is most stable at a given temperature and pressure. The conditions under which two or three phases of a pure substance can coexist in equilibrium are also discussed. *P–T*, *P–V*, and *P–V–T* phase diagrams summarize all of this information in a form that is very useful to chemists.

8.1 What Determines the Relative Stability of the Solid, Liquid, and Gas Phases?

Substances are found in solid, liquid, and gaseous phases. **Phase** refers to a form of matter that is uniform with respect to chemical composition and the state of aggregation on both microscopic and macroscopic length scales. For example, liquid water in a beaker is a single-phase system, but a mixture of ice and liquid water consists of two distinct phases, each of which is uniform on microscopic and macroscopic length scales. Although a substance may exist in several different solid phases, it can only exist in a single gaseous state. Most substances have a single liquid state, although there are exceptions such as helium, which can be a normal liquid or a superfluid. In this section, the conditions under which a pure substance spontaneously forms a solid, liquid, or gas are discussed.

Experience demonstrates that as *T* is lowered from 300. to 250. K at atmospheric pressure, liquid water is converted to a solid phase. Similarly, as liquid water is heated to 400. K at atmospheric pressure, it vaporizes to form a gas. Experience also shows that if a solid block of carbon dioxide is placed in an open container at 1 bar, it sublimes over time without passing through the liquid phase. Because of this property, solid CO_2 is known as dry ice. These observations can be generalized to state that the solid phase is the most stable state of a substance at sufficiently low temperatures, and that the gas phase is the most stable state of a substance at sufficiently high temperatures. The liquid state is stable at intermediate temperatures if it exists at the pressure of interest. What determines which of the solid, liquid, or gas phases exists and which phase is most stable at a given temperature and pressure?

As discussed in Chapter 6, the criterion for stability at constant temperature and pressure is that the Gibbs energy, $G(T, P, n)$, be minimized. Because for a pure substance the chemical potential μ is defined as

$$\mu = \left(\frac{\partial G}{\partial n}\right)_{T,P} = \left(\frac{\partial[nG_m]}{\partial n}\right)_{T,P} = G_m$$

where n designates the number of moles of substance in the system, $d\mu = dG_m$, and we can express the differential $d\mu$ as

$$d\mu = -S_m\, dT + V_m\, dP \tag{8.1}$$

From this equation, how μ varies with changes in P and T can be determined:

$$\left(\frac{\partial\mu}{\partial T}\right)_P = -S_m \quad\text{and}\quad \left(\frac{\partial\mu}{\partial P}\right)_T = V_m \tag{8.2}$$

Because S_m and V_m are always positive, μ decreases as the temperature increases, and it increases as the pressure increases. Section 5.4 demonstrated that S varies slowly with T (approximately as $\ln T$). Therefore, over a limited range in T, a plot of μ versus T at constant P is a curve with a slowly increasing negative slope, approximating a straight line.

It is also known from experience that heat is absorbed as a solid melts to form a liquid and as a liquid vaporizes to form a gas. Both processes are endothermic with $\Delta H > 0$. Because at the melting and boiling temperature the entropy increases in a stepwise fashion by $\Delta S = \Delta H_{transition}/T$, the entropy of the three phases follows this order:

$$S_m^{gas} > S_m^{liquid} > S_m^{solid} \tag{8.3}$$

The functional relation between μ and T for the solid, liquid, and gas phases is graphed at a given value of P in Figure 8.1. The molar entropy of a phase is the negative of the slope of the μ versus T curve, and the relative entropies of the three phases are given by Equation (8.3). The stable state of the system at any given temperature is that phase that has the lowest μ.

Assume that the initial state of the system is described by the dot in Figure 8.1. The most stable phase is the solid phase, because μ for the liquid and gas phases is much larger than that for the solid. As the temperature is increased, the chemical potential falls as μ remains on the solid curve. However, because the slope of the liquid and gas curves are greater than that of the solid, each of these μ versus T curves will intersect the solid curve for some value of T. In Figure 8.1, the liquid curve intersects the solid curve at T_m, which is called the melting temperature. At this temperature, the solid and liquid phases coexist and are in thermodynamic equilibrium. However, if the temperature is raised by an infinitesimal amount dT, the solid will melt completely because the liquid phase has the lower chemical potential at $T_m + dT$. Similarly, the liquid and gas phases are in thermodynamic equilibrium at the boiling temperature T_b. For $T > T_b$, the system is entirely in the gas phase. Note that the progression of solid $\rightarrow$ liquid $\rightarrow$ gas as T increases at this value of P can be explained with no other information than that $(\partial\mu/\partial T)_P = -S_m$ and that $S_m^{gas} > S_m^{liquid} > S_m^{solid}$.

If the temperature is changed too quickly, the equilibrium state of the system may not be reached. For example, it is possible to form a superheated liquid in which the liquid phase is metastable above T_b. Superheated liquids are dangerous because of the large volume increase that occurs if the system suddenly converts to the stable vapor phase. Boiling chips are often used in chemical laboratories to avoid the formation of superheated liquids. Similarly, it is possible to form a supercooled liquid, in which case the liquid is metastable below T_m. Glasses are made by cooling a viscous liquid fast

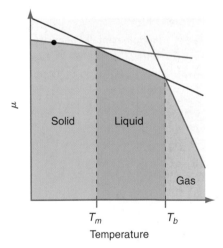

FIGURE 8.1

The chemical potential of a substance in the solid (green curve), liquid (blue curve), and gaseous (brown curve) states is plotted as a function of the temperature for a given value of pressure. The substance melts at the temperature T_m, corresponding to the intersection of the solid and liquid curves. It boils at the temperature T_b, corresponding to the intersection of the liquid and gas curves. The temperature ranges in which the different phases are the most stable are indicated by shaded areas. The three curves shown are actually curved slightly downward but have been approximated as straight lines.

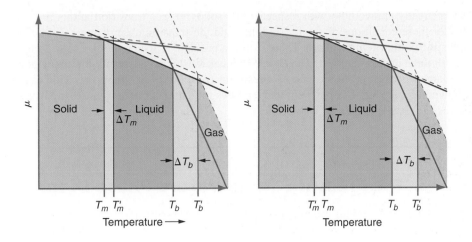

FIGURE 8.2
The left diagram applies if $V_m^{liquid} > V_m^{solid}$. The right diagram applies if $V_m^{liquid} < V_m^{solid}$. The solid curves show μ as a function of temperature for all three phases at $P = P_1$. The dashed curves show the same information for $P = P_2$, where $P_2 > P_1$. The unprimed temperatures refer to $P = P_1$, and the primed temperatures refer to $P = P_2$. The shifts in the solid and liquid curves are greatly exaggerated. The colored areas correspond to the temperature range in which the phases are most stable. The shaded area between T_m and T'_m is either solid or liquid, depending on P. The shaded area between T_b and T'_b is either liquid or gas, depending on P.

enough to avoid crystallization. These disordered materials lack the periodicity of crystals but behave mechanically like solids. Seed crystals can be used to increase the rate of crystallization if the viscosity of the liquid is not too high and the cooling rate is sufficiently slow. Liquid crystals, which can be viewed as a state of matter intermediate between a solid and a liquid, are discussed in Section 8.10.

In Figure 8.1, we consider changes with T at constant P. How is the relative stability of the three phases affected if P is changed at constant T? From Equation (8.2), $(\partial \mu / \partial P)_T = V_m$ and $V_m^{gas} \gg V_m^{liquid}$. For most substances, $V_m^{liquid} > V_m^{solid}$. Therefore, the μ versus T curve for the gas changes much more rapidly with pressure (by a factor of ~1000) than the liquid and solid curves. This behavior is illustrated in Figure 8.2, where it can be seen that the temperature at which the solid and the liquid curves intersect shifts as the pressure is increased. Because $V_m^{gas} \gg V_m^{liquid} > 0$, an increase in P always leads to a **boiling point elevation.** An increase in P leads to a **freezing point elevation** if $V_m^{liquid} > V_m^{solid}$ and to a **freezing point depression** if $V_m^{liquid} < V_m^{solid}$, as is the case for water. Few substances obey the relation $V_m^{liquid} < V_m^{solid}$; the consequences of this unusual behavior for water will be discussed in Section 8.2.

The μ versus T curve for a gas shifts along the T axis much more rapidly with P than the liquid and solid curves. As a consequence, changes in P can change the way in which a system progresses through the phases with increasing T from the "normal" order solid $\rightarrow$ liquid $\rightarrow$ gas shown in Figure 8.1. For example, the sublimation of dry ice at 298 K and 1 bar can be explained using Figure 8.3a. For CO_2 at the given pressure, the μ versus T curve for the gas intersects the corresponding curve for the solid at a lower temperature than the liquid curve. Therefore, the solid $\rightarrow$ liquid transition is

FIGURE 8.3
The chemical potential of a substance in the solid (green curve), liquid (blue curve), and gaseous (brown curve) states is plotted as a function of temperature for a fixed value of pressure. **(a)** The pressure lies below the triple point pressure, and the solid sublimes. **(b)** The pressure corresponds to the triple point pressure. At T_{tp}, all three phases coexist in equilibrium. The colored areas correspond to the temperature range in which the phases are the most stable. The liquid phase is not stable in part (a), and is only stable at the single temperature T_{tp} in part (b).

energetically unfavorable with respect to the solid $\rightarrow$ gas transition at this pressure. Under these conditions, the solid sublimes and the transition temperature T_s is called the **sublimation temperature.** There is also a pressure at which the μ versus T curves for all three phases intersect. The P, V_m, T values for this point specify the **triple point,** so named because all three phases coexist in equilibrium at this point. This case is shown in Figure 8.3b. Triple point temperatures for a number of substances are listed in Table 8.1 (see Appendix B, Data Tables).

8.2 The Pressure–Temperature Phase Diagram

As shown in the previous section, at a given value of pressure and temperature a system containing a pure substance may consist of a single phase, two phases in equilibrium, or three phases in equilibrium. The usefulness of a **phase diagram** is that it displays this information graphically. Although any two of the macroscopic system variables P, V, and T can be used to construct a phase diagram, the $P-T$ diagram is particularly useful. In this section, the features of a $P-T$ phase diagram that are common to pure substances are discussed. Phase diagrams must generally be determined experimentally because material-specific forces between atoms determine the temperatures and pressures at which different phases are stable. Increasingly, calculations have become sufficiently accurate that major features of phase diagrams can be obtained using microscopic theoretical models. However, as shown in Section 8.4, thermodynamics can say a great deal about the phase diagram without considering the microscopic properties of the system.

The **$P-T$ phase diagram,** a sample of which is shown in Figure 8.4, displays stability regions for a pure substance as a function of pressure and temperature. Most P, T points correspond to a single solid, liquid, or gas phase. At the triple point, all three phases coexist. The triple point of water is 273.16 K and 611 Pa. All P, T points for which the same two phases coexist at equilibrium fall on a curve. Such a curve is called a **coexistence curve.** Three separate coexistence curves are shown in Figure 8.4, corresponding to solid–gas, solid–liquid, and gas–solid coexistence. As shown in Section 8.5, the slopes of the solid-gas and liquid–gas curves are always positive. The slope of the solid–liquid curve can be either positive or negative.

The boiling point of a substance is defined as the temperature at which the vapor pressure of the substance is equal to the external pressure. The **standard boiling**

FIGURE 8.4
A $P-T$ phase diagram displays single-phase regions, coexistence curves for which two phases coexist at equilibrium, and a triple point. The processes corresponding to paths a, b, c, and the two processes labeled d are described in the text. Two solid-liquid coexistence curves are shown. For most substances, the solid curve, which has a positive slope, is observed. For water or any other substance for which the volume increases in a transition from the liquid to the solid, the red dashed curve corresponding to a negative slope is observed.

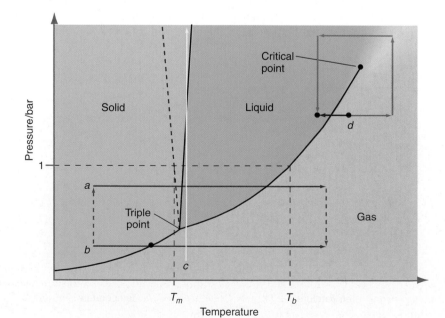

temperature is the temperature at which the vapor pressure of the substance is 1 bar. The **normal boiling temperature** is the temperature at which the vapor pressure of the substance is 1 atm. Values of the normal boiling and freezing temperatures for a number of substances are shown in Table 8.2. Because 1 bar is slightly less than 1 atm, the standard boiling temperature is slightly less than the normal boiling temperature. Along two-phase curves in which one of the coexisting phases is the gas, P refers to the **vapor pressure** of the substance. In all regions, P refers to the external pressure

TABLE 8.2	Melting and Boiling Temperatures and Enthalpies of Transition at 1 atm Pressure				
Substance	Name	T_m (K)	ΔH_{fusion} (kJ mol^{-1}) at T_m	T_b (K)	$\Delta H_{vaporization}$ (kJ mol^{-1}) at T_b
Ar	Argon	83.8	1.12	87.3	6.43
Cl$_2$	Chlorine	171.6	6.41	239.18	20.41
Fe	Iron	1811	13.81	3023	349.5
H$_2$	Hydrogen	13.81	0.12	20.4	0.90
H$_2$O	Water	273.15	6.010	373.15	40.65
He	Helium	0.95	0.021	4.22	0.083
I$_2$	Iodine	386.8	14.73	457.5	41.57
N$_2$	Nitrogen	63.5	0.71	77.5	5.57
Na	Sodium	370.87	2.60	1156	98.0
NO	Nitric oxide	109.5	2.3	121.41	13.83
O$_2$	Oxygen	54.36	0.44	90.7	6.82
SO$_2$	Sulfur dioxide	197.6	7.40	263.1	24.94
Si	Silicon	1687	50.21	2628	359
W	Tungsten	3695	52.31	5933	422.6
Xe	Xenon	161.4	1.81	165.11	12.62
CCl$_4$	Carbon tetrachloride	250	3.28	349.8	29.82
CH$_4$	Methane	90.68	0.94	111.65	8.19
CH$_3$OH	Methanol	175.47	3.18	337.7	35.21
CO	Carbon monoxide	68	0.83	81.6	6.04
C$_2$H$_4$	Ethene	103.95	3.35	169.38	13.53
C$_2$H$_6$	Ethane	90.3	2.86	184.5	14.69
C$_2$H$_5$OH	Ethanol	159.0	5.02	351.44	38.56
C$_3$H$_8$	Propane	85.46	3.53	231.08	19.04
C$_5$H$_5$N	Pyridine	231.65	8.28	388.38	35.09
C$_6$H$_6$	Benzene	278.68	9.95	353.24	30.72
C$_6$H$_5$OH	Phenol	314.0	11.3	455.02	45.69
C$_6$H$_5$CH$_3$	Toluene	178.16	6.85	383.78	33.18
C$_{10}$H$_8$	Naphthalene	353.3	17.87	491.14	43.18

Sources: Data from Lide, D. R., ed. *Handbook of Chemistry and Physics*; 83rd ed. Boca Raton, FL: CRC Press, 2002; Lide, D. R., ed. *CRC Handbook of Thermophysical and Thermochemical Data*. Boca Raton, FL: CRC Press, 1994; and Blachnik, R., ed. *D'Ans Lax Taschenbuch für Chemiker und Physiker*, 4th ed. Berlin: Springer, 1998.

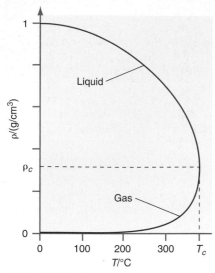

FIGURE 8.5

The density of the liquid and gaseous phases of water are shown as a function of temperature. At the critical point, the densities are equal.

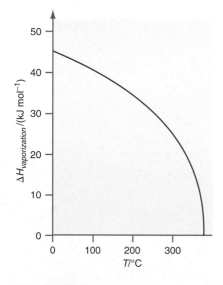

FIGURE 8.6

The enthalpy of vaporization for water is shown as a function of temperature. At the critical point, $\Delta H_{vaporization}$ approaches zero.

that would be exerted on the pure substance if it were confined in a piston and cylinder assembly.

The solid–liquid coexistence curve traces out the melting point as a function of pressure. The magnitude of the slope of this curve is large, as proven in Section 8.5. Therefore, T_m is a weak function of the pressure. If the solid is more dense than the liquid, the slope of this curve is positive, and the melting temperature increases with pressure. This is the case for most substances. If the solid is less dense than the liquid, the slope is negative and the melting temperature decreases with pressure. Water is one of the few substances that exhibits this behavior. Imagine the fate of aquatic plants and animals in climate zones where the temperature routinely falls below 0°C in the winter if water behaved "normally." Lakes would begin to freeze over at the water–air interface, and the ice formed would fall to the bottom of the lake. This would lead to more ice formation until the whole lake was full of ice. In cool climate zones, it is likely that ice at the bottom of the lakes would remain throughout the summer. The aquatic life that we are familiar with would not survive under such a scenario.

The slope of the liquid–gas coexistence curve is much smaller than that of the solid–liquid coexistence curve, as proven in Section 8.5. Therefore, the boiling point is a much stronger function of the pressure than the freezing point. The boiling point always increases with pressure. This property is utilized in a pressure cooker, where increasing the pressure by 1 bar increases the boiling temperature of water by approximately 20°C. The rate of the chemical processes involved in cooking increase exponentially with T. Therefore, a pressure cooker operating at $P = 2$ bar can cook food in 20% to 40% of the time required for cooking at atmospheric pressure. By contrast, a mountain climber in the Himalayas would find that the boiling temperature of water is reduced by approximately 25°C relative to sea level. Cooking takes significantly longer under these conditions.

The solid–gas coexistence curve ends at the triple point. Whereas the liquid–solid coexistence curve extends indefinitely, the liquid–gas curve ends at the **critical point,** characterized by $T = T_c$ and $P = P_c$. For $T > T_c$ and $P > P_c$, the liquid and gas phases have the same density as shown in Figure 8.5, so it is not meaningful to refer to distinct phases. Because the liquid and gas phases are indistinguishable at the critical point, $\Delta H_{vaporization}$ approaches zero as the critical point is reached as shown in Figure 8.6.

Substances for which $T > T_c$ and $P > P_c$ are called **supercritical fluids.** As discussed in Section 8.9, supercritical fluids have unusual properties that make them useful in chemical technologies.

Each of the paths labeled a, b, c, and d in Figure 8.4 corresponds to a process that demonstrates the usefulness of the $P–T$ phase diagram. In the following, each process is considered individually. Process a follows a constant pressure (isobaric) path. An example of this path is heating one mole of ice. The system is initially in the solid single–phase region. Assume that heat is added to the system at a constant rate using current flow through a resistive heater. Because the pressure is constant, $q_P = \Delta H$. Furthermore, as discussed in Section 2.4, $\Delta H \approx C_P \Delta T$ in a single–phase region. Combining these equations, $\Delta T = q_P / C_P^{solid}$. Along path a, the temperature increases linearly with q_P in the single–phase solid region as shown in Figure 8.7. At the melting temperature T_m, heat continues to be absorbed by the system as the solid is transformed into a liquid. This system now consists of two distinct phases, solid and liquid. As heat continues to flow into the system, the temperature will not increase until the system consists entirely of liquid. The heat taken up per mole of the system at the constant temperature T_m is ΔH_{fusion}.

The temperature again increases linearly with q_P and $\Delta T = q_P / C_P^{liquid}$ until the boiling point is reached. At this temperature, the system consists of two phases, liquid and gas. The temperature remains constant until all the liquid has been converted into gas. The heat taken up per mol of the system at the constant temperature T_b is $\Delta H_{vaporization}$. Finally, the system enters the single–phase gas region. Along path b, the pressure is less than the triple point pressure. Therefore, the liquid phase is not stable,

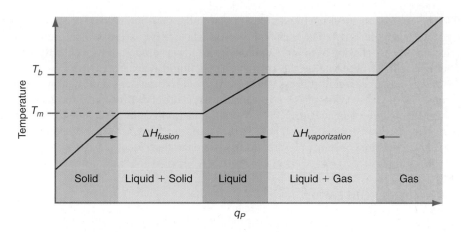

FIGURE 8.7
The trends in the temperature versus heat curve are shown schematically for the process corresponding to path a in Figure 8.4. The temperature rises linearly with q_P in single-phase regions and remains constant along the two-phase curves as the relative amounts of the two phases in equilibrium change (not to scale).

and the solid is converted directly into the gaseous form. As Figure 8.4 shows, there is only one two-phase interval along this path. A diagram indicating the relationship between temperature and heat flow is shown in Figure 8.8.

Note that the initial and final states in process b can be reached by an alternative route described by the vertical dashed arrows in Figure 8.4. The pressure of the system in process b is increased to the value of the initial state of process a at constant temperature. The process follows that described for process a, after which the pressure is returned to the final pressure of process b. Invoking Hess's law, the enthalpy change for this pathway and for pathway b are equal. Now imagine that the constant pressures for processes a and b differ only by an infinitesimal amount, although that for a is higher than the triple point pressure, and that for b is lower than the triple point pressure. We can express this mathematically by setting the pressure for process a equal to $P_{tp} + dP$, and that for process b equal to $P_{tp} - dP$. We examine the limit $dP \rightarrow 0$. In this limit, $\Delta H \rightarrow 0$ for the two steps in the process indicated by the dashed arrows because $dP \rightarrow 0$ Therefore, ΔH for the transformation solid $\rightarrow$ liquid $\rightarrow$ gas in process a and for the transformation solid $\rightarrow$ gas in process b must be identical. We conclude that

$$\Delta H = \Delta H_{sublimation} = \Delta H_{fusion} + \Delta H_{vaporization} \qquad (8.4)$$

This statement is strictly true at the triple point. Far from the triple point, the temperature–dependent heat capacities of the different phases and the temperature dependence of the transition enthalpies must be taken into account.

Path c indicates an isothermal process in which the pressure is increased. The initial state of the system is the single–phase gas region. As the gas is compressed, it is liquefied as it crosses the gas–liquid coexistence curve. As the pressure is increased further, the sample freezes as it crosses the liquid–solid coexistence curve. Freezing is exothermic, and heat must flow to the surroundings as the liquid solidifies. If the process is reversed, heat must flow into the system to keep T constant as the solid melts.

If T is below the triple point temperature, the liquid exists at equilibrium only if the slope of the liquid–solid coexistence curve is negative, as is the case for water. Solid water below the triple point temperature can melt at constant T if the pressure is increased sufficiently to cross the liquid–solid coexistence curve. In an example of such a process, a thin wire to which a heavy weight is attached on each end is stretched over a block of ice. With time, it is observed that the wire lies within the ice block and eventually passes through the block. There is no visible evidence of the passage of the wire in the form of a narrow trench. What happens in this process? Because the wire is thin, the force on the wire results in a high pressure in the area of the ice block immediately below the wire. This high pressure causes local melting of the ice below the wire. Melting allows the wire to displace the liquid water, which flows to occupy the volume immediately above the wire. Because in this region water no longer experiences a high pressure, it freezes again and hides the passage of the wire.

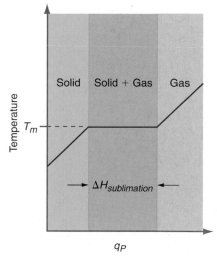

FIGURE 8.8
The temperature versus heat curve is shown for the process corresponding to path b in Figure 8.4. The temperature rises linearly with q_P in single-phase regions and remains constant along the two-phase curves as the relative amounts of the two phases in equilibrium change (not to scale).

The consequences of having a critical point in the gas–liquid coexistence curve are illustrated in Figure 8.4. The gas–liquid coexistence curve is crossed in the constant pressure process starting at point *d* and moving to lower temperatures. A clearly visible interface will be observed as the two-phase gas–liquid coexistence curve is crossed. However, the same overall process can be carried out in four steps indicated by the blue arrows. In this case, two–phase coexistence is not observed, because the gas–liquid coexistence curve ends at $P_{critical}$. The overall transition is the same along both paths, namely, gas is transformed into liquid. However, no interface will be observed if P for the higher pressure constant pressure process is greater than $P_{critical}$.

EXAMPLE PROBLEM 8.1

Draw a generic $P-T$ phase diagram like that shown in Figure 8.4. Draw pathways in the diagram that correspond to the processes described here:

a. You hang wash out to dry for a temperature below the triple point value. Initially, the water in the wet clothing is frozen. However, after a few hours in the sun, the clothing is warmer, dry, and soft.

b. A small amount of ethanol is contained in a thermos bottle. A test tube is inserted into the neck of the thermos bottle through a rubber stopper. A few minutes after filling the test tube with liquid nitrogen, the ethanol is no longer visible at the bottom of the bottle.

c. A transparent cylinder and piston assembly contains only a pure liquid in equilibrium with its vapor phase. An interface is clearly visible between the two phases. When you increase the temperature by a small amount, the interface disappears.

Solution

The phase diagram with the paths is shown here:

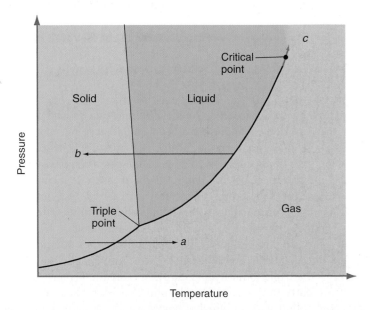

Paths *a*, *b*, and *c* are not unique. Path *a* must occur at a pressure lower than the triple point pressure. Process *b* must occur at a pressure greater than the triple point pressure and originate on the liquid–gas coexistence line. Path *c* will lie on the liquid–gas coexistence line up to the critical point, but it can deviate once $T > T_c$ and $P > P_c$.

A $P-T$ phase diagram for water at high P values is shown in Figure 8.9. Because we plot $\ln P$ rather than P, the curvature of the solid–liquid and liquid–gas coexistence curves decreases with increasing temperature rather than increases as shown in Figure 8.4. Water has a number of solid phases that are stable in different pressure ranges because they have

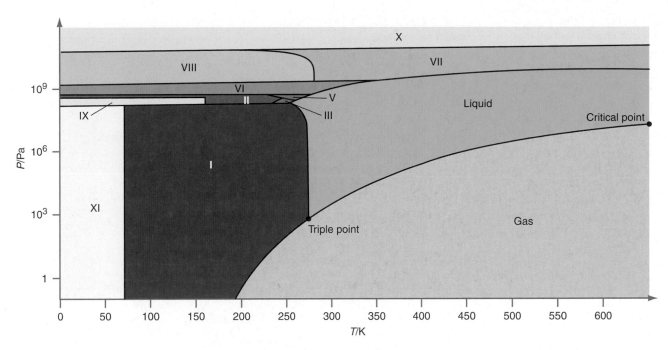

FIGURE 8.9
The $P-T$ phase diagram for H_2O is shown for pressures up to $P = 10^{12}$ Pa and $T = T_c$.
Source: Adapted from "Water Structure and Science," http://www.lsbu.ac.uk/water/phase.html, courtesy of Martin Chaplin.

different densities. Eleven different crystalline forms of ice have been identified up to a pressure of 10^{12} Pa. Note for example in Figure 8.9 that ice VI does not melt until the temperature is raised to ~300 K for $P \approx$ 1000 MPa. For a comprehensive collection of material on the phase diagram of water, see http://www.lsbu.ac.uk/water/phase.html.

Hexagonal ice (ice I) is the normal form of ice and snow. The structure shown in Figure 8.10 may be thought of as consisting of a set of parallel sheets connected to one another through hydrogen bonding. Hexagonal ice has a fairly open structure with a density of 0.931 g cm^{-3} near the triple point. Figure 8.10 also shows the crystal structure of ice VI. All water molecules in this structure are hydrogen bonded to four other molecules. Ice VII is much more closely packed than hexagonal ice and does not float on liquid water.

As shown in Figure 8.9, phase diagrams can be quite complex for simple substances because a number of solid phases can exist as P and T are varied. A further example is sulfur, which can also exist in several different solid phases. A portion of the phase diagram for sulfur is shown in Figure 8.11, and the solid phases are described by

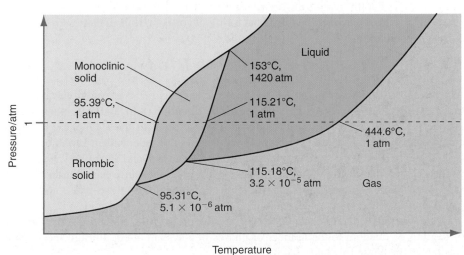

FIGURE 8.11
The $P-T$ phase diagram for sulfur (not to scale).

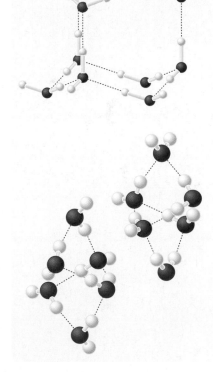

FIGURE 8.10
Two different crystal structures are shown. Hexagonal ice (top) is the stable form of ice under atmospheric conditions. Ice VI (bottom) is only stable at elevated pressures as shown in the phase diagram of Figure 8.9. The dashed lines indicate hydrogen bonds.

FIGURE 8.12
The P–T phase diagram for CO_2 (not to scale).

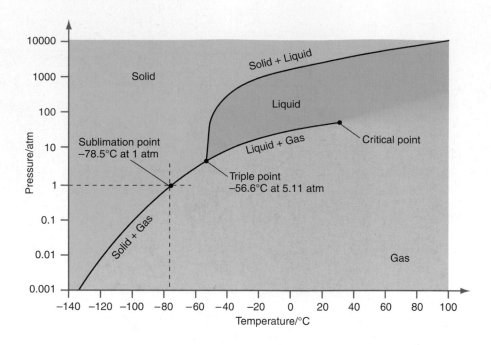

the symmetry of their unit cells. Note that several points correspond to three-phase equilibria. By contrast, the CO_2 phase diagram shown in Figure 8.12 is simpler. It is similar in structure to that of water, but the solid–liquid coexistence curve has a positive slope. Several of the end-of-chapter problems and questions refer to the phase diagrams in Figures 8.9, 8.11, and 8.12.

8.3 The Phase Rule

As we have seen in Section 8.1, the coexistence of two phases, α and β, of a substance requires that the chemical potentials are equal

$$\mu_\alpha(T, P) = \mu_\beta(T, P) \tag{8.5}$$

The state variables T and P, which are independent in a single-phase region, are linked through Equation (8.5) for a two-phase coexistence region. Consequently, only one of T and P is needed to describe the system, and the other variable is determined by Equation (8.5). For this reason, the two-phase coexistence regions in Figures 8.4, 8.9, 8.11, and 8.12 are curves that have the functional form $P(T)$ or $T(P)$. If three phases, α, β, and γ coexist in equilibrium,

$$\mu_\alpha(T, P) = \mu_\beta(T, P) = \mu_\gamma(T, P) \tag{8.6}$$

which imposes another requirement on the variables T and P. Now only one combination of T and P satisfies the requirement posed by Equation (8.6), and three-phase coexistence for a pure substance occurs only at a single point in the P, T phase diagram called the triple point. These restrictions on the number of independent variables accessible to a system can be summarized as follows. Because T and P can be varied independently in a single-phase region, a system of a pure substance has two **degrees of freedom.** The system has one and zero degrees of freedom in a two-phase and three-phase coexistence region, respectively.

The American chemist J. W. Gibbs derived the **phase rule,** which links the number of degrees of freedom to the number of phases in a system at equilibrium. For a pure substance, the phase rule takes the form

$$F = 3 - p \tag{8.7}$$

where F is the number of degrees of freedom, and p is the number of phases. Gibbs proved that no more than three phases of a pure substance can be in equilibrium as F cannot be a negative number. As we will see in Chapter 9, the number of degrees of freedom increases if a system contains several chemically independent species, for example in a system consisting of ethanol and water.

8.4 The Pressure–Volume and Pressure–Volume–Temperature Phase Diagrams

In Section 8.2, the regions of stability and equilibrium among the solid, liquid, and gas phases were described using a $P–T$ phase diagram. Any phase diagram that includes only two of the three state variables is limited because it does not contain information on the third variable. We first complement the information contained in the $P–T$ phase diagram with a **$P–V$ phase diagram,** and then combine these two representations into a $P–V–T$ phase diagram. Figure 8.13 shows a $P–V$ phase diagram for a substance for which $V_m^{liquid} > V_m^{solid}$.

Significant differences are seen in the way that two- and three-phase coexistence are represented in the two-phase diagrams. The two-phase coexistence curves of the $P–T$ phase diagram become two-phase regions in the $P–V$ phase diagram because the volume of a system in which two phases coexist varies with the relative amounts of the material in each phase. For pressures well below the critical point, the range in V over which the gas and liquid coexist is large compared to the range in V over which the solid and liquid coexist, because $V_m^{solid} < V_m^{liquid} \ll V_m^{gas}$. Therefore, the gas–liquid coexistence region is broader than the solid–liquid coexistence region. Note that the triple point in the $P–T$ phase diagram becomes a triple line in the $P–V$ diagram. Although P and T have unique values at the triple point, V can range between a maximum value for which the system consists almost entirely of gas with traces of the liquid and solid phases and a minimum value for which the system consists almost entirely of solid with traces of the liquid and gas phases.

The usefulness of the $P–V$ diagram is illustrated by tracing several processes in Figure 8.13. In process a, a solid is converted to a gas by increasing the temperature in an isobaric process for which P is greater than the triple point pressure. This same process was depicted in Figure 8.4. In the $P–V$ phase diagram, it is clear that this process involves large changes in volume in the two-phase coexistence region over which the temperature remains constant, which was not obvious in Figure 8.4. Process b shows an isobaric transition from solid to gas for P below the triple point pressure, for which the system has only one two-phase coexistence region. Process c shows a constant volume transition from a system consisting of solid and vapor in equilibrium to a supercritical fluid. How does the temperature change along this path?

Process b in Figure 8.13 is known as **freeze drying.** Assume that the food to be freeze dried is placed in a vessel at $-10°C$, and the system is allowed to equilibrate at this temperature. The partial pressure of $H_2O(g)$ in equilibrium with the ice crystals in the food under these conditions is 260 Pa. A vacuum pump is now started, and the gas phase is pumped away. The temperature reaches a steady-state value determined by heat conduction into the sample and heat loss through sublimation. The solid has an equilibrium vapor pressure determined by the steady-state temperature. As the pump removes water from the gas phase, ice spontaneously sublimes in order to keep the gas-phase water partial pressure at its equilibrium value. After the sublimation is complete, the food has been freeze dried. The advantage of freeze drying over boiling off water at 1 atm is that the food can be dehydrated at low temperatures at which the food is not cooked.

All the information on the values of P, V, and T corresponding to single-phase regions, two-phase regions, and the triple point displayed in the $P–T$ and $P–V$ phase diagrams is best displayed in a three-dimensional **$P–V–T$ phase diagram.** Such a diagram is shown in

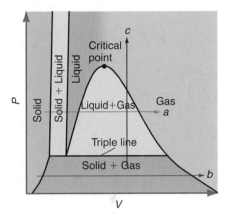

FIGURE 8.13
A $P–V$ phase diagram displays single- and two-phase coexistence regions, a critical point, and a triple line. The two-phase coexistence areas are colored.

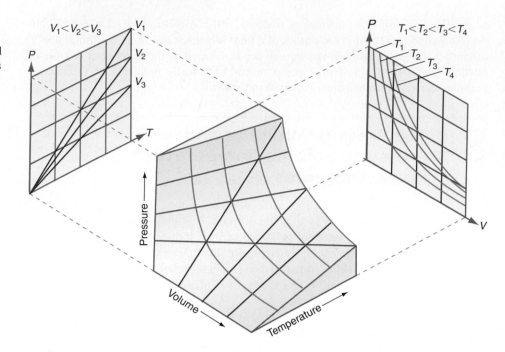

Figure 8.14 for an ideal gas, which does not exist in the form of condensed phases. It is easy to obtain the *P–T* and *P–V* phase diagrams from the *P–V–T* phase diagram. The *P–T* phase diagram is a projection of the three-dimensional surface on the *P–T* plane, and the *P–V* phase diagram is a projection of the three-dimensional surface on the *P–V* plane.

Figure 8.15 shows a *P–V–T* diagram for a substance that expands upon melting. The usefulness of the *P–V–T* phase diagram can be illustrated by revisiting the isobaric conversion of a solid to a gas at a temperature above the triple point shown as process *a* in Figure 8.4. This process is shown as the path $a \rightarrow b \rightarrow c \rightarrow d \rightarrow e \rightarrow f$ in Figure 8.15. We can see now that the temperature increases along the segments $a \rightarrow b, c \rightarrow d$, and $e \rightarrow f$, all of which lie within single-phase regions, and it remains constant along the segments $b \rightarrow c$ and $d \rightarrow e$, which lie within two-phase regions. Similarly, process *c* in Figure 8.4 is shown as the path $g \rightarrow h \rightarrow i \rightarrow k \rightarrow l \rightarrow m$ in Figure 8.15.

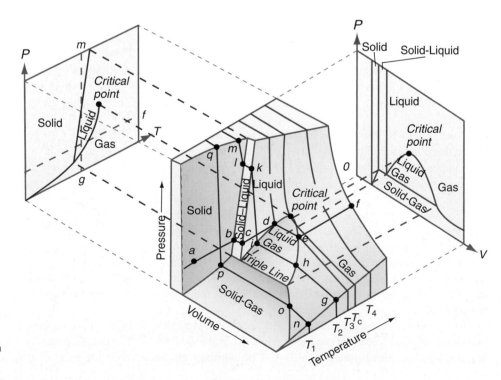

8.5 Providing a Theoretical Basis for the *P–T* Phase Diagram

In this section, a theoretical basis is provided for the coexistence curves that separate different single-phase regions in the *P–T* phase diagram. Along the coexistence curves, two phases are in equilibrium. From Section 5.6, we know that if two phases, α and β, are in equilibrium at a pressure P and temperature T, their chemical potentials must be equal:

$$\mu_\alpha(P, T) = \mu_\beta(P, T) \tag{8.8}$$

If the macroscopic variables are changed by a small amount, $P, T \rightarrow P + dP$, $T + dT$ such that the system pressure and temperature still lie on the coexistence curve, then

$$\mu_\alpha(P, T) + d\mu_\alpha = \mu_\beta(P, T) + d\mu_\beta \tag{8.9}$$

In order for the two phases to remain in equilibrium,

$$d\mu_\alpha = d\mu_\beta \tag{8.10}$$

Because $d\mu$ can be expressed in terms of dT and dP,

$$d\mu_\alpha = -S_{m\alpha}\, dT + V_{m\alpha}\, dP \quad \text{and} \quad d\mu_\beta = -S_{m\beta}\, dT + V_{m\beta}\, dP \tag{8.11}$$

The expressions for $d\mu$ can be equated, giving

$$-S_{m\alpha}\, dT + V_{m\alpha}\, dP = -S_{m\beta}\, dT + V_{m\beta}\, dP \quad \text{or}$$
$$(S_{m\beta} - S_{m\alpha})\, dT = (V_{m\beta} - V_{m\alpha})\, dP \tag{8.12}$$

Assume that as $P, T \rightarrow P + dP, T + dT$ an incremental amount of phase α is transformed to phase β. In this case, $\Delta S_m = S_{m\beta} - S_{m\alpha}$ and $\Delta V_m = V_{m\beta} - V_{m\alpha}$. Rearranging Equation (8.12) gives the **Clapeyron equation:**

$$\frac{dP}{dT} = \frac{\Delta S_m}{\Delta V_m} \tag{8.13}$$

The Clapeyron equation allows us to calculate the slope of the coexistence curves in a *P–T* phase diagram if ΔS_m and ΔV_m for the transition are known. This information is necessary when constructing a phase diagram. We next use the Clapeyron equation to estimate the slope of the solid–liquid coexistence curve. At the melting temperature,

$$\Delta G_{fusion} = \Delta H_{fusion} - T\Delta S_{fusion} = 0 \tag{8.14}$$

Therefore, the ΔS_m values for the fusion transition can be calculated from the enthalpy of fusion and the fusion temperature. Values of the normal fusion and vaporization temperatures, as well as ΔH_m for fusion and vaporization, are shown in Table 8.2 for a number of different elements and compounds. Although there is a significant variation in these values, for our purposes it is sufficient to use the average value of $\Delta S_{fusion} = 22\ \text{J mol}^{-1}\,\text{K}^{-1}$ calculated from the data in Table 8.2 in order to estimate the slope of the solid–liquid coexistence curve.

For the fusion transition, ΔV is small because the densities of the solid and liquid states are quite similar. The average ΔV_{fusion} for Ag, AgCl, Ca, CaCl$_2$, K, KCl, Na, NaCl, and H$_2$O is $+4.0 \times 10^{-6}\ \text{m}^3\,\text{mol}^{-1}$. Of these substances, only H$_2$O has a negative value for ΔV_{fusion}. We next use the average values of ΔS_{fusion} and ΔV_{fusion} to estimate the slope of the solid–liquid coexistence curve:

$$\left(\frac{dP}{dT}\right)_{fusion} = \frac{\Delta S_{fusion}}{\Delta V_{fusion}} \approx \frac{22\ \text{J mol}^{-1}\,\text{K}^{-1}}{\pm 4.0 \times 10^{-6}\ \text{m}^3\,\text{mol}^{-1}}$$

$$= \pm 5.5 \times 10^6\ \text{Pa K}^{-1} = \pm 55\ \text{bar K}^{-1} \tag{8.15}$$

Inverting this result, $(dT/dP)_{fusion} \approx \pm 0.018\,\text{K bar}^{-1}$. An increase of P by ~ 50 bar is required to change the melting temperature by one degree. This result explains the very steep solid–liquid coexistence curve shown in Figure 8.4.

The same analysis applies to the liquid–gas coexistence curve. Because $\Delta H_m^{vaporization}$ and $\Delta V_{vaporization} = V_m^{gas} - V_m^{liquid}$ are always positive, $(dP/dT)_{vaporization}$ is always positive. The average of $\Delta S_{vaporization}$ for the substances shown in Table 8.2 is $95\,\text{J mol}^{-1}\,\text{K}^{-1}$. This value is in accord with **Trouton's rule,** which states that $\Delta S_{vaporization} \approx 90\,\text{J mol}^{-1}\,\text{K}^{-1}$ for liquids. The rule fails for liquids in which there are strong interactions between molecules such as $-\text{OH}$ or $-\text{NH}_2$ groups capable of forming hydrogen bonds. For these substances, $\Delta S_{vaporization} > 90\,\text{J mol}^{-1}\,\text{K}^{-1}$.

The molar volume of an ideal gas is approximately 20. L mol^{-1} in the temperature range in which many liquids boil. Because $V_m^{gas} \gg V_m^{liquid}$, $\Delta V_{vaporization} \approx 20. \times 10^{-3}\,\text{m}^3\,\text{mol}^{-1}$. The slope of the liquid–gas coexistence curve is given by

$$\left(\frac{dP}{dT}\right)_{vaporization} = \frac{\Delta S_{vaporization}}{\Delta V_{vaporization}} \approx \frac{95\,\text{J mol}^{-1}\,\text{K}^{-1}}{2.0 \times 10^{-2}\,\text{m}^3\,\text{mol}^{-1}} \approx 4.8 \times 10^3\,\text{Pa K}^{-1}$$

$$= 4.8 \times 10^{-2}\,\text{bar K}^{-1} \qquad (8.16)$$

This slope is a factor of 10^3 smaller than the slope for the liquid–solid coexistence curve. Inverting this result, $(dT/dP)_{vaporization} \approx 21\,\text{K bar}^{-1}$. This result shows that it takes only a modest increase in the pressure to increase the boiling point of a liquid by a significant amount. For this reason, a pressure cooker or automobile radiator does not need to be able to withstand high pressures. Note that the slope of the liquid–gas coexistence curve in Figure 8.4 is much less than that of the solid–liquid coexistence curve. The slope of both curves increases with T because ΔS increases with increasing T and ΔV decreases with increasing P.

The solid–gas coexistence curve can also be analyzed using the Clapeyron equation. Because entropy is a state function, the entropy change for the processes solid $(P, T) \rightarrow$ gas (P, T) and solid $(P, T) \rightarrow$ liquid $(P, T) \rightarrow$ gas (P, T) must be the same. Therefore, $\Delta S_{sublimation} = \Delta S_{fusion} + \Delta S_{vaporization} > \Delta S_{vaporization}$. However, because the molar volume of the gas is so much larger than that of the solid or liquid, $\Delta V_{sublimation} \approx \Delta V_{vaporization}$. We conclude that $(dP/dT)_{sublimation} > (dP/dT)_{vaporization}$. Therefore, the slope of the solid–gas coexistence curve will be greater than that of the liquid–gas coexistence curve. Because this comparison applies to a common value of the temperature, it is best made for temperatures just above and just below the triple point temperature. This difference in slope of these two coexistence curves is exaggerated in Figure 8.4.

8.6 Using the Clausius–Clapeyron Equation to Calculate Vapor Pressure as a Function of *T*

From watching a pot of water as it is heated on a stove, it is clear that the vapor pressure of a liquid increases rapidly with increasing temperature. The same conclusion holds for a solid below the triple point. To calculate the vapor pressure at different temperatures, the Clapeyron equation must be integrated. Consider the solid–liquid coexistence curve:

$$\int_{P_i}^{P_f} dP = \int_{T_i}^{T_f} \frac{\Delta S_{fusion}}{\Delta V_{fusion}}\, dT = \int_{T_i}^{T_f} \frac{\Delta H_{fusion}}{\Delta V_{fusion}} \frac{dT}{T} \approx \frac{\Delta H_{fusion}}{\Delta V_{fusion}} \int_{T_i}^{T_f} \frac{dT}{T} \qquad (8.17)$$

where the integration is along the solid–liquid coexistence curve. In the last step, it has been assumed that ΔH_{fusion} and ΔV_{fusion} are independent of T over the temperature range of interest. Assuming that $(T_f - T_i)/T_i$ is small, the previous equation can be simplified to give

$$P_f - P_i = \frac{\Delta H_{fusion}}{\Delta V_{fusion}} \ln \frac{T_f}{T_i} = \frac{\Delta H_{fusion}}{\Delta V_{fusion}} \ln \frac{T_i + \Delta T}{T_i} \approx \frac{\Delta H_{fusion}}{\Delta V_{fusion}} \frac{\Delta T}{T_i} \qquad (8.18)$$

The last step uses the result $\ln(1 + x) \approx x$ for $x \ll 1$, obtained by expanding $\ln(1 + x)$ in a Taylor series about $x = 0$. We see that the vapor pressure of a solid varies linearly with ΔT in this limit. The value of the slope dP/dT was discussed in the previous section.

For the liquid–gas coexistence curve, we have a different result because $\Delta V \approx V^{gas}$. Assuming that the ideal gas law holds, we obtain the **Clausius-Clapeyron equation,**

$$\frac{dP}{dT} = \frac{\Delta S_{vaporization}}{\Delta V_{vaporization}} \approx \frac{\Delta H_{vaporization}}{TV_m^{gas}} = \frac{P\Delta H_{vaporization}}{RT^2}$$

$$\frac{dP}{P} = \frac{\Delta H_{vaporization}}{R}\frac{dT}{T^2} \qquad (8.19)$$

Assuming that $\Delta H_{vaporization}$ remains constant over the range of temperature of interest, the variation of the vapor pressure of the liquid with temperature is given by

$$\int_{P_i}^{P_f}\frac{dP}{P} = \frac{\Delta H_{vaporization}}{R} \times \int_{T_i}^{T_f}\frac{dT}{T^2}$$

$$\ln\frac{P_f}{P_i} = -\frac{\Delta H_{vaporization}}{R} \times \left(\frac{1}{T_f} - \frac{1}{T_i}\right) \qquad (8.20)$$

We see that the vapor pressure of a liquid rises exponentially with temperature. The same procedure is followed for the solid–gas coexistence curve. The result is the same as Equation (8.20) with $\Delta H_{sublimation}$ substituted for $\Delta H_{vaporization}$. Equation (8.20) provides a way to determine the enthalpy of vaporization for a liquid by measuring its vapor pressure as a function of temperature, as shown in Example Problem 8.2. In this discussion, it has been assumed that $\Delta H_{vaporization}$ is independent of temperature. More accurate values of the vapor pressure as a function of temperature can be obtained by fitting experimental data. This leads to an expression for the vapor pressure as a function of temperature. These functions for selected liquids and solids are listed in Tables 8.3 and 8.4 (see Appendix B, Data Tables).

EXAMPLE PROBLEM 8.2

The normal boiling temperature of benzene is 353.24 K, and the vapor pressure of liquid benzene is 1.19×10^4 Pa at 20.0°C. The enthalpy of fusion is 9.95 kJ mol^{-1}, and the vapor pressure of solid benzene is 137 Pa at -44.3°C. Calculate the following:

a. $\Delta H_{vaporization}$

b. $\Delta S_{vaporization}$

c. Triple point temperature and pressure

Solution

a. We can calculate $\Delta H_{vaporization}$ using the Clausius-Clapeyron equation because we know the vapor pressure at two different temperatures:

$$\ln\frac{P_f}{P_i} = -\frac{\Delta H_{vaporization}}{R}\left(\frac{1}{T_f} - \frac{1}{T_i}\right)$$

$$\Delta H_{vaporization} = -\frac{R\ln\dfrac{P_f}{P_i}}{\left(\dfrac{1}{T_f} - \dfrac{1}{T_i}\right)} = -\frac{8.314\,\text{J}\,\text{mol}^{-1}\text{K}^{-1} \times \ln\dfrac{101,325\,\text{Pa}}{1.19 \times 10^4\,\text{Pa}}}{\left(\dfrac{1}{353.24\,\text{K}} - \dfrac{1}{273.15\,\text{K} + 20.0\,\text{K}}\right)}$$

$$= 30.7\,\text{kJ}\,\text{mol}^{-1}$$

b. $\Delta S_{vaporization} = \dfrac{\Delta H_{vaporization}}{T_b} = \dfrac{30.7 \times 10^3\,\text{J}\,\text{mol}^{-1}}{353.24\,\text{K}} = 86.9\,\text{J}\,\text{mol}^{-1}\text{K}^{-1}$

c. At the triple point, the vapor pressures of the solid and liquid are equal:

$$\ln \frac{P_{tp}^{liquid}}{P^{\circ}} = \ln \frac{P_i^{liquid}}{P^{\circ}} - \frac{\Delta H_{vaporization}}{R}\left(\frac{1}{T_{tp}} - \frac{1}{T_i^{liquid}}\right)$$

$$\ln \frac{P_{tp}^{solid}}{P^{\circ}} = \ln \frac{P_i^{solid}}{P^{\circ}} - \frac{\Delta H_{sublimation}}{R}\left(\frac{1}{T_{tp}} - \frac{1}{T_i^{solid}}\right)$$

$$\ln \frac{P_i^{liquid}}{P^{\circ}} - \ln \frac{P_i^{solid}}{P^{\circ}} - \frac{\Delta H_{sublimation}}{RT_i^{solid}} + \frac{\Delta H_{vaporization}}{RT_i^{liquid}}$$

$$= \frac{\Delta H_{vaporization} - \Delta H_{sublimation}}{RT_{tp}}$$

$$T_{tp} = \frac{\Delta H_{vaporization} - \Delta H_{sublimation}}{R\left(\ln \dfrac{P_i^{liquid}}{P^{\circ}} - \ln \dfrac{P_i^{solid}}{P^{\circ}} - \dfrac{\Delta H_{sublimation}}{RT_i^{solid}} + \dfrac{\Delta H_{vaporization}}{RT_i^{liquid}}\right)}$$

$$= \frac{-9.95 \times 10^3 \text{ J mol}^{-1}}{8.314\, \text{J K}^{-1}\text{mol}^{-1} \times \left(\begin{array}{c}\ln \dfrac{1.19 \times 10^4 \text{ Pa}}{10^5 \text{ Pa}} - \ln \dfrac{137\,\text{Pa}}{10^5 \text{ Pa}} - \dfrac{(30.7 \times 10^3 + 9.95 \times 10^3)\,\text{J mol}^{-1}}{8.314\,\text{J K}^{-1}\text{mol}^{-1} \times 228.9\,\text{K}} \\[2mm] + \dfrac{30.7 \times 10^3 \text{mol}^{-1}}{8.314\,\text{J K}^{-1}\text{mol}^{-1} \times 293.15\,\text{K}}\end{array}\right)}$$

$$= 277 \text{ K}$$

We calculate the triple point pressure using the Clapeyron equation:

$$\ln \frac{P_f}{P_i} = -\frac{\Delta H_{vaporization}}{R}\left(\frac{1}{T_f} - \frac{1}{T_i}\right)$$

$$\ln \frac{P_{tp}}{101,325} = -\frac{30.7 \times 10^3 \text{ J mol}^{-1}}{8.314 \text{ J mol}^{-1}\text{K}^{-1}} \times \left(\frac{1}{278 \text{ K}} - \frac{1}{353.24 \text{ K}}\right)$$

$$\ln \frac{P_{tp}}{P^{\circ}} = 8.70483$$

$$P_{tp} = 6.03 \times 10^3 \text{ Pa}$$

8.7 The Vapor Pressure of a Pure Substance Depends on the Applied Pressure

Consider the piston and cylinder assembly containing water at 25°C shown in Figure 8.16. The equilibrium vapor pressure of water at this temperature is $P^* = 3.16 \times 10^3$ Pa, or 0.0316 bar. Therefore, if the weightless piston is loaded with a mass sufficient to generate a pressure of 1 bar, the system is in the single–phase liquid region of the phase diagram in Figure 8.4. This state of the system is shown in Figure 8.16a. The mass is reduced so that the pressure is exactly equal to the vapor pressure of water. The system now lies in the two-phase liquid–gas region of the phase diagram described by the liquid–gas coexistence curve. The piston can be pulled outward or pushed inward while maintaining this pressure. This action leads to a larger or smaller volume for the gas phase, but the pressure will remain constant at 3.16×10^3 Pa as long as the temperature of the system remains constant. This state of the system is shown in Figure 8.16b.

Keeping the temperature constant, enough argon gas is introduced into the cylinder such that the sum of the argon and H_2O partial pressures is 1 bar. This state of the system is shown in Figure 8.16c. What is the vapor pressure of water in this case, and does it differ from that for the system shown in Figure 8.16b? The vapor pressure P is used to denote the partial pressure of water in the gas phase, and **P** is used to denote the sum of the argon and water partial pressures.

To calculate the partial pressure of water in the argon-water mixture, the following equilibrium condition holds:

$$\mu_{liquid}(T, \mathbf{P}) = \mu_{gas}(T, P) \tag{8.21}$$

Differentiating this expression with respect to **P**, we obtain

$$\left(\frac{\partial \mu_{liquid}(T, \mathbf{P})}{\partial \mathbf{P}}\right)_T = \left(\frac{\partial \mu_{gas}(T, P)}{\partial P}\right)_T \left(\frac{\partial P}{\partial \mathbf{P}}\right)_T \tag{8.22}$$

Because $d\mu = -S_m\, dT + V_m\, dP$, $(d\mu/dP)_T = V_m$, and the previous equation becomes

$$V_m^{liquid} = V_m^{gas}\left(\frac{\partial P}{\partial \mathbf{P}}\right)_T \quad \text{or} \quad \left(\frac{\partial P}{\partial \mathbf{P}}\right)_T = \frac{V_m^{liquid}}{V_m^{gas}} \tag{8.23}$$

This equation shows that the vapor pressure P increases if the total pressure **P** increases. However, the rate of increase is small because the ratio $V_m^{liquid}/V_m^{gas} \ll 1$. It is reasonable to replace V_m^{gas} with the ideal gas value RT/P in Equation (8.23). This leads to the equation

$$\frac{RT}{P}\, dP = V_m^{liquid}\, d\mathbf{P} \quad \text{or} \quad RT\int_{P^*}^{P}\frac{dP'}{P'} = V_m^{liquid}\int_{P^*}^{\mathbf{P}} d\mathbf{P}' \tag{8.24}$$

Integrating Equation (8.24) gives

$$RT\ln\left(\frac{P}{P*}\right) = V_m^{liquid}(\mathbf{P} - P*) \tag{8.25}$$

For the specific case of water, $P^* = 0.0316$ bar, $\mathbf{P} = 1$ bar, and $V_m^{liquid} = 1.81 \times 10^{-5}\, \text{m}^3\, \text{mol}^{-1}$:

$$\ln\left(\frac{P}{P*}\right) = \frac{V_m^{liquid}\left(\mathbf{P} - P*\right)}{RT}$$

$$= \frac{1.81 \times 10^{-5}\,\text{m}^3\,\text{mol}^{-1} \times (1 - 0.0316) \times 10^5\,\text{Pa}}{8.314\,\text{J}\,\text{mol}^{-1}\,\text{K}^{-1} \times 298\,\text{K}} = 7.04 \times 10^{-4}$$

$$P = 1.0007\, P* \approx 0.0316\,\text{bar}$$

For an external pressure of 1 bar, the effect is negligible. However, for $\mathbf{P} = 100.$ bar, $P = 0.0339$ bar, amounting to an increase in the vapor pressure of 7%.

$P_{external} = 1.00$ bar
(a)

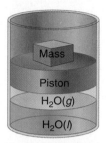

$P_{external} = 0.0316$ bar
(b)

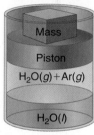

$P_{external} = 1.00$ bar
(c)

FIGURE 8.16
A piston and cylinder assembly at 298 K is shown with the contents being pure water **(a)** at a pressure greater than the vapor pressure, **(b)** at a pressure equal to the vapor pressure, and **(c)** at 1 bar for a mixture of argon and water.

8.8 Surface Tension

In discussing the liquid phase, the effect of the boundary surface on the properties of the liquid has been neglected. In the absence of a gravitational field, a liquid droplet will assume a spherical shape, because in this geometry the maximum number of molecules is surrounded by neighboring molecules. Because the interaction between molecules in a liquid is attractive, minimizing the surface-to-volume ratio minimizes the energy. How does the energy of the droplet depend on its surface area? Starting with the equilibrium spherical shape, assume that the droplet is distorted to create more

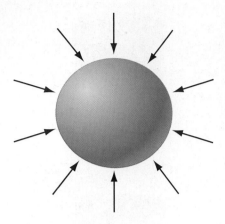

FIGURE 8.17
The forces acting on a spherical droplet that arise from surface tension.

area while keeping the volume constant. The work associated with the creation of additional surface area at constant V and T is

$$dA = \gamma d\sigma \qquad (8.26)$$

where A is the Helmholtz energy, γ is the surface tension, and σ is the unit element of area. The **surface tension** has the units of energy/area or J m^{-2}, which is equivalent to N m^{-1} (Newtons per meter). Equation (8.26) predicts that a liquid, a bubble, or a liquid film suspended in a wire frame will tend to minimize its surface area because $dA < 0$ for a spontaneous process at constant V and T.

Consider the spherical droplet depicted in Figure 8.17. There must be a force acting on the droplet in the radially inward direction for the liquid to assume a spherical shape. An expression for the force can be generated as follows. If the radius of the droplet is increased from r to $r + dr$, the area increases by $d\sigma$.

$$\sigma = 4\pi r^2 \quad \text{so} \quad d\sigma = 8\pi r \, dr \qquad (8.27)$$

From Equation (8.26), the work done in the expansion of the droplet is $8\pi\gamma r \, dr$. The force, which is normal to the surface of the droplet, is the work divided by the distance or

$$F = 8\pi\gamma r \qquad (8.28)$$

The net effect of this force is to generate a pressure differential across the droplet surface. At equilibrium, there is a balance between the inward and outward acting forces. The inward acting force is the sum of the force exerted by the external pressure and the force arising from the surface tension, whereas the outward acting force arises solely from the pressure in the liquid:

$$4\pi r^2 P_{outer} + 8\pi\gamma r = 4\pi r^2 P_{inner} \quad \text{or}$$

$$P_{inner} = P_{outer} + \frac{2\gamma}{r} \qquad (8.29)$$

Note that $P_{inner} - P_{outer} \rightarrow 0$ as $r \rightarrow \infty$. Therefore, the pressure differential exists only for a curved surface. From the geometry in Figure 8.17, it is apparent that the higher pressure is always on the concave side of the interface. Values for the surface tension for a number of liquids are listed in Table 8.5.

Equation (8.29) has interesting implications for the relative stabilities of bubbles with different curvatures $1/r$. Consider two air-filled bubbles of a liquid with the same surface tension γ, one with a large radius R_1, and one with a smaller radius R_2. Assume there is a uniform pressure outside both bubbles. From Equation (8.29) we obtain the difference between the pressures P_1 and P_2 within each bubble:

$$P_1 - P_2 = \frac{2\gamma}{R_1} - \frac{2\gamma}{R_2} = 2\gamma\left(\frac{1}{R_1} - \frac{1}{R_2}\right) \qquad (8.30)$$

Now suppose the two bubbles come into contact as shown at the top of Figure 8.18. Because $R_1 > R_2$, from Equation (8.30) the pressure P_2 inside bubble 2 (with respect

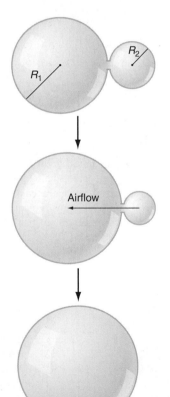

FIGURE 8.18
Two bubbles with unequal radii R_1 and R_2 make contact (top). Because the pressure within a bubble varies inversely with the bubble radius, air flows from the smaller into the larger bubble until a single large bubble remains (bottom).

TABLE 8.5	Surface Tension of Selected Liquids at 298 K				
Formula	Name	γ (mN m^{-1})	Formula	Name	(mN m^{-1})
Br$_2$	Bromine	40.95	CS$_2$	Carbon disulfide	31.58
H$_2$O	Water	71.99	C$_2$H$_5$OH	Ethanol	21.97
Hg	Mercury	485.5	C$_6$H$_5$N	Pyridine	36.56
CCl$_4$	Carbon tetrachloride	26.43	C$_6$H$_6$	Benzene	28.22
CH$_3$OH	Methanol	22.07	C$_8$H$_{18}$	Octane	21.14

Source: Data from Lide, D. R., ed. *Handbook of Chemistry and Physics.* 83rd ed. Boca Raton, FL: CRC Press, 2002.

to some common exterior pressure) is greater than the pressure P_1 in bubble 1, so air will flow from bubble 2 to bubble 1 until the smaller bubble disappears entirely as shown at the bottom of Figure 8.18. In foams this process is called "coarsening" and in crystals it is called "Ostwald ripening."

A biologically relevant issue that follows from Equations (8.29) and (8.30) is the stability of lung tissue. Lung tissue is composed of small water-lined, air-filled chambers called alveoli. According to Equation (8.29), the pressure difference across the surface of a spherical air-filled cavity is proportional to the surface tension and inversely proportional to the radius of the cavity. Therefore, if two air-filled alveoli, approximated as spheres with radii r and R and for which $r < R$ are interconnected, the smaller alveolus will collapse and the larger alveolus will expand because the pressure within the smaller alveolus is greater than the pressure within the larger alveolus. In theory, due to the surface tension of water that lines the alveoli, these cavities will "coarsen" as smaller alveoli collapse, leaving an ever diminishing number of alveoli of increasing size. Coarsening of the alveoli according to Equations (8.29) will result eventually in the collapse of the lungs.

In the alveoli, a lipid surfactant called phosphatidylcholine lowers the surface tension of the water lining the alveoli, thus stabilizing the lungs. But in persons lacking sufficient surfactant in the lining of their lungs, such as prematurely born infants and heavy smokers, the surface tension of the water in the alveoli is not lowered, and as a consequence lung tissue is destabilized by coarsening.

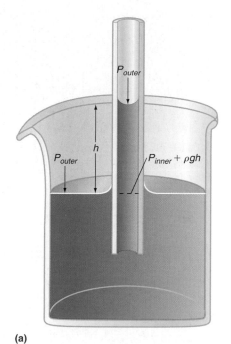

(a)

Another consequence of the pressure differential across a curved surface is that the vapor pressure of a droplet depends on its radius. By substituting numbers in Equation (8.29) and using Equation (8.25) to calculate the vapor pressure, we find that the vapor pressure of a 10^{-7} m water droplet is increased by 1%, that of a 10^{-8} m droplet is increased by 11%, and that of a 10^{-9} m droplet is increased by 270%. [At such a small diameter, the application of Equation (8.29) is questionable because the size of an individual water molecule is comparable to the droplet diameter. Therefore, a microscopic theory is needed to describe the forces within the droplet.] This effect plays a role in the formation of liquid droplets in a condensing gas such as fog. Small droplets evaporate more rapidly than large droplets, and the vapor condenses on the larger droplets, allowing them to grow at the expense of small droplets.

Capillary rise and **capillary depression** are other consequences of the pressure differential across a curved surface. Assume that a capillary of radius r is partially immersed in a liquid. When the liquid comes in contact with a solid surface, there is a natural tendency to minimize the energy of the system. If the surface tension of the liquid is lower than that of the solid, the liquid will wet the surface, as shown in Figure 8.19a. However, if the surface tension of the liquid is higher than that of the solid, the liquid will avoid the surface, as shown in Figure 8.19b. In either case, there is a pressure differential in the capillary across the gas–liquid interface because the interface is curved. If we assume that the liquid–gas interface is tangent to the interior wall of the capillary at the solid–liquid interface, the radius of curvature of the interface is equal to the capillary radius.

The difference in the pressure across the curved interface $2\gamma/r$ is balanced by the weight of the column in the gravitational field ρgh. Therefore, $2\gamma/r = \rho gh$ and the capillary rise or depression is given by

$$h = \frac{2\gamma}{\rho gr} \qquad (8.31)$$

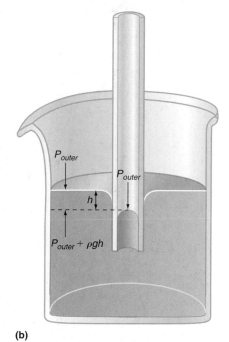

(b)

FIGURE 8.19
(a) If the liquid wets the interior wall of the capillary, a capillary rise is observed. The combination Pyrex–water exhibits this behavior. (b) If the liquid does not wet the capillary surface, a capillary depression is observed. The combination Pyrex–mercury exhibits this behavior.

In the preceding discussion, it was assumed that either (1) the liquid completely wets the interior surface of the capillary, in which case the liquid coats the capillary walls but does not fill the core, or (2) the liquid is completely nonwetting, in which case the liquid does not coat the capillary walls but fills the core. In a more realistic model, the interaction is intermediate between these two extremes. In this case, the liquid–surface is characterized by the **contact angle** θ, as shown in Figure 8.20.

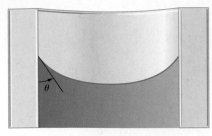

FIGURE 8.20
For cases intermediate between wetting and nonwetting, the contact angle θ lies in the range $0° < \theta < = 180°$.

Complete **wetting** corresponds to $\theta = 0°$, and complete **nonwetting** corresponds to $\theta = 180°$. For intermediate cases,

$$P_{inner} = P_{outer} + \frac{2\gamma \cos \theta}{r} \quad \text{and} \quad h = \frac{2\gamma \cos \theta}{\rho g r} \qquad (8.32)$$

The measurement of the contact angle is one of the main experimental methods used to measure the difference in surface tension at the solid–liquid interface.

EXAMPLE PROBLEM 8.3

The six-legged water strider supports itself on the surface of a pond on four of its legs. Each of these legs causes a depression to be formed in the pond surface. Assume that each depression can be approximated as a hemisphere of radius 1.2×10^{-4} m and that θ (as in Figure 8.20) is $0°$. Calculate the force that one of the insect's legs exerts on the pond.

Solution

$$\Delta P = \frac{2\gamma \cos \theta}{r} = \frac{2 \times 71.99 \times 10^{-3}\,\text{N m}^{-1}}{1.2 \times 10^{-4}\,\text{m}} = 1.20 \times 10^3\,\text{Pa}$$

$$F = PA = P \times \pi r^2 = 1.20 \times 10^3\,\text{Pa} \times \pi (1.2 \times 10^{-4}\,\text{m})^2 = 5.4 \times 10^{-5}\,\text{N}$$

EXAMPLE PROBLEM 8.4

Water is transported upward in trees through channels in the trunk called xylem. Although the diameter of the xylem channels varies from species to species, a typical value is 2.0×10^{-5} m. Is capillary rise sufficient to transport water to the top of a redwood tree that is 100 m high? Assume complete wetting of the xylem channels.

Solution

From Equation (8.31),

$$h = \frac{2\gamma}{\rho g r \cos \theta} = \frac{2 \times 71.99 \times 10^{-3}\,\text{N m}^{-1}}{997\,\text{kg m}^{-3} \times 9.81\,\text{m s}^{-2} \times 2.0 \times 10^{-5}\,\text{m} \times 1} = 0.74\,\text{m}$$

No, capillary rise is not sufficient to account for water supply to the top of a redwood tree.

As Example Problem 8.4 shows, capillary rise is insufficient to account for water transport to the leaves in all but the smallest plants. The property of water that accounts for water supply to the top of a redwood is its high **tensile strength.** Imagine pulling on a piston and cylinder containing only liquid water to create a negative pressure. How hard can a person pull on the water without "breaking" the water column? The answer to this question depends on whether bubbles are nucleated in the liquid. This phenomenon is called cavitation. If cavitation occurs, the bubbles will grow rapidly as the piston is pulled outward. The bubble pressure is given by Equations (8.31), where $P_{external}$ is the vapor pressure of water. The height of the water column in this case is limited to about 9.7 m. However, bubble nucleation is a kinetic phenomenon initiated at certain sites at the wall surrounding the water, and under the conditions present in xylem tubes it is largely suppressed. In the absence of bubble nucleation, theoretical calculations predict that negative pressure in excess of 1000 atm can be generated. The pressure is negative because the water is under tension rather than compression. Experiments on water inclusions in very small cracks in natural rocks have verified these estimates. However, bubble nucleation occurs at much lower negative pressures in capillaries similar in diameter to xylem tubes. Even in these capillaries, negative pressures of more than 50 atm have been observed.

How does the high tensile strength of water explain the transport of water to the top of a redwood? If one cuts into a tree near its base, the sap oozes rather than spurts out, showing that the pressure in the xylem tubes is ~ 1 atm at the base of a tree, whereas the pressure at the base of a tall water column would be much greater than 1 atm. Imagine the redwood in its infancy as a seedling. Capillary rise is sufficient to fill the xylem tubes to the top of the plant. As the tree grows higher, the water is pulled upward because of its high tensile strength. As the height of the tree increases, the pressure at the top becomes increasingly negative. As long as cavitation does not occur, the water column remains intact. As water evaporates from the leaves, it is resupplied from the roots through the pressure gradient in the xylem tubes that arises from the weight of the column. If the tree (and each xylem tube) grows to a height of ~ 100 m, and $P = 1$ atm at the base, the pressure at the top must be ~ -9 atm, from $\Delta P = \rho g h$. We again encounter a negative pressure because the water is under tension. If water did not have a sufficiently high tensile strength, gas bubbles would form in the xylem tubes. This would disrupt the flow of sap, and tall trees could not exist.

S U P P L E M E N T A L

8.9 Chemistry in Supercritical Fluids

Chemical reactions can take place in the gas phase, in solution (homogeneous reactions) or on the surfaces of solids (heterogeneous reactions). Solvents with suitable properties can influence both the yield and the selectivity of reactions in solution. It has been found that the use of supercritical fluids as solvents increases the number of parameters that chemists have at their disposal to tune a reaction system to meet their needs.

Supercritical fluids (SCFs) near the critical point with reduced temperature and pressures in the range $T_r \sim 1.0–1.1$ and $P_r \sim 1–2$ have a density that is an appreciable fraction of the liquid–phase density. They are unique in that they exhibit favorable properties of liquids and gases. Because the density is high, the solubility of solid substances is quite high, but the diffusion of solutes in the fluid is higher than in the normal liquid. This is the case because the density of the SCF is lower than that in normal liquids. For similar reasons, the viscosity of SCFs is lower than that in normal liquids. As a result, mass transfer is faster, and the overall reaction rate can be increased. Because an SCF is more gas-like than a liquid, the solubility of gases can be much higher in the SCF. This property is of particular usefulness in enhancing the reactivity of reactions in which one of the reactants is a gas, such as oxidation or hydrogenation. For example, hydrogen generally has a low solubility in organic solvents but is quite soluble in organic SCFs.

Carbon dioxide and water exhibit the unusual properties of SCFs. Because the values for the critical pressure and temperature of CO_2 ($P_c = 73.74$ bar and $T_c = 304$ K) are easily attainable, it is possible to use chemically inert supercritical CO_2 to replace toxic organic solvents in the dry cleaning industry. Although supercritical CO_2 is a good solvent for nonpolar molecules, it must be mixed with other substances to achieve a required minimum solubility for polar substances. This can be done without increasing the toxicity of the process greatly. Supercritical CO_2 is also used in the decaffeination of coffee and tea.

The critical constants of H_2O are considerably higher ($P_c = 220.64$ bar and $T_c = 647$ K), so the demands on the vessel used to contain the supercritical fluid are higher than for CO_2. However, as supercritical H_2O is formed, many of the hydrogen bonds in the normal liquid are broken. As a result, the dielectric constant can be varied between the normal value for the liquid of 80 and 5. At the lower end of this range, supercritical water acts like a nonpolar solvent, and is effective in dissolving organic materials. Supercritical water has considerable potential for use at the high temperatures at which organic solvents begin to decompose. For example, it can be used to destroy toxic substances through oxidation in the decontamination of contaminated groundwater. One challenge in using supercritical water is that it is highly corrosive, placing demands on the materials used to construct a practical facility.

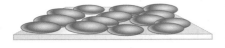

FIGURE 8.21
Liquid crystals are generally formed from polar organic molecules with a rod-like shape.

8.10 Liquid Crystal Displays

Liquid crystals are an exception to the general statement (superfluid He is another) that there are three equilibrium states of matter, namely solids, liquids, and gases. **Glasses** are liquids of such high viscosity that they cannot achieve equilibrium on the timescale of a human lifetime. Glasses are a commonly encountered nonequilibrium state of matter. An example is SiO_2 in the form of window glass. The properties of **liquid crystals** are intermediate between liquids and solids. Molecules that form liquid crystals typically contain rod-shaped structural elements. Several such molecules are shown in Figure 8.21.

How do liquid crystals differ from the other states of matter? The ordering in solid, liquid, and liquid crystal phases of such molecules is shown schematically in Figure 8.22. Whereas the crystalline solid phase is perfectly ordered and the liquid phase has no residual order, the liquid crystal phase retains some order. The axes of all molecules deviate only by a small amount from their value in the liquid crystal structure shown.

The **twisted nematic phase** shown in Figure 8.23 is of particular importance because liquid crystals with this structure are the basis of the multibillion-dollar liquid crystal display (LCD) industry and also the basis for sensor strips that change color with temperature. The twisted nematic phase consists of parallel planes in which the angle of the preferred orientation direction increases in a well-defined manner as the number of layers increases. If light is incident on such a crystal, it is partially reflected and partially transmitted from a number of layers. A maximum in the reflection occurs if the angle of incidence and the spacing between layers is such that constructive interference occurs in the light reflected from successive layers. Because this occurs only in a narrow range of wavelengths, the film appears colored, with the color determined by the wavelength. As the temperature increases or decreases, the layer spacing and, therefore, the color of the reflected light changes. For this reason, the color of the liquid crystal film changes with temperature.

The way in which an **LCD display** functions is illustrated in Figure 8.24. A twisted nematic liquid crystal film is sandwiched between two transparent conducting electrodes. The thickness of the film is such that the orientational direction rotates by a total of 90° from the bottom to the top of the film. Ambient light is incident on an upper polarizing filter that allows only one direction of polarization to pass. The rod-like molecules act as a waveguide and rotate the plane of polarization as the light passes through the liquid crystal film. Therefore, the light passes through the lower polarizer, which is rotated by 90° with respect to the first.

If an electric field is applied to the electrodes, the rotation of the orientational direction from plane to plane no longer occurs, because the polar molecules align along the

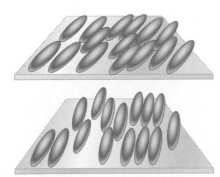

FIGURE 8.23
In a twisted nematic phase, the direction of preferential orientation of the rod-like molecules increases in a well-defined manner from one plane to the next.

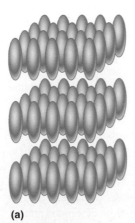

(a)

(b)

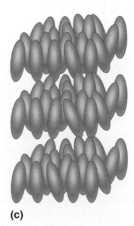

(c)

FIGURE 8.22
Solid **(a)** liquid **(b)** and liquid crystal **(c)** phases are shown.

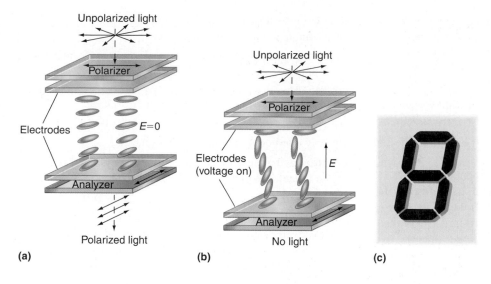

(a) **(b)** **(c)**

FIGURE 8.24
An LCD consists of a twisted nematic liquid film enclosed between parallel transparent conducting electrodes. Polarizers whose transmission direction is rotated 90° with respect to one another are mounted on the electrodes. **(a)** Light passing through the first polarizer is transmitted by the second polarizer because the plane of polarization of the light is rotated by the crystal. **(b)** The orientational ordering of the twisted nematic phase is destroyed by application of the electric field. No light is passed. **(c)** Arrangement of electrodes in an LCD alphanumeric display.

electric field. Therefore, the plane of polarization of the light transmitted by the upper polarizer is not rotated as it passes through the film. As a result, the light is not able to pass through the second polarizer. Now imagine the lower electrode to be in the form of a mirror. In the absence of the electric field (Figure 8.24a), the display will appear bright because the plane of polarization of the light reflected from the mirror is again rotated 90° as it passes back through the film. Consequently, it passes through the upper polarizer. By contrast, no light is reflected if the field is on (Figure 8.24b), and the display appears dark. Now imagine each of the electrodes to be patterned as shown in Figure 8.24c, and it will be clear why in a liquid crystal watch display dark numbers are observed on a light background.

Vocabulary

boiling point elevation	freezing point elevation	P–V–T phase diagram
capillary depression	glass	standard boiling temperature
capillary rise	LCD display	sublimation temperature
Clapeyron equation	liquid crystals	supercritical fluids
Clausius-Clapeyron equation	nonwetting	surface tension
coexistence curve	normal boiling temperature	tensile strength
contact angle	phase	triple point
critical point	phase diagram	Trouton's rule
degrees of freedom	phase rule	twisted nematic phase
freeze drying	P–T phase diagram	vapor pressure
freezing point depression	P–V phase diagram	wetting

Conceptual Problems

Q8.1 Why is it reasonable to show the μ versus T segments for the three phases as straight lines as is done in Figure 8.1? More realistic curves would have some curvature. Is the curvature upward or downward on a μ versus T plot? Explain your answer.

Q8.2 Why do the temperature versus heat curves in the solid, liquid, and gas regions of Figure 8.7 have different slopes?

Q8.3 Figure 8.7 is not drawn to scale. What would be the relative lengths on the q_P axis of the liquid + solid, liquid, and liquid + gas segments for water if the drawing were to scale and the system consisted of H_2O?

Q8.4 Show the paths $n \rightarrow o \rightarrow p \rightarrow q$ and $a \rightarrow b \rightarrow c \rightarrow d \rightarrow e \rightarrow f$ of the P–V–T phase diagram of Figure 8.15 in the P–T phase diagram of Figure 8.4.

Q8.5 At a given temperature, a liquid can coexist with its gas at a single value of the pressure. However, you can sense the presence of $H_2O(g)$ above the surface of a lake by the humidity, and it is still there if the barometric pressure rises or falls at constant temperature. How is this possible?

Q8.6 Why are the triple point temperature and the normal freezing point very close in temperature for most substances?

Q8.7 Give a molecular level explanation as to why the surface tension of $Hg(l)$ is not zero.

Q8.8 A vessel containing a liquid is opened inside an evacuated chamber. Will you see a liquid–gas interface if the volume of the initially evacuated chamber is (a) less than the critical volume, (b) a factor of 10 larger than the critical volume, and (c) a factor of 1.05 larger than the critical volume?

Q8.9 Why are there no points in the phase diagram for sulfur in Figure 8.11 that show rhombic and monoclinic solid phases in equilibrium with liquid and gaseous sulfur?

Q8.10 What is the physical origin of the pressure difference across a curved liquid–gas interface?

Q8.11 A triple point refers to a point in a phase diagram for which three phases are in equilibrium. Do all triple points correspond to a gas–liquid–solid equilibrium?

Q8.12 Why does the liquid–gas coexistence curve in a $P-T$ phase diagram end at the critical point?

Q8.13 How can you get a $P-T$ phase diagram from a $P-V-T$ phase diagram?

Q8.14 Why does the triple point in a $P-T$ diagram become a triple line in a $P-V$ diagram?

Q8.15 Why does water have several different solid phases but only one liquid and one gaseous phase?

Q8.16 As the pressure is increased at $-45°C$, ice I is converted to ice II. Which of these phases has the lower density?

Q8.17 Why is $\Delta H_{sublimation} = \Delta H_{fusion} + \Delta H_{vaporization}$?

Q8.18 What can you say about the density of liquid and gaseous water as the pressure approaches the critical value from lower values? Assume that the temperature is constant at the critical value.

Q8.19 What can you say about $\Delta H_{vaporization}$ of a liquid as the temperature approaches the critical temperature?

Q8.20 Is the following statement correct? *Because dry ice sublimes at 298 K, carbon dioxide has no liquid phase.* Explain your answer.

Q8.21 Redraw Figure 8.13 indicating the four-step process d (blue arrows) in Figure 8.4.

Numerical Problems

Problem numbers in red indicate that the solution to the problem is given in the *Student's Solutions Manual.*

P8.1 Use the vapor pressures of tetrachloromethane given in the following table to calculate the enthalpy of vaporization using a graphical method or a least squares fitting routine.

T (K)	P (Pa)	T (K)	P (Pa)
280.	6440.	320.	37130.
290.	10540.	330.	53250.
300.	16580.	340.	74520.
310.	25190.		

P8.2 The vapor pressure of a liquid can be written in the empirical form known as the Antoine equation, where $A(1)$, $A(2)$, and $A(3)$ are constants determined from measurements:

$$\ln \frac{P(T)}{Pa} = A(1) - \frac{A(2)}{\dfrac{T}{K} + A(3)}$$

Starting with this equation, derive an equation giving $\Delta H_{vaporization}$ as a function of temperature.

P8.3 In Section 8.8, it is stated that the maximum height of a water column in which cavitation does not occur is 9.7 m. Show that this is the case at 298 K.

P8.4 Use the vapor pressures for tetrachloromethane given in the following table to estimate the temperature and pressure of the triple point and also the enthalpies of fusion, vaporization, and sublimation.

Phase	T (K)	P (Pa)
Solid	230.	225.7
Solid	250.	905
Liquid	280.	6440.
Liquid	340.	62501

P8.5 Within what range can you restrict the values of P and T if the following information is known about CO_2? Use Figure 8.12 to answer this question.

a. As the temperature is increased, the solid is first converted to the liquid and subsequently to the gaseous state.

b. An interface delineating liquid and gaseous phases is observed throughout the pressure range between 6 and 65 atm.

c. Solid, liquid, and gas phases coexist at equilibrium.

d. Only a liquid phase is observed in the pressure range from 10. to 50. atm.

e. An increase in temperature from $-80.°$ to $20.°C$ converts a solid to a gas with no intermediate liquid phase.

P8.6   A *P–T* phase diagram for potassium is shown next.

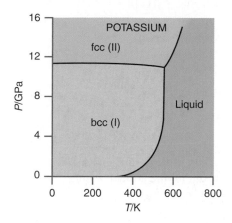

Source: Phase Diagrams of the Elements by David A. Young. © 1991 Regents of the University of California. Published by the University of California Press.

a. Which phase has the higher density, the fcc or the bcc phase? Explain your answer.

b. Indicate the range of *P* and *T* in the phase diagram for which fcc and liquid potassium are in equilibrium. Does fcc potassium float on or sink in liquid potassium? Explain your answer.

c. Redraw this diagram for a different pressure range and indicate where you expect to find the vapor phase. Explain how you chose the slope of your liquid–vapor coexistence line.

P8.7   A cell is roughly spherical with a radius of 20.0×10^{-6} m. Calculate the work required to expand the cell surface against the surface tension of the surroundings if the radius increases by a factor of three. Assume the cell is surrounded by pure water and that $T = 298.15$ K.

P8.8   It has been suggested that the surface melting of ice plays a role in enabling speed skaters to achieve peak performance. Carry out the following calculation to test this hypothesis. At 1 atm pressure, ice melts at 273.15 K, $\Delta H_{fusion} = 6010.$ J mol^{-1}, the density of ice is 920. kg m^{-3}, and the density of liquid water is 997. kg m^{-3}.

a. What pressure is required to lower the melting temperature by 4.0°C?

b. Assume that the width of the skate in contact with the ice has been reduced by sharpening to 19×10^{-3} cm, and that the length of the contact area is 18 cm. If a skater of mass 78 kg is balanced on one skate, what pressure is exerted at the interface of the skate and the ice?

c. What is the melting point of ice under this pressure?

d. If the temperature of the ice is −4.0°C, do you expect melting of the ice at the ice-skate interface to occur?

P8.9   Answer the following questions using the *P–T* phase diagram for carbon shown next.

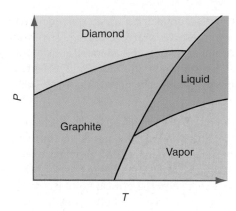

a. Which substance is more dense, graphite or diamond? Explain your answer.

b. Which phase is more dense, graphite or liquid carbon? Explain your answer.

c. Why does the phase diagram have two triple points? Explain your answer.

P8.10   You have a compound dissolved in chloroform and need to remove the solvent by distillation. Because the compound is heat sensitive, you hesitate to raise the temperature above 5.00°C and decide on vacuum distillation. What is the maximum pressure at which the distillation takes place?

P8.11   Use the vapor pressures for hexane given in the following table to estimate the temperature and pressure of the triple point and also the enthalpies of fusion, vaporization, and sublimation.

Phase	*T* (K)	*P* (Pa)
Solid	168	0.1296
Solid	178	1.111
Liquid	290.	1.400×10^4
Liquid	320.	4.822×10^4

P8.12   Are the following two *P–T* phase diagrams likely to be observed for a pure substance? If not, explain all features of the diagram that will not be observed.

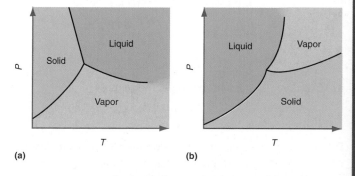

P8.13   Autoclaves that are used to sterilize surgical tools require a temperature of 120.°C to kill some bacteria. If water is used for this purpose, at what pressure must the autoclave operate?

P8.14 You have collected a tissue specimen that you would like to preserve by freeze drying. To ensure the integrity of the specimen, the temperature should not exceed $-5.00\,°C$. The vapor pressure of ice at 273.16 K is 624 Pa. What is the maximum pressure at which the freeze drying can be carried out?

P8.15 The phase diagram of NH_3 can be characterized by the following information. The normal melting and boiling temperatures are 195.2 and 239.82 K, respectively, and the triple point pressure and temperature are 6077 Pa and 195.41 K, respectively. The critical point parameters are 112.8×10^5 Pa and 405.5 K. Make a sketch of the $P-T$ phase diagram (not necessarily to scale) for NH_3. Place a point in the phase diagram for the following conditions. State which and how many phases are present.

a. 195.41 K, 9506 Pa

b. 195.41 K, 6077 Pa

c. 245.45 K, 101,325 Pa

d. 415 K, 105×10^5 Pa

e. 185.5 K, 6077 Pa

P8.16 Use the vapor pressures of ice given here to calculate the enthalpy of sublimation using a graphical method or a least squares fitting routine.

T (K)	P (Torr)
200.	0.1676
210.	0.7233
220.	2.732
230.	9.195
240.	27.97
250.	77.82
260.	200.2

P8.17 Calculate the vapor pressure for a mist of spherical water droplets of radius (a) 1.95×10^{-8} m and (b) 2.25×10^{-6} m surrounded by water vapor at 298 K. The vapor pressure of water at this temperature is 25.2 Torr.

P8.18 The vapor pressure of ethanol(l) is given by

$$\ln\left(\frac{P}{\text{Pa}}\right) = 23.58 - \frac{3.6745 \times 10^3}{\dfrac{T}{K} - 46.702}$$

a. Calculate the standard boiling temperature.

b. Calculate $\Delta H_{vaporization}$ at 298 K and at the standard boiling temperature.

P8.19 Use the following vapor pressures of propane given here to calculate the enthalpy of vaporization using a graphical method or a least squares fitting routine.

P8.20 The vapor pressure of liquid benzene is 20170 Pa at 298.15 K, and $\Delta H_{vaporization} = 30.72$ kJ mol^{-1} at 1 atm

T (K)	P (Pa)
100.	0.01114
120.	2.317
140.	73.91
160.	838.0
180.	5054
200.	2.016×10^4
220.	6.046×10^4

pressure. Calculate the normal and standard boiling points. Does your result for the normal boiling point agree with that in Table 8.3? If not, suggest a possible cause.

P8.21 Benzene (l) has a vapor pressure of 0.1269 bar at 298.15 K and an enthalpy of vaporization of 30.72 kJ mol^{-1}. The $C_{P,m}$ of the vapor and liquid phases at that temperature are 82.4 and 136.0 J K^{-1} mol^{-1}, respectively. Calculate the vapor pressure of $C_6H_6(l)$ at 340.0 K assuming

a. that the enthalpy of vaporization does not change with temperature.

b. that the enthalpy of vaporization at temperature T can be calculated from the equation $\Delta H_{vaporization}(T) = \Delta H_{vaporization}(T_0) + \Delta C_P(T - T_0)$ assuming that ΔC_P does not change with temperature.

P8.22 Use the values for ΔG_f° (CCl_4, l) and ΔG_f° (CCl_4, g) from Appendix B to calculate the vapor pressure of CCl_4 at 298.15 K.

P8.23 Calculate the vapor pressure of water droplets of radius 1.00×10^{-8} m at 360. K in equilibrium with water vapor. Use the tabulated value of the density and the surface tension at 298 K from Appendix B for this problem. (*Hint*: You need to calculate the vapor pressure of water at this temperature.)

P8.24 The vapor pressure of an unknown solid is given by $\ln(P/\text{Torr}) = 22.413 - 2035(K/T)$, and the vapor pressure of the liquid phase of the same substance is approximately given by $\ln(P/\text{Torr}) = 18.352 - 1736(K/T)$.

a. Calculate $\Delta H_{vaporization}$ and $\Delta H_{sublimation}$.

b. Calculate ΔH_{fusion}.

c. Calculate the triple point temperature and pressure.

P8.25 For water, $\Delta H_{vaporization}$ is 40.656 kJ mol^{-1}, and the normal boiling point is 373.12 K. Calculate the boiling point for water on the top of Mt. Everest (elevation 8848 m), where the barometric pressure is 253 Torr.

P8.26 Calculate the difference in pressure across the liquid–air interface for a (a) mercury and (b) methanol droplet of radius 125 nm.

P8.27 Calculate the vapor pressure of $CH_3OH(l)$ at 298.15 K if He is added to the gas phase at a partial pressure of 200. bar using the data tables. By what factor does the vapor pressure change?

P8.28 Use the vapor pressures of $SO_2(l)$ given in the following table to calculate the enthalpy of vaporization using a graphical method or a least squares fitting routine.

T (K)	P (Pa)	T (K)	P (Pa)
190.	824.1	240.	3.211×10^4
200.	2050	250.	5.440×10^4
210.	4591	260.	8.793×10^4
220.	9421		
230.	1.795×10^4		

P8.29 Prove that a substance for which the solid–liquid coexistence curve has a negative slope contracts upon melting.

P8.30 Are the following two P–T phase diagrams likely to be observed for a pure substance? If not, explain all features of the diagram that will not be observed.

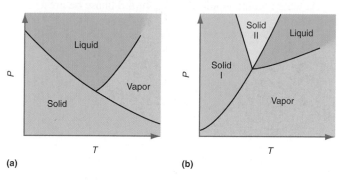

(a) (b)

P8.31 The vapor pressure of methanol(l) is 16.94×10^3 Pa at 298.15 K. Use this value to calculate ΔG_f° (CH_3OH, g) $- \Delta G_f^\circ$ (CH_3OH, l). Compare your result with those in Table 4.1.

P8.32 Within what range can you restrict the values of P and/or T if the following information is known about sulfur? Use Figure 8.11 to answer this problem.

a. Only the rhombic solid phase is observed for $P = 1$ atm.

b. When the pressure on the vapor is increased, the monoclinic solid phase is formed.

c. Solid, liquid, and gas phases coexist at equilibrium.

d. As the temperature is increased, the rhombic solid phase is converted to the monoclinic solid phase.

e. As the temperature is increased at 1 atm, the rhombic solid phase is converted to the liquid directly.

P8.33 The normal melting point of H_2O is 273.15 K, and $\Delta H_{fusion} = 6010.$ J mol^{-1}. Calculate the change in the normal freezing point at 100. and 500. bar compared to that at 1 bar assuming that the density of the liquid and solid phases remains constant at 997 and 917 kg m^{-3}, respectively. Explain why your answer is positive (or negative).

P8.34 Carbon tetrachloride melts at 250. K. The vapor pressure of the liquid is 10,539 Pa at 290. K and 74,518 Pa at 340. K. The vapor pressure of the solid is 270. Pa at 232 K and 1092 Pa at 250. K.

a. Calculate $\Delta H_{vaporization}$ and $\Delta H_{sublimation}$.

b. Calculate ΔH_{fusion}.

c. Calculate the normal boiling point and $\Delta S_{vaporization}$ at the boiling point.

d. Calculate the triple point pressure and temperature.

P8.35 In Equations (8.16), $(dP/dT)_{vaporization}$ was calculated by assuming that $V_m^{gas} \gg V_m^{liquid}$. In this problem, you will test the validity of this approximation. For water at its normal boiling point of 373.12 K, $\Delta H_{vaporization} = 40.656 \times 10^3$ J mol^{-1}, $\rho_{liquid} = 958.66$ kg m^{-3}, and $\rho_{gas} = 0.58958$ kg m^{-3}. Compare the calculated values for $(dP/dT)_{vaporization}$ with and without the approximation of Equation (8.16). What is the relative error in making the approximation?

P8.36 The densities of a given solid and liquid of molar mass 122.5 g mol^{-1} at its normal melting temperature of 427.15 K are 1075 and 1012 kg m^{-3}, respectively. If the pressure is increased to 120. bar, the melting temperature increases to 429.35 K. Calculate ΔH_{fusion}° and ΔS_{fusion}° for this substance.

P8.37 The variation of the vapor pressure of the liquid and solid forms of a pure substance near the triple point are given by $\ln \dfrac{P_{solid}}{Pa} = -8750 \dfrac{K}{T} + 34.143$ and

$\ln \dfrac{P_{liquid}}{Pa} = -4053 \dfrac{K}{T} + 21.10$. Calculate the temperature and pressure at the triple point.

P8.38 Use the vapor pressures of n-butane given in the following table to calculate the enthalpy of vaporization using a graphical method or a least squares fitting routine.

T (°C)	P (Pa)	T (°C)	P (Pa)
−134.3	1.000	−49.10	1.000×10^4
−121.0	10.00	−0.800	1.000×10^5
−103.9	100.0		
−81.10	1000.		

P8.39 At 298.15 K, ΔG_f° (HCOOH, g) $= -351.0$ kJ mol^{-1} and ΔG_f° (HCOOH, l) $= -361.4$ kJ mol^{-1}. Calculate the vapor pressure of formic acid at this temperature.

P8.40 In this problem, you will calculate the differences in the chemical potentials of ice and supercooled water and of steam and superheated water, all at 1 atm pressure shown schematically in Figure 8.1. For this problem,

$S^{\circ}_{H_2O,\,s} = 48.0$ J mol^{-1} K^{-1}, $S^{\circ}_{H_2O,\,l} = 70.0$ J mol^{-1} K^{-1}, and $S^{\circ}_{H_2O,\,g} = 188.8$ J mol^{-1} K^{-1}.

a. By what amount does the chemical potential of water exceed that of ice at $-2.25°C$?

b. By what amount does the chemical potential of water exceed that of steam at $102.25°C$?

P8.41 Calculate the vapor pressure of a droplet of benzene of radius 1.25×10^{-8} m at $38.0°C$ in equilibrium with its vapor. Use the tabulated value of the density and the surface tension at 298 K from Appendix B for this problem. (*Hint*: You need to calculate the vapor pressure of benzene at this temperature.)

P8.42 Solid iodine, $I_2(s)$, at $25.0°C$ has an enthalpy of sublimation of 56.30 kJ mol^{-1}. The C_{Pm} of the vapor and solid phases at that temperature are 36.9 and 54.4 J K^{-1} mol^{-1}, respectively. The sublimation pressure at $25.0°C$ is 0.30844 Torr. Calculate the sublimation pressure of the solid at the melting point ($113.6°C$) assuming

a. that the enthalpy of sublimation does not change with temperature.

b. that the enthalpy of sublimation at temperature T can be calculated from the equation $\Delta H_{sublimation}(T) = \Delta H_{sublimation}(T_0) + \Delta C_P(T - T_0)$ and ΔC_P does not change with T.

P8.43 Consider the transition between two forms of solid tin, $Sn(s, gray) \rightarrow Sn(s, white)$. The two phases are in equilibrium at 1 bar and $18°C$. The densities for gray and white tin are 5750 and 7280 kg m^{-3}, respectively, and the molar entropies for gray and white tin are 44.14 and 51.18 J K^{-1} mol^{-1}, respectively. Calculate the temperature at which the two phases are in equilibrium at 350. bar.

P8.44 A reasonable approximation to the vapor pressure of krypton is given by $\log_{10}(P/\text{Torr}) = b - 0.05223(a/T)$. For solid krypton, $a = 10{,}065$ and $b = 7.1770$. For liquid krypton, $a = 9377.0$ and $b = 6.92387$. Use these formulas to estimate the triple point temperature and pressure and also the enthalpies of vaporization, fusion, and sublimation of krypton.

P8.45 20.0 g of water is in a container of 20.0 L at 298.15 K. The vapor pressure of water at this temperature is 23.76 Torr.

a. What phases are present?

b. At what volume would only the gas phase be present?

c. At what volume would only the liquid phase be present?

Ideal and Real Solutions

In an ideal solution of A and B, the A–B interactions are the same as the A–A and B–B interactions. In this case, the relationship between the solution concentration and gas-phase partial pressure is described by Raoult's law for each component. In an ideal solution, the vapor over a solution is enriched in the most volatile component, allowing a separation into its components through fractional distillation. Nonvolatile solutes lead to a decrease of the vapor pressure above a solution. Such solutions exhibit a freezing point depression and a boiling point elevation. These properties depend only on the concentration and not on the identity of the nonvolatile solutes for an ideal solution. Real solutions are described by a modification of the ideal dilute solution model. In an ideal dilute solution, the solvent obeys Raoult's law, and the solute obeys Henry's law. This model is limited in its applicability. To quantify deviations from the ideal dilute solution model, we introduce the concept of the *activity*. The activity of a component of the solution is defined with respect to a standard state. Knowledge of the activities of the various components of a reactive mixture is essential in modeling chemical equilibrium. The thermodynamic equilibrium constant for a reaction in solution is calculated by expressing the reaction quotient Q in terms of activities rather than concentrations.

9.1 Defining the Ideal Solution

In an ideal gas, the atoms or molecules do not interact with one another. Clearly, this is not a good model for a liquid, because without attractive interactions, a gas will not condense. The attractive interaction in liquids varies greatly, as shown by the large variation in boiling points among the elements; for instance, helium has a normal boiling point of 4.2 K, whereas hafnium has a boiling point of 5400 K.

In developing a model for solutions, the vapor phase that is in equilibrium with the solution must be taken into account. Consider pure liquid benzene in a beaker placed in a closed room. Because the liquid is in equilibrium with the vapor phase above it, there is a nonzero partial pressure of benzene in the air surrounding the beaker. This pressure is called the vapor pressure of benzene at the temperature of the liquid. What happens when toluene is added to the beaker? It is observed that the partial pressure of benzene is reduced, and the vapor phase now contains both benzene and toluene. For this particular mixture, the partial pressure of each component (i) above the liquid is given by

$$P_i = x_i P_i^* \quad i = 1, 2 \tag{9.1}$$

where x_i is the mole fraction of that component in the liquid. This equation states that the partial pressure of each of the two components is directly proportional to the vapor pressure of the corresponding pure substance P_i^* and that the proportionality constant is x_i. Equation (9.1) is known as **Raoult's law** and is the definition of an **ideal solution.** Raoult's law holds for each substance in an ideal solution over the range from $0 \leq x_i \leq 1$. In binary solutions, one refers to the component that has the higher value of x_i as the **solvent** and the component that has the lower value of x_i as the **solute.**

Few solutions satisfy Raoult's law. However, it is useful to study the thermodynamics of ideal solutions and to introduce departures from ideal behavior later. Why is Raoult's law not generally obeyed over the whole concentration range of a binary solution consisting of molecules A and B? Equation (9.1) is only obeyed if the A–A, B–B, and A–B interactions are all equally strong. This criterion is satisfied for a mixture of benzene and toluene because the two molecules are very similar in size, shape, and chemical properties. However, it is not satisfied for arbitrary molecules. Raoult's law is an example of a **limiting law;** the solvent in a real solution obeys Raoult's law as the solution becomes highly dilute.

Raoult's law is derived in Example Problem 9.1 and can be rationalized using the model depicted in Figure 9.1. In the solution, molecules of solute are distributed in the solvent. The solution is in equilibrium with the gas phase, and the gas-phase composition is determined by a dynamic balance between evaporation from the solution and condensation from the gas phase, as indicated for one solvent and one solute molecule in Figure 9.1.

FIGURE 9.1

Schematic model of a solution. The white and black spheres represent solvent and solute molecules, respectively.

EXAMPLE PROBLEM 9.1

Assume that the rates of evaporation R_{evap} and condensation R_{cond} of the solvent from the surface of pure liquid solvent are given by the expressions

$$R_{evap} = A k_{evap}$$

$$R_{cond} = A k_{cond} P_{solvent}^*$$

where A is the surface area of the liquid and k_{evap} and k_{cond} are the rate constants for evaporation and condensation, respectively. Derive a relationship between the vapor pressure of the solvent above a solution and above the pure solvent.

Solution

For the pure solvent, the equilibrium vapor pressure is found by setting the rates of evaporation and condensation equal:

$$R_{evap} = R_{cond}$$

$$A k_{evap} = A k_{cond} P_{solvent}^*$$

$$P_{solvent}^* = \frac{k_{evap}}{k_{cond}}$$

Next, consider the ideal solution. In this case, the rate of evaporation is reduced by the factor $x_{solvent}$ because only that fraction of the surface is available for evaporation of the solvent.

$$R_{evap} = A k_{evap} x_{solvent}$$

$$R_{cond} = A k_{cond} P_{solvent}$$

and at equilibrium

$$R_{evap} = R_{cond}$$

$$Ak_{evap}x_{solvent} = Ak_{cond}P_{solvent}$$

$$P_{solvent} = \frac{k_{evap}}{k_{cond}}x_{solvent} = P^*_{solvent}x_{solvent}$$

The derived relationship is Raoult's law.

9.2 The Chemical Potential of a Component in the Gas and Solution Phases

If the liquid and vapor phases are in equilibrium, the following equation holds for each component of the solution, where μ_i is the chemical potential of species i:

$$\mu_i^{solution} = \mu_i^{vapor} \tag{9.2}$$

Recall from Section 6.3 that the chemical potential of a substance in the gas phase is related to its partial pressure P_i by

$$\mu_i^{vapor} = \mu_i^{\circ} + RT \ln \frac{P_i}{P^{\circ}} \tag{9.3}$$

where μ_i° is the chemical potential of pure component i in the gas phase at the standard state pressure $P^{\circ} = 1$ bar. Because at equilibrium $\mu_i^{solution} = \mu_i^{vapor}$, Equation (9.3) can be written in the form

$$\mu_i^{solution} = \mu_i^{\circ} + RT \ln \frac{P_i}{P^{\circ}} \tag{9.4}$$

For pure liquid i in equilibrium with its vapor, $\mu_i^*(\text{liquid}) = \mu_i^*(\text{vapor}) = \mu_i^*$. Therefore, the chemical potential of the pure liquid is given by

$$\mu_i^* = \mu_i^{\circ} + RT \ln \frac{P_i^*}{P^{\circ}} \tag{9.5}$$

Subtracting Equation (9.5) from (9.4) gives

$$\mu_i^{solution} = \mu_i^* + RT \ln \frac{P_i}{P_i^*} \tag{9.6}$$

For an ideal solution, $P_i = x_i P_i^*$. Combining Equations (9.6) and (9.1), the central equation describing ideal solutions is obtained:

$$\mu_i^{solution} = \mu_i^* + RT \ln x_i \tag{9.7}$$

This equation relates the chemical potential of a component in an ideal solution to the chemical potential of the pure liquid form of component i and the mole fraction of that component in the solution. This equation is most useful in describing the thermodynamics of solutions in which all components are volatile and miscible in all proportions.

Keeping in mind that $\mu_i = G_{i,m}$, the form of Equation (9.7) is identical to that derived for the Gibbs energy of a mixture of gases in Section 6.5. Therefore, one can derive relations for the thermodynamics of mixing to form ideal solutions as was done for ideal gases in Section 6.6. These relations are shown in Equation (9.8). Note in particular that $\Delta H_{mixing} = \Delta V_{mixing} = 0$ for an ideal solution:

$$\Delta G_{mixing} = nRT \sum_i x_i \ln x_i$$

$$\Delta S_{mixing} = -\left(\frac{\partial \Delta G_{mixing}}{\partial T}\right)_{P,n_1,n_2} = -nR \sum_i x_i \ln x_i$$

$$\Delta V_{mixing} = \left(\frac{\partial \Delta G_{mixing}}{\partial P}\right)_{T,n_1,n_2} = 0 \quad \text{and}$$

$$\Delta H_{mixing} = \Delta G_{mixing} + T\Delta S_{mixing}$$

$$= nRT \sum_i x_i \ln x_i - T\left(nR \sum_i x_i \ln x_i\right) = 0 \tag{9.8}$$

EXAMPLE PROBLEM 9.2

An ideal solution is made from 5.00 mol of benzene and 3.25 mol of toluene. Calculate ΔG_{mixing} and ΔS_{mixing} at 298 K and 1 bar pressure. Is mixing a spontaneous process?

Solution

The mole fractions of the components in the solution are $x_{benzene} = 0.606$ and $x_{toluene} = 0.394$.

$$\Delta G_{mixing} = nRT \sum_i x_i \ln x_i$$

$$= 8.25 \text{ mol} \times 8.314 \text{ J mol}^{-1} \text{ K}^{-1}$$
$$\times 298 \text{ K} \times (0.606 \ln 0.606 + 0.394 \ln 0.394)$$

$$= -13.7 \times 10^3 \text{ J}$$

$$\Delta S_{mixing} = -nR \sum_i x_i \ln x_i$$

$$= -8.25 \text{ mol} \times 8.314 \text{ J mol}^{-1} \text{ K}^{-1} \times (0.606 \ln 0.606 + 0.394 \ln 0.394)$$

$$= 46.0 \text{ J K}^{-1}$$

Mixing is spontaneous because $\Delta G_{mixing} < 0$ for an ideal solution. If two liquids are miscible, it is always true that $\Delta G_{mixing} < 0$.

9.3 Applying the Ideal Solution Model to Binary Solutions

Although the ideal solution model can be applied to any number of components, to simplify the mathematics, the focus in this chapter is on binary solutions, which consist of only two components. Because Raoult's law holds for both components of the mixture, $P_1 = x_1 P_1^*$ and $P_2 = x_2 P_2^* = (1 - x_1)P_2^*$. The total pressure in the gas phase varies linearly with the mole fraction of each of its components in the liquid:

$$P_{total} = P_1 + P_2 = x_1 P_1^* + (1 - x_1)P_2^* = P_2^* + (P_1^* - P_2^*)x_1 \tag{9.9}$$

The individual partial pressures as well as P_{total} above a benzene–1,2 dichloroethane ($C_2H_4Cl_2$) solution are shown in Figure 9.2. Small deviations from Raoult's law are seen. Such deviations are typical because few solutions obey Raoult's law exactly. Nonideal solutions, which generally exhibit large deviations from Raoult's law, are discussed in Section 9.9.

The concentration unit used in Equation (9.9) is the mole fraction of each component in the liquid phase. The mole fraction of each component in the gas phase can also be calculated. Using the definition of the partial pressure and the symbols y_1 and y_2 to denote the gas-phase mole fractions, we can write

$$y_1 = \frac{P_1}{P_{total}} = \frac{x_1 P_1^*}{P_2^* + (P_1^* - P_2^*)x_1} \tag{9.10}$$

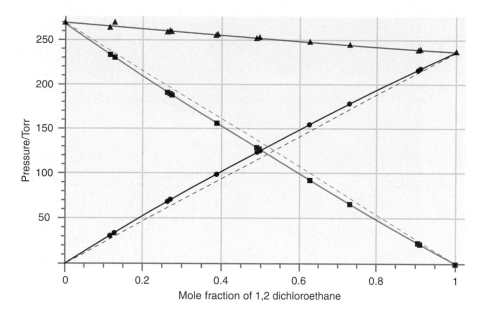

FIGURE 9.2
The vapor pressure of benzene (blue),
1,2 dichloroethane (red), and the total
vapor pressure (purple) above the solution
is shown as a function of the mole fraction
of 1,2 dichloroethane. The symbols are
data points [Zawidski, J. v. *Zeitschrift für
Physikalische ChePmie* 35 (1900): 129].
The solid curves are polynomial fits to the
data. The dashed lines are calculated using
Raoult's law.

To obtain the pressure in the vapor phase as a function of y_1, we first solve Equation (9.10) for x_1:

$$x_1 = \frac{y_1 P_2^*}{P_1^* + (P_2^* - P_1^*)y_1} \tag{9.11}$$

and obtain P_{total} from $P_{total} = P_2^* + (P_1^* - P_2^*)x_1$

$$P_{total} = \frac{P_1^* P_2^*}{P_1^* + (P_2^* - P_1^*)y_1} \tag{9.12}$$

Equation (9.12) can be rearranged to give an equation for y_1 in terms of the vapor pressures of the pure components and the total pressure:

$$y_1 = \frac{P_1^* P_{total} - P_1^* P_2^*}{P_{total}(P_1^* - P_2^*)} \tag{9.13}$$

The variation of the total pressure with x_1 and y_1 is not the same, as is seen in Figure 9.3. In Figure 9.3a, the system consists of a single-phase liquid for pressures above the curve and of a two-phase vapor–liquid mixture for points lying on the curve in the $P-x_1$ diagram. Only points lying above the curve are meaningful, because points lying below the curve do not correspond to equilibrium states at which liquid is present. In Figure 9.3b, the system consists of a single-phase vapor for pressures below the curve and of a two-phase vapor–liquid mixture for points lying on the curve in the $P-y_1$ diagram. Points lying above the curve do not correspond to equilibrium states at which vapor is present. The excess vapor would condense to form liquid.

Note that the pressure is plotted as a function of different variables in the two parts of Figure 9.3. To compare the gas phase and liquid composition at a given total pressure, both are graphed as a function of $Z_{benzene}$, which is called the **average composition** of benzene in the whole system in Figure 9.4. The average composition Z is defined by

$$Z_{benzene} = \frac{n_{benzene}^{liquid} + n_{benzene}^{vapor}}{n_{toluene}^{liquid} + n_{toluene}^{vapor} + n_{benzene}^{liquid} + n_{benzene}^{vapor}} = \frac{n_{benzene}}{n_{total}}$$

In the region labeled "Liquid" in Figure 9.4, the system consists entirely of a liquid phase, and $Z_{benzene} = x_{benzene}$. In the region labeled "Vapor," the system consists entirely of a gaseous phase and $Z_{benzene} = y_{benzene}$. The area separating the single-phase liquid and vapor regions corresponds to the two-phase liquid–vapor coexistence region.

We next apply the phase rule to the benzene–toluene ideal solution for which there are two components (benzene and toluene) and three independent state variables,

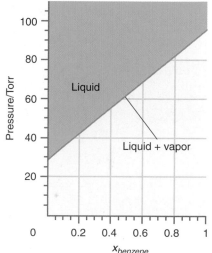

(a)

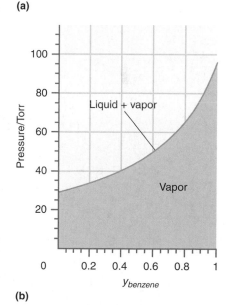

(b)

FIGURE 9.3
The total pressure above a
benzene–toluene ideal solution is shown
for different values of **(a)** the mole fraction of benzene in the solution and **(b)** the
mole fraction of benzene in the vapor.
Points on the curves correspond to
vapor–liquid coexistence. Only the curves
and the shaded areas are of physical
significance as explained in text.

A P–Z phase diagram is shown for a benzene-toluene ideal solution. The upper curve shows the vapor pressure as a function of $x_{benzene}$. The lower curve shows the vapor pressure as a function of $y_{benzene}$. Above the two curves, the system is totally in the liquid phase, and below the two curves, the system is totally in the vapor phase. The area intermediate between the two curves shows the liquid–vapor coexistence region. The horizontal lines connecting the curves are called tie lines.

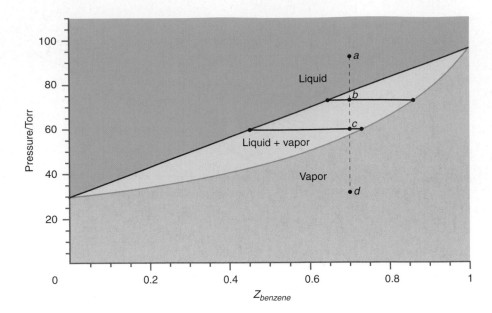

namely P, T, and $Z_{benzene}$ (or $Z_{toluene}$). Because $Z_{benzene} = 1 - Z_{toluene}$, these two variables are not independent. For a system containing C components, the phase rule discussed in Section 8.3 takes the form

$$F = C - p + 2 \qquad (9.14)$$

For the benzene–toluene ideal solution, $C = 2$ and $F = 4 - p$. Therefore, in the single phase regions labeled liquid and vapor in Figure 9.4, the system has three degrees of freedom. One of these degrees of freedom is already fixed because the temperature is held constant in Figure 9.4. Therefore, the two remaining degrees of freedom are P and $Z_{benzene}$. In the two-phase region labeled liquid + vapor, the system has only one degree of freedom. Once P is fixed, the Z values at the ends of the horizontal line fix $x_{benzene}$ and $y_{benzene}$ and therefore $Z_{benzene}$. We note that the number of components can be less than the number of chemical species. For example, a solution containing the three species A, B, and C linked by the chemical equilibrium $A + B \rightleftharpoons C$ contains only two independent components because only two of the three concentrations are independent variables.

To demonstrate the usefulness of this **pressure–average composition (P–Z) diagram,** consider a constant temperature process in which the pressure of the system is decreased from the value corresponding to point a in Figure 9.4. Because $P > P_{total}$, initially the system is entirely in the liquid phase. As the pressure is decreased at constant composition, the system remains entirely in the liquid phase until the constant composition line intersects the P versus $x_{benzene}$ curve. At this point, the system enters the two-phase vapor–liquid coexistence region. As point b is reached, what are the values for $x_{benzene}$ and $y_{benzene}$, the mole fractions of benzene in the liquid and vapor phases? These values can be determined by constructing a tie line in the two-phase coexistence region.

A **tie line** (see Figure 9.4) is a horizontal line at the pressure of interest that connects the P versus $x_{benzene}$ and P versus $y_{benzene}$ curves. Note that for all values of the pressure, $y_{benzene}$ is greater than $x_{benzene}$, showing that the vapor phase is always enriched in the more volatile or higher vapor pressure component in comparison with the liquid phase.

EXAMPLE PROBLEM 9.3

An ideal solution of 5.00 mol of benzene and 3.25 mol of toluene is placed in a piston and cylinder assembly. At 298 K, the vapor pressure of the pure substances are $P^*_{benzene} = 96.4$ Torr and $P^*_{toluene} = 28.9$ Torr.

a. The pressure above this solution is reduced from 760. Torr. At what pressure does the vapor phase first appear?

b. What is the composition of the vapor under these conditions?

Solution

a. The mole fractions of the components in the solution are $x_{benzene} = 0.606$ and $x_{toluene} = 0.394$. The vapor pressure above this solution is

$$P_{total} = x_{benzene} P_{benzene}^* + x_{toluene} P_{toluene}^*$$

$$= 0.606 \times 96.4 \text{ Torr} + 0.394 \times 28.9 \text{ Torr}$$

$$= 69.8 \text{ Torr}$$

No vapor will be formed until the pressure has been reduced to this value.

b. The composition of the vapor at a total pressure of 69.8 Torr is given by

$$y_{benzene} = \frac{P_{benzene}^* P_{total} - P_{benzene}^* P_{toluene}^*}{P_{total}\left(P_{benzene}^* - P_{toluene}^*\right)}$$

$$= \frac{96.4 \text{ Torr} \times 69.8 \text{ Torr} - 96.4 \text{ Torr} \times 28.9 \text{ Torr}}{69.8 \text{ Torr} \times (96.4 \text{ Torr} - 28.9 \text{ Torr})} = 0.837$$

$$y_{toluene} = 1 - y_{benzene} = 0.163$$

Note that the vapor is enriched relative to the liquid in the more volatile component, which has the higher vapor pressure and lower boiling temperature.

To calculate the relative amount of material in each of the two phases in a coexistence region, we derive the lever rule for a binary solution of the components A and B. Figure 9.5 shows a magnified portion of Figure 9.4 centered at the tie line that passes through point *b*. We derive the lever rule using the following geometrical argument. The lengths of the line segments *lb* and *bv* are given by

$$lb = Z_B - x_B = \frac{n_B^{tot}}{n^{tot}} - \frac{n_B^{liq}}{n_{liq}^{tot}} \qquad \textbf{(9.15)}$$

$$bv = y_B - Z_B = \frac{n_B^{vapor}}{n_{vapor}^{tot}} - \frac{n_B^{tot}}{n^{tot}} \qquad \textbf{(9.16)}$$

The superscripts on *n* indicate whether we are referring to the moles of component B in the vapor or liquid phases or the total number of moles of B in the system. If Equation (9.15) is multiplied by n_{liq}^{tot}, Equation (9.16) is multiplied by n_{vapor}^{tot}, and the two equations are subtracted, we find that

$$lb\, n_{liq}^{tot} - bv\, n_{vapor}^{tot} = \frac{n_B^{tot}}{n^{tot}}\left(n_{liq}^{tot} + n_{vapor}^{tot}\right) - \left(n_B^{liq} + n_B^{vapor}\right) = n_B^{tot} - n_B^{tot} = 0$$

We conclude that

$$\frac{n_{liq}^{tot}}{n_{vap}^{tot}} = \frac{bv}{lb} \qquad \textbf{(9.17)}$$

It is convenient to restate this result as

$$n_{liq}^{tot}(Z_B - x_B) = n_{vapor}^{tot}(y_B - Z_B) \qquad \textbf{(9.18)}$$

Equation (9.18) is called the **lever rule** by analogy with the torques acting on a lever of length $lb + bv$ with the fulcrum positioned at point *b*. For the specific case shown in Figure 9.5, $n_{liq}^{tot}/n_{vapor}^{tot} = 2.34$. Therefore, 70.1% of the total number of moles in the system is in the liquid phase, and 29.9% is in the vapor phase.

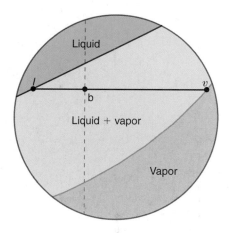

FIGURE 9.5
An enlarged region of the two-phase coexistence region of Figure 9.4 is shown. The vertical line through point *b* indicates a transition at constant system composition. The lever rule (see text) is used to determine what fraction of the system is in the liquid and vapor phases.

EXAMPLE PROBLEM 9.4

For the benzene–toluene solution of Example Problem 9.3, calculate

a. the total pressure

b. the liquid composition

c. the vapor composition

when 1.50 mol of the solution has been converted to vapor.

Solution

The lever rule relates the average composition, $Z_{benzene} = 0.606$, and the liquid and vapor compositions:

$$n_{vapor}^{tot}\left(y_{benzene} - Z_{benzene}\right) = n_{liq}^{tot}\left(Z_{benzene} - x_{benzene}\right)$$

Entering the parameters of the problem, this equation simplifies to

$$6.75x_{benzene} + 1.50y_{benzene} = 5.00$$

The total pressure is given by

$$P_{total} = x_{benzene}P_{benzene}^* + (1 - x_{benzene})P_{toluene}^*$$

$$= [96.4x_{benzene} + 28.9(1 - x_{benzene})]\text{Torr}$$

and the vapor composition is given by

$$y_{benzene} = \frac{P_{benzene}^* P_{total} - P_{benzene}^* P_{toluene}^*}{P_{total}(P_{benzene}^* - P_{toluene}^*)} = \left[\frac{96.4\dfrac{P_{total}}{\text{Torr}} - 2786}{67.5\dfrac{P_{total}}{\text{Torr}}}\right]$$

These three equations in three unknowns can be solved by using an equation solver or by eliminating the variables by combining equations. For example, the first equation can be used to express $y_{benzene}$ in terms of $x_{benzene}$. This result can be substituted into the second and third equations to give two equations in terms of $x_{benzene}$ and P_{total}. The solution for $x_{benzene}$ obtained from these two equations can be substituted in the first equation to give $y_{benzene}$. The relevant answers are $x_{benzene} = 0.561$, $y_{benzene} = 0.810$, and $P_{total} = 66.8$ Torr.

9.4 The Temperature–Composition Diagram and Fractional Distillation

The enrichment of the vapor phase above a solution in the more volatile component is the basis for fractional distillation, an important separation technique in chemistry and in the chemical industry. It is more convenient to discuss fractional distillation using a **temperature–composition diagram** than using the pressure–composition diagram discussed in the previous section. The temperature-composition diagram gives the temperature of the solution as a function of the average system composition for a predetermined total vapor pressure, P_{total}. It is convenient to use the value $P_{total} = 1$ atm so the vertical axis is the normal boiling point of the solution. Figure 9.6 shows a boiling temperature–composition diagram for a benzene–toluene solution. Neither the $T_b - x_{benzene}$ nor the $T_b - y_{benzene}$ curves are linear in a temperature–composition diagram. Note that because the more volatile component has the lower boiling point, the vapor and liquid regions are inverted when compared with the pressure–composition diagram.

The principle of **fractional distillation** can be illustrated using the sequence of lines labeled a through k in Figure 9.6. The vapor above the solution at point a is enriched in benzene. If this vapor is separated from the original solution and condensed by lowering the temperature, the resulting liquid will have a higher mole fraction of benzene than the original solution. As for the original solution, the vapor above this separately collected liquid is enriched in benzene. As this process is repeated, the successively collected vapor samples become more enriched in benzene. In the limit of a very large number of steps, the last condensed samples are essentially pure benzene. The multistep procedure described earlier is very cumbersome because it requires the collection and evaporation of many different samples. In practice, the separation into pure benzene and toluene is accomplished using a distillation column, shown schematically in Figure 9.7.

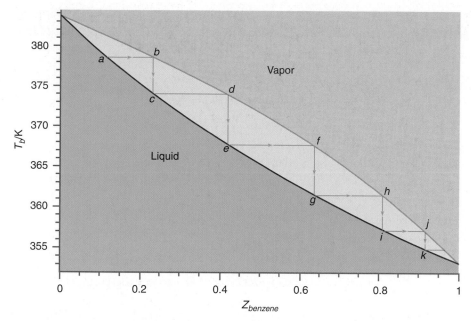

FIGURE 9.6
The boiling temperature of an ideal solution of the components benzene and toluene is plotted versus the average system composition, $Z_{benzene}$. The upper curve shows the boiling temperature as a function of $y_{benzene}$, and the lower curve shows the boiling temperature as a function of $x_{benzene}$. The area intermediate between the two curves shows the vapor–liquid coexistence region.

Rather than have the whole system at a uniform temperature, a distillation column operates with a temperature gradient so that the top part of the column is at a lower temperature than the solution being boiled off. The horizontal segments ab, cd, ef, and so on in Figure 9.6 correspond to successively higher (cooler) portions of the column. Each of these segments corresponds to one of the distillation stages in Figure 9.7. Because the vapor is moving upward through the condensing liquid, heat and mass exchange facilitates equilibration between the two phases.

Distillation plays a major role in the separation of crude oil into its various components. The lowest boiling liquid is gasoline, followed by kerosene, diesel fuel and heating oil, and gas oil, which is further broken down into lower molecular weight components by catalytic cracking. It is important in distillation to keep the temperature of the boiling liquid as low as possible to avoid the occurrence of thermally induced reactions. A useful way to reduce the boiling temperature is to distill the liquid under a partial vacuum, which is done by pumping away the gas phase. The boiling point of a typical liquid mixture can be reduced by approximately 100°C if the pressure is reduced from 760 to 20 Torr.

Although the principle of fractional distillation is the same for real solutions, it is not possible to separate a binary solution into its pure components if the nonideality is strong enough. If the A–B interactions are more attractive than the A–A and B–B interactions, the boiling point of the solution will go through a maximum at a concentration intermediate between $x_A = 0$ and $x_A = 1$. An example of such a case is an acetone–chloroform mixture. The hydrogen on the chloroform forms a hydrogen bond with the oxygen in acetone as shown in Figure 9.8, leading to stronger A–B than A–A and B–B interactions.

A boiling point diagram for this case is shown in Figure 9.9a. At the maximum boiling temperature, the liquid and vapor composition lines are tangent to one another. Fractional distillation of such a solution beginning with an initial value of x_A greater than that corresponding to the maximum boiling point is shown schematically in the figure. The component with the lowest boiling point will initially emerge at the top of the distillation column. In this case, it will be pure component A. However, the liquid left in the heated flask will not be pure component B, but rather the solution corresponding to the concentration at which the maximum boiling point is reached. Continued boiling of the solution at this composition will lead to evaporation at constant composition. Such a mixture is called an azeotrope, and because $T_{b,azeotrope} > T_{b,A}, T_{b,B}$, it is called a maximum boiling azeotrope. An example for a maximum boiling azeotrope is a mixture of H_2O ($T_b = 100°C$) and HCOOH ($T_b = 100°C$) at $x_{H_2O} = 0.427$, which boils at 107.2°C. Other commonly occurring azeotropic mixtures are listed in Table 9.1.

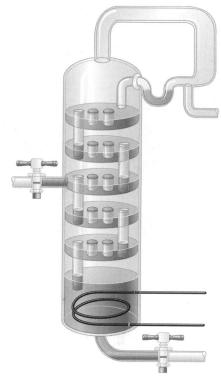

FIGURE 9.7
Schematic of a fractional distillation column. The solution to be separated into its components is introduced at the bottom of the column. The resistive heater provides the energy needed to vaporize the liquid. It can be assumed that the liquid and vapor are at equilibrium at each level of the column. The equilibrium temperature decreases from the bottom to the top of the column.

H₃C
   \
    C=O·····H—CCl₃
   /
H₃C

FIGURE 9.8
Hydrogen bond formation between acetone and chloroform leads to the formation of a maximum boiling azeotrope.

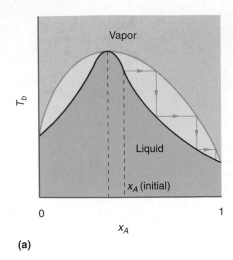

(a)

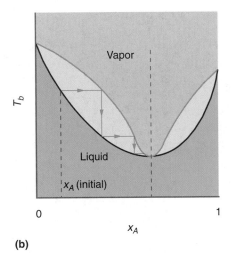

(b)

FIGURE 9.9
A boiling point diagram is shown for (a) maximum and (b) minimum boiling point azeotropes. The dashed lines indicate the initial composition of the solution and the composition of the azeotrope. The sequence of horizontal segments with arrows corresponds to successively higher (cooler) portions of the column.

TABLE 9.1 Composition and Boiling Temperatures of Selected Azeotropes

Azeotropic Mixture	Boiling Temperature of Components (°C)	Mole Fraction of First Component	Azeotrope Boiling Point (°C)
Water–ethanol	100./78.5	0.096	78.2
Water–trichloromethane	100./61.2	0.160	56.1
Water–benzene	100./80.2	0.295	69.3
Water–toluene	100./111	0.444	84.1
Ethanol–hexane	78.5/68.8	0.332	58.7
Ethanol–benzene	78.5/80.2	0.440	67.9
Ethyl acetate–hexane	78.5/68.8	0.394	65.2
Carbon disulfide–acetone	46.3/56.2	0.608	39.3
Toluene-acetic acid	111/118	0.625	100.7

Source: Lide, D. R., ed. *Handbook of Chemistry and Physics.* 83rd ed. Boca Raton, FL: CRC Press, 2002.

If the A–B interactions are less attractive than the A–A and B–B interactions, a minimum boiling azeotrope can be formed. A schematic boiling point diagram for such an azeotrope is also shown in Figure 9.9b. Fractional distillation of such a solution beginning with an initial value of x_A less than that corresponding to the minimum boiling point leads to a liquid with the azeotropic composition initially emerging at the top of the distillation column. An example for a minimum boiling azeotrope is a mixture of CS_2 ($T_b = 46.3°C$) and acetone ($T_b = 56.2°C$) at $x_{CS_2} = 0.608$, which boils at $39.3°C$.

It is still possible to collect one component of an azeotropic mixture using the property that the azeotropic composition is a function of the total pressure as shown in Figure 9.10. The mixture is first distilled at atmospheric pressure and the volatile distillate is collected. This vapor is condensed and subsequently distilled at a reduced pressure for which the azeotrope contains less A (more B). If this mixture is distilled, the azeotrope evaporates, leaving some pure B behind.

9.5 The Gibbs–Duhem Equation

In this section, we show that the chemical potentials of the two components in a binary solution are not independent. This is an important result, because it allows the chemical potential of a nonvolatile solute such as sucrose in a volatile solvent such as water to be determined. Such a solute has no measurable vapor pressure; therefore, its chemical potential cannot be measured directly. As shown later, its chemical potential can be determined knowing only the chemical potential of the solvent as a function of concentration.

From Chapter 6, the differential form of the Gibbs energy is given by

$$dG = -S \, dT + V \, dP + \sum_i \mu_i dn_i \tag{9.19}$$

For a binary solution at constant T and P, this equation reduces to

$$dG = \mu_1 dn_1 + \mu_2 dn_2 \tag{9.20}$$

Imagine starting with an infinitesimally small amount of a solution at constant T and P. The amount is gradually increased at constant composition. Because of this restriction, the chemical potentials are unchanged as the size of the system is changed. Therefore, the μ_i can be taken out of the integral:

$$\int_0^G dG' = \mu_1 \int_0^{n_1} dn'_1 + \mu_2 \int_0^{n_2} dn'_2 \quad \text{or}$$

$$G = \mu_1 n_1 + \mu_2 n_2 \tag{9.21}$$

The primes have been introduced to avoid using the same symbol for the integration variable and the upper limit. The total differential of the last equation is

$$dG = \mu_1 dn_1 + \mu_2 dn_2 + n_1 d\mu_1 + n_2 d\mu_2 \tag{9.22}$$

The previous equation differs from Equation (9.20) because, in general, we have to take changes of the composition of the solution into account. Therefore, μ_1 and μ_2 must be regarded as variables. Equating the expressions for dG in Equations (9.20) and (9.22), one obtains the **Gibbs–Duhem equation** for a binary solution, which can be written in either of two forms:

$$n_1 d\mu_1 + n_2 d\mu_2 = 0 \quad \text{or} \quad x_1 d\mu_1 + x_2 d\mu_2 = 0 \tag{9.23}$$

This equation states that the chemical potentials of the components in a binary solution are not independent. If the change in the chemical potential of the first component is $d\mu_1$, the change of the chemical potential of the second component is given by

$$d\mu_2 = -\frac{n_1 d\mu_1}{n_2} \tag{9.24}$$

The use of the Gibbs–Duhem equation is illustrated in Example Problem 9.5.

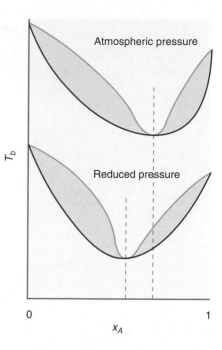

FIGURE 9.10
Because the azeotropic composition depends on the total pressure, pure B can be recovered from the A–B mixture by first distilling the mixture under atmospheric pressure and, subsequently, under a reduced pressure. Note that the boiling temperature is lowered as the pressure is reduced.

EXAMPLE PROBLEM 9.5

One component in a solution follows Raoult's law, $\mu_1^{solution} = \mu_1^* + RT \ln x_1$ over the entire range $0 \le x_1 \le 1$. Using the Gibbs–Duhem equation, show that the second component must also follow Raoult's law.

Solution

From Equation (9.24),

$$d\mu_2 = -\frac{x_1 d\mu_1}{x_2} = -\frac{n_1}{n_2} d(\mu_1^* + RT \ln x_1) = -RT \frac{x_1}{x_2} \frac{dx_1}{x_1}$$

Because $x_1 + x_2 = 1$, then $dx_2 = -dx_1$ and $d\mu_2 = RT \, dx_2/x_2$. Integrating this equation, one obtains $\mu_2 = RT \ln x_2 + C$, where C is a constant of integration. This constant can be evaluated by examining the limit $x_2 \to 1$. This limit corresponds to the pure substance 2 for which $\mu_2 = \mu_2^* = RT \ln 1 + C$. We conclude that $C = \mu_2^*$ and, therefore, $\mu_2^{solution} = \mu_2^* + RT \ln x_2$.

9.6 Colligative Properties

Many solutions consist of nonvolatile solutes that have limited solubility in a volatile solvent. Examples are solutions of sucrose or sodium chloride in water. Important properties of these solutions, including boiling point elevation, freezing point depression, and osmotic pressure are found to depend only on the solute concentration—not on the nature of the solute. These properties are called **colligative properties.** In this section, colligative properties are discussed using the model of the ideal solution. Corrections for nonideality are made in Section 9.13.

As discussed in Section 9.1, the vapor pressure above a solution containing a solute is lowered with respect to the pure solvent. Generally, the solute does not crystallize out with the solvent during freezing because the solute cannot be easily integrated into the solvent crystal structure. For this case, the change in the solvent chemical potential on dissolution of a nonvolatile solute can be understood using Figure 9.11a.

Only the liquid chemical potential is affected through formation of the solution. Although the gas pressure above the solution is lowered by the addition of the solute, the chemical potential of the gas is unaffected. This is the case because the comparison in Figure 9.11a is made at constant pressure. The chemical potential of the solid is

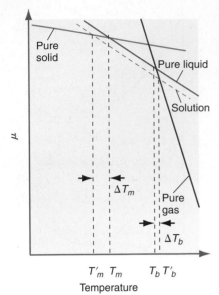

(a)

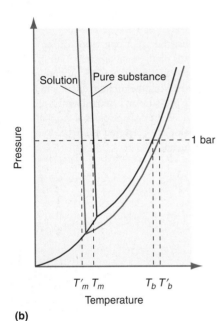

(b)

FIGURE 9.11
Illustration of the boiling point elevation and freezing point depression in two different ways: **(a)** These effects arise because the chemical potential of the solution is lowered through addition of the nonvolatile solute, while the chemical potential of the vapor and liquid is unaffected at constant pressure. **(b)** The same information from part (a) is shown using a P–T phase diagram.

unaffected because of the assumption that the solute does not crystallize out with the solvent. As shown in the figure, the melting temperature T_m, defined as the intersection of the solid and liquid μ versus T curves, is lowered. Similarly, the boiling temperature T_b is raised by dissolution of a nonvolatile solute in the solvent.

The same information is shown in a P–T phase diagram in Figure 9.11b. Because the vapor pressure above the solution is lowered by the addition of the solute, the liquid–gas coexistence curve intersects the solid–gas coexistence curve at a lower temperature than for the pure solvent. This intersection defines the triple point, and it must also be the origin of the solid–liquid coexistence curve. Therefore, the solid–liquid coexistence curve is shifted to lower temperatures through the dissolution of the nonvolatile solute. The overall effect of the shifts in the solid–gas and solid–liquid coexistence curves is a **freezing point depression** and a **boiling point elevation,** both of which arise because of the lowering of the vapor pressure of the solvent. This effect depends only on the concentration, and not on the identity, of the nonvolatile solute. The preceding discussion is qualitative in nature. In the next two sections, quantitative relationships are developed between the colligative properties and the concentration of the nonvolatile solute.

9.7 The Freezing Point Depression and Boiling Point Elevation

If the solution is in equilibrium with the pure solid solvent, the following relation must be satisfied:

$$\mu_{solution} = \mu_{solid}^* \tag{9.25}$$

Recall that $\mu_{solution}$ refers to the chemical potential of the solvent in the solution, and μ_{solid}^* refers to the chemical potential of the pure solvent in solid form. From Equation (9.7), we can express $\mu_{solution}$ in terms of the chemical potential of the pure solvent and its concentration and rewrite Equation (9.25) in the form

$$\mu_{solvent}^* + RT \ln x_{solvent} = \mu_{solid}^* \tag{9.26}$$

This equation can be solved for $\ln x_{solvent}$:

$$\ln x_{solvent} = \frac{\mu_{solid}^* - \mu_{solvent}^*}{RT} \tag{9.27}$$

The difference in chemical potentials $\mu_{solid}^* - \mu_{solvent}^* = -\Delta G_{fusion}$ so that

$$\ln x_{solvent} = \frac{-\Delta G_{fusion}}{RT} \tag{9.28}$$

Because we are interested in how the freezing temperature is related to $x_{solvent}$ at constant pressure, we need the partial derivative $(\partial T / \partial x_{solvent})_P$. This quantity can be obtained by differentiating Equation (9.28) with respect to $x_{solvent}$:

$$\left(\frac{\partial \ln x_{solvent}}{\partial x_{solvent}} \right)_P = \frac{1}{x_{solvent}} = -\frac{1}{R} \left(\frac{\partial \frac{\Delta G_{fusion}}{T}}{\partial T} \right)_P \left(\frac{\partial T}{\partial x_{solvent}} \right)_P \tag{9.29}$$

The first partial derivative on the right side of Equation (9.29) can be simplified using the Gibbs–Helmholtz equation (see Section 6.3), giving

$$\frac{1}{x_{solvent}} = \frac{\Delta H_{fusion}}{RT^2} \left(\frac{\partial T}{\partial x_{solvent}} \right)_P \quad \text{or}$$

$$\frac{dx_{solvent}}{x_{solvent}} = d \ln x_{solvent} = \frac{\Delta H_{fusion}}{R} \frac{dT}{T^2} \quad (\text{constant } P) \tag{9.30}$$

This equation can be integrated between the limits given by the pure solvent ($x_{solvent} = 1$) for which the fusion temperature is T_{fusion}, and an arbitrary small concentration of solute for which the fusion temperature is T:

$$\int_{1}^{x_{solvent}} \frac{dx}{x} = \int_{T_{fusion}}^{T} \frac{\Delta H_{fusion}}{R} \frac{dT'}{T'^2} \qquad (9.31)$$

For $x_{solvent}$, not very different from 1, ΔH_{fusion} is assumed to be independent of T, and Equation (9.31) simplifies to

$$\frac{1}{T} = \frac{1}{T_{fusion}} - \frac{R \ln x_{solvent}}{\Delta H_{fusion}} \qquad (9.32)$$

It is more convenient in discussing solutions to use the molality of the solute expressed in moles of solute per kg of solvent rather than the mole fraction of the solvent as the concentration unit. For dilute solutions $\ln x_{solvent} = \ln(n_{solvent}/(n_{solvent} + m_{solute}M_{solvent}n_{solvent})) = -\ln(1 + M_{solvent}m_{solute})$ and Equation (9.32) can be rewritten in terms of the molality (m) rather than the mole fraction. The result is

$$\Delta T_f = -\frac{RM_{solvent}T_{fusion}^2}{\Delta H_{fusion}} m_{solute} = -K_f m_{solute} \qquad (9.33)$$

where $M_{solvent}$ is the molar mass of the solvent. In going from Equation 9.32 to Equation 9.33 we have made the approximation $\ln(1 + M_{solvent}m_{solute}) \approx M_{solvent}m_{solute}$ and $1/T - 1/T_{fusion} \approx -\Delta T_f/T_{fusion}^2$. Note that K_f depends only on the properties of the solvent and is primarily determined by the molecular mass and the enthalpy of fusion. For most solvents, the magnitude of K_f lies between 1.5 and 10 as shown in Table 9.2. However, K_f can reach unusually high values, for example 30. (K kg)/mol for carbon tetrachloride and 40. (K kg)/mol for camphor.

The **boiling point elevation** can be calculated using the same argument with $\Delta G_{vaporization}$ and $\Delta H_{vaporization}$ substituted for ΔG_{fusion} and ΔH_{fusion}. The result is

$$\left(\frac{\partial T}{\partial m_{solute}}\right)_{P, m \to 0} = \frac{RM_{solvent}T_{vaporization}^2}{\Delta H_{vaporization}} \qquad (9.34)$$

TABLE 9.2 Freezing Point Depression and Boiling Point Elevation Constants

Substance	Standard Freezing Point (K)	K_f (K kg mol^{-1})	Standard Boiling Point (K)	K_b (K kg mol^{-1})
Acetic acid	289.6	3.59	391.2	3.08
Benzene	278.6	5.12	353.3	2.53
Camphor	449	40.	482.3	5.95
Carbon disulfide	161	3.8	319.2	2.40
Carbon tetrachloride	250.3	30.	349.8	4.95
Cyclohexane	279.6	20.0	353.9	2.79
Ethanol	158.8	2.0	351.5	1.07
Phenol	314	7.27	455.0	3.04
Water	273.15	1.86	373.15	0.51

Source: Lide, D. R., ed. *Handbook of Chemistry and Physics.* 83rd ed. Boca Raton, FL: CRC Press, 2002.

and

$$\Delta T_b = \frac{RM_{solvent}T^2_{vaporization}}{\Delta H_{vaporization}}m_{solute} = K_b\,m_{solute} \qquad (9.35)$$

Because $\Delta H_{vaporization} > \Delta H_{fusion}$, it follows that $K_f > K_b$. Typically, K_b values range between 0.5 and 5.0 K/(mol kg^{-1}) as shown in Table 9.2. Note also that, by convention, both K_f and K_b are positive; hence the negative sign in Equation (9.33) is replaced by a positive sign in Equation (9.35).

EXAMPLE PROBLEM 9.6

In this example, 4.50 g of a substance dissolved in 125 g of CCl_4 leads to an elevation of the boiling point of 0.650 K. Calculate the freezing point depression, the molar mass of the substance, and the factor by which the vapor pressure of CCl_4 is lowered.

Solution

$$\Delta T_f = \left(\frac{K_f}{K_b}\right)\Delta T_b = -\frac{30.\ \text{K/(mol kg}^{-1})}{4.95\ \text{K/(mol kg}^{-1})} \times 0.650\ \text{K} = -3.9\ \text{K}$$

To avoid confusion, we use the symbol m for molality and $\mathbf{m}$ for mass. We solve for the molar mass M_{solute} using Equation (9.35):

$$\Delta T_b = K_b m_{solute} = K_b \times \left(\frac{\mathbf{m}_{solute}/M_{solute}}{\mathbf{m}_{solvent}}\right)$$

$$M_{solute} = \frac{K_b\,\mathbf{m}_{solute}}{\mathbf{m}_{solvent}\,\Delta T_b}$$

$$M_{solute} = \frac{4.95\ \text{K kg mol}^{-1} \times 4.50\ \text{g}}{0.125\ \text{kg} \times 0.650\ \text{K}} = 274\ \text{g mol}^{-1}$$

We solve for the factor by which the vapor pressure of the solvent is reduced by using Raoult's law:

$$\frac{P_{solvent}}{P^*_{solvent}} = x_{solvent} = 1 - x_{solute} = 1 - \frac{n_{solute}}{n_{solute} + n_{solvent}}$$

$$= 1 - \frac{\dfrac{4.50\ \text{g}}{274\ \text{g mol}^{-1}}}{\left(\dfrac{4.50\ \text{g}}{274\ \text{g mol}^{-1}}\right) + \left(\dfrac{125\ \text{g}}{153.8\ \text{g mol}^{-1}}\right)} = 0.980$$

FIGURE 9.12

An osmotic pressure arises if a solution containing a solute that cannot pass through the membrane boundary is immersed in the pure solvent.

9.8 The Osmotic Pressure

Some membranes allow the passage of small molecules like water, yet do not allow larger molecules like sucrose to pass through them. Such a **semipermeable membrane** is an essential component in medical technologies such as kidney dialysis, which is described later. If a sac of such a membrane containing a solute that cannot pass through the membrane is immersed in a beaker containing the pure solvent, then initially the solvent diffuses into the sac. Diffusion ceases when equilibrium is attained, and at equilibrium the pressure is higher in the sac than in the surrounding solvent. This result is shown schematically in Figure 9.12. The process in which the solvent diffuses through a membrane and dilutes a solution is known as **osmosis**. The amount by which the pressure in the solution is raised is known as the **osmotic pressure**.

To understand the origin of the osmotic pressure, denoted by π, the equilibrium condition is applied to the contents of the sac and the surrounding solvent:

$$\mu_{solvent}^{solution}(T, P + \pi, x_{solvent}) = \mu_{solvent}^{*}(T, P) \quad \textbf{(9.36)}$$

Using Raoult's law to express the concentration dependence of $\mu_{solvent}$,

$$\mu_{solvent}^{solution}(T, P + \pi, x_{solvent}) = \mu_{solvent}^{*}(T, P + \pi) + RT \ln x_{solvent} \quad \textbf{(9.37)}$$

Because μ for the solvent is lower in the solution than in the pure solvent, only an increased pressure in the solution can raise its μ sufficiently to achieve equilibrium with the pure solvent. The dependence of μ on pressure and temperature is given by $d\mu = dG_m = V_m dP - S_m dT$. At constant T we can write

$$\mu_{solvent}^{*}(T, P + \pi, x_{solvent}) - \mu_{solvent}^{*}(T, P) = \int_{P}^{P+\pi} V_m^{*} dP' \quad \textbf{(9.38)}$$

where V_m^{*} is the molar volume of the pure solvent and P is the pressure in the solvent outside the sac. Because a liquid is nearly incompressible, it is reasonable to assume that V_m^{*} is independent of P to evaluate the integral in the previous equation. Therefore, $\mu_{solvent}^{*}(T, P + \pi, x_{solvent}) - \mu_{solvent}^{*}(T, P) = V_m^{*}\pi$, and Equation (9.37) reduces to

$$\pi V_m^{*} + RT \ln x_{solvent} = 0 \quad \textbf{(9.39)}$$

For a dilute solution, $n_{solvent} \gg n_{solute}$, and

$$\ln x_{solvent} = \ln(1 - x_{solute}) \approx -x_{solute} = -\frac{n_{solute}}{n_{solute} + n_{solvent}} \approx -\frac{n_{solute}}{n_{solvent}} \quad \textbf{(9.40)}$$

Equation (9.39) can be simplified further by recognizing that for a dilute solution, $V \approx n_{solvent} V_m^{*}$. With this substitution, Equation (9.39) becomes

$$\pi = \frac{n_{solute} RT}{V} \quad \textbf{(9.41)}$$

which is known as the **van't Hoff equation.** Note the similarity in form between this equation and the ideal gas law.

An important application of the selective diffusion of the components of a solution through a membrane is dialysis. In healthy individuals, the kidneys remove waste products from the bloodstream, whereas individuals with damaged kidneys use a dialysis machine for this purpose. Blood from the patient is shunted through tubes made of a selectively porous membrane surrounded by a flowing sterile solution made up of water, sugars, and other components. Blood cells and other vital components of blood are too large to fit through the pores in the membranes, but urea and salt flow out of the bloodstream through membranes into the sterile solution and are removed as waste.

EXAMPLE PROBLEM 9.7

Calculate the osmotic pressure generated at 298 K if a cell with a total solute concentration of 0.500 mol L^{-1} is immersed in pure water. The cell wall is permeable to water molecules, but not to the solute molecules.

Solution

$$\pi = \frac{n_{solute} RT}{V} = 0.500 \text{ mol L}^{-1} \times 8.206 \times 10^{-2} \text{ L atm K}^{-1} \text{ mol}^{-1} \times 298 \text{ K}$$

$$= 12.2 \text{ atm}$$

As this calculation shows, the osmotic pressure generated for moderate solute concentrations can be quite high. Hospital patients have died after pure water has accidentally been injected into their blood vessels, because the osmotic pressure is sufficient to burst the walls of blood cells.

Plants use osmotic pressure to achieve mechanical stability in the following way. A plant cell is bounded by a cellulose cell wall that is permeable to most components of the aqueous solutions that it encounters. A semipermeable cell membrane through which water, but not solute molecules, can pass is located just inside the cell wall. When the plant has sufficient water, the cell membrane expands, pushing against the cell wall, which gives the plant stalk a high rigidity. However, if there is a drought, the cell membrane is not totally filled with water and it moves away from the cell wall. As a result, only the cell wall contributes to the rigidity, and plants droop under these conditions.

Another important application involving osmotic pressure is the desalination of sea-water using reverse osmosis. Seawater is typically 1.1 molar in NaCl. Equation (9.41) shows that an osmotic pressure of 27 bar is needed to separate the solvated Na^+ and Cl^- ions from the water. If the seawater side of the membrane is subjected to a pressure greater than 27 bar, H_2O from the seawater will flow through the membrane, resulting in a separation of pure water from the seawater. This process is called reverse osmosis. The challenge in carrying out reverse osmosis on the industrial scale needed to provide a coastal city with potable water is to produce robust membranes that accommodate the necessary flow rates without getting fouled by algae and also effectively separate the ions from the water. The mechanism that prevents passage of the ions through suitable membranes is not fully understood. It is not based on the size of pores within the membrane alone and involves charged surfaces within the hydrated pores. These membrane-anchored ions repel the mobile Na^+ and Cl^- ions, while allowing the passage of the neutral water molecule.

9.9 Real Solutions Exhibit Deviations from Raoult's Law

In Sections 9.1 through 9.8, the discussion has been limited to ideal solutions. However, in general, if two volatile and miscible liquids are combined to form a solution, Raoult's law is not obeyed. This is the case because the A–A, B–B, and A–B interactions in a binary solution of A and B are in general not equal. If the A–B interactions are less (more) attractive than the A–A and B–B interactions, positive (negative) deviations from Raoult's law will be observed. An example of a binary solution with positive deviations from Raoult's law is CS_2—acetone. Experimental data for this system are shown in Table 9.3 and the data are plotted in Figure 9.13. How can a thermodynamic framework analogous to that presented for the ideal solution in Section 9.2 be developed for real solutions? This issue is addressed throughout the rest of this chapter.

Figure 9.13 shows that the partial and total pressures above a real solution can differ substantially from the behavior predicted by Raoult's law. Another way that ideal and real solutions differ is that the set of equations denoted in Equation (9.8), which describes the change in volume, entropy, enthalpy, and Gibbs energy that results from mixing, are not applicable to real solutions. For real solutions, these equations can only be written in a much less explicit form. Assuming that A and B are miscible,

$$\Delta G_{mixing} < 0$$
$$\Delta S_{mixing} > 0$$
$$\Delta V_{mixing} \neq 0$$
$$\Delta H_{mixing} \neq 0 \tag{9.42}$$

Whereas $\Delta G_{mixing} < 0$ and $\Delta S_{mixing} > 0$ always hold for miscible liquids, ΔV_{mixing} and ΔH_{mixing} can be positive or negative, depending on the nature of the A–B interaction in the solution.

TABLE 9.3 Partial and Total Pressures above a CS_2–Acetone Solution

x_{CS_2}	P_{CS_2} (Torr)	$P_{acetone}$ (Torr)	P_{total} (Torr)	x_{CS_2}	P_{CS_2} (Torr)	$P_{acetone}$ (Torr)	P_{total} (Torr)
0	0	343.8	343.8	0.4974	404.1	242.1	646.2
0.0624	110.7	331.0	441.7	0.5702	419.4	232.6	652.0
0.0670	119.7	327.8	447.5	0.5730	420.3	232.2	652.5
0.0711	123.1	328.8	451.9	0.6124	426.9	227.0	653.9
0.1212	191.7	313.5	505.2	0.6146	427.7	225.9	653.6
0.1330	206.5	308.3	514.8	0.6161	428.1	225.5	653.6
0.1857	258.4	295.4	553.8	0.6713	438.0	217.0	655.0
0.1991	271.9	290.6	562.5	0.6713	437.3	217.6	654.9
0.2085	283.9	283.4	567.3	0.7220	446.9	207.7	654.6
0.2761	323.3	275.2	598.5	0.7197	447.5	207.1	654.6
0.2869	328.7	274.2	602.9	0.8280	464.9	180.2	645.1
0.3502	358.3	263.9	622.2	0.9191	490.7	123.4	614.1
0.3551	361.3	262.1	623.4	0.9242	490.0	120.3	610.3
0.4058	379.6	254.5	634.1	0.9350	491.9	109.4	601.3
0.4141	382.1	253.0	635.1	0.9407	492.0	103.5	595.5
0.4474	390.4	250.2	640.6	0.9549	496.2	85.9	582.1
0.4530	394.2	247.6	641.8	0.9620	500.8	73.4	574.2
0.4933	403.2	242.8	646.0	0.9692	502.0	62.0	564.0
				1	512.3	0	512.3

Source: Zawidski, J. v. *Zeitschrift für Physikalische Chemie* 35 (1900): 129.

As indicated in Equation (9.42), the volume change upon mixing is not generally 0. Therefore, the volume of a solution will not be given by

$$V_m^{ideal} = x_A V_{m,A}^* + (1 - x_A)V_{m,B}^* \qquad (9.43)$$

as expected for 1 mol of an ideal solution, where $V_{m,A}^*$ and $V_{m,B}^*$ are the molar volumes of the pure substances A and B. Figure 9.14 shows $\Delta V_m = V_m^{real} - V_m^{ideal}$ for an acetone–chloroform solution as a function of $x_{chloroform}$. Note that ΔV_m can be positive or negative for this solution, depending on the value of $x_{chloroform}$. The deviations from ideality are small but are clearly evident.

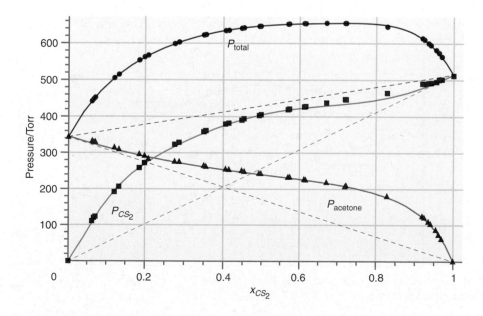

FIGURE 9.13
The data in Table 9.3 are plotted versus x_{CS_2}. The dashed lines show the expected behavior if Raoult's law were obeyed.

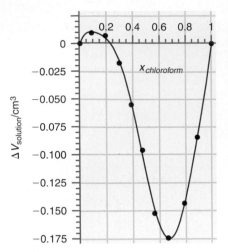

FIGURE 9.14

Deviations in the volume from the behavior expected for 1 mol of an ideal solution [Equation (9.43)] are shown for the acetone–chloroform system as a function of the mole fraction of chloroform.

The deviation of the volume from ideal behavior can best be understood by defining the concept of **partial molar quantities.** This concept is illustrated by discussing the **partial molar volume.** The volume of 1 mol of pure water at 25°C is 18.1 cm^3. However, if 1 mol of water is added to a large volume of an ethanol–water solution with $x_{H_2O} = 0.75$, the volume of the solution increases by only 16 cm^3. This is the case because the local structure around a water molecule in the solution is more compact than in pure water. The partial molar volume of a component in a solution is defined as the volume by which the solution changes if 1 mol of the component is added to such a large volume that the solution composition can be assumed constant. This statement is expressed mathematically in the following form:

$$\overline{V}_1(P, T, n_1, n_2) = \left(\frac{\partial V}{\partial n_1} \right)_{P, T, n_2} \tag{9.44}$$

With this definition, the volume of a binary solution is given

$$V = n_1 \overline{V}_1(P, T, n_1, n_2) + n_2 \overline{V}_2(P, T, n_1, n_2) \tag{9.45}$$

Note that because the partial molar volumes depend on the concentration of all components, the same is true of the total volume.

One can form partial molar quantities for any extensive property of a system (for example U, H, G, A, and S). Partial molar quantities (other than the chemical potential, which is the partial molar Gibbs energy) are usually denoted by the appropriate symbol topped by a horizontal bar. The partial molar volume is a function of P, T, n_1, and n_2, and $\overline{V}_i$ can be greater than or less than the molar volume of the pure component. Therefore, the volume of a solution of two miscible liquids can be greater than or less than the sum of the volumes of the pure components of the solution. Figure 9.15 shows data for the partial volumes of acetone and chloroform in an acetone–chloroform binary solution at 298 K. Note that the changes in the partial molar volumes with concentration are small, but not negligible.

In Figure 9.15, we can see that $\overline{V}_1$ increases if $\overline{V}_2$ decreases and vice versa. This is the case because partial molar volumes are related in the same way as the chemical potentials are related in the Gibbs–Duhem equation [Equation (9.23)]. In terms of the partial molar volumes, the Gibbs–Duhem equation takes the form

$$x_1 \, d\overline{V}_1 + x_2 \, d\overline{V}_2 = 0 \quad \text{or} \quad d\overline{V}_1 = -\frac{x_2}{x_1} d\overline{V}_2 \tag{9.46}$$

Therefore, as seen in Figure 9.15, if $\overline{V}_2$ changes by an amount $d\overline{V}_2$ over a small concentration interval, $\overline{V}_1$ will change in the opposite direction. The Gibbs–Duhem equation is applicable to both ideal and real solutions.

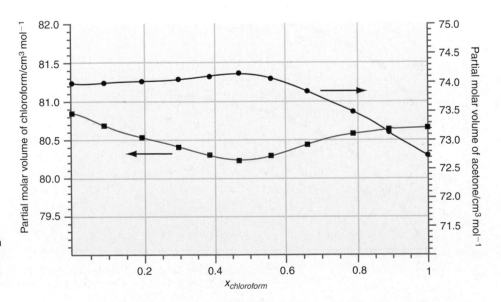

FIGURE 9.15

The partial molar volumes of chloroform (blue curve) and acetone (red curve) in a chloroform–acetone binary solution are shown as a function of $x_{chloroform}$.

9.10 The Ideal Dilute Solution

Although no simple model exists that describes all real solutions, there is a limiting law that merits further discussion. In this section, we describe the model of the ideal dilute solution, which provides a useful basis for describing the properties of real solutions if they are dilute. Just as for an ideal solution, at equilibrium the chemical potentials of a component in the gas and solution phase of a real solution are equal. As for an ideal solution, the chemical potential of a component in a real solution can be separated into two terms, a standard state chemical potential and a term that depends on the partial pressure.

$$\mu_i^{solution} = \mu_i^* + RT \ln \frac{P_i}{P_i^*} \tag{9.47}$$

Recall that for a pure substance $\mu_i^*(\text{vapor}) = \mu_i^*(\text{liquid}) = \mu_i^*$. Because the solution is not ideal, $P_i \neq x_i P_i^*$.

First consider only the solvent in a dilute binary solution. To arrive at an equation for $\mu_i^{solution}$ that is similar to Equation (9.7) for an ideal solution, we define the dimensionless **activity,** a_{solvent}, of the solvent by

$$a_{solvent} = \frac{P_{solvent}}{P_{solvent}^*} \tag{9.48}$$

Note that $a_{solvent} = x_{solvent}$ for an ideal solution. For a nonideal solution, the activity and the mole fraction are related through the **activity coefficient** $\gamma_{solvent}$, defined by

$$\gamma_{solvent} = \frac{a_{solvent}}{x_{solvent}} \tag{9.49}$$

The activity coefficient quantifies the degree to which the solution is nonideal. The activity plays the same role for a component of a solution that the fugacity plays for a real gas in expressing deviations from ideal behavior. In both cases, ideal behavior is observed in the appropriate limit, namely, $P \rightarrow 0$ for the gas, and $x_{solute} \rightarrow 0$ for the solution. To the extent that there is no atomic-scale model that tells us how to calculate γ, it should be regarded as a correction factor that exposes the inadequacy of our model, rather than as a fundamental quantity. As will be discussed in Chapter 10, there is such an atomic-scale model for dilute electrolyte solutions.

How is the chemical potential of a component related to its activity? Combining Equations (9.47) and (9.48), one obtains a relation that holds for all components of a real solution:

$$\mu_i^{solution} = \mu_i^* + RT \ln a_i \tag{9.50}$$

Equation (9.50) is a central equation in describing real solutions. It is the starting point for the discussion in the rest of this chapter.

The preceding discussion focused on the solvent in a dilute solution. However, the ideal dilute solution is defined by the conditions $x_{solute} \rightarrow 0$ and $x_{solvent} \rightarrow 1$. Because the solvent and solute are considered in different limits, we use different expressions to relate the mole fraction of a component and the partial pressure of the component above the solution.

Consider the partial pressure of acetone as a function of x_{CS_2} shown in Figure 9.16. Although Raoult's law is not obeyed over the whole concentration range, it is obeyed in the limit that $x_{acetone} \rightarrow 1$ and $x_{CS_2} \rightarrow 0$. In this limit, the average acetone molecule at the surface of the solution is surrounded by acetone molecules. Therefore, to a good approximation, $P_{acetone} = x_{acetone} P_{acetone}^*$ as $x_{acetone} \rightarrow 1$. Because the majority species is defined to be the solvent, we see that Raoult's law is obeyed for the solvent in a dilute solution. This limiting behavior is also observed for CS_2 *in the limit in which it is the solvent,* as seen in Figure 9.13.

Consider the opposite limit in which $x_{acetone} \rightarrow 0$. In this case, the average acetone molecule at the surface of the solution is surrounded by CS_2 molecules. Therefore, the

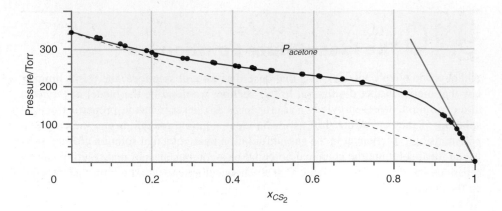

molecule experiences very different interactions with its neighbors than if it were surrounded by acetone molecules. For this reason, $P_{acetone} \neq x_{acetone} P^{*}_{acetone}$ as $x_{acetone} \rightarrow 0$. However, it is apparent from Figure 9.16 that $P_{acetone}$ also varies linearly with $x_{acetone}$ in this limit. This behavior is described by the following equation:

$$P_{acetone} = x_{acetone} k_H^{acetone} \text{ as } x_{acetone} \rightarrow 0 \qquad (9.51)$$

This relationship is known as **Henry's law,** and the constant k_H is known as the **Henry's law constant.** The value of the constant depends on the nature of the solute and solvent and quantifies the degree to which deviations from Raoult's law occur. As the solution approaches ideal behavior, $k_H^i \rightarrow P_i^{*}$. For the data shown in Figure 9.13, $k_H^{CS_2} = 1750$ Torr and $k_H^{acetone} = 1950$ Torr. Note that these values are substantially greater than the vapor pressures of the pure substances, which are 512.3 and 343.8 Torr, respectively. The Henry's law constants are less than the vapor pressures of the pure substances if the system exhibits negative deviations from Raoult's law. Henry's law constants for aqueous solutions are listed for a number of solutes in Table 9.4.

Based on these results, the **ideal dilute solution** is defined. *An ideal dilute solution is a solution in which the solvent is described using Raoult's law and the solute is described using Henry's law.* As shown by the data in Figure 9.13, the partial pressures above the CS_2–acetone mixture are consistent with this model in either of two limits, $x_{acetone} \rightarrow 1$ or $x_{CS_2} \rightarrow 1$. In the first of these limits, we consider acetone to be the solvent and CS_2 to be the solute. Acetone is the solute and CS_2 is the solvent in the second limit.

TABLE 9.4   Henry's Law Constants for Aqueous Solutions Near 298 K

Substance	k_H (Torr)	k_H (bar)
Ar	2.80×10^7	3.72×10^4
C_2H_6	2.30×10^7	3.06×10^4
CH_4	3.07×10^7	4.08×10^4
CO	4.40×10^6	5.84×10^3
CO_2	1.24×10^6	1.65×10^3
H_2S	4.27×10^5	5.68×10^2
He	1.12×10^8	1.49×10^6
N_2	6.80×10^7	9.04×10^4
O_2	3.27×10^7	4.95×10^4

Source: Alberty, R. A., and Silbey, R. S. *Physical Chemistry.* New York: John Wiley & Sons, 1992.

9.11 Activities Are Defined with Respect to Standard States

The ideal dilute solution model's predictions that Raoult's law is obeyed for the solvent and Henry's law is obeyed for the solute are not valid over a wide range of concentration. The concept of the activity coefficient introduced in Section 9.10 is used to quantify these deviations. In doing so, it is useful to define the activities in such a way that the solution approaches ideal behavior in the limit of interest, which is generally $x_A \to 0$, or $x_A \to 1$. With this choice, the activity approaches the concentration, and it is reasonable to set the activity coefficient equal to 1. Specifically, $a_i \to x_i$ as $x_i \to 1$ for the solvent, and $a_i \to x_i$ as $x_i \to 0$ for the solute. The reason for this choice is that numerical values for activity coefficients are generally not known. Choosing the standard state as described earlier ensures that the concentration (divided by the unit concentration to make it dimensionless), which is easily measured, is a good approximation to the activity.

In Section 9.10 the activity and activity coefficient for the solvent in a dilute solution ($x_{solvent} \to 1$) were defined by the relations

$$a_i = \frac{P_i}{P_i^*} \quad \text{and} \quad \gamma_i = \frac{a_i}{x_i} \tag{9.52}$$

As shown in Figure 9.13, the activity approaches unity as $x_{solvent} \to 1$. We refer to an activity calculated using Equation (9.52) as being based on a **Raoult's law standard state.** The standard state chemical potential based on Raoult's law is $\mu_{solvent}^*$, which is the chemical potential of the pure solvent.

However, this definition of the activity and choice of a standard state is not optimal for the solute at a low concentration, because the solute obeys Henry's law rather than Raoult's law and, therefore, the activity coefficient will differ appreciably from one. In this case,

$$\mu_{solute}^{solution} = \mu_{solute}^* + RT \ln \frac{k_H^{solute} x_{solute}}{P_{solute}^*} = \mu_{solute}^{*H} + RT \ln x_{solute} \text{ as } x_{solute} \to 0 \tag{9.53}$$

The standard state chemical potential is the value of the chemical potential when $x_i = 1$. We see that the Henry's law standard state chemical potential is given by

$$\mu_{solute}^{*H} = \mu_{solute}^* + RT \ln \frac{k_H^{solute}}{P_{solute}^*} \tag{9.54}$$

The activity and activity coefficient based on Henry's law are defined, respectively, by

$$a_i = \frac{P_i^H}{k_i^H} \text{ and } \gamma_i = \frac{a_i}{x_i} \tag{9.55}$$

We note that Henry's law is still obeyed for a solute that has such a small vapor pressure that we refer to it as a nonvolatile solute.

The **Henry's law standard state** is a state in which the pure solute has a vapor pressure $k_{H, solute}$ rather than its actual value P_{solute}^*. It is a hypothetical state that does not exist. Recall that the value $k_{H, solute}$ is obtained by extrapolation from the low coverage range in which Henry's law is obeyed. Although this definition may seem peculiar, only in this way can we ensure that $a_{solute} \to x_{solute}$ and $\gamma_{solute} \to 1$ as $x_{solute} \to 0$. We reiterate the reason for this choice. If the preceding conditions are satisfied, the concentration (divided by the unit concentration to make it dimensionless) is, to a good approximation, equal to the activity. Therefore, equilibrium constants for reactions in solution can be calculated without having numerical values for activity coefficients. We emphasize the difference in the Raoult's law and Henry's law standard states because it is the standard state chemical potentials μ_{solute}^* and μ_{solute}^{*H} that are used to calculate $\Delta G°$ and the thermodynamic equilibrium constant K. Different choices for standard states will result in different numerical values for K. The standard chemical potential μ_{solute}^{*H} refers to the hypothetical standard state in which $x_{solute} = 1$, and each solute species is in an environment characteristic of the infinitely dilute solution.

We now consider a less well-defined situation. For solutions in which the components are miscible in all proportions, such as the CS_2−acetone system, either a Raoult's law or a Henry's law standard state can be defined, as we show with sample calculations in Example Problems 9.8 and 9.9. This is the case because there is no unique choice for the standard state over the entire concentration range in such a system. Numerical values for the activities and activity coefficients will differ, depending on whether the Raoult's law or the Henry's law standard state is used.

EXAMPLE PROBLEM 9.8

Calculate the activity and activity coefficient for CS_2 at $x_{CS_2} = 0.3502$ using data from Table 9.3. Assume a Raoult's law standard state.

Solution

$$a_{CS_2}^R = \frac{P_{CS_2}}{P_{CS_2}^*} = \frac{358.3 \text{ Torr}}{512.3 \text{ Torr}} = 0.6994$$

$$\gamma_{CS_2}^R = \frac{a_{CS_2}^R}{x_{CS_2}} = \frac{0.6994}{0.3502} = 1.997$$

The activity and activity coefficients for CS_2 are concentration dependent. Results calculated as in Example Problem 9.8 using a Raoult's law standard state are shown in Figure 9.17 as a function of x_{CS_2}. For this solution, $\gamma_{CS_2}^R > 1$ for all values of the concentration for which $x_{CS_2} < 1$. Note that $\gamma_{CS_2}^R \rightarrow 1$ as $x_{CS_2} \rightarrow 1$ as the model requires. The activity and activity coefficients for CS_2 using a Henry's law standard state are shown in Figure 9.18 as a function of x_{CS_2}. For this solution, $\gamma_{CS_2}^H < 1$ for all values of the concentration for which $x_{CS_2} > 0$. Note that $\gamma_{CS_2}^H \rightarrow 1$ as $x_{CS_2} \rightarrow 0$ as the model requires. Which of these two possible standard states should be chosen? There is a good answer to this question only in the limits $x_{CS_2} \rightarrow 0$ or $x_{CS_2} \rightarrow 1$. For intermediate concentrations, either standard state can be used.

EXAMPLE PROBLEM 9.9

Calculate the activity and activity coefficient for CS_2 at $x_{CS_2} = 0.3502$ using data from Table 9.3. Assume a Henry's law standard state.

Solution

$$a_{CS_2}^H = \frac{P_{CS_2}}{k_{H,CS_2}} = \frac{358.3 \text{ Torr}}{1750 \text{ Torr}} = 0.204$$

$$\gamma_{CS_2}^H = \frac{a_{CS_2}^H}{x_{CS_2}} = \frac{0.204}{0.3502} = 0.584$$

The Henry's law standard state just discussed is defined with respect to concentration measured in units of the mole fraction. This is not a particularly convenient scale, and either the molarity or molality concentration scales are generally used in laboratory experiments. The mole fraction of the solute can be converted to the molality scale by dividing the first expression in Equation (9.56) by $n_{solvent} M_{solvent}$:

$$x_{solute} = \frac{n_{solute}}{n_{solvent} + n_{solute}} = \frac{m_{solute}}{\dfrac{1}{M_{solvent}} + m_{solute}} \quad (9.56)$$

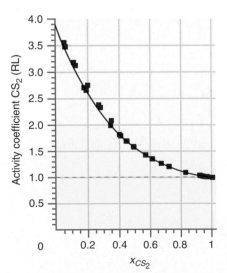

FIGURE 9.17

The activity and activity coefficient for CS_2 in a CS_2−acetone solution based on a Raoult's law standard state are shown as a function of x_{CS_2}.

where m_{solute} is the molality of the solute, and $M_{solvent}$ is the molar mass of the solvent in kg mol^{-1}. We see that $m_{solute} \rightarrow x_{solute}/M_{solvent}$ as $x_{solute} \rightarrow 0$. Using molality as the concentration unit, the activity and activity coefficient of the solute are defined by

$$a_{solute}^{molality} = \frac{P_{solute}}{k_H^{molality}} \quad \text{with} \quad a_{solute}^{molality} \rightarrow m_{solute} \quad \text{as} \quad m_{solute} \rightarrow 0 \quad \text{and} \quad \textbf{(9.57)}$$

$$\gamma_{solute}^{molality} = \frac{a_{solute}^{molality}}{m_{solute}} \quad \text{with} \quad \gamma_{solute}^{molality} \rightarrow 1 \quad \text{as} \quad m_{solute} \rightarrow 0 \qquad \textbf{(9.58)}$$

The Henry's law constants and activity coefficients determined on the mole fraction scale must be recalculated to describe a solution characterized by its molality or molarity. Example Problem 9.10 shows the conversion from mole fraction to molarity; similar conversions can be made if the concentration of the solution is expressed in terms of molality.

What is the standard state if concentrations rather than mole fractions are used? The standard state in this case is the hypothetical state in which Henry's law is obeyed by a solution that is 1.0 molar (or 1.0 molal) in the solute concentration. It is a hypothetical state because at this concentration, substantial deviations from Henry's law will be observed. Although this definition may seem peculiar at first reading, only in this way can we ensure that the activity becomes equal to the molarity (or molality), and the activity coefficient approaches 1, as the solute concentration approaches 0.

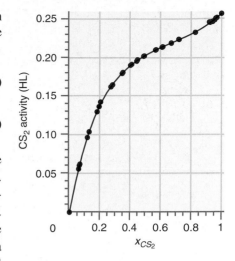

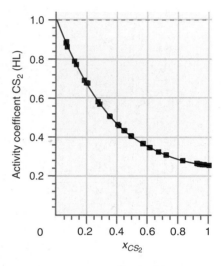

FIGURE 9.18

The activity and activity coefficient for CS_2 in a CS_2−acetone solution based on a Henry's law standard state are shown as a function of x_{CS_2}.

EXAMPLE PROBLEM 9.10

a. Derive a general equation, valid for dilute solutions, relating the Henry's law constants for the solute on the mole fraction and molarity scales.

b. Determine the Henry's law constants for acetone on the molarity scale. Use the results for the Henry's law constants on the mole fraction scale cited in Section 9.10. The density of acetone is 789.9 g L^{-1}.

Solution

a. We use the symbol c_{solute} to designate the solute molarity, and $c°$ to indicate a 1 molar concentration.

$$\frac{dP}{d\left(\dfrac{c_{solute}}{c°}\right)} = \frac{dP}{dx_{solute}} \frac{dx_{solute}}{d\left(\dfrac{c_{solute}}{c°}\right)}$$

To evaluate this equation, we must determine

$$\frac{dx_{solute}}{d\left(\dfrac{c_{solute}}{c°}\right)}$$

$$x_{solute} = \frac{n_{solute}}{n_{solute} + n_{solvent}} \approx \frac{n_{solute}}{n_{solvent}} = \frac{n_{solute}M_{solvent}}{V_{solution}\rho_{solution}} = c_{solute}\frac{M_{solvent}}{\rho_{solution}}$$

Therefore,

$$\frac{dP}{d\left(\dfrac{c_{solute}}{c°}\right)} = \frac{c°M_{solvent}}{\rho_{solution}} \frac{dP}{dx_{solute}}$$

b. $$\frac{dP}{d\left(\dfrac{c_{solute}}{c°}\right)} = \frac{1 \text{ mol L}^{-1} \times 58.08 \text{ g mol}^{-1}}{789.9 \text{ g L}^{-1}} \times 1950 \text{ Torr} = 143.4 \text{ Torr}$$

The colligative properties discussed for an ideal solution in Sections 9.7 and 9.8 refer to the properties of the solvent in a dilute solution. The Raoult's law standard state applies to this case and in an ideal dilute solution, Equations (9.33), (9.35) and (9.41) can be used with activities replacing concentrations:

$$\Delta T_f = -K_f \gamma m_{solute}$$

$$\Delta T_b = K_b \gamma m_{solute}$$

$$\pi = \gamma c_{solute} RT \tag{9.59}$$

The activity coefficients are defined with respect to molality for the boiling point elevation ΔT_b and freezing point depression ΔT_f, and with respect to molarity for the osmotic pressure π. Equations (9.59) provide a useful way to determine activity coefficients as shown in Example Problem 9.11.

EXAMPLE PROBLEM 9.11

In 500. g of water, 24.0 g of a nonvolatile solute of molecular weight 241 g mol^{-1} is dissolved. The observed freezing point depression is 0.359°C. Calculate the activity coefficient of the solute.

Solution

$$\Delta T_f = -K_f \gamma m_{solute}; \quad \gamma = -\frac{\Delta T_f}{K_f m_{solute}}$$

$$\gamma = \frac{0.359 \text{ K}}{1.86 \text{ K kg mol}^{-1} \times \dfrac{24.0}{241 \times 0.500} \text{ mol kg}^{-1}} = 0.969$$

9.12 Henry's Law and the Solubility of Gases in a Solvent

The ideal dilute solution model can be applied to the solubility of gases in liquid solvents. An example for this type of solution equilibrium is the amount of N_2 absorbed by water at sea level, which is considered in Example Problem 9.12. In this case, one of the components of the solution is a liquid and the other is a gas. The equilibrium of interest between the solution and the vapor phase is

$$N_2(\text{aqueous}) \rightleftharpoons N_2(\text{vapor}) \tag{9.60}$$

The chemical potential of the dissolved N_2 is given by

$$\mu_{N_2}^{solution} = \mu_{N_2}^{*H}(\text{vapor}) + RT \ln a_{solute} \tag{9.61}$$

In this case, a Henry's law standard state is the appropriate choice because the nitrogen is sparingly soluble in water. The mole fraction of N_2 in solution, x_{N_2}, is given by

$$x_{N_2} = \frac{n_{N_2}}{n_{N_2} + n_{H_2O}} \approx \frac{n_{N_2}}{n_{H_2O}} \tag{9.62}$$

The amount of dissolved gas is given by

$$n_{N_2} = n_{H_2O} x_{N_2} = n_{H_2O} \frac{P_{N_2}}{k_H^{N_2}} \tag{9.63}$$

Example Problem 9.12 shows how Equation (9.63) is used to model the dissolution of a gas in a liquid.

EXAMPLE PROBLEM 9.12

The average human with a body weight of 70. kg has a blood volume of 5.00 L. The Henry's law constant for the solubility of N_2 in H_2O is 9.04×10^4 bar at 298 K. Assume that this is also the value of the Henry's law constant for blood and that the density of blood is 1.00 kg L^{-1}.

a. Calculate the number of moles of nitrogen absorbed in this amount of blood in air of composition 80.% N_2 at sea level, where the pressure is 1 bar, and at a pressure of 50. bar.

b. Assume that a diver accustomed to breathing compressed air at a pressure of 50. bar is suddenly brought to sea level. What volume of N_2 gas is released as bubbles in the diver's bloodstream?

Solution

$$n_{N_2} = n_{H_2O} \frac{P_{N_2}}{k_H^{N_2}}$$

a.
$$= \frac{5.0 \times 10^3 \text{g}}{18.02 \text{ g mol}^{-1}} \times \frac{0.80 \text{ bar}}{9.04 \times 10^4 \text{ bar}}$$

$$= 2.5 \times 10^{-3} \text{ mol at 1 bar total pressure}$$

At 50. bar, $n_{N_2} = 50. \times 2.5 \times 10^{-3}$ mol $= 0.13$ mol.

b. $V = \dfrac{nRT}{P}$

$$= \frac{(0.13 \text{ mol} - 2.5 \times 10^{-3} \text{ mol}) \times 8.314 \times 10^{-2} \text{ L bar mol}^{-1} \text{ K}^{-1} \times 300. \text{ K}}{1.00 \text{ bar}}$$

$$= 3.2 \text{ L}$$

The symptoms induced by the release of air into the bloodstream are known to divers as the bends. The volume of N_2 just calculated is far more than is needed to cause the formation of arterial blocks due to gas-bubble embolisms.

9.13 Chemical Equilibrium in Solutions

The concept of activity can be used to express the thermodynamic equilibrium constant in terms of activities for real solutions. Consider a reaction between solutes in a solution. At equilibrium, the following relation must hold:

$$\left(\sum_j \nu_j \mu_j (\text{solution}) \right)_{equilibrium} = 0 \qquad \textbf{(9.64)}$$

where the subscript states that the individual chemical potentials must be evaluated under equilibrium conditions. Each of the chemical potentials in Equation (9.64) can be expressed in terms of a standard state chemical potential and a concentration-dependent term. Assume a Henry's law standard state for each solute. Equation (9.64) then takes the form

$$\sum_j \nu_j \mu_j^{*H}(\text{solution}) + RT \sum_j \ln (a_i^{eq})^{\nu_j} = 0 \qquad \textbf{(9.65)}$$

Using the relation between the Gibbs energy and the chemical potential, the previous equation can be written in the form

$$\Delta G_{reaction}^\circ = -RT \sum_j \ln (a_i^{eq})^{\nu_j} = -RT \ln K \qquad \textbf{(9.66)}$$

The equilibrium constant in terms of activities is given by

$$K = \prod_i (a_i^{eq})^{\nu_j} = \prod_i (\gamma_i^{eq})^{\nu_j} \left(\frac{c_i^{eq}}{c^\circ} \right)^{\nu_j} \tag{9.67}$$

where the symbol $\prod$ indicates that the terms following the symbol are multiplied with one another. It can be viewed as a generalization of K_P, defined in Equation (6.60) and can be applied to equilibria involving gases, liquids, dissolved species, and solids. For gases, the fugacities divided by f° (see Section 7.5) are the activities.

To obtain a numerical value for K, the standard state Gibbs reaction energy ΔG_R° must be known. As for gas-phase reactions, the ΔG_R° must be determined experimentally. This can be done by measuring the individual activities of the species in solution and calculating K from these results. After a series of ΔG_R° for different reactions has been determined, they can be combined to calculate the ΔG_R° for other reactions, as discussed for reaction enthalpies in Chapter 4. Because of the significant interactions between the solutes and the solvent, K values depend on the nature of the solvent, and for electrolyte solutions to be discussed in Chapter 10, they additionally depend on the ionic strength.

An equilibrium constant in terms of molarities or molalities can also be defined starting from Equation (9.67) and setting all activity coefficients equal to 1. This is most appropriate for a dilute solution of a nonelectrolyte, using a Henry's law standard state:

$$K = \prod_i (\gamma_i^{eq})^{\nu_j} \left(\frac{c_i^{eq}}{c^\circ} \right)^{\nu_j} \approx \prod_i \left(\frac{c_i^{eq}}{c^\circ} \right)^{\nu_j} \tag{9.68}$$

EXAMPLE PROBLEM 9.13

a. Write the equilibrium constant for the reaction $N_2(aq, m) \rightleftharpoons N_2(g, P)$ in terms of activities at 25°C, where m is the molality of $N_2(aq)$.

b. By making suitable approximations, convert the equilibrium constant of part (a) into one in terms of pressure and molality only.

Solution

a. $K = \prod_i (a_i^{eq})^{\nu_j} = \prod_i (\gamma_i^{eq})^{\nu_j} \left(\frac{c_i^{eq}}{c^\circ} \right)^{\nu_j} = \dfrac{\left(\dfrac{\gamma_{N_2,g} P}{P^\circ} \right)}{\left(\dfrac{\gamma_{N_2,aq} m}{m^\circ} \right)}$

$= \dfrac{\gamma_{N_2,g} \left(\dfrac{P}{P^\circ} \right)}{\gamma_{N_2,aq} \left(\dfrac{m}{m^\circ} \right)}$

b. Using a Henry's law standard state for dissolved N_2, $\gamma_{N_2,aq} \approx 1$, because the concentration is very low. Similarly, because N_2 behaves like an ideal gas up to quite high pressures at 25°C, $\gamma_{N_2,g} \approx 1$. Therefore,

$$K \approx \dfrac{\left(\dfrac{P}{P^\circ} \right)}{\left(\dfrac{m}{m^\circ} \right)}$$

Note that in this case, the equilibrium constant is simply the Henry's law constant in terms of molality.

The numerical values for the dimensionless thermodynamic equilibrium constant depend on the choice of the standard states for the components involved in the reaction. The same is true for ΔG_R°. Therefore, it is essential to know which standard state has been assumed before an equilibrium constant is used. The activity coefficients of most neutral solutes are close to 1 with the appropriate choice of standard state. Therefore, example calculations of chemical equilibrium using activities will be deferred until electrolyte solutions are discussed in Chapter 10. For such solutions, γ_{solute} differs substantially from 1, even for dilute solutions.

An important example of equilibrium processes in solution is the physical interaction of a molecule with a binding site. Such **binding equilibria** processes are ubiquitous in chemistry, being central to the function of oxygen transport proteins such as hemoglobin or myoglobin, the action of enzymes, and the adsorption of molecules to surfaces. All of these processes involve an equilibrium between the free molecule (R) and the molecule absorbed or "bound" to another species (protein, solid, etc.) containing a binding site (M). In the limit where there is a single binding site per species, the equilibrium can be written as

$$R + M \rightleftharpoons RM \qquad (9.69)$$

The equilibrium constant for this reaction in terms of the concentrations of the species of interest is

$$K = \frac{c_{RM}}{c_R \times c_M} \qquad (9.70)$$

(It is common in biochemistry that K is expressed in terms of concentrations, and therefore has units.) Assuming that there is a single binding site for the molecule, then the average number of bound molecules per species ($\bar{\nu}$) is given by

$$\bar{\nu} = \frac{c_{RM}}{c_M + c_{RM}} \qquad (9.71)$$

Expressing this fraction in terms of the equilibrium constant yields

$$\bar{\nu} = \frac{c_{RM}}{c_M + c_{RM}}\left(\frac{1/(c_R c_M)}{1/(c_R c_M)}\right) = \frac{K}{1/c_R + K} = \frac{Kc_R}{1 + Kc_R} \qquad (9.72)$$

This expression demonstrates that $\bar{\nu}$ depends on both the molecular concentration and the value of the equilibrium constant. The variation of $\bar{\nu}$ with c_R for different values of K are presented in Figure 9.19. With an increase in K (consistent with the bound form of the molecule being lower in Gibbs energy relative to the free form), the average number of bound molecules per species approaches unity (that is, the binding sites are completely occupied) even at low values of c_R.

The previous development was limited to the existence of a single molecular binding site; however, what species absorbing the molecules has more than one binding site? Two issues must be considered to address this case. First, binding to one site may have an effect on binding at other sites, and this effect is important in the binding of oxygen to hemoglobin. However, in this development we restrict ourselves to **independent site binding** where the binding at one site has no effect on binding to other sites. In this case the average number of molecules bound to a specific site (i) is

$$\bar{\nu}_i = \frac{K_i c_R}{1 + K_i c_R} \qquad (9.73)$$

If there are N binding sites, then the average number of molecules adsorbed per species is simply the sum of the average values for each specific binding site:

$$\bar{\nu} = \sum_{i=1}^{N} \bar{\nu}_i = \sum_{i=1}^{N} \frac{K_i c_R}{1 + K_i c_R} \qquad (9.74)$$

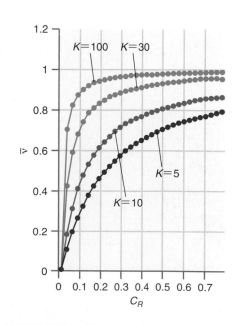

FIGURE 9.19

The average number of bound molecules per species $\bar{\nu}$ is plotted against the concentration of free species c_R for K values of 5, 10, 30, and 100.

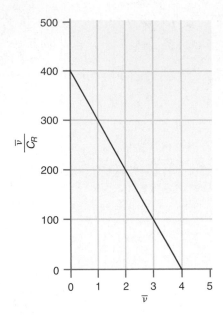

FIGURE 9.20
Scatchard plot of independent binding for $K = 100$ and $N = 4$. Note that the y intercept is $K \times N = 400$, the slope is $-K = -100$, and the x intercept is 4.

The second issue to address is the equivalency of the binding sites. Assuming that all of the binding sites are equivalent $\bar{\nu}$ becomes

$$\bar{\nu} = \sum_{i=1}^{N} \frac{K_i c_R}{1 + K_i c_R} = \frac{N K c_R}{1 + K c_R} \tag{9.75}$$

This expression is almost identical to our previous expression for $\bar{\nu}$ for a single binding site, but with the addition of a factor of N. Therefore, at large values of c_R the average number of bound molecules per species will approach N instead of 1. To determine K and N for specific binding equilibrium, the expression for $\bar{\nu}$ is recast in a linear form referred to as the **Scatchard equation:**

$$\bar{\nu} = \frac{N K c_R}{1 + K c_R}$$

$$\frac{\bar{\nu}}{c_R} = -K\bar{\nu} + NK \tag{9.76}$$

The Scatchard equation demonstrates that a plot of $\bar{\nu}/c_R$ versus $\bar{\nu}$ should yield a straight line having a slope of $-K$ and a y-intercept of NK from which N can be readily determined. A **Scatchard plot** for $K = 100$ and $N = 4$ is presented in Figure 9.20. Deviations from linear behavior in a Scatchard plot can occur if the binding is not independent or if the binding sites are not equivalent.

EXAMPLE PROBLEM 9.14

A variety of molecules physically bind to DNA by intercalation or insertion between the base pairs. The intercalation of chlorobenzylidine (CB) to calf thymus DNA was studied by circular dichroism spectroscopy [Zhong, W., Yu, J.-S., Liang, Y., Fan, K., and Lai, L., *Spectrochimica Acta*, Part A, 60A (2004): 2985]. Using the following representative binding data, determine the binding constant (K) and the number of binding sites per base (N).

c_{CB} ($\times 10^6$ M)	1.0	2.0	3.0	5.0	10.0
$\bar{\nu}$	0.006	0.012	0.018	0.028	0.052

Solution

A plot of $\bar{\nu}/c_{CB}$ versus $\bar{\nu}$. using binding data provided allows for a determination of K and N. Using the data, the following Scatchard plot is constructed:

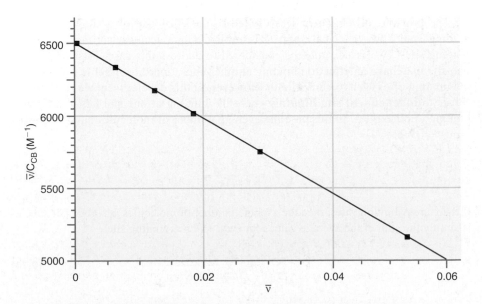

A best fit to a straight line results in a slope of $-26{,}000 \ \mathrm{M}^{-1}$ and y-intercept of 6500. Comparison to the Scatchard equation reveals that the slope is equal to $-K$ so that $K = 26{,}000 \ \mathrm{M}^{-1}$. In addition, the y-intercept is equal to the product of N and K; therefore, $N = 0.25$. This value of N is consistent with a binding site consisting of four bases, or two base pairs consistent with intercalation between base pairs.

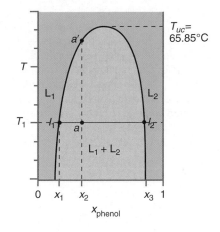

FIGURE 9.21

A T–x phase diagram is shown for the partially miscible liquids water and phenol. The pressure is held constant at 1 bar.

9.14 Solutions Formed from Partially Miscible Liquids

Two liquids need not be miscible in all proportions. Over a limited range of relative concentration, a homogeneous solution can be formed, and outside of this range, two liquid phases separated by a phase boundary can be observed. An example is the water-phenol system for which a T–x phase diagram is shown in Figure 9.21. The pressure is assumed to be sufficiently high that no gas phase is present.

As the phenol concentration increases from 0 at the temperature T_1, a single liquid phase of variable composition is formed. However, at x_1, the appearance of a second liquid phase is observed, separated from the first phase by a horizontal boundary with the denser phase below the less dense phase. If we start with pure phenol and add water at T_1, separation into two phases is observed at x_3. In the phenol-water system, the solubility increases with temperature so that the extent in x of the two-phase region decreases with increasing T, and phase separation is observed only at a single composition at the **upper consolate temperature** T_{uc}. For $T > T_{uc}$, phenol and water are completely miscible.

The composition of the two phases in equilibrium at a given temperature is given by the x values at the ends of the tie line $l_1 l_2$, and the relative amount of the two phases present is given by the lever rule:

$$\frac{\text{moles } \mathrm{L}_1}{\text{moles } \mathrm{L}_2} = \frac{al_2}{al_1} = \frac{x_3 - x_2}{x_2 - x_1} \qquad (9.77)$$

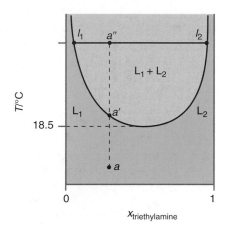

FIGURE 9.22

A T–x phase diagram showing a lower consolate temperature is shown for the partially miscible liquids water and triethylamine. The pressure is held constant at 1 bar.

Although less common, liquid–liquid systems are known in which the solubility decreases as the temperature increases. In this case, a **lower consolate temperature** is observed below which the two liquids are completely miscible. An example is the triethylamine–water system shown in Figure 9.22. The water–nicotine system has both a lower and upper consolate temperature, and phase separation is only observed between 61°C and 210°C as shown in Figure 9.23.

We next consider solutions of partially miscible liquids at pressures for which a vapor phase is in equilibrium with one or more liquid phases. The composition of the vapor for a given value of the average composition of the liquid depends on the relative values of the upper consolate temperature and the minimum boiling temperature for the azeotrope. For the system shown in Figure 9.24a, the vapor phase is only present above the upper consolate temperature. In this case, the vapor and liquid composition is the same as that described in Section 9.4 for minimum boiling point azeotropes. However, if the upper consolate temperature lies above the boiling temperature of the minimum boiling azeotrope, the T–x diagram shows new features as illustrated for the water–butanol system in Figure 9.24b.

First, consider the system at point a in the phase diagram corresponding to a single liquid phase enriched in butanol. If the temperature is increased to T_b, the gas-phase composition is given by point b. If the vapor of this composition is removed from the system and cooled to the original temperature T_a, it is at point c in the phase diagram, which corresponds to two liquid phases of composition d and e in equilibrium with the vapor phase. This system consists of two components and, therefore, $F = 4 - p$.

Next, consider heating the liquid described by point f in the two-liquid-phase region in a piston and cylinder assembly at constant pressure. Vapor first appears when the temperature is increased to 94°C and has the composition given by point i,

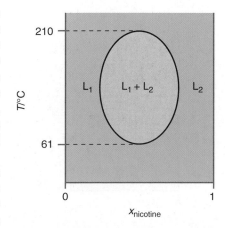

FIGURE 9.23

A T–x phase diagram showing an upper and lower consolate temperature is shown for the partially miscible liquids water and nicotine. The pressure is held constant at 1 bar.

T–x composition phase diagrams are shown for partially miscible liquids. **(a)** The upper consolate temperature is lower in temperature than the low boiling azeotrope. **(b)** The upper consolate temperature is higher in temperature than the minimum boiling azeotrope. The pressure is held constant at 1 bar.

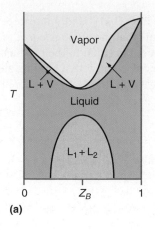

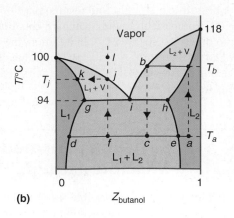

corresponding to a minimum boiling point azeotrope. The vapor is in equilibrium with two liquid phases that have the composition given by points *g* and *h*. Because the pressure is constant in Figure 9.24b and $p = 3$ at point *i*, the system has zero degrees of freedom. Distillation of the homogeneous solution results in two separate liquid phases of fixed composition *g* and *h*. As more heat is put into the system, liquid is converted to vapor, but the composition of the two liquid phases and the vapor phase remains constant as described by points *g*, *h*, and *i*. Because the vapor is enriched in butanol with respect to the liquid, the butanol-rich phase L_2 is converted to vapor before the water-rich phase L_1. At the point when the L_2 phase disappears, the temperature increases beyond 94°C and the vapor composition changes along the *i-j* curve. As the vapor approaches the composition corresponding to *j*, the entire system is converted to vapor and the last drop of L_1, which disappears at T_j, has the composition given by point *k*. Further heat input results in an increase in the temperature of the vapor with no change in its composition.

Summarizing these changes, heating increases the amount of material in the vapor phase. As long as the two phases of composition *g* and *h* are present, the temperature is constant and the amount of vapor increases, but its composition is fixed at *i*. After the last drop of L_2 of composition *h* boils off, the vapor composition continuously changes along the curve *i-j*. As the last drop of L_1 of composition *k* evaporates, the vapor composition approaches the average composition of the system given by *f*, *j*, and *l*.

9.15 The Solid-Solution Equilibrium

If a dilute liquid solution of B in A is cooled sufficiently, pure solvent A will crystallize out in a solid phase as discussed in Section 9.7 in connection with the freezing point depression. Point *a* in Figure 9.25a represents a solution of composition *b* in equilibrium with pure solid A (point *c*). The relative number of moles of solution and solid can be found using the lever rule. The temperature *T* at which freezing occurs is given by

$$\frac{1}{T} = \frac{1}{T_{fusion}} - \frac{R \ln x_{solvent}}{\Delta H_{fusion}} \qquad (9.78)$$

where T_{fusion} and ΔH_{fusion} are the freezing temperature and enthalpy of fusion of the pure solvent. *T* is shown schematically as a function of x_B in Figure 9.25a, and the dashed range indicates the region beyond which Equation (9.78) is unlikely to be valid.

The same argument holds for a dilute solution of A in B whereby we have interchanged the solvent and solute roles. Therefore, the appropriate phase diagram for all values of Z_A and Z_B is that shown in Figure 9.25b. We use the variables Z_A and Z_B rather than x_A and x_B because the solid is not homogeneous. In Figure 9.25b point *a* represents solid B in equilibrium with a solution of composition *b*, and point *c* represents pure solid A in equilibrium with the solution of composition *d*. Point *f* is a special point called the

T–x phase diagrams showing solid-liquid equilibria in a two-component system. The pressure is held constant at 1 bar.

eutectic point that is the minimum melting point observed in the A–B system. The temperature T_e is called the **eutectic temperature** (from the Greek, easily melted).

Because the tie line at T_e extends across the entire phase diagram, a solution at T_e is in equilibrium with both pure A (point e) and pure B (point g). As more heat is put into the system, the volume of the solution phase increases but its composition remains constant at Z_B until one of the pure solids is completely melted because $F = 0$.

Eutectic materials are widely used in making plumbing and electrical connections using a solder. Until the 1980s, copper plumbing components were connected using Pb–Sn solders but as even small amounts of lead present a health hazard, Pb–Sn solders have been replaced by lead-free solders which may contain tin, copper, silver, indium, zinc, antimony, or bismuth. Another important application of eutectic materials is making electrical connections to silicon chips in the microelectronics industry. Gold is deposited onto the Si surface using masks to establish an appropriate pattern. Subsequently, the chip is heated to near 400°C. As the temperature increases, gold atoms diffuse rapidly into the silicon, and a very thin liquid film forms at the interface once enough gold has diffused to reach the eutectic composition. As more heat is put into the chip, a larger volume of eutectic alloy is produced, which is ultimately the electrical connection. The Au–Si phase diagram is shown in Figure 9.26, and it is seen that the eutectic temperature is approximately 700°C and 950°C below the melting point of pure gold and Si, respectively.

The eutectic mixture is useful in bonding gold to silicon to form an electrical connection. Imagine that gold wire is placed in contact with a Si chip on which Au has been deposited. If the chip is heated, the eutectic liquid will form before the Si or the Au melts. Therefore, the eutectic will connect the gold wire to the chip without melting the Au or the Si.

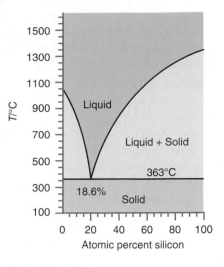

FIGURE 9.26
The T–x phase diagram is shown for the gold–silicon system. Note that the eutectic melting point is much lower than the melting points of the pure substances Au and Si. The pressure is constant at 1 bar.

Vocabulary

activity	Henry's law standard state	pressure–average composition diagram
activity coefficient	ideal dilute solution	Raoult's law
average composition	ideal solution	Raoult's law standard state
binding equilibria	independent site binding	Scatchard equation
boiling point elevation	lever rule	Scatchard plot
colligative properties	limiting law	semipermeable membrane
eutectic temperature	lower and upper consolate temperature	solute
fractional distillation	osmosis	solvent
freezing point depression	osmotic pressure	temperature–composition diagram
Gibbs–Duhem equation	partial molar quantities	tie line
Henry's law	partial molar volume	van't Hoff equation
Henry's law constant		

Conceptual Problems

Q9.1 Why is the magnitude of the boiling point elevation less than that of the freezing point depression?

Q9.2 Fractional distillation of a particular binary liquid mixture leaves behind a liquid consisting of both components in which the composition does not change as the liquid is boiled off. Is this behavior characteristic of a maximum or a minimum boiling point azeotrope?

Q9.3 In the description of Figure 9.24b, the following sentence appears: "At the point when the L_2 phase disappears,

the temperature increases beyond 94°C and the vapor composition changes along the i-j curve." Why does the vapor composition change along the i-j curve?

Q9.4 Explain why chemists doing quantitative work using liquid solutions prefer to express concentration in terms of molality rather than molarity.

Q9.5 Explain the usefulness of a tie line on a P–Z phase diagram such as that of Figure 9.4.

Q9.6   For a pure substance, the liquid and gaseous phases can only coexist for a single value of the pressure at a given temperature. Is this also the case for an ideal solution of two volatile liquids?

Q9.7   Using the differential form of G, $dG = V\,dP - S\,dT$, show that if $\Delta G_{mixing} = nRT \sum_i x_i \ln x_i$, then $\Delta H_{mixing} = \Delta V_{mixing} = 0$.

Q9.8   What information can be obtained from a tie line in a P–Z phase diagram?

Q9.9   You boil an ethanol–benzene mixture with $x_{ethanol} = 0.35$. What is the composition of the vapor phase that first appears? What is the composition of the last liquid left in the vessel?

Q9.10   What can you say about the composition of the solid below the eutectic temperature in Figure 9.26 on a microscopic scale?

Q9.11   Is a whale likely to get the bends when it dives deep into the ocean and resurfaces? Answer this question by considering the likelihood of a scuba diver getting the bends if he or she dives and resurfaces on one lung full of air as opposed to breathing air for a long time at the deepest point of the dive.

Q9.12   Why is the preferred standard state for the solvent in an ideal dilute solution the Raoult's law standard state? Why is the preferred standard state for the solute in an ideal dilute solution the Henry's law standard state? Is there a preferred standard state for the solution in which $x_{solvent} = x_{solute} = 0.5$?

Q9.13   The entropy of two liquids is increased if they mix. How can immiscibility be explained in terms of thermodynamic state functions?

Q9.14   Explain why colligative properties depend only on the concentration and not on the identity of the molecule.

Q9.15   The statement "The boiling point of a typical liquid mixture can be reduced by approximately 100°C if the pressure is reduced from 760 to 20 Torr" is found in Section 9.4. What figure(s) in Chapter 9 can you identify to support this statement in a qualitative sense?

Numerical Problems

Problem numbers in **red** indicate that the solution to the problem is given in the *Student's Solutions Manual.*

P9.1   Ratcliffe and Chao [*Canadian Journal of Chemical Engineering* 47 (1969): 148] obtained the following tabulated results for the variation of the total pressure above a solution of isopropanol ($P_1^* = 1008$ Torr) and n-decane ($P_2^* = 48.3$ Torr) as a function of the mole fraction of the n-decane in the solution and vapor phases. Using these data, calculate the activity coefficients for both components using a Raoult's law standard state.

P (Torr)	x_2	y_2
942.6	0.1312	0.0243
909.6	0.2040	0.0300
883.3	0.2714	0.0342
868.4	0.3360	0.0362
830.2	0.4425	0.0411
786.8	0.5578	0.0451
758.7	0.6036	0.0489

P9.2   At a given temperature, a nonideal solution of the volatile components A and B has a vapor pressure of 795 Torr. For this solution, $y_A = 0.375$. In addition, $x_A = 0.310$, $P_A^* = 610$ Torr, and $P_B^* = 495$ Torr. Calculate the activity and activity coefficient of A and B.

P9.3   Two liquids, A and B, are immiscible for $T < 75.0°C$ and for $T > 45.0°C$ and are completely miscible outside of this temperature range. Sketch the phase diagram, showing as much information as you can from these observations.

P9.4   At 350. K, pure toluene and hexane have vapor pressures of 3.57×10^4 Pa and 1.30×10^5 Pa, respectively.

a. Calculate the mole fraction of hexane in the liquid mixture that boils at 350. K at a pressure of 1 atm.

b. Calculate the mole fraction of hexane in the vapor that is in equilibrium with the liquid of part (a).

P9.5   The partial molar volumes of water and ethanol in a solution with $x_{H_2O} = 0.45$ at 25°C are 17.0 and 57.5 cm^3 mol^{-1}, respectively. Calculate the volume change upon mixing sufficient ethanol with 3.75 mol of water to give this concentration. The densities of water and ethanol are 0.997 and 0.7893 g cm^{-3}, respectively, at this temperature.

P9.6   A solution is made up of 222.9 g of ethanol and 130.8 g of H_2O. If the volume of the solution is 403.4 cm^3 and the partial molar volume of H_2O is 17.0 cm^3, what is the partial molar volume of ethanol under these conditions?

P9.7   The osmotic pressure of an unknown substance is measured at 298 K. Determine the molecular weight if the concentration of this substance is 31.2 kg m^{-3} and the osmotic pressure is 5.30×10^4 Pa. The density of the solution is 997 kg m^{-3}.

P9.8   At 303. K, the vapor pressure of benzene is 120. Torr and that of hexane is 189 Torr. Calculate the vapor pressure of a solution for which $x_{benzene} = 0.28$ assuming ideal behavior.

P9.9   An ideal solution is made up of the volatile liquids A and B, for which $P_A^* = 165$ Torr and $P_B^* = 85.1$ Torr. As the pressure is reduced, the first vapor is observed at a total pressure of 110. Torr. Calculate x_A.

P9.10   At high altitudes, mountain climbers are unable to absorb a sufficient amount of O_2 into their bloodstreams to maintain a high activity level. At a pressure of 1 bar, blood is typically 95% saturated with O_2, but near 18,000 feet where the pressure is 0.50 bar, the corresponding degree of saturation is 71%. Assuming that the Henry's law constant for blood is the same as for water, calculate the amount of O_2 dissolved in 1.00 L of blood for pressures of 1 bar and 0.500 bar. Air contains 20.99% O_2 by volume. Assume that the density of blood is 998 kg m^{-3}.

P9.11 At $-47.0°C$, the vapor pressure of ethyl bromide is 10.0 Torr and that of ethyl chloride is 40.0 Torr. Assume that the solution is ideal. Assume there is only a trace of liquid present and the mole fraction of ethyl chloride in the vapor is 0.80 and answer these questions:

a. What is the total pressure and the mole fraction of ethyl chloride in the liquid?

b. If there are 5.00 mol of liquid and 3.00 mol of vapor present at the same pressure as in part (a), what is the overall composition of the system?

P9.12 A and B form an ideal solution at 298 K, with $x_A = 0.320$, $P_A^* = 84.3$ Torr, and $P_B^* = 41.2$ Torr.

a. Calculate the partial pressures of A and B in the gas phase.

b. A portion of the gas phase is removed and condensed in a separate container. Calculate the partial pressures of A and B in equilibrium with this liquid sample at 298 K.

P9.13 Describe what you would observe if you heated the liquid mixture at the composition corresponding to point i in Figure 9.24b from a temperature below T_a to 118°C.

P9.14 The heat of fusion of water is 6.008×10^3 J mol^{-1} at its normal melting point of 273.15 K. Calculate the freezing point depression constant K_f.

P9.15 At 39.9°C, a solution of ethanol ($x_1 = 0.9006$, $P_1^* = 130.4$ Torr) and isooctane ($P_2^* = 43.9$ Torr) forms a vapor phase with $y_1 = 0.6667$ at a total pressure of 185.9 Torr.

a. Calculate the activity and activity coefficient of each component.

b. Calculate the total pressure that the solution would have if it were ideal.

P9.16 Calculate the solubility of H_2S in 1 L of water if its pressure above the solution is 2.75 bar. The density of water at this temperature is 997 kg m^{-3}.

P9.17 The binding of NADH to human liver mitochondrial isozyme was studied [*Biochemistry* 28 (1989): 5367] and it was determined that only a single binding site is present with $K = 2.0 \times 10^7$ M^{-1}. What concentration of NADH is required to occupy 10% of the binding sites?

P9.18 Given the vapor pressures of the pure liquids and the overall composition of the system, what are the upper and lower limits of pressure between which liquid and vapor coexist in an ideal solution?

P9.19 A and B form an ideal solution. At a total pressure of 0.720 bar, $y_A = 0.510$ and $x_A = 0.420$. Using this information, calculate the vapor pressure of pure A and of pure B.

P9.20 The partial pressures of Br_2 above a solution containing CCl_4 as the solvent at 25°C are found to have the values listed in the following table as a function of the mole fraction of Br_2 in the solution [Lewis G. N., and Storch, H. *J. American Chemical Society* 39 (1917): 2544]. Use these data and a graphical method to determine the Henry's law constant for Br_2 in CCl_4 at 25°C.

x_{Br_2}	P (Torr)	x_{Br_2}	P (Torr)
0.00394	1.52	0.0130	5.43
0.00420	1.60	0.0236	9.57
0.00599	2.39	0.0238	9.83
0.0102	4.27	0.0250	10.27

P9.21 The data from Problem P9.20 can be expressed in terms of the molality rather than the mole fraction of Br_2. Use the data from the following table and a graphical method to determine the Henry's law constant for Br_2 in CCl_4 at 25°C in terms of molality.

m_{Br_2}	P (Torr)	m_{Br_2}	P (Torr)
0.026	1.52	0.086	5.43
0.028	1.60	0.157	9.57
0.039	2.39	0.158	9.83
0.067	4.27	0.167	10.27

P9.22 The densities of pure water and ethanol are 997 and 789 kg m^{-3}, respectively. For $x_{ethanol} = 0.35$, the partial molar volumes of ethanol and water are 55.2 and 17.8×10^{-3} L mol^{-1}, respectively. Calculate the change in volume relative to the pure components when 2.50 L of a solution with $x_{ethanol} = 0.35$ is prepared.

P9.23 Two liquids, A and B, are immiscible for $x_A = x_B = 0.5$, for $T < 75.0°C$ and completely miscible for $T > 75.0°C$. Sketch the phase diagram, showing as much information as you can from these observations.

P9.24 An ideal solution is formed by mixing liquids A and B at 298 K. The vapor pressure of pure A is 151 Torr and that of pure B is 84.3 Torr. If the mole fraction of A in the vapor is 0.610, what is the mole fraction of A in the solution?

P9.25 A solution is prepared by dissolving 45.2 g of a non-volatile solute in 119 g of water. The vapor pressure above the solution is 22.51 Torr and the vapor pressure of pure water is 23.76 Torr at this temperature. What is the molecular weight of the solute?

P9.26 A sample of glucose ($C_6H_{12}O_6$) of mass 13.2 g is placed in a test tube of radius 1.25 cm. The bottom of the test tube is a membrane that is semipermeable to water. The tube is partially immersed in a beaker of water at 298 K so that the bottom of the test tube is only slightly below the level of the water in the beaker. The density of water at this temperature is 997 kg m^{-3}. After equilibrium is reached, how high is the water level of the water in the tube above that in the beaker? What is the value of the osmotic pressure? You may find the approximation $\ln(1/(1 + x)) \approx -x$ useful.

P9.27 A volume of 5.50 L of air is bubbled through liquid toluene at 298 K, thus reducing the mass of toluene in the beaker by 2.38 g. Assuming that the air emerging from the beaker is saturated with toluene, determine the vapor pressure of toluene at this temperature.

P9.28 The vapor pressures of 1-bromobutane and 1-chlorobutane can be expressed in the form

$$\ln \frac{P_{bromo}}{Pa} = 17.076 - \frac{1584.8}{\frac{T}{K} - 111.88}$$

and

$$\ln \frac{P_{chloro}}{Pa} = 20.612 - \frac{2688.1}{\frac{T}{K} - 55.725}$$

Assuming ideal solution behavior, calculate x_{bromo} and y_{bromo} at 305 K and a total pressure of 9750. Pa.

P9.29 In an ideal solution of A and B, 3.00 mol are in the liquid phase and 5.00 mol are in the gaseous phase. The overall composition of the system is $Z_A = 0.375$ and $x_A = 0.250$. Calculate y_A.

P9.30 Assume that 1-bromobutane and 1-chlorobutane form an ideal solution. At 273 K, $P^*_{chloro} = 3790$ Pa and $P^*_{bromo} = 1394$ Pa. When only a trace of liquid is present at 273 K, $y_{chloro} = 0.750$.

a. Calculate the total pressure above the solution.

b. Calculate the mole fraction of 1-chlorobutane in the solution.

c. What value would Z_{chloro} have in order for there to be 4.86 mol of liquid and 3.21 mol of gas at a total pressure equal to that in part (a)? [*Note:* This composition is different from that of part (a).]

P9.31 DNA is capable of forming complex helical structures. An unusual triple-helix structure of poly(dA).2poly(dT) DNA was studied by P. V. Scaria and R. H. Shafer [*Journal of Biological Chemistry* 266 (1991): 5417] where the intercalation of ethidium bromide was studied using UV absorption and circular dichroism spectroscopy. The following representative data were obtained using the results of this study:

c_{Eth} 1(μ M)	0.63	1.7	5.0	10.0	14.7
$\overline{\nu}$		0.00683	0.01675	0.0388	0.0590	0.0705

Using the data, determine K and N for the binding of ethidium bromide to the DNA triple-helical structure.

P9.32 Calculate the activity and activity coefficient for CS_2 at $x_{CS_2} = 0.722$ using the data in Table 9.3 for both a Raoult's law and a Henry's law standard state.

P9.33 The dissolution of 7.75 g of a substance in 825 g of benzene at 298 K raises the boiling point by 0.575°C. Note that $K_f = 5.12$ K kg mol^{-1}, $K_b = 2.53$ K kg mol^{-1}, and the density of benzene is 876.6 kg m^{-3}. Calculate the freezing point depression, the ratio of the vapor pressure above the solution to that of the pure solvent, the osmotic pressure, and the molecular weight of the solute. $P^*_{benzene} = 103$ Torr at 298 K.

P9.34 Describe what you would observe if you heated the solid at the composition 40. atomic percent Si in Figure 9.26 from 300.°C to 1300.°C.

P9.35 An ideal dilute solution is formed by dissolving the solute A in the solvent B. Write expressions equivalent to Equations (9.9) through (9.13) for this case.

P9.36 Describe the changes you would observe as the temperature of a mixture of phenol and water at point a in Figure 9.21 is increased until the system is at point a'. How does the relative amount of separate phases of phenol and water change along this path?

P9.37 Describe the changes you would observe as the temperature of a mixture of triethylamine and water at point a in Figure 9.22 is increased until the system is at point $a"$. How does the relative amount of separate phases of triethylamine and water change along this path?

P9.38 Describe the changes in a beaker containing water and butanol that you would observe along the path $a \rightarrow b \rightarrow c$ in Figure 19.24b. How would you calculate the relative amounts of different phases present along the path?

P9.39 Describe the changes in a beaker containing water and butanol that you would observe along the path $f \rightarrow j \rightarrow i$ in Figure 19.24b. How would you calculate the relative amounts of different phases present along the path?

P9.40 Describe the changes in a beaker containing water and butanol that you would observe along the path $f \rightarrow j \rightarrow k$ in Figure 19.24b. How would you calculate the relative amounts of different phases present along the path?

P9.41 Describe the system at points a and c in Figure 19.25b. How would you calculate the relative amounts of different phases present at these points?

Electrolyte Solutions

E lectrolyte solutions are quite different from the ideal and real solutions of neutral solutes discussed in Chapter 9. The fundamental reason for this difference is that solutes in electrolyte solutions exist as solvated positive and negative ions. In Chapter 4, the formation enthalpy and Gibbs energy of a pure element in its standard state at 1 bar of pressure were set equal to zero. These assumptions allow the formation enthalpies and Gibbs energies of compounds to be calculated using the results of thermochemical experiments. In electrolyte solutions, an additional assumption, $\Delta G_f^\circ(H^+, aq) = 0$, is made to allow the formation enthalpies, Gibbs energies, and entropies of individual ions to be determined. Why are electrolyte and nonelectrolyte solutions so different? The Coulomb interactions between ions in an electrolyte solution are of much longer range than the interactions between neutral solutes. For this reason, electrolyte solutions deviate from ideal behavior at much lower concentrations than do solutions of neutral solutes. Although a formula unit of an electrolyte dissociates into positive and negative ions, only the mean activity and activity coefficient of the ions produced is accessible through experiments. The Debye–Hückel limiting law provides a useful way to calculate activity coefficients for dilute electrolyte solutions.

10.1 The Enthalpy, Entropy, and Gibbs Energy of Ion Formation in Solutions

In this chapter, substances called **electrolytes,** which dissociate into positively and negatively charged mobile solvated ions when dissolved in an appropriate solvent, are discussed. Consider the following overall reaction in water:

$$1/2 \ H_2(g) + 1/2 \ Cl_2(g) \rightarrow H^+(aq) + Cl^-(aq) \qquad (10.1)$$

in which $H^+(aq)$ and $Cl^-(aq)$ represent mobile solvated ions. Although similar in structure, Equation (10.1) represents a reaction that is quite different than the gas-phase dissociation of an HCl molecule to give $H^+(g)$ and $Cl^-(g)$. For the reaction in solution, ΔH_R is -167.2 kJ mol^{-1}. The shorthand notation $H^+(aq)$ and $Cl^-(aq)$ refers to positive and negative ions as well as their associated hydration shell. The **solvation shell** is essential in lowering the energy of the ions, thereby making the previous reaction spontaneous. Although energy flow into the system is required to dissociate and ionize hydrogen and chlorine, even more energy is gained in the reorientation of the dipolar water molecules around the ions in the solvation shell. Therefore, the reaction is exothermic.

The standard state enthalpy for this reaction can be written in terms of formation enthalpies:

$$\Delta H_R^\circ = \Delta H_f^\circ(H^+, aq) + \Delta H_f^\circ(Cl^-, aq) \tag{10.2}$$

There is no contribution of $H_2(g)$ and $Cl_2(g)$ to ΔH_R° in Equation (10.2) because ΔH_f° for a pure element in its standard state is zero.

Unfortunately, no direct calorimetric experiment can measure only the heat of formation of the solvated anion or cation. This is the case because the solution remains electrically neutral; any dissociation reaction of a neutral solute must produce both anions and cations. As we have seen in Chapter 4, tabulated values of formation enthalpies, entropies, and Gibbs energies for various chemical species are very useful. How can this information be obtained for individual solvated cations and anions?

The discussion in the rest of this chapter is restricted to aqueous solutions, for which water is the solvent, but can be generalized to other solvents. Values of thermodynamic functions for anions and cations in aqueous solutions can be obtained by making an appropriate choice for the zero of ΔH_f°, ΔG_f°, and S_m°. By convention, the formation Gibbs energy for $H^+(aq)$ at unit activity is set equal to zero at all temperatures:

$$\Delta G_f^\circ(H^+, aq) = 0 \quad \text{for all } T \tag{10.3}$$

With this choice,

$$S_f^\circ(H^+, aq) = -\left(\frac{\partial \Delta G_f^\circ(H^+, aq)}{\partial T}\right)_P = 0 \quad \text{and}$$

$$\Delta H_f^\circ(H^+, aq) = \Delta G_f^\circ(H^+, aq) + TS_m^\circ(H^+, aq) = 0 \tag{10.4}$$

Using the convention of Equation (10.3), which has the consequences shown in Equation (10.4), the values of ΔH_f°, ΔG_f°, and S_m° for an individual ion can be assigned numerical values, as shown next.

As discussed earlier, ΔH_R° for the reaction $1/2\,H_2(g) + 1/2\,Cl_2(g) \rightarrow H^+(aq) + Cl^-(aq)$ can be directly measured. The value of ΔG_R° can be determined from $\Delta G_R^\circ = -RT \ln K$ by measuring the degree of dissociation in the reaction, using the solution conductivity, and ΔS_R° can be determined from the relation

$$\Delta S_R^\circ = \frac{\Delta H_R^\circ - \Delta G_R^\circ}{T}$$

Using the conventions stated in Equations (10.3) and (10.4) together with the conventions regarding ΔH_f° and ΔG_f° for pure elements, $\Delta H_R^\circ = \Delta H_f^\circ(Cl^-, aq)$, $\Delta G_R^\circ = \Delta G_f^\circ(Cl^-, aq)$ and

$$\Delta S_R^\circ = S_m^\circ(Cl^-, aq) - \frac{1}{2} S_m^\circ(H_2, g) - \frac{1}{2} S_m^\circ(Cl_2, g)$$

for the reaction of Equation (10.1). In this way, the numerical values $\Delta H_f(Cl^-, aq) = -167.2$ kJ mol^{-1}, $S_m^\circ(Cl^-, aq) = 56.5$ J K^{-1} mol^{-1}, and $\Delta G_f^\circ(Cl^-, aq) = -131.2$ kJ mol^{-1} can be obtained.

These values can be used to determine the formation functions of other ions. To illustrate how this is done, consider the following reaction:

$$NaCl(s) \rightarrow Na^+(aq) + Cl^-(aq) \tag{10.5}$$

for which ΔH_R° is found to be $+3.90$ kJ mol^{-1}. For this reaction,

$$\Delta H_R^\circ = \Delta H_f^\circ(Cl^-, aq) + \Delta H_f^\circ(Na^+, aq) - \Delta H_f^\circ(NaCl, s) \tag{10.6}$$

We use the tabulated value of $\Delta H_f^\circ(NaCl, s) = -411.2 \text{ kJ mol}^{-1}$ and the value for $\Delta H_f^\circ(Cl^-, aq)$ just determined to obtain a value for $\Delta H_f^\circ(Na^+, aq) = -240.1 \text{ kJ mol}^{-1}$. Proceeding to other reactions that involve either $Na^+(aq)$ or $Cl^-(aq)$, the enthalpies of formation of the counter ions can be determined. This procedure can be extended to include other ions. Values for ΔG_f° and S_m° can be determined in a similar fashion. Values for ΔH_f°, ΔG_f°, and S_m° for aqueous ionic species are tabulated in Table 10.1. These thermodynamic quantities are called **conventional formation enthalpies, conventional Gibbs energies of formation,** and **conventional formation entropies** because of the convention described earlier.

Note that ΔH_f°, ΔG_f°, and S_m° for ions are defined relative to $H^+(aq)$. Negative values for ΔH_f° indicate that the formation of the solvated ion is more exothermic than the formation of $H^+(aq)$. A similar statement can be made for ΔG_f°. Generally speaking, ΔH_f° for multiply charged ions is more negative than that of singly charged ions, and ΔH_f° for a given charge is more negative for smaller ions because of the stronger electrostatic attraction between the multiply charged or smaller ion and the water in the solvation shell.

Recall from Section 5.8 that the entropy of an atom or molecule was shown to be always positive. This is not the case for solvated ions because the entropy is measured relative to $H^+(aq)$. The entropy decreases as the hydration shell is formed because liquid water molecules are converted to relatively immobile molecules. Ions with a negative value for the conventional standard entropy such as $Mg^{2+}(aq)$, $Zn^{2+}(aq)$, and $PO_4^{3-}(aq)$ have a larger charge-to-size ratio than $H^+(aq)$. For this reason, the solvation shell is more tightly bound. Conversely, ions with a positive value for the standard entropy such as $Na^+(aq)$, $Cs^+(aq)$, and $NO_3^-(aq)$ have a smaller charge-to-size ratio than $H^+(aq)$ and a less tightly bound solvation shell.

TABLE 10.1 Conventional Formation Enthalpies, Gibbs Energies, and Entropies of Selected Aqueous Anions and Cations

Ion	ΔH_f° (kJ mol^{-1})	ΔG_f° (kJ mol^{-1})	S_m° (J K^{-1} mol^{-1})
$Ag^+(aq)$	105.6	77.1	72.7
$Br^-(aq)$	−121.6	−104.0	82.4
$Ca^{2+}(aq)$	−542.8	−553.6	−53.1
$Cl^-(aq)$	−167.2	−131.2	56.5
$Cs^+(aq)$	−258.3	−292.0	133.1
$Cu^+(aq)$	71.7	50.0	40.6
$Cu^{2+}(aq)$	64.8	65.5	−99.6
$F^-(aq)$	−332.6	−278.8	−13.8
$H^+(aq)$	0	0	0
$I^-(aq)$	−55.2	−51.6	111.3
$K^+(aq)$	−252.4	−283.3	102.5
$Li^+(aq)$	−278.5	−293.3	13.4
$Mg^{2+}(aq)$	−466.9	−454.8	−138.1
$NO_3^-(aq)$	−207.4	−111.3	146.4
$Na^+(aq)$	−240.1	−261.9	59.0
$OH^-(aq)$	−230.0	−157.2	−10.9
$PO_4^{3-}(aq)$	−1277.4	−1018.7	−220.5
$SO_4^{2-}(aq)$	−909.3	−744.5	20.1
$Zn^{2+}(aq)$	−153.9	−147.1	−112.1

Source: Lide, D. R., ed. *Handbook of Chemistry and Physics.* 83rd ed. Boca Raton, FL: CRC Press 2002.

10.2 Understanding the Thermodynamics of Ion Formation and Solvation

As discussed in the preceding section, ΔH_f°, ΔG_f°, and S_m° can be determined for a formula unit but not for an individual ion in a calorimetric experiment. However, as seen next, values for thermodynamic functions associated with individual ions can be calculated with a reasonable level of confidence using a thermodynamic model. This result allows the conventional values of ΔH_f°, ΔG_f°, and S_m° to be converted to absolute values for individual ions. In the following discussion, the focus is on ΔG_f°.

We first discuss the individual contributions to ΔG_f°, and do so by analyzing the following sequence of steps that describe the formation of $H^+(aq)$ and $Cl^-(aq)$:

$1/2\ H_2(g) \rightarrow H(g)$	$\Delta G^\circ = 203.3\ \text{kJ mol}^{-1}$
$1/2\ Cl_2(g) \rightarrow Cl(g)$	$\Delta G^\circ = 105.7\ \text{kJ mol}^{-1}$
$H(g) \rightarrow H^+(g) + e^-$	$\Delta G^\circ = 1312\ \text{kJ mol}^{-1}$
$Cl(g) + e^- \rightarrow Cl^-(g)$	$\Delta G^\circ = -349\ \text{kJ mol}^{-1}$
$Cl^-(g) \rightarrow Cl^-(aq)$	$\Delta G^\circ = \Delta G_{solvation}^\circ(Cl^-, aq)$
$H^+(g) \rightarrow H^+(aq)$	$\Delta G^\circ = \Delta G_{solvation}^\circ(H^+, aq)$

$1/2\ H_2(g) + 1/2\ Cl_2(g) \rightarrow H^+(aq) + Cl^-(aq)$	$\Delta G_R^\circ = -131.2\ \text{kJ mol}^{-1}$

This pathway is shown pictorially in Figure 10.1. Because G is a state function, both the black and red paths must have the same ΔG value. The first two reactions in this sequence are the dissociation of the molecules in the gas phase, and the second two reactions are the formation of gas phase ions from the neutral gas phase atoms. ΔG° can be determined experimentally for these four reactions. Substituting the known values for ΔG° for these four reactions in ΔG° for the overall process,

$$\Delta G_R^\circ = \Delta G_{solvation}^\circ(Cl^-, aq) + \Delta G_{solvation}^\circ(H^+, aq) + 1272\ \text{kJ mol}^{-1} \quad \textbf{(10.7)}$$

Equation 10.7 allows us to relate the $\Delta G_{solvation}^\circ$ of the H^+ and Cl^- ions with ΔG_R° for the overall reaction.

As Equation (10.7) shows, $\Delta G_{solvation}^\circ$ plays a critical role in the determination of the Gibbs energies of ion formation. Although $\Delta G_{solvation}^\circ$ of an individual cation or anion cannot be determined experimentally, it can be estimated using a model developed by Max Born. In this model, the solvent is treated as a uniform fluid with the appropriate dielectric constant, and the ion is treated as a charged sphere. How can $\Delta G_{solvation}^\circ$ be calculated with these assumptions? At constant T and P, the nonexpansion work for a reversible process equals ΔG for the process. Therefore, if the reversible work associated with solvation can be calculated, ΔG for the process is known. Imagine a process in which a neutral atom A gains the charge Q, first in a vacuum and secondly in a uniform dielectric medium. The value of $\Delta G_{solvation}^\circ$ of an ion with a charge q is the reversible work for the process $(A(g) \rightarrow A^Q(aq))_{solvation}$ minus that for the reversible process $(A(g) \rightarrow A^Q(g))_{vacuum}$.

FIGURE 10.1
ΔG° is shown pictorially for two different paths starting with $1/2\ H_2(g)$ and $1/2\ Cl_2(g)$ and ending with $H^+(aq) + Cl^-(aq)$. The units for the numbers are kJ mol^{-1}. Because ΔG is the same for both paths, $\Delta G_{solvation}^\circ(H^+, aq)$ can be expressed in terms of gas-phase dissociation and ionization energies.

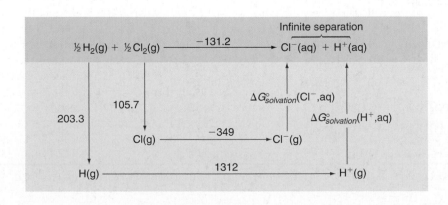

The electrical potential around a sphere of radius r with the charge Q' is given by $\phi = Q'/4\pi\varepsilon r$. From electrostatics, the work in charging the sphere by the additional amount dQ is $\phi\, dQ$. Therefore, the work in charging a neutral sphere in vacuum to the charge Q is

$$w = \int_0^Q \frac{Q'\, dQ'}{4\pi\varepsilon_0 r} = \frac{1}{4\pi\varepsilon_0 r}\int_0^Q Q'\, dQ' = \frac{Q^2}{8\pi\varepsilon_0 r} \tag{10.8}$$

where ε_0 is the permittivity of free space. The work of the same process in a solvent is $Q^2/8\pi\varepsilon_0\varepsilon_r r$, where ε_r is the relative permittivity (dielectric constant) of the solvent. Consequently, $\Delta G^{\circ}_{solvation}$ for an ion of charge $Q = ze$ is given by

$$\Delta G^{\circ}_{solvation} = \frac{z^2 e^2 N_A}{8\pi\varepsilon_0 r}\left(\frac{1}{\varepsilon_r} - 1\right) \tag{10.9}$$

Because $\varepsilon_r > 1$, $\Delta G^{\circ}_{solvation} < 0$, showing that solvation is a spontaneous process. Values for ε_r for a number of solvents are listed in Table 10.2 (see Appendix B, Data Tables).

To test the Born model, we need to compare absolute values of $\Delta G^{\circ}_{solvation}$ for ions of different radii with the functional form proposed in Equation (10.9). However, this comparison requires knowledge of $\Delta G^{\circ}_{solvation}(H^+, aq)$ to convert experimentally determined values of $\Delta G^{\circ}_{solvation}$ referenced to $H^+(aq)$ to absolute values. It turns out that $\Delta G^{\circ}_{solvation}(H^+, aq)$, and $\Delta H^{\circ}_{solvation}(H^+, aq)$, and $S^{\circ}_{solvation}(H^+, aq)$ can be calculated. Because the calculation is involved, the results are simply stated here.

$$\Delta H^{\circ}_{solvation}(H^+, aq) \approx -1090\,\text{kJ mol}^{-1}$$

$$\Delta G^{\circ}_{solvation}(H^+, aq) \approx -1050\,\text{kJ mol}^{-1}$$

$$S^{\circ}_{solvation}(H^+, aq) \approx -130\,\text{J mol}^{-1}\text{K}^{-1} \tag{10.10}$$

The values listed in Equation (10.10) can be used to calculate absolute values of $\Delta H^{\circ}_{solvation}$, $\Delta G^{\circ}_{solvation}$, and $S^{\circ}_{solvation}$ for other ions from the conventional values referenced to $H^+(aq)$. These calculated absolute values can be used to test the validity of the Born model. If the model is valid, a plot of $\Delta G^{\circ}_{solvation}$ versus z^2/r will give a straight line as shown by Equation (10.9), and the data points for individual ions should lie on the line. The results are shown in Figure 10.2, where r is the ionic radius obtained from crystal structure determinations.

The first and second clusters of data points in Figure 10.2 are for singly and doubly charged ions, respectively. The data are compared with the result predicted by Equation (10.9) in Figure 10.2a. As can be seen from the figure, the trends are reproduced, but there is no quantitative agreement. The agreement can be considerably improved by using an effective radius for the solvated ion rather than the ionic radius from crystal structure determinations. The effective radius is defined as the distance from the center of the ion to the center of charge in the dipolar water molecule in the solvation shell. Latimer, Pitzer, and Slansky [*J. Chemical Physics*, 7 (1939) 109] found the best agreement with the Born equation by adding 0.085 nm to the crystal radius of positive ions, and 0.100 nm to the crystal radius for negative ions to account for the fact that the H_2O molecule is not a point dipole. This difference is explained by the fact that the center of charge in the water molecule is closer to positive ions than to negative ions. Figure 10.2b shows that the agreement obtained between the predictions of Equation (10.9) and experimental values is very good if this correction to the ionic radii is made.

Figure 10.2 shows good agreement between the predictions of the Born model and calculated values for $\Delta G^{\circ}_{solvation}$, and justifies the approach used to calculate absolute enthalpies, Gibbs energies and entropies for solvated ions. However, because of uncertainties about the numerical values of the ionic radii and for the dielectric constant of the immobilized water in the solvation shell of an ion, the uncertainty is $\pm 50\,\text{kJ mol}^{-1}$ for the absolute solvation enthalpy and Gibbs energy and $\pm 10\,\text{J K}^{-1}\text{mol}^{-1}$ for the absolute solvation entropy. Because these uncertainties are large compared to the uncertainty of the thermodynamic functions using the convention described in Equations (10.3) and (10.4),

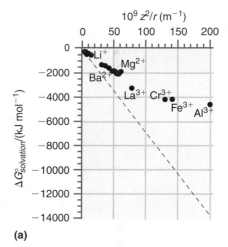

(a)

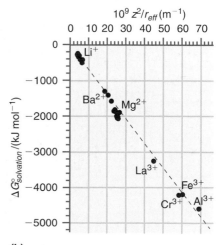

(b)

FIGURE 10.2

(a) The solvation energy calculated using the Born model is shown as a function of z^2/r. (b) The same results are shown as a function of z^2/r_{eff}. (See text.) The dashed line shows the behavior predicted by Equation (10.9).

conventional rather than absolute values of ΔH_f°, ΔG_f°, and S_m° for solvated ions in aqueous solutions are generally used by chemists.

10.3 Activities and Activity Coefficients for Electrolyte Solutions

The ideal dilute solution model presented for the activity and activity coefficient of components of real solutions in Chapter 9 is not valid for electrolyte solutions. This is the case because solute–solute interactions are dominated by the long-range electrostatic forces present between ions in electrolyte solutions. For example, the reaction that occurs when NaCl is dissolved in water is

$$NaCl(s) + H_2O(l) \rightarrow Na^+(aq) + Cl^-(aq) \tag{10.11}$$

In dilute solutions, NaCl is completely dissociated. Therefore, all solute–solute interactions are electrostatic in nature.

Although the concepts of activity and activity coefficients introduced in Chapter 9 are applicable to electrolytes, these concepts must be formulated differently for electrolytes to include the Coulomb interactions among ions. We next discuss a model for electrolyte solutions that is stated in terms of the chemical potentials, activities, and activity coefficients of the individual ionic species. However, only mean activities and activity coefficients (to be defined later) can be determined through experiments.

Consider the Gibbs energy of the solution, which can be written as

$$G = n_{solvent}\,\mu_{solvent} + n_{solute}\,\mu_{solute} \tag{10.12}$$

For the general electrolyte $A_{v_+}B_{v_-}$ that dissociates completely, one can also write an equivalent expression for G:

$$G = n_{solvent}\,\mu_{solvent} + n_+\mu_+ + n_-\mu_- = n_{solvent}\,\mu_{solvent} + n_{solute}(v_+\mu_+ + v_-\mu_-) \tag{10.13}$$

where v_+ and v_- are the stoichiometric coefficients of the cations and anions, respectively, produced upon dissociation of the electrolyte. In shorthand notation, an electrolyte is called a 1–1 electrolyte if $v_+ = 1$ and $v_- = 1$. Similarly, for a 2–3 electrolyte, $v_+ = 2$ and $v_- = 3$. Because Equations (10.12) and (10.13) describe the same solution, they are equivalent. Therefore,

$$\mu_{solute} = v_+\mu_+ + v_-\mu_- \tag{10.14}$$

Although this equation is formally correct for a strong electrolyte, one can never make a solution of either cations or anions alone, because any solution is electrically neutral. Therefore, it is useful to define a **mean ionic chemical potential** $\mu_\pm$ for the solute

$$\mu_\pm = \frac{\mu_{solute}}{v} = \frac{v_+\mu_+ + v_-\mu_-}{v} \tag{10.15}$$

where $v = v_+ + v_-$. The reason for doing so is that $\mu_\pm$ can be determined experimentally, whereas μ_+ and μ_- are not accessible through experiments.

The next task is to relate the chemical potentials of the solute and its individual ions to the activities of these species. For the individual ions,

$$\mu_+ = \mu_+^\circ + RT \ln a_+ \quad \text{and} \quad \mu_- = \mu_-^\circ + RT \ln a_- \tag{10.16}$$

where the standard chemical potentials of the ions, μ_+° and μ_-°, are based on a Henry's law standard state. Substituting Equation (10.16) in (10.15), an equation for the mean ionic chemical potential is obtained that is similar in structure to the expressions that we derived for the ideal dilute solution:

$$\mu_\pm = \mu_\pm^\circ + RT \ln a_\pm \tag{10.17}$$

The **mean ionic activity** $a_\pm$ is related to the individual ion activities by

$$a_\pm^v = a_+^{v_+} a_-^{v_-} \quad \text{or} \quad a_\pm = (a_+^{v_+} a_-^{v_-})^{1/v} \tag{10.18}$$

EXAMPLE PROBLEM 10.1

Write the mean ionic activities of NaCl, K_2SO_4, and H_3PO_4 in terms of the ionic activities of the individual anions and cations. Assume complete dissociation.

Solution

$$a_{NaCl}^2 = a_{Na^+}a_{Cl^-} \quad \text{or} \quad a_{NaCl} = \sqrt{a_{Na^+}a_{Cl^+}}$$

$$a_{K_2SO_4}^3 = a_{K^+}^2 a_{SO_4^{2-}} \quad \text{or} \quad a_{K_2SO_4} = (a_{K^+}^2 a_{SO_4^{2-}})^{1/3}$$

$$a_{H_3PO_4}^4 = a_{H^+}^3 a_{PO_4^{3-}} \quad \text{or} \quad a_{H_3PO_4} = (a_{H^+}^3 a_{PO_4^{3-}})^{1/4}$$

If the ionic activities are referenced to the concentration units of molality, then

$$a_+ = \frac{m_+}{m^\circ}\gamma_+ \quad \text{and} \quad a_- = \frac{m_-}{m^\circ}\gamma_- \tag{10.19}$$

where $m_+ = v_+m$ and $m_- = v_-m$. Because the activity is unitless, the molality must be referenced to a standard state concentration chosen to be $m^\circ = 1 \, mol \, kg^{-1}$. As in Chapter 9, a hypothetical standard state based on molality is defined. In this standard state, Henry's law, which is valid in the limit $m \rightarrow 0$, is obeyed up to a concentration of $m = 1$ molal. Substitution of Equation (10.19) in Equation (10.18) shows that

$$a_\pm^v = \left(\frac{m_+}{m^\circ}\right)^{v_+}\left(\frac{m_-}{m^\circ}\right)^{v_-}\gamma_+^{v_+}\gamma_-^{v_-} \tag{10.20}$$

To simplify this notation, we define the **mean ionic molality** $m_\pm$ and **mean ionic activity coefficient** $\gamma_\pm$ by

$$m_\pm^v = m_+^{v_+}m_-^{v_-}$$

$$m_\pm = (v_+^{v_+}v_-^{v_-})^{1/v}m \quad \text{and}$$

$$\gamma_\pm^v = \gamma_+^{v_+}\gamma_-^{v_-}$$

$$\gamma_\pm = (\gamma_+^{v_+}\gamma_-^{v_-})^{1/v} \tag{10.21}$$

With these definitions, the mean ionic activity is related to the mean ionic activity coefficient and mean ionic molality as follows:

$$a_\pm^v = \left(\frac{m_\pm}{m^\circ}\right)^v\gamma_\pm^v \quad \text{or} \quad a_\pm = \left(\frac{m_\pm}{m^\circ}\right)\gamma_\pm \tag{10.22}$$

Equations (10.19) through (10.22) relate the activities, activity coefficients, and molalities of the individual ionic species to mean ionic quantities and measurable properties of the system such as the molality and activity of the solute. With these definitions, Equation (10.17) defines the chemical potential of the electrolyte solute in terms of its activity:

$$\mu_{solute} = \mu_{solute}^\circ + RT \ln a_\pm^v \tag{10.23}$$

Equations (10.20) and (10.21) can be used to express the chemical potential of the solute in terms of measurable or easily accessible quantities:

$$\mu_{solute} = \mu_\pm = \left[v\mu_\pm^\circ + RT \ln(v_+^{v_+}v_-^{v_-})\right] + vRT \ln\left(\frac{m}{m^\circ}\right) + vRT \ln \gamma_\pm \tag{10.24}$$

The first term in the square bracket is defined by the "normal" standard state, which is usually taken to be a Henry's law standard state. The second term is obtained from the chemical formula for the solute. These two terms can be combined to create a new standard state $\mu_\pm^{\circ\circ}$ defined by the terms in the square brackets in Equation (10.24):

$$\mu_{solute} = \mu_\pm = \mu_\pm^{\circ\circ} + vRT \ln\left(\frac{m}{m^\circ}\right) + vRT \ln \gamma_\pm \tag{10.25}$$

The first two terms in Equation (10.25) correspond to the "ideal" ionic solution, which is associated with $\gamma_\pm = 1$.

The last term in Equation (10.25), which is the most important term in this discussion, contains the deviations from ideal behavior. The mean activity coefficient $\gamma_\pm$ can be obtained through experiment. For example, the activity coefficient of the solvent can be determined by measuring the boiling point elevation, the freezing point depression, or the lowering of the vapor pressure above the solution upon solution formation. The activity of the solute is obtained from that of the solvent using the Gibbs–Duhem equation. As shown in Section 11.8, $\gamma_\pm$ can also be determined through measurements on electrochemical cells. In addition, a very useful theoretical model allows $\gamma_\pm$ to be calculated for dilute electrolytic solutions. This model is discussed in the next section.

10.4 Calculating $\gamma_\pm$ Using the Debye–Hückel Theory

There is no model that adequately explains the deviations from ideality for the solutions of neutral solutes discussed in Sections 9.1 through 9.4. This is the case because the deviations arise through A–A, B–B, and A–B interactions that are specific to components A and B. This precludes a general model that holds for arbitrary A and B. However, the situation for solutions of electrolytes is different.

Deviations from ideal solution behavior occur at a much lower concentration for electrolytes than for nonelectrolytes, because the dominant interaction between the ions in an electrolyte is a long-range electrostatic Coulomb interaction rather than a short-range van der Waals or chemical interaction. Because of its long range, the Coulomb interaction among the ions cannot be neglected even for very dilute solutions of electrolytes. The Coulomb interaction allows a model of electrolyte solutions to be formulated independent of the identity of the solute for the following reason. The attractive or repulsive interaction of two ions depends only on their charge and separation, and not on their chemical identity. Therefore, the solute–solute interactions can be modeled knowing only the charge on the ions, and the model becomes independent of the identity of the solute species.

Measurements of activity coefficients in electrolyte solutions show that $\gamma_\pm \rightarrow 1$ for dilute solutions in the limit $m \rightarrow 0$. Because $\gamma_\pm < 1$, the chemical potential of the solute in a dilute solution is lower than that for a solution of uncharged solute species. Why is this the case? The lowering of μ_{solute} arises because the net electrostatic interaction among the ions surrounding an arbitrarily chosen central ion is attractive rather than repulsive. The model that describes the lowering of the energy of electrolytic solutions is due to Peter Debye and Erich Hückel. Rather than derive their results, we describe the essential features of their model next.

The solute ions in the solvent give rise to a spatially dependent electrostatic potential, ϕ, which can be calculated if the spatial distribution of ions is known. In dilute electrolyte solutions, the energy increase or decrease experienced by an ion of charge $\pm ze$ if the potential ϕ could be turned on suddenly is small compared to the thermal energy. This condition can be expressed in the form

$$|\pm ze\phi| \ll k_B T \tag{10.26}$$

In Equation (10.26), e is the charge on a proton, and k is Boltzmann's constant. In this limit, the dependence of ϕ on the spatial coordinates and the spatial distribution of the ions around an arbitrary central ion can be calculated. In contrast to the potential around an isolated ion in a dielectric medium, which is described by

$$\phi_{isolated\ ion}(r) = \frac{\pm ze}{4\pi\varepsilon_r\varepsilon_0 r} \tag{10.27}$$

the potential in the dilute electrolyte solution has the form

$$\phi_{solution}(r) = \frac{\pm ze}{4\pi\varepsilon_r\varepsilon_0 r}\exp(-\kappa r) \tag{10.28}$$

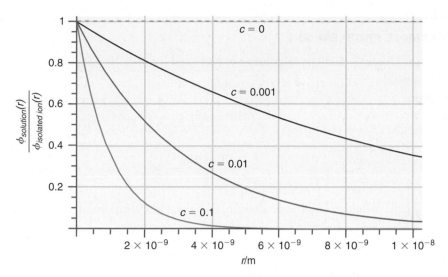

FIGURE 10.3
The ratio of the falloff in the electrostatic potential in the electrolyte solution to that for an isolated ion is shown as a function of the radial distance for three different molarities of a 1–1 electrolyte such as NaCl.

In Equations (10.27) and (10.28), ε_0 and ε_r are the permittivity of free space and the relative permittivity (dielectric constant) of the dielectric medium or solvent, respectively. Because of the exponential term, $\phi_{solution}(r)$ falls off much more rapidly with r than $\phi_{isolated\ ion}(r)$. We say that an individual ion experiences a **screened potential** from the other ions.

The Debye–Hückel theory shows that κ is related to the individual charges on the ions and to the solute molality m by

$$\kappa^2 = e^2 N_A \left(1000\ \text{L m}^{-3}\right) m \left(\frac{\nu_+ z_+^2 + \nu_- z_-^2}{\varepsilon_0 \varepsilon_r k_B T}\right) \rho_{solvent} \qquad \textbf{(10.29)}$$

From this formula, we can see that screening becomes more effective as the concentration of the ionic species increases. Screening is also more effective for multiply charged ions and for larger values of ν_+ and ν_-.

The ratio

$$\frac{\phi_{solution}(r)}{\phi_{isolated\ ion}(r)} = e^{-\kappa r}$$

is plotted in Figure 10.3 for different values of m for an aqueous solution of a 1–1 electrolyte. Note that the potential falls off much more rapidly with the radial distance r in the electrolyte solution than in the uniform dielectric medium. Note also that the potential falls off more rapidly with increasing concentration of the electrolyte. The origin of this effect is that ions of sign opposite to the central ion are more likely to be found close to the central ion. These surrounding ions form a diffuse ion cloud around the central ion, as shown pictorially in Figure 10.4. If a spherical surface is drawn centered at the central ion, the net charge within the surface can be calculated. The results show that the net charge has the same sign as the central charge, falls off rapidly with distance, and is close to zero for $\kappa r \sim 8$. For larger values of κr, the central ion is completely screened by the diffuse ion cloud, meaning that the net charge in the sphere around the central ion is zero. The net effect of the diffuse ion cloud is to screen the central ion from the rest of the solution, and the quantity $1/\kappa$ is known as the **Debye–Hückel screening length.** Larger values of κ correspond to a smaller diffuse cloud, and a more effective screening.

It is convenient to combine the concentration-dependent terms that contribute to κ in the **ionic strength** I, which is defined by

$$I = \frac{m}{2} \sum_i \left(\nu_{i+} z_{i+}^2 + \nu_{i-} z_{i-}^2\right) = \frac{1}{2} \sum_i \left(m_{i+} z_{i+}^2 + m_{i-} z_{i-}^2\right) \qquad \textbf{(10.30)}$$

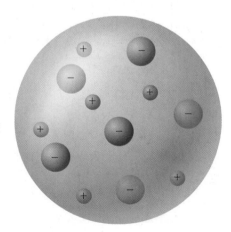

FIGURE 10.4
Pictorial rendering of the arrangement of ions about an arbitrary ion in an electrolyte solution. The central ion is more likely to have oppositely charged ions as neighbors. The large circle represents a sphere of radius $r \sim 8/\kappa$. From a point outside of this sphere, the charge on the central ion is essentially totally screened.

EXAMPLE PROBLEM 10.2

Calculate I for (a) a 0.050 molal solution of NaCl and for (b) a Na_2SO_4 solution of the same molality.

Solution

a. $$I_{NaCl} = \frac{m}{2}(v_+z_+^2 + v_-z_-^2) = \frac{0.050\,mol\,kg^{-1}}{2} \times (1 + 1) = 0.050\,mol\,kg^{-1}$$

b. $$I_{Na_2SO_4} = \frac{m}{2}(v_+z_+^2 + v_-z_-^2) = \frac{0.050\,mol\,kg^{-1}}{2} \times (2 + 4) = 0.15\,mol\,kg^{-1}$$

Using the definition of the ionic strength, Equation (10.29) can be written in the form

$$\kappa = \sqrt{\left(\frac{2e^2N_A}{\varepsilon_0 k_B T}\right)(1000L\,m^{-3})}\sqrt{\left(\frac{I}{\varepsilon_r}\right)\rho_{solvent}}$$

$$= 2.91 \times 10^8 \sqrt{\frac{I/mol\,kg^{-1}}{\varepsilon_r}\frac{\rho_{solvent}}{kg\,L^{-1}}}\,m^{-1} \quad \text{at 298 K} \qquad \textbf{(10.31)}$$

The first factor in this equation contains only fundamental constants that are independent of the solvent and solute as well as the temperature. The second factor contains the ionic strength of the solution and the unitless relative permittivity of the solvent. For the more conventional units of mol L^{-1}, and for water, for which $\varepsilon_r = 78.5$, $\kappa = 3.29 \times 10^9\sqrt{I}\,m^{-1}$ at 298 K.

By calculating the charge distribution of the ions around the central ion and the work needed to charge these ions up to their charges z_+ and z_- from an initially neutral state, Debye and Hückel were able to obtain an expression for the mean ionic activity coefficient. It is given by

$$\ln \gamma_\pm = -|z_+z_-|\frac{e^2\kappa}{8\pi\varepsilon_0\varepsilon_r k_B T} \qquad \textbf{(10.32)}$$

This equation is known as the **Debye–Hückel limiting law.** It is called a limiting law because Equation (10.32) is only obeyed for small values of the ionic strength. Note that because of the negative sign in Equation (10.32), $\gamma_\pm < 1$. From the concentration dependence of κ shown in Equation (10.31), the model predicts that $\ln \gamma_\pm$ decreases with the ionic strength as $\sqrt{I}$. This dependence is shown in Figure 10.5. Although all three solutions have the same solute concentration, they have different values for z^+ and z^-. For this reason, the three lines have a different slope.

Equation (10.32) can be simplified for a particular choice of solvent and temperature. For aqueous solutions at 298.15 K, the result is

$$\log \gamma_\pm = -0.5092|z_+z_-|\sqrt{I} \quad \text{or} \quad \ln\gamma_\pm = -1.173|z_+z_-|\sqrt{I} \qquad \textbf{(10.33)}$$

How well does the Debye–Hückel limiting law agree with experimental data? Figure 10.6 shows a comparison of the model with data for aqueous solutions of $AgNO_3$ and $CaCl_2$. In each case, $\ln \gamma_\pm$ is plotted versus $\sqrt{I}$. The Debye–Hückel limiting law predicts that the data will fall on the line indicated in each figure. The data points deviate from the predicted behavior above $\sqrt{I} = 0.1$ for $AgNO_3$ ($m = 0.01$), and above $\sqrt{I} = 0.006$ for $CaCl_2$ ($m = 0.004$). In the limit that $I \to 0$, the limiting law is obeyed. However, the deviations are significant at a concentration for which a neutral solute would exhibit ideal behavior.

The deviations continue to increase with increasing ionic strength. Figure 10.7 shows experimental data for $ZnBr_2$ out to $\sqrt{I} = 5.5$, corresponding to $m = 10$. Note that, although the Debye–Hückel limiting law is obeyed as $I \to 0$, $\ln \gamma_\pm$ goes through a minimum and begins to increase with increasing ionic strength. At the highest value of the ionic strength, $\gamma_\pm = 2.32$, which is significantly greater than one. Although the deviations from ideal behavior are less pronounced in Figure 10.6, the trend is the same for all the solutes. The mean ionic activity coefficient $\gamma_\pm$ falls off more slowly with the ionic strength than predicted by the Debye–Hückel limiting law. The behavior shown in Figure 10.7 is typical for most electrolytes; after passing through a minimum, $\gamma_\pm$ rises

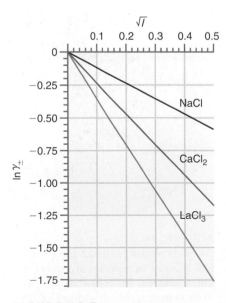

FIGURE 10.5
The decrease in the Debye–Hückel mean activity coefficient with the square root of the ionic strength is shown for a 1–1, a 1–2, and a 1–3 electrolyte, all of the same molality in the solute.

with increasing ionic strength, and for high values of I, $\gamma_\pm > 1$. Experimental values for $\gamma_\pm$ for a number of solutes at different concentrations in aqueous solution are listed in Table 10.3 (see Appendix B, Data Tables).

There are a number of reasons why the experimental values of $\gamma_\pm$ differ at high ionic strength from those calculated from the Debye–Hückel limiting law. They mainly involve the assumptions made in the model. It has been assumed that the ions can be treated as point charges with zero volume, whereas ions and their associated solvation shells occupy a finite volume. As a result, there is an increase in the repulsive interaction among ions in an electrolyte over that predicted for point charges, which becomes more important as the concentration increases. Repulsive interactions raise the energy of the solution and, therefore, increase $\gamma_\pm$. The Debye–Hückel model also assumes that the solvent can be treated as a structureless dielectric medium. However, the ion is surrounded by a relatively ordered primary solvation shell, as well as by more loosely bound water molecules. The atomic level structure of the solvation shell is not adequately represented by using the dielectric strength of bulk solvent. Another factor that has not been taken into account is that as the concentration increases, some ion pairing occurs such that the concentration of ionic species is less than would be calculated assuming complete dissociation.

Additionally, consider the fact that the water molecules in the solvation shell have effectively been removed from the solvent. For example, in an aqueous solution of H_2SO_4, approximately nine H_2O molecules are tightly bound per dissolved H_2SO_4 formula unit. Therefore, the number of moles of H_2O as solvent in 1 L of a one molar H_2SO_4 solution is reduced from 55 for pure H_2O to 46 in the solution. Consequently, the actual solute molarity is larger than that calculated by assuming that all the H_2O is in the form of solvent. Because the activity increases linearly with the actual molarity, $\gamma_\pm$ increases as the solute concentration increases. If there were no change in the enthalpy of solvation with concentration, all the H_2O molecules would be removed from the solvent at a concentration of six molar H_2SO_4. Clearly, this assumption is unreasonable. What actually happens is that solvation becomes energetically less favorable as the H_2SO_4 concentration increases. This corresponds to a less negative value of $\ln \gamma_\pm$, or equivalently to an increase in $\gamma_\pm$. Summing up, many factors explain why the Debye–Hückel limiting law is only valid for small concentrations. Because of the complexity of these different factors, there is no simple formula based on theory that can replace the Debye–Hückel limiting law. However, the main trends exhibited in Figures 10.6 and 10.7 are reproduced in more sophisticated theories of electrolyte solutions.

Because none of the usual models are valid at high concentrations, empirical models that "improve" on the Debye–Hückel model by predicting an increase in $\gamma_\pm$ for high concentrations are in widespread use. An empirical modification of the Debye–Hückel limiting law that has the form

$$\log_{10} \gamma_\pm = -0.51|z_+ z_-| \left[\frac{\left(\dfrac{I}{m^\circ}\right)^{1/2}}{1 + \left(\dfrac{I}{m^\circ}\right)^{1/2}} - 0.30\left(\dfrac{I}{m^\circ}\right) \right] \qquad \textbf{(10.34)}$$

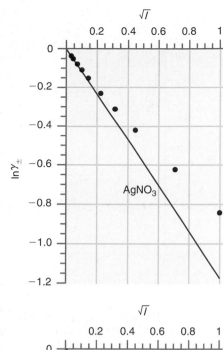

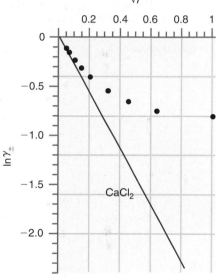

FIGURE 10.6

The experimentally determined activity coefficients for $AgNO_3$ and $CaCl_2$ are shown as a function of the square root of the ionic strength. The solid lines are the prediction of the Debye–Hückel theory.

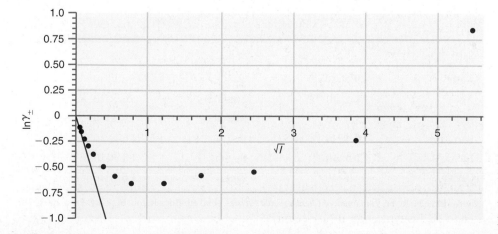

FIGURE 10.7

Experimentally determined values for the mean activity coefficient for $ZnBr_2$ are shown as a function of the square root of the ionic strength. The solid line is the prediction of the Debye–Hückel theory.

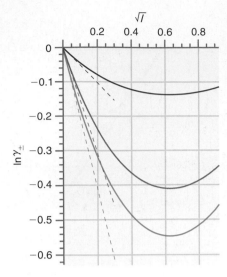

FIGURE 10.8

Comparison between the predictions of the Debye–Hückel limiting law (dashed lines) and the Davies equation (solid curves) for 1–1 (red), 1–2 (purple), and 1–3 (blue) electrolytes.

is known as the **Davies equation.** As seen in Figure 10.8, this equation for $\gamma_\pm$ shows the correct limiting behavior for low I values, and the trend at higher values of I is in better agreement with the experimental results shown in Figures 10.6 and 10.7. However, unlike the Debye–Hückel limiting law, there is no theoretical basis for the Davies equation.

10.5 Chemical Equilibrium in Electrolyte Solutions

As discussed in Section 9.13, the equilibrium constant in terms of activities is given by Equation (9.66):

$$K = \prod_i (a_i^{eq})^{\nu_j} \tag{10.35}$$

It is convenient to define the activity of a species relative to its molarity. In this case,

$$a_i = \gamma_i \frac{c_i}{c^\circ} \tag{10.36}$$

where γ_i is the activity coefficient of species i. We next specifically consider chemical equilibrium in electrolyte solutions, illustrating that activities rather than concentrations must be taken into account to accurately model equilibrium concentrations. We first restrict our considerations to the range of ionic strengths for which the Debye–Hückel limiting law is valid. As an example, we calculate the degree of dissociation of MgF_2 in water. The equilibrium constant in terms of molarities for ionic salts is usually given the symbol K_{sp}, where the subscript refers to the solubility product. The equilibrium constant K_{sp} is unitless and has the value of 6.4×10^{-9} for the reaction shown in Equation (10.37). Values for K_{sp} are generally tabulated for reduced concentration units of molarity (c/c°) rather than molality (m/m°), and values for selected substances are listed in Table 10.4. Because the mass of 1 L of water is 0.998 kg, the numerical value of the concentration is the same on both scales for dilute solutions.

We next consider dissociation of MgF_2 in an aqueous solution:

$$MgF_2(s) \rightarrow Mg^{2+}(aq) + 2F^-(aq) \tag{10.37}$$

Because the activity of the pure solid can be set equal to one,

$$K_{sp} = a_{Mg^{2+}}a_{F^-}^2 = \left(\frac{c_{Mg^{2+}}}{c^\circ}\right)\left(\frac{c_{F^-}}{c^\circ}\right)^2 \gamma_\pm^3 = 6.4 \times 10^{-9} \tag{10.38}$$

From the stoichiometry of the overall equation, we know that $c_{F^-} = 2c_{Mg^{2+}}$, but Equation (10.38) still contains two unknowns, $\gamma_\pm$ and c_{F^-}, that we solve for iteratively as shown in Example Problem 10.3.

TABLE 10.4 Solubility Product Constants (Molarity Based) for Selected Salts

Salt	K_{sp}	Salt	K_{sp}
AgBr	4.9×10^{-13}	$CaSO_4$	4.9×10^{-6}
AgCl	1.8×10^{-10}	$Mg(OH)_2$	5.6×10^{-11}
AgI	8.5×10^{-17}	$Mn(OH)_2$	1.9×10^{-13}
$Ba(OH)_2$	5.0×10^{-3}	$PbCl_2$	1.6×10^{-5}
$BaSO_4$	1.1×10^{-10}	$Pb\,SO_4$	1.8×10^{-8}
$CaCO_3$	3.4×10^{-9}	ZnS	1.6×10^{-23}

Source: Lide, D. R., ed. *Handbook of Chemistry and Physics.* 83rd ed. Boca Raton, FL: CRC Press 2002.

EXAMPLE PROBLEM 10.3

Calculate the solubility of MgF_2 in an aqueous solution. Use the Debye–Hückel limiting law to obtain $\gamma_\pm$. $K_{sp} = 6.4 \times 10^{-9}$.

Solution

The equilibrium expression is stated in Equation 10.38.

$$K_{sp} = a_{Mg^{2+}}a_{F^-}^2 = \left(\frac{c_{Mg^{2+}}}{c^\circ}\right)\left(\frac{c_{F^-}}{c^\circ}\right)^2 \gamma_\pm^3 = 6.4 \times 10^{-9}$$

First assume that $\gamma_\pm = 1$, and solve Equation (10.38) for $c_{Mg^{2+}}$.

$$K_{sp} = a_{Mg^{2+}}a_{F^-}^2 = \left(\frac{c_{Mg^{2+}}}{c^\circ}\right)\left(\frac{c_{F^-}}{c^\circ}\right)^2 = 6.4 \times 10^{-9}$$

$$4x^3 = 6.4 \times 10^{-9}; x = \frac{c_{Mg^{2+}}}{c^\circ} = 1.17 \times 10^{-3}\,\text{mol L}^{-1} = 1.17 \times 10^{-3}\,\text{mol kg}^{-1}$$

We next calculate the ionic strength of the solution using the concentrations just obtained.

$$I_{MgF_2} = \frac{m}{2}\left(v_+z_+^2 + v_-z_-^2\right) = \frac{1.17 \times 10^{-3}\,\text{mol kg}^{-1}}{2} \times (2^2 + 2)$$

$$= 3.51 \times 10^{-3}\,\text{mol kg}^{-1}$$

We next calculate $\gamma_\pm$ from the Debye–Hückel limiting law of Equation (10.33).

$$\ln \gamma_\pm = -1.173|z_+z_-|\sqrt{I}$$

$$\ln \gamma_\pm = -1.173 \times 2 \times \sqrt{3.51 \times 10^{-3}} = -0.1390$$

$$\gamma_\pm = 0.870$$

We use this value of $\gamma_\pm$ in the equilibrium expression and recalculate the solubility.

$$K_{sp} = a_{Mg^{2+}}a_{F^-}^2 = \left(\frac{c_{Mg^{2+}}}{c^\circ}\right)\left(\frac{c_{F^-}}{c^\circ}\right)^2 \gamma_\pm^3 = 6.4 \times 10^{-9}$$

$$4x^3 = \frac{6.4 \times 10^{-9}}{(0.870)^3}; x = \frac{c_{Mg^{2+}}}{c^\circ} = 1.34 \times 10^{-3}\,\text{mol L}^{-1} = 1.34 \times 10^{-3}\,\text{mol kg}^{-1}$$

A second iteration gives $\gamma_\pm = 0.862$ and $c_{Mg^{2+}} = 1.36 \times 10^{-3}\,\text{mol L}^{-1}$, and a third iteration gives $\gamma_\pm = 0.861$ and $c_{Mg^{2+}} = 1.36 \times 10^{-3}\,\text{mol L}^{-1}$, showing that the iteration has converged. These results show that assuming that $\gamma_\pm = 1$ leads to an unacceptably large error of 14% in $c_{Mg^{2+}}$.

Another effect of the ionic strength on solubility is described by the terms *salting in* and *salting out*. The behavior shown in Figure 10.7, in which the activity coefficient first decreases and subsequently increases with concentration, affects the solubility of a salt in the following way. For a salt such as MgF_2, the product $\left[Mg^{2+}\right]\left[F^-\right]^2\gamma_\pm^3 = K_{sp}$ is constant as the concentration varies for constant T, because the thermodynamic equilibrium constant K depends only on T. Therefore, the concentrations $\left[Mg^{2+}\right]$ and $\left[F^-\right]$ change in an opposite way to $\gamma_\pm$. At small values of the ionic strength, $\gamma_\pm < 1$, and the solubility increases as $\gamma_\pm$ decreases with concentration until the minimum in a plot of $\gamma_\pm$ versus I is reached. This effect is known as **salting in**. For high values of the ionic strength, $\gamma_\pm > 1$ and the solubility is less than at low values of I. This effect is known as **salting out**. Salting out can be used to purify proteins by selective precipitation from solution. For example, the blood clotting protein fibrinogen is precipitated in a 0.8 molar ammonium sulfate solution, whereas a 2.4 molar concentration of ammonium sulfate is required to precipitate albumin.

We next consider equilibria in electrolyte solutions at concentrations where the Debye–Hückel limiting law is no longer valid. In this case, activity coefficients must be estimated as there is no theoretical model for calculating them. The Davies [Equation (10.34)] can be used in such a case. An example is a buffer solution that consists of a weak acid and its conjugate base.

EXAMPLE PROBLEM 10.4

Calculate the pH of a buffer solution that is 0.100 molar in CH_3COOH and 0.100 molar in CH_3COONa. The molality based equilibrium constant K for the dissociation of acetic acid is 1.75×10^{-5}. Compare your value with what you would have calculated assuming $\gamma_\pm = 1$.

Solution

We write the equilibrium constant in the form

$$K = \frac{\gamma_\pm m(H_3O^+)\gamma_\pm m(CH_3COO^-)}{m(CH_3COOH)}$$

$$1.75 \times 10^{-5} = \frac{(\gamma_\pm)^2 x \times (0.100 + x)}{(0.100 - x)} \approx (\gamma_\pm)^2 x$$

$$m(H_3O^+) = \frac{1.75 \times 10^{-5}}{(\gamma_\pm)^2}$$

$$pH = \log\left\{m(H_3O^+)\right\}$$

To calculate $\gamma_\pm$ from the Davies equation, we must first calculate the ionic strength of the solution. The small value of K tells us that the degree of ionization of the acetic acid is small so that the ionic strength can be calculated from the concentration of CH_3COONa alone.

$$I = \frac{m}{2}(v_+z_+^2 + v_-z_-^2) = \frac{0.100\,\text{mol kg}^{-1}}{2} \times (1 + 1) = 0.100\,\text{mol kg}^{-1}$$

Using the Davies equation

$$\log_{10}\gamma_\pm = -0.51|z_+z_-|\left[\frac{\left(\frac{I}{m^\circ}\right)^{1/2}}{1 + \left(\frac{I}{m^\circ}\right)^{1/2}} - 0.30\left(\frac{I}{m^\circ}\right)\right]$$

$$= -0.51\left[\frac{(0.100)^{1/2}}{1 + (0.100)^{1/2}} - 0.30 \times 0.100\right] = -0.1270$$

$$\gamma_\pm = 0.746$$

$$m(H_3O^+) = \frac{1.75 \times 10^{-5}}{(0.746)^2} = 3.14 \times 10^{-5}\,\text{mol kg}^{-1}$$

$$pH = -\log\left\{3.14 \times 10^{-5}\right\} = 4.50$$

If we had assumed $\gamma_\pm = 1$

$$m(H_3O^+) = \frac{1.75 \times 10^{-5}}{(\gamma_\pm)^2} = 1.75 \times 10^{-5}$$

This value for $m(H_3O^+)$ gives a pH value of 4.76, which is significantly different from the value calculated earlier.

Vocabulary

conventional formation enthalpies	electrolyte	mean ionic molality
conventional formation entropies	ionic strength	salting in
conventional Gibbs energies of formation	mean ionic activity	salting out
Davies equation	mean ionic activity coefficient	screened potential
Debye–Hückel limiting law	mean ionic chemical potential	solvation shell
Debye–Hückel screening length		

Conceptual Problems

Q10.1 Discuss how the Debye–Hückel screening length changes as the (a) temperature, (b) dielectric constant, and (c) ionic strength of an electrolyte solution are increased.

Q10.2 Why is it not possible to measure the Gibbs energy of solvation of Cl^- directly?

Q10.3 Why are activity coefficients calculated using the Debye–Hückel limiting law always less than one?

Q10.4 How is the mean ionic chemical potential of a solute related to the chemical potentials of the anion and cation produced when the solute is dissolved in water?

Q10.5 How is the chemical potential of a solute related to its activity?

Q10.6 Tabulated values of standard entropies of some aqueous ionic species are negative. Why is this statement not inconsistent with the third law of thermodynamics?

Q10.7 Why is it not possible to measure the activity coefficient of $Na^+(aq)$?

Q10.8 Why is it possible to formulate a general theory for the activity coefficient for electrolyte solutions, but not for nonelectrolyte solutions?

Q10.9 Why does an increase in the ionic strength in the range where the Debye–Hückel law is valid lead to an increase in the solubility of a weakly soluble salt?

Q10.10 What is the correct order of the following inert electrolytes in their ability to increase the degree of dissociation of acetic acid?

a. 0.001 m NaCl

b. 0.001 m KBr

c. 0.10 m $CuCl_2$

Q10.11 How does salting in affect solubility?

Q10.12 Why is it not appropriate to use ionic radii from crystal structures to calculate $\Delta G^\circ_{solvation}$ of ions using the Born model?

Q10.13 Why do deviations from ideal behavior occur at lower concentrations for electrolyte solutions than for solutions in which the solute species are uncharged?

Q10.14 Why is the value for the dielectric constant for water in the solvation shell around ions less than that for bulk water?

Q10.15 What can you conclude about the interaction between ions in an electrolyte solution if the mean ionic activity coefficient is greater than one?

Q10.16 Why is the inequality $\gamma_\pm < 1$ always satisfied in dilute electrolyte solutions?

Q10.17 Under what conditions does $\gamma_\pm \to 1$ for electrolyte solutions?

Q10.18 How do you expect S°_m for an ion in solution to change as the ionic radius increases at constant charge?

Q10.19 How do you expect S°_m for an ion in solution to change as the charge increases at constant ionic radius?

Q10.20 It takes considerable energy to dissociate NaCl in the gas phase. Why does this process occur spontaneously in an aqueous solution? Why does it not occur spontaneously in CCl_4?

Numerical Problems

Problem numbers in **red** indicate that the solution to the problem is given in the *Student's Solutions Manual*.

P10.1 Calculate $\Delta S^\circ_{reaction}$ for the reaction $Ba(NO_3)_2(aq) + 2KCl(aq) \rightarrow BaCl_2(s) + 2KNO_3(aq)$.

P10.2 Calculate $\Delta S^\circ_{reaction}$ for the reaction $AgNO_3(aq) + KCl(aq) \rightarrow AgCl(s) + KNO_3(aq)$.

P10.3 Using the Debye–Hückel limiting law, calculate the value of $\gamma_\pm$ in (a) a 7.2×10^{-3} m solution of NaBr, (b) a

7.50×10^{-4} m solution of $SrCl_2$, and (c) a 2.25×10^{-3} m solution of $CaHPO_4$. Assume complete dissociation.

P10.4 Calculate the mean ionic molality $m_\pm$ in 0.0750 m solutions of (a) $Ca(NO_3)_2$, (b) NaOH, (c) $MgSO_4$, and (d) $AlCl_3$.

P10.5 A weak acid has a dissociation constant of $K_a = 2.50 \times 10^{-2}$. (a) Calculate the degree of dissociation for a 0.093 m solution of this acid using the Debye-Hückel limiting law. (b) Calculate the degree of dissociation for a 0.093 m solution of this acid that is also 0.200 m in KCl from the Debye–Hückel limiting law using an iterative calculation until the answer is constant in the second decimal place. (c) Repeat the calculation in (b) using the mean activity coefficient for KCl in Table 10.3. Is the use of the Debye–Hückel limiting law advisable at the given KCl concentration? Do you need to repeat the iterative calculation of (a) to solve (b) and (c)?

P10.6 Calculate the mean ionic activity of a 0.0350 m Na_3PO_4 solution for which the mean activity coefficient is 0.685.

P10.7 At 25°C, the equilibrium constant for the dissociation of acetic acid K_a is 1.75×10^{-5}. Using the Debye–Hückel limiting law, calculate the degree of dissociation in 0.150 m and 1.50 m solutions using an iterative calculation until the answer is constant to within $+/-2$ in the second decimal place. Compare these values with what you would obtain if the ionic interactions had been ignored. Compare your results with the degree of dissociation of the acid assuming $\gamma_\pm = 1$.

P10.8 From the data in Table 10.3 (see Appendix B, Data Tables), calculate the activity of the electrolyte in 0.200 m solutions of
a. KCl
b. H_2SO_4
c. $MgCl_2$

P10.9 Estimate the degree of dissociation of a 0.200 m solution of nitrous acid ($K_a = 4.00 \times 10^{-4}$) that is also 0.500 m in the strong electrolyte given in parts (a) through (c). Use the data tables to obtain $\gamma_\pm$, as the electrolyte concentration is too high to use the Debye–Hückel limiting law.
a. $Ba(Cl)_2$
b. KOH
c. $AgNO_3$

Compare your results with the degree of dissociation of the acid in the absence of other electrolytes.

P10.10 Calculate ΔH_R° and ΔG_R° for the reaction $Ba(NO_3)_2(aq) + 2\,KCl(aq) \rightarrow BaCl_2(s) + 2KNO_3(aq)$.

P10.11 Express $a_\pm$ in terms of a_+ and a_- for (a) Li_2CO_3, (b) $CaCl_2$, (c) Na_3PO_4, and (d) $K_4Fe(CN)_6$. Assume complete dissociation.

P10.12 Calculate $\Delta G_{solvation}^\circ$ in an aqueous solution for $Rb^+(aq)$ using the Born model. The radius of the Rb^+ ion is 161 pm.

P10.13 Calculate the ionic strength in a solution that is 0.0750 m in K_2SO_4, 0.0085 m in Na_3PO_4, and 0.0150 m in $MgCl_2$.

P10.14 Calculate I, $\gamma_\pm$, and $a_\pm$ for a 0.0120 m solution of Na_3PO_4 at 298 K. Assume complete dissociation.

P10.15 Express $\mu_\pm$ in terms of μ_+ and μ_- for (a) NaCl, (b) $MgBr_2$, (c) Li_3PO_4, and (d) $Ca(NO_3)_2$. Assume complete dissociation.

P10.16 In the Debye–Hückel theory, the counter charge in a spherical shell of radius r and thickness dr around the central ion of charge $+Q$ is given by $-Q\kappa^2 re^{-\kappa r}dr$. Calculate the radius at which the counter charge has its maximum value

r_{max}, from this expression. Evaluate r_{max} for a 0.090 m solution of Na_3PO_4 at 298 K.

P10.17 Calculate the solubility of $CaCO_3$ ($K_{sp} = 3.4 \times 10^{-9}$) (a) in pure H_2O and (b) in an aqueous solution with $I = 0.0250$ mol kg^{-1}. For part (a), do an iterative calculation of $\gamma_\pm$ and the solubility until the answer is constant in the second decimal place. Do you need to repeat this procedure in part (b)?

P10.18 Calculate the probability of finding an ion at a distance greater than $1/\kappa$ from the central ion.

P10.19 Express $\gamma_\pm$ in terms of γ_+ and γ_- for (a) $SrSO_4$, (b) $MgBr_2$, (c) K_3PO_4, and (d) $Ca(NO_3)_2$. Assume complete dissociation.

P10.20 Calculate the mean ionic molality and mean ionic activity of a 0.105 m K_3PO_4 solution for which the mean ionic activity coefficient is 0.225.

P10.21 The base dissociation constant of dimethylamine,

$$(CH_3)_2NH(aq) + H_2O(aq) \rightarrow CH_3NH_3{}^+(aq) + OH^-(aq)$$

is 5.12×10^{-4}. Calculate the extent of hydrolysis for (a) a 0.210 m solution of $(CH_3)_2NH$ in water using an iterative calculation until the answer is constant to within $+/-2$ in the second decimal place. (b) Repeat the calculation for a solution that is also 0.500 m in $NaNO_3$. Do you need to use an iterative calculation in this case?

P10.22 Dichloroacetic acid has a dissociation constant of $K_a = 3.32 \times 10^{-2}$. Calculate the degree of dissociation for a 0.105 m solution of this acid (a) from the Debye–Hückel limiting law using an iterative calculation until the answer is constant to within $+/-2$ in the second decimal place. (b) Repeat the calculation assuming that the mean ionic activity coefficient is one.

P10.23 Calculate the Debye–Hückel screening length $1/\kappa$ at 298 K in a 0.0075 m solution of K_3PO_4.

P10.24 Calculate I, $\gamma_\pm$, and $a_\pm$ for a 0.0215 m solution of K_2SO_4 at 298 K. How confident are you that your calculated results will agree with experimental results?

P10.25 Calculate I, $\gamma_\pm$, and $a_\pm$ for a 0.0175 m solution of Na_3PO_4 at 298 K. Assume complete dissociation. How confident are you that your calculated results will agree with experimental results?

P10.26 Calculate the ionic strength of each of the solutions in Problem P10.4.

P10.27 Calculate the value of $m_\pm$ in 5.5×10^{-3} molal solutions of (a) KCl, (b) $Ca(NO_3)_2$, and (c) $ZnSO_4$. Assume complete dissociation.

P10.28 Calculate ΔH_R° and ΔG_R° for the reaction $AgNO_3(aq) + KCl(aq) \rightarrow AgCl(s) + KNO_3(aq)$.

P10.29 Calculate the pH of a buffer solution that is 0.200 molal in CH_3COOH and 0.15 molal in CH_3COONa using the Davies equation to calculate $\gamma_\pm$. What pH value would you have calculated if you had assumed that $\gamma_\pm = 1$?

P10.30 Use the Davies equation to calculate $\gamma_\pm$ for a 1.00 molar solution of KOH. Compare your answer with the value in Table 10.3.

Electrochemical Cells, Batteries, and Fuel Cells

If a metal electrode is immersed in an aqueous solution containing cations of that metal, an equilibrium that leads to negative charge formation on the electrode is established. This configuration of electrode and solution is called a half-cell. Two half-cells can be combined to form an electrochemical cell. The equilibrium condition in an electrochemical cell is that the electrochemical potential, rather than the chemical potential, of a species is the same in all parts of the cell. The electrochemical potential can be changed through the application of an electrical potential external to the cell. This allows the direction of spontaneous change in the cell reaction to be reversed. Electrochemical cells can be used to determine the equilibrium constant for the cell reaction and to determine the mean activity coefficient of a solute. Electrochemical cells can also be used to provide power, in which case they are called batteries. Electrochemical cells in which the reactants can be supplied continuously are called fuel cells.

11.1 The Effect of an Electrical Potential on the Chemical Potential of Charged Species

If a Zn electrode is partially immersed in an aqueous solution of $ZnSO_4$, an equilibrium is established between $Zn(s)$ and $Zn^{2+}(aq)$ as a small amount of the Zn goes into solution as $Zn^{2+}(aq)$ ions as depicted in Figure 11.1. However, the electrons remain on the Zn electrode. Therefore, a negative charge builds up on the Zn electrode, and a corresponding positive charge builds up in the surrounding solution. This charging leads to a difference in the electrical potential ϕ between the electrode and the solution, which we call the half-cell potential. As we will see in Section 11.2, two half-cells are combined to form an electrochemical cell. The charge separation in the system arises through the dissociation equilibrium

$$Zn(s) \rightleftharpoons Zn^{2+}(aq) + 2e^- \tag{11.1}$$

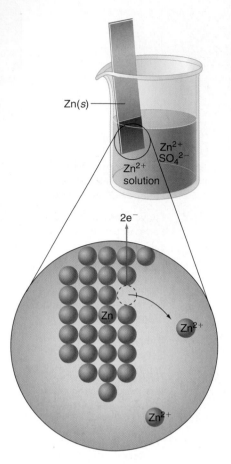

FIGURE 11.1

When a Zn electrode is immersed in an aqueous solution containing $Zn^{2+}(aq)$ ions, a very small amount of the Zn goes into solution as Zn^{2+} (aq), leaving two electrons behind on the Zn electrode per ion formed.

The equilibrium position in this reaction lies far toward Zn(s). At equilibrium, fewer than 10^{-14} mol of the Zn(s) dissolves in 1 liter of solution to form $Zn^{2+}(aq)$. However, this minuscule amount of charge transfer between the electrode and the solution is sufficient to create a difference of approximately 1 V in the electrical potential between the Zn electrode and the electrolyte solution. A similar dissociation equilibrium is established for other metal electrodes. Because the value of the equilibrium constant depends on ΔG_f° of the solvated metal ion, the equilibrium constant for the dissociation reaction and ϕ depends on the identity of the metal.

Can ϕ be measured directly? Let us assume that we can carry out the measurement using two chemically inert Pt wires as probes. One Pt wire is placed on the Zn electrode, and the second Pt wire is placed in the $ZnSO_4$ solution. However, the measured voltage is the difference in electrical potential between a Pt wire connected to a Zn electrode in a $ZnSO_4$ solution and a Pt electrode in a $ZnSO_4$ solution, which is not what we want. A difference in electrical potential can only be measured between one phase and a second phase *of identical composition*. For example, the difference in electrical potential across a resistor is measured by contacting the metal wire at each end of the resistor with two metal probes of identical composition connected to the terminals of a voltmeter. Although we can not measure the half-cell potential ϕ directly, half-cell potentials can be determined relative to a reference half-cell as will be shown in Section 11.2.

How are chemical species affected by the electrical potential ϕ? To a very good approximation, the chemical potential of a **neutral** atom or molecule is not affected if a small electrical potential is applied to the environment containing the species. However, this is not the case for a **charged** species such as an Na^+ ion in an electrolyte solution. The work required to transfer dn moles of charge reversibly from a chemically uniform phase at an electrical potential ϕ_1 to a second, otherwise identical phase at an electrical potential ϕ_2 is equal to the product of the charge and the difference in the electrical potential between the two locations:

$$dw_{rev} = (\phi_2 - \phi_1)\, dQ \qquad (11.2)$$

In this equation, $dQ = -zF\, dn$ is the charge transferred through the potential, z is the charge in units of the electron charge $(+1, -1, +2, -2, \ldots)$, and the **Faraday constant** F is the absolute magnitude of the charge associated with 1 mol of a singly charged species. The Faraday constant has the numerical value $F = 96,485$ Coulombs $mole^{-1}$ $(C\ mol^{-1})$.

Because the work being carried out in this reversible process is nonexpansion work, $dw_{rev} = dG$, which is the difference in the **electrochemical potential** $\tilde{\mu}$ of the charged particle in the two phases:

$$dG = \tilde{\mu}_2\, dn - \tilde{\mu}_1\, dn \qquad (11.3)$$

The electrochemical potential is a generalization of the chemical potential to include the effect of an electrical potential on a charged particle. It is the sum of the normal chemical potential μ and a term that results from the nonzero value of the electrical potential:

$$\tilde{\mu} = \mu + z\phi F \qquad (11.4)$$

Note that with this definition $\tilde{\mu} \to \mu$ as $\phi \to 0$.

Combining Equations (11.2) and (11.4) gives

$$\tilde{\mu}_2 - \tilde{\mu}_1 = +z(\phi_2 - \phi_1)F \quad \text{or} \quad \tilde{\mu}_2 = \tilde{\mu}_1 + z(\phi_2 - \phi_1)F \qquad (11.5)$$

Because only the difference in the electrical potential between two points can be measured, one can set $\phi_1 = 0$ in Equation (11.5) to obtain the result

$$\tilde{\mu}_2 = \tilde{\mu}_1 + z\phi F \qquad (11.6)$$

This result shows that charged particles in two otherwise identical phases have different values for the electrochemical potential if the phases are at different electrical potentials. Because the particles will flow in a direction that decreases their electrochemical potential, the flow of negatively charged particles in a conducting phase is toward a region of more positive electric potential. The opposite is true for positively charged particles.

Equation (11.6) is the basis for understanding all electrochemical reactions. In an electrochemical environment, the equilibrium condition is

$$\Delta G_R = \sum_i \nu_i \widetilde{\mu}_i = 0, \text{ rather than } \Delta G_R = \sum_i \nu_i \mu_i = 0 \qquad \textbf{(11.7)}$$

Chemists have a limited ability to change ΔG_R by varying P, T, and concentration. However, because the electrochemical potential can be varied through the application of an electrical potential, ΔG_R can be changed easily because $z_i \phi F$ can be larger in magnitude than $\sum_i \nu_i \mu_i$ and can have either the same or the opposite sign. Because a change in the electrical potential can lead to a change in the sign of ΔG_R, the direction of spontaneous change in an electrochemical reaction system can be changed simply by applying suitable electrical potentials within the system. A practical example of changing the sign of ΔG_R is the recharging of a battery by the application of an external potential.

11.2 Conventions and Standard States in Electrochemistry

How can we assign value to $\widetilde{\mu}$ for the individual species in the equilibrium $\text{Zn}(s) \rightleftharpoons$ $\text{Zn}^{2+}(aq) + 2e^-$? Because $\text{Zn}(s)$ is in its standard reference state and is uncharged, $\widetilde{\mu}(\text{Zn}(s)) = \mu(\text{Zn}(s)) = \Delta G_f(\text{Zn}(s)) = 0$. We next consider $\text{Zn}^{2+}(aq)$. We consider only the case that the overall chemical reaction in an electrochemical cell takes place in a single phase, and therefore ϕ has the same value for all solvated ionic reactants and products. In this case the equilibrium condition becomes

$$\Delta G_R = \sum_i \nu_i \widetilde{\mu}_i = \sum_i \nu_i \mu_i + F\varphi \sum_i \nu_i z_i = \sum_i \nu_i \mu_i = 0 \qquad \textbf{(11.7a)}$$

because $\sum_i \nu_i z_i = 0$. We see that the value of ϕ does not influence the equilibrium if the overall reaction takes place in a single phase. Therefore we can set $\phi = 0$ and adopt the convention

$$\widetilde{\mu}_i = \mu_i \text{ (ions in solution)} \qquad \textbf{(11.8)}$$

Adopting this standard state simplifies calculations because μ_i can be calculated from the solute concentration at low concentrations using the Debye–Hückel limiting law discussed in Section 10.4 or at higher concentrations if the activity coefficients are known.

Next, consider the appropriate standard state for an electron in a metal electrode. We cannot set $\phi = 0$ as we did for the ions in solution because there is a potential difference between the solution and the electrode. As shown in Equation (11.6), the electrochemical potential consists of two parts, a chemical component and a component that depends on the electrical potential. For an electron in a metal, there is no way to determine the relative magnitude of the two components. Therefore, it is convenient to choose the standard state μ_i (electrons in electrode) $= 0$ so that

$$\widetilde{\mu}_{e^-} = -\phi F \text{ (electrons in metal electrode)} \qquad \textbf{(11.9)}$$

As we discussed earlier, half-cell potentials cannot be measured directly. They are measured relative to one another rather than absolutely. To understand how this is done, it is useful to consider an **electrochemical cell,** which consists of two half-cells, such as the one shown in Figure 11.2 This particular cell is known as the **Daniell cell,** after its inventor. On the left, a Zn electrode is immersed in a solution of ZnSO_4. The solute is completely dissociated to form $\text{Zn}^{2+}(aq)$ and $\text{SO}_4^{2-}(aq)$. On the right, a Cu electrode

FIGURE 11.2
Schematic diagram of the Daniell cell.
Zn^{2+}/Zn and Cu^{2+}/Cu half-cells are connected through a salt bridge in the internal circuit. A voltmeter is shown in the external circuit. The inset shows the atomic level processes that occur at each electrode.

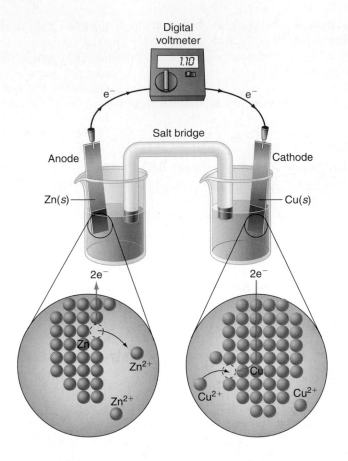

is immersed in a solution of $CuSO_4$, which is completely dissociated to form Cu^{2+} (aq) and SO_4^{2-} (aq). The two half-cells are connected by an ionic conductor known as a salt bridge. The **salt bridge** consists of an electrolyte such as KCl suspended in a gel. A salt bridge allows current to flow between the half-cells while preventing the mixing of the solutions. A metal wire fastened to each electrode allows the electron current to flow through the external part of the circuit. Note that because the wire is connected on one end to a Cu electrode and on the other end to a Zn electrode, the two phases between which we are measuring the electrical potential are not identical.

Using the experimental setup of Figure 11.2, the electrical potential difference between two half-cells can be measured, rather than the absolute electrical potential of each half-cell. However, we need potentials of individual half-cells. Therefore, it is convenient to choose one half-cell as a reference and arbitrarily assign an electrical potential of zero to this half-cell. Once this is done, the electrical potential associated with any other half-cell can be determined by combining it with the reference half-cell. The measured potential difference across the cell is associated with the half-cell of interest. It is next shown that the standard hydrogen electrode fulfills the role of a reference half-cell of zero potential. The measurement of electrical potentials is discussed using this cell. The reaction in the standard hydrogen electrode is

$$H^+(aq) + e^- \rightarrow \frac{1}{2} H_2(g) \qquad \textbf{(11.10)}$$

and the equilibrium in the half-cell is described by

$$\mu_{H^+}(aq) + \widetilde{\mu}_{e^-} = \frac{1}{2}\mu_{H_2(g)} \qquad \textbf{(11.11)}$$

Half-cell reactions such as Equation (11.10) are generally written as reduction reactions by convention as is done in Tables 11.1 and 11.2 (see Appendix B, Data Tables) even though equilibrium is established in the half-cell. It is useful to separate $\mu_{H_2(g)}$ and $\mu_{H^+}(aq)$ into a standard state portion and a portion that depends on the activity and use Equation (11.9) for the electrochemical potential of the electron. The preceding equation then takes the form

$$\mu_{H^+}^\circ + RT \ln a_{H^+} - F\phi_{H^+/H_2} = \frac{1}{2}\mu_{H_2}^\circ + \frac{1}{2}RT \ln f_{H_2} \qquad (11.12)$$

where f_{H_2} is the fugacity of the hydrogen gas. Solving Equation (11.12) for ϕ_{H^+/H_2}

$$\phi_{H^+/H_2} = \frac{\mu_{H^+}^\circ - \frac{1}{2}\mu_{H_2}^\circ}{F} - \frac{RT}{F} \ln \frac{f_{H_2}^{1/2}}{a_{H^+}} \qquad (11.13)$$

For unit activities of all species, the cell has its standard state potential, designated ϕ_{H^+/H_2}°. Because $\mu_{H_2}^\circ = \Delta G_f^\circ(H_2, g) = 0$,

$$\phi_{H^+/H_2}^\circ = \frac{\mu_{H^+}^\circ}{F} \qquad (11.14)$$

In Section 10.1 the convention that $\Delta G_f^\circ(H^+, aq) = \mu_{H^+}^\circ = 0$ was introduced. Therefore, we find that

$$\phi_{H^+/H_2}^\circ = 0 \qquad (11.15)$$

We have shown that the standard hydrogen electrode is a convenient **reference electrode** with zero potential against which the potentials of all other half-cells can be measured. A schematic drawing of this electrode is shown in Figure 11.3. To achieve equilibrium on a short timescale, this reaction $H^+(aq) + e^- \rightleftharpoons 1/2\ H_2(g)$ is carried out over a Pt catalyst electrode. It is also necessary to establish a standard state for the activity of $H^+(aq)$. It is customary to use a Henry's law standard state based on molarity. Therefore, $a_i \rightarrow c_i$ and $\gamma_i \rightarrow 1$ as $c_i \rightarrow 0$. The standard state is a (hypothetical) aqueous solution of $H^+(aq)$ that shows ideal solution behavior at a concentration of $c^\circ = 1\ \text{mol L}^{-1}$.

The usefulness of the result $\phi_{H^+/H_2}^\circ = 0$ is that values for the electrical potential can be assigned to individual half-cells by measuring their potential relative to the H^+/H_2 half-cell. For example, the cell potential of the electrochemical cell in Figure 11.4 is assigned to the Zn/Zn^{2+} half-cell if the $H^+(aq)$ and $H_2(g)$ activities both have the value 1. Although not directly measurable, absolute values for half-cell potentials can be determined. In Supplemental Section 11.16 it is shown that the absolute potential of the standard hydrogen electrode is -4.44 ± 0.02 V. Because only changes in energy rather than absolute energies can be measured, chemists generally use half-cell potentials relative to the standard hydrogen electrode assuming that $\phi_{H^+/H_2}^\circ = 0$.

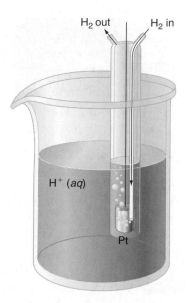

FIGURE 11.3

The standard hydrogen electrode consists of a solution of an acid such as HCl, H_2 gas, and a Pt catalyst electrode that allows the equilibrium in the half-cell reaction to be established rapidly. The activities of H_2 and H^+ are equal to one.

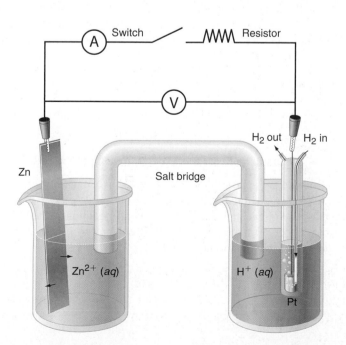

FIGURE 11.4

In a cell consisting of a half-cell and the standard hydrogen electrode, the entire cell voltage is assigned to the half-cell.

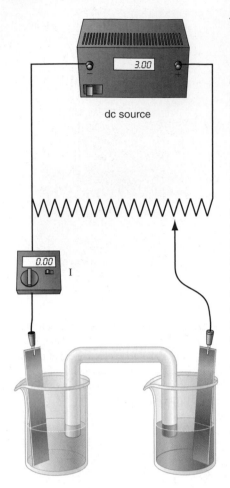

FIGURE 11.5
Schematic diagram showing how the reversible cell potential is measured.

11.3 Measurement of the Reversible Cell Potential

The cell potential measured under reversible conditions is directly related to the state functions G, H, and S. The reversible cell potential, also called **electromotive force (emf),** is determined in an experiment depicted in Figure 11.5. The dc source provides a voltage to a potentiometer circuit with a sliding contact. The sliding contact is attached to the positive cell terminal as shown, and the slider is adjusted until the current-sensing device labeled I shows a null current. At this position of the potentiometer, the voltage applied through the potentiometer exactly opposes the cell potential. The voltage measured in this way is the reversible cell potential. If the sliding contact is moved to a position slightly to the left of this position, the electron current will flow through the external circuit in one direction. However, if the sliding contact is moved to a position slightly to the right of this position, the electron current will flow through the external circuit in the opposite direction, showing that the direction of the cell reaction has been reversed. Because a small variation of the applied voltage can reverse the direction of spontaneous change, the criterion for reversibility is established. This discussion also demonstrates that the direction of spontaneous change in the cell can be reversed by changing the electrochemical potential of the electrons in one of the electrodes relative to that in the other electrode using an external voltage source.

11.4 Chemical Reactions in Electrochemical Cells and the Nernst Equation

What reactions occur in the Daniell cell shown in Figure 11.2? If the half-cells are connected through the external circuit, Zn atoms leave the Zn electrode to form Zn^{2+} in solution, and Cu^{2+} ions are deposited as Cu atoms on the Cu electrode. In the external circuit, it is observed that electrons flow through the wires and the resistor in the direction from the Zn electrode to the Cu electrode. These observations are consistent with the following electrochemical reactions:

$$\text{Left half-cell: } Zn(s) \rightarrow Zn^{2+}(aq) + 2e^- \tag{11.16}$$

$$\text{Right half-cell: } Cu^{2+}(aq) + 2e^- \rightarrow Cu(s) \tag{11.17}$$

$$\text{Overall: } Zn(s) + Cu^{2+}(aq) \rightleftharpoons Zn^{2+}(aq) + Cu(s) \tag{11.18}$$

In the left half-cell, Zn is being oxidized to Zn^{2+}, and in the right half-cell, Cu^{2+} is being reduced to Cu. By convention, the electrode at which oxidation occurs is called the **anode,** and the electrode at which reduction occurs is called the **cathode.** Each half-cell in an electrochemical cell must contain a species that can exist in an oxidized and a reduced form. For a general redox reaction, the reactions at the anode and cathode and the overall reaction can be written as follows:

$$\text{Anode: } Red_1 \rightarrow Ox_1^{n_1^+} + n_1 e^- \tag{11.19}$$

$$\text{Cathode: } Ox_2 + n_2 e^- \rightarrow Red_2^{n_2^-} \tag{11.20}$$

$$\text{Overall: } n_2 Red_1 + n_1 Ox_2 \rightleftharpoons n_2 Ox_1^{n_1^+} + n_1 Red_2^{n_2^-} \tag{11.21}$$

Note that electrons do not appear in the overall reaction because the electrons produced at the anode are consumed at the cathode.

How are the cell voltage and ΔG_R for the overall reaction related? This important relationship can be determined from the electrochemical potentials of the species involved in the overall reaction of the Daniell cell:

$$\Delta G_R = \widetilde{\mu}_{Zn^{2+}} + \widetilde{\mu}_{Cu} - \widetilde{\mu}_{Cu^{2+}} - \widetilde{\mu}_{Zn} = \mu^{\circ}_{Zn^{2+}} - \mu^{\circ}_{Cu^{2+}} + RT \ln\frac{a_{Zn^{2+}}}{a_{Cu^{2+}}}$$

$$= \Delta G_R^{\circ} + RT \ln\frac{a_{Zn^{2+}}}{a_{Cu^{2+}}} \tag{11.22}$$

If this reaction is carried out reversibly, the electrical work done is equal to the product of the charge and the potential difference through which the charge is moved. However, the reversible work at constant pressure is also equal to ΔG. Therefore, we can write the following equation:

$$\Delta G_R = -nF\Delta\phi \tag{11.23}$$

In Equation (11.23), $\Delta\phi$ is the measured potential difference generated by the spontaneous chemical reaction for particular values of $a_{Zn^{2+}}$ and $a_{Cu^{2+}}$, and n is the number of moles of electrons involved in the redox reaction. The measured cell voltage is directly proportional to ΔG. For a reversible reaction, the symbol E is used in place of $\Delta\phi$, and E is referred to as the **electromotive force (emf).** Using this definition, we rewrite Equation (11.23) as follows:

$$-2FE = \Delta G_R^\circ + RT \ln \frac{a_{Zn^{2+}}}{a_{Cu^{2+}}} \tag{11.24}$$

For standard state conditions, $a_{Zn^{2+}} = a_{Cu^{2+}} = 1$, and Equation (11.24) takes the form $\Delta G_R^\circ = -2FE^\circ$. This definition of E° allows Equation (11.24) to be rewritten as

$$E = E^\circ - \frac{RT}{2F} \ln\frac{a_{Zn^{2+}}}{a_{Cu^{2+}}} \tag{11.25}$$

For a general overall electrochemical reaction involving the transfer of n moles of electrons,

$$E = E^\circ - \frac{RT}{nF} \ln Q \tag{11.26}$$

where Q is the familiar reaction quotient. The preceding equation is known as the **Nernst equation.** At 298.15 K, the Nernst equation can be written in the form

$$E = E^\circ - \frac{0.05916\text{ V}}{n} \log_{10} Q \tag{11.27}$$

This function is graphed in Figure 11.6. The Nernst equation allows the emf for an electrochemical cell to be calculated if the activity is known for each species and if E° is known.

The Nernst equation has been derived on the basis of the overall cell reaction. For a half-cell, an equation of a similar form can be derived. The equilibrium condition for the half-cell reaction

$$Ox^{n+} + ne^- \rightleftharpoons Red \tag{11.28}$$

is given by

$$\mu_{Ox^{n+}} + n\tilde{\mu}_{e-} = \mu_{Red} \tag{11.29}$$

Using the convention for the electrochemical potential of an electron in a metal electrode [Equation (11.9)], Equation (11.29) can be written in the form

$$\mu_{Ox^{n+}}^\circ + RT \ln a_{Ox^{n+}} - nF\phi_{Ox/Red} = \mu_{Red}^\circ + RT \ln a_{Red}$$

$$\phi_{Ox/Red} = -\frac{\mu_{Red}^\circ - \mu_{Ox^{n+}}^\circ}{nF} - \frac{RT}{nF} \ln\frac{a_{Red}}{a_{Ox^{n+}}}$$

$$E_{Ox/Red} = E_{Ox/Red}^\circ - \frac{RT}{nF} \ln\frac{a_{Red}}{a_{Ox^{n+}}} \tag{11.30}$$

The last line in Equation (11.30) has the same form as the Nernst equation, but the activity of the electrons does not appear in Q. An example of the application of Equation (11.30) to a half-cell reaction is shown in Example Problem 11.1.

EXAMPLE PROBLEM 11.1

Calculate the potential of the H^+/H_2 half-cell when $a_{H^+} = 0.770$ and $f_{H_2} = 1.13$.

Solution

$$E = E^\circ - \frac{0.05916\text{ V}}{n} \log_{10}\frac{a_{H^+}}{\sqrt{f_{H_2}}} = 0 - \frac{0.05916\text{ V}}{1} \log_{10}\frac{0.770}{\sqrt{1.13}} = 0.0083\text{ V}$$

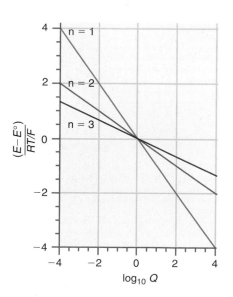

FIGURE 11.6
The cell potential E varies linearly with log Q. The slope of a plot of $(E - E^\circ)/(RT/F)$ is inversely proportional to the number of electrons transferred in the redox reaction.

11.5 Combining Standard Electrode Potentials to Determine the Cell Potential

A representative set of standard potentials is listed in Tables 11.1 and 11.2 (see Appendix B, Data Tables). By convention, half-cell emfs are always tabulated as reduction potentials. However, whether the reduction or the oxidation reaction is spontaneous in a half-cell is determined by the relative emfs of the two half-cells that make up the electrochemical cell. Because $\Delta G_R^{\circ} = -nFE^{\circ}$, and ΔG_R° for the oxidation and reduction reactions are equal in magnitude and opposite in sign,

$$E_{reduction}^{\circ} = -E_{oxidation}^{\circ} \tag{11.31}$$

How is the cell potential related to the potentials of the half-cells? The standard potential of a half-cell is given by $E^{\circ} = -\Delta G_R^{\circ}/nF$ and is an intensive property because although both ΔG_R° and n are extensive quantities, the ratio $\Delta G_R^{\circ}/n$ is an intensive quantity. In particular, E° is not changed if all the stoichiometric coefficients are multiplied by any integer, because both ΔG_R° and n are changed by the same factor. Therefore,

$$E_{cell}^{\circ} = E_{reduction}^{\circ} + E_{oxidation}^{\circ} \tag{11.32}$$

even if the balanced reaction for the overall cell is multiplied by an arbitrary number. In Equation (11.32), the standard potentials on the right refer to the half-cells.

EXAMPLE PROBLEM 11.2

An electrochemical cell is constructed using a half-cell for which the reduction reaction is given by

$$Fe(OH)_2(s) + 2e^- \rightarrow Fe(s) + 2\,OH^-(aq) \qquad E^{\circ} = -0.877 \text{ V}$$

It is combined with a half-cell for which the reduction reaction is given by

a. $Al^{3+}(aq) + 3e^- \rightarrow Al(s)$ $E^{\circ} = -1.66$ V

b. $AgBr(s) + e^- \rightarrow Ag(s) + Br^-(aq)$ $E^{\circ} = +0.071$ V

The activity of all species is one for both reactions. Write the overall reaction for the cells in the direction of spontaneous change. Is the Fe reduced or oxidized in the spontaneous reaction?

Solution

The emf for the cell is the sum of the emfs for the half-cells, one of which is written as an oxidation reaction, and the other of which is written as a reduction reaction. For the direction of spontaneous change, $E^{\circ} = E_{reduction}^{\circ} + E_{oxidation}^{\circ} > 0$. Note that the two half-cell reactions are combined with the appropriate stoichiometry so that electrons do not appear in the overall equation. Recall that the half-cell emfs are not changed in multiplying the overall reactions by the integers necessary to eliminate the electrons in the overall equation, because E° is an intensive quantity.

a. $3\,Fe(OH)_2(s) + 2\,Al(s) \rightleftharpoons 3\,Fe(s) + 6\,OH^-(aq) + 2\,Al^{3+}(aq)$

$E^{\circ} = -0.877 \text{ V} + 1.66 \text{ V} = +0.783$ V

In this cell, Fe is reduced.

b. $Fe(s) + 2\,OH^-(aq) + 2\,AgBr(s) \rightleftharpoons Fe(OH)_2(s) + 2\,Ag(s) + 2\,Br^-(aq)$

$E^{\circ} = 0.071 \text{ V} + 0.877 \text{ V} = +0.95$ V

In this cell, Fe is oxidized.

Because ΔG is a state function, the cell potential for a third half-cell can be obtained from the cell potentials of two half-cells if they have an oxidation or reduction reaction in common. The procedure is analogous to the use of Hess's law in Chapter 4 and is illustrated in Example Problem 11.3.

EXAMPLE PROBLEM 11.3

You are given the following reduction reactions and $E°$ values:

$$Fe^{3+}(aq) + e^- \rightarrow Fe^{2+}(aq) \qquad E° = +0.771 \text{ V}$$

$$Fe^{2+}(aq) + 2e^- \rightarrow Fe(s) \qquad E° = -0.447 \text{ V}$$

Calculate $E°$ for the half-cell reaction $Fe^{3+}(aq) + 3e^- \rightarrow Fe(s)$.

Solution

We calculate the desired value of $E°$ by converting the given $E°$ values to $\Delta G°$ values, and combining these reduction reactions to obtain the desired equation.

$$Fe^{3+}(aq) + e^- \rightarrow Fe^{2+}(aq)$$

$$\Delta G° = -nFE° = -1 \times 96485 \text{C mol}^{-1} \times 0.771 \text{ V} = -74.39 \text{ kJ mol}^{-1}$$

$$Fe^{2+}(aq) + 2e^- \rightarrow Fe(s)$$

$$\Delta G° = -nFE° = -2 \times 96485 \text{ C mol}^{-1} \times (-0.447 \text{ V}) = 86.26 \text{ kJ mol}^{-1}$$

We next add the two equations as well as their $\Delta G°$ to obtain

$$Fe^{3+}(aq) + 3e^- \rightarrow Fe(s)$$

$$\Delta G° = -74.39 \text{ kJ mol}^{-1} + 86.26 \text{ kJ mol}^{-1} = 11.87 \text{ kJ mol}^{-1}$$

$$E°_{Fe^{3+}/Fe} = -\frac{\Delta G°}{nF} = \frac{-11.87 \times 10^3 \text{ J mol}^{-1}}{3 \times 96485 \text{ C mol}^{-1}} = -0.041 \text{ V}$$

The $E°$ values cannot be combined directly, because they are intensive rather than extensive quantities.

The preceding calculation can be generalized as follows. Assume that n_1 electrons are transferred in the reaction with the potential $E°_{A/B}$, and n_2 electrons are transferred in the reaction with the potential $E°_{B/C}$. If n_3 electrons are transferred in the reaction with the potential $E°_{A/C}$, then $n_3 E°_{A/C} = n_1 E°_{A/B} + n_2 E°_{B/C}$.

11.6 Obtaining Reaction Gibbs Energies and Reaction Entropies from Cell Potentials

In Section 11.4, we demonstrated that $\Delta G_R = -nF\Delta\phi$. Therefore, if the cell potential is measured under standard conditions,

$$\Delta G_R° = -nFE° \tag{11.33}$$

If $E°$ is known, $\Delta G_R°$ can be determined using Equation (11.33). For example, $E°$ for the Daniell cell is $+1.10$ V. Therefore, $\Delta G_R°$ for the reaction $Zn(s) + Cu^{2+}(aq) \rightleftharpoons Zn^{2+}(aq) + Cu(s)$ is

$$\Delta G_R° = -nFE° = -2 \times 96,485 \text{ C mol}^{-1} \times 1.10 \text{ V} = -212 \text{ kJ mol}^{-1} \tag{11.34}$$

The reaction entropy is related to $\Delta G_R°$ by

$$\Delta S_R° = -\left(\frac{\partial \Delta G_R°}{\partial T}\right)_P = nF\left(\frac{\partial E°}{\partial T}\right)_P \tag{11.35}$$

Therefore, a measurement of the temperature dependence of $E°$ can be used to determine $\Delta S_R°$, as shown in Example Problem 11.4.

EXAMPLE PROBLEM 11.4

The standard potential of the cell formed by combining the $Cl_2(g)/Cl^-(aq)$ half-cell with the standard hydrogen electrode is $+1.36$ V, and $(\partial E°/\partial T)_P = -1.20 \times 10^{-3}$ V K^{-1}. Calculate $\Delta S_R°$ for the reaction $H_2(g) + Cl_2(g) \rightleftharpoons 2H^+(aq) + 2Cl^-(aq)$. Compare your result with the value that you obtain by using the values of $\Delta S°$ in the data tables from Appendix B.

Solution

From Equation (11.35),

$$\Delta S_R° = -\left(\frac{\partial \Delta G_R°}{\partial T}\right)_P = nF\left(\frac{\partial E°}{\partial T}\right)_P$$

$$= 2 \times 96480 \, C \, mol^{-1} \times (-1.2 \times 10^{-3} \, V \, K^{-1})$$

$$= -2.3 \times 10^2 \, J \, K^{-1} \, mol^{-1}$$

From Table 10.1 (see Appendix B, Data Tables),

$$\Delta S_R° = 2S_m°(H^+, aq) + 2S_m°(Cl^-, aq) - S_m°(H_2, g) - S_m°(Cl_2, g)$$

$$= 2 \times 0 + 2 \times 56.5 \, J \, K^{-1} mol^{-1} - 130.7 \, J \, K^{-1} mol^{-1} - 223.1 \, J \, K^{-1} mol^{-1}$$

$$= -2.408 \times 10^2 \, J \, K^{-1} mol^{-1}$$

The limited precision of the temperature dependence of $E°$ limits the precision in the determination of $\Delta S_R°$.

11.7 The Relationship between the Cell EMF and the Equilibrium Constant

If the redox reaction is allowed to proceed until equilibrium is reached, $\Delta G = 0$, so that $E = 0$. For the equilibrium state, the reaction quotient $Q = K$. Therefore,

$$E° = \frac{RT}{nF} \ln K \qquad (11.36)$$

Equation (11.36) shows that a measurement of the standard state cell potential, for which $a_i = 1$ for all species in the redox reaction, allows K for the overall reaction to be determined. Although this statement is true, it is not practical to adjust all activities to the value 1. The experimental determination of $E°$ is discussed in the next section.

Tabulated half-cell potentials provide a powerful way to determine the equilibrium constant in an electrochemical cell using Equation (11.36). To determine K, the overall reaction must be separated into the oxidation and reduction half-reactions, and the number of electrons transferred must be determined. For example, consider the reaction

$$2\,MnO_4^-(aq) + 6\,H^+(aq) + 5\,HOOCCOOH(aq) \rightleftharpoons$$
$$2\,Mn^{2+}(aq) + 8\,H_2O(l) + 10\,CO_2(g) \quad (11.37)$$

The reduction and oxidation half-reactions are

$$MnO_4^-(aq) + 8\,H^+(aq) + 5e^- \rightarrow Mn^{2+}(aq) + 4\,H_2O(l) \quad E° = +1.51 \text{ V} \quad (11.38)$$

$$HOOCCOOH(aq) \rightarrow 2\,H^+(aq) + 2\,CO_2(g) + 2e^- \quad E° = +0.49 \text{ V} \quad (11.39)$$

To eliminate the electrons in the overall equation, the first equation must be multiplied by 2, and the second equation must be multiplied by 5; $E°$ is unchanged by doing so. However, n is affected by the multipliers. In this case $n = 10$. Therefore,

$$\Delta G° = -nFE° = -10 \times 96485 \, C \, mol^{-1} \times (1.51 \text{ V} + 0.49 \text{ V})$$

$$= -1.93 \times 10^2 \, kJ \, mol^{-1}$$

$$\ln K = -\frac{\Delta G°}{RT} = \frac{1.93 \times 10^2 \, kJ \, mol^{-1}}{8.314 \, J \, K^{-1} \, mol^{-1} \times 298.15 \, K} = 778 \qquad (11.40)$$

As this result shows, the equilibrium corresponds to essentially complete conversion of reactants to products. Example Problem 11.5 shows the same calculation for the Daniell cell.

EXAMPLE PROBLEM 11.5

For the Daniell cell $E° = 1.10$ V. Calculate K for the reaction at 298.15 K
$Zn(s) + Cu^{2+}(aq) \rightleftharpoons Zn^{2+}(aq) + Cu(s)$.

Solution

$$\ln K = \frac{nF}{RT}E° = \frac{2 \times 96485 \text{ C mol}^{-1} \times 1.10 \text{ V}}{8.314 \text{ J K}^{-1} \text{ mol}^{-1} \times 298.15 \text{ K}}$$

$$= 85.63$$

$$K = 1.55 \times 10^{37}$$

Note that the equilibrium constant calculated in Example Problem 11.5 is so large that it could not have been measured by determining the activities of $a_{Zn^{2+}}$ and $a_{Cu^{2+}}$ by spectroscopic methods. This would require a measurement technique that is accurate over more than 30 orders of magnitude in the activity. By contrast, the equilibrium constant in an electrochemical cell can be determined with high accuracy using only a voltmeter.

A further example of the use of electrochemical measurements to determine equilibrium constants is the solubility constant for a weakly soluble salt. If the overall reaction corresponding to dissolution can be generated by combining half-cell potentials, then the solubility constant can be calculated from the potentials. For example, the following half-cell reactions can be combined to calculate the solubility product of AgBr.

$$AgBr(s) + e^- \rightarrow Ag(s) + Br^-(aq) \qquad E° = 0.07133 \text{ V} \quad \text{and}$$

$$\underline{Ag(s) \rightarrow Ag^+(aq) + e^- \qquad\qquad\qquad E° = -0.7996 \text{ V}}$$

$$AgBr(s) \rightleftharpoons Ag^+(aq) + Br^-(aq) \qquad E° = -0.7283 \text{ V}$$

$$\ln K_{sp} = \frac{nF}{RT}E° = \frac{1 \times 96485 \text{ C mol}^{-1} \times (-0.7283 \text{ V})}{8.314 \text{ J K}^{-1} \text{mol}^{-1} \times 298.15 \text{ K}} = -28.35$$

The value of the solubility constant is $K_{sp} = 4.88 \times 10^{-13}$.

EXAMPLE PROBLEM 11.6

A concentration cell consists of two half-cells that are identical except for the activities of the redox components. Consider two half-cells based on the $Ag^+(aq) + e^- \rightarrow Ag(s)$ reaction. The left half-cell contains $AgNO_3$ at unit activity, and the right half-cell initially had the same concentration of $AgNO_3$, but just enough $NaCl(aq)$ has been added to precipitate the $Ag^+(aq)$ as AgCl. Write an equation for the overall cell reaction. If the emf of this cell is 0.29 V, what is K_{sp} for AgCl?

Solution

The overall reaction is $Ag^+(aq, a = 1) \rightleftharpoons Ag^+(aq, a)$ with a concurrent transport of Cl^- in the opposite direction through the salt bridge between the half-cells. In the right half-cell we have the equilibrium $AgCl(s) \rightleftharpoons Ag^+(aq, a_\pm) + Cl^-(aq, a_\pm)$, so that $a_{Ag^+} a_{Cl^-} = a_\pm^2 = K_{sp}$.

Because the half-cell reactions are the same, $E° = 0$ and

$$E = -\frac{0.05916 \text{ V}}{n} \log_{10} \frac{a_\pm}{1} = -\frac{0.05916}{1} \log_{10} \sqrt{K_{sp}}$$

$$= -\frac{0.05916 \text{ V}}{2} \log_{10} K_{sp}$$

$$\log_{10} K_{sp} = -\frac{2E}{0.05916} = -\frac{2 \times (0.29 \text{ V})}{0.05916 \text{ V}} = -9.804$$

$$K_{sp} = 1.57 \times 10^{-10}$$

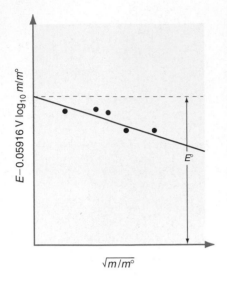

FIGURE 11.7
The value of $E°$ and the activity coefficient can be measured by plotting the left-hand side of Equation (11.42) against the square root of the molality.

11.8 Determination of $E°$ and Activity Coefficients Using an Electrochemical Cell

The main problem in determining standard potentials lies in knowing the value of the activity coefficient $\gamma_\pm$ for a given solute concentration. The best strategy is to carry out measurements of the cell potential at low concentrations, where $\gamma_\pm \rightarrow 1$, rather than near unit activity, where $\gamma_\pm$ differs appreciably from 1. Consider an electrochemical cell consisting of the Ag^+/Ag and standard hydrogen electrode half-cells at 298 K. The cell reaction is $Ag^+(aq) + 1/2\,H_2(g) \rightleftharpoons Ag(s) + H^+(aq)$ and $Q = (a_{Ag^+})^{-1}$. Because the activities of $H_2(g)$ and $H^+(aq)$ are 1, they do not appear in Q. Assume that the Ag^+ arises from the dissociation of $AgNO_3$. Recall that the activity of an individual ion cannot be measured directly. It must be calculated from the measured activity $a_\pm$ and the definition $a_\pm^\nu = a_+^{\nu^+} a_-^{\nu^-}$. In this case, $a_\pm^2 = a_{Ag^+} a_{NO_3^-}$ and $a_\pm = a_{Ag^+} = a_{NO_3^-}$. Similarly, $\gamma_\pm = \gamma_{Ag^+} = \gamma_{NO_3^-}$ and $m_{Ag^+} = m_{NO_3^-} = m_\pm = m$, and E is given by

$$E = E°_{Ag^+/Ag} + \frac{RT}{F}\ln a_{Ag^+} = E°_{Ag^+/Ag} + \frac{RT}{F}\ln(m/m°) + \frac{RT}{F}\ln\gamma_\pm \quad (11.41)$$

At low enough concentrations, the Debye–Hückel limiting law is valid and $\log\gamma_\pm = -0.5092\sqrt{m_\pm/m°}$ at 298 K as discussed in Section 10.4. Using this relation, Equation (11.41) can be rewritten in the form

$$E - 0.05916\log_{10}(m/m°) = E°_{Ag^+/Ag} - 0.05916 \times 0.5090\sqrt{(m/m°)}$$

$$= E°_{Ag^+/Ag} - 0.03011\sqrt{(m/m°)} \quad (11.42)$$

The left-hand side of this equation can be calculated from measurements and plotted as a function of $\sqrt{(m/m°)}$. The results will resemble the graph shown in Figure 11.7. An extrapolation of the line that best fits the data to $m = 0$ gives $E°$ as the intercept with the vertical axis. Once $E°$ has been determined, Equation (11.41) can be used to calculate $\gamma_\pm$.

Electrochemical cells provide a powerful method of determining activity coefficients because cell potentials can be measured more accurately and more easily than colligative properties such as freezing point depression or boiling point elevation. Note that although the Debye–Hückel limiting law was used to determine $E°$, it is not necessary to use the limiting law to calculate activity coefficients once $E°$ is known.

11.9 Cell Nomenclature and Types of Electrochemical Cells

It is useful to use an abbreviated notation to describe an electrochemical cell. This notation includes all species involved in the cell reaction and phase boundaries within the cell, which are represented by a vertical line. As will be seen later in this section, the metal electrodes appear at the ends of this notation, the half-cell in which oxidation occurs is written on the left, and the electrode is called the anode. The half-cell in which reduction occurs is written on the right, and the electrode is called the cathode.

We briefly discuss an additional small contribution to the cell potential that arises from the differing diffusion rates of large and small ions in an electrical field. As an electrochemical reaction proceeds, ions that diffuse rapidly across a liquid–liquid junction, such as H^+, will travel farther than ions that diffuse slowly, such as Cl^-, in a given time. At steady state, a dipole layer is built up across this junction, and the rates of ion transfer through this dipole layer become equal. This kinetic effect will give rise to a small **junction potential** between two liquids of different composition or concentration. Such a junction potential is largely eliminated by a salt bridge. An interface for which the junction potential has been eliminated is indicated by a pair of vertical lines. The separation of different phases that are in contact and allow electron transfer is shown by a solid vertical line. A single dashed line is used to indicate a liquid–liquid interface across which charge transfer can occur.

For example, the abbreviated notation for the Daniell cell containing a salt bridge is

$$Zn(s)|ZnSO_4(aq)\|CuSO_4(aq)|Cu(s) \tag{11.43}$$

and a cell made up of the Zn/Zn^{2+} half-cell and the standard hydrogen electrode is described by

$$Zn(s)|ZnSO_4(aq)\|H^+(aq)|H_2(g)|Pt(s) \tag{11.44}$$

The overall reaction in this cell is $Zn(s) + 2\,H^+(aq) \rightleftharpoons Zn^{2+}(aq) + H_2(g)$. In general, in such a cell the solutions are physically separated by a porous membrane to prevent mixing of the solutions. In this case, the junction potential has not been eliminated.

The half-cell and overall reactions can be determined from the abbreviated notation in the following way. An electron is transferred from the electrode at the far left of the abbreviated notation to the electrode on the far right through the external circuit. The number of electrons is then adjusted to fit the half-cell reaction. This procedure is illustrated in Example Problem 11.7.

EXAMPLE PROBLEM 11.7

Determine the half-cell reactions and the overall reaction for the cell designated

$$Ag(s)|AgCl(s)|Cl^-(aq, a_\pm = 0.0010)\|Fe^{2+}(aq, a_\pm = 0.50)$$
$$Fe^{3+}(aq, a_\pm = 0.10)|Pt(s)$$

Solution

The anode and cathode reactions are

$$Ag(s) + Cl^-(aq) \rightarrow AgCl(s) + e^-$$
$$Fe^{3+}(aq) + e^- \rightarrow Fe^{2+}(aq)$$

The overall reaction is

$$Ag(s) + Cl^-(aq) + Fe^{3+}(aq) \rightleftharpoons AgCl(s) + Fe^{2+}(aq)$$

Only after the cell potential is calculated is it clear whether the reaction or the reverse reaction is spontaneous.

We have already discussed several specific half-cells and next discuss different types of half-cells. The standard hydrogen electrode involves the equilibrium between a gas and a dissolved species. A second such electrode is the Cl_2/Cl^- electrode, for which the reduction reaction is

$$Cl_2(g) + 2e^- \rightarrow 2\,Cl^-(aq) \quad E^\circ = +1.36\text{ V} \tag{11.45}$$

Another type of half-cell that is frequently encountered involves a metal and a metal ion in solution. Both half-cells in the Daniell cell fall into this category.

$$Zn^{2+}(aq) + 2e^- \rightarrow Zn(s) \quad E^\circ = -0.76\text{ V} \tag{11.46}$$

A number of half-cells consist of a metal, an insoluble salt containing the metal, and an aqueous solution containing the anion of the salt. Two examples of this type of half-cell are the Ag–AgCl half-cell for which the reduction reaction is

$$AgCl(s) + e^- \rightarrow Ag(s) + Cl^-(aq) \quad E^\circ = +0.22\text{ V} \tag{11.47}$$

and the calomel (mercurous chloride) electrode, which is frequently used as a reference electrode in electrochemical cells:

$$Hg_2Cl_2(s) + 2e^- \rightarrow 2\,Hg(l) + 2\,Cl^-(aq) \quad E^\circ = +0.27\text{ V} \tag{11.48}$$

In a further type of half-cell, both species are present in solution, and the electrode is an inert conductor such as Pt, which allows an electrical connection to be made to the solution. For example, in the Fe^{3+}/Fe^{2+} half-cell, the reduction reaction is

$$Fe^{3+}(aq) + e^- \rightarrow Fe^{2+}(aq) \quad E^\circ = 0.771\text{ V} \tag{11.49}$$

**Most Strongly Reducing
(The metal is least easily oxidized.)**

Gold

(most positive reduction potential)

Platinum

Palladium

Silver

Rhodium

Copper

Mercury
(Hydrogen; zero reduction potential
by convention)

Lead

Tin

Nickel

Iron

Zinc

Chromium

Vanadium

Manganese

Magnesium

Sodium

Calcium

Potassium

Rubidium

Cesium

Lithium
(most negative reduction potential)

**Least Strongly Reducing
(The metal is most easily oxidized.)**

11.10 The Electrochemical Series

Tables 11.1 and 11.2 (see Appendix B, Data Tables) list the reduction potentials of commonly encountered half-cells. The emf of a cell constructed from two of these half-cells with standard reduction potentials E_1° and E_2° is given by

$$E_{cell}^\circ = E_1^\circ - E_2^\circ \qquad \textbf{(11.50)}$$

The potential E_{cell}° will be positive and, therefore, $\Delta G < 0$, if the reduction potential for reaction 1 is more positive than that of reaction 2. Therefore, the relative strength of a species as an oxidizing agent follows the order of the numerical value of its reduction potential in Table 11.2. The **electrochemical series** shown in Table 11.3 is obtained if the oxidation of neutral metals to their most common oxidation state is considered. For example, the entry for gold in Table 11.3 refers to the reduction reaction

$$Au^{3+}(aq) + 3e^- \rightarrow Au(s) \quad E^\circ = 1.498 \text{ V}$$

In a redox couple formed from two entries in the list shown in Table 11.3, the species lying higher in the list will be reduced, and the species lying lower in the list will be oxidized in the spontaneous reaction. For example, the table predicts that the spontaneous reaction in the copper–zinc couple is $Zn(s) + Cu^{2+}(aq) \rightarrow Zn^{2+}(aq) + Cu(s)$ and not the reverse reaction.

EXAMPLE PROBLEM 11.8

For the reduction of the permanganate ion MnO_4^- to Mn^{2+} in an acidic solution, $E^\circ = +1.51$ V. The reduction reactions and standard potentials for Zn^{2+}, Ag^+, and Au^+ are given here:

$$Zn^{2+}(aq) + 2e^- \rightarrow Zn(s) \qquad E^\circ = -0.7618 \text{ V}$$

$$Ag^+(aq) + e^- \rightarrow Ag(s) \qquad E^\circ = 0.7996 \text{ V}$$

$$Au^+(aq) + e^- \rightarrow Au(s) \qquad E^\circ = 1.692 \text{ V}$$

Which of these metals will be oxidized by the MnO_4^- ion?

Solution

The cell potentials assuming the reduction of the permanganate ion and oxidation of the metal are

$$\text{Zn:} \quad 1.51 \text{ V} + 0.761 \text{ V} = 2.27 \text{ V} > 0$$

$$\text{Ag:} \quad 1.51 \text{ V} - 0.7996 \text{ V} = 0.710 \text{ V} > 0$$

$$\text{Au:} \quad 1.51 \text{ V} - 1.692 \text{ V} = -0.18 \text{ V} < 0$$

If $E^\circ > 0$, $\Delta G < 0$. On the basis of the sign of the cell potential, we conclude that only Zn and Ag will be oxidized by the MnO_4^- ion.

11.11 Thermodynamics of Batteries and Fuel Cells

Batteries and fuel cells are electrochemical cells that are designed to maximize the ratio of output power to the cell weight or volume. **Batteries** contain the reactants needed to support the overall electrochemical reaction, whereas **fuel cells** are designed to accept a continuous flow of reactants from the surroundings. Batteries that cannot be recharged are called primary batteries, whereas rechargeable batteries are called secondary batteries.

It is useful to compare the relative amount of work that can be produced through an electrochemical reaction with the work that a heat engine could produce using the same overall reaction. The maximum electrical work is given by

$$w_{electrical} = -\Delta G = -\Delta H\left(1 - \frac{T\Delta S}{\Delta H}\right) \tag{11.51}$$

whereas the maximum work available through a reversible heat engine operating between T_h and T_c is

$$w_{thermal} = q_{hot}\varepsilon = -\Delta H\left(\frac{T_h - T_c}{T_h}\right) \tag{11.52}$$

where ε is the efficiency of a reversible heat engine (see Section 5.2). To compare the maximum thermal and electrical work, we use the overall reaction for the familiar lead-acid battery used in cars. For this reaction, $\Delta G_R^\circ = -376.97\,\text{kJ mol}^{-1}$, $\Delta H_R^\circ = -227.58\,\text{kJ mol}^{-1}$, and $\Delta S_R^\circ = 501.1\,\text{J K}^{-1}\,\text{mol}^{-1}$. Assuming $T_h = 600.\,\text{K}$ and $T_c = 300.\,\text{K}$ and that the battery operates at 300. K, then

$$\frac{w_{electrical}}{w_{thermal}} = 3.31 \tag{11.53}$$

This calculation shows that much more work can be produced in the electrochemical reaction than in the thermal reaction. This comparison does not even take into account that the lead-acid battery can be recharged, whereas the thermal reaction can only be run once.

11.12 The Electrochemistry of Commonly Used Batteries

The lead-acid battery was invented in 1859 and is still widely used in automobiles. Because the power required to start an automobile engine is on the order of a kilowatt, the current capacity of such a battery must be on the order of a hundred amperes. Additionally, such a battery must be capable of 500 to 1500 recharging cycles from a deep discharge. In recharging batteries, the reaction product in the form of a solid must be converted back to the reactant, also in the form of a solid. Because the solids in general have a different crystal structure and density, the conversion induces mechanical stress in the anode and cathode, which ultimately leads to a partial disintegration of these electrodes. This is the main factor that limits the number of charge–discharge cycles that a battery can tolerate.

The electrodes in the lead-acid battery consist of Pb powder and finely divided PbO and $PbSO_4$ supported on a Pb frame. The electrodes are supported in a container containing concentrated H_2SO_4. In use, a battery is discharged. Some batteries, including the lead-acid battery, can be recharged by applying an external voltage to convert products back to reactants. In that case, the role of anode and cathode are reversed. To avoid confusion, we write half cell and overall reactions for the discharge mode. In the lead-acid battery, the cell reactions at the cathode and anode are

$$PbO_2(s) + 4\,H^+(aq) + SO_4^{2-}(aq) + 2e^- \rightarrow PbSO_4(s) + 2\,H_2O(l)$$

$$E^\circ = 1.685\,\text{V} \tag{11.54}$$

$$Pb(s) + SO_4^{2-}(aq) \rightarrow PbSO_4(s) + 2e^- \quad E^\circ = -0.356\,\text{V} \tag{11.55}$$

respectively, and the overall reaction is

$$PbO_2(s) + Pb(s) + 2\,H_2SO_4(aq) \rightleftharpoons 2\,PbSO_4(s) + 2\,H_2O(l) \quad E^\circ = 2.04\,\text{V} \tag{11.56}$$

The arrows would point in the opposite direction and the sign of the emfs would be reversed for the charging mode.

Six such cells connected in series are required for a battery that provides a nominal potential of 12 V. The lead-acid battery is very efficient in that more than 90% of the electrical charge used to charge the battery is available in the discharge part of the cycle.

FIGURE 11.8

A number of different batteries are classified with their specific energy density per unit volume and per unit mass.

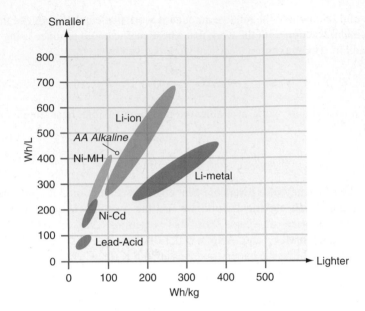

This means that side reactions such as the electrolysis of water play a minimal role in charging the battery. However, only about 50% of the lead in the battery is converted between PbO_2 and $PbSO_4$. Because Pb has a large atomic mass, this limited convertibility decreases the power per unit mass figure of merit for the battery. Parasitic side reactions also lead to a self-discharge of the cell without current flowing in the external circuit. For the lead-acid battery, the capacity is diminished by approximately 0.5% per day through self-discharge.

As batteries have become more common in portable devices such as cell phones and laptop computers, energy density is a major criterion in choosing the most suitable battery chemistry for a specific application. Figure 11.8 shows a comparison of different battery types. The lead-acid battery has the lowest specific energy either in terms of volume or mass. Next we discuss the chemistry of three commonly used rechargeable batteries: the alkaline, nickel metal hydride, and lithium ion batteries.

The individual elements of the alkaline cell are shown in Figure 11.9. The anode in this cell is powdered zinc, and the cathode is in the form of a MnO_2 paste mixed with

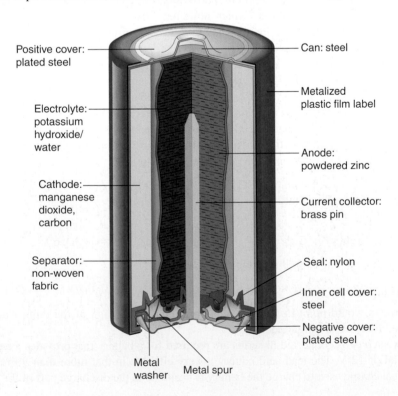

FIGURE 11.9

Schematic diagram of an alkaline cell.

powdered carbon to impart conductivity. KOH is used as the electrolyte. The anode and cathode reactions are

Anode: $Zn(s) + 2\,OH^-(aq) \rightarrow ZnO(s) + H_2O(l) + 2e^-$ $E° = 1.1\ V$ **(11.57)**

Cathode: $2\,MnO_2(s) + H_2O(l) + 2e^- \rightarrow Mn_2O_3(s) + 2\,OH^-(aq)$ $E° = -0.76\ V$
(11.58)

Nickel metal hydride batteries are currently used in hybrid vehicles that rely on dc motors to drive the vehicle in city traffic and use a gasoline engine for higher speed driving. The Toyota Prius uses 28 modules of 6 cells, each with a nominal voltage of 1.2 V and a total voltage of 201.6 V to power the vehicle. The capacity of the battery pack is ~1100 Wh. The anode and cathode reactions are

Anode: $MH(s) + OH^-(aq) \rightarrow M + H_2O(l) + e^-$ $E° = 0.83\ V$ **(11.59)**

Cathode: $NiOOH(s) + H_2O(l) + 2e^- \rightarrow Ni(OH)_2(s) + OH^-(aq)$
$E° = 0.52\ V$ **(11.60)**

The electrolyte is $KOH(aq)$, and the overall reaction is

$$MH(s) + NiOOH(s) \rightleftharpoons M + Ni(OH)_2(s)\quad E° = 1.35\ V \quad \textbf{(11.61)}$$

where M designates an alloy that can contain V, Ti, Zr, Ni, Cr, Co, and Fe.

Lithium ion batteries find applications as diverse as cell phones, where a high energy density per unit volume is required, and electric vehicles, where a high energy density per unit mass is required. The electrodes in lithium ion batteries contain Li^+ ions, and the cell voltage reflects the difference in the binding strength of Li^+ in the two materials. The structure of a cylindrical lithium ion battery is shown schematically in Figure 11.10. The two electrodes are separated by an electrolyte–saturated polymer membrane through which the Li^+ ions move in the internal circuit. The electrolyte is a lithium salt dissolved in an organic solvent, for example, 1M $LiPF_6$ in a mixture of ethylene carbonate and diethyl carbonate. Aqueous electrolytes would limit the cell voltage to 1.2 V because at larger potentials, water is reduced or oxidized. Figure 11.11 shows a number of materials that can be used as electrodes. Materials that fall outside of the band gap of the electrolyte are unsuitable because their use initiates reduction or oxidation of the solvent. It would appear that carbon is unsuitable, but the formation of a thin solid/electrolyte interface layer stabilizes carbon with respect to solvent reactions, and it is the

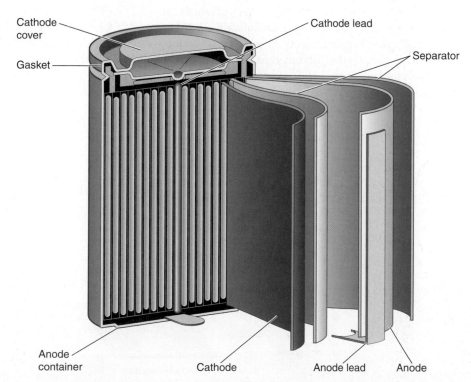

Cathode cover

Gasket

Cathode lead

Separator

Anode container

Cathode

Anode lead

Anode

FIGURE 11.10

Schematic structure of a cylindrical lithium ion battery. The anode and cathode material are formed of thin sheets to optimize the transport kinetics of Li^+ ions.

FIGURE 11.11
The cell potential of a lithium ion battery versus the energy density per unit mass for a number of electrode materials is shown. The dashed lines indicate the voltage range in which 1M LiPF$_6$ in a 1:1 mixture of ethylene carbonate and diethyl carbonate is stable with respect to reduction or oxidation.
Source: Goodenough J. B. and Kim Y. *Chemistry of Materials* 22 (2010), 587.

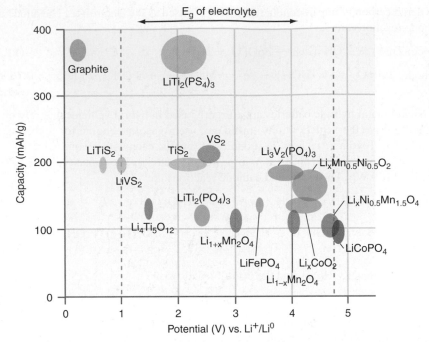

most widely used anode electrode. Using a carbon anode and LiCoO$_2$(s) as the cathode allows a cell potential of 3.7 V to be achieved. As Figure 11.11 shows, higher potentials are possible, but batteries with long life cycles using materials other than LiCoO$_2$(s) have not yet been developed.

Rechargeable lithium batteries have the following half-cell reactions while discharging the battery:

$$\text{positive electrode:} \quad Li_{1-x}CoO_2(s) + xLi^+(solution) + xe^- \rightarrow LiCoO_2(s) \tag{11.62}$$

$$\text{negative electrode:} \quad CLi_x \rightarrow C(s) + xLi^+(solution) + xe^- \tag{11.63}$$

The right arrows indicate the discharge directions. In these equations, x is a small positive number. The overall cell reaction is

$$Li_{1-x}CoO_2(s) + CLi_x \rightleftharpoons LiCoO_2(s) + C(graphite) \quad E° \sim 3.7 \text{ V} \tag{11.64}$$

and the fully charged battery has a cell potential of ~3.7 V. The structures of LiCoO$_2$(s) and CLi$_x$ are shown schematically in Figure 11.12. CLi$_x$ designates Li atoms intercalated between sheets of graphite; it is not a stoichiometric compound. In a lithium-ion battery the lithium ions are transported to and from the cathode or anode, with the transition metal, cobalt (Co), in Li$_x$CoO$_2$ being oxidized from Co^{3+} to Co^{4+} during charging, and reduced from Co^{4+} to Co^{3+} during discharge.

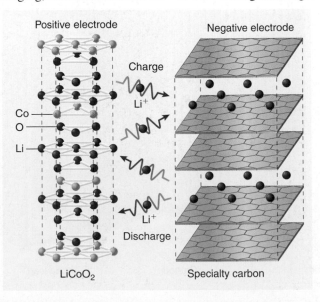

FIGURE 11.12
The cell voltage in a lithium battery is generated by moving the lithium between a lattice site in LiCoO$_2$ and an intercalation position between sheets of graphite.

11.13 Fuel Cells

The primary advantage of fuel cells over batteries is that they can be continually refueled and do not require a downtime for recharging. A number of different technologies are used in fuel cells. Most are still in the research and development stage, and only phosphoric acid fuel cells are available as off-the-shelf technology. Figure 11.13 shows a number of different types of fuel cells. All use O_2 or air as the oxidant, and the fuel can be either pure H_2 or H_2 produced on board by reforming hydrocarbons, methanol, or glucose. The different operating temperature and available power for the types of fuel cells shown in Figure 11.13 are shown in Figure 11.14.

The following discussion is restricted to the most widely understood technology, the **proton exchange membrane fuel cell** (**PEMFC**) using $H_2(g)$ as a fuel.

Proton exchange membrane fuel cells using H_2 and O_2 as the reactants were originally used in the NASA *Gemini* space flights of the 1960s. The principles underlying this technology have not changed in the intervening years. However, advances in technology have significantly increased the power generated per unit weight as well as the power generated per unit area of the electrodes. A schematic drawing of a single PEMFC and the relevant half-cell reactions is shown in Figure 11.15. The anode and cathode have channels that transport H_2 and O_2 to the catalyst and membrane. An intermediate diffusion layer ensures uniform delivery of reactants and separation of gases from the water formed in the reaction. The membrane is coated with a catalyst (usually Pt or Pt alloys) that facilitates the H^+ formation and the reaction between O_2 and H^+. Individual units can be arranged back to back to form a stack so that the potential across the stack is a multiple of the individual cell potential.

The heart of this fuel cell is the proton exchange membrane, which functions as a solid electrolyte. This membrane facilitates the passage of H^+ from the anode to the cathode in the internal circuit, and it also prevents electrons and negative ions from moving in the opposite direction. The membrane must be thin ($\sim 10 - 100 \; \mu m$) to allow rapid charge transport at reasonably high current densities and must be unreactive under the potentials present in the cell. The most widely used membranes are polymeric forms of perfluorosulfonic acids. These membranes are quite conductive if fully hydrated. The membrane structure consists of spherical cavities, or inverted micelles ~4 nm in diameter, connected by cylindrical channels ~1 nm in diameter. The interior of the cavities and channels is lined with SO_3^- groups. The $-CF_2-$ backbone of the polymer provides rigidity, makes the membrane unreactive, and increases the

FIGURE 11.13

Working principles of some fuel cells: EFC, enzymatic fuel cell; AFC, alkaline fuel cell; DMFC, direct methanol fuel cell; PEFC, polymer electrolyte fuel cell; PAFC, phosphoric acid fuel cell; MCFC, molten carbonate fuel cell; and SOFC, solid oxide fuel cell. (Legend of components: A, anode; E, electrolyte; PEM proton exchange membrane and C, cathode (YSZ) yttria-stabilized zirconia.)

Source: Sundmacher, K. Fuel Cell Engineering: Toward the Design of Efficient Electrochemical Power Plants. *Industrial Engineering Chemical Research 49* (2010): 10159–10182, fig 4.

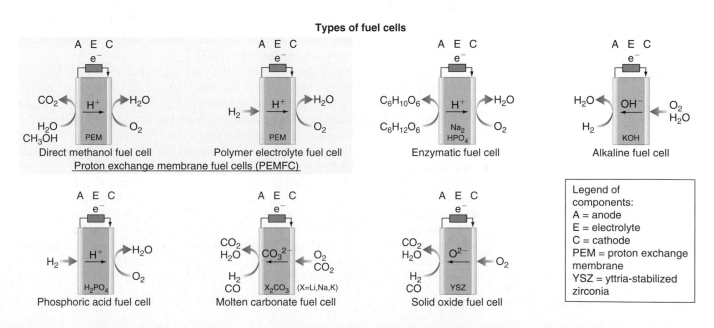

Types of fuel cells

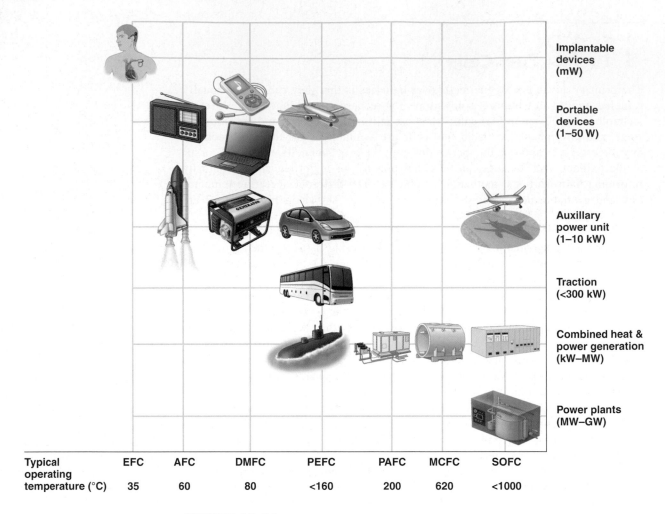

Typical operating temperature (°C)	EFC	AFC	DMFC	PEFC	PAFC	MCFC	SOFC
	35	60	80	<160	200	620	<1000

FIGURE 11.14

The optimal temperature of operation and the application potential of the different types of fuels cells is shown.

Source: Fuel Cell Engineering: Toward the Design of Efficient Electrochemical Power Plants. Sundermacher, K., *Industrial Engineering Chemical Research*, *49* (2010): 10159–10182, fig 5.

acidity of the terminal SO_3^- groups. The positively charged H^+ can migrate through this network under the influence of the electrical potential gradient across the membrane that arises through the half-cell reactions. By contrast, negatively charged species in a cavity cannot migrate through the network, because they are repelled by the negatively charged SO_3^- groups that are covalently bonded to long polymer chains in the narrow channels that connect adjacent cavities.

The electrochemistry of the two half-cell reactions shown in Figure 11.15 is well understood. The main challenges in manufacturing a fuel cell based on the overall reaction $H_2 + 1/2\,O_2 \rightarrow H_2O$ lie in maintaining a high cell potential at the necessary current output. A number of strategies have been pursued to improve the performance of proton exchange membrane fuel cells. Optimizing the interface between the gas diffusion element and the membrane has been the most important area of progress. The use of porous and finely divided materials allows the electrode surface area to be maximized per unit volume of the fuel cell. Careful integration of the catalyst and membrane surface also allows the amount of the expensive Pt-based catalysts to be minimized. A future goal is to develop operating conditions that do not require highly pure H_2. Currently, high purity is required because trace amounts (~50 ppm) of CO poison the Pt catalyst surface and render it inactive for H_2 dissociation. Fuel cells that operate at temperatures above 100°C are much

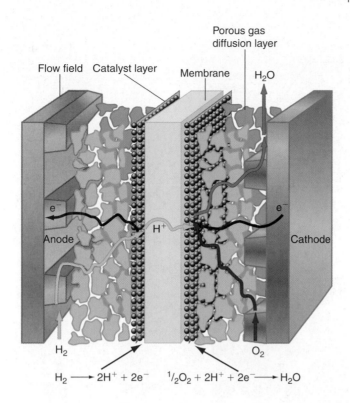

Porous gas
diffusion layer

Flow field   Catalyst layer   Membrane H_2O

e^- e^-

H^+

Anode   Cathode

H_2 O_2

$H_2 \longrightarrow 2H^+ + 2e^-$ $\frac{1}{2}O_2 + 2H^+ + 2e^- \longrightarrow H_2O$

FIGURE 11.15

Schematic diagram of a proton exchange membrane fuel cell. The half-cell reactions are shown. The channels in the anode and cathode facilitate the supply of O_2 and H_2 to the cell and carry away the H_2O reaction product. The gas diffusion layer ensures that the reactants are uniformly distributed over the membrane surface.

less sensitive to poisoning by CO. However, operation at these temperatures requires proton exchange membranes that use ionic liquids other than water to maintain a high conductivity.

The goals of much current research on fuel cells are to develop inexpensive catalysts not based on noble metals and using fuels other than pure hydrogen. Because gaseous H_2 has a low energy density per unit volume, it is not an economically useful fuel unless it is produced at the site of the fuel cell using renewable techniques. Reforming fuels such as methane to produce hydrogen or direct methanol oxidation are the most promising avenues to pursue. A recent review of the many types of fuel cells is Sundmacher, K. Fuel Cell Engineering: Toward the Design of Efficient Electrochemical Power Plants. *Industrial Engineering Chemical Research 49* (2010): 10159–10182.

SUPPLEMENTAL

11.14 Electrochemistry at the Atomic Scale

The half-cell reaction $Cu^{2+}(aq) + 2e^- \rightarrow Cu(s)$ only describes an overall process. What is known about this reaction at an atomic scale? In addressing this question, we draw extensively on a review article by Dieter Kolb [*Surface Science* 500 (2002), 722–740].

We first describe the structure of the interface between the solution and a metal electrode in an electrochemical cell, which is known as the electrical double layer. To a good approximation, the interface can be modeled as a parallel plate capacitor as depicted in Figure 11.16a. In this figure, the positively charged plate of the capacitor is the metal electrode, and the negative plate is made of negative ions that are surrounded by their solvation shells. The two types of negative ions at the interface are distinguished by the forces that hold them in this region. Specifically bound ions are those that form a chemical bond with one or more of the metal atoms at the surface of the electrode. Examples of specifically bound ions are

FIGURE 11.16

(a) The electrical double layer, showing specifically and nonspecifically adsorbed ions and water molecules, and the inner and outer Helmholtz planes. (b) Variation of the electrical potential in the electrical double layer, which has a thickness ~3 nm. The red solid line passes through the center of a nonspecifically adsorbed ion, and the dashed line passes through the center of a specifically adsorbed ion. (c) The electrochemical cell. The region that is shown greatly magnified in parts (a) and (b) is indicated.

[Kolb, D.M. "An Atomistic View of Electrochemistry." *Surface Science* 500 (2002): 722–740, Figure 1.]

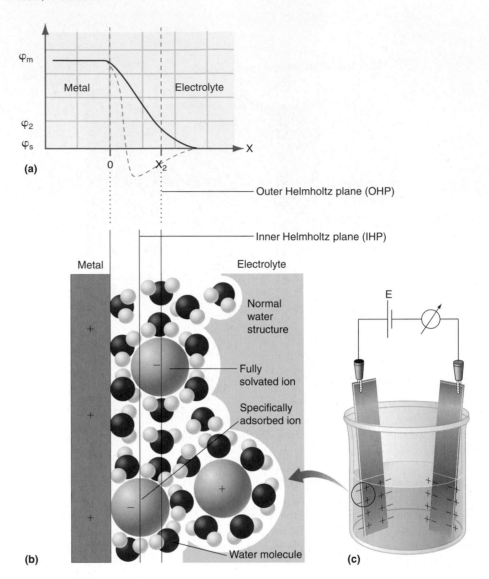

Cl^- and Br^-. It is advantageous for these ions to form bonds with the surface, because the water molecules in their solvation shell are not as strongly bound as in positive ions such as Na^+ and K^+. The part of the solvation shell directed toward the electrode atoms to which the ion is bonded is missing. The plane that goes through the center of the specifically adsorbed ions is known as the inner Helmholtz plane. The second type of ion that is found close to the positively charged electrode consists of fully solvated ions, which are called nonspecifically adsorbed ions. The plane that goes through the center of the nonspecifically adsorbed ions is known as the outer Helmholtz plane. Fully solvated positive ions are found outside the electrical double layer.

As shown in Figure 11.16b, the falloff in the electrical potential between the metal electrode and the electrolyte solution occurs in a very small distance of approximately 3 nm, resulting in an electrical field in the electrical double layer as large as 3×10^7 V cm^{-1}. Because the charge that can be accommodated at the surface of the electrode by the specifically and nonspecifically adsorbed ions is on the order of 0.1–0.2 electron per atom of metal at the surface, the electrical double layer has a very large capacitance of 20–50 μC cm^{-2}. This property makes the metal-electrolyte interface useful in supercapacitors, which can provide energy storage at a very high energy density. For a typical difference in potential between the electrode and the solution of ~1 V, the energy stored in the interfacial capacitance can be as large as 150 kJ kg^{-1}. Because this density is so high, electrochemical capacitors can be used to provide backup electrical energy in the event of a power failure in electronic devices. Note in Figure 11.16b that the electrical potential has fallen to the value in the middle of the solution at a distance

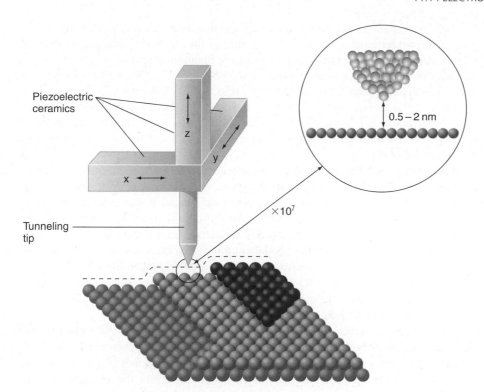

FIGURE 11.17
Schematic picture of an STM. The piezo elements labeled *x* and *y* are used to scan the tip parallel to the surface, and the *z* piezo element is used to vary the tip–surface distance. The inset shows a greatly magnified image of the region between the end of the tip and the surface. The dashed line is a cut through a contour map of the surface resulting from the measurement.

of approximately 5 nm. All important aspects of an electrochemical reaction occur in this region immediately adjacent to the electrode surface.

Having described the structure of the solution adjacent to the electrode surface, what can be said about the structure of the electrode surface at an atomic scale? Much of what is known about the structure of the electrode surface at an atomic scale has been obtained using the scanning tunneling microscope (STM) discussed in Chapter 16, and the reader is encouraged to review this material before proceeding. In an STM, a metal tip is positioned within 0.5–2 nm of a conducting surface, and at these distances, electrons can tunnel between the tip and surface. The tunneling current falls off exponentially with the tip–surface distance. Using piezoelectric elements, the tip height is varied as the tip is scanned over the surface to keep the tunneling current at a constant value, typically 1 nA. The voltage applied to the piezoelectric element normal to the surface is a direct measure of the height of the surface at a point parallel to the surface. These voltage values are used to construct a topographical map of the surface. A schematic picture of the essential elements of an STM is shown in Figure 11.17.

In an electrochemical STM, the tip is coated with an insulating material except in the immediate vicinity of the tunneling region, as illustrated in Figure 11.18, to avoid current passing from the tip to the surface through the conducting electrolyte solution.

The structure of an electrode surface in an electrochemical environment depends on the possible reactions that the electrode can undergo in the solution and the applied potential. For a material such as platinum, which is not easily oxidized, it is possible

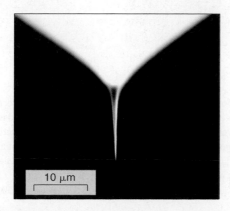

FIGURE 11.18
Very sharp STM tips can be prepared using electrochemical etching. In an electrochemical STM, the shank of the tip is coated with an insulating material to suppress ohmic conduction between the tip and surface through the electrolyte solution.
[Courtesy of D. M. Kolb, University of Ulm.]

by careful preparation to obtain an electrode that is virtually identical to what is expected from terminating the crystal structure of the electrode material. An example of this ideal case is shown in Figure 11.19a for a platinum electrode surface. The great majority of Pt atoms are in their ideal lattice positions. A much smaller number are located at the steps that form the edges of large terraces that make up the long-range structure of the surface. A detail showing the arrangement of the Pt atoms at the edge of a terrace is shown in Figure 11.19b. Although the step atoms are a small minority of the total number of surface atoms, they play a significant role in electrochemical processes, as shown later.

For the case of the platinum electrode, the surface corresponds to the most densely packed plane of the face-centered cubic lattice, which has the lowest surface energy. What is the atomic level structure of an electrode surface if the surface plane is not that of lowest surface energy? An example for this case is shown in Figure 11.20. The gold surface consists of a square array of Au atoms rather than the lowest energy close-packed layer, which has hexagonal symmetry. To minimize its energy, the topmost surface layer undergoes a reconstruction to a close-packed layer with the same structure as shown in Figure 11.19b. Because the hexagonal surface layer and the underlying square lattice do not have the same symmetry, the topmost layer of the electrode is periodically buckled.

The process in which the surface of the electrode undergoes the reconstruction described earlier can be imaged with the electrochemical STM, as shown in Figure 11.21. The light areas in this image are gold islands, around which the reconstruction must detour. Because of the square symmetry of the underlying layers, the reconstruction proceeds in two domains along two directions oriented at 90° with respect to one another. Note the defects in the reconstructed layer that arise from a meandering of a reconstructed part of a domain, defects that arise through the intersection of two domains of different orientation, as well as defects due to the Au islands present on the surface.

We now return to the influence of surface defects in electrochemical reactions. If an Au electrode is immersed in a $CuSO_4$ solution, and the potential is adjusted appropriately, Cu will be deposited on the electrode, as described by the reaction $Cu^{2+}(aq) + 2e^- \rightarrow Cu(s)$. How does the Cu layer grow on the Au surface? The answer is provided by the STM image shown in Figure 11.22. The initial stage of the Cu film is the formation of small Cu islands that are exclusively located at step edges of the underlying Au surface. This result can be understood by realizing that

10 nm × 10 nm

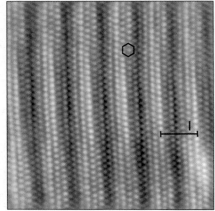

$E_{SCE} = -0.2$ V

FIGURE 11.20
A 10-nm × 10-nm image of a well-prepared gold electrode. Each dot corresponds to an individual Au atom. The topmost layer undergoes a reconstruction to form a close-packed layer of hexagonal symmetry as indicated by the hexagon. Because this hexagonal layer is not in registry with the underlying square lattice, the surface plane is buckled with the repeat unit labeled *l*. At the bottom, E_{SCE} indicates that the potential is measured relative to a standard calomel electrode, for which the half-cell reaction is
$Hg_2Cl_2(s) + 2e^- \rightarrow$
$\qquad 2 Hg(l) + 2 Cl^-(aq)$.
[Kolb, D.M. "An Atomistic View of Electrochemistry." *Surface Science* 500 (2002): 722–740, Figure 6.]

(a)   1000 nm × 1000 nm   **(b)**

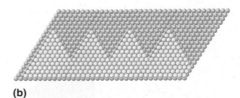

FIGURE 11.19
(a) A 1000-nm × 1000-nm STM image of a well-prepared platinum electrode exposing a close-packed layer of Pt atoms in the surface. The image shows several stacked terraces, which differ in height by one atomic layer. (b) Ball model showing the atomic layer structure underlying the image of part (a). Only step edges in which the step edge of a terrace consists of a close-packed row of atoms is observed in part (a). Note the lateral offset in successively higher terraces, which are characteristic of the stacking in planes of a face-centered cubic lattice.
[Part (a) Kolb, D.M. "An Atomistic View of Electrochemistry." *Surface Science* 500 (2002): 722–740, Figure 3.]

80 nm × 80 nm

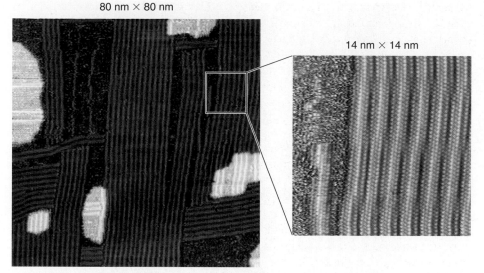

$E_{SCE} = -250$ mV

14 nm × 14 nm

FIGURE 11.21
STM images of a gold electrode in which the reconstruction of the topmost layer is not complete. The light areas are higher-lying Au islands. Note that the reconstruction proceeds along two perpendicular directions. This occurs because the underlying lattice structure has square symmetry. [Kolb, D.M. "An Atomistic View of Electrochemistry." *Surface Science* 500 (2002): 722–740, Figure 7.]

the coordination number of a Cu atom at a gold site at the step edge is larger than on a flat portion of the surface. As more copper is deposited, the initially formed islands grow laterally and eventually merge to form a uniform layer.

Several important conclusions can be made concerning the growth of an electrochemically deposited metal film from these studies. The fact that island growth is observed rather than random deposition of Cu atoms indicates that the initially adsorbed Cu atoms have a high mobility parallel to the surface. They diffuse across a limited region of the surface until they encounter the edge of an island. Because these edge sites have a higher coordination number than a site on the terrace, they are strongly bound and diffuse no further. The other conclusion that can be drawn is that the metal film grows one layer at a time. A second layer is not nucleated until the underlying layer is nearly completed. Layer-by-layer growth ensures that a compact and crystalline film is formed.

As the deposition of Cu on a gold electrode shows, step edges play a central role in determining the formation of electrochemically deposited layers. Although small in number, surface defects such as step edges or vacancies play a major role in the chemistry of a wide range of chemical processes including catalytic cracking of crude oil and the corrosion of metals.

Electrochemical scanning tunneling microscopy has also provided new insight into the atomic level processes underlying the electrochemical dissolution of a metal electrode. To carry out these studies, an STM was constructed that allowed 25 images to be obtained per second. A comparison of images obtained at closely spaced times allows researchers to follow the ongoing processes [Magnussen O. et al. *Electrochimica Acta* 46 (2001):3725–3733]. A mode of growth was observed in which individual atoms were added to a single row at the edge of a terrace as shown in Figure 11.23.

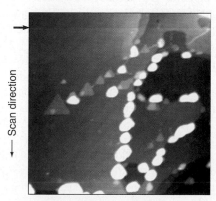

← Scan direction

170 nm × 170 nm

FIGURE 11.22
Image of a gold electrode in which a small amount of copper has been electrochemically deposited. Note that the Cu is deposited in the form of small islands that are anchored at step edges of the underlying gold electrode. The onset of the deposition is initiated by the horizontal arrow in the figure. At that point, the electrode potential was increased to a value that allows the reaction $Cu^{2+}(aq) + 2e^- \rightarrow Cu(s)$ to proceed. [Kolb, D.M. Engelmann, G.E., Ziegler, J.C. "On the Unusual Electochemical Stability of Nanofabricated Copper Clusters." *Angewandte Chemie International Edition* 39 (2000): 1123–1125.]

FIGURE 11.23
A series of images taken at the indicated time intervals is shown as Cu is deposited electrochemically on a copper electrode. Note that an individual row at the edge of the upper terrace grows from the top to the bottom of the image.

[Photo courtesy of R. J. Behm, University of Ulm. Magnussen et al., "In-Situ Atomic-Scale Studies of the Mechanisms and Dynamics of Metal STM." *Electrochemica Acta* 46 (2001): 3725–3733, Figure 1.]

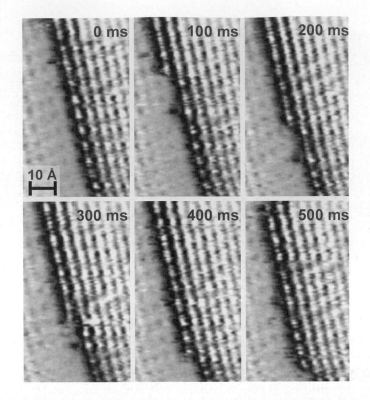

FIGURE 11.24
The arrows indicate examples of a concerted removal of whole or multiple steps for a copper electrode: (a) dissolution and (b) growth. The times at which the images were acquired relative to the first image are shown.

[Photo courtesy of R. J. Behm, University of Ulm. Magnussen et al., "In-Situ Atomic-Scale Studies of the Mechanisms and Dynamics of Metal STM." *Electrochemica Acta* 46 (2001); 3725–3733, Figure 2.]

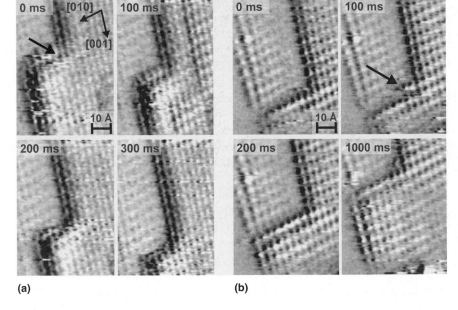

(a)　　　　　　(b)

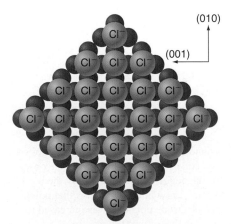

FIGURE 11.25
Because they are larger, the Cl^- ions cannot form the same structure as the underlying Cu surface. Instead, they form a square lattice that is rotated by 45° with respect to the copper unit cell.

However, a second mode of growth and dissolution is observed in which entire segments of a row can grow or dissolve collectively. By adjusting the electrode potential slightly, either Cu growth or dissolution can be initiated. Figure 11.24a shows a sequence of images for electrode dissolution, and Figure 11.24b shows a sequence for Cu deposition on the electrode. In both cases, entire rows or row segments are deposited at once rather than in the atom-by-atom fashion depicted in Figure 11.23.

The Cu electrode has a square symmetry in this case, and one might assume that growth and dissolution occur with equal probability in either of two directions oriented 90° to one another. However, the images in Figure 11.24 indicate that this is not the case. Growth or dissolution occurs only in the crystallographic direction labeled (001) and not along the direction labeled (010). This behavior is a consequence of the atomic level structure of the Cl^- specifically adsorbed ions of the HCl electrolyte. This structure is shown in Figure 11.25, and consists of a square lattice that has a larger unit cell length than the underlying Cu lattice and is rotated 45° with respect to the Cu lattice.

The adsorbed Cl^- layer removes the equivalence of the (010) and (001) directions as is shown in Figure 11.26. The kinks formed along step edges along these two directions have a different structure. Therefore, they are expected to have a different reactivity in the dissolution/process. As these experiments show, specifically adsorbed ions play a central role in an electrochemical reaction.

The ability to verify that an electrode is both crystalline and has a low defect density using STM has made the interpretation of experimental results obtained through classical electrochemical techniques significantly easier. An example is cyclic voltammetry, in which the electrode potential is varied linearly with time and then the potential is changed in the opposite direction and returned to the initial value. As the potential is varied, the electrode current is measured. Because the area in a plot of current versus time is the electrical charge, cyclic voltammetry provides a way to identify regions of potential in which significant amounts of charge are transferred to or from the electrode. These are the regions in which electrochemical reactions proceed. An example of a cyclic voltammogram for the deposition of less than one monolayer of Cu on a gold electrode is shown in Figure 11.27.

STM studies show that Cu can form a low-density ordered structure on the surface that has a saturated coverage of 0.67 monolayer, and a monolayer structure that has the same structure as the underlying Au electrode. Metal deposition begins to occur at a potential of 0.4 V and the surface is fully covered as the potential is lowered to 0 V. Deposition corresponds to a positive current in Figure 11.27 (the scan direction is from right to left in the figure), and dissolution corresponds to a negative current (left to right in the figure). Two well-defined peaks are observed in both the growth and dissolution directions. The emf associated with these peaks can be used to calculate the ΔG_f° associated with the formation of a chemically bonded Cu atom at the site on the Au surface characteristic of the different geometries of the two ordered Cu phases. Both of these peaks occur at a potential less than that needed to deposit copper on a copper electrode, because Cu atoms are more strongly bound to surface sites on the Au surface than on the copper surface. One refers to underpotential deposition to emphasize that the potential is less than that expected for a material deposited on its own lattice. The combination of a structural determination at the atomic level provided by STM with the *I–V* relation provided by cyclic voltammetry is well suited to understanding electrochemical processes at surfaces.

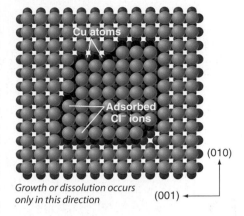

FIGURE 11.26

The kinks formed in step edges along the (010) and (001) directions have a different structure. This explains the different reactivity observed along these directions as shown in Figure 11.24. The small light and dark red spheres denote Cu atoms, and the large green spheres denote specifically adsorbed Cl^- ions.

Source: Reprinted from O.M. Magnussen et al., "In-Situ Atomic-Scale Studies of the Mechanisms and Dynamics of Metal STM," *Electrochimica Acta* 46: 3725–3733 (2001). Copyright 2001, with permission from Elsevier.

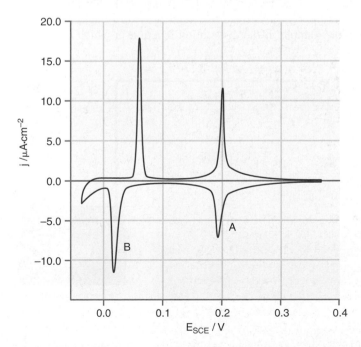

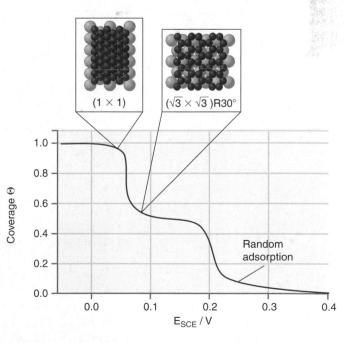

FIGURE 11.27

(a) A cyclic voltammogram is shown for the deposition/dissolution of Cu on a well-ordered single-crystal gold electrode. By measuring the area under the peaks, the total amount of Cu deposited or dissolved can be determined. This result is plotted in part (b). The amount by which peaks A and B are displaced in voltage in the dissolution process from their values in the deposition process is a measure of the rate of electron transfer in the processes. Cu and Au atoms are colored red and orange, respectively.

[Kolb, D.M. "An Atomistic View of Electrochemistry." *Surface Science* 500 (2002): 722–740, Figure 11.]

11.15 Using Electrochemistry for Nanoscale Machining

Many current technologies require fabrication of miniature devices with dimensional control in the micron to nanometer scale. For example, lithographic methods are used to make the masks employed in the fabrication of integrated circuits. Until recently, electrochemical machining techniques have not been able to provide dimensional control better than 10 μm for a reason related to the capacitance of the electrical double layer. If the external voltage is changed in an electrochemical cell, how long does it take for the cell to reach equilibrium with respect to the potential distribution? As discussed in the previous section, the electrical double region can be modeled as a large capacitor that must be charged up to the value of the new potential as the external voltage is changed. The current must flow through the electrolyte solution, which has an ohmic resistance, and the cell can be modeled as a series R-C circuit with R equal to the electrical resistance of the solution, and C equal to the capacitance of the electrical double layer.

If a metal tool is placed very close to an electrode, the resistance associated with a current path between the tool and the electrode depends on the path, as shown in Figure 11.28. Those paths immediately below the tool correspond to those of smallest electrical resistance. It can be shown that the time-dependent voltage across the capacitor in an R-C series circuit is given by

$$V(t) = V_0\left(1 - e^{-t/RC}\right) \tag{11.65}$$

The rate at which V/V_0 increases for a point on the surface directly below the tool and a point well to one side of the tool is shown in Figure 11.29. Below the tool, the potential is built up to its full value at much shorter times than outside of this region. Until the potential is built up to the value required for an electrochemical reaction of interest, the reaction will not proceed. Rolf Schuster and coworkers [*Electrochimica Acta* 48 (2003), 20] have used this fact to supply the external voltage in the form of a short pulse, rather than as a dc voltage. As a consequence, the reaction of interest occurs directly under the tool, and only to a negligible extent outside of this region. For example, if the dissolution reaction in which material is

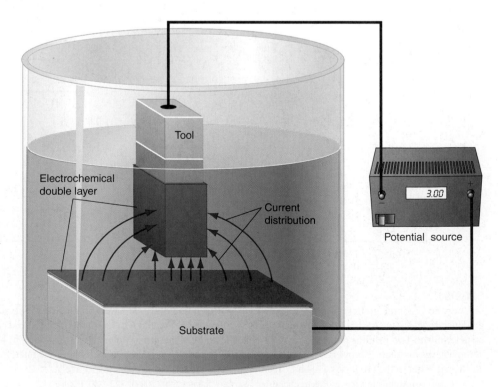

FIGURE 11.28

Possible current paths between a tool and an electrode immersed in an electrolyte solution. Before a reaction can proceed, the capacitance of the electrical double layer must be built up to its full value.

Source: Reprinted with permission from A.L. Trimmer et al., "Single-Step Electochemical Machining of Complex Nanostructures with Ultrashort Voltage Pulses," *Applied Physics Letters* 82 (19): 3327 (2003). Copyright 2003 American Institute of Physics.

removed from the work piece only takes place at potentials higher than that indicated by the horizontal bar in Figure 11.29, it will only take place under the tool for a short enough pulse. Note that if a dc potential is applied instead of a pulse, the reaction will not be localized.

Examples of the patterns that have been generated in this way by metal dissolution are shown in Figure 11.30. A resolution of ~20 nm has been obtained by using pulses of 200-ps duration.

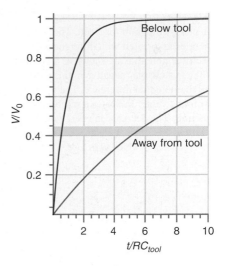

FIGURE 11.29
The ratio V/V_0 is shown as a function of t/RC_{tool} for a point immediately below the tool, and away from the tool at a point where the resistance along the path is higher by a factor of 10.

11.16 Absolute Half-Cell Potentials

The cell potentials listed in Tables 11.1 and 11.2 (see Appendix B, Data Tables) are all measured relative to the standard hydrogen electrode. Because an electrochemical cell consists of two half-cells, any common reference for the half-cell potentials cancels when calculating the cell potential. For this reason, equilibrium constants, activity coefficients, and the direction of spontaneous change can all be determined without absolute half-cell potentials. However, there is no reason why a half-cell potential cannot be determined in principle. The difficulty is in devising an appropriate experiment to measure this quantity. Recall that putting one lead of a voltmeter on the electrode and the second lead in the solution will not give the half-cell potential as discussed in Section 11.1. Although not necessary for discussing electrochemical reactions, absolute half-cell potentials are useful in formulating a microscopic model of electrochemical processes. For example, absolute values for half-cell potentials allow the solvation Gibbs energy of the ion involved in the redox reaction to be calculated directly, rather than to rely on a calculation as outlined in Section 10.2. We next describe a method for the determination of half-cell potentials that was formulated by R. Gomer and G. Tryson [*J. Chem. Phys.* 66 (1977), 4413–4424]. The following discussion assumes familiarity with the particle in a box model of the conduction electrons in a metal (see Chapters 15 and 16).

Consider the electrochemical cell shown schematically in Figure 11.31. The conduction electrons in each metal electrode are described using the particle in a box model, and all energy levels up to the highest occupied level, called the Fermi level, are filled. The vertical distance corresponds to energy for a negative test charge, and the horizontal axis corresponds to distance. The double layer acts as a capacitor and shifts the potential

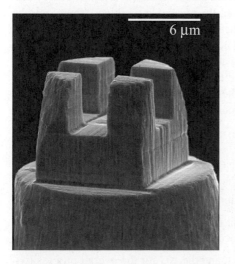

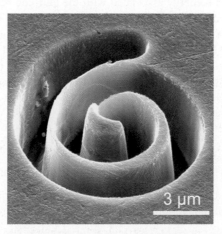

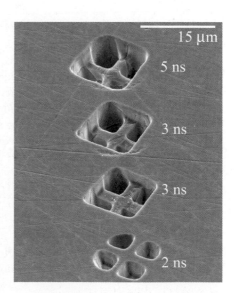

FIGURE 11.30
Results obtained using electrochemical machining of nickel in an HCl solution using the tool shown in the left image. Note the dependence of the resolution on pulse length in the middle image.
[Left and middle images reprinted with permission from Trimmer A.L. et al. *Applied Physics Letters* 82 (2003):3327–3329; right image reprinted with permission from Kock, M. et al. *Electrochemica Acta* 48, 20–22, 2003:3213–3219.]

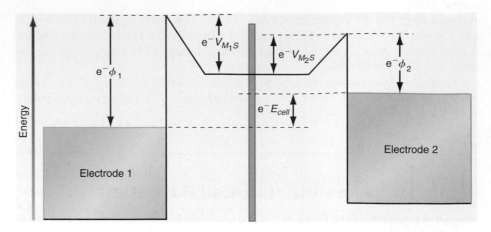

in the solution relative to the Fermi level by the amount $e^-V_{M_1S}$ or $e^-V_{M_2S}$ as shown in Figure 11.31. The change in potential within the double layer is depicted as being linear rather than as shown correctly in Figure 11.14. To remove an electron from the metal, it is necessary to give an electron at the Fermi level an amount of energy equal to the work function ϕ. The **work function** of a metal is analogous to the ionization energy of an atom, and different metals have different values of the work function. Because of the presence of the double layer at each electrode, a negatively charged ion or an electron has a lower energy in the bulk of the solution than at the surface of an electrode.

How do the two half-cells line up in an energy diagram? This question can be answered by recognizing that the electrochemical potential of an electron in a metal is the Fermi energy. If two metals are placed in contact, electrons flow from one metal to the other until the electrochemical potential of an electron is the same in each metal. This is equivalent to saying that the Fermi levels of the two electrodes lie at the same energy. The equilibrium

$$M^{n+}(aq) + ne^- \rightarrow M(s) \tag{11.66}$$

that is established separately in each half-cell determines the drop in electrical potential across the double layer and thereby shifts the Fermi level of each metal electrode relative to the energy of the bulk solution. Therefore, the Fermi levels of the two electrodes are exactly offset by the Gibbs energy:

$$\Delta G = -e^-E_{cell} \tag{11.67}$$

How can an absolute half-cell potential be defined that is consistent with the energy diagram in Figure 11.31? To answer this question, the reversible work and, therefore, the Gibbs energy associated with the half-cell reaction $M^{n+}(aq) + ne^- \rightarrow M(s)$ is calculated. For simplicity, it is assumed that $n = 1$. Because G is a state function, any convenient path between the initial states of the metal atom in the solid and the final state of the ion in the bulk solution and the electron at the Fermi level of the electrode can be chosen. Breaking this overall process down into the individual steps shown in the following equations simplifies the analysis. The ΔG associated with each step is indicated.

$$M_1(s) \rightarrow M_1(g) \qquad\qquad\qquad \Delta G = \Delta G_{vaporization} \tag{11.68}$$

$$M_1(g) \rightarrow M_1^+(g) + e^-(M_1) \qquad \Delta G = e^-(I_1 - \phi_1) \tag{11.69}$$

$$M_1^+(g) \rightarrow M_1^+\,(aq, \text{at } M_1) \qquad\quad \Delta G = \Delta G_{solvation} \tag{11.70}$$

$$M_1^+(aq, \text{at } M_1) \rightarrow M_1^+(aq, \text{bulk solution}) \quad \Delta G = e^-V_{M_1S} \tag{11.71}$$

Note that in the final step, the solvated ion at the electrode is transferred through the double layer into the solution. The overall process is

$M_1(s) \rightarrow M_1^+(aq, \text{bulk solution}) + e^-(M_1)$, and the ΔG for the overall process is

$$\Delta G_1 = \Delta G_{vaporization, 1} + e^-(I_1 - \phi_1) + \Delta G_{solvation, 1} + e^-V_{M_1S} \tag{11.72}$$

The analogous equation for electrode 2 is

$$\Delta G_2 = \Delta G_{vaporization, 2} + e^-(I_2 - \phi_2) + \Delta G_{solvation, 2} + e^-V_{M_2S} \tag{11.73}$$

If the two half-cells are in equilibrium, $\Delta G_1 = \Delta G_2$ and

$$(\Delta G_{vaporization,1} - \Delta G_{vaporization,2}) + (\Delta G_{solvation,1} - \Delta G_{solvation,2}) + e^-(I_1 - I_2)$$
$$= e^-(V_{M_2S} - \phi_2) - e^-(V_{M_1S} - \phi_1) = e^- E_{cell} \qquad \textbf{(11.74)}$$

Equating $e^-(V_{M_1S} - \phi_1) - e^-(V_{M_2S} - \phi_2)$ with $e^- E_{cell}$ follows from the energy diagram shown in Figure 11.31.

Equation (11.74) shows that a plausible definition of each of the half-cell potentials in terms of the physical parameters of the problem is given by

$$E_1 = V_{M_1S} - \phi_1 \quad \text{and} \quad E_2 = V_{M_2S} - \phi_2 \qquad \textbf{(11.75)}$$

Note that the half-cell potential is not the difference in the electrical potential between the electrode and the bulk solution. Only if the potential is defined in this way will $E = E_1 + E_2$, as can be seen in Figure 11.31.

Although Equation (11.75) defines an absolute half-cell potential, numerical values for the half-cell can only be determined if V_{MS} and ϕ are known. Note that the work function in Equation (11.75) is not the work function in air or in vacuum, but the work function in solution. The value of ϕ will vary for these different environments because of molecular layers absorbed on the metal surface. As discussed in Section 11.2, V_{MS} cannot be measured directly using a voltmeter and a probe inserted in the solution. However, V_{MS} can be measured without contacting the solution using the experimental setup shown schematically in Figure 11.32.

The electrode of interest is immersed in the electrolyte solution. A reference electrode (in this case gold) in the form of a flat plate is oriented parallel to the surface of the electrolyte solution. Gold is chosen because the work function in air is known. The Au and metal electrodes are connected by a metal wire, and a variable dc voltage source is inserted in the wire. The gold electrode is mounted in such a way that its distance to the electrolyte surface can be varied in a periodic fashion. The energy diagram shown in Figure 11.33 is used to understand how the experimental setup can be used to measure V_{MS}. By connecting the electrode in the solution with the gold electrode in air, their Fermi levels are equalized as shown in Figure 11.33a. Note that there is no double-layer potential on the electrode in air. The difference in the electrical potential of the gold electrode and the surface of the electrolyte surface is designated as V_{M_2Au}.

Because of the connecting wire shown in Figure 11.32, the Au electrode and the electrochemical double layer of the immersed electrode form a capacitor that charges up because of the potential difference between the two plates that arises through equalizations of the Fermi levels in the electrodes. First consider the case in which the adjustable voltage in Figure 11.32 is set equal to zero. The potential difference between the surface of the electrolyte and the parallel gold electrode is shown in Figure 11.33a. As the Au plate is vibrated, the capacitance changes and, therefore, an alternating current flows in the external circuit as the capacitor is alternately charged and discharged. Next carefully adjust the variable voltage in the wire as the Au electrode vibrates until no current can be detected in the external circuit. If no current flows, the voltage difference between the Au electrode and the electrolyte surface must be equal to zero. This case corresponds to the energy diagram of Figure 11.33b. The null current must correspond to the condition

$$\Delta V = V_{M_2Au} \qquad \textbf{(11.76)}$$

Note that in this experiment, V_{M_2Au} can be determined experimentally without inserting a probe into the liquid. Because ϕ_{Au} is known, $E_2 = V_{M_2Au} - \phi_{Au}$ can be calculated, and therefore the absolute half-cell potential can be determined.

Using this method, Gomer and Tryson first determined the absolute half-cell potential of the standard hydrogen electrode. More recent measurements use the more reliable value of the work function of the Hg surface of 4.50 eV to obtain the value $E^\circ_{SHE} = -4.44 \pm 0.02$ V [S. Trassatti, Pure & Applied Chemistry, 58 (1986) 955]. The absolute half-cell potentials of all other half-cell potentials can be determined by adding -4.44 to the values in Tables 11.1 and 11.2 (see Appendix B, Data Tables).

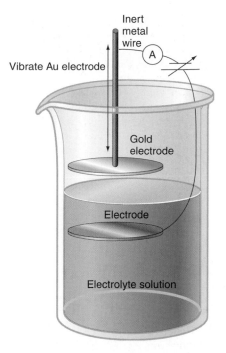

FIGURE 11.32
Apparatus used to determine absolute half-cell potentials.

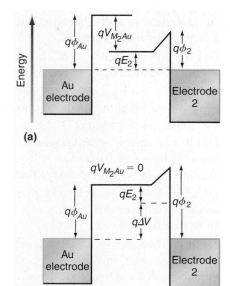

FIGURE 11.33
These energy diagrams are used to explain how the absolute half-cell potential can be determined using the experiment illustrated in Figure 11.32. (a) The variable voltage in series with the wire is set equal to zero. (b) The variable voltage is adjusted so that the current flow in the external circuit is equal to zero.

Once the absolute half-cell potential is known, the important (and otherwise experimentally inaccessible) quantity $\Delta G_{solvation}$ can be calculated. From Equations (11.68) through (11.71), the relation between the half-cell potential and the energetic parameters in these equations is

$$E_{half\text{-}cell} = -\frac{1}{e^-}(\Delta G_{vaporization} + \Delta G_{solvation} + Ie^-) \qquad \textbf{(11.77)}$$

This equation allows $\Delta G_{solvation}$ of an individual ion to be determined from the appropriate half-cell potential. The solvation Gibbs energy calculated in this way is in good agreement with the values calculated from the Born model if the ionic radii obtained from crystal structures are corrected for the water of solvation as discussed in Section 10.2.

Vocabulary

anode

battery

cathode

charged

Daniell cell

electrochemical cell

electrochemical potential

electrochemical series

electromotive force (emf)

Faraday constant

fuel cell

junction potential

Nernst equation

neutral

proton exchange membrane fuel cell (PEMFC)

reference electrode

salt bridge

work function

Conceptual Problems

Q11.1   To determine standard cell potentials, measurements are carried out in very dilute solutions rather than at unit activity. Why is this the case?

Q11.2   Show that if $\Delta G_f^\circ(H^+, aq) = 0$ for all T, the potential of the standard hydrogen electrode is zero.

Q11.3   How is it possible to deposit Cu on an Au electrode at a potential lower than that corresponding to the reaction $Cu^{2+}(aq) + 2e^- \rightarrow Cu(s)$?

Q11.4   Explain why the magnitude of the maximum work available from a battery can be greater than the magnitude of the reaction enthalpy of the overall cell reaction.

Q11.5   How can one conclude from Figure 11.22 that Cu atoms can diffuse rapidly over a well-ordered Au electrode in an electrochemical cell?

Q11.6   The temperature dependence of the potential of a cell is vanishingly small. What does this tell you about the thermodynamics of the cell reaction?

Q11.7   Why is the capacitance of an electrolytic capacitor so high compared with conventional capacitors of similar size?

Q11.8   What is the difference in the chemical potential and the electrochemical potential for an ion and for a neutral species in solution? Under what conditions is the electrochemical potential equal to the chemical potential for an ion?

Q11.9   Why is it possible to achieve high-resolution electrochemical machining by applying a voltage pulse rather than a dc voltage to the electrode being machined?

Q11.10   Can specifically adsorbed ions in the electrochemical double layer influence electrode reactions?

Q11.11   Why is it not necessary to know absolute half-cell potentials to determine the emf of an electrochemical cell?

Q11.12   What is the voltage between the terminals of a battery in which the contents are in chemical equilibrium?

Q11.13   By convention, the anode of a battery is where oxidation takes place. Is this true when the battery is charged, discharged, or both?

Q11.14   You wish to maximize the emf of an electrochemical cell. To do so, should the concentrations of the products in the overall reaction be high or low relative to those of the reactants? Explain your answer.

Q11.15   If you double all the coefficients in the overall chemical reaction in an electrochemical cell, the equilibrium constant changes. Does the emf change? Explain your answer.

Q11.16   Why can batteries only be recharged a limited number of times?

Q11.17   Why can more work be extracted from a fuel cell than a combustion engine for the same overall reaction?

Q11.18   What is the function of a salt bridge in an electrochemical cell?

Q11.19   How does the emf of an electrochemical cell change if you increase the temperature?

Q11.20   What thermodynamic quantity that cannot be measured directly can be calculated from absolute half-cell potentials?

Numerical Problems

Problem numbers in red indicate that the solution to the problem is given in the *Student's Solutions Manual*.

P11.1 You are given the following half-cell reactions:

$$Pd^{2+}(aq) + 2e^- \rightarrow Pd(s) \qquad E° = 0.83 \text{ V}$$

$$PdCl_4^{2-}(aq) + 2e^- \rightarrow Pd(s) + 4\,Cl^-(aq) \quad E° = 0.64 \text{ V}$$

a. Calculate the equilibrium constant for the reaction

$$Pd^{2+}(aq) + 4\,Cl^-(aq) \rightleftharpoons PdCl_4^{2-}(aq)$$

b. Calculate $\Delta G°$ for this reaction.

P11.2 For the half-cell reaction $AgBr(s) + e^- \rightarrow Ag(s) + Br^-(aq)$, $E° = +0.0713$ V. Using this result and $\Delta G_f°(AgBr, s) = -96.9$ kJ mol^{-1}, determine $\Delta G_f°(Br^-, aq)$.

P11.3 For the half-cell reaction $Hg_2Cl_2(s) + 2e^- \rightarrow 2Hg(l) + 2Cl^-(aq)$, $E° = +0.27$ V. Using this result and $\Delta G_f°(Hg_2Cl_2, s) = -210.7$ kJ mol^{-1}, determine $\Delta G_f°(Cl^-, aq)$.

P11.4 Determine the half-cell reactions and the overall cell reaction, calculate the cell potential, and determine the equilibrium constant at 298.15 K for the cell

$$Cd^{2+}(aq, a_{Cd^{2+}} = 0.150)|Cd(s)$$
$$|Cl^-(aq, a_{Cl^-} = 0.0100)|Ag(s)|AgCl(s)$$

Is the cell reaction spontaneous as written?

P11.5 The standard half-cell potential for the reaction $O_2(g) + 4\,H^+(aq) + 4e^- \rightarrow 2\,H_2O(l)$ is $+1.229$ V at 298.15 K. Calculate E for a 0.300-molal solution of H_2SO_4 for $a_{O_2} = 1.00$ (a) assuming that the a_{H^+} is equal to the molality and (b) using the measured mean ionic activity coefficient for this concentration from the data tables. How large is the relative error if the concentrations, rather than the activities, are used?

P11.6 Determine the half-cell reactions and the overall cell reaction, calculate the cell potential, and determine the equilibrium constant at 298.15 K for the cell

$$H_2(g)|Pt(s)\,|H^+(aq, a_{H^+} = 0.250)|Cu^{2+}$$
$$(aq, a_{Cu^{2+}} = 0.100)|Cu(s)$$

Is the cell reaction spontaneous as written?

P11.7 Consider the Daniell cell for the indicated molalities: $Zn(s)|ZnSO_4(aq, 0.200\ m)||CuSO_4(aq, 0.400\ m)|Cu(s)$. The activity coefficient $\gamma_\pm$ for the indicated concentrations can be found in the data tables. Calculate E by a) setting the activity equal to the molality and b) by using the correct values for $\gamma_\pm$. How large is the relative error if the concentrations, rather than the activities, are used?

P11.8 Determine the half-cell reactions and the overall cell reaction, calculate the cell potential, and determine the equilibrium constant at 298.15 K for the cell

$$Zn(s)|Zn^{2+}(aq, a_\pm = 0.0120)|$$
$$|Mn^{3+}(aq, a_\pm = 0.200), Mn^{2+}(aq, a_\pm = 0.0250)\,|Pt(s)$$

Is the cell reaction spontaneous as written?

P11.9 Consider the half-cell reaction $AgCl(s) + e^- \rightarrow Ag(s) + Cl^-(aq)$. If $\mu°(AgCl, s) = -109.71$ kJ mol^{-1}, and if $E° = +0.222$ V for this half-cell, calculate the standard Gibbs energy of formation of $Cl^-(aq)$.

P11.10 For a given overall cell reaction, $\Delta S_R° = 16.5$ J mol^{-1}K^{-1} and $\Delta H_R° = -270.0$ kJ mol^{-1}. Calculate $E°$ and $(\partial E°/\partial T)_P$. Assume that $n = 2$.

P11.11 Consider the cell $Fe(s)|FeSO_4(aq, a_\pm = 0.0250)|Hg_2SO_4(s)|Hg(l)$.

a. Write the cell reaction.

b. Calculate the cell potential, the equilibrium constant for the cell reaction, and $\Delta G_R°$ at 25°C.

P11.12 Between 0° and 90°C, the potential of the cell $Pt(s)|H_2(g, f = 1.00\text{ atm})|HCl(aq, m = 0.100)|AgCl(s)|Ag(s)$ is described by the equation $E(V) = 0.35510 - 0.3422 \times 10^{-4}t - 3.2347 \times 10^{-6}t^2 + 6.314 \times 10^{-9}t^3$, where t is the temperature on the Celsius scale. Write the cell reaction and calculate $\Delta G_R°$, $\Delta H_R°$, and $\Delta S_R°$ for the cell reaction at 50.°C.

P11.13

a. Calculate $\Delta G_R°$ and the equilibrium constant K at 298.15 K for the reaction $2\,Hg(l) + Cl_2(g) \rightleftharpoons Hg_2Cl_2(s)$.

b. Calculate K using Table 4.1. What value of ΔG_R would make the value of K the same as calculated from the half-cell potentials?

P11.14 Consider the couple $Ox + e^- \rightarrow Red$ with the oxidized and reduced species at unit activity. What must be the value of $E°$ for this half-cell if the reductant Red is to liberate hydrogen at 1.00 atm from

a. an acidic solution with $a_{H^+} = 2.50$ and $a_{H_2} = 1.00$?

b. a basic solution with pH = 9.00?

c. Is hydrogen a better reducing agent in acid or basic solution? Explain your answer.

P11.15 By finding appropriate half-cell reactions, calculate the equilibrium constant at 298.15 K for the following reactions:

a. $4\,NiOOH(s) + 2\,H_2O(l) \rightleftharpoons 4\,Ni(OH)_2(s) + O_2(g)$

b. $4\,NO_3^-(aq) + 4\,H^+(aq) \rightleftharpoons 4\,NO(g) + 2\,H_2O(l) + 3\,O_2(g)$

P11.16 The cell potential E for the cell $Pt(s)|H_2(g, a_{H_2} = 1.00)|H^+(aq, a_{H^+} = 1.00)||NaCl(aq, m = 0.300)|AgCl(s)|Ag(s)$ is $+0.260$ V. Determine γ_{Cl^-} assuming that $\gamma_\pm = \gamma_{Na^+} = \gamma_{Cl^-}$.

P11.17 The Edison storage cell is described by $Fe(s)|FeO(s)|KOH(aq, a_{KOH})|Ni_2O_3(s)|NiO(s)|Ni(s)$ and the half-cell reactions are as follows:

$$Ni_2O_3(s) + H_2O(l) + 2e^- \rightarrow 2\,NiO(s) + 2\,OH^-(aq)$$
$$E° = 0.40 \text{ V}$$

$$FeO(s) + H_2O(l) + 2e^- \rightarrow Fe(s) + 2\,OH^-(aq)$$
$$E° = -0.87 \text{ V}$$

a. What is the overall cell reaction?

b. How does the cell potential depend on the activity of the KOH?

c. How much electrical work can be obtained per kilogram of the active materials in the cell?

P11.18 Consider the Daniell cell, for which the overall cell reaction is $Zn(s) + Cu^{2+}(aq) \rightleftharpoons Zn^{2+}(aq) + Cu(s)$. The concentrations of $CuSO_4$ and $ZnSO_4$ are $2.50 \times 10^{-3}\ m$ and $1.10 \times 10^{-3}\ m$, respectively.

a. Calculate E setting the activities of the ionic species equal to their molalities.

b. Calculate $\gamma_\pm$ for each of the half-cell solutions using the Debye–Hückel limiting law.

c. Calculate E using the mean ionic activity coefficients determined in part (b).

P11.19 The standard potential E° for a given cell is 1.135 V at 298.15 K and $(\partial E^\circ/\partial T)_P = -4.10 \times 10^{-5}\ \mathrm{V\ K^{-1}}$. Calculate ΔG_R°, ΔS_R°, and ΔH_R°. Assume that $n = 2$.

P11.20 Determine E° for the reaction $Cr^{2+}(aq) + 2e^- \rightarrow Cr(s)$ from the one-electron reduction potential for $Cr^{3+}(aq)$ and the three-electron reduction potential for $Cr^{3+}(aq)$ given in Table 11.1 (see Appendix B).

P11.21 Harnet and Hamer [*J. American Chemical Society* 57 (1935): 33] report values for the potential of the cell $Pt(s)|PbSO_4(s)|H_2SO_4(aq, a)|PbSO_4(s)|PbO_2(s)|Pt(s)$ over a wide range of temperature and H_2SO_4 concentrations. In $1.00\ m\ H_2SO_4$, their results were described by $E(V) = 1.91737 + 56.1 \times 10^{-6}\ t + 108 \times 10^{-8}\ t^2$, where t is the temperature on the Celsius scale. Calculate ΔG_R°, ΔH_R°, and ΔS_R° for the cell reaction at 11° and $35°C$.

P11.22 Consider the reaction $Sn(s) + Sn^{4+}(aq) \rightleftharpoons 2Sn^{2+}(aq)$. If metallic tin is in equilibrium with a solution of $Sn^{2+}(aq)$ in which $a_{Sn^{2+}} = 0.250$, what is the activity of $Sn^{4+}(aq)$ at equilibrium at 298.15 K?

P11.23 Consider the half-cell reaction $O_2(g) + 4H^+(aq) + 4e^- \rightarrow 2H_2O(l)$. By what factor are n, Q, E, and E° changed if all the stoichiometric coefficients are multiplied by the factor two? Justify your answers.

P11.24 Calculate ΔG_R° and the equilibrium constant at 298.15 K for the reaction $Cr_2O_7^{2-}(aq) + 3H_2(g) + 8H^+(aq) \rightarrow 2Cr^{3+}(aq) + 7H_2O(l)$.

P11.25 The half-cell potential for the reaction $O_2(g) + 4H^+(aq) + 4e^- \rightarrow 2H_2O(l)$ is +1.03 V at 298.15 K when $a_{O_2} = 1.00$. Determine a_{H^+}.

P11.26 Using half-cell potentials, calculate the equilibrium constant at 298.15 K for the reaction $2H_2O(l) \rightleftharpoons 2H_2(g) + O_2(g)$. Compare your answer with that calculated using ΔG_f° values from Table 4.1 (see Appendix B). What is the value of E° for the overall reaction that makes the two methods agree exactly?

P11.27 The data in the following table have been obtained for the potential of the cell $Pt(s)|H_2(g, f = 1.00\ atm)|HCl(aq, m)|AgCl(s)|Ag(s)$ as a function of m at 25°C.

m/mol $\mathrm{kg^{-1}}$	E/V	m/mol $\mathrm{kg^{-1}}$	E/V	m/mol $\mathrm{kg^{-1}}$	E/V
0.00100	0.57915	0.0200	0.43024	0.500	0.27231
0.00200	0.54425	0.0500	0.38588	1.000	0.23328
0.00500	0.49846	0.100	0.35241	1.500	0.20719
0.0100	0.46417	0.200	0.31874	2.000	0.18631

a. Determine E° using a graphical method.

b. Calculate $\gamma_\pm$ for HCl at $m = 0.00100$, 0.0100, and 0.100 $\mathrm{mol\ kg^{-1}}$.

P11.28 Consider the cell $Pt(s)|H_2(g, 1\ atm)|H^+(aq, a = 1.00)\ |Fe^{3+}(aq), Fe^{2+}(aq)|Pt(s)$ given that $Fe^{3+} + e^- \rightarrow Fe^{2+}$ and $E^\circ = 0.771$ V.

a. If the cell potential is 0.712 V, what is the ratio of $Fe^{2+}(aq)$ to $Fe^{3+}(aq)$?

b. What is the ratio of these concentrations if the cell potential is 0.830 V?

c. Calculate the fraction of the total iron present as $Fe^{3+}(aq)$ at cell potentials of 0.650, 0.700, 0.750, 0.771, 0.800, and 0.900 V. Graph the result as a function of the cell potential.

P11.29 Determine K_{sp} for AgBr at 298.15 K using the electrochemical cell described by

$$Ag(s)|Ag^+(aq, a_{Ag^+})\|Br^-(aq, a_{Br^-})|AgBr(s)|Ag(s)$$

P11.30 By finding appropriate half-cell reactions, calculate the equilibrium constant at 298.15 K for the following reactions:

a. $2Cd(s) + O_2(g) + 2H_2O(l) \rightleftharpoons 2Cd(OH)_2(s)$

b. $2MnO_2(s) + 4OH^-(aq) + O_2(g)$
$\rightleftharpoons 2MnO_4^{2-}(aq) + 2H_2O(l)$

12

From Classical to Quantum Mechanics

As scientists became able to investigate the atomic realm, they obtained results that were inconsistent with classical physics. Classical physics predicted that all bodies at a temperature other than zero kelvin radiate an infinite amount of energy. It incorrectly predicted that the kinetic energy of electrons produced upon illuminating a metal surface in vacuum with light is proportional to the light intensity, and it could not explain the diffraction of an electron by a crystalline solid. Rutherford's laboratory showed that atoms consist of a small, positively charged nucleus surrounded by a diffuse cloud of electrons. Classical physics, however, predicted that such an atom was unstable and that the electrons would spiral into the nucleus while radiating energy to the environment. These inconsistencies between classical theory and experimental observations provided the stimulus for the development of quantum mechanics.

12.1 Why Study Quantum Mechanics?

Imagine how difficult it would be for humans to function in a world governed by underlying principles without knowing what they were. If we could not calculate the trajectory of a projectile, we could not launch a satellite. Without understanding how energy is transformed into work, we could not design an automobile that gets more mileage for a given amount of fuel. Technology arises from an understanding of matter and energy, which argues for a broad understanding of scientific principles.

Chemistry is a molecular science; the goal of chemists is to understand macroscopic behavior in terms of the properties of individual atoms and molecules. In the first decade of the 20th century, scientists learned that an atom consisted of a small, positively charged nucleus surrounded by a diffuse electron cloud. However, this structure was not compatible with classical physics (the physics of pre-1900), which predicted that the electrons would follow a spiral trajectory and end in the nucleus. Classical physics was also unable to explain why graphite conducts electricity and diamond does not or why the light emitted by a hydrogen discharge lamp appears at only a small number of wavelengths.

These deficiencies in classical physics made it clear that another physical model was needed to describe matter at the microscopic scale of atoms and molecules. Over a period of about 20 years, quantum mechanics was developed, and scientists found that the puzzling phenomena just cited can be explained using quantum mechanics. The central feature that distinguishes quantum and classical mechanics is wave-particle duality. At the atomic level, electrons, protons, and light all behave as wave/particles as opposed to waves or particles. It is the experiment that determines whether wave or particle behavior will be observed.

Although few people may know it, we are already users of quantum mechanics. We take the stability of the atom with its central positively charged nucleus and surrounding electron cloud, the laser in our CD players, the integrated circuit in our computers, and the chemical bond between atoms to form molecules for granted. We know that infrared spectroscopy provides a useful way to identify chemical compounds and that nuclear magnetic resonance spectroscopy provides a powerful tool to image internal organs. However, these spectroscopies would not be possible if atoms and molecules could have *any* value of energy as predicted by classical physics. Quantum mechanics predicts that atoms and molecules can only have discrete energies and provides a common basis for understanding all spectroscopies.

Many areas of modern technology such as microfabrication of integrated circuits are based on quantum mechanics. Quantum mechanical calculations of chemical properties of pharmaceutical molecules are now sufficiently accurate that molecules can be designed for a specific application before they are tested at the laboratory bench. Quantum computing, in which a logic state can be described by zero *and* one rather than zero *or* one, is a very active area of research. If quantum computers can ultimately be realized, they will be much more powerful than current computers. As many sciences such as biology become increasingly focused on the molecular level, more scientists will need to be able to think in terms of quantum mechanical models. Therefore, a basic understanding of quantum mechanics is an essential part of the chemist's knowledge base.

12.2 Quantum Mechanics Arose out of the Interplay of Experiments and Theory

Scientific theories gain acceptance if they help us to understand the world around us. A key feature of validating theories is to compare the result of new experiments with the prediction of currently accepted theories. If the experiment and the theory agree, we gain confidence in the model underlying the theory; if not, the model needs to be modified. At the end of the 19th century, Maxwell's electromagnetic theory unified existing knowledge in the areas of electricity, magnetism, and waves. This theory, combined with the well-established field of Newtonian mechanics, ushered in a new era of maturity for the physical sciences. Many scientists of that era believed that there was little left in the natural sciences to learn. However, the growing ability of scientists to probe natural phenomena at an atomic level soon showed that this presumption was incorrect. The field of quantum mechanics arose in the early 1900s as scientists became able to investigate natural phenomena at the newly accessible atomic level. A number of key experiments showed that the predictions of classical physics were inconsistent with experimental outcomes. Several of these experiments are described in more detail in this chapter in order to show the important role that experiments have had—and continue to have—in stimulating the development of theories to describe the natural world.

In the rest of this chapter, experimental evidence is presented for two key properties that have come to distinguish classical and quantum physics. The first of these is **quantization.** Energy at the atomic level is not a continuous variable, but it comes in discrete packets called *quanta*. The second key property is **wave-particle duality.** At the atomic level, light waves have particle-like properties, and atoms as well as subatomic particles such as electrons have wave-like properties. Neither quantization nor wave-particle duality were known concepts until the experiments described in Sections 12.3 through 12.7 were conducted.

12.3 Blackbody Radiation

Think of the heat that a person feels from the embers of a fire. The energy that the body absorbs is radiated from the glowing coals. An idealization of this system that is more amenable to theoretical study is a red-hot block of metal with a spherical cavity in its interior that can be observed through a hole small enough that the conditions inside the block are not perturbed. An **ideal blackbody** is shown in Figure 12.1. Under the condition of equilibrium between the radiation field inside the cavity and the glowing piece of matter, classical electromagnetic theory can predict what frequencies ν of light are radiated and their relative magnitudes. The result is

$$\rho(\nu,T)\ d\nu = \frac{8\pi\nu^2}{c^3}\ \overline{E}_{osc}\ d\nu \qquad \textbf{(12.1)}$$

In this equation, ρ is the **spectral density,** which has the units of energy $\times$ (volume)$^{-1}$ $\times$ (frequency)$^{-1}$. The spectral density is a function of the temperature T and the frequency ν. The speed of light is c, and $\overline{E}_{osc}$ is the average energy of an oscillating dipole in the solid. In words, the spectral density is the energy stored in the electromagnetic field of the blackbody radiator at frequency ν per unit volume and unit frequency.

The factor $d\nu$ is used on both sides of this equation because we are asking for the energy density observed within the frequency interval of width $d\nu$ centered at the frequency ν. Classical theory further predicts that the average energy of an oscillator is simply related to the temperature by

$$\overline{E}_{osc} = k_B T \qquad \textbf{(12.2)}$$

in which k is the Boltzmann constant. Combining these two equations results in an expression for $\rho(\nu,T)\ d\nu$, the amount of energy per unit volume in the frequency range between ν and $\nu + d\nu$ in equilibrium with a blackbody at temperature T:

$$\rho(\nu,T)\ d\nu = \frac{8\pi k_B T \nu^2}{c^3}\ d\nu \qquad \textbf{(12.3)}$$

It is possible to measure the spectral density of the radiation emitted by a blackbody, and the results are shown in Figure 12.2 for several temperatures together with the result predicted by classical theory. The experimental curves have a common behavior. The spectral density is peaked in a broad maximum and falls off to both lower and higher frequencies. The shift of the maxima to higher frequencies with increasing temperatures is consistent with our experience that if more power is put into an electrical heater, its color will change from dull red to yellow (increasing frequency).

The comparison of the spectral density distribution predicted by classical theory with that observed experimentally for $T = 6000.$ K is particularly instructive. The two curves show similar behavior at low frequencies, but the theoretical curve keeps on increasing with frequency as Equation (12.3) predicts. Because the area under the $\rho(\nu,T)$ versus ν curves gives the total energy per unit volume of the field of the blackbody, classical theory predicts that a blackbody will emit an infinite amount of energy at all temperatures above absolute zero! It is clear that this prediction is incorrect, but scientists at the beginning of the 20th century were greatly puzzled about where the theory went wrong.

In looking at data such as that shown in Figure 12.2, the German physicist Max Planck was able to develop some important insights that ultimately led to an understanding of **blackbody radiation.** It was understood at the time that the origin of blackbody radiation was the vibration of electric dipoles formed by atomic nuclei and their associated electrons that emit radiation at the frequency at which they oscillate. Planck saw that the discrepancy between experiment and classical theory occurred at high and not at low frequencies. The absence of high-frequency radiation at low temperatures showed that the high-frequency dipole oscillators emitted radiation

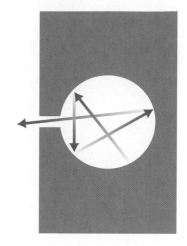

FIGURE 12.1
An idealized blackbody. A cubical solid at a high temperature emits photons from an interior spherical surface. The photons reflect several times before emerging through a narrow channel. The reflections ensure that the radiation is in thermal equilibrium with the solid.

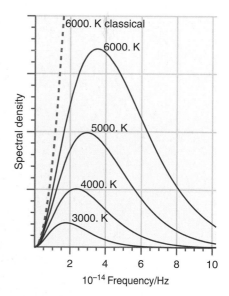

FIGURE 12.2
The red curves show the light intensity emitted from an ideal blackbody as a function of the frequency for different temperatures. The dashed curve shows the predictions of classical theory for $T = 6000.$ K.

only at high temperatures. Unless a large amount of energy is put into the blackbody (high temperature) it will not be possible to excite the high-energy (high-frequency) oscillators.

Planck found that he could obtain agreement between theory and experiment only if he assumed that the energy radiated by the blackbody was related to the frequency by

$$E = nh\nu \tag{12.4}$$

Planck's constant h was initially an unknown proportionality constant and n is a positive integer ($n = 1, 2, \ldots$). The frequency ν is continuous, but for a given ν, the energy is *quantized*. Equation (12.4) was a radical departure from classical theory, in which the energy stored in electromagnetic radiation is proportional to the square of the amplitude but independent of the frequency. This relationship between energy and frequency ushered in a new era of physics. Energy in classical theory is a *continuous* quantity, which means that it can take on all values. Equation (12.4) states that the energy radiated by a blackbody can take on only a set of *discrete* values for each frequency. Its main justification was that agreement between theory and experiment could be obtained. Using Equation (12.4) and some classical physics, Planck obtained the following relationship:

$$\overline{E}_{osc} = \frac{h\nu}{e^{h\nu/k_BT} - 1} \tag{12.5}$$

It is useful to obtain an approximate value for $\overline{E}_{osc}$ from this equation in two limits: at high temperatures, where $h\nu/k_BT \ll 1$, and at low temperatures, where $h\nu/k_BT \gg 1$. At high temperatures, the exponential function in Equation (12.5) can be expanded in a Taylor-Maclaurin series (see the Math Supplement, Appendix A), giving

$$\overline{E}_{osc} = \frac{h\nu}{(1 + h\nu/k_BT + \ldots) - 1} \approx k_BT \tag{12.6}$$

just as classical theory had predicted. However, for low temperatures corresponding to $h\nu/k_BT \gg 1$, the denominator in Equation (12.5) becomes very large, and $\overline{E}_{osc}$ approaches zero. The high-frequency oscillators do not contribute to the radiated energy at low and moderate temperatures.

Using Equation (12.5), in 1901 Planck obtained the following general formula for the spectral radiation density from a blackbody:

$$\rho(\nu,T)d\nu = \frac{8\pi h\nu^3}{c^3} \frac{1}{e^{h\nu/k_BT} - 1} d\nu \tag{12.7}$$

The value of the constant h was not known and Planck used it as a parameter to fit the data. He was able to reproduce the experimental data at all temperatures with the single adjustable parameter h which through more accurate measurements currently has the value $h = 6.626\,069\,57 \times 10^{-34}$ J s. In calculations in this book, we use only four significant figures. Obtaining this degree of agreement using a single adjustable parameter was a remarkable achievement. However, Planck's explanation, which relied on the assumption that the energy of the radiation came in discrete packets or quanta, was not accepted initially. Soon afterward, Einstein's explanation of the photoelectric effect gave support to Planck's hypothesis.

12.4 The Photoelectric Effect

Imagine a copper plate in a vacuum. Light incident on the plate can be absorbed, leading to the excitation of electrons to unoccupied energy levels. Sufficient energy can be transferred to the electrons such that some leave the metal and are ejected into the vacuum. The electrons that have been emitted from the copper upon illumination can be collected by another electrode in the vacuum system, called the collector.

This process of electron ejection by light is called the **photoelectric effect.** A schematic apparatus is shown in Figure 12.3. The absorbed light energy must be balanced by the energy required to eject an electron at equilibrium and the kinetic energy of the emitted electrons, because the energy of the system is constant. Classical theory makes the following predictions:

- Light is incident as a plane wave over the whole copper plate. Therefore, the light is absorbed by many electrons in the solid. Any one electron can absorb only a small fraction of the incident light.
- Electrons are emitted to the collector for all light frequencies, provided that the light is sufficiently intense.
- The kinetic energy per electron increases with the light intensity.

The results of the experiment can be summarized as follows:

- The number of emitted electrons is proportional to the light intensity, but their kinetic energy is independent of the light intensity.
- No electrons are emitted unless the frequency ν is above a threshold frequency ν_0 even for high light intensities.
- The kinetic energy of the emitted electrons depends on the frequency in the manner depicted in Figure 12.4.
- Electrons are emitted even at such low light intensities that all the light absorbed by the entire copper plate is barely enough to eject a single electron, based on energy conservation considerations.

Just as for blackbody radiation, the inability of classical theory to correctly predict experimental results stimulated a new theory. In 1905, Albert Einstein hypothesized that the energy of light was proportional to its frequency:

$$E = \beta\nu \qquad (12.8)$$

where β is a constant to be determined. This is a marked departure from classical electrodynamics, in which there is no relation between the energy of a light wave and its frequency. Invoking energy conservation, the energy of the electron E_e is related to that of the light by

$$E_e = \beta\nu - \phi \qquad (12.9)$$

The binding energy of the electron in the solid, which is analogous to the ionization energy of an atom, is designated by ϕ in this equation and is called the **work function.** In words, this equation says that the kinetic energy of the photoelectron that has escaped from the solid is smaller than the photon energy by the amount with which the electron is bound to the solid. Einstein's theory gives a prediction of the dependence of the kinetic energy of the photoelectrons as a function of the light frequency that can be compared directly with experiment. Because ϕ can be determined independently, only β is unknown. It can be obtained by fitting the data points in Figure 12.4 to Equation (12.9). The results shown by the red line in Figure 12.4 not only reproduce the data very well, but they yield the striking result that β, the slope of the line, is identical to Planck's constant h. The equation that relates the energy of light to its frequency

$$E = h\nu \qquad (12.10)$$

is one of the most widely used equations in quantum mechanics and earned Albert Einstein a Nobel Prize in physics. A calculation involving the photoelectric effect is carried out in Example Problem 12.1.

The agreement between the theoretical prediction and the experimental data validates Einstein's fundamental assumption that the energy of light is proportional to its frequency. This result also suggested that h is a "universal constant" that appears in seemingly unrelated phenomena. Its appearance in this context gained greater acceptance for the assumptions Planck used to explain blackbody radiation.

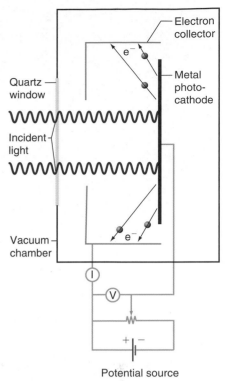

FIGURE 12.3
The electrons emitted by the surface upon illumination are incident on the collector, which is at an appropriate electrical potential to attract them. The experiment is carried out in a vacuum chamber to avoid collisions and capture of electrons by gas molecules.

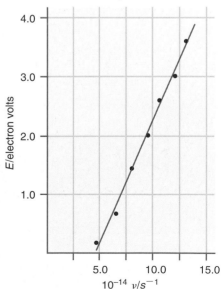

FIGURE 12.4
The energy of photo-ejected electrons is shown as a function of the light frequency. The individual data points are well fit by a straight line as shown.

EXAMPLE PROBLEM 12.1

Light with a wavelength of 300. nm is incident on a potassium surface for which the work function ϕ is 2.26 eV. Calculate the kinetic energy and speed of the ejected electrons.

Solution

Using Equation (12.9), we write $E_e = h\nu - \phi = (hc/\lambda) - \phi$ and convert the units of ϕ from electron-volts to joules: $\phi = (2.26\,\text{eV})(1.602 \times 10^{-19}\,\text{J/eV}) = 3.62 \times 10^{-19}\,\text{J}$. Electrons will only be ejected if the photon energy $h\nu$ is greater than ϕ. The photon energy is calculated to be

$$\frac{hc}{\lambda} = \frac{(6.626 \times 10^{-34}\,\text{J s})(2.998 \times 10^8\,\text{m s}^{-1})}{300. \times 10^{-9}\,\text{m}} = 6.62 \times 10^{-19}\,\text{J}$$

which is sufficient to eject electrons.

Using Equation (12.9), we obtain $E_e = (hc/\lambda) - \phi = 3.00 \times 10^{-19}\,\text{J}$. Using $E_e = 1/2\,mv^2$, we calculate that

$$v = \sqrt{\frac{2E_e}{m}} = \sqrt{\frac{2\,(3.00 \times 10^{-19}\,\text{J})}{9.109 \times 10^{-31}\,\text{kg}}} = 8.12 \times 10^5\,\text{m s}^{-1}$$

Another important conclusion can be drawn from the observation that even at very low light intensities, photoelectrons are emitted from the solid. More precisely, photoelectrons are detected even at intensities so low that all the energy incident on the solid surface is only slightly more than the threshold energy required to yield a single photoelectron. This means that the light that liberates the photoelectron is not uniformly distributed over the surface. If this were true, no individual electron could receive enough energy to escape into the vacuum. The surprising conclusion of this experiment is that all of the incident light energy can be concentrated in a single electron excitation. This led to the coining of the term **photon** to describe a spatially localized packet of light. Because this spatial localization is characteristic of particles, the conclusion that light can exhibit particle-like behavior under some circumstances was inescapable.

Many experiments have shown that light exhibits wave-like behavior. It has long been known that light can be diffracted by an aperture or slit. However, the photon in the photoelectric effect that exhibits particle-like properties and the photon in a diffraction experiment that exhibits wave-like properties are one and the same. This recognition forces us to conclude that light has a wave-particle duality, and depending on the experiment, it can manifest as a wave or as a particle. This important recognition leads us to the third fundamental experiment to be described: the diffraction of electrons by a crystalline solid. Because diffraction is proof of wave-like behavior, if particles can be diffracted, they exhibit a particle-wave duality just as light does.

12.5 Particles Exhibit Wave-Like Behavior

In 1924, Louis de Broglie suggested that a relationship that had been derived to relate momentum and wavelength for light should also apply to particles. The **de Broglie relation** states that

$$\lambda = \frac{h}{p} \tag{12.11}$$

in which h is the by now familiar Planck constant and p is the particle momentum given by $p = mv$, in which the momentum is expressed in terms of the particle mass and velocity. This proposed relation was confirmed in 1927 by Davisson and Germer, who carried out a diffraction experiment. Diffraction is the change in the directions and intensities of waves after passing by or through an aperture or grating whose characteristic size is approximately the same as the wavelength of the waves. Diffraction by a double slit is discussed in more detail in Section 12.6. Putting numbers in Equation (12.11) will demonstrate that it is difficult to obtain wavelengths much longer than 1 nm even with particles as light as the electron, as shown in Example Problem 12.2. Therefore, diffraction requires a grating with atomic dimensions, and an ideal candidate is a crystalline solid. Davisson and Germer observed diffraction of electrons from crystalline NiO in their classic experiment to verify the de Broglie relation. Figure 12.5 shows a scan through a diffraction pattern obtained by diffracting a beam of He atoms from a crystal surface of nickel.

EXAMPLE PROBLEM 12.2

Electrons are used to determine the structure of crystal surfaces. To have diffraction, the wavelength λ of the electrons should be on the order of the lattice constant, which is typically 0.30 nm. What energy do such electrons have, expressed in electron-volts and joules?

Solution

Using Equation (12.11) and the expression $E = p^2/2m$ for the kinetic energy, we obtain

$$E = \frac{p^2}{2m} = \frac{h^2}{2m\lambda^2} = \frac{(6.626 \times 10^{-34}\ \text{J s})^2}{2(9.109 \times 10^{-31}\ \text{kg})(3.0 \times 10^{-10}\ \text{m})^2}$$

$$= 2.7 \times 10^{-18}\ \text{J or } 17\ \text{eV}$$

The Davisson-Germer experiment was critical in the development of quantum mechanics in that it showed that particles exhibit wave behavior. If this is the case, there must be a wave equation that relates the spatial and time dependencies of the wave amplitude for the (wave-like) particle. This equation could be used to describe an atomic scale system rather than Newton's second law $F = ma$. It was Erwin Schrödinger who formulated this wave equation, which will be discussed in Chapter 13.

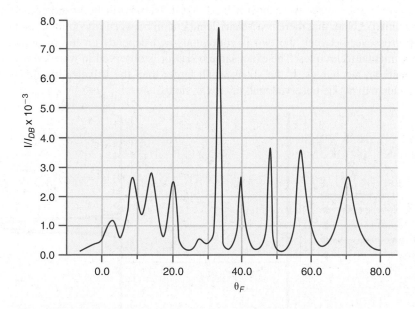

FIGURE 12.5
Diffraction scan obtained by rotating a mass spectrometer around a nickel single crystal surface on which a collimated He beam was incident. Each peak corresponds to a different diffraction maximum.
Source: Reprinted with permission from "A helium diffraction study of the structure of the Ni(115) surface," by D.S. Kaufman et al., from *The Journal of Chemical Physics,* Vol. 86, issue 6, pp. 3682 (1987). Copyright 1987, American Institute of Physics.

12.6 Diffraction by a Double Slit

There is probably no single experiment that exhibits the surprising nature of quantum mechanics as well as the diffraction of particles by a double slit. An idealized version of this experiment is described next, but everything in the explanation has been confirmed by experiments carried out with particles such as neutrons, electrons, and He atoms. We first briefly review classical diffraction of waves.

Diffraction is a phenomenon that is widely exploited in science. For example, the atomic level structure of DNA was in large part determined by analyzing the diffraction of X rays from crystalline DNA samples. Figure 12.6 illustrates diffraction of light from a thin slit in an otherwise opaque wall.

It turns out that the analysis of this problem is much simpler if the screen on which the image is projected is far away from the slit. Mathematically, this requires that $b \gg a$. In ray optics, which is used to determine the focusing effect of a lens on light, the light incident on the slit from the left in Figure 12.6 would give a sharp image of the slit on the screen. In this case parallel light is assumed to be incident on the slit and, therefore, the image and slit dimensions are identical. The expected intensity pattern is that shown by the blue lines in the figure. Instead, an intensity distribution like that shown by the red curve is observed if the light wavelength is comparable in magnitude to the slit width.

The origin of this pattern of alternating maxima and minima (which lies well outside the profile expected from ray optics) is wave interference. Its origin can be understood by treating each point in the plane of the slit as a source of cylindrical waves (Huygens' construction). Maxima and minima arise as a result of a path difference between the sources of the cylindrical waves and the screen, as shown in Figure 12.7. The condition that the minima satisfy is

$$\sin \theta = \frac{n\lambda}{a}, \quad n = \pm 1, \pm 2, \pm 3, \pm \ldots \qquad (12.12)$$

This equation helps us to understand under what conditions we might observe diffraction. The wavelength of light in the middle of the visible spectrum is about 600. nm or 6.00×10^{-4} mm. If this light is allowed to pass through a 1.00-mm-wide slit and the angle calculated at which the first minimum will appear, the result is $\theta = 0.03°$ for $n = 1$. This minimum is not easily observable because it lies so close to the maximum, and we expect to see a sharp image of the slit on our screen, just as in ray optics. However, if the slit width is decreased to 1.00×10^{-2} mm, then $\theta = 3.4°$. This minimum is easily observable and successive bands of light and darkness will be observed instead of a sharp image. Note that there is no clear demarcation between ray optics and diffraction. The crossover between the two is continuous and depends on the resolution of the experimental techniques. The exact same behavior is observed in wave-particle duality in quantum mechanics. If the slit is much larger than the wavelength, diffraction by particles will not be observed, and ray optics holds.

FIGURE 12.6
Diffraction of light of wavelength λ from a slit whose long axis is perpendicular to the page. The arrows from the left indicate parallel rays of light incident on an opaque plate containing the slit. Instead of seeing a sharp image of the slit on the screen, a diffraction pattern will be seen. This is schematically indicated in a plot of intensity versus distance. In the absence of diffraction, the intensity versus distance indicated by the blue lines would be observed.

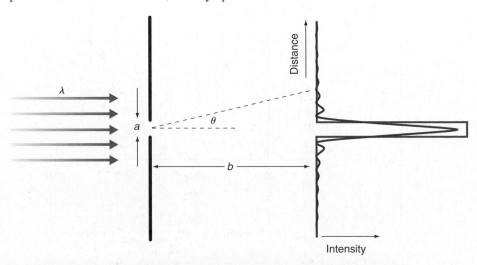

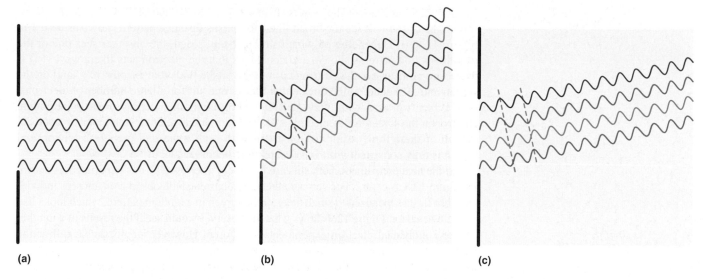

(a)   (b)   (c)

Consider the experimental setup designed to detect the diffraction of particles shown in Figure 12.8. The essential feature of the apparatus is a metal plate in which two rectangular slits of width *a* have been cut. The long axis of the rectangles is perpendicular to the plane of the page. Why two slits? Rather than detecting the diffraction from the individual slits, the apparatus is designed to detect diffraction from the *combination of the two slits*. Diffraction will only be observed for case 2 if the particle passes through both slits simultaneously which is hard to imagine from the vantage point of classical physics.

First, we need a source of particles, for instance, an electron gun. By controlling the energy of the electron, the wavelength is varied. Each electron has a **random phase angle** with respect to every other electron. Consequently, two electrons can never interfere with one another to produce a diffraction pattern. One electron gives rise to the diffraction pattern, but many electrons are needed to amplify the signal so that we can see the pattern. A more exact way to say this is that the intensities of the electron waves add together rather than the amplitudes.

A phosphorescent screen that lights up when energy from an incident wave or particle is absorbed (as in a television picture tube) is mounted behind the plate with the slits. The electron energy is adjusted so that diffraction by the single slits of width *a* results in broad maxima with the first intensity minimum at a large diffraction angle. The distance between the two slits, *b*, has been chosen such that we will observe a number of intensity oscillations for small diffraction angles. The diffraction patterns in Figure 12.8 (case 2) were calculated for the ratio $b/a = 5$.

FIGURE 12.7
Each segment of a slit through which light is diffracted can be viewed as a source of waves that interfere with one another. **(a)** The waves that emerge perpendicular to the slit are all in phase and give rise to the principal maximum in the diffraction pattern. **(b)** Successive waves that emerge at the angle shown are exactly out of phase. They will interfere destructively and a minimum intensity will be observed. **(c)** Every other wave is out of phase and destructive interference with a minimum intensity will be observed. The wavelength and slit width are not drawn to scale.

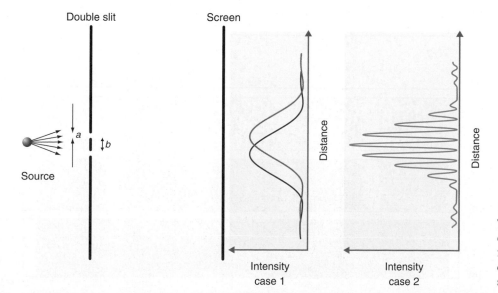

FIGURE 12.8
The double-slit diffraction experiment. case 1 describes the outcome of the diffraction when one of the slits is blocked. case 2 describes the outcome when both slits are open.

Now let's talk about the results. If one slit is closed and an observer looks at the screen, he or she will see the broad intensity versus distance pattern shown as case 1 in Figure 12.8. (i) Depending on which slit has been closed, the observer sees one of the two diffraction patterns shown, and concludes that the electron acts like a wave. (ii) If the observer measures the electron current, he or she finds that exactly 50% land on the screen and 50% land on the device that blocks one slit for a large number of electrons. (iii) When working with very sensitive phosphors, the arrival of each individual electron at the screen is detected by a flash of light localized to a small area of a screen. Which of these three results is consistent with both wave *and* particle behavior, and which is only consistent with wave *or* particle behavior?

In the next experiment, both slits are left open. The result of this experiment is shown in Figure 12.8 as case 2. For very small electron currents, the observer again sees individual light flashes localized to small areas of the screen in a random pattern, which looks like particle behavior. Figure 12.9 shows what the observer would see if the results of a number of these individual electron experiments are stored. However, unmistakable diffraction features are seen if we accumulate the results of many individual light flashes. This shows that a wave (a single electron) is incident on *both* of the slits simultaneously.

How can the results of this experiment be understood? The fact that diffraction is seen from a single slit as well as from the double slit shows the wave-like behavior of the electron. Yet individual light flashes are observed on the screen, which is what we expect from particle trajectories. To add to the complexity, the spatial distribution of the individual flashes on the screen is what we expect from waves rather than from particles. The measurement of the electron current to the slit blocker seems to indicate that the electron *either* went through one slit *or* through the other. However, this conclusion is inconsistent with the appearance of a diffraction pattern because a diffraction pattern only arises if one and the same electron goes through both slits!

Regardless of how we turn these results around, we will find that all the results are inconsistent with the logic of classical physics, namely, the electron goes through one slit or the other. This either/or logic cannot explain the results. In a quantum mechanical description, the electron wave function is a superposition of wave functions for going through the top slit and the bottom slit, which is equivalent to saying that the electron can go through both slits. We will have much more to say about wave functions in the next few chapters. The act of measurement, such as blocking one slit, changes the wave function such that the electron goes through either the top slit or the bottom one. The results represent a mixture of particle and wave behavior. Individual electrons move through the slits and generate points of light on the screen. This behavior is particle-like. However, the location of the points of light on the screen is not what is expected from classical

FIGURE 12.9

Simulation of the diffraction pattern observed in the double-slit experiment for (top to bottom) (**a**) 60, (**b**) 250, (**c**) 1000, and (**d**) 3000 particles. The bottom panel shows what would be expected for a wave incident on the apparatus. Bright red corresponds to high intensity and blue corresponds to low intensity. Note that the diffraction pattern only becomes obvious after a large number of particles have passed through the apparatus, although intensity minima are evident even for 60 incident particles.

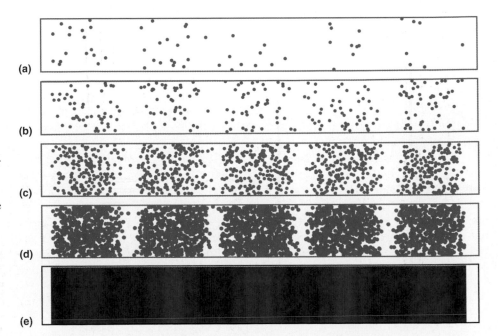

trajectories; it is governed by the diffraction pattern. This behavior is wave-like. Whereas in classical mechanics the operative word concerning several possible modes of behavior is *or*, in quantum mechanics it is *and*. If all of this seems strange at first sight, welcome to the crowd! Although particle diffraction has been observed directly only for atomic and molecular masses up to 20 amu (neon), the de Broglie relation has been verified for molecules as heavy as tetraphenylporphyrin, which has a molecular mass of 614 amu.

12.7 Atomic Spectra and the Bohr Model of the Hydrogen Atom

The most direct evidence of energy quantization comes from the analysis of the light emitted from highly excited atoms in a plasma. The structure of the atom was not known until fundamental studies using the scattering of alpha particles were carried out in Ernest Rutherford's laboratory beginning in 1910. These experiments showed that the positive and negative charges in an atom were separated. The positive charge is contained in the nucleus, whereas the negative charge of the electrons occupies a much greater volume that is centered at the nucleus. In analogy to our solar system, the first picture that emerged of the atom was of electrons orbiting the nucleus.

However, this picture of the atom is inconsistent with electrodynamic theory. An electron orbiting the nucleus is constantly accelerating and must therefore radiate energy. In a classical picture, the electron would continually radiate away its kinetic energy and eventually fall into the nucleus as depicted in Figure 12.10.

This clearly was not happening, but why? We will answer this question when we discuss the hydrogen atom in Chapter 20. Even before Rutherford's experiments, it was known that if an electrical arc is placed across a vacuum tube with a small partial pressure of hydrogen, light is emitted. Our present picture of this phenomenon is that the atom takes up energy from the electromagnetic field and makes a transition to an excited state. The excited state has a limited lifetime, and when the transition to a state of lower energy occurs, light is emitted. An apparatus used to obtain atomic spectra and a typical spectrum are shown schematically in Figure 12.11.

How did scientists working in the 1890s explain these spectra? The most important experimental observation made is that over a wide range of wavelengths, light emitted from atoms is only observed at certain discrete wavelengths; that is, it is quantized. This result was not understandable on the basis of classical theory because in classical physics, energy is a continuous variable. Even more baffling to these first spectroscopists was that they could derive a simple relationship to explain all of the frequencies that appeared in the hydrogen emission spectrum. For the emission spectra observed, the inverse of the wavelength $1/\lambda = \tilde{\nu}$ of all lines in an atomic hydrogen spectrum is given by equations of the type

$$\tilde{\nu}(\text{cm}^{-1}) = R_H \ (\text{cm}^{-1}) \left(\frac{1}{n_1^2} - \frac{1}{n^2} \right), n > n_1 \qquad (12.13)$$

in which only a single parameter n_1 appears. In this equation, n is an integer that takes on the values $n_1 + 1, n_1 + 2, n_1 + 3, \ldots$, and R_H is called the **Rydberg constant,** which has the value 109,677.581 cm^{-1}. What gives rise to such a simple relationship and why does n take on only integral values?

In 1911 Niels Bohr, who played a seminal role in the development of quantum mechanics, proposed a model for the hydrogen atom that explained its emission spectrum. Even though Bohr's model was superseded by the Schrödinger model described in Chapter 20, it offered the first explanation of how quantized energy levels arise in atoms as a result of wave-particle duality. Bohr assumed a simple model of the hydrogen atom in which an electron revolved around the nucleus in a circular orbit. The orbiting electron experiences two forces: a Coulombic attraction to the nucleus, and a centrifugal force that is opposite in direction. In a stable orbit, these two forces are equal.

$$\frac{e^2}{4\pi\varepsilon_0 r^2} = \frac{m_e v^2}{r} \qquad (12.14)$$

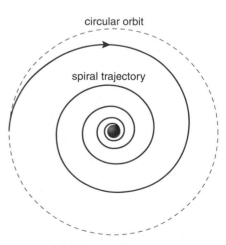

circular orbit

spiral trajectory

FIGURE 12.10

Classical particle-based physics predicts that an electron in a circular orbit will lose energy by radiation and spiral into the nucleus.

FIGURE 12.11

Light emitted from a hydrogen discharge lamp is passed through a narrow slit and separated into its component wavelengths by a dispersing element. As a result, multiple images of the slit, each corresponding to a different wavelength, are seen on the photographic film. One of the different series of spectral lines for H is shown. $\tilde{\nu}$ represents the inverse wavelength [see Equation (12.13)].

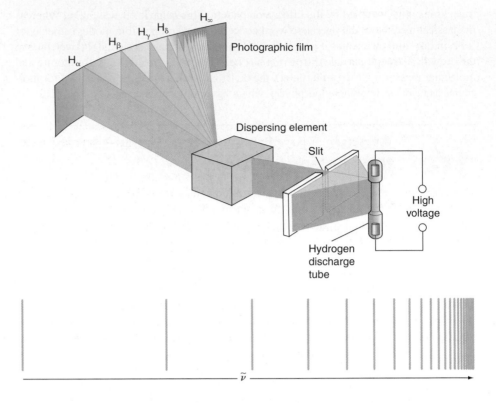

In Equation (12.14), e is the charge on the electron, m_e and v are its mass and speed, and r is the orbit radius.

Bohr next introduced wave-particle duality by asserting that the electron had the de Broglie wavelength $\lambda = h/p$. He made a new assumption that the length of an orbit had to be an integral number of wavelengths.

$$2\pi r = n\lambda = n\frac{h}{p} \qquad (12.15)$$

Which leads to the condition

$$m_e v r = n\hbar, \text{ where } n = 1,2,3\ldots \qquad (12.16)$$

We have introduced the symbol $\hbar$ for $h/2\pi$. The rationale for Equation (12.15) is shown in Figure 12.12.

Bohr reasoned that unless the orbit length is an integral number of wavelengths, the wave will destructively interfere with itself, and the amplitude will decrease to zero in a

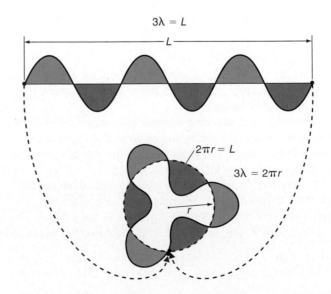

FIGURE 12.12

In analogy to a wave on a string (upper image), Bohr postulated a wave traveling on a circular orbit (lower image). Unless the circumference of the orbit is an integral number of wavelengths, the wave will cancel itself out.

few orbits. The assertion that there is a stable orbit for the electron goes beyond classical physics as shown in Figure 12.10.

Solving Equation (12.16) for v and substituting the result in Equation (12.14) gives the following expression for the orbit radius r:

$$r = \frac{\varepsilon_0 h^2 n^2}{\pi m_e e^2} = \frac{4\pi \varepsilon_0 \hbar^2 n^2}{m_e e^2} \tag{12.17}$$

Equation (12.17) shows that the electron can only have certain discrete values for the orbit radii, each corresponding to a different value of n. We next show that the discrete set of orbit radii gives rise to a discrete set of energy levels.

EXAMPLE PROBLEM 12.3

Calculate the radius of the electron in H in its lowest energy state, corresponding to $n = 1$.

Solution

$$r = \frac{4\pi\varepsilon_0\hbar^2 n^2}{m_e e^2}$$

$$= \frac{4\pi \times 8.85419 \times 10^{-12}\,C^2\,N^{-1}\,m^{-2} \times (1.0545 \times 10^{-34}\,J\,s)^2 \times 1^2}{9.109 \times 10^{-31}\,kg \times (1.6022 \times 10^{-19}\,C)^2}$$

$$= 5.292 \times 10^{-11}\,m$$

The total energy of the electron in the hydrogen atom is the sum of its kinetic and potential energies.

$$E_{total} = E_{kinetic} + E_{potential} = \frac{1}{2} m_e v^2 - \frac{e^2}{4\pi\varepsilon_0 r} \tag{12.18}$$

We transform Equation (12.18) into a more useful form by first eliminating v using Equation (12.14).

$$E_{total} = \frac{1}{2}\left(\frac{e^2}{4\pi\varepsilon_0 r}\right) - \left(\frac{e^2}{4\pi\varepsilon_0 r}\right) = -\frac{e^2}{8\pi\varepsilon_0 r} \tag{12.19}$$

We next eliminate r using Equation (12.17), obtaining Equation (12.20), which shows that the energy levels in the Bohr model are discrete.

$$E_n = -\frac{m_e e^4}{8\varepsilon_0^2 h^2 n^2} \quad n = 1,2,3,\dots \tag{12.20}$$

All energy values have negative values because the zero of energy, which is arbitrary, corresponds to $n \to \infty$, corresponding to a proton and an electron at infinite separation. The ground state energy is the lowest energy that a hydrogen atom can have that corresponds to $n = 1$.

Because the energy of the electron can have only certain discrete values, the light emitted when an electron makes a transition from a higher to a lower energy level has a discrete set of frequencies:

$$\nu_{n_2 \to n_1} = \frac{m_e e^4}{8\varepsilon_0^2 h^3}\left(\frac{1}{n_1^2} - \frac{1}{n_2^2}\right) \quad n_2 > n_1 \tag{12.21}$$

We see that Equation (12.21) provides a rationale for the empirical formula given by Equation (12.13), and the calculated and measured frequencies for the hydrogen atom are in quantitative agreement. This agreement between theory and experiment appears to justify the assumptions made. A number of possible energy transitions in the Bohr model are shown in Figure 12.13.

Although the Bohr model predicts the absorption and emission frequencies observed in the hydrogen atom, it does not give quantitative agreement with spectra

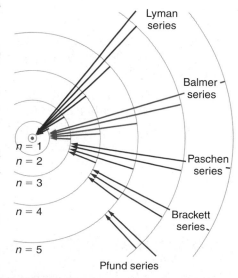

FIGURE 12.13

Transitions in the Bohr model giving rise to light emission are shown. The series differ in the quantum number of the final state.

observed for any atom containing more than one electron, for reasons that will become clear in Chapters 21 and 22. There is also a fundamental flaw in the Bohr model that was discovered by Werner Heisenberg 14 years after the model was introduced. Heisenberg showed that it is not possible to simultaneously know the electron orbit radius and its momentum as Bohr had assumed. We discuss Heisenberg's work on the uncertainty principle in Chapter 17.

Vocabulary

blackbody radiation

de Broglie relation

ideal blackbody

photoelectric effect

photon

Planck's constant

quantization

random phase angle

Rydberg constant

spectral density

wave-particle duality

work function

Conceptual Problems

Q12.1   Why is there an upper limit to the photon energy that can be observed in the discrete emission spectrum of the hydrogen atom?

Q12.2   Why was the wave nature of particles not discovered until atomic level experiments became possible?

Q12.3   Classical physics predicts that there is no stable orbit for an electron moving around a proton. The Bohr model of the hydrogen atom preceded quantum mechanics. Justify the criterion that Niels Bohr used to define special orbits that he assumed were stable.

Q12.4   You observe light passing through a slit of width a and pass from the range $\lambda \gg a$ to $\lambda \ll a$. Will you observe a sharp transition between ray optics and diffraction? Explain why or why not.

Q12.5   Which of the experimental results for the photoelectric effect suggests that light can display particle-like behavior?

Q12.6   Is the intensity observed from the diffraction experiment depicted in Figure 12.7 the same for the angles shown in parts (b) and (c)?

Q12.7   What feature of the distribution depicted as case 1 in Figure 12.8 tells you that it arises from diffraction?

Q12.8   Why does the analysis of the photoelectric effect based on classical physics predict that the kinetic energy of electrons will increase with increasing light intensity?

Q12.9   In the double-slit experiment, researchers found that an equal number of electrons pass through each slit. Does this result allow you to distinguish between particle-like and wave-like behavior?

Q12.10   The inability of classical theory to explain the spectral density distribution of a blackbody was called the *ultraviolet catastrophe*. Why is this name appropriate?

Q12.11   In the diffraction of electrons by crystals, the volume sampled by the diffracting electrons is on the order of 3 to 10 atomic layers. If He atoms are diffracted from the surface, only the topmost atomic layer is sampled. Can you explain this difference?

Q12.12   Why is a diffraction pattern generated by an electron gun formed by electrons interfering with themselves rather than with one another?

Q12.13   How can data from photoelectric effect experiments be used to obtain numerical values for the Planck constant h?

Q12.14   How did Planck conclude that the discrepancy between experiments and classical theory for blackbody radiation was at high and not low frequencies?

Q12.15   Write down formulas relating the wave number with the frequency, wavelength, and energy of a photon.

Q12.16   Planck's explanation of blackbody radiation was met by skepticism by his colleagues because Equation (12.4) seemed like a mathematical trick rather than being based on a microscopic model. Justify this equation using Einstein's explanation of the photoelectric effect that came five years later.

Numerical Problems

Problem numbers in red indicate that the solution to the problem is given in the *Student's Solutions Manual*.

P12.1   When a molecule absorbs a photon, momentum is conserved. If an H_2 molecule at 500. K absorbs an ultraviolet photon of wavelength 175 nm, what is the change in its velocity Δv?

Given that its average speed is $v_{rms} = \sqrt{3k_BT/m}$, what is $\Delta v/v_{rms}$?

P12.2   A more accurate expression for $\overline{E}_{osc}$ would be obtained by including additional terms in the Taylor-Maclaurin series. The Taylor-Maclaurin series expansion of $f(x)$ in the vicinity of x_0 is given by (see Math Supplement)

$$f(x) = f(x_0) + \left(\frac{df(x)}{dx}\right)_{x=x_0} (x - x_0)$$
$$+ \frac{1}{2!}\left(\frac{d^2 f(x)}{dx^2}\right)_{x=x_0} (x - x_0)^2$$
$$+ \frac{1}{3!}\left(\frac{d^3 f(x)}{dx^3}\right)_{x=x_0} (x - x_0)^3 + \ldots$$

Use this formalism to better approximate $\overline{E}_{osc}$ by expanding $e^{h\nu/k_BT}$ in powers of $h\nu/k_BT$ out to $(h\nu/k_BT)^3$ in the vicinity of $h\nu/k_BT = 0$. Calculate the relative error $(\overline{E}_{osc} - k_BT)/\overline{E}_{osc}$ if you had not included the additional terms for $\nu = 9.00 \times 10^{11}$ s^{-1} at temperatures of 1000., 600., and 200. K.

P12.3 The observed lines in the emission spectrum of atomic hydrogen are given by

$$\tilde{\nu}(\text{cm}^{-1}) = R_H(\text{cm}^{-1}) \left(\frac{1}{n_1^2} - \frac{1}{n^2}\right) \text{cm}^{-1}, \ n > n_1$$

In the notation favored by spectroscopists, $\tilde{\nu} = 1/\lambda = E/hc$ and $R_H = 109,677$ cm^{-1}. The Lyman, Balmer, and Paschen series refers to $n_1 = 1, 2,$ and 3, respectively, for emission from atomic hydrogen. What is the highest value of $\tilde{\nu}$ and E in each of these series?

P12.4 Calculate the speed that a gas-phase fluorine molecule would have if it had the same energy as an infrared photon ($\lambda = 1.00 \times 10^4$ nm), a visible photon ($\lambda = 500.$ nm), an ultraviolet photon ($\lambda = 100.$ nm), and an X-ray photon ($\lambda = 0.100$ nm). What temperature would the gas have if it had the same energy as each of these photons? Use the root mean square speed, $v_{rms} = \langle v^2 \rangle^{1/2} = \sqrt{3k_BT/m}$, for this calculation.

P12.5 Calculate the highest possible energy of a photon that can be observed in the emission spectrum of H.

P12.6 What is the maximum number of electrons that can be emitted if a potassium surface of work function 2.40 eV absorbs 5.00×10^{-3} J of radiation at a wavelength of 325 nm? What is the kinetic energy and velocity of the electrons emitted?

P12.7 Show that the energy density radiated by a blackbody

$$\frac{E_{total}(T)}{V} = \int_0^\infty \rho(\nu,T)d\nu = \int_0^\infty \frac{8\pi h\nu^3}{c^3} \frac{1}{e^{h\nu/k_BT} - 1} d\nu$$

depends on the temperature as T^4. (*Hint:* Make the substitution of variables $x = h\nu/k_BT$.) The definite integral $\int_0^\infty [x^3/(e^x - 1)]dx = \pi^4/15$. Using your result, calculate the energy density radiated by a blackbody at 1100. and 6000. K.

P12.8 What speed does an F_2 molecule have if it has the same momentum as a photon of wavelength 225 nm?

P12.9 A newly developed substance that emits 250. W of photons with a wavelength of 325 nm is mounted in a small rocket initially at rest in outer space such that all of the radiation is released in the same direction. Because momentum is conserved, the rocket will be accelerated in the opposite direction. If the total mass of the rocket is 14.2 kg, how fast will it be traveling at the end of 30 days in the absence of frictional forces?

P12.10 In the discussion of blackbody radiation, the average energy of an oscillator $\overline{E}_{osc} = h\nu/(e^{h\nu/k_BT} - 1)$ was approximated as $\overline{E}_{osc} = h\nu/[(1 + h\nu/k_BT) - 1] = k_BT$ for $h\nu/k_BT \ll 1$. Calculate the relative error $= (E - E_{approx})/E$ in making this approximation for $\nu = 7.5 \times 10^{12}$ s^{-1} at temperatures of 5000., 1500., and 300. K. Can you predict what the sign of the relative error will be without a detailed calculation?

P12.11 Using the root mean square speed, $v_{rms} = \langle v^2 \rangle^{1/2} = \sqrt{3k_BT/m}$, calculate the gas temperatures of He and Ar for which $\lambda = 0.25$ nm, a typical value needed to resolve diffraction from the surface of a metal crystal. On the basis of your result, explain why Xe atomic beams are not suitable for atomic diffraction experiments.

P12.12 Electrons have been used to determine molecular structure by diffraction. Calculate the speed and kinetic energy of an electron for which the wavelength is equal to a typical bond length, namely, 0.125 nm.

P12.13 For a monatomic gas, one measure of the "average speed" of the atoms is the root mean square speed, $v_{rms} = \langle v^2 \rangle^{1/2} = \sqrt{3k_BT/m}$, in which m is the molecular mass and k is the Boltzmann constant. Using this formula, calculate the de Broglie wavelength for H_2 and Ar at 200. and at 900. K.

P12.14 The distribution in wavelengths of the light emitted from a radiating blackbody is a sensitive function of the temperature. This dependence is used to measure the temperature of hot objects, without making physical contact with those objects, in a technique called *optical pyrometry*. In the limit $(hc/\lambda k_BT) \gg 1$, the maximum in a plot of $\rho(\lambda,T)$ versus λ is given by $\lambda_{max} = hc/5k_BT$. At what wavelength does the maximum in $\rho(\lambda,T)$ occur for $T = 675, 1150.,$ and 6200. K?

P12.15 A beam of electrons with a speed of 5.25×10^4 m/s is incident on a slit of width 200. nm. The distance to the detector plane is chosen such that the distance between the central maximum of the diffraction pattern and the first diffraction minimum is 0.300 cm. How far is the detector plane from the slit?

P12.16 If an electron passes through an electrical potential difference of 1 V, it has an energy of 1 electron-volt. What potential difference must it pass through in order to have a wavelength of 0.300 nm?

P12.17 Calculate the longest and the shortest wavelength observed in the Balmer series.

P12.18 X rays can be generated by accelerating electrons in a vacuum and letting them impact on atoms in a metal surface. If the 1250. eV kinetic energy of the electrons is completely converted to the photon energy, what is the wavelength of the

X rays produced? If the electron current is 3.50×10^{-5} A, how many photons are produced per second?

P12.19 The following data were observed in an experiment on the photoelectric effect from potassium:

10^{19} Kinetic Energy (J)	4.49	3.09	1.89	1.34	0.700	0.311
Wavelength (nm)	250.	300.	350.	400.	450.	500.

Graphically evaluate these data to obtain values for the work function and Planck's constant.

P12.20 The power (energy per unit time) radiated by a blackbody per unit area of surface expressed in units of W m^{-2} is given by $P = \sigma T^4$ with $\sigma = 5.67 \times 10^{-8}$ W m^{-2} K^{-4}. The radius of the sun is approximately 7.00×10^5 km and the surface temperature is 5800. K. Calculate the total energy radiated per second by the sun. Assume ideal blackbody behavior.

P12.21 The work function of palladium is 5.22 eV. What is the minimum frequency of light required to observe the photoelectric effect on Pd? If light with a 200. nm wavelength is absorbed by the surface, what is the velocity of the emitted electrons?

P12.22 Assume that water absorbs light of wavelength 4.20×10^{-6} m with 100% efficiency. How many photons are required to heat 5.75 g of water by 1.00 K? The heat capacity of water is 75.3 J mol^{-1} K^{-1}.

P12.23 Calculate the longest and the shortest wavelength observed in the Lyman series.

P12.24 A 1000. W gas discharge lamp emits 4.50 W of ultraviolet radiation in a narrow range centered near 275 nm. How many photons of this wavelength are emitted per second?

P12.25 The power per unit area emitted by a blackbody is given by $P = \sigma T^4$ with $\sigma = 5.67 \times 10^{-8}$ W m^{-2} K^{-4}. Calculate the power radiated per second by a spherical blackbody of radius 0.500 m at 925 K. What would the radius of a blackbody at 3000. K be if it emitted the same power as the spherical blackbody of radius 0.500 m at 925 K?

P12.26 A ground state H atom absorbs a photon and makes a transition to the $n = 4$ energy level. It then emits a photon of frequency 1.598×10^{14} s^{-1}. What is the final energy and n value of the atom?

P12.27 Pulsed lasers are powerful sources of nearly monochromatic radiation. Lasers that emit photons in a pulse of 5.00 ns duration with a total energy in the pulse of 0.175 J at 875 nm are commercially available.

a. What is the average power (energy per unit time) in units of watts (1 W = 1 J/s) associated with such a pulse?

b. How many photons are emitted in such a pulse?

Web-Based Simulations, Animations, and Problems

W12.1 The maximum in a plot of the spectral density of blackbody radiation versus T is determined for a number of values of T using numerical methods. Using these results, the validity of the approximation $\lambda_{max} = hc/5k_B T$ is tested graphically.

W12.2 The total radiated energy of blackbody radiation is calculated numerically for the temperatures of W12.1. Using these results, the exponent in the relation $E = CT^\alpha$ is determined graphically.

W12.3 Diffraction of visible light from a single slit is simulated. The slit width and light wavelength are varied using sliders. The student is asked to draw conclusions about how the diffraction pattern depends on these parameters.

W12.4 Diffraction of a particle from single and double slits is simulated. The intensity distribution on the detector plane is updated as each particle passes through the slits. The slit width and light wavelength are varied using sliders. The student is asked to draw conclusions about how the diffraction pattern depends on these parameters.

The Schrödinger Equation

T he key to understanding why classical mechanics does not provide an appropriate framework for understanding phenomena at the atomic level is the recognition that wave-particle duality needs to be integrated into the physics. Rather than solving Newton's equations of motion for a particle, an appropriate wave equation needs to be solved for the wave-particle. Erwin Schrödinger was the first to formulate such an equation successfully. Operators, eigenfunctions, wave functions, and eigenvalues are key concepts that arise in a viable framework to solve quantum mechanical wave equations. The eigenvalues correspond to the possible values of measured results, or observables, in an experiment. These new concepts are introduced in this chapter, and will be discussed in Chapters 14 and 15.

13.1 What Determines If a System Needs to Be Described Using Quantum Mechanics?

Quantum mechanics was viewed as a radically different way of looking at matter at the molecular, atomic, and subatomic levels in the 1920s. However, the historical distance we have from what was a revolution at the time makes the quantum view much more familiar today. It is important to realize that classical and quantum mechanics are not two competing ways to describe the world around us. Each has its usefulness in a different regime of physical properties that describe reality. Quantum mechanics merges seamlessly into classical mechanics in moving from atoms to masses the size of baseballs. Classical mechanics can be derived from quantum mechanics in the limit that allowed energy values are continuous rather than discrete. Some of these complexities will require differentiated thinking as you gain an understanding of quantum mechanics. For instance, it is not correct to say that in dealing with atoms, a quantum mechanical description must always be used.

To illustrate this point, consider a container filled with argon gas at a low pressure. At the atomic level, the origin of pressure is the collision of rapidly and randomly moving argon atoms with the container walls. Classical mechanics gives a perfectly good description of the origin of pressure in this case. However, if we pass ultraviolet light through hydrogen gas and ask how much energy can be taken up by an H_2 molecule, we must use a quantum mechanical description. At first, this seems puzzling—why do we need quantum mechanics in one case but not the other? On further consideration, we discover that a very few important relationships govern whether a classical description suffices in a given case. We next discuss these relationships in order to develop an understanding of when to use a classical description and when to use a quantum description for a given system.

The essence of quantum mechanics is that particles and waves are not really separate and distinct entities. Waves can show particle-like behavior as illustrated by the photoelectric effect. Particles can also show wave-like properties as shown by the diffraction of atomic beams from surfaces. How can we develop criteria that tell us when a particle description (classical) of an atomic or molecular system is sufficient and when we need to use a wave description (quantum mechanical)? Two criteria are used: the magnitude of the wavelength of the particle relative to the dimensions of the problem and the degree to which the allowed energy values form a **continuous energy spectrum.**

A good starting point is to think about diffraction of light of wavelength λ passing through a slit of width a. Ray optics is a good description as long as $\lambda \ll a$. Diffraction is only observed when the wavelength is comparable to the slit width. How big is the wavelength of a molecule? Of a macroscopic mass like a baseball? By putting numbers into Equation (12.11), we will find that the wavelength for a room temperature H_2 molecule is about 10^{-10} m and that for a baseball is about 10^{-34} m. Keep in mind that because p rather than v appears in the denominator of Equation (12.11) which defines the wavelength of a particle, the wavelength of a toluene molecule with the same velocity as an H_2 molecule is about a factor of 50 smaller. As we learned in discussing the Davisson–Germer experiment in Chapter 12, crystalline solids have regular spacings that are appropriate for the diffraction of electrons as well as light atoms and molecules. Particle diffraction is a demonstration of wave-particle duality. To see the wave character of a baseball, we need to come up with a diffraction experiment. We will not see diffraction of a baseball because we cannot construct an opening whose size is ~1×10^{-34} m. This does not mean that wave-particle duality breaks down for macroscopic masses; it simply means that the wave character of a baseball does not manifest. There is no sharp boundary such that above a certain value for the momentum we are dealing with a particle and below it we are dealing with a wave. The degree to which each of these properties is exhibited flows smoothly from one extreme to the other. Consider the second example cited earlier. Adding energy to hydrogen molecules using UV light cannot be treated classically, because energy is taken up by the electrons in H_2. The localization of the electrons to a small volume around the nuclei brings out their wave-like character, and therefore the process must be described using quantum mechanics.

We next discuss the second criterion for determining when we need a quantum mechanical description of a system. It is based on the energy spectrum of the system. Because all values of the energy are allowed for a classical system, it is said to have a **continuous energy spectrum.** In a bounded quantum mechanical system, only certain values of the energy are allowed, and such a system has a **discrete energy spectrum.** To make this criterion quantitative, we need to discuss the Boltzmann distribution.

We will cover more of Boltzmann's work in statistical thermodynamics. At this point, we attempt to make his most important result plausible so that we can apply it in our studies of quantum mechanics. Consider a one-liter container filled with an ideal atomic gas at the standard conditions of 1 bar and a temperature of 298.15 K. Because the atoms have no rotational or vibrational degrees of freedom, all of their energy is in the form of translational kinetic energy. At equilibrium, not all of the atoms have the same kinetic energy. In fact, the atoms exhibit a broad range of energies. To define the distribution of atoms having a given energy, descriptors, such as the mean, the median, or the most probable energy per atom are used. For the atoms under consideration, the root mean square energy is simply related to the absolute temperature T by

$$E_{rms} = \frac{3}{2}k_B T \tag{13.1}$$

The Boltzmann constant k_B is the familiar ideal gas law constant R divided by Avogadro's number.

We said that there is a broad distribution of kinetic energy in the gas for the individual atoms. What governs the probability of observing one value of the energy as opposed to another? This question led Ludwig Boltzmann to one of the most important equations in physics and chemistry. Looking specifically at our case, it relates the number of atoms n_i that have energy ε_i to the number of atoms n_j that have energy ε_j by the equation

$$\frac{n_i}{n_j} = \frac{g_i}{g_j} e^{-[\varepsilon_i - \varepsilon_j]/k_B T} \qquad (13.2)$$

This formula is called the **Boltzmann distribution.** An important concept to keep in mind is that a formula is just a shorthand way of describing phenomena that occur in the real world. It is critical to understand what lies behind the formula. Take a closer look at this equation. It says that the ratio of the number of atoms having the energy ε_i to the number having the energy ε_j depends on three things. It depends exponentially on the difference in the energies and the reciprocal of the temperature. This means that this ratio varies rapidly with temperature and $\varepsilon_i - \varepsilon_j$. The equation also states that it is the ratio of the energy difference to $k_B T$ that is important. What is $k_B T$? It has the units of energy and is approximately the average energy that an atom has at temperature T. We can understand this exponential term as telling us that the larger the temperature, the closer the ratio n_i/n_j will be to unity; the probability of an atom having a given energy falls off exponentially with increasing energy.

The third factor that influences the ratio n_i/n_j is the ratio g_i/g_j. The quantities g_i and g_j are the degeneracies of the energy levels i and j. The **degeneracy** of an energy level counts the number of ways that an atom can have an energy ε within the interval $\varepsilon_i - \Delta\varepsilon < \varepsilon < \varepsilon_i + \Delta\varepsilon$. The degeneracy can depend on the energy. In our example, degeneracy can be illustrated as follows. The energy of an atom $\varepsilon_i = \frac{1}{2}mv_i^2$ is determined by $v_i^2 = v_{xi}^2 + v_{yi}^2 + v_{zi}^2$. We have explicitly written that the energy depends only on the speed of the atom and not on its individual velocity components. For a fixed value of $\Delta\varepsilon$, there are many more ways of combining different individual velocity components to give the same speed at large speeds than there are for low speeds. Therefore, the degeneracy corresponding to a particular energy ε_i increases with the speed.

The importance of these considerations will become clearer as we continue to apply quantum mechanics to atoms and molecules. As already stated, a quantum mechanical system has a discrete rather than continuous energy spectrum. If $k_B T$ is small compared to the spacing between allowed energies, the distribution of states in energy will be very different from a classical system, which has a continuous energy spectrum. On the other hand, if $k_B T$ is much larger than the energy spacing, classical and quantum mechanics will give the same result for the relative numbers of atoms or molecules of different energy. This scenario occurs in either of two limits: large T or small $\varepsilon_i - \varepsilon_j$, illustrating how it is possible to have a continuous transition between classical and quantum mechanics. A large increase in T could cause a system that exhibited quantum behavior at low temperatures to exhibit classical behavior at high temperatures. A calculation using the Boltzmann distribution for a two-level system is carried out in Example Problem 13.1.

EXAMPLE PROBLEM 13.1

Consider a system of 1000. particles that can only have two energies, ε_1 and ε_2, with $\varepsilon_2 > \varepsilon_1$. The difference in the energy between these two values is $\Delta\varepsilon = \varepsilon_2 - \varepsilon_1$. Assume that $g_1 = g_2 = 1$.

a. Graph the number of particles, n_1 and n_2, in states ε_1 and ε_2 as a function of $k_B T/\Delta\varepsilon$. Explain your result.

b. At what value of $k_B T/\Delta\varepsilon$ do 750. of the particles have the energy ε_1?

Solution

Using information from the problem and Equation (13.2), we can write down the following two equations: $n_2/n_1 = e^{-\Delta\varepsilon/k_BT}$ and $n_1 + n_2 = 1000$. We solve these two equations for n_2 and n_1 to obtain

$$n_2 = \frac{1000.\,e^{-\Delta\varepsilon/k_BT}}{1 + e^{-\Delta\varepsilon/k_BT}} \quad \text{and} \quad n_1 = \frac{1000.}{1 + e^{-\Delta\varepsilon/k_BT}}$$

If these functions are plotted as a function of $kT/\Delta\varepsilon$, the following graphs result:

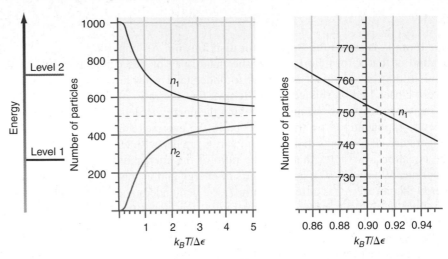

We see from the left graph that as long as $k_BT/\Delta\varepsilon$ is small, the vast majority of the particles have the lower energy. How can we interpret this result? As long as the thermal energy of the particle, which is approximately k_BT, is much less than the difference in energy between the two allowed values, the particles with the lower energy are unable to gain energy through collisions with other particles. However, as $k_BT/\Delta\varepsilon$ increases (which is equivalent to a temperature increase for a fixed energy difference between the two values), the random thermal energy available to the particles enables some of them to jump up to the higher energy value. Therefore, n_1 decreases and n_2 increases. For all finite temperatures, $n_1 > n_2$. As T approaches infinity, n_1 becomes equal to n_2.

Part (b) is solved graphically. n_1 is shown as a function of $k_BT/\Delta\varepsilon$ on an expanded scale on the right side of the preceding graphs, and we see that $n_1 = 750$. for $k_BT/\Delta\varepsilon = 0.91$.

Example Problem 13.1 shows that the population of states associated with the energy values ε_i and ε_j are very different if

$$\frac{(\varepsilon_i - \varepsilon_j)}{k_BT} \gg 1 \tag{13.3}$$

and very similar if the inequality is reversed. What are the consequences of this result?

Consider a quantum mechanical system, which, unlike a classical system, has a discrete energy spectrum. The allowed values of energy are called **energy levels.** Anticipating a system that we will deal with in Chapter 18, we refer specifically to the vibrational energy levels of a molecule. The allowed levels are equally spaced with an interval ΔE. These discrete energy levels are numbered with integers, beginning with one. Under what conditions will this quantum mechanical system *appear* to follow classical behavior? It will do so if the discrete energy spectrum *appears* to be continuous. How can this occur? In a gas at equilibrium, the total energy of an individual molecule fluctuates within a range $\Delta E \approx k_BT$ through collisions of molecules with one another. Therefore, the energy of a molecule with a particular vibrational quantum number fluctuates within a range of width k_BT centered at the particular energy level. A plot of the relative number of molecules having a vibrational energy E as a function of E is shown in Figure 13.1 for sharp energy levels and

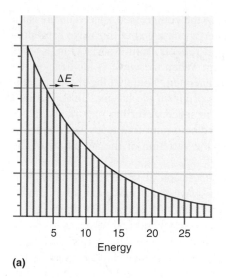

(a)

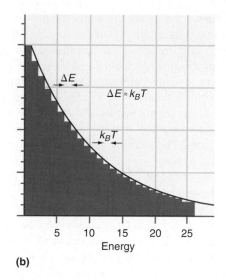

(b)

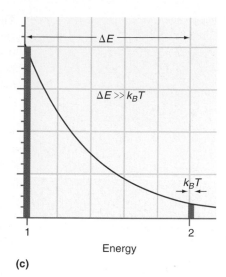

(c)

for the two indicated limits, $\Delta E \approx k_B T$ and $\Delta E \gg k_B T$. The plot is generated using the Boltzmann distribution.

For $\Delta E \approx k_B T$, each discrete energy level is sufficiently broadened by energy fluctuations that adjacent energy levels can no longer be distinguished. This is indicated by the overlap of the purple bars representing individual states shown in Figure 13.1b. In this limit, any energy that we choose in the range shown lies in the purple area. It corresponds to an allowed value and therefore the discrete energy spectrum *appears* to be continuous. Classical behavior will be observed under these conditions. However, if $\Delta E \gg k_B T$, an arbitrarily chosen energy in the range lies in the green area with high probability, because the purple bars of width $k_B T$ are widely separated. The range between the two purple bars corresponds to forbidden energies and, therefore, the discontinuous nature of the energy spectrum is observable. Quantum mechanical behavior is observed under these conditions.

Summarizing this discussion, if the allowed energies form a continuum, classical mechanics is sufficient to describe *that feature* of the system. If the allowed energies are discrete, a quantum mechanical description is needed. The words "that feature" require emphasis. The pressure exerted by the H_2 molecules in the box arises from momentum transfer governed by the molecules' translational energy spectrum, which *appears* to be continuous, as we will learn in Chapter 15. Therefore, we do not need quantum mechanics to discuss the pressure in the box. However, if we discuss light absorption by the same H_2 molecules, a quantum mechanical description of light absorption is required. This is so because light absorption involves an electronic excitation of the molecule, with electronic energy level spacings much larger than $k_B T$. Therefore, these levels remain discrete at all reasonable temperatures.

FIGURE 13.1
The relative population in the different energy levels designated by the integers 1–25 (vertical axis) is plotted at constant T as a function of energy for **(a)** sharp energy levels, **(b)** for $\Delta E \approx k_B T$, and **(c)** for $\Delta E \gg k_B T$. In parts (a), (b), and (c), the Boltzmann distribution describes the relative populations. However, the system behaves as if it has a continuous energy spectrum only if $\Delta E \approx k_B T$ or if $\Delta E < k_B T$.

13.2 Classical Waves and the Nondispersive Wave Equation

In Chapter 12, we learned that particles exhibit wave character under certain conditions. This suggests that there is a wave equation that should be used to describe particles. This equation is called the Schrödinger equation, and it is the fundamental equation used to describe atoms and molecules. However, before discussing the Schrödinger equation, we briefly review classical waves and the classical wave equation.

What characteristics differentiate waves and particles? Think about the collision between two billiard balls. We can treat the balls as point masses (any pool player will recognize this as an idealization) and apply Newton's laws of motion to calculate trajectories, momenta, and energies as a function of time if we know all the forces acting on the balls. Now think of a person shouting. Often there is an echo. What is happening here? The vocal cords create a local compression of the air in the larynx. This compression zone propagates away from its source as a wave with the speed of sound.

The louder the sound, the larger the pressure is in the compressed zone. The pressure variation is the amplitude of the wave and the energy contained in this wave is proportional to the square of the amplitude. The sound reflects from a surface and comes back as a weakened local compression of the air. When the wave is incident on the eardrum, a signal that we recognize as sound is generated. Note that the energy associated with the sound wave is only localized at its origin in the larynx. Unlike billiard balls, waves are not just located at a single point in space. A further important characteristic of a wave is that it has a characteristic velocity and frequency with which it propagates. The velocity and frequency govern the variation of the amplitude of the wave with time.

A wave can be represented pictorially by a succession of **wave fronts,** corresponding to surfaces over which the amplitude of the wave has a maximum or minimum value. A point source emits **spherical waves** as shown in Figure 13.2b, and the light passing through a rectangular slit can be represented by cylindrical waves as shown in Figure 13.2c. The waves sent out from a faraway source such as the sun when viewed from Earth are spherical waves with such little curvature that they can be represented as **plane waves** as shown in Figure 13.2a.

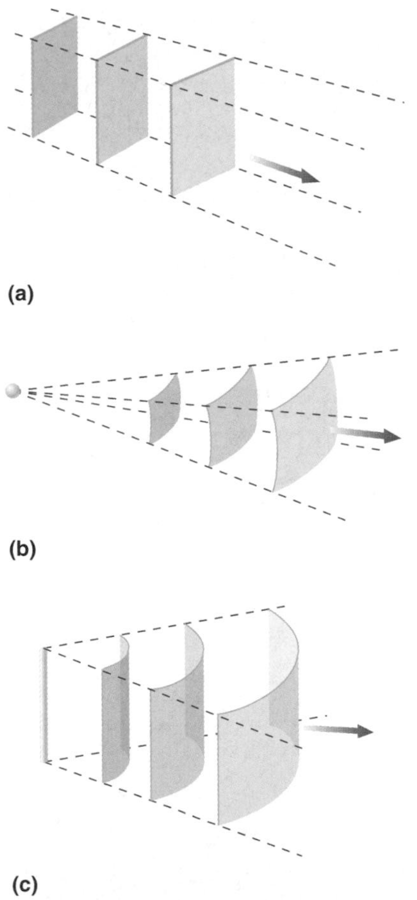

Mathematically, the amplitude of a wave can be described by a **wave function.** The wave function describes how the amplitude of the wave depends on the variables x and t. The variable x is measured along the direction of propagation. For convenience, only sinusoidal waves of wavelength λ and the single **frequency** $\nu = 1/T$, where T is the **period,** are considered. The velocity v, frequency ν, and **wavelength** λ, are related by $v = \lambda\nu$. The peak-to-peak amplitude of the wave is $2A$:

$$\Psi(x, t) = A \sin 2\pi\left(\frac{x}{\lambda} - \frac{t}{T}\right) \tag{13.4}$$

In this equation we have arbitrarily chosen our zero of time and distance such that $\Psi(0,0) = 0$. This equation represents a wave that is moving in the direction of positive x. We can prove this by considering how a specific feature of this wave changes with time. The wave amplitude is zero for

$$2\pi\left(\frac{x}{\lambda} - \frac{t}{T}\right) = n\pi \tag{13.5}$$

where n is an integer. Solving for x, the location of the nodes is obtained:

$$x = \lambda\left(\frac{n}{2} + \frac{t}{T}\right) \tag{13.6}$$

Note that x increases as t increases, showing that the wave is moving in the direction of positive x. Figure 13.3 shows a graph of the wave function given in Equation (13.4). To graph this function in two dimensions, one of the variables is kept constant.

The functional form in Equation (13.4) appears so often that it is convenient to combine some of the constants and variables to write the wave amplitude as

$$\Psi(x, t) = A \sin(kx - \omega t) \tag{13.7}$$

The quantity k is called the **wave vector** and is defined by $k = 2\pi/\lambda$. The quantity $\omega = 2\pi\nu$ is called the **angular frequency.**

Because the wave amplitude is a simple sine function in our case, it has the same value as the argument changes by 2π. The choice of a zero in position or time is arbitrary and is chosen at our convenience. To illustrate this, consider Equation (13.7) rewritten in the form

$$\Psi(x, t) = A \sin(kx - \omega t + \phi) \tag{13.8}$$

in which the quantity ϕ has been added to the argument of the sin function. This is appropriate when $\Psi(0,0) \neq 0$. The argument of the wave function is called the **phase,** and a change in the initial phase ϕ shifts the wave function to the right or left relative to the horizontal axes in Figure 13.3 depending on the sign of ϕ.

(a)

(b)

(c)

FIGURE 13.2
Waves can be represented by a succession of surfaces over which the amplitude of the wave has its maximum or minimum value. The distance between successive surfaces is the wavelength. Representative surfaces are shown for **(a)** plane waves, **(b)** spherical waves, and **(c)** cylindrical waves. The direction of propagation of the waves is perpendicular to the surfaces as indicated by the blue arrows.

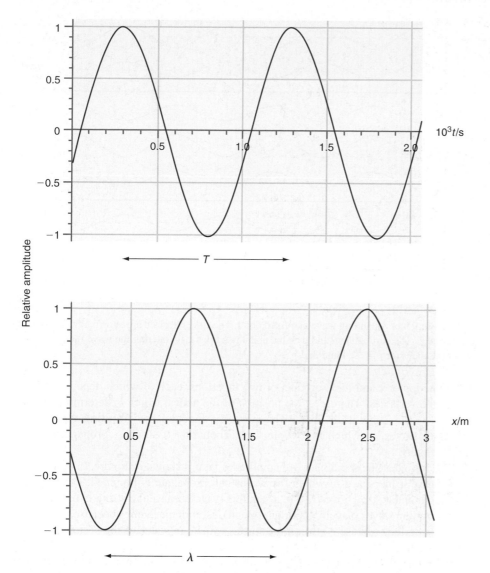

FIGURE 13.3
The upper panel shows the wave amplitude as a function of time at a fixed point. The wave is completely defined by the period, the maximum amplitude, and the amplitude at $t = 0$. The lower panel shows the analogous information when the wave amplitude is plotted as a function of distance for a fixed time.
$\lambda = 1.46$ m and $T = 1.00 \times 10^{-3}$ s.

When two or more waves are present in the same region of space, their time-dependent amplitudes add together, and the waves are said to interfere with one another. The **interference** between two waves gives rises to an enhancement in a region of space (**constructive interference**) if the wave amplitudes are both positive or both negative. It can also lead to a cancellation of the wave amplitude in a region of space (**destructive interference**) if the wave amplitudes are opposite in sign and equal in amplitude. At the constructive interference condition, maxima of the waves from the two sources line up (constructive interference) because the phases of the two waves are the same to within an integral multiple of 2π. They are out of phase at the destructive interference condition where the phases differ by $(2n + 1)\pi$, where n is an integer.

Interference can also result in a very different time dependence of the wave amplitude than was discussed for a traveling wave, namely, the formation of spatially fixed nodes where the amplitude is zero at all times. Consider the superposition of two waves of the same frequency and amplitude that are moving in opposite directions. The resultant wave amplitude is the sum of the individual amplitudes:

$$\Psi(x, t) = A[\sin(kx - \omega t) + \sin(kx + \omega t)] \tag{13.9}$$

Using the standard trigonometric identity $\sin(\alpha \pm \beta) = \sin \alpha \cos \beta \pm \cos \alpha \sin \beta$, Equation (13.9) can be simplified to

$$\Psi(x, t) = 2A \sin kx \cos \omega t = \psi(x) \cos \omega t \tag{13.10}$$

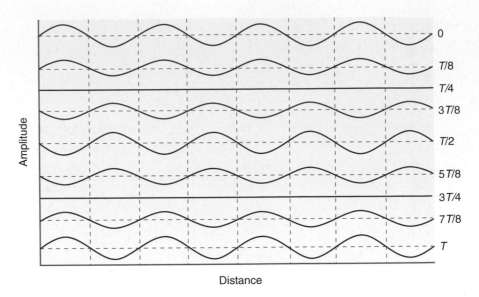

FIGURE 13.4
Time evolution of a standing wave at a fixed point. The time intervals are shown as a function of the period T. The vertical lines indicate the nodal positions x_0. Note that the wave function has temporal nodes for $t = T/4$ and $3T/4$.

This function of x and t is a product of two functions, each of which depends only on one of the variables. Therefore, the position of the nodes, which is determined by $\sin kx = 0$, is the same at all times. This property distinguishes **standing waves** from **traveling waves,** in which the whole wave, including the nodes, propagates at the same velocity.

The form that the standing wave amplitudes take is shown in Figure 13.4. Standing waves arise if the space in which the waves can propagate is bounded. For instance, plucking a guitar string gives rise to a standing wave because the string is fixed at both ends. Standing waves play an important role in quantum mechanics because, as demonstrated later, they represent **stationary states,** which are states of the system in which the measurable properties of the system do not change with time.

We return to the functional dependencies of the wave amplitude on time and distance for a traveling wave. For wave propagation in a medium for which all frequencies move with the same velocity (a nondispersive medium), the variation of the amplitude with time and distance are related by

$$\frac{\partial^2 \Psi(x, t)}{\partial x^2} = \frac{1}{v^2} \frac{\partial^2 \Psi(x, t)}{\partial t^2} \qquad \textbf{(13.11)}$$

Equation (13.11) is known as the **classical nondispersive wave equation** and v designates the velocity at which the wave propagates. This equation provides a starting point in justifying the Schrödinger equation, which is the fundamental quantum mechanical wave equation. (See the Math Supplement, Appendix A, for a discussion of partial differentiation.) Example Problem 13.2 demonstrates that the traveling wave that we have used in Equation (13.8) is a solution of the nondispersive wave equation.

EXAMPLE PROBLEM 13.2

The nondispersive wave equation in one dimension is given by

$$\frac{\partial^2 \Psi(x, t)}{\partial x^2} = \frac{1}{v^2} \frac{\partial^2 \Psi(x, t)}{\partial t^2}$$

Show that the traveling wave $\Psi(x, t) = A \sin(kx - \omega t + \phi)$ is a solution of the nondispersive wave equation. How is the velocity of the wave related to k and ω in this case?

Solution

$$\frac{\partial^2 \Psi(x,t)}{\partial x^2} = \frac{1}{v^2} \frac{\partial^2 \Psi(x,t)}{\partial t^2}$$

$$\frac{\partial^2 A \sin(kx - \omega t + \phi)}{\partial x^2} = -k^2 A \sin(kx - \omega t + \phi)$$

$$\frac{1}{v^2} \frac{\partial^2 A \sin(kx - \omega t + \phi)}{\partial t^2} = \frac{-\omega^2}{v^2} A \sin(kx - \omega t + \phi)$$

Because $v = \omega/k$, the nondispersive wave equation is obeyed.

13.3 Waves Are Conveniently Represented as Complex Functions

The mathematics of dealing with wave functions is much simpler if they are represented in the complex number plane. As discussed, a wave traveling in the positive x direction can be described by the function

$$\Psi(x,t) = A \sin(kx - \omega t + \phi) = A \cos(kx - \omega t + \phi - \pi/2)$$

$$= A \cos(kx - \omega t + \phi') \qquad \text{(13.12)}$$

where $\phi' = \phi - \pi/2$. Using Euler's formula, $e^{i\alpha} = \exp(i\alpha) = \cos \alpha + i \sin \alpha$, Equation (13.12) can be written as

$$\Psi(x,t) = \text{Re}\left(A e^{i(kx - \omega t + \phi')}\right)$$

$$= \text{Re}\left(A \exp i(kx - \omega t + \phi')\right) \qquad \text{(13.13)}$$

in which the notation Re indicates that we are considering only the real part of the complex function that follows.

Whereas the wave functions considered previously (for example, sound waves) have real amplitudes, quantum mechanical wave functions can have complex amplitudes. Working with only the real part of the functions makes some of the mathematical treatment more cumbersome, so it is easier to work with the whole complex function knowing that we can always extract the real part if we wish to do so.

The wave function of Equation (13.13) can then be written in the form

$$\Psi(x,t) = A \exp i(kx - \omega t + \phi') \qquad \text{(13.14)}$$

where A is a constant. All quantities that fully characterize the wave, namely the maximum amplitude, the wavelength, the period of oscillation, and the phase angle for $t = 0$ and $x = 0$, are contained in Equation (13.14). Operations such as differentiation, integration, and adding two waves to form a superposition are much easier when working with complex exponential notation than with real trigonometric wave functions.

The following bullets list the properties of complex numbers that will be used frequently in this book. Example Problems 13.3 and 13.4 show how to work with complex numbers. See the Math Supplement (Appendix A) for a more detailed discussion of complex numbers.

- A complex number or function can be written as $a + ib$, where a and b are real numbers or real functions and $i = \sqrt{-1}$, or equivalently in the form $re^{i\theta}$, where $r = \sqrt{a^2 + b^2}$ and $\theta = \sin^{-1}(b/r)$.

- The **complex conjugate** of a complex number or function f is denoted f^*. The complex conjugate is obtained by substituting $-i$ in f wherever i occurs. The complex conjugate of the number or function $a + ib$, is $a - ib$, and the complex conjugate of $re^{i\theta}$ is $re^{-i\theta}$.

- The magnitude of a complex number or function f is a real number or function denoted $|f|$, where $|f| = \sqrt{f^* f}$. For $f = a + ib$ or $re^{-i\theta}$, $|f| = \sqrt{a^2 + b^2}$, or r, respectively.

EXAMPLE PROBLEM 13.3

a. Express the complex number $4 - 4i$ in the form $re^{i\theta}$.

b. Express the complex number $3e^{i3\pi/2}$ in the form $a + ib$.

Solution

a. The magnitude of $4 + 4i$ is $[(4 + 4i)(4 - 4i)]^{1/2} = 4\sqrt{2}$. The phase is given by

$$\sin\theta = \frac{-4}{4\sqrt{2}} = -\frac{1}{\sqrt{2}} \text{ or } \theta = \sin^{-1}\left(-\frac{1}{\sqrt{2}}\right) = -\frac{\pi}{4} \text{ or } \frac{7\pi}{4}$$

Therefore, $4 - 4i$ can be written as $4\sqrt{2}\,e^{-i(\pi/4)}$.

b. Using the relation $e^{i\alpha} = \exp(i\alpha) = \cos\alpha + i\sin\alpha$, $3e^{i3\pi/2}$ can be written as

$$3\left(\cos\frac{3\pi}{2} + i\sin\frac{3\pi}{2}\right) = 3(0 - i) = -3i$$

EXAMPLE PROBLEM 13.4

Determine the magnitude of the following complex numbers:

a. $(1 + i)(\sqrt{2} + 5i)$

c. $\dfrac{e^{\sqrt{2}i\pi}\,e^{-3i\pi}}{4e^{i\pi/4}}$

b. $\dfrac{1 + \sqrt{3}i}{11 - 2i}$

d. $\dfrac{1 + 6i}{i}$

Solution

The magnitude of a complex number or function f is $\sqrt{f^*f}$. Note that the magnitude of a complex number is a real number.

a. $\sqrt{(1 + i)(\sqrt{2} + 5i)(1 - i)(\sqrt{2} - 5i)} = 3\sqrt{6}$

b. $\sqrt{\dfrac{1 + \sqrt{3}i}{11 - 2i}\dfrac{1 - \sqrt{3}i}{11 + 2i}} = \dfrac{2}{5\sqrt{5}}$

c. $\sqrt{\dfrac{e^{\sqrt{2}i\pi}\,e^{-3i\pi}}{4e^{i\pi/4}}\dfrac{e^{-\sqrt{2}i\pi}\,e^{+3i\pi}}{4e^{-i\pi/4}}} = \dfrac{1}{4}$

d. $\sqrt{\dfrac{1 + 6i}{i}\dfrac{1 - 6i}{-i}} = \sqrt{37}$

13.4 Quantum Mechanical Waves and the Schrödinger Equation

In this section, we justify the time-independent Schrödinger equation by combining the classical nondispersive wave equation and the de Broglie relation. For classical standing waves, we showed in Equation (13.10) that the wave function is a product of two functions, one of which depends only on spatial coordinates, and the other of which depends only on time:

$$\Psi(x, t) = \psi(x)\cos\omega t \qquad (13.15)$$

If this function is substituted in Equation (13.11), we obtain

$$\frac{d^2\psi(x)}{dx^2} + \frac{\omega^2}{v^2}\psi(x) = 0 \qquad (13.16)$$

The time-dependent part $\cos \omega t$ cancels because it appears on both sides of the equation after the derivative with respect to time is taken. Using the relations $\omega = 2\pi\nu$ and $\nu\lambda = \text{v}$, Equation (13.16) becomes

$$\frac{d^2\psi(x)}{dx^2} + \frac{4\pi^2}{\lambda^2}\psi(x) = 0 \qquad (13.17)$$

To this point, everything that we have written is for a classical wave. We introduce quantum mechanics by using the de Broglie relation, $\lambda = h/p$, for the wavelength. The momentum is related to the total energy E and the potential energy $V(x)$ by

$$\frac{p^2}{2m} = E - V(x) \quad \text{or} \quad p^2 = 2m(E - V(x)) \qquad (13.18)$$

Introducing this expression for the momentum into the de Broglie relation, and substituting the expression obtained for λ into Equation (13.17), we obtain

$$\frac{d^2\psi(x)}{dx^2} + \frac{8\pi^2 m}{h^2}[E - V(x)]\psi(x) = 0 \qquad (13.19)$$

Using the abbreviation $\hbar = h/2\pi$ and rewriting Equation (13.19), we obtain the **time-independent Schrödinger equation** in one dimension:

$$-\frac{\hbar^2}{2m}\frac{d^2\psi(x)}{dx^2} + V(x)\psi(x) = E\psi(x) \qquad (13.20)$$

This is the fundamental equation used to study the stationary states of quantum mechanical systems. The familiar $1s$ and $2p_z$ orbitals of the hydrogen atom are examples of stationary states obtained from the time-independent Schrödinger equation.

There is an analogous quantum mechanical form of the time-dependent classical nondispersive wave equation. It is called the **time-dependent Schrödinger equation** and has the following form:

$$-\frac{\hbar^2}{2m}\frac{\partial^2\Psi(x,t)}{\partial x^2} + V(x,t)\Psi(x,t) = i\hbar\frac{\partial\Psi(x,t)}{\partial t} \qquad (13.21)$$

This equation relates the temporal and spatial derivatives of $\Psi(x, t)$ with the potential energy function $V(x, t)$. It is applied in systems in which the energy changes with time. For example, the time-dependent equation is used to model transitions, in which the energy of a molecule changes as it absorbs a photon.

These two equations that Schrödinger formulated are the basis of all quantum mechanical calculations. Their validity has been confirmed by countless experiments carried out during the last 80 years. The equations look very different from Newton's equations of motion. The mass of the particle appears in both forms of the Schrödinger equation, but what meaning can be attached to $\Psi(x, t)$ and $\psi(x)$? This question will be discussed in some detail in Chapter 14. For now, we say that $\Psi(x, t)$ and $\psi(x)$ represent the amplitude of the wave that describes the particle or system of particles under consideration. To keep the notation simple, we have considered a one-dimensional system, but, in general, the spatial part of Ψ, denoted ψ, depends on all spatial coordinates.

Our main focus is on the stationary states of a quantum mechanical system. For these states, both the time-dependent and time-independent Schrödinger equations are satisfied. In this case,

$$i\hbar\frac{\partial\Psi(x,t)}{\partial t} = E\Psi(x,t) \qquad (13.22)$$

For stationary states, $\Psi(x, t) = \psi(x)f(t)$. Substituting this expression in Equation (13.22) gives

$$i\hbar\frac{df(t)}{dt} = Ef(t) \quad \text{or} \quad \frac{df(t)}{dt} = -i\frac{E}{\hbar}f(t) \qquad (13.23)$$

Solving this equation, we obtain $f(t) = e^{-i(E/\hbar)t}$. We have shown that wave functions that describe states whose energy is independent of time have the form

$$\Psi(x, t) = \psi(x)e^{-i(E/\hbar)t} \tag{13.24}$$

Note that $\Psi(x, t)$ is the product of two functions, each of which depends on only one variable. A standing wave as in Equation (13.15) has the same form. That is not a coincidence, because stationary states in quantum mechanics are represented by standing waves.

13.5 Solving the Schrödinger Equation: Operators, Observables, Eigenfunctions, and Eigenvalues

Now that we have introduced the quantum mechanical wave equation, we need to learn how to work with it. In this section, we develop this topic by introducing the language used in solving the Schrödinger equation. The key concepts introduced are those of *operators, observables, eigenfunctions,* and *eigenvalues*. These terms are defined later. A good understanding of these concepts is necessary to understand the quantum mechanical postulates in Chapter 14.

Both forms of the Schrödinger equation are differential equations whose solutions depend on the potential energy $V(x)$. Our emphasis in the next chapters will be on using the solutions of the time-independent equation for various problems such as the harmonic oscillator or the H atom to enhance our understanding of chemistry. We do not focus on methods to solve differential equations. However, it is very useful to develop a general understanding of the formalism used to solve the time-independent Schrödinger equation. This initial introduction is brief because our primary goal is to obtain a broad overview of the language of quantum mechanics. As we work with these new concepts in successive chapters, they will become more familiar.

We begin by illustrating the meaning of the term *operator* in the context of classical mechanics. Think about how we would describe the time evolution of a system consisting of a particle on which a force is acting. The velocity at time t_1 is known and we wish to know the velocity at a later time t_2. We write down Newton's second law

$$m\frac{d^2x}{dt^2} = F(x, t) \tag{13.25}$$

and integrate it to give

$$v(t_2) = v(t_1) + \frac{1}{m}\int_{t_1}^{t_2} F(x, t)\, dt \tag{13.26}$$

In words, one could describe this process as the series of operations:

- Integrate the force acting on the particle over the interval t_1 to t_2.
- Multiply by the inverse of the mass.
- Add this quantity to the velocity at time t_1.

These actions have the names *integrate, form the inverse, multiply,* and *add*, and they are all called operators. Note that we started at the right-hand side of the equation and worked our way to the left.

How are operators used in quantum mechanics? To every measurable quantity (**observable**), such as energy, momentum, or position, there is a corresponding **operator** in quantum mechanics. Quantum mechanical operators usually involve differentiation with respect to a variable such as x or multiplication by x or a function of the energy such as $V(x)$. Operators are denoted by a caret: $\hat{O}$.

Just as a differential equation has a set of solutions, an operator $\hat{O}$ has a set of eigenfunctions and eigenvalues. This means that there is a set of wave functions ψ_n with the index n such that

$$\hat{O}\psi_n = a_n\psi_n \qquad (13.27)$$

The operator acting on these special wave functions returns the wave function multiplied by a number. These special functions are called the **eigenfunctions** of the operator and the a_n are called the **eigenvalues.** The eigenvalues for quantum mechanical operators are always real numbers because they correspond to the values of observables that are measured in an experiment. There are in general an infinite number of eigenfunctions for a given operator for the specific system under consideration. For example, the eigenfunctions for the total energy operator (kinetic plus potential energy) for the hydrogen atom are the wave functions that describe the orbitals that we know as $1s$, $2s$, $2p_x$, ... The set of these eigenfunctions is infinite in size. The corresponding eigenvalues are the $1s$, $2s$, $2p_x$, . . . orbital energies.

We can now recognize that the time-independent Schrödinger equation is an eigenvalue equation for the total energy, E

$$\left\{\frac{-\hbar^2}{2m}\frac{\partial^2}{\partial x^2} + V(x)\right\}\psi_n(x) = E_n\psi_n(x) \qquad (13.28)$$

where the expression in the curly brackets { } is the total energy operator. This operator is given the $\hat{H}$ symbol in quantum mechanics and is called the Hamiltonian for historical reasons. With this notation, Equation (13.28) can be written in the form

$$\hat{H}\psi_n(x) = E_n\psi_n(x) \qquad (13.29)$$

The operator acting on one of its eigenfunctions returns the eigenfunction multiplied by the corresponding eigenvalue. In Example Problem 13.5, this formalism is applied for two operators. Solving the time-independent Schrödinger equation is equivalent to finding the set of eigenfunctions and eigenvalues that are the solutions to the eigenvalue problem of Equation (13.29). In this chapter, we consider only a single operator acting on a function. In Chapter 17, we will show that the outcome of two sequential operations on a wave function can depend on the order in which the operations occur. This fact has important implications for the measurement process in quantum mechanics.

EXAMPLE PROBLEM 13.5

Consider the operators d/dx and d^2/dx^2. Is $\psi(x) = Ae^{ikx} + Be^{-ikx}$ an eigenfunction of these operators? If so, what are the eigenvalues? A, B, and k are real numbers.

Solution

To test if a function is an eigenfunction of an operator, we carry out the operation and see if the result is the same function multiplied by a constant:

$$\frac{d(Ae^{ikx} + Be^{-ikx})}{dx} = ik\,Ae^{ikx} - ik\,Be^{-ikx} = ik(Ae^{ikx} - Be^{-ikx})$$

In this case, the result is not $\psi(x)$ multiplied by a constant, so $\psi(x)$ is not an eigenfunction of the operator d/dx unless either A or B is zero. We consider the second operator.

$$\frac{d^2(Ae^{ikx} + Be^{-ikx})}{dx^2} = (ik)^2 Ae^{ikx} + (-ik)^2 Be^{-ikx}$$

$$= -k^2(Ae^{ikx} + Be^{-ikx}) = -k^2\psi(x)$$

This result shows that $\psi(x)$ is an eigenfunction of the operator d^2/dx^2 with the eigenvalue $-k^2$.

In general, a quantum mechanical operator such as $\hat{H}$ has an infinite number of eigenfunctions. How are the eigenfunctions of a quantum mechanical operator related to one another? We discuss two of the most important properties in the next sections, namely, *orthogonality* and *completeness*.

13.6 The Eigenfunctions of a Quantum Mechanical Operator Are Orthogonal

We are familiar with the concept of orthogonal vectors. For example, orthogonality in three-dimensional Cartesian coordinate space is defined by

$$\mathbf{x} \cdot \mathbf{y} = \mathbf{x} \cdot \mathbf{z} = \mathbf{y} \cdot \mathbf{z} = 0 \qquad (13.30)$$

in which the scalar product between the unit vectors along the x, y, and z axes is zero. In function space, the analogous expression that defines **orthogonality** between the eigenfunctions $\psi_i(x)$ and $\psi_j(x)$ of a quantum mechanical operator is

$$\int_{-\infty}^{\infty} \psi_i^*(x)\psi_j(x)\, dx = 0 \quad \text{unless } i = j \qquad (13.31)$$

Example Problem 13.6 shows that graphical methods can be used to determine if two functions are orthogonal.

EXAMPLE PROBLEM 13.6

Show graphically that $\sin x$ and $\cos 3x$ are orthogonal functions over the interval $[-2j\pi, 2j\pi]$ where for the purposes of our discussion j is a very large integer. Also show graphically that $\int_{-2j\pi}^{2j\pi} (\sin mx)(\sin nx)\, dx \neq 0$ for $n = m = 1$.

Solution

The functions are shown in the following graphs. The vertical axes have been offset to avoid overlap and the horizontal line indicates the zero for each plot. Because the functions are periodic, we can draw conclusions about their behavior in a very large interval that is an integral multiple of 2π by considering their behavior in any interval that is an integral multiple of the period.

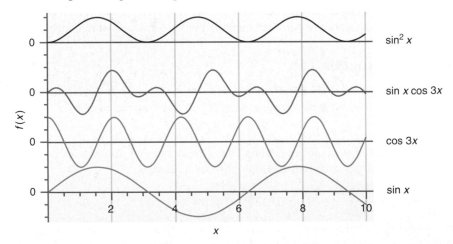

The integral of these functions equals the sum of the areas between the curves and the zero line. Areas above and below the line contribute with positive and negative signs, respectively, and indicate that $\int_{-\infty}^{\infty} \sin x \cos 3x\, dx = 0$ and $\int_{-\infty}^{\infty} \sin x \sin x\, dx > 0$. By similar means, we could show that any two functions of the type $\sin mx$ and $\sin nx$ or $\cos mx$ and $\cos nx$ are orthogonal unless $n = m$. Are the functions $\cos mx$ and $\sin mx$ ($m = n$) orthogonal?

Recall that the superscript * on a function indicates the complex conjugate. The product $\psi_i^*(x)\psi_j(x)$ rather than $\psi_i(x)\psi_j(x)$ occurs in Equation (13.31) because wave functions in quantum mechanics can be complex functions of x and t. If in addition to Equation (13.31) the integral has the value one for $i = j$, we say that the functions are **normalized** and form an **orthonormal** set. As we will see in Chapter 14, wave functions must be normalized so that they can be used to calculate probabilities. We show how to normalize wave functions in Example Problems 13.7 and 13.8.

EXAMPLE PROBLEM 13.7

Normalize the function $a(a - x)$ over the interval $0 \leq x \leq a$.

Solution

To normalize a function $\psi(x)$ over the given interval, we multiply it by a constant N, and then calculate N from the equation $N^2 \int_0^a \psi^*(x)\psi(x)\, dx = 1$.

In this particular case,

$$N^2 \int_0^a [a(a - x)]^2\, dx = 1$$

$$N^2 a^2 \int_0^a [a^2 - 2ax + x^2]\, dx = 1$$

$$N^2 \left(a^4 x - a^3 x^2 + a^2 \frac{x^3}{3} \right)_0^a = 1$$

$$N^2 \frac{a^5}{3} = 1 \quad \text{so that} \quad N = \sqrt{\frac{3}{a^5}}$$

The normalized wave function is $\sqrt{\dfrac{3}{a^5}} a(a - x)$

Up until now, we have considered functions of a single variable. This restricts us to dealing with a single spatial dimension. As we will see in Chapter 18, important problems such as the harmonic oscillator can be solved in a one-dimensional framework. The extension to three independent variables becomes important in describing three-dimensional systems. The three-dimensional system of most importance to us is the atom. Closed-shell atoms are spherically symmetric, and atomic wave functions are best described by spherical coordinates, shown in Figure 13.5. Therefore, it is helpful to become familiar with integrations in these coordinates. The Math Supplement (Appendix A) provides a more detailed discussion of working with

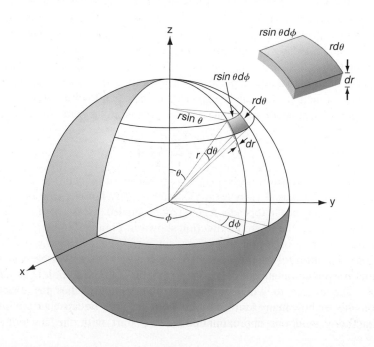

FIGURE 13.5

Defining variables and the volume element in spherical coordinates.

spherical coordinates. Note in particular that the volume element in spherical coordinates is $r^2 \sin \theta \, dr \, d\theta \, d\phi$ and not $dr \, d\theta \, d\phi$. A function is normalized in spherical coordinates in Example Problem 13.8.

EXAMPLE PROBLEM 13.8

Normalize the function e^{-r} over the interval $0 \le r \le \infty; 0 \le \theta \le \pi; \ 0 \le \phi \le 2\pi$.

Solution

We proceed as in Example Problem 13.7, remembering that the volume element in spherical coordinates is $r^2 \sin \theta \, dr \, d\theta \, d\phi$:

$$N^2 \int_0^{2\pi} d\phi \int_0^\pi \sin \theta \, d\theta \int_0^\infty r^2 e^{-2r} \, dr = 1$$

$$4\pi N^2 \int_0^\infty r^2 e^{-2r} dr = 1$$

Using the standard integral $\int_0^\infty x^n e^{-ax} \, dx = n!/a^{n+1}$ ($a > 0$, n is a positive integer), we obtain

$$4\pi N^2 \frac{2!}{2^3} = 1 \ \text{ so that } N = \sqrt{\frac{1}{\pi}}. \text{ The normalized wave function is } \sqrt{\frac{1}{\pi}} e^{-r}.$$

Note that the integration of any function involving r where $r = \sqrt{x^2 + y^2 + z^2}$ requires integration over all three variables, even if the function does not explicitly involve θ or ϕ.

13.7 The Eigenfunctions of a Quantum Mechanical Operator Form a Complete Set

The eigenfunctions of a quantum mechanical operator have another very important property that we will use frequently in later chapters, namely that the eigenfunctions of a quantum mechanical operator form a **complete set.** The idea of a complete set is familiar from the three-dimensional Cartesian coordinate system. Because any three-dimensional vector can be expressed as a linear combination of the three mutually perpendicular unit vectors **x, y,** and **z,** we say that these three unit vectors form a complete set.

Completeness is also an important concept in function space. To say that the eigenfunctions of any quantum mechanical vector form a complete set means that any well-behaved wave function $f(x)$ can be expanded in the eigenfunctions $\phi_n(x)$ of any of the quantum mechanical operators of interest to us defined in the same space, x in this case:

$$\psi(x) = \sum_{n=1}^\infty b_n \phi_n(x) \tag{13.32}$$

Before we expand wave functions in a complete set of functions in later chapters, we first illustrate how to expand a simple sawtooth function in a **Fourier sine and cosine series.** See the Math Supplement (Appendix A) for a more detailed discussion of Fourier series.

We approximate the sawtooth function shown in Figure 13.6 by a linear combination of the mutually orthogonal functions $\sin(n\pi x/b)$ and $\cos(n\pi x/b)$. These functions form an infinitely large complete set for $n = 1, 2, 3, \ldots \infty$. Because these functions form a complete set only as $n \to \infty$, the approximation becomes exact as $n \to \infty$. The degree to which the approximation approaches the exact function depends only on how many terms we include in the sum. Because we are interested in knowing how good our approximation is, we start with the sawtooth function,

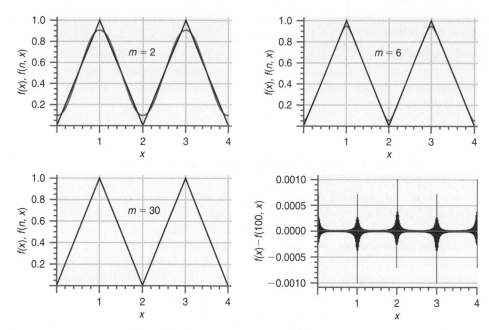

FIGURE 13.6
The sawtooth function (red curve) is compared with the finite Fourier series defined by Equation (13.33) (blue curve) containing m terms for $m = 2$, 6, and 30. The difference between the sawtooth function and the finite Fourier series for $m = 100$ is shown in the bottom right panel as a function of x.

approximate it by the finite sum in Equation (13.33), and evaluate how well we succeed for different values of m.

$$f(x) = d_0 + \sum_{n=1}^{m}\left[c_n \sin\left(\frac{n\pi x}{b}\right) + d_n \cos\left(\frac{n\pi x}{b}\right) \right] \tag{13.33}$$

We first need a way to calculate the coefficients d_0, c_m, and d_m. In our case, the function is even, $f(x) = f(-x)$, so that all the coefficients c_m are identically zero. For the Fourier series, the values for the d_m are easily obtained using the mutual orthogonality property of the sine and cosine functions demonstrated in Example Problem 13.6.

To obtain the d_m, we multiply both sides of Equation (13.33) by one of the expansion functions, for example $\cos(m\pi x/b)$, and integrate over the interval $-b$, b:

$$\int_{-b}^{b} f(x) \cos\left(\frac{m\pi x}{b}\right) dx = \int_{-b}^{b} \cos\left(\frac{m\pi x}{b}\right)$$

$$\times \left(d_0 + \sum_{n} d_n \cos\left(\frac{n\pi x}{b}\right) \right) dx$$

$$= \int_{-b}^{b} \left(\cos\left(\frac{m\pi x}{b}\right) \right) d_m \cos\left(\frac{m\pi x}{b}\right) dx = b d_m$$

Only one term in the summation within the integral gives a nonzero contribution, because all cosine functions for which $m \neq n$ are orthogonal. We have used one of the standard integrals listed in the Math Supplement (Appendix A) to obtain this result. We conclude that the optimal values for the coefficients are given by

$$d_m = \frac{1}{b}\int_{-b}^{b} f(x) \cos\left(\frac{m\pi x}{b}\right) dx, \quad m \neq 0 \quad \text{and}$$

$$d_0 = \frac{1}{2b}\int_{-b}^{b} f(x)\, dx$$

Using these equations to obtain the optimal coefficients will make the finite sum of Equation (13.33) nearly exact if enough terms can be included in the sum. How good is the approximation if m is finite? This question is answered in Figure 13.6, which shows that the sawtooth function can be described reasonably well for $m < 30$. To make this statement more quantitative, we graph $f(x) - \left(d_0 + \sum_{n=1}^{100}\left[c_n \sin\left(\frac{n\pi x}{b}\right) + d_n \cos\left(\frac{n\pi x}{b}\right)\right]\right)$ versus x in Figure 13.6 for a 101 term series. We see that the difference is less than 0.1% of the maximum amplitude for the sawtooth function. Generalizing this result to other functions, the maximum error occurs at the points for which the slope of the function is discontinuous.

13.8 Summing Up the New Concepts

In this chapter we introduced a number of the key tools in quantum mechanics that are used to solve the Schrödinger equation. The time-dependent and time-independent Schrödinger equations play the role in solving quantum mechanical problems that Newton's laws play in classical mechanics. Operators, eigenfunctions, and observables form the framework for solving the time-independent Schrödinger equation. All of these concepts will be applied to problems of chemical interest in the next few chapters. However, we will first introduce and discuss the five postulates of quantum mechanics in Chapter 14.

Vocabulary

angular frequency
Boltzmann distribution
classical nondispersive wave equation
complete set
completeness
complex conjugate
constructive interference
continuous energy spectrum
degeneracy
destructive interference
discrete energy spectrum
eigenfunction

eigenvalue
energy levels
Fourier sine and cosine series
frequency
interference
normalized
observable
operator
orthogonality
orthonormal
period
phase

plane wave
spherical wave
standing wave
stationary state
time-dependent Schrödinger equation
time-independent Schrödinger equation
traveling wave
wave front
wave function
wave vector
wavelength

Conceptual Problems

Q13.1 One source emits spherical waves and another emits plane waves. For which source does the intensity measured by a detector of fixed size fall off more rapidly with distance? Why?

Q13.2 What is the relationship between evaluating an integral and graphing the integrand?

Q13.3 A traveling wave with arbitrary phase ϕ can be written as $\psi(x, t) = A \sin(kx - \omega t + \phi)$. What are the units of ϕ? Show that ϕ could be used to represent a shift in the origin of time or distance.

Q13.4 Why is it true for any quantum mechanical problem that the set of wave functions is larger than the set of eigenfunctions?

Q13.5 By discussing the diffraction of a beam of particles by a single slit, justify the statement that there is no sharp boundary between particle-like and wave-like behavior.

Q13.6 Redraw Figure 13.2 to show surfaces corresponding to both minimum and maximum values of the amplitude.

Q13.7 Give three examples of properties of a gas phase molecule of H_2 that are quantized and three properties that are not quantized.

Q13.8 Why is it necessary in normalizing the function re^{-r} in spherical coordinates to integrate over θ and ϕ even though it is not a function of θ and ϕ?

Q13.9 If $\psi(x, t) = A \sin(kx - \omega t)$ describes a wave traveling in the plus x direction, how would you describe a wave traveling in the minus x direction? Justify your answer.

Q13.10 In Figure 13.6 the extent to which the approximate and true functions agree was judged visually. How could you quantify the quality of the fit?

Q13.11   Why does a quantum mechanical system with discrete energy levels behave as if it has a continuous energy spectrum if the energy difference between energy levels ΔE satisfies the relationship $\Delta E \ll k_B T$?

Q13.12   Distinguish between the following terms applied to a set of functions: *orthogonal, normalized,* and *orthonormal.*

Q13.13   Why can we conclude that the wave function $\psi(x, t) = \psi(x)e^{-i(E/\hbar)t}$ represents a standing wave?

Q13.14   What is the usefulness of a set of complete functions?

Q13.15   Can the function $\sin kx$ be normalized over the interval $-\infty < x < \infty$? Explain your answer.

Q13.16   A linear operator satisfies the condition $\hat{A}(f(x) + g(x)) = \hat{A}f(x) + \hat{A}g(x)$. Are $\hat{A}$ or $\hat{B}$ linear operators if $\hat{A}f(x) = (f(x))^2$ and $\hat{B}f(x) = df(x)/dx$?

Q13.17   Is $\hat{A}$ a linear operator if $\hat{A}f(x) = d^2f(x)/dx^2 + xf(x)$?

Q13.18   Two operators can be applied to a function in succession. By definition, $\hat{A}\hat{B}f(x) = \hat{A}[\hat{B}f(x)]$. Evaluate $\hat{A}\hat{B}$ $f(x)$ if $\hat{A} = d/dx$, $\hat{B} = x$, and $f(x) = \cos x$.

Q13.19   Is $\cos x$ an eigenfunction of the operator $\hat{A}$ if $\hat{A}f(x) = xf(x)$?

Q13.20   Which of the following functions are eigenfunctions of the operator $\hat{B}$ if $\hat{B}f(x) = d^2f(x)/dx^2$: x^2, $\cos x$, e^{-3ix}? State the eigenvalue if applicable.

Numerical Problems

Problem numbers in **red** indicate that the solution to the problem is given in the *Student's Solutions Manual.*

P13.1   A wave traveling in the z direction is described by the wave function $\Psi(z,t) = A_1 \mathbf{x} \sin(kz - \omega t + \phi_1) + A_2 \mathbf{y} \sin(kz - \omega t + \phi_2)$, where $\mathbf{x}$ and $\mathbf{y}$ are vectors of unit length along the x and y axes, respectively. Because the amplitude is perpendicular to the propagation direction, $\Psi(z, t)$ represents a transverse wave.

a. What requirements must A_1 and A_2 satisfy for a plane polarized wave in the *x-z* plane? The amplitude of a plane polarized wave is non-zero only in one plane.

b. What requirements must A_1 and A_2 satisfy for a plane polarized wave in the *y-z* plane?

c. What requirements must A_1 and A_2 and ϕ_1 and ϕ_2 satisfy for a plane polarized wave in a plane oriented at 45° to the *x-z* plane?

d. What requirements must A_1 and A_2 and ϕ_1 and ϕ_2 satisfy for a circularly polarized wave? The phases of the two components of a circularly polarized wave differ by $\pi/2$.

P13.2   Because $\int_{-d}^{d} \cos(n\pi x/d) \cos(m\pi x/d)\, dx = 0$, $m \neq n$, the functions $\cos(n\pi x/d)$ for $n = 1, 2, 3, \ldots$ form an orthogonal set in the interval $(-d, d)$. What constant must these functions be multiplied by to form an orthonormal set?

P13.3   Determine in each of the following cases if the function in the first column is an eigenfunction of the operator in the second column. If so, what is the eigenvalue?

a. x^2 $\dfrac{x^2}{8}d^2/dx^2$

b. $x^3 + y^3$ $x^3(\partial^3/\partial x^3) + y^3(\partial^3/\partial y^3)$

c. $\sin 2\theta \cos \phi$ $\partial^4/\partial\theta^4$

P13.4   If two operators act on a wave function as indicated by $\hat{A}\hat{B}f(x)$, it is important to carry out the operations in succession with the first operation being that nearest to the function. Mathematically, $\hat{A}\hat{B}f(x) = \hat{A}(\hat{B}f(x))$ and $\hat{A}^2f(x) = \hat{A}(\hat{A}f(x))$. Evaluate the following successive operations $\hat{A}\hat{B}f(x)$. The operators $\hat{A}$ and $\hat{B}$ are listed in the first two columns and $f(x)$ is listed in the third column.

a. $\dfrac{d}{dy}$ y ye^{-2y^3}

b. y $\dfrac{d}{dy}$ ye^{-2y^3}

c. $y\dfrac{\partial}{\partial x}$ $x\dfrac{\partial}{\partial y}$ $e^{-2(x+y)}$

d. $x\dfrac{\partial}{\partial y}$ $y\dfrac{\partial}{\partial x}$ $e^{-2(x+y)}$

Are your answers to parts (a) and (b) identical? Are your answers to parts (c) and (d) identical? As we will learn in Chapter 17, switching the order of the operators can change the outcome of the operation $\hat{A}\hat{B}f(x)$.

P13.5   Let (1, 0) and (0, 1) represent the unit vectors along the x and y directions, respectively. The operator

$$\begin{pmatrix} \cos \theta & -\sin \theta \\ \sin \theta & \cos \theta \end{pmatrix}$$

effects a rotation in the *x-y* plane. Show that the length of an arbitrary vector

$$\begin{pmatrix} a \\ b \end{pmatrix} = a\begin{pmatrix} 1 \\ 0 \end{pmatrix} + b\begin{pmatrix} 0 \\ 1 \end{pmatrix}$$

which is defined as $\sqrt{a^2 + b^2}$, is unchanged by this rotation. See the Math Supplement (Appendix A) for a discussion of matrices.

P13.6 Carry out the following coordinate transformations:

a. Express the point $x = 3$, $y = 1$, and $z = 1$ in spherical coordinates.

b. Express the point $r = 5$, $\theta = \dfrac{\pi}{4}$, and $\phi = \dfrac{3\pi}{4}$ in Cartesian coordinates.

P13.7 Operators can also be expressed as matrices and wave functions as column vectors. The operator matrix

$$\begin{pmatrix} a & \beta \\ \delta & \varepsilon \end{pmatrix}$$

acts on the wave function $\begin{pmatrix} a \\ b \end{pmatrix}$ according to the rule

$$\begin{pmatrix} \alpha & \beta \\ \delta & \varepsilon \end{pmatrix}\begin{pmatrix} a \\ b \end{pmatrix} = \begin{pmatrix} \alpha a + \beta b \\ \delta a + \varepsilon b \end{pmatrix}$$

In words, the 2×2 matrix operator acting on the two-element column wave function generates another two-element column wave function. If the wave function generated by the operation is the original wave function multiplied by a constant, the wave function is an eigenfunction of the operator. What is the effect of the operator

$$\begin{pmatrix} 0 & 1 \\ 1 & 0 \end{pmatrix}$$

on the column vectors $(1, 0)$, $(0, 1)$, $(1, 1)$, and $(-1, 1)$? Are these wave functions eigenfunctions of the operator? See the Math Supplement (Appendix A) for a discussion of matrices.

P13.8 Show that

$$\frac{a + ib}{c + id} = \frac{ac + bd + i(bc - ad)}{c^2 + d^2}$$

P13.9 Express the following complex numbers in the form $re^{i\theta}$.

a. $5 + 6i$ **d.** $\dfrac{5 + i}{3 - 4i}$

b. $2i$ **e.** $\dfrac{2 - i}{1 + i}$

c. 4

P13.10 Show that the set of functions $\phi_n(\theta) = e^{in\theta}$, $0 \leq \theta \leq 2\pi$, is orthogonal if n is an integer. To do so, you need to show that the integral $\int_0^{2\pi} \phi_m^*(\theta)\phi_n(\theta)\, d\theta = 0$ for $m \neq n$ if n and m are integers.

P13.11 Operate with (a) $\dfrac{\partial}{\partial x} + \dfrac{\partial}{\partial y} + \dfrac{\partial}{\partial z}$ and

(b) $\dfrac{\partial^2}{\partial x^2} + \dfrac{\partial^2}{\partial y^2} + \dfrac{\partial^2}{\partial z^2}$ on the function

$Ae^{-ik_1 x}\, e^{-ik_2 y}\, e^{-ik_3 z}$. Is the function an eigenfunction of either operator? If so, what is the eigenvalue?

P13.12 Which of the following wave functions are eigenfunctions of the operator d^2/dx^2? If they are eigenfunctions, what is the eigenvalue?

a. $a(e^{-3x} + e^{-3ix})$ **d.** $\cos\dfrac{ax}{\pi}$

b. $\sin\dfrac{2\pi x}{a}$ **e.** e^{-ix^2}

c. e^{-2ix}

P13.13 Does the superposition $\psi(x, t) = A\sin(kx - \omega t) + 2A\sin(kx + \omega t)$ generate a standing wave? Answer this question by using trigonometric identities to combine the two terms.

P13.14 Determine in each of the following cases if the function in the first column is an eigenfunction of the operator in the second column. If so, what is the eigenvalue?

a. $\cos\theta$ $\dfrac{1}{\sin\theta}\dfrac{d}{d\theta}\left(\sin\theta\dfrac{d}{d\theta}\right)$

b. $e^{-(2ix^2)}$ $\dfrac{d^2}{dx^2} + 16x^2$

c. $\cos x \sin x$ $\dfrac{d^2}{dx^2} - 2$

P13.15 Show by carrying out the integration that $\sin(m\pi x/a)$ and $\cos(m\pi x/a)$, where m is an integer, are orthogonal over the interval $0 \leq x \leq a$. Would you get the same result if you used the interval $0 \leq x \leq 3a/4$? Explain your result.

P13.16 To plot $\Psi(x, t) = A\sin(kx - vt)$ as a function of one of the variables x and t, the other variable needs to be set at a fixed value, x_0 or t_0. If $\Psi(x_0, 0)/\Psi_{max} = -0.280$, what is the constant value of x_0 in the upper panel of Figure 13.3? If $\Psi(0, t_0)/\psi_{max} = -0.309$, what is the constant value of t_0 in the lower panel of Figure 13.3? (*Hint:* The inverse sine function has two solutions within an interval of 2π. Make sure that you choose the correct one.)

P13.17 Determine in each of the following cases if the function in the first column is an eigenfunction of the operator in the second column. If so, what is the eigenvalue?

a. $e^{-i(7x+y)}$ $\dfrac{\partial^2}{\partial x^2}$

b. $\sqrt{3x^2 + 2y^2}$ $(1/3x)(3x^2 + 2y^2)\dfrac{\partial}{\partial x}$

c. $\sin\theta\cos\theta$ $\dfrac{1}{\sin\theta}\dfrac{d}{d\theta}\left(\sin\theta\dfrac{d}{d\theta}\right)$

P13.18 Assume that a system has a very large number of energy levels given by the formula $\varepsilon = \varepsilon_0 l^2$ with $\varepsilon_0 = 1.75 \times 10^{-22}$ J, where l takes on the integral values 1, 2, 3, ... Assume further that the degeneracy of a level is given by $g_l = 2l$. Calculate the ratios n_4/n_1 and n_8/n_1 for $T = 125$ K and $T = 750.$ K.

P13.19 Is the function $2x^2 - 1$ an eigenfunction of the operator $-(3/2 - x^2)(d^2/dx^2) + 2x(d/dx)$? If so, what is the eigenvalue?

P13.20 Find the result of operating with $d^2/dx^2 - 2x^2$ on the function e^{-ax^2}. What must the value of a be to make this function an eigenfunction of the operator? What is the eigenvalue?

P13.21 Determine in each of the following cases if the function in the first column is an eigenfunction of the operator in the second column. If so, what is the eigenvalue?

a. $\sin\theta\cos\phi$ $\partial^4/\partial\phi^4$

b. $e^{-(3x^2/\sqrt{2})}$ $(1/x)\,d/dx$

c. $\cos\dfrac{2\pi x}{a}$ $\left(1/\tan\dfrac{2\pi x}{a}\right)d/dx$

P13.22 Find the result of operating with $d^2/dx^2 + d^2/dy^2 + d^2/dz^2$ on the function $x^2 + y^2 + z^2$. Is this function an eigenfunction of the operator?

P13.23 Using the exponential representation of the sine and cosine functions, $\cos\theta = \frac{1}{2}(e^{i\theta} + e^{-i\theta})$ and $\sin\theta = \frac{1}{2i}(e^{i\theta} - e^{-i\theta})$, show that

a. $\cos^2\theta + \sin^2\theta = 1$

b. $d(\cos\theta)/d\theta = -\sin\theta$

c. $\sin\left(\theta + \dfrac{\pi}{2}\right) = \cos\theta$

P13.24 If two operators act on a wave function as indicated by $\hat{A}\hat{B}f(x)$, it is important to carry out the operations in succession with the first operation being that nearest to the function. Mathematically, $\hat{A}\hat{B}f(x) = \hat{A}(\hat{B}f(x))$ and $\hat{A}^2f(x) = \hat{A}(\hat{A}f(x))$. Evaluate the following successive operations $\hat{A}\hat{B}f(x)$. The operators $\hat{A}$ and $\hat{B}$ are listed in the first and second columns and $f(x)$ is listed in the third column. Compare your answers to parts (a) and (b), and to (c) and (d).

a. $\dfrac{d}{dx}$ x $x^2 + e^{ax^2}$

b. x $\dfrac{d}{dx}$ $x^2 + e^{ax^2}$

c. $\dfrac{\partial^2}{\partial y^2}$ y^2 $(\cos 3y)\sin^2 x$

d. y^2 $\dfrac{\partial^2}{\partial y^2}$ $(\cos 3y)\sin^2 x$

P13.25 Make the three polynomial functions a_0, $a_1 + b_1 x$, and $a_2 + b_2 x + c_2 x^2$ orthonormal in the interval $-1 \le x \le +1$ by determining appropriate values for the constants a_0, a_1, b_1, a_2, b_2, and c_2.

P13.26 Consider a two-level system with $\varepsilon_1 = 2.25 \times 10^{-22}$ J and $\varepsilon_2 = 4.50 \times 10^{-21}$ J. If $g_2 = 2g_1$, what value of T is required to obtain $n_2/n_1 = 0.175$? What value of T is required to obtain $n_2/n_1 = 0.750$?

P13.27 Find the result of operating with $(1/r^2)(d/dr)(r^2\,d/dr) + 2/r$ on the function $A e^{-br}$. What must the values of A and b be to make this function an eigenfunction of the operator?

P13.28 Normalize the set of functions $\phi_n(\theta) = e^{in\theta}$, $0 \le \theta \le 2\pi$. To do so, you need to multiply the functions by a normalization constant N so that the integral $N N^* \int_0^{2\pi} \phi_m^*(\theta)\phi_n(\theta)\,d\theta = 1$ for $m = n$.

P13.29 In normalizing wave functions, the integration is over all space in which the wave function is defined.

a. Normalize the wave function $x(a - x)y(b - y)$ over the range $0 \le x \le a$, $0 \le y \le b$. The element of area in two-dimensional Cartesian coordinates is $dx\,dy$; a and b are constants.

b. Normalize the wave function $e^{-(2r/b)}\sin\theta\sin\phi$ over the interval $0 \le r < \infty$, $0 \le \theta \le \pi$, $0 \le \phi \le 2\pi$. The volume element in three-dimensional spherical coordinates is $r^2\sin\theta\,dr\,d\theta\,d\phi$, and b is a constant.

P13.30 Operate with (a) $\dfrac{\partial}{\partial x} + \dfrac{\partial}{\partial y} + \dfrac{\partial}{\partial z}$ and

(b) $\dfrac{\partial^2}{\partial x^2} + \dfrac{\partial^2}{\partial y^2} + \dfrac{\partial^2}{\partial z^2}$ on the function $A\,e^{-ik_1x}e^{-ik_1y}e^{-ik_1z}$.

Under what conditions is the function an eigenfunction of one or both operators? What is the eigenvalue?

P13.31 Form the operator $\hat{A}^2$ if $\hat{A} = x - d/dx$. Be sure to include an arbitrary function on which the operator acts.

P13.32 Use a Fourier series expansion to express the function $f(y) = y^2$, $-a \le y \le a$, in the form

$$f(x) = d_0 + \sum_{n=1}^{m} c_n \sin\left(\frac{n\pi y}{a}\right) + d_n \cos\left(\frac{n\pi y}{a}\right)$$

Obtain d_0 and the first five pairs of coefficients c_n and d_n.

P13.33 Is the function $e^{-(ax^2/2)}$ an eigenfunction of the operator $d^2/dy^2 - a^2y^2$? If so, what is the eigenvalue?

P13.34 Show that the following pairs of wave functions are orthogonal over the indicated range.

a. e^{-ax^2} and $x(x^2 - 1)e^{-ax^2}$, $-\infty \le x < \infty$ where α is a constant that is greater than zero

b. $(6r/a_0 - r^2/a_0^2)e^{-r/3a_0}$ and $(r/a_0)e^{-r/2a_0}\cos\theta$ over the interval $0 \le r < \infty$, $0 \le \theta \le \pi$, $0 \le \phi \le 2\pi$

P13.35 Express the following complex numbers in the form $a + ib$.

a. $2e^{3i\pi/2}$ c. $e^{i\pi}$

b. $4\sqrt{3}\,e^{i\pi/4}$ d. $\dfrac{\sqrt{5}}{1 + \sqrt{2}}e^{i\pi/4}$

P13.36 Which of the following wave functions are eigenfunctions of the operator d/dx? If they are eigenfunctions, what is the eigenvalue?

a. $ae^{-3x} + be^{-3ix}$ d. $\cos ax$

b. $\sin^2 x$ e. e^{-ix^2}

c. e^{-ix}

P13.37 Form the operator $\hat{A}^2$ if $\hat{A} = d^2/dy^2 + 3y(d/dy) - 5$. Be sure to include an arbitrary function on which the operator acts.

Web-Based Simulations, Animations, and Problems

W13.1 The motion of transverse, longitudinal, and surface traveling waves is analyzed by varying the frequency and amplitude.

W13.2 Two waves of the same frequency traveling in opposite directions are combined. The relative amplitude is changed with sliders and the relative phase of the waves is varied. The effect of these changes on the superposition wave is investigated.

W13.3 Two waves, both of which are standing waves, are combined. The effect of varying the wavelength, period, and phase of the waves on the resulting wave using sliders is investigated.

W13.4 Several functions are approximated by a Fourier series in which the number of sine and cosine terms is varied. The degree to which the approximate function differs from the exact function is assessed.

The Quantum Mechanical Postulates

Quantum mechanics can be formulated in terms of six postulates. A postulate is a claim or an assumption of truth, especially as a basis for reasoning. Postulates cannot be proven, but they can be tested. The five postulates discussed in this chapter provide a framework for summarizing the basic concepts of quantum mechanics. The quantum mechanical postulates have been extensively tested since they were proposed in the early 1930s. No case has been found in which they predict an outcome that is in conflict with the result of an experiment.

The previous chapters focused on the classical mechanics of particles and on the mathematical description of waves. Wave-particle duality and the conditions under which the wave character of a particle (which is always present) becomes evident were discussed. We briefly discussed quantum mechanical wave functions and quantum mechanical operators and showed that values for the **observables** are obtained by operating on the wave function with the relevant operator. The rules for how information is obtained from wave functions can be summarized in a few **postulates.** The test of any set of postulates is their ability to explain the world around us.

In this chapter, five postulates are stated and explained. In the following chapters, we apply these postulates to model systems and compare the results with those obtained from classical mechanics. In Chapter 21, a sixth postulate is introduced.

14.1 The Physical Meaning Associated with the Wave Function Is Probability

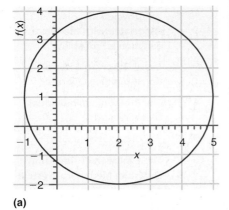

> **POSTULATE 1**
> The state of a quantum mechanical particle is completely specified by a **wave function** $\Psi(x, t)$. To simplify the notation, only one spatial coordinate is considered. The probability that the particle will be found at time t_0 in a spatial interval of width dx centered at x_0 is given by $\Psi^*(x_0, t_0)\Psi(x_0, t_0)\, dx$.

For a sound wave, the wave function $\Psi(x, t)$ is the pressure at a time t and position x. For a water wave, $\Psi(x, t)$ is the height of the wave as a function of position and time. What physical meaning can we associate with $\Psi(x, t)$ for quantum systems? For a particle (which also has wave character), the probability $P(x_0, t_0)$ of finding the particle at position x_0 at time t_0 within an interval dx is

$$P(x_0, t_0) = \Psi^*(x_0, t_0)\Psi(x_0, t_0)\, dx = |\Psi(x_0, t_0)|^2\, dx \qquad (14.1)$$

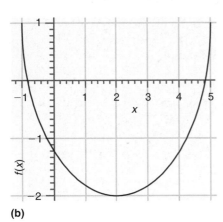

Unlike classical waves such as sound waves, the wave amplitude $\Psi(x, t)$ itself has no direct physical meaning in quantum mechanics. Because the probability is related to the *square of the magnitude* of $\Psi(x, t)$, given by $\Psi^*(x, t)\Psi(x, t)$, the wave function can be complex or negative and still be associated with a probability that lies between zero and one. The wave amplitude $\Psi(x, t)$ can be multiplied by -1, or its phase can be changed by multiplying it by a complex function of magnitude one such as $e^{i\theta(x,t)}$, without changing $\Psi^*(x, t)\Psi(x, t)$. Therefore, all wave functions that are identical except for a phase angle $\theta(x, t)$ are indistinguishable in that they generate the same observables. The wave function is a complete description of the system in that any measurable property (observable) can be obtained from the wave function as will be described later.

The association of the wave function with the probability places an important requirement on a wave function called **normalization.** The probability that the particle is found in an interval of width dx centered at the position x must lie between zero and one. The sum of the probabilities over all intervals accessible to the particle is one because the particle is somewhere in its range. Consider a particle that is confined to a one-dimensional space of infinite extent. The requirement that the particle is somewhere in the interval leads to the following normalization condition:

$$\int_{-\infty}^{\infty} \Psi^*(x, t)\Psi(x, t)\, dx = 1 \qquad (14.2)$$

FIGURE 14.1
(a) $f(x)$ has two values for nearly all values of x, and (b) $f(x)$ has only one value for each value of x.

Such a definition is obviously meaningless if the integral does not exist. Therefore, $\Psi(x, t)$ must satisfy several mathematical conditions to ensure that it represents a possible physical state. These conditions are as follows:

- The wave function must be a **single-valued function** of the spatial coordinates. If this were not the case, a particle would have more than one probability of being found in the same interval. For example, for the ellipse depicted in Figure 14.1a, $f(x)$ has two values for each value of x except the two points at which the tangent line is vertical. If only the part of the ellipse is considered for which $f(x) < 1$, as in Figure 14.1b, $f(x)$ has only one value for each value of x.

- The second derivative must exist and be well behaved. If this were not the case, we could not set up the Schrödinger equation. This is not the case if the wave function and/or its first derivative are discontinuous. As shown in Figure 14.2, $\sin x$ is a

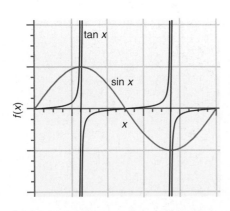

FIGURE 14.2
Examples of continuous and discontinuous functions.

continuous function of x. A function $f(x)$ is continuous at the point a if the following conditions hold:

- $f(x)$ is defined at a.
- $\lim_{x \to a} f(x)$ exists.
- $\lim_{x \to a} f(x) = f(a)$

The function $\tan x$ is not continuous, because $\tan \pi/2$ is not defined and $\lim_{x \to \pi/2} f(x)$ does not exist.

- The wave function cannot have an infinite amplitude over a finite interval. If this were the case, the wave function could not be normalized. For example, the function

$$\Psi(x, t) = e^{-i(E/\hbar)t} \frac{1}{x^2} \sin \frac{2\pi x}{a}, \, 0 \le x \le a \text{ cannot be normalized.}$$

14.2 Every Observable Has a Corresponding Operator

POSTULATE 2

For every measurable property of a system such as position, momentum, and energy, there exists a corresponding operator in quantum mechanics. An experiment in the laboratory to measure a value for such an observable is simulated in the theory by operating on the wave function of the system with the corresponding operator.

All quantum mechanical **operators** belong to a mathematical class called **Hermitian operators** that have real eigenvalues. For a Hermitian operator $\hat{A}$, $\int \psi^*(x)[\hat{A}\psi(x)] \, dx = \int \psi(x)[\hat{A} \, \psi(x)]^* dx$. The most important observables in classical mechanics, the corresponding quantum mechanical operators, and the symbols for these operators are listed in Table 14.1. To simplify the notation, only one spatial coordinate is

TABLE 14.1	Observables and Their Quantum Mechanical Operators	
Observable	Operator	Symbol for Operator
Momentum	$-i\hbar \dfrac{\partial}{\partial x}$	$\hat{p}_x$
Kinetic energy	$-\dfrac{\hbar^2}{2m} \dfrac{\partial^2}{\partial x^2}$	$\hat{E}_{kinetic} = \dfrac{1}{2m}(\hat{p}_x)(\hat{p}_x)$
Position	x	$\hat{x}$
Potential energy	$V(x)$	$\hat{E}_{potential}$
Total energy	$-\dfrac{\hbar^2}{2m} \dfrac{\partial^2}{\partial x^2} + V(x)$	$\hat{H}$
Angular momentum	$-i\hbar\left(y\dfrac{\partial}{\partial z} - z\dfrac{\partial}{\partial y} \right)$	$\hat{l}_x$
	$-i\hbar\left(z\dfrac{\partial}{\partial x} - x\dfrac{\partial}{\partial z} \right)$	$\hat{l}_y$
	$-i\hbar\left(x\dfrac{\partial}{\partial y} - y\dfrac{\partial}{\partial x} \right)$	$\hat{l}_z$

considered, except for angular momentum. Partial derivatives have been retained because the wave function depends on both position and time. The total energy operator is called the Hamiltonian for historical reasons and is given the symbol $\hat{H}$. For the position and potential energy operators, the operation is "multiply on the left by the position or potential energy." Operators act on a wave function from the left, and the order of operation is important. For example, $\hat{p}_x\hat{x}$ operating on the wave function $\sin x$ gives $-i\hbar(\sin x + x\cos x)$, whereas $\hat{x}\hat{p}_x$ operating on the same wave function gives $-i\hbar x\cos x$. As discussed in Chapter 17, operators for which the interchanging of the order does not change the result have a particular role in quantum mechanics.

14.3 The Result of an Individual Measurement

POSTULATE 3
In any single measurement of the observable that corresponds to the operator $\hat{A}$, the only values that will ever be measured are the eigenvalues of that operator.

This postulate states, for example, that if the energy of the hydrogen atom is measured, the only values obtained are the energies that are the **eigenvalues** E_n of the time-independent Schrödinger equation:

$$\hat{H}\Psi_n(x, t) = E_n\Psi_n(x, t) \tag{14.3}$$

This makes sense because the energy levels of the hydrogen atom are discrete and, therefore, only those energies are allowed. What gives pause for thought is that the wave function need not be an **eigenfunction** of $\hat{H}$, because the eigenfunctions are a subset of the infinite number of functions that satisfy all the requirements to be an acceptable wave function. We address this issue in the following postulate.

14.4 The Expectation Value

POSTULATE 4
If the system is in a state described by the wave function $\Psi(x, t)$, and the value of the observable a is measured once on each of many identically prepared systems, the average value (also called the **expectation value**) of all of these measurements is given by

$$\langle a \rangle = \frac{\int_{-\infty}^{\infty}\Psi^*(x, t)\hat{A}\Psi(x, t)\, dx}{\int_{-\infty}^{\infty}\Psi^*(x, t)\Psi(x, t)\, dx} \tag{14.4}$$

For the case in which $\Psi(x, t)$ is normalized, the denominator in this expression has the value 1. Wave functions are usually normalized, and in Equations (14.5) through (14.8), this is assumed to be the case. This postulate requires some explanation. As we know, two cases apply with regard to $\Psi(x, t)$: it either is or is not an eigenfunction of the operator $\hat{A}$. These two cases need to be examined separately.

In the first case, $\Psi(x, t)$ is a normalized eigenfunction of $\hat{A}$, for example, $\phi_j(x, t)$. Because $\hat{A}\phi_j(x, t) = a_j \phi_j(x, t)$,

$$\langle a \rangle = \int_{-\infty}^{\infty} \phi_j^*(x, t)\hat{A}\phi_j(x, t) \, dx = a_j \int_{-\infty}^{\infty} \phi_j^*(x, t)\phi_j(x, t) \, dx$$

$$= a_j \tag{14.5}$$

If $\Psi(x, t)$ is $\phi_j(x, t)$, all measurements will give the same answer, namely, a_j.

Now consider the second case, in which $\Psi(x, t)$ is not an eigenfunction of the operator $\hat{A}$. Because the eigenfunctions of $\hat{A}$ form a complete set, $\Psi(x, t)$ can be expanded in terms of these eigenfunctions:

$$\Psi(x, t) = \sum_n b_n \phi_n(x, t) \tag{14.6}$$

Because $\Psi(x, t)$ is normalized, $\sum_m b_m^* b_m = \sum_m |b_m|^2 = 1$. The expression for $\Psi(x, t)$ in Equation (14.6) can be inserted in Equation (14.4), giving

$$\langle a \rangle = \int \Psi^*(x, t)\hat{A}\Psi(x, t)dx$$

$$= \int_{-\infty}^{\infty} \left[\sum_{m=1}^{\infty} b_m^*\phi_m^*(x, t) \right]\left[\sum_{n=1}^{\infty} a_n b_n \phi_n(x, t) \right] dx$$

$$= \sum_{m=1}^{\infty} \sum_{n=1}^{\infty} b_m^* b_n a_n \int_{-\infty}^{\infty} \phi_m^*(x, t) \phi_n(x, t) \, dx \tag{14.7}$$

This expression can be greatly simplified by making use of the property that the eigenfunctions of a quantum mechanical operator are orthogonal. Because the eigenfunctions of $\hat{A}$ form an **orthonormal set,** the only terms in this double sum for which the integral is nonzero are those for which $m = n$. The integral has the value 1 for these terms, because the eigenfunctions of $\hat{A}$ are normalized. Therefore,

$$\langle a \rangle = \sum_{m=1}^{\infty} a_m b_m^* b_m = \sum_{m=1}^{\infty} |b_m|^2 a_m \tag{14.8}$$

What are the b_m? They are the **expansion coefficients** of the wave function in the complete set of the eigenfunctions of the operator $\hat{A}$. The coefficient b_m is a measure of the extent to which the wave function "looks like" the mth eigenfunction of the operator $\hat{A}$. To illustrate this point, consider the case in which $\Psi(x, t) = \phi_n(x, t)$. In this case, all of the b_m except the one value corresponding to $m = n$ are zero, $b_n = 1$, and $\langle a \rangle = a_n$. So if $\phi(x, t)$ is one of the eigenfunctions of $\hat{A}$, only one of the b_m is nonzero and the average value is just the eigenvalue corresponding to that eigenfunction. If only three of the b_m are nonzero, for example b_2, b_8, and b_{11}, then $b_2^2 + b_8^2 + b_{11}^2 = 1$ and $\langle a \rangle$ is given by

$$\langle a \rangle = |b_2|^2 a_2 + |b_8|^2 a_8 + |b_{11}|^2 a_{11} \tag{14.9}$$

Note that $\langle a \rangle$ is not simply an average of these three eigenvalues; instead, it is a **weighted average.** The weighting factor $|b_m|^2$ is directly related to the contribution of the mth eigenfunction to the wave function $\Psi(x, t)$.

The fourth postulate allows us to calculate the result of a large number of measurements, each carried out only once on a large number of identically prepared systems. What will be measured in each of these individual measurements? The third postulate says that the only possible result of a single measurement is one of the eigenvalues a_n. However, it does not tell us which of the a_n will be measured. The answer is that there is no way of knowing the outcome of an **individual measurement,** and that the outcomes from identically prepared systems are not the same. This is a sharp break with the predictability we have come to depend on in classical mechanics.

FIGURE 14.3

A large number of identically prepared systems consist of a single hydrogen atom in a three-dimensional box. The atom is described by the superposition state $\Psi_{electronic} = b_1\Psi_{1s} + b_2\Psi_{2s} + b_3\Psi_{2p_x} + b_4\Psi_{3s}$. Consider a hypothetical experiment that measures the total energy and is completed in such a short time that transitions to the ground state can be neglected. The result of the first measurement on each system is probabilistic, whereas successive measurements are deterministic.

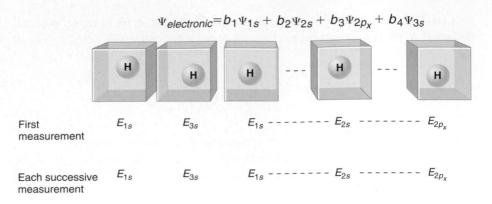

$$\Psi_{electronic} = b_1\Psi_{1s} + b_2\Psi_{2s} + b_3\Psi_{2p_x} + b_4\Psi_{3s}$$

| First measurement | E_{1s} | E_{3s} | E_{1s} -------- E_{2s} -------- E_{2p_x} |
| Each successive measurement | E_{1s} | E_{3s} | E_{1s} -------- E_{2s} -------- E_{2p_x} |

Consider a hypothetical example. Suppose that a single hydrogen atom could be isolated in a box and the electronic wave function prepared such that it is in a superposition of the ground state, in which the electron is in the $1s$ orbital, and the excited states, in which the electron is in the $2s$, $2p_x$, and $3s$ orbitals. Assume that the wave function for this **superposition state** is

$$\Psi_{electronic} = b_1\Psi_{1s} + b_2\Psi_{2s} + b_3\Psi_{2p_x} + b_4\Psi_{3s} \qquad (14.10)$$

An example of a superposition state is the particle in the double-slit experiment going through both slits simultaneously. We now prepare a large number of these systems, each of which has the same wave function, and carry out a measurement of the total energy of the atom. The results that would be obtained are illustrated in Figure 14.3. Even though the systems are identical, the same value is not obtained for the energy of the atom in each measurement.

More generally, the particular value observed in one measurement will be any one of the eigenvalues a_n for which the corresponding b_n is nonzero. This is a **probabilistic outcome,** similar to asking what the chance is of rolling a six with one throw of a die. In this more familiar case, there is no way to predict the outcome of a single throw. However, if the die is thrown a large number of times, the six will land facing up a proportion of times that almost always approaches 1/6. The equivalent case to the die for the wave function is that all of the coefficients b_m have the same value. In the particular case under consideration, we have only four nonzero coefficients and, therefore, we will only measure one of the values E_{1s}, E_{2s}, E_{2p}, or E_{3s} in an individual measurement, but we have no way of knowing which of these values we will obtain. The certainty that we are familiar with from classical mechanics—that identically prepared systems all have the same outcomes in a measurement—is replaced in quantum mechanics by the probabilistic outcome just described.

More can be said about the outcome of a large number of measurements than about the outcome of a single measurement. Consider the more general result stated in Equation (14.8): the average value of a large number of measurements carried out once on identically prepared systems is given by a sum containing the possible eigenvalues of the operator weighted by $|b_m|^2$, the square of the expansion coefficient. The bigger the contribution of an eigenfunction $\phi_m(x, t)$ of $\hat{A}$ to $\Psi(x, t)$ $\left(\text{larger }|b_m|^2\right)$, the more probable it is that the outcome of an individual measurement will be a_m and the more a_m will influence the average value $\langle a \rangle$. There is no way to predict which of the a_m will be found in an individual measurement. However, if this same experiment is repeated many times on identical systems, the average value can be predicted with very high precision. It is important to realize that this is not a shortcoming of how the "identical" systems were prepared. These systems are identical in every way and there is no reason to believe that we have left something out that resulted in this probabilistic result.

To illustrate the preceding discussion, consider the three different normalized superposition wave functions shown in Figure 14.4. They are made of the normalized

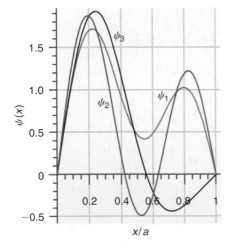

FIGURE 14.4

The three normalized wave functions $\psi_1(x), \psi_2(x),$ and $\psi_3(x)$ (blue, purple, and red curves, respectively) are defined over the interval $0 \leq x \leq a$. The amplitude of the wave functions is zero at both ends of the interval.

eigenfunctions $\phi_1(x), \phi_2(x)$, and $\phi_3(x)$ of the operator $\hat{A}$ with eigenvalues a_1, $4a_1$, and $9a_1$, respectively. The superposition wave functions are the following combinations of $\phi_1(x), \phi_2(x)$, and $\phi_3(x)$:

$$\psi_1(x) = \frac{\sqrt{11}}{4}\phi_1(x) + \frac{1}{4}\phi_2(x) + \frac{1}{2}\phi_3(x)$$

$$\psi_2(x) = \frac{1}{2}\phi_1(x) + \frac{1}{4}\phi_2(x) + \frac{\sqrt{11}}{4}\phi_3(x)$$

$$\psi_3(x) = \frac{1}{2}\phi_1(x) + \frac{\sqrt{11}}{4}\phi_2(x) + \frac{1}{4}\phi_3(x) \tag{14.11}$$

An individual measurement of the observable a gives only one of the values a_1, $4a_1$, and $9a_1$ regardless of which wave function describes the system. However, the probability of observing these values depends on the system wave function. For example, the probability of observing the value $9a_1$ is given by the square of the magnitude of the coefficient of $\phi_3(x)$ and is $1/4$, $11/16$, and $1/16$, respectively, depending on whether the state is described by $\psi_1(x), \psi_2(x)$, or $\psi_3(x)$.

Notice that the postulate specified that the measurement was to be carried out only once on each of a large number of identically prepared systems. What lies behind this requirement? We have just learned that the first measurement will give one of the eigenvalues of the operator corresponding to the observable being measured. We have also learned that we have no way to predict the outcome of a single measurement. What is expected if a second measurement of the same observable were carried out on the same system? The experimentally established answer to this question is illustrated in Figure 14.3. In successive measurements on the same system, exactly the same result will be obtained that was obtained in the first experiment. If further successive measurements are carried out, all of the results will be the same. The probabilistic result is obtained only on the first measurement; after that, the result is deterministic.

How can this transition from a probabilistic to a **deterministic outcome** be understood? Note that the second and all successive results are exactly what would be expected if the system were in a particular eigenstate of the operator for which only one coefficient b_m is nonzero, namely, Ψ_{1s}, or Ψ_{2s}, or Ψ_{2p_x}, or Ψ_{3s}, and not in the original superposition state $\Psi_{electronic} = b_1\Psi_{1s} + b_2\Psi_{2s} + b_3\Psi_{2p_x} + b_4\Psi_{3s}$. In fact, this is the key to understanding this very puzzling result. The act of carrying out a quantum mechanical measurement appears to convert the wave function of a system to an eigenfunction of the operator corresponding to the measured quantity! We are accustomed to thinking of our role in carrying out a measurement in classical mechanics as being passive. We simply note what the system is doing and it is not influenced by us. The **measurement process** in quantum mechanics is radically different. In fact, the standard interpretation of quantum mechanics attributed to the school of Niels Bohr gives the measurement process a central role in the outcome of the experiment. This has vexed many scientists, most notably Albert Einstein. Applying this reasoning to the macroscopic world, he remarked to a colleague, "Do you really think that the moon is not there when we are not looking at it?" However strange this may all seem, no one has devised an experiment to show that the view of the measurement process in quantum mechanics stated in this postulate is incorrect.

Assume now that the superposition state that describes the system is not known. This is generally the case for a real system. Can we determine the wave function from measurements like those shown in Figure 14.3? By measuring the frequency with which a particular eigenvalue is measured, the various $|b_m|^2$ can be determined. However, this only allows b_m to be determined to within a multiplicative factor $e^{i\theta(x,t)}$. Unfortunately, this does not provide enough information to reconstruct the wave function from experimental measurements. This is a general result; the wave function of a superposition state cannot be determined by any experimental means.

14.5 The Evolution in Time of a Quantum Mechanical System

POSTULATE 5

The evolution in time of a quantum mechanical system is governed by the time-dependent Schrödinger equation:

$$\hat{H}\ \Psi(x, t) = i\hbar \frac{\partial \Psi(x, t)}{\partial t} \tag{14.12}$$

In this case, the total energy operator is given by $\hat{H} = (-\hbar^2/2m)(\partial^2/\partial x^2) + V(x, t)$. This looks like more familiar territory in that the equation has a unique solution for a set of given initial conditions. We call this behavior *deterministic* (like Newton's second law) in contrast to the probabilistic nature of Postulate 4. The fourth and fifth postulates are not contradictory. If a measurement is carried out at time t_0, Postulate 4 applies. If we ask what state the system will be in for a time $t_1 > t_0$, *without carrying out a measurement in this time interval*, Postulate 5 applies. If at time t_1 we carry out a measurement again, Postulate 4 will apply.

Note that for wave functions that are solutions of the time-independent Schrödinger equation, $\Psi(x, t) = \psi(x)e^{-i(E/\hbar)t}$. In this case, in solving the eigenvalue equation for any operator $\hat{A}$ that is not a function of time, we can write

$$\hat{A}(x)\Psi_n(x, t) = a_n\Psi_n(x, t)$$

$$\hat{A}(x)\psi_n(x)e^{-i(E/\hbar)t} = a_n\psi_n(x)e^{-i(E/\hbar)t} \quad \text{or}$$

$$\hat{A}(x)\psi_n(x) = a_n\psi_n(x) \tag{14.13}$$

This means that eigenvalue equations can be written for $\hat{A}(x)$ using only the spatial part of the wave function $\psi(x)$, knowing that $\psi(x)$ and $\Psi(x, t)$ are related by $\Psi(x, t) = \psi(x)e^{-i(E/\hbar)t}$.

14.6 Do Superposition Wave Functions Really Exist?

A simple demonstration of superposition can be made using a beam of light that passes through a polarizing filter. Such a filter, which is widely used in sunglasses, is made up of long polyvinyl alcohol molecules embedded in a polymer film that become oriented parallel to one another by stretching the film. Elemental iodine molecules that are added to the film before stretching also become oriented, leading to a much enhanced electrical conductivity parallel to the PVA molecules. Light incident on the polarization filter has random polarization, and we can express the polarization as having a component parallel to the molecular axis of the filter and a second component perpendicular to the axis. When a light beam is incident on the filter, the oscillating electric field associated with electromagnetic radiation can set the electrons associated with the I_2 molecules in motion if the electric field is parallel to the PVA molecules. In this case, the light is absorbed by the filter. If the electric field is perpendicular to the axis of the oriented molecules, the electrons cannot couple to the oscillating field, and the light passes through the filter. The light that passes through the filter will have its electric field oriented perpendicular to the molecular axis and is said to be plane polarized.

It is also possible to construct a polarization filter that will reflect one direction of the polarization and allow the other direction to pass through as shown in Figure 14.5.

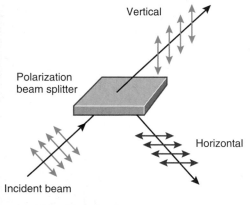

FIGURE 14.5

Schematic representation of the action of a polarization beam splitter on a plane polarized beam of photons. A plane polarized beam of photons is incident on the beam splitter. If a measurement of the polarization of the exiting photon beams is carried out, we find that the photons whose direction is unchanged are vertically polarized relative to the beam splitter, and those that are reflected are horizontally polarized. The incident beam has a plane of polarization midway between the horizontal and vertical directions.

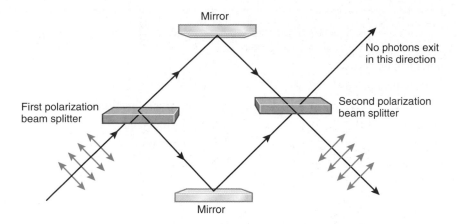

FIGURE 14.6

A second splitter and two mirrors are added to the experiment shown in Figure 14.15 in order to recombine the two beams of photons exiting the first splitter. Photons exit the apparatus only in the direction shown, and their polarization is identical to that incident on the first splitter. The polarization between the two splitters is not known because it is not measured. See text.

Such a filter is called a polarization beam splitter. We direct plane polarized light on the splitter where we choose the plane of polarization to be midway between parallel and perpendicular so that the outgoing beams have equal amplitude. So far we have considered light only in the wave picture. However, we also know that light consists of photons in a particle picture. We can reduce the light intensity to a level such that only one photon at a time is in the splitter. What do we observe in this case? We observe that an individual photon is either reflected or transmitted by the splitter and consequently has either parallel or perpendicular polarization with respect to the splitter. An interpretation of this result is that each photon is forced into an eigenfunction of the polarization operator that has only two possible eigenvalues, namely parallel and perpendicular.

We now combine two polarization beam splitters as shown in Figure 14.6. Using the mirrors, we recombine the two beams of photons exiting the first splitter. We observe that photons exit the second splitter in only one of the possible directions and the plane of polarization of the emerging light is identical to that incident on the first splitter. No photons are observed in the second possible direction. How can this result be explained? We note that there is an important difference between the two experiments shown in Figures 14.5 and 14.6.

In the experiment with a single splitter, we carry out a measurement immediately after the splitter and thereby force the photons into one of the two possible eigenfunctions of the polarization operator. In the second case, we do not carry out a measurement until after the second splitter. We consider two possibilities for the polarization of the photons that exit the first splitter in Figure 14.6. The first is that each photon passing through the first splitter in Figure 14.6 has a well-defined plane of polarization just as in Figure 14.5. In this case, the photons reflected from the upper mirror have vertical polarization and pass through the second splitter in the downward direction of Figure 14.6. The photons reflected from the lower mirror have horizontal polarization and are reflected from the second splitter. Both photons emerge in the observed direction. However, if this is the case, we should observe a stream of exiting photons each of which has either vertical or horizontal polarization. This is not consistent with the observed result.

The second possibility for the polarization of the photons that exit the first splitter is a superposition of parallel and perpendicular polarization in each of the two directions. This possibility is equivalent to each photon taking both possible paths through the first splitter rather than one path or the other. In this case, recombining the two paths at the second splitter recreates the original polarization direction midway between parallel and perpendicular, just as observed. Therefore, this experiment demonstrates that each photon exiting the first splitter is in a superposition state of the two possible polarizations rather than being either vertically or horizontally polarized.

Vocabulary

continuous function

deterministic outcome

eigenfunction

eigenvalue

expansion coefficient

expectation value

Hermitian operator

individual measurement

measurement process

normalization

observable

operator

orthonormal set

postulate

probabilistic outcome

single-valued function

superposition state

wave function

weighted average

Conceptual Problems

Q14.1 The amplitude of a standing wave function representing a moving particle changes from positive to negative values in the domain $(0, a)$ over which the wave function is defined. It must therefore pass through a stationary node at some value x_0, where $0 < x_0 < a$. Therefore the probability of the particle being at x_0 is zero and the particle cannot get from a position $x < x_0$ to a position $x > x_0$. Is this reasoning correct?

Q14.2 According to the 3rd postulate, in any single measurement of the total energy, the only values that will ever be measured are the eigenvalues of the total energy operator. Apart from the discrete energy values characteristic of a quantum mechanical system, is the result of an individual measurement of the total energy identical to the result obtained on a classical system?

Q14.3 Why must an acceptable wave function be single-valued?

Q14.4 Why must the first derivative of an acceptable wave function be continuous?

Q14.5 Why must a quantum mechanical operator $\hat{A}$ satisfy the relation, $\int \psi^*(x)[\hat{A}\psi(x)]\,dx = \int \psi(x)[\hat{A}\psi(x)]^*\,dx$?

Q14.6 If you flip a coin, what prediction can you make about it coming up heads in a single event?

Q14.7 If you flip a coin 1000. times, what prediction can you make about the number of times it comes up heads?

Q14.8 A superposition wave function can be expanded in the eigenfunctions of the operator corresponding to an observable to be measured. In analogy to rolling a single die, each of the infinite number of eigenvalues of the operator is equally likely to be measured. Is this statement correct?

Q14.9 If a system is in an eigenstate of the operator of interest, the wave function of the system can be determined. Explain this assertion. How could you know that the system is in an eigenstate of the operator of interest?

Q14.10 If the wave function for a system is a superposition wave function, the wave function of the system cannot be determined. Explain this assertion.

Q14.11 If hair color were a quantum mechanical observable, you would not have a hair color until you looked in the mirror or someone else looked at you. Is this reasoning consistent with the discussion of quantum mechanics in this chapter?

Q14.12 What did Einstein mean in his famous remark "Do you really think that the moon is not there when we are not looking at it?"

Q14.13 Would the outcome of the experiment shown in Figure 14.6 change if you carried out a measurement of the polarization between the two polarization beam splitters?

Numerical Problems

P14.1 Which of the following functions are single-valued functions of the variable x?

a. x^2

b. $\sqrt{x}$

c. $\sqrt{x} + 3x$

d. $\cos^{-1}\dfrac{2\pi x}{a}$

P14.2 Which of the following functions are single-valued functions of the variable x?

a. $\sin\dfrac{2\pi x}{a}$

b. $e^{3\sqrt{x}}$

c. $1 - 3\sin^2 x$

d. $e^{2\pi i x}$

P14.3 Graph

$$f(x) = \begin{cases} x^2 + 1, -1 \le x \le 1 \text{ except } x = 0 \\ 0, x = 0 \end{cases}$$ over the inter-

val $-4 \le x \le 4$. Is $f(x)$ a continuous function of x?

P14.4 Graph $f(x) = |x|$ and its first derivative over the interval $-4 \le x \le 4$. Are $f(x)$ and $df(x)/dx$ continuous functions of x?

P14.5 Is the function $(x^2 - 1)/(x - 1)$ continuous at $x = 1$? Answer this question by evaluating $f(1)$ and $\lim_{x \to 1} f(x)$.

P14.6 Is the function $1/(1 - x)^2$ continuous at $x = 1$? Answer this question using the criteria listed in Section 14.1.

P14.7 Consider the function $f(x) = \sin \pi x/2, -1 < x < 1$ and $f(x) = -1, x \le -1, f(x) = 1, x \ge 1$.

Graph $f(x), \dfrac{df}{dx}$, and $\dfrac{d^2f}{dx^2}$ over the interval $-2 < x < 2$.

Which, if any, of the functions is continuous over the interval?

P14.8 Which of the following functions are acceptable wave functions over the indicated interval?

a. e^{-x} $0 < x < \infty$

b. e^{-x} $-\infty < x < \infty$

c. $e^{-2\pi i x}$ $-100 < x < 100$

d. $\dfrac{1}{x}$ $1 < x < \infty$

Explain your answers.

P14.9 Which of the following functions are acceptable wave functions over the indicated interval?

a. $e^{-x^2/2}$ $-\infty < x < \infty$

b. e^{-ix} $0 < x < 2\pi$

c. $x^2 e^{-2\pi i x}$ $0 < x < \infty$

d. xe^{-x} $0 < x < \infty$

Explain your answers.

P14.10 In combining operators sequentially, it is useful to insert an arbitrary function after the operator to avoid errors. For example if the operators $\hat{A}$ and $\hat{B}$ are x and d/dx, then $\hat{A}\hat{B} \, f(x) = x df(x)/dx$. Derive the operator for kinetic energy using the classical relation for the kinetic energy, $p^2/2m$, and the operator for linear momentum listed in Table 14.1.

P14.11 For a Hermetian operator $\hat{A}$, $\int \psi(x)[\hat{A}\psi(x)] \, dx = \int \psi(x)[\hat{A}\psi(x)]^* dx$. Assume that $\hat{A} \, f(x) = (a + ib) f(x)$ where a and b are constants. Show that if $\hat{A}$ is a Hermetian operator, $b = 0$ so that the eigenvalues of $f(x)$ are real.

P14.12 Show that if $\psi_n(x)$ and $\psi_m(x)$ are solutions of the time independent Schrödinger equation, $\Psi(x,t) = \psi_n(x)e^{-(iE_n/\hbar)t} + \psi_m(x)e^{-(iE_m/\hbar)t}$ is a solution of the time dependent Schrödinger equation.

P14.13 Is the relation $(\hat{A}f(x))/f(x) = \hat{A}$ always obeyed? If not, give an example to support your conclusion.

P14.14 Is the relation $\hat{A}[f(x) + g(x)] = \hat{A}f(x) + \hat{A}g(x)$ always obeyed? If not, give an example to support your conclusion.

P14.15 In classical mechanics, the angular momentum vector $\mathbf{L}$ is defined by $\mathbf{L} = \mathbf{r} \times \mathbf{p}$. Determine the x component of $\mathbf{L}$. Substitute quantum mechanical operators for the components of $\mathbf{r}$ and $\mathbf{p}$ to prove that $\hat{l}_x = -i\hbar \left(y \dfrac{\partial}{\partial z} - z \dfrac{\partial}{\partial y} \right)$.

P14.16 Show that the three wave functions in Equation 14.11 are normalized.

15

Using Quantum Mechanics on Simple Systems

The framework described in Chapters 13 and 14 is used to solve two problems in a quantum mechanical framework that are familiar from classical mechanics. Both involve the motion of a particle on which no forces are acting. In the first case, the particle is not constrained. In the second, it is constrained to move within the confines of a box but has no other forces acting on it. We find that unlike classical mechanics, where the energy spectrum is continuous and the particle is equally likely to be found anywhere the box, the quantum mechanical particle in the box has a discrete energy spectrum and is more likely to be found in locations within the box that depend on the quantum mechanical state.

15.1 The Free Particle

The simplest classical system imaginable is the free particle, a particle in a one-dimensional space on which no forces are acting. We begin with

$$F = ma = m\frac{d^2x}{dt^2} = 0 \tag{15.1}$$

This differential equation can be solved to obtain

$$x = x_0 + v_0 t \tag{15.2}$$

Verify that this is a solution by substitution in Equation (15.1). We can calculate the position at any time if the boundary conditions of the problem, namely the initial position and velocity, are known.

How do we calculate the position of the wave-particle using quantum mechanics? The condition that no forces can be acting on the particle means that the potential energy is constant and independent of x and t. Therefore, we use the time-independent Schrödinger equation in one dimension,

$$-\frac{\hbar^2}{2m}\frac{d^2\psi(x)}{dx^2} + V(x)\psi(x) = E\psi(x) \tag{15.3}$$

to solve for the dependence of the wave function $\psi(x)$ on x. Whenever the potential energy $V(x)$ is constant, we can choose to make it zero because there is no fixed reference point for the zero of potential energy, and only changes in this quantity are measurable. The Schrödinger equation for this problem reduces to

$$\frac{d^2\psi(x)}{dx^2} = -\frac{2m}{\hbar^2}E\psi(x) \tag{15.4}$$

In words, $\psi(x)$ is a function that when differentiated twice returns the same function multiplied by a constant. Equation (15.4) has two solutions and the most appropriate form of these solutions (trigonometric or exponential) for our purposes is

$$\psi^+(x) = A_+e^{+i\sqrt{(2mE/\hbar^2)}\,x} = A_+e^{+ikx}$$

$$\psi^-(x) = A_-e^{-i\sqrt{(2mE/\hbar^2)}\,x} = A_-e^{-ikx} \tag{15.5}$$

in which the constants in the exponent have been combined using

$$k = 2\pi/\lambda = \sqrt{2mE/\hbar^2} \tag{15.6}$$

Note that the last equality is consistent with the definition of the classical kinetic energy, $E = 1/2\ mv^2$ using the de Broglie relation [Equation (12.11)]. We have been working with $\psi(x)$ rather than $\Psi(x, t)$. To obtain $\Psi(x, t)$, these two solutions are multiplied by $e^{-i(E/\hbar)t}$ or equivalently $e^{-i\omega t}$, where the relation $E = \hbar\omega$ has been used.

These solutions are plane waves, one moving to the right (positive x direction), the other moving to the left (negative x direction). The eigenvalues for the total energy can be found by substituting the wave functions of Equation (15.5) into Equation (15.4). For both solutions, $E = \hbar^2 k^2/2m$. Because k is a constant, these wave functions represent waves moving at a constant velocity determined by their initial velocity. Therefore, the quantum mechanical solution of this problem contains the same information as the classical problem, namely, motion with a constant velocity. One other important similarity between the classical and quantum mechanical free particle is that both can take on all values of energy because k is a continuous variable. The quantum mechanical free particle has a continuous energy spectrum. Why is this the case? We will learn the answer to this question in the next section of this chapter.

Because a plane wave is not localized in space, we cannot speak of its position as we did for the particle. However, the **probability** of finding the particle in an interval of length dx can be calculated. The free-particle wave functions cannot be normalized over the interval $-\infty < x < \infty$, but if x is restricted to the interval $-L \leq x \leq L$ where L can be very large, the probability of finding the particle described by $\psi^+(x)$ at position x in the interval dx is

$$P(x)\,dx = \frac{\psi^{+*}(x)\,\psi^+(x)\,dx}{\displaystyle\int_{-L}^{L}\psi^{+*}(x)\,\psi^+(x)\,dx} = \frac{A_+^* A_+ e^{-ikx}\,e^{+ikx}\,dx}{A_+^* A_+ \displaystyle\int_{-L}^{L} e^{-ikx}\,e^{+ikx}\,dx} = \frac{dx}{2L} \tag{15.7}$$

The same result is found for $\psi^-(x)$. The coefficients A_+ and A_- cancel because they appear in both the numerator and the denominator. We find that $P(x)\,dx$ is independent of x, which means that the particle is equally likely to be anywhere in the interval. This result is equivalent to saying that nothing is known about the position of the particle. As will be shown in Chapter 17, this result is linked to the fact that the momentum of the particle has been precisely specified to have the values $\hbar k$ and $-\hbar k$ for the wave functions

$\psi^+(x)$ and $\psi^-(x)$, respectively. We can verify that the eigenfunctions of the total energy operator are also eigenfunctions of the momentum operator by applying the momentum operator to these total energy eigenfunctions.

15.2 The Particle in a One-Dimensional Box

The next case to be considered is the **particle in a box.** To keep the mathematics simple, the box is one dimensional; that is, it is the one-dimensional analog of a single atom moving freely in a cube that has impenetrable walls. Two- and three-dimensional boxes are dealt with in Section 15.3 and in the problems at the end of this chapter. The impenetrable walls are modeled by making the potential energy infinite outside of a region of width a. The potential is depicted in Figure 15.1.

$$V(x) = 0, \quad \text{for } a \geq x \geq 0$$
$$V(x) = \infty, \text{ for } x > a, x < 0 \tag{15.8}$$

How does this change in the potential affect the eigenfunctions that were obtained for the free particle? To answer this question, the Schrödinger equation is written in the following form:

$$\frac{d^2\psi(x)}{dx^2} = \frac{2m}{\hbar^2}[V(x) - E]\,\psi(x) \tag{15.9}$$

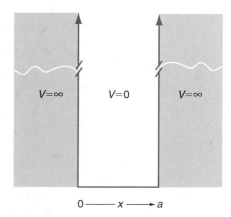

Outside of the box, where the potential energy is infinite, the second derivative of the wave function would be infinite if $\psi(x)$ were not zero for all x values outside the box. Because $d^2\psi(x)/dx^2$ must exist and be well behaved, $\psi(x)$ must be zero everywhere outside of the box as well as at $x = 0$ and $x = a$.

FIGURE 15.1
The potential described by Equation (15.8) is depicted. Because the particle is confined to the range $0 \leq x \leq a$, we say that it is confined to a one-dimensional box.

$$\psi(0) = \psi(a) = 0, \ \psi(x) = 0 \ \text{for } x > a, x < 0 \tag{15.10}$$

Equation (15.10) lists **boundary conditions** that any well-behaved wave function for the one-dimensional box must satisfy.

Inside the box, where $V(x) = 0$, the Schrödinger equation is identical to that for a free particle [Equation (15.4)], so the solutions must be the same. For ease in applying the boundary conditions, the solution is written in a trigonometric form equivalent to that of Equation (15.5):

$$\psi(x) = A \sin kx + B \cos kx \tag{15.11}$$

Now the boundary conditions given by Equation (15.10) are applied. Putting the values $x = 0$ and $x = a$ in Equation (15.11), we obtain

$$\psi(0) = 0 + B = 0$$
$$\psi(a) = A \sin ka = 0 \tag{15.12}$$

The first condition can only be satisfied by the condition that $B = 0$. The second condition can be satisfied if either $A = 0$ or $ka = n\pi$ with n being an integer. Setting A equal to zero would mean that the wave function is always zero, which is unacceptable because then there is no particle in the box. Therefore, we conclude that

$$\psi_n(x) = A \sin\left(\frac{n\pi x}{a}\right), \quad \text{for } n = 1, 2, 3, 4, \ldots \tag{15.13}$$

The requirement that $ka = n\pi$ will turn out to have important consequences for the energy spectrum of the particle in the box. Each different value of n corresponds to a different eigenfunction. To use operator language, we have found the infinite set of eigenfunctions of the total energy operator for the potential energy defined by Equation (15.8).

Note the undefined constant A in these equations. This constant can be determined by normalization, that is, by realizing that $\psi^*(x)\psi(x)\,dx$ represents the probability of finding the particle in the interval of width dx centered at x. Because the probability of finding the particle somewhere in the entire interval is 1,

$$\int_0^a \psi^*(x)\,\psi(x)\,dx = A^*A \int_0^a \sin^2\left(\frac{n\pi x}{a}\right) dx = 1 \qquad (15.14)$$

This integral is evaluated using the standard integral

$$\int \sin^2(by)\,dy = \frac{y}{2} - \frac{\sin(2by)}{4b}$$

resulting in $A = \sqrt{2/a}$, so the normalized eigenfunctions are

$$\psi_n(x) = \sqrt{\frac{2}{a}}\sin\left(\frac{n\pi x}{a}\right) \qquad (15.15)$$

What are the energy eigenvalues that go with these eigenfunctions? Applying the total energy operator to the eigenfunctions will give back the eigenfunction multiplied by the eigenvalue. We find that

$$-\frac{\hbar^2}{2m}\frac{d^2\psi_n(x)}{dx^2} = \frac{\hbar^2}{2m}\left(\frac{n\pi}{a}\right)^2\sqrt{\frac{2}{a}}\sin\left(\frac{n\pi x}{a}\right) \qquad (15.16)$$

Because

$$-\frac{\hbar^2}{2m}\frac{d^2\psi_n(x)}{dx^2} = E_n\psi_n(x)$$

the following result is obtained:

$$E_n = \frac{\hbar^2}{2m}\left(\frac{n\pi}{a}\right)^2 = \frac{h^2 n^2}{8ma^2}, \quad \text{for } n = 1, 2, 3, \ldots \qquad (15.17)$$

An important difference is seen when this result is compared to that obtained for the free particle. The energy for the particle in the box can only take on discrete values. We say that the energy of the particle in the box is **quantized** and the integer n is a **quantum number.** Another important result of this calculation is that the lowest allowed energy is greater than zero. The particle has a nonzero minimum energy, known as a **zero point energy.**

Why are quite different results obtained for the free and the confined particle? A comparison of these two problems reveals that quantization entered through the confinement of the particle. Because the particle is confined to the box, the amplitude of all allowed wave functions must be zero everywhere outside the box. By considering the limit $a \rightarrow \infty$, the confinement condition is removed. Example Problem 15.1 shows that the discrete energy spectrum becomes continuous in this limit.

The lowest four energy levels for the particle in the box are shown in Figure 15.2 superimposed on an energy-versus-distance-diagram. The eigenfunctions are also shown in this figure. Keep in mind that the time-independent part of the wave function is graphed. The full wave function is obtained by multiplying the wave functions shown in Figure 15.2 by $e^{-i(E/\hbar)t}$. If this is done, the variation of the total wave function with time is exactly what was shown in Figure 13.4 for a **standing wave,** if the real and imaginary parts of

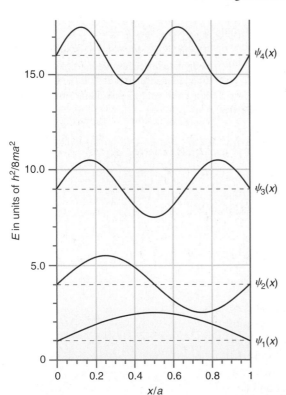

FIGURE 15.2

The first few eigenfunctions for the particle in a box are shown together with the corresponding energy eigenvalues. The energy scale is shown on the left. The wave function amplitude is shown on the right with the zero for each level indicated by the dashed line.

$\Psi(x, t)$ are considered separately. This result turns out to be general: the wave function for a state whose energy is independent of time is a standing and not a **traveling wave.** A standing wave has **nodes** that are at fixed distances independent of time, whereas the nodes move in time for a traveling wave. For this reason, the boundary conditions of Equation (15.10) cannot be satisfied for a traveling wave. A node is a point where a wave function goes through zero. An end point is not a node.

The particle in the box is also useful for showing that the quantization of the energy ultimately has its origin in the coupling of wave properties and boundary conditions. In moving from $\psi_n(x)$ to $\psi_{n+1}(x)$, the number of half-wavelengths, and therefore the number of nodes, has been increased by one. There is no way to add anything other than an integral number of half-wavelengths and still have $\psi_n(x) = 0$ at the ends of the box. Therefore, the wave vector k will increase in discrete increments rather than continuously in going from one stationary state to another. Because

$$k = \frac{2\pi}{\lambda} = \frac{p}{\hbar} = \frac{\sqrt{2mE}}{\hbar}$$

the allowed energies E also increase in jumps rather than in a continuous fashion as in classical mechanics. Thinking in this way also helps in understanding the origin of the zero point energy. Because $E = h^2/2m\lambda^2$, zero energy corresponds to an infinite wavelength, but the longest wavelength for which $\psi_n(x) = 0$ at the ends of the box is $\lambda = 2a$. Substituting this value in the equation for E gives exactly the zero point energy. Note that the zero point energy approaches zero as a approaches infinity. In this limit, the particle becomes free.

Looking at Equation (15.17), which shows the dependence of the total energy eigenvalues on the quantum number n, it is not immediately obvious that the energy spectrum will become continuous in the **classical limit** of very large n because the spacing between adjacent levels increases with n. This issue is addressed in Example Problem 15.1.

The total energy is one example of an observable that can be calculated once the eigenfunctions of the time-independent Schrödinger equation are known. Another observable that comes directly from solving this equation is the quantum mechanical analogue of position. Recall that the probability of finding the particle in any interval of width dx in the one-dimensional box is given by $\psi^*(x)\psi(x)\,dx$. The **probability density** $\psi^*(x)\psi(x)$ at a given point is shown in Figure 15.3 for the first few eigenfunctions.

How can these results be understood? Looking back at the discussion of waves in Chapter 13, recall that to ask for the position of a wave is not meaningful because the wave is not localized at a point. Wave-particle duality modifies the classical picture of being able to specify the location of a particle. Figure 15.3 shows the probability density of finding the particle in the vicinity of a given value of x rather than the position of that particle. We see that the probability of finding the particle outside of the box is zero, but that the probability of finding the particle within an interval dx in the box depends on the position and the quantum number. Although $|\psi(x)|^2$ can be zero at nodal positions, $\int_{x-\Delta x}^{x+\Delta x} \psi^*(x')\psi(x')\,dx'$ is never zero for a finite interval Δx inside the box. This means that there is no finite length interval inside the box in which the particle is not found. For the ground state, it is much more likely that the particle is found near the center of the box than at the edges. A classical particle would be found with the same probability everywhere. Does this mean that quantum mechanics and classical mechanics are in conflict? No, because we need to consider large values of n to compare with the classical limit. However, a feature in Figure 15.3 that appears hard to understand is the oscillations in $\psi^2(x)$. They will not disappear for large n; they will just be spaced more closely together. Because there are no such oscillations for the classical case, we need to make the quantum oscillations disappear for very large n.

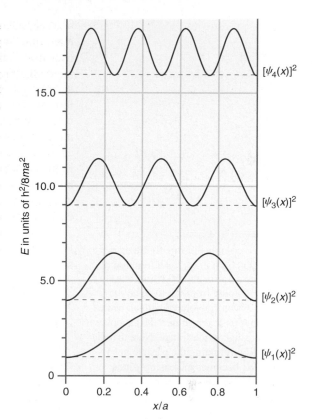

FIGURE 15.3

The square of the magnitude of the wave function, or probability density, is shown as a function of distance together with the corresponding energy eigenvalues. The energy scale is shown on the left. The square of the wave function amplitude is shown on the right with the zero for each level indicated by the dashed line.

The way to understand the convergence to the classical limit is to consider the measurement process. Any measurement has a certain resolution that averages data over the resolution range. What effect will the limited resolution have on a measurement like probability? The result is shown in Figure 15.4. The probability density $\psi^2(x)$ is shown for the 1st, 30th, and 50th eigenstates of the particle in the box for three different limits of resolution. The probability density for the ground state is unaffected by including a resolution limit. However, as the resolution of the measurement decreases, we see that the probability density for the 50th state is beginning to approach the classical behavior of a constant probability everywhere. The classical limit is closer to $n = 1 \times 10^{10}$ rather than 50 for $E \approx kT$ at realistic temperatures and box dimensions on the order of centimeters. The difference between the quantum and classical results disappears as n becomes large. This is a general result known as the **correspondence principle.**

This first attempt to apply quantum rather than classical mechanics to two familiar problems has led to several useful insights. By representing a wave-particle as a wave, familiar questions that can be asked in classical mechanics become inappropriate. An example is "Where is the particle at time t_0?" The appropriate question in quantum mechanics is "What is the probability of finding the particle at time t_0 in an interval of length dx centered at the position x_0?" For the free particle, we found that the relationship between momentum and

FIGURE 15.4
The three columns in this figure each show $\psi_n^2(x)/[\psi_1^2(x)]_{\text{max}}$ as a function of x/a for $n = 1$, 30, and 50 (from bottom to top). In going from left to right, the data have been convoluted with an instrument function that averages the data over an increasingly wider range. Note that the probability of finding the particle in an interval dx becomes increasingly independent of position as n increases for lower resolution.

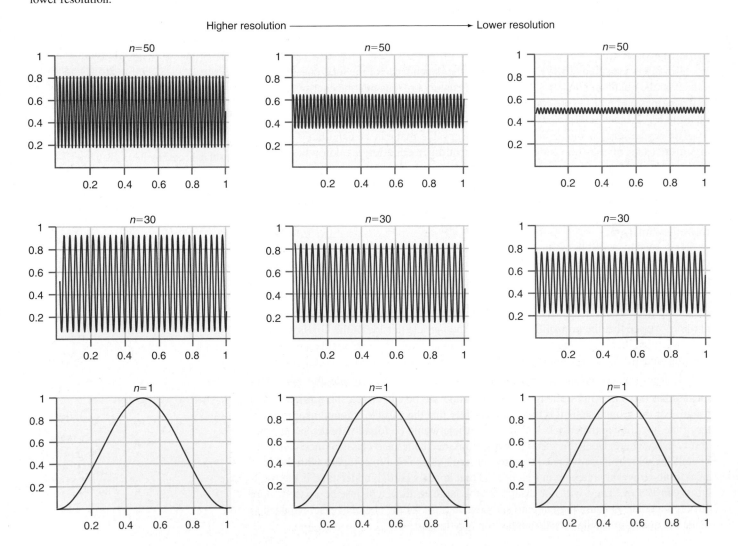

Higher resolution ⟶ Lower resolution

energy is the same as in classical mechanics and that there are no restrictions on the allowed energy. Restricting the motion of a particle to a finite region on the order of its wavelength has a significant effect on many observables associated with the particle. We saw that the origin of the effect is the requirement that the amplitude of the wave function be zero at the ends of the box for all times. This requirement changes the eigenfunctions of the Schrödinger equation from the traveling waves of the free particle to standing waves. Only discrete values of the particle momentum are allowed because of the condition $ka = n\pi$, $n = 1, 2, 3, \ldots$. Because $E = h^2 n^2/8ma^2$, the particle can only have certain values for the energy, and these values are determined by the length of the box. Wave-particle duality also leads to a nonuniform probability for finding the particle in the box.

EXAMPLE PROBLEM 15.1

From the formula given for the energy levels for the particle in the box, $E_n = h^2 n^2/8ma^2$ for $n = 1, 2, 3, \ldots$, we can see that the spacing between adjacent levels increases with n. This appears to indicate that the energy spectrum does not become continuous for large n, which must be the case for the quantum mechanical result to be identical to the classical result in the high-energy limit. A better way to look at the spacing between levels is to consider the ratio $(E_{n+1} - E_n)/E_n$. Form this ratio and show that $\Delta E/E$ becomes a smaller fraction of the energy as $n \to \infty$. This result shows that the energy spectrum becomes continuous for large n.

Solution

$$\frac{E_{n+1} - E_n}{E_n} = \left(\frac{h^2[(n+1)^2 - n^2]/8ma^2}{h^2 n^2/8ma^2} \right) = \frac{2n+1}{n^2}$$

which approaches zero as $n \to \infty$. Both the level spacing and the energy increase with n, but the energy increases faster (as n^2), making the energy spectrum appear to be continuous as $n \to \infty$. This is another example of the correspondence principle.

15.3 Two- and Three-Dimensional Boxes

The one-dimensional box is a useful model system because the conceptual simplicity allows the focus to be on the quantum mechanics rather than on the mathematics. The extension of the formalism developed for the one-dimensional problem to two and three dimensions has several aspects that are of use in understanding topics such as the rotation of molecules and the electronic structure of atoms, which cannot be reduced to one-dimensional problems.

Our focus here is on the three-dimensional box because the reduction in dimensionality from three to two is straightforward. The potential energy is given by

$$V(x, y, z) = 0 \quad \text{for } 0 \le x \le a; \quad 0 \le y \le b; \quad 0 \le z \le c$$

$$= \infty \quad \text{otherwise} \tag{15.18}$$

As before, the amplitude of the eigenfunctions of the total energy operator is identically zero outside the box. Inside the box, the Schrödinger equation can be written as

$$-\frac{\hbar^2}{2m} \left(\frac{\partial^2}{\partial x^2} + \frac{\partial^2}{\partial y^2} + \frac{\partial^2}{\partial z^2} \right) \psi(x, y, z) = E\psi(x, y, z) \tag{15.19}$$

This differential equation is solved assuming that $\psi(x, y, z)$ has the form

$$\psi(x, y, z) = X(x)Y(y)Z(z) \tag{15.20}$$

in which $\psi(x, y, z)$ is the product of three functions, each of which depends on only one of the variables. The assumption is valid in this case because $V(x, y, z)$ is independent of

x, y, and z inside the box. This process is referred to as a **separation of variables.** It is also valid for a potential of the form $V(x, y, z) = V_x(x) + V_y(y) + V_z(z)$. Substituting Equation (15.20) in Equation (15.19), we obtain

$$-\frac{\hbar^2}{2m}\left(Y(y)Z(z)\frac{d^2 X(x)}{dx^2} + X(x)Z(z)\frac{d^2 Y(y)}{dy^2} + X(x)Y(y)\frac{d^2 Z(z)}{dz^2}\right)$$
$$= EX(x)Y(y)Z(z) \tag{15.21}$$

Note that Equation (15.21) no longer contains partial derivatives, because each of the three functions X, Y, and Z depends on only one variable. Dividing by the product $X(x)Y(y)Z(z)$ results in

$$-\frac{\hbar^2}{2m}\left[\frac{1}{X(x)}\frac{d^2 X(x)}{dx^2} + \frac{1}{Y(y)}\frac{d^2 Y(y)}{dy^2} + \frac{1}{Z(z)}\frac{d^2 Z(z)}{dz^2}\right] = E \tag{15.22}$$

The form of this equation shows that E can be viewed as having independent contributions from the three coordinates, $E = E_x + E_y + E_z$, and the original differential equation in three variables reduces to three differential equations, each in one variable:

$$-\frac{\hbar^2}{2m}\frac{d^2 X(x)}{dx^2} = E_x X(x); \quad -\frac{\hbar^2}{2m}\frac{d^2 Y(y)}{dy^2} = E_y Y(y); \quad -\frac{\hbar^2}{2m}\frac{d^2 Z(z)}{dz^2} = E_z Z(z)$$
$$\tag{15.23}$$

Each of these equations has the same form as the equation that was solved for the one-dimensional problem. Therefore, the total energy eigenfunctions have the form

$$\psi_{n_x n_y n_z}(x, y, z) = N \sin\frac{n_x \pi x}{a} \sin\frac{n_y \pi y}{b} \sin\frac{n_z \pi z}{c} \tag{15.24}$$

and the total energy has the form

$$E = \frac{h^2}{8m}\left(\frac{n_x^2}{a^2} + \frac{n_y^2}{b^2} + \frac{n_z^2}{c^2}\right) \tag{15.25}$$

This is a general result. *If the total energy can be written as a sum of independent terms corresponding to different degrees of freedom, then the wave function is a product of individual terms, each corresponding to one of the degrees of freedom.*

Because this is a three-dimensional problem, the eigenfunctions depend on three quantum numbers. Because more than one set of the three quantum numbers may have the same energy [for example, $(1, 2, 1)$, $(2, 1, 1)$, and $(1, 1, 2)$ if $a = b = c$], several distinct eigenfunctions of the total energy operator may have the same energy. In this case, we say that the energy level is **degenerate,** and the number of states, each represented by a distinct eigenfunction, that have the same energy is the **degeneracy** of the level.

What form do ψ and E take for the two-dimensional box? How many quantum numbers are needed to characterize ψ and E for the two-dimensional problem? Additional issues related to the functional form, degeneracy, and normalization of the total energy eigenfunctions are covered in the end-of-chapter problems.

We have made a considerable effort to understand the particle in the box, because this model is very useful in understanding properties that can be measured for real systems. Some of these systems will be discussed in Chapter 16. However, we first return to the postulates introduced in Chapter 14, now that the Schrödinger equation has been solved for an interesting system.

15.4 Using the Postulates to Understand the Particle in the Box and Vice Versa

Because of its simplicity, the particle in a box is an excellent teaching tool for learning how to apply quantum mechanics to a specific system. In this section, each of the postulates is applied to this problem using the eigenvalues and eigenfunctions calculated earlier. We begin with the first postulate.

POSTULATE 1:

The state of a quantum mechanical system is completely specified by a wave function $\Psi(x, t)$. The probability that a particle will be found at time t in a spatial interval of width dx centered at x_0 is given by $\Psi^*(x_0, t)\Psi(x_0, t)\, dx$.

This postulate states that all the information that can ever be obtained about the system is contained in the wave function. At this point it is useful to review the distinction between a wave function and an eigenfunction. A wave function is any mathematically well-behaved function that satisfies the boundary conditions and that can be normalized to allow a meaningful definition of probability. An eigenfunction must satisfy these and one more criterion. A wave function is an eigenfunction of an operator $\hat{A}$ only if it satisfies the relationship $\hat{A}\psi_n(x) = a_n\psi_n(x)$. These criteria are illustrated in Example Problem 15.2.

EXAMPLE PROBLEM 15.2

Consider the superposition wave function $\psi(x) = c\sin(\pi x/a) + d\sin(2\pi x/a)$.

a. Is $\psi(x)$ an acceptable wave function for the particle in the box?

b. Is $\psi(x)$ an eigenfunction of the total energy operator $\hat{H}$?

c. Is $\psi(x)$ normalized?

Solution

a. If $\psi(x)$ is to be an acceptable wave function, it must satisfy the boundary conditions $\psi(x) = 0$ at $x = 0$ and $x = a$. The first and second derivatives of $\psi(x)$ must also be well-behaved functions between $x = 0$ and $x = a$. This is the case for $\psi(x)$. We conclude that $\psi(x) = c\sin(\pi x/a) + d\sin(2\pi x/a)$ is an acceptable wave function for the particle in the box.

b. Although $\psi(x) = c\sin(\pi x/a) + d\sin(2\pi x/a)$ may be an acceptable wave function, it need not be an eigenfunction of a given operator. To see if $\psi(x)$ is an eigenfunction of the total energy operator, the operator is applied to the wave function:

$$-\frac{\hbar^2}{2m}\frac{d^2}{dx^2}\left(c\sin\left(\frac{\pi x}{a}\right) + d\sin\left(\frac{2\pi x}{a}\right)\right)$$

$$= \frac{\hbar^2\pi^2}{2ma^2}\left(c\sin\left(\frac{\pi x}{a}\right) + 4d\sin\left(\frac{2\pi x}{a}\right)\right)$$

The result of this operation is not $\psi(x)$ multiplied by a constant. Therefore, $\psi(x)$ is not an eigenfunction of the total energy operator.

c. To see if $\psi(x)$ is normalized, the following integral is evaluated:

$$\int_0^a \left| c\sin\left(\frac{\pi x}{a}\right) + d\sin\left(\frac{2\pi x}{a}\right) \right|^2 dx$$

$$= \int_0^a \left[c^*c\sin^2\left(\frac{\pi x}{a}\right) + d^*d\sin^2\left(\frac{2\pi x}{a}\right) + (cd^* + c^*d)\sin\left(\frac{\pi x}{a}\right)\sin\left(\frac{2\pi x}{a}\right) \right] dx$$

$$= \int_0^a |c|^2\sin^2\left(\frac{\pi x}{a}\right) dx + \int_0^a |d|^2\sin^2\left(\frac{2\pi x}{a}\right) dx$$

$$+ \int_0^a (cd^* + c^*d)\sin\left(\frac{\pi x}{a}\right)\sin\left(\frac{2\pi x}{a}\right) dx$$

Using the standard integral $\int \sin^2(by)dy = y/2 - (1/4b)\sin(2\,by)$ and recognizing that the third integral is zero because all $\sin nx$ functions with different n are orthogonal,

$$\int_0^a \left[|c|^2 \sin^2\left(\frac{\pi x}{a}\right) + |d|^2 \sin^2\left(\frac{2\pi x}{a}\right) \right] dx$$

$$= |c|^2 \left[\frac{a}{2} - \frac{a(\sin 2\pi - \sin 0)}{4\pi} \right] + |d|^2 \left[\frac{a}{2} - \frac{a(\sin 4\pi - \sin 0)}{8\pi} \right]$$

$$= \frac{a}{2}(|c|^2 + |d|^2)$$

Therefore, $\psi(x)$ is not normalized in general, but the wave function

$$\sqrt{\frac{2}{a}} \left[c \sin\left(\frac{\pi\,x}{a}\right) + d \sin\left(\frac{2\pi\,x}{a}\right) \right]$$

is normalized for the condition that $|c|^2 + |d|^2 = 1$.

Note that a superposition wave function has a more complicated dependence on time than does an eigenfunction of the total energy operator. For instance, $\psi(x, t)$ for the wave function under consideration is given by

$$\Psi(x, t) = \sqrt{\frac{2}{a}} \left[ce^{-iE_1t/\hbar} \sin\left(\frac{\pi x}{a}\right) + de^{-iE_2t/\hbar} \sin\left(\frac{2\pi x}{a}\right) \right] \neq \psi(x)f(t)$$

This wave function cannot be written as a product of a function of x and a function of t. It is not a standing wave and does not describe a state whose properties are, in general, independent of time.

All of the particle in the box eigenfunctions, $\psi_n(x) = \sqrt{2/a} \sin(n\pi x/a)$, for $n = 1, 2, 3, \ldots$ are normalized, meaning that the total probability of finding the particle somewhere between $x = 0$ and $x = a$ is one. In other words, the particle is somewhere in the box. We cannot predict with certainty the outcome of a single measurement in which the position of the particle is determined, because these eigenfunctions of the total energy operator are not eigenfunctions of the position operator. In Chapter 17, we will discuss why the eigenvalues of $\hat{H}$ and $\hat{x}$ cannot be determined simultaneously. We can, however, predict the average value determined in a large number of independent measurements of the particle position. This is equivalent to asking for the probability density of finding the particle at a given position. The formula for calculating this probability is stated in the first postulate. The total probability of finding the particle in a finite length interval is obtained by integrating the probability density, as shown in Example Problem 15.3.

EXAMPLE PROBLEM 15.3

What is the probability P of finding the particle in the central third of the box if it is in its ground state?

Solution

For the ground state, $\psi_1(x) = \sqrt{2/a} \sin(\pi x/a)$. From the postulate, P is the sum of all the probabilities of finding the particle in intervals of width dx within the central third of the box. This probability is given by the integral

$$P = \frac{2}{a} \int_{a/3}^{2a/3} \sin^2\left(\frac{\pi x}{a}\right) dx$$

Solving this integral as in Example Problem 15.2,

$$P = \frac{2}{a}\left[\frac{a}{6} - \frac{a}{4\pi}\left(\sin\frac{4\pi}{3} - \sin\frac{2\pi}{3}\right)\right] = 0.609$$

Although we cannot predict the outcome of a single measurement, we can predict that for 60.9% of a large number of individual measurements, the particle is found in the central third of the box. What is the probability of finding a classical particle in this interval?

Postulate 2 is a recipe for associating classical observables with quantum mechanical operators and need not be considered further. Postulates 3 and 4 are best understood by considering them together.

POSTULATE 3:

In any single measurement of the observable that corresponds to the operator $\hat{A}$, the only values that will ever be measured are the eigenvalues of that operator.

POSTULATE 4:

If the system is in a state described by the wave function $\Psi(x, t)$, and the value of the observable a is measured once each on many identically prepared systems, the average value of all of these measurements is given by

$$\langle a \rangle = \frac{\displaystyle\int_{-\infty}^{\infty} \Psi^*(x, t)\hat{A}\Psi(x, t)\, dx}{\displaystyle\int_{-\infty}^{\infty} \Psi^*(x, t)\Psi(x, t)\, dx} \tag{15.26}$$

The wave function for particle in its ground state is $\psi(x) = \sqrt{2/a}\,\sin(\pi x/a)$, which is a normalized eigenfunction of the total energy operator. Applying the operator to this wave function returns the function multiplied by the constant. This is the value of the energy that is determined in any single measurement and, therefore, it is also the average of all values for the energy that are measured on many particles prepared in the same state.

Now consider a measurement of the total energy for a case in which the wave function of the system is not an eigenfunction of this operator. As you convinced yourself in Example Problem 15.2, the normalized superposition wave function

$$\psi(x) = \sqrt{\frac{2}{a}}\left(c\sin\frac{\pi x}{a} + d\sin\frac{2\pi x}{a}\right)$$

where $|c|^2 + |d|^2 = 1$ is not an eigenfunction of $\hat{H}$. Postulate 4 says that the average value of the energy for a large number of identical measurements on a system whose state is described by a normalized wave function is

$$\langle E \rangle = \int_0^a \psi^*(x)\left[-\frac{\hbar^2}{2m}\frac{d^2}{dx^2} + V(x)\right]\psi(x)\, dx \tag{15.27}$$

We now substitute the expression for $\psi(x)$ into Equation (15.27):

$$\langle E \rangle = \frac{2}{a}\int_0^a \left(c^*\sin\frac{\pi x}{a} + d^*\sin\frac{2\pi x}{a}\right)\left[-\frac{\hbar^2}{2m}\frac{d^2}{dx^2}\right]\left(c\sin\frac{\pi x}{a} + d\sin\frac{2\pi x}{a}\right) dx$$

$$\tag{15.28}$$

Multiplying out the terms in the brackets, and recognizing that each of the individual terms in the parentheses is an eigenfunction of the operator, $\langle E \rangle$ reduces to

$$\langle E \rangle = \frac{2}{a}\left[|c|^2 E_1 \int_0^a \sin^2 \frac{\pi x}{a}\, dx + |d|^2 E_2 \int_0^a \sin^2 \frac{2\pi x}{a}\, dx\right]$$

$$+ \frac{2}{a}\left[c^* d\, E_2 \int_0^a \sin \frac{\pi x}{a} \sin \frac{2\pi x}{a}\, dx + d^* c\, E_1 \int_0^a \sin \frac{\pi x}{a} \sin \frac{2\pi x}{a}\, dx\right] \qquad \textbf{(15.29)}$$

We know the value of each of the first two integrals is $a/2$ from our efforts to normalize the functions. Each of the last two integrals is identically zero because the sine functions with different arguments are mutually orthogonal. Therefore, the result of these calculations is

$$\langle E \rangle = |c|^2 E_1 + |d|^2 E_2 \qquad \textbf{(15.30)}$$

where $E_n = n^2 h^2/8ma^2$. Because $|c|^2 + |d|^2 = 1$, $\langle E \rangle$ is a weighted average of E_1 and E_2. As seen in Example Problem 15.2, the superposition wave function does not describe a stationary state, and the average values of observables such as $\langle p \rangle$ and $\langle x \rangle$ are functions of time as shown in Problem W15.6. However, the average energy is independent of time because the energy is conserved.

Note that this result is exactly what was derived for a more general case in discussing Postulate 4 (see Chapter 14). We next discuss in more detail what will be obtained for an individual measurement of the total energy and relate it to the result that was just derived for the average of many individual measurements. Postulate 3 says that in an individual measurement, only one of the eigenvalues of the operator can be measured. In this case, it means that only one of the infinite set of E_n given by $E_n = n^2 h^2/8ma^2$, $n = 1, 2, 3, \ldots$, is a possible result of an individual measurement. What is the likelihood that the value E_2 will be measured? Postulate 4 gives a recipe for answering this question. It tells us to expand the system wave function in the complete set of functions that are the eigenfunctions of the operator of interest. The probability that an individual measurement will give E_n is given by the square of the expansion coefficient of that eigenfunction in the expression for the wave function. In the particular case under consideration, the wave function can be written as follows:

$$\psi(x) = c\,\psi_1(x) + d\,\psi_2(x) + 0 \times (\psi_3(x) + \ldots + \psi_n(x) + \ldots) \qquad \textbf{(15.31)}$$

in which it has been made explicit that the coefficients of all the eigenfunctions other than $\psi_1(x)$ and $\psi_2(x)$ are zero. Therefore, given the wave function for the system, individual measurements on identically prepared systems will never give anything other than E_1 or E_2. The probability of obtaining E_1 is c^2 and the probability of obtaining E_2 is d^2. From this result, it is clear that the average value for the energy determined from a large number of measurements is $c^2 E_1 + d^2 E_2$.

A more detailed discussion of causality in quantum mechanics would lead us to a number of conclusions that differ significantly from our experience with classical mechanics. For instance, it is not possible to predict whether E_1 or E_2 would be measured in an individual measurement any more than the outcome of a single throw of a die can be predicted. However, if the energy is measured again on the same system (rather than carrying out a second measurement on an identically prepared system), the same result will be obtained as in the initial measurement. This conclusion also holds for all subsequent measurements. This last result is particularly intriguing because it suggests that through the measurement process, the system has been forced into an eigenfunction of the operator corresponding to the quantity being measured.

Now consider a measurement of the momentum or the position. As shown earlier, we need to know the wave function that describes the system to carry out such a calculation. For this calculation, assume that the system is in one of the eigenstates of the total energy operator, for which $\psi_n(x) = \sqrt{2/a}\,\sin(n\pi x/a)$. From the second postulate, and Table 14.1, the quantum mechanical operator associated with momentum is $-i\hbar(d/dx)$. Although $\psi(x)$ is an eigenfunction of the total energy operator, it is not clear if it is an eigenfunction of the momentum operator. Verify that it is not an eigenfunction of this operator by operating on the wave function with the momentum operator. We will return to the significance of this result in Chapter 17, but we first proceed in applying the postulates. Postulate 4 defines how the average value of the momentum obtained in a large number of individual measurements on an identically prepared system can be calculated. The result is given by

$$\langle p \rangle = \int_0^a \psi^*(x)\hat{p}\psi(x)dx$$

$$= \frac{2}{a} \int_0^a \sin\left(\frac{n\pi x}{a}\right)\left[-i\hbar \frac{d}{dx}\sin\left(\frac{n\pi x}{a}\right)\right]dx \qquad \textbf{(15.32)}$$

$$= \frac{-2i\hbar n\pi}{a^2} \int_0^a \sin\left(\frac{n\pi x}{a}\right)\cos\left(\frac{n\pi x}{a}\right)dx = \frac{-i\hbar}{a}\left[\sin^2 n\pi - \sin^2 0\right] = 0$$

Note that the result is the same for all values of n. We know that the energy of the lowest state is greater than zero and that all the energy is in the form of kinetic energy. Because $E = p^2/2m \neq 0$, the magnitude of p must be greater than zero for an individual measurement. How can the result that the average value of the momentum is zero be understood?

Keep in mind that, classically, the particle is bouncing back and forth between the two walls of the one-dimensional box with a constant velocity. Therefore, it is equally likely that the particle is moving in the $+x$ and $-x$ directions and that its momentum is positive or negative. For this reason, the average momentum is zero. This result holds up in a quantum mechanical picture. However, a major difference exists between the quantum and classical pictures. In classical mechanics, the magnitude of the momentum of the particle is known to be $\sqrt{2m\,E_{kin}}$ exactly. In quantum mechanics, a consequence of confining the particle to a box of length a is that an uncertainty has been introduced in its momentum that is proportional to $1/a$. This issue will be discussed in depth in Chapter 17. The calculation for the average value of position is carried out in Example Problem 15.4.

EXAMPLE PROBLEM 15.4

Assume that a particle is confined to a box of length a, and that the system wave function is $\psi(x) = \sqrt{2/a}\,\sin(\pi x/a)$.

a. Is this state an eigenfunction of the position operator?

b. Calculate the average value of the position $\langle x \rangle$ that would be obtained for a large number of measurements. Explain your result.

Solution

a. The position operator $\hat{x} = x$. Because $x\,\psi(x) = \sqrt{2/a}\,x\,\sin(\pi x/a) \neq c\psi(x)$, where c is a constant, the wave function is not an eigenfunction of the position operator.

b. The expectation value is calculated using the fourth postulate:

$$\langle x \rangle = \frac{2}{a}\int_0^a \left\{\sin\left(\frac{\pi x}{a}\right)\right\}x\,\sin\left(\frac{\pi x}{a}\right)dx = \frac{2}{a}\int_0^a x\left\{\sin\left(\frac{\pi x}{a}\right)\right\}^2 dx$$

Using the standard integral $\displaystyle\int x(\sin bx)^2\,dx = \frac{x^2}{4} - \frac{\cos 2bx}{8b^2} - \frac{x\sin 2bx}{4b}$

$$\langle x \rangle = \frac{2}{a}\left[\frac{x^2}{4} - \frac{\cos\left(\dfrac{2\pi x}{a}\right)}{8\left(\dfrac{\pi}{a}\right)^2} - \frac{x\sin\left(\dfrac{2\pi x}{a}\right)}{4\left(\dfrac{\pi}{a}\right)}\right]_0^a$$

$$= \frac{2}{a}\left[\left(\frac{a^2}{4} - \frac{a^2}{8\pi^2} - 0\right) + \frac{a^2}{8\pi^2}\right] = \frac{a}{2}$$

The average position is midway in the box. This is exactly what we would expect, because the particle is equally likely to be in each half of the box.

Vocabulary

boundary condition	particle in a box	standing wave
classical limit	probability	traveling wave
correspondence principle	probability density	wave vector
degeneracy	quantized	zero point energy
degenerate	quantum number	
node	separation of variables	

Conceptual Problems

Q15.1 We set the potential energy in the particle in the box equal to zero and justified it by saying that there is no absolute scale for potential energy. Is this also true for kinetic energy?

Q15.2 Discuss why a quantum mechanical particle in a box has a zero point energy in terms of its wavelength.

Q15.3 How does an expectation value for an observable differ from an average of all possible eigenvalues?

Q15.4 Is the probability distribution for a free particle consistent with a purely particle picture, a purely wave picture, or both?

Q15.5 Show that it is not possible to normalize the free-particle wave functions over the whole range of motion of the particle.

Q15.6 The probability density for a particle in a box is an oscillatory function even for very large energies. Explain how the classical limit of a constant probability density that is independent of position is achieved for large quantum numbers.

Q15.7 Explain using words, rather than equations, why if $V(x, y, z) \neq V_x(x) + V_y(y) + V_z(z)$, the total energy eigenfunctions cannot be written in the form $\psi(x, y, z) = X(x)Y(y)Z(z)$.

Q15.8 Can a guitar string be in a superposition of states or is such a superposition only possible for a quantum mechanical system?

Q15.9 Show that for the particle in the box total energy eigenfunctions, $\psi_n(x) = \sqrt{2/a} \sin(n\pi x/a)$, $\psi(x)$ is a continuous function at the edges of the box. Is $d\psi/dx$ a continuous function of x at the edges of the box?

Q15.10 Why are standing-wave solutions for the free particle not compatible with the classical result $x = x_0 + v_0 t$?

Q15.11 What is the difference between probability and probability density?

Q15.12 Why are traveling-wave solutions for the particle in the box not compatible with the boundary conditions?

Q15.13 Can the particles in a one-dimensional box, a square two-dimensional box, and a cubic three-dimensional box all have degenerate energy levels?

Q15.14 Invoke wave-particle duality to address the following question: How does a particle get through a node in a wave function to get to the other side of the box?

Q15.15 Why is the zero point energy lower for a He atom in a box than for an electron?

Q15.16 What are the units of the probability density for the particle in a one-dimensional box?

Q15.17 What are the units of the probability density for the particle in a three-dimensional box?

Q15.18 What is the relationship between the zero point energy for a H atom and a H_2 molecule in a one-dimensional box?

Q15.19 Show that the correct energy eigenvalues for the particle in a one-dimensional box are obtained even if the total energy eigenfunctions are not normalized.

Q15.20 What are the possible results for the energy that would be obtained in a measurement on the particle in a one-dimensional box if the wave function is $\psi_n(x) = \sqrt{2/a} \sin(7\pi x/a)$?

Numerical Problems

Problem numbers in **red** indicate that the solution to the problem is given in the *Student's Solutions Manual*.

P15.1 This problem explores under what conditions the classical limit is reached for a macroscopic cubic box of edge length a.

A nitrogen molecule of average translational energy $3/2 \, k_B T$ is confined in a cubic box of volume $V = 1.250 \text{ m}^3$ at 298 K. Use the result from Equation (15.25) for the dependence of the energy levels on a and on the quantum numbers n_x, n_y, and n_z.

a. What is the value of the "reduced quantum number" $\alpha = \sqrt{n_x^2 + n_y^2 + n_z^2}$ for $T = 298$ K?

b. What is the energy separation between the levels α and $\alpha + 1$? (*Hint:* Subtract $E_{\alpha+1}$ from E_α *before* plugging in numbers.)

c. Calculate the ratio $(E_{\alpha+1} - E_\alpha)/k_B T$ and use your result to conclude whether a classical or quantum mechanical description is appropriate for the particle.

P15.2 Calculate the expectation values $\langle x \rangle$ and $\langle x^2 \rangle$ for a particle in the state $n = 5$ moving in a one-dimensional box of length 2.50×10^{-10}. Is $\langle x^2 \rangle = \langle x \rangle^2$? Explain your answer.

P15.3 Normalize the total energy eigenfunctions for the three-dimensional box in the interval $0 \leq x \leq a$, $0 \leq y \leq b, 0 \leq z \leq c$.

P15.4 Is the superposition wave function for the free particle $\psi(x) = A_+ e^{+i\sqrt{(2mE/\hbar^2)}x} + A_- e^{-i\sqrt{(2mE/\hbar^2)}x}$ an eigenfunction of the momentum operator? Is it an eigenfunction of the total energy operator? Explain your result.

P15.5 Suppose that the wave function for a system can be written as

$$\psi(x) = \frac{\sqrt{3}}{4}\phi_1(x) + \frac{\sqrt{3}}{2\sqrt{2}}\phi_2(x) + \frac{2 + \sqrt{3}\,i}{4}\phi_3(x)$$

and that $\phi_1(x)$, $\phi_2(x)$, and $\phi_3(x)$ are normalized eigenfunctions of the operator $\hat{E}_{kinetic}$ with eigenvalues E_1, $2E_1$, and $4E_1$, respectively.

a. Verify that $\psi(x)$ is normalized.

b. What are the possible values that you could obtain in measuring the kinetic energy on identically prepared systems?

c. What is the probability of measuring each of these eigenvalues?

d. What is the average value of $E_{kinetic}$ that you would obtain from a large number of measurements?

P15.6 Consider a free particle moving in one dimension whose probability of moving in the positive x direction is four times that for moving in the negative x direction. Give as much information as you can about the wave function of the particle.

P15.7 Are the eigenfunctions of $\hat{H}$ for the particle in the one-dimensional box also eigenfunctions of the momentum operator $\hat{p}_x$? Calculate the average value of p_x for the case $n = 3$. Repeat your calculation for $n = 5$ and, from these two results, suggest an expression valid for all values of n. How does your result compare with the prediction based on classical physics?

P15.8 Evaluate the normalization integral for the eigenfunctions of $\hat{H}$ for the particle in the box $\psi_n(x) = A\sin(n\pi x/a)$ using the trigonometric identity $\sin^2 y = (1 - \cos 2y)/2$.

P15.9 Is the function $\psi(y) = A(y/b)[1 - (y/b)]$ an acceptable wave function for the particle in the one-dimensional infinite depth box of length b? Calculate the normalization constant A and the expectation values $\langle y \rangle$ and $\langle y^2 \rangle$.

P15.10 What is the solution of the time-dependent Schrödinger equation $\Psi(x, t)$ for the total energy eigenfunction $\psi_4(x) = \sqrt{2/a}\sin(3\pi x/a)$ for an electron in a one-dimensional box of length 1.00×10^{-10} m? Write explicitly in terms of the parameters of the problem. Give numerical values for the angular frequency ω and the wavelength of the particle.

P15.11 Derive an equation for the probability that a particle characterized by the quantum number n is in the first 25% ($0 \leq x \leq 0.25a$) of an infinite depth box. Show that this probability approaches the classical limit as $n \to \infty$.

P15.12 It is useful to consider the result for the energy eigenvalues for the one-dimensional box $E_n = h^2 n^2/8ma^2$, $n = 1, 2, 3, \ldots$ as a function of n, m, and a.

a. By what factor do you need to change the box length to decrease the zero point energy by a factor of 50 for a fixed value of m?

b. By what factor would you have to change n for fixed values of a and m to increase the energy by a factor of 600?

c. By what factor would you have to increase a at constant n to have the zero point energies of a Ne atom be equal to the zero point energy of a hydrogen atom in the box?

P15.13 Show that the energy eigenvalues for the free particle, $E = \hbar^2 k^2/2m$, are consistent with the classical result $E = (1/2)mv^2$.

P15.14 **a.** Show by substitution into Equation (15.19) that the eigenfunctions of $\hat{H}$ for a box with lengths along the x, y, and z directions of a, b, and c, respectively, are

$$\psi_{n_x, n_y, n_z}(x, y, z) = N\sin\left(\frac{n_x \pi x}{a}\right)\sin\left(\frac{n_y \pi y}{b}\right)\sin\left(\frac{n_z \pi z}{c}\right)$$

b. Obtain an expression for E_{n_x, n_y, n_z} in terms of n_x, n_y, n_z, and a, b, and c.

P15.15 Calculate the wavelength of the light emitted when an electron in a one-dimensional box of length 5.0 nm makes a transition from the $n = 7$ state to the $n = 6$ state.

P15.16 A bowling ball has a weight of 12 lb and the length of the lane is approximately 60. ft. Treat the ball in the lane as a particle in a one-dimensional box. What quantum number corresponds to a velocity of 7.5 miles per hour?

P15.17 For a particle in a two-dimensional box, the total energy eigenfunctions are $\psi_{n_x n_y}(x, y) = N\sin\dfrac{n_x \pi x}{a}\sin\dfrac{n_y \pi y}{b}$

a. Obtain an expression for E_{n_x, n_y} in terms of n_x, n_y, a, and b by substituting this wave function into the two-dimensional analog of Equation (15.19).

b. Contour plots of several eigenfunctions are shown here. The x and y directions of the box lie along the horizontal and vertical directions, respectively. The amplitude has been displayed as a gradation in colors. Regions of positive and negative amplitude are indicated. Identify the values of the quantum numbers n_x and n_y for plots a–f.

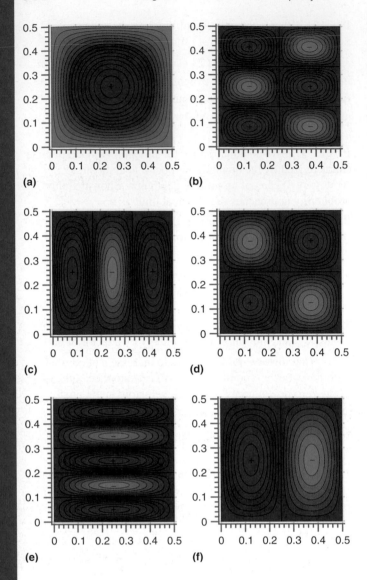

(a)

(b)

(c)

(d)

(e)

(f)

P15.18 Consider the contour plots of Problem P15.17.

a. What are the most likely area or areas $\Delta x \, \Delta y$ to find the particle for each of the eigenfunctions of $\hat{H}$ depicted in plots a–f?

b. For the one-dimensional box, the nodes are points. What form do the nodes take for the two-dimensional box? Where are the nodes located in plots a–f? How many nodes are there in each contour plot?

P15.19 Using your result from P15.17, how many energy levels does a particle of mass m in a two-dimensional box of edge length a have with $E \leq 29h^2/8ma^2$? What is the degeneracy of each level?

P15.20 Calculate (a) the zero point energy of a He atom in a one-dimensional box of length 1.00 cm and (b) the ratio of the zero point energy to kT at 300. K.

P15.21 Normalize the total energy eigenfunction for the rectangular two-dimensional box,

$$\psi_{n_x, n_y}(x, y) = N \sin\left(\frac{n_x \pi x}{a}\right) \sin\left(\frac{n_y \pi y}{b}\right)$$

in the interval $0 \leq x \leq a, 0 \leq y \leq b$.

P15.22 Generally, the quantization of translational motion is not significant for atoms because of their mass. However, this conclusion depends on the dimensions of the space to which they are confined. Zeolites are structures with small pores that we describe by a cube with edge length 1 nm. Calculate the energy of a H_2 molecule with $n_x = n_y = n_z = 10$. Compare this energy to $k_B T$ at $T = 300.$ K. Is a classical or a quantum description appropriate?

P15.23 Are the eigenfunctions of $\hat{H}$ for the particle in the one-dimensional box also eigenfunctions of the position operator $\hat{x}$? Calculate the average value of x for the case where $n = 3$. Explain your result by comparing it with what you would expect for a classical particle. Repeat your calculation for $n = 5$ and, from these two results, suggest an expression valid for all values of n. How does your result compare with the prediction based on classical physics?

P15.24 What is the zero point energy and what are the energies of the lowest seven energy levels in a three-dimensional box with $a = b = c$? What is the degeneracy of each level?

P15.25 In discussing the Boltzmann distribution in Chapter 13, we used the symbols g_i and g_j to indicate the degeneracies of the energy levels i and j. By degeneracy, we mean the number of distinct quantum states (different quantum numbers) all of which have the same energy.

a. Using your answer to Problem P15.17a, what is the degeneracy of the energy level $9h^2/4ma^2$ for the square two-dimensional box of edge length a?

b. Using your answer to Problem P15.14b, what is the degeneracy of the energy level $17h^2/8ma^2$ for a three-dimensional cubic box of edge length a?

P15.26 Show by examining the position of the nodes that $\text{Re}[A_+ e^{i(kx - \omega t)}]$ and $\text{Re}[A_- e^{i(-kx - \omega t)}]$ represent plane waves moving in the positive and negative x directions, respectively. The notation Re[ ] refers to the real part of the function in the brackets.

P15.27 Two wave functions are distinguishable if they lead to a different probability density. Which of the following wave functions are distinguishable from $\sin kx$?

a. $(e^{ikx} - e^{-ikx})/2$

b. $e^{i\theta} \sin kx$, θ a constant

c. $\cos(kx - \pi/2)$

d. $i \cos(kx + \pi/2)(\sin \theta + i \cos \theta)\left(-\dfrac{\sqrt{2}}{2} + i\dfrac{\sqrt{2}}{2}\right)$,

θ is a constant

P15.28 Is the superposition wave function $\psi(x) = \sqrt{2/a}[\sin(n\pi x/a) + \sin(m\pi x/a)]$ an eigenfunction of the total energy operator for the particle in the box?

P15.29 The smallest observed frequency for a transition between states of an electron in a one-dimensional box is $3.0 \times 10^{13} \, s^{-1}$. What is the length of the box?

P15.30 Are the total energy eigenfunctions for the free particle in one dimension, $\psi^+(x) = A_+ e^{+i\sqrt{(2mE/\hbar^2)}x}$ and

$\psi^-(x) = A_- e^{-i\sqrt{(2mE/\hbar^2)}x}$, eigenfunctions of the one-dimensional linear momentum operator? If so, what are the eigenvalues?

P15.31   Use the eigenfunction $\psi(x) = A'e^{+ikx} + B'e^{-ikx}$ rather than $\psi(x) = A \sin kx + B \cos kx$ to apply the boundary conditions for the particle in the box.

a. How do the boundary conditions restrict the acceptable choices for A' and B' and for k?

b. Do these two functions give different probability densities if each is normalized?

P15.32   Consider a particle in a one-dimensional box defined by $V(x) = 0, a > x > 0$ and $V(x) = \infty, x \geq a, x \leq 0$. Explain why each of the following unnormalized functions is or is not an acceptable wave function based on criteria such as being consistent with the boundary conditions, and with the association of $\psi^*(x)\psi(x)\,dx$ with probability.

a. $A \cos \dfrac{n\pi x}{a} + B \sin \dfrac{n\pi x}{a}$

b. $C\left(1 - \sin \dfrac{n\pi x}{a}\right)$

c. $Cx^3(x - a)$

d. $D(a - x)x$

e. $\dfrac{E}{\cos(n\pi x/a)}$

P15.33   Use your result from Problem P15.17 and make an energy level diagram for the first five energy levels of a square two-dimensional box of edge length b. Indicate which of the energy levels are degenerate and the degeneracy of these levels.

P15.34   Calculate the probability that a particle in a one-dimensional box of length a is found between $0.32a$ and $0.35a$ when it is described by the following wave functions:

a. $\sqrt{\dfrac{2}{a}} \sin\left(\dfrac{\pi x}{a}\right)$

b. $\sqrt{\dfrac{2}{a}} \sin\left(\dfrac{3\pi x}{a}\right)$

What would you expect for a classical particle? Compare your results for (a) and (b) with the classical result.

Web-Based Simulations, Animations, and Problems

W15.1   The motion of a classical particle in a box potential is simulated. The particle energy and the potential in the two halves of the box are varied using sliders. The kinetic energy is displayed as a function of the position x, and the result of measuring the probability of detecting the particle at x is displayed as a density plot. The student is asked to use the information gathered to explain the motion of the particle.

W15.2   Wave functions for $n = 1–5$ are shown for the particle in the infinite depth box, and the energy levels are calculated. Sliders are used to vary the box length and the mass of the particle. The student is asked questions that clarify the relationship between the level energy, the mass, and the box length.

W15.3   The probability is calculated for finding a particle in the infinite depth box in the interval $0 \rightarrow 0.1a$, $0.1a \rightarrow 0.2a, \ldots, 0.9a \rightarrow 1.0a$ for $n = 1, n = 2$, and $n = 50$. The student is asked to explain these results.

W15.4   Contour plots are generated for the total energy eigenfunctions of the particle in the two-dimensional infinite depth box,

$$\psi_{n_x n_y}(x, y) = N \sin \frac{n_x \pi x}{a} \sin \frac{n_y \pi y}{b}$$

The student is asked questions about the nodal structure of these eigenfunctions and asked to assign quantum numbers n_x and n_y to each contour plot.

W15.5   The student is asked to determine if the normalized wave function

$$\psi(x) = \sqrt{\frac{105}{a^7}} x^2(x - a)^2$$

is an acceptable wave function for the particle in the infinite depth box based on graphs of $\psi(x)$ and $d\psi(x)/dx$ as a function of x. The wave function $\psi(x)$ is expanded in eigenfunctions of the total energy operator. The student is asked to determine the probability of observing certain values of the total energy in a measurement on the system.

W15.6   The normalized wave function,

$$\Psi(x, t) = \sqrt{\frac{2}{a}} \left[ce^{-iE_1 t/\hbar} \sin\left(\frac{\pi x}{a}\right) + de^{-iE_2 t/\hbar} \sin\left(\frac{2\pi x}{a}\right) \right]$$

with $|c|^2 + |d|^2 = 1$ is a superposition of the ground state and first excited state for the particle in the infinite depth box. Simulations are carried out to determine if $\langle E \rangle$, $\langle p \rangle$, and $\langle x \rangle$ are independent of time for this superposition state.

The Particle in the Box and the Real World

W‌hy have we spent so much time trying to understand the quantum mechanical particle in a box? The particle in a box is a simple model that can be used to explore concepts such as why core electrons are not involved in chemical bonds, the stabilizing effect of delocalized π electrons in aromatic molecules, and the ability of metals to conduct electrons. It also provides a framework for understanding the tunneling of quantum mechanical particles through (not over!) barriers and size quantization, both of which find applications in quantum wells and quantum dots.

16.1 The Particle in the Finite Depth Box

Before applying the particle in a box model to the "real world," the box must be modified to make it more realistic. This is done by letting the box have a finite depth, which allows the particle to escape. This modification is necessary to model problems such as the ionization of an atom. The potential is defined by

$$V(x) = 0, \quad \text{for } -a/2 \le x \le a/2$$

$$V(x) = V_0, \quad \text{for } x > a/2, x < -a/2 \qquad \textbf{(16.1)}$$

The origin of the x coordinate has been changed from one end of the box (Chapter 15) to the center of the box to simplify the mathematics of solving the Schrödinger equation. The shift of the origin changes the functional form of the total energy eigenfunctions, as you will see in the end-of-chapter problems. However, it has no physical consequences in that eigenvalues and graphs of the eigenfunctions superimposed on the potential are identical for both choices of the point $x = 0$.

How do the eigenfunctions and eigenvalues for the Schrödinger equation for the **finite depth box** differ from those for the infinitely deep potential? For $E > V(x)$ (inside the box), the eigenfunctions have the oscillatory behavior that was exhibited for the infinitely deep box. However, because $V_0 < \infty$, the reasoning following Equation (15.8) no longer holds; the amplitude of the eigenfunctions need not be zero at the ends of the box. For $E < V(x)$ (outside of the box), the eigenfunctions decay exponentially with distance

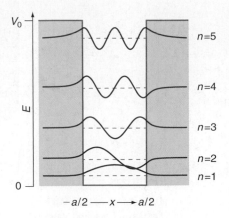

FIGURE 16.1
Eigenfunctions and allowed energy levels are shown for an electron in a well of depth $V_0 = 1.20 \times 10^{-18}$ J and width $a = 1.00 \times 10^{-9}$ m.

from the box, as we will show next. These two regions are considered separately. Inside the box, $V(x) = 0$, and

$$\frac{d^2\psi(x)}{dx^2} = -\frac{2mE}{\hbar^2}\psi(x) \tag{16.2}$$

Outside of the box, the Schrödinger equation has the form

$$\frac{d^2\psi(x)}{dx^2} = \frac{2m(V_0 - E)}{\hbar^2}\psi(x) \tag{16.3}$$

The difference in sign on the right-hand side makes a big difference in the eigenfunctions! Inside the box, the solutions have the same general form as discussed in Chapter 15, but outside the box, they have the form

$$\psi(x) = A\,e^{-\kappa x} + Be^{+\kappa x} \quad \text{for } \infty \geq x \geq a/2 \quad \text{and}$$

$$\psi(x) = A'\,e^{-\kappa x} + B'e^{+\kappa x} \quad \text{for } -\infty \leq x \leq -a/2$$

$$\text{where} \quad \kappa = \sqrt{\frac{2m(V_0 - E)}{\hbar^2}} \tag{16.4}$$

The functions of Equation (16.4) are solutions to Equation (16.3). The coefficients (A, B and A', B') are different on each side of the box. Because $\psi(x)$ must remain finite for very large positive and negative values of x, $B = A' = 0$. By requiring that $\psi(x)$ and $d\psi(x)/dx$ are continuous at the box boundaries and imposing a normalization condition, the Schrödinger equation can be solved for the eigenfunctions and eigenvalues in the potential for given values of m, a, and V_0. If $\psi(x)$ was not continuous at the boundaries, the probability density would have two different values at the same point, which makes no sense. If $d\psi(x)/dx$ was not continuous at the boundaries, $d^2\psi(x)/dx^2$ would not exist and we could not solve the Schrödinger equation. If the wave functions were not normalized, we could not associate $\psi^*(x)\psi(x)$ with a probability density. The details of the solution are left to the end-of-chapter problems. The allowed energy levels and the corresponding eigenfunctions for a finite depth potential are shown in Figure 16.1. The yellow areas correspond to the region for which $E_{potential} > E_{total}$. Because $E_{total} = E_{kinetic} + E_{potential}$, $E_{kinetic} < 0$ in this region. For a particle, $E_{kinetic} = p^2/2m$ and a negative value for $E_{kinetic}$ implies that the momentum is imaginary. For this reason, $E_{kinetic} < 0$ defines what is called the **classically forbidden region.**

Two major differences in the solutions between the finite and the infinite depth box are immediately apparent. First, the potential has only a finite number of total energy eigenvalues, which correspond to bound states. The number depends on m, a, and V_0. Second, the amplitude of the wave function does not go to zero at the edge of the box. We explore the consequences of this second difference when discussing tunneling later. As seen in Figure 16.2, the falloff of the wave function outside of the box is not the same for all eigenfunctions: $\psi(x)$ falls off most rapidly with distance for the most strongly bound state ($V_0 \gg E$) in the potential and most slowly for the least strongly bound state in the potential ($V_0 \sim E$). Equation (16.4) predicts this trend.

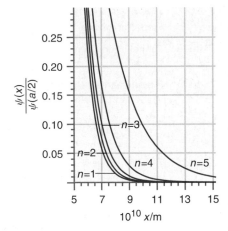

FIGURE 16.2
Decrease of the amplitude of the eigenfunctions as a function of distance from the center of the box. All eigenfunctions have been normalized to the value one at $x = a/2$ for purposes of comparison.

16.2 Differences in Overlap between Core and Valence Electrons

Figure 16.2 shows that weakly bound states have wave functions that leak quite strongly into the region outside of the box. What are the consequences of this behavior? Take this potential as a crude model for electrons in an atom. Strongly bound levels correspond to **core electrons** and weakly bound levels correspond to **valence electrons.** What happens when a second atom is placed close enough to the first atom that a chemical bond is formed? The results in Figure 16.3 show that the falloff of the wave functions for the weakly bound states in the box is gradual enough that both wave functions have a nonzero amplitude in the region between the wells. *These wave functions have a* *significant overlap.* Note that this is not the case for the strongly bound levels; these energy eigenfunctions have a small overlap.

We conclude that a correlation exists between the nonzero overlap required for chemical bond formation and the position of the energy level in the potential. This is our first application of the particle in the box model. It provides an understanding of why chemical bonds involve the least strongly bound, or valence, electrons and not the more strongly bound, or core, electrons. We will have more to say on this topic when the chemical bond is discussed in Chapter 23.

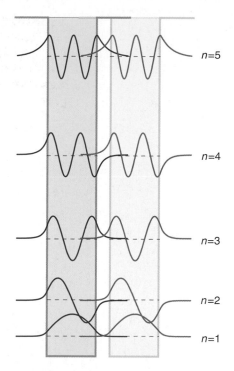

FIGURE 16.3
Overlap of wave functions from two closely spaced finite depth wells. The vertical scale has been expanded relative to Figure 16.1 to better display the overlap.

16.3 Pi Electrons in Conjugated Molecules Can Be Treated as Moving Freely in a Box

The absorption of light in the visible and ultraviolet (UV) part of the electromagnetic spectrum in molecules is a result of the excitation of electrons from occupied to unoccupied energy levels. If the electrons are delocalized as in an organic molecule with a **π-bonded network,** the maximum in the absorption spectrum shifts from the UV into the visible range. The greater the degree of **delocalization,** the more the absorption maximum shifts toward the red end of the visible spectrum. The energy levels for such a conjugated system can be described quite well with a one-dimensional particle in a box model. The series of dyes, 1,4-diphenyl–1,3-butadiene, 1,6-diphenyl–1,3,5-hexatriene, and 1,8-diphenyl–1,3,5,7-octatetraene consist of a planar backbone of alternating C—C and C=C bonds and have phenyl groups attached to the ends. The phenyl groups serve the purpose of decreasing the volatility of the compound. The π-bonded network does not include the phenyl groups, but does include the terminal carbon–phenyl group bond length. Only the π-bonded electrons are modeled using the particle in the box. Because each energy level can be occupied by two electrons, the highest occupied energy level corresponds to $n = 2$, 3, and 4 for the series of molecules considered.

The longest wavelength at which light is absorbed occurs when one of the electrons in the highest occupied energy level is promoted to the lowest lying unoccupied level. As Equation (15.17) shows, the energy level spacing depends on the length of the π-bonded network. For 1,4-diphenyl–1,3-butadiene, 1,6-diphenyl–1,3,5-hexatriene, and 1,8-diphenyl–1,3,5,7-octatetraene, the maximum wavelength at which absorption occurs is 345, 375, and 390 nm, respectively. From these data, and taking into account the quantum numbers corresponding to the highest occupied and lowest unoccupied levels, the apparent network length can be calculated. We demonstrate the calculation for 1,6-diphenyl–1,3,5-hexatriene, for which the transition corresponds to $n_i = 3 \rightarrow n_f = 4$ as indicated in Figure 16.4.

$$
\begin{aligned}
a &= \sqrt{\frac{(n_f^2 - n_i^2)h^2}{8m\Delta E}} = \sqrt{\frac{(n_f^2 - n_i^2)h\lambda_{max}}{8mc}} \\
&= \sqrt{\frac{(4^2 - 3^2)(6.626 \times 10^{-34}\,\text{J s})(375 \times 10^{-9}\,\text{m})}{8(9.109 \times 10^{-31}\,\text{kg})(2.998 \times 10^8\,\text{m s}^{-1})}} \\
&= 892\,\text{pm}
\end{aligned}
\tag{16.5}
$$

The apparent and calculated network length has been compared for each of the three molecules by B. D. Anderson [*J. Chemical Education* 74 (1997): 985]. Values are shown in Table 16.1. The agreement is reasonably good, given the simplicity of the model. Most importantly, the model correctly predicts that because λ is proportional to a^2, shorter π-bonded networks show absorption at smaller wavelengths. This trend is confirmed by experiment.

For 1,6-diphenyl–1,3,5-hexatriene in the **ground state,** the highest occupied energy level corresponds to $n = 3$. Does this mean that in a large number of molecules there will be very few molecules for which the $n = 4$ level is occupied at 300. K? This question can be answered with the help of the Boltzmann distribution.

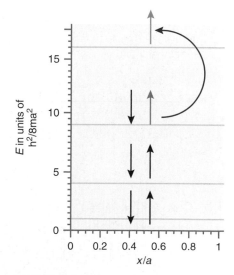

FIGURE 16.4
The transition from the highest occupied energy level to the lowest unoccupied energy level is shown for the particle in the box model of the π electrons in 1,6-diphenyl–1,3,5-hexatriene.

TABLE 16.1 Calculated Network Length for Conjugated Molecules

Compound	Apparent Network Length (pm)	Calculated Network Length (pm)
1,4-diphenyl–1,3-butadiene	723	695
1,6-diphenyl–1,3,5-hexatriene	892	973
1,8-diphenyl–1,3,5,7-octatetraene	1030	1251

The energy difference between the two levels is given by

$$\Delta E = \frac{h^2(n_f^2 - n_i^2)}{8ma^2} = \frac{7 \times (6.626 \times 10^{-34}\,\text{J s})^2}{8 \times 9.109 \times 10^{-31}\,\text{kg} \times (973 \times 10^{-12}\,\text{m})^2}$$

$$= 4.45 \times 10^{-19}\,\text{J} \tag{16.6}$$

Because there are two quantum states for each value of n, $g_4 = g_3 = 2$. The ratio of the population in the $n = 4$ level to that in the $n = 3$ level is given by

$$\frac{n_4}{n_3} = \frac{g_4}{g_3} e^{-\Delta E/k_B T} = \exp\left[-\frac{4.45 \times 10^{-19}\,\text{J}}{1.381 \times 10^{-23}\,\text{J K}^{-1} \times 300.\,\text{K}}\right]$$

$$= 2.0 \times 10^{-47} \tag{16.7}$$

Therefore, the $n = 3 \rightarrow n = 4$ transition cannot be achieved by the exchange of translational energy in the collision between molecules at 300. K, and essentially all molecules are in their electronic ground state.

16.4 Why Does Sodium Conduct Electricity and Why Is Diamond an Insulator?

As discussed earlier, valence electrons on adjacent atoms in a molecule or a solid can have an appreciable overlap. This means that the electrons can "hop" from one atom to the next. Consider Na, which has one valence electron per atom. If two Na atoms are bonded to form a dimer, the valence level that was localized on each atom will be delocalized over both atoms as is illustrated in Figure 16.5. Now add additional Na atoms to form a one-dimensional Na crystal. A crystalline metal can be thought of as a box with a periodic corrugated potential at the bottom. To illustrate the relationship to a box model, the potential of a one-dimensional periodic array of Na^+ potentials arising from the atomic cores at lattice sites is shown in Figure 16.6. Because the Na $3s$ valence electrons can be found with equal probability at any Na atom, one electron per atom is delocalized over the whole metal sample. This is exactly the model of the particle in the box.

The potential of Figure 16.6 can be idealized to a box as shown in Figure 16.7. This box differs from the simple boxes discussed earlier in an essential way. There are many

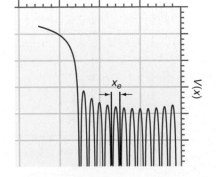

FIGURE 16.6
The potential energy resulting from a one-dimensional periodic array of Na^+ ions. One valence electron per Na is delocalized over this box. The quantity x_e represents the lattice spacing.

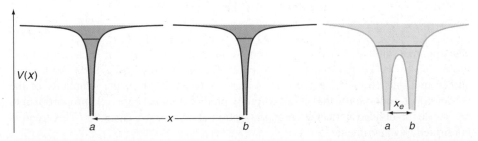

FIGURE 16.5
At large distances, the valence level on each Na atom is localized on that atom. When they are brought close enough together to form the dimer, the barrier between them is lowered, and the level is delocalized over both atoms. The quantity x_e represents the bond length of the dimer.

atoms in the atomic chain under consideration (large *a*), such that the energy levels for the delocalized electrons are very closely spaced in what is called the conduction band as discussed next. What is the energy-level spacing for the delocalized electrons in a 1.00-cm-long box? About 2×10^7 Na atoms will fit into the box. If each atom donates one electron to the band, we can easily show that at the highest filled level,

$$E_{n+1} - E_n = \frac{(n+1)^2 h^2}{8ma^2} - \frac{n^2 h^2}{8ma^2} = \frac{h^2}{8ma^2}(2n+1)$$

$$= (2n+1)(6.02 \times 10^{-34} \text{ J}) \qquad \textbf{(16.8)}$$

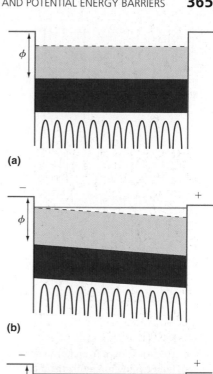

(a)

This spacing between levels is at most only $\approx 10^{-6} k_B T$ for $T = 300$. K and, therefore, the energy spectrum is essentially continuous. All energies within the range bounded by the bottom of the red shaded area in Figure 16.7 for low energies and the dashed line for high energies are accessible. This set of continuous energy levels is referred to as an **energy band.** The band shown in Figure 16.7a extends up to the dashed line, beyond which there are no allowed energy levels until the energy has increased by ΔE. An energy range, ΔE, in which there are no allowed states is called a **band gap** (See Figure 16.15). For Na, not all available states in the band are filled, as shown in Figure 16.7. The range of energies between the top of the red area and the dashed line corresponds to unfilled **conduction band** states. The fact that the band is only partially filled is critical in making Na an **electrical conductor,** as explained next.

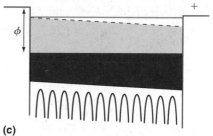

(b)

What happens when an electrical potential is applied between the two ends of the box? The field gives rise to a gradient of potential energy along the box superimposed on the original potential as shown in Figure 16.7b. The unoccupied states on the side of the metal with the more positive electrical potential have a lower energy than the occupied states with the more negative electrical potential. This makes it energetically favorable for the electrons to move toward the end of the box with the more positive voltage as shown in Figure 16.7c. This flow of electrons through the metal is the current that flows through the "wire." It occurs because of the **overlap of wave functions** on adjacent atoms, which leads to hopping, and because the energy levels are so close together that they form a continuous energy spectrum.

(c)

What makes diamond an **insulator** in this picture? Bands are separated from one another by band gaps, in which there are no allowed eigenfunctions of the total energy operator. In diamond, all quantum states in the band accessible to the delocalized valence electrons are filled. The highest filled energy band in semiconductors and insulators is called the **valence band.** In Figure 16.7, this corresponds to extending the red area up to the dashed line. As the energy increases, a range is encountered in which there are no allowed states of the system until the conduction band of allowed energy levels is reached. This means that, although we could draw diagrams just like the upper two panels of Figure 16.7 for diamond, the system cannot respond as shown in the lower panel. There are no unoccupied states in the valence band that can be used to transport electrons through the crystal. Therefore, diamond is an insulator. **Semiconductors** also have a band gap separating the fully occupied valence and the empty conduction band. However, in semiconductors, the band gap is smaller than for insulators, allowing them to become conductors at elevated temperatures. The band structure of solids is discussed in more detail in Chapter 24 after the chemical bond has been discussed.

FIGURE 16.7

Idealization of a metal in the particle in the box model. The horizontal scale is greatly expanded to show the periodic potential. Actually, more than 10^7 Na atoms will fit into a 1-cm-long box. The red shaded band shows the range of energies filled by the valence electrons of the individual atoms. The highest energy that can be occupied in this band is indicated by the dashed line. The energy required to remove an electron from the highest occupied state is the *work function*, ϕ. **(a)** The metal without an applied potential. **(b)** The effect on the energy levels of applying an electric field. **(c)** The response of the metal to the change in the energy levels induced by the electric field. The thin solid line at the top of the band in parts (b) and (c) indicates where the energy of the highest level would lie in the absence of an electric field.

16.5 Traveling Waves and Potential Energy Barriers

In the previous sections, we focused on the energy of bound states for a finite depth box. We next investigate how a traveling wave-particle is affected by a sudden change in the potential energy in the form of a step potential in order to develop a framework for discussing quantum mechanical tunneling in Section 16.6.

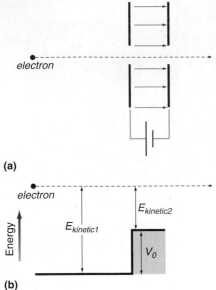

(a)

(b)

FIGURE 16.8
(a) An electron moves from left to right from a field free region into a parallel plate capacitor in which it is slowed down.
(b) The energetics of the electron is shown in the regions before entering the capacitor and in the capacitor.

Figure 16.8a shows an electron that moves from a region in which there is no electric field and enters a parallel plate capacitor in which the electric field opposes the motion of the particle. The energetics of this event are shown in Figure 16.8b. The electron is slowed down abruptly as it enters the region of higher potential energy. Energy is conserved in this event so that the increase in the potential energy at the step leads to a decrease in kinetic energy. Classically, we expect that the probability that the electron passes the step is one for $E > V_0$, and zero if $E < V_0$.

To solve this problem quantum mechanically, we need to solve the time-independent Schrödinger equation for the following potential:

$$V(x) = \begin{cases} 0, & x < 0 \\ V_0, & x \geq 0 \end{cases} \tag{16.9}$$

Just as we did for the free particle in Section 15.1, we write the Schrödinger equation in the form

$$\frac{d^2\psi(x)}{dx^2} = \left[\frac{2m(V(x) - E)}{\hbar^2}\right]\psi(x) \tag{16.10}$$

and solve it in the two separate regions for $x < 0$ and $x > 0$. We then combine these two solutions while requiring the continuity of $\psi(x)$ and $d\psi(x)/dx$ at $x = 0$ where the electron passes from one region to the other.

In the region $x < 0$, we consider the possibility that the electron will be reflected by the sudden change in the potential energy. Classically, reflection would not occur for $E > V_0$, but we know that light waves incident on an interface between two materials of differing refractive index can be reflected or transmitted. An appropriate wave function for the electron that includes reflection is

$$\psi(x) = Ae^{+ik_1x} + Be^{-ik_1x}, \quad x < 0 \tag{16.11}$$

where the first term is the wave incident on the barrier and the second term is the reflected wave. The wave vector k_1 is related to the kinetic energy by

$$k_1 = \sqrt{\frac{2mE_{kinetic1}}{\hbar^2}} \tag{16.12}$$

In the region for $x > 0$, we write the wave function in the form

$$\psi(x) = Ce^{+ik_2x}, \quad x > 0 \quad \text{where} \quad k_2 = \sqrt{\frac{2mE_{kinetic2}}{\hbar^2}} \tag{16.13}$$

There is no wave moving in the direction of decreasing x values in this region because the electron experiences no forces that could turn it around.

We next require that $\psi(x)$ and $d\psi(x)/dx$ in the two regions have the same values at $x = 0$. If the wave functions did not have the same value, the probability density at $x = 0$ would have two different values at the same point, which makes no sense. If $d\psi(x)/dx$ for the two regions did not have the same value at $x = 0$, $d^2\psi(x)/dx^2$ would not exist and we could not solve the Schrödinger equation. Applying the continuity condition on $\psi(x)$, we obtain

$$Ae^{+ik_1 0} + Be^{-ik_1 0} = Ce^{+ik_2 0}$$

$$A + B = C \tag{16.14}$$

Applying the continuity condition on $d\psi(x)/dx$, we obtain

$$ik_1 Ae^{+ik_1 0} - ik_1 Be^{-ik_1 0} = ik_2 Ce^{+ik_2 0}$$

$$k_1(A - B) = k_2 C \tag{16.15}$$

To obtain the probabilities for reflection and transmission, we must take the speeds of the particle in the two regions into account. The number of particles reflected or transmitted per unit time is given by the product of the probability density and the speed. The units

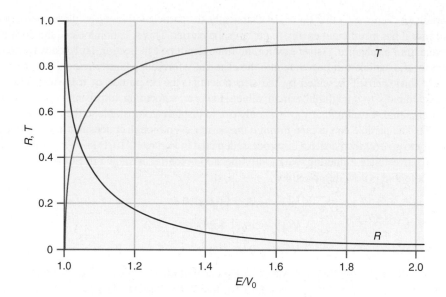

FIGURE 16.9
The probability for reflection R and the probability for transmission T from a step potential of height V_0 are shown as a function of the ratio E/V_0.

of this product are probability/length $\times$ length/time = probability/time as required. The transmission probability T and the reflection probability R are given by

$$R = \frac{\text{number reflected /unit time}}{\text{number incident /unit time}} = \frac{B^*B \times k_1}{A^*A \times k_1} = \frac{B^*B}{A^*A}$$

$$T = \frac{\text{number transmitted /unit time}}{\text{number incident /unit time}} = \frac{C^*C \times k_2}{A^*A \times k_1} \qquad (16.16)$$

We can use Equations (16.14) and (16.15) to express B in terms of A to calculate R and to express C in terms of A to calculate T. The results are

$$R = \frac{\left(\dfrac{k_1 - k_2}{k_1 + k_2}A\right)^*\left(\dfrac{k_1 - k_2}{k_1 + k_2}A\right)}{A^*A} = \left(\frac{k_1 - k_2}{k_1 + k_2}\right)^2$$

$$T = \frac{\left(\dfrac{2k_1}{k_1 + k_2}A\right)^*\left(\dfrac{2k_1}{k_1 + k_2}A\right)}{A^*A} \times \frac{k_2}{k_1} = \frac{4k_1 k_2}{(k_1 + k_2)^2} \qquad (16.17)$$

These results can be expressed in terms of E and V_0 where $E = E_{kinetic1}$.

$$R = \frac{\left(\sqrt{E} - \sqrt{E - V_0}\right)^2}{\left(\sqrt{E} + \sqrt{E - V_0}\right)^2}$$

$$T = \frac{4\sqrt{E(E - V_0)}}{\left(\sqrt{E} + \sqrt{E - V_0}\right)^2} \qquad (16.18)$$

Note that we are only considering energies for which $E > V_0$. The probabilities for reflection and transmission as a function of the dimensionless parameter E/V_0 are shown in Figure 16.9. Classical physics would predict that $T = 1$ and $R = 0$ for $E \geq V_0$ and that $T = 0$ and $R = 1$ for $E < V_0$. Quantum mechanics predicts that $R \rightarrow 1$ as $E \rightarrow V_0$, but R only approaches zero asymptotically for $E \gg V_0$.

16.6 Tunneling through a Barrier

In the preceding section, we saw that a wave-particle approaching a step potential can be reflected even if its energy is greater than the barrier height. For the step potential, the barrier is present for all positive values of x. We next consider a barrier of finite width for a particle of energy such that $V_0 > E$. Classically, the particle will not pass the barrier region because it has insufficient energy to get over the barrier. This situation looks quite different in quantum mechanics. As we saw for the finite depth box in Section 16.1, the wave function for the particle can penetrate into the classically forbidden barrier region. For the infinitely thick barrier presented by the step potential, the amplitude of the wave function decays rapidly to a negligibly small value. However, something surprising happens if the barrier is thin, meaning that $V_0 > E$ only over a distance comparable to the particle wavelength. The particle can escape *through* the barrier even though it does not have sufficient energy to go *over* the barrier. This process, depicted in Figure 16.10, is known as **tunneling.**

To investigate tunneling, we modify the step potential as shown in Equation (16.19). The potential is now described by

$$V(x) = 0, \quad \text{for } x < 0$$

$$V(x) = V_0, \quad \text{for } 0 \le x \le a$$

$$V(x) = 0, \quad \text{for } x > a \qquad \textbf{(16.19)}$$

The oscillating wave function for two incident particle energies is shown to the left of the barrier where $E > V_0$ in Figure 16.10. Inside the barrier where $E < V_0$, the wave function decays exponentially with distance. If the **barrier width** a is small enough that $\psi(x)$ has not decayed to a negligibly small value by the time it arrives at the end of the barrier at $x = a$, the wave function in the region $x > a$ will have a finite amplitude. Because $V(x) = 0$ for $x > a$, the wave function in this region is again a traveling wave. If the amplitude is greater than zero for $x > a$, the particle has a finite probability of escaping from the well even though its energy is less than the height of the barrier.

Figure 16.10 shows that tunneling is much more likely for particles with energies near the top of the barrier. This is due to the degree to which the wave function in the barrier falls off with distance as $e^{-\kappa x}$, in which the **decay length** $1/\kappa$ is given by $\sqrt{\hbar^2/2m(V_0 - E)}$. Because the wave function decays more slowly as $E \rightarrow V_0$, the amplitude of the wave function at $x = a$ is greater and tunneling is more likely to occur.

Rather than derive expressions for R and T as for the step potential, we leave the details of the calculations for the end-of-chapter problems. To illustrate the extreme sensitivity of the tunneling probability on the barrier width and particle energy, the results of a calculation for electron tunneling through a barrier of height 10. eV are shown in Figure 16.11. Note that for the most narrow barrier, the tunneling probability approaches 0.5 as $E \rightarrow V_0$, and that for wide barriers, the tunneling probably is very small unless $E \rightarrow V_0$.

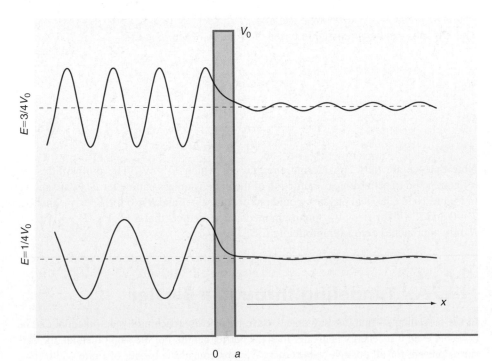

FIGURE 16.10

Wave-particles corresponding to the indicated energy are incident from the left on a barrier of height $V_0 = 1.60 \times 10^{-19}$ J and width 9.00×10^{-10} m. The exponentially decaying wave function is shown inside the barrier, and the incident and transmitted wave functions are shown for $x < 0$ and $x > a$, respectively.

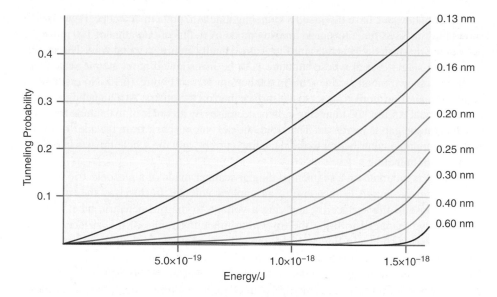

FIGURE 16.11
The tunneling probability for an electron is shown for seven barrier widths as a function of the particle energy for a barrier height of 1.6×10^{-18} J. $E = V_0$ for $E = 1.6 \times 10^{-18}$ J.

16.7 The Scanning Tunneling Microscope and the Atomic Force Microscope

It was not known that particles could tunnel through a barrier until the advent of quantum mechanics. In the early 1980s, the tunneling of electrons between two solids was used to develop an atomic resolution microscope. Gerd Binnig and Heinrich Rohrer received a Nobel Prize for the invention of the **scanning tunneling microscope (STM)** in 1986.

The STM allows the imaging of solid surfaces with atomic resolution with surprisingly simple instrumentation. The STM and a closely related device called the **atomic force microscope (AFM)** have been successfully used to study phenomena at atomic and near atomic resolution in a wide variety of areas including chemistry, physics, biology, and engineering. The invention of the STM and AFM played a significant role in enabling the development of nanotechnology. The essential elements of an STM are a sharp metallic tip and a conducting sample over which the tip is scanned to create an image of the sample surface. In an STM, the barrier between these two conductors is usually vacuum, and electrons are made to tunnel across this barrier, as discussed later. As might be expected, the barrier width needs to be on the order of atomic dimensions to observe tunneling. Electrons with an energy of typically 5 eV are used to tunnel from the metal tip to the surface. This energy corresponds to the **work function** as well as to the barrier height $V_0 - E$ in Figure 16.10. The decay length $\sqrt{\hbar^2/2m(V_0 - E)}$ for such an electron in the barrier is about 0.1 nm. Therefore, if the tip and sample are brought to within a nanometer of one another, electron tunneling will be observed between them.

How does a scanning tunneling microscope work? We address this question first in principle and then from a practical point of view. Because the particle in a box is a good model for the conduction of electrons in the metal solid, the tip and surface can be represented by boxes as shown in Figure 16.12. For convenience, the part of the box below the lowest energy that can be occupied by the core electrons has been omitted, and only the part of the box immediately adjacent to the tip–sample gap is shown. The tip and sample in general have different work functions as indicated. If they are not connected in an external circuit, their energy diagrams line up as in Figure 16.12a. When they are connected in an external circuit, charge flows between the tip and sample until the highest occupied level is the same everywhere as shown in Figure 16.12b.

Tunneling takes place at constant energy, which in Figure 16.12 corresponds to the horizontal dashed line. However, for the configuration shown in Figure 16.12b, there is no empty state on the sample into which an electron from the tip can tunnel. To allow tunneling to occur, a small (0.01–1 V) electrical potential is placed between the two metals. This raises the highest filled energy level of the tip relative to that of the sample. Now tunneling of electrons can take place from tip to sample, resulting in a net current flow.

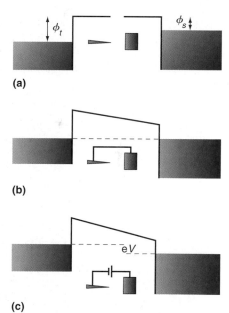

(a)

(b)

(c)

FIGURE 16.12
(a) If the conducting tip and surface are electrically isolated from one another, their energy diagrams line up. (b) If they are connected by a wire in an external circuit, charge flows from the lower work function material into the higher work function material until the highest occupied states have the same energy in both materials. (c) By applying a voltage V between the two materials, the highest occupied levels have an offset of energy eV. This allows tunneling to occur from left to right. The subscripts t and s refer to tip and surface.

Up until now, we have discussed a tunneling junction, not a microscope. Figure 16.13 shows how an STM functions in an imaging mode. A radius of curvature of 100 nm at the apex of the tip is routinely achievable by electrolytically etching a metal wire. The sample could be a single crystal whose structure is to be investigated at an atomic scale. This junction is shown on an atomic scale in the bottom part of Figure 16.13. No matter how blunt the tip is, one atom is closer to the surface than all the others. At a tunneling gap distance of about 0.5 nm, the tunneling current decreases by an order of magnitude for every 0.1 nm that the gap is increased. Therefore, the next atoms back from the apex of the tip make a negligible contribution to the tunneling current, and the whole tip acts like a single atom for tunneling.

The tip is mounted on a segmented tubular scanner made of a piezoelectric material that changes its length in response to an applied voltage. In this way the tip can be brought close to the surface by applying a voltage to the piezoelectric tube. Assume that we have managed to bring the tip within tunneling range of the surface. On the magnified scale shown in Figure 16.13, the individual atoms in the tip and surface are seen at a tip–surface spacing of about 0.5 nm. Keep in mind that the wave functions for the tunneling electrons in the tip decay rapidly in the region between tip and sample, as shown earlier in Figure 16.2. If the tip is directly over a surface atom, the amplitude of the wave function is large at the surface atom and the tunneling current is high. If the tip is between surface atoms, the amplitude of the wave functions is smaller and the tunneling current will be lower. To scan over the surface, different voltages are applied to the four segmented electrodes on the piezo tube. This allows a topographical image of the surface to be obtained. Because the tunneling current varies exponentially with the tip–surface distance, the microscope provides a very high sensitivity to changes in the height of the surface that occur on an atomic scale.

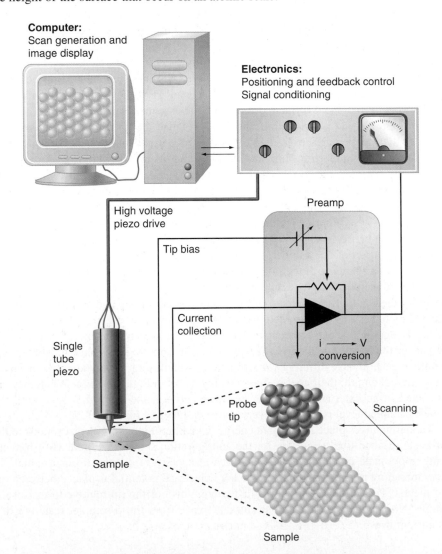

FIGURE 16.13

Schematic representation of a scanning tunneling microscope.

Source: Used by permission of Kevin E. Johnson (Pacific University), from University of Washington thesis, 1991.

In this abbreviated description, some details have been glossed over. The current is usually kept constant as the tip is scanned over the surface using a feedback circuit to keep the tip–surface distance constant. This is done by changing the voltage to the piezo tube electrodes as the tip scans over the surface. Additionally, a vibrational isolation system is required for the STM to prevent the tip from crashing into the surface as a result of vibrations always present in a laboratory. Figure 16.14 provides an example of the detail that can be seen with a scanning tunneling microscope. The individual planes, which are stacked together to make the silicon crystal, and the 0.3 nm height change between planes are clearly seen. Defects in the crystal structure are also clearly resolved. Researchers are using this microscope in many new applications aimed at understanding the structure of solid surfaces and modifying surfaces atom by atom.

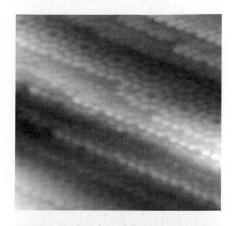

EXAMPLE PROBLEM 16.1

As was found for the finite depth well, the wave function amplitude decays in the barrier according to $\psi(x) = A \exp[-\sqrt{2m(V_0 - E)/\hbar^2}\, x]$. This result will be used to calculate the sensitivity of the scanning tunneling microscope. Assume that the tunneling current through a barrier of width a is proportional to $|A|^2 \exp[-2\sqrt{2m(V_0 - E)/\hbar^2}\, a]$.

a. If $V_0 - E$ is 4.50 eV, how much larger would the current be for a barrier width of 0.20 nm than for 0.30 nm?

b. A friend suggests to you that a proton tunneling microscope would be equally effective as an electron tunneling microscope. For a 0.20 nm barrier width, by what factor is the tunneling current changed if protons are used instead of electrons?

Solution

a. Putting the numbers into the formula given, we obtain

$$\frac{I(a = 2.0 \times 10^{-10}\,\text{m})}{I(a = 3.0 \times 10^{-10}\,\text{m})} = \exp\left[-2\sqrt{\frac{2m(V_0 - E)}{\hbar^2}}\right.$$
$$\left. \times (2.0 \times 10^{-10}\,\text{m} - 3.0 \times 10^{-10}\,\text{m})\right]$$

$$= \exp\left[-2\sqrt{\frac{2 \times 9.109 \times 10^{-31}\,\text{kg} \times 4.50\,\text{eV} \times 1.602 \times 10^{-19}\,\text{J/eV}}{(1.055 \times 10^{-34}\,\text{J s})^2}} \times (-1.0 \times 10^{-10}\,\text{m})\right]$$

$$= 8.8$$

Even a small distance change results in a substantial change in the tunneling current.

b. We find that the tunneling current for protons is appreciably smaller than that for electrons.

$$\frac{I(proton)}{I(electron)} = \frac{\exp\left[-2\sqrt{\frac{2m_{proton}(V_0 - E)}{\hbar^2}}\, a\right]}{\exp\left[-2\sqrt{\frac{2m_{electron}(V_0 - E)}{\hbar^2}}\, a\right]}$$

$$= \exp\left[-2\sqrt{\frac{2(V_0 - E)}{\hbar^2}}\left(\sqrt{m_{proton}} - \sqrt{m_{electron}}\right)a\right]$$

$$= \exp\left[-2\sqrt{\frac{2 \times 4.50\,\text{eV} \times 1.602 \times 10^{-19}\,\text{J/eV}}{(1.055 \times 10^{-34}\,\text{J s})^2}} \times \left(\sqrt{1.67 \times 10^{-27}\,\text{kg}} - \sqrt{9.11 \times 10^{-31}\,\text{kg}}\right) \times 2.0 \times 10^{-10}\,\text{m}\right]$$

$$= 1.2 \times 10^{-79}$$

This result does not make the proton tunneling microscope look very promising.

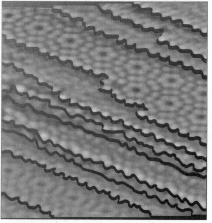

FIGURE 16.14
STM images of the (111) surface of Si. The upper image shows a 200. × 200. nm region with a high density of atomic steps, and the light dots correspond to individual Si atoms. The lower image shows how the image is related to the structure of parallel crystal planes separated by steps of one atom height. The step edges are shown as dark ribbons.
[Courtesy of Johnson, Kevin. "The Thermal Decomposition and Desorption Mechanism of Ultra-Thin Oxide on Silicon Studied by Scanning Tunneling Microscopy." PhD thesis, University of Washington, 1991.]

FIGURE 16.15

Schematic diagram of an atomic force microscope. **(a)** A tip mounted on a microfabricated cantilever is scanned over a surface in the *xy* plane by applying dc voltages to a segmented piezoelectric tube. If the tip experiences an attractive or repulsive force from the surface, the cantilever is deflected from its horizontal position. As a result, the laser light reflected from the back of the cantilever onto a segmented photodetector is differently distributed on the segments, giving rise to a difference current that is the input to a feedback controller. The controller changes the length of the piezoelectric tube in such a way to keep the cantilever deflection constant as the tip scans across the surface. Therefore, the surface image obtained corresponds to a constant force that can be varied using the feedback circuit. **(b)** The AFM can be modified to allow measurements in a liquid or controlled atmosphere using an o-ring seal mounted on the piezoelectric tube.

Source: Engel, T., Drobny, G., Reid, P. *Physical Chemistry for the Life Sciences*. Prentice Hall.

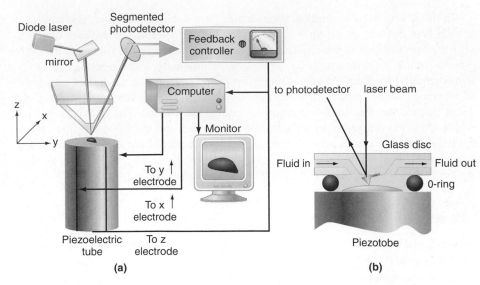

(a) (b)

The STM is limited to studies on conductive surfaces because although the tunneling current is small, the current density is very high. The AFM extends the range of the STM by allowing studies to be carried out on nonconductive surfaces. It does so by probing the force between the tip and the surface without any current passing across the junction. Although the AFM is not based on tunneling, we discuss it here because the atomic force microscope and the scanning tunneling microscope complement one another in structural studies of solid surfaces at the molecular and atomic level as we show later. The German physicist Gerd Binnig is a co-inventor of the AFM as well as the STM.

In an AFM, a tip attached to a flexible cantilever is scanned over the surface of a sample using the same feedback circuitry as for an STM as shown in Figure 16.15. The tip and cantilever shown in Figure 16.16 are generally microfabricated from Si, and the deflection of the cantilever from its horizontal position is given by

$$x = -\frac{F}{k} \qquad (16.20)$$

Where F is the force exerted on the cantilever, and k is its spring constant, which can have values in the range 0.01–100 N m^{-1}, depending on the application. The tip has a radius of curvature of $\sim$10–20 nm, and the force of interaction between the tip and the surface is primarily determined by those few atoms on the tip closest to the surface. The force is attractive and van der Waals in nature except for very small tip–surface distances, in which case repulsive electron–electron forces dominate. The deflection of the

FIGURE 16.16

Scanning electron micrograph of a micro-machined silicon cantilever with an integrated tip. This is a Pointprobe sensor made by Nanosensors GmbH und Co. KG, D-25870 Norderfriedrichskoog, Germany. Photo courtesy of Nanosensors GmbH & Co. KG.

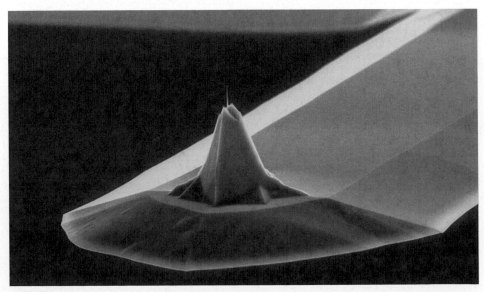

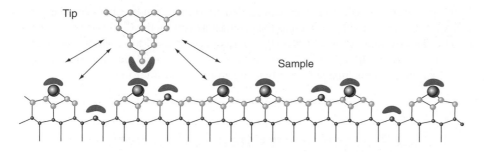

Tip

Sample

FIGURE 16.17

If an AFM tip is scanned over the surface in the more distant attractive part of the tip–surface potential that probes van der Waals forces, the atomic scale structure is averaged out because the tip senses many atoms as indicated by the blue arrows. If the tip is sufficiently close to the surface, the repulsive force between the occupied orbitals at the end of the tip and the occupied orbitals of the surface atoms (shown in red) is probed. Because these forces are very localized, atomic resolution is possible.

Source: Reprinted fig. 6 with permission from F. Giessibl, "Advances in atomic force microscopy," *Reviews of Modern Physics* 75: 949 (2003). Copyright 2003 by the American Physical Society. http://link.aps.org/doi/10.1103/RevModPhys.75.949.

cantilever is measured using a laser similar to that in a CD player. The light reflected from the back of the cantilever is incident on a segmented photodetector, and the deflection of the cantilever can be determined by comparing the signal from the segments of the photodetector. The feedback circuit keeps the cantilever deflection, and therefore the tip–surface force, constant as the tip is scanned across the surface. Whereas in an STM, an image corresponds to a surface contour at constant tunneling current, in an AFM, an image corresponds to a surface contour at constant force. Image acquisition is sufficiently fast that many kinetic processes can be imaged in real time.

The AFM design described is suitable for high resolution work in a controlled environment including liquids. To obtain ultrahigh molecular and atomic resolution, the device must be operated in ultrahigh vacuum in order to avoid contamination of the surface under study. The highest resolution images have been obtained in microscopes cooled to a temperature of ~5 K to minimize drifts in the area being scanned by thermal gradients. For atomic resolution studies, the tip must be scanned close enough to the surface that the repulsive rather than the attractive forces between the tip and the surface are sensed as shown in Figure 16.17. The cantilever is oscillated at an eigenfrequency determined by its geometry at an amplitude of ~1 nm. The repulsive interaction between the tip and the surface leads to small shifts in the cantilever eigenfrequency that can be used as the feedback parameter to keep the force constant as the tip is scanned over the surface. If the tip and surface are conductive, both STM and AFM images of the surface can be obtained with the same instrument. With an appropriate choice of cantilever and tip, AFM can be used to measure friction, conductivity, temperature and variations of chemical composition on surfaces with high resolution. For more details, see Giessibl, F. J. "Advances in Atomic Force Microscopy." *Reviews in Modern Physics* 75 (2003): 949, and Gross, L., Mohn, F., Moll, N., Liljeroth, P., and Meyer, G. "The Chemical Structure of a Molecule Resolved by Atomic Force Microscopy." *Science* 325 (2009): 1110.

Figure 16.18 shows images of pentacene molecules bound to a metal surface obtained using both STM and AFM modes of operation. The STM image arises from tunneling out of the highest filled molecular orbitals of the molecule which are delocalized over the molecule. For this reason, atomic scale resolution is not obtained. Note the nodal structure seen in the image. The AFM image in Figure 16.18b arises from the repulsive electrostatic force between the filled orbitals on CO and the electron density

FIGURE 16.18

STM and AFM imaging of pentacene on a copper surface. The tip has been prepared with a CO molecule at the apex with the oxygen atom pointing to the surface being scanned. **(a)** Ball-and-stick model of the pentacene molecule. **(b)** Constant-current STM image of a single pentacene molecule. The scale on the right shows the correspondence between the gray scale and the height within the molecule. **(c)** Constant-height AFM image of a single pentacene molecule. The scale on the right shows the correspondence between the gray scale and the frequency shift observed in scanning over the molecule. **(d)** Constant height image showing six pentacene molecules. The scale on the right shows the correspondence between the gray scale and the frequency shift observed in scanning over the molecule.

Source: Gross. L., Mohn, F., Moll, N., Liljeroth, P., and Meyer, G. "The Chemical Structure of a Molecule Resolved by Atomic Force Microscopy." *Science* 325 (2009): 1110.

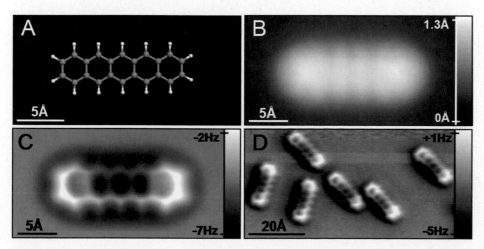

in the pentacene molecule. The enhanced density due to bonding between the carbon atoms of the benzene rings as well as the electron density in the C–H bonds can be seen clearly. This example shows both how tunneling can be used to image the structure of molecular orbitals directly (See also Section 24.6) and how the electron density in chemical bonds can be imaged directly using atomic force microscopy.

16.8 Tunneling in Chemical Reactions

Most chemical reactions are thermally activated; they proceed faster as the temperature of the reaction mixture is increased. This behavior is typical of reactions for which an energy barrier must be overcome in order to transform reactants into products. This barrier is referred to as the **activation energy** for the reaction. By increasing the temperature of the reactants, the fraction that has an energy that exceeds the activation energy is increased, allowing the reaction to proceed.

Tunneling provides another mechanism to convert reactants to products that does not require an increase in energy of the reactants for the reaction to proceed. It is well known that hydrogen transfer reactions can involve tunneling. An example is the reaction $R_1OH + R_2O^- \rightarrow R_2OH + R_1O^-$, where R_1 and R_2 are two different organic groups. The test for tunneling in this case is to substitute deuterium for hydrogen. If the reaction is thermally activated, the change in reaction rate is small and can be attributed to the different ground-state vibrational frequency of $-OH$ and $-OD$ bonds (see Chapter 19). However, if tunneling occurs, the rate decreases greatly because the tunneling rate depends exponentially on the decay length $\sqrt{\hbar^2/2m(V_0 - E)}$. However, it is not widely appreciated that tunneling can be important for heavier atoms such as C and O. A report by Zuev *et al.* [*Science* 299 (2003): 867] shows that the rate of the ring expansion reaction depicted in Figure 16.19 is faster than the predicted thermally activated rate by the factor 10^{152} at 10 K! This increase is due to the tunneling pathway. Because the tunneling rate depends exponentially on $\sqrt{\hbar^2/2m(V_0 - E)}$, heavier atom tunneling is only appreciable if $(V_0 - E)$ is very small. However, in a number of reactions, particularly in the fields of chemical catalysis and enzymology, this condition is met.

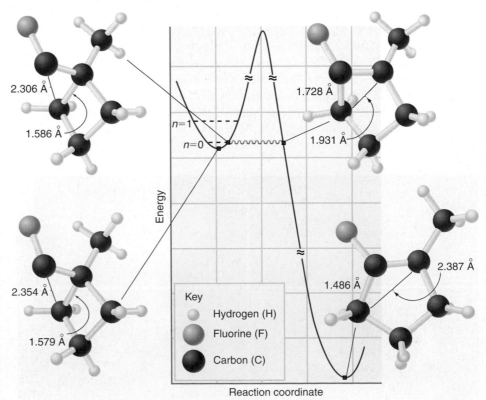

FIGURE 16.19
The structures of four species along the reaction path from reactant to product are shown together with a schematic energy diagram. The reaction occurs not by surmounting the barrier, but by tunneling through the barrier at the energy indicated by the wavy line.
Source: Figure 3 from Peter S. Zuev, et al., "Carbon Tunneling from a Single Quantum State," *Science*, New Series, Vol. 299: 867–870, Feb. 7, 2003. Copyright © 2003, The American Association for the Advancement of Science. Reprinted with permission from AAAS.

16.9 Quantum Wells and Quantum Dots

Just as not all atoms have the same ionization energy, not all solids have the same work function. The width and energetic position of the bands of allowed states are also not the same. These facts can be used to engineer some very useful devices. One good example is a device called a **quantum well structure.** Gallium arsenide is a widely used semiconductor in microelectronics applications. $Al_\alpha Ga_{1-\alpha}As$ is a substitutional alloy in which some of the Ga atoms are replaced by Al atoms. It can be combined with GaAs to form crystalline **heterostructures** that consist of alternating layers of GaAs and $Al_\alpha Ga_{1-\alpha}As$. Both substances are semiconductors that, like insulators, have a fully occupied energy band derived from their valence electrons. The fully occupied band is referred to as the valence band. As the energy increases, a band gap evolves that has no states, followed by an empty band that can be occupied by electrons, called the **conduction band.** However, there are only enough electrons in the electrically neutral crystal to fill the valence band. This is analogous to the H atom in which the $1s$ state is occupied and the $2s$ state is empty. This band structure is shown in Figure 16.20.

By means of a technique called molecular beam epitaxy in which materials are slowly evaporated onto a growing crystal under extremely low pressures, one can grow a crystalline structure in which a 0.1 to 1 nm layer of GaAs is sandwiched between two macroscopically thick (several micrometers) $Al_\alpha Ga_{1-\alpha}As$ layers. Such a heterostructure is depicted on the left in Figure 16.21. When this GaAs layer is considered as a three-dimensional (3D) box, it has energy levels that depend on three quantum numbers because this is a 3D problem:

$$E_{n_x n_y n_z} = \frac{h^2}{8m} \left(\frac{n_z^2 + n_y^2}{b^2} + \frac{n_x^2}{a^2} \right) \qquad \textbf{(16.21)}$$

The length b is on the order of 1000. nm, whereas a is 0.1 to 1.0 nm. Therefore, the energy spectrum is essentially continuous in n_z and n_y, but discrete in n_x. What does the band-gap region in such an alternating layer structure look like? This can be deduced from Figure 16.20 and is shown in Figure 16.21.

In this very thin layer of GaAs, the empty conduction band has lower energy states in the GaAs region than elsewhere in the heterostructure. The $Al_\alpha Ga_{1-\alpha}As$ layers have macroscopic dimensions in all three directions, so that the particle in the box states form a continuous energy spectrum. By contrast, the GaAs layer has relatively large dimensions parallel to the layer, but atomic scale dimensions along the x direction perpendicular to the interface between the substances. Along this direction, the quantization conditions are those expected from a particle in a finite well, leading to discrete energy levels as shown in Figure 16.21. Along the other two directions, the energy-level spectrum is continuous. By choosing this unusual geometry for the box, the system has a continuous energy spectrum along the y and z directions and a discrete energy

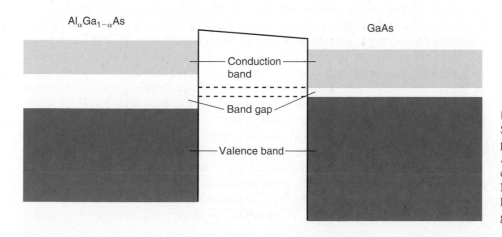

FIGURE 16.20
Schematic representation of relative positions of the bands in GaAs and $Al_\alpha Ga_{1-\alpha}As$ connected in an external circuit as in Figure 16.12b (not to scale). Note that the smaller band gap in GaAs lines up with the center of the larger band gap in $Al_\alpha Ga_{1-\alpha}As$.

FIGURE 16.21
Schematic depiction of the heterostructure (left) and the band and band-gap structure in the immediate vicinity of the GaAs layer (right). Not to scale.

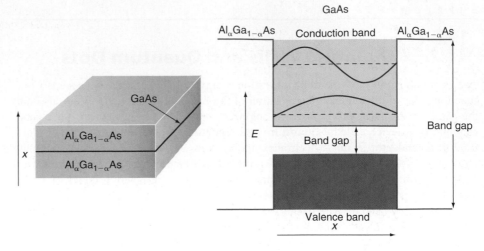

spectrum along the x direction. As discussed later, it is possible to selectively change the discrete energy spectrum.

This is certainly a novel structure, and it is also useful because it can be made to function as a very efficient laser. In the ground state, the valence band is fully occupied and the conduction band is empty. The lowest energy excitation from the valence band into the conduction band lies in the GaAs layer. Therefore, it is possible to efficiently excite these transitions by putting an amount of energy into the system that is equal to or larger than the band-gap energy in GaAs but less than the band-gap energy in $Al_\alpha Ga_{1-\alpha}As$. When the system decays to the ground state, a photon is emitted with frequency $\nu = \Delta E/h$, in which ΔE is the difference in energy between the excited energy level in the conduction and the empty states in the valence band. A laser of this type has two advantages over more conventional solid-state lasers. The first is that such lasers can be very efficient in producing photons. The second is that the energy levels in the GaAs layer can be changed by varying the layer thickness, as predicted by Equation (16.21). This allows for tuning of the laser frequency through a limited range. Devices based on the principles outlined here are called *quantum well devices*.

The technique used to manufacture heterostructures like those just discussed is molecular beam epitaxy (MBE). Because the materials must be deposited in a very high vacuum, MBE is an expensive technique. New techniques involving size-controlled crystallization in solution offer a less expensive way to synthesize nanoscale particles. Such techniques can produce crystalline spherical particles of compound semiconductors such as CdSe with uniform diameters in the range of 1 to 10. nm. This results in the energy levels being quantized in all three directions and opens up new possibilities for these structures, which are called **quantum dots.** Quantum dots have a band-gap energy that strongly depends on their diameter for the reasons discussed earlier.

Assume that all states below the band gap are filled and all states above the band gap of width E_{bg} are empty in the ground state of the quantum dot, making it a semiconductor. Transitions from states below to those above the band gap can occur through absorption of visible light. Subsequently, the electron in the excited state can drop to an empty state below the band gap, emitting a photon in a process called fluorescence with a wavelength $\lambda = hc/E_{bg}$. Because the energy levels and E_{bg} depend on the length b, λ also depends on b. This property is illustrated in Figure 16.22a. For CdSe quantum dots, the emission wavelength increases from 450 nm (blue light) to 650 nm (red light) as the dot diameter increases from 2 to 8 nm. Figure 16.22b shows another important property of quantum dots. Although they absorb light over a wide range of wavelengths, they emit light in a much smaller range of wavelengths. This occurs because electrons excited from occupied states just below the band gap to states well above the band gap in absorption rapidly lose energy and relax to states just above the band gap. Therefore the light emitted in fluorescence is in a narrow frequency range determined by the band-gap energy of the semiconducting quantum dot.

Quantum dots are currently being used in bioanalytical methods. The usefulness of these quantum dots is their ability to act as tags for biologically interesting substrates

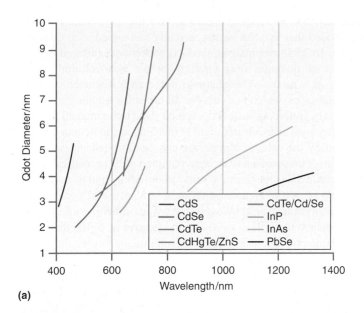

(a)

Absorption and emission spectra of four CdSe/ZnS quantum dots of different diameters

(b)

FIGURE 16.22

(a) The dependence of the wavelength of the light emitted in a transition from just above to just below the band gap is shown as a function of the quantum dot diameter for a number of materials. (b) The top panel shows the absorption spectrum of four CdSe/ZnS quantum dots of different diameters, and the bottom panel shows the corresponding emission spectrum. Note that absorption occurs over a much larger range of wavelengths than emission. The vertical bar indicates the wavelength of a 488 nm argon ion laser that can be used to excite electrons from below to above the band gap for all four diameters. Using this laser ensures that absorption and emission occur at distinctly different wavelengths.

Source: **Figure 16.22a** Figure 1A from X. Michalat, et al., "Quantum Dots for Live Cells, in vivo Imaging, and Diagnostics" from *Science*, New Series, Vol. 307: 538–544, Jan. 28, 2005. Copyright © 2005, The American Association for the Advancement of Science. Reprinted with permission from AAAS.

such as proteins, as shown in Figure 16.23. By functionalizing such quantum dots with an appropriate molecular layer, they can be made soluble in aqueous solutions and tethered to the protein of interest.

The following example illustrates the usefulness of a protein with a fluorescent label. After letting the tagged proteins enter a cell and attach to their receptors, the cell is illuminated with light and the quantum dots act as point sources of fluorescent light whose location can be imaged using optical microscopy. Because the light used for excitation and the fluorescent light have different wavelengths, it is easy to distinguish between them using optical filters. The same excitation wavelength can be used for quantum dots of different size, so that several different ligand–receptor combinations can be probed simultaneously if the individual ligands are tethered to quantum dots of differing diameter. It might appear that the number of possible different fluorescent tags is limited by the overlap in the wavelengths at which they fluoresce. However, one can also tether different combinations of a few different quantum dots to a protein, creating a barcode. For instance, the intensity versus wavelength distribution of the fluorescent signal from a tagged protein to which two 1 nm, one 3 nm, and two 5 nm quantum dots have been attached is different from all other distinct possible permutations of five quantum dots. This analysis method, which is based on size quantization, offers new analytical techniques for measuring the spatial distribution of molecules in inherently heterogeneous biological environments.

Because a quantum dot absorbs strongly over a wide range of wavelengths but fluoresces in a narrow range of wavelengths, it can be used as an internal light source for imaging the interior of semitransparent specimens. Figure 16.24 shows an image obtained by projecting the capillary structure of adipose tissue in a 250-μm-thick specimen surrounding a surgically exposed ovary of a living mouse on a plane [Larson *et al., Science* 300 (2003): 1434]. Additionally, the rate of blood flow and the differences in systolic and diastolic pressure can be directly observed in these experiments. It is not possible to obtain such images with X-ray-based techniques, because of the absence of a contrast mechanism.

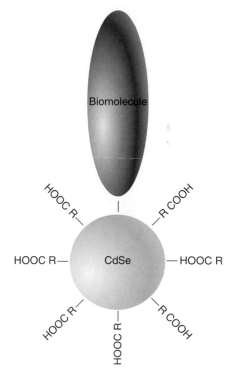

FIGURE 16.23

A CdSe quantum dot can be made soluble in an aqueous solution by coating it with a single molecular layer of an organic acid. When tethered to a biomolecule of interest, it can be used as a fluorescent tag to locate the biomolecule when the biomolecule is bound to a receptor in a heterogeneous environment such as a cell.

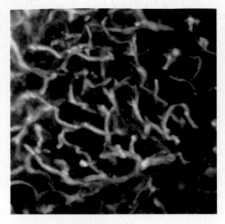

FIGURE 16.24

This 160. × 160. μm image was obtained by projecting the capillary structure in a 250-μm-thick specimen of adipose tissue in the skin of a living mouse using CdSe quantum dots that fluoresce at 550 nm.

[Reproduced with permission from Larson et al. "Water-Soluble Quantum Dots for Multiphoton Fluorescence Imaging in Vico." *Science* 300 (2003): 1434. © 2003 American Association for the Advancement of Science.]

FIGURE 16.25

Quantum dots were injected into the paw of a mouse. (**a**) The middle panel shows a video image taken 5 minutes after the quantum dots were introduced. The right panel is a fluorescence image that shows localization of the quantum dots in the lymph node. The left panel shows that without the quantum dots, no fluorescence is observed. (**b**) Surgery after injection of a chemical mapping agent that is known to localize in lymph nodes confirms that the quantum dots are localized in the lymph nodes.

[From S. Kim et al. "Near-Infrared Fluorescent Type II Quantum Dots for Sentinel Lymph Node Mapping." *Nature Biotechnology* 22 (2004): 93–97.]

The usefulness of quantum dots in imaging tissues in vivo has significant potential applications in surgery provided that toxicity issues currently associated with quantum dots can be resolved. Figure 16.25 shows images obtained with a near-infrared camera resulting from the injection of quantum dots emitting in the near-infrared region (840–860 nm) into the paw of a mouse. The quantum dots had a diameter ~15 nm including the functionalization layer, which is a suitable diameter for trapping in lymph nodes. Fluorescence from the quantum dots allow the lymph node to be imaged through the overlying tissue layers. Because near-infrared light is invisible to the human eye and visible light is not registered by the camera, the mouse can be simultaneously illuminated with both types of light. A comparison of images obtained with near-infrared and visible light can be used to guide the surgeon in the removal of tumors and to verify that all affected tissues have been removed. Near-infrared light is also useful because this wavelength minimizes the absorption of light by overlying tissues, allowing tissues containing quantum dots to be imaged through an overlying tissue layer of 6–10 cm.

Quantum dots have several applications that are in developmental stages. It may be possible to use them to couple electrical signal amplification, currently based on charge carrier conduction in semiconductors, with light amplification, in an application known as optoelectronics. Additionally, the reduced dimensions of quantum dots utilized as wavelength-tunable lasers allow them to be integrated into conventional silicon-based microelectronics.

a

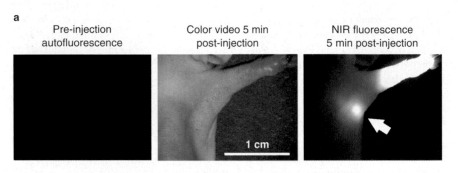

Pre-injection autofluorescence   Color video 5 min post-injection   NIR fluorescence 5 min post-injection

1 cm

b

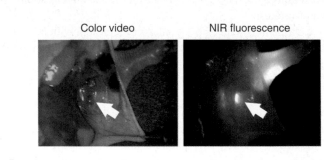

Color video   NIR fluorescence

Vocabulary

activation energy	electrical conductor	quantum well structure
atomic force microscope (AFM)	energy band	scanning tunneling microscope (STM)
band gap	finite depth box	semiconductor
barrier width	ground state	tunneling
classically forbidden region	heterostructure	valence band
conduction band	insulator	valence electrons
core electrons	overlap of wave functions	work function
decay length	π-bonded network	
delocalization	quantum dot	

Conceptual Problems

Q16.1 Why is it necessary to apply a bias voltage between the tip and surface in a scanning tunneling microscope?

Q16.2 The amplitude of the wave on the right side of the barrier in Figure 16.10 is much smaller than that of the wave incident on the barrier. What happened to the "rest of the wave"?

Q16.3 Why is a tunneling current not observed in an STM when the tip and the surface are 1 mm apart?

Q16.4 Redraw Figure 16.7 for an insulator.

Q16.5 Explain how it is possible to create a three-dimensional electron conductor that has a continuous energy spectrum in two dimensions and a discrete energy spectrum in the third dimension.

Q16.6 Explain, without using equations, why tunneling is more likely for the particle with $E = 3/4V_0$ than for $E = 1/4V_0$ in Figure 16.10.

Q16.7 What is the advantage of using quantum dots that fluoresce in the near infrared for surgical applications?

Q16.8 The overlap between wave functions can either be constructive or destructive, just as for waves. Can you distinguish between constructive and destructive overlap for the various energy levels in Figure 16.3?

Q16.9 Explain how you can use size-quantized quantum dots to create a protein with a barcode that can be read using light.

Q16.10 An STM can also be operated in a mode in which electrons tunnel from the surface into the tip. Use Figure 16.12 to explain how you would change the experimental setup to reverse the tunneling current.

Q16.11 For CdSe quantum dots, the emission wavelength increases from 450. nm to 650. nm as the dot diameter increases from 2 to 8 nm. Calculate the band gap energy for these two particle diameters.

Q16.12 Why is it necessary to functionalize CdSe quantum dots with groups such as organic acids to make them useful in bioanalytical applications?

Q16.13 Why must the amplitudes of the first derivatives of the energy eigenfunctions in the finite depth box and in the adjoining barrier regions have the same value at the boundary?

Q16.14 Why must the amplitudes of the energy eigenfunctions in the finite depth box and in the adjoining barrier regions have the same value at the boundary?

Q16.15 Explain how a quantum dot can absorb light over a range of wavelengths and emit light over a much smaller range of wavelengths.

Q16.16 Explain why the speed of the particle needs to be taken into account in calculating the probability for transmission over a step potential.

Q16.17 The reflection probability from a step potential was calculated for $E > V_0$ in Section 16.5. Is Equation (16.18) valid for $E < V_0$? What information can you extract from Figure 16.1 that will allow you to state the value of R for a step potential if $E < V_0$?

Q16.18 Figure 16.17 shows that atomic level resolution is only attainable in the repulsive portion of the tip–surface potential. What does this tell you about the range of the attractive and repulsive parts of the potential?

Q16.19 Why is atomic level resolution obtained on pentacene in the AFM mode as shown in Figure 16.18, but not in the STM mode?

Q16.20 Why were quantum dots emitting in the near-infrared region used for the surgery experiment shown in Figure 16.25?

Numerical Problems

P16.1 In this problem, you will calculate the transmission probability through the barrier illustrated in Figure 16.10. We first go through the mathematics leading to the solution. You will then carry out further calculations.

The domain in which the calculation is carried out is divided into three regions for which the potentials are

$$V(x) = 0 \quad \text{for } x \leq 0 \qquad \text{Region I}$$

$$V(x) = V_0 \quad \text{for } 0 < x < a \qquad \text{Region II}$$

$$V(x) = 0 \quad \text{for } x \geq a \qquad \text{Region III}$$

The spatial part of the wave functions must have the following form in the three regions if $E < V_0$:

$$\psi(x) = A \exp\left[+i\sqrt{\frac{2mE}{\hbar^2}}\, x\right] + B \exp\left[-i\sqrt{\frac{2mE}{\hbar^2}}\, x\right]$$

$$= A e^{+ikx} + B e^{-ikx} \qquad \text{Region I}$$

$$\psi(x) = C \exp\left[-\sqrt{\frac{2m(V_0 - E)}{\hbar^2}}\, x\right]$$

$$+ D \exp\left[+\sqrt{\frac{2m(V_0 - E)}{\hbar^2}}\, x\right]$$

$$= C e^{-\kappa x} + D e^{+\kappa x} \qquad \text{Region II}$$

$$\psi(x) = F \exp\left[+i\sqrt{\frac{2mE}{\hbar^2}}\, x\right] + G \exp\left[-i\sqrt{\frac{2mE}{\hbar^2}}\, x\right]$$

$$= F e^{+ikx} + G e^{-ikx} \qquad \text{Region III}$$

Assume that the wave approaches the barrier from the negative x direction. The coefficient B cannot be set equal to zero because $Be^{-i\sqrt{(2mE/\hbar^2)}x}$ represents reflection from the barrier. However, G can be set equal to zero because there is no wave incident on the barrier from the positive x direction.

a. The wave functions and their derivatives must be continuous at $x = 0$ and $x = a$. Show that the coefficients must satisfy the following conditions:

$$A + B = C + D \qquad Ce^{-\kappa a} + De^{+\kappa a} = Fe^{+ika}$$

$$A - B = -\frac{i\kappa}{k}(-C + D) \quad -Ce^{-\kappa a} + De^{+\kappa a} = \frac{ik}{\kappa}Fe^{+ika}$$

b. Because the transmission probability is given by $|F/A|^2$, it is useful to manipulate these equations to get a relationship between F and A. By adding and subtracting the first pair of equations, A and B can be expressed in terms of C and D. The second pair of equations can be combined in the same way to give equations for D and C in terms of F. Show that

$$D = \frac{ik\,e^{+ika} + \kappa e^{+ika}}{2\kappa\,e^{+\kappa a}}F$$

$$C = \frac{-ik\,e^{+ika} + \kappa e^{+ika}}{2\kappa\,e^{-\kappa a}}F, \text{ and}$$

$$A = \frac{(ik - \kappa)C + (ik + \kappa)D}{2ik}$$

c. Substitute these results for C and D in terms of F into

$$A = \frac{(ik - \kappa)C + (ik + \kappa)D}{2ik}$$

to relate A and F. Show that

$$2ikA = \frac{e^{+ika}}{2\kappa}\big[(ik - \kappa)(-ik + \kappa)e^{+\kappa a}$$

$$+ (ik + \kappa)(ik + \kappa)e^{-\kappa a}\big]F$$

d. Using the hyperbolic trigonometric functions

$$\sinh x = \frac{e^x - e^{-x}}{2} \text{ and } \cosh x = \frac{e^x + e^{-x}}{2}$$

and the relationship $\cosh^2 x - \sinh^2 x = 1$, show that

$$\left|\frac{F}{A}\right|^2 = \frac{16(\kappa k)^2}{16(\kappa k)^2 + \big(4(k^2 - \kappa^2)^2 + 16(\kappa k)^2\big)\sinh^2(\kappa a)}$$

$$= \frac{1}{1 + [(k^2 + \kappa^2)^2 \sinh^2(\kappa a)]/4(\kappa k)^2}$$

e. Plot the transmission probability for an electron as a function of energy for $V_0 = 1.6 \times 10^{-19}$ J and $a = 9.0 \times 10^{-10}$ m up to an energy of 8×10^{-19} J. At what energy is the tunneling probability 0.1? At what energy is the tunneling probability 0.02?

f. Plot the transmission probability for an electron of energy 0.50×10^{-19} J as a function of the barrier width for $V_0 = 1.6 \times 10^{-19}$ J between 2×10^{-10} and 8×10^{-10} m. At what barrier width is the transmission probability 0.2?

P16.2   Semiconductors can become conductive if their temperature is raised sufficiently to populate the (empty) conduction band from the highest filled levels in the valence band. The ratio of the populations in the highest level of the conduction band to that of the lowest level in the valence band is

$$\frac{n_{conduction}}{n_{valence}} = \frac{g_{conduction}}{g_{valence}}e^{-\Delta E/k_B T}$$

where ΔE is the band gap, which is 1.12 eV for Si and 5.5 eV for diamond. Assume for simplicity that the ratio of the degeneracies is one and that the semiconductor becomes sufficiently conductive when

$$\frac{n_{conduction}}{n_{valence}} = 5.5 \times 10^{-7}$$

At what temperatures will silicon and diamond become sufficiently conductive? Given that diamond sublimates near 3000. K, could you heat diamond enough to make it conductive and not sublimate it?

P16.3   For the π network of β carotene modeled using the particle in the box, the position-dependent probability density of finding 1 of the 22 electrons is given by

$$P_n(x) = |\psi_n(x)|^2 = \frac{2}{a}\sin^2\left(\frac{n\pi x}{a}\right)$$

The quantum number n in this equation is determined by the energy level of the electron under consideration. As we saw in Chapter 15, this function is strongly position dependent. The question addressed in this problem is as follows: Would you also expect the total probability density defined by $P_{total}(x) = \sum_n |\psi_n(x)|^2$ to be strongly position dependent? The sum is over all the electrons in the π-nework.

a. Calculate the total probability density $P_{total}(x) = \sum_n |\psi_n(x)|^2$ using the box length $a = 0.29$ nm, and plot your results as a function of x. Does $P_{total}(x)$ have the same value near the ends and at the middle of the molecule?

b. Determine $\Delta P_{total}(x)/\langle P_{total}(x)\rangle$, where $\Delta P_{total}(x)$ is the peak-to-peak amplitude of $P_{total}(x)$ in the interval between 0.12 and 0.16 nm.

c. Compare the result of part (b) with what you would obtain for an electron in the highest occupied energy level.

d. What value would you expect for $P_{total}(x)$ if the electrons were uniformly distributed over the molecule? How does this value compare with your result from part (a)?

P16.4   Calculate the energy levels of the π-network in hexatriene, C_6H_8, using the particle in the box model. To calculate the box length, assume that the molecule is linear and use the values 135 and 154 pm for C$=$C and C$-$C bonds. What is the wavelength of light required to induce a transition from the ground state to the first excited state? How does this compare with the experimentally observed value of 240 nm? What does the comparison made suggest to you about estimating the length of the π-network by adding bond lengths for this molecule?

P16.5 Calculate the energy levels of the π-network in octatetraene, C_8H_{10}, using the particle in the box model. To calculate the box length, assume that the molecule is linear and use the values 135 and 154 pm for $C=C$ and $C-C$ bonds. What is the wavelength of light required to induce a transition from the ground state to the first excited state?

P16.6 The maximum safe current in a copper wire with a diameter of 3.0 mm is about 20. amperes. In an STM, a current of 1.0×10^{-9} A passes from the tip to the surface in a filament of diameter ~1.0 nm. Compare the current density in the copper wire with that in the STM.

P16.7 In this problem, you will solve for the total energy eigenfunctions and eigenvalues for an electron in a finite depth box. We first go through the calculation for the box parameters used in Figure 16.1. You will then carry out the calculation for a different set of parameters.

We describe the potential in this way:

$$V(x) = V_0 \quad \text{for } x \le -\frac{a}{2} \qquad \text{Region I}$$

$$V(x) = 0 \quad \text{for } -\frac{a}{2} < x < \frac{a}{2} \quad \text{Region II}$$

$$V(x) = V_0 \quad \text{for } x \ge \frac{a}{2} \qquad \text{Region III}$$

The eigenfunctions must have the following form in these three regions:

$$\psi(x) = B \exp\left[+\sqrt{\frac{2m(V_0 - E)}{\hbar^2}} \, x \right]$$
$$+ B' \exp\left[-\sqrt{\frac{2m(V_0 - E)}{\hbar^2}} \, x \right]$$
$$= Be^{+\kappa x} + B'e^{-\kappa x} \qquad \text{Region I}$$

$$\psi(x) = C \sin\sqrt{\frac{2mE}{\hbar^2}} \, x + D \cos\sqrt{\frac{2mE}{\hbar^2}} \, x$$
$$= C \sin kx + D \cos kx \qquad \text{Region II}$$

$$\psi(x) = A \exp\left[-\sqrt{\frac{2m(V_0 - E)}{\hbar^2}} \, x \right]$$
$$+ A' \exp\left[+\sqrt{\frac{2m(V_0 - E)}{\hbar^2}} \, x \right]$$
$$= Ae^{-\kappa x} + A'e^{+\kappa x} \qquad \text{Region III}$$

So that the wave functions remain finite at large positive and negative values of x, $A' = B' = 0$. An additional condition must also be satisfied. To arrive at physically meaningful solutions for the eigenfunctions, the wave functions in the separate regions must have the same amplitude and derivatives at the values of $x = a/2$ and $x = -a/2$ bounding the regions. This restricts the possible values for the coefficients

$A, B, C,$ and D. Show that applying these conditions gives the following equations:

$$Be^{-\kappa(a/2)} = -C \sin k\frac{a}{2} + D \cos k\frac{a}{2}$$

$$B\kappa e^{-\kappa(a/2)} = Ck \cos k\frac{a}{2} + Dk \sin k\frac{a}{2}$$

$$Ae^{-\kappa(a/2)} = C \sin k\frac{a}{2} + D \cos k\frac{a}{2}$$

$$-A\kappa e^{-\kappa(a/2)} = Ck \cos k\frac{a}{2} - Dk \sin k\frac{a}{2}$$

These two pairs of equations differ on the right side only by the sign of one term. We can obtain a set of equations that contain fewer coefficients by adding and subtracting each pair of equations to give

$$(A + B)e^{-\kappa(a/2)} = 2D \cos\left(k\frac{a}{2} \right)$$

$$(A - B)e^{-\kappa(a/2)} = 2C \sin\left(k\frac{a}{2} \right)$$

$$(A + B)\kappa e^{-\kappa(a/2)} = 2Dk \sin\left(k\frac{a}{2} \right)$$

$$-(A - B)\kappa e^{-\kappa(a/2)} = 2Ck \cos\left(k\frac{a}{2} \right)$$

At this point we notice that by dividing the equations in each pair, the coefficients can be eliminated to give

$$\kappa = k \tan\left(k\frac{a}{2} \right) \text{ or } \sqrt{\frac{2m(V_0 - E)}{\hbar^2}}$$

$$= \sqrt{\frac{2mE}{\hbar^2}} \tan\left(\sqrt{\frac{2mE}{\hbar^2}} \frac{a}{2} \right) \text{ and }$$

$$-\kappa = k \cot\left(k\frac{a}{2} \right) \text{ or } -\sqrt{\frac{2m(V_0 - E)}{\hbar^2}}$$

$$= \sqrt{\frac{2mE}{\hbar^2}} \cot\left(\sqrt{\frac{2mE}{\hbar^2}} \frac{a}{2} \right)$$

Multiplying these equations on both sides by $a/2$ gives dimensionless parameters, and the final equations are

$$\sqrt{\frac{m(V_0 - E)a^2}{2\hbar^2}} = \sqrt{\frac{mEa^2}{2\hbar^2}} \tan\sqrt{\frac{mEa^2}{2\hbar^2}} \text{ and }$$

$$-\sqrt{\frac{m(V_0 - E)a^2}{2\hbar^2}} = \sqrt{\frac{mEa^2}{2\hbar^2}} \cot\sqrt{\frac{mEa^2}{2\hbar^2}}$$

The allowed energy values E must satisfy these equations. They can be obtained by graphing the two sides of each equation against E. The intersections of the two curves are the allowed energy eigenvalues. For the parameters in the caption of

Figure 16.1, $V_0 = 1.20 \times 10^{-18}$ J and $a = 1.00 \times 10^{-9}$ m, the following two graphs are obtained:

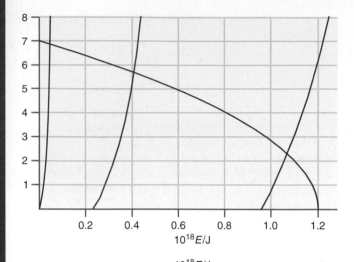

$10^{18} E/J$

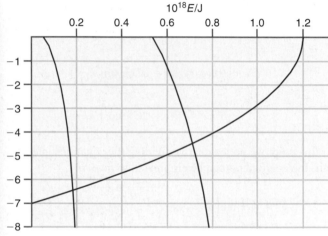

The five allowed energy levels are at 4.61×10^{-20}, 4.09×10^{-19}, and 1.07×10^{-18} J (top figure), and 1.84×10^{-19} and 7.13×10^{-19} J (bottom figure).

a. Given these values, calculate λ for each energy level. Is the relation $\lambda = 2a/n$ (for n an integer) that arose from the calculations on the infinitely deep box still valid? Compare the values with the corresponding energy level in the infinitely deep box. Explain why the differences arise.

b. Repeat this calculation for $V_0 = 5.00 \times 10^{-19}$ J and $a = 0.900 \times 10^{-9}$ m. Do you think that there will be fewer or more bound states than for the problem just worked out? How many allowed energy levels are there for this well depth, and what is the energy corresponding to each level?

P16.8 An electron of energy 5.0 eV approaches a step potential of height 2.0 eV. Calculate the probabilities that the electron will be reflected and transmitted.

Computational Problems

More detailed instructions on carrying out this calculation using Spartan Student are found on the book website at *www.masteringchemistry.com*.

C16.1 Build (a) ethylene, (b) the trans conformation for 1,3 butadiene, and (c) all trans hexatriene and calculate the ground-state (singlet) energy of these molecules using the B3LYP method with the $6\text{-}311\text{+}G^{**}$ basis set. Repeat your calculation for the triplet state, which corresponds to the excitation of a π electron from the highest filled energy level to the lowest unoccupied energy level. Use a nonplanar input geometry for the triplet states. Compare the energy difference from these calculations to literature values of the maximum in the UV-visible absorption spectrum.

Web-Based Simulations, Animations, and Problems

W16.1 The Schrödinger equation is solved numerically for the particle in the finite height box. Using the condition that the wave function must approach zero amplitude in the classically forbidden region, the energy levels are determined for a fixed particle mass, box depth, and box length. The particle mass and energy and the box depth and length are varied with sliders to demonstrate how the number of bound states varies with these parameters.

W16.2 The Schrödinger equation is solved numerically to calculate the tunneling probability for a particle through a thin finite barrier. Sliders are used to vary the barrier width and height and the particle energy and mass. The dependence of the tunneling probability on these variables is investigated.

Commuting and Noncommuting Operators and the Surprising Consequences of Entanglement

Classical physics predicts that there is no limit to the amount of information (observables) that can be known about a system at a given instant of time. This is not the case in quantum mechanics. Two observables can be known simultaneously only if the outcome of the measurements is independent of the order in which they are conducted. An uncertainty relation limits the degree to which observables of other operators can be known simultaneously. Although this result is counterintuitive from a classical perspective, the Stern–Gerlach experiment clearly demonstrates that this prediction of quantum mechanics is obeyed at the atomic level. Because a quantum state can be a superposition of individual states, two particles can be entangled. Entanglement is the basis of both teleportation and quantum computing.

17.1 Commutation Relations

In classical mechanics, a system under consideration can in principle be described completely. For instance, for a mass falling in a gravitational field, its position, momentum, kinetic energy, and potential energy can be determined simultaneously at any point on its trajectory. The uncertainty in the measurements is only limited by the capabilities of the measurement technique. All of these observables (and many more) can be known simultaneously. This is not generally true from a quantum mechanical perspective. In the quantum world, in some cases two observables can be known simultaneously with high accuracy. However, in other cases, two observables have a fundamental uncertainty that cannot be removed through any measurement techniques. However, as will be shown later, in the classical limit of very large quantum numbers, the fundamental uncertainty for such observables is less than the uncertainty associated with experimental techniques. This result shows that quantum mechanics is consistent with classical mechanics for large quantum numbers.

The values of two different observables a and b, which correspond to the operators $\hat{A}$ and $\hat{B}$, can be simultaneously determined only if the measurement process used does not change the state of the system. Otherwise, the system on which the second measurement is carried out is not the same as for the first measurement. Let $\psi_n(x)$ be the wave function that characterizes the system. How can the measurements of the observables corresponding to the operators $\hat{A}$ and $\hat{B}$ be described? Carrying out a measurement of the observables corresponding first to the operator $\hat{A}$ and subsequently to the operator $\hat{B}$ is equivalent to evaluating $\hat{B}[\hat{A}\psi_n(x)]$. If $\psi_n(x)$ is an eigenfunction of $\hat{A}$, then $\hat{B}[\hat{A}\psi_n(x)] = \alpha_n\hat{B}\psi_n(x)$. The only case in which the second measurement does not change the state of the system is if $\psi_n(x)$ is also an eigenfunction of $\hat{B}$. In this case, $\hat{B}[\hat{A}\psi_n(x)] = \beta_n\alpha_n\psi_n(x)$. Reversing the order of the two operations gives $\hat{A}[\hat{B}\psi_n(x)] = \alpha_n\beta_n\psi_n(x)$. Because the eigenvalues β_n and α_n are simply constants, $\beta_n\alpha_n\psi_n(x) = \alpha_n\beta_n\psi_n(x)$ and, therefore, $\hat{B}[\hat{A}\psi_n(x)] = \hat{A}[\hat{B}\psi_n(x)]$.

We have just shown that the act of measurement changes the state of the system unless the system wave function is an eigenfunction of the two different operators. Therefore, this is a condition for being able to simultaneously know the observables corresponding to these operators. How can one know if two operators have a common set of eigenfunctions? The example just discussed suggests a simple test that can be applied. Only if

$$\hat{A}[\hat{B} f(x)] - \hat{B}[\hat{A} f(x)] = 0 \tag{17.1}$$

for $f(x)$, an arbitrary function, will $\hat{A}$ and $\hat{B}$ have a common set of eigenfunctions, and only then can the corresponding observables be known simultaneously.

If two operators have a common set of eigenfunctions, we say that they **commute.** The difference $\hat{A}[\hat{B} f(x)] - \hat{B}[\hat{A} f(x)]$ is abbreviated $[\hat{A}, \hat{B}]f(x)$ and the expression in the square brackets is called the **commutator** of the operators $\hat{A}$ and $\hat{B}$. If the value of the commutator is not zero for an arbitrary function $f(x)$, the corresponding observables cannot be determined simultaneously and exactly. We will have more to say about what is meant by *exactly* later in this chapter.

EXAMPLE PROBLEM 17.1

Determine whether the momentum and (a) the kinetic energy and (b) the total energy can be known simultaneously.

Solution

We determine whether two operators $\hat{A}$ and $\hat{B}$ commute by evaluating the commutator $\hat{A}[\hat{B} f(x)] - \hat{B}[\hat{A} f(x)]$. If the commutator is zero, the two observables can be determined simultaneously and exactly.

a. For momentum and kinetic energy, we evaluate

$$-i\hbar\frac{d}{dx}\left(-\frac{\hbar^2}{2m}\frac{d^2}{dx^2}\right)f(x) - \left(-\frac{\hbar^2}{2m}\frac{d^2}{dx^2}\right)\left(-i\hbar\frac{d}{dx}\right)f(x)$$

In calculating the third derivative, it does not matter if the function is first differentiated twice and then once or the other way around. Therefore, the momentum and the kinetic energy can be determined simultaneously and exactly.

b. For momentum and total energy, we evaluate

$$-i\hbar \frac{d}{dx}\left(-\frac{\hbar^2}{2m}\frac{d^2}{dx^2} + V(x)\right)f(x) - \left(-\frac{\hbar^2}{2m}\frac{d^2}{dx^2} + V(x)\right)\left(-i\hbar \frac{d}{dx}\right)f(x)$$

Because the kinetic energy and momentum operators commute, per part (a), this expression is equal to

$$-i\hbar \frac{d}{dx}(V(x)f(x)) + i\hbar V(x)\frac{d}{dx}f(x)$$

$$= -i\hbar V(x)\frac{d}{dx}f(x) - i\hbar f(x)\frac{d}{dx}V(x) + i\hbar V(x)\frac{d}{dx}f(x)$$

$$= -i\hbar f(x)\frac{d}{dx}V(x)$$

We conclude the following:

$$\left[V(x), -i\hbar \frac{d}{dx}\right] = -i\hbar \frac{d}{dx}V(x) \neq 0$$

Therefore, the momentum and the total energy cannot be known simultaneously and exactly. Note that the arbitrary function $f(x)$ is not present in the final expression for the commutator. Note also that the momentum and the total energy can be known simultaneously if $[dV(x)]/dx = 0$. This corresponds to a constant potential energy for all values of x, in other words, the free particle of Section 15.1.

Now apply the formalism just discussed to the particle in the box in its lowest energy state. In Chapter 15, we found that although the wave function is an eigenfunction of the total energy operator, it is not an eigenfunction of the momentum operator. Therefore, these two operators do not commute. If the total energy of the particle is measured, the value $E = h^2/8ma^2$ is obtained. If the average momentum is subsequently determined from a number of individual measurements, the result is $\langle p_x \rangle = 0$. This result merely states that it is equally likely that positive and negative values will be obtained. There is no way of knowing what the magnitude and sign of the momentum will be for an individual measurement. Because the energy is known precisely, nothing is known about the momentum. This result is consistent with the fact that the two operators do not commute.

17.2 The Stern–Gerlach Experiment

Consider next a real experiment in a simple quantum mechanical framework that illustrates some of the concepts discussed in the preceding section in more concrete terms. This experiment also illustrates how quantum mechanical concepts of measurement arose out of analyzing results obtained in the laboratory. In this experiment, a beam of silver atoms having a well-defined direction passes through a magnetic field that has a constant value in the xy plane and varies linearly with the z coordinate, which is chosen to be perpendicular to the path of the atoms. We say that the magnetic field has a gradient in the z direction. An atomic beam of silver atoms can be made in a vacuum system by heating solid silver in an oven to a temperature at which the vapor pressure of Ag is in the range of 10^{-2} torr. Letting the atoms escape through a series of collimating apertures in the wall of the oven results in a beam of Ag atoms, all traveling in the same direction, which is chosen to be the y direction. The atoms pass through the magnetic field and are detected some distance beyond the magnet. The forces acting on the magnetic dipoles representing individual Ag atoms is depicted in Figure 17.1, and the **Stern–Gerlach experiment** is shown schematically in Figure 17.2.

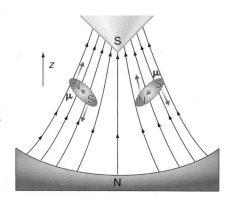

FIGURE 17.1
The effect of an inhomogeneous magnetic field on magnetic dipoles is to orient and deflect them in opposite directions, depending on the sign of the component of the magnetic moment along the z direction.

FIGURE 17.2
Schematic representation of the
Stern–Gerlach experiment. The inhomogeneous magnetic field separates the silver beam into two, and only two,
components.

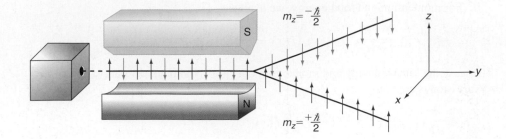

Silver atoms have a single unpaired electron that has an intrinsic magnetic moment. We return to the consequences of this fact later. The magnetic moment is associated with what is called the *electron spin,* although the picture of a spherical electron spinning around an axis through its center is incorrect. It turns out that the spin emerges naturally in relativistic quantum mechanics. (View these remarks as an aside because none of this was known at the time the Stern–Gerlach experiment was conducted.)

Because each atom has a magnetic moment associated with the unpaired electron, the atom is deflected in the z direction as it passes through the inhomogeneous magnetic field. The atom is not deflected along the x and y directions, because the magnetic field is constant along these directions. What outcome is expected in this experiment? Consider the classical system of a beam of magnetic dipoles. We expect that the magnetic dipoles are randomly oriented in space and that only their z component is affected by the magnet. Because the z component takes on all possible values between $+|\boldsymbol{\mu}|$ and $-|\boldsymbol{\mu}|$, where $\boldsymbol{\mu}$ is the magnetic moment of the atom, the silver atoms will be equally distributed along a range of z values at the detector. The z values can be predicted from the geometry of the experiment and the strength of the field gradient if the magnetic moment is known.

What are the results of the experiment? Silver atoms are deflected only in the z direction, but only two z values are observed. One corresponds to an upward deflection and the other to a downward deflection of the same magnitude. *What conclusions can be drawn from this experiment?* We conclude that the operator called "measure the z component of the magnetic moment," denoted by $\hat{A}$, has only two eigenfunctions with eigenvalues that are equal in magnitude but opposite in sign. We call the two eigenfunctions α and β and assume that they are normalized. Because the experiment shows that these two eigenfunctions form a complete set (only two deflection angles are observed), any acceptable wave function can be written as a linear combination of α and β. Therefore, the initial normalized wave function that describes a single silver atom is

$$\psi = c_1\alpha + c_2\beta \quad \text{with } |c_1|^2 + |c_2|^2 = 1 \tag{17.2}$$

We cannot specify the values of c_1 and c_2, because they refer to individual measurements, and only the total number of silver atoms in the two deflected beams at the detector is measured. However, the relative number of Ag atoms that was deflected upward and downward can be measured for a large number of atoms. This ratio is one, and therefore $|c_1|^2_{average} = |c_2|^2_{average} = 1/2$. The average is over all the atoms that have landed on the detector.

Now carry this experiment a step further. We follow the path of the downwardly deflected atoms, which have the wave function $\psi = \alpha$ and deflect them once again. However, this time the magnet has been turned 90° so that the magnetic field gradient is in the x direction. Note that now there is an inhomogeneity in the x direction, such that the atoms are separated along this direction. The operator is now "measure the x component of the magnetic moment," which is denoted $\hat{B}$. The experiment shows that this operator also has two and only two eigenfunctions that we call δ and γ. They have the same eigenvalues as α and β, respectively. If the relative number of Ag atoms deflected in the $+x$ and $-x$ directions is measured, the ratio is determined to be one. We conclude that the wave function prior to entering the second magnet was

$$\psi = c_3\delta + c_4\gamma \quad \text{with } |c_3|^2 + |c_4|^2 = 1 \tag{17.3}$$

As before, $|c_3|^2_{average} = |c_4|^2_{average} = 1/2$.

Now comes the punch line. We ask the question "Do the operators $\hat{A}$ and $\hat{B}$ commute?" This question is answered by repeating the first measurement to see if the state of the system has been changed by carrying out the second measurement. Experimentally, a third magnet that has the same alignment as the first magnet is added. This third magnet acts on one of the two separated beams that have emerged from the second magnet, as shown in Figure 17.3. If the operators commute, a single downwardly deflected beam of Ag atoms corresponding to $\psi = \alpha$ will be observed. If they do not commute, the wave function for the atoms entering the third magnet will no longer be an eigenfunction of $\hat{A}$, and two beams will be observed. Why is this? If the wave function that describes the Ag atom emerging from the second magnet is not an eigenfunction of $\hat{A}$, it still can be represented as a linear combination of the two eigenfunctions of $\hat{A}$. A state whose wave function is a linear combination of α and β will give rise to two deflected beams of Ag atoms.

The result of the experiment with the third magnet is that two beams emerge, just as was seen from the first magnet! *We conclude that the operators $\hat{A}$, "measure the z component of the magnetic moment," and $\hat{B}$, "measure the x component of the magnetic moment," do not commute.* This means that a silver atom does not simultaneously have well-defined values for both μ_z and μ_x. This is, of course, not the conclusion reached by applying classical mechanics to a classical magnetic moment. The experiment is a good illustration of how the quantum mechanical postulates arose from consideration of the outcomes of experiments.

Because the magnetic moment and the angular momentum of a charged particle differ only by a multiplicative constant, we have also shown that the operators for the individual components of the angular momentum vector do not commute. The consequences of this result will be discussed in Chapter 18.

17.2.1 The History of the Stern–Gerlach Experiment

This classic experiment, carried out in 1921, was designed to distinguish between the quantum mechanical model of the atom proposed by Niels Bohr and classical planetary models. A silver beam generated by an oven in a vacuum chamber was collimated by two narrow slits of 0.03-mm width. The beam passed through an inhomogeneous magnet 3.5 cm in length and impinged on a glass plate. After about an hour of operation, the plate was removed and examined visually. Only about one atomic layer of Ag was deposited on the plate in this time, making the detection of the spatial distribution of the silver atoms very difficult. The key to their successful detection was that both Stern and Gerlach smoked cheap cigars with a high sulfur content. The sulfur-containing smoke reacted with the Ag atoms, producing Ag_2S, which was clearly visible under a microscope, even though the amount deposited was less than 10^{-7} mol. Upon successful completion of the experiment, Gerlach sent Bohr the following postcard, which shows the result obtained without the magnetic field (left) and with the magnetic field (right). The splitting of the beam into two distinct components is clearly visible. The handwritten notes explain the experiment and congratulate Bohr, saying that the results confirm his theory.

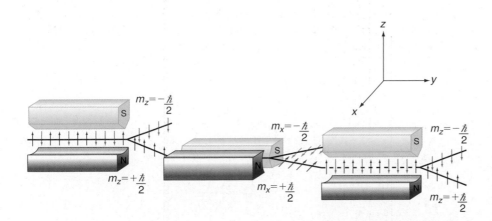

FIGURE 17.3

One of the beams exiting from the first magnet has been passed through a second magnet rotated by 90°. Again the beam is split into two components. The third magnet gives a result that is different than what would have been expected from classical physics.

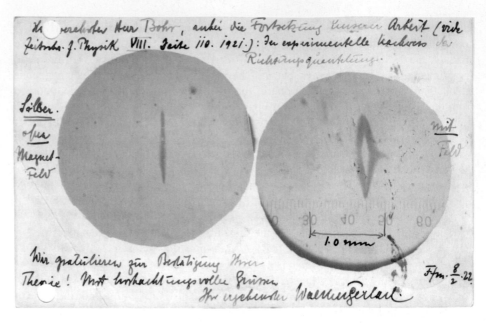

Courtesy of the Niels Bohr Archive, Copenhagen.

Although the results did not confirm the classical model of the atom, the agreement with the Bohr model turned out to be fortuitous and incorrect. Several years later, researchers discovered that the electron has an intrinsic angular moment (spin). This angular moment—and not a magnetic moment produced by electrons orbiting around the nucleus—is the basis for the deflection observed. A more detailed account of this experiment can be found in an article by B. Friedrich and D. Herschbach in the December 2003 issue of *Physics Today*.

17.3 The Heisenberg Uncertainty Principle

The best-known case of noncommuting operators concerns position and momentum and is associated with the **Heisenberg uncertainty principle.** This principle quantifies the uncertainty in the position and momentum of a quantum mechanical particle that arises from the fact that $[\hat{x}, \hat{p}_x] \neq 0$.

The uncertainty principle can be nicely illustrated with the free particle. As discussed in Section 15.1, the free-particle total energy eigenfunctions have the form $\Psi(x, t) = A \exp i(kx - \omega t - \phi)$. What can be said about the position and momentum of states described by this wave function? It is convenient to set $\phi = 0$ and $t = 0$ so that we can focus on the spatial variation of $\psi(x)$.

By operating on this wave function with the momentum operator, it can be easily shown that it is an eigenfunction of the momentum operator with the eigenvalue $\hbar k = hk/2\pi$. To discuss probability, this wave function must be normalized. As shown in Section 15.1, a plane wave cannot be normalized over an interval that is infinite, but it can be normalized over the finite interval $-L \leq x \leq L$:

$$\int_{-L}^{L} \Psi^*(x)\Psi(x)\, dx = 1$$

$$A^*A \int_{-L}^{L} e^{-ikx}e^{ikx}\, dx = 1 \qquad (17.4)$$

$$|A| = \frac{1}{\sqrt{2L}}$$

Now that the function is normalized, we calculate the probability of finding the particle near $x = x_0$:

$$P(x_0) \, dx = \psi^*(x_0) \psi(x_0) \, dx \qquad (17.5)$$

We see that the probability is $P(x) \, dx = dx/2L$ independent of position. This means that it is equally probable that the particle will be found anywhere. Now let the interval length L become arbitrarily large. The probability of finding the particle within the interval dx centered at $x = x_0$ approaches zero! *We conclude that if a particle is prepared in a state in which the momentum is exactly known, then its position is completely unknown.* It turns out that if a particle is prepared such that its position is exactly known (the wave function is an eigenfunction of the position operator), then its momentum is completely unknown.

This result is completely at variance with expectations based on classical mechanics, because a simultaneous knowledge of position and momentum is essential to calculating trajectories of particles subject to forces. How can this counterintuitive result be understood?

The uncertainty in position arises because the momentum is precisely known. Is it possible to construct a wave function for which the momentum is not precisely known? Will such a wave function give more information about the position of the particle than the plane wave $\Psi(x, t) = A \exp i(kx - \omega t - \phi)$ does? These questions can be answered by constructing a wave function that is a superposition of several plane waves and then examining its properties. Consider the superposition of plane waves of very similar wave vectors given by

$$\psi(x) = \frac{1}{2} A e^{ik_0 x} + \frac{1}{2} A \sum_{n=-m}^{m} e^{i(k_0 + n\Delta k)x}, \quad \text{with } \Delta k \ll k \qquad (17.6)$$

This superposition wave function is not an eigenfunction of the momentum operator. The upper portion of Figure 17.4 shows the real part of each of the 21 individual terms in an interval of approximately seven wavelengths about an arbitrarily chosen zero of distance for $m = 10$. We also choose to make the amplitude of the wave function zero outside of the range of distances shown. This ensures that the particle is somewhere in the interval.

How does the amplitude of $\psi(x)$ vary over the interval? At $x = 0$, all 21 waves constructively interfere, but at $x = \pm 3.14 \times 10^{-10}$ m they undergo destructive interference. Consequently, the wave function, which is a superposition of these waves, has a maximum amplitude at $x = 0$ and a value of zero at $x = \pm 3.14 \times 10^{-10}$ m. The amplitude oscillates about zero at intermediate values of x. How does the probability density vary over the interval? Evaluating Equation (17.6) for $m = 10$ and

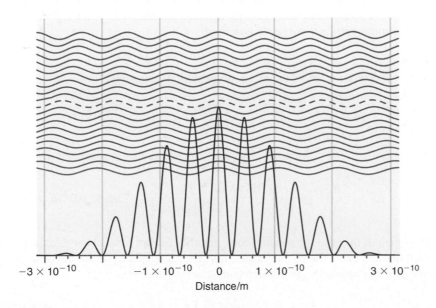

−3 × 10⁻¹⁰   −1 × 10⁻¹⁰   0   1 × 10⁻¹⁰   3 × 10⁻¹⁰

Distance/m

FIGURE 17.4

The top part of the figure shows 21 waves, each of which has zero amplitude outside the range of distances shown. They have been displaced vertically for purposes of display. The bottom part shows the probability density $\psi^*(x)\psi(x)$ resulting from adding all 21 waves. The wave vector k_0 has the value 7.00×10^{10} m⁻¹.

forming $|\psi(x)|^2$, the function shown in the lower part of Figure 17.4 is obtained. Because $|\psi(x)|^2$ is strongly peaked at the center of the interval, by superposing these 21 waves we see that the particle has been localized. The oscillations shown in Figure 17.4 are a result of having taken only 21 terms in the superposition. They would disappear, leaving a broad smooth curve that is the envelope of the red curve, if an infinite number of waves of intermediate wavelengths had been included in the superposition.

What does this calculation show? Because $\psi(x)$ is not an eigenfunction of the momentum operator, an uncertainty is connected with the momentum of the particle. In going from a single plane wave to the superposition function $\psi(x)$, the uncertainty in momentum has increased. As the curve for $|\psi(x)|^2$ in Figure 17.4 shows, increasing the uncertainty in momentum has decreased the uncertainty in position. Such a superposition wave function is referred to as a **wave packet** because it has wave character but is localized to a finite interval.

Because 21 waves of differing momentum have been superposed to construct the wave function, the momentum is no longer exactly known. Can we make this statement more quantitative? The value of p is known fairly well if $\Delta k \ll k_0$, because an individual measurement of the momentum for a state described by $\psi(x)$ gives values in the following range:

$$\hbar(k_0 - m\Delta k) \leq p \leq \hbar(k_0 + m\Delta k) \tag{17.7}$$

Comparing the results just obtained for $\psi(x)$ with those for a single plane wave of precisely determined momentum allows the following conclusion to be made: *as a result of the superposition of many plane waves, the position of the particle is no longer completely unknown, and the momentum of the particle is no longer exactly known.* Figure 17.4 shows that the approximate position of the particle can be known as long as an uncertainty in its momentum can be tolerated. The lesson of this discussion is that both position and momentum cannot be known *exactly and simultaneously* in quantum mechanics. We must accept a trade-off between the uncertainty in p and that of x. This result was quantified by Heisenberg in his famous uncertainty principle:

$$\Delta p \Delta x \geq \frac{\hbar}{2} \tag{17.8}$$

EXAMPLE PROBLEM 17.2

Assume that the double-slit experiment could be carried out with electrons using a slit spacing of $b = 10.0$ nm. To be able to observe diffraction, we choose $\lambda = b$, and because diffraction requires reasonably monochromatic radiation, we choose $\Delta p/p = 0.0100$. Show that with these parameters, the uncertainty in the position of the electron is greater than the slit spacing b.

Solution

Using the de Broglie relation, the mean momentum is given by

$$\langle p \rangle = \frac{h}{\lambda} = \frac{6.626 \times 10^{-34} \text{ J s}}{1.00 \times 10^{-8} \text{ m}} = 6.626 \times 10^{-26} \text{ kg m s}^{-1}$$

and $\Delta p = 6.626 \times 10^{-28}$ kg m s^{-1}. The minimum uncertainty in position is given by

$$\Delta x = \frac{\hbar}{2\Delta p} = \frac{1.055 \times 10^{-34} \text{ J s}}{2 \times 6.626 \times 10^{-28} \text{ kg m s}^{-1}} = 7.96 \times 10^{-8} \text{ m}$$

which is greater than the slit spacing. Note that the concept of an electron trajectory is not well defined under these conditions. This offers an explanation for the observation that the electron appears to go through both slits simultaneously!

The trajectory of a particle for which the momentum and energy are exactly known is not a well-defined concept in quantum mechanics. However, a good approximation for a "trajectory" in quantum mechanical systems is obtained by using wave packets.

What is the practical effect of the uncertainty principle? Does this mean that we have no idea what trajectories the electrons in a TV picture tube will follow or where a baseball thrown by a pitcher will pass a waiting batter? As mentioned earlier, this gets down to what is meant by *exact*. An exact trajectory could be calculated if $\hbar$ were equal to zero, rather than being a small number. Because $\hbar$ is a very small number, the uncertainty principle does not affect the calculation of the trajectories of baseballs, rockets, or other macroscopic objects. Although the uncertainty principle holds for both electrons and for baseballs, the effect is so small that it is not detectable for large masses.

EXAMPLE PROBLEM 17.3

The electrons in a TV picture tube have an energy of about 1.00×10^4 eV. If $\Delta p/p = 0.010$ in the direction of the electron trajectory for this case, calculate the minimum uncertainty in the position that defines where the electrons land on the phosphor in the picture tube.

Solution

Using the relation $\langle p \rangle = \sqrt{2mE}$, the momentum is calculated as follows:

$$\langle p \rangle = \sqrt{2 \times 9.109 \times 10^{-31} \text{ kg} \times 1.00 \times 10^4 \text{ eV} \times 1.602 \times 10^{-19} \text{ J/eV}}$$

$$= 5.41 \times 10^{-23} \text{ kg m s}^{-1}$$

Proceeding as in Example Problem 17.2,

$$\Delta x = \frac{\hbar}{2\Delta p} = \frac{1.055 \times 10^{-34} \text{ J s}}{2 \times 5.41 \times 10^{-25} \text{ kg m s}^{-1}} = 9.8 \times 10^{-11} \text{ m}$$

This distance is much smaller than could be measured and, therefore, the uncertainty principle has no effect in this instance.

EXAMPLE PROBLEM 17.4

An (over)educated baseball player tries to convince his manager that he cannot hit a 100 mile per hour (44.7 m s^{-1}) baseball that has a mass of 140. g and relative momentum uncertainty of 1.00% because the uncertainty principle does not allow him to estimate its position within 0.1 mm. Is his argument valid?

Solution

The momentum is calculated using the following equation:

$$p = mv = 0.140 \text{ kg} \times 44.7 \text{ m s}^{-1} = 6.26 \text{ kg m s}^{-1}, \text{ and } \Delta p = 0.0626 \text{ kg m s}^{-1}$$

Substituting in the uncertainty principle,

$$\Delta x = \frac{\hbar}{2\Delta p} = \frac{1.055 \times 10^{-34} \text{ J s}}{2 \times 0.0626 \text{ kg m s}^{-1}} = 8.43 \times 10^{-34} \text{ m}$$

The uncertainty is not zero, but it is well below the experimental sensitivity. Sorry, back to the minor leagues.

This result—that it is not possible to know the exact values of two observables simultaneously—is not restricted to position and momentum. It applies to any two observables whose corresponding operators do not commute. Energy and time are another example of two observables that are linked by an uncertainty principle. The energy of the H atom with the electron in the 1s state can only be known to high accuracy because it has a very long lifetime. This is the case because there is no lower state to which it can decay. Excited states that rapidly decay to the ground state have an uncertainty in their energy. Evaluation of the commutator is the means used to test whether any two observables can be determined simultaneously and exactly.

SUPPLEMENTAL

17.4 The Heisenberg Uncertainty Principle Expressed in Terms of Standard Deviations

This section addresses the topic of how to use the Heisenberg uncertainty principle in a quantitative fashion. This inequality can be written in the form

$$\sigma_x \sigma_p \geq \frac{\hbar}{2} \tag{17.9}$$

In this equation, σ_p and σ_x are the standard deviations that would be obtained by analyzing the distribution of a large number of measured values of position and momentum. The **standard deviations, σ_p and σ_x,** are related to observables by the relations

$$\sigma_p^2 = \langle p^2 \rangle - \langle p \rangle^2 \quad \text{and} \quad \sigma_x^2 = \langle x^2 \rangle - \langle x \rangle^2 \tag{17.10}$$

where σ_p^2 is called the variance in the momentum.

EXAMPLE PROBLEM 17.5

Starting with the definition for the standard deviation in position,

$\sigma_x = \sqrt{(1/N) \sum_{i=1}^{N} (x_i - \langle x \rangle)^2}$, derive the expression for σ_x^2 in Equation (17.10).

Solution

$$\sigma_x^2 = \frac{1}{N} \sum_{i=1}^{N} (x_i - \langle x \rangle)^2 = \frac{1}{N} \sum_{i=1}^{N} (x_i^2 - 2x_i\langle x \rangle + \langle x \rangle^2)$$

$$= \langle x^2 \rangle - 2\langle x \rangle\langle x \rangle + \langle x \rangle^2$$

$$= \langle x^2 \rangle - \langle x \rangle^2$$

The fourth postulate of quantum mechanics tells how to calculate these observables from the normalized wave functions:

$$\langle p^2 \rangle = \int \psi^*(x)\hat{p}^2 \psi(x)dx \quad \text{and}$$

$$\langle p \rangle^2 = \left(\int \psi^*(x)\hat{p}\psi(x)dx \right)^2$$

Similarly,

$$\langle x^2 \rangle = \int \psi^*(x)\hat{x}^2 \psi(x)dx \quad \text{and}$$

$$\langle x \rangle^2 = \left(\int \psi^*(x)\hat{x}\psi(x)dx \right)^2 \tag{17.11}$$

To illustrate how to use the Heisenberg uncertainty principle, we carry out a calculation for σ_p and σ_x using the particle in the box as an example. The normalized wave functions are given by $\psi_n(x) = \sqrt{2/a}\,\sin(n\pi x/a)$ and the operators needed are $\hat{p} = -i\hbar(\partial/\partial x)$ and $\hat{x} = x$.

Using the standard integrals

$$\int x \sin^2 bx \, dx = \frac{x^2}{4} - \frac{1}{4b}x \sin 2bx - \frac{1}{8b^2}\cos 2bx \quad \text{and}$$

$$\int x^2 \sin^2 bx \, dx = \frac{1}{6}x^3 - \left(\frac{1}{4b}x^2 - \frac{1}{8b^3}\right)\sin 2bx - \frac{1}{4b^2}x\cos 2bx$$

it is found that

$$\langle x \rangle = \int_0^a \sqrt{\frac{2}{a}}\sin\left(\frac{n\pi x}{a}\right) x \sqrt{\frac{2}{a}}\sin\left(\frac{n\pi x}{a}\right) dx = \frac{2}{a}\int_0^a x \sin^2\left(\frac{n\pi x}{a}\right) dx = \frac{1}{2}a$$

$$\langle x^2 \rangle = \int_0^a \sqrt{\frac{2}{a}}\sin\left(\frac{n\pi x}{a}\right) x^2 \sqrt{\frac{2}{a}}\sin\left(\frac{n\pi x}{a}\right) dx = \frac{2}{a}\int_0^a x^2 \sin^2\left(\frac{n\pi x}{a}\right) dx$$

$$= a^2\left(\frac{1}{3} - \frac{1}{2\pi^2 n^2}\right)$$

$$\langle p \rangle = \int_0^a \sqrt{\frac{2}{a}}\sin\left(\frac{n\pi x}{a}\right)\left(-i\hbar\frac{\partial}{\partial x}\sqrt{\frac{2}{a}}\sin\left(\frac{n\pi x}{a}\right)\right) dx$$

$$= -i\hbar\frac{2\pi n}{a^2}\int_0^a \sin\left(\frac{n\pi x}{a}\right)\cos\left(\frac{n\pi x}{a}\right) dx = 0$$

$$\langle p^2 \rangle = \int_0^a \sqrt{\frac{2}{a}}\sin\left(\frac{n\pi x}{a}\right)\left(-\hbar^2\frac{\partial^2}{\partial x^2}\sqrt{\frac{2}{a}}\sin\left(\frac{n\pi x}{a}\right)\right) dx$$

$$= \frac{2\pi^2 n^2 \hbar^2}{a^3}\int_0^a \sin^2\left(\frac{n\pi x}{a}\right) dx = \frac{n^2\pi^2\hbar^2}{a^2}$$

With these results, σ_p becomes

$$\sigma_p = \sqrt{\frac{n^2\pi^2\hbar^2}{a^2}} = \frac{n\pi\hbar}{a} \quad \text{and} \quad \sigma_x = a\sqrt{\left(\frac{1}{12} - \frac{1}{2\pi^2 n^2}\right)} \qquad \textbf{(17.12)}$$

Next, these results are verified as being compatible with the uncertainty principle for $n = 1$:

$$\sigma_p \sigma_x = \frac{n\pi\hbar}{a}\sqrt{a^2\left(\frac{1}{12} - \frac{1}{2\pi^2 n^2}\right)} = \hbar\sqrt{\left(\frac{\pi^2 n^2}{12} - \frac{1}{2}\right)}$$

$$= 0.57\hbar > \frac{\hbar}{2} \quad \text{for } n = 1 \qquad \textbf{(17.13)}$$

Because this function has its minimum value for $n = 1$, the uncertainty principle is satisfied for all values of n.

In evaluating a quantum mechanical result, it is useful to make sure that it converges to the classical result as $n \to \infty$. To do so, the relative uncertainties in x and p are evaluated. The quantity $\sqrt{\langle p^2 \rangle}$ rather than $\langle p \rangle$ is used for this calculation because $\langle p \rangle = 0$. The following result is obtained:

$$\frac{\sigma_x}{\langle x \rangle} = \frac{a\sqrt{\left(\frac{1}{12} - \frac{1}{2\pi^2 n^2}\right)}}{a/2} = \sqrt{\frac{1}{3} - \frac{2}{\pi^2 n^2}} \quad \text{and} \quad \frac{\sigma_p}{\sqrt{\langle p^2 \rangle}} = \frac{n\pi\hbar/a}{n\pi\hbar/a} = 1 \quad \textbf{(17.14)}$$

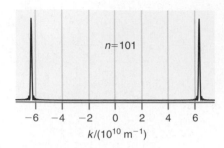

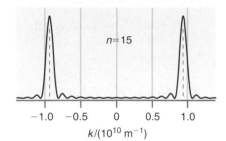

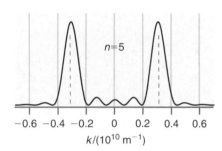

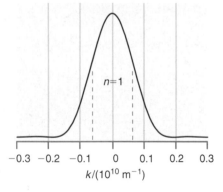

FIGURE 17.5

The relative probability density of observing a particular value of k, $A_k^* A_k$ (vertical axis), is graphed versus k for a 5.00-nm-long box for several values of n. The dashed lines for $n = 1$, 5, and 15 show the classically expected values $p = \pm \sqrt{2mE}$.

The interesting result is obtained that the relative uncertainty $\sigma_x / \langle x \rangle$ increases as $n \to \infty$. How can this result be understood? Looking back at the probability density in Figure 17.4, we see that the particle is most likely to be found near the center of the box for $n = 1$, whereas it is equally likely to be anywhere in the box for large n. The fact that the ground-state particle is more confined than the classical particle is at first surprising, but is consistent with the discussion in Chapter 15.

The result that the relative uncertainty in momentum is independent of the momentum is counterintuitive because in the classical limit, the uncertainty in the momentum is expected to be negligible. It turns out that the result for $\sigma_p / \sqrt{\langle p^2 \rangle}$ in Equation (17.14) is misleading because there are two values of p for a given value of p^2. The variance calculated earlier is characteristic of the set of the two p values, and what we want to know is $\sigma_p / \sqrt{\langle p^2 \rangle}$ for each value of p individually. How can the desired result be obtained?

The result is obtained by expanding the eigenfunctions $\psi_n(x)$ in the eigenfunctions of the momentum operator. In a fashion similar to that used to generate the data in Figure 17.4, we ask what values of k and what relative amplitudes A_k are required to represent the wave functions

$$\psi_n(x) = \sqrt{\frac{2}{a}} \sin\left(\frac{n\pi x}{a}\right), \quad \text{for } n = 1, 2, 3, 4, \ldots \quad a > x > 0 \text{ and}$$

$$\psi_n(x) = 0, \quad \text{for } 0 \geq x, \quad x \geq a \tag{17.15}$$

in the form

$$\psi_n(x) = \sum_{k=-\infty}^{\infty} A_k e^{ikx} \tag{17.16}$$

Expressing the eigenfunctions in this way allows the probability density of observing a particular value of p for a particle whose wave function is an eigenfunction of the total energy operator to be calculated. As outlined in the discussion of the fourth postulate in Chapter 14, the probability density of measuring a given momentum is proportional to $A_k^* A_k$. This quantity is shown as a function of k for several values of n in Figure 17.5, where, for $n = 101$, the result looks quite classical in that the observed values are sharply peaked at the two classically predicted values $p = \pm \sqrt{2mE}$. However, as n becomes smaller, quantum effects become much clearer. The most probable values of p are still given by $p = \pm \sqrt{2mE}$ for $n = 5$ and 15, but subsidiary maxima are seen, and the width of the peaks (which is a measure of the uncertainty in p) is substantial. For $n = 1$, the distribution is peaked at $p = 0$, rather than the classical values. For this lowest energy state, quantum and classical mechanics give very different results.

Figure 17.5 demonstrates that the relative uncertainty $\sigma_p / \sqrt{\langle p^2 \rangle}$ decreases as p increases. You will explore this issue more quantitatively in the end-of-chapter problems. The counterintuitive result of Equation (17.14)—that the relative uncertainty in the momentum is constant—is an artifact of characterizing the distribution consisting of two widely separated peaks by one variance, rather than looking at each of the peaks individually.

SUPPLEMENTAL

17.5 A Thought Experiment Using a Particle in a Three-Dimensional Box

Think of the following experiment: one particle is put in an opaque box, and the top is securely fastened. From the outside, a partition is slid into the box, dividing it into two equal leak-tight volumes. This partition allows the initial box to be separated into two separate leak-tight boxes, each with half the volume. These two boxes are separated by sending one of them to the moon. Finally, an observer opens one of the boxes. The observer finds that the box he has opened is either empty or that it contains the particle. From the viewpoint of classical mechanics, this is a straightforward experiment. If the box that was opened is empty, then that half of the box was empty when the partition was initially inserted. What does this problem look like from a quantum mechanical point of view? The individual steps are illustrated in Figure 17.6.

Initially, we know only that the particle is somewhere in the box before the partition is inserted. Because it exhibits wave-particle duality, the position of the particle cannot be determined exactly. If two eigenstates of the position operator, ψ_{left} and ψ_{right}, are defined, then the initial wave function is given by

$$\psi = a\psi_{left} + b\psi_{right}, \quad \text{with } |a|^2 + |b|^2 = 1 \qquad (17.17)$$

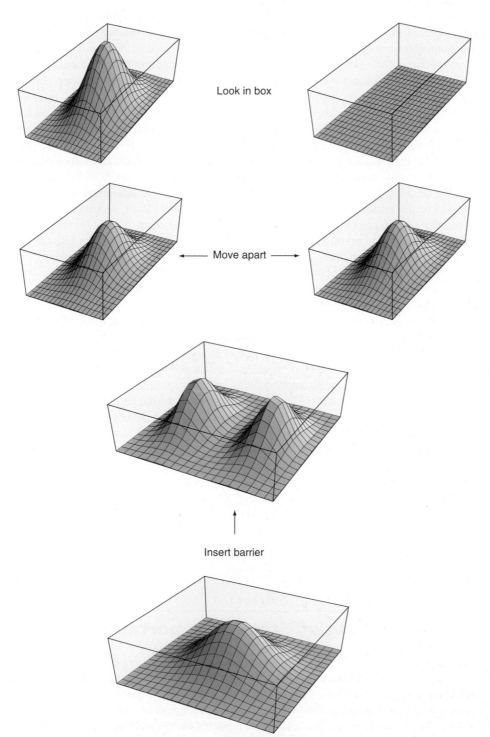

Look in box

← Move apart →

Insert barrier

FIGURE 17.6
Thought experiment using a particle in a box. The square of the wave function is plotted along the x and y coordinates of the box.

In the figure, it has been assumed that $a = b$. The square of the wave function is nonzero everywhere in the box and goes to zero at the walls. When the partition is inserted, what we have just said is again true, except that now the wave function also goes to zero along the partition. Classically, the particle is either in the left- or the right-hand side of the combined box, although it is not known which of these possibilities applies.

From a quantum mechanical perspective, such a definitive statement cannot be made. We can merely say that there is an equal probability of finding the particle in each of the two parts of the original box. Therefore, when the two halves of the box are separated, the integral of the square of the wave function is one-half in each of the smaller boxes.

Now the box is opened. This is equivalent to applying the position operator to the wave function of Equation (17.17). According to the discussion in Chapter 14, the wave function becomes either ψ_{left} or ψ_{right}. We do not know which of these will be the final wave function of the system, but we do know that in a large number of measurements, the probability of finding it on the left is a^2. Assume the case shown in the top part of Figure 17.4 in which the particle is found in the left box. *In that case, the integral of the square of the wave function in that box instantaneously changes from 0.5 to 1.0 at the moment we look into the box, and the integral of the square of the wave function in the other box drops from 0.5 to zero!* Because this result does not depend on the distance of separation between the boxes, this distance can be made large enough that the boxes are not coupled by a physical force. Even so, the one box "knows" instantaneously what has been learned about the other box. This is the interpretation of quantum mechanics attributed to the Copenhagen school of Niels Bohr, which gives the act of measurement a central role in the outcome of an experiment. Nearly 80 years after the formulation of quantum theory, the search for an "observer-free" theory has not yet led to a widely accepted alternative to the interpretation of the Copenhagen school.

Before dismissing this scenario as unrealistic, and accepting the classical view that the particle really is in one part of the box *or* the other, have another look at Figure 14.3. The results shown there demonstrate clearly that the outcome of an experiment on *identically prepared* quantum mechanical systems is inherently probabilistic. Therefore, the wave function for an individual system must be formulated in such a way that it includes all possible outcomes of an experiment. This means that, in general, it describes a superposition state. The result that measurements on identically prepared systems lead to different outcomes has been amply documented by experiments at the atomic level, and this precludes the certainty in the classical assertion that the particle really is in one part of the box *or* the other. Where does the classical limit appear in this case? For instance, one might ask why the motion of a human being is not described by the Schrödinger equation rather than Newton's second law if every atom in our body is described by quantum mechanics. This topic is an active area of research, and the current view is that the superposition wave function of a macroscopic system is unstable because of interactions with the environment. The superposition state decays very rapidly to a single term. This decay has the consequence that the strange behavior characteristic of quantum mechanical superposition states is no longer observed in large "classical systems."

SUPPLEMENTAL

17.6 Entangled States, Teleportation, and Quantum Computers

Erwin Schrödinger first noted a prediction of quantum mechanics that was very much at variance with classical physics. It is that two quantum particles can be coupled in such a way that their properties are no longer independent of one another no matter how far apart they may be. We say that the particles are **entangled.** This consequence of entanglement was pointed out by Einstein and called a "spooky action at a distance" to indicate what he believed to be a serious flaw in quantum mechanics. Definitive experiments to determine whether entanglement could be observed were not possible until the 1970s, when it was shown that Einstein was wrong in this instance.

Consider the following example of entanglement. A particle with no magnetic moment decays, giving two identical particles whose z component of the magnetic moment (which we call m_z) can take on the values $\pm 1/2$. Each of these particles is sent through a Stern–Gerlach analyzer as described in Section 17.2. A series of measurements of m_z for particle one gives $\pm 1/2$ in a random pattern; it is not possible to predict the outcome of a single measurement. However, because angular momentum is conserved, if one of the particles is found to have m_z equal to $+1/2(\uparrow)$, the other must have $m_z = -1/2(\downarrow)$. There are only two possibilities for the two particles, $\uparrow\downarrow$ or $\downarrow\uparrow$ where the left arrow in each case indicates m_z for particle one. Because the combinations $\uparrow\downarrow$ or $\downarrow\uparrow$ occur with equal probability, the two particles must be described by a single superposition wave function, which we write schematically as $\uparrow\downarrow + \downarrow\uparrow$. Note that neither particle can be described by its own wave function as a result of the entanglement.

This result implies that the second particle has no well-defined value of m_z until a measurement is carried out on the first particle. Because the roles of particles one and two can be reversed, quantum mechanics tells us that neither of the particles has a well-defined value of m_z until a measurement is carried out. This result violates a basic principle of classical physics called local realism. **Local realism** asserts that (1) Measured results correspond to elements of reality. For example, if I determine that a person's hair is black, according to local realism, that person's hair was black before I make the measurement and is black regardless of whether a measurement is ever made. (2) Measured results are independent of any action that might be taken at a distant location at the same time. If a person has an identical twin on the other side of the planet, a measurement of the twin's hair color has no influence on a measurement of the other twin's hair color made at the same time.

The experiment just described shows that local realism is not valid because there is no value for m_z until a measurement is carried out and because the m_z values of the two particles remain coupled no matter how far apart they are when the first measurement is made. Another experiment that illustrates this surprising result is depicted in Figure 17.7. Two entangled photons are passed through optical fibers to locations spaced 10 km apart. Photon 1 is passed through a double slit and exhibits a diffraction pattern. If the profile of the light intensity corresponding to photon 2 is determined, it corresponds to that of a photon that has passed through a double slit, even though it has not! If a person and his or her identical twin were quantum mechanically entangled, neither twin's hair color would be known before a measurement was made. Any possible hair color would be equally likely to be determined for one twin in a measurement, and the other twin would be found to have the same hair color.

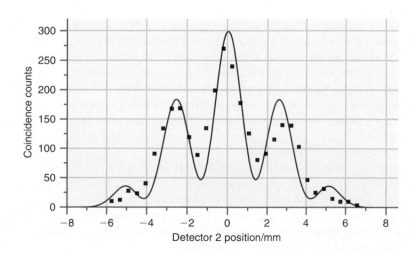

FIGURE 17.7

The spatial distribution of the light intensity for photon 2 shows a diffraction pattern (black squares) even though it has not passed through a slit. This result arises because photons 1 and 2 are entangled. The red curve is the diffraction pattern calculated using the experimental parameters.

Source: Reprinted fig. 2 with permission from D.V. Strekalov *et al.*, "Observation of Two-Photon 'Ghost' Interference and Diffraction," *Physical Review Letters*, 74 (18): 3600–3603 (May 1995). Copyright 1995 by the American Physical Society. http://link.aps.org/doi/10.1103/PhysRevLett. 74.3600

Does entanglement suggest that information can be transmitted instantaneously over an arbitrarily large distance? To answer this question, we consider how information about a system can be transmitted to a distant location, first for a classical system and then for a quantum mechanical system. Classically, a copy of the original information or object is created at the distant location. A classical system can be copied as often as desired, and the accuracy of the copy is limited only by the quality of the tools used. In principle, the copies can be so well made that they are indistinguishable from the original. The speed with which information is transferred is limited by the speed of light. By contrast, the information needed to make a copy of a quantum mechanical system cannot be obtained, because it is impossible to determine the state of the system exactly by measurement. If the system wave function is given by

$$\psi = \sum_m b_m \phi_m \tag{17.18}$$

in which the ϕ_m are the eigenfunctions of an appropriate quantum mechanical operator, experiments can only determine the absolute magnitudes $|b_m|^2$. This is not enough information to determine the wave function. Therefore, the information needed to make a copy is not available. Making a copy of a quantum mechanical system is also in violation of the Heisenberg uncertainty principle. If a copy could be made, one could easily measure the momentum of one of the copies and measure the position of the other copy. If this were possible, both the momentum and position could be known simultaneously.

Given these limitations of knowledge of quantum mechanical systems, how can a quantum mechanical system be transported to a distant location, and how is this transfer related to entanglement? Consider the following experiment described by Anton Zeilinger in *Scientific American*, April 2000, in which a photon at one location was recreated at a second location. Although photons were used in this experiment, there is no reason in principle why atoms or molecules could not be transferred from one location to another in the same way.

Bob and Alice are at distant locations and share an entangled photon pair, of which Bob has photon B and Alice has photon A as shown in Figure 17.8. Each of them carefully stores his or her photon so that the entanglement is maintained. At a later time, Alice has another photon that we call X, which she would like to send to Bob. How can this be done? She cannot measure the polarization state directly and send this information to Bob, because the act of measurement would change the state of the photon. Instead, she entangles X and A.

What are the consequences of the entanglement of A and X on B? We know that whatever state X has, A must have the orthogonal state. If X is vertically (horizontally) polarized, then A must be horizontally (vertically) polarized. However, the same logic must apply to A and B because they are also entangled. Whatever state A has, B must have the orthogonal state. If the state of B is orthogonal to that of A and the state of A is orthogonal to that of X, *then the states of B and X must be identical.* This follows from the fact that there are only two possible eigenfunctions of the polarization operator.

What has been accomplished by this experiment? Photon B acquires the original polarization of Alice's photon X and is therefore identical in every way to the original state of X. However, the state of X has been irreversibly changed at Alice's location, because in order to know that photons A and X have been successfully entangled, Alice has to pass both her photons through A detector. Therefore, the properties of X have been changed at Alice's location and transferred to Bob's location. This process is called **teleportation,** defined as the transfer of A quantum state from one location to another. Note that the uncertainty principle has not been violated because the photon has been teleported rather than copied.

Maintaining the entanglement of pairs A and B and A and X is the crucial ingredient of teleportation. Neither Bob nor Alice knows the state of X at the start or the end of the experiment. This is the case because neither of them has measured the state of the photon directly. Had they done so, the state of the photon would have been irreversibly changed. It is only because they did not determine the state of the photon that the recreation of photon X at Bob's location was possible.

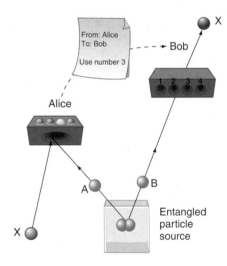

FIGURE 17.8

Teleportation of photon X from Alice to Bob. Note the classical communication channel that Alice uses to communicate the outcome of her measurement on A and X to Bob.

If the preceding outcome were the only possible outcome of Alice's entanglement of A and X, the transmission of information from Alice to Bob would be instantaneous, regardless of the distance between them. Therefore, it would be faster than the speed of light. Unfortunately, it turns out that Alice's entanglement of A and X has four possible outcomes, which we will not discuss other than to say that each is equally probable in the entanglement of an individual photon pair. Although there is no way to predict which of the four outcomes will occur, Alice has detectors that will tell her *after the fact* which outcome occurred.

In each of these outcomes, the entanglement of A and X is transferred to B, but in three of the four, Bob must carry out an operation on B, such as to rotate its polarization by a fixed angle, in order to make B identical to X. How does this affect what Bob knows about B? Without knowing which of the four outcomes Alice detected, Bob does not know how his photon has been transformed. Only if Alice sends him the result of her measurement does Bob know what he must do to B to make it indistinguishable from X. It is the need for this additional information that limits the speed of quantum information transfer through teleportation. Although the state of Bob's photon B is instantaneously transformed as Alice entangles A and X, he cannot interpret his results without additional information from her. Because Alice's information must be sent to Bob using conventional methods such as phone, fax, or e-mail, the overall process of teleportation is limited by the speed of light. Although the state of entangled particles changes instantaneously, information transmission utilizing entanglement cannot proceed faster than the speed of light.

In principle, the same technique could be used to teleport an atom, a molecule, or even an organism. The primary requirement is that it must be possible to create entangled pairs of the object to be teleported. The initial experiment was carried out with photons because experimental methods to entangle photons are available. As discussed earlier, entangled states are fragile and can decay to a single eigenfunction of the operator rapidly through interactions with the environment. This is especially true of systems containing a large number of atoms. However, it is possible to entangle atoms, and it seems within reach to entangle small molecules.

Entanglement has a further interesting application. It provides the basis for the **quantum computer,** which currently exists only as a concept. Such a computer would be far more powerful than the largest supercomputers currently available. How does a quantum computer differ from a classical computer? In a classical computer, information is stored in bits. A **bit** generally takes the form of a macroscopic object like a wire or a memory element that can be described in terms of a property such as a voltage. For example, two different ranges of voltage are used to represent the numbers 0 and 1. Within this binary system, an n bit memory can have 2^n possible states that range between 00000...00 and 11111...11. A three-bit memory can have the eight states 100, 010, 001, 110, 101, 011, 111, and 000. Information such as text and images can be stored in the form of such states. Mathematical or logical operations can be represented as transformations between such states. Logic gates operate on binary strings to carry out mathematical operations. Software provides an instruction set to route the data through the logic gates that are the heart of the computer hardware. This is the basis on which classical computers operate.

The quantum analog of the bit, in which two numbers characterize the entity, is the **qubit,** which has the property that it is *simultaneously* a linear combination of 0 and 1, rather than being either 1 or 0. The advantage of a qubit over a bit can be illustrated with the following example. A conventional three-bit array can only have one of eight possible states at a given time. By contrast, qubits can be entangled with one another so that a 3-qubit entangled array is in a superposition of all eight possible binary strings of length 3 at a given time. More generally, an M-qubit entangled array is in a superposition of all 2^M possible binary strings of length M. If this input signal can be processed using quantum gates without destroying the entanglement, 2^M simultaneous calculations ($\sim 10^{30}$ for $M = 100$) could be done in parallel by an M-qubit quantum computer. One of the most interesting applications of quantum computing is data encryption. Shor's algorithm, which allows the rapid factorization of very large numbers, would allow modestly sized quantum computers to outperform the largest classical supercomputers in the area of data encryption.

FIGURE 17.9
A combination of beam splitters and mirrors with a phase shifter is used to illustrate how a qubit can be generated. Partial reflection occurs because a semi-transparent silver layer is evaporated onto a glass substrate. The reflecting layer is on the top of the left beam splitter and on the bottom of the right beam splitter.

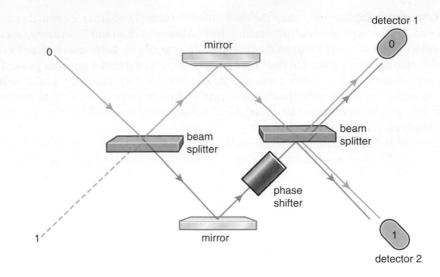

The generation of a qubit using photons is explained using Figure 17.9. A single photon is incident from the top left on a beam splitter, which has a probability of 0.5 for reflection and for transmission of a photon. We assign a photon with this direction the value 0. Just as for a particle incident on a double slit, each photon follows *both* pathways of reflection and transmission and reflection, rather than being either reflected or transmitted as was discussed in Section 14.6. The two mirrors are used to combine the two pathways at a second beam splitter where each incident photon again follows both transmission and reflection pathways. On the final part of the path to the detectors, the initial reflection and transmission pathways are combined and interference of the two beams occurs. To understand what the detectors register, three simple rules must be followed: (1) If a photon is reflected at an interface for which the refractive index behind the mirror is larger than in front of the mirror, a phase shift of π (half a wavelength) occurs. (2) If a photon is reflected at an interface for which the refractive index behind the mirror is smaller than in front of the mirror, no phase shift occurs. (3) In passing through the higher-index glass making up the beam splitters, a phase shift ϕ occurs that is proportional to the path length.

We first remove the phase shifter. Along the orange pathway leading to detector 1, phase shifts of π, π, and 2ϕ occur for a total of $2\pi + 2\phi$. Along the green pathway also leading to detector 1, phase shifts of ϕ, π, and ϕ occur for a total of $\pi + 2\phi$. Therefore, the two pathways are out of phase by π, meaning that destructive interference occurs, and no signal is registered at detector 1. We next carry out the same analysis for detector 2. Along the orange pathway leading to detector 2, phase shifts of π, π, and ϕ occur for a total of $2\pi + \phi$. Along the green pathway also leading to detector 1, phase shifts of ϕ, π, and π occur for a total of $2\pi + \phi$. There is no phase difference between the two pathways. Therefore, constructive interference occurs, and a signal is registered at detector 1. Had we considered the incident photon indicated by the dashed path given the value 1, we would have found that a signal is registered at detector 1 but not at detector 2. This device forms a NOT gate, because it makes the transformations $0 \rightarrow 1$ and $1 \rightarrow 0$.

If a phase shifter in the form of a piece of glass of variable thickness is inserted into one leg of the interferometer, the relative phase of the two pathways can be changed to any value from 0 to π. This means that the incoming signals 0 and 1 are transformed into a superposition of 0 and 1. In this way, a bit has been transformed into a qubit.

Three major hurdles must be overcome to construct a quantum computer: the entanglement of real qubits, the maintenance of entanglement over a long enough time to allow calculations to be carried out, and the extraction of the desired result from the superposition of all possible outcomes. The most critical task in realizing quantum computers is the requirement that a quantum computer's internal operation must be isolated from the rest of the universe. Small amounts of information leakage from the computer can destroy the fragile entangled states on which the quantum computer depends. Ions trapped in electromagnetic fields, arrays of trapped neutral atoms, and nuclear spins on different atoms in a molecule have been successfully entangled and are useful models for quantum computers. However, the ultimate goal is a solid-state device that is compatible with current microelectronic technology. Some progress in this direction has been achieved by using quantum dots or trapped dopant atoms embedded in semiconductors. For a recent review of the various physical systems that have been investigated to generate qubits and to entangle them, see Ladd, T.D., Jelezko, F., Laflamme, R., Nakamura, Y., Monroe, C., and O'Brien, J.L. "Quantum Computers." *Nature*, 464 (2010): 45.

Vocabulary

bit	Heisenberg uncertainty principle	standard deviation
commutator	local realism	Stern–Gerlach experiment
commute	quantum computer	teleportation
entangled	qubit	wave packet

Conceptual Problems

Q17.1 How did Stern and Gerlach conclude that the operator "measure the z component of the magnetic moment of an Ag atom" has only two eigenfunctions with eigenvalues that have the same magnitude and opposite sign?

Q17.2 Have a closer look at Equation (17.6) and Figure 17.4. How would Figure 17.4 change if Δk decreases for constant m? How well is the momentum known if $\Delta k \rightarrow 0$?

Q17.3 Why is maintaining the entanglement of pairs A and B and A and X the crucial ingredient of teleportation?

Q17.4 Why is it not possible to reconstruct the wave function of a quantum mechanical superposition state from experiments?

Q17.5 Why does the relative uncertainty in x for the particle in the box increase as $n \rightarrow \infty$?

Q17.6 Why is the statistical concept of variance a good measure of uncertainty in a quantum mechanical measurement?

Q17.7 Derive a relationship between $[\hat{A}, \hat{B}]$ and $[\hat{B}, \hat{A}]$.

Q17.8 How does a study of the eigenfunctions for the particle in the box lead to the conclusion that the position uncertainty has its minimum value for $n = 1$?

Q17.9 What is the difference between a bit and a qubit?

Q17.10 Why does it follow from the Heisenberg uncertainty principle that it is not possible to make exact copies of quantum mechanical objects?

Q17.11 Which result of the Stern–Gerlach experiment leads to the conclusion that the operators for the z and x components of the magnetic moment do not commute?

Q17.12 The Heisenberg uncertainty principle says that the momentum and position of a particle cannot be known simultaneously and exactly. Can that information be obtained by measuring the momentum and quickly following up with a measurement of the position?

Q17.13 Why isn't the motion of a human being described by the Schrödinger equation rather than Newton's second law if every atom in our body is described by quantum mechanics?

Q17.14 Explain the following statement: if $\hbar = 0$, it would be possible to measure the position and momentum of a particle exactly and simultaneously.

Q17.15 Why is $\sqrt{\langle p^2 \rangle}$ rather than $\langle p \rangle$ used to calculate the relative uncertainty for the particle in the box?

Q17.16 How would the results of the Stern and Gerlach be different if they had used a Mg beam instead of Ag beam?

Q17.17 How would the results of the Stern and Gerlach be different if they had used a homogeneous magnetic field instead of an inhomogeneous field?

Q17.18 Discuss whether the results shown in Figure 17.7 are consistent with local realism.

Q17.19 An electron and a He atom have the same uncertainty in their speed. What can you say about the relative uncertainty in position for the two particles?

Q17.20 Describe the trends in Figure 17.5 that you expect to see as the quantum number n increases.

Numerical Problems

Problem numbers in **red** indicate that the solution to the problem is given in the *Student's Solutions Manual*.

P17.1 In this problem, we consider the calculations for σ_p and σ_x for the particle in the box shown in Figure 17.5 in more detail. In particular, we want to determine how the absolute uncertainty Δp_x and the relative uncertainty $\Delta p_x/p_x$ of a single peak corresponding to either the most probable positive or negative momentum depends on the quantum number n.

a. First we must relate k and p_x. From $E = p_x^2/2m$ and $E = n^2h^2/8ma^2$, show that $p_x = nh/2a$.

b. Use the result from part (a) together with the relation linking the length of the box and the allowed wavelengths to obtain $p_x = \hbar k$.

c. Relate Δp_x and $\Delta p_x/p_x$ with k and Δk.

d. The following graph shows $|A_k|^2$ versus $k-k_{peak}$. By plotting the results of Figure 17.5 in this way, all peaks appear at the same value of the abscissa. Successive curves have been shifted upward to avoid overlap. Use the width of the $|A_k|^2$ peak at half height as a measure of Δk. What can you conclude from this graph about the dependence of Δp_x on n?

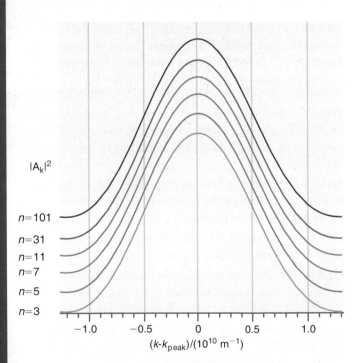

e. The following graph shows $|A_k|^2$ versus k/n for $n = 3$, $n = 5$, $n = 7$, $n = 11$, $n = 31$, and $n = 101$. Use the width of the $|A_k|^2$ peak at half height as a measure of $\Delta k/n$. Using the graphs, determine the dependence of $\Delta p_x/p_x$ on n. One way to do this is to assume that the width depends on n like $(\Delta p_x/p_x) = n^\alpha$ where α is a constant to be determined. If this relationship holds, a plot of $\ln(\Delta p_x/p_x)$ versus $\ln n$ will be linear and the slope will give the constant α.

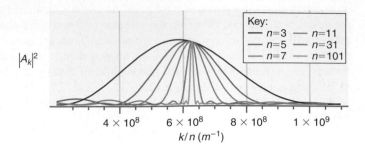

P17.2 Consider the results of Figure 17.5 more quantitatively. Describe the values of x and k by $x \pm \Delta x$ and $k_0 \pm \Delta k$. Evaluate Δx from the zero of distance to the point at which the envelope of $\psi^*(x)\psi(x)$ is reduced to one-half of its peak value. Evaluate Δk from $\Delta k = |1/2(k_0 - k_{min})|$ where k_0 is the average wavevector of the set of 21 waves (11th of 21) and k_{min} corresponds to the 21st of the 21 waves. Is your estimated value of $\Delta p \, \Delta x = \hbar \, \Delta k \, \Delta x$ in reasonable agreement with the Heisenberg uncertainty principle?

P17.3 Evaluate the commutator $[d/dy, 1/y^2]$ by applying the operators to an arbitrary function $f(y)$.

P17.4 Show

a. that $\psi(x) = e^{-x^2/2}$ is an eigenfunction of $\hat{A} = x^2 - \partial^2/\partial x^2$; and

b. that $\hat{B}\psi(x)$ (where $\hat{B} = x - \partial/\partial x$) is another eigenfunction of $\hat{A}$.

P17.5 Another important uncertainty principle is encountered in time-dependent systems. It relates the lifetime of a state Δt with the measured spread in the photon energy ΔE associated with the decay of this state to a stationary state of the system. "Derive" the relation $\Delta E \, \Delta t \geq \hbar/2$ in the following steps.

a. Starting from $E = p_x^2/2m$ and $\Delta E = (dE/dp_x)\Delta p_x$, show that $\Delta E = v_x \Delta p_x$.

b. Using $v_x = \Delta x/\Delta t$, show that $\Delta E \, \Delta t = \Delta p_x \Delta x \geq \hbar/2$.

c. Estimate the width of a spectral line originating from the decay of a state of lifetime 1.0×10^{-9} s and 1.0×10^{-11} s in inverse seconds and inverse centimeters.

P17.6 Evaluate the commutator $[x(\partial/\partial y), y(\partial/\partial x)]$ by applying the operators to an arbitrary function $f(x,y)$.

P17.7 Evaluate $[\hat{A}, \hat{B}]$ if $\hat{A} = x + d/dx$ and $\hat{B} = x - d/dx$.

P17.8 Consider the entangled wave function for two photons,

$$\psi_{12} = \frac{1}{\sqrt{2}}(\psi_1(H)\psi_2(V) + \psi_1(V)\psi_2(H))$$

Assume that the polarization operator $\hat{P}_i$ has the properties $\hat{P}_i\psi_i(H) = -\psi_i(H)$ and $\hat{P}_i\psi_i(V) = +\psi_i(V)$ where $i = 1$ or $i = 2$. H and V designate horizontal and vertical polarization, respectively.

a. Show that ψ_{12} is not an eigenfunction of $\hat{P}_1$ or $\hat{P}_2$.

b. Show that each of the two terms in ψ_{12} is an eigenfunction of the polarization operator $\hat{P}_1$.

c. What is the average value of the polarization P_1 that you will measure on identically prepared systems?

P17.9 Evaluate the commutator $[\hat{p}_x + \hat{p}_x^2, \hat{p}_x^2]$ by applying the operators to an arbitrary function $f(x)$.

P17.10 Revisit the double-slit experiment of Example Problem 17.2. Using the same geometry and relative uncertainty in the momentum, what electron momentum would give a position uncertainty of 2.50×10^{-10} m? What is the ratio of the wavelength and the slit spacing for this momentum? Would you expect a pronounced diffraction effect for this wavelength?

P17.11 Evaluate the commutator $[y^2, d^2/dy^2]$ by applying the operators to an arbitrary function $f(y)$.

P17.12 Revisit the TV picture tube of Example Problem 17.3. Keeping all other parameters the same, what electron energy would result in a position uncertainty of 1.00×10^{-8} m along the direction of motion?

P17.13 Evaluate the commutator $[(d^2/dy^2)y]$ by applying the operators to an arbitrary function $f(y)$.

P17.14 If the wave function describing a system is not an eigenfunction of the operator $\hat{B}$, measurements on identically prepared systems will give different results. The variance of this set of results is defined in error analysis as $\sigma_B^2 = \langle (B - \langle B \rangle)^2 \rangle$, where B is the value of the observable in a single measurement and $\langle B \rangle$ is the average of all measurements. Using the definition of the average value from the quantum mechanical postulates, $\langle A \rangle = \int \psi^*(x)\hat{A}\psi(x)dx$, show that $\sigma_B^2 = \langle B^2 \rangle - \langle B \rangle^2$.

P17.15 Apply the Heisenberg uncertainty principle to estimate the zero point energy for the particle in the box.

a. First, justify the assumption that $\Delta x \le a$ and that, as a result, $\Delta p \ge \hbar/2a$. Justify the statement that, if $\Delta p \ge 0$, we cannot know that $E = p^2/2m$ is identically zero.

b. Make this application more quantitative. Assume that $\Delta x = 0.35a$ and $\Delta p = 0.35p$ where p is the momentum in the lowest energy state. Calculate the total energy of this state based on these assumptions and compare your result with the ground-state energy for the particle in the box.

c. Compare your estimates for Δp and Δx with the more rigorously derived uncertainties σ_p and σ_x of Equation (17.13).

P17.16 Evaluate the commutator $[d/dx, x^2]$ by applying the operators to an arbitrary function $f(x)$.

P17.17 Evaluate the commutator $[\hat{x}, \hat{p}_x]$ by applying the operators to an arbitrary function $f(x)$. What value does the commutator $[\hat{p}_x, \hat{x}]$ have?

P17.18 In this problem, you will carry out the calculations that describe the Stern–Gerlach experiment shown in Figure 17.2. Classically, a magnetic dipole $\boldsymbol{\mu}$ has the potential energy $E = -\boldsymbol{\mu} \cdot \mathbf{B}$. If the field has a gradient in the z direction, the magnetic moment will experience a force, leading it to be deflected in the z direction. Because classically $\boldsymbol{\mu}$ can take on any value in the range $-|\boldsymbol{\mu}| \le \mu_z \le |\boldsymbol{\mu}|$, a continuous range of positive and negative z deflections of a beam along the y direction will be observed. From a quantum mechanical perspective, the forces are the same as in the classical picture, but μ_z can only take on a discrete set of values. Therefore, the incident beam will be split into a discrete set of beams that have different deflections in the z direction.

a. The geometry of the experiment is shown here. In the region of the magnet indicated by d_1, the Ag atom experiences a constant force. It continues its motion in the force-free region indicated by d_2.

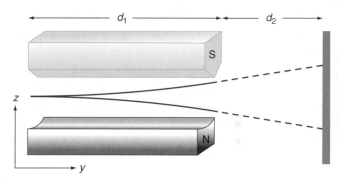

If the force inside the magnet is F_z, show that $|z| = 1/2(F_z/m_{Ag})t_1^2 + t_2 v_z(t_1)$. The flight times t_1 and t_2 correspond to the regions d_1 and d_2.

b. Show that assuming a small deflection,

$$|z| = F_z \left(\frac{d_1 d_2 + \tfrac{1}{2}d_1^2}{m_{Ag} v_y^2} \right)$$

c. The magnetic moment of the electron is given by $|\boldsymbol{\mu}| = g_s \mu_B/2$. In this equation, μ_B is the Bohr magneton and has the value 9.274×10^{-24} J/T. The gyromagnetic ratio of the electron g_s has the value 2.00231. If $\partial B_z/\partial z = 750.$ T m^{-1}, and d_1 and d_2 are 0.175 and 0.225 m, respectively, and $v_y = 475$ m s^{-1}, what values of z will be observed?

P17.19 Evaluate the commutator $[(d^2/dy^2) - y, (d^2/dy^2) + y]$ by applying the operators to an arbitrary function $f(y)$.

P17.20 Evaluate the commutator $[\hat{x}, \hat{p}_x^2]$ by applying the operators to an arbitrary function $f(x)$.

P17.21 What is wrong with the following argument? We know that the functions $\psi_n(x) = \sqrt{2/a} \sin(n\pi x/a)$ are eigenfunctions of the total energy operator for the particle in the infinitely deep box. We also know that in the box, $E = p_x^2/2m + V(x) = p_x^2/2m$. Therefore, the operator for E_{total} is proportional to the operator for p_x^2. Because the operators for p_x^2 and p_x commute as you demonstrated in Problem P17.8, the functions $\psi_n(x) = \sqrt{2/a} \sin(n\pi x/a)$

are eigenfunctions of both the total energy and momentum operators.

P17.22   For linear operators $\hat{A}$, $\hat{B}$, and $\hat{C}$, show that $[\hat{A}, \hat{B}\,\hat{C}] = [\hat{A}, \hat{B}]\,\hat{C} + \hat{B}\,[\hat{A}, \hat{C}]$.

P17.23   The muzzle velocity of a rifle bullet is 890. m s^{-1} along the direction of motion. If the bullet weighs 35 g, and the uncertainty in its momentum is 0.20%, how accurately can the position of the bullet be measured along the direction of motion?

Web-Based Simulations, Animations, and Problems

W17.1   The simulation of particle diffraction from a single slit is used to illustrate the dependence between the uncertainty in the position and momentum. The slit width and particle velocity are varied using sliders.

W17.2   The Heisenberg uncertainty principle states that $\Delta p \Delta x > \hbar/2$. In an experiment, it is more likely that λ is varied rather than p, where λ is the de Broglie wavelength of the particle. The relationship between Δx and $\Delta \lambda$ will be determined using a simulation. Δx will be measured as a function of $\Delta \lambda$ at a constant value of λ, and as a function of λ for a constant value of $\Delta \lambda$.

W17.3   The uncertainty in momentum will be determined for the total energy eigenfunctions for the particle in the infinite depth box for several values of the quantum number n. The function describing the distribution in k,

$$g_n(k) = \frac{1}{\sqrt{2\pi}} \int_{-\infty}^{\infty} f_n(x)\, e^{-ikx}\, dx = \frac{1}{\sqrt{2\pi}} \int_{0}^{a} \sin\frac{n\pi x}{a}\, e^{-ikx}\, dx$$

will be determined. The values of k for which this function has maxima will be compared with that expected for a classical particle of momentum $p = \sqrt{2mE}$. The width in k of the function $g_n(k)$ on n will be investigated.

A Quantum Mechanical Model for the Vibration and Rotation of Molecules

A molecule has translational, vibrational, and rotational types of motion. Each of these can be separately described by its own energy spectrum and energy eigenfunctions. As shown in Chapter 15, the particle in the box is a useful model for exploring the translational degree of freedom. In this chapter, quantum mechanics is used to study the vibration and rotation of a diatomic molecule. We first consider the vibrational degree of freedom, modeled by the harmonic oscillator. Like the particle in the box, the quantum mechanical harmonic oscillator has a discrete energy spectrum. We then formulate and solve a quantum mechanical model for rotational motion. This model provides a basis for understanding the orbital motion of electrons around the nucleus of an atom as well as the rotation of a molecule about its principal axes.

18.1 The Classical Harmonic Oscillator

The harmonic oscillator is reviewed from the perspective of classical mechanics in this section. Consider two masses m_1 and m_2 that are connected by a coiled spring. When at rest, the spring is at its equilibrium length. If the masses are pushed together, the spring is compressed, and if the masses are pulled apart, the spring is extended. In each case, the spring resists any attempt to move the masses away from their equilibrium positions. If the deviation of the spacing between the masses from its rest position is denoted by x, then

$$x = [x_{m_1}(t) + x_{m_2}(t)] - [x_{m_1} + x_{m_2}]_{equilibrium} \qquad \textbf{(18.1)}$$

Positive and negative values of x correspond to stretching and compression of the spring, respectively, as shown in Figure 18.1. Experimentally, it is found that to double x, the

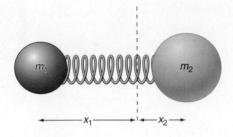

FIGURE 18.1
Two unequal masses are shown connected by a spring of force constant k. The intersection of the vertical line with the spring indicates the center of mass.

force exerted on the system must be doubled. This means that a linear relationship exists between the force and x given by

$$F = -kx \tag{18.2}$$

In this equation, k is called the **force constant.** The negative sign shows that the force and the displacement are in opposite directions.

Before developing a mathematical model of the **harmonic oscillator,** it is useful to make a mental image of what happens when the spring is either stretched or compressed and then let go. In either case, the force that the spring exerts on each of the masses will be in the direction opposite that of the applied force. As soon as the spacing of the masses reaches its equilibrium distance, the direction of the spring's force changes. This causes the direction of motion to reverse; an initial stretch becomes a compression and vice versa. In the absence of dissipative forces, the masses continue through alternate half cycles of being farther apart and closer together than their equilibrium distance. This system exhibits **oscillatory behavior.**

Although the masses move in opposite directions, the magnitudes of their displacements are not equal if their masses are unequal. This makes it hard to develop a simple picture of the time evolution of this system. However, somewhere between the masses is a point that does not move. This is called the center of mass, and a transformation to **center of mass coordinates** gives us a simpler description of the oscillatory motion of the harmonic oscillator.

Before going on, it is useful to summarize what information is needed to describe the harmonic oscillator and what information can be derived using classical mechanics. The oscillator is described by the two masses m_1 and m_2 and the force constant k, which allows the force F acting on each of the masses to be calculated. To solve Newton's second law of motion, two independent pieces of information are needed that describe the state of the system at a given initial time. The value of x and the kinetic energy of the oscillator at a given time will do. From this information, the positions x_1 and x_2, the velocities $\mathbf{v}_1$ and $\mathbf{v}_2$, and the kinetic and potential energies of each mass can be determined as a function of time. This is more information than necessary because we are more interested in the potential and kinetic energies associated with the entire oscillator as a unit than with the values for each of the masses separately. This is the main reason for working with the center of mass coordinates.

In the center of mass coordinates, the physical picture of the system changes from two masses connected by a spring of force constant k to a single mass, called the reduced mass μ, connected by a spring of the same force constant to an immovable wall. Why is this transformation used? We do this because only the *relative* motion of these two masses with respect to one another and not their individual motions is of interest. This change of coordinates also reduces the description of the periodic motion to a single coordinate.

The location of the center of mass x_{cm} and the reduced mass μ are given by the equations

$$x_{cm} = \frac{m_1 x_1 + m_2 x_2}{m_1 + m_2} \tag{18.3}$$

and

$$\mu = \frac{m_1 m_2}{m_1 + m_2} \tag{18.4}$$

We now use Newton's second law of motion to investigate the dynamics of the harmonic oscillator. Because the motion is in one dimension, the scalar magnitude of the force and acceleration can be used in what follows. Recall that the variable x denotes the deviation of the spring extension from its equilibrium position. Starting with

$$F = \mu a = \mu \frac{d^2 x}{dt^2} \tag{18.5}$$

where a is the acceleration, and using Equation (18.2) for the force, the differential equation

$$\mu \frac{d^2 x}{dt^2} + kx = 0 \tag{18.6}$$

describes the time dependence of the distance between the masses relative to its equilibrium value.

The general solution to this differential equation is

$$x(t) = c_1 e^{+i\sqrt{(k/\mu)}t} + c_2 e^{-i\sqrt{(k/\mu)}t} \quad \textbf{(18.7)}$$

in which c_1 and c_2 are arbitrary coefficients. At this point the **Euler formula,** $r e^{\pm i\theta} = r\cos\theta \pm ir\sin\theta$, is used to write Equation (18.7) in the form

$$x(t) = c_1\left(\cos\sqrt{\frac{k}{\mu}}t + i\sin\sqrt{\frac{k}{\mu}}t\right) + c_2\left(\cos\sqrt{\frac{k}{\mu}}t - i\sin\sqrt{\frac{k}{\mu}}t\right) \quad \textbf{(18.8)}$$

The last equation can be further simplified to

$$x(t) = b_1\cos\sqrt{\frac{k}{\mu}}t + b_2\sin\sqrt{\frac{k}{\mu}}t \quad \textbf{(18.9)}$$

with $b_1 = c_1 + c_2$ and $b_2 = i(c_1 - c_2)$. This is the general solution of the differential equation with no restrictions on b_1 and b_2. Because the amplitude of oscillation is real, a boundary condition that requires that b_1 and b_2 be real is imposed. The general solution contains two constants of integration that can be determined for a specific solution through the boundary conditions, $x(0)$, and $v(0) = [dx(t)/dt]_{t=0}$. For instance, if $x(0) = 0$ and $v(0) = v_0$, then

$$x(0) = b_1\cos\left(\sqrt{\frac{k}{\mu}} \times 0\right) = b_1 = 0$$

$$v(0) = \left(\frac{dx(t)}{dt}\right)_{t=0} = b_2\sqrt{\frac{k}{\mu}}\cos\left(\sqrt{\frac{k}{\mu}} \times 0\right) = b_2\sqrt{\frac{k}{\mu}} \quad \text{and}$$

$$b_2 = \sqrt{\frac{\mu}{k}}v_0 \quad \textbf{(18.10)}$$

The specific solution takes the form

$$x(t) = \sqrt{\frac{\mu}{k}}v_0\sin\sqrt{\frac{k}{\mu}}t \quad \textbf{(18.11)}$$

Note that only the second term in Equation (18.9) remains. This is because we arbitrarily choose $x(0) = 0$ and $v(0) = v_0$. Other boundary conditions could lead to solutions in which both b_1 and b_2 are nonzero.

Because the sine and cosine functions are periodic functions of the variable t, x exhibits oscillatory motion. The period of oscillation T is defined by the relation

$$\sqrt{\frac{k}{\mu}}(t + T) - \sqrt{\frac{k}{\mu}}t = 2\pi \quad \textbf{(18.12)}$$

giving

$$T = 2\pi\sqrt{\frac{\mu}{k}} \quad \textbf{(18.13)}$$

The inverse of T is called the frequency ν:

$$\nu = \frac{1}{2\pi}\sqrt{\frac{k}{\mu}} \quad \textbf{(18.14)}$$

These definitions of ν and T allow x to be written in the form

$$x(t) = b_1\cos 2\pi\frac{t}{T} + b_2\sin 2\pi\frac{t}{T} \quad \textbf{(18.15)}$$

Often, the angular frequency, $\omega = 2\pi\nu$, is introduced, giving

$$x(t) = b_1\cos\omega t + b_2\sin\omega t, \quad \text{or equivalently,} \quad x(t) = A\sin(\omega t + \alpha) \quad \textbf{(18.16)}$$

where the phase shift α is explored in Example Problem 18.1. The oscillatory periodic motion of the harmonic oscillator is depicted in Figure 18.2.

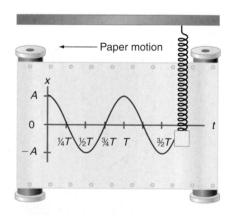

FIGURE 18.2
The periodic motion of a harmonic oscillator is revealed if the vertical motion is displayed on a moving piece of paper.

With a mathematical description of the motion, our mental picture of the oscillatory behavior can be tested. Because the potential energy $E_{potential}$ and the kinetic energy $E_{kinetic}$ of the oscillator are related to the magnitudes of **x** and **v** by the equations

$$E_{potential} = \frac{1}{2}kx^2 \quad \text{and} \quad E_{kinetic} = \frac{1}{2}\mu v^2 \qquad (18.17)$$

$E_{potential}$ and $E_{kinetic}$ can be expressed in terms of x, as is done in Example Problem 18.1.

Visualize the harmonic oscillator in terms of the potential and kinetic energies. Energy can be pumped into a harmonic oscillator at rest by stretching or compressing the spring. The maximum displacement from the equilibrium position depends on the force constant and the amount of energy taken up. The kinetic and potential energies also oscillate with time. As the amount of energy increases, its maximum amplitude of vibration and its maximum velocity increase. The harmonic oscillator can have any positive value for the total energy. Because there are no constraints on what value of the energy is allowed, the classical harmonic oscillator has a **continuous energy spectrum.**

EXAMPLE PROBLEM 18.1

For a harmonic oscillator described by $x(t) = A\sin(\omega t + \alpha)$, $\omega = (k/\mu)^{1/2}$, answer the following questions.

a. What are the units of A? What role does α have in this equation?

b. Graph the kinetic and potential energies given by the following equations as a function of time:

$$E_{kinetic} = \frac{1}{2}mv^2 \quad \text{and} \quad E_{potential} = \frac{1}{2}kx^2$$

c. Show that the sum of the kinetic and potential energies is independent of time.

Solution

a. Because $x(t)$ has the units of length and the sine function is dimensionless, A must have the units of length. The quantity α sets the value of x at $t = 0$, because $x(0) = A\sin(\alpha)$.

b. We begin by expressing the kinetic and potential energies in terms of $x(t)$:

$$E_{kinetic} = \frac{1}{2}\mu v^2 = \frac{1}{2}\mu\left(\frac{dx}{dt}\right)^2$$

$$= \frac{\mu}{2}\left(A\omega\cos(\omega t + \alpha)\right)^2$$

$$= \frac{1}{2}\mu\omega^2 A^2\cos^2(\omega t + \alpha)$$

$$E_{potential} = \frac{1}{2}kx^2 = \frac{1}{2}kA^2\sin^2(\omega t + \alpha)$$

$$= \frac{1}{2}\mu\omega^2 A^2\sin^2(\omega t + \alpha) \text{ because } \omega = \sqrt{\frac{k}{\mu}} \text{ and } k = \mu\omega^2$$

In the following figure, the energy is expressed in increments of $(1/2)\mu\omega^2 A^2$ and we have arbitrarily chosen $\alpha = \pi/6$. Note that the kinetic and potential energies are out of phase. Why is this the case?

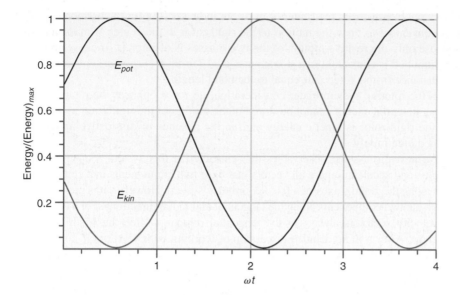

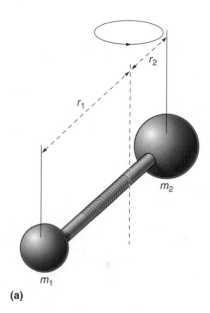

c. The dashed line in the preceding figure is the sum of the kinetic and potential energies, which is a constant. This can be verified algebraically by adding the expressions for $E_{kinetic}$ and $E_{potential}$:

$$E_{total} = \frac{1}{2}\mu\omega^2 A^2 \cos^2(\omega t + \alpha) + \frac{1}{2}\mu\omega^2 A^2 \sin^2(\omega t + \alpha)$$

$$= \frac{1}{2}\mu\omega^2 A^2 \left[\cos^2(\omega t + \alpha) + \sin^2(\omega t + \alpha)\right]$$

$$= \frac{1}{2}\mu\omega^2 A^2$$

Note that the sum of the kinetic and potential energies is independent of time, as must be the case, because no energy is added to the system after the initial stretching of the spring and there is no mechanism such as frictional forces for losing energy.

18.2 Angular Motion and the Classical Rigid Rotor

The harmonic oscillator is a good example of linear motion. In this system, the vectors for the velocity, momentum, and acceleration are all parallel to the direction of motion. However, not all motion is linear, making it necessary to analyze the motion induced if the applied force is not along the initial direction of motion. Why is rotational motion of interest to chemists? Energy can be taken up by a molecule in any of several ways. The first of these is translational kinetic energy, which is associated with the collective motion of all atoms in the molecule or with the center of mass. A second way to store energy, in the form of vibrational energy, was just discussed. It was shown that vibrational energy is both kinetic and potential and can be taken up by stretching bonds within a molecule. Now set the molecule spinning in addition to having it vibrate and undergo translational motion. Additional energy is taken up in this collective rotational motion. The **rigid rotor** is a simple example of angular motion. It is a good model for thinking about rotation of a diatomic molecule. The term *rigid* stems from the assumption that the rotational motion does not result in a stretching of the bond.

Consider the rigid rotor shown in Figure 18.3. The axis of rotation is perpendicular to the plane of rotation and passes through the center of mass. The distance of the

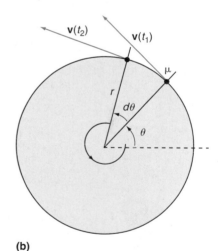

(b)

FIGURE 18.3
(a) The rigid rotor consists of two masses separated by a fixed distance. The dashed vertical line is the axis of rotation. It is perpendicular to the plane of rotation and passes through the center of mass. **(b)** The rigid rotor in the center of mass coordinates is a single particle of reduced mass μ rotating on a ring of radius equal to the bond length. The position and velocity of the reduced mass μ are shown at two times.

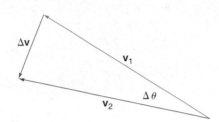

individual masses from the center of mass are indicated. As for the harmonic oscillator, it is convenient to view the motion of the rigid rotor in the center of mass coordinates because only the relative motion of the two masses is of interest. In these coordinates, the dumbbell is equivalent to a single mass $\mu = (m_1 m_2)/(m_1 + m_2)$ moving on a ring at a distance from a fixed axis equal to the bond length.

In the rotating system under consideration, no force opposes the rotation. For this reason, potential energy cannot be stored in the motion of the rigid rotor. All energy put into the rigid rotor is kinetic energy and, in the absence of dissipative losses, will be retained indefinitely.

We next discuss the observables that characterize this system. Force, momentum, velocity, and acceleration are all vectors that have two components that could be chosen to lie along the x and y axes of a fixed coordinate system. However, it is more convenient to take the two components along the tangential and radial directions. For a circular orbit, the velocity vector is always in the tangential direction. (See the Math Supplement, Appendix A for a more detailed discussion of working with vectors.) If no acceleration occurs along this direction, the magnitude of the velocity is constant in time. Figure 18.4 shows that $\Delta\mathbf{v} = \mathbf{v}_2 - \mathbf{v}_1$ is not zero because the particle experiences an acceleration on this orbit. Because the acceleration is given by $\mathbf{a} = \lim_{\Delta t \to 0} (\mathbf{v}_2 - \mathbf{v}_1)/(t_2 - t_1)$ the acceleration is not zero for circular motion. Because $\mathbf{v}_1$ and $\mathbf{v}_2$ have the same magnitude, only the radial component of the acceleration is nonzero. This component is called the **centripetal acceleration** $a_{centripetal}$ and has the magnitude

$$a_{centripetal}(t) = \frac{|\mathbf{v}(t)|^2}{r} \tag{18.18}$$

In circular motion, the total accumulated rotation angle θ is analogous to the distance variable in linear motion. The angle is typically measured in **radians.** A radian is the angle for which the arc length is equal to the radius; radians are related to degrees by 2π radians $= 360°$ or one radian $\approx 57.3°$. **Angular velocity** and **angular acceleration,** which are analogous to $\mathbf{v}$ and $\mathbf{a}$ in linear motion, are defined by

$$|\boldsymbol{\omega}| = \frac{d\theta}{dt} \quad \text{and} \quad \alpha = \frac{d|\boldsymbol{\omega}|}{dt} = \frac{d^2\theta}{dt^2} \tag{18.19}$$

The directions of both $\boldsymbol{\omega}$ and $\boldsymbol{\alpha}$ are determined by the right-hand rule and point along the axis of rotation. The application of the right-hand rule in determining the direction of $\boldsymbol{\omega}$ is illustrated in Figure 18.5. The angular acceleration is nonzero if the particle is not moving at constant speed on its circular orbit. Keep in mind that $\boldsymbol{\omega}$ is a vector perpendicular to the plane of rotation. Because the velocity is also defined by the expression

$$\mathbf{v} = \frac{\Delta \mathbf{s}}{\Delta t} = \frac{\mathbf{r}\Delta\boldsymbol{\theta}}{\Delta t} \quad \text{in the limit as} \quad \Delta t \to 0, \mathbf{v} = \frac{\mathbf{r}d\theta}{dt} = r\omega \tag{18.20}$$

the magnitudes of the angular and linear velocities can be related. In the case under consideration, the acceleration along the direction of motion is zero, and the expression for $d\theta/dt$ in Equation (18.19) can be integrated to obtain

$$\theta = \theta_0 + \omega t \tag{18.21}$$

For a constant acceleration along the direction of motion,

$$\omega = \omega_0 + \alpha t$$

$$\theta = \theta_0 + \omega_0 t + \frac{1}{2}\alpha t^2 \tag{18.22}$$

The kinetic energy can be expressed in the form

$$E_{kinetic} = \frac{1}{2}\mu v^2 = \frac{1}{2}\mu r^2 \omega^2 = \frac{1}{2}I\omega^2 \tag{18.23}$$

The quantity μr^2 is called the **moment of inertia** and given the symbol I. With this definition, the kinetic energy takes on a form similar to that in linear motion with the moment of inertia and the angular velocity taking on the role of the mass and linear velocity.

We next develop a relationship similar to $\mathbf{F} = m\mathbf{a} = d\mathbf{p}/dt$ for angular motion. The angular momentum, $\mathbf{l}$, is defined by

$$\mathbf{l} = \mathbf{r} \times \mathbf{p} \tag{18.24}$$

in which $\times$ indicates the vector cross product between $\mathbf{r}$ and $\mathbf{p}$. The use of the right-hand rule to determine the orientation of $\mathbf{l}$ relative to $\mathbf{r}$ and $\mathbf{p}$ is shown in Figure 18.6.

The magnitude of $\mathbf{l}$ is given by

$$l = pr \sin \phi = \mu vr \sin \phi \tag{18.25}$$

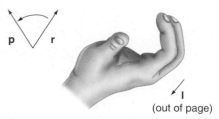
(a)

in which ϕ is the angle between the vectors $\mathbf{r}$ and $\mathbf{p}$. For circular motion, $\mathbf{r}$ and $\mathbf{p}$ are perpendicular so that $l = \mu vr$. The equation $E = p^2/2m$ and the definition of angular momentum can be used to express the kinetic energy in terms of l:

$$E = \frac{p^2}{2\mu} = \frac{l^2}{2\mu r^2} = \frac{l^2}{2I} \tag{18.26}$$

Classical mechanics does not place any restrictions on the direction or magnitude of $\mathbf{l}$. As for any observable in a classical system, the magnitude of $\mathbf{l}$ can change by an incrementally small amount. Therefore, any amount of energy can be stored in the rigid rotor, and an increase in the energy appears as an increase in the angular frequency. Because the amount of energy can be increased by an infinitesimally small amount, the classical rigid rotor has a continuous energy spectrum, just like the classical harmonic oscillator.

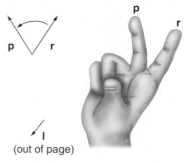
(b)

18.3 The Quantum Mechanical Harmonic Oscillator

In this and the next two sections, we develop quantum mechanical models for the harmonic oscillator and rigid rotor. The free particle and the particle in the box discussed in Chapters 15 and 16 were useful for understanding how **translational motion** in various potentials is described in the context of wave-particle duality. In applying quantum mechanics to molecules, two other types of motion that molecules can undergo require discussion: **vibration** and **rotation.** The energy needed to stretch the chemical bond can be described by a simple potential function such as that shown in Figure 18.7. The existence of a stable chemical bond implies that a minimum energy exists at the equilibrium **bond length.** The position of atoms in a molecule is dynamic rather than static. Think of the chemical bond as a spring rather than a rigid bar connecting the two atoms. Thermal energy increases the vibrational amplitude of the atoms about their equilibrium positions but does not change the vibrational frequency to a good approximation. The potential becomes steeply repulsive at short distances as the electron clouds of the atoms interpenetrate. It levels out at large distances because the overlap of electrons between the atoms required for chemical bond formation falls to zero.

(c)

The exact form of $V(x)$ as a function of x depends on the molecule under consideration. However, as will be shown in Chapter 19, only the lowest one or two vibrational energy levels are occupied for most molecules for $T \sim 300$ K. Therefore, it is a good approximation to say that the functional form of the potential energy near the equilibrium bond length can be approximated by the harmonic potential

$$V(x) = \frac{1}{2}kx^2 \tag{18.27}$$

In Equation (18.27), k is the force constant. For weakly bound molecules or high temperatures, the more realistic Morse potential (red curve in Figure 18.7) discussed in Section 19.3 should be used.

We expect the wave-particle of mass μ vibrating around its equilibrium distance to be described by a set of wave functions $\psi_n(x)$. To find these wave functions and the corresponding allowed vibrational energies, the Schrödinger equation with the appropriate potential energy function must be solved:

$$-\frac{\hbar^2}{2\mu}\frac{d^2\psi_n(x)}{dx^2} + \frac{kx^2}{2}\psi_n(x) = E_n\psi_n(x) \tag{18.28}$$

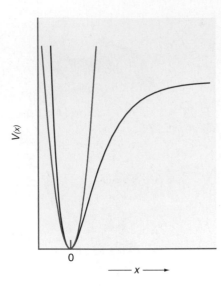

FIGURE 18.7
Potential energy $V(x)$ as a function of the displacement x from the equilibrium bond length for a diatomic molecule. The zero of energy is chosen to be the bottom of the potential. The red curve depicts a realistic potential in which the molecule dissociates for large values of x. The purple curve shows a harmonic potential, $V(x) = (1/2)kx^2$, which is a good approximation to the realistic potential near the bottom of the well.

The solution of this second-order differential equation was well known in the mathematical literature from other contexts well before the development of quantum mechanics. We simply state that the normalized wave functions are

$$\psi_n(x) = A_n H_n\left(\alpha^{1/2} x\right) e^{-\alpha x^2/2}, \quad \text{for } n = 0, 1, 2, \ldots \tag{18.29}$$

EXAMPLE PROBLEM 18.2

Show that the function $e^{-\beta x^2}$ satisfies the Schrödinger equation for the quantum harmonic oscillator. What conditions does this place on β? What is E?

Solution

$$-\frac{\hbar^2}{2\mu}\frac{d^2\psi_n(x)}{dx^2} + V(x)\psi_n(x) = E_n\psi_n(x)$$

$$-\frac{\hbar^2}{2\mu}\frac{d^2(e^{-\beta x^2})}{dx^2} + V(x)(e^{-\beta x^2}) = -\frac{\hbar^2}{2\mu}\frac{d(-2\beta x e^{-\beta x^2})}{dx} + \frac{1}{2}kx^2(e^{-\beta x^2})$$

$$= -\frac{\hbar^2}{2\mu}(-2\beta\, e^{-\beta x^2}) + \frac{\hbar^2}{2\mu}(-4\beta^2 x^2 e^{-\beta x^2})$$

$$+ \frac{1}{2}kx^2(e^{-\beta x^2})$$

The function is an eigenfunction of the total energy operator only if the last two terms cancel:

$$\hat{H}_{total}\, e^{-\beta x^2} = \frac{\hbar^2\beta}{\mu}e^{-\beta x^2} \quad \text{if } \beta^2 = \frac{1}{4}\frac{k\mu}{\hbar^2}$$

$$\text{Finally, } E = \frac{\hbar^2\beta}{\mu} = \frac{\hbar^2}{\mu}\sqrt{\frac{1}{4}\frac{k\mu}{\hbar^2}} = \frac{\hbar}{2}\sqrt{\frac{k}{\mu}}$$

In the preceding equation, several constants have been combined to give $\alpha = \sqrt{k\mu/\hbar^2}$, and the normalization constant A_n is given by

$$A_n = \frac{1}{\sqrt{2^n n!}}\left(\frac{\alpha}{\pi}\right)^{1/4} \tag{18.30}$$

The solution is written in this manner because the set of functions $H_n(\alpha^{1/2}x)$ is well known in mathematics as **Hermite polynomials.** The first few eigenfunctions $\psi_n(x)$ are given by

$$\psi_0(x) = \left(\frac{\alpha}{\pi}\right)^{1/4} e^{-(1/2)\alpha x^2}$$

$$\psi_1(x) = \left(\frac{4\alpha^3}{\pi}\right)^{1/4} x e^{-(1/2)\alpha x^2}$$

$$\psi_2(x) = \left(\frac{\alpha}{4\pi}\right)^{1/4}(2\alpha x^2 - 1)e^{-(1/2)\alpha x^2}$$

$$\psi_3(x) = \left(\frac{\alpha^3}{9\pi}\right)^{1/4}(2\alpha x^3 - 3x)e^{-(1/2)\alpha x^2} \tag{18.31}$$

$\psi_0, \psi_2, \psi_4, \ldots$ are even functions of x, $[\psi(x) = \psi(-x)]$, whereas $\psi_1, \psi_3, \psi_5, \ldots$ are odd functions of x $[\psi(x) = -\psi(-x)]$.

A necessary boundary condition is that the amplitude of the wave functions remains finite at large values of x. As for the particle in the box, this boundary condition gives rise to quantization. In this case, the quantization condition is not easy to derive.

However, it can be shown that the amplitude of the wave functions approaches zero for large x values only if the following condition is met:

$$E_n = \hbar \sqrt{\frac{k}{\mu}} \left(n + \frac{1}{2} \right) = h\nu \left(n + \frac{1}{2} \right) \quad \text{with } n = 0, 1, 2, 3, \ldots \quad \textbf{(18.32)}$$

Once again, we see that the imposition of boundary conditions has led to a discrete energy spectrum. Unlike the classical analogue, the energy stored in the quantum mechanical harmonic oscillator can only take on discrete values. As for the particle in the box, the lowest state accessible to the system still has an energy greater than zero, referred to as a **zero point energy**. The **frequency of oscillation** is given by

$$\nu = \frac{1}{2\pi} \sqrt{\frac{k}{\mu}} \quad \textbf{(18.33)}$$

just as for the classical harmonic oscillator.

EXAMPLE PROBLEM 18.3

a. Is $\psi_1(x) = (4\alpha^3/\pi)^{1/4} x e^{-(1/2)\alpha x^2}$ an eigenfunction of the kinetic energy operator? Is it an eigenfunction of the potential energy operator?

b. What are the average values of the kinetic and potential energies for a quantum mechanical oscillator in this state?

Solution

a. As discussed in Chapter 17, neither the potential energy operator nor the kinetic energy operator commutes with the total energy operator. Therefore, because $\psi_1(x) = (4\alpha^3/\pi)^{1/4} x e^{-(1/2)\alpha x^2}$ is an eigenfunction of the total energy operator, it is not an eigenfunction of the potential or kinetic energy operators.

b. The fourth postulate states how the average value of an observable can be calculated. Because

$$\hat{E}_{potential}(x) = V(x) \text{ and } \hat{E}_{kinetic}(x) = -\frac{\hbar^2}{2\mu} \frac{d^2}{dx^2}$$

then

$$\langle E_{potential} \rangle = \int \psi_1^*(x) V(x) \psi_1(x) \, dx$$

$$= \int_{-\infty}^{\infty} \left(\frac{4\alpha^3}{\pi} \right)^{1/4} x e^{-(1/2)\alpha x^2} \left(\frac{1}{2} k x^2 \right) \left(\frac{4\alpha^3}{\pi} \right)^{1/4} x e^{-(1/2)\alpha x^2} \, dx$$

$$= \frac{1}{2} k \left(\frac{4\alpha^3}{\pi} \right)^{1/2} \int_{-\infty}^{\infty} x^4 e^{-\alpha x^2} \, dx = k \left(\frac{4\alpha^3}{\pi} \right)^{1/2} \int_0^{\infty} x^4 e^{-\alpha x^2} \, dx$$

The limits can be changed as indicated in the last integral because the integrand is an even function of x. To obtain the solution, the following standard integral found in the Math Supplement is used:

$$\int_0^{\infty} x^{2n} e^{-ax^2} \, dx = \frac{1 \times 3 \times 5 \cdots (2n-1)}{2^{n+1} a^n} \sqrt{\frac{\pi}{a}}$$

The calculated values for the average potential and kinetic energy are

$$\langle E_{potential} \rangle = \frac{1}{2} k \left(\frac{4\alpha^3}{\pi} \right)^{1/2} \left(\sqrt{\frac{\pi}{\alpha}} \right) \frac{3}{4\alpha^2}$$

$$= \frac{3k}{4\alpha} = \frac{3}{4} \hbar \sqrt{\frac{k}{\mu}}$$

$$\langle E_{kinetic} \rangle = \int \psi_1^*(x)\left(-\frac{\hbar^2}{2\mu}\frac{d^2}{dx^2}\right)\psi_1(x)\,dx$$

$$= \int_{-\infty}^{\infty}\left(\frac{4\alpha^3}{\pi}\right)^{1/4} xe^{-(1/2)\alpha x^2}\left(-\frac{\hbar^2}{2\mu}\frac{d^2}{dx^2}\right)\left(\frac{4\alpha^3}{\pi}\right)^{1/4} xe^{-(1/2)\alpha x^2}\,dx$$

$$= -\frac{\hbar^2}{2\mu}\left(\frac{4\alpha^3}{\pi}\right)^{1/2}\int_{-\infty}^{\infty}(\alpha^2 x^4 - 3\alpha x^2)e^{-\alpha x^2}\,dx$$

$$= -\frac{\hbar^2}{\mu}\left(\frac{4\alpha^3}{\pi}\right)^{1/2}\int_{0}^{\infty}(\alpha^2 x^4 - 3\alpha x^2)e^{-\alpha x^2}\,dx$$

$$= -\frac{\hbar^2}{2\mu}\left(\frac{4\alpha^3}{\pi}\right)^{1/2}\left(\alpha^2\left[\sqrt{\frac{\pi}{\alpha}}\frac{3}{4\alpha^2}\right] - 3\alpha\left[\sqrt{\frac{\pi}{\alpha}}\frac{1}{2\alpha}\right]\right)$$

$$= \frac{3}{4}\frac{\hbar^2\alpha}{\mu} = \frac{3}{4}\hbar\sqrt{\frac{k}{\mu}}$$

Note that just as for the classical harmonic oscillator discussed in Section 18.1, the average values of the kinetic and potential energies are equal. When the kinetic energy has its maximum value, the potential energy is zero and vice versa. In general, we find that for the nth state,

$$\langle E_{kinetic,n} \rangle = \langle E_{potential,n} \rangle = \frac{\hbar}{2}\sqrt{\frac{k}{\mu}}\left(n + \frac{1}{2}\right)$$

As was done for the particle in the box, it is useful to plot $\psi(x)$ and $\psi^2(x)$ against x. They are shown superimposed on the potential energy function in Figures 18.8 and 18.9.

It is instructive to compare the quantum mechanical with the classical results. In quantum mechanics the value of x cannot be known if the system is in an eigenstate of the total energy operator, because these two operators do not commute. This issue arose earlier in considering the particle in the box. What can one say about x, the amplitude of the vibration?

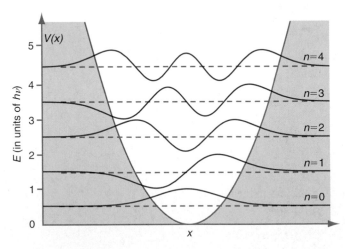

FIGURE 18.8
The first few eigenfunctions of the quantum mechanical harmonic oscillator are shown together with the potential function. The amplitude of the corresponding eigenfunction is shown superimposed on each energy level, with the zero of amplitude for the eigenfunctions indicated by the dashed lines. The yellow area indicates the classically forbidden region for which $E < V$.

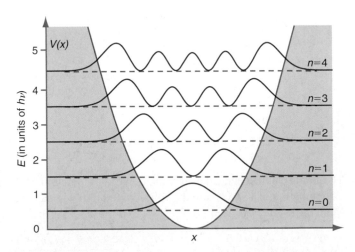

FIGURE 18.9
The square of the first few eigenfunctions of the quantum mechanical harmonic oscillator (the probability density) is superimposed on the energy spectrum and shown together with the potential function. The yellow area indicates the classically forbidden region.

Only the probability of the vibrational amplitude having a particular value of x within an interval dx can be calculated, and this probability is given by $\psi^2(x)\,dx$. For the classical harmonic oscillator, the probability of the vibrational amplitude having a particular value of x within the interval dx can also be calculated. Because the probability density varies inversely with the velocity, the maximum values are found at the turning points and its minimum value is found at $x = 0$. To visualize this behavior, imagine a frictionless ball rolling on a parabolic track under the influence of gravity. The ball moves fastest at the lowest point on the track and stops momentarily as it reverses its direction at the highest points on either side of the track. Figure 18.10 shows a comparison of $\psi_{12}^2(x)$ and the probability density of finding a particular amplitude for a classical oscillator with the same total energy as a function of x. A large quantum number has been used for comparison because in the limit of high energies (very large quantum numbers), classical and quantum mechanics give the same result as required by the correspondence principle.

The main difference between the classical and quantum mechanical results are the behavior in the classically forbidden region and the oscillations in $\psi_{12}^2(x)$, which are absent in the classical result. In calculating the probability of finding the value x for the oscillation amplitude in the interval Δx, it is necessary to evaluate

$$\int_{-\Delta x/2}^{\Delta x/2} \psi^2(x)\,dx$$

rather than the probability density $\psi^2(x)$. For large quantum numbers, Δx is large in comparison to the distance between neighboring oscillations in $\psi^2(x)$. Therefore, the oscillations in the probability density $\psi^2(x)$ are averaged out in performing the integration, so that the quantum and classical results agree well. The argument is the same as that used in calculating the results for the particle in the box shown in Figure 15.4.

We have been working with the time-independent Schrödinger equation, whose eigenfunctions allow the probability density to be calculated. To describe the time dependence of the oscillation amplitude, the total wave function, $\Psi_n(x, t) = e^{-i\omega t}\psi_n(x)$, is constructed. The spatial amplitude shown in Figure 18.8 is modulated by the factor $e^{-i\omega t}$, which has a frequency given by $\omega = \sqrt{k/\mu}$. Because $\psi_n(x, t)$ is a standing wave, the nodal positions shown in Figures 18.8 and 18.9 do not move with time.

In looking at Figures 18.8 and 18.9, several similarities are seen with Figures 15.2 and 15.3, in which the equivalent results were shown for the particle in the box. The eigenfunctions are again standing waves, but they are now in a box with a more complicated shape. Successive eigenfunctions add one more oscillation within the "box," and the amplitude of the wave function is small at the edge of the "box." The reason why it is small rather than zero follows the same lines as the discussion of the particle in the finite depth box in Chapter 16. The quantum mechanical harmonic oscillator also has a zero point energy, meaning that the lowest possible energy state still has vibrational energy. The origin of this zero point energy is similar to that for the particle in the box. By attaching a spring to the particle, its motion has been constrained. As the spring is made stiffer (larger k), the particle is more constrained and the zero point energy increases. This is the same trend observed for the particle in the box as the length is decreased.

Note, however, that important differences exist in the two systems that are a result of the more complicated shape of the harmonic oscillator "box." Although the harmonic oscillator wave functions show oscillatory behavior, they are no longer represented by simple sine functions because the classical probability density is not independent of x. The energy spacing is the same between adjacent energy levels; that is, it does not increase with the quantum number as was the case for the particle in the box. These differences show the sensitivity of the eigenfunctions and eigenvalues to the functional form of the potential.

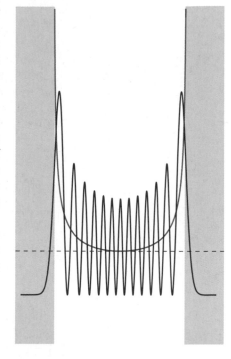

FIGURE 18.10
The calculated probability density for the vibrational amplitude is shown for the 12th eigenstate of the quantum mechanical oscillator (red curve). The classical result is shown by the purple curve. The yellow area indicates the classically forbidden region.

18.4 Quantum Mechanical Rotation in Two Dimensions

Quantum mechanical models were developed for translation in Chapter 15 and for vibration in Section 18.3. We now consider rotation to complete the description of the fundamental types of motion available to a molecule. To a good approximation, the three types of motion—translation, vibration, and rotation—can be dealt with independently. This treatment is exact rather than approximate (1) if the translational part of the total energy operator depends only on the translational coordinates of the center of mass, (2) if the rotational part depends only on the angular coordinates of the center of mass, and (3) if the vibrational part depends only on the internal coordinates of the molecule. This condition cannot be exactly satisfied, because the types of motion are not totally decoupled. For example, the average bond length of a rapidly rotating molecule is slightly longer than for a molecule that is not rotating because of the centrifugal forces acting on the atoms. However, although the coupling between the types of motion can be measured using sensitive spectroscopic techniques, it is small for most molecules.

Neglecting this coupling, the total energy operator can be written as a sum of individual operators for the types of motion for the molecule:

$$\hat{H}_{total} = \hat{H}_{trans}(r_{cm}) + \hat{H}_{vib}(\tau_{internal}) + \hat{H}_{rot}(\theta_{cm}, \phi_{cm}) \qquad \textbf{(18.34)}$$

In this equation, r_{cm}, θ_{cm}, and ϕ_{cm} refer to the spatial coordinates of the center of mass in spherical coordinates (see the Math Supplement, Appendix A). The symbol $\tau_{internal}$ refers collectively to the vibrational amplitudes of all atoms in the molecule around their equilibrium position. Because different variables appear in $\hat{H}_{trans}(r_{cm})$, $\hat{H}_{vib}(\tau_{internal})$, and $\hat{H}_{rot}(\theta_{cm}, \phi_{cm})$, it is possible to solve the Schrödinger equation for each type of motion separately. In this approximation, the total energy is given by the sum of the individual contributions,

$$E_{total} = E_{trans}(r_{cm}) + E_{vib}(\tau_{internal}) + E_{rot}(\theta_{cm}, \phi_{cm}) \qquad \textbf{(18.35)}$$

and the system wave function is a product of the eigenfunctions for the three types of motion:

$$\psi_{total} = \psi_{trans}(r_{cm}) \psi_{vib}(\tau_{internal}) \psi_{rot}(\theta_{cm}, \phi_{cm}) \qquad \textbf{(18.35a)}$$

Because the wave function is a product of individual terms that depend on different variables, what has been accomplished in Equations (18.35) and (18.35a) is a **separation of variables.**

Whereas for a diatomic molecule translation can be considered in one to three independent dimensions and vibration in one dimension, rotation requires at least a two-dimensional description. We restrict our considerations to diatomic molecules because the motion is easy to visualize. However, the process outlined next can be generalized to any molecule if several angular coordinates are included. Rotation in a two-dimensional space is discussed first because the mathematical formalism needed to describe such a problem is less complicated. The formalism is extended in Section 18.5 to rotation in three dimensions.

Rotation in two dimensions occurs only in a constrained geometry. An example is a molecule adsorbed on a smooth surface. Consider a diatomic molecule with masses m_1 and m_2 and a fixed bond length r_0 freely rotating in the xy plane. Because the bond length is assumed to remain constant as the molecule rotates, this model is often referred to as the rigid rotor. By transforming to the center of mass coordinate system, this problem becomes equivalent to a single reduced mass $\mu = (m_1 m_2)/(m_1 + m_2)$ rotating in the xy plane on a ring of radius r_0, just as for the classical rotor.

EXAMPLE PROBLEM 18.4

The bond length for $H^{19}F$ is 91.68×10^{-12} m. Where does the axis of rotation intersect the molecular axis?

Solution

The position of the center of mass is given by $x_{cm} = (m_H x_H + m_F x_F)/(m_H + m_F)$. We choose the origin of our coordinate system to be at the F atom, so $x_F = 0$ and

$x_H = 91.68 \times 10^{-12}$ m. Substituting $m_F = 18.9984$ amu and $m_H = 1.008$ amu, we find that $x_{cm} = 4.62 \times 10^{-12}$ m. Therefore $x_F = 4.62 \times 10^{-12}$ m and $x_H = 87.06 \times 10^{-12}$ m. We see that the axis of rotation is very close to the F atom. This effect is even more pronounced for HI or HCl.

Because it has been assumed that the particle experiences no hindrance to rotation, the potential energy is constant everywhere. Therefore, we can conveniently set $V(x, y) = 0$ everywhere without affecting the eigenfunctions of the total energy operator. The Schrödinger equation in Cartesian coordinates for this problem is

$$-\frac{\hbar^2}{2\mu}\left(\frac{\partial^2\psi(x, y)}{\partial x^2} + \frac{\partial^2\psi(x, y)}{\partial y^2}\right)_{r=r_0} = E\psi(x, y) \qquad (18.36)$$

The subscript after the bracket makes it clear that the radius is constant. Although Equation (18.36) is correct, it is always best to choose a coordinate system that reflects the symmetry of the system being considered. In this case, two-dimensional polar coordinates with the variables r and ϕ are the logical choice. In these coordinates, with r fixed at r_0, the operator $(\partial^2/\partial x^2) + (\partial^2/\partial y^2)$ becomes $(1/r_0^2)(\partial^2/\partial\phi^2)$. Therefore, the Schrödinger equation takes the simple form

$$-\frac{\hbar^2}{2\mu r_0^2}\frac{d^2\Phi(\phi)}{d\phi^2} = E\Phi(\phi) \qquad (18.37)$$

where the eigenfunction $\Phi(\phi)$ depends only on the angle ϕ. We have changed the symbol for the wave function to emphasize the change in the variables. This equation has the same form as the Schrödinger equation for a free particle, which was solved in Chapter 15. You should verify that the two linearly independent solutions to this equation are

$$\Phi_+(\phi) = A_{+\phi}e^{i|m_l|\phi} \quad \text{and} \quad \Phi_-(\phi) = A_{-\phi}e^{-i|m_l|\phi} \qquad (18.38)$$

The two solutions correspond to counterclockwise and clockwise rotation.

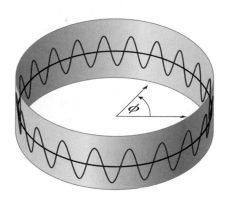

$m_l = \pm$integer

EXAMPLE PROBLEM 18.5

Determine the normalization constant $A_{\pm\phi}$ in Equation (18.38).

Solution

The variable ϕ can take on values between 0 and 2π. The following result is obtained:

$$\int_0^{2\pi} \Phi_{m_l}^*(\phi)\,\Phi_{m_l}(\phi)\,d\phi = 1$$

$$(A_{\pm\phi})^2\int_0^{2\pi} e^{\mp im_l\phi}e^{\pm im_l\phi}\,d\phi = (A_{\pm\phi})^2\int_0^{2\pi} d\phi = 1$$

$$A_{\pm\phi} = \frac{1}{\sqrt{2\pi}}$$

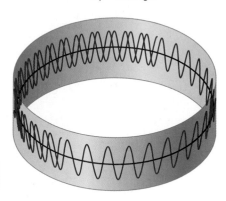

$m_l \neq \pm$integer

FIGURE 18.11
If the condition $m_l =$ integer is not met, the wave function does not have the same value for $\phi + 2\pi$ as for ϕ. The real part of the wave function is plotted as a function of ϕ in each case.

To obtain solutions of the Schrödinger equation that describe this physical problem, it is necessary to introduce the boundary condition $\Phi(\phi + 2\pi) = \Phi(\phi)$. This condition states that there is no way to distinguish the particle that has rotated n times around the circle from one that has rotated $n + 1$ times around the circle. Without this condition, the probability density would have multiple values for ϕ and $\phi + 2n\pi$, as shown in Figure 18.11, which is unacceptable. Applying the single-value condition to the

FIGURE 18.12
The real part of the second through seventh eigenfunctions for the rigid rotor with rotation confined to a plane is plotted as a function of ϕ. In the center of mass coordinates, this problem is equivalent to the particle on a ring. What does the first eigenfunction look like?

eigenfunction, $e^{im_l[\phi+2\pi]} = e^{im_l\phi}$ or $e^{2\pi im_l} = 1$. Using Euler's relation, this expression is equivalent to

$$\cos 2\pi m_l + i \sin 2\pi m_l = 1 \qquad (18.39)$$

To satisfy this condition, m_l must equal $0, \pm 1, \pm 2, \pm 3, \dots$. The boundary condition generates the quantization rules for the quantum number m_l. The motivation for using the subscript l on the quantum number m will become clear when rotation in three dimensions is considered.

What do these eigenfunctions look like? Because they are complex functions of the angle ϕ, only the real part of the function is shown in Figure 18.12. The imaginary part is identical in shape but is shifted in phase by the angle $\pi/2$. Note that, as for the particle in the box and the harmonic oscillator, the lowest energy state has no nodes, and the number of nodes, which is twice the quantum number, increases with m_l.

Putting the eigenfunctions back into Equation (18.37) allows the corresponding eigenvalues E_{m_l} to be calculated. The energy-level spectrum is discrete and is given by

$$E_{m_l} = \frac{\hbar^2 m_l^2}{2\mu r_0^2} = \frac{\hbar^2 m_l^2}{2I} \quad \text{for } m_l = 0, \pm 1, \pm 2, \pm 3, \dots \qquad (18.40)$$

In Equation (18.40), $I = \mu r_0^2$ is the moment of inertia. Note that states with $+m_l$ and $-m_l$ have the same energy, although the wave functions corresponding to these states are orthogonal to one another. We say that the energy levels with $m_l \neq 0$ are *twofold degenerate*.

The origin of the energy quantization is again a boundary condition. In this case, imagine the ring as a box of length 2π defined by the variable ϕ. The boundary condition given in Equation (18.39) states that an integral number of wavelengths must fit into this "box." For a classical rigid rotor,

$$E = \frac{|\mathbf{l}|^2}{2\mu r_0^2} = \frac{|\mathbf{l}|^2}{2I} = \frac{1}{2}I\omega^2 \qquad (18.41)$$

where, throughout this chapter, the symbol $\mathbf{l}$ is used for the angular momentum vector, $|\mathbf{l}|$ for its magnitude, and $\hat{l}$ for the angular momentum operator. Equation (18.41) also holds for the quantum mechanical rigid rotor, with the association $\omega = m_l\hbar/I$. Therefore, the quantization of energy means that only a discrete set of rotational frequencies is allowed.

One aspect of the eigenvalues for free rotation in two dimensions is different from what was encountered with the particle in the box or the harmonic oscillator: no zero point energy is associated with free rotational motion; $E_{m_l} = 0$ when $m_l = 0$. Why is this the case? A zero point energy appears only if the potential confines the motion to a limited region. In free rotation, there is no confinement and no zero point energy. Of course, a gas phase diatomic molecule also moves and vibrates. Therefore, the rotating molecule has a zero point energy associated with these degrees of freedom.

The angular momentum can also be calculated for the two-dimensional rigid rotor. For rotation in the xy plane, the angular momentum vector lies on the z axis. The angular momentum operator in these coordinates takes the simple form $\hat{l}_z = -i\hbar(\partial/\partial\phi)$. Applying this operator to an eigenfunction,

$$\hat{l}_z\Phi_{\pm}(\phi) = \frac{-i\hbar}{\sqrt{2\pi}}\frac{d\,e^{\pm im_l\phi}}{d\phi} = \frac{\pm\,m_l\hbar}{\sqrt{2\pi}}e^{\pm im_l\phi} = \pm m_l\hbar\,\Phi_{\pm}(\phi) \qquad (18.42)$$

A similar equation can be written for $\Phi_{-}(\phi)$. This result shows that the angular momentum is quantized. We see that $\Phi_{+}(\phi)$ and $\Phi_{-}(\phi)$ are eigenfunctions of both the total energy and the angular momentum operators for the two-dimensional rigid rotor. As we will see, this is not the case for rotation in three dimensions. Because the angular momentum has the values $+\hbar m_l$ and $-\hbar m_l$, Equation (18.40) can be written in the form

$$E_{m_l} = \frac{\hbar^2 m_l^2}{2I} = \frac{|\mathbf{l}|^2}{2I}$$

just as in classical mechanics.

What can be said about the value of the rotation angle with respect to a fixed direction in the xy plane? We know that the probability of finding a particular angle ϕ in the interval $d\phi$ is

$$P(\phi)\,d\phi = \Phi^*(\phi)\Phi(\phi)\,d\phi = \left(\frac{1}{\sqrt{2\pi}}\right)^2 e^{\pm im_l\phi}e^{\mp im_l\phi}\,d\phi = \frac{d\phi}{2\pi} \quad \textbf{(18.43)}$$

The probability of finding the particle in a given interval $d\phi$ is the same for all values of ϕ. Just as for the position of a free particle whose linear momentum is precisely defined, nothing is known about the angular position of the molecule whose angular momentum is precisely defined. The origin of this result is that the operators $\hat{\phi}$ and $\hat{l}_z$ do not commute, just as $\hat{x}$ and $\hat{p}_x$ do not commute.

18.5 Quantum Mechanical Rotation in Three Dimensions

In the case just considered, the motion has been constrained to two dimensions. Now imagine the more familiar case of a diatomic molecule freely rotating in three-dimensional space. This problem is not more difficult, but the mathematics is more cumbersome than the two-dimensional case just considered. Again, we transform to the center of mass coordinate system, and the rotational motion is transformed to the motion of a particle on the surface of a sphere of radius r_0. As before, it is advantageous to express the kinetic and potential energy operators in an appropriate coordinate system, which in this case is spherical coordinates. Because there is no hindrance to rotation, the potential energy is constant and can be set equal to zero. In this coordinate system, which is depicted in Figure 13.5, the Schrödinger equation is

$$-\frac{\hbar^2}{2\mu r_0^2}\left[\frac{1}{\sin\theta}\frac{\partial}{\partial\theta}\left(\sin\theta\frac{\partial Y(\theta,\phi)}{\partial\theta}\right) + \frac{1}{\sin^2\theta}\frac{\partial^2 Y(\theta,\phi)}{\partial\phi^2}\right] = EY(\theta,\phi) \quad \textbf{(18.44)}$$

Figure 13.5 defines the relationship between x, y, and z in Cartesian coordinates, and r, θ, and ϕ in spherical coordinates.

Our task is to find the eigenfunctions $Y(\theta,\phi)$ and the corresponding eigenvalues that are the solutions of this equation. Although the solution of this partial differential equation is not discussed in detail here, the first few steps are outlined because they provide some important physical insights. Combining constants in the form

$$\beta = \frac{2\mu r_0^2 E}{\hbar^2} \quad \textbf{(18.45)}$$

multiplying through on the left by $\sin^2\theta$, and rearranging this equation results in Equation (18.46)

$$\sin\theta\frac{\partial}{\partial\theta}\left(\sin\theta\frac{\partial Y(\theta,\phi)}{\partial\theta}\right) + [\beta\sin^2\theta]Y(\theta,\phi) = -\frac{\partial^2 Y(\theta,\phi)}{\partial\phi^2} \quad \textbf{(18.46)}$$

On the right side of the equation, the differentiation is with respect to ϕ only. On the left side of the equation, the differentiation is with respect to θ only. If this equality is to hold for all ϕ and θ, $Y(\theta,\phi)$ must be the product of two functions, each of which depends on only one of the two independent variables:

$$Y(\theta,\phi) = \Theta(\theta)\Phi(\phi) \quad \textbf{(18.47)}$$

This separation of variables leads to a major simplification in solving the Schrödinger equation Equation (18.44).

The functions $Y(\theta,\phi)$ are known as the **spherical harmonic functions** and are discussed in detail later in this chapter. Substituting Equation (18.47) into Equation (18.46) and dividing through by $\Theta(\theta)\Phi(\phi)$, we obtain

$$\frac{1}{\Theta(\theta)}\sin\theta\frac{d}{d\theta}\left(\sin\theta\frac{d\Theta(\theta)}{d\theta}\right) + \beta\sin^2\theta = -\frac{1}{\Phi(\phi)}\frac{d^2\Phi(\phi)}{d\phi^2} \quad \textbf{(18.48)}$$

Note that this equation no longer contains partial derivatives. Because each side of the equation depends on only one of the variables and the equality exists for all values of the variables, it must be true that both sides of the equation are equal to the same constant:

$$\frac{1}{\Theta(\theta)} \sin\theta \frac{d}{d\theta}\left(\sin\theta \frac{d\Theta(\theta)}{d\theta}\right) + \beta\sin^2\theta = m_l^2 \quad \text{and}$$

$$\frac{1}{\Phi(\phi)} \frac{d^2\Phi(\phi)}{d\phi^2} = -m_l^2 \qquad \qquad (18.49)$$

Looking back at the differential equation for rotation in two dimensions, it is clear why the constant is written in this way. The solutions for the second equation can be obtained immediately because the same equation was solved for the molecule rotating in two dimensions:

$$\Phi_+(\phi) = A_{+\phi}e^{i|m_l|\phi} \text{ and } \Phi_-(\phi) = A_{-\phi}e^{-i|m_l|\phi}, \quad \text{for } m_l = 0, 1, 2, 3, \ldots \quad (18.50)$$

where the part of $Y(\theta, \phi)$ that depends on ϕ is associated with the quantum number m_l.

The first equation in Equation (18.49) allows the part of $Y(\theta, \phi)$ that depends on θ to be determined. It can be solved to give a set of eigenfunctions and their corresponding eigenvalues. Rather than work through the solution, the results are summarized with a focus on the eigenvalues. A discussion of the spherical harmonics is postponed until Section 18.7. Two boundary conditions must be satisfied to solve Equation (18.49). To ensure that the functions $Y(\theta, \phi)$ are single-valued functions of θ and ϕ and that the amplitude of these functions remains finite everywhere, the following conditions must be met. We state rather than derive these conditions:

$$\beta = l(l+1), \quad \text{for } l = 0, 1, 2, 3, \ldots \quad \text{and}$$

$$m_l = -l, -(l-1), -(l-2), \ldots, 0, \ldots, (l-2), (l-1), l \quad (18.51)$$

Both l and m_l must be integers. Note that l and m_l are the quantum numbers for the three-dimensional rigid rotor. To emphasize this result, the spherical harmonic functions are written in the form

$$Y(\theta, \phi) = Y_l^{m_l}(\theta, \phi) = \Theta_l^{m_l}(\theta)\Phi_{m_l}(\phi) \quad (18.52)$$

The function $\Theta_l^{m_l}(\theta)$ is associated with both quantum numbers l and m_l, and the function $\Phi_{m_l}(\phi)$ is associated only with the quantum number m_l. For a given value of l, there are $2l + 1$ different values of m_l ranging from $-l$ to $+l$. We next consider the origin of these quantum numbers more closely.

Why are there two quantum numbers for rotation in three dimensions, whereas there is only one for rotation in two dimensions? The answer is related to the dimensionality of the problem. For rotation in two dimensions, r was held constant. Therefore, ϕ is the only variable in the problem and there is only one boundary condition. For rotation in three dimensions, r is again held constant and, therefore, only the two boundary conditions on θ and ϕ generate quantum numbers. For the same reason, the particle in the one-dimensional box is characterized by a single quantum number, whereas three quantum numbers are required to characterize the particle in the three-dimensional box.

What observables of the rotating molecule are associated with the quantum numbers l and m_l? From the equation

$$\beta = \frac{2\mu r_0^2 E}{\hbar^2} = \frac{2I}{\hbar^2}E = l(l+1)$$

the energy eigenvalues for rotation in three dimensions can be obtained. This shows that the quantum number l is associated with the total energy observable,

$$E_l = \frac{\hbar^2}{2I}l(l+1), \quad \text{for } l = 0, 1, 2, 3, \ldots \quad (18.53)$$

and that the total energy eigenfunctions $Y_l^{m_l}(\theta, \phi)$ satisfy the eigenvalue equation

$$\hat{H}_{total} Y_l^{m_l}(\theta, \phi) = \frac{\hbar^2}{2I}l(l+1)Y_l^{m_l}(\theta, \phi), \quad \text{for } l = 0, 1, 2, 3, \ldots \quad (18.54)$$

Note that the rotational energy values are quantized and that, once again, the quantization arises through a boundary condition. Note that the energy levels depend differently on the quantum number than the energy levels for rotation in two dimensions for which

$$E_{m_l} = \frac{\hbar^2 m_l^2}{2\mu r_0^2} = \frac{\hbar^2 m_l^2}{2I}, \quad \text{for } m_l = 0, \pm 1, \pm 2, \pm 3, \ldots \quad (18.55)$$

For rotation in three dimensions the energy depends on the quantum number l, but not on m_l. Why is this the case? As will be shown in Section 18.7, the quantum number m_l determines the z component of the vector $\mathbf{l}$. Because $\mathbf{E}_{total} = |\mathbf{l}|^2/2\mu r_0^2$, the energy of rotation depends only on the magnitude of the angular momentum and not its direction. Therefore, all $2l + 1$ total energy eigenfunctions that have the same l value but different m_l values have the same energy. This means that the **degeneracy** of each energy level is $2l + 1$. Recall that for rotation in two dimensions, the degeneracy of each energy level is two, except for the $m_l = 0$ level, which is nondegenerate.

18.6 The Quantization of Angular Momentum

We continue our discussion of three-dimensional rotation, although now it is discussed in the context of **angular momentum** rather than energy as was done earlier. Why is angular momentum important in quantum chemistry? Consider a familiar example from introductory chemistry, namely, the s, p, and d **orbitals** associated with atoms of the periodic table. This notation will be discussed in more detail in Chapter 20. We know that the bonding behavior of s, p, and d electrons is quite different. Why is an s orbital spherically symmetrical, whereas a p orbital has a dumbbell structure? Why are three energetically degenerate p orbitals directed along the x, y, and z directions? The origin of these chemically important properties is the particular value of l or m_l associated with these orbitals.

As discussed earlier, the spherical harmonic functions $Y_l^{m_l}(\theta, \phi)$, are eigenfunctions of the total energy operator for a molecule that rotates freely in three dimensions. Are these functions also eigenfunctions of other operators of interest to us? Because the potential energy is zero for a free rotor, the total energy stored in rotational motion is given by the kinetic energy $E_{total} = |\mathbf{l}|^2/2I$, in which $\mathbf{l}$ is the angular momentum and $I = \mu r_0^2$. Note that E_{total} and $|\mathbf{l}^2|$ differ only by the constant $1/2I$. Therefore, the corresponding operators $\hat{H}_{total}$ and $\hat{l}^2$ also satisfy this relationship. Because they differ only by a multiplicative constant, these two operators commute with one another and have a common set of eigenfunctions. Furthermore, because E_{total} is quantized, it can be concluded that $|\mathbf{l}^2|$ is also quantized. Using the proportionality of E_{total} and $|\mathbf{l}^2|$, the eigenvalue equation for the operator $\hat{l}^2$ can immediately be written from Equation (18.54):

$$\hat{l}^2 Y_l^{m_l}(\theta, \phi) = \hbar^2 l(l + 1) Y_l^{m_l}(\theta, \phi) \quad (18.56)$$

The notation explicitly shows that the quantum numbers l and m_l are defining indices for the eigenfunctions of $\hat{H}_{total}$ and $\hat{l}^2$. Because the eigenvalues for $\hat{l}^2$ are given by $\hbar^2 l(l + 1)$, the magnitude of the angular momentum takes on the quantized values $|\mathbf{l}| = \hbar\sqrt{l(l + 1)}$.

Note that it is $\hat{l}^2$ and not $\hat{l}$ that commutes with $\hat{H}_{total}$. We now focus our attention on the angular momentum $\mathbf{l}$ and the corresponding operator $\hat{l}^2$. How many components does $\mathbf{l}$ have? For rotation in the xy plane, the angular momentum vector has only a single component that lies on the z axis. For rotation in three dimensions, the angular momentum vector has the three components l_x, l_y, and l_z, which are obtained from the vector cross product $\mathbf{l} = \mathbf{r} \times \mathbf{p}$. See the Math Supplement for a more detailed discussion of the cross product and angular motion. As might be expected from the discussion of the Stern–Gerlach experiment in Chapter 17, the operators $\hat{l}_x$, $\hat{l}_y$, and $\hat{l}_z$ do not commute.

As you will see when working the end-of-chapter problems, the operators $\hat{l}_x$, $\hat{l}_y$, and $\hat{l}_z$ have the following form in Cartesian coordinates:

$$\hat{l}_x = -i\hbar\left(y\frac{\partial}{\partial z} - z\frac{\partial}{\partial y}\right)$$

$$\hat{l}_y = -i\hbar\left(z\frac{\partial}{\partial x} - x\frac{\partial}{\partial z}\right)$$

$$\hat{l}_z = -i\hbar\left(x\frac{\partial}{\partial y} - y\frac{\partial}{\partial x}\right) \tag{18.57}$$

Although not derived here, the operators have the following form in spherical coordinates:

$$\hat{l}_x = -i\hbar\left(-\sin\phi\frac{\partial}{\partial\theta} - \cot\theta\cos\phi\frac{\partial}{\partial\phi}\right)$$

$$\hat{l}_y = -i\hbar\left(\cos\phi\frac{\partial}{\partial\theta} - \cot\theta\sin\phi\frac{\partial}{\partial\phi}\right)$$

$$\hat{l}_z = -i\hbar\left(\frac{\partial}{\partial\phi}\right) \tag{18.58}$$

As you will verify in the end-of-chapter problems for the operators in Cartesian coordinates, the commutators relating the operators $\hat{l}_x$, $\hat{l}_y$, and $\hat{l}_z$ are given by

$$[\hat{l}_x, \hat{l}_y] = i\hbar\hat{l}_z$$

$$[\hat{l}_y, \hat{l}_z] = i\hbar\hat{l}_x$$

$$[\hat{l}_z, \hat{l}_x] = i\hbar\hat{l}_y \tag{18.59}$$

Note that the order of the commutator is important, that is, $[\hat{l}_x, \hat{l}_y] = -[\hat{l}_y, \hat{l}_x]$.

What are the consequences of the fact that the operators corresponding to the components of the angular momentum do not commute with one another? Because the commutators are not zero, the direction of the angular momentum vector cannot be specified for rotation in three dimensions. To do so, it would be necessary to know all three components simultaneously, which would require that the three commutators in Equation (18.59) are zero. Given that $\hat{l}_x$, $\hat{l}_y$, and $\hat{l}_z$ do not commute, what can be known about the components of the angular momentum for a molecule whose wave function is an eigenfunction of the total energy operator?

To answer this question, we look more closely at the operators for the individual components of the angular momentum. In spherical coordinates, $\hat{l}_x$ and $\hat{l}_y$ depend on both θ and ϕ, but as Equation (18.58) shows, $\hat{l}_z$ depends only on ϕ. As shown earlier, the spherical harmonics, $Y_l^{m_l}(\theta, \phi) = \Theta_l^{m_l}(\theta)\Phi_{m_l}(\phi)$, are eigenfunctions of the total energy operator and of $\hat{l}^2$. We now show that the spherical harmonics are also eigenfunctions of $\hat{l}_z$. Applying $\hat{l}_z$ to the functions $Y_l^{m_l}(\theta, \phi)$, we obtain

$$\hat{l}_z(Y_l^{m_l}(\theta, \phi)) = \Theta(\theta)\left[-i\hbar\frac{\partial}{\partial\phi}\left(\frac{1}{\sqrt{2\pi}}e^{im_l\phi}\right)\right] = m_l\hbar\Theta(\theta)\Phi(\phi),$$

for $m_l = 0, \pm1, \pm2, \pm3, \ldots, \pm l$ \hspace{1cm} (18.60)

showing that the $Y_l^{m_l}(\theta, \phi)$ are eigenfunctions of $\hat{l}_z$. What can we conclude from Equation (18.60)? Because the spherical harmonics are eigenfunctions of both $\hat{l}^2$ and $\hat{l}_z$, both the magnitude of $|\mathbf{l}|$ and its z component can be known simultaneously. In other words, one can know the length of the vector $\mathbf{l}$ and one of its components, but it is not possible to simultaneously know the other two components of $\mathbf{l}$.

Why has $\hat{l}_z$ rather than $\hat{l}_x$ or $\hat{l}_y$ been singled out, and what makes the z component special? There is nothing special about the z direction, and one could have just as easily chosen another direction. The way in which the variables are defined in spherical coordinates makes $\hat{l}_z$ take on a simple form. Therefore, when a direction is chosen, it is convenient to make it the z direction. The essence of the preceding discussion is that one can know the magnitude of $\mathbf{l}$ and only one of its components simultaneously.

The consequences of the different commutation relations among $\hat{H}, \hat{l}^2, \hat{l}_x, \hat{l}_y,$ and $\hat{l}_z$ are explored in Section 18.8, which deals with spatial quantization.

18.7 The Spherical Harmonic Functions

Until now, only the eigenvalues for $\hat{l}^2$, $\hat{H}$, and $\hat{l}_z$ for rotation in three dimensions have been discussed. We now discuss the spherical harmonic functions, $Y_l^{m_l}(\theta, \phi)$, which are the eigenfunctions common to these three operators. They are listed here for the first few values of l and m_l:

$$Y_0^0(\theta, \phi) = \frac{1}{(4\pi)^{1/2}}$$

$$Y_1^0(\theta, \phi) = \left(\frac{3}{4\pi}\right)^{1/2} \cos\theta$$

$$Y_1^{\pm 1}(\theta, \phi) = \left(\frac{3}{8\pi}\right)^{1/2} \sin\theta\, e^{\pm i\phi}$$

$$Y_2^0(\theta, \phi) = \left(\frac{5}{16\pi}\right)^{1/2} (3\cos^2\theta - 1)$$

$$Y_2^{\pm 1}(\theta, \phi) = \left(\frac{15}{8\pi}\right)^{1/2} \sin\theta \cos\theta\, e^{\pm i\phi}$$

$$Y_2^{\pm 2}(\theta, \phi) = \left(\frac{15}{32\pi}\right)^{1/2} \sin^2\theta\, e^{\pm 2i\phi} \tag{18.61}$$

As seen earlier in Equation (18.50), the ϕ dependence is a simple exponential function. The θ dependence enters as a polynomial in $\sin\theta$ and $\cos\theta$. The numerical factor in front of these functions ensures that they are normalized over the intervals $0 \le \theta \le \pi$ and $0 \le \phi \le 2\pi$. Because the spherical harmonics are eigenfunctions of the time-independent Schrödinger equation, they represent standing waves on the surface of a sphere in which the nodal positions are independent of time.

For $l = 0$, the eigenfunction is equal to a constant determined by the normalization condition. What does this mean? Remember that the square of the wave function gives the probability density for finding the particle at the coordinates θ and ϕ within the interval $d\theta$ and $d\phi$. These coordinates specify the angle defining the internuclear axis in a diatomic molecule. If the wave function is independent of θ and ϕ, any orientation of the internuclear axis in the rotation of a molecule is equally likely. This must be the case for a state in which the angular momentum is zero. A net angular momentum, corresponding to $l > 0$, requires that the wave function and the probability density distribution not have spherical symmetry.

The spherical harmonics are complex functions unless $m_l = 0$. Graphing complex functions requires double the number of dimensions as for real functions, so that it is customary to instead form appropriate linear combinations of the $Y_l^{m_l}(\theta, \phi)$ to generate real functions. These functions, which still form an orthonormal set, are given in the following equations. Equation (18.62) lists the p functions, and Equation (18.63) lists the d functions. We recognize the abbreviations in connection with the orbital designations for the hydrogen atom. As shown in Chapter 20, the functions shown in Figures 18.13 and 18.14 appear in the solutions of the Schrödinger equation for the hydrogen atom. Because of this, they merit more discussion.

$$p_x = \frac{1}{\sqrt{2}}\left(Y_1^1 + Y_1^{-1}\right) = \sqrt{\frac{3}{4\pi}} \sin\theta \cos\phi$$

$$p_y = \frac{1}{\sqrt{2}i}\left(Y_1^1 - Y_1^{-1}\right) = \sqrt{\frac{3}{4\pi}} \sin\theta \sin\phi \tag{18.62}$$

$$p_z = Y_1^0 = \sqrt{\frac{3}{4\pi}} \cos\theta$$

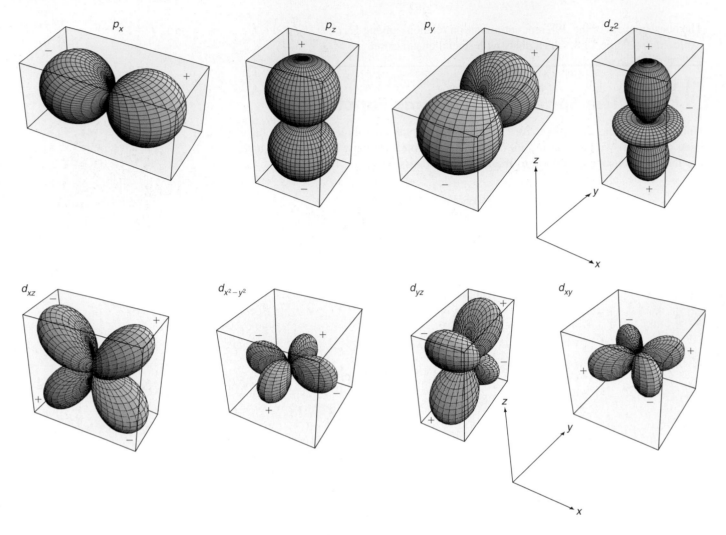

FIGURE 18.13
Three-dimensional perspective plots of the p and d linear combinations of the spherical harmonics. The plots show three-dimensional surfaces in which the relationship of the angles θ and ϕ to the Cartesian axes is defined in Figure 13.5. The distance from the origin to a point on the surface (θ, ϕ) represents the absolute magnitude of the functions defined by Equations (18.62) and (18.63). The sign of the functions in the different lobes is indicated by plus and minus signs.

$$d_{z^2} = Y_2^1 = \sqrt{\frac{5}{16\pi}}\,(3\cos^2\theta - 1)$$

$$d_{xz} = \frac{1}{\sqrt{2}}\left(Y_2^1 + Y_2^{-1}\right) = \sqrt{\frac{15}{4\pi}}\,\sin\theta\cos\theta\cos\phi$$

$$d_{yz} = \frac{1}{\sqrt{2}\,i}\left(Y_2^1 - Y_2^{-1}\right) = \sqrt{\frac{15}{4\pi}}\,\sin\theta\cos\theta\sin\phi \qquad \textbf{(18.63)}$$

$$d_{x^2-y^2} = \frac{1}{\sqrt{2}}\left(Y_2^2 + Y_2^{-2}\right) = \sqrt{\frac{15}{16\pi}}\,\sin^2\theta\cos 2\phi$$

$$d_{xy} = \frac{1}{\sqrt{2}\,i}\left(Y_2^2 - Y_2^{-2}\right) = \sqrt{\frac{15}{16\pi}}\,\sin^2\theta\sin 2\phi$$

These functions depend on two variables, θ and ϕ, and the way in which they are named refers them back to Cartesian coordinates. In graphing the functions, spherical coordinates have been used, whereby the radial coordinate is used to display the value of the amplitude, $r = f(\theta, \phi)$. All the functions generate lobular patterns in which the amplitude of the function in a lobe is either positive or negative. These signs are indicated in the plots.

The p functions form a set of three mutually perpendicular dumbbell structures. The wave function has the same amplitude but a different sign in the two lobes, and each function has a nodal plane passing through the origin. Four of the five d functions have a more complex four-lobed shape with nodal planes separating lobes in which the function has opposite signs. Because l is larger for the d than for the p functions, more nodes are seen in both angles. As for the particle in the box wave functions, an increase in the number of nodes corresponds to an increase in the energy of the quantum state. For the particle in the box, an increase in the number of nodes over a fixed interval corresponds to a shorter wavelength and, through the de Broglie relation, to a higher linear momentum. For the rigid rotor, an increase in the number of nodes over a fixed interval corresponds to a higher angular momentum. We return to the spherical harmonic functions when discussing the orbitals for the H atom in Chapter 20.

Up to this point, questions have been asked about the energy and the momentum. What can be learned about the angular orientation of the internuclear axis for the rotating molecule? This information is given by the probability density, defined by the first postulate as the square of the magnitude of the wave function. The probability density for the p and d functions is very similar in shape to the wave function amplitude shown in Figure 18.13, although the amplitude in all lobes is positive. Taking the p_z plot as an example, Figure 18.13 shows that the maximum amplitude of $|Y_1^0|^2$ is found along the positive and negative z axis. A point on the z axis corresponds to the probability density for finding the molecular axis parallel to the z axis.

An alternate graphical representation can be used that recognizes that spherical harmonics can be used to represent waves on the surface of a sphere. This can be done by displaying the amplitude of the desired function on the sphere at the location θ, ϕ using a color scale. This is done in Figure 18.15, where the square of the amplitude of the p_z and p_y functions is plotted as a color scale on the surface of a sphere. Black and red regions correspond to high and low probability densities, respectively. For the p_z function, there is a much higher probability density of finding the particle near the z axis than in the $z = 0$ plane. This means that the molecular axis is much more likely to be parallel to the z axis than to lie in the xy plane. For a state whose wave function is p_y, the internuclear axis is much more likely to be parallel to the y axis than to lie in the $y = 0$ plane. This is consistent with the angular orientation of the maxima of these functions shown in Figure 18.13. Why is the probability density not more sharply peaked in a small angular region near the z or y axis? If the wave function is the p_z function, E_{total}, $|l^2|$, and l_z are well defined. However, the operators for the angular coordinates ϕ and θ do not commute with the operators for E_{total}, $|l^2|$, and l_z. As a consequence, the angular position coordinates are not known exactly and only average values can be determined for these observables.

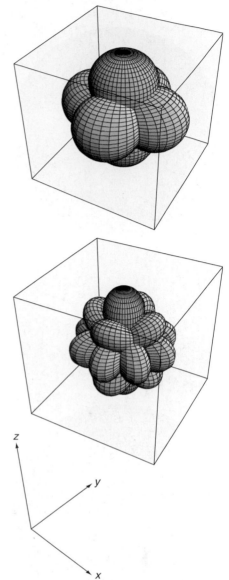

FIGURE 18.14
Three-dimensional perspective plots show the three p and the five d linear combinations of the spherical harmonics superimposed. The convention used in displaying the functions is explained in the text and in the caption for Figure 18.13.

18.8 Spatial Quantization

The fact that the operators $\hat{H}$, $\hat{l}^2$, and $\hat{l}_z$ commute whereas $\hat{l}_x$, $\hat{l}_y$, and $\hat{l}_z$ do not commute with one another states that the energy, the magnitude of the angular momentum vector, and the value of any one of its components can be known simultaneously but that the other two components of the angular momentum cannot be known. Contrast this with classical mechanics in which all three components of an angular momentum vector can be specified simultaneously. In that case, both the length of the vector and its direction can be known.

We summarize what can be known about the angular momentum vector associated with a molecule rotating in three dimensions pictorially. In doing so, classical and quantum mechanical descriptions are mixed. For this reason, the following is a **semiclassical** description. The one component that is known is chosen to be along the z direction. In

p_z

p_y

FIGURE 18.15
The absolute magnitude of the amplitude of the p_z and p_y functions is plotted on the surface of a unit sphere. Black and red regions correspond to high and low probability densities, respectively.

Figure 18.16, we show what can be known about $\mathbf{l}$ and l_z. The magnitude of $\mathbf{l}$ is $\sqrt{l(l+1)}\,\hbar$ and that of $l_z = m_l\hbar$. The vector $\mathbf{l}$ cannot lie on the z axis because $|\sqrt{m_l(m_l+1)}| \geq |m_l|$ so that $|m_l| \leq l$. From another point of view, $\mathbf{l}$ cannot lie on the z axis because the commutators in Equation (18.59) are not zero. If $\mathbf{l}$ did lie on the z axis, then l_x and l_y would both be zero and, therefore, all three components of the vector $\mathbf{l}$ could be known simultaneously. The fact that only $|\mathbf{l}|$ and one of its components can be known simultaneously is a direct manifestation of the fact that the operators $\hat{l}_x$, $\hat{l}_y$, and $\hat{l}_z$ do not commute with one another.

Although the picture in Figure 18.16 is useful, it does not depict $\mathbf{l}$ as a three-dimensional vector. We modify this figure to take the three-dimensional nature of $\mathbf{l}$ into account in Figure 18.17 for the case where $l = 2$ and $m_l = 2$. The vector $\mathbf{l}$ is depicted as a line on the surface of the cone beginning at its apex. The magnitude of $\mathbf{l}$ and its projection on the z axis are known exactly and can be determined from the figure. However, the components of the angular momentum vector along the x and y axes, l_x and l_y, cannot be known exactly and simultaneously. All that is knowable about them is that $l^2 - l_z^2 = l_x^2 + l_y^2 = l(l+1)\hbar^2 - m_l^2\hbar^2$. This equation defines the circle terminating the cone at its open end. Figure 18.17 depicts all that can be known simultaneously about the components of the angular momentum. To give a more physical picture to Figure 18.17, a classical rigid rotor for which the z component of the angular momentum vector is the same as for the quantum mechanical case is also shown. Do not take this comparison literally, because the rotor can be depicted as shown only because all three components of the angular momentum can be known simultaneously. This is not possible for a quantum mechanical rigid rotor.

Figure 18.18 combines the information about all possible values of m_l consistent with $l = 2$ in one figure. Such a depiction is often referred to as a **vector model of angular momentum.** Only the orientations of $\mathbf{l}$ for which the vector lies on one of the cones are allowed. A surprising result emerges from these considerations. Not only are the possible magnitudes of the angular momentum quantized, but the vector can only have certain orientations in space! This result is referred to as **spatial quantization.**

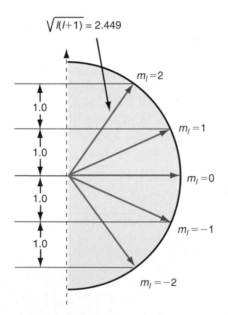

FIGURE 18.16
Possible orientations of the angular momentum vector $|\mathbf{l}| = \sqrt{l(l+1)}\,\hbar$ and $l_z = m_l\hbar, l \geq |m_l|$ for $l = 2$. The lengths indicated are in units of $\hbar$.

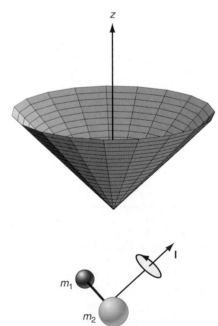

FIGURE 18.17
Example of an angular momentum vector for which only l^2, l_z, and $l_x^2 + l_y^2$ are known. In this case, $l = 2$ and $m_l = +2$. The right side of the figure illustrates a classical rigid rotor for which the angular momentum vector has the same l_z component.

What is the analogous situation in classical mechanics? Because l_x, l_y, and l_z can be known simultaneously for a classical system, and because their values are not quantized, the possible orientations of **l** map out a continuous spherical surface. The contrast between classical and quantum mechanical behavior is clearly evident! It is also apparent how quantum and classical results merge for high energies (large quantum numbers) as required by the correspondence principle. For a given l value, there are $2l + 1$ conical surfaces on a vector diagram like that shown in Figure 18.18. For large values of l, the individual cones are so close together that they merge into a sphere, and the angular momentum vector no longer seems to exhibit spatial quantization.

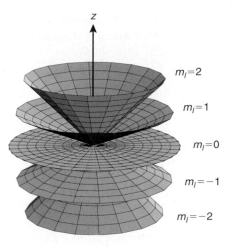

FIGURE 18.18
All possible orientations of an angular momentum vector with $l = 2$. The z component of the angular momentum is shown in units of $\hbar$.

EXAMPLE PROBLEM 18.6

How many cones of the type shown in Figure 18.18 will there be for $l = 1000$? What is the closest allowed angle between **l** and the z axis?

Solution

There will be $2l + 1$ or 2001 cones. The smallest allowed angle is for $l_z = 1000\hbar$, and is given by

$$\cos \phi = \frac{l}{\sqrt{l(l + 1)}} = \frac{1000}{1000.50}$$

$$\phi = 0.03 \text{ radians} = 1.7°$$

Vocabulary

angular acceleration	force constant	rotation
angular momentum	frequency of oscillation	semiclassical
angular velocity	harmonic oscillator	separation of variables
bond length	Hermite polynomial	spatial quantization
center of mass coordinates	moment of inertia	spherical harmonic functions
centripetal acceleration	orbital	translational motion
continuous energy spectrum	oscillatory behavior	vector model of angular momentum
degeneracy	radians	vibration
Euler formula	rigid rotor	zero point energy

Conceptual Problems

Q18.1 Why is the probability of finding the harmonic oscillator at its maximum extension or compression larger than that for finding it at its rest position?

Q18.2 Why does the energy of a rotating molecule depend on l but not on m_l?

Q18.3 Are the real functions listed in Equations (18.62) and (18.63) eigenfunctions of $\hat{l}_z$? Justify your answer.

Q18.4 Spatial quantization was discussed in Section 18.8. Suppose that we have a gas consisting of atoms, each of which has a nonzero angular momentum. Are all of their angular momentum vectors aligned?

Q18.5 Does the average length of a quantum harmonic oscillator depend on its energy? Answer this question by referring to the harmonic potential function shown in Figure 18.7.

The average length is the midpoint of the horizontal line connecting the two parts of $V(x)$.

Q18.6 Does the bond length of a real molecule depend on its energy? Answer this question by referring to Figure 18.7. The bond length is the midpoint of the horizontal line connecting the two parts of $V(x)$.

Q18.7 Why can the angular momentum vector lie on the z axis for two-dimensional rotation in the xy plane but not for rotation in three-dimensional space?

Q18.8 How does the total energy of the quantum harmonic oscillator depend on its maximum extension?

Q18.9 Explain why the amplitude of the total energy eigenfunctions for the quantum mechanical harmonic oscillator increases with $|x|$ as shown in Figure 18.10.

Q18.10   Why is it possible to write the total energy eigenfunctions for rotation in three dimensions in the form $Y_l^{m_l}(\theta, \phi) = \Theta(\theta)\Phi(\phi)$?

Q18.11   The two linearly independent total energy eigenfunctions for rotation in two dimensions are

$$\Phi_+(\phi) = \frac{1}{\sqrt{2\pi}}e^{i|m_l|\phi} \quad \text{and} \quad \Phi_-(\phi) = \frac{1}{\sqrt{2\pi}}e^{-i|m_l|\phi}.$$

What is the evolution in time of ϕ for each of these solutions?

Q18.12   Why is only one quantum number needed to characterize the eigenfunctions for rotation in two dimensions, whereas two quantum numbers are needed for rotation in three dimensions?

Q18.13   What makes the z direction special such that $\hat{l}^2$, $\hat{H}$, and $\hat{l}_z$ commute, whereas $\hat{l}_y$, $\hat{l}_z$, and $\hat{l}_x$ do not commute?

Q18.14   How are the spherical harmonics combined to form real p and d functions? What is the advantage in doing so?

Q18.15   The zero point energy of the particle in the box goes to zero as the length of the box approaches infinity. What is the appropriate analogue for the quantum harmonic oscillator?

Q18.16   Figure 18.12 shows the solutions to the time independent Schrödinger equation for the rigid rotor in two dimensions. Describe the corresponding solutions for the time dependent Schrödinger equation.

Q18.17   Use the anharmonic potential function in Figure 18.7 to demonstrate that rotation and vibration are not separable degrees of freedom for large quantum numbers.

Q18.18   Conservation of energy requires that the variation of the potential and kinetic energies with the oscillator extension be exactly out of phase. Explain this statement.

Q18.19   What is the degeneracy of the energy levels for the rigid rotor in two dimensions? If it is not 1, explain why.

Q18.20   For a two-dimensional harmonic oscillator, $V(x,y) = k_x x^2 + k_y y^2$. Write an expression for the energy levels of such an oscillator in terms of k_x and k_y.

Numerical Problems

Problem numbers in **red** indicate that the solution to the problem is given in the *Student's Solutions Manual*.

P18.1   A gas-phase $^1\text{H}^{127}\text{I}$ molecule, with a bond length of 160.92 pm, rotates in three-dimensional space.

a. Calculate the zero point energy associated with this rotation.

b. What is the smallest quantum of energy that can be absorbed by this molecule in a rotational excitation?

P18.2   In this problem you will derive the commutator $[\hat{l}_x, \hat{l}_y] = i\hbar \hat{l}_z$.

a. The angular momentum vector in three dimensions has the form $\mathbf{l} = \mathbf{i}l_x + \mathbf{j}l_y + \mathbf{k}l_z$ where the unit vectors in the x, y, and z directions are denoted by $\mathbf{i}$, $\mathbf{j}$, and $\mathbf{k}$. Determine l_x, l_y, and l_z by expanding the 3×3 cross product $\mathbf{l} = \mathbf{r} \times \mathbf{p}$. The vectors $\mathbf{r}$ and $\mathbf{p}$ are given by $\mathbf{r} = \mathbf{i}x + \mathbf{j}y + \mathbf{k}z$ and $\mathbf{p} = \mathbf{i}p_x + \mathbf{j}p_y + \mathbf{k}p_z$.

b. Substitute the operators for position and momentum in your expressions for l_x and l_y. Always write the position operator to the left of the momentum operator in a simple product of the two.

c. Show that $[\hat{l}_x, \hat{l}_y] = i\hbar \hat{l}_z$.

P18.3   In discussing molecular rotation, the quantum number J is used rather than l. Using the Boltzmann distribution, calculate n_J/n_0 for $^1\text{H}^{35}\text{Cl}$ for $J = 0, 5, 10$, and 20 at $T = 1025$ K. Does n_J/n_0 go through a maximum as J increases? If so, what can you say about the value of J corresponding to the maximum?

P18.4   Draw a picture (to scale) showing all angular momentum cones consistent with $l = 5$. Calculate the half angles for each of the cones.

P18.5 $^1\text{H}^{19}\text{F}$ has a force constant of 966 N m^{-1} and a bond length of 91.68 pm. Calculate the frequency of the light corresponding to the lowest energy pure vibrational and pure rotational transitions. In what regions of the electromagnetic spectrum do the transitions lie?

P18.6   The wave functions p_x and d_{xz} are linear combinations of the spherical harmonic functions, which are eigenfunctions of the operators $\hat{H}$, $\hat{l}^2$, and $\hat{l}_z$ for rotation in three dimensions. The combinations have been chosen to yield real functions. Are these functions still eigenfunctions of $\hat{l}_z$? Answer this question by applying the operator to the functions.

P18.7   At what values of θ does $Y_2^0(\theta, \phi) = (5/16\pi)^{1/2}(3\cos^2\theta - 1)$ have nodes? Are the nodes points, lines, planes, or other surfaces?

P18.8   The vibrational frequency for D$_2$ expressed in wave numbers is 3115 cm^{-1}. What is the force constant associated with the bond? How much would a classical spring with this force constant be elongated if a mass of 1.50 kg were attached to it? Use the gravitational acceleration on Earth at sea level for this problem.

P18.9   In discussing molecular rotation, the quantum number J is used rather than l. Calculate E_{rot}/k_BT for $^1\text{H}^{81}$ Br for $J = 0, 5, 10$, and 20 at 298 K. For which of these values of J is $E_{rot}/k_BT \geq 10$.?

P18.10   Show by carrying out the necessary integration that the eigenfunctions of the Schrödinger equation for rotation in two dimensions, $\frac{1}{\sqrt{2\pi}}e^{im_l\phi}$ and $\frac{1}{\sqrt{2\pi}}e^{in_l\phi}$, $m_l \neq n_l$ are orthogonal.

P18.11   Evaluate the average of the square of the linear momentum of the quantum harmonic oscillator $\langle p_x^2 \rangle$ for the ground state ($n = 0$) and first two excited states ($n = 1$ and $n = 2$). Use the hint about evaluating integrals in Problem P18.12.

P18.12 Show by carrying out the appropriate integration that the total energy eigenfunctions for the harmonic oscillator $\psi_0(x) = (\alpha/\pi)^{1/4}e^{-(1/2)\alpha x^2}$ and $\psi_2(x) = (\alpha/4\pi)^{1/4}(2\alpha x^2 - 1)e^{-(1/2)\alpha x^2}$ are orthogonal over the interval $-\infty < x < \infty$ and that $\psi_2(x)$ is normalized over the same interval. In evaluating integrals of this type,

$$\int_{-\infty}^{\infty} f(x)\,dx = 0 \text{ if } f(x) \text{ is an odd function of } x \text{ and}$$

$$\int_{-\infty}^{\infty} f(x)\,dx = 2\int_{0}^{\infty} f(x)\,dx \text{ if } f(x) \text{ is an even function of } x.$$

P18.13 Two 3.25 g masses are attached by a spring with a force constant of $k = 450.$ kg s^{-2}. Calculate the zero point energy of the system and compare it with the thermal energy kT at 298 K. If the zero point energy were converted to translational energy, what would be the speed of the masses?

P18.14 Calculate the frequency and wavelength of the radiation absorbed when a quantum harmonic oscillator with a frequency of 3.15×10^{13} s^{-1} makes a transition from the $n = 2$ to the $n = 3$ state.

P18.15 Evaluate the average kinetic and potential energies, $\langle E_{kinetic}\rangle$ and $\langle E_{potential}\rangle$, for the ground state $(n = 0)$ of the harmonic oscillator by carrying out the appropriate integrations.

P18.16 The vibrational frequency of $^{35}Cl_2$ is 1.68×10^{13} s^{-1}. Calculate the force constant of the molecule. How large a mass would be required to stretch a classical spring with this force constant by 2.25 cm? Use the gravitational acceleration on Earth at sea level for this problem.

P18.17 Evaluate $\langle x^2\rangle$ for the ground state $(n = 0)$ and first two excited states $(n = 1$ and $n = 2)$ of the quantum harmonic oscillator. Use the hint about evaluating integrals in Problem P18.12.

P18.18 A coin with a mass of 8.31 g suspended on a rubber band has a vibrational frequency of 7.50 s^{-1}. Calculate (a) the force constant of the rubber band, (b) the zero point energy, (c) the total vibrational energy if the maximum displacement is 0.725 cm, and (d) the vibrational quantum number corresponding to the energy in part (c).

P18.19 Calculate the position of the center of mass of (a) $^{1}H^{19}F$, which has a bond length of 91.68 pm, and (b) HD, which has a bond length of 74.15 pm.

P18.20 Show that the function $Y_2^0(\theta, \phi) = (5/16\pi)^{1/2}(3\cos^2\theta - 1)$ is normalized over the interval $0 \le \theta \le \pi$ and $0 \le \phi \le 2\pi$.

P18.21 Is it possible to simultaneously know the angular orientation of a molecule rotating in a two-dimensional space and its angular momentum? Answer this question by evaluating the commutator $[\phi, -i\hbar(\partial/\partial\phi)]$.

P18.22 The force constant for the $^{35}Cl_2$ molecule is 323 N m^{-1}. Calculate the vibrational zero point energy of this molecule. If this amount of energy were converted to translational energy, how fast would the molecule be moving? Compare this speed to the root mean square

speed from the kinetic gas theory, $|\mathbf{v}|_{rms} = \sqrt{3k_BT/m}$ for $T = 300.$ K.

P18.23 The force constant for a $^{1}H^{127}I$ molecule is 314 N m^{-1}.

a. Calculate the zero point vibrational energy for this molecule for a harmonic potential.

b. Calculate the light frequency needed to excite this molecule from the ground state to the first excited state.

P18.24 At 300. K, most molecules are not in their ground rotational state. Is this also true for their vibrational degree of freedom? Calculate $N_{n=1}/N_{n=0}$ and $N_{n=2}/N_{n=0}$ for the $^{127}I_2$ molecule for which the force constant is 172 N m^{-1}. At what temperature is $N_{n=2}/N_{n=0} = 0.500$? Repeat the calculation for H_2 and explain the difference in the results.

P18.25 An $^{1}H^{19}F$ molecule, with a bond length of 91.68 pm, absorbed on a surface rotates in two dimensions.

a. Calculate the zero point energy associated with this rotation.

b. What is the smallest quantum of energy that can be absorbed by this molecule in a rotational excitation?

P18.26 Verify that $\psi_1(x)$ in Equation 18.31 is a solution of the Schrödinger equation for the quantum harmonic oscillator. Determine the energy eigenvalue.

P18.27 Evaluate the average kinetic and potential energies, $\langle E_{kinetic}\rangle$ and $\langle E_{potential}\rangle$, for the second excited state $(n = 2)$ of the harmonic oscillator by carrying out the appropriate integrations.

P18.28 By substituting in the Schrödinger equation for the harmonic oscillator, show that the ground-state vibrational wave function is an eigenfunction of the total energy operator. Determine the energy eigenvalue.

P18.29 Evaluate the average linear momentum of the quantum harmonic oscillator $\langle p_x\rangle$ for the ground state $(n = 0)$ and first two excited states $(n = 1$ and $n = 2)$. Use the hint about evaluating integrals in Problem P18.12.

P18.30 By substituting in the Schrödinger equation for rotation in three dimensions, show that the rotational wave function $(5/16\pi)^{1/2}(3\cos^2\theta - 1)$ is an eigenfunction of the total energy operator. Determine the energy eigenvalue.

P18.31 Use $\sqrt{\langle x^2\rangle}$ as calculated in Problem P18.17 as a measure of the vibrational amplitude for a molecule. What fraction is $\sqrt{\langle x^2\rangle}$ of the 127.5 pm bond length of the $^{1}H^{35}Cl$ molecule for $n = 0$, 1, and 2? The force constant for H^{35}Cl is 516 N m^{-1}.

P18.32 Evaluate $\langle x\rangle$ for the ground state $(n = 0)$ and first two excited states $(n = 1$ and $n = 2)$ of the quantum harmonic oscillator. Use the hint about evaluating integrals given in Problem P18.12.

P18.33 Using your results for Problems P18.11, 17, 29, and 32, calculate the uncertainties in the position and momentum $\sigma_p^2 = \langle p^2\rangle - \langle p\rangle^2$ and $\sigma_x^2 = \langle x^2\rangle - \langle x\rangle^2$ for the ground state $(n = 0)$ and first two excited states $(n = 1$ and $n = 2)$ of the quantum harmonic oscillator. Compare your results with the predictions of the Heisenberg uncertainty principle.

P18.34 An $H^{35}Cl$ molecule has the rotational quantum number $J = 8$ and vibrational quantum number $n = 0$.

a. Calculate the rotational and vibrational energy of the molecule. Compare each of these energies with k_BT at 300. K.

b. Calculate the period for vibration and rotation. How many times does the molecule rotate during one vibrational period?

P18.35 Calculate the first five energy levels for a $^{35}Cl_2$ molecule, which has a bond length of 198.8 pm, (a) if it rotates freely in three dimensions and (b) if it is adsorbed on a surface and forced to rotate in two dimensions.

P18.36 Calculate the constants b_1 and b_2 in Equation (18.9) for the condition $x(0) = x_{max}$, the maximum extension of the oscillator. What is $v(0)$ for this condition?

P18.37 Calculate the reduced mass, the moment of inertia, the angular momentum, and the energy in the $J = 1$ rotational level for H_2, which has a bond length of 74.14 pm.

Web-Based Simulations, Animations, and Problems

W18.1 The motion of a particle in a harmonic potential is investigated, and the particle energy and force constant k are varied using sliders. The potential and kinetic energy are displayed as a function of the position x, and the result of measuring the probability of detecting the particle at x is displayed as a density plot. The student is asked to use the information gathered to explain the motion of the particle.

W18.2 The allowed energy levels for the harmonic oscillator are determined by numerical integration of the Schrödinger equation, starting in the classically forbidden region to the left of the potential. The criterion that the energy is an eigenvalue for the problem is that the wave function decays to zero in the classically forbidden region to the right of the potential. The zero point energy is determined for different values of k. The results are graphed to obtain a functional relationship between the zero point energy and k.

W18.3 The probability of finding the harmonic oscillator in the classically forbidden region P_n is calculated. The student generates a set of values for P_n for $n = 0, 1, 2, ..., 20$ and graphs them.

The Vibrational and Rotational Spectroscopy of Diatomic Molecules

C hemists have a wide range of spectroscopic techniques available to them. With these techniques, unknown molecules can be identified, bond lengths can be measured, and the force constants associated with chemical bonds can be determined. Spectroscopic techniques are based on transitions that occur between different energy states of molecules when they interact with electromagnetic radiation. In this chapter, we describe how light interacts with molecules to induce transitions between states. In particular, we discuss the absorption of electromagnetic radiation in the infrared and microwave regions of the spectrum. Light of these wavelengths induces transitions between eigenstates of vibrational and rotational energy.

19.1 An Introduction to Spectroscopy

The various forms of **spectroscopy** are among the most powerful tools that chemists have at their disposal to probe the world at an atomic and molecular level. In this chapter, we begin a discussion of molecular spectroscopy that will be taken up again in later chapters. Atomic spectroscopy will be discussed separately in Chapter 22. The information that is accessible through molecular spectroscopy includes bond lengths (rotational spectroscopy) and the vibrational frequencies of molecules (vibrational spectroscopy). In addition, the allowed energy levels for electrons in molecules can be determined with electronic spectroscopy, which is discussed in Chapter 25. This spectroscopic information is crucial for a deeper understanding of the chemical bonding and the reactivity of molecules. In most spectroscopies, atoms or molecules absorb electromagnetic radiation and undergo transitions between allowed quantum states.

In most experiments, the attenuation or enhancement of the incident radiation resulting from absorption or emission of radiation is measured as a function of the incident wavelength or frequency. Because quantum mechanical systems have a discrete energy

FIGURE 19.1

The electromagnetic spectrum depicted on a logarithmic wavelength scale.

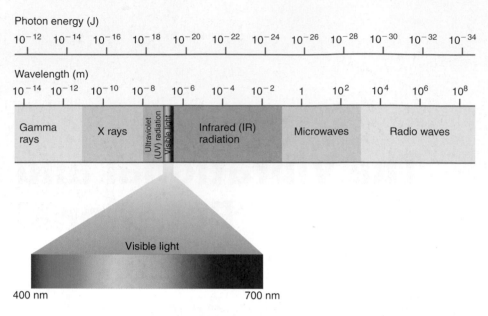

spectrum, an absorption or emission spectrum consists of individual peaks, each of which is associated with a transition between two allowed energy levels of the system. As we show later in Supplemental Section 19.9, the frequency at which energy is absorbed or emitted is related to the energy levels involved in the transitions by

$$h\nu = |E_2 - E_1| \tag{19.1}$$

The photon energy that is used in chemical spectroscopies spans more than 16 orders of magnitude in going from the radio frequency to the X-ray region. This is an indication of the very different energy-level spacings probed by these techniques. The energy-level spacing is smallest in nuclear magnetic resonance (NMR) spectroscopy, which is discussed in Chapter 28, and largest for electronic spectroscopy. Transitions between rotational and vibrational energy levels are intermediate between these two extremes, with rotational energy levels being more closely spaced in energy than vibrational energy levels. The electromagnetic spectrum is depicted schematically in Figure 19.1. Note that visible light is a very small part of this spectrum.

The spectral regions associated with various spectroscopies are shown in Table 19.1. Spectroscopists commonly use the quantity **wave number** $\tilde{\nu} = 1/\lambda$ which has units of inverse centimeters, rather than the wavelength λ or frequency ν to designate spectral transitions for historical reasons. The relationship between ν and $\tilde{\nu}$ is given by $\nu = \tilde{\nu}c$, where c is the speed of light. It is important to use consistent units when calculating the energy difference between states associated with a frequency in units of inverse seconds and wave numbers in units of inverse centimeters. Equation (19.1) expressed for both units is $|E_2 - E_1| = h\nu = hc\,\tilde{\nu}$.

The fact that atoms and molecules possess a set of discrete energy levels is an essential feature of all spectroscopies. If molecules had a continuous energy spectrum, it would be very difficult to distinguish them on the basis of their absorption spectra.

TABLE 19.1	**Important Spectroscopies and Their Spectral Range**				
Spectral Range	λ (m)	ν (Hz)	$\tilde{\nu}$ (cm^{-1})	Energy (J)	Spectroscopy
Radio	>0.1	$<3 \times 10^9$	>0.1	$<2 \times 10^{-24}$	NMR
Microwave	$0.001 - 0.1$	$3 \times 10^9 - 3 \times 10^{11}$	$0.1 - 10$	$2 \times 10^{-24} - 2 \times 10^{-22}$	Rotational
Infrared	$7 \times 10^{-7} - 1 \times 10^{-3}$	$3 \times 10^{11} - 4 \times 10^{14}$	$10 - 1 \times 10^4$	$2 \times 10^{-22} - 3 \times 10^{-19}$	Vibrational
Visible	$4 \times 10^{-7} - 7 \times 10^{-7}$	$4 \times 10^{14} - 7 \times 10^{14}$	$1 \times 10^4 - 3 \times 10^4$	$3 \times 10^{-19} - 5 \times 10^{-19}$	Electronic
Ultraviolet	$1 \times 10^{-8} - 4 \times 10^{-7}$	$7 \times 10^{14} - 3 \times 10^{16}$	$3 \times 10^4 - 1 \times 10^6$	$5 \times 10^{-19} - 2 \times 10^{-17}$	Electronic

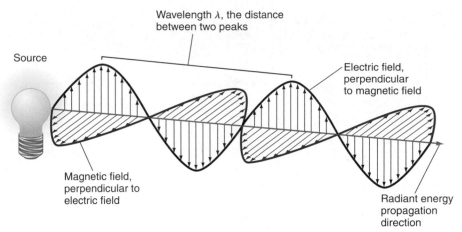

Wavelength λ, the distance between two peaks

Source

Electric field, perpendicular to magnetic field

Magnetic field, perpendicular to electric field

Radiant energy propagation direction

FIGURE 19.2
The electric and magnetic fields associated with a traveling light wave.

However, as discussed in Section 19.4, not all transitions between arbitrarily chosen states occur. **Selection rules** tell us which transitions will be experimentally observed. Because spectroscopies involve transitions between quantum states, we must first describe how electromagnetic radiation interacts with molecules.

We begin with a qualitative description of energy transfer from the electromagnetic field to a molecule leading to vibrational excitation. Light is a traveling electromagnetic wave that has magnetic and electric field components that are perpendicular to the propagation direction as shown in Figure 19.2. Consider the effect of a time-dependent electric field on a classical dipolar diatomic "molecule" constrained to move in one dimension. Such a molecule is depicted in Figure 19.3. If the spring were replaced by a rigid rod, the molecule could not take up energy from the field. However, the spring allows the two masses to oscillate about their equilibrium distance, thereby generating a periodically varying dipole moment. If the electric field and oscillation of the dipole moment have the same frequency, the molecule can absorb energy from the field. For a classical "molecule," any amount of energy can be taken up and the absorption spectrum is continuous.

For a real quantum mechanical molecule, the interaction with the electromagnetic field is similar. The electric field acts on a dipole moment within the molecule that can be of two types: permanent and dynamic. Polar molecules like HCl have a **permanent dipole moment.** As molecules vibrate, an additional induced **dynamic dipole moment** can be generated. How does the dynamic dipole arise? The magnitude of the dipole moment depends on the bond length and the degree to which charge is transferred from one atom to another. In turn, the charge transfer depends on the overlap of the electron densities of the atoms, which in turn depends on the internuclear distances. As the molecule vibrates, its dipole moment changes because of these effects, generating a dynamic dipole moment. Because the vibrational amplitude is a small fraction of the bond distance, the dynamic dipole moment is generally small compared to the permanent dipole moment.

As will be seen in the next section, it is the dynamic rather than the permanent dipole moment that determines if a molecule will absorb energy in the infrared region. By contrast, it is the permanent dipole moment that determines if a molecule will undergo rotational transitions by absorbing energy in the microwave region. Homonuclear diatomic molecules have neither permanent nor dynamic dipole moments and cannot absorb infrared radiation. However, vibrational spectroscopy on these molecules can be carried out using the Raman effect as discussed in Section 19.8 or in electronic spectroscopy as discussed in Chapter 25.

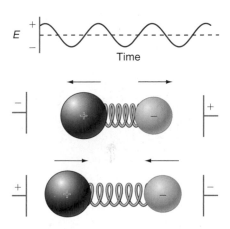

FIGURE 19.3
Schematic of the interaction of a classical harmonic oscillator constrained to move in one dimension under the influence of an electric field. The sinusoidally varying electric field shown at the top of the figure is applied between a pair of capacitor plates. The arrows indicate the direction of force on each of the two charged masses. If the phases of the field and vibration are as shown and the frequencies are equal, the oscillator will absorb energy in both the stretching and compression half cycles.

19.2 Absorption, Spontaneous Emission, and Stimulated Emission

We now move from a classical picture to a quantum mechanical description involving discrete energy levels. The basic processes by which photon-assisted transitions between energy levels occur are **absorption, spontaneous emission,** and **stimulated**

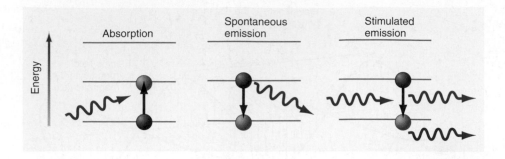

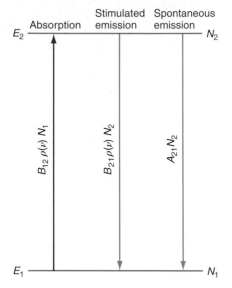

FIGURE 19.5
The rate at which transitions occur between two levels. It is in each case proportional to the product of the appropriate rate coefficient A_{21}, B_{12}, or B_{21} for the process and the population in the originating state, N_1 or N_2. For absorption and stimulated emission, the rate is additionally proportional to the radiation density $\rho(\nu)$.

emission. For simplicity, only transitions in a two-level system are considered as shown in Figure 19.4.

In absorption, the incident photon induces a transition to a higher level, and in emission, a photon is emitted as an excited state relaxes to one of lower energy. Absorption and stimulated emission are initiated by a photon incident on the molecule of interest. As the name implies, spontaneous emission is a random event and its rate is related to the lifetime of the excited state. These three processes are not independent in a system at equilibrium, as can be seen by considering Figure 19.5. At equilibrium, the overall transition rate from level 1 to 2 must be the same as that from 2 to 1. This means that

$$B_{12}\rho(\nu)N_1 = B_{21}\rho(\nu)N_2 + A_{21}N_2 \tag{19.2}$$

Whereas spontaneous emission is independent of the radiation density at a given frequency $\rho(\nu)$, the rates of absorption and stimulated emission are directly proportional to $\rho(\nu)$. The proportionality constants for the three processes are A_{21}, B_{12}, and B_{21}, respectively. Each of these rates is directly proportional to the number of molecules (N_1 or N_2) in the state from which the transition originates. This means that unless the lower state is populated, a signal will not be observed in an absorption experiment. Similarly, unless the upper state is populated, a signal will not be observed in an emission experiment.

The appropriate function to use for $\rho(\nu)$ in Equation (19.2) is the blackbody spectral density function of Equation (12.7), because $\rho(\nu)$ is the distribution of frequencies at equilibrium for a given temperature. Following this reasoning, Einstein concluded that

$$B_{12} = B_{21} \quad \text{and} \quad \frac{A_{21}}{B_{21}} = \frac{16\pi^2\hbar\nu^3}{c^3} \tag{19.3}$$

This result is derived in Example Problem 19.1.

EXAMPLE PROBLEM 19.1

Derive the equations $B_{12} = B_{21}$ and $A_{21}/B_{21} = 16\pi^2\hbar\nu^3/c^3$ using these two pieces of information: (1) the overall rate of transition between levels 1 and 2 (see Figure 19.5) is zero at equilibrium, and (2) the ratio of N_2 to N_1 is governed by the Boltzmann distribution.

Solution

The rate of transitions from level 1 to level 2 is equal and opposite to the transitions from level 2 to level 1. This gives the equation $B_{12}\rho(\nu)N_1 = B_{21}\rho(\nu)N_2 + A_{21}N_2$. The Boltzmann distribution function states that

$$\frac{N_2}{N_1} = \frac{g_2}{g_1}e^{-h\nu/k_BT}$$

In this case, $g_2 = g_1$. These two equations can be solved for $\rho(\nu)$, giving $\rho(\nu) = A_{21}/(B_{12}e^{h\nu/k_BT} - B_{21})$. As Planck showed, $\rho(\nu)$ has the form shown in Equation (12.7) so that

$$\rho(\nu) = \frac{A_{21}}{B_{12}e^{h\nu/k_BT} - B_{21}} = \frac{8\pi h\nu^3}{c^3}\frac{1}{e^{h\nu/k_BT} - 1}$$

For these two expressions to be equal, $B_{12} = B_{21}$ and $A_{21}/B_{21} = 8\pi h\nu^3/c^3 = 16\pi^2\hbar\nu^3/c^3$.

Spontaneous emission and stimulated emission differ in an important respect. Spontaneous emission is a completely random process, and the emitted photons are incoherent, by which we mean that their phases are random. In stimulated emission, the phase and direction of propagation are the same as that of the incident photon. This is referred to as coherent photon emission. A lightbulb is an **incoherent photon source.** The phase relation between individual photons is random, and because the propagation direction of the photons is also random, the intensity of the source falls off as the square of the distance. A laser is a **coherent source** of radiation. All photons are in phase, and because they have the same propagation direction, the divergence of the beam is very small. This explains why a laser beam that is reflected from the moon still has a measurable intensity when it returns to Earth. We will have more to say about lasers when atomic spectroscopy is discussed in Chapter 22.

19.3 An Introduction to Vibrational Spectroscopy

We now have a framework with which we can discuss spectroscopy as a chemical tool. Two features have enabled vibrational spectroscopy to achieve the importance that it has as a tool in chemistry. The first is that the vibrational frequency depends primarily on the identity of the two vibrating atoms on either end of the bond and to a much lesser degree on the presence of atoms farther away from the bond. This property generates characteristic frequencies for atoms joined by a bond known as **group frequencies.** We discuss group frequencies further in Section 19.5. The second feature is that a particular vibrational mode in a molecule has only one characteristic frequency of appreciable intensity. We discuss this feature next.

In any spectroscopy, transitions occur from one energy level to another. As discussed in Section 19.2, the energy level from which the transition originates must be occupied in order to generate a spectral signal. Which of the infinite set of vibrational levels has a substantial probability of being occupied? Table 19.2 shows the number of diatomic molecules in the first excited vibrational state (N_1) relative to those in the ground state (N_0) at 300. and 1000. K. The calculations have been carried out using the Boltzmann distribution. We see that nearly all the molecules in a macroscopic sample are in their ground vibrational state at room temperature because $N_1/N_0 \ll 1$. Even at 1000. K, N_1/N_0 is very small except for Br_2. This means that for these molecules, absorption of light at the characteristic frequency will occur from molecules in the $n = 0$ state. What final states are possible? As shown in the next section, for absorption by a quantum mechanical harmonic oscillator, $\Delta n = n_{final} - n_{initial} = +1$. Because only the $n = 0$ state has an appreciable population, with few exceptions only the $n = 0 \rightarrow n = 1$ transition is observed in vibrational spectroscopy.

TABLE 19.2 Vibrational State Populations for Selected Diatomic Molecules

Molecule	$\tilde{\nu}$ (cm^{-1})	ν (s^{-1})	N_1/N_0 for 300. K	N_1/N_0 for 1000. K
H—H	4400	1.32×10^{14}	6.88×10^{-10}	1.78×10^{-3}
H—F	4138	1.24×10^{14}	2.42×10^{-9}	2.60×10^{-3}
H—Br	2649	7.94×10^{13}	3.05×10^{-6}	2.21×10^{-2}
N—N	2358	7.07×10^{13}	1.23×10^{-5}	3.36×10^{-2}
C—O	2170	6.51×10^{13}	3.03×10^{-5}	4.41×10^{-2}
Br—Br	323	9.68×10^{12}	0.213	0.628

EXAMPLE PROBLEM 19.2

A strong absorption of infrared radiation is observed for $^1H^{35}Cl$ at 2991 cm^{-1}.

- Calculate the force constant k for this molecule.
- By what factor do you expect this frequency to shift if deuterium is substituted for hydrogen in this molecule? The force constant is unaffected by this substitution.

Solution

a. We first write $\Delta E = h\nu = hc/\lambda = \hbar\sqrt{k/\mu}$. Solving for k,

$$k = 4\pi^2\left(\frac{c}{\lambda}\right)^2 \mu$$

$$= 4\pi^2(2.998 \times 10^8 \text{ m s}^{-1})^2 \left(\frac{2991}{\text{cm}} \times \frac{100 \text{ cm}}{1 \text{ m}}\right)^2 \frac{(1.008)(34.969) \text{ amu}}{35.977}$$

$$\times \left(\frac{1.661 \times 10^{-27} \text{ kg}}{1 \text{ amu}}\right)$$

$$= 516.3 \text{ N m}^{-1}$$

b. $\dfrac{\nu_{DCl}}{\nu_{HCl}} = \sqrt{\dfrac{\mu_{HCl}}{\mu_{DCl}}} = \sqrt{\dfrac{m_H m_{Cl}(m_D + m_{Cl})}{m_D m_{Cl}(m_H + m_{Cl})}} = \sqrt{\left(\dfrac{1.0078}{2.0140}\right)\left(\dfrac{36.983}{35.977}\right)}$

$$= 0.717$$

The vibrational frequency for DCl is lower by a substantial amount. Would the shift be as great if ^{37}Cl were substituted for ^{35}Cl? The fact that vibrational frequencies are so strongly shifted by isotopic substitution of deuterium for hydrogen makes infrared spectroscopy a valuable tool for determining the presence of hydrogen atoms in molecules.

Note that the high sensitivity available in modern instrumentation to carry out vibrational spectroscopy does make it possible in favorable cases to see vibrational transitions originating from the $n = 0$ state for which $\Delta n = +2, +3, \ldots$. These **overtone** transitions are much weaker than the $\Delta n = +1$ absorption but are possible because the selection rule $\Delta n = +1$ is not rigorously obeyed for an anharmonic potential, as discussed later. This more advanced topic is explored in Problem P19.22 at the end of the chapter.

The overtone transitions are useful because they allow us to determine the degree to which real molecular potentials differ from the simple **harmonic potential,** $V(x) = (1/2)kx^2$. To a good approximation, a realistic **anharmonic potential** can be described in analytical form by the **Morse potential:**

$$V(x) = D_e\left[1 - e^{-\alpha(x-x_e)}\right]^2 \tag{19.4}$$

in which D_e is the dissociation energy relative to the bottom of the potential and $\alpha = \sqrt{k/(2D_e)}$. The force constant k for the Morse potential is defined by $k = (d^2V/dx^2)_{x=x_e}$ just as for the harmonic potential. The **bond energy** D_0 is defined with respect to the lowest allowed level, rather than to the bottom of the potential, as shown in Figure 19.6.

The energy levels for this potential are given by

$$E_n = h\nu\left(n + \frac{1}{2}\right) - \frac{(h\nu)^2}{4D_e}\left(n + \frac{1}{2}\right)^2 \tag{19.5}$$

The second term gives the anharmonic correction to the energy levels. Measurements of the frequencies of the overtone vibrations allow the parameter D_e in the Morse potential to be determined for a specific molecule. This provides a useful method for determining the details of the interaction potential in a molecule. CO is an example of a diatomic molecule for which overtone vibrations are easily observed.

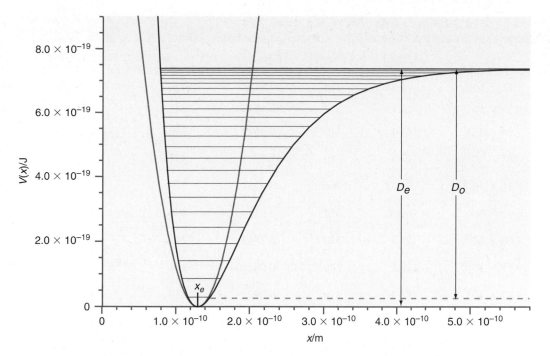

FIGURE 19.6
Morse potential, $V(x)$ (red curve), as a function of the bond length x for HCl, using the parameters from Example Problem 19.3. The zero of energy is chosen to be the bottom of the potential. The purple curve shows a harmonic potential, which is a good approximation to the Morse potential near the bottom of the well. The horizontal lines indicate allowed energy levels in the Morse potential. D_e and D_0 represent the bond energies defined with respect to the bottom of the potential and the lowest state, respectively, and x_e is the equilibrium bond length.

EXAMPLE PROBLEM 19.3

The Morse potential can be used to model dissociation as illustrated in this example. The $^1H^{35}Cl$ molecule can be described by a Morse potential with $D_e = 7.41 \times 10^{-19}$ J. The force constant k for this molecule is 516.3 N m^{-1} and $\nu = 8.97 \times 10^{13}$ s^{-1}. Calculate the number of allowed vibrational states in this potential and the bond energy for the $^1H^{35}Cl$ molecule.

Solution

We solve the equation

$$E_n = h\nu\left(n + \frac{1}{2}\right) - \frac{(h\nu)^2}{4D_e}\left(n + \frac{1}{2}\right)^2 = D_e$$

to obtain the highest value of n consistent with the potential. Using the parameters given earlier, we obtain the following equation for n:

$$-1.1918 \times 10^{-21}\, n^2 + 5.8243 \times 10^{-20}\, n + 2.942 \times 10^{-20} = 7.41 \times 10^{-19}$$

Both solutions to this quadratic equation give $n = 24.4$, so we conclude that the potential has 24 allowed levels. If the left side of the equation is graphed versus n, we obtain the results shown in the following figure.

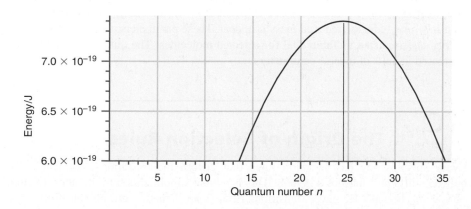

TABLE 19.3 Values of Molecular Constants for Selected Diatomic Molecules

	$\widetilde{\nu}$ (cm^{-1})	$\widetilde{\nu}$ (s^{-1})	x_e (pm)	k (Nm^{-1})	B (cm^{-1})	D_0 (kJ mol^{-1})	D_0 (J molecule^{-1})
H_2	4401	1.32×10^{14}	74.14	575	60.853	436	7.24×10^{-19}
D_2	3115	9.33×10^{13}	74.15	577	30.444	443	7.36×10^{-19}
$^1H^{81}Br$	2649	7.94×10^{13}	141.4	412	8.4649	366	6.08×10^{-19}
$^1H^{35}Cl$	2991	8.97×10^{13}	127.5	516	10.5934	432	7.17×10^{-19}
$^1H^{19}F$	4138	1.24×10^{14}	91.68	966	20.9557	570	9.46×10^{-19}
$^1H^{127}I$	2309	6.92×10^{13}	160.92	314	6.4264	298	4.95×10^{-19}
$^{35}Cl_2$	559.7	1.68×10^{13}	198.8	323	0.2440	243	4.03×10^{-19}
$^{79}Br_2$	325.3	9.75×10^{12}	228.1	246	0.082107	194	3.22×10^{-19}
$^{19}F_2$	916.6	2.75×10^{13}	141.2	470	0.89019	159	2.64×10^{-19}
$^{127}I_2$	214.5	6.43×10^{12}	266.6	172	0.03737	152	2.52×10^{-19}
$^{14}N_2$	2359	7.07×10^{13}	109.8	2295	1.99824	945	1.57×10^{-18}
$^{16}O_2$	1580.	4.74×10^{13}	120.8	1177	1.44563	498	8.27×10^{-19}
$^{12}C^{16}O$	2170.	6.51×10^{13}	112.8	1902	1.9313	1076	1.79×10^{-18}

Source: Lide, D. R., ed. *CRC Handbook of Chemistry and Physics.* 83rd ed. Boca Raton, FL: CRC Press, 2003.

Note that E_n decreases for $n > 25$. This is mathematically correct, but unphysical because for $n > 24$, the molecule has a continuous energy spectrum, and Equation (19.5) is no longer valid.

The bond energy D_0 is not D_e but $D_e - E_0$ where

$$E_0 = \frac{h\nu}{2} - \frac{(h\nu)^2}{16D_e}$$

from Equation (19.5), because the molecule has a zero point vibrational energy. Using the parameters given earlier, the bond energy is 7.11×10^{-19} J. The Morse and harmonic potentials as well as the allowed energy levels for this molecule are shown in Figure 19.6.

The material-dependent parameters that determine the frequencies observed in vibrational spectroscopy for diatomic molecules are the force constant k and the reduced mass μ. The corresponding parameters for rotational spectroscopy (see Section 19.6) are the rotational constant $B = h/(8\pi^2 c \mu r_0^2)$ in which the bond length, r_0 or x_e, and the reduced mass μ appear. These parameters, along with the bond energy D_0, are listed in Table 19.3 for selected molecules. The quantities B and $\widetilde{\nu}$ are expressed in units of inverse centimeters.

19.4 The Origin of Selection Rules

Every spectroscopy has selection rules that govern the transitions that can occur between different states of a system. This is a great simplification in the interpretation of spectra, because far fewer transitions occur than if there were no selection rules. How do these selection rules arise? We next derive the selection rules for vibrational spectroscopy based on the quantum mechanical harmonic oscillator.

As discussed later in Supplemental Section 19.9, the transition probability from state n to state m is only nonzero if the **transition dipole moment** μ_x^{mn} satisfies the following condition:

$$\mu_x^{mn} = \int \psi_m^*(x)\mu_x(x_e + x)\psi_n(x)\, dx \neq 0 \qquad (19.6)$$

In this equation, x is the vibrational amplitude and μ_x is the dipole moment along the electric field direction, which we take to be the x axis.

In the following discussion, we show how selection rules for vibrational excitation arise from Equation (19.6). As discussed in Section 19.1, the dipole moment μ_x will change slightly as the molecule vibrates. Because the amplitude of vibration x is an oscillatory function of t, the molecule has a time-dependent dynamic dipole moment. We take this into account by expanding μ_x in a Taylor series about the equilibrium bond length. Because x is the amplitude of vibration, the equilibrium bond length x_e corresponds to $x = 0$:

$$\mu_x(x_e + x(t)) = \mu_{0x_e} + x(t)\left(\frac{d\mu_x}{dx}\right)_{x=0} + \dots \qquad (19.7)$$

in which the values of μ_{0x} and $(d\mu_x/dx)_{x=0}$ depend on the molecule under consideration. Note that because $x = x(t)$, $\mu_x(x_e + x(t))$ is a function of time. The first term in Equation (19.7) is the permanent dipole moment at the equilibrium position, and the second term is the dynamic dipole moment. As we saw earlier, for absorption experiments, it is reasonable to assume that only the $n = 0$ state is populated. Using Equation (18.29), which gives explicit expressions for the eigenfunctions m,

$$\mu_x^{m0} = A_m A_0 \mu_{0x} \int_{-\infty}^{\infty} H_m(\alpha^{1/2}x)\, H_0(\alpha^{1/2}x)e^{-\alpha x^2}\, dx$$

$$+ A_m A_0\left[\left(\frac{d\mu_x}{dx}\right)_{x=0}\right]\int_{-\infty}^{\infty} H_m(\alpha^{1/2}x)\, x\, H_0(\alpha^{1/2}x)e^{-\alpha x^2}\, dx \qquad (19.8)$$

The first integral is zero because different eigenfunctions are orthogonal. To solve the second integral, we need to use the specific functional form of $H_m(\alpha^{1/2}x)$. However, because the integration is over the symmetric interval $-\infty < x < \infty$, this integral is zero if the integrand is an odd function of x. As Equation (18.31) shows, the Hermite polynomials $H_m(\alpha^{1/2}x)$ are odd functions of x if m is odd and even functions of x if m is even. The term $x\, H_0(\alpha^{1/2}x)e^{-\alpha x^2}$ in the integrand is an odd function of x and, therefore, μ_x^{m0} is zero if $H_m(\alpha^{1/2}x)$ is even. This simplifies the problem because only transitions of the type

$$n = 0 \rightarrow m = 2b + 1, \quad \text{for } b = 0, 1, 2, \dots \qquad (19.9)$$

can have nonzero values for μ_x^{m0}.

Do all the transitions indicated in Equation (19.9) lead to nonzero values of μ_x^{m0}? To answer this question, the integrand $H_m(\alpha^{1/2}x)x\, H_0(\alpha^{1/2}x)\, e^{-\alpha x^2}$ is graphed against x for the transitions $n = 0 \rightarrow m = 1$, $n = 0 \rightarrow m = 3$, and $n = 0 \rightarrow m = 5$ in Figure 19.7. Whereas the integrand is positive everywhere for the $n = 0 \rightarrow n = 1$ transition, the areas above the dashed line exactly cancel those below the line for the $n = 0 \rightarrow m = 3$ and $n = 0 \rightarrow m = 5$ transitions, showing that $\mu_x^{mn} = 0$. Therefore, $\mu_x^{m0} \neq 0$ only for the first of the three transitions shown and $\mu_x^{m0} = 0$ for $\Delta n \neq +1$. It can be shown more generally that in the dipole approximation, the selection rule for absorption is $\Delta n = +1$, and for emission, it is $\Delta n = -1$. Selection rules are different for different spectroscopies. However, within the dipole approximation, the selection rules for any spectroscopy are calculated using Equation (19.6) and the appropriate total energy eigenfunctions.

Note that because we found that the first integral in Equation (19.8) was zero, the absence or presence of a permanent dipole moment μ_{0x} is not relevant for the absorption of infrared radiation. For vibrational excitation to occur, the dynamic dipole moment must be nonzero. Because of this condition, homonuclear diatomic molecules do not absorb light in the infrared. This has important consequences for our environment. The temperature of Earth is determined primarily by an energy balance between visible and ultraviolet (UV) radiation absorbed from the sun and infrared radiation emitted by

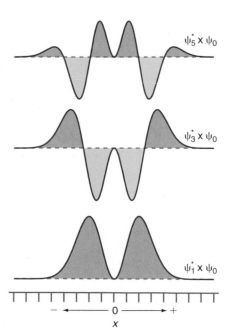

FIGURE 19.7
The integrand $H_m(\alpha^{1/2}x) \times H_0(\alpha^{1/2}x)e^{-\alpha x^2}$ is graphed as a function of x for the transitions $n = 0 \rightarrow m = 1$, $n = 0 \rightarrow m = 3$, and $n = 0 \rightarrow m = 5$. The dashed line shows the zero level for each graph.

FIGURE 19.8
The atmospheric CO_2 concentration is shown as a function of time since the industrial revolution began.

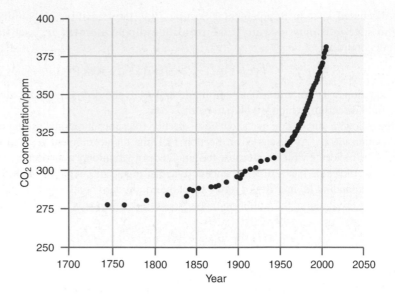

the planet. The molecules N_2, O_2, and H_2, which have no permanent or transient dipole moment, together with the rare gases make up 99.93% of the atmosphere. These gases do not absorb the infrared radiation emitted by Earth. Therefore, almost all of the emitted infrared radiation passes through the atmosphere and escapes into space. By contrast, greenhouse gases such as CO_2, NO_x, H_2O, and hydrocarbons absorb the infrared radiation emitted by Earth and radiate a portion of it back to Earth. However, as you will conclude in Problem P19.15 at the end of the chapter, not all the vibrational modes of CO_2 are infrared active. The concentration of CO_2 in the atmosphere has risen significantly since the beginning of the industrial revolution as shown in Figure 19.8. The result is an increase in Earth's temperature and global warming.

19.5 **Infrared Absorption Spectroscopy**

The most basic result of quantum mechanics is that atoms and molecules possess a discrete energy spectrum and that energy can only be absorbed or emitted in amounts that correspond to the difference between two energy levels. Because the energy spectrum for each chemical species is unique, the allowed transitions between these levels provide a "fingerprint" for that species. Using such a fingerprint to identify and quantify the species is a primary role of all chemical spectroscopies. For a molecule of known composition, the vibrational spectrum can also be used to determine the symmetry of the molecule and the force constants associated with the characteristic vibrations.

In absorption spectroscopies in general, electromagnetic radiation from a source of the appropriate wavelength is incident on a sample that is confined in a cell. The chemical species in the sample undergo transitions that are allowed by the appropriate selection rules among rotational, vibrational, or electronic states. The incident light of intensity $I_0(\lambda)$ is attenuated in passing a distance dl through the sample as described by the differential form of the **Beer–Lambert law** in which M is the concentration of the absorber; $I(\lambda)$ is the intensity of the transmitted light leaving the cell. Units of moles per liter are commonly used for M in liquid solutions, and partial pressure is used for gas mixtures:

$$dI(\lambda) = -\varepsilon(\lambda)M\,I(\lambda)\,dl \qquad (19.10)$$

This equation can be integrated to give

$$\frac{I(\lambda)}{I_0(\lambda)} = e^{-\varepsilon(\lambda)Ml} \qquad (19.11)$$

The information on the discrete energy spectrum of the chemical species in the cell is contained in the wavelength dependence of the **molar absorption coefficient** $\varepsilon(\lambda)$. It is evident that the strength of the absorption is proportional to $I(\lambda)/I_0(\lambda)$, which increases with $\varepsilon(\lambda)$, M, and with path length l. Because $\varepsilon(\lambda)$ is a function of the wavelength, absorption spectroscopy experiments typically consist of the elements shown in Figure 19.9. In the most basic form of this spectroscopy, a **monochromator** is used to separate the broadband radiation from the source into its constituent wavelengths. After passing through the sample, the transmitted light impinges on the detector. With this setup, only one wavelength can be measured at a time. This form of absorption spectroscopy is unnecessarily time consuming in comparison with Fourier transform techniques, which are discussed in Supplemental Section 19.7.

Light source

EXAMPLE PROBLEM 19.4

The molar absorption coefficient $\varepsilon(\lambda)$ for ethane is 40. $(\text{cm bar})^{-1}$ at a wavelength of 12 µm. Calculate $I(\lambda)/I_0(\lambda)$ in a 1.0-cm-long absorption cell if ethane is present at a contamination level of 2.0 ppm in one bar of air. What cell length is required to make $I(\lambda)/I_0(\lambda) = 0.90$?

Solution

Using $\dfrac{I(\lambda)}{I_0(\lambda)} = e^{-\varepsilon(\lambda)Ml}$

$$\frac{I(\lambda)}{I_0(\lambda)} = \exp\left\{-\left[40.(\text{cm bar})^{-1}(2.0 \times 10^{-6}\,\text{bar})(1.0\,\text{cm})\right]\right\} = 0.9992 \approx 1.0$$

This result shows that for this cell length, light absorption is difficult to detect. Rearranging the Beer–Lambert equation, we have

$$l = -\frac{1}{M\varepsilon(\lambda)}\ln\left(\frac{I(\lambda)}{I_0(\lambda)}\right) = -\frac{1}{40.(\text{cm bar})^{-1}(2.0 \times 10^{-6}\,\text{bar})}\ln(0.90)$$

$$= 1.3 \times 10^3\,\text{cm}$$

Path lengths of this order are possible in sample cells in which the light undergoes multiple reflections from mirrors outside of the cell. Even longer path lengths are possible in cavity ringdown spectroscopy. In this method, the absorption cell is mounted between two highly focusing mirrors with a reflectivity greater than 99.99%. Because of the many reflections that take place between the mirrors without appreciable attenuation of the light, the effective length of the cell is very large. The detection sensitivity to molecules such as NO_2 is less than 10 parts per billion using this technique.

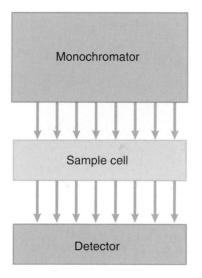

FIGURE 19.9
In an absorption experiment, the dependence of the sample absorption on wavelength is determined. A monochromator is used to filter out a particular wavelength from the broadband light source.

How does $\varepsilon(\lambda)$ depend on the wavelength or frequency? We know that for a harmonic oscillator, $\nu = (1/2\pi)\sqrt{k/\mu}$ so that the masses of the atoms and the force constant of the bond determine the resonant frequency. Now consider a molecule such as

$$\begin{array}{c} \text{O} \\ \parallel \\ \text{R}-\text{C}-\text{R}' \end{array}$$

The vibrational frequency of the C and O atoms in the carbonyl group is determined by the force constant for the $C{=}O$ bond. This force constant is primarily determined by the chemical bond between these atoms and to a much lesser degree by the adjacent R and R′ groups. For this reason, the carbonyl group has a characteristic frequency at which it absorbs infrared radiation that varies in a narrow range for different molecules. These group frequencies are very valuable in determining the structure of molecules, and an illustrative set is shown in Table 19.4.

TABLE 19.4	Selected Group Frequencies		
Group	Frequency (cm^{-1})	Group	Frequency (cm^{-1})
O—H stretch	3450–3650	C=O stretch	1650–1750
N—H stretch	3300–3500	C=C stretch	1620–1680
C—H stretch	2800–3000	C—C stretch	1200–1300
C—H bend	1450–1480	C—Cl stretch	600–800

We have shown that a diatomic molecule has a single vibrational peak of appreciable intensity because the overtone frequencies have very low intensities. How many vibrational peaks are observed for larger molecules in an infrared absorption experiment? A molecule consisting of n atoms has three translational degrees of freedom, and two or three rotational degrees of freedom depending on whether it is a linear or nonlinear molecule. The remaining $3n - 6$ (nonlinear molecule) or $3n - 5$ (linear) degrees of freedom are vibrational modes. For example, benzene has 30 vibrational modes. However, some of these modes have the same frequency (they are degenerate in energy), so that benzene has only 20 distinct vibrational frequencies.

We now examine some experimental data. Vibrational spectra for gas-phase CO and CH_4 are shown in Figure 19.10. Because CO and CH_4 are linear and nonlinear molecules, we expect one and nine vibrational modes, respectively. However, the spectrum for CH_4 shows two rather than nine peaks that we might associate with vibrational transitions. We also see several unexpected broad peaks in the CH_4 spectrum. The single peak in the CO spectrum is much broader than would be expected for a vibrational peak, and it has a deep minimum at the central frequency.

These spectra look different than expected for two reasons. The broadening in the CO absorption peak and the broad envelopes of additional peaks for CH_4 result from transitions between different rotational energy states that occur simultaneously with the $n = 0 \rightarrow n = 1$ transition between vibrational energy levels. We discuss transitions between rotational energy levels and analyze a high-resolution infrared absorption spectrum for a diatomic molecule in some detail in Section 19.6. At this point we simply note that absorption of infrared radiation results in both rotational and vibrational rather than just vibrational transitions.

The second unexpected feature in Figure 19.10 is that two and not nine peaks are observed in the CH_4 spectrum. Why is this? To discuss the vibrational modes of polyatomic species in more detail, the information about molecular symmetry and group theory discussed in Chapter 27 is needed. At this point, we simply state the results. In applying group theory to the CH_4 molecule, the 1306 cm^{-1} peak can be associated with three degenerate C—H bending modes, and the 3020 cm^{-1} peak can be associated with three degenerate C—H stretching modes. This still leaves three vibrational modes unaccounted for. Again applying group theory to the CH_4 molecule, one finds that these modes are symmetric and do not satisfy the condition $d\mu_x/dx \neq 0$. Therefore, they are infrared inactive. However, the stretching mode for CO and all modes for CH_4 are active in Raman spectroscopy, which we discuss in Supplemental Section 19.8.

Of the 30 vibrational modes for benzene, four peaks (corresponding to 7 of the 30 modes) are observed in infrared spectroscopy, and seven peaks (corresponding to 12 of the 30 modes) are observed in Raman spectroscopy. None of the frequencies is observed in both Raman and infrared spectroscopy. Eleven vibrational modes are neither infrared nor Raman active.

Although the discussion to this point might lead us to believe that each bonded pair of atoms in a molecule vibrates independently of the others, this is not the case. For example, we might think that the linear CO_2 molecule has a single C=O stretching

FIGURE 19.10

Infrared absorption spectra of gaseous CO and CH_4. The curves are offset vertically for clarity.

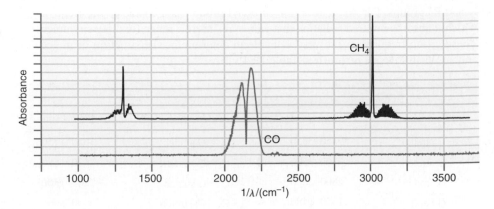

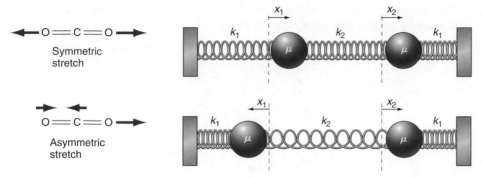

frequency, because the two C=O bonds are equivalent. However, experiments show that this molecule has two distinct C=O stretching frequencies. Why is this the case? When one C=O bond vibrates, the atomic positions and electron distribution throughout the molecule are changed, thereby influencing the other C=O bond. In other words, we can view the CO_2 molecule as consisting of two coupled harmonic oscillators. In the center of mass coordinates, each of the two C=O groups is modeled as a mass coupled to a wall by a spring with force constant k_1 (see Section 18.1). We model the coupling as a second spring with force constant k_2 that connects the two oscillators. The model is depicted in Figure 19.11.

This coupled system has two vibrational frequencies: the symmetrical and antisymmetric modes. In the symmetrical mode, the vibrational amplitude is equal in both magnitude and sign for the individual oscillators. In this case, the C atom does not move. This is equivalent to the coupling spring in Figure 19.11 having the same length during the whole vibrational period. Therefore, the vibrational frequency is unaffected by the coupling and is given by

$$\nu_{symmetric} = \frac{1}{2\pi}\sqrt{\frac{k_1}{\mu}} \qquad (19.12)$$

In the antisymmetric mode, the C atom does move. This is equivalent to the vibrational amplitude being equal in magnitude and opposite in sign for the individual oscillators. In this mode, the spring representing the coupling is doubly stretched, once by each of the oscillators. The resulting force on the reduced mass representing each oscillator is

$$F = -(k_1 + 2k_2)x \qquad (19.13)$$

and the resulting frequency of this antisymmetric mode is

$$\nu_{antisymmetric} = \frac{1}{2\pi}\sqrt{\frac{k_1 + 2k_2}{\mu}} \qquad (19.14)$$

We see that the C=O bond coupling gives rise to two different vibrational stretching frequencies and that the antisymmetric mode has the higher frequency as illustrated in Figure 19.12 for H_2O.

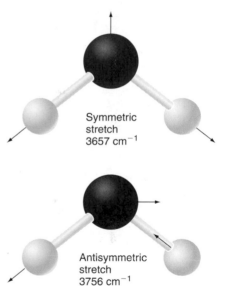

Symmetric
stretch
3657 cm^{-1}

Antisymmetric
stretch
3756 cm^{-1}

19.6 Rotational Spectroscopy

As for the harmonic oscillator, a selection rule governs the absorption of electromagnetic energy for a molecule to change its rotational energy, namely, $\Delta J = J_{final} - J_{initial} = \pm 1$. Although we do not derive this selection rule, Example Problem 19.5 shows that it holds for a specific case.

Note that we have just changed the symbol for the angular momentum quantum number from l to J. The quantum number l is usually used for orbital angular momentum (for example, the electron orbiting around the nucleus), and J is usually used for rotating molecules.

EXAMPLE PROBLEM 19.5

Using the following total energy eigenfunctions for the three-dimensional rigid rotor, show that the $J = 0 \rightarrow J = 1$ transition is allowed and that the $J = 0 \rightarrow J = 2$ transition is forbidden:

$$Y_0^0(\theta, \phi) = \frac{1}{(4\pi)^{1/2}}$$

$$Y_1^0(\theta, \phi) = \left(\frac{3}{4\pi}\right)^{1/2} \cos \theta$$

$$Y_2^0(\theta, \phi) = \left(\frac{5}{16\pi}\right)^{1/2} (3\cos^2 \theta - 1)$$

The notation $Y_J^{M_J}$ is used for the preceding functions.

Solution

Assuming the electromagnetic field to lie along the z axis, $\mu_z = \mu \cos \theta$, and the transition dipole moment takes the form

$$\mu_z^{J0} = \mu \int_0^{2\pi} d\phi \int_0^\pi Y_J^0(\theta, \phi)(\cos \theta) Y_0^0(\theta, \phi) \sin \theta \, d\theta$$

For the $J = 0 \rightarrow J = 1$ transition,

$$\mu_z^{10} = \mu \frac{\sqrt{3}}{4\pi} \int_0^{2\pi} d\phi \int_0^\pi \cos^2 \theta \sin \theta \, d\theta = \frac{\mu\sqrt{3}}{2}\left[-\frac{\cos^3 \theta}{3}\right]_{\theta=0}^{\theta=\pi} = \frac{\mu\sqrt{3}}{3} \neq 0$$

For the $J = 0 \rightarrow J = 2$ transition,

$$\mu_z^{20} = \mu \frac{\sqrt{5}}{8\pi} \int_0^{2\pi} d\phi \int_0^\pi (3\cos^2 \theta - 1) \cos \theta \sin \theta \, d\theta$$

$$= \mu \frac{\sqrt{5}}{8\pi}\left[-\frac{3\cos^4 \theta}{4} + \frac{\cos^2 \theta}{2}\right]_{\theta=0}^{\theta=\pi}$$

$$= \mu \frac{\sqrt{5}}{8\pi}\left[-\frac{1}{4} + \frac{1}{4}\right] = 0$$

The preceding calculations show that the $J = 0 \rightarrow J = 1$ transition is allowed and that the $J = 0 \rightarrow J = 2$ transition is forbidden.

E

Time

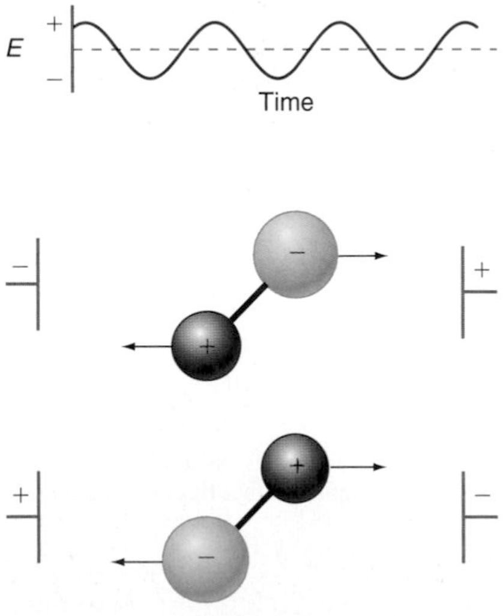

FIGURE 19.13

The interaction of a rigid rotor with an electric field. Imagine the sinusoidally varying electric field shown at the top of the figure applied between a pair of capacitor plates. The arrows indicate the direction of force on each of the two charged masses. If the frequencies of the field and rotation are equal, the rotor will absorb energy from or emit energy into the electric field.

In discussing vibrational spectroscopy, we learned that a molecule must have a nonzero dynamic dipole moment to absorb infrared radiation. By contrast, a molecule must have a permanent dipole moment to absorb energy in the microwave frequency range in which rotational transitions occur. As was the case for vibrational spectroscopy, the dominant interaction with the electric field is through the dipole moment. This is shown schematically in Figure 19.13.

As shown in Section 18.5, the dependence of the rotational energy on the quantum number is given by

$$E = \frac{\hbar^2}{2\mu r_0^2} J(J + 1) = \frac{h^2}{8\pi^2 \mu r_0^2} J(J + 1) = hcBJ(J + 1) \qquad (19.15)$$

In this equation, the constants specific to a molecule are combined in the so-called **rotational constant** $B = h/(8\pi^2 c\mu r_0^2)$. The factor c is included in B so that it has the units of inverse centimeters rather than inverse seconds. The energy levels and transitions allowed by the selection rule $\Delta J = J_{final} - J_{initial} = \pm 1$ as well as a simulated rotational spectrum are shown in Figure 19.14.

We can calculate the energy corresponding to rotational transitions for $\Delta J = +1$ and $\Delta J = -1$ originating from energy level J. $\Delta J = +1$ corresponds to absorption and $\Delta J = -1$ corresponds to emission of a photon. In the following equations, J is the quantum number of the state from which the transition occurs.

$$\Delta E = E(J_{final}) - E(J_{initial})$$

for $\Delta J = +1$

$$\Delta E_+ = \frac{\hbar^2}{2\mu r_0^2}(J+1)(J+2) - \frac{\hbar^2}{2\mu r_0^2}J(J+1)$$

$$= \frac{\hbar^2}{2\mu r_0^2}\left[(J^2 + 3J + 2) - (J^2 + J)\right] = \frac{\hbar^2}{\mu r_0^2}(J+1)$$

$$= 2hcB(J+1) \quad \text{and for } \Delta J = -1$$

$$\Delta E_- = \frac{\hbar^2}{2\mu r_0^2}(J-1)J - \frac{\hbar^2}{2\mu r_0^2}J(J+1)$$

$$= \frac{\hbar^2}{2\mu r_0^2}\left[(J^2 - J) - (J^2 + J)\right] = -\frac{\hbar^2}{\mu r_0^2}J = -2hcBJ \quad \textbf{(19.16)}$$

Note that $|\Delta E_+| \neq |\Delta E_-|$ because the energy levels are not equally spaced. We see that the larger the J value of the originating energy level, the more energetic the photon must be to promote excitation to the next highest energy level. Because the rotational energy does not depend on m_J, each energy level is $2J + 1$-fold degenerate.

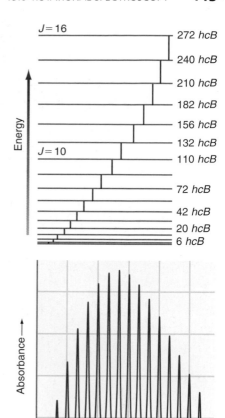

FIGURE 19.14
The energy levels for a rigid rotor are shown in the top panel and the spectrum observed through absorption of microwave radiation is shown in the bottom panel. The allowed transitions between levels are shown as vertical bars.

EXAMPLE PROBLEM 19.6

Because of the very high precision of frequency measurements, bond lengths can be determined with a correspondingly high precision, as illustrated in this example. From the rotational microwave spectrum of $^1H^{35}Cl$, we find that $B = 10.59342$ cm^{-1}. Given that the masses of ^1H and ^{35}Cl are 1.0078250 and 34.9688527 amu, respectively, determine the bond length of the $^1H^{35}Cl$ molecule.

Solution

$$B = \frac{h}{8\pi^2 \mu c r_0^2}$$

$$r_0 = \sqrt{\frac{h}{8\pi^2 \mu c B}}$$

$$= \sqrt{\frac{6.62606957 \times 10^{-34} \text{ J s}}{8\pi^2 c \left(\dfrac{(1.0078250)(34.9688527) \text{ amu}}{1.0078250 + 34.9688527}\right)(1.66054 \times 10^{-27} \text{ kg amu}^{-1})(10.59342 \text{ cm}^{-1})}}$$

$$= 1.274551 \times 10^{-10} \text{ m}$$

The structure of a rotational spectrum becomes more apparent when we consider the energy-level spacing in more detail. Table 19.5 shows the frequencies needed to excite various transitions consistent with the selection rule $\Delta J = J_{final} - J_{initial} = +1$ in general and also for $^1H^{35}Cl$. Each of these transitions can lead to absorption of electromagnetic radiation. We see that for successive initial values of J, the ΔE associated with

TABLE 19.5 Rotational Frequencies, $\Delta(\Delta\nu)$ and $\Delta E/k_B T$ at 300. K for $^1H^{35}Cl$

$J \to J'$	$\Delta\nu$	$\Delta\nu$ HCl/s^{-1}	$\Delta(\Delta\nu)$	$\Delta(\Delta\nu)$ HCl/s^{-1}	$\Delta E/k_B T$ at 300. K
$0 \to 1$	$2cB$	6.3158×10^{11}	$2cB$	6.3158×10^{11}	0.102
$1 \to 2$	$4cB$	1.27036×10^{12}	$2cB$	6.3158×10^{11}	0.203
$2 \to 3$	$6cB$	1.90554×10^{12}	$2cB$	6.3158×10^{11}	0.305
$3 \to 4$	$8cB$	2.54072×10^{12}	$2cB$	6.3158×10^{11}	0.406
$4 \to 5$	$10cB$	3.1759×10^{12}	$2cB$	6.3158×10^{11}	0.508

the transition increases in such a way that the difference between these $\Delta\nu$, which we call $\Delta(\Delta\nu)$, is constant. This means that the spectrum for a molecule immersed in a microwave field with a broad range of frequencies shows a series of equally spaced lines, separated in frequency by $2cB$ as seen in Figure 19.14.

How many absorption peaks will be observed? For vibrational spectroscopy, we expect only one intense peak for the following reasons. The energy-level spacing between adjacent levels is the same for all values of the quantum number in the harmonic approximation so that given the selection rule $\Delta n = +1$, all transitions have the same frequency. Also, in general only the $n = 0$ energy level has a significant population so that even taking anharmonicity into account will not generate additional peaks originating from peaks with $n > 0$. However, the situation is different for rotational transitions. Note that because the rotational energy levels are not equally spaced in energy, different transitions give rise to separate peaks. Additionally, $\Delta E_{rotation} < k_B T$ under most conditions so that many rotational energy levels will be populated. Therefore, many peaks are generally observed in a rotational spectrum.

Up to this point, we have considered rotation and vibration separately. In the microwave region of the electromagnetic spectrum, the photon energy is sufficient to excite rotational transitions but not to excite vibrational transitions. However, this is not the case for infrared radiation. Diatomic molecules that absorb infrared radiation can make transitions in which both n and J change according to the selection rules $\Delta n = +1$ and $\Delta J = \pm 1$. Therefore, an infrared absorption spectrum contains both vibrational and rotational transitions. What does a rotational-vibrational spectrum look like? To answer this question, first consider the relative photon energies associated with rotational and vibrational excitation. The energy levels for both degrees of freedom are indicated schematically in Figure 19.15. The ratio of the smallest value of ΔE in a rotational transition to that in a vibrational transition is

$$\frac{\Delta E_{rot}}{\Delta E_{vib}} = \frac{\hbar^2/\mu r_0^2}{\hbar\sqrt{k/\mu}} = \frac{\hbar}{r_0^2\sqrt{k\mu}} \quad (19.17)$$

This ratio is molecule specific, but we consider two extremes. For H_2 and I_2, $\Delta E_{rot}/\Delta E_{vib}$ is 0.028 and 0.00034, respectively, where ΔE_{rot} is for the $J = 0 \to J = 1$ transition. In both cases, there are many rotational levels between adjacent vibrational levels. Large moments of inertia (large atomic masses and/or long bonds) and large force constants (strong bonds) lead to a smaller value of $\Delta E_{rot}/\Delta E_{vib}$. It is largely the difference in the **moment of inertia** $I = \mu r_0^2$ that makes the ratio so different for I_2 and H_2.

On the basis of the previous discussion, what will be seen in an infrared absorption experiment on a diatomic molecule in which both rotational and vibrational transitions occur? As discussed in Section 19.3, the dominant vibrational transition is $n = 0 \to n = 1$. All transitions must now satisfy two selection rules, $\Delta n = +1$ and $\Delta J = \pm 1$. As discussed earlier, a vibrational-rotational spectrum will exhibit many different rotational transitions. What can one predict about the relative intensities of the peaks? Recall that the intensity of a spectral line in an absorption experiment is determined by the number of molecules in the energy level from which the transition originates. (This rule holds as long as the upper state population is small compared to the lower state population.) How many molecules are there in states for a given value of J

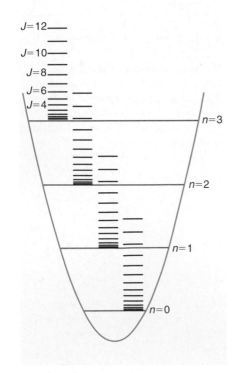

FIGURE 19.15
Schematic representation of rotational and vibrational levels. Each vibrational level has a set of rotational levels associated with it. Therefore, vibrational transitions usually also involve rotational transitions. The rotational levels are shown on an expanded energy scale and are much more closely spaced for real molecules.

relative to the number in the ground state for which $J = 0$? This ratio can be calculated using the Boltzmann distribution:

$$\frac{n_J}{n_0} = \frac{g_J}{g_0} e^{-(E_J - E_0)/k_B T} = (2J + 1) e^{-\hbar^2 J(J+1)/2I k_B T} \qquad \text{(19.18)}$$

The term in front of the exponential is the ratio of the degeneracy of the energy level J to that for $J = 0$. It generally dominates n_J/n_0 for small J and sufficiently large T. However, as J increases, the exponential term causes n_J/n_0 to decrease rapidly with increasing J. For molecules with a large moment of inertia, the exponential term does not dominate until J is quite large. As a result, many rotational energy levels are occupied; this behavior is seen for CO in Figure 19.16. Because many levels are occupied, a large number of peaks are observed in a rotational spectrum. For a molecule with a small moment of inertia, the rotational levels can be far enough apart that few rotational states are populated. This behavior is shown in Figure 19.16 for HD. At 100. K, only the $J = 0$, 1, and 2 states have an appreciable population. Increasing the temperature raises this upper value of J to approximately 4 and 7 for 300. and 700. K, respectively. The corresponding J values for CO are 13, 23, and 33.

Therefore, as long as n_J/n_0 increases with J, the intensity of the spectral peaks originating from states with those J values will increase. Beyond the J values for which n_J/n_0 increases, the intensity of the peaks decreases.

A simulated rotational-vibrational infrared absorption spectrum for HCl is shown in Figure 19.17. Such a spectrum consists of two nearly symmetric parts. The higher frequency part of the spectrum corresponds to transitions in which $\Delta J = +1$ and is called the **R branch.** The lower frequency part of the spectrum corresponds to transitions in which $\Delta J = -1$ and is called the **P branch.** Note that the gap in the center of the spectrum corresponds to $\Delta J = 0$, which is a forbidden transition in the dipole approximation for a linear molecule. Without going into more detail, note that Raman spectroscopy (see Supplemental Section 19.8) also shows both rotational and vibrational transitions. However, the selection rules are different. For rotational Raman spectra, the selection rule is $\Delta J = 0, \pm 2$, and not $\Delta J = \pm 1$ as it is for absorption spectra in the infrared or microwave ranges.

Based on this discussion of rotational-vibrational spectroscopy and the results shown in Figures 19.16 and 19.17, it is useful to revisit the infrared spectra of CO and CH_4 shown in Figure 19.10. The broad unresolved peaks seen for CO between 2000 and 2250 cm^{-1} are the P and R branches corresponding to rotational-vibrational excitations. The minimum near 2150 cm^{-1} corresponds to the forbidden $\Delta J = 0$ transition. The broad and only partially resolved peaks for CH_4 seen around the sharp peaks centered near 1300 and 3000 cm^{-1} are again the P and R branches. The $\Delta J = 0$ transition is allowed for methane and is the reason why the sharp central peaks are observed in the methane spectrum seen in Figure 19.10. To demonstrate the origin of the broad CO peaks in Figure 19.10, a high-resolution infrared absorption spectrum for this molecule is shown in Figure 19.18. It is apparent that the envelopes of the P and R branches in this figure correspond to the broad unresolved peaks in Figure 19.10.

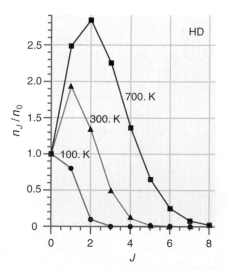
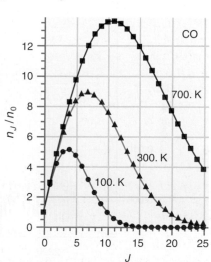

FIGURE 19.16
The number of molecules in energy levels corresponding to the quantum number J relative to the number in the ground state is shown as a function of J for two molecules at three different temperatures.

FIGURE 19.17
Simulated 300. K infrared absorption spectrum and energy diagram for $H^{35}Cl$. The two indices above the peak refer to the initial (first) and final (second) J values. The region of the spectrum with $\Delta J = +1$ (higher frequency) is called the R branch, and the region of the spectrum with $\Delta J = -1$ (lower frequency) is called the P branch. The energy levels in the upper part of the figure are not drawn to scale.

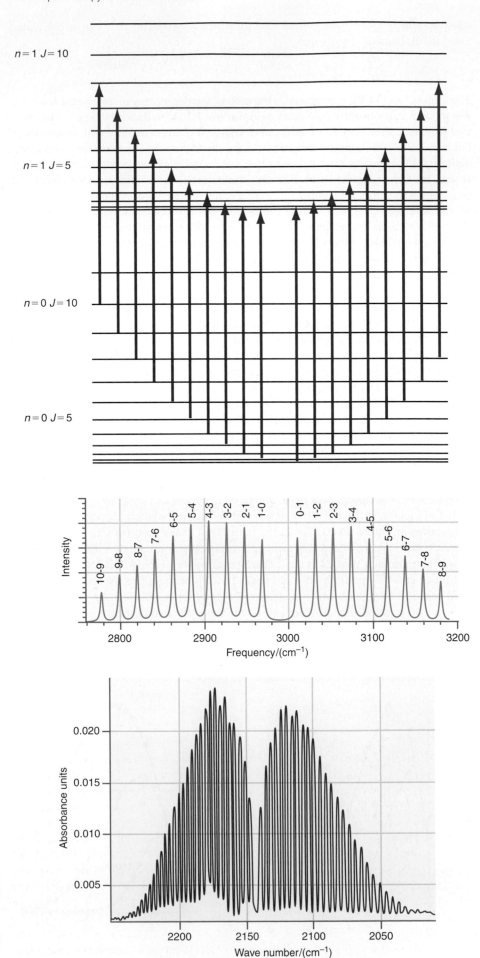

FIGURE 19.18
A high-resolution spectrum is shown for CO in which the P and R branches are resolved into the individual rotational transitions.

19.7 Fourier Transform Infrared Spectroscopy

How are infrared absorption spectra obtained in practice? We now turn to a discussion of Fourier transform infrared (FTIR) spectroscopy, which is the most widely used technique for obtaining vibrational absorption spectra. FTIR spectroscopy improves on the schematic absorption experiment shown in Figure 19.9 by eliminating the monochromator and by using a broadband blackbody radiation source. By simultaneously analyzing the absorption throughout the spectral range of the light source, it achieves a **multiplex advantage** that is equivalent to carrying out many single-wavelength experiments in parallel. This technique allows a spectrum to be obtained in a short time and has led to a revolution in the field of vibrational spectroscopy. We describe how **FTIR spectroscopy** works in this section.

The multiplex advantage in FTIR is gained by using a **Michelson interferometer** to determine the frequencies at which radiation is absorbed by molecules. A schematic drawing of this instrument is shown in Figure 19.19. The functioning of a Michelson interferometer is first explained by analyzing its effect on monochromatic radiation. An incoming traveling plane wave of amplitude $A_0 \exp\{i(kx - \omega t)\}$ and intensity I_0 impinges on a beam splitter S that both transmits and reflects 50% of the incident light. Each of these two waves is reflected back from a mirror (M_1 or M_2) and is incident on the beam splitter S. The wave that is reflected back from the movable mirror M_2 and transmitted by S interferes with the wave that is reflected from the fixed mirror M_1 and reflected from S. The recombined wave resulting from this interference travels in the negative y direction and has an amplitude at the detector plane $y = y_D$ given by

$$A(t) = \frac{A_0}{\sqrt{2}} \Big[\exp\{i(ky_D - \omega t)\} + \exp\{i(k[y_D + \Delta d(t)] - \omega t)\} \Big]$$

$$= \frac{A_0}{\sqrt{2}} \left(1 + e^{i\delta(t)} \right) \exp\{i(ky_D - \omega t)\} \qquad \textbf{(19.19)}$$

The phase difference $\delta(t)$ results from the path difference Δd that the two interfering waves have traveled Δd. It arises because mirrors M_1 and M_2 are not equidistant from the beam splitter:

$$\delta(t) = \frac{2\pi}{\lambda}(2SM_1 - 2SM_2) = \frac{2\pi}{\lambda}\Delta d(t) \qquad \textbf{(19.20)}$$

In this equation, SM_1 and SM_2 are the distances between the beam splitter and mirrors 1 and 2, respectively. The intensity of the resultant wave at the detector plane I is proportional to the product $A(t)A^*(t)$:

$$I(t) = \frac{I_0}{2}(1 + \cos\delta(t)) = \frac{I_0}{2}\left(1 + \cos\frac{2\pi\Delta d(t)}{\lambda}\right) \qquad \textbf{(19.21)}$$

where $I_0 = A_0^2$. The intensity varies periodically with distance as mirror M_2 is moved toward the beam splitter. Whenever $\Delta d = n\lambda$, the interference is constructive and the maximum intensity is transmitted to the detector. Whenever $\Delta d = (2n + 1)(\lambda/2)$, the interference is destructive and the wave is fully reflected back into the source.

The signal measured by the detector is called an **interferogram** because it results from the interference of the two waves. In this case the interferogram is described by a single sine wave, so that a frequency analysis of the intensity gives a single frequency corresponding to the incident plane wave. The output of the interferometer for a single incident frequency is shown in Figure 19.20. This simple example illustrates how the frequency of the radiation that enters the interferometer can be determined from the experimentally obtained interferogram.

We now consider the more interesting case encountered when the incident wave is composed of a number of different frequencies. This case describes a realistic situation in which a blackbody source of infrared light passes through a sample and enters the

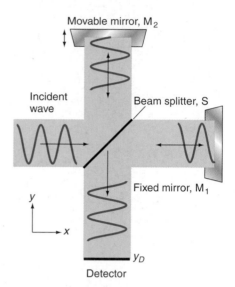

FIGURE 19.19
Schematic diagram of a Michelson interferometer.

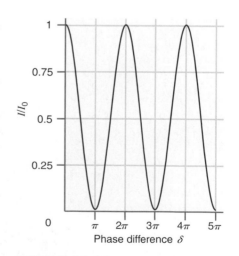

FIGURE 19.20
Intensity measured at the detector as a function of the phase difference in the two arms of a Michelson interferometer for a single frequency.

interferometer. Only certain frequencies from the source are absorbed by vibrational excitations of the molecules. The interferometer sees the blackbody distribution of frequencies of the source from which certain frequencies have been attenuated through absorption. What can we expect for the case of several incident frequencies? We can write the amplitude of the wave resulting from the interference of the two reflections from mirrors M_1 and M_2 as follows:

$$A(t) = \sum_j \frac{A_j}{\sqrt{2}}\left(1 + \exp\left[i\left(\frac{2\pi\,\Delta d(t)}{\lambda_j}\right)\right]\right) \exp\left[i\left(\frac{2\pi}{\lambda_j}y_D - \omega_j t\right)\right] \quad (19.22)$$

In this equation, the subscript j refers to the individual frequencies incident in the beam entering the interferometer. As you will see in the problems at the end of this chapter, if the mirror is moving with a velocity v, the measured intensity at the detector is

$$I(t) = \frac{1}{2}\sum_j I(\omega_j)\left(1 + \cos\left[\omega_j \frac{2v}{c}t\right]\right) \quad (19.23)$$

The interferogram $I(t)$ is determined by the distribution of frequencies entering the interferometer. Figure 19.21b shows interferograms that result from the sample spectra shown in Figure 19.21a. In practice, the opposite path is followed, in which the measured interferogram is converted to a spectrum using Fourier transform techniques.

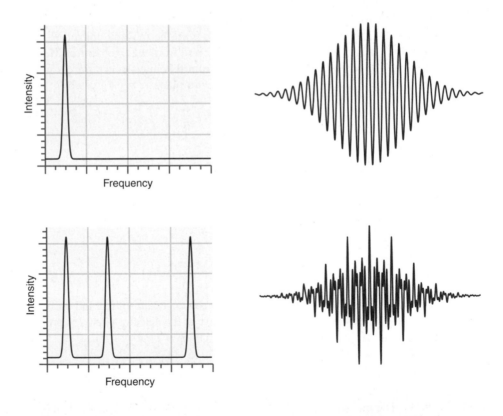

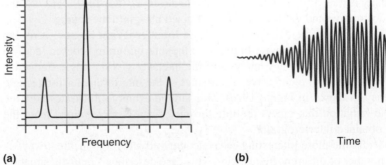

FIGURE 19.21
(a) Simulated sample spectra and (b) the resulting interferograms calculated using Equation (19.23).

(a)

(b)

Note the contrast between the interferogram of Figure 19.21 and the interferogram in Figure 19.20. Whereas the interferogram in Figure 19.20 is calculated for a single frequency, a real spectral line has a finite width in frequency. The effect of this finite width is to dampen the amplitude of the interferogram for longer and shorter times relative to the central value of $t = 0$. For our purposes, it is sufficient to note that the interferograms for the different sample spectra are clearly different. Although the characteristic absorption frequencies cannot be obtained directly by inspection of the interferogram, they are readily apparent after the data have been Fourier transformed from the time domain into the frequency domain.

Because the information about absorption at all frequencies is determined simultaneously, an FTIR spectrum can be obtained quickly with high sensitivity. For example, the components of automobile gas exhaust are typically (percent by volume) N_2 (71%), CO_2 (18%), H_2O (9.2%), CO (0.85%), O_2 and noble gases (0.7%), NO_x (0.08%), and hydrocarbons (0.05%). The concentration of these components other than N_2, O_2, and the rare gases can be determined in well under a minute by recording a single FTIR spectrum.

S U P P L E M E N T A L

19.8 Raman Spectroscopy

As discussed in the previous sections, absorption of light in the infrared portion of the spectrum can lead to transitions between eigenstates of the vibrational-rotational energy. Another interaction between a molecule and an electromagnetic field can also lead to vibrational and rotational excitation. It is called the **Raman effect** after its discoverer and involves scattering of a photon by the molecule. We can think of scattering as the collision between a molecule and a photon in which energy and momentum are transferred between the two collision partners. Raman spectroscopy complements infrared absorption spectroscopy because it obeys different selection rules. For instance, the stretching mode in a homonuclear diatomic molecule is Raman active but infrared inactive. The reasons for this difference will become clear after molecular symmetry and group theory are discussed in Chapter 27.

Consider a molecule with a characteristic vibrational frequency ν_{vib} in an electromagnetic field that has a time-dependent electric field given by

$$E = E_0 \cos (2\pi \nu t) \qquad \textbf{(19.24)}$$

The electric field distorts the molecule slightly because the negatively charged valence electrons and the positive nuclei and their tightly bound core electrons experience forces in opposite directions. This induces a time-dependent dipole moment of magnitude $\mu_{induced}(t)$ in the molecule of the same frequency as the field. The dipole moment is linearly proportional to the magnitude of the electric field, and the proportionality constant is the **polarizability** α. The polarizability is an anisotropic quantity and its value depends on the direction of the electric field relative to the molecular axes:

$$\mu_{induced}(t) = \alpha E_0 \cos (2\pi \nu t) \qquad \textbf{(19.25)}$$

The polarizability depends on the bond length $x_e + x(t)$, where x_e is the equilibrium value. The polarizability α can be expanded in a Taylor-Maclaurin series (see the Math Supplement, Appendix A) in which terms beyond the first order have been neglected:

$$\alpha(x_e + x) = \alpha(x_e) + x \left(\frac{d\alpha}{dx} \right)_{x=x_e} + \ldots \qquad \textbf{(19.26)}$$

Due to the vibration of the molecule, $x(t)$ is time dependent and is given by

$$x(t) = x_{max} \cos (2\pi \nu_{vib} t) \qquad \textbf{(19.27)}$$

Combining this result with Equation (19.26), we can rewrite Equation (19.25) in the form

$$\mu_{induced}(t) = \alpha E = E_0 \cos (2\pi \nu t) \left[\alpha(x_e) + \left\{ \left(\frac{d\alpha}{dx} \right)_{x=x_e} \right\} x_{max} \cos (2\pi \nu_{vib} t) \right] \textbf{(19.28)}$$

which can be simplified using the trigonometric identity $\cos x \cos y = \frac{1}{2}[\cos(x - y) + \cos(x + y)]$ to

$$\mu_{induced}(t) = \alpha E = \alpha(x_e)E_0 \cos(2\pi\nu t)$$
$$+ \left[\left(\frac{d\alpha}{dx}\right)_{x=x_e}\right] x_{max} E_0 [\cos\{2\pi(\nu + \nu_{vib})t\} + \cos\{2\pi(\nu - \nu_{vib})t\}] \qquad \textbf{(19.29)}$$

The time-varying dipole moment radiates light of the same frequency as the dipole moment, and at the frequencies ν, $(\nu - \nu_{vib})$, and $(\nu + \nu_{vib})$. These three frequencies are referred to as the **Rayleigh, Stokes,** and **anti–Stokes frequencies,** respectively. We see that in addition to scattered light at the incident frequency, light will also be scattered at frequencies corresponding to vibrational excitation and de-excitation. Higher order terms in the expansion for the polarizability [Equation (19.26)] also lead to scattered light at the frequencies $\nu \pm 2\nu_{vib}, \nu \pm 3\nu_{vib}, \ldots$, but the scattered intensity at these frequencies is much weaker than at the primary frequencies.

Equation (19.29) illustrates that the intensity of the Stokes and anti–Stokes peaks is zero unless $d\alpha/dx \neq 0$. We conclude that for vibrational modes to be Raman active, the polarizability of the molecule must change as it vibrates. This condition is satisfied for many vibrational modes and, in particular, it is satisfied for the stretching vibration of a homonuclear molecule, although $d\mu_x/dx = 0$ for these molecules, making them infrared inactive. Not all vibrational modes that are infrared active are Raman active and vice versa. This is why infrared and Raman spectroscopies provide a valuable complement to one another.

A schematic picture of the scattering event in Raman spectroscopy on an energy scale is shown in Figure 19.22. This diagram is quite different from that considered earlier in depicting a transition between two states. The initial and final states are the $n = 0$ and $n = 1$ states at the bottom of the figure. To visualize the interaction of the molecule with the photon of energy $h\nu$, which is much greater than the vibrational energy spacing, we imagine the scattered photon to be "absorbed" by the molecule, resulting in a much higher intermediate energy "state." This very short-lived "state" quickly decays to the final state. Whereas the initial and final states are eigenfunctions of the time-independent Schrödinger equation, the upper "state" in this energy diagram need not satisfy this condition. Therefore, it is referred to as a virtual state.

Are the intensities of the Stokes and anti–Stokes peaks equal? We know that their relative intensity is governed by the relative number of molecules in the originating states. For the Stokes line, the transition originates from the $n = 0$ state, whereas for the anti–Stokes line, the transition originates from the $n = 1$ state. Therefore, the relative intensity of the Stokes and anti–Stokes peaks can be calculated using the Boltzmann distribution:

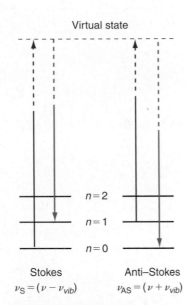

Virtual state

$n = 2$

$n = 1$

$n = 0$

Stokes

$\nu_S = (\nu - \nu_{vib})$

Anti–Stokes

$\nu_{AS} = (\nu + \nu_{vib})$

FIGURE 19.22

Schematic depiction of the Raman scattering event. The spectral peak resulting in vibrational excitation is called the Stokes peak, and the spectral peak originating from vibrational de-excitation is called the anti–Stokes peak.

$$\frac{I_{anti\text{-}Stokes}}{I_{Stokes}} = \frac{n_{excited}}{n_{ground}} = \frac{e^{-3h\nu/2k_BT}}{e^{-h\nu/2k_BT}} = e^{-h\nu/k_BT} \qquad \textbf{(19.30)}$$

For vibrations for which $\tilde{\nu}$ is in the range of 1000 to 3000 cm^{-1}, $\frac{I_{anti\text{-}Stokes}}{I_{Stokes}}$ ranges between 8×10^{-3} and 5×10^{-7} at 300 K. This calculation shows that the intensities of the Stokes and anti–Stokes peaks will be quite different. In this discussion of the Raman effect, we have only considered vibrational transitions. However, just as for infrared absorption spectra, Raman spectra show peaks originating from both vibrational and rotational transitions.

Raman and infrared spectroscopy are complementary and both can be used to study the vibrations of molecules. Both techniques can be used to determine the identities of molecules in a complex mixture by comparing the observed spectral peaks with characteristic group frequencies. The most significant difference between these two spectroscopies is the light source needed to implement the technique. For infrared absorption spectroscopy, the light source is in the infrared. Because Raman spectroscopy is a scattering technique, the frequency of the light used need not match the frequency of the transition being studied. Therefore, a source in the visible part of the spectrum is generally used to study rotational and vibrational modes. This has several advantages over infrared sources. By shifting the vibrational spectrum from the infrared into the visible part of the spectrum, commonly available lasers can be used to obtain Raman spectra. Intense lasers are necessary because the probability for Raman scattering is generally on the order of 10^{-6} or less. Furthermore, shifting the frequency of the source from the infrared into the visible part of the spectrum can reduce interference with absorbing species that are not of primary interest. For instance, infrared spectra of aqueous solutions always contain strong water peaks that may mask other peaks of interest. By shifting the source frequency to the visible part of the spectrum, such interferences can be eliminated.

Another interesting application of the Raman effect is the Raman microscope or microprobe. Because Raman spectroscopy is done in the visible part of the light spectrum, it can be combined with optical microscopy to obtain spectroscopic information with a spatial resolution of better than 0.01 mm. An area in which this technique has proved particularly useful is as a nondestructive probe of the composition of gas inclusions such as CH_4, CO, H_2S, N_2, and O_2 in mineral samples. Raman microscopy has also been used in biopsy analyses to identify mineral particles in the lung tissues of silicosis victims and to analyze the composition of gallstones.

SUPPLEMENTAL

19.9 How Does the Transition Rate between States Depend on Frequency?

Now that we have some familiarity with the terms *absorption, spontaneous emission,* and *stimulated emission,* the frequency dependence of the interaction of molecules with light can be examined. Until now, we have only dealt with potential energy functions that are independent of time. In any spectroscopic method, transitions occur from one state to another. Transitions cannot be induced by a time-independent potential, because the eigenfunctions of the time-independent Schrödinger equation are stationary states and have a constant energy. We now outline how a time-dependent electromagnetic field of light incident on molecules with a discrete set of energy levels can induce transitions between these levels. To make the mathematics more tractable, we consider a two-state system in which the states are denoted by 1 and 2 and the normalized solutions to the time-independent Schrödinger equation are $\psi_1(x)$ and $\psi_2(x)$ with eigenvalues E_1 and E_2, respectively. We assume that $E_2 > E_1$. The corresponding wave functions including the time dependence are $\Psi_1 = \psi_1(x)e^{-i(E_1/\hbar)t}$ and $\Psi_2 = \psi_2(x)e^{-i(E_2/\hbar)t}$. We assume that the system is in the ground state (state 1) at time $t = 0$.

When the light is turned on, the molecule interacts with the electric field of the light through its permanent or induced dipole moment, and the time-dependent potential energy is given by

$$\hat{H}_{field}(t) = -\boldsymbol{\mu} \cdot \mathbf{E} = -\mu_x E_0 \cos(2\pi\nu t) = -\frac{\mu_x E_0}{2}\left(e^{2\pi i\nu t} + e^{-2\pi i\nu t}\right) \quad \textbf{(19.31)}$$

We have assumed that the electric field E_0 lies along the x axis. Writing the operator $\hat{H}_{field}(t)$ in this way is called the **dipole approximation,** because much smaller terms involving higher order multipoles are neglected. What change will the system undergo under the influence of the light? We can expect transitions from the ground state to the first excited state to occur.

Because this is a time-dependent system, we must solve the time-dependent Schrödinger equation $(\hat{H}_0 + \hat{H}_{field})\Phi(x, t) = i\hbar(\partial\Phi(x, t)/\partial t)$. In this two-level system, Ψ_1 and Ψ_2 form a complete set, and therefore the eigenfunctions of the operator $\hat{H}_0 + \hat{H}_{field}$ must be a linear combination of Ψ_1 and Ψ_2:

$$\Phi(x, t) = a_1(t)\Psi_1 + a_2(t)\Psi_2 \quad \textbf{(19.32)}$$

$a_1^*a_1$ and $a_2^*a_2$ are the probabilities that the state is in level 1 and 2, respectively. At $t = 0$, $a_1^*a_1 = 1$ and $a_2^*a_2 = 0$. At later times, $a_1^*a_1 < 1$ and $a_2^*a_2 > 0$ as the transition to state 2 occurs. Our goal is to derive an expression for $a_2^*(t)a_2(t)$ and to determine how it depends on the frequency of the electric field.

Substituting Equation (19.32) in the time-dependent Schrödinger equation we obtain

$$a_1(t)\hat{H}_0\Psi_1 + a_2(t)\hat{H}_0\Psi_2 + a_1(t)\hat{H}_{field}\Psi_1 + a_2(t)\hat{H}_{field}\Psi_2$$

$$= i\hbar\Psi_1\frac{da_1(t)}{dt} + i\hbar\Psi_2\frac{da_2(t)}{dt} + i\hbar a_1(t)\frac{d\Psi_1(t)}{dt} + i\hbar a_2(t)\frac{d\Psi_2(t)}{dt} \quad (19.33)$$

This equation can be simplified by evaluating $i\hbar a_1(t)\dfrac{d\Psi_1(t)}{dt} + i\hbar a_2(t)\dfrac{d\Psi_2(t)}{dt}$:

$$i\hbar a_1(t)\frac{d\Psi_1(t)}{dt} + i\hbar a_2(t)\frac{d\Psi_2(t)}{dt} = i\hbar a_1(t)\psi_1(x)\frac{de^{-i(E_1/\hbar)t}}{dt} + i\hbar a_2(t)\psi_2(x)\frac{de^{-i(E_2/\hbar)t}}{dt}$$

$$= a_1(t)E_1\Psi_1(x) + a_2(t)E_2\Psi_2(x) \quad (19.34)$$

Because $\hat{H}_0\Psi_1 = E_1\Psi_1$ and $\hat{H}_0\Psi_2 = E_2\Psi_2$, $i\hbar a_1(t)\dfrac{d\Psi_1(t)}{dt} + i\hbar a_2(t)\dfrac{d\Psi_2(t)}{dt}$ cancels out $a_1(t)\hat{H}_0\Psi_1 + a_2(t)\hat{H}_0\Psi_2$ on the left side of Equation (19.33) and Equation (19.34) takes the simpler form

$$a_1(t)\hat{H}_{field}\Psi_1 + a_2(t)\hat{H}_{field}\Psi_2 = i\hbar\Psi_1\frac{da_1(t)}{dt} + i\hbar\Psi_2\frac{da_2(t)}{dt} \quad (19.35)$$

In order to obtain an equation for $da_2(t)/dt$, we next multiply Equation 19.35 on the left by Ψ_2^* and integrate over the spatial coordinate x to obtain

$$a_1(t)\int \Psi_2^*\hat{H}_{field}\Psi_1\,dx + a_2(t)\int \Psi_2^*\hat{H}_{field}\Psi_2\,dx$$

$$= i\hbar\frac{da_1(t)}{dt}\int \Psi_2^*\Psi_1\,dx + i\hbar\frac{da_2(t)}{dt}\int \Psi_2^*\Psi_2\,dx \quad (19.36)$$

Because Ψ_1 and Ψ_2 are orthonormal, the last two integrals can be evaluated and Equation (19.36) can be simplified to

$$i\hbar\frac{da_2(t)}{dt} = a_1(t)\int \Psi_2^*\hat{H}_{field}\Psi_1\,dx + a_2(t)\int \Psi_2^*\hat{H}_{field}\Psi_2\,dx \quad (19.37)$$

Equation (19.37) can be simplified further if only changes in the coefficients $a_1(t)$ and $a_2(t)$ for small values of t are considered. In this limit, we can replace $a_1(t)$ and $a_2(t)$ on the right side of this equation by their initial values, $a_1(t) = 1$ and $a_2(t) = 0$. Therefore, only one term remains on the right side of Equation (19.37). It turns out that imposing this limit does not affect the general conclusions drawn next. We also replace Ψ_1 and Ψ_2 by the complete form $\Psi_1 = \psi_1(x)e^{-i(E_1/\hbar)t}$ and $\Psi_2 = \psi_2(x)e^{-i(E_2/\hbar)t}$. After doing so, the following equations are obtained:

$$i\hbar\frac{da_2(t)}{dt} = \exp\left[\frac{i}{\hbar}(E_2 - E_1)t\right]\int \psi_2^*(x)\,\hat{H}_{field}\,\psi_1(x)dx$$

$$= -\frac{E_0}{2}\exp\left[\frac{i}{\hbar}(E_2 - E_1)t\right](\exp[2\pi i\nu t] + \exp[-2\pi i\nu t])$$

$$\times \int \psi_2^*(x)\,\mu_x\psi_1(x)\,dx$$

$$= -\frac{E_0}{2}\left(\exp\left[\frac{i}{\hbar}(E_2 - E_1 + h\nu)t\right] + \exp\left[\frac{i}{\hbar}(E_2 - E_1 - h\nu)t\right]\right)$$

$$\times \int \psi_2^*(x)\,\mu_x\psi_1(x)\,dx$$

$$i\hbar\frac{da_2(t)}{dt} = -\mu_x^{21}\frac{E_0}{2}\left(\exp\left[\frac{i}{\hbar}(E_2 - E_1 + h\nu)t\right] + \exp\left[\frac{i}{\hbar}(E_2 - E_1 - h\nu)t\right]\right) \quad (19.38)$$

In the last equation we have introduced the transition dipole moment μ_x^{21} defined in Equation (19.6) by $\mu_x^{mn} = \int \psi_m^*(x)\mu_x(x_e + x)\psi_n(x)\,dx$. The transition dipole moment is important because it generates the selection rules for any spectroscopy, as discussed in Section 19.4. Next, the last equation in Equation (19.38) is integrated with respect to time, using the dummy variable t' to obtain $a_2(t)$:

$$a_2(t) = \frac{i}{\hbar}\mu_x^{21}\frac{E_0}{2}\int_0^t\left(e^{\frac{i}{\hbar}(E_2-E_1+h\nu)t'} + e^{\frac{i}{\hbar}(E_2-E_1-h\nu)t'}\right)dt' \qquad \textbf{(19.39)}$$

$$= \mu_x^{21}\frac{E_0}{2}\left(\frac{-1 + e^{\frac{i}{\hbar}(E_2-E_1+h\nu)t}}{E_2 - E_1 + h\nu} + \frac{-1 + e^{-\frac{i}{\hbar}(E_2-E_1-h\nu)t}}{E_2 - E_1 - h\nu}\right) \qquad \textbf{(19.40)}$$

This expression looks complicated, but it contains a great deal of useful information that can be extracted fairly easily. Most importantly, it is seen that $a_2(t) = 0$ for all times unless the transition dipole moment $\mu_z^{21} \neq 0$. Next, we look at the terms in the parentheses. The numerator in each of the terms is an oscillating function of time. The period of oscillation approaches zero in the first and second terms as $E_1 - E_2 \rightarrow h\nu$ and $E_2 - E_1 \rightarrow h\nu$, respectively. In these limits, the denominator approaches zero. These are the conditions that lead a_2 to grow rapidly with time. The second term corresponds to absorption of a photon because we have chosen $E_2 > E_1$. The first term corresponds to stimulated emission of a photon. Stimulated emission is of importance in understanding lasers; this process is discussed in more detail in Chapter 22. However, because the current topic is absorption, we focus on the narrow range of energy around $E_2 - E_1 = h\nu$ in which only the absorption peak appears. The behavior of $a_2(t)$ at the resonance is not easy to discern, because $\lim a_2(t) = 0/0$ as $E_2 - E_1 \rightarrow h\nu$. We use L'Hôpital's rule,

$$\lim\left[\frac{f(x)}{g(x)}\right]_{x\rightarrow 0} = \lim\left[\frac{df(x)/dx}{dg(x)/dx}\right]_{x\rightarrow 0} \qquad \textbf{(19.41)}$$

which in this case takes the form

$$F(t) = \lim_{E_2-E_1-h\nu\rightarrow 0}\left[\frac{-1 + \exp\left[-\frac{i}{\hbar}(E_2 - E_1 - h\nu)t\right]}{(E_2 - E_1 - h\nu)}\right]$$

$$= \lim_{E_2-E_1-h\nu\rightarrow 0}\left[\frac{\left[d\left(-1 + \exp\left[-\frac{i}{\hbar}(E_2 - E_1 - h\nu)t\right]\right)\right]/d(E_2 - E_1 - h\nu)}{d(E_2 - E_1 - h\nu)/d(E_2 - E_1 - h\nu)}\right]$$

$$= -\frac{it}{\hbar}\left[\exp\left(\frac{-i}{\hbar}[E_2 - E_1 - h\nu]t\right)\right]_{E_2-E_1-h\nu=0} = -\frac{it}{\hbar} \qquad \textbf{(19.42)}$$

The important result that emerges from this calculation is that at the resonance condition $E_2 - E_1 = h\nu$, the magnitude of $a_2(t)$ increases linearly with t. How does $a_2(t)$ change with t near but not at the resonance condition? We can get this information if $a_2(t)$ is graphed versus t for the $h\nu$ values near the resonance as shown in Figure 19.23.

Figure 19.23 shows that $|F(t)|$ and therefore $|a_2(t)|$ increases nearly linearly with time for small values of t if $E_2 - E_1$ is extremely close to $h\nu$. This means that the probability of finding the atom or molecule in the excited state increases with time. However, for photon energies that deviate even by 1 ppm from this limit, $|a_2(t)|$ will oscillate and remain small. The oscillations will be more frequent and smaller in amplitude the more $h\nu$ differs from $E_2 - E_1$. The probability of finding the atom or molecule in the excited state remains small if $h\nu$ differs even slightly from $E_2 - E_1$ and the atom or molecule is unable to take up energy from the electromagnetic field. *We conclude that the rate of transition from the ground to the excited state is appreciable only if $h\nu$ is equal to $E_2 - E_1$.*

Our final goal is to find an expression for $a_2^*(t)a_2(t)$ that represents the probability of finding the molecule in the excited state with energy E_2 after it has been exposed to

FIGURE 19.23

The change in the magnitude of $|F(t)|$ with time for the three photon energies indicated. The curves are calculated for $E_2 - E_1 = 5.00 \times 10^{-19}$ J.

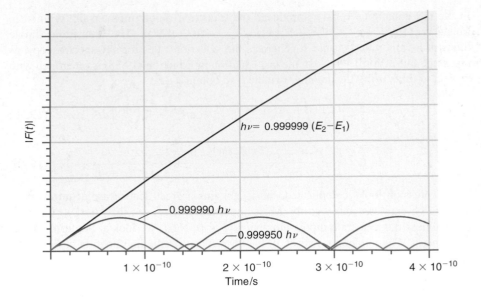

the light for the time t. We leave this part of the derivation for the end-of-chapter problems and simply state the result here:

$$a_2^*(t)a_2(t) = E_0^2[\mu_x^{21}]^2 \frac{\sin^2[(E_2 - E_1 - h\nu)t/2\hbar]}{(E_2 - E_1 - h\nu)^2} \tag{19.43}$$

Figure 19.24 shows a graph of $a_2^*(t)a_2(t)$ against $(E_2 - E_1 - h\nu)/(E_2 - E_1)$ for 40, 120, and 400 ps. As expected, the probability of finding the molecule in the excited state is sharply peaked if the photon energy satisfies the condition $E_2 - E_1 = h\nu$. Because $|a_2(t)|$ increases linearly with time for $E_2 - E_1 = h\nu$, $a_2^*(t)a_2(t)$ increases as t^2 at resonance. The different curves in Figure 19.24 have been normalized to the same maximum value to allow a direct comparison of their widths in energy. The relative amplitudes of $a_2^*(t)a_2(t)$ at resonance for 40, 120, and 400 ps are 1, 9, and 100, respectively. Because the peak height varies with time as t^2 and the width decreases as $1/t$, the total area under the resonance varies as t. This shows that the probability of finding the molecule in the upper state increases linearly with time.

As Figure 19.24 shows, the photon energy range over which absorption occurs becomes narrower as the time t increases. What is the origin of this effect? Small values of t are equivalent to short light pulses. If the time profile of the pulse is expanded as a Fourier series in the frequency, the 40 ps pulse contains a broader range of frequencies than the 400 ps pulse. For this reason, the range of energy over which energy is taken up by the system is larger for the 40 ps pulse than for the 400 ps pulse.

The probability density $a_2^*(t)a_2(t)$ is closely related to the intensity observed in an absorption spectrum. How is the broadening that was just discussed related to the linewidth observed in an experimentally determined spectrum? To answer this question, we must distinguish between an intrinsic and a measured linewidth. By intrinsic linewidth, we mean the linewidth that would be measured if the spectrometer were perfect. However, a real spectrometer is defined by an instrument function, which is the output of the spectrometer for a very narrow spectral peak. The observed spectrum results from the convolution of the instrument function with the intrinsic linewidth.

Based on theoretical calculations, the intrinsic linewidth for vibrational spectra is less than ~10^{-3} cm^{-1}. This is very small compared with the resolution of conventional infrared spectrometers, which is typically no better than 0.1 cm^{-1}. Therefore, the width of peaks in a spectrum is generally determined by the instrumental function as shown in the top panel of Figure 19.25 and gives no information about the intrinsic linewidth. However, peaks that are broader than the instrument function are obtained if a sample contains many different local environments for the entity generating the peak. For example, the O—H stretching region in an infrared spectrum in liquid water is very broad. This is the case because of the many different local geometries that arise from

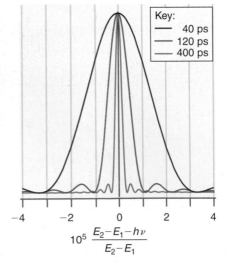

FIGURE 19.24

Graph of $a_2^*(t)a_2(t)$ against $(E_2 - E_1 - h\nu)/(E_2 - E_1)$. In the calculations $E_2 - E_1 = 5.00 \times 10^{-19}$ J. The range of $E_2 - E_1 - h\nu$ shown in the graph is only 80 parts per million of $E_2 - E_1$. The broadest and narrowest resonances are observed for $t = 40$ and 400 ps, respectively. The intermediate resonance is for $t = 120$ ps. All curves have been normalized to the same amplitude so that the peak widths can be directly compared.

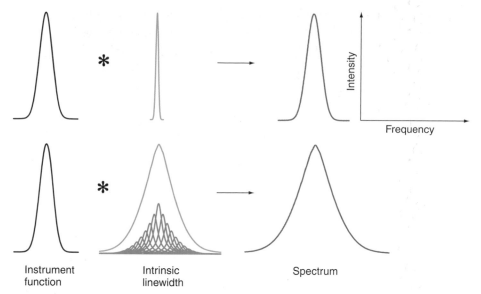

FIGURE 19.25
An experimental spectrum, as shown, arises from the convolution (indicated by the symbol *) of the instrument function and the intrinsic linewidth of the transition. For a homogeneous sample (top panel) the width of the spectrum is often determined by the instrument function. For an inhomogeneous sample, the intrinsic linewidth is the sum of the linewidths of the many different local environments. For inhomogeneous samples, the width of the spectrum can be determined by the intrinsic linewidth, rather than the instrument function.

hydrogen bonding between H_2O molecules, and each of them gives rise to a slightly different O—H stretching frequency. This effect is referred to as **inhomogeneous broadening** and is illustrated in the bottom panel of Figure 19.25.

Vocabulary

absorption	inhomogeneous broadening	R branch
anharmonic potential	interferogram	Raman effect
anti–Stokes frequency	Michelson interferometer	Rayleigh frequency
Beer–Lambert law	molar absorption coefficient	rotational constant
bond energy	moment of inertia	selection rule
coherent source	monochromator	spectroscopy
dipole approximation	Morse potential	spontaneous emission
dynamic dipole moment	multiplex advantage	stimulated emission
FTIR spectroscopy	overtone	Stokes frequency
group frequencies	P branch	transition dipole moment
harmonic potential	permanent dipole moment	wave number
incoherent photon source	polarizability	

Conceptual Problems

Q19.1 Why would you observe a pure rotational spectrum in the microwave region and a rotational-vibrational spectrum rather than a pure vibrational spectrum in the infrared region?

Q19.2 Solids generally expand as the temperature increases. Such an expansion results from an increase in the bond length between adjacent atoms as the vibrational amplitude increases. Will a harmonic potential lead to thermal expansion? Will a Morse potential lead to thermal expansion?

Q19.3 How can you observe vibrational transitions in Raman spectroscopy using visible light lasers where the photon energy is much larger than the vibrational energy spacing?

Q19.4 A molecule in an excited state can decay to the ground state either by stimulated emission or spontaneous emission. Use the Einstein coefficients to predict how the relative probability of these processes changes as the frequency of the transition doubles.

Q19.5 In Figure 19.16, n_J/n_0 increases initially with J for all three temperatures for CO but only for the two highest temperatures for HD. Explain this difference.

Q19.6 What is the difference between the transition dipole moment and the dynamic dipole moment?

Q19.7 Nitrogen and oxygen do not absorb infrared radiation and are therefore not greenhouse gases. Why is this the case?

Q19.8 Does the initial excitation in Raman spectroscopy result in a stationary state of the system? Explain your answer.

Q19.9 What feature of the Morse potential makes it suitable for modeling dissociation of a diatomic molecule?

Q19.10 If the rotational levels of a diatomic molecule were equally spaced and the selection rule remained unchanged, how would the appearance of the rotational-vibrational spectrum in Figure 19.17 change?

Q19.11 If a spectral peak is broadened, can you always conclude that the excited state has a short lifetime?

Q19.12 What is the difference between a permanent and a dynamic dipole moment?

Q19.13 What is the explanation for the absence of a peak in the rotational-vibrational spectrum near 3000 cm^{-1} in Figure 19.15?

Q19.14 What is the advantage in acquiring a vibrational spectrum using a FTIR spectrometer over a spectrometer in which the absorption is measured separately at each wavelength?

Q19.15 The number of molecules in a given energy level is proportional to $e^{-\Delta E/k_B T}$ where ΔE is the difference in energy between the level in question and the ground state. How is it possible that a higher lying rotational energy level can have a higher population than the ground state?

Q19.16 The square of a number of vibrational energy eigenfunctions are shown superimposed on a Morse potential in the following figure. Assign quantum numbers to the levels shown. Explain the differences in the shape of the eigenfunctions compared to those for a harmonic potential.

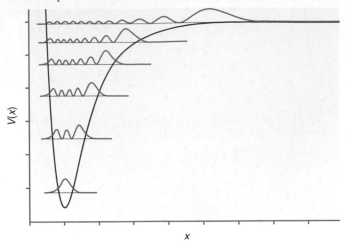

Q19.17 As a diatomic molecule rotates, the centrifugal force leads to a small change in the bond length. Do you expect the bond length to increase or decrease? Do you expect the difference between adjacent rotational energy peaks $\Delta(\Delta\nu)$ to increase or decrease?

Q19.18 For a harmonic potential, the vibrational force constant (a) is independent of the quantum number n and (b) independent of x-x_e for the molecule. Do you expect the same behavior for a Morse potential?

Q19.19 Use your answer from Q19.18 to compare the force constants for compression and stretching at the classical turning points for the levels shown in Q19.16. What trend do you see as n increases?

Q19.20 How many vibrational degrees of freedom do each of the following molecules have? NH_3, HCN, C_2H_6, C_{60}?

Numerical Problems

Problem numbers in **red** indicate that the solution to the problem is given in the *Student's Solutions Manual*.

P19.1 The $^1H^{35}Cl$ molecule can be described by a Morse potential with $D_e = 7.41 \times 10^{-19}$ J. The force constant k for this molecule is 516 N m^{-1} and $\nu = 8.97 \times 10^{13}$ s^{-1}.

a. Calculate the lowest four energy levels for a Morse potential.

b. Calculate the fundamental frequency ν_0 corresponding to the transition $n = 0 \rightarrow n = 1$ and the frequencies of the first three overtone vibrations. How large would the relative error be if you assume that the first three overtone frequencies are $2\nu_0$, $3\nu_0$, and $4\nu_0$?

P19.2 The infrared spectrum of $^7Li^{19}F$ has an intense line at 910.57 cm^{-1}. Calculate the force constant and period of vibration of this molecule.

P19.3 Purification of water for drinking using UV light is a viable way to provide potable water in many areas of the world. Experimentally, the decrease in UV light of wavelength 250 nm follows the empirical relation $I/I_0 = e^{-\varepsilon' l}$ where l is the distance that the light passed through the water and ε' is an effective absorption coefficient. $\varepsilon' = 0.070$ cm^{-1} for pure water and 0.30 cm^{-1} for water exiting a wastewater treatment plant. What distance corresponds to a decrease in I of 15% from its incident value for (a) pure water and (b) waste water?

P19.4 A simulated infrared absorption spectrum of a gas-phase organic compound is shown in the following figure. Use the characteristic group frequencies listed in Section 19.6 to decide whether this compound is more likely to be Cl_2CO, $(CH_3)_2CO$, CH_3OH, CH_3COOH, CH_3CN, CCl_4, or C_3H_8. Explain your reasoning.

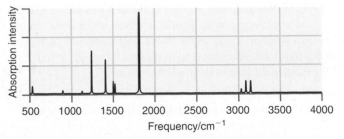

P19.5 The molecules $^{16}O^{12}C^{32}S$ and $^{16}O^{12}C^{34}S$ have values for $h/8\pi^2 I$ of 6081.490×10^6 s^{-1} and 5932.816×10^6 s^{-1}, respectively. Calculate the C—O and C—S bond distances.

P19.6 A simulated infrared absorption spectrum of a gas-phase organic compound is shown in the following figure. Use the characteristic group frequencies listed in Section 19.6 to decide whether this compound is more likely to be Cl_2CO, $(CH_3)_2CO$, CH_3OH, CH_3COOH, CH_3CN, CCl_4, or C_3H_8. Explain your reasoning.

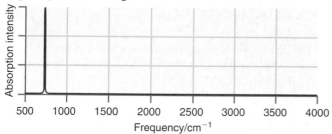

P19.7 The rotational constant for $^{14}N_2$ determined from microwave spectroscopy is 1.99824 cm^{-1}. The atomic mass of ^{14}N is 14.003074007 amu. Calculate the bond length in $^{14}N_2$ to the maximum number of significant figures consistent with this information.

P19.8 An infrared absorption spectrum of an organic compound is shown in the following figure. Use the characteristic group frequencies listed in Section 19.6 to decide whether this compound is more likely to be ethyl amine, pentanol, or acetone.

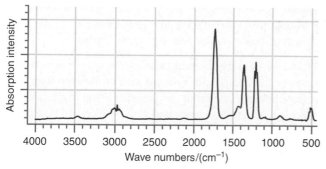

P19.9 Calculate the zero point energies for $^1H^{19}F$ and $^2D^{19}F$. Compare the difference in the zero point energies to $k_B T$ at 298 K.

P19.10 Write an expression for the moment of inertia of the acetylene molecule in terms of the bond distances. Does this molecule have a pure rotational spectrum?

P19.11 Show that the selection rule for the two-dimensional rotor in the dipole approximation is $\Delta m_l = \pm 1$. Use $A_{+\phi}e^{im_1\phi}$ and $A'_{+\phi}e^{im_2\phi}$ for the initial and final states of the rotor and $\mu\cos\phi$ as the dipole moment element.

P19.12 Following Example Problem 19.5, show that the $J = 1 \rightarrow J = 2$ rotational transition is allowed.

P19.13 Selection rules in the dipole approximation are determined by the integral $\mu_x^{mn} = \int \psi_m^*(\tau)\mu_x(\tau)\psi_n(\tau)\,d\tau$. If this integral is nonzero, the transition will be observed in an absorption spectrum. If the integral is zero, the transition is "forbidden" in the dipole approximation. It actually occurs with low probability because the dipole approximation is not exact. Consider the particle in the one-dimensional box and set $\mu_x = -ex$.

a. Calculate μ_x^{12} and μ_x^{13} in the dipole approximation. Can you see a pattern and discern a selection rule? You may need to evaluate a few more integrals of the type μ_x^{1m}. The standard integral

$$\int x \sin\left(\frac{\pi x}{a}\right)\sin\left(\frac{n\pi x}{a}\right)dx$$

$$= \frac{1}{2}\left(\frac{a^2\cos\dfrac{(n-1)\pi x}{a}}{(n-1)^2\pi^2} + \frac{ax\sin\dfrac{(n-1)\pi x}{a}}{(n-1)\pi}\right)$$

$$- \frac{1}{2}\left(\frac{a^2\cos\dfrac{(n+1)\pi x}{a}}{(n+1)^2\pi^2} + \frac{ax\sin\dfrac{(n+1)\pi x}{a}}{(n+1)\pi}\right)$$

is useful for solving this problem.

b. Determine the ratio μ_x^{12}/μ_x^{14}. On the basis of your result, would you modify the selection rule that you determined in part (a)?

P19.14 The bond length of $^7Li^1H$ is 159.49 pm. Calculate the value of B and the spacing between lines in the pure rotational spectrum of this molecule in units of s^{-1}.

P19.15 Calculating the motion of individual atoms in the vibrational modes of molecules (called normal modes) is an advanced topic. Given the normal modes shown in the following figure, decide which of the normal modes of CO_2 and H_2O have a nonzero dynamical dipole moment and are therefore infrared active. The motion of the atoms in the second of the two doubly degenerate bend modes for CO_2 is identical to the first but is perpendicular to the plane of the page.

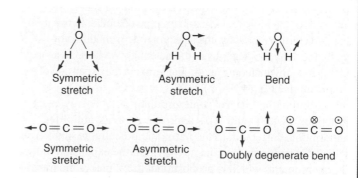

P19.16 The force constants for F_2 and I_2 are 470. and 172 N m^{-1}, respectively. Calculate the ratio of the vibrational state populations n_1/n_0 and n_2/n_0 at $T = 300.$ and at 1000. K.

P19.17 The rigid rotor model can be improved by recognizing that in a realistic anharmonic potential, the bond length increases with the vibrational quantum number n. Therefore, the rotational constant depends on n, and it can be shown that $B_n = B - (n + 1/2)\alpha$, where B is the rigid rotor value. The constant α can be obtained from experimental spectra. For $^1H^{81}Br$, $B = 8.46488$ cm^{-1} and $\alpha = 0.23328$ cm^{-1}. Using this more accurate formula for B_n, calculate the bond length for HBr in the ground state and for $n = 3$.

P19.18 Greenhouse gases generated from human activity absorb infrared radiation from Earth and keep it from being dispersed outside our atmosphere. This is a major cause of global warming. Compare the path length required to absorb 90.% of Earth's radiation near a wavelength of 7 μm for CH_3CCl_3 $[\varepsilon(\lambda) = 1.8 \text{ (cm atm)}^{-1}]$ and the chlorofluorocarbon CFC-14 $[\varepsilon(\lambda) = 4.1 \times 10^3 \text{ (cm atm)}^{-1}]$ assuming that each of these gases has a partial pressure of 1.5×10^{-6} atm.

P19.19 Show that the Morse potential approaches the harmonic potential for small values of the vibrational amplitude. (*Hint*: Expand the Morse potential in a Taylor-Maclaurin series.)

P19.20 The rotational constant for $^7Li^{19}F$ determined from microwave spectroscopy is 1.342583 cm^{-1}. The atomic masses of ^7Li and ^{19}F are 7.00160041 and 18.9984032 amu, respectively. Calculate the bond length in $^7Li^{19}F$ to the maximum number of significant figures consistent with this information.

P19.21 A simulated infrared absorption spectrum of a gas-phase organic compound is shown in the following figure. Use the characteristic group frequencies listed in Section 19.6 to decide whether this compound is more likely to be Cl_2CO, $(CH_3)_2CO$, CH_3OH, CH_3COOH, CH_3CN, CCl_4, or C_3H_8. Explain your reasoning.

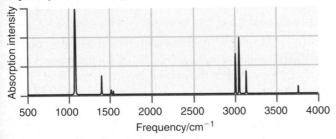

P19.22 Overtone transitions in vibrational absorption spectra for which $\Delta n = +2, +3, \ldots$ are forbidden for the harmonic potential $V = (1/2)kx^2$ because $\mu_x^{mn} = 0$ for $|m - n| \neq 1$ as shown in Section 19.4. However, overtone transitions are allowed for the more realistic anharmonic potential. In this problem, you will explore how the selection rule is modified by including anharmonic terms in the potential. We do so in an indirect manner by including additional terms in the expansion of the dipole moment $\mu_x(x_e + x) = \mu_{0x} + x(d\mu_x/dx)_{r_e} + \ldots$ but assuming that the harmonic oscillator total energy eigenfunctions are still valid. This approximation is valid if the anharmonic correction to the harmonic potential is small. You will show that including the next term in the expansion of the dipole moment, which is proportional to x^2, makes the transitions $\Delta n = \pm 2$ allowed.

a. Show that Equation (19.8) becomes

$$\mu_x^{m0} = A_m A_0 \mu_{0x} \int_{-\infty}^{\infty} H_m(\alpha^{1/2}x) \, H_0(\alpha^{1/2}x)e^{-\alpha x^2} \, dx$$

$$+ A_m A_0 \left(\frac{d\mu_x}{dx}\right)_{x=0} \int_{-\infty}^{\infty} H_m(\alpha^{1/2}x) \, x \, H_0(\alpha^{1/2}x)e^{-\alpha x^2} \, dx$$

$$+ \frac{A_m A_0}{2!} \left(\frac{d^2\mu_x}{dx^2}\right)_{x=0} \int_{-\infty}^{\infty} H_m(\alpha^{1/2}x) \, x^2 H_0(\alpha^{1/2}x)e^{-\alpha x^2} \, dx$$

b. Evaluate the effect of adding the additional term to μ_x^{mn}. You will need the recursion relationship

$$\alpha^{1/2}x \, H_n(\alpha^{1/2}x) = nH_{n-1}(\alpha^{1/2}x) + \tfrac{1}{2}H_{n+1}(\alpha^{1/2}x).$$

c. Show that both the transitions $n = 0 \to n = 1$ and $n = 0 \to n = 2$ are allowed in this case.

P19.23 The fundamental vibrational frequencies for $^1H_2$ and $^2D_2$ are 4401 and 3115 cm^{-1}, respectively, and D_e for both molecules is 7.677×10^{-19} J. Using this information, calculate the bond energy of both molecules.

P19.24 A simulated infrared absorption spectrum of a gas-phase organic compound is shown in the following figure. Use the characteristic group frequencies listed in Section 19.6 to decide whether this compound is more likely to be Cl_2CO, $(CH_3)_2CO$, CH_3OH, CH_3COOH, CH_3CN, CCl_4, or C_3H_8. Explain your reasoning.

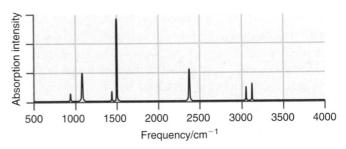

P19.25 Isotopic substitution is used to identify characteristic groups in an unknown compound using vibrational spectroscopy. Consider the C—C bond in ethane ($^{12}C_2\,^1H_6$). By what factor would the frequency change if deuterium were substituted for all the hydrogen atoms? Treat the H and D atoms as being rigidly attached to the carbon.

P19.26 A simulated infrared absorption spectrum of a gas-phase organic compound is shown in the following figure. Use the characteristic group frequencies listed in Section 19.6 to decide whether this compound is more likely to be Cl_2CO, $(CH_3)_2CO$, CH_3OH, CH_3COOH, CH_3CN, CCl_4, or C_3H_8. Explain your reasoning.

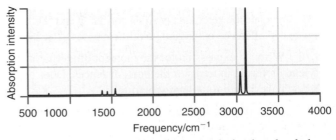

P19.27 Fill in the missing step in the derivation that led to the calculation of the spectral line shape in Figure 19.23. Starting from

$$a_2(t) = \mu_x^{21} \frac{E_0}{2}\left(\frac{1 - e^{\frac{i}{\hbar}(E_2 - E_1 + h\nu)t}}{E_2 - E_1 + h\nu} + \frac{1 - e^{-\frac{i}{\hbar}(E_2 - E_1 - h\nu)t}}{E_2 - E_1 - h\nu}\right)$$

and neglecting the first term in the parentheses, show that

$$a_2^*(t)a_2(t) = E_0^2[\mu_x^{21}]^2 \frac{\sin^2[(E_2 - E_1 - h\nu)t/2\hbar]}{(E_2 - E_1 - h\nu)^2}$$

P19.28   The force constant for $^7Li_2$ is 26.0 N m^{-1}. Calculate the vibrational frequency and zero point energy of this molecule.

P19.29   Because the intensity of a transition to first order is proportional to the population of the originating state, the J value for which the maximum intensity is observed in a rotational-vibrational spectrum is not generally $J = 0$. Treat J in the equation

$$\frac{n_J}{n_0} = \frac{g_J}{g_0}e^{-(E_J-E_0)/k_BT} = (2J + 1)e^{-\hbar^2 J(J+1)/(2Ik_BT)}$$

as a continuous variable.

a. Show that

$$\frac{d\left(\frac{n_J}{n_0}\right)}{dJ} = 2e^{-\hbar^2 J(J+1)/(2I\,k_BT)} - \frac{(2J + 1)^2\hbar^2}{2Ik_BT}e^{-\hbar^2 J(J+1)/(2I\,k_BT)}$$

b. Show that setting $d(n_J/n_0)/dJ = 0$ gives the equation

$$2 - \frac{(2J_{max} + 1)^2\hbar^2}{2I\,k_BT} = 0$$

c. Show that the solution of this quadratic equation is

$$J_{max} = \frac{1}{2}\left[\sqrt{\frac{4Ik_BT}{\hbar^2}} - 1\right]$$

In this problem, we assume that the intensity of the individual peaks is solely determined by the population in the originating state and that it does not depend on the initial and final J values.

P19.30   A strong absorption band in the infrared region of the electromagnetic spectrum is observed at $\tilde{\nu} = 1298$ cm^{-1} for $^{40}Ca^1H$. Assuming that the harmonic potential applies, calculate the fundamental frequency ν in units of inverse seconds, the vibrational period in seconds, and the zero point energy for the molecule in joules and electron-volts.

P19.31   The spacing between lines in the pure rotational spectrum of $^{11}B^2D$ is 3.9214×10^{11} s^{-1}. Calculate the bond length of this molecule.

P19.32   In this problem, you will derive the equations used to explain the Michelson interferometer for incident light of a single frequency.

a. Show that the expression

$$A(t) = \frac{A_0}{\sqrt{2}}(1 + e^{i\delta(t)}) \exp\left[i(ky_D - \omega t)\right]$$

represents the sum of two waves of the form $A_0/\sqrt{2}\exp\left[i(kx - \omega t)\right]$, one of which is phase shifted by the amount $\delta(t)$ evaluated at the position y_D.

b. Show using the definition $I(t) = A(t)A^*(t)$ that $I(t) = I_0/[2(1 + \cos\delta(t))]$.

c. Expressing $\delta(t)$ in terms of $\Delta d(t)$, show that

$$I(t) = \frac{I_0}{2}\left(1 + \cos\frac{2\pi\,\Delta d(t)}{\lambda}\right)$$

d. Expressing $\Delta d(t)$ in terms of the mirror velocity v, show that

$$I(t) = \frac{I_0}{2}\left(1 + \cos\left[\frac{2v}{c}\omega t\right]\right)$$

P19.33   Calculate the moment of inertia, the magnitude of the rotational angular momentum, and the energy in the $J = 4$ rotational state for $^{14}N_2$.

P19.34   A simulated infrared absorption spectrum of a gas-phase organic compound is shown in the following figure. Use the characteristic group frequencies listed in Section 19.6 to decide whether this compound is more likely to be Cl_2CO, $(CH_3)_2CO$, CH_3OH, CH_3COOH, CH_3CN, CCl_4, or C_3H_8. Explain your reasoning.

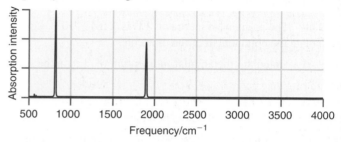

P19.35   A measurement of the vibrational energy levels of $^{12}C^{16}O$ gives the relationship

$$\tilde{\nu}(n) = 2170.21\left(n + \frac{1}{2}\right)cm^{-1} - 13.461\left(n + \frac{1}{2}\right)^2 cm^{-1}$$

where n is the vibrational quantum number. The fundamental vibrational frequency is $\tilde{\nu}_0 = 2170.21$ cm^{-1}. From these data, calculate the depth D_e of the Morse potential for $^{12}C^{16}O$. Calculate the bond energy of the molecule.

P19.36   Using the formula for the energy levels for the Morse potential,

$$E_n = h\nu\left(n + \frac{1}{2}\right) - \frac{(h\nu)^2}{4D_e}\left(n + \frac{1}{2}\right)^2$$

show that the energy spacing between adjacent levels is given by

$$E_{n+1} - E_n = h\nu - \frac{(h\nu)^2}{2D_e}(n + 1)$$

P19.37   Use your results from Problem P19.36 to solve the following problem. For $^1H^{35}Cl$, $D_e = 7.41 \times 10^{-19}$ J and $\nu = 8.97 \times 10^{13}$ s^{-1}. As n increases, the energy difference between adjacent vibrational levels decreases and approaches zero, corresponding to dissociation. Assuming a Morse potential, calculate all discrete vibrational energy values for $^1H^{35}Cl$. What value of n corresponds to dissociation?

P19.38   In Problem P19.29 you obtained the result

$$J_{max} = (1/2)\left[\sqrt{4Ik_BT/\hbar^2} - 1\right]$$

Using this result, estimate T for the simulated $^1H^{35}Cl$ rotational spectra shown in the following figure. Give realistic estimates of the precision with which you can determine T from the spectra. In generating the simulation, we assumed that the intensity of the individual peaks is solely determined by the population in the originating state and that it does not depend on the initial and final J values.

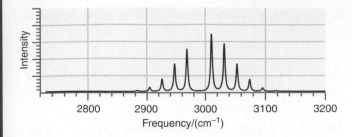

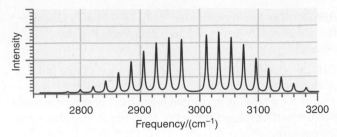

P19.39   Of the 190 nm wavelength light incident on a 15.0-mm-thick piece of fused silica quartz glass, 35% passes through the glass and the remainder is absorbed. What percentage of the light will pass through a 35.0-mm-thick piece of the same glass?

P19.40   The moment of inertia of $^7Li_2$ is 4.161×10^{-46} kg m^2. Calculate the bond length of the molecule.

P19.41   Calculate the angular momentum of $^7Li_2$ in the $J = 5$ state.

P19.42   The rotational energy of $^7Li_2$ in the $J = 5$ state is 4.0126×10^{-22} J. Calculate the bond length of the molecule.

P19.43   A simulated infrared absorption spectrum of a gas-phase organic compound is shown in the following figure. Use the characteristic group frequencies listed in Section 19.6 to decide whether this compound is more likely to be Cl_2CO, $(CH_3)_2CO$, CH_3OH, CH_3COOH, CH_3CN, CCl_4, or C_3H_8. Explain your reasoning.

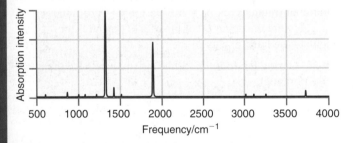

P19.44   An infrared absorption spectrum of an organic compound is shown in the following figure. Use the characteristic group frequencies listed in Section 19.6 to decide whether this compound is more likely to be hexene, hexane, or hexanol.

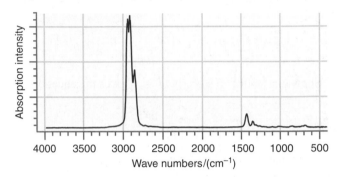

P19.45   If the vibrational potential is not harmonic, the force constant is not independent of degree of stretching or compression of a molecule. Using the relation $k_{effective} = (d^2V(x)/dx^2)$, derive an expression for the vibrational force constant for a Morse potential as a function of $x–x_e$. Using the parameters from Table 19.3, plot $k_{effective}$ as a function of x over a ± 5.0 pm range from x_e for $^1H^{35}Cl$. What is the variation of $k_{effective}$ over this range of x?

Computational Problems

More detailed instructions on carrying out these calculations using Spartan Physical Chemistry are found on the book website at *www.masteringchemistry.com*.

C19.1   Build structures for the gas-phase (a) hydrogen fluoride ($^1H^{19}F$), (b) hydrogen chloride ($^1H^{35}Cl$), (c) carbon monoxide ($^{12}C^{16}O$), and (d) sodium chloride ($^{23}Na^{35}Cl$) molecules. (For Spartan, these are the default isotopic masses.) Calculate the equilibrium geometry and the IR spectrum using the B3LYP method with the 6-311+G** basis set.

a. Compare your result for the vibrational frequency with the experimental value listed in Table 19.3. What is the relative error in the calculation?

b. Calculate the force constant from the vibrational frequency and reduced mass. Determine the relative error using the experimental value in Table 19.3.

c. Calculate the values for the rotational constant B using the calculated bond length. Determine the relative error using the experimental value in Table 19.3.

C19.2   Calculate the bond energy in gaseous (a) hydrogen fluoride ($^1H^{19}F$), (b) hydrogen chloride ($^1H^{35}Cl$), (c) carbon monoxide ($^{12}C^{16}O$), and (d) sodium chloride ($^{23}Na^{35}Cl$) molecules by comparing the total energies of the species in the dissociation reactions [e.g., $HF(g) \rightarrow H(g) + F(g)$]. Use the B3LYP method with the 6-31G* basis set. Determine the relative error of the calculation using the experimental value in Table 19.3.

C19.3 Build structures for the gas-phase (a) NF_3, (b) PCl_3 and (c) SO_3 molecules. Calculate the equilibrium geometry and the IR spectrum using the B3LYP method with the 6-31G* basis set. Animate the vibrational normal modes and classify them as symmetrical stretch, symmetrical deformation, degenerate stretch, and degenerate deformation.

C19.4 Build structures for the gas-phase (a) F_2CO, (b) Cl_2CO, and (c) O_2NF molecules of the structural form X_2YZ. Calculate the equilibrium geometry and the IR spectrum using the B3LYP method with the 6-311+G** basis set.

Animate the vibrational normal modes and classify them as Y-Z stretch, YX_2 scissors, antisymmetric X-Y stretch, YX_2 rock, and $Y-X_2$ wag.

C19.5 Build structures for the bent gas-phase (a) HOF, (b) ClOO, and (c) HSO molecules of the structural form XYZ. Calculate the equilibrium geometry and the IR spectrum using the B3LYP method with the 6-31G* basis set. Animate the vibrational normal modes and classify them as Y-Z stretch, X-Y stretch, and X-Y-Z bend.

Web-Based Simulations, Animations, and Problems

W19.1 A pair of emission spectra, one from an unknown (hypothetical) atom and one resulting from the electron energy levels entered using sliders, is displayed. The student adjusts the displayed energy levels in order to replicate the atomic spectrum and, hence, determine the actual electron energy levels in the atom.

W19.2 The number of allowed energy levels in a Morse potential is determined for variable values of the vibrational frequency and the well depth.

W19.3 The normal modes for H_2O are animated. Each normal mode is associated with a local motion from a list displayed in the simulation.

W19.4 The normal modes for CO_2 are animated. Each normal mode is associated with a local motion from a list displayed in the simulation.

W19.5 The normal modes for NH_3 are animated. Each normal mode is associated with a local motion from a list displayed in the simulation.

W19.6 The normal modes for formaldehyde are animated. Each normal mode is associated with a local motion from a list displayed in the simulation.

W19.7 Simulated rotational (microwave) spectra are generated for one or more of the diatomic molecules $^{12}C^{16}O$, $^{1}H^{19}F$, $^{1}H^{35}Cl$, $^{1}H^{79}Br$, and $^{1}H^{127}I$. Using a slider, the temperature is varied. The J value corresponding to the maximum intensity peak is determined and compared with the prediction from the formula

$$J_{max} = \frac{1}{2}\left[\sqrt{\frac{4I\,k_BT}{\hbar^2}} - 1\right]$$

The number of peaks that have an intensity greater than half of that for the largest peak is determined at different temperatures. The frequencies of the peaks are then used to generate the rotational constants B and α_e.

W19.8 Simulated rotational-vibrational (infrared) spectra are generated for one or more diatomic molecules including $^{12}C^{16}O$, $^{1}H^{19}F$, $^{1}H^{35}Cl$, $^{1}H^{79}Br$, or $^{1}H^{127}I$ for predetermined temperatures. The frequencies of the peaks are then used to generate the rotational constants B and α_e, and the force constant k.

The Hydrogen Atom

Classical physics is unable to explain the stability of atoms. In this chapter, we solve the Schrödinger equation for the motion of an electron in a spherically symmetric Coulomb potential and show that an atom consisting of an electron orbiting around a central positively charged nucleus is stable. To emphasize the similarities and differences between quantum mechanical and classical models, a comparison is made between the quantum mechanical picture of the hydrogen atom and the popularly depicted shell picture of the atom.

20.1 Formulating the Schrödinger Equation

After having applied quantum mechanics to a number of simple problems, we turn to one of the triumphs of quantum mechanics: the understanding of atomic structure and spectroscopy. As discussed in Chapter 21, for atoms with more than one electron, the Schrödinger equation cannot be solved exactly. However, for the hydrogen atom, the Schrödinger equation can be solved exactly, and many of the results we obtain from that solution can be generalized to many-electron atoms.

To set the stage historically, experiments by Rutherford had established that the positive charge associated with an atom was localized at the center of the atom and that the electrons were spread out over a large volume (relative to nuclear dimensions) centered at the nucleus. The **shell model** in which the electrons are confined in spherical shells centered at the nucleus had a major flaw when viewed from the vantage point of classical physics. An electron orbiting around the nucleus undergoes accelerated motion and radiates energy. Therefore, it will eventually fall into the nucleus. Atoms are not stable according to classical mechanics. The challenge for quantum mechanics was to provide a framework within which the stability of atoms could be understood.

We model the hydrogen atom as made up of an electron moving about a proton located at the origin of the coordinate system. The two particles attract one another and the interaction potential is given by a simple **Coulomb potential:**

$$V(\mathbf{r}) = -\frac{e^2}{4\pi\varepsilon_0|\mathbf{r}|} = -\frac{e^2}{4\pi\varepsilon_0 r} \qquad (20.1)$$

In this equation, e is the electron charge, and ε_0 is the permittivity of free space. In the text that follows, we abbreviate the magnitude of the vector $\mathbf{r}$ as r, the distance between the nucleus, and the electron. Because the potential is spherically symmetrical, we choose spherical polar coordinates (r, θ, ϕ) to formulate the Schrödinger equation for this problem. In doing so, it takes on the formidable form

$$-\frac{\hbar^2}{2m_e}\left[\frac{1}{r^2}\frac{\partial}{\partial r}\left(r^2\frac{\partial\psi(r,\theta,\varphi)}{\partial r}\right) + \frac{1}{r^2\sin\theta}\frac{\partial}{\partial\theta}\left(\sin\theta\frac{\partial\psi(r,\theta,\varphi)}{\partial\theta}\right) + \frac{1}{r^2\sin^2\theta}\frac{\partial^2\psi(r,\theta,\varphi)}{\partial\varphi^2}\right]$$

$$-\frac{e^2}{4\pi\varepsilon_0 r}\psi(r,\theta,\phi) = E\psi(r,\theta,\phi) \tag{20.2}$$

In this equation m_e is the electron mass.

20.2 Solving the Schrödinger Equation for the Hydrogen Atom

Because $V(r)$ depends only on r and not on the angles θ and ϕ, we can achieve a **separation of variables,** as discussed in Sections 15.3 and 18.4, and write the wave function as a product of three functions, each of which depends on only one of the variables:

$$\psi(r,\theta,\phi) = R(r)\Theta(\theta)\Phi(\phi) \tag{20.3}$$

This simplifies the solution of the partial differential equation greatly. We also recognize that, apart from constants, the angular part of Equation (20.2), the last two terms in the brackets, is the operator $\hat{l}^2$ discussed in Section 18.6. Therefore, the angular part of $\psi(r, \theta, \phi)$ is the product $\Theta(\theta)\Phi(\phi)$ that we encountered in solving the Schrödinger equation for the rigid rotor, namely, the normalized spherical harmonic functions $Y_l^{m_l}(\theta, \phi)$. Therefore, the only part of $\psi(r, \theta, \phi)$ that remains unknown is the radial function $R(r)$.

Equation (20.2) can be reduced to a radial equation in the following way. Substituting the product function $\psi(r,\theta,\phi) = R(r)\Theta(\theta)\Phi(\phi)$ into Equation (20.2), and taking out those parts not affected by the partial derivative in front of each term, we obtain

$$-\frac{\hbar^2}{2m_e r^2}\Theta(\theta)\Phi(\phi)\frac{d}{dr}\left[r^2\frac{d\,R(r)}{dr}\right] + \frac{1}{2m_e r^2}R(r)\hat{l}^2\Theta(\theta)\Phi(\phi)$$

$$-\Theta(\theta)\Phi(\phi)\left[\frac{e^2}{4\pi\varepsilon_0 r}\right]R(r) = ER(r)\Theta(\theta)\Phi(\phi) \tag{20.4}$$

We know that $\hat{l}^2\Theta(\theta)\Phi(\phi) = \hbar^2 l(l+1)\Theta(\theta)\Phi(\phi)$. Putting this result into Equation (20.4), and canceling the product $\Theta(\theta)\Phi(\phi)$ that appears in each term, a differential equation is obtained for $R(r)$:

$$-\frac{\hbar^2}{2m_e r^2}\frac{d}{dr}\left[r^2\frac{dR(r)}{dr}\right] + \left[\frac{\hbar^2 l(l+1)}{2m_e r^2} - \frac{e^2}{4\pi\varepsilon_0 r}\right]R(r) = ER(r) \tag{20.5}$$

Before continuing, we summarize the preceding discussion. The Schrödinger equation was formulated for the hydrogen atom. It differs from the rigid rotor problem, where r has a fixed value, in that the potential is not zero; instead, it depends inversely on r. Because the potential is not dependent on the angular coordinates, the solutions to the Schrödinger equation for θ and ϕ are the same as those obtained for the rigid rotor. In the rigid rotor, r was fixed at a constant value that is appropriate for a diatomic molecule with a stiff bond. For the electron–proton distance in the hydrogen atom, this is clearly not appropriate, and the wave function will depend on r. We have been able to separate out the dependence of the wave function on the radial coordinate r from that on the angles θ and ϕ. We now take a closer look at the eigenvalues and eigenfunctions for Equation (20.5).

Note that the second term on the left-hand side of Equation (20.5) can be viewed as an **effective potential,** $V_{eff}(r)$. It is made up of the **centrifugal potential,** which varies as $+1/r^2$, and the Coulomb potential, which varies as $-1/r$:

$$V_{eff}(r) = \frac{\hbar^2 l(l+1)}{2m_e r^2} - \frac{e^2}{4\pi\varepsilon_0 r} \tag{20.6}$$

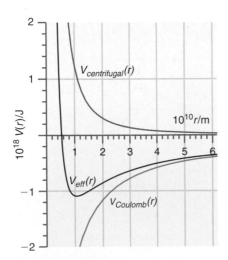

FIGURE 20.1

The individual contributions to the effective potential and their sum are plotted as a function of distance. The centrifugal potential used is for $l = 1$; larger values of l make the effective potential more repulsive at small r.

Each of the terms that contribute to $V_{eff}(r)$ and their sums are graphed as a function of distance in Figure 20.1.

Because the first term is repulsive and varies more rapidly with r than the Coulomb potential, it dominates at small distances if $l \neq 0$. Both terms approach zero for large values of r. The resultant potential is repulsive at short distances for $l > 0$ and is more repulsive the greater the value of l. The net result of this repulsive centrifugal potential is to force the electrons in orbitals with $l > 0$ (looking ahead, p, d, and f electrons) on average farther from the nucleus than s electrons for which $l = 0$.

20.3 Eigenvalues and Eigenfunctions for the Total Energy

Equation (20.5) can be solved using standard mathematical methods, so we concern ourselves only with the results. Note that the energy E only appears in the radial equation and not in the angular equation. Because only one variable is involved in this equation, the energy is expected to depend on a single quantum number. The quantization condition that results from the restriction that $R(r)$ be well behaved at large values of r $[R(r) \rightarrow 0$ as $r \rightarrow \infty]$ is

$$E_n = -\frac{m_e e^4}{8\varepsilon_0^2 h^2 n^2} \quad , \text{ for } n = 1, 2, 3, 4, \dots \quad \textbf{(20.7)}$$

This formula is usually simplified by combining a number of constants in the form $a_0 = \varepsilon_0 h^2/\pi m_e e^2$. The quantity a_0 has the value 0.529×10^{-10} m and is called the **Bohr radius.** Use of this definition leads to the following formula:

$$E_n = -\frac{e^2}{8\pi\varepsilon_0 a_0 n^2} = -\frac{2.179 \times 10^{-18} \text{ J}}{n^2} = -\frac{13.60 \text{ eV}}{n^2} \quad n = 1, 2, 3, 4, \dots \quad \textbf{(20.8)}$$

Note that E_n goes to zero as $n \rightarrow \infty$. As previously emphasized, the zero of energy is a matter of convention rather than being a quantity that can be determined. As n approaches infinity, the electron is on average farther and farther from the nucleus, and the zero of energy corresponds to the electron at infinite separation from the nucleus. All negative energies correspond to bound states of the electron in the Coulomb potential. Positive energies correspond to states in which the atom is ionized.

As has been done previously for the particle in the box and the harmonic oscillator, the energy eigenvalues can be superimposed on a potential energy diagram, as shown in Figure 20.2. The potential forms a "box" that acts to confine the particle. This box has a peculiar form in that it is infinitely deep at the center of the atom, and the depth falls off inversely with distance from the proton. Figure 20.2 shows that the two lowest energy levels have an appreciable separation in energy and that the separation for adjacent energy levels becomes rapidly smaller as $n \rightarrow \infty$. All states for which $5 < n < \infty$ have energies in the narrow range between $\sim -1 \times 10^{-19}$ J and zero. Although this seems strange at first, it is exactly what is expected based on the results for the particle in the box. Because of the shape of the potential, the H atom box is very narrow for the first few energy eigenstates but becomes very wide for large n. The particle in the box formula [Equation (15.17)] predicts that the energy spacing varies as the inverse of the square of box length. This is the trend seen in Figure 20.2. Note also that the wave functions penetrate into the classically forbidden region just as for the particle in the finite depth box and the harmonic oscillator.

Although the energy depends on a single quantum number n, the eigenfunctions $\psi(r, \theta, \phi)$ are associated with three quantum numbers because three boundary conditions arise in a three-dimensional problem. The other two quantum numbers are l and m_l, which arise from the angular coordinates. As for the rigid rotor, these quantum numbers are not independent. Their relationship is given by

$$n = 1, 2, 3, 4, \dots$$
$$l = 0, 1, 2, 3, \dots, n - 1$$
$$m_l = 0, \pm 1, \pm 2, \pm 3, \dots \pm l \quad \textbf{(20.9)}$$

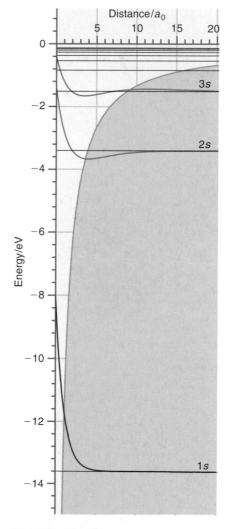

FIGURE 20.2

The border of the yellow classically forbidden region is the Coulomb potential, which is shown together with E_n for $n = 1$ through $n = 10$ and the $1s$, $2s$, and $3s$ wave functions.

The relationship between l and m_l was discussed in Section 18.3. Although we do not present a justification of the relationship between n and l here, all the conditions in Equation (20.9) emerge naturally out of the boundary conditions in the solution of the differential equations.

The radial functions $R(r)$ are products of an exponential function with a polynomial in the dimensionless variable r/a_0. Their functional form depends on the quantum numbers n and l. The first few normalized radial functions $R_{nl}(r)$ are as follows:

$$n = 1, l = 0 \quad R_{10}(r) = 2\left(\frac{1}{a_0}\right)^{3/2} e^{-r/a_0}$$

$$n = 2, l = 0 \quad R_{20}(r) = \frac{1}{\sqrt{8}}\left(\frac{1}{a_0}\right)^{3/2}\left(2 - \frac{r}{a_0}\right) e^{-r/2a_0}$$

$$n = 2, l = 1 \quad R_{21}(r) = \frac{1}{\sqrt{24}}\left(\frac{1}{a_0}\right)^{3/2}\frac{r}{a_0} e^{-r/2a_0}$$

$$n = 3, l = 0 \quad R_{30}(r) = \frac{2}{81\sqrt{3}}\left(\frac{1}{a_0}\right)^{3/2}\left(27 - 18\frac{r}{a_0} + 2\frac{r^2}{a_0^2}\right) e^{-r/3a_0}$$

$$n = 3, l = 1 \quad R_{31}(r) = \frac{4}{81\sqrt{6}}\left(\frac{1}{a_0}\right)^{3/2}\left(6\frac{r}{a_0} - \frac{r^2}{a_0^2}\right) e^{-r/3a_0}$$

$$n = 3, l = 2 \quad R_{32}(r) = \frac{4}{81\sqrt{30}}\left(\frac{1}{a_0}\right)^{3/2}\frac{r^2}{a_0^2} e^{-r/3a_0}$$

To form the hydrogen atom eigenfunctions, we combine $R_{nl}(r)$ with the spherical harmonics and list here the first few of the infinite set of normalized wave functions $\psi(r,\theta,\phi) = R(r)\Theta(\theta)\Phi(\phi)$ for the hydrogen atom. Note that, in general, the eigenfunctions depend on r, θ, and ϕ, but are not functions of θ and ϕ for $l = 0$. The quantum numbers are associated with the wave functions using the notation ψ_{nlm_l}:

$$n = 1, l = 0, m_l = 0 \quad \psi_{100}(r) = \frac{1}{\sqrt{\pi}}\left(\frac{1}{a_0}\right)^{3/2} e^{-r/a_0}$$

$$n = 2, l = 0, m_l = 0 \quad \psi_{200}(r) = \frac{1}{4\sqrt{2\pi}}\left(\frac{1}{a_0}\right)^{3/2}\left(2 - \frac{r}{a_0}\right) e^{-r/2a_0}$$

$$n = 2, l = 1, m_l = 0 \quad \psi_{210}(r, \theta, \phi) = \frac{1}{4\sqrt{2\pi}}\left(\frac{1}{a_0}\right)^{3/2}\frac{r}{a_0} e^{-r/2a_0}\cos\theta$$

$$n = 2, l = 1, m_l = \pm 1 \quad \psi_{21\pm1}(r, \theta, \phi) = \frac{1}{8\sqrt{\pi}}\left(\frac{1}{a_0}\right)^{3/2}\frac{r}{a_0} e^{-r/2a_0}\sin\theta\, e^{\pm i\phi}$$

$$n = 3, l = 0, m_l = 0 \quad \psi_{300}(r) = \frac{1}{81\sqrt{3\pi}}\left(\frac{1}{a_0}\right)^{3/2}\left(27 - 18\frac{r}{a_0} + 2\frac{r^2}{a_0^2}\right) e^{-r/3a_0}$$

$$n = 3, l = 1, m_l = 0 \quad \psi_{310}(r, \theta, \phi) = \frac{1}{81}\left(\frac{2}{\pi}\right)^{1/2}\left(\frac{1}{a_0}\right)^{3/2}\left(6\frac{r}{a_0} - \frac{r^2}{a_0^2}\right) e^{-r/3a_0}\cos\theta$$

$$n = 3, l = 1, m_l = \pm 1 \quad \psi_{31\pm1}(r, \theta, \phi) = \frac{1}{81\sqrt{\pi}}\left(\frac{1}{a_0}\right)^{3/2}\left(6\frac{r}{a_0} - \frac{r^2}{a_0^2}\right) e^{-r/3a_0}\sin\theta\, e^{\pm i\phi}$$

$$n = 3, l = 2, m_l = 0 \quad \psi_{320}(r, \theta, \phi) = \frac{1}{81\sqrt{6\pi}}\left(\frac{1}{a_0}\right)^{3/2}\frac{r^2}{a_0^2} e^{-r/3a_0}(3\cos^2\theta - 1)$$

$$n = 3, l = 2, m_l = \pm 1 \quad \psi_{32\pm1}(r, \theta, \phi) = \frac{1}{81\sqrt{\pi}}\left(\frac{1}{a_0}\right)^{3/2}\frac{r^2}{a_0^2}e^{-r/3a_0}\sin\theta\cos\theta\,e^{\pm i\phi}$$

$$n = 3, l = 2, m_l = \pm 2 \quad \psi_{32\pm2}(r, \theta, \phi) = \frac{1}{162\sqrt{\pi}}\left(\frac{1}{a_0}\right)^{3/2}\frac{r^2}{a_0^2}e^{-r/3a_0}\sin^2\theta\,e^{\pm 2i\phi}$$

These functions are referred to both as the H atom eigenfunctions and the H atom **orbitals.** A shorthand notation for the quantum numbers is to give the numerical value of n followed by a symbol indicating the values of l and m_l. The letters s, p, d, and f are used to denote $l = 0, 1, 2,$ and 3, respectively, $\psi_{100}(r)$ is referred to as the 1s orbital or wave function, and all three wave functions with $n = 2$ and $l = 1$ are referred to as 2p orbitals. The wave functions are real functions if $m_l = 0$, and complex functions otherwise. The angular and radial portions of the wave functions have nodes that are discussed in more detail later in this chapter. These functions have been normalized in keeping with the association between probability density and $\psi(r, \theta, \phi)$ stated in the first postulate (see Chapter 14).

EXAMPLE PROBLEM 20.1

Normalize the functions $e^{-r/2a_0}$ and $(r/a_0)\,e^{-r/2a_0}\sin\theta\,e^{+i\phi}$ in three-dimensional spherical coordinates.

Solution

In general, a wave function $\psi(\tau)$ is normalized by multiplying it by a constant N defined by $N^2\int\psi^*(\tau)\psi(\tau)\,d\tau = 1$. In three-dimensional spherical coordinates, $d\tau = r^2\sin\theta\,dr\,d\theta\,d\phi$, as discussed in Section 13.6. The normalization integral becomes $N^2\int_0^\pi \sin\theta\,d\theta\int_0^{2\pi}d\phi\int_0^\infty \psi^*(r,\theta,\phi)\psi(r,\theta,\phi)r^2\,dr = 1$.

For the first function,

$$N^2\int_0^\pi \sin\theta\,d\theta\int_0^{2\pi}d\phi\int_0^\infty e^{-r/2a_0}e^{-r/2a_0}r^2\,dr = 1$$

We use the standard integral

$$\int_0^\infty x^n e^{-ax}\,dx = \frac{n!}{a^{n+1}}$$

Integrating over the angles θ and ϕ, we obtain $4\pi N^2\int_0^\infty e^{-r/2a_0}e^{-r/2a_0}r^2\,dr = 1$. Evaluating the integral over r,

$$4\pi N^2\frac{2!}{1/a_0^3} = 1 \quad \text{or} \quad N = \frac{1}{2\sqrt{2\pi}}\left(\frac{1}{a_0}\right)^{3/2}$$

For the second function,

$$N^2\int_0^\pi \sin\theta\,d\theta\int_0^{2\pi}d\phi\int_0^\infty \left(\frac{r}{a_0}e^{-r/2a_0}\sin\theta\,e^{-i\phi}\right)\left(\frac{r}{a_0}e^{-r/2a_0}\sin\theta\,e^{+i\phi}\right)r^2\,dr = 1$$

This simplifies to

$$N^2\int_0^\pi \sin^3\theta\,d\theta\int_0^{2\pi}d\phi\int_0^\infty \left(\frac{r}{a_0}\right)^2 e^{-r/a_0}r^2\,dr = 1$$

Integrating over the angles θ and ϕ using the result $\int_0^\pi \sin^3\theta\,d\theta = 4/3$, we obtain

$$\frac{8\pi}{3}N^2\int_0^\infty \left(\frac{r}{a_0}\right)^2 e^{-r/a_0}r^2\,dr = 1$$

Using the same standard integral as in the first part of the problem,

$$\frac{8\pi}{3} N^2 \frac{1}{a_0^2}\left(\frac{4!}{1/a_0^5}\right) = 1 \quad \text{or} \quad N = \frac{1}{8\sqrt{\pi}}\left(\frac{1}{a_0}\right)^{3/2}$$

Each eigenfunction listed here describes a separate state of the hydrogen atom. However, as we have seen, the energy depends only on the quantum number n. Therefore, all states with the same value for n, but different values for l and m_l, have the same energy and we say that the energy levels are degenerate. Using the formulas given in Equation (20.9), we can see that the **degeneracy** of a given level is n^2. Therefore, the $n = 2$ level has a fourfold degeneracy and the $n = 3$ level has a ninefold degeneracy.

The angular part of each hydrogen atom total energy eigenfunction is a spherical harmonic function. As discussed in Section 18.5, these functions are complex unless $m_l = 0$. To facilitate making graphs, it is useful to form combinations of those hydrogen orbitals $\psi_{nlm_l}(r, \theta, \phi)$ for which $m_l \neq 0$ are real functions of r, θ, and ϕ. As discussed in Section 18.5, this is done by forming linear combinations of $\psi_{nlm_l}(r, \theta, \phi)$ and $\psi_{nl-m_l}(r, \theta, \phi)$. The first few of these combinations, resulting in the $2p$, $3p$, and $3d$ orbitals, are shown here:

$$\psi_{2p_x}(r, \theta, \phi) = \frac{1}{4\sqrt{2\pi}}\left(\frac{1}{a_0}\right)^{3/2}\frac{r}{a_0}e^{-r/2a_0}\sin\theta\cos\phi$$

$$\psi_{2p_y}(r, \theta, \phi) = \frac{1}{4\sqrt{2\pi}}\left(\frac{1}{a_0}\right)^{3/2}\frac{r}{a_0}e^{-r/2a_0}\sin\theta\sin\phi$$

$$\psi_{2p_z}(r, \theta, \phi) = \frac{1}{4\sqrt{2\pi}}\left(\frac{1}{a_0}\right)^{3/2}\frac{r}{a_0}e^{-r/2a_0}\cos\theta$$

$$\psi_{3p_x}(r, \theta, \phi) = \frac{\sqrt{2}}{81\sqrt{\pi}}\left(\frac{1}{a_0}\right)^{3/2}\left(6\frac{r}{a_0} - \frac{r^2}{a_0^2}\right)e^{-r/3a_0}\sin\theta\cos\phi$$

$$\psi_{3p_y}(r, \theta, \phi) = \frac{\sqrt{2}}{81\sqrt{\pi}}\left(\frac{1}{a_0}\right)^{3/2}\left(6\frac{r}{a_0} - \frac{r^2}{a_0^2}\right)e^{-r/3a_0}\sin\theta\sin\phi$$

$$\psi_{3p_z}(r, \theta, \phi) = \frac{\sqrt{2}}{81\sqrt{\pi}}\left(\frac{1}{a_0}\right)^{3/2}\left(6\frac{r}{a_0} - \frac{r^2}{a_0^2}\right)e^{-r/3a_0}\cos\theta$$

$$\psi_{3d_{z^2}}(r, \theta, \phi) = \frac{1}{81\sqrt{6\pi}}\left(\frac{1}{a_0}\right)^{3/2}\frac{r^2}{a_0^2}e^{-r/3a_0}(3\cos^2\theta - 1)$$

$$\psi_{3d_{xz}}(r, \theta, \phi) = \frac{\sqrt{2}}{81\sqrt{\pi}}\left(\frac{1}{a_0}\right)^{3/2}\frac{r^2}{a_0^2}e^{-r/3a_0}\sin\theta\cos\theta\cos\phi$$

$$\psi_{3d_{yz}}(r, \theta, \phi) = \frac{\sqrt{2}}{81\sqrt{\pi}}\left(\frac{1}{a_0}\right)^{3/2}\frac{r^2}{a_0^2}e^{-r/3a_0}\sin\theta\cos\theta\sin\phi$$

$$\psi_{3d_{x^2-y^2}}(r, \theta, \phi) = \frac{1}{81\sqrt{2\pi}}\left(\frac{1}{a_0}\right)^{3/2}\frac{r^2}{a_0^2}e^{-r/3a_0}\sin^2\theta\cos2\phi$$

$$\psi_{3d_{xy}}(r, \theta, \phi) = \frac{1}{81\sqrt{2\pi}}\left(\frac{1}{a_0}\right)^{3/2}\frac{r^2}{a_0^2}e^{-r/3a_0}\sin^2\theta\sin2\phi$$

When is it appropriate to use these functions as opposed to the complex functions $\psi_{nlm_l}(r)$? The real functions are more useful in visualizing chemical bonds, so those will generally be used throughout this book. However, both representations are useful in different applications, and we note that although the real functions are eigenfunctions of $\hat{H}$ and $\hat{l}^2$, they are not eigenfunctions of $\hat{l}_z$.

The challenge we posed for quantum mechanics at the beginning of this chapter was to provide an understanding for the stability of atoms. By verifying that there is a set of eigenfunctions and eigenvalues of the time-independent Schrödinger equation for a system consisting of a proton and an electron, we have demonstrated that there are states whose energy is independent of time. Because the energy eigenvalues are all negative numbers, all of these states are more stable than the reference state of zero energy that corresponds to the proton and electron separated by an infinite distance. Because $n \geq 1$, the energy cannot approach $-\infty$, corresponding to the electron falling into the nucleus. These results show that when the wave nature of the electron is taken into account, the H atom is stable.

As with any new theory, the true test is consistency with experimental data. Although the wave functions are not directly observable, we know that the spectral lines from a hydrogen arc lamp (measured as early as 1885) must involve transitions between two stable states of the hydrogen atom. Therefore, the frequencies measured by the early experimentalists in emission spectra must be given by

$$\nu = \left| \frac{1}{h} (E_{initial} - E_{final}) \right| \qquad (20.10)$$

In a more exact treatment, the origin of the coordinate system describing the H atom is placed at the center of mass of the proton and electron rather than at the position of the proton. Using Equation (20.7) with the reduced mass of the atom in place of m_e and Equation (20.10), quantum theory predicts that the frequencies of all the spectral lines are given by

$$\nu = \left| \frac{\mu e^4}{8\varepsilon_0^2 h^3} \left(\frac{1}{n_{initial}^2} - \frac{1}{n_{final}^2} \right) \right| \qquad (20.11)$$

where $\mu = m_e m_p/(m_e + m_p)$ is the reduced mass of the atom, which is 0.05% less than m_e. Spectroscopists commonly refer to spectral lines in units of wave numbers. Rather than reporting values of ν, they use the units $\tilde{\nu} = \nu/c = 1/\lambda$. The combination of constants $m_e e^4/8\varepsilon_0^2 h^3 c$ is called the **Rydberg constant.** It has the value 109,677.581 cm^{-1} in quantitative agreement with the experimental value.

Equation (20.11) quantitatively predicts all observed spectral lines for the hydrogen atom. It also correctly predicts the very small shifts in frequency observed for the isotopes of hydrogen, which have slightly different reduced masses. The agreement between theory and experiment verifies that the quantum mechanical model for the hydrogen atom is valid and accurate. We discuss the selection rules for transitions between electronic states in atoms in Chapter 22. Some of these transitions are shown superimposed on a set of energy levels in Figure 20.3.

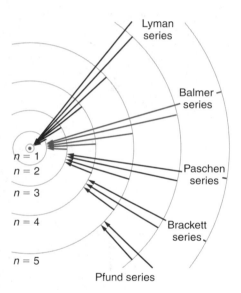

FIGURE 20.3
Energy-level diagram for the hydrogen atom showing the allowed transitions for $n < 6$. Because for $E > 0$ the energy levels are continuous, the absorption spectrum will be continuous above an energy that depends on the initial n value. The different sets of transitions are named after the scientists who first investigated them.

EXAMPLE PROBLEM 20.2

Consider an excited state of the H atom with the electron in the 2s orbital.

a. Is the wave function that describes this state,

$$\psi_{200}(r) = \frac{1}{\sqrt{32\pi}} \left(\frac{1}{a_0} \right)^{3/2} \left(2 - \frac{r}{a_0} \right) e^{-r/2a_0}$$

an eigenfunction of the kinetic energy? Of the potential energy?

b. Calculate the average values of the kinetic and potential energies for an atom described by this wave function.

Solution

a. We know that this function is an eigenfunction of the total energy operator because it is a solution of the Schrödinger equation. You can convince yourself that the total energy operator does not commute with either the kinetic energy operator or the potential energy operator by extending the discussion of Example Problem 20.1. Therefore, this wave function cannot be an eigenfunction of either of these operators.

b. The average value of the kinetic energy is given by

$$\langle E_{kinetic} \rangle = \int \psi^*(\tau)\, \hat{E}_{kinetic}\, \psi(\tau)\, d\tau$$

$$= -\frac{\hbar^2}{2m_e}\frac{1}{32\pi a_0^3}\int_0^{2\pi} d\phi \int_0^{\pi} \sin\theta\, d\theta \int_0^{\infty}\left(2 - \frac{r}{a_0}\right) \times$$

$$e^{-r/2a_0}\left(\frac{1}{r^2}\frac{d}{dr}\left[r^2 \frac{d}{dr}\left\{\left(2 - \frac{r}{a_0}\right)e^{-r/2a_0}\right\}\right]\right)r^2\, dr$$

$$= -\frac{\hbar^2}{2m_e}\frac{1}{8a_0^3}\int_0^{\infty}\left(2 - \frac{r}{a_0}\right)e^{-r/2a_0}\left(-\frac{e^{-r/2a_0}}{4a_0^3 r}\right)(16a_0^2 - 10a_0 r + r^2)r^2\, dr$$

$$= -\frac{\hbar^2}{2m_e}\frac{1}{8a_0^3}\left(\frac{9}{a_0^2}\int_0^{\infty}r^2 e^{-r/a_0}\, dr - \frac{8}{a_0}\int_0^{\infty}r e^{-r/a_0}\, dr - \frac{3}{a_0^3}\int_0^{\infty}r^3 e^{-r/a_0}\, dr\right.$$

$$\left. + \frac{1}{4a_0^4}\int_0^{\infty}r^4 e^{-r/a_0}\, dr\right)$$

We use the standard integral, $\int_0^{\infty} x^n e^{-ax}\, dx = n!/a^{n+1}$:

$$\langle E_{kinetic} \rangle = \hbar^2/8m_e a_0^2$$

Using the relationship $a_0 = \varepsilon_0 h^2/\pi m_e e^2$,

$$\langle E_{kinetic} \rangle = \frac{e^2}{32\pi\varepsilon_0 a_0} = -E_n, \quad \text{for } n = 2$$

The average potential energy is given by

$$\langle E_{potential} \rangle = \int \psi^*(\tau)\, \hat{E}_{potential}\, \psi(\tau)\, d\tau$$

$$= -\frac{e^2}{4\pi\varepsilon_0}\frac{1}{32\pi a_0^3}\int_0^{2\pi} d\phi \int_0^{\pi} \sin\theta\, d\theta \int_0^{\infty}\left[\left(2 - \frac{r}{a_0}\right)e^{-r/2a_0}\right] \times$$

$$\left(\frac{1}{r}\right)\left[\left(2 - \frac{r}{a_0}\right)e^{-r/2a_0}\right]r^2\, dr$$

$$= -\frac{e^2}{4\pi\varepsilon_0}\frac{1}{8a_0^3}\left(4\int_0^{\infty}r e^{-r/a_0}\, dr - \frac{4}{a_0}\int_0^{\infty}r^2 e^{-r/a_0}\, dr + \frac{1}{a_0^2}\int_0^{\infty}r^3 e^{-r/a_0}\, dr\right)$$

$$= -\frac{e^2}{4\pi\varepsilon_0}\frac{1}{8a_0^3}(2a_0^2)$$

$$= -\frac{e^2}{16\pi\varepsilon_0 a_0} = 2E_n \quad \text{for } n = 2$$

We see that $\langle E_{potential} \rangle = 2\langle E_{total} \rangle$ and $\langle E_{potential} \rangle = -2\langle E_{kinetic} \rangle$. The relationship of the kinetic and potential energies is a specific example of the **virial theorem** and holds for any system in which the potential is Coulombic.

20.4 The Hydrogen Atom Orbitals

We now turn to the total energy eigenfunctions (or orbitals) of the hydrogen atom. What insight can be gained from them? Recall the early quantum mechanics shell model of atoms proposed by Niels Bohr. It depicted electrons as orbiting around the nucleus and associated orbits of small radius with more negative energies. Only certain orbits were allowed in order to give rise to a discrete energy spectrum. This model was discarded because defining orbits exactly is inconsistent with the Heisenberg uncertainty principle. The model postulated by Schrödinger and other pioneers of quantum theory replaced knowledge of the location of the electron in the hydrogen atom with knowledge of the probability of finding it in a small volume element at a specific location. As we have seen in considering the particle in the box and the harmonic oscillator, this probability is proportional to $\psi^*(r, \theta, \phi)\psi(r, \theta, \phi)\, d\tau$.

To what extent does the exact quantum mechanical solution resemble the shell model? To answer this question, information must be extracted from the H atom orbitals. A new concept, the radial distribution function, is introduced for this purpose. We begin our discussion by focusing on the wave functions $\psi_{nlm_l}(r, \theta, \phi)$. Next we discuss what can be learned about the probability of finding the electron in a particular region in space, $\psi^2_{nlm_l}(r, \theta, \phi)\, r^2 \sin\theta\, dr\, d\theta\, d\phi$. Finally, we define the radial distribution function and look at the similarities and differences between quantum mechanical and shell models of the hydrogen atom.

The initial step is to look at the ground-state (lowest energy state) wave function for the hydrogen atom, and to find a good way to visualize this function. Because

$$\psi_{100}(r) = \frac{1}{\sqrt{\pi}}\left(\frac{1}{a_0}\right)^{3/2} e^{-r/a_0}$$

is a function of the three spatial coordinates x, y, and z, we need a four-dimensional space to plot ψ_{100} as a function of all its variables. Because such a space is not readily available, the number of variables will be reduced. The dimensionality of the representation can be reduced by evaluating $r = \sqrt{x^2 + y^2 + z^2}$ in one of the xy, xz, or yz planes by setting the third coordinate equal to zero. Three common ways of depicting $\psi_{100}(r)$ are shown in Figures 20.4 through 20.6. In Figure 20.4a, a three-dimensional plot of $\psi_{100}(r)$ evaluated in the xy half-plane ($z = 0$, $y \geq 0$) is shown in perspective. Although it is difficult to extract quantitative information from such a plot directly, it allows a good visualization of the function. We clearly see that the wave function has its maximum value at $r = 0$ (the nuclear position) and that it falls off rapidly with increasing distance from the nucleus.

More quantitative information is available in a contour plot shown in Figure 20.4b in which $\psi_{100}(r)$ is evaluated in the xy plane from a vantage point on the z axis. In this case, the outermost contour represents 10% of the maximum value, and successive contours are spaced at equal intervals. The shading indicates the value of the function, with darker colors representing larger values of the amplitude. This way of depicting $\psi_{100}(r)$ is more quantitative than that of Figure 20.4a in that we can recognize that the contours of constant amplitude are circles and that the contour spacing becomes smaller as r approaches zero. A third useful representation is to show the value of the function with two variables set equal to zero. This represents a cut through $\psi_{100}(r)$ in a plane perpendicular to the

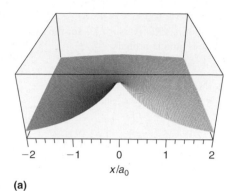

(a)

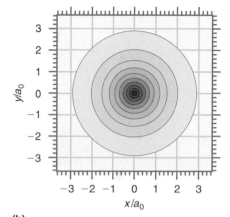

(b)

FIGURE 20.4
(a) 3D perspective and (b) contour plot of $\psi_{100}(r)$. Darker contour colors indicate larger values for the magnitude of the amplitude.

FIGURE 20.5
Three-dimensional perspective plots of the $1s$, $2s$, and $3s$ orbitals. The dashed lines indicate the zero of amplitude for the wave functions. The "×2" refers to the fact that the amplitude of the wave function has been multiplied by 2 to make the subsidiary maxima apparent. The horizontal axis shows radial distance in units of a_0.

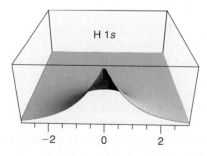

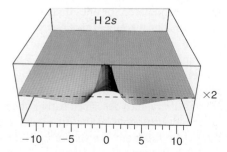

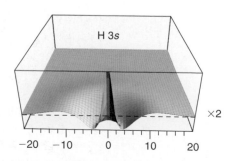

FIGURE 20.6
Plot of $a_0^{3/2} R(r)$ versus r/a_0 for the first few H atomic orbitals.

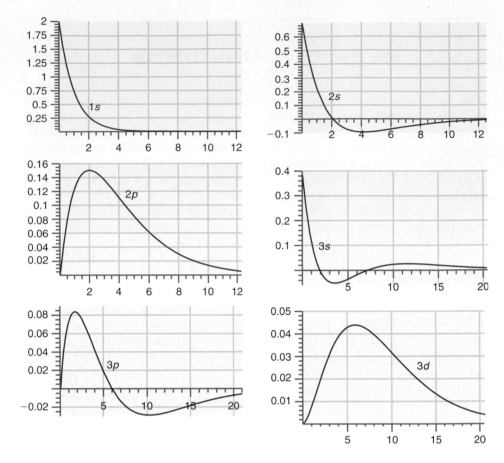

$x-y$ plane. Because ψ_{100} is independent of the angular coordinates, the same result is obtained for all planes containing the z axis. As we will see later, this is only true for orbitals for which $l = 0$. This way of depicting $\psi_{100}(r)$ is shown as the front edge of the three-dimensional plot in Figure 20.4a and in Figures 20.5 and 20.6. All of these graphical representations contain exactly the same information.

Because $\psi_{100}(r)$ is a function of the single variable r the function can be graphed directly. However, it is important to keep in mind that r is a three-dimensional function of x, y, and z. For $l = 0$, the wave function does not depend on θ and ϕ. For $l > 0$, the wave function does depend on θ and ϕ, and a plot of the amplitude of a wave function versus r assumes that θ and ϕ are being held constant at values that need to be specified. These values generally correspond to a maximum in the angular part of the wave function.

In Figure 20.6, the radial wave function amplitude $R(r)$ is graphed versus r. What should we expect having solved the particle in the box and harmonic oscillator problems? Because the eigenfunctions of the Schrödinger equation are standing waves, the solutions should be oscillating functions that have nodes. There should be no nodes in the ground state, and the number of nodes should increase as the quantum number increases.

First consider the eigenfunctions with $l = 0$, namely, the $1s$, $2s$, and $3s$ orbitals. From Figure 20.5, we clearly see that ψ_{100} has no nodes as expected. The $2s$ and $3s$ orbitals have one and two nodes, respectively. Because these nodes correspond to constant values of r, they are spherical **nodal surfaces,** rather than the nodal points previously encountered for one-dimensional potentials.

Now consider the eigenfunctions with $l > 0$. Why do the $2p$ and $3d$ functions in Figure 20.6 appear not to have nodal surfaces? This is related to the fact that the function is graphed for particular values of θ and ϕ. To see the nodes, the angular part of these eigenfunctions must be displayed.

Whereas the spherically symmetric s orbitals are equally well represented by the three forms of graphics described earlier, the p and d orbitals can best be visualized with a contour plot analogous to that of Figure 20.4b. Contour plots for the $2p_y$, $3p_y$, $3d_{xy}$, and $3d_{z^2}$ wave functions are shown in Figure 20.7. This nomenclature was defined in Section 20.3. We can now see that the $2p_y$ wave function has a nodal plane defined by

$y = 0$; however, it appears in the angular rather than the radial part of the wave function. It can be shown that the radial part of the energy eigenfunctions has $n - l - 1$ nodal surfaces. There are l nodal surfaces in the angular part of the energy eigenfunctions, making a total of $n - 1$ nodes, just as was obtained for the particle in the box and the harmonic oscillator. As can be seen in Figure 20.7, the $3p_y$ wave function has a second nodal surface in addition to the nodal plane at $y = 0$. This second node comes from the radial part of the energy eigenfunction and is a spherical surface. The d orbitals have a more complex nodal structure that can include spheres, planes, and cones. The $3d_{xy}$ orbital has two nodal planes that intersect in the z axis. The $3d_{z^2}$ orbital has two conical nodal surfaces, whose axis of rotation is the z axis.

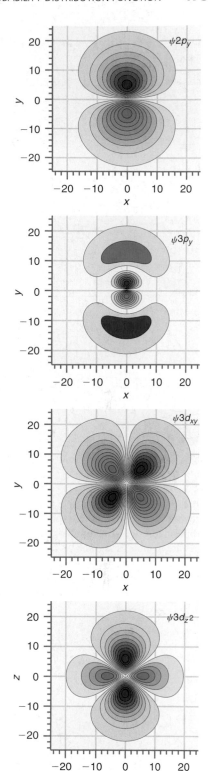

EXAMPLE PROBLEM 20.3

Locate the nodal surfaces in

$$\psi_{310}(r, \theta, \phi) = \frac{1}{81}\left(\frac{2}{\pi}\right)^{1/2}\left(\frac{1}{a_0}\right)^{3/2}\left(6\frac{r}{a_0} - \frac{r^2}{a_0^2}\right)e^{-r/3a_0}\cos\theta$$

Solution

We consider the angular and radial nodal surfaces separately. The angular part, $\cos\theta$, is zero for $\theta = \pi/2$. In three-dimensional space, this corresponds to the plane $z = 0$. The radial part of the equations is zero for finite values of r/a_0 for $(6r/a_0 - r^2/a_0^2) = 0$. This occurs at $r = 0$, which is not a node and at $r = 6a_0$. The first value is a point in three-dimensional space and the second is a spherical surface. This wave function has one angular and one radial node. In general, an orbital characterized by n and l has l angular nodes and $n - 1 - 1$ radial nodes.

20.5 The Radial Probability Distribution Function

To continue the discussion of the similarities and differences between the quantum mechanical and shell models of the hydrogen atom, let us see what information can be obtained from $\psi^2_{nlm_l}(r, \theta, \phi)$, which is the probability density of finding the electron at a particular point in space. We again consider the s orbitals and the p and d orbitals separately. We first show $\psi^2_{n00}(r, \theta, \phi)$ as a three-dimensional graphic in Figure 20.8 for $n = 1$, 2, and 3. Subsidiary maxima are seen in addition to the main maximum at $r = 0$. Figure 20.9 shows a graph of $R^2_{nlm_l}(r)$ as a function of r.

Now consider $\psi^2_{nlm_l}(r, \theta, \phi)$ for $l > 0$. As expected from the effect of the centrifugal potential (see Figure 20.1), the electron is pushed away from the nucleus, so that $\psi^2_{nlm_l}(r, \theta, \phi)$ goes to zero as r approaches zero. Because a nonzero angular momentum is associated with these states, $\psi^2_{nlm_l}(r, \theta, \phi)$ is not spherically symmetric. All of this makes sense in terms of our picture of p and d orbitals.

EXAMPLE PROBLEM 20.4

a. At what point does the probability density for the electron in a 2s orbital have its maximum value?

b. Assume that the nuclear diameter for H is 2×10^{-15} m. Using this assumption, calculate the total probability of finding the electron in the nucleus if it occupies the 2s orbital.

Solution

a. The point at which $\psi^*(\tau)\psi(\tau)\,d\tau$ and, therefore, $\psi(\tau)$ has its greatest value is found from the wave function:

$$\psi_{200}(r) = \frac{1}{\sqrt{32\pi}}\left(\frac{1}{a_0}\right)^{3/2}\left(2 - \frac{r}{a_0}\right)e^{-r/2a_0}$$

which has its maximum value at $r = 0$, or at the nucleus as seen in Figure 20.6.

FIGURE 20.7

Contour plots for the orbitals indicated. Positive and negative amplitudes are shown as red and blue, respectively. Darker colors indicate larger values for the magnitude of the amplitude. Distances are in units of a_0.

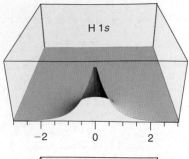

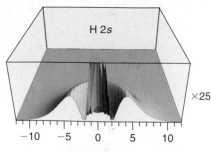

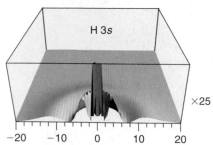

FIGURE 20.8
Three-dimensional perspective plots of the square of the wave functions for the orbitals indicated. The numbers on the axes are in units of a_0. The "×25" refers to the fact that the amplitude of the wave function has been multiplied by 25 to make the subsidiary maxima apparent.

b. The result obtained in part (a) seems unphysical, but is a consequence of wave-particle duality in describing electrons. It is really only a problem if the total probability of finding the electron within the nucleus is significant. This probability is given by

$$P = \frac{1}{32\pi}\left(\frac{1}{a_0}\right)^3 \int_0^{2\pi} d\phi \int_0^{\pi} \sin\theta \, d\theta \int_0^{r_{nucleus}} r^2\left(2 - \frac{r}{a_0}\right)^2 e^{-r/a_0} \, dr$$

Because $r_{nucleus} \ll a_0$, we can evaluate the integrand by assuming that $(2 - r/a_0)^2 e^{-r/a_0} \sim 2$ over the interval $0 \le r \le r_{nucleus}$:

$$P = \frac{1}{32\pi}\left(\frac{1}{a_0}\right)^3 4\pi\left[\left(2 - \frac{r_{nucleus}}{a_0}\right)^2 e^{-r_{nucleus}/a_0}\right]\int_0^{r\,nucleus} r^2 \, dr$$

$$= \frac{1}{32\pi}\left(\frac{1}{a_0}\right)^3\left[\left(2 - \frac{r_{nucleus}}{a_0}\right)^2 e^{-r_{nucleus}/a_0}\right]\frac{4\pi}{3}r_{nucleus}^3$$

Because $2 - (r_{nucleus}/a_0) \approx 2$ and $e^{-r_{nucleus}/a_0} \approx 1$,

$$P = \frac{1}{6}\left(\frac{r_{nucleus}}{a_0}\right)^3 = 9.0 \times 10^{-15}$$

Because this probability is vanishingly small, even though the wave function has its maximum amplitude at the nucleus the probability of finding the electron in the nucleus is essentially zero.

At this point, we ask a different question involving probability. What is the most probable distance from the nucleus at which the electron will be found? For the $1s$, $2s$, and $3s$ orbitals, the maximum probability density is at the nucleus. This result seems to

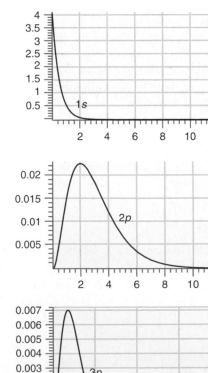

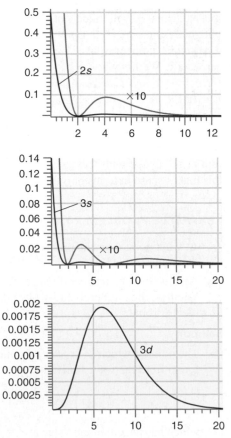

FIGURE 20.9
Plot of $a_0^3 R^2(r)$ versus r/a_0 for the first few H atomic orbitals. The numbers on the horizontal axis are in units of a_0.

predict that the most likely orbit for the electron has a radius of zero. Clearly, we are missing something, because this result is inconsistent with a shell model. It turns out that we are not asking the right question. The probability as calculated from the probability density is correct, but it gives the likelihood of finding the particle in the vicinity of a *particular point* for a given value of r, θ, and ϕ. Why is this not the information we are looking for? Imagine that a planet has a circular orbit and we want to determine the radius of the orbit. To do so, we must find the planet. If we looked at only one point on a spherical shell of a given radius for different values of the radius, we would be unlikely to find the planet. To find the planet, we need to look everywhere on a shell of a given radius simultaneously.

How do we apply this reasoning to finding the electron on the hydrogen atom? The question we need to ask is "What is the probability of finding the electron at a particular value of r, regardless of the values of θ and ϕ?" This probability is obtained by integrating the probability density $\psi^2_{nlm_l}(r, \theta, \phi)\, r^2 \sin \theta\, dr\, d\theta\, d\phi$ over all values of θ and ϕ. This gives the probability of finding the electron in a spherical shell of radius r and thickness dr rather than the probability of finding the electron near a given point on the spherical shell of thickness dr with the particular coordinates r_0, θ_0, ϕ_0. For example, for the $1s$ orbital the probability of finding the electron in a spherical shell of radius r and thickness dr is

$$P_{1s}(r)\, dr = \frac{1}{\pi a_0^3} \int_0^{2\pi} d\phi \int_0^\pi \sin \theta\, d\theta\, r^2 e^{-2r/a_0}\, dr$$

$$= \frac{4}{a_0^3} r^2 e^{-2r/a_0}\, dr \qquad (20.12)$$

EXAMPLE PROBLEM 20.5

Consider an excited hydrogen atom with the electron in the $2s$ orbital.

a. Calculate the probability of finding the electron in the volume about a point defined by

$$5.22 \times 10^{-10}\,\text{m} \le r \le 5.26 \times 10^{-10}\,\text{m}, \frac{\pi}{2} - 0.01 \le \phi \le \frac{\pi}{2} + 0.01,$$

$$\frac{\pi}{2} - 0.01 \le \theta \le \frac{\pi}{2} + 0.01$$

b. Calculate the probability of finding the electron in the spherical shell defined by

$$5.22 \times 10^{-10}\,\text{m} \le r \le 5.26 \times 10^{-10}\,\text{m}$$

Solution

a. We numerically solve the integral $\iiint \psi^*(r, \theta, \phi)\psi(r, \theta, \phi)\, r^2 \sin \theta\, dr\, d\theta\, d\phi$. The result is

$$P = \frac{1}{32\pi}\left(\frac{1}{a_0}\right)^3 \int_{\frac{\pi}{2}-0.01}^{\frac{\pi}{2}+0.01} d\phi \int_{\frac{\pi}{2}-0.01}^{\frac{\pi}{2}+0.01} \sin \theta\, d\theta \int_{5.22\times10^{-10}}^{5.26\times10^{-10}} r^2\left(2 - \frac{r}{a_0}\right)^2 e^{-r/a_0}\, dr$$

$$= 0.00995 \times 6.76 \times 10^{30}\,\text{m}^{-3} \times 0.020 \times 0.0200 \times 3.43 \times 10^{-33}\,\text{m}^3 = 9.23 \times 10^{-8}$$

b. In this case, we integrate over all values of the angles:

$$P = \frac{1}{32\pi}\left(\frac{1}{a_0}\right)^3 \int_0^{2\pi} d\phi \int_0^\pi \sin \theta\, d\theta \int_{5.22\times10^{-10}}^{5.26\times10^{-10}} r^2\left(2 - \frac{r}{a_0}\right)^2 e^{-r/a_0}\, dr$$

$$= 0.00995 \times 6.76 \times 10^{30}\,\text{m}^{-3} \times 4\pi \times 3.43 \times 10^{-33}\,\text{m}^3 = 2.90 \times 10^{-3}$$

This probability is greater than that calculated in part (a) by a factor of 3.1×10^4 because we have integrated the probability density over the whole spherical shell of thickness 4×10^{-12} m.

Because the integration of the probability density over the angles θ and ϕ amounts to an averaging of $\psi(r, \theta, \phi)^2$ over all angles, it is most meaningful for the s orbitals whose amplitudes are independent of the angular coordinates. However, to arrive at a uniform definition for all orbitals, a new function, the **radial distribution function, $P_{nl}(r)$,** is defined.

$$P_{nl}(r)\,dr = \left\{ \int_0^{2\pi} d\phi \int_0^{\pi} [Y_l^{m_l}(\theta, \phi)]^* \, [Y_l^{m_l}(\theta, \phi)] \sin\theta \, d\theta \right\} r^2 R_{nl}^2(r)\,dr = r^2 R_{nl}^2(r)\,dr$$

(20.13)

The radial distribution function is the probability function of choice to determine the most likely radius to find the electron for a given orbital. Understanding the difference between the radial distribution function $P_{nl}(r)\,dr$ and the probability density $\psi_{nlm_l}^*(r)\psi_{nlm_l}(r)\, r^2 \sin\theta \, dr \, d\theta \, d\phi$ is very important in working with the hydrogen atom wave functions.

EXAMPLE PROBLEM 20.6

Calculate the maxima in the radial probability distribution for the $2s$ orbital. What is the most probable distance from the nucleus for an electron in this orbital? Are there subsidiary maxima?

Solution

The radial distribution function is

$$P_{20}(r) = r^2 R_{20}^2(r) = \frac{1}{8}\left(\frac{1}{a_0}\right)^3 r^2 \left(2 - \frac{r}{a_0}\right)^2 e^{-r/a_0}$$

To find the maxima, we plot $P(r)$ and

$$\frac{dP_{20}(r)}{dr} = \frac{r}{8a_0^6}(8a_0^3 - 16a_0^2 r + 8a_0 r^2 - r^3)e^{-r/a_0}$$

versus r/a_0 and look for the nodes in this function. These functions are plotted as a function of r/a_0 in the following figure:

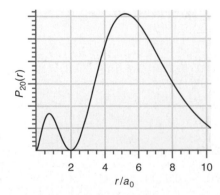
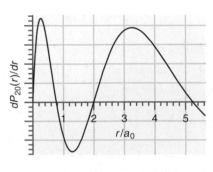

We see that the principal maximum in $P_{20}(r)$ is at $5.24\, a_0$. This corresponds to the most probable distance of a $2s$ electron from the nucleus. The subsidiary maximum is at $0.76\, a_0$. The minimum is at $2\, a_0$.

The resulting radial distribution function only depends on r, and not on θ and ϕ. Therefore, we can display $P_{nl}(r)\,dr$ versus r in a graph as shown in Figure 20.10.

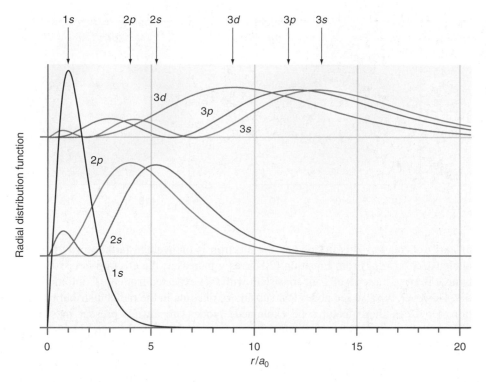

FIGURE 20.10
Plot of $r^2 R_{nl}^2(r)$ versus r/a_0 for the first few H atomic orbitals. The curves for $n = 2$ and $n = 3$ have been displaced vertically as indicated. The position of the principal maxima for each orbital is indicated by an arrow.

20.6 The Validity of the Shell Model of an Atom

What can we conclude from Figure 20.10 regarding a shell model for the hydrogen atom? By now, we have become accustomed to the idea of wave-particle duality. Waves are not sharply localized, so a shell model like that shown in Figure 20.11 with electrons as point masses orbiting around the nucleus is not viable in quantum mechanics. If there are some remnants of a shell model in the hydrogen atom, there is a greater likelihood of finding the electron at some distance from the nucleus than others.

The quantum mechanical analogue of the shell model can be generated in the following way. Imagine that three-dimensional images of the shell model for hydrogen with the electron in the 1s, 2s, or 3s levels were taken at a large number of random times. A cut through the resulting images at the $z = 0$ plane would reveal sharply defined circles with a different radius for each orbital. The quantum mechanical analogue of this process is depicted in Figure 20.12. The principal maxima seen in Figure 20.10 are the source of the darkest rings in each part of Figure 20.12. The rings are broad in comparison to the sharp circle of the classical model. The subsidiary maxima seen in Figure 20.10 appear as less intense rings for the 2s and 3s orbitals.

The radial distribution function gives results that are more in keeping with our intuition and with a shell model than what we saw in the plots for the probability density $\psi^*(r, \theta, \phi)\psi(r, \theta, \phi)$. For the 1s orbital, the radial distribution function is peaked at a value of a_0. However, the peak has a considerable width, whereas a shell model would give a sharp peak of nearly zero width. This contradiction is reminiscent of our discussion of the double-slit diffraction experiment. Because wave-particle duality is well established, it is not useful to formulate models that are purely particle-like or purely wave-like. The broadening of the orbital shell over what we would expect in a particle picture is a direct manifestation of the wave nature of the electron, and the existence of an orbit is what we would expect in a particle picture. Both aspects of wave-particle duality are evident.

It is useful to summarize the main features that appear in Figure 20.10 and 20.12 for the radial probability distribution. We see broad maxima that move to greater values

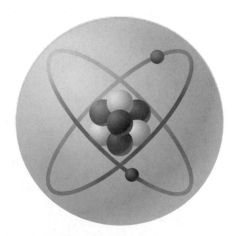

FIGURE 20.11
A shell model of an atom with electrons moving on spherical shells around the nucleus.

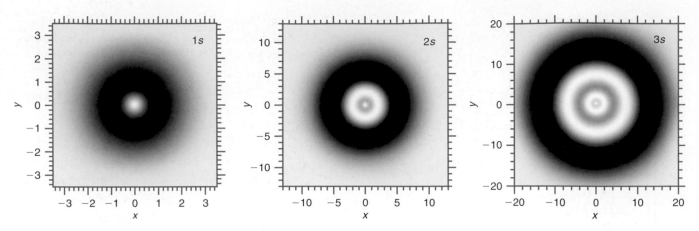

FIGURE 20.12
The radial probability distribution evaluated for $z = 0$ is plotted in the $x-y$ plane with lengths in units of a_0. Darker regions correspond to greater values of the function. The sharp circle in a classical shell model becomes a broad ring in a quantum mechanical model over which the probability of finding the electron varies. Less intense subsidiary rings are also observed for the $2s$ and $3s$ orbitals.

of r as n increases. This means that the electron is on average farther away from the nucleus for large n. From Equation (20.8), as n increases, the electron is less strongly bound. Both of these results are consistent with that expected from the Coulomb potential. However, we also see nodes and subsidiary maxima in the radial distribution function. How can these features be explained? Nodes are always present in standing waves, and eigenfunctions of the time-independent Schrödinger equation are standing waves. The nodes are directly analogous to the nodes observed for the particle in the box wave functions and are a manifestation of wave-particle duality. The subsidiary maxima are another manifestation of the wave character of the electron and occur whenever wave interference occurs. Recall that such subsidiary maxima are also observed in diffraction experiments. It is tempting to assign orbital radii to the H atomic orbitals with values corresponding to the positions of the principal maxima. The maxima are indicated by arrows in Figure 20.10. However, this amounts to reducing a function to a single number and is unwise.

Vocabulary

Bohr radius	degeneracy	orbital	separation of variables
centrifugal potential	effective potential	radial distribution function	shell model
Coulomb potential	nodal surface	Rydberg constant	virial theorem

Conceptual Problems

Q20.1   What possible geometrical forms can the nodes in the angular function for p and d orbitals in the H atom have? What possible geometrical forms can the nodes in the radial function for s, p, and d orbitals in the H atom have?

Q20.2   What transition gives rise to the highest frequency spectral line in the Lyman series?

Q20.3   Is it always true that the probability of finding the electron in the H atom is greater in the interval $r - dr < r < r + dr$ than in the interval $r - dr < r < r + dr, \theta - d\theta < \theta < \theta + d\theta$, $\phi - d\phi < \phi < \phi + d\phi$?

Q20.4   Why are the total energy eigenfunctions for the H atom not eigenfunctions of the kinetic energy?

Q20.5   How do the results shown in Figure 20.10 differ from the predictions of the Bohr model of the H atom?

Q20.6   What effect does the centrifugal potential have in determining the maximum in the radial function for the $3s$, $3p$, and $3d$ orbitals?

Q20.7   How does the effective potential differ for p and d electrons?

Q20.8   Why does the centrifugal potential dominate the effective potential for small values of r?

Q20.9   If the probability density of finding the electron in the $1s$ orbital in the H atom has its maximum value for $r = 0$, does this mean that the proton and electron are located at the same point in space?

Q20.10   Explain the different degree to which the $1s$, $2s$, and $3s$ total energy eigenfunctions penetrate into the classically forbidden region.

Q20.11 What are the units of the H atom total energy eigenfunctions? Why is $a_0^{3/2}R(r)$ graphed in Figure 20.6 rather than $R(r)$?

Q20.12 Why is the radial probability function rather than $\psi^*(r)\psi(r)\,r^2\sin\theta\,dr\,d\theta\,d\phi$ the best measure of the probability of finding the electron at a distance r from the nucleus?

Q20.13 Use an analogy with the particle in the box to explain why the energy levels for the H atom are more closely spaced as n increases.

Q20.14 Explain why the radial distribution function rather than the square of the magnitude of the wave function should be used to make a comparison with the shell model of the atom.

Q20.15 What is the difference between an angular and a radial node? How can you distinguish the two types of nodes in a contour diagram such as Figure 20.7?

Q20.16 What is the minimum photon energy needed to ionize a hydrogen atom in the ground state?

Q20.17 To what physical state does a hydrogen atom energy of $+1.0 \times 10^{-19}$ J correspond?

Q20.18 Why does the centrifugal potential force the $3d$ electrons further from the nucleus than the $3s$ electrons?

Q20.19 Why is the radial probability density rather than the probability density used to calculate the most probable distance of the electron from the nucleus?

Numerical Problems

Problem numbers in red indicate that the solution to the problem is given in the *Student's Solutions Manual*.

P20.1 Calculate the wave number corresponding to the most and least energetic spectral lines in the Lyman, Balmer, and Paschen series for the hydrogen atom.

P20.2 Show that the function $(r/a_0)e^{-r/2a_0}$ is a solution of the following differential equation for $l = 1$

$$-\frac{\hbar^2}{2m_e r^2}\frac{d}{dr}\left[r^2\frac{dR(r)}{dr}\right]$$

$$+\left[\frac{\hbar^2 l(l+1)}{2m_e r^2} - \frac{e^2}{4\pi\varepsilon_0 r}\right]R(r) = ER(r)$$

What is the eigenvalue? Using this result, what is the value for the principal quantum number n for this function?

P20.3 Determine the probability of finding the electron in the region for which the ψ_{320} wavefunction is negative (the toroidal region).

P20.4 Calculate the expectation value for the potential energy of the H atom with the electron in the $1s$ orbital. Compare your result with the total energy.

P20.5 Calculate the probability that the $1s$ electron for H will be found between $r = a_0$ and $r = 2a_0$.

P20.6 Calculate the distance from the nucleus for which the radial distribution function for the $2p$ orbital has its main and subsidiary maxima.

P20.7 Calculate the expectation value of the radius $\langle r \rangle$ at which you would find the electron if the H atom wave function is $\psi_{100}(r)$.

P20.8 Calculate the expectation value for the kinetic energy of the H atom with the electron in the $2s$ orbital. Compare your result with the total energy.

P20.9 Ions with a single electron such as He$^+$, Li^{2+}, and Be^{3+} are described by the H atom wave functions with Z/a_0 substituted for $1/a_0$, where Z is the nuclear charge. The $1s$ wave function

becomes $\psi(r) = 1/\sqrt{\pi}(Z/a_0)^{3/2}e^{-Zr/a_0}$. Using this result, calculate the total energy for the $1s$ state in H, He$^+$, Li^{2+}, and Be^{3+} by substitution in the Schrödinger equation.

P20.10 Ions with a single electron such as He$^+$, Li^{2+}, and Be^{3+} are described by the H atom wave functions with Z/a_0 substituted for $1/a_0$, where Z is the nuclear charge. The $1s$ wave function becomes $\psi(r) = 1/\sqrt{\pi}(Z/a_0)^{3/2}e^{-Zr/a_0}$. Using this result, compare the mean value of the radius $\langle r \rangle$ at which you would find the $1s$ electron in H, He$^+$, Li^{2+}, and Be^{3+}.

P20.11 As the principal quantum number n increases, the electron is more likely to be found far from the nucleus. It can be shown that for H and for ions with only one electron such as He$^+$, $\langle r \rangle_{nl} = \dfrac{n^2 a_0}{Z}\left[1 + \dfrac{1}{2}\left(1 - \dfrac{l(l+1)}{n^2}\right)\right]$

Calculate the value of n for an s state in the hydrogen atom such that $\langle r \rangle = 500.\ a_0$. Round up to the nearest integer. What is the ionization energy of the H atom in this state in electron-volts? Compare your answer with the ionization energy of the H atom in the ground state.

P20.12 In this problem, you will calculate the probability of finding an electron within a sphere of radius r for the H atom in its ground state.

a. Show using integration by parts, $\int u\,dv = uv - \int v\,du$, that
$$\int r^2 e^{-r/\alpha}\,dr = e^{-r/\alpha}(-2\alpha^3 - 2\alpha^2 r - \alpha r^2).$$

b. Using this result, show that the probability of finding the electron within a sphere of radius r for the hydrogen atom in its ground state is
$$1 - e^{-2r/a_0} - \frac{2r}{a_0}\left(1 + \frac{r}{a_0}\right)e^{-2r/a_0}$$

c. Evaluate this probability for $r = 0.25\ a_0$, $r = 2.25\ a_0$, and $r = 5.5\ a_0$.

P20.13 The radius of an atom r_{atom} can be defined as that value for which 90% of the electron charge is contained within a sphere of radius r_{atom}. Use the formula in P20.12b to calculate the radius of the H atom.

P20.14 Use the result of P20.13.

a. Calculate the mass density of the H atom.

b. Compare your answer with the nuclear density assuming a nuclear radius of 1.0×10^{-15} m.

c. Calculate the mass density of the H atom outside of the nucleus.

P20.15 Calculate the expectation value $\langle r - \langle r \rangle \rangle^2$ if the H atom wave function is $\psi_{100}(r)$.

P20.16 In spherical coordinates, $z = r\cos\theta$. Calculate $\langle z \rangle$ and $\langle z^2 \rangle$ for the H atom in its ground state. Without doing the calculation, what would you expect for $\langle x \rangle$ and $\langle y \rangle$, and $\langle x^2 \rangle$ and $\langle y^2 \rangle$? Why?

P20.17 The force acting between the electron and the proton in the H atom is given by $F = -e^2/4\pi\varepsilon_0 r^2$. Calculate the expectation value $\langle F \rangle$ for the 1s and $2p_z$ states of the H atom in terms of e, ε_0, and a_0.

P20.18 The d orbitals have the nomenclature d_{z^2}, d_{xy}, d_{xz}, d_{yz}, and $d_{x^2-y^2}$. Show how the d orbital

$$\psi_{3d_{yz}}(r, \theta, \phi) = \frac{\sqrt{2}}{81\sqrt{\pi}} \left(\frac{1}{a_0}\right)^{3/2} \frac{r^2}{a_0^2} e^{-r/3a_0} \sin\theta \cos\theta \sin\phi$$

can be written in the form $yzF(r)$.

P20.19 Calculate the expectation value of the moment of inertia of the H atom in the 2s and $2p_z$ states in terms of μ and a_0.

P20.20 The energy levels for ions with a single electron such as He^+, Li^{2+}, and Be^{3+} are given by $E_n = -Z^2 e^2/(8\pi\varepsilon_0 a_0 n^2)$, $n = 1, 2, 3, 4, \ldots$. Calculate the ionization energies of H, He^+, Li^{2+}, and Be^{3+} in their ground states in units of electron-volts (eV).

P20.21 Calculate the mean value of the radius $\langle r \rangle$ at which you would find the electron if the H atom wave function is $\psi_{210}(r, \theta, \phi)$.

P20.22 The total energy eigenvalues for the hydrogen atom are given by $E_n = -e^2/(8\pi\varepsilon_0 a_0 n^2)$, $n = 1, 2, 3, 4, \ldots$, and the three quantum numbers associated with the total energy eigenfunctions are related by $n = 1, 2, 3, 4, \ldots$; $l = 0, 1, 2, 3, \ldots, n-1$; and $m_l = 0, \pm 1, \pm 2, \pm 3, \ldots \pm l$.

Using the nomenclature ψ_{nlm_l}, list all eigenfunctions that have the following total energy eigenvalues:

a. $E = -\dfrac{e^2}{32\pi\varepsilon_0 a_0}$

b. $E = -\dfrac{e^2}{72\pi\varepsilon_0 a_0}$

c. $E = -\dfrac{e^2}{128\pi\varepsilon_0 a_0}$

d. What is the degeneracy of each of these energy levels?

P20.23 Locate the radial and angular nodes in the H orbitals $\psi_{3p_x}(r, \theta, \phi)$ and $\psi_{3p_z}(r, \theta, \phi)$.

P20.24 Calculate the average value of the kinetic and potential energies for the H atom in its ground state.

P20.25 Show by substitution that $\psi_{100}(r, \theta, \phi) = 1/\sqrt{\pi}(1/a_0)^{3/2}e^{-r/a_0}$ is a solution of

$$-\frac{\hbar^2}{2m_e}\left[\frac{1}{r^2}\frac{\partial}{\partial r}\left(r^2\frac{\partial\psi(r,\theta,\phi)}{\partial r}\right) + \frac{1}{r^2\sin\theta}\frac{\partial}{\partial\theta}\left(\sin\theta\frac{\partial\psi(r,\theta,\phi)}{\partial\theta}\right) + \frac{1}{r^2\sin^2\theta}\frac{\partial^2\psi(r,\theta,\phi)}{\partial\phi^2}\right]$$
$$-\frac{e^2}{4\pi\varepsilon_0 r}\psi(r,\theta,\phi) = E\psi(r,\theta,\phi)$$

What is the eigenvalue for the total energy? Use the relation $a_0 = \varepsilon_0 h^2/(\pi m_e e^2)$ to simplify your answer.

P20.26 Show that the total energy eigenfunctions $\psi_{100}(r)$ and $\psi_{200}(r)$ are orthogonal.

P20.27 As will be discussed in Chapter 21, core electrons shield valence electrons so that they experience an effective nuclear charge Z_{eff} rather than the full nuclear charge. Given that the first ionization energy of Li is 5.39 eV, use the formula in Problem P20.20 to estimate the effective nuclear charge experienced by the 2s electron in Li.

P20.28 Is the total energy wave function

$$\psi_{310}(r, \theta, \phi) = \frac{1}{81}\left(\frac{2}{\pi}\right)^{1/2}\left(\frac{1}{a_0}\right)^{3/2}\left(6\frac{r}{a_0} - \frac{r^2}{a_0^2}\right)e^{-r/3a_0}\cos\theta$$

an eigenfunction of any other operators? If so, which ones? What are the eigenvalues?

P20.29 Show that the total energy eigenfunctions $\psi_{210}(r, \theta, \phi)$ and $\psi_{211}(r, \theta, \phi)$ are orthogonal. Do you have to integrate over all three variables to show that the functions are orthogonal?

P20.30 Calculate $\langle r \rangle$ and the most probable value of r for the H atom in its ground state. Explain why they differ with a drawing.

P20.31 How many radial and angular nodes are there in the following H orbitals?

a. $\psi_{2p_x}(r, \theta, \phi)$

b. $\psi_{2s}(r)$

c. $\psi_{3d_{xz}}(r, \theta, \phi)$

d. $\psi_{3d_{x^2-y^2}}(r, \theta, \phi)$

P20.32 Show that $\psi_{2p_x}(r, \theta, \phi)$ and $\psi_{2p_y}(r, \theta, \phi)$ can be written in the form $Nxe^{-r/2a_0}$ and $N'ye^{-r/2a_0}$ where N and N' are normalization constants.

P20.33 Using the result of Problem P20.12, calculate the probability of finding the electron in the 1s state outside a sphere of radius $0.75a_0$, $2.5a_0$, and $4.5a_0$.

Many-Electron Atoms

The Schrödinger equation cannot be solved analytically for atoms containing more than one electron because of the electron–electron repulsion term in the potential energy. Instead, approximate numerical methods can be used to obtain the eigenfunctions and eigenvalues of the Schrödinger equation for many-electron atoms. Having more than one electron in an atom also raises new issues that we have not considered, including the indistinguishability of electrons, the electron spin, and the interaction between orbital and spin magnetic moments. The Hartree–Fock method provides a way to calculate total energies and orbital energies for many-electron atoms in the limit that the motion of individual electrons is assumed to be uncorrelated.

21.1 Helium: The Smallest Many-Electron Atom

The Schrödinger equation for the hydrogen atom can be solved analytically because this atom has only one electron. The complexity of solving the Schrödinger equation for systems that have more than one electron can be illustrated using the He atom. Centering the coordinate system at the nucleus and neglecting the kinetic energy of the nucleus, the Schrödinger equation takes the form

$$\left(-\frac{\hbar^2}{2m_e}\nabla_{e1}^2 - \frac{\hbar^2}{2m_e}\nabla_{e2}^2 - \frac{2e^2}{4\pi\varepsilon_0 r_1} - \frac{2e^2}{4\pi\varepsilon_0 r_2} + \frac{e^2}{4\pi\varepsilon_0 r_{12}}\right)\psi(\mathbf{r}_1, \mathbf{r}_2)$$

$$= E\psi(\mathbf{r}_1, \mathbf{r}_2) \tag{21.1}$$

In this equation, $r_1 = |\mathbf{r}_1|$ and $r_2 = |\mathbf{r}_2|$ are the distances of electrons 1 and 2 from the nucleus, $r_{12} = |\mathbf{r}_1 - \mathbf{r}_2|$, and ∇_{e1}^2 is shorthand for

$$\frac{1}{r_1^2}\frac{\partial}{\partial r_1}\left(r_1^2\frac{\partial}{\partial r_1}\right) + \frac{1}{r_1^2\sin^2\theta_1}\frac{\partial^2}{\partial\phi_1^2} + \frac{1}{r_1^2\sin\theta_1}\frac{\partial}{\partial\theta_1}\left(\sin\theta_1\frac{\partial}{\partial\theta_1}\right)$$

This is the part of the operator that is associated with the kinetic energy of electron 1, expressed in spherical coordinates. The last three terms in Equation (21.1) are the potential energy operators for the electron–nucleus attraction and the electron–electron repulsion. The variables $r_1 = |\mathbf{r}_1|$, $r_2 = |\mathbf{r}_2|$, and $r_{12} = |\mathbf{r}_1 - \mathbf{r}_2|$ are shown in Figure 21.1.

The eigenfunctions of the Schrödinger equation depend on the coordinates of both electrons. If this formalism is applied to argon, each many-electron eigenfunction

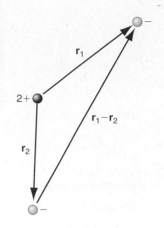

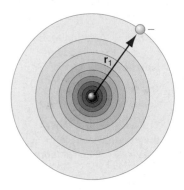

FIGURE 21.1
The top image shows the proton and two electrons in He. The bottom image shows that if the position of electron 2 is averaged over its orbit, electron 1 sees a spherically symmetric charge distribution due to the proton and electron 2.

depends simultaneously on the coordinates of 18 electrons! However, we also know that electrons in different atomic orbitals have quite different properties. For instance, valence electrons are involved in chemical bonds, and core electrons are not. Therefore, it seems reasonable to express a many-electron eigenfunction in terms of individual electron orbitals, each of which depends only on the coordinates of one electron. This is called the **orbital approximation,** in which the many-electron eigenfunctions of the Schrödinger equation are expressed as a product of one-electron orbitals:

$$\psi(\mathbf{r}_1, \mathbf{r}_2, \ldots, \mathbf{r}_n) = \phi_1(\mathbf{r}_1)\phi_2(\mathbf{r}_2)\ldots\phi_n(\mathbf{r}_n) \qquad \textbf{(21.2)}$$

This is not equivalent to saying that all of the electrons are independent of one another because, as we will see, the functional form for each $\phi_n(\mathbf{r}_n)$ is influenced by all the other electrons. The one-electron orbitals $\phi_n(\mathbf{r}_n)$ turn out to be quite similar to the functions $\psi_{nlm_l}(r, \theta, \phi)$ obtained for the hydrogen atom in Chapter 20, and they are labeled with indices such as 1*s,* 2*p,* and 3*d.* Each of the $\phi_n(\mathbf{r}_n)$ is associated with a one-electron **orbital energy** ε_n.

The orbital approximation allows an *n*-electron Schrödinger equation to be written as *n* one-electron Schrödinger equations, one for each electron. However, a further problem arises in solving these *n* equations. Because of the form of the **electron–electron repulsion** term, $e^2/(4\pi\varepsilon_0 r_{12})$, the potential energy operator no longer has spherical symmetry, so that the potential no longer has the form $V = V(r)$. This is evident from Figure 21.1 because the vector $\mathbf{r}_1 - \mathbf{r}_2$ does not start at the nucleus. Therefore, the Schrödinger equation cannot be solved analytically, and numerical methods must be used. For these methods to be effective, further approximations beyond the orbital approximation have to be made. Perhaps the most serious of these approximations is that one cannot easily include what electrons do naturally in a many-electron atom, namely, stay out of each other's way by undergoing a correlated motion. Whereas **electron correlation** ensures that the repulsion among electrons is minimized, the numerical methods introduced in this chapter to solve the Schrödinger equation assume that the electrons move independently of one another. As discussed in Chapter 26, corrections can be made that largely eliminate the errors generated through this assumption.

A schematic illustration of how a neglect of electron correlation simplifies solving the Schrödinger equation is shown in Figure 21.1 for the He atom. We know from introductory chemistry that both electrons occupy what we call the 1*s* **orbital**, implying that the wave functions are similar to

$$\frac{1}{\sqrt{\pi}}\left(\frac{\zeta}{a_0}\right)^{3/2} e^{-\zeta r/a_0}$$

Zeta (ζ) is the **effective nuclear charge** felt by the electron. The importance of ζ in determining chemical properties is discussed later in this chapter. If the assumption is made that the motion of electrons 1 and 2 is uncorrelated, electron 1 can interact with the nucleus and the spatially averaged charge distribution arising from electron 2. This spatially averaged charge distribution is determined by $\phi^*(\mathbf{r}_2)\phi(\mathbf{r}_2)$. Think of electron 2 as being smeared out in a distribution that is spherically symmetrical about the nucleus, with a negative charge in the volume element $d\tau$ proportional to $-e\phi^*(\mathbf{r}_2)\phi(\mathbf{r}_2)\,d\tau$.

The advantage of this approximation becomes apparent in Figure 21.1, because the effective charge distribution that electron 1 experiences is spherically symmetrical. Because the potential energy V depends only on r, each one-electron wave function can be written as a product of radial and angular functions, $\phi(\mathbf{r}) = \phi(r, \theta, \phi) = R(r)\Theta(\theta)\Phi(\phi)$. Although the radial functions differ from those for the hydrogen atom, the angular functions are the same so that the *s, p, d, f,* . . . nomenclature used for the hydrogen atom also applies to many-electron atoms.

This quick look at the helium atom illustrates the approach that we take in the rest of this chapter. The Schrödinger equation is solved for many-electron atoms by approximating the true wave function by products of orbitals, each of which depends only on the coordinates of one electron. This approximation reduces the *n*-electron Schrödinger equation to *n* one-electron Schrödinger equations. The set of *n* equations is solved to obtain the one-electron energies and orbitals ε_i and ϕ_i. The solutions are approximate because of the orbital approximation and because electron correlation is neglected. However, before this approach is implemented, two important concepts must be introduced, namely, electron spin and the indistinguishability of electrons.

21.2 Introducing Electron Spin

Electron spin plays an important part in formulating the Schrödinger equation for many-electron atoms. In discussing the Stern–Gerlach experiment in Chapter 17, we focused on commutation relations rather than the other surprising result of this experiment, which is that two and only two deflected beams are observed. In order for a silver atom to be deflected in an inhomogeneous field, it must have a magnetic moment and an associated angular momentum. What is the origin of this moment? An electric current passing through a loop of wire produces a magnetic field and, therefore, the loop has a magnetic moment. An electron in an orbit around a nucleus for which $l > 0$ has a nonzero angular momentum because of the nonspherical electron charge distribution. However, Ag has a closed-shell configuration plus a single $5s$ valence electron. A closed shell has a spherical electron charge distribution and no net angular momentum. Therefore, the magnetic moment must be associated with the $5s$ electron, which has no orbital angular momentum because $l = 0$. If this electron has an intrinsic angular momentum, which we call s, it will be split into $2s + 1$ components in passing through the magnet. The fact that two components are observed in the Stern–Gerlach experiment shows that $s = 1/2$. Therefore, there is a z component of angular momentum $s_z = m_s \hbar = \pm \hbar/2$ associated with the $5s$ electron. The origin of this effect cannot be an orbital angular momentum because for an s electron $l = 0$ and because orbital angular momentum comes in quanta twice that size. This intrinsic electron spin angular momentum is a vector called $\mathbf{s}$, and its z component is called s_z to distinguish it from orbital angular momentum. The term *intrinsic* refers to the fact that the spin is independent of the environment of the electron. The use of the term spin implies that the electron is spinning about an axis. Although the nomenclature is appealing, there is no physical basis for this association.

How does the existence of spin change what has been discussed up to now? As we show later, each of the orbitals in a many-electron atom can be doubly occupied; one electron has $m_s = +1/2$, and the other has $m_s = -1/2$. This adds a fourth quantum number to the H atom eigenfunctions that is now labeled $\psi_{nlm_lm_s}(r, \theta, \phi)$. Because electron spin is an intrinsic property of the electron, it does not depend on the spatial variables r, θ, and ϕ.

How can this additional quantum number be incorporated in the formalism described for the hydrogen atom? This can be done by defining spin wave functions called α and β, which are eigenfunctions of the spin angular momentum operators $\hat{s}^2$ and $\hat{s}_z$. Because all angular momentum operators have the same properties, the spin operators follow the commutation rules listed in Equation (18.57). As for the orbital angular momentum, only the magnitude of the spin angular momentum and one of its components can be known simultaneously. The spin operators $\hat{s}^2$ and $\hat{s}_z$ have the following properties:

$$\hat{s}^2 \alpha = \hbar^2 s (s + 1)\alpha = \frac{\hbar^2}{2}\left(\frac{1}{2} + 1\right)\alpha$$

$$\hat{s}^2 \beta = \hbar^2 s(s + 1)\beta = \frac{\hbar^2}{2}\left(\frac{1}{2} + 1\right)\beta$$

$$\hat{s}_z \alpha = m_s \hbar \alpha = \frac{\hbar}{2}\alpha \,, \quad \hat{s}_z \beta = m_s \hbar \beta = -\frac{\hbar}{2}\beta$$

$$\int \alpha^* \beta \, d\sigma = \int \beta^* \alpha \, d\sigma = 0$$

$$\int \alpha^* \alpha \, d\sigma = \int \beta^* \beta \, d\sigma = 1 \qquad \textbf{(21.3)}$$

In these equations, σ is called the spin variable. It is not a spatial variable and the "integration" over σ exists only formally so that we can define orthogonality. The H atom eigenfunctions are redefined by multiplying them by α and β and including a quantum number for spin. For example, the H atom 1s eigenfunctions take the form

$$\psi_{100\frac{1}{2}}(r) = \frac{1}{\sqrt{\pi}} \left(\frac{1}{a_0}\right)^{3/2} e^{-r/a_0} \alpha \quad \text{and}$$

$$\psi_{100-\frac{1}{2}}(r) = \frac{1}{\sqrt{\pi}} \left(\frac{1}{a_0}\right)^{3/2} e^{-r/a_0} \beta \tag{21.4}$$

The eigenfunctions remain orthonormal because with this formalism

$$\iiiint \psi^*_{100\frac{1}{2}}(r, \sigma)\psi_{100-\frac{1}{2}}(r, \sigma) \, dr \, d\theta \, d\phi \, d\sigma$$

$$= \iiint \psi^*_{100}(r)\psi_{100}(r) \, dr \, d\theta \, d\phi \int \alpha^*\beta \, d\sigma = 0$$

and

$$\iiiint \psi^*_{100\frac{1}{2}}(r, \sigma)\psi_{100\frac{1}{2}}(r, \sigma) \, dr \, d\theta \, d\phi \, d\sigma$$

$$= \iiint \psi^*_{100}(r)\psi_{100}(r) \, dr \, d\theta \, d\phi \int \alpha^*\alpha \, d\sigma = 1 \tag{21.5}$$

These two eigenfunctions have the same energy because the total energy operator of Equation (21.1) does not depend on the spin. Having discussed how to include electron spin in a wave function, we now take on the issue of keeping track of electrons in a many-electron atom.

21.3 Wave Functions Must Reflect the Indistinguishability of Electrons

In discussing He in Section 21.1, the electrons were numbered 1 and 2. Macroscopic objects can be distinguished from one another, but in an atom we have no way to distinguish between any two electrons. This fact needs to be taken into account in the formulation of a wave function. How can **indistinguishability** be introduced into the orbital approximation? Consider an n-electron wave function written as the product of n one-electron wave functions, which we describe using the notation $\psi(1, 2, \ldots, n) = \psi(r_1\theta_1\phi_1\sigma_1, r_2\theta_2\phi_2\sigma_2, \ldots, r_n\theta_n\phi_n\sigma_n)$. The position variables are suppressed in favor of keeping track of the electrons. How does indistinguishability affect how the wave function is written? We know that the wave function itself is not an observable, but the square of the magnitude of the wave function is proportional to the electron density and is an observable. Because the two electrons in He are indistinguishable, no observable of the system can be changed if the electron labels 1 and 2 are interchanged. Therefore, $\psi^2(1,2) = \psi^2(2,1)$. This equation can be satisfied either by $\psi(1,2) = \psi(2,1)$ or $\psi(1,2) = -\psi(2,1)$. We refer to the wave function as being a **symmetric wave function** if $\psi(1,2) = \psi(2,1)$ or an **antisymmetric wave function** if $\psi(1,2) = -\psi(2,1)$. For a ground-state He atom, examples of symmetric and antisymmetric wave functions are as follows:

$$\psi_{symmetric}(1,2) = \phi_{1s}(1)\alpha(1)\phi_{1s}(2)\beta(2) + \phi_{1s}(2)\alpha(2)\phi_{1s}(1)\beta(1) \quad \text{and}$$

$$\psi_{antisymmetric}(1,2) = \phi_{1s}(1)\alpha(1)\phi_{1s}(2)\beta(2) - \phi_{1s}(2)\alpha(2)\phi_{1s}(1)\beta(1) \tag{21.6}$$

where $\phi(1) = \phi(\mathbf{r}_1)$. Wolfgang Pauli showed that only an antisymmetric wave function is allowed for electrons, a result that can be formulated as a further fundamental postulate of quantum mechanics.

> **POSTULATE 6:**
> Wave functions describing a many-electron system must change sign (be antisymmetric) under the exchange of any two electrons.

This postulate is also known as the **Pauli principle.** This principle states that different product wave functions of the type $\psi(1, 2, 3, \ldots, n) = \phi_1(1)\phi_2(2)\ldots\phi_n(n)$ must be combined such that the resulting wave function changes sign when any two electrons are interchanged. A combination of such terms is required because a single-product wave function cannot be made antisymmetric in the interchange of two electrons. For example, $\phi_{1s}(1)\alpha(1)\phi_{1s}(2)\beta(2) \neq -\phi_{1s}(2)\alpha(2)\phi_{1s}(1)\beta(1)$.

How can antisymmetric wave functions be constructed? Fortunately, there is a simple way to do so using determinants. They are known as **Slater determinants** and have the form

$$\psi(1,2,3,\ldots,n) = \frac{1}{\sqrt{n!}} \begin{vmatrix} \phi_1(1)\alpha(1) & \phi_1(1)\beta(1) & \ldots & \phi_m(1)\beta(1) \\ \phi_1(2)\alpha(2) & \phi_1(2)\beta(2) & \ldots & \phi_m(2)\beta(2) \\ \ldots & \ldots & \ldots & \ldots \\ \phi_1(n)\alpha(n) & \phi_1(n)\beta(n) & \ldots & \phi_m(n)\beta(n) \end{vmatrix} \quad (21.7)$$

where $m = n/2$ if n is even and $m = (n + 1)/2$ if n is odd. The one-electron orbitals in which the n electrons are sequentially filled are listed going across each row with one row for each electron. The factor in front of the determinant takes care of the normalization if the one-electron orbitals are individually normalized. The Slater determinant is simply a recipe for constructing an antisymmetric wave function, and none of the individual entries in the determinant has a separate reality. For the ground state of He, the antisymmetric wave function is the 2×2 determinant:

$$\psi(1,2) = \frac{1}{\sqrt{2}} \begin{vmatrix} 1s(1)\alpha(1) & 1s(1)\beta(1) \\ 1s(2)\alpha(2) & 1s(2)\beta(2) \end{vmatrix}$$

$$= \frac{1}{\sqrt{2}}[1s(1)\alpha(1)1s(2)\beta(2) - 1s(1)\beta(1)1s(2)\alpha(2)]$$

$$= \frac{1}{\sqrt{2}}1s(1)1s(2)[\alpha(1)\beta(2) - \beta(1)\alpha(2)] \quad (21.8)$$

The shorthand notation $\phi_{1s}(1)\alpha(1) = \phi_{100+\frac{1}{2}}(r_1, \theta_1, \phi_1, \sigma_1) = 1s(1)\alpha(1)$ has been used in the preceding determinant.

Determinants are used in constructing antisymmetric wave functions because their value automatically changes sign when two rows (which refer to individual electrons) are interchanged. This can easily be verified by comparing the values of the following determinants:

$$\begin{vmatrix} 3 & 6 \\ 4 & 2 \end{vmatrix} \quad \text{and} \quad \begin{vmatrix} 4 & 2 \\ 3 & 6 \end{vmatrix}$$

Writing the wave function as a determinant also demonstrates another formulation of the Pauli principle. The value of a determinant is zero if two rows are identical. *This is equivalent to saying that the wave function is zero if all quantum numbers of any two electrons are the same.* Example Problem 21.1 illustrates how to work with determinants. Further information on determinants can be found in the Math Supplement (see Appendix A).

EXAMPLE PROBLEM 21.1

Consider the determinant

$$\begin{vmatrix} 3 & 1 & 5 \\ 4 & -2 & 1 \\ 3 & 2 & 7 \end{vmatrix}$$

a. Evaluate the determinant by expanding it in the cofactors of the first row.

b. Show that the value of the related determinant

$$\begin{vmatrix} 4 & -2 & 1 \\ 4 & -2 & 1 \\ 3 & 2 & 7 \end{vmatrix}$$

in which the first two rows are identical, is zero.

c. Show that exchanging the first two rows changes the sign of the value of the determinant.

Solution

The value of a 2×2 determinant

$$\begin{vmatrix} a & b \\ c & d \end{vmatrix} = ad - bc$$

We reduce a higher order determinant to a 2×2 determinant by expanding it in the cofactors of a row or column (see the Math Supplement). Any row or column can be used for this reduction, and all will yield the same result. The cofactor of an element a_{ij}, where i is the index of the row and j is the index of the column, is the $(n - 1) \times (n - 1)$ determinant that is left by ignoring the elements in the ith row and in the jth column. In our case, we reduce the 3×3 determinant to a sum of 2×2 determinants by adding the first row cofactors, each of which is multiplied by $(-1)^{i+j} a_{ij}$. For the given determinant,

a. $\begin{vmatrix} 3 & 1 & 5 \\ 4 & -2 & 1 \\ 3 & 2 & 7 \end{vmatrix} = 3(-1)^{1+1}\begin{vmatrix} -2 & 1 \\ 2 & 7 \end{vmatrix} + 1(-1)^{1+2}\begin{vmatrix} 4 & 1 \\ 3 & 7 \end{vmatrix} + 5(-1)^{1+3}\begin{vmatrix} 4 & -2 \\ 3 & 2 \end{vmatrix}$

$= 3(-14 - 2) - 1(28 - 3) + 5(8 + 6) = -3$

b. $\begin{vmatrix} 4 & -2 & 1 \\ 4 & -2 & 1 \\ 3 & 2 & 7 \end{vmatrix} = 4(-1)^{1+1}\begin{vmatrix} -2 & 1 \\ 2 & 7 \end{vmatrix} + (-2)(-1)^{1+2}\begin{vmatrix} 4 & 1 \\ 3 & 7 \end{vmatrix}$

$+ 1(-1)^{1+3}\begin{vmatrix} 4 & -2 \\ 3 & 2 \end{vmatrix}$

$= 4(-14 - 2) + 2(28 - 3) + 1(8 + 6) = 0$

c. $\begin{vmatrix} 4 & -2 & 1 \\ 3 & 1 & 5 \\ 3 & 2 & 7 \end{vmatrix} = 4(-1)^{1+1}\begin{vmatrix} 1 & 5 \\ 2 & 7 \end{vmatrix} - 2(-1)^{1+2}\begin{vmatrix} 3 & 5 \\ 3 & 7 \end{vmatrix}$

$+ 1(-1)^{1+3}\begin{vmatrix} 3 & 1 \\ 3 & 2 \end{vmatrix}$

$= 4(7 - 10) + 2(21 - 15) + 1(6 - 3) = +3$

For ground-state helium, both electrons have the same values of n, l, and m_l, but the values of m_s are $+1/2$ for one electron and $-1/2$ for the other. We now describe the way in which electrons are assigned to orbitals by a configuration. A **configuration** specifies the values of n and l for each electron. For example, the configuration for ground-state He is $1s^2$ and that for ground state F is $1s^2 2s^2 2p^5$. The quantum numbers m_l and m_s are not specified in a configuration. Describing the quantum state of an atom requires this information, as will be discussed in Chapter 22.

EXAMPLE PROBLEM 21.2

This problem illustrates how determinantal wave functions can be associated with putting α and β spins in a set of orbitals. The first excited state of the helium atom can be described by the configuration $1s^1 2s^1$. However, four different spin orientations are consistent with this notation, as shown pictorially in the following figure. Do not take these pictures too literally, because they imply that one electron can be distinguished from another.

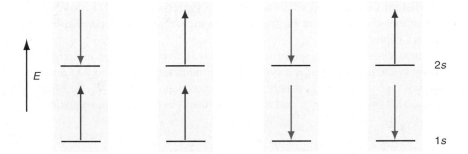

Keep in mind that the ↑ and ↓ notation commonly used for α and β spins is shorthand for the more accurate vector model depiction discussed in Chapter 18 and shown here:

Write determinantal wave functions that correspond to these pictures.

Solution

$$\psi_1(1, 2) = \frac{1}{\sqrt{2}} \begin{vmatrix} 1s(1)\alpha(1) & 2s(1)\beta(1) \\ 1s(2)\alpha(2) & 2s(2)\beta(2) \end{vmatrix}; \quad \psi_2(1, 2) = \frac{1}{\sqrt{2}} \begin{vmatrix} 1s(1)\alpha(1) & 2s(1)\alpha(1) \\ 1s(2)\alpha(2) & 2s(2)\alpha(2) \end{vmatrix};$$

$$\psi_3(1, 2) = \frac{1}{\sqrt{2}} \begin{vmatrix} 1s(1)\beta(1) & 2s(1)\beta(1) \\ 1s(2)\beta(2) & 2s(2)\beta(2) \end{vmatrix}; \quad \psi_4(1, 2) = \frac{1}{\sqrt{2}} \begin{vmatrix} 1s(1)\beta(1) & 2s(1)\alpha(1) \\ 1s(2)\beta(2) & 2s(2)\alpha(2) \end{vmatrix}$$

The neutral atom that has three electrons is Li. If the third electron is put in the $1s$ orbital, the determinantal wave function

$$\psi(1, 2, 3) = \frac{1}{\sqrt{3!}} \begin{vmatrix} 1s(1)\alpha(1) & 1s(1)\beta(1) & 1s(1)\alpha(1) \\ 1s(2)\alpha(2) & 1s(2)\beta(2) & 1s(2)\alpha(2) \\ 1s(3)\alpha(3) & 1s(3)\beta(3) & 1s(3)\alpha(3) \end{vmatrix}$$

is obtained, where the third electron can have either α or β spin. However, the first and third columns in this determinant are identical, so that $\psi(1, 2, 3) = 0$. Therefore, the third electron must go into the next higher energy orbital with $n = 2$. *This example shows that the Pauli exclusion principle requires that each orbital have a maximum occupancy of two electrons.* The configuration of ground-state Li is $1s^2 2s^1$. For $n = 2$, l can take on the value 0 with the only possible m_l value of 0, or 1 with the possible m_l values of 0, +1, and −1. Each of the possible sets of n, l, and m_l can be combined with $m_s = \pm 1/2$. Therefore, there are eight different sets of quantum numbers for $n = 2$.

The set of orbitals with the same values of n and l comprises a **subshell,** and the set of orbitals with the same n value comprises a **shell.** The connotation of a shell is demonstrated pictorially in Figure 20.12.

21.4 Using the Variation Method to Solve the Schrödinger Equation

In Section 21.1, we concluded that electron–electron repulsion terms in the total energy operator for many-electron atoms preclude an analytical solution to the Schrödinger equation. However, numerical methods are available for calculating one-electron energies and the orbitals ε_i and ϕ_i that include electron–electron repulsion. The goal is to obtain as good an approximation as possible to the total energy eigenfunctions and eigenvalues for the many-electron atom. Only one of these methods, the **Hartree–Fock self-consistent field method** combined with the **variation method,** is discussed here. Other methods that go beyond Hartree–Fock by including electron correlation are discussed in Chapter 26.

We next discuss the variation method, which is frequently used in computational chemistry calculations. Consider a system in its ground state with energy E_0 and the corresponding eigenfunction ψ_0, which satisfies the equation $\hat{H}\psi_0 = E_0\psi_0$. Multiplying this expression on the left by ψ_0^* and integrating results in the following equation:

$$E_0 = \frac{\int \psi_0^* \hat{H} \psi_0 \, d\tau}{\int \psi_0^* \psi_0 \, d\tau} \tag{21.9}$$

The denominator takes into account that the wave function may not be normalized. For a many-electron atom, the total energy operator can be formulated, but the exact total energy eigenfunctions are unknown. How can the energy be calculated in this case? The **variation theorem** states that no matter what approximate wave function Φ is substituted for the ground-state eigenfunction in Equation (21.9), the energy is always greater than or equal to the true energy. Expressed mathematically, the theorem says that

$$E = \frac{\int \Phi^* \hat{H} \Phi \, d\tau}{\int \Phi^* \Phi \, d\tau} \geq E_0 \tag{21.10}$$

The proof of this theorem is included as an end-of-chapter problem. How can this method be implemented to obtain good approximate wave functions and energies? We parameterize the **trial wave function** Φ and find the optimal values for the parameters by minimizing the energy with respect to each parameter. This procedure gives the best energy that can be obtained for that particular choice of a trial wave function. The better the choice made for the trial function, the closer the calculated energy will be to the true energy.

We illustrate this formalism using the particle in the box as a specific example. Any trial function used must satisfy a number of general conditions (single valued, normalizable, the function and its first derivative are continuous) and also the boundary condition that the wave function goes to zero at the ends of the box. We use the trial function of Equation (21.11) to approximate the ground-state wave function. This wave function satisfies the criteria just listed. This function contains a single parameter α that is used to minimize the energy:

$$\Phi(x) = \left(\frac{x}{a} - \frac{x^3}{a^3}\right) + \alpha\left(\frac{x^5}{a^5} - \frac{1}{2}\left(\frac{x^7}{a^7} + \frac{x^9}{a^9}\right)\right), \quad 0 < x < a \tag{21.11}$$

We first calculate the energy for $\alpha = 0$ and obtain

$$E = \frac{-\dfrac{\hbar^2}{2m}\displaystyle\int_0^a \left(\frac{x}{a} - \frac{x^3}{a^3}\right)\frac{d^2}{dx^2}\left(\frac{x}{a} - \frac{x^3}{a^3}\right)dx}{\displaystyle\int_0^a \left(\frac{x}{a} - \frac{x^3}{a^3}\right)^2 dx} = 0.133\frac{h^2}{ma^2} \tag{21.12}$$

Because the trial function is not the exact ground-state wave function, the energy is higher than the exact value $E_0 = 0.125(h^2/ma^2)$. How similar is the trial function to the ground-state eigenfunction? A comparison between the exact solution and the trial function with $\alpha = 0$ is shown in Figure 21.2a.

To find the optimal value for α, E is first expressed in terms of h, m, a, and α, and then minimized with respect to α. The energy E is given by

$$
E = \frac{-\dfrac{\hbar^2}{2m}\displaystyle\int_0^a \left[\begin{array}{c} \left[\left(\dfrac{x}{a} - \dfrac{x^3}{a^3}\right) + \alpha\left(\dfrac{x^5}{a^5} - \dfrac{1}{2}\left(\dfrac{x^7}{a^7} + \dfrac{x^9}{a^9}\right)\right)\right] \times \\[2ex] \dfrac{d^2}{dx^2}\left[\left(\dfrac{x}{a} - \dfrac{x^3}{a^3}\right) + \alpha\left(\dfrac{x^5}{a^5} - \dfrac{1}{2}\left(\dfrac{x^7}{a^7} + \dfrac{x^9}{a^9}\right)\right)\right] \end{array}\right] dx}{\displaystyle\int_0^a \left[\left(\dfrac{x}{a} - \dfrac{x^3}{a^3}\right) + \alpha\left(\dfrac{x^5}{a^5} - \dfrac{1}{2}\left(\dfrac{x^7}{a^7} + \dfrac{x^9}{a^9}\right)\right)\right]^2 dx}
\tag{21.13}
$$

Carrying out this integration gives E in terms of $\hbar$, m, a, and α:

$$
E = \frac{\hbar^2}{2ma^2} \frac{\left(\dfrac{4}{5} + \dfrac{116\alpha}{231} + \dfrac{40247\alpha^2}{109395}\right)}{\left(\dfrac{8}{105} + \dfrac{8\alpha}{273} + \dfrac{1514\alpha^2}{230945}\right)}
\tag{21.14}
$$

To minimize the energy with respect to the variational parameter, we differentiate this function with respect to α, set the resulting equation equal to zero, and solve for α. The solutions are $\alpha = -5.74$ and $\alpha = -0.345$. The second of these solutions corresponds to the minimum in E. Substituting this value in Equation (21.14) gives $E = 0.127(h^2/ma^2)$, which is very close to the true value of $0.125(h^2/ma^2)$. The optimized trial function is shown in Figure 21.2b. We can see that, by choosing the optimal value of α, $E \rightarrow E_0$ and $\Phi \rightarrow \psi_0$. No better value for the energy can be obtained with this particular choice of a trial wave function, and this illustrates a limitation of the variation method. The "best" energy obtained depends on the choice of the trial function. For example, a lower energy is obtained if a function of the type $\Phi(x) = x^\alpha(a - x)^\alpha$ is minimized with respect to α. This example shows how the variation method can be implemented by optimizing approximate solutions to the Schrödinger equation.

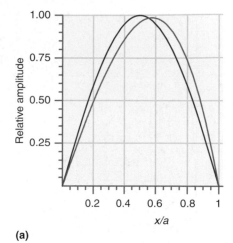

(a)

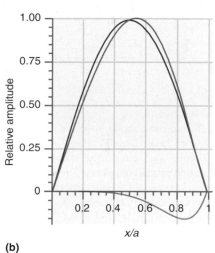

(b)

FIGURE 21.2
Exact (red curve) and approximate (purple curve) wave functions for the ground state of the particle in the box. (a) The approximate wave function contains only the first term in Equation (21.11). (b) The optimal approximate wave function contains both terms of Equation (21.11). The light blue curve shows the contribution of the second term in Equation (21.11) to the approximate wave function.

21.5 The Hartree–Fock Self-Consistent Field Method

We now return to the problem at hand, namely, solving the Schrödinger equation for many-electron atoms. The starting point is to use the orbital approximation and to take the Pauli exclusion principle into account. Antisymmetry of the wave function with respect to electron exchange is accomplished by expressing the wave function as a Slater determinant

$$
\psi(1,2,3,\ldots,n) = \frac{1}{\sqrt{n!}} \begin{vmatrix} \phi_1(1)\alpha(1) & \phi_1(1)\beta(1) & \cdots & \phi_m(1)\beta(1) \\ \phi_1(2)\alpha(2) & \phi_1(2)\beta(2) & \cdots & \phi_m(2)\beta(2) \\ \cdots & \cdots & \cdots & \cdots \\ \phi_1(n)\alpha(n) & \phi_1(n)\beta(n) & \cdots & \phi_m(n)\beta(n) \end{vmatrix}
\tag{21.15}
$$

in which the individual entries ϕ_j are modified H atom orbitals as described later. The Hartree–Fock method is a prescription for finding the single Slater determinant that gives the lowest energy for the ground-state atom in the absence of electron correlation. (More correctly, configurations with more than one unpaired electron require more than one Slater determinant.)

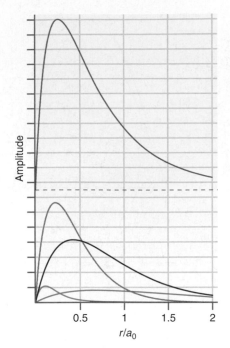

FIGURE 21.3

The top curve shows the radial function for the 2p orbital in Ne determined in a Hartree–Fock calculation. It has been shifted upward for clarity. The bottom four curves are the individual terms in the four-element basis set.

As for the helium atom discussed in Section 21.1, it is assumed that the electrons are uncorrelated and that a particular electron feels the spatially averaged electron charge distribution of the remaining $n - 1$ electrons. These approximations reduce the radial part of the n-electron Schrödinger equation to n one-electron Schrödinger equations that have the form

$$\left(\frac{\hbar^2}{2m} \nabla_i^2 + V_i^{eff}(r) \right) \phi_i(r) = \varepsilon_i \phi_i(r), \quad i = 1, \ldots, n \qquad \textbf{(21.16)}$$

in which the effective potential energy felt by the first electron, $V_1^{eff}(r)$, takes into account the electron-nuclear attraction and the repulsion between electron 1 and all other electrons. The Hartree–Fock method allows the best (in a variational sense) one-electron orbitals $\phi_i(r)$ and the corresponding orbital energies ε_i to be calculated.

Because of the neglect of electron correlation, the effective potential is spherically symmetrical and, therefore, the angular part of the wave functions is identical to the solutions for the hydrogen atom. *This means that the s, p, d, ... orbital nomenclature derived for the hydrogen atom remains intact for the one-electron orbitals for all atoms.* What remains to be found are solutions to the radial part of the Schrödinger equation.

To optimize the radial part of the determinantal wave function, the variational method outlined in Section 21.4 is used. What functions should be used for the individual entries $\phi_j(r)$ in the determinant? Each $\phi_j(r)$ is expressed as a linear combination of suitable **basis functions** $f_i(r)$ as shown in Equation (21.17).

$$\phi_j(r) = \sum_{i=1}^{m} c_i f_i(r) \qquad \textbf{(21.17)}$$

What do we mean by a set of suitable functions? Recall that a well-behaved function can be expanded in a Fourier series as a sum of sine and cosine functions, which in this context are basis functions. There are many other choices for individual members of a basis set. The criterion for a "good" basis set is that the number of terms in the sum m representing $\phi_j(r)$ is as small as possible and that the basis functions enable the Hartree–Fock calculations to be carried out rapidly. Two examples of basis set expansions for atomic orbitals are shown in Figures 21.3 and 21.4.

In Figure 21.3, the 2p atomic orbital of Ne obtained in a Hartree–Fock calculation is shown together with the individual contributions to Equation (21.17) where each member of the $m = 4$ basis set is of the form $f_i(r) = N_i r \exp[-\zeta_i r/a_0]$ and N_i is a normalization constant. In a second example, the H 1s AO and the contributions of each member of the $m = 3$ basis set to Equation (21.17) are shown in Figure 21.4, where the

FIGURE 21.4

The left panel shows a fit to a H 1s orbital with a single Gaussian function. The agreement is not good. The right panel shows a best fit (purple curve) using a basis set of three Gaussian functions which are also shown. Except very near the nucleus, the three basis function fit is very good.

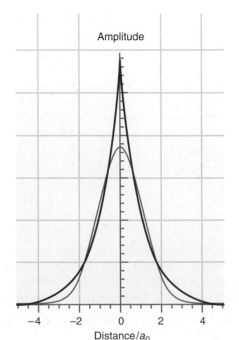

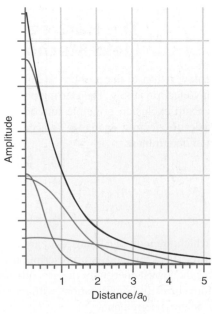

basis set functions are of the form $f_i(r) = N_i \exp[-\zeta_i(r/a_0)^2]$ (Gaussian functions). In both cases, the coefficients c_i in Equation (21.17) are used as variational parameters to optimize $\phi_j(r)$ and the ζ_i values are optimized separately. Although the Gaussian functions do not represent the H $1s$ function accurately near the nucleus, they are well suited to Hartree–Fock calculations and are the most widely used basis functions in computational chemistry. (See Chapter 26 for a more detailed discussion of Gaussian basis functions.)

The preceding discussion describes the input to a calculation of the orbital energies, but there is a problem in proceeding with the calculation. To solve the Schrödinger equation for electron 1, $V_1^{eff}(r)$ must be known, and this means that we must know the functional form of all the other orbitals $\phi_2(\mathbf{r}_2), \phi_3(\mathbf{r}_3), \ldots, \phi_n(\mathbf{r}_n)$. This is also the case for the remaining $n-1$ electrons. In other words, the answers must be known in order to solve the problem.

The way out of this quandary is to use an iterative approach. A reasonable guess is made for an initial set of $\phi_j(r)$. Using these orbitals, an effective potential is calculated, and the energy and improved orbital functions, $\phi'_j(r)$, for each of the n electrons are calculated. The $\phi'_j(r)$ are used to calculate a new effective potential, which is used to calculate a further improved set of orbitals, $\phi''_j(r)$, and this procedure is repeated for all electrons until the solutions for the energies and orbitals are self-consistent, meaning that they do not change significantly in a further iteration. This procedure, coupled with the variation method in optimizing the parameters in the orbitals, is very effective in giving the best one-electron orbitals and energies available for a many-electron atom in the absence of electron correlation. More accurate calculations that include electron correlation are discussed in Chapter 26.

The accuracy of a Hartree–Fock calculation depends primarily on the size of the basis set. This dependence is illustrated in Table 21.1 in which the calculated total energy of He and the 1s orbital energy are shown for three different basis sets. In each case, $\phi_{1s}(r)$ has the form

$$\phi_{1s}(r) = \sum_{i=1}^{m} c_i N_i e^{-\zeta_i r/a_0} \tag{21.18}$$

where N_i is a normalization constant for the ith basis function and m is the number of basis functions. It is seen that there is almost no change in going from two to five basis functions, which represents the Hartree–Fock limit of a complete basis set in this case. The one element or single zeta basis set gives an energy that differs significantly from the Hartree–Fock limit. We return to this basis set in discussing the effective nuclear charge later. The He 1s orbital cannot be accurately represented by a single exponential function as was the case for the hydrogen atom.

One might think that the total energy of an atom is the sum of the orbital energies, or for helium, $\varepsilon_{total} = 2\varepsilon_{1s}$. As shown in Table 21.1, $\varepsilon_{total} - 2\varepsilon_{1s} < 0$, and this result

TABLE 21.1 Total Energy and 1s Orbital Energy for He for Three Different Basis Sets Used to Represent the 1s Orbital

Number of Basis Functions, m	Exponents, ζ_i	Total Energy of He, ε_{total} (eV)	1s Orbital Energy, ε_{1s} (eV)	$\varepsilon_{total} - 2\varepsilon_{1s}$ (eV)
5	1.41714, 2.37682, 4.39628, 6.52699, 7.94252	−77.8703	−24.9787	−27.9129
2	2.91093, 1.45363	−77.8701	−24.9787	−27.9133
1	1.68750	−77.4887	−24.3945	−28.6998

The data is taken from E. Clementi and C. Roetti. "Roothaan-Hartree-Fock Atomic Wavefunctions: Basis Functions and Their Coefficients for Ground and Certain Excited States of Neutral and Ionized Atoms, $Z \leq 54$." *Atomic Data and Nuclear Data Tables* 14 (1974): 177.

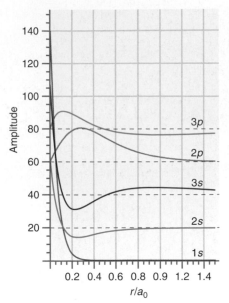

FIGURE 21.5

Hartree–Fock radial functions are shown for Ar. The curves are offset vertically to allow individual functions to be compared. [Calculated from data in E. Clementi and C. Roetti. "Roothaan-Hartree-Fock Atomic Wavefunctions: Basis Functions and Their Coefficients for Ground and Certain Excited States of Neutral and Ionized Atoms, $Z \leq 54$." *Atomic Data and Nuclear Data Tables* 14 (1974): 177.]

can be understood by considering how electron–electron repulsion is treated in a Hartree–Fock calculation. The $1s$ orbital energy is calculated using an effective potential in which repulsion between the two electrons in the orbital is included. Therefore, assuming that $\varepsilon_{total} = 2\varepsilon_{1s}$ counts the repulsion between the two electrons twice and gives a value for ε_{total} that is more positive than the true total energy.

Radial functions for Ar in the Hartree–Fock limit of a large basis set are shown in Figure 21.5. It is seen that they have the same nodal structure as the orbitals for the hydrogen atom.

The Hartree–Fock radial functions can be used to obtain the radial probability distribution for many-electron atoms from

$$P(r) = \sum_i n_i r^2 R_i^2(r) \qquad (21.19)$$

where $R_i(r)$ is the radial function corresponding to the ith subshell, for example $2s$, $3p$, or $4d$, and n_i is the number of electrons in the subshell. $P(r)$ is shown for Ne, Ar, and Kr in Figure 21.6. Note that the radial distribution exhibits a number of maxima, one for each occupied shell and that the contributions from different shells overlap. The width of $P(r)$ for a given shell increases with n; it is smallest for $n = 1$ and largest for the largest n value.

Hartree–Fock orbital energies ε_i are shown in Figure 21.7 for the first 36 elements in the periodic table. An important result of these calculations is that the ε_i for many-electron atoms depend on both the principal quantum number n and on the angular momentum quantum number l. Within a shell of principal quantum number n, $\varepsilon_{ns} < \varepsilon_{np} < \varepsilon_{nd} < \dots$. This was not the case for the H atom. This result can be understood by considering the radial distribution functions for Kr shown in Figure 21.8. As discussed in Chapter 20, this function gives the probability of finding an electron at a given distance from the nucleus. The subsidiary maximum near $r = 0.02\,a_0$ in the $3s$ radial distribution function indicates that there is a higher probability of finding the $3s$ electron close to the nucleus than is the case for the

FIGURE 21.6

Radial distribution functions calculated from Hartree–Fock wave functions are shown for Ne, Ar, and Kr. The colored curves show the contributions from the individual shells and the purple curve shows the total radial distribution function. The $n = 1$ curve for Kr is not shown for clarity.

[Calculated from data in E. Clementi and C. Roetti. "Roothaan-Hartree-Fock Atomic Wavefunctions: Basis Functions and Their Coefficients for Ground and Certain Excited States of Neutral and Ionized Atoms, $Z \leq 54$." *Atomic Data and Nuclear Data Tables* 14 (1974): 177.]

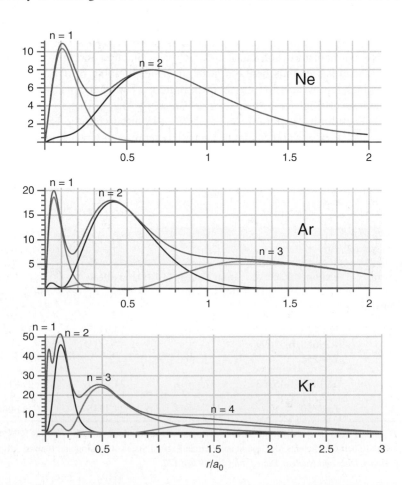

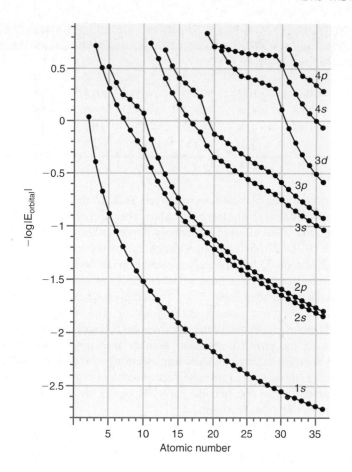

FIGURE 21.7
The one-electron orbital energies obtained from Hartree–Fock calculations are shown on a logarithmic scale for the first 36 elements.
[The data are taken from E. Clementi and C. Roetti. "Roothaan-Hartree-Fock Atomic Wavefunctions: Basis Functions and Their Coefficients for Ground and Certain Excited States of Neutral and Ionized Atoms, $Z \leq 54$." *Atomic Data and Nuclear Data Tables* 14 (1974): 177.]

$3p$ and $3d$ electrons. The potential energy associated with the attraction between the nucleus and the electron falls off as $1/r$ so that its magnitude increases substantially as the electron comes closer to the nucleus. As a result, the $3s$ electron is bound more strongly to the nucleus and, therefore, the orbital energy is more negative than for the $3p$ and $3d$ electrons. The same argument can be used to understand why the $3p$ orbital energy is lower than that for the $3d$ orbital. Figure 21.7 shows that the energy of a given orbital decreases strongly with the atomic number. This is a result of the increase in the attractive force between the nucleus and an electron as the charge on the nucleus increases.

It is important to realize that ε_i for a many-electron atom depends on the electron configuration and on the atomic charge because ε_i is determined in part by the average distribution of all other electrons. For example, the Hartree–Fock limiting value for ε_{1s} is -67.4 eV for neutral Li and -76.0 eV for Li$^+$ for which the 2s electron has been removed.

A further useful result from Hartree–Fock calculations are values for the effective nuclear charge, ζ. The effective nuclear charge takes into account that an electron farther from the nucleus experiences a smaller nuclear charge than that experienced by an inner

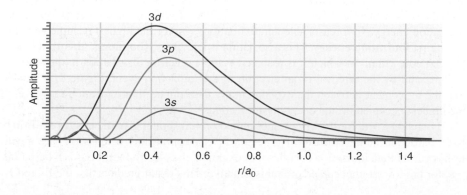

FIGURE 21.8
The contributions of the $3s$, $3p$, and $3d$ subshells to the radial distribution function of krypton obtained from Hartree–Fock calculations are shown.
[Calculated from data in E. Clementi and C. Roetti. "Roothaan-Hartree-Fock Atomic Wavefunctions: Basis Functions and Their Coefficients for Ground and Certain Excited States of Neutral and Ionized Atoms $Z \leq 54$." *Atomic Data and Nuclear Data Tables* 14 (1974): 177.]

TABLE 21.2 Effective Nuclear Charges for Selected Atoms

	H (1)							He (2)
$1s$	1.00							1.69
	Li (3)	Be (4)	B (5)	C (6)	N (7)	O (8)	F (9)	Ne (10)
$1s$	2.69	3.68	4.68	5.67	6.66	7.66	8.65	9.64
$2s$	1.28	1.91	2.58	3.22	3.85	4.49	5.13	5.76
$2p$			2.42	3.14	3.83	4.45	5.10	5.76

electron. This can be seen by referring to Figure 21.1. To the electron in question, it looks as though the nuclear charge has been reduced because of the presence of the other smeared-out electrons. This effect is particularly important for valence electrons and we say that they are *shielded* from the full nuclear charge by the core electrons closer to the nucleus. Table 21.2 shows ζ for all occupied orbitals in the first 10 atoms in the periodic table. The zeta values are obtained from a Hartree–Fock calculation using the single zeta basis set discussed earlier (See Table 21.1). The difference between the true and effective nuclear charge is a direct measure of the shielding. The effective nuclear charge is nearly equal to the nuclear charge for the $1s$ orbital but falls off quite rapidly for the outermost electron as the principal quantum number increases. Whereas electrons of smaller n value are quite effective in **shielding** electrons with greater n values from the full nuclear charge, those in the same shell are much less effective. Therefore $Z - \zeta$ increases in moving across the periodic table. However, as Example Problem 21.3 shows, some subtle effects are involved.

EXAMPLE PROBLEM 21.3

The effective nuclear charge seen by a $2s$ electron in Li is 1.28. We might expect this number to be 1.0 rather than 1.28. Why is ζ larger than 1? Similarly, explain the effective nuclear charge seen by a $2s$ electron in carbon.

Solution

The effective nuclear charge seen by a $2s$ electron in Li will be only 1.0 if all the charge associated with the $1s$ electrons is located between the nucleus and the $2s$ shell. As Figure 20.10 shows, a significant fraction of the charge is located farther from the nucleus than the $2s$ shell, and some of the charge is quite close to the nucleus. Therefore, the effective nuclear charge seen by the $2s$ electrons is reduced by a number smaller than 2. On the basis of the argument presented for Li, we expect the shielding by the $1s$ electrons in carbon to be incomplete and we might expect the effective nuclear charge felt by the $2s$ electrons in carbon to be more than 4. However, carbon has four electrons in the $n = 2$ shell, and although shielding by electrons in the same shell is less effective than shielding by electrons in inner shells, the total effect of all four $n = 2$ electrons reduces the effective nuclear charge felt by the $2s$ electrons to 3.22.

We now turn our attention to the orbital energies ε_i. What observables can be associated with the orbital energies? The most meaningful link of ε_i to physical properties is to the ionization energy. To a reasonable approximation, $-\varepsilon_i$ for the highest occupied orbital is the first **ionization energy.** This association is known as **Koopmans' theorem** in the "frozen core" limit, in which it is assumed that the electron distribution in the atom is not affected by the removal of an electron in the ionization event. Figure 21.9 shows that the agreement between the experimentally determined first ionization energy and the highest occupied orbital energy is quite good.

By analogy, $-\varepsilon_i$ for the lowest unoccupied orbital should give the **electron affinity** for a particular atom. However, Hartree–Fock electron affinity calculations are much less accurate than ionization energies. For example, the electron affinity for F based on $-\varepsilon_i$ for the lowest unoccupied orbital is negative. This result predicts that the F^- ion is

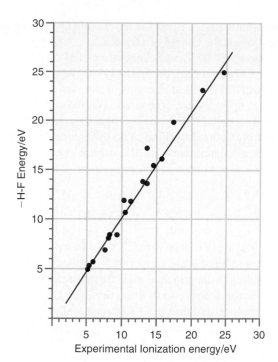

FIGURE 21.9
The negative of the highest occupied Hartree–Fock orbital is graphed against experimentally determined first ionization energies of the first 18 elements. If the two values were identical, all points would lie on the red line.

less stable than the neutral F atom, contrary to experiment. A better estimate of the electron affinity of F is obtained by comparing the total energies of F and F^-. This gives a value for the electron affinity of 0.013 eV, which is still much smaller than the experimental value of 3.34 eV. More accurate calculations, including electron correlation as discussed in Chapter 26, are necessary to obtain accurate results for the ionization energy and electron affinity of atoms.

The electron configuration of most atoms can be obtained by using Figure 21.10, which shows the order in which the atomic orbitals are generally filled based on the orbital energy sequence of Figure 21.7. Filling orbitals in this sequential order is known as the **Aufbau principle,** and it is often asserted that the relative order of orbital energies explains the electron configurations of the atoms in the periodic table. However, this assertion is not always true.

To illustrate this point, consider the known configurations of the first transition series shown in Table 21.3. Figure 21.7 shows that the $4s$ orbital is lower in energy than the $3d$ orbital for K and Ca but that the order is reversed for higher atomic numbers. Is the order in which the s and d subshells are filled in the 4th period explained by the relative energy of the orbitals? If this were the case, the configuration $[Ar]4s^03d^n$ with $n = 3, \ldots, 10$ would be predicted for the sequence scandium-nickel where [Ar] is an abbreviation for the configuration of Ar. However, with the exception of Cr and Cu, the experimentally determined configurations are given by $[Ar]4s^23d^n$, with $n = 1, \ldots, 10$

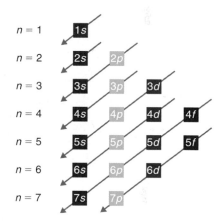

FIGURE 21.10
The order in which orbitals in many-electron atoms are filled for most atoms is described by the gray lines, starting from the top of the figure. Twelve of the forty transition elements show departures from this order.

TABLE 21.3 Configurations for Fourth Row Atoms

Nuclear Charge	Element	Electron Configuration	Nuclear Charge	Element	Electron Configuration
19	K	$[Ar]4s^1$	25	Mn	$[Ar]4s^23d^5$
20	Ca	$[Ar]4s^2$	26	Fe	$[Ar]4s^23d^6$
21	Sc	$[Ar]4s^23d^1$	27	Co	$[Ar]4s^23d^7$
22	Ti	$[Ar]4s^23d^2$	28	Ni	$[Ar]4s^23d^8$
23	V	$[Ar]4s^23d^3$	29	Cu	$[Ar]4s^13d^{10}$
24	Cr	$[Ar]4s^13d^5$	30	Zn	$[Ar]4s^23d^{10}$

for the sequence scandium-zinc. Cr and Cu have a single $4s$ electron because a half-filled or filled d shell lowers the energy of an atom.

As has been shown by L. G. Vanquickenbourne et al. "Transition Metals and the Aufbau Principle." *J. Chemical Education* 71 (1994): 469–471, the observed configurations can be explained if the total energies of the various possible configurations rather than the orbital energies are compared. We show that it is favorable in the neutral atom to fill the s orbital before the d orbital by considering the energetic cost of moving a $4s$ electron to the $3d$ orbital. The difference in the total energy of the two configurations is a balance between the orbital energies and the electrostatic repulsion of the electrons involved in the promotion. ΔE for a $4s^2 3d^n \rightarrow 4s^1 3d^{n+1}$ promotion is given by

$$\Delta E(4s \rightarrow 3d) \cong (\varepsilon_{3d} - \varepsilon_{4s}) + [E_{repulsive}(3d, 3d) - E_{repulsive}(3d, 4s)] \quad \textbf{(21.20)}$$

The second term in Equation (21.20) represents the difference in the repulsive energies of the two configurations. What is the sign of the second term? Figure 21.6 shows that the distance corresponding to the principal maxima in the radial probability distribution for a typical many-electron atom follows the order $3s > 3p > 3d$. We conclude that the d electrons are more localized than the s electrons, and therefore the repulsive energies follow the order $E_{repulsive}(3d, 3d) > E_{repulsive}(3d, 4s) > E_{repulsive}(4s, 4s)$. Therefore, the sign of the second term in Equation (21.20) is positive. For this transition metal series, the magnitude of the repulsive term is greater than the magnitude of the difference in the orbital energies. Therefore, even though $(\varepsilon_{3d} - \varepsilon_{4s}) < 0$ for scandium, the promotion $4s^2 3d^1 \rightarrow 4s^0 3d^3$ does not occur because $(\varepsilon_{3d} - \varepsilon_{4s}) + [E_{repulsive}(3d, 3d) - E_{repulsive}(3d, 4s)] > 0$ and is larger than $|(\varepsilon_{3d} - \varepsilon_{4s})|$. The energy lowering from promotion to the lower orbital energy is more than offset by the energy increase resulting from electron repulsion. Therefore, Sc has the configuration $[Ar]4s^2 3d^1$ rather than $[Ar]4s^0 3d^3$.

These calculations also explain the seemingly anomalous configurations for the doubly charged positive ions in the sequence scandium-zinc, which are $[Ar]4s^0 3d^n$ with $n = 1, \dots, 21$. The removal of two electrons significantly increases the effective nuclear charge felt by the remaining electrons. As a result, both ε_{4s} and ε_{3d} are lowered substantially, but ε_{3d} is lowered more. Therefore, $\varepsilon_{3d} - \varepsilon_{4s}$ becomes more negative. For the doubly charged ions, the magnitude of the repulsive term is less than the magnitude of the difference in the orbital energies. As a consequence, the doubly ionized configurations are those that would be predicted by filling the lower lying $3d$ orbital before the $4d$ orbital.

Recall that Hartree–Fock calculations neglect electron correlation. Therefore, the total energy is larger than the true energy by an amount called the **correlation energy.** For example, the correlation energy for He is 110 kJ mol^{-1}. This amount increases somewhat faster than the number of electrons in the atom. Although the correlation energy is a small percentage of the total energy of the atom and decreases with the atomic number (1.4% for He and 0.1% for K), it presents a problem in the application of Hartree–Fock calculations to chemical reactions for the following reason. In chemical reactions, we are not interested in the total energies of the reactants and products but rather in ΔG_R and ΔH_R. These changes are on the order of 100 kJ mol^{-1} so that errors in quantum chemical calculations resulting from the neglect of the electron correlation can lead to significant errors in thermodynamic calculations. However, the neglect of correlation is often less serious than might be expected. The resulting error in the total energy is often similar for the reactants and products if the number of unpaired electrons is the same for reactants and products. For such reactions, the neglect of electron correlation largely cancels in thermodynamic calculations. Additionally, the coordinated work of many quantum chemists over decades has led to computational methods that go beyond Hartree–Fock by including electron correlation. These advances make it possible to calculate thermodynamic functions and activation energies for many reactions for which it would be very difficult to obtain experimental data. These computational methods will be discussed in Chapter 26.

21.6 Understanding Trends in the Periodic Table from Hartree–Fock Calculations

We briefly summarize the main results of Hartree–Fock calculations for atoms:

- The orbital energy depends on both n and l. Within a shell of principal quantum number n, $\varepsilon_{ns} < \varepsilon_{np} < \varepsilon_{nd} < \dots$.

- Electrons in a many-electron atom are shielded from the full nuclear charge by other electrons. Shielding can be modeled in terms of an effective nuclear charge. Core electrons are more effective in shielding outer electrons than electrons in the same shell.

- The ground-state configuration for an atom results from a balance between orbital energies and electron–electron repulsion.

In addition to the orbital energies, two parameters that can be calculated using the Hartree–Fock method are very useful in understanding chemical trends in the periodic table. They are the atomic radius and the electronegativity. Values for atomic radii are obtained by calculating the radius of the sphere that contains ~90% of the electron charge. This radius is determined by the effective charge felt by valence shell electrons.

The degree to which atoms accept or donate electrons to other atoms in a reaction is closely related to the first ionization energy and the electron affinity, which we associate with the HOMO and LUMO orbitals. For example, the energy of these orbitals allows us to predict whether the ionic NaCl species is better described by Na^+Cl^- or Na^-Cl^+. Formation of Na^+ and Cl^- ions at infinite separation requires

$$\Delta E = E_{ionization}^{Na} - E_{electron\ affinity}^{Cl} = 5.14\ \text{eV} - 3.61\ \text{eV} = 1.53\ \text{eV} \quad \textbf{(21.21)}$$

Formation of oppositely charged ions requires

$$\Delta E = E_{ionization}^{Cl} - E_{electron\ affinity}^{Na} = 12.97\ \text{eV} - 0.55\ \text{eV} = 12.42\ \text{eV} \quad \textbf{(21.22)}$$

In each case, additional energy is gained by bringing the ions together. Clearly the formation of Na^+Cl^- is favored over Na^-Cl^+. The concept of **electronegativity,** which is given the symbol χ, quantifies this tendency of atoms to either accept or donate electrons to another atom in a chemical bond. Because the noble gases in group VIII do not form chemical bonds (with very few exceptions), they are not generally assigned values of χ.

Several definitions of electronegativity (which has no units) exist, but all lead to similar results when scaled to the same numerical range. For instance, χ as defined by Mulliken is given by

$$\chi = 0.187(IE + EA) + 0.17 \quad \textbf{(21.23)}$$

where IE is the first ionization energy and EA is the electron affinity. It is basically the average of the first ionization energy and the electron affinity with the parameters 0.187 and 0.17 chosen to optimize the correlation with the earlier electronegativity scale of Pauling, which is based on bond energies. The Mulliken definition of χ can be understood using Figure 21.11.

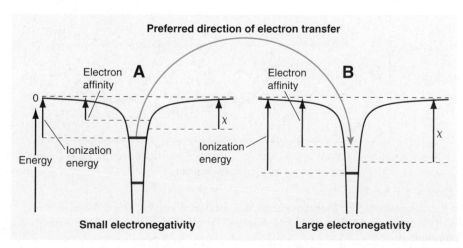

FIGURE 21.11
The energy of the molecule AB is lowered if electron charge is transferred from A to B rather than from B to A.

Assume that an atom with a small ionization energy and electron affinity (A) forms a bond with an atom that has a larger ionization energy and electron affinity (B). Partial charge transfer from A to B lowers the energy of the system and is therefore favored over the reverse process, which increases the energy of the system. Chemical bonds between atoms with large differences in χ have a strong ionic character because significant electron transfer occurs. Chemical bonds between atoms that have similar χ values are largely covalent, because the driving force for electron transfer is small, and valence electrons are shared nearly equally by the atoms.

Figure 21.12 compares values for the atomic radius, first ionization energy, and χ as a function of atomic number up to $Z = 55$. This range spans one period in which only the $1s$ orbital is filled, two short periods in which only s and p orbitals are filled, and two longer periods in which d orbitals are also filled. Beginning with the covalent radius, we see the trends predicted from calculated ζ values for the valence electrons that increase in going across a period and down a group as shown for the main group elements in Figure 21.13.

The radii decrease continuously in going across a period but increase abruptly as n increases by one in moving to the next period. Moving down a group of the periodic table, the radius increases with n because ζ increases more slowly with the nuclear charge than in moving across a period. Small radii are coupled with large ζ, and this combination leads to a large ionization energy. Therefore, changes in the ionization energy follow the opposite trend to that for the atomic radii. The ionization energy falls in moving down a column, because the atomic radius increases more rapidly than ζ increases. The electronegativity follows the same pattern as the ionization energy because in general the ionization energy is larger than the electron affinity.

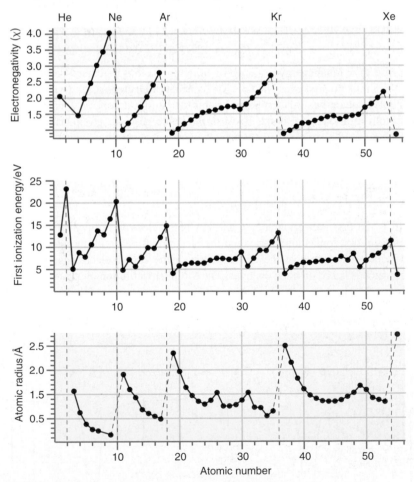

FIGURE 21.12

The electronegativity, first ionization energy, and covalent atomic radius are plotted as a function of the atomic number for the first 55 elements. Dashed vertical lines mark the completion of each period.

H 1s 1							He 1s 1.69
Li 2s 1.28	Be 2s 1.91	B 2s 2.58 2p 2.42	C 2s 3.22 2p 3.14	N 2s 3.85 2p 3.83	O 2s 4.49 2p 4.45	F 2s 5.13 2p 5.10	Ne 2s 5.76 2p 5.76
Na 3s 2.51	Mg 3s 3.31	Al 3s 4.12 3p 4.07	Si 3s 4.90 3p 4.29	P 3s 5.64 3p 4.89	S 3s 6.37 3p 5.48	Cl 3s 7.07 3p 6.12	Ar 3s 7.76 3p 6.76
K 4s 3.50	Ca 4s 4.40	Ga 4s 7.07 4p 6.22	Ge 4s 8.04 4p 6.78	As 4s 8.94 4p 7.45	Se 4s 9.76 4p 8.29	Br 4s 10.55 4p 9.03	Kr 4s 11.32 4p 9.77
Rb 5s 4.98	Sr 5s 6.07	In 5s 9.51 5p 8.47	Sn 5s 10.63 5p 9.10	Sb 5s 10.61 5p 9.99	Te 5s 12.54 5p 10.81	I 5s 13.40 5p 11.61	Xe 5s 14.22 5p 12.42

FIGURE 21.13
Effective nuclear charges are shown for valence shell electrons of main group elements in the first five periods in the periodic table.

Vocabulary

antisymmetric wave function

Aufbau principle

basis functions

configuration

correlation energy

effective nuclear charge

electron affinity

electron correlation

electron–electron repulsion

electron spin

electronegativity

Hartree–Fock self-consistent field method

indistinguishability

ionization energy

Koopmans' theorem

orbital

orbital approximation

orbital energy

Pauli exclusion principle

shell

shielding

Slater determinant

subshell

symmetric wave function

trial wave function

variational method

variational theorem

Conceptual Problems

Q21.1 Why does the effective nuclear charge for the $1s$ orbital increase by 0.99 in going from oxygen to fluorine but only increases by 0.65 for the $2p$ orbital?

Q21.2 There are more electrons in the $n = 4$ shell than for the $n = 3$ shell in krypton. However, the peak in the radial distribution in Figure 21.6 is smaller for the $n = 4$ shell than for the $n = 3$ shell. Explain this fact.

Q21.3 How is the effective nuclear charge related to the size of the basis set in a Hartree–Fock calculation?

Q21.4 The angular functions, $\Theta(\theta)\Phi(\phi)$, for the one-electron Hartree–Fock orbitals are the same as for the hydrogen atom, and the radial functions and radial probability functions are similar to those for the hydrogen atom. The contour coloring is explained in the caption to figure 20.7. The following figure shows (a) a contour plot in the xy plane with the y axis being the vertical axis, (b) the radial function, and (c) the radial probability distribution for a one-electron orbital. Identify the orbital ($2s$, $4d_{xz}$, and so on).

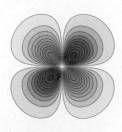

(a)

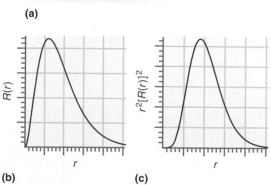

(b)   (c)

Q21.5 What is the functional dependence of the $1s$ orbital energy on Z in Figure 21.7? Check your answer against a few data points.

Q21.6 See Question Q21.4.

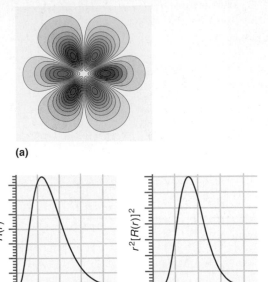

(a)

(b)   (c)

Q21.7 Explain why shielding is more effective by electrons in a shell of lower principal quantum number than by electrons having the same principal quantum number.

Q21.8 Are the elements of a basis set observable in an experiment? Explain your reasoning.

Q21.9 Show using an example that the following two formulations of the Pauli exclusion principle are equivalent:

a. Wave functions describing a many-electron system must change sign under the exchange of any two electrons.

b. No two electrons may have the same values for all four quantum numbers.

Q21.10 See Question Q21.4.

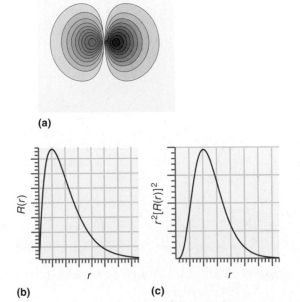

(a)

(b)   (c)

Q21.11 See Question Q21.4.

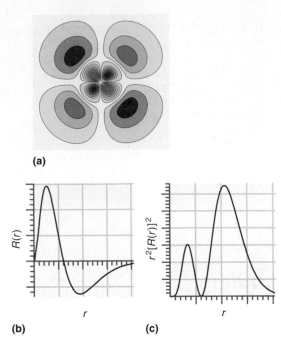

(a)

(b)   (c)

Q21.12 Why is the total energy of a many-electron atom not equal to the sum of the orbital energies for each electron?

Q21.13 See Question Q21.4.

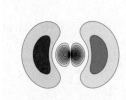

(a)

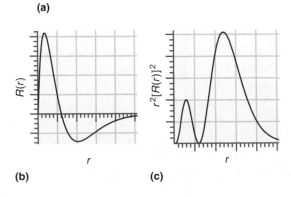

(b)   (c)

Q21.14 See Question Q21.4.

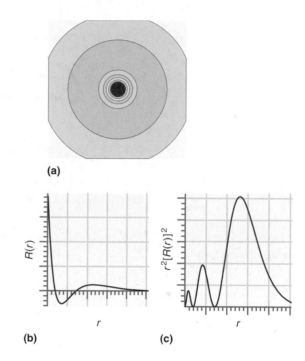

(a)

(b)

(c)

Q21.15 See Question Q21.4.

(a)

(b)

(c)

Q21.16 Show that the Slater determinant formalism automatically incorporates the Pauli exclusion principle by evaluating the He ground-state wave function of Equation (21.8), giving both electrons the same quantum numbers.

Q21.17 Is there a physical reality associated with the individual entries of a Slater determinant?

Q21.18 See Question Q21.4.

(a)

(b)

(c)

Q21.19 How can you tell if one basis set is better than another in calculating the total energy of an atom?

Q21.20 Why is the *s, p, d,* ... nomenclature derived for the H atom also valid for many-electron atoms?

Q21.21 Would the trial wave function

$$\Phi(x) = \left(\frac{x}{a} - \frac{x^3}{a^3}\right) + \alpha\left(\frac{x^5}{a^5} - \frac{1}{2}\left(\frac{x^7}{a^7}\right)\right), \quad 0 < x < a$$

have been a suitable choice for the calculations carried out in Section 21.4? Justify your answer.

Numerical Problems

Problem numbers in **red** indicate that the solution to the problem is given in the *Student's Solutions Manual*.

P21.1 Is $\psi(1, 2) = 1s(1)\alpha(1)\,1s(2)\beta(2) + 1s(2)\alpha(2)1s(1)\beta(1)$ an eigenfunction of the operator $\hat{S}_z$? If so, what is its eigenvalue M_S?

P21.2 Calculate the angles that a spin angular momentum vector for an individual electron can make with the *z* axis.

P21.3 In this problem we represent the spin eigenfunctions and operators as vectors and matrices.

a. The spin eigenfunctions are often represented as the column vectors

$$\alpha = \begin{pmatrix} 1 \\ 0 \end{pmatrix} \quad \text{and} \quad \beta = \begin{pmatrix} 0 \\ 1 \end{pmatrix}$$

Show that α and β are orthogonal using this representation.

b. If the spin angular momentum operators are represented by the matrices

$$\hat{s}_x = \frac{\hbar}{2}\begin{pmatrix} 0 & 1 \\ 1 & 0 \end{pmatrix}, \hat{s}_y = \frac{\hbar}{2}\begin{pmatrix} 0 & -i \\ i & 0 \end{pmatrix}, \hat{s}_z = \frac{\hbar}{2}\begin{pmatrix} 1 & 0 \\ 0 & -1 \end{pmatrix}$$

show that the commutation rule $[\hat{s}_x, \hat{s}_y] = i\hbar\hat{s}_z$ holds.

c. Show that

$$\hat{s}^2 = \hat{s}_x^2 + \hat{s}_y^2 + \hat{s}_z^2 = \frac{\hbar^2}{4}\begin{pmatrix} 3 & 0 \\ 0 & 3 \end{pmatrix}$$

d. Show that α and β are eigenfunctions of $\hat{s}_z$ and $\hat{s}^2$. What are the eigenvalues?

e. Show that α and β are not eigenfunctions of $\hat{s}_x$ and $\hat{s}_y$.

P21.4 In this problem you will prove that the ground-state energy for a system obtained using the variational method is greater than the true energy.

a. The approximate wave function Φ can be expanded in the true (but unknown) eigenfunctions ψ_n of the total energy operator in the form $\Phi = \sum_n c_n \psi_n$. Show that by substituting $\Phi = \sum_n c_n \psi_n$ in the equation

$$E = \frac{\displaystyle\int \Phi^* \hat{H} \Phi \, d\tau}{\displaystyle\int \Phi^* \Phi \, d\tau}$$

you obtain the result

$$E = \frac{\displaystyle\sum_n \sum_m \int (c_n^* \psi_n^*)\hat{H}(c_m \psi_m)\, d\tau}{\displaystyle\sum_n \sum_m \int (c_n^* \psi_n^*)(c_m \psi_m)\, d\tau}$$

b. Because the ψ_n are eigenfunctions of $\hat{H}$, they are orthogonal and $\hat{H}\psi_n = E_n\psi_n$. Show that this information allows us to simplify the expression for E from part (a) to

$$E = \frac{\displaystyle\sum_m E_m c_m^* c_m}{\displaystyle\sum_m c_m^* c_m}$$

c. Arrange the terms in the summation such that the first energy is the true ground-state energy E_0 and the energy increases with the summation index m. Why can you conclude that $E - E_0 \geq 0$?

P21.5 In this problem you will show that the charge density of the filled $n = 2, l = 1$ subshell is spherically symmetrical and that therefore $\mathbf{L} = 0$. The angular distribution of the electron charge is simply the sum of the squares of the magnitude of the angular part of the wave functions for $l = 1$ and $m_l = -1, 0$, and 1.

a. Given that the angular part of these wave functions is

$$Y_1^0(\theta, \phi) = \left(\frac{3}{4\pi}\right)^{1/2}\cos\theta$$

$$Y_1^1(\theta, \phi) = \left(\frac{3}{8\pi}\right)^{1/2}\sin\theta\, e^{i\phi}$$

$$Y_1^{-1}(\theta, \phi) = \left(\frac{3}{8\pi}\right)^{1/2}\sin\theta\, e^{-i\phi}$$

write an expression for $|Y_1^0(\theta, \phi)|^2 + |Y_1^1(\theta, \phi)|^2 + |Y_1^{-1}(\theta, \phi)|^2$.

b. Show that $|Y_1^0(\theta, \phi)|^2 + |Y_1^1(\theta, \phi)|^2 + |Y_1^{-1}(\theta, \phi)|^2$ does not depend on θ and ϕ.

c. Why does this result show that the charge density for the filled $n = 2, l = 1$ subshell is spherically symmetrical?

P21.6 The operator for the square of the total spin of two electrons is $\hat{S}_{total}^2 = (\hat{S}_1 + \hat{S}_2)^2 = \hat{S}_1^2 + \hat{S}_2^2 + 2(\hat{S}_{1x}\hat{S}_{2x} + \hat{S}_{1y}\hat{S}_{2y} + \hat{S}_{1z}\hat{S}_{2z})$. Given that

$$\hat{S}_x\alpha = \frac{\hbar}{2}\beta, \quad \hat{S}_y\alpha = \frac{i\hbar}{2}\beta, \quad \hat{S}_z\alpha = \frac{\hbar}{2}\alpha,$$

$$\hat{S}_x\beta = \frac{\hbar}{2}\alpha, \quad \hat{S}_y\beta = \frac{i\hbar}{2}\alpha, \quad \hat{S}_z\beta = \frac{\hbar}{2}\beta,$$

show that $\alpha(1)\,\alpha(2)$ and $\beta(1)\,\beta(2)$ are eigenfunctions of the operator $\hat{S}_{total}^2$. What is the eigenvalue in each case?

P21.7 Show that the functions $[\alpha(1)\beta(2) + \beta(1)\alpha(2)]/\sqrt{2}$ and $[\alpha(1)\beta(2) - \beta(1)\alpha(2)]/\sqrt{2}$ are eigenfunctions of $\hat{S}_{total}^2$. What is the eigenvalue in each case?

P21.8 In this problem, you will use the variational method to find the optimal $1s$ wave function for the hydrogen atom starting from the trial function $\Phi(r) = e^{-\alpha r}$ with α as the variational parameter. You will minimize

$$E(\alpha) = \frac{\displaystyle\int \Phi^* \hat{H} \Phi \, d\tau}{\displaystyle\int \Phi^* \Phi \, d\tau}$$

with respect to α.

a. Show that

$$\hat{H}\Phi = -\frac{\hbar^2}{2m_e}\frac{1}{r^2}\frac{\partial}{\partial r}\left(r^2\frac{\partial\Phi(r)}{\partial r}\right) - \frac{e^2}{4\pi\varepsilon_0 r}\Phi(r)$$

$$= \frac{\alpha\hbar^2}{2m_e r^2}(2r - \alpha r^2)e^{-\alpha r} - \frac{e^2}{4\pi\varepsilon_0 r}e^{-\alpha r}$$

b. Obtain the result $\int \Phi^* \hat{H}\Phi\, d\tau = 4\pi\int_0^\infty r^2\Phi^*\hat{H}\Phi\, dr = \pi\hbar^2/(2m_e\alpha) - e^2/(4\varepsilon_0\alpha^2)$ using the standard integrals in the Math Supplement.

c. Show that $\int \Phi^*\Phi\, d\tau = 4\pi\int_0^\infty r^2\Phi^*\Phi\, dr = \pi/\alpha^3$ using the standard integrals in the Math Supplement.

d. You now have the result $E(\alpha) = \hbar^2\alpha^2/(2m_e) - e^2\alpha/(4\pi\varepsilon_0)$. Minimize this function with respect to α and obtain the optimal value of α.

e. Is $E(\alpha_{optimal})$ equal to or greater than the true energy? Why?

P21.9 You have commissioned a measurement of the second ionization energy from two independent research teams. You find that they do not agree and decide to plot the data together with known values of the first ionization energy. The results are shown here:

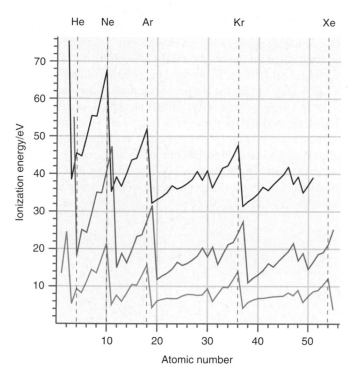

The lowest curve is for the first ionization energy and the upper two curves are the results for the second ionization energy from the two research teams. The uppermost curve has been shifted vertically to avoid an overlap with the other new data set. On the basis of your knowledge of the periodic table, you suddenly know which of the two sets of data is correct and the error that one of the teams of researchers made. Which data set is correct? Explain your reasoning.

P21.10 Classify the following functions as symmetric, antisymmetric, or neither in the exchange of electrons 1 and 2:

a. $[1s(1)2s(2) + 2s(1)1s(2)] \times [\alpha(1)\beta(2) - \beta(1)\alpha(2)]$

b. $[1s(1)2s(2) + 2s(1)1s(2)]\alpha(1)\alpha(2)$

c. $[1s(1)2s(2) + 2s(1)1s(2)][\alpha(1)\beta(2) + \beta(1)\alpha(2)]$

d. $[1s(1)2s(2) - 2s(1)1s(2)][\alpha(1)\beta(2) + \beta(1)\alpha(2)]$

e. $[1s(1)2s(2) + 2s(1)1s(2)] \times [\alpha(1)\beta(2) - \beta(1)\alpha(2) + \alpha(1)\alpha(2)]$

P21.11 Write the Slater determinant for the ground-state configuration of Be.

P21.12 The exact energy of a ground state He atom is -79.01 eV. Calculate the correlation energy and the ratio of the correlation energy to the total energy for He using the results in Table 21.1.

P21.13 The ground state wave function of Li^{2+} is $\pi^{-1/2}(Z/a_0)e^{-Zr/a_0}$ where Z is the nuclear charge. Calculate the expectation value of the potential energy for Li^{2+}.

P21.14 Calculate the position of the maximum in the radial distribution function for Li^{2+} in its ground state using the wave function in P21.13.

P21.15 – P21.20 refer to the first ionization energies and electron affinities of the first 11 elements (units of eV) shown in the following table.

Element	H	He	Li	Be	B	C	N	O	F	Ne	Na
First Ionization Energy (eV)	13.6	24.6	5.4	9.3	8.3	11.3	14.5	13.6	17.4	21.6	5.1
Electron Affinity (eV)	0.8	<0	0.6	<0	0.3	1.3	−0.1	1.5	3.4	<0	0.5

P21.15 Why is the magnitude of the electron affinity for a given element smaller than the magnitude of the first ionization energy?

P21.16 The electron affinities of He, Be, and Ne are negative, meaning that the negative ion is less stable than the neutral atom. Give an explanation of why this is so for these three elements.

P21.17 Are the effective nuclear charges listed in Figure 21.13 helpful in explaining the trend in the first ionization energy with increasing atomic number? Explain your answer.

P21.18 Are the effective nuclear charges listed in Figure 21.13 helpful in explaining the trend in the electron affinity with increasing atomic number? Explain your answer.

P21.19 Explain why the electron affinity of N is negative.

P21.20 Explain why the first ionization energy and electron affinity for F are larger than for O.

Computational Problems

More detailed instructions on carrying out these calculations using Spartan Physical Chemistry are found on the book website at *www.masteringchemistry.com*. Gaussian basis sets are discussed in Chapter 26.

C21.1 Calculate the total energy and $1s$ orbital energy for Ne using the Hartree–Fock method and the (a) 3-21G, (b) 6-31G*, and (c) 6-311+G** basis sets. Note the number of basis functions used in the calculations. Calculate the relative error of your result compared with the Hartree–Fock limit of -128.854705 hartree for each basis set. Rank the basis sets in terms of their approach to the Hartree–Fock limit for the total energy.

C21.2   Calculate the total energy and $4s$ orbital energy for K using the Hartree–Fock method and the (a) 3-21G and (b) 6-31G* basis sets. Note the number of basis functions used in the calculations. Calculate the percentage deviation from the Hartree–Fock limits, which are -16245.7 eV for the total energy and -3.996 eV for the $4s$ orbital energy. Rank the basis sets in terms of their approach to the Hartree–Fock limit for the total energy. What percentage error in the Hartree–Fock limit to the total energy corresponds to a typical reaction enthalpy change of 100. $kJ\ mol^{-1}$?

C21.3   Calculate the ionization energy for (a) Li, (b) F, (c) S, (d) Cl, and (e) Ne using the Hartree–Fock method and the 6-311+G** basis set. Carry out the calculation in two different ways: (a) Use Koopmans' theorem and (b) compare the total energy of the neutral and singly ionized atom. Compare your answers with literature values.

C21.4   Calculate the electron affinity for (a) Li, (b) F, (c) S, and (d) Cl using the Hartree–Fock method and the 6-311+G** basis set by comparing the total energy of the neutral and singly ionized atom. Compare your answers with literature values.

C21.5   Using your results from C21.3 and C21.4, calculate the Mulliken electronegativity for (a) Li, (b) F, (c) S, and (d) Cl. Compare your results with literature values.

C21.6   To assess the accuracy of the Hartree–Fock method for calculating energy changes in reactions, calculate the total energy change for the reaction $CH_3OH \rightarrow CH_3 + OH$ by calculating the difference in the total energy of reactants and products (ΔU) using the Hartree–Fock method and the 6-31G* basis set. Compare your result with a calculation using the B3LYP method and the same basis set and with the experimental value of 410. $kJ\ mol^{-1}$. As discussed in Chapter 26, the B3LYP method takes electron correlation into account. What percentage error in the Hartree–Fock total energy for CH_3OH would account for the difference between the calculated and experimental value of (ΔU)?

Quantum States for Many-Electron Atoms and Atomic Spectroscopy

Having more than one electron in an atom raises the issues of the indistinguishability of electrons, the electron spin, and the interaction between orbital and spin magnetic moments. Taking these issues into consideration leads to a new set of quantum numbers for the states of many-electron atoms and the grouping of these states into levels and terms. Atomic spectroscopies give information on the discrete energy levels of an atom and provide the basis for understanding the coupling of the individual spin and orbital angular momentum vectors in a many-electron atom. Because the discrete energy levels for atoms differ, atomic spectroscopies give information on the identity and concentration of atoms in a sample. For this reason, atomic spectroscopies are widely used in analytical chemistry. The discrete energy spectra of atoms and the difference in the rates of transition between quantum states can be used to construct lasers that provide an intense and coherent source of monochromatic radiation. Atomic spectroscopies can also provide elemental identification specific to the first few atomic layers of a solid. The reactions of electronically excited atoms can differ dramatically from their ground-state counterparts, as evidenced by reactions in Earth's atmosphere.

22.1 Good Quantum Numbers, Terms, Levels, and States

How are quantum numbers assigned to many-electron atoms? The quantum numbers n, l, m_l, and m_s that were used to characterize total energy eigenfunctions for the H atom are associated with the eigenvalues of the operators $\hat{H}$, $\hat{l}^2$, $\hat{l}_z$, and $\hat{s}_z$. It can be shown

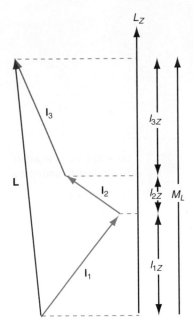

FIGURE 22.1
The sum of three classical angular momentum vectors is depicted. Whereas it is necessary to know the direction of each vector to calculate **L,** this is not necessary to calculate M_L. As discussed in Section 18.8, each angular momentum vector would need to be represented by a cone to be consistent with the commutation relations among $\hat{L}_x$, $\hat{L}_y$, and $\hat{L}_z$.

that the eigenvalues of a given operator are independent of time only if the operator commutes with $\hat{H}$. The H atom quantum numbers are **good quantum numbers** because the set of operators $\hat{l}^2$, $\hat{l}_z$, $\hat{s}^2$, and $\hat{s}_z$ commutes with the total energy operator $\hat{H}$. Operators that generate good quantum numbers are of particular interest to us in obtaining the values of time-independent observables for atoms and molecules.

However, n, l, m_l, and m_s are not good quantum numbers for any many-electron atom or ion. Therefore, another set of quantum numbers whose corresponding operators do commute with $\hat{H}$ must be found. Our primary focus is on a model that adequately describes atoms with $Z < 40$, and we extend this model to atoms for which $Z > 40$ in Section 22.3, where the reason for this restriction on the value of the atomic number is explained. Good quantum numbers are generated by forming vector sums of the electron orbital and spin angular momenta separately, **L** and **S,** which have the z components M_L and M_S, respectively. Only electrons in unfilled subshells contribute to these sums:

$$\mathbf{L} = \sum_i \mathbf{l}_i, \qquad \mathbf{S} = \sum_i \mathbf{s}_i \qquad (22.1)$$

where the summation is over the electrons in unfilled subshells. As discussed in Chapter 18 for **l,** the magnitudes of **L** and **S** are $\sqrt{L(L + 1)}\hbar$ and $\sqrt{S(S + 1)}\hbar$ respectively.

Figure 22.1 illustrates vector addition in classical physics. Note that in order to carry out the vector summations, all three vector components must be known. However, it follows from the commutation rules between $\hat{l}_x$, $\hat{l}_y$, and $\hat{l}_z$ [see Equation (18.57)] that only the length of an angular momentum vector and one of its components (which we choose to be on the z axis) can be known in quantum mechanics. This means that the summation shown in Figure 22.1 cannot actually be carried out. By contrast, it is easy to form the sum M_L because the known components l_{zi} add as scalars, $M_L = \sum_i l_{zi}$. As we will see later, it is sufficient to know M_L and M_S in order to determine the good quantum numbers L and S.

We next discuss many electron-atom operators $\hat{L}^2$, $\hat{L}_z$, $\hat{S}^2$, and $\hat{S}_z$, which are formed from one-electron operators. These operators commute with $\hat{H}$ for a many-electron atom with $Z < 40$. The capitalized form of the operators refers to the resultant for all electrons in unfilled subshells of the atom. These operators are defined by

$$\hat{S}_z = \sum_i \hat{s}_{z,i} \quad \text{and} \quad \hat{S}^2 = \left(\sum_i \hat{s}_i \right)^2$$

$$\hat{L}_z = \sum_i \hat{l}_{z,i} \quad \text{and} \quad \hat{L}^2 = \left(\sum_i \hat{l}_i \right)^2 \qquad (22.2)$$

in which the index i refers to the individual electrons in unfilled subshells. The good quantum numbers for many-electron atoms for $Z < 40$ are L, S, M_L, and M_S.

As can be inferred from Equation (22.2), the calculation for $\hat{S}^2$ is somewhat complex and is not discussed here. By contrast, $\hat{S}_z$ can be calculated easily as shown in Example Problem 22.1.

FIGURE 22.2
The top line shows the level of approximation, the second line shows the group of states that are degenerate in energy, and the bottom line shows the good quantum numbers in each level of approximation.

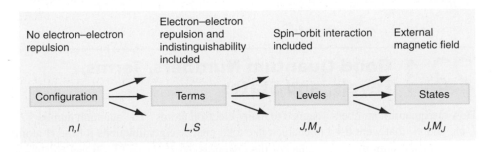

EXAMPLE PROBLEM 22.1

Is $\psi(1, 2) = 1s(1)\alpha(1)1s(2)\beta(2) - 1s(2)\alpha(2)1s(1)\beta(1)$ an eigenfunction of the operator $\hat{S}_z$? If so, what is its eigenvalue M_S?

Solution

$$\hat{S}_z = \hat{s}_z(1) + \hat{s}_z(2) \text{ where } \hat{s}_z(i) \text{ acts only on electron } i$$

$$\hat{S}_z\psi(1, 2) = (\hat{s}_z(1) + \hat{s}_z(2))\psi(1, 2)$$

$$= (\hat{s}_z(1) + \hat{s}_z(2))[1s(1)\alpha(1)1s(2)\beta(2) - 1s(2)\alpha(2)1s(1)\beta(1)]$$

$$= (\hat{s}_z(1))[1s(1)\alpha(1)1s(2)\beta(2) - 1s(2)\alpha(2)1s(1)\beta(1)]$$

$$+ (\hat{s}_z(2))[1s(1)\alpha(1)1s(2)\beta(2) - 1s(2)\alpha(2)1s(1)\beta(1)]$$

$$= \frac{\hbar}{2}[1s(1)\alpha(1)1s(2)\beta(2)] + \frac{\hbar}{2}[1s(2)\alpha(2)1s(1)\beta(1)]$$

$$- \frac{\hbar}{2}[1s(1)\alpha(1)1s(2)\beta(2)] - \frac{\hbar}{2}[1s(2)\alpha(2)1s(1)\beta(1)]$$

$$= \left(\frac{\hbar}{2} - \frac{\hbar}{2}\right)[1s(1)\alpha(1)1s(2)\beta(2) - 1s(2)\alpha(2)1s(1)\beta(1)] = 0 \times \psi(1, 2)$$

This result shows that the wave function is an eigenfunction of $\hat{S}_z$ with $M_S = 0$.

The occupied orbitals of an atom are specified in a **configuration.** For example, the electron configuration of neon is $1s^2 2s^2 2p^6$. Although a configuration is a very useful way to describe the electronic structure of atoms, it does not completely specify the quantum state of a many-electron atom because it is based on the one-electron quantum numbers n and l. Taking electron–electron repulsion into account and invoking the Pauli exclusion principle splits a configuration into terms as shown in Figure 22.2. A **term** is a group of states that has the same L and S values. Describing the states of many-electron atoms by terms is appropriate for atoms with a nuclear charge of $Z < 40$ because L and S are "good enough" quantum numbers for these atoms, meaning that the difference in energy between quantum states in a term is very small compared to the energy separation of the terms. Levels will be discussed in Section 22.3.

22.2 The Energy of a Configuration Depends on Both Orbital and Spin Angular Momentum

As proved in Supplemental Section 22.11, the energy of an atom depends on the value of the quantum number S. If an atom has at least two unpaired electrons (electrons in orbitals that are singly occupied), then the atom can have more than one value for S. Consider the excited state of He with the configuration $1s^1 2s^1$. Because both electrons have $l = 0$, $|\mathbf{L}| = 0$. We next show that there are two different values of $|\mathbf{S}|$ consistent with the $1s^1 2s^1$ configuration and formulate antisymmetric wave functions for each value of S.

Recall that an individual electron can be characterized by a spin angular momentum vector $\mathbf{s}$ of magnitude $|\mathbf{s}| = \sqrt{s(s + 1)}\hbar$ where the quantum number s can only have the single value $s = 1/2$. The vector $\mathbf{s}$ has $2s + 1 = 2$ possible orientations with the z component $s_z = \pm1/2\hbar$. We say that two spins can only be **parallel,** $\alpha(1)\alpha(2)$ and $\beta(1)\beta(2)$, or **antiparallel,** $\alpha(1)\beta(2)$ and $\beta(1)\alpha(2)$.

Figure 22.3 shows that adding the scalar components m_s for the two electrons in each of the four possible combinations gives the values $M_S = m_{s1} + m_{s2} = 0$ twice, as well as $M_S = m_{s1} + m_{s2} = +1$ and -1. Surprisingly, the possible values of S for He in the $1s^1 2s^1$ configuration can be deduced using only this information about M_S. We know that $S \geq |M_S|$ because the spin angular momentum follows the same rules

FIGURE 22.3
Possible alignment of the spins in the He configuration $1s^1 2s^1$. An upward-pointing arrow corresponds to $m_s = +1/2$ and a downward-pointing arrow corresponds to $m_s = -1/2$.

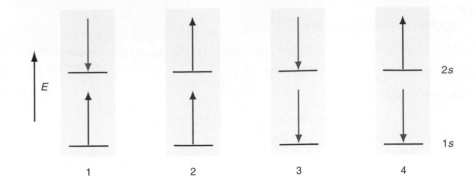

as the orbital angular momentum. Because there is no value for $M_S > 1$ among these four possible spin combinations, $M_S = \pm 1$ is only consistent with $S = 1$. Because M_S takes on all integral values between $+S$ and $-S$, the $S = 1$ group must include $M_S = 0, +1$, and -1. This accounts for three of the four values of M_S listed earlier. The one remaining combination has $M_S = 0$, which is only consistent with $S = 0$.

We have just shown that three of the four possible spin combinations are characterized by $S = 1$ with $M_S = \pm 1$ and 0 and that the fourth has $S = 0$ with $M_S = 0$. Because of the number of possible M_s values, the $S = 0$ spin combination is called a **singlet** and the $S = 1$ spin combination is called a **triplet**. Singlet and triplet states are encountered frequently in chemistry and are associated with **paired** and **unpaired electrons,** respectively.

Now that we know the S values for the four spin combinations, we can write antisymmetric wave functions for He $1s^1 2s^1$ that are eigenfunctions of $\hat{S}^2$ with $S = 0$ and $S = 1$.

$$S = 0 \quad \psi_{singlet} = \frac{1}{\sqrt{2}}[1s(1)2s(2) + 2s(1)1s(2)] \frac{1}{\sqrt{2}}[\alpha(1)\beta(2) - \beta(1)\alpha(2)]$$

$$S = 1 \quad \psi_{triplet} = \frac{1}{\sqrt{2}}[1s(1)2s(2) - 2s(1)1s(2)] \times$$

$$\begin{cases} \alpha(1)\alpha(2) \quad \text{or} \\ \beta(1)\beta(2) \quad \text{or} \\ \frac{1}{\sqrt{2}}[\alpha(1)\beta(2) + \beta(1)\alpha(2)] \end{cases} \qquad \textbf{(22.3)}$$

For the wave functions that describe the three different states for the triplet, $S = 1$, $|\mathbf{S}| = \sqrt{2}\hbar$, and (from top to bottom) $M_S = 1, -1$, and 0. The singlet consists of a single state with $S = 0$ and $M_S = 0$. Note that the antisymmetry of the total wave function is achieved by making the spatial part symmetric and the spin part antisymmetric for the singlet wave function and the other way around for the triplet wave functions.

The vector model of angular momentum can be used to depict singlet and triplet states, as shown in Figure 22.4. Although the individual spins cannot be located on the cones, their motion is coupled so that $M_S = 0$ and $S = 0$ for the singlet state. For a triplet state, there is a similar coordinated precession, but in this case, the vectors add rather than cancel and $S = 1$. Because $S = 1$, there must be three different cones corresponding to $M_S = -1, 0$, and 1.

We next make it plausible that the total energy for many-electron atoms also depends on $|\mathbf{L}|$. In the Hartree–Fock self-consistent field method, the actual positions of the electrons are approximated by their average positions. This results in a spherically symmetric charge distribution for closed subshells. As discussed in Chapter 21, this approximation greatly simplifies the calculations of orbital energies and wave functions for many-electron atoms. However, by looking at the angular part of the hydrogen atom wave functions (Figure 20.7), we can see that if l is not zero (for example, the $2p$ electrons in carbon), the electron probability distribution is not spherically symmetrical. Electrons in states characterized by $l = 1$ that have different values of m_l ($-1, 0$, or $+1$) have different orientations of the same spatial probability distributions. Two such electrons, therefore, have different repulsive interactions depending on their m_l values. By looking at Figure 18.13, we can see

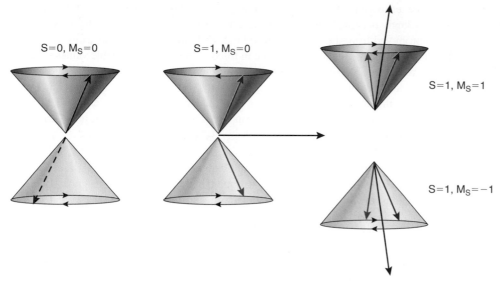

FIGURE 22.4
Vector model of the singlet and triplet states. The individual spin angular momentum vectors and their vector sum **S** (black arrow) are shown for the triplet states. For the singlet state (left image), $|\mathbf{S}| = 0$ and $M_S = 0$. The dashed arrow in the left image indicates that the vector on the yellow cone is on the opposite side of the cone from the vector on the purple cone.

that two electrons in the p_x orbital repel each other more strongly than if one of the electrons is in the p_x orbital and the other is in the p_z or p_y orbital.

Because the m_s value constrains the choices for m_l for electrons in the same orbital through the Pauli principle, the repulsive interactions between these electrons are determined by both l and s. Recall that a configuration specifies only the n and l values for the electrons and not the m_l and m_s values. For many atoms, the configuration does not completely define the quantum state. When is this the case and how does the angular momentum affect the orbital energies of the atom? As you will verify in the end-of-chapter problems, only partially filled subshells contribute to **L** and **S**. Under what conditions do the values of m_l and m_s for a given configuration lead to different spatial distributions of electrons and therefore to a different electron–electron repulsion? This occurs when there are at least two electrons in the valence shell and when there are multiple possible choices in m_l and m_s for these electrons consistent with the Pauli principle and the configuration. This is not the case for the ground states of the rare gases, the alkali metals, the alkaline earth metals, group III, and the halogens. Atoms in all of these groups have either a filled shell or subshell or only one electron or one electron fewer than the maximum number of electrons in a subshell. None of these atoms has more than one unpaired electron in its ground state and all are uniquely described by their configuration. However, the ground states for carbon, nitrogen, and oxygen are not completely described by a configuration. Several quantum states, all of which are consistent with the configuration, have significantly different values for the total energy as well as different chemical reactivities.

For atoms with $Z < 40$, the total energy is essentially independent of M_S and M_L. Therefore, a group of different quantum states that have the same values for L and S but different values of M_L and M_S is degenerate in energy. Such a group of states is called a term, and the L and S values for the term are indicated by the **term symbol** $^{(2S+1)}L$. Terms with $L = 0, 1, 2, 3, 4,...$ are given the symbols S, P, D, F, G,..., respectively. Because there are $2L + 1$ quantum states (different M_L values) for a given value of L and the $2S + 1$ states (different M_S values) for a given value of S, a term will include $(2L + 1)(2S + 1)$ quantum states, all of which have the same energy to a good approximation. This is the **degeneracy of a term.** The superscript $2S + 1$ is called the **multiplicity,** and the words *singlet* and *triplet* refer to $2S + 1 = 1$ and 3, respectively. Extending this formalism, $2S + 1 = 2$ and 4 are associated with doublets and quartets. For a filled subshell or shell,

$$M_L = \sum_i m_{li} = M_S = \sum_i m_{si} = 0 \qquad (22.4)$$

and $M_L = 0$ and $M_S = 0$ are only consistent with $L = 0$ and $S = 0$. Therefore all atoms with no unpaired electrons that have either a filled valence subshell or shell are characterized by the term ^1S. Note that the term symbols do not depend on the principal quantum number of the valence shell. Carbon, which has the $1s^2 2s^2 2p^2$ configuration, has the same set of terms as silicon, which has the $1s^2 2s^2 2p^6 3s^2 3p^2$ configuration.

How are terms generated for a given configuration? The simplest case is for a configuration with singly occupied subshells. An example is C $1s^2 2s^2 2p^1 3d^1$, in which an electron has been promoted from the $2p$ to the $3d$ orbital. Only the $2p$ and $3d$ electrons need to be considered, because the other electrons are in filled subshells. The possible values of L and S are given by the **Clebsch–Gordon series.** When applied to the two-electron case, allowed L values are given by $l_1 + l_2, l_1 + l_2 - 1, l_1 + l_2 - 2, \ldots, |l_1 - l_2|$. Using the same rule, the allowed S values are $s_1 - s_2$ and $s_1 + s_2$. For our example, $l_1 = 1$, $l_2 = 2$, and $s_1 = s_2 = 1/2$. Therefore, L can have the values 3, 2, and 1, and S can have the values 1 and 0. We conclude that the $1s^2 2s^2 2p^1 3d^1$ configuration generates ^3F, ^3D, ^3P, ^1F, ^1D, and ^1P terms. The degeneracy of these terms, $(2L + 1)(2S + 1)$, is 21, 15, 9, 7, 5, and 3, respectively, which corresponds to a total of 60 quantum states. Looking back at the configuration, the $2p$ electron can have $m_l = \pm 1$ and 0 and $m_s = \pm 1/2$. This gives six possible combinations of m_l and m_s. The $3d$ electron can have $m_l = \pm 1, \pm 2,$ and 0, and $m_s = \pm 1/2$. This gives 10 possible combinations of m_l and m_s. Because any combination for the $2p$ electron can be used with any combination of the $3d$ electron, there are a total of $6 \times 10 = 60$ combinations of m_l and m_s consistent with the $1s^2 2s^2 2p^1 3d^1$ configuration. These combinations generate the 60 states that belong to the ^3F, ^3D, ^3P, ^1F, ^1D, and ^1P terms.

The same method can be extended to more than two electrons by first calculating L and S for two electrons and adding in the remaining electrons one by one. For example, consider the L values associated with the C $1s^2 2s^1 2p^1 3p^1 3d^1$ configuration. Combining the $2s$ and $2p$ electrons gives only $L = 1$. Combining this L value with the $3p$ electron gives 2, 1, and 0. Combining these values with the $3d$ electron gives possible L values of 4, 3, 2, 1, and 0. The maximum value of S is $n/2$, where n is the number of different singly filled subshells. The minimum value of S is 0 if n is even, and $1/2$ if n is odd. For our example, the possible S values are 2, 1, and 0. Which terms are generated by these values of L and S?

Assigning terms to a configuration is more complicated if subshells contain more than one electron, because the Pauli exclusion principle must be obeyed. To illustrate such a case, consider the ground state of carbon, which has the configuration $1s^2 2s^2 2p^2$. We need only consider the $2p$ electrons. Because m_l can have any of the values $-1, 0$, or $+1$, and m_s can have the values $+1/2$ and $-1/2$ for p electrons, six combinations of the quantum numbers m_s and m_l for the first electron are possible. The second electron will have one fewer possible combination because of the Pauli principle. This appears to give a total of $6 \times 5 = 30$ combinations of quantum numbers for the two electrons. However, this assumes that the electrons are distinguishable, which overcounts the possible number of combinations by a factor of 2. Taking this into account, there are 15 possible quantum states of the carbon atom consistent with the configuration $1s^2 2s^2 2p^2$, which are shown schematically in Figure 22.5.

To determine the possible terms consistent with a p^2 configuration, it is convenient to display the information in Figure 22.5 in tabular form, as shown in Table 22.1. In setting up Table 22.1, we have relied only on the z components m_{si} and m_{li}. Using these components, $M_S = \sum_i m_{si}$ and $M_L = \sum_i m_{li}$ can be easily calculated because no vector addition is involved. To derive terms from this table, it is necessary to determine what values for L and S are consistent with the tabulated M_S and M_L values. How can this be done knowing only M_L and M_S? We first determine which values of L and S are consistent with the entries for M_L and M_S in the table given that $-S \leq M_S \leq +S$ and $-L \leq M_L \leq +L$. A good way to start is to look at the highest value for $|M_L|$ first. This requires careful bookkeeping.

m_l

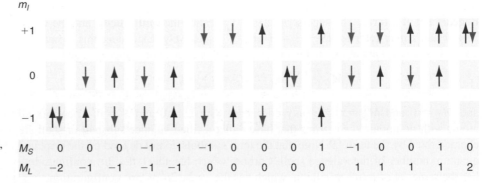

FIGURE 22.5

The different ways in which two electrons can be placed in p orbitals is shown. Upward- and downward-pointing arrows correspond to $m_s = +1/2$ and $m_s = -1/2$, respectively, and M_S and M_L are the scalar sums of the m_s and m_l, respectively.

| M_S | 0 | 0 | 0 | −1 | 1 | −1 | 0 | 0 | 0 | 1 | −1 | 0 | 0 | 1 | 0 |
| M_L | −2 | −1 | −1 | −1 | −1 | 0 | 0 | 0 | 0 | 0 | 1 | 1 | 1 | 1 | 2 |

TABLE 22.1 States and Terms for the np^2 Configuration

m_{l1}	m_{l2}	$M_L = m_{l1} + m_{l2}$	m_{s1}	m_{s2}	$M_s = m_{s1} + m_{s2}$	Term
-1	-1	-2	$1/2$	$-1/2$	0	^1D
			$-1/2$	$-1/2$	-1	^3P
0	-1	-1	$-1/2$	$1/2$	0	$^1D, {}^3P$
			$1/2$	$-1/2$	0	
			$1/2$	$1/2$	1	^3P
0	0	0	$1/2$	$-1/2$	0	
			$-1/2$	$1/2$	0	$^1D, {}^3P, {}^1S$
1	-1	0	$1/2$	$-1/2$	0	
			$-1/2$	$-1/2$	-1	^3P
			$1/2$	$1/2$	1	^3P
			$-1/2$	$-1/2$	-1	^3P
1	0	1	$-1/2$	$1/2$	0	$^1D, {}^3P$
			$1/2$	$-1/2$	0	
			$1/2$	$1/2$	1	^3P
1	1	2	$1/2$	$-1/2$	0	^1D

The top and bottom entries in the table have the largest M_L values of -2 and $+2$, respectively. They must belong to a term with $L = 2$ (a D term), because $|M_L|$ can be no greater than L. All states with M_L values of -2 and $+2$ have $M_S = 0$ because the set of quantum numbers for each electron must differ. Stated differently, because $m_{l1} = m_{l2}$, $m_{s1} \neq m_{s2}$, and therefore $M_S = 0$. We conclude that $S = 0, 2S + 1 = 1$, and the D term must be ^1D. This term has $(2S + 1)(2L + 1) = 5$ states associated with it. It includes states with $M_L = -2, -1, 0, +1$, and $+2$, all of which have $M_S = 0$. These 5 states are mentally removed from the table, which leaves us with 10 states. Of those remaining, the next highest value of $|M_L|$ is $+1$, which must belong to a P term. Because there is a combination with $M_L = 1$ and $M_S = 1$, the P term must be ^3P. This term has $(2S + 1)$ $(2L + 1) = 9$ states associated with it and by mentally removing these 9 states from the table, a single state is left with $M_L = M_S = 0$. This is a complete ^1S term. By a process of elimination, we have found that the 15 combinations of m_l and m_s consistent with the configuration $1s^2 2s^2 2p^2$ separate into ^1D, ^1S, and ^3P terms. This conclusion is true for any np^2 configuration. Because the ^1D, ^1S, and ^3P terms have 5, 1, and 9 states associated with them, a total of 15 states are associated with the terms of the $1s^2 2s^2 2p^2$ configuration just as for the classification scheme based on the individual quantum numbers n, l, m_s, and m_l.

EXAMPLE PROBLEM 22.2

What terms result from the configuration $ns^1 d^1$? How many quantum states are associated with each term?

Solution

Because the electrons are not in the same subshell, the Pauli principle does not limit the combinations of m_l and m_s. Using the guidelines formulated earlier, $S_{min} = 1/2 - 1/2 = 0$, $S_{max} = 1/2 + 1/2 = 1$, $L_{min} = 2 - 0 = 2$, and $L_{max} = 2 + 0 = 2$. Therefore, the terms that arise from the configuration $ns^1 d^1$ are ^3D and ^1D. Table 22.2 shows how these terms arise from the individual quantum numbers. In setting up the table, we have relied only on the z components of the vectors m_{si} and m_{li}. Using these components, $M_S = \sum_i m_{si}$ and $M_L = \sum_i m_{li}$ can be easily calculated because no vector addition is involved. Because each term has $(2S + 1)(2L + 1)$ states, the ^3D term consists of 15 states, and the ^1D term consists of 5 states as shown in Table 22.2.

TABLE 22.2 States and Terms for the ns^1d^1 Configuration

m_{l1}	m_{l2}	$M_L = m_{l1} + m_{l2}$	m_{s1}	m_{s2}	$M_S = m_{s1} + m_{s2}$	Term
0	−2	−2	−1/2	−1/2	−1	^3D
			−1/2	1/2	0	$^1D, {}^3D$
			1/2	−1/2	0	
			1/2	1/2	1	^3D
0	−1	−1	−1/2	−1/2	−1	^3D
			−1/2	1/2	0	$^1D, {}^3D$
			1/2	−1/2	0	
			1/2	1/2	1	^3D
0	0	0	−1/2	−1/2	−1	^3D
			−1/2	1/2	0	$^1D, {}^3D$
			1/2	−1/2	0	
			1/2	1/2	1	^3D
0	1	1	−1/2	−1/2	−1	^3D
			−1/2	1/2	0	$^1D, {}^3D$
			1/2	−1/2	0	
			1/2	1/2	1	^3D
0	2	2	−1/2	−1/2	−1	^3D
			−1/2	1/2	0	$^1D, {}^3D$
			1/2	−1/2	0	
			1/2	1/2	1	^3D

The preceding discussion has demonstrated how to generate the terms associated with a particular configuration. The same procedure can be followed for any configuration and a few examples are shown in Table 22.3 for electrons in the same shell. The numbers in parentheses behind the term symbol indicate the number of different terms of that type that belong to the configuration. A simplifying feature in generating terms is that the same results are obtained for a given number of electrons or "missing electrons" (sometimes called holes) in a subshell. For example, d^1 and d^9 configurations result in the same terms. Note that configurations with a single electron or hole in the unfilled shell or subshell give only a single term as discussed earlier. In filled shells or subshells, $M_L = M_S = 0$ because m_l and m_s take on all possible values between their maximum positive and negative values. For this reason the term symbol for s^2, p^6, and d^{10} is ^1S.

TABLE 22.3 Possible Terms for Indicated Configurations

Electron Configuration	Term Symbol
s^1	^2S
p^1, p^5	^2P
p^2, p^4	$^1S, {}^1D, {}^3P$
p^3	$^2P, {}^2D, {}^4S$
d^1, d^9	^2D
d^2, d^8	$^1S, {}^1D, {}^1G, {}^3P, {}^3F$
d^3, d^7	$^4F, {}^4P, {}^2H, {}^2G, {}^2F, {}^2D\,(2), {}^2P$
d^4, d^6	$^5D, {}^3H, {}^3G, {}^3F\,(2), {}^3D, {}^3P\,(2), {}^1I, {}^1G\,(2), {}^1F, {}^1D\,(2), {}^1S\,(2)$
d^5	$^6S, {}^4G, {}^4F, {}^4D, {}^4P, {}^2I, {}^2H, {}^2G\,(2), {}^2F\,(2), {}^2D\,(3), {}^2P, {}^2S$

EXAMPLE PROBLEM 22.3

How many states are consistent with a d^2 configuration? What L values result from this configuration?

Solution

The first electron can have any of the m_l values ± 2, ± 1, and 0, and either of the m_s values $\pm 1/2$. This gives 10 combinations. The second electron can have 9 combinations, and the total number of combinations for both electrons is $10 \times 9 = 90$. However, because the electrons are not distinguishable, we must divide this number by 2 and obtain 45 states. Using the formula $L = l_1 + l_2, l_1 + l_2 - 1, \ldots, |l_1 - l_2|$, we conclude that L values of 4, 3, 2, 1, and 0 are allowed. Therefore, this configuration gives rise to G, F, D, P, and S terms. Table 22.3 shows that the allowed terms for these L values are ^{1}S, ^{1}D, ^{1}G, ^{3}P, and ^{3}F. The degeneracy of each term is given by $(2L + 1)(2S + 1)$ and is 1, 5, 9, 9, and 21, respectively. Therefore, the d^2 configuration gives rise to 45 distinct quantum states, just as was calculated based on the possible combinations of m_l and m_s.

The relative energy of the different terms has not been discussed yet. From the examination of a large body of spectroscopic data, Friedrich Hund deduced **Hund's rules,** which state that for a given configuration the following are true:

RULE 1:

The lowest energy term is that which has the greatest spin multiplicity. For example, the ^{3}P term of an np^2 configuration is lower in energy than the ^{1}D and ^{1}S terms.

RULE 2:

For terms that have the same spin multiplicity, the term with the greatest orbital angular momentum lies lowest in energy. For example, the ^{1}D term of an np^2 configuration is lower in energy than the ^{1}S term.

Hund's rules predict that in placing electrons in one-electron orbitals, the number of unpaired electrons should be maximized. This is why Cr has the configuration $[Ar]4s^1 3d^5$ rather than $[Ar]4s^2 3d^4$. Hund's rules imply that the energetic consequences of electron–electron repulsion are greater for spin than for orbital angular momentum. As we will see in Section 22.10, atoms in quantum states described by different terms can have substantially different chemical reactivity.

Although some care is needed to establish the terms that belong to a particular configuration such as p^n or d^n, it is straightforward to predict the lowest energy term among the possible terms using the following recipe. Create boxes, one for each of the possible values of m_l. Place the electrons specified by the configuration in the boxes in such a way that $M_L = \sum_i m_{li}$ is maximized and that the number of unpaired spins is maximized. L and S for the lowest energy term are given by $L = M_{L, \text{max}}$ and $S = M_{S, \text{max}}$. This procedure is illustrated for the p^2 and d^6 configurations in Example Problem 22.4.

EXAMPLE PROBLEM 22.4

Determine the lowest energy term for the p^2 and d^6 configurations.

Solution

The placement of the electrons is as shown here:

For the p^2 configuration, $M_{L, max} = 1$ and $M_{S, max} = 1$. Therefore, the lowest energy term is ^3P. For the d^6 configuration, $M_{L, max} = 2$ and $M_{S, max} = 2$. Therefore, the lowest energy term is ^5D. It is important to realize that this procedure only provides a recipe for finding the lowest energy term. The picture used in the recipe has no basis in reality, because no association of a term with particular values of m_s and m_l can be made.

22.3 Spin-Orbit Coupling Breaks Up a Term into Levels

Up until now, we have said that all states in a term have the same energy. This is a good approximation for atoms with $Z < 40$. However, even for these atoms, the terms are split into closely spaced levels. What is this splitting due to? We know that electrons have nonzero magnetic moments if $L > 0$ and $S > 0$. The separate spin and orbital magnetic moments can interact through **spin-orbit coupling,** just as two bar magnets interact. As a result of this interaction, the total energy operator contains an extra term proportional to $\mathbf{L} \cdot \mathbf{S}$. Under these conditions, the operators $\hat{L}^2$, $\hat{L}_z$, $\hat{S}^2$, and $\hat{S}_z$ no longer commute with $\hat{H}$, but the operators $\hat{J}^2$ and $\hat{J}_z$ where $\mathbf{J}$ is the **total angular momentum** defined by

$$\mathbf{J} = \mathbf{L} + \mathbf{S} \qquad (22.5)$$

do commute with $\hat{H}$. If the coupling is sufficiently large as in atoms for which $Z > 40$, the only good quantum numbers are J and M_J, the projection of J on the z axis. The magnitude of $\mathbf{J}$ can take on all values given by $J = L + S, L + S - 1, L + S - 2, \ldots,$ $|L - S|$ and M_J can take on all values between zero and J that differ by one.

For example, the ^3P term has J values of 2, 1, and 0. All quantum states with the same J value have the same energy and belong to the same **level.** The additional quantum number J is included in the nomenclature for a level as a subscript in the form $^{(2S+1)}L_J$. In counting states, $2J + 1$ states with different M_J values are associated with each J value. This gives five states associated with $^3P_2$, three states associated with $^3P_1$, and one state associated with $^3P_0$. The total of nine states in the three levels is the same as the number of states in the ^3P term, as deduced from the formula $(2L + 1)(2S + 1)$.

Taking spin-orbit coupling into account gives Hund's third rule:

> **RULE 3:**
> The order in energy of levels in a term is given by the following:
>
> - If the unfilled subshell is exactly or more than half full, the level with the highest J value has the lowest energy.
> - If the unfilled subshell is less than half full, the level with the lowest J value has the lowest energy.

Therefore, the $^3P_0$ level has the lowest energy for an np^2 configuration. The $^3P_2$ level has the lowest energy for an np^4 configuration, which describes O.

In a magnetic field, states with the same J, but different M_J, have different energies. For atoms with $Z < 40$, this energy splitting is less than the energy separation between levels, which is in turn less than the energy separation between terms. However, all of these effects are observable in spectroscopies as shown for carbon in Figure 22.6, and many of them have practical implications in analytical chemistry. Clearly the energy levels of many-electron atoms have a higher level of complexity than those for the hydrogen atom. This complexity gives more detailed information about atoms through spectroscopic experiments that can be used to better understand the quantum mechanics of many-electron atoms.

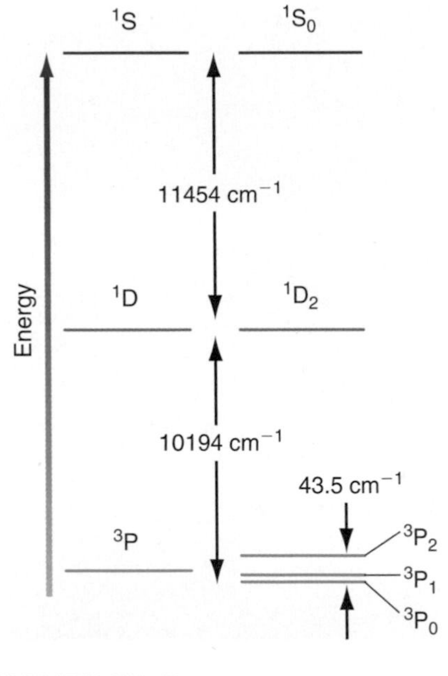

FIGURE 22.6

Assuming a spherically symmetric electron distribution, there would be a single energy for a configuration of the carbon atom. Taking the dependence of the electron repulsion on the directions of L and S into account splits the configuration into terms of different energy as shown. Taking the coupling of L and S into account leads to a further splitting of the terms into levels according to the J values as shown on the right. The separation of the levels for the ^3P term has been multiplied by a factor of 25 to make it visible.

EXAMPLE PROBLEM 22.5

What values of J are consistent with the terms ^2P and ^3D? How many states with different values of M_J correspond to each?

Solution

The quantum number J can take on all values given by $J = L + S, L + S - 1,$ $L + S - 2, \ldots, |L - S|$. For the ^2P term, $L = 1$ and $S = 1/2$. Therefore, J can have the values 3/2 and 1/2. There are $2J + 1$ values of M_J, or 4 and 2 states, respectively.

For the ^3D term, $L = 2$ and $S = 1$. Therefore, J can have the values 3, 2, and 1. There are $2J + 1$ values of M_J or 7, 5, and 3 states, respectively.

22.4 The Essentials of Atomic Spectroscopy

With an understanding of the quantum states of many-electron atoms, we turn our attention to atomic spectroscopy. All spectroscopies involve the absorption or emission of electromagnetic radiation that induces transitions between states of a quantum mechanical system. In this chapter, we discuss transitions between electronic states in atoms. Whereas the energies involved in rotational and vibrational transitions are on the order of 1 and 10 kJ mol^{-1}, respectively, photon energies associated with electronic transitions are on the order of 200 to 1000 kJ mol^{-1}. Typically, such energies are associated with visible, UV, or X-ray photons.

The information on atomic energy levels discussed in previous sections is derived from atomic spectra. The interpretation of spectra requires knowledge of the selection rules for the spectroscopy being used. **Selection rules** can be derived based on the **dipole approximation** (Section 19.4). Although transitions that are forbidden in the dipole approximation may be allowed in a higher level theory, the absorption or emission peaks are very weak. In Chapter 19, the dipole selection rule $\Delta n = \pm 1$ was derived for vibrational transitions, and it was stated without proof that the selection rule for rotational transitions in diatomic molecules is $\Delta J = \pm 1$. What selection rules apply for transitions between atomic levels? If the **L–S** coupling scheme outlined in Section 22.2 applies (atomic numbers less than ~40), the dipole selection rules for atomic transitions are $\Delta l = \pm 1$, and $\Delta L = 0, \pm 1$, and $\Delta J = 0, \pm 1$. There is an additional selection rule, $\Delta S = 0$, for the spin angular momentum. Note that the first selection rule refers to the angular momentum of an electron involved in the transition, whereas the other rules refer to the vector sums for all electrons in the atom. Keep in mind that aside from the rotational spectroscopy selection rule cited earlier, the quantum number J in this chapter refers to the total electron angular momentum and not to the rotational angular momentum.

Atomic spectroscopy is important in many practical applications such as analytical chemistry and lasers, which we discuss in this chapter. At a fundamental level, the relative energy of individual quantum states can be measured to high precision using spectroscopic techniques. An application of such high-precision measurements is the standard for the time unit of a second, which is based on a transition between states in the cesium atom that has the frequency 9.192631770×10^9 s^{-1}.

Because the energy levels of the hydrogen atom can be written as

$$E_n = -\frac{\mu e^4}{8\varepsilon_0^2 h^2 n^2} \tag{22.6}$$

where n is the principal quantum number, the frequency for absorption lines in the hydrogen spectrum is given by

$$\tilde{\nu} = \frac{\mu e^4}{8\varepsilon_0^2 h^3 c}\left(\frac{1}{n_{initial}^2} - \frac{1}{n_{final}^2}\right) = R_H\left(\frac{1}{n_{initial}^2} - \frac{1}{n_{final}^2}\right) \tag{22.7}$$

where R_H is the Rydberg constant and μ is the reduced mass of the atom. The derivation of this formula was one of the early major triumphs of quantum mechanics. The Rydberg constant is one of the most precisely known fundamental constants, and it has the value 109677.581 cm^{-1}. The series of spectral lines associated with $n_{initial} = 1$ is called the Lyman series, and the series associated with $n_{initial} = 2, 3, 4,$ and 5 are called the Balmer, Paschen, Brackett, and Pfund series, respectively, after the spectroscopists who identified them.

EXAMPLE PROBLEM 22.6

The absorption spectrum of the hydrogen atom shows lines at 82,258; 97,491; 102,823; 105,290; and 106,631 cm^{-1}. There are no lower frequency lines in the spectrum. Use graphical methods to determine $n_{initial}$ and the ionization energy of the hydrogen atom in this state.

Solution

The knowledge that frequencies for transitions follow a formula like that of Equation (22.7) allows $n_{initial}$ and the ionization energy to be determined from a limited number of transitions between bound states. The plot of $\tilde{\nu}$ versus assumed values of $1/n_{final}^2$ has a slope of $-R_H$ and an intercept with the frequency axis of $R_H/n_{initial}^2$. However, both $n_{initial}$ and n_{final} are unknown, so that in plotting the data, n_{final} values have to be assigned to the observed frequencies. For the lowest energy transition, n_{final} is $n_{initial} + 1$. We try different combinations of n_{final} and $n_{initial}$ values to see if the slope and intercept are consistent with the expected values of $-R_H$ and $R_H/n_{initial}^2$. In this case, the sequence of spectral lines is assumed to correspond to $n_{final} = 2, 3, 4, 5,$ and 6 for an assumed value of $n_{initial} = 1$; $n_{final} = 3, 4, 5, 6,$ and 7 for an assumed value of $n_{initial} = 2$; and $n_{final} = 4, 5, 6, 7,$ and 8 for an assumed value of $n_{initial} = 3$. The plots are shown in the following figure:

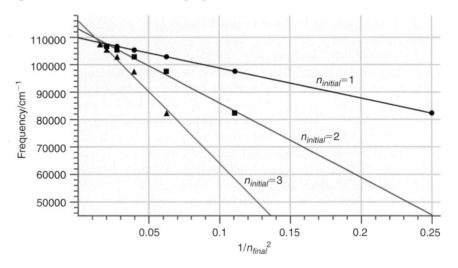

The slopes and intercepts calculated for these assumed values of $n_{initial}$ are

Assumed $n_{initial}$	Slope (cm^{-1})	Intercept (cm^{-1})
1	-1.10×10^5	1.10×10^5
2	-2.71×10^5	1.13×10^5
3	-5.23×10^5	1.16×10^5

Because the slope is $-R_H$, for only one of the three assumed values, we conclude that $n_{initial} = 1$. The ionization energy of the hydrogen atom in this state is hcR_H, corresponding to $n_{final} \longrightarrow \infty$, or 2.18×10^{-18} J. The appropriate number of significant figures for the slope and intercept is approximate in this example and must be based on an error analysis of the data.

Information from atomic spectra is generally displayed in a standard format called a **Grotrian diagram.** An example is shown in Figure 22.7 for He, for which **L–S** coupling is a good model. The figure shows the configuration information next to the energy level, and the configurations are arranged according to their energy and term symbols. The triplet and singlet states are shown in separate parts of the diagram because

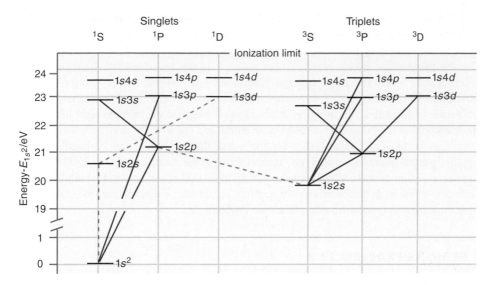

FIGURE 22.7
The ground and the first few excited states
of the He atom are shown on an energy
scale. All terms for which $l > 2$ and
$n > 4$ have been omitted to simplify the
presentation. The top horizontal line indicates the ionization energy of He. Below
this energy, all states are discrete. Above
this level, the energy spectrum is continuous. Several, but not all, allowed (solid
lines) and forbidden (dashed lines) transitions are shown. Which selection rule do
the forbidden transitions violate?

transitions between these states do not occur as a consequence of the $\Delta S = 0$ selection
rule. The ^{3}P and ^{3}D He terms are split into levels with different J values, but because the
spin-orbit interaction is so small for He, the splitting is not shown in Figure 22.7.
For example, the $^3P_0$ and $^3P_2$ levels arising from the $1s2p$ configuration differ in energy
by only 0.0006%.

22.5 Analytical Techniques Based on Atomic Spectroscopy

The absorption and emission of light that occurs in transitions between different atomic
levels provides a powerful tool for qualitative and quantitative analysis of samples of
chemical interest. For example, the concentration of lead in human blood and the presence of toxic metals in drinking water are routinely determined using **atomic emission**
and **atomic absorption spectroscopy.** Figure 22.8 illustrates how these two spectroscopic techniques are implemented. A sample, ideally in the form of very small droplets
($\sim 1-10 \ \mu$m in diameter) of a solution or suspension, is injected into the heated zone of
the spectrometer. The heated zone may take the form of a flame, an electrically heated

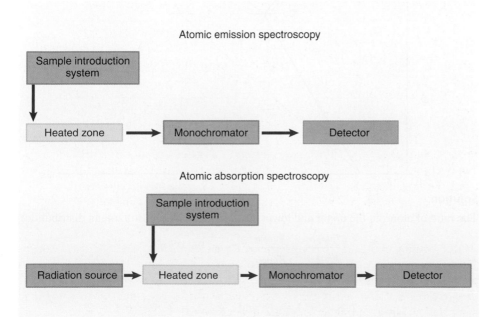

FIGURE 22.8
Schematic diagram of atomic emission
and atomic absorption spectroscopies.

graphite furnace, or a plasma arc source. The main requirement of the heated zone is that it must convert a portion of the molecules in the sample of interest into atoms in their ground and excited states.

We first discuss atomic emission spectroscopy. In this technique, the light emitted by excited-state atoms as they undergo transitions back down to the ground state is dispersed into its component wavelengths by a monochromator and the intensity of the radiation is measured as a function of wavelength. Because the emitted light intensity is proportional to the number of excited-state atoms and because the wavelengths at which emission occurs are characteristic for the atom, the technique can be used for both qualitative and quantitative analysis. Temperatures in the range of 1800 to 3500 K can be achieved in flames and carbon furnaces and up to 10,000 K can be reached in plasma arc sources. These high temperatures are required to produce sufficient excited-state atoms that emit light as demonstrated in Example Problem 22.7.

EXAMPLE PROBLEM 22.7

The $^2S_{1/2} \longrightarrow {}^2P_{3/2}$ transition in sodium has a wavelength of 589.0 nm. This is one of the lines characteristic of the sodium vapor lamps used for lighting streets, and it gives the lamps their yellow-orange color. Calculate the ratio of the number of atoms in these two states at 1500., 2500., and 3500. K. The following figure is a Grotrian diagram for Na (not to scale) in which the transition of interest is shown as a blue line.

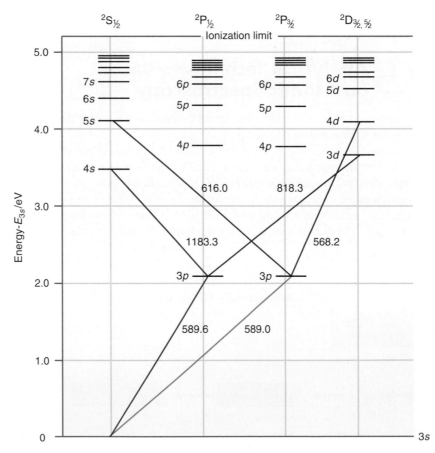

Solution

The ratio of atoms in the upper and lower levels is given by the Boltzmann distribution:

$$\frac{n_{upper}}{n_{lower}} = \frac{g_{upper}}{g_{lower}} e^{-(\varepsilon_{upper}-\varepsilon_{lower})/k_B T}$$

The degeneracies g are given by $2J + 1$, which is the number of states in each level:

$$g_{upper} = 2 \times \frac{3}{2} + 1 = 4 \quad \text{and} \quad g_{lower} = 2 \times \frac{1}{2} + 1 = 2$$

From the Boltzmann distribution,

$$\frac{n_{upper}}{n_{lower}} = \frac{g_{upper}}{g_{lower}} \exp\left(-hc/\lambda_{transition}k_B T\right)$$

$$= \frac{4}{2}\exp\left[-\frac{(6.626 \times 10^{-34}\,\text{J s})(2.998 \times 10^8\,\text{m s}^{-1})}{(589.0 \times 10^{-9}\,\text{m})(1.381 \times 10^{-23}\,\text{J K}^{-1}) \times T}\right]$$

$$= 2\exp\left(-\frac{24420}{T}\right) = 1.699 \times 10^{-7} \text{ at 1500. K}, 1.145 \times 10^{-4} \text{ at 2500. K},$$

and 1.866×10^{-3} at 3500. K

As seen in the preceding example problem, the fraction of atoms in the excited state is quite small, but it increases rapidly with temperature. The very high temperature plasma arc sources are widely used because they allow light emission from both more highly excited states and from ions to be observed. This greatly increases the sensitivity of the technique. However, because photons can be detected with very high efficiency, measurements can be obtained from systems for which n_{upper}/n_{lower} is quite small. For instance, a temperature of ~3000 K is reached in an oxygen-natural gas flame. If a small amount of NaCl is put into the flame, $n_{upper}/n_{lower} \sim 6 \times 10^{-4}$ for Na as shown in Example Problem 22.7. Even with this rather low degree of excitation, a bright yellow emission resulting from the 589.0 and 589.6 nm emission lines in the flame is clearly visible with the naked eye. The sensitivity of the technique can be greatly enhanced using photomultipliers, and spectral transitions for which $n_{upper}/n_{lower} < 10^{-10}$ are routinely used in analytical chemistry.

Atomic absorption spectroscopy differs from atomic emission spectroscopy in that light is passed through the heated zone and the absorption associated with transitions from the lower to the upper state is detected. Because this technique relies on the population of low-lying rather than highly excited atomic states, it has some advantages in sensitivity over atomic emission spectroscopy. It became a very widely used technique when researchers realized that the sensitivity would be greatly enhanced if the light source were nearly monochromatic with a wavelength centered at $\lambda_{transition}$. The advantage of this arrangement can be seen from Figure 22.9.

Only a small fraction of the broadband light that passes through the heated zone is absorbed in the transition of interest. To detect the absorption, the light needs to be dispersed with a grating and the intensity of the light must be measured as a function of frequency. Because the monochromatic source matches the transition both in frequency and in linewidth, detection is much easier. Only a simple monochromator is needed to remove background light before the light is focused on the detector. The key to the implementation of this technique was the development of hollow cathode gas discharge

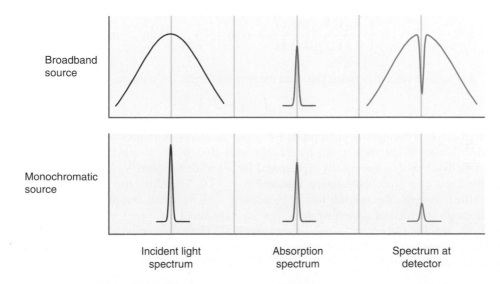

Broadband source

Monochromatic source

Incident light spectrum · Absorption spectrum · Spectrum at detector

FIGURE 22.9
The intensity of light as a function of its frequency is shown at the entrance to the heated zone and at the detector for broadband and monochromatic sources. The absorption spectrum of the atom to be detected is shown in the middle column.

lamps that emit light at the characteristic frequencies of the cathode materials. By using an array of these relatively inexpensive lamps on a single spectrometer, analyses for a number of different elements of interest can be carried out.

The sensitivity of atomic emission and absorption spectroscopy depends on the element and ranges from $10^{-4}\,\mu g/mL$ for Mg to $10^{-2}\,\mu g/mL$ for Pt. These techniques are used in a wide variety of applications, including drinking water analysis and engine wear, by detecting trace amounts of abraded metals in lubricating oil.

22.6 The Doppler Effect

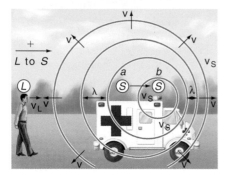

A further application of atomic spectroscopy results from the **Doppler effect.** If a source is radiating light and moving relative to an observer, the observer sees a change in the frequency of the light as shown in Figure 22.10 for sound.

The shift in frequency is given by the formula

$$\omega = \omega_0 \sqrt{(1 \pm v_z/c)/(1 \mp v_z/c)} \qquad (22.8)$$

In this formula, v_z is the velocity component in the observation direction, c is the speed of light, and ω_0 is the light frequency in the frame in which the source is stationary. The upper and lower signs refer to the object approaching and receding from the observer, respectively. Note that the frequency shift is positive for objects that are approaching (a so-called "blue shift") and negative for objects that are receding (a so-called "red shift"). The **Doppler shift** is used to measure the speed at which stars and other radiating astronomical objects are moving relative to Earth.

FIGURE 22.10
The frequency of light or sound at the position of the observer L depends on v_S, the relative velocity of the source S and v_L, the velocity of the observer L.

EXAMPLE PROBLEM 22.8

A line in the Lyman emission series for atomic hydrogen ($n_{final} = 1$), for which the wavelength is at 121.6 nm for an atom at rest, is seen for a particular quasar at 445.1 nm. Is the source approaching toward or receding from the observer? What is the magnitude of the velocity?

Solution

Because the frequency observed is less than that which would be observed for an atom at rest, the object is receding. The relative velocity is given by

$$\left(\frac{\omega}{\omega_0}\right)^2 = \left(1 - \frac{v_z}{c}\right) \Big/ \left(1 + \frac{v_z}{c}\right), \text{ or}$$

$$\frac{v_z}{c} = \frac{1 - (\omega/\omega_0)^2}{1 + (\omega/\omega_0)^2} = \frac{1 - (\lambda_0/\lambda)^2}{1 + (\lambda_0/\lambda)^2}$$

$$= \frac{1 - (121.6/445.1)^2}{1 + (121.6/445.1)^2} = 0.8611; \quad v_z = 2.581 \times 10^8\,\mathrm{m\,s^{-1}}$$

For source velocities much less than the speed of light, the nonrelativistic formula

$$\omega = \omega_0\left(1/1 \mp \frac{v_z}{c}\right) \qquad (22.9)$$

applies. This formula is appropriate for a gas of atoms or molecules for which the distribution of speeds is given by the Maxwell–Boltzmann distribution. Because all velocity directions are equally represented for a particular speed, v_z has a large range for a gas at a given temperature, centered at $v_z = 0$. Therefore, the frequency is not shifted; instead, the spectral line is broadened. This is called **Doppler broadening.** Because atomic and molecular velocities are very small compared with the speed of light, the broadening of a line of frequency ω_0 is on the order of 1 part in 10^6. This effect is not as dramatic as the shift in frequency for the quasar, but it is still of importance in determining the linewidth of a laser, as we will see in the next section.

22.7 **The Helium-Neon Laser**

In this section, we demonstrate the relevance of the basic principles discussed earlier to the functioning of a laser. We focus on the He-Ne laser. To understand this laser, the concepts of absorption, spontaneous emission, and stimulated emission introduced in Chapter 19 are used. All three processes obey the same selection rules for an atom: $\Delta l = \pm 1$ for an electron and $\Delta L = 0$ or ± 1.

Spontaneous and stimulated emission differ in an important respect. **Spontaneous emission** is a completely random process in time, and the photons that are emitted are incoherent, meaning that their phases are random. A light bulb is an **incoherent photon source.** Because all propagation directions are equally likely, the intensity of the source falls off as the square of the distance. In **stimulated emission,** the phase and direction of propagation are the same as that of the incident photon. This is referred to as coherent photon emission. A **laser** is a coherent photon source. All photons are in phase, and because they have the same propagation direction, the divergence of the beam is very small. This explains why a laser beam that is reflected from the moon still has a measurable intensity when it returns to Earth. This discussion makes it clear that a **coherent photon source** must be based on stimulated rather than spontaneous emission. However, $B_{12} = B_{21}$, as was shown in Section 19.2. Therefore, the rates of absorption and stimulated emission are equal for $N_1 = N_2$. Stimulated emission will only dominate over absorption if $N_2 > N_1$. This condition is called a **population inversion** because for equal level degeneracies, the higher energy state has the higher population. The key to making a practical laser is to create a stable population inversion. Although a population inversion is not possible under equilibrium conditions, it is possible to maintain such a distribution under steady-state conditions if the relative rates of the transitions between levels are appropriate. This is illustrated in Figure 22.11.

Figure 22.11 can be used to understand how the population inversion between the levels involved in the lasing transition is established and maintained. The lengths of the horizontal lines representing the levels are proportional to the level populations N_1 to N_4. The initial step involves creating a significant population in level 4 by transitions from level 1. This is accomplished by an external source, which for the He-Ne laser is an electrical discharge in a tube containing the gas mixture. Relaxation to level 3 can occur through spontaneous emission of a photon as indicated by the wavy arrow. Similarly, relaxation from level 2 to level 1 can also occur through spontaneous emission of a photon. If this second relaxation process is fast compared to the first, N_3 will be maintained at a higher level than N_2. In this way, a population inversion is established between levels two and three. The advantage of having the lasing transition between levels 3 and 2 rather than 2 and 1 is that N_2 can be kept low if relaxation to level 1 from level 2 is fast. It is not possible to keep N_1 at a low level because atoms in the ground state cannot decay to a lower state.

This discussion shows how a population inversion can be established. How can a continuous lasing transition based on stimulated emission be maintained? This is made possible by carrying out the process indicated in Figure 22.11 in an **optical resonator** as shown in Figure 22.12.

The He-Ne mixture is put into a glass tube with carefully aligned parallel mirrors on each end. Electrodes are inserted to maintain the electrical discharge that pumps level 4 from level 1. Light reflected back and forth in the optical cavity between the two mirrors interferes constructively only if $n\lambda = n(c/\nu) = 2d$, where d is the distance between mirrors and n is an integer. The next constructive interference occurs when $n \rightarrow n + 1$. The difference in frequency between these two modes is $\Delta \nu = c/2d$, which defines the bandwidth of the cavity. The number of modes that contribute to laser action is determined by two factors: the frequencies of the **resonator modes** and the width in frequency of the stimulated emission transition. The width of the transition is determined by Doppler broadening, which arises through the thermal motion of gas-phase Ne atoms. A schematic diagram of a He-Ne laser, including the anode, cathode, and power supply needed to maintain the electrical discharge as well as the optical resonator, is shown in Figure 22.13.

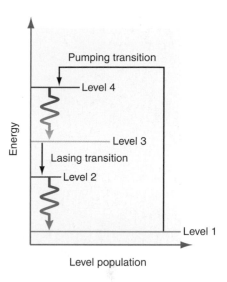

FIGURE 22.11
Schematic representation of a four-state laser. The energy is plotted vertically, and the level population is plotted horizontally.

FIGURE 22.12
Schematic representation of a He-Ne laser operated as an optical resonator. The parallel lines in the resonator represent coherent stimulated emission that is amplified by the resonator, and the red waves represent incoherent spontaneous emission events.

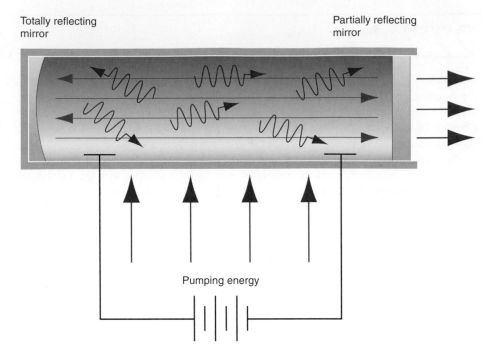

FIGURE 22.13
Schematic diagram of a He-Ne laser.

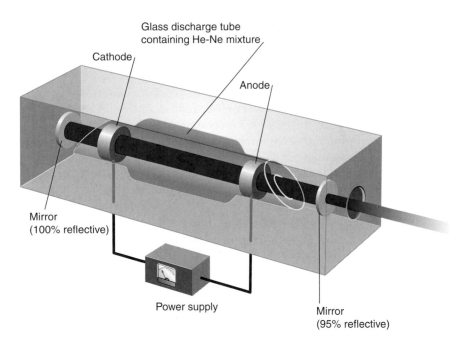

Six of the possible resonator modes are indicated in Figure 22.14. The curve labeled "Doppler linewidth" gives the relative number of atoms in the resonator as a function of the frequency at which they emit light. The product of these two functions gives the relative intensities of the stimulated emission at the different frequencies supported by the resonator. This product is shown as a function of frequency in Figure 22.14b. Because of losses in the cavity, the number of atoms in level 3 is continuously depleted. A laser transition can only be sustained if enough atoms in the cavity are in the excited state at a supported resonance. In Figure 22.14, only two resonator modes lead to a sufficient intensity to sustain the laser. The main function of the optical resonator is to decrease the $\Delta\nu$ associated with the frequency of the lasing transition to less than the Doppler limit. Example Problem 22.9 shows how the number of supported modes varies with the gas temperature.

EXAMPLE PROBLEM 22.9

As shown in Chapter 33, the distribution function that describes the probability of finding a particular value of magnitude of the velocity along one dimension v in a gas at temperature T is given by

$$f(v)dv = \sqrt{\frac{m}{2\pi kT}} \, e^{-mv^2/2k_BT} dv$$

This velocity distribution leads to the broadening of a laser line in frequency given by

$$I(\nu)d\nu = \sqrt{\frac{mc^2}{2\pi k_BT\nu_0^2}} \exp\left\{-\frac{mc^2}{2kT}\left(\frac{\nu - \nu_0}{\nu_0}\right)^2\right\} d\nu$$

The symbol c stands for the speed of light, and k is the Boltzmann constant. We next calculate the broadening of the 632.8 nm line in the He-Ne laser as a function of T.

a. Plot $I(\nu)$ for $T = 100.0, 300.0$, and 1000. K, using the mass appropriate for a Ne atom, and determine the width in frequency at half the maximum amplitude of $I(\nu)$ for each of the three temperatures.

b. Assuming that the amplification threshold is 50% of the maximum amplitude, how many modes could lead to amplification in a cavity of length 100. cm?

Solution

a. This function is of the form of a normal or Gaussian distribution given by

$$f_{\nu_0,\sigma}(\nu) \, d\nu = \frac{1}{\sigma\sqrt{2\pi}} e^{-(\nu-\nu_0)/2\sigma^2} d\nu$$

The full width at half height is 2.35 σ, or for this case, $2.35\sqrt{kT\nu_0^2/mc^2}$. This gives half widths of $7.554 \times 10^8 \text{ s}^{-1}$, $1.308 \times 10^9 \text{ s}^{-1}$, and $2.388 \times 10^9 \text{ s}^{-1}$ at temperatures of 100.0, 300.0, and 1000. K, respectively. The functions $I(\nu)$ are plotted here:

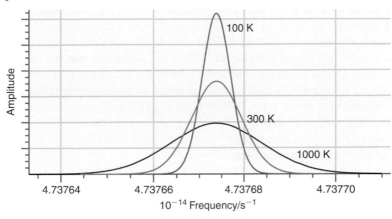

b. The frequency spacing between two modes is given by

$$\Delta\nu = \frac{c}{2d} = \frac{2.998 \times 10^8 \text{ m s}^{-1}}{2 \times 1.00 \text{ m}} = 1.50 \times 10^8 \text{ s}^{-1}$$

The width of the velocity distribution will support 5 modes at 100.0 K, 8 modes at 300.0 K, and 15 modes at 1000. K. The smaller Doppler broadening at low temperatures reduces the number of possible modes considerably.

By adding a further optical filter in the laser tube, it is possible to have only one mode enhanced by multiple reflections in the optical resonator. In addition, all light of the correct frequency but with a propagation direction that is not perpendicular to the mirrors, is not enhanced through multiple reflections. In this way, the resonator establishes a standing

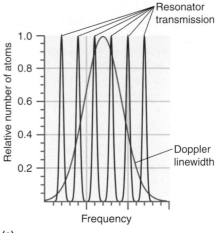

(a)

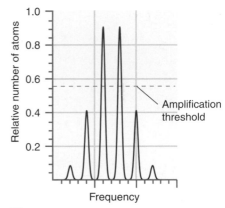

(b)

FIGURE 22.14
The linewidth of a transition in a He-Ne laser is Doppler broadened through the Maxwell–Boltzmann velocity distribution. **(a)** The resonator transmission decreases the linewidth of the lasing transition to less than the Doppler limit. **(b)** The amplification threshold further reduces the number of frequencies supported by the resonator.

FIGURE 22.15

Transitions in the He-Ne laser. The slanted solid lines in the upper right side of the figure show three possible lasing transitions.

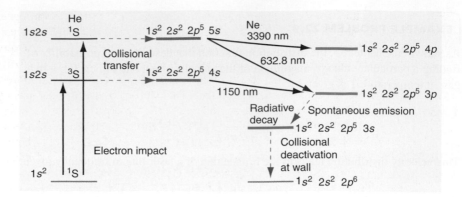

wave at the lasing frequency that has a propagation direction aligned along the laser tube axis. This standing wave causes further photons to be emitted from the lasing medium (He-Ne mixture) through stimulated emission. As discussed previously, these photons are exactly in phase with the photons that stimulate the emission and have the same propagation direction. These photons amplify the standing wave, which because of its greater intensity causes even more stimulated emission. Allowing one of the end mirrors to be partially transmitting lets some of the light escape, and the result is a coherent, well-collimated laser beam.

To this point, the laser has been discussed at a schematic level. How does this discussion relate to the atomic energy levels of He and Ne shown schematically in Figure 22.15? The electrical discharge in the laser tube produces electrons and positively charged ions. The electrons are accelerated in the electric field and can excite the He atoms from states in the ^1S term of the $1s^2$ configuration to states in the ^1S and ^3S terms of the $1s2s$ configuration. This is the **pumping transition** in the scheme shown in Figure 22.11. This transition occurs through a collision rather than through the absorption of a photon and, therefore, the normal selection rules do not apply. Because the selection rules $\Delta S = 0$ and $\Delta l = \pm 1$ prohibit transitions to the ground state, these states are long lived. The excited He atoms efficiently transfer their energy through collisions to states in the $2p^5 5s$ and $2p^5 4s$ configurations of Ne. This creates a population inversion relative to Ne states in the $2p^5 4p$ and $2p^5 3p$ configuration. These levels are involved in the **lasing transition** through stimulated emission. Spontaneous emission to states in the $2p^5 3s$ configuration and collisional deactivation at the inner surface of the optical resonator depopulate the lower state of the lasing transitions and ensure that the population inversion is maintained. The initial excitation is to excited states of He, which consist of a single term. However, the excited-state configurations of Ne give rise to several terms (^3P and ^1P for $2p^5 4s$ and $2p^5 5s$, and ^3D, ^1D, ^3P, ^1P, ^3S, and ^1S for $2p^5 3p$ and $2p^5 4p$). The manifold of these states is indicated in the figure by thicker lines, indicating a range of energies.

Note that a number of wavelengths can lead to lasing transitions. Coating the mirrors in the optical resonator ensures that they are reflective only in the range of interest. The resonator is usually configured to support the 632.8 nm transition in the visible part of the spectrum. This corresponds to the red light characteristic of He-Ne lasers.

22.8 Laser Isotope Separation

A number of ways are available for separating atoms and molecules into their different isotopes. Separation by diffusion in the gas phase is possible because the speed of molecules depends on the molecular weight M as $M^{-1/2}$. To fabricate fuel rods for nuclear reactors, uranium fuel, which contains the isotopes ^{234}U, ^{235}U, and ^{238}U must be enriched in the fissionable isotope ^{235}U. This has been done on a large scale by reacting uranium with fluorine to produce the gas-phase molecule UF_6. This gas is enriched in the ^{235}U isotope by centrifugation.

It is feasible, although not practical on an industrial scale, to create a much higher degree of enrichment by using selective laser ionization of the ^{235}U isotope. The principle is shown schematically in Figure 22.16. A tunable copper vapor laser with a very narrow linewidth is used to excite ground-state uranium atoms to an excited state involving the $7s$ electrons. As Figure 22.16b indicates, the electronic states of the different isotopes have slightly different energies, and the bandwidth of the laser is sufficiently small that only one isotope is excited. A second laser pulse is used to ionize the selectively excited isotope. The ions can be collected by electrostatic attraction to a metal electrode at an appropriate electrical potential. Because neither of the lasers produce photons of energy sufficient to ionize the uranium atoms directly, only those atoms selectively excited by the copper vapor laser by the first pulse are ionized by the second pulse.

Why do the different isotopes have slightly different atomic energy levels? The $-1/r$ Coulomb potential that attracts the electron to the nucleus is valid outside the nucleus, but the potential levels off inside the nucleus, which has a diameter of about 1.6×10^{-14} m or $3 \times 10^{-4} a_0$. The distance at which the Coulomb potential is no longer valid depends on the nuclear diameter and, therefore, on the number of neutrons in the nucleus. This effect is negligible for states with $l > 0$, because the effective potential discussed in Section 20.2 keeps these electrons away from the nucleus. Only s states, for which the wave function has its maximum amplitude at the center of the nucleus, exhibit energy-level splitting. The magnitude of the splitting depends on the nuclear diameter. The splitting for the uranium isotopes is only about 2×10^{-3} percent of the ionization energy. However, the very small bandwidth that is attainable in lasers allows selective excitation to a single level even in cases for which the energy-level spacing is very small.

22.9 Auger Electron and X-Ray Photoelectron Spectroscopies

Most of the atomic spectroscopies that we have discussed have been illustrated with gas-phase examples. Another useful application of spectroscopic methods is in the analysis of the elemental composition of surfaces. This capability is important in such fields as corrosion and heterogeneous catalysis in which a chemical reaction takes place at the interface between a solid phase and a gaseous or liquid phase. Sampling of a surface in a way that is relevant to the localization of the reaction at the surface requires that the method be sensitive only to the first few atomic layers of the solid. The two spectroscopies described in this section satisfy this requirement for reasons to be discussed next. They are also applicable in other environments such as the gas phase.

Both of these spectroscopies involve the ejection of an electron from individual atoms in a solid and the measurement of the electron energy. To avoid energy losses due to collisions of the ejected electron with gas-phase molecules, the solid sample is examined in a vacuum chamber. If the electron has sufficient energy to escape from the solid into the vacuum, it will have a characteristic energy simply related to the energy level from which it originated. To escape into the vacuum, it must travel from its point of origin to the surface of the solid. This process is analogous to a gas-phase atom traveling through a gas. The atom will travel a certain distance (that depends on the gas pressure) before it collides with another atom. In the collision, it exchanges energy and momentum with its collision partner and thereby loses memory about its previous momentum and energy. Similarly, an electron generated in the solid traveling toward the surface suffers collisions with other electrons and loses memory of the energy levels from which it originated if its path is too long. Only those atoms within one **inelastic mean free path** of the surface eject electrons into the gas phase whose energy is simply related to the atomic energy levels. Electrons emitted from other atoms simply contribute to the background signal. The mean free path for electrons depends on the energy of the electrons but is relatively material independent. It has its minimum value of about 2 atomic layers near 40 eV and increases slowly to about 10 atomic layers at 1000 eV. Electrons in this energy range that have been ejected from atoms can provide information that is highly surface sensitive.

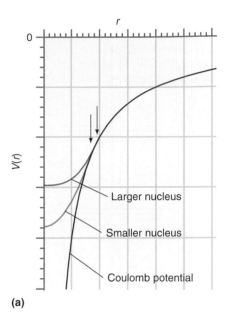

(a)

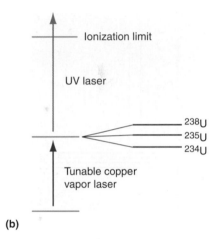

(b)

FIGURE 22.16
The principle of laser isotope excitation is shown schematically (not to scale). The electron-nucleus potential deviates from a Coulomb potential at the nuclear radius, which is indicated by the vertical arrows in part (**a**). Because this distance depends on the nuclear volume, it depends on the number of neutrons in the nucleus for a given atomic number. (**b**) This very small variation in $V(r)$ for isotopes ^{234}U through ^{238}U gives rise to an energy splitting for states involving s electrons. By means of a combined two-photon excitation and ionization, this splitting can be utilized to selectively ionize a particular isotope.

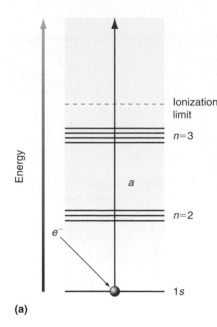

(a)

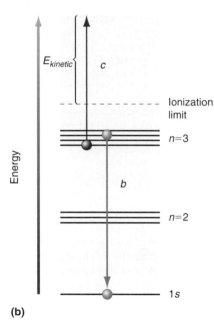

(b)

FIGURE 22.17
The principle of Auger electron spectroscopy is illustrated schematically.
(a) A core level hole is formed by energy transfer from an incident photon or electron. **(b)** The core hole is filled through relaxation from a higher level, and a third electron is emitted to conserve energy (c). The energy of the emitted electron can be measured and is characteristic of the particular element.

EXAMPLE PROBLEM 22.10

Upon impingement of X rays from a laboratory source, titanium atoms near the surface of bulk TiO_2 emit electrons into a vacuum with energy of 790 eV. The finite mean free path of these electrons leads to an attenuation of the signal for Ti atoms beneath the surface according to $[I(d)]/[I(0)] = e^{-d/\lambda}$. In this equation, d is the distance to the surface and λ is the mean free path. If λ is 2.0 nm, what is the sensitivity of Ti atoms 10.0 nm below the surface relative to those at the surface?

Solution

Substituting in the equation $[I(d)]/[I(0)] = e^{-d/\lambda}$, we obtain
$\dfrac{I(10.0)}{I(0)} = e^{-10.0/2.0} = 6.7 \times 10^{-3}$. This result illustrates the surface sensitivity of the technique.

Auger electron spectroscopy (AES) is schematically illustrated in Figure 22.17. An electron (or photon) ejects an electron from a low-lying level in an atom. This hole is quickly filled by a transition from a higher state. This event alone, however, does not conserve energy. Energy conservation is accomplished by the simultaneous ejection of a second electron into the gas phase. It is the kinetic energy of this electron that is measured. Although three different energy levels are involved in this spectroscopy, the signatures of different atoms are quite easy to distinguish. The main advantage of using electrons rather than photons to create the initial hole is to gain spatial resolution. Electron beams can be focused to a spot size on the order of 10–100 nm and, therefore, Auger spectroscopy with electron excitation is routinely used in many industries to map out elemental distribution at the surfaces of solids with very high lateral resolution.

Results using scanning Auger spectroscopy to study the growth of Cu_2O nanodots on the surface of a $SrTiO_3$ crystal are shown in Figure 22.18. The scanning electron microscopy (SEM) image shows the structure of the surface but gives no information on the elemental composition. Scanning Auger images of the same area are shown for Cu, Ti, and O. These images show that Cu does not uniformly coat the surface but instead forms three-dimensional crystallites. This conclusion can be drawn from the absence of Cu and the presence of Ti in the areas between the nanodots. The Cu_2O nanodots appear darker (lower O content) than the underlying $SrTiO_3$ surface in the oxygen image because Cu_2O has only one O for every two Cu cations, whereas the surface has three O in a formula unit containing two cations. These results show that the nanodot deposition process can be understood using a surface-sensitive spectroscopy.

X-ray photoelectron spectroscopy (XPS) is simpler than AES in that only one level is involved. A photon of energy is absorbed by an atom, and to conserve energy, an electron is ejected with kinetic energy

$$E_{kinetic} = h\nu - E_{binding} \tag{22.10}$$

A small correction term that involves the work functions of the solid (defined in Chapter 12) and the detector has been omitted. A schematic picture of the process that gives rise to an ejected electron is shown in Figure 22.19. Currently, no off-the-shelf X-ray lasers are available, so sources cannot be made with very small bandwidths. However, using monochromatized X-ray sources, distinctly different peaks are observed for substances in which the same atom is present in chemically nonequivalent environments. This **chemical shift** is also illustrated in Figure 22.19. A positive value for the chemical shift indicates a higher binding energy for the electron in the atom than would be measured for the free atom. The origin of the chemical shift can be understood in a simple model, although accurate calculations require a more detailed treatment.

Consider the different binding environment of the carbon atom in ethyltrifluoroacetate, whose structure is shown in Figure 22.19b. The carbon atom in the CF_3 group experiences a net electron withdrawal to the much more electronegative F atoms.

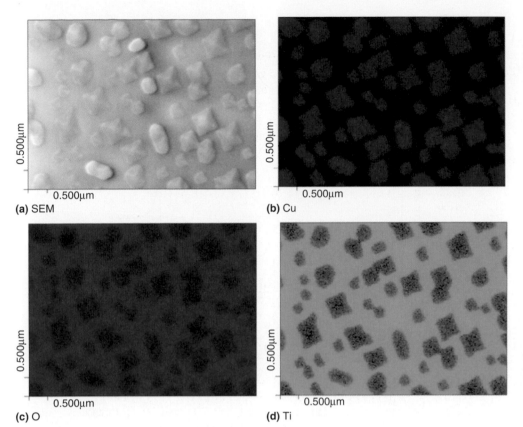

(a) SEM   **(b)** Cu

(c) O   **(d)** Ti

FIGURE 22.18
Scanning electron microscopy and scanning Auger spectroscopy images for copper, oxygen, and titanium are shown for a 0.5×0.5-μm area of a $SrTiO_3$ crystal surface on which Cu_2O nanodots have been deposited by Cu evaporation in an oxygen containing plasma. **(a)** The SEM image obtained without energy analysis shows structure but gives no information on the elemental distribution. **(b)**, **(c)**, and **(d)** are obtained using energy analysis of backscattered electrons. Light and dark areas correspond to high and low values respective of Cu **(b)**, O **(c)**, and Ti **(d)**. An analysis of the data shows that the light areas in **(b)** contain Cu and O, but no Ti, and that the dark areas in **(d)** contain no Ti.] "Synthesis and Characterization of Self-Assembled Cu_2O Nano-Dots."
Source: Liang, Y.; Lea, A. S.; McCready, D. E.; Meethunkij, P.; *Proceedings - Electrochemical Society* (2001): 125.

Therefore, the $1s$ electrons lose some of the shielding effect they had from the $2p$ electrons and, as a result, the C $1s$ electron experiences a slightly greater nuclear charge. This leads to an increase in the binding energy or a positive chemical shift. The carbon atom with double and single bonds to oxygen experiences an electron withdrawal, although to a lesser degree. The carbon of the methyl group has little electron transfer, and the methylene carbon experiences a larger electron withdrawal because it is directly bonded to an oxygen. Although these effects are small, they are easily measurable. Therefore, a photoelectron spectrum gives information on the oxidation state as well as on the identity of the element.

An example in which the surface sensitivity of XPS is used is shown in Figure 22.20. The growth of iron and iron oxide films on a crystalline magnesium oxide surface are monitored under different conditions. The goal is to determine the oxidation state of the iron in the film. The X-ray photon ejects an electron from the $2p$ level of Fe species near the surface, and the signal is dominated by those species within ~1 nm of the surface. The spin angular momentum **s** of the remaining electron in the Fe $2p$ level couples with its orbital angular momentum **l** to form the total angular momentum vector **j** with two possible values for the quantum number, $j = 1/2$ and $3/2$. These two states are of different energy and, therefore, two peaks are observed in the spectrum. The ratio of the measured photoemission signal from these states is given by the ratio of their degeneracy, or

$$\frac{I(2p_{3/2})}{I(2p_{1/2})} = \frac{2 \times \left(\frac{3}{2}\right) + 1}{2 \times \left(\frac{1}{2}\right) + 1} = 2$$

The binding energy corresponding to the Fe peaks clearly shows that deposition in vacuum leads to metallic or zero valent Fe species. Different crystalline phases of the iron oxide formed by exposing the film to oxygen gas while depositing iron have different ratios of Fe(II) and Fe(III).

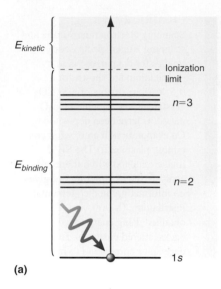

(a)

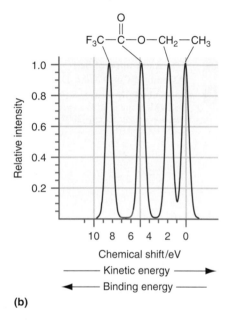

(b)

FIGURE 22.19

(a) Principle of X-ray photoelectron spectroscopy. (b) A spectrum exhibiting chemical shifts for the carbon $1s$ level of the ethyltrifluoroacetate molecule is shown.

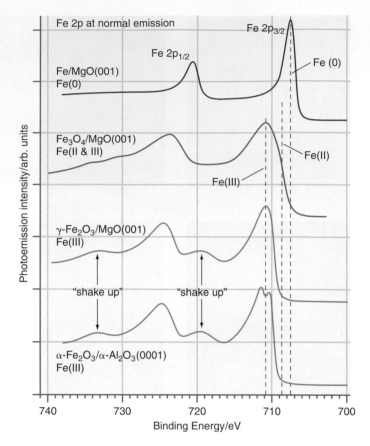

FIGURE 22.20

XP spectra are shown for the deposition of Fe films on a crystalline MgO surface. Note the splitting of the peaks originating from the $2p$ core level as a result of spin-orbit interaction. Shake-up features originate when valence electrons are promoted to higher levels in the photoemission event. This promotion reduces the kinetic energy of the ejected electron.

Source: Graph courtesy of Scott A. Chambers/Pacific Northwest National Laboratory.

22.10 Selective Chemistry of Excited States: O(^{3}P) and O(^{1}D)

The interaction of sunlight with molecules in the atmosphere leads to an interconnected set of chemical reactions, which in part determines the composition of Earth's atmosphere. Oxygen is a major species involved in these reactions. Solar radiation in the ultraviolet range governs the concentrations of O·, O_2, and O_3 according to

$$O_2 + h\nu \longrightarrow O\cdot + O\cdot$$

$$O\cdot + O_2 + M \longrightarrow O_3$$

$$O_3 + h\nu \longrightarrow O\cdot + O_2$$

$$O\cdot + O_3 \longrightarrow 2O_2 \qquad (22.11)$$

where M is another molecule in the atmosphere that takes up the energy released in forming O_3. For wavelengths less than 300 nm, ^{1}D oxygen atoms are produced in the stratosphere by the reaction

$$O_3 + h\nu \longrightarrow O_2 + O\cdot(^1D) \qquad (22.12)$$

For longer wavelengths, ground-state ^{3}P oxygen atoms are produced. Importantly, O·(^{1}D) has an excess energy of 190 kJ mol^{-1} relative to O·(^{3}P). This energy can be used to overcome an activation barrier to reaction. For example, the reaction

$$O\cdot(^3P) + H_2O \longrightarrow \cdot OH + \cdot OH \qquad (22.13)$$

is endothermic by 70 kJ mol^{-1}, whereas the reaction

$$O \cdot (^1D) + H_2O \longrightarrow \cdot OH + \cdot OH \qquad (22.14)$$

is exothermic by 120 kJ mol^{-1}. Because a radiative transition from $O \cdot (^1D)$ to $O \cdot (^3P)$ is forbidden by the selection rule $\Delta S = 0$, the $O \cdot (^1D)$ atoms are long lived and their concentration is predominantly depleted by reactions with other species.

The $O \cdot (^1D)$ atoms are primarily responsible for generating reactive hydroxyl and methyl radicals through the reactions

$$O \cdot (^1D) + H_2O \longrightarrow \cdot OH + \cdot OH$$

$$O \cdot (^1D) + CH_4 \longrightarrow \cdot OH + \cdot CH_3 \qquad (22.15)$$

As before, the reactivity for $O \cdot (^1D)$ is much higher than that for $O \cdot (^3P)$ largely because the excess energy in electronic excitation can be used to overcome the activation barrier for the reaction. $O \cdot (^1D)$ is also involved in generating the reactive NO intermediate from N_2O and $Cl \cdot$ from chlorofluorocarbons.

SUPPLEMENTAL

22.11 Configurations with Paired and Unpaired Electron Spins Differ in Energy

In this section, we show that the energy of a configuration depends on whether the spins are paired for a specific case, namely, the excited states of He with the configuration $1s^1 2s^1$. Antisymmetric wave functions for each value of S were formulated in Section 22.2. In the following, the energy for the singlet and triplet states is calculated, and we show that the triplet state lies lower in energy than the singlet state.

The Schrödinger equation for the singlet wave function can be written as

$$\left(\hat{H}_1 + \hat{H}_2 + \frac{e^2}{4\pi\varepsilon_0 r_{12}} \right) \psi(1, 2) = E_{singlet} \, \psi(1, 2) \qquad (22.16)$$

where $\hat{H}_1$ and $\hat{H}_2$ are the total energy operators neglecting electron–electron repulsion and the subscripts refer to the electron involved. $\psi(1, 2)$ is the unknown exact wave function. The spin part of the wave function is not included because the total energy operator does not contain terms that depend on spin. Because we do not know the exact wave function, we approximate it by the simple singlet wave function of Equation (22.3). Keep in mind that the singlet wave function is not an eigenfunction of the total energy operator. To obtain the expectation value for the total energy using this approximate wave function, one multiplies on the left by the complex conjugate of the wave function and integrates over the spatial coordinates. Because the $1s$ and $2s$ functions are real, the function and its complex conjugate are identical.

$$E_{singlet} = \frac{1}{2} \iint [1s(1)2s(2) + 2s(1)1s(2)] \left(\hat{H}_1 + \hat{H}_2 + \frac{e^2}{4\pi\varepsilon_0 r_{12}} \right)$$
$$\times [1s(1)2s(2) + 2s(1)1s(2)] d\tau_1 d\tau_2 \qquad (22.17)$$

As you will see when you work the end-of-chapter problems, the two integrals arising from $\hat{H}_1$ and $\hat{H}_2$ give $E_{1s} + E_{2s}$ where $E_{1s} = -e^2/(2\pi\varepsilon_0 a_0)$ and $E_{2s} = -e^2/(8\pi\varepsilon_0 a_0)$ are the H atom eigenvalues for $\zeta = 2$. Using this result,

$$E_{singlet} = E_{1s} + E_{2s}$$
$$+ \frac{1}{2} \iint [1s(1)2s(2) + 2s(1)1s(2)] \left(\frac{e^2}{4\pi\varepsilon_0 r_{12}} \right)$$
$$\times [1s(1)2s(2) + 2s(1)1s(2)] d\tau_1 d\tau_2 \qquad (22.18)$$

The remaining integral can be simplified as you will also see in the end-of-chapter problems to yield $E_{singlet}$:

$$E_{singlet} = E_{1s} + E_{2s} + J_{12} + K_{12}, \text{ where} \tag{22.19}$$

$$J_{12} = \frac{e^2}{8\pi\varepsilon_0} \iint [1s(1)]^2 \left(\frac{1}{r_{12}}\right)[2s(2)]^2 \, d\tau_1 d\tau_2 \text{ and}$$

$$K_{12} = \frac{e^2}{8\pi\varepsilon_0} \iint [1s(1)2s(2)] \left(\frac{1}{r_{12}}\right)[1s(2)2s(1)] \, d\tau_1 d\tau_2 \tag{22.20}$$

If the calculation is carried out for the triplet state, the corresponding result is

$$E_{triplet} = E_{1s} + E_{2s} + J_{12} - K_{12} \tag{22.21}$$

Focus on the results rather than the mathematics. In the absence of the repulsive interaction between the two electrons, the total energy is simply $E_{1s} + E_{2s}$. Including the Coulomb repulsion between the electrons and making the wave function antisymmetric give rise to the additional terms J_{12} and K_{12}. The energy shift relative to $E_{1s} + E_{2s}$ is $J_{12} + K_{12}$ for the singlet state and $J_{12} - K_{12}$ for the triplet state. This shows that the triplet and singlet states of $He(1s^1 2s^1)$ have energies that differ by $2K_{12}$. Because all the terms that appear in the first integral of Equation (22.20) are positive, $J_{12} > 0$. It can also be shown that K_{12} is positive. *Therefore, it has been shown that the triplet state for the first excited state of He lies lower in energy than the singlet state. This is a general result for singlet and triplet states.*

Looking back, this result is based on a purely mathematical argument. Can a physical meaning be attached to J_{12} and K_{12}? Imagine that electrons 1 and 2 were point charges. In that case, the integral J_{12} simplifies to $e^2/(4\pi\varepsilon_0|r_1 - r_2|)$. The electrons of He can be thought of as diffuse charge clouds. The integral J_{12} is simply the electrostatic interaction between the diffuse charge distributions $\rho(1)$ and $\rho(2)$ where $\rho(2) = [2s(2)]^2 d\tau_2$ and $\rho(1) = [1s(1)]^2 d\tau_1$. Because J_{12} can be interpreted in this way, it is called the **Coulomb integral.** Unlike J_{12}, the integral K_{12} has no classical physical interpretation. The product $[1s(1)2s(2)][1s(2)2s(1)] d\tau_1 d\tau_2$ does not fit the definition of charge because it does not have the form $|\psi(1)|^2 |\psi(2)|^2 d\tau_1 d\tau_2$. Because the electrons have been exchanged between the two parts of this product, K_{12} is referred to as the **exchange integral.** It has no classical analogue and arises from the fact that the singlet and triplet wave functions are written as a superposition of two parts in order to satisfy the Pauli principle.

The singlet and triplet wave functions also differ in the degree to which they include electron correlation. We know that electrons avoid one another because of their Coulomb repulsion. If we let electron 2 approach electron 1, the spatial part of the singlet wave function

$$\frac{1}{\sqrt{2}}[1s(1)2s(2) + 2s(1)1s(2)] \longrightarrow \frac{1}{\sqrt{2}}[1s(1)2s(1) + 2s(1)1s(1)]$$

$$= \frac{2}{\sqrt{2}} 1s(1)2s(1)$$

because $r_2, \theta_2, \phi_2 \longrightarrow r_1, \theta_1, \phi_1$, but the spatial part of the triplet wave function

$$\frac{1}{\sqrt{2}}[1s(1)2s(2) - 2s(1)1s(2)] \longrightarrow \frac{1}{\sqrt{2}}[1s(1)2s(1) - 2s(1)1s(1)] = 0$$

This shows that the triplet wave function has a greater degree of electron correlation built into it than the singlet wave function, because the probability of finding both electrons in a given region falls to zero as the electrons approach one another.

Why is the energy of the triplet state lower than that of the singlet state? One might think that the electron–electron repulsion is lower in the triplet state because of the electron correlation and that this is the origin of the lower total energy. In fact, this is not correct. A more detailed analysis shows that the electron–electron repulsion is actually greater in the triplet than in the singlet state. However, on average the electrons are slightly closer to the nucleus in the triplet state. The increased electron–nucleus attraction outweighs the electron–electron repulsion and, therefore, the triplet state has a lower energy. Note that spin influences the energy even though the total energy operator does not contain any terms involving spin. Spin enters the calculation through the antisymmetrization required by the Pauli principle. *Generalizing this result, it can be concluded that for a given configuration, a state in which the spins are unpaired has a lower energy than a state in which the spins are paired.*

Vocabulary

antiparallel	exchange integral	pumping transition
atomic absorption spectroscopy	good quantum number	resonator modes
atomic emission spectroscopy	Grotrian diagram	selection rule
Auger electron spectroscopy	Hund's rules	singlet
chemical shift	incoherent photon source	spin-orbit coupling
Clebsch–Gordon series	inelastic mean free path	spontaneous emission
coherent photon source	laser	stimulated emission
configuration	lasing transition	term
Coulomb integral	level	term symbol
degeneracy of a term	multiplicity	total angular momentum
dipole approximation	optical resonator	triplet
Doppler broadening	paired electrons	unpaired electrons
Doppler effect	parallel spins	X-ray photoelectron spectroscopy
Doppler shift	population inversion	

Conceptual Problems

Q22.1 Justify the statement that the Coulomb integral J defined in Equation (22.20) is positive by explicitly formulating the integral that describes the interaction between two negative classical charge clouds.

Q22.2 Without invoking equations, explain why the energy of the triplet state is lower than that of the singlet state for He in the $1s^1 2s^1$ configuration.

Q22.3 How can the width of a laser line be less than that determined by Doppler broadening?

Q22.4 Why is an electronically excited atom more reactive than the same ground-state atom?

Q22.5 Why is atomic absorption spectroscopy more sensitive in many applications than atomic emission spectroscopy?

Q22.6 Why does the Doppler effect lead to a shift in the wavelength of a star but to a broadening of a transition in a gas?

Q22.7 Why are n, l, m_l, and m_s not good quantum numbers for many-electron atoms?

Q22.8 Write an equation giving the relationship between the Rydberg constant for H and for Li^{2+}.

Q22.9 Can the individual states in Table 22.1 be distinguished experimentally?

Q22.10 How is it possible to determine the L and S value of a term knowing only the M_L and M_S values of the states?

Q22.11 What is the origin of the chemical shift in XPS?

Q22.12 Why are two medium-energy photons rather than one high-energy photon used in laser isotope separation?

Q22.13 Why does one need to put a sample in a vacuum chamber to study it with XPS or AES?

Q22.14 Why is XPS a surface-sensitive technique?

Q22.15 Explain the direction of the chemical shifts for Fe(0), Fe(II), and Fe(III) in Figure 22.20.

Numerical Problems

Problem numbers in **red** indicate that the solution to the problem is given in the *Student's Solutions Manual*.

P22.1 The principal line in the emission spectrum of potassium is violet. On close examination, the line is seen to be a doublet with wavelengths of 393.366 and 396.847 nm. Explain the source of this doublet.

P22.2 The absorption spectrum of the hydrogen atom shows lines at 5334, 7804, 9145, 9953, and 10,478 cm^{-1}. There are no lower frequency lines in the spectrum. Use the graphical methods discussed in Example Problem 22.6 to determine $n_{initial}$ and the ionization energy of the hydrogen atom in this state. Assume values for $n_{initial}$ of 1, 2, and 3.

P22.3 Using Table 22.3, which lists the possible terms that arise from a given configuration, and Hund's rules, write the term symbols for the ground state of the atoms H through F in the form $^{(2S+1)}L_J$.

P22.4 In this problem, you will supply the missing steps in the derivation of the formula $E_{singlet} = E_{1s} + E_{2s} + J + K$ for the singlet level of the $1s^1 2s^1$ configuration of He.

a. Expand Equation (22.17) to obtain

$$E_{singlet} = \frac{1}{2} \iint \frac{[1s(1)2s(2) + 2s(1)1s(2)](\hat{H}_1)}{[1s(1)2s(2) + 2s(1)1s(2)]d\tau_1 d\tau_2}$$

$$+ \frac{1}{2} \iint \frac{[1s(1)2s(2) + 2s(1)1s(2)](\hat{H}_2)}{[1s(1)2s(2) + 2s(1)1s(2)]d\tau_1 d\tau_2}$$

$$+ \frac{1}{2} \iint [1s(1)2s(2) + 2s(1)1s(2)] \times$$
$$\left(\frac{e^2}{4\pi\varepsilon_0 |r_1 - r_2|}\right) \times$$
$$[1s(1)2s(2) + 2s(1)1s(2)]d\tau_1 d\tau_2$$

b. Starting from the equations $\hat{H}_i 1s(i) = E_{1s} 1s(i)$ and $\hat{H}_i 2s(i) = E_{2s} 2s(i)$, show that $E_{singlet} = E_{1s} + E_{2s}$

$$+ \frac{1}{2} \iint [1s(1)2s(2) + 2s(1)1s(2)]\left(\frac{e^2}{4\pi\varepsilon_0 |r_1 - r_2|}\right) \times$$
$$[1s(1)2s(2) + 2s(1)1s(2)]d\tau_1 d\tau_2$$

c. Expand the previous equation using the definitions

$$J = \frac{e^2}{8\pi\varepsilon_0} \iint [1s(1)]^2 \left(\frac{1}{|r_1 - r_2|}\right)[2s(2)]^2 d\tau_1 d\tau_2 \text{ and}$$

$$K = \frac{e^2}{8\pi\varepsilon_0} \iint [1s(1)2s(2)]\left(\frac{1}{|r_1 - r_2|}\right)[1s(2)2s(1)] \times$$
$$d\tau_1 d\tau_2$$

to obtain the desired result, $E_{singlet} = E_{1s} + E_{2s} + J + K$.

P22.5 What J values are possible for a ^6H term? Calculate the number of states associated with each level and show that the total number of states is the same as that calculated from the term symbol.

P22.6 Using Table 22.3, which lists the possible terms that arise from a given configuration, and Hund's rules, write the configurations and term symbols for the ground state of the ions F^- and Ca^{2+} in the form $^{(2S+1)}L_J$.

P22.7 The Doppler broadening in a gas can be expressed as $\Delta\nu = (2\nu_0/c)\sqrt{2\ln 2(RT/M)}$, where M is the molar mass. For the sodium $3p\ ^2P_{3/2} \longrightarrow 3s\ ^2S_{1/2}$ transition, $\nu_0 = 5.0933 \times 10^{14}\,s^{-1}$. Calculate $\Delta\nu$ and $\Delta\nu/\nu_0$ at 500.0 K.

P22.8 Calculate the transition dipole moment, $\mu_z^{mn} = \int \psi_m^*(\tau)\,\mu_z\,\psi_n(\tau)\,d\tau$ where $\mu_z = -er\cos\theta$ for a transition from the $1s$ level to the $2p_z$ level in H. Show that this transition is allowed. The integration is over r, θ, and ϕ. Use

$$\psi_{210}(r, \theta, \phi) = \frac{1}{\sqrt{32\pi}}\left(\frac{1}{a_0}\right)^{3/2}\frac{r}{a_0}e^{-r/2a_0}\cos\theta$$

for the $2p_z$ wave function.

P22.9 Consider the $1s\,np\ ^3P \to 1s\,nd\ ^3D$ transition in He. Draw an energy-level diagram, taking the spin-orbit coupling that splits terms into levels into account. Into how many levels does each term split? The selection rule for transitions in this case is $\Delta J = 0, \pm 1$. How many transitions will be observed in an absorption spectrum? Show the allowed transitions in your energy diagram.

P22.10 Atomic emission experiments of a mixture show a calcium line at 422.673 nm corresponding to a $^1P_1 \to ^1S_0$ transition and a doublet due to potassium $^2P_{3/2} \to ^2S_{1/2}$ and $^2P_{1/2} \to ^2S_{1/2}$ transitions at 764.494 and 769.901 nm, respectively.

a. Calculate the ratio g_{upper}/g_{lower} for each of these transitions.

b. Calculate n_{upper}/n_{lower} for a temperature of 1600°C for each transition.

P22.11 How many ways are there to place three electrons into an f subshell? What is the ground-state term for the f^3 configuration, and how many states are associated with this term? See Problem P22.36.

P22.12 Calculate the wavelengths of the first three lines of the Lyman, Balmer, and Paschen series, and the series limit (the shortest wavelength) for each series.

P22.13 The Lyman series in the hydrogen atom corresponds to transitions that originate from the $n = 1$ level in absorption or that terminate in the $n = 1$ level for emission. Calculate the energy, frequency (in inverse seconds and inverse centimeters), and wavelength of the least and most energetic transition in this series.

P22.14 The inelastic mean free path of electrons in a solid, λ, governs the surface sensitivity of techniques such as AES and XPS. The electrons generated below the surface must make their way to the surface without losing energy in order to give elemental and chemical shift information. An empirical expression for elements that give λ as a function of the kinetic energy of the electron generated in AES or XPS is $\lambda = 538E^{-2} + 0.41(lE)^{0.5}$. The units of λ are monolayers, E is the kinetic energy of the electron in eV, and l is the monolayer thickness in nanometers. On the basis of this equation, what kinetic energy maximizes the surface sensitivity for a monolayer thickness of 0.3 nm? An equation solver would be helpful in obtaining the answer.

P22.15 The effective path length that an electron travels before being ejected into the vacuum is related to the depth below the surface at which it is generated and the exit angle by $d = \lambda \cos \theta$, where λ is the inelastic mean free path and θ is the angle between the surface normal and the exit direction.

a. Justify this equation based on a sketch of the path that an electron travels before exiting into the vacuum.

b. The XPS signal from a thin layer on a solid surface is given by $I = I_0(1 - e^{-d/(\lambda \cos \theta)})$, where I_0 is the signal that would be obtained from an infinitely thick layer, and λ is defined in Problem P22.14. Calculate the ratio I/I_0 at $\theta = 0$ for $\lambda = 2d$. Calculate the exit angle required to increase I/I_0 to 0.50.

P22.16 List the allowed quantum numbers m_l and m_s for the following subshells and determine the maximum occupancy of the subshells:

a. $2p$ **b.** $3d$ **c.** $4f$ **d.** $5g$

P22.17 What are the levels that arise from the following terms? How many states are there in each level?

a. ^4F **b.** ^2D **c.** ^2S **d.** ^4P

P22.18 As discussed in Chapter 20, in a more exact solution of the Schrödinger equation for the hydrogen atom, the coordinate system is placed at the center of mass of the atom rather than at the nucleus. In that case, the energy levels for a one-electron atom or ion of nuclear charge Z are given by

$$E_n = -\frac{Z^2 \mu e^4}{32\pi^2 \varepsilon_0^2 \hbar^2 n^2}$$

where μ is the reduced mass of the atom. The masses of an electron, a proton, and a tritium (^3H or T) nucleus are given by 9.1094×10^{-31} kg, 1.6726×10^{-27} kg, and 5.0074×10^{-27} kg, respectively. Calculate the frequency of the $n = 1 \rightarrow n = 4$ transition in H and T to five significant figures. Which of the transitions, $1s \rightarrow 4s$, $1s \rightarrow 4p$, $1s \rightarrow 4d$, could the frequencies correspond to?

P22.19 Derive the ground-state term symbols for the following configurations:

a. $s^1 d^5$ **b.** f^3 **c.** g^2

P22.20 Calculate the terms that can arise from the configuration $np^1n'p^1$, $n \neq n'$. Compare your results with those derived in the text for np^2. Which configuration has more terms and why?

P22.21 For a closed-shell atom, an antisymmetric wave function can be represented by a single Slater determinant.

For an open-shell atom, more than one determinant is needed. Show that the wave function for the $M_S = 0$ triplet state of He $1s^1 2s^1$ is a linear combination of two of the Slater determinants of Example Problem 22.1. Which of the two are needed and what is the linear combination?

P22.22 Calculate the transition dipole moment, $\mu_z^{mn} = \int \psi_m^*(\tau) \mu_z \psi_n(\tau) d\tau$ where $\mu_z = -er \cos \theta$ for a transition from the $1s$ level to the $2s$ level in H. Show that this transition is forbidden. The integration is over r, θ, and ϕ.

P22.23 Use the transition frequencies shown in Example Problem 22.7 to calculate the energy (in joules and electron-volts) of the six levels relative to the $3s\,^2S_{1/2}$ level. State your answers with the correct number of significant figures.

P22.24 Derive the ground-state term symbols for the following atoms or ions:

a. H   **b.** F⁻   **c.** Na⁺   **d.** Sc

P22.25 The spectrum of the hydrogen atom reflects the splitting of the $1s^2$ S and $2p^2$ P terms into levels. The energy difference between the levels in each term is much smaller than the difference in energy between the terms. Given this information, how many spectral lines are observed in the $1s^2$ S $\rightarrow$ $2p^2$ P transition? Are the frequencies of these transitions very similar or quite different?

P22.26 Using Table 22.3, which lists the possible terms that arise from a given configuration, and Hund's rules, write the term symbols for the ground state of the atoms K through Cu, excluding Cr, in the form $^{(2S+1)}L_J$.

P22.27 What atomic terms are possible for the following electron configurations? Which of the possible terms has the lowest energy?

a. $ns^1 np^1$ **b.** $ns^1 nd^1$ **c.** $ns^2 np^1$ **d.** $ns^1 np^2$

P22.28 Two angular momenta with quantum numbers $j_1 = 3/2$ and $j_2 = 5/2$ are added. What are the possible values of J for the resultant angular momentum states?

P22.29 Derive the ground-state term symbols for the following configurations:

a. d^2 **b.** f^9 **c.** f^{14}

P22.30 The first ionization potential of ground-state He is 24.6 eV. The wavelength of light associated with the $1s2p$ ^1P term is 58.44 nm. What is the ionization energy of the He atom in this excited state?

P22.31 In the Na absorption spectrum, the following transitions are observed:

$$4p\,^2P \rightarrow 3s\,^2S \quad \lambda = 330.26 \text{ nm}$$

$$3p\,^2P \rightarrow 3s\,^2S \quad \lambda = 589.593 \text{ nm, } 588.996 \text{ nm}$$

$$5s\,^2S \rightarrow 3p\,^2P \quad \lambda = 616.073 \text{ nm, } 615.421 \text{ nm}$$

Calculate the energies of the $4p\,^2P$ and $5s\,^2S$ states with respect to the $3s\,^2S$ ground state.

P22.32 The Grotrian diagram in Figure 22.7 shows a number of allowed electronic transitions for He. Which of the following transitions shows multiple spectral peaks due to a splitting of terms into levels? How many peaks are observed in each case? Are any of the following transitions between energy levels forbidden by the selection rules?

a. $1s^2\ {}^1S \rightarrow 1s2p\ {}^1P$

b. $1s2p\ {}^1P \rightarrow 1s3s\ {}^1S$

c. $1s2s\ {}^3S \rightarrow 1s2p\ {}^3P$

d. $1s2p\ {}^3P \rightarrow 1s3d\ {}^3D$

P22.33   List the quantum numbers L and S that are consistent with the following terms:

a. ${}^4S$ **b.** ${}^4G$ **c.** ${}^3P$ **d.** ${}^2D$

P22.34   The transition $Al[Ne](3s)^2(3p)^1 \rightarrow Al[Ne](3s)^2(4s)^1$ has two lines given by $\widetilde{\nu} = 25354.8\ \text{cm}^{-1}$ and $\widetilde{\nu} = 25242.7\ \text{cm}^{-1}$. The transition $Al[Ne](3s)^2(3p)^1 \rightarrow Al[Ne](3s)^2(3d)^1$ has three lines given by $\widetilde{\nu} = 32444.8\ \text{cm}^{-1}$, $\widetilde{\nu} = 32334.0\ \text{cm}^{-1}$, and $\widetilde{\nu} = 32332.7\ \text{cm}^{-1}$. Sketch an energy-level diagram of the states involved and explain the source of all lines. *[Hint: The lowest energy levels are P levels and the highest are D levels. The energy spacing between the D levels is less than for the P levels.]*

P22.35   Given that the levels in the ${}^3P$ term for carbon have the relative energies (expressed in wave numbers) of

${}^3P_1 - {}^3P_0 = 16.4\ \text{cm}^{-1}$ and ${}^3P_2 - {}^3P_1 = 27.1\ \text{cm}^{-1}$, calculate the ratio of the number of C atoms in the ${}^3P_2$ and ${}^3P_0$ levels at 200.0 and 1000. K.

P22.36   A general way to calculate the number of states that arise from a given configuration is as follows. Calculate the combinations of m_l and m_s for the first electron, and call that number n. The number of combinations used is the number of electrons, which we call m. The number of unused combinations is $n - m$. According to probability theory, the number of distinct permutations that arise from distributing the m electrons among the n combinations is $n!/[m!(n - m)!]$.

For example, the number of states arising from a p^2 configuration is $6!/[2!4!] = 15$, which is the result obtained in Section 22.2. Using this formula, calculate the number of possible ways to place five electrons in a d subshell. What is the ground-state term for the d^5 configuration and how many states does the term include?

P22.37   The ground-state level for the phosphorus atom is ${}^4S_{3/2}$. List the possible values of L, M_l, S, M_S, J, and M_J consistent with this level.

P22.38   Derive the ground-state term symbols for the following atoms:

a. F   **b.** Na   **c.** P

Web-Based Simulations, Animations, and Problems

W22.1   The individual processes of absorption, spontaneous emission, and stimulated emission are simulated in a two-level system. The level of pumping needed to sustain lasing is

experimentally determined by comparing the population of the upper and lower levels in the lasing transition.

The Chemical Bond in Diatomic Molecules

The chemical bond is at the heart of chemistry. We begin with a qualitative molecular orbital model for chemical bonding using the H_2^+ molecule as an example. We show that H_2^+ is more stable than widely separated H and H^+ because of delocalization of the electron over the molecule and localization of the electron in the region between the two nuclei. The molecular orbital model provides a good understanding of the electronic structure of diatomic molecules and is used to understand the bond order, bond energy, and bond length of homonuclear diatomic molecules. The formalism is extended to describe bonding in strongly polar molecules such as HF.

23.1   Generating Molecular Orbitals from Atomic Orbitals

Because the essence of chemistry is bonds between atoms, chemists need to have a firm understanding of the theory of the chemical bond. In this chapter, the origin of the chemical bond is explored using the H_2^+ molecule as an example. We then discuss chemical bonding in first and second row diatomic molecules. In Chapter 24, localized and delocalized bonding models will be used to understand and predict the shape of small molecules. The discussion in this chapter and Chapter 24 is largely qualitative in character. In Chapter 26, numerical methods for quantum chemical calculations on molecules are discussed. It may be useful to work on Chapter 26 in parallel with Chapters 23 and 24.

A chemical bond is formed between two atoms if the energy of the molecule passes through a minimum at an equilibrium distance that is smaller than the energy of the separated atoms. How does the electron distribution around the nuclei change when a chemical bond is formed? In answering this question, we consider the relative energies of two H atoms compared to the H_2 molecule. Two H atoms are more stable than the four infinitely separated charges by 2624 kJ mol^{-1}. The H_2 molecule is more stable than two infinitely separated H atoms by 436 kJ mol^{-1}. Therefore, the chemical bond lowers the total energy of the two protons and two electrons by 17%. Although appreciable, the **bond energy** is a small fraction of the total energy of the widely separated electrons and nuclei. This result suggests that the charge distribution in a molecule is quite similar to a superposition of the charge distribution of the individual atoms.

However, as we will see later, valence electrons that are localized on an individual atom for large internuclear distances are **delocalized,** meaning that they have a finite probability of being found anywhere in the molecule. Core electrons generally remain localized on individual atoms.

As discussed in Chapter 21, the introduction of a second electron vastly complicates the task of finding solutions to the Schrödinger equation for atoms. This is also true for molecules. The exact **molecular wave functions** for a molecule with n electrons and m nuclei are functions of the positions of all the electrons and nuclei

$$\psi_i^{molecule} = \psi(\mathbf{r}_1, \mathbf{r}_2, ..., \mathbf{r}_n, \mathbf{R}_1, \mathbf{R}_2, ..., \mathbf{R}_m) \tag{23.1}$$

where the $\mathbf{r}$ and $\mathbf{R}$ are positions of the electrons and nuclei respectively. In order to be able to solve the Schrodinger equation for a molecule, an approximate wave function with fewer variables is needed. The parts of $\psi_i^{molecule}$ for the motion of the nuclei and the electrons, both of which appear in Equation (23.1), can be separated using the **Born–Oppenheimer approximation.** Because the electron is lighter than the proton by a factor of nearly 2000, the electron charge quickly rearranges in response to the slower periodic motion of the nuclei in molecular vibrations. Because of the very different timescales for nuclear and electron motion, the two motions can be decoupled and we can write Equation (23.1) in the form

$$\psi_i^{molecule} \approx \psi_{BO}(\mathbf{r}_1, \mathbf{r}_2, ..., \mathbf{r}_n)\psi_{nuclear}(\mathbf{R}_1, \mathbf{R}_2, ..., \mathbf{R}_m) \tag{23.2}$$

$\psi_{nuclear}$ describes the motions of the nuclei in vibration and rotation of the molecule and ψ_{BO} describes the electrons for an instantaneous fixed positions of the nuclei. We next solve the Schrödinger equation for ψ_{BO} and calculate the total energy of the molecule at a fixed set of nuclear positions using further approximations that will be discussed later. If this procedure is repeated for many values of $\mathbf{R}_1, \mathbf{R}_2, ..., \mathbf{R}_m$, we can determine an energy function, $E_{total}(\mathbf{R}_1, \mathbf{R}_2, ..., \mathbf{R}_m)$. The values for $\mathbf{R}_1, \mathbf{R}_2, ..., \mathbf{R}_m$ at the minimum in E determine the equilibrium nuclear positions.

The total energy operator for a diatomic molecule in the Born–Oppenheimer approximation is given by

$$\hat{H} = -\frac{\hbar^2}{2m_e}\sum_{i=1}^{n}\nabla_i^2 - \sum_{i=1}^{n}\left(\frac{Z_Ae^2}{4\pi\varepsilon_0 r_{iA}} + \frac{Z_Be^2}{4\pi\varepsilon_0 r_{iB}}\right) + \sum_{i=1}^{n}\sum_{j>i}^{n}\frac{e^2}{4\pi\varepsilon_0 r_{ij}} + \frac{Z_AZ_Be^2}{4\pi\varepsilon_0 R_{AB}}$$

$$\tag{23.3}$$

The first term is the kinetic energy of the electrons, the second term is the Coulomb attraction between the n electrons and two nuclei, the third term is the electron–electron repulsion, and the last term is the nuclear–nuclear repulsion. The restriction $j > i$ on the summation in the third term ensures that the electron–electron repulsion between electrons i and j is not counted twice.

The last term in $\hat{H}$ is a constant because we are assuming that the nuclei do not move. It is convenient to separate out this term and to write an electronic total energy operator

$$\hat{H}_{el} = -\frac{\hbar^2}{2m_e}\sum_{i=1}^{n}\nabla_i^2 - \sum_{i=1}^{n}\left(\frac{Z_Ae^2}{4\pi\varepsilon_0 r_{iA}} + \frac{Z_Be^2}{4\pi\varepsilon_0 r_{iB}}\right) + \sum_{i=1}^{n}\sum_{j>i}^{n}\frac{e^2}{4\pi\varepsilon_0 r_{ij}} \tag{23.4}$$

The eigenvalues for the electronic Schrödinger equation $\hat{H}_{el}\psi_{el} = E_{el}\psi_{el}$ are related to the total energy eigenvalues by

$$E_{total} = E_{el} + \frac{Z_AZ_Be^2}{4\pi\varepsilon_0 R_{AB}} \tag{23.5}$$

The reason for separating out the nuclear repulsion term will become clear when we discuss the molecular orbital energy diagram. The energy eigenfunctions are identical for $\hat{H}_{el}$ and $\hat{H}$. Only the eigenvalues are affected by separating out the nuclear repulsion [see Equation (23.5)].

The goal in this chapter is to develop a qualitative model of chemical bonding in diatomic molecules. Quantitative computational chemistry models discussed in Chapter 26 are required to determine accurate bond lengths and bond energies. The qualitative model that we discuss assumes that electrons in molecules occupy **molecular orbitals (MOs)** that extend over the molecule similar to how electrons in an atom occupy atomic orbitals. A given MO, ψ_{el}, can be written as a linear combination of the **atomic orbitals (AOs)** on individual atoms in the molecule. This is called the **LCAO-MO model.** The justification for this assumption is that the linear combination of AOs is the simplest wave function we can write that leads to the electron delocalization over the molecule. In the rest of this chapter, we drop the subscript *el* to simplify the notation. Keep in mind that we are calculating only the electronic part of $\psi_i^{molecule}$ and that we are doing so at a fixed set of nuclear positions. To simplify the mathematics, we consider only a diatomic molecule AB and assume that each MO is generated by combining only one AO on each atom, ϕ_a and ϕ_b on atoms A and B, respectively. The AOs are the **basis functions** for the MO. Such a small basis set is inadequate for quantitative calculations, and in solving the computational problems at the end of the chapter, you will use much larger basis sets.

We next write an approximate MO in terms of the atomic orbitals, $\psi_1 = c_a\phi_a + c_b\phi_b$, and minimize the MO energy with respect to the values of the AO coefficients c_1 and c_2. The expectation value of the MO energy ε for this approximate wave function is given by

$$
\langle \varepsilon \rangle = \frac{\int \psi_1^* \hat{H}_{el} \psi_1 \, d\tau}{\int \psi_1^* \psi_1 \, d\tau}
$$

$$
= \frac{\int (c_a\phi_a + c_b\phi_b)^* \hat{H}_{el}(c_a\phi_a + c_b\phi_b) \, d\tau}{\int (c_a\phi_a + c_b\phi_b)^*(c_a\phi_a + c_b\phi_b) \, d\tau}
$$

$$
= \frac{(c_a)^2 \int \phi_a^* \hat{H}_{el}\phi_a \, d\tau + (c_b)^2 \int \phi_b^* \hat{H}_{el}\phi_b \, d\tau + 2c_a c_b \int \phi_a^* \hat{H}_{el}\phi_b \, d\tau}{(c_a)^2 \int \phi_a^* \phi_a \, d\tau + (c_b)^2 \int \phi_b^* \phi_b \, d\tau + 2c_a c_b \int \phi_a^* \phi_b \, d\tau} \quad \textbf{(23.6)}
$$

Because the AOs are normalized, the first two integrals in the denominator of the last line of Equation (23.6) have the value 1.

$$
\langle \varepsilon \rangle = \frac{(c_a)^2 H_{aa} + (c_b)^2 H_{bb} + 2c_a c_b H_{ab}}{(c_a)^2 + (c_b)^2 + 2c_a c_b S_{ab}} \quad \textbf{(23.7)}
$$

In the preceding equation, the symbol H_{ij} is a shorthand notation for the integrals involving $\hat{H}_{el}$ and the AOs i and j as follows:

$$
H_{ij} = \int \phi_i^*(\tau)\hat{H}_{el}\phi_j(\tau) \, d\tau \quad \textbf{(23.8)}
$$

S_{ab} is called the **overlap integral** and is an abbreviation for $S_{ab} = \int \phi_a^* \phi_b \, d\tau$. The overlap is a new concept that was not encountered in atomic systems. The meaning of S_{ab} is indicated pictorially in Figure 23.1. In words, it is a measure of the degree to which both of the AOs have nonzero values in the same region. S_{ab} can have values between zero and one. It has the value zero for widely separated atoms and increases as the atoms approach one another. As we will see later, in order to have chemical bond formation it is necessary that $S_{ab} > 0$.

To minimize ε with respect to the coefficients, ε is first differentiated with respect to c_a and c_b. We then set the two resulting expressions equal to zero and

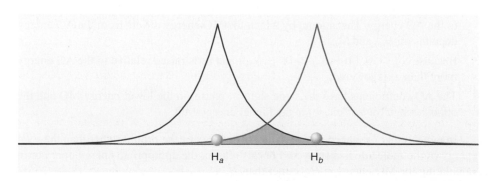

FIGURE 23.1
The amplitude of two $1s$ atomic orbitals is shown along an axis connecting the atoms. The overlap is appreciable only for regions in which the amplitude of both AOs is significantly different from zero. Such a region is shown schematically in orange. In reality, the overlap occurs in three-dimensional space.

H_a H_b

solve for c_a and c_b. By multiplying both sides of the equation by the denominator before differentiating, the following two equations are obtained:

$$(2c_a + 2c_b S_{ab})\varepsilon + \frac{\partial \varepsilon}{\partial c_a}\left((c_a)^2 + (c_b)^2 + 2c_a c_b S_{ab}\right) = 2c_a H_{aa} + 2c_b H_{ab}$$

$$(2c_b + 2c_a S_{ab})\varepsilon + \frac{\partial \varepsilon}{\partial c_b}\left((c_a)^2 + (c_b)^2 + 2c_a c_b S_{ab}\right) = 2c_b H_{bb} + 2c_a H_{ab} \quad \textbf{(23.9)}$$

Setting $\partial \varepsilon / \partial c_a$ and $\partial \varepsilon / \partial c_b = 0$ and rearranging these two equations results in the following two linear equations for c_a and c_b, which are called the **secular equations:**

$$c_a(H_{aa} - \varepsilon) + c_b(H_{ab} - \varepsilon S_{ab}) = 0$$

$$c_a(H_{ab} - \varepsilon S_{ab}) + c_b(H_{bb} - \varepsilon) = 0 \quad \textbf{(23.10)}$$

As shown in the Math Supplement (Appendix A), these equations have a solution other than $c_a = c_b = 0$ only if the **secular determinant** satisfies the condition

$$\begin{vmatrix} H_{aa} - \varepsilon & H_{ab} - \varepsilon S_{ab} \\ H_{ab} - \varepsilon S_{ab} & H_{bb} - \varepsilon \end{vmatrix} = 0 \quad \textbf{(23.11)}$$

The secular determinant is a 2×2 determinant because the basis set consists of only one AO on each atom.

Expanding the determinant generates a quadratic equation for the MO energy ε. The two solutions are

$$\varepsilon = \frac{1}{2 - 2S_{ab}^2}[H_{aa} + H_{bb} - 2S_{ab}H_{ab}] \pm \frac{1}{2 - 2S_{ab}^2}$$

$$\times \left[\sqrt{\left(H_{aa}^2 + 4H_{ab}^2 + H_{bb}^2 - 4S_{ab}H_{ab}H_{bb} - 2H_{aa}(H_{bb} + 2S_{ab}H_{ab} - 2S_{ab}^2 H_{bb})\right)} \right]$$

$$\textbf{(23.12)}$$

For homonuclear diatomic molecules $H_{aa} = H_{bb}$. In this case, Equation (23.12) simplifies to

$$\varepsilon_1 = \frac{H_{aa} + H_{ab}}{1 + S_{ab}} \quad \text{and} \quad \varepsilon_2 = \frac{H_{aa} - H_{ab}}{1 - S_{ab}} \quad \textbf{(23.13)}$$

We return to hetereonuclear diatomic molecules in Section 23.8. Using the H_2^+ molecule as an example, we show later that H_{aa} and H_{ab} are both negative and since $S_{ab} > 0$, $\varepsilon_2 > \varepsilon_1$. Substituting ε_1 in Equations (23.10), we find that $c_a = c_b$, whereas if ε_2 is substituted in the same equations, we obtain $c_a = -c_b$.

Figure 23.2 summarizes the following results of this discussion pictorially in a **molecular orbital energy diagram** using H_2 as an example:

- Two localized AOs combine to form two delocalized MOs provided that S_{ab} is nonzero. This is the case if there are regions in space in which the amplitudes of both AOs are nonzero.

- The energy of one MO is lowered and the energy of the other MO is raised relative to the AO energy. The amount by which the MO energy differs from the AO energy depends on H_{ab} and S_{ab}.

- Because $S_{ab} > 0$, $(1 + S_{ab}) > (1 - S_{ab})$ and ε_2 is raised relative to the AO energy more than ε_1 is lowered.

- The AO coefficients have the same sign (in-phase) in the lower energy MO and the opposite sign (out-of-phase) in the higher energy MO.

In a molecular orbital energy diagram, the energy of the orbital rather than the total energy of the molecule is displayed. For this reason, the appropriate energy operator in calculating the MO energy is $\hat{H}_{el}$ rather than $\hat{H}$.

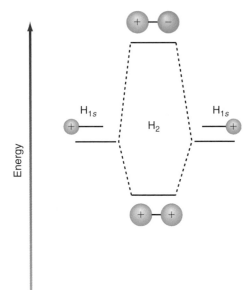

FIGURE 23.2

Molecular orbital energy diagram for a qualitative description of bonding in H_2. The atomic orbitals are shown to the left and right, and the molecular orbitals are shown in the middle. Dashed lines connect the MO with the AOs from which it was constructed. Shaded circles have a diameter proportional to the coefficients c_a and c_b. Red and blue shading signifies positive and negative signs of the AO coefficients, respectively. Interchanging red and blue does not generate a different MO.

EXAMPLE PROBLEM 23.1

Show that substituting $\varepsilon_1 = \dfrac{H_{aa} + H_{ab}}{1 + S_{ab}}$ in Equations (23.10) gives the result $c_a = c_b$.

Solution

$$c_a\left(H_{aa} - \frac{H_{aa} + H_{ab}}{1 + S_{ab}}\right) + c_b\left(H_{ab} - \frac{H_{aa} + H_{ab}}{1 + S_{ab}}S_{ab}\right) = 0$$

$$c_a([1 + S_{ab}]H_{aa} - [H_{aa} + H_{ab}]) + c_b([1 + S_{ab}]H_{ab} - [H_{aa} + H_{ab}]S_{ab}) = 0$$

$$c_a(H_{aa} + S_{ab}H_{aa} - H_{aa} - H_{ab}) + c_b(H_{ab} + H_{ab}S_{ab} - H_{aa}S_{ab} - H_{ab}S_{ab}) = 0$$

$$c_a(H_{aa}S_{ab} - H_{ab}) - c_b(H_{aa}S_{ab} - H_{ab}) = 0$$

$$c_a = c_b$$

Substitution in the second of the two Equations (23.10) gives the same result.

23.2 The Simplest One-Electron Molecule: H₂⁺

In the previous section, we outlined a formalism to generate MOs from AOs. We next apply this formalism to the only molecule for which the electronic Schrödinger equation can be solved exactly, the one-electron H_2^+ molecular ion. Just as for atoms, the Schrödinger equation cannot be solved exactly for any molecule containing more than one electron. Rather than discuss the exact solution, we approach H_2^+ using the LCAO-MO model, which gives considerable insight into chemical bonding and, most importantly, can be extended easily to many-electron molecules.

We begin by setting up the electronic Schrödinger equation for H_2^+ in the Born–Oppenheimer approximation. Figure 23.3 shows the relative positions of the two protons and the electron in H_2^+ at a particular instant in time. The total energy operator for this molecule has the form

$$\hat{H} = -\frac{\hbar^2}{2m_e}\nabla_e^2 - \frac{e^2}{4\pi\varepsilon_0}\left(\frac{1}{r_a} + \frac{1}{r_b}\right) + \frac{e^2}{4\pi\varepsilon_0}\frac{1}{R} \tag{23.14}$$

The first term is the electron kinetic energy, the second term is the attractive Coulombic interaction between the electron and each of the nuclei, and the last term is the nuclear–nuclear repulsion. We again separate out the nuclear repulsion term and write an electronic energy operator

$$\hat{H}_{el} = -\frac{\hbar^2}{2m_e}\nabla_e^2 - \frac{e^2}{4\pi\varepsilon_0}\left(\frac{1}{r_a} + \frac{1}{r_b}\right) \tag{23.15}$$

From experimental results, we know that H_2^+ is a stable species, so that solving the Schrödinger equation for H_2^+ must give at least one bound state. We define the zero of total energy as an H atom and an H^+ ion that are infinitely separated. Given this choice of the zero, a stable molecule has a negative energy. The energy function $E_{total}(R)$ has a minimum value for a distance R_e, which is the equilibrium bond length.

We next discuss the approximate wave functions for the H_2^+ molecule in the LCAO-MO model. Imagine slowly bringing together a H atom and a H^+ ion. At infinite separation, the electron is in a $1s$ orbital on either one nucleus or the other. However, as the internuclear distance approaches R_e, the potential energy wells for the two species overlap, and the barrier between them is lowered. Consequently, the electron can move back and forth between the Coulomb wells on the two nuclei. It is equally likely to be on nucleus a as on nucleus b so that the molecular wave function looks like the superposition of a $1s$ orbital on each nucleus as shown pictorially in Figure 23.4.

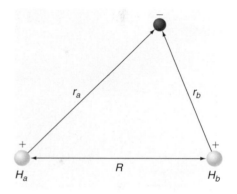

FIGURE 23.3
The two protons and the electron are shown at one instant in time. The quantities R, r_a, and r_b represent the distances between the charged particles.

FIGURE 23.4
The potential energy of the H_2^+ molecule is shown for two different values of R (red curves). At large distances, the electron will be localized in a $1s$ orbital either on nucleus a or b. However, at the equilibrium bond length R_e, the two Coulomb potentials overlap, allowing the electron to be delocalized over the whole molecule. The purple curve represents the amplitude of the atomic (top panel) and molecular (bottom panel) wave functions, and the solid horizontal lines represent the corresponding energy eigenvalues.

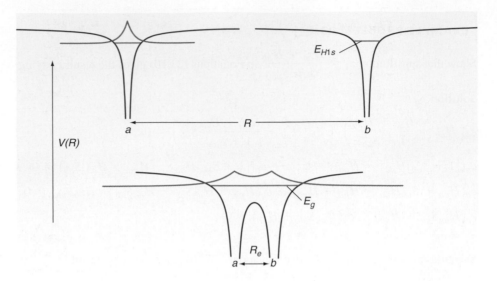

The AOs used to form the MOs are the $1s$ orbitals, ϕ_{H1s}. To allow the electron distribution around each nucleus to change as the bond is formed, a **variational parameter** ζ is inserted in each AO:

$$\phi_{H1s} = \frac{1}{\sqrt{\pi}}\left(\frac{\zeta}{a_0}\right)^{3/2} e^{-\zeta r/a_0} \tag{23.16}$$

This parameter looks like an effective nuclear charge. You will see in the end-of-chapter problems that varying ζ allows the size of the orbital to change.

In the previous section, we showed that $c_a = \pm c_b$. Although the signs of c_a and c_b can differ, the magnitude of the coefficients is the same. Using this result, the two MOs are

$$\psi_g = c_g(\phi_{H1s_a} + \phi_{H1s_b})$$
$$\psi_u = c_u(\phi_{H1s_a} - \phi_{H1s_b}) \tag{23.17}$$

The wave functions for a homonuclear diatomic molecule are classified as g or u based on whether they change signs upon undergoing inversion through the center of the molecule. If the origin of the coordinate system is placed at the center of the molecule, inversion corresponds to $\psi(x, y, z) \rightarrow \psi(-x, -y, -z)$. If this operation leaves the wave function unchanged, that is, $\psi(x, y, z) = \psi(-x, -y, -z)$, it has **g symmetry.** If $\psi(x, y, z) = -\psi(-x, -y, -z)$, the wave function has **u symmetry.** The subscripts g and u refer to the German words *gerade* and *ungerade,* which can be translated as even and odd and are also referred to as **symmetric** and **antisymmetric.** See Figures 23.8 and 23.12 for illustrations of g and u MOs. We will see later that only ψ_g describes a stable, chemically bonded H_2^+ molecule.

The values of c_g and c_u can be determined by normalizing ψ_g and ψ_u. Note that the integrals used in the normalization are over all three spatial coordinates. Normalization requires that

$$1 = \int c_g^*(\phi_{H1s_a}^* + \phi_{H1s_b}^*)c_g(\phi_{H1s_a} + \phi_{H1s_b})\, d\tau$$
$$= c_g^2\left(\int \phi_{H1s_a}^*\phi_{H1s_a}\, d\tau + \int \phi_{H1s_b}^*\phi_{H1s_b}\, d\tau + 2\int \phi_{H1s_b}^*\phi_{H1s_a}\, d\tau\right) \tag{23.18}$$

The first two integrals in the second line have the value 1 because the H$1s$ orbitals are normalized, and we obtain the result

$$c_g = \frac{1}{\sqrt{2 + 2S_{ab}}} \tag{23.19}$$

The coefficient c_u has a similar form, as you will see in the end-of-chapter problems.

$$c_u = \frac{1}{\sqrt{2 - 2S_{ab}}} \tag{23.20}$$

23.3 The Energy Corresponding to the H_2^+ Molecular Wave Functions ψ_g and ψ_u

Keep in mind that the molecular wave functions we are using are approximate rather than exact eigenfunctions of the total energy operator of Equation (23.15). Therefore, we can only calculate the expectation value of the electronic energy for the state corresponding to ψ_g:

$$E_g = \frac{\int \psi_g^* \hat{H}_{el} \psi_g \, d\tau}{\int \psi_g^* \psi_g \, d\tau} = \frac{H_{aa} + H_{ab}}{1 + S_{ab}} \tag{23.21}$$

This result was derived in Section 23.1 where we also showed that $E_u = \dfrac{H_{aa} - H_{ab}}{1 - S_{ab}}$.

Looking ahead, we will find that the total energy corresponding to ψ_g is lower than that corresponding to ψ_u and that only ψ_g describes a stable H_2^+ molecule. To understand the difference between ψ_g and ψ_u, we must look in more detail at the integrals H_{aa} and H_{ab}.

To evaluate H_{aa}, we use $\hat{H}_{el}$ from Equation (23.15):

$$H_{aa} = \int \phi_{H1s_a}^* \left(-\frac{\hbar^2}{2m} \nabla^2 - \frac{e^2}{4\pi\varepsilon_0 r_a} \right) \phi_{H1s_a} \, d\tau$$

$$- \int \phi_{H1s_a}^* \left(\frac{e^2}{4\pi\varepsilon_0 r_b} \right) \phi_{H1s_a} \, d\tau \tag{23.22}$$

Assume initially that $\zeta = 1$, in which case ϕ_{H1s_a} is an eigenfunction of the operator in parentheses

$$\left(-\frac{\hbar^2}{2m} \nabla^2 - \frac{e^2}{4\pi\varepsilon_0 r_a} \right) \phi_{H1s} = E_{1s} \phi_{H1s} \tag{23.23}$$

Because the atomic wave functions are normalized, the first integral is equal to E_{1s} and H_{aa} is given by

$$H_{aa} = E_{1s} - J, \text{ where } J = \int \phi_{H1s_a}^* \left(\frac{e^2}{4\pi\varepsilon_0 r_b} \right) \phi_{H1s_a} \, d\tau \tag{23.24}$$

J represents the energy of interaction of the electron viewed as a negative diffuse charge cloud on atom a with the positively charged nucleus b. This result is exactly what would be calculated in classical electrostatics for a diffuse negative charge of density $\phi_{H1s_a}^* \phi_{H1s_a}$. What is the physical meaning of the energy H_{aa}? The quantity H_{aa} represents the total energy of an undisturbed hydrogen atom separated from a bare proton by the distance R excluding the nuclear repulsion. As $R \rightarrow \infty$, $H_{aa} \rightarrow E_{1s}$. What is the sign of H_{aa}? We know that $E_{1s} < 0$ and because all the terms in the integrand for J are positive, $J > 0$. Therefore, $H_{aa} < 0$.

Next, the energy $H_{ab} = H_{ba}$ is evaluated. Substituting as before, we find that

$$H_{ba} = \int \phi_{H1s_b}^* \left(-\frac{\hbar^2}{2m} \nabla^2 - \frac{e^2}{4\pi\varepsilon_0 r_a} \right) \phi_{H1s_a} \, d\tau$$

$$- \int \phi_{H1s_b}^* \left(\frac{e^2}{4\pi\varepsilon_0 r_b} \right) \phi_{H1s_a} \, d\tau \tag{23.25}$$

Evaluating the first integral gives $S_{ab} E_{1s}$ and

$$H_{ab} = S_{ab} E_{1s} - K \quad \text{where } K = \int \phi_{H1s_b}^* \left(\frac{e^2}{4\pi\varepsilon_0 r_b} \right) \phi_{H1s_a} \, d\tau \tag{23.26}$$

In this model, K plays a central role in the lowering of the energy that leads to the formation of a bond. However, it has no simple physical interpretation. It is a direct consequence of writing the MO as a superposition of two AOs, which leads to an interference term in $\psi_g^* \psi_g$ as seen in Equation (23.18). Both J and K are positive because all terms that appear in the integrals are positive over the entire range of the integration. Quantitative calculations show that near the equilibrium distance $R = R_e$, both H_{aa} and H_{ab} are negative, and $|H_{ab}| > |H_{aa}|$. For many electron atoms, integrals similar to J and K are generated and are referred to as Coulomb and exchange integrals, respectively.

The differences ΔE_g and ΔE_u between the electronic energy of the molecule in the states described by ψ_g and ψ_u and the energy of the H1s AO are calculated in Example Problem 23.2.

EXAMPLE PROBLEM 23.2

Using Equation (23.13) and (23.21), express the change in the MO energies resulting from bond formation, $\Delta E_g = E_g - E_{1s}$ and $\Delta E_u = E_u - E_{1s}$, in terms of J, K, and S_{ab}.

Solution

$$E_g = \frac{H_{aa} + H_{ab}}{1 + S_{ab}} = \frac{E_{1s} - J + S_{ab}E_{1s} - K}{1 + S_{ab}} = \frac{(1 + S_{ab})E_{1s} - J - K}{1 + S_{ab}}$$

$$= E_{1s} - \frac{J + K}{1 + S_{ab}}$$

$$\Delta E_g = E_g - E_{1s} = -\frac{J + K}{1 + S_{ab}}$$

$$E_u = \frac{H_{aa} - H_{ab}}{1 - S_{ab}} = \frac{E_{1s} - J - S_{ab}E_{1s} + K}{1 - S_{ab}} = \frac{(1 - S_{ab})E_{1s} - J + K}{1 - S_{ab}}$$

$$= E_{1s} - \frac{J - K}{1 - S_{ab}}$$

$$\Delta E_u = E_u - E_{1s} = -\frac{J - K}{1 - S_{ab}}$$

As discussed earlier, both J and K are positive. Quantitative calculations show that near the equilibrium distance $|K| > |J|$, so that ΔE_u is positive and ΔE_g is negative, meaning that the u state is raised and the g state is lowered in energy with respect to the H1s AO. These calculations also show that $|\Delta E_u| > |\Delta E_g|$ in agreement with Figure 23.2.

To assess the stability of the molecule with respect to its dissociation products, we must include the nuclear repulsion term and calculate E_{total} rather than E_{el} as a function of R. Using the approximate wave function of Equation (23.17), an analytical expression can be obtained for $E_{total}(R, \zeta)$. For details, see I. N. Levine, *Quantum Chemistry*. The energy is minimized with respect to ζ at each of the R values in a variational calculation. The resulting $E_{total}(R)$ curves are shown schematically in Figure 23.5. The value of the energy as $R \rightarrow \infty$ is the total energy of a H atom and a proton at infinite separation, or E_{1s}. For the H atom as for any atom, $E_{total} = E_{el}$. The most important

FIGURE 23.5
Schematic energy functions $E_{total}(R)$ are shown for the g and u states in the approximate solution discussed. The reference energy (-13.6 eV) corresponds to infinitely separated H and H$^+$ species E_{1s}, which is the limit of H_{aa} as $R \rightarrow \infty$.

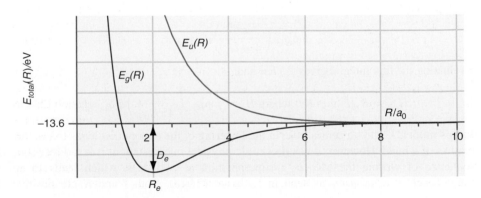

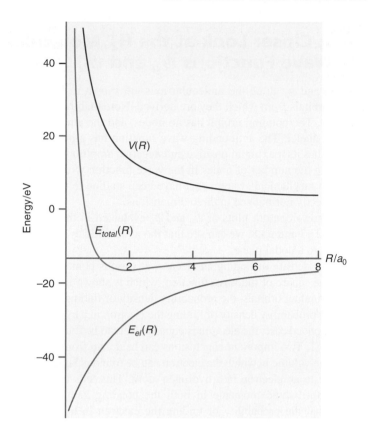

FIGURE 23.6

E_{total}, E_{el}, and the nuclear repulsion energy $V(R)$ obtained in an exact calculation are shown as a function of R. The horizontal line corresponds to -13.6 eV. The data for E_{el} is taken from "Electron Energy for H_2^+ in the Ground State," H. Wind, *Journal of Chemical Physics* 42(1965): 2371.

conclusions that can be drawn from this figure are that ψ_g describes a stable H_2^+ molecule because the energy has a well-defined minimum at $R = R_e$ and that ψ_u does not describe a bound state of H and H^+ because $E_u(R) > 0$ for all R, which makes the molecule unstable with respect to dissociation. *Therefore, we conclude that only a H_2^+ molecule described by ψ_g is a stable molecule.* The ψ_g and ψ_u wave functions are referred to as **bonding** and **antibonding molecular orbitals,** respectively, to emphasize their relationship to the chemical bond.

The equilibrium distance R_e and the bond energy D_e are of particular interest and have the values $R_e = 1.98\ a_0$ and 2.36 eV. ζ has the value 1.24 for ψ_g and 0.90 for ψ_u at R_e. The result that $\zeta > 1$ for ψ_g shows that the optimal H1s AO to use in constructing ψ_g is contracted relative to a free H atom. This means that the electron in H_2^+ in the ψ_g state is pulled in closer to each of the nuclei than it would be in a free hydrogen atom. The opposite is true for the ψ_u state.

Values of E_{total}, E_{el} and the nuclear repulsion energy $V(R)$ obtained in an exact calculation are shown as a function of R in Figure 23.6. As $R \rightarrow \infty$, $V(R) \rightarrow 0$, and $E_{el} \rightarrow -13.6$ eV, which is the electronic energy (and total energy) of a H atom. As $R \rightarrow 0$, $E_{el} \rightarrow -54.4$ eV, which is the electronic energy of a He^+ ion. At large R values, $E_{total}(R)$ is dominated by $E_{el}(R)$ and is negative. However, at small R values, $E_{total}(R)$ is dominated by $V(R)$ and is positive. This crossover results in a minimum in the total energy at $R = 1.98\ a_0$ and a bond energy of 2.79 eV or 269 kJ mol^{-1}. The calculated binding energy D_e in the simple model is 2.36 eV, which is reasonably close to the exact value, and the exact and calculated R_e values are both 1.98 a_0. The fact that the approximate values are quite close to the exact values validates the assumption that the exact molecular wave function is quite similar to ψ_g.

What have we learned so far about the origin of the chemical bond? It is tempting to attribute the binding to H_{ab} or K and, within the LCAO-MO formalism that we have used, this is correct. However, other formalisms for solving the Schrödinger equation for the H_2^+ molecule do not give rise to these integrals. We should, therefore, look for an explanation of chemical binding that is independent of the formalism used. For this reason, we seek the origin of the chemical bond in the differences between ψ_g and ψ_u as the wave functions are essentially independent of the method used to obtain them. This statement is true without the caveat for sufficiently accurate calculations using different methods.

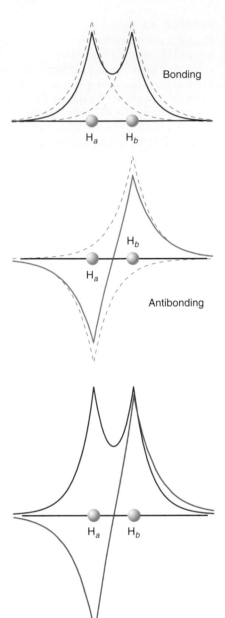

FIGURE 23.7

Molecular wave functions ψ_g and ψ_u (solid lines), evaluated along the internuclear axis are shown in the top two panels. The unmodified ($\zeta = 1$) H1s orbitals from which they were generated are shown as dashed lines. The bottom panel shows a direct comparison of ψ_g and ψ_u.

23.4 A Closer Look at the H_2^+ Molecular Wave Functions ψ_g and ψ_u

The values of ψ_g and ψ_u along the molecular axis are shown in Figure 23.7 together with the atomic orbitals from which they are derived. Note that the two wave functions are quite different. The bonding orbital has no nodes, and the amplitude of ψ_g is quite high between the nuclei. The antibonding wave function has a node midway between the nuclei and ψ_u has its maximum positive and negative amplitudes at the nuclei. Note that the increase in the number of nodes in the wave function with energy is similar to the other quantum mechanical systems that have been studied to this point. Both wave functions are correctly normalized in three dimensions.

Figure 23.8 shows contour plots of ψ_g and ψ_u evaluated in the $z = 0$ plane. If we compare Figures 23.7 and 23.8, we can see that the node midway between the H atoms in ψ_u corresponds to a nodal plane.

The probability density of finding an electron at various points along the molecular axis is given by the square of the wave function, which is shown in Figure 23.9. For the antibonding and bonding orbitals, the probability density of finding the electron in H_2^+ is compared with the probability density of finding the electron in a hypothetical nonbonded case. For the nonbonded case, the electron is equally likely to be found on each nucleus in H1s AOs and $\zeta = 1$. Two important conclusions can be drawn from this figure. First, for both ψ_g and ψ_u, the volume in which the electron can be found is large compared with the volume accessible to an electron in a hydrogen atom. This tells us that the electron is delocalized over the whole molecule in both the bonding and antibonding orbitals. Second, we see that the probability of finding the electron in the region between the nuclei is quite different for ψ_g and ψ_u. For the antibonding orbital, the probability is zero midway between the two nuclei, but for the bonding orbital, it is quite high. This difference is what makes the g state a bonding state and the u state an antibonding state.

This pronounced difference between ψ_g^2 and ψ_u^2 is explored further in Figure 23.10. The *difference* between the probability density for these orbitals and the hypothetical nonbonding state is shown in this figure. This difference tells us how the electron density would change if we could suddenly switch on the interaction at the equilibrium geometry. We see that for the antibonding state, electron density would move from the region between the two nuclei to the outer regions of the molecule. For the bonding state, electron density would move both to the region between the nuclei and closer to each nucleus. The origin of the density increase between the nuclei for the bonding orbital is the interference term $2\phi_{H1s_a}\phi_{H1s_b}$ in $(\phi_{H1s_a} + \phi_{H1s_b})^2$. The origin of the density increase near each nucleus is the increase in ζ from 1.00 to 1.24 in going from the free atom to the H_2^+ molecule.

The probability density is increased relative to the nonbonding case in the region between the nuclei and decreased by the same amount outside of this region. The opposite is true for ψ_u. Although it may not be apparent in Figures 23.7 to 23.10, the wave functions satisfy this requirement. Only small changes in the probability density outside

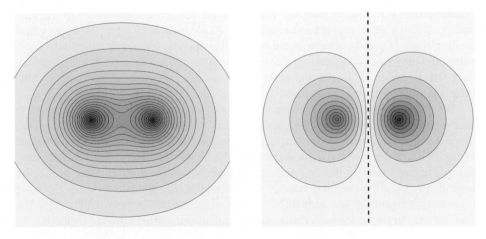

FIGURE 23.8

Contour plots of ψ_g (left) and ψ_u (right). Positive and negative amplitudes are shown as red and blue respectively. Darker colors indicate larger values for the magnitude of the amplitude. The dashed line indicates the position of the nodal plane in ψ_u.

of the region between the nuclei are needed to balance larger changes in this region, because the integration volume outside of the region between the nuclei is much larger. The data shown as a line plot in Figure 23.10 are shown as a contour plot in Figure 23.11. Red and blue correspond to the most positive and least positive values for $\Delta\psi_g^2$ and $\Delta\psi_u^2$, respectively. The outermost contour for $\Delta\psi_g^2$ in Figure 23.11 corresponds to a negative value, and it is seen that the corresponding area is large. The product of the small negative charge in $\Delta\psi_g^2$ with the large volume corresponding to the contour area is equal in magnitude and opposite in sign to the increase in $\Delta\psi_g^2$ in the bonding region.

The comparison of the electron charge densities associated with ψ_g and ψ_u helps us to understand the important ingredients in chemical bond formation. For both states, the electronic charge undergoes a **delocalization** over the whole molecule. However, charge is also localized in the molecular orbitals, and this **localization** is different in the bonding and antibonding states. In the bonding state, the electronic charge redistribution relative to the nonbonded state leads to a charge buildup both near the nuclei and between the nuclei. In the antibonding state, the electronic charge redistribution leads to a charge buildup outside of the region between the nuclei. We conclude that electronic charge buildup between the nuclei is an essential ingredient of a chemical bond.

We now look at how this charge redistribution affects the kinetic and potential energy of the H_2^+ molecule. A more detailed account is given by N. C. Baird [*J. Chemical Education* 63 (1986): 660]. The **virial theorem** is very helpful in this context. The virial theorem applies to atoms or molecules described either by exact wave functions or by approximate wave functions if these wave functions have been optimized with respect to all possible parameters. This theorem says that for a Coulomb potential, the average kinetic and potential energies are related by

$$\langle E_{potential} \rangle = -2\langle E_{kinetic} \rangle \qquad (23.27)$$

Because $E_{total} = E_{potential} + E_{kinetic}$, it follows that

$$\langle E_{total} \rangle = -\langle E_{kinetic} \rangle = \frac{1}{2}\langle E_{potential} \rangle \qquad (23.28)$$

Because this equation applies both to the nonbonded case and to the H_2^+ molecule at its equilibrium geometry, the change in total, kinetic, and potential energies associated with bond formation is given by

$$\langle \Delta E_{total} \rangle = -\langle \Delta E_{kinetic} \rangle = \frac{1}{2}\langle \Delta E_{potential} \rangle \qquad (23.29)$$

For the molecule to be stable, $\langle \Delta E_{total} \rangle < 0$ and, therefore, $\langle \Delta E_{kinetic} \rangle > 0$ and $\langle \Delta E_{potential} \rangle < 0$. Bond formation must lead to an increase in the kinetic energy and a decrease in the potential energy. How does this result relate to the competing effects of charge localization and delocalization that we saw for ψ_g and ψ_u?

Imagine that we could break down the change in the electron charge distribution as the bond is formed into two separate steps. First, we bring the proton and H atom to a distance R_e and let them interact, keeping the effective nuclear charge at the value $\zeta = 1$. In this step, the kinetic energy of the electron decreases, and it can be shown that the potential energy changes little. Therefore, the total energy will decrease. Why is the kinetic

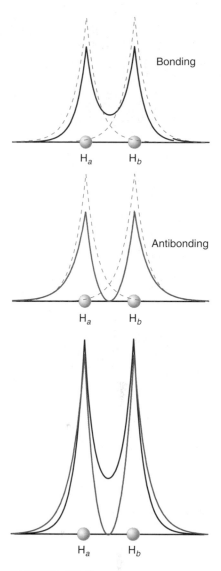

FIGURE 23.9

The upper two panels show the probability densities ψ_g^2 and ψ_u^2 along the internuclear axis for the bonding and antibonding wave functions. The dashed lines show $\frac{1}{2}\psi_{H1s_a}^2$ and $\frac{1}{2}\psi_{H1s_b}^2$, which are the probability densities for unmodified ($\zeta = 1$) H1s orbitals on each nucleus. The lowest panel shows a direct comparison of ψ_g^2 and ψ_u^2. Both molecular wave functions are correctly normalized in three dimensions.

FIGURE 23.10

The red curve shows ψ_g^2 (left panel) and the purple curve shows ψ_u^2 (right panel). The light blue curves show the differences $\Delta\psi_g^2 = \psi_g^2 - 1/2(\psi_{H1s_a})^2 - 1/2(\psi_{H1s_b})^2$ (left panel) and $\Delta\psi_u^2 = \psi_u^2 - 1/2(\psi_{H1s_a})^2 - 1/2(\psi_{H1s_b})^2$ (right panel). These differences are a measure of the change in electron density near the nuclei due to bond formation. A charge buildup occurs for the bonding orbital and a charge depletion occurs for the antibonding orbital in the region between the nuclei.

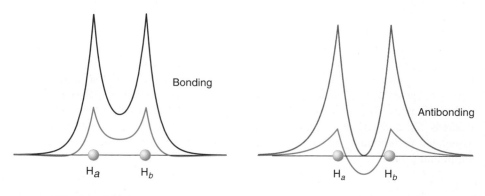

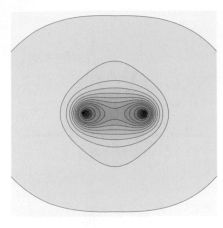

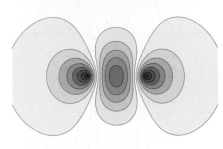

FIGURE 23.11
Contour plots of $\Delta\psi_g^2$ (top) and $\Delta\psi_u^2$ (bottom). Positive and negative amplitudes are shown as blue red and blue respectively. Darker colors indicate larger values for the magnitude of the amplitude.

energy lower? The explanation follows directly from our analysis of the particle in the one-dimensional box: as the box length increases, the kinetic energy decreases. Similarly, as the electron is delocalized over the whole space of the molecule, the kinetic energy decreases. By looking only at this first step, we see that electron delocalization alone will lead to bond formation. However, the total energy of the molecule can be reduced further at the fixed internuclear distance R_e by optimizing ζ. At the optimal value of $\zeta = 1.24$, some of the electron charge is withdrawn from the region between the nuclei and redistributed around the two nuclei. Because the size of the "box" around each atom is decreased, the kinetic energy of the molecule is increased. This increase is sufficiently large that $\langle \Delta E_{kinetic} \rangle > 0$ for the overall two-step process.

However, increasing ζ from 1.0 to 1.24 decreases the potential energy of the molecule because of the increased Coulombic interaction between the electron and the two protons. The result is that $\langle \Delta E_{potential} \rangle$ is lowered more than $\langle \Delta E_{kinetic} \rangle$ is raised. Therefore, the total energy of the molecule decreases further in this second step. Although the changes in $\langle \Delta E_{potential} \rangle$ and $\langle \Delta E_{kinetic} \rangle$ are both quite large, $\langle \Delta E_{total} \rangle$ changes very little as ζ increases from 1.0 to 1.24. Although $\langle \Delta E_{kinetic} \rangle > 0$ for the two step process, the dominant driving force for bond formation is electron delocalization, which is associated with $\langle \Delta E_{kinetic} \rangle < 0$. This result holds for bond formation in general.

At this point, we summarize what has been learned about the chemical bond. We have carried out an approximate solution of the Schrödinger equation for the simplest molecule imaginable and have developed a formalism based on delocalized molecular orbitals derived from atomic orbitals. We conclude that both charge delocalization and localization play a role in chemical bond formation. Delocalization promotes bond formation because the kinetic energy is lowered as the electron occupies a larger region in the molecule than it would in the atom. However, localization through the contraction of atomic orbitals and the accumulation of electron density between the atoms in the state described by ψ_g lowers the total energy even further. Both localization and delocalization play a role in bond formation, and it is this complex interplay between opposites that leads to a strong chemical bond.

23.5 Homonuclear Diatomic Molecules

In this section, we develop a qualitative picture of the shape and spatial extent of molecular orbitals for homonuclear diatomic molecules. Following the same path used in going from the H atom to many-electron atoms, we construct MOs for many-electron molecules on the basis of the excited states of the H_2^+ molecule. These MOs are useful in describing bonding in first and second row homonuclear diatomic molecules. Heteronuclear diatomic molecules are discussed in Section 23.8.

All MOs for homonuclear diatomics can be divided into two groups with regard to each of two **symmetry operations.** The first of these is rotation about the molecular axis which is taken to be along the z axis. If this rotation leaves the MO unchanged, it has no nodes that contain this axis, and the MO has $\boldsymbol{\sigma}$ **symmetry.** Combining s AOs always gives rise to σ MOs for diatomic molecules. If the MO has one nodal plane containing the molecular axis, the MO has $\boldsymbol{\pi}$ **symmetry.** All diatomic MOs have either σ or π and either g or u symmetry. Combining p_x or p_y AOs always gives rise to π MOs if the AOs have a common nodal plane. The second operation is inversion through the center of the molecule. Placing the origin at the center of the molecule, inversion corresponds to $\psi(x, y, z) \rightarrow \psi(-x, -y, -z)$. If this operation leaves the MO unchanged, the MO has g symmetry. If $\psi(x, y, z) \rightarrow -\psi(-x, -y, -z)$, the MO has u symmetry. All MOs are constructed using $n = 1$ and $n = 2$ AOs. Molecular orbitals for H_2^+ of g and u symmetry are shown in Figure 23.12. Note that $1\sigma_g$ and $1\pi_u$ are bonding MOs, whereas $1\sigma_u^*$ and $1\pi_g^*$ are antibonding MOs showing that u and g cannot be uniquely associated with bonding and antibonding. The symbol * is usually used to indicate an antibonding MO.

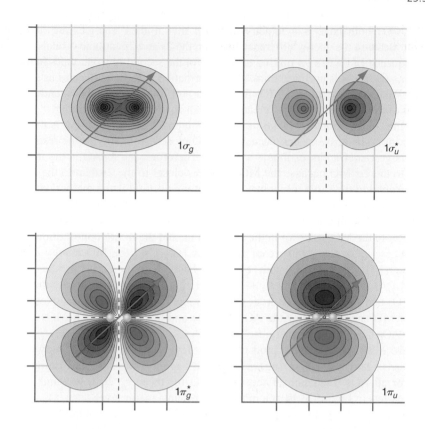

FIGURE 23.12
Contour plots of several bonding and antibonding orbitals of H_2^+. Positive and negative amplitudes are shown as red and blue respectively. Darker colors indicate larger values for the magnitude of the amplitude. The green arrows show the transformation $(x, y, z) \rightarrow (-x, -y, -z)$ for each orbital. If the amplitude of the wave function changes sign under this transformation, it has u symmetry. If it is unchanged, it has g symmetry.

Only atomic orbitals of the same symmetry can combine with one another to form a molecular orbital. For this example, we consider only s and p electrons. Figure 23.13 shows that a net nonzero overlap between two atomic orbitals occurs only if both AOs are either cylindrically symmetric with respect to the molecular axis (σ MOs) or if both have a common nodal plane that coincides with the molecular axis (π MOs).

Two different notations are commonly used to describe MOs in homonuclear diatomic molecules. In the first, the MOs are classified according to symmetry and increasing energy. For instance, a $2\sigma_g$ orbital has the same symmetry but a higher energy than the $1\sigma_g$ orbital. In the second notation, the integer indicating the relative energy is omitted, and the AOs from which the MOs are generated are listed instead. For instance, the $\sigma_g(2s)$ MO has a higher energy than the $\sigma_g(1s)$ MO. The superscript * is used to designate antibonding orbitals. Two types of MOs can be generated by combining $2p$ AOs. If the axis of the $2p$ orbital lies on the intermolecular axis (by convention the z axis), two σ MOs are generated. These MOs are called $3\sigma_g$ and $3\sigma_u^*$ depending on the relative phase of the AOs. Adding $2p_x$ (or $2p_y$) orbitals on each atom gives π MOs because of the nodal plane containing the molecular axis. These MOs are called $1\pi_u$ and $1\pi_g^*$ MOs.

In principle, we should take linear combinations of all the basis functions of the same symmetry (either σ or π) when constructing MOs. However, little mixing occurs between AOs of the same symmetry if they have greatly different orbital energies. For example, the mixing between $1s$ and $2s$ AOs for the second row homonuclear diatomics can be neglected at our level of discussion. However, for these same molecules, the $2s$

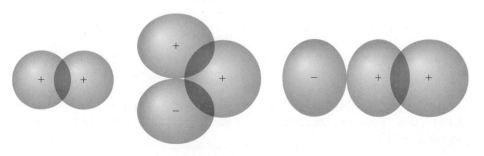

FIGURE 23.13
The overlap between two $1s$ orbitals $(\sigma + \sigma)$, a $1s$ and a $2p_x$ or $2p_y$ $(\sigma + \pi)$, and a $1s$ and a $2p_z$ $(\sigma + \sigma)$ are depicted from left to right. Note that the two shaded areas in the middle panel have opposite signs, so the net overlap of these two atomic orbitals of different symmetry is zero.

and $2p_z$ AOs both have σ symmetry and will mix if their energies are not greatly different. Because the energy difference between the $2s$ and $2p_z$ atomic orbitals increases in the sequence Li $\rightarrow$ F, **$s-p$ mixing** decreases for the second row diatomics in the order Li_2, B_2, ..., O_2, F_2. It is useful to think of MO formation in these molecules as a two-step process. We first create separate MOs from the $2s$ and $2p$ AOs, and subsequently combine the MOs of the same symmetry to create new MOs that include $s-p$ mixing.

Are the contributions from the s and p AOs equally important in MOs that exhibit $s-p$ mixing? The answer is no because the AO closest in energy to the resulting MO has the largest coefficient c_{ij} in $\psi_j = \Sigma_i c_{ij} \phi_i$. Therefore, the $2s$ AO is the major contributor to the 2σ MO because the MO energy is closer to the $2s$ than to the $2p$ orbital energy. Applying the same reasoning, the $2p_z$ atomic orbital is the major contributor to the 3σ MO. The MOs used to describe chemical bonding in first and second row homonuclear diatomic molecules are shown in Table 23.1. The AO that is the major contributor to the MO is shown in the last column, and the minor contribution is shown in parentheses. For the sequence of molecules $H_2 \rightarrow N_2$, the MO energy calculated using higher level methods with extended basis sets increases in the sequence $1\sigma_g < 1\sigma_u^* < 2\sigma_g < 2\sigma_u^* < 1\pi_u < 3\sigma_g < 1\pi_g^* < 3\sigma_u^*$. Moving across the periodic table to O_2 and F_2, the relative order of the $1\pi_u$ and $3\sigma_g$ MOs changes. Note that the first four MO energies follow the AO sequence and that the σ and π MOs generated from $2p$ AOs have different energies.

Figure 23.14 shows contour plots of the first few H_2^+ MOs, including only the major AO in each case (no $s-p$ mixing). The orbital exponent has not been optimized and $\zeta = 1$ for all AOs. Inclusion of the minor AO for the $2\sigma_g$, $2\sigma_u^*$, $3\sigma_g$, and $3\sigma_u^*$ MOs alters the plots in Figure 23.14 at a minor rather than a major level.

We next discuss the most important features of these plots. As might be expected, the $1\sigma_g$ orbital has no nodes, whereas the $2\sigma_g$ orbital has a nodal surface and the $3\sigma_g$ orbital has two nodal surfaces. All σ_u^* orbitals have a nodal plane perpendicular to the internuclear axis. The π orbitals have a nodal plane containing the internuclear axis. The amplitude for all the antibonding σ MOs is zero midway between the atoms on the molecular axis. This means that the probability density for finding electrons in this region will be small. The antibonding $1\sigma_u^*$ and $3\sigma_u^*$ orbitals have a nodal plane, and the $2\sigma_u^*$ orbital has both a nodal plane and a nodal surface. The $1\pi_u$ orbital has no nodal plane other than on the intermolecular axis, whereas the $1\pi_g^*$ orbital has one nodal plane in the bonding region.

Note that the MOs made up of AOs with $n = 1$ do not extend as far away from the nuclei as the MOs made up of AOs with $n = 2$. In other words, electrons that occupy valence AOs are more likely to overlap with their counterparts on neighboring atoms than are electrons in core AOs. This fact is important in understanding which electrons participate in making bonds in molecules, as well as in understanding reactions between molecules. The MOs shown in Figure 23.14 are specific to H_2^+ and have been calculated using $R = 2.00\ a_0$ and $\zeta = 1$. The detailed shape of these MOs varies from molecule to molecule and depends primarily on the effective nuclear charge ζ and the bond length. We can get a qualitative idea of what the MOs look like for other molecules by using the

TABLE 23.1	**Molecular Orbitals Used to Describe Chemical Bonding in Homonuclear Diatomic Molecules**		
MO Designation	Alternate	Character	Atomic Orbitals
$1\sigma_g$	$\sigma_g(1s)$	Bonding	$1s$
$1\sigma_u^*$	$\sigma_u^*(1s)$	Antibonding	$1s$
$2\sigma_g$	$\sigma_g(2s)$	Bonding	$2s\ (2p_z)$
$2\sigma_u^*$	$\sigma_u^*(2s)$	Antibonding	$2s\ (2p_z)$
$3\sigma_g$	$\sigma_g(2p_z)$	Bonding	$2p_z\ (2s)$
$3\sigma_u^*$	$\sigma_u^*(2p_z)$	Antibonding	$2p_z\ (2s)$
$1\pi_u$	$\pi_u(2p_x, 2p_y)$	Bonding	$2p_x, 2p_y$
$1\pi_g^*$	$\pi_g^*(2p_x, 2p_y)$	Antibonding	$2p_x, 2p_y$

H_2^+ MOs with the effective nuclear charge obtained from Hartree–Fock calculations for the molecule of interest.

For example, the bond length for F_2 is ~35% greater than that for H_2^+ and $\zeta = 8.65$ and 5.1 for the $1s$ and $2p$ orbitals, respectively. Because $\zeta > 1$, the amplitude of the fluorine AOs falls off much more rapidly with the distance from the nucleus than is the case for the H_2^+ molecule. Figure 23.15 shows $1\sigma_g$, $3\sigma_u^*$, and $1\pi_u$ MOs for these ζ values generated using the H_2^+ AOs. Note how much more compact the AOs and MOs are compared with $\zeta = 1$. The overlap between the $1s$ orbitals used to generate the lowest energy MO in F_2 is very small. For this reason, electrons in this MO do not contribute to the chemical bond in F_2. Note also that the $3\sigma_u^*$ orbital for F_2 exhibits three nodal surfaces between the atoms rather than one node shown in Figure 23.15 for H_2^+ with $\zeta = 1$. Unlike the $1\pi_u$ MO for H_2^+, the F_2 $1\pi_u$ MO shows distinct contributions from each atom because the amplitude of the $2p$ AOs falls off rapidly along the internuclear axis. However, apart from these differences, the general features shown in Figure 23.15 are common to the MOs of all first and second row homonuclear diatomics.

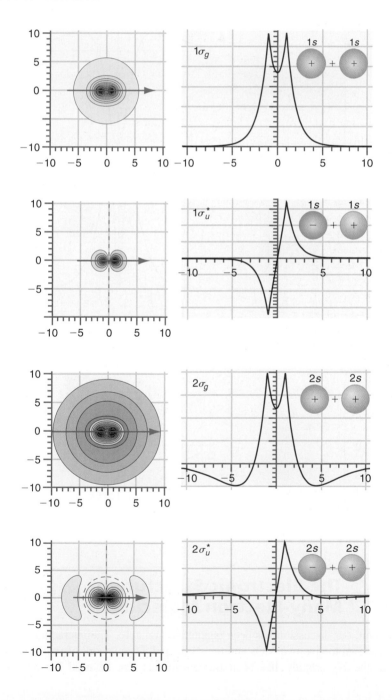

FIGURE 23.14

MOs based on the ground and excited states for H_2^+ generated from $1s$, $2s$, and $2p$ atomic orbitals. Contour plots are shown on the left and line scans along the path indicated by the green arrow are shown on the right. Positive and negative amplitudes are shown as red and blue respectively. Darker colors indicate larger values for the magnitude of the amplitude. Dashed lines and curves indicate nodal surfaces. Lengths are in units of a_0, and $R_e = 2.00\ a_0$.

FIGURE 23.14
(continued)

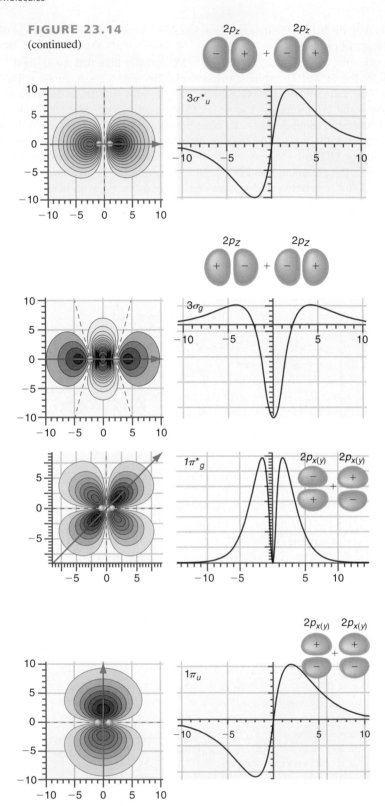

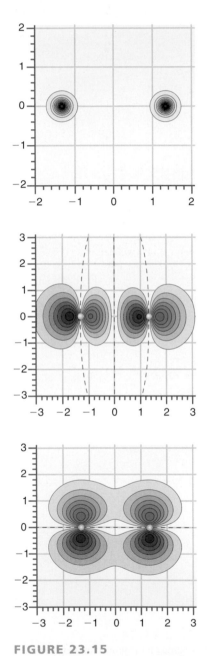

FIGURE 23.15
Contour plots for the $1\sigma_g$ (top), $3\sigma_u^*$ (center), and $1\pi_u$ (bottom) H_2^+ MOs with ζ values appropriate to F_2. Positive and negative amplitudes are shown as red and blue respectively. Darker colors indicate larger values for the magnitude of the amplitude. Dashed lines indicate nodal surfaces. Light circles indicate position of nuclei. Lengths are in units of a_0, and $R_e = 2.66\ a_0$.

23.6 The Electronic Structure of Many-Electron Molecules

To this point, our discussion has been qualitative in nature. The interaction of two AOs has been shown to give two MOs and the shape and a framework of molecular orbitals, based on the H_2^+ orbitals, has been introduced that can be used for many-electron diatomic molecules. To calculate aspects of diatomic molecules such as the MO energies, the bond length, and the dipole moment, the Schrödinger equation must be solved numerically. As for many-electron atoms, the starting point for quantitative molecular calculations is the Hartree–Fock model. As the formulation of the model is more complex for molecules than for atoms, we refer the interested reader to Chapter 26 and to sources such as I. N. Levine, *Quantum Chemistry*. As was discussed for many electron atoms in Chapter 21, the crucial input for a calculation is the expansion of the one-electron molecular orbitals ψ_j in a basis set of the N basis functions ϕ_i, and a variety of basis sets is available in commercially available computational chemistry software.

$$\psi_j = \sum_{i=1}^{N} c_{ij}\phi_i \qquad (23.30)$$

Although calculations using the Hartree–Fock model generally give sufficiently accurate values for bond lengths in diatomic molecules and bond angles in polyatomic molecules, accurate energy level calculations require electron correlation to be taken into account as discussed in Chapter 26.

Once the MO energy levels have been calculated, a **molecular configuration** is obtained by putting two electrons in each MO, in order of increasing orbital energy, until all electrons have been accommodated. If the degeneracy of an energy level is greater than one, Hund's first rule is followed and the electrons are placed in the MOs in such a way that the total number of unpaired electrons is maximized.

We first discuss the molecular configurations for H_2 and He_2. The MO energy diagrams in Figure 23.16 show the number and spin of the electrons rather than the magnitude and sign of the AO coefficients as was the case in Figure 23.2. What can we say about the magnitude and sign of the AO coefficients for each of the four MOs in Figure 23.16?

The interaction of $1s$ orbitals on each atom gives rise to a bonding and an antibonding MO as shown schematically in Figure 23.16. Each MO can hold two electrons of opposite spin. The configurations for H_2 and He_2 are $(1\sigma_g)^2$ and $(1\sigma_g)^2(1\sigma_u^*)^2$, respectively. We should consider two cautionary remarks about the interpretation of molecular orbital energy diagrams. First, just as for the many-electron atom, the total energy of a molecule is not the sum of the MO energies. Therefore, it is not always valid to draw conclusions about the stability or bond strength of a molecule solely on the basis of the orbital energy diagram. Secondly, the words *bonding* and *antibonding* give information about the relative signs of the AO coefficients in the MO, but they do not convey whether the electron is bound to the molecule. The total energy for any stable molecule is lowered by adding electrons to any orbital for which the energy is less than zero. For example, O_2^- is a stable species compared to O_2 and an electron at infinity, even though the additional electron is placed in an antibonding MO.

For H_2, both electrons are in the $1\sigma_g$ MO, which is lower in energy than the $1s$ AOs. Calculations show that the $1\sigma_u^*$ MO energy is greater than zero. In this case, the total energy is lowered by putting electrons in the $1\sigma_g$ orbital and rises if electrons are additionally put into the $1\sigma_u^*$ as would be the case for H_2^-. In the MO model, He_2 has two electrons in each of the $1\sigma_g$ and $1\sigma_u^*$ orbitals. Because the energy of the $1\sigma_u^*$ orbital is greater than zero, He_2 is not a stable molecule in this model. In fact, He_2 is stable only below ~ 5 K as a result of a very weak van der Waals interaction, rather than chemical bond formation.

The preceding examples used a single $1s$ orbital on each atom to form molecular orbitals. We now discuss the molecules F_2 and N_2, for which both s and p AOs contribute to the MOs. Combining n AOs generates n MOs, so combining the $1s$, $2s$, $2p_x$, $2p_y$, and $2p_z$ AOs on N and F generates 10 MOs for F_2 and N_2. Although MOs with contributions from the $1s$ and $2s$ AOs are in principle possible, mixing does not occur for either molecule because the AOs have very different energies. Mixing between $2s$, $2p_x$ and $2p_y$, or $2p_x$ and $2p_y$ AOs does not occur, because the net overlap is zero. We next consider mixing between the $2s$ and $2p_z$ AOs. For F_2, $s-p$ mixing can be neglected because the $2s$ AO lies 21.6 eV below the $2p$ AO. The F_2 MOs, in order of increasing energy, are $1\sigma_g < 1\sigma_u^* < 2\sigma_g < 2\sigma_u^* < 3\sigma_g < 1\pi_u = 1\pi_u < 1\pi_g^* = 1\pi_g^*$ and the configuration for F_2 is $(1\sigma_g)^2(1\sigma_u^*)^2(2\sigma_g)^2(2\sigma_u^*)^2(3\sigma_g)^2(1\pi_u)^2(1\pi_u)^2(1\pi_g^*)^2(1\pi_g^*)^2$. For this molecule, the 2σ MOs are quite well described by a single $2s$ AO on each atom,

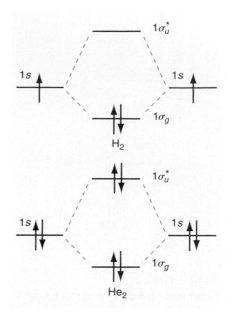

FIGURE 23.16

Atomic and molecular orbital energies and occupation for H_2 and He_2. Upward- and downward-pointing arrows indicate α and β spins respectively. The energy splitting between the MO levels is not to scale.

and the 3σ MOs are quite well described by a single $2p_z$ AO on each atom. Because the $2p_x$ and $2p_y$ AOs have a net zero overlap with each other, each of the doubly degenerate $1\pi_u$ and $1\pi_g^*$ molecular orbitals originates from a single AO on each atom. Figure 23.17 shows a molecular orbital energy diagram for F_2. Note that the $3\sigma_g$ and $3\sigma_u^*$ MOs have a greater energy separation than the $1\pi_u$ and $1\pi_g^*$ MOs. This is the case because the overlap of the $2p_z$ AOs is greater than the overlap of the $2p_x$ or $2p_y$ AOs.

For N_2, the $2s$ AO lies below the $2p$ AO by only 12.4 eV, and in comparison to F_2, $s-p$ mixing is not negligible. The MOs, in order of increasing energy, are $1\sigma_g < 1\sigma_u^* < 2\sigma_g < 2\sigma_u^* < 1\pi_u = 1\pi_u < 3\sigma_g < 1\pi_g^* = 1\pi_g^*$ and the configuration is $(1\sigma_g)^2(1\sigma_u^*)^2(2\sigma_g)^2(2\sigma_u^*)^2(1\pi_u)^2(1\pi_u)^2(3\sigma_g)^2$. Because of $s-p$ mixing, the 2σ and 3σ MOs have significant contributions from both $2s$ and $2p_z$ AOs with the result that the $3\sigma_g$ MO is higher in energy than the $1\pi_u$ MO. A MO energy diagram for N_2 is shown in Figure 23.18. The shape of the 2σ and 3σ N_2 MOs schematically indicates $s-p$ mixing. The $2\sigma_g$ MO has more bonding character because the probability of finding the electron between the atoms is higher than it was without $s-p$ mixing.

FIGURE 23.17
Schematic MO energy diagram for the valence electrons in F_2. The degenerate p and π orbitals are shown slightly offset in energy. The dominant atomic orbital contributions to the MOs are shown as solid lines. Minor contributions due to $s-p$ mixing have been neglected. The MOs are schematically depicted to the right of the figure. The $1\sigma_g$ and $1\sigma_u^*$ MOs are not shown.

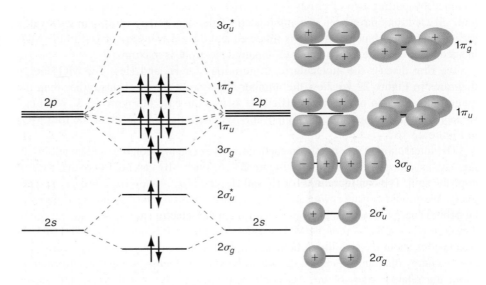

FIGURE 23.18
Schematic MO energy diagram for the valence electrons in N_2. The degenerate p and π orbitals are shown slightly offset in energy. The dominant AO contributions to the MOs are shown as solid lines. Lesser contributions arising from $s-p$ mixing are shown as dashed lines. The MOs are schematically depicted to the right of the figure. The $1\sigma_g$ and $1\sigma_u^*$ MOs are not shown.

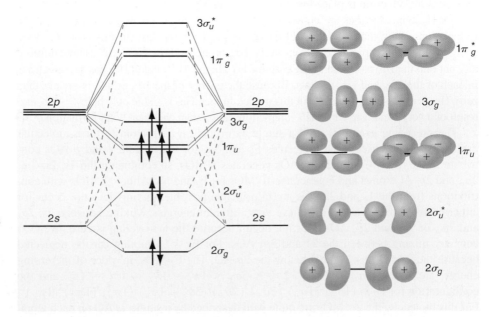

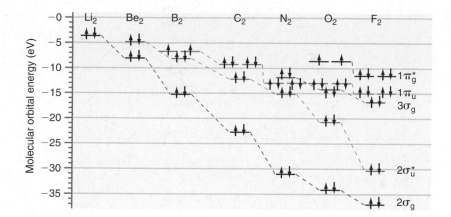

FIGURE 23.19
Molecular orbital energy levels for occupied MOs of the second row diatomics. The $1\sigma_g$ and $1\sigma_u^*$ orbitals lie at much lower values of energy and are not shown.
[From calculations by E. R. Davidson, unpublished]

Applying the same reasoning, the $2\sigma_u^*$ MO has become less antibonding and the $3\sigma_g$ MO has become less bonding for N_2 in comparison with F_2. We can see from the overlap in the AOs that the triple bond in N_2 arises from electron occupation of the $3\sigma_g$ and the pair of $1\pi_u$ MOs.

On the basis of this discussion of H_2, He_2, N_2, and F_2, the MO formalism is extended to all first and second row homonuclear diatomic molecules. After the relative energies of the molecular orbitals are established from numerical calculations, the MOs are filled in the sequence of increasing energy, and the number of unpaired electrons for each molecule can be predicted. The results for the second row are shown in Figure 23.19. Using Hund's first rule, we see that both B_2 and O_2 are predicted to have two unpaired electrons; therefore, these molecules should have a net magnetic moment (they are paramagnetic), whereas all other homonuclear diatomics should have a zero net magnetic moment (they are diamagnetic). These predictions are in good agreement with experimental measurements, which provides strong support for the validity of the MO model.

Figure 23.19 shows that the energy of the molecular orbitals tends to decrease with increasing atomic number in this series. This is a result of the increase in ζ in going across the periodic table. The larger effective nuclear charge and the smaller atomic size leads to a lower AO energy, which in turn leads to a lower MO energy. However, the $3\sigma_g$ orbital energy falls more rapidly across this series than the $1\pi_u$ orbital. This occurs because of a number of factors, including the decrease of $s-p$ mixing when going from Li_2 to F_2 and the change in overlap of the AOs resulting from changes in the bond length and effective nuclear charge. As a result, an inversion occurs in the order of molecular orbital energies between the $1\pi_u$ and $3\sigma_g$ orbitals for O_2 and F_2 relative to the other molecules in this series.

23.7 Bond Order, Bond Energy, and Bond Length

Molecular orbital theory has shown its predictive power by providing an explanation of the observed net magnetic moment in B_2 and O_2 and the absence of a net magnetic moment in the other second row diatomic molecules. We now show that the theory can also provide an understanding of trends in the binding energy and the vibrational force constant for these molecules. Figure 23.20 shows data for these observables for the series $H_2 \rightarrow Ne_2$. As the number of electrons in the diatomic molecule increases, the bond energy has a pronounced maximum for N_2 and a smaller maximum for H_2. The vibrational force constant k shows the same trend. The bond length increases as the bond energy and force constant decrease in the series $Be_2 \rightarrow N_2$, but it exhibits a more complicated trend for the lighter molecules. All of these data can be qualitatively understood using molecular orbital theory.

Consider the MO energy diagrams for H_2 and He_2 in Figure 23.16. For simplicity, we assume that the total energy of the molecules is proportional to the sum of the orbital energies. Because the bonding orbital is lower in energy than the atomic orbitals from which it was created, putting electrons into a bonding orbital leads to an energy

FIGURE 23.20
Bond energy, bond length, and vibrational force constant of the first 10 diatomic molecules as a function of the number of electrons in the molecule. The upper panel shows the calculated bond order for these molecules. The dashed line indicates the dependence of the bond length on the number of electrons if the He$_2$ data point is omitted.

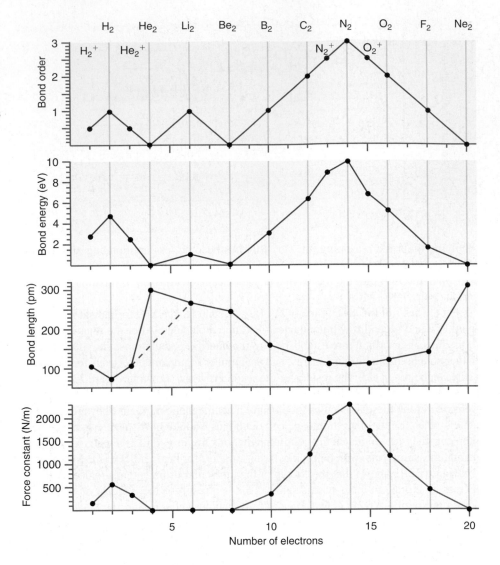

lowering with respect to the atoms. This makes the molecule more stable than the separated atoms, which is characteristic of a chemical bond. Similarly, putting two electrons into each of the bonding and antibonding orbitals leads to a total energy that is greater than that of the separated molecules. Therefore, the molecule is unstable with respect to dissociation into two atoms. This result suggests that stable bond formation requires more electrons to be in bonding than in antibonding orbitals. We introduce the concept of **bond order,** which is defined as

$$\text{Bond order} = 1/2[(\text{total bonding electrons}) - (\text{total antibonding electrons})]$$

We expect the bond energy to be very small for a bond order of zero and to increase with increasing bond order. As shown in Figure 23.20, the bond order shows the same trend as the bond energies. The bond order also tracks the vibrational force constant very well. Again, we can explain the data by associating a stiffer bond with a higher bond order. This agreement is a good example of how a model becomes validated and useful when it provides an understanding for different sets of experimental data.

The relationship between the bond length and the number of electrons in the molecule is influenced both by the bond order and by the variation of the atomic radius with the effective nuclear charge. For a given atomic radius, the bond length is expected to vary inversely with the bond order. This trend is approximately followed for the series Be$_2 \rightarrow$ N$_2$ in which the atomic radii are not constant but decrease steadily. The bond length increases in going from He$_2^+$ to Li$_2$ because the valence electron in Li is in the 2s rather than the 1s AO. The correlation between bond order and bond length also breaks down for He$_2$ because the atoms are not really chemically bonded. On balance, the trends shown in Figures 23.19 and 23.20 provide significant support for the concepts underlying molecular orbital theory.

EXAMPLE PROBLEM 23.3

Arrange the following in terms of increasing bond energy and bond length on the basis of their bond order: N_2^+, N_2, N_2^-, and N_2^{2-}.

Solution

The ground-state configurations for these species are

$$N_2^+: (1\sigma_g)^2(1\sigma_u^*)^2(2\sigma_g)^2(2\sigma_u^*)^2(1\pi_u)^2(1\pi_u)^2(3\sigma_g)^1$$

$$N_2: (1\sigma_g)^2(1\sigma_u^*)^2(2\sigma_g)^2(2\sigma_u^*)^2(1\pi_u)^2(1\pi_u)^2(3\sigma_g)^2$$

$$N_2^-: (1\sigma_g)^2(1\sigma_u^*)^2(2\sigma_g)^2(2\sigma_u^*)^2(1\pi_u)^2(1\pi_u)^2(3\sigma_g)^2(1\pi_g^*)^1$$

$$N_2^{2-}: (1\sigma_g)^2(1\sigma_u^*)^2(2\sigma_g)^2(2\sigma_u^*)^2(1\pi_u)^2(1\pi_u)^2(3\sigma_g)^2(1\pi_g^*)^1(1\pi_g^*)^1$$

In this series, the bond order is 2.5, 3, 2.5, and 2. Therefore, the bond energy is predicted to follow the order $N_2 > N_2^+$, $N_2^- > N_2^{2-}$ using the bond order alone. However, because of the extra electron in the antibonding $1\pi_g^*$ MO, the bond energy in N_2^- will be less than that in N_2^+. Because bond lengths decrease as the bond strength increases, the bond length will follow the opposite order.

Looking back at what we have learned about homonuclear diatomic molecules, several important concepts stand out. Combining atomic orbitals on each atom to form molecular orbitals provides a way to generate molecular configurations for molecules. Although including many AOs on each atom (that is, using a larger basis set) is necessary to calculate accurate MO energies, important trends can be predicted using a minimal basis set of one or two AOs per atom. The symmetry of atomic orbitals is important in predicting whether they contribute to a given molecular orbital. The concept of bond order allows us to understand why He_2, Be_2, and Ne_2 are not stable and why the bond in N_2 is so strong.

23.8 Heteronuclear Diatomic Molecules

We extend the discussion of Section 23.1 on generating molecular orbitals to heteronuclear diatomic molecules for which the AO energies are not equal. We again consider only one AO on each atom. To be specific, let ϕ_1 be a hydrogen $1s$ orbital and let ϕ_2 be a fluorine $2p_z$ orbital in the molecule HF. The bonding and antibonding MOs have the form

$$\psi_1 = c_{1H}\phi_{H1s} + c_{1F}\phi_{F2p_z} \quad \text{and} \quad \psi_2 = c_{2H}\phi_{H1s} + c_{2F}\phi_{F2p_z} \quad \textbf{(23.31)}$$

where the coefficients are to be determined. The MOs labeled 1 and 2 are the in-phase and out-of-phase combinations of the AOs, respectively. Normalization requires that

$$(c_{1H})^2 + (c_{1F})^2 + 2c_{1H}c_{1F}S_{HF} = 1 \quad \text{and}$$

$$(c_{2H})^2 + (c_{2F})^2 + 2c_{2H}c_{2F}S_{HF} = 1 \quad \textbf{(23.32)}$$

To calculate ε_1, ε_2, c_{1H}, c_{2H}, c_{1F}, and c_{2F}, we need numerical values for H_{HH}, H_{FF}, H_{HF}, and S_{HF}. To a good approximation, H_{HH} and H_{FF} correspond to the first ionization energies of H and F, respectively, and fitting experimental data gives the approximate empirical relation $H_{HF} = -1.75S_{HF}\sqrt{H_{HH}H_{FF}}$. We assume that $S_{HF} = 0.30$, $H_{HH} = -13.6$ eV, and $H_{FF} = -18.6$ eV, so that $H_{HF} = -8.35$ eV. We are looking for trends rather than striving for accuracy, so these approximate values are sufficiently good for our purposes. Substituting these values in Equation (23.12) gives the MO energy levels shown next. Example Problem 23.4 shows how to obtain the corresponding values of the coefficients.

$$\varepsilon_1 = -19.6 \text{ eV} \qquad \psi_1 = 0.34\phi_{H1s} + 0.84\phi_{F2p_z}$$

$$\varepsilon_2 = -10.3 \text{ eV} \qquad \psi_2 = 0.99\phi_{H1s} - 0.63\phi_{F2p_z} \qquad \textbf{(23.33)}$$

Note that the magnitudes of the coefficients in the MOs are not equal. The coefficient of the lower energy AO has the larger magnitude in the in-phase (bonding) MO and the smaller magnitude in the out-of-phase (antibonding) MO. The MO energy results for HF are shown in a molecular orbital energy diagram in Figure 23.21. The relative size of the AO coeffcients are indicated by the size of the AO, and the sign of the coefficient is indicated by the color of the symbol.

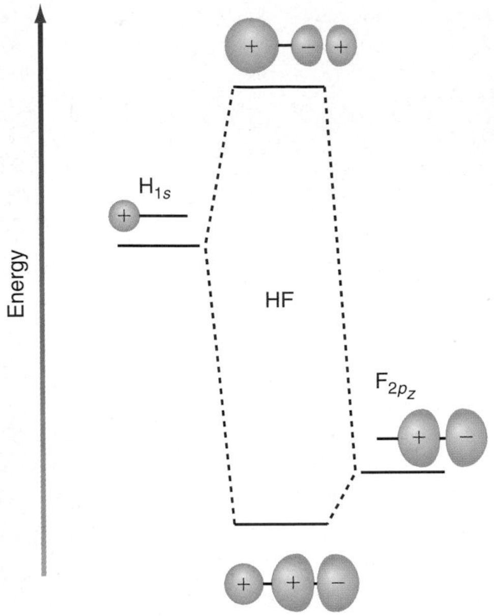

FIGURE 23.21
Molecular orbital energy diagram for a qualitative description of bonding in HF. The atomic orbitals are shown to the left and right, and the molecular orbitals are shown in the middle. Dashed lines connect the MO with the AOs from which it was constructed. Shaded circles have a diameter proportional to the coefficients c_{ij}. Red and blue shading signifies positive and negative signs of the AO coefficients, respectively.

EXAMPLE PROBLEM 23.4

Calculate c_{2H} and c_{2F} for the antibonding HF MO for which $\varepsilon_2 = -10.3$ eV. Calculate c_{1H} and c_{1F} for the HF bonding MO for which $\varepsilon_1 = -19.6$ eV. Assume that $S_{HF} = 0.30$.

Solution

We first obtain the result $H_{HF} = -1.75\,S_{HF}\sqrt{H_{HH}H_{FF}} = -8.35$ eV. We calculate c_{1H}/c_{1F} and c_{2H}/c_{2F} by substituting the values for ε_1 and ε_2 in the first equation in Equation (23.10)

$$c_{2H}(H_{HH} - \varepsilon_2) + c_{2F}(H_{HF} - \varepsilon_2 S_{HF}) = 0.$$

For $\varepsilon_2 = -10.3\ eV,\ c_{2H}(-13.6 + 10.3) + c_{2F}(-8.35 + 0.30 \times 10.3) = 0$

$$\frac{c_{2H}}{c_{2F}} = -1.58$$

Using this result in the normalization equation $c_{2H}^2 + c_{2F}^2 + 2c_{2H}c_{2F}S_{HF} = 1$

$$c_{2H} = 0.99,\ c_{2F} = -0.63,\ \text{and}\ \psi_2 = 0.99\phi_{H1s} - 0.63\phi_{F2p_z}$$

For $\varepsilon_1 = -19.6\ eV,\ c_{1H}(-13.6 + 19.6) + c_{1F}(-8.35 + 0.3 \times 19.6) = 0$

$$\frac{c_{1H}}{c_{1F}} = 0.41$$

Using this result in the normalization equation $c_{1H}^2 + c_{1F}^2 + 2c_{1H}c_{1F}S_{HF} = 1$

$$c_{1H} = 0.34,\ c_{1F} = 0.84,\ \text{and}\ \psi_1 = 0.34\phi_{H1s} + 0.84\phi_{F2p_z}$$

Just as for H_2^+, in the bonding MO, the coefficients of the AOs have the same sign (in-phase). In the antibonding MO, they have the opposite sign (out-of-phase). However, because the AO energies are not equal, the magnitude of the coefficient of the lower energy AO is *larger* in the bonding orbital and *smaller* in the antibonding orbital.

The relative magnitude of the coefficients of the AOs gives information about the charge distribution in the molecule, within the framework of the following simple model. Consider an electron in the HF bonding MO described by $\psi_1 = 0.34\phi_{H1s} + 0.84\phi_{F2p_z}$. The molecular dipole moment is greater as the difference between the coefficients increases. Because of the association made in the first postulate between $|\psi|^2$ and probability, the individual terms in $\int \psi_1^* \psi_1\, d\tau = (c_{1H})^2 + (c_{1F})^2 + 2c_{1H}c_{1F}S_{HF} = 1$ can be interpreted in the following way. We associate $(c_{1H})^2 = 0.12$ with the probability of finding the electron around the H atom, $(c_{1F})^2 = 0.71$ with the probability of finding the electron around the F atom, and $2c_{1H}c_{1F}S_{HF} = 0.17$ with the probability of finding the electron shared by the F and H atoms. We divide the shared probability equally between the atoms. This gives the probabilities of $(c_{1H})^2 + c_{1H}c_{1F}S_{HF} = 0.21$ and $(c_{1F})^2 + c_{1H}c_{1F}S_{HF} = 0.79$ for finding the electron on the H and F atoms, respectively. This result is reasonable given the known electronegativities of F and H. By comparison, Hartree–Fock calculations using a 28-member basis set give a charge of $+0.48$ and -0.48 on the H and F atoms, respectively. These calculated charges give rise to a dipole moment of 2.03 debye (1 debye $= 3.34 \times 10^{-30}$ C m), which is in good agreement with the experimental value of 1.91 debye.

Note that, although this method of assigning charge due to Robert Mulliken is reasonable, there is no unique way to distribute the electron charge in an MO among atoms because the charge on an atom is not a quantum mechanical observable. For a pictorial explanation of this assertion, see Figure 26.23. Note, however, that the charge transfer is in the opposite direction for the antibonding MO. We find that the shared probability has a positive sign for a bonding orbital and a negative sign for an antibonding orbital. This is a useful criterion for distinguishing between bonding and antibonding MOs.

The results for HF show that the bonding MO has a greater amplitude on F, which has the lower energy AO. In other words, the bonding MO is more localized on F than on H. We generalize this result to a molecule HX where the AO energy of X lies significantly lower than that of H by calculating ε_1, c_{1H}, and c_{1X} for different AO energies of X. The results are shown in Table 23.2 where $H_{HH} = -13.6$ eV and $S_{HX} = 0.30$.

Note that as the X AO energy becomes more negative, the X AO coefficient $\rightarrow 1$ and the H AO coefficient $\rightarrow 0$ in the bonding MO. It is also seen that the MO energy approaches the lower AO energy as the X AO energy becomes more negative, which means that the MO is essentially identical to the AO. This result shows that although we have assumed that MOs are delocalized over the molecule, a MO formed from AOs that differ substantially in energy is essentially localized on the atom with the lower AO energy.

We next discuss the nomenclature for MOs for heteronuclear diatomics. Because the two atoms are dissimilar, the u and g symmetries do not apply since inversion interchanges the nuclei. However, the MOs will still have either σ or π symmetry. Therefore, the MOs on a heteronuclear diatomic molecule are numbered differently than for the molecules Li_2-N_2:

Homonuclear	$1\sigma_g$	$1\sigma_u^*$	$2\sigma_g$	$2\sigma_u^*$	$1\pi_u$	$3\sigma_g$	$1\pi_g^*$	$3\sigma_u^*$	....
Heteronuclear	1σ	2σ	3σ	4σ	1π	5σ	2π	6σ	....

For larger molecules, the bonding and antibonding character can become difficult to discern. In these cases the symbol * is often not used. A common numbering system is to assign the 1σ MO to the lowest-energy valence MO rather than including, for example, the $1s$ electrons on F, which are localized on the F atom.

To illustrate the differences between homonuclear and heteronuclear diatomic molecules, we consider HF and construct MOs using the $1s$ AO on H and the $2s$ and $2p$ AOs on F. The molecular orbital energy diagram for HF is shown in Figure 23.22. The AOs on the two atoms that give rise to the MOs are shown on the right side of the diagram, with the size of the orbital proportional to its coefficient in the MO. Numerical calculations show that the $1s$ electrons are almost completely localized on the F atom. The 1π electrons are completely localized on the F atom because the $2p_x$ and $2p_y$

TABLE 23.2   AO Coefficients and MO Energies for Different Values of H_{xx}

H_{xx}(eV)	c_{1H}	c_{1F}	ε_1(eV)
-18.6	0.345	0.840	-19.9
-23.6	0.193	0.925	-24.1
-33.6	0.055	0.982	-33.7
-43.6	0.0099	1.00	-43.6

FIGURE 23.22
Schematic energy diagram showing the relationship between the atomic and molecular orbital energy levels for the valence electrons in HF. The degenerate p and π orbitals are shown slightly offset in energy. The dominant atomic orbital contributions to the MOs are shown as solid lines. Lesser contributions are shown as dashed lines. The MOs are depicted to the right of the figure. We assign the $1s$ electrons on F to the 1σ MO, which is localized on the F atom.

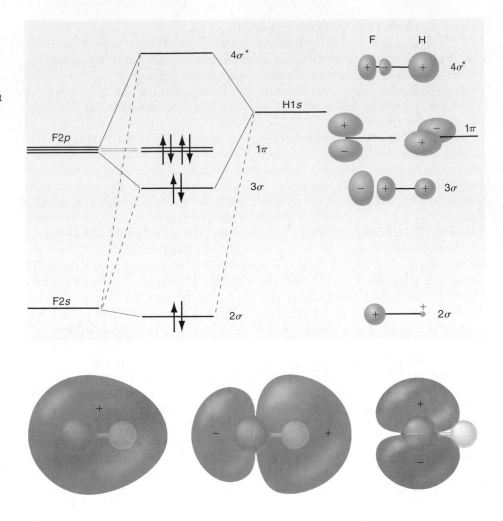

FIGURE 23.23
The 2σ, 3σ, and 1π MOs for HF are shown from left to right.

orbitals on F have a zero net overlap with the $1s$ orbital on H. Electrons in MOs localized on a single atom are referred to as nonbonding electrons. The mixing of $2s$ and $2p$ AOs in the 3σ and $4\sigma^*$ MOs changes the electron distribution in the HF molecule somewhat when compared with a homonuclear diatomic molecule. The 3σ MO has less bonding character and the $4\sigma^*$ MO has less antibonding character. Note that the total bond order is approximately one because the 3σ MO is largely localized on the F atom, the 3σ MO is not totally bonding, and the 1π MOs are completely localized on the F atom. The MO energy diagram depicts the MOs in terms of their constituent AOs. MOs 2 through 4 obtained in calculations using a 28-member basis set are shown in Figure 23.23.

As expected, in the 2σ bonding orbital the electron density is much greater on the more electronegative fluorine than on the hydrogen. However, in the antibonding $4\sigma^*$ orbital, this polarity is reversed. As you will see in the end-of-chapter problems, the estimated dipole moment is smaller in the excited state than in the ground state.

23.9 The Molecular Electrostatic Potential

As discussed in Section 23.8, the charge on an atom in a molecule is not a quantum mechanical observable and, consequently, atomic charges cannot be assigned uniquely. However, we know that the electron charge is not uniformly distributed in a polar molecule. For example, the region around the oxygen atom in H_2O has a net negative charge, whereas the region around the hydrogen atoms has a net positive charge.

How can this non-uniform charge distribution be discussed? To do so, we introduce the **molecular electrostatic potential,** which is the electrical potential felt by a test charge at various points in the molecule.

The molecular electrostatic potential is calculated by considering the contribution of the valence electrons and the atomic nuclei separately. Consider the nuclei first. For a point charge of magnitude q, the electrostatic potential $\phi(r)$ at a distance r from the charge, is given by

$$\phi(r) = \frac{q}{4\pi\varepsilon_0 r} \tag{23.34}$$

Therefore, the contribution to the molecular electrostatic potential from the atomic nuclei is given by

$$\phi_{nuclei}(x_1, y_1, z_1) = \sum_i \frac{q_i}{4\pi\varepsilon_0 r_i} \tag{23.35}$$

where q_i is the atomic number of nucleus i, and r_i is the distance of nucleus i from the observation point with the coordinates (x_1, y_1, z_1). The sum extends over all atoms in the molecule.

The electrons in the molecule can be considered as a continuous charge distribution with a density at a point with the coordinates (x, y, z) that is related to the n-electron wave function by

$$\rho(x, y, z) = -e \int \ldots \int (\psi(x, y, z; x_1, y_1, z_1; \ldots; x_n, y_n, z_n))^2$$

$$\times \, dx_1 \, dy_1 \, dz_1 \ldots dx_n \, dy_n \, dz_n \tag{23.36}$$

The integration is over the position variables of all n electrons. Combining the contributions of the nuclei and the electrons, the molecular electrostatic potential is given by

$$\phi(x_1, y_1, z_1) = \sum_i \frac{q_i}{4\pi\varepsilon_0 r_i} - e \iiint \frac{\rho(x, y, z)}{4\pi\varepsilon_0 r_e} \, dx \, dy \, dz \tag{23.37}$$

where r_e is the distance of an infinitesimal volume element of electron charge from the observation point with the coordinates (x_1, y_1, z_1).

The molecular electrostatic potential must be calculated numerically using the Hartree–Fock method or other methods discussed in Chapter 26. To visualize the polarity in a molecule, it is convenient to display a contour of constant electron density around the molecule and then display the values of the molecular electrostatic potential on the density contour using a color scale, as shown for HF in Figure 23.24. Negative values of the electrostatic potential, shown in red, are found near atoms to which electron charge transfer occurs. For HF, this is the region around the fluorine atom. Positive values of the molecular electrostatic potential, shown in blue, are found around atoms from which electron transfer occurs, as for the hydrogen atom in HF.

The calculated molecular electrostatic potential function identifies regions of a molecule that are either electron rich or depleted in electrons. We can use this function to predict regions of a molecule that are susceptible to nucleophilic or electrophilic attack as in enzyme–substrate reactions. The molecular electrostatic potential is particularly useful because it can also be used to obtain a set of atomic charges that is more reliable than the Mulliken model discussed in Section 23.8. This is done by initially choosing a set of atomic charges and calculating an approximate molecular electrostatic potential around a molecule using the set of charges in Equation (23.35). These atomic charges are varied systematically, subject to the constraint that the total charge is zero for a neutral molecule, until optimal agreement is obtained between the approximate and the accurate molecular electrostatic potential calculated from Equation (23.37). The atomic charges obtained in computational chemistry software such as *Spartan* are calculated in this way as discussed in Chapter 26.

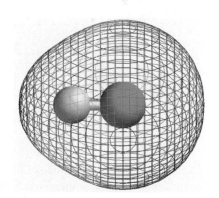

FIGURE 23.24
The grid shows a surface of constant electron density for the HF molecule. The fluorine atom is shown in green. The color shading on the grid indicates the value of the molecular electrostatic potential. Red and blue correspond to negative and positive values, respectively.

Vocabulary

π symmetry	bonding molecular orbital	molecular configuration	secular determinant
σ symmetry	Born–Oppenheimer	molecular electrostatic	secular equations
antibonding molecular orbital	approximation	potential	$s-p$ mixing
antisymmetric wave function	delocalization	molecular orbital (MO)	symmetric wave function
atomic orbital (AO)	delocalized	molecular orbital energy	symmetry operation
basis functions	g symmetry	diagram	u symmetry
bond energy	LCAO-MO model	molecular wave function	variational parameter
bond order	localization	overlap integral	virial theorem

Conceptual Problems

Q23.1   The following images show contours of constant electron density for H_2 calculated using the methods described in Chapter 26. The values of electron density are (a) 0.10, (b) 0.15, (c) 0.20, (d) 0.25, and (e) 0.30 electron/a_0^3.

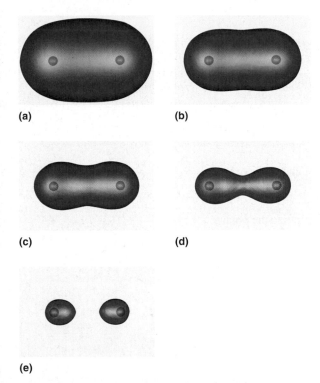

(a)

(b)

(c)

(d)

(e)

a. Explain why the apparent size of the H_2 molecule as approximated by the volume inside the contour varies in the sequence a–e.

b. Notice the neck that forms between the two hydrogen atoms in contours c and d. What does neck formation tell you about the relative density in the bonding region and in the region near the nuclei?

c. Explain the shape of the contours in image e by comparing this image with Figures 23.9 and 23.10.

d. Estimate the electron density in the bonding region midway between the H atoms by estimating the value of the electron density at which the neck disappears.

Q23.2   Consider the molecular electrostatic potential map for the NH_3 molecule shown here. Is the hydrogen atom (shown as a white sphere) an electron acceptor or an electron donor in this molecule?

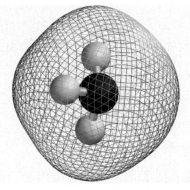

Q23.3   Give examples of AOs for which the overlap reaches its maximum value only as the internuclear separation approaches zero in a diatomic molecule. Also give examples of AOs for which the overlap goes through a maximum value and then decreases as the internuclear separation approaches zero.

Q23.4   Why is it reasonable to approximate H_{11} and H_{22} by the appropriate ionization energy of the corresponding neutral atom?

Q23.5   Identify the molecular orbitals for F_2 in the images shown here in terms of the two designations discussed in Section 23.7. The molecular axis is the z axis, and the y axis is tilted slightly out of the plane of the image.

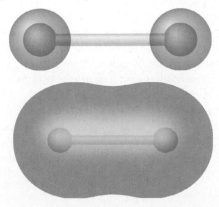

Q23.6 The molecular electrostatic potential maps for LiH and HF are shown here. Does the apparent size of the hydrogen atom (shown as a white sphere) tell you whether it is an electron acceptor or an electron donor in these molecules?

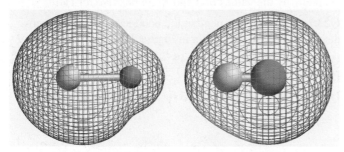

Q23.7 For H_2^+, explain why H_{aa} is the total energy of an undisturbed hydrogen atom separated from a bare proton by the distance R.

Q23.8 Distinguish between the following concepts used to describe chemical bond formation: basis set, minimal basis set, atomic orbital, molecular orbital, and molecular wave function.

Q23.9 Consider the molecular electrostatic potential map for the BH_3 molecule shown here. Is the hydrogen atom (shown as a white sphere) an electron acceptor or an electron donor in this molecule?

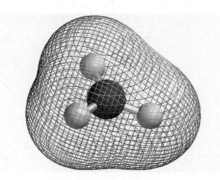

Q23.10 Using Figures 23.7 and 23.8, explain why $\Delta\psi_g^2 < 0$ and $\Delta\psi_u^2 > 0$ outside of the bonding region of H_2^+.

Q23.11 Consider the molecular electrostatic potential map for the BeH_2 molecule shown here. Is the hydrogen atom (shown as a white sphere) an electron acceptor or an electron donor in this molecule?

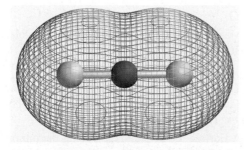

Q23.12 Why are MOs on heteronuclear diatomic molecules not labeled with g and u subscripts?

Q23.13 See Question Q23.5 for the images shown here.

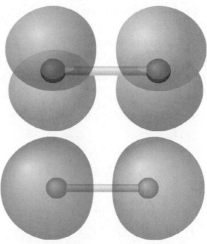

Q23.14 What is the justification for saying that, in expanding MOs in terms of AOs, the equality $\psi_j(1) = \sum_i c_{ij}\phi_i(1)$ can in principle be satisfied?

Q23.15 Why are the magnitudes of the coefficients c_a and c_b in the H_2^+ wave functions ψ_g and ψ_u equal?

Q23.16 Explain why $s-p$ mixing is more important in Li_2 than in F_2.

Q23.17 Justify the Born–Oppenheimer approximation based on vibrational frequencies and the timescale for electron motion.

Q23.18 Why can you conclude that the energy of the antibonding MO in H_2^+ is raised more than the energy of the bonding MO is lowered?

Q23.19 Does the total energy of a molecule rise or fall when an electron is put in an antibonding orbital?

Q23.20 Consider the molecular electrostatic potential map for the LiH molecule shown here. Is the hydrogen atom (shown as a white sphere) an electron acceptor or an electron donor in this molecule?

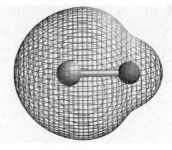

Q23.21 Consider the molecular electrostatic potential map for the H_2O molecule shown here. Is the hydrogen atom (shown as a white sphere) an electron acceptor or an electron donor in this molecule?

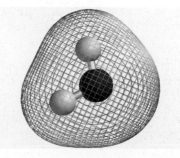

Q23.22   For the case of two H$1s$ AOs, the value of the overlap integral S_{ab} is never exactly zero even at very large separation of the H atoms. Explain this statement.

Q23.23   If there is a node in ψ_u, is the electron in this wave function really delocalized? How does it get from one side of the node to the other?

Q23.24   See Question Q23.5 for the images shown here.

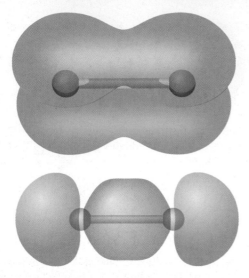

Q23.25   By considering each term in

$$K = \int \phi^*_{H1s_b}\left(\frac{e^2}{4\pi\varepsilon_0 r_b}\right)\phi_{H1s_a}\,d\tau$$

and

$$J = \int \phi^*_{H1s_a}\left(\frac{e^2}{4\pi\varepsilon_0 r_b}\right)\phi_{H1s_a}\,d\tau$$

explain why the values of J and K are positive for H_2^+.

Q23.26   Why do we neglect the bond length in He_2 when discussing the trends shown in Figure 23.20?

Q23.27   Explain why the nodal structures of the $1\sigma_g$ MOs in H_2 and F_2 differ.

Q23.28   See Question Q23.5 for the images shown here.

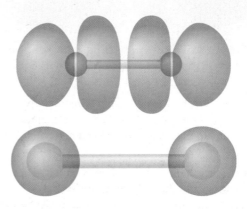

Q23.29   In discussing Figure 23.2, the following statement is made: *Interchanging red and blue does not generate a different MO.* Justify this statement.

Numerical Problems

Problem numbers in **red** indicate that the solution to the problem is given in the *Student's Solutions Manual.*

P23.1   Using ζ as a variational parameter in the normalized function $\psi_{H1s} = 1/\sqrt{\pi}(\zeta/a_0)^{3/2}e^{-\zeta r/a_0}$ allows one to vary the size of the orbital. Show this by calculating the probability of finding the electron inside a sphere of radius a_0 for different values of ζ using the standard integral

$$\int x^2 e^{-ax}\,dx = -e^{-ax}\left(\frac{2}{a^3} + 2\frac{x}{a^2} + \frac{x^2}{a}\right)$$

a. Obtain an expression for the probability as a function of ζ.

b. Evaluate the probability for $\zeta = 1.5, 2.5,$ and 3.5.

P23.2   The overlap integral for ψ_g and ψ_u as defined in Section 23.3 is given by

$$S_{ab} = e^{-\zeta R/a_0}\left(1 + \zeta\frac{R}{a_0} + \frac{1}{3}\zeta^2\frac{R^2}{a_0^2}\right)$$

Plot S_{ab} as a function of R/a_0 for $\zeta = 0.8$, 1.0, and 1.2. Estimate the value of R/a_0 for which $S_{ab} = 0.4$ for each of these values of ζ.

P23.3   Sketch out a molecular orbital energy diagram for CO and place the electrons in the levels appropriate for the ground state. The AO ionization energies are O$2s$: 32.3 eV; O$2p$: 15.8 eV; C$2s$: 19.4 eV; and C$2p$: 10.9 eV. The MO energies follow the sequence (from lowest to highest) 1σ, 2σ, 3σ, 4σ, 1π, 5σ, 2π, 6σ. Connect each MO level with the level of the major contributing AO on each atom.

P23.4   Explain the difference in the appearance of the MOs in Problem P23.13 with those for HF. Based on the MO energies, do you expect LiH$^+$ to be stable? Do you expect LiH$^-$ to be stable?

P23.5   Calculate the bond order in each of the following species. Predict which of the two species in the following pairs has the higher vibrational frequency:

a. Li$_2$ or Li$_2^+$ **b.** C$_2$ or C$_2^+$

c. O$_2$ or O$_2^+$ **d.** F$_2$ or F$_2^-$

P23.6   Make a sketch of the highest occupied molecular orbital (HOMO) for the following species:

a. N$_2^+$ **b.** Li$_2^+$ **c.** O$_2^-$ **d.** H$_2^-$ **e.** C$_2^+$

P23.7   The ionization energy of CO is greater than that of NO. Explain this difference based on the electron configuration of these two molecules.

P23.8   A Hartree–Fock calculation using the minimal basis set of the $1s$, $2s$, $2p_x$, $2p_y$, and $2p_z$ AOs on each of N and O generated the energy eigenvalues and AO coefficients listed in the following table:

MO	$\varepsilon(eV)$	c_{N1s}	c_{N2s}	c_{N2p_z}	c_{N2p_x}	c_{N2p_y}	c_{O1s}	c_{O2s}	c_{O2p_z}	c_{O2p_x}	c_{O2p_y}
3	−41.1	−0.13	+0.39	+0.18	0	0	−0.20	+0.70	+0.18	0	0
4	−24.2	−0.20	0.81	−0.06	0	0	0.16	−0.71	−0.30	0	0
5	−18.5	0	0	0	0	0.70	0	0	0	0	0.59
6	−15.2	+0.09	−0.46	+0.60	0	0	+0.05	−0.25	−0.60	0	0
7	−15.0	0	0	0	0.49	0	0	0	0	0.78	0
8	−9.25	0	0	0	0	0.83	0	0	0	0	−0.74

a. Designate the MOs in the table as σ or π symmetry and as bonding or antibonding. Assign the MOs to the following images, in which the O atom is red. The molecular axis is the z axis.

b. This calculation gives incorrect results for the shape and energies of MOs 5 and 7. Based on how these MOs arise, what energies and shapes would you expect for them?

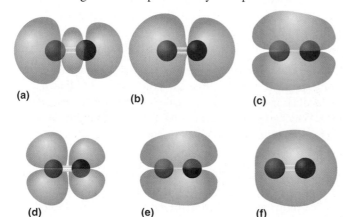

(a)   (b)   (c)

(d)   (e)   (f)

P23.9 Calculate the value for the coefficients of the AOs in Example Problem 23.4 for $S_{12} = 0.45$. How are they different from the values calculated in that problem for $S_{12} = 0.3$? Can you offer an explanation for the changes?

P23.10 Using the method of Mulliken, calculate the probabilities of finding an electron involved in the chemical bond on the H and F atoms for the bonding and antibonding MOs for Problem P23.9.

P23.11 Arrange the following in terms of decreasing bond energy and bond length: O_2^+, O_2, O_2^-, and O_2^{2-}.

P23.12 Predict the bond order in the following species:

a. N_2^+ b. Li_2^+ c. O_2^- d. H_2^- e. C_2^+

P23.13 Images of molecular orbitals for LiH calculated using the minimal basis set are shown here. In these images, the smaller atom is H. The H $1s$ AO has a lower energy than the Li $2s$ AO. The energy of the MOs is (left to right) -63.9 eV, -7.92 eV, and $+2.14$ eV. Make a molecular orbital diagram for this molecule, associate the MOs with the images, and designate the MOs in the following images as filled or empty. Which MO is the HOMO? Which MO is the LUMO? Do you expect the dipole moment in this molecule to have the negative end on H or Li?

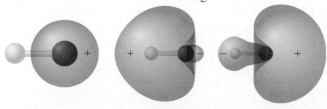

P23.14 What is the electron configuration corresponding to O_2, O_2^-, and O_2^+? What do you expect the relative order of bond strength to be for these species? Which, if any, have unpaired electrons?

P23.15 Calculate the dipole moment of HF for the bonding MO in Equation (23.33). Use the method outlined in Section 23.8 to calculate the charge on each atom. The bond length in HF is 91.7 pm. The experimentally determined dipole moment of ground-state HF is 1.91 debye, where 1 debye $= 3.33 \times 10^{-30}$ C m. Compare your result with this value. Does the simple theory give a reliable prediction of the dipole moment?

P23.16 Evaluate the energy for the two MOs generated by combining two H1s AOs. Carry out the calculation for $S_{12} = 0.15, 0.30$, and 0.45 to mimic the effect of decreasing the atomic separation in the molecule. Use the parameters $H_{11} = H_{22} = -13.6$ eV and $H_{12} = -1.75 S_{12}\sqrt{H_{11}H_{22}}$. Explain the trend that you observe in the results.

P23.17 Show that calculating E_u in the manner described by Equation (23.21) gives the result $E_u = (H_{aa} - H_{ab})/(1 - S_{ab})$.

P23.18 A surface displaying a contour of the total charge density in LiH is shown here. The molecular orientation is the same as in Problem P23.13. What is the relationship between this surface and the MOs displayed in Problem P23.13? Why does this surface closely resemble one of the MOs?

P23.19 Sketch the molecular orbital energy diagram for the radical OH based on what you know about the corresponding diagram for HF. How will the diagrams differ? Characterize the HOMO and LUMO as antibonding, bonding, or nonbonding.

P23.20 The bond dissociation energies of the species NO, CF^-, and CF^+ follow the trend $CF^+ > NO > CF^-$. Explain this trend using MO theory.

P23.21 Evaluate the energy for the two MOs generated by combining a H1s and a F2p AO. Use Equation (23.12) and carry out the calculation for $S_{HF} = 0.075, 0.18$, and 0.40 to mimic the effect of increasing the atomic separation in the molecule. Use the parameters $H_{11} = -13.6$ eV, $H_{22} = -18.6$ eV, and $H_{12} = -1.75 S_{12}\sqrt{H_{11}H_{22}}$. Explain the trend that you observe in the results.

P23.22 The expressions $(c_{11})^2 + c_{11}c_{21}S_{12}$ and $(c_{12})^2 + c_{11}c_{21}S_{12}$ for the probability of finding an electron on the H and F atoms in HF, respectively, were derived in Section 23.8. Use your results from Problem P23.21 and these expressions to calculate the probability of finding an electron

in the bonding orbital on the F atom for $S_{HF} = 0.075, 0.18,$ and 0.40. Explain the trend shown by these results.

P23.23 Follow the procedure outlined in Section 23.2 to determine c_u in Equation (23.17).

P23.24 Calculate the bond order in each of the following species. Which of the species in parts (a–d) do you expect to have the shorter bond length?
a. Li_2 or Li_2^+ **b.** C_2 or C_2^+ **c.** O_2 or O_2^+ **d.** F_2 or F_2^-

Computational Problems

More detailed instructions on carrying out these calculations using Spartan Physical Chemistry are found on the book website at *www.masteringchemistry.com*.

C23.1 According to Hund's rules, the ground state of O_2 should be a triplet because the last two electrons are placed in a doubly degenerate set of π MOs. Calculate the energy of the singlet and triplet states of O_2 using the B3LYP method and the 6-31G* basis set. Does the singlet or triplet have the lower energy? Both states will be populated if the energy difference $\Delta E \sim k_B T$. For which temperature is this the case?

C23.2 If the ground state of oxygen is a diradical, you might think that O_2 would dimerize to form square planar O_4 to achieve a molecule in which all electrons are paired. Optimize the geometry and calculate the energies of triplet O_2 and singlet O_4 using the B3LYP method and the 6-31G* basis set. Do you predict O_4 to be more or less stable than 2 O_2 molecules? Use a nonplanar shape in building your O_4 molecule. Is the geometry optimized molecule planar or nonplanar?

C23.3 O_6 might be more stable than O_4 because the bond angle is larger, leading to less steric strain. Optimize the geometry and compare the energy of O_6 with 1.5 times the energy of O_4 using the B3LYP method and the 6-31G* basis set. Is O_6 more stable than O_4? Use a nonplanar shape in building your O_6 molecule. Is the geometry optimized molecule planar or nonplanar?

C23.4 In a LiF crystal, both the Li and F are singly ionized species. Optimize the geometry and calculate the charge on Li and F in a single LiF molecule using the B3LYP method and the 6-31G* basis set. Are the atoms singly ionized? Compare the value of the bond length with the distance between Li^+ and F^- ions in the crystalline solid.

C23.5 Does LiF dissociate into neutral atoms or into Li^+ and F^-? Answer this question by comparing the energy difference between reactants and products for the reactions $LiF(g) \rightarrow Li(g) + F(g)$ and $LiF(g) \rightarrow Li^+(g) + F^-(g)$ using the B3LYP method and the 6-31G* basis set.

C23.6 Calculate Hartree–Fock MO energy values for HF using the MP2 method and the 6-31G* basis set. Make a molecular energy diagram to scale omitting the lowest energy MO. Why can you neglect this MO? Characterize the other MOs as bonding, antibonding, or nonbonding.

C23.7

a. Based on the molecular orbital energy diagram in Problem C23.6, would you expect triplet neutral HF in which an electron is promoted from the 1π to the $4\sigma^*$ MO to be more or less stable than singlet HF?

b. Calculate the equilibrium bond length and total energy for singlet and triplet HF using the MP2 method and the 6-31G* basis set. Using the frequency as a criterion, are both stable molecules? Compare the bond lengths and vibrational frequencies.

c. Calculate the bond energy of singlet and triplet HF by comparing the total energies of the molecules with the total energy of F and H. Are your results consistent with the bond lengths and vibrational frequencies obtained in part (b)?

C23.8 Computational chemistry allows you to carry out calculations for hypothetical molecules that do not exist in order to see trends in molecular properties. Calculate the charge on the atoms in singlet HF and in triplet HF for which the bond length is fixed at 10% greater than the bond length for singlet HF. Are the trends that you see consistent with those predicted by Figure 23.22? Explain your answer.

Web-Based Simulations, Animations, and Problems

W23.1 Two atomic orbitals are combined to form two molecular orbitals. The energy levels of the molecular orbitals and the coefficients of the atomic orbitals in each MO are

calculated by varying the relative energy of the AOs and the overlap, S_{12}, using sliders.

Molecular Structure and Energy Levels for Polyatomic Molecules

For diatomic molecules, the only structural element is the bond length, whereas in polyatomic molecules, bond lengths, bond angles, and the arrangement of the atoms determine the energy of the molecule. In this chapter, we discuss both localized and delocalized bonding models that enable the structures of small molecules to be predicted. We also discuss the usefulness of computational chemistry in determining the structure and energy levels of small molecules.

24.1 Lewis Structures and the VSEPR Model

In Chapter 23, we discussed chemical bonding and the electronic structure of diatomic molecules. Molecules with more than two atoms introduce a new aspect to our discussion of chemical bonding, namely, bond angles. In this chapter, the discussion of bonding is expanded to include the structure of small molecules. This will allow us to answer questions such as "Why is the bond angle 104.5° in H_2O and 92.2° in H_2S?" The most straightforward answer to this question is that the angles 104.5° and 92.2° in H_2O and in H_2S minimize the total energy of these molecules. As will be shown in Chapter 26, numerical quantum mechanical calculations of bond angles are in very good agreement with experimentally determined values. This result confirms that the approximations made in the calculation are valid and gives confidence that bond angles in molecules for which there are no data can be calculated. An interpretation of the numerical calculations is required to provide an understanding about *why* a bond angle of 104.5° minimizes the energy for H_2O, whereas a bond angle of 92.2° minimizes the energy for H_2S.

Results from calculations can be used to formulate useful qualitative theoretical models. For example, Walsh's Rules, which are discussed in Section 24.5, predict how the bond angle in a class of molecules H_2X with X equal to O, S, Se, or Te depends on X. Gaining a qualitative understanding of why small molecules have a particular structure is the primary goal of this chapter.

Molecular structure is addressed from two different vantage points. The significant divide between these points of view is their description of the electrons in a molecule as being localized, as in the valence bond (VB) model, or delocalized, as in the molecular orbital (MO) model. As discussed in Chapter 23, MO theory is based on electron orbitals that are delocalized over the entire molecule. By contrast, a **Lewis structure** represents molecular fluorine as $\ddot{\text{F}}$—$\ddot{\text{F}}$, which is a description in terms of localized bonds and lone pairs. These two viewpoints seem to be irreconcilable at first glance. However, as shown in Section 24.6, each point of view can be reformulated in the language of the other.

We first discuss molecular structure using **localized bonding models.** We do so because there is a long tradition in chemistry of describing chemical bonds in terms of the interaction between neighboring atoms. A great deal was known about the thermo-chemical properties, stoichiometry, and structure of molecules before the advent of quantum mechanics. For instance, scientists knew that a set of two atom bond enthalpies could be extracted from experimental measurements. Using these bond enthalpies, the enthalpy of formation of molecules can be calculated with reasonable accuracy as the sum of the bond enthalpies associated with the reactants minus the sum of the bond enthalpies for the product. Similarly, the bond length between two specific atoms, O—H, is found to be nearly the same in many different compounds. As discussed in Chapter 19, the characteristic vibrational frequency of a group such as —OH is largely independent of the composition of the rest of the molecule. Results such as these give strong support for the idea that a molecule can be described by a set of coupled but nearly independent chemical bonds between adjacent atoms. The molecule can be assembled by linking these chemical bonds.

Figure 24.1 shows how a structural formula is used to describe ethanol. This structural formula provides a pictorial statement of a localized bonding model. However, a picture like this raises a number of questions. Do the spokes in a ball and spoke model have any reality? Localized bonding models imply that bonding electrons are localized between adjacent atoms. However, we know from studying the particle in the box, the hydrogen atom, and many-electron atoms that localizing electrons results in a high energy cost. We also know that it is not possible to distinguish one electron from another, so does it make sense to assign some electrons in F_2 to lone pairs and others to the bond? At first glance, a localized model of bonding seems to be at odds with what we have learned about quantum mechanics. Yet, the preceding discussion provides credibility for a local model of chemical bonding. Comparing and contrasting the localized and delocalized models of chemical bonding and molecular structure is a major theme of this chapter.

A useful place to start a discussion of localized bonding is with Lewis structures, which emphasize the pairing of electrons as the basis for chemical bond formation. Bonds are shown as connecting lines, and electrons not involved in the bonds are indicated by dots. Lewis structures for a few representative small molecules are shown here:

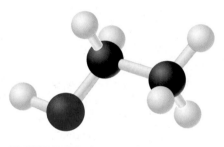

FIGURE 24.1
Ethanol depicted in the form of a ball-and-stick model.

$$\text{H—H} \qquad \text{H—}\ddot{\underset{\cdot\cdot}{\text{Cl}}}\text{:} \qquad \overset{\text{H}}{\underset{\text{H}}{\diagdown}}\text{C}\overset{\text{H}}{\underset{\text{H}}{\diagup}} \qquad \text{H—}\overset{\cdot\cdot}{\underset{|}{\text{N}}}\text{—H} \qquad \text{:}\overset{\cdot\cdot}{\text{O}}\text{—H}$$
$$\qquad\qquad\qquad\qquad\qquad\qquad\qquad\qquad\qquad \text{H} \qquad\qquad \text{H}$$

Lewis structures are useful in understanding the stoichiometry of a molecule and in emphasizing the importance of nonbonding electron pairs, also called **lone pairs.** Lewis structures are less useful in predicting the geometrical structure of molecules.

The **valence shell electron pair repulsion (VSEPR) model** provides a qualitative rationalization of molecular structures using the Lewis concepts of localized bonds and lone pairs. The basic assumptions of the model can be summarized in the following statements about a central atom that may have lone pairs and is bonded to several atomic ligands:

• The ligands and lone pairs around a central atom act as if they repel one another. They adopt an arrangement in three dimensional space that maximizes their angular separation.

• A lone pair occupies more angular space than a ligand.

- The amount of angular space occupied by a ligand decreases as its electronegativity increases and increases as the electronegativity of the central atom increases.

- A multiply bonded ligand occupies more angular space than a singly bonded ligand.

As Figure 24.2 shows, the structure of a large number of molecules can be understood using the VSEPR model. For example, the decrease in the bond angle in the molecules CH_4, NH_3, and H_2O can be explained on the basis of the greater angular space occupied by lone pairs than by ligands. The tendency of lone pairs to maximize their angular separation also explains why XeF_2 is linear, SO_2 is bent, and IF_4^- is planar. However, in some cases the model is inapplicable or does not predict the correct structure. For instance, a radical, such as CH_3, that has an unpaired electron is planar and, therefore, does not fit into the VSEPR model. Alkaline earth dihalides such as CaF_2 and $SrCl_2$ are angular rather than linear as would be predicted by the model. SeF_6^{2-} and $TeCl_6^{2-}$ are octahedral even though they each have a lone pair in addition to the six ligands. This result indicates that lone pairs do not always exert an influence on molecular shape. In addition, lone pairs do not play as strong a role as the model suggests in transition metal complexes.

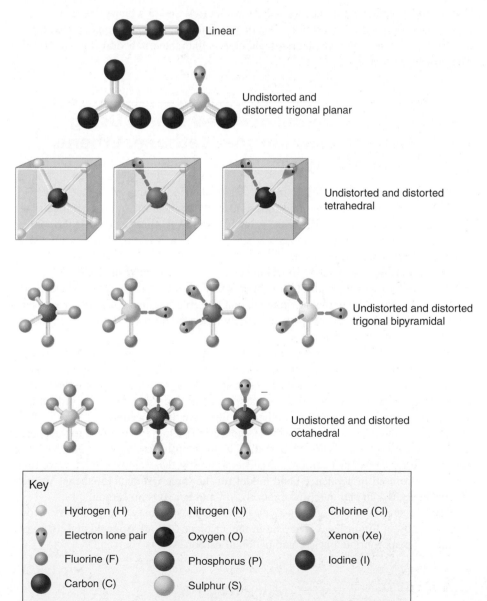

Linear

Undistorted and distorted trigonal planar

Undistorted and distorted tetrahedral

Undistorted and distorted trigonal bipyramidal

Undistorted and distorted octahedral

Key

Hydrogen (H)	Nitrogen (N)	Chlorine (Cl)
Electron lone pair	Oxygen (O)	Xenon (Xe)
Fluorine (F)	Phosphorus (P)	Iodine (I)
Carbon (C)	Sulphur (S)	

FIGURE 24.2
Examples of correctly predicted molecular shapes using the VSEPR model.

EXAMPLE PROBLEM 24.1

Using the VSEPR model, predict the shape of NO_3^- and OCl_2.

Solution

The following Lewis structure shows one of the three resonance structures of the nitrate ion:

Because the central nitrogen atom has no lone pairs and the three oxygens are equivalent, the nitrate ion should be planar with a 120° bond angle. This is the observed structure.

The Lewis structure for OCl_2 is

The central oxygen atom is surrounded by two ligands and two lone pairs. The ligands and lone pairs are described by a distorted tetrahedral arrangement, leading to a bent molecule. The bond angle should be less than the tetrahedral angle of 109.5°. The observed bond angle is 111°.

24.2 Describing Localized Bonds Using Hybridization for Methane, Ethene, and Ethyne

As discussed earlier, the VSEPR model is useful in predicting the shape of a wide variety of molecules. Although the rules used in its application do not specifically use the vocabulary of quantum mechanics, **valence bond (VB) theory** does use the concept of localized orbitals to explain molecular structure. In the VB model, AOs on the same atom are combined to generate a set of directed orbitals in a process called **hybridization.** The combined orbitals are referred to as **hybrid orbitals.** The hybrid orbitals are assumed to contribute independently to the electron density and to the energy of the molecule to the maximum extent possible because this allows the assembly of the molecule out of separate and largely independent parts. This requires the set of hybrid orbitals to be orthogonal.

How is hybridization used to describe molecular structure? Consider the sequence of molecules methane, ethene, and ethyne. From previous chemistry courses, we know that carbon in these molecules is characterized by the sp^3, sp^2, and sp **hybridizations,** respectively. What is the functional form associated with these different hybridizations? We construct the hybrid orbitals for ethene to illustrate the procedure.

To model the three σ bonds in ethene, the carbon AOs are hybridized to the configuration $1s^2 2p_y^1(\psi_a)^1(\psi_b)^1(\psi_c)^1$ rather than to the configuration $1s^2 2s^2 2p^2$, which is appropriate for an isolated carbon atom. The orbitals ψ_a, ψ_b, and ψ_c are the wave functions that are used in a valence bond model for the three σ bonds in ethene. We next formulate ψ_a, ψ_b, and ψ_c in terms of the $2s$, $2p_x$, and $2p_z$ AOs on carbon.

The three sp^2-hybrid orbitals ψ_a, ψ_b, and ψ_c must satisfy the geometry shown schematically in Figure 24.3. They lie in the xz plane and are oriented at 120° to one another. The appropriate linear combination of carbon AOs is

$$\psi_a = c_1\phi_{2p_z} + c_2\phi_{2s} + c_3\phi_{2p_x}$$

$$\psi_b = c_4\phi_{2p_z} + c_5\phi_{2s} + c_6\phi_{2p_x}$$

$$\psi_c = c_7\phi_{2p_z} + c_8\phi_{2s} + c_9\phi_{2p_x} \qquad \textbf{(24.1)}$$

How can c_1 through c_9 be determined? A few aspects of the chosen geometry simplify the task of determining the coefficients. Because the $2s$ orbital is spherically symmetrical, it will contribute equally to each of the hybrid orbitals. Therefore, $c_2 = c_5 = c_8$. These three coefficients must satisfy the equation $\sum_i (c_{2si})^2 = 1$, where the subscript $2s$ refers to the $2s$ AO. This equation states that all of the $2s$ contributions to the hybrid orbitals must be accounted for. We choose $c_2 < 0$ in the preceding equations to make the $2s$ orbital

$$\psi_{200}(r) = \frac{1}{\sqrt{32\pi}}\left(\frac{1}{a_0}\right)^{3/2}\left(2 - \frac{r}{a_0}\right)e^{-r/2a_0}$$

have a positive amplitude in the bonding region. (For graphs of the $2s$ AO amplitude versus r, see Figures 20.5 and 20.6.) Therefore, we conclude that

$$c_2 = c_5 = c_8 = -\frac{1}{\sqrt{3}}$$

From the orientation of the orbitals seen in Figure 24.3, $c_3 = 0$ because ψ_a is oriented on the z axis. Because the hybrid orbital points along the positive z axis, $c_1 > 0$. We can also conclude that $c_4 = c_7$, that both are negative, and that $-c_6 = c_9$ with $c_9 > 0$. Based on these considerations, Equation (24.1) simplifies to

$$\psi_a = c_1\phi_{2p_z} - \frac{1}{\sqrt{3}}\phi_{2s}$$

$$\psi_b = c_4\phi_{2p_z} - \frac{1}{\sqrt{3}}\phi_{2s} - c_6\phi_{2p_x}$$

$$\psi_c = c_4\phi_{2p_z} - \frac{1}{\sqrt{3}}\phi_{2s} + c_6\phi_{2p_x} \qquad \textbf{(24.2)}$$

As shown in Example Problem 24.2, the remaining unknown coefficients can be determined by normalizing and orthogonalizing ψ_a, ψ_b, and ψ_c.

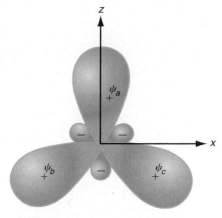

FIGURE 24.3
Geometry of the sp^2-hybrid orbitals used in Equation (24.1). In this and in most of the figures in this chapter, we use a "slimmed down" picture of hybrid orbitals to separate individual orbitals. A more correct form for $s-p$ hybrid orbitals is shown in Figure 24.5.

EXAMPLE PROBLEM 24.2

Determine the three unknown coefficients in Equation (24.2) by normalizing and orthogonalizing the hybrid orbitals.

Solution

We first normalize ψ_a. Terms such as $\int \phi_{2p_x}^* \phi_{2p_z} \, d\tau$ and $\int \phi_{2s}^* \phi_{2p_x} \, d\tau$ do not appear in the following equations because all of the AOs are orthogonal to one another. Evaluation of the integrals is simplified because the individual AOs are normalized.

$$\int \psi_a^* \psi_a \, d\tau = (c_1)^2 \int \phi_{2p_z}^* \phi_{2p_z} \, d\tau + \left(-\frac{1}{\sqrt{3}}\right)^2 \int \phi_{2s}^* \phi_{2s} \, d\tau = 1$$

$$= (c_1)^2 + \frac{1}{3} = 1$$

which tells us that $c_1 = \sqrt{2/3}$. Orthogonalizing ψ_a and ψ_b, we obtain

$$\int \psi_a^* \psi_b \, d\tau = c_4\sqrt{\frac{2}{3}} \int \phi_{2p_z}^* \phi_{2p_z} \, d\tau + \left(-\frac{1}{\sqrt{3}}\right)^2 \int \phi_{2s}^* \phi_{2s} \, d\tau = 0$$

$$= c_4\sqrt{\frac{2}{3}} + \frac{1}{3} = 0 \quad \text{and}$$

$$c_4 = -\sqrt{\frac{1}{6}}$$

Normalizing ψ_b, we obtain

$$\int \psi_b^* \psi_b \, d\tau = \left(-\frac{1}{\sqrt{6}}\right)^2 \int \phi_{2p_z}^* \phi_{2p_z} \, d\tau$$

$$+ \left(-\frac{1}{\sqrt{3}}\right)^2 \int \phi_{2s}^* \phi_{2s} \, d\tau + (-c_6)^2 \int \phi_{2p_x}^* \phi_{2p_x} \, d\tau$$

$$= (c_6)^2 + \frac{1}{3} + \frac{1}{6} = 1 \quad \text{and}$$

$$c_6 = +\frac{1}{\sqrt{2}}$$

We have chosen the positive root so that the coefficient of ϕ_{2px} in ψ_b is negative. Using these results, the normalized and orthogonal set of hybrid orbitals is

$$\psi_a = \sqrt{\frac{2}{3}}\phi_{2p_z} - \frac{1}{\sqrt{3}}\phi_{2s}$$

$$\psi_b = -\frac{1}{\sqrt{6}}\phi_{2p_z} - \frac{1}{\sqrt{3}}\phi_{2s} - \frac{1}{\sqrt{2}}\phi_{2p_x}$$

$$\psi_c = -\frac{1}{\sqrt{6}}\phi_{2p_z} - \frac{1}{\sqrt{3}}\phi_{2s} + \frac{1}{\sqrt{2}}\phi_{2p_x}$$

Note that ψ_c is normalized and orthogonal to ψ_a and ψ_b.

How can the $2s$ and $2p$ character of the hybrids be quantified? Because the sum of the squares of the coefficients for each hybrid orbital equals 1, the p and s character of the hybrid orbital can be calculated. The fraction of $2p$ character in ψ_b is $1/6 + 1/2 = 2/3$. The fraction of $2s$ character is $1/3$. Because the ratio of the $2p$ to $2s$ character is 2:1, we refer to sp^2 hybridization.

How do we know that these hybrid orbitals are oriented with respect to one another as shown in Figure 24.3? Because ψ_a has no component of the $2p_x$ orbital, it must lie on the z axis, corresponding to a value of zero for the polar angle θ. To demonstrate that the ψ_b orbital is oriented as shown, we find its maximum value with respect to the variable θ, which is measured from the z axis.

EXAMPLE PROBLEM 24.3

Demonstrate that the hybrid orbital ψ_b has the orientation shown in Figure 24.3.

Solution

To carry out this calculation, we have to explicitly include the θ dependence of the $2p_x$ and $2p_z$ orbitals from Chapter 20. In doing so, we set the azimuthal angle ϕ, discussed in Section 20.4, equal to zero:

$$\frac{d\psi_b}{d\theta} = \left[\frac{1}{\sqrt{32\pi}}\left(\frac{\zeta}{a_0}\right)^{3/2} e^{-\zeta r/2a_0}\right]$$

$$\times \frac{d}{d\theta}\left(-\frac{1}{\sqrt{6}}\frac{\zeta r}{a_0}\cos\theta - \frac{1}{\sqrt{3}}\left[2 - \frac{\zeta r}{a_0}\right] - \frac{1}{\sqrt{2}}\frac{\zeta r}{a_0}\sin\theta\right) = 0$$

which simplifies to

$$\frac{1}{\sqrt{6}}\sin\theta - \frac{1}{\sqrt{2}}\cos\theta = 0 \quad \text{or} \quad \tan\theta = \sqrt{3}$$

This value for $\tan\theta$ is satisfied by $\theta = 60°$ and $240°$. Applying the condition that $d^2\psi_b/d\theta^2 < 0$ for the maximum, we conclude that $\theta = 240°$ corresponds to the maximum and $\theta = 60°$ corresponds to the minimum. Similarly, it can be shown that ψ_c has its maximum value at $120°$ and a minimum at $300°$.

TABLE 24.1 C—C Bond Types

Carbon—Carbon Single Bond Types	σ Bond Hybridization	s-to-p Ratio	Angle between Equivalent σ Bonds (°)	Carbon—Carbon Single Bond Length (pm)
$\geq$C—C$\leq$	sp^3	1:3	109.4	154
$\geq$C—C$<$	sp^2	1:2	120	146
$\equiv$C—C$\equiv$	sp	1:1	180	138

Example Problem 24.3 shows that sp^2 hybridization generates three equivalent hybrid orbitals that are separated by an angle of 120°. By following the procedure outlined earlier, it can be shown that the set of orthonormal sp-hybrid orbitals that are oriented 180° apart is

$$\psi_a = \frac{1}{\sqrt{2}}(-\phi_{2s} + \phi_{2p_z})$$

$$\psi_b = \frac{1}{\sqrt{2}}(-\phi_{2s} - \phi_{2p_z}) \tag{24.3}$$

and that the set of tetrahedrally oriented orthonormal hybrid orbitals for sp^3 hybridization that are oriented 109.4° apart is

$$\psi_a = \frac{1}{2}(-\phi_{2s} + \phi_{2p_x} + \phi_{2p_y} + \phi_{2p_z})$$

$$\psi_b = \frac{1}{2}(-\phi_{2s} - \phi_{2p_x} - \phi_{2p_y} + \phi_{2p_z})$$

$$\psi_c = \frac{1}{2}(-\phi_{2s} + \phi_{2p_x} - \phi_{2p_y} - \phi_{2p_z})$$

$$\psi_d = \frac{1}{2}(-\phi_{2s} - \phi_{2p_x} + \phi_{2p_y} - \phi_{2p_z}) \tag{24.4}$$

By combining s and p orbitals, at most four hybrid orbitals can be generated. To describe bonding around a central atom with coordination numbers greater than four, d orbitals need to be included in forming the hybrids. Although hybrid orbitals with d character are not discussed here, the principles used in constructing them are the same as those outlined earlier.

The properties of C—C single bonds depend on the hybridization of the carbon atoms, as shown in Table 24.1. The most important conclusion that can be drawn from this table for the discussion in the next section is that increasing the s character in $s-p$ hybrids increases the bond angle. Note also that the C—C single bond length becomes shorter as the s character of the hybridization increases, and that the C—C single bond energy increases as the s character of the hybridization increases.

24.3 Constructing Hybrid Orbitals for Nonequivalent Ligands

In the preceding section, the construction of hybrid orbitals for equivalent ligands was considered. However, in general, molecules contain nonequivalent ligands as well as non-bonding electron lone pairs. How can hybrid orbitals be constructed for such molecules if the bond angles are not known? By considering the experimentally determined structures of a wide variety of molecules, Henry Bent formulated the following guidelines:

- Central atoms that obey the octet rule can be classified into three structural types. Central atoms that are surrounded by a combination of four single bonds or electron pairs are to a first approximation described by a tetrahedral geometry and sp^3

hybridization. Central atoms that form one double bond and a combination of two single bonds or electron pairs are to a first approximation described by a trigonal geometry and sp^2 hybridization. Central atoms that form two double bonds or one triple bond and either a single bond or an electron pair are to a first approximation described by a linear geometry and sp hybridization.

• The presence of different ligands is taken into account by assigning a different hybridization to all nonequivalent ligands and lone pairs. The individual hybridization is determined by the electronegativity of each ligand. A nonbonding electron pair can be considered to be electropositive or, equivalently, to have a small electronegativity. Bent's rule states that atomic s character concentrates in hybrid orbitals directed toward electropositive ligands and that p character concentrates in hybrid orbitals directed toward electronegative ligands.

We now apply these guidelines to H_2O. The oxygen atom in H_2O is to a first approximation described by a tetrahedral geometry and sp^3 hybridization. However, because the H atoms are more electronegative than the electron pairs, the p character of the hybrid orbitals directed toward the hydrogen atoms will be greater than that of sp^3 hybridization. Because Table 24.1 shows that increasing the p character decreases the bond angle, Bent's rule says that the H—O—H bond angle will be less than 109.4°. Note that the effect of Bent's rule is the same as the effect of the VSEPR rules listed in Section 24.1. However, the hybridization model provides a basis for the rules.

Although useful in predicting bond angles, Bent's rule is not quantitative. To make it predictive, a method is needed to assign a hybridization to a specific combination of two atoms that is independent of the other atoms in the molecule. Several authors have developed methods that meet this need, for example, D. M. Root *et al.* in *J. American Chemical Society* 115 (1993): 4201–4209.

EXAMPLE PROBLEM 24.4

a. Use Bent's rule to decide if the X—C—X bond angle in F_2CO is larger or smaller than in H_2CO.

$$\begin{array}{c} X \\ \diagdown \\ \diagup \\ X \end{array} C=O$$

b. Use Bent's rule to estimate whether the H—C—H bond angle in FCH_3 and $ClCH_3$ differ from 109.5°.

Solution

a. To first order, the carbon atom exhibits sp^2 hybridization. Because F is more electronegative than H, the hybridization of the C—F ligand contains more p character than does the C—H ligand. Therefore, the F—C—F bond angle will be smaller than the H—C—H bond angle.

b. For both FCH_3 and $ClCH_3$, H is more electropositive than the halogen atom so that the C—H bonds have greater s character than the C-halogen bond. This makes the H—C—H bond angle greater than 109.5° in both molecules.

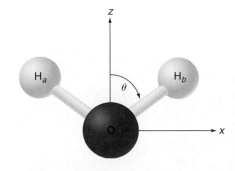

FIGURE 24.4
Coordinate system used to generate the hybrid orbitals on the oxygen atom that are suitable for describing the structure of H_2O.

To test the predictions of Bent's rule, the hybrid orbitals for water are constructed using the known bond angle, and their individual hybridizations are determined. Constructing the lone pair orbitals is left as an end-of-chapter problem. The valence electron configuration of the water molecule can be written in the form $1s^2_{oxygen}(\psi_{OH})^2$ $(\psi_{OH})^2(\psi_{lone\ pair})^2(\psi_{lone\ pair})^2$. Each ψ_{OH} and $\psi_{lone\ pair}$ describes localized hybrid orbitals. From the known geometry, the two bonding orbitals are oriented at 104.5° with respect to one another, as shown in Figure 24.4. Starting with this input, how do we construct ψ_{OH} and $\psi_{lone\ pair}$ from the atomic orbitals on hydrogen and oxygen? To describe the H_2O molecule, a pair of orthogonal equivalent s–p hybrids, called ψ_a and ψ_b, is constructed on the oxygen atom. The calculation is initially carried out for an arbitrary bond angle.

The hybrid orbitals are described by

$$\psi_a = N\left[(\cos\theta)\,\phi_{2p_z} + (\sin\theta)\,\phi_{2p_x} - \alpha\phi_{2s}\right]$$

$$\psi_b = N\left[(\cos\theta)\,\phi_{2p_z} - (\sin\theta)\,\phi_{2p_x} - \alpha\phi_{2s}\right] \qquad \textbf{(24.5)}$$

where N is a normalization constant and α is the relative amplitude of the $2s$ and $2p$ orbitals.

To derive Equation (24.5), visualize ϕ_{2p_x} and ϕ_{2p_z} as vectors along the x and z directions. Because the $2s$ orbital has one radial node, the $2s$ orbital coefficient in Equation (24.5) is negative, which generates a positive amplitude at the position of the H atom. The two hybrid orbitals are orthogonal only if

$$\int \psi_a^* \psi_b \, d\tau = N^2 \int \left[(\cos\theta)\,\phi_{2p_z} + (\sin\theta)\,\phi_{2p_x} - \alpha\phi_{2s}\right]$$

$$\times \left[(\cos\theta)\,\phi_{2p_z} - (\sin\theta)\,\phi_{2p_x} - \alpha\phi_{2s}\right] d\tau$$

$$= N^2\left[\cos^2\theta \int \phi_{2p_z}^* \phi_{2p_z} \, d\tau - \sin^2\theta \int \phi_{2p_x}^* \phi_{2p_x} \, d\tau + \alpha^2 \int \phi_{2s}^* \phi_{2s} \, d\tau\right] = 0 \qquad \textbf{(24.6)}$$

Terms such as $\int \phi_{2p_x}^* \phi_{2p_z} \, d\tau$ and $\int \phi_{2s}^* \phi_{2p_x} \, d\tau$ do not appear in this equation because all of the atomic orbitals are orthogonal to one another. Because each of the AOs is normalized, Equation (24.6) reduces to

$$N^2[\cos^2\theta - \sin^2\theta + \alpha^2] = N^2[\cos 2\theta + \alpha^2] = 0 \quad \text{or}$$

$$\cos 2\theta = -\alpha^2 \qquad \textbf{(24.7)}$$

In simplifying this equation, we have used the identity $\cos^2 x - \sin^2 y = \cos(x+y)\cos(x-y)$. Because $\alpha^2 > 0$, $\cos 2\theta < 0$ and the bond angle $180° \geq 2\theta \geq 90°$. What has this calculation shown? We have demonstrated that it is possible to create two hybrid orbitals separated by a bonding angle in this angular range simply by varying the relative contributions of the $2s$ and $2p$ orbitals to the hybrid.

The hybrid orbitals in Equation (24.5) are not specific to a particular molecule other than that the two atoms that bond to the central oxygen atom are identical. We now calculate the value of α that generates the correct bond angle in H_2O. Calculating α by substituting the known value $\theta = 52.25°$ in Equation (24.7), we find that the unnormalized hybrid orbitals that describe bonding in water are

$$\psi_a = N[0.61\,\phi_{2p_z} + 0.79\,\phi_{2p_x} - 0.50\,\phi_{2s}]$$

$$\psi_b = N[0.61\,\phi_{2p_z} - 0.79\,\phi_{2p_x} - 0.50\,\phi_{2s}] \qquad \textbf{(24.8)}$$

EXAMPLE PROBLEM 24.5

Normalize the hybrid orbitals given in Equation (24.8).

Solution

$$\int \psi_a^* \psi_a \, d\tau = N^2 (0.61)^2 \int \phi_{2p_z}^* \phi_{2p_z} \, d\tau$$

$$+ N^2 (0.79)^2 \int \phi_{2p_x}^* \phi_{2p_x} \, d\tau + N^2 (0.50)^2 \int \phi_{2s}^* \phi_{2s} \, d\tau = 1$$

Other terms do not contribute because the atomic orbitals are orthogonal to one another.

$$\int \psi_a^* \psi_a \, d\tau = N^2 (0.61)^2 + N^2 (0.79)^2 + N^2 (0.50)^2 = 1.25\, N^2 = 1$$

$$N = 0.89$$

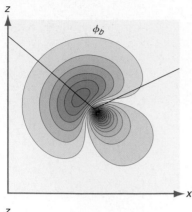

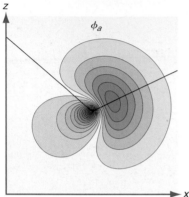

FIGURE 24.5
Directed hybrid bonding orbitals for H_2O. The black lines show the bond angle and orbital orientation. Positive and negative amplitudes are shown as red and blue respectively. Darker colors indicate larger values for the magnitude of the amplitude.

Using the result of Example Problem 24.5, the normalized hybrid orbitals can be written as follows:

$$\psi_a = 0.55\phi_{2p_z} + 0.71\phi_{2p_x} - 0.45\phi_{2s}$$
$$\psi_b = 0.55\phi_{2p_z} - 0.71\phi_{2p_x} - 0.45\phi_{2s} \qquad (24.9)$$

Because the sum of the squares of the coefficients for each hybrid orbital equals 1, we can calculate their p and s character. The fraction of $2p$ character is $(0.55)^2 + (0.71)^2 = 0.80$. The fraction of $2s$ character is $(-0.45)^2 = 0.20$. Therefore, the hybridization of the bonding hybrid orbitals is described as sp^4. These hybrid orbitals have more p character than the first approximation sp^3, as predicted by Bent's rule.

The two hybrid orbitals are shown in Figure 24.5. Note that each of the directed hybrid orbitals lies along one of the bonding directions and has little amplitude along the other bonding direction. These hybrid orbitals could be viewed as the basis for the line connecting bonded atoms in the Lewis structure for water. Figure 24.5 shows a realistic representation of the hybrid orbitals to compare with the "slimmed down" version of Figure 24.3.

This calculation for H_2O illustrates how to construct bonding hybrid orbitals with a desired relative orientation. To this point, the energetics of this process have not been discussed. In many-electron atoms, the $2p$ orbital energy is greater than that for the $2s$ orbital. How can these orbitals be mixed in all possible proportions without putting energy into the atom? To create the set of occupied hybrid orbitals on an isolated ground-state oxygen atom would indeed require energy; however, the subsequent formation of bonds to the central atom lowers the energy, leading to an overall decrease in the energy of the molecule relative to the isolated atoms after bond formation. In the language of the hybridization model, the energy cost of promoting the electrons from the $1s^2 2s^2 2p^4$ configuration to the $1s^2 \psi_c^2 \psi_a^2 \psi_a^1 \psi_b^1$ configuration is more than offset by the energy gained in forming two O—H bonds. Keep in mind that the individual steps in the formation of the H_2O molecule such as promotion of the O atom, followed by the creation of O—H bonds, are only an aid in describing the formation of H_2O, rather than a series of actual events. The language of the hybridization model should not be taken too literally because neither the promotion process nor hybrid orbitals are observables, and it is important to distinguish between a model and reality. The reality of orbitals is discussed in Section 24.6 in more detail.

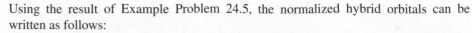

24.4 Using Hybridization to Describe Chemical Bonding

By using the hybridization model to create local bonding orbitals, the concepts inherent in Lewis structures can be given a quantum mechanical basis. As an example, consider BeH_2, which is not observed as an isolated molecule because it forms a solid through polymerization of BeH_2 units stabilized by hydrogen bonds. We consider only a single BeH_2 unit. Be has the configuration $1s^2 2s^2 2p^0$, and because it has no unpaired electrons, it is not obvious how bonding to the H atoms can be explained in the Lewis model. In the VB model, the $2s$ and $2p$ orbitals are hybridized to create bonding hybrids on the Be atom. Because the bond angle is known to be 180°, two equivalent and orthogonal sp-hybrid orbitals are constructed as given by Equation (24.3). This allows Be to be described as $1s^2 (\psi_a)^1 (\psi_b)^1$. In this configuration, Be has two unpaired electrons and, therefore, the hybridized atom is divalent. The orbitals are depicted schematically in Figure 24.6. To make a connection to Lewis structures, the bonding electron pair is placed in the overlap region between the Be and H orbitals as indicated by the dots. In reality, the bonding electron density is distributed over the entire region in which the orbitals have a nonzero amplitude. We return to BeH_2 in Section 24.6 where we compare localized and delocalized bonding models for this molecule.

The chemically most important use of the hybridization model is in describing bonding in molecules containing carbon. Figure 24.7 depicts valence bond hybridization in

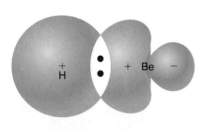

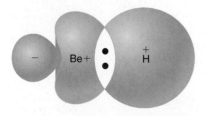

FIGURE 24.6
Bonding in BeH_2 using two sp-hybrid orbitals on Be. The two Be—H hybrid bonding orbitals are shown separately.

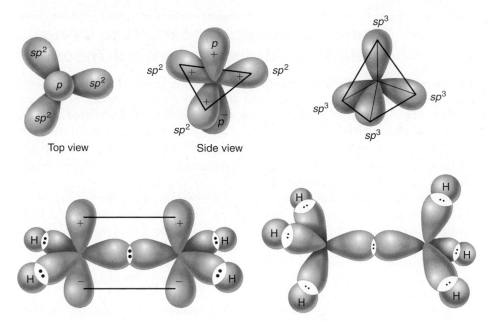

FIGURE 24.7
The top panel shows the arrangement of the hybrid orbitals for sp^2 and sp^3 carbon. The bottom panel shows a schematic depiction of bonding in ethene (left) and ethane (right) using hybrid bonding orbitals.

ethene and ethane. For ethene, each carbon atom is promoted to the $1s^2 2p_y^1(\psi_a)^1(\psi_b)^1(\psi_c)^1$ configuration before forming four C—H σ bonds, a C—C σ bond, and a C—C π bond. The maximal overlap between the p orbitals to create a π bond in ethene occurs when all atoms lie in the same plane. The double bond is made from one σ and one π bond. For ethane, each carbon atom is promoted to the $1s^2(\psi_a)^1(\psi_b)^1(\psi_c)^1(\psi_d)^1$ configuration before forming six C—H σ bonds and a C—C σ bond.

In closing this discussion of hybridization, we emphasize the positive aspects of the model and point out some of its shortcomings. The main usefulness of hybridization is that it is an easily understandable model with considerable predictive power using, for instance, Bent's Rule. It retains the main features of Lewis structures in describing local orthogonal bonds between adjacent atoms in terms of electron pairing, and it justifies Lewis structures in the language of quantum mechanics. Hybridization also provides a theoretical basis for the VSEPR rules.

Hybridization also offers more than a useful framework for understanding bond angles in molecules. Because the $2s$ AO is lower in energy than the $2p$ level in many-electron atoms, the electronegativity of a hybridized atom increases with increasing s character. Therefore, the hybridization model predicts that an sp-hybridized carbon atom is more electronegative than an sp^3-hybridized carbon atom. Evidence for this effect is that the positive end of the dipole moment in N≡C—Cl is on the Cl atom. We conclude that the carbon atom in the cyanide group is more electronegative than a chlorine atom. Because increased s character leads to shorter bond lengths, and because shorter bonds generally have a greater bond strength, the hybridization model provides a correlation of s character and bond strength.

The hybridization model also has a few shortcomings. For known bond angles, the hybridization can be calculated as was done for ethane and H_2O. However, semiempirical prescriptions must be used to estimate the s and p character of a hybrid orbital for a molecule in the absence of structural information. It is also more straightforward to construct an appropriate hybridization for symmetric molecules such as methane than for molecules with electron lone pairs and several different ligands bonded to the central atom. Additionally, the depiction of bonding hybrids in Figure 24.7 seems to imply that the electron density is highly concentrated along the bonding directions. This is not true, as can be seen by looking at the realistic representation of hybrid orbitals in Figure 24.5. Finally, the conceptual formalism used in creating hybrid orbitals—in particular, promotion followed by hybridization—assumes much more detail than can be verified by experiments.

24.5 Predicting Molecular Structure Using Qualitative Molecular Orbital Theory

We now consider a **delocalized bonding model** of the chemical bond. MO theory approaches the structure of molecules quite differently than local models of chemical bonding. The electrons involved in bonding are assumed to be delocalized over the molecule. Each one-electron molecular orbital σ_j is expressed as a linear combination of atomic orbitals such as $\sigma_j(k) = \sum_i c_{ij} \phi_i(k)$, which refers to the jth molecular orbital for electron k. The many-electron wave function ψ is written as a Slater determinant in which the individual entries are the $\sigma_j(k)$.

In quantitative molecular orbital theory, which will be discussed in Chapter 26, structure emerges naturally as a result of solving the Schrödinger equation and determining the atomic positions for which the energy has its minimum value. Although this concept can be formulated in a few words, carrying out this procedure is a complex exercise in numerical computing. In this section, our focus is on a more qualitative approach that conveys the spirit of molecular orbital theory, but that can be written down without extensive mathematics.

To illustrate this approach, we use qualitative MO theory to understand the bond angle in triatomic molecules of the type H_2A, where A is one of the atoms in the sequence Be → O, and show that a qualitative picture of the optimal bond angle can be obtained by determining how the energy of the individual occupied molecular orbitals varies with the bond angle. In doing so, we assume that the total energy of the molecule is proportional to the sum of the orbital energies. This assumption can be justified, although we do not do so here. Experimentally, we know that BeH_2 is a linear molecule and H_2O has a bond angle of 104.5°. How can this difference be explained using MO theory?

The minimal basis set used here to construct the MOs consists of the $1s$ orbitals on each of the H atoms and the $1s$, $2s$, $2p_x$, $2p_y$, and $2p_z$ orbitals on atom A. Seven MOs can be generated using these seven AOs. We omit the two lowest MOs generated from the $1s$ oxygen AO from the following discussion because the corresponding electrons are localized on the oxygen atom. Water has eight valence electrons that occupy four of the five remaining MOs.

Recall that the orbital energy increases with the number of nodes for the particle in the box, the harmonic oscillator, and the H atom. We also know that the lower the AO energies, the lower the MO energy will be. The occupied valence MOs for water are shown in Figure 24.8 in terms of the AOs from which they are constructed. The relative MO orbital energies are discussed later. The MOs are labeled according to their symmetry with respect to a set of rotation and reflection operations that leave the water molecule unchanged. We will discuss the importance of molecular symmetry in constructing MOs from AOs at some length in Chapter 27. However, in the present context it is sufficient to think of these designations simply as labels. Because the oxygen $2s$ AO is lower in energy than the $2p$ AOs, the MO with no nodes designated $1a_1$ in Figure 24.8 is expected to have the lowest energy of all possible valence MOs. The next higher MOs involve $2p$ AOs on the O atom and the $1s$ AO on the H atoms.

The three oxygen $2p$ orbitals are differently oriented with respect to the plane containing the H atoms. As a result, the MOs that they generate are quite different in energy. Assume that the H_2O molecule lies in the xz plane with the z axis bisecting the H—O—H angle as shown in Figure 24.4. The $1b_2$ MO, generated using the $2p_x$ AO, and the $2a_1$ MO, generated using the $2p_z$ AO, each have no nodes in the O—H region and, therefore, have binding character. However, because each has one node, both MOs have a higher energy than the $1a_1$ MO. Calculations show that the MO generated using the $2p_x$ AO has a lower orbital energy than that generated using the $2p_z$ AO. Note that some s−p mixing has been incorporated in the $2a_1$ and $3a_1$ MO generated from the $2s$ and $2p_z$ AO. Having discussed the MOs formed using the $2p_x$ and $2p_z$ AOs, we turn to the $2p_y$ AO. The $2p_y$ orbital has no net overlap with the H atoms and gives rise to the $1b_1$ nonbonding MO that is localized on the O atom. Because this MO is not stabilized through interaction with the H $1s$ AOs, it has the highest energy of all the occupied

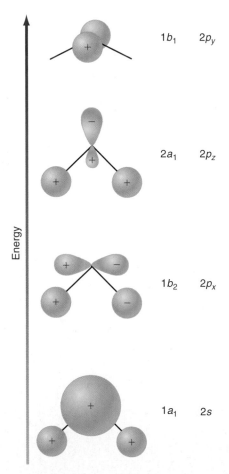

$1b_1$	$2p_y$
$2a_1$	$2p_z$
$1b_2$	$2p_x$
$1a_1$	$2s$

Energy

FIGURE 24.8
The valence MOs occupied in the ground state of water are shown in order of increasing orbital energy. The MOs are depicted in terms of the AOs from which they are constructed. The second column gives the MO symmetry, and the third column lists the dominant AO orbital on the oxygen atom.

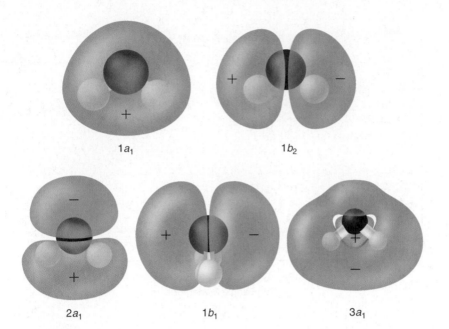

FIGURE 24.9
The first five valence MOs for H_2O are depicted. The $1b_1$ and $3a_1$ MOs are the HOMO and LUMO, respectively. Note that the $1b_1$ MO is the AO corresponding to the nonbonding $2p_y$ electrons on oxygen. The plane of the molecule has been rotated for the $1b_1$ MO to better display the nodal structure.

MOs. Numerically calculated molecular orbitals including the $3a_1$ LUMO are depicted in Figure 24.9.

The preceding discussion about the relative energy of the MOs is sufficient to allow us to draw the MO energy diagram shown in Figure 24.10. The MO energy levels in this figure are drawn for a particular bond angle near 105°, but the energy levels vary with 2θ, as shown in Figure 24.11, in what is known as a **Walsh correlation diagram.** It is important to understand the trends shown in this figure and are discussed next because the variation of the MO energies with angle is ultimately responsible for BeH_2 being linear and H_2O being bent.

How does the $1a_1$ energy vary with 2θ? The overlap between the s orbitals on A and H is independent of 2θ, but as this angle decreases from 180°, the overlap between the H atoms increases. This stabilizes the molecule and, therefore, the $1a_1$ energy decreases. By contrast, the overlap between the $2p_y$ orbital and the H $1s$ orbitals is a maximum at 180° and, therefore, the $1b_2$ energy increases as 2θ decreases. Because the effect of the H—H overlap on the $1a_1$ energy is a secondary effect, the $1b_2$ energy falls more rapidly with increasing 2θ than the $1a_1$ energy increases.

We now consider the $2a_1$ and $1b_1$ energies. The $2p_y$ and $2p_z$ orbitals are nonbonding and degenerate for a linear H_2A molecule. However, as 2θ decreases from 180°, the O $2p_z$ orbital has a net overlap with the H $1s$ AOs and has increasingly more bonding

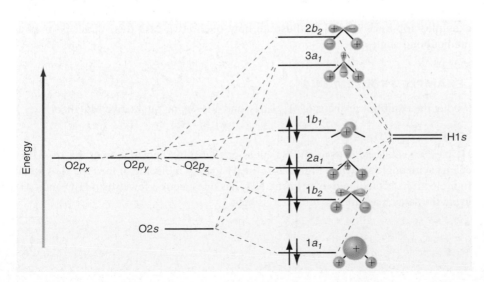

FIGURE 24.10
Molecular orbital energy-level diagram for H_2O at its equilibrium geometry.

FIGURE 24.11
Schematic variation of the MO energies
for water with bond angle. The symbols
used on the left to describe the MOs are
based on symmetry considerations and are
valid for $2\theta < 180°$. This nomenclature is
discussed in Chapter 27.

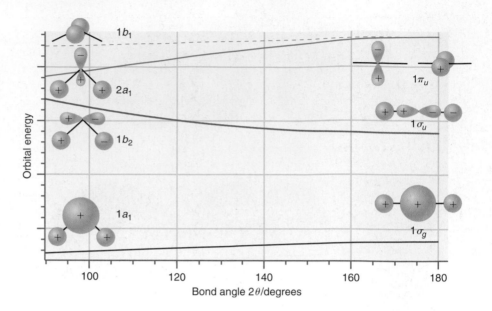

FIGURE 24.11
Schematic variation of the MO energies
for water with bond angle. The symbols
used on the left to describe the MOs are
based on symmetry considerations and are
valid for $2\theta < 180°$. This nomenclature is
discussed in Chapter 27.

character. Therefore, the $2a_1$ MO energy decreases as 2θ decreases from $180°$. The $1b_1$ MO remains nonbonding as 2θ decreases from $180°$, but electron repulsion effects lead to a slight decrease of the MO energy. These variations in the MO energies are depicted in Figure 24.11.

The MO energy diagram of Figure 24.10 is equally valid for H_2A molecules with A being 2nd period elements other than oxygen. Consider the molecules BeH_2 and H_2O. BeH_2 has four valence electrons that are placed in the two lowest-lying valence MOs, $1a_1$ and $1b_2$. Because the $1b_2$ orbital energy decreases with increasing 2θ more than the $2a_1$ orbital energy increases, the total energy of the molecule is minimized if $2\theta = 180°$. This qualitative argument predicts that BeH_2, as well as any other four-valence electron H_2A molecule, is linear and has the valence electron configuration $(1\sigma_g)^2(1\sigma_u)^2$. Note that the description of H_2A in terms of σ MOs with g or u symmetry applies only to a linear molecule, whereas a description in terms of $1a_1$ and $1b_2$ applies to all bent molecules.

We now consider H_2O, which has eight valence electrons. In this case, the lowest four MOs are doubly occupied. At what angle is the total energy of the molecule minimized? For water, the decrease in the energy of the $1a_1$ and $2a_1$ MOs as 2θ decreases more than offsets the increase in energy for the $1b_2$ MO. Therefore, H_2O is bent rather than linear and has the valence electron configuration $(1a_1)^2(1b_2)^2(2a_1)^2(1b_1)^2$. The degree of bending depends on how rapidly the energy of the MOs changes with angle. Numerical calculations for water using this approach predict a bond angle that is very close to the experimental value of $104.5°$. These examples for BeH_2 and H_2O illustrate how qualitative MO theory can be used to predict bond angles.

EXAMPLE PROBLEM 24.6

Predict the equilibrium shape of H_3^+, LiH_2, and NH_2 using qualitative MO theory.

Solution

H_3^+ has two valence electrons and is bent as predicted by the variation of the $1a_1$ MO energy with angle shown in Figure 24.11. LiH_2 or any molecule of the type H_2A with four electrons is predicted to be linear. NH_2 has one electron fewer than H_2O and, using the same reasoning as for water, is bent.

24.6 How Different Are Localized and Delocalized Bonding Models?

Molecular orbital theory and hybridization-based valence bond theory have been developed using delocalized and localized bonding, respectively. These models approach the chemical bond from very different starting points. However, it is instructive to compare the molecular wave functions generated by these models using BeH_2 as an example. We have already discussed BeH_2 using hybridization in Section 24.4 and now formulate the many-electron wave function using the MO model. To minimize the size of the determinant in Equation (24.11), we assume that the Be $1s$ electrons are not delocalized over the molecule. With this assumption, BeH_2 has the configuration $(1s_{Be})^2(1\sigma_g)^2(1\sigma_u)^2$. On the basis of the symmetry requirements posed on the MOs by the linear geometry (see Chapter 27), the two lowest energy MOs are

$$\sigma_g = c_1(\phi_{H1sA} + \phi_{H1sB}) + c_2\phi_{Be2s}$$

$$\sigma_u = c_3(\phi_{H1sA} - \phi_{H1sB}) - c_4\phi_{Be2p_z} \qquad (24.10)$$

The many-electron determinantal wave function that satisfies the Pauli requirement is

$$\psi(1,2,3,4) = \frac{1}{\sqrt{4!}}\begin{vmatrix} \sigma_g(1)\alpha(1) & \sigma_g(1)\beta(1) & \sigma_u(1)\alpha(1) & \sigma_u(1)\beta(1) \\ \sigma_g(2)\alpha(2) & \sigma_g(2)\beta(2) & \sigma_u(2)\alpha(2) & \sigma_u(2)\beta(2) \\ \sigma_g(3)\alpha(3) & \sigma_g(3)\beta(3) & \sigma_u(3)\alpha(3) & \sigma_u(3)\beta(3) \\ \sigma_g(4)\alpha(4) & \sigma_g(4)\beta(4) & \sigma_u(4)\alpha(4) & \sigma_u(4)\beta(4) \end{vmatrix}$$

$$(24.11)$$

Each entry in the determinant is an MO multiplied by a spin function.

We now use a property of a determinant that you will prove in the end-of-chapter problems for a 2×2 determinant, namely,

$$\begin{vmatrix} a & c \\ b & d \end{vmatrix} = \begin{vmatrix} a & \gamma a + c \\ b & \gamma b + d \end{vmatrix} \qquad (24.12)$$

This equation says that one can add a column of the determinant multiplied by an arbitrary constant γ to another column *without changing the value of the determinant*. For reasons that will become apparent shortly, we replace the MOs σ_g and σ_u with the new MOs $\sigma' = \sigma_g + (c_1/c_3)\sigma_u$ and $\sigma'' = \sigma_g - (c_1/c_3)\sigma_u$. These hybrid MOs are related to the AOs by

$$\sigma' = 2c_1\phi_{H1sA} + \left(c_2\phi_{Be2s} - \frac{c_1c_4}{c_3}\phi_{Be2p_z}\right)$$

$$\sigma'' = 2c_1\phi_{H1sB} + \left(c_2\phi_{Be2s} + \frac{c_1c_4}{c_3}\phi_{Be2p_z}\right) \qquad (24.13)$$

Transforming from σ_g and σ_u to σ' and σ'' requires two steps like the one in Equation (24.12). Note that with this transformation, ϕ_{H1sB} no longer appears in σ' and ϕ_{H1sA} no longer appears in σ''. Because of the property of determinants cited earlier, neither $\psi(1,2,3,4)$—*nor any molecular observable*—will be affected by this change in the MOs. Therefore, the configurations $(1s_{Be})^2(1\sigma_g)^2(1\sigma_u)^2$ and $(1s_{Be})^2$ $[1\sigma_g + (c_1/c_3)1\sigma_u]^2 [1\sigma_g - (c_1/c_3)1\sigma_u]^2$ are completely equivalent, and no experiment can distinguish between them.

Why have we made this particular change? Equation (24.13) and Figure 24.12 show that the new MOs σ' and σ'' are localized bonding MOs, one combining the $1s$ orbital on H_A with an $s-p$ hybrid AO on Be, and the other combining the $1s$ orbital on H_B with an $s-p$ hybrid AO on Be. In other words, the two delocalized MOs σ_g and σ_u have been transformed into two localized MOs *without changing the molecular wave function* $\psi(1,2,3,4)$. This result can be generalized to the statement that for any closed-shell molecular configuration, the set of delocalized MOs can be transformed into a set of localized orbitals predominantly involving two neighboring atoms. Such a transformation is not possible for open-shell molecules or conjugated and aromatic molecules in which at least some of the electrons are delocalized over the molecule.

As the BeH_2 example shows, the distinction between localized and delocalized orbitals is not as clear-cut as it seemed to be at the beginning of this chapter. Working with σ' and σ'' has some disadvantages, because they are not eigenfunctions of the total energy operator. This means that we cannot assign orbital energies to these functions or draw energy-level diagrams as can be done for the delocalized MOs that are solutions to the molecular Hartree–Fock equations. Additionally, computional algorithms used to solve the Schrödinger equation are more efficient when formulated in terms of delocalized MOs than for localized orbitals. Because of these advantages, delocalized rather than localized MOs are generally used to calculate wave functions and energy levels in molecules.

The preceding discussion implies that there is no unique set of one-electron MOs for a molecule and raises the question "How 'real' are the molecular orbitals depicted in this and the previous chapter?" It is useful to distinguish between observables and elements of a model that are not amenable to measurement. Although a many-electron wave function $\psi(1, \ldots, n)$ cannot be determined experimentally, the electron density is proportional to $\sum_i |\psi_i(1, \ldots, n)|^2$ and can be measured using techniques such as X-ray diffraction. Because of the summation over all occupied orbitals, X-ray diffraction does not give information directly about individual MOs.

Although the individual one-electron MOs are also not amenable to direct measurement, experimental measurements can be made that strongly reflect the spatial distribution of the true many-electron wave function over a molecule for a given energy. An example is the use of scanning tunneling microscopy to measure the variation of the tunneling current over a molecule for different values of the energy of the tunneling electrons. The principle of the measurement is shown in Figure 24.13.

As discussed in Section 16.6, the difference between the highest occupied energy levels in the tip and sample can be varied by applying a voltage between the two elements of the STM. With the voltage polarity shown in Figure 24.13, the tunneling current flows from the surface to the tip, both of which are metals. As discussed in

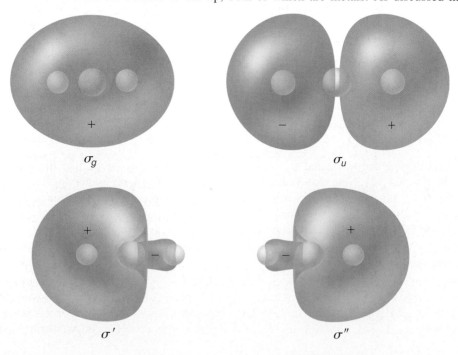

FIGURE 24.12

Schematic representation of the delocalized MOs σ_g and σ_u and the localized bonding orbitals σ' and σ''.

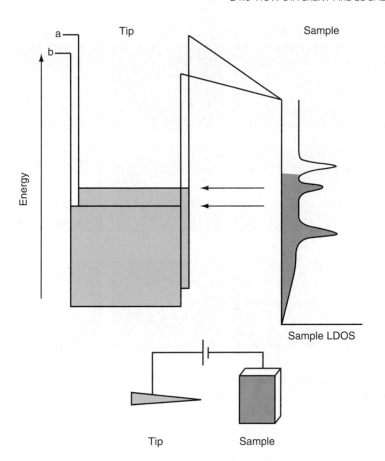

FIGURE 24.13
In scanning tunneling spectroscopy, a voltage applied between the tip and sample aligns different energy levels of the molecule adsorbed on the surface with the highest occupied energy level in the tip. Scanning of the tip over the molecule gives an image of the local density of states in the molecule at a fixed energy corresponding to the fixed voltage. To a good approximation, this is the magnitude of the MO, in case (a) the HOMO. If an appropriate voltage with the reverse polarity is applied, tunneling from the tip to the empty states of the sample can occur and the LUMO can be imaged.

Section 16.4, a metal is characterized by a continuum of states in the conduction band. If a molecule with discrete energy levels is adsorbed on the surface, the number of states at a given energy, called the local density of states, has a multi-peaked structure, where the peaks correspond to the discrete MO energy levels. Tunneling occurs into the highest occupied level in the tip and the tunneling current is proportional to the local density of states. If the highest occupied level of the tip is aligned with a peak in the local density of states of the adsorbed molecule as shown in case *a* of Figure 24.13, scanning of the tip over the molecule will give an image of the magnitude of the molecular orbital corresponding to the peak energy. However, if the highest occupied level of the tip is aligned with a gap between peaks in the local density of states of the adsorbed molecule as shown in case *b* of Figure 24.13, scanning of the tip over the molecule will give an image of the geometry of the molecule because the density of states on the metal surface is approximately constant over the region being scanned.

Figure 24.14 shows the results of carrying out the experiment just described on individual pentacene molecules separated from a copper surface by an atomically thin layer of NaCl. This ultrathin insulator layer prevents coupling between the electronic states of pentacene with the delocalized states of the metal. The preceding discussion, we have assumed that the density of states in the tip has no influence on the measurement. Figure 24.14 shows results for two tip configurations: a bare metal tip and a tip terminated in a single pentacene molecule. Although the general features of the results are similar for both tips, the details differ. A comparison of the results obtained with the pentacene terminated tip with calculations using density functional theory (see Chapter 26) for a gas phase pentacene molecule show very good agreement between the calculated and observed HOMO and LUMO probability densities.

What does this experiment tell us about the reality of molecular orbitals? Adsorption of the molecule on the surface is necessary to immobilize it. This weak "bonding" influences the electronic structure of the pentacene. Additionally, the local density of states in the tip and surface influence the results, so that the images shown in Figure 24.14 are not exact images of the HOMO and LUMO probability densities for an isolated pentacene molecule. However, the high degree of correspondence between the calculation and

FIGURE 24.14

Images of the local density of states of a pentacene molecule adsorbed on a silver surface are shown in the first two rows of columns one and three. Calculated probability densities for the HOMO and LUMO are shown in the third row. The first two rows of the middle column show results obtained in an energy gap for which the local density of states for pentacene are negligible.

Source: Repp, J., Meyer, G., Stojkovic, S., Gourdon, A., Joachim, C. *Molecules on Insulating Films: Scanning-Tunneling Microscopy Imaging of Individual Molecular Orbitals.* Physical Review Letters, 94, no.2 (2005): 026803.

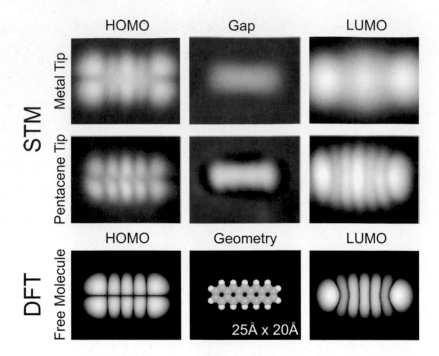

experimental results imply that orbital approximation that leads to the picture of one-electron MOs is valid in this case. Because the HOMO and LUMO are occupied by the π electrons in this aromatic molecule, a delocalized model is necessary for their description.

These experimental results do not contradict the assertion made in Section 17.6 that it is not possible to reconstruct a wave function from experimental results because at best only the magnitude of the wave function can be determined.

24.7 Molecular Structure and Energy Levels from Computational Chemistry

Solving the Schrödinger equation analytically is possible only for one-electron atoms and molecules. MOs and energy levels for many electron atoms and molecules must be obtained using numerical methods to solve the Schrödinger equation. Fortunately, readily available software can be used to solve structures such as those discussed in this chapter on standard personal computers in a few minutes. Inexpensive or free versions of such software are available. At this time, there is little need for oversimplified, nonquantitative models that were developed before the advent of computers. The methods and approximations used, as well as the accuracy of these calculations, are discussed in detail in Chapter 26. At this point, we present a few results from computational chemistry that support other aspects discussed in Chapters 23 and 24.

Lewis structures represent bonding and lone pair electrons differently. Can this picture be supported by rigorous calculations? Figure 24.15 shows negative electrostatic potential surfaces for molecules that Lewis structures would assign one, two, and three lone pairs. It is seen that the shape and extent of the surfaces is consistent

FIGURE 24.15

Electrostatic potential surfaces for ammonia (left), water (center), and hydrogen fluoride (right) support the concept of localized lone pairs.

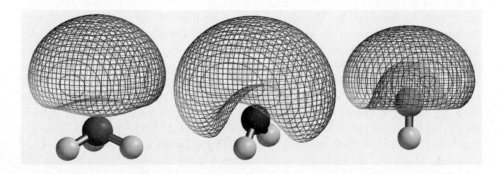

with the number of lone pairs as well as their distribution in space to minimize repulsive interactions.

Figure 24.16 shows calculated symmetry adapted MOs for the 8 occupied MOs on formaldehyde, which has 16 electrons. The MO designations indicate whether the MO changes its sign when the molecule undergoes a rotation about a symmetry axis or a reflection in a symmetry plane. This topic is discussed in Chapter 27. Consider them labels for this discussion.

The lowest energy MO, designated $1a_1$, is localized on O and corresponds to the $1s$ AO. The second-lowest MO, designated $2a_1$, is localized on C and corresponds to the $1s$ AO. Note that although these two MOs were assumed to be delocalized over the molecule, the calculation shows that they are localized as expected for core electrons. The next lowest MO, designated $3a_1$, has contributions from the O $2s$ and C $2s$ AOs and a small contribution from the H atoms. This MO contributes to bonding in the C—O region. Participation of the $2p$ electrons is first observed in the $4a_1$ MO. It has contributions from the C sp^2 hybrid AO directed in the negative x direction toward the H atoms and the $2p_x$ lone pair on O. This MO is bonding in the C—H regions and antibonding in the C—O region. The $1b_2$ MO has contributions from the $2p_y$ lone pair AO on O, mixing in-phase with C—H bonding orbitals formed from $2p_y$ on C and the out-of-phase combination of H $1s$ AOs. This MO is bonding in both the C—O and C—H regions. The $5a_1$ MO has contributions from the $2p_x$ lone pair on O, which mixes in phase with the sp^2 hybrid AOs on C. These hybrid orbitals are directed toward the H atoms. This MO is bonding

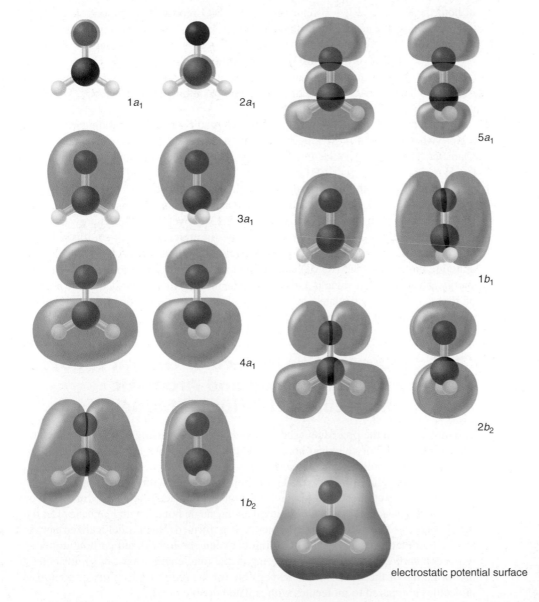

$1a_1$ $2a_1$

$5a_1$

$3a_1$

$1b_1$

$4a_1$

$2b_2$

$1b_2$

electrostatic potential surface

FIGURE 24.16
The symmetry-adapted MOs are shown for formaldehyde. For all but the localized MOs, two perpendicular orientations of the molecule are shown. The last image shows a charge density contour enclosing 90% of the electron charge with a superimposed electrostatic potential map in which red and blue correspond to negative and positive regions of the molecule.

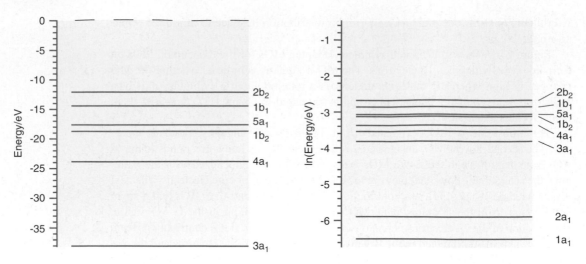

FIGURE 24.17
The formaldehyde valence MO energy levels are shown on the left on a linear energy scale. As the $1a_1$ and $2a_1$ MO levels are -560.2 and -308.7 eV, respectively, they are not shown. All MO energy levels are shown on the right on a logarithmic energy scale. Values were obtained from an MP2 calculation using the 6-31G* basis set (see Chapter 26).

in both the C—O and C—H regions. The $1b_1$ MO represents a π bond in the C—O region. Because the net overlap with the H 1s AOs is zero, there is no contribution from H to this MO. This MO is bonding in the C—O region and nonbonding in the C—H regions. The $2b_2$ MO is the HOMO. It is the antibonding version of the $1b_2$ MO. The $2p_y$ lone pair AO on O interacts in an antibonding fashion with the C—H bonds that are formed from the $2p_y$ AO on C and the out-of-phase combination of H 1s AOs. This MO is nonbonding in CO and bonding in CH regions.

The electrons in the $3a_1$, $1b_2$, $5a_1$ and $1b_1$ MOs all contribute to bonding in the C—O region whereas the $4a_1$ MO decreases the bond order because of its antibonding character. The electrons in the $4a_1$, $1b_2$, $5a_1$ and $2b_2$ MOs all contribute to bonding in the C—H region.

The charges on the individual atoms calculated as discussed in Section 23.8 are H: -0.064, C: $+0.60$, O: -0.47, and the calculated dipole moment is 2.35 debye, which compares well with the measured value of 2.33 debye. The MO energy levels are shown in Figure 24.17. As AOs from three types of atoms are involved, it is not possible to draw lines linking the MO energy levels to AO energy levels as was done for diatomic molecules. The HOMO energy level should correspond to the first ionization energy. The calculated value is 12.0 eV, and the measured value is 11 eV. The accuracy of these and other calculated quantities depend significantly on the computational model used, as will be discussed in Chapter 26.

24.8 Qualitative Molecular Orbital Theory for Conjugated and Aromatic Molecules: The Hückel Model

The molecules in the preceding sections can be discussed using either a localized or a delocalized model of chemical bonding. This is not the case for conjugated and aromatic molecules, in which a delocalized model must be used. **Conjugated molecules** such as 1,3-butadiene have a planar carbon backbone with alternating single and double bonds. Butadiene has single and double bond lengths of 147 and 134 pm, respectively. The single bonds are shorter than the single bond length in ethane (154 pm), which suggests that a delocalized π network is formed. Such a delocalized network can be modeled in terms of the coupling between sp^2-hybridized carbon atoms in a σ-bonded carbon backbone. The lowering of the total energy that can be attributed to the formation of the π network is responsible for the reduced reactivity of conjugated molecules compared to molecules with isolated double bonds.

Aromatic molecules are a special class of conjugated molecules. They are based on ring structures that are particularly stable in chemical reactions. The presence of "closed circuits" of mobile electrons is required for a molecule to be aromatic. Because such currents imply electron delocalization, bonding in aromatic molecules cannot be explained by electron pairing in localized bonds. For example, benzene has six C—C bonds of equal length, 139 pm, a value between the single and double bond lengths in 1,3-butadiene. This suggests that the six π electrons are distributed over all six carbon atoms. Therefore, a delocalized model is required to discuss aromatic molecules.

Erich Hückel formulated a useful application of qualitative MO theory to calculate the energy levels of the delocalized π electrons in conjugated and aromatic molecules. Despite its simplicity, the **Hückel model** correctly predicts the stabilization that arises from delocalization and predicts which of many possible cyclic polyenes will be aromatic. In the Hückel model, the π network of MOs can be treated separately from the σ network of the carbon backbone. The Hückel model uses hybridization and the localized valence bond model to describe the σ-bonded skeleton and MO theory to describe the delocalized π electrons.

In the Hückel theory, the p atomic orbitals that combine to form π MOs are treated separately from the sp^2 σ-bonded carbon backbone. For the four-carbon π network in butadiene, the π MO can be written in the form

$$\psi_\pi = c_1\phi_{2pz1} + c_2\phi_{2pz2} + c_3\phi_{2pz3} + c_4\phi_{2pz4} \qquad \textbf{(24.14)}$$

As was done in Section 23.1, the variational method is used to calculate the coefficients that give the lowest energy for the four MOs that result from combining four AOs. We obtain the following secular equations:

$$c_1(H_{11} - \varepsilon S_{11}) + c_2(H_{12} - \varepsilon S_{12}) + c_3(H_{13} - \varepsilon S_{13}) + c_4(H_{14} - \varepsilon S_{14}) = 0$$

$$c_1(H_{21} - \varepsilon S_{21}) + c_2(H_{22} - \varepsilon S_{22}) + c_3(H_{23} - \varepsilon S_{23}) + c_4(H_{24} - \varepsilon S_{24}) = 0$$

$$c_1(H_{31} - \varepsilon S_{31}) + c_2(H_{32} - \varepsilon S_{32}) + c_3(H_{33} - \varepsilon S_{33}) + c_4(H_{34} - \varepsilon S_{34}) = 0$$

$$c_1(H_{41} - \varepsilon S_{41}) + c_2(H_{42} - \varepsilon S_{42}) + c_3(H_{43} - \varepsilon S_{43}) + c_4(H_{44} - \varepsilon S_{44}) = 0$$

$$\textbf{(24.15)}$$

Similar to the discussion in Chapter 23, integrals of the type H_{aa} are called Coulomb integrals, integrals of the type H_{ab} are called resonance integrals, and integrals of the type S_{ab} are called overlap integrals. Rather than evaluate these integrals, in the Hückel model thermodynamic and spectroscopic data obtained from different conjugated molecules are used to obtain their values. Because it relies on both theoretical and experimental input, the Hückel model is a **semiempirical theory.**

In the Hückel model, the Coulomb and resonance integrals are assumed to be the same for all conjugated hydrocarbons and are given the symbols α and β, respectively, where α is the negative of the ionization energy of the $2p$ orbital, and β, which is negative, is usually left as an adjustable parameter. Do not confuse this notation with spin up and spin down. The **secular determinant** that is used to obtain the MO energies and the coefficients of the AOs for 1,3-butadiene is

$$\begin{vmatrix} H_{11} - \varepsilon S_{11} & H_{12} - \varepsilon S_{12} & H_{13} - \varepsilon S_{13} & H_{14} - \varepsilon S_{14} \\ H_{21} - \varepsilon S_{21} & H_{22} - \varepsilon S_{22} & H_{23} - \varepsilon S_{23} & H_{24} - \varepsilon S_{24} \\ H_{31} - \varepsilon S_{31} & H_{32} - \varepsilon S_{32} & H_{33} - \varepsilon S_{33} & H_{34} - \varepsilon S_{34} \\ H_{41} - \varepsilon S_{41} & H_{42} - \varepsilon S_{42} & H_{43} - \varepsilon S_{43} & H_{44} - \varepsilon S_{44} \end{vmatrix} \qquad \textbf{(24.16)}$$

Several simplifying assumptions are made in the Hückel model to make it easier to solve secular determinants. The first is $S_{ii} = 1$ and $S_{ij} = 0$ unless $i = j$. This is a rather drastic simplification, because if the overlap between adjacent atoms were zero, no bond formation would occur. It is also assumed that $H_{ij} = \beta$ if i and j are on adjacent C atoms, $H_{ij} = \alpha$ if $i = j$, and $H_{ij} = 0$ otherwise. Setting $H_{ij} = 0$ for nonadjacent carbon atoms amounts to saying that the primary interaction is between the neighboring $2p_z$ orbitals. The result of the simplifying assumptions is that all elements of the determinant that are more than one position removed from the diagonal are zero for non-cyclic polyenes.

FIGURE 24.18
Energy levels and molecular orbitals for butadiene in the Hückel approximation. The sizes of the $2p_z$ AOs in the left column are proportional to their coefficients in the MO. Calculated MOs (see text) are shown in the right column. Red and blue lobes refer to positive and negative amplitudes, respectively. The vertical dashed lines indicate nodal planes.

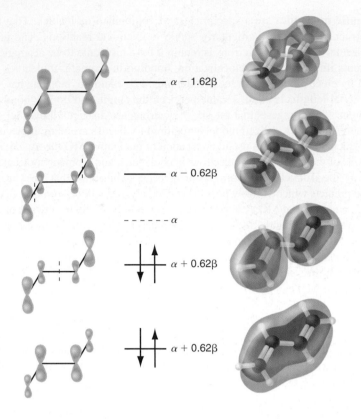

$$\alpha - 1.62\beta$$

$$\alpha - 0.62\beta$$

$$- - - - - - \alpha$$

$$\alpha + 0.62\beta$$

$$\alpha + 0.62\beta$$

With these assumptions, the secular determinant for butadiene is

$$\begin{vmatrix} \alpha - \varepsilon & \beta & 0 & 0 \\ \beta & \alpha - \varepsilon & \beta & 0 \\ 0 & \beta & \alpha - \varepsilon & \beta \\ 0 & 0 & \beta & \alpha - \varepsilon \end{vmatrix} = 0 \qquad \textbf{(24.17)}$$

As shown in Example Problem 24.7, this determinant has the solutions for the π orbital energies shown in Figure 24.18.

Consider several of the results shown in Figure 24.18. First, the coefficients of the AOs in the different MOs are not the same. Secondly, a pattern of nodes is seen that is identical to the other quantum mechanical systems that have been solved. The ground state has no nodes perpendicular to the plane of the molecule, and successively higher MOs have an increasing number of nodes. Recall that nodes correspond to regions in which the probability of finding the electron is zero. Because the nodes appear between the carbon atoms, they add an antibonding character to the MO, which increases the MO energy.

EXAMPLE PROBLEM 24.7

Solve the secular determinant for butadiene to obtain the MO energies.

Solution

The 4 × 4 secular determinant can be expanded to yield (see the Math Supplement, Appendix A) the following equation:

$$\begin{vmatrix} \alpha - \varepsilon & \beta & 0 & 0 \\ \beta & \alpha - \varepsilon & \beta & 0 \\ 0 & \beta & \alpha - \varepsilon & \beta \\ 0 & 0 & \beta & \alpha - \varepsilon \end{vmatrix}$$

$$= (\alpha - \varepsilon) \begin{vmatrix} \alpha - \varepsilon & \beta & 0 \\ \beta & \alpha - \varepsilon & \beta \\ 0 & \beta & \alpha - \varepsilon \end{vmatrix} - \beta \begin{vmatrix} \beta & \beta & 0 \\ 0 & \alpha - \varepsilon & \beta \\ 0 & \beta & \alpha - \varepsilon \end{vmatrix}$$

$$= (\alpha - \varepsilon)^2 \begin{vmatrix} \alpha - \varepsilon & \beta \\ \beta & \alpha - \varepsilon \end{vmatrix} - \beta(\alpha - \varepsilon) \begin{vmatrix} \beta & \beta \\ 0 & \alpha - \varepsilon \end{vmatrix} - \beta^2 \begin{vmatrix} \alpha - \varepsilon & \beta \\ \beta & \alpha - \varepsilon \end{vmatrix} + \beta^2 \begin{vmatrix} 0 & \beta \\ 0 & \alpha - \varepsilon \end{vmatrix}$$

$$= (\alpha - \varepsilon)^4 - (\alpha - \varepsilon)^2 \beta^2 - (\alpha - \varepsilon)^2 \beta^2 - (\alpha - \varepsilon)^2 \beta^2 + \beta^4$$

$$= (\alpha - \varepsilon)^4 - 3(\alpha - \varepsilon)^2 \beta^2 + \beta^4 = \frac{(\alpha - \varepsilon)^4}{\beta^4} - \frac{3(\alpha - \varepsilon)^2}{\beta^2} + 1 = 0$$

This equation can be written in the form of a quadratic equation:

$$\frac{(\alpha - \varepsilon)^2}{\beta^2} = \frac{3 \pm \sqrt{5}}{2}$$

which has the four solutions $\varepsilon = \alpha \pm 1.62\beta$ and $\varepsilon = \alpha \pm 0.62\beta$.

The effort required to obtain the energy levels and the AO coefficients for a monocyclic polyene can be greatly simplified by using the following geometrical construction: inscribe a regular polygon with the shape of the polyene in a circle of radius 2β with one vertex of the polygon pointing directly downward. Draw a horizontal line at each point for which the polygon touches the circle. These lines correspond to the energy levels, with the center of the circle corresponding to the energy α. This method is illustrated in Example Problem 24.8.

EXAMPLE PROBLEM 24.8

Use the inscribed polygon method to calculate the Hückel MO energy levels for benzene.

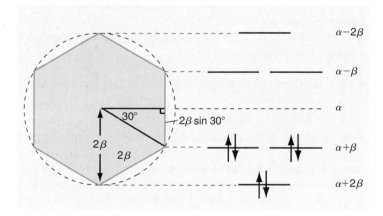

Solution

The geometrical construction shows that the energy levels are $\alpha + 2\beta$, $\alpha + \beta$, $\alpha - \beta$, and $\alpha - 2\beta$. The sum of the orbital energies for the six π electrons is $6\alpha + 8\beta$.

The benzene MOs and their energies are shown in Figure 24.19. Note that the energy levels for $\alpha + \beta$ and $\alpha - \beta$ are doubly degenerate. As was the case for butadiene, the lowest MO has no nodes perpendicular to the molecular plane, and the energy of the MO increases with the number of additional nodal planes. The average orbital energy of a π electron in benzene is

$$\frac{1}{6} \left[2(\alpha + 2\beta) + 4(\alpha + \beta) \right] = \alpha + 1.33\beta$$

FIGURE 24.19
Energy levels and molecular orbitals for benzene in the Hückel approximation. The sizes of the $2p_z$ AOs are proportional to their coefficients in the MO. Calculated MOs (see text) are shown in a three-dimensional perspective for the filled MOs and as an on-top view for the unfilled MOs. Red and blue lobes refer to positive and negative amplitudes. Thin dashed lines indicate nodal planes.

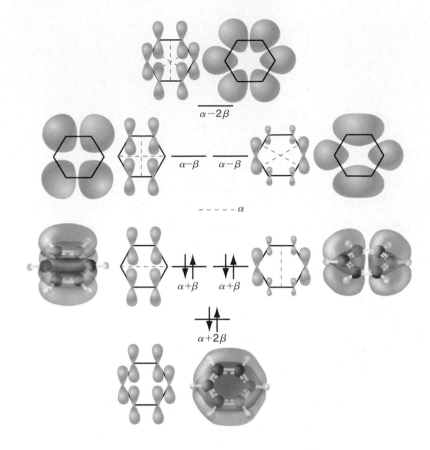

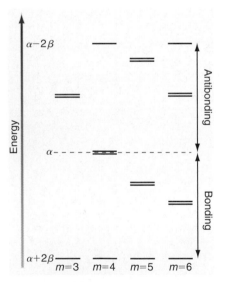

FIGURE 24.20
The energy of the π MOs is shown for cyclic polyenes described by the formula $(CH)_m$ with $m = 3$ to 6 π bonded carbons. The doubly degenerate pairs are shown slightly separated in energy for clarity.

The energy levels for the smaller monocyclic polyenes $(CH)_m$, where m is the number of π bonded carbons, exhibit the pattern shown in Figure 24.20. The energy value α separates the bonding and antibonding MOs.

This figure provides a justification for the following **Hückel rules** for a monocyclic conjugated system with N π electrons:

- If $N = 4n + 2$, where n is an integer 0, 1, 2, ..., the molecule is stabilized through the π delocalization network.
- If $N = 4n + 1$ or $4n + 3$, the molecule is a free radical.
- If $N = 4n$, the molecule has two unpaired electrons and is very reactive.

The justification for these rules can be understood from Figure 24.20. For each cyclic polyene, the lowest energy level is nondegenerate and has the energy $\alpha + 2\beta$. All other levels are doubly degenerate, with the exception of the highest level if m is even. The maximum stabilization is attained if $N = 4n + 2$, because all π electrons are paired and in bonding orbitals for which $\varepsilon < \alpha$. For $n = 1$, six π electrons correspond to the maximal stabilization. Benzene, for which $m = 6$, is an example of this case. Next consider benzene with one fewer or one more π electron. Because $N = 5$ or 7, both species are radicals because the highest occupied energy level is not filled, and both are less stable than benzene. What happens to a system of maximum stabilization if two electrons are removed? Because all energy levels, except the lowest (and if m is even, the highest), are doubly degenerate, each of the degenerate levels has an occupancy of one for $N = 4n$, and the molecule is a diradical.

These rules can be used to make useful predictions. For example, C_3H_3, which is formed from cyclopropene by the removal of one H atom, should be more stable as $C_3H_3^+ (N = 2)$ than as neutral $C_3H_3(N = 3)$ or $C_3H_3^- (N = 4)$. Undistorted cyclobutadiene $(N = 4)$ with four π electrons will be a diradical and, therefore, very reactive. The maximum stabilization for C_5H_5 is for $N = 6$, as is seen in Figure 24.20. Therefore, $C_5H_5^-$ is predicted to be more stable than C_5H_5 or $C_5H_5^+$. These predictions have been verified by experiment and show that the Hückel model has considerable predictive power, despite its significant approximations.

At present, the Hückel model is primarily useful for explaining the $4n + 2$ rule. Readily available computational chemistry software can rapidly calculate MOs and their corresponding energy levels on personal computers, making the determination of α and β from experimental data unnecessary. The calculated MOs shown in Figures 24.18 and 24.19 were obtained in this way using the B3LYP method of density functional theory and the 6-31G* basis set (see Chapter 26).

We now discuss the **resonance stabilization energy** that arises in aromatic compounds through the presence of closed circuits of mobile electrons. No unique method is available for calculating this stabilization energy. However, a reasonable way to determine this energy is to compare the π network energy of the cyclic polyene with that of a linear polyene that consists of alternating double and single bonds with the same number and arrangement of hydrogen atoms. In some cases, this may be a hypothetical molecule whose π network energy can be calculated using the method outlined earlier. As has been shown by L. Schaad and B. Hess [*J. Chemical Education* 51 (1974): 640–643], meaningful results for the resonance stabilization energy can be obtained only if the reference molecule is similar in all aspects except one: it is a linear rather than a cyclic polyene. For benzene, the reference molecule has the total π energy $6\alpha + 7.608\beta$. From Figure 24.20, note that the corresponding value for benzene is $6\alpha + 8\beta$. Therefore, the resonance stabilization energy per π electron in benzene is $(8.000\beta - 7.608\beta)/6 = 0.065\beta$. By considering suitable reference compounds, these authors have calculated the resonance stabilization energy for a large number of compounds, some of which are shown in Figure 24.21.

Figure 24.21 indicates that benzene and benzocyclobutadiene have the greatest resonance delocalization energy on this basis. Molecules with negative values for the resonance stabilization energy are predicted to be more stable as linear polyenes than as cyclic polyenes and are referred to as **antiaromatic molecules.** Note that these calculations only give information on the π network energy and that the total energy of the molecule is assumed to be proportional to the sum of the occupied π orbital energies. We have also ignored the possible effect of strain energy that arises if the C—C—C bonding angles are significantly different from 120°, which is optimal for sp^2 hybridization. For example, cyclobutadiene, for which the bond angle is 90°, has an appreciable strain

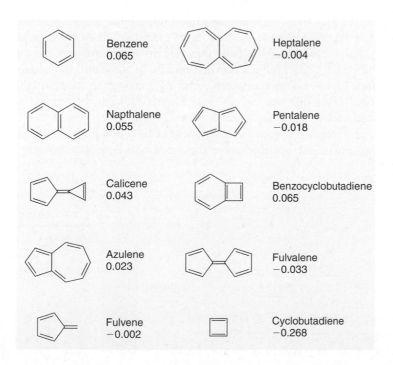

FIGURE 24.21

The resonance delocalization energy per π electron of a number of cyclic polyenes is shown in units of β.

energy associated with the σ-bonded backbone, which destabilizes the molecule relative to a linear polyene. A ranking of the degree of aromaticity based on experimental data such as thermochemistry, reactivity, and chemical shifts using nuclear magnetic resonance spectroscopy (see Chapter 28) is in good agreement with the predictions of the Hückel model if the appropriate reference molecule is used to calculate the resonance stabilization energy.

Although the examples used here to illustrate aromaticity are planar compounds, this is not a requirement for aromaticity. Sandwich compounds such as ferrocene, as well as the fullerenes, also show aromatic behavior. For these molecules, the closed circuits of mobile electrons extend over all three dimensions. These calculations for conjugated and aromatic molecules show the power of the Hückel model in obtaining useful results with minimal computational effort and without evaluating any integrals or even using numerical values for α and β. Fewer simplifying assumptions are made in the extended Hückel model, which treats the σ and π electrons similarly.

24.9 From Molecules to Solids

The Hückel model is also useful for understanding the energy levels in a solid, which can be thought of as a giant molecule. In discussing the application of the particle in the box model to solids in Chapter 16, we learned that a solid has an energy spectrum that has both continuous and discrete aspects. Within a range of energies called a band, the energy spectrum is continuous. However, the energy bands are separated by **band gaps** in which no quantum states are allowed. The Hückel model (Figure 24.22) is useful in developing an understanding of how this energy spectrum is generated.

Consider a one-dimensional chain of atoms in which π bonds are formed. Combining N $2p_x$ atomic orbitals creates the same number of π MOs as was seen for ethene, butadiene, and benzene. Hückel theory predicts that the difference in energy between the lowest- and highest-energy MO depends on the length of the conjugated chain but approaches the value 4β as the chain becomes infinitely long. All N energy levels still must lie in the range between $\alpha + 2\beta$ and $\alpha - 2\beta$. Therefore, as $N \rightarrow \infty$, the spacing between adjacent levels becomes vanishingly small, and the energy spectrum becomes continuous, generating a band.

The wave functions of the long one-dimensional chain are schematically indicated in Figure 24.23. At the bottom of the band, all AOs are in phase (fully bonding), but at the top of the band the AOs on adjacent atoms are out of phase (fully antibonding). At energies near the middle of the band, the nodal spacing is intermediate between N and one atomic spacing, making the state partially bonding. The energy versus distance curves from Figure 23.4 can be applied to the one-dimensional chain. This has been done in Figure 24.23. For a two-atom solid (diatomic molecule), the wave function is either fully bonding or fully antibonding. For a long chain, all possible wave functions between fully bonding and fully

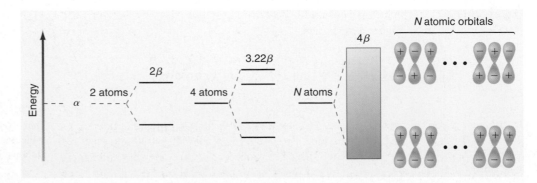

FIGURE 24.22
MOs generated in an atom chain using the Hückel model. As N becomes very large, the energy spectrum becomes continuous. The energy range of the MOs is shown in units of β.

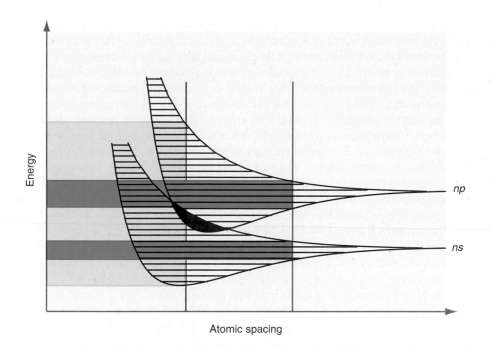

Energy

Atomic spacing

np

ns

FIGURE 24.23
Bands generated from two different AOs are shown. The width in energy of the band depends on the atomic spacing. For the equilibrium spacing indicated by the red line, the two bands overlap and all energy values between the top and bottom of the yellow-shaded area are allowed. This is not true for significantly larger or shorter atomic spacings, and the solid would exhibit a band gap at the spacing indicated by the blue line. In this case the two narrow bands indicated by the green areas do not overlap.

antibonding are possible. Therefore, the entire energy range indicated by horizontal black lines in Figure 24.23 is allowed.

We consider a specific example that demonstrates the contrasts among a conductor, a semiconductor, and an insulator. In solids, separate bands are generated from different AOs, such as the $3s$ and $3p$ AOs on Mg. If there is sufficient overlap between AOs to generate bonding, the bands are wide in energy. If this is not the case, the bands are narrow in energy. Magnesium, with the $[Ne]3s^2$ atomic configuration, has two $3s$ valence electrons that go into a band generated from the overlap of the $3s$ electrons on neighboring Mg atoms. Because N Mg atoms generate N MOs, each of which can be doubly occupied, the $2N$ Mg valence electrons completely fill the $3s$-generated band (lower band in Figure 24.23). If there were a gap between this and the next-highest band (upper band in Figure 24.23), which is generated from the $3p$ electrons, Mg would be an insulator. This corresponds to the atomic spacing indicated by the vertical blue line. However, in this case, the $3s$ and $3p$ bands overlap, corresponding to the atomic spacing indicated by the vertical red line. As a result, the unoccupied states in the overlapping bands are only infinitesimally higher in energy than the highest filled state. For this reason, Mg is a conductor.

If there is a gap between a completely filled band and the empty band of next higher energy, the solid is either an insulator or a semiconductor. The distinction between a semiconductor and an insulator is the width of the energy gap. If $E_{gap} \gg kT$ at temperatures below the melting point of the solid, the material is an insulator. Diamond is an insulator even at very high temperatures because it has a large band gap. However, if at elevated temperatures $E_{gap} \sim k_B T$, the Boltzmann distribution [Equation (13.2)] predicts that it will be easy to promote an electron from the filled valence band to the empty conduction band. In this case, the highest filled state is infinitesimally lower in energy than the lowest unfilled state, and the solid is a conductor. Silicon and germanium are called semiconductors because they behave like insulators at low temperatures and like conductors at higher temperatures.

24.10 Making Semiconductors Conductive at Room Temperature

In its pure state, silicon is conductive to an appreciable extent only at temperatures greater than 900 K because it has a band gap of 1.1 eV. Yet computers and other devices that are based on silicon technology function at room temperature. For this to happen, these devices must transmit electrical currents at 300. K. What enables silicon to be

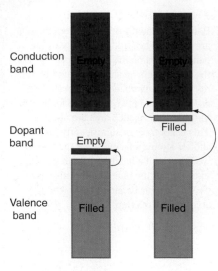

FIGURE 24.24

Modification of the silicon band structure generated by the introduction of dopants. The excitation that leads to conduction is from the valence band to the dopant band (*p*-type), as shown on the left, or dopant band to conduction band (*n*-type), as shown on the right, rather than across the Si band gap as indicated by the right-most curved arrow. Occupied and unoccupied bands are indicated by blue and red coloring, respectively. Energy increases vertically in the figure.

conductive at such low temperatures? The key to changing the properties of Si is the introduction of other atoms that occupy Si sites in the silicon crystal structure. The introduction of these foreign atoms in the Si crystal lattice is called **doping.**

Silicon is normally doped using atoms such as boron or phosphorus, which have one fewer or one more valence electron than silicon, respectively. Typically, the dopant concentration is on the order of a few parts per million relative to the Si concentration. How does this make Si conductive at lower temperatures? The Coulomb potential associated with the phosphorus atoms overlaps with the Coulomb potentials of the neighboring Si atoms, and the valence electrons of the P atom become delocalized throughout the crystal and form a separate band as discussed in Section 16.4. Because P has one more valence electron than Si, this band is only partially filled. As indicated in Figure 24.24, this band is located ~0.04 eV below the bottom of the empty conduction band. Electrons can be thermally excited from the dopant band to populate the empty Si conduction band. Importantly, it is the 0.04 eV rather than the Si band gap of 1.1 eV that must be comparable to k_BT to produce delocalized electrons in the conduction band. Therefore, phosphorus-doped silicon is conductive at 300. K, where $k_BT \approx 0.04$ eV. Because the dominant charge carriers are negative, one refers to an *n*-type semiconductor.

Boron can be introduced as a dopant at a ppm concentration. The Si crystal site that a boron atom occupies has one valence electron fewer than the neighboring sites and acts like a positive charge, which is referred to as a hole. The hole is delocalized throughout the lattice and acts like a mobile positive charge because electrons from adjacent Si atoms can fill it, leaving the B atom with an extra negative charge while the hole jumps from Si to Si atom. In this case, the empty dopant band is located ~0.045 eV above the top of the filled valence band. Thermal excitations of electrons from the filled valence band into the empty dopant band make the *p*-type semiconductor conductive. Because the dominant charge carriers in this case are positive, one refers to a *p*-type semiconductor.

For both *n*-type and *p*-type semiconductors, the activation energy to promote charge carriers and to induce conduction is much less than the Si band gap. The modifications to the Si **band structure** introduced by dopants are illustrated in Figure 24.24.

Vocabulary

antiaromatic molecules	Hückel rules	semiempirical theory
aromatic molecules	hybrid orbital	sp, sp^2, and sp^3 hybridization
band gap	hybridization	valence bond (VB) theory
band structure	Lewis structure	valence shell electron pair repulsion (VSEPR) model
conjugated molecules	localized bonding model	
delocalized bonding model	lone pair	Walsh correlation diagram
doping	resonance stabilization energy	
Hückel model	secular determinant	

Conceptual Problems

Q24.1   Why can it be unclear whether a material is a semiconductor or an insulator?

Q24.2   How do the values of the AO coefficients in a MO differ for a delocalized and a localized bond?

Q24.3   What experimental evidence can you cite in support of the hypothesis that the electronegativity of a hybridized atom increases with increasing *s* character?

Q24.4   Explain why all possible wave functions between the fully bonding and the fully antibonding are possible for the bands shown in Figure 24.22.

Q24.5   On the basis of what you know about the indistinguishability of electrons and the difference between the wave functions for bonding electrons and lone pairs, discuss the validity and usefulness of the Lewis structure for the fluorine molecule (:F̈—F̈:).

Q24.6 What evidence can you find in Table 24.1 that C—C sp bonds are stronger than sp^3 bonds?

Q24.7 How is it possible that a semiconductor would become metallic if the nearest neighbor spacing could be changed sufficiently?

Q24.8 Why are localized and delocalized models equally valid for describing bonding in closed-shell molecules? Why can't experiments distinguish between these models?

Q24.9 The hybridization model assumes that atomic orbitals are recombined to prepare directed orbitals that have the bond angles appropriate for a given molecule. What aspects of the model can be tested by experiments, and what aspects are conjectures that are not amenable to experimental verification?

Q24.10 Why can't localized orbitals be represented in an MO energy diagram?

Q24.11 In using the sum of the occupied MO energies to predict the bond angle in H_2A molecules, the total energy of the molecule is assumed to be proportional to the sum of the occupied MO energies. This assumption can be justified. Do you expect this sum to be greater than or smaller than the total energy?

Q24.12 In explaining molecular structure, the MO model uses the change in MO energy with bond angle. Explain why the decrease in energy of the $1a_1$ and $2a_1$ MOs as 2θ decreases more than offsets the increase in energy for the $1b_2$ MO for water.

Q24.13 What is the in-plane amplitude of the wave functions describing the π network in the conjugated molecules shown in Figures 24.18 and 24.19?

Q24.14 What is the rationale for setting $H_{ij} = 0$ for nonadjacent atoms in the Hückel model?

Q24.15 A certain cyclic polyene is known to be nonplanar. Are the MO energy levels of this molecule well described by the Hückel model? Justify your answer.

Numerical Problems

Problem numbers in **red** indicate that the solution to the problem is given in the *Student's Solutions Manual*.

P24.1 Show that the determinantal property

$$\begin{vmatrix} a & c \\ b & d \end{vmatrix} = \begin{vmatrix} a & \gamma a + c \\ b & \gamma b + d \end{vmatrix}$$

used in the discussion of localized and delocalized orbitals in Section 24.6 is correct.

P24.2 Predict whether LiH_2^+ and NH_2^- should be linear or bent based on the Walsh correlation diagram in Figure 24.11. Explain your answers.

P24.3 Use the framework described in Section 24.3 to construct normalized hybrid bonding orbitals on the central oxygen in O_3 that are derived from $2s$ and $2p$ atomic orbitals. The bond angle in ozone is 116.8°.

P24.4 Are the localized bonding orbitals in Equation (24.13) defined by

$$\sigma' = 2c_1\phi_{H1sA} + \left(c_2\phi_{Be2s} - \frac{c_1c_4}{c_3}\phi_{Be2p_z}\right) \text{ and }$$

$$\sigma'' = 2c_1\phi_{H1sB} + \left(c_2\phi_{Be2s} + \frac{c_1c_4}{c_3}\phi_{Be2p_z}\right)$$

orthogonal? Answer this question by evaluating the integral $\int(\sigma')^*\sigma''\, d\tau$.

P24.5 Use the method described in Example Problem 24.3 to show that the sp-hybrid orbitals $\psi_a = 1/\sqrt{2}(-\phi_{2s} + \phi_{2p_z})$ and $\psi_b = 1/\sqrt{2}(-\phi_{2s} - \phi_{2p_z})$ are oriented 180° apart.

P24.6 Use the formula $\cos 2\theta = -\alpha^2$ and the method in Section 24.2 to derive the formula $\psi_a = 1/\sqrt{2}(-\phi_{2s} + \phi_{2p_z})$ and $\psi_b = 1/\sqrt{2}(-\phi_{2s} - \phi_{2p_z})$ for two sp-hybrid orbitals directed 180° apart. Show that these hybrid orbitals are orthogonal.

P24.7 Show that two of the set of four equivalent orbitals appropriate for sp^3 hybridization,

$$\psi_a = \frac{1}{2}(-\phi_{2s} + \phi_{2p_x} + \phi_{2p_y} + \phi_{2p_z}) \text{ and }$$

$$\psi_b = \frac{1}{2}(-\phi_{2s} - \phi_{2p_x} - \phi_{2p_y} + \phi_{2p_z})$$

are orthogonal.

P24.8 Show that the water hybrid bonding orbitals given by $\psi_a = 0.55\phi_{2p_z} + 0.71\phi_{2p_x} - 0.45\phi_{2s}$ and $\psi_b = 0.55\phi_{2p_z} - 0.71\phi_{2p_x} - 0.45\phi_{2s}$ are orthogonal.

P24.9 Predict which of the bent molecules, BH_2 or NH_2, should have the larger bond angle on the basis of the Walsh correlation diagram in Figure 24.11. Explain your answer.

P24.10 Derive two additional mutually orthogonal hybrid orbitals for the lone pairs on oxygen in H_2O, each of which is orthogonal to ψ_a and ψ_b, by following these steps:

a. Starting with the following formulas for the lone pair orbitals

$$\psi_c = d_1\phi_{2p_z} + d_2\phi_{2p_y} + d_3\phi_{2s} + d_4\phi_{2p_x}$$

$$\psi_d = d_5\phi_{2p_z} + d_6\phi_{2p_y} + d_7\phi_{2s} + d_8\phi_{2p_x}$$

use symmetry conditions to determine d_2 and d_4 and to determine the ratio of d_3 to d_7 and of d_4 to d_8.

b. Use the condition that the sum of the square of the coefficients over all the hybrid orbitals and lone pair orbitals is 1 to determine the unknown coefficients.

P24.11 Use the Boltzmann distribution to answer parts (a) and (b):

a. Calculate the ratio of the number of electrons at the bottom of the conduction band to those at the top of the valence band for pure Si at 300. K. The Si band gap is 1.1 eV.

b. Calculate the ratio of the number of electrons at the bottom of the conduction band to those at the top of the dopant band for P-doped Si at 300. K. The top of the dopant band lies 0.040 eV below the bottom of the Si conduction band.

Assume for these calculations that the ratio of the degeneracies is unity. What can you conclude about the room temperature conductivity of these two materials on the basis of your calculations?

P24.12 Use the VSEPR method to predict the structures of the following:

a. PCl_5 **b.** SO_2 **c.** XeF_2 **d.** XeF_6

P24.13 Use the framework described in Section 24.3 to derive the normalized hybrid lone pair orbital on the central oxygen in O_3 that is derived from $2s$ and $2p$ atomic orbitals. The bond angle in ozone is 116.8°.

P24.14 Using your results from Problem P24.10, do the following:

a. Calculate the s and p character of the water lone pair hybrid orbitals.

b. Show that the lone pair orbitals are orthogonal to each other and to the hybrid bonding orbitals.

P24.15 Use the VSEPR model to predict the structures of the following:

a. PF_3 **b.** CO_2 **c.** BrF_5 **d.** SO_3^{2-}

P24.16 The occupied MOs of ammonia are shown next along with the MO energies. Indicate which AOs are most important in each MO and indicate the relative phases of the AOs. Classify the MOs as localized or delocalized, and bonding, nonbonding or antibonding.

P24.17 Predict whether the ground state or the first excited state of CH_2 should have the larger bond angle on the basis of the Walsh correlation diagram shown in Figure 24.11. Explain your answer.

P24.18 The occupied MOs of ethene are shown next along with the MO energies. Indicate which AOs are most important in each MO and indicate the relative phases of the AOs. Classify the MOs as localized or delocalized, σ or π bonds, and bonding, nonbonding or antibonding.

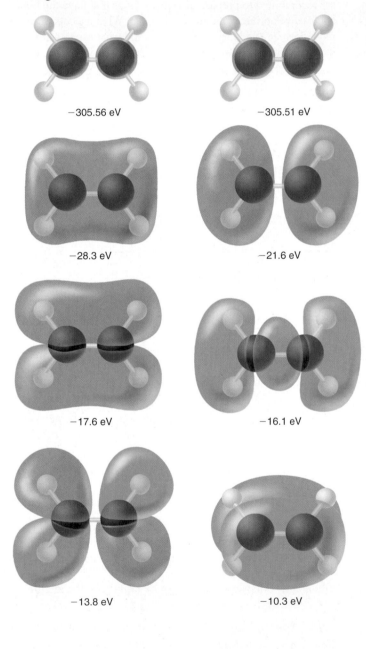

−305.56 eV   −305.51 eV

−28.3 eV   −21.6 eV

−17.6 eV   −16.1 eV

−13.8 eV   −10.3 eV

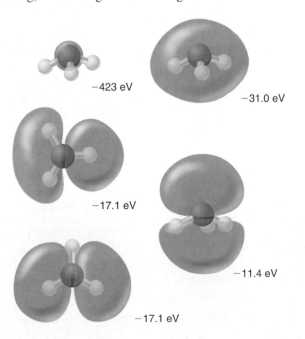

−423 eV

−31.0 eV

−17.1 eV

−11.4 eV

−17.1 eV

P24.19 The occupied MOs of hydrogen cyanide are shown next along with the MO energies. Indicate which AOs are most important in each MO and indicate the relative phases of the AOs. Classify the MOs as localized or delocalized, σ or π bonds, and bonding, nonbonding or antibonding.

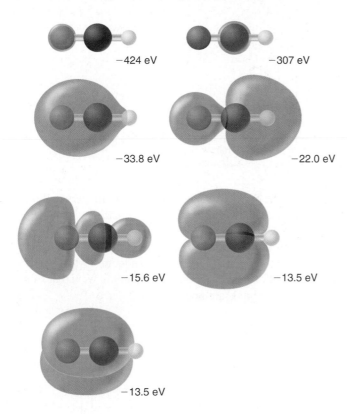

P24.20 Use the geometrical construction shown in Example Problem 24.8 to derive the π electron MO levels for cyclobutadiene. What is the total π energy of the molecule? How many unpaired electrons will the molecule have?

P24.21 Determine the AO coefficients for the lowest energy Hückel π MO for butadiene.

P24.22 Use the geometrical construction shown in Example Problem 24.8 to derive the π electron MO levels for the cyclopentadienyl radical. What is the total π energy of the molecule? How many unpaired electrons will the molecule have?

P24.23 The allyl cation $CH_2 = CH - CH_2^+$ has a delocalized π network that can be described by the Hückel method. Derive the MO energy levels of this species and place the electrons in the levels appropriate for the ground state. Using the butadiene MOs as an example, sketch what you would expect the MOs to look like. Classify the MOs as bonding, antibonding, or nonbonding.

P24.24 Write down and solve the secular determinant for the π system of ethylene in the Hückel model. Determine the coefficients for the $2p_z$ AOs on each of the carbons and make a sketch of the MOs. Characterize the MOs as bonding and antibonding.

P24.25 Use the geometrical construction shown in Example Problem 24.8 to derive the energy levels of the cycloheptatrienyl cation. What is the total π energy of the molecule? How many unpaired electrons will the molecule have? Would you expect this species, the neutral species, or the anion to be aromatic? Justify your answer.

P24.26 One of the low-energy geometries of digermane, Ge_2H_2, is ethene-like. The Lewis-dot structure shown is one of three Lewis-dot resonant forms. The order of the Ge — Ge bond has been described as between two and three. Bond orders above two rely on the lone electrons participating in bonding. The five highest-energy occupied valence molecular orbitals are shown next. Classify the MOs as localized or delocalized, σ or π bonds, and bonding, nonbonding or antibonding.

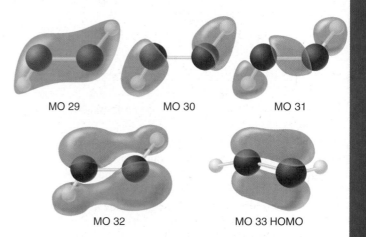

P24.27 $S-p$ hybridization on each Ge atom in planar *trans*-digermane has been described as $sp^{1.5}$ for the Ge—Ge sigma bond and $sp^{1.8}$ for the Ge—H bond. Suppose that the Ge lone electron (in terms of Lewis-dot valence electrons) is in an sp^n-hybrid orbital, where n is the hybrid's p character. The three in-plane $s-p$ hybrid orbitals on germanium must be normalized and mutually orthogonal. Assume that the molecule lies in the xz plane with the Ge—Ge bond on the z axis. Express the hybrid orbitals as linear combinations of the p and s AOs and calculate the coefficients. Use these values to calculate n.

P24.28 $S-p$ hybridization on each Ge atom in planar *trans*-digermane has been described as $sp^{1.5}$ for the Ge—Ge sigma bond and $sp^{1.8}$ for the Ge—H bond. Calculate the H—Ge—Ge bond angle based on this information. Note that the $4p_x$ and $4p_y$ orbitals are proportional to $\cos(\theta)$ and $\sin(\theta)$, respectively and use the coefficients determined in Problem P24.27 to solve this problem.

P24.29 The energy of the occupied valence MOs of H_2S are shown as a function of the H—S—H bond angle. Compared to the analogous diagram, Figure 24.11, for H_2O, the $2a_1$ MO energy decreases more as the bond angle approaches 90 degrees. Explain, based on the MO diagram, why H_2S is bent, and why its bond angle (92°) is smaller than the bond angle in water.

For comparison, also offer a bond-angle explanation based on hybridization rather than on the MO diagram.

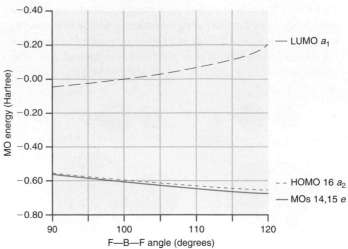

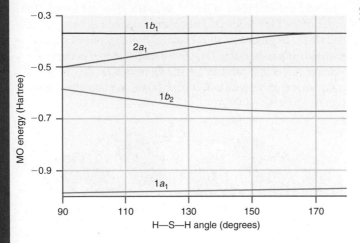

P24.30 The following diagram shows the energies of valence molecular orbitals of boron trifluoride. The energies of three occupied orbitals (the a_2 HOMO and doubly degenerate e orbitals) are shown. The energy of the unoccupied LUMO is also shown. The angle on the abscissa is the F—B—F bond angle. Based on the MO diagram, is boron trifluoride planar or pyramidal? Which structure does the VSEPR model predict?

P24.31 The density of states (DOS) of pyrite, crystalline FeS_2, as calculated by Eyert, *et al., Physical Review B* 55 (6350): 1998 is shown next. The highest occupied energy level corresponds to zero energy. Based on the DOS, is pyrite an insulator, a conductor, or a semiconductor? Also, how does the DOS support the localized-bonding view that some iron valence orbitals are non-bonding?

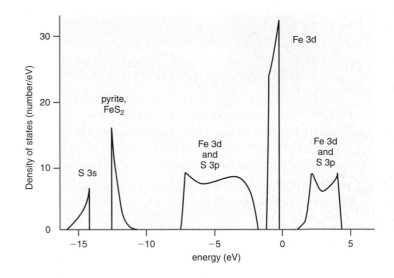

Computational Problems

More detailed instructions on carrying out these calculations using Spartan Physical Chemistry are found on the book website at *www.masteringchemistry.com.*

C24.1 Calculate the bond angles in NH_3 and in NF_3 using the density functional method with the B3LYP functional and the 6-31G* basis set. Compare your result with literature values. Do your results agree with the predictions of the VSEPR model and Bent's rule?

C24.2 Calculate the bond angles in H_2O and in H_2S using the density functional method with the B3LYP functional and the 6-31G* basis set. Compare your result with literature values. Do your results agree with the predictions of the VSEPR model and Bent's rule?

C24.3 Calculate the bond angle in ClO_2 using the density functional method with the B3LYP functional and the 6-31G* basis set. Compare your result with literature values. Does your result agree with the predictions of the VSEPR model?

C24.4 SiF_4 has four ligands and one lone pair on the central S atom. Which of the following structures do you expect to be the equilibrium form based on a calculation using the density functional method with the B3LYP functional and the 6-31G* basis set? In (a) the structure is a trigonal bipyramid, (b) is a square planar structure, and (c) is a see-saw structure.

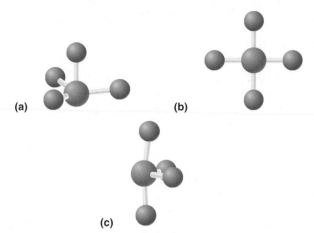

(a) (b) (c)

C24.5 Calculate the bond angles in singlet BeH_2, doublet NH_2, and singlet BH_2 using the Hartree–Fock method and the 6-31G* basis set. Explain your results using the Walsh diagram of Figure 24.11.

C24.6 Calculate the bond angle in singlet LiH_2^+ using the Hartree–Fock method and the 6-31G* basis set. Can you explain your results using the Walsh diagram of Figure 24.11? (Hint: Determine the calculated bond lengths in the LiH_2^+ molecule.)

C24.7 Calculate the bond angle in singlet and triplet CH_2 and doublet CH_2^+ using the Hartree–Fock method and the 6-31G* basis set. Can you explain your results using the Walsh diagram of Figure 24.11?

C24.8 Calculate the bond angle in singlet NH_2^+, doublet NH_2, and singlet NH_2^- using the Hartree–Fock method and the 6-31G* basis set. Can you explain your results using the Walsh diagram of Figure 24.11?

C24.9 How essential is coplanarity to conjugation? Answer this question by calculating the total energy of butadiene using the Hartree–Fock method and the 6-31G* basis set for dihedral angles of 0, 45, and 90 degrees.

C24.10 Calculate the equilibrium structures for singlet and triplet formaldehyde using the density functional method with the B3LYP functional and the 6-311+G** basis set. Choose (a) planar and (b) pyramidal starting geometries. Calculate vibrational frequencies for both starting geometries. Are any of the frequencies imaginary? Explain your results.

C24.11 Calculate the equilibrium structure for Cl_2O using the density functional method with the B3LYP functional and the 6-31G* basis set. Obtain an infrared spectrum and activate the normal modes. What are the frequencies corresponding to the symmetric stretch, the asymmetric stretch, and the bending modes?

C24.12 Calculate the equilibrium structures for PF_3 using the density functional method with the B3LYP functional and the 6-31G* basis set. Obtain an infrared spectrum and activate the normal modes. What are the frequencies corresponding to the symmetric stretch, the symmetric deformation, the degenerate stretch, and the degenerate deformation modes?

C24.13 Calculate the equilibrium structures for C_2H_2 using the density functional method with the B3LYP functional and the 6-31G* basis set. Obtain an infrared spectrum and activate the normal modes. What are the frequencies corresponding to the symmetric $C-H$ stretch, the antisymmetric $C-H$ stretch, the $C-C$ stretch, and the two bending modes?

C24.14 Calculate the structure of $N\equiv C-Cl$ using the density functional method with the B3LYP functional and the 6-31G* basis set. Which is more electronegative, the Cl or the cyanide group? What result of the calculation did you use to answer this question?

Electronic Spectroscopy

Absorption of visible or ultraviolet light can lead to transitions between the ground state and excited electronic states of atoms and molecules. Vibrational transitions that occur together with electronic transitions are governed by the Franck-Condon factors rather than the $\Delta n = \pm 1$ dipole selection rule. The excited state can relax to the ground state through a combination of fluorescence, internal conversion, intersystem crossing, and phosphorescence. Fluorescence is very useful in analytical chemistry and can detect as little as 2×10^{-13} mol/L of a strongly fluorescing species. Ultraviolet photoemission can be used to obtain information about the orbital energies of molecules. Linear and circular dichroism spectroscopy can be used to determine the secondary and tertiary structure of biomolecules in solution.

25.1 The Energy of Electronic Transitions

In Chapter 19 spectroscopy and the basic concepts relevant to transitions between energy levels of a molecule were introduced. Recall that the energy spacing between rotational levels is much less than the spacing between vibrational levels. Extending this comparison to electronic states, $\Delta E_{electronic} \gg \Delta E_{vibrational} > \Delta E_{rotational}$. Whereas rotational and vibrational transitions are induced by microwave and infrared radiation, electronic transitions are induced by visible and ultraviolet (UV) radiation. Just as an absorption spectrum in the infrared exhibits both rotational and vibrational transitions, an absorption spectrum in the visible and UV range exhibits a number of electronic transitions, and a specific electronic transition will contain vibrational and rotational fine structure.

Electronic excitations are responsible for giving color to the objects we observe because the human eye is sensitive to light only in the limited range of wavelengths in which some electronic transitions occur. Either the reflected or the transmitted light is observed, depending on whether the object is opaque or transparent. Transmitted and reflected light complement the absorbed light. For example, a leaf is green because chlorophyll absorbs in the blue (450 nm) and red (650 nm) regions of the visible light spectrum. Electronic excitations can be detected (at a limited resolution) without the aid of a spectrometer because the human eye is a very sensitive detector of radiation.

At a wavelength of 500 nm, the human eye can detect one part in 10^6 of the intensity of sunlight on a bright day. This corresponds to as few as 500 photons per second incident on an area of 1 mm^2.

Because the electronic spectroscopy of a molecule is directly linked to its energy levels, which are in turn determined by its structure and chemical composition, UV-visible spectroscopy provides a very useful qualitative tool for identifying molecules. In addition, for a given molecule, electronic spectroscopy can be used to determine energy levels in molecules. However, the UV and visible photons that initiate an electronic excitation perturb a molecule far more than rotational or vibrational excitation. For example, the bond length in electronically excited states of O_2 is as much as 30% longer than that in the ground state. Whereas in its ground state, formaldehyde is a planar molecule, it is pyramidal in its lowest two excited states. As might be expected from such changes in geometry, the chemical reactivity of excited-state species can be quite different from the reactivity of the ground-state molecule.

25.2 Molecular Term Symbols

We begin our discussion of electronic excitations by introducing **molecular term** symbols, which describe the electronic states of molecules in the same way that atomic term symbols describe atomic electronic states. The following discussion is restricted to diatomic molecules. A quantitative discussion of electronic spectroscopy requires a knowledge of molecular term symbols. However, electronic spectroscopy can be discussed at a qualitative level without discussing molecular term symbols. To do so, move directly to Section 25.4.

The component of **L** and **S** along the molecular axis (M_L and M_S), which is chosen to be the z axis, and S are the only good quantum numbers (see Section 22.1) by which to specify individual states in diatomic molecules. Therefore, term symbols for molecules are defined using these quantities. As for atoms, only unfilled subshells need to be considered to obtain molecular term symbols. As discussed in Chapter 23, in the first and second row diatomic molecules, the molecular orbitals (MOs) are either of the σ or π type. Just as for atoms, the quantum numbers m_{li} and m_{si} can be added to generate M_L and M_S for the molecule because they are scalars rather than vectors. The addition process is described by the equations

$$M_L = \sum_{i=1}^{n} m_{li} \quad \text{and} \quad M_S = \sum_{i=1}^{n} m_{si} \tag{25.1}$$

in which m_{li} and m_{si} are the z components of orbital and spin angular momentum for the ith electron in its molecular orbital and the summation is over unfilled subshells. The molecular orbitals of diatomic molecules have either σ symmetry, in which the orbital is unchanged by rotation around the molecular axis, or π symmetry, in which the MO has a nodal plane passing through the molecular axis. For a σ orbital, $m_l = 0$, and for a π orbital, $m_l = \pm 1$. Note that the $m_l = 0$ value does not occur for a π MO because this value corresponds to the $2p_z$ AO, which forms a σ MO. M_S is calculated from the individual spin angular momentum vector components $m_{si} = \pm 1/2$ in the same way for molecules as for atoms (see Chapter 22). The allowed values of the quantum numbers S and L can be calculated from $-L \le M_L \le L$ and $-S \le M_S \le S$ to generate a molecular term symbol of the form $^{2S+1}\Lambda$, where $\Lambda = |M_L|$. For molecules, the following symbols are used for different Λ values to avoid confusion with atomic terms:

$$
\begin{array}{ccccc}
\Lambda & 0 & 1 & 2 & 3 \\
\text{Symbol} & \Sigma & \Pi & \Delta & \Phi
\end{array}
\tag{25.2}
$$

A g or u right subscript is added to the molecular term symbol for homonuclear diatomics as illustrated in Example Problem 25.1. Because heteronuclear diatomics do not possess an inversion center, they do not have g or u symmetry. This formalism will become clearer with a few examples.

EXAMPLE PROBLEM 25.1

What is the molecular term symbol for the H_2 molecule in its ground state? In its first two excited states?

Solution

In the ground state, the H_2 molecule is described by the $(1\sigma_g)^2$ configuration. For both electrons, $m_l = 0$. Therefore, $\Lambda = 0$, and we are dealing with a Σ term. Because of the Pauli principle, one electron has $m_s = +1/2$ and the other has $m_s = -1/2$. Therefore, $M_S = 0$ and it follows that $S = 0$. It remains to be determined whether the MO has g or u symmetry. Each term in the antisymmetrized MO is of the form $\sigma_g \times \sigma_g$ (see Section 23.6). Recall that the products of two even or odd functions is even, and the product of an odd and an even function is odd. Therefore, the product of two g (or two u) functions is a g function, and the ground state of the H_2 molecule is $^1\Sigma_g$.

After promotion of an electron, the configuration is $(1\sigma_g)^1(1\sigma_u^*)^1$, and because the electrons are in separate MOs, this configuration leads to both **singlet states** and **triplet states.** Again, because $m_l = 0$ for both electrons, we are dealing with a Σ term. Because the two electrons are in different MOs, $m_s = \pm 1/2$ for each electron, giving m_s values of $-1, 0$ (twice), and $+1$. This is consistent with $S = 1$ and $S = 0$. Because the product of a u and a g function is a u function, both singlet and triplet states are u functions. Therefore, the first two excited states are described by the terms $^3\Sigma_u$ and $^1\Sigma_u$. Using Hund's first rule, we conclude that the triplet state is lower in energy than the singlet state.

In a more complete description, an additional subscript $+$ or $-$ is added to Σ terms only, depending on whether the antisymmetrized molecular wave function changes sign $(-)$ or remains unchanged $(+)$ in a reflection through any plane containing the molecular axis. The assignment of $+$ or $-$ to the terms is an advanced topic that is discussed in Supplemental Section 25.14. For our purposes, the following guidelines are sufficient for considering the ground state of second row homonuclear diatomic molecules:

- If all MOs are filled, $+$ applies.
- If all partially filled MOs have σ symmetry, $+$ applies.
- For partially filled MOs of π symmetry (for example, B_2 and O_2), if Σ terms arise, the triplet state is associated with $-$, and the singlet state is associated with $+$.

These guidelines do not apply to excited states. We conclude that the term corresponding to the $(\pi^*)^2$ ground-state configuration of O_2 (see Figure 23.18) is designated by $^3\Sigma_g^-$. The other terms that arise from the ground-state configuration are discussed in Example Problem 25.2.

EXAMPLE PROBLEM 25.2

Determine the possible molecular terms for O_2, which has the following configuration: $(1\sigma_g)^2(1\sigma_u^*)^2(2\sigma_g)^2(2\sigma_u^*)^2(3\sigma_g)^2(1\pi_u)^2(1\pi_u)^2(1\pi_g^*)^1(1\pi_g^*)^1$

Solution

Only the last two electrons contribute to nonzero net values of M_L and M_S because the other subshells are filled. The various possibilities for combining the orbital and spin angular momenta of these two electrons in a way consistent with the Pauli principle are given in the following table. The Λ values are determined as discussed for atomic terms in Chapter 22. Because $M_L \le L$, the first two entries in the table belong to a Δ term. Because $M_S = 0$ for both entries, it is a $^1\Delta$ term. Of the remaining four entries, two have $|M_S| = 1$, corresponding to a triplet term. One of the two other entries with $M_S = 0$ must also belong to this term. Because $M_L = 0$ for all four entries, it is a $^3\Sigma$ term. The remaining entry corresponds to a $^1\Sigma$ term.

m_{l1}	m_{l2}	$M_L = m_{l1} + m_{l2}$	m_{s1}	m_{s2}	$M_S = m_{s1} + m_{s2}$	Term
1	1	2	+1/2	−1/2	0	$\left.\right\}{}^1\Delta$
−1	−1	−2	+1/2	−1/2	0	
1	−1	0	+1/2	+1/2	1	$\left.\right\}{}^3\Sigma$
1	−1	0	−1/2	−1/2	−1	
1	−1	0	+1/2	−1/2	0	$\left.\right\}{}^1\Sigma, {}^3\Sigma$
1	−1	0	−1/2	+1/2	0	

The next task is the assignment of the g or u label to these molecular terms. Because both of the electrons are in an MO of g symmetry, the overall symmetry of the term will be g in all cases.

The $+$ and $-$ symbols are assigned in Supplemental Section 25.14. We show there that the singlet term is ${}^1\Sigma_g^+$ and the triplet term is ${}^3\Sigma_g^-$. By Hund's first rule, the ${}^3\Sigma_g^-$ term is lowest in energy and is the ground state. Experimentally, the ${}^1\Delta_g$ and ${}^1\Sigma_g^+$ terms are found to lie 0.98 and 1.62 eV higher in energy, respectively, than the ground state.

In terms of arrows indicating the spin orientations, the allowed combinations of m_l and m_s in the table can be represented in a shorthand notation by

$$(1\sigma_g)^2 \ (1\sigma_u^*)^2 \ (2\sigma_g)^2 \ (2\sigma_u^*)^2 \ (3\sigma_g)^2 \ (1\pi_u)^2 \ (1\pi_u)^2 \ (1\pi_g^*)^1 \ (1\pi_g^*)^1$$

$$(\uparrow\downarrow) \ (\uparrow\downarrow) \ (\uparrow\downarrow) \ (\uparrow\downarrow) \ (\uparrow\downarrow) \ (\uparrow\downarrow) \ (\uparrow\downarrow) \quad (\uparrow) \quad\quad (\downarrow) \quad {}^1\Sigma_g^+, {}^1\Delta_g$$

$$\left\{
\begin{array}{l}
(1\sigma_g)^2 \ (1\sigma_u^*)^2 \ (2\sigma_g)^2 \ (2\sigma_u^*)^2 \ (3\sigma_g)^2 \ (1\pi_u)^2 \ (1\pi_u)^2 \ (1\pi_g^*)^1 \ (1\pi_g^*)^1 \\
\left[
\begin{array}{l}
(\uparrow\downarrow) \ (\uparrow\downarrow) \ (\uparrow\downarrow) \ (\uparrow\downarrow) \ (\uparrow\downarrow) \ (\uparrow\downarrow) \ (\uparrow\downarrow) \quad (\uparrow) \quad (\uparrow) \\
(\uparrow\downarrow) \ (\uparrow\downarrow) \ (\uparrow\downarrow) \ (\uparrow\downarrow) \ (\uparrow\downarrow) \ (\uparrow\downarrow) \ (\uparrow\downarrow) \quad (\uparrow) \quad (\downarrow) \\
\qquad\qquad\qquad\qquad\qquad + \\
(\uparrow\downarrow) \ (\uparrow\downarrow) \ (\uparrow\downarrow) \ (\uparrow\downarrow) \ (\uparrow\downarrow) \ (\uparrow\downarrow) \ (\uparrow\downarrow) \quad (\downarrow) \quad (\uparrow) \\
(\uparrow\downarrow) \ (\uparrow\downarrow) \ (\uparrow\downarrow) \ (\uparrow\downarrow) \ (\uparrow\downarrow) \ (\uparrow\downarrow) \ (\uparrow\downarrow) \quad (\downarrow) \quad (\downarrow)
\end{array}
\right]
\end{array}
\right\} {}^3\Sigma_g^-$$

Note that this notation with arrows pointing up and down to indicate α and β spins is inadequate because it is not possible to represent the different values of m_{l1} and m_{l2}.

On the basis of this discussion, the **molecular configuration** and the ground-state terms for the first row homonuclear diatomic molecules are listed in Table 25.1. The procedure for heteronuclear diatomics is similar, but it differs in that the numbering of the MOs is different and the g and u symmetries do not apply.

TABLE 25.1 Terms for Ground-State Second Row Diatomics

Molecule	Electron Configuration	Ground-State Term
H_2^+	$(1\sigma_g)^1$	${}^2\Sigma_g^+$
H_2	$(1\sigma_g)^2$	${}^1\Sigma_g^+$
He_2^+	$(1\sigma_g)^2(1\sigma_u^*)^1$	${}^2\Sigma_u^+$
Li_2	$(1\sigma_g)^2(1\sigma_u^*)^2(2\sigma_g)^2$	${}^1\Sigma_g^+$
B_2	$(1\sigma_g)^2(1\sigma_u^*)^2(2\sigma_g)^2(2\sigma_u^*)^2(1\pi_u)^1(1\pi_u)^1$	${}^3\Sigma_g^-$
C_2	$(1\sigma_g)^2(1\sigma_u^*)^2(2\sigma_g)^2(2\sigma_u^*)^2(1\pi_u)^2(1\pi_u)^2$	${}^1\Sigma_g^+$
N_2^+	$(1\sigma_g)^2 \ (1\sigma_u^*)^2 \ (2\sigma_g)^2 \ (2\sigma_u^*)^2 \ (1\pi_u)^2 \ (1\pi_u)^2 \ (3\sigma_g)^1$	${}^2\Sigma_g^+$
N_2	$(1\sigma_g)^2(1\sigma_u^*)^2(2\sigma_g)^2(2\sigma_u^*)^2(1\pi_u)^2(1\pi_u)^2(3\sigma_g)^2$	${}^1\Sigma_g^+$
O_2^+	$(1\sigma_g)^2(1\sigma_u^*)^2(2\sigma_g)^2(2\sigma_u^*)^2(3\sigma_g)^2(1\pi_u)^2(1\pi_u)^2(1\pi_g^*)^1$	${}^2\Pi_g$
O_2	$(1\sigma_g)^2(1\sigma_u^*)^2(2\sigma_g)^2(2\sigma_u^*)^2(3\sigma_g)^2(1\pi_u)^2(1\pi_u)^2(1\pi_g^*)^1(1\pi_g^*)^1$	${}^3\Sigma_g^-$
F_2	$(1\sigma_g)^2(1\sigma_u^*)^2(2\sigma_g)^2(2\sigma_u^*)^2(3\sigma_g)^2(1\pi_u)^2(1\pi_u)^2(1\pi_g^*)^2(1\pi_g^*)^2$	${}^1\Sigma_g^+$

25.3 Transitions between Electronic States of Diatomic Molecules

Diatomic molecules have the most easily interpretable electronic spectra because the spacing between the various rotational-vibrational-electronic states is sufficiently large to allow individual states to be resolved. Potential energy curves for the five lowest lying bound states of O_2 are shown in Figure 25.1. Vibrational energy levels are indicated schematically in the figure, but rotational levels are not shown. Note that the lowest four states all dissociate to give two ground-state ^3P oxygen atoms, whereas the highest energy state shown dissociates to give one ^3P and one ^1D oxygen atom. The letter X before $^3\Sigma_g^-$ indicates that the term symbol refers to the ground state. Electronic states of higher energy are designated by A, B, C, ... if they have the same multiplicity, $2S + 1$, as the ground state, and a, b, c, ... if they have a different multiplicity.

The bond length of excited-state molecules is generally greater and the binding energy generally less than that for the ground state. This is the case because the excited states generally have a greater antibonding character than the ground states. The decrease in bond order leads to a smaller bond energy, a larger bond length, and a lower vibrational frequency for the excited-state species. You will address the fact that the bond lengths for the first two excited states are similar to that for the ground state in the end-of-chapter questions.

Although a symbol such as $^3\Sigma_g^-$ completely describes the quantum state for a ground-state O_2 molecule, it is also useful to associate a molecular configuration with the state. Starting with a configuration makes it easier to visualize a transition in terms of promoting an electron from an occupied to an unoccupied level. To what configurations do the excited states shown in Figure 25.1 correspond? The $X^3\Sigma_g^-$, $a^1\Delta_g$, and $b^1\Sigma_g^+$ states all belong to the ground-state configuration $(1\sigma_g)^2(1\sigma_u^*)^2(2\sigma_g)^2(2\sigma_u^*)^2$ $(3\sigma_g)^2(1\pi_u)^2(1\pi_u)^2(1\pi_g^*)^1(1\pi_g^*)^1$ but are associated with different M_L and M_S values as was shown in Example Problem 25.2. The $A^3\Sigma_u^+$ and $B^3\Sigma_u^-$ states are associated with the $(1\sigma_g)^2(1\sigma_u^*)^2(2\sigma_g)^2(2\sigma_u^*)^2(3\sigma_g)^2(1\pi_u)^1(1\pi_u)^2(1\pi_g^*)^1(1\pi_g^*)^2$ configuration. Keep in mind that although a molecular term can be associated with a configuration, in general, several molecular terms are generated from the same configuration.

Spectroscopy involves transitions between molecular states. What selection rules govern transitions between different electronic states? The selection rules for molecular electronic transitions are most well defined for lower molecular weight diatomic molecules in which spin-orbit coupling is not important. This is the case if the atomic number of the atoms, Z, is less than 40. For these molecules, the selection rules are

$$\Delta\Lambda = 0, \pm 1, \text{ and } \Delta S = 0 \quad (25.3)$$

Recall that Λ is the component of the total orbital angular momentum $\mathbf{L}$ along the molecular axis. The value $\Delta\Lambda = 0$ applies to a $\Sigma \leftrightarrow \Sigma$ transition, and $\Delta\Lambda = \pm 1$ applies to $\Sigma \leftrightarrow \Pi$ transitions. Further selection rules are associated with the $+/-$ and g/u parities. For homonuclear diatomics, $u \leftrightarrow g$ transitions are allowed, but $u \leftrightarrow u$ and $g \leftrightarrow g$ transitions are forbidden. The transitions $\Sigma^- \leftrightarrow \Sigma^-$ and $\Sigma^+ \leftrightarrow \Sigma^+$ are allowed, but $\Sigma^+ \leftrightarrow \Sigma^-$ transitions are forbidden. All of these selection rules can be derived by calculating the transition dipole element defined in Section 19.5.

With these selection rules in mind, we consider the possible transitions among the states shown in Figure 25.1 for O_2. The $X^3\Sigma_g^- \rightarrow a^1\Delta_g$ and $X^3\Sigma_g^- \rightarrow b^1\Sigma_g^+$ transitions are forbidden because of the $\Delta S = 0$ selection rule and because $g \leftrightarrow g$ transitions are forbidden. The $X^3\Sigma_g^- \rightarrow A^3\Sigma_u^+$ transition is forbidden because $\Sigma^+ \leftrightarrow \Sigma^-$ transitions are forbidden. Therefore, the lowest allowed transition originating from the ground state is $X^3\Sigma_g^- \rightarrow B^3\Sigma_u^-$. Absorption from the ground state into various vibrational levels of the $B^3\Sigma_u^-$ excited state occurs in a band between 175 and 200 nm wavelengths. An interesting consequence of these selection rules is that if transitions from the ground state to the first two excited states were allowed, O_2 would absorb light in the visible part of the spectrum, and Earth's atmosphere would not be transparent.

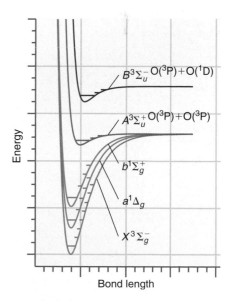

FIGURE 25.1

Potential energy curves for the ground state of O_2 and for the four lowest excited states. The spectroscopic designation of the states is explained in the text. Horizontal lines indicate vibrational levels for each state.

If sufficient energy is taken up by the molecule, dissociation can occur through the pathway

$$O_2 + h\nu \rightarrow 2\,O\cdot \tag{25.4}$$

The maximum wavelength consistent with this reaction is 242 nm. This reaction is an example of a **photodissociation** reaction. This particular reaction is of great importance in the stratosphere because it is the only significant pathway for forming the atomic oxygen needed for ozone production through the reaction

$$O\cdot + O_2 + M \rightarrow O_3 + M^* \tag{25.5}$$

where M designates a gas-phase spectator species that takes up energy released in the O_3 formation reaction. Because O_3 absorbs UV radiation strongly over the 220 to 350 nm range, it plays a vital role in filtering out UV radiation from the sunlight incident on the planet. The ozone layer located in the stratosphere 10–50 km above Earth's surface absorbs 97–99% of the sun's high frequency ultraviolet light, light that is potentially damaging to life on Earth.

25.4 The Vibrational Fine Structure of Electronic Transitions in Diatomic Molecules

Each of the molecular bound states shown in Figure 25.1 has well-defined vibrational and rotational energy levels. As discussed in Chapter 19, changes in the vibrational state can occur together with a change in the rotational state. Similarly, the vibrational and rotational quantum numbers can change during electronic excitation. We next discuss the vibrational excitation and de-excitation associated with electronic transitions but do not discuss the associated rotational transitions. We will see that the $\Delta n = \pm 1$ selection rule for vibrational transitions within a given electronic state does not hold for transitions between two electronic states.

What determines Δn in a vibrational transition between electronic states? This question can be answered by looking more closely at the **Born–Oppenheimer approximation,** which was introduced in Chapter 23. This approximation stated mathematically says that the total wave function for the molecule can be factored into two parts. The part that depends only on the position of the nuclei $(\mathbf{R}_1, \ldots, \mathbf{R}_m)$ is associated with vibration of the molecule. The second part depends only on the position of the electrons $(\mathbf{r}_1, \ldots, \mathbf{r}_n)$ at a fixed position of all the nuclei. This part describes electron "motion" in the molecule:

$$\psi(\mathbf{r}_1, \ldots, \mathbf{r}_n, \mathbf{R}_1, \ldots, \mathbf{R}_m) = \psi^{electronic}(\mathbf{r}_1, \ldots, \mathbf{r}_n, \mathbf{R}_1^{fixed}, \ldots, \mathbf{R}_m^{fixed})$$
$$\times \phi^{vibrational}(\mathbf{R}_1, \ldots, \mathbf{R}_m) \tag{25.6}$$

As discussed in Section 19.5, the spectral line corresponding to an electronic transition (initial $\rightarrow$ final) has a measurable intensity only if the value of the transition dipole moment is different from zero:

$$\mu^{fi} = \int \psi_f^*(\mathbf{r}_1, \ldots, \mathbf{r}_n, \mathbf{R}_1, \ldots, \mathbf{R}_m)\,\hat{\mu}\,\psi_i(\mathbf{r}_1, \ldots, \mathbf{r}_n, \mathbf{R}_1, \ldots, \mathbf{R}_m)\,d\tau \neq 0 \tag{25.7}$$

The superscripts and subscripts f and i refer to the final and initial states in the transition. In Equation (25.7), the dipole moment operator $\hat{\mu}$ is given by

$$\hat{\mu} = -e\sum_{j=1}^{n}\mathbf{r}_j \tag{25.8}$$

where the summation is over the positions of the electrons.

Because the total wave function can be written as a product of electronic and vibrational parts, Equation (25.7) becomes

$$\mu^{fi} = \int (\phi_f^{vibrational}(\mathbf{R}_1, \ldots, \mathbf{R}_m))^* \phi_i^{vibrational}(\mathbf{R}_1, \ldots, \mathbf{R}_m)\, d\tau$$

$$\times \int (\psi_f^{electronic}(\mathbf{r}_1, \ldots, \mathbf{r}_n, \mathbf{R}_1^{fixed}, \ldots, \mathbf{R}_m^{fixed}))^* \hat{\mu}\, \psi_i^{electronic}(\mathbf{r}_1, \ldots, \mathbf{r}_n, \mathbf{R}_1^{fixed}, \ldots, \mathbf{R}_m^{fixed})\, d\tau$$

$$= S \int \psi_f^*(\mathbf{r}_1, \ldots, \mathbf{r}_n, \mathbf{R}_1^{fixed}, \ldots, \mathbf{R}_m^{fixed}) \hat{\mu}\, \psi_i(\mathbf{r}_1, \ldots, \mathbf{r}_n, \mathbf{R}_1^{fixed}, \ldots, \mathbf{R}_m^{fixed})\, d\tau \qquad \textbf{(25.9)}$$

Note that the first of the two product integrals in Equation (25.9) represents the overlap S between the vibrational wave functions in the ground and excited states. The magnitude of the square of this integral for a given transition is known as the **Franck-Condon factor** and is a measure of the expected intensity of an electronic transition. The Franck-Condon factor replaces the selection rule $\Delta n = \pm 1$ obtained for pure vibrational transitions derived in Section 19.4 as a criterion for the intensity of a transition:

$$S^2 = \left| \int (\phi_f^{vibrational})^* \phi_i^{vibrational}\, d\tau \right|^2 \qquad \textbf{(25.10)}$$

The **Franck-Condon principle** states that transitions between electronic states correspond to vertical lines on an energy versus internuclear distance diagram. The basis of this principle is that electronic transitions occur on a timescale that is very short compared to the vibrational period of a molecule. Therefore, the atoms do not move during the transition. As Equation (25.10) shows, the intensity of a vibrational-electronic transition is governed by the overlap between the final and initial vibrational wave functions at fixed values of the internuclear distances. Is it necessary to consider all vibrational levels in the ground state as an initial state for an electronic transition? As discussed in Chapter 19, nearly all of the molecules in the ground state have the vibrational quantum number $n = 0$, for which the maximum amplitude of the wave function is at the equilibrium bond length. As shown in Figure 25.2, vertical transitions predominantly occur from this ground vibrational state to several vibrational states in the upper electronic state.

How does the Franck-Condon principle determine the n values in the excited state that give the most intense spectral lines? The most intense electronic transitions are to vibrational levels in the upper electronic state that have the largest overlap with the ground vibrational level in the lower electronic state. As Figure 18.10 shows, the vibrational wave functions have their largest amplitude near the R value at which the energy level meets the potential curve, because this corresponds to the classical turning point. For the example shown in Figure 25.2, the overlap $\left| \int (\phi_f^{vibrational})^* \phi_i^{vibrational}\, d\tau \right|$ is greatest between the $n = 0$ vibrational state of the ground electronic state and the $n = 4$ vibrational state of the excited electronic state. Although this transition has the maximum overlap and generates the most intense spectral line, other states close in energy to the most probable state will also give rise to spectral lines. Their intensity is lower because S is smaller.

The fact that a number of vibrational transitions are observed in an electronic transition is very useful in obtaining detailed information about both the ground electronic state potential energy surface and that of the electronic state to which the transition occurs. For example, vibrational transitions are observed in the electronic spectra of O_2 and N_2, although neither of these molecules absorbs energy in the infrared. Because multiple vibrational peaks are often observed in electronic spectra, the bond strength of the molecule in the excited states can be determined by fitting the observed frequencies of the transitions to a model potential such as the Morse potential discussed in Section 19.3. Because the excited state can also correspond to a photodissociation product, electronic spectroscopy can be used to determine the vibrational force constant and bond energy of highly reactive species such as the CN radical that cannot be studied with conventional IR absorption techniques.

For the example shown in Figure 25.2, the molecule will exhibit a discrete energy spectrum in the visible or UV region of the spectrum. However, for some conditions the electronic absorption spectrum for a diatomic molecule is continuous. A continuous spectrum is observed if the photon energy is sufficiently high that excitation occurs to an unbound region of an excited state. This is illustrated in Figure 25.3. In this case, a discrete

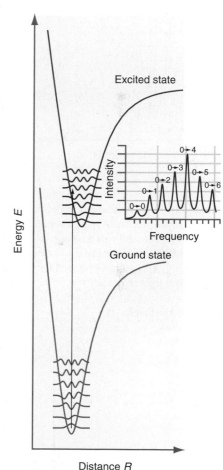

FIGURE 25.2

The relation between energy and bond length is shown for two electronic states. Only the lowest vibrational energy levels and the corresponding wave functions are shown. The vertical line shows the most probable transition predicted by the Franck-Condon principle. The inset shows the relative intensities of different vibrational lines in an absorption spectrum for the potential energy curves shown.

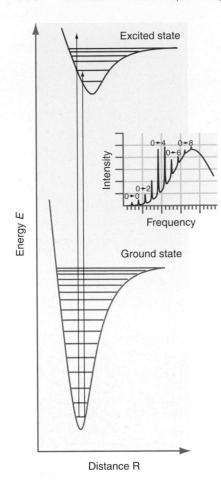

FIGURE 25.3

For absorption from the ground vibrational state of the ground electronic state to the excited electronic state, a continuous energy spectrum will be observed for sufficiently high photon energy. A discrete energy spectrum is observed for an incident light frequency $\nu < E/h$. A continuous spectrum is observed for higher frequencies.

energy spectrum is observed for low photon energy and a **continuous energy spectrum** is observed for incident light frequencies $\nu > E/h$, where E corresponds to the energy of the transition to the highest bound state in the excited state potential. A purely continuous energy spectrum at all energies is observed if the excited state is a nonbinding state, such as that corresponding to the first excited state for H_2^+.

The preceding discussion briefly summarizes the most important aspects of the electronic spectroscopy of diatomic molecules. In general, the vibrational energy levels for these molecules are sufficiently far apart that individual transitions can be resolved. We next consider polyatomic molecules, for which this is not usually the case.

25.5 UV-Visible Light Absorption in Polyatomic Molecules

Many rotational and vibrational transitions are possible if an electronic transition occurs in polyatomic molecules. Large molecules have large moments of inertia, and as Equation (19.15) shows, this leads to closely spaced rotational energy levels. A large molecule may have ~1000 rotational levels in an interval of 1 cm^{-1}. For this reason, individual spectral lines overlap so that broad bands are often observed in UV-visible absorption spectroscopy. This is schematically indicated in Figure 25.4. An electronic transition in an atom gives a sharp line. An electronic transition in a diatomic molecule has additional structure resulting from vibrational and rotational transitions that can often be resolved into individual peaks. However, the many rotational and vibrational transitions possible in a polyatomic molecule generally overlap, giving rise to a broad, nearly featureless band. This overlap makes it difficult to extract information on the initial and final states involved in an electronic transition in polyatomic molecules. In addition, there are no good angular momentum quantum numbers for triatomic and larger molecules. Therefore, the main selection rule that applies is $\Delta S = 0$, together with selection rules based on the symmetry of the initial and final states.

The number of transitions observed can be reduced dramatically by obtaining spectra at low temperatures. Low-temperature spectra for individual molecules can be obtained either by embedding the molecule of interest in a solid rare gas matrix at cryogenic temperatures or by expanding gaseous He containing the molecules of interest in dilute concentration through a nozzle into a vacuum. The He gas as well as the molecules of interest are cooled to very low temperatures in the expansion. An example of the elimination of spectral congestion through such a gas expansion is shown in Figure 25.5. The temperature of 9 K is reached by simply expanding the 300 K gas mixture into a vacuum using a molecular beam apparatus.

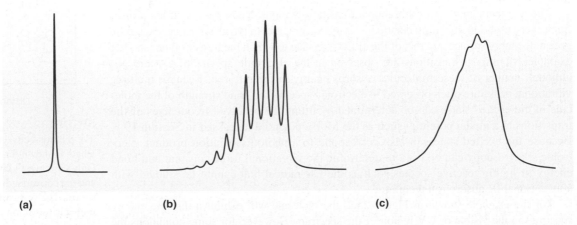

(a)   (b)   (c)

FIGURE 25.4

The intensity of absorption in a small part of the UV-visible range of the electromagnetic spectrum is shown schematically for **(a)** an atom, **(b)** a diatomic molecule, and **(c)** a polyatomic molecule.

The concept of chromophores is particularly useful for discussing the electronic spectroscopy of polyatomic molecules. As discussed in Chapter 19, characteristic vibrational frequencies are associated with neighboring atoms in larger molecules. Similarly, the absorption of UV and visible light in larger molecules can be understood by visualizing the molecule as a system of coupled entities, such as $-C=C-$ or $-O-H$, that are called **chromophores.** A chromophore is a chemical entity embedded within a molecule that absorbs radiation at nearly the same wavelength in different molecules. Common chromophores in electronic spectroscopy are $C=C$, $C=O$, $C\equiv N$, or $C=S$ groups. Each chromophore has one or several characteristic absorption frequencies in the UV, and the UV absorption spectrum of the molecule, to a first approximation, can be thought of as arising from the sum of the absorption spectra of its chromophores. The wavelengths and absorption strengths associated with specific chromophores are discussed in Section 25.6.

As discussed in Chapter 24, the electronic structure of molecules can be viewed in either a localized or delocalized framework. In viewing the transitions involved in electronic spectroscopy, it is often useful to work from a localized bonding model. However, electrons in radicals and those in delocalized π bonds in conjugated and aromatic molecules need to be described in a delocalized rather than a localized binding model.

What transitions are most likely to be observed in electronic spectroscopy? Electronic excitation involves the promotion of an electron from an occupied MO to a higher energy unfilled or partially filled MO. Consider the electronic ground-state configuration of formaldehyde, H_2CO, and those of its lowest lying electronically excited states. In a localized bonding model, the carbon $2s$ and $2p$ electrons combine to form sp^2-hybrid orbitals on the carbon atom as shown in Figure 25.6.

We write the ground-state configuration in the localized orbital notation $(1s_O)^2(1s_C)^2(2s_O)^2(\sigma_{CH})^2(\sigma'_{CH})^2(\sigma_{CO})^2(\pi_{CO})^2(n_O)^2(\pi^*_{CO})^0$ to emphasize that the $1s$ and $2s$ electrons on oxygen and the $1s$ electrons on carbon remain localized on the atoms and are not involved in the bonding. There is also an electron lone pair in a nonbonding MO, designated by n_O, localized on the oxygen atom. Bonding orbitals are primarily localized on adjacent $C-H$ or $C-O$ atoms as indicated in the configuration. The $C-H$ bonds and one of the $C-O$ bonds are σ bonds, and the remaining $C-O$ bond is a π bond.

What changes in the occupation of the MO energy levels can be associated with the electronic transitions observed for formaldehyde? To answer this question, it is useful to generalize the results obtained for MO formation in diatomic molecules to the CO chromophore in formaldehyde. In a simplified picture of this molecule, we expect that the σ_{CO} orbital formed primarily from the $2p_z$ orbital on O and one of the sp^2-hybrid orbitals on C has the lowest energy and that the antibonding combination of the same orbitals has the highest energy. The orbital formed from the $2p$ levels on each atom has the next lowest energy, and the antibonding π^* combination has the next highest energy. The lone pair electrons that occupy the $2p$ orbital on O have an energy intermediate between the π and π^* levels. The very approximate molecular orbital energy diagram shown in Figure 25.7 is sufficient to discuss the transitions that formaldehyde undergoes in the UV-visible region.

From the MO energy diagram, we conclude that the nonbonding orbital on O derived from the $2p$ AO is the HOMO, and the empty π^* orbital is the LUMO. The lowest excited state is reached by promoting an electron from the n_O to the π^*_{CO} orbital and is called an $n \rightarrow \pi^*$ **transition.** The resulting state is associated with the configuration $(1s_O)^2(1s_C)^2(2s_O)^2(\sigma_{CH})^2(\sigma'_{CH})^2(\sigma_{CO})^2(\pi_{CO})^2(n_O)^1(\pi^*_{CO})^1$. The next excited state is reached by promoting an electron from the π_{CO} to the π^*_{CO} MO and is called a $\pi \rightarrow \pi^*$ **transition.** The resulting state is associated with the configuration $(1s_O)^2(1s_C)^2(2s_O)^2(\sigma_{CH})^2(\sigma'_{CH})^2(\sigma_{CO})^2(\pi_{CO})^1(n_O)^2(\pi^*_{CO})^1$.

However, as was the case for atoms, these configurations do not completely describe the quantum states because the alignment of the spins in the unfilled orbitals is not specified by the configuration. Because each of the excited-state configurations just listed has two half-filled MOs, both singlet and triplet states arise from each configuration. The relative energy of these states is indicated in Figure 25.8. Just as for diatomic molecules, for the same configuration, triplet states lie lower in energy than singlet states. The difference in energy between the singlet and triplet states is specific to a molecule but typically lies between 2 and 10 eV.

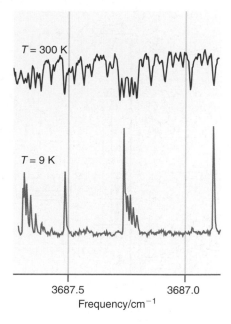

FIGURE 25.5

A small portion of the electronic absorption spectrum of methanol is shown at 300 and 9 K using expansion of a dilute mixture of methanol in He through a nozzle into a vacuum. At 300 K, the molecule absorbs almost everywhere in the frequency range. At 9 K, very few rotational and vibrational states are populated, and individual spectral features corresponding to rotational fine structure are observed.
Source: Reprinted from P. Carrick, et al., "The OH Stretching Fundamental of Methanol," *Journal of Molecular Structure* 223: 171–184, (June 1990), Copyright 1990, with permission from Elsevier.

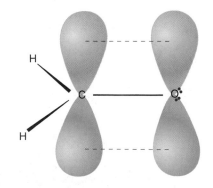

FIGURE 25.6

Valence bond picture of the formaldehyde molecule. The solid lines indicate σ bonds and the dashed lines indicate a π bond. The nonequivalent lone pairs on oxygen are also shown.

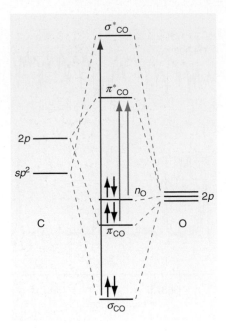

FIGURE 25.7

A simplified MO energy diagram is shown for the C—O bonding interaction in formaldehyde. The most important allowed transitions between these levels are shown. Only one of the sp^2 orbitals on carbon is shown, because the other two hybrid orbitals form σ_{CH} bonds.

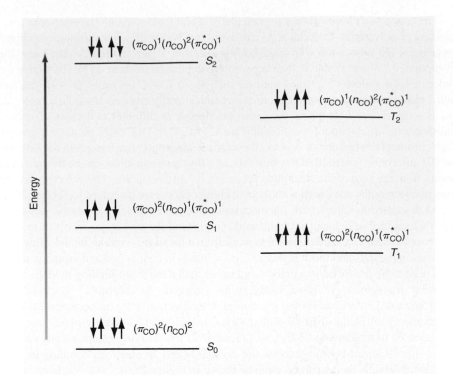

FIGURE 25.8

The ground state of formaldehyde is a singlet and is designated S_0. Successively higher energy singlet and triplet states are designated S_1, S_2, T_1, and T_2. The electron configurations and the alignment of the unpaired spins for the states involved in the most important transitions are also shown. The energy separation between the singlet and triplet states is not to scale in this figure.

The energy difference between the initial and final states determines the frequency of the spectral line. Although large variations can occur among different molecules for a given type of transition, generally the energy increases in the sequence $n \rightarrow \pi^*$, $\pi \rightarrow \pi^*$, and $\sigma \rightarrow \sigma^*$. The $\pi \rightarrow \pi^*$ transitions require multiple bonds, and occur in alkenes, alkynes, and aromatic compounds. The $n \rightarrow \pi^*$ transitions require both a nonbonding electron pair and multiple bonds and occur in molecules containing carbonyls, thiocarbonyls, nitro, azo, and imine groups and in unsaturated halocarbons. The $\boldsymbol{\sigma \rightarrow \sigma^*}$ **transitions** are seen in many molecules, particularly in alkanes, in which none of the other transitions is possible.

25.6 Transitions among the Ground and Excited States

We next generalize the preceding discussion for formaldehyde to an arbitrary molecule. What transitions can take place among ground and excited states? Consider the energy levels for such a molecule shown schematically in Figure 25.9. The ground state is, in general, a singlet state, and the excited states can be either a singlet or triplet state. We include only one excited singlet and triplet state in addition to the ground state and consider the possible transitions among these states. The restriction is justified because an initial excitation to higher-lying states will rapidly decay to the lowest-lying state of the same multiplicity through a process called internal conversion, which is discussed later. The diagram also includes vibrational levels associated with each of the electronic levels. Rotational levels are omitted to simplify the diagram. The fundamental rule governing transitions is that all transitions must conserve energy and angular momentum. For transitions within a molecule, this condition can be satisfied by transferring energy between electronic, vibrational, and rotational states. Alternatively, energy can be conserved by transferring energy between a molecule and its surroundings.

Four types of transitions are indicated in Figure 25.9. **Radiative transitions,** in which a photon is absorbed or emitted, are indicated by solid lines. **Nonradiative transitions,** in

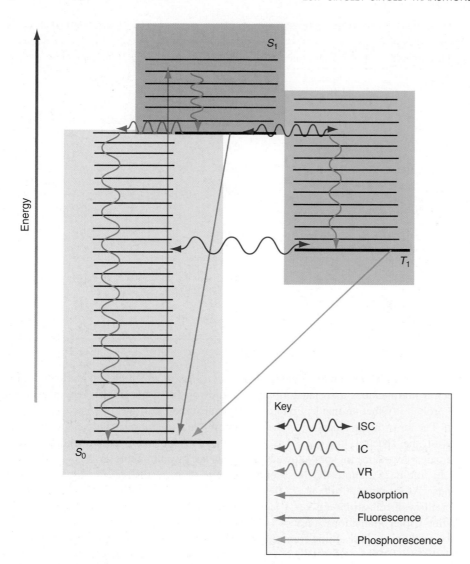

FIGURE 25.9
A Jablonski diagram depicting various photophysical processes, where S_0 is the ground electronic singlet state, S_1 is the first excited singlet state, and T_1 is the first excited triplet state. Radiative processes are indicated by the straight lines. The nonradiative processes of intersystem crossing (ISC), internal conversion (IC), and vibrational relaxation (VR) are indicated by the wavy lines.

which energy is transferred between different degrees of freedom of a molecule or to the surroundings, are indicated by wavy vertical lines. **Internal conversion** occurs without a change in energy between states of the same multiplicity and is shown as a horizontal wavy line. An **intersystem crossing** differs from an internal conversion in that a change in multiplicity($\Delta S \neq 0$) occurs. The pathway by which a molecule in an excited state decays to the ground state depends on the rates of a number of competing processes. In the next two sections, these processes are discussed individually.

25.7 Singlet–Singlet Transitions: Absorption and Fluorescence

As discussed in Section 25.5, an absorption band in an electronic spectrum can be associated with a specific chromophore. Whereas in atomic spectroscopy the selection rule $\Delta S = 0$ is strictly obeyed, in molecular spectroscopy one finds instead that spectral lines for transitions corresponding to $\Delta S = 0$ are much stronger than those for which this condition is not fulfilled. It is useful to quantify what is meant by strong and weak absorption. If I_0 is the incident light intensity at the frequency of interest and I_t is the intensity of transmitted light, the dependence of I_t/I_0 on the concentration c and the path length l is described by **Beer's law:**

$$\log\left(\frac{I_t}{I_0}\right) = -\varepsilon l c \qquad (25.11)$$

The **molar extinction coefficient** ε is a measure of the strength of the transition. It is independent of the path length and concentration and is characteristic of the chromophore.

TABLE 25.2 Characteristic Parameters for Common Chromophores

Chromophore	Transition	λ_{max} (nm)	ε_{max} (dm^3 mol^{-1} cm^{-1})
N=O	$n \rightarrow \pi^*$	660	200
N=N	$n \rightarrow \pi^*$	350	100
C=O	$n \rightarrow \pi^*$	280	20
NO$_2$	$n \rightarrow \pi^*$	270	20
C$_6$H$_6$ (benzene)	$\pi \rightarrow \pi^*$	260	200
C=N	$n \rightarrow \pi^*$	240	150
C=C—C=O	$\pi \rightarrow \pi^*$	220	2×10^5
C=C—C=C	$\pi \rightarrow \pi^*$	220	2×10^5
S=O	$n \rightarrow \pi^*$	210	1.5×10^3
C=C	$\pi \rightarrow \pi^*$	180	1×10^3
C—C	$\sigma \rightarrow \sigma^*$	<170	1×10^3
C—H	$\sigma \rightarrow \sigma^*$	<170	1×10^3

The **integral absorption coefficient,** $A = \int \varepsilon(\nu)\, d\nu$, in which the integration over the spectral line includes associated vibrational and rotational transitions, is a measure of the probability that an incident photon will be absorbed in a specific electronic transition. The terms A and ε depend on the frequency, and ε measured at the maximum intensity of the spectral line ε_{max} has been tabulated for many chromophores. Some characteristic values for spin-allowed transitions are given in Table 25.2.

In Table 25.2, note the large enhancement of ε_{max} that occurs for conjugated bonds. As a general rule, ε_{max} lies between 10 and 5×10^4 dm^3 mol^{-1} cm^{-1} for spin-allowed transitions ($\Delta S = 0$), and between 1×10^{-4} and 1 dm^3 mol^{-1}cm^{-1} for singlet–triplet transitions ($\Delta S = 1$). Therefore, the attenuation of light passing through the sample resulting from singlet–triplet transitions will be smaller by a factor of ~10^4 to 10^7 than the attenuation from singlet–singlet transitions. This illustrates that in an absorption experiment, transitions for which $\Delta S = 1$ are not totally forbidden if spin-orbit coupling is not negligible but are typically too weak to be of much importance. However, as discussed in Section 25.8, singlet–triplet transitions are important for phosphorescence.

The excited-state molecule in S_1 can return to the ground state S_0 through radiative or nonradiative transitions involving collisions with other molecules. What determines which of these two pathways will be followed? An isolated excited-state molecule (for instance, in interstellar space) cannot exchange energy with other molecules through collisions and, therefore, nonradiative transitions (other than isoenergetic internal electronic-to-vibrational energy transfer) will not occur. However, excited-state molecules in a crystal, in solution, or in a gas undergo frequent collisions with other molecules in which they lose energy and return to the lowest vibrational state of S_1 through vibrational relaxation. This process generally occurs much faster than a radiative transition directly from a vibrationally excited state in S_1 to a vibrational state S_0. Once in the lowest vibrational state of S_1, either of three events can occur. The molecule can undergo a radiative transition to a vibrational state in S_0 in a process called **fluorescence,** or it can make a nonradiative transition to an excited vibrational state of T_1 through intersystem crossing. Intersystem crossing violates the $\Delta S = 0$ selection rule and, therefore, occurs at a very low rate in comparison with the other processes depicted in Figure 25.9.

Because vibrational relaxation is generally fast in comparison with fluorescence the vibrationally excited-state molecule will relax to the ground vibrational state of S_1 before undergoing fluorescence to S_0. As a result of the relaxation, the fluorescence spectrum is shifted to lower energies relative to the absorption spectrum, as shown in Figure 25.10. When comparing absorption and fluorescence spectra, it is

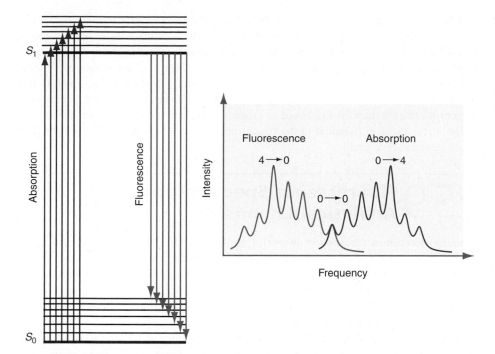

FIGURE 25.10
Illustration of the absorption and fluorescence bands expected if internal conversion is fast relative to fluorescence. The relative intensities of individual transitions within the absorption and fluorescence bands are determined by the Franck-Condon principle.

often seen that for potentials that are symmetric about the minimum (for example, a harmonic potential), the band of lines corresponding to absorption and fluorescence are mirror images of one another. This relationship is shown in Figure 25.10.

25.8 Intersystem Crossing and Phosphorescence

Although intersystem crossing between singlet and triplet electronic states is forbidden by the $\Delta S = 0$ selection rule, the probability of this happening is high for many molecules. The probability of intersystem crossing transitions is enhanced by two factors: a very similar molecular geometry in the excited singlet and triplet states, and a strong spin-orbit coupling, which allows the spin flip associated with a singlet–triplet transition to occur. The processes involved in phosphorescence are illustrated in a simplified fashion in Figure 25.11 for a diatomic molecule.

Imagine that a molecule is excited from S_0 to S_1. This is a dipole-allowed transition, so it has a high probability of occurring. Through collisions with other molecules, the excited-state molecule loses vibrational energy and decays to the lowest vibrational state of S_1. As shown in Figure 25.11, the potential energy curves can overlap such that an excited vibrational state in S_1 can have the same energy as an excited vibrational state in T_1. In this case, the molecule has the same geometry and energy in both singlet and triplet states. In Figure 25.11, this occurs for $n = 4$ in the state S_1. If the spin-orbit coupling is strong enough to initiate a spin flip, the molecule can cross over to the triplet state without a change in geometry or energy. Through vibrational relaxation it will rapidly relax to the lowest vibrational state of T_1. At this point, it can no longer make a transition back to S_1 because the ground vibrational state of T_1 is lower than any state in S_1.

However, from the ground vibrational state of T_1, the molecule can decay radiatively to the ground state in the dipole transition forbidden process called **phosphorescence.** This is not a high-probability event because nonradiative processes involving collisions between molecules or with the walls of the reaction vessel can compete effectively with phosphorescence. Therefore, the probability for a $T_1 \rightarrow S_0$ phosphorescence transition is generally much lower than for fluorescence. It usually lies in the range of 10^{-2} to 10^{-5}. The relative probabilities of fluorescence and phosphorescence if collisional relaxation can occur is determined by the lifetime of the excited state. Fluorescence is an allowed transition, and the excited-state lifetime is short, typically less than 10^{-7} s. By contrast, phosphorescence is a forbidden transition and the excited-state lifetime is typically longer than 10^{-3} s. On this time scale collisional relaxation has a high probability.

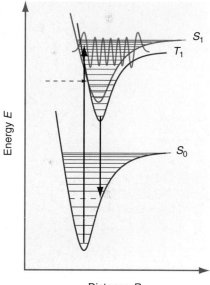

FIGURE 25.11
Process giving rise to phosphorescence illustrated for a diatomic molecule. Absorption from S_0 leads to a population of excited vibrational states in S_1. The molecule has a finite probability of making a transition to an excited vibrational state of T_1 if it has the same geometry in both states and if there are vibrational levels of the same energy in both states. The dashed arrow indicates the coincidence of vibrational energy levels in T_1 and S_1. For reasons of clarity, only the lowest vibrational levels in T_1 are shown. The initial excitation to S_1 occurs to a vibration state of maximum overlap with the ground state of S_0 as indicated by the blue vibrational wave function.

Fluorescence can be induced using broadband radiation or highly monochromatic laser light. Fluorescence spectroscopy is well suited for detecting very small concentrations of a chemical species if the wavelength of the emission lies in the visible-UV part of the electromagnetic spectrum where there is little background noise near room temperature. As shown in Figure 25.10, relaxation to lower vibrational levels within the excited electronic state has the consequence that the fluorescence signal occurs at a longer wavelength than the light used to create the excited state. Therefore, the contribution of the incident radiation to the background at the wavelength used to detect the fluorescence is very small.

25.9 Fluorescence Spectroscopy and Analytical Chemistry

We now describe a particularly powerful application of fluorescence spectroscopy, namely, the sequencing of the human genome. The goal of the human genome project was to determine the sequence of the four bases, A, C, T, and G, in DNA that encode all the genetic information necessary for propagating the human species. A sequencing technique based on laser-induced fluorescence spectroscopy that has been successfully used in this effort can be divided into three parts.

In the first part, a section of DNA is cut into small lengths of 1000 to 2000 base pairs using mechanical shearing. Each of these pieces is replicated to create many copies, and these replicated pieces are put into a solution with a mixture of the four bases, A, C, T, and G. A reaction is set in motion that leads to the strands growing in length through replication. A small fraction of each of the A, C, T, and G bases in solution that are incorporated into the pieces of DNA has been modified in two ways. The modified base terminates the replication process. It also contains a dye chosen to fluoresce strongly at a known wavelength. The initial segments continue to grow if they incorporate unmodified bases, and no longer grow if they incorporate one of the modified bases. As a result of these competing processes involving the incorporation of modified or unmodified bases, a large number of partial replicas of the whole DNA are created, each of which is terminated in the base that has a fluorescent tag built into it. The ensemble of these partial replicas contains all possible lengths of the original DNA segment that terminate in the particular base chosen. If the lengths of these segments can be measured, then the positions of the particular base in the DNA segment can be determined.

The lengths of the partial replicas are measured using capillary electrophoresis coupled with detection using laser-induced fluorescence spectroscopy. In this method, a solution containing the partial replicas is passed through a glass capillary filled with a gel. An electrical field along the capillary causes the negatively charged DNA partial replicas to travel down the column with a speed that depends inversely on their length. Because of the different migration speeds, a separation in length occurs as the partial replicas pass through the capillary. At the end of the capillary, the partial replicas emerge from the capillary into a buffer solution that flows past the capillary, forming a sheath. The flow pattern of the buffer solution is carefully controlled to achieve a focusing of the emerging stream containing the partial replicas to a diameter somewhat smaller than the inner diameter of the capillary. A schematic diagram of such a sheath flow cuvette electrophoresis apparatus is shown in Figure 25.12. An array of capillaries is used rather than a single capillary in order to obtain the multiplexing advantage of carrying out several experiments in parallel.

The final part of the sequencing procedure is to measure the time that each of the partial replicas spent in transit through the capillary, which determines its length, and to identify the terminating base. The latter task is accomplished by means of laser-induced fluorescence spectroscopy. A narrow beam of visible laser light is passed through all the capillaries in series. Because of the very dilute solutions involved, the attenuation of the laser beam by each successive capillary is very small. The fluorescent light emitted from each of the capillaries is directed to light-sensing photodiodes by means of a microscope objective and individual focusing lenses. A rotating filter wheel between the microscope objective and the focusing lenses allows a discrimination to be made among the four

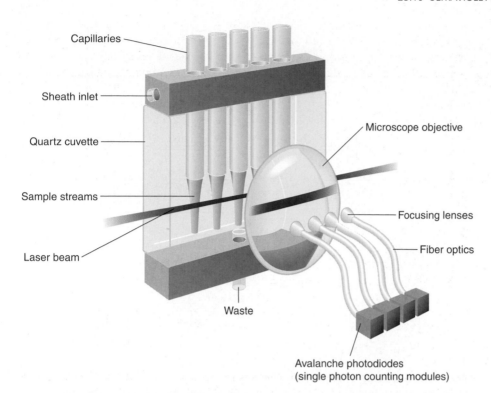

Capillaries

Sheath inlet

Quartz cuvette

Sample streams

Laser beam

Microscope objective

Focusing lenses

Fiber optics

Waste

Avalanche photodiodes
(single photon counting modules)

FIGURE 25.12
Schematic diagram of the application of fluorescence spectroscopy in the sequencing of the human genome. From: Dovichi, Norm. *Development of DNA Sequencer. Science* 285 (1999): 1016.

different fluorescent dyes with which the bases were tagged. The sensitivity of the system shown in Figure 25.12 is 130 ± 30 molecules in the volume illuminated by the laser that corresponds to a concentration of 2×10^{-13} mol/L! This extremely high sensitivity is a result of coupling the sensitive fluorescence technique to a sample cell designed with a very small sampling volume. Matching the laser beam diameter to the sample size and reducing the size of the cuvette result in a significant reduction in background noise. Commercial versions of this approach utilizing 96 parallel capillaries played a major part in the first phase of the sequencing of the human genome.

25.10 Ultraviolet Photoelectron Spectroscopy

Spectroscopy in general, and electronic spectroscopy in particular, gives information on the energy difference between the initial and final states rather than the energy levels involved in the transition. However, the energy of both occupied and unoccupied molecular orbitals is of particular interest to chemists. Information at this level of detail cannot be obtained directly from a UV absorption spectrum, because only a difference between energy levels is measured. However, information about the orbitals involved in the transition can be extracted from an experimentally obtained spectrum using a model. For example, the molecular orbital model described in Chapter 23 can be used to calculate the orbital energy levels for a molecule. With these results, an association can be made between energy-level differences calculated from observed spectral peaks and orbital energy levels obtained from the model.

Of all the possible forms of electronic spectroscopy, **UV photoelectron spectroscopy** comes closest to the goal of directly identifying the orbital energy level from which an electronic transition originates. What is the principle of this spectroscopy? As in the photoelectric effect discussed in Chapter 12, an incident photon of sufficiently high energy ejects an electron from one of the filled valence orbitals of the molecule, creating a positive ion as shown in Figure 25.13, using O_2 as an example. The kinetic energy of the ejected electron is related to the total energy required to form the positive ion via **photoionization,**

$$E_{kinetic} = h\nu - \left[E_f + \left(n_f + \frac{1}{2} \right) h\nu_{vibration} \right]$$ (25.12)

FIGURE 25.13
The ground-state molecular orbital diagram of O_2 is shown on the left. An incident UV photon can eject an electron from one of the occupied MOs, generating an O_2^+ ion and an unbound electron whose kinetic energy can be measured as shown for two different MOs in the center and right of the figure. Electrons ejected from different MOs will differ in their kinetic energy.

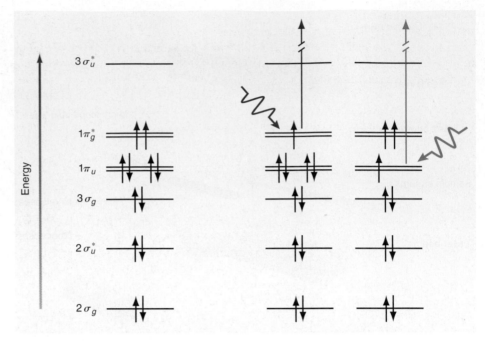

FIGURE 25.14

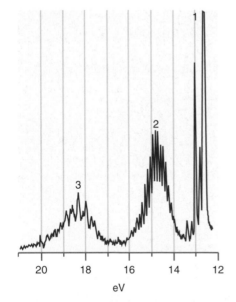

UV photoelectron spectrum of gas-phase H_2O. Three groups of peaks are seen. The structure within each group results from vibrational excitation of the cation formed in the photoionization process.
Source: From "High resolution molecular photoelectron spectroscopy II. Water and deuterium oxide" by C.R. Brundle and D.W. Turner. Proc. Roy. Soc. A. 307: 27–36 (1968), fig. 1, p. 27. Used by permission of The Royal Society and the author.

where E_f is the energy of the cation, which is formed by the removal of the electron, in its ground state. Equation (25.12) takes vibrational excitation of the cation into account, which by conservation of energy leads to a lower kinetic energy for the photoejected electron. Because, in general, either the initial or final state is a radical, a delocalized MO model must be used to describe UV photoelectron spectroscopy.

Under the assumptions to be discussed next, the measured value of $E_{kinetic}$ can be used to obtain the energy of the orbital $\varepsilon_{orbital}$ from which the electron originated. The energy of the cation, E_f, which can be determined directly from a photoelectron spectrum, is equal to $\varepsilon_{orbital}$ if the following assumptions are valid:

- The nuclear positions are unchanged in the transition (Born–Oppenheimer approximation).
- The orbitals for the atom and ion are the same (**frozen orbital approximation**). This assumes that the electron distribution is unchanged in the ion, even though the ion has one fewer electron.
- The total electron correlation energy in the molecule and ion are the same.

The association of E_f with $\varepsilon_{orbital}$ for the neutral molecule under these assumptions is known as **Koopmans' theorem.** In comparing spectra obtained for a large number of molecules with high-level numerical calculations, the measured and calculated orbital energies are often found to differ by approximately 1 to 3 eV. The difference results primarily from the last two assumptions not being entirely satisfied.

This discussion suggests that a photoelectron spectrum consists of a series of peaks, each of which can be associated with a particular molecular orbital of the molecule. Figure 25.14 shows a photoelectron spectrum obtained for gas-phase water molecules for a photon energy $h\nu = 21.8$ eV, corresponding to a strong UV emission peak from a helium discharge lamp. Each of the three groups of peaks can be associated with a particular molecular orbital of H_2O, and the approximately equally spaced peaks within a group correspond to vibrational excitations of the cation formed in the photoionization process.

An analysis of this spectrum offers a good opportunity to compare and contrast localized and delocalized models of chemical bonding in molecules. It turns out that the assignment of peaks in a molecular photoelectron spectrum to individual localized orbitals is not valid. This is the case because the molecular wave function must exhibit the symmetry of the molecule. This important topic will be discussed in some detail in Chapter 27. Using the photoelectron spectrum of H_2O as an example, we show that the correct assignment of peaks in photoelectron spectra is to delocalized linear combinations of the localized orbitals, rather than to individual localized orbitals.

In a localized bonding model, water has two lone pairs and two O—H bonding orbitals. Because the lone pairs and the bonding orbitals are identical except for their orientation, one might expect to observe one group of peaks associated with the lone pair and one group of peaks associated with the bonding orbital in the photoelectron spectrum. In fact, four rather than two groups are observed if the photon energy is significantly higher than that used to obtain the data shown in Figure 25.14. In the localized bonding model, this discrepancy can be understood in terms of the coupling between the lone pairs and between the bonding orbitals. The coupling leads to **symmetric combinations** and **antisymmetric combinations,** just as was observed for vibrational spectroscopy in Section 19.5. In the molecular orbital model, the result can be understood by solving the Hartree–Fock equations, which generates four distinct MOs. We refer to these MOs as symmetric (S) or antisymmetric (A) and as having lone pair (or nonbonding) character (n) or sigma character (σ).

We now return to the photoelectron spectrum of Figure 25.14. The group of peaks below 13eV can be attributed to ε_{nA}. The corresponding MO wave function is the $1b_1$ orbital of Figure 24.8, which can be associated with the antisymmetric combination of the lone pairs. The group of peaks between 14 and 16 eV can be attributed to ε_{nS}. The corresponding wave function is the $2a_1$ orbital, which can be associated with the symmetric combination of the lone pairs. The group between 17 and 20 eV can be attributed to $\varepsilon_{\sigma A}$. The corresponding wave function is the $1b_2$ orbital, which can be associated with the antisymmetric combination of the localized O—H bonding orbitals. The group attributed to $\varepsilon_{\sigma S}$ lies at higher ionization energies than were accessible in the experiment and, therefore, is not observed. The corresponding wave function is the $1a_1$ orbital, which can be associated with the symmetric combination of the localized O—H bonding orbitals. The nomenclature used for these MOs, which was introduced in Chapter 24, will be explained in Chapter 27.

The preceding analysis leaves us with the following question: Why do **equivalent bonds** or lone pairs give rise to several different orbital energies? A nonmathematical explanation follows. Although the localized bonding orbitals are equivalent and orthonormal, the electron distribution in one O—H bond is not independent of the electron distribution in the other O—H bond because of Coulombic interactions between the two bonding regions. Therefore, an electronic excitation in one local bonding orbital changes the potential energy felt by the electrons in the region of the other local bonding orbital. This interaction leads to a coupling between the two localized bonds. By forming symmetric and antisymmetric combinations of the local orbitals, the coupling is removed. However, the local character of the bonding orbitals has also been removed. Therefore, the decoupled molecular wave functions cannot be identified with a state that is localized in only one of the two O—H regions. Only the decoupled wave functions, and not the localized orbitals, are consistent with the symmetry of the molecule.

In the case of water, the two equivalent localized O—H bonds give rise to two distinct orbital energies. However, in highly symmetric molecules, the number of distinct orbital energies can be less than the number of equivalent localized bonds. For instance, the three equivalent localized N—H bonds in NH_3 give rise to two distinct orbital energies, and the four equivalent localized C—H bonds in CH_4 give rise to two distinct orbital energies. The reason for these differences will become apparent after molecular symmetry is discussed in Chapter 27.

25.11 Single Molecule Spectroscopy

Spectroscopic measurements as described earlier are generally carried out in a sample cell in which a very large number of the molecules of interest, called an ensemble, are present. In general, the local environment of the molecules in an ensemble is not identical, which leads to inhomogeneous broadening of an absorption line as discussed in Section 19.9. Figure 25.15 shows how a broad absorption band arises if the corresponding narrow bands

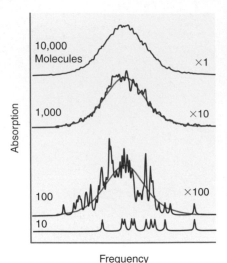

FIGURE 25.15

The absorption spectrum of an individual molecule is narrow, but the peak occurs over a range of frequencies for different molecules as shown in the lowest curve for 10 molecules in the sampling volume. As the number of molecules in the sampling volume is increased, the observed peak shows inhomogeneous broadening and is characteristic of the ensemble average (purple curves) rather than of an individual molecule.

for individual molecules in the ensemble are slightly shifted in frequency because of variations in the immediate environment of a molecule.

Clearly, more information is obtained from the spectra of the individual molecules than from the inhomogeneously broadened band. The "true" absorption spectrum for an individual molecule is observed only if the number of molecules in the volume being sampled is very small, for example, the bottom spectrum in Figure 25.15.

Single molecule spectroscopy is particularly useful in understanding the structure-function relationship for biomolecules. The **conformation** of a biomolecule refers to the arrangement of its constituent atoms in space and can be discussed in terms of primary, secondary, and tertiary structure. The **primary structure** is determined by the backbone of the molecule, for example, peptide bonds in a polypeptide. The term **secondary structure** refers to the local conformation of a part of the polypeptide. Two common secondary structures of polypeptides are the α-helix and the β-sheet as shown in Figure 25.16. **Tertiary structure** refers to the overall shape of the molecule; globular proteins are folded into a spherical shape, whereas fibrous proteins have polypeptide chains that arrange into parallel strands or sheets.

Keep in mind that the conformation of a biomolecule in solution is not static. Collisions with solvent and other solute molecules continuously change the energy and the conformation of a dissolved biomolecule with time. What are the consequences of such conformational changes for an enzyme? Because the activity is intimately linked to structure, conformational changes lead to fluctuations in activity, making an individual enzyme molecule alternately active and inactive as a function of time. Spectroscopic measurements carried out on an ensemble of enzyme molecules give an average over all possible conformations and hence over all possible activities for the enzyme. Such measurements are of limited utility in understanding how structure and chemical

FIGURE 25.16

The **(a)** α-helix and **(b)** β-sheet are two important forms in which proteins are found in aqueous solution. In both structures, hydrogen bonds form between imino (—NH—) groups and carbonyl groups.

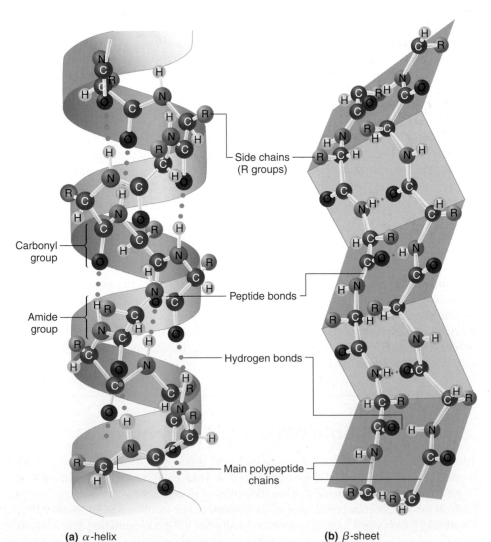

(a) α-helix   **(b)** β-sheet

activity are related. As we show in the next section, single molecule spectroscopy can go beyond the ensemble limit and gives information on the possible conformations of biomolecules and on the timescales on which transitions to different conformations take place.

To carry out single molecule spectroscopy, the number of molecules in the sampling volume must be reduced to approximately one. How is a spectrum of individual biomolecules in solution obtained? Only molecules in the volume that is both illuminated by the light source and imaged by the detector contribute to the measured spectrum. For this number to be approximately one, a laser focused to a small diameter is used to excite the molecules of interest, and a confocal microscope is used to collect the photons emitted in fluorescence from a small portion of the much larger cylindrical volume of the solution illuminated by the laser. In a confocal microscope, the sampling volume is at one focal point and the detector is behind a pinhole aperture located at the other focal point of the microscope imaging optics. Because photons that originate outside of a volume of approximately $1 \ \mu m \times 1 \ \mu m \times 1 \ \mu m$ centered at the focal point are not imaged on the pinhole aperture, they cannot reach the detector. Therefore, the sampling volume is much smaller than the illuminated volume. To ensure that no more than a few molecules are likely to be found in the sampling volume, the concentration of the biomolecule must be less than $\sim 1 \times 10^{-6}$ M. Fluorescence spectroscopy is well suited for single molecule studies because as discussed in Section 25.8, vibrational relaxation ensures that the emitted photons have a lower frequency than the laser used to excite the molecule. Therefore, optical filters can be used to ensure that scattered laser light or photons from Raman scattering outside of sampling volume do not reach the detector. If the molecules being investigated are immobilized, they can also be imaged using the same experimental techniques.

25.12 Fluorescent Resonance Energy Transfer (FRET)

FRET is a form of single molecule spectroscopy that has proved to be very useful in studying biochemical systems. An electronically excited molecule can lose energy by either radiative or nonradiative events as discussed in Section 25.9. We refer to the molecule that loses energy as the **donor** and the molecule that accepts the energy as the **acceptor.** If the emission spectrum of the donor overlaps the absorption spectrum of the acceptor as shown in Figure 25.17, then we refer to **resonance energy transfer** as shown in Figure 25.18. Under resonance conditions, the energy transfer from the donor to the acceptor can occur with a high efficiency.

The probability for resonant energy transfer is strongly dependent on the distance between the two molecules. It was shown by Theodor Förster that the rate at which resonant energy transfer occurs decreases as the sixth power of the donor-acceptor distance.

$$k_{ret} = \frac{1}{\tau_D^{\circ}} \left(\frac{R_0}{r} \right)^6 \tag{25.13}$$

In Equation (25.13), τ_D° is the lifetime of the donor in its excited state and R_0, the critical Förster radius, is the distance at which the resonance transfer rate and the rate for spontaneous decay of the excited state donor are equal. Both these quantities can be determined experimentally. The sensitive dependence of the resonance energy transfer on the donor-acceptance distance makes it possible to use FRET as a **spectroscopic ruler** to measure donor-acceptor distances in the 10–100 nm range.

Figure 25.19 illustrates how FRET can be used to determine the conformation of a biomolecule. Schuler et al. attached dyes acting as donor and acceptor molecules to the ends of polyproline peptides of defined length containing between 6 and 40 proline residues. The donor absorbed a photon from a laser, and the efficiency with which the photon was transferred to the acceptor was measured. The efficiency is defined as the fraction of excitations of the donor that result in excitation of the acceptor. As Equation (25.13) shows, the efficiency falls off as the sixth power of the donor-acceptor distance. The results for a large number of measurements are shown in Figure 25.19b. A range of values for the efficiency is seen for each polypeptide. The variation of the maximum efficiency with the length

FIGURE 25.17
The red curves show the range in wavelength over which absorption and emission occur for the donor and acceptor. Note that emission occurs at greater wavelengths than absorption as discussed in Section 25.7. If the emission band for the donor overlaps the absorption band for the acceptor, resonant energy transfer can occur.

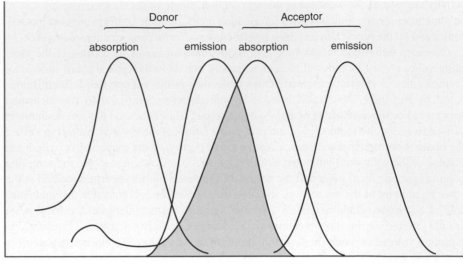

demonstrates that the peptide is not a rigid rod. The width of the distribution in efficiency shows that molecules of the same length can have a number of possible conformations, each of which has a different donor-acceptor distance. The width increases with the peptide length because more twists and turns can occur in a longer strand.

An interesting application of single molecule FRET is in probing the conformational flexibility of single-stranded DNA in solution. The conformational flexibility of single-stranded DNA plays an important role in many DNA processes such as replication, repair, and transcription. Such a strand can be viewed as a flexible rod, approximately 2 nm in diameter. To put the dimensions of a strand in perspective, if it were a rubber tube of 1 cm in diameter, its length would be nearly 1 kilometer. A single molecule of DNA can be as long as ~1 cm in length, yet must fit in the nucleus of a cell, which is typically ~1 μm in diameter. To do so, the conformation of a DNA strand might take the form of a very long piece of spaghetti coiled upon itself as shown in Figure 25.20.

Such a complex conformation is best described by statistical models, one of which is called the worm-like chain model.

In the **worm-like chain model,** the strand takes the form of a flexible rod that is continuously and randomly curved in all possible directions. However, there is an

FIGURE 25.18
Individual events in the emission spectrum of the excited donor and the absorption spectrum of the acceptor are shown. The donor emission photon energies labeled 1, 2, and 3 give rise to the absorption transitions labeled 1', 2', and 3', respectively. Note that vibrational relaxation in the excited acceptor state will lead to a shift in the wavelength of the light emitted by the acceptor. This shift allows the use of optical filters to detect acceptor emission in the presence of scattered light from the laser used to excite the donor.

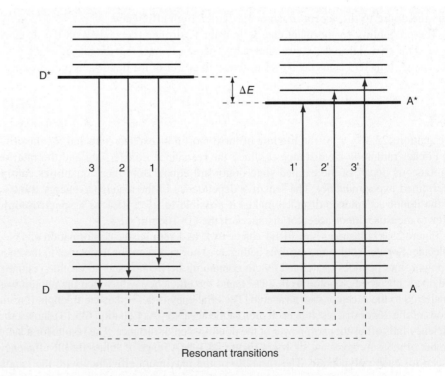

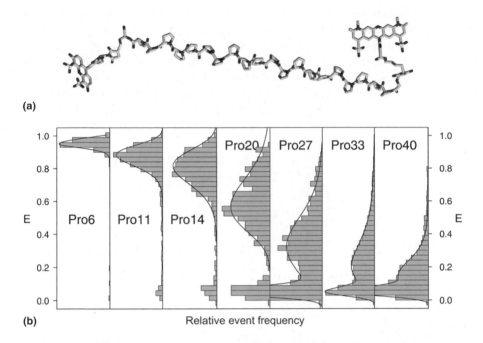

(a)

(b)   Relative event frequency

FIGURE 25.19
(a) Donor (left) and acceptor (right) dyes are attached to a polyproline peptide, which becomes increasingly flexible as its length is increased. (b) The efficiency E of resonant energy transfer from the donor to acceptor is shown for peptides of different lengths. The length of the bars represents the relative event frequency for a large number of measurements on individual molecules. Note that the width of the distribution in E increases with the length of the peptide. The peak near zero efficiency is an experimental artifact due to inactive acceptors.
Source: From B. Schuler et al., "Polyproline and the "Spectroscopic Ruler" Revisited with Single-Molecule Fluorescence," *Proceedings of the National Academy of Sciences of the United States of America* 102: 2754–2759 (2005). Copyright 2005 National Academy of Sciences, U.S.A. Used by permission.

energetic cost of bending the strand, which is available to the strand through the energy transferred in collisions with other species in solution. This energy depends linearly on the temperature. The energy required to bend the rod depends on the radius of curvature; a very gentle bend with a large radius of curvature requires much less energy than a sharp hairpin turn. In the limit of zero kelvin, the collisional energy transfer approaches zero, and the strand takes the form of a rigid rod. As the temperature increases, fluctuations in the radius of curvature increasingly occur along the rod. This behavior is described in the worm-like chain model by the **persistence length,** which is the length that can be traveled along the rod in a straight line before the rod bends in a different direction. As the temperature increases from zero kelvin to room temperature, a worm-like chain changes in conformation from a rigid rod of infinite persistence length to the tangle depicted in Figure 25.20, which has a very small persistence length.

How well does the worm-like chain model describe single-stranded DNA? This was tested by M. C. Murphy et al., who attached flexible single-stranded DNAs to a rigid tether, which was immobilized by bonding the biotin at the end of the tether to a streptavidin-coated quartz surface as shown in Figure 25.21.

A donor fluorophore was attached to the free end of the flexible strand and an acceptor was attached to the rigid end. The length of the strand was varied between 10 and 70 nucleotides corresponding to distances from ~60 nm to ~420 nm between the donor and acceptor. After measuring τ_D° and R_0, the efficiency of resonant energy transfer from the donor to the acceptor was measured as a function of the strand length in NaCl solution whose concentration ranged from 2.5×10^{-3} M to 2 M. The results are shown in Figure 25.22 and compared with calculations in which the persistence length was used as a parameter.

The peak in Figure 25.22a shifts to lower efficiencies as the strand length increases, because the donor and acceptor are farther apart. The width of the distribution in efficiency shows that molecules of the same length can have a number of possible conformations, each of which has a different donor-acceptor distance. Figure 25.22b shows a comparison between measured efficiencies as a function of strand length with predictions of the worm-like chain model. It is seen that the model represents the data well and that the persistence length decreases from 3 nm at low NaCl concentration to 1.5 nm at the highest concentration. The decrease in persistence length as the NaCl concentration increases can be attributed to a reduction in the repulsive interaction between the charged phosphate groups on the DNA through screening of the charge by the ionic solution (see Section 10.4). In this case, FRET measurements have provided a validation of the worm-like chain model for the conformation of DNA.

FIGURE 25.20
The conformation of a long rod-like molecule in solution can be highly tangled.

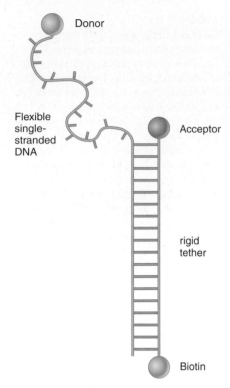

FIGURE 25.21

A donor and acceptor are attached to opposite ends of flexible single strands of DNA. The strands are attached to a silica substrate. The rigid tether has the function of moving the acceptor away from the quartz surface into the solution.

Source: Reprinted from M.C. Murphy et al., "Probing Single-Stranded DNA Conformational Flexibility Using Fluorescence Spectroscopy," *Biophysical Journal* 86 (4): 2530-2537 (April 2004), fig.1, p. 2531, fig. 3, p. 2533, Copyright 2004, with permission from Elsevier.

FIGURE 25.22

(a) The FRET efficiency is shown for Poly dT ssDNA of length 40, 27, and 17 nucleotides (top to bottom panel). The peak at zero efficiency is an experimental artifact due to inactive acceptors. The length of the bars represents the relative event frequency for a large number of measurements on individual molecules.
(b) The FRET efficiency is shown as a function of N, the number of nucleotides, for various concentrations of NaCl. The various curves are calculated curves using the persistence length as a parameter. The best fit curves and the corresponding persistence length are shown for each salt concentration.

Source: Reprinted from M.C. Murphy et al., "Probing Single-Stranded DNA Conformational Flexibility Using Fluorescence Spectroscopy," *Biophysical Journal* 86 (4): 2530-2537 (April 2004), fig.1, p. 2531, fig. 3, p. 2533, Copyright 2004, with permission from Elsevier.

Similar studies have been carried out using electron transfer reactions rather than resonant energy transfer between the donor and acceptor in order to probe the time scale of the conformational fluctuations of a single protein molecule. Yang et al. [*Science* 302 (2003): 262] found that conformational fluctuations occur over a wide range of timescales ranging from hundreds of microseconds to seconds. This result suggests that there are many different pathways that lead from one conformer to another and provides valuable data to researchers who model protein folding and other aspects of the conformational dynamics of biomolecules.

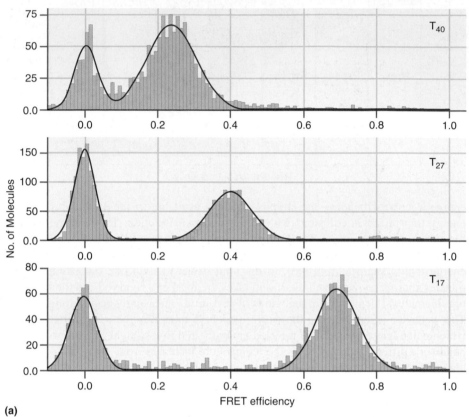

(a)

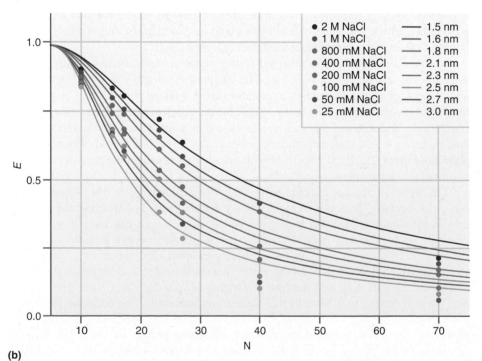

(b)

25.13 Linear and Circular Dichroism

Because the structure of a molecule is closely linked to its reactivity, it is a goal of chemists to understand the structure of a molecule of interest. This is a major challenge in the case of biomolecules because the larger the molecule, the more challenging it is to determine the structure. However, there are techniques available to determine aspects of the molecular structure of biomolecules, although they do not give the positions of all atoms in the molecule. Linear and circular dichroism are particularly useful in giving information on the secondary structure of biomolecules.

As discussed in Section 19.1, light is a transverse electromagnetic wave that interacts with molecules through a coupling of the electric field **E** of the light to the permanent or transient dipole moment $\boldsymbol{\mu}$ of the molecule. Both **E** and $\boldsymbol{\mu}$ are vectors, and in classical physics the strength of the interaction is proportional to the scalar product **E**•$\boldsymbol{\mu}$. In quantum mechanics, the strength of the interaction is proportional to **E**•$\boldsymbol{\mu}^{fi}$, where the **transition dipole moment** is defined by

$$\boldsymbol{\mu}^{fi} = \int \psi_f^*(\tau)\,\mu(\tau)\psi_i(\tau)\,d\tau \qquad \textbf{(25.14)}$$

In Equation (25.14), $d\tau$ is the infinitesimal three dimensional volume element and ψ_i and ψ_f refer to the initial and final states in the transition in which a photon is absorbed or emitted. The spatial orientation of $\boldsymbol{\mu}^{fi}$ is determined by evaluating an integral such as Equation (25.14), which goes beyond the level of this text. We show the orientation of $\boldsymbol{\mu}^{fi}$ for the amide group, which is the building block for the backbone of a polypeptide for a given $\pi \rightarrow \pi^*$ transition in Figure 25.23.

Many biomolecules have a long rod-like shape and can be oriented by embedding them in a film and then stretching the film. For such a sample, the molecule and therefore $\boldsymbol{\mu}^{fi}$ has a well-defined orientation in space. The electric field **E** can also be oriented in a plane with any desired orientation using a polarization filter, in which **linearly polarized light** is generated, as shown in Figure 25.24.

If the plane of polarization is varied with respect to the molecular orientation, the measured absorbance A will vary. It has a maximum value if **E** and $\boldsymbol{\mu}^{fi}$ are parallel and is zero if **E** and $\boldsymbol{\mu}^{fi}$ are perpendicular. In **linear dichroism spectroscopy,** the variation of the absorbance with the orientation of plane-polarized light is measured. It is useful because it allows the direction of $\boldsymbol{\mu}^{fi}$ to be determined for an oriented molecule whose secondary structure is not known. One measures the absorbance with **E** parallel and perpendicular to the molecular axis. The difference $A_\parallel - A_\perp$ relative to the absorbance for randomly polarized light is the quantity of interest.

We illustrate the application of linear dichroism spectroscopy in determining the secondary structure of a polypeptide in the following discussion. The amide groups shown in Figure 25.23 interact with one another because of their close spacing, and the interaction can give rise to a splitting of the transition into two separate peaks. The orientation of $\boldsymbol{\mu}^{fi}$ for each peak depends on the polypeptide secondary structure. For the case of an α-helix, a transition near 208 nm with $\boldsymbol{\mu}^{fi}$ parallel to the helix axis and a transition near 190 nm with $\boldsymbol{\mu}^{fi}$ perpendicular to the helix axis is predicted from theory. Figure 25.25 shows the absorbance for randomly polarized light and $A_\parallel - A_\perp$ for a polypeptide. The data show that $A_\parallel < A_\perp$ for the transition near 190 nm and that $A_\parallel > A_\perp$ near 208 nm. This shows that the secondary structure of this polypeptide is an α-helix. By contrast, the absorption for randomly polarized light gives no structural information.

Because molecules must be oriented in space for linear dichroism spectroscopy, the technique cannot be used for biomolecules in a static solution. For solutions, circular dichroism spectroscopy is widely used to obtain secondary structural information. In this spectroscopy, circularly polarized light, which is depicted in Figure 25.26, is passed through the solution.

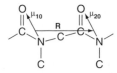

FIGURE 25.23
The amide bonds in a polypeptide chain are shown. The transition dipole moment is shown for a $\pi \rightarrow \pi^*$ transition.

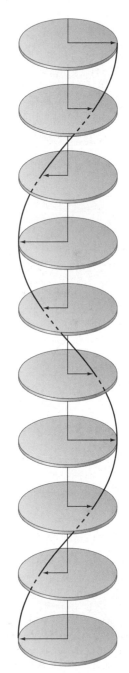

FIGURE 25.24
The arrows in successive images indicate the direction of the electric field vector as a function of time or distance. For linearly polarized light, the amplitude of the electric field vector changes periodically, but is confined to the plane of polarization.

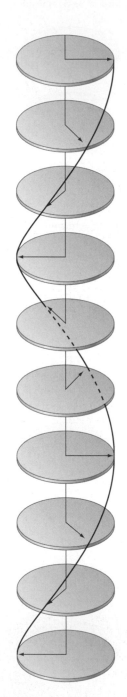

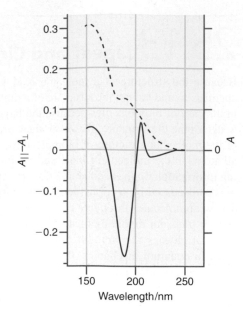

FIGURE 25.25

The normal isotropic absorbance, A (dashed line), and $A_{\parallel} - A_{\perp}$ (solid line) are shown as a function of the wavelength for an oriented film of poly(γ-ethyl-L-glutamate) in which it has the conformation of an α-helix.

Source: Adapted from data from J. Brahms et al., "Application of a New Modulation Method for Linear Dichroism Studies of Oriented Biopolymers in The Vacuum Ultraviolet." *Proceedings of the National Academy of Sciences USA* 60 (1968): 1130.

Biomolecules are optically active, meaning that they do not possess a center of inversion. For an optically active molecule, the absorption for circularly polarized light in which the direction of rotation is clockwise (R) differs from that in which the direction of rotation is counterclockwise (L). This difference in A can be expressed as a difference in the extinction coefficient ε.

$$\Delta A(\lambda) = A_L(\lambda) - A_R(\lambda) = [\varepsilon_L(\lambda) - \varepsilon_R(\lambda)]lc = \Delta\varepsilon lc \qquad (25.15)$$

In Equation (25.15), l is the path length in the sample cell, and c is the concentration. In practice, the difference between $A_L(\lambda)$ and $A_R(\lambda)$ is usually expressed as the molar residual ellipticity, which is the shift in the phase angle θ between the components of the circularly polarized light in the form

$$\theta = 2.303 \times (A_L - A_R) \times 180/(4\pi) \text{ degrees} \qquad (25.16)$$

Circular dichroism can only be observed if ε is nonzero and is usually observed in the visible part of the light spectrum.

As in the case of linear dichroism, $\Delta A(\lambda)$ for a given transition is largely determined by the secondary structure and is much less sensitive to other aspects of the conformation. A discussion of how $\Delta A(\lambda)$ depends on the secondary structure is beyond the level of this text, but it can be shown that common secondary structures such as the α-helix, the β-sheet, a single turn, and a random coil have a distinctly different $\varepsilon(\lambda)$ dependence as shown in Figure 25.27. In this range of wavelengths, the absorption corresponds to $\pi \rightarrow \pi^*$ transitions of the amide group.

The absorbance curves for the different secondary structures in Figure 25.25 are sufficiently different that the extracted $\Delta\varepsilon(\lambda)$ curve obtained for a protein of unknown secondary structure in solution can be expressed in the form

$$\Delta\varepsilon_{observed}(\lambda) = \sum_i F_i \Delta\varepsilon_i(\lambda) \qquad (25.17)$$

where $\Delta\varepsilon_i(\lambda)$ is the curve corresponding to one of the possible secondary structures of the biomolecule and F_i is the fraction of the peptide chromophores in that particular secondary structure. A best fit of the data to Equation (25.17) using widely available software allows a determination of the F_i to be made.

Figure 25.28 shows the results of an application of circular dichroism in determining the secondary structure of α-synuclein bound to unilamellar phospholipid vesicles, which were used as a model for cell membranes. α-Synuclein is a small soluble protein of 140–143 amino acids that is found in high concentration in presynaptic nerve terminals. A mutation in this protein has been linked to Parkinson's disease and it is believed to be a precursor in the formation of extracellular plaques in Alzheimer's disease.

FIGURE 25.26

The arrows in successive images indicate the direction of the electric field vector as a function of time or distance. For circularly polarized light, the amplitude of the electric field vector is constant, but its plane of polarization undergoes a periodic variation.

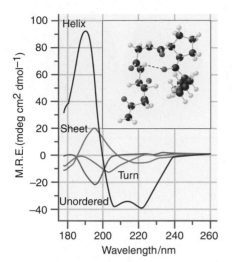

FIGURE 25.27
The mean residual ellipticity θ is shown as a function of wavelength for biomolecules having different secondary structures. Because the curves are distinctly different, circular dichroism spectra can be used to determine the secondary structure for optically active molecules. The inset shows the hydrogen bonding between different amide groups that generates different secondary structures.
Source: From John T. Pelton, "Secondary Considerations," *Science* 291: 2175–2176, March 16, 2001. Copyright © 2001, The American Association for the Advancement of Science. Reprinted with permission from AAAS.

As can be seen by comparing the spectra in Figure 25.28 with those of Figure 25.27, the conformation of α-synuclein in solution is that of a random coil. However, upon binding to unilamellar phospholipid vesicles, the circular dichroism spectrum is dramatically changed and is characteristic of an α-helix. These results show that the binding of α-synuclein requires a conformational change. This conformational change can be understood from the known sequence of amino acids in the protein. By forming an α-helix, the polar and nonpolar groups in the protein are shifted to opposite sides of the helix. This allows the polar groups to associate with the acidic phospholipids, leading to a stronger binding than would be the case for a random coil.

S U P P L E M E N T A L

25.14 Assigning + and − to Σ Terms of Diatomic Molecules

In this section we illustrate how the + and − symmetry designations are applied to Σ terms for homonuclear diatomic molecules. A more complete discussion can be found in *Quantum Chemistry*, sixth edition, by I. Levine, or in *Atoms and Molecules* by M. Karplus and R. N. Porter.

Recall that only partially filled MOs need to be considered in generating term symbols from a molecular configuration. The + and − designations refer to the change in sign of the molecular wave function on reflection in a plane that contains the molecular axis. If there is no change in sign, the + designation applies; if the wave function does change sign, the − designation applies. In the simplest case, all MOs are filled or the unpaired electrons are all in σ MOs. For such states, the + sign applies because there is no change in the sign of the wave function as a result of the reflection operation, as can be seen in Figure 25.29.

We next discuss molecular terms that do not fit into these categories, using O_2 as an example. The configuration for ground-state O_2 is $(1\sigma_g)^2(1\sigma_u^*)^2(2\sigma_g)^2(2\sigma_u^*)^2(3\sigma_g)^2$ $(1\pi_u)^2(1\pi_u)^2(1\pi_g^*)^1(1\pi_g^*)^1$, where we associate the partially filled MOs with the out-of-phase combinations of the $2p_x$ and $2p_y$ AOs as shown in Figure 25.30. Because filled MOs can be ignored, O_2 has a $(\pi^*)^2$ configuration, with one electron on each of the two degenerate π^* MOs. Recall that, in general, a configuration gives rise to several quantum states. Because the two electrons are in different $1\pi_g^*$ MOs, all six combinations of ± 1 for m_l and $\pm 1/2$ for m_s are possible. For example, the Σ terms for which $M_L = m_{l1} + m_{l2} = 0$ occur as singlet and triplet terms. To satisfy the Pauli exclusion principle, the overall wave function (which is a product of spin and spatial parts) must be antisymmetric in the exchange of two electrons.

However, just as the $2p_x$ and $2p_y$ AOs are not eigenfunctions of the operator $\hat{l}_z$, as discussed in Section 18.5, the MOs depicted in Figure 25.30 are not eigenfunctions

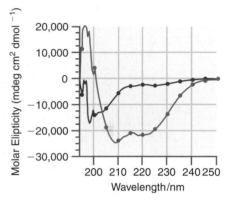

FIGURE 25.28
The molar ellipticity is shown as a function of the wavelength for α-synuclein in solution (red circles) and for α-synuclein bound to unilamellar phospholipid vesicles (blue circles).
Source: Figure 4B from W.Sean Davidson, et al., "Stabilization of a-Synuclein Secondary Structure upon Binding to Synthetic Membranes," *The Journal of Biological Chemistry,* 273 (16): 9443–9449, April 1998. Copyright © 1998, by the American Society for Biochemistry and Molecular Biology. Reprinted with permission.

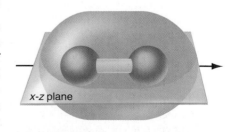

x-z plane

FIGURE 25.29
Reflection of a σ MO in a plane passing through the molecular axis, leaving the wave function unchanged.

FIGURE 25.30
The two degenerate $1\pi_g^*$ wave functions are depicted.

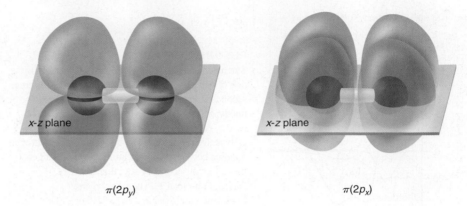

$\pi(2p_y)$ $\pi(2p_x)$

of the operator $\hat{L}_z$. To discuss the assignment of $+$ and $-$ to molecular terms, we can only use wave functions that are eigenfunctions of $\hat{L}_z$. In the cylindrical coordinates appropriate for a diatomic molecule, $\hat{L}_z = -i\hbar(\partial/\partial\phi)$, where ϕ is the angle of rotation around the molecular axis, and the eigenfunctions of this operator have the form $\psi(\phi) = Ae^{-i\Lambda\phi}$, as shown in Section 18.4. We cannot depict these complex functions, because this requires a six-dimensional space, rather than the three-dimensional space required to depict real functions.

The O_2 molecule has a $(\pi^*)^2$ configuration, and antisymmetric molecular wave functions can be formed either by combining symmetric spatial functions with antisymmetric spin functions or vice versa. All possible combinations are shown in the following equations. The subscript $+1$ or -1 on the spatial function indicates the value of m_l.

$$\psi_1 = \pi_{+1}\pi_{+1}(\alpha(1)\beta(2) - \beta(1)\alpha(2))$$

$$\psi_2 = \pi_{-1}\pi_{-1}(\alpha(1)\beta(2) - \beta(1)\alpha(2))$$

$$\psi_3 = (\pi_{+1}\pi_{-1} + \pi_{-1}\pi_{+1})(\alpha(1)\beta(2) - \beta(1)\alpha(2))$$

$$\psi_4 = (\pi_{+1}\pi_{-1} - \pi_{-1}\pi_{+1})\alpha(1)\alpha(2)$$

$$\psi_5 = (\pi_{+1}\pi_{-1} - \pi_{-1}\pi_{+1})(\alpha(1)\beta(2) + \beta(1)\alpha(2))$$

$$\psi_6 = (\pi_{+1}\pi_{-1} - \pi_{-1}\pi_{+1})\beta(1)\beta(2)$$

As shown in Section 22.2, the first three wave functions are associated with singlet states, and the last three are associated with triplet states. Because $\Lambda = |M_L|$, ψ_1 and ψ_2 belong to a Δ term, and ψ_3 through ψ_6 belong to Σ terms.

We next determine how these six wave functions are changed on reflection through a plane containing the molecular axis. As shown in Figure 25.31, reflection through such a plane changes the rotation angle $+\phi$ into $-\phi$. As a consequence, each eigenfunction of $\hat{L}_z$, $Ae^{-i\Lambda\phi}$ is transformed into $Ae^{+i\Lambda\phi}$, which is equivalent to changing the sign of M_L. Therefore, $\pi_{+1} \rightarrow \pi_{-1}$ and $\pi_{-1} \rightarrow \pi_{+1}$. Note that reflection does not change the sign of the wave function for ψ_1 through ψ_3 because $(-1) \times (-1) = 1$. Therefore, the plus sign applies and the term corresponding to ψ_3 is $^1\Sigma_g^+$. However, reflection does change the sign of the wave function for ψ_4 through ψ_6 because $(-1) \times (+1) = -1$; therefore, the minus sign applies. Because these three wave functions belong to a triplet term, the term symbol is $^3\Sigma_g^-$. A similar analysis can be carried out for other configurations.

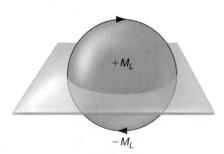

FIGURE 25.31
The rotation angle ϕ is transformed into $-\phi$ through reflection in a plane that contains the molecular axis. This is equivalent to changing $+M_L$ into $-M_L$.

Vocabulary

acceptor	chromophore	equivalent bonds
antisymmetric combination	conformation	fluorescence
Beer's law	continuous energy spectrum	Franck-Condon factor
Born–Oppenheimer approximation	donor	Franck-Condon principle

frozen orbital approximation

integral absorption coefficient

internal conversion

intersystem crossing

Koopmans' theorem

linear dichroism spectroscopy

linearly polarized light

molar extinction coefficient

molecular configuration

molecular term

$n \rightarrow \pi^*$ transition

nonradiative transition

$\pi \rightarrow \pi^*$ transition

persistence length

phosphorescence

photodissociation

photoionization

primary structure

radiative transition

resonance energy transfer

$\sigma \rightarrow \sigma^*$ transition

secondary structure

singlet state

spectroscopic ruler

symmetric combination

tertiary structure

transition dipole moment

triplet state

UV photoelectron spectroscopy

worm-like chain model

Conceptual Problems

Q25.1 Predict the number of unpaired electrons and the ground-state term for the following:

a. BO **b.** LiO

Q25.2 How can FRET give information about the tertiary structure of a biological molecule in solution?

Q25.3 Photoionization of a diatomic molecule produces a singly charged cation. For the molecules listed here, calculate the bond order of the neutral molecule and the lowest energy cation. For which of the molecules do you expect the $n = 0 \rightarrow n' = 1$ vibrational peak to have a higher intensity than the $n = 0 \rightarrow n' = 0$ vibrational peak? The term n refers to the vibrational quantum number in the ground state, and n' refers to the vibrational quantum number in the excited state.

a. H_2 **b.** O_2 **c.** F_2 **d.** NO

Q25.4 What would the intensity versus frequency plot in Figure 25.10 look like if fluorescence were fast with respect to internal conversion?

Q25.5 What aspect of the confocal microscope makes single molecule spectroscopy in solutions possible?

Q25.6 Explain why the fluorescence and absorption groups of peaks in Figure 25.10 are shifted and show mirror symmetry for idealized symmetrical ground-state and excited-state potentials.

Q25.7 The rate of fluorescence is in general higher than that for phosphorescence. Can you explain this fact?

Q25.8 Can linear dichroism spectroscopy be used for molecules in a static solution or in a flowing solution? Explain your answer.

Q25.9 Predict the number of unpaired electrons and the ground-state term for the following:

a. NO **b.** CO

Q25.10 How many distinguishable states belong to the following terms:

a. $^1\Sigma_g^+$ **b.** $^3\Sigma_g^-$ **c.** $^2\Pi$ **d.** $^2\Delta$

Q25.11 Explain why the spectator species M in Equation (25.5) is needed to make the reaction proceed.

Q25.12 Because internal conversion is in general very fast, the absorption and fluorescence spectra are shifted in frequency as shown in Figure 25.10. This shift is crucial in making

fluorescence spectroscopy capable of detecting very small concentrations. Can you explain why?

Q25.13 Make a sketch, like that in the inset of Figure 25.2 of what you might expect the electronic spectrum to look like for the ground and excited states shown in the figure below.

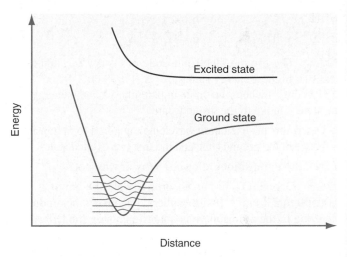

Q25.14 Why are the spectra of the individual molecules shown in the bottom trace of Figure 25.15 shifted in frequency?

Q25.15 Suppose you obtain the UV photoelectron spectrum shown here for a gas-phase molecule. Each of the groups corresponds to a cation produced by ejecting an electron from a different MO. What can you conclude about the bond length of the cations in the three states formed relative to the ground-state neutral molecule? Use the relative intensities of the individual vibrational peaks in each group to answer this question.

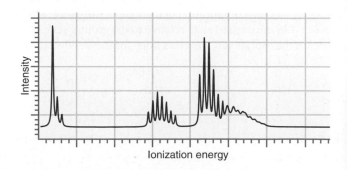

Q25.16 The ground state of O_2^+ is $X^2\Pi_g$, and the next few excited states, in order of increasing energy, are $a^4\Pi_u$, $A^2\Pi_u$, $b^4\Sigma_g^-$, $^2\Delta_g$, $^2\Sigma_g^-$, and $c^4\Sigma_u^-$. On the basis of selection rules, which of the excited states can be accessed from the ground state by absorption of UV light?

Q25.17 The relative intensities of vibrational peaks in an electronic spectrum are determined by the Franck-Condon factors. How would the potential curve for the excited state in Figure 25.2 need to be shifted along the distance axis for the $n = 0 \rightarrow n' = 0$ transition to have the highest intensity? The term n refers to the vibrational quantum number in the ground state, and n' refers to the vibrational quantum number in the excited state.

Q25.18 Calculate the bond order for O_2 in the $X^3\Sigma_g^-$, $a^1\Delta_g$, $b^1\Sigma_g^+$, $A^3\Sigma_u^+$, and $B^3\Sigma_u^-$ states. Arrange these states in order

of increasing bond length on the basis of bond order. Do your results agree with the potential energy curves shown in Figure 25.1?

Q25.19 How can circular dichroism spectroscopy be used to determine the secondary structure of a biomolecule?

Q25.20 What does the word *resonance* in FRET refer to?

Q25.21 In a simple model used to analyze UV photoelectron spectra, the orbital energies of the neutral molecule and the cation formed by ejection of an electron are assumed to be the same. In fact, some relaxation occurs to compensate for the reduction in the number of electrons by one. Would you expect the orbital energies to increase or decrease in the relaxation? Explain your answer.

Numerical Problems

P25.1 Determine whether the following transitions are allowed or forbidden:

a. $^3\Pi_u \rightarrow {}^3\Sigma_g^-$ b. $^1\Sigma_g^+ \rightarrow {}^1\Pi_g$

c. $^3\Sigma_g^- \rightarrow {}^3\Pi_g$ d. $^1\Pi_g \rightarrow {}^1\Delta_u$

P25.2 The ground electronic state of O_2 is $^3\Sigma_g^-$, and the next two highest energy states are $^1\Delta_g$ (7918 cm^{-1}), and $^1\Sigma_g^+$ (13195 cm^{-1}) where the value in parentheses is the energy of the state relative to the ground state.

a. Determine the excitation wavelength required for a transition between the ground state and the first two excited states.

b. Are these transitions allowed? Why or why not?

P25.3 Ozone (O_3) has an absorptivity at 300. nm of 0.00500 $torr^{-1}$ cm^{-1}. In atmospheric chemistry the amount of ozone in the atmosphere is quantified using the Dobson unit (DU), where 1 DU is equivalent to a 10^{-2} mm thick layer of ozone at 1 atm and 273.15 K.

a. Calculate the absorbance of the ozone layer at 300. nm for a typical coverage of 300. DU.

b. Seasonal stratospheric ozone depletion results in a decrease in ozone coverage to values as low as 120. DU. Calculate the absorbance of the ozone layer at this reduced coverage.

In each part, also calculate the transmission from the absorbance using Beer's Law.

P25.4 Consider a diatomic molecule for which the bond force constant in the ground and excited electronic states is the same, but the equilibrium bond length is shifted by an amount δ in the excited state relative to the ground state. For this case the vibrational wavefunctions for the $n = 0$ state are

$$\psi_{g,0} = \left(\frac{\alpha}{\pi}\right)^{1/4} e^{-\frac{1}{2}ar^2}, \psi_{e,0} = \left(\frac{\alpha}{\pi}\right)^{1/4} e^{-\frac{1}{2}a(r-\delta)^2}, \alpha = \sqrt{\frac{k\mu}{\hbar^2}}$$

Calculate the Franck-Condon factor for the 0-0 transition for this molecule by evaluating the following expression:

$$\left| \int_{-\infty}^{\infty} \psi_{g,0}^* \psi_{e,0} dr \right|^2$$

In evaluating this expression, the following integral will be useful:

$$\int_{-\infty}^{\infty} e^{-ax^2 - bx} dx = \left(\frac{\pi}{a}\right)^{1/2} e^{b^2/4a} \quad (a > 0)$$

P25.5 One method for determining Franck-Condon factors between the $n = 0$ vibrational state of the ground electronic state and the n_{th} vibrational level of an electronic excited state is:

$$FC_{0-n} = \frac{1}{n!}\left(\frac{\delta^2}{2}\right)^n \exp\left(-\frac{\delta^2}{2}\right)$$

Where δ is the dimensionless displacement of the excited state relative to the ground state and can be related to atomic displacements through

$$\delta = \left(\frac{\mu\omega}{\hbar}\right)^{1/2}(r_e - r_g)$$

a. Determine the Franck-Condon factors for $n = 0$ to $n = 5$ when $\delta = 0.20$ corresponding to the excited-state potential surface being slightly displaced from that of the ground state.

b. How would you expect the Franck-Condon factors to change if the excited-state displacement increases to $\delta = 2.0$? Verify your expectation by calculating the Franck-Condon factors from $n = 0$ to $n = 5$ for this displacement.

P25.6 When vibrational transitions are observed in an electronic absorption spectra, these transitions can be used to determine dissociation energies. Specifically, a Birge-Sponer plot is constructed where the energy difference between successive vibrational transitions n and $n+1$ ($\Delta G_{n+1/2}$) (see figure) is plotted versus the vibrational level number. Note that G does not refer to the Gibbs energy in this context.

The central idea behind the approach is that the dissociation energy is equal to the sum of these energy differences from $n = 0$ to the dissociation limit:

$$D_0 = \Delta G_{1/2} + \Delta G_{3/2} + \Delta G_{5/2} + \cdots = \sum_{n=0}^{n_{diss}} \Delta G_{n+1/2}$$

a. For the ground state of I_2 the following values for G versus n were determined (*J. Chem. Phys 32* (1960): 738):

n	ΔG (cm^{-1})
0	213.31
1	212.05
2	210.80
3	209.66
4	208.50
5	207.20
6	205.80
7	204.55
8	203.18
9	201.93
10	199.30
11	198.05
12	196.73
13	195.36
14	194.36
15	192.73
16	191.31
17	189.96
18	188.47
19	187.07

If the potential function can be described by a Morse potential [Equation (19.5)], G will be a linear function of $n + 1/2$. Construct a Birge-Sponer plot (ΔG versus $n + 1/2$) using the given data, and using the best fit to a straight line to determine the value of n where $\Delta G = 0$. This is the I_2 ground-state vibrational quantum number at dissociation.

b. The area under the Birge-Sponer plot is equal to the dissociation energy D_0. This area can be determined by summing the ΔG values from $n = 0$ to n at dissociation (determined in part a). Perform this summation to determine D_0 for ground state I_2. You can also integrate the best fit equation to determine D_0.

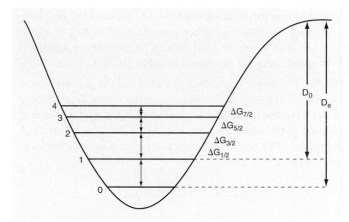

P25.7 Birge-Sponer plots are generally made using ΔG for n values far from the dissociation limit. If the data includes n values close to the dissociation limit, deviations from a linear relationship between ΔG and $n + 1/2$ are observed. Taking these deviations into account allows a more accurate determination of the n value corresponding to dissociation and D_0 to be made. A student determined the following values for ΔG versus n for the ground state of H_2:

n	ΔG (cm^{-1})
0	4133
1	3933
2	3733
3	3533
4	3233
5	3000
6	2733
7	2533
8	2267
9	2000
10	1733
11	1400
12	1067
13	633

a. Construct a Birge-Sponer plot (ΔG versus $n + 1/2$) using the given data, and fit the data assuming a linear relationship between ΔG and n. Determine the value of n where $\Delta G = 0$. This is the H_2 ground-state vibrational quantum number at dissociation.

b. The area under the Birge-Sponer plot is equal to the dissociation energy D_0. This area can be determined by summing the ΔG values from $n = 0$ to n at dissociation (determined in part a). Perform this summation to determine D_0 for ground state I_2. You can also integrate your best fit expression $\Delta G(n)$ from zero to the n value corresponding to dissociation. Compare your result with the value shown in Table 19.3.

c. Determine the value of n where $\Delta G = 0$ and D_0, assuming a quadratic relationship between ΔG and n, $\Delta G = a + b$ $(n + {}^1/_2) + c(n + {}^1/_2)^2$. Which fit gives better agreement with the value for D_0 shown in Table 19.3?

P25.8 Electronic spectroscopy of the Hg-Ar van der Waals complex was performed to determine the dissociation energy of the complex in the first excited state (Quayle, C. J. K. et al. *Journal of Chemical Physics* 99 (1993): 9608). As described in Problem P25.6, the following data regarding ΔG versus n was obtained:

n	G (cm^{-1})
1	37.2
2	34
3	31.6
4	29.2
5	26.8

a. Construct a Birge-Sponer plot (ΔG versus $n + {}^1/_2$) using the given data, and using a best fit to a straight line determine the value of n where $\Delta G = 0$. This is the vibrational quantum number at dissociation in the excited state. $\Delta G_{{}^1/_2}$ can also be determined from the plot.

b. The area under the Birge-Sponer plot is equal to the dissociation energy D_0 for this van der Waals complex in the excited state. The area can be determined by summing the ΔG values from $n = 0$ to n at dissociation (determined in part a). You can also integrate the best fit equation to determine D_0.

P25.9 Green fluorescent protein (GFP) and variants of this protein have been developed for in vivo FRET studies (Pollok B. and Heim R. *Trends in Cell Biology* 9 (1999): 57). Two variants of GFP, cyan fluorescent protein (CFP) and yellow fluorescent protein (YFP), form a FRET pair where $R_0 = 4.72$ nm. The excited-state lifetime of the CFP donor in the absence of YFP is 2.7 ns.

a. At what distance will the rate of energy transfer be equal to the excited-state decay rate for isolated CFP, equal to the inverse of the excited-state lifetime?

b. Determine the distance at which the energy transfer rate will be five times the excited-state decay rate.

P25.10 Structural changes in proteins have been measured using FRET with the amino acid tryptophan as the donor and dansyl as the acceptor where dansyl is attached to the protein through addition to amino acids with aliphatic amine groups such as lysine. For this pair $R_0 = 2.1$ nm, and the excited state lifetime of tryptophan is ~1.0 ns. Determine the rate of energy transfer for $r = 0.50, 1.0, 2.0, 3.0,$ and 5.0 nm.

P25.11 In the polyproline "spectroscopic ruler" experiment shown in Figure 25.19, the FRET pair employed is comprised of the fluorescent dyes Alexa Fluor 488 (excited-state lifetime of 4.1 ns) and Alexa Fluor 594. For this FRET pair $R_0 = 5.4$ nm. The distance between the FRET pair ranges from 2.0 nm for Pro6 to 12.5 nm for Pro40. Calculate the variation in the energy-transfer rate for $r = 2.0, 7.0,$ and 12.0 nm. Do your results agree with the trend evident in Figure 25.19?.

Computational Chemistry

Warren J. Hehre, CEO, Wavefunction, Inc.

To the memory of Sir John Pople, 1925–2004

The Schrödinger equation can be solved exactly only for atoms or molecules containing one electron. For this reason, numerical methods that allow us to calculate approximate wave functions and values for observables such as energy, equilibrium bond lengths and angles, and dipole moments are at the heart of computational chemistry. The starting point for our discussion is the Hartree–Fock molecular orbital model. Although this model gives good agreement with experiment for some variables such as bond lengths and angles, it is inadequate for calculating many other observables. By extending the model to include electron correlation in a more realistic manner, and by judicious choice of a basis set, more accurate calculations can be made. The configuration interaction, Møller-Plesset, and density functional methods are discussed in this chapter, and the trade-off between computational cost and accuracy is emphasized. The 37 problems provided with this chapter are designed to give the student a working, rather than a theoretical, knowledge of computational chemistry.

26.1 The Promise of Computational Chemistry

Calculations on molecules based on quantum mechanics, once a mere novelty, are now poised to complement experiments as a means to uncover and explore new chemistry. The most important reason for this is that the theories underlying the calculations have now evolved to the point at which a variety of important quantities, among them molecular equilibrium geometry and reaction energetics, can be obtained with sufficient accuracy to actually be of use. Also important are the spectacular advances in computer hardware that have been made during the past decade. Taken together, this means that good theories can now be routinely applied to real systems. Finally, current computer software can be easily and productively used with little special training.

In making these quantum mechanics calculations, however, significant obstacles remain. For one, the chemist is confronted with many choices to make and few guidelines on how to make these choices. The fundamental problem is that the mathematical equations that arise from the application of quantum mechanics to chemistry—and that ultimately govern molecular structure and properties—cannot be solved analytically. Approximations need to be made in order to realize equations that can actually be solved. Severe approximations may lead to methods that can be widely applied, but may not yield accurate information. Less severe approximations may lead to methods that are more accurate, but too costly to apply routinely. In short, no one method of calculation is likely to be ideal for all applications, and the ultimate choice of specific methods rests on a balance between accuracy and cost. We equate cost with the computational time required to carry out the calculation.

The purpose of this chapter is to guide the student past the point of merely thinking about quantum mechanics as one of several components of a physical chemistry course and to instead have the student actually use quantum mechanics to address real chemical problems. The chapter starts with the many-electronic Schrödinger equation and then outlines the approximations that need to be made to transform this equation into what is now commonly known as Hartree–Fock theory. In the spirit of emphasizing the concepts rather than the theoretical framework, mathematical descriptions of the theoretical models discussed appear in boxes. A detailed understanding of this framework, however desirable, is not necessary to apply quantum mechanics to chemistry.

A focus on the limitations of Hartree–Fock theory leads to ways to improve on it and to a range of practical quantum chemical models. A few of these models are examined in detail and their performance and cost discussed. Finally, a series of graphical techniques is presented to portray the results of quantum chemical calculations. Aside from its practical focus, what sets this chapter apart from the remainder of this text is the problems. None of these are of the pencil-and-paper type; instead they require use of a quantum chemical program[1] on a digital computer. For the most part, the problems are open ended (as is an experimental laboratory) meaning that the student is free to explore. Problems that use the quantum chemical models under discussion are referenced throughout the chapter. Working problems as they are presented, before proceeding to the next section, is strongly recommended.

This Icon Indicates That Relevant Computational Problems are Available in the End-of-Chapter Problems.

26.2 Potential Energy Surfaces

Chemists are familiar with the plot of energy versus the torsion angle involving the central carbon–carbon bond in *n*-butane. Figure 26.1 reveals three energy minima, corresponding to staggered structures, and three energy maxima, corresponding to eclipsed structures. One of the minima is given by a torsion angle of 180° (the so-called *anti* structure), and it is lower in energy and distinct from the other two minima with torsion angles of approximately 60° and 300° (so-called *gauche* structures), which are identical. Similarly, one of the energy maxima corresponding to a torsion angle of 0° is distinct from the other two maxima with torsion angles of approximately 120° and 240°, which are identical.

Eclipsed forms of *n*-butane are not stable molecules; instead they correspond only to hypothetical structures between *anti* and *gauche* minima. Thus, any sample of *n*-butane is made up of only two distinct compounds, *anti* *n*-butane and *gauche* *n*-butane. The relative abundance of the two compounds as a function of temperature is given by the Boltzmann distribution (see the discussion in Section 13.1).

The important geometrical coordinate in the example of Figure 26.1 can be clearly identified as a torsion involving one particular carbon–carbon bond. More generally, the

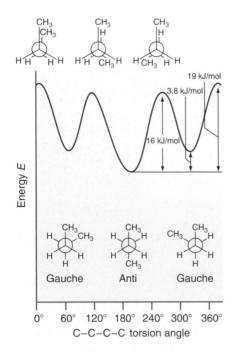

FIGURE 26.1

The energy of *n*-butane is shown as a function of the CCCC torsion angle, which is the reaction coordinate.

[1]The problems have been designed with the capabilities of the Student Edition of the Spartan molecular modeling program in mind. Other programs that allow equilibrium and transition-state geometry optimization, conformational searching, energy, property, and graphical calculations using Hartree–Fock, and density functional and MP2 models can also be used. The only exceptions are problems that appear early in the chapter before calculation models have been fully introduced. These problems make use of precalculated Spartan files that will be made available to students. See *www.masteringchemistry.com*.

important coordinate will be some combination of bond distances and angles and will be referred to simply as the **reaction coordinate**. This leads to a general type of plot in which the energy is given as a function of the reaction coordinate. Diagrams like this are commonly referred to as **reaction coordinate diagrams** or **potential energy surfaces** and provide essential connections between important chemical observables—structure, stability, reactivity, and selectivity—and energy.

26.2.1 Potential Energy Surfaces and Geometry

The positions of the energy minima along the reaction coordinate give the equilibrium structures of the reactants and products as shown in Figure 26.2. Similarly, the position of the energy maximum defines the transition state. For example, where the reaction involves *gauche n*-butane going to the more stable *anti* conformer, the reaction coordinate may be thought of as a simple torsion about the central carbon–carbon bond, and the individual reactant, transition-state, and product structures in terms of this coordinate are depicted in Figure 26.3.

Equilibrium structure (geometry) can be determined from experiments as long as the molecule can be prepared and is sufficiently long lived to be subject to measurement. On the other hand, the geometry of a transition state cannot be established from measurement. This is simply because the transition state does not exist in terms of a sufficiently large population of molecules on which measurements can be performed.

Both equilibrium and transition-state structure can be determined from calculations. The former requires a search for an energy minimum on a potential energy surface, whereas the latter requires a search for an energy maximum along the reaction coordinate (and a minimum along each of the remaining coordinates). To see what is actually involved, the qualitative picture provided earlier must be replaced by a rigorous mathematical treatment. Reactants, products, and transition states are all stationary points on the potential energy diagram. In the one-dimensional case (the reaction coordinate diagram alluded to previously), this means that the first derivative of the potential energy with respect to the reaction coordinate is zero:

$$\frac{dV}{dR} = 0 \tag{26.1}$$

The same must be true in dealing with a many-dimensional potential energy diagram (a potential energy surface). Here all partial derivatives of the energy with respect to each of the $3N - 6$ (N atoms) independent geometrical coordinates (R_i) are zero:

$$\frac{\partial V}{\partial R_i} = 0 \quad i = 1, 2, \ldots, 3N - 6 \tag{26.2}$$

In the one-dimensional case, reactants and products are energy minima and are characterized by a positive second energy derivative:

$$\frac{d^2V}{dR^2} > 0 \tag{26.3}$$

The transition state is an energy maximum and is characterized by a negative second energy derivative:

$$\frac{d^2V}{dR^2} < 0 \tag{26.4}$$

In the many-dimensional case, each independent coordinate, R_i, gives rise to $3N - 6$ second derivatives:

$$\frac{\partial^2V}{\partial R_i R_1}, \frac{\partial^2V}{\partial R_i R_2}, \frac{\partial^2V}{\partial R_i R_3}, \ldots, \frac{\partial^2V}{\partial R_i R_{3N-6}} \tag{26.5}$$

Problem P26.1

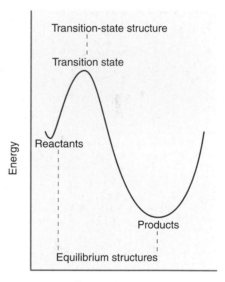

FIGURE 26.2
A reaction coordinate diagram shows the energy as a function of the reaction coordinate. Reactants and products correspond to minima, and the transition state corresponds to a maximum along this path.

Gauche
n-butane
"reactant"

"Transition state"

Anti
n-butane
"product"

FIGURE 26.3
The structure of the reactant, product, and transition state in the "reaction" of *gauche n*-butane to *anti n*-butane.

This leads to a matrix of second derivatives (the so-called Hessian):

$$
\begin{bmatrix}
\dfrac{\partial^2 V}{\partial R_1^2} & \dfrac{\partial^2 V}{\partial R_1 R_2} & \cdots & \\
\dfrac{\partial^2 V}{\partial R_2 R_1} & \dfrac{\partial^2 V}{\partial R_2^2} & \cdots & \\
\cdots & \cdots & & \\
\cdots & \cdots & & \dfrac{\partial^2 V}{\partial R_{3N-6}^2}
\end{bmatrix}
\tag{26.6}
$$

In this form, it is not possible to say whether any given coordinate corresponds to an energy minimum, an energy maximum, or neither. To see the correspondence, the original set of geometrical coordinates (R_i) is replaced by a new set of coordinates (ξ_i), which leads to a matrix of second derivatives that is diagonal:

$$
\begin{bmatrix}
\dfrac{\partial^2 V}{\partial \xi_1^2} & 0\cdots & & 0 \\
0 & \dfrac{\partial^2 V}{\partial \xi_2^2}\cdots & & 0 \\
\cdots & \cdots & & \cdots \\
0 & 0\cdots & & \dfrac{\partial^2 V}{\partial \xi_{3N-6}^2}
\end{bmatrix}
\tag{26.7}
$$

The ξ_i are unique and referred to as **normal coordinates.** Stationary points for which all second derivatives (in normal coordinates) are positive are energy minima:

$$
\frac{\partial^2 V}{\partial \xi_i^2} > 0 = 1, 2, \ldots, 3N - 6
\tag{26.8}
$$

These correspond to equilibrium forms (reactants and products). Stationary points for which all but one of the second derivatives are positive are so-called (first-order) saddle points and may correspond to transition states. If they do, the coordinate for which the second derivative is negative is referred to as the reaction coordinate (ξ_p):

$$
\frac{\partial^2 V}{\partial \xi_p^2} < 0
\tag{26.9}
$$

26.2.2 Potential Energy Surfaces and Vibrational Spectra

The vibrational frequency for a diatomic molecule A-B is given by Equation (26.10) as discussed in Section 18.1:

$$
\nu = \frac{1}{2\pi}\sqrt{\frac{k}{\mu}}
\tag{26.10}
$$

In this equation, k is the force constant, which is in fact the second energy derivative of the potential energy, V, with respect to the bond length, R, at its equilibrium position

$$
k = \frac{d^2 V(R)}{dR^2}
\tag{26.11}
$$

and μ is the reduced mass,

$$
\mu = \frac{m_A m_B}{m_A + m_B}
\tag{26.12}
$$

where m_A and m_B are masses of atoms A and B.

Polyatomic systems are treated in a similar manner. Here, the force constants are the elements in the diagonal representation of the Hessian [Equation (26.7)]. Each vibrational mode is associated with a particular motion of atoms away from their equilibrium positions on the potential energy surface. Low frequencies correspond to motions in shallow regions of the surface, whereas high frequencies correspond to motions in steep regions. Note that one of the elements of the Hessian for a transition state will be a negative number, meaning that the corresponding frequency will be imaginary [the square root of a negative number as in Equation (26.10)]. This normal coordinate refers to motion along the reaction coordinate.

26.2.3 Potential Energy Surfaces and Thermodynamics

The relative stability of reactant and product molecules is indicated on the potential energy surface by their energies. The thermodynamic state functions internal energy, U, and enthalpy, H, can be obtained from the energy of a molecule calculated by quantum mechanics, as discussed in Section 26.8.4.

The most common case is, as depicted in Figure 26.4, the one in which energy is released in the reaction. This kind of reaction is said to be **exothermic,** and the difference in stabilities of reactant and product is simply the enthalpy difference ΔH. For example, the reaction of *gauche* n-butane to *anti* n-butane is exothermic, and $\Delta H = -3.8 \text{ kJ/mol}$ as shown in Figure 26.1.

Thermodynamics tells us that if we wait long enough the amount of products in an exothermic reaction will be greater than the amount of reactants. The actual ratio of the number of molecules of products ($n_{products}$) to reactants ($n_{reactants}$) also depends on the temperature and follows from the Boltzmann distribution:

$$\frac{n_{products}}{n_{reactants}} = \exp\left[-\frac{E_{products} - E_{reactants}}{k_B T}\right] \quad (26.13)$$

where $E_{products}$ and $E_{reactants}$ are the energies per molecule of the products and reactants, respectively, T is the temperature, and k is the Boltzmann constant. The Boltzmann distribution tells us the relative amounts of the products and reactants at equilibrium. Even small energy differences between major and minor products lead to large product ratios, as shown in Table 26.1. The product formed in greatest abundance is that with the lowest energy, irrespective of the reaction pathway. In this case, the product is referred to as the **thermodynamic product** and the reaction is said to be **thermodynamically controlled.**

26.2.4 Potential Energy Surfaces and Kinetics

A potential energy surface also reveals information about the rate at which a reaction occurs. The difference in energy between reactants and the transition state as shown in Figure 26.5 determines the kinetics of the reaction. The absolute reaction rate depends both on the concentrations of the reactants, $[A]^a$, $[B]^b$, ..., where $a, b, \ldots$ are typically integers or half integers, and a quantity k', called the **rate constant:**

$$\text{Rate} = k'[A]^a[B]^b[C]^c \ldots \quad (26.14)$$

The rate constant is given by the Arrhenius equation and depends on the temperature:

$$k' = A \exp\left[-\frac{(E_{transition\ state} - E_{reactants})}{k_B T}\right] \quad (26.15)$$

Here, $E_{transition\ state}$ and $E_{reactants}$ are the energies per molecule of the transition state and the reactants, respectively. Note that the rate constant and the overall rate do not depend on the energies of reactants and products, but only on the difference in energies between reactants and the transition state. This difference is commonly referred to

Problems P26.2–P26.3

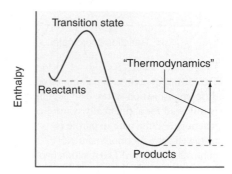

FIGURE 26.4
The energy difference between the reactants and products determines the thermodynamics of a reaction.

TABLE 26.1 The Ratio of the Major to Minor Product Is shown as a Function of the Energy Difference between these Products	
Energy Difference kJ/mol	Major: Minor (at Room Temperature)
2	~80 : 20
4	~90 : 10
8	~95 : 5
12	~99 : 1

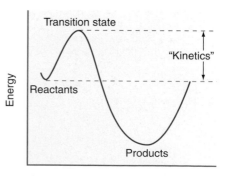

FIGURE 26.5
The energy difference between the reactants and transition state determines the rate of a reaction.

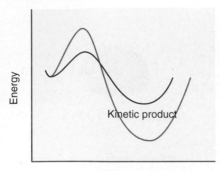

Energy

Kinetic product

Reaction coordinate

FIGURE 26.6
Two different pathways passing through different transition states. The path in red is followed for a kinetically controlled reaction, and the path in purple is followed for a thermodynamically controlled reaction. The reaction coordinate differs for the two pathways.

Problem P26.4

as the **activation energy** and is usually given the symbol $\Delta E^{\ddagger}$. Other factors such as the likelihood of encounters between molecules and the effectiveness of these encounters in promoting reactions are taken into account by way of the **preexponential factor,** A, which is generally assumed to have the same value for reactions involving a single set of reactants going to different products or for reactions involving closely related reactants.

In general, the lower the activation energy, the faster the reaction. In the limit $\Delta E^{\ddagger} = 0$, the reaction rate will be limited entirely by how rapidly the molecules can move. Such limiting reactions are known as **diffusion-controlled reactions.** The product formed in greatest amount in a kinetically controlled reaction is that proceeding via the lowest energy transition state, irrespective of whether or not this is the thermodynamically stable product. For example, a **kinetically controlled** reaction will proceed along the red pathway in Figure 26.6, and the product formed is different than that corresponding to equilibrium in the system. The **kinetic product** ratio shows a dependence on activation energy differences that is analogous to that of Equation (26.13) with $E_{transition\ state}$ in place of $E_{product}$.

26.3 Hartree–Fock Molecular Orbital Theory: A Direct Descendant of the Schrödinger Equation

The Schrödinger equation is deceptive in that, although it is remarkably easy to write down for any collection of nuclei and electrons, it has proven to be insolvable except for the one-electron case (the hydrogen atom). This situation was elaborated as early as 1929 by Dirac, one of the early founders of quantum mechanics:

> The underlying physical laws necessary for the mathematical theory of a large part of physics and the whole of chemistry are thus completely known, and the difficulty is only that the exact application of these laws leads to equations much too complicated to be solvable.
>
> P. A. M. Dirac, 1902–1984

To realize a practical quantum mechanical theory, it is necessary to make three approximations to the general multinuclear, multielectron Schrödinger equation:

$$\hat{H}\Psi = E\Psi \tag{26.16}$$

where E is the total energy of the system and Ψ is the n-electron wave function that depends both on the identities and positions of the nuclei and on the total number of electrons. The Hamiltonian $\hat{H}$ provides the recipe for specifying the kinetic and potential energies for each of the particles:

$$\hat{H} = -\frac{\hbar^2}{2m_e}\sum_{i}^{electrons}\nabla_i^2 - \frac{\hbar^2}{2}\sum_{A}^{nuclei}\frac{1}{M_A}\nabla_A^2 - \frac{e^2}{4\pi\varepsilon_0}\sum_{i}^{electrons}\sum_{A}^{nuclei}\frac{Z_A}{r_{iA}}$$

$$+ \frac{e^2}{4\pi\varepsilon_0}\sum_{i>j}^{electrons}\sum_{j}^{electrons}\frac{1}{r_{ij}} + \frac{e^2}{4\pi\varepsilon_0}\sum_{A>B}^{nuclei}\sum_{B}^{nuclei}\frac{Z_A Z_B}{R_{AB}} \tag{26.17}$$

where Z_A is the nuclear charge, M_A is the mass of nucleus A, m_e is the mass of the electron, R_{AB} is the distance between nuclei A and B, r_{ij} is the distance between electrons i and j, r_{iA} is the distance between electron i and nucleus A, ε_0 is the permittivity of free space, and $\hbar$ is the Planck constant divided by 2π.

The first approximation takes advantage of the fact that nuclei move much more slowly than do electrons. We assume that the nuclei are stationary from the perspective of the electrons (see Section 23.2), which is known as the **Born–Oppenheimer approximation.** This assumption leads to a nuclear kinetic energy term in Equation (26.17), the second term, which is zero, and a nuclear–nuclear Coulombic energy term, the last term, which is constant. What results is the **electronic Schrödinger equation:**

$$\hat{H}^{el}\Psi^{el} = E^{el}\Psi^{el} \qquad (26.18)$$

$$\hat{H}^{el} = -\frac{\hbar^2}{2m_e}\sum_{i}^{electrons}\nabla_i^2 - \frac{e^2}{4\pi\varepsilon_0}\sum_{i}^{electrons}\sum_{A}^{nuclei}\frac{Z_A}{r_{iA}} + \frac{e^2}{4\pi\varepsilon_0}\sum_{j}^{electrons}\sum_{i}^{electrons}\frac{1}{r_{ij}} \quad (26.19)$$

The (constant) nuclear–nuclear Coulomb energy, the last term in Equation (26.17) needs to be added to E^{el} to get the total energy. Note that nuclear mass does not appear in the electronic Schrödinger equation. To the extent that the Born–Oppenheimer approximation is valid, this means that isotope effects on molecular properties must have a different origin.

Equation (26.18), like Equation (26.16), is insolvable for the general (many-electron) case and further approximations need to be made. The most obvious thing to do is to assume that electrons move independently of each other, which is what is done in the **Hartree–Fock approximation.** In practice, this can be accomplished by assuming that individual electrons are confined to functions called spin orbitals, χ_i. Each of the N electrons feels the presence of an average field made up of all of the other $(N-1)$ electrons. To ensure that the total (many-electron) wave function Ψ is antisymmetric upon interchange of electron coordinates, it is written in the form of a single determinant called the **Slater determinant** (see Section 21.3):

$$\Psi = \frac{1}{\sqrt{n!}}\begin{vmatrix} \chi_1(1) & \chi_2(1)\cdots & \chi_n(1) \\ \chi_1(2) & \chi_2(2)\cdots & \chi_n(2) \\ \cdots & \cdots & \cdots \\ \chi_1(n) & \chi_2(n) & \chi_n(n) \end{vmatrix} \qquad (26.20)$$

Individual electrons are represented by different rows in the determinant, which means that interchanging the coordinates of two electrons is equivalent to interchanging two rows in the determinant, multiplying its value by -1. Spin orbitals are the product of spatial functions or molecular orbitals, ψ_i, and spin functions, α or β. The fact that there are only two kinds of spin functions (α and β) leads to the conclusion that two electrons at most may occupy a given molecular orbital. Were a third electron to occupy the orbital, two rows in the determinant would be the same, as was shown in Section 21.3. Therefore, the value of the determinant would be zero. Thus, the notion that electrons are paired is a consequence of the Hartree–Fock approximation through the use of a determinant for the wave function. The set of molecular orbitals leading to the lowest energy is obtained by a process referred to as a **self-consistent-field (SCF) procedure,** which was discussed in Section 21.5 for atoms and in Section 24.1 for molecules.

The Hartree–Fock approximation leads to a set of differential equations, the **Hartree–Fock equations,** each involving the coordinates of a single electron. Although they can be solved numerically, it is advantageous to introduce an additional approximation in order to transform the Hartree–Fock equations into a set of algebraic equations. The basis for this approximation is the expectation that the one-electron solutions for many-electron molecules will closely resemble the one-electron wave functions for the hydrogen atom. After all, molecules are made up of atoms, so why should molecular solutions not be made up of atomic solutions? As discussed in Section 24.2, the molecular orbitals ψ_i are expressed as linear combinations of a basis set of prescribed functions known as basis functions, ϕ:

$$\psi_i = \sum_{\mu}^{basis\ functions} c_{\mu i}\phi_{\mu} \qquad (26.21)$$

In this equation, the coefficients $c_{\mu i}$ are the (unknown) molecular orbital coefficients. Because the ϕ are usually centered at the nuclear positions, they are referred to as atomic orbitals, and Equation (26.21) is called the **linear combination of atomic orbitals**

(LCAO) approximation. Note, that in the limit of a complete (infinite) basis set, the LCAO approximation is exact at the Hartree–Fock level.

MATHEMATICAL FORMULATION OF THE HARTREE–FOCK METHOD

The Hartree–Fock and LCAO approximations, taken together and applied to the electronic Schrödinger equation, lead to a set of matrix equations now known as the **Roothaan–Hall equations:**

$$\mathbf{Fc} = \varepsilon \mathbf{Sc} \tag{26.22}$$

where $\mathbf{c}$ are the unknown molecular orbital coefficients [see Equation (26.21)], ε are orbital energies, $\mathbf{S}$ is the overlap matrix, and $\mathbf{F}$ is the Fock matrix, which is analogous to the Hamiltonian in the Schrödinger equation:

$$F_{\mu\nu} = H_{\mu\nu}^{core} + J_{\mu\nu} - K_{\mu\nu} \tag{26.23}$$

where H^{core} is the so-called core Hamiltonian, the elements of which are given by

$$H_{\mu\nu}^{core} = \int \phi_\mu(1) \left[-\frac{\hbar^2}{2m_e}\nabla^2 - \frac{e^2}{4\pi\varepsilon_0}\sum_A^{nuclei}\frac{Z_A}{r_{1A}} \right] \phi_\nu(1)\, d\tau \tag{26.24}$$

Coulomb and exchange elements are given by:

$$J_{\mu\nu} = \sum_\lambda^{basis\ functions} \sum_\sigma P_{\lambda\sigma}(\mu\nu|\lambda\sigma) \tag{26.25}$$

$$K_{\mu\nu} = \frac{1}{2}\sum_\lambda^{basis\ functions} \sum_\sigma P_{\lambda\sigma}(\mu\lambda|\nu\sigma) \tag{26.26}$$

where $\mathbf{P}$ is called the density matrix, the elements of which involve a product of two molecular orbital coefficients summed over all occupied molecular orbitals (the number of which is simply half the total number of electrons for a closed-shell molecule):

$$P_{\lambda\sigma} = 2\sum_i^{occupied\ molecular\ orbitals} c_{\lambda i}c_{\sigma i} \tag{26.27}$$

and $(\mu\nu|\lambda\sigma)$ are two-electron integrals, the number of which increases as the fourth power of the number of basis functions. Therefore, the cost of a calculation rises rapidly with the size of the basis set:

$$(\mu\nu|\lambda\sigma) = \iint \phi_\mu(1)\phi_\nu(1)\left[\frac{1}{r_{12}}\right]\phi_\lambda(2)\phi_\sigma(2)\, d\tau_1\, d\tau_2 \tag{26.28}$$

Methods resulting from solution of the Roothaan–Hall equations are called **Hartree–Fock models.** The corresponding energy in the limit of a complete basis set is called the **Hartree–Fock energy.**

26.4 Properties of Limiting Hartree–Fock Models

As discussed earlier in Section 21.5, total energies obtained from limiting (complete basis set) Hartree–Fock calculations will be too large (positive). This can be understood by recognizing that the Hartree–Fock approximation leads to replacement of instantaneous interactions between individual pairs of electrons with a picture in which each electron interacts with a charge cloud formed by all other electrons. The loss of flexibility causes electrons to get in each other's way to a greater extent than would actually be the case, leading to an overall electron repulsion energy that is too large and, hence, a total energy that is too large. The direction of the error in the total energy is also a direct consequence of the fact that Hartree–Fock models are variational. The limiting Hartree–Fock energy must be larger than (or at best equal to) the energy that would result from the solution of the exact Schrödinger equation.

The difference between the limiting Hartree–Fock energy and the exact Schrödinger energy is called the **correlation energy.** The name *correlation* stems from the idea that the motion of one electron necessarily adjusts to or correlates with the motions of all other electrons. Any restriction on the freedom of electrons to move independently will, therefore, reduce their ability to correlate with other electrons.

The magnitude of the correlation energy may be quite large in comparison with typical bond energies or reaction energies. However, a major part of the total correlation energy may be insensitive to molecular structure, and Hartree–Fock models, which provide an incomplete account of correlation, may provide acceptable accounts of the energy change in some types of chemical reactions. It is also often the case that other properties, such as equilibrium geometries and dipole moments, are less influenced by correlation effects than are total energies. The sections that follow explore to what extent these conclusions are valid.

It is important to realize that calculations cannot actually be carried out at the Hartree–Fock limit. Presented here under the guise of limiting Hartree–Fock quantities are the results of calculations performed with a relatively large and flexible basis set, specifically the 6-311+G** basis set. (Basis sets are discussed at length in Section 26.7.) Although such a treatment leads to total energies that are higher than actual limiting Hartree–Fock energies by several tens to several hundreds of kilojoules per mole (depending on the size of the molecule), it is expected that errors in relative energies as well as in geometries, vibrational frequencies, and properties such as dipole moments will be much smaller.

26.4.1 Reaction Energies

The most easily understood problem with limiting Hartree–Fock models is uncovered in comparisons of **homolytic bond dissociation** energies. In such a reaction, a bond is broken leading to two radicals, for example, in methanol:

$$CH_3 - OH \longrightarrow \cdot CH_3 + \cdot OH$$

As seen from the data in Table 26.2, Hartree–Fock dissociation energies are too small. In fact, limiting Hartree–Fock calculations suggest an essentially zero O—O bond energy in hydrogen peroxide and a negative F—F "bond energy" in the fluorine molecule! Something is seriously wrong. To see what is going on, consider the analogous bond dissociation reaction in the hydrogen molecule:

$$H - H \longrightarrow \cdot H + \cdot H$$

Each of the hydrogen atoms that make up the product contains only a single electron, and its energy is given exactly by the (limiting) Hartree–Fock model. On the other hand, the reactant contains two electrons and, according to the variation principle, its energy must be too high (too positive). Therefore, the bond dissociation energy must be too low

TABLE 26.2 Homolytic Bond Dissociation Energies (kJ/mol)

Molecule (bond)	Hartree–Fock Limit	Experiment	Δ
$CH_3 - CH_3 \longrightarrow \cdot CH_3 + \cdot CH_3$	276	406	−130
$CH_3 - NH_2 \longrightarrow \cdot CH_3 + \cdot NH_2$	238	389	−151
$CH_3 - OH \longrightarrow \cdot CH_3 + \cdot OH$	243	410	−167
$CH_3 - F \longrightarrow \cdot CH_3 + \cdot F$	289	477	−188
$NH_2 - NH_2 \longrightarrow \cdot NH_2 + \cdot NH_2$	138	289	−151
$HO - OH \longrightarrow \cdot OH + \cdot OH$	−8	230	−238
$F - F \longrightarrow \cdot F + \cdot F$	−163	184	−347

TABLE 26.3 Relative Energies of Structural Isomers (kJ/mol)

Reference Compound	Isomer	Hartree–Fock Limit	Experiment	Δ
Acetonitrile	Methyl isocyanide	88	88	0
Acetaldehyde	Oxirane	134	113	21
Acetic acid	Methyl formate	71	75	−4
Ethanol	Dimethyl ether	46	50	−4
Propyne	Allene	8	4	4
	Cyclopropene	117	92	25
Propene	Cyclopropane	42	29	13
1,3-Butadiene	2-Butyne	29	38	−9
	Cyclobutene	63	46	17
	Bicyclo[1.1.0]butane	138	109	29

(too negative). To generalize, because the products of a homolytic bond dissociation reaction will contain one fewer electron pair than the reactant, the products would be expected to have lower correlation energy. The correlation energy associated with an electron pair is greater than that for a separated pair of electrons.

The poor results seen for homolytic bond dissociation reactions do not necessarily carry over into other types of reactions as long as the total number of electron pairs is maintained. A good example is found in energy comparisons among structural isomers (see Table 26.3). Although bonding may be quite different in going from one isomer to another, for example, one single and one double bond in propene versus three single bonds in cyclopropane, the total number of bonds is the same in reactants and products:

$$CH_3CH=CH_2 \longrightarrow \overset{CH_2}{\underset{H_2C-CH_2}{\triangle}}$$

The errors noted here are an order of magnitude less than those found for homolytic bond dissociation reactions, although in some of the comparisons they are still quite large, in particular, where small-ring compounds are compared with (unsaturated) acyclics, for example, propene with cyclopropane.

The performance of limiting Hartree–Fock models for reactions involving even more subtle changes in bonding is better still. For example, the data in Table 26.4 show that calculated energies of protonation of nitrogen bases relative to the energy of

TABLE 26.4 Proton Affinities of Nitrogen Bases Relative to the Proton Affinity of Methylamine (kJ/mol)

Base	Hartree–Fock Limit	Experiment	Δ
Ammonia	−50	−38	−12
Aniline	−25	−10	−15
Methylamine	0	0	—
Dimethylamine	29	27	2
Pyridine	29	29	0
Trimethylamine	50	46	4
Diazabicyclooctane	75	60	15
Quinuclidine	92	75	17

protonation of methylamine as a standard, for example, pyridine relative to methylamine.

The results are typically in reasonable accord with their respective experimental values.

26.4.2 Equilibrium Geometries

Systematic discrepancies are also noted in comparisons involving limiting Hartree–Fock and experimental equilibrium geometries. Two comparisons are provided. The first (Table 26.5) involves the geometries of the hydrogen molecule, lithium hydride, methane, ammonia, water, and hydrogen fluoride, whereas the second (Table 26.6) involves AB bond distances in two-heavy-atom hydrides, H_mABH_n. Most evident is the fact that, aside from lithium hydride, all calculated bond distances are shorter than experimental values. In the case of bonds to hydrogen, the magnitude of the error increases with the electronegativity of the heavy atom. In the case of the two-heavy-atom hydrides, the error increases

TABLE 26.5 Structures of One-Heavy-Atom Hydrides (bond distances, Å; bond angles, °)

Molecule	Geometrical Parameter	Hartree–Fock Limit	Experiment	Δ
H_2	r(HH)	0.736	0.742	−0.006
LiH	r(LiH)	1.607	1.596	+0.011
CH_4	r(CH)	1.083	1.092	−0.009
NH_3	r(NH)	1.000	1.012	−0.012
	<(HNH)	107.9	106.7	−1.2
H_2O	r(OH)	0.943	0.958	−0.015
	<(HOH)	106.4	104.5	+1.9
HF	r(FH)	0.900	0.917	−0.017

TABLE 26.6 Bond Distances in Two Heavy Metal Hydrides (Å)

Molecule (Bond)	Hartree–Fock Limit	Experiment	Δ
Ethane (H_3C—CH_3)	1.527	1.531	−0.004
Methylamine (H_3C—NH_2)	1.453	1.471	−0.018
Methanol (H_3C—OH)	1.399	1.421	−0.022
Methyl fluoride (H_3C—F)	1.364	1.383	−0.019
Hydrazine (H_2N—NH_2)	1.412	1.449	−0.037
Hydrogen peroxide (HO—OH)	1.388	1.452	−0.064
Fluorine (F—F)	1.330	1.412	−0.082
Ethylene (H_2C=CH_2)	1.315	1.339	−0.024
Formaldimine (H_2C=NH)	1.247	1.273	−0.026
Formaldehyde (H_2C=O)	1.178	1.205	−0.027
Diimide (NH=NH)	1.209	1.252	−0.043
Oxygen (O=O)	1.158	1.208	−0.050
Acetylene (HC≡CH)	1.185	1.203	−0.018
Hydrogen cyanide (HC≡N)	1.124	1.153	−0.029
Nitrogen (N≡N)	1.067	1.098	−0.031

TABLE 26.7 Symmetric Stretching Frequencies in Diatomic and Small Polyatomic Molecules (cm^{-1})

Molecule	Hartree–Fock Limit	Experiment	Δ
Lithium fluoride	927	914	13
Fluorine	1224	923	301
Lithium hydride	1429	1406	23
Carbon monoxide	2431	2170	261
Nitrogen	2734	2360	374
Methane	3149	3137	12
Ammonia	3697	3506	193
Water	4142	3832	310
Hydrogen fluoride	4490	4139	351
Hydrogen	4589	4401	188

substantially when two electronegative elements are involved in the bond. Thus, although errors in bond distances for methylamine, methanol, and methyl fluoride are fairly small, those for hydrazine, hydrogen peroxide, and fluorine molecule are much larger.

The reason for this trend—limiting Hartree–Fock bond distances being shorter than experimental values—as well as the reason that lithium hydride is an exception will become evident when we examine how Hartree–Fock models can be extended to treat electron correlation in Section 26.6.

26.4.3 Vibrational Frequencies

A few comparisons of limiting Hartree–Fock and experimental symmetric stretching frequencies for diatomic and small polyatomic molecules are provided in Table 26.7. (Note that the experimentally measured frequencies have been corrected for anharmonic behavior before being compared with calculated **harmonic frequencies.**) The systematic error in equilibrium bond distances for limiting Hartree–Fock models (calculated distances are shorter than experimental lengths) seems to be paralleled by a systematic error in stretching frequencies (calculated frequencies are larger than experimental frequencies). This is not unreasonable: too short a bond implies too strong a bond, which translates to a frequency that is too large. Note, however, that homolytic bond dissociation energies from limiting Hartree–Fock models are actually smaller (not larger) than experimental values, an observation that might imply that frequencies should be smaller (not larger) than experimental values. The reason for the apparent contradiction is that the Hartree–Fock model does not dissociate to the proper limit of two radicals as a bond is stretched.

26.4.4 Dipole Moments

Electric dipole moments for a few simple molecules from limiting Hartree–Fock calculations are compared with experimental moments in Table 26.8. The calculations reproduce the overall ordering of dipole moments. Although the sample is too small

TABLE 26.8 Electric Dipole Moments (debyes)

Molecule	Hartree–Fock Limit	Experiment	Δ
Methylamine	1.5	1.31	0.2
Ammonia	1.7	1.47	0.2
Methanol	1.9	1.70	0.2
Hydrogen fluoride	2.0	1.82	0.2
Methyl fluoride	2.2	1.85	0.3
Water	2.2	1.85	0.3

to generalize, the calculated values are consistently larger than the corresponding experimental quantities. This might seem to be at odds with the notion that limiting Hartree–Fock bond lengths in these same molecules are smaller than experimental distances (which would imply dipole moments should be smaller than experimental values). We address this issue later in Section 26.8.8.

26.5 Theoretical Models and Theoretical Model Chemistry

As discussed in the preceding sections, limiting Hartree–Fock models do not provide results that are identical to experimental results. This is, of course, a direct consequence of the Hartree–Fock approximation, which replaces instantaneous interactions between individual electrons by interactions between a particular electron and the average field created by all other electrons. Because of this, electrons get in each other's way to a greater extent than they should. This leads to an overestimation of the electron–electron repulsion energy and too high a total energy.

At this point it is instructive to introduce the idea of a **theoretical model,** that is, a detailed recipe starting from the electronic Schrödinger equation and ending with a useful scheme, as well as the notion that any given theoretical model necessarily leads to a set of results, a **theoretical model chemistry.** At the outset, we might anticipate that the less severe the approximations that make up a particular theoretical model, the closer will be its results to experiment. The terms *theoretical model* and *theoretical model chemistry* were introduced by Sir John Pople, who in 1998 received the Nobel Prize in chemistry for his work in bringing quantum chemistry into widespread use.

All possible theoretical models may be viewed in the context of the two-dimensional diagram shown in Figure 26.7. The horizontal axis relates the extent to which the motions of electrons in a many-electron system are independent of each other or, alternately, the degree to which electron correlation is taken into account. At the extreme left are Hartree–Fock models. The vertical axis designates the basis set, which is used to represent the individual molecular orbitals. At the top is a so-called minimal basis set, which involves the fewest possible functions discussed in Section 26.7.1, while at the very bottom is the hypothetical complete basis set. The bottom of the column of Hartree–Fock models (at the far left) is called the Hartree–Fock limit.

Proceeding all the way to the right in Figure 26.7 (electron correlation fully taken into account) and then all the way to the bottom (both complete basis set and electron

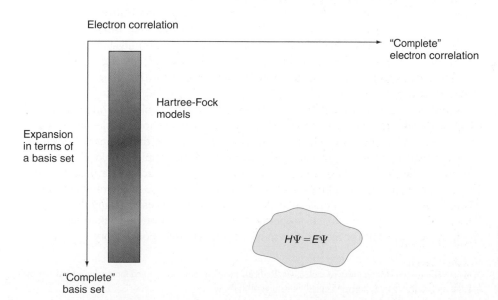

FIGURE 26.7
Different theoretical models can be classified by the degree to which electron correlation is taken into account and by the size of the basis set used.

correlation taken into account) on this diagram is functionally equivalent to solving the Schrödinger equation exactly—something that, as stated earlier, cannot be realized. Note, however, that starting from some position on the diagram, that is, some level of treatment of electron correlation and some basis set, if moving down and to the right produces no significant change in a particular property of interest, then we can reasonably conclude that further motion would also not result in change in this property. In effect, this would signal that the exact solution has been achieved.

To the extent that it is possible, any theoretical model should satisfy a number of conditions. Most importantly, it should yield a unique energy, among other molecular properties, given only the kinds and positions of the nuclei, the total number of electrons, and the number of unpaired electrons. A model should not appeal in any way to chemical intuition. Also important is that, if at all possible or practical, the magnitude of the error of the calculated energy should increase roughly in proportion to molecular size, that is, the model should be **size consistent.** Only then is it reasonable to anticipate that reaction energies can be properly described. Less important, but highly desirable, is that the model energy should represent a bound to the exact energy, that is, the model should be **variational.** Finally, a model needs to be practical, that is, able to be applied not only to very simple or idealized systems, but also to problems that are actually of interest. Were this not an issue, then it would not be necessary to move beyond the Schrödinger equation itself.

Hartree–Fock models, which have previously been discussed, are well defined and yield unique properties. They are both size consistent and variational. Most importantly, Hartree–Fock models are presently applicable to molecules comprising upward of 50 to 100 atoms. We have already seen that limiting Hartree–Fock models also provide excellent descriptions of a number of important chemical observables, most important among them, equilibrium geometry and the energies of some kinds of reactions. We shall see in Section 26.8 that "practical" Hartree–Fock models are also quite successful in similar situations.

26.6 Moving Beyond Hartree–Fock Theory

We next discuss improvements to the Hartree–Fock model that have the effect of moving down and to the right in Figure 26.7. Because these improvements increase the cost of a calculation, it is important to ask if they are necessary for a given calculation. This question must be answered by determining the extent to which the value of the observable of interest has the desired accuracy. Sections 26.8.1 through 26.8.11 explicitly address this question for a number of important observables, among them equilibrium geometries, reaction energies, and dipole moments.

Two fundamentally different approaches for moving beyond Hartree–Fock theory have received widespread attention. The first increases the flexibility of the Hartree–Fock wave function (associated with the electronic ground state) by combining it with wave functions corresponding to various excited states. The second introduces an explicit term in the Hamiltonian to account for the interdependence of electron motions.

Solution of the Roothaan-Hall equations results in a set of molecular orbitals, each of which is doubly occupied,[2] and a set of higher energy unoccupied molecular orbitals. The number of occupied molecular orbitals is equal to half of the number of electrons for closed-shell molecules, whereas the number of unoccupied molecular orbitals depends on the choice of basis set. Typically this number is much larger than the number of occupied molecular orbitals, and for the hypothetical case of a complete basis set it is infinite. The unoccupied molecular orbitals play no part in establishing the Hartree–Fock energy nor any ground-state properties obtained from Hartree–Fock models. They are, however, the basis for models that move beyond Hartree–Fock theory.

---

[2]This is valid for the vast majority of molecules. Radicals have one singly occupied molecular orbital and the oxygen molecule has two singly occupied molecular orbitals.

26.6.1 **Configuration Interaction Models**

It can be shown that in the limit of a complete basis set, the energy resulting from the optimum linear combination of the ground-state electronic configuration (that obtained from Hartree–Fock theory) and all possible excited-state electronic configurations formed by promotion of one or more electrons from occupied to unoccupied molecular orbitals is the same as would result from solution of the full many-electron Schrödinger equation. An example of such a promotion is shown in Figure 26.8.

This result, referred to as **full configuration interaction,** while interesting, is of no practical value simply because the number of excited-state electronic configurations is infinite. Practical configuration interaction models may be realized first by assuming a finite basis set and then by restricting the number of excited-state electronic configurations included in the mixture. Because of these two restrictions, the final energy is not the same as would result from solution of the exact Schrödinger equation. Operationally, what is required is first to obtain the Hartree–Fock wave function, and then to write a new wave function as a sum, the leading term of which, Ψ_0, is the Hartree–Fock wave function, and remaining terms, Ψ_s, are wave functions derived from the Hartree–Fock wave function by electron promotions:

$$\Psi = a_0\Psi_0 + \sum_{s>0} a_s\Psi_s \tag{26.29}$$

The unknown linear coefficients, a_s, are determined by solving Equation (26.30):

$$\sum_s (H_{st} - E\delta_{st})a_s = 0 \tag{26.30}$$

where the matrix elements are given by

$$H_{st} = \int \cdots \int \Psi_s \hat{H} \Psi_t \, d\tau_1 \, d\tau_2 \ldots d\tau_n \tag{26.31}$$

The lowest energy wave function obtained from solution of Equation (26.30) corresponds to the energy of the electronic ground state.

One approach for limiting the number of electron promotions is referred to as the **frozen core approximation.** In effect, this eliminates any promotions from molecular orbitals that correspond essentially to (combinations of) inner-shell or core electrons. Although the total contribution to the energy arising from inner-shell promotions is not insignificant, experience suggests that this contribution is nearly identical for the same types of atoms in different molecules. A more substantial approximation is to limit the number of promotions based on the total number of electrons involved, that is, **single-electron promotions, double-electron promotions,** and so on. Configuration interaction based on single-electron promotions only, the so-called **CIS method,** leads to no improvement of the (Hartree–Fock) energy or wave function. The simplest procedure to use that actually leads to improvement over Hartree–Fock is the so-called **CID method,** which is restricted to double-electron promotions:

$$\Psi_{CID} = a_0\Psi_0 + \overset{\substack{molecular\ orbitals\\occ}}{\sum_{i<j}} \overset{unocc}{\sum_{a<b}} a_{ij}^{ab}\Psi_{ij}^{ab} \tag{26.32}$$

A somewhat less restricted recipe, the so-called CISD method, considers both single- and double-electron promotions:

$$\Psi_{CISD} = a_0\Psi_0 + \overset{\substack{molecular\ orbitals\\occ\quad unocc}}{\sum_i \sum_a} a_i^a\Psi_i^a + \overset{\substack{molecular\ orbitals\\occ\quad unocc}}{\sum_{i<j} \sum_{a<b}} a_{ij}^{ab}\Psi_{ij}^{ab} \tag{26.33}$$

Solution of Equation (26.30) for either CID or CISD methods is practical for reasonably large systems. Both methods are obviously well defined and they are variational. However, neither method (or any limited configuration interaction method) is size consistent. This can easily be seen by considering the CISD description of a two-electron system, for example, a helium atom as shown in Figure 26.9, using just two basis

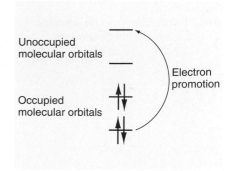

FIGURE 26.8
Electron promotion from occupied to unoccupied molecular orbitals.

FIGURE 26.9
The CISD description of He.

FIGURE 26.10
The CISD description of He$_2$ restricted to
one- and two-electron promotions.

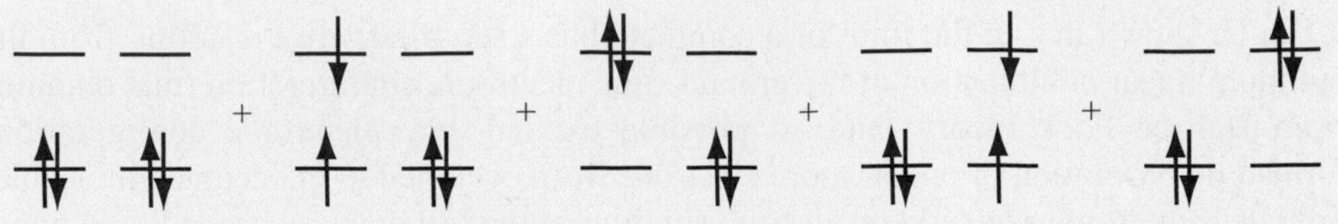

functions, which leads to one occupied and one unoccupied molecular orbital. In this case the CISD description for the isolated atom is exact (within the confines of the basis set), meaning that all possible electron promotions have been explicitly considered. Similarly, the description of two helium atoms treated independently is exact.

Next consider the corresponding CISD treatment of two helium atoms together but at infinite separation, as shown in Figure 26.10. This description is not exact because three- and four-electron promotions have not been taken into account. Thus, the calculated energies of two helium atoms treated separately and two helium atoms at infinite separation will be different. Size consistency is a very important attribute for any quantum chemical model, and its absence for any practical configuration interaction models makes them much less appealing than they otherwise might be.

26.6.2 Møller-Plesset Models

Practical size-consistent alternatives to configuration interaction models are **Møller-Plesset models,** in particular, the second-order Møller-Plesset model (MP2). Møller-Plesset models are based on the recognition that, while the Hartree–Fock wave function Ψ_0 and ground-state energy E_0 are approximate solutions to the Schrödinger equation, they are exact solutions to an analogous problem involving the Hartree–Fock Hamiltonian, $\hat{H}_0$, in place of the exact Hamiltonian, $\hat{H}$. Assuming that the Hartree–Fock wave function Ψ and energy are, in fact, very close to the exact wave function and ground-state energy E, the exact Hamiltonian can then be written in the following form:

$$\hat{H} = \hat{H}_0 + \lambda\hat{V} \tag{26.34}$$

In Equation (26.34), $\hat{V}$ is a small perturbation and λ is a dimensionless parameter. Using perturbation theory the exact wave function and energy are expanded in terms of the Hartree–Fock wave function and energy yields

$$E = E^{(0)} + \lambda E^{(1)} + \lambda^2 E^{(2)} + \lambda^3 E^{(3)} + \ldots \tag{26.35}$$

$$\Psi = \Psi_0 + \lambda\Psi^{(1)} + \lambda^2\Psi^{(2)} + \lambda^3\Psi^{(3)} + \ldots \tag{26.36}$$

MATHEMATICAL FORMULATION OF MØLLER-PLESSET MODELS
Substituting the expansions of Equations (26.34) to (26.36) into the Schrödinger equation and gathering terms in λ^n yields

$$\hat{H}_0\Psi_0 = E^{(0)}\Psi_0 \tag{26.37a}$$

$$\hat{H}_0\Psi^{(1)} + \hat{V}\Psi_0 = E^{(0)}\Psi^{(1)} + E^{(1)}\Psi_0 \tag{26.37b}$$

$$\hat{H}_0\Psi^{(2)} + \hat{V}\Psi^{(1)} = E^{(0)}\Psi^{(2)} + E^{(1)}\Psi^{(1)} + E^{(2)}\Psi_0 \tag{26.37c}$$

$$\ldots$$

Multiplying each of the Equations (26.37) by Ψ_0 and integrating over all space yields the following expression for the nth-order (MPn) energy:

$$E^{(0)} = \int \ldots \int \Psi_0\hat{H}_0\Psi_0 \, d\tau_1 \, d\tau_2 \ldots d\tau_n \tag{26.38a}$$

$$E^{(1)} = \int \ldots \int \Psi_0 \hat{V} \Psi_0 \, d\tau_1 \, d\tau_2 \ldots d\tau_n \qquad \text{(26.38b)}$$

$$E^{(2)} = \int \ldots \int \Psi_0 \hat{V} \Psi^{(1)} d\tau_1 \, d\tau_2 \ldots d\tau_n \qquad \text{(26.38c)}$$

$$\ldots$$

In this framework, the Hartree–Fock energy is the sum of the zero- and first-order Møller-Plesset energies:

$$E^{(0)} + E^{(1)} = \int \ldots \int \Psi_0 (\hat{H}_0 + \hat{V}) \Psi_0 \, d\tau_1 \, d\tau_2 \ldots d\tau_n \qquad \text{(26.39)}$$

The first correction, $E^{(2)}$ can be written as follows:

$$E^{(2)} = \sum_{i<j}^{\substack{\text{molecular orbitals} \\ occ}} \sum_{a<b}^{unocc} \frac{[(ij\|ab)]^2}{(\varepsilon_a + \varepsilon_b - \varepsilon_i - \varepsilon_j)} \qquad \text{(26.40)}$$

where ε_i and ε_j are energies of occupied molecular orbitals, and ε_a and ε_b are energies of unoccupied molecular orbitals. The integrals $(ij\|ab)$ over filled (i and j) and empty (a and b) molecular orbitals account for changes in electron–electron interactions as a result of electron promotion,

$$(ia\|jb) = -(ib|ja) \qquad \text{(26.41)}$$

in which the integrals $(ij|ab)$ and $(ib|ja)$ involve molecular orbitals rather than basis functions, for example,

$$(ia|jb) = \int \psi_i(1) \psi_a(1) \left[\frac{1}{r_{12}} \right] \psi_j(2) \psi_b(2) \, d\tau_1 \, d\tau_2 \qquad \text{(26.42)}$$

The two integrals are related by a simple transformation,

$$(ij|ab) = \sum_{\mu} \sum_{\nu} \sum_{\lambda}^{\substack{\text{basis functions}}} \sum_{\sigma} c_{\mu i} c_{\nu j} c_{\lambda a} c_{\sigma b} (\mu\nu|\lambda\sigma) \qquad \text{(26.43)}$$

where $(\mu\nu|\lambda\sigma)$ are given by Equation (26.28).

The MP2 model is well defined and leads to unique results. As mentioned previously, MP2 is size consistent, although (unlike configuration interaction models) it is not variational. Therefore, the calculated energy may be lower than the exact value.

26.6.3 Density Functional Models

The second approach for moving beyond the Hartree–Fock model is now commonly known as **density functional theory.** It is based on the availability of an exact solution for an idealized many-electron problem, specifically an electron gas of uniform density. The part of this solution that relates only to the exchange and correlation contributions is extracted and then directly incorporated into an SCF formalism much like Hartree–Fock formalism. Because the new exchange and correlation terms derive from idealized problems, density functional models, unlike configuration interaction and Møller-Plesset models, do not limit to the exact solution of the Schrödinger equation. In a sense, they are empirical in that they incorporate external data (the form of the solution of the idealized problem). What makes density functional models of great interest is their significantly lower computation cost than either configuration interaction or Møller-Plesset models. For his discovery, leading up to the development of practical density functional models, Walter Kohn was awarded the Nobel Prize in chemistry in 1998.

The Hartree–Fock energy may be written as a sum of the kinetic energy, E_T, the electron–nuclear potential energy, E_V, Coulomb, E_J, and exchange, E_K, components of the electron–electron interaction energy:

$$E^{HF} = E_T + E_V + E_J + E_K \tag{26.44}$$

The first three of these terms carry over directly to density functional models, whereas the Hartree–Fock exchange energy is replaced by a so-called exchange/correlation energy, E_{XC}, the form of which follows from the solution of the idealized electron gas problem:

$$E^{DFT} = E_T + E_V + E_J + E_{XC} \tag{26.45}$$

Except for E_T, all components depend on the total electron density, $\rho(\mathbf{r})$:

$$\rho(\mathbf{r}) = 2 \sum_i^{orbitals} \left| \psi_i(\mathbf{r}) \right|^2 \tag{26.46}$$

The ψ_i are orbitals, strictly analogous to molecular orbitals in Hartree–Fock theory.

MATHEMATICAL FORMULATION OF DENSITY FUNCTIONAL THEORY

Within a finite basis set (analogous to the LCAO approximation for Hartree–Fock models), the components of the density functional energy, E^{DFT}, can be written as follows:

$$E_T = \sum_\mu^{basis\ functions} \sum_\nu \int \phi_\mu(\mathbf{r}) \left[-\frac{\hbar^2 e^2}{2m_e} \nabla^2 \right] \phi_\nu(\mathbf{r})\, d\mathbf{r} \tag{26.47}$$

$$E_V = \sum_\mu^{basis\ functions} \sum_\nu P_{\mu\nu} \sum_A^{nuclei} \int \phi_\mu(\mathbf{r}) \left[-\frac{Z_A e^2}{4\pi\varepsilon_0 |\mathbf{r} - \mathbf{R}_A|} \right] \phi_\nu(\mathbf{r})\, d\mathbf{r} \tag{26.48}$$

$$E_J = \frac{1}{2} \sum_\mu \sum_\nu \sum_\lambda^{basis\ functions} \sum_\sigma P_{\mu\nu} P_{\lambda\sigma} (\mu\nu|\lambda\sigma) \tag{26.49}$$

$$E_{XC} = \int f(\rho(\mathbf{r}), \nabla\rho(\mathbf{r})\ldots)\, d\mathbf{r} \tag{26.50}$$

where Z is the nuclear charge, $|\mathbf{r} - \mathbf{R}_A|$ is the distance between the nucleus and the electron density, $\mathbf{P}$ is the density matrix [Equation (26.27)], and the $(\mu\nu|\lambda\sigma)$ are two-electron integrals [Equation (26.28)]. The $f(\rho(\mathbf{r}), \nabla\rho(\mathbf{r}), \ldots)$ is the so-called exchange/correlation functional, which depends on the electron density. In the simplest form of the theory, it is obtained by fitting the density resulting from the idealized electron gas problem to a function. Better models result from also fitting the gradient of the density. Minimizing E^{DFT} with respect to the unknown orbital coefficients yields a set of matrix equations, the Kohn-Sham equations, analogous to the Roothaan–Hall equations [Equation (26.22)]:

$$\mathbf{Fc} = \varepsilon\mathbf{Sc} \tag{26.51}$$

Here the elements of the Hartree–Fock matrix are given by

$$F_{\mu\nu} = H_{\mu\nu}^{core} + J_{\mu\nu} - F_{\mu\nu}^{XC} \tag{26.52}$$

and are defined analogously to Equations (26.25) and (26.26), respectively, and $\mathbf{F}^{XC}$ is the exchange/correlation part, the form of which depends on the particular exchange/correlation functional used. Note that substitution of the Hartree–Fock exchange, $\mathbf{K}$, for $\mathbf{F}^{XC}$ yields the Roothaan–Hall equations.

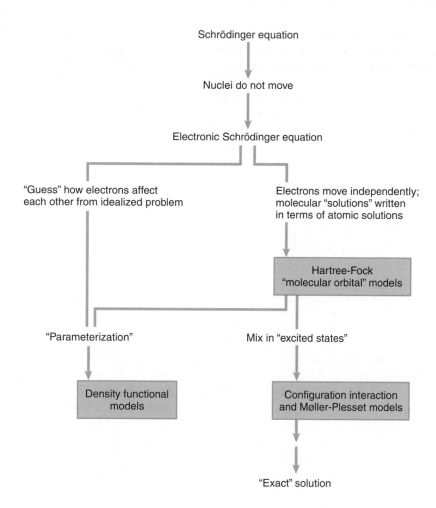

Density functional models are well defined and yield unique results. They are neither size consistent nor variational. Note that if the exact exchange/correlation functional had been known for the problem at hand (rather than only for the idealized many-electron gas problem), then the density functional approach would be exact. Although better forms of such functionals are constantly being developed, at present, there is no systematic way to improve the functional to achieve an arbitrary level of accuracy.

26.6.4 Overview of Quantum Chemical Models

An overview of quantum chemical models, starting with the Schrödinger equation, and including Hartree–Fock models, configuration interaction and Møller-Plesset models, and density functional models, is provided in Figure 26.11.

26.7 Gaussian Basis Sets

The LCAO approximation requires the use of a basis set made up of a finite number of well-defined functions centered on each atom. The obvious choice for the functions would be those corresponding closely to the exact solution of the hydrogen atom, that is, a polynomial in the Cartesian coordinates multiplying an exponential in r. However, the use of these functions was not cost effective, and early numerical calculations were carried out using nodeless **Slater-type orbitals** (STOs), defined by

$$\phi(r, \theta, \phi) = \frac{(2\zeta/a_0)^{n+1/2}}{[(2n)!]^{1/2}} r^{n-1} e^{-\zeta r/a_0} Y_l^m(\theta, \phi) \text{ß} \qquad (26.53)$$

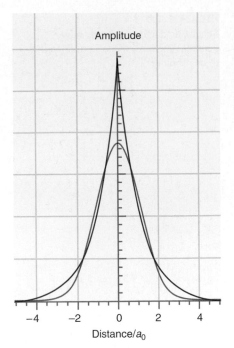

FIGURE 26.12

A cut through the hydrogen atom $1s$ AO (red curve) is compared with a single Gaussian function (purple curve). Note that the AO has a cusp at the nucleus, whereas the Gaussian function has zero slope at the nucleus. Note also that the Gaussian function falls off more rapidly with distance because of the r^2 dependence in the exponent.

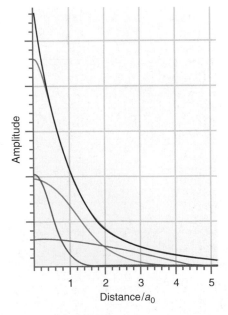

FIGURE 26.13

A hydrogen atom $1s$ AO (red curve) can be fit with the three Gaussian functions with different α values shown in green, or brown, and purple. Both the α values and the coefficients multiplying the Gaussian functions are optimized in the best fit function, shown in blue.

The symbols n, m, and l denote the usual quantum numbers and ζ is the effective nuclear charge. Use of these so-called Slater functions was entertained seriously in the years immediately following the introduction of the Roothaan–Hall equations, but soon abandoned because they lead to integrals that are difficult if not impossible to evaluate analytically. Further work showed that the cost of calculations can be further reduced if the AOs are expanded in terms of **Gaussian functions,** which have the form

$$g_{ijk}(r) = N\,x^i y^j z^k e^{-\alpha r^2} \qquad (26.54)$$

In this equation, x, y, and z are the position coordinates measured from the nucleus of an atom; i, j, and k are nonnegative integers, and α is an orbital exponent. An s-type function (zeroth order Gaussian) is generated by setting $i = j = k = 0$; a p-type function (first order Gaussian) is generated if one of i, j, and k is 1 and the remaining two are 0; and a d-type function (second order Gaussian) is generated by all combinations that give $i + j + k = 2$. Note that this recipe leads to six rather than five d-type functions, but appropriate combinations of these six functions give the usual five d-type functions and a sixth function that has s symmetry.

Gaussian functions lead to integrals that are easily evaluated. With the exception of so-called semi-empirical models, which do not actually entail evaluation of large numbers of difficult integrals, all practical quantum chemical models now make use of Gaussian functions.

Given the different radial dependence of STOs and Gaussian functions, it is not obvious at first glance that Gaussian functions are appropriate choices for AOs. Figure 26.12 shows a comparison of the two functional forms. The solution to this problem is to approximate the STO by a linear combination of Gaussian functions having different α values, rather than by a single Gaussian function. For example, a best fit to a $1s$-type STO using three Gaussians is shown in Figure 26.13. We can see that, although the region near the nucleus is not fit well, in the bonding region beyond 0.5 a_0, the fit is very good. The fit near the nucleus can be improved by using more Gaussian functions.

In practice, instead of taking individual Gaussian functions as members of the basis set, a normalized linear combination of Gaussian functions with fixed coefficients is constructed to provide a best fit to an AO. The value of each coefficient is optimized either by seeking minimum atom energies or by comparing calculated and experimental results for "representative" molecules. These linear combinations are called **contracted functions.** The contracted functions become the elements of the basis set. Although the coefficients in the contracted functions are fixed, the coefficients $c_{\mu i}$ in Equation (26.21) are variable and are optimized in the solution of the Schrödinger equation.

26.7.1 Minimal Basis Sets

Although there is no limit to the number of functions that can be placed on an atom, there is a minimum number. The minimum number is the number of functions required to hold all the electrons of the atom while still maintaining its overall spherical nature. This simplest representation or **minimal basis set** involves a single ($1s$) function for hydrogen and helium, a set of five functions ($1s$, $2s$, $2p_x$, $2p_y$, $2p_z$) for lithium to neon, and a set of nine functions ($1s$, $2s$, $2p_x$, $2p_y$, $2p_z$, $3s$, $3p_x$, $3p_y$, $3p_z$) for sodium to argon. Note that although $2p$ functions are not occupied in the lithium or beryllium atoms (and $3p$ functions are not occupied in the sodium or magnesium atoms), they are needed to provide proper descriptions of the bonding in molecular systems. For example, the bonding in a molecule such as lithium fluoride involves electron donation from a lone pair on fluorine to an appropriate (p-type) empty orbital on lithium (back bonding) as shown in Figure 26.14.

Of the minimal basis sets that have been devised, perhaps the most widely used and extensively documented is the **STO-3G basis set.** Here, each of the basis functions is expanded in terms of three Gaussian functions, where the values of the Gaussian exponents and the linear coefficient have been determined by least squares as best fits to Slater-type (exponential) functions.

The STO-3G basis set and all minimal basis sets have two obvious shortcomings: the first is that all basis functions are either themselves spherical or come in sets that, when taken together, describe a sphere. This means that atoms with spherical molecular environments or nearly spherical molecular environments will be better described than atoms with aspherical environments. This suggests that comparisons among different molecules will be biased in favor of those incorporating the most spherical atoms. The second shortcoming follows from the fact that basis functions are atom centered. This restricts their ability to describe electron distributions between nuclei, which are a critical element of chemical bonds. Minimal basis sets such as STO-3G are primarily of historical interest and have largely been replaced in practical calculations by split-valence basis sets and polarization basis sets, which have been formulated to address these two shortcomings. These basis sets are discussed in the following two subsections.

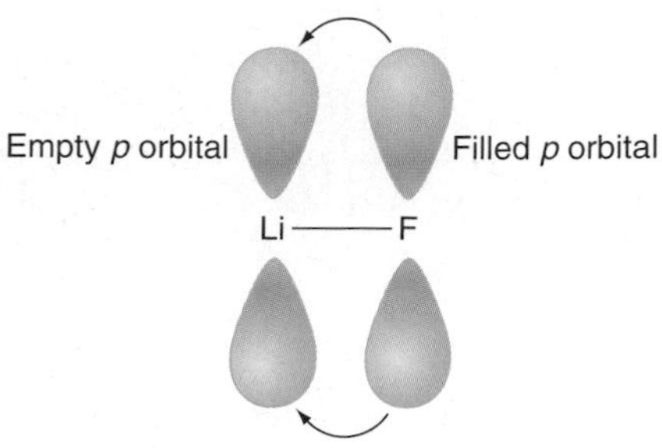

26.7.2 Split-Valence Basis Sets

The first shortcoming of a minimal basis set, namely, a bias toward atoms with spherical environments, can be addressed by providing two sets of valence basis functions: an inner set, which is more tightly held and an outer set, which is more loosely held. The iterative process leading to solution of the Roothaan–Hall equations adjusts the balance of the two parts independently for the three Cartesian directions, by adjusting the individual molecular orbital coefficients. For example, the proper linear combination to produce a molecular orbital suitable for σ bonding might involve a large coefficient (σ_{inner}) multiplying the inner basis function (in the σ direction) and a small coefficient (σ_{outer}) multiplying the outer basis function, whereas that to produce a molecular orbital suitable for π bonding might involve a small coefficient (π_{inner}) multiplying the inner basis function and a large coefficient (π_{outer}) multiplying the outer basis function as shown in Figure 26.15. The fact that the three Cartesian directions are treated independently of each other means that the atom (in the molecule) may be nonspherical.

A **split-valence basis set** represents core atomic orbitals by one set of functions and valence atomic orbitals by two sets of functions, $1s$, $2s^i$, $2p_x^i$, $2p_y^i$, $2p_z^i$, $2s^o$, $2p_x^o$, $2p_y^o$, $2p_z^o$ for lithium to neon and $1s$, $2s$, $2p_x$, $2p_y$, $2p_z$, $3s^i$, $3p_x^i$, $3p_y^i$, $3p_z^i$, $3s^o$, $3p_x^o$, $3p_y^o$, $3p_z^o$ for sodium to argon. Note that the valence $2s$ ($3s$) functions are also split into inner (superscript i) and outer (superscript o) components, and that hydrogen atoms are also represented by inner and outer valence ($1s$) functions. Among the simplest split-valence basis sets are 3-21G and 6-31G. Each core atomic orbital in the 3-21G basis set is expanded in terms of three Gaussians, whereas basis functions representing inner and outer components of valence atomic orbitals are expanded in terms of two and one Gaussians, respectively. The 6-31G basis sets are similarly constructed, with core orbitals represented in terms of six Gaussians and valence orbitals split into three and one Gaussian components. Expansion coefficients and Gaussian exponents for 3-21G and 6-31G basis sets have been determined by Hartree–Fock energy minimization on atomic ground states.

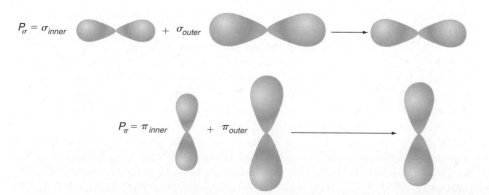

$P_\sigma = \sigma_{inner}$ $+$ σ_{outer} $\longrightarrow$

$P_\pi = \pi_{inner}$ $+$ π_{outer} $\longrightarrow$

FIGURE 26.14
$2p$ orbitals need to be included in the Li basis set to allow back bonding.

FIGURE 26.15
A split-valence basis set provides a way to allow the electron distribution about an atom to be nonspherical.

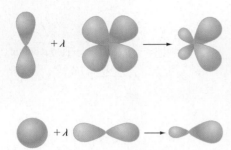

26.7.3 Polarization Basis Sets

The second shortcoming of a minimal (or split-valence) basis set, namely, that the basis functions are centered on atoms rather than between atoms, can be addressed by providing d-type functions on main-group elements (where the valence orbitals are of s and p type), and (optionally) p-type functions on hydrogen (where the valence orbital is of s type). This allows displacement of electron distributions away from the nuclear positions, as depicted in Figure 26.16.

The inclusion of **polarization functions** can be thought about either in terms of hybrid orbitals, for example, pd and sp hybrids, or alternatively in terms of a Taylor series expansion of a function (d functions are the first derivatives of p functions and p functions are the first derivatives of s functions). Although the first way of thinking is quite familiar to chemists (Pauling hybrids), the second offers the advantage of knowing what steps might be taken next to effect further improvement, that is, adding second and third derivatives.

Among the simplest polarization basis sets is 6-31G*, constructed from 6-31G by adding a set of d-type polarization functions written in terms of a single Gaussian for each heavy (non-hydrogen) atom. A set of six second-order Gaussians is added in the case of 6-31G*. Gaussian exponents for polarization functions have been chosen to give the lowest energies for representative molecules. Polarization of the s orbitals on hydrogen atoms is necessary for an accurate description of the bonding in many systems (particularly those in which hydrogen is a bridging atom). The 6-31G** basis set is identical to 6-31G*, except that it also provides three p-type polarization functions for hydrogen.

26.7.4 Basis Sets Incorporating Diffuse Functions

Calculations involving anions, for example, absolute acidity calculations, and calculations of molecules in excited states and of UV absorption spectra often pose special problems. This is because the highest energy electrons for such species may only be loosely associated with specific atoms (or pairs of atoms). In these situations, basis sets may need to be supplemented by **diffuse functions,** such as diffuse s- and p-type functions, on heavy (non-hydrogen) atoms (designated with a plus sign as in 6-31 + G* and 6-31 + G**). It may also be desirable to provide hydrogens with diffuse s-type functions (designated by two plus signs as in 6-31++G* and 6-31++G**.

26.8 Selection of a Theoretical Model

By now, it should be apparent to the reader that many different models are available and useful in describing molecular geometry, reaction energies, and other properties. All of these models ultimately stem from the electronic Schrödinger equation, and they differ from each other both in the manner in which they treat electron correlation and in the nature of the atomic basis set. Each distinct combination (a theoretical model) leads to a scheme with its own particular characteristics (a theoretical model chemistry).

Hartree–Fock models may be seen as the parent model in that they treat electron correlation in the simplest possible manner, in effect, replacing instantaneous electron–electron interactions with average interactions. Despite their simplicity, Hartree–Fock models have proven to be remarkably successful in a large number of situations and remain a mainstay of computational chemistry.

As discussed earlier, **correlated models** can be broadly divided into two categories: density functional models, which provide an explicit empirical term in the Hamiltonian to account for electron correlation, and configuration interaction and Møller-Plesset models, which start from the Hartree–Fock description and then optimally mix together wave functions corresponding to the ground and various excited states. Each of these models exhibits its own particular characteristics.

Of course, no single theoretical model is likely to be ideal for all applications. A great deal of effort has gone into defining the limits of different models and judging the degree of success and the pitfalls of each. Most simply, success depends on the ability of a model to consistently reproduce known (experimental) data. This assumes that reliable experimental data are available or, at least, that errors in the data have been quantified. These include data on the geometries and conformations (shapes) of stable molecules, the enthalpies of chemical reactions (thermodynamics), and on such properties as vibrational frequencies (infrared spectra) and dipole moments. Quantum mechanical models may also be applied to high-energy molecules (reactive intermediates) for which reliable experimental data may be difficult to come by, and to reaction transition states, which may not even be directly observed, much less characterized. Although no experimental transition-state structures are available with which to compare the results of the calculations, experimental kinetic data may be interpreted to provide information about activation energies. As an alternative, transition-state geometries can instead be compared with the results of high-level quantum chemical calculations.

The success of a quantum chemical model is not an absolute. Different properties and certainly different problems may require different levels of confidence to actually be of value. Neither is success sufficient. A model also needs to be practical for the task at hand. The nature and size of the system needs to be taken into account, as do the available computational resources and the experience and patience of the practitioner. Practical models usually do share one feature in common, in that they are not likely to be the best possible treatments to have been formulated. Compromise is almost always an essential component of model selection.

Oddly enough, the main problem faced by those who wish to apply computation to investigate chemistry is not the lack of suitable models but rather the excess of models. Quite simply, there are too many choices. In this spirit, consideration from this point on will be limited to just four theoretical models: Hartree–Fock models with 3-21G split-valence and 6-31G* polarization basis sets, the B3LYP/6-31G* density functional model, and the MP2/6-31G* model. Although all of these models can be routinely applied to molecules of considerable size, they differ by two orders of magnitude in the amount of computer time they require. Thus, it is quite important to know where the less time-consuming models perform satisfactorily and where the more time-consuming models are needed. Note that although this set of models has been successfully applied to a wide range of chemical problems, some problems may require more accurate and more time-consuming models.

It is difficult to quantify the overall computation time of a calculation, because it depends not only on the specific system and task at hand, but also on the sophistication of the computer program and the experience of the user. For molecules of moderate size (say, 10 atoms other than H), the HF/6-31G*, B3LYP/6-31G*, and MP2/6-31G* models would be expected to exhibit overall computation times in a ratio of roughly 1:1.5:10. The HF/3-21G model will require a third to half of the computation time required by the corresponding HF/6-31G* model, whereas the computation time of Hartree–Fock, B3LYP, and MP2 models with basis sets larger than 6-31G* will increase roughly as the cube (HF and B3LYP) and the fifth power (MP2) of the total number of basis functions. Geometry optimizations and frequency calculations are typically an order of magnitude more time-consuming than energy calculations, and the ratio will increase with increasing complexity (number of independent geometrical variables) of the system. Transition-state geometry optimizations are likely to be even more time-consuming than equilibrium geometry optimizations, due primarily to a poorer initial guess of the geometry.

Only a few calculated properties are examined in this discussion: equilibrium bond distances, reaction energies, conformational energy differences, and dipole moments. Comparisons between the results of the calculations and experimental data are few for each of these, but sufficient to establish meaningful trends.

26.8.1 Equilibrium Bond Distances

A comparison of calculated and experimentally determined carbon–carbon bond distances in hydrocarbons is provided in Table 26.9. Whereas errors in measured bond distances are typically on the order of ± 0.02 Å, experimental data for hydrocarbons and other small molecules presented here are better, and comparisons with the results of calculations to 0.01 Å are meaningful. In terms of mean absolute error, all four models perform admirably. The B3LYP/6-31G* and MP2/6-31G* models perform better than the two Hartree–Fock models, due in most part to a sizable systematic error in carbon–carbon double bond lengths. With one exception, Hartree–Fock double bond lengths are shorter than

		Hartree–Fock		B3LYP	MP2	
TABLE 26.9 Bond Distances in Hydrocarbons (Å)						
Bond	Hydrocarbon	3-21G	6-31G*	6-31G*	6-31G*	Experiment
C—C	But-1-yne-3-ene	1.432	1.439	1.424	1.429	1.431
	Propyne	1.466	1.468	1.461	1.463	1.459
	1,3-Butadiene	1.479	1.467	1.458	1.458	1.483
	Propene	1.510	1.503	1.502	1.499	1.501
	Cyclopropane	1.513	1.497	1.509	1.504	1.510
	Propane	1.541	1.528	1.532	1.526	1.526
	Cyclobutane	1.543	1.548	1.553	1.545	1.548
C=C	Cyclopropene	1.282	1.276	1.295	1.303	1.300
	Allene	1.292	1.296	1.307	1.313	1.308
	Propene	1.316	1.318	1.333	1.338	1.318
	Cyclobutene	1.326	1.322	1.341	1.347	1.332
	But-1-yne-ene	1.320	1.322	1.341	1.344	1.341
	1,3-Butadiene	1.320	1.323	1.340	1.344	1.345
	Cyclopentadiene	1.329	1.329	1.349	1.354	1.345
Mean absolute error		0.011	0.011	0.006	0.007	—

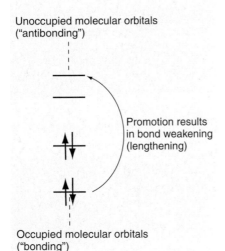

Unoccupied molecular orbitals
("antibonding")

Promotion results
in bond weakening
(lengthening)

Occupied molecular orbitals
("bonding")

FIGURE 26.17
The promotion of electrons to unfilled orbitals reduces the bond strength and leads to bond lengthening.

experimental distances. This is easily rationalized. Treatment of electron correlation (for example, in the MP2 model) involves the promotion of electrons from occupied molecular orbitals (in the Hartree–Fock wave function) to unoccupied molecular orbitals. Because occupied molecular orbitals are (generally) net bonding in character, and because unoccupied molecular orbitals are (generally) net antibonding in character, any promotions should result in bond weakening (lengthening) as illustrated in Figure 26.17. This in turn suggests that bond lengths from limiting Hartree–Fock models are necessarily shorter than exact values. Apparently, Hartree–Fock models with 3-21G and 6-31G* basis set are close enough to the limit for this behavior to be seen.

Consistent with such an interpretation, B3LYP/6-31G* and MP2/6-31G* double bond lengths do not show a systematic trend and are both smaller and larger than experimental values.

Similar comments can be made regarding CN and CO bond distances (Table 26.10). In terms of mean absolute error, the performance of the B3LYP/6-31G* and MP2/6-31G* models is similar to that previously noted for CC bonds in hydrocarbons, but the two Hartree–Fock models do not fare as well. Note that although bond distances from the HF/6-31G* model are constantly smaller than measured values, in accord with the picture presented for hydrocarbons, HF/3-21G bond lengths do not show such a trend. It appears that the 3-21G basis set is not large enough to closely mirror the Hartree–Fock limit in this instance. Most bond distances obtained from the B3LYP/6-31G* and MP2/6-31G* models are actually slightly larger than experimental distances. (The CN bond length in formamide is the only significant exception.) Bond lengthening from the corresponding (6-31G* basis set) Hartree–Fock model is a direct consequence of treatment of electron correlation.

In summary, all four models provide a plausible account of equilibrium bond lengths. Similar comments also apply to bond angles and more generally to the structures of larger molecules.

26.8.2 Finding Equilibrium Geometries

As detailed at the start of this chapter, an equilibrium structure is a point on a multidimensional potential energy surface for which all first energy derivatives with respect to the individual geometrical coordinates are zero, and for which the diagonal representation of the matrix of second energy derivatives has all positive elements. In simple terms, an equilibrium structure corresponds to the bottom of a well on the overall potential energy surface.

TABLE 26.10	Bond Distances in Molecules with Heteroatoms (Å)					
		Hartree–Fock		B3LYP	MP2	
Bond	Hydrocarbon	3-21G	6-31G*	6-31G*	6-31G*	Experiment
C—N	Formamide	1.351	1.349	1.362	1.362	1.376
	Methyl isocyanide	1.432	1.421	1.420	1.426	1.424
	Trimethylamine	1.471	1.445	1.456	1.455	1.451
	Aziridine	1.490	1.448	1.473	1.474	1.475
	Nitromethane	1.497	1.481	1.499	1.488	1.489
C—O	Formic acid	1.350	1.323	1.347	1.352	1.343
	Furan	1.377	1.344	1.364	1.367	1.362
	Dimethyl ether	1.435	1.392	1.410	1.416	1.410
	Oxirane	1.470	1.401	1.430	1.439	1.436
Mean absolute error		0.017	0.018	0.005	0.005	—

Not all equilibrium structures correspond to (kinetically) stable molecules, meaning that not all equilibrium structures will correspond to detectable (let alone characterizable) molecules. Stability also implies that the well is deep enough to preclude the molecule from being transformed into other molecules. Equilibrium structures that no doubt exist but cannot be detected easily are commonly referred to as **reactive intermediates.**

Geometry optimization does not guarantee that the final geometry will have a lower energy than any other geometry of the same molecular formula. All that it guarantees is that the geometry will correspond to a local minimum, that is, a geometry the energy of which is lower than that of any similar geometry. However, the resulting structure may still not be the lowest energy structure possible for the molecule. Other local minima that are actually lower in energy may exist and be accessible via low-energy rotations about single bonds or puckering of rings (see Figure 26.1). The full collection of local minima are referred to as **conformers.** Finding the lowest energy conformer or global minimum requires repeated geometry optimization starting with different initial geometries as discussed in Section 26.8.6.

Finding an equilibrium structure is not as difficult a chore as it might first appear. For one, chemists know a great deal about what molecules look like and can usually provide an excellent starting structure. Also, optimization to a minimum is an important task in many fields of science and engineering, and very good algorithms exist with which to accomplish it.

Geometry optimization is an iterative process. The energy and energy gradient (first derivatives with respect to all geometrical coordinates) are calculated for the initial geometry, and this information is then used to project a new geometry. This process needs to continue until the lowest energy or optimized geometry is reached. Three criteria must be satisfied before a geometry is accepted as optimized. First, successive geometry changes must not lower the energy by more than a specified (small) value. Second, the energy gradient must closely approach zero. Third, successive iterations must not change any geometrical parameter by more than a specified (small) value.

In principle, geometry optimization carried out in the absence of symmetry must result in a local energy minimum. On the other hand, the imposition of symmetry may result in a geometry that is not an energy minimum. The most conservative tactic is always to optimize geometry in the absence of symmetry. If this is not practical, and if there is any doubt whatsoever that the symmetrical structure actually corresponds to an energy minimum, then it is always possible to verify that the geometry located indeed corresponds to a local minimum by calculating vibrational frequencies for the final (optimized) geometry. These should all be real numbers. The presence of an imaginary frequency indicates that the corresponding coordinate is not an energy minimum.

Problems P26.5–P26.12

TABLE 26.11 Homolytic Bond Dissociation Energies (kJ/mol)

Bond Dissociation Reaction	Hartree–Fock 3-21G	Hartree–Fock 6-31G*	B3LYP 6-31G*	MP2 6-31G*	Experiment
$CH_3—CH_3 \rightarrow \cdot CH_3 + \cdot CH_3$	285	293	406	414	406
$CH_3—NH_2 \rightarrow \cdot CH_3 + \cdot NH_2$	247	243	372	385	389
$CH_3—OH \rightarrow \cdot CH_3 + \cdot OH$	222	247	402	410	410
$CH_3—F \rightarrow \cdot CH_3 + \cdot F$	247	289	473	473	477
$NH_2—NH_2 \rightarrow \cdot NH_2 + \cdot NH_2$	155	142	293	305	305
$HO—OH \rightarrow \cdot OH + \cdot OH$	13	0	226	230	230
$F—F \rightarrow \cdot F + \cdot F$	−121	−138	176	159	159
Mean absolute error	190	186	9	2	—

26.8.3 Reaction Energies

Reaction energy comparisons are divided into three parts: bond dissociation energies, energies of reactions relating structural isomers, and relative proton affinities. Bond dissociation reactions are the most disruptive, because they lead to a change in the number of electron pairs. Structural isomer comparisons maintain overall electron pair count, but swap bonds of one kind for those of another. Relative proton affinity comparisons are least disruptive in that they maintain the numbers of each kind of formal chemical bond and lead only to subtle changes in the molecular environment.

A comparison of homolytic bond dissociation energies based on calculation and on experimental thermochemical data is provided in Table 26.11. Hartree–Fock models with the 3-21G and 6-31G* basis set turn in a very poor performance, paralleling the poor performance of limiting Hartree–Fock models (see discussion in Section 26.4.1). Bond energies are far too small, consistent with the fact that the total correlation energy for the radical products is smaller than that for the reactant due to a decrease in the number of electron pairs. B3LYP/6-31G* and especially MP2/6-31G* models fare much better (results for the latter are well inside the experimental error bars).

"Which of several possible structural isomers is most stable?" and "What are the relative energies of any reasonable alternatives?" are without doubt two of the most commonly asked questions relating to thermochemistry. The ability to pick out the lowest energy isomer and at least rank the energies of higher energy isomers is essential to the success of any model. A few comparisons of this kind are found in Table 26.12.

TABLE 26.12 Relative Energies Isomer - Reference Compound of Structural Isomers (kJ/mol)

Reference Compound	Isomer	Hartree–Fock 3-21G	Hartree–Fock 6-31G*	B3LYP 6-31G*	MP2 6-31G*	Experiment
Acetonitrile	Methyl isocyanide	88	100	113	121	88
Acetaldehyde	Oxirane	142	130	117	113	113
Acetic acid	Methyl formate	54	54	50	59	75
Ethanol	Dimethyl ether	25	29	21	38	50
Propyne	Allene	13	8	−13	21	4
	Cyclopropene	167	109	92	96	92
Propene	Cyclopropane	59	33	33	17	29
1,3-butadiene	2-Butyne	17	29	33	17	38
	Cyclobutane	75	54	50	33	46
	Bicyclo [1.1.0] butane	192	126	117	88	109
Mean absolute error		32	13	12	15	—

TABLE 26.13	Proton Affinities of Nitrogen Bases Relative to the Proton Affinity of Methylamine (kJ/mol)				
	Hartree–Fock		B3LYP	MP2	
Base	3-21G	6-31G*	6-31G*	6-31G*	Experiment
Ammonia	−42	−46	−42	−42	−38
Aniline	−38	−17	−21	−13	−10
Methylamine	0	0	0	0	0
Dimethylamine	29	29	25	25	27
Pyridine	17	29	25	13	29
Trimethylamine	46	46	38	38	46
Diazabicyclooctane	67	71	59	54	60
Quinuclidine	79	84	75	71	75
Mean absolute error	8	5	4	6	—

In terms of mean absolute error, three of the four models provide similar results. The HF/3-21G model is inferior. None of the models is actually up to the standard that would make it a useful reliable replacement for experimental data (<5 kJ/mol). More detailed comparisons provide insight. For example, Hartree–Fock models consistently disfavor small-ring cyclic structures over their unsaturated cyclic isomers whereas neither the B3LYP/6-31G* nor the MP2/6-31G* methods show a consistent preference.

The final comparison (Table 26.13) is between proton affinities of a variety of nitrogen bases and that of methylamine as a standard, that is,

$$BH^+ + NH_3 \longrightarrow B + NH_4^+$$

This type of comparison is important not only because proton affinity (basicity) is an important property in its own right, but also because it typifies property comparisons among sets of closely related compounds. The experimental data derive from equilibrium measurements in the gas phase and are accurate to ± 4 kJ/mol. In terms of mean absolute error, all four models turn in similar and respectable accounts over what is a considerable range (>100 kJ/mol) of experimental proton affinities. The HF/3-21G model is clearly the poorest performer, due primarily to underestimation of the proton affinities of aniline and pyridine.

Problems P26.13–P26.16

26.8.4 Energies, Enthalpies, and Gibbs Energies

Quantum chemical calculations account for reaction thermochemistry by combining the energies of reactant and product molecules at 0 K. Additionally, the residual energy of vibration (the so-called zero point energy discussed in Section 18.1) is ignored. On the other hand, experimental thermochemical comparisons are most commonly based on enthalpies or Gibbs energies of 1 mol of real (vibrating) molecules at some finite temperature (typically 298.15 K). The connection between the various quantities involves the mass, equilibrium geometry, and set of vibrational frequencies for each of the molecules in the reaction. Calculating thermodynamic quantities is straightforward but, because it requires frequencies, consumes significant computation time, and is performed only where necessary.

We start with two familiar thermodynamic relationships:

$$\Delta G = \Delta H - T\Delta S \qquad (26.55)$$

$$\Delta H = \Delta U + \Delta(PV) \approx \Delta U \qquad (26.56)$$

where G is the Gibbs energy, H is the enthalpy, S is the entropy, U is the internal energy, and T, P, and V are the temperature, pressure, and volume, respectively.

For most cases, the $\Delta(PV)$ term can be ignored, meaning that the $\Delta U = \Delta H$ at 0 K. Three steps are required to obtain ΔG, the first two to relate the quantum mechanical energy at 0 K to the internal energy at 298 K, and the third to calculate the Gibbs energy.

1. *Correction of the internal energy for finite temperature.* The change in internal energy from 0 K to a finite temperature, T, $\Delta U(T)$, is given by

$$\Delta U(T) = \Delta U_{trans}(T) + \Delta U_{rot}(T) + \Delta U_{vib}(T) \tag{26.57}$$

$$\Delta U_{trans}(T) = \frac{3}{2}RT \tag{26.58}$$

$$\Delta U_{rot}(T) = \frac{3}{2}RT \ (RT \text{ for a linear molecule}) \tag{26.59}$$

$$\Delta U_{vib}(T) = U_{vib}(T) - U_{vib}(0K) = N_A \sum_i^{\nu_i} \frac{h\nu_i}{e^{h\nu_i/kT} - 1} \tag{26.60}$$

The ν_i are vibrational frequencies, N_A is Avogadro's number, and R, k, and h are the gas constant, the Boltzmann constant, and the Planck constant, respectively.

2. *Correction for zero point vibrational energy.* The zero point vibrational energy, $U_{vib}(0)$, of n moles of a molecule at 0 K is given by

$$U_{vib}(0) = nN_A E_{zero\ point} = \frac{1}{2}nN_A \sum_i^{\nu_i} h\nu_i \tag{26.61}$$

where N_A is Avogadro's number. This calculation also requires knowledge of the vibrational frequencies.

3. *Entropy.* The absolute entropy, S, of n moles of a molecule may be written as a sum of terms:

$$S = S_{trans} + S_{rot} + S_{vib} + S_{el} - nR[\ln(nN_A) - 1] \tag{26.62}$$

$$S_{trans} = nR\left[\frac{3}{2} + \ln\left(\left(\frac{nRT}{P}\right)\left(\frac{2\pi mkT}{h^2}\right)^{3/2}\right)\right] \tag{26.63}$$

$$S_{rot} = nR\left[\frac{3}{2} + \ln\left(\left(\frac{\sqrt{\pi}}{\sigma}\right)\left(\frac{kT}{hcB_A}\right)^{1/2}\left(\frac{kT}{hcB_B}\right)^{1/2}\left(\frac{kT}{hcB_C}\right)^{1/2}\right)\right] \tag{26.64}$$

$$S_{vib} = nR\sum_i^{\nu_i}\left[\left(\frac{\mu_i}{e^{\mu_i} - 1}\right) + \ln\left(\frac{1}{1 - e^{-\mu_i}}\right)\right] \tag{26.65}$$

$$S_{el} = nR \ln g_0 \tag{26.66}$$

In these equations, m is the molecular mass, B_i is the rotational constant, σ is the symmetry number, $\mu_i = h\nu_i/kT$, c is the speed of light, and g_0 is the degeneracy of the electronic ground state (normally equal to one).

Note that molecular structure enters into the rotational entropy through B, and the vibrational frequencies enter into the vibrational entropy. The translational entropy cancels in a (mass) balanced reaction, and the electronic entropy is usually zero because for most molecules $g_0 = 1$. Note also that the expression provided for the vibrational contribution to the entropy goes to infinity as the vibrational frequency goes to zero. This is clearly wrong and has its origin in the use of the linear harmonic oscillator approximation to derive the expression. Unfortunately, low-frequency modes are the major contributors to the vibrational entropy, and caution must be exercised when using the preceding formulas for the case of frequencies below approximately 300 cm^{-1}. In this case, the molecular partition function must be evaluated term by term rather than assuming the classical limit.

Problem P26.17

26.8.5 Conformational Energy Differences

Rotation around single bonds may give rise to rotational isomers (conformers). Because bond rotation is almost always a very low energy process, this means that more than one conformer may be present at equilibrium. For example, *n*-butane exists as a mixture of *anti* and *gauche* conformers, as shown in Figures 26.1 and 26.18. The same reasoning carries over to molecules incorporating flexible rings, where conformer interconversion may be viewed in terms of a process involving restricted rotation about the bonds in the ring.

Knowledge of the conformer of lowest energy and, more generally, the distribution of conformers is important because many molecular properties depend on detailed molecular shape. For example, whereas *gauche n*-butane is a polar molecule (albeit very weakly polar), *anti n*-butane is nonpolar, and the value of the dipole moment for an actual sample of *n*-butane would depend on how much of each species was actually present.

Experimentally, a great deal is known about the conformational preferences of molecules in the solid state (from X-ray crystallography). Far less is known about the conformations of isolated (gas-phase) molecules, although there are sufficient data to allow gross assessment of practical quantum chemical models. Experimental conformational energy differences are somewhat more scarce, but accurate data are available for a few very simple (two-conformer) systems. Comparison of these data with the results of calculations for hydrocarbons is provided in Table 26.14. These are expressed in terms of the energy of the high-energy conformer relative to that of the low-energy conformer.

All models correctly assign the ground-state conformer in all molecules. In terms of mean absolute error, the MP2/6-31G* model provides the best description of conformational energy differences and the HF/6-31G* model the worst description. Hartree–Fock models consistently overestimate differences (the sole exception is for the *trans/gauche* energy difference in 1,3-butadiene from the 3-21G model), in some cases by large amounts (nearly 5 kJ/mol for the *equatorial/axial* energy difference in *tert*-butylcyclohexane from the 3-21G model). Correlated models also typically (but not always) overestimate energy differences, but the magnitudes of the errors are much smaller than those seen for Hartree–Fock models.

26.8.6 Determining Molecular Shape

Many molecules can (and do) exist in more than one shape, arising from different arrangements around single bonds and/or flexible rings. The problem of identifying the lowest energy conformer (or the complete set of conformers) in simple molecules such as *n*-butane and cyclohexane is straightforward, but rapidly becomes difficult as the number of conformational degrees of freedom increases, due simply to the large number of arrangements that need to be examined. For example, a systematic search on a molecule with N single bonds and step size of $360°/M$, would need to examine M^N

FIGURE 26.18
Structures of two *n*-butane conformers.

Anti n-butane   *Gauche n*-butane

Problems P26.18–P26.20

TABLE 26.14   Conformational Energy in Hydrocarbons (kJ/mol)

Hydrocarbon	Low-Energy/ High-Energy Conformer	Hartree–Fock		B3LYP	MP2	
		3-21G	6-31G*	6-31G*	6-31G*	Experiment
n-Butane	*anti/gauche*	3.3	4.2	3.3	2.9	2.80
1-Butene	*skew/cis*	3.3	2.9	1.7	2.1	0.92
1,3-Butadiene	*trans/gauche*	11.3	13.0	15.1	10.9	12.1
Cyclohexane	*chair/twist-boat*	27.2	28.5	26.8	27.6	19.7–25.9
Methylcyclohexane	*equatorial/axial*	7.9	9.6	8.8	7.9	7.32
tert-Butylcyclohexane	*equatorial/axial*	27.2	25.5	22.2	23.4	22.6
cis-1,3-Dimethylcyclohexane	*equatorial/axial*	26.4	27.2	25.1	23.8	23.0
Mean absolute error		1.9	2.3	1.3	0.9	—

FIGURE 26.19
Inversion of NH_3 leads to its mirror image.

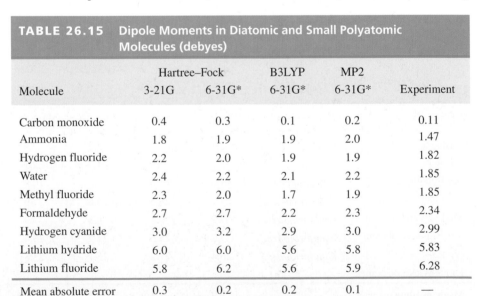

FIGURE 26.20
Pseudorotation leads to exchange of *equatorial* and *axial* positions at a trigonal bipyramidal phosphorus center.

conformers. For a molecule with three single bonds and a step size of 120° ($M = 3$), this leads to 27 conformers; for a molecule with eight single bonds, more than 6500 conformers would need to be considered. It is clear that it will not always be possible to look everywhere, and sampling techniques will need to replace systematic procedures for complex molecules. The most common of these are so-called Monte Carlo methods (which randomly sample different conformations) and molecular dynamics techniques (which follow motion among different conformers in time).

26.8.7 Alternatives to Bond Rotation

Single-bond rotation (including restricted bond rotation in flexible rings) is the most common mechanism for conformer interconversion, but it is by no means the only mechanism. At least two other processes are known: inversion and pseudorotation. Inversion is normally associated with pyramidal nitrogen or phosphorus and involves a planar (or nearly planar) transition state, for example, in ammonia, as shown in Figure 26.19. Note that the starting and ending molecules are mirror images. Were the nitrogen to be bonded to three different groups and were the nitrogen lone pair to be counted as a fourth group, inversion would result in a change in chirality at this center. Pseudorotation, which is depicted in Figure 26.20, is normally associated with trigonal bipyramidal phosphorus and involves a square-based-pyramidal transition state. Note that pseudorotation interconverts *equatorial* and *axial* positions on phosphorus.

Both inversion of pyramidal nitrogen and pseudorotation around trigonal bipyramidal phosphorus are very low energy processes ($<20-30$ kJ/mol) and generally proceed rapidly at 298 K. On the other hand, inversion of pyramidal phosphorus is more difficult (100 kJ/mol) and is inhibited at 298 K.

Problem P26.21

26.8.8 Dipole Moments

Calculated dipole moments for a selection of diatomic and small polyatomic molecules are compared with experimental values in Table 26.15. The experimental data cover a wide spectrum of molecules, from carbon monoxide, which is close to being

	Hartree–Fock		B3LYP	MP2	
TABLE 26.15 Dipole Moments in Diatomic and Small Polyatomic Molecules (debyes)					
Molecule	3-21G	6-31G*	6-31G*	6-31G*	Experiment
Carbon monoxide	0.4	0.3	0.1	0.2	0.11
Ammonia	1.8	1.9	1.9	2.0	1.47
Hydrogen fluoride	2.2	2.0	1.9	1.9	1.82
Water	2.4	2.2	2.1	2.2	1.85
Methyl fluoride	2.3	2.0	1.7	1.9	1.85
Formaldehyde	2.7	2.7	2.2	2.3	2.34
Hydrogen cyanide	3.0	3.2	2.9	3.0	2.99
Lithium hydride	6.0	6.0	5.6	5.8	5.83
Lithium fluoride	5.8	6.2	5.6	5.9	6.28
Mean absolute error	0.3	0.2	0.2	0.1	—

nonpolar, to lithium fluoride, which is close to being fully ionic. All models provide a good overall account of this range. In terms of mean absolute error, the HF/3-21G model fares worst and the MP2/6-31G* model fares best, but the differences are not large. Note that dipole moments from the two Hartree–Fock models are consistently larger than experimental values, the only exception being for lithium fluoride. This is in accord with the behavior of the limiting Hartree–Fock model (see discussion earlier in Section 26.4.4) and may now easily be rationalized. Recognize that electron promotion from occupied to unoccupied molecular orbitals (either implicit or explicit in all electron correlation models) takes electrons from "where they are" (negative regions) to "where they are not" (positive regions), as illustrated in Figure 26.21. In formaldehyde, for example, the lowest energy promotion is from a nonbonded lone pair localized on oxygen into a π^* orbital principally concentrated on carbon. As a result, electron correlation acts to reduce overall charge separation and to reduce the dipole moment in comparison with the Hartree–Fock value. This is supported by the fact that dipole moments from correlated (B3LYP/6-31G* and MP2/6-31G*) calculations are not consistently larger than experimental values.

26.8.9 Atomic Charges: Real or Make Believe?

Charges are part of the everyday language of chemistry and, aside from geometries and energies, are certainly the most commonly demanded quantities from quantum chemical calculations. Charge distributions not only assist chemists in assessing overall molecular structure and stability, but they also tell them about the chemistry that molecules can undergo. Consider, for example, the two resonance structures that a chemist would draw for acetate anion, $CH_3CO_2^-$, as shown in Figure 26.22. This figure indicates that the two CO bonds are equivalent and should be intermediate in length between single and double linkages, and that the negative charge is evenly distributed on the two oxygens. Taken together, these two observations suggest that the acetate ion is delocalized and therefore particularly stable.

Despite their obvious utility, atomic charges are not measurable properties, nor can they be determined uniquely from calculations. Although the total charge on a molecule (the total nuclear charge and the sum of the charges on all of the electrons) is well defined, and although overall charge distribution may be inferred from such observables as the dipole moment, it is not possible to assign discrete atomic charges. To do this would require accounting both for the nuclear charge and for the charge of any electrons uniquely associated with the particular atom. Although it is reasonable to assume that the nuclear contribution to the total charge on an atom is simply the atomic number, it is not at all obvious how to partition the total electron distribution by atoms. Consider, for example, the electron distribution for the heteronuclear diatomic molecule hydrogen fluoride, shown in Figure 26.23. Here, the surrounding contour is a particular electron density surface that, for example, corresponds to a van der Waals surface and encloses a large fraction of the total electron density. In this picture, the surface has been drawn to suggest that more electrons are associated with fluorine than with hydrogen. This is entirely reasonable, given the known polarity of the molecule, that is, $^{\delta+}H\text{—}F^{\delta-}$, as evidenced experimentally by the direction of its dipole moment. It is, however, not at all apparent how to divide this surface between the two nuclei. Are any of the divisions shown in Figure 26.23 better than the others? No! Atomic charges are not molecular properties, and it is not possible to provide a unique definition (or even a definition that

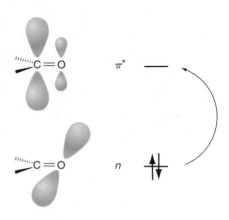

FIGURE 26.21
Accounting for electron correlation involves excitations such as the $n \rightarrow \pi^*$ transition in formaldehyde, which moves charge from oxygen to carbon and reduces the dipole moment in the carbonyl group.

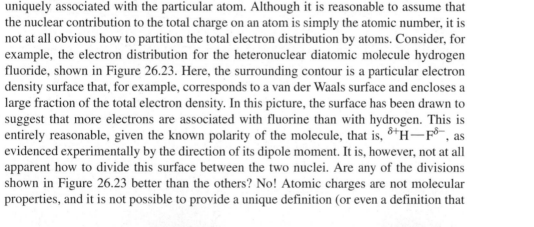

FIGURE 26.22
The two Lewis structures of the acetate ion.

FIGURE 26.23
Three different ways of partitioning the electrons in hydrogen fluoride between hydrogen and fluorine.

will satisfy all). We can calculate (and measure using X-ray diffraction) molecular charge distributions, that is, the number of electrons in a particular volume of space, but it is not possible to uniquely partition them among the atomic centers.

Despite the obvious problem with their definition, atomic charges are still useful, and several recipes have been formulated to calculate them. The simplest of these, now referred to as **Mulliken population analysis,** was discussed in Section 23.8.

MATHEMATICAL DESCRIPTION OF THE MULLIKEN POPULATION ANALYSIS

The Mulliken population analysis starts from the definition of the electron density, $\rho(\mathbf{r})$, in the framework of the Hartree–Fock model:

$$\rho(\mathbf{r}) = \sum_{\mu}^{basis\ functions} \sum_{\nu} P_{\mu\nu}\phi_{\mu}(\mathbf{r})\phi_{\nu}(\mathbf{r}) \tag{26.67}$$

where $P_{\mu\nu}$ is an element of the density matrix [see Equation (26.27)], and the summations are carried out over all atom-centered basis functions, ϕ_{μ}. Summing over basis functions and integrating over all space leads to an expression for the total number of electrons, n:

$$\int \rho(\mathbf{r})\,d\mathbf{r} = \sum_{\mu}^{basis\ functions} \sum_{\nu} P_{\mu\nu} \int \phi_{\mu}(\mathbf{r})\phi_{\nu}(\mathbf{r})\,d\mathbf{r}$$

$$= \sum_{\mu}^{basis\ functions} \sum_{\nu} P_{\mu\nu} S_{\mu\nu} = n \tag{26.68}$$

where $S_{\mu\nu}$ are elements of the overlap matrix:

$$S_{\mu\nu} = \int \phi_{\mu}(\mathbf{r})\phi_{\nu}(\mathbf{r})\,d\mathbf{r} \tag{26.69}$$

Analogous expressions can be constructed for correlated models. The important point is that it is possible to equate the total number of electrons in a molecule to a sum of products of density matrix and overlap matrix elements as follows:

$$\sum_{\mu}^{basis\ functions} \sum_{\nu} P_{\mu\nu}S_{\mu\nu} = \sum_{\mu}^{basis\ functions} P_{\mu\mu} + 2\sum_{\mu\neq\nu}^{basis\ functions}\sum P_{\mu\nu}S_{\mu\nu} = n \tag{26.70}$$

It is reasonable (but not necessarily correct) to assign any electrons associated with a particular diagonal element, $\mu\mu$, to that atom on which the basis function ϕ_{μ} is located. It is also reasonable to assign electrons associated with off-diagonal elements, $\mu\nu$, where both ϕ_{μ} and ϕ_{ν} reside on the same atom, to that atom. However, it is not apparent how to partition electrons from density matrix elements, $\mu\nu$, where ϕ_{μ} and ϕ_{ν} reside on different atoms. Mulliken provided a recipe. Give each atom half of the total, which is very simple but completely arbitrary! According to Mulliken's scheme, the gross electron population, q_{μ}, for basis function ϕ_{μ} is given by:

$$q_{\mu} = P_{\mu\mu} + \sum_{\nu}^{basis\ functions} P_{\mu\nu}S_{\mu\nu} \tag{26.71}$$

Atomic electron populations, q_A, and atomic charges, Q_A, follow, where Z_A is the atomic number of atom A:

$$q_A = \sum_{\mu}^{\substack{basis\ functions \\ on\ atom\ A}} q_{\mu} \tag{26.72}$$

$$Q_A = Z_A - q_A \tag{26.73}$$

An entirely different approach to providing atomic charges is to fit the value of some property that has been calculated based on the exact wave function with that obtained from representation of the electronic charge distribution in terms of a collection of atom-centered charges. One choice of property is the electrostatic potential, ε_p. This represents the energy of interaction of a unit positive charge at some point in space, p, with the nuclei and the electrons of a molecule:

$$\varepsilon_p = \sum_{A}^{nuclei} \frac{Z_A e^2}{4\pi\varepsilon_0 R_{Ap}} - \frac{e^2}{4\pi\varepsilon_0} \sum_{\mu}^{basis\ functions} \sum_{\nu} P_{\mu\nu} \int \frac{\phi_\mu(\mathbf{r})\phi_\nu(\mathbf{r})}{r_p} d\mathbf{r} \quad \textbf{(26.74)}$$

Z_A are atomic numbers, $P_{\mu\nu}$ are elements of the density matrix, and R_{Ap} and r_p are distances separating the point charges from the nuclei and electrons, respectively. The first summation is over nuclei and the second pair of summations is over basis functions.

Operationally, electrostatic-fit charges are obtained by first defining a grid of points surrounding the molecule, then calculating the electrostatic potential at each of these grid points, and finally providing a best (least-squares) fit of the potential at the grid points to an approximate electrostatic potential, ε_p^{approx}, based on replacing the nuclei and electron distribution by a set of atom-centered charges, Q_A, subject to overall charge balance:

$$\varepsilon_p^{approx} = \sum_{A}^{nuclei} \frac{e^2 Q_A}{4\pi\varepsilon_0 R_{Ap}} \quad \textbf{(26.75)}$$

The lack of uniqueness of the procedure results from selection of the grid points.

Problem P26.22

26.8.10 **Transition-State Geometries and Activation Energies**

Quantum chemical calculations need not be limited to the description of the structures and properties of stable molecules, that is, molecules that can actually be observed and characterized experimentally. They may as easily be applied to molecules that are highly reactive (reactive intermediates) and, even more interesting, to transition states, which cannot be observed let alone characterized. However, activation energies (the energy difference between the reactants and the transition state) can be inferred from experimental kinetic data. The complete absence of experimental data on transition-state geometries complicates assessment of the performance of different models. However, it is possible to get around this by assuming that some particular (high-level) model yields reasonable geometries for the transition state, and then to compare the results of the other models with this standard. The MP2/6-311+G** model has been selected as the standard.

The most conspicuous difference between the structure data presented in Table 26.16 and previous comparisons involving equilibrium bond distances is the much larger variation among different models. This should not come as a surprise. Transition states represent a compromise situation in which some bonds are being broken while others are being formed, and the potential energy surface around the transition state would be expected to be flat, meaning that large changes in geometry are expected to lead only to small changes in the energy. In terms of mean absolute deviations from the standard, the MP2/6-31G* model fares best and the two Hartree–Fock models fare worst, but all models give reasonable results. In terms of individual comparisons, the largest deviations among different models correspond to making and breaking single bonds. In such situations, the potential energy surface is expected to be quite flat.

As discussed in Section 26.2.4, an experimental activation energy can be obtained from the temperature dependence of the measured reaction rate by way of the Arrhenius equation, Equation (26.15). This first requires that a rate law be postulated [Equation (26.14)]. Association of the activation energy with the difference in energies between reactants and transition state (as obtained from quantum chemical calculations) requires the further assumption that all reacting molecules pass through the

TABLE 26.16   Key Bond Distances in Transition States for Organic Reactions (Å)

Reaction/Transition State	Bond Length	Hartree–Fock 3-21G	Hartree–Fock 6-31G*	B3LYP 6-31G*	MP2 6-31G*	MP2 6-311+G**
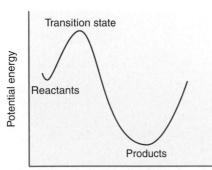	a	1.88	1.92	1.90	1.80	1.80
	b	1.29	1.26	1.29	1.31	1.30
	c	1.37	1.37	1.38	1.38	1.39
	d	2.14	2.27	2.31	2.20	2.22
	e	1.38	1.38	1.38	1.39	1.39
	f	1.39	1.39	1.40	1.41	1.41
	a	1.40	1.40	1.42	1.43	1.43
	b	1.37	1.38	1.39	1.39	1.39
	c	2.11	2.12	2.11	2.02	2.07
	d	1.40	1.40	1.41	1.41	1.41
	e	1.45	1.45	1.48	1.55	1.53
	f	1.35	1.36	1.32	1.25	1.25
	a	1.39	1.38	1.40	1.40	1.40
	b	1.37	1.37	1.38	1.38	1.38
	c	2.12	2.26	2.18	2.08	2.06
	d	1.23	1.22	1.24	1.25	1.24
	e	1.88	1.74	1.78	1.83	1.83
	f	1.40	1.43	1.42	1.41	1.41
Mean absolute deviation from MP2/6-311+G**		0.05	0.05	0.03	0.01	—

transition state. In effect, this implies that all reactants have the same energy, or that none of the reactants has energy in excess of that needed to reach the transition state. This is the essence of transition-state theory.

Absolute activation energies for a small series of organic reactions are provided in Table 26.17. As with transition-state geometries, results from the practical models are compared with those of the standard, MP2/6-311+G**. Overall, the performance of Hartree–Fock models is very poor. In most cases, the activation energies are over-estimated by large amounts. This is not surprising in view of previous comparisons involving homolytic bond dissociation energies (see Table 26.11), which were too small. The argument that might be given here is that a transition state is typically more tightly bound than the reactants, meaning that correlation effects will be greater. The B3LYP/6-31G* and MP2/6-31G* perform much better, and lead to errors (relative to the standard) that are comparable to those previously noted for reaction energy comparisons.

26.8.11 Finding a Transition State

The usual picture of a chemical reaction in terms of a one-dimensional potential energy (or reaction coordinate) diagram is shown in Figure 26.24. The vertical axis corresponds to the energy of the system, and the horizontal axis (reaction coordinate) corresponds to the geometry of the system. The starting point on the diagram (reactants) is an energy minimum, as is the ending point (products). Motion along the reaction coordinate is assumed to be continuous and to pass through a single energy maximum called the transition state. As described in Section 26.2.1, a transition state on a real many-dimensional potential energy surface corresponds to a point that is actually an energy minimum in all but one dimension and an energy maximum along the reaction coordinate. The obvious

FIGURE 26.24
The potential energy surface for a reaction is typically represented by a one-dimensional representation of the energy as a function of the reaction coordinate.

TABLE 26.17 Absolute Activation Energies for Organic Reactions (kJ/mol)

Reaction	Hartree–Fock 3-21G	Hartree–Fock 6-31G*	B3LYP 6-31G*	MP2 6-31G*	MP2 6-311+G**	Experiment
$CH_3NC \longrightarrow CH_3CN$	238	192	172	180	172	159
$HCO_2CH_2CH_3 \longrightarrow HCO_2H + C_2H_4$	259	293	222	251	234	167, 184
(cyclohexadiene → benzene type)	192	238	142	117	109	151
(oxygen-containing ring rearrangement)	176	205	121	109	105	130
(cyclopentadiene + ethylene → norbornene)	126	167	84	50	38	84
(cyclohexene → + C_2H_4)	314	356	243	251	230	—
$HCNO + C_2H_2 \longrightarrow$ (isoxazole)	105	146	50	33	38	—
(diene rearrangement)	230	247	163	159	142	—
(cyclobutene → butadiene)	176	197	151	155	142	—
(lactone → + CO_2)	247	251	167	184	172	—
(SO_2 ring → + SO_2)	205	205	92	105	92	—
Mean absolute deviation from MP2/6-311+G**	71	100	17	13	—	—

analogy is to the crossing of a mountain range, the goal of which is simply to get from one side of the range to the other side with minimal effort.

Crossing over the top of a "mountain" (pathway A), which corresponds to crossing through an energy maximum on a (two-dimensional) potential energy surface, accomplishes the goal, as shown in Figure 26.25. However, it is not likely to be the chosen pathway. This is because less effort (energy) will be expended by going through a "pass" between two "mountains" (pathway B), a maximum in one dimension but a minimum in the other dimension. This is referred to as a saddle point and corresponds to a transition state.

A single molecule may have many transition states (some corresponding to real chemical reactions and others not), and merely finding *a* transition state does not guarantee that it is *the* transition state, meaning that it is at the top of the lowest energy pathway that smoothly connects reactants and products. Although it is possible to verify the smooth connection of reactants and products, it will generally not be possible to know with complete certainty that what has been identified as the transition state is in fact the lowest energy structure over which the reaction might proceed, or whether in fact the actual reaction proceeds over a transition state that is not the lowest energy structure.

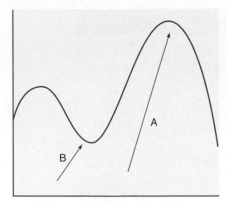

FIGURE 26.25
A reaction in two dimensions is analogous to crossing a mountain range. Pathway A, which goes over the top of a mountain, requires more effort to traverse than pathway B, which goes through a pass between two mountains.

Problems P26.23–P26.25

FIGURE 26.26
The calculated transition state for the isomerization of methyl isocyanide to acetonitrile is consistent with a three-membered ring in the reaction scheme shown.

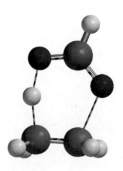

FIGURE 26.27
The calculated transition state in the pyrolysis of ethyl formate is consistent with a six-membered ring in the reaction scheme shown.

The fact that transition states, like the reactants and products of a chemical reaction, correspond to well-defined structures, means that they can be fully characterized from calculation. However, this is one area where the results of calculation cannot be tested, except with reference to chemical intuition. For example, it is reasonable to expect that the transition state for the unimolecular isomerization of methyl isocyanide to acetonitrile takes the form of a three-membered ring,

$$H_3C-N\equiv C \longrightarrow \underset{N=C}{\overset{CH_3}{\diagup}} \longrightarrow N\equiv C-CH_3$$

in accord with the structure actually calculated, which is shown in Figure 26.26.

It is also reasonable to expect that the transition state for pyrolysis of ethyl formate leading to formic acid and ethylene will take the form of a six-membered ring:

This expectation agrees with the result of the calculation, as shown in Figure 26.27.

26.9 Graphical Models

In addition to numerical quantities (bond lengths and angles, energies, dipole moments, and so on), quantum chemical calculations furnish a wealth of information that is best displayed in the form of images. Among the results of calculations that have proven to be of value are the molecular orbitals themselves, the electron density, and the electrostatic potential. These can all be expressed as three-dimensional functions of the coordinates. One way to display them on a two-dimensional video screen (or on a printed page) is to define a surface of constant value, a so-called isovalue surface or, more simply, isosurface:

$$f(x, y, z) = constant \tag{26.76}$$

The value of the constant may be chosen to reflect a particular physical observable of interest, for example, the "size" of a molecule in the case of display of electron density.

26.9.1 Molecular Orbitals

As detailed in Section 26.3, molecular orbitals, ψ, are written in terms of linear combinations of basis functions, ϕ, which are centered on the individual nuclei:

$$\psi_i = \sum_{\mu}^{basis\ functions} c_{\mu i}\phi_\mu \tag{26.21}$$

Although it is tempting to associate a molecular orbital with a particular bond, more often than not this is inappropriate. Molecular orbitals will generally be spread out (delocalized) over the entire molecule, whereas bonds are normally associated with a pair of atoms. Also, molecular orbitals, unlike bonds, show the symmetry of the molecule. For example, the equivalence of the two OH bonds in water is revealed by the two molecular orbitals best describing OH bonding as shown in Figure 26.28.

Molecular orbitals, in particular, the highest energy occupied molecular orbital (the **HOMO**) and the lowest energy unoccupied molecular orbital (the **LUMO**), are often quite familiar to chemists. The former holds the highest energy (most available) electrons and should be subject to attack by electrophiles, whereas the latter provides the lowest energy space for additional electrons and should be subject to attack by nucleophiles. For example, the HOMO in formaldehyde is in the heavy-atom plane of the

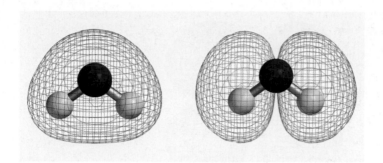

FIGURE 26.28
These molecular orbitals can be identified
with the O—H bonds in water.

molecule, indicating that attack by an electrophile, for example, a proton, will occur
here, as shown in Figure 26.29, whereas the LUMO is out of plane on the carbonyl carbon, consistent with the known nucleophilic chemistry.

26.9.2 Orbital Symmetry Control of Chemical Reactions

Woodward and Hoffmann, building on the earlier ideas of Fukui, first clearly pointed
out how the symmetries of the HOMO and LUMO (together referred to as the **frontier
molecular orbitals**) could be used to rationalize why some chemical reactions proceed easily whereas others do not. For example, the fact that the HOMO in *cis*-1,3-
butadiene is able to interact favorably with the LUMO in ethene suggests that the two
molecules should readily combine in a concerted manner to form cyclohexene in a
process called Diels-Alder cycloaddition. This process is depicted in Figure 26.30. On
the other hand, interaction between the HOMO on one ethene and the LUMO on
another ethene is not favorable, as illustrated in Figure 26.31, and concerted addition
to form cyclobutane would not be expected. Reactions which are allowed or forbidden
because of orbital symmetry have been collected under what is now known as the
"Woodward-Hoffmann" rules. For their work, Hoffmann and Fukui shared the Nobel
Prize in chemistry in 1981.

Problems P26.26–P26.31

26.9.3 Electron Density

The electron density, $\rho(\mathbf{r})$, is a function of the coordinates $\mathbf{r}$, defined such that $\rho(\mathbf{r})d\mathbf{r}$ is
the number of electrons inside a small volume $d\mathbf{r}$. This is what is measured in an X-ray
diffraction experiment. Electron density $\rho(\mathbf{r})$ is written in terms of a sum of products of
basis functions, ϕ_μ:

$$\rho(\mathbf{r}) = \sum_{\mu}^{basis\ functions} \sum_{\nu} P_{\mu\nu}\phi_\mu(\mathbf{r})\phi_\nu(\mathbf{r}) \qquad (26.77)$$

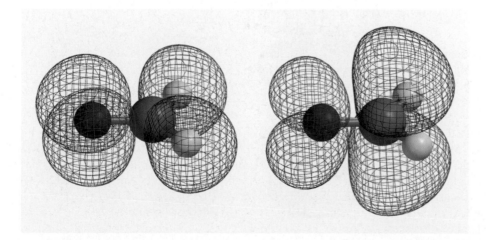

FIGURE 26.29
The HOMO (left) and LUMO (right)
for formaldehyde identify regions where
electrophilic and nucleophilic attack,
respectively, are likely to occur.

FIGURE 26.30
The HOMO of butadiene (bottom) is able to interact with the LUMO of ethene (top), resulting in cycloaddition, in agreement with experiment.

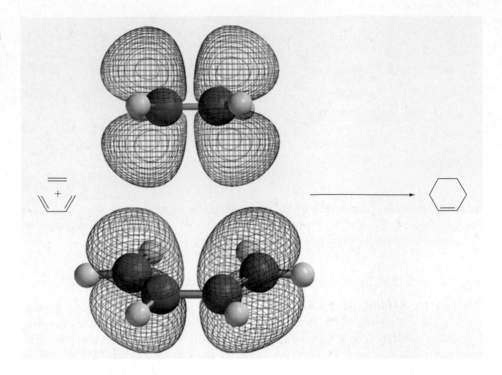

where $P_{\mu\nu}$ are elements of the density matrix [Equation (26.27)]. The electron density can be portrayed in terms of a surface (an **electron density surface**) with the size and shape of the surface being given by the value of the density, for example, in cyclohexanone in Figure 26.32.

Depending on the value, isodensity surfaces can either serve to locate atoms (left image in Figure 26.32), to delineate chemical bonds (center image), or to indicate overall molecular size and shape (right image). The regions of highest electron density surround the heavy (non-hydrogen) atoms in a molecule. This is the basis of X-ray crystallography, which locates atoms by identifying regions of high electron density. Also interesting are regions of lower electron density. For example, a 0.1 electrons/a_0^3 isodensity surface for cyclohexanone conveys essentially the same information as a conventional skeletal structure model; that is, it depicts the locations of bonds. A surface of 0.002 electrons/a_0^3 provides a good fit to conventional space-filling models and, hence, serves to portray overall molecular size and shape. As is the case with the space-filling model, this definition of molecular size is completely arbitrary (except that it closely

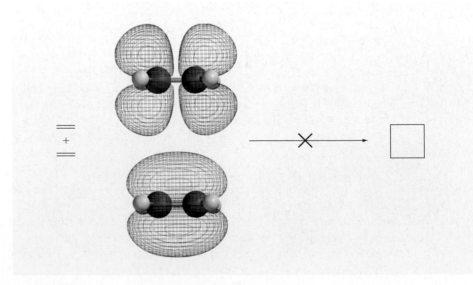

FIGURE 26.31
The HOMO of ethene (bottom) is not able to interact with the corresponding LUMO (top), suggesting that cycloaddition is not likely to occur, in agreement with experiment.

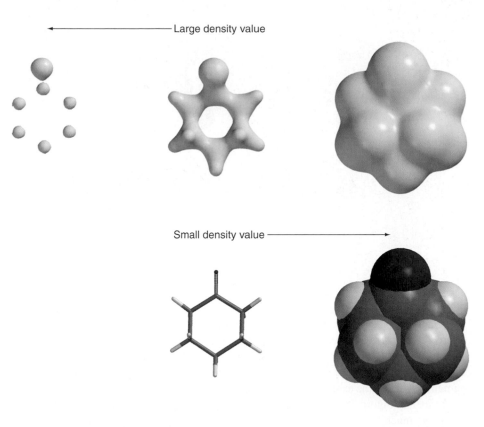

Large density value

Small density value

FIGURE 26.32
Electron density surfaces for cyclo-hexanone corresponding to three different values of the electron density: 0.4 electron/a_0^3 (left), 0.1 electrons/a_0^3 (center), and 0.002 electrons/a_0^3 (right). Conventional skeletal and space-filling models appear underneath the last two electron density surfaces.

matches experimental data on how closely atoms fit together in crystalline solids). A single parameter, namely, the value of the electron density at the surface, has replaced the set of atomic radii used for space-filling models. These latter two electron density surfaces are examined in more detail in the following section.

26.9.4 Where Are the Bonds in a Molecule?

An electron density surface can be employed to reveal the location of bonds in a molecule. Of course, chemists routinely employ a variety of tactics to depict chemical bonding, ranging from pencil sketches (Lewis structures) to physical models such as Dreiding models. The most important advantage of electron density surfaces is that they can be applied to elucidate bonding and not only to portray bonding in cases where the location of bonds is known. For example, the electron density surface for diborane (Figure 26.33) clearly shows a molecule with very little electron density concentrated between the two borons. This fact suggests that the appropriate Lewis structure of the two shown in Figure 26.34 is the one that lacks a boron–boron bond, rather than the one that shows the two borons directly bonded.

Another important application of electron density surfaces is to the description of the bonding in transition states. An example is the pyrolysis of ethyl formate, leading to formic acid and ethylene, which is illustrated in Figure 26.35. The electron density surface offers clear evidence of a **late transition state,** meaning that the CO bond is nearly fully cleaved and the migrating hydrogen is more tightly bound to oxygen (as in the product) than to carbon (as in the reactant).

26.9.5 How Big Is a Molecule?

The size of a molecule can be defined according to the amount of space that it takes up in a liquid or solid. The so-called space-filling or CPK model has been formulated to portray molecular size, based on fitting the experimental data to a set of atomic radii (one for each atom type). Although this simple model is remarkably satisfactory over-all, some problematic cases do arise, in particular for atoms that may adopt different oxidation states, for example, Fe^0 in $FeCO_5$ versus Fe^{II} in $FeCl_4^{2-}$.

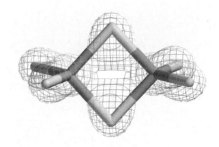

FIGURE 26.33
An electron density surface for diborane shows that there is no boron–boron bond.

Problem P26.32

FIGURE 26.34
Two possible Lewis structures of diborane differ in that only one has a boron–boron bond.

FIGURE 26.35
An electron density structure for the
transition state in the pyrolysis of ethyl
formate shows a six-membered ring
consistent with the conventional Lewis
picture shown in the reaction scheme.

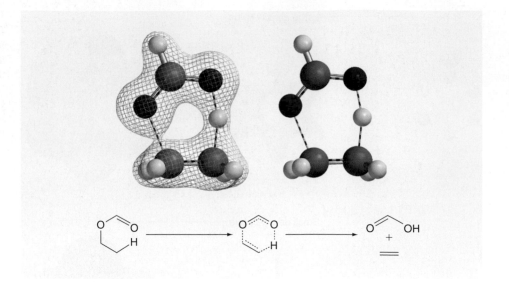

Problems P26.33–P26.34

Because the electrons—not the underlying nuclei—dictate overall molecular size, the electron density provides an alternate measure of how much space molecules actually take up. Unlike space-filling models, electron density surfaces respond to changes in the chemical environment and allow atoms to adjust their sizes in response to different environments. An extreme example concerns the size of hydrogen in main-group hydrides.

As seen in Figure 26.36, electron density surfaces reveal that the hydrogen in lithium hydride is much larger than that in hydrogen fluoride, consistent with the fact that the former serves as a base (hydride donor), whereas the latter serves as an acid (proton donor). Hydrogen sizes in beryllium hydride, borane, methane, ammonia, and water are intermediate and parallel the ordering of the electronegativities of the heavy atom.

26.9.6 Electrostatic Potential

The **electrostatic potential,** ε_p, is defined as the energy of interaction of a positive point charge located at p with the nuclei and electrons of a molecule:

$$\varepsilon_p = \sum_{A}^{nuclei} \frac{e^2 Z_A}{4\pi\varepsilon_0 R_{Ap}} - \sum_{\mu}^{basis\ functions} \sum_{\nu} P_{\mu\nu} \int \frac{\phi_\mu^*(\mathbf{r})\phi_\nu(\mathbf{r})}{r_p} d\tau \qquad \textbf{(26.78)}$$

Notice that the electrostatic potential represents a balance between repulsion of the point charge by the nuclei (first summation) and attraction of the point charge by the

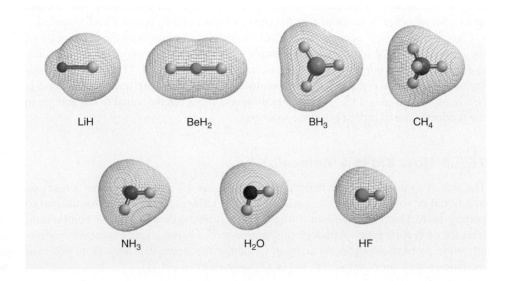

FIGURE 26.36
Electron density surfaces are shown for
hydrides of lithium to fluorine.

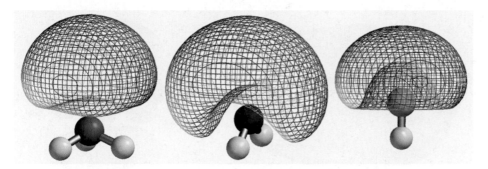

FIGURE 26.37
Electrostatic potential surfaces for ammonia (left), water (center), and hydrogen fluoride (right) are useful in depicting lone pairs.

electrons (second summation). $P_{\mu\nu}$ are elements of the density matrix [see Equation (26.27)] and the ϕ are atomic basis functions.

26.9.7 Visualizing Lone Pairs

The octet rule dictates that each main-group atom in a molecule will be surrounded by eight valence electrons. These electrons can either be tied up in bonds (two electrons for a single bond, four electrons for a double bond, 6 electrons for a triple bond), or can remain with the atom as a nonbonded or lone pair of electrons. Although you cannot actually see bonds, you can see their consequence (the atoms to which bonds are made). On this basis, lone pairs would seem to be completely invisible, because there are no telltale atoms. However, the fact that the electrons in lone pairs should be highly accessible suggests another avenue. Regions of space around a molecule where the potential is negative suggest an excess of electrons. To the extent that lone pairs represent electron-rich environments, they should be revealed by electrostatic potential surfaces. A good example is provided by negative electrostatic potential surfaces for ammonia, water, and hydrogen fluoride, as shown in Figure 26.37.

The electron-rich region in ammonia is in the shape of a lobe pointing in the fourth tetrahedral direction, whereas that in water takes the form of a crescent occupying two tetrahedral sites. At first glance, the electrostatic potential surface for hydrogen fluoride is nearly identical to that in ammonia. Closer inspection reveals that rather than pointing away from the fluorine (as it points away from ammonia), the surface encloses the atom. All in all, these three surfaces are entirely consistent with conventional Lewis structures for the three hydrides shown in Figure 26.38.

A related comparison between electrostatic potential surfaces for ammonia in both the observed pyramidal and unstable trigonal planar geometries is shown in Figure 26.39. As previously mentioned, the former depicts a lobe pointing in the fourth tetrahedral direction, and the electrostatic potential surface for the planar trigonal arrangement shows two equal out-of-plane lobes. This is, of course, consistent

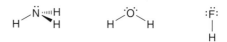

FIGURE 26.38
Lewis structures for ammonia, water, and hydrogen fluoride.

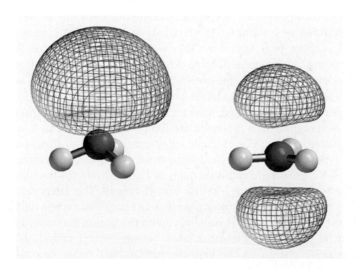

FIGURE 26.39
Electrostatic potential surfaces show that the lone pair is directed to one side of pyramidal ammonia, but is equally distributed on both sides of planar ammonia.

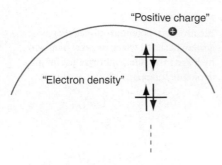

"Positive charge"

"Electron density"

FIGURE 26.40

An electrostatic potential map shows the value of the electrostatic potential at all locations on a surface of electron density (corresponding to overall size and shape).

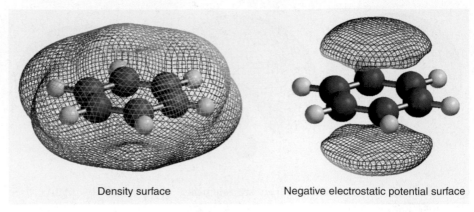

Density surface

Negative electrostatic potential surface

FIGURE 26.41

An electron density surface and a negative electrostatic potential surface are shown for benzene.

with the fact that pyramidal ammonia has a dipole moment (with the negative end pointing in the direction of the lone pair), whereas planar ammonia does not have a dipole moment.

26.9.8 Electrostatic Potential Maps

Graphical models need not be restricted to portraying a single quantity. Additional information can be presented in terms of a property map on top of an isosurface, where different colors can be used to portray different property values. Most common are maps on electron density surfaces. Here the surface can be used to designate overall molecular size and shape, and the colors to represent the value of some property at various locations on the surface. The most commonly used property map is the **electrostatic potential map,** schematically depicted in Figure 26.40. This gives the value of the electrostatic potential at locations on a particular surface, most commonly a surface of electron density corresponding to overall molecular size.

To see how an electrostatic potential map (and by implication any property map) is constructed, first consider both a density surface and a particular (negative) electrostatic potential surface for benzene, as shown in Figure 26.41. Both of these surfaces convey structure. The density surface reveals the size and shape of benzene, and the negative electrostatic potential surface delineates in which regions surrounding benzene a particular (negative) electrostatic potential will be felt.

Next, consider making a map of the value of the electrostatic potential on the density surface (an electrostatic potential map), using colors to designate values of the potential. This leaves the density surface unchanged (insofar as it represents the size and shape of benzene), but replaces the gray-scale image (conveying only structural information) with a color image (conveying the value of the electrostatic potential *in addition to* structure). An electrostatic map for benzene is presented in Figure 26.42. Colors near red represent large negative values of the potential, whereas colors near blue represent large positive values (orange, yellow, and green represent intermediate values of the potential). Note that the π system is red, consistent with the (negative) potential surface previously shown.

Electrostatic potential maps are used for a myriad of purposes other than rapidly conveying which regions of a molecule are likely to be electron rich and which are likely to be electron poor. For example, they can be used to distinguish between molecules in which charge is localized from those where it is delocalized.

Compare the electrostatic potential maps in Figure 26.43 for the planar (top) and perpendicular (bottom) structures of the benzyl cation. The latter reveals a heavy concentration of positive charge (blue color) on the benzylic carbon and perpendicular to the plane of the ring. This is consistent with the notion that only a single Lewis structure can be drawn. On the other hand, planar benzyl cation shows no such buildup of positive charge on the benzylic carbon, but rather delocalization onto

Problem P26.35

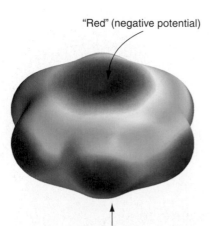

"Red" (negative potential)

"Blue" (positive potential)

FIGURE 26.42

An electrostatic potential map of benzene.

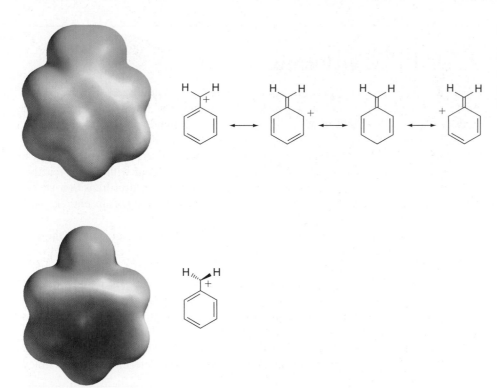

FIGURE 26.43
An electrostatic potential map for planar benzyl cation (top) shows delocalization of positive charge, whereas that for perpendicular benzyl cation (bottom) shows charge localization.

ortho and *para* ring carbons, consistent with the fact that several Lewis structures can be drawn.

Electrostatic potential maps can also be employed to characterize transition states in chemical reactions. A good example is pyrolysis of ethyl formate (leading to formic acid and ethylene):

Here, the electrostatic potential map shown in Figure 26.44 (based on an electron density surface appropriate to identify bonds) clearly shows that the hydrogen being transferred (from carbon to oxygen) is positively charged; that is, it is an electrophile.

Problems P26.36–P26.37

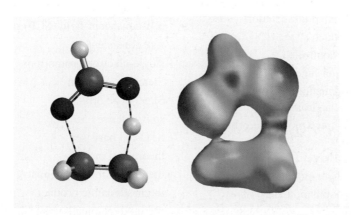

FIGURE 26.44
An electrostatic potential map is useful in depicting the charge distribution in the transition state for the pyrolysis of ethyl formate.

26.10 Conclusion

Quantum chemical calculations are rapidly becoming a viable alternative to experiments as a means to investigate chemistry. Continuing rapid advances in computer hardware and software technology will only further this trend and lead to even wider adoption among mainstream chemists. Calculations are already able to properly account for molecular structure and energetics, among other important quantities. Perhaps most intriguing is the ability of the calculations to deal with highly reactive molecules, which may be difficult to synthesize, and with reaction transition states, which cannot be observed at all. In this regard, calculations open up entirely new avenues for chemical research.

Quantum chemical calculations do have limitations. Most conspicuous of these is the trade-off between accuracy and cost. Practical quantum chemical models do not always yield results that are sufficiently accurate to actually be of value, and models that are capable of yielding accurate results may not yet be practical for the system of interest. Second, a number of quantities important to chemists cannot yet be routinely and reliably obtained from calculations. The most important limitation, however, is that, for the most part, calculations apply strictly to isolated molecules (gas phase), whereas much if not most chemistry is carried out in solution. Practical models that take solvent into account are being developed and tested to determine their accuracy and limitations.

The prognosis is very bright. Your generation will have at its disposal a whole range of powerful tools for exploring and understanding chemistry, just like the generations before you were given technologies such as the laser to exploit. It is all part of the natural evolution of the science.

Vocabulary

activation energy
Born–Oppenheimer approximation
CID method
CIS method
conformer
contracted functions
correlated models
correlation energy
density functional theory
diffuse functions
diffusion-controlled reactions
double-electron promotions
electron density surface
electronic Schrödinger equation
electrostatic potential
electrostatic potential map
exothermic
frontier molecular orbitals
frozen core approximation
full configuration interaction

Gaussian functions
harmonic frequencies
Hartree–Fock approximation
Hartree–Fock energy
Hartree–Fock equations
Hartree–Fock model
highest-occupied molecular orbital (HOMO)
homolytic bond dissociation
kinetic product
kinetically controlled
late transition state
linear combination of atomic orbitals (LCAO) approximation
lowest-occupied molecular orbital (LUMO)
minimal basis set
Møller-Plesset models
Mulliken population analysis
normal coordinates
polarization basis set

polarization functions
potential energy surface
preexponential factor
rate constant
reaction coordinate
reaction coordinate diagrams
reactive intermediates
Roothaan–Hall equations
self-consistent-field (SCF) procedure
size consistent
single-electron promotions
Slater determinant
Slater-type orbitals
split-valence basis set
STO-3G basis set
theoretical model
theoretical model chemistry
thermodynamic product
thermodynamically controlled
variational

Numerical Problems

P26.1 The assumption that the reaction coordinate in going from *gauche* to *anti* n-butane is a simple torsion is an oversimplification, because other geometrical changes no doubt also occur during rotation around the carbon–carbon bond, for example, changes in bond lengths and angles. Examine the energy profile for *n*-butane ("n-butane" on the precalculated Spartan file) and plot the change in distance of the central CC bond and CCC bond angle as a function of the torsion angle. Are the bond length and bond angle nearly identical or significantly different (>0.02 Å and $>2°$) for the two equilibrium forms of *n*-butane? Are the two parameters nearly identical or significantly different between the *anti* form and either or both of the transition states? Explain your results.

P26.2 Ammonia provides a particularly simple example of the dependence of vibrational frequencies on the atomic masses and of the use of vibrational frequencies to distinguish between a stable molecule and a transition state. First examine the vibrational spectrum of pyramidal ammonia ("ammonia" on the precalculated Spartan file).

a. How many vibrational frequencies are there? How does this number relate to the number of atoms? Are all frequencies real numbers or are one or more imaginary numbers? Describe the motion associated with each frequency and characterize each as being primarily bond stretching, angle bending, or a combination of the two. Is bond stretching or angle bending easier? Do the stretching motions each involve a single NH bond or do they involve combinations of two or three bonds?

b. Next, consider changes to the vibrational frequencies of ammonia as a result of substituting deuteriums for hydrogens ("perdeuteroammonia" on the precalculated Spartan file). Are the frequencies in ND_3 larger, smaller, or unchanged from those in NH_3? Are any changes greater for motions that are primarily bond stretching or motions that are primarily angle bending?

c. Finally, examine the vibrational spectrum of an ammonia molecule that has been constrained to a planar geometry ("planar ammonia" on the Spartan download). Are all the frequencies real numbers? If not, describe the motions associated with any imaginary frequencies and relate them to the corresponding motion(s) in the pyramidal equilibrium form.

P26.3 The presence of the carbonyl group in a molecule is easily confirmed by an intense line in the infrared spectrum around 1700 cm^{-1} that corresponds to the C$=$O stretching vibration. Locate this line in the calculated infrared spectrum of acetone ("acetone" on the precalculated Spartan file) and note its position in the overall spectrum (relative to the positions of the other lines) and the intensity of the absorption.

a. Speculate why this line is a reliable diagnostic for carbonyl functionality.

b. Examine the lowest frequency mode for acetone and then the highest frequency mode. Describe each and relate to the relative ease of difficulty of the associated motion.

P26.4 Chemists recognize that the cyclohexyl radical is likely to be more stable than the cyclopentylmethyl radical, because they know that six-membered rings are more stable than five-membered rings and, more importantly, that secondary radicals are more stable than primary radicals. However, much important chemistry is not controlled by what is most stable (thermodynamics) but rather by what forms most readily (kinetics). For example, loss of bromine from 6-bromohexene leading initially to hex-5-enyl radical, results primarily in product from cyclopentylmethyl radical.

The two possible interpretations for the experimental result are that the reaction is thermochemically controlled but that our understanding of radical stability is wrong or that the reaction is kinetically controlled.

a. First, see if you can rule out the first possibility. Examine structures and total energies for cyclohexyl and cyclopentylmethyl radicals ("cyclohexyl and cyclopentylmethyl radicals" on the precalculated Spartan file). Which radical, cyclohexyl or cyclopentylmethyl, is more stable (lower in energy)? Is the energy difference large enough such that only the more stable radical is likely to be observed? (Recall that at room temperature an energy difference of 12 kJ/mol corresponds to a product ratio of >99:1.) Do you conclude that ring closure is under thermodynamic control?

b. The next objective is to establish which ring closure, to cyclohexyl radical or to cyclopentylmethyl radical, is easier; that is, which product, cyclohexane or methylcyclopentane, is the kinetic product? Examine structures and total energies for the transition states for the two ring closures ("to cyclohexyl and cyclopentylmethyl radicals" on the Spartan download). Which radical, cyclohexyl or cyclopentylmethyl, is more easily formed?

c. Consider the following relationships between transition-state energy difference, $\Delta E^{\ddagger}$, and the ratio of major to minor (kinetic) products, calculated from the Boltzmann distribution:

$\Delta E^{\ddagger}$ (kJ/mol)	Major : Minor (room temperature)
4	~90:10
8	~95:5
12	~99:1

What is the approximate ratio of products suggested by the calculations? How does this compare with what is observed? Do you conclude that ring closure is under kinetic control?

P26.5 VSEPR (valence state electron pair repulsion) theory was formulated to anticipate the local geometry about an atom in a molecule (see discussion in Section 25.1). All that is required is the number of electron pairs surrounding the atom, broken down into bonded pairs and nonbonded (lone) pairs. For example, the carbon in carbon tetrafluoride is surrounded by four electron pairs, all of them tied up in CF bonds, whereas the sulfur in sulfur tetrafluoride is surrounded by five electron pairs, four of which are tied up in SF bonds with the fifth being a lone pair.

VSEPR theory is based on two simple rules. The first is that electron pairs (either lone pairs or bonds) will seek to avoid each other as much as possible. Thus, two electron pairs will lead to a linear geometry, three pairs to a trigonal planar geometry, four pairs to a tetrahedral geometry, five pairs to a trigonal bipyramidal geometry, and six pairs to an octahedral geometry. Although this knowledge is sufficient to assign a geometry for a molecule such as carbon tetrafluoride (tetrahedral), it is not sufficient to specify the geometry of a molecule such as sulfur tetrafluoride. Does the lone pair assume an *equatorial* position on the trigonal bipyramid leading to a seesaw geometry, or an *axial* position leading to a trigonal pyramidal geometry?

Seesaw   **Trigonal pyramidal**

The second rule, that lone pairs take up more space than bonds, clarifies the situation. The seesaw geometry in which the lone pair is 90° to two of the SF bonds and 120° to the other two bonds is preferable to the trigonal pyramidal geometry in which three bonds are 90° to the lone pair.

Although VSEPR theory is easy to apply, its results are strictly qualitative and often of limited value. For example, although the model tells us that sulfur tetrafluoride adopts a seesaw geometry, it does not reveal whether the trigonal pyramidal structure (or any other structure) is an energy minimum, and if it is, what its energy is relative to the seesaw form. Also it has little to say when more than six electron pairs are present. For example, VSEPR theory tells us that xenon hexafluoride is not octahedral, but it does not tell us what geometry the molecule actually assumes. Hartree–Fock molecular orbital calculations provide an alternative.

a. Optimize the structure of SF_4 in a seesaw geometry (C_{2v} symmetry) using the HF/3-21G model and calculate vibrational frequencies (the infrared spectrum). This calculation is necessary to verify that the energy is at a minimum. Next, optimize the geometry of SF_4 in a trigonal pyramidal geometry and calculate its vibrational frequencies. Is the seesaw structure an energy minimum?

What leads you to your conclusion? Is it lower in energy than the corresponding trigonal pyramidal structure in accordance with VSEPR theory? What is the energy difference between the two forms? Is it small enough that both might actually be observed at room temperature? Is the trigonal pyramidal structure an energy minimum?

b. Optimize the geometry of XeF_6 in an octahedral geometry (O_h symmetry) using the HF/3-21G model and calculate vibrational frequencies. Next, optimize XeF_6 in a geometry that is distorted from octahedral (preferably a geometry with C_1 symmetry) and calculate its vibrational frequencies. Is the octahedral form of XeF_6 an energy minimum? What leads you to your conclusion? Does distortion lead to a stable structure of lower energy?

P26.6 Each of the carbons in ethane is surrounded by four atoms in a roughly tetrahedral geometry; each carbon in ethene is surrounded by three atoms in a trigonal planar geometry and each carbon in acetylene by two atoms in a linear geometry. These structures can be rationalized by suggesting that the valence $2s$ and $2p$ orbitals of carbon are able to combine either to produce four equivalent sp^3 hybrids directed toward the four corners of a tetrahedron, or three equivalent sp^2 hybrids directed toward the corners of an equilateral triangle with a p orbital left over, or two equivalent sp hybrids directed along a line with two p orbitals left over. The $2p$ atomic orbitals extend farther from carbon than the $2s$ orbital. Therefore, sp^3 hybrids will extend farther than sp^2 hybrids, which in turn will extend farther than sp hybrids. As a consequence, bonds made with sp^3 hybrids should be longer than those made with sp^2 hybrids, which should in turn be longer than those made with sp hybrids.

a. Obtain equilibrium geometries for ethane, ethene, and acetylene using the HF/6-31G* model. Is the ordering in CH bond lengths what you expect on the basis of the hybridization arguments? Using the CH bond length in ethane as a standard, what is the percent reduction in CH bond lengths in ethene? In acetylene?

b. Obtain equilibrium geometries for cyclopropane, cyclobutane, cyclopentane, and cyclohexane using the HF/6-31G* model. Are the CH bond lengths in each of these molecules consistent with their incorporating sp^3-hybridized carbons? Note any exceptions.

c. Obtain equilibrium geometries for propane, propene, and propyne using the HF/6-31G* model. Is the ordering of bond lengths the same as that observed for the CH bond lengths in ethane, ethene, and acetylene? Are the percent reductions in bond lengths from the standard (propane) similar ($\pm10\%$) to those seen for ethene and acetylene (relative to ethane)?

P26.7 The bond angle about oxygen in alcohols and ethers is typically quite close to tetrahedral ($109.5°$), but opens up significantly in response to extreme steric crowding, for example, in going from *tert*-butyl alcohol to di-*tert*-butyl ether:

This is entirely consistent with the notion that while lone pairs take up space, they can be "squeezed" to relieve crowding. Another way to relieve unfavorable steric interactions (without changing the position of the lone pairs) is to increase the CO bond distance.

a. Build *tert*-butyl alcohol and di-*tert*-butyl ether and optimize the geometry of each using the HF/6-31G* model. Are the calculated bond angles involving oxygen in accord with the values given earlier, in particular with regard to the observed increase in bond angle? Do you see any lengthening of the CO bond in the ether over that in the alcohol? If not, or if the effect is very small (<0.01 Å), speculate why not.

b. Next, consider the analogous trimethylsilyl compounds Me_3SiOH and $Me_3SiOSiMe_3$. Calculate their equilibrium geometries using the HF/6-31G* model. Point out any similarities and any differences between the calculated structures of these compounds and their *tert*-butyl analogues. In particular, do you see any widening of the bond angle involving oxygen in response to increased steric crowding? Do you see lengthening of the SiO bond in the ether over that of the alcohol? If not, rationalize what you do see.

P26.8 Water contains two acidic hydrogens that can act as hydrogen-bond donors and two lone pairs that can act as hydrogen-bond acceptors:

Given that all are tetrahedrally disposed around oxygen, this suggests two reasonable structures for the hydrogen-bonded dimer of water, $(H_2O)_2$, one with a single hydrogen bond and one with two hydrogen bonds:

Whereas the second seems to make better use of water's attributes, in doing so, it imposes geometrical restrictions on the dimer.

Build the two dimer structures. Take into account that the hydrogen-bond distance $(O \cdot \cdot \cdot H)$ is typically on the order of 2 Å. Optimize the geometry of each using the HF/6-31G* model and, following this, calculate vibrational frequencies.

Which structure, singly or doubly hydrogen bonded, is more stable? Is the other (higher energy) structure also an energy minimum? Explain how you reached your conclusion. If the dimer with the single hydrogen bond is more stable, speculate what this has told you about the geometric requirements of hydrogen bonds. Based on your experience with water dimers, suggest a "structure" for liquid water.

P26.9 For many years, a controversy raged concerning the structures of so-called "electron-deficient" molecules, that is, molecules with insufficient electrons to make normal two-atom, two-electron bonds. Typical is ethyl cation, $C_2H_5^+$, formed from protonation of ethene.

Is it best represented as an open Lewis structure with a full positive charge on one of the carbons, or as a hydrogen-bridged structure in which the charge is dispersed onto several atoms? Build both open and hydrogen-bridged structures for ethyl cation. Optimize the geometry of each using the B3LYP/6-31G* model and calculate vibrational frequencies. Which structure is lower in energy, the open or hydrogen-bridged structure? Is the higher energy structure an energy minimum? Explain your answer.

P26.10 One of the most powerful attractions of quantum chemical calculations over experiments is their ability to deal with any molecular system, stable or unstable, real or imaginary. Take as an example the legendary (but imaginary) kryptonite molecule. Its very name gives us a formula, KrO_2^{2-}, and the fact that this species is isoelectronic with the known linear molecule, KrF_2, suggests that it too should be linear.

a. Build KrF_2 as a linear molecule $(F—Kr—F)$, optimize its geometry using the HF/6-31G* model, and calculate vibrational frequencies. Is the calculated KrF bond distance close to the experimental value (1.89 Å)? Does the molecule prefer to be linear or does it want to bend? Explain how you reached this conclusion.

b. Build KrO_2^{2-} as a linear molecule (or as a bent molecule if the preceding analysis has shown that KrF_2 is not linear), optimize its structure using the HF/6-31G* model, and calculate vibrational frequencies. What is the structure of KrO_2^{2-}?

P26.11 Discussion of the VSEPR model in Section 25.1 suggested a number of failures, in particular, in CaF_2 and $SrCl_2$, which (according to the VSEPR) should be linear but which are apparently bent, and in SeF_6^{2-} and $TeCl_6^{2-}$, which should not be octahedral but apparently are. Are these really failures or does the discrepancy lie with the fact that the experimental structures correspond to the solid rather than the gas phase (isolated molecules)?

a. Obtain equilibrium geometries for linear CaF_2 and $SrCl_2$ and also calculate vibrational frequencies (infrared spectra). Use the HF/3-21G model, which has actually proven to be quite successful in describing the structures of main-group inorganic molecules. Are the linear structures for CaF_2 and $SrCl_2$ actually energy minima? Elaborate. If one or both are not, repeat your optimization starting with a bent geometry.

b. Obtain equilibrium geometries for octahedral SeF_6^{2-} and $TeCl_6^{2-}$ and also calculate vibrational frequencies. Use the HF/3-21G model. Are the octahedral structures for SeF_6^{2-} and $TeCl_6^{2-}$ actually energy minima? Elaborate. If one or both are not, repeat your optimization starting with distorted structures (preferably with C_1 symmetry).

P26.12 Benzyne has long been implicated as an intermediate in nucleophilic aromatic substitution, for example,

Although the geometry of benzyne has yet to be conclusively established, the results of a ^{13}C labeling experiment leave little doubt that two (adjacent) positions on the ring are equivalent:

There is a report, albeit controversial, that benzyne has been trapped in a low-temperature matrix and its infrared spectrum recorded. Furthermore, a line in the spectrum at 2085 cm^{-1} has been assigned to the stretching mode of the incorporated triple bond.

Optimize the geometry of benzyne using the HF/6-31G* model and calculate vibrational frequencies. For reference, perform the same calculations on 2-butyne. Locate the C≡C stretching frequency in 2-butyne and determine an appropriate scaling factor to bring it into agreement with the corresponding experimental frequency (2240 cm^{-1}). Then, identify the vibration corresponding to the triple-bond stretch in benzyne and apply the same scaling factor to this frequency. Finally, plot the calculated infrared spectra of both benzyne and 2-butyne.

Does your calculated geometry for benzyne incorporate a fully formed triple bond? Compare with the bond in 2-butyne as a standard. Locate the vibrational motion in benzyne corresponding to the triple bond stretch. Is the corresponding (scaled) frequency significantly different (>100 cm^{-1}) from the frequency assigned in the experimental investigation? If it is, are you able to locate any frequencies from your calculation that would fit with the assignment of a benzyne mode at 2085 cm^{-1}? Elaborate. Does the calculated infrared spectrum provide further evidence for or against the experimental observation? (*Hint*: Look at the intensity of the triple-bond stretch in 2-butyne.)

P26.13 All chemists know that benzene is unusually stable, that is, it is aromatic. They are also well aware that many other similar molecules are stabilized by aromaticity to some extent and, more often than not, can recognize aromatic molecules as those with delocalized bonding. What most chemists are unable to do, however, is to "put a number" on the aromatic stabilization afforded benzene or to quantify aromatic stabilization among different molecules. This is not to say that methods have not been proposed (for a discussion see Section 25.7), but rather that these methods have rarely been applied to real molecules.

Assigning a value to aromatic stabilization is actually quite straightforward. Consider a hypothetical reaction in which a molecule of hydrogen is added to benzene to yield 1,3-cyclohexadiene. Next, consider analogous hydrogenation reactions of 1,3-cyclohexadiene (leading to cyclohexene) and of cyclohexene (leading to cyclohexane):

Addition of H$_2$ to benzene trades an H—H bond and a C—C π bond for two C—H bonds, but in so doing destroys the aromaticity, whereas H$_2$ addition to either 1,3-cyclohexadiene or cyclohexene trades the same bonds but does not result in any loss of aromaticity (there is nothing to lose). Therefore, the difference in the heats of hydrogenation (134 kJ/mol referenced to 1,3-cyclohexadiene and 142 kJ/mol referenced to cyclohexene) is a measure of the aromaticity of benzene.

Reliable quantitative comparisons require accurate experimental data (heats of formation). These will generally be available only for very simple molecules and will almost never be available for novel interesting compounds. As a case in point, consider to what extent, if any, the 10 π-electron molecule 1,6-methanocyclodeca-1,3,5,7,9-pentaene ("bridged naphthalene") is stabilized by aromaticity. Evidence provided by the X-ray crystal structure suggests a fully delocalized π system. The 10 carbons that make up the base are very nearly coplanar and all CC bonds are intermediate in length between normal single and double linkages, just as they are in naphthalene:

1,6-Methanocyclodeca-1,3,5,7,9-pentaene   Naphthalene

Calculations provide a viable alternative to experiment for thermochemical data. Although absolute hydrogenation energies may be difficult to describe with currently practical models, hydrogenation energies relative to a closely related standard compound are much easier to accurately describe. In this case, the natural standard is benzene.

a. Optimize the geometries of benzene, 1,3-cyclohexadiene, naphthalene, and 1,2-dihydronaphthalene using the HF/6-31G* model. Evaluate the energy of the following reaction, relating the energy of hydrogenation of naphthalene to that of benzene (as a standard):

naphthalene   1,3-cyclohexadiene   1,2-dihydronaphthalene   benzene

On the basis of relative hydrogenation energies, would you say that naphthalene is stabilized (by aromaticity) to about the same extent as is benzene or to a lesser or greater extent? Try to explain your result.

b. Optimize the geometries of 1,6-methanocyclodeca-1,3,5,7,9-pentaene and its hydrogenation product using the HF/6-31G* model. Evaluate the energy of hydrogenation relative to that of naphthalene. On the basis of relative hydrogenation energies, would you say that the bridged naphthalene is stabilized to about the same extent as is naphthalene or to a lesser or greater extent? Try to explain your result.

P26.14 Singlet and triplet carbenes exhibit different properties and show markedly different chemistry. For example, a singlet carbene will add to a *cis*-disubstituted alkene to produce only *cis*-disubstituted cyclopropane products (and to a *trans*-disubstituted alkene to produce only *trans*-disubstituted cyclopropane products), whereas a triplet carbene will add to produce a mixture of *cis* and *trans* products.

The origin of the difference lies in the fact that triplet carbenes are biradicals (or diradicals) and exhibit chemistry similar to that exhibited by radicals, whereas singlet carbenes incorporate both a nucleophilic site (a low-energy unfilled molecular orbital) and an electrophilic site (a high-energy filled molecular orbital); for example, for singlet and triplet methylene:

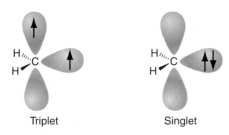

It should be possible to take advantage of what we know about stabilizing radical centers versus stabilizing empty orbitals and use that knowledge to design carbenes that will either be singlets or triplets. Additionally, it should be possible to say with confidence that a specific carbene of interest will either be a singlet or a triplet and, thus, to anticipate its chemistry.

The first step is to pick a model and then to establish the error in the calculated singlet–triplet energy separation in methylene where the triplet is known experimentally to be approximately 42 kJ/mol lower in energy than the singlet. This can then be applied as a correction for calculated singlet–triplet separations in other systems.

a. Optimize the structures of both the singlet and triplet states of methylene using both Hartree–Fock and B3LYP density functional models with the 6-31G* basis set. Which state (singlet or triplet) is found to be of lower energy according to the HF/6-31G* calculations? Is the singlet or the triplet unduly favored at this level of calculation? Rationalize your result. (*Hint*: Triplet methylene contains one fewer electron pair than singlet methylene.) What energy correction needs to be applied to calculated singlet–triplet energy separations? Which state (singlet or triplet) is found to be of lower energy according to the B3LYP/6-31G* calculations? What energy correction needs to be applied to calculated energy separations?

b. Proceed with either the HF/6-31G* or B3LYP/6-31G* model, depending on which leads to better agreement

for the singlet–triplet energy separation in methylene. Optimize singlet and triplet states for cyanomethylene, methoxymethylene, and cyclopentadienylidene:

Cyanomethylene	Methoxymethylene	Cyclopentadienylidene
NC—C—H	MeO—C—H	HC=C / HC—CH with CH=CH

Apply the correction obtained in the previous step to estimate the singlet–triplet energy separation in each. For each of the three carbenes, assign the ground state as singlet or triplet. Relative to hydrogen (in methylene), has the cyano substituent in cyanomethylene and the methoxy substituent in methoxymethylene led to favoring of the singlet or the triplet? Rationalize your result by first characterizing cyano and methoxy substituents as π donors or π acceptors, and then speculating about how a donor or acceptor would stabilize or destabilize singlet and triplet methylene. Has incorporation into a cyclopentadienyl ring led to increased preference for a singlet or triplet ground state (relative to the preference in methylene)? Rationalize your result. (*Hint*: Count the number of π electrons associated with the rings in both singlet and triplet states.)

P26.15 Electron-donating groups on benzene promote electrophilic aromatic substitution and lead preferentially to so-called *ortho* and *para* products over *meta* products, whereas electron-withdrawing groups retard substitution and lead preferentially to *meta* products (over *ortho* and *para* products), for example, for electrophilic alkylation:

We can expect the first step in the substitution to be addition of the electrophile, leading to a positively charged adduct:

So-called benzenium ions have been characterized spectroscopically and X-ray crystal structures for several are known. Will the stabilities of benzenium ion intermediates anticipate product distribution?

a. Optimize the geometries of benzene, aniline, and nitrobenzene using the HF/3-21G model. You will need their energies to ascertain the relative reactivities of the three substituted benzenes. Also, optimize the geometry of the benzenium ion using the HF/3-21G model. A good guess is a planar six-membered ring comprising five sp^2 carbons and an sp^3 carbon with bond distances between sp^2 carbons intermediate in length between single and double bonds. It should have C_{2v} symmetry. In terms of ring bond distances, how does your calculated structure compare with the experimental X-ray geometry of heptamethylbenzenium ion?

CH₃ CH₃ 1.49
H₃C CH₃
H₃C CH₃ 1.37
CH₃
1.42

b. Optimize the geometries of methyl cation adducts of benzene, aniline (*meta* and *para* isomers only), and nitrobenzene (*meta* and *para* isomers only) using the HF/3-21G model. Use the calculated structure of the parent benzenium ion as a template. Which isomer, *meta* or *para*, of the aniline adduct is more stable? Which isomer of the nitrobenzene adduct is more stable? Considering only the lower energy isomer for each system, order the binding energies of methyl cation adducts of benzene, aniline, and nitrobenzene, that is: E (substituted benzene methyl cation adduct) – E (substituted benzene) – E (methyl cation). You will need to calculate the energy of the methyl cation using the HF/3-21G model. Which aromatic compound should be most reactive? Which should be least reactive? Taken as a whole, do your results provide support for the involvement of benzenium ion adducts in electrophilic aromatic substitution? Explain.

P26.16 Aromatic molecules such as benzene typically undergo substitution when reacted with an electrophile such as Br_2, whereas alkenes such as cyclohexene most commonly undergo addition:

Addition   **Substitution**

What is the reason for the change in preferred reaction in moving from the alkene to the arene? Use the Hartree–Fock 6-31G* model to obtain equilibrium geometries and energies for reactants and products of both addition and substitution reactions of both cyclohexene and benzene (four reactions in total). Assume *trans* addition products (1,2-dibromocyclohexane and 5,6-dibromo-1,3-cyclohexadiene). Is your result consistent with what is actually observed? Are all four reactions exothermic? If one or more are not exothermic, provide a rationale as to why.

P26.17 Evaluate the difference between change in energy at 0 K in the absence of zero point vibration and both change in

enthalpy and in free energy for real molecules at 298 K. Consider both a unimolecular isomerization that does not lead to a net change in the number of molecules and a thermal decomposition reaction that leads to an increase in the number of molecules.

a. Calculate ΔU, $\Delta H(298)$, and $\Delta G(298)$ for the following isomerization reaction:

$$CH_3N{\equiv}C \longrightarrow CH_3C{\equiv}N$$

Obtain equilibrium geometries for both methyl isocyanide and acetonitrile using the B3LYP/6-31G* density functional model. Do the calculated values for ΔU and ΔH (298) differ significantly (by more than 10%)? If so, is the difference due primarily to the temperature correction or to the inclusion of zero point energy (or to a combination of both)? Is the calculated value for ΔG (298) significantly different from that of ΔH (298)?

b. Repeat your analysis (again using the B3LYP/6-31G* model) for the following pyrolysis reaction:

$$HCO_2CH_2CH_3 \longrightarrow HCO_2H + H_2C{\equiv}CH_2$$

Do these two reactions provide a similar or a different picture as to the importance of relating experimental thermochemical data to calculated ΔG values rather than ΔU values? If different, explain your result.

P26.18 Hydrazine would be expected to adopt a conformation in which the NH bonds stagger. There are two likely candidates, one with the lone pairs on nitrogen *anti* to each other and the other with the lone pairs *gauche*:

Anti hydrazine   *Gauche* hydrazine

On the basis of the same arguments made in VSEPR theory (electron pairs take up more space than bonds) you might expect that *anti* hydrazine would be the preferred structure.

a. Obtain energies for the *anti* and *gauche* conformers of hydrazine using the HF/6-31G* model. Which is the more stable conformer? Is your result in line with what you expect from VSEPR theory?

You can rationalize your result by recognizing that when electron pairs interact they form combinations, one of which is stabilized (relative to the original electron pairs) and one of which is destabilized. The extent of destabilization is greater than that of stabilization, meaning that overall interaction of two electron pairs is unfavorable energetically:

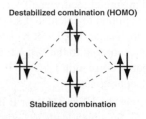

Destabilized combination (HOMO)

Stabilized combination

b. Measure the energy of the highest occupied molecular orbital (the HOMO) for each of the two hydrazine conformers. This corresponds to the higher energy (destabilized) combination of electron pairs. Which hydrazine conformer (*anti* or *gauche*) has the higher HOMO energy? Is this also the higher energy conformer? If so, is the difference in HOMO energies comparable to the difference in total energies between the conformers?

P26.19 Diels-Alder cycloaddition of 1,3-butadiene with acrylonitrile requires that the diene be in a *cis* (or *cis*-like) conformation:

In fact, 1,3-butadiene exists primarily in a *trans* conformation, the *cis* conformer being approximately 9 kJ/mol less stable and separated from the *trans* conformer by a low-energy barrier. At room temperature, only about 5% of butadiene molecules will be in a *cis* conformation. Clearly, rotation into a *cis* conformation is required before reaction can proceed.

Conduct a search for a substituted 1,3-butadiene that actually prefers to exist in a *cis* (or *cis*-like) conformation as opposed to a *trans* conformation. The only restriction you need to be aware of is that the diene needs to be electron rich in order to be reactive. Restrict your search to alkyl and alkoxy substituents as well as halogen. Use the HF/3-21G model. Report your successes and provide rationales.

P26.20 The energy of rotation about a single bond is a periodic function of the torsion angle, ϕ, and is, therefore, appropriately described in terms of a truncated Fourier series, the simplest acceptable form of which is given by

$$V(\phi) = \frac{1}{2}V_1(1 - \cos \phi) + \frac{1}{2}V_2(1 - \cos 2\phi)$$
$$+ \frac{1}{2}V_3(1 - \cos 3\phi)$$
$$= V_1(\phi) + V_2(\phi) + V_3(\phi)$$

Here, V_1 is the onefold component (periodic in 360°), V_2 is the twofold component (periodic in 180°), and V_3 is the threefold component (periodic in 120°).

A Fourier series is an example of an orthogonal polynomial, meaning that the individual terms which it comprises are independent of each other. It should be possible, therefore, to dissect a complex rotational energy profile into a series of *N*-fold components and to interpret each of these components independent of all others. The one-fold component is quite easy to rationalize. For example, the onefold term for rotation about the central bond in *n*-butane no doubt reflects the crowding of methyl groups,

whereas the onefold term in 1,2-difluoroethane probably reflects differences in electrostatic interactions as represented by bond dipoles:

The threefold component represents the difference in energy between eclipsed and staggered arrangements about a single bond. However, the twofold component is perhaps the most interesting of the three and is what concerns us here. It relates to the difference in energy between planar and perpendicular arrangements.

Optimize the geometry of dimethyl peroxide (CH_3OOCH_3) subject to the COOC dihedral angle being held at 0°, 20°, 40°, ..., 180° (10 optimizations in total). Use the B3LYP/6-31G* density functional model. Construct a plot of energy versus dihedral angle and fit this to a three-term Fourier series. Does the Fourier series provide a good fit to your data? If so, what is the dominant term? Rationalize it. What is the second most important term? Rationalize your result.

P26.21 Pyramidal inversion in the cyclic amine aziridine is significantly more difficult than inversion in an acyclic amine, for example, requiring 80 kJ/mol versus 23 kJ/mol in dimethylamine according to HF/6-31G* calculations. One plausible explanation is that the transition state for inversion needs to incorporate a planar trigonal nitrogen center, which is obviously more difficult to achieve in aziridine, where one bond angle is constrained to a value of around 60°, than it is in dimethylamine. Such an interpretation suggests that the barriers to inversion in the corresponding four- and five-membered ring amines (azetidine and pyrrolidine) should also be larger than normal and that the inversion barrier in the six-membered ring amine (piperidine) should be quite close to that for the acyclic.

Optimize the geometries of aziridine, azetidine, pyrrolidine, and piperidine using the HF/6-31G* model. Starting from these optimized structures, provide guesses at the respective inversion transition states by replacing the tetrahedral nitrogen center with a trigonal center. Obtain transition states using the same Hartree–Fock model and calculate inversion barriers. Calculate vibrational frequencies to verify that you have actually located the appropriate inversion transition states.

Do the calculated inversion barriers follow the order suggested in the preceding figure? If not, which molecule(s) appear to be anomalous? Rationalize your observations by considering other changes in geometry from the amine to the transition state.

P26.22 Molecules such as dimethylsulfoxide and dimethyl-sulfone can either be represented as *hypervalent*, that is, with more than the normal complement of eight valence electrons around sulfur, or as *zwitterions*, in which sulfur bears a positive charge:

$$
\begin{array}{ccc}
CH_3 & & CH_3 \\
\quad S=O & vs. & \quad S^{\pm}-O^{-} \\
CH_3 & & CH_3
\end{array}
\qquad
\begin{array}{ccc}
CH_3 & & CH_3 \\
\quad S{\lesseqgtr}^{O}_{O} & vs. & \quad S^{2\pm}{\lesseqgtr}^{O^{-}}_{O^{-}} \\
CH_3 & & CH_3
\end{array}
$$

Atomic charges obtained from quantum chemical calculations can help to decide which representation is more appropriate.

a. Obtain equilibrium geometries for dimethylsulfide, $(CH_3)_2S$, and dimethylsulfoxide using the HF/3-21G model and obtain charges at sulfur based on fits to the electrostatic potential. Is the charge on sulfur in dimethylsulfoxide about the same as that on sulfur in dimethylsulfide (normal sulfur), or has it increased by one unit, or is it somewhere between? Would you conclude that dimethylsulfoxide is best represented as a hypervalent molecule, as a zwitterion, or something between? See if you can support your conclusion with other evidence (geometries, dipole moments, and so on).

b. Repeat your analysis for dimethylsulfone. Compare your results for the charge at sulfur to those for dimethylsulfide and dimethylsulfoxide.

P26.23 Hydroxymethylene has never actually been observed, although it is believed to be an intermediate both in the photofragmentation of formaldehyde to hydrogen and carbon monoxide,

$$ H_2CO \xrightarrow{h\nu} [H\ddot{C}OH] \longrightarrow H_2 + CO $$

and in the photodimerization of formaldehyde in an argon matrix:

$$ H_2CO \xrightarrow{h\nu} [H\ddot{C}OH] \xrightarrow{H_2CO} HOCH_2CHO $$

Does hydroxymethylene actually exist? To have a chance "at life," it must be separated from both its rearrangement product (formaldehyde) and from its dissociation product (hydrogen and carbon monoxide) by a sizable energy barrier (>80 kJ/mol). Of course, it must also actually be a minimum on the potential energy surface.

a. First calculate the energy difference between formalde-hyde and hydroxymethylene and compare your result to the indirect experimental estimate of 230 kJ/mol. Try two different models, B3LYP/6-31G* and MP2/6-31G*. Following calculation of the equilibrium geometry for hydroxymethylene, obtain vibrational frequencies. Is hydroxymethylene an energy minimum? How do you know? Is the energy difference inferred from experiment reasonably well reproduced with one or both of the two models?

b. Proceed with the model that gives the better energy difference and try to locate transition states both for isomerization of hydroxymethylene to formaldehyde and for dissociation to hydrogen and carbon monoxide. Be certain to calculate vibrational frequencies for the two transition states. On the basis of transition states you have located, would you expect that both isomerization and dissociation reactions are available to hydroxymethylene? Explain. Do both suggest that hydroxymethylene is in a deep enough energy well to actually be observed?

P26.24 The three vibrational frequencies in H_2O (1595, 3657, and 3756 cm^{-1}) are all much larger than the corresponding frequencies in D_2O (1178, 1571, and 2788 cm^{-1}). This follows from the fact that vibrational frequency is given by the square root of a (mass-independent) quantity, which relates to the curvature of the energy surface at the minima, divided by a quantity that depends on the masses of the atoms involved in the motion.

As discussed in Section 26.8.4, vibrational frequencies enter into both terms required to relate the energy obtained from a quantum chemical calculation (stationary nuclei at 0 K) to the enthalpy obtained experimentally (vibrating nuclei at finite temperature), as well as the entropy required to relate enthalpies to free energies. For the present purpose, focus is entirely on the so-called zero point energy term, that is, the energy required to account for the latent vibrational energy of a molecule at 0 K.

The zero point energy is given simply as the sum over individual vibrational energies (frequencies). Thus, the zero point energy for a molecule in which isotopic substitution has resulted in an increase in mass will be reduced from that in the unsubstituted molecule:

A direct consequence of this is that enthalpies of bond disso-ciation for isotopically substituted molecules (light to heavy) are smaller than those for unsubstituted molecules.

a. Perform B3LYP/6-31G* calculations on HCl and on its dissociation products, chlorine atom and hydrogen atom. Following geometry optimization on HCl, calculate the vibrational frequency for both HCl and DCl and evaluate the zero point energy for each. In terms of a percentage of the total bond dissociation energy, what is the change noted in going from HCl to DCl?

d_1-Methylene chloride can react with chlorine atoms in either of two ways: by hydrogen abstraction (producing HCl) or by deuterium abstraction (producing DCl):

Which pathway is favored on the basis of thermodynamics and which is favored on the basis of kinetics?

b. Obtain the equilibrium geometry for dichloromethyl radical using the B3LYP/6-31G* model. Also obtain vibrational frequencies for both the unsubstituted and the deuterium-substituted radical and calculate zero point energies for the two abstraction pathways (you already have zero point energies for HCl and DCl). Which pathway is favored on the basis of thermodynamics? What would you expect the (thermodynamic) product ratio to be at room temperature?

c. Obtain the transition state for hydrogen abstraction from methylene chloride using the B3LYP/6-31G* model. A reasonable guess is shown here:

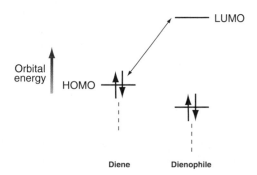

Calculate vibrational frequencies for the two possible structures with one deuterium and evaluate the zero point energies for these two structures. (For the purpose of zero point energy calculation, ignore the imaginary frequency corresponding to the reaction coordinate.) Which pathway is favored on the basis of kinetics? Is it the same or different from the thermodynamic pathway? What would you expect the (kinetic) product ratio to be at room temperature?

P26.25 Diels-Alder reactions commonly involve electron-rich dienes and electron-deficient dienophiles:

The rate of these reactions generally increases with the π-donor ability of the diene substituent, Y, and with the π-acceptor ability of the dienophile substituent, X. The usual interpretation is that electron donors will push up the energy of the HOMO on the diene and that electron acceptors will push down the energy of the LUMO on the dienophile:

The resulting decrease in the HOMO–LUMO gap leads to a stronger interaction between diene and dienophile and to a decrease in the activation barrier.

a. Obtain equilibrium geometries for acrylonitrile, 1,1 dicyanoethylene, *cis*- and *trans*-1,2-dicyanoethylene, tricyanoethylene, and tetracyanoethylene using the HF/3-21G model.

0   4.64   1.94
Acrylonitrile   1, 1-Dicyanoethene   *cis*-1,2-Dicyanoethene

1.89   5.66   7.61
trans-1, 2-Dicyanoethene   Tricyanoethene   Tetracyanoethene

Plot the LUMO energy for each dienophile versus the log of the observed relative rate for its addition to cyclopentadiene (listed below the structures in the preceding figure). Is there a reasonable correlation between LUMO energy and relative rate?

b. Obtain transition-state geometries for Diels-Alder cyclo-additions of acrylonitrile and cyclopentadiene and tetracyanoethylene and cyclopentadiene using the HF/3-21G model. Also obtain a geometry for cyclopentadiene. Calculate activation energies for the two reactions.

How does the calculated difference in activation energies compare with the experimental difference (based on a value of 7.61 for the difference in the log of the rates and assuming 298 K)?

P26.26 It is well known that cyanide acts as a "carbon" and not a "nitrogen" nucleophile in S_N2 reactions, for example,

$$:N\equiv C:\curvearrowright CH_3 \overset{\curvearrowright}{-}I \longrightarrow :N\equiv C-CH_3 + I^-$$

How can this behavior be rationalized with the notion that nitrogen is in fact more electronegative than carbon and, therefore, would be expected to hold any excess electrons?

a. Optimize the geometry of cyanide using the HF/3-21G model and examine the HOMO. Describe the shape of the HOMO of cyanide. Is it more concentrated on carbon or nitrogen? Does it support the picture of cyanide acting as a carbon nucleophile? If so, explain why your result is not at odds with the relative electronegativities of carbon and nitrogen.

Why does iodide leave following nucleophilic attack by cyanide on methyl iodide?

b. Optimize the geometry of methyl iodide using the HF/3-21G model and examine the LUMO. Describe the shape of the LUMO of methyl iodide. Does it anticipate the loss of iodide following attack by cyanide? Explain.

P26.27 At first glance, the structure of diborane would seem unusual. Why shouldn't the molecule assume the same geometry as ethane, which after all has the same number of heavy atoms and the same number of hydrogens?

Diborane   Ethane

The important difference between the two molecules is that diborane has two fewer electrons than ethane and is not able to make the same number of bonds. In fact, it is ethene which

684 **CHAPTER 26** Computational Chemistry

has the same number of electrons, to which diborane is structurally related.

Obtain equilibrium geometries for both diborane and ethene using the HF/6-31G* model and display the six valence molecular orbitals for each. Associate each valence orbital in ethene with its counterpart in diborane. Focus on similarities in the structure of the orbitals and not on their position in the lists of orbitals. To which orbital in diborane does the π orbital in ethene (the HOMO) best relate? How would you describe this orbital in diborane? Is it B—B bonding, B—H bonding, or both?

P26.28 Molecular orbitals are most commonly delocalized throughout the molecule and exhibit distinct bonding or antibonding character. Loss of an electron from a specific molecular orbital from excitation by light or by ionization would, therefore, be expected to lead to distinct changes in bonding and changes in molecular geometry.

a. Obtain equilibrium geometries for ethene, formaldimine, and formaldehyde using the HF/6-31G* model and display the highest occupied and lowest unoccupied molecular orbitals (HOMO and LUMO, respectively) for each. What would happen to the geometry around carbon (remain planar versus pyramidalize), to the C=X bond length, and (for formaldimine) to the C=NH bond angle if an electron were to be removed from the HOMO of ethene, formaldimine, and formaldehyde?

b. Obtain equilibrium geometries for radical cations of ethene, formaldimine, and formaldehyde using the HF/6-31G* model. Are the calculated geometries of these species, in which an electron has been removed from the corresponding neutral molecule, in line with your predictions based on the shape and nodal structure of the HOMO?

Unoccupied molecular orbitals are also delocalized and also show distinct bonding or antibonding character. Normally, this is of no consequence. However, were these orbitals to become occupied (from excitation or from capture of an electron), then changes in molecular geometry would also be expected. What would happen to the geometry around carbon, to the C=X bond length, and (for formaldimine) to the C=NH bond angle, if an electron were to be added to the LUMO of ethene, formaldimine, and formaldehyde?

c. Obtain equilibrium geometries for the radical anions of ethene, formaldimine, and formaldehyde using the HF/6-31G* model. Are the calculated geometries of these species, in which an electron has been added to the corresponding neutral molecule, in line with your predictions based on the shape and nodal structure of the LUMO?

The first excited state of formaldehyde (the so-called $n \rightarrow \pi^*$ state) can be thought of as arising from the promotion of one electron from the HOMO (in the ground state of formaldehyde) to the LUMO. The experimental equilibrium geometry of the molecule shows lengthening of the CO bond and a pyramidal carbon (ground-state values are shown in parentheses):

$$1.32\text{Å } (1.21\text{Å})$$
$$H\text{\tiny{\textbar\textbar\textbar}}C\text{\Large{=}}O$$
$$H \quad 154° \ (180°)$$

d. Rationalize this experimental result on the basis of what you know about the HOMO and LUMO in formaldehyde and your experience with calculations on the radical cation and radical anion of formaldehyde.

P26.29 BeH_2 is linear, whereas CH_2 with two additional electrons and H_2O with four additional electrons are both bent to a similar degree. Could these changes in geometry have been anticipated by examining the shapes of the bonding molecular orbitals?

a. Perform a series of geometry optimizations on BeH_2 with the bond angle constrained at 90°, 100°, 110°, ..., 180° (10 optimizations in total). Use the HF/6-31G* model. Plot the total energy, along with the HOMO and LUMO energies versus bond angle. Also, display the HOMO and LUMO for one of your structures of intermediate bond angle.

Does the energy of the HOMO of BeH_2 increase (more positive) or decrease in going from a bent to a linear structure, or does it remain constant, or is the energy at a minimum or maximum somewhere between? Would this result have been anticipated by examining the shape and nodal structure of the HOMO?

Does the energy of the LUMO of BeH_2 increase or decrease with increase in bond angle, or does it remain constant, or is the energy at a minimum or maximum somewhere between? Rationalize your result by reference to the shape and nodal structure of the LUMO. What do you anticipate would happen to the geometry of BeH_2 as electrons are added to the LUMO? Take a guess at the structure of $BH_2^\bullet$ (one electron added to the LUMO) and singlet CH_2 (two electrons added to the LUMO).

b. Optimize the geometries of (singlet) $BH_2^\bullet$ and singlet CH_2 using the HF/6-31G* model. Are the results of the quantum chemical calculations in line with your qualitative arguments?

c. Perform a series of geometry optimizations on singlet CH_2 with the bond angle constrained to 90°, 100°, 110°, ..., . Plot the total energy as a function of the angle as well as the HOMO and LUMO energies.

Display the LUMO for some intermediate structure. Does the plot of HOMO energy versus angle in CH_2 mirror the plot of LUMO energy versus angle in BeH_2? Rationalize your answer. Does the energy of the LUMO in CH_2 increase, decrease, or remain constant with increase in bond angle (or is it at a minimum or maximum somewhere between)? Is the change in LUMO energy smaller, larger, or about the same as the change in the energy of the HOMO over the same range of bond angles? Rationalize these two observations by reference to the shape and nodal structure of the LUMO. What do you anticipate would happen to the geometry of CH_2 as electrons are added to the LUMO? Take a guess at the structure of $NH_2^\bullet$ (one electron added to the LUMO) and H_2O (two electrons added to the LUMO).

d. Optimize the geometries of $NH_2^\bullet$ and H_2O using the HF/6-31G* model. Are the results of the quantum chemical calculations in line with your qualitative arguments?

P26.30 Olefins assume planar (or nearly planar) geometries wherever possible. This ensures maximum overlap between p orbitals and maximum π-bond strength. Any distortion away from planarity should reduce orbital overlap and bond strength. In principle, π-bond strength can be determined experimentally, by measuring the activation energy required for *cis-trans* isomerization, for example, in *cis*-1,2-dideuteroethylene:

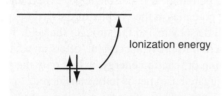

Another measure of π-bond strength, at least π-bond strength relative to a standard, is the energy required to remove an electron from the π orbital, or the ionization energy:

Ionization energy

Non-planar olefins might be expected to result from incorporation to a *trans* double bond into a small ring. Small-ring cycloalkenes prefer *cis* double bonds, and the smallest *trans* cycloalkene to actually have been isolated is cyclooctene. It is known experimentally to be approximately 39 kJ/mol less stable than *cis*-cyclooctene. Is this a measure of reduction in π bond strength?

Optimize the geometries of both *cis*- and *trans*-cyclooctene using the HF/3-21G model. (You should first examine the possible conformers available to each of the molecules.) Finally, calculate and display the HOMO for each molecule.

Is the double bond in *trans*-cyclooctene significantly distorted from its ideal planar geometry? If so, would you characterize the distortion as puckering of the double bond carbons or as twisting around the bond, or both? Does the HOMO in *trans*-cyclooctene show evidence of distortion? Elaborate. Is the energy of the HOMO in *trans*-cyclooctene significantly higher (less negative) than that in *cis*-cyclooctene? How does the energy difference compare to the experimentally measured difference in ionization potentials between the two isomers (0.29 eV)? How does the difference in HOMO energies (ionization potentials) relate to the calculated (measured) difference in isomer energies?

P26.31 Singlet carbenes add to alkenes to yield cyclopropanes. Stereochemistry is maintained, meaning that *cis*- and *trans*-substituted alkenes give *cis*- and *trans*-substituted cyclopropanes, respectively; for example:

This implies that the two σ bonds are formed more or less simultaneously, without the intervention of an intermediate that would allow *cis-trans* isomerization.

Locate the transition state for addition of singlet difluorocarbene and ethene using the HF/3-21G model and, following this, calculate vibrational frequencies. When completed, verify that you have in fact found a transition state and that it appears to be on the way to the correct product.

What is the orientation of the carbene relative to ethene in your transition state? Is it the same orientation as adopted in the product (1,1-difluorocyclopropane)? If not, what is the reason for the difference? (*Hint*: Consider that the π electrons on ethylene need to go into a low-lying unoccupied molecular orbital on the carbene. Build difluorocarbene and optimize its geometry using the HF/3-21G model and display the LUMO.)

P26.32 Further information about the mechanism of the ethyl formate pyrolysis reaction can be obtained by replacing the static picture with a movie, that is, an animation along the reaction coordinate. Bring up "ethyl formate pyrolysis" (on the Spartan download) and examine the change in electron density as the reaction proceeds. Do hydrogen migration and CO bond cleavage appear to occur in concert or is one leading the other?

P26.33 Do related molecules with the same number of electrons occupy the same amount of space, or are other factors (beyond electron count) of importance when dictating overall size requirements? Obtain equilibrium geometries for methyl anion, ammonia, and hydronium cation using the HF/6-31G* model and compare electron density surfaces corresponding to enclosure of 99% of the total electron density. Do the three molecules take up the same amount of space? If not, why not?

P26.34 Lithium provides a very simple example of the effect of oxidation state on overall size. Perform HF/6-31G* calculations on lithium cation, lithium atom, and lithium anion, and compare the three electron density surfaces corresponding to enclosure of 99% of the total electron density. Which is smallest? Which is largest? How does the size of lithium relate to the number of electrons? Which surface most closely resembles a conventional space-filling model? What, if anything does this tell you about the kinds of molecules that were used to establish the space-filling radius for lithium?

P26.35 A surface for which the electrostatic potential is negative delineates regions in a molecule that are subject to electrophilic attack. It can help you to rationalize the widely different chemistry of molecules that are structurally similar.

Optimize the geometries of benzene and pyridine using the HF/3-21G model and examine electrostatic potential surfaces corresponding to -100 kJ/mol. Describe the potential surface for each molecule. Use it to rationalize the following experimental observations: (1) Benzene and its derivatives undergo electrophilic aromatic substitution far more readily than do pyridine and its derivatives;

(2) protonation of perdeuterobenzene (C_6D_6) leads to loss of deuterium, whereas protonation of perdeuteropyridine (C_5D_5N) does not lead to loss of deuterium; and (3) benzene typically forms π-type complexes with transition models, whereas pyridine typically forms σ-type complexes.

P26.36 Hydrocarbons are generally considered to be nonpolar or weakly polar at best, characterized by dipole moments that are typically only a few tenths of a debye. For comparison, dipole moments for molecules of comparable size with heteroatoms are commonly several debyes. One recognizable exception is azulene, which has a dipole moment of 0.8 debye:

Azulene   Naphthalene

Optimize the geometry of azulene using the HF/6-31G* model and calculate an electrostatic potential map. For reference, perform the same calculations on naphthalene, a nonpolar isomer of azulene. Display the two electrostatic potential maps side by side and on the same (color) scale. According to its electrostatic potential map, is one ring in azulene more negative (relative to naphthalene as a standard) and one ring more positive? If so, which is which? Is this result consistent with the direction of the dipole moment in azulene? Rationalize your result. (*Hint*: Count the number of π electrons.)

As written, this is a highly *endothermic* process, because not only is a bond broken but two charged molecules are created from the neutral acid. It occurs readily in solution only because the solvent acts to disperse charge.

Acid strength can be calculated simply as the difference in energy between the acid and its conjugate base (the energy of the proton is 0). In fact, acid strength comparisons among closely related systems, for example, carboxylic acids, are quite well described with practical quantum chemical models. This is consistent with the ability of the same models to correctly account for relative base strengths (see discussion in Section 26.8.3).

Another possible measure of acid strength is the degree of positive charge on the acidic hydrogen as measured by the electrostatic potential. It is reasonable to expect that the more positive the potential in the vicinity of the hydrogen, the more easily it will dissociate and the stronger the acid. This kind of measure, were it to prove successful, offers an advantage over the calculation of reaction energy, in that only the acid (and not the conjugate base) needs to be considered.

a. Obtain equilibrium geometries for nitric acid, sulfuric acid, acetic acid, and ethanol using the HF/3-21G model, and compare electrostatic potential maps. Be certain to choose the same (color) scale for the four acids. For which acid is the electrostatic potential in the vicinity of (the acidic) hydrogen most positive? For which is it least positive? Do electrostatic potential maps provide a qualitatively correct account of the relative acid strength of these four compounds?

Acid	pKa	Acid	pKa
Cl_3CCO_2H	0.7	HCO_2H	3.75
HO_2CCO_2H	1.23	*trans*-$ClCH{=}CHCO_2H$	3.79
Cl_2CHCO_2H	1.48	$C_6H_5CO_2H$	4.19
$NCCH_2CO_2H$	2.45	*p*-$ClC_6H_4CH{=}CHCO_2H$	4.41
$ClCH_2CO_2H$	2.85	*trans*-$CH_3CH{=}CHCO_2H$	4.70
trans-$HO_2CCH{=}CHCO_2H$	3.10	CH_3CO_2H	4.75
p-$HO_2CC_6H_4CO_2H$	3.51	$(CH_3)_3CCO_2H$	5.03

P26.37 Chemists know that nitric and sulfuric acids are strong acids and that acetic acid is a weak acid. They would also agree that ethanol is at best a very weak acid. Acid strength is given directly by the energetics of deprotonation (heterolytic bond dissociation); for example, for acetic acid:

$$CH_3CO_2H \longrightarrow CH_3CO_2^- + H^+$$

b. Obtain equilibrium geometries for several of the carboxylic acids found in the following table using the HF/3-21G model and display an electrostatic potential map for each.

"Measure" the most positive value of the electrostatic potential associated with the acidic hydrogen in each of these compounds and plot this against experimental pKa (given in the preceding table). Is there a reasonable correlation between acid strengths and electrostatic potential at hydrogen in this closely related series of acids?

Molecular Symmetry

The combination of group theory and quantum mechanics provides a powerful tool for understanding the consequences of molecular symmetry. In this chapter, after a brief description of the most important aspects of group theory, several applications are discussed. They include using molecular symmetry to decide which atomic orbitals contribute to molecular orbitals, understanding the origin of spectroscopic selection rules, identifying the normal modes of vibration for a molecule, and determining if a particular molecular vibration is infrared active and/or Raman active.

27.1 Symmetry Elements, Symmetry Operations, and Point Groups

An individual molecule has an inherent symmetry based on the spatial arrangement of its atoms. For example, after a rotation of benzene by 60° about an axis that is perpendicular to the plane of the molecule and that passes through the center of the molecule, the molecule cannot be distinguished from the original configuration. Solid benzene in a crystalline form has additional symmetries that arise from the way in which individual benzene molecules are arranged in the crystal structure. These symmetry elements are essential in discussing diffraction of X rays. However, in this chapter, the focus is on the symmetry of an individual molecule.

Why is molecular symmetry useful to chemists? The symmetry of a molecule determines a number of its important properties. For example, CF_4 has no dipole moment, but H_2O has a dipole moment because of the symmetry of these molecules. All molecules have vibrational modes. However, the number of vibrational modes that are infrared and Raman active and the degeneracy of a given vibrational frequency depend on the molecular symmetry. Symmetry also determines the selection rules for transitions between states of the molecule in all forms of spectroscopy, and symmetry determines which atomic orbitals contribute to a given molecular orbital.

The focus in this chapter is on applying the predictive power of group theory to problems of interest in quantum chemistry, rather than on formally developing the mathematical framework. Therefore, results from group theory that are needed for specific applications are introduced without their derivations. These results are highlighted in shaded text boxes. Readers who wish to see these results derived or discussed in more detail are referred to standard texts such as *Symmetry and Structure* by

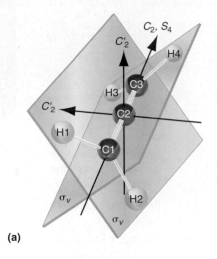

(a)

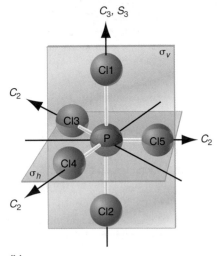

(b)

FIGURE 27.1
The symmetry elements of allene
(CH_2CCH_2) are shown in **(a)** and those
for PCl_5 are shown in **(b)**. Only one of the
three σ_v planes for PCl_5 is shown.

S. F. A. Kettle, *Molecular Symmetry and Group Theory* by R. L. Carter, and *Chemical Applications of Group Theory*, by F. A. Cotton. In Sections 27.1 through 27.5, the essentials of group theory that are needed to address problems of chemical interest are discussed. With a working knowledge of reducible and irreducible representations and character tables, several applications of group theory to chemistry are presented in the rest of the chapter. In Section 27.6, group theory is used to construct molecular orbitals (MOs) that incorporate the symmetry of the molecule under consideration from atomic orbitals (AOs). In Section 27.7, we discuss the normal modes for the vibration of molecules, and in Section 27.8, we show that symmetry determines whether a given vibrational mode of a molecule is infrared or Raman active. We will also show that symmetry determines the number of normal modes that have the same vibrational frequency.

We begin our discussion of molecular symmetry by discussing symmetry elements and symmetry operations. **Symmetry elements** are geometric entities such as axes, planes, or points with respect to which operations can be carried out. **Symmetry operations** are actions with respect to the symmetry elements that leave the molecule in a configuration that cannot be distinguished from the original configuration. There are only five different types of symmetry elements for an isolated molecule, although a molecule may require several elements of each type—n-fold rotation axes, n-fold rotation-reflection axes, or mirror planes—to fully define its symmetry. These elements and operations are listed in Table 27.1. Operators are indicated by a caret above the symbol.

Whereas other symmetry elements generate a single operation, C_n and S_n axes generate n operations. We choose the direction of rotation to be counterclockwise. However, if carried through consistently, either direction can be used. Examples of these symmetry elements are illustrated in Figure 27.1 for allene and PCl_5. Consider first the following symmetry elements for allene:

- A rotation of $360°/2 = 180°$ about the C_2 **rotation axes** passing through the carbon atoms leaves the molecule in a position that is indistinguishable from its initial position.
- A rotation of $360°/4 = 90°$ about the twofold axis discussed in the previous point, followed by a reflection through a plane perpendicular to the axis that passes through the central carbon atom also leaves the allene molecule unchanged. The combined operation is called an S_4 fourfold **rotation-reflection axis.** This axis and the C_2 rotation axis of the previous bullet are collinear.
- Two further C_2 rotation axes exist in this molecule. Both pass through the central carbon atom (C2). Consider the two planes shown in Figure 27.1a. One contains H1 and H2, and the other contains H3 and H4. The two C_2 axes bisect the angle between the two planes and, therefore, are perpendicular to one another.
- The molecule contains two mirror planes, as shown. Because they contain the main twofold axis, which is referred to as the vertical axis, they are designated with σ_v.

These symmetry elements for allene are shown in Figure 27.1. Consider next the PCl_5 molecule, which has the following symmetry elements:

- A threefold rotation axis C_3 that passes through Cl1, Cl2, and the central P atom.

TABLE 27.1   Symmetry Elements and Their Corresponding Operations

Symmetry Elements		Symmetry Operations	
E	Identity	$\hat{E}$	leave molecule unchanged
C_n	n-Fold rotation axis	$\hat{C}_n, \hat{C}_n^2, \ldots, \hat{C}_n^n$	rotate about axis by $360°/n$ 1, 2, ..., n times (indicated by superscript)
σ	Mirror plane	$\hat{\sigma}$	reflect through the mirror plane
i	Inversion center	$\hat{\imath}$	$(x, y, z) \rightarrow (-x, -y, -z)$
S_n	n-Fold rotation-reflection axis	$\hat{S}_n$	rotate about axis by $360°/n$, and reflect through a plane perpendicular to the axis.

- A **mirror plane,** σ_h, that passes through the centers of the three equatorial Cl atoms. Reflection through this plane leaves the equatorial Cl atoms in their original location and exchanges the axial Cl atoms.
- Three C_2 axes that pass through the central P atom and one of the equatorial Cl atoms.
- Three mirror planes, σ_v, that contain Cl1, Cl2, and P as well as one of Cl3, Cl4, or Cl5.

One of these planes is shown in Figure 27.1b. As we will see in Section 27.2, allene and PCl_5 can each be assigned to a group on the basis of symmetry elements of the molecule.

What is the relationship between symmetry elements, the symmetry operators, and the group? A set of symmetry elements forms a **group** if the following statements are true about their corresponding operators:

- The successive application of two operators is equivalent to one of the operations of the group. This guarantees that the group is closed.
- An **identity operator,** $\hat{E}$, exists that commutes with any other operator and leaves the molecule unchanged. Although this operator seems trivial, it plays an important role as we will see later. The identity operator has the property that $\hat{A}\hat{E} = \hat{E}\hat{A} = \hat{A}$ where $\hat{A}$ is an arbitrary element of the group.

- The group contains an **inverse operator** for each element in the group. If $\hat{B}^{-1}$ is the inverse operator of $\hat{B}$, then $\hat{B}\hat{B}^{-1} = \hat{B}^{-1}\hat{B} = \hat{E}$. If $\hat{A} = \hat{B}^{-1}$, then $\hat{A}^{-1} = \hat{B}$. In addition, $\hat{E} = \hat{E}^{-1}$.
- The operators are **associative,** meaning that $\hat{A}(\hat{B}\hat{C}) = (\hat{A}\hat{B})\hat{C}$.

The groups of interest in this chapter are called **point groups** because the set of symmetry elements intersects in a point or set of points. To utilize the power of group theory in chemistry, molecules are assigned to point groups on the basis of the symmetry elements characteristic of the particular molecule. Each point group has its own set of symmetry elements and corresponding operations. We work with several of these groups in more detail in the following sections.

27.2 Assigning Molecules to Point Groups

How is the point group to which a molecule belongs determined? The assignment is made using the logic diagram of Figure 27.2. To illustrate the use of this logic diagram, we assign NF_3, CO_2, and $Au(Cl_4)^-$ to specific point groups. In doing so, it is useful to first identify the major symmetry elements. After a tentative assignment of a point group is made based on these symmetry elements, it is necessary to verify that the other symmetry elements of that group are also present in the molecule. We start at the top of the diagram and follow the branching points.

NF_3 is a pyramidal molecule that has a threefold axis (C_3) passing through the N atom and a point in the plane of the F atoms that is equidistant from all three F atoms. NF_3 has no other rotation axes. The molecule has three mirror planes in which the C_3 axis, the N atom, and one F atom lie. These planes are perpendicular to the line connecting the other two fluorine atoms. Because the C_3 axis lies in the mirror plane, we conclude that NF_3 belongs to the C_{3v} group. The pathway through the logic diagram of Figure 27.2 is shown as a red line.

Carbon dioxide is a linear molecule with an **inversion center.** These symmetry characteristics uniquely specify CO_2 as belonging to the $D_{\infty h}$ group. The ∞ appears rather than the subscript n because any rotation about the molecular axis leaves the molecule unchanged.

$Au(Cl_4)^-$ is a square planar complex with a C_4 axis. It has C_2 axes perpendicular to the C_4 axis, but no other C_n axis with $n > 2$. It has mirror planes, one of which is perpendicular to the C_4 axis. Therefore, this complex belongs to the D_{4h} group. Trace the

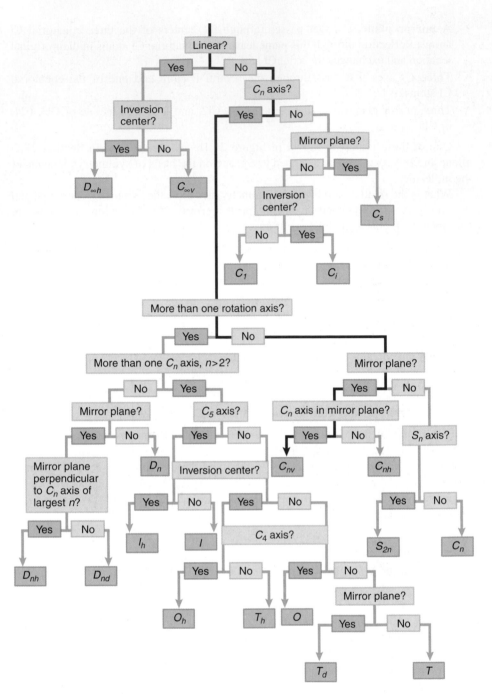

FIGURE 27.2

Logic diagram indicating how molecules are assigned to point groups. The red line indicates how NF_3 is assigned to the C_{3v} point group.

paths through the logic diagram for these molecules to see if you would have made the same assignments.

These examples illustrate how a given molecule can be assigned to a point group, but have only utilized a few of the symmetry operations of a given group. A number of point groups applicable to small molecules are listed in Table 27.2. All symmetry elements of the group are listed. Note that several groups have different categories or **classes** of symmetry elements such as C_n and σ, which are indicated by single and double primes. Classes are defined in Section 27.3.

The preceding discussion of the symmetry elements of a group has been of a general nature. In the following section, we discuss the symmetry elements of the C_{2v} group, to which water belongs, in greater detail.

TABLE 27.2   Selected Point Groups and Their Elements

Point Group	Symmetry Elements	Example Molecule
C_s	E, σ	BFClBr (planar)
C_2	E, C_2	H_2O_2
C_{2v}	E, C_2, σ, σ'	H_2O
C_{3v}	$E, C_3, C_3^2, 3\sigma$	NF_3
$C_{\infty v}$	$E, C_\infty, \infty\sigma$	HCl
C_{2h}	E, C_2, σ, i	$trans$-$C_2H_2F_2$
D_{2h}	$E, C_2, C_2', C_2'', \sigma, \sigma', \sigma'', i$	C_2F_4
D_{3h}	$E, C_3, C_3^2, 3C_2, S_3, S_3^2, \sigma, 3\sigma'$	SO_3
D_{4h}	$E, C_4, C_4^3, C_2, 2C_2', 2C_2'', i, S_4, S_4^3, \sigma, 2\sigma', 2\sigma''$	XeF_4
D_{6h}	$E, C_6, C_6^5, C_3, C_3^2, C_2, 3C_2', 3C_2'', i, S_3, S_3^2,$ $S_6, S_6^5, \sigma, 3\sigma', 3\sigma''$	C_6H_6 (benzene)
$D_{\infty h}$	$E, C_\infty, S_\infty, \infty C_2, \infty\sigma, \sigma', i$	H_2, CO_2
T_d	$E, 4C_3, 4C_3^2, 3C_2, 3S_4, 3S_4^3, 6\sigma$	CH_4
O_h	$E, 4C_3, 4C_3^2, 6C_2, 3C_4, 3C_2, i, 3S_4, 3S_4^3,$ $4S_6, 4S_6^5, 3\sigma, 6\sigma'$	SF_6

27.3 The H₂O Molecule and the C_{2v} Point Group

To gain practice in working with the concepts introduced in the preceding section, we next consider a specific molecule, express the symmetry operators mathematically, and show that the symmetry elements form a group. We do so by representing the operators as matrices and showing the requirements that the elements of any group must meet for this particular group.

Figure 27.3 shows all the symmetry elements for the water molecule. By convention, the rotation axis of highest symmetry (principal rotation axis), C_2, is oriented

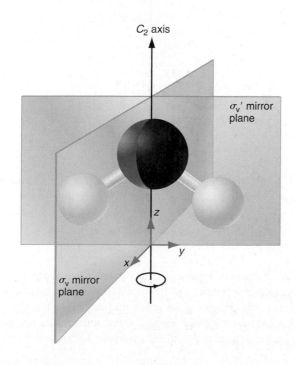

FIGURE 27.3
The water molecule is shown together with its symmetry elements. Convince yourself that the two mirror planes are in different classes.

along the z axis. The C_2 axis passes through the O atom. The molecule has two mirror planes oriented at 90° to one another, and their line of intersection is the C_2 axis. Because the mirror planes contain the principal rotation axis, the symmetry planes are referred to as vertical planes and designated by the subscript v. Mirror planes perpendicular to the principal rotation axis are referred to as horizontal and are designated by the subscript h. The molecule lies in the plane designated σ_v', and the second mirror plane, designated σ_v, bisects the H—O—H bond angle. As shown in Example Problem 27.1, these two mirror planes belong to different classes and, therefore, have different symbols.

> Elements that belong to the same class can be transformed into one another by other symmetry operations of the group. For example, the operators $\hat{C}_n, \hat{C}_n^2, \ldots, \hat{C}_n^n$, belong to the same class.

EXAMPLE PROBLEM 27.1

a. Are the three mirror planes for the NF_3 molecule in the same or in different classes?

b. Are the two mirror planes for H_2O in the same or in different classes?

Solution

a. NF_3 belongs to the C_{3v} group, which contains the rotation operators $\hat{C}_3, \hat{C}_3^2 = (\hat{C}_3)^{-1}$, and $\hat{C}_3^3 = \hat{E}$ and the vertical mirror planes $\hat{\sigma}_v(1)$, $\hat{\sigma}_v(2)$, and $\hat{\sigma}_v(3)$. These operations and elements are illustrated by this figure:

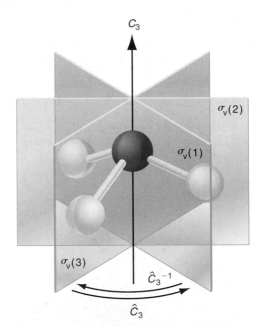

We see that $\hat{C}_3$ converts $\sigma_v(1)$ to $\sigma_v(3)$, and $\hat{C}_3^2 = (\hat{C}_3)^{-1}$ converts $\sigma_v(1)$ to $\sigma_v(2)$. Therefore, all three mirror planes belong to the same class.

b. Figure 27.3 shows that neither the $\hat{C}_2$ nor the $\hat{E}$ operation converts σ_v to σ_v'. Therefore, these two mirror planes are in different classes.

Using the logic diagram of Figure 27.2, we conclude that H_2O belongs to the C_{2v} group. This point group is given the shorthand notation C_{2v} because it has a C_2 axis and vertical mirror planes. The C_{2v} group has four symmetry elements: the identity element, a C_2 rotation axis, and two mutually perpendicular mirror planes. The corresponding operators are the identity operator $\hat{E}$ and the operators $\hat{C}_2, \hat{\sigma}$, and $\hat{\sigma}'$.

To understand how these operators act, we must introduce mathematical representations of the operators and then carry out the operations. To do so, the operators of the C_{2v} group are represented by 3×3 matrices, which act on a vector in three-dimensional space. See the Math Supplement (Appendix A) for an introduction to working with matrices.

Consider the effect of the symmetry operators on an arbitrary vector $\mathbf{r} = (x_1, y_1, z_1)$, originating at the intersection of the mirror planes and the C_2 axis. The vector $\mathbf{r}$ is converted to the vector (x_2, y_2, z_2) through the particular symmetry operation. We begin with a counterclockwise rotation by the angle θ about the z axis. As Example Problem 27.2 shows, the transformation of the components of the vector is described by Equation (27.1):

$$\begin{pmatrix} x_2 \\ y_2 \\ z_2 \end{pmatrix} = \begin{pmatrix} \cos\theta & -\sin\theta & 0 \\ \sin\theta & \cos\theta & 0 \\ 0 & 0 & 1 \end{pmatrix} \begin{pmatrix} x_1 \\ y_1 \\ z_1 \end{pmatrix} \tag{27.1}$$

EXAMPLE PROBLEM 27.2

Show that a rotation about the z axis can be represented by the matrix

$$\begin{pmatrix} \cos\theta & -\sin\theta & 0 \\ \sin\theta & \cos\theta & 0 \\ 0 & 0 & 1 \end{pmatrix}$$

Show that for a rotation of $180°$ this matrix takes the form

$$\begin{pmatrix} -1 & 0 & 0 \\ 0 & -1 & 0 \\ 0 & 0 & 1 \end{pmatrix}$$

Solution

The z coordinate is unchanged in a rotation about the z axis, so we need only consider the vectors $\mathbf{r}_1 = (x_1, y_1)$ and $\mathbf{r}_2 = (x_2, y_2)$ in the xy plane. These equations can be derived from the figure that follows:

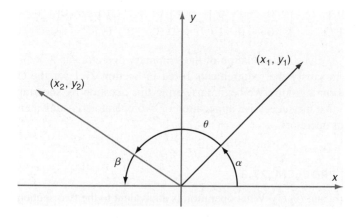

$$\theta = 180° - \alpha - \beta$$
$$x_1 = r\cos\alpha, \quad y_1 = r\sin\alpha$$
$$x_2 = -r\cos\beta, \quad y_2 = r\sin\beta$$

Using the identities $\cos(\phi \pm \delta) = \cos\phi\cos\delta \mp \sin\phi\sin\delta$ and $\sin(\phi \pm \delta) = \sin\phi\cos\delta \pm \cos\phi\sin\delta$, x_2 and y_2 can be expressed in terms of θ and α.

$$x_2 = -r\cos\beta = -r\cos(180° - \alpha - \theta)$$
$$= r\sin 180° \sin(-\theta - \alpha) - r\cos 180° \cos(-\theta - \alpha)$$
$$= r\cos(-\theta - \alpha) = r\cos(\theta + \alpha) = r\cos\theta\cos\alpha - r\sin\theta\sin\alpha$$
$$= x_1\cos\theta - y_1\sin\theta$$

Using the same procedure, it can be shown that $y_2 = x_1 \sin \theta + y_1 \cos \theta$.

The coordinate z is unchanged in the rotation, so that $z_2 = z_1$. The three equations

$$x_2 = x_1 \cos \theta - y_1 \sin \theta$$
$$y_2 = x_1 \sin \theta + y_1 \cos \theta \quad \text{and}$$
$$z_2 = z_1$$

can be expressed in the matrix form

$$\begin{pmatrix} x_2 \\ y_2 \\ z_2 \end{pmatrix} = \begin{pmatrix} \cos \theta & -\sin \theta & 0 \\ \sin \theta & \cos \theta & 0 \\ 0 & 0 & 1 \end{pmatrix} \begin{pmatrix} x_1 \\ y_1 \\ z_1 \end{pmatrix}$$

Because $\cos(180°) = -1$ and $\sin(180°) = 0$, the matrix for $180°$ rotation around the z axis takes the form

$$\begin{pmatrix} -1 & 0 & 0 \\ 0 & -1 & 0 \\ 0 & 0 & 1 \end{pmatrix}$$

The effect of the four operators, $\hat{E}$, $\hat{C}_2$, $\hat{\sigma}_v$, and $\hat{\sigma}_v'$ on **r** can also be deduced from Figure 27.4. Convince yourself, using Example Problem 27.1 and Figure 27.4, that the symmetry operators of the C_{2v} group have the following effect on the vector (x, y, z):

$$\hat{E} \begin{pmatrix} x \\ y \\ z \end{pmatrix} \Rightarrow \begin{pmatrix} x \\ y \\ z \end{pmatrix}, \quad \hat{C}_2 \begin{pmatrix} x \\ y \\ z \end{pmatrix} \Rightarrow \begin{pmatrix} -x \\ -y \\ z \end{pmatrix}, \quad \hat{\sigma}_v \begin{pmatrix} x \\ y \\ z \end{pmatrix} \Rightarrow \begin{pmatrix} x \\ -y \\ z \end{pmatrix}, \quad \hat{\sigma}_v' \begin{pmatrix} x \\ y \\ z \end{pmatrix} \Rightarrow \begin{pmatrix} -x \\ y \\ z \end{pmatrix} \quad (27.2)$$

Given these results, the operators $\hat{E}$, $\hat{C}_2$, $\hat{\sigma}_v$, and $\hat{\sigma}_v'$ can be described by the following 3×3 matrices:

$$\hat{E}: \begin{pmatrix} 1 & 0 & 0 \\ 0 & 1 & 0 \\ 0 & 0 & 1 \end{pmatrix} \quad \hat{C}_2: \begin{pmatrix} -1 & 0 & 0 \\ 0 & -1 & 0 \\ 0 & 0 & 1 \end{pmatrix} \quad \hat{\sigma}_v: \begin{pmatrix} 1 & 0 & 0 \\ 0 & -1 & 0 \\ 0 & 0 & 1 \end{pmatrix} \quad \hat{\sigma}_v': \begin{pmatrix} -1 & 0 & 0 \\ 0 & 1 & 0 \\ 0 & 0 & 1 \end{pmatrix} (27.3)$$

Equation (27.3) gives a formulation of the symmetry operators as 3×3 matrices. Do these operators satisfy the requirements listed in Section 27.1 for the corresponding elements to form a group? We begin answering this question by showing in Example Problem 27.3 that the successive application of two operators is equivalent to applying one of the four operators.

EXAMPLE PROBLEM 27.3

Evaluate $\hat{C}_2 \hat{\sigma}_v$ and $\hat{C}_2 \hat{C}_2$. What operation is equivalent to the two sequential operations?

Solution

$$\hat{C}_2 \hat{\sigma}_v = \begin{pmatrix} -1 & 0 & 0 \\ 0 & -1 & 0 \\ 0 & 0 & 1 \end{pmatrix} \begin{pmatrix} 1 & 0 & 0 \\ 0 & -1 & 0 \\ 0 & 0 & 1 \end{pmatrix} = \begin{pmatrix} -1 & 0 & 0 \\ 0 & 1 & 0 \\ 0 & 0 & 1 \end{pmatrix} = \hat{\sigma}_v'$$

$$\hat{C}_2 \hat{C}_2 = \begin{pmatrix} -1 & 0 & 0 \\ 0 & -1 & 0 \\ 0 & 0 & 1 \end{pmatrix} \begin{pmatrix} -1 & 0 & 0 \\ 0 & -1 & 0 \\ 0 & 0 & 1 \end{pmatrix} = \begin{pmatrix} 1 & 0 & 0 \\ 0 & 1 & 0 \\ 0 & 0 & 1 \end{pmatrix} = \hat{E}$$

We see that the product of the two operators is another operator of the group.

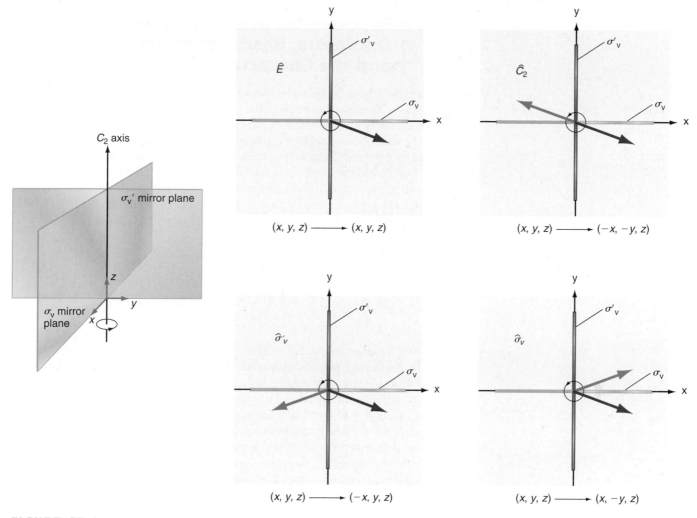

FIGURE 27.4
Schematic of the effect of the four symmetry operations of the C_{2v} group on an arbitrary vector (x, y, z). The symmetry elements are shown on the left. Because z is unchanged through any of the operations, it is sufficient to determine the changes in the xy coordinates through the symmetry operations. This is shown on the right side of the figure viewed along the C_2 axis. The wide lines along the x and y axis represent the σ_v and σ'_v mirror planes, respectively. The red vector is transformed into the green vector in each case.

By repeating the procedure from Example Problem 27.3 with all possible combinations of operators, Table 27.3 can be generated. This table shows that, as required, the result of any two successive operations is another of these four symmetry operations. The table also shows that $\hat{C}_2\hat{C}_2 = \hat{\sigma}_v\hat{\sigma}_v = \hat{\sigma}'_v\hat{\sigma}'_v = \hat{E}$. Each operator has an inverse operator in the group, and in this particular case, each operator is its own inverse operator. The operations are also associative, which can be shown by evaluating an arbitrary combination of three operators such as $\hat{\sigma}_v(\hat{C}_2\hat{\sigma}'_v) - (\hat{\sigma}_v\hat{C}_2)\hat{\sigma}'_v$. If the operators are associative, this expression will equal zero. Using the multiplication table to evaluate the products in parentheses in the following equation, the result is

$$\hat{\sigma}_v(\hat{C}_2\hat{\sigma}'_v) - (\hat{\sigma}_v\hat{C}_2)\hat{\sigma}'_v = \hat{\sigma}_v\hat{\sigma}_v - \hat{\sigma}'_v\hat{\sigma}'_v = \hat{E} - \hat{E} = 0 \qquad (27.4)$$

You can convince yourself that any other combination of three operators will give the same result. We have now shown that the four symmetry elements characteristic of the water molecule satisfy the requirements of a group.

In this section, it was useful to express the operators of the C_{2v} group as 3×3 matrices in order to generate the group multiplication table. It turns out that these operators can be expressed in many different ways. This important topic is discussed in the following section.

TABLE 27.3 Multiplication Table for Operators of the C_{2v} Group

Second Operation	First Operation			
	$\hat{E}$	$\hat{C}_2$	$\hat{\sigma}_v$	$\hat{\sigma}'_v$
$\hat{E}$	$\hat{E}$	$\hat{C}_2$	$\hat{\sigma}_v$	$\hat{\sigma}'_v$
$\hat{C}_2$	$\hat{C}_2$	$\hat{E}$	$\hat{\sigma}'_v$	$\hat{\sigma}_v$
$\hat{\sigma}_v$	$\hat{\sigma}_v$	$\hat{\sigma}'_v$	$\hat{E}$	$\hat{C}_2$
$\hat{\sigma}'_v$	$\hat{\sigma}'_v$	$\hat{\sigma}_v$	$\hat{C}_2$	$\hat{E}$

27.4 Representations of Symmetry Operators, Bases for Representations, and the Character Table

The matrices derived in the previous section are called **representations** of that group, meaning that the multiplication table of the group can be reproduced with the matrices. For this group the symmetry operators can be represented by numbers, and these numbers obey the multiplication table of a group. How can the operators of the C_{2v} group be represented by numbers? Surprisingly, each operation can be represented by either the number $+1$ or -1 and the multiplication table is still satisfied. As shown later, this is far from a trivial result. You will show in the end-of-chapter problems that the following four sets of $+1$ and -1, denoted Γ_1 through Γ_4, each satisfy the C_{2v} multiplication table and, therefore, are individual representations of the C_{2v} group:

Representation	E	C_2	σ_v	σ_v'
Γ_1	1	1	1	1
Γ_2	1	1	-1	-1
Γ_3	1	-1	1	-1
Γ_4	1	-1	-1	1

Other than the trivial set in which the value zero is assigned to all operators, no other set of numbers satisfies the multiplication table. The fact that a representation of the group can be constructed using only the numbers $+1$ and -1 means that 1×1 matrices are sufficient to describe all operations of the C_{2v} group. This conclusion can also be reached by noting that all four 3×3 matrices derived in the previous section are diagonal, meaning that x, y, and z transform independently in Equation (27.2).

It is useful to regard the set of numbers for an individual representation as a row vector, which we designate Γ_1 through Γ_4 for the C_{2v} group. Each group has an infinite number of different representations. For example, had we considered a Cartesian coordinate system at the position of each atom in water, we could have used 9×9 matrices to describe the operators. However, a much smaller number of representations, called **irreducible representations,** play a fundamental role in group theory. The irreducible representations are the matrices of smallest dimension that obey the multiplication table of the group. We cite the following theorem from group theory:

> A group has as many irreducible representations as it has classes of symmetry elements.

Irreducible representations play a central role in discussing molecular symmetry. We explore irreducible representations in greater depth in the next section.

Because the C_{2v} group has four classes of symmetry elements, only four different irreducible representations of this group are possible. This is an important result that we will return to in Section 27.5. The usefulness of these representations in quantum chemistry can be seen by considering the effect of symmetry operations on the oxygen AOs in H_2O. Consider the three different $2p$ atomic orbitals on the oxygen atom shown in Figure 27.5.

How are the three oxygen $2p$ orbitals transformed under the symmetry operations of the C_{2v} group? Numbers are assigned to the transformation of the $2p$ orbitals in the following way. If the sign of each lobe is unchanged by the operation, $+1$ is assigned to the transformation. If the sign of each lobe is changed, -1 is assigned to the transformation. These are the only possible outcomes for the symmetry operations of the C_{2v} group. We consider the $2p_z$ AO first. Figure 27.5 shows that the sign of each lobe remains the same after each operation. Therefore, we assign $+1$ to each operation. For the $2p_x$ AO, the $\hat{C}_2$ rotation and the $\hat{\sigma}_v'$ reflection change the sign of each lobe, but the sign of each lobe is unchanged after the $\hat{E}$ and $\hat{\sigma}_v$ operations. Therefore, we assign $+1$ to the $\hat{E}$ and $\hat{\sigma}_v$ operators and -1 to the $\hat{C}_2$ and $\hat{\sigma}_v'$ operators. Similarly, for the $2p_y$ AO, we assign $+1$ to the $\hat{E}$ and $\hat{\sigma}_v'$ operators and -1 to the $\hat{C}_2$ and $\hat{\sigma}_v$ operators. Note that if

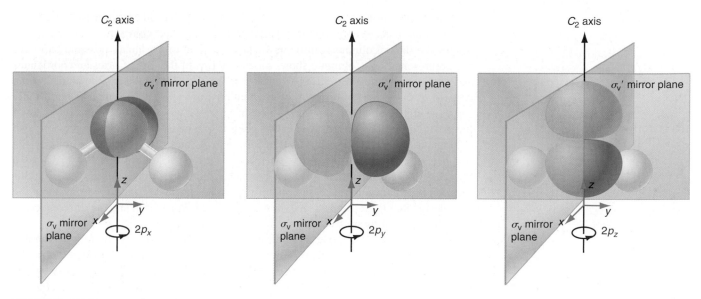

FIGURE 27.5
The three p orbitals on the oxygen atom transform differently under the symmetry operations of the C_{2v} group.

you arrange the numbers $+1$ and -1 obtained separately for the $2p_z$, $2p_x$, and $2p_y$ orbitals in the order $\hat{E}$, $\hat{C}_2$, $\hat{\sigma}_v$, and $\hat{\sigma}'_v$, the sequences that we have just derived are identical to the first, third, and fourth representations for the C_{2v} group.

Because each of the $2p_z$, $2p_x$, and $2p_y$ AOs can be associated with a different representation, each of these AOs **forms a basis** for one of the representations. Had we considered an unoccupied $3d_{xy}$ AO on the oxygen, we would have found that it forms a basis for the second representation. In the nomenclature used in group theory, one says that an AO, or any other function, **belongs to a particular representation** if it forms a basis for that representation. To this point, we have shown that 3×3 matrices, an appropriate set of the numbers $+1$ and -1, and the AOs of oxygen all form a basis for the C_{2v} point group.

The information on the possible representations discussed can be assembled in a form known as a **character table.** Each point group has its unique character table. The character table for the C_{2v} group is as follows:

	E	C_2	σ_v	σ'_v			
A_1	1	1	1	1	z	x^2, y^2, z^2	$2p_z(O)$
A_2	1	1	-1	-1	R_z	xy	$3d_{xy}(O)$
B_1	1	-1	1	-1	x, R_y	xz	$2p_x(O)$
B_2	1	-1	-1	1	y, R_x	yz	$2p_y(O)$

Much of the information in this character table was derived in order to make the origin of the individual entries clear. However, this task is not necessary, because character tables for point groups in this standard format are widely accessible and are listed in Appendix D.

The character table is the single most important result of group theory for chemists. Therefore, the structure and individual entries in the character table are now discussed in detail. The leftmost column in a character table shows the symbol for each irreducible representation. By convention, a representation that is symmetric $(+1)$ with respect to rotation about the principal axis, C_2 in this case, is given the symbol A. A representation that is antisymmetric (-1) with respect to rotation about the principal axis is given the symbol B. The subscript 1 (2) is used for representations that are symmetric (antisymmetric) with respect to a C_2 axis perpendicular to the principal axis. If such an axis is not an element of the group, the symmetry with respect to a vertical mirror plane, $\hat{\sigma}_v$ in this case, is used. The representation in which all entries are $+1$ is called the **totally symmetric representation.** Every group has a totally symmetric representation.

The next section of the table (columns 2 through 5) has an entry for each operation of the group in each representation. These entries are called **characters.** The right section of the table (columns 6 through 8) shows several of the many possible bases for each representation. Column 6 shows bases in terms of the three Cartesian coordinates and rotations about the three axes. Column 8 shows the AOs on the oxygen atom that can be used as bases for the different representations; note that this column is not usually shown in character tables. It is shown here because we will work further with this set of **basis functions.** The information in this column can be inferred from the previous two columns as the p_x, p_y, and p_z AOs transform as x, y, and z, respectively. Similarly, the d_{z^2}, d_{xy}, d_{yz}, $d_{x^2-y^2}$, and d_{xy} AOs transform as their subscript indices. The s AOs are a basis for A_1 because of their spherical symmetry. Next consider columns 6 and 7 in this section, which have entries based on the x, y, and z coordinates and rotations about the axes designated R_x, R_y, and R_z. We show later that the R_z rotation and the different coordinate combinations are bases for the indicated representations.

How can it be shown that the indicated functions are bases for the four irreducible representations? Equation (27.2) shows that the effect of any of the C_{2v} operators on the components x, y, and z of an arbitrary three-dimensional vector are $x \rightarrow \pm x$, $y \rightarrow \pm y$, and $z \rightarrow z$. Because z does not change sign under any of the operators, all characters for the representation have the value $+1$. Therefore, z is a basis for the A_1 representation. Similarly, because x^2, y^2, and z^2 do not change under any of the operations, these functions are also bases for the A_1 representation. Equation (27.2) shows that the product $xy \rightarrow xy$ for $\hat{E}$ and $\hat{C}_2$ and $xy \rightarrow -xy$ for $\hat{\sigma}_v$ and $\hat{\sigma}'_v$. Therefore, the product xy is a basis for the A_2 representation. Because z does not change sign under any operation, xz and yz transform as x and y. Therefore, Equation (27.2) shows that the functions x and xz are bases for the B_1 representation, and y and yz are bases for the B_2 representation.

Example Problem 27.1 demonstrated that in the operation R_z (C_2 in this case), $x \rightarrow -x$, $y \rightarrow -y$, and $z \rightarrow z$. Therefore, the product xy is unchanged because $xy \rightarrow (-x)(-y) = xy$. This shows that both R_z and xy are bases for the A_2 representation. We will not prove that R_x and R_y are bases for the B_1 and B_2 representations, but the procedure to do so is the same as for the other representations. As we saw in Section 27.3, the rotation operators are three-dimensional matrices. Therefore, in contrast to the coordinate bases, the rotation operators are bases for **reducible representations,** because their dimension is greater than one.

As shown earlier, all irreducible representations of the C_{2v} group are one dimensional. However, it is useful to consider reducible representations for this group such as R_x, R_y, and R_z, all of which are three dimensional, to visualize how individual operators act on an arbitrary vector. Some of the groups discussed in this chapter also have irreducible representations whose dimensionality is two or three. Therefore, before we begin to work on problems of chemical interest using character tables, it is necessary to discuss the dimensionality of irreducible representations.

27.5 The Dimension of a Representation

The bases for the different representations of the C_{2v} group include either x or y or z, but not a linear combination of two coordinates such as $x + y$. This is the case because under any transformation $(x, y, z) \rightarrow (x', y', z')$, x' is only a function of x as opposed to being a function of x and y or x and z or x, y, and z. Similar statements can be made for y' and z'. As a consequence, all of the matrices that describe the operators for the C_{2v} group have a diagonal form, as shown in Equation (27.3).

The matrix generated by two successive operations of diagonal matrices, which is denoted by $\hat{R}''' = \hat{R}'\hat{R}''$, is also a diagonal matrix whose elements are given by

$$\hat{R}'''_{ii} = \hat{R}'_{ii}\hat{R}''_{ii} \tag{27.5}$$

The **dimension of a representation** is defined as the size of the matrix used to represent the symmetry operations. As discussed earlier, the matrices of Equation (27.5) form a three-dimensional representation of the C_{2v} group. However, because all of the matrices are diagonal, the 3×3 matrix operations can be reduced to three 1×1 matrix operations, which consist of the numbers $+1$ and -1. Therefore, the three-dimensional reducible representation of Equation (27.5) can be reduced to three one-dimensional representations.

Point groups can also have **two-dimensional** and **three-dimensional irreducible representations.** If x' and/or $y' = f(x, y)$ for a representation, then the basis will be (x, y) and the dimension of that irreducible representation is two. At least one of the matrices representing the operators will have the form

$$\begin{pmatrix} a & b & 0 \\ c & d & 0 \\ 0 & 0 & e \end{pmatrix}$$

in which entries a through e are in general nonzero. If x' and/or y' and/or $z' = f(x, y, z)$ the dimension of the representation is three and at least one of the operators will have the form

$$\begin{pmatrix} a & b & c \\ d & e & f \\ g & h & j \end{pmatrix}$$

in which entries a through j are in general nonzero.

How does one know how many irreducible representations a group has and what their dimension is? The following result of group theory is used to answer this question:

The dimension of the different irreducible representations, d_j, and the **order of the group,** h, defined as the number of symmetry elements in the group, are related by the equation

$$\sum_{j=1}^{N} d_j^2 = h \qquad (27.6)$$

This sum is over the irreducible representations of the group.

Because every point group contains the one-dimensional totally symmetric representation, at least one of the $d_j = 1$. We apply this formula to the C_{2v} representations. This group has four elements, and all belong to different classes. Therefore, there are four different representations. The only set of nonzero integers that satisfies the equation

$$d_1^2 + d_2^2 + d_3^2 + d_4^2 = 4 \qquad (27.7)$$

is $d_1 = d_2 = d_3 = d_4 = 1$. We conclude that all of the irreducible representations of the C_{2v} group are one dimensional. Because a 1×1 matrix cannot be reduced to one of lower dimensionality, all one-dimensional representations are irreducible.

For the C_{2v} group, the number of irreducible representations is equal to the number of elements and classes. More generally, the number of irreducible representations is equal to the number of classes for any group. Recall that all operators generated from a single symmetry element and successive applications of other operators of the group belong to the same class. For example, consider NF_3, which belongs to the C_{3v} group. As shown in Example Problem 27.1, the C_3 and C_3^2 rotations of the C_{3v} group belong to the same class. The three σ_v mirror planes also belong to the same class because the second and third planes are generated from the first by applying $\hat{C}_3$ and $\hat{C}_3^2$. Therefore, the C_{3v} group has six elements, but only three classes.

We next show that the C_{3v} point group has one representation that is not one dimensional. Using the result from Example Problem 27.2, the matrix that describes a $120°$ rotation is

$$\hat{C}_3 = \begin{pmatrix} \cos\theta & -\sin\theta & 0 \\ \sin\theta & \cos\theta & 0 \\ 0 & 0 & 1 \end{pmatrix} = \begin{pmatrix} -1/2 & -\sqrt{3}/2 & 0 \\ \sqrt{3}/2 & -1/2 & 0 \\ 0 & 0 & 1 \end{pmatrix} \qquad (27.8)$$

The other operators in this group have a diagonal form. The $\hat{C}_3$ operator does not have a diagonal form, and $\hat{C}_3$ acting on the vector (x, y, z) mixes x and y. However, z' depends only on z and not on x or y. Therefore, it is possible to reduce the 3×3 matrix operator for $\hat{C}_3$ into separate irreducible 2×2 and 1×1 matrix operators. We conclude that the C_{3v} point group contains a two-dimensional irreducible representation. Example Problem 27.4 shows how to determine the number and dimension of the remaining irreducible representations for the C_{3v} group.

EXAMPLE PROBLEM 27.4

The C_{3v} group has the elements $\hat{E}$, $\hat{C}_3$, and $\hat{C}_3^2$ and three σ_v mirror planes. How many different irreducible representations does this group have, and what is the dimensionality of each irreducible representation?

Solution

The order of the group is the number of elements, so $h = 6$. The number of representations is the number of classes. As discussed earlier, $\hat{C}_3$ and $\hat{C}_3^2$ belong to one class, and the same is true of the three σ_v reflections. Although the group has six elements, it has only three classes. Therefore, the group has three irreducible representations. The equation $l_1^2 + l_2^2 + l_3^2 = 6$ is solved to find the dimension of the representations, and one of the values must be 1. The only possible solution is $l_1 = l_2 = 1$ and $l_3 = 2$. We see that the C_{3v} group contains one two-dimensional representation and two one-dimensional representations.

To gain practice in working with irreducible representations of more than one dimension, the matrices for the individual operations that describe the two-dimensional representation in the C_{3v} group are derived next. Example Problem 27.2 shows how to set up the matrices for rotation operators. Figure 27.6 shows how the $x-y$ coordinate system is transformed by a mirror plane, σ.

The values x' and y' are related to x and y by

$$x' = -x \cos 2\theta - y \sin 2\theta$$
$$y' = -x \sin 2\theta + y \cos 2\theta \qquad (27.9)$$

Equation (27.9) is used to evaluate the 2×2 matrices for the mirror planes $\hat{\sigma}$, $\hat{\sigma}'$, and $\hat{\sigma}''$ at 0, $\pi/3$, and $2\pi/3$, and Equation (27.1) is used to evaluate the 2×2 matrices for $\hat{C}_3$ and $\hat{C}_3^2$. The resulting operators for the two-dimensional representation of the C_{3v} group are shown in Equation (27.10). Remember that $\hat{\sigma}$, $\hat{\sigma}'$, and $\hat{\sigma}''$ all belong to the one class, as do $\hat{C}_3$ and $\hat{C}_3^2$.

$$\hat{E} = \begin{pmatrix} 1 & 0 \\ 0 & 1 \end{pmatrix}$$

$$\hat{\sigma} = \begin{pmatrix} -\cos 0 & -\sin 0 \\ -\sin 0 & \cos 0 \end{pmatrix} = \begin{pmatrix} -1 & 0 \\ 0 & 1 \end{pmatrix}$$

$$\hat{\sigma}' = \begin{pmatrix} -\cos(2\pi/3) & -\sin(2\pi/3) \\ -\sin(2\pi/3) & \cos(2\pi/3) \end{pmatrix} = \begin{pmatrix} 1/2 & \sqrt{3}/2 \\ \sqrt{3}/2 & -1/2 \end{pmatrix}$$

$$\hat{\sigma}'' = \begin{pmatrix} -\cos(4\pi/3) & -\sin(4\pi/3) \\ -\sin(4\pi/3) & \cos(4\pi/3) \end{pmatrix} = \begin{pmatrix} 1/2 & -\sqrt{3}/2 \\ -\sqrt{3}/2 & -1/2 \end{pmatrix}$$

$$\hat{C}_3 = \begin{pmatrix} \cos(2\pi/3) & -\sin(2\pi/3) \\ \sin(2\pi/3) & \cos(2\pi/3) \end{pmatrix} = \begin{pmatrix} -1/2 & -\sqrt{3}/2 \\ \sqrt{3}/2 & -1/2 \end{pmatrix}$$

$$\hat{C}_3^2 = \begin{pmatrix} \cos(4\pi/3) & -\sin(4\pi/3) \\ \sin(4\pi/3) & \cos(4\pi/3) \end{pmatrix} = \begin{pmatrix} -1/2 & \sqrt{3}/2 \\ -\sqrt{3}/2 & -1/2 \end{pmatrix} \qquad (27.10)$$

How is the character table for the C_{3v} group constructed? In particular, how are characters assigned to the two-dimensional representation, which is generally called E? (Do not

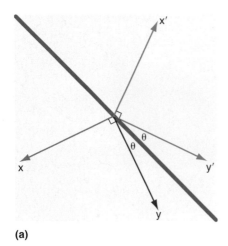

(a)

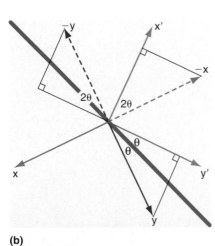

(b)

FIGURE 27.6
Schematic depiction of the transformation of $x-y$ coordinates effected by reflection through a mirror plane, σ, containing the z axis. **(a)** The $x-y$ coordinate system for which the y axis is rotated by θ relative to the mirror plane is reflected through the mirror plane (purple line). This operation generates the $x'-y'$ coordinate system. **(b)** The geometry used to derive Equation (27.9). is shown.

confuse this symbol for a two-dimensional representation with the operator $\hat{E}$.) The following theorem of group theory is used:

> The character for an operator in a representation of dimension higher than one is given by the sum of the diagonal elements of the matrix.

Using this rule, we see that the character of $\hat{\sigma}$, $\hat{\sigma}'$, and $\hat{\sigma}''$ is 0, and the character of $\hat{C}_3$ and $\hat{C}_3^2$ is -1. As expected, the character of all elements in a class is the same. Recall also that every group has a totally symmetric representation in which all characters are $+1$.

Because the C_{3v} group contains three classes, it must have three irreducible representations. We enter the information that we obtained earlier for A_1 and E in the following partially completed character table. All of the symmetry operators of a class are grouped together in a character table. For example, in the following listing, the elements C_3 and C_3^2 are listed as $2C_3$ to make the notation compact.

	E	$2C_3$	$3\sigma_v$
A_1	1	1	1
?	a	b	c
E	2	-1	0

How can the values for the characters a, b, and c be obtained? We use another result from group theory:

> If the set of characters associated with a representation of the group is viewed as a vector, $\Gamma_i = \chi_i(\hat{R}_j)$, with one component for each element of the group, the following condition holds:
>
> $$\Gamma_i \Gamma_k = \sum_{j=1}^{h} \chi_i(\hat{R}_j)\chi_k(\hat{R}_j) = h\delta_{ik}, \text{ where } \delta_{ik} = 0 \text{ if } i \neq k \text{ and } 1 \text{ if } i = k \quad \textbf{(27.11)}$$
>
> or, equivalently, $\Gamma_i \Gamma_k = \boldsymbol{\chi}_i(\hat{R}_j) \cdot \boldsymbol{\chi}_k(\hat{R}_j) = h\delta_{ik}$. The sum is over all elements of the group.

EXAMPLE PROBLEM 27.5

Determine the unknown coefficients a, b, and c for the preceding partially completed character table and assign the appropriate symbol to the irreducible representation.

Solution

From Example Problem 27.4, we know that the unknown representation is one dimensional. From Equation (27.11), we know that the $\chi_i(\hat{R}_j)$ for different values of the index i are orthogonal. Therefore,

$$\boldsymbol{\chi}_? \cdot \boldsymbol{\chi}_{A_1} = a + b + b + c + c + c = a + 2b + 3c = 0$$

$$\boldsymbol{\chi}_? \cdot \boldsymbol{\chi}_E = 2a - b - b = 2a - 2b = 0$$

We could also have taken the sum over classes and multiplied each term by the number of elements in the class, because all elements in a class have the same character. We also know that $a = 1$ because it is the character of the identity operator. Solving the equations gives the results of $b = 1$ and $c = -1$. Because the character of C_3 is $+1$, and the character of σ_v is -1, the unknown representation is designated A_2. Table 27.4 shows the completed C_{3v} character table.

Note that the two-dimensional basis functions occur in pairs. You will be asked to verify that z and R_z are bases for the A_1 and A_2 representations, respectively, in the end-of-chapter problems.

TABLE 27.4   The C_{3v} Character Table

	E	$2C_3$	$3\sigma_v$		
A_1	1	1	1	z	$x^2 + y^2, z^2$
A_2	1	1	-1	R_z	
E	2	-1	0	$(x, y), (R_x, R_y)$	$(x^2 - y^2, xy), (xz, yz)$

27.6 Using the C_{2v} Representations to Construct Molecular Orbitals for H₂O

A number of aspects of group theory have been discussed in the preceding sections. In particular, the structure of character tables, which are the most important result of group theory for chemists, has been explained. We now illustrate the usefulness of character tables to solve a problem of chemical interest, namely, the construction of MOs that incorporate the symmetry of a molecule. Why is this necessary?

To answer this question, consider the relationship among the total energy operator, the molecular wave functions ψ_j, and the symmetry of the molecule. A molecule that has undergone one of its symmetry operations, $\hat{A}$, is indistinguishable from the original molecule. Therefore, $\hat{H}$ must also be unchanged under this and any other symmetry operation of the group, because the total energy of the molecule is the same in any of its equivalent positions. If this is the case, then $\hat{H}$ belongs to the totally symmetric representation.

Because the order of applying $\hat{H}$ and $\hat{A}$ to the molecule is immaterial, it follows that $\hat{H}$ and $\hat{A}$ commute. Therefore, as discussed in Chapter 17, eigenfunctions of $\hat{H}$ can be found that are simultaneously eigenfunctions of $\hat{A}$ and of all other operators of the group. These **symmetry-adapted MOs** are of central importance in quantum chemistry. In this section, we illustrate how to generate symmetry-adapted MOs from AOs. Not all AOs contribute to a particular symmetry-adapted MO. Invoking the symmetry of a molecule results in a set of MOs consisting of fewer AOs than would have been obtained had the molecular symmetry been neglected.

Consider a specific example. Which of the AOs on oxygen contribute to the symmetry-adapted MOs on water? We begin by asking which of the four oxygen valence AOs can be combined with the hydrogen AOs to form symmetry-adapted MOs. All possible combinations are shown in Figure 27.7. In order to take the symmetry of the water molecule into account, the hydrogen AOs will appear as in-phase or out of phase combinations. Consider first the in-phase combination, $\phi_+ = \phi_{H1sA} + \phi_{H1sB}$.

The overlap integral S_{+j} between the orbital ϕ_+ and an oxygen AO ϕ_j is defined by

$$S_{+j} = \int \phi_+^* \phi_j \, d\tau \tag{27.12}$$

Only the oxygen AOs that have a nonzero overlap with the hydrogen AOs are useful in forming chemical bonds. Because S_{+j} is just a number, it cannot change upon applying any of the operators of the C_{2v} group to the integral. In other words, S_{+j} belongs to the A_1 representation. The same must be true of the integrand and, therefore, the integrand must also belong to the A_1 representation. If ϕ_+ belongs to one representation and ϕ_j belongs to another, what can be said about the symmetry of the direct product $\phi_+ \cdot \phi_j$? A result of group theory is used to answer this question:

The character for an operator $\hat{R}$ ($\hat{E}$, $\hat{C}_2$, $\hat{\sigma}_v$, or $\hat{\sigma}'_v$ for the C_{2v} group) of the direct product of two representations is given by

$$\chi_{product}(\hat{R}) = \chi_i(\hat{R})\chi_j(\hat{R}) \tag{27.13}$$

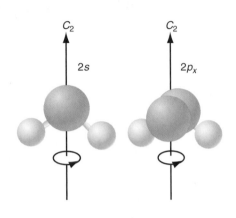
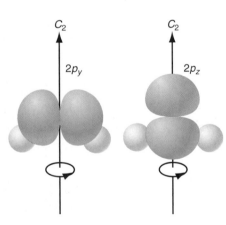

FIGURE 27.7
Depiction of the oxygen atomic orbitals that will be considered as contributors to the MO formed using $\phi_+ = \phi_{H1sA} + \phi_{H1sB}$.

For example, if ϕ_+ belongs to A_2, and ϕ_j belongs to B_2, $\Gamma_{product}$ can be calculated from the $\chi_{product}$ terms as follows:

$$\Gamma_{product} = \boldsymbol{\chi}_{A_2} \cdot \boldsymbol{\chi}_{B_2} = [1 \times 1 \quad 1 \times (-1) \quad (-1) \times (-1) \quad (-1) \times 1]$$
$$= (1 \quad -1 \quad 1 \quad -1) \tag{27.14}$$

Looking at the C_{2v} character table, we can see that the direct product $A_2 \cdot B_2$ belongs to B_1.

How is this result useful in deciding which of the oxygen AOs contribute to the symmetry-adapted water MOs? Because the integrand must belong to the A_1 representation, each character of the representation of $\phi_+^* \phi_j$ must be equal to one. We conclude that

$$\sum_{k=1}^{h} \chi_+(\hat{R}_k)\chi_j(\hat{R}_k) = h \tag{27.15}$$

However, according to Equation (27.11), this equation is never satisfied if the two representations to which the orbitals belong denoted $+$ and j are different. We conclude that *the overlap integral between two combinations of AOs is nonzero only if the combinations belong to the same representation.*

Using this result, which of the oxygen AOs in Figure 27.7 form symmetry-adapted MOs with the combination $\phi_+ = \phi_{H1sA} + \phi_{H1sB}$? The orbital ϕ_+ is unchanged by any of the symmetry operators, so it must belong to the A_1 representation. The $2s$ AO on oxygen is spherically symmetrical, so that it transforms as $x^2 + y^2 + z^2$. As the C_{2v} character table shows, the $2s$ AO belongs to the A_1 representation, as does the $2p_z$ orbital. By contrast, the $2p_x$ and $2p_y$ AOs on oxygen belong to the B_1 and B_2 representations, respectively. Therefore, only the oxygen $2s$ and $2p_z$ AOs belong to the same irreducible representation as ϕ_+, and only these AOs will contribute to MOs involving ϕ_+. The combinations of the $2s$ AO and $2p_z$ oxygen AO with $\phi_+ = \phi_{H1sA} + \phi_{H1sB}$ result in the $1a_1$, the $2a_1$, and the $3a_1$, MOs for water shown in Figure 27.8. We now have an explanation for the nomenclature introduced in Figure 27.7 for the symmetry-adapted water MOs. The a_1 refers to the particular irreducible representation of the C_{2v} group, and the integer 1, 2, ..., refers to the lowest, next lowest, ..., energy MO belonging to the A_1 representation.

EXAMPLE PROBLEM 27.6

Which of the oxygen AOs shown in Figure 27.7 will participate in forming symmetry-adapted water MOs with the antisymmetric combination of hydrogen AOs defined by $\phi_- = \phi_{H1sA} - \phi_{H1sB}$?

Solution

The antisymmetric combination of the H AOs is given by $\phi_- = \phi_{H1sA} - \phi_{H1sB}$, shown in the margin. By considering the C_{2v} operations shown in Figure 27.4, convince yourself that the characters for the different operations are $\hat{E}:+1$, $\hat{C}_2:-1$, $\hat{\sigma}_v:-1$, and $\hat{\sigma}'_v:+1$. Therefore, ϕ_- belongs to the B_2 representation. Of the valence oxygen AOs, only the $2p_y$ orbital belongs to the B_2 representation. Therefore, the only symmetry adapted MOs formed from ϕ_- and the $2s$ and $2p$ orbitals that have a nonzero overlap among the AOs are the MOs denoted $1b_2$ and $2b_2$, which are shown in Figure 27.8. This nomenclature indicates that they are the lowest and next lowest energy MOs of B_2 symmetry.

We now make an important generalization of the result just obtained. The same symmetry considerations used for the overlap integral apply in evaluating integrals of the type $H_{ab} = \int \psi_a^* \hat{H} \psi_b \, d\tau$. As shown in Chapters 23 and 24, such integrals appear whenever the total energy is calculated. The value of H_{ab} is zero unless $\psi_a^* \hat{H} \psi_b$ belongs to the A_1 representation. Because $\hat{H}$ belongs to the A_1 representation, H_{ab} will be zero unless ψ_a and ψ_b belong to the same representation (not necessarily the A_1 representation). Only then will the integrand $\psi_a^* \hat{H} \psi_b$ contain the A_1 representation. This important result

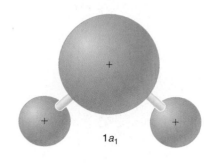

$1a_1$

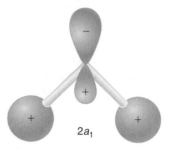

$2a_1$

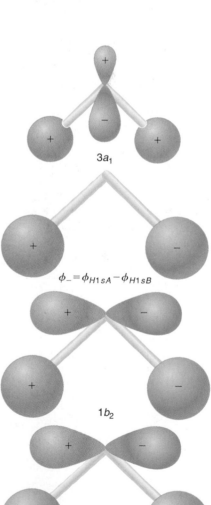

$3a_1$

$\phi_- = \phi_{H1sA} - \phi_{H1sB}$

$1b_2$

$2b_2$

FIGURE 27.8

Water molecular orbitals formed from the symmetric and anti-symmetric combinations of the hydrogen AOs are shown.

is of great help in evaluating entries in a secular determinant such as those encountered in Chapter 23.

In the preceding discussion, symmetry-adapted MOs for H_2O were generated from AOs that belong to the different irreducible representations of the C_{2v} group in an ad hoc manner. In Supplemental Section 27.9, we discuss a powerful method, called the projection operator method, that allows the symmetry-adapted MOs to be constructed for arbitrary molecules.

27.7 The Symmetries of the Normal Modes of Vibration of Molecules

The vibrational motions of individual atoms in a molecule might appear to be chaotic and independent of one another. However, the selection rules for infrared vibrational and Raman spectroscopy are characteristic of the normal modes of a molecule, which can be described in the following way. In a normal mode vibration, each atom is displaced from its equilibrium position by a vector that can but need not lie along the bond direction (for example in a bending mode). The directions and magnitudes of the displacements are not the same for all atoms. The following can be said of the motion of the atoms in the normal modes:

- During a vibrational period, the center of mass of the molecule remains fixed and all atoms in the molecule undergo in-phase periodic motion about their equilibrium positions.
- All atoms in a molecule reach their minimum and maximum amplitudes at the same time.
- These collective motions are called **normal modes,** and the frequencies are called the **normal mode frequencies.**
- The frequencies measured in vibrational spectroscopy are the normal mode frequencies.
- All normal modes are independent in the harmonic approximation, meaning that excitation of one normal mode does not transfer vibrational energy into another normal mode.
- Any seemingly random motion of the atoms in a molecule can be expressed as a linear combination of the normal modes of that molecule.

How many normal modes does a molecule have? An isolated atom has three translational degrees of freedom; therefore, a molecule consisting of n atoms has $3n$ degrees of freedom. Three of these are translations of the molecule and are not of interest here. A nonlinear molecule with n atoms has three degrees of rotational freedom, and the remaining $3n - 6$ internal degrees of freedom correspond to normal modes of vibration. Because a linear molecule has only two degrees of rotational freedom, it has $3n - 5$ normal modes of vibration. For a diatomic molecule, there is only one vibrational mode, and the motion of the atoms is directed along the bond. In the harmonic approximation,

$$V(x) = \frac{1}{2}kx^2 = \frac{1}{2}\left(\frac{d^2V}{dx^2}\right)x^2$$

where x is the displacement from the equilibrium position in the center of mass coordinates. For a molecule with N vibrational degrees of freedom, the potential energy is given by

$$V(q_1, q_2, \ldots, q_N) = \frac{1}{2}\sum_{i=1}^{N}\sum_{j=1}^{N}\frac{\partial^2V}{\partial q_i\partial q_j}q_iq_j \tag{27.16}$$

where the q_i designates the individual normal mode displacements. Classical mechanics allows us to find a new set of vibrational coordinates $Q_j(q_1, q_2, \ldots, q_N)$ that simplify Equation (27.16) to the form

$$V(Q_1, Q_2, \ldots, Q_N) = \frac{1}{2}\sum_{i=1}^{N}\left(\frac{\partial^2V}{\partial Q_i^2}\right)Q_i^2 \tag{27.17}$$

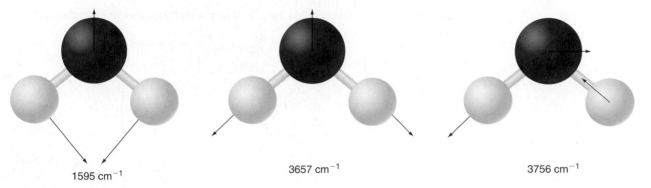

FIGURE 27.9

The normal modes of H_2O are depicted, with the vectors indicating atomic displacements (not to scale). From left to right, the modes correspond to a bond bending, an O—H symmetric stretch, and an O—H asymmetric stretch. The experimentally observed frequencies are indicated.

The $Q_j(q_1, q_2, \ldots, q_N)$ are known as the **normal coordinates** of the molecule. This transformation has significant advantages in describing vibrational motion. Because there are no cross terms of the type Q_iQ_j in the potential energy, the vibrational modes are independent in the harmonic approximation, meaning that

$$\psi_{vibrational}(Q_1, Q_2, \ldots, Q_N) = \psi_1(Q_1)\psi_2(Q_2)\ldots\psi_N(Q_N) \quad \text{and}$$

$$E_{vibrational} = \sum_{i=1}^{N}\left(n_j + \frac{1}{2}\right)h\nu_j \qquad \textbf{(27.18)}$$

Because of the transformation to normal coordinates, each of the normal modes contributes independently to the energy, and the vibrational motions of different normal modes are not coupled, consistent with the properties in the first paragraph of this section. Finding the normal modes is a nontrivial but straightforward exercise that can be done most easily using numerical methods. The calculated normal modes of H_2O are shown in Figure 27.9. The arrows show the displacement of each atom at a given time. After half the vibrational period, each arrow has the same magnitude, but the direction is opposite to that shown in the figure. We do not carry out a normal mode calculation, but focus instead on the symmetry properties of the normal modes.

Just as the $2p$ atomic orbitals on the oxygen atom belong to individual representations of the C_{2v} group, the normal modes of a molecule belong to individual representations. The next task is to identify the symmetry of the three different normal modes of H_2O. To do so, a coordinate system is set up at each atom and a matrix representation formed that is based on the nine x, y, and z coordinates of the atoms in the molecule. Figure 27.10 illustrates the geometry under consideration.

Consider the C_2 operation. By visualizing the motion of the coordinate systems on the three atoms, convince yourself that the individual coordinates are transformed as follows under this operation:

$$\begin{pmatrix} x_1 \\ y_1 \\ z_1 \\ x_2 \\ y_2 \\ z_2 \\ x_3 \\ y_3 \\ z_3 \end{pmatrix} \Rightarrow \begin{pmatrix} -x_3 \\ -y_3 \\ z_3 \\ -x_2 \\ -y_2 \\ z_2 \\ -x_1 \\ -y_1 \\ z_1 \end{pmatrix} \qquad \textbf{(27.19)}$$

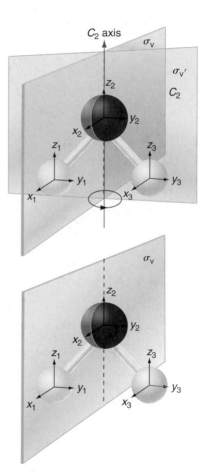

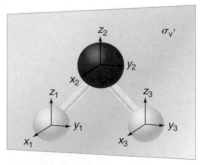

FIGURE 27.10

Transformations of coordinate systems on the atom in water. Each is considered to be a separate entity under symmetry operations of the C_{2v} group.

You can also convince yourself that the 9×9 matrix that describes this transformation is

$$
\hat{C}_2 = \begin{pmatrix}
0 & 0 & 0 & 0 & 0 & 0 & -1 & 0 & 0 \\
0 & 0 & 0 & 0 & 0 & 0 & 0 & -1 & 0 \\
0 & 0 & 0 & 0 & 0 & 0 & 0 & 0 & 1 \\
0 & 0 & 0 & -1 & 0 & 0 & 0 & 0 & 0 \\
0 & 0 & 0 & 0 & -1 & 0 & 0 & 0 & 0 \\
0 & 0 & 0 & 0 & 0 & 1 & 0 & 0 & 0 \\
-1 & 0 & 0 & 0 & 0 & 0 & 0 & 0 & 0 \\
0 & -1 & 0 & 0 & 0 & 0 & 0 & 0 & 0 \\
0 & 0 & 1 & 0 & 0 & 0 & 0 & 0 & 0
\end{pmatrix}
\qquad \textbf{(27.20)}
$$

Note that this matrix has a simple structure. It consists of identical 3×3 subunits shown in boxes. More importantly, the diagonal elements of the subunits lie along the diagonal of the 9×9 matrix only if the atom is not shifted to another position through the transformation. Because H atoms 1 and 3 exchange places under the C_2 operation, they do not contribute to the character of the operator, which is the sum of the diagonal elements of the matrix. This result leads to the following guidelines for calculating the character of each element of the group in the 9×9 matrix representation:

- If the atom remains in the same position under the transformation, and the sign of x, y, or z is not changed, the value $+1$ is associated with each unchanged coordinate.

- If the sign of x, y, or z is changed, the value -1 is associated with each changed coordinate.

- If the coordinate system is exchanged with the position of another coordinate system, the value zero is associated with each of the three coordinates.

- Recall that only the diagonal elements contribute to the character. Therefore, only atoms that are not shifted by an operation contribute to the character.

This procedure is applied to the water molecule. Because nothing changes under the E operation, the character of $\hat{E}$ is 9. Under the rotation of 180°, the two H atoms are interchanged, so that none of the six coordinates contributes to the character of the $\hat{C}_2$ operator. On the oxygen atom, $x \rightarrow -x$, $y \rightarrow -y$, and $z \rightarrow z$. Therefore, the character of $\hat{C}_2$ is -1. For the $\hat{\sigma}_v$ operation, the H atoms are again interchanged so that they do not contribute to the character of $\hat{\sigma}_v$. On the oxygen atom, $x \rightarrow x$, $y \rightarrow -y$, and $z \rightarrow z$. Therefore, the character of $\hat{\sigma}_v$ is $+1$. For the $\hat{\sigma}_v'$ operation, on the H atoms, $x \rightarrow -x$, $y \rightarrow y$, and $z \rightarrow z$, so that the two H atoms contribute 2 to the $\hat{\sigma}_v'$ character. On the oxygen atom, $x \rightarrow -x$, $y \rightarrow y$, and $z \rightarrow z$ so that the O atom contributes $+1$ to the $\hat{\sigma}_v'$ character. Therefore, the total character of $\hat{\sigma}_v'$ is $+3$. These considerations show that the reducible representation formed using the coordinate systems on the three atoms as a basis is

E	C_2	σ_v	σ_v'	
9	-1	1	3	**(27.21)**

This is a reducible representation because it is a nine-dimensional representation, whereas all irreducible representations of the C_{2v} group are one dimensional. To use this result to characterize the symmetry of the normal modes of water, it is necessary to decompose this reducible representation into the irreducible representations that it contains, as follows:

The general method for decomposing a reducible representation into its irreducible representations utilizes the vector properties of the representations introduced in Section 27.3. Take the scalar product between the reducible representation $\Gamma_{reducible}(\hat{R}_j)$ and each of the irreducible representations $\Gamma_i(\hat{R}_j)$ in turn, and divide by the order of the group. The result of this procedure is a positive integer n_i that is the number of times each representation appears in the irreducible representation. This statement is expressed by the equation

$$n_i = \frac{1}{h} \Gamma_i \Gamma_{reducible} = \frac{1}{h} \chi_i(\hat{R}_j) \cdot \chi_{reducible}(\hat{R}_j) = \frac{1}{h} \sum_{j=1}^{h} \chi_i(\hat{R}_j) \chi_{reducible}(\hat{R}_j),$$

for $i = 1, 2, \ldots, N$

(27.22)

We calculate the contribution of the individual irreducible representations to this reducible presentation using Equation (27.22):

$$n_{A_1} = \frac{1}{h} \sum_{j=1}^{h} \chi_{A_1}(\hat{R}_j) \chi_{reducible}(\hat{R}_j) = \frac{1 \times 9 + 1 \times (-1) + 1 \times 1 + 1 \times 3}{4} = 3$$

$$n_{A_2} = \frac{1}{h} \sum_{j=1}^{h} \chi_{A_2}(\hat{R}_j) \chi_{reducible}(\hat{R}_j) = \frac{1 \times 9 + 1 \times (-1) + (-1) \times 1 + (-1) \times 3}{4} = 1$$

$$n_{B_1} = \frac{1}{h} \sum_{j=1}^{h} \chi_{B_1}(\hat{R}_j) \chi_{reducible}(\hat{R}_j) = \frac{1 \times 9 + (-1) \times (-1) + 1 \times 1 + (-1) \times 3}{4} = 2$$

$$n_{B_2} = \frac{1}{h} \sum_{j=1}^{h} \chi_{B_2}(\hat{R}_j) \chi_{reducible}(\hat{R}_j) = \frac{1 \times 9 + (-1) \times (-1) + (-1) \times 1 + 1 \times 3}{4} = 3$$

(27.23)

This calculation shows that $\Gamma_{reducible} = 3A_1 + A_2 + 2B_1 + 3B_2$. However, not all of these representations describe vibrational normal modes. The translation of the molecules along the x, y, and z axes as well as their rotation about the same axes must be separated out to obtain the representations of the vibrational normal modes. This can be done by subtracting the representations belonging to x, y, and z as well as to R_x, R_y, and R_z. Representations for these degrees of freedom can be determined from the C_{2v} character table. Eliminating them gives the representations of the three vibrational modes as

$$\Gamma_{reducible} = 3A_1 + A_2 + 2B_1 + 3B_2 - (B_1 + B_2 + A_1) - (B_2 + B_1 + A_2)$$
$$= 2A_1 + B_2$$

(27.24)

This calculation has shown that the symmetry of the H_2O molecule dictates the symmetry of the normal modes. Of the three normal modes, one belongs to B_2 and two belong to A_1. The normal mode calculations outlined here give the modes shown in Figure 27.9.

How can these modes be assigned to different irreducible representations of the C_{2v} group? The arrows on each atom in Figure 27.9 show the direction and magnitude of the displacement at a given time. All displacement vectors are reversed after half a period. If the set of displacement vectors is to be a basis for a representation, they must transform as the characters of the particular representation. Consider first the 1595-cm^{-1} normal mode. The direction and magnitude of each vector is unaffected by each of the operations E, C_2, σ_v, and σ'_v. Therefore, this mode must belong to the A_1 representation. The same is true of the 3657-cm^{-1} normal mode. By contrast, the displacement vector on the O atom is reversed upon carrying out the C_2 operation for the 3756-cm^{-1} normal mode. Because the H atoms are interchanged, their displacement vectors do not contribute to the character of the C_2 operation, which is -1. Therefore, this mode must belong to either the B_1 or B_2 representations. Which of these is appropriate can be decided by examining the effect of the σ'_v operation on the individual displacement vectors. Because the vectors lie in the mirror plane, they are unchanged in the reflection, corresponding to a character of $+1$. Therefore, the 3756-cm^{-1} normal mode belongs to the B_2 representation.

The water molecule is small enough that the procedure described can be carried out without a great deal of effort. For larger molecules, the effort is significantly greater, but the normal modes and the irreducible representations to which they belong can be calculated using widely available quantum chemistry software. Many of these programs allow an animation of the vibration to be displayed, which is helpful in assigning the dominant motion to a stretch or a bend. Normal mode animations for several molecules are explored in the Web-based problems of Chapter 19 and in the computational problems for Chapter 24.

27.8 Selection Rules and Infrared versus Raman Activity

We next show that the selection rule for infrared absorption spectroscopy, $\Delta n = +1$, can be derived using group theory. More importantly, we show that for allowed transitions, $\Gamma_{reducible}$ as calculated in the previous section must contain the A_1 representation. As discussed in Section 19.3, for most molecules, only the $n = 0$ vibrational state is populated to a significant extent at 300 K. The molecule can be excited to a state with $n_j > 0$ through the absorption of infrared energy if the dipole matrix element satisfies the condition given by

$$\mu_{Q_j}^{m \leftarrow 0} = \left(\frac{\partial \mu}{\partial Q_j}\right) \int \psi_m^*(Q_j)\hat{\mu}(Q_j)\psi_0(Q_j)dQ_j \neq 0,$$

where $j = $ one of $1, 2, \ldots, 3N - 6$ **(27.25)**

We have modified Equation (19.6), which is applicable to a diatomic molecule, to the more general case of a polyatomic molecule and expressed the position variable in terms of the normal coordinate. To simplify the mathematics, the electric field is oriented along the normal coordinate. From Chapter 18, ψ_0, ψ_1, and ψ_2 are given by

$$\psi_0(Q_j) = \left(\frac{\alpha_j}{\pi}\right)^{1/4} e^{-(\alpha_j Q_j^2)/2}$$

$$\psi_1(Q_j) = \left(\frac{4\alpha_j^3}{\pi}\right)^{1/4} Q_j e^{-(\alpha_j Q_j^2)/2}$$

$$\psi_2(Q_j) = \left(\frac{\alpha_j}{4\pi}\right)^{1/4} (2\alpha_j Q_j^2 - 1)e^{-(\alpha_j Q_j^2)/2}$$ **(27.26)**

and the dipole moment operator is given by

$$\hat{\mu}(Q_j) = \mu_e + \left[\left(\frac{\partial \mu}{\partial Q_j}\right)Q_j + \cdots\right]$$

where μ_e is the static dipole moment. Higher terms are neglected in the harmonic approximation.

For what final states ψ_f will Equation (27.25) for the transition dipole moment be satisfied? Section 27.6 demonstrated that for the integral to be nonzero, the integrand $\psi_m^*(Q_j)\mu(Q_j)\psi_0(Q_j)$ must belong to the A_1 representation. The C_{2v} character table shows that Q_j^2 is a basis for this representation, so the integrand must be a function of Q_j^2 only. We know that ψ_0 is an even function of Q_j, $\psi_0(Q_j) = \psi_0(-Q_j)$, and that μ is an odd function of Q_j, $\mu(Q_j) = -\mu(-Q_j)$. Under what condition will the integrand be an even function of Q_j? It will be an even function only if ψ_m^* is an odd function of Q_j, $\psi_m^*(Q_j) = f(Q_j)$. Because of this restriction, $n = 0 \rightarrow n = 1$ is an allowed transition, but $n = 0 \rightarrow n = 2$ is not allowed in the dipole approximation. This is the same conclusion that was reached in Section 19.4, using a different line of reasoning.

The preceding discussion addressed the selection rule, but did not address the symmetry requirements for the normal modes that satisfy Equation (27.25). Because $\hat{\mu}(Q_j) = f(Q_j)$ and transforms as x, y, or z and $\psi_0(Q_j) = f(Q_j^2)$ transform as x^2, y^2, or z^2, $\psi_m(Q_j)$ must transform as x, y, or z, in order for $\psi_m^*(Q_j)\mu(Q_j)\psi_0(Q_j)$ to transform as x^2, y^2, or z^2. This gives us the requirement that a normal mode is infrared active; it must have x, y, or z as a basis. For H_2O, this means that the normal modes must belong to A_1, B_1, or B_2. Because, as shown in Equation (27.24), the three normal modes belong to A_1 and B_2, we conclude that all are infrared active. As discussed in more advanced texts, normal modes of a molecule are Raman active if the bases of the representation to which the normal mode belongs are the x^2, y^2, z^2, xy, yz, or xz functions. By looking at the C_{2v} character table, we can see that all three normal modes of water are Raman active. It is not generally the case that all normal modes are both infrared and Raman active for a molecule.

Based on this discussion, recall the infrared absorption spectrum for CH_4 shown in Figure 19.10. Although CH_4 has $3n - 6 = 9$ normal modes, only two peaks are observed. Methane belongs to the T_d point group, and an analysis equivalent to that which led to Equation (27.24) shows that

$$\Gamma_{reducible} = A_1 + E + 2T_2 \qquad (27.27)$$

The dimensions of the representations are one for A_1, two for E, and three for T_2. Therefore, all nine normal modes are accounted for. An examination of the character table for the T_d group shows that only the T_2 representations have x, y, or z as a basis. Therefore, only six of the nine normal modes of methane are infrared active. Why are only two peaks observed in the spectrum? The following result of group theory is used:

All normal modes that belong to a particular representation have the same frequency.

Therefore, each of the T_2 representations has three degenerate vibrational frequencies. For this reason, only two vibrational frequencies are observed in the infrared absorption spectrum of CH_4 shown in Figure 19.10. However, each frequency corresponds to three distinct but degenerate normal modes.

SUPPLEMENTAL

27.9 Using the Projection Operator Method to Generate MOs That Are Bases for Irreducible Representations

In Section 27.6, symmetry-adapted MOs for H_2O were generated from AOs that belong to the different irreducible representations of the C_{2v} group in an ad hoc manner. We next discuss a powerful method, called the **projection operator method,** that allows the same end to be achieved for arbitrary molecules. The method is applied to ethene, which belongs to the D_{2h} point group.

The symmetry elements for ethene and the D_{2h} character table are shown in Table 27.5 and Figure 27.11. Aside from the identity element, the group contains three C_2 axes and three mirror planes, all of which form separate classes, as well as an inversion center. Irreducible representations in groups with an inversion center have the subscript g or u denoting that they are symmetric $(+1)$ or antisymmetric (-1) with respect to the inversion center.

TABLE 27.5 The Character Table for the D_{2h} Point Group

	E	$C_2(z)$	$C_2(y)$	$C_2(x)$	i	$\sigma(xy)$	$\sigma(xz)$	$\sigma(yz)$		
A_g	1	1	1	1	1	1	1	1		x^2, y^2, z^2
B_{1g}	1	1	-1	-1	1	1	-1	-1	R_z	xy
B_{2g}	1	-1	1	-1	1	-1	1	-1	R_y	xz
B_{3g}	1	-1	-1	1	1	-1	-1	1	R_x	yz
A_u	1	1	1	1	-1	-1	-1	-1		
B_{1u}	1	1	-1	-1	-1	-1	1	1	z	
B_{2u}	1	-1	1	-1	-1	1	-1	1	y	
B_{3u}	1	-1	-1	1	-1	1	1	-1	x	

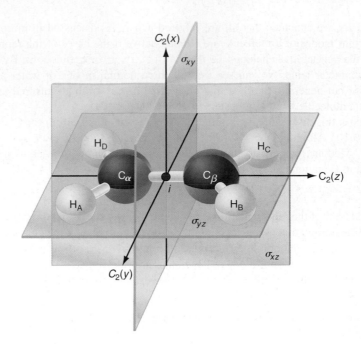

FIGURE 27.11
The symmetry elements of the D_{2h} group are shown using ethene as an example. The symbol at the intersection of the C_2 axes indicates the inversion center. The molecule lies in the yz plane.

Next consider how the individual H atoms are affected by the symmetry operations of the group. Convince yourself that the result of applying a symmetry operation to the molecule shifts the atom H_A as listed here:

$$
\begin{array}{cccc}
\hat{E} & \hat{C}_2(z) & \hat{C}_2(y) & \hat{C}_2(x) \\
\hline
H_A \to H_A & H_A \to H_D & H_A \to H_B & H_A \to H_C \\
\hat{i} & \hat{\sigma}(xy) & \hat{\sigma}(xz) & \hat{\sigma}(yz) \\
\hline
H_A \to H_C & H_A \to H_B & H_A \to H_D & H_A \to H_A
\end{array}
\qquad (27.28)
$$

Next consider the atom C_α and follow the same procedure used earlier for the $2s$, $2p_x$, $2p_y$, and $2p_z$ AOs. Convince yourself that the results shown in Table 27.6 are correct. These results are used to generate symmetry-adapted MOs for ethene using the method described here:

> The following procedure, based on the projection operator method, can be used to generate a symmetry-adapted MO from AOs that forms a basis for a given representation. The recipe consists of the following steps:
>
> • Choose an AO on an atom and determine into which AO it is transformed by each symmetry operator of the group.
>
> • Multiply the AO of the transformed species by the character of the operator in the representation of interest for each symmetry operator.
>
> • The resulting linear combination of these AOs forms a MO that is a basis for that representation.

The use of the projection operator method is illustrated in the following two example problems.

TABLE 27.6   Effect of the Symmetry Operations on the Carbon Atom Orbitals

	$\hat{E}$	$\hat{C}_2(z)$	$\hat{C}_2(y)$	$\hat{C}_2(x)$	$\hat{i}$	$\hat{\sigma}(xy)$	$\hat{\sigma}(xz)$	$\hat{\sigma}(yz)$
$2s$	$C_\alpha \to C_\alpha$	$C_\alpha \to C_\alpha$	$C_\alpha \to C_\beta$	$C_\alpha \to C_\beta$	$C_\alpha \to C_\beta$	$C_\alpha \to C_\beta$	$C_\alpha \to C_\alpha$	$C_\alpha \to C_\alpha$
$2p_x$	$C_\alpha \to C_\alpha$	$C_\alpha \to -C_\alpha$	$C_\alpha \to -C_\beta$	$C_\alpha \to C_\beta$	$C_\alpha \to -C_\beta$	$C_\alpha \to C_\beta$	$C_\alpha \to C_\alpha$	$C_\alpha \to -C_\alpha$
$2p_y$	$C_\alpha \to C_\alpha$	$C_\alpha \to -C_\alpha$	$C_\alpha \to C_\beta$	$C_\alpha \to -C_\beta$	$C_\alpha \to -C_\beta$	$C_\alpha \to C_\beta$	$C_\alpha \to -C_\alpha$	$C_\alpha \to C_\alpha$
$2p_z$	$C_\alpha \to C_\alpha$	$C_\alpha \to C_\alpha$	$C_\alpha \to -C_\beta$	$C_\alpha \to -C_\beta$	$C_\alpha \to -C_\beta$	$C_\alpha \to -C_\beta$	$C_\alpha \to C_\alpha$	$C_\alpha \to C_\alpha$

EXAMPLE PROBLEM 27.7

Form a linear combination of the H atomic orbitals in ethene that is a basis for the B_{1u} representation. Show that there is no combination of the H atomic orbitals in ethene that is a basis for the B_{3u} representation.

Solution

We take ϕ_{H_A} as the initial orbital, and multiply the AO into which ϕ_{H_A} is transformed by the character of the B_{1u} representation for each operator and sum these terms. The result is

$$
\begin{aligned}
\psi^H_{B_{1u}} &= 1 \times \phi_{H_A} + 1 \times \phi_{H_D} - 1 \times \phi_{H_B} - 1 \times \phi_{H_C} - 1 \times \phi_{H_C} \\
&\quad - 1 \times \phi_{H_B} + 1 \times \phi_{H_D} + 1 \times \phi_{H_A} \\
&= \phi_{H_A} + \phi_{H_D} - \phi_{H_B} - \phi_{H_C} - \phi_{H_C} - \phi_{H_B} + \phi_{H_D} + \phi_{H_A} \\
&= 2(\phi_{H_A} - \phi_{H_B} - \phi_{H_C} + \phi_{H_D})
\end{aligned}
$$

This molecular wave function has not yet been normalized. Pictorially, this combination looks like this:

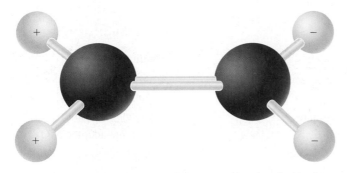

Follow the same procedure to generate the linear combination for the B_{3u} representation:

$$
\psi^H_{B_{3u}} = \phi_{H_A} - \phi_{H_D} - \phi_{H_B} + \phi_{H_C} - \phi_{H_C} + \phi_{H_B} + \phi_{H_D} - \phi_{H_A} = 0
$$

This result shows that there is no linear combination of the H AOs that is a basis for B_{3u}.

EXAMPLE PROBLEM 27.8

Use the same procedure as in Example Problem 27.7 to form a linear combination of the C atomic orbitals in ethene that is a basis for the B_{1u} representation.

Solution

Follow the same procedure outlined in Example Problem 27.7 and apply it to each of the carbon valence AOs:

$$
\begin{aligned}
2s:\ & 1 \times \phi_{C_\alpha} + 1 \times \phi_{C_\alpha} - 1 \times \phi_{C_\beta} - 1 \times \phi_{C_\beta} - 1 \times \phi_{C_\beta} - 1 \times \phi_{C_\beta} \\
& + 1 \times \phi_{C_\alpha} + 1 \times \phi_{C_\alpha} = 4\phi_{C_\alpha} - 4\phi_{C_\beta} \\
2p_x:\ & 1 \times \phi_{C_\alpha} - 1 \times \phi_{C_\alpha} + 1 \times \phi_{C_\beta} - 1 \times \phi_{C_\beta} + 1 \times \phi_{C_\beta} - 1 \times \phi_{C_\beta} \\
& + 1 \times \phi_{C_\alpha} - 1 \times \phi_{C_\alpha} = 0 \\
2p_y:\ & 1 \times \phi_{C_\alpha} - 1 \times \phi_{C_\alpha} - 1 \times \phi_{C_\beta} + 1 \times \phi_{C_\beta} + 1 \times \phi_{C_\beta} - 1 \times \phi_{C_\beta} \\
& - 1 \times \phi_{C_\alpha} + 1 \times \phi_{C_\alpha} = 0 \\
2p_z:\ & 1 \times \phi_{C_\alpha} + 1 \times \phi_{C_\alpha} + 1 \times \phi_{C_\beta} + 1 \times \phi_{C_\beta} + 1 \times \phi_{C_\beta} + 1 \times \phi_{C_\beta} \\
& + 1 \times \phi_{C_\alpha} + 1 \times \phi_{C_\alpha} = 4\phi_{C_\alpha} + 4\phi_{C_\beta}
\end{aligned}
$$

This result shows that the appropriate linear combination of carbon AOs to construct the symmetry-adapted MO that is a basis for the B_{1u} representation is

$$
\psi^C_{B_{1u}} = c_1(\phi_{C_{2s\alpha}} - \phi_{C_{2s\beta}}) + c_2(\phi_{C_{2pz\alpha}} + \phi_{C_{2pz\beta}})
$$

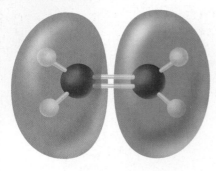

FIGURE 27.12
The ethene MO that is a basis of the B_{1u} representation.

Combining the results of the last two example problems, we find that the symmetry-adapted MO that includes AOs on all atoms and is also a representation of the B_{1u} representation is

$$\psi_{B1u} = c_1(\phi_{C_{2s\alpha}} - \phi_{C_{2s\beta}}) + c_2(\phi_{C_{2pz\alpha}} + \phi_{C_{2pz\beta}}) + c_3(\phi_{H_A} - \phi_{H_B} - \phi_{H_C} + \phi_{H_D})$$

An image of this molecular orbital is shown in Figure 27.12.

The values of the AO coefficients in the MOs cannot be obtained from symmetry considerations, but we can determine which coefficients are zero, equal in magnitude, and equal or opposite in sign. For the case of interest, c_1 through c_3 must be determined in a variational calculation in which the total energy of the molecule is minimized. Note that without taking symmetry into consideration, 12 coefficients would have been required to specify the wave function (one AO on each H, and four AOs on each C). We see that forming the symmetry-adapted MO significantly reduces the number of coefficients required in the calculation from 12 to just 3. This example shows the simplification of the molecular wave function that is obtained by forming symmetry-adapted MOs.

Vocabulary

associative operation

basis function

belongs to a particular representation

character

character table

class

dimension of a representation

forms a basis

group

identity operator

inverse operator

inversion center

irreducible representation

mirror plane

normal coordinate

normal mode

normal mode frequency

order of a group

point group

projection operator method

reducible representation

representation

rotation axis

rotation-reflection axis

symmetry elements

symmetry operations

symmetry-adapted MO

three-dimensional irreducible representation

totally symmetric representation

two-dimensional irreducible representation

Conceptual Problems

Q27.1 Can a molecule with an inversion center have a dipole moment? Give an example of a molecule with this symmetry element and explain your reasoning.

Q27.2 Which of the three normal modes of H_2O in Figure 27.9 is best described as a bending mode? Does the bond angle remain unchanged in any of the modes? Which requires less energy, bond bending or bond stretching?

Q27.3 Why does the list of elements for the D_{6h} group in Table 27.2 not list the elements C_6^2, C_6^3, and C_6^4?

Q27.4 Why does the list of elements for the D_{6h} group in Table 27.2 not list the elements S_6^2, S_6^3, and S_6^4?

Q27.5 How are quantum mechanical calculations in the LCAO-MO model simplified through the construction of symmetry-adapted MOs?

Q27.6 Some symmetry operations can be carried out physically using a ball-and-stick model of a molecule and others can only be imagined. Give two examples of each category.

Q27.7 Why does the C_{3v} group have a two-dimensional irreducible representation? Answer this question by referring to the form of the matrices that represent the operations of the group.

Q27.8 Can NH_3 have molecular orbitals that are triply degenerate in energy?

Q27.9 Can a molecule with an inversion center be superimposed on its mirror image and therefore be chiral? Give an example of a molecule with this symmetry element and explain your reasoning.

Q27.10 Why are all one-dimensional representations irreducible?

Q27.11 What is the difference between a symmetry element and a symmetry operation?

Q27.12 Can a molecule with D_{2h} symmetry have a dipole moment? Give an example of a molecule with this symmetry and explain your reasoning.

Q27.13 Can a molecule with C_{3h} symmetry have a dipole moment? Give an example of a molecule with this symmetry and explain your reasoning.

Q27.14 Explain why only two peaks are observed in the infrared spectrum of methane although six of the nine normal modes are infrared active.

Q27.15 Explain why the overlap integral between two combinations of AOs is nonzero only if the combinations belong to the same representation.

Numerical Problems

Problem numbers in **red** indicate that the solution to the problem is given in the *Student's Solutions Manual*.

P27.1 Show that a molecule with an inversion center implies the presence of an S_2 element.

P27.2 Use the 3×3 matrices for the C_{2v} group in Equation (27.2) to verify the group multiplication table for the following successive operations:

a. $\hat{\sigma}_v \hat{\sigma}_v'$ **b.** $\hat{\sigma}_v \hat{C}_2$ **c.** $\hat{C}_2 \hat{C}_2$

P27.3 Use the logic diagram of Figure 27.2 to determine the point group for the planar molecule *trans*$-$HBrC$=$CBrH. Indicate your decision-making process as was done in the text for NH_3.

P27.4 The D_3 group has the following classes: E, $2C_3$, and $3C_2$. How many irreducible representations does this group have and what is the dimensionality of each?

P27.5 Benzene, C_6H_6, belongs to the D_{6h} group. The reducible representation for the vibrational modes is

$$\Gamma_{reducible} = 2A_{1g} + A_{2g} + A_{2u} + 2B_{1u} + 2B_{2g}$$
$$+ 2B_{2u} + E_{1g} + 3E_{1u} + 4E_{2g} + 2E_{2u}$$

a. How many vibrational modes does benzene have?

b. How many of these modes are infrared active and to which representation do they belong?

c. Which of the infrared active modes are degenerate in energy and what is the degeneracy for each?

d. How many of these modes are Raman active and to which representation do they belong?

e. Which of the Raman active modes are degenerate in energy and what is the degeneracy for each?

f. Which of the infrared modes are also Raman active?

P27.6 NH_3 belongs to the C_{3v} group. The reducible representation for the vibrational modes is $\Gamma_{reducible} = 2A_1 + 2E$

a. How many vibrational modes does NH_3 have?

b. How many of these modes are infrared active and to which representation do they belong?

c. Are any of the infrared active modes degenerate in energy?

d. How many of these modes are Raman active and to which representation do they belong?

e. Are any of the Raman active modes degenerate in energy?

f. How many modes are both infrared and Raman active?

P27.7 XeF$_4$ belongs to the D_{4h} point group with the following symmetry elements: E, C_4, C_4^2, C_2, C_2', C_2'', i, S_4, S_4^2,

σ, $2\sigma'$, and $2\sigma''$. Make a drawing similar to Figure 27.1 showing these elements.

P27.8 Methane belongs to the T_d group. The reducible representation for the vibrational modes is $\Gamma_{reducible} = A_1 + E + 2T_2$.

a. Show that the A_1 and T_2 representations are orthogonal to each other and to the other representations in the table.

b. What is the symmetry of each of the vibrational modes that gives rise to Raman activity? Are any of the Raman active modes degenerate in energy?

P27.9 Use the 3×3 matrices for the C_{2v} group in Equation (27.2) to verify the associative property for the following successive operations:

a. $\hat{\sigma}_v(\hat{\sigma}_v' \hat{C}_2) = (\hat{\sigma}_v \hat{\sigma}_v')\hat{C}_2$

b. $(\hat{\sigma}_v \hat{E})\hat{C}_2 = \hat{\sigma}_v(\hat{E}\hat{C}_2)$

P27.10 Use the logic diagram of Figure 27.2 to determine the point group for allene. Indicate your decision-making process as was done in the text for NH_3.

P27.11 To determine the symmetry of the normal modes of methane, an analysis of the transformation of individual coordinate systems on the five atoms is carried out, as shown in Figure 27.10 for H_2O. After the rotational and translational representations are removed, the following reducible representation $\chi_{reducible}$ is obtained for the vibrational modes:

$\hat{E}$	$8\hat{C}_3$	$3\hat{C}_2$	$6\hat{C}_4$	$6\hat{\sigma}_d$
9	0	1	-1	3

Using the character table for the T_d group, verify that $\Gamma_{reducible} = A_1 + E + 2T_2$.

P27.12 Use the logic diagram of Figure 27.2 to determine the point group for the planar molecule *cis*$-$HBrC$=$CCIH. Indicate your decision-making process as was done in the text for NH_3.

P27.13 Decompose the following reducible representation into irreducible representations of the C_{2v} group:

$\hat{E}$	$\hat{C}_2$	$\hat{\sigma}_v$	$\hat{\sigma}_v'$
4	0	0	0

P27.14 Show that z is a basis for the A_1 representation and that R_z is a basis for the A_2 representation of the C_{3v} group.

P27.15 Use the logic diagram of Figure 27.2 to determine the point group for PCl$_5$. Indicate your decision-making process as was done in the text for NH_3.

P27.16 Use the method illustrated in Example Problem 27.2 to generate a 3×3 matrix for the following:

a. $\hat{C}_6$ operator

b. $\hat{S}_4$ operator

c. $\hat{i}$ operator

P27.17 Consider the function $f(x, y) = xy$ integrated over a square region in the xy plane centered at the origin.

a. Draw contours of constant f values (positive and negative) in the plane and decide whether the integral can have a nonzero value.

b. Use the information that the square has D_{4h} symmetry and determine which representation the integrand belongs to. Decide whether the integral can have a nonzero value from this information.

P27.18 Use the logic diagram of Figure 27.2 to determine the point group for CH_3Cl. Indicate your decision-making process as was done in the text for NH_3.

P27.19 CH_4 belongs to the T_d point group with the following symmetry elements: E, $4C_3$, $4C_3^2$, $3C_2$, $3S_4$, $3S_4^3$, and 6σ. Make drawings similar to Figure 27.1 showing these elements.

P27.20 Show that the presence of a C_2 axis and a mirror plane perpendicular to the rotation axis imply the presence of a center of inversion.

P27.21 Decompose the following reducible representation into irreducible representations of the C_{3v} group:

$\hat{E}$	$2\hat{C}_3$	$3\hat{\sigma}_v$
5	2	-1

P27.22 Assume that a central atom in a molecule has ligands with C_{4v} symmetry. Decide by evaluating the appropriate transition dipole element if the transition $p_x \rightarrow p_z$ is allowed with the electric field in the z direction.

P27.23 Show that a molecule with a C_n axis cannot have a dipole moment perpendicular to the axis.

P27.24 The C_{4v} group has the following classes: E, $2C_4$, C_2, $2\sigma_v$ and $2\sigma_d$. How many irreducible representations does this group have and what is the dimensionality of each? σ_d refers to a dihedral mirror plane. For example in the molecule BrF_5, the σ_v mirror planes each contain two of the equatorial F atoms, whereas the dihedral mirror planes do not contain the equatorial F atoms.

P27.25 Use the 2×2 matrices of Equation (27.10) to derive the multiplication table for the C_{3v} group.

Nuclear Magnetic Resonance Spectroscopy

Although the nuclear magnetic moment interacts only weakly with an external magnetic field, this interaction provides a very sensitive probe of the local electron distribution in a molecule. A nuclear magnetic resonance (NMR) spectrum can distinguish between inequivalent nuclei such as ^1H at different sites in a molecule. Individual spins, such as nearby ^1H nuclei, can couple to generate a multiplet splitting of NMR peaks. This splitting can be used to determine the structure of small organic molecules. NMR can also be used as a nondestructive imaging technique that is widely used in medicine and in the study of materials. Pulsed NMR and 2D Fourier transform techniques provide a powerful combination to determine the structure of large molecules of biological interest.

28.1 Intrinsic Nuclear Angular Momentum and Magnetic Moment

Recall that the electron has an intrinsic magnetic moment. Some, but not all, nuclei also have an intrinsic magnetic moment. Because the **nuclear magnetic moment** of the proton is about 2000 times weaker than that of the electron magnetic moment, it has an insignificant effect on the one-electron energy levels in the hydrogen atom. The nuclear magnetic moment does not generate chemical effects in that the reactivity of a molecule containing ^{12}C with zero nuclear spin is no different than the reactivity of a molecule containing ^{13}C with nuclear spin 1/2. However, the nuclear magnetic moment gives rise to an important spectroscopy. As shown later, a nucleus with a nonzero nuclear spin is an extremely sensitive probe of the local electron distribution within a molecule. Because of this sensitivity, nuclear magnetic resonance spectroscopy is arguably the single most important spectroscopic technique used by chemists today. NMR spectroscopy can be used to determine the structure of complex biomolecules, to map out the electron distribution in molecules, to study the kinetics of chemical transformations, and to nondestructively image internal organs in the human body. What is the basis for this spectroscopy?

Whereas electrons only have the spin quantum number 1/2, nuclear spins can take on integral multiples of 1/2. For example, ^{12}C and ^{16}O have spin 0, ^1H and ^{19}F have spin 1/2, and ^2H and ^{14}N have spin 1. The nuclear magnetic moment $\boldsymbol{\mu}$ and the nuclear angular momentum $\mathbf{I}$ are proportional to one another according to

$$\boldsymbol{\mu} = g_N \frac{e\hbar}{2m_{proton}}\mathbf{I} = g_N \beta_N \mathbf{I} = \gamma \hbar \mathbf{I} \tag{28.1}$$

In the SI system of units, $\boldsymbol{\mu}$ has the units of ampere $(\text{meter})^2 = \text{joule}(\text{tesla})^{-1}$, and $\mathbf{I}$ has the units of joule second. In these equations, the quantity $\beta_N = e\hbar/2m_{proton}$, which has the value $5.0507866 \times 10^{-27}$ J T^{-1}, is called the **nuclear magneton** and $\gamma = g_N \beta_N/\hbar$ is called the **magnetogyric ratio.** Just as for the orbital angular momentum (see Chapter 18), the z component of the intrinsic nuclear angular momentum can take on the values $m_z \hbar$ with $-I \leq m_z \leq I$, where $|\mathbf{I}| = \hbar\sqrt{I(I+1)}$. Because m_{proton} is greater than m_e by about a factor of 2000, the nuclear magnetic moment is much smaller than the electron magnetic moment for the same value of I. The **nuclear g factor** g_N, which is a dimensionless number, is characteristic of a particular nucleus. Values of these quantities for the nuclei most commonly used in NMR spectroscopy are shown in Table 28.1. Because the abundantly occurring nuclei ^{12}C and ^{16}O have no nuclear magnetic moment, they do not have a signature in an NMR experiment. In the rest of this chapter, we focus our attention on ^1H. However, this formalism can be applied to other spin-active nuclei in a straightforward manner.

As we learned when considering the electron spin, the quantum mechanical operators for orbital and spin angular momentum have the same commutation relations. The same relations also apply to nuclear spin. Therefore, we can immediately conclude that we can only know the magnitude of the nuclear angular momentum and one of its components simultaneously. The other two components remain unknown. As for electron spin, the nuclear angular momentum is quantized in units of $\hbar/2$. For ^1H, which has the spin quantum number 1/2, the operator $\hat{I}^2$ has two eigenfunctions that are usually called α and β. They correspond to $I_z = +(1/2)\hbar$ and $I_z = -(1/2)\hbar$, respectively. These functions α and β satisfy the relations

$$\hat{I}^2\alpha = \frac{1}{2}\left(\frac{1}{2}+1\right)\hbar^2\alpha; \qquad \hat{I}_z\alpha = +\frac{1}{2}\hbar\alpha$$

$$\hat{I}^2\beta = \frac{1}{2}\left(\frac{1}{2}+1\right)\hbar^2\beta; \qquad \hat{I}_z\beta = -\frac{1}{2}\hbar\beta \tag{28.2}$$

Note that by convention the same nomenclature is used for the electron spin and nuclear spin eigenfunctions. Although this presents a possibility for confusion, it emphasizes the fact that both sets of eigenfunctions have the same relationship to their angular momentum operators.

TABLE 28.1	Parameters for Spin-Active Nuclei			
Nucleus	Isotopic Abundance (%)	Spin	Nuclear g Factor g_N	Magnetogyric Ratio $\gamma/10^7$ (rad T^{-1} s^{-1})
^1H	99.985	1/2	5.5854	26.75
^{13}C	1.108	1/2	1.4042	6.73
^{31}P	100	1/2	2.2610	10.84
^2H	0.015	1	0.8574	4.11
^{14}N	99.63	1	0.4036	1.93

28.2 The Energy of Nuclei of Nonzero Nuclear Spin in a Magnetic Field

Classically, a magnetic moment or dipole can have any orientation in a magnetic field $\mathbf{B}_0$, and its energy in a particular orientation relative to the field (which we choose to lie along the z direction) is given by

$$E = -\boldsymbol{\mu} \cdot \mathbf{B}_0 = -\gamma B_0 m_z \hbar \tag{28.3}$$

However, we know that I_z for an atom of nuclear spin $1/2$ like ^{1}H can only have two values. Additionally, the magnetic moment of a single spin cannot be oriented parallel to the quantization axis because the components of the angular momentum operator do not commute (see Section 18.4). Therefore, the only allowed energy values for spin $1/2$ are

$$E = -\left(\pm \frac{1}{2} \right) g_N \beta_N B_0 = -\left(\pm \frac{1}{2} \right) \gamma B_0 \tag{28.4}$$

Although only two discrete energy levels are possible, the energy of these levels is a continuous function of the magnetic field, as shown in Figure 28.1.

Equation (28.4) shows that the two orientations of the magnetic moment have different potential energies. Additionally, a magnetic moment that is not parallel to the magnetic field experiences a force. For a classical magnetic moment, the torque is given by

$$\boldsymbol{\Gamma} = \boldsymbol{\mu} \times \mathbf{B}_0 \tag{28.5}$$

The torque is perpendicular to the plane containing $\mathbf{B}_0$ and $\boldsymbol{\mu}$ and, therefore, leads to a movement of $\boldsymbol{\mu}$ on the surface of a cone about the magnetic field direction. This motion is called **precession** and is analogous to the motion of a spinning top in a gravitational field. The precession of individual spins is shown in Figure 28.2. In NMR spectroscopy, one deals with a finite volume that contains many individual spins. Therefore, it is useful to define the **macroscopic magnetic moment M,** which is the vector sum of the individual magnetic moments $\mathbf{M} = \sum_i \boldsymbol{\mu}_i$. Whereas classical mechanics is not appropriate for describing individual nuclear magnetic moments, it is useful for describing the behavior of **M.** In Figure 28.2, all the individual $\boldsymbol{\mu}_i$ yellow cones have the same magnitude for the z component, but their transverse components are randomly oriented in the xy plane. Therefore, in a macroscopic sample containing on the order of Avogadro's number of nuclear spins, the transverse component of **M** is zero. We conclude that **M** lies on the z axis, which corresponds to the field direction.

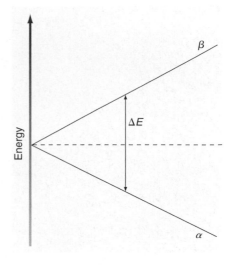

FIGURE 28.1
Energy of a nuclear spin of quantum number $1/2$ as a function of the magnetic field.

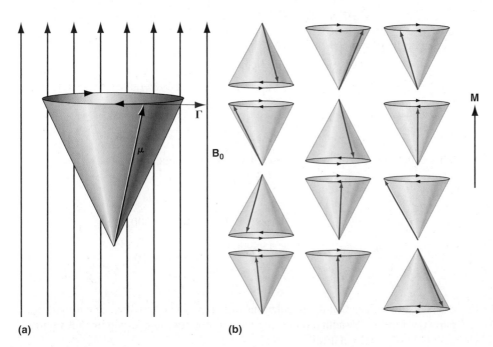

(a) (b)

FIGURE 28.2
(a) Precession of an individual nuclear spin about the magnetic field direction for an $\boldsymbol{\alpha}$ spin. (b) The magnetization vector **M** resulting from summing the individual spin magnetic moments (yellow cones) is oriented parallel to the magnetic field. It has no transverse component.

The frequency with which an individual magnetic moment precesses about the magnetic field direction is given by

$$\nu = \frac{1}{2\pi}\gamma B_0 \quad \text{or} \quad \omega = 2\pi\nu = \gamma B_0 \qquad (28.6)$$

and is called the **Larmor frequency.** The Larmor frequency increases linearly with the magnetic field and has characteristic values for different nuclei. For instance, ^{1}H has a resonance frequency of 500 MHz at a field of approximately 12 T.

In NMR spectroscopy, as in any spectroscopy, a transition must be induced between two different energy levels so that the absorption or emission of the electromagnetic energy that occurs can be detected. As we saw earlier, a spin $1/2$ system has only two levels, and their separation increases linearly with $\mathbf{B}_0$. As shown in Example Problem 28.1, the energy separation of these two levels is very small compared to kT. This makes the energy absorption difficult to detect, because the levels are nearly equally populated. Therefore, a major focus within the technology supporting NMR spectroscopy has been the development of very high magnetic fields to increase the energy separation of these two levels. Currently, by means of superconducting magnets, fields of up to approximately 21.1 T can be generated. This is a factor of nearly 10^6 higher than the Earth's magnetic field.

EXAMPLE PROBLEM 28.1

  a. Calculate the two possible energies of the ^{1}H nuclear spin in a uniform magnetic field of 5.50 T.
  b. Calculate the energy ΔE absorbed in making a transition from the α to the β state. If a transition is made between these levels by the absorption of electromagnetic radiation, what region of the spectrum is used?
  c. Calculate the relative populations of these two states in equilibrium at 300. K.

Solution

  a. The two energies are given by

$$E = \pm\frac{1}{2}g_N\beta_N B_0$$

$$= \pm\frac{1}{2} \times 5.5854 \times 5.051 \times 10^{-27}\,\text{J/T} \times 5.50\,\text{T}$$

$$= \pm7.76 \times 10^{-26}\,\text{J}$$

  b. The energy difference is given by

$$\Delta E = 2(7.76 \times 10^{-26}\,\text{J})$$

$$= 1.55 \times 10^{-25}\,\text{J}$$

$$\nu = \frac{\Delta E}{h} = \frac{1.55 \times 10^{-25}}{6.626 \times 10^{-34}} = 2.34 \times 10^8\,\text{s}^{-1}$$

This is in the range of frequencies called radio frequencies.

  c. The relative populations of the two states are given by

$$\frac{n_\beta}{n_\alpha} = \exp\left(-\frac{E_\beta - E_\alpha}{k_BT}\right) = \exp\left(\frac{-2 \times 7.76 \times 10^{-26}\,\text{J}}{1.381 \times 10^{-23}\,\text{J K}^{-1} \times 300.\,\text{K}}\right) = 0.999963$$

$$\frac{n_\alpha - n_\beta}{\frac{1}{2}(n_\beta + n_\alpha)} \approx \frac{(1 - 0.999963)\,n_\alpha}{n_\alpha} = 3.7 \times 10^{-5}$$

From this result, we see that the populations of the two states are the same to within a few parts per million. Note that observing the appropriate rules for significant figures, we would obtain a ratio of 1.00.

The solution to part (c) of Example Problem 28.1 shows that because $E_\beta - E_\alpha \ll k_B T, n_\alpha \approx n_\beta$. This result has important consequences for implementing NMR spectroscopy. As we learned in Chapter 19, if a system with only two energy levels is exposed to radiation of frequency $\nu = (E_\beta - E_\alpha)/h$, and if $n_\alpha \approx n_\beta$, the rate of upward transitions is nearly equal to the rate of downward transitions. Therefore, only a very small fraction of the nuclear spins contributes to the NMR signal. More generally, the energy absorbed is proportional to the product of $E_\beta - E_\alpha$ and $n_\alpha - n_\beta$. Both of these quantities increase as the magnetic field B_0 increases, and this is a major reason for carrying out NMR experiments at high magnetic fields.

The energy level diagram of Figure 28.1 indicates that under the condition

$$\nu_0 = \frac{E_\beta - E_\alpha}{h} = \frac{g_N \beta_N B_0}{2\pi} = \frac{\gamma B_0}{2\pi} \qquad (28.7)$$

energy can be absorbed by a sample containing atoms with a nonzero nuclear spin.

How can transitions be induced? As Figure 28.2 shows, the net magnetization induced by the static field $\mathbf{B}_0$ is parallel to the field. Inducing transitions is equivalent to rotating $\mathbf{M}$ away from the direction of $\mathbf{B}_0$. The torque acting on $\mathbf{M}$ is $\mathbf{\Gamma} = \mathbf{M} \times \mathbf{B}_{rf}$, where $\mathbf{B}_{rf}$ is the radiofrequency field inducing the transitions. To obtain the maximum effect, the time-dependent electromagnetic field $\mathbf{B}_{rf}$ should lie in a plane perpendicular to the static field $\mathbf{B}_0$. Equation (28.7) can be satisfied either by tuning the monochromatic radiofrequency input to the resonance value at a constant magnetic field, or vice versa. As we discuss in Supplemental Sections 28.12 through 28.14, modern NMR spectroscopy uses neither of these methods; instead it utilizes radiofrequency pulse techniques.

If no more information than that outlined in the preceding paragraphs could be obtained with this technique, NMR spectroscopy would simply be an expensive tool for quantitatively analyzing the elemental composition of compounds. However, two important aspects of this technique make it very useful for obtaining additional chemical information at the molecular level. The first of these is that the magnetic field in Equation (28.7) is not the applied external field, but rather the local field. As we will see, the local field is influenced by the electron distribution on the atom of interest as well as by the electron distribution on nearby atoms. This difference between the external and induced magnetic fields is the origin of the **chemical shift.** The H atoms in methane and chloroform have a different Larmor frequency because of this chemical shift. The origin of the chemical shift is discussed in Sections 28.3 through 28.6. The second important aspect is that individual magnetic dipoles interact with one another. This leads to a splitting of the energy levels of a two-spin system and the appearance of multiplet spectra in NMR. As discussed in Sections 28.7 and 28.8, the multiplet structure of a NMR resonance absorption gives direct structural information about the molecule.

28.3 The Chemical Shift for an Isolated Atom

When an atom is placed in a magnetic field, a circulation current is induced in the electron charge around the nucleus that generates a secondary magnetic field. The direction of the induced magnetic field at the position of the nucleus of interest opposes the external field; this phenomenon is referred to as a **diamagnetic response.** The origin of this response is shown in Figure 28.3. At distances from the center of the distribution that are large compared to an atomic diameter, the field is the same as that of a magnetic dipole. The z component of the induced magnetic field is given by

$$B_z = \frac{\mu_0}{4\pi} \frac{|\boldsymbol{\mu}|}{r^3} (3\cos^2\theta - 1) \qquad (28.8)$$

In Equation (28.8), μ_0 is the vacuum permeability, $\boldsymbol{\mu}$ is the induced magnetic moment, and θ and r define the coordinates of the observation point relative to the center of the charge distribution. Note that the induced field falls off rapidly with distance.

FIGURE 28.3
The shaded spherical volume represents a negatively charged classical continuous charge distribution. When placed in a magnetic field, the distribution will circulate as indicated by the horizontal orbit, viewed from the perspective of classical electromagnetic theory. The motion will induce a magnetic field at the center of the distribution that opposes the external field. This classical picture is not strictly applicable at the atomic level, but the outcome is the same as a rigorous quantum mechanical treatment.

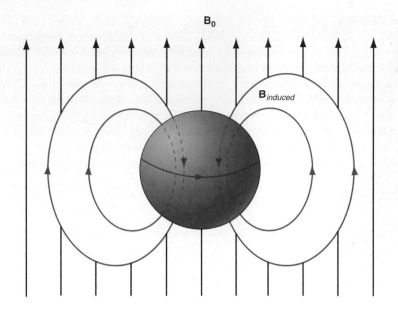

Depending on whether $3\cos^2\theta > 1$ or $3\cos^2\theta < 1$, B_z will add to or subtract from, the external applied magnetic field at the point r, θ. As we will see later, the angular dependence of B_z is important in averaging out the induced magnetic field of freely tumbling molecules in a solution.

For a diamagnetic response, the induced field at the nucleus of interest is opposite in direction and is linearly proportional to the external field in magnitude. Therefore, we can write $\mathbf{B}_{induced} = -\sigma\,\mathbf{B}_0$, which defines the **shielding constant** σ. The total field at the nucleus is given by the sum of the external and induced fields,

$$\mathbf{B}_{total} = (1 - \sigma)\,\mathbf{B}_0 \qquad (28.9)$$

and the resonance frequency taking the shielding into account is given by

$$\nu_0 = \frac{\gamma B_0 (1 - \sigma)}{2\pi} \quad \text{or} \quad \omega_0 = \gamma B_0 (1 - \sigma) \qquad (28.10)$$

Because $\sigma > 0$ for a diamagnetic response, the resonance frequency of a nucleus in an atom is lower than would be expected for the bare nucleus. The frequency shift is given by

$$\Delta\nu = \frac{\gamma B_0}{2\pi} - \frac{\gamma B_0 (1 - \sigma)}{2\pi} = \frac{\sigma \gamma B_0}{2\pi} \qquad (28.11)$$

which shows that the electron density around the nucleus reduces the resonance frequency of the nuclear spin. This effect is the basis for the chemical shift in NMR. The shielding constant σ increases with the electron density around the nucleus and, therefore, with the atomic number. Although ^{1}H is the most utilized nuclear probe in NMR, it has the smallest shielding constant of all atoms because it has only one electron orbiting around the nucleus. By comparison, σ for ^{13}C and ^{31}P are a factor of 15 and 54 greater, respectively.

28.4 The Chemical Shift for an Atom Embedded in a Molecule

We now consider the effect of neighboring atoms in a molecule on the chemical shift of a ^{1}H atom. As we have seen, the frequency shift for an atom depends linearly on the shielding constant σ. Because σ depends on the electron density around the nuclear spin of interest, it will change as neighboring atoms or groups either withdraw or increase electron density from the hydrogen atom of interest. This leads to a shift in frequency that makes NMR a sensitive probe of the chemical environment around a nucleus with nonzero nuclear spin. For ^{1}H, σ is typically in the range of 10^{-5} to 10^{-6},

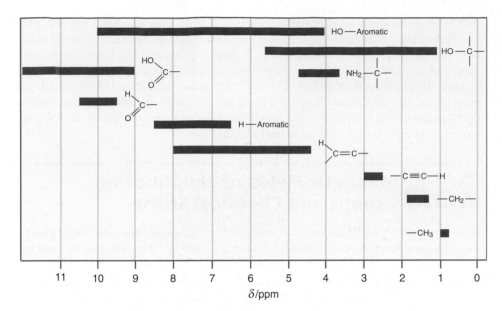

FIGURE 28.4
Chemical shifts δ as defined by Equation (28.12) for ^{1}H in different classes of chemical compounds. Extensive compilations of chemical shifts are available in the chemical literature.

so that the change in the resonance frequency due to the chemical shift is quite small. It is convenient to define a dimensionless quantity δ to characterize this frequency shift, with δ defined relative to a reference compound by

$$\delta = 10^6 \frac{(\nu - \nu_{ref})}{\nu_{ref}} = 10^6 \frac{\gamma B_0 (\sigma_{ref} - \sigma)}{\gamma B_0 (1 - \sigma_{ref})} \approx 10^6 (\sigma_{ref} - \sigma) \qquad \textbf{(28.12)}$$

For ^{1}H NMR, tetramethylsilane, $(CH_3)_4Si$, is usually used as a reference compound. Defining the chemical shift in this way has the advantage that δ is independent of the frequency, so that all measurements using spectrometers with different magnetic fields will give the same value of δ.

Figure 28.4 illustrates the observed ranges of δ for hydrogen atoms in different types of chemical compounds. The figure shows that the chemical shift for the OH group in alcohols is quite different than the chemical shift for H atoms in a methyl group. It also shows that the range in observed chemical shifts for a class of compounds can be quite large, as is seen for the aromatic alcohols. How can these chemical shifts be understood?

Although a quantitative understanding of these shifts requires the consideration of many factors, two factors are responsible for the major part of the chemical shift: the electronegativity of the neighboring group and the induced magnetic field of the neighboring group at the position of the nucleus of interest. We discuss each of these effects in the next two sections.

28.5 Electronegativity of Neighboring Groups and Chemical Shifts

Rather than consider individual atoms near the nuclear spin of interest, we consider groups of atoms such as —OH or —CH_2—. If a neighboring group is more electronegative than hydrogen, it will withdraw electron density from the region around the ^{1}H nucleus. Therefore, the nucleus is less shielded, and the NMR resonance frequency appears at a larger value of δ. For example, the chemical shift for ^{1}H in the methyl halides follows the sequence $CH_3I < CH_3Br < CH_3Cl < CH_3F$. The range of this effect is limited to about three or four bond lengths as can be shown by considering the chemical shifts in 1-chlorobutane. In this molecule, δ for the ^{1}H on the CH_2 group closest to the Cl is almost 3 ppm larger than the ^{1}H on the terminal CH_3 group, which has nearly the same δ as in propane.

As Figure 28.4 shows, the chemical shifts for different classes of molecules are strongly correlated with their electron-withdrawing ability. Carboxyl groups are very effective in withdrawing charge from around the ^{1}H nucleus; therefore, the chemical

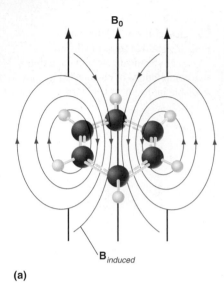

(a)

(b)

FIGURE 28.5

(a) The induced magnetic field generated by a circulating ring current in benzene. Note that in the plane of the molecule, the induced field is in the same direction as the external field outside the ring, and in the opposite direction inside the ring.

(b) 18-Annulene provides a confirmation of this model, because δ has the opposite sign for interior and exterior ^{1}H.

shift is large and positive. Aldehydes, alcohols, and amines are somewhat less effective in withdrawing electron charge. Aromatic rings are somewhat more effective than double and triple bonds in withdrawing charge. A methyl group attached to an electron-rich atom such as Li or Al will have a negative chemical shift, indicating that the ^{1}H nucleus is more shielded than in $(CH_3)_4Si$. However, the spread in chemical shifts for any of these classes of compounds can be quite large, and the ranges for different classes overlap. The spread and overlap arise because of the induced magnetic field of neighboring groups. This topic is discussed in the next section.

28.6 Magnetic Fields of Neighboring Groups and Chemical Shifts

The magnetic field at a ^{1}H nucleus is a superposition of the external field, the local field induced by the diamagnetic response of the electrons around the ^{1}H nucleus, and the local induced magnetic fields from neighboring atoms or groups. The value of σ is small for the H atom because an isolated H atom has only one electron; therefore, the magnetic field at a ^{1}H nucleus in a molecules is often dominated by the local induced magnetic fields from neighboring atoms or groups. We focus on neighboring groups rather than on individual neighboring atoms, because groups can have a high diamagnetic or paramagnetic response. The stronger the magnetic field induced by a diamagnetic or paramagnetic response in a group, the greater the effect it will have at the neighboring ^{1}H nucleus under consideration. It is helpful to think of the neighboring group as a magnetic dipole $\boldsymbol{\mu}$ whose strength and direction are determined by the magnitude and sign of its shielding constant σ. Groups containing delocalized electrons such as aromatic groups, carbonyl, and other groups containing multiple bonds give rise to large values of $\boldsymbol{\mu}$. Aromatic rings generate a large $\boldsymbol{\mu}$ value because the delocalized electrons can give rise to a ring current, as illustrated in Figure 28.5a, just as a current is induced in a macroscopic wire loop by a time-dependent magnetic field. This model of the ring current predicts that the chemical shift of the interior and exterior ^{1}H atoms attached to an aromatic system should be in the opposite direction. In fact, δ for an exterior ^{1}H of 18-annulene is +9.3 and that for an interior ^{1}H is −3.0.

An aromatic ring has a strong **magnetic anisotropy.** A sizable ring current is induced when the magnetic field is perpendicular to the plane of the ring, but the current is negligible when the magnetic field lies in the plane of the ring. This is true of many neighboring groups; the magnitude of $\boldsymbol{\mu}$ depends on the orientation of the group relative to the field.

Although we have considered individual molecules, the NMR signal of a solution sample is generated by the large number of molecules contained in the sampling volume. Therefore, the observed σ is an average over all possible orientations of the molecule, $\sigma_{average} = \langle \sigma_{individual} \rangle$, where $\sigma_{individual}$ applies to a particular orientation. In a gas or a solution, the molecules in the sample have all possible orientations with respect to the magnetic field. To determine how this random orientation affects the spectrum, we must ask if the shielding or deshielding of a ^{1}H by a neighboring group depends on the orientation of the molecule relative to the field.

Consider the induced magnetic field of a neighboring group at the ^{1}H of interest. Because the direction of the induced magnetic field is linked to the external field, rather than to the molecular axis, it retains its orientation relative to the field as the molecule tumbles in a gas or a solution. For the case of an isotropic neighboring group, $\langle B_z \rangle$ is obtained by averaging the induced magnetic field given by Equation (28.8) over all possible angles. Because the induced magnetic field for an isotropic neighboring group is independent of θ, $|\boldsymbol{\mu}|$ can be taken out of the integral, leading to

$$\langle B_z \rangle = \frac{\mu_0}{4\pi} \frac{|\boldsymbol{\mu}|}{r^3} \int_0^{2\pi} d\phi \int_0^{\pi} (3\cos^2\theta - 1)\sin\theta d\theta = \frac{\mu_0}{2} \frac{|\boldsymbol{\mu}|}{r^3} \left[-\cos^3\theta + \cos\theta \right]_0^{\pi} = 0 \quad \textbf{(28.13)}$$

Equation (28.13) shows that as a result of the tumbling, $\langle B_z \rangle = 0$ unless the neighboring group has a magnetic anisotropy. In this case, $\boldsymbol{\mu}$ depends on θ and ϕ, and $\mu(\theta, \phi)$ must remain inside the integral.

If the neighboring groups are magnetically isotropic and the molecule is tumbling freely, the NMR spectrum of a sample in solution is greatly simplified because $\sigma_{average} = 0$. For a solid in which tumbling cannot occur, there is another way to eliminate dipolar interactions from neighboring groups. This is done by orienting the sample in the static magnetic field, choosing the angle $\theta = 54.74°$ at which $\langle B_z \rangle$ goes to zero. Solid-state NMR spectra and the technique of magic angle spinning are discussed in Section 28.10.

This and the previous section have provided a brief introduction to the origin of the chemical shift in NMR spectroscopy. For ^1H, the range of observed values for δ among different chemical compounds is about 10 ppm. For nuclei in atoms that can exhibit both paramagnetic and diamagnetic behavior, δ can vary over a much wider range. For example, δ for ^{19}F can vary by 1000 ppm for different chemical compounds. Vast libraries of ^1H NMR spectra for different compounds have been assembled and provide chemists with a valuable tool for identifying chemical compounds on the basis of chemical shifts in their NMR spectra.

28.7 Multiplet Splitting of NMR Peaks Arises through Spin–Spin Coupling

What might one expect the spectrum of a molecule of ethanol, with three different types of hydrogens, to look like? A good guess is that each group of chemically equivalent protons resonates in a separate frequency range, one corresponding to the methyl group, another to the methylene group, and the third to the OH group. The OH proton is most strongly deshielded (largest δ) because it is directly bound to the electronegative oxygen atom. It is found near 5 ppm. Because the methylene group is closer to the electronegative OH group, the protons are more deshielded and appear at larger values of δ (near 3.5 ppm) than the methyl protons, which are found near 1 ppm. Furthermore, we expect that the areas of the peaks have the ratio $CH_3:CH_2:OH = 3:2:1$ because the NMR signal is proportional to the number of spins. A simulated NMR spectrum for ethanol is shown in Figure 28.6.

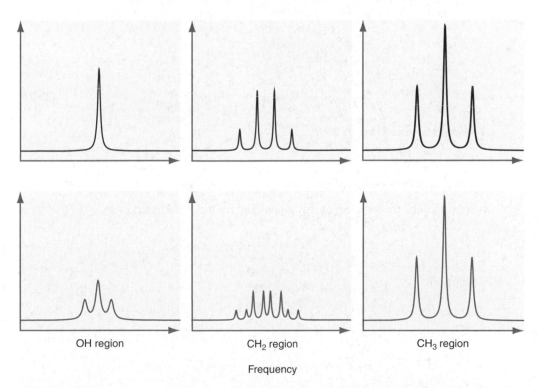

OH region   CH$_2$ region   CH$_3$ region

Frequency

FIGURE 28.6

Simulated NMR spectrum showing the intensity (vertical axis) as a function of frequency for ethanol. The top panel shows the multiplet structure at room temperature. The lower panel shows the multiplet structure observed at lower temperature in acid-free water. The different portions of the spectrum are not to scale, but have the relative areas discussed in the text.

A very important feature that has not been discussed yet is shown in Figure 28.6: the individual peaks are split into **multiplets.** At low temperature and in the absence of acidic protons, the OH proton resonance is a triplet, whereas the CH_3 proton resonance is a triplet and the CH_2 resonance is an octet. At higher temperature, a change in the NMR spectrum is observed. The OH proton resonance is a singlet, the CH_3 proton resonance is a triplet, and the CH_2 resonance is a quartet. How can this splitting be understood? The higher temperature spectrum is the result of rapid transfer of the OH proton between ethanol and water, as discussed later in Section 28.9. For now, we turn our attention to the origin of **multiplet splitting.**

Multiplets arise as a result of spin–spin interactions among different nuclei. We first consider the case of two distinguishable noninteracting spins such as the ^1H nuclei of the CH_3 and CH_2 groups in ethanol. We give these spins the labels 1 and 2 and subsequently introduce the interaction. The spin energy operator for the noninteracting spins is

$$\hat{H} = -\gamma B_0 (1 - \sigma_1)\hat{I}_{z_1} - \gamma B_0 (1 - \sigma_2)\hat{I}_{z_2} \qquad (28.14)$$

and the eigenfunctions of this operator are products of the eigenfunctions of the individual operators $\hat{I}_{z_1}$ and $\hat{I}_{z_2}$:

$$\psi_1 = \alpha(1)\alpha(2)$$
$$\psi_2 = \beta(1)\alpha(2)$$
$$\psi_3 = \alpha(1)\beta(2)$$
$$\psi_4 = \beta(1)\beta(2) \qquad (28.15)$$

We solve the Schrödinger equation for the corresponding eigenvalues, which are as follows (see Example Problem 28.2):

$$E_1 = -\hbar\gamma B_0 \left(1 - \frac{\sigma_1 + \sigma_2}{2}\right)$$

$$E_2 = -\frac{\hbar\gamma B_0}{2}(\sigma_1 - \sigma_2)$$

$$E_3 = \frac{\hbar\gamma B_0}{2}(\sigma_1 - \sigma_2)$$

$$E_4 = \hbar\gamma B_0 \left(1 - \frac{\sigma_1 + \sigma_2}{2}\right) \qquad (28.16)$$

We have assumed that $\sigma_1 > \sigma_2$.

EXAMPLE PROBLEM 28.2

Show that the total nuclear energy eigenvalue for the wave function $\psi_2 = \beta(1)\alpha(2)$ is

$$E_2 = -\frac{\hbar\gamma B_0}{2}(\sigma_1 - \sigma_2)$$

Solution

$$\hat{H}\psi_2 = [-\gamma B_0 (1 - \sigma_1)\hat{I}_{z_1} - \gamma B_0 (1 - \sigma_2)\hat{I}_{z_2}]\beta(1)\alpha(2)$$

$$= [-\gamma B_0 (1 - \sigma_1)\hat{I}_{z_1}]\beta(1)\alpha(2) + [-\gamma B_0 (1 - \sigma_2)\hat{I}_{z_2}]\beta(1)\alpha(2)$$

$$= \frac{\hbar}{2}\gamma B_0 (1 - \sigma_1)\beta(1)\alpha(2) - \frac{\hbar}{2}\gamma B_0 (1 - \sigma_2)\beta(1)\alpha(2)$$

$$= -\frac{\hbar\gamma B_0}{2}(\sigma_1 - \sigma_2)\beta(1)\alpha(2)$$

$$= E_2\psi_2$$

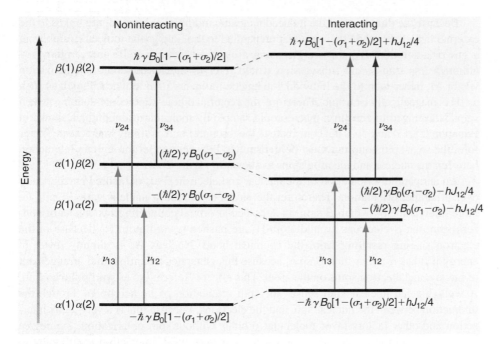

FIGURE 28.7
The energy levels for two noninteracting spins and the allowed transitions between these levels are shown on the left. The same information is shown on the right for interacting spins. The splitting between levels 2 and 3 and the energy shifts of all four levels for interacting spins are greatly magnified to emphasize the spin–spin interactions.

You will calculate the energy eigenvalues for the other eigenfunctions in Equation (28.15) in the end-of-chapter problems. All four energy eigenvalues are plotted in the energy diagram of Figure 28.7. We initially focus on the left half of this figure, which shows the energy levels for noninteracting spins. The selection rule for NMR spectroscopy is that only one of the spins can change in a transition. The four allowed transitions are indicated in the figure. For the noninteracting spin case, $E_2 - E_1 = E_4 - E_3$ and $E_3 - E_1 = E_4 - E_2$. Therefore, the NMR spectrum contains only two peaks corresponding to the frequencies

$$\nu_{12} = \nu_{34} = \frac{E_2 - E_1}{h} = \frac{\gamma B_0(1 - \sigma_1)}{2\pi}$$

$$\nu_{13} = \nu_{24} = \frac{E_4 - E_2}{h} = \frac{\gamma B_0(1 - \sigma_2)}{2\pi} \qquad (28.17)$$

You will calculate the allowed frequencies in the end-of-chapter problems. This result shows that the splitting of a single peak into multiplets is not observed for noninteracting spins.

We next consider the case of interacting spins. Because each of the nuclear spins acts like a small bar magnet, they interact with one another through **spin–spin coupling.** There are two different types of spin–spin coupling: through-space vectorial dipole–dipole coupling, which is important in the NMR of solids (see Section 28.10), and through-bond, or scalar, dipole–dipole coupling, which is considered next.

The spin energy operator that takes scalar dipole–dipole coupling into account is

$$\hat{H} = -\gamma B_0(1 - \sigma_1)\hat{I}_{z_1} - \gamma B_0(1 - \sigma_2)\hat{I}_{z_2} + \frac{hJ_{12}}{\hbar^2}\hat{I}_1 \cdot \hat{I}_2 \qquad (28.18)$$

In this equation, J_{12}, is called the **coupling constant** and is a measure of the strength of the interaction between the individual magnetic moments. The factor $h/\hbar^2$ in the last term of Equation (28.18) is included to make the units of J_{12} be s^{-1}. What is the origin of this through-bond coupling interaction? Two possibilities are considered: vectorial dipole–dipole coupling and the interaction between nuclear and electron spins.

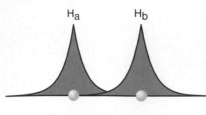

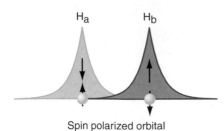

FIGURE 28.8

Schematic illustration of how spin polarized orbitals couple nuclear spins even though they are highly shielded from one another through the electron density. The upper and lower arrows in the lower part of the figure indicate the electron and nuclear spin, respectively.

Because the directions of the induced magnetic moments $\boldsymbol{\mu}_1$ and $\boldsymbol{\mu}_2$ are linked to the external field, they retain their orientation parallel to the field as the molecule tumbles in a gas or a solution. (Again, we are using a classical picture, and a more rigorous— although less transparent—discussion would refer to the macroscopic magnetization vector **M,** rather than to the individual magnetic moments.) An individual nucleus such as ^{1}H is magnetically isotropic. Therefore, the vectorial dipole–dipole interaction between spins is averaged to zero in a macroscopic sample by molecular tumbling, as shown in Equation (28.13) and does not contribute to the through-bond coupling interaction. Therefore, the spin–spin coupling must be transmitted between nuclei through an interaction between the nuclear and electron spins as shown in Figure 28.8.

An antiparallel orientation of the nuclear and electron spins is favored energetically over a parallel orientation. Therefore, the electrons around a nucleus with β spin are more likely to be of α than β spin. In a molecular orbital connecting two nuclei of non-zero spin, the β electrons around atom H_a are pushed toward atom H_b because of the electron sharing resulting from the chemical bond. Nucleus H_b is slightly lower in energy if it has α rather than β spin, because this generates an antiparallel arrangement of nuclear and electron spins on the atom. This effect is referred to as **spin polarization.** A well-shielded nuclear spin senses the spin orientation of its neighbors through the interaction between the nuclear spin and the electrons. Because this is a very weak interaction and other factors favor molecular orbitals without spin polarization, the degree of spin polarization is very small. However, this very weak interaction is sufficient to account for the parts per million changes in the frequency of NMR transitions.

At this point, we discuss the spin energy operator for interacting spins and use an approximation method to determine the spin energy eigenvalues of this operator. The eigenfunctions are linear combinations of the eigenfunctions for noninteracting spins and need not concern us further. The approximation method is called first-order perturbation theory. It is applicable when we know how to solve the Schrödinger equation for a problem that is very similar to the one of interest. In this case, the problem we know how to solve is for noninteracting spins. If the change in the energy levels brought about by an additional interaction term in the spin energy operator, $\hat{H}_{interaction}$, is small, then we state without proof that the first-order correction to the energy for the case of two interacting spins is given by

$$\Delta E_j = \iint \psi_j^* \hat{H}_{interaction} \psi_j \, d\tau_1 \, d\tau_2 = \frac{4\pi^2}{h} J_{12} \iint \psi_j^* \hat{I}_1 \cdot \hat{I}_2 \psi_j \, d\tau_1 \, d\tau_2 \quad \textbf{(28.19)}$$

In this equation, the wave functions are those for the problem in the absence of $\hat{H}_{interaction}$, and the integration is over the two spin variables.

To evaluate this integral, we write $\hat{I}_1 \cdot \hat{I}_2 = \hat{I}_{1x}\hat{I}_{2x} + \hat{I}_{1y}\hat{I}_{2y} + \hat{I}_{1z}\hat{I}_{2z}$ and must solve equations of the type

$$\Delta E_j = \frac{4\pi^2}{h} J_{12} \iint \alpha^*(1)\alpha^*(2)\hat{I}_{1x}\hat{I}_{2x}\alpha(1)\alpha(2) \, d\tau_1 \, d\tau_2$$

We know that α and β are eigenfunctions of $\hat{I}_z$ and that they are not eigenfunctions of $\hat{I}_x$ and $\hat{I}_y$. The following relations, which are not proved, are used to solve the necessary integrals as shown in Example Problem 28.3:

$$\hat{I}_x\alpha = \frac{\hbar}{2}\beta; \qquad \hat{I}_y\alpha = \frac{i\hbar}{2}\beta; \qquad \hat{I}_z\alpha = \frac{\hbar}{2}\alpha$$

$$\hat{I}_x\beta = \frac{\hbar}{2}\alpha; \qquad \hat{I}_y\beta = -\frac{i\hbar}{2}\alpha; \qquad \hat{I}_z\beta = -\frac{\hbar}{2}\beta \qquad \textbf{(28.20)}$$

EXAMPLE PROBLEM 28.3

Show that the energy correction to $\psi_2 = \alpha(1)\beta(2)$ is $\Delta E_2 = -(h J_{12}/4)$

Solution

We evaluate

$$\Delta E_2 = \frac{4\pi^2}{h} J_{12} \iint \alpha^*(1)\beta^*(2)[\hat{I}_{1x}\hat{I}_{2x} + \hat{I}_{1y}\hat{I}_{2y} + \hat{I}_{1z}\hat{I}_{2z}]\alpha(1)\beta(2)\, d\tau_1\, d\tau_2$$

$$= \frac{4\pi^2}{h} J_{12} \left[\begin{array}{l} \iint \alpha^*(1)\beta^*(2)[\hat{I}_{1x}\hat{I}_{2x}]\alpha(1)\beta(2)\, d\tau_1\, d\tau_2 \\[6pt] + \iint \alpha^*(1)\beta^*(2)[\hat{I}_{1y}\hat{I}_{2y}]\alpha(1)\beta(2)\, d\tau_1\, d\tau_2 \\[6pt] + \iint \alpha^*(1)\beta^*(2)[\hat{I}_{1z}\hat{I}_{2z}]\alpha(1)\beta(2)\, d\tau_1\, d\tau_2 \end{array} \right]$$

$$= \frac{4\pi^2}{h} J_{12} \left[\begin{array}{l} \iint \alpha^*(1)\beta^*(2)\left[\dfrac{\hbar^2}{4}\right]\beta(1)\alpha(2)\, d\tau_1\, d\tau_2 \\[6pt] + \iint \alpha^*(1)\beta^*(2)\left[-\dfrac{i^2\hbar^2}{4}\right]\beta(1)\alpha(2)\, d\tau_1\, d\tau_2 \\[6pt] + \iint \alpha^*(1)\beta^*(2)\left[-\dfrac{\hbar^2}{4}\right]\alpha(1)\beta(2)\, d\tau_1\, d\tau_2 \end{array} \right]$$

Because of the orthogonality of the spin functions, the first two integrals are zero and

$$\Delta E_2 = \frac{4\pi^2}{h}\left(-\frac{\hbar^2}{4}\right) J_{12} \iint \alpha^*(1)\beta^*(2)\alpha(1)\beta(2)\, d\tau_1 d\tau_2 = \frac{4\pi^2}{h} J_{12}\left(-\frac{\hbar^2}{4}\right) = -\frac{h J_{12}}{4}$$

Note that because J_{12} has the units of s^{-1}, hJ has the unit joule.

You will use the procedure of Example Problem 28.3 in the end-of-chapter problems to show that the spin energy eigenvalue for a given state is changed relative to the case of noninteracting spins by the amount

$$\Delta E = m_1 m_2 h J_{12} \text{ with } m_1 \text{ and } m_2 = +\frac{1}{2} \text{ for } \alpha \text{ and } -\frac{1}{2} \text{ for } \beta \qquad \textbf{(28.21)}$$

A given energy level is shifted to a higher energy if both spins are of the same orientation, and to a lower energy if the orientations are different. As you will see in the end-of-chapter problems, the frequencies of the allowed transitions including the spin–spin coupling are

$$\nu_{12} = \frac{\gamma B_0(1 - \sigma_1)}{2\pi} - \frac{J_{12}}{2}$$

$$\nu_{34} = \frac{\gamma B_0(1 - \sigma_1)}{2\pi} + \frac{J_{12}}{2}$$

$$\nu_{13} = \frac{\gamma B_0(1 - \sigma_2)}{2\pi} - \frac{J_{12}}{2}$$

$$\nu_{24} = \frac{\gamma B_0(1 - \sigma_2)}{2\pi} + \frac{J_{12}}{2} \qquad \textbf{(28.22)}$$

The energy levels and transitions corresponding to these frequencies are shown on the right side of Figure 28.7. This calculation shows that spin–spin interactions result in the appearance of multiplet splitting in NMR spectra. Each of the two peaks that appeared in the spectrum in the absence of spin–spin interactions is now split into a doublet in

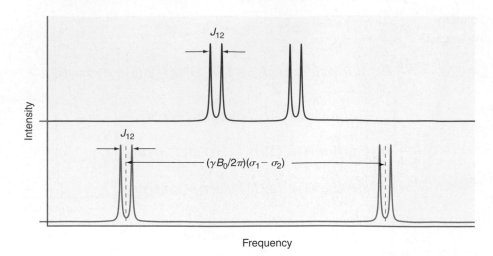

FIGURE 28.9

Splitting of a system of two interacting spins into doublets for two values of B_0. The spacing within the doublet is independent of the magnetic field strength, but the spacing of the doublets increases linearly with B_0.

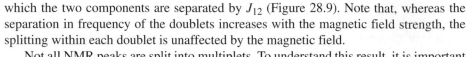

CH$_2$CF$_2$

CH$_2$F$_2$

FIGURE 28.10

The H atoms in CH_2CF_2 are chemically equivalent, but magnetically inequivalent. The H atoms in CH_2F_2 are chemically and magnetically equivalent.

which the two components are separated by J_{12} (Figure 28.9). Note that, whereas the separation in frequency of the doublets increases with the magnetic field strength, the splitting within each doublet is unaffected by the magnetic field.

Not all NMR peaks are split into multiplets. To understand this result, it is important to distinguish between **chemically** as opposed to **magnetically equivalent nuclei.** Consider the two molecules shown in Figure 28.10. In both cases, the two H atoms and the two F atoms are chemically equivalent. The nuclei of chemically equivalent atoms are also magnetically equivalent *only* if the interactions that they experience with other nuclei of nonzero spin are identical. Because the two F nuclei in CH_2F_2 are equidistant from each H atom, the two H—F couplings are identical and the ^{1}H are magnetically equivalent. However, the two H—F couplings in CH_2CF_2 are different because the spacing between the H and F nuclei is different. Therefore, the ^{1}H nuclei in this molecule are magnetically inequivalent. Multiplet splitting only arises through the interaction of magnetically inequivalent nuclei and is observed in CH_2CF_2, but not in CH_2F_2 or the reference compound $(CH_3)_4Si$. Because the derivation of this result is somewhat lengthy, it is omitted in this chapter.

28.8 Multiplet Splitting When More Than Two Spins Interact

For simplicity, we have considered only the case of two coupled spins in the previous sections. However, many organic molecules have more than two inequivalent protons that are close enough to one another to generate multiplet splittings. In this section, several different coupling schemes are considered. The frequencies for transitions in such a system involving the nuclear spin A can be written as

$$\nu_A = \frac{\gamma_A B(1 - \sigma_A)}{2\pi} - \sum_{X \neq A} J_{AX} m_X m_A \tag{28.23}$$

where the summation is over all other spin-active nuclei. The strength of the interaction that leads to peak splitting is weak because J_{AX} falls off rapidly with distance. Therefore, the neighboring spins must be rather close in order to generate peak splitting. Experiments have shown that generally only those atoms within three or four bond lengths of the nucleus of interest have a sufficiently strong interaction to generate peak splitting. In strongly coupled systems such as those with conjugated bonds, the coupling can still be strong when the spins are farther apart.

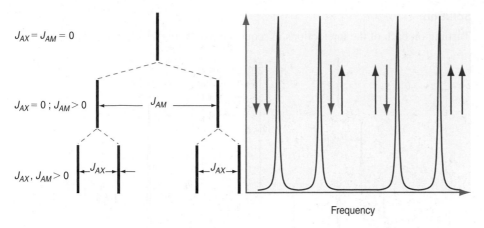

FIGURE 28.11
Coupling scheme and expected NMR spectrum for spin A coupled to spins M and X with different coupling constants J_{AX} and J_{AM}. The vertical axis shows the spectrum intensity.

To illustrate the effect of spin–spin interaction in generating multiplet splittings, we consider the coupling of three distinct spin 1/2 nuclei that we label A, M, and X. The two coupling constants are J_{AM} and J_{AX} with $J_{AM} > J_{AX}$. The effect of these couplings can be determined by turning on the couplings individually as indicated in Figure 28.11. The result is that each of the lines in the doublet that arises from turning on the interaction J_{AM} is again split into a second doublet when the interaction J_{AX} is turned on as shown in Figure 28.11.

A special case occurs when A and M are identical so that $J_{AM} = J_{AX}$. The middle two lines for the AMX case now lie at the same frequency, giving rise to the AX_2 pattern shown in Figure 28.12. Because the two lines lie at the same frequency, the resulting spectrum is a triplet with the intensity ratio $1:2:1$. Such a spectrum is observed for the methylene protons in the molecule $CHCl_2\text{---}CH_2\text{---}CHCl_2$.

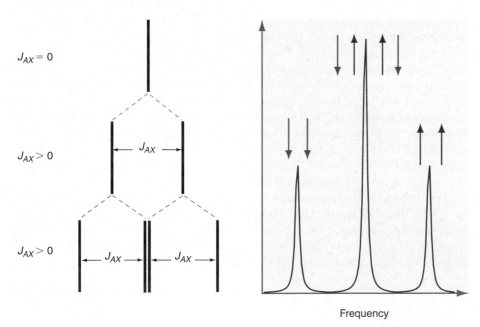

FIGURE 28.12
Coupling scheme and expected NMR spectrum for spin A coupled to two spins X. In this case, there is only one coupling constant J_{AX}. The closely spaced pair of lines in the lower part of the left figure actually coincide. They have been shown separated to make their origin clear. The vertical axis shows the spectrum intensity.

EXAMPLE PROBLEM 28.4

Using the same reasoning as that applied to the AX_2 case, predict the NMR spectrum for an AX_3 spin system. Such a spectrum is observed for the methylene protons in the molecule $CH_3\text{---}CH_2\text{---}CCl_3$ where the coupling is to the methyl group hydrogens.

Solution

Turning on each of the interactions in sequence results in the following diagram:

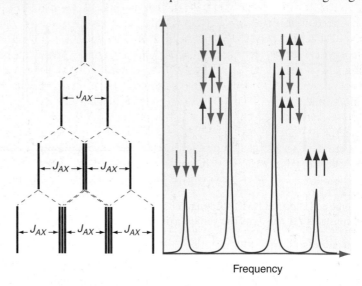

The end result is a quartet with the intensity ratios $1:3:3:1$. These results can be generalized to the rule that if a ^1H nucleus has n equivalent ^1H neighbors, its NMR spectral line will be split into $n+1$ peaks. The relative intensity of these peaks is given by the coefficients in the expansion of $(1 + x)^n$, the binomial expression. The closely spaced pair of lines in the left figure actually coincide. They have been shown separated to make their origin clear.

Given the results of the last two sections, we are (almost) at the point of being able to understand the fine structure in the NMR spectrum of ethanol shown in Figure 28.6. As discussed in Section 28.7, the resonance near 5 ppm can be attributed to the OH proton, the resonance near 3.5 ppm can be attributed to the CH_2 protons, and the resonance near 1 ppm can be attributed to the CH_3 protons. This is consistent with the integrated intensities of the peaks, which from high to low δ are in the ratio $1:2:3$. We now consider the multiplet splitting. Invoking the guideline that spins located more than three bonds away do not generate a peak splitting, we conclude that the CH_3 resonance is a triplet because it is split by the two CH_2 protons. The OH proton is too distant to generate a further splitting of the CH_3 group. We conclude that the CH_2 resonance is an octet (two pairs of quartets) because it is split by the three equivalent CH_3 protons and the OH proton. We predict that the OH resonance is a triplet because it is split by the two equivalent CH_2 protons. In fact, this is exactly what is observed for the NMR spectrum of ethanol at low temperatures. This example shows the power of NMR spectroscopy in obtaining structural information at the molecular level.

For ethanol at room temperature, these predictions are correct for the CH_3 group, but not for the other groups. The CH_2 hydrogen resonance is a quartet and the OH proton resonance is a singlet. This tells us that there is something that we have overlooked regarding the OH group. What has been overlooked is the rapid exchange of the OH proton with water, a topic that is discussed in the next section.

28.9 Peak Widths in NMR Spectroscopy

The ability of any spectroscopic technique to deliver useful information is limited by the width of the peaks in frequency. If two different NMR active nuclei in a sample have characteristic frequencies that are significantly closer than the width of the peaks, it is difficult to distinguish them. For samples in solutions, NMR spectra can exhibit peak widths of as little as 0.1 Hz, whereas for solid samples, peak widths of 10 kHz are not atypical. What are the reasons for such a large variation of peak widths in NMR spectra?

To answer this question, the change in the **magnetization vector M** with time must be considered. The vector **M** has two components: M_z or $\mathbf{M}_{\parallel}$, which is parallel to the static field $\mathbf{B}_0$, and M_{xy} or $\mathbf{M}_{\perp}$, which is perpendicular to the field. Assume that the system has been perturbed so that **M** is not parallel to $\mathbf{B}_0$. How does the system of spins return to equilibrium? Note first that M_z decays at a different rate than $M_{x\text{-}y}$. It is not surprising that these two processes have different rates. To relax M_z, energy must be transferred to the surroundings, which is usually referred to as the lattice. The characteristic time associated with this process is called the longitudinal or **spin-lattice relaxation time T_1.** The relaxation of $M_{x\text{-}y}$ occurs through a randomization or **dephasing** of the spins and does not involve energy transfer to the surroundings because this component of the magnetization vector is perpendicular to $\mathbf{B}_0$. The characteristic time associated with this process is called the transverse or **spin–spin relaxation time T_2.** Because M_z will return to its initial value only after $\mathbf{M}_{\perp} \rightarrow 0$, we conclude that $T_1 \geq T_2$.

The relaxation time T_1 determines the rate at which the energy absorbed from the radiofrequency field is dissipated to the surroundings. If T_1 is not sufficiently small, energy is not lost quickly enough to the surroundings, and the population of the excited state becomes as large as that of the ground state. If the populations of the ground and excited states are equal, the net absorption at the transition frequency is zero, and we say that the transition is a **saturated transition.** In obtaining NMR spectra, the radiofrequency power is kept low in order to avoid saturation.

How is the rate of relaxation of **M** related to the NMR linewidth? In discussing this issue, it is useful to view the experiment from two vantage points, time and frequency. As shown later in Section 28.12, the NMR signal is proportional to $M_{x\text{-}y}$, which decreases with increasing time with the functional form e^{-t/T_2} in the time domain. In a measurement of the peak width as a function of the frequency, we look at the same process in the frequency domain because the signal in the frequency domain is the Fourier transform of the time-domain signal. Because of this relationship between the two domains, T_2 determines the spectral linewidth. The linewidth can be estimated with the Heisenberg uncertainty principle. The lifetime of the excited state, Δt, and the width in frequency of the spectral line corresponding to the transition to the ground state, $\Delta \nu$, are inversely related by

$$\frac{\Delta E \, \Delta t}{h} \approx 1 \quad \text{or} \quad \Delta \nu \approx \frac{1}{\Delta t} \qquad \textbf{(28.24)}$$

In the NMR experiment, T_2 is equivalent to Δt and, therefore, it determines the width of the spectral line, $\Delta \nu$. For this reason, narrow spectral features correspond to large values of T_2. In solution, T_2 can be orders of magnitude greater than for an ordered or disordered solid of the same substance. Therefore, NMR spectra in solution, in which through-space vectorial dipole–dipole coupling is averaged to zero through the tumbling of molecules resulting in large T_2 values, consist of narrow lines. By contrast, solid-state spectra exhibit broad lines because T_2 is small. The vectorial dipole–dipole coupling is not averaged to zero in this case because the molecules are fixed at their lattice sites.

The lifetime of the excited state in NMR spectroscopy can be significantly changed relative to the preceding discussion if the spins are strongly coupled to their surroundings. For example, this occurs if a proton on a tumbling molecule in solution undergoes a chemical exchange between two different sites. Consider the proton exchange reaction for ethanol:

$$\text{CH}_3\text{CH}_2\text{OH} + \text{H}_3\text{O}^+ \; \rightleftharpoons \; \text{CH}_3\text{CH}_2\text{OH}_2^+ + \text{H}_2\text{O} \qquad \textbf{(28.25)}$$

The exchange decreases the lifetime of the excited state, or T_2, leading to a broadening of the NMR peak. It turns out that the peaks become significantly broader only if the site exchange time is in the range of 10^{-4} to 10 s. This effect is referred to as **motional broadening.**

For a significantly faster exchange, only a single sharp peak is observed, and this effect is referred to as **motional narrowing.** Because the exchange occurs in times faster than 10^{-4} s, motional narrowing is observed for ethanol at room temperature. For this reason, the portion of the ethanol NMR spectrum shown in Figure 28.6 corresponding to the OH proton is a singlet rather than a triplet. However, at low temperatures and under acid-free conditions, the exchange rate can be sufficiently reduced so that the exchange can be ignored. In this case, the OH ^{1}H signal is a triplet. We now understand

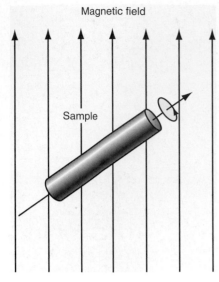

FIGURE 28.13
In magic angle spinning, the sample is rapidly spun about its axis, which is tilted 54.74° with respect to the static magnetic field.

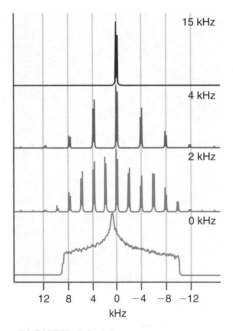

FIGURE 28.14
The ^{13}C NMR spectrum of a powder in which the unit cell contains a molecule with two inequivalent —C═O groups. The green spectrum shows the broad and nearly featureless solid-state spectrum. The 15-kHz spectrum (red) shows only two sharp peaks that can be attributed to the two chemically inequivalent —C═O groups. The other spectra are taken for different spinning frequencies. The spinning sidebands seen at 2 and 4 kHz are experimental artifacts that arise if the spinning frequency is not sufficiently high.

Source: Published by permission of Gary Drobny, University of Washington.

why the 300 K CH$_2$ hydrogen resonance in ethanol is a quartet rather than an octet and why the OH hydrogen resonance is a singlet rather than a triplet.

28.10 Solid-State NMR

Whereas NMR spectra with well-separated narrow peaks are generally observed in solution, this is not the case for solids because direct dipole–dipole coupling between spins is not averaged to zero in solids as it is through molecular tumbling in solution. As we saw in Section 28.3, the magnetic field of a neighboring dipole can increase or decrease the external field $\mathbf{B_0}$ at the position of a spin, leading to a shift in the resonance frequency. The frequency shift resulting from direct coupling between two dipoles i and j is

$$\Delta\nu_{d-d} \propto \frac{3\mu_i\mu_j}{hr_{ij}^3}(3\cos^2\theta_{ij} - 1) \tag{28.26}$$

In this equation, r_{ij} is the distance between the dipoles and θ_{ij} is the angle between the magnetic field direction and the vector connecting the dipoles. Why did we not consider direct dipole–dipole coupling in discussing NMR spectra of solutions? Because molecules in a solution are rapidly tumbling, the time-averaged value of $\cos^2\theta_{ij}$, rather than the instantaneous value, determines $\Delta\nu_{d-d}$. As shown in Section 28.6, $\langle\cos^2\theta_{ij}\rangle = 1/3$ and, therefore, $\Delta\nu_{d-d} = 0$ for rapidly tumbling molecules in solution. By contrast, in solids the relative orientation of all the spin-active nuclei is frozen because of the crystal structure. For this reason, $\Delta\nu_{d-d}$ can be as large as several hundred kilohertz. This leads to very broad spectral features in the NMR spectra of solids. Given this situation, why carry out NMR experiments on solids?

This question can be answered in several ways. First, many materials are only available as solids, so that the option of obtaining solution spectra is not available. Second, useful information about the molecular anisotropy of the chemical shift can be obtained from solid-state NMR spectra. Finally, the technique of magic angle spinning can be used to transform broad solid-state spectra into spectra with linewidths comparable to those obtained in solution, as discussed next.

In general, a sample used in solid-state NMR experiments consists of many individual solid particles that are randomly oriented with respect to one another rather than a single crystal. Now imagine that a molecule in the unit cell is rotating about an axis rather than tumbling freely. Although not derived here, the time average of $3\cos^2\theta_{ij} - 1$ in this case is given by

$$\langle 3\cos^2\theta_{ij} - 1\rangle = (3\cos^2\theta' - 1)\left(\frac{3\cos^2\gamma_{ij} - 1}{2}\right) \tag{28.27}$$

In this equation, θ' is the angle that the sample rotation axis makes with B_0, and γ_{ij} is the angle between the vector $\mathbf{r}_{ij}$ that connects the magnetic dipoles i and j and the rotation axis. If the whole solid sample is rotated rapidly, then all pairs of coupled dipoles in the entire sample have the same value of θ' even though they have different values of γ_{ij}. If we choose to make $\theta' = 54.74°$, then $\langle 3\cos^2\theta' - 1\rangle = 0$, $\Delta\nu_{d-d} = 0$, and the broadening introduced by direct dipole coupling vanishes. Because this choice of θ' has such a dramatic effect, it is referred to as the **magic angle,** and the associated technique is referred to as **magic angle spinning** (Figure 28.13). An example of how a broad solid-state NMR spectrum can be transformed into a sharp spectrum through magic angle spinning is shown in Figure 28.14.

28.11 NMR Imaging

One of the most important applications of NMR spectroscopy is its use in imaging the interior of solids. In the health sciences, **NMR imaging** has proved to be the most powerful and least invasive technique for obtaining information on soft tissue such as internal organs in humans. How is the spatial resolution needed for imaging obtained using NMR? For imaging, a **magnetic field gradient** is superimposed onto the constant magnetic field normally used in NMR. In this way, the resonance frequency of a given spin depends not only on the identity of the spin (that is, ^1H or ^{13}C), but also on the local magnetic field, which is determined by the location of the spin relative to the poles of the magnet. Figure 28.15 illustrates how the addition of a field gradient to the constant magnetic field allows the spatial mapping of spins to be carried out. Imagine a sphere and a cube containing ^1H_2O immersed in a background that contains no spin-active nuclei. In the absence of the field gradient, all spins in the structures resonate at the same frequency, giving rise to a single NMR peak. However, with the field gradient present, each volume element of the structure along the gradient has a different resonance frequency. The intensity of the NMR peak at each frequency is proportional to the total number of spins in the volume. A plot of the NMR peak intensity versus field strength gives a projection of the volume of the structures along the gradient direction. If a number of scans corresponding to different directions of the gradient are obtained, the three-dimensional structure of the specimen can be reconstructed, provided that the scans cover a range of at least 180°.

The particular usefulness of NMR for imaging biological samples relies on the different properties that can be used to create contrast in an image. In X-ray radiography, the image contrast is determined by the differences in electron density in various parts of the structure. Because carbon has a lower atomic number than oxygen, it does not scatter X rays as strongly as oxygen. Therefore, fatty tissue appears lighter in a transmission image than tissues with a high density of water. However, this difference in scattering power is small and often gives insufficient contrast. To obtain a higher contrast, material that strongly scatters X rays is injected or ingested. For NMR spectroscopy, several different properties can be utilized to provide image contrast without adding foreign substances.

The properties include the relaxation times T_1 and T_2, as well as chemical shifts and flow rates. The relaxation time offers the most useful contrast mechanism. The relaxation times T_1 and T_2 for water can vary in biological tissues from 0.1 s to several seconds. The more strongly bound the water is to a biological membrane, the greater the change in its relaxation time relative to freely tumbling water molecules. For example, the brain can be imaged with high contrast because the relaxation times of ^1H in gray matter, white matter, and spinal fluid are quite different. Data acquisition methods have been developed to enhance the signal amplitude for a particular range of relaxation times, enabling the contrast to be optimized for the problem of interest. Figure 28.16 shows an NMR image of a human brain.

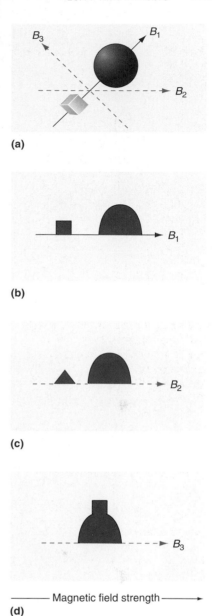

(a)

(b)

(c)

————— Magnetic field strength —————→
(d)

FIGURE 28.15
(a) Two structures are shown along with the three gradient directions indicated along which NMR spectra will be taken. In each case, spins within a thin volume element slice along the gradient resonate at the same frequency. This leads to a spectrum that is a projection of the volume onto the gradient axis. Image reconstruction techniques originally developed for X rays can be used to determine the three-dimensional structure.
(b–d) NMR spectra that would be observed along the B_1, B_2, and B_3 directions indicated in part (a).

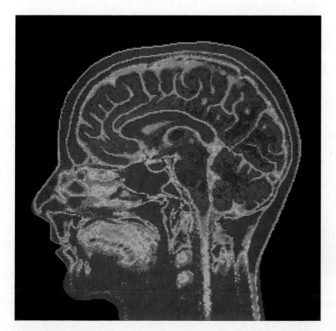

FIGURE 28.16
NMR image of a human brain. The section shown is from a noninvasive scan of the patient's head. The contrast has its origin in the dependence of the relaxation time on the strength of binding of the water molecule to different biological tissues.
[© M. Kulyk/Photo Researchers, Inc.]

Chemical shift imaging can be used to localize metabolic processes and to follow signal transmission in the brain through chemical changes that occur at nerve synapses. One variation of flow imaging is based on the fact that it takes times that are several multiples of T_1 for the local magnetization to achieve its equilibrium value. If, for instance, blood flows into the region under investigation on shorter timescales, it will not have the full magnetization of the spins that have been exposed to the field for much longer times. In such a case, the ^1H_2O in the blood resonates at a different frequency than the surrounding ^1H spins.

NMR imaging also has many applications in materials science, for example, in the measurement of the chemical cross-link density in polymers, the appearance of heterogeneities in elastomers such as rubber through vulcanization or aging, and the diffusion of solvents into polymers. Voids and defects in ceramics and the porosity of ceramics can be detected by nondestructive NMR imaging.

SUPPLEMENTAL

28.12 The NMR Experiment in the Laboratory and Rotating Frames

As discussed at the beginning of this chapter, NMR peaks can be observed by varying either the magnetic field strength or the frequency of the applied ac field. However, modern NMR spectrometers utilize Fourier transform techniques because they greatly enhance the rate at which information can be acquired. In this and the next section, we describe the principles underlying Fourier transform NMR experiments.

A schematic diagram of the main components of an NMR experiment is shown in Figure 28.17. A sample is placed in a strong static magnetic field $\mathbf{B_0}$ that is directed along the z axis. A coil wound around the sample generates a much weaker oscillating radio-frequency (rf) magnetic field $\mathbf{B_1}$ of frequency ω that is directed along the y axis. A third detector coil used to detect the signal (not shown) is also wound around the sample. The sample under consideration has a single characteristic frequency, $\omega = \omega_0$. Additional frequencies that arise from chemical shifts are considered later. Why are two separate magnetic fields needed for the experiment? The static magnetic field $\mathbf{B_0}$ gives rise to the two energy levels shown as a function of the magnetic field strength in Figure 28.1. It does not induce transitions between the two states. However, the rf field $\mathbf{B_1}$ induces transitions between the two levels if the resonance condition $\omega = \omega_0$ is met.

To see how $\mathbf{B_1}$ induces a transition, we consider an alternative way of representing this rf field. The linearly polarized field $\mathbf{B_1}$ is mathematically equivalent to the superposition of two circularly polarized fields rotating in opposite directions. This can be seen by writing the two circularly polarized fields as

$$\mathbf{B}_1^{cc} = B_1(\mathbf{x}\ \cos \omega t\ + \mathbf{y}\ \sin \omega t)$$

$$\mathbf{B}_1^{c} = B_1(\mathbf{x}\ \cos \omega t\ - \mathbf{y}\ \sin \omega t) \qquad (28.28)$$

In these equations, $\mathbf{x}$ and $\mathbf{y}$ are unit vectors along the x and y directions, and the superscripts c and cc refer to clockwise and counterclockwise rotation, respectively, as shown in Figure 28.18. The sum of these fields has zero amplitude in the y direction and an oscillatory amplitude in the x direction. This is analogous to the superposition

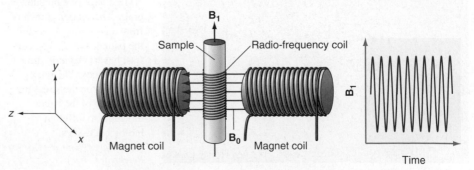

FIGURE 28.17
Schematic of the NMR experiment showing the static field and the rf field coil.

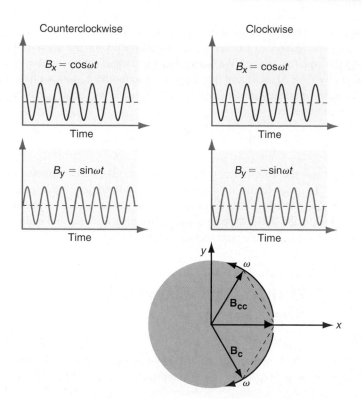

FIGURE 28.18
The superposition of two circularly polarized magnetic fields rotating in opposite directions leads to a linearly polarized magnetic field.

of two traveling waves to create a standing wave, a topic that was considered in Section 13.2. Of the two rotating components, only the counterclockwise component that is rotating in the same direction as the magnetic dipole will induce transitions; therefore, a linearly polarized magnetic field $\mathbf{B}_1$ has the same effect as a circularly polarized field that is rotating counterclockwise in the xy plane. For this reason, we can associate the part of the linearly polarized field that is effective for NMR spectroscopy with $\mathbf{B}_1^{cc} = B_1(\mathbf{x} \cos \omega t + \mathbf{y} \sin \omega t)$.

At this point, we discuss the precession of $\mathbf{M}$ about the total magnetic field. We consider the precession in the frame of reference rotating about the external magnetic field axis at the frequency ω of the rf field. The resultant magnetic field that is experienced by the nuclear spins is the vector sum of $\mathbf{B}_0$ and $\mathbf{B}_1$ and is depicted in Figure 28.19.

An observer in the laboratory frame sees a static field in the z direction, a circularly polarized field rotating at the frequency ω in the xy plane, and a resultant field that precesses around the z axis at the frequency ω. The resultant field is the vector sum of the static and rf fields. The total nuclear magnetic moment precesses around the resultant field, and this precession about a vector, which is itself precessing about the z axis, is difficult to visualize. The geometry becomes simpler if we view the motion of the magnetic moment from a frame of reference that is rotating about the z axis at the frequency ω. We choose the zero of time such that $\mathbf{B}_1$ lies along the x axis. According to classical mechanics, in the **rotating frame,** the rf field and the static field are stationary, and the

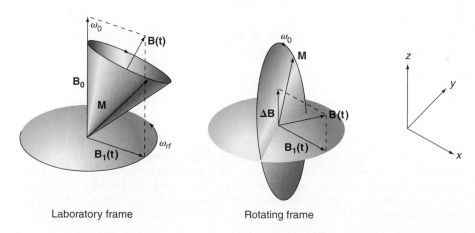

Laboratory frame   Rotating frame

FIGURE 28.19
The NMR experiment as viewed from the laboratory and the rotating frame of reference.

magnetic moment precesses about the resultant field $\Delta\mathbf{B}$, where $\Delta\mathbf{B} = \mathbf{B} - \mathbf{B}_1$, with a frequency $\omega_0 - \omega$. What can we say about the magnitude of the static field $\Delta\mathbf{B}$ along the z axis in the rotating frame? We know that the torque acting on the magnetization vector is given by $\mathbf{\Gamma} = \mathbf{M} \times \mathbf{B}$ and that the magnetic moment has not changed. In order for the precession frequency to decrease from ω to $\omega_0 - \omega$, the apparent static field in the rotating frame must be

$$\Delta\mathbf{B} = \mathbf{B}_0 - \frac{\omega}{\gamma} = \frac{1}{\gamma}(\omega_0 - \omega) \tag{28.29}$$

As ω_{rf} approaches the resonance condition $\omega_0 = \gamma\mathbf{B}_0$, $\Delta\mathbf{B}$ approaches zero and $\mathbf{B} = \mathbf{B}_1$. In the rotating frame at resonance, the half-angle of the precession cone increases to 90°, and $\mathbf{M}$ now precesses in the yz plane at the resonance frequency $\boldsymbol{\omega}$. The usefulness of viewing the NMR experiment in the rotating frame is that it allows the NMR pulse sequences described in the next section to be visualized easily.

S U P P L E M E N T A L

28.13 Fourier Transform NMR Spectroscopy

NMR spectra can be obtained by scanning the static magnetic field or the frequency of the rf magnetic field. In these methods, data are only obtained at one particular frequency at any one instant of time. Because a sample typically contains different molecules with multiple resonance frequencies ω_0, obtaining data in this way is slow. If instead the rf signal is applied in the form of short pulses in a controlled sequence, information about a wide spectrum of resonance frequencies can be obtained simultaneously. In the following, we illustrate how this method, called **Fourier transform NMR spectroscopy**, is implemented. The procedure is depicted in Figure 28.20.

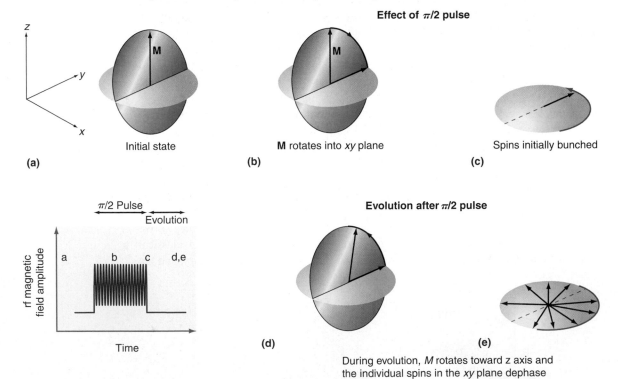

FIGURE 28.20
RF pulse timing and the effect on $\mathbf{M}$ as viewed from the rotating frame. At time a, $\mathbf{M}$ points along the z axis. As the $\pi/2$ pulse is applied (time b), $\mathbf{M}$ precesses in the yz plane and rotates into the xy plane. At the end of the $\pi/2$ pulse (time c) $\mathbf{M}$ points along the y axis and precesses in the xy plane. During the evolution time (d, e) after the pulse is turned off, $\mathbf{M}$ relaxes to its initial orientation along the z axis. The z component increases with the relaxation time T_1. Simultaneously, the $x-y$ component of $\mathbf{M}$ decays with the relaxation time T_2 as the individual spins dephase.

At resonance in the rotating frame, the magnetic moment **M** is stationary and is aligned along the z axis before the rf pulse is applied. As soon as the pulse is applied, **M** begins to precess in the yz plane. The angle through which **M** precesses is given by

$$\alpha = \gamma B_1 t_p = \omega t_p \qquad (28.30)$$

in which t_p is the length of time that the rf field B_1 is on. The pulse length can be chosen so that **M** rotates 90°, after which time it lies in the xy plane. This is called a **$\pi/2$ pulse.** In the xy plane, the individual spins precess at slightly different frequencies because of their differing local fields which may be caused by different chemical shifts or field inhomogeneities. Immediately after the $\pi/2$ pulse, the spins are bunched together in the xy plane and **M** is aligned along the y axis. However, this is not the lowest energy configuration of the system because **M** is perpendicular rather than parallel to the static field. With increasing time, the magnetic moment returns to its equilibrium orientation parallel to the z axis by undergoing spin-lattice relaxation with the characteristic relaxation time T_1.

What happens to the component of **M** in the xy plane? The vector sum of the individual spin magnetic moments in the xy plane is the transverse magnetic moment component. Because the individual spins precess at different frequencies in the xy plane, they will fan out, leading to a dephasing of the spins. This process occurs with the spin–spin relaxation time T_2. As the spins dephase, the magnitude of the transverse component decays to its equilibrium value of zero, as shown in Figure 28.21. Three major mechanisms lead to dephasing: unavoidable inhomogeneities in $\mathbf{B_0}$, chemical shifts, and **transverse relaxation** due to spin–spin interactions.

How is the NMR spectrum generated using the Fourier transform technique? This process is indicated in Figure 28.21. The variation of **M** with time traces a spiral in which M_z increases and the **transverse magnetization** M_{xy} decreases with time. Because the detector coil has its axis along the y axis, it is not sensitive to changes in M_z. However, changes in M_{xy} induce a time-dependent voltage in the coil and, for that reason, the evolution of M_{xy} with time is shown separately in Figure 28.21. Because M_{xy} is a periodic function with the angular frequency ω, the induced voltage in the detector coil is alternately positive and negative. Because of the damping from spin–spin relaxation, its amplitude decays with time as e^{-t/T_2}. The process by which M_{xy} decays to its equilibrium value after the rf pulse is turned off is called **free induction decay.** This experiment provides a way to measure T_2.

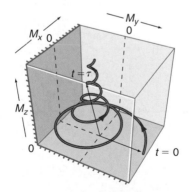

Evolution of **M** in three dimensions

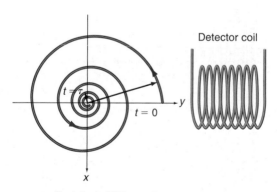

Evolution of **M** in xy plane

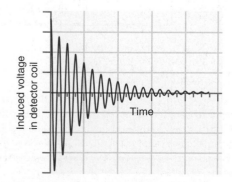

FIGURE 28.21

Evolution of the magnetization vector **M** in three dimensions and M_{xy} as a function of time. The variation of M_{xy} with time leads to exponentially decaying induced rf voltage in the detector coil.

Recall that because of the chemical shifts and unavoidable heterogeneities in the static magnetic field, not all spins have the same resonance frequency. Each group of spins with the same chemical shift gives rise to a different magnetization vector **M** that induces an ac voltage in the detector coil with a frequency equal to its characteristic precession frequency. Because all of these frequencies are contained in the signal, it contains the spectral information in a form that is not easily interpreted. However, by taking the **Fourier transform** of the detector coil signal,

$$I(\omega) = \int_0^\infty I(t)[\cos \omega t + i \sin \omega t] \, dt \qquad (28.31)$$

which is readily accomplished on a laboratory computer, the spectrum can be obtained as a function of frequency rather than time. Examples of the relationship between the free induction decay curves and the spectrum obtained through Fourier transformation are shown in Figure 28.22 for one, two, and three different frequencies.

What is the advantage of the Fourier transform technique over scanning either the magnetic field strength or the rf field strength to obtain an NMR spectrum? By using the Fourier transform technique, the whole spectral range is accessed at all times in which the data are collected. By contrast, in the scanning techniques, the individual frequencies are accessed serially. In any experiment as insensitive as NMR spectroscopy, it is difficult to extract useful signal from a background of noise. Therefore, any method in which more data are collected in a given time is to be preferred. Two arguments may be useful in gaining an understanding of how the method works. The first of these is an analogy with a mechanical resonator. If a bell is struck with a hammer, it will ring with its characteristic frequencies no matter what kind of hammer is used and how it is hit. Similarly, a solution containing precessing spins also has its collection of resonant frequencies and the "right hammer," in this case an rf pulse, excites the spins *at their resonant frequencies* regardless of which additional frequencies are contained in the pulse. The second argument is mathematical in nature. In analogy to the discussion in Section 13.7, many frequency components are required to describe a time-dependent function that changes rapidly over a small time interval. To write the $\pi/2$ pulse of Figure 28.20 as a sum of sine and cosine terms, $f(t) = d_0 + \sum_{n=1}^{m}(c_n \sin n\omega t + d_n \cos n\omega t)$ requires many terms. In this sense, the rf pulse consists of many individual frequencies. Therefore, the pulse

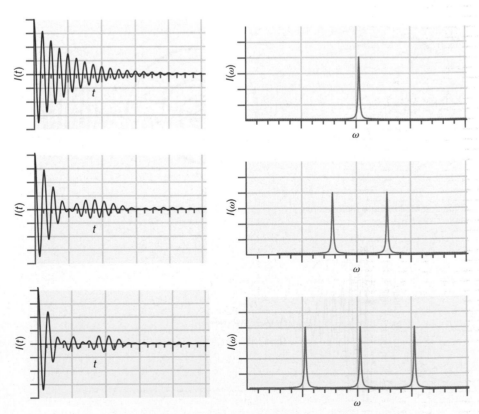

FIGURE 28.22

Free induction decay curves on the left for one, two, and three equal amplitude frequency components. The NMR spectrum on the right is the Fourier transform of the free induction decay curves.

experiment is equivalent to carrying out many parallel experiments with rf magnetic fields of different frequency.

Fourier transform NMR provides the opportunity to manipulate the evolution of **M** by the application of successive rf pulses with varying length, intensity, frequency, and phase. Such a succession of pulses is called a **pulse sequence.** Pulse sequences are designed to manipulate the evolution of spins and reveal interactions between them or to selectively detect certain relaxation pathways. Pulse sequences are the foundation of modern NMR and constitute the basis of multidimensional NMR. The usefulness of these techniques can be understood by describing the spin–echo experiment.

To obtain the frequency spectrum from the free induction decay curve, T_2 must be known. The **spin–echo technique** uses a pulse sequence of particular importance to measure the transverse relaxation time T_2. The experiment is schematically outlined in Figure 28.23. After an initial $\pi/2$ pulse from a coil along the x axis, the spins begin to fan out in the xy plane as a result of unavoidable inhomogeneities in $\mathbf{B_0}$ and because of the presence of chemical shifts. The decay of the signal that results from that part of the dephasing which originates from chemical shifts and field inhomogeneities can be eliminated in the following way. Rather than considering the resultant transverse magnetization component M_{xy}, we consider two populations of spins A and B. Spins A and B correspond to a Larmor frequency slightly higher and slightly lower than the frequency of the rf field, respectively. There is no coupling between the spins in this example; the spin–echo experiment for coupled spins is discussed in the next section. In a frame that is rotating at the average Larmor frequency, one of these spins will move clockwise and the other counterclockwise, as shown in Figure 28.23.

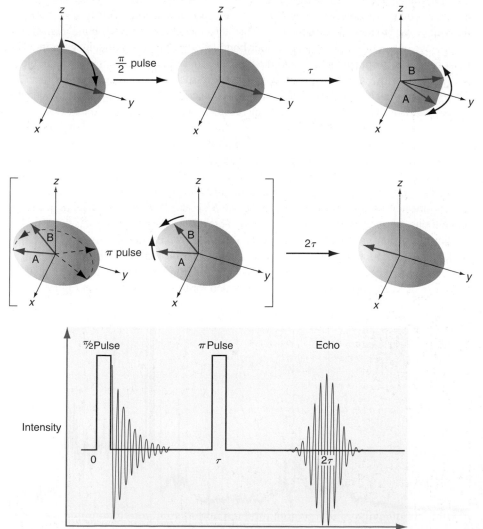

FIGURE 28.23
Schematic representation of the spin–echo experiment. The $\pi/2$ pulse applied along the $-x$ axis rotates **M** into the xy plane. After an evolution time τ in which free induction decay occurs, a π pulse is applied along the x axis. The effect of the π pulse on spin A is shown by the red arc. As a result of the π pulse, the fanning out process resulting from field inhomogeneities and chemical shifts is reversed. An echo will be observed in the detector coil at time 2τ and successive echoes will be observed at further integral multiples of τ. The amplitude of the successive echoes decreases with time because of transverse spin relaxation. The images in the square brackets show the effect of the π pulse on spins A and B. The left image shows the transformation of the spins, and the right image shows the resulting direction of precession.

After a time τ, a π pulse is applied, again along the x axis. This pulse causes the transformation $M_x \rightarrow M_x$ and $M_y \rightarrow -M_y$. As a result, the spins are flipped with respect to the x axis. The direction of precession of spins A and B is unchanged, but after the π pulse the angle between spins A and B now decreases with time. Therefore, the trend toward dephasing is reversed, and the spins will be in phase again after a second time interval τ equal to the initial evolution time. An echo of the original free induction decay signal is observed at 2τ. The amplitude of the echo is smaller than the original signal by the factor e^{-t/T_2} because of the dephasing resulting from transverse relaxation. Therefore, by measuring the amplitude of the echo, T_2 can be determined. Note that only the dephasing that occurs because of the field homogeneities and chemical shifts can be reversed using this technique. The spin–echo experiment is the most accurate method of determining the transverse relaxation time T_2.

SUPPLEMENTAL

28.14 Two-Dimensional NMR

A ^{1}H NMR spectrum for a given molecule in solution contains a wealth of information. For large molecules, the density of spectral peaks can be very high as shown in Figure 28.24. Because of the high density, it is difficult to assign individual peaks to a particular ^{1}H in the molecule. One of the major uses of NMR is to determine the structure of molecules in their natural state in solution. To identify the molecule, it is necessary to know which peaks belong to equivalent ^{1}H that are split into a multiplet through coupling to other spins. Similarly, it would be useful to identify those peaks corresponding to ^{1}H that are coupled by through-bond interactions as opposed to through-space interactions. This type of information can be used to identify the structure of the molecule because through-bond interactions only occur over a distance of three to four bond lengths, whereas through-space interactions can identify spins that are more than three to four bond lengths apart, but are close to one another by virtue of a secondary structure, such as a folding of the molecule. **Two-dimensional NMR (2D-NMR)** allows such experiments to be carried out by separating the overlapped spectra of chemically inequivalent spins in multiple dimensions.

What is meant by 2D-NMR? We answer this question by describing how 2D-NMR is used to extract information from the five-peak one-dimensional spectrum shown in Figure 28.25. On the basis of the information contained in this NMR spectrum alone, there is no way to distinguish between peaks that arise from a chemical shift alone and peaks that arise from a chemical shift plus spin–spin coupling. The goal of the following 2D example is to outline how such a separation among the five peaks can be accomplished. For pedagogical reasons, we apply the analysis to a case for which the origin of each peak is known. This spectrum results from two nonequivalent ^{1}H, separated by a chemical shift δ, in which one peak is split into a doublet and the second is split into a triplet through spin–spin coupling.

The key to the separation between peaks corresponding to coupled and uncoupled ^{1}H is the use of an appropriate pulse sequence, which for this case is shown in Figure 28.26.

FIGURE 28.24

One-dimensional ^{1}H NMR spectrum of a small protein (molecular weight: ~17 kDa) in aqueous solution. The large number of overlapping broad peaks precludes a structural determination on the basis of the spectrum.

Source: Published by permission of Rachel Klevit, University of Washington.

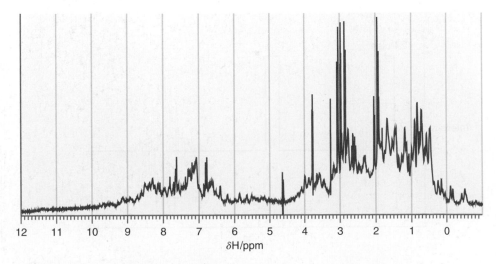

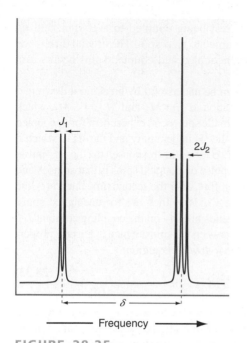

FIGURE 28.25
Illustration of a conventional one-dimensional NMR spectrum consisting of a doublet and a triplet separated by a chemical shift δ.

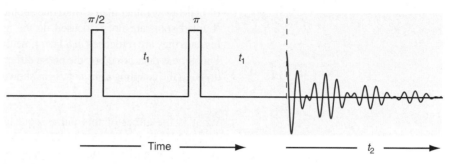

FIGURE 28.26
An initial $\pi/2$ pulse is applied along the x axis to initiate the experiment. After time t_1, a π pulse is applied, again along the x axis. After a second time interval t_1, the detector is turned on at the time indicated by the dashed line, and the signal is measured by the detector coil along the y axis as a function of the time t_2.

We recognize this pulse sequence as that used in the spin–echo experiment, and the effect of this pulse sequence on uncoupled chemically shifted ^{1}H nuclei was discussed in the previous section. In that case, we learned that the spins are refocused into an echo if the first and second time intervals are of equal length. However, we now consider the case of coupled spins. What is changed in the outcome of this experiment through the coupling? This question is answered in Figure 28.27.

As for the spin–echo experiment without coupling, we consider two spins, one higher and one lower in frequency than ν_0. In this case, the frequency difference between the spins is a result of the coupling, as opposed to a chemical shift. From Equation (28.21), the two frequencies are given by

$$\nu_B = \nu_0 - \frac{J}{2} \quad \text{and} \quad \nu_A = \nu_0 + \frac{J}{2} \qquad (28.32)$$

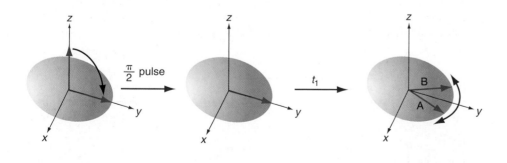

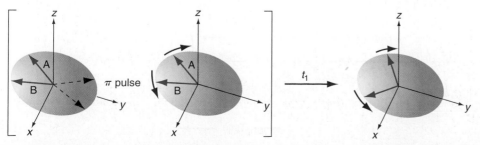

FIGURE 28.27
Illustration of the effect of the pulse sequence in Figure 28.26 on two coupled ^{1}H nuclei. Note the different effect of the π pulse illustrated in the square brackets compared with Figure 28.23. Spin A (B) is rotated as in Figure 28.27, but also converted to spin B (A).

where ν_A originates from the ^1H spin of interest being coupled to a β spin, and ν_B originates from the ^1H spin of interest being coupled to an α spin. The crucial difference between the effect of the pulse sequence on uncoupled and coupled spins occurs as a result of the π pulse.

The effect of the π pulse on coupled spins can be understood by breaking it down into two steps. Initially, the pulse causes the transformation $M_x \rightarrow M_x$ and $M_y \rightarrow -M_y$, which would make the spins rotate toward each other as in the spin–echo experiment for uncoupled spins (see Figure 28.23). However, because both the ^1H under study and the ^1H to which it is coupled make the transitions $\alpha \rightarrow \beta$ and $\beta \rightarrow \alpha$ in response to the π pulse, $\nu_B \rightarrow \nu_A$ and $\nu_A \rightarrow \nu_B$. The total effect of the π pulse on coupled spins is that spins A and B rotate away rather than toward one another. Therefore, after the second time interval t_1, the coupled spins are not refocused on the negative y axis as they are for uncoupled spins. Instead, they are refocused at a later time that depends linearly on the coupling constant J_{12}. This time is determined by the phase difference between the spins, which is linearly proportional to the coupling constant J_{12} as shown in the following equation:

$$\phi = 2\pi J_{12} t_1 \qquad (28.33)$$

What is the effect of this pulse sequence on the uncoupled spins in the sample? All uncoupled chemically shifted ^1H will give a pronounced echo after the second time interval t_1 shown in Figure 28.23, regardless of the value of δ. Therefore, coupled and uncoupled spins behave quite differently in response to the pulse sequence of Figure 28.26.

How can the values for J_{12} and δ contained in the spectrum of Figure 28.25 be separately determined? First, a series of experiments is carried out for different values of t_1. The evolution of M_{x-y} in this time interval depends on both J and δ. The free induction decay curves $A(t_1, t_2)$ are obtained as a function of t_2 for each value of t_1. Note that the chemically shifted spins are refocused through the spin echo at the zero of t_2. Therefore, the value of $A(t_1, t_2 = 0)$ depends only on J, and not on δ. For all times $t_2 > 0$, the evolution of M_{x-y} once again depends on both J and δ. The set of $A(t_1, t_2)$ are shown in Figure 28.28 in the time interval denoted t_2. Next, each of these signals $A(t_1, t_2)$ is Fourier transformed with respect to t_2 to give $C(t_1, \omega_2)$. The sign of $C(t_1, \omega_2)$ is determined by $A(t_1, t_2 = 0)$, and can be either positive or negative. Each of the $C(t_1, \omega_2)$ for given values

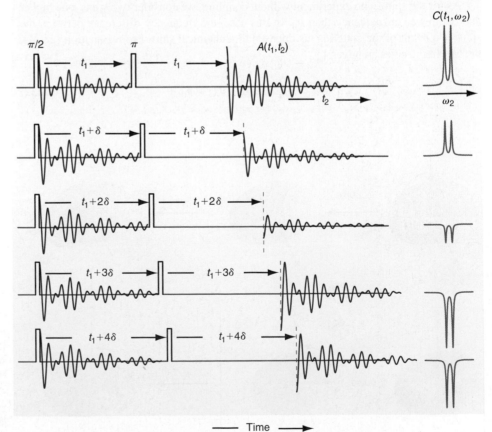

FIGURE 28.28

A series of NMR experiments corresponding to the pulse sequence of Figure 28.26 is shown for different values of t_1 for two coupled spins with a single coupling constant J. The Fourier transformed signal $C(t_1, \omega_2)$ obtained from a Fourier transformation of $A(t_1, t_2)$ with respect to t_2 is shown on the far right side of the figure.

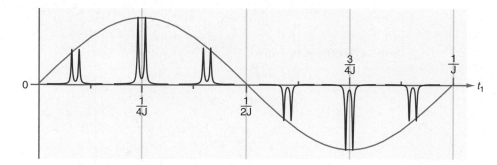

FIGURE 28.29
The function $C(t_1, \omega_2)$ is shown for different values of t_1. The periodicity in time evident in the figure can be expressed as a frequency by carrying out a Fourier transformation with respect to t_1.

of t_1 corresponds to the doublet shown in the rightmost column of Figure 28.28. Although $C(t_1, \omega_2)$, in general, exhibits a number of peaks, only two are shown for illustrative purposes. The dependence of $C(t_1, \omega_2)$ on t_1 is shown in Figure 28.29.

A periodic variation of $C(t_1, \omega_2)$ is observed, with the period T given by $T = 1/J$ as shown in Figure 28.29. We conclude that $C(t_1, \omega_2)$ is an amplitude-modulated periodic function whose period is determined by the coupling constant J. The periodicity in time can be converted to a frequency by a further Fourier transformation of $C(t_1, \omega_2)$, this time with respect to the time t_1, to give the function $G(\omega_1, \omega_2)$. Because the experiment has two characteristic frequencies, they can be used to define the two dimensions of the 2D technique. Function $G(\omega_1, \omega_2)$ is closely related to the desired 2D-NMR spectrum. As shown earlier, Fourier transformation with respect to the frequency ω_1 allows us to extract information from the data set on both δ and J_{12}, whereas the frequency ω_2 depends only on δ. Therefore, the information on δ and J_{12} can be independently obtained.

The J_{12} dependence can be separated from $C(t_1, \omega_2)$ to obtain a function $F(\omega_1, \omega_2)$ in which ω_2 depends only on δ and ω_1 depends only on J_{12}. Function $F(\omega_1, \omega_2)$ is referred to as the 2D J-δ spectrum and is shown as a contour plot in Figure 28.30. This function has two maxima along the ω_2 axis corresponding to the two multiplets of the 1D spectrum in Figure 28.25 that are separated by δ. At each of the δ values, further peaks will be observed along the ω_1 axis, with one peak for each member of the multiplet. The measured separation allows the value of J_{12} to be determined. As can be seen from the figure, the pulse sequence of Figure 28.26 allows a clear separation to be made between peaks in the 1D spectrum arising from a chemical shift and those arising from spin–spin coupling. It is clear that the information content of a 2D-NMR spectrum is much higher than that of a 1D spectrum.

The power of 2D-NMR is further illustrated for structural studies with another example. For this example, a pulse sequence is used that reveals the through-bond coupling of two ^1H. This particular 2D technique is called COSY (an acronym for *CO*rrelated *S*pectroscop*Y*). We illustrate the information that can be obtained from a COSY experiment for the molecule 1-bromobutane. The 1D-NMR spectrum of this

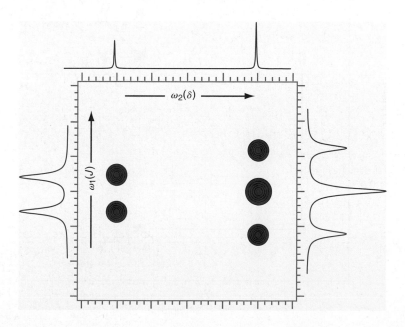

FIGURE 28.30
The two-dimensional function $F(\omega_1, \omega_2)$, corrected mathematically to separate δ and J on the ω_2 axis, is displayed as a contour plot. The horizontal scan above the contour plot shows the δ contribution to the 1D-NMR curve, and the two vertical plots on either side of the contour plot show the spin–spin coupling contribution to the 1D-NMR curve of Figure 28.25.
Source: Published by permission of Tom Pratum, Western Washington University.

FIGURE 28.31

1D-NMR spectrum of 1-bromobutane. The multiplet splitting is not clearly seen because of the large range of δ used in the plot. The assignment of the individual peaks to equivalent ^{1}H spins in the molecule is indicated.

Source: Published by permission of Tom Pratum, Western Washington University.

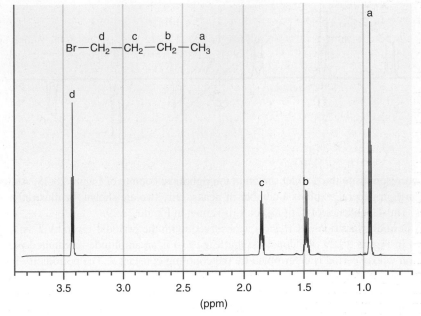

molecule is shown in Figure 28.31. It consists of four peaks with multiplet splittings. On the basis of the discussion in Section 28.5, the peak assignments can be made readily by considering the effect of the electronegative Br atom on the different carbon atoms. The ^{1}H in the CH_2 group attached to the Br (d) generate a triplet, those in the adjacent CH_2 group (c) generate a five-peak multiplet, those in the CH_2 group (b) generate a six-peak multiplet, and those in the terminal CH_3 group generate a triplet. The integrated peak areas are in the ratio a:b:c:d = 3:2:2:2. These are the results expected on the basis of the discussion in Sections 28.7 and 28.8.

We now show that 2D-NMR can be used to find out which of the ^{1}H spins are coupled to one another. By applying the COSY pulse sequence, the 2D-NMR spectrum can be obtained as a function of ω_1 and ω_2. The results are shown in Figure 28.32 in the form of a contour plot. The 1D spectrum shown in this figure corresponds to the diagonal in the 2D spectrum representation. Four peaks are seen corresponding to different

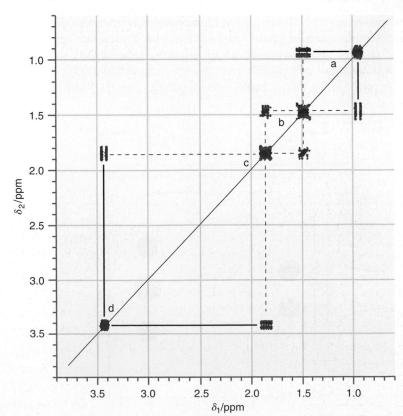

FIGURE 28.32

2D-NMR data for 1-bromobutane in the form of a contour plot. Dashed lines indicate coupling of a set of spins to two other groups. Solid lines indicate coupling of a set of spins to only one other group.

Source: Published by permission of Tom Pratum, Western Washington University.

δ values. We also see off-diagonal peaks at positions that are symmetrical with respect to the diagonal. These peaks identify spins that are coupled. The strength of the coupling can be determined from the intensity of each peak. We can determine which spins are coupled by moving vertically and horizontally from the off-diagonal peaks until the diagonal is reached. It is seen that spins (d) couple only with spins (c), spins (c) couple with both spins (d) and (b), spins (b) couple with both spins (c) and (a), and spins (a) couple only with spins (b). Therefore, the 2D COSY experiment allows us to determine which spins couple with one another. Note that these results are exactly what we would expect for the structural model shown in Figure 28.31 if coupling is ineffective for spins that are separated by more than three bond lengths.

Again, for pedagogical purposes we have chosen to analyze a simple spin system. This particular 2D-NMR experiment actually gives no more information than could have been deduced from the observed multiplet splitting. However, for large molecules with a molecular weight of several thousand Daltons and many inequivalent ^1H, COSY spectra give detailed information on the through-bond coupling of chemically inequivalent ^1H.

A classically based discussion of the effect of a pulse sequence on a sample containing spin-active nuclei analogous to the spin–echo experiment is not adequate to describe the COSY experiment. The higher level quantum mechanical description of this experiment is discussed in advanced texts. An analogous technique, called NOESY, gives information on the through-space coupling of inequivalent ^1H. These two techniques are just a small subset of the many powerful techniques available to NMR spectroscopists. Because of this diversity of experiments achievable through different pulse sequences, 2D-NMR is a powerful technique for the structural determination of biomolecules.

Vocabulary

chemical shift
chemical shift imaging
chemically equivalent nuclei
coupling constant
dephasing
diamagnetic response
Fourier transform
Fourier transform NMR spectroscopy
free induction decay
Larmor frequency
macroscopic magnetic moment
magic angle
magic angle spinning
magnetic anisotropy

magnetic field gradient
magnetically equivalent nuclei
magnetization vector
magnetogyric ratio
motional broadening
motional narrowing
multiplet
multiplet splitting
nuclear g factor
nuclear magnetic moment
nuclear magneton
NMR imaging
$\pi/2$ pulse
precession

pulse sequence
rotating frame
saturated transition
shielding constant
spin–echo technique
spin-lattice relaxation time T_1
spin polarization
spin–spin coupling
spin–spin relaxation time T_2
transverse magnetization
transverse relaxation
two-dimensional NMR (2D-NMR)

Conceptual Problems

Q28.1 Why can the signal loss resulting from spin dephasing caused by magnetic field inhomogeneities and chemical shift be recovered in the spin–echo experiment?

Q28.2 Why do neighboring groups lead to a net induced magnetic field at a given spin in a molecule in the solid state, but not for the same molecule in solution?

Q28.3 Why is it useful to define the chemical shift relative to a reference compound as follows?

$$\delta = 10^6 \frac{(\nu - \nu_{ref})}{\nu_{ref}}$$

Q28.4 What is the advantage of a 2D-NMR experiment over a 1D-NMR experiment?

Q28.5 Why do magnetic field inhomogeneities of only a few parts per million pose difficulties in NMR experiments?

Q28.6 Why does NMR lead to a higher contrast in the medical imaging of soft tissues than X ray techniques?

Q28.7 Why is the multiplet splitting for coupled spins independent of the static magnetic field?

Q28.8 Why does the H atom on the OH group not lead to a multiplet splitting of the methyl hydrogens of ethanol?

Q28.9 Why are the multiplet splittings in Figure 28.9 not dependent on the static magnetic field?

Q28.10 Redraw Figure 28.2 for β spins. What is the direction of precession for the spins and for the macroscopic magnetic moment?

Q28.11 Why is the measurement time in NMR experiments reduced by using Fourier transform techniques?

Q28.12 Order the molecules CH_3I, CH_3Br, CH_3Cl, and CH_3F in terms of increasing chemical shift for ^1H. Explain your answer.

Q28.13 Explain why $T_1 \geq T_2$.

Q28.14 Explain why two magnetic fields, a static field and a radiofrequency field, are needed to carry out NMR experiments. Why must the two field directions be perpendicular?

Q28.15 Explain the difference in the mechanism that gives rise to through-space dipole–dipole coupling and through-bond coupling.

Numerical Problems

P28.1 Predict the number of chemically shifted ^1H peaks and the multiplet splitting of each peak that you would observe for diethyl ether. Justify your answer.

P28.2 Using your results from the previous problems, show that there are four possible transitions between the energy levels of two interacting spins and that the frequencies are given by

$$\nu_{12} = \frac{\gamma B(1 - \sigma_1)}{2\pi} - \frac{J_{12}}{2}$$

$$\nu_{34} = \frac{\gamma B(1 - \sigma_1)}{2\pi} + \frac{J_{12}}{2}$$

$$\nu_{13} = \frac{\gamma B(1 - \sigma_2)}{2\pi} - \frac{J_{12}}{2}$$

$$\nu_{24} = \frac{\gamma B(1 - \sigma_2)}{2\pi} + \frac{J_{12}}{2}$$

P28.3 For a fixed frequency of the radiofrequency field, ^1H, ^{13}C, and ^{31}P will be in resonance at different values of the static magnetic field. Calculate the value of $\mathbf{B_0}$ for these nuclei to be in resonance if the radiofrequency field has a frequency of 250. MHz.

P28.4 Using the matrix representation of the operators and spin eigenfunctions of Problem P28.7, show that the relationships listed in Equation (28.20) are obeyed.

P28.5 Predict the number of chemically shifted ^1H peaks and the multiplet splitting of each peak that you would observe for bromoethane. Justify your answer.

P28.6 A 250 MHz ^1H spectrum of a compound shows two peaks. The frequency of one peak is 510 Hz higher than that of the reference compound (tetramethylsilane) and the second peak is at a frequency 170 Hz lower than that of the reference compound. What chemical shift should be assigned to these two peaks?

P28.7 The nuclear spin operators can be represented as 2×2 matrices in the form and α and β can be represented as column vectors in the form

$$\alpha = \begin{pmatrix} 1 \\ 0 \end{pmatrix} \text{ and } \beta = \begin{pmatrix} 0 \\ 1 \end{pmatrix}$$

Given that

$$\hat{I}_x = \frac{\hbar}{2}\begin{pmatrix} 0 & 1 \\ 1 & 0 \end{pmatrix}, \quad \hat{I}_y = \frac{\hbar}{2}\begin{pmatrix} 0 & -i \\ i & 0 \end{pmatrix}, \quad \hat{I}_z = \frac{\hbar}{2}\begin{pmatrix} 1 & 0 \\ 0 & -1 \end{pmatrix}$$

and

$$\hat{I}^2 = \left(\frac{\hbar}{2}\right)^2 \begin{pmatrix} 3 & 0 \\ 0 & 3 \end{pmatrix}$$

show that

$$\hat{I}^2\alpha = \frac{1}{2}\left(\frac{1}{2} + 1\right)\hbar^2\alpha, \quad \hat{I}_z\alpha = +\frac{1}{2}\hbar\alpha, \quad \hat{I}^2\beta = \frac{1}{2}\left(\frac{1}{2} + 1\right)\hbar^2\beta,$$

and $\hat{I}_z\beta = -\frac{1}{2}\hbar\beta$

P28.8 Predict the number of chemically shifted ^1H peaks and the multiplet splitting of each peak that you would observe for 1,1,1,2-tetrachloroethane. Justify your answer.

P28.9 Predict the number of chemically shifted ^1H peaks and the multiplet splitting of each peak that you would observe for 1,1,2,2-tetrachloroethane. Justify your answer.

P28.10 Predict the number of chemically shifted ^1H peaks and the multiplet splitting of each peak that you would observe for nitroethane. Justify your answer.

P28.11 Predict the number of chemically shifted ^1H peaks and the multiplet splitting of each peak that you would observe for nitromethane. Justify your answer.

P28.12 Predict the number of chemically shifted ^1H peaks and the multiplet splitting of each peak that you would observe for 1,1,2-trichloroethane. Justify your answer.

P28.13 Calculate the spin energy eigenvalues for the wave functions $\psi_1 = \alpha(1)\alpha(2)$, $\psi_3 = \alpha(1)\beta(2)$, and $\psi_4 = \beta(1)\beta(2)$ [Equation (28.15)] for noninteracting spins.

P28.14 Predict the number of chemically shifted ^1H peaks and the multiplet splitting of each peak that you would observe for 1-chloropropane. Justify your answer.

P28.15 Consider the first-order correction to the energy of interacting spins illustrated in Example Problem 28.3 for ψ_2. Calculate the energy correction to the wave functions $\psi_1 = \alpha(1)\alpha(2)$, $\psi_2 = \beta(1)\alpha(2)$, and $\psi_4 = \beta(1)\beta(2)$. Show that your results are consistent with $\Delta E = m_1 m_2 h J_{12}$ with m_1 and $m_2 = +1/2$ for α and $-1/2$ for β.

29 Probability

The concept of probability is central to many areas of chemistry. The characterization of large assemblies of atoms and molecules, from experimental observations to theoretical descriptions, relies on the concepts from statistics and probability. Given the utility of these concepts in chemistry, the central ideas of probability theory are presented in this chapter, including permutations, configurations, probability distribution functions, and the use of these functions to determine benchmark values that characterize the probability distribution.

29.1 Why Probability?

Physical chemistry can be partitioned into two perspectives of matter. One perspective is the microscopic viewpoint utilized by quantum mechanics, in which matter is described through a detailed analysis of its atomic and molecular components. This approach is elegant in its detail and was triumphant in describing many experimental observations that escaped classical descriptions of matter. For example, the observation of discrete emission from the hydrogen atom can only be explained using quantum theory. So successful is this approach that it stands as one of the greatest human accomplishments of the 20th century, and the ramifications of the quantum perspective continue to be explored.

Given the success of quantum theory in describing aspects of nature that are beyond the reach of classical mechanics, some might be tempted to dismiss classical, macroscopic descriptions of matter as irrelevant. However, this macroscopic perspective utilized by thermodynamics is extremely powerful in its ability to predict the outcome of chemical events. Thermodynamics involves the numerous relationships between macroscopic observables and relates experimental measurements of macroscopic properties to predictions of chemical behavior. Perhaps the most impressive aspect of thermodynamics is its ability to predict reaction spontaneity. By simply considering the Gibbs or Helmholtz energy difference between reactants and products, it is possible to state with certainty if a reaction will occur spontaneously. Even though it has impressive predictive utility, thermodynamics is of limited help if we want to know why a reaction occurs in the first place. What are the molecular details that give rise to Gibbs energy, and why should this quantity vary from one species to the next? Unfortunately, answers to these questions are beyond the descriptive bounds of thermodynamics. Can the detailed molecular descriptions available from quantum mechanics be used to formulate an answer to these questions? This type of approach demands that the quantum perspective converge with that of thermodynamics. The link between these perspectives is developed in the following four chapters.

Statistical mechanics allows for the translation between the microscopic properties of matter and its macroscopic behavior. In this approach, a thermodynamic system is described as a collection of smaller units, a reduction in scale that can be taken to the atomic or molecular level. Starting from the microscopic perspective, statistical mechanics allows one to take detailed quantum descriptions of atoms and molecules and determine the corresponding thermodynamic properties. For example, consider a system consisting of 1 mol of gaseous HCl. We can take our knowledge of the quantum energetics of HCl and use this information in combination with statistical mechanics to determine thermodynamic properties of the system such as internal energy, heat capacity, entropy, and other properties described earlier in this book.

However, the question remains as to how this statistical bridge will be built. The task at hand is to consider a single atom or molecule and scale this perspective up to assemblies on the order of 10^{23}! Such an approach necessitates a quantitative description of chemistry as a collection of events or observables, a task that is readily met by probability theory. Therefore, the mathematical tools of probability theory are required before proceeding with a statistical development. Probability is a concept of central utility in discussing chemical systems; the mathematical tools developed in this section will find wide application in subsequent chapters.

29.2 Basic Probability Theory

Probability theory was initially developed in the late 1600s as a mathematical formalism for describing games of chance. Consistent with these origins, the majority of illustrative examples employed in this chapter involve games of chance. Central to probability theory are random **variables,** or quantities that can change in value throughout the course of an experiment or series of events. A simple example of a variable is the outcome of a coin toss, with the variable being the side of the coin observed after tossing the coin. The variable can assume one of two values—heads or tails—and the value the variable assumes may change from one coin toss to the next. Variables can be partitioned into two categories: discrete variables and continuous variables.

Discrete variables assume only a number of specific values. The outcome of a coin toss is an excellent example of a discrete variable in that the outcome can be only one of two values: heads or tails. For another example, imagine a classroom with 100 desks in it, each of which is numbered. If we define the variable *chair number* as being the number on the chair, then this variable can assume values ranging from 1 to 100 in integer values. The possible values a variable can assume are collectively called the **sample space** of the variable. In the chair example, the sample space is equal to the collection of integers from 1 to 100, that is, $\{1, 2, 3, \ldots, 100\}$.

Continuous variables can assume any value within a set of limits, for example, the variable X that can have any value in the range of $1 \leq X \leq 100$. Thermodynamics provides another well-known example of a continuous variable: temperature. The absolute temperature scale ranges from 0 K to infinity, with the variable, temperature, able to assume any value between these two limits. For continuous variables, the sample space is defined by the limiting values of the variable.

The treatment of probability differs depending on whether the variable of interest is discrete or continuous. Probability for discrete variables is mathematically simpler to describe; therefore, we focus on the discrete case first and then generalize to the continuous case later in Section 29.5.

Once a variable and its corresponding sample space have been defined, the question becomes "To what extent will the variable assume any individual value from the sample space?" In other words, we are interested in the **probability** that the variable will assume a certain value. Imagine a lottery where balls numbered 1 to 50 are mixed inside a machine, and a single ball is selected. What is the probability that the ball chosen will have the value one (①)? Since the value chosen is random, the probability of selecting ① is simply 1/50. Does this mean that we will select ① only once every 50 selections? Consider each ball selection as an individual experiment, and after each experiment the selected ball is thrown back into the machine and another experiment is performed. The probability of selecting ① in any experiment is 1/50, but if the outcome of each experiment is independent of other outcomes, then it is not inconceivable that ① will not be selected after 50 trials, *or* that ① will be retrieved more than once. However, if the experiment is performed many times, the end result will be the retrieval of ① a proportion of times that approaches 1/50th of the total number of trials. This simple example illustrates a very important point: probabilities dictate the likelihood that the variable will assume a given value as determined from an infinite number of experiments. As scientists, we are not able to perform an infinite number of experiments; therefore, the extension of probabilities to situations involving a limited number of experiments is done with the understanding that the probabilities provide an approximate expectation of an experimental outcome.

In the lottery example, the probability of selecting any single ball is 1/50. Because there are 50 balls total, the sum of the probabilities for selecting each individual ball must be equal to 1. Consider a variable X for which the sample space consists of M values denoted as $\{x_1, x_2, \ldots, x_M\}$. The probability that the variable X will assume one of these values (p_i) is

$$0 \le p_i \le 1 \tag{29.1}$$

where the subscript indicates one of the values contained in the set space ($i = 1, 2, \ldots, M$). Furthermore, X must assume some value from the sample set in a given experiment dictating that the sum of all probabilities will equal unity:

$$p_1 + p_2 + \cdots + p_M = \sum_{i=1}^{M} p_i = 1 \tag{29.2}$$

In Equation (29.2), the sum of probabilities has been indicated by the summation sign, with the limits of summation indicating that the sum is taken over the entire sample space, from $i = 1$ to M. The combination of the sample space, $S = \{x_1, x_2, \ldots, x_M\}$, and corresponding probabilities, $P = \{p_1, p_2, \ldots, p_M\}$, is known as the **probability model** for the experiment.

EXAMPLE PROBLEM 29.1

What is the probability model for the lottery experiment described in the preceding text?

Solution

The value of the ball that is retrieved in an individual experiment is the variable of interest, and it can take on integer values from 1 to 50. Therefore, the sample space is

$$S = \{1, 2, 3, \ldots, 50\}.$$

If the probability of retrieving any individual ball is equal, and there are 50 balls total, then the probabilities are given by

$$P = \{p_1, p_2, \ldots, p_{50}\} \text{ with all } p_i = 1/50$$

Finally, we note that the sum of all probabilities is equal to 1:

$$p_{total} = \sum_{i=1}^{50} p_i = \left(\frac{1}{50}\right)_1 + \left(\frac{1}{50}\right)_2 + \cdots + \left(\frac{1}{50}\right)_{50} = 1$$

The preceding discussion described the probability associated with a single experiment; however, there are times when one is more interested in the probability associated with a given outcome for a series of experiments, that is, the **event probability.**

FIGURE 29.1
Potential outcomes after tossing a coin four times. Red signifies heads and blue signifies tails.

For example, imagine tossing a coin four times. What is the probability that at least two heads are observed after four tosses? All of the potential outcomes for this series of experiments are given in Figure 29.1. Of the 16 potential outcomes, 11 have at least two heads. Therefore, the probability of obtaining at least two heads after tossing a coin four times is 11/16, or the number of outcomes of interest divided by the total number of outcomes.

Let the sample space associated with a particular variable be S where $S = \{s_1, s_2, \ldots, s_N\}$, and let the probability that the outcome or event of interest E occurs be equal to P_E. Finally, there are j values in S corresponding to the outcome of interest. If the probability of observing an individual value is equivalent, then P_E is given by

$$P_E = \left(\frac{1}{N}\right)_1 + \left(\frac{1}{N}\right)_2 + \cdots + \left(\frac{1}{N}\right)_j = \frac{j}{N} \tag{29.3}$$

This expression states that the probability of the event outcome of interest occurring is equal to the sum of the probabilities for each individual value in sample space corresponding to the desired outcome. Alternatively, if there are N values in sample space, and N_E of these values correspond to the event of interest, then P_E simply becomes

$$P_E = \frac{N_E}{N} \tag{29.4}$$

EXAMPLE PROBLEM 29.2

What is the probability of selecting a heart from a standard deck of 52 cards?

Solution

In a standard deck of cards, each suit has 13 cards and there are 4 suits total (hearts, spades, clubs, and diamonds). The sample space consists of the 52 cards, of which 13 correspond to the event of interest (selecting a heart). Therefore,

$$P_E = \frac{N_E}{N} = \frac{13}{52} = \frac{1}{4}$$

29.2.1 The Fundamental Counting Principle

In the preceding examples, the number of ways a given event could be accomplished was determined by counting. This "brute force" approach can be used when dealing with just a few experiments, but what if we toss a coin 50 times and are interested in the probability of the coin landing heads 20 times out of the 50 tosses? Clearly, writing down every possible outcome and counting would be a long and tedious process. A more efficient method for determining the number of arrangements is illustrated by the following example. Imagine that the instructor of a class consisting of 30 students needs to assemble these students into a line. How many arrangements of students are possible? There are

30 possibilities for the selection of the first student in line, 29 possibilities for the second, and so on until the last student is placed in line. If the probabilities for picking any given student are equal, then the total number of ways to arrange the students (W) is

$$W = (30)(29)(28)\ldots(2)(1) = 30! = 2.65 \times 10^{32}$$

The exclamation point symbol (!) in this expression is referred to as **factorial**, with $n!$ indicating the product of all values from 1 to n, and $0! = 1$. The preceding result in its most general form is known as the **fundamental counting principle.**

FUNDAMENTAL COUNTING PRINCIPLE: For a series of manipulations $\{M_1, M_2 \ldots, M_j\}$ having n_i ways to accomplish M_i, the total number of ways to perform the entire series of manipulations ($Total_M$) is the product of the number of ways to perform each manipulation under the assumption that the ways are independent:

$$Total_M = (n_1)(n_2)\ldots(n_j) \tag{29.5}$$

EXAMPLE PROBLEM 29.3

How many 5-card arrangements are possible from a standard deck of 52 cards?

Solution

Employing the fundamental counting principle, each manipulation is receiving a card in your hand; therefore, five manipulations are required. There are 52 possible cards we could receive as our first card, or 52 ways to accomplish the first manipulation ($n_1 = 52$). Next, there are 51 possible cards we could receive as the second card in our hand, or 51 ways to accomplish the second manipulation ($n_2 = 51$). Following this logic

$$Total_M = (n_1)(n_2)(n_3)(n_4)(n_5)$$
$$= (52)(51)(50)(49)(48) = 311{,}875{,}200$$

EXAMPLE PROBLEM 29.4

The electron configuration for the first excited state of He is $1s^12s^1$. Using the fundamental counting principle, how many possible spin states are expected for this excited-state electron configuration?

Solution

Because the electrons are in different orbitals, they do not have to be spin paired. Therefore, there are two choices for the spin state of the first electron, and two choices for the spin state of the second electron such that

$$Total_M = (n_1)(n_2) = (2)(2) = 4$$

29.2.2 Permutations

In the example of a classroom with 30 students, we found that there are $30!$ unique ways of arranging the students into a line, or $30!$ **permutations.** The total number of objects that are arranged is known as the order of the permutation, denoted as n, such that there are $n!$ total permutations of n objects. Thus far we have assumed that the entire set of n objects is used, but how many permutations are possible if only a subset of objects is employed in constructing the permutation? Let $P(n, j)$ represent the number of permutations possible using a subset of j objects from the total group of n; $P(n, j)$ is equal to

$$P(n, j) = n(n - 1)(n - 2)\ldots(n - j + 1) \tag{29.6}$$

Equation (29.6) can be rewritten by noting that

$$n(n-1)\ldots(n-j+1) = \frac{n(n-1)\ldots(1)}{(n-j)(n-j-1)\ldots(1)} = \frac{n!}{(n-j)!} \quad (29.7)$$

Therefore, $P(n, j)$ is given by the following relationship:

$$P(n, j) = \frac{n!}{(n-j)!} \quad (29.8)$$

EXAMPLE PROBLEM 29.5

The coach of a basketball team has 12 players on the roster but can only play 5 players at one time. How many 5-player arrangements are possible using the 12-player roster?

Solution

For this problem the permutation order (n) is 12, and the subset (j) is 5 such that

$$P(n, j) = P(12,5) = \frac{12!}{(12-5)!} = 95,040$$

29.2.3 Configurations

The previous section discussed the number of ordered arrangements or permutations possible using a given number of objects. However, many times one is instead interested in the number of *unordered* arrangements that are possible. The basketball team example of Example Problem 29.5 is an excellent illustration of this point. In a basketball game, the coach is generally more concerned with which five players are in the game at any one time rather than the order in which they entered the game. An unordered arrangement of objects is referred to as a **configuration.** Similar to permutations, configurations can also be constructed using all objects in the set being manipulated (n), or just a subset of objects with size j. We will refer to configurations using the nomenclature $C(n, j)$.

For a conceptual example of configurations, consider the four colored balls shown in Figure 29.2. How many three-ball configurations and associated permutations can be

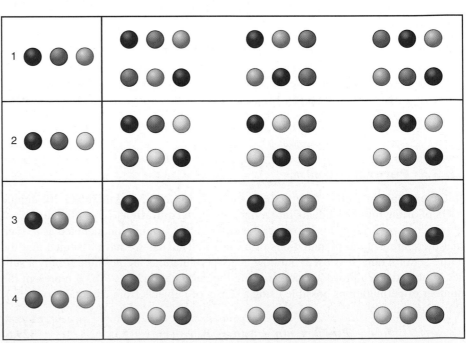

FIGURE 29.2
Illustration of configurations and permutations using four colored balls. The left-hand column presents the four possible three-color configurations, and the right-hand column presents the six permutations corresponding to each configuration.

made using these balls? The possibilities are illustrated in the figure. Notice that a configuration is simply a collection of three colored balls, and that a permutation corresponds to an ordered arrangement of these same balls such that there are four possible configurations with each configuration having six associated permutations. This observation can be used to develop the following mathematical relationship between configurations and permutations:

$$C(n, j) = \frac{P(n, j)}{j!}$$

$C(n, j)$ is the number of configurations that are possible using a subset of j objects from a total number of n objects. Substituting the definition of $P(n, j)$ from Equation (29.8) into the preceding expression results in the following relationship:

$$C(n, j) = \frac{P(n, j)}{j!} = \frac{n!}{j!(n - j)!} \qquad \textbf{(29.9)}$$

EXAMPLE PROBLEM 29.6

If you are playing cards with a standard 52-card deck, how many possible 5-card combinations or "hands" are there?

Solution

Each 5-card hand is a configuration or subset taken from the 52-card deck. Therefore, $n = 52$, $j = 5$, and

$$C(n, j) = C(52, 5) = \frac{52!}{5!(52 - 5)!} = \frac{52!}{(5!)(47!)} = 2{,}598{,}960$$

This result should be contrasted with the 311,875,200 permutations obtained in Example Problem 29.3. Generally, card players are more interested in the five cards in their possession at any time and not on the order in which those cards arrived.

29.2.4 A Counting Example: Bosons and Fermions (Advanced)

The concepts of permutation and configuration will be exceedingly important in the next chapter, but we can introduce a simple counting problem here to demonstrate the connection between probability theory and chemistry. Consider the following question: how many ways can a set of n indistinguishable particles be arranged into x equally accessible states, with each state capable of holding any number of particles? This counting problem is encountered when describing particles known as **bosons,** in which multiple particles can occupy the same state. Photons and particles of integer spin such as ^{4}He are bosons, and these particles follow **Bose–Einstein statistics.** To describe the arrangement of such particles over a collection of states, we first consider a manageable example consisting of four particles and three states, as illustrated in Figure 29.3. In the figure, each particle is shown as a red circle, and the three states are shown by rectangles. The four possible configurations have a collection of associated permutations given by the numerals on the right side of each configuration. For example, the second configuration with three particles in one state, one in a second state, and no particles in the third state has six associated permutations. The total number of possible arrangements is equal to the total number of permutations, or 15 in this example.

Another way to envision the possible configurations in this example is shown in Figure 29.4. Here, the particles are confined to a single box with two movable walls allowing for three separate partitions. The figure demonstrates that there are again four possible configurations, identical to the result shown in Figure 29.3. The advantage of this depiction is that we can envision this problem as counting the number of permutations associated with a collection of six indistinguishable objects: four particles and two

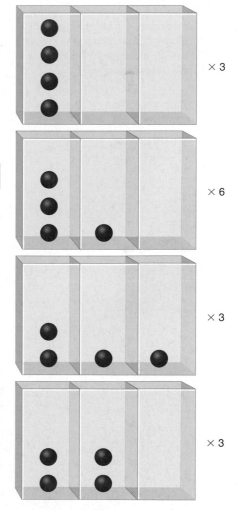

FIGURE 29.3

Configurations associated with arranging four identical particles (red balls) in three states (rectangles). The number of permutations associated with each configuration is given by the numerals to the right of each configuration.

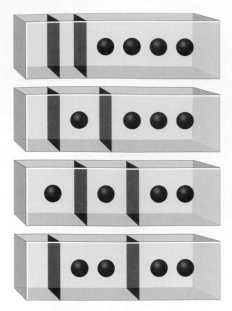

FIGURE 29.4
Second model for the four-particle/three-state arrangement depicted in Figure 29.3. In this model, states correspond to regions inside a single rectangle created by movable walls (black rectangles).

movable walls. For a given number of states x there will be $x - 1$ movable walls. Combining the walls with the n particles, the total number of permutations is

$$P_{BE} = \frac{(n + x - 1)!}{n!(x - 1)!} \qquad (29.10)$$

The subscript BE refers to Bose–Einstein, reflecting the fact that we are discussing bosons. Employing Equation (29.10) to our illustrative example with $n = 4$ and $x = 3$, we find that $P_{BE} = 15$ consistent with our earlier result from counting. This result can be understood using the probability concepts introduced to this point. Specifically, the total number of objects being manipulated is $n + x - 1$ corresponding to $(n + x - 1)!$ permutations if the particles and walls were distinguishable. To account for the fact that the particles and walls are indistinguishable, we divided by $n!$ and $(x - 1)!$ resulting in the final expression for P_{BE}.

The second type of particles are **fermions,** which are species of noninteger spin such as electrons or ^{3}He. In a collection of fermions, each particle has a unique set of quantum numbers such that no two particles will occupy the same state. Fermions follow **Fermi-Dirac statistics.** Suppose we have n fermions to distribute over x states. How many possible arrangement permutations are there? There will be x choices for placement of the first particle, $x - 1$ choices for the second, and so forth so that the total number of state arrangements is given by

$$x(x - 1)(x - 2)\ldots(x - n + 1) = \frac{x!}{(x - n)!}$$

However, as in the boson example, the particles are indistinguishable, requiring that we divide the above expression by $n!$ resulting in

$$P_{FD} = \frac{x!}{n!(x - n)!} \qquad (29.11)$$

EXAMPLE PROBLEM 29.7

How many quantum states are possible for a carbon atom with the configuration $1s^2 2s^2 2p^2$?

Solution

Since this problem involves the arrangement of electrons, Fermi-Dirac statistics apply. Since only the two electrons in the p orbitals contribute to defining the number of states (recall, electrons in the s orbitals must be spin paired) we have $n = 2$. Next, there are three p orbitals and two possible spin orientations in each orbital such that $x = 6$. Therefore, the total number of quantum states, or arrangement permutations, is

$$P_{FD} = \frac{x!}{n!(x - n)!} = \frac{6!}{2!(4!)} = 15$$

29.2.5 Binomial Probabilities

The probability of an event E occurring out of a total number of possible outcomes was denoted as P_E in Equation (29.4). We can also define the **complement** of P_E as the probability of an outcome other than that associated with the event of interest and denote this quantity as P_{EC}. With this definition, the sum of P_E and P_{EC} are related as follows:

$$P_E + P_{EC} = 1 \qquad (29.12)$$

Equation (29.12) states that the probability of the event of interest occurring combined with the event not occurring must be equal to unity and provides a definition for experiments known as **Bernoulli trials.** In such a trial, the outcome of a given experiment will be a success (i.e., the outcome of interest) or a failure (i.e., not the outcome of interest). The tossing of a coin is one example of a Bernoulli trial where the outcome "heads" can be considered a success and "tails" a failure (or vice versa). A collection of Bernoulli trials is known as a **binomial experiment,** and these simple experiments provide a framework in which to explore probability distributions. It is critical to note that the outcome of each trial is independent of the outcome of any other trial in the experiment. That is, previous or the subsequent outcomes have no effect on the outcome of the current trial. Consider a binomial experiment in which a coin is tossed four times and this question is asked: "What is the probability of observing heads (or a successful outcome) every time?" Because the probability of success for each trial is $1/2$, the total probability is the product of the success probabilities for each trial:

$$P_E = \left(\frac{1}{2}\right)\left(\frac{1}{2}\right)\left(\frac{1}{2}\right)\left(\frac{1}{2}\right) = \frac{1}{16}$$

Note that this is the answer one would reach by considering all of the possible permutations encountered when flipping a coin four times as illustrated in Figure 29.1.

For a series of Bernoulli trials in which the probability of success for a single trial is P_E, the probability of obtaining j successes in a trial consisting of n trials is given by

$$P(j) = C(n, j)(P_E)^j(1 - P_E)^{n-j} \qquad \textbf{(29.13)}$$

The $(P_E)^j$ term in the preceding expression is the product of probabilities for the j successful trials. Because n total trials were performed, $(n - j)$ trials must have failed, and the probability of these trials occurring is given by $(1 - P_E)^{n-j}$. But why does the total number of configurations appear in Equation (29.13)? The answer to this question lies in the difference between permutations and configurations. Again, consider a series of four coin tosses where the outcome of interest is the exact permutation {H, T, T, H}. If H is a successful trial occurring with probability P_E, then the probability of observing this permutation is

$$P = (P_E)(1 - P_E)(1 - P_E)(P_E) = (P_E)^2(1 - P_E)^2 = \left(\frac{1}{2}\right)^2\left(\frac{1}{2}\right)^2 = \frac{1}{16} \qquad \textbf{(29.14)}$$

but this is also the probability for observing {H, H, H, H}. That is, the probability of observing a specific order of trial outcomes, or a single permutation, is equivalent. If the outcome of interest is two successful trials that can occur in any order, then the probabilities for all possible permutations corresponding to two successful trials must be added, and this is accomplished by the inclusion of $C(n, j)$ in Equation (29.13).

EXAMPLE PROBLEM 29.8

Imagine tossing a coin 50 times. What are the probabilities of observing heads 25 times (i.e., 25 successful experiments) and just 10 times?

Solution

The trial of interest consists of 50 separate experiments; therefore, $n = 50$. Considering the case of 25 successful experiments where $j = 25$. The probability (P_{25}) is

$$P_{25} = C(n, j)(P_E)^j(1 - P_E)^{n-j}$$

$$= C(50, 25)(P_E)^{25}(1 - P_E)^{25}$$

$$= \left(\frac{50!}{(25!)(25!)}\right)\left(\frac{1}{2}\right)^{25}\left(\frac{1}{2}\right)^{25} = (1.26 \times 10^{14})(8.88 \times 10^{-16}) = 0.11$$

For the case of 10 successful experiments, $j = 10$ such that

$$P_{10} = C(n, j)(P_E)^j(1 - P_E)^{n-j}$$

$$= C(50, 10)(P_E)^{10}(1 - P_E)^{40}$$

$$= \left(\frac{50!}{(10!)(40!)}\right)\left(\frac{1}{2}\right)^{10}\left(\frac{1}{2}\right)^{40} = (1.03 \times 10^{10})(8.88 \times 10^{-16}) = 9.1 \times 10^{-6}$$

29.3 Stirling's Approximation

When calculating $P(n, j)$ and $C(n, j)$, it is necessary to evaluate factorial quantities. In the examples encountered so far, n and j were sufficiently small such that these quantities could be evaluated on a calculator. However, this approach to evaluating factorial quantities is limited to relatively small numbers. For example, 100! is equal to 9.3×10^{157}, which is an extremely large number and beyond the range of many calculators. Furthermore, we are interested in extending the probability concepts we have developed up to chemical systems for which $n \approx 10^{23}$! The factorial of such a large number is simply beyond the computational ability of most calculators.

Fortunately, approximation methods are available that will allow us to calculate the factorial of large numbers. The most famous of these methods is known as **Stirling's approximation,** which provides a simple method by which to calculate the natural log of $N!$. A simplified version of this approximation is

$$\ln N! = N \ln N - N \tag{29.15}$$

Equation (29.15) is readily derived as follows:

$$\ln(N!) = \ln\left[(N)(N-1)(N-2)\ldots(2)(1)\right]$$

$$= \ln(N) + \ln(N-1) + \ln(N-2) + \cdots + \ln(2) + \ln(1)$$

$$= \sum_{n=1}^{N} \ln(n) \approx \int_{1}^{N} \ln(n)\, dn$$

$$= N \ln N - N - (1 \ln 1 - 1) \approx N \ln N - N \tag{29.16}$$

The replacement of the summation by an integral is appropriate when N is large. The final result is obtained by evaluating the integral over the limits indicated. Note that the main assumption inherent in this approximation is that N is a large number. The central concern in applying Stirling's approximation is whether N is sufficiently large to justify its application. Example Problem 29.9 illustrates this point.

EXAMPLE PROBLEM 29.9

Evaluate $\ln(N!)$ for $N = 10, 50$, and 100 using a calculator, and compare the result to that obtained using Stirling's approximation.

Solution

For $N = 10$, using a calculator we can determine that $N! = 3.63 \times 10^{6}$ and $\ln(N!) = 15.1$ Using Stirling's approximation

$$\ln(N!) = N \ln N - N = 10 \ln(10) - 10 = 13.0$$

This value represents a 13.9% error relative to the exact result, a substantial difference. The same procedure for $N = 50$ and 100 results in the following:

N	$\ln(N!)$ **Calculated**	$\ln(N!)$ **Stirling**	**Error (%)**
50	148.5	145.6	2.0
100	363.7	360.5	0.9

This problem demonstrates that there are significant differences between the exact and approximate results even for $N = 100$ but that the magnitude of this error decreases as N increases. For the chemical systems encountered in subsequent chapters, N will be $\sim 10^{23}$, many orders of magnitude larger than the values studied in this example. Therefore, for our purposes Stirling's approximation represents an elegant and sufficiently accurate method by which to evaluate the factorial of large quantities.

29.4 Probability Distribution Functions

Returning to the coin-tossing experiment, we now ask what is the probability of obtaining a given outcome (for example, the number of heads) after tossing a coin 50 times. Using Equation (29.13), a table of probability as a function of the number of heads can be constructed with $n = 50$ (total number of tosses) and $j = $ the number of heads (i.e., the number of successful tosses):

Number of Heads	Probability	Number of Heads	Probability
0	8.88×10^{-16}	30	0.042
1	4.44×10^{-14}	35	2.00×10^{-3}
2	1.09×10^{-12}	40	9.12×10^{-6}
5	1.88×10^{-9}	45	1.88×10^{-9}
10	9.12×10^{-6}	48	1.09×10^{-12}
15	2.00×10^{-3}	49	4.44×10^{-14}
20	0.042	50	8.88×10^{-16}
25	0.112		

Rather than reading all of the probability values from a table, this same information can be presented graphically by plotting the probability as a function of outcome. This plot for the case where $P_E = 0.5$ (i.e., the coin landing heads or tails is equally likely) is shown as the red line in Figure 29.5. Notice that the maximum probability is predicted to be 25 heads corresponding to the most probable outcome (as intuition suggests). A second feature of this distribution of probabilities is that the actual value of the probability corresponding to 25 successful trials is not unity, but 0.112. However, summing all of the outcome probabilities reveals that

$$P_0 + P_1 + \cdots + P_{50} = \sum_{j=0}^{50} P_j = 1 \qquad \textbf{(29.17)}$$

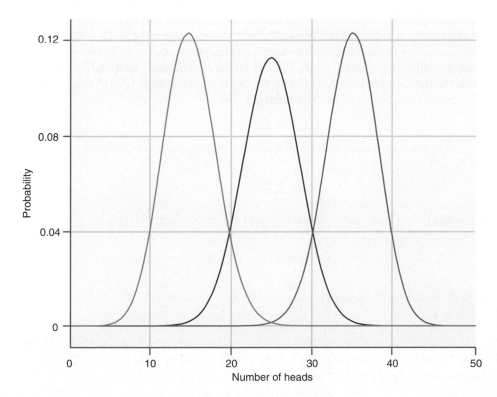

FIGURE 29.5
Plot of the probability of the number of heads being observed after flipping a coin 50 times. The red curve represents the distribution of probabilities for $P_E = 0.5$, the blue curve for $P_E = 0.3$, and the purple curve for $P_E = 0.7$.

Figure 29.5 presents the variation in probability of event outcome as a function of the number of heads observed after flipping a coin 50 times. This plot also demonstrates that the probability formula for Bernoulli trials can be used to describe the variation or distribution of probability versus event outcome. Therefore, the probability expression we have employed can also be thought of as the distribution function for a binomial experiment. From Equation (29.13), the probability of observing j successful trials following n total trials is given by

$$P(j) = \frac{n!}{j!(n-j)!}(P_E)^j(1-P_E)^{n-j} \quad \text{for} \quad j = 0, 1, 2, \ldots, n \quad \textbf{(29.18)}$$

Equation (29.18) is valid when $0 \le P_E \le 1$. In the coin-flipping experiment just discussed, the assumption was made that the coin can land either heads or tails with equal probability such that $P_E = 0.5$. What would happen if a coin were used where the probability of observing heads is 0.7? First, we would expect the probability distribution of trial outcomes to vary with respect to the case where $P_E = 0.5$. For the experiment in which $P_E = 0.7$ and $(1 - P_E) = 0.3$, the distribution function $P(j)$ can be calculated and compared to the previous case, as illustrated in Figure 29.5. Comparison of the probability distributions demonstrates that the most probable outcome has shifted to a greater number of heads, H = 35 in the case of $P_E = 0.7$. The outcome H = 15 is most probable when $P_E = 0.3$. A final aspect to note is that in addition to the maximum probability changing from H = 25 as P_E is changed from 0.5 to some other value, the probability for the most likely outcome also increases. For example, for $P_E = 0.5$, the maximum probability is 0.112 for H = 25. If $P_E = 0.7$, the maximum probability now becomes 0.128 for H = 35. That is, not only does the most probable outcome depend on P_E, but the probability of observing the most probable outcome also changes with P_E.

The coin-tossing experiment provides an excellent example of a probability distribution function and motivates a more formal definition of such functions. A **probability distribution function** represents the probability of a variable (X) having a given value, with the probability described by a function

$$P(X_i) \propto f_i \quad \textbf{(29.19)}$$

In Equation (29.19), $P(X_i)$ is the probability that the variable X will have some value X_i in the sample space. This equation states that the set of probabilities $\{P(X_1), P(X_2), \ldots, P(X_M)\}$ will be proportional to the value of the distribution function evaluated for the corresponding value of the variable given by $\{f_1, f_2, \ldots, f_M\}$. In the coin toss example, the variable *number of heads* could assume a value between 0 and 50, and Equation (29.13) was employed to determine the corresponding probability $P(X_i)$ for each possible value of the variable. For a binomial experiment, Equation (29.13) represents the function f in Equation (29.19). We can express Equation (29.19) as an equality by introducing a proportionality constant (C), and

$$P(X_i) = Cf_i \quad \textbf{(29.20)}$$

Imposing the requirement that the total probability be equal to unity results in

$$\sum_{i=1}^{M} P(X_i) = 1 \quad \textbf{(29.21)}$$

Equation (29.21) is readily evaluated to yield the proportionality constant:

$$1 = \sum_{i=1}^{M} Cf_i = Cf_1 + Cf_2 + \cdots + Cf_M$$

$$1 = C(f_1 + f_2 + \cdots + f_M) = C\sum_{i=1}^{M} f_i$$

$$C = \frac{1}{\sum_{i=1}^{M} f_i} \quad \textbf{(29.22)}$$

Substitution into our original expression for probability provides the final result of interest:

$$P(X_i) = \frac{f_i}{\sum\limits_{i=1}^{M} f_i} \tag{29.23}$$

Equation (29.23) states that the probability of a variable having a given value from the sample space is given by the value of the probability function for this outcome divided by the sum of the probabilities for all possible outcomes. Equation (29.23) is a general expression for probability, and we will use this construct in the upcoming chapter when defining the Boltzmann distribution, one of the central results of statistical thermodynamics.

EXAMPLE PROBLEM 29.10

What is the probability of receiving any 1 card from a standard deck of 52 cards?

Solution

The variable of interest is the card received, which can be any one of 52 cards (that is, the sample space for the variable consists of the 52 cards in the deck). The probability of receiving any card is equal so that $f_i = 1$ for $i = 1$ to 52. With these definitions, the probability becomes

$$P(X_i) = \frac{f_i}{\sum\limits_{i=1}^{52} f_i} = \frac{1}{\sum\limits_{i=1}^{52} f_i} = \frac{1}{52}$$

29.5 Probability Distributions Involving Discrete and Continuous Variables

To this point we have assumed that the variable of interest is discrete. As such, probability distributions can be constructed by calculating the corresponding probability when X assumes each value from the sample set. However, what if the variable X is continuous? In this case the probability must be determined for the variable having a value within a portion of the domain of X, denoted as dX. In this case we employ the **probability density** $P(X)$, and $P(X)\,dX$ is the probability that the variable X has a value in the range of dX. By analogy with the development for discrete variables, the probability is then given by

$$P(X)\,dX = Cf(X)\,dX \tag{29.24}$$

which states that the probability $P(X)\,dX$ is proportional to some function $f(X)\,dX$ as yet undefined. Applying the normalization condition to ensure that the total probability is unity over the domain of the variable $(X_1 \leq X \leq X_2)$:

$$\int_{X_1}^{X_2} P(X)\,dX = C \int_{X_1}^{X_2} f(X)\,dX = 1 \tag{29.25}$$

The second equality in Equation (29.25) dictates that

$$C = \frac{1}{\displaystyle\int_{X_1}^{X_2} f(X)\,dX} \tag{29.26}$$

Equation (29.26) is identical to the discrete variable result [Equation (29.22)] except that summation has been replaced by integration since we are now interested in continuous variables. With the preceding definition of the proportionality constant, the probability is defined as

$$P(X)\,dX = \frac{f(X)\,dX}{\displaystyle\int_{X_1}^{X_2} f(X)\,dX} \tag{29.27}$$

The similarity of this expression to the corresponding expression for discrete variables [Equation (29.23)] illustrates an important point. When working with probability distributions for continuous variables, integration over the domain of the variable is performed. In contrast, summation is performed when the variable is discrete. Although the mathematics changes slightly between discrete and continuous variables, it is important to realize that the conceptual description of probability is unchanged. Continuous probability distributions are common in quantum mechanics. For example, the normalization condition is applied to the spatial domain of the wave function through the following relationship:

$$\int \psi^*(x, t)\psi(x, t)\,dx = 1$$

The product of the wave function with its complex conjugate represents the probability density for the spatial location of the particle, and it is equivalent to $P(X)$ in Equation (29.24). By integrating this product over all space, all possible locations of the particle are included such that the probability of the particle being somewhere must be one.

For another example of probability distribution functions involving continuous variables, consider the following probability distribution function associated with the translational kinetic energy of ideal gas particles (E), a topic that will be discussed in detail in Chapter 33:

$$P(E)\,dE = 2\pi \left(\frac{1}{\pi RT}\right)^{3/2} E^{1/2} e^{-E/RT}\,dE \tag{29.28}$$

In Equation (29.28), T is temperature and R is the ideal gas constant. The variable of interest E is continuous in the domain from 0 to infinity. Figure 29.6 presents an illustration of this distribution function for three temperatures: 300., 400., and 500. K.

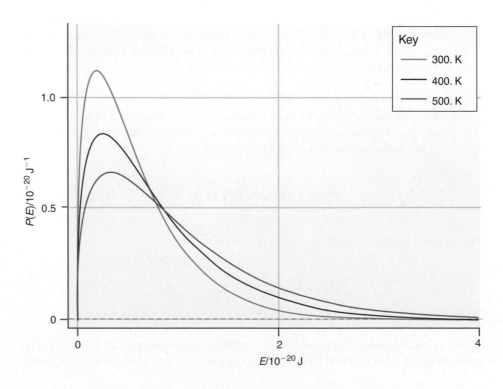

FIGURE 29.6
Probability distributions for the translational kinetic energy of an ideal gas.

Notice that energy corresponding to maximum probability changes as a function of temperature and that the probability of finding a particle at larger kinetic energies also increases with temperature. This example demonstrates that a substantial amount of information regarding the behavior of a chemical system can be succinctly presented using probability distributions.

29.5.1 Continuous Representation of Discrete Variables

We have seen that it is possible to evaluate probabilities involving discrete variables by summation; however, we will encounter situations in later chapters where summation involves a prohibitively large number of terms. In such cases, treating a discrete variable as continuous can simplify matters greatly. However, one concern with this approach is the error introduced by including values for the discrete variable that are not contained in the sample set. Under what conditions are such errors acceptable? Consider the following unnormalized probability distribution function:

$$P(X) = e^{-0.3X} \tag{29.29}$$

We are interested in normalizing this distribution by determining the total probability P_{total}. First, X is treated as a discrete variable with the sample space consisting of integer values ranging from 0 to 100, and P_{total} is

$$P_{total} = \sum_{X=0}^{100} e^{-0.3X} = 3.86 \tag{29.30}$$

Next, X is treated as continuous with the domain of the sample set equal to $0 \le X \le 100$. The corresponding expression for the total probability is

$$P_{total} = \int_0^{100} e^{-0.3X} dX = 3.33 \tag{29.31}$$

The only difference between these last two equations is that the summation over discrete values of the variable X has been replaced by integration over the range of the variable. The preceding comparison demonstrates that the continuous approximation is close to the exact result given by summation. In general, if the differences between values the function can assume are small relative to the domain of interest, then treating a discrete variable as continuous is appropriate. This issue will become critical when the various energy levels of an atom or molecule are discussed. Specifically, the approximation of Equation (29.31) will be used to treat translational and rotational states from a continuous perspective where direct summation is impractical. In the remainder of this text, situations in which the continuous approximation is not valid will be carefully noted.

EXAMPLE PROBLEM 29.11

Using the following distribution function

$$P(X) = e^{-0.05X}$$

Determine the total probability if the variable X is discrete with the sample space consisting of integer values ranging from 0 to 100. Compare this result to that if X is a continuous variable with a range of $0 \le X \le 100$.

Solution

This is the same analysis performed in the previous experiment, but now the differences between values are smaller in comparison to the previous distribution. Therefore, we would expect the discrete and continuous results to be closer in value. First, treating X as a discrete variable, we get

$$P_{total} = \sum_{X=0}^{100} e^{-0.05X} = 20.4$$

Next, treating X as continuous

$$P_{total} = \int_0^{100} e^{-0.05X}\, dX = 19.9$$

Comparison of these results demonstrates that the difference between the discrete and continuous treatments is smaller for this more "fine-grain" distribution.

29.6 Characterizing Distribution Functions

Distribution functions provide all the probability information available for a given system; however, there will be times when a full description of the distribution function is not required. For example, imagine a series of experiments in which only certain aspects of a molecular distribution are studied, and the aspects of the distribution function of interest are those addressed by experiment. Consider the translational kinetic energy distribution depicted in Figure 29.6. What if all we were interested in was the kinetic energy where the probability distribution function was at a maximum? In this case, full knowledge of the probability distribution is not needed. Instead, a single quantity for comparison to experiment, or a "benchmark" value, is all that is needed. In this section, benchmark values of substantial utility when characterizing distribution functions are presented.

29.6.1 Average Values

The **average value** of a quantity is perhaps the most useful way to characterize a distribution function. Consider a function $g(X)$ whose value is dependent on variable X. The average value of this function is dependent on the probability distribution associated with the variable X. If the probability distribution describing the likelihood of X assuming some value is known, this distribution can be employed to determine the average value for the function as follows:

$$\langle g(X) \rangle = \sum_{i=1}^{M} g(X_i) P(X_i) = \frac{\displaystyle\sum_{i=1}^{M} g(X_i) f_i}{\displaystyle\sum_{i=1}^{M} f_i} \tag{29.32}$$

Equation (29.32) states that in order to determine the average value of the function $g(X)$, one simply sums the values of this function determined for each value of the sample set, $\{X_1, X_2, \ldots, X_M\}$, multiplied by the probability of the variable assuming this value. The summation in the denominator provides for normalization of the probability distribution. The angle brackets around $g(X)$ denotes the average value of the function, a quantity that is also referred to as an **expectation value.**

EXAMPLE PROBLEM 29.12

Imagine that you are at a carnival and observe a dart game in which you throw a single dart at the following target:

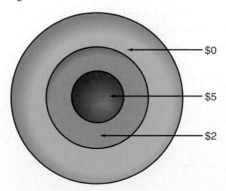

The dollar amounts in the figure indicate the amount of money you will win if you hit the corresponding part of the target with your dart. The geometry of the target is such that the radii of the circles increase linearly between successive areas. Finally, you are a sufficiently accomplished dart thrower that you will not miss the target altogether. If it costs you $1.50 each time you throw a dart, is this carnival game something you should play?

Solution

We will assume that the probability of hitting each section of the target is directly proportional to the area. If the radius of the inner circle is r, then the radii of the second and third circles are $2r$ and $3r$, respectively. The area of the inner curve A_1 is πr^2 and the areas of the outer two circles are

$$A_2 = \pi(2r)^2 - \pi r^2 = 3A_1$$

$$A_3 = \pi(3r)^2 - \pi(2r)^2 = 5A_1$$

Finally, total area is given by the sum of the individual areas, or $9A_1$, such that the probability of hitting each part of the target is

$$f_1 = \frac{A_1}{A_1 + A_2 + A_3} = \frac{A_1}{9A_1} = \frac{1}{9}$$

$$f_2 = \frac{A_2}{A_1 + A_2 + A_3} = \frac{3A_1}{9A_1} = \frac{3}{9}$$

$$f_3 = \frac{A_3}{A_1 + A_2 + A_3} = \frac{5A_1}{9A_1} = \frac{5}{9}$$

The sum of probabilities equals 1 so that the probability distribution is normalized. The quantity of interest is the average payout, or $\langle n_\$ \rangle$, given by

$$\langle n_\$ \rangle = \frac{\sum_{i=1}^{3} n_{\$, i}\, f_i}{\sum_{i=1}^{3} f_i} = \frac{(\$5)\frac{1}{9} + (\$2)\frac{3}{9} + (\$0)\frac{5}{9}}{1} = \$1.22$$

Comparison of the average payout to the amount you pay to throw a dart ($1.50) suggests that this game is a money-losing activity (from the dart-thrower's perspective). Notice in this example that the "experimental" quantities you are comparing are the amount of money you have to pay and the amount of money you can expect to receive on average, which is the benchmark of the distribution of primary interest when you make a decision to play.

29.6.2 Distribution Moments

Some of the most widely used benchmark values for distributions involve functions of the form $\langle x^n \rangle$ where n is an integer. If $n = 1$, the corresponding function $\langle x \rangle$ is referred to as the **first moment** of the distribution function, and it is equal to the average value of the distribution as discussed earlier. If $n = 2$, $\langle x^2 \rangle$ is referred to as the **second moment** of the distribution. Finally, the square root of $\langle x^2 \rangle$ is referred to as the root-mean-squared or "rms" value. The first and second moments, as well as the rms value, are extremely useful quantities for characterizing distribution functions of atomic and molecular properties. For example, in the upcoming discussion of molecular motion in Chapter 33 we will find that a collection of molecules will have a distribution of speeds. Rather than discussing the entire speed distribution, the moments of the distribution are instead used to describe the system of interest. **Distribution moments** are readily calculated for both discrete and continuous distributions as Example Problem 29.13 illustrates.

EXAMPLE PROBLEM 29.13

Consider the following distribution function:

$$P(x) = Cx^2e^{-ax^2}$$

Probability distributions of this form are encountered when describing speed distributions of ideal gas particles. In this probability distribution function, C is a normalization constant that ensures the total probability is equal to one, and a is also a constant that depends on atomic or molecular properties and temperature. The domain of interest is given by $\{0 \leq x \leq \infty\}$. Are the mean and rms values for this distribution the same?

Solution

First, the normalization condition is applied to the probability:

$$1 = \int_0^\infty Cx^2e^{-ax^2}\,dx = C\int_0^\infty x^2e^{-ax^2}\,dx$$

The integral is easily evaluated using the integral tables provided in Appendix A, Math Supplement:

$$\int_0^\infty x^2e^{-ax^2}\,dx = \frac{1}{4a}\sqrt{\frac{\pi}{a}}$$

resulting in the following definition for the normalization constant:

$$C = \frac{1}{\dfrac{1}{4a}\sqrt{\dfrac{\pi}{a}}} = \frac{4a^{3/2}}{\sqrt{\pi}}$$

With this normalization constant, the mean and rms values of the distribution function can be determined. The mean is equal to $\langle x \rangle$, which is determined as follows:

$$\langle x \rangle = \int_0^\infty xP(x)\,dx = \int_0^\infty x\left(\frac{4a^{3/2}}{\sqrt{\pi}}x^2e^{-ax^2}\right)dx$$

$$= \frac{2}{\sqrt{\pi a}}$$

Proceeding in a similar fashion for the second moment ($\langle x^2 \rangle$),

$$\langle x^2 \rangle = \int_0^\infty x^2 P(x)\,dx = \int_0^\infty x^2\left(\frac{4a^{3/2}}{\sqrt{\pi}}x^2e^{-ax^2}\right)dx$$

$$= \frac{3}{2a}$$

The resulting rms value is:

$$\text{rms} = \sqrt{\langle x^2 \rangle} = \sqrt{\frac{3}{2a}}$$

The mean and rms values for the distribution are not equivalent. The normalized distribution and the location of the mean and rms values for the case where $a = 0.3$ are shown here:

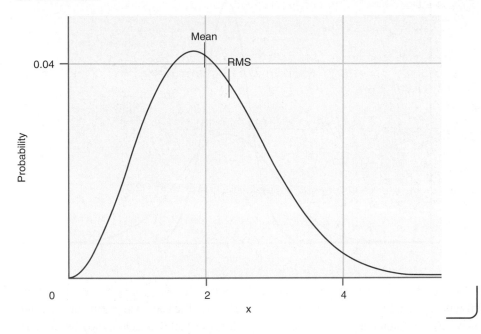

29.6.3 Variance

The **variance** (denoted as σ^2) provides a measure of the width of a distribution. The variance is defined as the average deviation squared from the mean of the distribution:

$$\sigma^2 = \langle (x - \langle x \rangle)^2 \rangle = \langle x^2 - 2x\langle x \rangle + \langle x \rangle^2 \rangle \qquad (29.33)$$

It is important to remember that the value for the mean (or first moment $\langle x \rangle$) is a constant and should not be confused with the variable x. Equation (29.33) can be simplified by using the following two general properties for the averages involving two functions, $b(x)$ and $d(x)$:

$$\langle b(x) + d(x) \rangle = \langle b(x) \rangle + \langle d(x) \rangle \qquad (29.34)$$

$$\langle cb(x) \rangle = c\langle b(x) \rangle \qquad (29.35)$$

In the second property, c is simply a constant. Using these properties the expression for variance becomes:

$$\sigma^2 = \langle x^2 - 2x\langle x \rangle + \langle x \rangle^2 \rangle = \langle x^2 \rangle - \langle 2x\langle x \rangle \rangle + \langle x \rangle^2$$

$$= \langle x^2 \rangle - 2\langle x \rangle\langle x \rangle + \langle x \rangle^2$$

$$\sigma^2 = \langle x^2 \rangle - \langle x \rangle^2 \qquad (29.36)$$

In other words, the variance of a distribution is equal to the difference between the second moment and the square of the first moment.

To illustrate the use of variance as a benchmark, consider the following distribution function, which is referred to as a **Gaussian distribution:**

$$P(X)\, dX = \frac{1}{(2\pi\sigma^2)^{1/2}} e^{-(X-\delta)^2/2\sigma^2} \qquad (29.37)$$

The Gaussian distribution is the "bell-shaped curve" of renown in the social sciences, employed widely through chemistry and physics and well known to any college students concerned about their course grade. It is the primary distribution function utilized to describe error in experimental measurements. The variable X is continuous in the domain

FIGURE 29.7
The influence of variance on Gaussian probability distribution functions. The figure presents the evolution in probability $P(X)$ as a function of the variable X. The distributions for two values of the variance of the distribution σ^2 are presented: 2.0 (red line) and 0.4 (purple line). Notice that an increase in the variance corresponds to an increase in the width of the distribution.

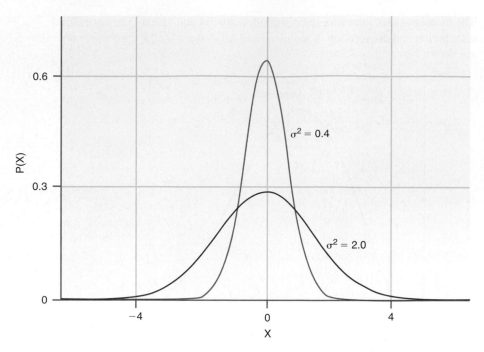

from negative infinity to infinity, $\{-\infty \leq X \leq \infty\}$. The Gaussian probability distribution has a maximum at $X = \delta$. The width of the distribution is determined by the variance σ^2 with an increase in the variance corresponding to increased width. The dependence of the Gaussian distribution on variance is illustrated in Figure 29.7 using Equation (29.37) with $\delta = 0$. As the variance of the distribution increases from $\sigma^2 = 0.4$ to 2.0, the width of the probability distribution increases.

EXAMPLE PROBLEM 29.14

Determine the variance for the distribution function $P(x) = Cx^2 e^{-ax^2}$.

Solution

This is the distribution function employed in Example Problem 29.13, and we have for the first and second moments

$$\langle x \rangle = \frac{2}{\sqrt{\pi a}}$$

$$\langle x^2 \rangle = \frac{3}{2a}$$

Using these values in the expression for variance, we find

$$\sigma^2 = \langle x^2 \rangle - \langle x \rangle^2 = \frac{3}{2a} - \left(\frac{2}{\sqrt{\pi a}}\right)^2$$

$$= \frac{3}{2a} - \frac{4}{\pi a} = \frac{0.23}{a}$$

Vocabulary

average values	bosons	discrete variable
Bernoulli trial	complement	distribution moments
binomial experiment	configuration	event probability
Bose–Einstein statistics	continuous variable	expectation value

factorial	permutation	sample space
Fermi–Dirac statistics	Poisson distribution	second moment
fermions	probability	Stirling's approximation
first moment	probability density	variable
fundamental counting principle	probability distribution function	variance
Gaussian distribution	probability model	

Conceptual Problems

Q29.1 What is the difference between a configuration and a permutation?

Q29.2 What are the elements of a probability model, and how do they differ for continuous and discrete variables?

Q29.3 How does Figure 29.2 change if one is concerned with two versus three colored-ball configurations and permutations?

Q29.4 What is Stirliing's Approximation? Why is it useful? When is it applicable?

Q29.5 What is a Bernoulli Trial?

Q29.6 What must the outcome of a binomial experiment be if $P_E = 1$?

Q29.7 Why is normalization of a probability distribution important? What would one have to consider when working with a probability distribution that was not normalized?

Q29.8 When can one make the approximation of treating a probability distribution involving discrete variables as continuous?

Q29.9 What properties of atomic and molecular systems could you imagine describing using probability distributions?

Q29.10 What is the difference between average and root-mean-squared?

Q29.11 When is the higher moment of a probability distribution more useful as a benchmark value as opposed to simply using the mean of the distribution?

Numerical Problems

Problem numbers in **red** indicate that the solution to the problem is given in the *Student's Solutions Manual*.

P29.1 Suppose that you draw a card from a standard deck of 52 cards. What is the probability of drawing

a. an ace of any suit?

b. the ace of spades?

c. How would your answers to parts (a) and (b) change if you were allowed to draw three times, replacing the card drawn back into the deck after each draw?

P29.2 You are dealt a hand consisting of 5 cards from a standard deck of 52 cards. Determine the probability of obtaining the following hands:

a. flush (five cards of the same suit)

b. a king, queen, jack, ten, and ace of the same suit (a "royal flush")

P29.3 A pair of standard dice are rolled. What is the probability of observing the following:

a. The sum of the dice is equal to 7.

b. The sum of the dice is equal to 9.

c. The sum of the dice is less than or equal to 7.

P29.4 Answer Problem P29.3 assuming that "shaved" dice are used so that the number 6 appears twice as often as any other number.

P29.5 **a.** Consider assigning phone numbers to an area where each number consists of seven digits, each of which can be between 0 and 9. How many phone numbers can be assigned to the area?

b. To serve multiple areas, you decide to introduce an area code that serves as an identifier. In the United States, area codes were initially three-digit numbers in which the first digit ranged from 2 to 9, the second digit was either 0 or 1, and the third digit ranged from 1 to 9. How many areas could be specified using this system?

c. Using area codes with individual phone numbers, how many unique phones could this system support?

P29.6 Proteins are made up of individual molecular units of unique structure known as amino acids. The order or "sequence" of amino acids is an important factor in determining protein structure and function. There are 20 naturally occurring amino acids.

a. How many unique proteins consisting of 8 amino acids are possible?

b. How does your answer change if a specific amino acid can only appear once in the protein?

P29.7 Atomic chlorine has two naturally occurring isotopes, ^{35}Cl and ^{37}Cl. If the molar abundance of these isotopes are 75.4% and 24.6%, respectively, what fraction of a mole of molecular chlorine (Cl_2) will have one of each isotope? What fraction will contain just the ^{35}Cl isotope?

P29.8 The natural molar abundance of ^{13}C is roughly 1%. What is the probability of having a single ^{13}C isotope in benzene (C_6H_6)? What is the probability that two ^{13}C isotopes will be adjacent to each other in benzene?

P29.9 Evaluate the following:

a. the number of permutations employing all objects in a six-object set

b. the number of permutations employing four objects from a six-object set

c. the number of permutations employing no objects from a six-object set

d. $P(50,10)$

P29.10 Determine the number of permutations of size 3 that can be made from the set $\{1, 2, 3, 4, 5, 6\}$. Write down all of the permutations.

P29.11 Determine the numerical values for the following:

a. the number of configurations employing all objects in a six-object set

b. the number of configurations employing four objects from a six-object set

c. the number of configurations employing no objects from a six-object set

d. $C(50,10)$

P29.12 Radio station call letters consist of four letters (for example, KUOW).

a. How many different station call letters are possible using the 26 letters in the English alphabet?

b. Stations west of the Mississippi River must use the letter K as the first call letter. Given this requirement, how many different station call letters are possible if repetition is allowed for any of the remaining letters?

c. How many different station call letters are possible if repetition is not allowed for any of the letters?

P29.13 Four bases (A, C, T, and G) appear in DNA. Assume that the appearance of each base in a DNA sequence is random.

a. What is the probability of observing the sequence AAGACATGCA?

b. What is the probability of finding the sequence GGGGGAAAAA?

c. How do your answers to parts (a) and (b) change if the probability of observing A is twice that of the probabilities used in parts (a) and (b) of this question when the preceding base is G?

P29.14 The natural abundance of ^{13}C is roughly 1%, and the abundance of deuterium (^2H or D) is 0.015%. Determine the probability of finding the following in a mole of acetylene:

a. H-^{13}C-^{13}C-H

b. D-^{12}C-^{12}C-D

c. H-^{13}C-^{12}C-D

P29.15 In the neck of the flask depicted in the following figure, five red balls rest on five blue balls. Suppose the balls are tipped back into the flask, shaken, and the flask is reinverted. What is the probability that the order depicted in the figure will be seen?

P29.16 The Washington State Lottery consists of drawing five balls numbered 1 to 43 and a single ball numbered 1 to 23 from a separate machine.

a. What is the probability of hitting the jackpot in which the values for all six balls are correctly predicted?

b. What is the probability of predicting just the first five balls correctly?

c. What is the probability of predicting the first five balls in the exact order they are picked?

P29.17 Fermions and bosons demonstrate different distribution statistics over a set of quantum states. However, in Chapter 30 we will encounter the Boltzmann distribution, in which we essentially ignore the differentiation between fermions and bosons. This is appropriate only in the "dilute limit" in which the number of available states far outnumbers the number of particles. To illustrate this convergence

a. determine the number of arrangement permutations possible for 3 bosons and 10 states, and repeat this calculation for fermions.

b. repeat the calculations from part (a) for 3 particles, but now 100 states. What do you notice about the difference between the two results?

P29.18 Consider the 25 players on a professional baseball team. At any point, 9 players are on the field.

a. How many 9-player batting orders are possible given that the order of batting is important?

b. How many 9-player batting orders are possible given that the all-star designated hitter must be batting in the fourth spot in the order?

c. How many 9-player fielding teams are possible under the assumption that the location of the players on the field is not important?

P29.19 Imagine an experiment in which you flip a coin 4 times. Furthermore, the coin is balanced fairly such that the probability of landing heads or tails is equivalent. After tossing the coin 10 times, what is the probability of observing the following specific outcomes:

a. no heads

b. two heads

c. five heads

d. eight heads

P29.20 Imagine performing the coin-flip experiment of Problem P29.19, but instead of using a fair coin, a weighted coin is employed for which the probability of landing heads is two-fold greater than landing tails. After tossing the coin 10 times, what is the probability of observing the following specific outcomes:

a. no heads

b. two heads

c. five heads

d. eight heads

P29.21 In Chapter 34 we will model particle diffusion as a random walk in one dimension. In such processes, the probability of moving an individual step in the $+x$ or $-x$ direction is equal to $1/2$. Imagine starting at $x = 0$ and performing a random walk in which 20 steps are taken.

a. What is the farthest distance the particle can possibly move in the $+x$ direction? What is the probability of this occurring?

b. What is the probability the particle will not move at all?

c. What is the probability of the particle moving half the maximum distance in the $+x$ direction?

d. Plot the probability of the particle moving a given distance versus distance. What does the probability distribution look like? Is the probability normalized?

P29.22 Simplify the following expressions:

a. $\dfrac{n!}{(n-2)!}$

b. $\dfrac{n!}{\left(\dfrac{n}{2}!\right)^2}$ (for even n)

P29.23 Another form of Stirling's approximation is

$$N! = \sqrt{2\pi N}\left(\frac{N}{e}\right)^N$$

Use this approximation for $N = 10, 50, 100$ and compare your results to those given in Example Problem 12.9.

P29.24 You are at a carnival and are considering playing the dart game described in Example Problem 29.12; however, you are confident of your dart-throwing skills such that the probability of hitting the center area of the target is three times greater than the probability determined by area. Assuming the confidence in your skills is warranted, is it a good idea to play?

P29.25 Radioactive decay can be thought of as an exercise in probability theory. Imagine that you have a collection of radioactive nuclei at some initial time (N_0) and are interested in how many nuclei will still remain at a later time (N). For first-order radioactive decay, $N/N_0 = e^{-kt}$. In this expression, k is known as the decay constant and t is time.

a. What is the variable of interest in describing the probability distribution?

b. At what time will the probability of nuclei undergoing radioactive decay be 0.50?

P29.26 First order decay processes as described in the previous problem can also be applied to a variety of atomic and molecular processes. For example, in aqueous solution the decay of singlet molecular oxygen ($O_2(^1\Delta_g)$) to the ground-state triplet configuration proceeds according to

$$\frac{[O_2(^1\Delta_g)]}{[O_2(^1\Delta_g)]_0} = e^{-(2.4 \times 10^5 \, \text{s}^{-1})t}$$

In this expression, $[O_2(^1\Delta_g)]$ is the concentration of singlet oxygen at a given time, and the subscript "0" indicates that this is the concentration of singlet oxygen present at the beginning of the decay process corresponding to $t = 0$.

a. How long does one have to wait until 90% of the singlet oxygen has decayed?

b. How much singlet oxygen remains after $t = (2.4 \times 10^5 \, \text{s}^{-1})^{-1}$?

P29.27 In a subsequent chapter we will encounter the energy distribution $P(\varepsilon) = Ae^{-\varepsilon/kT}$, where $P(\varepsilon)$ is the probability of a molecule occupying a given energy state, ε is the energy of the state, k is a constant equal to $1.38 \times 10^{-23} \, \text{J K}^{-1}$, and T is temperature. Imagine that there are three energy states at 0, 100., and 500. J mol^{-1}.

a. Determine the normalization constant for this distribution.

b. What is the probability of occupying the highest energy state at 298 K?

c. What is the average energy at 298 K?

d. Which state makes the largest contribution to the average energy?

P29.28 Assume that the probability of occupying a given energy state is given by the relationship provided in Problem P29.27.

a. Consider a collection of three total states with the first state located at $\varepsilon = 0$ and others at kT and $2kT$, respectively, relative to this first state. What is the normalization constant for the probability distribution?

b. How would your answer change if there are five states with $\varepsilon = kT$ in addition to the single states at $\varepsilon = 0$ and $\varepsilon = 2kT$?

c. Determine the probability of occupying the energy level $\varepsilon = kT$ for the cases in which one and five states exist at this energy.

P29.29 Consider the following probability distribution corresponding to a particle located between point $x = 0$ and $x = a$:

$$P(x)\,dx = C\sin^2\left[\frac{\pi x}{a}\right]dx$$

a. Determine the normalization constant C.

b. Determine $\langle x \rangle$.

c. Determine $\langle x^2 \rangle$.

d. Determine the variance.

P29.30 Consider the probability distribution for molecular velocities in one dimension (v_x) given by
$$P(v_x)\, dv_x = Ce^{-mv_x^2/2kT}\, dv_x.$$

a. Determine the normalization constant C.

b. Determine $\langle v_x \rangle$.

c. Determine $\langle v_x^2 \rangle$.

d. Determine the variance.

P29.31 One classic problem in quantum mechanics is the "harmonic oscillator." In this problem a particle is subjected to a one-dimensional potential (taken to be along x) of the form $V(x) \propto x^2$ where $-\infty \leq x \leq \infty$. The probability distribution function for the particle in the lowest-energy state is

$$P(x) = Ce^{-ax^2/2}$$

Determine the expectation value for the particle along x (that is, $\langle x \rangle$). Can you rationalize your answer by considering the functional form of the potential energy?

P29.32 A crude model for the molecular distribution of atmospheric gases above Earth's surface (denoted by height h) can be obtained by considering the potential energy due to gravity:

$$P(h) = e^{-mgh/kT}$$

In this expression m is the per-particle mass of the gas, g is the acceleration due to gravity, k is a constant equal to 1.38×10^{-23} J K^{-1}, and T is temperature. Determine $\langle h \rangle$ for methane (CH$_4$) using this distribution function.

P29.33 Another use of distribution functions is determining the most probable value, which is done by realizing that at the distribution maximum the derivative of the distribution function with respect to the variable of interest is zero. Using this concept, determine the most probable value of x ($0 \leq x \leq \infty$) for the following function:

$$P(x) = Cx^2 e^{-ax^2}$$

Compare your result to $\langle x \rangle$ and x_{rms} when $a = 0.3$ (see Example Problem 29.13).

P29.34 In nonlinear optical switching devices based on dye-doped polymer systems, the spatial orientation of the dye molecules in the polymer is an important parameter. These devices are generally constructed by orienting dye molecules with a large dipole moment using an electric field. Imagine placing a vector along the molecular dipole moment such that the molecular orientation can be described by the orientation of this vector in space relative to the applied field (z direction) as illustrated here:

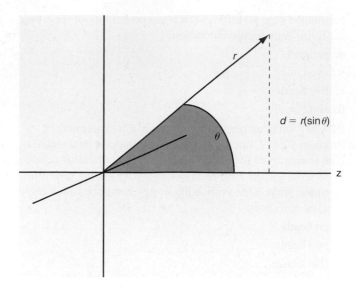

For random molecular orientation about the z axis, the probability distribution describing molecular orientation along the z axis is given by $P(\theta) = \sin\theta\, d\theta / \displaystyle\int_0^\pi \sin\theta\, d\theta$. Orientation is quantified using moments of $\cos\theta$.

a. Determine $\langle \cos\theta \rangle$ for this probability distribution.

b. Determine $\langle \cos^2\theta \rangle$ for this probability distribution.

P29.35 The **Poisson distribution** is a widely used discrete probability distribution in science:

$$P(x) = \frac{e^{-\lambda}\lambda^x}{x!}$$

This distribution describes the number of events (x) occurring in a fixed period of time. The events occur with a known average rate corresponding to λ, and event occurrence does not depend on when other events occur. This distribution can be applied to describe the statistics of photon arrival at a detector as illustrated by the following:

a. Assume that you are measuring a light source with an average output of 5 photons per second. What is the probability of measuring 5 photons in any 1-second interval?

b. For this same source, what is the probability of observing 8 photons in any 1-second interval?

c. Assume a brighter photon source is employed with an average output of 50 photons per second. What is the probability of observing 50 photons in any 1-second interval?

For Further Reading

Bevington, P., and D. Robinson. *Reduction and Error Analysis for the Physical Sciences*. New York: McGraw-Hill, 1992.

Dill, K., and S. Bromberg. *Molecular Driving Forces*. New York: Garland Science, 2003.

McQuarrie, D. *Mathematical Methods for Scientists and Engineers*. Sausalito, CA: University Science Books, 2003.

Nash, L. K. *Elements of Statistical Thermodynamics*. San Francisco: Addison-Wesley, 1972.

Ross, S. *A First Course in Probability Theory*, 3rd ed. New York: Macmillan, 1988.

Taylor, J. R. *An Introduction to Error Analysis*. Mill Valley, CA: University Science Books, 1982.

The Boltzmann Distribution

Employing statistical concepts, one can determine the most probable distribution of energy in a chemical system. This distribution, referred to as the Boltzmann distribution, represents the most probable configuration of energy for a chemical system at equilibrium and also gives rise to the thermodynamic properties of the system. In this chapter, the Boltzmann distribution is derived starting with the probability concepts introduced in the previous chapter. This distribution is then applied to some elementary examples to demonstrate how the distribution of energy in a chemical system depends on both the available energy and the energy-level spacings that characterize the system. The concepts outlined here provide the conceptual framework required to apply statistical mechanics to atomic and molecular systems.

30.1 Microstates and Configurations

We begin by extending the concepts of probability theory introduced in the previous chapter to chemical systems. Although such an extension of probability theory may appear difficult, the sheer size of chemical systems makes the application of these statistical concepts straightforward. Recall the discussion of permutations and configurations illustrated by tossing a coin four times. The possible outcomes for this experiment are presented in Figure 30.1.

Figure 30.1 illustrates that there are five possible outcomes for the trial: from no heads to all heads. Which of these trial outcomes is most likely? Using probability theory, we found that the most probable outcome will be the one with the most possible ways to achieve that outcome. In the coin toss example, the configuration "2 Head" has the greatest number of ways to achieve this configuration; therefore, the "2 Head" configuration represents the most likely outcome. In the language of probability theory, the configuration with the largest number of corresponding permutations will be the most likely trial outcome. As discussed in the previous chapter, the probability P_E of this configuration representing the trial outcome is given by

$$P_E = \frac{E}{N} \tag{30.1}$$

FIGURE 30.1
Possible configurations and permutations for a Bernoulli trial consisting of flipping a coin four times. Blue indicates tails and red indicates heads.

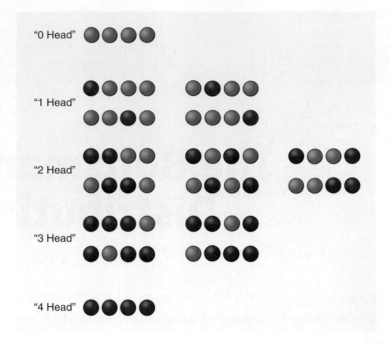

where E is the number of permutations associated with the event of interest, and N is the total number of possible permutations. This relatively simple equation provides the key idea for this entire chapter: *the most likely configurational outcome for a trial is the configuration with the greatest number of associated permutations.* The application of this idea to macroscopic chemical systems is aided by the fact that for systems containing a large number of units, one configuration will have vastly more associated permutations than any other configuration. As such, this configuration will be the only one that is observed to an appreciable extent.

The application of probability theory to chemical systems is achieved by considering system energy. Specifically, we are interested in developing a formalism that is capable of identifying the most likely configuration or distribution of energy in a chemical system. To begin, consider a simple "molecular" system consisting of three quantum harmonic oscillators that share a total of three quanta of energy. The energy levels for the oscillators are given by

$$E_n = h\nu\left(n + \frac{1}{2}\right) \quad \text{for } n = 0, 1, 2, \dots, \infty \tag{30.2}$$

In this equation, h is Planck's constant, ν is the oscillator frequency, and n is the quantum number associated with a given energy level of the oscillator. This quantum number can assume integer values starting at 0 and increasing to infinity; therefore, the energy levels of the quantum harmonic oscillator consist of a ladder or "manifold" of equally spaced levels. The lowest level ($n = 0$) has an energy of $h\nu/2$ referred to as the zero point energy. Employed here is a modified version of the harmonic oscillator in which the ground-state energy ($n = 0$) is zero so that the level energies are given by

$$E_n = h\nu n \quad \text{for } n = 0, 1, 2, \dots, \infty \tag{30.3}$$

When considering atomic and molecular systems in the upcoming chapters, the relative difference in energy levels will prove to be the relevant quantity of interest, and a similar modification will be employed. The final important point to note is that the oscillators are distinguishable. That is, each oscillator in the three-oscillator collection can be readily identified. This assumption must be relaxed when describing molecular systems such as an ideal gas because atomic or molecular motion prohibits identification of each gaseous particle. The extension of this approach to indistinguishable particles is readily accomplished; therefore, starting with the distinguishable case will not prove problematic.

In this example system, the three oscillators are of equal frequency such that the energy levels for all three are identical. Three quanta of energy are placed into this system, and we ask the question "What is the most probable distribution of energy?" From the preceding discussion, probability theory dictates that the most likely distribution of energy corresponds to the configuration with the largest number of associated permutations. How can the concepts of configuration and permutations be connected to oscillators and energy? Two central definitions are introduced here that will prove useful in making this connection: a **configuration** is a general arrangement of total energy available to the system, and a **microstate** is a specific arrangement of energy that describes the energy contained by each individual oscillator. This definition of a configuration is equivalent to the definition of configuration from Chapter 29, and microstates are equivalent to permutations. To determine the most likely configuration of energy in the example system, we will simply count all of the possible microstates and arrange them with respect to their corresponding configurations.

In the first configuration depicted in Figure 30.2, each oscillator has one quantum of energy such that all oscillators populate the $n = 1$ energy level. Only one permutation is associated with this configuration. In terms of the nomenclature just introduced, there is one microstate corresponding to this energy configuration. In the next configuration illustrated in the figure, one oscillator contains two quanta of energy, a second contains one quantum of energy, and a third contains no energy. Six potential arrangements correspond to this general distribution of energy; that is, six microstates correspond to this configuration. The last configuration depicted is one in which all three quanta of energy reside on a single oscillator. Because there are three choices for which oscillator will have all three quanta of energy, there are three corresponding microstates for this configuration. It is important to note that the total energy of all of the arrangements just mentioned is the same and that the only difference is the distribution of the energy over the oscillators.

Which configuration of energy would we expect to observe? Just like the coin tossing example, we expect to see the energy configuration that has the largest number of microstates. In this example, that configuration is the second one discussed, or the "2, 1, 0" configuration. If all microstates depicted have an equal probability of being observed, the probability of observing the 2, 1, 0 configuration is simply the number of microstates associated with this configuration divided by the total number of microstates available, or

$$P_E = \frac{E}{N} = \frac{6}{6 + 3 + 1} = \frac{6}{10} = 0.6$$

Note that although this example involves a "molecular" system, the concepts encountered can be generalized to probability theory. Whether tossing a coin or distributing energy among distinguishable oscillators, the ideas are the same.

30.1.1 Counting Microstates and Weight

The three-oscillator example provides an approach for finding the most probable configuration of energy for a chemical system: determine all of the possible configurations of energy and corresponding microstates and identify the configuration with the greatest number of microstates. Clearly, this would be an extremely laborious task for a chemical system of interesting size. Fortunately, there are ways to obtain a quantitative count of all of the microstates associated with a given configuration without actually "counting" them. First, recall from Chapter 29 that the total number of possible permutations given N objects is $N!$ For the most probable 2, 1, 0 configuration described earlier, there are three objects of interest (i.e., three oscillators) such that $N! = 3! = 6$. This is exactly the same number of microstates associated with this configuration. But what of the other configurations? Consider the 3, 0, 0 configuration in which one oscillator has all three quanta of energy. In assigning quanta of energy to this system to construct each microstate, there are three choices of where to place the three quanta of energy, and two remaining choices for zero quanta. However, this latter choice is redundant in that it does not matter which oscillator receives zero quanta first. The two

FIGURE 30.2
Configurations and associated permutations involving the distribution of three quanta of energy over three distinguishable oscillators.

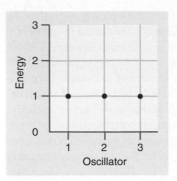

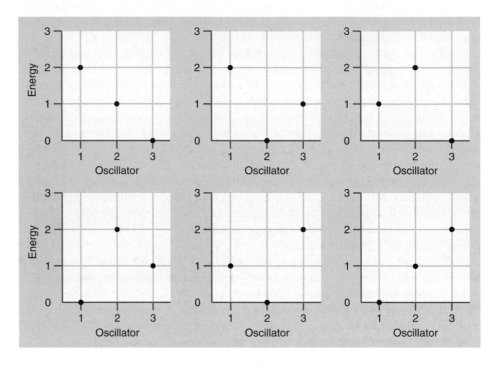

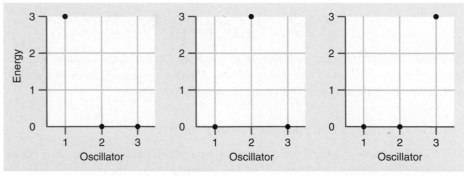

conceptually different arrangements correspond to exactly the same microstate and, thus, are indistinguishable. To determine the number of microstates associated with such distributions of energy, the total number of possible permutations is divided by a factor that corrects for overcounting, which for the '3, 0, 0' configuration is accomplished as follows:

$$\text{Number of microstates} = \frac{3!}{2!} = 3$$

This expression is simply the probability for the number of expression permutations available using a subgroup from an overall group of size N. Therefore, if no two oscillators reside in the same energy level, then the total number of microstates available is given by $N!$, where N is the number of oscillators. However, if two or more oscillators occupy the same energy state (including the zero-energy state), then we need to divide by a term that corrects for overcounting of identical permutations. The total number of microstates associated with a given configuration of energy is referred to as the **weight** of the configuration, W, which is given by

$$W = \frac{N!}{a_0! \, a_1! \, a_2! \ldots a_n!} = \frac{N!}{\prod_j a_j!} \tag{30.4}$$

In Equation (30.4), W is the weight of the configuration of interest, N is the number of units over which energy is distributed, and the a_n terms represent the number of units occupying the nth energy level. The a_n quantities are referred to as **occupation numbers** because they describe how many units occupy a given energy level. For example, in the 3, 0, 0 configuration presented in Figure 30.2, $a_0 = 2$, $a_3 = 1$, and all other $a_n = 0$ (with $0! = 1$). The denominator in our expression for weight is evaluated by taking the product of the factorial of the occupation numbers ($a_n!$), with the product of these values denoted by the $\prod$ symbol (which is analogous to the Σ symbol denoting summation). Equation (30.4) is not limited to our specific example but is a general relationship that applies to any collection of distinguishable units for which only one state is available at a given energy level. The situation in which multiple states exist at a given energy level is discussed later in this chapter.

EXAMPLE PROBLEM 30.1

What is the weight associated with the configuration corresponding to observing 40 heads after flipping a coin 100 times? How does this weight compare to that of the most probable outcome?

Solution

Using the expression for weight in Equation (30.4), the coin flip can be envisioned as a system in which two states can be populated: heads or tails. In addition, the number of distinguishable units (N) is 100, the number of coin tosses. Using these definitions,

$$W = \frac{N!}{a_H! \, a_T!} = \frac{100!}{40! \, 60!} = 1.37 \times 10^{28}$$

The most probable outcome corresponds to the configuration where 50 heads are observed such that

$$W = \frac{N!}{a_H! \, a_T!} = \frac{100!}{50! \, 50!} = 1.01 \times 10^{29}$$

30.1.2 The Dominant Configuration

The three-oscillator example from the preceding section illustrates a few key ideas that will guide us toward our development of the Boltzmann distribution. Specifically, weight is the total number of permutations corresponding to a given configuration. The probability of observing a configuration is given by the weight of that configuration divided by the total weight:

$$P_i = \frac{W_i}{W_1 + W_2 + \ldots + W_N} = \frac{W_i}{\sum_{j=1}^{N} W_j} \tag{30.5}$$

where P_i is the probability of observing configuration i, W_i is the weight associated with this configuration, and the denominator represents the sum of weights for all possible configurations. Equation (30.5) predicts that the configuration with the largest weight will have the greatest probability of being observed. The configuration with the largest weight is referred to as the **dominant configuration.**

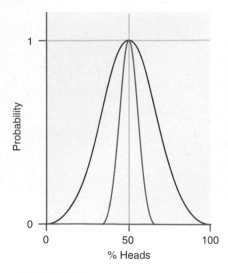

FIGURE 30.3
Comparison of relative probability (probability/maximum probability) for outcomes of a coin-flip trial in which the number of tosses is 10 (red line) and 100 (purple line). The x axis is the percentage of tosses that are heads. Notice that all trials have a maximum probability at 50% heads; however, as the number of tosses increases, the probability distribution becomes more centered about this value as evidenced by the decrease in distribution width.

Given the definition of the dominant configuration, the question arises as to how dominant this configuration is relative to other configurations. A conceptual answer to this question is provided by an experiment in which a coin is tossed 10 times. How probable is the outcome of four heads relative to five heads? Although the outcome of $n_H = 5$ has the largest weight, the weight of $n_H = 4$ is of comparable value. Therefore, observing the $n_H = 4$ configuration would not be at all surprising. But what if the coin were flipped 100 times? How likely would the outcome of $n_H = 40$ be relative to $n_H = 50$? This question was answered in Example Problem 30.1, and the weight of $n_H = 50$ was significantly greater than $n_H = 40$. As the number of tosses increases, the probability of observing heads 50% of the time becomes greater relative to any other outcome. In other words, the configuration associated with observing 50% heads should become the more dominant configuration as the number of coin tosses increases. An illustration of this expectation is presented in Figure 30.3 in which the relative weights associated with observing heads a certain percentage of the time after tossing a coin 10 and 100 times are presented. The figure demonstrates that after 10 tosses, the relative probability of observing something other than five heads is appreciable. However, as the number of coin tosses increases, the probability distribution narrows such that the final result of 50% heads becomes dominant. Taking this argument to sizes associated with molecular assemblies, imagine performing Avogadro's number of coin tosses! Given the trend illustrated in Figure 30.3, one would expect the probability to be sharply peaked at 50% heads. In other words, as the number of experiments in the trial increases, the 50% heads configuration will not only become the most probable, but also the weight associated with this configuration will become so large that the probability of observing another outcome is minuscule. Indeed, the most probable configuration evolves into the dominant configuration as the size of the system increases.

EXAMPLE PROBLEM 30.2

Consider a collection of 10,000 particles with each particle capable of populating one of three energy levels having energies 0, ε, and 2ε with a total available energy of 5000 ε. Under the constraint that the total number of particles and total energy be constant, determine the dominant configuration.

Solution

With constant total energy, only one of the energy-level populations is independent. Treating the number of particles in the highest energy level (N_3) as the independent variable, the number of particles in the intermediate (N_2) and lowest (N_1) energy levels is given by

$$N_2 = 5000 - 2N_3$$
$$N_1 = 10,000 - N_2 - N_3$$

Because the number of particles in a given energy level must be greater than or equal to 0, the preceding equations demonstrate that N_3 can range from 0 to 2500. Given the size of the state populations, it is more convenient to calculate the natural log of the weight associated with each configuration of energy as a function of N_3 by using Equation (30.4) in the following form:

$$\ln W = \ln \left(\frac{N!}{N_1!N_2!N_3!} \right) = \ln N! - \ln N_1! - \ln N_2! - \ln N_3!$$

Each term can be readily evaluated using Stirling's approximation, and results of this calculation are presented in Figure 30.4(a). This figure demonstrates that ln W has a maximum value at $N_3 \approx 1200$ (or 1162 to be precise). The dominance of this configuration is also illustrated in Figure 30.4(b), where the weight of the configurations corresponding to the allowed values of N_3 are compared with that of the dominant configuration. Even for this relatively simple system having only 10,000 particles, the weight is sharply peaked at the dominant configuration.

30.2 Derivation of the Boltzmann Distribution

As the size of the system increases, a single configurational outcome will have such a large relative weight that only this configuration will be observed. In this limit, it becomes pointless to define all possible configurations, and only the outcome associated with the dominant configuration is of interest. A method is needed by which to identify this configuration directly. Inspection of Figures 30.3 and 30.4 reveals that the dominant configuration can be determined as follows. Because the dominant configuration has the largest associated weight, any change in outcome corresponding to a different configuration will be reflected by a reduction in weight. Therefore, the dominant configuration can be identified by locating the peak of the curve corresponding to weight as a function of **configurational index,** denoted by χ. Because W will be large for molecular systems, it is more convenient to work with $\ln W$, and the search criterion for the dominant configuration becomes

$$\frac{d \ln W}{d\chi} = 0 \tag{30.6}$$

This expression is a mathematical definition of the dominant configuration, and it states that if a configuration space is searched by monitoring the change in $\ln W$ as a function of configurational index, a maximum will be observed that corresponds to the dominant configuration. A graphical description of the search criterion is presented in Figure 30.5.

The distribution of energy associated with the dominant configuration is known as the **Boltzmann distribution.** We begin our derivation of this distribution by taking the natural log of the weight using the expression for weight developed previously and applying Stirling's approximation:

$$\ln W = \ln N! - \ln \prod_n a_n!$$

$$= N \ln N - \sum_n a_n \ln a_n \tag{30.7}$$

In obtaining Equation 30.7 the following equality was employed:

$$N = \sum_n a_n \tag{30.8}$$

This equation makes intuitive sense. The objects in our collection must be in one of the available energy levels; therefore, summation over the occupation numbers is equivalent to counting all objects.

The criterion for the dominant configuration requires differentiation of $\ln W$ by some relevant configurational index, but what is this index? We are interested in the distribution of energy among a collection of molecules, or the number of molecules that resides in a given energy level. Because the number of molecules residing in a given energy level is the occupation number a_n, the occupation number provides a relevant configurational index. Recognizing this, differentiation of $\ln W$ with respect to a_n yields the following:

$$\frac{d \ln W}{da_n} = \frac{dN}{da_n} \ln N + N\frac{d \ln N}{da_n} - \frac{d}{da_n} \sum_n (a_n \ln a_n) \tag{30.9}$$

The following mathematical relationship was employed in obtaining the previous relationship:

$$\frac{d \ln x}{dx} = \frac{1}{x}$$

such that

$$\frac{d \ln W}{da_n} = \ln N + N\left(\frac{1}{N}\right) - (\ln a_n + 1) = -\ln\left(\frac{a_n}{N}\right) \tag{30.10}$$

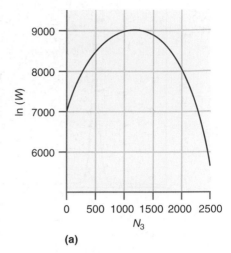

(a)

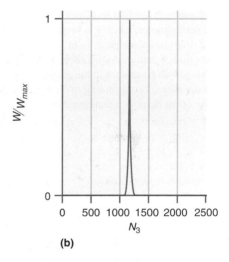

(b)

FIGURE 30.4
Illustration of the dominant configuration for a system consisting of 10,000 particles with each particle having three energy levels at energies of 0, ε, and 2ε as discussed in Example Problem 30.2. The number of particles populating the higher energy level is N_3, and the energy configurations are characterized by the population in this level. **(a)** Variation in the natural log of the weight $\ln W$ for energy configurations as a function of N_3, demonstrating that $\ln W$ has a maximum at $N_3 \approx 1200$. **(b)** Variation in the weight associated with a given configuration to that of the dominant configuration. The weight is sharply peaked around $N_3 \approx 1200$ corresponding to the dominant configuration of energy.

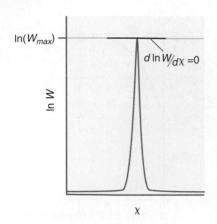

FIGURE 30.5
Mathematical definition of the dominant configuration. The change in the natural log of weight ln W as a function of configuration index χ is presented. If we determine the change in ln W as a function of configuration index, the change will equal zero at the maximum of the curve corresponding to the location of the dominant configuration.

Comparison of Equation (30.10) and the search criteria for the dominant configuration suggests that if $\ln(a_n/N) = 0$ our search is complete. However, this assumes that the occupation numbers are independent, yet the sum of the occupation numbers must equal N, dictating that reduction of one occupation number must be balanced by an increase in another. In other words, one object in the collection is free to gain or lose energy, but a corresponding amount of energy must be lost or gained elsewhere in the system. We can ensure that both the number of objects and total system energy are constant by requiring that

$$\sum_n da_n = 0 \quad \text{and} \quad \sum_n \varepsilon_n da_n = 0 \qquad (30.11)$$

The da_n terms denote change in occupation number a_n. In the expression on the right, ε_n is the energy associated with the nth energy level. Because the conservation of N and energy has not been required to this point, these conditions are now included by introducing Lagrange multipliers, α and β, as weights for the corresponding constraints into the differential as follows:

$$d \ln W = 0 = \sum_n - \ln\left(\frac{a_n}{N}\right) da_n + \alpha \sum_n da_n - \beta \sum_n \varepsilon_n da_n$$

$$= \sum_n \left(-\ln\left(\frac{a_n}{N}\right) + \alpha - \beta\varepsilon_n\right) da_n \qquad (30.12)$$

The Lagrange method of undetermined multipliers is described in the Math Supplement. Formally, this technique allows for maximization of a function that is dependent on many variables that are constrained among themselves. The key step in this approach is to determine the identity of α and β by noting that in Equation (30.12) the equality is only satisfied when the terms in parentheses are equal to zero:

$$0 = -\ln\left(\frac{a_n}{N}\right) + \alpha - \beta\varepsilon_n$$

$$\ln\left(\frac{a_n}{N}\right) = \alpha - \beta\varepsilon_n$$

$$\frac{a_n}{N} = e^\alpha e^{-\beta\varepsilon_n}$$

$$a_n = Ne^\alpha e^{-\beta\varepsilon_n} \qquad (30.13)$$

At this juncture the Lagrange multipliers can be defined. First, α is defined by summing both sides of the preceding equality over all energy levels. Recognizing that $\Sigma a_n = N$,

$$N = \sum_n a_n = Ne^\alpha \sum_n e^{-\beta\varepsilon_n}$$

$$1 = e^\alpha \sum_n e^{-\beta\varepsilon_n}$$

$$e^\alpha = \frac{1}{\sum_n e^{-\beta\varepsilon_n}} \qquad (30.14)$$

This last equality is a central result in statistical mechanics. The denominator in Equation (30.14) is referred to as the **partition function** q and is defined as

$$q = \sum_n e^{-\beta\varepsilon_n} \qquad (30.15)$$

The value of β will be determined next. The partition function represents the sum over all terms that describes the probability associated with the variable of interest, in this case ε_n, or the energy of level n. Using the partition function with Equation (30.13), the probability of occupying a given energy level p_n becomes

$$p_n = \frac{a_n}{N} = e^\alpha e^{-\beta\varepsilon_n} = \frac{e^{-\beta\varepsilon_n}}{q} \qquad (30.16)$$

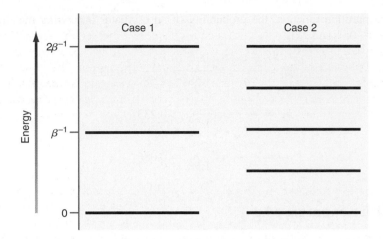

FIGURE 30.6
Two example oscillators. In case 1, the energy spacing is β^{-1}, and the energy spacing is $\beta^{-1}/2$ in case 2.

Equation (30.16) is the final result of interest. It quantitatively describes the probability of occupying a given energy for the dominant configuration of energy. This well-known and important result is referred to as the Boltzmann distribution. We can compare Equation (30.16) to the expression from Chapter 29 for probability involving discrete variables, where probability was defined as follows:

$$P(X_i) = \frac{f_i}{\displaystyle\sum_{j=1}^{M} f_j} \qquad (30.17)$$

This comparison demonstrates that the Boltzmann distribution is nothing more than a statement of probability, with the partition function serving to normalize the probability distribution.

Quantitative application of Equation (30.16) requires that we determine β, the second of our Lagrange multipliers. However, a conceptual discussion at this juncture can provide some insight into how this distribution describes molecular systems. Imagine a collection of harmonic oscillators as in the previous example, but instead of writing down all microstates and identifying the dominant configuration, the dominant configuration is instead given by the Boltzmann distribution law of Equation (30.16). This law establishes that the probability of observing an oscillator in a given energy level is dependent on level energy (ε_n) as $\exp(-\beta\varepsilon_n)$. Because the exponent must be unitless, β must have units of inverse energy.[1] Recall that the energy levels of the harmonic oscillator (neglecting zero point energy) are $\varepsilon_n = nh\nu$ for $n = 0, 1, 2, \ldots, \infty$. Therefore, our conceptual example will employ oscillators where $h\nu = \beta^{-1}$ as illustrated in Figure 30.6.

Using this value for the energy spacings, the exponential terms in the Boltzmann distribution are easily evaluated:

$$e^{-\beta\varepsilon_n} = e^{-\beta(n/\beta)} = e^{-n} \qquad (30.18)$$

The partition function is evaluated by performing the summation over the energy levels:

$$q = \sum_{n=0}^{\infty} e^{-n} = 1 + e^{-1} + e^{-2} + \ldots$$

$$= \frac{1}{1 - e^{-1}} = 1.58 \qquad (30.19)$$

In the last step of this example, we have used the following series expression (where $|x| < 1$):

$$\frac{1}{1 - x} = 1 + x + x^2 + \ldots$$

---

[1]This simple unit analysis points to an important result: that β is related to energy. As will be shown, $1/\beta$ provides a measure of the energy available to the system.

With the partition function, the probability of an oscillator occupying the first three levels ($n = 0, 1$, and 2) is

$$p_0 = \frac{e^{-\beta\varepsilon_0}}{q} = \frac{e^{-0}}{1.58} = \frac{1}{1.58} = 0.633$$

$$p_1 = \frac{e^{-\beta\varepsilon_1}}{q} = \frac{e^{-1}}{1.58} = 0.233$$

$$p_2 = \frac{e^{-\beta\varepsilon_2}}{q} = \frac{e^{-2}}{1.58} = 0.086 \tag{30.20}$$

EXAMPLE PROBLEM 30.3

For the example just discussed, what is the probability of finding an oscillator in energy levels $n \geq 3$?

Solution

The Boltzmann distribution is a normalized probability distribution. As such, the sum of all probabilities equals unity:

$$p_{total} = 1 = \sum_{n=0}^{\infty} p_n$$

$$1 = p_0 + p_1 + p_2 + \sum_{n=3}^{\infty} p_n$$

$$1 - (p_0 + p_1 + p_2) = 0.048 = \sum_{n=3}^{\infty} p_n$$

In other words, only 4.8% of the oscillators in our collection will be found in levels $n \geq 3$.

We continue with our conceptual example by asking the following question: "How will the probability of occupying a given level vary with a change in energy separation between levels?" In the first example, the energy spacings were equal to β^{-1}. A reduction in energy-level spacings to half this value requires that $h\nu = \beta^{-1}/2$. It is important to note that β has not changed relative to the previous example; only the separation in energy levels has changed. With this new energy separation, the exponential terms in the Boltzmann distribution become

$$e^{-\beta\varepsilon_n} = e^{-\beta(n/2\beta)} = e^{-n/2} \tag{30.21}$$

Substituting this equation into the expression for the partition function,

$$q = \sum_{n=0}^{\infty} e^{-n/2} = 1 + e^{-1/2} + e^{-1} + \dots$$

$$= \frac{1}{1 - e^{-1/2}} = 2.54 \tag{30.22}$$

Using this value for the partition function, the probability of occupying the first three levels ($n = 0, 1$, and 2) corresponding to this new spacing is

$$p_0 = \frac{e^{-\beta\varepsilon_0}}{q} = \frac{e^{-0}}{2.54} = \frac{1}{2.54} = 0.394$$

$$p_1 = \frac{e^{-\beta\varepsilon_1}}{q} = \frac{e^{-1/2}}{2.54} = 0.239$$

$$p_2 = \frac{e^{-\beta\varepsilon_2}}{q} = \frac{e^{-1}}{2.54} = 0.145 \tag{30.23}$$

Comparison with the previous system probabilities in Equation (30.20) illustrates some interesting results. First, with a decrease in energy-level spacings, the probability of occupying the lowest energy level ($n = 0$) decreases, whereas the probability of occupying the other energy levels increases. Reflecting this change in probabilities, the value of the partition function has also increased. Since the partition function represents the sum of the probability terms over all energy levels, an increase in the magnitude of the partition function reflects an increase in the probability of occupying higher energy levels. That is, the partition function provides a measure of the number of energy levels that are occupied for a given value of β.

EXAMPLE PROBLEM 30.4

For the preceding example with decreased energy-level spacings, what is the probability of finding an oscillator in energy states $n \geq 3$?

Solution

The calculation from the previous example is used to find that

$$\sum_{n=3}^{\infty} p_n = 1 - (p_0 + p_1 + p_2) = 0.222$$

Consistent with the discussion, the probability of occupying higher energy levels has increased substantially with a reduction in level spacings.

30.2.1 Degeneracy

To this point, the assumption has been that only one state is present at a given energy level; however, when discussing atomic and molecular systems, more than a single state may be present at a given energy. The presence of multiple states at a given energy level is referred to as **degeneracy,** and degeneracy is incorporated into the expression for the partition function as follows:

$$q = \sum_n g_n e^{-\beta \varepsilon_n} \tag{30.24}$$

Equation (30.24) is identical to the previous definition of q from Equation (30.15) with the exception that the term g_n has been included. This term represents the number of states present at a given energy level, or the degeneracy of the level. The corresponding expression for the probability of occupying energy level ε_i is

$$p_i = \frac{g_i e^{-\beta \varepsilon_i}}{q} \tag{30.25}$$

In Equation (30.25), g_i is the degeneracy of the level with energy-level ε_i, and q is as defined in Equation (30.24).

How does degeneracy influence probability? Consider Figure 30.7 in which a system with single states at energy 0 and β^{-1} is shown with a similar system in which two states are present at energy β^{-1}. The partition function for the first system is

$$q_{system1} = \sum_n g_n e^{-\beta \varepsilon_n} = 1 + e^{-1} = 1.37 \tag{30.26}$$

For the second system, q is

$$q_{system2} = \sum_n g_n e^{-\beta \varepsilon_n} = 1 + 2e^{-1} = 1.74 \tag{30.27}$$

The corresponding probability of occupying a state at energy β^{-1} for the two systems is given by

$$p_{system1} = \frac{g_i e^{-\beta \varepsilon_i}}{q} = \frac{e^{-1}}{1.37} = 0.27$$

$$p_{system2} = \frac{g_i e^{-\beta \varepsilon_i}}{q} = \frac{2e^{-1}}{1.74} = 0.42 \tag{30.28}$$

FIGURE 30.7
Illustration of degeneracy. In system 1, one state is present at energies 0 and β^{-1}. In system 2, the energy spacing is the same, but at energy β^{-1} two states are present such that the degeneracy at this energy is 2.

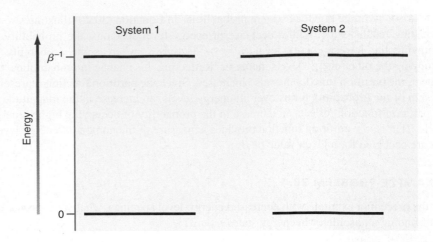

Notice that the probability of occupying a state at energy β^{-1} is greater for system 2, the system with degeneracy. This increase reflects the fact that two states are now available for population at this energy. However, this increase is not simply twice that of the nondegenerate case (system 1) since the value of the partition function also changes due to degeneracy.

30.3 Dominance of the Boltzmann Distribution

The search criterion for the Boltzmann distribution as illustrated in Figure 30.5 dictates that there will be an energy distribution for which the number of corresponding microstates is greatest. However, the question remains as to just how dominant the Boltzmann distribution is. Another way to approach this question is to ask "What exactly is the width of the curve presented in Figure 30.5?" Is there sufficient breadth in this distribution that a slight change in configuration results in another energy distribution, different from that of the dominant configuration but with a reasonable probability of being observed?

Consider an isolated macroscopic assembly of distinguishable units.[2] The dominant configuration will be that having the largest number of microstates corresponding to weight – W_{max}. In addition, consider a slightly different configuration having weight – W, and $W < W_{max}$. Let a_n be the fractional change in the number of units present in the nth energy state:

$$\alpha_n = \frac{a'_n - a_n}{a_n} = \frac{\delta_n}{a_n} \tag{30.29}$$

In this expression, δ_n is simply the change in occupation number a_n to some new configuration (denoted as a'_n) relative to the occupation numbers corresponding to the dominant configuration. The system is isolated, dictating that both the number of particles and the total amount of energy are conserved, resulting in $\Delta N = 0$ and $\Delta E = 0$. With these definitions, the ratio of W_{max}/W is given by the following equation:

$$\frac{W_{max}}{W} = \frac{\dfrac{N!}{\prod\limits_n a_n!}}{\dfrac{N!}{\prod\limits_n (a_n + \delta_n)!}} = \prod_n \left(a_n + \frac{\delta_n}{2} \right)^{\delta_n} \tag{30.30}$$

The term $(a_n + \delta_n)$ consists of δ factors from $(a + 1)$ to $(a + \delta)$. The last equality in Equation (30.30) is satisfied when δ is sufficiently small such that $|\delta_n| \ll a_n$. The ratio of weights provides a measure of the number of microstates associated with the

---

[2]A full presentation of the following derivation appears in L. K. Nash, "On the Boltzmann Distribution Law," *J. Chemical Education* 59 (1982): 824.

modified configuration versus the Boltzmann distribution. Because W will be quite large for an assembly of molecules, the natural log of this ratio is used:

$$\ln\left(\frac{W_{max}}{W}\right) = \sum_n \ln\left(a_n + \frac{\delta_n}{2}\right)^{\delta_n}$$

$$= \sum_n \delta_n \ln\left(a_n + \frac{\delta_n}{2}\right)$$

$$= \sum_n \delta_n \ln a_n\left(1 + \frac{\delta_n}{2a_n}\right)$$

$$= \sum_n \delta_n \ln a_n + \sum_n \delta_n \ln\left(1 + \frac{\delta_n}{2a_n}\right) \qquad (30.31)$$

The last equality in Equation (30.31) can be simplified by noting that $\ln(1 \pm z) = \pm z$ for small z. Because δ_n represents a fractional change in occupation number corresponding to the dominant configuration, $\delta_n/2a_n \ll 1$. Therefore,

$$\ln\left(\frac{W_{max}}{W}\right) = \sum_n \delta_n \ln a_n + \sum_n \frac{\delta_n^2}{2a_n} \qquad (30.32)$$

Recall the following definition, which related the fractional change in occupation number to the occupation numbers themselves:

$$\alpha_n = \frac{\delta_n}{a_n} \qquad (30.33)$$

Substitution of the preceding relationship for δ_n results in

$$\ln\left(\frac{W_{max}}{W}\right) = \sum_n a_n\alpha_n \ln a_n + \sum_n \frac{a_n}{2}(\alpha_n^2)$$

$$= \sum_n a_n\alpha_n(\ln a_0 - \beta\varepsilon_n) + \sum_n \frac{a_n}{2}(\alpha_n^2)$$

$$= \sum_n a_n\alpha_n(\ln a_0) - \sum_n \beta\varepsilon_n a_n\alpha_n + \sum_n \frac{a_n}{2}(\alpha_n^2)$$

$$= \ln a_0 \sum_n a_n\alpha_n - \beta \sum_n \varepsilon_n a_n\alpha_n + \sum_n \frac{a_n}{2}(\alpha_n^2)$$

$$= \sum_n \frac{a_n}{2}(\alpha_n^2) \qquad (30.34)$$

The last step in this derivation was performed by realizing that $\Delta N = 0$ (and the first summation is equal to zero) and $\Delta E = 0$ (so that the second summation is also equal to zero). The root-mean-squared deviation in occupation number relative to those of the dominant configuration is defined as follows:

$$\alpha_{rms} = \left[\frac{\sum_n a_n\alpha_n^2}{N}\right]^{1/2} \qquad (30.35)$$

Given this equation, the ratio of weights corresponding to the Boltzmann distribution W_{max} and the slightly modified distribution W reduces to

$$\frac{W_{max}}{W} = e^{N\alpha_{rms}^2/2} \qquad (30.36)$$

To apply this relationship to a chemical system, imagine that the system of interest is a mole of molecules such that $N = 6.022 \times 10^{23}$. In addition, the fractional change in occupation number will be exceedingly small corresponding to $\alpha_{rms} = 10^{-10}$ (that is, one part in 10^{10}). Using these values,

$$\frac{W_{max}}{W} = e^{\frac{(6.022 \times 10^{23}) \times (10^{-20})}{2}} \approx e^{3000} = 10^{1302}$$

The ratio of weights is an extremely large number, and it demonstrates that a minute change in configuration will result in a significant reduction in weight. Clearly, the width of the curve illustrated in Figure 30.5 is extremely small in a system where N is on the order of Avogadro's number, and the most probable distribution is virtually the only distribution that will be observed for a macroscopic assembly of units. Of the total number of microstates available to a large assembly of units, the vast majority of these microstates correspond to the dominant configuration and a subset of configurations that differ from the dominant configuration by exceedingly small amounts such that the macroscopic properties of the assembly will be identical to those of the dominant configuration. In short, the macroscopic properties of the assembly are defined by the dominant configuration.

30.4 Physical Meaning of the Boltzmann Distribution Law

How does one know other configurations exist if the dominant configuration is all one expects to see for a system at equilibrium? Furthermore, are the other nondominant configurations of the system of no importance? Modern experiments are capable of displacing systems from equilibrium and monitoring the system as it relaxes back toward equilibrium. Therefore, the capability exists to experimentally prepare a nondominant configuration so that these configurations can be studied. An illuminating, conceptual answer to this question is provided by the following logical arguments. First, consider the central postulate of statistical mechanics:

> Every possible microstate of an isolated assembly of units occurs with equal probability.

How does one know that this postulate is true? Imagine a collection of 100 oscillators having 100 quanta of energy. The total number of microstates available to this system is on the order of 10^{200}, which is an extremely large number. Now, imagine performing an experiment in which the energy content of each oscillator is measured such that the corresponding microstate can be established. Also assume that a measurement can be performed every 10^{-9} s (1 nanosecond) such that microstates can be measured at a rate of 10^9 microstates per second. Even with such a rapid determination of microstates, it would take us 10^{191} s to count every possible microstate, a period of time that is much larger than the age of the universe! In other words, the central postulate cannot be verified experimentally. However, we will operate under the assumption that the central postulate is true because statistical mechanical descriptions of chemical systems have provided successful and accurate descriptions of macroscopic systems.

Even if the validity of the central postulate is assumed, the question of its meaning remains. To gain insight into this question, consider a large or macroscopic collection of distinguishable and identical oscillators. Furthermore, the collection is isolated, resulting in both the total energy and the number of oscillators being constant. Finally, the oscillators are free to exchange energy such that any configuration of energy (and, therefore, any microstate) can be achieved. The system is set free to evolve, and the following features are observed:

1. All microstates are equally probable; however, one has the greatest probability of observing a microstate associated with the dominant configuration.

2. As demonstrated in the previous section, configurations having a significant number of microstates will be only infinitesimally different from the dominant configuration. The macroscopic properties of the system will be identical to that of the dominant configuration. Therefore, with overwhelming probability, one will observe a macroscopic state of the system characterized by the dominant configuration.

3. Continued monitoring of the system will result in the observation of macroscopic properties of the systems that appear unchanging, although energy is still being

exchanged between the oscillators in our assembly. This macroscopic state of the system is called the **equilibrium state.**

4. Given items 1 through 3, the equilibrium state of the system is characterized by the dominant configuration.

This logical progression brings us to an important conclusion: *the Boltzmann distribution law describes the energy distribution associated with a chemical system at equilibrium.* In terms of probability, the fact that all microstates have equal probability of being observed does not translate into an equal probability of observing all configurations. As illustrated in Section 30.3, the vast majority of microstates correspond to the Boltzmann distribution, thereby dictating that the most probable configuration that will be observed is the one characterized by the Boltzmann distribution.

30.5 The Definition of β

Use of the Boltzmann distribution requires an operative definition for β, preferably one in which this quantity is defined in terms of measurable system variables. Such a definition can be derived by considering the variation in weight W as a function of total energy contained by an assembly of units E. To begin, imagine an assembly of 10 oscillators having only three quanta of total energy. In this situation, the majority of the oscillators occupy the lowest energy states, and the weight corresponding to the dominant configuration should be small. However, as energy is deposited into the system, the oscillators will occupy higher energy states and the denominator in Equation (30.4) will be reduced, resulting in an increase in W. Therefore, one would expect E and W to be correlated.

The relationship between E and W can be determined by taking the natural log of Equation (30.4):

$$\ln W = \ln N! - \ln \prod_n a_n!$$

$$= \ln N! - \sum_n \ln a_n! \tag{30.37}$$

Interest revolves around the change in W with respect to E, a relationship that requires the total differential of W:

$$d \ln W = -\sum_n \ln a_n!$$

$$= -\sum_n \ln a_n \, da_n \tag{30.38}$$

The result provided by Equation (30.38) was derived using Stirling's approximation to evaluate $\ln(a_n!)$. Simplification of Equation (30.38) is accomplished using the Boltzmann relationship to define the ratio between the occupation number for an arbitrary energy level ε_n versus the lowest or ground energy level ($\varepsilon_0 = 0$):

$$\frac{a_n}{a_0} = \frac{\dfrac{Ne^{-\beta\varepsilon_n}}{q}}{\dfrac{Ne^{-\beta\varepsilon_0}}{q}} = e^{-\beta\varepsilon_n} \tag{30.39}$$

$$\ln a_n = \ln a_0 - \beta\varepsilon_n \tag{30.40}$$

In the preceding steps, the partition function q and N simply cancel. Taking this expression for $\ln(a_n)$ and substituting into Equation (30.38) yields

$$d \ln W = -\sum_n (\ln a_0 - \beta\varepsilon_n) da_n$$

$$= -\ln a_0 \sum_n da_n + \beta \sum_n \varepsilon_n \, da_n \tag{30.41}$$

The first summation in Equation (30.41) represents the total change in occupation numbers, and it is equal to the change in the total number of oscillators in the system. Because the system is closed with respect to the number of oscillators, $dN = 0$ and the first summation is also equal to zero. The second term represents the change in total energy of the system (dE) accompanying the deposition of energy into the system:

$$\sum_n \varepsilon_n \, da_n = dE \tag{30.42}$$

With this last equality, the relationship between β, weight, and total energy is finally derived:

$$d \ln W = \beta dE \tag{30.43}$$

This last equality is quite remarkable and provides significant insight into the physical meaning of β. We began by recognizing that weight increases in proportion with the energy available to the system, and β is simply the proportionality constant in this relationship. Unit analysis of Equation (30.43) also demonstrates that β must have units of inverse energy as inferred previously.

Associating β with measurable system variables is the last step in deriving a full definition for the Boltzmann distribution. This step can be accomplished through the following conceptual experiment. Imagine two separate systems of distinguishable units at equilibrium having associated weights W_x and W_y. These assemblies are brought into thermal contact, and the composite system is allowed to evolve toward equilibrium, as illustrated in Figure 30.8. The composite system is then isolated from the surroundings such that the total energy available to the composite system is the sum of energy contained in the individual assemblies. The total weight of the combined system immediately after establishing thermal contact is the product of W_x and W_y. If the two systems are initially at different equilibrium conditions, the instantaneous composite system weight will be less than the weight of the composite system at equilibrium. Since the composite weight will increase as equilibrium is approached

$$d(W_x \cdot W_y) \geq 0 \tag{30.44}$$

This inequality can be simplified as follows (see the Math Supplement):

$$W_y dW_x + W_x dW_y \geq 0$$

$$\frac{dW_x}{W_x} + \frac{dW_y}{W_y} \geq 0$$

$$d \ln W_x + d \ln W_y \geq 0 \tag{30.45}$$

Substitution of Equation (30.43) into the last expression of Equation (30.45) results in

$$\beta_x dE_x + \beta_y dE_y \geq 0 \tag{30.46}$$

where β_x and β_y are the corresponding β values associated with the initial assemblies x and y. Correspondingly, dE_x and dE_y refer to the change in total energy for the individual assemblies. Because the composite system is isolated from the surroundings, any change in energy for assembly x must be offset by a corresponding change in assembly y:

$$dE_x + dE_y = 0$$

$$dE_x = -dE_y \tag{30.47}$$

Now, if dE_x is positive, then by Equation (30.46),

$$\beta_x \geq \beta_y \tag{30.48}$$

Can the preceding result be interpreted in terms of system variables? This question can be answered by considering the following. If dE_x is positive, energy flows into assembly x from assembly y. Thermodynamics dictates that because temperature is a measure of internal kinetic energy, an increase in the energy will be accompanied by an increase in the temperature of assembly x. A corresponding decrease in the temperature of

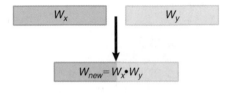

FIGURE 30.8
Two assemblies of distinguishable units, denoted x and y, are brought into thermal contact.

assembly y will also occur. Therefore, before equilibrium is established, thermodynamic considerations dictate that

$$T_y \geq T_x \qquad (30.49)$$

In order for Equations (30.48) and Equations (30.49) to be true, β must be inversely related to T. Furthermore, from unit analysis of Equation (30.43), we know that β must have units of inverse energy. This requirement is met by including a proportionality constant in the relationship between β and T, resulting in the final expression for β:

$$\beta = \frac{1}{kT} \qquad (30.50)$$

The constant in Equation (30.50), k, is referred to as **Boltzmann's constant** and has a numerical value of $1.3806488 \times 10^{-23}$ J K^{-1}. The product of k and Avogadro's number is equal to R, the ideal gas constant (8.3144621 J mol^{-1} K^{-1}). Although the joule is the SI unit for energy, much of the information regarding molecular energy levels is derived from spectroscopic measurements. These spectroscopic quantities are generally expressed in units of wavenumbers (cm^{-1}). The **wavenumber** is simply the number of waves in an electromagnetic field per centimeter. Conversion from wavenumbers to joules is performed by multiplying the quantity in wavenumbers by Planck's constant h and the speed of light c. In Example Problem 30.5, the vibrational energy levels for I_2 are given by the vibrational frequency of the oscillator $\tilde{\nu}$. Using this spectroscopic information, the vibrational level energies in joules are

$$E_n = nhc\tilde{\nu} = n(6.626 \times 10^{-34} \text{ J s}) (3.00 \times 10^{10} \text{ cm s}^{-1}) (208 \text{ cm}^{-1})$$
$$= n(4.13 \times 10^{-21} \text{J}) \qquad (30.51)$$

At times the conversion from wavenumbers to joules will prove inconvenient. In such cases, Boltzmann's constant can be expressed in units of wavenumbers instead of joules where $k = 0.69503476$ cm^{-1} K^{-1}. In this case, the spectroscopic quantities in wavenumbers can be used directly when evaluating partition functions and other statistical-mechanical expressions.

EXAMPLE PROBLEM 30.5

The vibrational frequency of I_2 is 208 cm^{-1}. What is the probability of I_2 populating the $n = 2$ vibrational level if the molecular temperature is 298 K?

Solution

Molecular vibrational energy levels can be modeled as harmonic oscillators; therefore, this problem can be solved by employing a strategy identical to the one just presented. To evaluate the partition function q, the "trick" used earlier was to write the partition function as a series and use the equivalent series expression:

$$q = \sum_n e^{-\beta \varepsilon_n} = 1 + e^{-\beta hc\tilde{\nu}} + e^{-2\beta hc\tilde{\nu}} + e^{-3\beta hc\tilde{\nu}} + \dots$$

$$= \frac{1}{1 - e^{-\beta hc\tilde{\nu}}}$$

Since $\tilde{\nu} = 208$ cm^{-1} and $T = 298$ K, the partition function is

$$q = \frac{1}{1 - e^{-\beta hc\tilde{\nu}}}$$

$$= \frac{1}{1 - e^{-hc\tilde{\nu}/kT}}$$

$$= \frac{1}{1 - \exp\left[-\left(\dfrac{(6.626 \times 10^{-34} \text{ Js})(3.00 \times 10^{10} \text{ cm s}^{-1})(208 \text{ cm}^{-1})}{(1.38 \times 10^{-23}\text{J K}^{-1})(298 \text{ K})}\right)\right]}$$

$$= \frac{1}{1 - e^{-1}} = 1.58$$

This result is then used to evaluate the probability of occupying the second vibrational state ($n = 2$) as follows:

$$p_2 = \frac{e^{-2\beta hc\tilde{\nu}}}{q}$$

$$= \frac{\exp\left[-2\left(\dfrac{(6.626 \times 10^{-34}\,\text{J s}^{-1})(3.00 \times 10^{10}\,\text{cm s}^{-1})(208\,\text{cm}^{-1})}{(1.38 \times 10^{-23}\,\text{J K}^{-1})(298\,\text{K})}\right)\right]}{1.58}$$

$$= 0.086$$

The last result in Example Problem 30.5 should look familiar. An identical example was worked earlier in this chapter where the energy-level spacings were equal to β^{-1} (case 1 in Figure 30.6) and the probability of populating states for which $n = 0, 1$, and 2 was determined. This previous example in combination with the molecular example just presented illustrates that the exponential term $\beta\varepsilon_n$ in the Boltzmann distribution and partition function can be thought of as a comparative term that describes the ratio of the energy needed to populate a given energy level versus the thermal energy available to the system, as quantified by kT. Energy levels that are significantly higher in energy than kT are not likely to be populated, whereas the opposite is true for energy levels that are small relative to kT.

Example Problem 30.5 is reminiscent of the development presented in the discussion of vibrational and rotational spectroscopy in which the use of the Boltzmann distribution to predict the relative population in vibrational and rotational states and the effect of these populations on vibrational and rotational transition intensities was presented. In addition, the role of the Boltzmann distribution in nuclear magnetic resonance spectroscopy (NMR) is explored in the following example problem.

EXAMPLE PROBLEM 30.6

In NMR spectroscopy, energy separation between spin states is created by placing nuclei in a magnetic field. Protons have two possible spin states: $+1/2$ and $-1/2$. The energy separation between these two states, ΔE, is dependent on the strength of the magnetic field and is given by

$$\Delta E = g_N \beta_N B = (2.82 \times 10^{-26}\,\text{J T}^{-1})B$$

where B is the magnetic field strength in tesla (T). Also, g_N and β_N are the nuclear g-factor and nuclear magneton for a proton, respectively. Early NMR spectrometers employed magnetic field strengths of approximately 1.45 T. What is the ratio of the population between the two spin states given this magnetic field strength and $T = 298$ K?

Solution

Using the Boltzmann distribution, the occupation number for energy levels is given by

$$a_n = \frac{Ne^{-\beta\varepsilon_n}}{q}$$

where N is the number of particles, ε_n is the energy associated with the level of interest, and q is the partition function. Using the preceding equation, the ratio of occupation numbers is given by

$$\frac{a_{+1/2}}{a_{-1/2}} = \frac{\dfrac{Ne^{-\beta\varepsilon_{+1/2}}}{q}}{\dfrac{Ne^{-\beta\varepsilon_{-1/2}}}{q}} = e^{-\beta(\varepsilon_{+1/2}-\varepsilon_{-1/2})} = e^{-\beta\Delta E}$$

Substituting for ΔE and β (and taking care that units cancel), the ratio of occupation numbers is given by

$$\frac{a_{+1/2}}{a_{-1/2}} = e^{-\beta \Delta E} = e - \Delta E/kt = \exp\left[\frac{-(2.82 \times 10^{-26} \text{ J T}^{-1})(1.45 \text{ T})}{(1.38 \times 10^{-23} \text{ J K}^{-1})(298 \text{ K})}\right]$$

$$= e^{-(9.94 \times 10^{-6})} = 0.999990$$

In other words, in this system the energy spacing is significantly smaller than the energy available (kT) such that the higher energy spin state is populated to a significant extent, and is nearly equal in population to that of the lower energy state.

Vocabulary

Boltzmann distribution

Boltzmann's constant

configuration

configurational index

degeneracy

dominant configuration

equilibrium state

microstate

occupation number

partition function

wavenumber

weight

Conceptual Problems

Q30.1 What is the difference between a configuration and a microstate?

Q30.2 What is meant by the "weight" of a configuration?

Q30.3 How does one calculate the number of microstates associated with a given configuration?

Q30.4 Describe what is meant by the phrase "the dominant configuration."

Q30.5 What is an occupation number? How is this number used to describe energy distributions?

Q30.6 What does a partition function represent? Can you describe this term using concepts from probability theory?

Q30.7 Explain the significance of the Boltzmann distribution. What does this distribution describe?

Q30.8 Why is the probability of observing a configuration of energy different from the Boltzmann distribution vanishingly small?

Q30.9 What is degeneracy? Can you conceptually relate the expression for the partition function without degeneracy to that with degeneracy?

Q30.10 How is β related to temperature? What are the units of kT?

Q30.11 How would you expect the partition function to vary with temperature? For example, what should the value of a partition function be at 0 K?

Numerical Problems

Problem numbers in **red** indicate that the solution to the problem is given in the *Student's Solutions Manual*.

P30.1

a. What is the possible number of microstates associated with tossing a coin N times and having it come up H times heads and T times tails?

b. For a series of 1000 tosses, what is the total number of microstates associated with 50% heads and 50% tails?

c. How much less probable is the outcome that the coin will land 40% heads and 60% tails?

P30.2 In Example Problem 30.1, the weights associated with observing 40 heads and 50 heads after flipping a coin 100 times were determined. Perform a similar calculation to determine the weights associated with observing 400 and 500 heads after tossing a coin 1000 times. (Note: Stirling's approximation will be useful in performing these calculations).

P30.3

a. Realizing that the most probable outcome from a series of N coin tosses is $N/2$ heads and $N/2$ tails, what is the expression for W_{max} corresponding to this outcome?

b. Given your answer for part (a), derive the following relationship between the weight for an outcome other than the most probable, W, and W_{max}:

$$\log\left(\frac{W}{W_{max}}\right) = -H \log\left(\frac{H}{N/2}\right) - T \log\left(\frac{T}{N/2}\right)$$

c. We can define the deviation of a given outcome from the most probable outcome using a "deviation index," $\alpha = (H - T)/N$. Show that the number of heads or tails can be expressed as $H = (N/2)(1 + \alpha)$ and $T = (N/2)(1 - \alpha)$.

d. Finally, demonstrate that $W/W_{max} = e^{-N\alpha^2}$.

P30.4 Consider the case of 10 oscillators and 8 quanta of energy. Determine the dominant configuration of energy for this system by identifying energy configurations and calculating the corresponding weights. What is the probability of observing the dominant configuration?

P30.5 Determine the weight associated with the following card hands:

a. Having any five cards

b. Having five cards of the same suit (known as a "flush")

P30.6 For a two-level system, the weight of a given energy distribution can be expressed in terms of the number of systems N and the number of systems occupying the excited state n_1. What is the expression for weight in terms of these quantities?

P30.7 The probability of occupying a given excited state p_i is given by $p_i = n_i/N = e^{-\beta\varepsilon_i}/q$, where n_i is the occupation number for the state of interest, N is the number of particles, and ε_i is the energy of the level of interest. Demonstrate that the preceding expression is independent of the definition of energy for the lowest state.

P30.8 Barometric pressure can be understood using the Boltzmann distribution. The potential energy associated with being a given height above Earth's surface is mgh, where m is the mass of the particle of interest, g is the acceleration due to gravity, and h is height. Using this definition of the potential energy, derive the following expression for pressure: $P = P_o e^{-mgh/kT}$. Assuming that the temperature remains at 298 K, what would you expect the relative pressures of N_2 and O_2 to be at the tropopause, the boundary between the troposphere and stratosphere roughly 11 km above Earth's surface? At Earth's surface, the composition of air is roughly 78% N_2, 21% O_2, and 1% other gases.

P30.9 Consider the following energy-level diagrams:

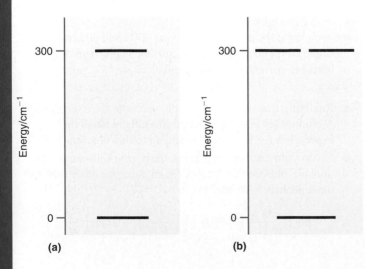

(a)   (b)

a. At what temperature will the probability of occupying the second energy level be 0.15 for the states depicted in part (a) of the figure?

b. Perform the corresponding calculation for the states depicted in part (b) of the figure. Before beginning the calculation, do you expect the temperature to be higher or lower than that determined in part (a) of this problem? Why?

P30.10 Consider the following energy-level diagrams, modified from Problem P30.9 by the addition of another excited state with energy of 600. cm^{-1}:

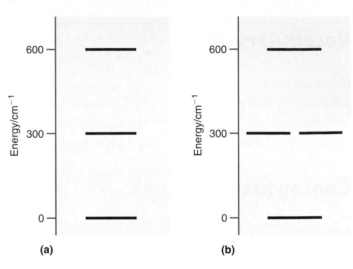

(a)   (b)

a. At what temperature will the probability of occupying the second energy level be 0.15 for the states depicted in part (a) of the figure?

b. Perform the corresponding calculation for the states depicted in part (b) of the figure.

(*Hint*: You may find this problem easier to solve numerically using a spreadsheet program such as Excel.)

P30.11 Consider the following sets of populations for four equally spaced energy levels:

ε/κ (K)	Set A	Set B	Set C
300.	5	3	4
200.	7	9	8
100.	1	1	16
0	33	31	32

a. Demonstrate that the sets have the same energy.

b. Determine which of the sets is the most probable.

c. For the most probable set, is the distribution of energy consistent with a Boltzmann distribution?

P30.12 A set of 13 particles occupies states with energies of 0, 100., and 200. cm^{-1}. Calculate the total energy and number of microstates for the following energy configurations:

a. $a_0 = 8$, $a_1 = 5$, and $a_2 = 0$

b. $a_0 = 9$, $a_1 = 3$, and $a_2 = 1$

c. $a_0 = 10$, $a_1 = 1$, and $a_2 = 2$

Do any of these configurations correspond to the Boltzmann distribution?

P30.13 For a set of nondegenerate levels with energy $\varepsilon/k = 0, 100.,$ and 200. K, calculate the probability of occupying each state when $T = 50., 500.,$ and 5000. K. As the temperature continues to increase, the probabilities will reach a limiting value. What is this limiting value?

P30.14 For two nondegenerate energy levels separated by an amount of energy $\varepsilon/k = 500.$ K, at what temperature will the population in the higher-energy state be $1/2$ that of the lower-energy state? What temperature is required to make the populations equal?

P30.15 Consider a collection of molecules where each molecule has two nondegenerate energy levels that are separated by 6000. cm^{-1}. Measurement of the level populations demonstrates that there are exactly 8 times more molecules in the ground state than in the upper state. What is the temperature of the collection?

P30.16 Consider a molecule having three energy levels as follows:

state	Energy (cm^{-1})	degeneracy
1	0	1
2	500.	3
3	1500.	5

What is the value of the partition function when $T = 300.$ and 3000. K?

P30.17 Consider the molecule described in the previous problem. Imagine a collection of N molecules all at $T = 300.$ K in which one of these molecules is selected. What is the probability that this molecule will be in the lowest-energy state? What is the probability that it will be in the higher-energy state?

P30.18 The emission from C can be used for wavelength calibration of instruments in the ultraviolet. This is generally performed by electron-impact initiated decomposition of a precursor (for example, CF_4) resulting in the production of electronically excited C, which relaxes to the ground electronic state by emitting a photon at 193.09 nm. Suppose one wanted to take another approach where C is heated until 5% of the atoms occupy the electronically excited state. Considering only the lowest-energy ground state and the electronically excited state, what temperature is required to achieve this excited-state population?

P30.19 The ^{13}C nucleus is a spin 1/2 particle as is a proton. However, the energy splitting for a given field strength is roughly 1/4 of that for a proton. Using a 1.45 T magnet as in Example Problem 30.6, what is the ratio of populations in the excited and ground spin states for ^{13}C at 298 K?

P30.20 ^{14}N is a spin 1 particle such that the energy levels are at 0 and $\pm\gamma B\hbar$, where γ is the magnetogyric ratio and B is the strength of the magnetic field. In a 4.8 T field, the energy splitting between any two spin states expressed as the resonance frequency is 14.45 MHz. Determine the occupation numbers for the three spin states at 298 K.

P30.21 When determining the partition function for the harmonic oscillator, the zero-point energy of the oscillator was ignored. Show that the expression for the probability of occupying a specific energy level of the harmonic oscillator with the inclusion of zero-point energy is identical to that ignoring zero-point energy.

P30.22 The vibrational frequency of I_2 is 208 cm^{-1}. At what temperature will the population in the first excited state be half that of the ground state?

P30.23 The vibrational frequency of Cl_2 is 525 cm^{-1}. Will the temperature at which the population in the first excited vibrational state is half that of the ground state be higher or lower relative to I_2 (see Problem P30.22)? What is this temperature?

P30.24 Determine the partition function for the vibrational degrees of freedom of Cl_2 ($\tilde{\nu} = 525$ cm^{-1}) and calculate the probability of occupying the first excited vibrational level at 300. and 1000. K. Determine the temperature at which identical probabilities will be observed for F_2 ($\tilde{\nu} = 917$ cm^{-1}).

P30.25 Hydroxyl radicals are of interest in atmospheric processes due to their oxidative ability. Determine the partition function for the vibrational degrees of freedom for OH ($\tilde{\nu} = 3735$ cm^{-1}) and calculate the probability of occupying the first excited vibrational level at 260. K. Would you expect the probability for occupying the first-excited vibrational level for OD ($\tilde{\nu} = 2721$ cm^{-1}) to be greater or less than for OH?

P30.26 Calculate the partition function for the vibrational energetic degree of freedom for $^1H_2$ where $\tilde{\nu} = 4401$ cm^{-1}. Perform this same calculation for D_2 (or $^2H_2$), assuming the force constant for the bond is the same as in $^1H_2$.

P30.27 A two-level system is characterized by an energy separation of 1.30×10^{-18} J. At what temperature will the population of the ground state be 5 times greater than that of the excited state?

P30.28 Rhodopsin is the pigment in the retina rod cells responsible for vision, and it consists of a protein and the co-factor retinal. Retinal is a π-conjugated molecule which absorbs light in the blue-green region of the visible spectrum, where photon absorption represents the first step in the visual process. Absorption of a photon results in retinal undergoing a transition from the ground- or lowest-energy state of the molecule to the first electronic excited state. Therefore, the wavelength of light absorbed by rhodopsin provides a measure of the ground and excited-state energy gap.

a. The absorption spectrum of rhodopsin is centered at roughly 500. nm. What is the difference in energy between the ground and excited state?

b. At a physiological temperature of 37°C, what is the probability of rhodopsin populating the first excited state? How susceptible do you think rhodopsin is to thermal population of the excited state?

P30.29 The lowest two electronic energy levels of the molecule NO are illustrated here:

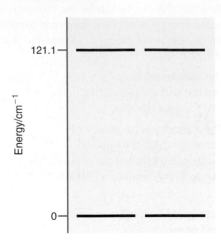

Determine the probability of occupying one of the higher energy states at 100., 500., and 2000. K.

P30.30 Molecular oxygen populating the excited-singlet state ($^1\Delta_g$) can relax to the ground triplet state ($^3\Sigma_g$, which is the lowest energy state) by emitting a 1270 nm photon.

a. Construct the partition function involving the ground and excited-singlet state of molecular oxygen.

b. What temperature is required to have 10% population in the excited-singlet state?

P30.31 The simplest polyatomic molecular ion is H_3^+, which can be thought of as molecular hydrogen with an additional proton. Infrared spectroscopic studies of interstellar space have identified this species in the atmosphere of Jupiter and other interstellar bodies. The rotational-vibrational spectrum of the ν_2 band of H_3^+ was first measured in the laboratory by T. Oka in 1980 [*Phys. Rev. Lett.* 45 (1980): 531]. The spectrum consists of a series of transitions extending from 2450 to 2950 cm^{-1}. Employing an average value of 2700. cm^{-1} for the energy of the first excited rotational-vibrational state of this molecule relative to the ground state, what temperature is required for 10% of the molecules to populate this excited state?

Web-Based Simulations, Animations, and Problems

W30.1 In this simulation the behavior of the partition function for a harmonic oscillator with temperature and oscillator frequency is explored. The variation in q with temperature for an oscillator where $\widetilde{\nu} = 1000$ cm^{-1} is studied, and variation in the individual level contributions to the partition function is studied. In addition, the change in level contributions as the oscillator frequency is varied is depicted. This simulation provides insight into the elements of the partition function and the variation of this function with temperature and energy-level spacings.

For Further Reading

Chandler, D. *Introduction to Modern Statistical Mechanics.* New York: Oxford, 1987.

Hill, T. *Statistical Mechanics. Principles and Selected Applications.* New York: Dover, 1956.

McQuarrie, D. *Statistical Mechanics.* New York: Harper & Row, 1973.

Nash, L. K. "On the Boltzmann Distribution Law." *J. Chemical Education* 59 (1982): 824.

Nash, L. K. *Elements of Statistical Thermodynamics.* San Francisco: Addison-Wesley, 1972.

Noggle, J. H. *Physical Chemistry.* New York: HarperCollins, 1996.

31

Ensemble and Molecular Partition Functions

The relationship between the microscopic description of individual molecules and the macroscopic properties of a collection of molecules is a central concept in statistical mechanics. In this chapter, the relationship between the partition function that describes a collection of noninteracting molecules and the partition function describing an individual molecule is developed. We demonstrate that the molecular partition function can be decomposed into the product of partition functions for each energetic degree of freedom and that the functional form of these partition functions is derived. The concepts outlined in this chapter provide the basic foundation on which statistical thermodynamics resides.

31.1 The Canonical Ensemble

An **ensemble** is defined as a large collection of identical units or replicas of a system. For example, a mole of water can be envisioned as an ensemble with Avogadro's number of identical units of water molecules. The ensemble provides a theoretical concept by which the microscopic properties of matter can be related to the corresponding thermodynamic system properties as expressed in the following postulate:

> The average value for a property of the ensemble corresponds to the time-averaged value for the corresponding macroscopic property of the system.

What does this postulate mean? Imagine the individual units of the ensemble sampling the available energy space; the energy content of each unit is measured at a single time, and the measured unit energies are used to determine the average energy for the ensemble. According to the postulate, this energy will be equivalent to the average energy of the ensemble as measured over time. This idea, first formulated by

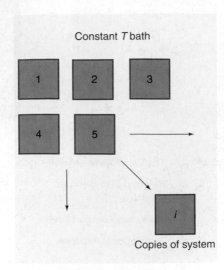

Constant *T* bath

Copies of system

FIGURE 31.1
The canonical ensemble is comprised of a collection of identical systems having fixed temperature, volume, and number of particles. The units are embedded in a constant *T* bath. The arrows indicate that an infinite number of copies of the system comprises the ensemble.

J. W. Gibbs in the late 1800s, lies at the heart of statistical thermodynamics and is explored in this chapter.

To connect ensemble average values and thermodynamic properties, we begin by imagining a collection of identical copies of the system as illustrated in Figure 31.1. These copies of the system are held fixed in space such that they are distinguishable. The volume *V*, temperature *T*, and number of particles in each system *N* are constant. An ensemble in which *V*, *T*, and *N* are constant is referred to as a **canonical ensemble.** The term *canonical* means "by common practice" because this is the ensemble employed unless the problem of interest dictates that other variables be kept constant. Note that other quantities can be constant to construct other types of ensembles. For example, if *N*, *V*, and energy are held fixed, the corresponding ensemble is referred to as *microcanonical.* However, for the purposes of this text, the canonical ensemble will prove sufficient.

In the canonical ensemble, each ensemble member is embedded in a temperature bath such that the total ensemble energy is constant. Furthermore, the walls that define the volume of the units can conduct heat, allowing for energy exchange with the surroundings. The challenge is to link the statistical development presented in the previous chapter to a similar statistical description for this ensemble. We begin by considering the total energy of the ensemble E_c which is given by

$$E_c = \sum_i a_{(c)i} E_i \tag{31.1}$$

In Equation (31.1), the terms $a_{(c)i}$ are the occupation numbers corresponding to the number of ensemble members having energy E_i. Proceeding exactly as in the previous chapter, the weight W_c associated with a specific configuration of energy among the N_c members of the ensemble is given by

$$W_c = \frac{N_c!}{\prod_i a_{(c)i}!} \tag{31.2}$$

This relationship can be used to derive the probability of finding an ensemble unit at energy E_i:

$$p(E_i) = \frac{W_i e^{-\beta E_i}}{Q} \tag{31.3}$$

Equation (31.3) looks very similar to the probability expression derived previously. In this equation, W_i can be thought of as the number of states present at a given energy E_i. The quantity Q in Equation (31.3) is referred to as the **canonical partition function** and is defined as follows:

$$Q = \sum_n e^{-\beta E_n} \tag{31.4}$$

In Equation (31.4), the summation is over all energy levels. The probability defined in Equation (31.3) is dependent on two factors: W_i, or the number of states present at a given energy that will increase with energy, and a Boltzmann term $e^{-\beta E_i}/Q$ that describes the probability of an ensemble unit having energy E_i that decreases exponentially with energy. The generic behavior of each term with energy is depicted in Figure 31.2. The product of these terms will reach a maximum corresponding to the average ensemble energy. The figure illustrates that an individual unit of the ensemble will have an energy that is equal to or extremely close to the average energy, and that units having energy far from this value will be exceedingly rare. We know this to be the case from experience. Imagine a swimming pool filled with water divided up into one-liter units. If the thermometer at the side of the pool indicates that the water temperature is 18°C, someone diving into the pool will not be worried that the liter of water immediately under his or her head will spontaneously freeze. That is, the temperature measured in one part of the pool is sufficient to characterize the temperature of the water in any part of the pool. Figure 31.2 provides an illustration of the statistical aspects underlying this expectation.

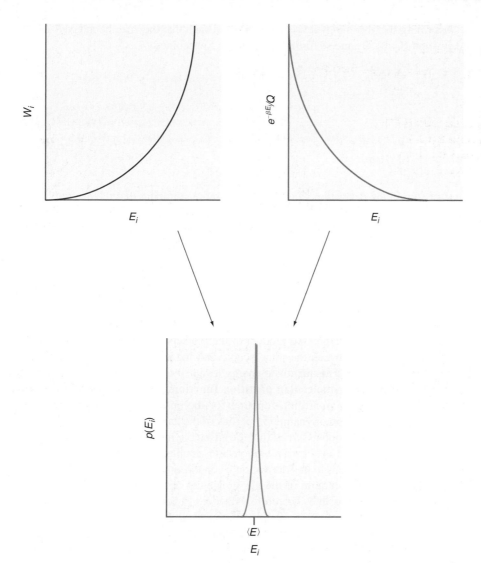

FIGURE 31.2
For the canonical ensemble, the probability of a member of the ensemble having a given energy is dependent on the product of W_i, the number of states present at a given energy, and the Boltzmann distribution function for the ensemble. The product of these two factors results in a probability distribution that is peaked about the average energy $\langle E \rangle$.

The vast majority of systems in the ensemble will have energy $\langle E \rangle$; therefore, the thermodynamic properties of the unit are representative of the thermodynamic properties of the ensemble, demonstrating the link between the microscopic unit and the macroscopic ensemble. To make this connection mathematically exact the canonical partition function Q must be related to the partition function describing the individual members of the ensemble q.

31.2 Relating Q to q for an Ideal Gas

In relating the canonical partition function Q to the partition function describing the members of the ensemble q our discussion is limited to systems consisting of independent "ideal" particles in which the interactions between particles is negligible (for example, an ideal gas). The relationship between Q and q is derived by considering an ensemble made up of two distinguishable units, A and B, as illustrated in Figure 31.3. For this simple ensemble, the partition function is

$$Q = \sum_n e^{-\beta E_n} = \sum_n e^{-\beta(\varepsilon_{A_n} + \varepsilon_{B_n})} \qquad \textbf{(31.5)}$$

In this expression, ε_{A_n} and ε_{B_n} refer to the energy levels associated with units A and B, respectively.

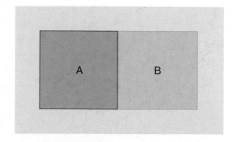

FIGURE 31.3
A two-unit ensemble. In this ensemble, the two units, A and B, are distinguishable.

Assuming that the energy levels are quantized such that they can be indexed A_0, A_1, A_2, etc. and B_0, B_1, B_2 and so forth, Equation (31.5) becomes

$$
\begin{aligned}
Q &= \sum_n e^{-\beta(\varepsilon_{A_n} + \varepsilon_{B_n})} = e^{-\beta(\varepsilon_{A_0} + \varepsilon_{B_0})} + e^{-\beta(\varepsilon_{A_0} + \varepsilon_{B_1})} + e^{-\beta(\varepsilon_{A_0} + \varepsilon_{B_2})} + \ldots \\
&\quad + e^{-\beta(\varepsilon_{A_1} + \varepsilon_{B_0})} + e^{-\beta(\varepsilon_{A_1} + \varepsilon_{B_1})} + e^{-\beta(\varepsilon_{A_1} + \varepsilon_{B_2})} + \ldots \\
&\quad + e^{-\beta(\varepsilon_{A_2} + \varepsilon_{B_0})} + e^{-\beta(\varepsilon_{A_2} + \varepsilon_{B_1})} + e^{-\beta(\varepsilon_{A_2} + \varepsilon_{B_2})} + \ldots \\
&= (e^{-\beta\varepsilon_{A_0}} + e^{-\beta\varepsilon_{A_1}} + e^{-\beta\varepsilon_{A_2}} + \ldots)(e^{-\beta\varepsilon_{B_0}} + e^{-\beta\varepsilon_{B_1}} + e^{-\beta\varepsilon_{B_2}} + \ldots) \\
&= (q_A)(q_B) \\
&= q^2
\end{aligned}
$$

The last step in the derivation is accomplished by recognizing that the ensemble units are identical such that the partition functions are also identical. Extending the preceding result to a system with N distinguishable units, the canonical partition function is found to be simply the product of unit partition functions

$$Q = q^N \quad \text{for } N \text{ distinguishable units} \tag{31.6}$$

Thus far no mention has been made of the size of the identical systems comprising the ensemble. The systems can be as small as desired, including just a single molecule. Taking the single-molecule limit, a remarkable conclusion is reached: the canonical ensemble is nothing more than the product of the molecular partition functions. This is the direct connection between the microscopic and macroscopic perspectives that we have been searching for. The quantized energy levels of the molecular (or atomic) system are embedded in the **molecular partition function q,** and this partition function can be used to define the partition function for the ensemble Q. Finally, Q can be directly related to the thermodynamic properties of the ensemble.

The preceding derivation assumed that the ensemble members were distinguishable. This might be the case for a collection of molecules coupled to a surface where they cannot move, but what happens to this derivation when the ensemble is in the gaseous state? Clearly, the translational motion of the gas molecules will make identification of each individual molecule impossible. Therefore, how does Equation (31.6) change if the units are indistinguishable? A simple counting example will help to answer this question. Consider three distinguishable oscillators (A, B, and C) with three total quanta of energy as described in the previous chapter. The dominant configuration of energy was with the oscillators in three separate energy states, denoted "2, 1, 0." The energy states relative to the oscillators can be arranged in six different ways:

A	B	C
2	1	0
2	0	1
1	2	0
0	2	1
1	0	2
0	1	2

However, if the three oscillators are indistinguishable, there is no difference among the arrangements listed. In effect, there is only one arrangement of energy that should be counted. This problem was encountered in the chapter on probability when discussing indistinguishable particles, and in such cases the total number of permutations was divided by $N!$ where N was the number of units in the collection. Extending this logic to a molecular ensemble dictates that the canonical partition function for indistinguishable particles have the following form:

$$Q = \frac{q^N}{N!} \quad \text{for } N \text{ indistinguishable units} \tag{31.7}$$

Equation (31.7) is correct in the limit for which the number of energy levels available is significantly greater than the number of particles. This discussion of statistical mechanics is limited to systems for which this is true, and the validity of this statement is demonstrated later in this chapter. It is also important to keep in mind that Equation (31.7) is limited to ideal systems of noninteracting particles such as an ideal gas.

31.3 Molecular Energy Levels

The relationship between the canonical and molecular partition functions provides the link between the microscopic and macroscopic descriptions of the system. The molecular partition function can be evaluated by considering molecular energy levels. For polyatomic molecules, there are four **energetic degrees of freedom** to consider in constructing the molecular partition function:

1. Translation
2. Rotation
3. Vibration
4. Electronic

Assuming the energetic degrees of freedom are not coupled, the total molecular partition function that includes all of these degrees of freedom can be decomposed into a product of partition functions corresponding to each degree of freedom. An equivalent approach is taken in quantum mechanics when separating the molecular Hamiltonian into translational, rotational, and vibrational components. Let ε_{Total} represent the energy associated with a given molecular energy level. This energy will depend on the translational, rotational, vibrational, and electronic level energies as follows:

$$\varepsilon_{Total} = \varepsilon_T + \varepsilon_R + \varepsilon_V + \varepsilon_E \tag{31.8}$$

Recall that the molecular partition function is obtained by summing over molecular energy levels. Using the expression for the total energy and substituting into the expression for the partition function, the following expression is obtained:

$$
\begin{aligned}
q_{Total} &= \sum g_{Total} e^{-\beta \varepsilon_{Total}} \\
&= \sum (g_T g_R g_V g_E) e^{-\beta(\varepsilon_T + \varepsilon_T + \varepsilon_T + \varepsilon_E)} \\
&= \sum (g_T e^{-\beta \varepsilon_T})(g_R e^{-\beta \varepsilon_R})(g_V e^{-\beta \varepsilon_V})(g_E e^{-\beta \varepsilon_E}) \\
&= q_T q_R q_V q_E
\end{aligned}
\tag{31.9}
$$

This relationship demonstrates that the total molecular partition function is simply the product of partition functions for each molecular energetic degree of freedom. Using this definition for the molecular partition function, the final relationships of interest are

$$Q_{Total} = q_{Total}^N \ (\text{distinguishable}) \tag{31.10}$$

$$Q_{Total} = \frac{1}{N!} q_{Total}^N \ (\text{indistinguishable}) \tag{31.11}$$

All that remains to derive are partition functions for each energetic degree of freedom, a task that is accomplished in the remainder of this chapter.

31.4 Translational Partition Function

Translational energy levels correspond to the translational motion of atoms or molecules in a container of volume V. Rather than work directly in three dimensions, a one-dimensional model is first employed and then later extended to three dimensions. From quantum mechanics, the energy levels of a molecule confined to a box were described by the "particle-in-a-box" model as illustrated in Figure 31.4. In this figure, a particle with mass m is free to move in the domain $0 \le x \le a$, where a is the length of the box. Using the expression for the energy levels provided in Figure 31.4, the partition function for translational energy in one dimension becomes

$$q_{T,1D} = \sum_{n=1}^{\infty} e^{\frac{-\beta n^2 h^2}{8ma^2}} \tag{31.12}$$

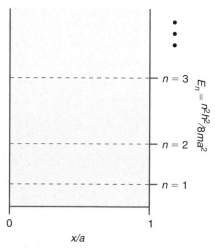

FIGURE 31.4
Particle-in-a-box model for translational energy levels.

Notice that the summation consists of an infinite number of terms. Furthermore, a closed-form expression for this series does not exist such that it appears one must evaluate the sum directly. However, a way around this apparently impossible task becomes evident when the spacing between energy translational energy states is considered, as illustrated in the following example.

EXAMPLE PROBLEM 31.1

What is the difference in energy between the $n = 2$ and $n = 1$ states for molecular oxygen constrained by a one-dimensional box having a length of 1.00 cm?

Solution

The energy difference is obtained by using the expression for the one-dimensional particle-in-a-box model as follows:

$$\Delta E = E_2 - E_1 = 3E_1 = \frac{3h^2}{8ma^2}$$

The mass of an O_2 molecule is 5.31×10^{-26} kg such that

$$\Delta E = \frac{3(6.626 \times 10^{-34}\ \text{J s})^2}{8(5.31 \times 10^{-26}\ \text{kg})(0.01\ \text{m})^2}$$

$$= 3.10 \times 10^{-38}\ \text{J}$$

Converting to units of cm^{-1}:

$$\Delta E = \frac{3.10 \times 10^{-38}\ \text{J}}{hc} = 1.56 \times 10^{-15}\ \text{cm}^{-1}$$

At 298 K, the amount of thermal energy available as given by the product of Boltzmann's constant and temperature kT is 207 cm^{-1}. Clearly, the spacings between translational energy levels are extremely small relative to kT at room temperature.

Because numerous translational energy levels are accessible at room temperature, the summation in Equation (31.12) can be replaced by integration with negligible error:

$$q_T = \sum e^{-\beta \alpha n^2} \approx \int_0^\infty e^{-\beta \alpha n^2}\,dn \tag{31.13}$$

In this expression, the following substitution was made to keep the collection of constant terms compact:

$$\alpha = \frac{h^2}{8ma^2} \tag{31.14}$$

The integral in Equation (31.13) is readily evaluated (see Appendix A, Math Supplement):

$$q_T \approx \int_0^\infty e^{-\beta \alpha n^2}\,dn = \frac{1}{2}\sqrt{\frac{\pi}{\beta \alpha}} \tag{31.15}$$

Substituting in for α, the **translational partition function** in one dimension becomes

$$q_{T,1D} = \left(\frac{2\pi m}{h^2 \beta}\right)^{1/2} a \tag{31.16}$$

This expression can be simplified by defining the **thermal de Broglie wavelength,** or simply the **thermal wavelength,** as follows:

$$\Lambda = \left(\frac{h^2 \beta}{2\pi m}\right)^{1/2} \tag{31.17}$$

such that

$$q_{T,1D} = \frac{a}{\Lambda} = \left(\frac{2\pi m}{\beta}\right)^{1/2}\frac{a}{h} = (2\pi mkT)^{1/2}\frac{a}{h} \tag{31.18}$$

Referring to Λ as the thermal wavelength reflects the fact that the average momentum of a gas particle p is equal to $(mkT)^{1/2}$. Therefore, Λ is essentially h/p, or the de Broglie wavelength of the particle. The translational degrees of freedom are considered separable; therefore, the three-dimensional translational partition function is the product of one-dimensional partition functions for each dimension:

$$q_{T,3D} = q_{T_x}q_{T_y}q_{T_z}$$

$$= \left(\frac{a_x}{\Lambda}\right)\left(\frac{a_y}{\Lambda}\right)\left(\frac{a_z}{\Lambda}\right)$$

$$= \left(\frac{1}{\Lambda}\right)^3 a_x a_y a_z$$

$$= \left(\frac{1}{\Lambda}\right)^3 V$$

$$q_{T,3D} = \frac{V}{\Lambda^3} = (2\pi mkT)^{3/2}\frac{V}{h^3} \tag{31.19}$$

where V is volume and Λ is the thermal wavelength [Equation (31.17)]. Notice that the translational partition function depends on both V and T. Recall the discussion from the previous chapter in which the partition function was described conceptually as providing a measure of the number of energy states available to the system at a given temperature. The increase in q_T with volume reflects the fact that as volume is increased, the translational energy-level spacings decrease such that more states are available for population at a given T. Given the small energy spacings between translational energy levels relative to kT at room temperature, we might expect that at room temperature a significant number of translational energy states are accessible. The following example provides a test of this expectation.

EXAMPLE PROBLEM 31.2

What is the translational partition function for Ar confined to a volume of 1.00 L at 298 K?

Solution

Evaluation of the translational partition function is dependent on determining the thermal wavelength [Equation (31.17)]:

$$\Lambda = \left(\frac{h^2\beta}{2\pi m}\right)^{1/2} = \frac{h}{(2\pi mkT)^{1/2}}$$

The mass of Ar is 6.63×10^{-26} kg. Using this value for m, the thermal wavelength becomes

$$\Lambda = \frac{6.626 \times 10^{-34}\,\text{J s}}{\left(2\pi(6.63 \times 10^{-26}\,\text{kg})(1.38 \times 10^{-23}\,\text{JK}^{-1})(298\,\text{K})\right)^{1/2}}$$

$$= 1.60 \times 10^{-11}\,\text{m}$$

The units of volume must be such that the partition function is unitless. Therefore, conversion of volume to units of cubic meters (m^3) is performed as follows:

$$V = 1.00\,\text{L} = 1000\,\text{mL} = 1000\,\text{cm}^3\left(\frac{1\,\text{m}}{100\,\text{cm}}\right)^3 = 0.001\,\text{m}^3$$

The partition function is simply the volume divided by the thermal wavelength cubed:

$$q_{T,3D} = \frac{V}{\Lambda^3} = \frac{0.001\,\text{m}^3}{(1.60 \times 10^{-11}\,\text{m})^3} = 2.44 \times 10^{29}$$

The magnitude of the translational partition function determined in Example Problem 31.2 illustrates that a vast number of translation energy states are available at room temperature. In fact, the number of accessible states is roughly 10^6 times larger than Avogadro's number, illustrating that the assumption that many more states are available relative to units in the ensemble (Section 31.2) is reasonable.

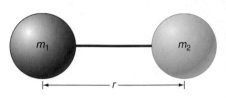

FIGURE 31.5
Schematic representation of a diatomic molecule, consisting of two masses (m_1 and m_2) joined by a chemical bond with the separation of atomic centers equal to the bond length r.

31.5 Rotational Partition Function: Diatomics

A **diatomic** molecule consists of two atoms joined by a chemical bond as illustrated in Figure 31.5. In treating rotational motion of diatomic molecules, the rigid rotor approximation is employed in which the bond length is assumed to remain constant during rotational motion and effects such as centrifugal distortion are neglected.

In deriving the rotational partition function, an approach similar to that used in deriving the translational partition function is employed. Within the rigid-rotor approximation, the quantum mechanical description of rotational energy levels for diatomic molecules dictates that the energy of a given rotational state E_J is dependent on the rotational quantum number J:

$$E_J = hcBJ(J + 1) \quad \text{for } J = 0, 1, 2, \ldots \quad (31.20)$$

where J is the quantum number corresponding to rotational energy level and can take on integer values beginning with zero. The quantity B is the **rotational constant** and is given by

$$B = \frac{h}{8\pi^2 cI} \quad (31.21)$$

where I is the moment of inertia, which is equal to

$$I = \mu r^2 \quad (31.22)$$

In the expression for the moment of inertia, r is the distance separating the two atomic centers and μ is the reduced mass, which for a diatomic molecule consisting of atoms having masses m_1 and m_2 is equal to

$$\mu = \frac{m_1 m_2}{m_1 + m_2} \quad (31.23)$$

Because diatomic molecules differ depending on the masses of atoms in the molecule and the bond length, the value of the rotational constant is molecule dependent. Using the preceding expression for the rotational energy, the rotational partition function can be constructed by simply substituting into the general form of the molecular partition function:

$$q_R = \sum_J g_J e^{-\beta E_J} = \sum_J g_J e^{-\beta hcBJ(J+1)} \quad (31.24)$$

In this expression, the energies of the levels included in the summation are given by $hcBJ(J + 1)$. However, notice that the expression for the rotational partition function contains an addition term, g_J, that represents the number of rotational states present at a given energy level, or the degeneracy of the rotational energy level. To determine the degeneracy, consider a rigid rotor and the time-independent Schrödinger equation:

$$H\psi = E\psi \quad (31.25)$$

For the rigid rotor, the Hamiltonian (H) is proportional to the square of the total angular momentum given by the operator $\hat{l}^2$. The eigenstates of this operator are the spherical harmonics with the following eigenvalues:

$$\hat{l}^2\psi = \hat{l}^2 Y_{l,m}(\theta, \phi) = \hbar^2 l(l + 1)Y_{l,m}(\theta, \phi) \quad (31.26)$$

In this expression, l is a quantum number corresponding to total angular momentum ranging from $0, 1, 2, \ldots$, to infinity. The spherical harmonics are also eigenfunctions of the $\hat{l}_z$ operator corresponding to the z component of the angular momentum. The corresponding eigenvalues employing the $\hat{l}_z$ operator are given by

$$\hat{l}_z Y_{l,m}(\theta, \phi) = \hbar m Y_{l,m}(\theta, \phi) \tag{31.27}$$

Possible values for the quantum number m in Equation (31.27) are dictated by the quantum number l:

$$m = -l \ldots 0 \ldots l \quad \text{or} \quad (2l + 1) \tag{31.28}$$

Thus, the degeneracy of the rotational energy levels originates from the quantum number m because all values of m corresponding to a given quantum number l will have the same total angular momentum and, therefore, the same energy. Using the value of $(2l + 1)$ or $(2J + 1)$ for the degeneracy, the rotational partition function is

$$q_R = \sum_J (2J + 1)e^{-\beta hcBJ(J+1)} \tag{31.29}$$

As written, evaluation of Equation (31.29) involves summation over all rotational states. A similar issue was encountered when the expression for the translational partition function was evaluated. The spacings between translational levels were very small relative to kT such that the partition function could be evaluated by integration rather than discrete summation. Are the rotational energy-level spacings also small relative to kT such that integration can be performed instead of summation?

To answer this question, consider the energy-level spacings for the rigid rotor presented in Figure 31.6 The energy of a given rotational state (in units of the rotational constant B) are presented as a function of the rotational quantum number J. The energy-level spacings are multiples of B. The value of B will vary depending on the molecule of interest, with representative values provided in Table 31.1. Inspection of the table reveals a few interesting trends. First, the rotational constant depends on the atomic mass, with an increase in atomic mass resulting in a reduction in the rotational constant. Second, the values for B are quite different; therefore, any comparison of rotational state energies to kT will depend on the diatomic of interest. For example, at 298 K, $kT = 207$ cm^{-1}, which is roughly equal to the energy of the $J = 75$ level of I_2. For this species, the energy-level spacings are clearly much smaller than kT and integration of the partition function is appropriate. However, for H_2 the $J = 2$ energy level is greater than kT so that integration would be inappropriate, and evaluation of the partition function by direct summation must be performed. In the remainder of this chapter, we assume that integration of the rotational partition function is appropriate unless stated otherwise.

With the assumption that the rotational energy-level spacings are small relative to kT, evaluation of the rotational partition function is performed with integration over the rotational states:

$$q_R = \int_0^\infty (2J + 1)e^{-\beta hcBJ(J+1)} dJ \tag{31.30}$$

Evaluation of the preceding integral is simplified by recognizing the following:

$$\frac{d}{dJ}e^{-\beta hcBJ(J+1)} = -\beta hcB(2J + 1)e^{-\beta hcBJ(J+1)}$$

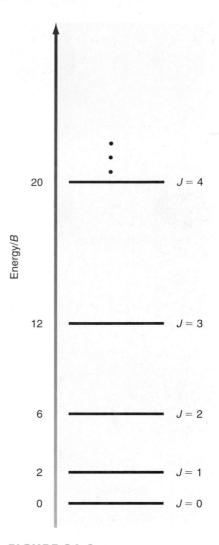

FIGURE 31.6
Rotational energy levels as a function of the rotational quantum number J. The energy of a given rotational state is equal to $BJ(J + 1)$.

TABLE 31.1	Rotational Constants for Some Representative Diatomic Molecules		
Molecule	B (cm^{-1})	Molecule	B (cm^{-1})
H^{35}Cl	10.595	H$_2$	60.853
H^{37}Cl	10.578	^{14}N^{16}O	1.7046
D^{35}Cl	5.447	^{127}I^{127}I	0.03735

Source: Herzberg, G., *Molecular Spectra and Molecular Structure, Volume 1: Spectra of Diatomic Molecules.* Melbourne, FL: Krieger Publishing, 1989.

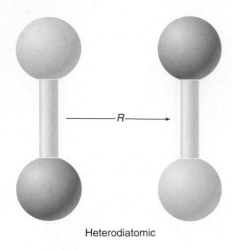

Heterodiatomic

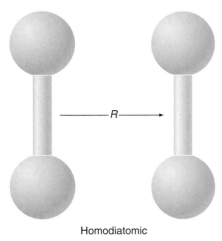

Homodiatomic

FIGURE 31.7
A 180° rotation of heterodiatomic and homodiatomic molecules.

Using this relationship, the expression for the **rotational partition function** can be rewritten and the result evaluated as follows:

$$q_R = \int_0^\infty (2J + 1)e^{-\beta hcBJ(J+1)}\,dJ = \int_0^\infty \frac{-1}{\beta hcB}\frac{d}{dJ}e^{-\beta hcBJ(J+1)}\,dJ$$

$$= \frac{-1}{\beta hcB}e^{-\beta hcBJ(J+1)}\Big|_0^\infty = \frac{1}{\beta hcB}$$

$$q_R = \frac{1}{\beta hcB} = \frac{kT}{hcB} \tag{31.31}$$

31.5.1 The Symmetry Number

The expression for the rotational partition function of a diatomic molecule provided in the previous section is correct for heterodiatomic species in which the two atoms comprising the diatomic are not equivalent. HCl is a heterodiatomic species because the two atoms in the diatomic, H and Cl, are not equivalent. However, the expression for the rotational partition function must be modified when applied to homodiatomic molecules such as N_2. A simple illustration of why such a modification is necessary is presented in Figure 31.7. In the figure, rotation of the heterodiatomic results in a species that is distinguishable from the molecule before rotation. However, the same 180° rotation applied to a homodiatomic results in a configuration that is equivalent to the prerotation form. This difference in behavior is similar to the differences between canonical partition functions for distinguishable and indistinguishable units. In the partition function case, the result for the distinguishable case was divided by $N!$ to take into account the "overcounting" of nonunique microstates encountered when the units are indistinguishable. In a similar spirit, for homodiatomic species the number of classical rotational states (i.e., distinguishable rotational configurations) is overcounted by a factor of 2.

To correct our rotational partition function for overcounting, we can simply divide the expression for the rotational partition function by the number of equivalent rotational configurations. This factor is known as the **symmetry number** σ and is incorporated into the partition function as follows:

$$q_R = \frac{1}{\sigma \beta hcB} = \frac{kT}{\sigma hcB} \tag{31.32}$$

The concept of a symmetry number can be extended to molecules other than diatomics. For example, consider a trigonal pyramidal molecule such as NH_3 as illustrated in Figure 31.8.

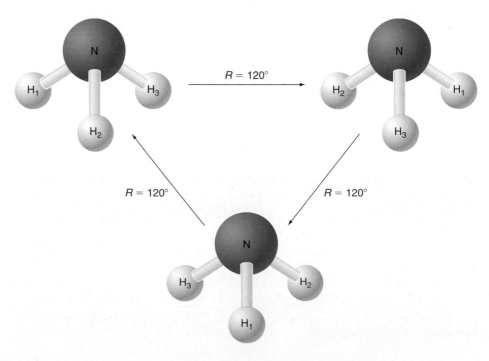

FIGURE 31.8
Rotational configurations for NH_3.

Imagine that performing a 120° rotation about an axis through the nitrogen atom and the center of the triangle made the three hydrogens. The resulting configuration would be exactly equivalent to the previous configuration before rotation. Furthermore, a second 120° rotation would produce a third configuration. A final 120° would result in the initial prerotation configuration. Therefore, NH_3 has three equivalent rotational configurations; therefore, $\sigma = 3$.

EXAMPLE PROBLEM 31.3

What is the symmetry number for methane (CH_4)?

Solution

To determine the number of equivalent rotational configurations, we will proceed in a fashion similar to that employed for NH_3. The tetrahedral structure of methane is shown in the following figure:

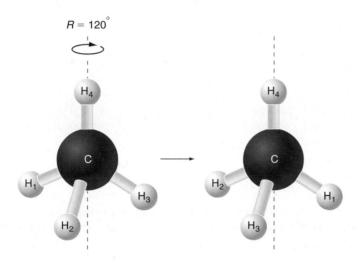

Similar to NH_3, three equivalent configurations can be generated by 120° rotation about the axis depicted by the dashed line in the figure. Furthermore, we can draw four such axes of rotation aligned with each of the four C—H bonds. Therefore, there are 12 total rotational configurations for CH_4 corresponding to $\sigma = 12$.

31.5.2 Rotational Level Populations and Spectroscopy

There is a direct relationship between the populations in various rotational energy levels and rotational-vibrational infrared absorption intensities. With the rotational partition function, we are in a position to explore this relationship in detail. The probability of occupying a given rotational energy level p_J is given by

$$p_J = \frac{g_J e^{-\beta hcBJ(J+1)}}{q_R} = \frac{(2J+1)e^{-\beta hcBJ(J+1)}}{q_R} \tag{31.33}$$

Previously $H^{35}Cl$ where $B = 10.595 \text{ cm}^{-1}$ was employed to illustrate the relationship between p_J and absorption intensity. At 300. K the rotational partition function for $H^{35}Cl$ is

$$q_R = \frac{1}{\sigma \beta hcB} = \frac{kT}{\sigma hcB} = \frac{(1.38 \times 10^{-23} \text{J K}^{-1})(300. \text{ K})}{(1)(6.626 \times 10^{-34} \text{J s})(3.00 \times 10^{10} \text{cm s}^{-1})(10.595 \text{ cm}^{-1})}$$
$$= 19.7 \tag{31.34}$$

With q_R, the level probabilities can be readily determined using Equation (31.33), and the results of this calculation are presented in Figure 31.9. The intensity of the P and R branch transitions in a rotational-vibrational infrared absorption spectrum are proportional to the probability of occupying a given J level. This dependence is reflected by the evolution in

FIGURE 31.9
Probability of occupying a rotational
energy level p_J as a function of rotational
quantum number J for H^{35}Cl at 300. K.

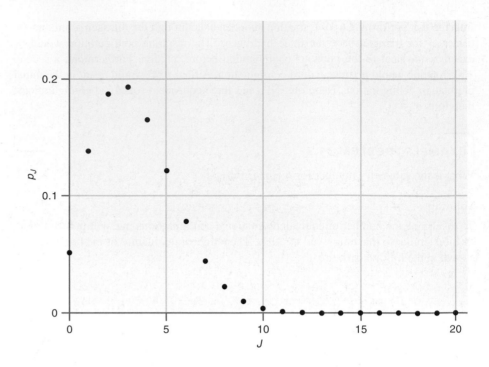

rotational-vibrational transition intensity as a function of J. The transition moment
demonstrates modest J dependence as well such that the correspondence between transition intensity and rotational level population is not exact.

EXAMPLE PROBLEM 31.4

In a rotational spectrum of HBr ($B = 8.46$ cm^{-1}), the maximum intensity transition in
the R-branch corresponds to the $J = 4$ to 5 transition. At what temperature was the
spectrum obtained?

Solution

The information provided for this problem dictates that the $J = 4$ rotational energy
level was the most populated at the temperature at which the spectrum was taken. To
determine the temperature, we first determine the change in occupation number for
the rotational energy level, a_J, versus J as follows:

$$a_J = \frac{N(2J + 1)e^{-\beta hcBJ(J+1)}}{q_R} = \frac{N(2J + 1)e^{-\beta hcBJ(J+1)}}{\left(\dfrac{1}{\beta hcB}\right)}$$

$$= N\beta hcB(2J + 1)e^{-\beta hcBJ(J+1)}$$

Next, we take the derivative of a_J with respect to J and set the derivative equal to zero
to find the maximum of the function:

$$\frac{da_J}{dJ} = 0 = \frac{d}{dJ}N\beta hcB(2J + 1)e^{-\beta hcBJ(J+1)}$$

$$0 = \frac{d}{dJ}(2J + 1)e^{-\beta hcBJ(J+1)}$$

$$0 = 2e^{-\beta hcBJ(J+1)} - \beta hcB(2J + 1)^2 e^{-\beta hcBJ(J+1)}$$

$$0 = 2 - \beta hcB(2J + 1)^2$$

$$2 = \beta hcB(2J + 1)^2 = \frac{hcB}{kT}(2J + 1)^2$$

$$T = \frac{(2J + 1)^2 hcB}{2k}$$

Substitution of $J = 4$ into the preceding expression results in the following temperature at which the spectrum was obtained:

$$T = \frac{(2J + 1)^2 hcB}{2k}$$

$$= \frac{(2(4) + 1)^2 (6.626 \times 10^{-34}\,\text{J s})(3.00 \times 10^{10}\,\text{cm s}^{-1})(8.46\,\text{cm}^{-1})}{2(1.38 \times 10^{-23}\,\text{J K}^{-1})}$$

$$= 4943\,\text{K}$$

31.5.3 An Advanced Topic: The Rotational States of H_2

The rotational-level distribution of H_2 provides an elegant example of the influence of molecular symmetry on the partition function. Molecular hydrogen exists in two forms, one in which the nuclear spins are paired (***para*-hydrogen**) and one in which the spins are aligned (***ortho*-hydrogen**). Because the hydrogen nuclei are spin $1/2$ particles, they are fermions. The Pauli exclusion principle dictates that when two identical fermions interchange position, the overall wave function describing the system must be antisymmetric (or change sign) with interchange. The wave function in this case can be separated into a product of spin and rotational components. The spin component of the wave function is considered first. For *para*-hydrogen, rotation results in the interchange of two nuclei (A and B) having opposite spin (α and β) such that the spin component of the wave function should be antisymmetric with nuclei interchange due to rotation. This requirement is accomplished using the following linear combination of nuclear spin states:

$$\psi_{spin,\,para} = \alpha(A)\beta(B) - \alpha(B)\beta(A)$$

Interchange of the nuclear labels A and B corresponding to rotation will result in the preceding wave function changing sign such that the spin component of the *para*-hydrogen wave function is antisymmetric with respect to interchange.

For *ortho*-hydrogen, rotation results in the interchange of two nuclei with the same spin; therefore, the spin wave function should be symmetric. Three combinations of nuclear spin states meet this requirement:

$$\psi_{spin,\,ortho} = \{\alpha(A)\alpha(B), \beta(A)\beta(B), \alpha(A)\beta(B) + \beta(A)\alpha(B)\}$$

In summary, the spin component of the wave function is antisymmetric with respect to nuclei interchange for *para*-hydrogen, but symmetric for *ortho*-hydrogen.

Next, consider the symmetry of the rotational component of the wave function. It can be shown that the symmetry of rotational states is dependent on the rotational quantum number J. If J is an even integer ($J = 0, 2, 4, 6, \ldots$) the corresponding rotational wave function is symmetric with respect to interchange, and if J is odd ($J = 1, 3, 5, 7, \ldots$) the wave function is antisymmetric. Because the wave function is the product of spin and rotational components, this product must be antisymmetric. Therefore, the rotational wave function for *para*-hydrogen is restricted to even-J levels, and for *ortho*-hydrogen is restricted to odd-J levels. Finally, the nuclear-spin-state degeneracy for *ortho*- and *para*-hydrogen is three and one, respectively. Therefore, the rotational energy levels for *ortho*-hydrogen have an additional threefold degeneracy.

A collection of molecular hydrogen will contain both *ortho*- and *para*-hydrogen so that the rotational partition function is

$$q_R = \frac{1}{4}\left[1 \sum_{J=0,2,4,6,\ldots} (2J + 1)e^{-\beta hcBJ(J+1)} + 3 \sum_{J=1,3,5,\ldots} (2J + 1)e^{-\beta hcBJ(J+1)} \right] \quad \textbf{(31.35)}$$

In Equation (31.35), the first term in brackets corresponds to *para*-hydrogen, and the second term corresponds to *ortho*-hydrogen. In essence, this expression for q_R represents average H_2 consisting of one part *para*-hydrogen and three parts *ortho*-hydrogen. Notice that the symmetry number is omitted in Equation (31.35) because overcounting of the allowed rotational levels has already been taken into account by restricting the summations to even or odd J. At high temperatures, the value of q_R determined using Equation (31.35) will, to good approximation, equal that obtained using Equation (31.32) with $\sigma = 2$, as the following example illustrates.

EXAMPLE PROBLEM 31.5

What is the rotational partition function for H_2 at 1000. K?

Solution

The rotational partition function for H_2, assuming that the high-temperature limit is valid, is given by

$$q_R = \frac{1}{\sigma \beta hcB} = \frac{1}{2\beta hcB}$$

With $B = 60.589 \text{ cm}^{-1}$ (Table 31.1):

$$q_R = \frac{1}{2\beta hcB} = \frac{kT}{2hcB}$$

$$= \frac{(1.38 \times 10^{-23} \text{ J K}^{-1})(1000. \text{ K})}{2(6.626 \times 10^{-34} \text{ J s})(3.00 \times 10^{10} \text{ cm s}^{-1})(60.589 \text{ cm}^{-1})} = 5.73$$

Evaluation of the rotational partition function by direct summation is performed as follows:

$$q_R = \frac{1}{4}\left[1 \sum_{J=0,2,4,6,\dots} (2J+1)e^{-\beta hcBJ(J+1)} \right.$$

$$\left. + 3 \sum_{J=1,3,5,\dots} (2J+1)e^{-\beta hcBJ(J+1)} \right] = 5.91$$

Comparison of these two expressions demonstrates that the high-T expression for q_R with $\sigma = 2$ provides a good estimate for the value of the rotational partition function of H_2.

31.5.4 The Rotational Temperature

Whether the rotational partition function should be evaluated by direct summation or integration is entirely dependent on the size of the rotational energy spacings relative to the amount of thermal energy available (kT). This comparison is facilitated through the introduction of the **rotational temperature** Θ_R defined as rotational constant divided by Boltzmann's constant:

$$\Theta_R = \frac{hcB}{k} \tag{31.36}$$

Unit analysis of Equation (31.36) dictates that Θ_R has units of temperature. We can rewrite the expression for the rotational partition function in terms of the rotational temperature as follows:

$$q_R = \frac{1}{\sigma \beta hcB} = \frac{kT}{\sigma hcB} = \frac{T}{\sigma \Theta_R}$$

A second application of the rotational temperature is as a comparative metric to the temperature at which the partition function is being evaluated. Figure 31.10 presents a comparison between q_R for $H^{35}Cl$ ($\Theta_R = 15.24 \text{ K}$) determined by summation [Equation (31.29)] and using the integrated form of the partition function [Equation (31.36)]. At low temperatures, significant differences between the summation and integrated results are evident. At higher temperatures, the summation result remains larger than the integrated result; however, both results predict that q_R will increase linearly with temperature. At high temperatures, the error in using the integrated result decreases such that for temperature where $T/\Theta_R \geq 10$, use of the integrated form of the rotational partition function is reasonable. The integrated form of the partition function is referred to as the **high-temperature or high-T limit** because it is applicable when kT is significantly greater than the rotational energy spacings. The following example illustrates the use of the rotational temperature in deciding which functional form of the rotational partition function to use.

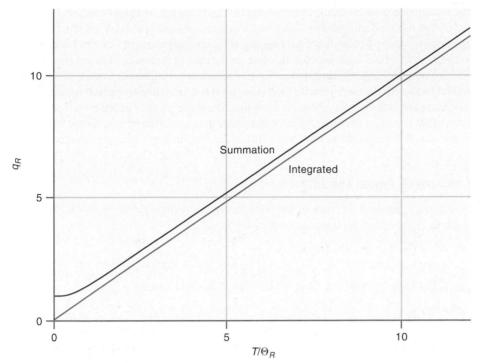

FIGURE 31.10
Comparison of q_R for $H^{35}Cl$ ($\Theta_R = 15.24$ K) determined by summation and by integration. Although the summation result remains greater than the integrated result at all temperatures, the fractional difference decreases with elevated temperatures such that the integrated form provides a sufficiently accurate measure of q_R when $T/\Theta_R > 10$.

EXAMPLE PROBLEM 31.6

Evaluate the rotational partition functions for I_2 at $T = 100.$ K.

Solution

Because $T = 100.$ K, it is important to ask how kT compares to the rotational energy-level spacings. Using Table 31.1, $B(I_2) = 0.0374$ cm^{-1} corresponding to rotational temperatures of

$$\Theta_R(I_2) = \frac{hcB}{k} = \frac{(6.626 \times 10^{-34}\,\text{J s})(3.00 \times 10^{10}\,\text{cm s}^{-1})(0.0374\,\text{cm}^{-1})}{1.38 \times 10^{-23}\,\text{J K}^{-1}} = 0.0538\,\text{K}$$

Comparison of these rotational temperatures to 100. K indicates that the high-temperature expression for the rotational partition function is valid for I_2:

$$q_R(I_2) = \frac{T}{\sigma\Theta_R} = \frac{100.\,\text{K}}{(2)(0.0539\,\text{K})} = 929$$

31.6 Rotational Partition Function: Polyatomics

In the diatomic systems described in the preceding section, there are two nonvanishing moments of inertia as illustrated in Figure 31.11. For **polyatomic** molecules (more than two atoms) the situation can become more complex.

If the polyatomic system is linear, there are again only two nonvanishing moments of inertia such that a linear polyatomic molecule can be treated using the same formalism as diatomic molecules. However, if the polyatomic molecule is not linear, then there are three nonvanishing moments of inertia. Therefore, the partition function that describes the rotational energy levels must take into account rotation about all three axes. Derivation of this partition function is not trivial; therefore, the result without derivation is stated here:

$$q_R = \frac{\sqrt{\pi}}{\sigma}\left(\frac{1}{\beta hcB_A}\right)^{1/2}\left(\frac{1}{\beta hcB_B}\right)^{1/2}\left(\frac{1}{\beta hcB_C}\right)^{1/2} \quad \textbf{(31.37)}$$

The subscript on B in Equation (31.37) indicates the corresponding moment of inertia as illustrated in Figure 31.11, and σ is the symmetry number as discussed earlier. In addition,

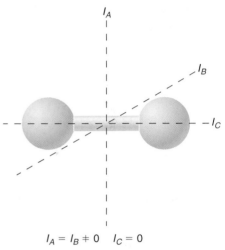

$I_A = I_B \neq 0 \quad I_C = 0$

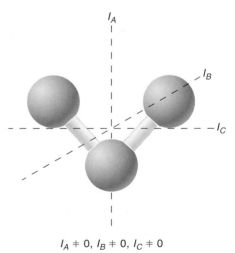

$I_A \neq 0, I_B \neq 0, I_C \neq 0$

FIGURE 31.11
Moments of inertia for diatomic and nonlinear polyatomic molecules. Note that in the case of the diatomic, $I_C = 0$ in the limit that the atomic masses are considered to be point masses that reside along the axis connecting the two atomic centers. Each moment of inertia will have a corresponding rotational constant.

the assumption is made that the polyatomic is "rigid" during rotational motion. The development of the rotational partition function for diatomic systems provides some intuition into the origin of this partition function. One can envision each moment of inertia contributing $(\beta hcB)^{-1/2}$ to the overall partition function. In the case of diatomics or linear polyatomics, the two nonvanishing moments of inertia are equivalent such that the product of the contribution from each moment results in the expression for the diatomic derived earlier. For the nonlinear polyatomic system, the partition function is the product of the contribution from each of the moments of inertia, which may or may not be equivalent as indicated by the subscripts on the corresponding rotational constants in the partition function presented earlier.

EXAMPLE PROBLEM 31.7

Evaluate the rotational partition functions for the following species at 298 K. You can assume that the high-temperature expression is valid.

  a. OCS ($B = 1.48 \text{ cm}^{-1}$)

  b. ONCl ($B_A = 2.84 \text{ cm}^{-1}$, $B_B = 0.191 \text{ cm}^{-1}$, $B_C = 0.179 \text{ cm}^{-1}$)

  c. CH_2O ($B_A = 9.40 \text{ cm}^{-1}$, $B_B = 1.29 \text{ cm}^{-1}$, $B_C = 1.13 \text{ cm}^{-1}$)

Solution

  a. OCS is a linear molecule as indicated by the single rotational constant. In addition, the molecule is asymmetric such that $\sigma = 1$. Using the rotational constant, the rotational partition function is

$$q_R = \frac{1}{\sigma \beta hcB} = \frac{kT}{hcB}$$

$$= \frac{(1.38 \times 10^{-23} \text{ J K}^{-1})(298 \text{ K})}{(6.626 \times 10^{-34} \text{ J s})(3.00 \times 10^{10} \text{ cm s}^{-1})(1.48 \text{ cm}^{-1})} = 140.$$

  b. ONCl is a nonlinear polyatomic. It is asymmetric such that $\sigma = 1$, and the partition function becomes

$$q_R = \frac{\sqrt{\pi}}{\sigma}\left(\frac{1}{\beta hcB_A}\right)^{1/2}\left(\frac{1}{\beta hcB_B}\right)^{1/2}\left(\frac{1}{\beta hcB_C}\right)^{1/2}$$

$$= \sqrt{\pi}\left(\frac{kT}{hc}\right)^{3/2}\left(\frac{1}{B_A}\right)^{1/2}\left(\frac{1}{B_B}\right)^{1/2}\left(\frac{1}{B_C}\right)^{1/2}$$

$$= \sqrt{\pi}\left(\frac{(1.38 \times 10^{-23} \text{ J K}^{-1})(298 \text{ K})}{(6.626 \times 10^{-34} \text{ J s})(3.00 \times 10^{10} \text{ cm s}^{-1})}\right)^{3/2}$$

$$\left(\frac{1}{2.84 \text{ cm}^{-1}}\right)^{1/2}\left(\frac{1}{0.191 \text{ cm}^{-1}}\right)^{1/2}\left(\frac{1}{0.179 \text{ cm}^{-1}}\right)^{1/2}$$

$$= 16{,}900$$

  c. CH_2O is a nonlinear polyatomic. However, the symmetry of this molecule is such that $\sigma = 2$. With this value for the symmetry number, the rotational partition function becomes

$$q_R = \frac{\sqrt{\pi}}{\sigma}\left(\frac{1}{\beta hcB_A}\right)^{1/2}\left(\frac{1}{\beta hcB_B}\right)^{1/2}\left(\frac{1}{\beta hcB_C}\right)^{1/2}$$

$$= \frac{\sqrt{\pi}}{2}\left(\frac{kT}{hc}\right)^{3/2}\left(\frac{1}{B_A}\right)^{1/2}\left(\frac{1}{B_B}\right)^{1/2}\left(\frac{1}{B_C}\right)^{1/2}$$

$$= \frac{\sqrt{\pi}}{2}\left(\frac{(1.38 \times 10^{-23} \text{ J K}^{-1})(298 \text{ K})}{(6.626 \times 10^{-34} \text{ J s})(3.00 \times 10^{10} \text{ cm s}^{-1})}\right)^{3/2}$$

$$\left(\frac{1}{9.40 \text{ cm}^{-1}}\right)^{1/2}\left(\frac{1}{1.29 \text{ cm}^{-1}}\right)^{1/2}\left(\frac{1}{1.13 \text{ cm}^{-1}}\right)^{1/2}$$

$$= 713$$

Note that the values for all three partition functions indicate that a substantial number of rotational states are populated at room temperature.

31.7 Vibrational Partition Function

The quantum mechanical model for vibrational degrees of freedom is the harmonic oscillator. In this model, each vibrational degree of freedom is characterized by a quadratic potential as illustrated in Figure 31.12. The energy levels of the harmonic oscillator are as follows:

$$E_n = hc\tilde{\nu}\left(n + \frac{1}{2}\right) \tag{31.38}$$

This equation demonstrates that the energy of a given level E_n is dependent on the quantum number n, which can take on integer values beginning with zero ($n = 0, 1, 2, \ldots$). The frequency of the oscillator, or vibrational frequency, is given by $\tilde{\nu}$ in units of cm^{-1}. Note that the energy of the $n = 0$ level is not zero, but $hc\tilde{\nu}/2$. This residual energy is known as the zero point energy and was discussed in detail during the quantum mechanical development of the harmonic oscillator. The expression for E_n provided in Equation 31.38 can be used to construct the vibrational partition function as follows:

$$q_V = \sum_{n=0}^{\infty} e^{-\beta E_n}$$

$$= \sum_{n=0}^{\infty} e^{-\beta hc\tilde{\nu}\left(n + \frac{1}{2}\right)}$$

$$= e^{-\beta hc\tilde{\nu}/2} \sum_{n=0}^{\infty} e^{-\beta hc\tilde{\nu}n} \tag{31.39}$$

The sum can be rewritten using the series identity:

$$\frac{1}{1 - e^{-\alpha x}} = \sum_{n=0}^{\infty} e^{-n\alpha x} \tag{31.40}$$

With this substitution, we arrive at the following expression for the **vibrational partition function:**

$$q_V = \frac{e^{-\beta hc\tilde{\nu}/2}}{1 - e^{-\beta hc\tilde{\nu}}} \qquad \text{(with zero point energy)} \tag{31.41}$$

Although this expression is correct as written, at times it is advantageous to redefine the vibrational energy levels such that $E_0 = 0$ with all levels decreased by an amount equal to the zero point energy. Why would this be an advantageous thing to do? Consider the calculation of the probability of occupying a given vibrational energy level p_n as follows:

$$p_n = \frac{e^{-\beta E_n}}{q_V} = \frac{e^{-\beta hc\tilde{\nu}\left(n + \frac{1}{2}\right)}}{\dfrac{e^{-\beta hc\tilde{\nu}/2}}{1 - e^{-\beta hc\tilde{\nu}}}} = \frac{e^{-\beta hc\tilde{\nu}/2}e^{-\beta hc\tilde{\nu}n}}{\dfrac{e^{-\beta hc\tilde{\nu}/2}}{1 - e^{-\beta hc\tilde{\nu}}}}$$

$$= e^{-\beta hc\tilde{\nu}n}\left(1 - e^{-\beta hc\tilde{\nu}}\right) \tag{31.42}$$

Notice that in Equation (31.42) the zero point energy contributions for both the energy level and the partition function cancel. Therefore, the relevant energy for determining p_n is not the absolute energy of a given level but the *relative* energy of the level. Given this, one can simply eliminate the zero point energy, resulting in the following expression for the vibrational partition function:

$$q_V = \frac{1}{1 - e^{-\beta hc\tilde{\nu}}} \qquad \text{(without zero point energy)} \tag{31.43}$$

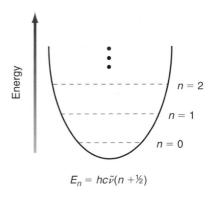

$$E_n = hc\tilde{\nu}(n + \tfrac{1}{2})$$

FIGURE 31.12
The harmonic oscillator model. Each vibrational degree of freedom is characterized by a quadratic potential. The energy levels corresponding to this potential are evenly spaced.

It is important to be consistent in including or not including zero point energy. For example, what if one were to perform the probability calculation presented earlier including zero point energy for the vibrational state of interest but not including zero point energy in the expression for the vibrational partition function? Proceeding as before, we arrive at the following *incorrect* result:

$$p_n = \frac{e^{-\beta E_n}}{q_V} = \frac{e^{-\beta hc\tilde{\nu}\left(n+\frac{1}{2}\right)}}{\dfrac{1}{1-e^{-\beta hc\tilde{\nu}}}} = \frac{e^{-\beta hc\tilde{\nu}/2}e^{-\beta hc\tilde{\nu}n}}{\dfrac{1}{1-e^{-\beta hc\tilde{\nu}}}}$$

$$= e^{-\beta hc\tilde{\nu}/2}e^{-\beta hc\tilde{\nu}n}\left(1-e^{-\beta hc\tilde{\nu}}\right)$$

Note that the zero point energy terms did not cancel as in the previous case, reflecting the fact that the energy of state n is defined differently relative to the expression for the partition function. In summary, once a decision has been made to include or ignore zero point energy, the approach taken must be consistently applied.

EXAMPLE PROBLEM 31.8

At what temperature will the vibrational partition function for I_2 ($\tilde{\nu} = 208 \text{ cm}^{-1}$) be greatest: 298 or 1000. K?

Solution

Because the partition function is a measure of the number of states that are accessible given the amount of energy available (kT), we would expect the partition function to be greater for $T = 1000$ K relative to $T = 298$ K. We can confirm this expectation by numerically evaluating the vibrational partition function at these two temperatures:

$$(q_V)_{298 \text{ K}} = \frac{1}{1-e^{-\beta hc\tilde{\nu}}} = \frac{1}{1-e^{-hc\tilde{\nu}/kT}}$$

$$= \frac{1}{1-\exp\left[-\dfrac{(6.626 \times 10^{-34}\,\text{J s})(3.00 \times 10^{10}\,\text{cm s}^{-1})(208\,\text{cm}^{-1})}{(1.38 \times 10^{-23}\,\text{J K}^{-1})(298\,\text{K})}\right]} = 1.58$$

$$(q_V)_{1000 \text{ K}} = \frac{1}{1-e^{-\beta hc\tilde{\nu}}}$$

$$= \frac{1}{1-\exp\left[-\dfrac{(6.626 \times 10^{-34}\,\text{J s})(3.00 \times 10^{10}\,\text{cm s}^{-1})(208\,\text{cm}^{-1})}{(1.38 \times 10^{-23}\,\text{J K}^{-1})(1000.\,\text{K})}\right]} = 3.86$$

Consistent with our expectation, the partition function increases with temperature, indicating that more states are accessible at elevated temperatures. The variation of q_v with temperature for I_2 is shown here:

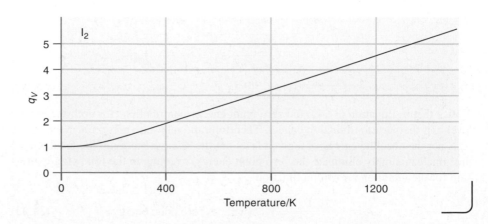

31.7.1 Beyond Diatomics: Multidimensional q_V

The expression for the vibrational partition function derived in the preceding subsection is for a single vibrational degree of freedom and is sufficient for diatomic molecules. However, triatomics and larger molecules (collectively referred to as polyatomics) require a different form for the partition function that takes into account all vibrational degrees of freedom. To define the vibrational partition function for polyatomics, we first need to know how many vibrational degrees of freedom there will be. A polyatomic molecule consisting of N atoms has $3N$ total degrees of freedom corresponding to three cartesian degrees of freedom for each atom. The atoms are connected by chemical bonds; therefore, the atoms are not free to move independently of each other. First, the entire molecule can translate through space; therefore, three of the $3N$ total degrees of freedom correspond to translational motion of the entire molecule. Next, a rotational degree of freedom will exist for each nonvanishing moment of inertia. As discussed in the section on rotational motion, linear polyatomics have two rotational degrees of freedom because there are two nonvanishing moments of inertia, and nonlinear polyatomic molecules have three rotational degrees of freedom. The remaining degrees of freedom are vibrational such that the number of vibrational degrees of freedom are

$$\text{Linear polyatomics: } 3N - 5 \qquad \textbf{(31.44)}$$

$$\text{Nonlinear polyatomics: } 3N - 6 \qquad \textbf{(31.45)}$$

Note that a diatomic molecule can be viewed as linear polyatomic with $N = 2$, and the preceding expressions dictate that there is only one vibrational degree of freedom [$3(2) - 5 = 1$] as stated earlier.

The final step in deriving the partition function for a polyatomic system is to recognize that within the harmonic approximation the vibrational degrees of freedom are separable and each vibration can be treated as a separate energetic degree of freedom. In Section 31.1, various forms of molecular energy were shown to be separable so that the total molecular partition function is simply the sum of the partition functions for each energetic degree of freedom. Similar logic applies to vibrational degrees of freedom where the total vibrational partition function is simply the product of vibrational partition functions for each vibrational degree of freedom:

$$(q_V)_{Total} = \prod_{i=1}^{3N-5 \text{ or } 3N-6} (q_V)_i \qquad \textbf{(31.46)}$$

In Equation (31.46), the total vibrational partition function is equal to the product of vibrational partition functions for each vibrational mode (denoted by the subscript i). There will be $3N - 5$ or $3N - 6$ mode-specific partition functions depending on the geometry of the molecule.

EXAMPLE PROBLEM 31.9

The triatomic chlorine dioxide (OClO) has three vibrational modes of frequency: 450, 945, and 1100 cm^{-1}. What is the value of the vibrational partition function for $T = 298$ K?

Solution

The total vibrational partition function is simply the product of the partition functions for each vibrational degree of freedom. Setting the zero point energy equal to zero, we find that

$$q_{450} = \frac{1}{1 - e^{-\beta hc(450 \text{ cm}^{-1})}}$$

$$= \frac{1}{1 - \exp\left[-\dfrac{(6.626 \times 10^{-34} \text{ J s})(3.00 \times 10^{10} \text{ cm s}^{-1})(450 \text{ cm}^{-1})}{(1.38 \times 10^{-23} \text{ J s})(298 \text{ K})}\right]}$$

$$= 1.13$$

$$q_{945} = \frac{1}{1 - e^{-\beta hc(945 \text{ cm}^{-1})}}$$

$$= \frac{1}{1 - \exp\left[-\dfrac{(6.626 \times 10^{-34} \text{ J s})(3.00 \times 10^{10} \text{ cm s}^{-1})(945 \text{ cm}^{-1})}{(1.38 \times 10^{-23} \text{ J s})(298 \text{ K})}\right]}$$

$$= 1.01$$

$$q_{1100} = \frac{1}{1 - e^{-\beta hc(1100 \text{ cm}^{-1})}}$$

$$= \frac{1}{1 - \exp\left[-\dfrac{(6.626 \times 10^{-34} \text{ J s})(3.00 \times 10^{10} \text{ cm s}^{-1})(1100 \text{ cm}^{-1})}{(1.38 \times 10^{-23} \text{ J s})(298 \text{ K})}\right]} = 1.00$$

$$(q_V)_{Total} = \prod_{i=1}^{3N-6} (q_V)_i = (q_{450})(q_{950})(q_{1100}) = (1.13)(1.01)(1.00) = 1.14$$

Note that the total vibrational partition function is close to unity. This is consistent with the fact that the vibrational energy spacings for all modes are significantly greater than kT such that few states other than $n = 0$ are populated.

31.7.2 High Temperature Approximation to q_V

Similar to the development of rotations, the **vibrational temperature** (Θ_V) is defined as the frequency of a given vibrational degree of freedom divided by k:

$$\Theta_V = \frac{hc\widetilde{\nu}}{k} \tag{31.47}$$

Unit analysis of Equation (31.47) dictates that Θ_V will have units of temperature (K). We can incorporate this term into our expression for the vibrational partition function as follows:

$$q_V = \frac{1}{1 - e^{-\beta hc\widetilde{\nu}}} = \frac{1}{1 - e^{-hc\widetilde{\nu}/kT}} = \frac{1}{1 - e^{-\Theta_V/T}} \tag{31.48}$$

The utility of this form of the partition function is that the relationship between vibrational energy and temperature becomes transparent. Specifically, as T becomes large relative to Θ_V, the exponent becomes smaller and the exponential term approaches 1. The denominator in Equation (31.48) will decrease such that the vibrational partition function will increase. If the temperature becomes sufficiently large relative to Θ_V, q_V can be reduced to a simpler form. The Math Supplement (Appendix A) provides the following series expression for exp $(-x)$:

$$e^{-x} = 1 - x + \frac{x^2}{2} - \cdots$$

For the vibrational partition function in Equation (31.48), $x = -\Theta_V/T$. When $T \gg \Theta_V$, x becomes sufficiently small that only the first two terms can be included in the series expression for $\exp(-x)$ since higher order terms are negligible. Substituting into the expression for the vibrational partition function:

$$q_V = \frac{1}{1 - e^{-\Theta_V/T}} = \frac{1}{1 - \left(1 - \dfrac{\Theta_V}{T}\right)} = \frac{T}{\Theta_V} \tag{31.49}$$

This result is the high-temperature (or high-T) limit for the vibrational partition function.

$$q_V = \frac{T}{\Theta_V} \quad \text{(high-T limit)} \tag{31.50}$$

When is Equation (31.50) appropriate for evaluating q_V as opposed to the exact expression? The answer to this question depends on both the vibrational frequency of interest

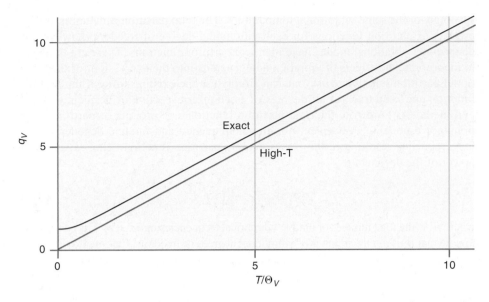

FIGURE 31.13
Comparison between the exact
[Equation (31.48)] and high-T
[Equation (31.50)] results for q_V.
Parameters employed in the calculation
correspond to I_2 ($\Theta_V = 299$ K). The frac-
tional difference between the two results is
0.5 for temperatures where $T/\Theta_V \geq 10$.
Exact agreement can be reached by
including the next term in the series
expansion of the exponential term in
Equation (31.49).

and the temperature. Figure 31.13 provides a comparison for I_2 ($\Theta_V = 299$ K) between
the exact expression without zero point energy [Equation (31.48)] and high-T expression [Equation (31.50)] for q_V. Similar to the case for rotations, the two results are significantly different at low temperature, but predict the same linear dependence on T at
elevated temperatures. When $T \geq 10\,\Theta_V$, the fractional difference between the high-T
and exact results is sufficiently small that the high-T result for q_V can be used. For the
majority of molecules, this temperature will be extremely high as shown in Example
Problem 31.10.

EXAMPLE PROBLEM 31.10

At what temperature is the high-T limit for q_V appropriate for F_2 ($\widetilde{\nu} = 917$ cm^{-1})?

Solution

The high-T limit is applicable when $T = 10\,\Theta_V$. The vibrational temperature for F_2 is

$$\Theta_V = \frac{hc\widetilde{\nu}}{k} = \frac{(6.626 \times 10^{-34}\,\text{J s})(3.00 \times 10^{10}\,\text{cm s}^{-1})(917\,\text{cm}^{-1})}{1.38 \times 10^{-23}\,\text{J K}^{-1}} = 1320\,\text{K}$$

Therefore, the high-T limit is applicable when $T = $ ~13,000 K. To make sure this is
indeed the case, we can compare the value for q_V determined by both the full expression for the partition function and the high-T approximation:

$$q_V = \frac{1}{1 - e^{-\Theta_V/T}} = \frac{1}{1 - e^{-1319\,\text{K}/13,000\,\text{K}}} = 10.4$$

$$\approx \frac{T}{\Theta_V} = \frac{13,000\,\text{K}}{1319\,\text{K}} = 9.9$$

Comparison of the two methods for evaluating the partition function demonstrates
that the high-T limit expression provides a legitimate estimate of the partition function at this temperature. However, the temperature at which this is true is exceedingly high.

31.7.3 Degeneracy and q_V

The total vibrational partition function for a polyatomic molecule is the product of the
partition functions for each vibrational degree of freedom. What if two or more of these
vibrational degrees of freedom have the same frequency such that the energy levels are
degenerate? It is important to keep in mind that there will always be $3N - 5$ or
$3N - 6$ vibrational degrees of freedom depending on geometry. In the case of degeneracy, two or more of these degrees of freedom will have the same vibrational energy

spacings, or the same vibrational temperature. The total partition function is still the product of partition functions for each vibrational degree of freedom; however, the degenerate vibrational modes have identical partition functions. There are two ways to incorporate the effects of vibrational degeneracy into the expression for the vibrational partition function. First, one can simply use the existing form of the partition function and keep track of all degrees of freedom irrespective of frequency. A second method is to rewrite the total partition function as a product of partition functions corresponding to a given vibrational frequency and include degeneracy at a that frequency resulting in the corresponding partition function being raised to the power of the degeneracy:

$$(q_V)_{Total} = \prod_{i=1}^{n'} (q_V)_i^{g_i} \tag{31.51}$$

where n' is the total number of unique vibrational frequencies indexed by i. It is important to note that n' is *not* the number of vibrational degrees of freedom! Carbon dioxide serves as a classic example of vibrational degeneracy, as Example Problem 31.11 illustrates.

EXAMPLE PROBLEM 31.11

CO_2 has the following vibrational degrees of freedom: 1388, 667.4 (doubly degenerate), and 2349 cm^{-1}. What is the total vibrational partition function for this molecule at 1000. K?

Solution

Evaluation of the partition function can be performed by calculating the individual vibrational partition functions for each unique frequency, then taking the product of these partition functions raised to the power of the degeneracy at a given frequency:

$$(q_V)_{1388} = \cfrac{1}{1 - \exp\left[-\cfrac{(6.626 \times 10^{-34}\,J\,s)(3.00 \times 10^{10}\,cm\,s^{-1})(1388\,cm^{-1})}{(1.38 \times 10^{-23}\,J\,K^{-1})(1000.\,K)}\right]} = 1.16$$

$$(q_V)_{667.4} = \cfrac{1}{1 - \exp\left[-\cfrac{(6.626 \times 10^{-34}\,J\,s)(3.00 \times 10^{10}\,cm\,s^{-1})(667.4\,cm^{-1})}{(1.38 \times 10^{-23}\,J\,K^{-1})(1000.\,K)}\right]} = 1.62$$

$$(q_V)_{2349} = \cfrac{1}{1 - \exp\left[-\cfrac{(6.626 \times 10^{-34}\,J\,s)(3.00 \times 10^{10}\,cm\,s^{-1})(2349\,cm^{-1})}{(1.38 \times 10^{-23}\,J\,K^{-1})(1000.\,K)}\right]} = 1.04$$

$$(q_V)_{Total} = \prod_{i=1}^{n'} (q_V)_i^{g_i} = (q_V)_{1388}(q_V)_{667.4}^2(q_V)_{2349}$$

$$= (1.16)(1.62)^2(1.04) = 3.17$$

31.8 The Equipartition Theorem

In the previous sections regarding rotations and vibrations, equivalence of the high-T and exact expressions for q_R and q_V, respectively, was observed when the temperature was sufficiently large that the thermal energy available to the system was significantly greater than the energy-level spacings. At these elevated temperatures, the quantum nature of the energy levels becomes unimportant, and a classical description of the energetics is all that is needed.

The definition of the partition function involves summation over quantized energy levels, and one might assume that there is a corresponding classical expression for the partition function in which a classical description of the system energetics is employed. Indeed there is such an expression; however, its derivation is beyond the scope of this text, so we simply state the result here. The expression for the three-dimensional partition function for a molecule consisting of N atoms is

$$q_{classical} = \frac{1}{h^{3N}} \int \cdots \int e^{-\beta H} dp^{3N} dx^{3N} \qquad (31.52)$$

In the expression for the partition function, the terms p and x represent the momentum and position coordinates for each particle, respectively, with three Cartesian dimensions available for each term. The integral is multiplied by h^{-3N}, which has units of $(\text{momentum} \times \text{distance})^{-3N}$ such that the partition function is unitless.

What does the term $e^{-\beta H}$ represent in $q_{classical}$? The H represents the classical Hamiltonian and, like the quantum Hamiltonian, is the sum of a system's kinetic and potential energy. Therefore, $e^{-\beta H}$ is equivalent to $e^{-\beta \varepsilon}$ in our quantum expression for the molecular partition function. Consider the Hamiltonian for a classical one-dimensional harmonic oscillator with reduced mass μ and force constant k:

$$H = \frac{p^2}{2\mu} + \frac{1}{2}kx^2 \qquad (31.53)$$

Using this Hamiltonian, the corresponding classical partition function for the one-dimensional harmonic oscillator is

$$q_{classical} = \frac{1}{h} \int dp \int dx\, e^{-\beta\left(\frac{p^2}{2\mu}+\frac{1}{2}kx^2\right)} = \frac{T}{\Theta_V} \qquad (31.54)$$

This result is in agreement with the high-T approximation to q_V derived using the quantum partition function of Equation (31.50). This example illustrates the applicability of classical statistical mechanics to molecular systems when the temperature is sufficiently high such that summation over the quantum states can be replaced by integration. Under these temperature conditions, knowledge of the quantum details of the system is not necessary because when evaluating Equation (31.54), nothing was implied regarding the quantization of the harmonic oscillator energy levels.

The applicability of classical statistical mechanics to molecular systems at high temperature finds application in an interesting theorem known as the **equipartition theorem.** This theorem states that any term in the classical Hamiltonian that is quadratic with respect to momentum or position (i.e., p^2 or x^2) will contribute $kT/2$ to the average energy. For example, the Hamiltonian for the one-dimensional harmonic oscillator [Equation (31.53)] has both a p^2 and x^2 term such that the average energy for the oscillator by equipartition should be kT (or NkT for a collection of N harmonic oscillators). In the next chapter, the equipartition result will be directly compared to the average energy determined using quantum statistical mechanics. At present, it is important to recognize that the concept of equipartition is a consequence of classical mechanics because, for a given energetic degree of freedom, the change in energy associated with passing from one energy level to the other must be significantly less than kT. As discussed earlier, this is true for translational and rotational degrees of freedom but is not the case for vibrational degrees of freedom except at relatively high temperatures.

31.9 Electronic Partition Function

Electronic energy levels correspond to the various arrangements of electrons in an atom or molecule. The hydrogen atom provides an excellent example of an atomic system where the orbital energies are given by

$$E_n = \frac{-m_e e^4}{8\varepsilon_o^2 h^2 n^2} = -109{,}737\,\text{cm}^{-1}\frac{1}{n^2} \quad (n = 1, 2, 3, \dots) \qquad (31.55)$$

This expression demonstrates that the energy of a given orbital in the hydrogen atom is dependent on the quantum number n. In addition, each orbital has a degeneracy of $2n^2$. Using Equation (31.55), the energy levels for the electron in the hydrogen atom can be determined as illustrated in Figure 31.14.

FIGURE 31.14
Orbital energies for the hydrogen atom.
(**a**) The orbital energies as dictated by
solving the Schrödinger equation for the
hydrogen atom. (**b**) The energy levels
shifted by the addition of 109,737 cm^{-1}
of energy such that the lowest orbital
energy is 0.

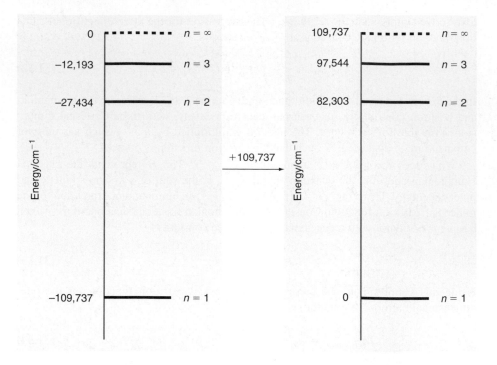

From the perspective of statistical mechanics, the energy levels of the hydrogen atom
represent the energy levels for the electronic energetic degree of freedom, with the corresponding partition function derived by summing over the energy levels. However, rather
than use the absolute energies as determined in the quantum mechanical solution to the
hydrogen atom problem, we will adjust the energy levels such that the energy associated
with the $n = 1$ orbital is zero, similar to the adjustment of the ground-state energy of the
harmonic oscillator to zero by elimination of the zero point energy. With this redefinition
of the orbital energies, the electronic partition for the hydrogen atom becomes

$$q_E = \sum_{n=1}^{\infty} g_n e^{-\beta hcE_n} = 2e^{-\beta hcE_1} + 8e^{-\beta hcE_2} + 18e^{-\beta hcE_3} + \dots$$

$$= 2e^{-\beta hc(0\ \text{cm}^{-1})} + 8e^{-\beta hc(82,303\ \text{cm}^{-1})} + 18e^{-\beta hc(97,544\ \text{cm}^{-1})} + \dots$$

$$= 2 + 8e^{-\beta hc(82,303\ \text{cm}^{-1})} + 18e^{-\beta hc(97,544\ \text{cm}^{-1})} + \dots \qquad \textbf{(31.56)}$$

The magnitude of the terms in the partition function corresponding to $n \geq 2$ will depend
on the temperature at which the partition function is being evaluated. However, note that
these energies are quite large. Consider defining the "electronic temperature," or Θ_E, in
exactly the same way as the rotational and vibrational temperatures were defined:

$$\Theta_E = \frac{hcE_n}{k} = \frac{E_n}{0.695\ \text{cm}^{-1}\,\text{K}^{-1}} \qquad \textbf{(31.57)}$$

With the definition of $E_1 = 0$, the energy of the $n = 2$ orbital, E_2, is 82,303 cm^{-1} corresponding to $\Theta_E = 118,421$ K! This is an extremely high temperature, and this simple calculation illustrates the primary difference between electronic energy levels and the other
energetic degrees of freedom discussed thus far. Electronic degrees of freedom are generally characterized by level spacings that are quite large relative to kT. Therefore, only the
ground electronic state is populated to a significant extent (although exceptions are known,
as presented in the problems at the end of this chapter). Applying this conceptual picture to
the hydrogen atom, the terms in the partition function corresponding to $n \geq 2$ should be
quite small at 298 K. For example, the term for the $n = 2$ state is as follows:

$$e^{-\beta hcE_2}$$

$$= \exp\left[\frac{-(6.626 \times 10^{-34}\ \text{J s})(3.00 \times 10^{10}\ \text{cm s}^{-1})(82,303\ \text{cm}^{-1})}{(1.38 \times 10^{-23}\ \text{J K}^{-1})(298\ \text{K})}\right] = e^{-397.5} \approx 0$$

Terms corresponding to higher energy orbitals will also be extremely small such that the electronic partition function for the hydrogen atom is ~ 2 at 298 K. In general, the contribution of each state to the partition function must be considered, resulting in the following expression for the **electronic partition function:**

$$q_E = \sum_n g_n e^{-\beta hc E_n} \qquad (31.58)$$

In the expression for the electronic partition function, the exponential term for each energy level is multiplied by the degeneracy of the level g_n. If the energy level spacings are very large compared to kT, then $q_E \approx g_o$, or the degeneracy of the ground state. However, certain atoms and molecules may have excited electronic states that are energetically accessible relative to kT, and the contribution of these states must be included in evaluating the partition function. The following problem provides an example of such a system.

EXAMPLE PROBLEM 31.12

The lowest nine energy levels for gaseous vanadium (V) have the following energies and degeneracies:

Level(n)	Energy(cm^{-1})	Degeneracy
0	0	4
1	137.38	6
2	323.46	8
3	552.96	10
4	2112.28	2
5	2153.21	4
6	2220.11	6
7	2311.36	8
8	2424.78	10

What is the value of the electronic partition function for V at 298 K?

Solution

Due to the presence of unpaired electrons in V, the electronic excited states are accessible relative to kT. Therefore, the partition function is not simply equal to the ground-state degeneracy and must instead be determined by writing out the summation explicitly, paying careful attention to the energy and degeneracy of each level:

$$q_E = \sum_n g_n e^{-\beta hc E_n} = g_o e^{-\beta hc E_0} + g_1 e^{-\beta hc E_1} + g_2 e^{-\beta hc E_2} + g_3 e^{-\beta hc E_0} + g_4 e^{-\beta hc E_0} + \dots$$

$$= 4\exp\left[\frac{-0 \text{ cm}^{-1}}{(0.695 \text{ cm}^{-1}\text{K}^{-1})(298 \text{ K})}\right] + 6\exp\left[\frac{-137.38 \text{ cm}^{-1}}{(0.695 \text{ cm}^{-1}\text{K}^{-1})(298 \text{ K})}\right]$$

$$+ 8\exp\left[\frac{-323.46 \text{ cm}^{-1}}{(0.695 \text{ cm}^{-1}\text{K}^{-1})(298 \text{ K})}\right] + 10\exp\left[\frac{-552.96 \text{ cm}^{-1}}{(0.695 \text{ cm}^{-1}\text{K}^{-1})(298 \text{ K})}\right]$$

$$+ 2\exp\left[\frac{-2112.28 \text{ cm}^{-1}}{(0.695 \text{ cm}^{-1}\text{K}^{-1})(298 \text{ K})}\right] + \dots$$

Notice in the preceding expression that the energy of state $n = 4$ (2112.28 cm^{-1}) is large with respect to kT (208 cm^{-1}). The exponential term for this state is approximately e^{-10}, or 4.5×10^{-5}. Therefore, the contribution of state $n = 4$ and higher energy states to the partition function will be extremely modest, and these terms can be disregarded when evaluating the partition function. Focusing on the lower energy states that make the dominant contribution to the partition function results in the following:

$$q_E \approx 4\exp\left[\frac{-0\ \text{cm}^{-1}}{(0.695\ \text{cm}^{-1}\,\text{K}^{-1})(298\ \text{K})}\right] + 6\exp\left[\frac{-137.38\ \text{cm}^{-1}}{(0.695\ \text{cm}^{-1}\,\text{K}^{-1})(298\ \text{K})}\right]$$

$$+ 8\exp\left[\frac{-323.46\ \text{cm}^{-1}}{(0.695\ \text{cm}^{-1}\,\text{K}^{-1})(298\ \text{K})}\right] + 10\exp\left[\frac{-552.96\ \text{cm}^{-1}}{(0.695\ \text{cm}^{-1}\,\text{K}^{-1})(298\ \text{K})}\right]$$

$$\approx 4 + 6(0.515) + 8(0.210) + 10(0.0693)$$

$$\approx 9.46$$

In summary, if the energy of an electronic excited state is sufficiently greater than kT, the contribution of the state to the electronic partition function will be minimal and the state can be disregarded in the numerical evaluation of the partition function.

Although the previous discussion focused on atomic systems, similar logic applies to molecules. Molecular electronic energy levels are described using molecular orbital (MO) theory. In MO theory, linear combinations of atomic orbitals are used to construct a new set of electronic orbitals known as molecular orbitals. The molecular orbitals differ in energy, and the electronic configuration of the molecule is determined by placing spin-paired electrons into the orbitals starting with the lowest energy orbital. The highest energy occupied molecular orbital is designated as the HOMO or highest occupied molecular orbital. The molecular orbital energy-level diagram for butadiene is presented in Figure 31.15.

Figure 31.15 presents both the lowest and next highest energy electronic energy states for butadiene, with the difference in energies corresponding to the promotion of an electron from the HOMO to the lowest unoccupied molecular orbital (LUMO). The separation in energy between these two states corresponds to the amount of energy it takes to excite the electron. The wavelength of the lowest energy electronic transition of butadiene is ~ 220 nm, demonstrating that the separation between the HOMO and LUMO is ~ 45,000 cm^{-1}, significantly greater than kT at 298 K. As such, only the lowest energy electronic state contributes to the electronic partition function, and $q_E = 1$ for butadiene since the degeneracy of the lowest energy level is one. Typically, the first

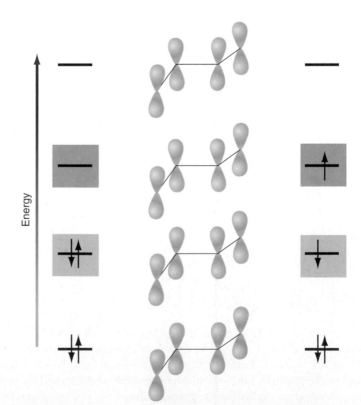

FIGURE 31.15
Depiction of the molecular orbitals for butadiene. The highest occupied molecular orbital (HOMO) is indicated by the orange rectangle, and the lowest unoccupied molecular orbital (LUMO) is indicated by the red rectangle. The lowest energy electron configuration is shown on the left, and the next highest energy configuration is shown on the right, corresponding to the promotion of an electron from the HOMO to the LUMO.

electronic excited level of molecules will reside 5000 to 50,000 cm^{-1} higher in energy than the lowest level such that at room temperature only this lowest level is considered in evaluating the partition function

$$q_E = \sum_{n=0} g_n e^{-\beta E_n} \approx g_0 \qquad (31.59)$$

In the absence of degeneracy of the ground electronic state, the electronic partition function will equal one.

31.10 Review

Given the numerous derivations, equations, and examples provided in this chapter, it is important for the reader to focus on the primary concepts developed throughout. There are two primary goals of this chapter: to relate the canonical partition Q to the molecular partition function q and to express the molecular partition function in terms of the individual energetic degrees of freedom. The relationship between the molecular and canonical partition function depends on whether the individual units comprising the ensemble are distinguishable or indistinguishable such that

$$Q_{Total} = q_{Total}^N \text{ (distinguishable)}$$

$$Q_{Total} = \frac{1}{N!} q_{Total}^N \text{ (indistinguishable)}$$

Evaluation of the canonical partition function requires knowledge of the total molecular partition function, which is equal to the product of the partition functions for each energetic degree of freedom:

$$q_{Total} = q_T q_R q_V q_E$$

An example will help to illustrate the approach that was developed in this chapter. Imagine that we are interested in the canonical partition function for a mole of a diatomic molecule at an arbitrary temperature. First, gaseous molecules are indistinguishable such that

$$Q = \frac{1}{N_A!} q_{Total}^{N_A}$$

Next, the expression for the total molecular partition function is simply given by

$$q_{Total} = q_T q_R q_V q_E$$

$$= \left(\frac{V}{\Lambda^3}\right)\left(\frac{1}{\sigma \beta h c B}\right)\left(\frac{1}{1 - e^{-\beta h c \tilde{\nu}}}\right)(g_0)$$

Notice that we have used the appropriate expressions for the rotational and vibrational partition functions for a diatomic molecule. Numerical evaluation of the preceding equation requires knowledge of the specific molecular parameters, the volume of the gas, and the temperature of the ensemble. However, the conceptual approach is universal and illustrates the connection between the ensemble Q and the microscopic properties of the individual units of the ensemble q. Using this approach, we can connect the microscopic description of molecules to the macroscopic behavior of molecular ensembles, an undertaking we explore in the next chapter.

Vocabulary

canonical ensemble	electronic partition function	equipartition theorem
canonical partition function	energetic degrees of freedom	high-temperature (high-T) limit
diatomic	ensemble	molecular partition function

ortho-hydrogen	rotational partition function	thermal wavelength
para-hydrogen	rotational temperature	translational partition function
polyatomic	symmetry number	vibrational partition function
rotational constant	thermal de Broglie wavelength	vibrational temperature

Conceptual Problems

Q31.1 What is the canonical ensemble? What properties are held constant in this ensemble?

Q31.2 What is the relationship between Q and q? How does this relationship differ if the particles are distinguishable versus indistinguishable?

Q31.3 List the atomic and/or molecular energetic degrees of freedom discussed in this chapter. For each energetic degree of freedom, briefly summarize the corresponding quantum mechanical model.

Q31.4 For which energetic degrees of freedom are the energy-level spacings small relative to kT at room temperature?

Q31.5 For the translational and rotational degrees of freedom, evaluation of the partition function involved replacement of the summation by integration. Why could integration be performed? How does this relate back to the discussion of probability distributions of discrete variables treated as continuous?

Q31.6 How many rotational degrees of freedom are there for linear and non-linear molecules?

Q31.7 Assuming $^{19}F_2$ and $^{35}Cl_2$ have the same bond length, which molecule do you expect to have the largest rotational constant?

Q31.8 Consider the rotational partition function for a non-linear polyatomic molecule. Can you describe the origin of each term in the partition function and why the partition function involves a product of terms?

Q31.9 What is the high-T approximation for rotations and vibrations? For which of these two degrees of freedom do you expect this approximation to be generally valid at room temperature?

Q31.10 In constructing the vibrational partition function, we found that the definition depended on whether zero-point energy was included in the description of the energy levels. However, the expression for the probability of occupying a specific vibrational energy level was independent of zero-point energy. Why?

Q31.11 Although the vibrational degrees of freedom are generally not in the high-T limit, is the vibrational partition function evaluated by discrete summation?

Q31.12 What is the form of the total vibrational partition function for a polyatomic molecule?

Q31.13 How does the presence of degeneracy affect the form of the total vibrational partition function?

Q31.14 What is the equipartition theorem? Why is this theorem inherently classical?

Q31.15 Why is the electronic partition function generally equal to the degeneracy of the ground electronic state?

Q31.16 What is q_{Total}, and how is it constructed using the partition functions for each energetic degree of freedom discussed in this chapter?

Q31.17 Why is it possible to set the energy of the ground vibrational and electronic energy level to zero?

Numerical Problems

Problem numbers in **red** indicate that the solution to the problem is given in the *Student's Solutions Manual*.

P31.1 Evaluate the translational partition function for H_2 confined to a volume of 100. cm^3 at 298 K. Perform the same calculation for N_2 under identical conditions. (*Hint:* Do you need to reevaluate the full expression for q_T?)

P31.2 Evaluate the translational partition function for $^{35}Cl_2$ confined to a volume of 1 L at 298 K. How does your answer change if the gas is $^{37}Cl_2$? (*Hint:* Can you reduce the ratio of translational partition functions to an expression involving mass only?)

P31.3 He has a normal boiling point of 4.2 K. For a mole of gaseous He at 4.2 K and 1 atm, is the high-temperature limit for translational degrees of freedom applicable?

P31.4 Evaluate the translational partition function for Ar confined to a volume of 1000. cm^3 at 298 K. At what temperature

will the translational partition function of Ne be identical to that of Ar at 298 K confined to the same volume?

P31.5 At what temperature are there Avogadro's number of translational states available for O_2 confined to a volume of 1000. cm^3?

P31.6 Imagine gaseous Ar at 298 K confined to move in a two-dimensional plane of area 1.00 cm^2. What is the value of the translational partition function?

P31.7 Researchers at IBM have used scanning tunneling microscopy to place atoms in nanoscale arrangements that they refer to as "quantum corrals" (*Nature* 363 [1993]: 6429). Imagine constructing a square corral with 4.00 nm sides. If CO is confined to move in the two-dimensional space defined by this corral, what is the value of the translational partition function?

P31.8 For N_2 at 77.3 K, 1.00 atm, in a 1.00-cm^3 container, calculate the translational partition function and ratio of this partition function to the number of N_2 molecules present under these conditions.

P31.9 What is the symmetry number for the following molecules?

a. $^{35}Cl^{37}Cl$

b. $^{35}Cl_2$

c. $^{16}O_2$

d. C_6H_6

e. CH_2Cl_2

P31.10 Determine the symmetry number for the following halogenated methanes: CCl_4, $CFCl_3$, CF_2Cl_2, CF_3Cl.

P31.11 Which species will have the largest rotational partition function at a given temperature: H_2, HD, or D_2? Which of these species will have the largest translational partition function assuming that volume and temperature are identical? When evaluating the rotational partition functions, you can assume that the high-temperature limit is valid.

P31.12 Consider *para*-H_2 ($B = 60.853$ cm^{-1}) for which only even-J levels are available. Evaluate the rotational partition function for this species at 50 K. Perform this same calculation for HD ($B = 45.655$ cm^{-1}).

P31.13 Calculate the rotational partition function for the interhalogen compound $F^{35}Cl$ ($B = 0.516$ cm^{-1}) at 298 K.

P31.14 Calculate the rotational partition function for $^{35}Cl_2$ (B $= 0.244$ cm^{-1}) at 298 K.

P31.15 For which of the following diatomic molecules is the high-temperature expression for the rotational partition function valid at 40. K?

a. DBr ($B = 4.24$ cm^{-1})

b. DI ($B = 3.25$ cm^{-1})

c. CsI ($B = 0.0236$ cm^{-1})

d. $F^{35}Cl$ ($B = 0.516$ cm^{-1})

P31.16 Calculate the rotational partition function for SO_2 at 298 K where $B_A = 2.03$ cm^{-1}, $B_B = 0.344$ cm^{-1}, and $B_C = 0.293$ cm^{-1}.

P31.17 Calculate the rotational partition function for ClNO at 500. K where $B_A = 2.84$ cm^{-1}, $B_B = 0.187$ cm^{-1}, and $B_C = 0.175$ cm^{-1}.

P31.18

a. In the rotational spectrum of $H^{35}Cl$ ($I = 2.65 \times 10^{-47}$ kg m^2), the transition corresponding to the $J = 4$ to $J = 5$ transition is the most intense. At what temperature was the spectrum obtained?

b. At 1000. K, which rotational transition of $H^{35}Cl$ would you expect to demonstrate the greatest intensity?

c. Would you expect the answers for parts (a) and (b) to change if the spectrum were of $H^{37}Cl$?

P31.19 What transition in the rotational spectrum of IF ($B = 0.280$ cm^{-1}) is expected to be the most intense at 298 K?

P31.20 For a rotational-vibrational spectrum of $H^{81}Br$ ($B = 8.46$ cm^{-1}) taken at 500. K, which R-branch transition to you expect to be the most intense?

P31.21 Calculate the rotational partition function for oxygen ($B = 1.44$ cm^{-1}) at its boiling point, 90.2 K, using the high-temperature approximation and by discrete summation. Why should only odd values of J be included in this summation?

P31.22 In microwave spectroscopy a traditional unit for the rotational constant is the Mc or "mega cycle" equal to 10^6 s^{-1}. For $^{14}N^{14}N^{16}O$ the rotational constant is 12,561.66 Mc.

a. Convert the above value for the rotational constant from Mc to cm^{-1}.

b. Determine the value of the rotational partition function at 298 K.

P31.23

a. Calculate the population of the first 10 rotational energy levels for HBr ($B = 8.46$ cm^{-1}) at 298 K.

b. Repeat this calculation for HF assuming that the bond length of this molecule is identical to that of HBr.

P31.24 In general, the high-temperature limit for the rotational partition function is appropriate for almost all molecules at temperatures above the boiling point. Hydrogen is an exception to this generality because the moment of inertia is small due to the small mass of H. Given this, other molecules with H may also represent exceptions to this general rule. For example, methane (CH_4) has relatively modest moments of inertia ($I_A = I_B = I_C = 5.31 \times 10^{-40}$ g cm^2) and has a relatively low boiling point of $T = 112$ K.

a. determine B_A, B_B, and B_C for this molecule.

b. Use the answer from part (a) to determine the rotational partition function. Is the high-temperature limit valid?

P31.25 When ^{4}He is cooled below 2.17 K it becomes a "superfluid" with unique properties such as a viscosity approaching zero. One way to learn about the superfluid environment is to measure the rotational-vibrational spectrum of molecules embedded in the fluid. For example, the spectrum of OCS in low-temperature ^{4}He droplet has been reported (*Journal of Chemical Physics* 112 [2000]: 4485). For $OC^{32}S$ the authors measured a rotational constant of 0.203 cm^{-1} and found that the intensity of the J = 0 to 1 transition was roughly 1.35 times greater than that of the J = 1 to 2 transition. Using this information, provide a rough estimate of the temperature of the droplet.

P31.26 Calculate the vibrational partition function for $H^{35}Cl$ ($\tilde{\nu} = 2990$ cm^{-1}) at 300. and 3000. K. What fraction of molecules will be in the ground vibrational state at these temperatures?

P31.27 Determine the rotational partition function for $I^{35}Cl$ (B $= 0.114$ cm^{-1}) at 298 K.

P31.28 For IF ($\tilde{\nu} = 610$. cm^{-1}) calculate the vibrational partition function and populations in the first three vibrational energy levels for $T = 300$. and 3000. K. Repeat this calculation for IBr ($\tilde{\nu} = 269$ cm^{-1}). Compare the probabilities for IF and IBr. Can you explain the differences between the probabilities of these molecules?

P31.29 Evaluate the vibrational partition function for H_2O at 2000. K where the vibrational frequencies are 1615, 3694, and 3802 cm^{-1}.

P31.30 Evaluate the vibrational partition function for SO_2 at 298 K where the vibrational frequencies are 519, 1151, and 1361 cm^{-1}.

P31.31 Evaluate the vibrational partition function for NH_3 at 1000. K where the vibrational frequencies are 950., 1627.5 (doubly degenerate), 3335, and 3414 cm^{-1} (doubly degenerate). Are there any modes that you can disregard in this calculation? Why or why not?

P31.32 Evaluate the vibrational partition function for $CFCl_3$ at 298 K where the vibrational frequencies are (with degeneracy in parenthesis) 1081, 847 (2), 535, 394 (2), 350., and 241(2) cm^{-1}.

P31.33 Determine the populations in $n = 0$ and 1 for $H^{81}Br$ ($\widetilde{\nu} = 2649$ cm^{-1}) at 298 K.

P31.34 Isotopic substitution is employed to isolate features in a vibrational spectrum. For example, the C=O stretch of individual carbonyl groups in the backbone of a polypeptide can be studied by substituting $^{13}C^{18}O$ for $^{12}C^{16}O$.

a. From quantum mechanics the vibrational frequency of a diatomic molecules depends on the bond force constant (κ) and reduced mass (μ) as follows:

$$\widetilde{\nu} = \sqrt{\frac{\kappa}{\mu}}$$

If the vibrational frequency of $^{12}C^{16}O$ is 1680 cm^{-1}, what is the expected frequency for $^{13}C^{18}O$?

b. Using the vibrational frequencies for $^{12}C^{16}O$ and $^{13}C^{18}O$, determine the value of the corresponding vibrational partition functions at 298 K. Does this isotopic substitution have a dramatic effect on q_V?

P31.35 In using statistical mechanics to describe the thermodynamic properties of molecules, high-frequency vibrations are generally not of importance under standard thermodynamic conditions since they are not populated to a significant extent. For example, for many hydrocarbons the C−H stretch vibrational degrees of freedom are neglected. Using cyclohexane as an example, the IR-absorption spectrum reveals that the C−H stretch transition are located at ~2850 cm^{-1}.

a. What is the value of the vibrational partition function for a mode of this frequency at 298 K?

b. At what temperature will this partition function reach a value of 1.1?

P31.36 In deriving the vibrational partition function, a mathematical expression for the series expression for the partition function was employed. However, what if one performed integration instead of summation to evaluate the partition function? Evaluate the following expression for the vibrational partition function:

$$q_V = \sum_{n=0}^{\infty} e^{-\beta hcn\widetilde{\nu}} \approx \int_0^{\infty} e^{-\beta hcn\widetilde{\nu}} \, dn$$

Under what conditions would you expect the resulting expression for q_V to be applicable?

P31.37 You have in your possession the first vibrational spectrum of a new diatomic molecule X_2 obtained at 1000. K. From the spectrum you determine that the fraction of molecules occupying a given vibrational energy state n is as follows:

n	0	1	2	3	>3
Fraction	0.352	0.184	0.0963	0.050.	0.318

What are the vibrational energy spacings for X_2?

P31.38

a. In this chapter, the assumption was made that the harmonic oscillator model is valid such that anharmonicity can be neglected. However, anharmonicity can be included in the expression for vibrational energies. The energy levels for an anharmonic oscillator are given by

$$E_n = hc\widetilde{\nu}\left(n + \frac{1}{2}\right) - hc\widetilde{\chi}\widetilde{\nu}\left(n + \frac{1}{2}\right)^2 + \dots$$

Neglecting zero point energy, the energy levels become $E_n = hc\widetilde{\nu}n - hc\widetilde{\chi}\widetilde{\nu}n^2 + \dots$. Using the preceding expression, demonstrate that the vibrational partition function for the anharmonic oscillator is

$$q_{V,\,anharmonic} = q_{V,\,harm}\left[1 + \beta hc\widetilde{\chi}\widetilde{\nu}q_{V,harm}^2(e^{-2\beta\widetilde{\nu}hc} + e^{-\beta\widetilde{\nu}hc})\right]$$

In deriving the preceding result, the following series relationship will prove useful:

$$\sum_{n=0}^{\infty} n^2 x^n = \frac{x^2 + x}{(1 - x)^3} \text{ for} |x| < 1$$

b. For H_2, $\widetilde{\nu} = 4401.2$ cm^{-1} and $\widetilde{\chi}\widetilde{\nu} = 121.3$ cm^{-1}. Use the result from part (a) to determine the percent error in q_V if anharmonicity is ignored.

P31.39 Consider a particle free to translate in one dimension. The classical Hamiltonian is $H = p^2/2m$.

a. Determine $q_{classical}$ for this system. To what quantum system should you compare it in order to determine the equivalence of the classical and quantum statistical mechanical treatments?

b. Derive $q_{classical}$ for a system with translational motion in three dimensions for which $H = (p_x^2 + p_y^2 + p_z^2)/2m$.

P31.40 A general expression for the classical Hamiltonian is

$$H = \alpha p_i^2 + H'$$

where p_i is the momentum along one dimension for particle i, α is a constant, and H' are the remaining terms in the Hamiltonian. Substituting this into the equipartition theorem yields

$$q = \frac{1}{h^{3N}} \iint e^{-\beta(\alpha p^2_i + H')} dp^{3N} dx^{3N}$$

a. Starting with this expression, isolate the term involving p_i and determine its contribution to q.

b. Given that the average energy $\langle \varepsilon \rangle$ is related to the partition function as follows,

$$\langle \varepsilon \rangle = \frac{-1}{q}\left(\frac{\delta q}{\delta \beta}\right)$$

evaluate the expression the contribution from p_i. Is your result consistent with the equipartiton theorem?

P31.41 Hydrogen isocyanide, HNC, is the tautomer of hydrogen cyanide (HCN). HNC is of interest as an intermediate species in a variety of chemical processes in interstellar space (T = 2.75 K).

a. For HCN the vibrational frequencies are 2041 cm^{-1} (CN stretch), 712 cm^{-1} (bend, doubly degenerate), and 3669 cm^{-1} (CH stretch). The rotational constant is 1.477 cm^{-1}. Calculate the rotational and vibrational partition functions for HCN in interstellar space. Before calculating the vibrational partition function, is there an approximation you can make that will simplify this calculation?

b. Perform the same calculations for HNC which has vibrational frequencies of 2024 cm^{-1} (NC stretch), 464 cm^{-1} (bend, doubly degenerate), and 3653 cm^{-1} v (NH stretch). The rotational constant is 1.512 cm^{-1}.

c. The presence of HNC in space was first established by Snyder and Buhl (*Bulletin of the American Astronomical Society* 3 [1971]: 388) through the microwave emission of the J = 1 to 0 transition of HNC at 90.665 MHz. Considering your values for the rotational partition functions, can you rationalize why this transition would be observed? Why not the J = 20-19 transition?

P31.42 Evaluate the electronic partition function for atomic Fe at 298 K given the following energy levels.

Level (n)	Energy (cm^{-1})	Degeneracy
0	0	9
1	415.9	7
2	704.0	5
3	888.1	3
4	978.1	1

P31.43

a. Evaluate the electronic partition function for atomic Si at 298 K given the following energy levels:

Level (n)	Energy(cm^{-1})	Degeneracy
0	0	1
1	77.1	3
2	223.2	5
3	6298	5

b. At what temperature will the $n = 3$ energy level contribute 0.100 to the electronic partition function?

P31.44 NO is a well-known example of a molecular system in which excited electronic energy levels are readily accessible at room temperature. Both the ground and excited electronic states are doubly degenerate and are separated by 121.1 cm^{-1}.

a. Evaluate the electronic partition function for this molecule at 298 K.

b. Determine the temperature at which $q_E = 3$.

P31.45 Rhodopsin is a biological pigment that serves as the primary photoreceptor in vision (*Science* 266 [1994]: 422). The chromophore in rhodopsin is retinal, and the absorption spectrum of this species is centered at roughly 500 nm. Using this information, determine the value of q_E for retinal. Do you expect thermal excitation to result in a significant excited-state population of retinal?

P31.46 Determine the total molecular partition function for I_2, confined to a volume of 1000. cm^3 at 298 K. Other information you will find useful is that $B = 0.0374$ cm^{-1}, ($\tilde{\nu} = 208$ cm^{-1}), and the ground electronic state is nondegenerate.

P31.47 Determine the total molecular partition function for gaseous H_2O at 1000. K confined to a volume of 1.00 cm^3. The rotational constants for water are $B_A = 27.8$ cm^{-1}, $B_B = 14.5$ cm^{-1}, and $B_C = 9.95$ cm^{-1}. The vibrational frequencies are 1615, 3694, and 3802 cm^{-1}. The ground electronic state is nondegenerate.

P31.48 The effect of symmetry on the rotational partition function for H_2 was evaluated by recognizing that each hydrogen is a spin 1/2 particle and is, therefore, a fermion. However, this development is not limited to fermions, but is also applicable to bosons. Consider CO_2 in which rotation by 180° results in the interchange of two spin 0 particles.

a. Because the overall wave function describing the interchange of two bosons must be symmetric with respect to exchange, to what J levels is the summation limited in evaluating q_R for CO_2?

b. The rotational constant for CO_2 is 0.390 cm^{-1}. Calculate q_R at 298 K. Do you have to evaluate q_R by summation of the allowed rotational energy levels? Why or why not?

Web-Based Simulations, Animations, and Problems

W31.1 In this Web-based simulation, the variation in q_T, q_R, and q_V with temperature is investigated for three diatomic molecules: HF, H^{35}Cl, and ^{35}ClF. Comparisons of q_T and q_R are performed to illustrate the mass and temperature dependence of these partition functions. Also, the expected dependence of q_T, q_R, and q_V on temperature in the high-temperature limit is investigated.

Computational Problems

More detailed instructions on carrying out these calculations using Spartan Physical Chemistry are found on the book website at *www.masteringchemistry.com*.

C31.1 Using the Hartree-Fock 6-31G* basis set, determine the vibrational frequencies for F_2O and Br_2O and calculate the vibrational partition function for these species at 500. K.

C31.2 Halogenated methanes are of interest as greenhouse gases. Perform a Hartree-Fock 6-31G* calculation of CFH_3 including determination of the IR spectrum. Does this molecule possess any strong transitions in the infrared? Calculate the vibrational partition function for the vibrational degree of freedom with the strongest predicted IR-absorption intensity.

C31.3 Nitryl chloride ($ClNO_2$) is an important compound in atmospheric chemistry. It serves as a reservoir compound for Cl and is produced in the lower troposphere through the reaction of N_2O_5 with Cl^- containing aerosols. Using the Hartree-Fock 3-21G basis set, determine the value of the vibrational partition function at 260. K. Does your answer change appreciably if you use the 6-31G* basis set?

C31.4 The rotational constant for 1,3-butadiyne (C_4H_2) is 4391.19 MHz. Performing a Hartree-Fock 6-31G* calculation, minimize the geometry of this species and compare the effective bond length determined using the rotational constant to the actual geometry of this species.

C31.5 Using Hartree-Fock with the 3-21G basis set, determine the value of the vibrational partition function for CF_2Cl_2 and CCl_4 at 298 K. What do you notice about the vibrational frequencies for the higher-symmetry species?

C31.6 The high-temperature limit was shown to be of limited applicability to vibrational degrees of freedom. As will be shown in the next chapter, this has a profound consequence when exploring the role of vibrations in statistical thermodynamics. For example, a coarse rule of thumb is that one usually does not need to worry about the contribution of CH, NH, and OH stretch vibrations. Perform a Hartree-Fock 6-31G* calculation on 1,3-cyclohexadiene, and calculate the IR intensities. Looking at the atomic displacements that make up the normal modes, identify those modes that are predominantly of C-H stretch character. In what frequency range are these modes located? Identify the lowest-energy C-H stretch and the lowest-frequency mode in general. Compare the value of the partition function for these two vibrational degrees of freedom at 298 K. How does this comparison support the "rule of thumb?"

C31.7 You are interested in designing a conducting polyene-based polymer. The design calls for maximizing the conjugation length, but as the conjugation length increases, the HOMO-LUMO energy gap corresponding to the lowest-energy electronic transition will decrease. If the gap becomes too small, then thermal excitation can result in population of the excited state. If thermal excitation is significant, it will degrade the performance of the polymer.

a. With a tolerance of 2% population in the first electronic excited state at 373 K, what is the smallest electronic energy gap that can be tolerated?

b. Using the Hartree-Fock 3-21G basis set, calculate the HOMO-LUMO energy gap for 1,3,5-hexatriene, 1,3,5,7-octatetraene, and 1,3,5,7,9-decapentaene. Using a plot of energy gap versus number of double bonds, determine the longest polyene structure that can be achieved while maintaining the 2% excited-state population tolerance, assuming that the energy gap varies linearly with conjugation length. (Note: This method provides a very rough estimate of the actual energy gap.)

c. (Advanced) In part (b) it was assumed that the HOMO-LUMO energy gap decreases linearly with an increase in conjugation length. Check this assumption by calculating this energy gap for the next two polyenes. Does the linear correlation continue?

For Further Reading

Chandler, D. *Introduction to Modern Statistical Mechanics*. New York: Oxford, 1987.

Hill, T. *Statistical Mechanics. Principles and Selected Applications*. New York: Dover, 1956.

McQuarrie, D. *Statistical Mechanics*. New York: Harper & Row, 1973.

Nash, L. K. "On the Boltzmann Distribution Law." *J. Chemical Education* 59 (1982): 824.

Nash, L. K. *Elements of Statistical Thermodynamics*. San Francisco: Addison-Wesley, 1972.

Noggle, J. H. *Physical Chemistry*. New York: HarperCollins, 1996.

Townes, C. H., and A. L. Schallow. *Microwave Spectroscopy*. New York: Dover, 1975. (This book contains an excellent appendix of spectroscopic constants.)

Widom, B. *Statistical Mechanics*. Cambridge: Cambridge University Press, 2002.

Statistical Thermodynamics

With the central concepts of statistical mechanics in hand, the relationship between statistical mechanics and classical thermodynamics can be explored. In this chapter, the microscopic viewpoint of matter is connected to fundamental thermodynamic quantities such as internal energy, entropy, and Gibbs free energy using statistical mechanics. As will be shown, the statistical perspective is not only capable of reproducing thermodynamic properties of matter, but it also provides critical insight into the microscopic details behind these properties. We will see that the insight into the behavior of chemical systems gained from the statistical perspective is remarkable.

32.1 Energy

We begin our discussion of statistical thermodynamics by returning to the canonical ensemble (Section 31.1) and considering the **average energy** content of an ensemble unit $\langle \varepsilon \rangle$, which is simply the **total energy** of the ensemble E divided by the number of units in the ensemble N:

$$\langle \varepsilon \rangle = \frac{E}{N} = \frac{\sum_n \varepsilon_n a_n}{N} = \sum_n \varepsilon_n \frac{a_n}{N} \quad (32.1)$$

In this equation, ε_n is the level energy and a_n is the occupation number for the level. Consistent with our development in Chapter 31, the ensemble is partitioned such that there is one atom or molecule in each unit. The Boltzmann distribution for a series of nondegenerate energy levels is

$$\frac{a_n}{N} = \frac{e^{-\beta \varepsilon_n}}{q} \quad (32.2)$$

In this expression, q is the molecular partition function and $\beta = (kT)^{-1}$. Substituting Equation (32.2) into Equation (32.1) yields

$$\langle \varepsilon \rangle = \sum_n \varepsilon_n \frac{a_n}{N} = \frac{1}{q} \sum_n \varepsilon_n e^{-\beta \varepsilon_n} \quad (32.3)$$

As a final step in deriving $\langle \varepsilon \rangle$, consider the derivative of the molecular partition function with respect to β, which is given by

$$\frac{-dq}{d\beta} = \sum_n \varepsilon_n e^{-\beta \varepsilon_n} \tag{32.4}$$

Using Equation (32.4), Equation (32.3) can be rewritten to obtain the following expressions for the average unit energy and total ensemble energy:

$$\langle \varepsilon \rangle = \frac{-1}{q}\left(\frac{dq}{d\beta}\right) = -\left(\frac{d\ln q}{d\beta}\right) \tag{32.5}$$

$$E = N\langle \varepsilon \rangle = \frac{-N}{q}\left(\frac{dq}{d\beta}\right) = -N\left(\frac{d\ln q}{d\beta}\right) \tag{32.6}$$

At times, Equation (32.5) and (32.6) are easier to evaluate by taking the derivative with respect to T rather than β. Using the definition $\beta = (kT)^{-1}$,

$$\frac{d\beta}{dT} = \frac{d}{dT}(kT)^{-1} = -\frac{1}{kT^2} \tag{32.7}$$

Using Equation (32.7), the expressions for average and total energy can be written as follows:

$$\langle \varepsilon \rangle = kT^2\left(\frac{d\ln q}{dT}\right) \tag{32.8}$$

$$E = NkT^2\left(\frac{d\ln q}{dT}\right) \tag{32.9}$$

Equation (32.9) demonstrates that E will change with temperature, a result we are familiar with from our study of thermodynamics. Example Problem 32.1 involves an ensemble comprised of particles with two energy levels, commonly referred to as a **two-level system** (see Figure 32.1), to illustrate the variation of E with T. To derive the functional form of E for this ensemble, q must be constructed and used to derive an expression for E, as shown next.

EXAMPLE PROBLEM 32.1

Determine the total energy of an ensemble consisting of N particles that have only two energy levels separated by energy $h\nu$.

Solution

The energy levels for the particles are illustrated in Figure 32.1. As mentioned, systems with only two energy levels are commonly referred to as two-level systems. To determine the average energy, the partition function describing this system must be evaluated. The partition function consists of a sum of two terms as follows:

$$q = 1 + e^{-\beta h\nu}$$

The derivative of the partition function with respect to β is

$$\frac{dq}{d\beta} = \frac{d}{d\beta}(1 + e^{-\beta h\nu})$$

$$= -h\nu e^{-\beta h\nu}$$

Using this result, the total energy is

$$E = \frac{-N}{q}\left(\frac{dq}{d\beta}\right) = \frac{-N}{(1 + e^{-\beta h\nu})}(-h\nu^{-\beta h\nu})$$

$$= \frac{Nh\nu e^{-\beta h\nu}}{1 + e^{-\beta h\nu}} = \frac{Nh\nu}{e^{\beta h\nu} + 1}$$

In the final step of this example, the expression for E was multiplied by unity in the form of $\exp(\beta h\nu)/\exp(\beta h\nu)$ to facilitate numerical evaluation.

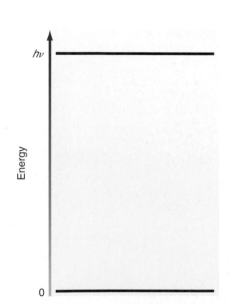

FIGURE 32.1
Depiction of the two-level system.

Figure 32.2 presents the evolution in E with T for the two-level system in Figure 32.1. The total energy as presented in the figure is divided by the number of particles in the ensemble N and the energy-level spacing $h\nu$. In addition, temperature is expressed as kT divided by the energy spacing between levels.

Two interesting trends are evident in Figure 32.2. First, at the lowest temperatures the total energy does not change appreciably as the temperature is increased until $kT/h\nu \approx 0.2$, at which point a significant increase in E is observed. Second, the total energy reaches a limiting value at high temperature, and further increases in T do not affect the total energy. Why does this occur? Recall that the change in energy of a system correlates with a change in occupation number, with higher energy levels being more readily populated as the temperature is increased. For the two-level system just presented, the probability of occupying the excited energy level from Equation (32.2) is

$$p_1 = \frac{a_1}{N} = \frac{e^{-\beta h\nu}}{q} = \frac{e^{-\beta h\nu}}{1 + e^{-\beta h\nu}} = \frac{1}{e^{\beta h\nu} + 1} \quad \textbf{(32.10)}$$

At $T = 0$, the exponential term in the denominator is infinite and $p_1 = 0$. The probability of occupying the excited state becomes finite when the amount of thermal energy available, kT, is comparable to the separation energy of the states, $h\nu$. At low temperatures $kT \ll h\nu$, and the probability of occupying the excited state is quite small such that the occupation number for the excited state a_1 is essentially zero. Inspection of Equation (32.10) demonstrates that p_1 will increase until $h\nu \ll kT$. At these elevated temperatures, the exponential term in the denominator approaches unity, and p_1 approaches its limiting value of 0.5. Because this is a two-level system, the probability of occupying the lower energy state must also be 0.5. When the probability of occupying all energy states is equal, the occupation numbers do not change. The exact result for the evolution of p_1 as a function of temperature is presented in Figure 32.3. Again, the evolution in temperature is described as $kT/h\nu$ such that kT is measured with respect to the energy gap between the two states. Notice that the change in p_1 exactly mimics the behavior observed for the total energy depicted in Figure 32.2, demonstrating that changes in energy correspond to changes in occupation number.

32.1.1 Energy and the Canonical Partition Function

Equation (32.9) allows one to calculate the ensemble energy, but to what thermodynamic quantity is this energy related? Recall that we are interested in the canonical ensemble in which N, V, and T are held constant. Because V is constant, there can be no P–V-type work, and by the first law of thermodynamics any change in internal energy must occur by heat flow, q_{Vol}. Using the first law, the change in heat is related to the change in system **internal energy** at constant volume by

$$U - U_o = q_{Vol} \quad \textbf{(32.11)}$$

In Equation (32.11), q_{Vol} is heat, not the molecular partition function. The energy as expressed by Equation (32.11) is the difference in internal energy at some finite temperature to that at 0 K. If there is residual, internal energy present at 0 K, it must be included to determine the overall energy of the system. However, by convention U_o is

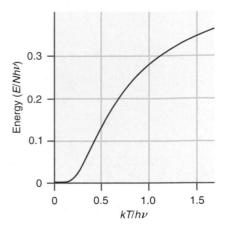
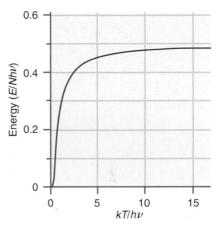

FIGURE 32.2
Total energy as a function of temperature is presented for an ensemble consisting of units that have two energy levels separated by an amount $h\nu$. Temperature is described relative to the energy gap through $kT/h\nu$. The plot on the top provides an expanded view of the low $kT/h\nu$ region plotted on the bottom.

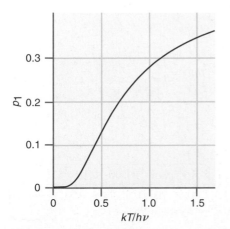
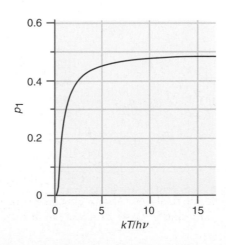

FIGURE 32.3
The probability of occupying the excited state in a two-level system is shown as a function of temperature. The evolution in temperature is plotted as $kT/h\nu$.

generally set to zero. For example, in the two-level example described earlier, the internal energy will be zero at 0 K since the energy of the ground state is defined as zero.

The second important relationship to establish is that between the internal energy and the canonical partition function. Fortunately, we have already encountered this relationship. For an ensemble of indistinguishable noninteracting particles, the canonical partition function is given by

$$Q = \frac{q^N}{N!} \tag{32.12}$$

In Equation (32.12), q is the molecular partition function. Taking the natural log of Equation (32.12) yields

$$\ln Q = \ln\left(\frac{q^N}{N!}\right) = N \ln q - \ln N! \tag{32.13}$$

Finally, taking the derivative of Equation (32.13) with respect to β and recognizing that $\ln(N!)$ is constant in the canonical ensemble,

$$\frac{d \ln Q}{d\beta} = \frac{d}{d\beta}(N \ln q) - \frac{d}{d\beta}(\ln N!)$$

$$= N\frac{d \ln q}{d\beta} \tag{32.14}$$

The last relationship is related to the total energy; therefore, the relationship between the canonical partition function and total internal energy U is simply

$$U = -\left(\frac{\partial \ln Q}{\partial \beta}\right)_V \tag{32.15}$$

EXAMPLE PROBLEM 32.2

For an ensemble consisting of 1.00 moles of particles having two energy levels separated by $hv = 1.00 \times 10^{-20}$ J, at what temperature will the internal energy of this system equal 1.00 kJ?

Solution

Using the expression for total energy and recognizing that $N = nN_A$,

$$U = -\left(\frac{\partial \ln Q}{\partial \beta}\right)_V = -nN_A\left(\frac{\partial \ln q}{\partial \beta}\right)_V$$

Evaluating the preceding expression and paying particular attention to units, we get

$$U = -nN_A\left(\frac{\partial}{\partial \beta} \ln q\right)_V = -\frac{nN_A}{q}\left(\frac{\partial q}{\partial \beta}\right)_V$$

$$\frac{U}{nN_A} = \frac{-1}{(1 + e^{-\beta hv})}\left(\frac{\partial}{\partial \beta}(1 + e^{-\beta hv})\right)_V$$

$$= \frac{hve^{-\beta hv}}{1 + e^{-\beta hv}} = \frac{hv}{e^{\beta hv} + 1}$$

$$\frac{nN_A hv}{U} - 1 = e^{\beta hv}$$

$$\ln\left(\frac{nN_A hv}{U} - 1\right) = \beta hv = \frac{hv}{kT}$$

$$T = \frac{hv}{k \ln\left(\dfrac{nN_A hv}{U} - 1\right)}$$

$$= \frac{1.00 \times 10^{-20} \text{ J}}{(1.38 \times 10^{-23} \text{ J K}^{-1}) \ln\left(\dfrac{(1.00 \text{ mol})(6.022 \times 10^{23} \text{ mol}^{-1})(1.00 \times 10^{-20} \text{ J})}{(1.00 \times 10^3 \text{ J})} - 1\right)}$$

$$= 449 \text{ K}$$

32.2 Energy and Molecular Energetic Degrees of Freedom

In the previous chapter, the relationship between the total molecular partition function (q_{Total}) and the partition functions corresponding to an individual energetic degree of freedom was derived under the assumption that the energetic degrees of freedom are not coupled:

$$q_{Total} = q_T q_R q_V q_E \qquad (32.16)$$

In this expression, the subscripts T, R, V, and E refer to translational, rotational, vibrational, and electronic energetic degrees of freedom, respectively. In a similar fashion, the internal energy can be decomposed into the contributions from each energetic degree of freedom:

$$
\begin{aligned}
U &= -\left(\frac{\partial \ln Q}{\partial \beta}\right)_V = -N\left(\frac{\partial \ln q}{\partial \beta}\right)_V \\
&= -N\left(\frac{\partial \ln (q_T q_R q_V q_E)}{\partial \beta}\right)_V \\
&= -N\left(\frac{\partial}{\partial \beta}(\ln q_T + \ln q_R + \ln q_V + \ln q_E)\right)_V \qquad (32.17) \\
&= -N\left[\left(\frac{\partial \ln q_T}{\partial \beta}\right)_V + \left(\frac{\partial \ln q_R}{\partial \beta}\right)_V + \left(\frac{\partial \ln q_V}{\partial \beta}\right)_V + \left(\frac{\partial \ln q_E}{\partial \beta}\right)_V\right] \\
&= U_T + U_R + U_V + U_E
\end{aligned}
$$

The last line in this expression demonstrates a very intuitive and important result—that the total internal energy is simply the sum of contributions from each molecular energetic degree of freedom. This result also illustrates the connection between the macroscopic property of the ensemble (internal energy) and the microscopic details of the units themselves (molecular energy levels). To relate the total internal energy to the energetic degrees of freedom, expressions for the energy contribution from each energetic degree of freedom (U_T, U_R, and so on) are needed. The remainder of this section is dedicated to deriving these relationships.

32.2.1 Translations

The contribution to the system internal energy from translational motion is

$$U_T = \frac{-N}{q_T}\left(\frac{\partial q_T}{\partial \beta}\right)_V \qquad (32.18)$$

In Equation (32.18), q_T is the translational partition function, which in three dimensions is given by

$$q_T = \frac{V}{\Lambda^3} \text{ with } \Lambda^3 = \left(\frac{h^2 \beta}{2\pi m}\right)^{3/2} \qquad (32.19)$$

In Equation (32.19), m is particle mass. With this partition function, the translational contribution to the internal energy becomes

$$
\begin{aligned}
U_T &= \frac{-N}{q_T}\left(\frac{\partial q_T}{\partial \beta}\right)_V = \frac{-N\Lambda^3}{V}\left(\frac{\partial}{\partial \beta}\frac{V}{\Lambda^3}\right)_V \\
&= -N\Lambda^3\left(\frac{\partial}{\partial \beta}\frac{1}{\Lambda^3}\right)_V \\
&= -N\Lambda^3\left(\frac{\partial}{\partial \beta}\left(\frac{2\pi m}{h^2 \beta}\right)^{3/2}\right)_V \\
&= -N\Lambda^3\left(\frac{2\pi m}{h^2}\right)^{3/2}\left(\frac{\partial}{\partial \beta}\beta^{-3/2}\right)_V \\
&= -N\Lambda^3\left(\frac{2\pi m}{h^2}\right)^{3/2}\frac{-3}{2}\beta^{-5/2}
\end{aligned}
$$

$$= \frac{3}{2} N \Lambda^3 \left(\frac{2\pi m}{h^2 \beta} \right)^{3/2} \beta^{-1}$$

$$= \frac{3}{2} N \beta^{-1}$$

$$U_T = \frac{3}{2} NkT = \frac{3}{2} nRT \tag{32.20}$$

Equation (32.20) should look familiar. Recall from Chapter 2 that the internal energy of an ideal monatomic gas is $nC_V \Delta T = 3/2(nRT)$ with $T_{initial} = 0$ K, identical to the result just obtained. The convergence between the thermodynamic and statistical mechanical descriptions of monatomic-gas systems is remarkable.

It is also interesting to note that the contribution of translational motion to the internal energy is equal to that predicted by the equipartition theorem (Section 31.8). The equipartition theorem states that any term in the classical Hamiltonian that is quadratic with respect to momentum or position will contribute $kT/2$ to the energy. The Hamiltonian corresponding to three-dimensional translational motion of an ideal monatomic gas is

$$H_{trans} = \frac{1}{2m}(p_x^2 + p_y^2 + p_z^2) \tag{32.21}$$

Each p^2 term in Equation (32.21) will contribute $1/2\, kT$ to the energy by equipartition such that the total contribution will be $3/2\, kT$, or $3/2\, RT$ for a mole of particles that is identical to the result derived using the quantum mechanical description of translational motion. This agreement is not surprising given the small energy gap between translational energy levels such that classical behavior is expected.

32.2.2 Rotations

Within the rigid rotor approximation, the rotational partition function for a diatomic molecule in the high-temperature limit is given by

$$q_R = \frac{1}{\sigma \beta hcB} \tag{32.22}$$

With this partition function, the contribution to the internal energy from rotational motion is

$$U_R = \frac{-N}{q_R} \left(\frac{\partial q_R}{\partial \beta} \right)_V = -N\sigma\beta hcB \left(\frac{\partial}{\partial \beta} \frac{1}{\sigma \beta hcB} \right)_V$$

$$= -N\beta \left(\frac{\partial}{\partial \beta} \beta^{-1} \right)_V$$

$$= -N\beta(-\beta^{-2})$$

$$= N\beta^{-1}$$

$$U_R = NkT = nRT \tag{32.23}$$

Recall from Chapter 31 that the rotational partition function employed in this derivation is for a diatomic in which the rotational temperature Θ_R is much less than kT. In this limit, a significant number of rotational states are accessible. If Θ_R is not small relative to kT, then full evaluation of the sum form of the rotational partition function is required to determine U_R.

In the high-temperature limit, the rotational energy can be thought of as containing contributions of $1/2\, kT$ from each nonvanishing moment of inertia. This partitioning of energy is analogous to the case of translational energy discussed earlier within the context of the equipartition theorem. The concept of equipartition can be used to extend the result for U_R obtained for a diatomic molecule to linear and nonlinear polyatomic molecules. Because each nonvanishing moment of inertia will provide $1/2\, kT$ to the rotational energy equipartition, we can state that

$$U_R = nRT \quad \text{(linear polyatomic)} \tag{32.24}$$

$$U_R = \frac{3}{2} nRT \quad \text{(nonlinear polyatomic)} \tag{32.25}$$

The result for nonlinear polyatomic molecules can be confirmed by evaluating the expression for the average rotational energy employing the partition function for a nonlinear polyatomic that was presented in the previous chapter.

32.2.3 Vibrations

Unlike translational and rotational degrees of freedom, vibrational energy-level spacings are typically greater than kT such that the equipartition theorem is usually not applicable to this energetic degree of freedom. Fortunately, within the harmonic oscillator model, the regular energy-level spacings provide for a relatively simple expression for the vibrational partition function neglecting zero-point energy:

$$q_V = (1 - e^{-\beta hc\tilde{\nu}})^{-1} \tag{32.26}$$

The term $\tilde{\nu}$ represents vibrational frequency in units of cm^{-1}. Using this partition function, the vibrational contribution to the average energy is

$$
\begin{aligned}
U_V &= \frac{-N}{q_V}\left(\frac{\partial q_V}{\partial \beta}\right)_V = -N(1 - e^{-\beta hc\tilde{\nu}})\left(\frac{\partial}{\partial \beta}(1 - e^{-\beta hc\tilde{\nu}})^{-1}\right)_V \\
&= -N\left(1 - e^{-\beta hc\tilde{\nu}}\right)\left(-hc\tilde{\nu}e^{-\beta hc\tilde{\nu}}\right)\left(1 - e^{-\beta hc\tilde{\nu}}\right)^{-2} \\
&= \frac{Nhc\tilde{\nu}e^{-\beta hc\tilde{\nu}}}{\left(1 - e^{-\beta hc\tilde{\nu}}\right)}
\end{aligned}
$$

$$U_V = \frac{Nhc\tilde{\nu}}{e^{\beta hc\tilde{\nu}} - 1} \tag{32.27}$$

The temperature dependence of U_V/Nhc for a vibrational degree of freedom with $\tilde{\nu} = 1000 \ cm^{-1}$ is presented in Figure 32.4. First, note that at lowest temperatures U_V is zero, which is reminiscent of the two-level example presented earlier in this chapter. At low temperatures, $kT \ll hc\tilde{\nu}$, such that there is insufficient thermal energy to populate the first excited vibrational state to an appreciable extent. However, for temperatures ≥ 1000 K the average energy increases linearly with temperature, identical to the behavior observed for translational and rotational energy. This observation suggests that a high-temperature expression for U_V also exists.

To derive the high-temperature limit expression for the vibrational energy, the exponential term in q_V is written using the following series expression:

$$e^x = 1 + x + \frac{x^2}{2!} + \cdots$$

With $x = \beta hc\tilde{\nu} = hc\tilde{\nu}/kT$, and when $kT \gg hc\tilde{\nu}$, the series is approximately equal to $1 + x$ yielding

$$
\begin{aligned}
U_V &= \frac{Nhc\tilde{\nu}}{e^{\beta hc\tilde{\nu}} - 1} = \frac{Nhc\tilde{\nu}}{(1 + \beta hc\tilde{\nu}) - 1} \\
&= \frac{N}{\beta}
\end{aligned}
$$

$$U_V = NkT = nRT \tag{32.28}$$

When the temperature is sufficiently high so that the high-temperature limit is applicable, the vibrational contribution to the internal energy is nRT, identical to the prediction of the equipartition theorem. Note that this is the contribution for a single vibrational degree of freedom and that the overall contribution will be the sum of contributions from all vibrational degrees of freedom. Furthermore, the applicability of the high-temperature approximation is dependent on the details of the vibrational energies for the system in question. In particular, the high-temperature approximation may be applicable for some low-energy vibrations but not applicable to higher energy vibrations.

The applicability of the high-temperature limit can be determined using the vibrational temperature defined previously:

$$\Theta_V = \frac{hc\tilde{\nu}}{k} \tag{32.29}$$

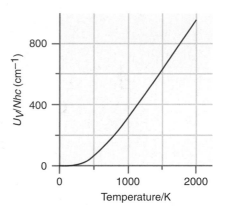

FIGURE 32.4

The variation in average vibrational energy as a function of temperature where $\tilde{\nu} = 1000 \ cm^{-1}$.

TABLE 32.1	Molecular Vibrational Temperatures		
Molecule	Θ_V (K)	Molecule	Θ_V (K)
I_2	309	N_2	3392
Br_2	468	CO	3121
Cl_2	807	O_2	2274
F_2	1329	H_2	6338

It was stated that the high-temperature limit is applicable when $T \geq 10\Theta_V$. Some examples of Θ_V for diatomic molecules are provided in Table 32.1. Inspection of the table demonstrates that relatively elevated temperatures must be reached before the high-temperature limit is applicable to the vast majority of vibrational degrees of freedom. If the high-temperature limit is not applicable, then the vibrational energy must be determined using Eq. 32.27.

32.2.4 Electronic

Because electronic energy-level spacings are generally quite large compared to kT, the partition function is simply equal to the ground-state degeneracy. Since the degeneracy is a constant, the derivative of this quantity with respect to β must be zero and

$$U_E = 0 \tag{32.30}$$

Exceptions to this result do exist. In particular, for systems in which electronic energy levels are comparable to kT, it is necessary to evaluate the full partition function.

EXAMPLE PROBLEM 32.3

The ground state of O_2 is $^3\Sigma_g^-$. When O_2 is electronically excited, emission from the excited state ($^1\Delta_g$) to the ground state is observed at 1263 nm. Calculate q_E and determine the electronic contribution to U for a mole of O_2 at 500. K.

Solution

The first step to solving this problem is construction of the electronic partition function. The ground state is threefold degenerate, and the excited state is nondegenerate. The energy of the excited state relative to the ground state is determined using the emission wavelength:

$$\varepsilon = \frac{hc}{\lambda} = \frac{(6.626 \times 10^{-34}\,\text{J s})(3.00 \times 10^8\,\text{m s}^{-1})}{1.263 \times 10^{-6}\,\text{m}} = 1.57 \times 10^{-19}\,\text{J}$$

Therefore, the electronic partition function is

$$q_E = g_0 + g_1 e^{-\beta\varepsilon} = 3 + e^{-\beta(1.57 \times 10^{-19}\,\text{J})}$$

With the electronic partition function, U_E is readily determined:

$$U_E = \frac{-nN_A}{q_E}\left(\frac{\partial q_E}{\partial \beta}\right)_V = \frac{(1\,\text{mol})\,6.022 \times 10^{23}\,\text{mol}^{-1}(1.57 \times 10^{-19}\,\text{J})e^{-\beta(1.57 \times 10^{-19}\,\text{J})}}{3 + e^{-\beta(1.57 \times 10^{-19}\,\text{J})}}$$

$$= \frac{94.5\,\text{kJ} \exp\left[-\dfrac{1.57 \times 10^{-19}\,\text{J}}{(1.38 \times 10^{-23}\,\text{J K}^{-1})(500.\,\text{K})}\right]}{3 + \exp\left[-\dfrac{1.57 \times 10^{-19}\,\text{J}}{(1.38 \times 10^{-23}\,\text{J K}^{-1})(500.\,\text{K})}\right]}$$

$$= 4.14 \times 10^{-6}\,\text{J}$$

Notice that U_E is quite small, reflecting the fact that even at 500. K the $^1\Delta_g$ excited state of O_2 is not readily populated. Therefore, the contribution to the internal energy from electronic degrees of freedom for O_2 at 500. K is negligible.

32.2.5 Review

At the beginning of this section, we established that the total average energy was simply the sum of average energies for each energetic degree of freedom. Applying this logic to a diatomic system, the total energy is given by

$$
\begin{aligned}
U_{Total} &= U_T + U_R + U_V + U_E \\
&= \frac{3}{2}NkT + NkT + \frac{Nhc\tilde{\nu}}{e^{\beta hc\tilde{\nu}} - 1} + 0 \\
&= \frac{5}{2}NkT + \frac{Nhc\tilde{\nu}}{e^{\beta hc\tilde{\nu}} - 1}
\end{aligned}
\tag{32.31}
$$

In arriving at Equation (32.31), the assumption has been made that the rotational degrees of freedom are in the high-temperature limit and that degeneracy of the ground electronic level makes the lone contribution to the electronic partition function. Although the internal energy is dependent on molecular details (B, $\tilde{\nu}$, and so on), it is important to realize that the total energy can be decomposed into contributions from each molecular degree of freedom.

32.3 Heat Capacity

As discussed in Chapter 2, the thermodynamic definition of the **heat capacity** at constant volume (C_V) is

$$
C_V = \left(\frac{\partial U}{\partial T}\right)_V = -k\beta^2\left(\frac{\partial U}{\partial \beta}\right)_V
\tag{32.32}
$$

Because the internal energy can be decomposed into contributions from each energetic degree of freedom, the heat capacity can also be similarly decomposed. In the previous section, the average internal energy contribution from each energetic degree of freedom was determined. Correspondingly, the heat capacity is given by the derivative of the internal energy with respect to temperature for a given energetic degree of freedom. In the remainder of this section, we use this approach to determine the translational, vibrational, rotational, and electronic contributions to the constant volume heat capacity.

EXAMPLE PROBLEM 32.4

Determine the heat capacity for an ensemble consisting of units that have only two energy levels separated by an arbitrary amount of energy, $h\nu$.

Solution

This is the same system discussed in Example Problem 32.1 where the partition function was determined to be $q = 1 + e^{-\beta h\nu}$. The corresponding average energy calculated using this partition function is $U = Nh\nu/(e^{\beta h\nu} + 1)$. Given the functional form of the average energy, the heat capacity is most easily determined by taking the derivative with respect to β as follows:

$$
\begin{aligned}
C_v &= -k\beta^2\left(\frac{\partial U}{\partial \beta}\right)_V = -Nk\beta^2\left(\frac{\partial}{\partial \beta}h\nu\,(e^{\beta h\nu} + 1)^{-1}\right)_V \\
&= \frac{Nk\beta^2(h\nu)^2 e^{\beta h\nu}}{(e^{\beta h\nu} + 1)^2}
\end{aligned}
$$

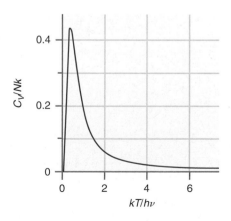

FIGURE 32.5
Constant volume heat capacity for a two-level system as a function of temperature. Note that the heat capacity has been divided by the product of Boltzmann's constant and the number of particles.

The functional form of the heat capacity is rather complex and is plotted in Figure 32.5. As observed previously, limiting behavior is observed at both low and high temperatures. At the lowest temperature, C_V is zero and then increases to a maximum value after which further increases in temperature result in a decrease in the heat capacity. This behavior is reminiscent of the evolution in energy as a function of temperature (Figure 32.2) and can be traced back to the same origin, the evolution in state populations as a function of temperature. Heat capacity is a measure of the ability of a system to absorb energy from the surroundings. At the lowest temperatures, there is insufficient energy to access the excited state such that the two-level system is incapable of absorbing energy. As the available thermal energy is increased, the excited state becomes accessible and the heat capacity increases. Finally, at the highest temperatures the populations in the ground and excited states reach their limiting values of 0.5. Here, the system is again incapable of absorbing energy, and the heat capacity approaches zero.

32.3.1 Translational Heat Capacity

Translational energy-level spacings are extremely small such that the high-temperature approximation is valid. The contribution of translational motion to the average molecular energy for an ideal gas is

$$U_T = \frac{3}{2}NkT \tag{32.33}$$

Therefore, the translational contribution to C_V is readily determined by taking the derivative of the preceding expression with respect to temperature:

$$(C_V)_T = \left(\frac{\partial U_T}{\partial T}\right)_V = \frac{3}{2}Nk \tag{32.34}$$

This result demonstrates that the contribution of translations to the overall constant volume heat capacity is simply a constant, with no temperature dependence. The constant value of the translational contribution to C_V down to very low temperatures is a consequence of the dense manifold of closely spaced translational states.

32.3.2 Rotational Heat Capacity

Assuming that the rotational energy spacings are small relative to kT, the high-temperature expression for the rotational contribution to the average energy can be used to determine the rotational contribution to C_V. The internal energy is dependent on molecular geometry:

$$U_R = NkT \text{ (linear)} \tag{32.35}$$

$$U_R = \frac{3}{2}NkT \text{ (nonlinear)} \tag{32.36}$$

Using these two equations, the rotational contribution to C_V becomes

$$(C_V)_R = Nk \text{ (linear)} \tag{32.37}$$

$$(C_V)_R = \frac{3}{2}Nk \text{ (nonlinear)} \tag{32.38}$$

Note that these expressions are correct in the high-temperature limit. What will occur at low temperature? At the lowest temperatures, there is insufficient thermal energy to provide for population of excited rotational energy levels. Therefore, the heat capacity approaches zero. As temperature is increased, the heat capacity increases until the high-temperature limit is reached. At these intermediate temperatures, the heat capacity must be determined by evaluating the summation form of q_R.

32.3.3 Vibrational Heat Capacity

In contrast to translations and rotations, the high-temperature limit is generally not applicable to the vibrational degrees of freedom. Therefore, the exact functional form of the energy must be evaluated to determine the vibrational contribution to C_V. The derivation is somewhat more involved, but still straightforward. First, recall that the vibrational contribution to U is [Equation (32.27)]

$$U_V = \frac{Nhc\tilde{\nu}}{e^{\beta hc\tilde{\nu}} - 1} \tag{32.39}$$

Given this equation, the vibrational contribution to the constant volume heat capacity is

$$(C_V)_{Vib} = \left(\frac{\partial U_V}{\partial T}\right)_V = -k\beta^2 \left(\frac{\partial U_V}{\partial \beta}\right)_V$$

$$= -Nk\beta^2 hc\tilde{\nu}\left(\frac{\partial}{\partial \beta}(e^{\beta hc\tilde{\nu}} - 1)^{-1}\right)_V$$

$$= -Nk\beta^2 hc\tilde{\nu}(-hc\tilde{\nu}e^{\beta hc\tilde{\nu}}(e^{\beta hc\tilde{\nu}} - 1)^{-2})V$$

$$(C_V)_{Vib} = Nk\beta^2(hc\tilde{\nu})^2\frac{e^{\beta hc\tilde{\nu}}}{(e^{\beta hc\tilde{\nu}} - 1)^2} \qquad (32.40)$$

The heat capacity can also be cast in terms of the vibrational temperature Θ_V as follows

$$(C_V)_{Vib} = Nk\left(\frac{\Theta_V}{T}\right)^2\frac{e^{\Theta_V/T}}{(e^{\Theta_V/T} - 1)^2} \qquad (32.41)$$

As mentioned in Chapter 31, a polyatomic molecule has $3N - 6$ or $3N - 5$ vibrational degrees of freedom, respectively. Each vibrational degree of freedom contributes to the overall vibrational constant volume heat capacity such that

$$(C_V)_{Vib,Total} = \sum_{m=1}^{3N-6 \text{ or } 3N-5}(C_V)_{Vib,m} \qquad (32.42)$$

How does the vibrational heat capacity change as a function of T? The contribution of a given vibrational degree of freedom depends on the spacing between vibrational levels relative to kT. Therefore, at lowest temperatures we expect the vibrational contribution to C_V to be zero. As the temperature increases, the lowest energy vibrational modes have spacings comparable to kT such that these modes will contribute to the heat capacity. Finally, the highest energy vibrational modes contribute only at high temperatures. In summary, we expect the vibrational contribution to C_V to demonstrate a significant temperature dependence.

Figure 32.6 presents the vibrational contribution to C_V as a function of temperature for a nonlinear triatomic molecule having three nondegenerate vibrational modes of frequency 100, 1000, and 3000 cm^{-1}. As expected, the lowest frequency mode contributes to the heat capacity at lowest temperatures, reaching a constant value of Nk for temperatures greatly in excess of Θ_V. As the temperature increases, each successively higher energy vibrational mode begins to contribute to the heat capacity and reaches the same limiting value of Nk at highest temperatures. Finally, the total vibrational heat capacity, which is simply the sum of the contributions from the individual modes, approaches a limiting value of $3Nk$.

The behavior illustrated in Figure 32.6 demonstrates that for temperatures where $kT \gg hc\tilde{\nu}$, the heat capacity approaches a constant value, suggesting that, like translations and rotations, there is a high-temperature limit for the vibrational contribution to C_V. Recall that the high-temperature approximation for the average energy for an individual vibrational degree of freedom is

$$U_V = \frac{N}{\beta} = NkT \qquad (32.43)$$

Differentiation of this expression with respect to temperature reveals that the vibrational contribution to the C_V is indeed equal to Nk per vibrational mode in the high-temperature limit as illustrated in Figure 32.6 and as expected from the equipartition theorem.

32.3.4 Electronic Heat Capacity

Because the partition function for the energetic degree of freedom is generally equal to the ground-state degeneracy, the resulting average energy is zero. Therefore, there is no contribution to the constant volume heat capacity from these degrees of freedom. However, for systems with electronic excited states that are comparable to kT, the contribution to C_V from electronic degrees of freedom can be finite and must be determined using the summation form of the partition function.

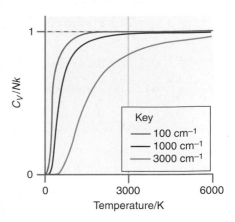

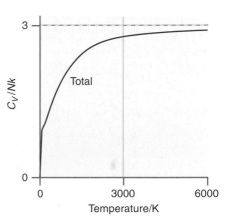

FIGURE 32.6

Evolution in the vibrational contribution to C_V as a function of temperature. Calculations are for a molecule with three vibrational degrees of freedom as indicated in the top panel. Contribution for each vibrational mode (top) and the total vibrational contribution (bottom).

32.3.5 Review

Similar to energy, the overall constant volume heat capacity is simply the sum of the contributions to the heat capacity from each individual degree of freedom. For example, for a diatomic molecule with translational and rotational degrees of freedom well described by the high-temperature approximation, we have

$$
\begin{aligned}
(C_V)_{Total} &= (C_V)_T + (C_V)_R + (C_V)_{Vib} \\
&= \frac{3}{2}Nk + Nk + Nk\beta^2(hc\widetilde{\nu})^2 \frac{e^{\beta hc\widetilde{\nu}}}{\left(e^{\beta hc\widetilde{\nu}} - 1\right)^2} \\
&= \frac{5}{2}Nk + Nk\beta^2(hc\widetilde{\nu})^2 \frac{e^{\beta hc\widetilde{\nu}}}{\left(e^{\beta hc\widetilde{\nu}} - 1\right)^2}
\end{aligned}
\tag{32.44}
$$

The theoretical prediction for the heat capacity for gaseous HCl as a function of temperature is presented in Figure 32.7. The figure illustrates that at the lowest temperatures the heat capacity is $3/2Nk$ due to the contribution of translational motion. As the temperature is increased, the rotational contribution to the heat capacity increases, reaching the high-temperature limit at ~150 K, which is approximately 10-fold greater than the rotational temperature of 15.2 K. The temperature dependence of the rotational contribution was determined by numerical evaluation, a tedious procedure that must be performed when the high-temperature limit is not applicable. Finally, the vibrational contribution to the heat capacity becomes significant at highest temperatures consistent with the high vibrational temperature of this molecule (~4000 K). The curve depicted in the figure is theoretical and does not take into account molecular dissociation or phase transitions.

32.3.6 The Einstein Solid

The **Einstein solid** model was developed to describe the thermodynamic properties of atomic crystalline systems. In this model, each atom is envisioned to occupy a lattice site where the restoring potential is described as a three-dimensional harmonic oscillator. All of the harmonic oscillators are assumed to be separable such that motion of the atom in one dimension does not affect the vibrational motion (i.e., energy) in orthogonal dimensions. Finally, the harmonic oscillators are assumed to be isoenergetic such that they are characterized by the same frequency. For a crystal containing N atoms, there are $3N$ vibrational degrees of freedom, and the total heat capacity will simply be the sum of contributions from each vibrational degree of freedom. Therefore,

$$
(C_V)_{Total} = 3Nk\left(\frac{\Theta_V}{T}\right)^2 \frac{e^{\Theta_V/T}}{\left(e^{\Theta_V/T} - 1\right)^2}
\tag{32.45}
$$

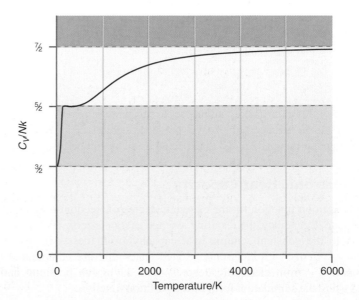

FIGURE 32.7

The constant volume heat capacity for gaseous HCl as a function of temperature. The contributions of translational (yellow), rotational (orange), and vibrational (light green) degrees of freedom to the heat capacity are shown.

This total heat capacity is identical to that of a collection of $3N$ harmonic oscillators. The important aspect to note about Equation (32.45) is that C_V of the crystal at a given temperature is predicted to depend only on the vibrational frequency. This suggests that measurement of the constant volume heat capacity of an atomic crystal as a function of temperature allows for a determination of the characteristic vibrational temperature and corresponding vibrational frequency. This was one of the earliest quantum mechanical models used to describe a thermodynamic observable. For example, a comparison of the Einstein model and experiment for diamond ($\Theta_V = 1320$ K) is presented in Figure 32.8, and the agreement is remarkably good. Note that the model predicts that the heat capacity should reach a limiting value of 24.91 J mol^{-1} K^{-1}, or $3R$, at high temperature. This limiting value is known as the **Dulong and Petit law,** and represents the high-temperature or classical prediction for the heat capacity of such systems. The elegance of the Einstein model is that it provides a reason for the failure of classical models at low temperatures where significant deviations from the Dulong and Petit law are observed. As has been seen, the quantum nature of vibrational degrees of freedom becomes exceedingly important when $\Theta_V > T$, and the Einstein model provided a critical view into the microscopic nature of matter manifesting itself in macroscopic behavior that classical mechanics cannot explain.

A second prediction of the Einstein model is that a plot of C_V versus T/Θ_V can describe the variation in heat capacity for all solids. Figure 32.9 presents a comparison of this prediction to the experimental values for three atomic crystals. The agreement is nothing short of astonishing and suggests that one only need measure C_V at a single temperature to determine the characteristic Θ_V for a given solid! Furthermore, with Θ_V in hand, the heat capacity at all other temperatures is defined! In short, starting with a quantum model for vibrations and the mathematics of probability theory, a unifying theory of heat capacity for atomic solids has been developed. This is quite an accomplishment.

The agreement between the Einstein model and experiment at high temperatures is excellent; however, discrepancies between this theory and experiment have been noted. First, the model is applicable only to atomic crystals, and it does not accurately reproduce the temperature dependence of C_V for molecular solids. Second, the Einstein model predicts that the heat capacity should demonstrate exponential dependence at low temperature; however, experimentally the heat capacity demonstrates T^3 dependence. The discrepancy between the Einstein model and experiment lies in the fact that the normal frequencies of the crystal are not due to vibrations of single atoms but are due to concerted harmonic motion of all the atoms. These lattice vibrations are not all characterized by the same frequency. More advanced treatments of atomic crystals, such as the Debye model, are capable of quantitatively reproducing crystalline heat capacities.

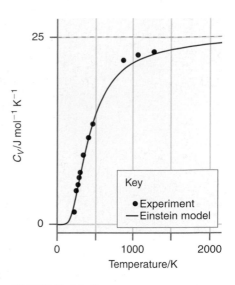

FIGURE 32.8
Comparison of C_V for diamond to the theoretical prediction of the Einstein solid model. The classical limit of 24.91 J mol^{-1} K^{-1} is shown as the dashed line.

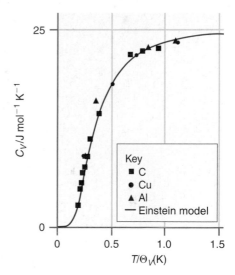

FIGURE 32.9
Comparison of C_V versus $T > \Theta_V$ for C, Cu, and Al.

32.4 Entropy

Entropy is perhaps the most misunderstood thermodynamic property of matter. Introductory chemistry courses generally describe the entropic driving force for a reaction as an inherent desire for the system to increase its "randomness" or "disorder." With statistical mechanics, we will see that the tendency of an isolated system to evolve toward the state of maximum entropy is a direct consequence of statistics. To illustrate this point, recall from Chapter 30 that a system approaches equilibrium by achieving the configuration of energy with the maximum weight. In other words, the configuration of energy that will be observed at equilibrium corresponds to W_{max}. The tendency of a system to maximize W and entropy S suggests that the relationship exists between these quantities. Boltzmann expressed this relationship in the following equation, known as **Boltzmann's formula:**

$$S = k \ln W \qquad (32.46)$$

This relationship states that entropy is directly proportional to $\ln(W)$, with the Boltzmann constant serving as the proportionality constant. Equation (32.46) makes it clear that a maximum in W will correspond to a maximum in S. Although at first view the Boltzmann formula appears to be a rather ad hoc statement, this formula is in fact equivalent to the thermodynamic definition of entropy discussed in Chapter 5. To illustrate this equivalence, consider the energy of an ensemble of particles. This energy is equal to the sum of the product of level energies and the occupation numbers for these levels such that

$$E = \sum_n \varepsilon_n a_n \tag{32.47}$$

The total differential of E is given by

$$dE = \sum_n \varepsilon_n \, da_n + \sum_n a_n \, d\varepsilon_n \tag{32.48}$$

Because volume is constant in the canonical ensemble, the level energies are also constant. Therefore, the second term in the preceding expression is zero, and the change in energy is related exclusively to changes in level occupation numbers:

$$dE = \sum_n \varepsilon_n \, da_n \tag{32.49}$$

The constraint of constant volume also dictates that there is no P–V-type work; therefore, by the first law of thermodynamics the change in energy must be due to heat flow:

$$dE = dq_{rev} = \sum_n \varepsilon_n \, da_n \tag{32.50}$$

where dq_{rev} is the reversible heat exchanged between system and surroundings. The thermodynamic definition of entropy is

$$dS = \frac{dq_{rev}}{T} \tag{32.51}$$

Comparison of the last two equations provides for the following definition of entropy:

$$dS = \frac{1}{T} \sum_n \varepsilon_n \, da_n = k\beta \sum_n \varepsilon_n \, da_n \tag{32.52}$$

We have rewritten T^{-1} as $k\beta$ anticipating our upcoming need for β. In the derivation of the Boltzmann distribution presented in Chapter 30, the following relationship was obtained:

$$\left(\frac{d \ln W}{da_n} \right) + \alpha - \beta \varepsilon_n = 0$$

$$\beta \varepsilon_n = \left(\frac{d \ln W}{da_n} \right) + \alpha \tag{32.53}$$

where α and β are constants (i.e., the Lagrange multipliers in the derivation of the Boltzmann distribution). Employing Equation (32.53), our expression for entropy becomes

$$
\begin{aligned}
dS = k\beta \sum_n \varepsilon_n \, da_n &= k \sum_n \left(\frac{d \ln W}{da_n} \right) da_n + k \sum_n \alpha \, da_n \\
&= k \sum_n \left(\frac{d \ln W}{da_n} \right) da_n + k\alpha \sum_n da_n \\
&= k \sum_n \left(\frac{d \ln W}{da_n} \right) da_n \\
&= k(d \ln W)
\end{aligned}
\tag{32.54}
$$

The third line above was obtained by recognizing that when the change in occupation number is summed over all possible states, this sum must equal zero since the gain in occupation number for a given state must be accompanied by the loss in another state. Equation (32.54) is simply an alternative form of the Boltzmann formula of Equation (32.46). The definition of entropy provided by the Boltzmann formula provides insight into the underpinnings of entropy, but how does one calculate the entropy for a molecular system? The answer to this question requires a bit more work. Because the partition function provides a measure of energy state accessibility, it can be assumed that a relationship between the partition function and entropy exists. This relationship can be derived by applying the Boltzmann formula as follows:

$$S = k \ln W = k \ln\left(\frac{N!}{\prod_n a_n!}\right) \tag{32.55}$$

which can be reduced as follows:

$$S = k \ln\left(\frac{N!}{\prod_n a_n!}\right)$$

$$= k \ln N! - k \ln \prod_n a_n!$$

$$= k \ln N! - k \sum_n \ln a_n!$$

$$= k(N \ln N - N) - k \sum_n (a_n \ln a_n - a_n)$$

$$= k\left(N \ln N - \sum_n a_n \ln a_n\right) \tag{32.56}$$

In the last step the following definition for N was used:

$$N = \sum_n a_n \tag{32.57}$$

Using this definition again:

$$S = k\left(N \ln N - \sum_n a_n \ln a_n\right)$$

$$= k\left(\sum_n a_n \ln N - \sum_n a_n \ln a_n\right)$$

$$= -k \sum_n a_n \ln \frac{a_n}{N}$$

$$= -k \sum_n a_n \ln p_n \tag{32.58}$$

In the last line of this equation, p_n is the probability of occupying energy level n as defined previously in Equation (32.10). The Boltzmann distribution law can be used to rewrite the probability term in Equation (32.58) as follows:

$$\ln p_n = \ln\left(\frac{e^{-\beta\varepsilon_n}}{q}\right) = -\beta\varepsilon_n - \ln q \tag{32.59}$$

The expression for entropy then becomes

$$S = -k \sum_n a_n \ln p_n$$

$$= -k \sum_n a_n(-\beta\varepsilon_n - \ln q)$$

$$= k\beta \sum_n a_n\varepsilon_n + k \sum_n a_n \ln q$$

$$= k\beta E + kN \ln q$$

$$= \frac{E}{T} + k \ln q^N$$

$$S = \frac{E}{T} + k \ln Q = \frac{U}{T} + k \ln Q \tag{32.60}$$

In this expression, U is the internal energy of the system and Q is the canonical partition function. In Equation (32.60), E is replaced by U consistent with the previous discussion regarding internal energy. The internal energy is related to the canonical partition function by

$$U = -\left(\frac{\partial \ln Q}{\partial \beta}\right)_V = kT^2\left(\frac{\partial \ln Q}{\partial T}\right)_V \qquad (32.61)$$

Using this relationship, we arrive at a very compact expression for entropy:

$$S = \frac{U}{T} + k \ln Q = kT\left(\frac{\partial \ln Q}{\partial T}\right)_V + k \ln Q$$

$$S = \left(\frac{\partial}{\partial T}(kT \ln Q)\right)_V \qquad (32.62)$$

32.4.1 Entropy of an Ideal Monatomic Gas

What is the general expression for the molar entropy of an ideal monatomic gas? A monatomic gas is a collection of indistinguishable particles. Assuming that the electronic partition function is unity (i.e., the ground electronic energy level is nondegenerate), only translational degrees of freedom remain to be evaluated, resulting in

$$S = \frac{U}{T} + k \ln Q$$

$$= \frac{1}{T}\left(\frac{3}{2}NkT\right) + k \ln \frac{q_{trans}^N}{N!}$$

$$= \frac{3}{2}Nk + Nk \ln q_{trans} - k(N \ln N - N)$$

$$= \frac{5}{2}Nk + Nk \ln q_{trans} - Nk \ln N$$

$$= \frac{5}{2}Nk + Nk \ln \frac{V}{\Lambda^3} - Nk \ln N$$

$$= \frac{5}{2}Nk + Nk \ln V - Nk \ln \Lambda^3 - Nk \ln N$$

$$= \frac{5}{2}Nk + Nk \ln V - Nk \ln \left(\frac{h^2}{2\pi mkT}\right)^{3/2} - Nk \ln N$$

$$= \frac{5}{2}Nk + Nk \ln V + \frac{3}{2}Nk \ln T - Nk \ln \left(\frac{N^{2/3}h^2}{2\pi mk}\right)^{3/2}$$

$$= \frac{5}{2}nR + nR \ln V + \frac{3}{2}nR \ln T - nR \ln \left(\frac{n^{2/3}N_A^{2/3}h^2}{2\pi mk}\right)^{3/2} \qquad (32.63)$$

The final line of Equation (32.63) is a version of the **Sackur–Tetrode equation,** which can be written in the more compact form:

$$S = nR \ln\left[\frac{e^{5/2} V}{\Lambda^3 N}\right] = nR \ln\left[\frac{RTe^{5/2}}{\Lambda^3 N_A P}\right] \quad \text{where } \Lambda^3 = \left(\frac{h^2}{2\pi mkT}\right)^{3/2} \qquad (32.64)$$

The Sackur–Tetrode equation reproduces many of the classical thermodynamics properties of ideal monatomic gases encountered previously. For example, consider the isothermal expansion of an ideal monatomic gas from an initial volume V_1 to a final volume V_2. Inspection of the expanded form of the Sackur–Tetrode equation [Equation (32.63)] demonstrates that all of the terms in this expression are unchanged except for the second term involving volume such that

$$\Delta S = S_{final} - S_{initial} = nR \ln\frac{V_2}{V_1} \qquad (32.65)$$

This is the same result obtained from classical thermodynamics. What if the entropy change were initiated by isochoric ($\Delta V = 0$) heating? Using the difference in temperature between initial (T_1) and final (T_2) states, Equation (32.63) yields

$$\Delta S = S_{final} - S_{initial} = \frac{3}{2}nR \ln\frac{T_2}{T_1} = nC_V \ln\frac{T_2}{T_1} \qquad (32.66)$$

Recognizing that $C_V = 3/2\, R$ for an ideal monatomic gas, we again arrive at a result first encountered in thermodynamics.

Does the Sackur–Tetrode equation provide any information not available from thermodynamics? Indeed, note the first and fourth terms in Equation (32.63). These terms are simply constants, with the latter varying with the atomic mass. Classical thermodynamics is entirely incapable of explaining the origin of these terms, and only through empirical studies could the presence of these terms be determined. However, their contribution to entropy appears naturally (and elegantly) when using the statistical perspective.

EXAMPLE PROBLEM 32.5

Determine the standard molar entropy of Ne and Kr under standard thermodynamic conditions.

Solution

Beginning with the expression for entropy [see Equation (32.63)]:

$$S = \frac{5}{2}R + R\ln\left(\frac{V}{\Lambda^3}\right) - R\ln N_A$$

$$= \frac{5}{2}R + R\ln\left(\frac{V}{\Lambda^3}\right) - 54.75\,R$$

$$= R\ln\left(\frac{V}{\Lambda^3}\right) - 52.25\,R$$

The conventional standard state is defined by $T = 298$ K and $V_m = 24.4$ L (0.0244 m^3). The thermal wavelength for Ne is

$$\Lambda = \left(\frac{h^2}{2\pi mkT}\right)^{1/2}$$

$$= \left(\frac{(6.626 \times 10^{-34}\,\text{J s})^2}{2\pi\left(\dfrac{0.02018\text{ kg mol}^{-1}}{N_A}\right)(1.38 \times 10^{-23}\,\text{J K}^{-1})(298\text{ K})}\right)^{1/2}$$

$$= 2.25 \times 10^{-11}\,\text{m}$$

Using this value for the thermal wavelength, the entropy becomes

$$S = R\ln\left(\frac{0.0244\text{ m}^3}{(2.25 \times 10^{-11}\,\text{m})^3}\right) - 52.25\,R$$

$$= 69.84\,R - 52.25\,R = 17.59\,R = 146\text{ J mol}^{-1}\text{K}^{-1}$$

The experimental value is 146.48 J mol^{-1} K^{-1}! Rather than determining the entropy of Kr directly, it is easier to determine the difference in entropy relative to Ne:

$$\Delta S = S_{Kr} - S_{Ne} = S = R\ln\left(\frac{V}{\Lambda_{Kr}^3}\right) - R\ln\left(\frac{V}{\Lambda_{Ne}^3}\right)$$

$$= R\ln\left(\frac{\Lambda_{Ne}}{\Lambda_{Kr}}\right)^3$$

$$= 3R\ln\left(\frac{\Lambda_{Ne}}{\Lambda_{Kr}}\right)$$

$$= 3R\ln\left(\frac{m_{Kr}}{m_{Ne}}\right)^{1/2}$$

$$= \frac{3}{2}R\ln\left(\frac{m_{Kr}}{m_{Ne}}\right) = \frac{3}{2}R\ln(4.15)$$

$$= 17.7\text{ J mol}^{-1}\text{K}^{-1}$$

Using this difference, the standard molar entropy of Kr becomes

$$S_{Kr} = \Delta S + S_{Ne} = 164 \text{ J mol}^{-1} \text{K}^{-1}$$

The experimental value is 163.89 J mol^{-1} K^{-1}, and the calculated value is again in excellent agreement!

When calculating the entropy for an ideal gas consisting of diatomic or polyatomic molecules, it is best to start with the general expression for entropy [Equation (32.62)] and express the canonical partition function in terms of the product of molecular partition functions for each energetic degree of freedom. In addition to the translational entropy term derived earlier for an ideal monatomic gas, contributions from rotational, vibrational, and electronic degrees of freedom will also be included when calculating the entropy.

32.5 Residual Entropy

As illustrated in Example Problem 32.5, when the entropy calculated using statistical mechanics is compared to experiment, good agreement is observed for a variety of atomic and molecular systems. However, for many molecular systems, this agreement is less than ideal. A famous example of such a system is carbon monoxide where the calculated entropy at thermodynamic standard temperature and pressure is 197.9 J mol^{-1} K^{-1} and the experimental value is only 193.3 J mol^{-1} K^{-1}. In this and other systems, the calculated entropy is always greater than that observed experimentally.

The reason for the systematic discrepancy between calculated and experimental entropies for such systems is **residual entropy,** or entropy associated with molecular orientation in the molecular crystal at low temperature. Using CO as an example, the weak electric dipole moment of the molecule dictates that dipole-dipole interactions do not play a dominant role in determining the orientation of one CO molecule relative to neighboring molecules in a crystal. Therefore, each CO can assume one of two orientations as illustrated in Figure 32.10. The solid corresponding to the possible orientations of CO will have an inherent randomness to it. Because each CO molecule can assume one of two possible orientations, the entropy associated with this orientational disorder is

$$S = k \ln W = k \ln 2^N = Nk \ln 2 = nR \ln 2 \qquad \textbf{(32.67)}$$

In Equation (32.67), W is the total number of CO arrangements possible, and it is equal to 2^N where N is the number of CO molecules. For a system consisting of 1 mol of CO, the residual entropy is predicted to be $R \ln 2$ or 5.76 J mol^{-1} K^{-1}, roughly equal to the difference between the experimental and calculated entropy values.

Finally, note that the concept of residual entropy sheds light on the origin of the third law of thermodynamics. As discussed in Chapter 5, the third law states that the entropy of a pure and crystalline substance is zero at 0 K. By "pure and crystalline," the third law means that the system must be pure with respect to both composition (i.e., a single component) and orientation in the solid at 0 K. For such a pure system, $W = 1$ and correspondingly $S = 0$ by Equation (32.67). Therefore, the definition of zero entropy provided by the third law is a natural consequence of the statistical nature of matter.

FIGURE 32.10

The origin of residual entropy for CO. Each CO molecule in the solid can have one of two possible orientations as illustrated by the central CO. Each CO will have two possible directions such that the total number of arrangements possible is 2^N where N is the number of CO molecules.

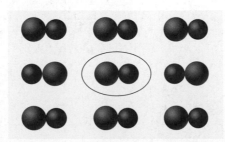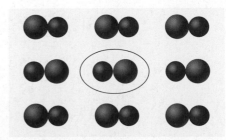

> **EXAMPLE PROBLEM 32.6**
>
> The van der Waals radii of H and F are similar such that steric effects on molecular ordering in the crystal are minimal. Do you expect the residual molar entropies for crystalline 1,2-difluorobenzene and 1,4-difluorobenzene to be the same?
>
> **Solution**
>
> The structures of 1,2-difluorobenzene and 1,4-difluorobenzene are shown here:
>
>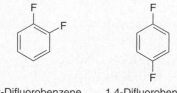
>
> 1,2-Difluorobenzene   1,4-Difluorobenzene
>
> In crystalline 1,2-difluorobenzene, there will be six possible arrangements that can be visualized by rotation of the molecule. Therefore, $W = 6^N$ and
>
> $$S = k \ln W = k \ln 6^{N_A} = N_A k \ln 6 = R \ln 6$$
>
> Similarly, for 1,4-difluorobenzene, there are three possible arrangements such that $W = 3^N$ and $S = R \ln 3$. The residual molar entropies are expected to differ for molecular crystals involving these species.

32.6 Other Thermodynamic Functions

The discussion thus far illustrates the convergence of the statistical and thermodynamics perspectives, and the utility of the statistical perspective in illustrating the underlying contributions to macroscopic properties of matter is impressive. The relationship between the canonical partition function and other thermodynamic quantities can be derived using the following familiar thermodynamic relationships for enthalpy H, Helmholtz energy A, and Gibbs energy G:

$$H = U + PV \tag{32.68}$$

$$A = U - TS \tag{32.69}$$

$$G = H - TS \tag{32.70}$$

Using these expressions, relationships between the canonical partition function and the previous thermodynamic quantities can be derived.

32.6.1 Helmholtz Energy

Beginning with the thermodynamic definition of A from Equation (32.69), the relationship of this quantity to the canonical partition function becomes

$$
\begin{aligned}
A &= U - TS \\
  &= U - T\left(\frac{U}{T} + k \ln Q\right) \\
A &= -kT \ln Q
\end{aligned}
\tag{32.71}
$$

The **Helmholtz energy** provides an interesting pathway to obtain a well-known relationship: the ideal gas law. From thermodynamics, pressure is related to the Helmholtz energy as described in Section 6.2 by

$$P = \left(\frac{-\partial A}{\partial V}\right)_T \tag{32.72}$$

Substituting the definition for A provided in Equation (32.71) into the preceding equation,

$$P = \left(\frac{-\partial}{\partial V}(-kT \ln Q) \right)_T$$

$$= kT \left(\frac{\partial}{\partial V} \ln Q \right)_T \tag{32.73}$$

The canonical partition function for an ideal gas is

$$Q = \frac{q^N}{N!}$$

Substituting this expression for the canonical partition function into the preceding equation,

$$P = kT \left(\frac{\partial}{\partial V} \ln \frac{q^N}{N!} \right)_T$$

$$= kT \left(\frac{\partial}{\partial V} \ln q^N - \frac{\partial}{\partial V} \ln N! \right)_T$$

$$= kT \left(\frac{\partial}{\partial V} \ln q^N \right)_T$$

$$= NkT \left(\frac{\partial}{\partial V} \ln q \right)_T \tag{32.74}$$

For a monatomic gas,

$$q = \frac{V}{\Lambda^3} \tag{32.75}$$

such that

$$P = NkT \left(\frac{\partial}{\partial V} \ln \frac{V}{\Lambda^3} \right)_T$$

$$= NkT \left(\frac{\partial}{\partial V} \ln V - \frac{\partial}{\partial V} \ln \Lambda^3 \right)_T$$

$$= NkT \left(\frac{\partial}{\partial V} \ln V \right)_T$$

$$= \frac{NkT}{V} = \frac{nRT}{V} \tag{32.76}$$

This result is nothing short of remarkable. This relationship was first obtained by empirical measurements, expressed as the laws of Boyle, Charles, Avogadro, and Gay-Lussac. However, in deriving this relationship we have not employed these laws. Instead, starting with a quantum mechanical model for translational motion, we have derived the ideal gas law from a purely theoretical perspective. Once again, statistical mechanics has provided microscopic insight into a macroscopic property. This example illustrates one of the major contributions of statistical mechanics to physical chemistry: the ability to predict relationships that are largely derived from empirical observations and subsequently stated as thermodynamic laws. As a final point, an objection that may be raised is that the preceding derivation was performed for a monatomic gas. However, the skeptical reader is encouraged to demonstrate that this result also holds for polyatomic systems, a task that is provided as a problem at the end of this chapter.

32.6.2 Enthalpy

Using the thermodynamic definition of enthalpy from Equation (32.68), this quantity can be expressed in terms of the canonical partition function as follows:

$$H = U + PV$$

$$= \left(\frac{-\partial}{\partial \beta} \ln Q \right)_V + V \left(\frac{-\partial A}{\partial V} \right)_T$$

$$= \left(\frac{-\partial}{\partial\beta}\ln Q\right)_V + V\left(\frac{-\partial}{dV}(-kT\ln Q)\right)_T$$

$$= kT^2\left(\frac{\partial}{\partial T}\ln Q\right)_V + VkT\left(\frac{\partial}{\partial V}\ln Q\right)_T$$

$$H = T\left[kT\left(\frac{\partial}{\partial T}\ln Q\right)_V + Vk\left(\frac{\partial}{\partial V}\ln Q\right)_T\right] \qquad \textbf{(32.77)}$$

Although Equation (32.77) is correct, it clearly requires a bit of work to implement. Yet, the statistical perspective has shown that one can relate enthalpy to microscopic molecular details through the partition function. When calculating enthalpy, it is sometimes easier to use a combination of thermodynamic and statistical perspectives, as the following example demonstrates.

EXAMPLE PROBLEM 32.7

What is the enthalpy of 1 mol of an ideal monatomic gas?

Solution

One approach to this problem is to start with the expression for the canonical partition function in terms of the molecular partition function for an ideal monatomic gas and evaluate the result. However, a more efficient approach is to begin with the thermodynamic definition of enthalpy:

$$H = U + PV$$

Recall that the translational contribution to U is $3/2\ RT$, and this is the only degree of freedom operative for the monatomic gas. In addition, we can apply the ideal gas law (because we have now demonstrated its validity from a statistical perspective) such that the enthalpy is simply

$$H = U + PV$$
$$= \frac{3}{2}RT + RT$$
$$= \frac{5}{2}RT$$

The interested reader is encouraged to obtain this result through a full evaluation of the statistical expression for enthalpy.

32.6.3 Gibbs Energy

Perhaps the most important state function to emerge from thermodynamics is the **Gibbs energy.** Using this quantity, one can determine if a chemical reaction will occur spontaneously. The statistical expression for the Gibbs energy is also derived starting with the thermodynamic definition of this quantity [Equation (32.70)]:

$$G = A + PV$$
$$= -kT\ln Q + VkT\left(\frac{\partial}{\partial V}\ln Q\right)_T$$
$$G = -kT\left[\ln Q - V\left(\frac{\partial\ln Q}{\partial V}\right)_T\right] \qquad \textbf{(32.78)}$$

Previously derived expressions for the Helmholtz energy and pressure were employed in arriving at the result of Equation (32.78). A more intuitive result can be derived by applying the preceding relationship to an ideal gas such that $PV = nRT = NkT$. With this relationship

$$G = A + PV$$
$$= -kT\ln Q + NkT$$
$$= -kT\ln\left(\frac{q^N}{N!}\right) + NkT$$

$$= -kT \ln q^N + kT \ln N! + NkT$$

$$= -NkT \ln q + kT(N \ln N - N) + NkT$$

$$= -NkT \ln q + NkT \ln N$$

$$G = -NkT \ln\left(\frac{q}{N}\right) = -nRT \ln\left(\frac{q}{N}\right) \tag{32.79}$$

This relationship is extremely important because it provides insight into the origin of the Gibbs energy. At constant temperature the nRT prefactor in the expression for G is equivalent for all species; therefore, differences in the Gibbs energy between chemical species must be due to the partition function. Because the Gibbs energy is proportional to $-\ln(q)$, an increase in the value for the partition function will result in a lower Gibbs energy. The partition function quantifies the number of states that are accessible at a given temperature; therefore, the statistical perspective dictates that species with a comparatively greater number of accessible energy states will have a lower Gibbs energy. This relationship will have profound consequences when discussing chemical equilibria in the next section.

EXAMPLE PROBLEM 32.8

Calculate the molar Gibbs energy of Ar at 298.15 K and 10^5 Pa, assuming that the gas demonstrates ideal behavior.

Solution

Argon is a monatomic gas; therefore, $q = q_{trans}$. Using Equation (32.79),

$$G° = -nRT \ln\left(\frac{q}{N}\right) = -nRT \ln\left(\frac{V}{N\Lambda^3}\right)$$

$$= -nRT \ln\left(\frac{kT}{P\Lambda^3}\right)$$

The superscript on G indicates standard thermodynamic conditions. In the last step, the ideal gas law was used to express V in terms of P, and the relationships $N = nN_A$ and $R = N_A k$ were employed. The units of pressure must be Pa = J m^{-3}. Solving for the thermal wavelength term Λ^3, we get

$$\Lambda^3 = \left(\frac{h^2}{2\pi mkT}\right)^{3/2}$$

$$= \left(\frac{(6.626 \times 10^{-34}\,\mathrm{J\,s})^2}{2\pi\left(\dfrac{0.040\,\mathrm{kg\,mol^{-1}}}{6.022 \times 10^{23}\,\mathrm{mol^{-1}}}\right)(1.38 \times 10^{-23}\,\mathrm{J\,K^{-1}})(298\,\mathrm{K})}\right)^{3/2}$$

$$= 4.09 \times 10^{-33}\,\mathrm{m^3}$$

With this result, $G°$ becomes

$$G° = -nRT \ln\left(\frac{kT}{P\Lambda^3}\right) = -(1\,\mathrm{mol})(8.314\,\mathrm{J\,mol^{-1}\,K^{-1}})$$

$$\times\,(298\,\mathrm{K}) \ln\left(\frac{(1.38 \times 10^{-23}\,\mathrm{J\,K^{-1}})(298\,\mathrm{K})}{(10^5\,\mathrm{Pa})(4.09 \times 10^{-33}\,\mathrm{m^3})}\right)$$

$$= -3.99 \times 10^4\,\mathrm{J} = -39.9\,\mathrm{kJ}$$

32.7 Chemical Equilibrium

Consider the following generic reaction:

$$a\text{J} + b\text{K} \rightleftharpoons c\text{L} + d\text{M} \qquad (32.80)$$

The change in Gibbs energy for this reaction is related to the Gibbs energy for the species involved in the reaction as follows:

$$\Delta G^\circ = cG_L^\circ + dG_M^\circ - aG_J^\circ - bG_K^\circ \qquad (32.81)$$

In this expression, the superscript indicates standard thermodynamic state. In addition, the equilibrium constant K is given by

$$\Delta G^\circ = -RT \ln K \qquad (32.82)$$

In the previous section of this chapter, the Gibbs energy was related to the molecular partition function. Therefore, it should be possible to define ΔG° and K in terms of partition functions for the various species involved. This development can be initiated by substituting into our expression for ΔG° from Equation (32.81) the definition for G given in Equation (32.79) and considering molar quantities such that $N = N_A$:

$$
\begin{aligned}
\Delta G^\circ &= c\left(-RT \ln\left(\frac{q_L^\circ}{N_A}\right)\right) + d\left(-RT \ln\left(\frac{q_M^\circ}{N_A}\right)\right) - a\left(-RT \ln\left(\frac{q_J^\circ}{N_A}\right)\right) \\
&\quad - b\left(-RT \ln\left(\frac{q_K^\circ}{N_A}\right)\right) \\
&= -RT \ln\left(\frac{\left(\dfrac{q_L^\circ}{N_A}\right)^c \left(\dfrac{q_M^\circ}{N_A}\right)^d}{\left(\dfrac{q_J^\circ}{N_A}\right)^a \left(\dfrac{q_K^\circ}{N_A}\right)^b}\right) \qquad (32.83)
\end{aligned}
$$

Comparison of the preceding relationship to the thermodynamics definition of ΔG° demonstrates that the equilibrium constant can be defined as follows:

$$K_P = \frac{\left(\dfrac{q_L^\circ}{N_A}\right)^c \left(\dfrac{q_M^\circ}{N_A}\right)^d}{\left(\dfrac{q_J^\circ}{N_A}\right)^a \left(\dfrac{q_K^\circ}{N_A}\right)^b} \qquad (32.84)$$

Although Equation (32.84) is correct as written, there is one final detail to consider. Specifically, imagine taking the preceding relationship to $T = 0\ \text{K}$ such that only the lowest energy states along all energetic degrees of freedom are populated. The translational and rotational ground states for all species are equivalent; however, the vibrational and electronic ground states are not. Figure 32.11 illustrates the

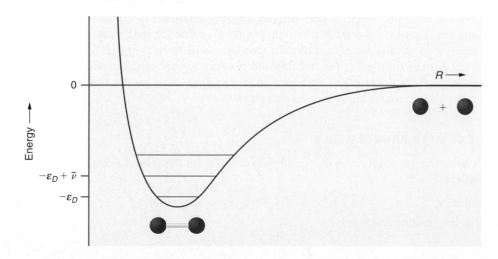

FIGURE 32.11
The ground-state potential energy curve for a diatomic molecule. The lowest three vibrational levels are indicated.

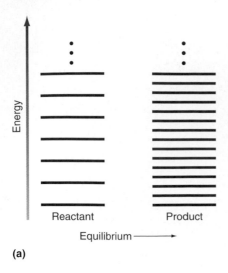

Energy

Reactant　　　　　Product

Equilibrium ⟶

(a)

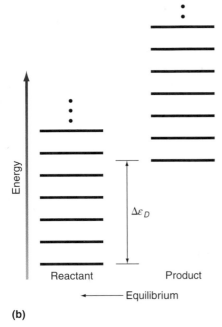

Energy

$\Delta\varepsilon_D$

Reactant　　　　　Product

⟵ Equilibrium

(b)

FIGURE 32.12
The statistical interpretation of equilibrium. (a) Reactant and product species having equal ground-state energies are depicted. However, the energy spacings of the product are less than the reactant such that more product states are available at a given temperature. Therefore, equilibrium will lie with the product. (b) Reactant and product species having equal state spacings are depicted. In this case, the product states are higher in energy than those of the reactant such that equilibrium lies with the reactant.

origin of this discrepancy. The figure illustrates the ground *vibronic* (vibrational and electronic) potential for a diatomic molecule. The presence of a bond between the two atoms in the molecule lowers the energy of the molecule relative to the separated atomic fragments. Because the energy of the atomic fragments is defined as zero, the ground vibrational state is lower than zero by an amount equal to the **dissociation energy** of the molecule ε_D. Furthermore, different molecules will have different values for the dissociation energy such that this offset from zero will be molecule specific.

Establishing a common reference state for vibrational and electronic degrees of freedom is accomplished as follows. First, differences in ε_D can be incorporated into the vibrational part of the problem such that the electronic partition function remains identical to our earlier definition. Turning to the vibrational problem, the general expression for the vibrational partition function can be written incorporating ε_D as follows:

$$q'_{vib} = \sum_n e^{-\beta\varepsilon_n} = e^{-\beta(-\varepsilon_D)} + e^{-\beta(-\varepsilon_D + \tilde{\nu})} + e^{-\beta(-\varepsilon_D + 2\tilde{\nu})} + \cdots$$

$$= e^{\beta\varepsilon_D}(1 + e^{-\beta\tilde{\nu}} + e^{-2\beta\tilde{\nu}} + \cdots)$$

$$= e^{\beta\varepsilon_D}q_{vib} \qquad (32.85)$$

In other words, the corrected form for the vibrational partition function is simply the product of our original q_{vib} (without zero point energy) times a factor that corrects for the offset in energy due to dissociation. With this correction factor in place, the ground vibrational states are all set to zero. Therefore, the final expression for the equilibrium constant is

$$K_P = \frac{\left(\dfrac{q_L^\circ}{N_A}\right)^c \left(\dfrac{q_M^\circ}{N_A}\right)^d}{\left(\dfrac{q_J^\circ}{N_A}\right)^a \left(\dfrac{q_K^\circ}{N_A}\right)^b} e^{\beta(c\varepsilon_L + d\varepsilon_M - a\varepsilon_J - b\varepsilon_K)} \quad K_P = \frac{\left(\dfrac{q_L^\circ}{N_A}\right)^c \left(\dfrac{q_M^\circ}{N_A}\right)^d}{\left(\dfrac{q_J^\circ}{N_A}\right)^a \left(\dfrac{q_K^\circ}{N_A}\right)^b} e^{-\beta\Delta\varepsilon} \quad (32.86)$$

The dissociation energies needed to evaluate Equation (32.86) are readily obtained using spectroscopic techniques.

What insight does this expression for the equilibrium constant provide? Equation (32.86) can be viewed as consisting of two parts. The first part is the ratio of partition functions for products and reactants. Because the partition functions quantify the number of energy states available, this ratio dictates that equilibrium will favor those species with the greatest number of available energy states at a given temperature. The second half is the dissociation energy. This term dictates that equilibrium will favor those species with the lowest energy states. This behavior is illustrated in Figure 32.12. The upper part of the figure shows a reactant and product species having equal ground-state energies. The only difference between the two is that the product species has more accessible energy states at a given temperature than the reactant. Another way to envision this relationship is that the partition function describing the product will be greater than that for the reactant at the same temperature such that, at equilibrium, products will be favored ($K > 1$). On the lower part of the figure, the reactant and product energy spacings are equivalent; however, the product states lie higher in energy than those of the reactant. In this case, equilibrium will lie with the reactant ($K < 1$).

EXAMPLE PROBLEM 32.9

What is the general form of the equilibrium constant for the dissociation of a diatomic molecule?

Solution

The dissociation reaction is

$$X_2(g) \rightleftharpoons 2X(g)$$

We first need to derive the partition functions that describe the reactants and products. The products are monatomic species such that only translations and electronic degrees of freedom are relevant. The partition function is then

$$q_X^\circ = q_T^\circ q_E = \left(\frac{V^\circ}{\Lambda_X^3}\right) g_o$$

The superscripts indicate standard thermodynamic conditions. With these conditions, $V^\circ = RT/P^\circ$ (for 1 mol) such that

$$q_X^\circ = q_T^\circ q_E = \left(\frac{RT}{\Lambda_X^3 P^\circ}\right) g_o$$

For molar quantities

$$\frac{q_X^\circ}{N_A} = \frac{g_o RT}{N_A \Lambda_X^3 P^\circ}$$

The partition function for X_2 will be equivalent to that for X, with the addition of rotational and vibrational degrees of freedom:

$$\frac{q_{X_2}^\circ}{N_A} = \frac{g_o RT}{N_A \Lambda_{X_2}^3 P^\circ} q_R q_V$$

Using the preceding expressions, the equilibrium constant for the dissociation of a diatomic becomes

$$K_P = \frac{\left(\dfrac{q_X^\circ}{N_A}\right)^2}{\left(\dfrac{q_{X_2}^\circ}{N_A}\right)} e^{-\beta \varepsilon_D} = \frac{\left(\dfrac{g_{o,X} RT}{N_A \Lambda_X^3 P^\circ}\right)^2}{\left(\dfrac{g_{o,X_2} RT}{N_A \Lambda_{X_2}^3 P^\circ}\right) q_R q_V} e^{-\beta \varepsilon_D}$$

$$= \left(\frac{g_{o,X}^2}{g_{o,X_2}}\right)\left(\frac{RT}{N_A P^\circ}\right)\left(\frac{\Lambda_{X_2}^3}{\Lambda_X^6}\right)\frac{1}{q_R q_V} e^{-\beta \varepsilon_D}$$

For a specific example, we can use the preceding expression to predict K_P for the dissociation of I_2 at 298 K given the following parameters:

$$g_{o,I} = 4 \quad \text{and} \quad g_{o,I_2} = 1$$
$$\Lambda_I = 8.98 \times 10^{-12} \text{ m} \quad \text{and} \quad \Lambda_{I_2} = 6.35 \times 10^{-12} \text{ m}$$
$$q_R = 2770$$
$$q_V = 1.58$$
$$\varepsilon_D = 12{,}461 \text{ cm}^{-1}$$

Evaluation of the thermal wavelengths, q_R, and q_V was performed as described in Chapter 31. Given these values, K_P becomes

$$K_P = \left(\frac{g_{o,I}^2}{g_{o,I_2}}\right)\left(\frac{RT}{N_A P^\circ}\right)\left(\frac{\Lambda_{I_2}^3}{\Lambda_I^6}\right)\frac{1}{q_R q_V} e^{-\beta \varepsilon_D}$$

$$= (16)(4.12 \times 10^{-26} \text{ m}^3)\left(\frac{(6.35 \times 10^{-12} \text{ m})^3}{(8.98 \times 10^{-12} \text{ m})^6}\right)\frac{1}{(2770)(1.58)}$$

$$\exp\left[\frac{-hc(12461 \text{ cm}^{-1})}{kT}\right]$$

$$= (7.81 \times 10^5)\exp\left[\frac{-(6.626 \times 10^{-34} \text{ J s})(3.00 \times 10^{10} \text{ cm s}^{-1})(12461 \text{ cm}^{-1})}{(1.38 \times 10^{-23} \text{ J K}^{-1})(298 \text{ K})}\right]$$

$$= 5.64 \times 10^{-22}$$

Using tabulated values for ΔG° provided in the back of the text and the relationship $\Delta G^\circ = -RT \ln K$, a value of $K = 5.92 \times 10^{-22}$ is obtained for this reaction.

Vocabulary

average energy	Gibbs energy	Sackur–Tetrode equation
Boltzmann's formula	heat capacity	total energy
dissociation energy	Helmholtz energy	translational heat capacity
Dulong and Petit law	internal energy	two-level system
Einstein solid	residual entropy	vibrational heat capacity
electronic heat capacity	rotational heat capacity	

Conceptual Problems

Q32.1 What is the relationship between ensemble energy and the thermodynamic concept of internal energy?

Q32.2 Why is the contribution of translational motion to the internal energy $3/2RT$ at 298 K?

Q32.3 List the energetic degrees of freedom expected to contribute to the internal energy of a diatomic molecule at 298 K. Given this list, what spectroscopic information do you need to numerically determine the internal energy?

Q32.4 List the energetic degrees of freedom for which the contribution to the internal energy determined by statistical mechanics is generally equal to the prediction of the equipartition theorem at 298 K.

Q32.5 Write down the contribution to the constant volume heat capacity from translations and rotations for an ideal monatomic, diatomic, and nonlinear polyatomic gas, assuming that the high-temperature limit is appropriate for the rotational degrees of freedom.

Q32.6 The constant volume heat capacity for all monatomic gases is 12.48 J mol^{-1} K^{-1}. Why?

Q32.7 When are rotational degrees of freedom expected to contribute R or $3/2R$ (linear and nonlinear, respectively) to the molar constant volume heat capacity? When will a vibrational degree of freedom contribute R to the molar heat capacity?

Q32.8 The molar constant volume heat capacity of N_2 is 20.8 J mol^{-1} K^{-1}. What is this value in terms of R? Can you make sense of this value?

Q32.9 Why do electronic degrees of freedom generally not contribute to the constant volume heat capacity?

Q32.10 Describe the model used to determine the heat capacity of atomic crystals.

Q32.11 What is the Boltzmann formula, and how can it be used to predict residual entropy?

Q32.12 How does the Boltzmann formula provide an understanding of the third law of thermodynamics?

Q32.13 For carbon monoxide, the calculated molar entropy was more negative than the experimental value. Why?

Q32.14 What thermodynamic property of what particular system does the Sackur–Tetrode equation describe?

Q32.15 Which thermodynamic quantity is used to derive the ideal gas law for a monatomic gas? What molecular partition function is employed in this derivation?

Q32.16 What is the definition of "zero" energy employed in constructing the statistical mechanical expression for the equilibrium constant? Why was this definition necessary?

Q32.17 Why should the equilibrium constant be dependent on the difference in Gibbs energy? How is this relationship described using statistical mechanics?

Q32.18 For the equilibrium involving the dissociation of a diatomic, what energetic degrees of freedom were considered for the diatomic, and for the atomic constituents?

Q32.19 The statistical mechanical expression for K_p consisted of two general parts. What are these parts, and what energetic degrees of freedom do they refer to?

Q32.20 Assume you have an equilibrium expression that involves monatomic species only. What difference in energy between reactants and products would you use in the expression for K_P?

Numerical Problems

Problem numbers in red indicate that the solution to the problem is given in the *Student's Solutions Manual*.

P32.1 In exploring the variation of internal energy with temperature for a two-level system with the ground and excited state separated by an energy of hv, the total energy was found to plateau at a value of $0.5Nhv$. Repeat this analysis for an

ensemble consisting of N particles having three energy levels corresponding to 0, hv, and $2hv$. What is the limiting value of the total energy for this system?

P32.2 For a two-level system where $v = 1.50 \times 10^{13}$ s^{-1}, determine the temperature at which the internal energy is equal to 0.25 Nhv, or 1/2 the limiting value of 0.50 Nhv.

P32.3 Consider two separate molar ensembles of particles characterized by the following energy-level diagrams:

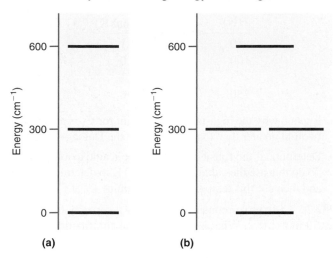

(a)   (b)

Derive expressions for the internal energy for each ensemble. At 298 K, which ensemble is expected to have the greatest internal energy?

P32.4 For an ensemble consisting of a mole of particles having two energy levels separated by 1000. cm^{-1}, at what temperature will the internal energy equal 3.00 kJ?

P32.5 What is the contribution to the internal energy from translations for an ideal monatomic gas confined to move on a surface? What is the expected contribution from the equipartition theorem?

P32.6 For a system of energy levels, $\varepsilon_m = m^2\alpha$, where α is a constant with units of energy and $m = 0, 1, 2, \ldots, \infty$. What is the internal energy and heat capacity of this system in the high-temperature limit?

P32.7 (Challenging) Building on the concept of equipartition, demonstrate that for any energy term of the form αx^2 where α is a constant, the contribution to the internal energy is equal to $kT/2$ by evaluating the following expression:

$$\varepsilon = \frac{\int_{-\infty}^{\infty} \alpha x^2 e^{-\alpha x^2/kT}\, dx}{\int_{-\infty}^{\infty} e^{-\alpha x^2/kT}\, dx}$$

P32.8 Consider the following table of diatomic molecules and associated rotational constants:

Molecule	B (cm^{-1})	$\widetilde{\nu}$ (cm^{-1})
$H^{35}Cl$	10.59	2886
$^{12}C^{16}O$	1.93	2170
^{39}KI	0.061	200.
CsI	0.024	120.

a. Calculate the rotational temperature for each molecule.

b. Assuming that these species remain gaseous at 100. K, for which species is the equipartition theorem prediction for the rotational contribution to the internal energy appropriate?

c. Calculate the vibrational temperature for each molecule.

d. If these species were to remain gaseous at 1000. K, for which species is the equipartition theorem prediction for the vibrational contribution to the internal energy appropriate?

P32.9 The four energy levels for atomic vanadium (V) have the following energies and degeneracies:

Level (n)	Energy (cm^{-1})	Degeneracy
0	0	4
1	137.38	6
2	323.46	8
3	552.96	10

What is the contribution to the average molar energy from electronic degrees of freedom for V when $T = 298$ K?

P32.10 The three lowest energy levels for atomic carbon (C) have the following energies and degeneracies:

Level (n)	Energy (cm^{-1})	Degeneracy
0	0	1
1	16.4	3
2	43.5	5

What is the contribution to the average molar energy from the electronic degrees of freedom for C when $T = 100.$ K?

P32.11 Consider an ensemble of units in which the first excited electronic state at energy ε_1 is m_1-fold degenerate, and the energy of the ground state is m_o-fold degenerate with energy ε_0.

a. Demonstrate that if $\varepsilon_0 = 0$, the expression for the electronic partition function is

$$q_E = m_o\left(1 + \frac{m_1}{m_o} e^{-\varepsilon_1/kT}\right)$$

b. Determine the expression for the internal energy U of an ensemble of N such units. What is the limiting value of U as the temperature approaches zero and infinity?

P32.12 Calculate the internal energy of He, Ne, and Ar under standard thermodynamic conditions. Do you need to redo the entire calculation for each species?

P32.13 Determine the molar internal energy of HCl ($B = 10.59$ cm^{-1} and $\widetilde{\nu} = 2886$ cm^{-1}) under standard thermodynamic conditions.

P32.14 How would you expect the molar internal energy of ^{79}BrF to compare to that of $^{79}Br^{35}Cl$ at 298 K? Check your answer by using the following data:

	B (cm^{-1})	$\widetilde{\nu}$ (cm^{-1})
^{79}BrF	0.356	671
$^{79}Br^{35}Cl$	0.153	420.

P32.15 Determine the vibrational contribution to C_V for HCl ($\widetilde{\nu} = 2886$ cm^{-1}) over a temperature range from 500. to 5000. K in 500. K intervals and plot your result. At what temperature do you expect to reach the high-temperature limit for the vibrational contribution to C_V?

P32.16 Determine the vibrational contribution to C_V for HCN where $\widetilde{\nu}_1 = 2041$ cm^{-1}, $\widetilde{\nu}_2 = 712$ cm^{-1} (doubly degenerate), and $\widetilde{\nu}_3 = 3369$ cm^{-1} at $T = 298, 500.$, and $1000.$ K.

P32.17 Carbon dioxide has attracted much recent interest as a greenhouse gas. Determine the vibrational contribution to C_V for CO_2 where $\widetilde{\nu}_1 = 2349$ cm^{-1}, $\widetilde{\nu}_2 = 667$ cm^{-1} (doubly degenerate), and $\widetilde{\nu}_3 = 1333$ cm^{-1} at $T = 260.$ K.

P32.18 The three lowest energy levels for atomic carbon (C) have the following energies and degeneracies:

Level (n)	Energy (cm^{-1})	Degeneracy
0	0	1
1	16.4	3
2	43.5	5

Determine the electronic contribution to C_V for atomic C at 100. K.

P32.19 Consider the following energy levels and associated degeneracies for atomic Fe:

Level (n)	Energy (cm^{-1})	Degeneracy
0	0	9
1	415.9	7
2	704.0	5
3	888.1	3
4	978.1	1

a. Determine the electronic contribution to C_V for atomic Fe at 150. K assuming that only the first two levels contribute to C_V.

b. How does your answer to part (a) change if the $n = 2$ level is included in the calculation of C_V? Do you need to include the other levels?

P32.20 The speed of sound is given by the relationship

$$c_{sound} = \left(\frac{\frac{C_p}{C_V} RT}{M} \right)^{1/2}$$

where C_p is the constant pressure heat capacity (equal to $C_V + R$), R is the ideal gas constant, T is temperature, and M is molar mass.

a. What is the expression for the speed of sound for an ideal monatomic gas?

b. What is the expression for the speed of sound of an ideal diatomic gas?

c. What is the speed of sound in air at 298 K, assuming that air is mostly made up of nitrogen ($B = 2.00$ cm^{-1} and $\widetilde{\nu} = 2359$ cm^{-1})?

P32.21 The measured molar heat capacities for crystalline KCl are as follows at the indicated temperatures:

T(K)	C_v (J/mol K)
50.	21.1
100.	39.0
175	46.1
250.	48.6

a. Explain why the high-temperature limit for C_V is apparently twofold greater than that predicted by the Dulong–Petit law.

b. Determine if the Einstein model is applicable to ionic solids. To do this use the value for C_V at 50.0 K to determine Θ_V, and then use this temperature to determine C_V at 175 K.

P32.22 The molar constant volume heat capacity for $I_2(g)$ is 28.6 J mol^{-1} K^{-1}. What is the vibrational contribution to the heat capacity? You can assume that the contribution from the electronic degrees of freedom is negligible.

P32.23 Inspection of the thermodynamic tables in the back of the text reveals that many molecules have quite similar constant volume heat capacities.

a. The value of $C_{V, M}$ for Ar(g) at standard temperature and pressure is 12.48 J mol^{-1} K^{-1}, identical to gaseous He(g). Using statistical mechanics, demonstrate why this equivalence is expected.

b. The value of $C_{V, M}$ for $N_2(g)$ is 20.8 J mol^{-1} K^{-1}. Is this value expected given your answer to part (a)? For N_2, $\widetilde{\nu} = 2359$ cm^{-1} and $B = 2.00$ cm^{-1}.

P32.24 Consider rotation about the C-C bond in ethane. A crude model for torsion about this bond is the "free rotor" model where rotation is considered unhindered. In this model the energy levels along the torsional degree of freedom are given by

$$E_j = \frac{\hbar^2 j^2}{2I} \quad \text{for } j = 0, \pm 1, \pm 2, ...$$

In this expression I is the moment of inertia. Using these energies, the summation expression for the corresponding partition function is

$$Q = \frac{1}{\sigma} \sum_{j=-\infty}^{\infty} e^{-E_j/kT}$$

where σ is the symmetry number.

a. Assuming that the torsional degree of freedom is in the high-temperature limit, evaluate the previous expression for Q.

b. Determine the contribution of the torsional degree of freedom to the molar constant-volume heat capacity.

c. The experimentally determined C_v for the torsional degree of freedom is approximately equal to R at 340. K. Can you rationalize why the experimental value is greater that than predicted using the free rotor model?

P32.25 Determine the molar entropy for 1 mol of gaseous Ar at 200., 300., and 500. K and $V = 1000.$ cm^3 assuming that Ar can be treated as an ideal gas. How does the result of this calculation change if the gas is Kr instead of Ar?

P32.26 The standard molar entropy of O_2 is 205.14 J mol^{-1} K^{-1} at $P = 1.00$ atm. Using this information, determine the bond length of O_2. For this molecule, $\tilde{\nu} = 1580.$ cm^{-1}, and the ground electronic state degeneracy is three.

P32.27 Determine the standard molar entropy of N_2O, a linear triatomic molecule at $P = 1.00$ atm. For this molecule, $B = 0.419$ cm^{-1} and $\tilde{\nu}_1 = 1285$ cm^{-1}, $\tilde{\nu}_2 = 589$ cm^{-1} (doubly degenerate), and $\tilde{\nu}_3 = 2224$ cm^{-1}.

P32.28 Determine the standard molar entropy of OClO, a nonlinear triatomic molecule where $B_A = 1.06$ cm^{-1}, $B_B = 0.31$ cm^{-1}, $B_C = 0.29$ cm^{-1}, $\tilde{\nu}_1 = 938$ cm^{-1}, $\tilde{\nu}_2 = 450.$ cm^{-1}, $\tilde{\nu}_3 = 1100.$ cm^{-1}, and $P = 1.00$ atm.

P32.29 Determine the standard molar entropy for the hydroxyl radical, OH, for which $\tilde{\nu} = 3735$ cm^{-1} and $B = 18.9$ cm^{-1}, the ground electronic state is doubly degenerate, and $P = 1.00$ atm.

P32.30 Determine the standard molar entropy of N_2 ($\tilde{\nu} = 2359$ cm^{-1}, $B = 2.00$ cm^{-1}, $g_0 = 1$) and the molar entropy when $P = 1.00$ atm but $T = 2500.$ K.

P32.31 Determine the standard molar entropy of H^{35}Cl at 298 K where $B = 10.58$ cm^{-1}, $\tilde{\nu} = 2886$ cm^{-1}, the ground-state electronic level degeneracy is 1, and $P = 1.00$ atm.

P32.32 Derive the expression for the standard molar entropy of a monatomic gas restricted to two-dimensional translational motion. (*Hint*: You are deriving the two-dimensional version of the Sackur–Tetrode equation.)

P32.33 The standard molar entropy of CO is 197.7 J mol^{-1} K^{-1}. How much of this value is due to rotational and vibrational motion of CO?

P32.34 The standard molar entropy of the tropospheric pollutant NO_2 is 240.1 J mol^{-1} K^{-1}. How much of this value is due to rotational motion? The vibrational frequencies of NO_2 are 1318, 750., and 1618 cm^{-1}, and the ground electronic state is doubly degenerate.

P32.35 Determine the standard molar entropy of CO_2 where $B = 0.39$ cm^{-1} and $P = 1.00$ atm. You can ignore the vibrational and electronic contributions to the standard molar entropy in this calculation.

P32.36 Consider the molecule NNO, which has a rotational constant nearly identical to CO_2. Would you expect the standard molar entropy for NNO to be greater or less than CO_2? Can you provide a rough estimation of how much greater?

P32.37 Entropy, heat, and temperature are related through the following expression:

$$\Delta S = \frac{\Delta q_{reversible}}{T}$$

This expression can be rearranged to provide an expression for T in terms of heat and weight (W, the number of microstates for a given configuration of energy):

$$T = \frac{\Delta q_{reversible}}{\Delta S} = \frac{\Delta q_{reversible}}{k \Delta \ln W}$$

a. Substitute the expression for W for a two-level system into the equation to define T in terms of heat, Boltzmann's

constant, the total number of two-state particles (N), and the number of particles in the excited state (n).

b. Determine the temperature for a two-level system where $N = 20$, the separation between energy levels is 100. cm^{-1}, and $q_{rev} = n(100.$ cm$^{-1})$. To perform this calculation, determine q_{rev} and $\ln W$ for n = 0 to 20 for integer values of n. The temperature can then be determined by expressing Δq_{rev} and $\Delta \ln W$ as

$$\Delta q_{reversible} = q_{rev,n+1} - q_{rev,n}$$
$$\Delta \ln W = \ln W_{n+1} - \ln W_n$$

c. Inspect the dependence of T on n. Can you explain this dependence?

P32.38 The molecule NO has a ground electronic level that is doubly degenerate, and a first excited level at 121.1 cm^{-1} that is also twofold degenerate. Determine the contribution of electronic degrees of freedom to the standard molar entropy of NO. Compare your result to $R\ln(4)$. What is the significance of this comparison?

P32.39 Determine the residual molar entropies for molecular crystals of the following:

a. ^{35}Cl^{37}Cl   **b.** $CFCl_3$

c. CF_2Cl_2 **d.** CO_2

P32.40 Using the Helmholtz energy, demonstrate that the pressure for an ideal polyatomic gas is identical to that derived for an ideal monatomic gas in the text.

P32.41 Derive an expression for the standard molar enthalpy of an ideal monatomic gas by evaluation of the statistical mechanical expression for enthalpy as opposed to the thermodynamic argument provided in Example Problem 32.7.

P32.42 Demonstrate that the molar enthalpy is equal to the molar energy for a collection of one-dimensional harmonic oscillators.

P32.43 Calculate the standard Helmholtz energy for molar ensembles of Ne and Kr at 298 K.

P32.44 What is the vibrational contribution to the Helmholtz and Gibbs energies from a molar ensemble of one-dimensional harmonic oscillators?

P32.45 Determine the molar standard Gibbs energy for ^{35}Cl^{35}Cl where $\tilde{\nu} = 560.$ cm^{-1}, $B = 0.244$ cm^{-1}, and the ground electronic state is nondegenerate.

P32.46 Determine the rotational and vibrational contributions to the molar standard Gibbs energy for N_2O (NNO), a linear triatomic molecule where $B = 0.419$ cm^{-1} and $\tilde{\nu}_1 = 1285$ cm^{-1}, $\tilde{\nu}_2 = 589$ cm^{-1} (doubly degenerate), and $\tilde{\nu}_3 = 2224$ cm^{-1}.

P32.47 Determine the equilibrium constant for the dissociation of sodium at 298 K: $Na_2(g) \rightleftharpoons 2\,Na(g)$. For Na_2, $B = 0.155$ cm^{-1}, $\tilde{\nu} = 159$ cm^{-1}, the dissociation energy is 70.4 kJ/mol, and the ground-state electronic degeneracy for Na is 2.

P32.48 The isotope exchange reaction for Cl_2 is as follows: ^{35}Cl^{35}Cl + ^{37}Cl^{37}Cl $\rightleftharpoons$ $2\,^{37}$Cl^{35}Cl. The equilibrium constant for this reaction is ~4. Furthermore, the equilibrium constant for similar isotope-exchange reactions is also close to this value. Demonstrate why this would be so.

P32.49 Consider the following isotope exchange reaction:

$$DCl(g) + HBr(g) \rightleftharpoons DBr(g) + HCl(g)$$

The amount of each species at equilibrium can be measured using proton and deuterium NMR (see *Journal of Chemical Education* 73 [1996]:99). Using the following spectroscopic information, determine K_p for this reaction at 298 K. For this reaction, $\Delta \varepsilon = 41 \text{ cm}^{-1}$ equal to the difference in zero-point energies between products versus reactants, and the ground-state electronic degeneracy is zero for all species.

	M (g mol^{-1})	B (cm^{-1})	$\tilde{\nu}$ (cm^{-1})
H^{35}Cl	35.98	10.59	2991
D^{35}Cl	36.98	5.447	2145
H^{81}Br	81.92	8.465	2649
D^{81}Br	82.93	4.246	1885

P32.50 The equilibrium between hydrogen cyanide (HCN) and its isomer hydrogen isocyanide (HNC) is important in interstellar chemistry:

$$HCN(g) \rightleftharpoons HNC(g)$$

A long-standing "puzzle" regarding this reaction is that in space (T = 2.75 K) surprisingly large amounts of HNC are observed. For example, HNC/HCN ratios approaching 20% have been observed in comets (*Advances in Space Research* 31 (2003):2577). Using the spectroscopic information provided in the following table and knowledge that the potential-energy surface minimum of HNC lies roughly 5200 cm^{-1} higher in energy relative to HCN, calculate the theoretical value for K_p for this reaction in interstellar space.

	M (g mol^{-1})	B (cm^{-1})	$\tilde{\nu}_1$ (cm^{-1})	$\tilde{\nu}_2$ (cm^{-1})	$\tilde{\nu}_3$ (cm^{-1})
HCN	27.03	1.477	2041	712	3669
HNC	27.03	1.512	2024	464	3653

P32.51 In "Direct Measurement of the Size of the Helium Dimer" by F. Luo, C. F. Geise, and W. R. Gentry (*J. Chemical Physics* 104 [1996]: 1151), evidence for the helium dimer is presented. As one can imagine, the chemical bond in the dimer is extremely weak, with an estimated value of only 8.3 mJ mol^{-1}.

a. An estimate for the bond length of 65 Å is presented in the paper. Using this information, determine the rotational constant for He$_2$. Using this value for the rotational constant, determine the location of the first rotational state. If correct, you will determine that the first excited rotational level is well beyond the dissociation energy of He$_2$.

b. Consider the following equilibrium between He$_2$ and its atomic constituents: $He_2(g) \rightleftharpoons 2He(g)$. If there are no rotational or vibrational states to consider, the equilibrium is determined exclusively by the translational degrees of freedom and the dissociation energy of He$_2$. Using the dissociation energy provided earlier and $V = 1000. \text{ cm}^3$, determine K_P assuming that $T = 10. \text{ K}$. The experiments were actually performed at 1 mK; why was such a low temperature employed?

Computational Problems

More detailed instructions on carrying out these calculations using Spartan Physical Chemistry are found on the book website at *www.masteringchemistry.com*.

C32.1 Perform a Hartree–Fock 6-31G* calculation on acetonitrile and determine the contribution of translations to the molar enthalpy and entropy under standard thermodynamic conditions. How does the calculation compare to the values determined by hand computation using the expressions for enthalpy and entropy presented in this chapter.

C32.2 Perform a Hartree–Fock 3-21G calculation on cyclohexane and determine the frequency of the lowest-frequency mode consistent with the reaction coordinate corresponding to the "boat-to-boat" conformational change. Using the calculated frequency of this vibrational degree of freedom, determine how much this mode contributes to the vibrational heat capacity at 298 K.

C32.3 In the treatment of C_V presented in this chapter the contribution of electronic degrees of freedom was ignored. To further explore this issue, perform a Hartree–Fock 6-31G* calculation on anthracene and determine the HOMO–LUMO energy gap, which provides a measure of the energy difference between the ground and first excited electronic states.

Using this energy gap, determine the temperature where 2% of the anthracene molecules populate the excited electronic state, and using this temperature determine the contribution of electronic degrees of freedom to C_V. Reflecting on this result, why can we generally ignore electronic degrees of freedom when calculating standard thermodynamic properties?

C32.4 In 2001 astronomers discovered the presence of vinyl alcohol (C_2H_3OH) in an interstellar cloud near the center of the Milky Way galaxy. This complex is a critical piece of the puzzle regarding the origin of complex organic molecules in space. The temperature of an interstellar cloud is ~20 K. Determine the vibrational frequencies of this compound using a Hartree–Fock 3-21G basis set, and determine if any of the vibrational modes of vinyl alcohol contribute more that $0.1R$ to C_V at this temperature.

C32.5 Nitrous acid (HONO) is of interest in atmospheric chemistry as a reservoir compound contributing to the NO$_x$ cycle. Perform a Hartree–Fock 6-31G* calculation on the cis conformer of this compound and determine the standard molar entropy of this compound. What are the contributions of vibrational and rotational degrees of freedom to the entropy? Using the techniques presented in this chapter and treating HONO as an ideal gas, determine the translational contribution to the standard molar entropy and compare to the computational result.

C32.6 For dibromine oxide (BrOBr), perform a Hartree–Fock 6-31G* calculation and determine the translational, rotational, and vibrational contribution to the standard molar entropy.

Web-Based Simulations, Animations, and Problems

W32.1 In this simulation, the temperature dependence of C_V for vibrational degrees of freedom is investigated for diatomic and polyatomic molecules. Diatomic molecules are studied, as are polyatomics with and without mode degeneracy. Comparisons of exact values to those expected for the high-temperature limit are performed.

For Further Reading

Chandler, D. *Introduction to Modern Statistical Mechanics.* New York, Oxford: 1987.

Hill, T. *Statistical Mechanics. Principles and Selected Applications.* New York, Dover: 1956.

Linstrom, P. J., and Mallard, W. G., Eds., *NIST Chemistry Webbook: NIST Standard Reference Database Number 69.* National Institute of Standards and Technology, Gaithersburg, MD, retrieved from http://webbook.nist.gov/chemistry. (This site contains a searchable database of thermodynamic and spectroscopic properties of numerous atomic and molecular species.)

McQuarrie, D. *Statistical Mechanics.* New York, Harper & Row: 1973.

Nash, L. K. "On the Boltzmann Distribution Law." *J. Chemical Education* 59 (1982): 824.

Nash, L. K. *Elements of Statistical Thermodynamics.* San Francisco, Addison-Wesley: 1972.

Noggle, J. H. *Physical Chemistry.* New York, HarperCollins: 1996.

Townes, C. H., and A. L. Schallow. *Microwave Spectroscopy.* New York, Dover: 1975. (This book contains an excellent appendix of spectroscopic constants.)

Widom, B. *Statistical Mechanics.* Cambridge, Cambridge University Press, 2002.

Kinetic Theory of Gases

Gas particle motion is of importance in many aspects of physical chemistry, from transport phenomena to chemical kinetics. In this chapter, the translational motion of gas particles is described. Gas particle motion is characterized by a distribution of velocities and speeds. These distributions, including the Maxwell speed distribution, are derived. Benchmark values for these distributions that provide insight into how the distributions change with temperature and particle mass are presented. Finally, molecular collisions are discussed, including the frequency of collisional events and the distance particles travel between collisions. The concepts presented in this chapter find wide application in the remainder of this text since they provide the first step in understanding gas-phase molecular dynamics.

33.1 Kinetic Theory of Gas Motion and Pressure

In this chapter, we expand on the microscopic viewpoint of matter by considering the translational motion of gas particles. **Gas kinetic theory** provides the starting point for this development and represents a central concept of physical chemistry. In this theory, gases are envisioned as a collection of atoms or molecules that we will refer to as *particles*. Gas kinetic theory is applicable when the particle density of the gas is such that the distance between particles is very large in comparison to their size. To illustrate this point, consider a mole of Ar at a temperature of 298 K and 1 atm pressure. Using the ideal gas law, the gas occupies a volume of 24.4 L or 0.0244 m³. Dividing this volume by Avogadro's number provides an average volume per Ar atom of 4.05×10^{-26} m³, or 40.5 nm³. The diameter of Ar is ~0.29 nm corresponding to a particle volume of 0.013 nm³. Comparison of the particle volume to the average volume of an individual Ar atom demonstrates that on average only 0.03% of the available volume is occupied by the particle. Even for particles with diameters substantially greater than Ar, the difference between the average volume per particle and the volume of the particle itself will be such that the distance between particles in the gas phase is substantial.

Given the large distance between particles in a gas, each particle is envisioned as traveling through space as a separate, unperturbed entity until a collision occurs with another particle or with the walls of the container confining the gas. In gas kinetic theory, particle motion is described using Newton's laws of motion. Although in previous chapters we have seen that classical descriptions fail to capture many of the microscopic properties of atoms and molecules, recall from Chapter 31 that the energy spacings between translational energy states are very small relative to kT such that a classical description of translational motion is appropriate.

One hallmark of kinetic theory is its ability to describe the pressure of an ideal gas as shown in Chapter 1. As opposed to the thermodynamic derivation of the ideal gas law, which relied on empirical relations between gas variables, the kinetic theory description of pressure relied on classical mechanics and a microscopic description of the system. The pressure exerted by a gas on the container confining the gas arose from collisions of gas particles with the container walls. By employing a classical description of a single molecular collision with the container wall and then scaling this result up to macroscopic proportions, one of the central results in the chemistry of gaseous systems was derived.

A second result derived from gas kinetic theory is the relationship between **root-mean-squared speed** and temperature. If the particle motion is random, the average velocities along all three Cartesian dimensions are equivalent. The average velocity along any dimension will be zero because there will be just as many particles traveling in both the positive and negative directions. In contrast, the root-mean-squared speed is given by the following using Equation (1.10):

$$\langle v^2 \rangle^{1/2} = \langle v_x^2 + v_y^2 + v_z^2 \rangle^{1/2}$$
$$= \langle 3v_x^2 \rangle^{1/2}$$
$$= \left(\frac{3kT}{m} \right)^{1/2} \tag{33.1}$$

Kinetic theory thus predicts that the root-mean-squared speed of the gas particles should increase as the square root of temperature and decrease as the square root of the particle mass.

The success of this approach is tempered by uncertainty regarding the assumptions made during the course of the derivation. For example, it has been assumed that the individual particle velocities can be characterized by some average value. Can we determine what this average value is given the distribution of molecular velocities that exist? Just what does the distribution of particle velocities or speed look like? In addition, the molecules collide with each other as well as the walls of the container. How frequent are such collisions? The frequency of molecular collisions will be important in subsequent chapters describing transport phenomena and the rates of chemical reactions. Therefore, a more critical look at atomic and molecular speed distributions and collisional dynamics is warranted.

33.2 Velocity Distribution in One Dimension

From the previous discussion of statistical thermodynamics, it is clear that a distribution of translational energies and, therefore, velocities will exist for a collection of gaseous particles. What does this distribution of velocities look like?

The variation in particle velocities is described by the **velocity distribution function.** In Chapter 29 the concept of a distribution function was presented. The velocity distribution function describes the probability of a gas particle having a velocity within a given range. In Section 31.4 we found that the translational energy-level spacings are sufficiently small that velocity can be treated as a continuous variable.

Therefore, the velocity distribution function describes the probability of a particle having a velocity in the range $v_x + dv_x$, $v_y + dv_y$, and $v_z + dv_z$.

To begin the derivation of the velocity distribution function, let $\Omega(v_x, v_y, v_z)$ represent the function that describes the distribution of velocity for an ensemble of gaseous particles. We assume that the distribution function can be decomposed into a product of distribution functions for each Cartesian dimension and the distribution of velocities in one dimension is independent of the distribution in the other two dimensions. With this assumption, $\Omega(v_x, v_y, v_z)$ is expressed as follows:

$$\Omega(v_x, v_y, v_z) = f(v_x)f(v_y)f(v_z) \tag{33.2}$$

In Equation (33.2), $f(v_x)$ is the velocity distribution for velocity in the x direction, and so forth. We assume that the gas is confined to an isotropic space such that the direction in which the particle moves does not affect the properties of the gas. In this case, the distribution function $\Omega(v_x, v_y, v_z)$ only depends on the magnitude of the velocity, or speed (ν). The natural log of Equation (33.2) yields

$$\ln \Omega(\nu) = \ln f(v_x) + \ln f(v_y) + \ln f(v_z)$$

To determine the velocity distribution along a single direction, the partial derivative of $\ln \Omega(\nu)$ is taken with respect to v_x while keeping the velocity along the other two directions constant:

$$\left(\frac{\partial \ln \Omega(\nu)}{\partial v_x} \right)_{v_y, v_z} = \frac{d \ln f(v_x)}{dv_x} \tag{33.3}$$

Equation (33.3) can be rewritten using the chain rule for differentiation (see Appendix A, Math Supplement) allowing for the derivative of $\ln \Omega(\nu)$ with respect to v_x to be written as

$$\left(\frac{d \ln \Omega(\nu)}{d\nu} \right)\left(\frac{\partial \nu}{\partial v_x} \right)_{v_y, v_z} = \frac{d \ln f(v_x)}{dv_x} \tag{33.4}$$

The second factor on the left-hand side of Equation (33.4) can be readily evaluated:

$$\left(\frac{\partial \nu}{\partial v_x} \right)_{v_y, v_z} = \left(\frac{\partial}{\partial v_x}(v_x^2 + v_y^2 + v_z^2)^{1/2} \right)_{v_y, v_z}$$

$$= \frac{1}{2}(2v_x)(v_x^2 + v_y^2 + v_z^2)^{-1/2}$$

$$= \frac{v_x}{\nu} \tag{33.5}$$

Substituting this result into Equation (33.4) and rearranging yields

$$\left(\frac{d \ln \Omega(\nu)}{d\nu} \right)\left(\frac{\partial \nu}{\partial v_x} \right)_{v_y, v_z} = \frac{d \ln f(v_x)}{dv_x}$$

$$\left(\frac{d \ln \Omega(\nu)}{d\nu} \right)\left(\frac{v_x}{\nu} \right) = \frac{d \ln f(v_x)}{dv_x}$$

$$\frac{d \ln \Omega(\nu)}{\nu \, d\nu} = \frac{d \ln f(v_x)}{v_x \, dv_x} \tag{33.6}$$

It is important to recall at this point that the velocity distributions along each direction are equivalent. Therefore, the preceding derivation could just as easily have been performed considering v_y or v_z, resulting in the following expressions analogous to Equation (33.6):

$$\frac{d \ln \Omega(\nu)}{\nu \, d\nu} = \frac{d \ln f(v_y)}{v_y \, dv_y} \tag{33.7}$$

$$\frac{d \ln \Omega(\nu)}{\nu \, d\nu} = \frac{d \ln f(v_z)}{v_z \, dv_z} \tag{33.8}$$

Comparison of Equations (33.6), (33.7), and (33.8) suggests that the following equality exists:

$$\frac{d \ln f(v_x)}{v_x dv_x} = \frac{d \ln f(v_y)}{v_y dv_y} = \frac{d \ln f(v_z)}{v_z dv_z} \tag{33.9}$$

In order for Equation (33.9) to be correct, each of the terms in Equations (33.6), (33.7), and (33.8) must be equal to a constant, γ, such that

$$\frac{d \ln f(v_j)}{v_j dv_j} = \frac{df(v_j)}{v_j f(v_j) dv_j} = -\gamma \quad \text{for } j = x, y, z \tag{33.10}$$

In this equation, the negative of γ has been employed, recognizing that γ must be a positive quantity to ensure that $f(v_j)$ does not diverge as v_j approaches infinity. Integration of Equation (33.10) results in the following expression for the velocity distribution along one direction:

$$\int \frac{df(v_j)}{f(v_j)} = -\int \gamma v_j \, dv_j$$

$$\ln f(v_j) = -\frac{1}{2}\gamma v_j^2$$

$$f(v_j) = Ae^{-\gamma v_j^2/2} \tag{33.11}$$

The last step remaining in the derivation is to determine the normalization constant A and γ. To determine A, we refer back to Chapter 29 and the discussion of normalized distribution functions and require that the velocity distribution be normalized. Because a particle can be traveling in either the $+j$ or $-j$ direction, the range of the velocity distribution is $-\infty \le v_j \le \infty$. Applying the normalization condition and integrating over this range,

$$\int_{-\infty}^{\infty} f(v_j) \, dv_j = 1 = \int_{-\infty}^{\infty} Ae^{-\gamma v_j^2/2} dv_j$$

$$1 = 2A \int_{0}^{\infty} e^{-\gamma v_j^2/2} dv_j$$

$$1 = A\sqrt{\frac{2\pi}{\gamma}}$$

$$\sqrt{\frac{\gamma}{2\pi}} = A \tag{33.12}$$

In evaluating this integral, we use the property of even integrands, which says that the integral from $-\infty$ to ∞ is equal to twice the integral from 0 to ∞. With the normalization factor, the velocity distribution in one dimension becomes

$$f(v_j) = \left(\frac{\gamma}{2\pi}\right)^{1/2} e^{-\gamma v_j^2/2} \tag{33.13}$$

All that remains is to evaluate γ. Earlier we encountered the following definition for $\langle v_x^2 \rangle$:

$$\langle v_x^2 \rangle = \frac{kT}{m} \tag{33.14}$$

Recall that angle brackets around v_x^2 indicate that this quantity is an average over the ensemble of particles. Furthermore, this quantity is equal to the second moment of the velocity distribution; therefore, γ can be determined as follows:

$$\langle v_x^2 \rangle = \frac{kT}{m} = \int_{-\infty}^{\infty} v_x^2 f(v_x) \, dv_x$$

$$= \int_{-\infty}^{\infty} v_x^2 \sqrt{\frac{\gamma}{2\pi}} e^{-\gamma v_x^2/2} dv_x$$

$$= \sqrt{\frac{\gamma}{2\pi}} \int_{-\infty}^{\infty} v_x^2 e^{-\gamma v_x^2/2} dv_x$$

$$= \sqrt{\frac{\gamma}{2\pi}} \left(\frac{1}{\gamma} \sqrt{\frac{2\pi}{\gamma}} \right)$$

$$= \frac{1}{\gamma}$$

$$\frac{m}{kT} = \gamma \qquad (33.15)$$

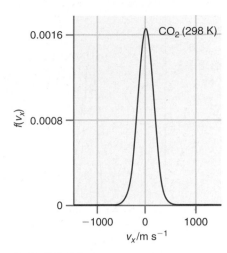

The integration was performed using the integral tables provided in the Math Supplement, Appendix A. With the definition of γ, the **Maxwell–Boltzmann velocity distribution** in one dimension becomes

$$f(v_j) = \left(\frac{m}{2\pi kT} \right)^{1/2} e^{(-mv_j^2/2kT)} = \left(\frac{M}{2\pi RT} \right)^{1/2} e^{(-Mv_j^2/2RT)} \qquad (33.16)$$

FIGURE 33.1
One-dimensional velocity distribution for CO_2 at 298 K.

In Equation (33.16), m is the particle mass in units of kilograms, and M is the molar mass in units of kg mol^{-1}, obtained from the expression involving m using the relationship $R = N_A k$. The velocity distribution is equal to the product of a preexponential factor that is independent of velocity, and an exponential factor that is velocity dependent, and this latter term is very reminiscent of the Boltzmann distribution. The one-dimensional velocity distribution in the x direction for CO_2 at 298 K is presented in Figure 33.1. Notice that the distribution maximum is at 0 m s^{-1}.

EXAMPLE PROBLEM 33.1

Compare $\langle v_x \rangle$ and $\langle v_x^2 \rangle$ for an ensemble of gaseous particles.

Solution

The average velocity is simply the first moment of the velocity distribution function:

$$\langle v_x \rangle = \int_{-\infty}^{\infty} v_x f(v_x) dv_x$$

$$= \int_{-\infty}^{\infty} v_x \left(\frac{m}{2\pi kT} \right)^{1/2} e^{-mv_x^2/2kT} dv_x$$

$$= \left(\frac{m}{2\pi kT} \right)^{1/2} \int_{-\infty}^{\infty} v_x e^{-mv_x^2/2kT} dv_x$$

$$= 0$$

The integral involves the product of odd (v_x) and even (exponential) factors so that the integral over the domain of v_x equals zero (a further description of even and odd functions is provided in the Math Supplement). The fact that the average value of $v_x = 0$ reflects the vectorial character of velocity with particles equally likely to be moving in the $+x$ or $-x$ direction.

The quantity $\langle v_x^2 \rangle$ was determined earlier:

$$\langle v_x^2 \rangle = \int_{-\infty}^{\infty} v_x^2 f(v_x) dv_x = \frac{kT}{m}$$

Notice that the average value for the second moment is greater than zero, reflecting the fact that the square of the velocity must be a positive quantity.

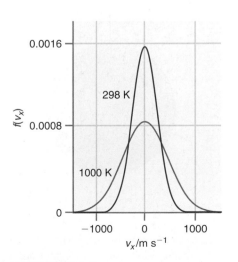

FIGURE 33.2
One-dimensional velocity distributions for Ar at 298 K (red line) and 1000 K (purple line).

Figure 33.2 presents the one-dimensional velocity distribution functions for Ar at two different temperatures. Notice how the width of the distribution increases with temperature consistent with the increased probability of populating higher energy translational states with correspondingly greater velocities.

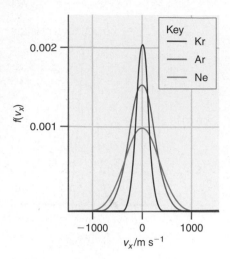

FIGURE 33.3
Velocity distributions for Kr (red line, molar mass = 83.8 g mol^{-1}), Ar (purple line, molar mass = 39.9 g mol^{-1}), and Ne (blue line, molar mass = 20.2 g mol^{-1}) at 298 K.

Figure 33.3 presents the distribution for Kr, Ar, and Ne at 298 K. The velocity distribution is narrowest for Kr and broadest for Ne reflecting the mass dependence of the distribution.

33.3 The Maxwell Distribution of Molecular Speeds

With one-dimensional velocity distributions in hand, the three-dimensional distribution of molecular speeds can be determined. First, it is important to recognize that speed is not a vector. The reason we concern ourselves with speed as opposed to velocity is that many physical properties of gases are dependent only on the speed of the gas particles and not on the direction of motion. Therefore, only the magnitude of the velocity, or speed, is generally of interest. As shown earlier, the **particle speed** ν is related to the one-dimensional velocity Cartesian components by the following:

$$\nu = (v_x^2 + v_y^2 + v_z^2)^{1/2} \tag{33.17}$$

We are interested in determining the particle speed distribution $F(\nu)$, but how can this distribution be derived using the velocity distributions derived in the previous section? We can connect these concepts using the geometric interpretation of velocity presented in Figure 33.4. The figure depicts **velocity space,** which can be understood in analogy to Cartesian space with linear distance (x, y, z) replaced by the Cartesian components of velocity (v_x, v_y, v_z). The figure demonstrates that the molecular velocity **v** is described by a vector with coordinates v_x, v_y, and v_z in velocity space with length equal to the speed [Equation (33.17)].

Particle speed distribution $F(\nu)$ is defined in terms of one-dimensional velocity distributions along each direction [Equation (33.15)] as follows:

$$F(\nu)\,d\nu = f(v_x)f(v_y)f(v_z)\,dv_x\,dv_y\,dv_z$$

$$= \left[\left(\frac{m}{2\pi kT}\right)^{1/2} e^{-mv_x^2/2kT}\right]\left[\left(\frac{m}{2\pi kT}\right)^{1/2} e^{-mv_y^2/2kT}\right]$$

$$\times \left[\left(\frac{m}{2\pi kT}\right)^{1/2} e^{-mv_z^2/2kT}\right]dv_x\,dv_y\,dv_z$$

$$= \left(\frac{m}{2\pi kT}\right)^{3/2} e^{[-m(v_x^2 + v_y^2 + v_z^2)]/2kT}\,dv_x\,dv_y\,dv_z \tag{33.18}$$

Notice that in Equation (33.18), the speed distribution of interest is defined with respect to the Cartesian components of velocity; therefore, the factors involving velocity need to be expressed in terms of speed to obtain $F(\nu)\,d\nu$. This transformation is accomplished as follows. First, the factor $dv_x\,dv_y\,dv_z$ is an infinitesimal volume element in velocity space (Figure 33.4). Similar to the transformation from Cartesian coordinates to spherical coordinates (see the Math Supplement), the velocity volume element can be written as $4\pi\nu^2\,d\nu$ after integration over angular dimensions leaving only a ν dependence. In addition, $v_x^2 + v_y^2 + v_z^2$ in the exponent of Equation (33.18) can be written as ν^2 [Equation (33.17)] such that

$$F(\nu)\,d\nu = 4\pi\left(\frac{m}{2\pi kT}\right)^{3/2}\nu^2 e^{-m\nu^2/2kT}\,d\nu \tag{33.19}$$

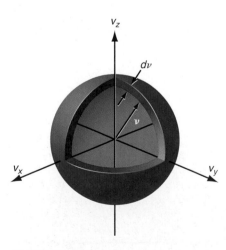

FIGURE 33.4
Illustration of velocity space. The Cartesian components of the particle velocity **v** are given by v_x, v_y, and v_z. The spherical shell represents a differential volume element of velocity space having volume $4\pi\nu^2\,d\nu$.

Equation (33.19) is written in terms of the mass of an individual gas particle. Because the molar mass M is equal to $N_A \times m$, where N_A is Avogadro's number and

m is the particle mass, and $R = N_A \times k$, Equation (33.19) can be written in terms of M as follows:

$$F(\nu)\, d\nu = 4\pi \left(\frac{M}{2\pi RT} \right)^{3/2} \nu^2 e^{-M\nu^2/2RT}\, d\nu \qquad \textbf{(33.20)}$$

The **Maxwell speed distribution** is given by Equation (33.19) or (33.20) and represents the probability distribution of a molecule having a speed between ν and $\nu + d\nu$. Comparison of this distribution to the one-dimensional velocity distribution of Equation (33.15) reveals many similarities, with the main difference being the ν^2 dependence that now appears in the preexponential factor in Equation (33.19) or (33.20). A second difference is the range of the distribution. Unlike velocity where negative values are possible since the particle is always free to move in the negative direction with respect to a given coordinate, particle speeds must be greater than or equal to zero so that the range of the distribution is from zero to infinity.

Figure 33.5 presents the Maxwell speed distribution for Ar at 298 and 1000 K and illustrates the dependence of the speed distribution on temperature. Notice that unlike the velocity distribution of Equation (33.15), the speed distribution is not symmetric. This is because the initial increase in probability is due to the ν^2 factor in Equation (33.19) or (33.20), but at higher speeds the probability decays exponentially. Also, as temperature increases two trends become evident. First, the maximum of the distribution shifts to higher speed as temperature increases. This is expected because an increase in temperature corresponds to an increase in kinetic energy and subsequently an increase in particle speed. Second, the entire distribution does not simply shift to higher velocity. Instead, the curvature of the distribution changes, with this behavior quite pronounced on the high-speed side of the distribution since an increase in kT will increase the probability of occupying higher energy translational states. Finally, note that the area under both curves is identical and equal to one expected for a normalized distribution.

Figure 33.6 presents a comparison of the particle speed distributions for Ne, Ar, and Kr at 298 K. The speed distribution peaks at lower speeds for heavier particles. This behavior can be understood since the average kinetic energy is $3/2\, kT$, a quantity that is only dependent on temperature. Because kinetic energy is also equal to $1/2\, m\nu^2$, an increase in mass must be offset by a reduction in the root-mean-squared speed. This expectation is reflected by the distributions presented in Figure 33.6.

One of the first detailed experimental verifications of the Maxwell distribution law was provided in 1955 by Miller and Kusch [*Phys. Rev.* 99 (1955): 1314]. A schematic drawing of the apparatus employed in this study is presented in Figure 33.7. An oven was used to create a gas of known temperature, and a hole was placed in the side of the oven through which the gas could emerge. The stream of gas molecules escaping the oven was then directed through spatial apertures to create a beam of particles, which was then directed toward a velocity selector. By changing the rotational speed of the velocity selector, the required gas speed necessary to pass through the cylinder is varied. The number of gas particles passing through the cylinder as a function of rotational speed is then measured to determine the distribution of gas speeds.

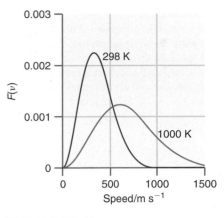

FIGURE 33.5
Speed distributions for Ar at 298 and 1000 K.

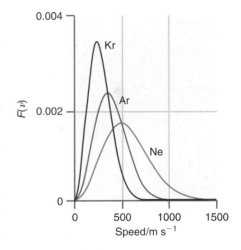

FIGURE 33.6
Speed distributions for Ne (blue line, molar mass = 20.2 g mol⁻¹), Ar (purple line, molar mass = 39.9 g mol⁻¹), and Kr (red line, molar mass = 83.8 g mol⁻¹) at 298 K.

FIGURE 33.7
Schematic of an experimental apparatus used to verify the Maxwell speed distribution. Molecules are emitted from the oven and pass through the slits to produce a beam of molecules. This beam reaches a velocity selector consisting of two rotating disks with slots in each disk. After passing through the first disk, the molecules will reach the second disk that will have rotated by angle θ. Only molecules for which the velocity equals $\omega x / \theta$, where ω is the angular rotational velocity of the disks and x is the separation between disks, will pass through the second disk and reach the detector.

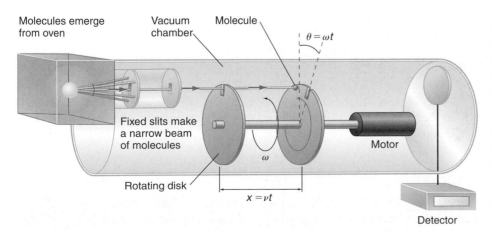

FIGURE 33.8
Experimentally determined distribution of particle speeds for gaseous potassium at 466 ± 2 K (circles). The expected Maxwellian distribution is presented as the red line, and demonstrates excellent agreement with experiment.

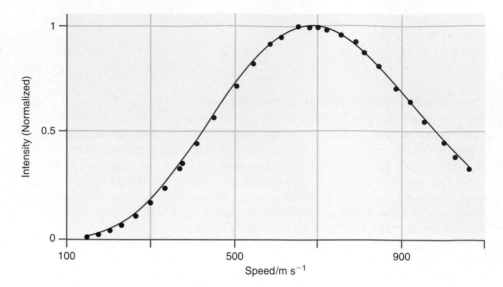

The results of this experiment for potassium at an oven temperature of 466 ± 2 K are presented in Figure 33.8. The comparison between the experimentally measured number of particles with a certain speed versus the theoretical prediction for a Maxwellian distribution is excellent. The interested reader is encouraged to read the Miller and Kusch manuscript for a fuller presentation of this elegant experiment.

33.4 Comparative Values for Speed Distributions: ν_{ave}, ν_{mp}, and ν_{rms}

The Maxwell speed distribution describes the probability of observing a particle within a given range of speeds; however, knowledge of the entire distribution is seldom required when comparing the properties of gases. Instead, representative quantities of this distribution that provide a metric as to how the distribution changes as a function of mass or temperature are sometimes more useful. For example, in Figure 33.6 it is clear that the speed distributions for Ne, Ar, and Kr are different, but can these differences be described without depicting the entire distribution? It would be much more convenient to compare only certain aspects of the distribution, such as the speed at which the distribution is maximized, or the average speed.

The first comparative value we consider is the **most probable speed,** or ν_{mp}, which is equal to the speed at which $F(\nu)$ is at a maximum. This quantity is determined by calculating the derivative of $F(\nu)$ with respect to speed:

$$
\begin{aligned}
\frac{dF(\nu)}{d\nu} &= \frac{d}{d\nu}\left(4\pi \left(\frac{m}{2\pi kT}\right)^{3/2} \nu^2 e^{-m\nu^2/2kT}\right) \\
&= 4\pi \left(\frac{m}{2\pi kT}\right)^{3/2} \frac{d}{d\nu}\left(\nu^2 e^{-m\nu^2/2kT}\right) \\
&= 4\pi \left(\frac{m}{2\pi kT}\right)^{3/2} e^{-m\nu^2/2kT}\left[2\nu - \frac{m\nu^3}{kT}\right]
\end{aligned}
$$

The most probable speed is the speed at which $dF(\nu)/d\nu$ is equal to zero, which will be the case when the factor contained in the square brackets in the preceding equation is equal to zero. Recognizing this, ν_{mp} is given by

$$
2\nu_{mp} - \frac{m\nu_{mp}^3}{kT} = 0
$$

$$
\nu_{mp} = \sqrt{\frac{2kT}{m}} = \sqrt{\frac{2RT}{M}} \tag{33.21}
$$

EXAMPLE PROBLEM 33.2

What is the most probable speed for Ne and Kr at 298 K?

Solution

First, ν_{mp} for Ne is readily determined using Equation (33.21):

$$\nu_{mp} = \sqrt{\frac{2RT}{M}} = \sqrt{\frac{2(8.314\ \text{J mol}^{-1}\ \text{K}^{-1})298\ \text{K}}{0.020\ \text{kg mol}^{-1}}} = 498\ \text{m s}^{-1}$$

The corresponding ν_{mp} for Kr can be determined in the same manner, or with reference to Ne as follows:

$$\frac{(\nu_{mp})_{Kr}}{(\nu_{mp})_{Ne}} = \sqrt{\frac{M_{Ne}}{M_{Kr}}} = \sqrt{\frac{0.020\ \text{kg mol}^{-1}}{0.083\ \text{kg mol}^{-1}}} = 0.491$$

With this result, ν_{mp} for Kr is readily determined:

$$(\nu_{mp})_{kr} = 0.491(\nu_{mp})_{Ne} = 244\ \text{m s}^{-1}$$

For the sake of comparison, the speed of sound in dry air at 298 K is 346 m/s, and a typical commercial airliner travels at about 500 miles/hour or 224 m/s.

Average speed can be determined using the Maxwell speed distribution and the definition of average value provided in Chapter 29.

$$\nu_{ave} = \langle \nu \rangle = \int_0^\infty \nu F(\nu)\, d\nu$$

$$= \int_0^\infty \nu \left(4\pi \left(\frac{m}{2\pi kT} \right)^{3/2} \nu^2 e^{-m\nu^2/2kT} \right) d\nu$$

$$= 4\pi \left(\frac{m}{2\pi kT} \right)^{3/2} \int_0^\infty \nu^3 e^{-m\nu^2/2kT}\, d\nu$$

$$= 4\pi \left(\frac{m}{2\pi kT} \right)^{3/2} \frac{1}{2} \left(\frac{2kT}{m} \right)^2$$

$$\nu_{ave} = \left(\frac{8kT}{\pi m} \right)^{1/2} = \left(\frac{8RT}{\pi M} \right)^{1/2} \qquad \textbf{(33.22)}$$

A solution to the integral in Equation (33.22) is presented in the Math Supplement.

The final comparative quantity is the **root-mean-squared speed,** or ν_{rms}. This quantity is equal to $[\langle \nu^2 \rangle]^{1/2}$, or simply the square root of the second moment of the distribution:

$$\nu_{rms} = \left[\langle \nu^2 \rangle \right]^{1/2} = \left(\frac{3kT}{m} \right)^{1/2} = \left(\frac{3RT}{M} \right)^{1/2} \qquad \textbf{(33.23)}$$

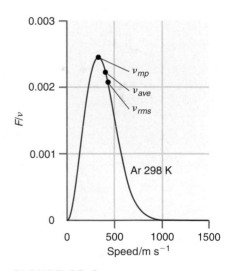

FIGURE 33.9
Comparison of $\nu_{mp}, \nu_{ave},$ and ν_{rms} for Ar at 298 K.

Notice that Equation (33.23) is equal to the prediction of kinetic theory of Equation (33.1). The locations of $\nu_{mp}, \nu_{ave},$ and ν_{rms} relative to the speed distribution for Ar at 298 K are presented in Figure 33.9. Comparison of Equation (33.21), (33.22), and (33.23) reveals that the only differences between the values are constants, which becomes evident when the ratios of these quantities are considered. Specifically, the ratio of $\nu_{rms}/\nu_{mp} = (3/2)^{1/2}$ and $\nu_{ave}/\nu_{mp} = (4/\pi)^{1/2}$ so that $\nu_{rms} > \nu_{ave} > \nu_{mp}$. Note also that all three **benchmark values** demonstrate the same dependence on T and particle mass: they increase as the square root of T and decrease as the square root of M.

EXAMPLE PROBLEM 33.3

Determine $\nu_{mp}, \nu_{ave},$ and ν_{rms} for Ar at 298 K.

Solution

Using Equations (33.21), (33.22), and (33.23), the benchmark speed values are as follows:

$$v_{mp} = \sqrt{\frac{2RT}{M}} = \sqrt{\frac{2(8.314 \text{ J mol}^{-1} \text{ K}^{-1})(298 \text{ K})}{0.040 \text{ kg mol}^{-1}}} = 352 \text{ m s}^{-1}$$

$$v_{ave} = \sqrt{\frac{8RT}{\pi M}} = \sqrt{\frac{8(8.314 \text{ J mol}^{-1} \text{ K}^{-1})(298 \text{ K})}{\pi(0.040 \text{ kg mol}^{-1})}} = 397 \text{ m s}^{-1}$$

$$v_{rms} = \sqrt{\frac{3RT}{M}} = \sqrt{\frac{3(8.314 \text{ J mol}^{-1} \text{ K}^{-1})(298 \text{ K})}{0.040 \text{ kg mol}^{-1}}} = 431 \text{ m s}^{-1}$$

33.5 Gas Effusion

As described earlier, the experiments that verified the accuracy of the Maxwell speed distribution were performed using a gas escaping through an aperture in the wall of the oven containing the gas (see Figure 33.7). In this technique, the gas confined to the box is at some finite pressure and is separated from a vacuum by a thin wall of the oven containing the aperture. The pressure of the gas and size of the aperture is such that molecules do not undergo collisions near or when passing through the aperture. The process by which a gas passes through an opening under these conditions is called **effusion** and is employed to produce a stream or "beam" of gas particles. For example, this technique is used to create atomic or molecular beams that can collide with beams of other molecules to study chemical reaction dynamics.

To derive the rate of gas effusion, we proceed in a fashion analogous to that used to derive pressure in Chapter 1. Let dN_c be the number of particles that hit the wall of the container. The collisional rate, dN_c/dt, is the number of collisions with the wall per unit time. This quantity will be proportional to the area being struck, A. In addition, the collisional rate will depend on particle velocity, with increased velocity resulting in an increased collisional rate. Finally, the collisional rate should be directly proportional to the particle density $\tilde{N}$ defined as the number of particles per unit volume. Taking these three ideas into account, we can write

$$\frac{dN_c}{dt} = \tilde{N}A \int_0^\infty v_x f(v_x)\, dv_x \qquad (33.24)$$

The integral in Equation (33.24) is simply the average particle velocity in the direction that will result in collision with the area of interest (taken as the positive x direction with corresponding limits of integration from zero to positive infinity). Evaluating this integral yields the following expression for the collision rate:

$$\frac{dN_c}{dt} = \tilde{N}A \int_0^\infty v_x \left(\frac{m}{2\pi kT}\right)^{1/2} e^{-mv_x^2/2kT}\, dv_x$$

$$= \tilde{N}A \left(\frac{m}{2\pi kT}\right)^{1/2} \int_0^\infty v_x e^{-mv_x^2/2kT}\, dv_x$$

$$= \tilde{N}A \left(\frac{m}{2\pi kT}\right)^{1/2} \left(\frac{kT}{m}\right)$$

$$= \tilde{N}A \left(\frac{kT}{2\pi m}\right)^{1/2}$$

$$\frac{dN_c}{dt} = \tilde{N}A \frac{1}{4} v_{ave} \qquad (33.25)$$

In the final step, the definition of average speed v_{ave} provided in Equation (33.22) has been used. The **collisional flux** Z_c is defined as the number of collisions per unit time and per unit area. This quantity is equal to collisional rate divided by the area of interest A:

$$Z_c = \frac{dN_c/dt}{A} = \frac{1}{4}\widetilde{N}v_{ave} \tag{33.26}$$

It is sometimes more convenient to express the collisional flux in terms of gas pressure. This is accomplished by rewriting $\widetilde{N}$ as follows:

$$\widetilde{N} = \frac{N}{V} = \frac{nN_A}{V} = \frac{P}{kT} \tag{33.27}$$

With this definition of $\widetilde{N}$, Z_c becomes

$$Z_c = \frac{P}{(2\pi m kT)^{1/2}} = \frac{PN_A}{(2\pi MRT)^{1/2}} \tag{33.28}$$

where m is the particle mass (in kilograms) and M is molar mass (in kg mol^{-1}). Evaluating the preceding expression requires careful attention to units, as Example Problem 33.4 illustrates.

EXAMPLE PROBLEM 33.4

How many collisions per second occur on a container wall with an area of 1 cm^2 for a collection of Ar particles at 1 atm and 298 K?

Solution

Using Equation (33.28):

$$Z_c = \frac{PN_A}{(2\pi MRT)^{1/2}} = \frac{(1.01325 \times 10^5\,\text{Pa})(6.022 \times 10^{23}\,\text{mol}^{-1})}{(2\pi(0.0400\,\text{kg mol}^{-1})(8.314\,\text{J mol}^{-1}\,\text{K}^{-1})(298\,\text{K}))^{1/2}}$$

$$= 2.45 \times 10^{27}\,\text{m}^{-2}\,\text{s}^{-1}$$

Notice that pressure is in units of Pa (kg m^{-1} s^{-2}) resulting in the appropriate units for Z_c. Then, multiplying the collisional flux by the area of interest yields the collisional rate:

$$\frac{dN_c}{dt} = Z_c A = (2.45 \times 10^{27}\,\text{m}^{-2}\,\text{s}^{-1})(10^{-4}\,\text{m}^2) = 2.45 \times 10^{23}\,\text{s}^{-1}$$

This quantity represents the number of collisions per second with a section of the container wall having an area of 1 cm^2. This is a rather large quantity, and it demonstrates the substantial number of collisions that occur in a container for a gas under standard temperature and pressure conditions.

Effusion will result in a decrease in gas pressure as a function of time. The change in pressure is related to the change in the number of particles in the container, N, as follows:

$$\frac{dP}{dt} = \frac{d}{dt}\left(\frac{NkT}{V}\right) = \frac{kT}{V}\frac{dN}{dt} \tag{33.29}$$

The quantity dN/dt can be related to the collisional flux [Equation (33.28)] by recognizing the following. First, if the space outside of the container is at a significantly lower pressure than the container and a particle escapes the container, it will not return. Second, each collision corresponds to a particle striking the aperture area so that the number of collisions with the aperture area is equal to the number of molecules lost, resulting in $N_c = N$ in Equation (33.26) and

$$\frac{dN}{dt} = -Z_c A = \frac{-PA}{(2\pi m kT)^{1/2}} \tag{33.30}$$

where the negative sign is consistent with the expectation that the number of particles in the container will decrease as effusion proceeds. Substituting the preceding result into Equation (33.29), the change in pressure as a function of time becomes

$$\frac{dP}{dt} = \frac{kT}{V}\left(\frac{-PA}{(2\pi m kT)^{1/2}}\right) \tag{33.31}$$

Integration of Equation (33.31) yields the following expression for container pressure as a function of time:

$$P = P_0 \exp\left[-\frac{At}{V}\left(\frac{kT}{2\pi m}\right)^{1/2}\right] \tag{33.32}$$

In Equation (33.32), P_0 is initial container pressure. This result demonstrates that effusion will result in an exponential decrease in container pressure as a function of time.

EXAMPLE PROBLEM 33.5

A 1 L container filled with Ar at 298 K and at an initial pressure of 1.00×10^{-2} atm is allowed to effuse through an aperture having an area of 0.01 μm^2. Will the pressure inside the container be significantly reduced after 1 hour of effusion?

Solution

In evaluating Equation (33.32), it is easiest to first determine the exponential factor and then determine the pressure:

$$\frac{At}{V}\left(\frac{kT}{2\pi m}\right)^{1/2} = \frac{10^{-14}\,m^2(3600\,s)}{10^{-3}\,m^3}\left(\frac{1.38 \times 10^{-23}\,J\,K^{-1}(298\,K)}{2\pi\left(\dfrac{0.0400\,kg\,mol^{-1}}{N_A}\right)}\right)^{1/2}$$

$$= 3.60 \times 10^{-8}\,s\,m^{-1}(99.3\,m\,s^{-1}) = 3.57 \times 10^{-6}$$

The pressure after 1 hour of effusion, therefore, is

$$P = P_0\,e^{-3.57\times 10^{-6}} = (1.00 \times 10^{-2}\,atm)e^{-3.57\times 10^{-6}} \approx 1.00 \times 10^{-2}\,atm$$

Given the large volume of the container and the relatively small aperture through which effusion occurs, the pressure inside the container is essentially unchanged.

33.6 Molecular Collisions

Kinetic theory can also be used to determine the collisional frequency between gaseous particles. Recall that one of the primary ideas behind kinetic theory is that the distance between gaseous particles is on average much greater than the actual particle volume. However, the particles translate through space, and collisions between particles will occur. How does one think about these collisions with respect to the intermolecular interactions that occur between particles? At high gas pressures, intermolecular forces are substantial and particle interactions become important. Even during collisions at low pressures, the intermolecular forces must be relevant during the collisions. Modeling collisions including the subtleties of intermolecular interactions is beyond the scope of this text. Instead, we adopt a limiting viewpoint of collisions in which we treat the particles as hard spheres. Billiard balls are an excellent example of a hard-sphere particle. Collisions occur when two billiard balls attempt to occupy the same region of space, and this is the only time the particles interact. We will see in upcoming chapters that an understanding of the frequency of molecular collisions is important in describing a variety of chemical phenomena, including the rates of chemical reactions.

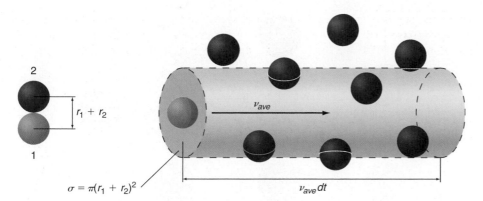

FIGURE 33.10
Schematic of the hard-sphere collisional process.

How frequent are molecular collisions? To answer this question, we assume that the particle of interest is moving and that all other molecules are stationary (we will relax this assumption shortly). In this picture, the particle of interest sweeps out a collisional cylinder, which determines the number of collisions the particle undergoes per unit time. A depiction of this cylinder is provided in Figure 33.10. Collisions occur between the particle of interest and other particles that are positioned within the cylinder. The area of the collisional cylinder base or **collisional cross section** σ is dependent on the radii of the gas particles r as follows:

$$\sigma = \pi (r_1 + r_2)^2 \tag{33.33}$$

The subscripts 1 and 2 in Equation (33.33) denote the two collisional partners, which may or may not be the same species. The length of the cylinder is given by the product of a given time interval (dt) and ν_{ave}, the average speed of the molecule. Therefore, the total cylinder volume is equal to $\sigma \nu_{ave} dt$ as depicted in Figure 33.10.

This derivation of the cylinder volume is not exactly correct because the other molecules are not stationary. To incorporate the motion of other molecules in the derivation, we now introduce the concept of **effective speed.** To illustrate this concept, imagine two particles moving with some average speed as depicted in Figure 33.11. The effective speed at which two particles approach each other is dependent on the relative direction of particle motion. The first case depicted in the figure is of two particles traveling in the same direction where $\langle \nu_{12} \rangle = \langle \nu_1 \rangle - \langle \nu_2 \rangle$. If the particles were identical, the effective speed would be zero since the molecules would travel with constant separation and never collide. The opposite case is depicted in the middle of the figure, where the effective speed is $\langle \nu_1 \rangle + \langle \nu_2 \rangle$ when the particles are traveling directly at each other. Again, if the particles were identical then $\langle \nu_{12} \rangle = 2\langle \nu_1 \rangle$.

A full derivation of the effective speed is quite involved and results in the intuitive result depicted as the third case in Figure 33.11. In this third case, the average approach angle is 90° and, using the Pythagorean theorem, the effective speed is equal to

$$\langle \nu_{12} \rangle = (\langle \nu_1 \rangle^2 + \langle \nu_2 \rangle^2)^{1/2} = \left[\left(\frac{8kT}{\pi m_1} \right) + \left(\frac{8kT}{\pi m_2} \right) \right]^{1/2}$$

$$= \left[\frac{8kT}{\pi} \left(\frac{1}{m_1} + \frac{1}{m_2} \right) \right]^{1/2}$$

$$= \left(\frac{8kT}{\pi \mu} \right)^{1/2} \tag{33.34}$$

where

$$\mu = \frac{m_1 m_2}{m_1 + m_2}$$

With the effective speed as defined by Equation (33.34), the collisional-cylinder volume is now defined. The number of collisional partners in the collisional cylinder is equal to

FIGURE 33.11
Depiction of effective speed in a two-
particle collision.

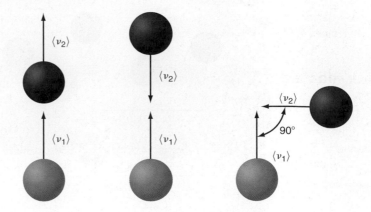

the product of the collisional-partner number density (N_2/V) and the volume of the cylinder, $V_{cyl} = \sigma v_{ave} dt$ (Figure 33.10). We define the individual **particle collisional frequency** z_{12} as the number of collisions an individual molecule (denoted by the subscript 1) undergoes with other collisional partners (denoted by the subscript 2) per unit time (dt). This quantity is equal to the number of collisional partners divided by dt:

$$z_{12} = \frac{N_2}{V}\left(\frac{V_{cyl}}{dt}\right) = \frac{N_2}{V}\left(\frac{\sigma v_{ave} dt}{dt}\right) = \frac{N_2}{V}\sigma\left(\frac{8kT}{\pi\mu}\right)^{1/2} \tag{33.35}$$

If the gas is comprised of one type of particle, $\mu = m_1/2$ and Equation (33.35) becomes

$$z_{11} = \frac{N_1}{V}\sigma\sqrt{2}\left(\frac{8kT}{\pi m_1}\right)^{1/2} = \frac{P_1 N_A}{RT}\sigma\sqrt{2}\left(\frac{8RT}{\pi M_1}\right)^{1/2} \tag{33.36}$$

The **total collisional frequency** is defined as the total number of collisions that occur for all gas particles. The total collisional frequency for a collection of two types of gas molecules Z_{12} is given by z_{12} times the number density of species 1:

$$Z_{12} = \frac{N_1}{V}z_{12} = \frac{N_1}{V}\frac{N_2}{V}\sigma\left(\frac{8kT}{\pi\mu}\right)^{1/2} = \left(\frac{P_1 N_A}{RT}\right)\left(\frac{P_2 N_A}{RT}\right)\sigma\left(\frac{8kT}{\pi\mu}\right)^{1/2} \tag{33.37}$$

The units of Z_{12} are collisions per cubic meter, or the total number of collisions per unit volume. The corresponding collisional frequency for a gas consisting of only one type of particle is

$$Z_{11} = \frac{1}{2}\frac{N_1}{V}z_{11} = \frac{1}{\sqrt{2}}\left(\frac{N_1}{V}\right)^2\sigma\left(\frac{8kT}{\pi m_1}\right)^{1/2} = \frac{1}{\sqrt{2}}\left(\frac{P_1 N_A}{RT}\right)^2\sigma\left(\frac{8RT}{\pi M_1}\right)^{1/2} \tag{33.38}$$

The factor of $1/2$ appears in Equation (33.38) to ensure that each collision is only counted once. Evaluation of Equations (33.35) through (33.38) requires knowledge of the collisional cross sections that are in turn dependent on the effective hard-sphere radii. As we will see in the next chapter, these values can be determined by the measure of various gas properties. Table 33.1 provides the hard-sphere radii for a variety of

TABLE 33.1	Collisional Parameters for Various Gases	
Species	**r (nm)**	**σ (nm^2)**
He	0.13	0.21
Ne	0.14	0.24
Ar	0.17	0.36
Kr	0.20	0.52
N_2	0.19	0.43
O_2	0.18	0.40
CO_2	0.20	0.52

common gases determined from such measurements. Generally, for monatomic gases and small molecules the radius is on the order of 0.2 nm.

EXAMPLE PROBLEM 33.6

What is z_{11} for CO_2 at 298 K and 1 atm?

Solution

The question asks for the single-particle collisional frequency of CO_2. Using Equation (33.36) and the collisional cross section provided in Table 33.1, we obtain

$$z_{CO_2} = \frac{P_{CO_2} N_A}{RT} \sigma \sqrt{2} \left(\frac{8RT}{\pi M_{CO_2}} \right)^{1/2}$$

$$= \frac{101{,}325 \text{ Pa } (6.022 \times 10^{23} \text{ mol}^{-1})}{8.314 \text{ J mol}^{-1} \text{ K}^{-1}(298 \text{ K})} (5.2 \times 10^{-19} \text{ m}^2) \sqrt{2}$$

$$\times \left(\frac{8(8.314 \text{ J mol}^{-1} \text{ K}^{-1})(298 \text{ K})}{\pi (0.044 \text{ kg mol}^{-1})} \right)^{1/2}$$

$$= 6.9 \times 10^9 \text{ s}^{-1}$$

This calculation demonstrates that a single CO_2 molecule undergoes roughly 7 billion collisions per second under standard temperature and pressure conditions! The inverse of the collisional frequency corresponds to the time between molecular collision or roughly 150 picoseconds (1 ps $= 10^{-12}$ s) between collisions.

EXAMPLE PROBLEM 33.7

What is the total collisional frequency (Z_{ArKr}) at 298 K for a collection of Ar and Kr confined to a 1 cm^3 container with partial pressures of 360. Torr for Ar and 400. Torr for Kr?

Solution

Evaluation of Equation (33.37) is best performed by evaluating each factor in the equation separately, then combining factors to calculate the total collisional frequency as follows:

$$\left(\frac{P_{Ar} N_A}{RT} \right) = \left(\frac{47{,}996 \text{ Pa } (6.022 \times 10^{23} \text{ mol}^{-1})}{8.314 \text{ J mol}^{-1} \text{ K}^{-1}(298 \text{ K})} \right) = 1.17 \times 10^{25} \text{ m}^{-3}$$

$$\left(\frac{P_{Kr} N_A}{RT} \right) = \left(\frac{53{,}328 \text{ Pa } (6.022 \times 10^{23} \text{ mol}^{-1})}{8.314 \text{ J mol}^{-1} \text{ K}^{-1}(298 \text{ K})} \right) = 1.30 \times 10^{25} \text{ m}^{-3}$$

$$\sigma = \pi (r_{Ar} + r_{Kr})^2 = \pi (0.17 \text{ nm} + 0.20 \text{ nm})^2 = 0.430 \text{ nm}^2 = 4.30 \times 10^{-19} \text{ m}^2$$

$$\mu = \frac{m_{Ar} m_{Kr}}{m_{Ar} + m_{Kr}} = \frac{(0.040 \text{ kg mol}^{-1})(0.084 \text{ kg mol}^{-1})}{(0.040 \text{ kg mol}^{-1}) + (0.084 \text{ kg mol}^{-1})} \times \frac{1}{N_A} = 4.50 \times 10^{-26} \text{ kg}$$

$$\left(\frac{8kT}{\pi \mu} \right)^{1/2} = \left(\frac{8(1.38 \times 10^{-23} \text{ J K}^{-1})(298 \text{ K})}{\pi (4.50 \times 10^{-26} \text{ kg})} \right)^{1/2} = 482 \text{ m s}^{-1}$$

$$Z_{ArKr} = \left(\frac{P_{Ar} N_A}{RT} \right) \left(\frac{P_{Kr} N_A}{RT} \right) \sigma \left(\frac{8kT}{\pi \mu} \right)^{1/2}$$

$$= (1.17 \times 10^{25} \text{ m}^{-3})(1.30 \times 10^{25} \text{ m}^{-3})(4.30 \times 10^{-19} \text{ m}^2)(482 \text{ m s}^{-1})$$

$$= 3.16 \times 10^{34} \text{ m}^{-3} \text{ s}^{-1} = 3.16 \times 10^{31} \text{ L}^{-1} \text{ s}^{-1}$$

Comparison of the last two example problems reveals that the total collisional frequency (Z_{12}) is generally much larger than the collisional frequency for an individual molecule (z_{12}). The total collisional frequency incorporates all collisions that occur for a collection of gas particles (consistent with the units of inverse volume); therefore, the magnitude of this value relative to the collisional frequency for a single particle is not unexpected.

33.7 The Mean Free Path

The **mean free path** is defined as the average distance a gas particle travels between successive collisions. In a given time interval dt, the distance a particle will travel is equal to $\nu_{ave}\, dt$ where ν_{ave} is the average speed of the particle. In addition, the number of collisions the particle undergoes is given by $(z_{11} + z_{12})\, dt$, where the frequency of collisions with either type of collisional partner in the binary mixture is included. Given these quantities, the mean free path λ is given by the average distance traveled divided by the number of collisions:

$$\lambda = \frac{\nu_{ave}\, dt}{(z_{11} + z_{12})\, dt} = \frac{\nu_{ave}}{(z_{11} + z_{12})} \tag{33.39}$$

If our discussion is limited to a gas with one type of particle, $N_2 = 0$ resulting in $z_{12} = 0$, and the mean free path becomes

$$\lambda = \frac{\nu_{ave}}{z_{11}} = \frac{\nu_{ave}}{\left(\dfrac{N_1}{V}\right)\sqrt{2}\sigma\nu_{ave}} = \left(\frac{RT}{P_1 N_A}\right)\frac{1}{\sqrt{2}\sigma} \tag{33.40}$$

Equation (33.40) demonstrates that the mean free path decreases if the pressure increases or if the collisional cross section of the particle increases. This behavior makes intuitive sense. As particle density increases (i.e., as pressure increases), we would expect the particle to travel a shorter distance between collisions. Also, as the particle size increases, we would expect the probability of collision to also increase, thereby reducing the mean free path.

What does the mean free path tell us about the length scale of collisional events relative to molecular size? Recall that one of the assumptions of kinetic theory is that the distance between particles is large compared to their size. Is the mean free path consistent with this assumption? To answer this question, we return to the first example provided in this chapter, Ar at a pressure of 1 atm and temperature of 298 K, for which the mean free path is

$$\lambda_{Ar} = \left(\frac{RT}{P_{Ar} N_A}\right)\frac{1}{\sqrt{2}\sigma}$$

$$= \left(\frac{(8.314\ \text{J mol}^{-1}\,\text{K}^{-1})(298\ \text{K})}{(101{,}325\ \text{Pa})(6.022 \times 10^{23}\ \text{mol}^{-1})}\right)\frac{1}{\sqrt{2}(3.6 \times 10^{-19}\ \text{m}^2)}$$

$$= 7.98 \times 10^{-8}\ \text{m} \approx 80\ \text{nm}$$

Compared to the 0.29 nm diameter of Ar, the mean free path demonstrates that an Ar atom travels an average distance equal to ~ 275 times its diameter between collisions. This difference in length scales is consistent with the assumptions of kinetic theory. To provide further insight into the behavior of a collection of gaseous particles, as indicated in Example Problem 33.6, the collisional frequency can be used to determine the timescale between collisions as follows:

$$\frac{1}{z_{11}} = \frac{\lambda}{\nu_{ave}} = \frac{7.98 \times 10^{-8}\ \text{m}}{397\ \text{m s}^{-1}} = 2.01 \times 10^{-10}\ \text{s}$$

A picosecond is equal to 10^{-12} s; therefore, an individual Ar atom undergoes on average a collision every 200 ps. As will be discussed in the following chapters, properties such as collisional frequency and mean free path are important in describing transport properties of gases and chemical reaction dynamics involving collisional processes. The physical picture of gas particle motion outlined here will prove critical in understanding these important aspects of physical chemistry.

Vocabulary

average speed

benchmark value

collisional cross section

collisional flux

effective speed

effusion

gas kinetic theory

Maxwell speed distribution

Maxwell–Boltzmann velocity
 distribution

mean free path

most probable speed

particle collisional frequency

particle speed

root-mean-squared speed

total collisional frequency

velocity distribution function

velocity space

Conceptual Problems

Q33.1 Why is probability used to describe the velocity and speed of gas molecules?

Q33.2 What is the most probable velocity for a one-dimensional gas velocity distribution? Why?

Q33.3 Provide a physical explanation as to why the Maxwell speed distribution approaches zero at high speeds. Why is $f(\nu) = 0$ at $\nu = 0$?

Q33.4 How would the Maxwell speed distributions for He versus Kr compare if the gases were at the same temperature?

Q33.5 Imagine that you are performing an experiment using a molecular beam using Ar at a given temperature. If you switch the gas to Kr, will you have to increase or decrease the temperature of the gas to achieve a distribution of speeds identical to Ar?

Q33.6 How does the average speed of a collection of gas particles vary with particle mass and temperature?

Q33.7 Does the average kinetic energy of a particle depend on particle mass?

Q33.8 Arrange ν_{mp}, ν_{ave}, and ν_{rms} in order from highest to lowest speed.

Q33.9 Why does the mean free path depend on σ? Would an increase in $\widetilde{N}$ increase or decrease the mean free path?

Q33.10 In effusion, how does the frequency of collisions with the opening scale with molecular mass?

Q33.11 What is the typical length scale for a molecular diameter?

Q33.12 What is the difference between z_{11} and z_{12}?

Q33.13 Define the mean free path. How does this quantity vary with number density, particle diameter, and average particle speed?

Numerical Problems

Problem numbers in red indicate that the solution to the problem is given in the *Student's Solutions Manual*.

P33.1 Consider a collection of gas particles confined to translate in two dimensions (for example, a gas molecule on a surface). Derive the Maxwell speed distribution for such a gas.

P33.2 Determine ν_{mp}, ν_{ave}, and ν_{rms} for the following species at 298 K:

a. Ne **b.** Kr **c.** CH_4 **d.** C_2H_6 **e.** C_{60}

P33.3 Compute ν_{mp}, ν_{ave}, and ν_{rms} for O_2 at 300. and 500. K. How would your answers change for H_2?

P33.4 Compute ν_{ave} for H_2O, HOD, and D_2O at 298 K. Do you need to perform the same calculation each time, or can you derive an expression that relates the ratio of average speeds for two gases to their respective masses?

P33.5 Compare the average speed and average translational kinetic energy of O_2 with that of CCl_4 at 298 K.

P33.6 How far, on average, does O_2 travel in 1 second at 298 K and 1 atm? How does this distance compare to that of Kr under identical conditions?

P33.7

a. What is the average time required for H_2 to travel 1.00 m at 298 K and 1 atm?

b. How much longer does it take N_2 to travel 1.00 m, on average, relative to H_2 under these same conditions?

c. (Challenging) What fraction of N_2 particles will require more than this average time to travel 1.00 m? Answering this question will require evaluating a definite integral of the speed distribution, which requires using numerical methods such as Simpson's rule.

P33.8 As mentioned in Section 33.3, the only differences between the quantities v_{mp}, v_{ave}, and v_{rms} involve constants.

a. Derive the expressions for v_{ave} and v_{rms} relative to v_{mp} provided in the text.

b. Your result from part (a) will involve quantities that are independent of gas-specific quantities such as mass or temperature. Given this, it is possible to construct a "generic" speed distribution curve for speed in reduced units of v/v_{mp}. Transform the Maxwell distribution into a corresponding expression involving reduced speed.

P33.9 At what temperature is the v_{rms} of Ar equal to that of SF_6 at 298 K? Perform the same calculation for v_{mp}.

P33.10 Determine the temperature at which v_{ave} for Kr is equal to that of Ne at 298 K.

P33.11 The probability that a particle will have a velocity in the x direction in the range of $-v_{x_0}$ and v_{x_0} is given by

$$f(-v_{x0} \le v_x \le v_{x0}) = \left(\frac{m}{2\pi kT}\right)^{1/2} \int_{v_{x_o}}^{v_{x_o}} e^{-mv_x^2/2kT} dv_x$$

$$= \left(\frac{2m}{\pi kT}\right)^{1/2} \int_{0}^{v_{x_o}} e^{-mv_x^2/2kT} dv_x$$

The preceding integral can be rewritten using the following substitution: $\xi^2 = mv_x^2/2kT$, resulting in $f(-v_{x0} \le v_x \le v_{x0}) = 2/\sqrt{\pi}\left(\int_0^{\xi_0} e^{-\xi^2} d\xi\right)$, which can be evaluated using the error function defined as $\text{erf}(z) = 2/\sqrt{\pi}\left(\int_0^z e^{-x^2} dx\right)$.

The complementary error function is defined as $\text{erfc}(z) = 1 - \text{erf}(z)$. Finally, a plot of both $\text{erf}(z)$ and $\text{erfc}(z)$ as a function of z is shown here:

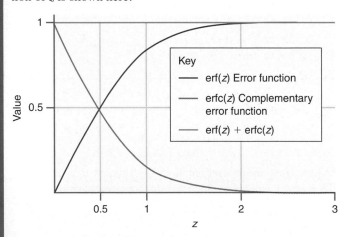

Using this graph of $\text{erf}(z)$, determine the probability that $|v_x| \le (2kT/m)^{1/2}$. What is the probability that $|v_x| > (2kT/m)^{1/2}$?

P33.12 The speed of sound is given by $v_{sound} = \sqrt{\gamma kT/m} = \sqrt{\gamma RT/M}$, where $\gamma = C_P/C_v$.

a. What is the speed of sound in Ne, Kr, and Ar at 1000. K?

b. At what temperature will the speed of sound in Kr equal the speed of sound in Ar at 1000. K?

P33.13 For O_2 at 1 atm and 298 K, what fraction of molecules has a speed that is greater than v_{rms}?

P33.14 The escape velocity from Earth's surface is given by $v_E = (2gR)^{1/2}$ where g is the gravitational acceleration (9.807 m s^{-2}) and R is the radius of Earth (6.37 × 10^6 m).

a. At what temperature will v_{mp} for N_2 be equal to the escape velocity?

b. How does the answer for part (a) change if the gas of interest is He?

c. What is the largest molecular mass that is capable of escaping Earth's surface at 298 K?

P33.15 For N_2 at 298 K, what fraction of molecules has a speed between 200. and 300. m/s? What is this fraction if the gas temperature is 500. K?

P33.16 A molecular beam apparatus employs supersonic jets that allow gas molecules to expand from a gas reservoir held at a specific temperature and pressure into a vacuum through a small orifice. Expansion of the gas results for achieving internal temperatures of roughly 10 K. The expansion can be treated as adiabatic, with the change in gas enthalpy accompanying expansion being converted to kinetic energy associated with the flow of the gas:

$$\Delta H = C_P T_R = \frac{1}{2} M v^2$$

The temperature of the reservoir (T_R) is generally greater than the final temperature of the gas, allowing one to consider the entire enthalpy of the gas to be converted into translational motion.

a. For a monatomic gas $C_P = 5/2\,R$. Using this information, demonstrate that the final flow velocity of the molecular beam is related to the initial temperature of the reservoir (T_R) by

$$v = \sqrt{\frac{5RT_R}{M}}$$

b. Using this expression, what is the flow velocity of a molecular beam of Ar where $T_R = 298$ K? Notice that this is remarkably similar to the average speed of the gas. Therefore, the molecular beam resulting can be described as a gas that travels with velocity v but with a very low internal energy. In other words, the distribution of molecular speeds around the flow velocity is significantly reduced in this process.

P33.17 Demonstrate that the Maxwell–Boltzmann speed distribution is normalized.

P33.18 (Challenging) Derive the Maxwell–Boltzmann distribution using the Boltzmann distribution introduced in statistical mechanics. Begin by developing the expression for the distribution in translational kinetic energy in one dimension and then extend it to three dimensions.

P33.19 Starting with the Maxwell speed distribution, demonstrate that the probability distribution for translational kinetic energy for $\varepsilon_T \gg kT$ is given by

$$f(\varepsilon_T)\,d\varepsilon_T = 2\pi\left(\frac{1}{\pi kT}\right)^{3/2} e^{-\varepsilon_T/kT} \varepsilon_T^{1/2}\,d\varepsilon_T$$

P33.20 Using the distribution of particle translational kinetic energy provided in Problem P33.19, derive expressions for the average and most probable translational kinetic energies for a collection of gaseous particles.

P33.21 (Challenging) Using the distribution of particle translational kinetic energy provided in Problem P33.19, derive an expression for the fraction of molecules that have energy greater than some energy ε^*. The rate of many chemical reactions is dependent on the thermal energy available kT versus some threshold energy. Your answer to this question will provide insight into why one might expect the rate of such chemical reactions to vary with temperature.

P33.22 As discussed in Chapter 29 the nth moment of a distribution can be determined as follows: $\langle x^n \rangle = \int x^n f(x) dx$, where integration is over the entire domain of the distribution. Derive expressions for the nth moment of the gas speed distribution.

P33.23 Imagine a cubic container with sides 1 cm in length that contains 1 atm of Ar at 298 K. How many gas–wall collisions are there per second?

P33.24 The vapor pressure of various substances can be determined using effusion. In this process, the material of interest is placed in an oven (referred to as a Knudsen cell) and the mass of material lost through effusion is determined. The mass loss (Δm) is given by $\Delta m = Z_c Am\Delta t$, where Z_c is the collisional flux, A is the area of the aperture through which effusion occurs, m is the mass of one atom, and Δt is the time interval over which the mass loss occurs. This technique is quite useful for determining the vapor pressure of nonvolatile materials. A 1.00 g sample of UF_6 is placed in a Knudsen cell equipped with a 100.-μm-radius hole and heated to 18.2°C where the vapor pressure is 100. Torr.

a. The best scale in your lab has an accuracy of ± 0.01 g. What is the minimum amount of time you must wait until the mass change of the cell can be determined by your balance?

b. How much UF_6 will remain in the Knudsen cell after 5.00 minutes of effusion?

P33.25 Imagine designing an experiment in which the presence of a gas is determined by simply listening to the gas with your ear. The human ear can detect pressures as low as 2×10^{-5} N m^{-2}. Assuming that the eardrum has an area of roughly 1 mm^2, what is the minimum collisional rate that can be detected by ear? Assume that the gas of interest is N_2 at 298 K.

P33.26

a. How many molecules strike a 1.00 cm$_2$ surface during 1 minute if the surface is exposed to O_2 at 1 atm and 298 K?

b. Ultrahigh vacuum studies typically employ pressures on the order of 10^{-10} Torr. How many collisions will occur at this pressure at 298 K?

P33.27 You are a NASA engineer faced with the task of ensuring that the material on the hull of a spacecraft can withstand puncturing by space debris. The initial cabin air pressure in the craft of 1 atm can drop to 0.7 atm before the safety of the crew is jeopardized. The volume of the cabin is 100. m^3,

and the temperature in the cabin is 285 K. Assuming it takes the space shuttle about 8 hours from entry into orbit until landing, what is the largest circular aperture created by a hull puncture that can be safely tolerated assuming that the flow of gas out of the spaceship is effusive? Can the escaping gas from the spaceship be considered as an effusive process? (You can assume that the air is adequately represented by N_2.)

P33.28 Many of the concepts developed in this chapter can be applied to understanding the atmosphere. Because atmospheric air is comprised primarily of N_2 (roughly 78% by volume), approximate the atmosphere as consisting only of N_2 in answering the following questions:

a. What is the single-particle collisional frequency at sea level, with $T = 298$ K and P = 1.0 atm? The corresponding single-particle collisional frequency is reported as 10^{10} s^{-1} in the *CRC Handbook of Chemistry and Physics* (62nd ed., p. F-171).

b. At the tropopause (11 km), the collisional frequency decreases to 3.16×10^9 s^{-1}, primarily due to a reduction in temperature and barometric pressure (i.e., fewer particles). The temperature at the tropopause is ~220 K. What is the pressure of N_2 at this altitude?

c. At the tropopause, what is the mean free path for N_2?

P33.29

a. The stratosphere begins at 11 km above Earth's surface. At this altitude $P = 22.6$ kPa and $T = -56.5$°C. What is the mean free path of N_2 at this altitude assuming N_2 is the only component of the stratosphere?

b. The stratosphere extends to 50.0 km where $P = 0.085$ kPa and $T = 18.3$°C. What is the mean free path of N_2 at this altitude?

P33.30

a. Determine the total collisional frequency for CO_2 at 1 atm and 298 K.

b. At what temperature would the collisional frequency be 10.% of the value determined in part (a)?

P33.31

a. A standard rotary pump is capable of producing a vacuum on the order of 10^{-3} Torr. What is the single-particle collisional frequency and mean free path for N_2 at this pressure and 298 K?

b. A cryogenic pump can produce a vacuum on the order of 10^{-10} Torr. What is the collisional frequency and mean free path for N_2 at this pressure and 298 K?

P33.32 Determine the mean free path for Ar at 298 K at the following pressures:

a. 0.5 atm

b. 0.005 atm

c. 5×10^{-6} atm

P33.33 Determine the mean free path at 500. K and 1 atm for the following:

a. Ne

b. Kr

c. CH_4

Rather than simply calculating the mean free path for each species separately, instead develop an expression for the ratio of mean free paths for two species and use the calculated value for one species to determine the other two.

P33.34   Consider the following diagram of a molecular beam apparatus:

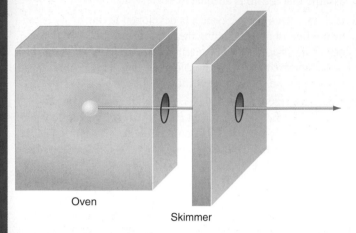

Oven

Skimmer

In the design of the apparatus, it is important to ensure that the molecular beam effusing from the oven does not collide with other particles until the beam is well past the skimmer, a device that selects molecules that are traveling in the appropriate direction, resulting in the creation of a molecular beam. The skimmer is located 10 cm in front of the oven so that a mean free path of 20. cm will ensure that the molecules are well past the skimmer before a collision can occur. If the molecular beam consists of O_2 at a temperature of 500. K, what must the pressure outside the oven be to ensure this mean free path?

P33.35   A comparison of ν_{ave}, ν_{mp}, and ν_{rms} for the Maxwell speed distribution reveals that these three quantities are not equal. Is the same true for the one-dimensional velocity distributions?

P33.36   At 30. km above Earth's surface (roughly in the middle of the stratosphere), the pressure is roughly 0.013 atm and the gas density is 3.74×10^{23} molecules/m³. Assuming N_2 is representative of the stratosphere, using the collisional diameter information provided in Table 33.1 determine

a. the number of collisions a single gas particle undergoes in this region of the stratosphere in 1.0 s.

b. the total number of particle collisions that occur in 1.0 s.

c. the mean free path of a gas particle in this region of the stratosphere.

Web-Based Simulations, Animations, and Problems

W33.1   In this simulation, the variations of gas particle velocity and speed distributions with particle mass and temperature are explored. Specifically, the variations in these distributions with mass are studied and compared to calculations performed by the student. Similar calculations are performed with respect to the distribution variation with temperature.

For Further Reading

Castellan, G. W. *Physical Chemistry*, 3rd ed. Reading, MA: Addison Wesley, 1983.

Hirschfelder, J. O., C. F. Curtiss, and R. B. Bird. *The Molecular Theory of Gases and Liquids*. New York: Wiley, 1954.

Liboff, R. L. *Kinetic Theory: Classical, Quantum, and Relativistic Descriptions*. New York: Springer, 2003.

McQuarrie, D. *Statistical Mechanics*. New York: Harper & Row, 1973.

Transport Phenomena

How will a system respond when it is not at equilibrium? The first steps toward answering this question are provided in this chapter. The study of system relaxation toward equilibrium is known as dynamics. In this chapter, transport phenomena involving the evolution of a system's physical properties such as mass or energy are described. All transport phenomena are connected by one central idea: the rate of change for a system's physical property is dependent on the spatial gradient of the property. In this chapter, this underlying idea is first described as a general concept, then applied to mass (diffusion), energy (thermal conduction), linear momentum (viscosity), and charge (ionic conductivity) transport. The timescale for mass transport is discussed and approached from both the macroscopic and microscopic perspective. It is important to note that although the various transport phenomena outlined here look different, the underlying concepts describing these phenomena have a common origin.

34.1 What Is Transport?

To this point we have been concerned with describing system properties at equilibrium. However, consider the application of an external perturbation to a system such that a property of the system is shifted away from equilibrium. Examples of such system properties are mass and energy. Once the external perturbation is removed, the system will evolve to reestablish the equilibrium distribution of the property. **Transport phenomena** involve the evolution of a system property in response to a nonequilibrium distribution of the property. The system properties of interest in this chapter are given in Table 34.1, and each property is listed with the corresponding transport process.

TABLE 34.1 Transported Properties and the Corresponding Transport Process

Property Transported	Transport Process
Matter	Diffusion
Energy	Thermal conductivity
Linear momentum	Viscosity
Charge	Ionic conductivity

In order for a system property to be transported, a spatial distribution of the property must be different from the distribution at equilibrium. Consider a collection of gas particles in which the equilibrium spatial distribution of the particles corresponds to the same particle number density throughout the container. What would happen if the particle number density were greater on one side of the container than the other? We would expect the gas particles to move in order to reestablish a homogeneous number density throughout the container. That is, the system evolves to reestablish a distribution of the system property that is consistent with equilibrium.

A central concept in transport phenomena is **flux,** defined as a quantity transferred through a given area in a given amount of time. Flux will occur when a spatial imbalance or gradient exists for a system property, and the flux will act in opposition to this gradient. In the example just discussed, imagine dividing the container into two parts with a partition and counting the number of particles that move from one side of the container to the other side, as illustrated in Figure 34.1. The flux in this case is equal to the number of particles that move through the partition per unit time. The theoretical underpinning of *all* of the transport processes listed in Table 34.1 involves flux and the fact that a spatial gradient in a system property will give rise to a corresponding flux. The most basic relationship between flux and the spatial gradient in transported property is as follows:

$$J_x = -\alpha \frac{d(\text{property})}{dx} \tag{34.1}$$

In Equation (34.1), J_x is the flux expressed in units of property area^{-1} time^{-1}. The derivative in Equation (34.1) represents the spatial gradient of the quantity of interest (mass, energy, etc.). The linear relationship between flux and the property spatial gradient is reasonable when the displacement away from equilibrium is modest. The limit of modest displacement of the system property away from equilibrium is assumed in the remainder of this chapter.

The negative sign in Equation (34.1) indicates that the flux occurs in the opposite direction of the gradient; therefore, flux will result in a reduction of the gradient if external action is not taken to maintain the gradient. If the gradient is externally maintained at a constant value, the flux will also remain constant. Again, consider Figure 34.1, which presents a graphical example of the relationship between gradient and flux. The gas density is greatest on the left-hand side of the container such that the particle density increases as one goes from the right side of the container to the left. According to Equation (34.1), particle flux occurs in opposition to the number density gradient in an attempt to make the particle density spatially homogeneous. The final quantity of interest in Equation (34.1) is the factor α. Mathematically, this quantity serves as the proportionality constant between the gradient and flux and is referred to as the **transport coefficient.** In the following sections, we will determine the transport coefficients for the processes listed in Table 34.1 and derive the expressions for flux involving these various transport phenomena. Although the derivations for each transport property will look different, it is important to note that all originate from Equation (34.1). That is, the underlying principle behind all transport phenomena is the relationship between flux and gradient.

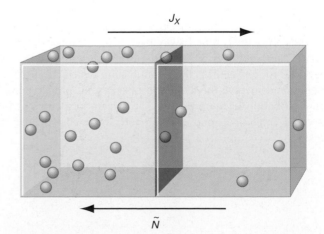

FIGURE 34.1
Illustration of flux. The flux J_x of gas particles is in opposition to the gradient in particle number density $\tilde{N}$.

34.2 **Mass Transport: Diffusion**

Diffusion is the process by which particle density evolves in response to a spatial gradient in concentration. With respect to thermodynamics, this spatial gradient represents a gradient in chemical potential, and the system will relax toward equilibrium by eliminating this gradient. The first case to be considered is diffusion in an ideal gas. Diffusion in liquids will be treated later in this chapter.

Consider a gradient in gas particle number density, $\widetilde{N}$, as depicted in Figure 34.2. According to Equation (34.1), there will be a flux of gas particles in opposition to the gradient. The flux is determined by quantifying the flow of particles per unit time through an imaginary plane located at $x = 0$ with area A. We will refer to this plane as the *flux plane*. Two other planes are located one mean free path $\pm\lambda$ away on either side of the flux plane, and the net flux arises from particles traveling from either of these planes to the flux plane. The mean free path is the distance a particle travels on average between collisions as defined in Equation (33.39).

Figure 34.2 demonstrates that a gradient in particle number density exists in the x direction such that

$$\frac{d\widetilde{N}}{dx} \neq 0$$

If the gradient in $\widetilde{N}$ were equal to zero, flux J_x would also equal zero by Equation (34.1). However, this does not mean that particles are now stationary. Instead, $J_x = 0$ indicates that the flow of particles through the flux plane from left to right is exactly balanced by the flow of particles from right to left. Therefore, the flux expressed in Equation (34.1) represents the net flux, or sum of flux in each direction through the flux plane.

Equation (34.1) provides the relationship between the flux and the spatial gradient in $\widetilde{N}$. Solution of this **mass transport** problem involves determining the proportionality constant α. This quantity is referred to as the diffusion coefficient for mass transport. To determine this constant, consider the particle number density at $\pm\lambda$:

$$\widetilde{N}(-\lambda) = \widetilde{N}(0) - \lambda\left(\frac{d\widetilde{N}}{dx}\right)_{x=0} \tag{34.2}$$

$$\widetilde{N}(\lambda) = \widetilde{N}(0) + \lambda\left(\frac{d\widetilde{N}}{dx}\right)_{x=0} \tag{34.3}$$

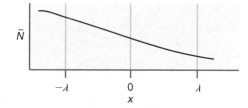

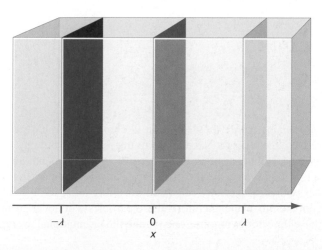

FIGURE 34.2
Model used to describe gas diffusion. The gradient in number density $\widetilde{N}$ results in particles diffusing from $-x$ to $+x$. The plane located at $x = 0$ is where the flux of particles in response to the gradient is calculated (the *flux plane*). Two planes, located one mean free path distance away ($\pm\lambda$) are considered with particles traveling from either of these planes to the flux plane. The total flux through the flux plane is equal to the difference in flux from the planes located at $\pm\lambda$.

These expressions state that the value of $\widetilde{N}$ away from $x = 0$ is equal to the value of $\widetilde{N}$ at $x = 0$ plus a second term representing the change in concentration as one moves toward the planes at $\pm\lambda$. Formally, Equation (34.2) and (34.3) are derived from a Taylor series expansion of the number density with respect to distance, and only the first two terms in the expansion are kept, consistent with λ being sufficiently small such that higher order terms of the expansion can be neglected. The diffusion process can be viewed as an effusion of particles through a plane or aperture of area A. Effusion was described in detail in Section 33.5. Proceeding in a similar fashion to the derivation of effusion presented earlier, the number of particles, N, striking a given area per unit time is equal to

$$\frac{dN}{dt} = J_x \times A \qquad (34.4)$$

Consider the flux of particles traveling from the plane at $-\lambda$ to the flux plane (Figure 34.2). We are interested in the number of particles striking the flux plane per unit time; therefore, we need to count only those particles traveling toward the flux plane. Again, this same problem was encountered in the section on gas effusion where the number of particles traveling toward the wall was equal to the product of number density and the average velocity in the $+x$ direction. Taking the identical approach, the flux in the $+x$ direction is given by

$$J_x = \widetilde{N} \int_0^\infty v_x f(v_x)\, dv_x$$

$$= \widetilde{N} \int_0^\infty v_x \left(\frac{m}{2\pi kT}\right)^{1/2} e^{-mv_x^2/2kT}\, dv_x$$

$$= \widetilde{N}\left(\frac{kT}{2\pi m}\right)^{1/2}$$

$$= \frac{\widetilde{N}}{4} v_{ave} \qquad (34.5)$$

In this equation, we have employed the definition of v_{ave} as defined in Equation (33.22). Substituting Equation (34.2) and (34.3) into the expression of Equation (34.5) for J_x, the flux from the planes located at $-\lambda$ and λ is given by

$$J_{-\lambda,0} = \frac{1}{4} v_{ave} \widetilde{N}(-\lambda) = \frac{1}{4} v_{ave}\left[\widetilde{N}(0) - \lambda\left(\frac{d\widetilde{N}}{dx}\right)_{x=0}\right] \qquad (34.6)$$

$$J_{\lambda,0} = \frac{1}{4} v_{ave} \widetilde{N}(\lambda) = \frac{1}{4} v_{ave}\left[\widetilde{N}(0) + \lambda\left(\frac{d\widetilde{N}}{dx}\right)_{x=0}\right] \qquad (34.7)$$

The total flux through the flux plane is simply the difference in flux from the planes at $\pm\lambda$:

$$J_{Total} = J_{-\lambda,0} - J_{\lambda,0} = \frac{1}{4} v_{ave}\left(-2\lambda\left(\frac{d\widetilde{N}}{dx}\right)_{x=0}\right)$$

$$= -\frac{1}{2} v_{ave} \lambda\left(\frac{d\widetilde{N}}{dx}\right)_{x=0} \qquad (34.8)$$

One correction remains before the derivation is complete. We have assumed that the particles move from the planes located at $\pm\lambda$ to the flux plane directly along the x axis. However, Figure 34.3 illustrates that if the particle trajectory is not aligned with the x axis, the particle will not reach the flux plane after traveling one mean free path. At this point, collisions with other particles can occur, resulting in postcollision particle trajectories away from the flux plane; therefore, these particles will not contribute to the flux. Inclusion of these trajectories requires one to take the orientational average of the mean

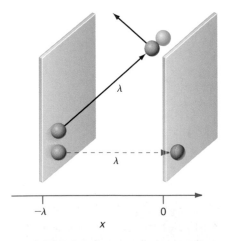

FIGURE 34.3
Particle trajectories aligned with the x axis (dashed line) result in the particle traveling between planes without collision. However, trajectories not aligned with the x axis (solid line) result in the particle not reaching the flux plane before a collision with another particle occurs. This collision may result in the particle being directed away from the flux plane.

free path, and this averaging results in a reduction in total flux as expressed by Equation (34.8) by a factor of $2/3$. With this averaging, the total flux becomes

$$J_{total} = -\frac{1}{3}v_{ave}\lambda\left(\frac{d\widetilde{N}}{dx}\right)_{x=0} \tag{34.9}$$

Equation (34.9) is identical to Equation (34.1), with the diffusion proportionality constant, or simply the **diffusion coefficient,** defined as follows:

$$D = \frac{1}{3}v_{ave}\lambda \tag{34.10}$$

The diffusion coefficient has units of $m^2\,s^{-1}$ in SI units. With this definition of the diffusion coefficient, Equation (34.9) becomes

$$J_{Total} = -D\left(\frac{d\widetilde{N}}{dx}\right)_{x=0} \tag{34.11}$$

Equation (34.11) is referred to as **Fick's first law** of diffusion. It is important to note that the diffusion coefficient is defined using parameters derived from gas kinetic theory first encountered in Chapter 33, namely, the average speed of the gas and the mean free path. Example Problem 34.1 illustrates the dependence of the diffusion coefficient on these parameters.

EXAMPLE PROBLEM 34.1

Determine the diffusion coefficient for Ar at 298 K and a pressure of 1.00 atm.

Solution

Using Equation (34.10) and the collisional cross section for Ar provided in Table 34.1,

$$D_{Ar} = \frac{1}{3}v_{ave,\,Ar}\lambda_{Ar}$$

$$= \frac{1}{3}\left(\frac{8RT}{\pi M_{Ar}}\right)^{1/2}\left(\frac{RT}{PN_A\sqrt{2}\sigma_{Ar}}\right)$$

$$= \frac{1}{3}\left(\frac{8(8.314\,\text{J mol}^{-1}\,\text{K}^{-1})298\,\text{K}}{\pi(0.040\,\text{kg mol}^{-1})}\right)^{1/2}\left(\frac{(8.314\,\text{J mol}^{-1}\,\text{K}^{-1})298\,\text{K}}{(101,325\,\text{Pa})(6.022\times10^{23}\,\text{mol}^{-1})}\times\frac{1}{\sqrt{2}(3.6\times10^{-19}\,\text{m}^2)}\right)$$

$$= \frac{1}{3}(397\,\text{m s}^{-1})(7.98\times10^{-8}\,\text{m})$$

$$= 1.1\times10^{-5}\,\text{m}^2\,\text{s}^{-1}$$

Transport properties of gases can be described using concepts derived from gas kinetic theory. For example, the diffusion coefficient is dependent on the mean free path, which is in turn dependent on the collisional cross section. One criticism of this approach is that parameters such as average velocity are derived using an equilibrium distribution, yet these concepts are now applied in a nonequilibrium context when discussing transport phenomena. The development presented here is performed under the assumption that the displacement of the system away from equilibrium is modest; therefore, equilibrium-based quantities remain relevant. That said, transport phenomena can be described using nonequilibrium distributions; however, the mathematical complexity of this approach is beyond the scope of this text.

With the expression for the diffusion coefficient in hand [Equation (34.10)], the relationship between this quantity and the details of the gas particles are clear. This relationship suggests that transport properties such as diffusion can be used to determine particle parameters such as effective size as described by the collisional cross section. Example Problem 34.2 illustrates the connection between the diffusion coefficient and particle size.

EXAMPLE PROBLEM 34.2

Under identical temperature and pressure conditions, the diffusion coefficient of He is roughly four times larger than that of Ar. Determine the ratio of the collisional cross sections.

Solution

Using Equation (34.10), the ratio of diffusion coefficients (after canceling the 1/3 constant term) can be written in terms of the average speed and mean free path as follows:

$$\frac{D_{He}}{D_{Ar}} = 4 = \frac{\nu_{ave,He}\lambda_{He}}{\nu_{ave,Ar}\lambda_{Ar}}$$

$$= \frac{\left(\dfrac{8RT}{\pi M_{He}}\right)^{1/2}\left(\dfrac{RT}{P_{He}N_A\sqrt{2}\sigma_{He}}\right)}{\left(\dfrac{8RT}{\pi M_{Ar}}\right)^{1/2}\left(\dfrac{RT}{P_{Ar}N_A\sqrt{2}\sigma_{Ar}}\right)}$$

$$= \left(\frac{M_{Ar}}{M_{He}}\right)^{1/2}\left(\frac{\sigma_{Ar}}{\sigma_{He}}\right)$$

$$\left(\frac{\sigma_{He}}{\sigma_{Ar}}\right) = \frac{1}{4}\left(\frac{M_{Ar}}{M_{He}}\right)^{1/2} = \frac{1}{4}\left(\frac{39.9 \text{ g mol}^{-1}}{4.00 \text{ g mol}^{-1}}\right)^{1/2} = 0.79$$

Recall from Section 33.6 that the collisional cross section for a pure gas is equal to πd^2 where d is the diameter of a gas particle. The ratio of collisional cross sections, determined using the diffusion coefficients, is consistent with a He diameter that is 0.89 smaller than that of Ar. However, the diameter of Ar as provided in tables of atomic radii is roughly 2.5 times greater than that of He. The origin of this discrepancy can be traced to the hard-sphere approximation for interparticle interactions.

34.3 The Time Evolution of a Concentration Gradient

As illustrated in the previous section, the existence of a concentration gradient results in particle diffusion. What is the timescale for diffusion, and how far can a particle diffuse in a given amount of time? These questions are addressed by the diffusion equation, which can be derived as follows. Beginning with Fick's first law, the particle flux is given by

$$J_x = -D\left(\frac{d\widetilde{N}(x)}{dx}\right) \tag{34.12}$$

The quantity J_x in Equation (34.12) is the flux through a plane located at x as illustrated in Figure 34.4. The flux at location $x + dx$ can also be written using Fick's first law:

$$J_{x+dx} = -D\left(\frac{d\widetilde{N}(x + dx)}{dx}\right) \tag{34.13}$$

The particle density at $(x + dx)$ is related to the corresponding value at x as follows:

$$\widetilde{N}(x + dx) = \widetilde{N}(x) + dx\left(\frac{d\widetilde{N}(x)}{dx}\right) \tag{34.14}$$

This equation is derived by keeping the first two terms in the Taylor series expansion of number density with distance equivalent to the procedure described earlier for obtaining

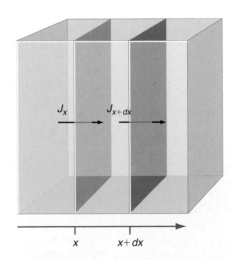

FIGURE 34.4
Depiction of flux through two separate planes. If $J_x = J_{x+dx}$, then the concentration between the planes will not change. However, if the fluxes are unequal, then the concentration will change with time.

Equation (34.2) and (34.3). Substituting Equation (34.14) into Equation (34.13), the flux through the plane at $(x + dx)$ becomes

$$J_{x+dx} = -D\left(\frac{d\widetilde{N}(x)}{dx} + \left(\frac{d^2\widetilde{N}(x)}{dx^2}\right)dx\right) \tag{34.15}$$

Consider the space between the two flux planes illustrated in Figure 34.4. The change in particle density in this region of space depends on the difference in flux through the two planes. If the fluxes are identical, the particle density will remain constant in time. However, a difference in flux will result in an evolution in particle number density as a function of time. The flux is equal to the number of particles that pass through a given area per unit time. If the area of the two flux planes is equivalent, then the difference in flux is directly proportional to the difference in number density. This relationship between the time dependence of the number density and the difference in flux is expressed as follows:

$$\frac{\partial \widetilde{N}(x,t)}{\partial t} = \frac{\partial(J_x - J_{x+dx})}{\partial x}$$

$$= \frac{\partial}{\partial x}\left[-D\left(\frac{\partial \widetilde{N}(x,t)}{\partial x}\right) - \left(-D\left(\left(\frac{\partial \widetilde{N}(x,t)}{\partial x}\right) + \left(\frac{\partial^2 \widetilde{N}(x,t)}{\partial x^2}\right)\partial x\right)\right)\right]$$

$$\frac{\partial \widetilde{N}(x,t)}{\partial t} = D\frac{\partial^2 \widetilde{N}(x,t)}{\partial x^2} \tag{34.16}$$

Equation (34.16) is called the **diffusion equation,** and is also known as **Fick's second law of diffusion.** Notice that $\widetilde{N}$ depends on both position x and time t. Equation (34.16) demonstrates that the time evolution of the concentration gradient is proportional to the second derivative of the spatial gradient in concentration. That is, the greater the "curvature" of the concentration gradient, the faster the relaxation will proceed. Equation (34.16) is a differential equation that can be solved using standard techniques and a set of initial conditions (see Crank entry in the Further Reading section at the end of the chapter) resulting in the following expression for $\widetilde{N}(x,t)$:

$$\widetilde{N}(x,t) = \frac{N_0}{2A(\pi Dt)^{1/2}}e^{-x^2/4Dt} \tag{34.17}$$

In this expression, N_0 represents the initial number of molecules confined to a plane at $t = 0$, A is the area of this plane, x is distance away from the plane, and D is the diffusion coefficient. Equation (34.17) can be viewed as a distribution function that describes the probability of finding a particle at time t at a plane located a distance x away from the initial plane at $t = 0$. An example of the spatial variation in $\widetilde{N}$ versus time is provided in Figure 34.5 for a species with $D = 10^{-5}$ m^2 s^{-1}, roughly equivalent to the diffusion coefficient of Ar at 298 K and 1 atm. The figure demonstrates that with an increase in time, $\widetilde{N}$ increases at distances farther away from the initial plane (located at 0 m in Figure 34.5).

Similar to other distribution functions encountered thus far, it is more convenient to use a metric or benchmark value that provides a measure of $\widetilde{N}(x,t)$ as opposed to describing the entire distribution. The primary metric employed to describe $\widetilde{N}(x,t)$ is the root-mean-square (rms) displacement, determined using what should by now be a familiar approach:

$$x_{rms} = \langle x^2 \rangle^{1/2} = \left[\frac{A}{N_0}\int_{-\infty}^{\infty}x^2\widetilde{N}(x,t)dx\right]^{1/2}$$

$$= \left[\frac{A}{N_0}\int_{-\infty}^{\infty}x^2\frac{N_0}{2A(\pi Dt)^{1/2}}e^{-x^2/4Dt}\,dx\right]^{1/2}$$

$$= \left[\frac{1}{2(\pi Dt)^{1/2}}\int_{-\infty}^{\infty}x^2 e^{-x^2/4Dt}\,dx\right]^{1/2}$$

$$x_{rms} = \sqrt{2Dt} \tag{34.18}$$

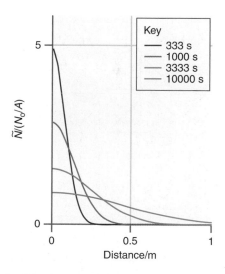

FIGURE 34.5
The spatial variation in particle number density $\widetilde{N}(x,t)$ as a function of time. The number density is defined with respect to N_0/A, the number of particles confined to a plane located at $x = 0$ of area A. In this example, $D = 10^{-5}$ m^{2}s^{-1}, a typical value for a gas at 1 atm and 298 K (see Example Problem 34.1). The corresponding diffusion time for a given concentration profile is indicated.

Notice that the rms displacement increases as the square root of both the diffusion coefficient and time. Equation (34.18) represents the rms displacement in a single dimension. For diffusion in three dimensions, the corresponding term r_{rms} can be determined using the Pythagorean theorem under the assumption that diffusion is equivalent in all three dimensions:

$$r_{rms} = \sqrt{6Dt} \qquad (34.19)$$

The diffusion relationships derived in this section and the previous section involved gases and employed concepts from gas kinetic theory. However, these relationships are also applicable to diffusion in solution, as will be demonstrated later in this chapter.

EXAMPLE PROBLEM 34.3

Determine x_{rms} for a particle where $D = 1.00 \times 10^{-5}\ \text{m}^2\ \text{s}^{-1}$ for diffusion times of 1000. and 10,000. s.

Solution

Employing Equation (34.18):

$$x_{rms,\ 1000\ s} = \sqrt{2\,Dt} = \sqrt{2(1.00 \times 10^{-5}\ \text{m}^2\ \text{s}^{-1})(1000\ \text{s})} = 0.141\ \text{m}$$

$$x_{rms,\ 10,000\ s} = \sqrt{2\,Dt} = \sqrt{2(1.00 \times 10^{-5}\ \text{m}^2\ \text{s}^{-1})(10,000\ \text{s})} = 0.447\ \text{m}$$

The diffusion coefficient employed in this example is equivalent to that used in Figure 34.5, and the rms displacements determined here can be compared to the spatial variation in $\tilde{N}(x, t)$ depicted in the figure to provide a feeling for the x_{rms} distance versus the overall distribution of particle diffusion distances versus time.

SUPPLEMENTAL

34.4 Statistical View of Diffusion

In deriving Fick's first law of diffusion, a gas particle was envisioned to move a distance equal to the mean free path before colliding with another particle. After this collision, memory of the initial direction of motion is lost and the particle is free to move in the same or a new direction until the next collision occurs. This conceptual picture of particle motion is mathematically described by the statistical approach to diffusion. In the statistical approach, illustrated in Figure 34.6a, particle diffusion is also modeled as a series of discrete displacements or steps, with the direction of one step being uncorrelated with that of the previous step. That is, once the particle has taken a step, the direction of the next step is random. A series of such steps is referred to as a **random walk.**

In the previous section we determined the probability of finding a particle at a distance x away from the origin after a certain amount of time. The statistical model of diffusion can be connected directly to this idea using the random walk model. Consider a particle undergoing a random walk along a single dimension x such that the particle moves one step in either the $+x$ or $-x$ direction (Figure 34.6b). After a certain number of steps, the particle will have taken Δ total steps with Δ_- steps in the $-x$ direction and Δ_+ steps in the $+x$ direction. The probability that the particle will have traveled a distance X from the origin is related to the weight associated with that distance, as given by the following expression:

$$W = \frac{\Delta!}{\Delta_+!\Delta_-!} = \frac{\Delta!}{\Delta_+!(\Delta - \Delta_+)!} \qquad (34.20)$$

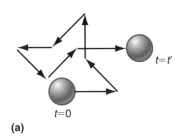

(a)

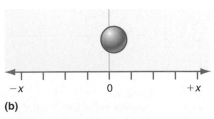

(b)

FIGURE 34.6

(a) Illustration of a random walk. Diffusion of the particle is modeled as a series of discrete steps (each arrow); the length of each step is the same, but the direction of the step is random. (b) Illustration of the one-dimensional random walk model.

Equation (34.20) is identical to the weight associated with observing a certain number of heads after tossing a coin Δ times, as discussed in Chapter 30. This similarity is not a coincidence—the one-dimensional random walk model is very much like tossing a coin. Each outcome of a coin toss is independent of the previous outcome, and only one of two outcomes is possible per toss. Evaluation of Equation (34.20) requires an expression for Δ_+ in terms of Δ. This relationship can be derived by recognizing that X is equal to the difference in the number of steps in the $+x$ and $-x$ direction:

$$X = \Delta_+ - \Delta_- = \Delta_+ - (\Delta - \Delta_+) = 2\Delta_+ - \Delta$$

$$\frac{X + \Delta}{2} = \Delta_+$$

With this definition for Δ_+, the expression for W becomes

$$W = \frac{\Delta!}{\left(\dfrac{\Delta + X}{2}\right)!\left(\Delta - \dfrac{\Delta + X}{2}\right)!} = \frac{\Delta!}{\left(\dfrac{\Delta + X}{2}\right)!\left(\dfrac{\Delta - X}{2}\right)!} \quad \textbf{(34.21)}$$

The probability of the particle being a distance X away from the origin is given by the weight associated with this distance divided by the total weight, 2^Δ, such that

$$P = \frac{W}{W_{Total}} = \frac{W}{2^\Delta} \propto e^{-X^2/2\Delta} \quad \textbf{(34.22)}$$

The final proportionality can be derived by evaluation of Equation (34.21) using Stirling's approximation. Recall from the solution to the diffusion equation in the previous section that the distance a particle diffuses away from the origin was also proportional to an exponential term:

$$\widetilde{N}(x, t) \propto e^{-x^2/4Dt} \quad \textbf{(34.23)}$$

In Equation (34.23), x is the actual diffusion distance, D is the diffusion coefficient, and t is time. For these two pictures of diffusion to converge on the same physical result, the exponents must be equivalent such that

$$\frac{x^2}{4Dt} = \frac{X^2}{2\Delta} \quad \textbf{(34.24)}$$

At this point, the random walk parameters X and Δ must be expressed in terms of the actual quantities of diffusion distance x and total diffusion time t. The total number of random walk steps is expressed as the total diffusion t time divided by the time per random walk step τ:

$$\Delta = \frac{t}{\tau} \quad \textbf{(34.25)}$$

In addition, x can be related to the random walk displacement X by using a proportionality constant x_o that represents the average distance in physical space a particle traverses between collisions such that

$$x = Xx_o \quad \textbf{(34.26)}$$

With these definitions for Δ and X, substitution into Equation (34.24) results in the following definition of D:

$$D = \frac{x_o^2}{2\tau} \quad \textbf{(34.27)}$$

Equation (34.27) is the **Einstein–Smoluchowski equation.** The importance of this equation is that it relates a macroscopic quantity D to microscopic aspects of the diffusion as described by the random walk model. For reactions in solution, x_o is generally taken to be the particle diameter. Using this definition and the experimental value for D, the timescale associated with each random walk event can be determined.

EXAMPLE PROBLEM 34.4

The diffusion coefficient of liquid benzene is 2.2×10^{-5} cm^2 s^{-1}. Given an estimated molecular diameter of 0.3 nm, what is the timescale for a random walk?

Solution

Rearranging Equation (34.27), the time per random walk step is

$$t = \frac{x_o^2}{2D} = \frac{(0.3 \times 10^{-9} \text{ m})^2}{2(2.2 \times 10^{-9} \text{ m}^2 \text{ s}^{-1})} = 2 \times 10^{-11} \text{ s}$$

This is an extremely short time, only 20 ps! This example illustrates that, on average, the diffusional motion of a benzene molecule in the liquid phase is characterized by short-range translational motion between frequent collisions with neighboring molecules.

The Einstein–Smoluchowski equation can also be related to gas diffusion described previously. If we equate x_o with the mean free path λ and define the time per step as the average time it takes a gas particle to translate one mean free path λ/ν_{ave}, then the diffusion coefficient is given by

$$D = \frac{\lambda^2}{2\left(\dfrac{\lambda}{\nu_{ave}}\right)} = \frac{1}{2}\lambda\nu_{ave}$$

This is exactly the same expression for D derived from gas kinetic theory in the absence of the 2/3 correction for particle trajectories as discussed earlier. That is, the statistical and kinetic theory viewpoints of diffusion provide equivalent descriptions of gas diffusion.

34.5 Thermal Conduction

Thermal conduction is the transport process in which energy migrates in response to a gradient in temperature. Figure 34.7 depicts a collection of identical gas particles for which a gradient in temperature exists. Note that the gradient is with respect to temperature only, and that the particle number density is the same throughout the box. Equilibrium is reached when the system has an identical temperature in all regions of the box. Because temperature and kinetic energy are related, relaxation toward equilibrium will involve the transport of kinetic energy from the high-temperature side of the box to the lower temperature side.

Thermal conduction occurs through a variety of mechanisms, depending on the phase of matter. In this derivation we assume that energy transfer occurs during particle collisions and that equilibrium with the energy gradient is established after each collision, thereby ensuring that the particles are at equilibrium with the gradient after each collision. This collisional picture of energy transfer is easy to envision for a gas; however, it can also be applied to liquids and solids. In these phases, molecules do not translate freely, yet the molecular energy can be transferred through collisions with nearby molecules, resulting in energy transfer. In addition to collisional energy transfer, energy can also be transferred through convection or radiative transfer. In convection, differences in density resulting from the temperature gradient can produce convection currents. Although energy is still transferred through collisional events when the particles in the currents collide with other particles, particle migration is not random such that transfer through convection is physically distinct from collisional transport as we have defined it. In radiative transfer, matter is treated as a blackbody that is capable of emitting and absorbing electromagnetic radiation. Radiation from higher temperature matter is absorbed by lower temperature matter, resulting in energy transfer. Both convection and radiative transfer are assumed to be negligible here.

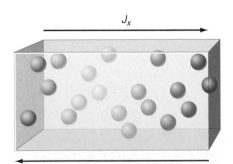

FIGURE 34.7

Temperature gradient in a collection of gas particles. Regions containing high kinetic energy particles are red and low kinetic energy regions are blue. The gradient in kinetic energy, and therefore temperature, is indicated, and the flux in energy in response to this gradient, J_x, is also shown.

The problem of thermal conduction is approached in exactly the same manner as for diffusion. The process begins by considering energy transfer for an ideal monatomic gas. Recall from statistical mechanics and thermodynamics that the average translational energy of a monatomic gas particle is $3/2\,kT$. In addition to translational energy, diatomic and polyatomic molecules can carry more energy in rotational and vibrational energetic degrees of freedom, and our treatment will be extended to these more complex molecules shortly. Kinetic energy is transferred when the flux plane is struck by particles from one side of the plane. In this collision, energy is transmitted from one side of the flux plane to the other. It is important to note that the particle number density is equivalent throughout the box such that mass transfer does not occur. Energy transport and the relaxation toward thermal equilibrium are accomplished through molecular collisions, not diffusion.

Using techniques identical to those employed in Section 34.2 to describe diffusion, the energy at planes located $\pm\lambda$ away from the flux plane (Figure 34.8) is

$$\varepsilon(-\lambda) = \varepsilon(0) - \lambda\left(\frac{d\varepsilon}{dx}\right)_{x=0} \tag{34.28}$$

$$\varepsilon(\lambda) = \varepsilon(0) + \lambda\left(\frac{d\varepsilon}{dx}\right)_{x=0} \tag{34.29}$$

Proceeding in a manner identical to the derivation in Section 34.2 for diffusion, the flux is defined as

$$J_x = \widetilde{N}\varepsilon(x)\int_0^{\infty} v_x f(v_x)\,dv_x = \frac{1}{4}\widetilde{N}\varepsilon(x)\nu_{ave} \tag{34.30}$$

In Equation (34.30), $\widetilde{N}$ is the number density of gas particles, which is constant (recall that there is no spatial gradient in number density, only energy). Substituting Equation (34.28) and (34.29) into Equation (34.30) yields the following expression for the flux in energy from planes located $\pm\lambda$ from the flux plane:

$$J_{-\lambda,0} = \frac{1}{4}\nu_{ave}\widetilde{N}\varepsilon(-\lambda) \tag{34.31}$$

$$J_{\lambda,0} = \frac{1}{4}\nu_{ave}\widetilde{N}\varepsilon(\lambda) \tag{34.32}$$

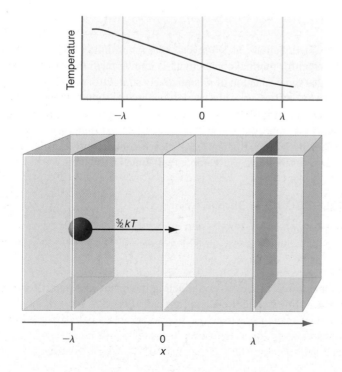

FIGURE 34.8
Model used to determine the thermal conductivity of a gas. The gradient in temperature will result in the transfer of kinetic energy from regions of high temperature (indicated by the red plane) to regions of lower temperature (indicated by the blue plane). The plane located at $x = 0$ is the location at which the flux of kinetic energy in response to the gradient is determined. Two planes, located one mean free path away ($\pm\lambda$) from the flux plane, are established, with particles traveling from one of these planes to the flux plane. Energy is transferred via collisional events at the flux plane.

The total energy flux is given by the difference between Equation (34.31) and (34.32):

$$J_{Total} = J_{-\lambda,0} - J_{\lambda,0}$$

$$= \frac{1}{4} v_{ave} \tilde{N} (\varepsilon(-\lambda) - \varepsilon(\lambda))$$

$$= \frac{1}{4} v_{ave} \tilde{N} \left(-2\lambda \left(\frac{d\varepsilon}{dx} \right)_{x=0} \right)$$

$$= -\frac{1}{2} v_{ave} \tilde{N} \lambda \left(\left(\frac{d\left(\frac{3}{2}kT \right)}{dx} \right)_{x=0} \right)$$

$$= -\frac{3}{4} k v_{ave} \tilde{N} \lambda \left(\frac{dT}{dx} \right)_{x=0} \tag{34.33}$$

After multiplying the total flux by 2/3 for orientational averaging of the particle trajectories, the expression for total flux through the flux plane becomes

$$J_{Total} = -\frac{1}{2} k v_{ave} \tilde{N} \lambda \left(\frac{dT}{dx} \right)_{x=0} \tag{34.34}$$

Equation (34.34) describes the total energy flux for an ideal monatomic gas, where the energy transferred per collision is $3/2\ kT$. The expression for total flux can be defined in terms of molar heat capacity, $C_{V,m}$, by recognizing that the constant volume molar heat capacity for a monatomic gas is $3/2\ R$; therefore,

$$\frac{C_{V,m}}{N_A} = \frac{3}{2}k \text{ (for an ideal monatomic gas)}$$

With this substitution, the flux can be expressed in terms of the $C_{V,m}$ for the species of interest and is not restricted to monatomic species. Employing this identity, the final expression for total energy flux is

$$J_{Total} = -\frac{1}{3} \frac{C_{V,m}}{N_A} v_{ave} \tilde{N} \lambda \left(\frac{dT}{dx} \right)_{x=0} \tag{34.35}$$

Comparison of Equation (34.35) to the general relationship between flux and gradient of Equation (34.1) dictates that the proportionality constant for energy transfer, referred to as the **thermal conductivity** κ be given by

$$\kappa = \frac{1}{3} \frac{C_{V,m}}{N_A} v_{ave} \tilde{N} \lambda \tag{34.36}$$

Unit analysis of Equation (34.36) reveals that κ has units of $J\ K^{-1}\ m^{-1}\ s^{-1}$. Every factor in the expression for thermal conductivity is known from thermodynamics or gas kinetic theory such that the evaluation of κ is relatively straightforward for gases. For fluids, thermal conductivity still serves as the proportionality constant between the net flux in energy and the gradient in temperature; however, a simple expression similar to Equation (34.36) does not exist. The appearance of the mean free path in Equation (34.36) demonstrates that similar to the diffusion coefficient, the thermal conductivity can be used to estimate the hard-sphere radius of gases, as Example Problem 34.5 illustrates.

EXAMPLE PROBLEM 34.5

The thermal conductivity of Ar at 300. K and 1 atm pressure is $0.0177\ J\ K^{-1}\ m^{-1}\ s^{-1}$. What is the collisional cross section of Ar assuming ideal gas behavior?

Solution

The collisional cross section is contained in the mean free path. Rearranging Equation (34.36) to isolate the mean free path, we obtain

$$\lambda = \frac{3\kappa}{\left(\dfrac{C_{V,m}}{N_A} \right) v_{ave} \tilde{N}}$$

The thermal conductivity is provided, and the other terms needed to calculate λ are as follows:

$$\frac{C_{V,m}}{N_A} = \frac{\frac{3}{2}R}{N_A} = \frac{3}{2}k = 2.07 \times 10^{-23} \text{ J K}^{-1}$$

$$v_{ave} = \left(\frac{8RT}{\pi M}\right)^{1/2} = 398 \text{ m s}^{-1}$$

$$\widetilde{N} = \frac{N}{V} = \frac{N_A P}{RT} = 2.45 \times 10^{25} \text{m}^{-3}$$

With these quantities, the mean free path is

$$\lambda = \frac{3(0.0177 \text{ J K}^{-1} \text{ m}^{-1} \text{ s}^{-1})}{(2.07 \times 10^{-23} \text{J K}^{-1})(398 \text{ m s}^{-1})(2.45 \times 10^{25} \text{m}^{-3})} = 2.63 \times 10^{-7} \text{m}$$

Using the definition of the mean free path (Section 33.7), the collisional cross section is

$$\sigma = \frac{1}{\sqrt{2}\widetilde{N}\lambda} = \frac{1}{\sqrt{2}(2.45 \times 10^{25}\text{m}^{-3})(2.63 \times 10^{-7}\text{m})}$$
$$= 1.10 \times 10^{-19} \text{ m}^2$$

This value for the collisional cross section corresponds to a particle diameter of 187 pm (pm $= 10^{-12}$ m), which is remarkably close to the 194 pm value provided by the tabulated atomic radii of Ar.

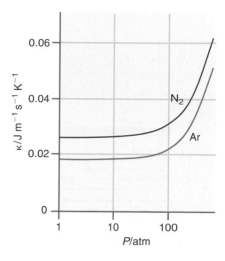

FIGURE 34.9
Thermal conductivity as a function of pressure for N_2 and Ar at 300 K. Note that the pressure scale is logarithmic. The figure demonstrates that κ is roughly independent of pressure up to 50 atm.

The thermal conductivity of a material is dependent on ν_{ave}, $\widetilde{N}$, and λ, quantities that are dependent on temperature and pressure. Therefore, κ will also demonstrate temperature and pressure dependence. Predictions regarding the dependence of κ on T and P can be made within the gas kinetic theory approach used to describe the gas particles. With respect to pressure, λ is inversely proportional to $\widetilde{N}$ such that the mean free path will decrease with pressure:

$$\lambda = \left(\frac{RT}{PN_A}\right)\frac{1}{\sqrt{2}\sigma} = \left(\frac{V}{N}\right)\frac{1}{\sqrt{2}\sigma} = \frac{1}{\widetilde{N}\sqrt{2}\sigma}$$

The product of the mean free path and $\widetilde{N}$ results in the absence of $\widetilde{N}$ dependence in Equation (34.36), leading to the surprising result that κ is predicted to be independent of pressure. Figure 34.9 illustrates the pressure dependence of κ for N_2 and Ar at 300 K. The figure demonstrates that κ is indeed essentially independent of pressure up to 50 atm! At elevated pressures, the intermolecular forces between gas molecules become appreciable, resulting in failure of the hard-sphere model, and the expressions derived using gas kinetic theory are not applicable. This region is indicated by the dramatic rise in κ at high pressures. The pressure dependence of κ is also observed at very low pressure, when the mean free path is greater than the dimensions of the container. In this case, the energy is transported from one side of the container to the other by wall–particle collisions such that κ increases with pressure in this low-pressure regime.

With respect to the temperature dependence of κ, the cancellation of $\widetilde{N}$ with the mean free path results in only ν_{ave} and $C_{V, m}$ carrying temperature dependence. For a monatomic gas, $C_{V, m}$ is expected to demonstrate minimal temperature dependence. Therefore, the ν_{ave} term in Equation (34.36) predicts that κ is proportional to $T^{1/2}$. Identical behavior is predicted for diatomic and polyatomic molecules in which the temperature dependence of $C_{V, m}$ is minimal. Figure 34.10 presents the variation in κ with temperature for N_2 and Ar at 1 atm. Also presented is the predicted $T^{1/2}$ dependence of κ. The comparison between the experimental and predicted temperature dependence demonstrates that κ increases more rapidly with temperature than the predicted $T^{1/2}$ dependence due to the presence of intermolecular interactions that are neglected in the hard-sphere approximation employed in gas kinetic theory. Notice that differences between the predicted and experimental κ are evident at low temperature

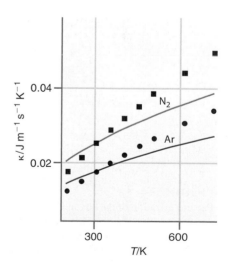

FIGURE 34.10
Variation of κ with temperature for N_2 and Ar at 1 atm. Experimental data are indicated by the squares and circles, and the predicted $T^{1/2}$ dependence as the solid lines. The calculated κ was set equal to the experimental value at 300 K, and then $T^{1/2}$ dependence was applied to generate the predicted variation in κ with temperature.

as well, as opposed to the good agreement between the predicted and experimental behavior of κ with pressure evident in Figure 34.9.

34.6 Viscosity of Gases

The third transport phenomenon considered is that of linear momentum. Practical experience provides an intuitive guide with respect to this area of transport. Consider the flow of a gas through a pipe under pressure. Some gases will flow more easily than others, and the property that characterizes resistance of flow is **viscosity,** represented by the symbol η (lowercase Greek eta). What does viscosity have to do with linear momentum? Figure 34.11 provides a cutaway view of a gas flowing between two plates. It can be shown experimentally that the velocity of the gas, v_x, is greatest midway between the plates and decreases as the gas approaches either plate with $v_x = 0$ at the fluid–plate boundary. Therefore, a gradient in v_x exists along the coordinate orthogonal to the direction of flow (z in Figure 34.11). Because the linear momentum in the x direction is mv_x, a gradient in linear momentum must also exist.

We assume that the gas flow is **laminar flow,** meaning that the gas can be decomposed into layers of constant speed as illustrated in Figure 34.11. This regime will exist for most gases and some liquids provided the flow rate is not too high (see Problem P34.18 at the end of the chapter). At high flow rates, the **turbulent flow** regime is reached where the layers are intermixed such that a clear dissection of the gas in terms of layers of the same speed cannot be performed. The discussion presented here is limited to conditions of laminar flow.

The analysis of linear momentum transport proceeds in direct analogy to diffusion and thermal conductivity. As illustrated in Figure 34.11, a gradient in linear momentum exists in the z direction; therefore, planes of similar linear momentum are defined parallel to the direction of fluid flow as illustrated in Figure 34.12. The transfer of linear momentum occurs by a particle from one momentum layer colliding with the flux plane and thereby transferring its momentum to the adjacent layer.

To derive the relationship between flux and the gradient in linear momentum, we proceed in a fashion analogous to that used to derive diffusion (Section 34.2) and thermal conductivity (Section 34.5). First, the linear momentum p at $\pm\lambda$ is given by

$$p(-\lambda) = p(0) - \lambda\left(\frac{dp}{dz}\right)_{z=0} \tag{34.37}$$

$$p(\lambda) = p(0) + \lambda\left(\frac{dp}{dz}\right)_{z=0} \tag{34.38}$$

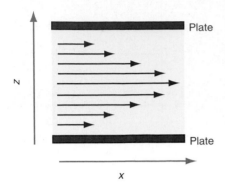

FIGURE 34.11
Cross section of a fluid flowing between two plates. The fluid is indicated by the green area between the plates, with arrow lengths representing the speed of the fluid.

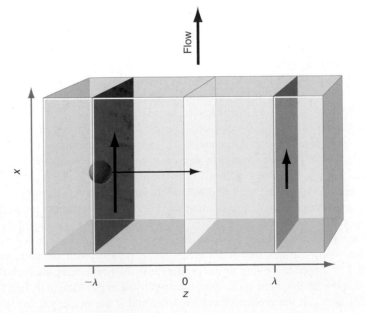

FIGURE 34.12
Parameterization of the box model used to derive viscosity. Planes of identical particle velocity (v_x) are given by the blue planes, with the magnitude of velocity given by the arrows. The gradient in linear momentum will result in momentum transfer from regions of high momentum (darker blue plane) to regions of lower momentum (indicated by the light blue plane). The plane located at $z = 0$ is the location at which the flux of linear momentum in response to the gradient is determined.

Proceeding just as before for diffusion and thermal conduction, the flux in linear momentum from each plane located at $\pm\lambda$ to the flux plane is

$$J_{-\lambda,0} = \frac{1}{4} \nu_{ave} \tilde{N} p(-\lambda) \tag{34.39}$$

$$J_{\lambda,0} = \frac{1}{4} \nu_{ave} \tilde{N} p(\lambda) \tag{34.40}$$

The total flux is the difference between Equations (34.39) and (34.40):

$$J_{Total} = \frac{1}{4} \nu_{ave} \tilde{N} \left(-2\lambda \left(\frac{dp}{dz} \right)_{z=0} \right)$$

$$= -\frac{1}{2} \nu_{ave} \tilde{N} \lambda \left(\frac{d(m\nu_x)}{dz} \right)_{z=0}$$

$$= -\frac{1}{2} \nu_{ave} \tilde{N} \lambda m \left(\frac{d\nu_x}{dz} \right)_{z=0} \tag{34.41}$$

Finally, Equation (34.41) is multiplied by 2/3 as a result of orientational averaging of the particle trajectories, resulting in the final expression for the total flux in linear momentum:

$$J_{Total} = -\frac{1}{3} \nu_{ave} \tilde{N} \lambda m \left(\frac{d\nu_x}{dz} \right)_0 \tag{34.42}$$

Comparison to Equation (34.1) indicates that the proportionality constant between flux and the gradient in velocity is defined as

$$\eta = \frac{1}{3} \nu_{ave} \tilde{N} \lambda m \tag{34.43}$$

Equation (34.43) represents the viscosity of the gas η given in terms of parameters derived from gas kinetic theory. The units of viscosity are the poise (P), or 0.1 kg m^{-1} s^{-1}. Notice the quantity 0.1 in the conversion to SI units. Viscosities are generally reported in μP (10^{-6} P) for gases and cP (10^{-2} P) for liquids.

EXAMPLE PROBLEM 34.6

The viscosity of Ar is 227 μP at 300. K and 1 atm. What is the collisional cross section of Ar assuming ideal gas behavior?

Solution

Because the collisional cross section is related to the mean free path, Equation (34.43) is first rearranged as follows:

$$\lambda = \frac{3\eta}{\nu_{ave} \tilde{N} m}$$

Next, we evaluate each term separately, then use these terms to calculate the mean free path:

$$\nu_{ave} = \left(\frac{8RT}{\pi M} \right)^{1/2} = 398 \text{ m s}^{-1}$$

$$\tilde{N} = \frac{P N_A}{RT} = 2.45 \times 10^{25} \text{m}^{-3}$$

$$m = \frac{M}{N_A} = \frac{0.040 \text{ kg mol}^{-1}}{6.022 \times 10^{23} \text{ mol}^{-1}} = 6.64 \times 10^{-26} \text{ kg}$$

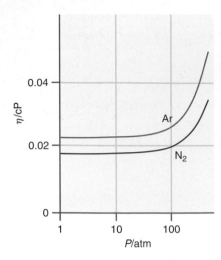

FIGURE 34.13
Pressure dependence of η for gaseous N_2 and Ar at 300 K.

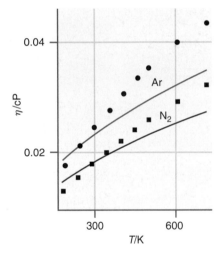

FIGURE 34.14
Temperature dependence of η for N_2 and Ar at 1 atm. Experimental values are given by the squares and circles, and the predicted $T^{1/2}$ dependence is given as the solid line.

$$\lambda = \frac{3\eta}{\nu_{ave}\tilde{N}m} = \frac{3(277 \times 10^{-6}\,P)}{(398\text{ m s}^{-1})(2.45 \times 10^{25}\text{ m}^{-3})(6.64 \times 10^{-26}\text{ kg})}$$

$$= \frac{3(227 \times 10^{-7}\text{ kg m}^{-1}\text{ s}^{-1})}{(398\text{ m s}^{-1})(2.45 \times 10^{25}\text{ m}^{-3})(6.64 \times 10^{-26}\text{ kg})} = 1.28 \times 10^{-7}\text{ m}$$

Note the conversion of poise to SI units in the last step. Using the definition of the mean free path, the collisional cross section can be determined:

$$\sigma = \frac{1}{\sqrt{2}\tilde{N}\lambda} = \frac{1}{\sqrt{2}(2.45 \times 10^{25}\text{ m}^{-3})\,(1.28 \times 10^{-7}\text{ m})} = 2.25 \times 10^{-19}\text{ m}^2$$

In Example Problem 34.5, the thermal conductivity of Ar provided an estimate for the collisional cross section that was roughly 1/2 this value. The discrepancy between collisional cross-section values determined using two different measured transport properties illustrates the approximate nature of the treatment presented here. Specifically, we have assumed that the intermolecular interactions are well modeled by the hard-sphere approximation. The difference in σ suggests that attractive forces are also important in describing the interaction of particles during collisions.

Similar to thermal conductivity, viscosity is dependent on ν_{ave}, $\tilde{N}$, and λ, quantities that are both temperature and pressure dependent. Both $\tilde{N}$ and λ demonstrate pressure dependence, but the product of $\tilde{N}$ and λ contains no net $\tilde{N}$ dependence; therefore, η is predicted to be independent of pressure. Figure 34.13 presents the pressure dependence of η for N_2 and Ar at 300 K. The behavior is almost identical to that observed for thermal conductivity (Figure 34.9), with η demonstrating little pressure dependence until $P \sim 50$ atm. At elevated pressures, intermolecular interactions become important, and the increased interaction between particles gives rise to a substantial increase in η for $P > 50$ atm, similar to the case for thermal conductivity.

With respect to the temperature dependence of η, the ν_{ave} term in Equation (34.43) dictates that η should increase as $T^{1/2}$, identical to the temperature dependence of thermal conductivity (Figure 34.10). This result is perhaps a bit surprising since liquids demonstrate a decrease in η with an increase in temperature. However, the predicted increase in η with temperature for a gas is borne out by experiment, as illustrated in Figure 34.14 where the variation in η with temperature for N_2 and Ar at 1 atm is presented. Also shown is the predicted $T^{1/2}$ dependence. The prediction of increased viscosity with temperature is a remarkable confirmation of gas kinetic theory. However, comparison of the experimental and predicted temperature dependence demonstrates that η increases more rapidly than predicted due to the presence of intermolecular interactions that are neglected in the hard-sphere model, similar to the discussion of the temperature dependence of κ provided earlier. The increase in η with temperature for a gas is consistent with the increase in velocity accompanying a rise in temperature and corresponding increase in momentum flux.

34.7 Measuring Viscosity

Viscosity is a measure of a fluid's resistance to flow; therefore, it is not surprising that the viscosity of gases and liquids is measured using flow. Viscosity is typically measured by monitoring the flow of a fluid through a tube with the underlying idea that the greater the viscosity, the smaller the flow through the tube. The following equation was derived by Poiseuille to describe flow of a liquid through a round tube under conditions of laminar flow:

$$\frac{\Delta V}{\Delta t} = \frac{\pi r^4}{8\eta}\left(\frac{P_2 - P_1}{x_2 - x_1}\right) \tag{34.44}$$

This equation is referred to as **Poiseuille's law.** In Equation (34.44), $\Delta V / \Delta t$ represents the volume of fluid ΔV that passes through the tube in a specific amount of time Δt, r is the radius of the tube through which the fluid flows, η is the fluid viscosity, and the factor in parentheses represents the macroscopic pressure gradient over the tube length. Notice that the fluid flow rate is dependent on the radius of the tube and is also inversely proportional to fluid viscosity. As anticipated, the more viscous the fluid, the smaller the flow rate. The flow of an ideal gas through a tube is given by

$$\frac{\Delta V}{\Delta t} = \frac{\pi r^4}{16 \eta L P_0} (P_2^2 - P_1^2) \qquad \textbf{(34.45)}$$

where L is the length of the tube, P_2 and P_1 are the pressures at the entrance and exit of the tube, respectively, and P_0 is the pressure at which the volume is measured (and is equal to P_1 if the volume is measured at the end of the tube).

EXAMPLE PROBLEM 34.7

Gas cylinders of CO_2 are sold in terms of weight of CO_2. A cylinder contains 50 lb (22.7 kg) of CO_2. How long can this cylinder be used in an experiment that requires flowing CO_2 at 293 K ($\eta = 146\ \mu P$) through a 1.00-m-long tube (diameter $= 0.75$ mm) with an input pressure of 1.05 atm and output pressure of 1.00 atm? The flow is measured at the tube output.

Solution

Using Equation (34.45), the gas flow rate $\Delta V / \Delta t$ is

$$\frac{\Delta V}{\Delta t} = \frac{\pi r^4}{16 \eta L P_0} (P_2^2 - P_1^2)$$

$$= \frac{\pi (0.375 \times 10^{-3}\ \text{m})^4}{16(1.46 \times 10^{-5}\ \text{kg m}^{-1}\ \text{s}^{-1})(1.00\ \text{m})(101{,}325\ \text{Pa})}$$

$$\times\ ((106{,}391\ \text{Pa})^2 - (101{,}325\ \text{Pa})^2)$$

$$= 2.76 \times 10^{-6}\ \text{m}^3\ \text{s}^{-1}$$

Converting the CO_2 contained in the cylinder to the volume occupied at 298 K and 1 atm pressure, we get

$$n_{CO_2} = 22.7\ \text{kg} \left(\frac{1}{0.044\ \text{kg mol}^{-1}} \right) = 516\ \text{mol}$$

$$V = \frac{nRT}{P} = 1.24 \times 10^4\ \text{L} \left(\frac{10^{-3}\ \text{m}^3}{\text{L}} \right) = 12.4\ \text{m}^3$$

Given the effective volume of CO_2 contained in the cylinder, the duration over which the cylinder can be used is

$$\frac{12.4\ \text{m}^3}{2.76 \times 10^{-6}\ \text{m}^3\ \text{s}^{-1}} = 4.49 \times 10^6\ \text{s}$$

This time corresponds to roughly 52 days.

A convenient tool for measuring the viscosity of liquids is an **Ostwald viscometer** (Figure 34.15). To determine the viscosity, one measures the time it takes for the liquid level to fall from the "high" level mark to the "low" level mark, and the fluid flows through a thin capillary that ensures laminar flow. The pressure driving the liquid through the capillary is $\rho g h$, where ρ is the fluid density, g is the acceleration due to gravity, and h is the difference in liquid levels in the two sections of the viscometer, as illustrated in the figure. Because the height difference will evolve as the fluid flows,

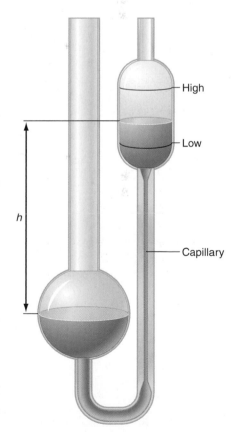

FIGURE 34.15
An Ostwald viscometer. The time for a volume of fluid (ΔV) to flow from the "High" level mark to the "Low" level mark is measured and then used to determine the viscosity of the fluid by means of Equation (34.47).

h represents an average height. Because $\rho g h$ is the pressure generating flow, Equation (34.44) can be rewritten as follows:

$$\frac{\Delta V}{\Delta t} = \frac{\pi r^4}{8\eta} \frac{\rho g h}{(x_2 - x_1)} = \frac{\pi r^4}{8\eta} \frac{\rho g h}{l} \tag{34.46}$$

where l is the length of the capillary. This equation can be rearranged:

$$\eta = \left(\frac{\pi r^4}{8} \frac{gh}{\Delta V l}\right)\rho\,\Delta t = A\rho\,\Delta t \tag{34.47}$$

In Equation (34.47), A is known as the viscometer constant and is dependent on the geometry of the viscometer. All of the viscometer parameters can be determined through careful measurement of the viscometer dimensions; however, the viscometer constant is generally determined by calibration using a fluid of known density and viscosity.

34.8 Diffusion in Liquids and Viscosity of Liquids

The concept of random motion resulting in particle diffusion was evidenced in the famous microscopy experiments of Robert Brown performed in 1827. (An excellent account of this work can be found in *The Microscope* 40 [1992]: 235–241.) In these experiments, Brown took ~5 μm-diameter pollen grains suspended in water, and using a microscope he was able to see that the particles were "very evidently in motion." After performing experiments to show that the motion was not from convection or evaporation, Brown concluded that the motion was associated with the particle itself. This apparently random motion of the pollen grain is referred to as **Brownian motion,** and this motion is actually the diffusion of a large particle in solution.

Consider Figure 34.16 where a particle of mass m is embedded in a liquid having viscosity η. The motion of the particle is driven by collisions with liquid particles, which will provide a time-varying force, $F(t)$. We decompose $F(t)$ into its directional components and focus on the component in the x direction, $F_x(t)$. Motion of the particle in the x direction will result in a frictional force due to the liquid's viscosity:

$$F_{fr,x} = -f\mathrm{v}_x = -f\left(\frac{dx}{dt}\right) \tag{34.48}$$

The negative sign in Equation (34.48) indicates that the frictional force is in opposition to the direction of motion. In Equation (34.48), f is referred to as the friction coefficient and is dependent on both the geometry of the particle and the viscosity of the fluid. The total force on the particle is simply the sum of the collisional and frictional forces:

$$F_{total,x} = F_x(t) + F_{fr,x}$$

$$m\left(\frac{d^2x}{dt^2}\right) = F_x(t) - f\left(\frac{dx}{dt}\right) \tag{34.49}$$

FIGURE 34.16

Illustration of Brownian motion. A spherical particle with radius r is embedded in a liquid of viscosity η. The particle undergoes collisions with solvent molecules (the red dot in the right-hand figure) resulting in a time-varying force, $F(t)$, that initiates particle motion. Particle motion is opposed by a frictional force that is dependent on the solvent viscosity.

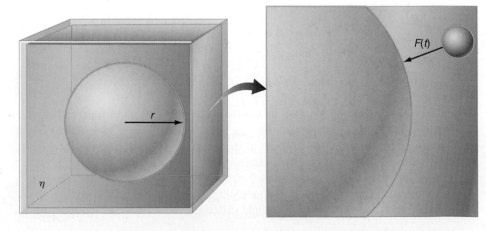

This differential equation was studied in 1905 by Einstein, who demonstrated that when averaged over numerous fluid–particle collisions the average square displacement of the particle in a specific amount of time t is given by

$$\langle x^2 \rangle = \frac{2kTt}{f} \tag{34.50}$$

where k is Boltzmann's constant and T is temperature. If the particle undergoing diffusion is spherical, the frictional coefficient is given by

$$f = 6\pi\eta r \tag{34.51}$$

such that

$$\langle x^2 \rangle = 2\left(\frac{kT}{6\pi\eta r}\right)t \tag{34.52}$$

Comparison of this result to the value of $\langle x^2 \rangle$ determined by the diffusion equation [Equation (34.18)] dictates that the term in parentheses is the diffusion coefficient D given by

$$D = \frac{kT}{6\pi\eta r} \tag{34.53}$$

Equation (34.53) is the **Stokes–Einstein equation** for diffusion of a spherical particle. This equation states that the diffusion coefficient is dependent on the viscosity of the medium, the size of the particle, and the temperature. This form of the Stokes–Einstein equation is applicable to diffusion in solution when the radius of the diffusing particle is significantly greater than the radius of a liquid particle. For diffusion of a particle in a fluid of similar size, experimental data demonstrate that Equation (34.53) should be modified as follows:

$$D = \frac{kT}{4\pi\eta r} \tag{34.54}$$

Equation (34.54) can also be used for **self-diffusion,** in which the diffusion of a single solvent molecule in the solvent itself occurs.

EXAMPLE PROBLEM 34.8

Hemoglobin is a protein responsible for oxygen transport. The diffusion coefficient of human hemoglobin in water at 298 K ($\eta = 0.891$ cP) is 6.90×10^{-11} m^2 s^{-1}. Assuming this protein can be approximated as spherical, what is the radius of hemoglobin?

Solution

Rearranging Equation (34.53) and paying close attention to units, the radius is

$$r = \frac{kT}{6\pi\eta D} = \frac{(1.38 \times 10^{-23} \text{ J K}^{-1})(298 \text{ K})}{6\pi(0.891 \times 10^{-3} \text{ kg m}^{-1} \text{ s}^{-1})(6.90 \times 10^{-11} \text{ m}^2 \text{ s}^{-1})}$$

$$= 3.55 \text{ nm}$$

X-ray crystallographic studies of hemoglobin have shown that this globular protein can be approximated as a sphere having a radius of 2.75 nm. One reason for this discrepancy is that in aqueous solution, hemoglobin will have associated water, which is expected to increase the effective size of the particle relative to the crystallographic measurement.

Unlike gases, the dependence of fluid viscosity on temperature is difficult to explain theoretically. The primary reason for this is that, on average, the distance between particles in a liquid is small, and intermolecular interactions become important when describing particle interactions. As a general rule, the stronger the intermolecular interactions in a liquid, the greater the viscosity of the liquid. With respect to temperature, liquid viscosities increase as the temperature decreases. The reason for this behavior is that as the temperature is reduced, the kinetic energy of the particles is also reduced, and the particles have less kinetic energy to overcome the potential energy arising from intermolecular interactions, resulting in the fluid being more resistant to flow.

SUPPLEMENTAL

34.9 Sedimentation and Centrifugation

An important application of transport phenomena in liquids is **sedimentation.** Sedimentation can be used with diffusion to determine the molecular weights of macromolecules. Figure 34.17 depicts a molecule with mass m undergoing sedimentation in a liquid of density ρ under the influence of Earth's gravitational field. Three distinct forces are acting on the particle:

1. The frictional force: $F_{fr} = -f\mathrm{v}_x$
2. The gravitational force: $F_{gr} = mg$
3. The buoyant force: $F_b = -m\overline{V}\rho g$

In the expression for the buoyant force, $\overline{V}$ is the **specific volume** of the solute, equal to the change in solution volume per mass of solute in units of $cm^3\ g^{-1}$.

Imagine placing the particle at the top of the solution and then letting it fall. Initially, the downward velocity (v_x) is zero, but the particle will accelerate and the velocity will increase. Eventually, a particle velocity will be reached where the frictional and buoyant forces are balanced by the gravitational force. This velocity is known as the terminal velocity, and when this velocity is reached the particle acceleration is zero. Using Newton's second law,

$$F_{Total} = ma = F_{fr} + F_{gr} + F_b$$

$$0 = -f\mathrm{v}_{x,ter} + mg - m\overline{V}\rho g$$

$$\mathrm{v}_{x,ter} = \frac{mg(1 - \overline{V}\rho)}{f}$$

$$\overline{s} = \frac{\mathrm{v}_{x,ter}}{g} = \frac{m(1 - \overline{V}\rho)}{f} \qquad \text{(34.55)}$$

The **sedimentation coefficient** $\overline{s}$ is defined as the terminal velocity divided by the acceleration due to gravity. Sedimentation coefficients are generally reported in the units of Svedbergs (S) with $1\ S = 10^{-13}$ s; however, we will use units of seconds to avoid confusion with other units in upcoming sections of this chapter.

Sedimentation is generally not performed using acceleration due to Earth's gravity. Instead, acceleration of the particle is accomplished using a **centrifuge,** with ultracentrifuges capable of producing accelerations on the order of 10^5 times the acceleration due to gravity. In centrifugal sedimentation, the acceleration is equal to $\varpi^2 x$ where ϖ is the angular velocity (radians s^{-1}) and x is the distance of the particle from the center of rotation. During centrifugal sedimentation, the particles will also reach a terminal velocity that depends on the acceleration due to centrifugation, and the sedimentation coefficient is expressed as

$$\overline{s} = \frac{\mathrm{v}_{x,ter}}{\varpi^2 x} = \frac{m(1 - \overline{V}\rho)}{f} \qquad \text{(34.56)}$$

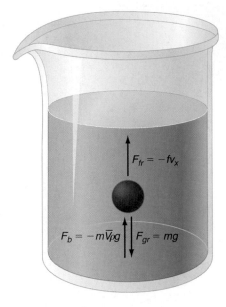

FIGURE 34.17
Illustration of the forces involved in sedimentation of a particle; F_{fr} is the frictional force, F_{gr} is the gravitational force, and F_b is the buoyant force.

EXAMPLE PROBLEM 34.9

The sedimentation coefficient of lysozyme (M = 14,100 g mol^{-1}) in water at 20°C is 1.91×10^{-13} s and the specific volume is 0.703 $cm^3\ g^{-1}$. The density of water at this temperature is 0.998 g cm^{-3} and $\eta = 1.002$ cP. Assuming lysozyme is spherical, what is the radius of this protein?

Solution

The frictional coefficient is dependent on molecular radius. Solving Equation (34.56) for f, we obtain

$$f = \frac{m(1 - \overline{V}\rho)}{\overline{s}} = \frac{\dfrac{(14{,}100 \text{ g mol}^{-1})}{(6.022 \times 10^{23} \text{ mol}^{-1})}\left(1 - (0.703 \text{ mL g}^{-1})(0.998 \text{ g mL}^{-1})\right)}{1.91 \times 10^{-13} \text{ s}}$$

$$= 3.66 \times 10^{-8} \text{ g s}^{-1}$$

The frictional coefficient is related to the radius for a spherical particle by Equation (34.51) such that

$$r = \frac{f}{6\pi\eta} = \frac{3.66 \times 10^{-8} \text{ g s}^{-1}}{6\pi(1.002 \text{ g m}^{-1} \text{ s}^{-1})} = 1.94 \times 10^{-9} \text{ m} = 1.94 \text{ nm}$$

One method for measurement of macromolecular sedimentation coefficients is by centrifugation, as illustrated in Figure 34.18. In this process, an initially homogeneous solution of macromolecules is placed in a centrifuge and spun. Sedimentation occurs, resulting in regions of the sample farther away from the axis of rotation experiencing an increase in macromolecule concentrations and a corresponding reduction in concentration for the sample regions closest to the axis of rotation. A boundary between these two concentration layers will be established, and this boundary will move away from the axis of rotation with time in the centrifuge. If we define the x_b as the midpoint of the **boundary layer** (Figure 34.18), the following relationship exists between the location of the boundary layer and centrifugation time:

$$\overline{s} = \frac{v_{x,ter}}{\varpi^2 x} = \frac{\dfrac{dx_b}{dt}}{\varpi^2 x_b}$$

$$\varpi^2 \overline{s} \int_0^t dt = \int_{x_{b,t=0}}^{x_{b,t}} \frac{dx_b}{x_b}$$

$$\varpi^2 \overline{s} t = \ln\left(\frac{x_{b,t}}{x_{b,t=0}}\right) \qquad (34.57)$$

Equation (34.57) suggests that a plot of $\ln(x_b/x_{b,t=0})$ versus time will yield a straight line with slope equal to ϖ^2 times the sedimentation coefficient. The determination of sedimentation coefficients by boundary centrifugation is illustrated in Example Problem 34.10.

EXAMPLE PROBLEM 34.10

The sedimentation coefficient of lysozyme is determined by centrifugation at 55,000 rpm in water at 20°C. The following data were obtained regarding the location of the boundary layer as a function of time:

Time (min)	x_b (cm)
0	6.00
30	6.07
60	6.14
90	6.21
120	6.28
150	6.35

Using these data, determine the sedimentation coefficient of lysozyme in water at 20°C.

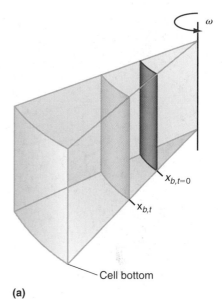

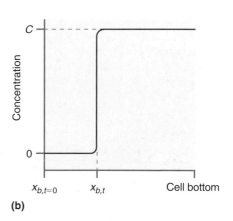

(a)

(b)

FIGURE 34.18
Determination of sedimentation coefficient by centrifugation. **(a)** Schematic drawing of the centrifuge cell, which is rotating with angular velocity ϖ. The blue plane at $x_{b,t=0}$ is the location of the solution meniscus before centrifugation. As the sample is centrifuged, a boundary between the solution with increased molecular concentration versus the solvent is produced. This boundary is represented by the yellow plane at $x_{b,t}$. **(b)** As centrifugation proceeds, the boundary layer will move toward the cell bottom. A plot of $\ln(x_{b,t}/x_{b,t=0})$ versus time will yield a straight line with slope equal to ϖ^2 times the sedimentation coefficient.

Solution

First, we transform the data to determine $\ln(x_b/x_{b,t=0})$ as a function of time:

Time (min)	x_b (cm)	$(x_b/x_{b,t=0})$	$\ln(x_b/x_{b,t=0})$
0	6.00	1	0
30	6.07	1.01	0.00995
60	6.14	1.02	0.01980
90	6.21	1.03	0.02956
120	6.28	1.04	0.03922
150	6.35	1.05	0.04879

The plot of $\ln(x_b/x_{b,t=0})$ versus time is shown here:

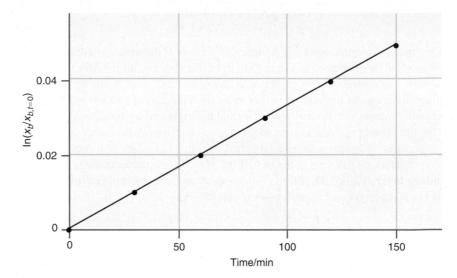

The slope of the line in the preceding plot is 3.75×10^{-4} min^{-1}, which is equal to ϖ^2 times the sedimentation coefficient:

$$3.75 \times 10^{-4} \text{ min}^{-1} = 6.25 \times 10^{-6} \text{ s}^{-1} = \varpi^2 \bar{s}$$

$$\bar{s} = \frac{6.25 \times 10^{-6} \text{ s}^{-1}}{\varpi^2} = \frac{6.25 \times 10^{-6} \text{ s}^{-1}}{((55{,}000 \text{ rev min}^{-1})(2\pi \text{ rad rev}^{-1})(0.0167 \text{ min s}^{-1}))^2}$$

$$\bar{s} = 1.88 \times 10^{-13} \text{ s}$$

Finally with knowledge of the sedimentation coefficient and diffusion coefficient, the molecular weight of a macromolecule can be determined. Equation (34.56) can be rearranged to isolate the frictional coefficient:

$$f = \frac{m(1 - \bar{V}\rho)}{\bar{s}} \tag{34.58}$$

The frictional coefficient can also be expressed in terms of the diffusion coefficient [Equations (34.51) and (34.53)]:

$$f = \frac{kT}{D} \tag{34.59}$$

Setting Equations (34.58) and Equation (34.59) equal, the weight of the molecule is given by

$$m = \frac{kT\bar{s}}{D(1 - \bar{V}\rho)} \quad \text{or} \quad M = \frac{RT\bar{s}}{D(1 - \bar{V}\rho)} \tag{34.60}$$

Equation (34.60) dictates that with the sedimentation and diffusion coefficients, as well as the specific volume of the molecule, the molecular weight of a macromolecule can be determined.

34.10 Ionic Conduction

Ionic conduction is a transport phenomenon in which electrical charge in the form of electrons or ions migrates under the influence of an electrical potential. The amount of charge that migrates is equal to the electrical current I, which is defined as the amount of charge Q migrating in a given time interval:

$$I = \frac{dQ}{dt} \qquad (34.61)$$

The unit used for current is the ampere (A), which is equal to one coulomb of charge per second:

$$1 \text{ A} = 1 \text{ C s}^{-1} \qquad (34.62)$$

Recall that the charge on an electron is 1.60×10^{-19} C. Therefore, 1.00 A of current corresponds to the flow of 6.25×10^{18} electrons in one second. Current can also be quantified by the current density j defined as the amount of current that flows through a conductor of cross-sectional area A (Figure 34.19):

$$j = \frac{I}{Area} \qquad (34.63)$$

In the remainder of this section, the cross-sectional area will be written as *Area* rather than A to avoid confusion with amperes.

The migration of charge is initiated by the presence of an electrical force, created by an electrical field. In Figure 34.19, the electric field across the conductor is generated using a battery. The current density is proportional to the magnitude of the electric field E, and the proportionality constant is known as the **electrical conductivity** κ:

$$j = \kappa E \qquad (34.64)$$

Unfortunately, by convention thermal conductivity and electrical conductivity are both denoted by the symbol κ. Therefore, we will use the terms *thermal* or *electrical* when discussing conductivity to avoid confusion. Electrical conductivity has units of siemens per meter, or S m^{-1}, where 1 S equals $1\Omega^{-1}$ (ohm^{-1}). Finally, the **resistivity** ρ of a material is the inverse of the electrical conductivity:

$$\rho = \frac{1}{\kappa} = \frac{E}{j} \qquad (34.65)$$

Resistivity is expressed in units of Ω m. Consider Figure 34.19 in which an electric field E is applied to a cylindrical conductor with a cross-sectional area and length l. If the cylinder is homogeneous in composition, the electric field and current density will be equivalent throughout the conductor, allowing us to write:

$$E = \frac{V}{l} \qquad (34.66)$$

$$j = \frac{I}{Area} \qquad (34.67)$$

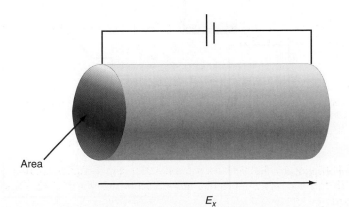

Area

E_x

FIGURE 34.19
Cross section of a current-carrying conductor. The conductor cross-sectional area is indicated by the shaded area. The direction of the applied electric field E_x created by the battery is also shown.

In Equation (34.66), V is the magnitude of the electric potential difference between the ends of the conductor, commonly referred to as the voltage. Substituting Equation (34.66) and (34.67) into the expression for resistivity, Equation (34.65), yields the following:

$$\rho = \frac{1}{\kappa} = \left(\frac{V}{I}\right)\frac{Area}{l} = R\frac{Area}{l} \tag{34.68}$$

In Equation (34.68), Ohm's law was used to rewrite V/I as R, the resistance. This equation demonstrates that the resistivity is proportional to the resistance of the material R in units of Ω with a proportionality constant that is equal to the area A of the conductor cross section divided by the conductor length l.

How is electrical conductivity related to transport? Again, consider the application of a battery to a conductor as shown in Figure 34.19. The battery will produce an electric field inside the conductor, and the direction of the electric field is taken to be the x direction. The electric field is related to the gradient in the electrical potential as follows:

$$E_x = -\frac{d\phi}{dx} \tag{34.69}$$

Using this definition, Equation (34.64) becomes

$$\frac{1}{Area}\frac{dQ}{dt} = -\kappa\frac{d\phi}{dx} \tag{34.70}$$

The left-hand side of Equation (34.70) is simply the flux of charge through the conductor in response to the gradient in electrical potential. Comparison of Equation (34.70) to Equation (34.1) demonstrates that ionic conductivity is described exactly as other transport processes. The proportionality constant between the potential gradient and flux in charge is the electrical conductivity.

Insight into the nature of charge transport in solution can be obtained by measuring the conductivity of ion-containing solutions. Measurements are usually performed using a conductivity cell, which is simply a container in which two electrodes of well-defined area and spacing are placed, and into which a solution of electrolyte is added (Figure 34.20). The resistance of the conductivity cell is generally measured using alternating currents to limit the buildup of electrolysis products at the electrodes. The resistance of the cell is measured using a Wheatstone bridge as illustrated in Figure 34.20. The operating principle behind this circuit is that when the two "arms" of

FIGURE 34.20
Schematic of the electrical circuit used to perform electrical conductivity measurements. The solution of interest is placed in the conductivity cell, which is placed in one arm of a Wheatstone bridge circuit, which includes R_3 and the cell. Resistor R_3 is adjusted until no current flows between points A and B. Under these conditions, the resistance of the cell can be determined.

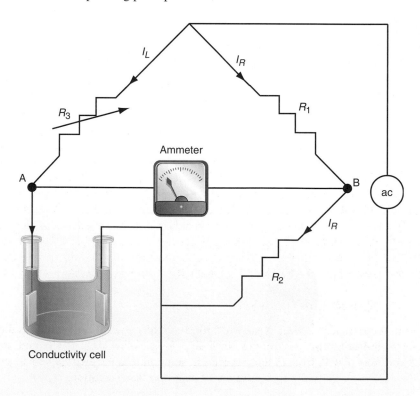

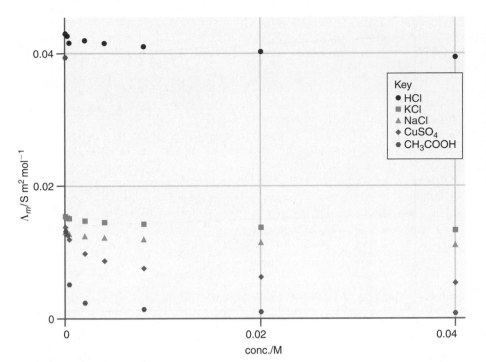

FIGURE 34.21
Molar conductivity Λ_m of various compounds as a function of concentration c. Measurements were performed in water at 25°C. Notice the mild reduction in Λ_m as a function of concentration for HCl, KCl, NaCl, and $CuSO_4$. This behavior is characteristic of a strong electrolyte. In contrast, CH_3COOH demonstrates a substantial reduction in Λ_m with increased concentration, characteristic of a weak electrolyte.

the bridge have equal resistance, no current will flow between points A and B. In the experiment, the resistance of R_3 is varied until this occurs. Under these conditions, the resistance of the cell is given by

$$R_{cell} = \frac{R_2 R_3}{R_1} \qquad (34.71)$$

If the resistance of the cell is measured, the electrical conductivity of the solution can be determined using Equation (34.68).

The electrical conductivity of a solution depends on the number of ions present. The conductivity of ionic solutions is expressed as the **molar conductivity** Λ_m, which is defined as follows:

$$\Lambda_m = \frac{\kappa}{c} \qquad (34.72)$$

where c is the molar concentration of electrolyte. The molar conductivity has units of $S\ m^2\ mol^{-1}$. If κ is linearly proportional to electrolyte concentration, then the molar conductivity will be independent of concentration. Figure 34.21 presents the measured molar conductivity for a variety of electrolytes as a function of concentration in water at 25°C. Two trends are immediately evident. First, none of the electrolytes demonstrates a concentration-independent molar conductivity. Second, the concentration dependence of the various species indicates that electrolytes can be divided into two categories. In the first category are **strong electrolytes,** which demonstrate a modest decrease in molar conductivity with an increase in concentration. In Figure 34.21, HCl, NaCl, KCl, and $CuSO_4$ are all strong electrolytes. In the second category are **weak electrolytes** represented by CH_3COOH in Figure 34.21. The molar conductivity of a weak electrolyte is comparable to a strong electrolyte at low concentration but decreases rapidly with increased concentration. Note that a given species can behave as either a weak or strong electrolyte depending on the solvent employed. In the remainder of this section, we discuss the underlying physics behind the conductivity of strong and weak electrolytes.

34.10.1 Strong Electrolytes

The central difference between weak and strong electrolytes is the extent of ion formation in solution. Strong electrolytes, including ionic solids (e.g., KCl) and strong acids/bases (e.g., HCl), exist as ionic species in solution, and the electrolyte concentration is directly related to the concentration of ionic species in solution.

FIGURE 34.22
Molar conductivity Λ_m versus the square root of concentration for three strong electrolytes: HCl, KCl, and NaCl. The solid lines are fits to Kohlrausch's law [Equation (34.73)].

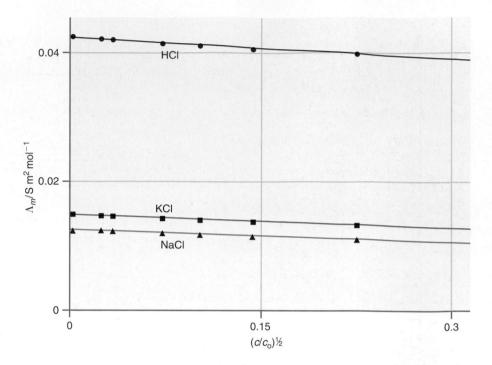

In 1900, Kohlrausch demonstrated that the concentration dependence of the molar conductivity for strong electrolytes is well described by the following equation:

$$\Lambda_m = \Lambda_m^0 - K\sqrt{\frac{c}{c_0}} \qquad (34.73)$$

This equation, known as **Kohlrausch's law,** states that the molar conductivity for strong electrolytes decreases as the square root of concentration, with slope given by the quantity K. The slope is found to depend more on the stoichiometry of the electrolyte (AB, AB_2, etc.) than composition. In Equation (34.73), Λ_m^0 is the molar conductivity at infinite dilution and represents the expected molar conductivity if interionic interactions were entirely absent. Clearly, any measurement at "infinite" dilution is impossible to perform; however, Kohlrausch's law demonstrates that Λ_m^0 can be readily determined by plotting Λ_m as a function of the square root of concentration (referenced to a standard concentration c_0, generally 1 M), and then extrapolating to $c = 0$. Figure 34.22 presents the molar conductivity of HCl, KCl, and NaCl originally presented in Figure 34.21, but now shown with Λ_m plotted as a function of $(c/c_0)^{1/2}$. The solid line is a fit to the data using Kohlrausch's law, and it demonstrates the ability of this equation to reproduce the concentration dependence of Λ_m for strong electrolytes.

Kohlrausch was also able to demonstrate that for strong electrolytes, the molar conductivity at infinite dilution can be described in terms of contributions from the individual ionic constituents as follows:

$$\Lambda_m^0 = \nu_+\lambda_+ + \nu_-\lambda_- \qquad (34.74)$$

In Equation (34.74), ν_+ and ν_- are the number of positively and negatively charged species, respectively, in the formula weight of the electrolyte. For example, $\nu_+ = \nu_- = 1$ for NaCl, corresponding to Na^+ and Cl^-. The **ionic equivalent conductance** of the cations and anions is denoted by λ_+ and λ_-, respectively, and represents the conductivity of an individual ionic constituent of the electrolyte. Equation (34.74) is known as the **law of independent migration of ions** and states that under dilute conditions the molecular conductivity of an electrolyte is equal to the sum of the conductivities of the ionic constituents. Table 34.2 provides the molar conductivities of various ions in aqueous solution.

TABLE 34.2	Ionic Equivalent Conductance Values for Representative Ions		
Ion	λ (S m^2 mol^{-1})	Ion	λ (S m^2 mol^{-1})
H^+	0.0350	OH^-	0.0199
Na^+	0.0050	Cl^-	0.0076
K^+	0.0074	Br^-	0.0078
Mg^{2+}	0.0106	F^-	0.0054
Cu^{2+}	0.0107	NO_3^-	0.0071
Ca^{2+}	0.0119	CO_3^{2-}	0.0139

EXAMPLE PROBLEM 34.11

What is the expected molar conductivity at infinite dilution for $MgCl_2$?

Solution

Using Equation (34.74) and the data provided in Table 34.2,

$$\Lambda_m^0(MgCl_2) = 1\lambda(Mg^{2+}) + 2\lambda(Cl^-)$$
$$= (0.0106 \text{ S m}^2 \text{ mol}^{-1}) + 2(0.0076 \text{ S m}^2 \text{ mol}^{-1})$$
$$= 0.0258 \text{ S m}^2 \text{ mol}^{-1}$$

As determined by a plot of conductivity versus $(c/c_0)^{1/2}$, Λ_m^0 for $MgCl_2$ is equal to 0.0212 S m^2 mol^{-1}. Comparison of this value to the expected molar conductivity calculated earlier demonstrates that $MgCl_2$ behaves as a strong electrolyte in aqueous solution.

What is the origin of the $c^{1/2}$ dependence stated in Kohlrausch's law? In Chapter 10 it was shown that the activity of electrolytes demonstrates a similar concentration dependence, and this dependence was related to dielectric screening as expressed by the Debye–Hückel limiting law. This law states that the natural log of the ion activity is proportional to the square root of the solution ionic strength, or ion concentration. An ion in solution is characterized not only by the ion itself, but also by the ionic atmosphere surrounding the ion as discussed in Chapter 10, and depicted in Figure 34.23. In the figure, a negative ion is surrounded by a few close-lying positive ions comprising the ionic atmosphere. In the absence of an electric field, the ionic atmosphere will be spherically symmetric, and the centers of positive and negative charge will be identical. However, application of a field will result in ion motion, and this motion will create two effects. First, consider the situation where the negative ion in Figure 34.23 migrates and the atmosphere follows this migrating ion. The ionic atmosphere is not capable of responding to the motion of the negative ion instantaneously, and the lag in response will result in displacement of the centers of positive and negative charge. This displacement will give rise to a local electric field in opposition to the applied field. Correspondingly, the migration rate of the ion, and subsequently the conductivity, will be reduced. This effect is called the **relaxation effect,** and this term implies that a time dependence is associated with this phenomenon. In this case, it is the time it takes the ionic atmosphere to respond to the motion of the negative ion.

Information regarding the timescale for ionic atmosphere relaxation can be gained by varying the frequency of the ac electric field to study ionic conductivity. A second effect that reduces the migration rate of the ion is the **electrophoretic effect.** The negatively and positively charged ions will migrate in opposite directions such that the viscous drag the ion experiences will increase. The reduction in ion mobility accompanying the increase in viscous drag will decrease the conductivity.

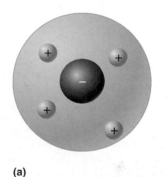
(a)

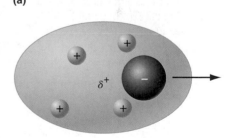
(b)

FIGURE 34.23
Negative ion in solution and associated ionic atmosphere. **(a)** In the absence of an electric field, the center of the positive and negative charge is identical. **(b)** However, when an electric field is present, the motion of the negative ion results in displacement of the center of the positive and negative charge. This local electric field is in opposition to the applied electric field such that the rate of ion migration is reduced.

34.10.2 Weak Electrolytes

Weak electrolytes undergo only fractional conversion into their ionic constituents in solution. Acetic acid, a weak acid, is a classic example of a weak electrolyte. Weak acids are characterized by incomplete conversion to H_3O^+ and the conjugate base. The dependence of Λ_m on concentration for a weak electrolyte can be understood by considering the extent of ionization. For the case of a weak acid (HA) with dissociation in water to form H_3O^+ and the conjugate base (A^-), incomplete ionization can be factored into the ionic and electrolyte concentrations as follows:

$$[H_3O^+] = [A^-] = \alpha c \qquad (34.75)$$

$$[HA] = (1 - \alpha)c \qquad (34.76)$$

In Equations (34.75) and (34.76), α is the degree of ionization (ranging in value from 0 to 1), and c is the initial concentration of electrolyte. Substituting Equations (34.75) and (34.76) into the expression for the equilibrium constant of a weak acid results in Equation (34.77):

$$K_a = \frac{\alpha^2 c}{(1 - \alpha)} \qquad (34.77)$$

Solving Equation (34.77) for α yields

$$\alpha = \frac{K_a}{2c}\left(\left(1 + \frac{4c}{K_a}\right)^{1/2} - 1\right) \qquad (34.78)$$

The molar conductivity of a weak electrolyte at a given electrolyte concentration Λ_m will be related to the molar conductivity at infinite dilution Λ_m^0 by

$$\Lambda_m = \alpha \Lambda_m^0 \qquad (34.79)$$

Using Equations (34.78) and (34.79), the concentration dependence of the molar conductivity can be determined and compared to experiment. The relationship between Λ_m and c for a weak electrolyte is described by the **Ostwald dilution law:**

$$\frac{1}{\Lambda_m} = \frac{1}{\Lambda_m^0} + \frac{c\Lambda_m}{K_a(\Lambda_m^0)^2} \qquad (34.80)$$

The comparison of the predicted behavior given by Equation (34.80) to experiment for acetic acid ($K_a = 1.8 \times 10^{-5}$, $\Lambda_m^0 = 0.03957$ S m^2 mol^{-1}) is presented in Figure 34.24, and excellent agreement is observed. The Ostwald dilution law can be used to determine the molar conductivity at infinite dilution for a weak acid.

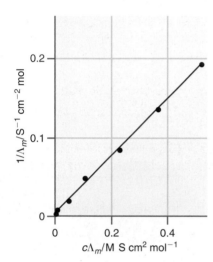

FIGURE 34.24
Comparison of experimental and predicted conductivity of acetic acid (CH_3COOH) as a function of concentration using the Ostwald dilution law of Equation (34.80). The y intercept on the graph is equal to $1/\Lambda_m^0$.

Vocabulary

boundary layer	ionic equivalent conductance	sedimentation coefficient
Brownian motion	Kohlrausch's law	self-diffusion
centrifuge	laminar flow	specific volume
diffusion	law of independent migration of ions	Stokes–Einstein equation
diffusion coefficient	mass transport	strong electrolytes
diffusion equation	molar conductivity	thermal conduction
Einstein–Smoluchowski equation	Ostwald dilution law	thermal conductivity
electrical conductivity	Ostwald viscometer	transport coefficient
electrophoretic effect	Poiseuille's law	transport phenomena
Fick's first law	random walk	turbulent flow
Fick's second law of diffusion	relaxation effect	viscosity
flux	resistivity	weak electrolytes
ionic conduction	sedimentation	

Conceptual Problems

Q34.1 What is the general relationship between the spatial gradient in a system property and the flux of that property?

Q34.2 What is the expression for the diffusion coefficient D in terms of gas kinetic theory parameters? How is D expected to vary with an increase in molecular mass or collisional cross section?

Q34.3 Would you expect the diffusion coefficient of H_2 to be greater or less than that of D_2? By how much? Assume that H_2 and D_2 are the same size.

Q34.4 Particles are confined to a plane and then allowed to diffuse. How does the number density vary with distance away from the initial plane?

Q34.5 How does the root-mean-square diffusion distance vary with the diffusion coefficient? How does this quantity vary with time?

Q34.6 What is the expression for thermal conductivity in terms of particle parameters derived from gas kinetic theory?

Q34.7 Why is the thermal conductivity for an ideal gas expected to be independent of pressure? Why does the thermal conductivity for an ideal gas increase as $T^{1/2}$?

Q34.8 In describing viscosity, what system quantity was transported? What is the expression for viscosity in

terms of particle parameters derived from gas kinetic theory?

Q34.9 What observable is used to measure the viscosity of a gas or liquid?

Q34.10 How does the viscosity of a gas vary with pressure?

Q34.11 What is Brownian motion?

Q34.12 Based on the Stokes–Einstein equation, how does the diffusion coefficient for a particle depend on the size of the particle?

Q34.13 Describe the random walk model of diffusion. How is this model related to Brownian motion?

Q34.14 In the Stokes–Einstein equation that describes particle diffusion for a spherical particle, how does the diffusion coefficient depend on fluid viscosity and particle size?

Q34.15 What forces are operative in particle sedimentation?

Q34.16 What is the difference between a strong and weak electrolyte?

Q34.17 According to Kohlrausch's law, how will the molar conductivity for a strong electrolyte change with concentration?

Numerical Problems

Problem numbers in red indicate that the solution to the problem is given in the *Student's Solutions Manual*.

P34.1 The diffusion coefficient for CO_2 at 273 K and 1 atm is 1.00×10^{-5} m^2 s^{-1}. Estimate the collisional cross section of CO_2 given this diffusion coefficient.

P34.2 The collisional cross section of N_2 is 0.43 nm^2. What is the diffusion coefficient of N_2 at a pressure of 1 atm and a temperature of 298 K?

P34.3

a. The diffusion coefficient for Xe at 273 K and 1.00 atm is 0.5×10^{-5} m^2 s^{-1}. What is the collisional cross section of Xe?

b. The diffusion coefficient of N_2 is threefold greater than that of Xe under the same pressure and temperature conditions. What is the collisional cross section of N_2?

P34.4

a. The diffusion coefficient of sucrose in water at 298 K is 0.522×10^{-9} m^2 s^{-1}. Determine the time it will take a sucrose molecule on average to diffuse an rms distance of 1 mm.

b. If the molecular diameter of sucrose is taken to be 0.8 nm, what is the time per random walk step?

P34.5

a. The diffusion coefficient of the protein lysozyme (MW = 14.1 kg/mol) is 0.104×10^{-5} cm^2 s^{-1}. How long

will it take this protein to diffuse an rms distance of 1 μm? Model the diffusion as a three-dimensional process.

b. You are about to perform a microscopy experiment in which you will monitor the fluorescence from a single lysozyme molecule. The spatial resolution of the microscope is 1 μm. You intend to monitor the diffusion using a camera that is capable of one image every 60 s. Is the imaging rate of the camera sufficient to detect the diffusion of a single lysozyme protein over a length of 1 μm?

c. Assume that in the microscopy experiment of part (b) you use a thin layer of water such that diffusion is constrained to two dimensions. How long will it take a protein to diffuse an rms distance of 1 μm under these conditions?

P34.6 A solution consisting of 1 g of sucrose in 10 mL of water is poured into a 1 L graduated cylinder with a radius of 2.5 cm. Then the cylinder is filled with pure water.

a. The diffusion of sucrose can be considered diffusion in one dimension. Derive an expression for the average distance of diffusion x_{ave}.

b. Determine x_{ave} and x_{rms} for sucrose for time periods of 1 s, 1 min, and 1 h.

P34.7 A thermopane window consists of two sheets of glass separated by a volume filled with air (which we will model as N_2 where $\kappa = 0.0240$ J K^{-1} m^{-1} s^{-1}). For a thermopane window that is 1 m^2 in area with a separation between glass sheets of 3 cm, what is the loss of energy when

a. the exterior of the window is at a temperature of 10°C and the interior of the window is at a temperature of 22°C?

b. the exterior of the window is at a temperature of −20°C and the interior of the window is at a temperature of 22°C?

c. the same temperature differential as in part (b) is used, but the window is filled with Ar ($\kappa = 0.0163$ J K^{-1} m^{-1} s^{-1}) rather than N$_2$?

P34.8 An advertisement for a thermopane window company touts Kr-filled windows and states that these windows provide ten times better insulation than conventional windows filled with Ar. Do you agree with this statement? What should the ratio of thermal conductivities be for Kr ($\sigma = 0.52$ nm^2) versus Ar ($\sigma = 0.36$ nm^2)?

P34.9 Two parallel metal plates separated by 1 cm are held at 300. and 298 K, respectively. The space between the plates is filled with N$_2$ ($\sigma = 0.430$ nm^2 and $C_{V,m} = 5/2\ R$). Determine the heat flux between the two plates in units of W cm^{-2}.

P34.10 Determine the thermal conductivity of the following species at 273 K and 1.00 atm:

a. Ar ($\sigma = 0.36$ nm^2)

b. Cl$_2$ ($\sigma = 0.93$ nm^2)

c. SO$_2$ ($\sigma = 0.58$ nm^2, geometry: bent)

You will need to determine $C_{V,m}$ for the species listed. You can assume that the translational and rotational degrees of freedom are in the high-temperature limit and that the vibrational contribution to $C_{V,m}$ can be ignored at this temperature.

P34.11 The thermal conductivity of Kr is 0.0087 J K^{-1} m^{-1} s^{-1} at 273 K and 1 atm. Estimate the collisional cross section of Kr.

P34.12 The thermal conductivity of N$_2$ at 298 K and 1 atm is 0.024 J K^{-1} m^{-1} s^{-1}. What is the collisional cross section of N$_2$ based on its thermal conductivity? For N$_2$ $C_{V,m} = 20.8$ J mol^{-1} K^{-1}.

P34.13 The thermal conductivity of Kr is roughly half that of Ar under identical pressure and temperature conditions. Both gases are monatomic such that $C_{V,m} = 3/2\ R$.

a. Why would one expect the thermal conductivity of Kr to be less than that of Ar?

b. Determine the ratio of collisional cross sections for Ar relative to Kr assuming identical pressure and temperature conditions.

c. For Kr at 273 K at 1 atm, $\kappa = 0.0087$ J K^{-1} m^{-1} s^{-1}. Determine the collisional cross section of Kr.

P34.14

a. Determine the ratio of thermal conductivity for N$_2$ ($\sigma = 0.43$ nm^2) at sea level ($T = 300.$ K, $P = 1.00$ atm) versus the lower stratosphere (T = 230. K, P = 0.25 atm).

b. Determine the ratio of thermal conductivity for N$_2$ at sea level if $P = 1$ atm, but the temperature is 100. K. Which energetic degrees of freedom will be operative at the lower temperature, and how will this affect $C_{V,m}$?

P34.15 The thermal conductivities of acetylene (C$_2$H$_2$) and N$_2$ at 273 K and 1 atm are 0.01866 and 0.0240 J K^{-1} m^{-1} s^{-1}, respectively. Based on these data, what is the ratio of the collisional cross section of acetylene relative to N$_2$?

P34.16

a. The viscosity of Cl$_2$ at 293 K and 1 atm is 132 μP. Determine the collisional cross section of this molecule based on the viscosity.

b. Given your answer in part (a), estimate the thermal conductivity of Cl$_2$ under the same pressure and temperature conditions.

P34.17

a. The viscosity of O$_2$ at 293 K and 1.00 atm is 204 μP. What is the expected flow rate through a tube having a radius of 2.00 mm, length of 10.0 cm, input pressure of 765 Torr, output pressure of 760. Torr, with the flow measured at the output end of the tube?

b. If Ar were used in the apparatus ($\eta = 223\ \mu$P) of part (a), what would be the expected flow rate? Can you determine the flow rate without evaluating Poiseuille's equation?

P34.18 The Reynolds' number (Re) is defined as Re $= \rho \langle v_x \rangle d / \eta$, where ρ and η are the fluid density and viscosity, respectively; d is the diameter of the tube through which the fluid is flowing; and $\langle v_x \rangle$ is the average velocity. Laminar flow occurs when Re < 2000, the limit in which the equations for gas viscosity were derived in this chapter. Turbulent flow occurs when Re > 2000. For the following species, determine the maximum value of $\langle v_x \rangle$ for which laminar flow will occur:

a. Ne at 293 K ($\eta = 313\ \mu$P, $\rho = $ that of an ideal gas) through a 2.00-mm-diameter pipe.

b. Liquid water at 293 K ($\eta = 0.891$ cP, $\rho = 0.998$ g mL^{-1}) through a 2.00-mm-diameter pipe.

P34.19 The viscosity of H$_2$ at 273 K at 1 atm is 84 μP. Determine the viscosities of D$_2$ and HD.

P34.20 An Ostwald viscometer is calibrated using water at 20°C ($\eta = 1.0015$ cP, $\rho = 0.998$ g mL^{-1}). It takes 15.0 s for the fluid to fall from the upper to the lower level of the viscometer. A second liquid is then placed in the viscometer, and it takes 37.0 s for the fluid to fall between the levels. Finally, 100. mL of the second liquid weighs 76.5 g. What is the viscosity of the liquid?

P34.21 How long will it take to pass 200. mL of H$_2$ at 273 K through a 10. cm-long capillary tube of 0.25 mm if the gas input and output pressures are 1.05 and 1.00 atm, respectively?

P34.22

a. Derive the general relationship between the diffusion coefficient and viscosity for a gas.

b. Given that the viscosity of Ar is 223 μP at 293 K and 1 atm, what is the diffusion coefficient?

P34.23

a. Derive the general relationship between the thermal conductivity and viscosity.

b. Given that the viscosity of Ar is 223 μP at 293 K and 1.00 atm, what is the thermal conductivity?

c. What is the thermal conductivity of Ne under these same conditions (the collisional cross section of Ar is 1.5 times that of Ne)?

P34.24 As mentioned in the text, the viscosity of liquids decreases with increasing temperature. The empirical equation $\eta(T) = Ae^{E/RT}$ provides the relationship between viscosity and temperature for a liquid. In this equation, A and E are constants, with E being referred to as the activation energy for flow.

a. How can one use the equation provided to determine A and E given a series of viscosity versus temperature measurements?

b. Use your answer in part (a) to determine A and E for liquid benzene given the following data:

T (°C)	η (cP)
5	0.826
40.	0.492
80.	0.318
120.	0.219
160.	0.156

P34.25 Poiseuille's law can be used to describe the flow of blood through blood vessels. Using Poiseuille's law, determine the pressure drop accompanying the flow of blood through 5.00 cm of the aorta ($r = 1.00$ cm). The rate of blood flow through the body is 0.0800 L s^{-1}, and the viscosity of blood is approximately 4.00 cP at 310 K.

P34.26 What is the flow of blood through a small vein (diameter = 3.00 mm) that is 1.00 cm long? The drop in blood pressure over this length is 40.0 Torr, and the viscosity of blood is approximately 4.00 cP at 310 K.

P34.27 Myoglobin is a protein that participates in oxygen transport. For myoglobin in water at 20°C, $\bar{s} = 2.04 \times 10^{-13}$ s, $D = 1.13 \times 10^{-11}$ m^2 s^{-1}, and $\bar{V} = 0.740$ cm^3 g^{-1}. The density of water is 0.998 g cm^{-3}, and the viscosity is 1.002 cP at this temperature.

a. Using the information provided, estimate the size of myoglobin.

b. What is the molecular weight of myoglobin?

P34.28 The molecular weight of bovine serum albumin (BSA) is 66,500 g mol^{-1} and has a specific volume of 0.717 cm^3 g^{-1}. Velocity centrifugation demonstrates that $\bar{s} = 4.31 \times 10^{-13}$ s for this protein. Assuming a solution density of 1.00 g cm^{-3} and viscosity of 1.00 cP, determine the friction coefficient and effective radius of BSA.

P34.29 You are interested in purifying a sample containing the protein alcohol dehydrogenase obtained from horse liver; however, the sample also contains a second protein, catalase. These two proteins have the following transport properties at 298 K:

	Catalase	Alcohol Dehydrogenase
$\bar{s}$ (s)	11.3×10^{-13}	4.88×10^{-13}
D(m^2 s^{-1})	4.1×10^{-11}	6.5×10^{-11}
$\bar{V}$(cm^3 g^{-1})	0.715	0.751

a. Determine the molecular weight of catalase and alcohol dehydrogenase.

b. You have access to a centrifuge that can provide angular velocities up to 35,000 rpm. For the species you expect to travel the greatest distance in the centrifuge tube, determine the time it will take to centrifuge until a 3 cm displacement of the boundary layer occurs relative to the initial 5 cm location of the boundary layer relative to the centrifuge axis.

c. To separate the proteins, you need a separation of at least 1.5 cm between the boundary layers associated with each protein. Using your answer to part (b), will it be possible to separate the proteins by centrifugation?

P34.30 Boundary centrifugation is performed at an angular velocity of 40,000 rpm to determine the sedimentation coefficient of cytochrome c ($M = 13,400$ g mol^{-1}) in water at 20°C ($\rho = 0.998$ g cm^{-3}, $\eta = 1.002$ cP). The following data are obtained on the position of the boundary layer as a function of time:

Time (h)	x_b (cm)
0	4.00
2.5	4.11
5.2	4.23
12.3	4.57
19.1	4.91

a. What is the sedimentation coefficient for cytochrome c under these conditions?

b. The specific volume of cytochrome c is 0.728 cm^3 g^{-1}. Estimate the size of cytochrome c.

P34.31 T. Svedberg measured the molecular weight of carbonyl hemoglobin (specific volume = 0.755 mL g^{-1}, $D = 7.00 \times 10^{-11}$ m^2 s^{-1}) by velocity centrifugation. In this experiment, 0.96 g of protein were dissolved in 100. mL of water at 303 K ($\rho = 0.998$ g cm^{-3}), and the solution was spun at 39,300 rpm. After 30.0 min, the concentration boundary advanced by 0.074 cm from an initial position of 4.525 cm. Calculate the molecular weight of carbonyl hemoglobin.

P34.32 A current of 2.00 A is applied to a metal wire for 30. s. How many electrons pass through a given point in the wire during this time?

P34.33 Use the following data to determine the conductance at infinite dilution for NaNO$_3$:

$$\Lambda_m^0(\text{KCl}) = 0.0149 \text{ S m}^2 \text{ mol}^{-1}$$
$$\Lambda_m^0(\text{NaCl}) = 0.0127 \text{ S m}^2 \text{ mol}^{-1}$$
$$\Lambda_m^0(\text{KNO}_3) = 0.0145 \text{ S m}^2 \text{ mol}^{-1}$$

P34.34 The following molar conductivity data are obtained for an electrolyte:

Concentration (M)	Λ_m (S m^2 mol^{-1})
0.0005	0.01245
0.001	0.01237
0.005	0.01207
0.01	0.01185
0.02	0.01158
0.05	0.01111
0.1	0.01067

Determine if the electrolyte is strong or weak, and determine the conductivity of the electrolyte at infinite dilution.

P34.35 The molar conductivity of sodium acetate, CH_3COONa, is measured as a function of concentration in water at 298 K, and the following data are obtained:

Concentration (M)	Λ_m (S m^2 mol^{-1})
0.0005	0.00892
0.001	0.00885
0.005	0.00857
0.01	0.00838
0.02	0.00812
0.05	0.00769
0.1	0.00728

Is sodium acetate a weak or strong electrolyte? Determine Λ_m^0 using appropriate methodology depending on your answer.

P34.36 Starting with Equations (34.47) and (34.48), derive the Ostwald dilution law.

P34.37 For a one-dimensional random walk, determine the probability that the particle will have moved six steps in either the $+x$ or $-x$ direction after 10, 20, and 100 steps.

P34.38 In the early 1990s, fusion involving hydrogen dissolved in palladium at room temperature, or *cold fusion,* was proposed as a new source of energy. This process relies on the diffusion of H_2 into palladium. The diffusion of hydrogen gas through a 0.005-cm-thick piece of palladium foil with a cross section of 0.750 cm^2 is measured. On one side of the foil, a volume of gas maintained at 298 K and 1 atm is applied, while a vacuum is applied to the other side of the foil. After 24 h, the volume of hydrogen has decreased by 15.2 cm^3. What is the diffusion coefficient of hydrogen gas in palladium?

P34.39 In the determination of molar conductivities, it is convenient to define the cell constant as $K = 1/A$, where l is the separation between the electrodes in the conductivity cell and A is the area of the electrodes.

a. A standard solution of KCl (conductivity or $\kappa = 1.06296 \times 10^{-6}$ S m^{-1} at 298 K) is employed to standardize the cell, and a resistance of 4.2156 Ω is measured. What is the cell constant?

b. The same cell is filled with a solution of HCl, and a resistance of 1.0326 Ω is measured. What is the conductivity of the HCl solution?

P34.40 Conductivity measurements were one of the first methods used to determine the autoionization constant of water. The autoionization constant of water is given by the following equation:

$$K_w = a_{H^+} a_{OH^-} = \left(\frac{[H^+]}{1\,M} \right) \left(\frac{[OH^-]}{1\,M} \right)$$

where a is the activity of the species, which is equal to the actual concentration of the species divided by the standard state concentration at infinite dilution. This substitution of concentrations for activities is a reasonable approximation given the small concentrations of H^+ and OH^- that result from autoionization.

a. Using the expression provided, show that the conductivity of pure water can be written as

$$\Lambda_m(H_2O) = (1\,M)K_w^{1/2}(\lambda(H^+) + \lambda(OH^-))$$

b. Kohlrausch and Heydweiller measured the conductivity of water in 1894 and determined that $\Lambda_m(H_2O) = 5.5 \times 10^{-6}$ S m^{-1} at 298 K. Using the information in Table 34.2, determine K_w.

Web-Based Simulations, Animations, and Problems

W34.1 In this problem, concentration time dependence in one dimension is depicted as predicted using Fick's second law of diffusion. Specifically, variation of particle number density as a function of distance with time and diffusion constant D is investigated. Comparisons of the full distribution to x_{rms} are performed to illustrate the behavior of x_{rms} with time and D.

For Further Reading

Bird, R. B., W. E. Stewart, and E. N. Lightfoot. *Transport Phenomena.* New York: Wiley, 1960.

Cantor, C. R., and P. R. Schimmel. *Biophysical Chemistry. Part II: Techniques for the Study of Biological Structure and Function.* San Francisco: W. H. Freeman, 1980.

Castellan, G. W. *Physical Chemistry.* Reading, MA: Addison-Wesley, 1983.

Crank, J. *The Mathematics of Diffusion.* Oxford: Clarendon Press, 1975.

Hirschfelder, J. O., C. F. Curtiss, and R. B. Bird. *The Molecular Theory of Gases and Liquids.* New York: Wiley, 1954.

Reid, R. C., J. M. Prausnitz, and T. K. Sherwood. *The Properties of Gases and Liquids.* New York: McGraw-Hill, 1977.

Vargaftik, N. B. *Tables on the Thermophysical Properties of Liquids and Gases.* New York: Wiley, 1975.

Welty, J. R., C. E. Wicks, and R. E. Wilson. *Fundamentals of Momentum, Heat, and Mass Transfer.* New York: Wiley, 1969.

Elementary Chemical Kinetics

In Chapters 35 and 36, the kinetics associated with chemical reactions are explored. The central question of interest in chemical kinetics is perhaps the first question one asks about any chemical reaction: Just how do the reactants become products? In chemical kinetics, this question is answered by determining the timescale and mechanism of a chemical reaction. In this chapter, the foundational tools of chemical kinetics are developed. These tools will be the central components of a kinetics "toolkit" to be used in describing complex reactions in Chapter 36. The importance of chemical kinetics is evidenced by its application in nearly every area of chemistry. Found in areas such as enzyme catalysis, materials processing, and atmospheric chemistry, chemical kinetics is clearly an important aspect of physical chemistry.

35.1 Introduction to Kinetics

In the previous chapter, transport processes involving the evolution of a system's physical properties toward equilibrium were discussed. In these processes, the system undergoes relaxation without a change in chemical composition. Transport phenomena are sometimes referred to as physical kinetics to indicate that the physical properties of the system are evolving, not the chemical composition. In this chapter and the next, we focus on chemical kinetics where the composition of the system evolves with time.

Chemical kinetics involves the study of the rates and mechanisms of chemical reactions. This area bridges an important gap in our discussion of chemical reactions to this point. Thermodynamic descriptions of chemical reactions involved the Gibbs or Helmholtz energy for a reaction and the corresponding equilibrium constant. These quantities are sufficient to predict the reactant and product concentrations at equilibrium but are of little use in determining the timescale over which the reaction occurs. That is, thermodynamics may dictate that a reaction is spontaneous, but it does not dictate the timescale over which the reaction will occur and reach equilibrium. Chemical kinetics provides information on the timescale of chemical reactions.

In the course of a chemical reaction, concentrations will change with time as "reactants" become "products." Figure 35.1 presents possibly the first chemical reaction

FIGURE 35.1
Concentration as a function of time for the conversion of reactant A into product B. The concentration of A at $t = 0$ is $[A]_0$, and the concentration of B is zero. As the reaction proceeds, the loss of A results in the production of B.

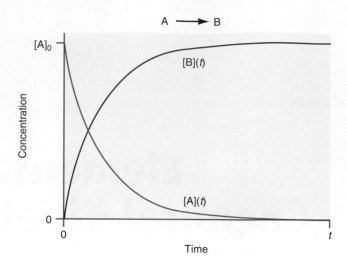

introduced in introductory chemistry: the conversion of reactant "A" into product "B." The figure illustrates that, as the reaction proceeds, a decrease in reactant concentration and a corresponding increase in product concentration are observed. One way to describe this process is to define the rate of concentration change with time, a quantity that is referred to as the reaction rate.

The central idea behind chemical kinetics is that by monitoring the rate at which chemical reactions occur and determining the dependence of this rate on system parameters such as temperature, concentration, and pressure, we can gain insight into the mechanism of the reaction. Experimental chemical kinetics includes the development of techniques that allow for the measurement and analysis of chemical reaction dynamics. In addition to experiments, theoretical work has been performed to understand reaction mechanisms and the underlying physics that govern the rates of chemical reactions. The synergy between experimental and theoretical chemical kinetics has provided for dramatic advances in this field.

35.2 Reaction Rates

Consider the following "generic" chemical reaction:

$$a\text{A} + b\text{B} + \ldots \rightarrow c\text{C} + d\text{D} + \ldots \tag{35.1}$$

In Equation (35.1), uppercase letters indicate a chemical species and lowercase letters represent the stoichiometric coefficient for the species in the balanced reaction. The number of moles of a species during any point of the reaction is given by

$$n_i = n_i^o + \nu_i \xi \tag{35.2}$$

where n_i is the number of moles of species i at any given time during the reaction, n_i^o is the number of moles of species i present before initiation of the reaction, ν_i is related to the stoichiometric coefficient of species i, and ξ the extent or advancement of the reaction equal to zero at the start of the reaction and one at completion. The advancement variable allows us to quantify the rate of the reaction with respect to all species, irrespective of stoichiometry (see the later discussion). For the reaction depicted in Equation (35.1), reactants will be consumed and products formed during the reaction. To ensure that this behavior is reflected in Equation (35.2), ν_i is set equal to -1 times the stoichiometric coefficient for reactants and is set equal to the stoichiometric coefficient for products.

The time evolution of the reactant and product concentrations is quantified by differentiating both sides of Equation (35.2) with respect to time:

$$\frac{dn_i}{dt} = \nu_i \frac{d\xi}{dt} \tag{35.3}$$

The **reaction rate** is defined as the change in the advancement of the reaction with time:

$$Rate = \frac{d\xi}{dt} \tag{35.4}$$

With this definition, the rate of reaction with respect to the change in the number of moles of a given species with time is

$$Rate = \frac{1}{\nu_i}\frac{dn_i}{dt} \tag{35.5}$$

As an example of how the rate of reaction is defined relative to the change in moles of reactant or product with time, consider the following reaction:

$$4\,NO_2(g) + O_2(g) \rightarrow 2\,N_2O_5(g) \tag{35.6}$$

The rate of reaction can be expressed with respect to any species in Equation (35.6):

$$Rate = -\frac{1}{4}\frac{dn_{NO_2}}{dt} = -\frac{dn_{O_2}}{dt} = \frac{1}{2}\frac{dn_{N_2O_5}}{dt} \tag{35.7}$$

Notice the sign convention of the coefficient with respect to reactants and products: negative for reactants and positive for products. Since the number of moles of reactant decreases with time, the negative sign for the coefficient ensures that the *Rate* is positive. Also, notice that the rate of reaction can be defined with respect to both reactants and products. In our example, 4 mol of NO_2 react with 1 mol of O_2 to produce 2 mol of N_2O_5 product. Therefore, the **rate of conversion** of NO_2 will be four times greater than the rate of O_2 conversion. Although the conversion rates are different, the reaction rate defined with respect to either species will be the same.

In applying Equation (35.5) to define the reaction rate, a set of stoichiometric coefficients must be employed; however, these coefficients are not unique. For example, if we multiply both sides of Equation (35.6) by a factor of 2, the expression for the rate of conversion must also change. Generally, one decides on a given set of coefficients for a balanced reaction and uses these coefficients consistently throughout a given kinetics problem.

In our present definition, the rate of reaction as written is an extensive property; therefore, it will depend on the system size. The rate can be made intensive by dividing Equation (35.5) by the volume of the system:

$$R = \frac{Rate}{V} = \frac{1}{V}\left(\frac{1}{\nu_i}\frac{dn_i}{dt}\right) = \frac{1}{\nu_i}\frac{d[i]}{dt} \tag{35.8}$$

In Equation (35.8), R is the intensive reaction rate. The last equality in Equation (35.8) is performed recognizing that moles of species i per unit volume is simply the molarity of species i, or $[i]$. Equation (35.8) is the definition for the rate of reaction at constant volume. For species in solution, the application of Equation (35.8) in defining the rate of reaction is clear, but it can also be used for gases, as Example Problem 35.1 illustrates.

EXAMPLE PROBLEM 35.1

The decomposition of acetaldehyde is given by the following balanced reaction:

$$CH_3COH(g) \rightarrow CH_4(g) + CO(g)$$

Define the rate of reaction with respect to the pressure of the reactant.

Solution

Beginning with Equation (35.2) and focusing on the acetaldehyde reactant, we obtain

$$n_{CH_3COH} = n^o_{CH_3COH} - \xi$$

Using the ideal gas law, the pressure of acetaldehyde is expressed as

$$P_{CH_3COH} = \frac{n_{CH_3COH}}{V}RT = [CH_3COH]RT$$

Therefore, the pressure is related to the concentration by the quantity RT. Substituting this result into Equation (35.8) with $\nu_i = 1$ yields

$$R = \frac{Rate}{V} = -\frac{1}{\nu_{CH_3COH}} \frac{d[CH_3COH]}{dt}$$

$$= -\frac{1}{RT} \frac{dP_{CH_3COH}}{dt}$$

35.3 Rate Laws

We begin our discussion of rate laws with a few important definitions. The rate of a reaction will generally depend on the temperature, pressure, and concentrations of species involved in the reaction. In addition, the rate may depend on the phase or phases in which the reaction occurs. Homogeneous reactions occur in a single phase, whereas heterogeneous reactions involve more than one phase. For example, reactions that occur on surfaces are classic examples of heterogeneous reactions. We will limit our initial discussion to homogeneous reactions, with heterogeneous reactivity discussed in Chapter 36. For the majority of homogeneous reactions, an empirical relationship between reactant concentrations and the rate of a chemical reaction can be written. This relationship is known as a **rate law**, and for the reaction shown in Equation (35.1) it is written as

$$R = k[A]^\alpha[B]^\beta \ldots \tag{35.9}$$

where [A] is the concentration of reactant A, [B] is the concentration of reactant B, and so forth. The constant α is known as the **reaction order** with respect to species A, β the reaction order with respect to species B, and so forth. The overall reaction order is equal to the sum of the individual reaction orders ($\alpha + \beta + \ldots$). Finally, the constant k is referred to as the **rate constant** for the reaction. The rate constant is independent of concentration but dependent on pressure and temperature, as discussed later in Section 35.9.

The reaction order dictates the concentration dependence of the reaction rate. The reaction order may be an integer or a fraction. *It cannot be overemphasized that reaction orders have no relation to stoichiometric coefficients, and they are determined by experiment.* For example, reconsider the reaction of nitrogen dioxide with molecular oxygen [Equation (35.6)]:

$$4\,NO_2(g) + O_2(g) \rightarrow 2\,N_2O_5(g)$$

The experimentally determined rate law expression for this reaction is

$$R = k[NO_2]^2[O_2] \tag{35.10}$$

The reaction is second order with respect to NO_2, first order with respect to O_2, and third order overall. Notice that the reaction orders are not equal to the stoichiometric coefficients. *All rate laws must be determined experimentally with respect to each reactant,* and the order of the reaction is not determined by the stoichiometry of the reaction.

In the rate law expression of Equation (35.9), the rate constant serves as the proportionality constant between the concentrations of the various species and the reaction rate. Inspection of Equation (35.8) demonstrates that the reaction rate will *always* have units of concentration time^{-1}. Therefore, the units of k must change with respect to the overall order of the reaction to ensure that the reaction rate has the correct units. The relationship between the rate law expression, order, and the units of k is presented in Table 35.1.

35.3.1 Measuring Reaction Rates

With the definitions for the reaction rate and rate law provided by Equations (35.8) and (35.9), the question of how one measures the rate of reaction becomes important. To illustrate this point, consider the following reaction:

$$A \xrightarrow{k} B \tag{35.11}$$

TABLE 35.1   Relationship between Rate Law, Order, and the Rate Constant k^*

Rate Law	Order	Units of k
$R = k$	Zero	$M\,s^{-1}$
$R = k[A]$	First order with respect to A	s^{-1}
	First order overall	
$R = k[A]^2$	Second order with respect to A	$M^{-1}\,s^{-1}$
	Second order overall	
$R = k[A][B]$	First order with respect to A	$M^{-1}\,s^{-1}$
	First order with respect to B	
	Second order overall	
$R = k[A][B][C]$	First order with respect to A	$M^{-2}\,s^{-1}$
	First order with respect to B	
	First order with respect to C	
	Third order overall	

*In the units of k, M represents mol L^{-1} or moles per liter.

The rate of this reaction in terms of [A] is given by

$$R = -\frac{d[A]}{dt} \qquad (35.12)$$

Furthermore, suppose experiments demonstrate that at a certain temperature and pressure the reaction is first order in A, first order overall, and $k = 40\ s^{-1}$ so that

$$R = k[A] = (40\ s^{-1})[A]$$

Equation (35.12) states that the rate of the reaction is equal to the negative of the time derivative of [A]. Imagine that we perform an experiment in which [A] is measured as a function of time as shown in Figure 35.2. The derivative in Equation (35.12) is simply the slope of the tangent for the concentration curve at a specific time. Therefore, the reaction rate will depend on the time at which the rate is determined. Figure 35.2 presents a measurement of the rate at two time points, $t = 0\ ms\,(1\ ms = 10^{-3}\ s)$ and $t = 30\ ms$. At $t = 0\ ms$, the reaction rate is given by the negative of the slope of the line corresponding to the change in [A] with time, per Equation (35.12):

$$R_{t=0} = -\frac{d[A]}{dt} = 40\ M\,s^{-1}$$

However, when measured at 30 ms the rate is

$$R_{t=30\,ms} = -\frac{d[A]}{dt} = 12\ M\,s^{-1}$$

Notice that the reaction rate is decreasing with time. This behavior is a direct consequence of the change of [A] as a function of time, as expected from the rate law of Equation (35.8). Specifically, at $t = 0$,

$$R_{t=0} = 40\ s^{-1}[A]_{t=0} = 40\ s^{-1}(1\ M) = 40\ M\,s^{-1}$$

However, by $t = 30\ ms$ the concentration of A has decreased to 0.3 M so that the rate is

$$R_{t=30\,ms} = 40\ s^{-1}[A]_{t=30\,ms} = 40\ s^{-1}(0.3\ M) = 12\ M\,s^{-1}$$

This difference in rates brings to the forefront an important issue in kinetics: how does one define a reaction rate if the rate changes with time? One convention is to define the rate before the reactant concentrations have undergone any substantial change from their initial values. The reaction rate obtained under such conditions is known as the **initial rate.** The initial rate in the previous example is that determined at $t = 0$. In the

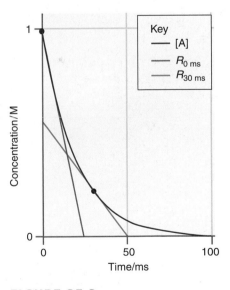

FIGURE 35.2

Measurement of the reaction rate. The concentration of reactant A as a function of time is presented. The rate R is equal to the slope of the tangent of this curve. This slope depends on the time at which the tangent is determined. The tangent determined 30 ms into the reaction is presented as the blue line, and the tangent at $t = 0$ is presented as the purple line.

remainder of our discussion of kinetics, the rate of reaction is taken to be synonymous with initial rate. However, the rate constant is independent of concentration; therefore, if the rate constant, concentrations, and order dependence of the reaction rate are known, the reaction rate can be determined at any time.

35.3.2 Determining Reaction Orders

Consider the following reaction:

$$A + B \xrightarrow{k} C \tag{35.13}$$

The rate law expression for this reaction is

$$R = k[A]^{\alpha}[B]^{\beta} \tag{35.14}$$

How can one determine the order of the reaction with respect to A and B? First, note that the measurement of the rate under a single set of concentrations for A and B will not by itself provide a measure of α and β because we have only one equation [Equation (35.14)], and two unknown quantities. Therefore, the determination of reaction order will involve the measurement of the reaction rate under various concentration conditions. The question then becomes "What set of concentrations should be used to determine the reaction rate?" One answer to this question is known as the **isolation method.** In this approach, the reaction is performed with all reactants but one in excess. Under these conditions, only the concentration of the reactant not in excess will vary to a significant extent during the reaction. For example, consider the A + B reaction shown in Equation (35.13). Imagine performing the experiment in which the initial concentration of A is 1.00 M and the concentration of B is 0.01 M. The rate of the reaction will be zero when all of reactant B has been used; however, the concentration of A will have been reduced to 0.99 M, only a slight reduction from the initial concentration. This simple example demonstrates that the concentration of species present in excess will be essentially constant with time. This time independence simplifies the reaction rate expression because the reaction rate will depend only on the concentration of the reactant not present in excess. In our example reaction where A is in excess, Equation (35.14) simplifies to

$$R = k'[B]^{\beta}, \text{ where } k' = k[A]^{\alpha} \tag{35.15}$$

The quantity k' is taken to be time-independent since k is constant in time, and $[A]^{\alpha}$ will change only slightly with time since A is in excess. In this limit the reaction rate depends on [B] exclusively, and the reaction order with respect to B is determined by measuring the reaction rate as [B] is varied. Of course, the isolation method could just as easily be applied to determine α by performing measurements with B in excess and varying [A].

A second strategy employed to determine reaction rates is referred to as the **method of initial rates.** In this approach, the concentration of a single reactant is changed while holding all other concentrations constant, and the initial rate of the reaction is determined. The variation in the initial rate as a function of concentration is then analyzed to determine the order of the reaction with respect to the reactant that is varied. Consider the reaction depicted by Equation (35.13). To determine the order of the reaction for each reactant, the reaction rate is measured as [A] is varied and the concentration of B is held constant. The reaction rate at two different values of [A] is then analyzed to determine the order of the reaction with respect to [A] as follows:

$$\frac{R_1}{R_2} = \frac{k[A]_1^{\alpha}[B]_0^{\beta}}{k[A]_2^{\alpha}[B]_0^{\beta}} = \left(\frac{[A]_1}{[A]_2}\right)^{\alpha}$$

$$\ln\left(\frac{R_1}{R_2}\right) = \alpha \ln\left(\frac{[A]_1}{[A]_2}\right) \tag{35.16}$$

Notice that [B] and k are constant in each measurement; therefore, they cancel when one evaluates the ratio of the measured reaction rates. Using Equation (35.16), the order of the reaction with respect to A is readily determined. A similar experiment to determine β is also performed where [A] is held constant and the dependence of the reaction rate on [B] is measured.

EXAMPLE PROBLEM 35.2

Using the following data for the reaction illustrated in Equation (35.13), determine the order of the reaction with respect to A and B, and the rate constant for the reaction:

[A] (M)	[B] (M)	Initial Rate (M s^{-1})
2.30×10^{-4}	3.10×10^{-5}	5.25×10^{-4}
4.60×10^{-4}	6.20×10^{-5}	4.20×10^{-3}
9.20×10^{-4}	6.20×10^{-5}	1.68×10^{-2}

Solution

Using the last two entries in the table, the order of the reaction with respect to A is

$$\ln\left(\frac{R_2}{R_3}\right) = \alpha \ln\left(\frac{[A]_2}{[A]_3}\right)$$

$$\ln\left(\frac{4.20 \times 10^{-3}}{1.68 \times 10^{-2}}\right) = \alpha \ln\left(\frac{4.60 \times 10^{-4}}{9.20 \times 10^{-4}}\right)$$

$$-1.386 = \alpha(-0.693)$$

$$2 = \alpha$$

Using this result and the first two entries in the table, the order of the reaction with respect to B is given by

$$\frac{R_1}{R_2} = \frac{k[A]_1^2[B]_1^\beta}{k[A]_2^2[B]_2^\beta} = \frac{[A]_1^2[B]_1^\beta}{[A]_2^2[B]_2^\beta}$$

$$\left(\frac{5.25 \times 10^{-4}}{4.20 \times 10^{-3}}\right) = \left(\frac{2.30 \times 10^{-4}}{4.60 \times 10^{-4}}\right)^2 \left(\frac{3.10 \times 10^{-5}}{6.20 \times 10^{-5}}\right)^\beta$$

$$0.500 = (0.500)^\beta$$

$$1 = \beta$$

Therefore, the reaction is second order in A, first order in B, and third order overall. Using any row from the table, the rate constant is readily determined:

$$R = k[A]^2[B]$$

$$5.2 \times 10^{-4}\,\text{M s}^{-1} = k(2.3 \times 10^{-4}\,\text{M})^2(3.1 \times 10^{-5}\,\text{M})$$

$$3.17 \times 10^8\,\text{M}^{-2}\,\text{s}^{-1} = k$$

Having determined k, the overall rate law is

$$R = (3.17 \times 10^8\,\text{M}^{-2}\,\text{s}^{-1})[A]^2[B]$$

The remaining question to address is how one actually measures the rate of a chemical reaction. Measurement techniques are usually separated into one of two categories: chemical and physical. As the name implies, **chemical methods** in kinetics studies rely on chemical processing to determine the progress of a reaction with respect to time. In this method, a chemical reaction is initiated and samples are removed from the reaction and manipulated such that the reaction in the sample is terminated. Termination of the reaction is accomplished by rapidly cooling the sample or by adding a chemical species that depletes one of the reactants. After stopping the reaction, the sample contents are analyzed. By performing this analysis on a series of samples removed from the original reaction container as a function of time after initiation of the reaction, the kinetics of the reaction can be determined. Chemical methods are generally cumbersome to use and are limited to reactions that occur on slow timescales.

The majority of modern kinetics experiments involve **physical methods.** In these methods, a physical property of the system is monitored as the reaction proceeds. For some reactions, the system pressure or volume provides a convenient physical property for monitoring the progress of a reaction. For example, consider the thermal decomposition of PCl_5:

$$PCl_5(g) \rightarrow PCl_3(g) + Cl_2(g)$$

As the reaction proceeds, for every gaseous PCl_5 molecule that decays, two gaseous product molecules are formed. Therefore, the total system pressure will increase as the reaction proceeds in a container with fixed volume. Measurement of this pressure increase as a function of time provides information on the reaction kinetics.

More complex physical methods involve techniques that are capable of monitoring the concentration of an individual species as a function of time. Many of the spectroscopic techniques described in this text are extremely useful for such measurements. For example, electronic absorption measurements can be performed in which the concentration of a species is monitored using the electronic absorption of a molecule and the Beer–Lambert law. Vibrational spectroscopic measurements using infrared absorption and Raman scattering can be employed to monitor vibrational transitions of reactants or products providing information on their consumption or production. Finally, NMR spectroscopy is a useful technique for following the reaction kinetics of complex systems.

The challenge in chemical kinetics is to perform measurements with sufficient time resolution to monitor the chemistry of interest. If the reaction is slow (seconds or longer), then the chemical methods just described can be used to monitor the kinetics. However, many chemical reactions occur on timescales as short as picoseconds (10^{-12} s) and femtoseconds (10^{-15} s). Reactions occurring on these short timescales are most easily studied using physical methods.

For reactions that occur on timescales as short as 1 ms (10^{-3} s), **stopped-flow techniques** provide a convenient method by which to measure solution phase reactions. These techniques are exceptionally popular for biochemical studies. A stopped-flow experiment is illustrated in Figure 35.3. Two reactants (A and B) are held in reservoirs connected to a syringe pump. The reaction is initiated by depressing the reactant syringes, and the reactants are mixed at the junction indicated in the figure. The reaction is monitored by observing the change in absorbance of the reaction mixture as a function of time. The temporal resolutions of stopped-flow techniques are generally limited by the time it takes for the reactants to mix.

Reactions that can be triggered by light are studied using **flash photolysis techniques.** In flash photolysis, the sample is exposed to a temporal pulse of light that initiates the reaction. Ultrafast light pulses as short as 10 femtoseconds $(10 \text{ fs} = 10^{-14} \text{ s})$ in the visible region of the electromagnetic spectrum are available such that reaction dynamics on this extremely short or ultrafast timescale can be studied. For reference, a 3000 cm^{-1} vibrational mode has a period of roughly 10 fs; therefore, using short optical pulses reactions can be initiated on the same timescale as vibrational molecular motion, and this capability has opened up many exciting fields in chemical kinetics, referred to as **femtochemistry.** This capability has been used to determine the ultrafast chemical kinetics associated with vision, photosynthesis, atmospheric processes, and charge-carrier dynamics in semiconductors. Very recently, femtochemical techniques have been extended to the X-ray region of the spectrum, allowing for the direct interrogation of photoinitiated structural changes using time-resolved scattering techniques. Recent references to some of this work are included in the "For Further Reading" section at the end of this chapter.

FIGURE 35.3

Schematic of a stopped-flow experiment. Two reactants are rapidly introduced into the mixing chamber by syringes. After mixing, the reaction kinetics are monitored by observing the change in sample concentration versus time, in this example by measuring the absorption of light as a function of time after mixing.

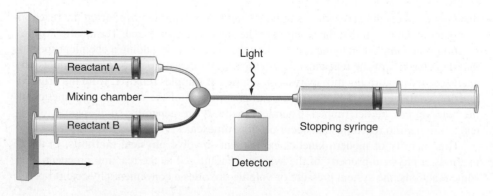

Short optical pulses can be used to perform vibrational spectroscopic measurements (infrared absorption or Raman) on the 100-fs timescale. Finally, NMR techniques, as well as optical absorption and vibrational spectroscopy, can be used to study reactions that occur on the microsecond (10^{-6} s) and longer timescale.

Another approach to studying chemical kinetics is that of **perturbation-relaxation methods.** In this approach, a chemical system initially at equilibrium is perturbed such that the system is no longer at equilibrium. By following the relaxation of the system toward the new equilibrium state, the rate constants for the reaction can be determined. Any system variable that affects the position of the equilibrium such as pressure, pH, or temperature can be used in a perturbation-relaxation experiment. Temperature perturbation or T-jump experiments are the most common type of perturbation experiment and are described in detail later in this chapter.

In summary, the measurement technique chosen for reaction rate determination will depend on both the specifics of the reaction as well as the timescale over which the reaction occurs. In any event, the determination of reaction rates is an experimental exercise and must be accomplished through careful measurements involving well-designed experiments.

35.4 Reaction Mechanisms

As discussed previously, the order of a reaction is not determined by the stoichiometry of the reaction. The reason for the inequivalence of the stoichiometric coefficient and reaction order is that the balanced chemical reaction provides no information with respect to the mechanism of the chemical reaction. A **reaction mechanism** is defined as the collection of individual kinetic processes or elementary steps involved in the transformation of reactants into products. The rate law expression for a chemical reaction, including the order of the reaction, is entirely dependent on the reaction mechanism. In contrast, the Gibbs energy for a reaction is dependent on the equilibrium concentration of reactants and products. Just as the study of concentrations as a function of reaction conditions provides information on the thermodynamics of the reaction, the study of reaction rates as a function of reaction conditions provides information on the reaction mechanism.

All reaction mechanisms consist of a series of **elementary reaction steps,** or chemical processes that occur in a single step. The **molecularity** of a reaction step is the stoichiometric quantity of reactants involved in the reaction step. For example, unimolecular reactions involve a single reactant species. An example of a unimolecular reaction step is the decomposition of a diatomic molecule into its atomic fragments:

$$I_2 \overset{k_d}{\to} 2\,I \tag{35.17}$$

Although Equation (35.17) is referred to as a unimolecular reaction, enthalpy changes accompanying the reaction generally involve the transfer of this heat through collisions with other, neighboring molecules. The role of collisional energy exchange with surrounding molecules will figure prominently in the discussion of unimolecular dissociation reactions in the following chapter (Section (36.3), but these energy-exchange processes are suppressed here. Bimolecular reaction steps involve the interaction of two reactants. For example, the reaction of nitric oxide with ozone is a bimolecular reaction:

$$NO + O_3 \overset{k_r}{\to} NO_2 + O_2 \tag{35.18}$$

The importance of elementary reaction steps is that the corresponding rate law expression for the reaction can be written based on the molecularity of the reaction. For the unimolecular reaction, the rate law expression is that of a first-order reaction. For the unimolecular decomposition of I_2 presented in Equation (35.17), the rate law expression for this elementary step is

$$R = -\frac{d[I_2]}{dt} = k_d[I_2] \tag{35.19}$$

Likewise, the rate law expression for the bimolecular reaction of NO and O_3 is

$$R = -\frac{d[NO]}{dt} = k_r[NO][O_3] \tag{35.20}$$

Comparison of the rate law expressions with their corresponding reactions demonstrates that for elementary reactions the order of the reaction is equal to the stoichiometric coefficient. It is important to keep in mind that *the equivalence of order and molecularity is only true for elementary reaction steps*.

A common problem in kinetics is identifying which of a variety of proposed reaction mechanisms is the "correct" mechanism. The design of kinetic experiments to differentiate between proposed mechanisms is quite challenging. Due to the complexity of many reactions, it is often difficult to experimentally differentiate between several potential mechanisms. A general rule of kinetics is that although it may be possible to rule out a proposed mechanism, it is never possible to prove unequivocally that a given mechanism is correct. The following example illustrates the origins of this rule. Consider the following reaction:

$$A \rightarrow P \tag{35.21}$$

As written, the reaction is a simple first-order transformation of reactant A into product P, and it may occur through a single elementary step. However, what if the reaction were to occur through two elementary steps as follows:

$$A \xrightarrow{k_1} I$$
$$I \xrightarrow{k_2} P \tag{35.22}$$

In this mechanism, the decay of reactant A results in the formation of an intermediate species I that undergoes subsequent decay to produce the reaction product P. One way to validate this mechanism is to observe the formation of the intermediate species. However, if the rate of the second reaction step is fast compared to the rate of the first step, the concentration of [I] will be quite small such that detection of the intermediate may be difficult. As will be seen later, in this limit the product formation kinetics will be consistent with the single elementary step mechanism, and verification of the two-step mechanism is not possible. It is usually assumed that the simplest mechanism consistent with the experimentally determined order dependence is correct until proven otherwise. In this example, a simple single-step mechanism would be considered "correct" until a clever chemist discovered a set of reaction conditions that demonstrates that the reaction must occur by a sequential mechanism.

In order for a reaction mechanism to be valid, the order of the reaction predicted by the mechanism must be in agreement with the experimentally determined rate law. In evaluating a reaction mechanism, one must express the mechanism in terms of elementary reaction steps. The remainder of this chapter involves an investigation of various elementary reaction processes and derivations of the rate law expressions for these elementary reactions. The techniques developed in this chapter can be readily employed in the evaluation of complex kinetic problems, as illustrated in Chapter 36.

35.5 Integrated Rate Law Expressions

The rate law determination methods described in Section 35.3 assume that one has a substantial amount of control over the reaction. Specifically, application of the initial rates method requires that the reactant concentrations be controlled and mixed in any proportion desired. In addition, this method requires that the rate of reaction be measured immediately after initiation of the reaction. Unfortunately, many reactions cannot be studied by this technique due to the instability of the reactants involved, or the timescale of the reaction of interest. In this case, other approaches must be employed.

One approach is to assume that the reaction occurs with a given order dependence and then determine how the concentrations of reactants and products will vary as a function of time. The predictions of the model are compared to experiment to determine if the model provides an appropriate description of the reaction kinetics. **Integrated rate law expressions** provide the predicted temporal evolution in reactant and product concentrations for reactions where the order is assumed. In this section these expressions are derived. For many elementary reactions, integrated rate law expressions can be derived, and some of those cases are considered in this section. However, more complex reactions may be difficult to approach using this technique, and one must resort to numerical methods to evaluate the kinetic behavior associated with a given reaction mechanism. Numerical techniques are discussed in Section 35.6.

35.5.1 First-Order Reactions

Consider the following elementary reaction step where reactant A decays, resulting in the formation of product P:

$$A \xrightarrow{k} P \tag{35.23}$$

If the reaction is first order with respect to [A], the corresponding rate law expression is

$$R = k[A] \tag{35.24}$$

where k is the rate constant for the reaction. The reaction rate can also be written in terms of the time derivative of [A]:

$$R = -\frac{d[A]}{dt} \tag{35.25}$$

Because the reaction rates given by Equations (35.24) and (35.25) are the same, we can write

$$\frac{d[A]}{dt} = -k[A] \tag{35.26}$$

Equation (35.26) is known as a differential rate expression. It relates the time derivative of A to the rate constant and concentration dependence of the reaction. It is also a standard differential equation that can be integrated as follows:

$$\int_{[A]_0}^{[A]} \frac{d[A]}{[A]} = \int_0^t -k\,dt$$

$$\ln\left(\frac{[A]}{[A]_0}\right) = -kt$$

$$[A] = [A]_0 e^{-kt} \tag{35.27}$$

The limits of integration employed in obtaining Equation (35.27) correspond to the initial concentration of reactant when the reaction is initiated ($[A] = [A]_0$ at $t = 0$) and the concentration of reactant at a given time after the reaction has started. If only the reactant is present at $t = 0$, the sum of reactant and product concentrations at any time must be equal to $[A]_0$. Using this idea, the concentration of product with time for this **first-order reaction** is

$$[P] + [A] = [A]_0$$

$$[P] = [A]_0 - [A]$$

$$[P] = [A]_0(1 - e^{-kt}) \tag{35.28}$$

Equation (35.27) demonstrates that for a first-order reaction, the concentration of A will undergo exponential decay with time. A graphically convenient version of Equation (35.27) for comparison to experiment is obtained by taking the natural log of the equation:

$$\ln[A] = \ln[A]_0 - kt \tag{35.29}$$

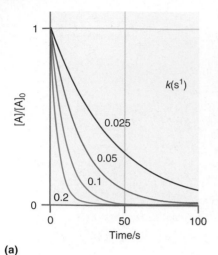

(a)

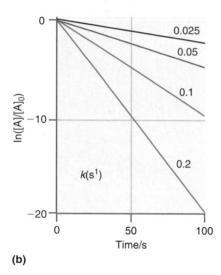

(b)

FIGURE 35.4
Reactant concentration as a function of time for a first-order chemical reaction as given by Equation (35.27). **(a)** Plots of [A] as a function of time for various rate constants k. The rate constant of a given curve is provided in the figure. **(b)** The natural log of reactant concentration as a function of time for a first-order chemical reaction as given by Equation (35.28).

Equation (35.29) predicts that for a first-order reaction, a plot of the natural log of the reactant concentration versus time will be a straight line of slope $-k$ and y intercept equal to the natural log of the initial concentration. Figure 35.4 provides a comparison of the concentration dependences predicted by Equations (35.27) and (35.29) for first-order reactions. It is important to note that the comparison of experimental data to an integrated rate law expression requires that the variation in concentration with time be accurately known over a wide range of reaction times to determine if the reaction indeed follows a certain order dependence.

35.5.2 Half-Life and First-Order Reactions

The time it takes for the reactant concentration to decrease to one-half of its initial value is called the **half-life** of the reaction and is denoted as $t_{1/2}$. For a first-order reaction, substitution of the definition for $t_{1/2}$ into Equation (35.29) results in the following:

$$-kt_{1/2} = \ln\left(\frac{[A]_0/2}{[A]_0}\right) = -\ln 2$$

$$t_{1/2} = \frac{\ln 2}{k} \tag{35.30}$$

Notice that the half-life for a first-order reaction is independent of the initial concentration, and only the rate constant of the reaction influences $t_{1/2}$.

EXAMPLE PROBLEM 35.3

The decomposition of N_2O_5 is an important process in tropospheric chemistry. The half-life for the first-order decomposition of this compound is 2.05×10^4 s. How long will it take for a sample of N_2O_5 to decay to 60% of its initial value?

Solution

Using Equation (35.29), the rate constant for the decay reaction is determined using the half-life as follows:

$$k = \frac{\ln 2}{t_{1/2}} = \frac{\ln 2}{2.05 \times 10^4 \text{ s}} = 3.38 \times 10^{-5} \text{ s}^{-1}$$

The time at which the sample has decayed to 60% of its initial value is then determined using Equation (35.27):

$$[N_2O_5] = 0.6[N_2O_5]_0 = [N_2O_5]_0 e^{-(3.38 \times 10^{-5} \text{ s}^{-1})t}$$

$$0.6 = e^{-(3.38 \times 10^{-5} \text{ s}^{-1})t}$$

$$\frac{-\ln(0.6)}{3.38 \times 10^{-5} \text{ s}^{-1}} = t = 1.51 \times 10^4 \text{ s}$$

Radioactive decay of unstable nuclear isotopes is an important example of a first-order process. The decay rate is usually stated as the half-life. Example Problem 35.4 demonstrates the use of radioactive decay in determining the age of a carbon-containing material.

EXAMPLE PROBLEM 35.4

Carbon-14 is a radioactive nucleus with a half-life of 5760 years. Living matter exchanges carbon with its surroundings (for example, through CO_2) so that a constant level of ^{14}C is maintained, corresponding to 15.3 decay events per minute. Once living matter has died, carbon contained in the matter is not exchanged with the surroundings, and the amount of ^{14}C that remains in the dead material decreases with time due to radioactive decay. Consider a piece of fossilized wood that demonstrates 2.4 ^{14}C decay events per minute. How old is the wood?

Solution

The ratio of decay events yields the amount of ^{14}C present currently versus the amount that was present when the tree died:

$$\frac{[^{14}C]}{[^{14}C]_0} = \frac{2.40 \text{ min}^{-1}}{15.3 \text{ min}^{-1}} = 0.157$$

The rate constant for isotope decay is related to the half-life as follows:

$$k = \frac{\ln 2}{t_{1/2}} = \frac{\ln 2}{5760 \text{ years}} = \frac{\ln 2}{1.82 \times 10^{11} \text{ s}} = 3.81 \times 10^{-12} \text{ s}^{-1}$$

With the rate constant and ratio of isotope concentrations, the age of the fossilized wood is readily determined:

$$\frac{[^{14}C]}{[^{14}C]_0} = e^{-kt}$$

$$\ln\left(\frac{[^{14}C]}{[^{14}C]_0}\right) = -kt$$

$$-\frac{1}{k}\ln\left(\frac{[^{14}C]}{[^{14}C]_0}\right) = -\frac{1}{3.81 \times 10^{-12} \text{ s}}\ln(0.157) = t$$

$$4.86 \times 10^{11} \text{ s} = t$$

This time corresponds to an age of roughly 15,400 years.

35.5.3 Second-Order Reaction (Type I)

Consider the following elementary reaction, which is second order with respect to the reactant A:

$$2\text{ A} \xrightarrow{k} \text{P} \tag{35.31}$$

Second-order reactions involving a single reactant species are referred to as **type I.** Another reaction that is second order overall involves two reactants, A and B, with a rate law that is first order with respect to each reactant. Such reactions are referred to as **second-order reactions of type II.** We focus first on the type I case. For this reaction, the corresponding rate law expression is

$$R = k[\text{A}]^2 \tag{35.32}$$

The rate as expressed as the derivative of reactant concentration is

$$R = -\frac{1}{2}\frac{d[\text{A}]}{dt} \tag{35.33}$$

The rates in the preceding two expressions are equivalent such that

$$-\frac{d[\text{A}]}{dt} = 2k[\text{A}]^2 \tag{35.34}$$

Generally, the quantity $2k$ is written as an effective rate constant, denoted as k_{eff}. With this substitution, integration of Equation (35.34) yields

$$-\int_{[\text{A}]_0}^{[\text{A}]}\frac{d[\text{A}]}{[\text{A}]^2} = \int_0^t k_{eff}dt$$

$$\frac{1}{[\text{A}]} - \frac{1}{[\text{A}]_0} = k_{eff}t$$

$$\frac{1}{[\text{A}]} = \frac{1}{[\text{A}]_0} + k_{eff}t \tag{35.35}$$

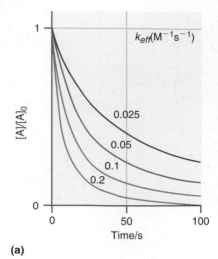

(a)

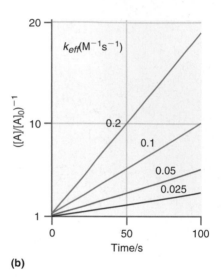

(b)

FIGURE 35.5
Reactant concentration as a function of time for a type I second-order chemical reaction. **(a)** Plots of [A] as a function of time for various rate constants. The rate constant of a given curve is provided in the figure. **(b)** The inverse of reactant concentration as a function of time as given by Equation (35.35).

Equation (35.35) demonstrates that for a second-order reaction, a plot of the inverse of reactant concentration versus time will result in a straight line having a slope of k_{eff} and y intercept of $1/[A]_0$. Figure 35.5 presents a comparison between [A] versus time for a second-order reaction and $1/[A]$ versus time. The linear behavior predicted by Equation (35.35) is evident.

35.5.4 Half-Life and Reactions of Second Order (Type I)

Recall that "half-life" refers to when the concentration of a reactant is half of its initial value. With this definition, the half-life for a type I second-order reaction is

$$t_{1/2} = \frac{1}{k_{eff}[A]_0} \tag{35.36}$$

In contrast to first-order reactions, the half-life for a second-order reaction is dependent on the initial concentration of reactant, with an increase in initial concentration resulting in a decrease in $t_{1/2}$. This behavior is consistent with a first-order reaction occurring through a unimolecular process, whereas the second-order reaction is a bimolecular process that involves the interaction of two species (for example, through collision). As such, the concentration dependence of the reaction rate is expected.

35.5.5 Second-Order Reaction (Type II)

Second-order reactions of type II involve two different reactants, A and B, as follows:

$$A + B \xrightarrow{k} P \tag{35.37}$$

Assuming that the reaction is first order in both A and B, the reaction rate is

$$R = k[A][B] \tag{35.38}$$

In addition, the rate with respect to the time derivative of the reactant concentrations is

$$R = -\frac{d[A]}{dt} = -\frac{d[B]}{dt} \tag{35.39}$$

The loss rate for the reactants is equal, so

$$[A]_0 - [A] = [B]_0 - [B]$$
$$[B]_0 - [A]_0 + [A] = [B]$$
$$\Delta + [A] = [B] \tag{35.40}$$

Equation (35.40) provides a definition for [B] in terms of [A] and the difference in initial concentration, $[B]_0 - [A]_0$, denoted as Δ. Beginning with the case where $\Delta \neq 0$ (that is, the initial concentrations of A and B are not the same) setting Equations (35.38) and (35.39) equal results in the following expression:

$$\frac{d[A]}{dt} = -k[A][B] = -k[A](\Delta + [A])$$

$$\int_{[A]_0}^{[A]} \frac{d[A]}{[A](\Delta + [A])} = -\int_0^t k \, dt$$

Next, the solution to an integral of the previous form is given by

$$\int \frac{dx}{x(c + x)} = -\frac{1}{c} \ln\left(\frac{c + x}{x}\right)$$

Using this solution, the integrated rate law expression becomes

$$-\frac{1}{\Delta} \ln\left(\frac{\Delta + [A]}{[A]}\right)\Big|_{[A]_0}^{[A]} = -kt$$

$$\frac{1}{\Delta}\left[\ln\left(\frac{\Delta + [A]}{[A]}\right) - \ln\left(\frac{\Delta + [A]_0}{[A]_0}\right)\right] = kt$$

$$\frac{1}{\Delta}\left[\ln\left(\frac{[B]}{[A]}\right) - \ln\left(\frac{[B]_0}{[A]_0}\right)\right] = kt$$

$$\frac{1}{[B]_0 - [A]_0}\ln\left(\frac{[B]/[B]_0}{[A]/[A]_0}\right) = kt \qquad (35.41)$$

Equation (35.41) is applicable when $[B]_0 \neq [A]_0$. For the case where $[B]_0 = [A]_0$, the concentrations of [A] and [B] reduce to the expression for a second-order reaction of type I with $k_{eff} = k$. The time evolution in reactant concentrations depends on the amount of each reactant present. Finally, the concept of half-life does not apply to second-order reactions of type II. Unless the reactants are mixed in stoichiometric proportions (1:1 for the case discussed in this section), the concentrations of both species will not be $1/2$ their initial concentrations at the identical time.

SUPPLEMENTAL

35.6 Numerical Approaches

For the simple reactions outlined in the preceding section, an integrated rate law expression can be readily determined. However, there is a wide variety of kinetic problems for which an integrated rate law expression cannot be obtained. How can one compare a kinetic model with experiment in the absence of an integrated rate law? In such cases, numerical methods provide another approach by which to determine the time evolution in concentrations predicted by a kinetic model. To illustrate this approach, consider the following first-order reaction:

$$A \xrightarrow{k} P \qquad (35.42)$$

The differential rate expression for this reaction is

$$\frac{d[A]}{dt} = -k[A] \qquad (35.43)$$

The time derivative corresponds to the change in [A] for a time duration that is infinitesimally small. Using this idea, we can state that for a finite time duration Δt, the change in [A] is given by

$$\frac{\Delta[A]}{\Delta t} = -k[A]$$

$$\Delta[A] = -\Delta t(k[A]) \qquad (35.44)$$

In Equation (35.44), [A] is the concentration of [A] at a specific time. Therefore, we can use this equation to determine the change in the concentration of A, or $\Delta[A]$, over a time period Δt and then use this concentration change to determine the concentration at the end of the time period. This new concentration can be used to determine the subsequent change in [A] over the next time period, and this process is continued until the reaction is complete. Mathematically,

$$[A]_{t+\Delta t} = [A]_t + \Delta[A]$$

$$= [A]_t - k\Delta t[A]_t \qquad (35.45)$$

In Equation (35.45), $[A]_t$ is the concentration at the beginning of the time interval, and $[A]_{t+\Delta t}$ is the concentration at the end of the time interval. This process is illustrated in Figure 35.6. In the figure, the initial concentration is used to determine $\Delta[A]$ over the time interval Δt. The concentration at this next time point, $[A]_1$, is used to determine

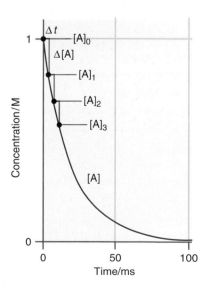

FIGURE 35.6
Schematic representation of the numerical evaluation of a rate law.

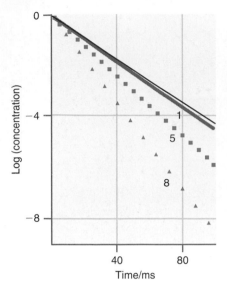

FIGURE 35.7

Comparison of the numerical approximation method to the integrated rate law expression for a first-order reaction. The rate constant for the reaction is 0.1 M s^{-1}. The time evolution in reactant concentration determined by the integrated rate law expression of Equation (35.27) is shown as the solid red line. Comparison to three numerical approximations is given, and the size of the time step (in ms) employed for each approximation is indicated. Notice the improvement in the numerical approximation as the time step is decreased.

$\Delta[A]$ over the next time interval, resulting in concentration $[A]_2$. This process is continued until the entire concentration profile is evaluated.

The specific example discussed here is representative of the general approach to numerically integrating differential equations, known as **Euler's method.** Application of Euler's method requires some knowledge of the timescale of interest, and then selection of a time interval Δt that is sufficiently small to capture the evolution in concentration. Figure 35.7 presents a comparison of the reactant concentration determined using the integrated rate law expression for a first-order reaction to that determined numerically for three different choices for Δt. The figure illustrates that the accuracy of this method is highly dependent on an appropriate choice for Δt. In practice, convergence of the numerical model is demonstrated by reducing Δt and observing that the predicted evolution in concentrations does not change.

The numerical method can be applied to any kinetic process for which differential rate expressions can be prescribed. Euler's method provides the most straightforward way by which to predict how reactant and product concentrations will vary for a specific kinetic scheme. However, this method is "brute force" in that a sufficiently small time step must be chosen to accurately capture the slope of the concentration, and the time steps may be quite small, requiring a large number of iterations in order to reproduce the full time course of the reaction. As such, Euler's method can be computationally demanding. More elegant approaches, such as the Runge–Kutta method, exist that allow for larger time steps to be performed in numerical evaluations, and the interested reader is encouraged to investigate these approaches.

35.7 Sequential First-Order Reactions

Many chemical reactions occur in a series of steps in which reactants are transformed into products through multiple sequential elementary reaction steps. For example, consider the following **sequential reaction** scheme:

$$A \xrightarrow{k_A} I \xrightarrow{k_I} P \tag{35.46}$$

In this scheme, the reactant A decays to form intermediate I, and this intermediate undergoes subsequent decay resulting in the formation of product P. Species I is known as an **intermediate.** The sequential reaction scheme illustrated in Equation (35.46) involves a series of elementary first-order reactions. Recognizing this, the differential rate expressions for each species can be written as follows:

$$\frac{d[A]}{dt} = -k_A[A] \tag{35.47}$$

$$\frac{d[I]}{dt} = k_A[A] - k_I[I] \tag{35.48}$$

$$\frac{d[P]}{dt} = k_I[I] \tag{35.49}$$

These expressions follow naturally from the elementary reaction steps in which a given species participates. For example, the decay of A occurs in the first step of the reaction. The decay is a first-order process, consistent with the differential rate expression in Equation (35.47). The formation of product P is also a first-order process per Equation (35.49). The expression of Equation (35.48) for intermediate I reflects the fact that I is involved in both elementary reaction steps, the decay of A ($k_A[A]$), and the formation of P ($-k_I[I]$). Correspondingly, the differential rate expression for [I] is the sum of the rates associated with these two reaction steps. To determine the concentrations of each species as a function of time, we begin with Equation (35.47), which can be readily integrated given a set of initial concentrations. Let only the reactant A be present at $t = 0$ such that

$$[A]_0 \neq 0; \ [I]_0 = 0; \ [P]_0 = 0 \tag{35.50}$$

With these initial conditions, the expression for [A] is exactly that derived previously:

$$[A] = [A]_0 e^{-k_A t} \qquad (35.51)$$

The expression for [A] given by Equation (35.51) can be substituted into the differential rate expression for I resulting in

$$\frac{d[I]}{dt} = k_A[A] - k_I[I]$$

$$= k_A[A]_0 e^{-k_A t} - k_I[I] \qquad (35.52)$$

Equation (35.52) is a differential equation that when solved yields the following expression for [I]:

$$[I] = \frac{k_A}{k_I - k_A}(e^{-k_A t} - e^{-k_I t})[A]_0 \qquad (35.53)$$

Finally, the expression for [P] is readily determined using the initial conditions of the reaction, with the initial concentration of A, $[A]_0$, equal to the sum of all concentrations for $t > 0$:

$$[A]_0 = [A] + [I] + [P]$$

$$[P] = [A]_0 - [A] - [I] \qquad (35.54)$$

Substituting Equations (35.51) and (35.53) into Equation (35.54) results in the following expression for [P]:

$$[P] = \left(\frac{k_A e^{-k_I t} - k_I e^{-k_A t}}{k_I - k_A} + 1\right)[A]_0 \qquad (35.55)$$

Although the expressions for [I] and [P] look complicated, the temporal evolution in concentration predicted by these equations is intuitive as shown in Figure 35.8. Figure 35.8a presents the evolution in concentration when $k_A = 2k_I$. Notice that A undergoes exponential decay resulting in the production of I. The intermediate in turn undergoes subsequent decay to form the product. The temporal evolution of [I] is extremely dependent on the relative rate constants for the production k_A and decay k_I. Figure 35.8b presents the case where $k_A > k_I$. Here, the maximum intermediate concentration is greater than in the first case. The opposite limit is illustrated in Figure 35.8c, where $k_A < k_I$ and the maximum in intermediate concentration is significantly reduced. This behavior is consistent with intuition: if the intermediate undergoes decay at a faster rate than the rate at which it is being formed, then the intermediate concentration will be small. Of course, the opposite logic holds as evidenced by the $k_A > k_I$ example presented in the Figure 35.8b.

FIGURE 35.8
Concentration profiles for a sequential reaction in which the reactant (A, blue line) forms an intermediate (I, purple line) that undergoes subsequent decay to form the product (P, red line) where (a) $k_A = 2k_I = 0.1$ s^{-1} and (b) $k_A = 8k_I = 0.4$ s^{-1}. Notice that both the maximal amount of I in addition to the time for the maximum is changed relative to the first panel. (c) $k_A = 0.025k_I = 0.0125$ s^{-1}. In this case, very little intermediate is formed, and the maximum in [I] is delayed relative to the first two examples.

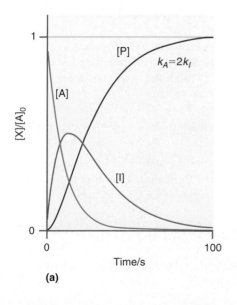

(a)

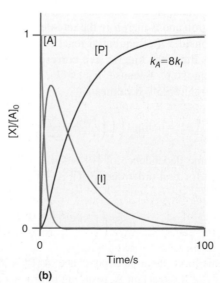

(b)

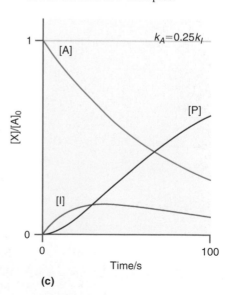

(c)

35.7.1 Maximum Intermediate Concentration

Inspection of Figure 35.8 demonstrates that the time at which the concentration of the intermediate species will be at a maximum depends on the rate constants for its production and decay. Can we predict when [I] will be at a maximum? The maximum intermediate concentration has been reached when the derivative of [I] with respect to time is equal to zero:

$$\left(\frac{d[I]}{dt}\right)_{t=t_{max}} = 0 \qquad (35.56)$$

Using the expression for [I] given in Equation (35.53) in the preceding equation, the time at which [I] is at a maximum, t_{max}, is

$$t_{max} = \frac{1}{k_A - k_I} \ln\left(\frac{k_A}{k_I}\right) \qquad (35.57)$$

EXAMPLE PROBLEM 35.5

Determine the time at which [I] is at a maximum for $k_A = 2k_I = 0.1 \text{ s}^{-1}$.

Solution

This is the first example illustrated in Figure 35.8 where $k_A = 0.1 \text{ s}^{-1}$ and $k_I = 0.05 \text{ s}^{-1}$. Using these rate constants and Equation (35.57), t_{max} is determined as follows:

$$t_{max} = \frac{1}{k_A - k_I} \ln\left(\frac{k_A}{k_I}\right) = \frac{1}{0.1 \text{ s}^{-1} - 0.05 \text{ s}^{-1}} \ln\left(\frac{0.1 \text{ s}^{-1}}{0.05 \text{ s}^{-1}}\right) = 13.9 \text{ s}$$

35.7.2 Rate-Determining Steps

In the preceding subsection, the rate of product formation in a sequential reaction was found to depend on the timescale for production and decay of the intermediate species. Two limiting situations can be envisioned for a sequential reaction. The first limit is where the rate constant for intermediate decay is much greater than the rate constant for production, that is, where $k_I \gg k_A$ in Equation (35.46). In this limit, any intermediate formed will rapidly go on to product, and the rate of product formation depends on the rate of reactant decay. The opposite limit occurs when the rate constant for intermediate production is significantly greater than the intermediate decay rate constant, that is, where $k_A \gg k_I$ in Equation (35.46). In this limit, reactants quickly produce intermediate, but the rate of product formation depends on the rate of intermediate decay. These two limits give rise to one of the most important approximations made in the analysis of kinetic problems, that of the **rate-determining step** or the rate-limiting step. The central idea behind this approximation is as follows: if one step in the sequential reaction is much slower than any other step, the slow step will control the rate of product formation and is therefore the rate-determining step.

Consider the sequential reaction illustrated in Equation (35.46) when $k_A \gg k_I$. In this limit, the kinetic step corresponding to the decay of intermediate I is the rate limiting step. Because $k_A \gg k_I$, $e^{-k_A t} \ll e^{-k_I t}$ and the expression for [P] of Equation (35.55) becomes

$$\lim_{k_A \gg k_I} [P] = \lim_{k_A \gg k_I} \left(\left(\left(\frac{k_A e^{-k_I t} - k_I e^{-k_A t}}{k_I - k_A}\right) + 1\right)[A]_0\right) = (1 - e^{-k_I t})[A]_0 \quad (35.58)$$

The time dependence of [P] when k_I is the rate-limiting step is identical to that predicted for first order decay of I resulting in product formation. The other limit occurs when $k_I \gg k_A$, where $e^{-k_I t} \ll e^{-k_A t}$, and the expression for [P] becomes

$$\lim_{k_I \gg k_A} [P] = \lim_{k_I \gg k_A} \left(\left(\left(\frac{k_A e^{-k_I t} - k_I e^{-k_A t}}{k_I - k_A}\right) + 1\right)[A]_0\right) = (1 - e^{-k_A t})[A]_0 \quad (35.59)$$

In this limit, the time dependence of [P] is identical to that predicted for the first-order decay of the reactant A, resulting in product formation.

When is the rate-determining step approximation appropriate? For the two-step reaction under consideration, 20-fold differences between rate constants are sufficient to ensure that the smaller rate constant will be rate determining. Figure 35.9 presents a comparison for [P] determined using the exact result from Equation (35.55) and the rate-limited prediction of Equations (35.58) and (35.59) for the case where $k_A = 20k_I = 1 \text{ s}^{-1}$ and where $k_A = 0.04k_I = 0.02 \text{ s}^{-1}$. In Figure 35.9a, decay of the intermediate is the rate-limiting step in product formation. Notice the rapid reactant decay, resulting in an appreciable intermediate concentration, with the subsequent decay of the intermediate reflected by a corresponding increase in [P]. The similarity of the exact and rate-limiting curves for [P] demonstrates the validity of the rate-limiting approximation for this ratio of rate constants. The opposite limit is presented in Figure 35.9b. In this case, decay of the reactant is the rate-limiting step in product formation. When reactant decay is the rate-limiting step, very little intermediate is produced. In this case, the loss of [A] is mirrored by an increase in [P]. Again, the agreement between the exact and rate-limiting descriptions of [P] demonstrates the validity of the rate-limiting approximation when a substantial difference in rate constants for intermediate production and decay exists.

35.7.3 The Steady-State Approximation

Consider the following sequential reaction scheme:

$$A \xrightarrow{k_A} I_1 \xrightarrow{k_1} I_2 \xrightarrow{k_2} P \tag{35.60}$$

In this reaction, product formation results from the formation and decay of two intermediate species, I_1 and I_2. The differential rate expressions for this scheme are as follows:

$$\frac{d[A]}{dt} = -k_A[A] \tag{35.61}$$

$$\frac{d[I_1]}{dt} = k_A[A] - k_1[I_1] \tag{35.62}$$

$$\frac{d[I_2]}{dt} = k_1[I_1] - k_2[I_2] \tag{35.63}$$

$$\frac{d[P]}{dt} = k_2[I_2] \tag{35.64}$$

A determination of the time-dependent concentrations for the species involved in this reaction by integration of the differential rate expressions is not trivial; therefore, how can the concentrations be determined? One approach is to use Euler's method (Section 35.6) and numerically determine the concentrations as a function of time. The result of this approach for $k_A = 0.02 \text{ s}^{-1}$ and $k_1 = k_2 = 0.2 \text{ s}^{-1}$ is presented in Figure 35.10. Notice that the relative magnitude of the rate constants results in only modest intermediate concentrations.

Inspection of Figure 35.10 illustrates that $[I_1]$ and $[I_2]$ undergo little change with time. As such, the time derivative of these concentrations is approximately equal to zero:

$$\frac{d[I]}{dt} = 0 \tag{35.65}$$

Equation (35.65) is known as the **steady-state approximation**. This approximation is used to evaluate the differential rate expressions by simply setting the time derivative of all intermediates to zero. This approximation is particularly good when the decay rate of the intermediate is greater than the rate of production so that the intermediates are present at very small concentrations during the reaction (as in the case illustrated in Figure 35.10). Applying the steady-state approximation to I_1 in our example reaction results in the following expression for $[I_1]$:

$$\frac{d[I_1]_{ss}}{dt} = 0 = k_A[A] - k_1[I_1]_{ss}$$

$$[I_1]_{ss} = \frac{k_A}{k_1}[A] = \frac{k_A}{k_1}[A]_0 e^{-k_A t} \tag{35.66}$$

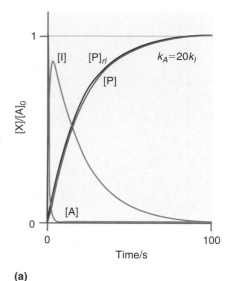

(a)

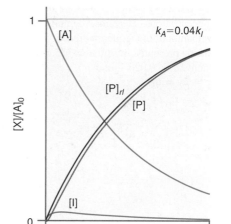

(b)

FIGURE 35.9
Rate-limiting step behavior in sequential reactions. **(a)** $k_A = 20k_I = 1 \text{ s}^{-1}$ such that the rate-limiting step is the decay of intermediate I. In this case, the reduction in [I] is reflected by the appearance of [P]. The time evolution of [P] predicted by the sequential mechanism is given by the purple line, and the corresponding evolution assuming rate-limiting step behavior, $[P]_{rl}$, is given by the red curve. **(b)** The opposite case from part (a) in which $k_A = 0.04k_I = 0.02 \text{ s}^{-1}$ such that the rate-limiting step is the decay of reactant A.

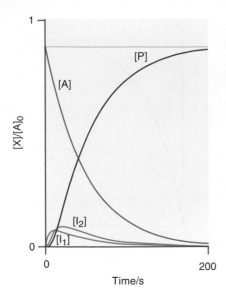

FIGURE 35.10
Concentrations determined by numerical evaluation of the sequential reaction scheme presented in Equation (35.60) where $k_A = 0.02$ s^{-1} and $k_1 = k_2 = 0.2$ s^{-1}.

where the subscript ss indicates that the concentration is that predicted using the steady-state approximation. The final equality in Equation (35.66) results from integration of the differential rate expression for [A] with the initial conditions that $[A]_0 \neq 0$ and all other initial concentrations are zero. The corresponding expression for $[I_2]$ under the steady-state approximation is

$$\frac{d[I_2]_{ss}}{dt} = 0 = k_1[I_1]_{ss} - k_2[I_2]_{ss}$$

$$[I_2]_{ss} = \frac{k_1}{k_2}[I_1]_{ss} = \frac{k_A}{k_2}[A]_0 e^{-k_A t} \tag{35.67}$$

Finally, the differential expression for P is

$$\frac{d[P]_{ss}}{dt} = k_2[I_2] = k_A[A]_0 e^{-k_A t} \tag{35.68}$$

Integration of Equation (35.68) results in the now familiar expression for [P]:

$$[P]_{ss} = [A]_0 (1 - e^{-k_A t}) \tag{35.69}$$

Equation (35.69) demonstrates that within the steady-state approximation, [P] is predicted to demonstrate appearance kinetics consistent with the first-order decay of A.

When is the steady-state approximation valid? The approximation requires that the concentration of intermediate be constant as a function of time. Consider the concentration of the first intermediate under the steady-state approximation. The time derivative of $[I_1]_{ss}$ is

$$\frac{d[I_1]_{ss}}{dt} = \frac{d}{dt}\left(\frac{k_A}{k_1}[A]_0 e^{-k_A t}\right) = -\frac{k_A^2}{k_1}[A]_0 e^{-k_A t} \tag{35.70}$$

The steady-state approximation is valid when Equation (35.70) is equal to zero, which is true when $k_1 \gg k_A^2[A]_0$. In other words, k_1 must be sufficiently large such that $[I_1]$ is small at all times. Similar logic applies to I_2 for which the steady-state approximation is valid when $k_2 \gg k_A^2[A]_0$.

Figure 35.11 presents a comparison between the numerically determined concentrations and those predicted using the steady-state approximation for the two-intermediate sequential reaction where $k_A = 0.02$ s^{-1} and $k_1 = k_2 = 0.2$ s^{-1}. Notice that even for these conditions where the steady-state approximation is expected to be valid, the discrepancy between [P] determined by numerical evaluation versus the steady-state approximation value, $[P]_{ss}$, is evident. For the examples presented here, the steady-state approximation is relatively easy to implement; however, for many reactions the approximation of constant intermediate concentration with time is not appropriate. In addition, the steady-state approximation is difficult to implement if the intermediate concentrations are not isolated to one or two of the differential rate expressions derived from the mechanism of interest.

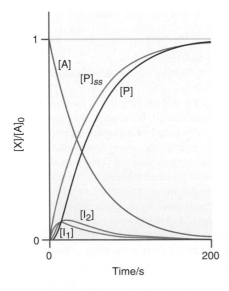

FIGURE 35.11
Comparison of the numerical and steady-state concentration profiles for the sequential reaction scheme presented in Equation (35.44) where $k_A = 0.02$ s^{-1} and $k_1 = k_2 = 0.2$ s^{-1}. Curves corresponding to the steady-state approximation are indicated by the subscript ss.

EXAMPLE PROBLEM 35.6

Consider the following sequential reaction scheme:

$$A \xrightarrow{k_A} I \xrightarrow{k_I} P$$

Assuming that only reactant A is present at $t = 0$, what is the expected time dependence of [P] using the steady-state approximation?

Solution

The differential rate expressions for this reaction were provided in Equations (35.47), (35.48), and (35.49):

$$\frac{d[A]}{dt} = -k_A[A]$$

$$\frac{d[I]}{dt} = k_A[A] - k_I[I]$$

$$\frac{d[P]}{dt} = k_I[I]$$

Applying the steady-state approximation to the differential rate expression for I and substituting in the integrated expression for [A] of Equation (35.51) yield

$$\frac{d[I]}{dt} = 0 = k_A[A] - k_I[I]$$

$$\frac{k_A}{k_I}[A] = \frac{k_A}{k_I}[A]_0 \, e^{-k_A t} = [I]$$

Substituting the preceding expression for [I] into the differential rate expression for the product and integrating yield

$$\frac{d[P]}{dt} = k_I[I] = \frac{k_A}{k_I}\left(k_I[A]_0 \, e^{-k_A t}\right)$$

$$\int_0^{[P]} d[P] = k_A[A]_0 \int_0^t e^{-k_A t} \, dt$$

$$[P] = k_A[A]_0\left[\frac{1}{k_A}\left(1 - e^{-k_A t}\right)\right]$$

$$[P] = [A]_0\left(1 - e^{-k_A t}\right)$$

This expression for [P] is identical to that derived when the decay of A is treated as the rate-limiting step in the sequential reaction [Equation (35.59)].

35.8 Parallel Reactions

In the reactions discussed thus far, reactant decay results in the production of only a single species. However, in many instances a single reactant can become a variety of products. Such reactions are referred to as **parallel reactions.** Consider the following reaction in which the reactant A can form one of two products, B or C:

$$\begin{array}{c}
\quad\quad\quad\quad B \\
\quad k_B \nearrow \\
A \\
\quad k_C \searrow \\
\quad\quad\quad\quad C
\end{array} \qquad (35.71)$$

The differential rate expressions for the reactant and products are

$$\frac{d[A]}{dt} = -k_B[A] - k_C[A] = -(k_B + k_C)[A] \qquad (35.72)$$

$$\frac{d[B]}{dt} = k_B[A] \qquad (35.73)$$

$$\frac{d[C]}{dt} = k_C[A] \qquad (35.74)$$

Integration of the preceding expression involving [A] with the initial conditions $[A]_0 \neq 0$ and $[B] = [C] = 0$ yields

$$[A] = [A]_0 \, e^{-(k_B + k_C)t} \qquad (35.75)$$

The product concentrations can be determined by substituting the expression for [A] into the differential rate expressions and integrating, which results in

$$[B] = \frac{k_B}{k_B + k_C}[A]_0 \left(1 - e^{-(k_B + k_C)t}\right) \qquad (35.76)$$

$$[C] = \frac{k_C}{k_B + k_C}[A]_0 \left(1 - e^{-(k_B + k_C)t}\right) \qquad (35.77)$$

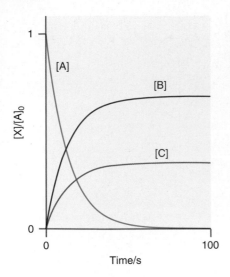

FIGURE 35.12

Concentrations for a parallel reaction where $k_B = 2k_C = 0.1 \text{ s}^{-1}$.

Figure 35.12 provides an illustration of the reactant and product concentrations for this branching reaction where $k_B = 2k_C = 0.1 \text{ s}^{-1}$. A few general trends demonstrated by branching reactions are evident in the figure. First, notice that the decay of A occurs with an apparent rate constant equal to $k_B + k_C$, the sum of rate constants for each reaction branch. Second, the ratio of product concentrations is independent of time. That is, at any time point the ratio $[B]/[C]$ is identical. This behavior is consistent with Equations (35.76) and (35.77) where this ratio of product concentrations is predicted to be

$$\frac{[B]}{[C]} = \frac{k_B}{k_C} \tag{35.78}$$

Equation (35.78) is a very interesting result. The equation states that as the rate constant for one of the reaction branches increases relative to the other, the final concentration of the corresponding product increases as well. Furthermore, notice that the ratio of the product concentrations in Equation (35.78) contains no time dependence; therefore, the product ratio will be the same throughout the course of the reaction.

Equation (35.78) demonstrates that the extent of product formation in a parallel reaction is dependent on the rate constants. Another way to view this behavior is with respect to probability; the larger the rate constant for a given process, the more likely that product will be formed. The **yield** Φ is defined as the probability that a given product will be formed by decay of the reactant:

$$\Phi_i = \frac{k_i}{\sum\limits_{n} k_n} \tag{35.79}$$

In Equation (35.79), k_i is the rate constant for the reaction leading to formation of the product of interest indicated by the subscript i. The denominator is the sum over all rate constants for the reaction branches. The total yield is the sum of the yields for forming each product, and it is normalized such that

$$\sum\limits_{i} \Phi_i = 1 \tag{35.80}$$

In the example reaction depicted in Figure 35.12 where $k_B = 2k_C$, the yield for the formation of product C is

$$\Phi_C = \frac{k_C}{k_B + k_C} = \frac{k_C}{(2k_C) + k_C} = \frac{1}{3} \tag{35.81}$$

Because there are only two branches in this reaction, $\Phi_B = 2/3$. Inspection of Figure 35.12 reveals that $[B] = 2[C]$, which is consistent with the calculated yields.

EXAMPLE PROBLEM 35.7

In acidic conditions, benzyl penicillin (BP) undergoes the following parallel reaction:

In the molecular structures, R_1 and R_2 indicate alkyl substituents. Imagine swallowing penicillin while in your stomach the pH is ~ 3. At this pH the rate constants for the processes at 22°C are $k_1 = 7.0 \times 10^{-4} \text{ s}^{-1}$, $k_2 = 4.1 \times 10^{-3} \text{ s}^{-1}$, and $k_3 = 5.7 \times 10^{-3} \text{ s}^{-1}$. What is the yield for P_1 formation?

Solution

Using Equation (35.79),

$$\Phi_{P_1} = \frac{k_1}{k_1 + k_2 + k_3} = \frac{7.0 \times 10^{-4} \text{ s}^{-1}}{7.0 \times 10^{-4} \text{ s}^{-1} + 4.1 \times 10^{-3} \text{ s}^{-1} + 5.7 \times 10^{-3} \text{ s}^{-1}}$$

$$= 0.067$$

Of the BP that undergoes acid-catalyzed dissociation, 6.7% will result in the formation of P_1.

35.9 Temperature Dependence of Rate Constants

As mentioned at the beginning of this chapter, rate constants (k) are generally temperature-dependent quantities. Experimentally, it is observed that for many reactions a plot of $\ln(k)$ versus T^{-1} demonstrates linear or close to linear behavior. The following empirical relationship between temperature and k, first proposed by Arrhenius in the late 1800s, is known as the **Arrhenius expression:**

$$k = Ae^{-E_a/RT} \tag{35.82}$$

In Equation (35.82), the constant A is referred to as the **frequency factor** or **Arrhenius preexponential factor,** and E_a is the **activation energy** for the reaction. The units of the preexponential factor are identical to those of the rate constant and will vary depending on the order of the reaction. The activation energy is in units of energy mol^{-1} (for example, kJ mol^{-1}). The natural log of Equation (35.82) results in the following expression:

$$\ln(k) = \ln(A) - \frac{E_a}{R}\frac{1}{T} \tag{35.83}$$

Equation (35.83) predicts that a plot of $\ln(k)$ versus T^{-1} will yield a straight line with slope equal to $-E_a/R$ and y intercept equal to $\ln(A)$. Example Problem 35.8 provides an application of Equation (35.83) to determine the Arrhenius parameters for a reaction.

EXAMPLE PROBLEM 35.8

The temperature dependence of the acid-catalyzed hydrolysis of penicillin (illustrated in Example Problem 35.7) is investigated, and the dependence of k_1 on temperature is given in the following table. What is the activation energy and Arrhenius preexponential factor for this branch of the hydrolysis reaction?

Temperature (°C)	k_1 (s^{-1})
22.2	7.0×10^{-4}
27.2	9.8×10^{-4}
33.7	1.6×10^{-3}
38.0	2.0×10^{-3}

Solution

A plot of $\ln(k_1)$ versus T^{-1} is shown here:

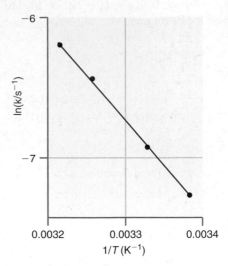

The data are indicated by the points, and the solid line corresponds to the linear least-squares fit to the data. The equation for the line is

$$\ln(k) = (-6300 \text{ K})\frac{1}{T} + 14.1$$

As shown in Equation (35.83), the slope of the line is equal to $-E_a/R$ such that

$$6300 \text{ K} = \frac{E_a}{R} \Rightarrow E_a = 52,400 \text{ J mol}^{-1} = 52.4 \text{ kJ mol}^{-1}$$

The y intercept is equal to $\ln(A)$ such that

$$A = e^{14.1} = 1.33 \times 10^6 \text{ s}^{-1}$$

The origin of the energy term in the Arrhenius expression can be understood as follows. The activation energy corresponds to the energy needed for the chemical reaction to occur. Conceptually, we envision a chemical reaction as occurring along an energy profile as illustrated in Figure 35.13. If the energy content of the reactants is greater than the activation energy, the reaction can proceed. In Figure 35.13 Boltzmann distributions

FIGURE 35.13
A schematic drawing of the energy profile for a chemical reaction. Reactants must acquire sufficient energy to overcome the activation energy E_a for the reaction to occur. The reaction coordinate represents the bonding and geometry changes that occur in the transformation of reactants into products. Shown here are Boltzmann distributions of molecular translational energy (E_{trans}) at 200 and 500 K. Notice that the fraction of molecules with translational energy greater than E_a increase with temperature.

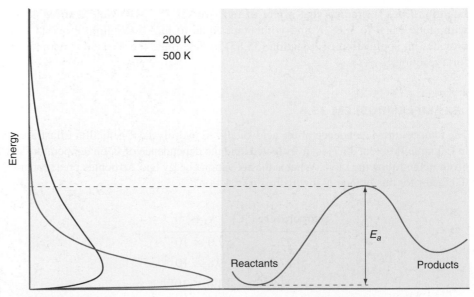

of translational energy at 200 K and 500 K are presented. Notice that the number of molecules with translational energy greater than E_a increases with temperature. This behavior is captured in the exponential dependence on the activation energy, given by $\exp(-E_a/RT)$. At fixed temperature, as the activation energy for a reaction increases the fraction of molecules with sufficient energy to react will decrease, and this will be reflected as a decrease in the reaction rate.

Not all chemical reactions demonstrate Arrhenius behavior. Specifically, the inherent assumption in Equation (35.83) is that both E_a and A are temperature-independent quantities. There are many reactions for which a plot of $\ln(k)$ versus T^{-1} does not yield a straight line, consistent with the temperature dependence of one or both of the Arrhenius parameters. Modern theories of reaction rates predict that the rate constant will demonstrate the following behavior:

$$k = aT^m e^{-E'/RT}$$

where a and E' are temperature-independent quantities, and m can assume values such as 1, $1/2$, and $-1/2$ depending on the details of the theory used to predict the rate constant. For example, in the upcoming section on activated complex theory (Section 35.14), a value of $m = 1$ is predicted. With this value for m, a plot of $\ln(k/T)$ versus T^{-1} should yield a straight line with slope equal to $-E'/R$ and the y intercept equal to $\ln(a)$. Although the limitations of the Arrhenius expression are well known, this relationship still provides an adequate description of the temperature dependence of reaction rate constants for a wide variety of reactions.

35.10 Reversible Reactions and Equilibrium

In the kinetic models discussed in earlier sections, it was assumed that once reactants form products, the opposite or "back" reaction does not occur. However, the **reaction coordinate** presented in Figure 35.14 suggests that, depending on the energetics of the reaction, such reactions can indeed occur. Specifically, the figure illustrates that reactants form products if they have sufficient energy to overcome the activation energy for the reaction. But what if the reaction coordinate is viewed from the product's perspective? Can the coordinate be followed in reverse, with products returning to reactants by overcoming the activation energy barrier from the product side E_a' of the coordinate? Such **reversible reactions** are discussed in this section.

Consider the following reaction in which the forward reaction is first order in A, and the back reaction is first order in B:

$$A \underset{k_B}{\overset{k_A}{\rightleftharpoons}} B \tag{35.84}$$

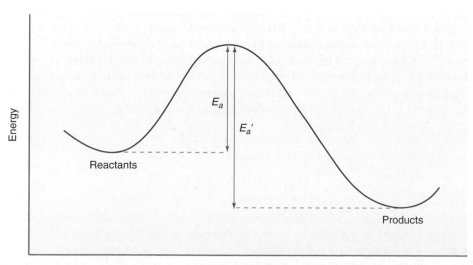

FIGURE 35.14
Reaction coordinate demonstrating the activation energy for reactants to form products E_a and the back reaction in which products form reactants E_a'.

The forward and back rate constants are k_A and k_B, respectively. Integrated rate law expressions can be obtained for this reaction starting with the differential rate expressions for the reactant and product:

$$\frac{d[A]}{dt} = -k_A[A] + k_B[B] \tag{35.85}$$

$$\frac{d[B]}{dt} = k_A[A] - k_B[B] \tag{35.86}$$

Equation (35.85) should be contrasted with the differential rate expression for first-order reactant decay given in Equation (35.26). Reactant decay is included through the $-k_A[A]$ term similar to first-order decay discussed earlier; however, a second term involving the formation of reactant by product decay $k_B[B]$ is now included. The initial conditions are identical to those employed in previous sections. Only reactant is present at the beginning of the reaction when $t = 0$. Also, the concentration of reactant and product for $t > 0$ must be equal to the initial concentration of reactant:

$$[A]_0 = [A] + [B] \tag{35.87}$$

With these initial conditions, Equation (35.85) can be integrated as follows:

$$\frac{d[A]}{dt} = -k_A[A] + k_B[B]$$
$$= -k_A[A] + k_B([A]_0 - [A])$$
$$= -[A](k_A + k_B) + k_B[A]_0$$

$$\int_{[A]_0}^{[A]} \frac{d[A]}{[A](k_A + k_B) - k_B[A]_0} = -\int_0^t dt \tag{35.88}$$

Equation (35.88) can be evaluated using the following standard integral:

$$\int \frac{dx}{(a + bx)} = \frac{1}{b} \ln(a + bx)$$

Using this relationship with the initial conditions specified earlier, the concentrations of reactant and products are

$$[A] = [A]_0 \frac{k_B + k_A e^{-(k_A + k_B)t}}{k_A + k_B} \tag{35.89}$$

$$[B] = [A]_0 \left(1 - \frac{k_B + k_A e^{-(k_A + k_B)t}}{k_A + k_B} \right) \tag{35.90}$$

Figure 35.15 presents the time dependence of [A] and [B] for the case where $k_A = 2k_B = 0.06 \text{ s}^{-1}$. Note that [A] undergoes exponential decay with an apparent rate constant equal to $k_A + k_B$, and [B] appears exponentially with an equivalent rate constant. If the back reaction were not present, [A] would be expected to decay to zero; however, the existence of the back reaction results in both [A] and [B] being nonzero at long times. The concentration of reactant and product at long times is defined as the equilibrium concentration. The equilibrium concentrations are equal to the limit of Equations (35.89) and (35.90) as time goes to infinity:

$$[A]_{eq} = \lim_{t \to \infty} [A] = [A]_0 \frac{k_B}{k_A + k_B} \tag{35.91}$$

$$[B]_{eq} = \lim_{t \to \infty} [B] = [A]_0 \left(1 - \frac{k_B}{k_A + k_B} \right) \tag{35.92}$$

Equations (35.91) and (35.92) demonstrate that the reactant and product concentrations reach a constant or equilibrium value that depends on the relative magnitude of the forward and back reaction rate constants.

In theory, one must wait an infinite amount of time before equilibrium is reached. Fortunately, in practice there will be a time after which the reactant and product concentrations are sufficiently close to equilibrium, and the change in these concentrations with time is so modest that approximating the system as having reached equilibrium is reasonable. This time is indicated by t_{eq} in Figure 35.15, where inspection of the figure demonstrates that the concentrations are at their equilibrium values for times $> t_{eq}$. After equilibrium has been established, the reactant and product concentrations are time independent such that

$$\frac{d[A]_{eq}}{dt} = \frac{d[B]_{eq}}{dt} = 0 \qquad (35.93)$$

The subscripts in Equation (35.93) indicate that equality applies only after equilibrium has been established. A common misconception is that Equation (35.93) states that at equilibrium the forward and back reaction rates are zero. Instead, at equilibrium the forward and back reaction rates are equal, but not zero, such that the macroscopic concentration of reactant or product does not evolve with time. That is, the forward and back reactions still occur, but they occur with equal rates at equilibrium. Using Equation (35.93) in combination with the differential rate expressions for the reactant [Equation (35.85)], we arrive at what is hopefully a familiar relationship:

$$\frac{d[A]_{eq}}{dt} = \frac{d[B]_{eq}}{dt} = 0 = -k_A[A]_{eq} + k_B[B]_{eq}$$

$$\frac{k_A}{k_B} = \frac{[B]_{eq}}{[A]_{eq}} = K_c \qquad (35.94)$$

In this equation, K_c is the equilibrium constant defined in terms of concentrations. This quantity is identical to that first encountered in thermodynamics (Chapter 6) and statistical mechanics (Chapter 32). We now have a definition of equilibrium from the kinetic perspective; therefore, Equation (35.94) is a remarkable result in which the concept of equilibrium as described by these three different perspectives is connected into one deceptively simple equation. From the kinetic standpoint, K_c is related to the ratio of forward and backward rate constants for the reaction. The greater the forward rate constant relative to that for the back reaction, the more equilibrium will favor products over reactants.

Figure 35.16 illustrates the methodology by which forward and backward rate constants can be determined. Specifically, measurement of the reactant decay kinetics (or equivalently the product formation kinetics) provides a measure of the apparent rate constant, $k_A + k_B$. The measurement of K_c, or the reactant and product concentrations at equilibrium, provides a measure of the ratio of the forward and backward rate constants.

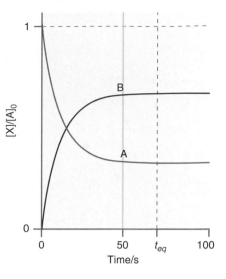

FIGURE 35.15
Time-dependent concentrations in which both forward and back reactions exist between reactant A and product B. In this example, $k_A = 2k_B = 0.06 \text{ s}^{-1}$. Note that the concentrations reach a constant value at longer times ($t \geq t_{eq}$) at which point the reaction reaches equilibrium.

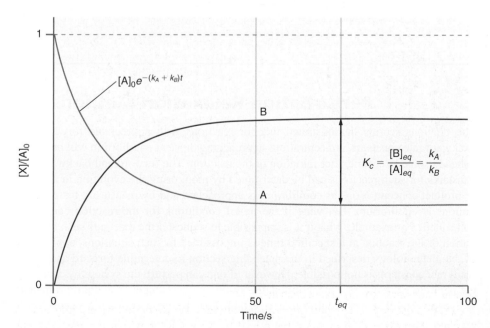

FIGURE 35.16
Methodology for determining forward and backward rate constants. The apparent rate constant for reactant decay is equal to the sum of forward k_A and backward k_B rate constants. The equilibrium constant is equal to k_A/k_B. These two measurements provide a system of two equations and two unknowns that can be readily evaluated to produce k_A and k_B.

Together, these measurements represent a system of two equations and two unknowns that can be readily solved to determine k_A and k_B.

EXAMPLE PROBLEM 35.9

Consider the interconversion of the "boat" and "chair" conformations of cyclohexane:

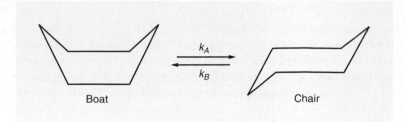

Boat   Chair

The reaction is first order in each direction, with an equilibrium constant of 10^4. The activation energy for the conversion of the chair conformer to the boat conformer is 42 kJ/mol. Assuming an Arrhenius preexponential factor of 10^{12} s^{-1}, what is the expected observed reaction rate constant at 298 K if one were to initiate this reaction starting with only the boat conformer?

Solution

Using the Arrhenius expression of Equation (35.82), k_B is given by

$$k_B = Ae^{-E_a/RT} = 10^{12} \text{ s}^{-1} \exp\left[\frac{-42,000 \text{ J mol}^{-1}}{(8.314 \text{ J mol}^{-1} \text{ K}^{-1})(298 \text{ K})}\right]$$

$$= 4.34 \times 10^4 \text{ s}^{-1}$$

Using the equilibrium constant, k_A can be determined as follows:

$$K_c = 10^4 = \frac{k_A}{k_B}$$

$$k_A = 10^4 k_B = 10^4 (4.34 \times 10^4 \text{ s}^{-1}) = 4.34 \times 10^8 \text{ s}^{-1}$$

Finally, the apparent rate constant is simply the sum of k_A and k_B:

$$k_{app} = k_A + k_B = 4.34 \times 10^8 \text{ s}^{-1}$$

SUPPLEMENTAL

35.11 Perturbation-Relaxation Methods

The previous section demonstrated that for reactions with appreciable forward and backward rate constants, concentrations approaching those at equilibrium will be established at some later time after initiation of the reaction. The forward and backward rate constants for such reactions can be determined by monitoring the evolution in reactant or product concentrations as equilibrium is approached and by measuring the concentrations at equilibrium. But what if the initial conditions for the reaction cannot be controlled? For example, what if it is impossible to sequester the reactants such that initiation of the reaction at a specified time is impossible? In such situations, application of the methodology described in the preceding section to determine forward and backward rate constants is not possible. However, if one can perturb the system by changing temperature, pressure, or concentration, the system will no longer be at equilibrium and will evolve until a new equilibrium is established. If the perturbation occurs on a timescale that is rapid compared to the system relaxation, the kinetics of the relaxation can be monitored and related to the forward and backward rate constants. This is the conceptual idea behind perturbation methods and their application to chemical kinetics.

There are many perturbation techniques; however, the focus here is on **temperature jump** (or T-jump) methods to illustrate the type of information available using perturbation techniques. Consider again the following reaction in which both the forward and back reactions are first order:

$$A \underset{k_B}{\overset{k_A}{\rightleftharpoons}} B \tag{35.95}$$

Next, a rapid change in temperature occurs such that the forward and backward rate constants are altered in accord with the Arrhenius expression of Equation (35.82), and a new equilibrium is established:

$$A \underset{k_B^+}{\overset{k_A^+}{\rightleftharpoons}} B \tag{35.96}$$

The superscript $+$ in this expression indicates that the rate constants correspond to the conditions after the temperature jump. As described next, one can jump to a final temperature of interest so that the reaction can be characterized at this temperature. Following the temperature jump, the concentrations of reactants and products will evolve until the new equilibrium concentrations are reached. At the new equilibrium, the differential rate expression for the reactant is equal to zero so that

$$\frac{d[A]_{eq}}{dt} = 0 = -k_A^+[A]_{eq} + k_B^+[B]_{eq}$$

$$k_A^+[A]_{eq} = k_B^+[B]_{eq} \tag{35.97}$$

The subscripts on the reactant and product concentrations represent the new equilibrium concentrations after the temperature jump. The evolution of reactant and product concentrations from the pre-temperature to post-temperature jump values can be expressed using a coefficient of reaction advancement (Section 35.2). Specifically, let the variable ξ represent the extent to which the pre-temperature jump concentration is shifted away from the concentration for the post-temperature jump equilibrium:

$$[A] - \xi = [A]_{eq} \tag{35.98}$$

$$[B] + \xi = [B]_{eq} \tag{35.99}$$

Immediately after the temperature jump, the concentrations will evolve until equilibrium is reached. Using this idea, the differential rate expression describing the extent of reaction advancement is as follows:

$$\frac{d\xi}{dt} = -k_A^+[A] + k_B^+[B]$$

Notice in this equation that the forward and backward rate constants are the post-temperature jump values. Substitution of Equations (35.98) and (35.99) into the differential rate expression yields the following:

$$\frac{d\xi}{dt} = -k_A^+(\xi + [A]_{eq}) + k_B^+(-\xi + [B]_{eq})$$

$$= -k_A^+[A]_{eq} + k_B^+[B]_{eq} - \xi(k_A^+ + k_B^+)$$

$$= -\xi(k_A^+ + k_B^+) \tag{35.100}$$

In the second step of the preceding equation, the first two terms cancel in accord with Equation (35.97). The relaxation time τ is defined as follows:

$$\tau = (k_A^+ + k_B^+)^{-1} \tag{35.101}$$

FIGURE 35.17

FIGURE 35.17
Example of a temperature-jump experiment for a reaction in which the forward and backward rate processes are first order. The orange and light green portions of the graph indicate times before and after the temperature jump, respectively. After the temperature jump, [A] decreases with a time constant related to the sum of the forward and backward rate constants. The change between the pre-jump and post-jump equilibrium concentrations is given by ξ_0.

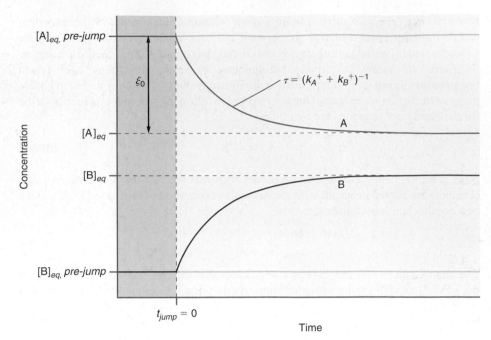

Employing the relaxation time, Equation (35.100) is readily evaluated:

$$\frac{d\xi}{dt} = -\frac{\xi}{\tau}$$

$$\int_{\xi_0}^{\xi} \frac{d\xi'}{\xi'} = -\frac{1}{\tau}\int_{0}^{t} dt'$$

$$\xi = \xi_0\, e^{-t/\tau} \qquad\qquad \textbf{(35.102)}$$

Equation (35.102) demonstrates that for this reaction the concentrations will change exponentially, and the relaxation time is the time it takes for the coefficient of reaction advancement to decay to e^{-1} of its initial value. The timescale for relaxation after the temperature jump is related to the sum of the forward and backward rate constants. This information in combination with the equilibrium constant (given by measurement of $[A]_{eq}$ and $[B]_{eq}$) can be used to determine the individual values for the rate constants. Figure 35.17 presents a schematic of this process.

SUPPLEMENTAL

35.12 The Autoionization of Water: A Temperature-Jump Example

In the autoionization of water, the equilibrium of interest is the following:

$$H_2O(aq) \underset{k_r}{\overset{k_f}{\rightleftharpoons}} H^+(aq) + OH^-(aq) \qquad\qquad \textbf{(35.103)}$$

The reaction is first order in the forward direction and second order in the reverse direction. The differential rate expressions that describe the temporal evolution of H_2O and H^+ concentrations are as follows:

$$\frac{d[H_2O]}{dt} = -k_f[H_2O] + k_r[H^+][OH^-] \qquad\qquad \textbf{(35.104)}$$

$$\frac{d[H^+]}{dt} = k_f[H_2O] - k_r[H^+][OH^-] \qquad\qquad \textbf{(35.105)}$$

Following a temperature jump to 298 K, the measured relaxation time constant was 37 μs. In addition, the pH of the solution is 7. Given this information, the forward and back rate constants can be determined as follows. First, the equilibrium constant after the temperature jump is

$$\frac{k_f^+}{k_r^+} = \frac{[H^+]_{eq}[OH^-]_{eq}}{[H_2O]_{eq}} = K_c \tag{35.106}$$

The differential rate expression for the extent of reaction advancement after the perturbation is given by

$$\begin{aligned}
\frac{d\xi}{dt} &= -k_f^+[H_2O] + k_r^+[H^+][OH^-] \\
&= -k_f^+(\xi + [H_2O]_{eq}) + k_r^+(\xi - [H^+]_{eq})(\xi - [OH^-]_{eq}) \\
&= -k_f^+\left(\xi + \frac{k_r^+}{k_f^+}[H^+]_{eq}[OH^-]_{eq}\right) + k_r^+(\xi - [H^+]_{eq})(\xi - [OH^-]_{eq}) \\
&= -k_f^+\xi - k_r^+\xi([H^+]_{eq} + [OH^-]_{eq}) + O(\xi^2)
\end{aligned} \tag{35.107}$$

The last term in Equation (35.107) represents terms on the order of ξ^2. If the extent of reaction advancement is small, corresponding to a small perturbation of the system temperature, then this term can be neglected, resulting in the following expression for the reaction advancement as a function of time:

$$\frac{d\xi}{dt} = -\xi(k_f^+ + k_r^+([H^+]_{eq} + [OH^-]_{eq})) \tag{35.108}$$

Proceeding as before, the relaxation time is defined as

$$\frac{1}{\tau} = (k_f^+ + k_r^+([H^+]_{eq} + [OH^-]_{eq})) \tag{35.109}$$

Substitution of Equation (35.109) into Equation (35.108) and integration yields an expression for the post-temperature jump evolution identical to that derived earlier in Equation (35.102). The parameters needed to determine the autoionization forward and backward rate constants are the expression for the relaxation time [Equation (35.109)] and the equilibrium constant. Recall that the experimental relaxation time was 37 μs (1 μs = 10^{-6} s) such that

$$\frac{1}{3.7 \times 10^{-5}\,s} = (k_f^+ + k_r^+([H^+]_{eq} + [OH^-]_{eq})) \tag{35.110}$$

In addition, the pH at equilibrium is 7.00 such that $[H^+] = [OH^-] = 1.0 \times 10^{-7}$ M. Finally, the concentration of water at 298 K is 55.5 M. Using this information, the ratio of the forward to back rate constants becomes

$$\frac{k_f^+}{k_r^+} = \frac{[H^+]_{eq}[OH^-]_{eq}}{[H_2O]_{eq}} = \frac{(1.0 \times 10^{-7}\,M)(1.0 \times 10^{-7}\,M)}{55.5\,M} = 1.8 \times 10^{-16}\,M \tag{35.111}$$

Substitution of Equation (35.111) into Equation (35.110) yields the following value for the reverse rate constant:

$$\begin{aligned}
\frac{1}{3.7 \times 10^{-5}\,s} &= (k_f^+ + k_r^+([H^+]_{eq} + [OH^-]_{eq})) \\
&= (1.8 \times 10^{-16}\,M(k_r^+) + k_r^+(2.0 \times 10^{-7}\,M))
\end{aligned}$$

$$\frac{1}{(3.7 \times 10^{-5}\,s)(2.0 \times 10^{-7}\,M)} = k_r^+$$

$$1.4 \times 10^{11}\,M^{-1}\,s^{-1} = k_r^+$$

Finally, the forward rate constant is

$$k^+ = (k_r^+)1.8 \times 10^{-16}\,M = (1.4 \times 10^{11}\,M^{-1}\,s^{-1})(1.8 \times 10^{-16}\,M)$$

$$= 2.5 \times 10^{-5}\,s^{-1}$$

Notice the substantial difference between the forward and backward rate constants, consistent with the modest amount of autoionized species in water. In addition, the forward and reverse rate constants are temperature dependent, and the autoionization constant also demonstrates temperature dependence.

35.13 Potential Energy Surfaces

In the discussion of the Arrhenius equation, the energetics of the reaction were identified as an important factor determining the rate of a reaction. This connection between reaction kinetics and energetics is central to the concept of the potential energy surface. To illustrate this concept, consider the following bimolecular reaction:

$$AB + C \rightarrow A + BC \tag{35.112}$$

The diatomic species AB and BC are stable, but we will assume that the triatomic species ABC and the diatomic species AC are not formed during the course of the reaction. This reaction can be viewed as the interaction of three atoms, and the potential energy of this collection of atoms can be defined with respect to the relative positions in space. The geometric relationship between these species is generally defined with respect to the distance between two of the three atoms (R_{AB} and R_{BC}) and the angle formed between these two distances, as illustrated in Figure 35.18.

The potential energy of the system can be expressed as a function of these coordinates. The variation of the potential energy with a change along these coordinates can then be presented as a graph or surface referred to as a **potential energy surface.** Formally, for our example reaction this surface would be four-dimensional (the three geometric coordinates and energy). The dimensionality of the problem can be reduced by considering the energetics of the reaction at a fixed value for one of the geometric coordinates. In our example reaction, the centers of A, B, and C must be aligned during the reaction such that $\theta = 180°$. With this constraint, the potential energy is reduced to a three-dimensional problem as shown in Figure 35.19. The graphs represent the variation in energy with displacement along R_{AB} and R_{BC} with the arrows indicating the direction of increased separation.

Figures 35.19a and 35.19b illustrate the three-dimensional potential energy surface, and the two minima in this surface corresponding to the stable diatomic molecules AB and BC. A more convenient way to view the potential energy surface is to use a two-dimensional **contour plot,** as illustrated in Figure 35.19c. One can think of this plot as a view straight down onto the three-dimensional surface presented in Figure 35.19. The lines on the contour plot connect regions of equal potential energy. On the lower left-hand region of the surface is a broad energetic plateau that corresponds to the energy when the three atoms are separated or the dissociated state A + B + C. The pathway corresponding to the reaction of B + C to form BC is indicated by the dashed line between points a and a'. The cross section of the potential energy surface along this line is presented in Figure 35.19d, and this contour is simply the potential energy diagram for the diatomic molecule BC. The depth of the potential is equal to the dissociation energy of the diatomic, $D_e(BC)$, and the minimum along R_{BC} corresponds to the equilibrium bond length of the diatomic. Figure 35.19e presents the corresponding diagram for the diatomic molecule AB, as indicated by the dashed line between points b and b' in Figure 35.19c.

The dashed line between points c and d in Figure 35.19c represents the system energy as C approaches AB and reacts to form BC and A under the constraint that $\theta = 180°$. This pathway represents the AB + C $\rightarrow$ A + BC reaction. The maximum in energy along this pathway is referred to as the **transition state** and is indicated by the double dagger symbol, ‡. The variation in energy as one proceeds from reactants to products along this reactive pathway can be plotted to construct a reaction coordinate diagram as presented in Figure 35.20. Note that the transition state corresponds to a maximum along the reaction coordinate; therefore, the activated complex is not a stable species (i.e., an intermediate) along the reaction coordinate.

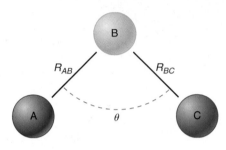

FIGURE 35.18
Definition of geometric coordinates for the AB + C $\rightarrow$ A + BC reaction.

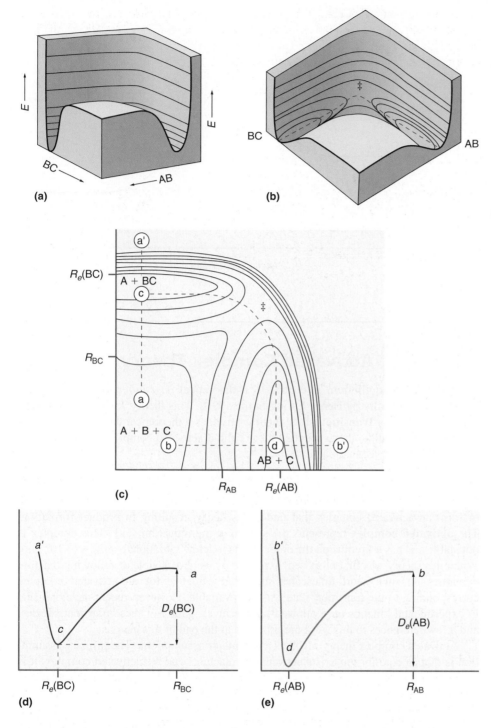

FIGURE 35.19
Illustration of a potential energy surface for the AB + C reaction at a colinear geometry ($\theta = 180°$ in Figure 35.18). **(a, b)** Three-dimensional views of the surface. **(c)** Contour plot of the surface with contours of equipotential energy. The curved dashed line represents one possible path of a reactive event, corresponding to its reaction coordinate. The transition state for this coordinate is indicated by the symbol ‡. **(d, e)** Cross sections of the potential energy surface along the lines $a'-a$ and $b'-b$, respectively. These two graphs correspond to the potential for two-body interactions of B with C, and A with B. [Adapted from J. H. Noggle, *Physical Chemistry*, 3rd Edition, © 1996. Reprinted and electronically reproduced by permission of Pearson Education, Inc., Upper Saddle River, New Jersey.]

The discussion of potential energy surfaces just presented suggests that the kinetics and product yields will depend on the energy content of the reactants and the relative orientation of reactants. This sensitivity can be explored using techniques of crossed-molecular beams. In this approach, reactants with well-defined energies are seeded into a molecular beam that intersects another beam of reactants at well-defined beam geometries. The products formed in the reaction can be analyzed in terms of their energetics, spatial distribution of the products, and beam geometry. This experimental information is then used to construct a potential energy surface (following a substantial amount of analysis). Crossed-molecular beam techniques have provided much insight into the nature of reactive pathways, and detailed, introductory references to this important area of research are provided at the end of this chapter.

FIGURE 35.20
Reaction coordinate diagram involving an activated complex and a reactive intermediate. The graph corresponds to the reaction-coordinate derived from the dashed line between points c and d on the contour plot of Figure 35.19c. The maximum in energy along this coordinate corresponds to the transition state, and the species at this maximum is referred to as an activated complex.

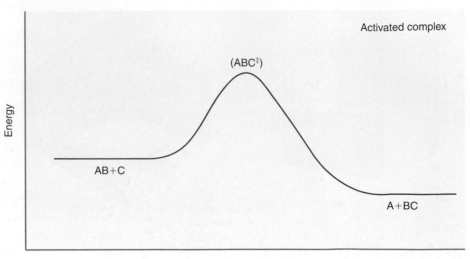

35.14 Activated Complex Theory

The concept of equilibrium is central to a theoretical description of reaction rates developed principally by Henry Eyring in the 1930s. This theory, known as **activated complex theory** or **transition state theory,** provides a theoretical description of reaction rates. To illustrate the conceptual ideas behind activated complex theory, consider the following bimolecular reaction:

$$A + B \xrightarrow{k} P \qquad (35.113)$$

Figure 35.21 illustrates the reaction coordinate for this process, where A and B react to form an activated complex that undergoes decay, resulting in product formation. The **activated complex** represents the system at the transition state. This complex is not stable and has a lifetime on the order of one to a few vibrational periods ($\sim 10^{-14}$ s). When this theory was first proposed, experiments were incapable of following reaction dynamics on such short timescales such that evidence for an activated complex corresponding to the transition state was not available. However, recent developments in experimental kinetics have allowed for the investigation of these transient species, and a few references to this work are provided at the end of this chapter.

Activated complex theory involves a few major assumptions. The primary assumption is that an equilibrium exists between the reactants and the activated complex. It is also assumed that the reaction coordinate describing decomposition of the activated complex can be mapped onto a single energetic degree of freedom of the activated complex. For example, if product formation involves the breaking of a bond, then the vibrational degree of freedom corresponding to bond stretching is taken to be the reactive coordinate.

With these approximations in mind, we can take the kinetic methods derived earlier in this chapter and develop an expression for the rate of product formation. For the example of the bimolecular reaction from Equation (35.113), the kinetic mechanism corresponding to the activated complex model described earlier is

$$A + B \underset{k_{-1}}{\overset{k_1}{\rightleftharpoons}} AB^{\ddagger} \qquad (35.114)$$

$$AB^{\ddagger} \xrightarrow{k_2} P \qquad (35.115)$$

Equation (35.114) represents the equilibrium between reactants and the activated complex, and Equation (35.115) represents the decay of the activated complex to form product. With the assumption of an equilibrium between the reactants and the activated complex, the

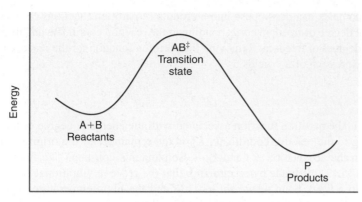

FIGURE 35.21
Illustration of transition state theory. Similar to reaction coordinates depicted previously, the reactants (A and B) and products (P) are separated by an energy barrier. The transition state is an activated reactant complex envisioned to exist at the free-energy maximum along the reaction coordinate.

differential rate expression for one of the reactants (A in this case) is set equal to zero consistent with equilibrium, and an expression for [AB‡] is obtained as follows:

$$\frac{d[A]}{dt} = 0 = -k_1[A][B] + k_{-1}[AB^{\ddagger}]$$

$$[AB^{\ddagger}] = \frac{k_1}{k_{-1}}[A][B] = \frac{K_c^{\ddagger}}{c^{\circ}}[A][B] \qquad (35.116)$$

In Equation (35.116), $K_c^{\ddagger}$ is the equilibrium constant involving the reactants and the activated complex, and it can be expressed in terms of the molecular partition functions of these species as described in Chapter 31. In addition, c° is the standard state concentration (typically 1 M), which appears in the following definition for $K_c^{\ddagger}$:

$$K_c^{\ddagger} = \frac{[AB^{\ddagger}]/c^{\circ}}{([A]/c^{\circ})([B]/c^{\circ})} = \frac{[AB^{\ddagger}]c^{\circ}}{[A][B]}.$$

The rate of the reaction is equal to the rate of product formation, which by Equation (35.115) is equal to

$$R = \frac{d[P]}{dt} = k_2[AB^{\ddagger}] \qquad (35.117)$$

Substitution into Equation (35.117) of the expression for [AB‡] provided in Equation (35.116) yields the following expression for the reaction rate:

$$R = \frac{d[P]}{dt} = \frac{k_2 K_c^{\ddagger}}{c^{\circ}}[A][B] \qquad (35.118)$$

Further evaluation of the reaction rate expression requires that k_2 be defined. This rate constant is associated with the rate of activated complex decay. Imagine that product formation requires the dissociation of a weak bond in the activated complex (since this bond eventually breaks, resulting in product formation). The activated complex is not stable, and the complex dissociates with motion along the corresponding bond-stretching coordinate. Therefore, k_2 is related to the vibrational frequency associated with bond stretching, ν. The rate constant is exactly equal to ν only if every time an activated complex is formed, it dissociates, resulting in product formation. However, it is possible that the activated complex will instead revert back to reactants in which case only a fraction of the activated complexes formed will result in product formation. To account for this possibility, a term referred to as the transmission coefficient κ is included in the definition of k_2:

$$k_2 = \kappa\nu \qquad (35.119)$$

With this definition of k_2, the reaction rate becomes

$$R = \frac{\kappa\nu K_c^{\ddagger}}{c^{\circ}}[A][B] \qquad (35.120)$$

One can express $K_c^{\ddagger}$ in terms of the partition function of reactants and the activated complex using the techniques outlined in Chapter 32. In addition, the partition function for the activated complex can decompose into a product of partition functions corresponding to vibrational degree of freedom corresponding to the reaction coordinate and the remaining energetic degrees of freedom. Removing the partition function for the reactive coordinate from the expression of $K_c^{\ddagger}$ yields

$$K_c^{\ddagger} = q_{rc}\overline{K_c^{\ddagger}} = \frac{k_B T}{h\nu}\overline{K_c^{\ddagger}} \tag{35.121}$$

where q_{rc} is the partition function associated with the energetic degree of freedom corresponding to the reactive coordinate, $\overline{K_c^{\ddagger}}$ is the remainder of the original equilibrium constant in the absence of q_{rc}, and k_B is Boltzmann's constant. The final equality in Equation (35.121) is made by recognizing that the reactive vibrational coordinate corresponds to a weak bond where $h\nu \ll kT$, and the high-temperature approximation for q_{rc} is valid. Substituting Equation (35.121) into Equation (35.120) yields the following expression for rate constant for product formation (k):

$$k = \kappa\frac{k_B T}{hc^{\circ}}\overline{K_c^{\ddagger}} \tag{35.122}$$

Equation (35.122) is the central result of activated complex theory, and it provides a connection between the rate constant for product formation and the molecular parameters for species involved in the reaction. Evaluation of this rate expression requires that one determine $\overline{K_c^{\ddagger}}$, which is related to the partition functions of the activated complex and reactants. The partition functions for the reactants can be readily determined; however, the partition function for the activated complex requires some thought.

The translational partition function for the complex can also be determined using the techniques described earlier, but determination of the rotational and vibrational partition functions requires some knowledge of the structure of the activated complex. The determination of the vibrational partition function is further complicated by the requirement that one of the vibrational degrees of freedom be designated as the reactive coordinate; however, identification of this coordinate may be far from trivial for an activated complex with more than one weak bond. At times, computational techniques can be used to provide insight into the structure of the activated complex and assist in determination of the partition function for this species. With these complications acknowledged, Equation (35.122) represents an important theoretical accomplishment in chemical reaction kinetics. Note that the presentation of activated complex theory provided here is a very rudimentary description of this field. Work continues to the present day to advance and refine this theory, and references are provided at the end of this chapter to review articles describing the significant advances in this field.

We end this discussion by connecting the results of activated complex theory to earlier thermodynamic descriptions of chemical reactions. Recall from thermodynamics that the equilibrium constant $K_c^{\ddagger}$ is related to the corresponding change in Gibbs energy using the following thermodynamic definition:

$$\Delta G^{\ddagger} = -RT \ln K_c^{\ddagger} \tag{35.123}$$

In this definition, $\Delta G^{\ddagger}$ is the difference in Gibbs energy between the transition state and reactants. With this definition for $K_c^{\ddagger}$, k becomes (setting $\kappa = 1$ for convenience)

$$k = \frac{k_B T}{hc^{\circ}}e^{-\Delta G^{\ddagger}/RT} \tag{35.124}$$

In addition, $\Delta G^{\ddagger}$ can be related to the corresponding changes in enthalpy and entropy using

$$\Delta G^{\ddagger} = \Delta H^{\ddagger} - T\Delta S^{\ddagger} \tag{35.125}$$

Substituting Equation (35.125) into Equation (35.124) yields

$$k = \frac{k_B T}{hc^{\circ}}e^{\Delta S^{\ddagger}/R}e^{-\Delta H^{\ddagger}/RT} \tag{35.126}$$

Equation (35.126) is known as the **Eyring equation.** Notice that the temperature dependence of the reaction rate constant predicted by transition state theory is different than that assumed by the Arrhenius expression of Equation (35.82). In particular, the preexponential term in the Eyring equation demonstrates temperature dependence as opposed to the temperature independence of the corresponding term in the Arrhenius expression. However, both the Eyring equation and the Arrhenius expression provide an expression for the temperature dependence of rate constants; therefore, one might expect that the parameters in the Eyring equation ($\Delta H^{\ddagger}$ and $\Delta S^{\ddagger}$) can be related to corresponding parameters in the Arrhenius expression (E_a and A). To derive this relationship, we begin with a modification of Equation (35.82) where the Arrhenius activation energy is written as

$$E_a = RT^2 \left(\frac{d \ln k}{dT} \right) \tag{35.127}$$

Substituting for k the expression given in Equation (35.122) yields

$$E_a = RT^2 \left(\frac{d}{dT} \ln \left(\frac{k_B T}{hc^{\circ}} \overline{K_c^{\ddagger}} \right) \right) = RT + RT^2 \left(\frac{d \ln \overline{K_c^{\ddagger}}}{dT} \right)$$

From thermodynamics the temperature derivative of $\ln(K_c)$ is equal to $\Delta U / RT^2$. Employing this definition to the previous equation results in the following:

$$E_a = RT + \Delta U^{\ddagger}$$

We also make use of the thermodynamic definition of enthalpy, $H = U + PV$, to write

$$\Delta U^{\ddagger} = \Delta H^{\ddagger} - \Delta(PV)^{\ddagger} \tag{35.128}$$

In Equation (35.128), the $\Delta(PV)^{\ddagger}$ term is related to the difference in the product PV with respect to the activated complex and reactants. For a solution-phase reaction, P is constant and the change in V is negligible such that $\Delta U^{\ddagger} \approx \Delta H^{\ddagger}$ and the activation energy in terms of $\Delta H^{\ddagger}$ becomes

$$E_a = \Delta H^{\ddagger} + RT \ \text{(solutions)} \tag{35.129}$$

Comparison of this result with Equation (35.126) demonstrates that the Arrhenius preexponential factor in this case is

$$A = \frac{ek_B T}{hc^{\circ}} e^{\Delta S^{\ddagger}/R} \ \text{(solutions, bimolecular)} \tag{35.130}$$

For solution-phase unimolecular reactions once again $\Delta U^{\ddagger} \approx \Delta H^{\ddagger}$ and the activation energy for a unimolecular solution phase reaction is identical to Equation (35.129). All that changes relative to the bimolecular case is the Arrhenius preexponential factor losing the factor of c° (see Equation 35.116), resulting in

$$A = \frac{ek_B T}{h} e^{\Delta S^{\ddagger}/R} \ \text{(solutions, unimolecular)} \tag{35.131}$$

For a gas-phase reaction, $\Delta(PV)^{\ddagger}$ in Equation (35.128) is proportional to the difference in the number of moles between the transition state and reactants. For a unimolecular ($\Delta n^{\ddagger} = 0$) and bimolecular ($\Delta n^{\ddagger} = -1$) reaction, E_a and A are given by

$$gas, uni \quad E_a = \Delta H^{\ddagger} + RT; \quad A = \frac{ek_B T}{h} e^{\Delta S^{\ddagger}/R} \tag{35.132}$$

$$gas, bi \quad E_a = \Delta H^{\ddagger} + 2RT; \quad A = \frac{e^2 k_B T}{hc^{\circ}} e^{\Delta S^{\ddagger}/R} \tag{35.133}$$

Notice now that both the Arrhenius activation energy and preexponential terms are expected to demonstrate temperature dependence. If $\Delta H^{\ddagger} \gg RT$, then the temperature dependence of E_a will be modest. Also notice that if the enthalpy of the transition state is lower than that of the reactants, the reaction rate may become faster as temperature is decreased! However, the entropy difference between the transition state and reactants is also important in determining the rate. If this entropy difference is positive and the activation energy is near zero, the reaction rate is determined by entropic rather than enthalpic factors.

EXAMPLE PROBLEM 35.10

The thermal decomposition reaction of nitrosyl halides is important in tropospheric chemistry. For example, consider the decomposition of NOCl:

$$2\ NOCl(g) \rightarrow 2\ NO(g) + Cl_2(g)$$

The Arrhenius parameters for this reaction are $A = 1.00 \times 10^{13}\ M^{-1}\ s^{-1}$ and $E_a = 104.0\ kJ\ mol^{-1}$. Calculate $\Delta H^{\ddagger}$ and $\Delta S^{\ddagger}$ for this reaction with $T = 300.\ K$.

Solution

This is a bimolecular reaction such that

$$\Delta H^{\ddagger} = E_a - 2RT = 104\ kJ\ mol^{-1} - 2(8.314\ J\ mol^{-1}\ K^{-1})(300.\ K)$$

$$= 104.0\ kJ\ mol^{-1} - (4.99 \times 10^3\ J\ mol^{-1})\left(\frac{1\ kJ}{1000\ J}\right) = 99.0\ kJ\ mol^{-1}$$

$$\Delta S^{\ddagger} = R \ln\left(\frac{Ahc^{\circ}}{e^2 kT}\right)$$

$$= (8.314\ J\ mol^{-1}\ K^{-1})\ln\left(\frac{(1.00 \times 10^{13}\ M^{-1}\ s^{-1})(6.626 \times 10^{-34}\ J\ s)(1\ M)}{e^2(1.38 \times 10^{-23}\ J\ K^{-1})(300.\ K)}\right)$$

$$= -12.7\ J\ mol^{-1}\ K^{-1}$$

One of the utilities of this calculation is that the sign and magnitude of $\Delta S^{\ddagger}$ provide information on the structure of the activated complex at the transition state relative to the reactants. The negative value in this example illustrates that the activated complex has a lower entropy (or is more ordered) than the reactants. This observation is consistent with a mechanism in which the two NOCl reactants form a complex that eventually decays to produce NO and Cl.

35.15 Diffusion Controlled Reactions

For bimolecular chemical reactions in solution, the presence of solvent molecules can result in reaction dynamics that differ significantly from those in the gas phase. For example, the activation energy and relative orientation of reacting species were identified as being key factors in defining the rate constant for the reaction in the previous section. Imagine a reaction occurring in solution as illustrated in Figure 35.22. Since the average kinetic energy of the reactants is $3/2\ RT$, the average translational velocity is the same as in the gas phase. However, in solution the presence of the solvent molecules results in a number of solvent–solute collisions before the reactants collide. Subsequently, the uninterrupted approach of the reactants characteristic of a gas-phase reaction is replaced by the reactants undergoing diffusion in solution until they encounter each other. In this case, the rate of diffusion can determine the rate of reaction.

The role of diffusion in solution-phase chemistry can be described using the following kinetic scheme:

$$A + B \xrightarrow{k_d} AB \tag{35.134}$$

$$AB \xrightarrow{k_r} A + B \tag{35.135}$$

$$AB \xrightarrow{k_p} P \tag{35.136}$$

In this scheme, reactants A and B diffuse with rate constant k_d until they make contact and form the intermediate complex AB. Once this complex is formed, dissociation can

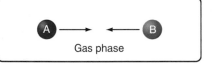

Gas phase

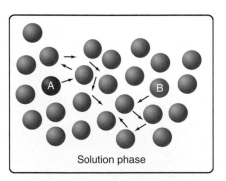

Solution phase

FIGURE 35.22

Top: Reactants A and B approach each other and collide in the gas phase. Bottom: In solution, the reactants undergo a series of collisions with the solvent. In this case, the approach of the reactants is dependent on the rate of reactant diffusion in solution.

occur to re-form the separate reactants (with rate constant k_r), or the reaction can continue resulting in product formation (with rate constant k_P). The expression for the reaction rate consistent with this scheme is

$$R = k_p[\text{AB}] \tag{35.137}$$

Since AB is an intermediate, the steady-state approximation is employed to express the concentration of this species in terms of the reactants:

$$\frac{d[\text{AB}]}{dt} = 0 = k_d[A][B] - k_r[AB] - k_p[AB]$$

$$[AB] = \frac{k_d[A][B]}{k_r + k_p} \tag{35.138}$$

Using this expression for [AB] the reaction rate becomes

$$R = \frac{k_p k_d}{k_r + k_p}[A][B] \tag{35.139}$$

If the rate constant for product formation is much greater than the decay of the intermediate complex to re-form the reactants ($k_p \gg k_r$), the rate expression becomes

$$R = k_d[A][B] \tag{35.140}$$

This is the **diffusion-controlled limit** where diffusion for the reactants limits the rate of product formation. The rate constant for diffusion can be related to the diffusion coefficients for the reactants by

$$k_d = 4\pi N_A(r_A + r_B)D_{AB} \tag{35.141}$$

In Equation (35.141), r_A and r_B are the radii of the reactants, and D_{AB} is the mutual diffusion coefficient equal to the sum of diffusion coefficients for the reactants ($D_{AB} = D_A + D_B$). For a spherical particle in solution the diffusion coefficient is related to the viscosity η by the Stokes–Einstein equation (Chapter 34):

$$D = \frac{k_B T}{6\pi \eta r} \tag{35.142}$$

In Equation (35.142) k is Boltzmann's constant, T is temperature, η is solvent viscosity, and r is the particle radius. This equation predicts that the reaction rate constant should decrease linearly with an increase in solvent viscosity.

To provide some insight into the magnitude of diffusion-controlled rate constants, consider the protonation of acetate (CH_3COO^-) in aqueous solution. The diffusion coefficient of CH_3COO^- is 1.1×10^{-5} cm^2 s^{-1} and of H^+ is 9.3×10^{-5} cm^2 s^{-1}, and $(r_A + r_B)$ is on the order of 5 Å for ions in solution. Assuming that this reaction is diffusion controlled, the rate constant for the reaction is 6.5×10^{-11} cm^3 s^{-1} per reactant pair. Using N_A to convert this quantity to a molar value, $k_d = 3.9 \times 10^{10}$ M^{-1} s^{-1}. Experimentally, the rate constants for acid/base neutralization reactions are greater than the diffusion-controlled limit due to Coulombic attraction of the ions accelerating the collisional rate beyond the diffusion-controlled limit. Another prediction of the diffusion-controlled limit is that the rate of reaction is dependent on the viscosity of solution.

The opposite limit for the reaction occurs when the rate constant for product formation is much smaller than the rate of complex dissociation. In this activation-controlled limit the expression for the reaction rate [(Equation (35.139)] is

$$R = \frac{k_p k_d}{k_r}[A][B] \tag{35.143}$$

In this limit, the rate of the reaction depends on the energetics of the reaction, which are contained in k_p.

EXAMPLE PROBLEM 35.11

In aqueous solution at 25°C and pH 7.4 the diffusion coefficient for hemoglobin (radius = 35 Å) is 7.6×10^{-7} cm^2 s^{-1}, and the diffusion coefficient for O_2 (radius = 2.0 Å) is 2.2×10^{-5} cm^2 s^{-1}. The rate constant for the binding of O_2 to hemoglobin is 4×10^7 M^{-1} s^{-1}. Is this a diffusion-controlled reaction?

Solution

Assuming the reaction is diffusion controlled, the rate constant is expected to be

$$k_d = 4\pi N_A (r_A + r_B) D_{AB}$$
$$= 4\pi N_A (35 \times 10^{-8} \text{ cm} + 2.0 \times 10^{-8} \text{ cm})$$
$$(7.6 \times 10^{-7} \text{ cm}^2 \text{ s}^{-1} + 2.2 \times 10^{-5} \text{ cm}^2 \text{ s}^{-1})$$
$$= 6.4 \times 10^{10} \text{ M}^{-1} \text{ s}^{-1}$$

The diffusion-controlled rate constant is significantly greater than the experimentally measured rate constant; therefore, the binding of O_2 to hemoglobin is not diffusion controlled.

Vocabulary

activated complex	frequency factor	rate of conversion
activated complex theory	half-life	reaction coordinate
activation energy	initial rate	reaction mechanism
Arrhenius expression	integrated rate law expression	reaction order
Arrhenius preexponential factor	intermediate	reaction rate
chemical kinetics	isolation method	reversible reaction
chemical methods	method of initial rates	second-order reaction (type I)
contour plot	molecularity	second-order reaction (type II)
diffusion-controlled limit	parallel reaction	sequential reaction
elementary reaction step	perturbation-relaxation methods	steady-state approximation
Euler's method	physical methods	stopped-flow techniques
Eyring equation	potential energy surface	temperature jump
femtochemistry	rate constant	transition state
first-order reaction	rate-determining step	transition state theory
flash photolysis techniques	rate law	yield

Conceptual Problems

Q35.1 Why is the stoichiometry of a reaction generally not sufficient to determine reaction order?

Q35.2 What is the difference between the order of a reaction with respect to a given species and the overall order?

Q35.3 What is an elementary chemical step, and how is one used in kinetics?

Q35.4 What is the difference between chemical and physical methods for studying chemical kinetics?

Q35.5 What are the fastest timescales on which chemical reactions can be investigated? Describe three experimental techniques for investigation of chemical reactions.

Q35.6 What is the method of initial rates, and why is it used in chemical kinetics studies?

Q35.7 What is a rate law expression, and how is it determined?

Q35.8 What is the difference between a first-order reaction and a second-order reaction?

Q35.9 What is a half-life? Is the half-life for a first-order reaction dependent on concentration?

Q35.10 In a sequential reaction, what is an intermediate?

Q35.11 What is meant by the rate-determining step in a sequential reaction?

Q35.12 What is the steady-state approximation, and when is this approximation employed?

Q35.13 In a parallel reaction in which two products can be formed from the same reactant, what determines the extent to which one product will be formed over another?

Q35.14 What is the kinetic definition of equilibrium?

Q35.15 In a temperature-jump experiment, why does a change in temperature result in a corresponding change in equilibrium?

Q35.16 What is a transition state? How is the concept of a transition state used in activated complex theory?

Q35.17 What is meant by a diffusion-controlled reaction?

Q35.18 What is a typical rate constant for a diffusion-controlled reaction in aqueous solution?

Q35.19 What is the relationship between the parameters in the Arrhenius equation and in the Eyring equation?

Numerical Problems

Problem numbers in **red** indicate that the solution to the problem is given in the *Student's Solutions Manual*.

P35.1 Express the rate of reaction with respect to each species in the following reactions:

a. $2 NO(g) + O_2(g) \rightarrow N_2O_4(g)$

b. $H_2(g) + I_2(g) \rightarrow 2 HI(g)$

c. $ClO(g) + BrO(g) \rightarrow ClO_2(g) + Br(g)$

P35.2 Consider the first-order decomposition of cyclobutane at 438°C at constant volume: $C_4H_8(g) \rightarrow 2C_2H_4(g)$

a. Express the rate of the reaction in terms of the change in total pressure as a function of time.

b. The rate constant for the reaction is $2.48 \times 10^{-4} s^{-1}$. What is the half-life?

c. After initiation of the reaction, how long will it take for the initial pressure of C_4H_8 to drop to 90% of its initial value?

P35.3 As discussed in the text, the total system pressure can be used to monitor the progress of a chemical reaction. Consider the following reaction: $SO_2Cl_2(g) \rightarrow SO_2(g) + Cl_2(g)$. The reaction is initiated, and the following data are obtained:

Time (h)	0	3	6	9	12	15
P_{Total} (kPa)	11.07	14.79	17.26	18.90	19.99	20.71

a. Is the reaction first or second order with respect to SO_2Cl_2?

b. What is the rate constant for this reaction?

P35.4 Consider the following reaction involving bromophenol blue (BPB) and OH^-: $HBPB(aq) + OH^-(aq) \rightarrow BPB^-(aq) + H_2O(l)$. The concentration of BPB can be monitored by following the absorption of this species and using the Beer–Lambert law. In this law, absorption A and concentration are linearly related.

a. Express the reaction rate in terms of the change in absorbance as a function of time.

b. Let A_o be the absorbance due to HBPB at the beginning of the reaction. Assuming that the reaction is first order with respect to both reactants, how is the absorbance of HBPB expected to change with time?

c. Given your answer to part (b), what plot would you construct to determine the rate constant for the reaction?

P35.5 For the following rate expressions, state the order of the reaction with respect to each species, the total order of the reaction, and the units of the rate constant k:

a. $R = k[ClO][BrO]$

b. $R = k[NO]^2[O_2]$

c. $R = k\dfrac{[HI]^2[O_2]}{[H^+]^{1/2}}$

P35.6 What is the overall order of the reaction corresponding to the following rate constants?

a. $k = 1.63 \times 10^{-4} M^{-1} s^{-1}$

b. $k = 1.63 \times 10^{-4} M^{-2} s^{-1}$

c. $k = 1.63 \times 10^{-4} M^{-1/2} s^{-1}$

P35.7 The reaction rate as a function of initial reactant pressures was investigated for the reaction $2 NO(g) + 2 H_2(g) \rightarrow N_2(g) + 2 H_2O(g)$, and the following data were obtained:

Run	P_o H$_2$ (kPa)	P_o NO (kPa)	Rate (kPa s^{-1})
1	53.3	40.0	0.137
2	53.3	20.3	0.033
3	38.5	53.3	0.213
4	19.6	53.3	0.105

What is the rate law expression for this reaction?

P35.8 One loss mechanism for ozone in the atmosphere is the reaction with the $HO_2 \cdot$ radical:

$$HO_2 \cdot (g) + O_3(g) \rightarrow OH \cdot (g) + 2O_2(g)$$

Using the following information, determine the rate law expression for this reaction:

Rate (cm^{-3} s^{-1})	[HO$_2 \cdot$] (cm^{-3})	[O$_3$] (cm^{-3})
1.9×10^8	1.0×10^{11}	1.0×10^{12}
9.5×10^8	1.0×10^{11}	5.0×10^{12}
5.7×10^8	3.0×10^{11}	1.0×10^{12}

P35.9 The disaccharide lactose can be decomposed into its constituent sugars galactose and glucose. This decomposition can be accomplished through acid-based hydrolysis or by the enzyme lactase. Lactose intolerance in humans is due to the lack of lactase production by cells in the small intestine. However, the stomach is an acidic environment; therefore, one might expect lactose hydrolysis to still be an efficient process. The following data were obtained on the rate of lactose decomposition as a function of acid and lactose concentration. Using this information, determine the rate law expression for the acid-based hydrolysis of lactose.

Initial Rate ($M^{-1} s^{-1}$)	[lactose]$_0$ (M)	$[H^+]$ (M)
0.00116	0.01	0.001
0.00232	0.02	0.001
0.00464	0.01	0.004

P35.10 (Challenging) The first-order thermal decomposition of chlorocyclohexane is as follows: $C_6H_{11}Cl(g) \rightarrow C_6H_{10}(g) + HCl(g)$. For a constant volume system the following total pressures were measured as a function of time:

Time (s)	P (Torr)	Time (s)	P (Torr)
3	237.2	24	332.1
6	255.3	27	341.1
9	271.3	30	349.3
12	285.8	33	356.9
15	299.0	36	363.7
18	311.2	39	369.9
21	322.2	42	375.5

a. Derive the following relationship for a first-order reaction: $P(t_2) - P(t_1) = (P(t_\infty) - P(t_0))e^{-kt_1}(1 - e^{-k(t_2-t_1)})$. In this relation, $P(t_1)$ and $P(t_2)$ are the pressures at two specific times; $P(t_0)$ is the initial pressure when the reaction is initiated, $P(t_\infty)$ is the pressure at the completion of the reaction, and k is the rate constant for the reaction. To derive this relationship do the following:

i. Given the first-order dependence of the reaction, write the expression for the pressure of chlorocyclohexane at a specific time t_1.

ii. Write the expression for the pressure at another time t_2, which is equal to $t_1 + \Delta$ where delta is a fixed quantity of time.

iii. Write expressions for $P(t_\infty) - P(t_1)$ and $P(t_\infty) - P(t_2)$.

iv. Subtract the two expressions from part (iii).

b. Using the natural log of the relationship from part (a) and the data provided in the table given earlier in this problem, determine the rate constant for the decomposition of chlorocyclohexane. (*Hint*: Transform the data in the table by defining t_2-t_1 to be a constant value, for example, 9 s.)

P35.11 You are given the following data for the decomposition of acetaldehyde:

Initial Concentration (M)	9.72×10^{-3}	4.56×10^{-3}
Half-Life (s)	328	685

Determine the order of the reaction and the rate constant for the reaction.

P35.12 Consider the schematic reaction $A \overset{k}{\rightarrow} P$.

a. If the reaction is one-half order with respect to A, what is the integrated rate law expression for this reaction?

b. What plot would you construct to determine the rate constant k for the reaction?

c. What would be the half-life for this reaction? Will it depend on initial concentration of the reactant?

P35.13 A certain reaction is first order, and 540 s after initiation of the reaction, 32.5% of the reactant remains.

a. What is the rate constant for this reaction?

b. At what time after initiation of the reaction will 10% of the reactant remain?

P35.14 The half-life of ^{238}U is 4.5×10^9 years. How many disintegrations occur in 1 min for a 10 mg sample of this element?

P35.15 You are performing an experiment using ^3H (half-life $= 4.5 \times 10^3$ days) labeled phenylalanine in which the five aromatic hydrogens are labeled. To perform the experiment, the initial activity cannot be lower than 10% of the initial activity when the sample was received. How long after receiving the sample can you wait before performing the experiment?

P35.16 One issue confronting the use of ^{14}C decay (half-life of 5760 years) to date materials is that of obtaining a standard. One approach to this issue is found in the field of dendrochronology or using tree rings to determine age. Using this approach, tree materials dating back 10,000 years have been identified. Assuming you had a sample of such a tree in which the number of decay events was 15.3 decays per minute before decomposition, what would the decays per minute be in the present day?

P35.17 A convenient source of gamma rays for radiation chemistry research is ^{60}Co, which undergoes the following decay process: $^{60}_{27}Co \overset{k}{\rightarrow} {}^{60}_{28}Ni + \beta^- + \gamma$. The half-life of ^{60}Co is 1.9×10^3 days.

a. What is the rate constant for the decay process?

b. How long will it take for a sample of ^{60}Co to decay to half of its original concentration?

P35.18 The growth of a bacterial colony can be modeled as a first-order process in which the probability of cell division is linear with respect to time such that $dN/N = \zeta dt$, where dN is the number of cells that divide in the time interval dt, and ζ is a constant.

a. Use the preceding expression to show that the number of cells in the colony is given by $N = N_0 e^{\zeta t}$, where N is the number of cells in the colony and N_0 is the number of cells present at $t = 0$.

b. The generation time is the amount of time it takes for the number of cells to double. Using the answer to part (a), derive an expression for the generation time.

c. In milk at 37°C, the bacterium *Lactobacillus acidophilus* has a generation time of about 75 min. Construct a plot of the *acidophilus* concentration as a function of time for time intervals of 15, 30, 45, 60, 90, 120, and 150 min after a colony of size N_0 is introduced to a container of milk.

P35.19 A technique for radioactively labeling proteins is electrophilic radioiodination in which an aromatic substitution of ^{131}I onto a tyrosine residue is performed as follows:

Using the activity of ^{131}I, one can measure protein lifetimes in a variety of biological processes. ^{131}I undergoes beta decay with a half-life of 8.02 days. Initially a protein labeled with ^{131}I has a specific activity of 1.0 μCi, which corresponds to 37,000 decay events every second. The protein is suspended in aqueous solution and exposed to oxygen for 5 days. After isolating the protein from solution, the protein sample is found to have a specific activity of 0.32 μCi. Is oxygen reacting with the tyrosine residues of the protein, resulting in the loss of ^{131}I?

P35.20 Molybdenum-99 decays to form "metastable" technetium-99 (^{99}Tc) through beta decay. This isotope has a very long lifetime (half-life of 6 hours) compared to other nuclear isomers that undergo gamma decay. The gamma rays emitted when ^{99}Tc decays are easily detected, which makes this species ideal for a variety of medical imaging applications such as single-photon emission computed tomography. At the beginning of this millennium, the decommissioning of two nuclear reactors created a critical shortage of ^{99}Tc, and this shortage remains to be addressed (*Science* 331 (2011): 227). Imagine you are performing an imaging experiment with ^{99}Tc in which you need 2 hours to obtain the image after injecting the isotope into the patient. How much of the ^{99}Tc will have decayed during this time?

P35.21 Show that the ratio of the half-life to the one-quarter life, $t_{1/2}/t_{1/4}$, for a reaction that is nth order ($n > 1$) in reactant A can be written as a function of n alone (that is, there is no concentration dependence in the ratio). (Note: The one-quarter life is defined as the time at which the concentration is $1/4$ of the initial concentration.)

P35.22 Given the following kinetic scheme and associated rate constants, determine the concentration profiles of all species using Euler's method. Assume that the reaction is initiated with only the reactant A present at an initial concentration of 1 M. To perform this calculation, you may want to use a spreadsheet program such as Excel.

$$A \xrightarrow{k=1.5 \times 10^{-3} \text{ s}^{-1}} B$$

$$A \xrightarrow{k=2.5 \times 10^{-3} \text{ s}^{-1}} C \xrightarrow{k=1.8 \times 10^{-3} \text{ s}^{-1}} D$$

P35.23 For the sequential reaction $A \xrightarrow{k_A} B \xrightarrow{k_B} C$, the rate constants are $k_A = 5 \times 10^6 \text{ s}^{-1}$ and $k_B = 3 \times 10^6 \text{ s}^{-1}$. Determine the time at which $[B]$ is at a maximum.

P35.24 For the sequential reaction $A \xrightarrow{k_A} B \xrightarrow{k_B} C$, $k_A = 1.00 \times 10^{-3} \text{ s}^{-1}$. Using a computer spreadsheet program such as Excel, plot the concentration of each species for cases where $k_B = 10k_A$, $k_B = 1.5k_A$, and $k_B = 0.1k_A$. Assume that only the reactant is present when the reaction is initiated.

P35.25 (Challenging) For the sequential reaction in Problem P35.21, plot the concentration of each species for the case where $k_B = k_A$. Can you use the analytical expression for [B] in this case?

P35.26 For a type II second-order reaction, the reaction is 60% complete in 60 seconds when $[A]_0 = 0.1$ M and $[B]_0 = 0.5$ M.

a. What is the rate constant for this reaction?

b. Will the time for the reaction to reach 60% completion change if the initial reactant concentrations are decreased by a factor of 2?

P35.27 Bacteriorhodopsin is a protein found in *Halobacterium halobium* that converts light energy into a transmembrane proton gradient that is used for ATP synthesis. After light is absorbed by the protein, the following initial reaction sequence occurs:

$$\text{Br} \xrightarrow{k_1=2.0 \times 10^{12} \text{ s}^{-1}} \text{J} \xrightarrow{k_2=3.3 \times 10^{11} \text{ s}^{-1}} \text{K}$$

a. At what time will the maximum concentration of the intermediate J occur?

b. Construct plots of the concentration of each species versus time.

P35.28 Bananas are somewhat radioactive due to the presence of substantial amounts of potassium. Potassium-40 decays by two different paths:

$$^{40}_{19}\text{K} \rightarrow {}^{40}_{20}\text{Ca} + \beta^- \ (89.3\%)$$

$$^{40}_{19}\text{K} \rightarrow {}^{40}_{18}\text{Ar} + \beta^+ \ (10.7\%)$$

The half-life for potassium decay is 1.3×10^9 years. Determine the rate constants for the individual channels.

P35.29 The bimolecular reaction of chlorine monoxide can result in the formation of three different combinations of products or product channels (the rate constant for each reaction is indicated):

$$\text{ClO} \cdot (g) + \text{ClO} \cdot (g) \rightarrow \text{Cl}_2(g) + \text{O}_2(g) \qquad k = 2.9 \times 10^6$$
$$\rightarrow \text{ClOO} \cdot (g) + \text{Cl} \cdot (g) \qquad k = 4.8 \times 10^6$$
$$\rightarrow \text{OClO} \cdot (g) + \text{Cl} \cdot (g) \qquad k = 2.1 \times 10^6$$

Determine the quantum yield for the three product channels.

P35.30 In the stratosphere, the rate constant for the conversion of ozone to molecular oxygen by atomic chlorine is $\text{Cl} \cdot (g) + \text{O}_3(g) \rightarrow \text{ClO} \cdot (g) + \text{O}_2(g)$ [(half-life of 5760 years) k = $(1.7 \times 10^{10} \text{ M}^{-1}\text{s}^{-1})e^{-260K/T}$].

a. What is the rate of this reaction at 20 km where $[\text{Cl}] = 5 \times 10^{-17}$ M, $[\text{O}_3] = 8 \times 10^{-9}$ M, and $T = 220$ K?

b. The actual concentrations at 45 km are $[\text{Cl}] = 3 \times 10^{-15}$ M and $[\text{O}_3] = 8 \times 10^{-11}$ M. What is the rate of the reaction at this altitude where $T = 270$ K?

c. (Optional) Given the concentrations in part (a), what would you expect the concentrations at 20. km to be assuming that the gravity represents the operative force defining the potential energy?

P35.31 An experiment is performed on the following parallel reaction:

Two things are determined: (1) The yield for B at a given temperature is found to be 0.3 and (2) the rate constants are described well by an Arrhenius expression with the activation to B and C formation being 27 and 34 kJ mol^{-1}, respectively, and with identical preexponential factors. Demonstrate that these two statements are inconsistent with each other.

P35.32 The reaction of atomic chlorine with ozone is the first step in the catalytic decomposition of stratospheric ozone by Cl$\cdot$:

$$Cl\cdot(g) + O_3(g) \rightarrow ClO\cdot(g) + O_2(g)$$

At 298 K the rate constant for this reaction is $6.7 \times 10^9 \, M^{-1} s^{-1}$. Experimentally, the Arrhenius pre-exponential factor was determined to be $1.4 \times 10^{10} \, M^{-1} s^{-1}$. Using this information determine the activation energy for this reaction.

P35.33 In addition to Cl$\cdot$, other halogens can potentially contribute to stratospheric ozone loss. For example, consider the reaction of Br$\cdot$ with ozone:

$$Br\cdot(g) + O_3(g) \rightarrow BrO\cdot(g) + O_2(g)$$

At 298 K the rate constant for this reaction is $1.2 \times 10^8 \, M^{-1} s^{-1}$, about a factor of 50 reduced from the Cl$\cdot$ value presented in the previous problem. Experimentally, the Arrhenius pre-exponential factor was determined to be $1.0 \times 10^{10} \, M^{-1} s^{-1}$.

a. Without doing any calculations, is the energy barrier for this reaction greater or less than that for Cl?

b. Calculate the activation energy for this reaction.

P35.34 A standard "rule of thumb" for thermally activated reactions is that the reaction rate doubles for every 10 K increase in temperature. Is this statement true independent of the activation energy (assuming that the activation energy is positive and independent of temperature)?

P35.35 Calculate the ratio of rate constants for two thermal reactions that have the same Arrhenius preexponential term but have activation energies that differ by 1.0, 10., and 30. kJ/mol for $T = 298$ K.

P35.36 The conversion of $NO_2(g)$ to $NO(g)$ and $O_2(g)$ can occur through the following reaction:

$$NO_2(g) \rightarrow 2NO(g) + O_2(g)$$

The activation energy for this reaction is 111 kJ mol^{-1} and the pre-exponential factor is $2.0 \times 10^{-9} \, M^{-1} s^{-1}$. Assume that these quantities are temperature independent.

a. What is the rate constant for this reaction at 298 K?

b. What is the rate constant for this reaction at the tropopause where T = 225 K.

P35.37 The activation energy for a reaction is 50. J mol^{-1}. Determine the effect on the rate constant for this reaction with a change in temperature from 273 K to 298 K.

P35.38 The rate constant for the reaction of hydrogen with iodine is $2.45 \times 10^{-4} \, M^{-1} s^{-1}$ at 302°C and 0.950 $M^{-1} s^{-1}$ at 508°C.

a. Calculate the activation energy and Arrhenius preexponential factor for this reaction.

b. What is the value of the rate constant at 400.°C?

P35.39 Consider the gas phase thermal decomposition of 1.0 atm of $(CH_3)_3COOC(CH_3)_3(g)$ to acetone $(CH_3)_2CO(g)$ and ethane $(C_2H_6)(g)$, which occurs with a rate constant of 0.0019 s^{-1}. After initiation of the reaction, at what time would you expect the pressure to be 1.8 atm?

P35.40 At 552.3 K, the rate constant for the thermal decomposition of SO_2Cl_2 is 1.02×10^{-6} s^{-1}. If the activation energy is 210. kJ mol^{-1}, calculate the Arrhenius preexponential factor and determine the rate constant at 600. K.

P35.41 The melting of double-strand DNA into two single strands can be initiated using temperature-jump methods. Derive the expression for the T-jump relaxation time for the following equilibrium involving double-strand (DS) and single-strand (SS) DNA:

$$DS \underset{k_r}{\overset{k_f}{\rightleftharpoons}} 2SS$$

P35.42 Consider the reaction

$$A + B \underset{k'}{\overset{k}{\rightleftharpoons}} P$$

A temperature-jump experiment is performed where the relaxation time constant is measured to be 310 μs, resulting in an equilibrium where $K_{eq} = 0.70$ with $[P]_{eq} = 0.20$ M. What are k and k'? (Watch the units!)

P35.43 In the limit where the diffusion coefficients and radii of two reactants are equivalent, demonstrate that the rate constant for a diffusion controlled reaction can be written as

$$k_d = \frac{8RT}{3\eta}$$

P35.44 In the following chapter, enzyme catalysis reactions will be extensively reviewed. The first step in these reactions involves the binding of a reactant molecule (referred to as a substrate) to a binding site on the enzyme. If this binding is extremely efficient (that is, equilibrium strongly favors the enzyme–substrate complex over separate enzyme and substrate) and the formation of product rapid, then the rate of catalysis could be diffusion limited. Estimate the expected rate constant for a diffusion controlled reaction using typical values for an enzyme ($D = 1.00 \times 10^{-7}$ cm^2 s^{-1} and $r = 40.0$ Å) and a small molecular substrate ($D = 1.00 \times 10^{-5}$ cm^2 s^{-1} and $r = 5.00$ Å).

P35.45 Imidazole is a common molecular species in biological chemistry. For example, it constitutes the side chain of the amino acid histidine. Imidazole can be protonated in solution as follows:

The rate constant for the protonation reaction is $5.5 \times 10^{10} \text{ M}^{-1}\text{ s}^{-1}$. Assuming that the reaction is diffusion controlled, estimate the diffusion coefficient of imidazole when $D(\text{H}^+) = 9.31 \times 10^{-5} \text{ cm}^2 \text{ s}^{-1}$, $r(\text{H}^+) \sim 1.0$ Å and r (imidazole) $= 6.0$ Å. Use this information to predict the rate of deprotonation of imidazole by OH^- ($D = 5.30 \times 10^{-5} \text{ cm}^2 \text{ s}^{-1}$ and $r = {\sim}1.5$ Å).

P35.46 Catalase is an enzyme that promotes the conversion of hydrogen peroxide (H_2O_2) into water and oxygen. The diffusion constant and radius for catalase are $6.0 \times 10^{-7} \text{ cm}^2 \text{ s}^{-1}$ and 51.2 Å. For hydrogen peroxide the corresponding values are $1.5 \times 10^{-5} \text{ cm}^2 \text{ s}^{-1}$ and $r \sim 2.0$ Å. The experimentally determined rate constant for the conversion of hydrogen peroxide by catalase is $5.0 \times 10^6 \text{ M}^{-1}\text{ s}^{-1}$. Is this a diffusion-controlled reaction?

P35.47 The unimolecular decomposition of urea in aqueous solution is measured at two different temperatures, and the following data are observed:

Trial Number	Temperature (°C)	k (s^{-1})
1	60.0	1.20×10^{-7}
2	71.5	4.40×10^{-7}

a. Determine the Arrhenius parameters for this reaction.
b. Using these parameters, determine $\Delta H^\ddagger$ and $\Delta S^\ddagger$ as described by the Eyring equation.

P35.48 The gas-phase decomposition of ethyl bromide is a first-order reaction, occurring with a rate constant that demonstrates the following dependence on temperature:

Trial Number	Temperature (K)	k (s^{-1})
1	800.	0.0360
2	900.	1.410

a. Determine the Arrhenius parameters for this reaction.
b. Using these parameters, determine $\Delta H^\ddagger$ and $\Delta S^\ddagger$ as described by the Eyring equation.

P35.49 Hydrogen abstraction from hydrocarbons by atomic chlorine is a mechanism for Cl • loss in the atmosphere. Consider the reaction of Cl • with ethane:

$$\text{C}_2\text{H}_6(g) + \text{Cl} \cdot (g) \rightarrow \text{C}_2\text{H}_5 \cdot (g) + \text{HCl}(g)$$

This reaction was studied in the laboratory, and the following data were obtained:

T (K)	k ($\times 10^{-10}$ M^{-2}s^{-1})
270	3.43
370	3.77
470	3.99
570	4.13
670	4.23

a. Determine the Arrhenius parameters for this reaction.
b. At the tropopause (the boundary between the troposphere and stratosphere located approximately 11 km above the surface of Earth), the temperature is roughly 220 K. What do you expect the rate constant to be at this temperature?
c. Using the Arrhenius parameters obtained in part (a), determine the Eyring parameters $\Delta H^\ddagger$ and $\Delta S^\ddagger$ for this reaction at 220 K.

P35.50 Consider the "unimolecular" isomerization of methylcyanide, a reaction that will be discussed in detail in Chapter 36: $\text{CH}_3\text{NC}(g) \rightarrow \text{CH}_3\text{CN}(g)$. The Arrhenius parameters for this reaction are $A = 2.5 \times 10^{16} \text{ s}^{-1}$ and $E_a = 272 \text{ kJ mol}^{-1}$. Determine the Eyring parameters $\Delta H^\ddagger$ and $\Delta S^\ddagger$ for this reaction with $T = 300.$ K.

P35.51 Reactions involving hydroxyl radical (OH •) are extremely important in atmospheric chemistry. The reaction of hydroxyl radical with molecular hydrogen is as follows:

$$\text{OH} \cdot (g) + \text{H}_2(g) \rightarrow \text{H}_2\text{O}(g) + \text{H} \cdot (g)$$

Determine the Eyring parameters $\Delta H^\ddagger$ and $\Delta S^\ddagger$ for this reaction where $A = 8.0 \times 10^{13} \text{ M}^{-1}\text{ s}^{-1}$ and $E_a = 42 \text{ kJ mol}^{-1}$.

P35.52 Chlorine monoxide (ClO •) demonstrates three bimolecular self-reactions:

$$Rxn_1: \text{ClO} \cdot (g) + \text{ClO} \cdot (g) \xrightarrow{k_1} \text{Cl}_2(g) + \text{O}_2(g)$$
$$Rxn_2: \text{ClO} \cdot (g) + \text{ClO} \cdot (g) \xrightarrow{k_2} \text{Cl} \cdot (g) + \text{ClOO} \cdot (g)$$
$$Rxn_3: \text{ClO} \cdot (g) + \text{ClO} \cdot (g) \xrightarrow{k_3} \text{Cl} \cdot (g) + \text{OClO} \cdot (g)$$

The following table provides the Arrhenius parameters for this reaction:

	A (M^{-1} s^{-1})	E_a (kJ/mol)
Rxn_1	6.08×10^8	13.2
Rxn_2	1.79×10^{10}	20.4
Rxn_3	2.11×10^8	11.4

a. For which reaction is $\Delta H^\ddagger$ greatest and by how much relative to the next closest reaction?
b. For which reaction is $\Delta S^\ddagger$ the smallest and by how much relative to the next closest reaction?

Computational Problems

More detailed instructions on carrying out these calculations using Spartan Physical Chemistry are found on the book website at *www.masteringchemistry.com*.

C35.1 Chlorofluorocarbons are a source of atomic chlorine in the stratosphere. In this problem the energy needed to dissociate the C—Cl bond in CF_3Cl will be determined.

a. Perform a Hartree–Fock 3-21G calculation on the freon CF_3Cl and determine the minimum energy of this compound.

b. Select the $C-Cl$ bond and calculate the ground-state potential energy surface along this coordinate by determining the energy of the compound for the following $C-Cl$ bond lengths:

r_{CCl} (Å)	E (Hartree)	r_{CCl} (Å)	E (Hartree)
1.40		2.40	
1.50		2.60	
1.60		2.80	
1.70		3.00	
1.80		4.00	
2.00		5.00	
2.20		6.00	

c. Using the difference between the minimum of the potential energy surface to the energy at 6.00 Å, determine the barrier to dissociation.

d. Assuming an Arrhenius preexponential factor of $10^{-12} s^{-1}$, what is the expected rate constant for dissociation based on this calculation at 220 K? Is thermal dissociation of the $C-Cl$ bond occurring to an appreciable extent in the stratosphere?

C35.2 Consider the dissociation of the $C-F$ bond in $CFCl_3$. Using a standard bond dissociation energy of 485 kJ mol^{-1}, what would be the effect on the predicted rate constant for dissociation if zero-point energy along the $C-F$ stretch coordinate were ignored? Performing a Hartree–Fock 6-31G* calculation, determine the frequency of the vibrational mode dominated by $C-F$ stretch character. Calculate the dissociation rate using the Arrhenius expression without consideration of zero-point energy by adding the zero-point energy to the standard dissociation energy. Assume $A = 10^{10} s^{-1}$ and $T = 298$ K. Perform the corresponding calculation using the standard dissociation energy only. Does zero-point energy make a significant difference in the rate constant for this dissociation?

Web-Based Simulations, Animations, and Problems

W35.1 In this problem, concentration profiles as a function of rate constant are explored for the following sequential reaction scheme:

$$A \xrightarrow{k_a} B \xrightarrow{k_b} C$$

Students vary the rate constants k_a and k_b and explore the following behavior:

a. The variation in concentrations as the rate constants are varied.

b. Comparison of the maximum intermediate concentration time determined by simulation and through computation.

c. Visualization of the conditions under which the rate-limiting step approximation is valid.

W35.2 In this simulation, the kinetic behavior of the following parallel reaction is studied:

$$A \xrightarrow{k_b} B; \quad A \xrightarrow{k_c} C$$

The variation in concentrations as a function of k_b and k_c is studied. In addition, the product yields for the reaction determined based on the relative values of k_b and k_c are compared to the simulation result.

For Further Reading

Brooks, P. R. "Spectroscopy of Transition Region Species." *Chemical Reviews* 87 (1987): 167.

Callender, R. H., R. B. Dyer, R. Blimanshin, and W. H. Woodruff. "Fast Events in Protein Folding: The Time Evolution of a Primary Process." *Annual Review of Physical Chemistry* 49 (1998): 173.

Castellan, G. W. *Physical Chemistry*. Reading, MA: Addison-Wesley, 1983.

Eyring, H., S. H. Lin, and S. M. Lin. *Basic Chemical Kinetics*. New York: Wiley, 1980.

Frost, A. A., and R. G. Pearson. *Kinetics and Mechanism*. New York: Wiley, 1961.

Gagnon, E., P. Ranitovic, X.-M. Tong, C. L. Cocke, M. M. Murnane, H. C. Kapteyn, and A. S. Sandhu. "Soft X-Ray-Driven Femtosecond Molecular Dynamics." *Science* 317 (2007): 1374.

Hammes, G. G. *Thermodynamics and Kinetics for the Biological Sciences*. New York: Wiley, 2000.

Laidler, K. J. *Chemical Kinetics*. New York: Harper & Row, 1987.

Martin, J.-L., and M. H. Vos. "Femtosecond Biology." *Annual Review of Biophysical and Biomolecular Structure* 21 (1992): 1999.

Pannetier, G., and P. Souchay. *Chemical Kinetics*. Amsterdam: Elsevier, 1967.

Schoenlein, R. W., L. A. Peteanu, R. A. Mathies, and C. V. Shank. "The First Step in Vision: Femtosecond Isomerization of Rhodopsin." *Science* 254 (1991): 412.

Steinfeld, J. I., J. S. Francisco, and W. L. Hase. *Chemical Kinetics and Dynamics*. Prentice-Hall, Upper Saddle River, NJ: Prentice-Hall, 1999.

Truhlar, D. G., W. L. Hase, and J. T. Hynes. "Current Status in Transition State Theory." *Journal of Physical Chemistry* 87 (1983): 2642.

Vos, M. H., F. Rappaport, J.-C. Lambry, J. Breton, and J.-L. Martin. "Visualization of Coherent Nuclear Motion in a Membrane Protein by Femtosecond Spectroscopy." *Nature* 363 (1993): 320.

Zewail, H. "Laser Femtochemistry." *Science* 242 (1988): 1645.

CHAPTER

36

Complex Reaction Mechanisms

In this chapter, the chemical kinetic tools developed in the previous chapter are applied to complex reactions. Reaction mechanisms and their use in predicting reaction rate law expressions are explored. The preequilibrium approximation is presented and used in the evaluation of catalytic reactions including enzyme catalysis. In addition, homogeneous and heterogeneous catalytic processes are described. Reactions involving radicals, including polymerization and radical-initiated explosions, are discussed. The chapter concludes with an introduction to photochemistry. The unifying theme behind these apparently different topics is that all of the reaction mechanisms for these phenomena can be developed using the techniques of elementary chemical kinetics. Seemingly complex reactions can be decomposed into a series of well-defined kinetic steps, thereby providing substantial insight into the underlying chemical reaction dynamics.

36.1 Reaction Mechanisms and Rate Laws

Reaction mechanisms are defined as the collection of individual kinetic processes or steps involved in the transformation of reactants into products. The rate law expression for a chemical reaction, including the order of the reaction, is entirely dependent on the reaction mechanism. For a reaction mechanism to be valid, the rate-law expression predicted by the mechanism must agree with experiment. Consider the following reaction:

$$2\,N_2O_5(g) \rightarrow 4\,NO_2(g) + O_2(g) \tag{36.1}$$

For the remainder of this section the phase of the species is suppressed for clarity. One possible mechanism for this reaction is a single step consisting of a bimolecular collision between two N_2O_5 molecules. A reaction mechanism that consists of a single elementary step is known as a **simple reaction.** The rate law predicted by this mechanism is second order with respect to N_2O_5. However, the experimentally determined rate law for this reaction is first order in N_2O_5, not second order. Therefore, the single-step mechanism cannot be correct. To explain the observed order dependence of the reaction rate, the following mechanism was proposed:

$$2\left\{N_2O_5 \underset{k_{-1}}{\overset{k_1}{\rightleftharpoons}} NO_2 + NO_3\right\} \tag{36.2}$$

$$NO_2 + NO_3 \xrightarrow{k_2} NO_2 + O_2 + NO \tag{36.3}$$

$$NO + NO_3 \xrightarrow{k_3} 2\,NO_2 \tag{36.4}$$

This mechanism is an example of a **complex reaction,** defined as a reaction that occurs in two or more elementary steps. In this mechanism, the first step, Equation (36.2), represents an equilibrium between N_2O_5 to NO_2 and NO_3. In the second step, Equation (36.3), the bimolecular reaction of NO_2 and NO_3 results in the dissociation of NO_3 to product NO and O_2. In the final step of the reaction, Equation (36.4), NO and NO_3 undergo a bimolecular reaction to produce $2\,NO_2$.

In addition to the reactant (N_2O_5) and overall reaction products (NO_2 and O_2), two other species appear in the mechanism (NO and NO_3) that are not in the overall reaction of Equation (36.1). These species are referred to as **reaction intermediates.** Reaction intermediates that are formed in one step of the mechanism must be consumed in a subsequent step. Given this requirement, step 1 of the reaction must occur twice in order to balance the NO_3 that appears in steps 2 and 3. Therefore, we have multiplied this reaction by 2 in Equation (36.2) to emphasize that it must occur twice for every occurrence of steps 2 and 3. The number of times a given step occurs in a reaction mechanism is referred to as the **stoichiometric number.** In the mechanism under discussion, step 1 has a stoichiometric number of 2, whereas the other two steps have stoichiometric numbers of 1. With correct stoichiometric numbers, the sum of the elementary reaction steps will produce an overall reaction that is stoichiometrically equivalent to the reaction of interest.

For a reaction mechanism to be considered valid, the mechanism must be consistent with the experimentally determined rate law. Using the mechanism depicted by Equations (36.2) through (36.4), the rate of the reaction is

$$R = -\frac{1}{2}\frac{d[N_2O_5]}{dt} = \frac{1}{2}\left(k_1[N_2O_5] - k_{-1}[NO_2][NO_3]\right) \tag{36.5}$$

Notice that the stoichiometric number of the reaction is not included in the differential rate expression. Equation (36.5) corresponds to the loss of N_2O_5 by unimolecular decay and production by the bimolecular reaction of NO_2 with NO_3. As discussed earlier, NO and NO_3 are reaction intermediates. Writing the differential rate expression for these species and applying the steady-state approximation to the concentrations of both intermediates (Section 35.7.3) yields

$$\frac{d[NO]}{dt} = 0 = k_2[NO_2][NO_3] - k_3[NO][NO_3] \tag{36.6}$$

$$\frac{d[NO_3]}{dt} = 0 = k_1[N_2O_5] - k_{-1}[NO_2][NO_3] - k_2[NO_2][NO_3] - k_3[NO][NO_3] \tag{36.7}$$

Equation (36.6) can be rewritten to produce the following expression for [NO]:

$$[NO] = \frac{k_2[NO_2]}{k_3} \tag{36.8}$$

Substituting this result into Equation (36.7) yields

$$0 = k_1[N_2O_5] - k_{-1}[NO_2][NO_3] - k_2[NO_2][NO_3] - k_3\left(\frac{k_2[NO_2]}{k_3}\right)[NO_3]$$

$$0 = k_1[N_2O_5] - k_{-1}[NO_2][NO_3] - 2k_2[NO_2][NO_3]$$

$$\frac{k_1[N_2O_5]}{k_{-1} + 2k_2} = [NO_2][NO_3] \tag{36.9}$$

Substituting Equation (36.9) into Equation (36.5) results in the following predicted rate law expression for this mechanism:

$$R = \frac{1}{2}(k_1[N_2O_5] - k_{-1}[NO_2][NO_3])$$

$$= \frac{1}{2}\left(k_1[N_2O_5] - k_{-1}\left(\frac{k_1[N_2O_5]}{k_{-1} + 2k_2}\right)\right)$$

$$= \frac{k_1 k_2}{k_{-1} + 2k_2}[N_2O_5] = k_{eff}[N_2O_5] \tag{36.10}$$

In Equation (36.10), the collection of rate constants multiplying $[N_2O_5]$ has been renamed k_{eff}. Equation (36.10) demonstrates that the mechanism is consistent with the experimentally observed first-order dependence on $[N_2O_5]$. However, as discussed in Chapter 35, the consistency of a reaction mechanism with the experimental order dependence of the reaction is not proof that the mechanism is absolutely correct but instead demonstrates that the mechanism is consistent with the experimentally determined order dependence.

The example just presented illustrates how reaction mechanisms are used to explain the order dependence of the reaction rate. A theme that reoccurs throughout this chapter is the relation between reaction mechanisms and elementary reaction steps. As we will see, the mechanisms for many complex reactions can be decomposed into a series of elementary steps, and the techniques developed in the previous chapter can be readily employed in the evaluation of these complex kinetic problems.

36.2 The Preequilibrium Approximation

The preequilibrium approximation is a central concept employed in the evaluation of reaction mechanisms. This approximation is used when equilibrium among a subset of species is established before product formation occurs. In this section, the preequilibrium approximation is defined. This approximation will prove to be extremely useful in subsequent sections.

36.2.1 General Solution

Consider the following reaction:

$$A + B \underset{k_r}{\overset{k_f}{\rightleftharpoons}} I \overset{k_p}{\rightarrow} P \tag{36.11}$$

In Equation (36.11), forward and back rate constants link the reactants A and B with an intermediate species, I. Decay of I results in the formation of product, P. If the forward and backward reactions involving the reactants and intermediate are more rapid than the decay of the intermediate to form products, then the reaction of Equation (36.11) can be envisioned as occurring in two distinct steps:

1. First, equilibrium between the reactants and the intermediate is maintained during the course of the reaction.

2. The intermediate undergoes decay to form product.

This description of events is referred to as the **preequilibrium approximation.** The application of the preequilibrium approximation in evaluating reaction mechanisms containing equilibrium steps is performed as follows. The differential rate expression for the product is

$$\frac{d[P]}{dt} = k_p[I] \tag{36.12}$$

In Equation (36.12), [I] can be rewritten by recognizing that this species is in equilibrium with the reactants; therefore,

$$\frac{[I]}{[A][B]} = \frac{k_f}{k_r} = K_c \tag{36.13}$$

$$[I] = K_c[A][B] \tag{36.14}$$

In the preceding equations, K_c is the equilibrium constant expressed in terms of reactant and product concentrations. Substituting the definition of [I] provided by Equation (36.14) into Equation (36.12), the differential rate expression for the product becomes

$$\frac{d[P]}{dt} = k_p[I] = k_pK_c[A][B] = k_{eff}[A][B] \tag{36.15}$$

Equation (36.15) demonstrates that with the preequilibrium approximation, the predicted rate law is second order overall and first order with respect to both reactants (A and B). Finally, the rate constant for product formation is not simply k_p but is instead the product of this rate constant with the equilibrium constant, which is in turn equal to the ratio of the forward and backward rate constants.

36.2.2 A Preequilibrium Example

The reaction of NO and O_2 to form product NO_2 provides an example in which the preequilibrium approximation provides insight into the mechanism of NO_2 formation. The specific reaction of interest is

$$2\,NO(g) + O_2(g) \rightarrow 2\,NO_2(g) \tag{36.16}$$

One possible mechanism for this reaction is that of a single elementary step corresponding to a trimolecular reaction of two NO molecules and one O_2 molecule. The experimental rate law for this reaction is second order in NO and first order in O_2, consistent with this mechanism. However, this mechanism was further evaluated by measuring the temperature dependence of the reaction rate. If correct, raising the temperature will increase the number of collisions, and the reaction rate should increase. However, as the temperature is increased, a *reduction* in the reaction rate is observed, proving that the trimolecular-collisional mechanism is incorrect. An alternative mechanism for this reaction was proposed:

$$2\,NO \underset{k_r}{\overset{k_f}{\rightleftarrows}} N_2O_2 \tag{36.17}$$

$$N_2O_2 + O_2 \xrightarrow{k_p} 2\,NO_2 \tag{36.18}$$

In the first step of this mechanism, Equation (36.17), an equilibrium between NO and the dimer N_2O_2 is established rapidly compared to the rate of product formation. In the second step, Equation (36.18), a bimolecular reaction involving the dimer and O_2 results in the production of the NO_2 product. The stoichiometric number for each step is 1. To evaluate this mechanism, the preequilibrium approximation is applied to step 1 of the mechanism, and the concentration of N_2O_2 is expressed as

$$[N_2O_2] = \frac{k_f}{k_r}[NO]^2 = K_c[NO]^2 \tag{36.19}$$

Using the second step of the mechanism, the reaction rate is written as

$$R = \frac{1}{2}\frac{d[NO_2]}{dt} = k_p[N_2O_2][O_2] \tag{36.20}$$

Substitution of Equation (36.19) into the differential rate expression for [NO_2] yields

$$R = k_p[N_2O_2][O_2] = k_pK_c[NO]^2[O_2] = k_{eff}[NO]^2[O_2] \tag{36.21}$$

The rate law predicted by this mechanism is second order in NO and first order in O_2, consistent with experiment. Furthermore, the preequilibrium approximation provides an explanation for the temperature dependence of product formation. Specifically, the formation of N_2O_2 is an exothermic process such that an increase in temperature shifts the equilibrium between NO and N_2O_2 toward NO. As such, there is less N_2O_2 to react with O_2 such that the rate of NO_2 formation decreases with increased temperature.

36.3 The Lindemann Mechanism

The **Lindemann mechanism** for **unimolecular reactions** provides an elegant example of the relationship between kinetics and reaction mechanisms. This mechanism was developed to describe the observed concentration dependence in unimolecular dissociation reactions of the form

$$A \rightarrow fragments \tag{36.22}$$

In this reaction, a reactant molecule undergoes decomposition when the energy content of one or more vibrational modes is sufficient for decomposition to occur. The question is "how does the reactant acquire sufficient energy to undergo decomposition?" One possibility is that the reactant acquires sufficient energy to react through a bimolecular collision. Experimentally, however, the rate of decomposition demonstrates only first-order behavior at high reactant concentrations, and not second order as expected for a single-step bimolecular mechanism. Frederick Lindemann proposed another mechanism to explain the order dependence of the reaction with respect to reactant concentration.

The Lindemann mechanism involves two steps. First, reactants acquire sufficient energy to undergo reaction through a bimolecular collision:

$$A + A \xrightarrow{k_1} A^* + A \tag{36.23}$$

In this reaction, A^* is the "activated" reactant that has received sufficient energy to undergo decomposition. The collisional partner of the activated reactant molecule leaves the collision with insufficient energy to decompose. In the second step of the Lindemann mechanism, the **activated reactant** undergoes one of two reactions: collision resulting in deactivation or decomposition resulting in product formation:

$$A^* + A \xrightarrow{k_{-1}} A + A \tag{36.24}$$

$$A^* \xrightarrow{k_2} P \tag{36.25}$$

The separation of the reaction into two steps is the key conceptual contribution of the Lindemann mechanism. Specifically, the mechanism implies that a separation in timescale exists between activation and deactivation/product formation. Inspection of the mechanism described by Equations (36.24) and (36.25) demonstrates that the only process resulting in product formation is the final decomposition step of Equation (36.25); therefore, the rate of product production is written as

$$\frac{d[P]}{dt} = k_2[A^*] \tag{36.26}$$

Evaluation of Equation (36.26) requires an expression for $[A^*]$ in terms of reactant concentration, $[A]$. Because A^* is an intermediate species, the relationship between $[A^*]$ and $[A]$ is obtained by writing the differential rate expression for $[A^*]$ and applying the steady-state approximation:

$$\frac{d[A^*]}{dt} = k_1[A]^2 - k_{-1}[A][A^*] - k_2[A^*] = 0$$

$$[A^*] = \frac{k_1[A]^2}{(k_{-1}[A] + k_2)} \tag{36.27}$$

Substituting Equation (36.27) into Equation (36.26) results in the final differential rate expression for [P]:

$$\frac{d[P]}{dt} = \frac{k_1 k_2 [A]^2}{k_{-1}[A] + k_2}$$ (36.28)

Equation (36.28) is the central result of the Lindemann mechanism. It states that the observed order dependence on [A] depends on the relative magnitude of $k_{-1}[A]$ versus k_2. At high reactant concentrations, $k_{-1}[A] > k_2$ and Equation (36.28) reduces to

$$\frac{d[P]}{dt} = \frac{k_1 k_2}{k_{-1}}[A]$$ (36.29)

Equation (36.29) demonstrates that at high reactant concentrations or pressures (recall that $P_A/RT = n_A/V = [A]$) the rate of product formation will be first order in [A], consistent with experiment. Mechanistically, at high pressures activated molecules will be produced faster than decomposition occurs such that the rate of decomposition is the rate-limiting step in product formation. At low reactant concentrations $k_2 > k_{-1}[A]$ and Equation (36.28) becomes

$$\frac{d[P]}{dt} = k_1 [A]^2$$ (36.30)

Equation (36.30) demonstrates that at low pressures the formation of activated complex becomes the rate-limiting step in the reaction and the rate of product formation is second order in [A].

The Lindemann mechanism can be generalized to describe a variety of unimolecular reactions through the following generic scheme:

$$A + M \underset{k_{-1}}{\overset{k_1}{\rightleftharpoons}} A^* + M$$ (36.31)

$$A^* \xrightarrow{k_2} P$$ (36.32)

In this mechanism, M is a collisional partner that can be the reactant itself (A) or some other species such as a nonreactive buffer gas added to the reaction. The rate of product formation can be written as follows:

$$\frac{d[P]}{dt} = \frac{k_1 k_2 [A][M]}{k_{-1}[M] + k_2} = k_{uni}[A]$$ (36.33)

In Equation (36.33), k_{uni} is the apparent rate constant for the reaction defined as

$$k_{uni} = \frac{k_1 k_2 [M]}{k_{-1}[M] + k_2}$$ (36.34)

In the limit of high M concentrations, $k_{-1}[M] \gg k_2$ and $k_{uni} = k_1 k_2 / k_{-1}$, resulting in an apparent rate constant that is independent of [M]. As [M] decreases, k_{uni} will decrease until $k_2 > k_{-1}[M]$, at which point $k_{uni} = k_1[M]$ and the apparent rate demonstrates first-order dependence on M. Figure 36.1 presents a plot of the observed rate constant for the isomerization of methyl isocyanide versus pressure measured at 230.4°C by Schneider and Rabinovitch. The figure demonstrates that the predicted linear relationship between k_{uni} and pressure at low pressure, which is consistent with the corresponding limiting behavior of Equation (36.34), is observed for this reaction. In addition, at high pressure k_{uni} reaches a constant value, which is also consistent with the limiting behavior of Equation (36.34).

The Lindemann mechanism provides a detailed prediction of how the rate constant for a unimolecular reaction will vary with pressure or concentration. Inverting Equation (36.34), the relationship between k_{uni} and reactant concentration becomes

$$\frac{1}{k_{uni}} = \frac{k_{-1}}{k_1 k_2} + \left(\frac{1}{k_1}\right)\frac{1}{[M]}$$ (36.35)

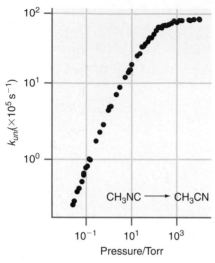

FIGURE 36.1

Pressure dependence of the observed rate constant for the unimolecular isomerization of methyl isocyanide.
[Data from Schneider and Rabinovitch, "Thermal Unimolecular Isomerization of Methyl Isocyanide - Fall-Off Behavior," *J. American Chemical Society* 84 (1962): 4225.]

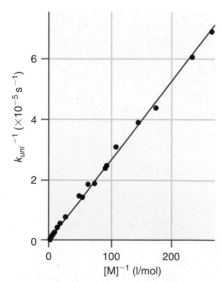

FIGURE 36.2

Plot of k_{uni}^{-1} versus $[M]^{-1}$ for the unimolecular isomerization of methyl isocyanide at 230.4°C. The solid line is the best fit to the data.

Equation (36.35) predicts that a plot of k_{uni}^{-1} versus $[M]^{-1}$ should yield a straight line with slope $1/k_1$ and y intercept of $k_{-1}/k_1 k_2$. A plot of the data presented in Figure 36.1 employing Equation (36.35) is presented in Figure 36.2. The figure demonstrates that the expected linear relationship between k_{uni}^{-1} and $[M]^{-1}$ that is observed for this reaction is consistent with the Lindemann mechanism. The solid line in the figure is the best fit to the data by a straight line. The slope of this line provides a value for k_1 of $4.16 \times 10^6 \text{ M}^{-1} \text{ s}^{-1}$, and the y intercept in combination with the value for k_1 dictates that $k_{-1}/k_2 = 1.76 \times 10^5 \text{ M}^{-1}$.

36.4 Catalysis

A **catalyst** is a substance that participates in chemical reactions by increasing the reaction rate, yet the catalyst itself remains intact after the reaction is complete. The general function of a catalyst is to provide an additional mechanism by which reactants are converted to products. The presence of a new reaction mechanism involving the catalyst results in a second reaction coordinate that connects reactants and products. The activation energy along this second reaction coordinate will be lower in comparison to the uncatalyzed reaction; therefore, the overall reaction rate will increase. For example, consider Figure 36.3 in which a reaction involving the conversion of reactant A to product B with and without a catalyst is depicted. In the absence of a catalyst, the rate of product formation is given by rate $= r_0$. In the presence of the catalyst, a second pathway is created, and the reaction rate is now the sum of the original rate plus the rate for the catalyzed reaction, or $r_0 + r_c$.

An analogy for a catalyzed reaction is found in the electrical circuits depicted in Figure 36.3. In the "catalyzed" electrical circuit, a second, parallel pathway for current flow has been added, allowing for increased total current when compared to the "uncatalyzed" circuit. By analogy, the addition of the second, parallel pathway is equivalent to the alternative reaction mechanism involving the catalyst.

To be effective, a catalyst must combine with one or more of the reactants or with an intermediate species involved in the reaction. After the reaction has taken place, the catalyst is freed and can combine with another reactant or intermediate in a subsequent reaction. The catalyst is not consumed during the reaction, so a small amount of catalyst can participate in numerous reactions. The simplest mechanism describing a catalytic process is as follows:

$$S + C \underset{k_{-1}}{\overset{k_1}{\rightleftharpoons}} SC \tag{36.36}$$

$$SC \overset{k_2}{\longrightarrow} P + C \tag{36.37}$$

where S represents the reactant or substrate, C is the catalyst, and P is the product. The **substrate–catalyst complex** is represented by SC and is an intermediate species in this mechanism. The differential rate expression for product formation is

$$\frac{d[P]}{dt} = k_2[SC] \tag{36.38}$$

Given that SC is an intermediate, we write the differential rate expression for this species and apply the steady-state approximation:

$$\frac{d[SC]}{dt} = k_1[S][C] - k_{-1}[SC] - k_2[SC] = 0$$

$$[SC] = \frac{k_1[S][C]}{k_{-1} + k_2} = \frac{[S][C]}{K_m} \tag{36.39}$$

In Equation (36.39), K_m is referred to as the **composite constant** and is defined as

$$K_m = \frac{k_{-1} + k_2}{k_1} \tag{36.40}$$

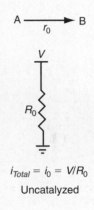

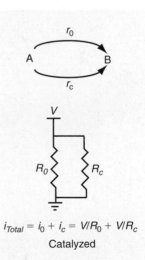

FIGURE 36.3
Illustration of catalysis. In the uncatalyzed reaction, the rate of reaction is given by r_0. In the catalyzed case, a new pathway is created by the presence of the catalyst with corresponding rate r_c. The total rate of reaction for the catalyzed case is $r_0 + r_c$. The analogous electrical circuits are also presented for comparison.

Substituting the expression for [SC] into Equation (36.38), the rate of product formation becomes

$$\frac{d[\mathrm{P}]}{dt} = \frac{k_2[\mathrm{S}][\mathrm{C}]}{K_m} \tag{36.41}$$

Equation (36.41) illustrates that the rate of product formation is expected to increase linearly with both substrate and catalyst concentrations. This equation is difficult to evaluate over the entire course of the reaction because the concentrations of substrate and catalyst given in Equation (36.41) correspond to species not in the SC complex, and these concentrations can be quite difficult to measure. A more convenient measurement is to determine how much substrate and catalyst are present at the beginning of the reaction. Conservation of mass dictates the following relationship between these initial concentrations and the concentrations of all species present after the reaction is initiated:

$$[\mathrm{S}]_0 = [\mathrm{S}] + [\mathrm{SC}] + [\mathrm{P}] \tag{36.42}$$

$$[\mathrm{C}]_0 = [\mathrm{C}] + [\mathrm{SC}] \tag{36.43}$$

Rearrangement of Equations (36.42) and (36.43) yields the following definitions for [S] and [C]:

$$[\mathrm{S}] = [\mathrm{S}]_0 - [\mathrm{SC}] - [\mathrm{P}] \tag{36.44}$$

$$[\mathrm{C}] = [\mathrm{C}]_0 - [\mathrm{SC}] \tag{36.45}$$

Substituting these expressions into Equation (36.39) yields

$$K_m[\mathrm{SC}] = [\mathrm{S}][\mathrm{C}] = ([\mathrm{S}]_0 - [\mathrm{SC}] - [\mathrm{P}])\,([\mathrm{C}]_0 - [\mathrm{SC}])$$

$$0 = [[\mathrm{C}]_0([\mathrm{S}]_0 - [\mathrm{P}])] - [\mathrm{SC}]([\mathrm{S}]_0 + [\mathrm{C}]_0 - [\mathrm{P}] + K_m) + [\mathrm{SC}]^2 \tag{36.46}$$

Equation (36.46) can be evaluated as a quadratic equation to determine [SC]. However, two assumptions are generally employed at this point to simplify matters. First, through control of the initial substrate and catalyst concentrations, conditions can be employed such that [SC] is small. Therefore, the $[\mathrm{SC}]^2$ term in Equation (36.46) can be neglected. Second, we confine ourselves to early stages of the reaction when little product has been formed; therefore, terms involving [P] can also be neglected. With these two approximations, Equation (36.46) is readily evaluated, providing the following expression for [SC]:

$$[\mathrm{SC}] = \frac{[\mathrm{S}]_0[\mathrm{C}]_0}{[\mathrm{S}]_0 + [\mathrm{C}]_0 + K_m} \tag{36.47}$$

Substituting Equation (36.47) into Equation (36.38), the rate of the reaction becomes

$$R_0 = \frac{d[\mathrm{P}]}{dt} = \frac{k_2[\mathrm{S}]_0[\mathrm{C}]_0}{[\mathrm{S}]_0 + [\mathrm{C}]_0 + K_m} \tag{36.48}$$

In Equation (36.48), the subscript on the rate indicates that this expression applies to the early-time or initial reaction rate. We next consider two limiting cases of Equation (36.48).

36.4.1 Case 1: $[\mathrm{C}]_0 \ll [\mathrm{S}]_0$

The most common case in catalysis is when there is much more substrate present in comparison to catalyst. In this limit $[\mathrm{C}]_0$ can be neglected in the denominator of Equation (36.48) and the rate becomes

$$R_0 = \frac{k_2[\mathrm{S}]_0[\mathrm{C}]_0}{[\mathrm{S}]_0 + K_m} \tag{36.49}$$

For substrate concentrations where $[S]_0 < K_m$, the reaction rate should increase linearly with substrate concentration, with a slope equal to $k_2[C]_0/K_m$. Parameters such as k_2 and K_m can be obtained by comparing experimental reaction rates to Equation (36.49). An alternative approach to determining these parameters is to invert Equation (36.49) to obtain the following relationship between the reaction rate and initial substrate concentration:

$$\frac{1}{R_0} = \left(\frac{K_m}{k_2[C]_0}\right)\frac{1}{[S]_0} + \frac{1}{k_2[C]_0} \tag{36.50}$$

Equation (36.50) demonstrates that a plot of the inverse of the initial reaction rate versus $[S]_0^{-1}$, referred to as a **reciprocal plot,** should yield a straight line. The y intercept and slope of this line provide a measure of K_m and k_2, assuming $[C]_0$ is known.

At elevated concentrations of substrate where $[S]_0 \gg K_m$, the denominator in Equation (36.49) can be approximated as $[S]_0$, resulting in the following expression for the reaction rate:

$$R_0 = k_2[C]_0 = R_{max} \tag{36.51}$$

In other words, the rate of reaction will reach a limiting value where the rate becomes zero order in substrate concentration. In this limit, the reaction rate can only be enhanced by increasing the amount of catalyst. An illustration of the variation in the reaction rate with initial substrate concentration predicted by Equations (36.49) and (36.50) is provided in Figure 36.4.

36.4.2 Case 2: $[C]_0 \gg [S]_0$

In this limit Equation (36.48) becomes

$$R_0 = \frac{k_2[S]_0[C]_0}{[C]_0 + K_m} \tag{36.52}$$

In this concentration limit, the reaction rate is first order in $[S]_0$ but can be first or zero order in $[C]_0$ depending on the magnitude of $[C]_0$ relative to K_m. In catalysis studies, this limit is generally avoided because the insight to be gained regarding the rate constants for the various reaction steps are more easily evaluated for the previously discussed Case 1. In addition, good catalysts can be expensive; therefore, employing excess catalyst in a reaction is not cost effective.

36.4.3 Michaelis–Menten Enzyme Kinetics

Enzymes are protein molecules that serve as catalysts in a wide variety of chemical reactions. Enzymes are noted for their reaction specificity, with nature having developed specific catalysts to facilitate the vast majority of biological reactions required for organism survival. An illustration enzyme with associated substrate is presented in Figure 36.5. The figure presents a space-filling model derived from a crystal structure of phospholipase A_2 (white) containing a bound substrate analogue (red). This enzyme catalyzes the hydrolysis of esters in phospholipids. The substrate analogue contains a stable phosphonate group in place of the enzyme-susceptible ester. The substrate analogue is resistant to enzymatic hydrolysis so that it does not suffer chemical breakdown during the structure determination process. With reactive substrate, ester hydrolysis occurs and the products of the reaction are released from the enzyme, resulting in regeneration of the free enzyme.

The kinetic mechanism of phospholipase A_2 catalysis can be described using the **Michaelis–Menten mechanism** of enzyme activity illustrated in Figure 36.6. The figure depicts the "lock-and-key" model for enzyme reactivity in which the substrate is bound to the active site of the enzyme where the reaction is catalyzed. The enzyme and substrate form the enzyme–substrate complex, which dissociates into product and uncomplexed enzyme. The interactions involved in creation of the enzyme–substrate

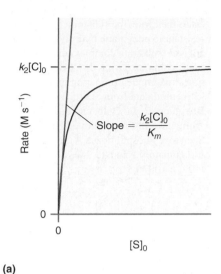

(a)

(b)

FIGURE 36.4
Illustration of the variation in the reaction rate with substrate concentration under Case 1 conditions as described in the text. **(a)** Plot of the initial reaction rate with respect to substrate concentration [Equation (36.49)]. At low substrate concentrations, the reaction rate increases linearly with substrate concentration. At high substrate concentrations, a maximum reaction rate of $k_2[C]_0$ is reached. **(b)** Reciprocal plot where the inverse of the reaction rate is plotted with respect to the inverse of substrate concentration [Equation (36.50)]. The y intercept of this line is equal to the inverse of the maximum reaction rate, or $(k_2[C]_0)^{-1}$. The slope of the line is equal to $K_m(k_2[C]_0)^{-1}$; therefore, with the slope and y intercept, K_m can be determined.

FIGURE 36.5
Space-filling model of the enzyme phospholipase A_2 (white) containing a bound substrate analogue (red). The substrate analogue contains a stable phosphonate group in place of the enzyme-susceptible ester; therefore, the substrate analogue is resistant to enzymatic hydrolysis and the enzyme–substrate complex remains stable in the complex during the X-ray diffraction structure determination process. [Structural data from Scott, White, Browning, Rosa, Gelb, and Sigler. "Structures of Free Inhibited Human Secretory Phospholipase A_2 from Inflammatory Exudate." *Science* 5034 (1991): 1007.]

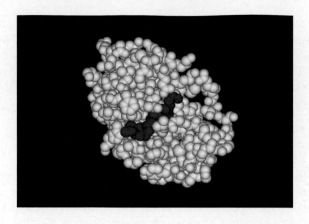

complex are enzyme specific. For example, the active site may bind the substrate in more than one location, thereby creating geometric strain that promotes product formation. The enzyme may orient the substrate so that the reaction geometry is optimized. In summary, the details of enzyme-mediated chemistry are highly dependent on the reaction of interest. Rather than an exhaustive presentation of enzyme kinetics, our motivation here is to describe enzyme kinetics within the general framework of catalyzed reactions.

A schematic description of the mechanism illustrated in Figure 36.6 is as follows:

$$E + S \underset{k_{-1}}{\overset{k_1}{\rightleftharpoons}} ES \overset{k_2}{\longrightarrow} E + P \tag{36.53}$$

In this mechanism, E is enzyme, S is substrate, ES is the complex, and P is product. Comparison of the mechanism of Equation (36.53) to the general catalytic mechanism described earlier in Equations (36.36) and (36.37) demonstrates that this mechanism is identical to the general catalysis mechanism except that the catalyst C is now the enzyme E. In the limit where the initial substrate concentration is substantially greater than that of the enzyme ($[S]_0 \gg [E]_0$ or Case 1 conditions as described previously), the rate of product formation is given by

$$R_0 = \frac{k_2[S]_0[E]_0}{[S]_0 + K_m} \tag{36.54}$$

In enzyme kinetics the composite constant K_m in Equation (36.54) is referred to as the **Michaelis constant** in enzyme kinetics, and Equation (36.54) is referred to as the **Michaelis–Menten rate law**. When $[S]_0 \gg K_m$, the Michaelis constant can be neglected, resulting in the following expression for the rate:

$$R_0 = k_2[E]_0 = R_{max} \tag{36.55}$$

Equation (36.55) demonstrates that the rate of product formation will plateau at some maximum value equal to the product of initial enzyme concentration and k_2, the rate constant for product formation, consistent with the behavior depicted in Figure 36.4. A reciprocal plot of the reaction rate can also be constructed by inverting Equation (36.54), which results in the **Lineweaver–Burk equation**:

$$\frac{1}{R_0} = \frac{1}{R_{max}} + \frac{K_m}{R_{max}} \frac{1}{[S]_0} \tag{36.56}$$

For the Michaelis–Menten mechanism to be consistent with experiment, a plot of the inverse of the initial rate with respect to $[S]_0^{-1}$ should yield a straight line from which the y intercept and slope can be used to determine the maximum reaction rate and the Michaelis constant. This reciprocal plot is referred to as the **Lineweaver–Burk plot.** In addition, because $[E]_0$ is readily determined experimentally, the maximum rate can be used to determine k_2, referred to as the **turnover number** of the enzyme [Equation (36.55)]. The turnover number can be thought of as the maximum number of

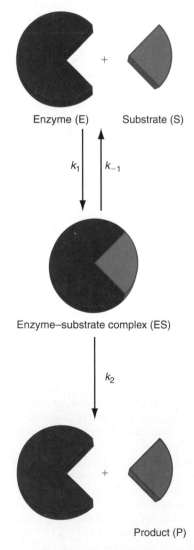

Enzyme (E)   Substrate (S)

k_1 k_{-1}

Enzyme–substrate complex (ES)

k_2

Product (P)

FIGURE 36.6
Schematic of enzyme catalysis.

substrate molecules per unit time that can be converted into product. Most enzymes demonstrate turnover numbers between 1 and 10^5 s^{-1} under physiological conditions.

EXAMPLE PROBLEM 36.1

DeVoe and Kistiakowsky [*J. American Chemical Society* 83 (1961): 274] studied the kinetics of CO_2 hydration catalyzed by the enzyme carbonic anhydrase:

$$CO_2 + H_2O \rightleftharpoons HCO_3^- + H^+$$

In this reaction, CO_2 is converted to bicarbonate ion. Bicarbonate is transported in the bloodstream and converted back to CO_2 in the lungs, a reaction that is also catalyzed by carbonic anhydrase. The following initial reaction rates for the hydration reaction were obtained for an initial enzyme concentration of 2.3 nM and temperature of 0.5°C:

Rate (M s^{-1})	[CO_2] (mM)
2.78×10^{-5}	1.25
5.00×10^{-5}	2.5
8.33×10^{-5}	5.0
1.67×10^{-4}	20.0

Determine K_m and k_2 for the enzyme at this temperature.

Solution

The Lineweaver–Burk plot of the rate^{-1} versus $[CO_2]^{-1}$ is shown here:

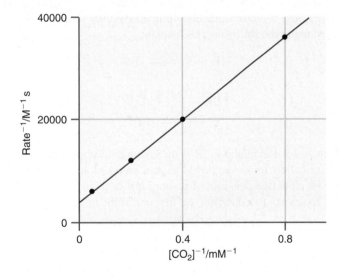

The y intercept for the best fit line to the data is 4000 M^{-1} s corresponding to $R_{max} = 2.5 \times 10^{-4}$ M s^{-1}. Using this value and $[E]_0 = 2.3$ nM, k_2 is

$$k_2 = \frac{R_{max}}{[E]_0} = \frac{2.5 \times 10^{-4} \text{ M s}^{-1}}{2.3 \times 10^{-9} \text{ M}} = 1.1 \times 10^5 \text{ s}^{-1}$$

Notice that the units of k_2, the turnover number, are consistent with a first-order process, in agreement with the Michaelis–Menten mechanism. The slope of the best fit line is 40 s such that, per Equation (36.56), K_m is given by

$$K_m = slope \times R_{max} = (40 \text{ s})(2.5 \times 10^{-4} \text{ M s}^{-1})$$

$$= 10 \text{ mM}$$

In addition to the Lineweaver–Burk plot, K_m can be estimated if the maximum rate is known. Specifically, if the initial rate is equal to one-half the maximum rate, Equation (36.54) reduces to

$$R_0 = \frac{k_2[S]_0[E]_0}{[S]_0 + K_m} = \frac{R_{max}[S]_0}{[S]_0 + K_m}$$

$$\frac{R_{max}}{2} = \frac{R_{max}[S]_0}{[S]_0 + K_m}$$

$$[S]_0 + K_m = 2[S]_0$$

$$K_m = [S]_0 \qquad (36.57)$$

Equation (36.57) demonstrates that when the initial rate is half the maximum rate, K_m is equal to the initial substrate concentration. Therefore, K_m can be determined by viewing a substrate saturation curve, as illustrated in Figure 36.7 for the carbonic-anhydrase catalyzed hydration of CO_2 discussed in Example Problem 36.1. The figure demonstrates that the initial rate is equal to half the maximum rate when $[S]_0 = 10$ mM. Therefore, the value of K_m determined in this relatively simple approach is in excellent agreement with that determined from the Lineweaver–Burk plot. Notice in Figure 36.7 that the maximum rate depicted was that employed using the Lineweaver–Burk analysis as shown in the example problem. When employing this method to determine K_m, the high-substrate-concentration limit must be carefully explored to ensure that the reaction rate is indeed at a maximum.

36.4.4 Competitive Inhibition in Enzyme Catalysis

The activity of an enzyme can be affected by the introduction of species that structurally resemble the substrate and that can occupy the enzyme active site; however, once bound to the active site the molecules are nonreactive. Such molecules are referred to as **competitive inhibitors.** The phosphonated substrate bound to phospholipase A_2 in Figure 36.5 is an example of a competitive inhibitor. Competitive inhibition can be described using the following mechanism:

$$E + S \underset{k_{-1}}{\overset{k_1}{\rightleftharpoons}} ES \qquad (36.58)$$

$$ES \xrightarrow{k_2} E + P \qquad (36.59)$$

$$E + I \underset{k_{-3}}{\overset{k_3}{\rightleftharpoons}} EI \qquad (36.60)$$

In this mechanism, I is the inhibitor, EI is the enzyme–inhibitor complex, and the other species are identical to those employed in the standard enzyme kinetic scheme of Equation (36.53). How does the rate of reaction differ from the noninhibited case discussed earlier? To answer this question, we first define the initial enzyme concentration:

$$[E]_0 = [E] + [EI] + [ES] \qquad (36.61)$$

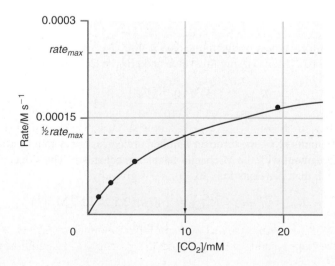

FIGURE 36.7

Determination of K_m for the carbonic-anhydrase catalyzed hydration of CO_2. The substrate concentration at which the rate of reaction is equal to half that of the maximum rate is equal to K_m.

Next, assuming that k_1, k_{-1}, k_3, and $k_{-3} \gg k_2$ the preequilibrium approximation is applied using Equations (36.58) and (36.60), yielding

$$K_s = \frac{[\text{E}][\text{S}]}{[\text{ES}]} \approx K_m \qquad (36.62)$$

$$K_i = \frac{[\text{E}][\text{I}]}{[\text{EI}]} \qquad (36.63)$$

In Equation (36.62), the constant describing the enzyme and substrate is written as K_m (Equation 36.40) when $k_{-1} \gg k_2$. With these relationships, Equation (36.61) can be written as

$$
\begin{aligned}
[\text{E}]_0 &= \frac{K_m[\text{ES}]}{[\text{S}]} + \frac{[\text{E}][\text{I}]}{K_i} + [\text{ES}] \\
&= \frac{K_m[\text{ES}]}{[\text{S}]} + \left(\frac{K_m[\text{ES}]}{[\text{S}]}\right)\frac{[\text{I}]}{K_i} + [\text{ES}] \\
&= [\text{ES}]\left(\frac{K_m}{[\text{S}]} + \frac{K_m[\text{I}]}{[\text{S}]K_i} + 1\right)
\end{aligned} \qquad (36.64)
$$

Solving Equation (36.64) for [ES] yields

$$[\text{ES}] = \frac{[\text{E}]_0}{1 + \dfrac{K_m}{[\text{S}]} + \dfrac{K_m[\text{I}]}{[\text{S}]K_i}} \qquad (36.65)$$

Finally, the rate of product formation is given by

$$
R = \frac{d[\text{P}]}{dt} = k_2[\text{ES}] = \frac{k_2[\text{E}]_0}{1 + \dfrac{K_m}{[\text{S}]} + \dfrac{K_m[\text{I}]}{[\text{S}]K_i}} = \frac{k_2[\text{S}][\text{E}]_0}{[\text{S}] + K_m\left(1 + \dfrac{[\text{I}]}{K_i}\right)}
$$

$$R \cong \frac{k_2[\text{S}]_0[\text{E}]_0}{[\text{S}]_0 + K_m\left(1 + \dfrac{[\text{I}]}{K_i}\right)} \qquad (36.66)$$

In Equation (36.66), the assumption that [ES] and [P] $\ll$ [S] has been employed so that [S] $\cong$ $[\text{S}]_0$, consistent with the previous treatment of uninhibited catalysis. Comparison of Equation (36.66) to the corresponding expression for the uninhibited case of Equation (36.54) illustrates that with competitive inhibition, a new apparent Michaelis constant can be defined:

$$K_m^* = K_m\left(1 + \frac{[\text{I}]}{K_i}\right) \qquad (36.67)$$

Notice that K_m^* reduces to K_m in the absence of inhibitor ([I] = 0). Next, using the definition of maximum reaction rate defined earlier in Equation (36.55), the reaction rate in the case of competitive inhibition can be written as

$$R_0 \cong \frac{R_{max}[\text{S}]_0}{[\text{S}]_0 + K_m^*} \qquad (36.68)$$

In the presence of inhibitor, $K_m^* \geq K_m$, and more substrate is required to reach half the maximum rate in comparison to the uninhibited case. The effect of inhibition can also be observed in a Lineweaver–Burk plot of the following form:

$$\frac{1}{R_0} = \frac{1}{R_{max}} + \frac{K_m^*}{R_{max}}\frac{1}{[\text{S}]_0} \qquad (36.69)$$

Because $K_m^* > K_m$, the slope of the Lineweaver–Burk plot will be greater with inhibitor compared to the slope without inhibitor. Figure 36.8 presents an illustration of this effect.

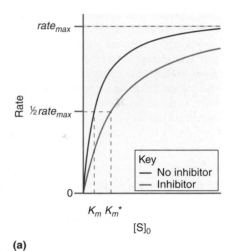

(a)

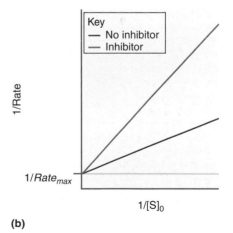

(b)

FIGURE 36.8

Comparison of enzymatic reaction rates in the presence and absence of a competitive inhibitor. **(a)** Plot of rate versus initial substrate concentration. The location of K_m and K_m^* is indicated. **(b)** Reciprocal plots ($1/R$ versus $1/[\text{S}]_0$). Notice that $1/R_{max}$ is identical in the presence and absence of a competitive inhibitor.

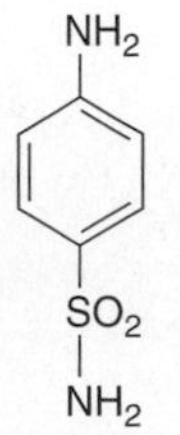

NH$_2$

SO$_2$

NH$_2$

Sulfanilamide

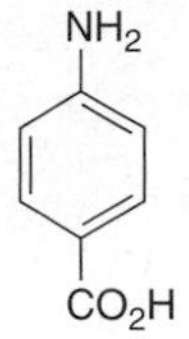

NH$_2$

CO$_2$H

p-Aminobenzoic acid

FIGURE 36.9
Structural comparison of the antibacterial drug sulfanilamide, a competitive inhibitor of the enzyme dihydropteroate synthetase, and the active substrate, p-aminobenzoic acid. The change in functional group from —CO$_2$H to —SO$_2$NH$_2$ is such that sulfanilamide cannot be used by bacteria to synthesize folate, and the bacterium starves.

Competitive inhibition has been used in drug design for antiviral, antibacterial, and antitumor applications. Many drugs are molecules that serve as competitive inhibitors for enzymes required for viral, bacterial, or cellular replication. For example sulfanilamide (Figure 36.9) is a powerful antibacterial drug. This compound is similar to p-aminobenzoic acid, the substrate for the enzyme dihydropteroate synthetase that participates in the production of folate. When present, the enzyme in bacteria cannot produce folate, and the bacteria die. However, humans do not possess this enzyme; they obtain folate from other sources. Therefore, sulfanilamide is not toxic.

36.4.5 Homogeneous and Heterogeneous Catalysis

A **homogeneous catalyst** is a catalyst that exists in the same phase as the species involved in the reaction, and a **heterogeneous catalyst** exists in a different phase. Enzymes serve as an example of a homogeneous catalyst; they exist in solution and catalyze reactions that occur in solution. A famous example of gas-phase catalysis is the catalytic depletion of stratospheric ozone by atomic chlorine. In the mid-1970s, F. Sherwood Rowland and Mario Molina proposed that Cl atoms catalyze the decomposition of stratospheric ozone by the following mechanism:

$$Cl + O_3 \xrightarrow{k_1} ClO + O_2 \tag{36.70}$$

$$\underline{ClO + O \xrightarrow{k_2} Cl + O_2} \tag{36.71}$$

$$O_3 + O \rightarrow 2\,O_2 \tag{36.72}$$

In this mechanism, Cl reacts with ozone to produce chlorine monoxide (ClO) and molecular oxygen. The ClO undergoes a second reaction with atomic oxygen, largely formed by O$_3$ photolysis, resulting in the reformation of Cl and the product of O$_2$. The sum of these reactions leads to the net conversion of O$_3$ and O to 2 O$_2$. Notice that the Cl is not consumed in the net reaction.

The catalytic efficiency of Cl can be determined using standard techniques in kinetics. The experimentally determined rate law expression for the uncatalyzed reaction of Equation (36.72) is

$$R_{nc} = k_{nc}[O][O_3] \tag{36.73}$$

The stratospheric temperature where this reaction occurs is roughly 220 K, at which temperature k_{nc} has a value of $3.30 \times 10^5 \text{ M}^{-1}\text{ s}^{-1}$. For the Cl catalyzed decomposition of ozone, the rate constants at this temperature are $k_1 = 1.56 \times 10^{10} \text{ M}^{-1}\text{ s}^{-1}$ and $k_2 = 2.44 \times 10^{10} \text{ M}^{-1}\text{ s}^{-1}$. To employ these rates in determining the overall rate of reaction, the rate law expression for the catalytic mechanism must be determined. Notice that both Cl and ClO are intermediates in this mechanism. Applying the steady-state approximation, the concentration of intermediates is taken to be a constant such that

$$[Cl]_{total} = [Cl] + [ClO] \tag{36.74}$$

where $[Cl]_{total}$ is defined as the sum of reaction intermediate concentrations, a definition that will prove useful in deriving the rate law. In addition, the steady-state approximation is applied in evaluating the differential rate expression for [Cl] as follows:

$$\frac{d[Cl]}{dt} = 0 = -k_1[Cl][O_3] + k_2[ClO][O]$$

$$k_1[Cl][O_3] = k_2[ClO][O]$$

$$\frac{k_1[Cl][O_3]}{k_2[O]} = [ClO] \tag{36.75}$$

Substituting Equation (36.75) into Equation (36.74) yields the following expression for [Cl]:

$$[\text{Cl}] = \frac{k_2[\text{Cl}]_{total}[\text{O}]}{k_1[\text{O}_3] + k_2[\text{O}]} \tag{36.76}$$

Using Equation (36.76), the rate law expression for the catalytic mechanism is determined as follows:

$$R_{cat} = -\frac{d[\text{O}_3]}{dt} = k_1[\text{Cl}][\text{O}_3] = \frac{k_1 k_2 [\text{Cl}]_{total}[\text{O}][\text{O}_3]}{k_1[\text{O}_3] + k_2[\text{O}]} \tag{36.77}$$

The composition of the stratosphere is such that $[\text{O}_3] \gg [\text{O}]$. Taken in combination with the numerical values for k_1 and k_2 presented earlier, the $k_2[\text{O}]$ term in the denominator of Equation (36.77) can be neglected, and the rate law expression for the catalyzed reaction becomes

$$R_{cat} = k_2[\text{Cl}]_{total}[\text{O}] \tag{36.78}$$

The ratio of catalyzed to uncatalyzed reaction rates is

$$\frac{\text{R}_{cat}}{\text{R}_{uc}} = \frac{k_2[\text{Cl}]_{total}}{k_{nc}[\text{O}_3]} \tag{36.79}$$

In the stratosphere $[\text{O}_3]$ is roughly 10^3 greater than $[\text{Cl}]_{total}$, and Equation (36.79) becomes

$$\frac{R_{cat}}{R_{uc}} = \frac{k_2}{k_{nc}} \times 10^{-3} = \frac{2.44 \times 10^{10}\,\text{M}^{-1}\text{s}^{-1}}{3.30 \times 10^5\,\text{M}^{-1}\text{s}^{-1}} \times 10^{-3} \approx 74$$

Therefore, through Cl-mediated catalysis, the rate of O_3 loss is roughly two orders of magnitude greater than the loss through the bimolecular reaction of O_3 and O directly.

Where does stratospheric Cl come from? Rowland and Molina proposed that a major source of Cl was from the photolysis of chlorofluorocarbons such as CFCl_3 and CF_2Cl_2, anthropogenic compounds that were common refrigerants at the time. These molecules are extremely robust, and when released into the atmosphere, they readily survive transport through the troposphere and into the stratosphere. Once in the stratosphere, these molecules can absorb a photon of light with sufficient energy to dissociate the C—Cl bond, and Cl is produced. This proposal served as the impetus to understand the details of stratospheric ozone depletion, and it led to the Montreal Protocol in which the vast majority of nations agreed to phase out the industrial use of chlorofluorocarbons.

Heterogeneous catalysts are extremely important in industrial chemistry. The majority of industrial catalysts are solids. For example, the synthesis of NH_3 from reactants N_2 and H_2 is catalyzed using Fe. This is an example of heterogeneous catalysis because the reactants and product are in the gas phase, but the catalyst is a solid. An important step in reactions involving solid catalysis is the adsorption of one or more of the reactants to the solid surface. First, we assume that the particles adsorb to the surface without changing their internal bonding, a process referred to as **physisorption.** A dynamic equilibrium exists between the free and surface-adsorbed species or adsorbate, and information regarding the kinetics of surface adsorption and desorption can be obtained by studying this equilibrium as a function of reactant pressure over the surface of the catalyst. A critical parameter in evaluating surface adsorption is the **fractional coverage** θ defined as

$$\theta = \frac{\text{Number of adsorption sites occupied}}{\text{Total number of adsorption sites}}$$

Figure 36.10 provides an illustration of a surface with a series of adsorption sites. Reactant molecules (given by the blue spheres) can exist in either the gas phase or be adsorbed to one of these sites. The fractional coverage is simply the fraction of adsorption sites occupied.

The fractional coverage can also be defined as $\theta = V_{adsorbed}/V_m$ where $V_{adsorbed}$ is the volume of adsorbate at a specific pressure and V_m is the volume of adsorbate in the high-pressure limit corresponding to monolayer coverage.

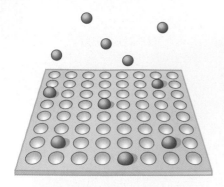

FIGURE 36.10
Illustration of fractional coverage θ. The surface (orange parallelogram) contains a series of adsorption sites (white circles). The reactant (blue spheres) exists in an equilibrium between free reactants and adsorbates. The fractional coverage is the number of occupied adsorption sites divided by the total number of sites on the surface.

Studies of adsorption involve measuring the extent of adsorption or θ as a function of reactant-gas pressure at a specific temperature. The variation in θ with pressure at fixed temperature is called an **adsorption isotherm.** The simplest kinetic model describing the adsorption process is known as the **Langmuir model,** where adsorption is described by the following mechanism:

$$R(g) + M \,(surface) \; \underset{k_d}{\overset{k_a}{\rightleftharpoons}} \; RM \,(surface) \tag{36.80}$$

In Equation (36.80), R is reagent, M (surface) is an unoccupied adsorption site on the surface of the catalyst, RM (surface) is an occupied adsorption site, k_a is the rate constant for adsorption, and k_d is the rate constant for desorption. Three approximations are employed in the Langmuir model:

1. Adsorption is complete once monolayer coverage has been reached.
2. All adsorption sites are equivalent, and the surface is uniform.
3. Adsorption and desorption are uncooperative processes. The occupancy state of the adsorption site will not affect the probability of adsorption or desorption for adjacent sites.

With these approximations, the rate of change in θ will depend on the rate constant for adsorption k_a, reagent pressure P, and the number of vacant sites, which is equal to $N(1 - \theta)$ or the total number of adsorption sites N times the fraction of sites that are open $(1 - \theta)$:

$$\left(\frac{d\theta}{dt}\right)_{abs} = k_a PN(1 - \theta) \tag{36.81}$$

The corresponding change in θ due to desorption is related to the rate constant for desorption k_d and the number of occupied adsorption sites $N\theta$ as follows:

$$\left(\frac{d\theta}{dt}\right)_{des} = -k_d N\theta \tag{36.82}$$

At equilibrium, the change in fractional coverage with time is equal to zero so we can write

$$\frac{d\theta}{dt} = 0 = k_a PN(1 - \theta) - k_d N\theta$$

$$(k_a PN + k_d N)\theta = k_a PN$$

$$\theta = \frac{k_a P}{k_a P + k_d}$$

$$\theta = \frac{KP}{KP + 1} \tag{36.83}$$

where K is the equilibrium constant defined as k_a/k_d. Equation (36.83) is the equation for the **Langmuir isotherm.** Figure 36.11 presents Langmuir isotherms for various values of k_a/k_d. Notice that as the rate constant for desorption increases relative to adsorption, higher pressures must be employed to reach $\theta = 1$, and this behavior can be understood based on the competition between the kinetics for adsorption and desorption. Correspondingly, if the rate constant for desorption is small, the coverage becomes independent of pressure for lower values of pressure.

In many instances adsorption is accompanied by dissociation of the adsorbate, a process that is referred to as **chemisorption** and described by the following mechanism:

$$R_2(g) + 2M \,(surface) \; \underset{k_d}{\overset{k_a}{\rightleftharpoons}} \; 2RM \,(surface)$$

Kinetic analysis of this mechanism (see the end-of-chapter problems) yields the following expression for θ:

$$\theta = \frac{(KP)^{1/2}}{1 + (KP)^{1/2}} \tag{36.84}$$

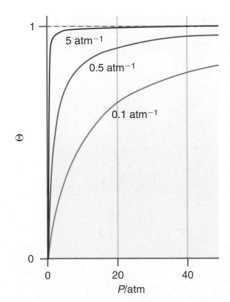

FIGURE 36.11
Langmuir isotherms for a range of k_a/k_d.

Inspection of Equation (36.84) reveals that the extent of surface coverage should demonstrate weaker pressure dependence compared to physisorption. Figure 36.12 provides a comparison of the isotherms predicted using nondissociative and dissociative mechanisms corresponding to the same ratio of k_a/k_d. Finally, different Langmuir isotherms can be collected and evaluated over a range of temperatures to determine K as a function of T. With this information, a van't Hoff plot of $\ln K$ versus $1/T$ should provide a straight line of slope $-\Delta H_{ads}/R$. Through this analysis, the enthalpy of adsorption ΔH_{ads} can be determined.

The assumptions employed in the Langmuir model may not be rigorously obeyed in real heterogeneous systems. First, surfaces are generally not uniform, resulting in the presence of more than one type of adsorption site. Second, the rate of adsorption and desorption may depend on the occupation state of nearby adsorption sites. Finally, it has been established that adsorbed molecules can diffuse on the surface and then adsorb corresponding to a kinetic process of adsorption that is more complicated than the Langmuir mechanism envisions.

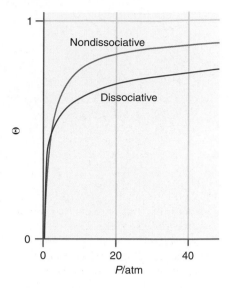

FIGURE 36.12
Comparison of Langmuir isotherms for nondissociative and dissociative adsorption with $k_a/k_d = 0.5 \text{ atm}^{-1}$.

EXAMPLE PROBLEM 36.2

The following data were obtained for the adsorption of Kr on charcoal at 193.5 K. Using the Langmuir model, construct the adsorption isotherm, and determine V_m and the equilibrium constant for adsorption/desorption.

V_{ads} (cm³ g⁻¹)	P (Torr)
5.98	2.45
7.76	3.5
10.1	5.2
12.35	7.2
16.45	11.2
18.05	12.8
19.72	14.6
21.1	16.1

Solution

The fractional coverage is related to the experimentally measured V_{ads}. The adsorption isotherm is given by a plot of V_{ads} versus P, which can be compared to the behavior predicted by Equation (36.83) as illustrated here:

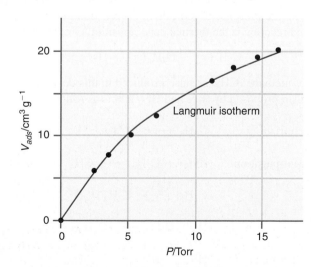

Although the comparison of the adsorption isotherm to Equation (36.83) illustrates that the Langmuir model is consistent with the adsorption of Kr on charcoal, determination of the Langmuir parameters is difficult because parameters such as V_m are unknown. This information is more readily determined by using the reciprocal of Equation (36.83):

$$\frac{1}{V_{ads}} = \left(\frac{1}{KV_m}\right)\frac{1}{P} + \frac{1}{V_m}$$

This equation demonstrates that a plot of $(V_{ads})^{-1}$ versus $(P)^{-1}$ should yield a straight line with slope equal to $(KV_m)^{-1}$ and y intercept equal to $(V_m)^{-1}$. A plot of the data in reciprocal form with the best fit line is shown next:

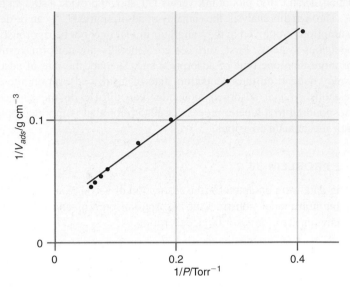

The y intercept obtained from the best fit line is 0.0293 g cm^{-3} such that $V_m = 34.1$ cm^3 g^{-1}. The slope of the best fit line is 0.3449 Torr g cm^{-3}. Using V_m determined from the y intercept, K is found to be 8.38×10^{-2} Torr^{-1}.

36.5 Radical-Chain Reactions

Radicals are chemical species that contain an unpaired electron. Due to the presence of the unpaired electron, radicals tend to be extremely reactive. In 1934, Rice and Herzfeld were able to demonstrate that the kinetic behavior of many organic reactions was consistent with the presence of radicals in the reaction mechanism. An example of a radical-mediated reaction is the thermal decomposition of ethane:

$$C_2H_6(g) \rightarrow C_2H_4(g) + H_2(g) \tag{36.85}$$

Small amounts of methane (CH_4) are also produced in this decomposition. The decomposition mechanism proposed by Rice and Herzfeld is as follows:

$$\text{Initiation} \qquad C_2H_6 \xrightarrow{k_1} 2\ CH_3\cdot \tag{36.86}$$

$$\text{Propagation} \qquad CH_3\cdot + C_2H_6 \xrightarrow{k_2} CH_4 + C_2H_5\cdot \tag{36.87}$$

$$C_2H_5\cdot \xrightarrow{k_3} C_2H_4 + H\cdot \tag{36.88}$$

$$H\cdot + C_2H_6 \xrightarrow{k_4} C_2H_5\cdot + H_2 \tag{36.89}$$

$$\text{Termination} \qquad H\cdot + C_2H_5\cdot \xrightarrow{k_5} C_2H_6 \tag{36.90}$$

In this section, we include a dot ($\cdot$) in the formula of a compound to indicate that the species is a radical. The first elementary step in the mechanism involves the creation of two methyl radicals, referred to as an **initiation step** [Equation (36.86)]. In an initiation step, radicals are produced from a precursor species. The next three steps in the mechanism [Equations (36.87) through (36.89)] are referred to as **propagation steps** in which a radical reacts with another species to produce radical and nonradical products, and the radical products go on to participate in subsequent reactions. The final step in the mechanism is a **termination step** in which two radicals recombine to produce a nonradical product.

Although the mechanism of Equations (36.86) through (36.90) is clearly complex, the rate law expression predicted by this mechanism is remarkably simple. To derive this rate law, we begin with the differential rate expression for the disappearance of ethane:

$$-\frac{d[C_2H_6]}{dt} = k_1[C_2H_6] + k_2[C_2H_6][CH_3\cdot] + k_4[C_2H_6][H\cdot] - k_5[C_2H_5\cdot][H\cdot]$$

$$(36.91)$$

Because the methyl radical is a reactive intermediate, the steady-state approximation is applied to this species, resulting in the following expression for $[CH_3]$:

$$\frac{d[CH_3\cdot]}{dt} = 0 = 2k_1[C_2H_6] - k_2[C_2H_6][CH_3\cdot] \qquad (36.92)$$

$$[CH_3\cdot] = \frac{2k_1}{k_2} \qquad (36.93)$$

The factor of 2 in Equation (36.92) originates from the relationship between the reaction rate and the rate of $CH_3\cdot$ appearance as discussed in Chapter 35. Next, the steady-state approximation is applied to the differential rate expressions for the ethyl radical and atomic hydrogen since they are also reaction intermediates:

$$\frac{d[C_2H_5\cdot]}{dt} = 0 = k_2[CH_3\cdot][C_2H_6] - k_3[C_2H_5\cdot]$$

$$+ k_4[C_2H_6][H\cdot] - k_5[C_2H_5\cdot][H\cdot] \qquad (36.94)$$

$$\frac{d[H\cdot]}{dt} = 0 = k_3[C_2H_5\cdot] - k_4[C_2H_6][H\cdot] - k_5[C_2H_5\cdot][H\cdot] \qquad (36.95)$$

Adding Equations (36.94), (36.95), and (36.92) yields the following expression for $[H\cdot]$:

$$0 = 2k_1[C_2H_6] - 2k_5[C_2H_5\cdot][H\cdot]$$

$$[H\cdot] = \frac{k_1[C_2H_6]}{k_5[C_2H_5\cdot]} \qquad (36.96)$$

Substituting Equation (36.96) into Equation (36.95) yields

$$0 = k_3[C_2H_5\cdot] - k_4[C_2H_6]\left(\frac{k_1[C_2H_6]}{k_5[C_2H_5\cdot]}\right) - k_5[C_2H_5\cdot]\left(\frac{k_1[C_2H_6]}{k_5[C_2H_5\cdot]}\right)$$

$$0 = k_3[C_2H_5\cdot] - \frac{k_4k_1[C_2H_6]^2}{k_5[C_2H_5\cdot]} - k_1[C_2H_6]$$

$$0 = [C_2H_5\cdot]^2 - \frac{k_4k_1[C_2H_6]^2}{k_5k_3} - \frac{k_1}{k_3}[C_2H_6][C_2H_5\cdot] \qquad (36.97)$$

The last expression is a quadratic equation in $[C_2H_5\cdot]$ for which the solution yields the following expression for $[C_2H_5\cdot]$:

$$[C_2H_5\cdot] = [C_2H_6]\left[\frac{k_1}{2k_3} + \left(\left(\frac{k_1}{2k_3}\right)^2 + \left(\frac{k_1k_4}{k_3k_5}\right)\right)^{1/2}\right] \qquad (36.98)$$

Experimentally, the rate constant for initiation k_1 is small such that only the lowest power term in Equation (36.98) is appreciable such that

$$[C_2H_5\cdot] = \left(\frac{k_1k_4}{k_3k_5}\right)^{1/2}[C_2H_6] \qquad (36.99)$$

With Equation (36.99), $[H\cdot]$ from Equation (36.96) becomes

$$[H\cdot] = \frac{k_1[C_2H_6]}{k_5[C_2H_5\cdot]} = \frac{k_1}{k_5}\left(\frac{k_3k_5}{k_1k_4}\right)^{1/2} = \left(\frac{k_1k_3}{k_4k_5}\right)^{1/2} \qquad (36.100)$$

With the preceding definitions for $[H\cdot]$ and $[C_2H_5\cdot]$ in hand, the differential rate expression for the disappearance of ethane [Equation (36.91)] becomes

$$-\frac{d[C_2H_6]}{dt} = \left(k_2[CH_3\cdot] + \left(\frac{k_1 k_3 k_4}{k_5}\right)^{1/2}\right)[C_2H_6] \qquad \textbf{(36.101)}$$

Finally, using the definition of $[CH_3\cdot]$ in Equation (36.93) and ignoring higher powers of k_1, the final differential rate expression for $[C_2H_6]$ is

$$-\frac{d[C_2H_6]}{dt} = \left(\frac{k_1 k_3 k_4}{k_5}\right)^{1/2}[C_2H_6] \qquad \textbf{(36.102)}$$

Equation (36.102) predicts that the decay of ethane should be first order with respect to ethane, as is observed experimentally. The remarkable aspect of this result is that from a very complex mechanism a relatively simple rate expression is derived. In general, even the most complex Rice–Herzfeld radical mechanisms will yield orders of 1/2, 1, 3/2, and 2.

EXAMPLE PROBLEM 36.3

Consider the following reaction of methane with molecular chlorine:

$$CH_4(g) + Cl_2(g) \rightarrow CH_3Cl(g) + HCl(g)$$

Experimental studies have shown that the rate law for this reaction is one-half order with respect to Cl_2. Is the following mechanism consistent with this behavior?

$$Cl_2 \xrightarrow{k_1} 2\,Cl\cdot$$

$$Cl\cdot + CH_4 \xrightarrow{k_2} HCl + CH_3\cdot$$

$$CH_3\cdot + Cl_2 \xrightarrow{k_3} CH_3Cl + Cl\cdot$$

$$Cl\cdot + Cl\cdot \xrightarrow{k_4} Cl_2$$

Solution

The rate of reaction in terms of product HCl is given by

$$R = \frac{d[HCl]}{dt} = k_2[Cl\cdot][CH_4]$$

Because $Cl\cdot$ is a reaction intermediate, it cannot appear in the final rate law expression, and $Cl\cdot$ must be expressed in terms of $[CH_4]$ and $[Cl_2]$. The differential rate expressions for $[Cl\cdot]$ and $[CH_3Cl\cdot]$ are

$$\frac{d[Cl\cdot]}{dt} = 2k_1[Cl_2] - k_2[Cl\cdot][CH_4] + k_3[CH_3\cdot][Cl_2] - 2k_4[Cl\cdot]^2$$

$$\frac{d[CH_3\cdot]}{dt} = k_2[Cl\cdot][CH_4] - k_3[CH_3\cdot][Cl_2]$$

Applying the steady-state approximation to the expression for $[CH_3\cdot]$ yields

$$[CH_3\cdot] = \frac{k_2[Cl\cdot][CH_4]}{k_3[Cl_2]}$$

Next, we substitute this definition of $[CH_3\cdot]$ into the differential rate expression for $[Cl\cdot]$ and apply the steady-state approximation:

$$0 = 2k_1[Cl_2] - k_2[Cl\cdot][CH_4] + k_3[CH_3\cdot][Cl_2] - 2k_4[Cl\cdot]^2$$

$$0 = 2k_1[Cl_2] - k_2[Cl\cdot][CH_4] + k_3\left(\frac{k_2[Cl\cdot][CH_4]}{k_3[Cl_2]}\right)[Cl_2] - 2k_4[Cl\cdot]^2$$

$$0 = 2k_1[Cl_2] - 2k_4[Cl\cdot]^2$$

$$[Cl\cdot] = \left(\frac{k_1}{k_4}[Cl_2]\right)^{1/2}$$

With this result, the predicted rate law expression becomes

$$R = k_2[\text{Cl}\cdot][\text{CH}_4] = k_2\left(\frac{k_1}{k_4}\right)^{1/2}[\text{CH}_4][\text{Cl}_2]^{1/2}$$

The mechanism is consistent with the experimentally observed one-half order dependence on $[\text{Cl}_2]$.

36.6 Radical-Chain Polymerization

A very important class of radical reactions are **radical polymerization reactions.** In these processes, a monomer is activated through the reaction with a radical initiator, creating a monomer radical. Next, the monomer radical reacts with another monomer to create a radical dimer. The radical dimer then reacts with another monomer, and the process continues, resulting in formation of a **polymer chain.** The mechanism for chain polymerization is as follows. First, the activated monomer must be created in an initiation step:

$$\text{I} \xrightarrow{k_i} 2\text{R}\cdot \tag{36.103}$$

$$\text{R}\cdot + \text{M} \xrightarrow{k_1} \text{M}_1\cdot \tag{36.104}$$

In this step, the initiator I is transformed into radicals $\text{R}\cdot$ that react with a monomer to form an activated monomer $\text{M}_1\cdot$. The next mechanistic step is propagation, in which the activated monomer reacts with another monomer to form activated dimer, and the dimer undergoes subsequent reactions as follows:

$$\text{M}_1\cdot + \text{M} \xrightarrow{k_p} \text{M}_2\cdot \tag{36.105}$$

$$\text{M}_2\cdot + \text{M} \xrightarrow{k_p} \text{M}_3\cdot \tag{36.106}$$

$$\text{M}_{n-1}\cdot + \text{M} \xrightarrow{k_p} \text{M}_n\cdot \tag{36.107}$$

In the preceding equations, the subscript indicates the number of monomers contained in the polymer chain, and the rate constant for propagation k_p is assumed to be independent of polymer size. The final step in the mechanism is termination in which two radical chains undergo reaction:

$$\text{M}_m\cdot + \text{M}_n\cdot \xrightarrow{k_t} \text{M}_{m+n} \tag{36.108}$$

The rate of activated monomer production is related to the rate of radical $(\text{R}\cdot)$ formation as follows:

$$\left(\frac{d[\text{M}\cdot]}{dt}\right)_{production} = 2\phi k_i[I] \tag{36.109}$$

where ϕ represents the probability that the initiator-generated radical $(\text{R}\cdot)$ will create a radical chain. Since active monomers combine to terminate polymerization [Equation (36.107)], the rate of activated monomer loss is equal to the rate of termination:

$$\left(\frac{d[\text{M}\cdot]}{dt}\right)_{decay} = -2k_t[\text{M}\cdot]^2 \tag{36.110}$$

The total differential rate expression for $[\text{M}\cdot]$ is given by the sum of Equations (36.109) and (36.110):

$$\frac{d[\text{M}\cdot]}{dt} = 2\phi k_i[I] - 2k_t[\text{M}\cdot]^2 \tag{36.111}$$

Because M· is an intermediate species, the steady-state approximation is applied to Equation (36.111), yielding the following:

$$[\text{M·}] = \left(\frac{\phi k_i}{k_t}\right)^{1/2}[\text{I}]^{1/2} \tag{36.112}$$

Finally, the monomer consumption is dominated by propagation such that the differential rate expression for [M] becomes

$$\frac{d[\text{M}]}{dt} = -k_p[\text{M·}][\text{M}]$$

$$= -k_p\left(\frac{\phi k_i}{k_t}\right)^{1/2}[\text{I}]^{1/2}[\text{M}] \tag{36.113}$$

Equation (36.113) demonstrates that the rate of monomer consumption is predicted to be overall 3/2 order, 1/2 order in initiator concentration, and first order in monomer concentration.

One measure of the efficiency of polymerization is the kinetic chain length ν. This quantity is defined as the average number of monomers in the polymer chain per active center monomer produced. Consistent with this definition, ν can be expressed as the rate of monomer unit consumption divided by the rate of active monomer production:

$$\nu = \frac{k_p[\text{M·}][\text{M}]}{2\phi k_i[\text{I}]} = \frac{k_p[\text{M·}][\text{M}]}{2k_t[\text{M·}]^2} = \frac{k_p[\text{M}]}{2k_t[\text{M·}]} \tag{36.114}$$

Substitution of Equation (36.112) into Equation (36.114) provides the final expression for the kinetic chain length:

$$\nu = \frac{k_p[\text{M}]}{2k_t\left(\dfrac{\phi k_i}{k_t}\right)^{1/2}[\text{I}]^{1/2}} = \frac{k_p[\text{M}]}{2(\phi k_i k_t)^{1/2}[\text{I}]^{1/2}} \tag{36.115}$$

Equation (36.115) predicts that the kinetic chain length will increase as the rate constants for chain initiation or termination or the concentration of initiator are reduced. Therefore, polymerization is usually carried out at minimal initiator concentrations such that the number of activated monomers is kept small.

36.7 Explosions

Consider a highly exothermic reaction in which a significant amount of heat is liberated during the reaction. The reaction will proceed with a certain initial rate, but if the heat liberated during the reaction is not dissipated, the system temperature will rise, as will the rate of reaction. The final result of this process is a thermal explosion. A second type of explosion involves chain-branching reactions. In the previous section, the concentration of radical intermediate species was determined by applying the steady-state approximation. However, what if the concentrations of radical intermediate species were not constant with time? Two limits can be envisioned: either a significant reduction or increase in radical intermediate concentration as time proceeds. If the concentration of radical intermediates decreases with time, then the reaction will terminate. What if the concentration of reactive intermediates increases rapidly with time? From the mechanisms discussed thus far, an increase in radical intermediate concentration would lead to the creation of more radical species. In this case, the number of radical chains increases exponentially with time, leading to explosion.

A standard introductory chemistry demonstration is the ignition of a balloon containing hydrogen and oxygen. The balloon is ignited, and if the two gases are present in the correct stoichiometric ratio, an explosion occurs as evidenced by a loud bang and a number of startled students. The specific reaction is

$$2\,\text{H}_2(g) + \text{O}_2(g) \rightarrow 2\,\text{H}_2\text{O}(g) \tag{36.116}$$

The reaction is deceptively simple, yet the mechanism of the reaction is not fully under-stood. Important mechanistic components of this reaction are

$$H_2 + O_2 \rightarrow 2\,OH\cdot \tag{36.117}$$

$$H_2 + OH\cdot \rightarrow H\cdot + H_2O \tag{36.118}$$

$$H\cdot + O_2 \rightarrow OH\cdot + \cdot O\cdot \tag{36.119}$$

$$\cdot O\cdot + H_2 \rightarrow OH\cdot + H\cdot \tag{36.120}$$

$$H\cdot + O_2 + M \rightarrow HO_2\cdot + M^* \tag{36.121}$$

The first step in the mechanism, Equation (36.117), is an initiation step in which two hydroxyl radicals (OH·) are created. The radical is propagated in the second step, Equation (36.118). Steps 3 and 4, Equations (36.119) and (36.120), are referred to as **branching reactions,** in which a single radical species reacts to produce two radical species. As such, the number of reactive radicals increases twofold in these branching steps. These branching steps lead to a chain-branching explosion because the concentration of the reactive species grows rapidly in time.

The occurrence of explosions for this reaction is dependent on temperature and pressure, as shown in Figure 36.13. First, explosion will occur only if the temperature is greater than 460°C. At lower temperatures, the rates for the various radical-producing reactions are insufficient to support appreciable chain branching. In addition to temperature, the pressure must also be sufficiently high to support chain branching. If the pressure is low, radicals that are produced can diffuse to the walls of the vessel where they are destroyed. Under these conditions, the rates of radical production and decay are balanced such that branching is not prevalent and explosion does not occur. As the pressure is increased, the first explosion regime is reached in which the radicals can participate in branching reactions before reaching the container walls. A further increase in pressure results in the reaction returning to a controlled regime where the pressure is so great that radical–radical reactions reduce the number of reactive species present. The final reaction in the mechanism under discussion is an example of such a reaction. In this step, H· and O_2 react to produce $HO_2\cdot$, which does not contribute to the reaction. The formation of this species requires a three-body collision, with the third species, M, taking away excess energy such that the $HO_2\cdot$ radical is stable. Such reactions will occur only at elevated pressures. Finally, at highest pressures another explosive regime is encountered. This is a thermal explosive regime in which the dissipation of heat is insufficient to keep the system temperature from increasing rapidly, providing for explosive behavior.

The likelihood of undergoing an explosion is highly dependent on radical concentration. A generic scheme for chain-branching reactions is as follows:

$$A + B \xrightarrow{k_i} R\cdot \tag{36.122}$$

$$R\cdot \xrightarrow{k_b} \phi R\cdot + P_1 \tag{36.123}$$

$$R\cdot \xrightarrow{k_t} P_2 \tag{36.124}$$

In this scheme, the first step involves the reaction of reactants A and B, resulting in the formation of radical R·. The second step is chain branching in which the radical undergoes reaction to produce other radicals with a branching efficiency given by ϕ. The final step of the mechanism is termination. Finally, the species P_1 and P_2 represent nonreactive products. The differential rate expression for [R·] consistent with this mechanism is as follows:

$$\begin{aligned}\frac{d[R\cdot]}{dt} &= k_i[A][B] - k_b[R\cdot] + \phi k_b[R\cdot] - k_t[R\cdot] \\ &= \Gamma + k_{eff}[R\cdot]\end{aligned} \tag{36.125}$$

In Equation (36.125), the following definitions have been employed:

$$\Gamma = k_i[A][B]$$

$$k_{eff} = k_b(\phi - 1) - k_t$$

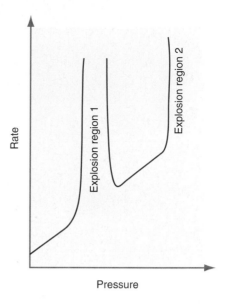

FIGURE 36.13
Schematic of an explosive reaction. As the pressure is increased, two explosive regimes are encountered. The region at lower pressures is due to chain-branching reactions, and the higher pressure region corresponds to a thermal explosion.

Equation (36.125) can be solved for $[R\cdot]$ to yield

$$[R\cdot] = \frac{\Gamma}{k_{eff}}(e^{k_{eff}\,t} - 1) \tag{36.126}$$

Equation (36.126) demonstrates that $[R\cdot]$ is dependent on k_{eff}. Two cases can be envisioned depending on the magnitude of k_t in comparison to $k_b(\phi-1)$. In the limit where $k_t \gg k_b(\phi-1)$, termination dominates and Equation (36.126) becomes

$$\lim_{k_t \gg k_b(\phi-1)} [R\cdot] = \frac{\Gamma}{k_t}(1 - e^{-k_t t}) \tag{36.127}$$

Equation (36.127) demonstrates that in this limit the radical concentration will reach a limiting value of Γ/k_t at $t = \infty$. The interpretation of this limiting behavior is that $[R\cdot]$ will never become large enough to support branching, and explosion will not occur. The second limit occurs when $k_b(\phi-1) \gg k_t$ and branching dominates. In this limit, Equation (36.126) becomes

$$\lim_{k_b(\phi-1) \gg k_t} [R\cdot] = \frac{\Gamma}{k_b(\phi - 1)}(e^{k_b(\phi-1)t} - 1) \tag{36.128}$$

Equation (36.128) demonstrates that $[R\cdot]$ is predicted to increase exponentially corresponding to explosion. This simple mechanism illustrates the importance of efficient propagation/branching in promoting explosions in chain-branching reactions.

36.8 Feedback, Nonlinearity, and Oscillating Reactions

The discussion of chemical kinetics thus far has largely involved systems in which reactants combine to produce products, and the reaction proceeds until equilibrium (defined as an absence of macroscopic variation in reactant of product concentrations with time) is reached. However, there is a class of reactions that demonstrate periodic variation in concentration in space or in time. These reactions are referred to as **oscillating reactions.** Oscillating reactions are generally open systems kept from reaching equilibrium by supplying reactants or removing products from the reaction. Under these conditions, concentrations of reactants, products, and intermediates can demonstrate macroscopic variation in concentrations. Once viewed as an oddity, oscillating reactions are found in a wide variety of systems in biology (cellular energy, chemotaxis) and chemistry (combustion, film growth). All oscillating reactions involve two kinetic features: nonlinearity and feedback.

The previous section describing explosions and radical chemistry provided an example of feedback. Specifically, in the conversion of hydrogen and oxygen to water [Equation (36.116)] the first step in the mechanism involved the formation of hydroxyl radical:

$$H_2 + O_2 \rightarrow 2\,OH\cdot$$

After formation, hydroxyl radical could in turn react with hydrogen:

$$H_2 + OH\cdot \rightarrow H\cdot + H_2O$$

Notice that once the $OH\cdot$ intermediate is formed, this species reacts with H_2, which results in an increase in the rate of H_2 consumption and correspondingly the reaction rate. This is an example of **feedback,** where products formed in one step of the mechanism can influence the rate of other steps. Acceleration of the reaction rate (as described previously) is an example of positive feedback, and negative feedback occurs when the formation of a species results in a reduction in the reaction rate.

What does it mean for a chemical reaction to be **nonlinear**? To explore this concept, we return to the coefficient of reaction advancement (ξ). For the following reaction:

$$A + BC \rightarrow AB + C \tag{36.129}$$

We define ξ with respect to A as follows:

$$\xi = \frac{[A]_0 - [A]}{[A]_0} \tag{36.130}$$

With this definition, ξ has a value of 1 at the beginning of the reaction and 0 when A has been consumed. If we measure the rate of the reaction with BC in excess, then the reaction will be first-order in A and the reaction rate defined with respect to ξ will be

$$\frac{d\xi}{dt} = k(1 - \xi) \tag{36.131}$$

This expression demonstrates that the reaction rate will initially be at a maximum and will decrease linearly as the reaction proceeds as shown in Figure 36.14. However, what if BC is not in excess and the reaction is second order? In this case the reaction rate becomes

$$\frac{d\xi}{dt} = k(1 - \xi)^2 \tag{36.132}$$

This expression dictates that the decrease in the reaction rate as ξ increases will be nonlinear as shown in Figure 36.14. Notice that nonlinear evolution of the reaction rate with ξ is a relatively common aspect of chemical kinetics. In this simple example, the reaction had to simply be second order for the reaction to be nonlinear.

What happens if feedback and nonlinearity are combined? Returning back to the reaction of molecular hydrogen and oxygen, we would expect the rate of the reaction to increase as hydroxyl radical is formed, and then decrease as the reactants are consumed. We can model this behavior in terms of ξ as follows:

$$\frac{d\xi}{dt} = k\xi(1 - \xi) \tag{36.133}$$

This expression can be directly related to the general relationship for quadratic **autocatalysis**:

$$A + B \xrightarrow{k} 2B \qquad R = k[A][B] \tag{36.134}$$

In this autocatalytic reaction we would expect the reaction rate to initially increase as the reaction proceeds due to the increase in [B]. As the reaction proceeds, [A] will eventually be depleted and the reaction rate will approach zero as ξ approaches one. This evolution in the reaction rate for this autocatalytic reaction is shown in Figure 36.14.

Autocatalytic reactions provide the opportunity for concentrations and reaction rates to evolve in time, including oscillatory variation in these quantities. Perhaps the most famous example of a chemical reaction demonstrating oscillations is the

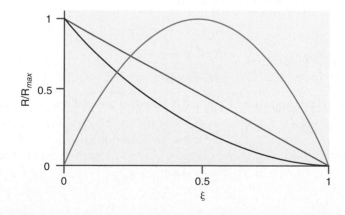

FIGURE 36.14
Illustration of the dependence of the reaction rate on the coefficient of reaction advancement (ξ). The purple curve represents linear dependence of the rate with ξ (Equation 36.131), and the red curve demonstrates nonlinear dependence for a second-order reaction (Equation 36.132). The blue curve represents quadratic autocatalysis (Equation 36.133) where both feedback and nonlinear behavior result in a maximum in the reaction rate after the reaction has been initiated.

Belousov-Zhabotinsky (BZ) reaction. The feedback steps in the BZ reaction are the following:

$$BrO_3^- + HBrO_2 + H^+ \underset{k_{-1}}{\overset{k_1}{\rightleftharpoons}} 2BrO_2{}^\bullet + H_2O \tag{36.135}$$

$$BrO_2{}^\bullet + Ce(III) + H^+ \xrightarrow{k_2} HBrO_2 + Ce(IV) \tag{36.136}$$

The feedback mechanism in this reaction involves $HBrO_2$, which reacts to yield two BrO_2 radicals, with these radicals forming the reactant $HBrO_2$ through the oxidation of Ce(III). If each of the BrO_2 radicals goes on to form $HBrO_2$, then the following net reaction occurs:

$$BrO_3^- + HBrO_2 + 3H^+ + 2Ce(III) \rightarrow 2HBrO_2 + 2Ce(IV) + H_2O \tag{36.137}$$

Comparison to Equation 36.134 reveals that this reaction is predicted to demonstrate quadratic autocatalysis. The dependence of the rate law on $HBrO_2$ can be further explored using basic kinetic tools. First, we can express the reaction rate with respect to $HBrO_2$ as follows:

$$R = \frac{d[HBrO_2]}{dt}$$
$$= -k_1[BrO_3^-][HBrO_2][H^+] + k_{-1}[BrO_2{}^\bullet]^2 + k_2[BrO_2{}^\bullet][Ce(III)][H^+] \tag{36.138}$$

Since the oxidation of Ce(III) is relatively slow, we can employ the preequilibrium approximation to derive an expression for $[BrO_2{}^\bullet]$:

$$[BrO_2{}^\bullet] = \left\{ \frac{k_1[BrO_3^-][HBrO_2][H^+]}{k_{-1}} \right\}^{1/2} \tag{36.139}$$

Substituting this into the rate expression yields

$$R = \frac{d[HBrO_2]}{dt} = k_2\left(\frac{k_1}{k_{-1}}\right)^{1/2}[BrO_3^-]^{1/2}[Ce(III)][H^+]^{3/2}[HBrO_2]^{1/2} \tag{36.140}$$

This rate expression is consistent with autocatalytic behavior as indicted by the positive exponent for $HBrO_2$, and half-order dependence. In contrast, if Fe(III)/Fe(IV) are used in place of Ce(III)/Ce(IV), the oxidation of Fe(III) is rapid and the steady-state approximation can be used to define $[BrO_2{}^\bullet]$. In this case the rate law expression for the reaction demonstrates autocatalytic behavior that is first order in $HBrO_2$:

$$R = \frac{d[HBrO_2]}{dt} = k_1[BrO_3^-][H^+][HBrO_2] \tag{36.141}$$

Autocatalysis can also lead to oscillatory behavior where concentrations of reactants and products demonstrate recurring maxima with time. Figure 36.15 presents a picture of a BZ reaction, where color variation is due to spatial variation in concentrations. The spatial patterns evolve with time since autocatalysis results in oscillations in concentrations with time and space.

The BZ reaction is rather complex. A simpler example of an oscillatory chemical reaction is the **Lotka-Volterra mechanism:**

$$A + X \xrightarrow{k_1} 2X \tag{36.142}$$

$$X + Y \xrightarrow{k_2} 2Y \tag{36.143}$$

$$Y \xrightarrow{k_3} B \tag{36.144}$$

The differential rate expressions for X and Y based on this mechanism are

$$\frac{d[X]}{dt} = 2k_1[A][X] - k_2[X][Y] \tag{36.145}$$

$$\frac{d[Y]}{dt} = 2k_2[X][Y] - k_3[Y] \tag{36.146}$$

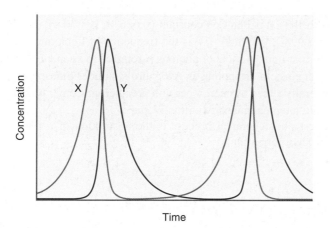

FIGURE 36.16
Oscillatory behavior of intermediate concentrations from the Lotka-Volterra mechanism.

Imagine that the reaction is carried out by keeping the concentration of A ([A]) constant, which can be achieved by continuously supplying A as the reaction proceeds. Furthermore, notice that the reaction rate is not dependent on the concentration of the product B. Under these conditions the only concentrations relevant to the reaction rate are the intermediates [X] and [Y]. Using numerical methods (such as Euler's method described in Section 35.6) [X] and [Y] can be determined using the differential rate expressions above for a given set of rate constants and $[A]_0$. The results of this calculation are shown in Figure 36.16. Notice that [X] and [Y] oscillate, demonstrating periodic recurrences in time. Also, notice that the maximum in [Y] occurs after the maximum in [X]. Inspection of the mechanism reveals why this would be the case. The first step of the mechanism is autocatalytic with respect to X, resulting in an increase in rate of X production. However, once X is produced it reacts with Y in the second step of the mechanism. Since this second step is autocatalytic with respect to Y, the rate of Y production will increase, resulting in a depletion of X. Finally, the decay in Y to form the product B given by the third step of the mechanism provides for a decrease in [Y]. Although we have explored the Lotka-Volterra mechanism in a chemical context, the differential rate expressions represent a general class of differential equations referred to as "predator-prey" equations. In our example, X is the prey and Y is the predator. The initial increase in prey population promotes an increase in the predator population. Ultimately, the predators consume the prey, resulting in a decrease in the prey population. With a decrease in the prey population, the predator populations will decline. Finally, a decline in predators allows the prey population to increase, and the cycle starts over again. These equations have been used to describe oscillations in economic activity, wildlife populations, and the spread of disease.

36.9 Photochemistry

Photochemical processes involve the initiation of a chemical reaction through the absorption of a photon by an atom or molecule. In these reactions, photons can be thought of as reactants, and initiation of the reaction occurs when the photon is absorbed. Photochemical reactions are important in a wide variety of areas. The primary event in vision involves the absorption of a photon by the visual pigment rhodopsin. Photosynthesis involves the conversion of light energy into chemical energy by plants and bacteria. Finally, numerous photochemical reactions occur in the atmosphere (e.g., ozone production and decomposition) that are critical to life on Earth. As illustrated by these examples, photochemical reactions are an extremely important area of chemistry, and they are explored in this section.

36.9.1 Photophysical Processes

When a molecule absorbs a photon of light, the energy contained in the photon is transferred to the molecule. The amount of energy contained by a photon is given by the Planck equation:

$$E_{photon} = h\nu = \frac{hc}{\lambda} \qquad \textbf{(36.147)}$$

In Equation (36.147), h is Planck's constant (6.626×10^{-34} J s), c is the speed of light in a vacuum (2.998×10^8 m s^{-1}), ν is the frequency of light, and λ is the corresponding wavelength of light. A mole of photons is referred to as an einstein, and the energy contained by an einstein of photons is Avogadro's number times E_{photon}. The intensity of light is generally stated as energy per unit area per unit time. Because one joule per second is a watt, a typical intensity unit is W cm^{-2}.

The simplest photochemical process is the absorption of a photon by a reactant resulting in product formation:

$$A \xrightarrow{h\nu} P \qquad (36.148)$$

The rate of reactant photoexcitation is given by

$$R = -\frac{d[A]}{dt} = \frac{I_{abs}1000}{l} \qquad (36.149)$$

In Equation (36.149), I_{abs} is the intensity of absorbed light in units of Einstein cm^{-2} s^{-1}, l is the path length of the sample in centimeters, and 1000 represents the conversion from cubic centimeters to liters such that the rate has appropriate units of M s^{-1}. In Equation (36.149), it is assumed that reactant excitation occurs through the absorption of a single photon. According to the Beer–Lambert law, the intensity of light transmitted through a sample (I_{trans}) is given by

$$I_{trans} = I_0 10^{-\varepsilon l[A]} \qquad (36.150)$$

where I_0 is the intensity of incident radiation, ε is the **molar absorptivity** of species A, and $[A]$ is the concentration of reactant. Recall that the molar absorptivity will vary with excitation wavelength. Because $I_{abs} = I_0 - I_{trans}$,

$$I_{abs} = I_0(1 - 10^{-\varepsilon l[A]}) \qquad (36.151)$$

The series expansion of the exponential term in Equation (36.151) is

$$10^{-\varepsilon l[A]} = 1 - 2.303\varepsilon l[A] + \frac{(2.303\varepsilon l[A])^2}{2!} - \cdots \qquad (36.152)$$

If the concentration of reactant is kept small, only the first two terms in Equation (36.152) are appreciable, and substitution into Equation (36.151) yields

$$I_{abs} = I_0(2.303)\varepsilon l[A] \qquad (36.153)$$

Substitution of Equation (36.153) into the rate expression for reactant photoexcitation of Equation (36.149) and integration yield the following expression for $[A]$:

$$[A] = [A]_0 e^{-I_0(2303)\varepsilon t} = [A]_0 e^{-kt} \qquad (36.154)$$

Equation (36.154) demonstrates that the absorption of light will result in the decay of reactant concentration consistent with first-order kinetic behavior. Most photochemical reactions are first order in reactant concentration such that Equation (36.154) describes the evolution in reactant concentration for the majority of photochemical processes. At times it is more useful to discuss photochemical processes with respect to the number of molecules as opposed to concentration. This is precisely the limit one encounters when considering the **photochemistry** of individual molecules as presented later. In this case, Equation (36.149) becomes

$$-\frac{dA}{dt} = I_0 \frac{2303\varepsilon}{N_A} A \qquad (36.155)$$

where A represents the number of molecules of reactant and N_A is Avogadro's number. Integrating Equation (36.155), we obtain

$$A = A_0 e^{-I_0(2303\varepsilon/N_A)t} = A_0 e^{-I_0\sigma_A t} \qquad (36.156)$$

where σ_A is known as the **absorption cross section** and the rate constant for excitation k_a is equal to $I_0\sigma_A$ with I_0 in units of photons cm^{-2} s^{-1}.

The absorption of light may occur when the photon energy is equal to the energy difference between two energy states of the molecule. A schematic of the processes that occur following photon absorption resulting in an electronic energy-level transition (or "electronic transition") is given in Figure 36.17. Such diagrams are referred to as **Jablonski diagrams** after Aleksander Jablonski, a Polish physicist who developed these diagrams for describing kinetic processes initiated by electronic transitions. In a Jablonski diagram, the vertical axis represents increasing energy. The electronic states depicted are the ground-state singlet S_0, first excited singlet S_1, and triplet T_1. In the singlet states, the electrons are spin paired such that the spin multiplicity is one (i.e., a "singlet"), and in the triplet state two electrons are unpaired such that the spin multiplicity is three (a "triplet"). The subscripts indicate the energy ordering of the states. Because triplets are generally formed by electronic excitation, the lowest energy triplet state is labeled T_1 as opposed to T_0 (the lowest energy spin configuration of molecular oxygen is a triplet, a famous exception to this generality). Finally, the lowest vibrational level for each electronic state is indicated by dark horizontal lines, with higher vibrational levels indicated by the lighter horizontal lines. In addition, a manifold of rotational states will exist for each vibrational level; however, the rotational energy levels have been suppressed for clarity in Figure 36.17.

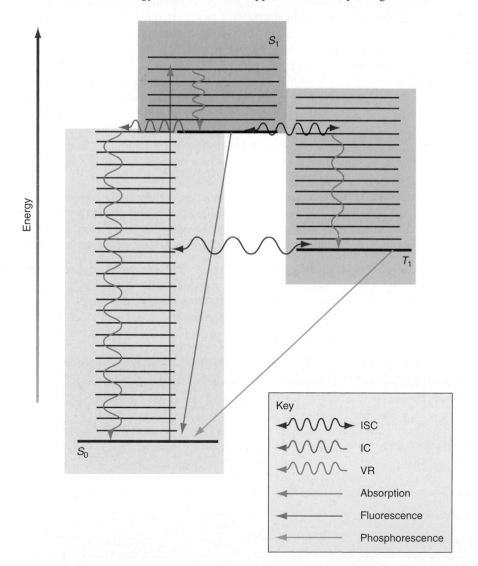

Key

~~~► ISC

~~~ IC

~~~ VR

◄─── Absorption

◄─── Fluorescence

◄─── Phosphorescence

FIGURE 36.17
A Jablonski diagram depicting various photophysical processes, where S_0 is the ground electronic singlet state, S_1 is the first excited singlet state, and T_1 is the first excited triplet state. Radiative processes are indicated by the straight lines. The nonradiative processes of intersystem crossing (ISC), internal conversion (IC), and vibrational relaxation (VR) are indicated by the wavy lines.

The solid and wavy lines in Figure 36.17 represent a variety of processes that couple the electronic states. These processes, including the absorption of light and subsequent energetic relaxation pathways, are referred to as **photophysical processes** because the structure of the molecule remains unchanged. In fact, many of the processes of interest in "photochemistry" do not involve photochemical transformation of the reactant at all but are instead photophysical in nature. The absorption of light decreases the population in the lowest energy singlet state S_0, referred to as a depletion. Correspondingly, the population in the first excited singlet S_1 is increased. The absorption transition depicted in Figure 36.17 is to a higher vibrational level in S_1, with the probability of transition to a specific vibrational level determined by the Franck–Condon factor between the lowest energy vibrational level in S_0 and the vibrational states in S_1.

After populating S_1, thermal equilibration of the vibrational energy will occur, a process referred to as vibrational relaxation. Vibrational relaxation is extremely rapid (~ 100 fs), and when complete, the vibrational state population in S_1 will be governed by the Boltzmann distribution. The vibrational energy-level spacings are assumed to be sufficiently large such that only the lowest vibrational level of S_1 is populated to a significant extent after equilibration. Decay of S_1 resulting in repopulation of S_0 can occur through one of three paths:

1. *Path 1*: Loss of excess electronic energy through the emission of a photon. Such processes are referred to as **radiative transitions.** The process by which photons are emitted in the radiative transitions between S_1 and S_0 is referred to as **fluorescence.** This process is equivalent to spontaneous emission.

2. *Path 2*: **Intersystem crossing** (ISC in Figure 36.14) resulting in population of T_1. This process involves a change in spin state, a process that is forbidden by quantum mechanics. As such, intersystem crossing is significantly slower than vibrational relaxation, but it is competitive with fluorescence in systems where the triplet state is populated to a significant extent. Following intersystem crossing, vibrational relaxation in the triplet vibrational manifold occurs, resulting in population of the lowest energy vibrational level. From this level, a second radiative transition can occur where S_0 is populated and the excess energy is released as a photon. This process is referred to as **phosphorescence.** Because the $T_1 - S_0$ transition also involves a change in spin, it is also forbidden by spin selection rules. Therefore, the rate for this process is slow, and phosphorescence occurs over longer timescales (10^{-6} s to seconds) as compared to fluorescence ($\sim 10^{-9}$ s).

3. *Path 3*: Rather than undergoing a radiative transition, decay from S_1 to a high vibrational level of S_0 can occur followed by rapid vibrational relaxation. This process is referred to as **internal conversion** or nonradiative decay. Nonradiative decay can also occur through the triplet state by intersystem crossing to S_0 followed by vibrational relaxation.

From the viewpoint of kinetics, the absorption of light and subsequent relaxation processes can be viewed as a collection of reactions with corresponding rates. Figure 36.18 presents a modified version of the Jablonski diagram that focuses on these processes and corresponding rate constants. The individual processes, reactions, and notation for the reaction rates are provided in Table 36.1.

TABLE 36.1   Photophysical Reactions and Corresponding Rate Expressions

Process	Reaction	Rate
Absorption/excitation	$S_0 + h\nu \rightarrow S_1$	$k_a[S_0]\ (k_a = I_0 \sigma_A)$
Fluorescence	$S_1 \rightarrow S_0 + h\nu$	$k_f[S_1]$
Internal conversion	$S_1 \rightarrow S_0$	$k_{ic}[S_1]$
Intersystem crossing	$S_1 \rightarrow T_1$	$k_{isc}^S[S_1]$
Phosphorescence	$T_1 \rightarrow S_0 + h\nu$	$k_p[T_1]$
Intersystem crossing	$T_1 \rightarrow S_0$	$k_{isc}^T[T_1]$

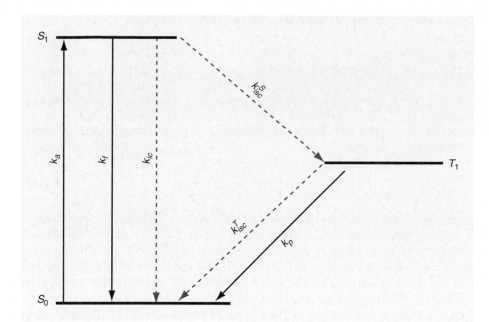

FIGURE 36.18
Kinetic description of photophysical processes. Rate constants are indicated for absorption (k_a), fluorescence (k_f), internal conversion (k_{ic}), intersystem crossing from S_1 to T_1 (k^S_{isc}), intersystem crossing from T_1 to S_0 (k^T_{isc}), and phosphorescence (k_p).

36.9.2 Fluorescence and Fluorescence Quenching

The photophysical processes outlined in Table 36.1 are present for any molecular system. To study excited state lifetimes, another photophysical process is introduced: **collisional quenching.** In this process, a collision occurs between a species Q and a molecule populating an excited electronic state. The result of the collision is the removal of energy from the molecule with the accompanying conversion of the molecule from S_1 to S_0:

$$S_1 + Q \xrightarrow{k_q} S_0 + Q \qquad (36.157)$$

The rate expression for this process is

$$R_q = k_q[S_1][Q] \qquad (36.158)$$

By studying the rate of collisional quenching as a function of [Q], it is possible to determine the k_f. To demonstrate this procedure, we begin by recognizing that in the kinetic scheme illustrated in Figure 36.18, S_1 can be considered an intermediate species. Under constant illumination, the concentration of this intermediate will not change. Therefore, we can write the differential rate expression for S_1 and apply the steady-state approximation:

$$\frac{d[S_1]}{dt} = 0 = k_a[S_0] - k_f[S_1] - k_{ic}[S_1] - k^S_{isc}[S_1] - k_q[S_1][Q] \qquad (36.159)$$

The **fluorescence lifetime** τ_f is defined as

$$\frac{1}{\tau_f} = k_f + k_{ic} + k^S_{isc} + k_q[Q] \qquad (36.160)$$

Using this definition of τ_f, Equation (36.159) becomes

$$\frac{d[S_1]}{dt} = 0 = k_a[S_0] - \frac{[S_1]}{\tau_f} \qquad (36.161)$$

Equation (36.161) is readily solved for $[S_1]$:

$$[S_1] = k_a[S_0]\tau_f \qquad (36.162)$$

The fluorescence intensity I_f depends on the rate of fluorescence given by

$$I_f = k_f[S_1] \tag{36.163}$$

Substituting Equation (36.162) into Equation (36.163) results in

$$I_f = k_a[S_0]k_f\tau_f \tag{36.164}$$

Inspection of the last two factors in Equation (36.164) illustrates the following relationship:

$$k_f\tau_f = \frac{k_f}{k_f + k_{ic} + k_{isc}^S + k_q[Q]} = \Phi_f \tag{36.165}$$

The product of the fluorescence rate constant and fluorescence lifetime is equivalent to the radiative rate constant divided by the sum of rate constants for all processes leading to the decay of S_1. In effect, S_1 decay can be viewed as a branching reaction, and the ratio of rate constants contained in Equation (36.165) can be rewritten as the quantum yield for fluorescence Φ_f, similar to the definition of reaction yield provided in Section 35.8. The fluorescence quantum yield is also defined as the number of photons emitted as fluorescence divided by the number of photons absorbed. Comparison of this definition to Equation (36.165) demonstrates that the fluorescence quantum yield will be large for molecules in which k_f is significantly greater than other rate constants corresponding to S_1 decay. Inverting Equation (36.164) and using the definition of τ_f, the following expression is obtained:

$$\frac{1}{I_f} = \frac{1}{k_a[S_0]}\left(1 + \frac{k_{ic} + k_{isc}^S}{k_f}\right) + \frac{k_q[Q]}{k_a[S_0]k_f} \tag{36.166}$$

For a fluorophore with a quantum yield approaching unity, $k_f \gg k_{ic}$ and k_{isc}^S. In fluorescence quenching experiments, fluorescence intensity is measured as a function of [Q]. Measurements are generally performed by referencing to the fluorescence intensity observed in the absence of quencher I_f^0 such that

$$\frac{I_f^0}{I_f} = 1 + \frac{k_q}{k_f}[Q] \tag{36.167}$$

Equation (36.167) reveals that a plot of the fluorescence intensity ratio as a function of [Q] will yield a straight line, with slope equal to k_q/k_f. Such plots are referred to as **Stern–Volmer plots,** an example of which is shown in Figure 36.19.

36.9.3 Measurement of τ_f

In the development presented in the preceding subsection, it was assumed that the system of interest was subjected to continuous irradiation so that the steady-state approximation could be applied to $[S_1]$. However, it is often more convenient to photoexcite the system with a temporally short burst of photons or pulse of light. If the temporal duration of the pulse is short compared to the rate of S_1 decay, the decay of this state can be measured directly by monitoring the fluorescence intensity as a function of time. Optical pulses as short as 4 fs (4×10^{-15} s) can be produced that provide excitation on a timescale that is significantly shorter than the decay time of S_1.

After excitation by a temporally short optical pulse, the concentration of molecules in $[S_1]$ will be finite. In addition, the rate constant for excitation is zero because $I_0 = 0$; therefore, the differential rate expression for S_1 becomes

$$\frac{d[S_1]}{dt} = -k_f[S_1] - k_{ic}[S_1] - k_{isc}^S[S_1] - k_q[Q][S_1]$$

$$\frac{d[S_1]}{dt} = -\frac{[S_1]}{\tau_f} \tag{36.168}$$

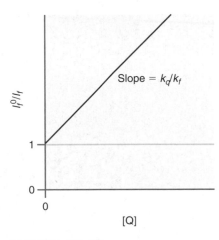

Slope = k_q/k_f

FIGURE 36.19
A Stern–Volmer plot. Intensity of fluorescence as a function of quencher concentration is plotted relative to the intensity in the absence of quencher. The slope of the line provides a measure of the quenching rate constant relative to the rate constant for fluorescence.

Equation (36.168) can be solved for $[S_1]$ resulting in

$$[S_1] = [S_1]_0 e^{-t/\tau_f} \tag{36.169}$$

Because the fluorescence intensity is linearly proportional to $[S_1]$ per Equation (36.163), Equation (36.169) predicts that the fluorescence intensity will undergo exponential decay with time constant τ_f. In the limit where $k_f \gg k_{ic}$ and $k_f \gg k_{isc}^S$, τ_f can be approximated as follows:

$$\lim_{k_f \gg k_{ic}, k_{isc}^S} \tau_f = \frac{1}{k_f + k_q[Q]} \tag{36.170}$$

In this limit, measurement of the fluorescence lifetime at a known quencher concentration combined with the slope from a Stern–Volmer plot is sufficient to uniquely determine k_f and k_q. Taking the reciprocal of Equation (36.170), we obtain

$$\frac{1}{\tau_f} = k_f + k_q[Q] \tag{36.171}$$

Equation (36.171) demonstrates that a plot of $(\tau_f)^{-1}$ versus $[Q]$ will yield a straight line with y intercept equal to k_f and slope equal to k_q.

EXAMPLE PROBLEM 36.4

Thomaz and Stevens (in *Molecular Luminescence*, New York: W. A. Benjamin Inc, 1969) studied the fluorescence quenching of pyrene in solution. Using the following information, determine k_f and k_q for pyrene in the presence of the quencher Br_6C_6.

[Br_6C_6] (M)	τ_f (s)
0.0005	2.66×10^{-7}
0.001	1.87×10^{-7}
0.002	1.17×10^{-7}
0.003	8.50×10^{-8}
0.005	5.51×10^{-8}

Solution

Using Equation (36.171), a plot of $(\tau_f)^{-1}$ versus $[Q]$ for this system is as follows:

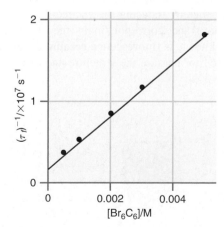

The best fit to the data by a straight line corresponds to a slope of $3.00 \times 10^9 \, s^{-1}$, which is equal to k_q by Equation (36.171), and a y intercept of $1.98 \times 10^6 \, s^{-1}$, which is equal to k_f.

FIGURE 36.20
Microscope image of single Rhodamine B dye molecules on glass. Image was obtained using a confocal scanning microscope with the bright spots in the image corresponding to molecular fluorescence. The image dimension is 5 μm by 5 μm.

36.9.4 Single-Molecule Fluorescence

Equation (36.169) describes how the population of molecules in S_1 will change with time, and the fluorescence intensity is predicted to demonstrate exponential decay. This predicted behavior is for a collection, or ensemble, of molecules; however, recent spectroscopic techniques and advances in light detection have allowed for the detection of fluorescence from a single molecule. Figure 36.20 presents an image of single molecules obtained using a confocal scanning microscope. In a confocal microscope, the excitation source and image occur at identical focal distances such that fluorescence from sample areas not directly in focus can be rejected. Using this technique in combination with laser excitation and efficient detectors, it is possible to observe the fluorescence from a single molecule. In Figure 36.20, the bright features represent fluorescence from single molecules. The spatial dimension of these features is determined by the diameter of the light beam at the sample ($\sim$300 nm).

What does the fluorescence from a single molecule look like as a function of time? Instead of a population of molecules in S_1 being responsible for the emission, the fluorescence is derived from a single molecule. Figure 36.21 presents the observed fluorescence intensity from a single molecule with continuous photoexcitation. Fluorescence is observed after the light field is turned on, and the molecule cycles between S_0 and S_1 due to photoexcitation and subsequent relaxation via fluorescence. This regime of constant fluorescence intensity continues until the fluorescence abruptly stops. At this point, depopulation of the S_1 state results in the production of T_1 or some other state of the molecule that does not fluoresce. Eventually, fluorescence is again observed at later times corresponding to the eventual recovery of S_0 by relaxation from these other states, with excitation resulting in the repopulation of S_1 followed by fluorescence. This pattern continues until a catastrophic event occurs in which the structural integrity of the molecule is lost. This catastrophic event is referred to as photodestruction, and it results in irreversible photochemical conversion of the molecule to another, nonemissive species.

Clearly, the fluorescence behavior observed in Figure 36.21 is dramatically different than the behavior predicted for an ensemble of molecules. Current interest in this field involves the application of single-molecule techniques to elucidate behavior that is not reflected by the ensemble. Such studies are extremely useful for isolating molecular dynamics from an ensemble of molecules having inherently inhomogeneous behavior. In addition, molecules can be studied in isolation of the bulk, thereby providing a window into the connection between molecular and ensemble behavior.

36.9.5 Fluorescence Resonance Energy Transfer

Another fluorescence quenching technique involves the transfer of excitation from one chromophore to another thereby reducing the excited-state population of the initially photoexcited chromophore and correspondingly the fluorescence from this chromophore. This process, known as **fluorescence resonance energy transfer,** or FRET, has been extensively used to measure the structure and dynamics of many biological

FIGURE 36.21
Fluorescence from a single Rhodamine B dye molecule. Steady illumination of the single molecule occurs at t_{on}, resulting in fluorescence, I_f. The fluorescence continues until decay of the S_1 state leads to population of a nonfluorescent state. At the end of the time axis, a brief period of fluorescence is observed corresponding to decay of the nonfluorescent state to populate S_0 followed by photoexcitation, resulting in the population of S_1 and fluorescence. However, this second period of fluorescence ends abruptly due to photodestruction of the molecule as evidenced by the absence of fluorescence after the decay event (t_{pd}).

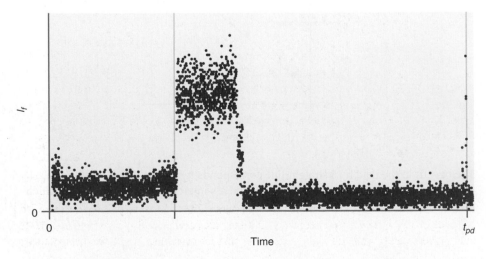

systems. The following mechanism can be employed to describe energy transfer between donor (D) and acceptor (A) chromophores:

$$D \xrightarrow{h\nu} D^* \tag{36.172}$$

$$D^* \xrightarrow{k_f} D \tag{36.173}$$

$$D^* \xrightarrow{k_{nr}} D \tag{36.174}$$

$$D^* + A \xrightarrow{k_{fret}} D + A^* \tag{36.175}$$

$$A^* \xrightarrow{k_{f'}} A \tag{36.176}$$

In this scheme, the donor is initially photoexcited, resulting in population of the first-excited singlet state. Decay from this state can occur through fluorescence, nonradiative decay (representing the sum of internal conversion and intersystem crossing), and resonant energy transfer to A resulting in this species populating the first-excited singlet state (A^*). Decay of the acceptor excited-state occurs through fluorescence of this species. Proceeding as in the previous section, in the absence of A the mechanism provides the following expression for the fluorescence quantum yield:

$$\Phi_f = \frac{k_f}{k_f + k_{nr}} \tag{36.177}$$

FRET experiments are generally performed with high-fluorescence quantum yield donors where $k_f \gg k_{nr}$. In the presence of A the expression for the fluorescence quantum yield becomes

$$\Phi_{f \, \text{w/fret}} = \frac{k_f}{k_f + k_{nr} + k_{fret}} \tag{36.178}$$

The efficiency of excitation transfer is related to the ratio of the fluorescence quantum yields as follows:

$$Eff = 1 - \frac{\Phi_{f \, w/fret}}{\Phi_f} \tag{36.179}$$

This expression illustrates that as k_{fret} becomes greater than k_f, the efficiency approaches unity.

What factors influence FRET efficiency? The theory for resonance energy transfer was first developed by T. Förster in the late 1940s. The central ideas inherent in Förster theory are that the efficiency of resonance energy transfer is dependent on the distance between the donor and acceptor, that the absorption band of the acceptor should overlap with the fluorescence band of the donor (that is, $S_0 - S_1$ energy gaps of the donor and acceptors are comparable), and that the relative orientation of the donor and acceptor pair will influence the efficiency of transfer. The distance dependence of the transfer efficiency predicted by Förster theory is

$$Eff = \frac{r_0^6}{r_0^6 + r^6} \tag{36.180}$$

In Equation (36.180), r is the separation distance between donor and acceptor, and r_0 is a pair-dependent quantity that defines the distance at which the transfer efficiency is 0.5. The value of r_0 depends on the spectral overlap of the donor fluorescence and acceptor absorption as well as the relative orientation between donor and acceptor:

$$r_0(\text{Å}) = 8.79 \times 10^{-5} \left(\frac{\kappa^2 J \Phi_f}{n^4} \right)^{1/6} \tag{36.181}$$

In this expression, κ depends on the relative orientation of the transition dipole moments, and is equal to 0 if they are perpendicular, 2 if they are parallel, and $1/3$ for random orientation. Since κ can be difficult to determine, the random-orientation value for this quantity is

TABLE 36.2 Values of R_0 for FRET Pairs

Donor	Acceptor	r_0 (Å)
EDANS	DABCYL	33
Pyrene	Coumarin	39
Dansyl	Octadecylrhodamine	43
IAEDANS	Fluorescein	46
Fluorescein	Tetramethylrhodamine	55

IAEDANS = 5-(((2-iodoacetyl)amino)ethyl)amino)naphthalene-1-sulfonic acid

EDANS = 5-((2-aminoethyl)amino)naphthalene-1-sulfonic acid

DABCYL = 4-((4-(dimethylamino)phenyl)azo)benzoic acid, succinimidyl ester

generally assumed. Also in the expression for r_0, Φ_f is the fluorescence quantum yield of the donor, n is the refractive index of the medium in which the transfer occurs, and J is the overlap integral between the donor fluorescence and acceptor absorption expressed as:

$$J = \int \varepsilon_A(\lambda)F_D(\lambda)\lambda^4 d\lambda \qquad (36.182)$$

In this expression, ε_A is the extinction coefficient of the acceptor, F_D is the fluorescence spectrum of the donor, and the integral is performed over all wavelengths. The value of this integral, and correspondingly the value for r_0, will vary as a function of donor–acceptor pair. To illustrate the connection between donor emission, acceptor absorption, and the overlap integral, Figure 36.22 presents the emission and absorption for the fluorescein/tetramethylrhodamine (TMR) FRET pair, and provides an illustration of J for this FRET pair.

A collection of FRET donor–acceptor pairs are presented in Table 36.2. When using FRET to measure distance, it is critical to choose a donor–acceptor pair whose r_0 is close to the length scale of interest. For distances where $r \gg r_0$, the FRET efficiency will be 0 so that the fluorescence quantum yield of the donor will be largely unaffected by the presence of the acceptor. In the other limit where $r_0 \gg r$, the quenching of the donor fluorescence from energy transfer will be extremely efficient and little emission from the donor will be observed. The overlap between the fluorescein emission and TMR absorption is shown.

EXAMPLE PROBLEM 36.5

You are designing a FRET experiment to determine the magnitude of the structural change introduced by substrate binding to an enzyme. Using site-specific mutagenesis, you have constructed a mutant form of the enzyme that possesses a single tyrosine residue and a single tryptophan residue, and these residues are separated by 11 Å. You would like to determine if the distance between these residues changes with substrate binding. The fluorescence of tyrosine overlaps with the tryptophan absorption; therefore, these two amino acids form a FRET pair for which $r_0 = 9$ Å, determined using the absorption and emission spectra in combination with Equation (36.181). Calculate the FRET efficiency at 11 Å separation and how much this distance must increase in order for the efficiency to decrease by 20%, the experimental detection limit.

Solution

Using the initial separation distance and r_0, the efficiency is determined as follows:

$$Eff = \frac{r_0^6}{r_0^6 + r^6} = \frac{(9\ \text{Å})^6}{(9\ \text{Å})^6 + (11\ \text{Å})^6} = 0.23$$

The detection limit corresponds to $Eff = 0.18$. Solving for r yields:

$$Eff = 0.18 = \frac{r_0^6}{r_0^6 + r^6} = \frac{(9\ \text{Å})^6}{(9\ \text{Å})^6 + r^6}$$

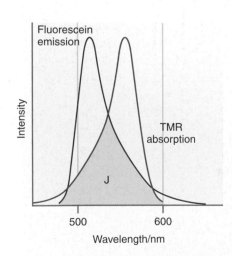

FIGURE 36.22
Illustration of the overlap integral ($J(\lambda)$) for the fluorescein/tetramethylrhodamine (TMR) FRET donor–acceptor pair.

$$\frac{(9\ \text{Å})^6}{0.18} = (9\ \text{Å})^6 + r^6$$

$$2.42 \times 10^6\ \text{Å}^6 = r^6$$

$$11.6\ \text{Å} = r$$

Notice that for this FRET pair the modification of the tyrosine–tryptophan separation accompanying substrate binding can be measured for a relatively limited change in r. This example illustrates the importance of choosing FRET pairs having r_0 values that are close to the length scale of interest.

An important application of resonant energy transfer involves light harvesting in photosynthetic pigments. The absorption of light is enabled by light-harvesting by antenna pigments contained in the thylakoid membranes of the chloroplast, the photosynthetic organelles of green plants and algae. These systems have evolved such that light primarily in the visible and near-infrared regions of the electromagnetic spectrum are absorbed corresponding to electronic transitions of the pigments. The most abundant plant pigments are chlorophylls a and b as illustrated in Figure 36.23. In addition, carotenoids such as β-carotene also serve as light-harvesting pigments. These light-harvesting pigments are organized into light-harvesting complexes. X-ray crystallographic studies have shown that the chlorophyll molecules are arranged in symmetric cyclical structures in the light-harvesting complexes. The structure of light-harvesting complex II consisting of chlorophyll and β-carotene is shown in Figure 36.23. This spatial arrangement of light-harvesting pigments provides for efficient inter-pigment transfer once a photon is absorbed and for transfer of this energy outside the light-harvesting complex.

The biological purpose of light-harvesting complexes is to absorb solar radiation and through resonance energy transfer deliver this energy to the reaction center. The intensity of sunlight is such that the probability of a chlorophyll molecule absorbing a photon is extremely modest. Incorporating numerous chlorophylls into a single light-harvesting complex provides for maximum collection efficiency. Resonance energy transfer results in the migration of absorbed photon energy from the light-harvesting complex to the reaction center as illustrated in Figure 36.24. Once the energy is transferred to the reaction center, it initiates an electron transfer process that is the start of

FIGURE 36.23
Left. Structures of photosynthetic light-harvesting pigments chlorophyll (a), β-carotene (b), and phycoerythrin (c). Chlorophylls a and b are the most abundant plant pigments. **Right.** X-ray crystal structure of the light-harvesting complex II of purple photosynthetic bacteria. The pigments, including bacteriochlorophyll (green) are held in this spatially complex arrangement by the surrounding protein (parts of which are shown in red and white).

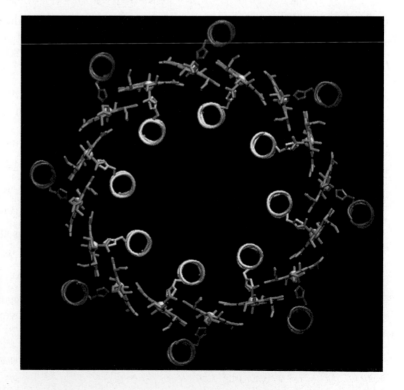

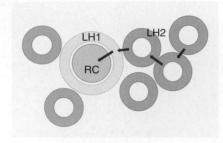

FIGURE 36.24
Schematic illustration of resonance energy transfer in photosynthesis where the energy initially obtained by photon absorption of a light-harvesting pigment is transferred to the reaction center through a series of resonance energy transfer steps.

the chemical transformations involved in photosynthesis. This electron transfer reaction will be further explored in the upcoming section on electron transfer.

36.9.6 Photochemical Processes

As discussed earlier, photochemical processes are distinct from photophysical processes in that the absorption of a photon results in chemical transformation of the reactant. For a photochemical process that occurs through the first excited singlet state S_1, a photochemical reaction can be viewed kinetically as another reaction branch, resulting in the decay of S_1. The corresponding expression for the rate corresponding to this photochemical reaction branch is

$$R_{photochem.} = k_{photo}^S [S_1] \qquad (36.183)$$

where k_{photo} is the rate constant for the photochemical reaction. For photochemical processes occurring through T_1, a rate expression similar to Equation (36.183) can be constructed as follows:

$$R_{photochem.} = k_{photo}^T [T_1] \qquad (36.184)$$

The absorption of a photon can also provide sufficient energy to initiate a chemical reaction. However, given the range of photophysical processes that occurs, absorption of a photon is not sufficient to guarantee that the photochemical reaction will occur. The extent of photochemistry is quantified by the overall **quantum yield** ϕ, which is defined as the number of reactant molecules consumed in photochemical processes per photon absorbed. The overall quantum yield can be greater than one, as demonstrated by the photoinitiated decomposition of $HI(g)$ that proceeds by the following mechanism:

$$HI + h\nu \rightarrow H\cdot + I\cdot \qquad (36.185)$$

$$H\cdot + HI \rightarrow H_2 + I\cdot \qquad (36.186)$$

$$I\cdot + I\cdot \rightarrow I_2 \qquad (36.187)$$

In this mechanism, absorption of a photon results in the loss of two HI molecules such that $\phi = 2$. In general, the overall quantum yield can be determined experimentally by comparing the molecules of reactant lost to the number of photons absorbed as illustrated in Example Problem 36.6.

EXAMPLE PROBLEM 36.6

The reactant 1,3-cyclohexadiene can be photochemically converted to *cis*-hexatriene. In an experiment, 2.5 mmol of cyclohexadiene are converted to *cis*-hexatriene when irradiated with 100. W of 280. nm light for 27.0 s. All of the light is absorbed by the sample. What is the overall quantum yield for this photochemical process?

Solution

First, the total photon energy absorbed by the sample, E_{abs}, is

$$E_{abs} = (power)\Delta t = (100.\ J\,s^{-1})(27.0\ s) = 2.70 \times 10^3\ J$$

Next, the photon energy at 280. nm is

$$E_{ph} = \frac{hc}{\lambda} = \frac{(6.626 \times 10^{-34}\ J\,s)(2.998 \times 10^8\ m\,s^{-1})}{2.80 \times 10^{-7}\ m} = 7.10 \times 10^{-19}\ J$$

The total number of photons absorbed by the sample is therefore

$$\frac{E_{abs}}{E_{ph}} = \frac{2.70 \times 10^3\ J}{7.10 \times 10^{-19}\ J\ photon^{-1}} = 3.80 \times 10^{21}\ photons$$

Dividing this result by Avogadro's number results in 6.31×10^{-3} Einsteins or moles of photons. Therefore, the overall quantum yield is

$$\phi = \frac{moles_{react}}{moles_{photon}} = \frac{2.50 \times 10^{-3}\ mol}{6.31 \times 10^{-3}\ mol} = 0.396 \approx 0.40$$

36.10 Electron Transfer

Electron transfer reactions involve the exchange of charge between chemical species. These reactions are ubiquitous in biological chemistry. For example, photosynthesis involves electron transfer in biological energy transduction. In plants, the net reaction from photosynthesis is the conversion of $CO_2(g)$ and $H_2O(l)$ to form carbohydrates and molecular oxygen:

$$6CO_2(g) + 6H_2O(l) \rightarrow C_6H_{12}O_6(s) + 6O_2(g) \qquad \Delta G^{\circ} = 2870 \text{ kJ}$$

Photosynthesis also occurs in certain bacteria. For example, in green sulfur bacteria H_2S serves as the reactant rather than water:

$$2H_2S(aq) + CO_2(g) \rightarrow CH_2O(g) + H_2O(l) + 2S(rh) \qquad \Delta G^{\circ} = 88 \text{ kJ}$$

In these reactions, the carbon in $CO_2(g)$ is reduced and H_2O/H_2S is oxidized such that net reactions involve the electron transfer. Although photosynthesis is based on a series of coupled reactions rather than the single-step reaction shown earlier, the net reactions do illustrate the importance of electron transfer in this process.

In the previous section we discussed the transfer of radiative energy from the light-harvesting complex to the photosynthetic reaction center. The radiative energy transferred to the reaction center is used to initiate an electron transfer reaction from a pair of chlorophyll molecules (known as the special pair) to a nearby pheophytin molecule (structurally similar to chlorophyll). This process is extremely fast, occurring in approximately 3 ps (3×10^{-12} s). A second electron transfer process occurs in which the electron on the pheophytin is transferred to a quinone in 200 ps. Ultimately, the electron is transferred to a second quinone in 100 μs. The electrons on the quinone are used to "split" water as follows:

$$2H_2O(l) \rightarrow O_2(g) + 4H^+(aq) + 4e^-$$

The electrons and hydrogen ions are transported in a series of sequential reactions ultimately resulting in the production of carbohydrate and molecular oxygen. The net effect of this transport is to create a transmembrane proton gradient that is used by the cell to drive ATP synthesis. Given the importance of electron transfer in photosynthesis and other areas of biological chemistry discussed earlier in the text, in this section we investigate simple kinetic models for charge transfer and explore **Marcus theory** for describing electron transfer processes.

36.10.1 Kinetic Model of Electron Transfer

Electron transfer involves the exchange of an electron from a donor molecule (D) to an acceptor molecule (A) resulting in donor oxidation (D^+) and reduction of the acceptor (A^-):

$$D + A \rightleftharpoons D^+ + A^- \qquad \text{(36.188)}$$

The reaction mechanism employed to describe bimolecular electron transfer in solution is as follows. First, the donor and acceptor form the donor–acceptor complex by diffusing in solution until these species come in contact. Formation of the donor–acceptor complex is modeled as a reversible process; therefore,

$$D + A \underset{k_{d'}}{\overset{k_d}{\rightleftharpoons}} DA \qquad \text{(36.189)}$$

In the next step of the mechanism, electron transfer from the donor to the acceptor occurs in the complex. In addition, the back-electron transfer process is also possible, corresponding to reformation of the donor–acceptor complex:

$$DA \underset{k_{b-et}}{\overset{k_{et}}{\rightleftharpoons}} D^+A^- \qquad \text{(36.190)}$$

In the final step of the mechanism, the post-electron transfer complex separates to form isolated oxidized donor and reduced acceptor:

$$D^+A^- \xrightarrow{k_{sep}} D^+ + A^- \tag{36.191}$$

The expression for the rate of reaction consistent with this mechanism is derived as follows. The last step of the mechanism results in the formation of products; therefore, the reaction rate can be written as

$$R = k_{sep}[D^+A^-] \tag{36.192}$$

Writing the differential-rate expression for D^+A^- and applying the steady-state approximation results in the following expression for $[D^+A^-]$:

$$\frac{d[D^+A^-]}{dt} = 0 = k_{et}[DA] - k_{b-et}[D^+A^-] - k_{sep}[D^+A^-]$$

$$[D^+A^-] = \frac{k_{et}[DA]}{k_{b-et} + k_{sep}} \tag{36.193}$$

The donor–acceptor complex (DA) is also an intermediate; therefore, writing the differential rate expression for this species and applying the steady-state approximation yields

$$\frac{d[DA]}{dt} = 0 = k_d[D][A] - k_{d'}[DA] - k_{et}[DA] + k_{b-et}[D^+A^-]$$

$$[DA] = \frac{k_d[D][A] + k_{b-et}[D^+A^-]}{k_{d'} + k_{et}} \tag{36.194}$$

Substituting this result into the expression for $[D^+A^-]$ results in the following:

$$[D^+A^-] = \frac{k_{et}[DA]}{k_{b-et} + k_{sep}} = \left(\frac{k_{et}}{k_{b-et} + k_{sep}}\right)\left(\frac{k_d[D][A] + k_{b-et}[D^+A^-]}{k_{d'} + k_{et}}\right)$$

$$[D^+A^-] = \frac{k_{et}k_d}{k_{b-et}k_{d'} + k_{sep}k_{d'} + k_{sep}k_{et}}[D][A] \tag{36.195}$$

Using this expression for $[D^+A^-]$, the reaction rate becomes

$$R = \frac{k_{sep}k_{et}k_d}{k_{b-et}k_{d'} + k_{sep}k_{d'} + k_{sep}k_{et}}[D][A] \tag{36.196}$$

The reaction is predicted to be first order with respect to both donor and acceptor concentrations (as expected), and the apparent microscopic rate constant is a composite of the microscopic rate constants for the various steps in the mechanism. If we assume that dissociation of the post-electron transfer complex is rapid compared to back-electron transfer ($k_{sep} \gg k_{b-et}$), then the reaction rate becomes

$$R = \frac{k_{et}k_d}{k_{d'} + k_{et}}[D][A] \tag{36.197}$$

Finally, if the rate constant for electron transfer is sufficiently large relative to dissociation of the donor–acceptor complex ($k_{et} \gg k_{d'}$), then formation of the donor acceptor becomes the rate-limiting step and electron transfer becomes a diffusion-controlled reaction as described in the previous chapter. In the opposite limit ($k_{et} \ll k_{d'}$), electron transfer becomes the rate-limiting step and the expression for the reaction rate becomes

$$R_{k_{d'} \gg k_{et}} = k_{et}K_{d,d'}[D][A] = k_{exp}[D][A] \tag{36.198}$$

In this expression, $K_{d,d'}$ is the equilibrium constant for the $D + A \rightleftharpoons DA$ step of the reaction mechanism described earlier, and k_{exp} is the experimentally measured rate constant. In this limit, the rate of electron transfer is determined by the rate constant for the transfer process itself. This later limit is also representative of systems in which the donor and acceptor are already in contact (for example, by covalently linking the donor and acceptor), and in many biological systems in which the electron donor and acceptor are held at a fixed distance by the surrounding protein matrix.

36.10.2 Marcus Theory

What factors determine the magnitude of the observed rate constant for electron transfer? The kinetic model for electron transfer presented earlier suggests two important factors in determining the value of k_{exp}. First, the model requires that donor and acceptor be in proximity before electron transfer can occur. Therefore, the rate constant should depend on the separation distance between the donor and acceptor. This distance dependence is expressed as

$$k_{exp} \propto e^{-\beta r} \tag{36.199}$$

In this expression, β is a constant that varies as a function of the system of interest and the medium in which electron transfer occurs, and r is the donor–acceptor separation distance. The second factor that influences the observed rate constant is thermodynamic in origin. If the charge transfer occurs over an energy barrier, we can refer back to transition state theory described in the previous chapter and model this barrier as the difference in free energy between the donor–acceptor complex and the activated complex corresponding to the free-energy maximum along the reaction coordinate connecting the donor–acceptor complex with the complex after electron transfer. Therefore, the experimentally observed rate constant for electron transfer depends on the Gibbs energy of activation $\Delta G^{\ddagger}$ as follows:

$$k_{exp} \propto e^{-\Delta G^{\ddagger}/kT} \tag{36.200}$$

Therefore, the rate constant for electron transfer can be written as

$$k_{exp} \propto e^{-\beta r} e^{-\Delta G^{\ddagger}/kT} \tag{36.201}$$

Rudolph Marcus received the Nobel Prize in Chemistry in 1992 for his contributions to defining this relationship and exploring the chemical factors that are important in determining the value of $\Delta G^{\ddagger}$. Specifically, Marcus noted that after electron transfer both the solvent and solute will undergo relaxation. For the solutes, the transfer of an electron will result in a change in bond order and the nuclei will relax to assume new equilibrium geometries. The solvent will rearrange in response to the change in charge distribution accompanying the formation of D^{+} and A^{-}. Taking these factors into account, Marcus defined the Gibbs energy of activation in terms of the standard change in Gibbs energy accompanying the charge transfer $(DA \rightleftharpoons D^{+}A^{-})$ and the change in Gibbs energy accompanying relaxation of the solvent and solute, referred to as the **reorganization energy** or λ:

$$\Delta G^{\ddagger} = \frac{(\Delta G^{\circ} + \lambda)^2}{4\lambda} \tag{36.202}$$

The reorganization energy can be thought of as the Gibbs energy accompanying rearrangement of bonds in the donor and acceptor as well as solvent rearrangement of accompanying evolution along the electron-transfer reaction coordinate. This expression demonstrates that the Gibbs energy of activation will be at a minimum when $-\Delta G^{\circ} = \lambda$. If the separation distance of the donor and acceptor is fixed, then the experimentally observed rate constant will be greatest at this value of ΔG°. A sketch of $\ln(k_{exp})$ versus $-\Delta G^{\circ}$ is presented in Figure 36.25. The maximum in the rate constant occurs when $-\Delta G^{\circ} = \lambda$, but notice that the rate constant is predicted to decrease as the charge-transfer state decreases in free energy relative to the neutral state. This is the so-called Marcus inverted regime of electron transfer.

The connection between the rate constant for reaction and the Gibbs energy for the reactants and products is presented in Figure 36.26. Two curves are shown, one representing the neutral donor and acceptor, and the other representing the donor cation and acceptor anion formed as a result of electron transfer. The potential energy surfaces representing donor and acceptor configurations are parabolic, consistent with the quadratic dependence of energy with displacement along nuclear vibrational coordinates (i.e., $V(x) = 1/2 \, kx^2$ where k is the force constant of the bond and x is displacement). The electron transfer occurs at the point along the reaction coordinate where the two

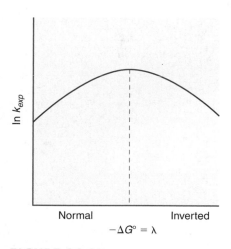

FIGURE 36.25

Illustration of the Marcus normal and inverted regimes. The experimentally measured rate constant is predicted to increase as the driving force for the reaction increases corresponding to the normal regime, and be at a maximum when $-\Delta G^{\circ} = \lambda$. As the driving force for the reaction increases beyond this point, the rate constant should decrease corresponding to the inverted regime.

FIGURE 36.26
Plots of the Gibbs energy versus the reaction coordinate for electron transfer. Three different values for $\Delta G°$ are presented corresponding to the normal, maximum, and inverted Marcus regimes.

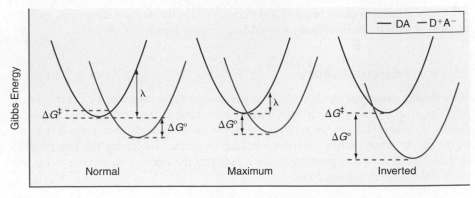

curves cross so that energy is conserved in the transfer. In addition, electron transfer is considered to be rapid compared to nuclear motion; therefore, the transfer occurs at one point along the reaction coordinate. In the normal regime, the value of $\Delta G°$ is such that there is a barrier to electron transfer that must be overcome. This barrier corresponds to $\Delta G^{\ddagger}$ as described earlier. As the Gibbs energy of the charge transfer state is reduced relative to that of the reactant state, the thermodynamic driving force for the reaction increases, and when $-\Delta G° = \lambda$, the electron transfer is predicted to be a barrierless process corresponding to a maximum in the rate constant for electron transfer. A further decrease in the Gibbs energy of the products relative to the reactants results in an increased barrier to electron transfer and corresponding reduction in the reaction rate constant, corresponding to the Marcus inverted regime.

A significant amount of experimental research has been performed to test the validity of Marcus theory, and in particular to obtain evidence for the inverted regime. One successful demonstration of the inverted regime is illustrated in Figure 36.27. Here, the biphenyl donor was connected to a variety of acceptors through a cyclohexane scaffold. As the acceptor is changed, $\Delta G°$ is altered. The data clearly illustrate a decrease in the observed rate constant for electron transfer with increased driving force for the reaction beyond ~ 1.1 eV. For other systems inverted-regime behavior has not been observed. In these systems, the participation of solute vibrational degrees of freedom in the electron transfer process decreases the barrier to formation of the activated complex in the inverted regime; therefore, the rate constant remains close to the maximum value even as the driving force for the reaction increases.

EXAMPLE PROBLEM 36.7

The experimental results presented earlier demonstrate that a maximum in the rate constant for electron transfer $(\sim 2.0 \times 10^9 \text{ s}^{-1})$ occurs when $-\Delta G° = 1.20$ eV. Given this observation, estimate the rate constant for electron transfer when 2-naphthoquinoyl is employed as the acceptor for which $-\Delta G° = 1.93$ eV.

Solution

The rate constant for electron transfer is predicted to be at a maximum when $-\Delta G° = \lambda$; therefore, $\lambda = 1.20$ eV. Using this information, the barrier to electron transfer is determined as follows:

$$\Delta G^{\ddagger} = \frac{(\Delta G° + \lambda)^2}{4\lambda} = \frac{(-1.93 \text{ eV} + 1.20 \text{ eV})^2}{4(1.20 \text{ eV})} = 0.111 \text{ eV}$$

With the barrier to electron transfer, the rate constant is estimated as follows:

$$\frac{k_{1.93\,\text{eV}}}{k_{\max}} = e^{-\Delta G^{\dagger}/kT} = exp\left(-0.111 \text{ eV} \times \frac{1.60 \times 10^{-19}\text{J}}{\text{eV}} \middle/ (1.38 \times 10^{-23}\text{J K}^{-1})(296 \text{ K})\right)$$

$$= 0.0129$$

$$k_{1.93\,\text{eV}} = k_{\max}(0.0129) = (2.0 \times 10^9 \text{ s}^{-1})(0.0129) = 2.6 \times 10^7 \text{ s}^{-1}$$

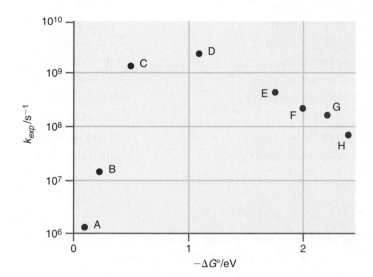

FIGURE 36.27
Experimental verification of the Marcus inverted regime in electron transfer. In this experiment, the electron transfer rates between the biphenyl donor and the various acceptors were measured in 2-methyltetrahydrofuran at 296 K. Notice that the rate constant reaches a maximum and then decreases as $-\Delta G°$ increases corresponding to an increase in driving force for the reaction. Figure adapted from Miller et al., *J. American Chemical Society*, 106 (1984): 3047.

Vocabulary

absorption cross section	Belousov-Zhabotinsky (BZ) reaction	collisional quenching
activated reactant	branching reaction	competitive inhibitor
adsorption isotherm	catalyst	complex reaction
autocatalysis	chemisorption	composite constant

electron transfer	Lineweaver–Burk equation	propagation step
enzymes	Lineweaver–Burk plot	quantum yield
feedback	Lotka-Volterra mechanism	radiative transitions
fluorescence	Marcus theory	radical
fluorescence lifetime	Michaelis constant	radical polymerization reaction
fluorescence resonance energy transfer	Michaelis–Menten mechanism	reaction intermediate
fractional coverage	Michaelis–Menten rate law	reaction mechanism
heterogeneous catalyst	molar absorptivity	reciprocal plot
homogeneous catalyst	nonlinear reaction	reorganization energy
initiation step	oscillating reaction	simple reaction
internal conversion	phosphorescence	Stern–Volmer plot
intersystem crossing	photochemistry	stoichiometric number
Jablonski diagram	photophysical processes	substrate–catalyst complex
Langmuir isotherm	physisorption	termination step
Langmuir model	polymer chain	turnover number
Lindemann mechanism	preequilibrium approximation	unimolecular reactions

Conceptual Problems

Q36.1 How is a simple reaction different from a complex reaction?

Q36.2 For a reaction mechanism to be considered correct, what property must it demonstrate?

Q36.3 What is a reaction intermediate? Can an intermediate be present in the rate law expression for the overall reaction?

Q36.4 What is the preequilibrium approximation, and under what conditions is it considered valid?

Q36.5 What is the main assumption in the Lindemann mechanism for unimolecular reactions?

Q36.6 How is a catalyst defined, and how does such a species increase the reaction rate?

Q36.7 What is an enzyme? What is the general mechanism describing enzyme catalysis?

Q36.8 What is the Michaelis–Menten rate law? What is the maximum reaction rate predicated by this rate law?

Q36.9 How is the standard enzyme kinetic scheme modified to incorporate competitive inhibition? What plot is used to establish competitive inhibition and to determine the kinetic parameters associated with inhibition?

Q36.10 What is the difference between a homogeneous and a heterogeneous catalyst?

Q36.11 What are the inherent assumptions in the Langmuir model of surface adsorption?

Q36.12 What is a radical? What elementary steps are involved in a reaction mechanism involving radicals?

Q36.13 In what ways are radical polymerization reactions similar to radical reactions in general?

Q36.14 What is autocatalysis?

Q36.15 What is an oscillating reaction?

Q36.16 What is photochemistry? How does one calculate the energy of a photon?

Q36.17 What depopulation pathways occur from the first excited singlet state? For the first excited triplet state?

Q36.18 What is the expected variation in excited state lifetime with quencher concentration in a Stern–Volmer plot?

Q36.19 What is the distance dependence for fluorescence resonance energy transfer predicted by the Förster model?

Q36.20 What two factors influence the electron transfer rate constant according to Marcus theory?

Numerical Problems

Problem numbers in **red** indicate that the solution to the problem is given in the *Student's Solutions Manual*.

P36.1 A proposed mechanism for the formation of $N_2O_5(g)$ from $NO_2(g)$ and $O_3(g)$ is

$$NO_2(g) + O_3(g) \xrightarrow{k_1} NO_3(g) + O_2(g)$$

$$NO_2(g) + NO_3(g) + M(g) \xrightarrow{k_2} N_2O_5(g) + M(g)$$

Determine the rate law expression for the production of $N_2O_5(g)$ given this mechanism.

P36.2 The reaction of nitric oxide ($NO(g)$) with molecular hydrogen ($H_2(g)$) results in the production of molecular nitrogen and water as follows:

$$2NO(g) + 2H_2(g) \rightarrow N_2O(g) + 2H_2O(g)$$

The experimentally-determined rate-law expression for this reaction is first order in $H_2(g)$ and second order in $NO(g)$.

a. Is the reaction as written consistent with the experimental order dependence for this reaction?

b. One potential mechanism for this reaction is as follows:

$$H_2(g) + 2NO(g) \xrightarrow{k_1} N_2O(g) + H_2O(g)$$

$$H_2(g) + N_2O(g) \xrightarrow{k_2} N_2(g) + H_2O(g)$$

Is this mechanism consistent with the experimental rate law?

c. An alternative mechanism for the reaction is

$$2NO(g) \underset{k_{-1}}{\overset{k_1}{\rightleftharpoons}} N_2O_2(g) \text{ (fast)}$$

$$H_2(g) + N_2O_2(g) \xrightarrow{k_2} N_2O(g) + H_2O(g)$$

$$H_2(g) + N_2O(g) \xrightarrow{k_3} N_2(g) + H_2O(g)$$

Show that this mechanism is consistent with the experimental rate law.

P36.3 In the troposphere carbon monoxide and nitrogen dioxide undergo the following reaction:

$$NO_2(g) + CO(g) \rightarrow NO(g) + CO_2(g)$$

Experimentally, the rate law for the reaction is second order in $NO_2(g)$, and $NO_3(g)$ has been identified as an intermediate in this reaction. Construct a reaction mechanism that is consistent with these experimental observations.

P36.4 The Rice–Herzfeld mechanism for the thermal decomposition of acetaldehyde ($CH_3CO(g)$) is

$$CH_3CHO(g) \xrightarrow{k_1} CH_3\cdot(g) + CHO\cdot(g)$$

$$CH_3\cdot(g) + CH_3CHO(g) \xrightarrow{k_2} CH_4(g) + CH_2CHO\cdot(g)$$

$$CH_2CHO\cdot(g) \xrightarrow{k_3} CO(g) + CH_3\cdot(g)$$

$$CH_3\cdot(g) + CH_3\cdot(g) \xrightarrow{k_4} C_2H_6(g)$$

Using the steady-state approximation, determine the rate of methane ($CH_4(g)$) formation.

P36.5 Consider the following mechanism for ozone thermal decomposition:

$$O_3(g) \underset{k_{-1}}{\overset{k_1}{\rightleftharpoons}} O_2(g) + O(g)$$

$$O_3(g) + O(g) \xrightarrow{k_2} 2O_2(g)$$

a. Derive the rate law expression for the loss of $O_3(g)$.

b. Under what conditions will the rate law expression for $O_3(g)$ decomposition be first order with respect to $O_3(g)$?

P36.6 Consider the formation of double-stranded (DS) DNA from two complementary single strands (S and S′) through the following mechanism involving an intermediate helix (IH):

$$S + S' \underset{k_{-1}}{\overset{k_1}{\rightleftharpoons}} IH$$

$$IH \xrightarrow{k_2} DS$$

a. Derive the rate law expression for this reaction employing the preequilibrium approximation.

b. What is the corresponding rate-law expression for the reaction employing the steady state approximation for the intermediate IH?

P36.7 The hydrogen–bromine reaction corresponds to the production of $HBr(g)$ from $H_2(g)$ and $Br_2(g)$ as follows: $H_2(g) + Br_2(g) \rightleftharpoons 2HBr(g)$. This reaction is famous for its complex rate law, determined by Bodenstein and Lind in 1906:

$$\frac{d[HBr]}{dt} = \frac{k[H_2][Br_2]^{1/2}}{1 + \dfrac{m[HBr]}{[Br_2]}}$$

where k and m are constants. It took 13 years for a likely mechanism of this reaction to be proposed, and this feat was accomplished simultaneously by Christiansen, Herzfeld, and Polyani. The mechanism is as follows:

$$Br_2(g) \underset{k_{-1}}{\overset{k_1}{\rightleftharpoons}} 2Br\cdot(g)$$

$$Br\cdot(g) + H_2(g) \xrightarrow{k_2} HBr(g) + H\cdot(g)$$

$$H\cdot(g) + Br_2(g) \xrightarrow{k_3} HBr(g) + Br\cdot(g)$$

$$HBr(g) + H\cdot(g) \xrightarrow{k_4} H_2(g) + Br\cdot(g)$$

Construct the rate law expression for the hydrogen–bromine reaction by performing the following steps:

a. Write down the differential rate expression for [HBr].

b. Write down the differential rate expressions for [Br·] and [H·].

c. Because $Br\cdot(g)$ and $H\cdot(g)$ are reaction intermediates, apply the steady-state approximation to the result of part (b).

d. Add the two equations from part (c) to determine [Br·] in terms of [Br_2].

e. Substitute the expression for [Br·] back into the equation for [H·] derived in part (c) and solve for [H·].

f. Substitute the expressions for [Br·] and [H·] determined in part (e) into the differential rate expression for [HBr] to derive the rate law expression for the reaction.

P36.8

a. For the hydrogen–bromine reaction presented in Problem P36.7 imagine initiating the reaction with only Br_2 and H_2 present. Demonstrate that the rate law expression at $t = 0$ reduces to

$$\left(\frac{d[HBr]}{dt}\right)_{t=0} = 2k_2\left(\frac{k_1}{k_{-1}}\right)^{1/2}[H_2]_0[Br_2]_0^{1/2}$$

b. The activation energies for the rate constants are as follows:

Rate Constant	ΔE_a (kJ/mol)
k_1	192
k_2	0
k_{-1}	74

What is the overall activation energy for this reaction?

c. How much will the rate of the reaction change if the temperature is increased to 400. K from 298 K?

P36.9 For the reaction $I^-(aq) + OCl^-(aq) \rightleftharpoons OI^-(aq) + Cl^-(aq)$ occurring in aqueous solution, the following mechanism has been proposed:

$$OCl^-(aq) + H_2O(l) \underset{k_{-1}}{\overset{k_1}{\rightleftharpoons}} HOCl(aq) + OH^-(aq)$$

$$I^-(aq) + HOCl(aq) \overset{k_2}{\longrightarrow} HOI(aq) + Cl^-(aq)$$

$$HOI(aq) + OH^-(aq) \overset{k_3}{\longrightarrow} H_2O(l) + OI^-(aq)$$

a. Derive the rate law expression for this reaction based on this mechanism. (*Hint:* [OH⁻] should appear in the rate law.)

b. The initial rate of reaction was studied as a function of concentration by Chia and Connick [*J. Physical Chemistry* 63 (1959): 1518], and the following data were obtained:

$[I^-]_0$ (M)	$[OCl^-]_0$ (M)	$[OH^-]_0$ (M)	Initial Rate (M s⁻¹)
2.0×10^{-3}	1.5×10^{-3}	1.00	1.8×10^{-4}
4.0×10^{-3}	1.5×10^{-3}	1.00	3.6×10^{-4}
2.0×10^{-3}	3.0×10^{-3}	2.00	1.8×10^{-4}
4.0×10^{-3}	3.0×10^{-3}	1.00	7.2×10^{-4}

Is the predicted rate law expression derived from the mechanism consistent with these data?

P36.10 Using the preequilibrium approximation, derive the predicted rate law expression for the following mechanism:

$$A_2 \underset{k_{-1}}{\overset{k_1}{\rightleftharpoons}} 2A$$

$$A + B \overset{k_2}{\longrightarrow} P$$

P36.11 Consider the following mechanism, which results in the formation of product P:

$$A \underset{k_{-1}}{\overset{k_1}{\rightleftharpoons}} B \underset{k_{-2}}{\overset{k_2}{\rightleftharpoons}} C$$

$$B \overset{k_3}{\rightarrow} P$$

If only the species A is present at $t = 0$, what is the expression for the concentration of P as a function of time? You can apply the preequilibrium approximation in deriving your answer.

P36.12 Consider the gas-phase isomerization of cyclopropane:

$$\underset{CH_2-CH_2}{\overset{CH_2}{\diagdown \diagup}} \longrightarrow CH_3CH=CH_2$$

Are the following data of the observed rate constant as a function of pressure consistent with the Lindemann mechanism?

P (Torr)	k (10⁴ s⁻¹)	P (Torr)	k (10⁴ s⁻¹)
84.1	2.98	1.36	1.30
34.0	2.82	0.569	0.857
11.0	2.23	0.170	0.486
6.07	2.00	0.120	0.392
2.89	1.54	0.067	0.303

P36.13 In the discussion of the Lindemann mechanism, it was assumed that the rate of activation by collision with another reactant molecule A was the same as collision with a nonreactant molecule M such as a buffer gas. What if the rates of activation for these two processes are different? In this case, the mechanism becomes

$$A + M \underset{k_{-1}}{\overset{k_1}{\rightleftharpoons}} A^* + M$$

$$A + A \underset{k_{-2}}{\overset{k_2}{\rightleftharpoons}} A^* + A$$

$$A^* \overset{k_3}{\longrightarrow} P$$

a. Demonstrate that the rate law expression for this mechanism is

$$R = \frac{k_3(k_1[A][M] + k_2[A]^2)}{k_{-1}[M] + k_{-2}[A] + k_{-3}}$$

b. Does this rate law reduce to the expected form when $[M] = 0$?

P36.14 In the unimolecular isomerization of cyclobutane to butylene, the following values for k_{uni} as a function of pressure were measured at 350 K:

P_0 (Torr)	110	210	390	760
k_{uni} (s⁻¹)	9.58	10.3	10.8	11.1

Assuming that the Lindemann mechanism accurately describes this reaction, determine k_1 and the ratio k_{-1}/k_2.

P36.15 Consider the collision-induced dissociation of $N_2O_5(g)$ via the following mechanism:

$$N_2O_5(g) + N_2O_5(g) \underset{k_{-1}}{\overset{k_1}{\rightleftharpoons}} N_2O_5(g)^* + N_2O_5(g)$$

$$N_2O_5(g)^* \overset{k_2}{\longrightarrow} NO_2(g) + NO_3(g)$$

The asterisk in the first reaction indicates that the reactant is activated through collision. Experimentally it is found that the reaction can be either first or second order in $N_2O_5(g)$ depending on the concentration of this species. Derive a rate law expression for this reaction consistent with this observation.

P36.16 The enzyme fumarase catalyzes the hydrolysis of fumarate: Fumarate(aq) + $H_2O(l) \rightarrow$ L-malate(aq). The turnover number for this enzyme is 2.5×10^3 s⁻¹, and the Michaelis constant is 4.2×10^{-6} M. What is the rate of fumarate conversion if the initial enzyme concentration is 1×10^{-6} M and the fumarate concentration is 2×10^{-4} M?

P36.17 The enzyme catalase catalyzes the decomposition of hydrogen peroxide. The following data are obtained regarding the rate of reaction as a function of substrate concentration:

$[H_2O_2]_0$ (M)	0.001	0.002	0.005
Initial Rate (M s^{-1})	1.38×10^{-3}	2.67×10^{-3}	6.00×10^{-3}

The concentration of catalase is 3.50×10^{-9} M. Use these data to determine R_{max}, K_m, and the turnover number for this enzyme.

P36.18 Peptide bond hydrolysis is performed by a family of enzymes known as serine proteases. The name is derived from a highly conserved serine residue that is critical for enzyme function. One member of this enzyme class is chymotrypsin, which preferentially cleaves proteins at residue sites with hydrophobic side chains such as phenylalanine, leucine, and tyrosine. For example, *N*-benzoyl-tyrosylamide (NBT) and *N*-acetyl-tyrosylamide (NAT) are cleaved by chymotrypsin:

N-benzoyl-tyrosylamide (NBT)   *N*-acetyl-tyrosylamide (NAT)

a. The cleavage of NBT by chymotrypsin was studied and the following reaction rates were measured as a function of substrate concentration:

[NBT] (mM)	1.00	2.00	4.00	6.00	8.00
R_0 (mM s^{-1})	0.040	0.062	0.082	0.099	0.107

Use these data to determine K_m and R_{max} for chymotrypsin with NBT as the substrate.

b. The cleavage of NAT is also studied and the following reaction rates versus substrate concentration were measured:

[NAT] (mM)	1.00	2.00	4.00	6.00	8.00
R_0 (mM s^{-1})	0.004	0.008	0.016	0.022	0.028

Use these data to determine K_m and R_{max} for chymotrypsin with NAT as the substrate.

P36.19 Protein tyrosine phosphatases (PTPases) are a general class of enzymes that are involved in a variety of disease processes including diabetes and obesity. In a study by Z.-Y. Zhang and coworkers [*J. Medicinal Chemistry* 43 (2000): 146], computational techniques were used to identify potential competitive inhibitors of a specific PTPase known as PTP1B. The

structure of one of the identified potential competitive inhibitors is shown here:

PTP1B inhibitor

The reaction rate was determined in the presence and absence of inhibitor I and revealed the following initial reaction rates as a function of substrate concentration:

[S] (μM)	R_0 (μM s^{-1}), $[I] = 0$	R_0 (μM s^{-1}), $[I] = 200\ \mu$M
0.299	0.071	0.018
0.500	0.100	0.030
0.820	0.143	0.042
1.22	0.250	0.070
1.75	0.286	0.105
2.85	0.333	0.159
5.00	0.400	0.200
5.88	0.500	0.250

a. Determine K_m and R_{max} for PTP1B.

b. Demonstrate that the inhibition is competitive, and determine K_i.

P36.20 The rate of reaction can be determined by measuring the change in optical rotation of the sample as a function of time if a reactant or product is chiral. This technique is especially useful for kinetic studies of enzyme catalysis involving sugars. For example, the enzyme invertase catalyzes the hydrolysis of sucrose, an optically active sugar. The initial reaction rates as a function of sucrose concentration are as follows:

[Sucrose]$_0$ (M)	R_0 (M s^{-1})
0.029	0.182
0.059	0.266
0.088	0.310
0.117	0.330
0.175	0.362
0.234	0.361

Use these data to determine the Michaelis constant for invertase.

P36.21 The enzyme glycogen synthase kinase 3β (GSK-3β) plays a central role in Alzheimer's disease. The onset of Alzheimer's disease is accompanied by the production of highly phosphorylated forms of a protein referred to as "τ." GSK-3β contributes to the hyperphosphorylation of τ such that inhibiting the activity of this enzyme represents a pathway for the development of an Alzheimer's drug. A compound known as Ro 31-8220 is a competitive inhibitor of GSK-3β. The following data were obtained for the rate of GSK-3β activity in the presence and absence of Ro 31-8220 [A. Martinez et al., *J. Medicinal Chemistry* 45 (2002): 1292]:

$[S](\mu M)$	$R_0\ (\mu M\ s^{-1}),\ [I] = 0$	$R_0\ (\mu M\ s^{-1}),\ [I] = 200\ \mu M$
66.7	4.17×10^{-8}	3.33×10^{-8}
40.0	3.97×10^{-8}	2.98×10^{-8}
20.0	3.62×10^{-8}	2.38×10^{-8}
13.3	3.27×10^{-8}	1.81×10^{-8}
10.0	2.98×10^{-8}	1.39×10^{-8}
6.67	2.31×10^{-8}	1.04×10^{-8}

Determine K_m and R_{max} for GSK-3β and, using the data with the inhibitor, determine K_m^* and K_I.

P36.22 In the Michaelis–Menten mechanism, it is assumed that the formation of product from the enzyme–substrate complex is irreversible. However, consider the following modified version in which the product formation step is reversible:

$$E + S \underset{k_{-1}}{\overset{k_1}{\rightleftharpoons}} ES \underset{k_{-2}}{\overset{k_2}{\rightleftharpoons}} E + P$$

Derive the expression for the Michaelis constant for this mechanism in the limit where $[S]_0 \gg [E]_0$.

P36.23 Reciprocal plots provide a relatively straightforward way to determine if an enzyme demonstrates Michaelis–Menten kinetics and to determine the corresponding kinetic parameters. However, the slope determined from these plots can require significant extrapolation to regions corresponding to low substrate concentrations. An alternative to the reciprocal plot is the Eadie–Hofstee plot in which the reaction rate is plotted versus the rate divided by the substrate concentration and the data are fit to a straight line.

a. Beginning with the general expression for the reaction rate given by the Michaelis–Menten mechanism:

$$R_0 = \frac{R_{max}[S]_0}{[S]_0 + K_m}$$

rearrange this equation to construct the following expression, which is the basis for the Eadie–Hofstee plot:

$$R_0 = R_{max} - K_m\left(\frac{R_0}{[S]_0}\right)$$

b. Using an Eadie–Hofstee plot, determine R_{max} and K_m for hydrolysis of sugar by the enzyme invertase using the following data:

$[Sucrose]_0$ (M)	R_0 (M s^{-1})
0.029	0.182
0.059	0.266
0.088	0.310
0.117	0.330
0.175	0.362
0.234	0.361

P36.24 Determine the predicted rate law expression for the following radical-chain reaction:

$$A_2 \xrightarrow{k_1} 2A\cdot$$

$$A\cdot \xrightarrow{k_2} B\cdot + C$$

$$A\cdot + B\cdot \xrightarrow{k_3} P$$

$$A\cdot + P \xrightarrow{k_4} B\cdot$$

P36.25 The overall reaction for the halogenation of a hydrocarbon (RH) using Br as the halogen is $RH(g) + Br_2(g) \rightarrow RBr(g) + HBr(g)$. The following mechanism has been proposed for this process:

$$Br_2(g) \xrightarrow{k_1} 2Br\cdot(g)$$

$$Br\cdot(g) + RH(g) \xrightarrow{k_2} R\cdot(g) + HBr(g)$$

$$R\cdot(g) + Br_2(g) \xrightarrow{k_3} RBr(g) + Br\cdot(g)$$

$$Br\cdot(g) + R\cdot(g) \xrightarrow{k_4} RBr(g)$$

Determine the rate law predicted by this mechanism.

P36.26 The chlorination of vinyl chloride, $C_2H_3Cl + Cl_2 \rightarrow C_2H_3Cl_3$, is believed to proceed by the following mechanism:

$$Cl_2 \xrightarrow{k_1} 2Cl\cdot$$

$$Cl\cdot + C_2H_3Cl \xrightarrow{k_2} C_2H_3Cl_2\cdot$$

$$C_2H_3Cl_2\cdot + Cl_2 \xrightarrow{k_3} C_2H_3Cl_3 + Cl\cdot$$

$$C_2H_3Cl_2\cdot + C_2H_3Cl_2\cdot \xrightarrow{k_4} \text{stable species}$$

Derive the rate law expression for the chlorination of vinyl chloride based on this mechanism.

P36.27 Determine the expression for fractional coverage as a function of pressure for the dissociative adsorption mechanism described in the text in which adsorption is accompanied by dissociation:

$$R_2(g) + 2M\ (surface) \underset{k_d}{\overset{k_a}{\rightleftharpoons}} 2RM\ (surface)$$

P36.28 The adsorption of ethyl chloride on a sample of charcoal at 0°C measured at several different pressures is as follows:

$P_{C_2H_5Cl}$ (Torr)	V_{ads} (mL)
20	3.0
50	3.8
100	4.3
200	4.7
300	4.8

Using the Langmuir isotherm, determine the fractional coverage at each pressure and V_m.

P36.29 Given the limitations of the Langmuir model, many other empirical adsorption isotherms have been proposed to better reproduce observed adsorption behavior. One of these empirical isotherms is the Temkin isotherm: $V_{adsorbed} = r \ln(sP)$, where V is the volume of gas adsorbed, P is pressure, and r and s are empirical constants.

a. Given the Temkin isotherm provided, what type of plot is expected to give a straight line?

b. Use your answer from part (a) to determine r and s for the data presented in Problem P36.24.

P36.30 Use the following data to determine the Langmuir adsorption parameters for nitrogen on mica:

V_{ads} (cm^3 g^{-1})	P (Torr)
0.494	2.1×10^{-3}
0.782	4.60×10^{-3}
1.16	1.30×10^{-2}

P36.31 Many surface reactions require the adsorption of two or more different gases. For the case of two gases, assuming that the adsorption of a gas simply limits the number of surface sites available for adsorption, derive expressions for the fractional coverage of each gas.

P36.32 DNA microarrays or "chips" first appeared on the market in 1996. These chips are divided into square patches, with each patch having strands of DNA of the same sequence attached to a substrate. The patches are differentiated by differences in the DNA sequence. One can introduce DNA or mRNA of unknown sequence to the chip and monitor to which patches the introduced strands bind. This technique has a wide variety of applications in genome mapping and other areas.

Modeling the chip as a surface with binding sites, and modeling the attachment of DNA to a patch using the Langmuir model, what is the required difference in the Gibbs energy of binding needed to modify the fractional coverage on a given patch from 0.90 to 0.10 for two different DNA strands at the same concentration at 298 K? In performing this calculation replace pressure (P) with concentration (c) in the fractional-coverage expression. Also, recall that $K = \exp(-\Delta G/RT)$.

P36.33 Another type of autocatalytic reaction is referred to as cubic autocatalytic corresponding to the following elementary process:

$$A + 2B \rightarrow 3B$$

Write the rate law expression for this elementary process. What would you expect the corresponding differential rate expression in terms of ξ (the coefficient of reaction advancement) to be?

P36.34 (Challenging) Cubic autocatalytic steps are important in a reaction mechanism referred to as the "brusselator" (named in honor of the research group in Brussels that initially discovered this mechanism):

$$A \xrightarrow{k_1} X$$
$$2X + Y \xrightarrow{k_2} 3X$$
$$B + X \xrightarrow{k_3} Y + C$$
$$X \xrightarrow{k_4} D$$

If [A] and [B] are held constant, this mechanism demonstrates interesting oscillatory behavior that we will explore in this problem.

a. Identify the autocatalytic species in this mechanism.

b. Write down the differential rate expressions for [X] and [Y].

c. Using these differential rate expressions, employ Euler's method (Section 35.6) to calculate [X] and [Y] versus time under the conditions $k_1 = 1.2$ s^{-1}, $k_2 = 0.5$ M^{-2} s^{-1}, $k_3 = 3.0$ M^{-1} s^{-1}, $k_4 = 1.2$ s^{-1}, and $[A]_0 = [B]_0 = 1$ M. Begin with $[X]_0 = 0.5$ M and $[Y]_0 = 0.1$ M. A plot of [Y] versus [X] should look like the top panel in the following figure.

d. Perform a second calculation identical to that in part (c), but with $[X]_0 = 3.0$ M and $[Y]_0 = 3.0$ M. A plot of [Y] versus [X] should look like the bottom panel in the following figure.

e. Compare the left and bottom panels in the following figure. Notice that the starting conditions for the reaction are different (indicated by the black spot). What DO the figures indicate regarding the oscillatory state the system evolves to?

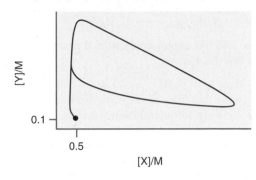

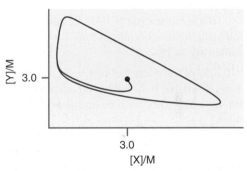

P36.35 The Kermack-McKendrick model was developed to explain the rapid rise and fall in the number of infected people during epidemics. This model involves the interaction of susceptible (S), infected (I), and recovered (R) people through the following mechanism:

$$S + I \xrightarrow{k_1} I + I$$
$$I \xrightarrow{k_2} R$$

a. Write down the differential rate expressions for S, I, and R.

b. The key quantity in this mechanism is called the epidemiological threshold defined as the ratio of $[S]k_1/k_2$. When this ratio is greater than 1 the epidemic will spread; however, when the threshold is less than 1 the epidemic will die out. Based on the mechanism, explain why this behavior is observed.

P36.36 A sunburn is caused primarily by sunlight in what is known as the UVB band, or the wavelength range from 290 to 320 nm. The minimum dose of radiation needed to create a sunburn (erythema) is known as a MED (minimum erythema dose). The MED for a person of average resistance to burning is 50.0 mJ cm^{-2}.

a. Determine the number of 290. nm photons corresponding to the MED, assuming each photon is absorbed. Repeat this calculation for 320. nm photons.

b. At 20° latitude, the solar flux in the UVB band at the surface of the earth is 1.45 mW cm^{-2}. Assuming that each photon is absorbed, how long would a person with unprotected skin be able to stand in the sun before acquiring one MED?

P36.37 A likely mechanism for the photolysis of acetaldehyde is

$$CH_3CHO(g) + h\nu \rightarrow CH_3{\cdot}(g) + CHO{\cdot}(g)$$

$$CH_3{\cdot}(g) + CH_3CHO(g) + \xrightarrow{k_1} CH_4(g) + CH_3CO{\cdot}(g)$$

$$CH_3CO{\cdot}(g) \xrightarrow{k_2} CO(g) + CH_3{\cdot}(g)$$

$$CH_3{\cdot}(g) + CH_3{\cdot}(g) \xrightarrow{k_3} C_2H_6(g)$$

Derive the rate law expression for the formation of $CO(g)$ based on this mechanism.

P36.38 If $\tau_f = 1 \times 10^{-10}$ s and $k_{ic} = 5 \times 10^8$ s^{-1}, what is ϕ_f? Assume that the rate constants for intersystem crossing and quenching are sufficiently small that these processes can be neglected.

P36.39 The quantum yield for $CO(g)$ production in the photolysis of gaseous acetone is unity for wavelengths between 250 and 320 nm. After 20.0 min of irradiation at 313 nm, 18.4 cm^3 of $CO(g)$ (measured at 1008 Pa and 22°C) is produced. Calculate the number of photons absorbed and the absorbed intensity in J s^{-1}.

P36.40 If 10% of the energy of a 100. W incandescent bulb is in the form of visible light having an average wavelength of 600. nm, how many photons are emitted per second from the light bulb?

P36.41 For phenanthrene, the measured lifetime of the triplet state τ_p is 3.3 s, the fluorescence quantum yield is 0.12, and the phosphorescence quantum yield is 0.13 in an alcohol-ether glass at 77 K. Assume that no quenching and no internal conversion from the singlet state occurs. Determine k_p, k_{isc}^T, and k_{isc}^S/k_f.

P36.42 In this problem you will investigate the parameters involved in a single-molecule fluorescence experiment. Specifically, the incident photon power needed to see a single molecule with a reasonable signal-to-noise ratio will be determined.

a. Rhodamine dye molecules are typically employed in such experiments because their fluorescence quantum yields are large. What is the fluorescence quantum yield for Rhodamine B (a specific rhodamine dye) where $k_r = 1 \times 10^9$ s^{-1} and $k_{ic} = 1 \times 10^8$ s^{-1}? You can ignore intersystem crossing and quenching in deriving this answer.

b. If care is taken in selecting the collection optics and detector for the experiment, a detection efficiency of 10% can be readily achieved. Furthermore, detector dark noise usually limits these experiments, and dark noise on the order of 10 counts s^{-1} is typical. If we require a signal-to-noise ratio of 10:1, then we will need to detect 100 counts s^{-1}. Given the detection efficiency, a total emission rate of 1000 fluorescence photons s^{-1} is required. Using the fluorescence quantum yield and a molar extinction coefficient for Rhodamine B of ~40,000 M^{-1} cm^{-1}, what is the intensity of light needed in this experiment in terms of photons cm^{-2} s^{-1}?

c. The smallest diameter focused spot one can obtain in a microscope using conventional refractive optics is approximately one-half the wavelength of incident light. Studies of Rhodamine B generally employ 532 nm light such that the focused-spot diameter is ~270 nm. Using this diameter, what incident power in watts is required for this experiment? Do not be surprised if this value is relatively modest.

P36.43 A central issue in the design of aircraft is improving the lift of aircraft wings. To assist in the design of more efficient wings, wind-tunnel tests are performed in which the pressures at various parts of the wing are measured generally using only a few localized pressure sensors. Recently, pressure-sensitive paints have been developed to provide a more detailed view of wing pressure. In these paints, a luminescent molecule is dispersed into an oxygen-permeable paint and the aircraft wing is painted. The wing is placed into an airfoil, and luminescence from the paint is measured. The variation in O_2 pressure is measured by monitoring the luminescence intensity, with lower intensity demonstrating areas of higher O_2 pressure due to quenching.

a. The use of platinum octaethylporphyrin (PtOEP) as an oxygen sensor in pressure-sensitive paints was described by Gouterman and coworkers [*Review of Scientific Instruments* 61 (1990): 3340]. In this work, the following relationship between luminescence intensity and pressure was derived: $I_0/I = A + B(P/P_0)$, where I_0 is the fluorescence intensity at ambient pressure P_0, and I is the fluorescence intensity at an arbitrary pressure P. Determine coefficients A and B in the preceding expression using the Stern–Volmer equation: $k_{total} = 1/\tau_l = k_l + k_q[Q]$. In this equation τ_l is the luminescence lifetime, k_l is the luminescent rate constant, and k_q is the quenching rate constant. In addition, the luminescent intensity ratio is equal to the ratio of luminescence quantum yields at ambient pressure Φ_0 and an arbitrary pressure Φ:

$$\Phi_0/\Phi = I_0/I.$$

b. Using the following calibration data of the intensity ratio versus pressure observed for PtOEP, determine A and B:

I_0/I	P/P_0	I_0/I	P/P_0
1.0	1.0	0.65	0.46
0.9	0.86	0.61	0.40
0.87	0.80	0.55	0.34
0.83	0.75	0.50	0.28
0.77	0.65	0.46	0.20
0.70	0.53	0.35	0.10

c. At an ambient pressure of 1 atm, $I_0 = 50,000$ (arbitrary units) and 40,000 at the front and back of the wing. The wind tunnel is turned on to a speed of Mach 0.36, and the measured luminescence intensity is 65,000 and 45,000 at the respective locations. What is the pressure differential between the front and back of the wing?

P36.44 Oxygen sensing is important in biological studies of many systems. The variation in oxygen content of sapwood trees was measured by del Hierro and coworkers

[*J. Experimental Biology* 53 (2002): 559] by monitoring the luminescence intensity of $[Ru(dpp)_3]^{2+}$ immobilized in a sol-gel that coats the end of an optical fiber implanted into the tree. As the oxygen content of the tree increases, the luminescence from the ruthenium complex is quenched. The quenching of $[Ru(dpp)_3]^{2+}$ by O_2 was measured by Bright and coworkers [*Applied Spectroscopy* 52 (1998): 750] and the following data were obtained:

I_0/I	$\%O_2$
3.6	12
4.8	20
7.8	47
12.2	100

a. Construct a Stern–Volmer plot using the data supplied in the table. For $[Ru(dpp)_3]^{2+}$ $k_r = 1.77 \times 10^5\ s^{-1}$, what is k_q?

b. Comparison of the Stern–Volmer prediction to the quenching data led the authors to suggest that some of the $[Ru(dpp)_3]^{2+}$ molecules are located in sol-gel environments that are not equally accessible to O_2. What led the authors to this suggestion?

P36.45 The pyrene/coumarin FRET pair ($r_0 = 39\ Å$) is used to study the fluctuations in enzyme structure during the course of a reaction. Computational studies suggest that the pair will be separated by 35 Å in one conformation and 46 Å in a second configuration. What is the expected difference in FRET efficiency between these two conformational states?

P36.46 In a FRET experiment designed to monitor conformational changes in T4 lysozyme, the fluorescence intensity fluctuates between 5000 and 10,000 counts per second. Assuming that 7500 counts represents a FRET efficiency of 0.5, what is the change in FRET pair separation distance during the reaction? For the tetramethylrhodamine/texas red FRET pair employed $r_0 = 50.\ Å$.

P36.47 One complication when using FRET is that fluctuations in the local environment can affect the $S_0 - S_1$ energy gap for the donor or acceptor. Explain how this fluctuation would impact a FRET experiment.

P36.48 One can use FRET to follow the hybridization of two complimentary strands of DNA. When the two strands bind to form a double helix, FRET can occur from the donor to the acceptor allowing one to monitor the kinetics of helix formation (for an example of this technique see *Biochemistry* 34 (1995): 285). You are designing an experiment using a FRET

pair with $R_0 = 59.6\ Å$. For B-form DNA the length of the helix increases 3.46 Å per residue. If you can accurately measure FRET efficiency down to 0.10, how many residues is the longest piece of DNA that you can study using this FRET pair?

P36.49 In Marcus theory for electron transfer, the reorganization energy is partitioned into solvent and solute contributions. Modeling the solvent as a dielectric continuum, the solvent reorganization energy is given by

$$\lambda_{sol} = \frac{(\Delta e)^2}{4\pi\varepsilon_0}\left(\frac{1}{d_1} + \frac{1}{d_2} - \frac{1}{r}\right)\left(\frac{1}{n^2} - \frac{1}{\varepsilon}\right)$$

where Δe is the amount of charge transferred, d_1 and d_2 are the ionic diameters of ionic products, r is the separation distance of the reactants, n^2 is the index of refraction of the surrounding medium, and ε is the dielectric constant of the medium. In addition, $(4\pi\varepsilon_0)^{-1} = 8.99 \times 10^9\ J\ m\ C^{-2}$.

a. For an electron transfer in water ($n = 1.33$ and $\varepsilon = 80.$), where the ionic diameters of both species are 6 Å and the separation distance is 15 Å, what is the expected solvent reorganization energy?

b. Redo the earlier calculation for the same reaction occurring in a protein. The dielectric constant of a protein is dependent on sequence, structure, and the amount of included water; however, a dielectric constant of 4 is generally assumed consistent with a hydrophobic environment. Using light-scattering measurements the dielectric constant of proteins has been estimated to be ~1.5.

P36.50 An experiment is performed in which the rate constant for electron transfer is measured as a function of distance by attaching an electron donor and acceptor to pieces of DNA of varying length. The measured rate constant for electron transfer as a function of separation distance is as follows:

$k_{exp}\ (s^{-1})$	2.10×10^8	2.01×10^7	2.07×10^5	204
Distance (Å)	14	17	23	32

a. Determine the value for β that defines the dependence of the electron transfer rate constant on separation distance.

b. It has been proposed that DNA can serve as an electron "π-way" facilitating electron transfer over long distances. Using the rate constant at 17 Å presented in the table, what value of β would result in the rate of electron transfer decreasing by only a factor of 10 at a separation distance of 23 Å?

Web-Based Simulations, Animations, and Problems

W36.1 In this problem, the Lindemann mechanism for unimolecular rearrangements is investigated. Students will investigate the dependence of the reaction rate on reactant concentration and on the relative rate constants for the reaction. The turnover of the rate law from second order to first order as the reactant concentration is

increased is explored. Also, dependence of this turnover on the relative magnitudes of $k_{-1}\ [A]$ and k_2 is investigated.

W36.2 In this problem, Michaelis–Menten enzyme kinetics are investigated, specifically, the variation in reaction rate with substrate concentration for three enzymes having significantly different Michaelis–Menten kinetic parameters. Students will investigate how the maximum reaction rate and overall kinetics depend on K_m and turnover number. Finally, hand calculations of enzyme kinetic parameters performed by the students are compared to the results obtained by simulation.

W36.3 In this problem, the Langmuir isotherms for nondissociative and dissociative adsorption to a surface are investigated. Students study the dependence of the isotherms for these two adsorption processes on the adsorption/desorption rate constants and on adsorbate pressure.

For Further Reading

CRC Handbook of Photochemistry and Photophysics. Boca Raton, FL: CRC Press, 1990.

Eyring, H., S. H. Lin, and S. M. Lin. *Basic Chemical Kinetics*. New York: Wiley, 1980.

Pannetier, G., and P. Souchay. *Chemical Kinetics*. Amsterdam: Elsevier, 1967.

Hammes, G. G. *Thermodynamics and Kinetics for the Biological Sciences*. New York: Wiley, 2000.

Fersht, A. *Enzyme Structure and Mechanism*. New York: W. H. Freeman, 1985.

Fersht, A. *Structure and Mechanism in Protein Science*. New York: W. H. Freeman, 1999.

Laidler, K. J. *Chemical Kinetics*. New York: Harper & Row, 1987.

Robinson, P. J., and K. A. Holbrook. *Unimolecular Reactions*. New York: Wiley, 1972.

Turro, N. J. *Modern Molecular Photochemistry*. Menlo Park, CA: Benjamin Cummings, 1978.

Simons, J. P. *Photochemistry and Spectroscopy*. New York Wiley, 1971.

Lim, E. G. *Molecular Luminescence*. Menlo Park, CA Benjamin Cummings, 1969.

Noggle, J. H. *Physical Chemistry*. New York: HarperCollins, 1996.

Scott, S. K. *Oscillations, Waves, and Chaos in Chemical Kinetics*. Oxford: Oxford University Press, 2004.

Math Supplement

A.1 Working with Complex Numbers and Complex Functions

Imaginary numbers can be written in the form

$$z = a + ib \qquad \text{(A.1)}$$

where a and b are real numbers and $i = \sqrt{-1}$. It is useful to represent complex numbers in the complex plane shown in Figure A.1. The vertical and horizontal axes correspond to the imaginary and real parts of z, respectively.

In the representation shown in Figure A.1, a complex number corresponds to a point in the complex plane. Note the similarity to the polar coordinate system. Because of this analogy, a complex number can be represented either as the pair (a, b), or by the radius vector r and the angle θ. From Figure A.1, it can be seen that

$$r = \sqrt{a^2 + b^2} \quad \text{and} \quad \theta = \cos^{-1}\frac{a}{r} = \sin^{-1}\frac{b}{r} = \tan^{-1}\frac{b}{a} \qquad \text{(A.2)}$$

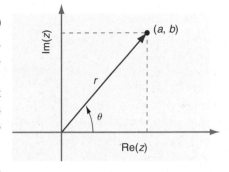

FIGURE A.1

Using the relations between a, b, and r as well as the Euler relation $e^{i\theta} = \cos\theta + i\sin\theta$, a complex number can be represented in either of two equivalent ways:

$$a + ib = r\cos\theta + r\sin\theta = re^{i\theta} = \sqrt{a^2 + b^2}\exp[i\tan^{-1}(b/a)] \qquad \text{(A.3)}$$

If a complex number is represented in one way, it can easily be converted to the other way. For example, we express the complex number $6 - 7i$ in the form $re^{i\theta}$. The magnitude of the radius vector r is given by $\sqrt{6^2 + 7^2} = \sqrt{85}$. The phase is given by $\tan\theta = (-7/6)$ or $\theta = \tan^{-1}(-7/6)$. Therefore, we can write $6 - 7i$ as $\sqrt{85}\exp[i\tan^{-1}(-7/6)]$.

In a second example, we convert the complex number $2e^{i\pi/2}$, which is in the $re^{i\theta}$ notation, to the $a + ib$ notation. Using the relation $e^{i\alpha} = \exp(i\alpha) = \cos\alpha + i\sin\alpha$, we can write $2e^{i\pi/2}$ as

$$2\left(\cos\frac{\pi}{2} + i\sin\frac{\pi}{2}\right) = 2(0 + i) = 2i$$

The complex conjugate of a complex number z is designated by z^* and is obtained by changing the sign of i, wherever it appears in the complex number. For example, if $z = (3 - \sqrt{5}i)e^{i\sqrt{2}\phi}$, then $z^* = (3 + \sqrt{5}i)e^{-i\sqrt{2}\phi}$. The magnitude of a complex number is defined by $\sqrt{zz^*}$ and is always a real number. This is the case for the previous example:

$$zz^* = (3 - \sqrt{5}i)e^{i\sqrt{2}\phi}(3 + \sqrt{5}i)e^{-i\sqrt{2}\phi}$$
$$= (3 - \sqrt{5}i)(3 + \sqrt{5}i)e^{i\sqrt{2}\phi - i\sqrt{2}\phi} = 14 \qquad \text{(A.4)}$$

Note also that $zz^* = a^2 + b^2$.

Complex numbers can be added, multiplied, and divided just like real numbers. A few examples follow:

$$(3 + \sqrt{2}i) + (1 - \sqrt{3}i) = [4 + (\sqrt{2} - \sqrt{3})i]$$
$$(3 + \sqrt{2}i)(1 - \sqrt{3}i) = 3 - 3\sqrt{3}i + \sqrt{2}i - \sqrt{6}i^2$$
$$= (3 + \sqrt{6}) + (\sqrt{2} - 3\sqrt{3})i$$
$$\frac{(3 + \sqrt{2}i)}{(1 - \sqrt{3}i)} = \frac{(3 + \sqrt{2}i)}{(1 - \sqrt{3}i)}\frac{(1 + \sqrt{3}i)}{(1 + \sqrt{3}i)}$$
$$= \frac{3 + 3\sqrt{3}i + \sqrt{2}i + \sqrt{6}i^2}{4}$$

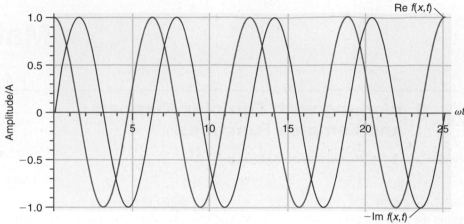

FIGURE A.2

$$= \frac{(3 - \sqrt{6}) + (3\sqrt{3} + \sqrt{2})i}{4}$$

Functions can depend on a complex variable. It is convenient to represent a plane traveling wave usually written in the form

$$\psi(x, t) = A \sin(kx - \omega t) \tag{A.5}$$

in the complex form

$$A e^{i(kx - \omega t)} = A \cos(kx - \omega t) - iA \sin(kx - \omega t) \tag{A.6}$$

Note that

$$\psi(x, t) = -\mathrm{Im}\, A e^{i(kx - \omega t)} \tag{A.7}$$

The reason for working with the complex form rather than the real form of a function is that calculations such as differentiation and integration can be carried out more easily. Waves in classical physics have real amplitudes, because their amplitudes are linked directly to observables. For example, the amplitude of a sound wave is the local pressure that arises from the expansion or compression of the medium through which the wave passes. However, in quantum mechanics, observables are related to $|\psi(x, t)|^2$ rather than $\psi(x, t)$. Because $|\psi(x, t)|^2$ is always real, $\psi(x, t)$ can be complex, and the observables associated with the wave function are still real.

For the complex function $f(x, t) = A e^{i(kx - \omega t)}$, $zz^* = \psi(x, t)\psi^*(x, t) = A e^{i(kx - \omega t)} A^* e^{-i(kx - \omega t)} = AA^*$, so that the magnitude of the function is a constant and does not depend on t or x. As Figure A.2 shows, the real and imaginary parts of $A e^{i(kx - \omega t)}$ depend differently on the variables x and t; they are phase shifted by $\pi/2$. The figure shows the amplitudes of the real and imaginary parts as a function of ωt for $x = 0$.

A.2 Differential Calculus

A.2.1 THE FIRST DERIVATIVE OF A FUNCTION

The derivative of a function has as its physical interpretation the slope of the function evaluated at the position of interest. For example, the slope of the function $y = x^2$ at the point $x = 1.5$ is indicated by the line tangent to the curve shown in Figure A.3.

Mathematically, the first derivative of a function $f(x)$ is denoted $f'(x)$ or $df(x)/dx$. It is defined by

$$\frac{df(x)}{dx} = \lim_{h \to 0} \frac{f(x + h) - f(x)}{h} \tag{A.8}$$

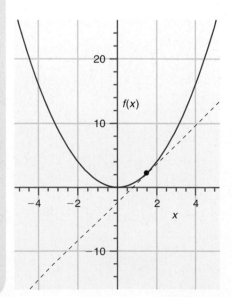

FIGURE A.3

For the function of interest,

$$\frac{df(x)}{dx} = \lim_{h \to 0} \frac{(x+h)^2 - (x)^2}{h}$$

$$= \lim_{h \to 0} \frac{2hx + h^2}{h} = \lim_{h \to 0} 2x + h = 2x \qquad \text{(A.9)}$$

In order for $df(x)/dx$ to be defined over an interval in x, $f(x)$ must be continuous over the interval.

Based on this example, $df(x)/dx$ can be calculated if $f(x)$ is known. Several useful rules for differentiating commonly encountered functions are listed next:

$$\frac{d(ax^n)}{dx} = anx^{n-1}, \quad \text{where } a \text{ is a constant and } n > 0 \qquad \text{(A.10)}$$

For example, $d(\sqrt{3}x^{4/3})/dx = (4/3)\sqrt{3}x^{1/3}$

$$\frac{d(ae^{bx})}{dx} = abe^{bx}, \quad \text{where } a \text{ and } b \text{ are constants} \qquad \text{(A.11)}$$

For example, $d(5e^{3\sqrt{2}x})/dx = 15\sqrt{2}e^{3\sqrt{2}x}$

$$\frac{d(ae^{bx})}{dx} = abe^{bx}, \quad \text{where } a \text{ and } b \text{ are constants}$$

$$\frac{d(a \sin x)}{dx} = a \cos x, \quad \text{where } a \text{ is a constant}$$

$$\frac{d(a \cos x)}{dx} = -a \sin x, \quad \text{where } a \text{ is a constant} \qquad \text{(A.12)}$$

Two useful rules in evaluating the derivative of a function that is itself the sum or product of two functions are as follows:

$$\frac{d[f(x) + g(x)]}{dx} = \frac{df(x)}{dx} + \frac{dg(x)}{dx} \qquad \text{(A.13)}$$

For example,

$$\frac{d(x^3 + \sin x)}{dx} = \frac{dx^3}{dx} + \frac{d \sin x}{dx} = 3x^2 + \cos x$$

$$\frac{d[f(x)g(x)]}{dx} = g(x)\frac{df(x)}{dx} + f(x)\frac{dg(x)}{dx} \qquad \text{(A.14)}$$

For example,

$$\frac{d[\sin(x) \cos(x)]}{dx} = \cos(x)\frac{d \sin(x)}{dx} + \sin(x)\frac{d \cos(x)}{dx}$$

$$= \cos^2 x - \sin^2 x$$

A.2.2 THE RECIPROCAL RULE AND THE QUOTIENT RULE

How is the first derivative calculated if the function to be differentiated does not have a simple form such as those listed in the preceding section? In many cases, the derivative can be found by using the product and quotient rules stated here:

$$\frac{d\left(\dfrac{1}{f(x)}\right)}{dx} = -\frac{1}{[f(x)]^2}\frac{df(x)}{dx} \qquad \text{(A.15)}$$

For example,

$$\frac{d\left(\dfrac{1}{\sin x}\right)}{dx} = -\frac{1}{\sin^2 x}\frac{d\sin x}{dx} = \frac{-\cos x}{\sin^2 x}$$

$$\frac{d\left[\dfrac{f(x)}{g(x)}\right]}{dx} = \frac{g(x)\dfrac{df(x)}{dx} - f(x)\dfrac{dg(x)}{dx}}{[g(x)]^2} \tag{A.16}$$

For example,

$$\frac{d\left(\dfrac{x^2}{\sin x}\right)}{dx} = \frac{2x\sin x - x^2\cos x}{\sin^2 x}$$

A.2.3 THE CHAIN RULE

In this section, we deal with the differentiation of more complicated functions. Suppose that $y = f(u)$ and $u = g(x)$. From the previous section, we know how to calculate $df(u)/du$. How do we calculate $df(u)/dx$? The answer to this question is stated as the chain rule:

$$\frac{df(u)}{dx} = \frac{df(u)}{du}\frac{du}{dx} \tag{A.17}$$

Several examples illustrating the chain rule follow:

$$\frac{d\sin(3x)}{dx} = \frac{d\sin(3x)}{d(3x)}\frac{d(3x)}{dx} = 3\cos(3x)$$

$$\frac{d\ln(x^2)}{dx} = \frac{d\ln(x^2)}{d(x^2)}\frac{d(x^2)}{dx} = \frac{2x}{x^2} = \frac{2}{x}$$

$$\frac{d\left(x + \dfrac{1}{x}\right)^{-4}}{dx} = \frac{d\left(x + \dfrac{1}{x}\right)^{-4}}{d\left(x + \dfrac{1}{x}\right)}\frac{d\left(x + \dfrac{1}{x}\right)}{dx} = -4\left(x + \frac{1}{x}\right)^{-5}\left(1 - \frac{1}{x^2}\right)$$

$$\frac{d\exp(ax^2)}{dx} = \frac{d\exp(ax^2)}{d(ax^2)}\frac{d(ax^2)}{dx} = 2ax\exp(ax^2), \quad \text{where } a \text{ is a constant}$$

A.2.4 HIGHER ORDER DERIVATIVES: MAXIMA, MINIMA, AND INFLECTION POINTS

A function $f(x)$ can have higher order derivatives in addition to the first derivative. The second derivative of a function is the slope of a graph of the slope of the function versus the variable. Mathematically,

$$\frac{d^2 f(x)}{dx^2} = \frac{d}{dx}\left(\frac{df(x)}{dx}\right) \tag{A.18}$$

For example,

$$\frac{d^2\exp(ax^2)}{dx^2} = \frac{d}{dx}\left[\frac{d\exp(ax^2)}{dx}\right] = \frac{d[2ax\exp(ax^2)]}{dx}$$

$$= 2a\exp(ax^2) + 4a^2x^2\exp(ax^2), \quad \text{where } a \text{ is a constant}$$

The second derivative is useful in identifying where a function has its minimum or maximum value within a range of the variable, as shown next.

Because the first derivative is zero at a local maximum or minimum, $df(x)/dx = 0$ at the values $x_{\max}$ and $x_{\min}$. Consider the function $f(x) = x^3 - 5x$ shown in Figure A.4 over the range $-2.5 \leq x \leq 2.5$.

By taking the derivative of this function and setting it equal to zero, we find the minima and maxima of this function in the range

$$\frac{d(x^3 - 5x)}{dx} = 3x^2 - 5 = 0, \quad \text{which has the solutions } x = \pm\sqrt{\frac{5}{3}} = 1.291$$

The maxima and minima can also be determined by graphing the derivative and finding the zero crossings as shown in Figure A.5.

Graphing the function clearly shows that the function has one maximum and one minimum in the range specified. What criterion can be used to distinguish between these extrema if the function is not graphed? The sign of the second derivative, evaluated at the point for which the first derivative is zero, can be used to distinguish between a maximum and a minimum:

$$\frac{d^2f(x)}{dx^2} = \frac{d}{dx}\left[\frac{df(x)}{dx}\right] < 0 \quad \text{for a maximum}$$

$$\frac{d^2f(x)}{dx^2} = \frac{d}{dx}\left[\frac{df(x)}{dx}\right] > 0 \quad \text{for a minimum} \qquad \text{(A.19)}$$

We return to the function graphed earlier and calculate the second derivative:

$$\frac{d^2(x^3 - 5x)}{dx^2} = \frac{d}{dx}\left[\frac{d(x^3 - 5x)}{dx}\right] = \frac{d(3x^2 - 5)}{dx} = 6x$$

By evaluating

$$\frac{d^2f(x)}{dx^2} \quad \text{at } x = \pm\sqrt{\frac{5}{3}} = \pm1.291$$

we see that $x = 1.291$ corresponds to the minimum, and $x = -1.291$ corresponds to the maximum.

If a function has an inflection point in the interval of interest, then

$$\frac{df(x)}{dx} = 0 \quad \text{and} \quad \frac{d^2f(x)}{dx^2} = 0 \qquad \text{(A.20)}$$

An example for an inflection point is $x = 0$ for $f(x) = x^3$. A graph of this function in the interval $-2 \leq x \leq 2$ is shown in Figure A.6. As you can verify,

$$\frac{dx^3}{dx} = 3x^2 = 0 \text{ at } x = 0 \quad \text{and} \quad \frac{d^2(x^3)}{dx^2} = 6x = 0 \text{ at } x = 0$$

A.2.5 MAXIMIZING A FUNCTION SUBJECT TO A CONSTRAINT

A frequently encountered problem is that of maximizing a function relative to a constraint. We first outline how to carry out a constrained maximization, and subsequently apply the method to maximizing the volume of a cylinder while minimizing its area. The theoretical framework for solving this problem originated with the French mathematician Lagrange, and the method is known as Lagrange's method of undetermined multipliers. We wish to maximize the function $f(x, y)$ subject to the constraint that $\phi(x, y) - C = 0$, where C is a constant. For example, you may want to maximize the area A of a rectangle while minimizing its circumference C. In this case, $f(x, y) = A(x, y) = xy$ and $\phi(x, y) = C(x, y) = 2(x + y)$, where x and y are the length and width of the rectangle. The total differentials of these functions are given by Equation (A.21):

$$df = \left(\frac{\partial f}{\partial x}\right)_y dx + \left(\frac{\partial f}{\partial y}\right)_x dy = 0 \quad \text{and} \quad d\phi = \left(\frac{\partial \phi}{\partial x}\right)_y dx + \left(\frac{\partial \phi}{\partial y}\right)_x dy = 0$$

$$\text{(A.21)}$$

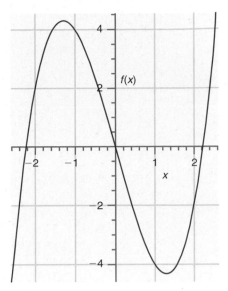

FIGURE A.4

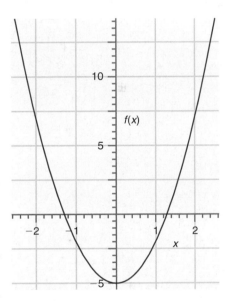

FIGURE A.5

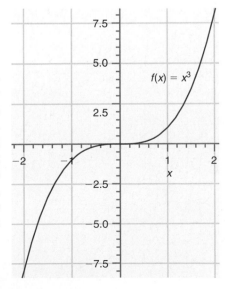

FIGURE A.6

If x and y were independent variables (there is no constraining relationship), the maximization problem would be identical to those dealt with earlier. However, because $d\phi = 0$ also needs to be satisfied, x and y are not independent variables. In this case, Lagrange found that the appropriate function to minimize is $f - \lambda\phi$, where λ is an undetermined multiplier. He showed that each of the expressions in the square brackets in the differential given by Equation (A.22) can be maximized independently. A separate multiplier is required for each constraint:

$$df = \left[\left(\frac{\partial f}{\partial x}\right)_y - \lambda\left(\frac{\partial \phi}{\partial x}\right)_y\right]dx + \left[\left(\frac{\partial f}{\partial y}\right)_x - \lambda\left(\frac{\partial \phi}{\partial y}\right)_x\right]dy \qquad \text{(A.22)}$$

We next use this method to maximize the volume V of a cylindrical can subject to the constraint that its exterior area A be minimized. The functions f and ϕ are given by

$$V = f(r,h) = \pi r^2 h \quad \text{and} \quad A = \phi(r,h) = 2\pi r^2 + 2\pi rh \qquad \text{(A.23)}$$

Calculating the partial derivatives and using Equation (A.22), we have

$$\left(\frac{\partial f(r,h)}{r}\right)_h = 2\pi rh \quad \left(\frac{\partial f(r,h)}{h}\right)_r = \pi r^2$$

$$\left(\frac{\partial \phi(r,h)}{r}\right)_h = 4\pi r + 2\pi h \quad \left(\frac{\partial \phi(r,h)}{h}\right)_r = 2\pi r$$

$$(2\pi rh - \lambda[4\pi r + 2\pi h])dr = 0 \quad \text{and} \quad (\pi r^2 - \lambda 2\pi r)dh = 0 \qquad \text{(A.24)}$$

Eliminating λ from these two equations gives

$$\frac{2\pi rh}{4\pi r + 2\pi h} = \frac{\pi r^2}{2\pi r} \qquad \text{(A.25)}$$

Solving for h in terms of r gives the result $h = 2r$. Note that there is no need to determine the value of multiplier λ. Perhaps you have noticed that beverage cans do not follow this relationship between r and h. Can you think of factors other than minimizing the amount of metal used in the can that might be important in this case?

A.3 Series Expansions of Functions

A.3.1 CONVERGENT INFINITE SERIES

Physical chemists often express functions of interest in the form of an infinite series. For this application, the series must converge. Consider the series

$$a_0 + a_1 x + a_2 x^2 + a_3 x^3 + \cdots a_n x^n + \cdots \qquad \text{(A.26)}$$

How can we determine if such a series converges? A useful convergence criterion is the ratio test. If the absolute ratio of successive terms (designated u_{n-1} and u_n) is less than 1 as $n \to \infty$, the series converges. We consider the series of Equation (A.26) with (a) $a_n = n!$ and (b) $a_n = 1/n!$, and apply the ratio test as shown in Equations (A.27a and b).

$$(a)\ \lim_{n\to\infty}\left|\frac{u_n}{u_{n-1}}\right| = \left|\frac{n!x^n}{(n-1)!x^{n-1}}\right| = \lim_{n\to\infty}|nx| > 1 \text{ unless } x = 0 \qquad \text{(A.27a)}$$

$$(b)\ \lim_{n\to\infty}\left|\frac{u_n}{u_{n-1}}\right| = \left|\frac{x^n/n!}{x^{n-1}/(n-1)!}\right| = \lim_{n\to\infty}\left|\frac{x}{n}\right| < 1 \text{ for all } x \qquad \text{(A.27b)}$$

We see that the infinite series converges if $a_n = 1/n!$ but diverges if $a_n = n!$.

The power series is a particularly important form of a series that is frequently used to fit experimental data to a functional form. It has the form

$$a_0 + a_1 x + a_2 x^2 + a_3 x^3 + a_1 x + a_4 x^4 + \cdots = \sum_{n=0}^{\infty} a_n x^n \qquad \text{(A.28)}$$

Fitting a data set to a series with a large number of terms is impractical, and to be useful, the series should contain as few terms as possible to satisfy the desired accuracy. For example, the function $\sin x$ can be fit to a power series over the interval $0 \leq x \leq 1.5$ by the following truncated power series

$$\sin x \approx -1.20835 \times 10^{-3} + 1.02102x - 0.0607398x^2 - 0.11779x^3$$

$$\sin x \approx -8.86688 \times 10^{-5} + 0.996755x + 0.0175769x^2$$
$$-0.200644x^3 - 0.027618x^4 \tag{A.29}$$

The coefficients in Equation (A.29) have been determined using a least squares fitting routine. The first series includes terms in x up to x^3, and is accurate to within 2% over the interval. The second series includes terms up to x^4, and is accurate to within 0.1% over the interval. Including more terms will increase the accuracy further.

A special case of a power series is the geometric series, in which successive terms are related by a constant factor. An example of a geometric series and its sum is given in Equation (A.30). Using the ratio criterion of Equation (A.27), convince yourself that this series converges for $|x| < 1$.

$$a(1 + x + x^2 + x^3 + \cdots) = \frac{a}{1 - x}, \quad \text{for } |x| < 1 \tag{A.30}$$

A.3.2 REPRESENTING FUNCTIONS IN THE FORM OF INFINITE SERIES

Assume that you have a function in the form $f(x)$ and wish to express it as a power series in x of the form

$$f(x) = a_0 + a_1x + a_2x^2 + a_3x^3 + \cdots \tag{A.31}$$

To do so, we need a way to find the set of coefficients $(a_0, a_1, a_2, a_3, \cdots)$. How can this be done?

If the functional form $f(x)$ is known, the function can be expanded about a point of interest using the Taylor-Maclaurin expansion. In the vicinity of $x = a$, the function can be expanded in the series

$$f(x) = f(a) + \left(\frac{df(x)}{dx}\right)_{x=a} (x - a) + \frac{1}{2!}\left(\frac{d^2f(x)}{dx^2}\right)_{x=a}(x - a)^2$$

$$+ \frac{1}{3!}\left(\frac{d^3f(x)}{dx^3}\right)_{x=a}(x - a)^3 + \cdots \tag{A.32}$$

For example, consider the expansion of $f(x) = e^x$ about $x = 0$. Because $(d^n e^x/dx^n)_{x=0} = 1$ for all values of n, the Taylor-Maclaurin expansion for e^x about $x = 0$ is

$$f(x) = 1 + x + \frac{1}{2!}x^2 + \frac{1}{3!}x^3 + \cdots \tag{A.33}$$

Similarly, the Taylor-Maclaurin expansion for $\ln(1 + x)$ is found by evaluating the derivatives in turn:

$$\frac{d \ln(1 + x)}{dx} = \frac{1}{1 + x}$$

$$\frac{d^2 \ln(1 + x)}{dx^2} = \frac{d}{dx}\frac{1}{(1 + x)} = -\frac{1}{(1 + x)^2}$$

$$\frac{d^3 \ln(1 + x)}{dx^3} = -\frac{d}{dx}\frac{1}{(1 + x)^2} = \frac{2}{(1 + x)^3}$$

$$\frac{d^4 \ln(1 + x)}{dx^4} = \frac{d}{dx}\frac{2}{(1 + x)^3} = \frac{-6}{(1 + x)^4}$$

Each of these derivatives must be evaluated at $x = 0$.

Using these results, the Taylor-Maclaurin expansion for $\ln(1 + x)$ about $x = 0$ is

$$f(x) = x - \frac{x^2}{2!} + \frac{2x^3}{3!} - \frac{6x^4}{4!} + \cdots = x - \frac{x^2}{2} + \frac{x^3}{3} - \frac{x^4}{4} + \cdots \tag{A.34}$$

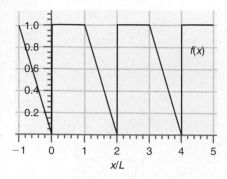

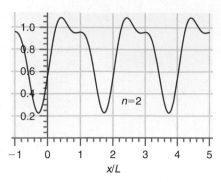

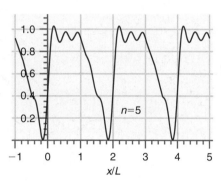

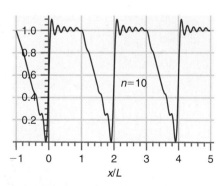

FIGURE A.7

The number of terms that must be included to adequately represent the function depends on the value of x. For $-1 \ll x \ll 1$, the series converges rapidly and, to a very good approximation, we can truncate the Taylor-Maclaurin series after the first one or two terms involving the variable. For the two functions just considered, it is reasonable to write $e^x \approx 1 + x$ and $\ln(1 \pm x) \approx \pm x$ if $-1 \ll x \ll 1$.

A second widely used series is the Fourier sine and cosine series. This series can be used to expand functions that are periodic over an interval $-L \leq x \leq L$ by the series

$$f(x) = \frac{1}{2}b_0 + \sum_n b_n \cos \frac{n\pi x}{L} + \sum_n a_n \sin \frac{n\pi x}{L} \tag{A.35}$$

A Fourier series is an infinite series, and the coefficients a_n and b_n can be calculated using the equations

$$a_n = \frac{1}{L} \int_{-L}^{+L} f(x) \sin \frac{n\pi x}{L} dx \quad \text{and} \quad b_n = \frac{1}{L} \int_{-L}^{+L} f(x) \cos \frac{n\pi x}{L} dx \tag{A.36}$$

The usefulness of the Fourier series is that a function can often be approximated by a few terms, depending on the accuracy desired.

For functions that are either even or odd with respect to the variable x, only either the sine or the cosine terms will appear in the series. For even functions, $f(-x) = f(x)$, and for odd functions, $f(-x) = -f(x)$. Because $\sin(-x) = -\sin(x)$ and $\cos(-x) = \cos(x)$, all coefficients a_n are zero for an even function, and all coefficients b_n are zero for an odd function. Note that Equation (A.29) is not an odd function of x because the function was only fit over the interval $0 \leq x \leq 1.5$.

Whereas the coefficients for the Taylor-Maclaurin series can be readily calculated, those for the Fourier series require more effort. To avoid mathematical detail here, the Fourier coefficients a_n and b_n are not explicitly calculated for a model function. The coefficients can be easily calculated using a program such as *Mathematica*. Our focus here is to show that periodic functions can be approximated to a reasonable degree by using the first few terms in a Fourier series, rather than to carry out the calculations.

To demonstrate the usefulness of expanding a function in a Fourier series, consider the function

$$f(x) = 1 \quad \text{for } 0 \leq x \leq L$$

$$f(x) = -x \quad \text{for } -L \leq x \leq 0 \tag{A.37}$$

which is periodic in the interval $-L \leq x \leq L$, in a Fourier series. This function is a demanding function to expand in a Fourier series because the function is discontinuous at $x = 0$ and the slope is discontinuous at $x = 0$ and $x = 1$. The function and the approximate functions obtained by truncating the series at $n = 2$, $n = 5$, and $n = 10$ are shown in Figure A.7. The agreement between the truncated series and the function is reasonably good for $n = 10$. The oscillations seen near $x/L = 0$ are due to the discontinuity in the function. More terms in the series are required to obtain a good fit if the function changes rapidly in a small interval.

A.4 Integral Calculus

A.4.1 DEFINITE AND INDEFINITE INTEGRALS

In many areas of physical chemistry, the property of interest is the integral of a function over an interval in the variable of interest. For example, the total probability of finding a particle within an interval $0 \leq x \leq a$ is the integral of the probability density $P(x)$ over the interval

$$P_{total} = \int_0^a P(x) \, dx \tag{A.38}$$

Geometrically, the integral of a function over an integral is the area under the curve describing the function. For example, the integral $\int_{-2.3}^{2.3} (x^3 - 5x) \, dx$ is the sum of the areas of the individual rectangles in Figure A.8 in the limit within which the width of the rectangles approaches zero. If the rectangles lie below the zero line, the incremental area is negative; if the rectangles lie above the zero line, the incremental area is positive. In this case the total area is zero because the total negative area equals the total positive area. This is the case because $f(x)$ is an odd function of x.

The integral can also be understood as an antiderivative. From this point of view, the integral symbol is defined by the relation

$$f(x) = \int \frac{df(x)}{dx} \, dx \qquad \text{(A.39)}$$

and the function that appears under the integral sign is called the integrand. Interpreting the integral in terms of area, we evaluate a definite integral, and the interval over which the integration occurs is specified. The interval is not specified for an indefinite integral.

The geometrical interpretation is often useful in obtaining an integral from experimental data when the functional form of the integrand is not known. For our purposes, the interpretation of the integral as an antiderivative is more useful. The value of the indefinite integral $\int (x^3 - 5x) \, dx$ is that function which, when differentiated, gives the integrand. Using the rules for differentiation discussed earlier, you can verify that

$$\int (x^3 - 5x) \, dx = \frac{x^4}{4} - \frac{5x^2}{2} + C \qquad \text{(A.40)}$$

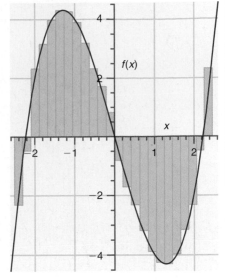

FIGURE A.8

Note the constant that appears in the evaluation of every indefinite integral. By differentiating the function obtained upon integration, you should convince yourself that any constant will lead to the same integrand. In contrast, a definite integral has no constant of integration. If we evaluate the definite integral

$$\int_{-2.3}^{2.3} (x^3 - 5x) \, dx = \left(\frac{x^4}{4} - \frac{5x^2}{2} + C \right)_{x=2.3} - \left(\frac{x^4}{4} - \frac{5x^2}{2} + C \right)_{x=-2.3} \qquad \text{(A.41)}$$

we see that the constant of integration cancels. Because the function obtained upon integration is an even function of x, $\int_{-2.3}^{2.3} (x^3 - 5x) \, dx = 0$, just as we saw in the geometric interpretation of the integral.

It is useful for the student of physical chemistry to commit the integrals listed next to memory, because they are encountered frequently. These integrals are directly related to the derivatives discussed in Section A.2:

$$\int df(x) = f(x) + C$$

$$\int x^n \, dx = \frac{x^{n+1}}{n+1} + C$$

$$\int \frac{dx}{x} = \ln x + C$$

$$\int e^{ax} = \frac{e^{ax}}{a} + C, \quad \text{where } a \text{ is a constant}$$

$$\int \sin x \, dx = -\cos x + C$$

$$\int \cos x \, dx = \sin x + C$$

However, the primary tool for the physical chemist in evaluating integrals is a good set of integral tables. The integrals that are most frequently used in elementary quantum mechanics are listed here; the first group lists indefinite integrals:

$$\int (\sin ax) \, dx = -\frac{1}{a} \cos ax + C$$

$$\int (\cos ax) \, dx = \frac{1}{a} \sin ax + C$$

$$\int (\sin^2 ax)\, dx = \frac{1}{2}x - \frac{1}{4a}\sin 2ax + C$$

$$\int (\cos^2 ax)\, dx = \frac{1}{2}x + \frac{1}{4a}\sin 2ax + C$$

$$\int (x^2 \sin^2 ax)\, dx = \frac{1}{6}x^3 - \left(\frac{1}{4a}x^2 - \frac{1}{8a^3}\right)\sin 2ax - \frac{1}{4a^2}x\cos 2ax + C$$

$$\int (x^2 \cos^2 ax)\, dx = \frac{1}{6}x^3 + \left(\frac{1}{4a}x^2 - \frac{1}{8a^3}\right)\sin 2ax + \frac{1}{4a^2}x\cos 2ax + C$$

$$\int (\sin^3 ax)\, dx = -\frac{3\cos ax}{4a} + \frac{\cos 3ax}{12a} + C$$

$$\int x^2 (\sin ax)\, dx = \frac{(a^2 x^2 - 2)\cos ax}{a^3} + \frac{2x\sin ax}{a^2} + C$$

$$\int x^2 (\cos ax)\, dx = \frac{(a^2 x^2 - 2)\sin ax}{a^3} + \frac{2x\cos ax}{a^2} + C$$

$$\int x^m e^{ax}\, dx = \frac{x^m e^{ax}}{a} - \frac{m}{a}\int x^{m-1} e^{ax}\, dx + C$$

$$\int \frac{e^{ax}}{x^m}\, dx = -\frac{1}{m-1}\frac{e^{ax}}{x^{m-1}} + \frac{a}{m-1}\int \frac{e^{ax}}{x^{m-1}}\, dx + C$$

The following group lists definite integrals.

$$\int_0^a \sin\left(\frac{n\pi x}{a}\right) \times \sin\left(\frac{m\pi x}{a}\right) dx = \int_0^a \cos\left(\frac{n\pi x}{a}\right) \times \cos\left(\frac{m\pi x}{a}\right) dx = \frac{a}{2}\delta_{mn}$$

$$\int_0^a \left[\sin\left(\frac{n\pi x}{a}\right)\right] \times \left[\cos\left(\frac{n\pi x}{a}\right)\right] dx = 0$$

$$\int_0^\pi \sin^2 mx\, dx = \int_0^\pi \cos^2 mx\, dx = \frac{\pi}{2}$$

$$\int_0^\infty \frac{\sin x}{\sqrt{x}}\, dx = \int_0^\infty \frac{\cos x}{\sqrt{x}}\, dx = \sqrt{\frac{\pi}{2}}$$

$$\int_0^\infty x^n e^{-ax}\, dx = \frac{n!}{a^{n+1}} \quad (a > 0,\, n \text{ positive integer})$$

$$\int_0^\infty x^{2n} e^{-ax^2}\, dx = \frac{1\cdot 3\cdot 5\cdots (2n-1)}{2^{n+1}a^n}\sqrt{\frac{\pi}{a}} \quad (a > 0,\, n \text{ positive integer})$$

$$\int_0^\infty x^{2n+1} e^{-ax^2}\, dx = \frac{n!}{2\,a^{n+1}} \quad (a > 0,\, n \text{ positive integer})$$

$$\int_0^\infty e^{-ax^2}\, dx = \left(\frac{\pi}{4a}\right)^{1/2}$$

In the first integral above, $\delta_{mn} = 1$ if $m = n$, and 0 if $m \neq n$.

A.4.2 MULTIPLE INTEGRALS AND SPHERICAL COORDINATES

In the previous section, integration with respect to a single variable was discussed. Often, however, integration occurs over two or three variables. For example, the wave functions for the particle in a two-dimensional box are given by

$$\psi_{n_x n_y}(x, y) = N \sin \frac{n_x \pi x}{a} \sin \frac{n_y \pi y}{b} \tag{A.42}$$

In normalizing a wave function, the integral of $|\psi_{n_x n_y}(x, y)|^2$ is required to equal 1 over the range $0 \le x \le a$ and $0 \le y \le b$. This requires solving the double integral

$$\int_0^b dy \int_0^a \left(N \sin \frac{n_x \pi x}{a} \sin \frac{n_y \pi y}{b} \right)^2 dx = 1 \tag{A.43}$$

to determine the normalization constant N. We sequentially integrate over the variables x and y or vice versa using the list of indefinite integrals from the previous section.

$$\int_0^b dy \int_0^a \left(N \sin \frac{n_x \pi x}{a} \sin \frac{n_y \pi y}{b} \right)^2 dx$$

$$= \left[\frac{1}{2} x - \frac{a}{4n\pi} \sin \frac{2n_x \pi x}{a} \right]_{x=0}^{x=a} \times N^2 \int_0^b \left(\sin \frac{n_y \pi y}{b} \right)^2 dy$$

$$1 = \left[\frac{1}{2} a - \frac{a}{4n\pi} (\sin 2n_x \pi - 0) \right] \times N^2 \int_0^b \left(\sin \frac{n_y \pi y}{b} \right)^2 dy$$

$$1 = N^2 \left[\frac{1}{2} a - \frac{a}{4n\pi} (\sin 2n_x \pi - 0) \right] \times \left[\frac{1}{2} b - \frac{a}{4n\pi} (\sin 2n_y \pi - 0) \right] = \frac{N^2 ab}{4}$$

$$N = \frac{2}{\sqrt{ab}}$$

Convince yourself that the normalization constant for the wave functions of the three-dimensional particle in the box

$$\psi_{n_x n_y n_z}(x, y, z) = N \sin \frac{n_x \pi x}{a} \sin \frac{n_y \pi y}{b} \sin \frac{n_z \pi z}{c} \tag{A.44}$$

has the value $N = 2\sqrt{2}/\sqrt{abc}$.

Up to this point, we have considered functions of a single variable. This restricts us to dealing with a single spatial dimension. The extension to three independent variables becomes important in describing three-dimensional systems. The three-dimensional system of most importance to us is the atom. Closed-shell atoms are spherically symmetric, so we might expect atomic wave functions to be best described by spherical coordinates. Therefore, you should become familiar with integrations in this coordinate system. In transforming from spherical coordinates r, θ, and ϕ to Cartesian coordinates x, y, and z, the following relationships are used:

$$x = r \sin \theta \cos \phi$$
$$y = r \sin \theta \sin \phi \tag{A.45}$$
$$z = r \cos \theta$$

These relationships are depicted in Figure A.9. For small increments in the variables r, θ, and ϕ, the volume element depicted in this figure is a rectangular solid of volume

$$dV = (r \sin \theta \, d\phi)(dr)(r \, d\theta) = r^2 \sin \theta \, dr \, d\theta \, d\phi \tag{A.46}$$

Note in particular that the volume element in spherical coordinates is not $dr \, d\theta \, d\phi$ in analogy with the volume element $dx \, dy \, dz$ in Cartesian coordinates.

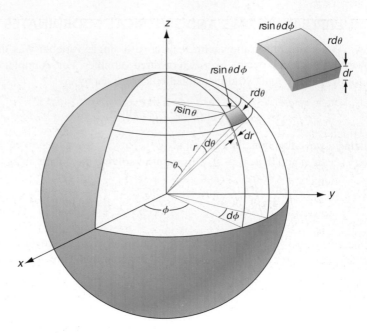

FIGURE A.9

In transforming from Cartesian coordinates x, y, and z to the spherical coordinates r, θ, and ϕ, these relationships are used:

$$r = \sqrt{x^2 + y^2 + z^2} \quad \theta = \cos^{-1}\frac{z}{\sqrt{x^2 + y^2 + z^2}} \text{ and } \phi = \tan^{-1}\frac{y}{x} \quad \text{(A.47)}$$

What is the appropriate range of variables to integrate over all space in spherical coordinates? If we imagine the radius vector scanning over the range $0 \le \theta \le \pi$; $0 \le \phi \le 2\pi$, the whole angular space is scanned. If we combine this range of θ and ϕ with $0 \le r \le \infty$, all of the three-dimensional space is scanned. Note that $r = \sqrt{x^2 + y^2 + z^2}$ is always positive.

To illustrate the process of integration in spherical coordinates, we normalize the function $e^{-r}\cos\theta$ over the interval $0 \le r \le \infty$; $0 \le \theta \le \pi$; $0 \le \phi \le 2\pi$:

$$N^2 \int_0^{2\pi} d\phi \int_0^{\pi} \sin\theta\,d\theta \int_0^{\infty} (e^{-r}\cos\theta)^2 r^2\,dr = N^2 \int_0^{2\pi} d\phi \int_0^{\pi} \cos^2\theta\sin\theta\,d\theta \int_0^{\infty} r^2 e^{-2r}\,dr = 1$$

It is most convenient to integrate first over ϕ, giving

$$2\pi N^2 \int_0^{\pi} \cos^2\theta\sin\theta\,d\theta \int_0^{\infty} r^2 e^{-2r}\,dr = 1$$

We next integrate over θ, giving

$$2\pi N^2 \left[\frac{-\cos^3\pi + \cos^3 0}{3}\right] \times \int_0^{\infty} r^2 e^{-2r}\,dr = \frac{4\pi N^2}{3} \int_0^{\infty} r^2 e^{-2r}\,dr = 1$$

We finally integrate over r using the standard integral

$$\int_0^{\infty} x^n e^{-ax}\,dx = \frac{n!}{a^{n+1}} \quad (a > 0, n \text{ positive integer})$$

The result is

$$\frac{4\pi N^2}{3} \int_0^{\infty} r^2 e^{-2r}\,dr = \frac{4\pi N^2}{3}\frac{2!}{8} = 1 \quad \text{or} \quad N = \sqrt{\frac{3}{\pi}}$$

We conclude that the normalized wave function is $\sqrt{3/\pi}\,e^{-r}\cos\theta$.

A.5 Vectors

The use of vectors occur frequently in physical chemistry. Consider circular motion of a particle at constant speed in two dimensions, as depicted in Figure A.10. The particle is moving in a counterclockwise direction on the ring-like orbit. At any instant in time, its position, velocity, and acceleration can be measured. The two aspects to these measurements are the magnitude and the direction of each of these observables. Whereas a scalar quantity such as speed has only a magnitude, a vector has both a magnitude and a direction.

For the particular case under consideration, the position vectors $\mathbf{r}_1$ and $\mathbf{r}_2$ extend outward from the origin and terminate at the position of the particle. The velocities $\mathbf{v}_1$ and $\mathbf{v}_2$ are related to the position vector as $\mathbf{v} = \lim_{\Delta t \to 0} [\mathbf{r}(t + \Delta t) - \mathbf{r}(t)]/\Delta t$. Therefore, the velocity vector is perpendicular to the position vector. The acceleration vector is defined by $\mathbf{a} = \lim_{\Delta t \to 0} [\mathbf{v}(t + \Delta t) - \mathbf{v}(t)]/\Delta t$. As we see in part (b) of Figure A.10, $\mathbf{a}$ is perpendicular to $\mathbf{v}$, and is antiparallel to $\mathbf{r}$. As this example of a relatively simple motion shows, vectors are needed to describe the situation properly by keeping track of both the magnitude and direction of each of the observables of interest. For this reason, it is important to be able to work with vectors.

In three-dimensional Cartesian coordinates, any vector can be written in the form

$$\mathbf{r} = x_1\mathbf{i} + y_1\mathbf{j} + z_1\mathbf{k} \tag{A.48}$$

where $\mathbf{i}$, $\mathbf{j}$, and $\mathbf{k}$ are the mutually perpendicular vectors of unit length along the x, y, and z axes, respectively, and x_1, y_1, and z_1 are numbers. The length of a vector is defined by the equation

$$|\mathbf{r}| = \sqrt{x_1^2 + y_1^2 + z_1^2} \tag{A.49}$$

This vector is depicted in the three-dimensional coordinate system shown in Figure A.11.

By definition, the angle θ is measured from the z axis, and the angle ϕ is measured in the xy plane from the x axis. The angles θ and ϕ are related to x_1, y_1, and z_1 by

$$\theta = \cos^{-1}\frac{z_1}{\sqrt{x_1^2 + y_1^2 + z_1^2}} \quad \text{and} \quad \phi = \tan^{-1}\frac{y_1}{x_1} \tag{A.50}$$

We next consider the addition and subtraction of two vectors. Two vectors $\mathbf{a} = x\mathbf{i} + y\mathbf{j} + z\mathbf{k}$ and $\mathbf{b} = x'\mathbf{i} + y'\mathbf{j} + z'\mathbf{k}$ can be added or subtracted according to the equations

$$\mathbf{a} \pm \mathbf{b} = (x \pm x')\mathbf{i} + (y \pm y')\mathbf{j} + (z \pm z')\mathbf{k} \tag{A.51}$$

The addition and subtraction of vectors can also be depicted graphically, as done in Figure A.12.

The multiplication of two vectors can occur in either of two forms. Scalar multiplication of $\mathbf{a}$ and $\mathbf{b}$, also called the dot product of $\mathbf{a}$ and $\mathbf{b}$, is defined by

$$\mathbf{a} \cdot \mathbf{b} = |\mathbf{a}||\mathbf{b}|\cos \alpha \tag{A.52}$$

where α is the angle between the vectors. For $\mathbf{a} = 3\mathbf{i} + 1\mathbf{j} - 2\mathbf{k}$ and $\mathbf{b} = 2\mathbf{i} + -1\mathbf{j} + 4\mathbf{k}$, the vectors in the previous equation can be expanded in terms of their unit vectors:

$$\mathbf{a} \cdot \mathbf{b} = (3\mathbf{i} + 1\mathbf{j} - 2\mathbf{k}) \cdot (2\mathbf{i} + -1\mathbf{j} + 4\mathbf{k})$$
$$= 3\mathbf{i} \cdot 2\mathbf{i} + 3\mathbf{i} \cdot (-1\mathbf{j}) + 3\mathbf{i} \cdot 4\mathbf{k} + 1\mathbf{j} \cdot 2\mathbf{i} + 1\mathbf{j} \cdot (-1\mathbf{j})$$
$$+ 1\mathbf{j} \cdot 4\mathbf{k} - 2\mathbf{k} \cdot 2\mathbf{i} - 2\mathbf{k} \cdot (-1\mathbf{j}) - 2\mathbf{k} \cdot 4\mathbf{k}$$

However, because $\mathbf{i}$, $\mathbf{j}$, and $\mathbf{k}$ are mutually perpendicular vectors of unit length, $\mathbf{i} \cdot \mathbf{i} = \mathbf{j} \cdot \mathbf{j} = \mathbf{k} \cdot \mathbf{k} = 1$ and $\mathbf{i} \cdot \mathbf{j} = \mathbf{i} \cdot \mathbf{k} = \mathbf{j} \cdot \mathbf{k} = 0$. Therefore, $\mathbf{a} \cdot \mathbf{b} = 3\mathbf{i} \cdot 2\mathbf{i} + 1\mathbf{j} \cdot (-1\mathbf{j}) - 2\mathbf{k} \cdot 4\mathbf{k} = -3$.

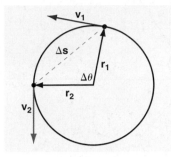

(a)

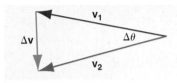

(b)

FIGURE A.10

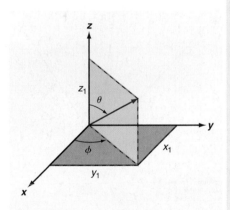

FIGURE A.11

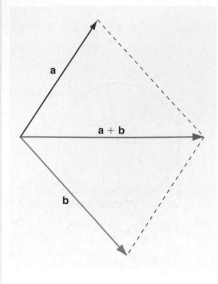

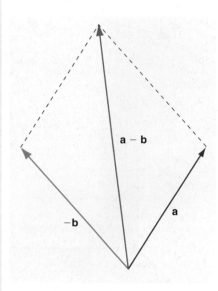

FIGURE A.12

The other form in which vectors are multiplied is the vector product, also called the cross product. The vector multiplication of two vectors results in a vector, whereas the scalar multiplication of two vectors results in a scalar. The cross product is defined by the equation

$$\mathbf{a} \times \mathbf{b} = \mathbf{c}|\mathbf{a}||\mathbf{b}| \sin \alpha \tag{A.53}$$

Note that $\mathbf{a} \times \mathbf{b} = -\mathbf{b} \times \mathbf{a}$ as shown in Figure A.13. By contrast, $\mathbf{a} \cdot \mathbf{b} = \mathbf{b} \cdot \mathbf{a}$.

In Equation (A.53), $\mathbf{c}$ is a vector of unit length that is perpendicular to the plane containing $\mathbf{a}$ and $\mathbf{b}$ and has a positive direction found by using the right-hand rule (see Chapter 18) and α is the angle between $\mathbf{a}$ and $\mathbf{b}$.

The cross product between two three-dimensional vectors $\mathbf{a}$ and $\mathbf{b}$ is given by

$$\begin{aligned}
\mathbf{a} \times \mathbf{b} &= (a_x\mathbf{i} + a_y\mathbf{j} + a_z\mathbf{k}) \times (b_x\mathbf{i} + b_y\mathbf{j} + b_z\mathbf{k}) \\
&= a_x\mathbf{i} \times b_x\mathbf{i} + a_x\mathbf{i} \times b_y\mathbf{j} + a_x\mathbf{i} \times b_z\mathbf{k} + a_y\mathbf{j} \times b_x\mathbf{i} + a_y\mathbf{j} \times b_y\mathbf{j} \\
&\quad + a_y\mathbf{j} \times b_z\mathbf{k} + a_z\mathbf{k} \times b_x\mathbf{i} + a_z\mathbf{k} \times b_y\mathbf{j} + a_z\mathbf{k} \times b_z\mathbf{k} \tag{A.54}
\end{aligned}$$

However, using the definition of the cross product in Equation (A.53),

$$\mathbf{i} \times \mathbf{i} = \mathbf{j} \times \mathbf{j} = \mathbf{k} \times \mathbf{k} = 0, \quad \mathbf{i} \times \mathbf{j} = \mathbf{k}, \quad \mathbf{i} \times \mathbf{k} = -\mathbf{j}$$
$$\mathbf{j} \times \mathbf{i} = -\mathbf{k}, \quad \mathbf{j} \times \mathbf{k} = \mathbf{i}, \quad \mathbf{k} \times \mathbf{i} = \mathbf{j}, \quad \mathbf{k} \times \mathbf{j} = -\mathbf{i}$$

Therefore, Equation (A.54) simplifies to

$$\mathbf{a} \times \mathbf{b} = (a_yb_z - a_zb_y)\mathbf{i} + (a_zb_x - a_xb_z)\mathbf{j} + (a_xb_y - a_yb_x)\mathbf{k} \tag{A.55}$$

As we will see in Section A.7, there is a simple way to calculate cross products using determinants.

The angular momentum $\mathbf{l} = \mathbf{r} \times \mathbf{p}$ is of particular interest in quantum chemistry, because s, p, and d electrons are distinguished by their orbital angular momentum. For the example of the particle rotating on a ring depicted at the beginning of this section, the angular momentum vector is pointing upward in a direction perpendicular to the plane of the page. In analogy to Equation (A.55),

$$\mathbf{l} = \mathbf{r} \times \mathbf{p} = (yp_z - zp_y)\mathbf{i} + (zp_x - xp_z)\mathbf{j} + (xp_y - yp_x)\mathbf{k}$$

A.6 Partial Derivatives

In this section, we discuss the differential calculus of functions that depend on several independent variables. Consider the volume of a cylinder of radius r and height h, for which

$$V = f(r, h) = \pi r^2 h \tag{A.56}$$

where V can be written as a function of the two variables r and h. The change in V with a change in r or h is given by the partial derivatives

$$\left(\frac{\partial V}{\partial r}\right)_h = lim_{\Delta r \to 0} \frac{V(r + \Delta r, h) - V(r, h)}{\Delta r} = 2\pi r h$$

$$\left(\frac{\partial V}{\partial h}\right)_r = lim_{\Delta h \to 0} \frac{V(r, h + \Delta h) - V(r, h)}{\Delta h} = \pi r^2 \tag{A.57}$$

The subscript h in $(\partial V / \partial r)_h$ reminds us that h is being held constant in the differentiation. The partial derivatives in Equation (A.57) allow us to determine how a function changes when one of the variables changes. How does V change if the values of both variables change? In this case, V changes to $V + dV$ where

$$dV = \left(\frac{\partial V}{\partial r}\right)_h dr + \left(\frac{\partial V}{\partial h}\right)_r dh \tag{A.58}$$

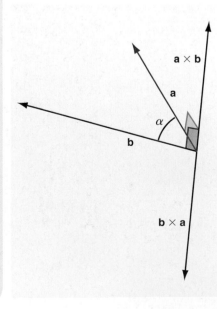

FIGURE A.13

These partial derivatives are useful in calculating the error in the function that results from errors in measurements of the individual variables. For example, the relative error in the volume of the cylinder is given by

$$\frac{dV}{V} = \frac{1}{V}\left[\left(\frac{\partial V}{\partial r}\right)_h dr + \left(\frac{\partial V}{\partial h}\right)_r dh\right] = \frac{1}{\pi r^2 h}[2\pi r h\, dr + \pi r^2 dh] = \frac{2dr}{r} + \frac{dh}{h}$$

This equation shows that a given relative error in r generates twice the relative error in V as a relative error in h of the same size.

We can also take second or higher derivatives with respect to either variable. The mixed second partial derivatives are of particular interest. The mixed partial derivatives of V are given by

$$\left(\frac{\partial}{\partial h}\left(\frac{\partial V}{\partial r}\right)_h\right)_r = \left(\partial\left(\frac{\partial[\pi r^2 h]}{\partial r}\right)_h \middle/ \partial h\right)_r = \left(\frac{\partial[2\pi r h]}{\partial h}\right)_r = 2\pi r$$

$$\left(\frac{\partial}{\partial r}\left(\frac{\partial V}{\partial h}\right)_r\right)_h = \left(\partial\left(\frac{\partial[\pi r^2 h]}{\partial h}\right)_r \middle/ \partial r\right)_h = \left(\frac{\partial[\pi r^2]}{\partial r}\right)_h = 2\pi r \qquad \textbf{(A.59)}$$

For the specific case of V, the order in which the function is differentiated does not affect the outcome. Such a function is called a state function. Therefore, for any state function f of the variables x and y,

$$\left(\frac{\partial}{\partial y}\left(\frac{\partial f(x,y)}{\partial x}\right)_y\right)_x = \left(\frac{\partial}{\partial x}\left(\frac{\partial f(x,y)}{\partial y}\right)_x\right)_y \qquad \textbf{(A.60)}$$

Because Equation (A.60) is satisfied by all state functions f it can be used to determine if a function f is a state function.

We demonstrate how to calculate the partial derivatives

$$\left(\frac{\partial f}{\partial x}\right)_y, \left(\frac{\partial f}{\partial y}\right)_x, \left(\frac{\partial^2 f}{\partial x^2}\right)_y, \left(\frac{\partial^2 f}{\partial y^2}\right)_x, \left(\partial\left(\frac{\partial f}{\partial x}\right)_y \middle/ \partial y\right)_x, \text{ and} \left(\partial\left(\frac{\partial f}{\partial y}\right)_x \middle/ \partial x\right)_y$$

for the function $f(x,y) = ye^{ax} + xy\cos x + y\ln xy$, where a is a real constant:

$$\left(\frac{\partial f}{\partial x}\right)_y = aye^{ax} + \frac{y}{x} + y\cos x - xy\sin x,$$

$$\left(\frac{\partial f}{\partial y}\right)_x = 1 + e^{ax} + x\cos x + \ln xy$$

$$\left(\frac{\partial^2 f}{\partial x^2}\right)_y = a^2 ye^{ax} - \frac{y}{x^2} - 2y\sin x - xy\cos x, \left(\frac{\partial^2 f}{\partial y^2}\right)_x = \frac{1}{y}$$

$$\left(\partial\left(\frac{\partial f}{\partial x}\right)_y \middle/ \partial y\right)_x = ae^{ax} + \frac{1}{x} + \cos x - x\sin x,$$

$$\left(\partial\left(\frac{\partial f}{\partial y}\right)_x \middle/ \partial x\right)_y = ae^{ax} + \frac{1}{x} + \cos x - x\sin x$$

Because we have shown that

$$\left(\partial\left(\frac{\partial f}{\partial x}\right)_y \middle/ \partial y\right)_x = \left(\partial\left(\frac{\partial f}{\partial y}\right)_x \middle/ \partial x\right)_y$$

$f(x, y)$ is a state function of the variables x and y.

Whereas the partial derivatives tell us how the function changes if the value of one of the variables is changed, the total differential tells us how the function changes when all of the variables are changed simultaneously. The total differential of the function $f(x, y)$ is defined by

$$df = \left(\frac{\partial f}{\partial x}\right)_y dx + \left(\frac{\partial f}{\partial y}\right)_x dy \qquad \textbf{(A.61)}$$

The total differential of the function used earlier is calculated as follows:

$$df = \left(aye^{ax} + \frac{y}{x} + y\cos x - xy\sin x \right) dx + (1 + e^{ax} + x\cos x + \log xy)\, dy$$

Two other important results from multivariate differential calculus are used frequently. For a function $z = f(x, y)$, which can be rearranged to $x = g(y, z)$ or $y = h(x, z)$,

$$\left(\frac{\partial x}{\partial y} \right)_z = \frac{1}{\left(\dfrac{\partial y}{\partial x} \right)_z} \tag{A.62}$$

The other important result that is used frequently is the cyclic rule:

$$\left(\frac{\partial x}{\partial y} \right)_z \left(\frac{\partial y}{\partial z} \right)_x \left(\frac{\partial z}{\partial x} \right)_y = -1 \tag{A.63}$$

Consider an additional example of calculating partial derivatives for a function encountered in quantum mechanics. The Schrödinger equation for the hydrogen atom takes the form

$$-\frac{\hbar^2}{2\mu}\left[\frac{1}{r^2}\frac{\partial}{\partial r}\left(r^2 \frac{\partial \psi(r,\theta,\phi)}{\partial r} \right) + \frac{1}{r^2 \sin\theta}\frac{\partial}{\partial \theta}\left(\sin\theta \frac{\partial \psi(r,\theta,\phi)}{\partial \theta} \right) + \frac{1}{r^2 \sin\theta}\frac{\partial^2 \psi(r,\theta,\phi)}{\partial \phi^2} \right]$$

$$-\frac{e^2}{4\pi\varepsilon_0 r}\psi(r,\theta,\phi) = E\psi(r,\theta,\phi)$$

Note that each of the first three terms on the left side of the equation involves partial differentiation with respect to one of the variables r, θ, and ϕ in turn. Two of the solutions to this differential equation are $(r/a_0)e^{-r/2a_0}\sin\theta e^{\pm i\phi}$. Each of these terms is evaluated separately to demonstrate how partial derivatives are taken in quantum mechanics. Although this is a more complex exercise than those presented earlier, it provides good practice in partial differentiation. For the first term, the partial derivative is taken with respect to r:

$$-\frac{\hbar^2}{2\mu}\frac{1}{\sqrt{64\pi}}\left(\frac{1}{a_0} \right)^{3/2}\left[\frac{1}{r^2}\frac{\partial}{\partial r}\left(r^2 \frac{\partial\left(\dfrac{r}{a_0}e^{-r/2a_0}\sin\theta e^{\pm i\phi} \right)}{\partial r} \right) \right]$$

$$= -\frac{\hbar^2}{2\mu}\frac{1}{\sqrt{64\pi}}\left(\frac{1}{a_0} \right)^{3/2}\sin\theta e^{\pm i\phi}\left[\frac{1}{r^2}\frac{\partial}{\partial r}\left(r^2 \frac{\partial\left(\dfrac{r}{a_0}e^{-r/2a_0} \right)}{\partial r} \right) \right]$$

$$= -\frac{\hbar^2}{2\mu}\frac{1}{\sqrt{64\pi}}\left(\frac{1}{a_0} \right)^{3/2}\sin\theta e^{\pm i\phi}\left[\frac{1}{r^2}\frac{\partial}{\partial r}\left(r^2 \left(\frac{1}{a_0}e^{-r/2a_0} - (r/2a_0^2)e^{-r/2a_0} \right) \right) \right]$$

$$= -\frac{\hbar^2}{2\mu}\frac{1}{\sqrt{64\pi}}\left(\frac{1}{a_0} \right)^{3/2}\sin\theta e^{\pm i\phi}\left[\frac{1}{r^2}\left(\begin{array}{c} -r^2\dfrac{e^{-r/2a_0}}{a_0^2} + r^3\dfrac{e^{-r/2a_0}}{4a_0^3} + 2r\dfrac{e^{-r/2a_0}}{a_0} \\ -2r^2\dfrac{e^{-r/2a_0}}{a_0^2} \end{array} \right) \right]$$

$$= -\frac{\hbar^2}{2\mu}\frac{1}{\sqrt{64\pi}}\left(\frac{1}{a_0} \right)^{3/2}\sin\theta e^{\pm i\phi}e^{-r/2a_0}\frac{(8a_0^2 - 8a_0 r + r^2)}{4a_0^3 r}$$

Partial differentiation with respect to θ is easier, because the terms that depend on r and ϕ are constant:

$$-\frac{\hbar^2}{2\mu}\frac{1}{\sqrt{64\pi}}\left(\frac{1}{a_0} \right)^{3/2}\left[\frac{1}{r^2 \sin\theta}\frac{\partial}{\partial \theta}\left(\sin\theta \frac{\partial\left(\dfrac{r}{a_0}e^{-r/2a_0}\sin\theta e^{\pm i\phi} \right)}{\partial \theta} \right) \right]$$

$$= -\frac{\hbar^2}{2\mu}\frac{1}{\sqrt{64\pi}}\left(\frac{1}{a_0} \right)^{3/2}\frac{r}{a_0}e^{-r/2a_0}e^{\pm i\phi}\left[\frac{1}{r^2 \sin\theta}\frac{\partial}{\partial \theta}\left(\sin\theta \frac{\partial(\sin\theta)}{\partial \theta} \right) \right]$$

$$= -\frac{\hbar^2}{2\mu}\frac{1}{\sqrt{64\pi}}\left(\frac{1}{a_0}\right)^{3/2}\frac{r}{a_0}e^{-r/2a_0}e^{\pm i\phi}\left[\frac{1}{r^2\sin\theta}\frac{\partial}{\partial\theta}(\sin\theta\cos\theta)\right]$$

$$= -\frac{\hbar^2}{2\mu}\frac{1}{\sqrt{64\pi}}\left(\frac{1}{a_0}\right)^{3/2}\frac{r}{a_0}e^{-r/2a_0}e^{\pm i\theta}\left[\frac{1}{r^2\sin\theta}(\cos{}^2\theta - \sin{}^2\theta)\right]$$

Partial differentiation with respect to ϕ is also not difficult, because the terms that depend on r and θ are constant:

$$-\frac{\hbar^2}{2\mu}\frac{1}{\sqrt{64\pi}}\left(\frac{1}{a_0}\right)^{3/2}\left[\frac{1}{r^2\sin\theta}\frac{\partial^2\frac{r}{a_0}e^{-r/2a_0}\sin\theta\, e^{\pm i\phi}}{\partial\phi^2}\right]$$

$$= -\frac{\hbar^2}{2\mu}\frac{1}{\sqrt{64\pi}}\left(\frac{1}{a_0}\right)^{3/2}\frac{r}{a_0}e^{-r/2a_0}\frac{1}{r^2}\left[\frac{\partial^2 e^{\pm i\phi}}{\partial\phi^2}\right]$$

$$= \frac{\hbar^2}{2\mu}\frac{1}{\sqrt{64\pi}}\left(\frac{1}{a_0}\right)^{3/2}\frac{r}{a_0}e^{-r/2a_0}\frac{1}{r^2}\left[e^{\pm i\phi}\right]$$

A.7 Working with Determinants

A determinant of nth order is a square $n \times n$ array of numbers symbolically enclosed by vertical lines. A fifth-order determinant is shown here with the conventional indexing of the elements of the array:

$$\begin{vmatrix} a_{11} & a_{12} & a_{13} & a_{14} & a_{15} \\ a_{21} & a_{22} & a_{23} & a_{24} & a_{25} \\ a_{31} & a_{32} & a_{33} & a_{34} & a_{35} \\ a_{41} & a_{42} & a_{43} & a_{44} & a_{45} \\ a_{51} & a_{52} & a_{53} & a_{54} & a_{55} \end{vmatrix} \qquad \textbf{(A.64)}$$

A 2×2 determinant has a value that is defined in Equation (A.65). It is obtained by multiplying the elements in the diagonal connected by a line with a negative slope and subtracting from this the product of the elements in the diagonal connected by a line with a positive slope.

$$\begin{vmatrix} a_{11} & a_{12} \\ a_{21} & a_{22} \end{vmatrix} = a_{11}a_{22} - a_{12}a_{21} \qquad \textbf{(A.65)}$$

The value of a higher order determinant is obtained by expanding the determinant in terms of determinants of lower order. This is done using the method of cofactors. We illustrate the use of method of cofactors by reducing a 3×3 determinant to a sum of 2×2 determinants. Any row or column can be used in the reduction process. We use the first row of the determinant in the reduction. The recipe is spelled out in this equation:

$$\begin{vmatrix} a_{11} & a_{12} & a_{13} \\ a_{21} & a_{22} & a_{23} \\ a_{31} & a_{32} & a_{33} \end{vmatrix} = (-1)^{1+1}a_{11}\begin{vmatrix} a_{22} & a_{23} \\ a_{32} & a_{33} \end{vmatrix} + (-1)^{1+2}a_{12}\begin{vmatrix} a_{21} & a_{23} \\ a_{31} & a_{33} \end{vmatrix}$$

$$+ (-1)^{1+3}a_{13}\begin{vmatrix} a_{21} & a_{22} \\ a_{31} & a_{32} \end{vmatrix}$$

$$= a_{11}\begin{vmatrix} a_{22} & a_{23} \\ a_{32} & a_{33} \end{vmatrix} - a_{12}\begin{vmatrix} a_{21} & a_{23} \\ a_{31} & a_{33} \end{vmatrix} + a_{13}\begin{vmatrix} a_{21} & a_{22} \\ a_{31} & a_{32} \end{vmatrix} \qquad \textbf{(A.66)}$$

Each term in the sum results from the product of one of the three elements of the first row, $(-1)^{m+n}$, where m and n are the indices of the row and column designating the element, respectively, and the 2×2 determinant obtained by omitting the entire row and column to which the element used in the reduction belongs. The product $(-1)^{m+n}$ and the 2×2

determinant are called the cofactor of the element used in the reduction. For example, the value of the following 3×3 determinant is found using the cofactors of the second row:

$$\begin{vmatrix} 1 & 3 & 4 \\ 2 & -1 & 6 \\ -1 & 7 & 5 \end{vmatrix} = (-1)^{2+1}2\begin{vmatrix} 3 & 4 \\ 7 & 5 \end{vmatrix} + (-1)^{2+2}(-1)\begin{vmatrix} 1 & 4 \\ -1 & 5 \end{vmatrix} + (-1)^{2+3}6\begin{vmatrix} 1 & 3 \\ -1 & 7 \end{vmatrix}$$

$$= -1 \times 2 \times (-13) + 1 \times (-1) \times 9 + (-1) \times 6 \times 10 = -43$$

If the initial determinant is of a higher order than 3, multiple sequential reductions as outlined earlier will reduce it in order by one in each step until a sum of 2×2 determinants is obtained.

The main usefulness for determinants is in solving a system of linear equations. Such a system of equations is obtained in evaluating the energies of a set of molecular orbitals obtained by combining a set of atomic orbitals. Before illustrating this method, we list some important properties of determinants that we will need in solving a set of simultaneous equations.

Property I The value of a determinant is not altered if each row in turn is made into a column or vice versa as long as the original order is kept. By this we mean that the nth row becomes the nth column. This property can be illustrated using 2×2 and 3×3 determinants:

$$\begin{vmatrix} 2 & 1 \\ 3 & -1 \end{vmatrix} = \begin{vmatrix} 2 & 3 \\ 1 & -1 \end{vmatrix} = -5 \quad \text{and} \quad \begin{vmatrix} 1 & 3 & 4 \\ 2 & -1 & 6 \\ -1 & 7 & 5 \end{vmatrix} = \begin{vmatrix} 1 & 2 & -1 \\ 3 & -1 & 7 \\ 4 & 6 & 5 \end{vmatrix} = -43$$

Property II If any two rows or columns are interchanged, the sign of the value of the determinant is changed. For example,

$$\begin{vmatrix} 2 & 1 \\ 3 & -1 \end{vmatrix} = -5, \text{ but } \begin{vmatrix} 1 & 2 \\ -1 & 3 \end{vmatrix} = +5 \quad \text{and}$$

$$\begin{vmatrix} 1 & 3 & 4 \\ 2 & -1 & 6 \\ -1 & 7 & 5 \end{vmatrix} = -43, \text{ but } \begin{vmatrix} 2 & -1 & 6 \\ 1 & 3 & 4 \\ -1 & 7 & 5 \end{vmatrix} = +43$$

Property III If two rows or columns of a determinant are identical, the value of the determinant is zero. For example,

$$\begin{vmatrix} 2 & 1 \\ 2 & 1 \end{vmatrix} = 2 - 2 = 0 \text{ and}$$

$$\begin{vmatrix} 1 & 1 & 4 \\ 2 & 2 & 6 \\ -1 & -1 & 5 \end{vmatrix} = (-1)^{2+1}2\begin{vmatrix} 1 & 4 \\ -1 & 5 \end{vmatrix} + (-1)^{2+2}2\begin{vmatrix} 1 & 4 \\ -1 & 5 \end{vmatrix}$$

$$+ (-1)^{2+3}6\begin{vmatrix} 1 & 1 \\ -1 & -1 \end{vmatrix}$$

$$= -1 \times 2 \times 9 + 1 \times 2 \times 9 + (-1) \times 6 \times 0 = 0$$

Property IV If each element of a row or column is multiplied by a constant, the value of the determinant is multiplied by that constant. For example,

$$\begin{vmatrix} 2 & 1 \\ 3 & -1 \end{vmatrix} = -5 \text{ and } \begin{vmatrix} 8 & 4 \\ 3 & -1 \end{vmatrix} = -20 \quad \text{and}$$

$$\begin{vmatrix} 1 & 2 & -1 \\ 3 & -1 & 7 \\ 4 & 6 & 5 \end{vmatrix} = -43 \text{ and } \begin{vmatrix} 1 & 3\sqrt{2} & 4 \\ 2 & -\sqrt{2} & 6 \\ -1 & 7\sqrt{2} & 5 \end{vmatrix} = -43\sqrt{2}$$

Property V   The value of a determinant is unchanged if a row or column multiplied by an arbitrary number is added to another row or column. For example,

$$\begin{vmatrix} 2 & 1 \\ 3 & -1 \end{vmatrix} = \begin{vmatrix} 2+1 & 1 \\ 3-1 & -1 \end{vmatrix} = \begin{vmatrix} 2 & 1 \\ 3 & -1 \end{vmatrix} = -5 \quad \text{and}$$

$$\begin{vmatrix} 1 & 3 & 4 \\ 2 & -1 & 6 \\ -1 & 7 & 5 \end{vmatrix} = \begin{vmatrix} 1 & 3 & 4 \\ 2-1 & -1+7 & 6+5 \\ -1 & 7 & 5 \end{vmatrix} = \begin{vmatrix} 1 & 3 & 4 \\ 2 & -1 & 6 \\ -1 & 7 & 5 \end{vmatrix} = -43$$

How are determinants useful? This question can be answered by illustrating how determinants can be used to solve a set of linear equations:

$$x + y + z = 10$$

$$3x + 4y - z = 12$$

$$-x + 2y + 5z = 26 \qquad \text{(A.67)}$$

This set of equations is solved by first constructing the 3×3 determinant that is the array of the coefficients of x, y, and z:

$$\mathbf{D}_{coefficients} = \begin{vmatrix} 1 & 1 & 1 \\ 3 & 4 & -1 \\ -1 & 2 & 5 \end{vmatrix} \qquad \text{(A.68)}$$

Now imagine that we multiply the first column by x. This changes the value of the determinant as stated in Property IV:

$$\begin{vmatrix} 1x & 1 & 1 \\ 3x & 4 & -1 \\ -1x & 2 & 5 \end{vmatrix} = x\mathbf{D}_{coefficients} \qquad \text{(A.69)}$$

We next add to the first column of $x\mathbf{D}_{coefficients}$ the second column of $\mathbf{D}_{coefficients}$ multiplied by y, and the third column multiplied by z. According to Properties IV and V, the value of the determinant is unchanged. Therefore,

$$\mathbf{D}_{c1} = \begin{vmatrix} 1 & 1 & 1 \\ 3 & 4 & -1 \\ -1 & 2 & 5 \end{vmatrix} = \begin{vmatrix} x+y+z & 1 & 1 \\ 3x+4y-z & 4 & -1 \\ -x+2y+5z & 2 & 5 \end{vmatrix} = \begin{vmatrix} 10 & 1 & 1 \\ 12 & 4 & -1 \\ 26 & 2 & 5 \end{vmatrix} = x\mathbf{D}_{coefficients}$$

$$\text{(A.70)}$$

To obtain the third determinant in the previous equation, the individual equations in Equation (A.67) are used to substitute the constants for the algebraic expression in the preceding determinants. From the previous equation, we conclude that

$$x = \frac{\mathbf{D}_{c1}}{\mathbf{D}_{coefficients}} = \frac{\begin{vmatrix} 10 & 1 & 1 \\ 12 & 4 & -1 \\ 26 & 2 & 5 \end{vmatrix}}{\begin{vmatrix} 1 & 1 & 1 \\ 3 & 4 & -1 \\ -1 & 2 & 5 \end{vmatrix}} = 3$$

To determine y and z, the exact same procedure can be followed, but we substitute instead in columns 2 and 3, respectively. The first step in each case is to multiply all elements of the second (third) row by $y(z)$. If we do so, we obtain the determinants $\mathbf{D}_{c2}$ and $\mathbf{D}_{c3}$:

$$\mathbf{D}_{c2} = \begin{vmatrix} 1 & 10 & 1 \\ 3 & 12 & -1 \\ -1 & 26 & 5 \end{vmatrix} \quad \text{and} \quad \mathbf{D}_{c3} = \begin{vmatrix} 1 & 1 & 10 \\ 3 & 4 & 12 \\ -1 & 2 & 26 \end{vmatrix}$$

and we conclude that

$$y = \frac{D_{c2}}{D_{coefficients}} = \frac{\begin{vmatrix} 1 & 10 & 1 \\ 3 & 12 & -1 \\ -1 & 26 & 5 \end{vmatrix}}{\begin{vmatrix} 1 & 1 & 1 \\ 3 & 4 & -1 \\ -1 & 2 & 5 \end{vmatrix}} = 2 \quad \text{and}$$

$$z = \frac{D_{c3}}{D_{coefficients}} = \frac{\begin{vmatrix} 1 & 1 & 10 \\ 3 & 4 & 12 \\ -1 & 2 & 26 \end{vmatrix}}{\begin{vmatrix} 1 & 1 & 1 \\ 3 & 4 & -1 \\ -1 & 2 & 5 \end{vmatrix}} = 5$$

This method of solving a set of simultaneous linear equations is known as Cramer's method.

If the constants in the set of equations are all zero, as in Equations A.71a and A.71b,

$$x + y + z = 0$$
$$3x + 4y - z = 0 \qquad\qquad \textbf{(A.71a)}$$
$$-x + 2y + 5z = 0$$

$$3x - y + 2z = 0$$
$$-x + y - z = 0$$
$$(1 + \sqrt{2})x + (1 - \sqrt{2})y + \sqrt{2}z = 0 \qquad \textbf{(A.71b)}$$

the determinants D_{c1}, D_{c2}, and D_{c3} all have the value zero. An obvious set of solutions is $x = 0$, $y = 0$ and $z = 0$. For most problems in physics and chemistry, this set of solutions is not physically meaningful and is referred to as the set of trivial solutions. A set of nontrivial solutions only exists if the equation $D_{coefficients} = 0$ is satisfied. There is no nontrivial solution to the set of Equation (A.71a) because $D_{coefficients} \neq 0$. There is a set of nontrivial solutions to the set of Equations (A.71b), because $D_{coefficients} = 0$ in this case.

Determinants offer a convenient way to calculate the cross product of two vectors, as discussed in Section A.5. The following recipe is used:

$$\mathbf{a} \times \mathbf{b} = \begin{vmatrix} \mathbf{i} & \mathbf{j} & \mathbf{k} \\ a_x & a_y & a_z \\ b_x & b_y & b_z \end{vmatrix} = \mathbf{i}\begin{vmatrix} a_y & a_z \\ b_y & b_z \end{vmatrix} - \mathbf{j}\begin{vmatrix} a_x & a_z \\ b_x & b_z \end{vmatrix} + \mathbf{k}\begin{vmatrix} a_x & a_y \\ b_x & b_y \end{vmatrix}$$
$$= (a_y b_z - a_z b_y)\mathbf{i} + (a_z b_x - a_x b_z)\mathbf{j} + (a_x b_y - a_y b_x)\mathbf{k} \quad \textbf{(A.72)}$$

Note that by referring to Property II, you can show that $\mathbf{b} \times \mathbf{a} = -\mathbf{a} \times \mathbf{b}$.

A.8 Working with Matrices

Physical chemists find widespread use for matrices. Matrices can be used to represent symmetry operations in the application of group theory to problems concerning molecular symmetry. They can also be used to obtain the energies of molecular orbitals formed through the linear combination of atomic orbitals. We next illustrate the use of matrices for representing the rotation operation that is frequently encountered in molecular symmetry considerations.

Consider the rotation of a three-dimensional vector about the z axis. Because the z component of the vector is unaffected by this operation, we need only consider the effect of the rotation operation on the two-dimensional vector formed by the projection of the three-dimensional vector on the xy plane. The transformation can be represented by $(x_1, y_1, z_1) \rightarrow (x_2, y_2, z_1)$. The effect of the operation on the x and y components of the vector is shown in Figure A.14.

Next, relationships are derived among (x_1, y_1, z_1), (x_2, y_2, z_1), the magnitude of the radius vector r, and the angles α and β, based on the preceding figure. The magnitude of the radius vector r is

$$r = \sqrt{x_1^2 + y_1^2 + z_1^2} = \sqrt{x_2^2 + y_2^2 + z_1^2} \qquad \text{(A.73)}$$

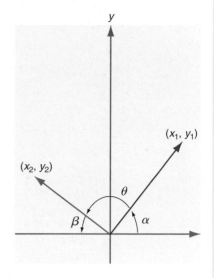

FIGURE A.14

Although the values of x and y change in the rotation, r is unaffected by this operation. The relationships between x, y, r, α, and β are given by

$$\theta = 180° - \alpha - \beta$$
$$x_1 = r \cos \alpha, \qquad y_1 = r \sin \alpha$$
$$x_2 = -r \cos \beta, \qquad y_1 = r \sin \beta \qquad \text{(A.74)}$$

In the following discussion, these identities are used:

$$\cos(\alpha \pm \beta) = \cos \alpha \cos \beta \pm \sin \alpha \sin \beta$$
$$\sin(\alpha \pm \beta) = \sin \alpha \cos \beta \pm \cos \alpha \sin \beta \qquad \text{(A.75)}$$

From Figure A.14, the following relationship between x_2 and x_1 and y_1 can be derived using the identities of Equation (A.75):

$$\begin{aligned}
x_2 &= -r \cos \beta = -r \cos(180° - \alpha - \theta) \\
&= r \sin 180° \sin(-\theta - \alpha) - r \cos 180° \cos(-\theta - \alpha) \\
&= r \cos(-\theta - \alpha) = r \cos(\theta + \alpha) = r \cos \theta \cos \alpha - r \sin \theta \sin \alpha \\
&= x_1 \cos \theta - y_1 \sin \theta \qquad \text{(A.76)}
\end{aligned}$$

Using the same procedure, the following relationship between y_2 and x_1 and y_1 can be derived:

$$y_2 = x_1 \sin \theta + y_1 \cos \theta \qquad \text{(A.77)}$$

Next, these results are combined to write the following equations relating x_2, y_2, and z_2 to x_1, y_1, and z_1:

$$\begin{aligned}
x_2 &= x_1 \cos \theta - y_1 \sin \theta \\
y_2 &= x_1 \sin \theta + y_1 \cos \theta \\
z_2 &= 0x_1 + 0y_1 + z_1 \qquad \text{(A.78)}
\end{aligned}$$

At this point, the concept of a matrix can be introduced. An $n \times m$ matrix is an array of numbers, functions, or operators that can undergo mathematical operations such as addition and multiplication with one another. The operation of interest to us in considering rotation about the z axis is matrix multiplication. We illustrate how matrices, which are designated in bold script, such as $\mathbf{A}$, are multiplied using 2×2 matrices as an example.

$$\mathbf{AB} = \begin{pmatrix} a_{11} & a_{12} \\ a_{21} & a_{22} \end{pmatrix} \begin{pmatrix} b_{11} & b_{12} \\ b_{21} & b_{22} \end{pmatrix} = \begin{pmatrix} a_{11}b_{11} + a_{12}b_{21} & a_{11}b_{12} + a_{12}b_{22} \\ a_{21}b_{11} + a_{22}b_{21} & a_{21}b_{12} + a_{22}b_{22} \end{pmatrix} \qquad \text{(A.79)}$$

Using numerical examples,

$$\begin{pmatrix} 2 & 1 \\ -3 & 4 \end{pmatrix} \begin{pmatrix} 1 & 6 \\ 2 & -1 \end{pmatrix} = \begin{pmatrix} 4 & 11 \\ 5 & -22 \end{pmatrix} \quad \text{and} \quad \begin{pmatrix} 1 & 6 \\ 2 & -1 \end{pmatrix} \begin{pmatrix} 1 \\ -1 \end{pmatrix} = \begin{pmatrix} -5 \\ 3 \end{pmatrix}$$

Now consider the initial and final coordinates (x_1, y_1, z_1) and (x_2, y_2, z_1) as 3×1 matrices (x_1, y_1, z_1) and (x_2, y_2, z_1). In that case, the set of simultaneous equations of Equation (A.78) can be written as

$$\begin{pmatrix} x_2 \\ y_2 \\ z_2 \end{pmatrix} = \begin{pmatrix} \cos \theta & -\sin \theta & 0 \\ \sin \theta & \cos \theta & 0 \\ 0 & 0 & 1 \end{pmatrix} \begin{pmatrix} x_1 \\ y_1 \\ z_1 \end{pmatrix} \qquad \text{(A.80)}$$

We see that we can represent the operator for rotation about the z axis, R_z, as the following 3×3 matrix:

$$R_z = \begin{pmatrix} \cos\theta & -\sin\theta & 0 \\ \sin\theta & \cos\theta & 0 \\ 0 & 0 & 1 \end{pmatrix} \tag{A.81}$$

The rotation operator for 180° and 120° rotation can be obtained by evaluating the sine and cosine functions at the appropriate values of θ. These rotation operators have the form

$$\begin{pmatrix} -1 & 0 & 0 \\ 0 & -1 & 0 \\ 0 & 0 & 1 \end{pmatrix} \quad \text{and} \quad \begin{pmatrix} 1/2 & -\sqrt{3}/2 & 0 \\ \sqrt{3}/2 & 1/2 & 0 \\ 0 & 0 & 1 \end{pmatrix}, \quad \text{respectively} \tag{A.82}$$

One special matrix, the identity matrix designated $\mathbf{I}$, deserves additional mention. The identity matrix corresponds to an operation in which nothing is changed. The matrix that corresponds to the transformation $(x_1, y_1, z_1) \rightarrow (x_2, y_2, z_2)$ expressed in equation form as

$$x_2 = x_1 + 0y_1 + 0z_1$$

$$y_2 = 0x_1 + y_1 + 0z_1$$

$$z_2 = 0x_1 + 0y_1 + z_1 \tag{A.83}$$

is the identity matrix

$$\mathbf{I} = \begin{pmatrix} 1 & 0 & 0 \\ 0 & 1 & 0 \\ 0 & 0 & 1 \end{pmatrix}$$

The identity matrix is an example of a diagonal matrix. It has this name because only the diagonal elements are nonzero. In the identity matrix of order $n \times n$ all diagonal elements have the value one.

The operation that results from the sequential operation of two individual operations represented by matrices $\mathbf{A}$ and $\mathbf{B}$ is the products of the matrices: $\mathbf{C} = \mathbf{AB}$. An interesting case illustrating this relationship is counterclockwise rotation through an angle θ followed by clockwise rotation through the same angle, which corresponds to rotation by $-\theta$. Because $\cos(-\theta) = \cos\theta$ and $\sin\theta = -\sin\theta$, the rotation matrix for $-\theta$ must be

$$R_{-z} = \begin{pmatrix} \cos\theta & \sin\theta & 0 \\ -\sin\theta & \cos\theta & 0 \\ 0 & 0 & 1 \end{pmatrix} \tag{A.84}$$

Because the sequential operations leave the vector unchanged, it must be the case that $\mathbf{R_z R_{-z}} = \mathbf{R_{-z} R_z} = \mathbf{I}$. We verify here that the first of these relations is obeyed:

$$\mathbf{R_z R_{-z}} = \begin{pmatrix} \cos\theta & -\sin\theta & 0 \\ \sin\theta & \cos\theta & 0 \\ 0 & 0 & 1 \end{pmatrix}\begin{pmatrix} \cos\theta & \sin\theta & 0 \\ -\sin\theta & \cos\theta & 0 \\ 0 & 0 & 1 \end{pmatrix}$$

$$= \begin{pmatrix} \cos^2\theta + \sin^2\theta + 0 & \sin\theta\cos\theta - \sin\theta\cos\theta + 0 & 0 \\ \sin\theta\cos\theta - \sin\theta\cos\theta + 0 & \cos^2\theta + \sin^2\theta + 0 & 0 \\ 0 & 0 & 1 \end{pmatrix}$$

$$= \begin{pmatrix} 1 & 0 & 0 \\ 0 & 1 & 0 \\ 0 & 0 & 1 \end{pmatrix} \tag{A.85}$$

Any matrix $\mathbf{B}$ that satisfies the relationship $\mathbf{AB} = \mathbf{BA} = \mathbf{I}$ is called the inverse matrix of $\mathbf{A}$ and is designated $\mathbf{A}^{-1}$. Inverse matrices play an important role in finding the energies of a set of molecular orbitals that is a linear combination of atomic orbitals.

Data tables referenced throughout the text are listed here and appear either on the page number in parentheses or, if no page number is listed, in this appendix.

Sources of Data

The most extensive databases for thermodynamic data (and the abbreviations listed with the tables) are as follows:

HCP Lide, D. R., Ed., *Handbook of Chemistry and Physics,* 83rd ed. CRC Press, Boca Raton, FL, 2002.

NIST Chemistry Webbook Linstrom, P. J., and W. G. Mallard, Eds., *NIST Chemistry Webbook: NIST Standard Reference Database Number 69.* National Institute of Standards and Technology, Gaithersburg, MD, retrieved from http://webbook.nist.gov.

Additional data sources used in the tables include the following:

Bard Bard, A. J., R. Parsons, and J. Jordan, *Standard Potentials in Aqueous Solution.* Marcel Dekker, New York, 1985.

DAL Blachnik, R., Ed., *D'Ans Lax Taschenbuch für Chemiker und Physiker*, 4th ed. Springer, Berlin, 1998.

HP Benenson, W., J. W. Harris, H. Stocker, and H. Lutz, *Handbook of Physics.* Springer, New York, 2002.

HTTD Lide, D. R., Ed., *CRC Handbook of Thermophysical and Thermochemical Data.* CRC Press, Boca Raton, FL, 1994.

TDOC Pedley, J. B., R. D. Naylor, and S. P. Kirby, *Thermochemical Data of Organic Compounds.* Chapman and Hall, London, 1977.

TDPS Barin, I., *Thermochemical Data of Pure Substances.* VCH Press, Weinheim, 1989.

AS Alberty, R. A., and R. S. Silbey, *Physical Chemistry.* John Wiley & Sons, New York, 1992.

TABLE 2.3   Physical Properties of Selected Elements

Densities are shown for nongaseous elements under standard conditions.

Substance	Atomic Weight	Melting Point (K)	Boiling Point (K)	$\rho°$ (kg m^{-3})	$C_{P,\,m}$ (J K^{-1} mol^{-1})	Oxidation States
Aluminum	26.982	933.47	2792.15	2698.9	25.4	3
Argon	39.948	83.79 tp (69 kPa)	87.30	—	20.79	
Barium	137.33	1000.15	2170.15	3620	28.07	2
Boron	10.811	2348.15	4273.15	2340	11.1	3
Bromine	79.904	265.95	331.95	3103	36.05	1, 3, 4, 5, 6
Calcium	40.078	1115.15	1757.15	1540	25.9	2
Carbon	12.011	4713.15		3513	6.113	2, 4
		(12.4 GPa)	4098.15	(diamond)	(diamond)	
		4762.15 tp	(graphite)	2250	8.527	
		(10.3 MPa)		(graphite)	(graphite)	
Cesium	132.91	301.65	944.15	1930	32.20	1
Chlorine	35.453	171.65	239.11	—	33.95	1, 3, 4, 5, 6, 7
Copper	63.546	1357.77	2835.15	8960	24.4	1, 2
Fluorine	18.998	53.48 tp	85.03	—	31.30	1
Gold	196.97	1337.33	3129.15	19320	25.42	1, 3
Helium	4.0026	0.95	4.22	—	20.79	
Hydrogen	1.0079	13.81	20.28	—	28.84	1
Iodine	126.90	386.85	457.55	4933	54.44	1, 3, 5, 7
Iron	55.845	1811.15	3134.15	7874	25.10	2, 3
Krypton	83.80	115.77 tp (73.2 kPa)	119.93	—	20.79	
Lead	207.2	600.61	2022.15	11343	26.44	2, 4
Lithium	6.941	453.65	1615.15	534	24.77	1
Magnesium	24.305	923.15	1363.15	1740	24.89	2
Manganese	54.938	1519.15	2334.15	7300	26.3	2, 3, 4, 6, 7
Mercury	200.59	234.31	629.88	13534	27.98	1, 2
Molybdenum	95.94	2896.15	4912.15	10222	23.90	2, 3, 4, 5, 6
Neon	20.180	24.54 tp (43 kPa)	27.07	—	20.79	
Nickel	58.693	1728.15	3186	8902	26.07	2, 3
Nitrogen	14.007	63.15	77.36	—	29.12	1, 2, 3, 4, 5
Oxygen	15.999	54.36	90.20	—	29.38	1, 2
Palladium	106.42	1828.05	3236.15	11995	25.98	2, 4
Phosphorus (white)	30.974	317.3	553.65	1823	23.84	3, 5
Platinum	195.08	2041.55	4098.15	21500	25.85	2, 4, 6
Potassium	39.098	336.65	1032.15	890	29.58	1
Rhenium	186.21	3459.15	5869.15	20800	25.31	2, 4, 5, 6, 7
Rhodium	102.91	2237.15	3968.15	12410	24.98	2, 3, 4
Ruthenium	101.07	2607.15	4423.15	12100	24.04	3, 4, 5, 6, 8
Silicon	28.086	1687.15	3538.15	2330	20.00	4
Silver	107.87	1234.93	2435.15	10500	25.35	1
Sodium	22.990	370.95	1156.15	971	28.24	1
Sulfur	32.066	388.36	717.75	1819	22.76	2, 4, 6
Tin	118.71	505.08	2879	7310	26.99	2, 4
Titanium	47.867	1941.15	3560.15	4506	25.05	2, 3, 4
Vanadium	50.942	2183.15	3680.15	6000	24.89	2, 3, 4, 5
Xenon	131.29	161.36 tp (81.6 kPa)	165.03	—	20.79	2, 4, 6, 8
Zinc	65.39	692.68	1180.15	7135	25.40	2

Sources: HCP and DAL.

TABLE 2.4   Physical Properties of Selected Compounds

Densities are shown for nongaseous compounds under standard conditions.

Formula	Name	Molecular Weight	Melting Point (K)	Boiling Point (K)	Density $\rho°$ (kg m^{-3})	Heat Capacity $C_{P,m}$ (J K^{-1} mol^{-1})
CO(g)	Carbon monoxide	28.01	68.13	81.65	—	29.1
COCl$_2$(g)	Phosgene	98.92	145.4	281	—	
CO$_2$(g)	Carbon dioxide	44.01	216.6 tp	194.75	—	37.1
D$_2$O(l)	Deuterium oxide	20.03	277	374.6	1108	
HCl(g)	Hydrogen chloride	36.46	158.98	188.15	—	29.1
HF(g)	Hydrogen fluoride	20.01	189.8	293.15	—	
H$_2$O	Water	18.02	273.15	373.15	998(l) 20°C	75.3(l)
						36.2(s)
					917(s) 0°C	
H$_2$O$_2$(l)	Hydrogen peroxide	34.01	272.72	423.35	1440	43.1
H$_2$SO$_4$(l)	Sulfuric acid	98.08	283.46	610.15	1800	
KBr(s)	Potassium bromide	119.00	1007.15	1708.15	2740	52.3
KCl(s)	Potassium chloride	74.55	1044.15		1988	51.3
KI(s)	Potassium iodide	166.0	954.15	1596.15	3120	52.9
NaCl(s)	Sodium chloride	58.44	1073.85	1738.15	2170	50.5
NH$_3$(g)	Ammonia	17.03	195.42	239.82	—	35.1
SO$_2$(g)	Sulfur dioxide	64.06	197.65	263.10	—	39.9
CCl$_4$(l)	Carbon tetrachloride	153.82	250.3	349.8	1594	131.3
CH$_4$(g)	Methane	16.04	90.75	111.65	—	35.7
HCOOH(l)	Formic acid	46.03	281.45	374.15	1220	99.04
CH$_3$OH(l)	Methanol	32.04	175.55	337.75	791.4	81.1
CH$_3$CHO(l)	Acetaldehyde	44.05	150.15	293.25	783.4	89.0
CH$_3$COOH(l)	Acetic acid	60.05	289.6	391.2	1044.6	123.1
CH$_3$COCH$_3$(l)	Acetone	58.08	178.5	329.2	789.9	125.45
C$_2$H$_5$OH(l)	Ethanol	46.07	158.8	351.5	789.3	112.3
C$_3$H$_7$OH(l)	1-Propanol	60.10	147.05	370.35	799.8	156.5
C$_4$H$_{11}$OH(l)	1-Butanol	74.12	183.35	390.85	809.8	176.9
C$_5$H$_5$NH$_2$(l)	Pyridine	79.10	231.55	388.35	981.9	193
C$_5$H$_{12}$(l)	Pentane	72.15	143.45	309.15	626.2	167.2
C$_5$H$_{11}$OH(l)	1-Pentanol	88.15	194.25	411.05	814.4	207.5
C$_6$H$_{12}$(l)	Cyclohexane	84.16	279.6	353.9	773.9	156.0
C$_6$H$_5$CHO(l)	Benzaldehyde	106.12	247.15	452.15	1041.5	172.0
C$_6$H$_5$COOH(s)	Benzoic acid	122.12	395.55	522.35	1265.9	147.8
C$_6$H$_5$CH$_3$(l)	Toluene	92.14	178.2	383.8	866.9	157.1
C$_6$H$_5$NH$_2$(l)	Aniline	93.13	267	457	1021.7	194.1
C$_6$H$_5$OH(s)	Phenol	94.11	314.05	454.95	1057.6	127.2
C$_6$H$_6$(l)	Benzene	78.11	278.6	353.3	876.5	135.7
1,2-(CH$_3$)$_2$C$_6$H$_5$(l)	o-Xylene	106.17	248	417.6	880.2	187.7
C$_8$H$_{18}$(l)	Octane	114.23	216.35	398.75	698.6	254.7

Sources: HCP and TDOC.

TABLE 2.5   Molar Heat Capacity, $C_{P,m}$, of Gases in the Range 298–800 K

Given by

$$C_{P,m} \ (\text{J K}^{-1} \text{ mol}^{-1}) = A(1) + A(2)\frac{T}{K} + A(3)\frac{T^2}{K^2} + A(4)\frac{T^3}{K^3}$$

Note that $C_{P,m}$ for solids and liquids at 298.15 K is listed in Tables 2.1 and 2.2.

Name	Formula	$C_{P,m}$ (298.15 K) in J K^{-1} mol^{-1}	A(1)	A(2)	A(3)	A(4)
All monatomic gases	He, Ne, Ar, Xe, O, H, among others	20.79	20.79			
Bromine	Br_2	36.05	30.11	0.03353	-5.5009×10^{-5}	3.1711×10^{-8}
Chlorine	Cl_2	33.95	22.85	0.06543	-1.2517×10^{-4}	1.1484×10^{-7}
Carbon monoxide	CO	29.14	31.08	-0.01452	3.1415×10^{-5}	-1.4973×10^{-8}
Carbon dioxide	CO_2	37.14	18.86	0.07937	-6.7834×10^{-5}	2.4426×10^{-8}
Fluorine	F_2	31.30	23.06	0.03742	-3.6836×10^{-5}	1.351×10^{-8}
Hydrogen	H_2	28.84	22.66	0.04381	-1.0835×10^{-4}	1.1710×10^{-7}
Water	H_2O	33.59	33.80	-0.00795	2.8228×10^{-5}	-1.3115×10^{-8}
Hydrogen bromide	HBr	29.13	29.72	-0.00416	7.3177×10^{-6}	
Hydrogen chloride	HCl	29.14	29.81	-0.00412	6.2231×10^{-6}	
Hydrogen fluoride	HF	29.14	28.94	0.00152	-4.0674×10^{-6}	3.8970×10^{-9}
Ammonia	NH_3	35.62	29.29	0.01103	4.2446×10^{-5}	-2.7706×10^{-8}
Nitrogen	N_2	29.13	30.81	-0.01187	2.3968×10^{-5}	-1.0176×10^{-8}
Nitric oxide	NO	29.86	33.58	-0.02593	5.3326×10^{-5}	-2.7744×10^{-8}
Nitrogen dioxide	NO_2	37.18	32.06	-0.00984	1.3807×10^{-4}	-1.8157×10^{-7}
Oxygen	O_2	29.38	32.83	-0.03633	1.1532×10^{-4}	-1.2194×10^{-7}
Sulfur dioxide	SO_2	39.83	26.07	0.05417	-2.6774×10^{-5}	
Methane	CH_4	35.67	30.65	-0.01739	1.3903×10^{-4}	8.1395×10^{-8}
Methanol	CH_3OH	44.07	26.53	0.03703	9.451×10^{-5}	-7.2006×10^{-8}
Ethyne	C_2H_2	44.05	10.82	0.15889	-1.8447×10^{-4}	8.5291×10^{-8}
Ethene	C_2H_4	42.86	8.39	0.12453	-2.5224×10^{-5}	-1.5679×10^{-8}
Ethane	C_2H_6	52.38	6.82	0.16840	-5.2347×10^{-5}	
Propane	C_3H_8	73.52	0.56	0.27559	-1.0355×10^{-4}	
Butane	C_4H_{10}	101.01	172.02	-1.08574	4.4887×10^{-3}	-6.5539×10^{-6}
Pentane	C_5H_{12}	120.11	2.02	0.44729	-1.7174×10^{-4}	
Benzene	C_6H_6	82.39	-46.48	0.53735	-3.8303×10^{-4}	1.0184×10^{-7}
Hexane	C_6H_{12}	142.13	-13.27	0.61995	-3.5408×10^{-4}	7.6704×10^{-8}

Sources: HCP and HTTD.

TABLE 2.6 Molar Heat Capacity, $C_{P,m}$, of Solids

$$C_{P,m} \text{ (J K}^{-1} \text{ mol}^{-1}) = A(1) + A(2)\frac{T}{K} + A(3)\frac{T^2}{K^2} + A(4)\frac{T^3}{K^3} + A(5)\frac{T^4}{K^4}$$

Formula	Name	$C_{P,m}$ (298.15 K) (J K^{-1} mol^{-1})	$A(1)$	$A(2)$	$A(3)$	$A(4)$	$A(5)$	Range (K)
Ag	Silver	25.35	26.12	-0.0110	3.826×10^{-5}	3.750×10^{-8}	1.396×10^{-11}	290–800
Al	Aluminum	24.4	6.56	0.1153	-2.460×10^{-4}	1.941×10^{-7}		200–450
Au	Gold	25.4	34.97	-0.0768	2.117×10^{-4}	-2.350×10^{-7}	9.500×10^{-11}	290–800
CsCl	Cesium chloride	52.5	43.38	0.0467	-8.973×10^{-5}	1.421×10^{-7}	-8.237×10^{-11}	200–600
CuSO$_4$	Copper sulfate	98.9	-13.81	0.7036	-1.636×10^{-3}	2.176×10^{-6}	-1.182×10^{-9}	200–600
Fe	Iron	25.1	-10.99	0.3353	-1.238×10^{-3}	2.163×10^{-6}	-1.407×10^{-9}	200–450
NaCl	Sodium chloride	50.5	25.19	0.1973	-6.011×10^{-4}	8.815×10^{-7}	-4.765×10^{-10}	200–600
Si	Silicon	20.0	-6.25	0.1681	-3.437×10^{-4}	2.494×10^{-7}	6.667×10^{-12}	200–450
C (graphite)	C	8.5	-12.19	0.1126	-1.947×10^{-4}	1.919×10^{-7}	-7.800×10^{-11}	290–600
C$_6$H$_5$OH	Phenol	127.4	-5.97	1.0380	-6.467×10^{-3}	2.304×10^{-5}	-2.658×10^{-8}	100–314
C$_{10}$H$_8$	Naphthalene	165.7	-6.16	1.0383	-5.355×10^{-3}	1.891×10^{-5}	-2.053×10^{-8}	100–353
C$_{14}$H$_{10}$	Anthracene	210.5	11.10	0.5816	2.790×10^{-4}			100–488

Sources: HCP and HTTD.

TABLE 4.1 Thermodynamic Data for Inorganic Compounds at 298.15 K

Substance	ΔH_f° (kJ mol^{-1})	ΔG_f° (kJ mol^{-1})	S_m° (J mol^{-1} K^{-1})	$C_{P,m}$ (J mol^{-1} K^{-1})	Atomic or Molecular Weight (amu)
Aluminum					
Al(s)	0	0	28.3	24.4	26.98
Al$_2$O$_3$(s)	-1675.7	-1582.3	50.9	79.0	101.96
Al^{3+}(aq)	-538.4	-485.0	-325		26.98
Antimony					
Sb(s)	0	0	45.7	25.2	121.75
Argon					
Ar(g)	0	0	154.8	20.8	39.95
Barium					
Ba(s)	0	0	62.5	28.1	137.34
BaO(s)	-548.0	-520.3	72.1	47.3	153.34
BaCO$_3$(s)	-1216.3	-1137.6	112.1	85.4	197.35
BaCl$_2$(s)	-856.6	-810.3	123.7	75.1	208.25
BaSO$_4$(s)	-1473.2	-1362.3	132.2	101.8	233.40
Ba^{2+}(aq)	-537.6	-560.8	9.6		137.34
Bromine					
Br$_2$(l)	0	0	152.2	75.7	159.82
Br$_2$(g)	30.9	3.1	245.5	36.0	159.82
Br(g)	111.9	82.4	175.0	20.8	79.91
HBr(g)	-36.3	-53.4	198.7	29.1	90.92
Br$^-$(aq)	-121.6	-104.0	82.4		79.91

Substance	ΔH_f° (kJ mol^{-1})	ΔG_f° (kJ mol^{-1})	S_m° (J mol^{-1} K^{-1})	$C_{P,\,m}$ (J mol^{-1} K^{-1})	Atomic or Molecular Weight (amu)
Calcium					
Ca(s)	0	0	41.6	25.9	40.08
CaCO$_3$(s) calcite	−1206.9	−1128.8	92.9	83.5	100.09
CaCl$_2$(s)	−795.4	−748.8	104.6	72.9	110.99
CaO(s)	−634.9	−603.3	38.1	42.0	56.08
CaSO$_4$(s)	−1434.5	−1322.0	106.5	99.7	136.15
Ca^{2+}(aq)	−542.8	−553.6	−53.1		40.08
Carbon					
Graphite(s)	0	0	5.74	8.52	12.011
Diamond(s)	1.89	2.90	2.38	6.12	12.011
C(g)	716.7	671.2	158.1	20.8	12.011
CO(g)	−110.5	−137.2	197.7	29.1	28.011
CO$_2$(g)	−393.5	−394.4	213.8	37.1	44.010
HCN(g)	135.5	124.7	201.8	35.9	27.03
CN$^-$(aq)	150.6	172.4	94.1		26.02
HCO$_3^-$(aq)	−692.0	−586.8	91.2		61.02
CO$_3^{2-}$(aq)	−675.2	−527.8	−50.0		60.01
Chlorine					
Cl$_2$(g)	0	0	223.1	33.9	70.91
Cl(g)	121.3	105.7	165.2	21.8	35.45
HCl(g)	−92.3	−95.3	186.9	29.1	36.46
ClO$_2$(g)	104.6	105.1	256.8	45.6	67.45
ClO$_4^-$(aq)	−128.1	−8.52	184.0		99.45
Cl$^-$(aq)	−167.2	−131.2	56.5		35.45
Copper					
Cu(s)	0	0	33.2	24.4	63.54
CuCl$_2$(s)	−220.1	−175.7	108.1	71.9	134.55
CuO(s)	−157.3	−129.7	42.6	42.3	79.54
Cu$_2$O(s)	−168.6	−146.0	93.1	63.6	143.08
CuSO$_4$(s)	−771.4	−662.2	109.2	98.5	159.62
Cu$^+$(aq)	71.7	50.0	40.6		63.54
Cu^{2+}(aq)	64.8	65.5	−99.6		63.54
Deuterium					
D$_2$(g)	0	0	145.0	29.2	4.028
HD(g)	0.32	−1.46	143.8	29.2	3.022
D$_2$O(g)	−249.2	−234.5	198.3	34.3	20.028
D$_2$O(l)	−294.6	−243.4	75.94	84.4	20.028
HDO(g)	−246.3	−234.5	199.4	33.8	19.022
HDO(l)	−289.9	−241.9	79.3		19.022
Fluorine					
F$_2$(g)	0	0	202.8	31.3	38.00
F(g)	79.4	62.3	158.8	22.7	19.00
HF(g)	−273.3	−275.4	173.8	29.1	20.01
F$^-$(aq)	−332.6	−278.8	−13.8		19.00
Gold					
Au(s)	0	0	47.4	25.4	196.97
Au(g)	366.1	326.3	180.5	20.8	197.97

(continued)

TABLE 4.1 Thermodynamic Data for Inorganic Compounds at 298.15 K (continued)

Substance	ΔH_f° (kJ mol^{-1})	ΔG_f° (kJ mol^{-1})	S_m° (J mol^{-1} K^{-1})	$C_{P,m}$ (J mol^{-1} K^{-1})	Atomic or Molecular Weight (amu)
Hydrogen					
$H_2(g)$	0	0	130.7	28.8	2.016
$H(g)$	218.0	203.3	114.7	20.8	1.008
$OH(g)$	39.0	34.2	183.7	29.9	17.01
$H_2O(g)$	−241.8	−228.6	188.8	33.6	18.015
$H_2O(l)$	−285.8	−237.1	70.0	75.3	18.015
$H_2O(s)$			48.0	36.2 (273 K)	18.015
$H_2O_2(g)$	−136.3	−105.6	232.7	43.1	34.015
$H^+(aq)$	0	0	0		1.008
$OH^-(aq)$	−230.0	−157.24	−10.9		17.01
Iodine					
$I_2(s)$	0	0	116.1	54.4	253.80
$I_2(g)$	62.4	19.3	260.7	36.9	253.80
$I(g)$	106.8	70.2	180.8	20.8	126.90
$I^-(aq)$	−55.2	−51.6	111.3		126.90
Iron					
$Fe(s)$	0	0	27.3	25.1	55.85
$Fe(g)$	416.3	370.7	180.5	25.7	55.85
$Fe_2O_3(s)$	−824.2	−742.2	87.4	103.9	159.69
$Fe_3O_4(s)$	−1118.4	−1015.4	146.4	150.7	231.54
$FeSO_4(s)$	−928.4	−820.8	107.5	100.6	151.92
$Fe^{2+}(aq)$	−89.1	−78.9	−137.7		55.85
$Fe^{3+}(aq)$	−48.5	−4.7	−315.9		55.85
Lead					
$Pb(s)$	0	0	64.8	26.4	207.19
$Pb(g)$	195.2	162.2	175.4	20.8	207.19
$PbO_2(s)$	−277.4	−217.3	68.6	64.6	239.19
$PbSO_4(s)$	−920.0	−813.20	148.5	86.4	303.25
$Pb^{2+}(aq)$	0.92	−24.4	18.5		207.19
Lithium					
$Li(s)$	0	0	29.1	24.8	6.94
$Li(g)$	159.3	126.6	138.8	20.8	6.94
$LiH(s)$	−90.5	−68.3	20.0	27.9	7.94
$LiH(g)$	140.6	117.8	170.9	29.7	7.94
$Li^+(aq)$	−278.5	−293.3	13.4		6.94
Magnesium					
$Mg(s)$	0	0	32.7	24.9	24.31
$Mg(g)$	147.1	112.5	148.6	20.8	24.31
$MgO(s)$	−601.6	−569.3	27.0	37.2	40.31
$MgSO_4(s)$	−1284.9	−1170.6	91.6	96.5	120.38
$MgCl_2(s)$	−641.3	−591.8	89.6	71.4	95.22
$MgCO_3(s)$	−1095.8	−1012.2	65.7	75.5	84.32
$Mg^{2+}(aq)$	−466.9	−454.8	−138.1		24.31
Manganese					
$Mn(s)$	0	0	32.0	26.3	54.94
$Mn(g)$	280.7	238.5	173.7	20.8	54.94
$MnO_2(s)$	−520.0	−465.1	53.1	54.1	86.94
$Mn^{2+}(aq)$	−220.8	−228.1	−73.6		54.94
$MnO_4^-(aq)$	−541.4	−447.2	191.2		118.94

Substance	ΔH_f° (kJ mol^{-1})	ΔG_f° (kJ mol^{-1})	S_m°(J mol^{-1} K^{-1})	$C_{P,\,m}$ (J mol^{-1} K^{-1})	Atomic or Molecular Weight (amu)
Mercury					
Hg(l)	0	0	75.9	28.0	200.59
Hg(g)	61.4	31.8	175.0	20.8	200.59
Hg$_2$Cl$_2$(s)	−265.4	−210.7	191.6	101.9	472.09
Hg^{2+}(aq)	170.2	164.4	−36.2		401.18
Hg$_2^{2+}$(aq)	166.9	153.5	65.7		401.18
Nickel					
Ni(s)	0	0	29.9	26.1	58.71
Ni(g)	429.7	384.5	182.2	23.4	58.71
NiCl$_2$(s)	−305.3	−259.0	97.7	71.7	129.62
NiO(s)	−239.7	−211.5	38.0	44.3	74.71
NiSO$_4$(s)	−872.9	−759.7	92.0	138.0	154.77
Ni^{2+}(aq)	−54.0	−45.6	−128.9		58.71
Nitrogen					
N$_2$(g)	0	0	191.6	29.1	28.013
N(g)	472.7	455.5	153.3	20.8	14.007
NH$_3$(g)	−45.9	−16.5	192.8	35.1	17.03
NO(g)	91.3	87.6	210.8	29.9	30.01
N$_2$O(g)	81.6	103.7	220.0	38.6	44.01
NO$_2$(g)	33.2	51.3	240.1	37.2	46.01
NOCl(g)	51.7	66.1	261.7	44.7	65.46
N$_2$O$_4$(g)	11.1	99.8	304.4	79.2	92.01
N$_2$O$_4$(l)	−19.5	97.5	209.2	142.7	92.01
HNO$_3$(l)	−174.1	−80.7	155.6	109.9	63.01
HNO$_3$(g)	−133.9	−73.5	266.9	54.1	63.01
NO$_3^-$(aq)	−207.4	−111.3	146.4		62.01
NH$_4^+$(aq)	−132.5	−79.3	113.4		18.04
Oxygen					
O$_2$(g)	0	0	205.2	29.4	31.999
O(g)	249.2	231.7	161.1	21.9	15.999
O$_3$(g)	142.7	163.2	238.9	39.2	47.998
OH(g)	39.0	34.22	183.7	29.9	17.01
OH$^-$(aq)	−230.0	−157.2	−10.9		17.01
Phosphorus					
P(s) white	0	0	41.1	23.8	30.97
P(s) red	−17.6	−12.1	22.8	21.2	30.97
P$_4$(g)	58.9	24.4	280.0	67.2	123.90
PCl$_5$(g)	−374.9	−305.0	364.6	112.8	208.24
PH$_3$(g)	5.4	13.5	210.2	37.1	34.00
H$_3$PO$_4$(l)	−1271.7	−1123.6	150.8	145.0	94.97
PO$_4^{3-}$(aq)	−1277.4	−1018.7	−220.5		91.97
HPO$_4^{2-}$(aq)	−1299.0	−1089.2	−33.5		92.97
H$_2$PO$_4^-$(aq)	−1302.6	−1130.2	92.5		93.97
Potassium					
K(s)	0	0	64.7	29.6	39.10
K(g)	89.0	60.5	160.3	20.8	39.10
KCl(s)	−436.5	−408.5	82.6	51.3	74.56
K$_2$O(s)	−361.5	−322.8	102.0	77.4	94.20
K$_2$SO$_4$(s)	−1437.8	−1321.4	175.6	131.5	174.27
K$^+$(aq)	−252.4	−283.3	102.5		39.10

(continued)

TABLE 4.1 Thermodynamic Data for Inorganic Compounds at 298.15 K (continued)

Substance	ΔH_f° (kJ mol^{-1})	ΔG_f° (kJ mol^{-1})	S_m° (J mol^{-1} K^{-1})	$C_{P,\,m}$ (J mol^{-1} K^{-1})	Atomic or Molecular Weight (amu)
Silicon					
Si(s)	0	0	18.8	20.0	28.09
Si(g)	450.0	405.5	168.0	22.3	28.09
SiCl$_4$(g)	−662.7	−622.8	330.9	90.3	169.70
SiO$_2$(quartz)	−910.7	−856.3	41.5	44.4	60.09
Silver					
Ag(s)	0	0	42.6	25.4	107.87
Ag(g)	284.9	246.0	173.0	20.8	107.87
AgCl(s)	−127.0	−109.8	96.3	50.8	143.32
AgNO$_2$(s)	−44.4	19.8	140.6	93.0	153.88
AgNO$_3$(s)	−44.4	19.8	140.9	93.1	169.87
Ag$_2$SO$_4$(s)	−715.9	−618.4	200.4	131.4	311.80
Ag$^+$(aq)	105.6	77.1	72.7		107.87
Sodium					
Na(s)	0	0	51.3	28.2	22.99
Na(g)	107.5	77.0	153.7	20.8	22.99
NaCl(s)	−411.2	−384.1	72.1	50.5	58.44
NaOH(s)	−425.8	−379.7	64.4	59.5	40.00
Na$_2$SO$_4$(s)	−1387.1	−1270.2	149.6	128.2	142.04
Na$^+$(aq)	−240.1	−261.9	59.0		22.99
Sulfur					
S(rhombic)	0	0	32.1	22.6	32.06
SF$_6$(g)	−1220.5	−1116.5	291.5	97.3	146.07
H$_2$S(g)	−20.6	−33.4	205.8	34.2	34.09
SO$_2$(g)	−296.8	−300.1	248.2	39.9	64.06
SO$_3$(g)	−395.7	−371.1	256.8	50.7	80.06
SO$_3^{2-}$(aq)	−635.5	−486.6	−29.3		80.06
SO$_4^{2-}$(aq)	−909.3	−744.5	20.1		96.06
Tin					
Sn(white)	0	0	51.2	27.0	118.69
Sn(g)	301.2	266.2	168.5	21.3	118.69
SnO$_2$(s)	−577.6	−515.8	49.0	52.6	150.69
Sn^{2+}(aq)	−8.9	−27.2	−16.7		118.69
Titanium					
Ti(s)	0	0	30.7	25.0	47.87
Ti(g)	473.0	428.4	180.3	24.4	47.87
TiCl$_4$(l)	−804.2	−737.2	252.4	145.2	189.69
TiO$_2$(s)	−944.0	−888.8	50.6	55.0	79.88
Xenon					
Xe(g)	0	0	169.7	20.8	131.30
XeF$_4$(s)	−261.5	−123	146	118	207.29
Zinc					
Zn(s)	0	0	41.6	25.4	65.37
ZnCl$_2$(s)	−415.1	−369.4	111.5	71.3	136.28
ZnO(s)	−350.5	−320.5	43.7	40.3	81.37
ZnSO$_4$(s)	−982.8	−871.5	110.5	99.2	161.43
Zn^{2+}(aq)	−153.9	−147.1	−112.1		65.37

Sources: HCP, HTTD, and TDPS.

TABLE 4.2   Thermodynamic Data for Selected Organic Compounds at 298.15 K

Substance	Formula	Molecular Weight	ΔH_f° (kJ mol^{-1})	$\Delta H_{combustion}^\circ$ (kJ mol^{-1})	ΔG_f° (kJ mol^{-1})	S_m° (J mol^{-1} K^{-1})	$C_{P,m}$ (J mol^{-1} K^{-1})
Carbon (graphite)	C	12.011	0	−393.5	0	5.74	8.52
Carbon (diamond)	C	12.011	1.89	−395.4	2.90	2.38	6.12
Carbon monoxide	CO	28.01	−110.5	−283.0	−137.2	197.7	29.1
Carbon dioxide	CO_2	44.01	−393.5		−394.4	213.8	37.1
Acetaldehyde(*l*)	C_2H_4O	44.05	−192.2	−1166.9	−127.6	160.3	89.0
Acetic acid(*l*)	$C_2H_4O_2$	60.05	−484.3	−874.2	−389.9	159.8	124.3
Acetone(*l*)	C_3H_6O	58.08	−248.4	−1790	−155.2	199.8	126.3
Benzene(*l*)	C_6H_6	78.12	49.1	−3268	124.5	173.4	136.0
Benzene(*g*)	C_6H_6	78.12	82.9	−3303	129.7	269.2	82.4
Benzoic acid(*s*)	$C_7H_6O_2$	122.13	−385.2	−3227	−245.5	167.6	146.8
1,3-Butadiene(*g*)	C_4H_6	54.09	110.0	−2541			79.8
n-Butane(*g*)	C_4H_{10}	58.13	−125.7	−2878	−17.0	310.2	97.5
1-Butene(*g*)	C_4H_8	56.11	−0.63	−2718	71.1	305.7	85.7
Carbon disulfide(*g*)	CS_2	76.14	116.9	−1112	66.8	238.0	45.7
Carbon tetrachloride(*l*)	CCl_4	153.82	−128.2	−360	−62.5	214.4	133.9
Carbon tetrachloride(*g*)	CCl_4	153.82	−95.7		−58.2	309.7	83.4
Cyclohexane(*l*)	C_6H_{12}	84.16	−156.4	−3920	26.8	204.5	154.9
Cyclopentane(*l*)	C_5H_{10}	70.13	−105.1	−3291	38.8	204.5	128.8
Cyclopropane(*g*)	C_3H_6	42.08	53.3	−2091	104.5	237.5	55.6
Dimethyl ether(*g*)	C_2H_6O	131.6	−184.1	−1460	−112.6	266.4	64.4
Ethane(*g*)	C_2H_6	30.07	−84.0	−1561	−32.0	229.2	52.5
Ethanol(*l*)	C_2H_6O	46.07	−277.6	−1367	−174.8	160.7	112.3
Ethanol(*g*)	C_2H_6O	46.07	−234.8	−1367	−167.9	281.6	65.6
Ethene(*g*)	C_2H_4	28.05	52.4	−1411	68.4	219.3	42.9
Ethyne(*g*)	C_2H_2	26.04	227.4	−1310	209.2	200.9	44
Formaldehyde(*g*)	CH_2O	30.03	−108.6	−571	−102.5	218.8	35.4
Formic acid(*l*)	CH_2O_2	46.03	−425.0	−255	−361.4	129.0	99.0
Formic acid(*g*)	CH_2O_2	46.03	−378.7	−256	−351.0	248.7	45.2
α-D-Glucose(*s*)	$C_6H_{12}O_6$	180.16	−1273.1	−2805	−910.6	209.2	219.2
n-Hexane(*l*)	C_6H_{14}	86.18	−198.7	−4163	−4.0	296.0	195.6
Hydrogen cyanide(*l*)	HCN	27.03	108.9		125.0	112.8	70.6
Hydrogen cyanide(*g*)	HCN	27.03	135.5		124.7	201.8	35.9
Methane(*g*)	CH_4	16.04	−74.6	−891	−50.5	186.3	35.7
Methanol(*l*)	CH_4O	32.04	−239.2	−726	−166.6	126.8	81.1
Methanol(*g*)	CH_4O	32.04	−201.0	−764	−162.3	239.9	44.1
Oxalic acid(*g*)	$C_2H_2O_4$	90.04	−731.8	−246	−662.7	320.6	86.2
n-Pentane(*g*)	C_5H_{12}	72.15	−146.9	−3509	−8.2	349.1	120.1
Phenol(*s*)	C_6H_6O	94.11	−165.1	−3054	−50.2	144.0	127.4
Propane(*g*)	C_3H_8	44.10	−103.8	−2219	−23.4	270.3	73.6
Propene(*g*)	C_3H_6	42.08	20.0	−2058	62.7	266.9	64.0
Propyne(*g*)	C_3H_4	40.07	184.9	−2058	194.5	248.2	60.7
Pyridine(*l*)	C_5H_5N	79.10	100.2	−2782		177.9	132.7
Sucrose(*s*)	$C_{12}H_{22}O_{11}$	342.3	−2226.1	−5643	−1544.6	360.2	424.3
Thiophene(*l*)	C_4H_4S	84.14	80.2	−2829		181.2	123.8
Toluene(*g*)	C_7H_8	92.14	50.5	−3910	122.3	320.8	104
Urea(*s*)	$C_2H_4N_2O$	60.06	−333.1	−635	−197.4	104.3	92.8

Sources: HCP, HTTD, TDPS, and TDOC.

TABLE 7.1 Second Virial Coefficients for Selected Gases in Units of cm^3 mol^{-1}

	200 K	300 K	400 K	500 K	600 K	700 K
Benzene		−1453	−712	−429	−291	−211
Cl$_2$		−299	−166	−97	−59	−36
CO$_2$		−126	−61.7	−30.5	−12.6	−1.18
H$_2$O		−1126	−356	−175	−104	−67
Heptane		−2782	−1233	−702	−452	−304
Kr	−117	−51.0	−23.0	−7.75	1.78	8.33
N$_2$	−34.4	−3.91	9.17	16.3	20.8	23.8
Octane		−4042	−1704	−936	−583	−375

Source: Calculated from HTTD.

TABLE 7.2 Critical Constants of Selected Substances

Substance	Formula	T_c (K)	P_c (bar)	$10^3 V_c$ (L)	$z_c = \dfrac{P_c V_c}{R T_c}$
Ammonia	NH$_3$	405.40	113.53	72.47	0.244
Argon	Ar	150.86	48.98	74.57	0.291
Benzene	C$_6$H$_6$	562.05	48.95	256.00	0.268
Bromine	Br$_2$	588.00	103.40	127.00	0.268
Carbon dioxide	CO$_2$	304.13	73.75	94.07	0.274
Carbon monoxide	CO	132.91	34.99	93.10	0.295
Ethane	C$_2$H$_6$	305.32	48.72	145.50	0.279
Ethanol	C$_2$H$_5$OH	513.92	61.37	168.00	0.241
Ethene	C$_2$H$_4$	282.34	50.41	131.1	0.281
Ethyne	C$_2$H$_2$	308.30	61.38	112.20	0.269
Fluorine	F$_2$	144.30	51.72	66.20	0.285
Hydrogen	H$_2$	32.98	12.93	64.20	0.303
Methane	CH$_4$	190.56	45.99	98.60	0.286
Methanol	CH$_3$OH	512.50	80.84	117.00	0.221
Nitrogen	N$_2$	126.20	33.98	90.10	0.292
Oxygen	O$_2$	154.58	50.43	73.37	0.288
Pentane	C$_5$H$_{12}$	469.70	33.70	311.00	0.268
Propane	C$_3$H$_8$	369.83	42.48	200.00	0.276
Pyridine	C$_5$H$_5$N	620.00	56.70	243.00	0.267
Tetrachloromethane	CCl$_4$	556.60	45.16	276.00	0.269
Water	H$_2$O	647.14	220.64	55.95	0.229
Xenon	Xe	289.74	58.40	118.00	0.286

Sources: HCP and DAL.

TABLE 7.4   van der Waals and Redlich–Kwong Parameters for Selected Gases

Substance	Formula	van der Waals		Redlich–Kwong	
		a (dm^6 bar mol^{-2})	b (dm^3 mol^{-1})	a (dm^6 bar mol^{-2} K$^{1/2}$)	b (dm^3 mol^{-1})
Ammonia	NH_3	4.225	0.0371	86.12	0.02572
Argon	Ar	1.355	0.0320	16.86	0.02219
Benzene	C_6H_6	18.82	0.1193	452.0	0.08271
Bromine	Br_2	9.75	0.0591	236.5	0.04085
Carbon dioxide	CO_2	3.658	0.0429	64.63	0.02971
Carbon monoxide	CO	1.472	0.0395	17.20	0.02739
Ethane	C_2H_6	5.580	0.0651	98.79	0.04514
Ethanol	C_2H_5OH	12.56	0.0871	287.7	0.06021
Ethene	C_2H_4	4.612	0.0582	78.51	0.04034
Ethyne	C_2H_2	4.533	0.05240	80.65	0.03632
Fluorine	F_2	1.171	0.0290	14.17	0.01993
Hydrogen	H_2	0.2452	0.0265	1.427	0.01837
Methane	CH_4	2.303	0.0431	32.20	0.02985
Methanol	CH_3OH	9.476	0.0659	217.1	0.04561
Nitrogen	N_2	1.370	0.0387	15.55	0.02675
Oxygen	O_2	1.382	0.0319	17.40	0.02208
Pentane	C_5H_{12}	19.09	0.1448	419.2	0.1004
Propane	C_3H_8	9.39	0.0905	182.9	0.06271
Pyridine	C_5H_5N	19.77	0.1137	498.8	0.07877
Tetrachloromethane	CCl_4	20.01	0.1281	473.2	0.08793
Water	H_2O	5.537	0.0305	142.6	0.02113
Xenon	Xe	4.192	0.0516	72.30	0.03574

Source: Calculated from critical constants.

TABLE 8.1   Triple Point Pressures and Temperatures of Selected Substances

Formula	Name	T_{tp} (K)	P_{tp} (Pa)
Ar	Argon	83.806	68950
Br_2	Bromine	280.4	5879
Cl_2	Chlorine	172.17	1392
HCl	Hydrogen chloride	158.8	
H_2	Hydrogen	13.8	7042
H_2O	Water	273.16	611.73
H_2S	Hydrogen sulfide	187.67	23180
NH_3	Ammonia	195.41	6077
Kr	Krypton	115.8	72920
NO	Nitrogen oxide	109.54	21916
O_2	Oxygen	54.36	146.33
SO_3	Sulfur trioxide	289.94	21130
Xe	Xenon	161.4	81590
CH_4	Methane	90.694	11696
CO	Carbon monoxide	68.15	15420
CO_2	Carbon dioxide	216.58	518500
C_3H_6	Propene	87.89	9.50×10^{-4}

Sources: HCP, HTTP, and DAL.

TABLE 8.3 Vapor Pressure and Boiling Temperature of Liquids

$$\ln\frac{P(T)}{Pa} = A(1) - \frac{A(2)}{\frac{T}{K} + A(3)} \qquad T_{vap}(P) = \frac{A(2)}{A(1) - \ln\frac{P}{Pa}} - A(3)$$

Molecular Formula	Name	T_{vap} (K)	$A(1)$	$A(2)$	$A(3)$	$10^{-3}P(298.15\ K)$ (Pa)	Range (K)
Ar	Argon	87.28	22.946	1.0325×10^3	3.130	—	73–90
Br_2	Bromine	331.9	20.729	2.5782×10^3	−51.77	28.72	268–354
HF	Hydrogen fluoride	292.65	22.893	3.6178×10^3	25.627	122.90	273–303
H_2O	Water	373.15	23.195	3.8140×10^3	−46.290		353–393
SO_2	Sulfur dioxide	263.12	21.661	2.3024×10^3	−35.960		195–280
CCl_4	Tetrachloromethane	349.79	20.738	2.7923×10^3	−46.6667	15.28	287–350
$CHCl_3$	Trichloromethane	334.33	20.907	2.6961×10^3	−46.926	26.24	263–335
HCN	Hydrogen cyanide	298.81	22.226	3.0606×10^3	−12.773	98.84	257–316
CH_3OH	Methanol	337.70	23.593	3.6971×10^3	−31.317	16.94	275–338
CS_2	Carbon disulfide	319.38	20.801	2.6524×10^3	−33.40	48.17	255–320
C_2H_5OH	Ethanol	351.45	23.58	3.6745×10^3	−46.702	7.87	293–366
C_3H_6	Propene	225.46	20.613	1.8152×10^3	−25.705	1156.6	166–226
C_3H_8	Propane	231.08	20.558	1.8513×10^3	−26.110	948.10	95–370
C_4H_9Br	1-Bromobutane	374.75	17.076	1.5848×10^3	−11.188	5.26	195–300
C_4H_9Cl	1-Chlorobutane	351.58	20.612	2.6881×10^3	−55.725	13.68	256–352
$C_5H_{11}OH$	1-Pentanol	411.133	20.729	2.5418×10^3	−134.93	0.29	410–514
C_6H_5Cl	Chlorobenzene	404.837	20.964	3.2969×10^3	−55.515	1.57	335–405
C_6H_5I	Iodobenzene	461.48	21.088	3.8136×10^3	−62.654	0.13	298–462
C_6H_6	Benzene	353.24	20.767	2.7738×10^3	−53.08	12.69	294–378
C_6H_{14}	Hexane	341.886	20.749	2.7081×10^3	−48.251	20.17	286–343
C_6H_5CHO	Benzaldehyde	451.90	21.213	3.7271×10^3	−67.156	0.17	311–481
$C_6H_5CH_3$	Toluene	383.78	21.600	3.6266×10^3	−23.778	3.80	360–580
$C_{10}H_8$	Naphthalene	491.16	21.100	4.0526×10^3	−67.866	0.01	353–453
$C_{14}H_{10}$	Anthracene	614.0	21.965	5.8733×10^3	−51.394		496–615

Sources: HCP and HTTP.

TABLE 8.4 Sublimation Pressure of Solids

$$\ln\frac{P(T)}{Pa} = A(1) - \frac{A(2)}{\frac{T}{K} + A(3)}$$

Molecular Formula	Name	$A(1)$	$A(2)$	$A(3)$	Range (K)
CCl_4	Tetrachloromethane	17.613	1.6431×10^3	−95.250	232–250
C_6H_{14}	Hexane	31.224	4.8186×10^3	−23.150	168–178
C_6H_5COOH	Benzoic acid	14.870	4.7196×10^3		293–314
$C_{10}H_8$	Naphthalene	31.143	8.5750×10^3		270–305
$C_{14}H_{10}$	Anthracene	31.620	1.1378×10^4		353–400

Sources: HCP and HTTP.

TABLE 10.2   Dielectric Constants, ε_r, of Selected Liquids

Substance	Dielectric Constant	Substance	Dielectric Constant
Acetic acid	6.2	Heptane	1.9
Acetone	21.0	Isopropyl alcohol	20.2
Benzaldehyde	17.8	Methanol	33.0
Benzene	2.3	Nitrobenzene	35.6
Carbon tetrachloride	2.2	o-Xylene	2.6
Cyclohexane	2.0	Phenol	12.4
Ethanol	25.3	Toluene	2.4
Glycerol	42.5	Water (273 K)	88.0
1-Hexanol	13.0	Water (373 K)	55.3

Source: HCP.

TABLE 10.3   Mean Activity Coefficients in Terms of Molalities at 298 K

Substance	0.1 m	0.2 m	0.3 m	0.4 m	0.5 m	0.6 m	0.7 m	0.8 m	0.9 m	1.0 m
$AgNO_3$	0.734	0.657	0.606	0.567	0.536	0.509	0.485	0.464	0.446	0.429
$BaCl_2$	0.500	0.444	0.419	0.405	0.397	0.391	0.391	0.391	0.392	0.395
$CaCl_2$	0.518	0.472	0.455	0.448	0.448	0.453	0.460	0.470	0.484	0.500
$CuCl_2$	0.508	0.455	0.429	0.417	0.411	0.409	0.409	0.410	0.413	0.417
$CuSO_4$	0.150	0.104	0.0829	0.0704	0.0620	0.0559	0.0512	0.0475	0.0446	0.0423
HCl	0.796	0.767	0.756	0.755	0.757	0.763	0.772	0.783	0.795	0.809
HNO_3	0.791	0.754	0.735	0.725	0.720	0.717	0.717	0.718	0.721	0.724
H_2SO_4	0.2655	0.2090	0.1826		0.1557		0.1417			0.1316
KCl	0.770	0.718	0.688	0.666	0.649	0.637	0.626	0.618	0.610	0.604
KOH	0.798	0.760	0.742	0.734	0.732	0.733	0.736	0.742	0.749	0.756
$MgCl_2$	0.529	0.489	0.477	0.475	0.481	0.491	0.506	0.522	0.544	0.570
$MgSO_4$	0.150	0.107	0.0874	0.0756	0.0675	0.0616	0.0571	0.0536	0.0508	0.0485
NaCl	0.778	0.735	0.710	0.693	0.681	0.673	0.667	0.662	0.659	0.657
NaOH	0.766	0.727	0.708	0.697	0.690	0.685	0.681	0.679	0.678	0.678
$ZnSO_4$	0.150	0.140	0.0835	0.0714	0.0630	0.0569	0.0523	0.0487	0.0458	0.0435

Source: HCP.

TABLE 11.1   Standard Reduction Potentials in Alphabetical Order

Reaction	$E°$ (V)	Reaction	$E°$ (V)
$Ag^+ + e^- \longrightarrow Ag$	0.7996	$Au^{3+} + 2e^- \longrightarrow Au^+$	1.401
$Ag^{2+} + e^- \longrightarrow Ag^+$	1.980	$Au^{3+} + 3e^- \longrightarrow Au$	1.498
$AgBr + e^- \longrightarrow Ag + Br^-$	0.07133	$AuBr_2^- + e^- \longrightarrow Au + 2Br^-$	0.959
$AgCl + e^- \longrightarrow Ag + Cl^-$	0.22233	$AuCl_4^- + 3e^- \longrightarrow Au + 4Cl^-$	1.002
$AgCN + e^- \longrightarrow Ag + CN^-$	-0.017	$Ba^{2+} + 2e^- \longrightarrow Ba$	-2.912
$AgF + e^- \longrightarrow Ag + F^-$	0.779	$Be^{2+} + 2e^- \longrightarrow Be$	-1.847
$Ag_4[Fe(CN)_6] + 4e^- \longrightarrow 4\,Ag + [Fe(CN)_6]^{4-}$	0.1478	$Bi^{3+} + 3e^- \longrightarrow Bi$	0.20
$AgI + e^- \longrightarrow Ag + I^-$	-0.15224	$Br_2(aq) + 2e^- \longrightarrow 2\,Br^-$	1.0873
$AgNO_2 + e^- \longrightarrow Ag + NO_2^-$	0.564	$BrO^- + H_2O + 2e^- \longrightarrow Br^- + 2\,OH^-$	0.761
$Al^{3+} + 3e^- \longrightarrow Al$	-1.662	$Ca^+ + e^- \longrightarrow Ca$	-3.80
$Au^+ + e^- \longrightarrow Au$	1.692	$Ca^{2+} + 2e^- \longrightarrow Ca$	-2.868

(continued)

TABLE 11.1   Standard Reduction Potentials in Alphabetical Order (continued)

Reaction	$E°$ (V)	Reaction	$E°$ (V)
$Cd^{2+} + 2e^- \longrightarrow Cd$	-0.4030	$In^{3+} + 3e^- \longrightarrow In$	-0.3382
$Cd(OH)_2 + 2e^- \longrightarrow Cd + 2\,OH^-$	-0.809	$K^+ + e^- \longrightarrow K$	-2.931
$CdSO_4 + 2e^- \longrightarrow Cd + SO_4^{2-}$	-0.246	$Li^+ + e^- \longrightarrow Li$	-3.0401
$Ce^{3+} + 3e^- \longrightarrow Ce$	-2.483	$Mg^{2+} + 2e^- \longrightarrow Mg$	-2.372
$Ce^{4+} + e^- \longrightarrow Ce^{3+}$	1.61	$Mg(OH)_2 + 2e^- \longrightarrow Mg + 2\,OH^-$	-2.690
$Cl_2(g) + 2e^- \longrightarrow 2\,Cl^-$	1.35827	$Mn^{2+} + 2e^- \longrightarrow Mn$	-1.185
$ClO_4^- + 2\,H^+ + 2e^- \longrightarrow ClO_3^- + H_2O$	1.189	$Mn^{3+} + e^- \longrightarrow Mn^{2+}$	1.5415
$ClO^- + H_2O + 2e^- \longrightarrow Cl^- + 2\,OH^-$	$.81$	$MnO_2 + 4\,H^+ + 2e^- \longrightarrow Mn^{2+} + 2\,H_2O$	1.224
$ClO_4^- + H_2O + 2e^- \longrightarrow ClO_3^- + 2\,OH^-$	0.36	$MnO_4^- + 4\,H^+ + 3e^- \longrightarrow MnO_2 + 2\,H_2O$	1.679
$Co^{2+} + 2e^- \longrightarrow Co$	-0.28	$MnO_4^{2-} + 2\,H_2O + 2e^- \longrightarrow MnO_2 + 4\,OH^-$	0.595
$Co^{3+} + e^- \longrightarrow Co^{2+}$	1.92	$MnO_4^- + 8\,H^+ + 5e^- \longrightarrow Mn^{2+} + 4\,H_2O$	1.507
$Cr^{2+} + 2e^- \longrightarrow Cr$	-0.913	$MnO_4^- + e^- \longrightarrow MnO_4^{2-}$	0.558
$Cr^{3+} + e^- \longrightarrow Cr^{2+}$	-0.407	$2\,NO + 2\,H^+ + 2e^- \longrightarrow N_2O + H_2O$	1.591
$Cr^{3+} + 3e^- \longrightarrow Cr$	-0.744	$HNO_2 + H^+ + e^- \longrightarrow NO + H_2O$	0.983
$Cr_2O_7^{2-} + 14\,H^+ + 6e^- \longrightarrow 2\,Cr^{3+} + 7\,H_2O$	1.232	$NO_2 + H_2O + 3e^- \longrightarrow NO + 2\,OH^-$	-0.46
$Cs^+ + e^- \longrightarrow Cs$	-2.92	$NO_3^- + 4\,H^+ + 3e^- \longrightarrow NO + 2\,H_2O$	0.957
$Cu^+ + e^- \longrightarrow Cu$	0.521	$NO_3^- + 2\,H^+ + e^- \longrightarrow NO_2^- + H_2O$	0.835
$Cu^{2+} + e^- \longrightarrow Cu^+$	0.153	$NO_3^- + H_2O + 2e^- \longrightarrow NO_2^- + 2\,OH^-$	0.10
$Cu(OH)_2 + 2e^- \longrightarrow Cu + 2\,OH^-$	-0.222	$Na^+ + e^- \longrightarrow Na$	-2.71
$F_2 + 2\,H^+ + 2e^- \longrightarrow 2\,HF$	3.053	$Ni^{2+} + 2e^- \longrightarrow Ni$	-0.257
$F_2 + 2e^- \longrightarrow 2\,F^-$	2.866	$NiOOH + H_2O + e^- \longrightarrow Ni(OH)_2 + OH^-$	0.52
$Fe^{2+} + 2e^- \longrightarrow Fe$	-0.447	$Ni(OH)_2 + 2e^- \longrightarrow Ni + 2\,OH^-$	-0.72
$Fe^{3+} + 3e^- \longrightarrow Fe$	-0.030	$NiO_2 + 2\,H_2O + 2e^- \longrightarrow Ni(OH)_2 + 2\,OH^-$	0.49
$Fe^{3+} + e^- \longrightarrow Fe^{2+}$	0.771	$NiO_2 + 4\,H^+ + 2e^- \longrightarrow Ni^{2+} + 2\,H_2O$	1.678
$[Fe(CN)_6]^{3-} + e^- \longrightarrow [Fe(CN)_6]^{4-}$	0.358	$O_2 + e^- \longrightarrow O_2^-$	-0.56
$2\,H^+ + 2e^- \longrightarrow H_2$	0	$O_2 + 2\,H^+ + 2e^- \longrightarrow H_2O_2$	0.695
$HBrO + H^+ + e^- \longrightarrow 1/2\,Br_2 + H_2O$	1.574	$O_2 + 4\,H^+ + 4e^- \longrightarrow 2\,H_2O$	1.229
$HClO + H^+ + e^- \longrightarrow 1/2\,Cl_2 + H_2O$	1.611	$O_2 + 2\,H_2O + 2e^- \longrightarrow H_2O_2 + 2\,OH^-$	-0.146
$HClO_2 + 3\,H^+ + 3e^- \longrightarrow 1/2\,Cl_2 + 2\,H_2O$	1.628	$O_2 + 2\,H_2O + 4e^- \longrightarrow 4\,OH^-$	0.401
$HO_2 + H^+ + e^- \longrightarrow H_2O_2$	1.495	$O_2 + H_2O + 2e^- \longrightarrow HO_2^- + OH^-$	-0.076
$HO_2 + H_2O + 2e^- \longrightarrow 3\,OH^-$	0.878	$O_3 + 2\,H^+ + 2e^- \longrightarrow O_2 + H_2O$	2.07
$2\,H_2O + 2e^- \longrightarrow H_2 + 2\,OH^-$	-0.8277	$O_3 + H_2O + 2e^- \longrightarrow O_2 + 2\,OH^-$	1.24
$H_2O_2 + 2\,H^+ + 2e^- \longrightarrow 2\,H_2O$	1.776	$Pb^{2+} + 2e^- \longrightarrow Pb$	-0.1262
$2\,H_2SO_3 + H^+ + 2e^- \longrightarrow H_2SO_4^- + 2\,H_2O$	-0.056	$Pb^{4+} + 2e^- \longrightarrow Pb^{2+}$	1.67
$H_2SO_3 + 4\,H^+ + 4e^- \longrightarrow S + 3\,H_2O$	0.449	$PbBr_2 + 2e^- \longrightarrow Pb + 2\,Br^-$	-0.284
$H_3PO_4 + 2\,H^+ + 2e^- \longrightarrow H_3PO_3 + H_2O$	-0.276	$PbCl_2 + 2e^- \longrightarrow Pb + 2\,Cl^-$	-0.2675
$Hg^{2+} + 2e^- \longrightarrow Hg$	0.851	$PbO + H_2O + 2e^- \longrightarrow Pb + 2\,OH^-$	-0.580
$Hg_2^{2+} + 2e^- \longrightarrow 2\,Hg$	0.7973	$PbO_2 + 4\,H^+ + 2e^- \longrightarrow Pb^{2+} + 2\,H_2O$	1.455
$Hg_2Cl_2 + 2e^- \longrightarrow 2\,Hg + 2\,Cl^-$	0.26808	$PbO_2 + SO_4^{2-} + 4\,H^+ + 2e^- \longrightarrow PbSO_4 + 2\,H_2O$	1.6913
$Hg_2SO_4 + 2e^- \longrightarrow 2\,Hg + SO_4^{2-}$	0.6125	$PbSO_4 + 2e^- \longrightarrow Pb + SO_4^{2-}$	-0.3505
$I_2 + 2e^- \longrightarrow 2\,I^-$	0.5355	$Pd^{2+} + 2e^- \longrightarrow Pd$	0.951
$I_3^- + 2e^- \longrightarrow 3\,I^-$	0.536	$Pt^{2+} + 2e^- \longrightarrow Pt$	1.118
$In^+ + e^- \longrightarrow In$	-0.14	$[PtCl_4]^{2-} + 2e^- \longrightarrow Pt + 4\,Cl^-$	0.755
$In^{2+} + e^- \longrightarrow In^+$	-0.40	$[PtCl_6]^{2-} + 2e^- \longrightarrow [PtCl_4]^{2-} + 2\,Cl^-$	0.68

Reaction	$E°$ (V)	Reaction	$E°$ (V)
$Pt(OH)_2 + 2e^- \longrightarrow Pt + 2\,OH^-$	0.14	$Sn^{2+} + 2e^- \longrightarrow Sn$	-0.1375
$Rb^+ + e^- \longrightarrow Rb$	-2.98	$Sn^{4+} + 2e^- \longrightarrow Sn^{2+}$	0.151
$Re^{3+} + 3e^- \longrightarrow Re$	0.300	$Ti^{2+} + 2e^- \longrightarrow Ti$	-1.630
$S + 2e^- \longrightarrow S^{2-}$	-0.47627	$Ti^{3+} + e^- \longrightarrow Ti^{2+}$	-0.9
$S + 2\,H^+ + 2e^- \longrightarrow H_2S(aq)$	0.142	$TiO_2 + 4\,H^+ + 2e^- \longrightarrow Ti^{2+} + 2\,H_2O$	-0.502
$S_2O_6^{2-} + 4\,H^+ + 2e^- \longrightarrow 2\,H_2SO_3$	0.564	$Zn^{2+} + 2e^- \longrightarrow Zn$	-0.7618
$S_2O_6^{2-} + 2e^- + 2\,H^+ \longrightarrow 2\,HSO_3^-$	0.464	$ZnO_2^{2-} + 2\,H_2O + 2e^- \longrightarrow Zn + 4\,OH^-$	-1.215
$S_2O_8^{2-} + 2e^- \longrightarrow 2\,SO_4^{2-}$	2.010	$Zr(OH)_2 + H_2O + 4e^- \longrightarrow Zr + 4\,OH^-$	-2.36

Source: HCP and Bard.

TABLE 11.2   Standard Reduction Potentials Ordered by Reduction Potential

Reaction	$E°$ (V)	Reaction	$E°$ (V)
$Ca^+ + e^- \longrightarrow Ca$	-3.80	$Cd^{2+} + 2e^- \longrightarrow Cd$	-0.4030
$Li^+ + e^- \longrightarrow Li$	-3.0401	$In^{2+} + e^- \longrightarrow In^+$	-0.40
$Rb^+ + e^- \longrightarrow Rb$	-2.98	$PbSO_4 + 2e^- \longrightarrow Pb + SO_4^{2-}$	-0.3505
$K^+ + e^- \longrightarrow K$	-2.931	$In^{3+} + 3e^- \longrightarrow In$	-0.3382
$Cs^+ + e^- \longrightarrow Cs$	-2.92	$PbBr_2 + 2e^- \longrightarrow Pb + 2\,Br^-$	-0.284
$Ba^{2+} + 2e^- \longrightarrow Ba$	-2.912	$Co^{2+} + 2e^- \longrightarrow Co$	-0.28
$Ca^{2+} + 2e^- \longrightarrow Ca$	-2.868	$H_3PO_4 + 2\,H^+ + 2e^- \longrightarrow H_3PO_3 + H_2O$	-0.276
$Na^+ + e^- \longrightarrow Na$	-2.71	$PbCl_2 + 2e^- \longrightarrow Pb + 2\,Cl^-$	-0.2675
$Mg(OH)_2 + 2e^- \longrightarrow Mg + 2\,OH^-$	-2.690	$Ni^{2+} + 2e^- \longrightarrow Ni$	-0.257
$Ce^{3+} + 3e^- \longrightarrow Ce$	-2.483	$CdSO_4 + 2e^- \longrightarrow Cd + SO_4^{2-}$	-0.246
$Mg^{2+} + 2e^- \longrightarrow Mg$	-2.372	$Cu(OH)_2 + 2e^- \longrightarrow Cu + 2\,OH^-$	-0.222
$Zr(OH)_2 + H_2O + 4e^- \longrightarrow Zr + 4\,OH^-$	-2.36	$AgI + e^- \longrightarrow Ag + I^-$	-0.15224
$Be^{2+} + 2e^- \longrightarrow Be$	-1.847	$O_2 + 2\,H_2O + 2e^- \longrightarrow H_2O_2 + 2\,OH^-$	-0.146
$Al^{3+} + 3e^- \longrightarrow Al$	-1.662	$In^+ + e^- \longrightarrow In$	-0.14
$Ti^{2+} + 2e^- \longrightarrow Ti$	-1.630	$Sn^{2+} + 2e^- \longrightarrow Sn$	-0.1375
$ZnO_2^{2-} + 2\,H_2O + 2e^- \longrightarrow Zn + 4\,OH^-$	-1.215	$Pb^{2+} + 2e^- \longrightarrow Pb$	-0.1262
$Mn^{2+} + 2e^- \longrightarrow Mn$	-1.185	$O_2 + H_2O + 2e^- \longrightarrow HO_2^- + OH^-$	-0.076
$Cr^{2+} + 2e^- \longrightarrow Cr$	-0.913	$2\,H_2SO_3 + H^+ + 2e^- \longrightarrow H_2SO_4^- + 2\,H_2O$	-0.056
$Ti^{3+} + e^- \longrightarrow Ti^{2+}$	-0.9	$Fe^{3+} + 3e^- \longrightarrow Fe$	-0.030
$2\,H_2O + 2e^- \longrightarrow H_2 + 2\,OH^-$	-0.8277	$AgCN + e^- \longrightarrow Ag + CN^-$	-0.017
$Cd(OH)_2 + 2e^- \longrightarrow Cd + 2\,OH^-$	-0.809	$2\,H^+ + 2e^- \longrightarrow H_2$	0
$Zn^{2+} + 2e^- \longrightarrow Zn$	-0.7618	$AgBr + e^- \longrightarrow Ag + Br^-$	0.07133
$Cr^{3+} + 3e^- \longrightarrow Cr$	-0.744	$NO_3^- + H_2O + 2e^- \longrightarrow NO_2^- + 2\,OH^-$	0.10
$Ni(OH)_2 + 2e^- \longrightarrow Ni + 2\,OH^-$	-0.72	$Pt(OH)_2 + 2e^- \longrightarrow Pt + 2\,OH^-$	0.14
$PbO + H_2O + 2e^- \longrightarrow Pb + 2\,OH^-$	-0.580	$S + 2\,H^+ + 2e^- \longrightarrow H_2S1aq2$	0.142
$O_2 + e^- \longrightarrow O_2^-$	-0.56	$Ag_4[Fe(CN)_6] + 4e^- \longrightarrow 4\,Ag + [Fe(CN)_6]^{4-}$	0.1478
$TiO_2 + 4\,H^+ + 2e^- \longrightarrow Ti^{2+} + 2\,H_2O$	-0.502	$Sn^{4+} + 2e^- \longrightarrow Sn^{2+}$	0.151
$S + 2e^- \longrightarrow S^{2-}$	-0.47627	$Cu^{2+} + e^- \longrightarrow Cu^+$	0.153
$NO_2 + H_2O + 3e^- \longrightarrow NO + 2\,OH^-$	-0.46	$Bi^{3+} + 3e^- \longrightarrow Bi$	0.20
$Fe^{2+} + 2e^- \longrightarrow Fe$	-0.447	$AgCl + e^- \longrightarrow Ag + Cl^-$	0.22233
$Cr^{3+} + e^- \longrightarrow Cr^{2+}$	-0.407	$Hg_2Cl_2 + 2e^- \longrightarrow 2\,Hg + 2\,Cl^-$	0.26808

(continued)

TABLE 11.2 **Standard Reduction Potentials Ordered by Reduction Potential (continued)**

Reaction	$E°$ (V)	Reaction	$E°$ (V)
$Re^{3+} + 3e^- \longrightarrow Re$	0.300	$AuCl_4^- + 3e^- \longrightarrow Au + 4Cl^-$	1.002
$[Fe(CN)_6]^{3-} + e^- \longrightarrow [Fe(CN)_6]^{4-}$	0.358	$Br_2(aq) + 2e^- \longrightarrow 2Br^-$	1.0873
$ClO_4^- + H_2O + 2e^- \longrightarrow ClO_3^- + 2OH^-$	0.36	$Pt^{2+} + 2e^- \longrightarrow Pt$	1.118
$O_2 + 2H_2O + 4e^- \longrightarrow 4OH^-$	0.401	$ClO_4^- + 2H^+ + 2e^- \longrightarrow ClO_3^- + H_2O$	1.189
$H_2SO_3 + 4H^+ + 4e^- \longrightarrow S + 3H_2O$	0.449	$MnO_2 + 4H^+ + 2e^- \longrightarrow Mn^{2+} + 2H_2O$	1.224
$S_2O_6^{2-} + 2e^- + 2H^+ \longrightarrow 2HSO_3^-$	0.464	$O_2 + 4H^+ + 4e^- \longrightarrow 2H_2O$	1.229
$NiO_2 + 2H_2O + 2e^- \longrightarrow Ni(OH)_2 + 2OH^-$	0.49	$Cr_2O_7^{2-} + 14H^+ + 6e^- \longrightarrow 2Cr^{3+} + 7H_2O$	1.232
$NiOOH + H_2O + e^- \longrightarrow Ni(OH)_2 + OH^-$	0.52	$O_3 + H_2O + 2e^- \longrightarrow O_2 + 2OH^-$	1.24
$Cu^+ + e^- \longrightarrow Cu$	0.521	$Cl_2(g) + 2e^- \longrightarrow 2Cl^-$	1.35827
$I_2 + 2e^- \longrightarrow 2I^-$	0.5355	$Au^{3+} + 2e^- \longrightarrow Au^+$	1.401
$I_3^- + 2e^- \longrightarrow 3I^-$	0.536	$PbO_2 + 4H^+ + 2e^- \longrightarrow Pb^{2+} + 2H_2O$	1.455
$MnO_4^- + e^- \longrightarrow MnO_4^{2-}$	0.558	$HO_2 + H^+ + e^- \longrightarrow H_2O_2$	1.495
$AgNO_2 + e^- \longrightarrow Ag + NO_2^-$	0.564	$Au^{3+} + 3e^- \longrightarrow Au$	1.498
$S_2O_6^{2-} + 4H^+ + 2e^- \longrightarrow 2H_2SO_3$	0.564	$MnO_4^- + 8H^+ + 5e^- \longrightarrow Mn^{2+} + 4H_2O$	1.507
$MnO_4^{2-} + 2H_2O + 2e^- \longrightarrow MnO_2 + 4OH^-$	0.595	$Mn^{3+} + e^- \longrightarrow Mn^{2+}$	1.5415
$Hg_2SO_4 + 2e^- \longrightarrow 2Hg + SO_4^{2-}$	0.6125	$HBrO + H^+ + e^- \longrightarrow 1/2Br_2 + H_2O$	1.574
$[PtCl_6]^{2-} + 2e^- \longrightarrow [PtCl_4]^{2-} + 2Cl^-$	0.68	$2NO + 2H^+ + 2e^- \longrightarrow N_2O + H_2O$	1.591
$O_2 + 2H^+ + 2e^- \longrightarrow H_2O_2$	0.695	$Ce^{4+} + e^- \longrightarrow Ce^{3+}$	1.61
$[PtCl_4]^{2-} + 2e^- \longrightarrow Pt + 4Cl^-$	0.755	$HClO + H^+ + e^- \longrightarrow 1/2Cl_2 + H_2O$	1.611
$BrO^- + H_2O + 2e^- \longrightarrow Br^- + 2OH^-$	0.761	$HClO_2 + 3H^+ + 3e^- \longrightarrow 1/2Cl_2 + 2H_2O$	1.628
$Fe^{3+} + e^- \longrightarrow Fe^{2+}$	0.771	$Pb^{4+} + 2e^- \longrightarrow Pb^{2+}$	1.67
$AgF + e^- \longrightarrow Ag + F^-$	0.779	$NiO_2 + 4H^+ + 2e^- \longrightarrow Ni^{2+} + 2H_2O$	1.678
$Hg_2^{2+} + 2e^- \longrightarrow 2Hg$	0.7973	$MnO_4^- + 4H^+ + 3e^- \longrightarrow MnO_2 + 2H_2O$	1.679
$Ag^+ + e^- \longrightarrow Ag$	0.7996	$PbO_2 + SO_4^{2-} + 4H^+ + 2e^- \longrightarrow PbSO_4 + 2H_2O$	1.6913
$ClO^- + H_2O + 2e^- \longrightarrow Cl^- + 2OH^-$	0.81	$Au^+ + e^- \longrightarrow Au$	1.692
$NO_3^- + 2H^+ + e^- \longrightarrow NO_2^- + H_2O$	0.835	$H_2O_2 + 2H^+ + 2e^- \longrightarrow 2H_2O$	1.776
$Hg^{2+} + 2e^- \longrightarrow Hg$	0.851	$Co^{3+} + e^- \longrightarrow Co^{2+}$	1.92
$HO_2 + H_2O + 2e^- \longrightarrow 3OH^-$	0.878	$Ag^{2+} + e^- \longrightarrow Ag^+$	1.980
$Pd^{2+} + 2e^- \longrightarrow Pd$	0.951	$S_2O_8^{2-} + 2e^- \longrightarrow 2SO_4^{2-}$	2.010
$NO_3^- + 4H^+ + 3e^- \longrightarrow NO + 2H_2O$	0.957	$O_3 + 2H^+ + 2e^- \longrightarrow O_2 + H_2O$	2.076
$AuBr_2^- + e^- \longrightarrow Au + 2Br^-$	0.959	$F_2 + 2e^- \longrightarrow 2F^-$	2.866
$HNO_2 + H^+ + e^- \longrightarrow NO + H_2O$	0.983	$F_2 + 2H^+ + 2e^- \longrightarrow 2HF$	3.053

Sources: HCP and Bard.

Point Group Character Tables

C.1 The Nonaxial Groups

C_1	E
A	1

C_s	E	σ_h		
A'	1	1	x, y, R_z	x^2, y^2, z^2, xy
A''	1	-1	z, R_x, R_y	yz, xz

C_i	E	i		
A_g	1	1	R_x, R_y, R_z	$x^2, y^2, z^2, xy, xz, yz$
A_u	1	-1	x, y, z	

C.2 The C_n Groups

C_2	E	C_2		
A	1	1	z, R_z	x^2, y^2, z^2, xy
B	1	-1	x, y, R_x, R_y	yz, xz

C_4	E	C_4	C_2	C_4^3		
A	1	1	1	1	z, R_z	$x^2 + y^2, z^2$
B	1	-1	1	-1		$x^2 - y^2, xy$
E	$\begin{cases} 1 \\ 1 \end{cases}$	$\begin{matrix} i \\ -i \end{matrix}$	$\begin{matrix} -1 \\ -1 \end{matrix}$	$\begin{matrix} -i \\ i \end{matrix}\Bigg\}$	$(x, y), (R_x, R_y)$	(yz, xz)

C.3 The D_n Groups

D_2	E	$C_2(z)$	$C_2(y)$	$C_2(x)$		
A	1	1	1	1		x^2, y^2, z^2
B_1	1	1	-1	-1	z, R_z	xy
B_2	1	-1	1	-1	y, R_y	xz
B_3	1	-1	-1	1	x, R_x	yz

D_3	E	$2C_3$	$3C_2$		
A_1	1	1	1		$x^2 + y^2, z^2$
A_2	1	1	-1	z, R_z	
E	2	-1	0	$(x, y), (R_x, R_y)$	$(x^2 - y^2, xy), (xz, yz)$

D_4	E	$2C_4$	$C_2(= C_4^2)$	$2C_2'$	$2C_2''$		
A_1	1	1	1	1	1		$x^2 + y^2, z^2$
A_2	1	1	1	-1	-1	z, R_z	
B_1	1	-1	1	1	-1		$x^2 - y^2$
B_2	1	-1	1	-1	1		xy
E	2	0	-2	0	0	$(x, y), (R_x, R_y)$	(xz, yz)

D_5	E	$2C_5$	$2C_5^2$	$5C_2$		
A_1	1	1	1	1		$x^2 + y^2, z^2$
A_2	1	1	1	-1	z, R_z	
E_1	2	$2 \cos 72°$	$2 \cos 144°$	0	$(x, y), (R_x, R_y)$	(xz, yz)
E_2	2	$2 \cos 144°$	$2 \cos 72°$	0		$(x^2 - y^2, xy)$

D_6	E	$2C_6$	$2C_3$	C_2	$3C_2'$	$3C_2''$		
A_1	1	1	1	1	1	1		$x^2 + y^2, z^2$
A_2	1	1	1	1	-1	-1	z, R_z	
B_1	1	-1	1	-1	1	-1		
B_2	1	-1	1	-1	-1	1		
E_1	2	1	-1	-2	0	0	$(x, y), (R_x, R_y)$	(xz, yz)
E_2	2	-1	-1	2	0	0		$(x^2 - y^2, xy)$

C.4 The C_{nv} Groups

C_{2v}	$\mathcal{E}$	C_2	$\sigma_v(xz)$	$\sigma_v'(yz)$		
A_1	1	1	1	1	z	x^2, y^2, z^2
A_2	1	1	-1	-1	R_z	xy
B_1	1	-1	1	-1	x, R_y	xz
B_2	1	-1	-1	1	y, R_x	yz

C_{3v}	E	$2C_3$	$3\sigma_v$		
A_1	1	1	1	z	$x^2 + y^2, z^2$
A_2	1	1	-1	R_z	
E	2	-1	0	$(x, y), (R_x, R_y)$	$(x^2 - y^2, xy), (xz, yz)$

C_{4v}	E	$2C_4$	C_2	$2\sigma_v$	$2\sigma_d$		
A_1	1	1	1	1	1	z	$x^2 + y^2, z^2$
A_2	1	1	1	-1	-1	R_z	
B_1	1	-1	1	1	-1		$x^2 - y^2$
B_2	1	-1	1	-1	1		xy
E	2	0	-2	0	0	$(x, y), (R_x, R_y)$	(xz, yz)

C_{5v}	E	$2C_5$	$2C_5^2$	$5\sigma_v$		
A_1	1	1	1	1	z	$x^2 + y^2, z^2$
A_2	1	1	1	-1	R_z	
E_1	2	$2\cos 72°$	$2\cos 144°$	0	$(x, y), (R_x, R_y)$	(xz, yz)
E_2	2	$2\cos 144°$	$2\cos 72°$	0		$(x^2 - y^2, xy)$

C_{6v}	E	$2C_6$	$2C_3$	C_2	$3\sigma_v$	$3\sigma_d$		
A_1	1	1	1	1	1	1	z	$x^2 + y^2, z^2$
A_2	1	1	1	1	-1	-1	R_z	
B_1	1	-1	1	-1	1	-1		
B_2	1	-1	1	-1	-1	1		
E_1	2	1	-1	-2	0	0	$(x, y), (R_x, R_y)$	(xz, yz)
E_2	2	-1	-1	2	0	0		$(x^2 - y^2, xy)$

C.5 The C_{nh} Groups

C_{2h}	E	C_2	i	σ_h		
A_g	1	1	1	1	R_z	x^2, y^2, z^2, xy
B_g	1	-1	1	-1	R_x, R_y	xz, yz
A_u	1	1	-1	-1	z	
B_u	1	-1	-1	1	x, y	

C_{4h}	E	C_4	C_2	C_4^3	i	S_4^3	σ_h	S_4		
A_g	1	1	1	1	1	1	1	1	R_z	$x^2 + y^2, z^2$
B_g	1	-1	1	-1	1	-1	1	-1		$x^2 - y^2, xy$
E_g	$\begin{cases} 1 \\ 1 \end{cases}$	$\begin{matrix} i \\ -i \end{matrix}$	$\begin{matrix} -1 \\ -1 \end{matrix}$	$\begin{matrix} -i \\ i \end{matrix}$	$\begin{matrix} 1 \\ 1 \end{matrix}$	$\begin{matrix} i \\ -i \end{matrix}$	$\begin{matrix} -1 \\ -1 \end{matrix}$	$\begin{matrix} -i \\ i \end{matrix}$	(R_x, R_y)	(xz, yz)
A_u	1	1	1	1	-1	-1	-1	-1	z	
B_u	1	-1	1	-1	-1	1	-1	1		
E_u	$\begin{cases} 1 \\ 1 \end{cases}$	$\begin{matrix} i \\ -i \end{matrix}$	$\begin{matrix} -1 \\ -1 \end{matrix}$	$\begin{matrix} -i \\ i \end{matrix}$	$\begin{matrix} -1 \\ -1 \end{matrix}$	$\begin{matrix} -i \\ i \end{matrix}$	$\begin{matrix} 1 \\ 1 \end{matrix}$	$\begin{matrix} -i \\ i \end{matrix}$	(x, y)	

C.6 The D_{nh} Groups

D_{2h}	E	$C_2(z)$	$C_2(y)$	$C_2(x)$	i	$\sigma(xy)$	$\sigma(xz)$	$\sigma(yz)$		
A_g	1	1	1	1	1	1	1	1		x^2, y^2, z^2
B_{1g}	1	1	-1	-1	1	1	-1	-1	R_z	xy
B_{2g}	1	-1	1	-1	1	-1	1	-1	R_y	xz
B_{3g}	1	-1	-1	1	1	-1	-1	1	R_x	yz
A_u	1	1	1	1	-1	-1	-1	-1		
B_{1u}	1	1	-1	-1	-1	-1	1	1	z	
B_{2u}	1	-1	1	-1	-1	1	-1	1	y	
B_{3u}	1	-1	-1	1	-1	1	1	-1	x	

D_{3h}	E	$2C_3$	$3C_2$	σ_h	$2S_3$	$3\sigma_v$		
A_1'	1	1	1	1	1	1		$x^2 + y^2, z^2$
A_2'	1	1	-1	1	1	-1	R_z	
E'	2	-1	0	2	-1	0	(x, y)	$(x^2 - y^2, xy)$
A_1''	1	1	1	-1	-1	-1		
A_2''	1	1	-1	-1	-1	1	z	
E''	2	-1	0	-2	1	0	(R_x, R_y)	(xz, yz)

D_{4h}	E	$2C_4$	C_2	$2C_2'$	$2C_2''$	i	$2S_4$	σ_h	$2\sigma_v$	$2\sigma_d$		
A_{1g}	1	1	1	1	1	1	1	1	1	1		$x^2 + y^2, z^2$
A_{2g}	1	1	1	-1	-1	1	1	1	-1	-1	R_z	
B_{1g}	1	-1	1	1	-1	1	-1	1	1	-1		$x^2 - y^2$
B_{2g}	1	-1	1	-1	1	1	-1	1	-1	1		xy
E_g	2	0	-2	0	0	2	0	-2	0	0	(R_x, R_y)	(xz, yz)
A_{1u}	1	1	1	1	1	-1	-1	-1	-1	-1		
A_{2u}	1	1	1	-1	-1	-1	-1	-1	1	1	z	
B_{1u}	1	-1	1	1	-1	-1	1	-1	-1	1		
B_{2u}	1	-1	1	-1	1	-1	1	-1	1	-1		
E_u	2	0	-2	0	0	-2	0	2	0	0	(x, y)	

D_{6h}	E	$2C_6$	$2C_3$	C_2	$3C_2'$	$3C_2''$	i	$2S_3$	$2S_6$	σ_h	$3\sigma_d$	$3\sigma_v$		
A_{1g}	1	1	1	1	1	1	1	1	1	1	1	1		x^2+y^2, z^2
A_{2g}	1	1	1	1	-1	-1	1	1	1	1	-1	-1	R_z	
B_{1g}	1	-1	1	-1	1	-1	1	-1	1	-1	1	-1		
B_{2g}	1	-1	1	-1	-1	1	1	-1	1	-1	-1	1		
E_{1g}	2	1	-1	-2	0	0	2	1	-1	-2	0	0	(R_x, R_y)	(xz, yz)
E_{2g}	2	-1	-1	2	0	0	2	-1	-1	2	0	0		(x^2-y^2, xy)
A_{1u}	1	1	1	1	1	1	-1	-1	-1	-1	-1	-1		
A_{2u}	1	1	1	1	-1	-1	-1	-1	-1	-1	1	1	z	
B_{1u}	1	-1	1	-1	1	-1	-1	1	-1	1	-1	1		
B_{2u}	1	-1	1	-1	-1	1	-1	1	-1	1	1	-1		
E_{1u}	2	1	-1	-2	0	0	-2	-1	1	2	0	0	(x, y)	
E_{2u}	2	-1	-1	2	0	0	-2	1	1	-2	0	0		

D_{8h}	E	$2C_8^3$	$2C_8$	$2C_4$	C_2	$4C_2'$	$4C_2''$	i	$2S_8^3$	$2S_8$	$2S_4$	σ_h	$4\sigma_d$	$4\sigma_v$		
A_{1g}	1	1	1	1	1	1	1	1	1	1	1	1	1	1		x^2+y^2, z^2
A_{2g}	1	1	1	1	1	-1	-1	1	1	1	1	1	-1	-1	R_z	
B_{1g}	1	-1	-1	1	1	1	-1	1	-1	-1	1	1	1	-1		
B_{2g}	1	-1	-1	1	1	-1	1	1	-1	-1	1	1	-1	1		
E_{1g}	2	$\sqrt{2}$	$-\sqrt{2}$	0	-2	0	0	2	$\sqrt{2}$	$-\sqrt{2}$	0	-2	0	0	(R_x, R_y)	(xz, yz)
E_{2g}	2	0	0	-2	2	0	0	2	0	0	-2	2	0	0		(x^2-y^2, xy)
E_{3g}	2	$-\sqrt{2}$	$\sqrt{2}$	0	-2	0	0	2	$-\sqrt{2}$	$\sqrt{2}$	0	-2	0	0		
A_{1u}	1	1	1	1	1	1	1	-1	-1	-1	-1	-1	-1	-1		
A_{2u}	1	1	1	1	1	-1	-1	-1	-1	-1	-1	-1	1	1	z	
B_{1u}	1	-1	-1	1	1	1	-1	-1	1	1	-1	-1	-1	1		
B_{2u}	1	-1	-1	1	1	-1	1	-1	1	1	-1	-1	1	-1		
E_{1u}	2	$\sqrt{2}$	$-\sqrt{2}$	0	-2	0	0	-2	$-\sqrt{2}$	$\sqrt{2}$	0	2	0	0	(x, y)	
E_{2u}	2	0	0	-2	2	0	0	-2	0	0	2	-2	0	0		
E_{3u}	2	$-\sqrt{2}$	$\sqrt{2}$	0	-2	0	0	-2	$\sqrt{2}$	$-\sqrt{2}$	0	2	0	0		

C.7 The D_{nd} Groups

D_{2d}	E	$2S_4$	C_2	$2C_2'$	$2\sigma_d$		
A_1	1	1	1	1	1		x^2+y^2, z^2
A_2	1	1	1	-1	-1	R_z	
B_1	1	-1	1	1	-1		x^2-y^2
B_2	1	-1	1	-1	1	z	xy
E	2	0	-2	0	0	$(x,y), (R_x, R_y)$	(xz, yz)

D_{3d}	E	$2C_3$	$3C_2$	i	$2S_6$	$3\sigma_d$		
A_{1g}	1	1	1	1	1	1		x^2+y^2, z^2
A_{2g}	1	1	-1	1	1	-1	R_z	
E_g	2	-1	0	2	-1	0	(R_x, R_y)	$(x^2-y^2, xy), (xz, yz)$
A_{1u}	1	1	1	-1	-1	-1		
A_{2u}	1	1	-1	-1	-1	1	z	
E_u	2	-1	0	-2	1	0	(x,y)	

D_{4d}	E	$2S_8$	$2C_4$	$2S_8^3$	C_2	$4C_2'$	$4\sigma_d$		
A_1	1	1	1	1	1	1	1		x^2+y^2, z^2
A_2	1	1	1	1	1	-1	-1	R_z	
B_1	1	-1	1	-1	1	1	-1		
B_2	1	-1	1	-1	1	-1	1	z	
E_1	2	$\sqrt{2}$	0	$-\sqrt{2}$	-2	0	0	(x,y)	
E_2	2	0	-2	0	2	0	0		(x^2-y^2, xy)
E_3	2	$-\sqrt{2}$	0	$\sqrt{2}$	-2	0	0	(R_x, R_y)	(xz, yz)

D_{6d}	E	$2S_{12}$	$2C_6$	$2S_4$	$2C_3$	$2S_{12}^5$	C_2	$6C_2'$	$6\sigma_d$		
A_1	1	1	1	1	1	1	1	1	1		x^2+y^2, z^2
A_2	1	1	1	1	1	1	1	-1	-1	R_z	
B_1	1	-1	1	-1	1	-1	1	1	-1		
B_2	1	-1	1	-1	1	-1	1	-1	1	z	
E_1	2	$\sqrt{3}$	1	0	-1	$-\sqrt{3}$	-2	0	0	(x,y)	
E_2	2	1	-1	-2	-1	1	2	0	0		(x^2-y^2, xy)
E_3	2	0	-2	0	2	0	-2	0	0		
E_4	2	-1	-1	2	-1	-1	2	0	0		
E_5	2	$-\sqrt{3}$	1	0	-1	$\sqrt{3}$	-2	0	0	(R_x, R_y)	(xz, yz)

C.8 The Cubic Groups

T_d	E	$8C_3$	$3C_2$	$6S_4$	$6\sigma_d$		
A_1	1	1	1	1	1		$x^2 + y^2 + z^2$
A_2	1	1	1	-1	-1		
E	2	-1	2	0	0		$(2z^2 - x^2 - y^2, x^2 - y^2)$
T_1	3	0	-1	1	-1	(R_x, R_y, R_z)	
T_2	3	0	-1	-1	1	(x, y, z)	(xy, xz, yz)

O	E	$8C_3$	$3C_2(= C_4^2)$	$6C_4$	$6C_2$		
A_1	1	1	1	1	1		$x^2 + y^2 + z^2$
A_2	1	1	1	-1	-1		
E	2	-1	2	0	0		$(2z^2 - x^2 - y^2, x^2 - y^2)$
T_1	3	0	-1	1	-1	$(R_x, R_y, R_z), (x, y, z)$	
T_2	3	0	-1	-1	1		(xy, xz, yz)

O_h	E	$8C_3$	$6C_2$	$6C_4$	$3C_2(= C_4^2)$	i	$6S_4$	$8S_6$	$3\sigma_h$	$6\sigma_d$		
A_{1g}	1	1	1	1	1	1	1	1	1	1		$x^2 + y^2 + z^2$
A_{2g}	1	1	-1	-1	1	1	-1	1	1	-1		
E_g	2	-1	0	0	2	2	0	-1	2	0		$(2z^2 - x^2 - y^2, x^2 - y^2)$
T_{1g}	3	0	-1	1	-1	3	1	0	-1	-1	(R_x, R_y, R_z)	
T_{2g}	3	0	1	-1	-1	3	-1	0	-1	1		(xz, yz, xy)
A_{1u}	1	1	1	1	1	-1	-1	-1	-1	-1		
A_{2u}	1	1	-1	-1	1	-1	1	-1	-1	1		
E_u	2	-1	0	0	2	-2	0	1	-2	0		
T_{1u}	3	0	-1	1	-1	-3	-1	0	1	1	(x, y, z)	
T_{2u}	3	0	1	-1	-1	-3	1	0	1	-1		

C.9 The Groups $C_{\infty v}$ and $D_{\infty h}$ for Linear Molecules

$C_{\infty v}$	E	$2C_\infty^\Phi$	...	$\infty\sigma_v$		
$A_1(\Sigma^+)$	1	1		1	z	$x^2 + y^2, z^2$
$A_2(\Sigma^-)$	1	1	...	-1	R_z	
$E_1(\Pi)$	2	$2\cos\Phi$	...	0	$(x, y), (R_x, R_y)$	(xz, yz)
$E_2(\Delta)$	2	$2\cos 2\Phi$	...	0		$(x^2 - y^2, xy)$
$E_3(\Phi)$	2	$2\cos 3\Phi$	...	0		
...	...	...	...	...		

$D_{\infty h}$	E	$2C_\infty^\Phi$	...	$\infty\sigma_v$	i	$2S_\infty^\Phi$	...	∞C_2		
Σ_g^+	1	1	...	1	1	1	...	1		$x^2 + y^2, z^2$
Σ_g^-	1	1	...	-1	1	1	...	-1	R_z	
Π_g	2	$2\cos\Phi$	...	0	2	$-2\cos\Phi$	...	0	(R_x, R_y)	(xy, yz)
Δ_g	2	$2\cos 2\Phi$	...	0	2	$2\cos 2\Phi$	...	0		$(x^2 - y^2, xy)$
...	...	...	...	...	...	...	...	...		
Σ_u^+	1	1	...	1	-1	-1	...	-1	z	
Σ_u^-	1	1	...	-1	-1	-1	...	1		
Π_u	2	$2\cos\Phi$	...	0	-2	$2\cos\Phi$	...	0	(x, y)	
Δ_u	2	$2\cos 2\Phi$	...	0	-2	$-2\cos 2\Phi$	...	o		
...	...	...	...	...	...	...	...	...		

Answers to Selected End-of-Chapter Problems

Numerical answers to problems are included here. Complete solutions to selected problems can be found in the *Student's Solutions Manual*.

Chapter 1

P1.1 1.27×10^6

P1.2 2.33×10^3 L

P1.3   26.9 bar

P1.4 $x_{CO_2} = 0.235,\ x_{H_2O} = 0.314,\ x_{O_2} = 0.451$

P1.5   32.0%

P1.6 1.11×10^{21}

P1.7   8.37 g

P1.8   37.9 L

P1.9   a. N_2 67.8%, O_2 18.2%, Ar 0.869%, and H_2O 13.1%

       b.   20.6 L

       c.   0.981

P1.10 3.7×10^3

P1.11   34.6 bar

P1.12 3.66×10^5 Pa

P1.13 $x^\circ_{H_2} = 0.366;\ x^\circ_{O_2} = 0.634$

P1.14   0.280 moles

P1.16   18.6 L

P1.17   0.27

P1.18   a. 2.18×10^{-2} bar

       b. $x_{O_2} = 0.0136,\ x_{N_2} = 0.805,\ x_{CO} = 0.181$

         $P_{O_2} = 2.97 \times 10^{-4}$ bar, $P_{N_2} = 1.76 \times 10^{-2}$ bar

         $P_{CO} = 3.94 \times 10^{-3}$ bar

P1.19   21.4 bar

P1.20   a. 3.86×10^4 Pa

       b. 6.68×10^4 without mixing

         1.49×10^5 with mixing

P1.21   0.149 L

P1.22 2.20×10^{-2} L

P1.23 7.95×10^4 Pa

P1.24   91

P1.25 2.69×10^{25}

P1.26   a. $P_{H_2} = 2.48 \times 10^6$ Pa; $P_{O_2} = 1.52 \times 10^5$ Pa

         $P_{total} = 2.64 \times 10^6$ Pa

         mol % H_2 = 94.2%; mol % O_2 = 5.78%

       b. $P_{N_2} = 1.34 \times 10^5$ Pa; $P_{O_2} = 8.23 \times 10^4$ Pa

         $P_{total} = 2.17 \times 10^5$ Pa

         mol % N_2 = 62.0%, mol % O_2 = 38.0%

       c. $P_{NH_3} = 1.63 \times 10^5$ Pa; $P_{CH_4} = 2.06 \times 10^5$ Pa

         $P_{total} = 3.69 \times 10^5$ Pa

         mol % NH_3 = 44.2%, mol % CH_4 = 55.8%

P1.27   0.557

P1.28 1.36×10^3 K

P1.29   34.8 L

P1.30   0.0040

P1.31 $x_{O_2} = 0.295,\quad x_{H_2} = 0.705$

P1.32   0.0919 atm mol^{-1} °C^{-1}, 265.0 °C

P1.33   2.37 L

P1.34 $V_{N_2} = 0.0521$ L, $V_{CO_2} = 0.312$ L

       $V_{total} = 0.364$ L

P1.35 6.44×10^8 L

P1.36   77.2 amu

Chapter 2

P2.1 7.82×10^3 J, 0.642 m

P2.2 $q = 0, w = \Delta U = 838$ J, $\Delta H = 1.40 \times 10^3$ J

P2.3   a. $w = -754$ J, ΔU and $\Delta H = 0$, $q = -w = 754$ J

       b. $\Delta U = -1.68 \times 10^3$ J,

         $w = 0, q = \Delta U = -1.68 \times 10^3$ J

         $\Delta H = -2.81 \times 10^3$ J

         $\Delta U_{total} = -1.68 \times 10^3$ J

         $w_{total} = -754$ J, $q_{total} = -930.$ J

         $\Delta H_{total} = -2.81 \times 10^3$ J

P2.4   15 g

P2.5   30. m

P2.6   a. -5.30×10^3 J,   b. -4.55×10^3 J

P2.7 $q = 0; \Delta U = w = -4.90 \times 10^3$ J, $\Delta H = -8.17 \times 10^3$ J

P2.8 $w = -4.68 \times 10^3$ J, 1.06 bar

P2.9 $w = -0.172$ J

       $q = 34.6 \times 10^3$ J

       $\Delta H = 34.6 \times 10^3$ J

       $\Delta U = 34.6 \times 10^3$ J

P2.10   0.46 J

P2.11   a.   418 K, $w = 4.63 \times 10^3$ J

       b.   954 K, 30.4×10^3 J

P2.12 $q = 0, \Delta U = w = -1300.$ J, $\Delta H = -2.17 \times 10^3$ J

P2.13   322 K

P2.14   a. -3.76×10^3 J,   b. -1.82×10^3 J

P2.15 134.0×10^3 Pa, 129.7×10^3 Pa

P2.16 1.12×10^3 K, $q = 0, w = \Delta U = 22.7 \times 10^3$ J

       $\Delta H = 3.79 \times 10^4$ J

P2.17   301 K

P2.19   step 1: 0, step 2: -22.0×10^3 J, step 3: 8.90×10^3 J,

       cycle: -13.1×10^3 J

P2.20 $w = 3.58 \times 10^3$ J

$q = 0$

$\Delta H = 5.00 \times 10^3$ J

$\Delta U = 3.58 \times 10^3$ J

P2.21 -35.6×10^3 J

P2.22 $T_f = 168$ K

$q = 0$

$w = \Delta U = -8.40 \times 10^3$ J

$\Delta H = -10.8 \times 10^3$ J

P2.23 $q = 0$

a. $T_f = 213$ K, b. 248 K

a. $w = -2.66 \times 10^3$ J, b. -3.17×10^3 J

a. $\Delta H = -4.44 \times 10^3$ J, b. -4.44×10^3 J

a. $\Delta U = -2.66 \times 10^3$ J, b. -3.17×10^3 J

P2.24 a. -17.393×10^3 J,

b. -17.675×10^3 J, -1.57%

P2.25 0.60 K

P2.26 a. $\Delta U = \Delta H = 0$, $w = -q = -1.27 \times 10^3$ J

b. $w = 0$,

$q = \Delta U = -1.14 \times 10^3$ J, $\Delta H = -1.91 \times 10^3$ J

For the overall process, $w = -1.27 \times 10^3$ J,

$q = 130.$ J, $\Delta U = -1.14 \times 10^3$ J, and

$\Delta H = -1.91 \times 10^3$ J

P2.27 a. $P_1 = 5.83 \times 10^5$ Pa, $w = -6.44 \times 10^3$ J

$\Delta U = 0$ and $\Delta H = 0$

$q = -w = 6.44 \times 10^3$ J

b. $P_2 = 7.35 \times 10^5$ Pa, $\Delta U = 2.29 \times 10^3$ J

$w = 0$, $q = 2.29 \times 10^3$ J

$\Delta H = 3.81 \times 10^3$ J

overall:

$q = 8.73 \times 10^3$ J, $w = -6.44 \times 10^3$ J

$\Delta U = 2.29 \times 10^3$ J, $\Delta H = 3.81 \times 10^3$ J

P2.28 $q = 0$, $\Delta U = w = -4.90 \times 10^3$ J, $\Delta H = -8.16 \times 10^3$ J

P2.29 408 K

P2.30 $\Delta U = -1.77 \times 10^3$ J, $\Delta H = q_P = -2.96 \times 10^3$ J

$w = 1.18 \times 10^3$ J

P2.31 5.6 K

P2.32 18.9×10^3 J, 7.30×10^3 J, 0

P2.33 251 K

P2.34 a. -19.2 K

b. 0 K

c. 18.0 K

P2.35 a. 158 K   b. 204 K

P2.36 -1190 J

P2.37 $\Delta H = 5.88 \times 10^4$ J, $\Delta U = 4.78 \times 10^4$ J

P2.38 $w = 0$, $\Delta U = q = 1.31 \times 10^4$ J, $\Delta H = 1.83 \times 10^4$ J

P2.39 c. 984 kg, 1.38×10^3 kg

P2.41 0.35 kg

P2.42 2.2×10^{-19} J

P2.43 $w = 1.22 \times 10^4$ J, $\Delta U = 0$ and $\Delta H = 0$

$q = -1.22 \times 10^4$ J

P2.44 0.99 J

Chapter 3

P3.3 $\Delta H = q = 8.58 \times 10^4$ J

$\Delta U = 7.18 \times 10^4$ J;

$w = -1.39 \times 10^4$ J

P3.5 292 K

P3.6 93.4 bar

P3.8 369 K

P3.9 1.78×10^3 J, 0

P3.11 314 K

P3.12 $q = \Delta H = 8.01 \times 10^4$ J;

$w = -8.85 \times 10^3$ J

$\Delta U = 7.62 \times 10^4$ J

Chapter 4

P4.1 SiF 596 kJ mol^{-1}, 593 kJ mol^{-1}

SiCl 398 kJ mol^{-1}, 396 kJ mol^{-1}

CF 489 kJ mol^{-1}, 487 kJ mol^{-1}

NF 279 kJ mol^{-1}, 276 kJ mol^{-1}

OF 215 kJ mol^{-1}, 213 kJ mol^{-1}

HF 568 kJ mol^{-1}, 565 kJ mol^{-1}

P4.2 -49.39 kJ mol^{-1}

P4.3 $\Delta U_f - 361$ kJ mol^{-1}; $\Delta H_f^\circ = -362$ kJ mol^{-1}

P4.4 91.7 kJ mol^{-1}

P4.5 $\Delta H = 5.80 \times 10^{17}$ kJ yr^{-1}

P4.7 a. 416 kJ mol^{-1}, 413 kJ mol^{-1}

b. 329 kJ mol^{-1}, 329 kJ mol^{-1}

c. 589 kJ mol^{-1}, 588 kJ mol^{-1}

P4.8 a. 428.22 kJ mol^{-1}, 425.74 kJ mol^{-1}

b. 926.98 kJ mol^{-1}, 922.02 kJ mol^{-1}

c. 498.76 kJ mol^{-1}, 498.28 kJ mol^{-1}

P4.9 -182.9 kJ mol^{-1}

P4.11 134.68 kJ mol^{-1}

P4.12 a. -73.0 kJ mol^{-1};

b. -804 kJ mol^{-1}

P4.13 $\Delta H_{combustion} = -3268$ kJ mol^{-1}

$\Delta U_{reaction} = -3264$ kJ mol^{-1}; 0.0122

P4.14 -59.8 kJ mol^{-1}

P4.15 6.64×10^3 J °C^{-1}

P4.16 $\Delta H = 7.89 \times 10^{17}$ kJ yr^{-1}

P4.17 a. -1816 kJ mol^{-1}, -1814 kJ mol^{-1}

 b. -116.2 kJ mol^{-1}, -113.7 kJ mol^{-1}

 c. 62.6 kJ mol^{-1}, 52.7 kJ mol^{-1}

 d. -111.6 kJ mol^{-1}, -111.6 kJ mol^{-1}

 e. 205.9 kJ mol^{-1}, 200.9 kJ mol^{-1}

 f. -172.8 kJ mol^{-1}, -167.8 kJ mol^{-1}

P4.18 -2.4×10^3 J mol^{-1}, 4.8%

P4.19 a. 696.8 kJ mol^{-1}

 b. -1165.1 kJ mol^{-1}

 c. -816.7 kJ mol^{-1}

P4.20 a. $\Delta H_{combustion} = -1.364 \times 10^3$ kJ mol^{-1}

 b. $\Delta H_f° = -280.0$. kJ mol^{-1}

P4.21 -812.2 kJ mol^{-1}

P4.22 a. $\Delta U = \Delta H = -5.635 \times 10^3$ kJ mol^{-1}

  $\Delta H = -5.635 \times 10^3$ kJ mol^{-1}

 b. $\Delta H_f° = -2.225 \times 10^3$ kJ mol^{-1}

 c. 1.19×10^3 J K^{-1}

P4.23 -1811 kJ mol^{-1}

P4.24 -20.6 kJ mol^{-1}, -178.2 kJ mol^{-1}

P4.25 180 kJ mol^{-1}

P4.26 -393.6 kJ mol^{-1}

P4.27 15.3 kJ mol^{-1}, 13.1%

P4.28 -266.3 kJ mol^{-1}, -824.2 kJ mol^{-1}

P4.29 415.8 kJ mol^{-1}, 1.2%

P4.30 -134 kJ mol^{-1}, $\approx 0\%$

P4.31 49.6 kJ mol^{-1}

P4.32 356.5 kJ mol^{-1}

P4.33 49 g

P4.34 1.8

P4.35 $6.5°C$

Chapter 5

P5.1 At 298.15 K,

$\Delta S = -262.4$ J K^{-1} mol^{-1}

$\Delta H = 2803$ kJ mol^{-1}

$\Delta S_{surroundings} = -9.40 \times 10^3$ J K^{-1} mol^{-1}

$\Delta S_{universe} = -9.66 \times 10^3$ J K^{-1} mol^{-1}

At 310. K,

$\Delta S = -273.4$ J K^{-1} mol^{-1}

$\Delta H = 2799$ kJ mol^{-1}

$\Delta S_{surroundings} = -9.03 \times 10^3$ J K^{-1} mol^{-1}

$\Delta S_{universe} = -9.30 \times 10^3$ J K^{-1} mol^{-1}

P5.2 a. 0.627, b. 0.398, c. 110.7

P5.3 983 kg

P5.4 4.85×10^8 J

P5.5 57.2 J K^{-1}

P5.6 a. $w = -6.83 \times 10^3$ J, $\Delta U = 10.2 \times 10^3$ J

  $q = \Delta H = 17.1 \times 10^3$ J, $\Delta S = 36.4$ J K^{-1}

 b. $w = 0$, $\Delta U = q = 10.2 \times 10^3$ J

  $\Delta H = 17.1 \times 10^3$ J, $\Delta S = 21.8$ J K^{-1}

 c. $\Delta U = \Delta H = 0$;

  $w_{reversible} = -q = -6.37 \times 10^3$ J

  $\Delta S = 20.6$ J K^{-1}

P5.7 a. $V_c = 113$ L; $V_d = 33.0$ L

 b. $w_{ab} = -9.44 \times 10^3$ J, $w_{bc} = -11.2 \times 10^3$ J

  $w_{cd} = 3.96 \times 10^3$ J, $w_{da} = 11.2 \times 10^3$ J

  $w_{total} = -5.49 \times 10^3$ J

 c. 0.581, 1.72 kJ

P5.8 $\Delta S = 256$ kJ k^{-1} week^{-1}

P5.9 a. $\Delta S_{surroundings} = 0$, $\Delta S = 0$, $\Delta S_{total} = 0$. not spontaneous.

 b. $\Delta S = 30.9$ J K^{-1}, $\Delta S_{surroundings} = 0$

  $\Delta S_{total} = 30.9$ J K^{-1}, spontaneous.

 c. $\Delta S = 30.9$ J K^{-1}

  $\Delta S_{surroundings} = -30.9$ J K^{-1}

  $\Delta S_{total} = 0$, not spontaneous

P5.10 0.538, 0.661

P5.11 $\Delta U = -17.9 \times 10^3$ J, $\Delta H = -25.0 \times 10^3$ J

$\Delta S = -80.5$ J K^{-1}

P5.12 27.9 J K^{-1}

P5.13 -0.0765 J K^{-1}, -0.0765 J K^{-1}

P5.14 -191.2 J K^{-1} mol^{-1}

P5.15 -2.5 K

P5.16 a. $q = 0$, $\Delta U = w = -4.96 \times 10^3$ J

  $\Delta H = -8.26 \times 10^3$ J, $\Delta S = 0$

 b. $q = 0$

  $\Delta U = w = -3.72 \times 10^3$ J, $\Delta H = -6.19 \times 10^3$ J

  $\Delta S = 10.1$ J K^{-1}

 c. $w = 0$, $\Delta U = \Delta H = 0$, $q = 0$;

  $\Delta S = 34.3$ J K^{-1}

P5.17 2.4

P5.18 $a \rightarrow b$: $\Delta U = \Delta H = 0$, $q = -w = 9.44 \times 10^3$ J

$b \rightarrow c$: $\Delta U = w = -11.2 \times 10^3$ J, $q = 0$

$\Delta H = -15.6 \times 10^3$ J

$c \rightarrow d$: $\Delta U = \Delta H = 0$, $q = -w = -3.96 \times 10^3$ J

$d \rightarrow a$: $\Delta U = w = 11.2 \times 10^3$ J, $q = 0$

$\Delta H = 15.6 \times 10^3$ J

$q_{total} = 5.49 \times 10^3$ J $= -w_{total}$

$\Delta U_{total} = \Delta H_{total} = 0$

P5.19 a. $1.0 \text{ J K}^{-1} \text{ mol}^{-1}$, b. $3.14 \text{ J K}^{-1} \text{ mol}^{-1}$

c. $\Delta S_{transition} = 8.24 \text{ J K}^{-1} \text{ mol}^{-1}$

$\Delta S_{fusion} = 25.1 \text{ J K}^{-1} \text{ mol}^{-1}$

P5.20 $\Delta U = 49.7 \text{ J}$, $w = -5.55 \times 10^3 \text{ J}$

$\Delta H = 85.2 \text{ J}$, $q = 5.59 \times 10^3 \text{ J}$, $\Delta S = 18.8 \text{ J K}^{-1}$

P5.21 $104.5 \text{ J K}^{-1} \text{ mol}^{-1}$

P5.22 $\Delta H = 3.75 \text{ kJ}$, $\Delta S = 12.1 \text{ J K}^{-1}$

P5.23 a. 105 J K^{-1}, b. 74.8 J K^{-1}

P5.25 $50.1 \text{ J K}^{-1} \text{ mol}^{-1}$

P5.26 3.0 J K^{-1}

P5.27 $\Delta S = 175 \text{ J K}^{-1} \text{ mol}^{-1}$

$\Delta S_{suroundings} = 225 \text{ J K}^{-1} \text{ mol}^{-1}$

and $\Delta S_{total} = 400. \text{ J K}^{-1} \text{ mol}^{-1}$

P5.28 $\Delta S = 53.0 \text{ J K}^{-1}$, $\Delta S_{suroundings} = -143 \text{ J K}^{-1}$

and $\Delta S_{universe} = -90.0 \text{ J K}^{-1}$

P5.29 $\Delta S = -21.48 \text{ J K}^{-1}$, $\Delta S_{surroundings} = 21.78 \text{ J K}^{-1}$

$\Delta S_{total} = 0.30 \text{ J K}^{-1}$

P5.30 $\Delta S = -1.66 \times 10^6 \text{ J K}^{-1}$

$\Delta S_{suroundings} = 2.03 \times 10^6 \text{ J K}^{-1}$

and $\Delta S_{universe} = 3.72 \times 10^5 \text{ J K}^{-1}$

P5.31 $135.1 \text{ J K}^{-1} \text{ mol}$

P5.32 a. $\Delta S_{total} = \Delta S + \Delta S_{surroundings} = 0 + 0 = 0.$ not spontaneous.

b. $\Delta S_{total} = \Delta S + \Delta S_{surroundings} = 10.1 \text{ J K}^{-1} + 0 = 10.1 \text{ J K}^{-1}.$ spontaneous.

P5.33 773 J s^{-1}

P5.34 $a \rightarrow b: \Delta S = -\Delta S_{surroundings} = 12.8 \text{ J K}^{-1}$

$\Delta S_{total} = 0$

$b \rightarrow c: \Delta S = -\Delta S_{surroundings} = 0, \Delta S_{total} = 0$

$c \rightarrow d: \Delta S = -\Delta S_{surroundings} = -12.8 \text{ J K}^{-1}$

$\Delta S_{total} = 0$

$d \rightarrow a: \Delta S = -\Delta S_{surroundings} = 0, \Delta S_{total} = 0$ to within the round-off error.

For the cycle, $\Delta S = \Delta S_{surroundings} = \Delta S_{total} = 0$ to within the round-off error.

P5.35 $\Delta H = 5.65 \times 10^3 \text{ J}$, $\Delta S = 17.9 \text{ J K}^{-1}$

P5.36 $\Delta S_{surroundings} = -20.6 \text{ J K}^{-1}$

$\Delta S_{total} = 0$, not spontaneous

P5.37 206 J K^{-1}

P5.38 $32.6 \text{ J K}^{-1} \text{ mol}^{-1}$

P5.39 a. 2.71 J K^{-1}; b. 154 J K^{-1}

P5.40 26.0 J K^{-1}

P5.41 $135.2 \text{ J K}^{-1} \text{ mol}^{-1}$

P5.42 19.4 m^2

P5.43 a. $q = 0$

$\Delta U = w = -3.56 \times 10^3 \text{ J}$, $\Delta H = -4.98 \times 10^3 \text{ J}$

$\Delta S = 0$, $\Delta S_{surroundings} = 0$, $\Delta S_{total} = 0$.

b. $w = 0$, $\Delta U = q = 3.56 \times 10^3 \text{ J}$, $\Delta H = 4.98 \times 10^3 \text{ J}$

$\Delta S = 16.0 \text{ J K}^{-1}$

c. $\Delta H = \Delta U = 0$, $w = -q = 4.40 \times 10^3 \text{ J}$

$\Delta S = -16.0 \text{ J K}^{-1}$

For the cycle,

$w_{cycle} = 840. \text{ J}$, $q_{cycle} = -840. \text{ J}$

$\Delta U_{cycle} = 0$, $\Delta H_{cycle} = 0$;

$\Delta S_{cycle} = 0$

P5.44 $\Delta S = 620 \text{ J K}^{-1} \text{ mol}^{-1}$

P5.45 a. $108.5 \text{ J mol}^{-1} \text{ K}^{-1}$

b. $25.7 \times 10^3 \text{ J mol}^{-1}$

Chapter 6

P6.1 $\Delta G_{combustion} = -3203 \times 10^3 \text{ kJ mol}^{-1}$

$\Delta A_{combustion} = -3199 \times 10^3 \text{ kJ mol}^{-1}$

P6.2 2.29×10^6 at 298 K and 2.37×10^2 at 490. K

P6.3 b. 2.11×10^{-2}

c. 1.96 moles of $N_2(g)$, 3.88 moles of $H_2(g)$, and 1.58 moles of $NH_3(g)$

P6.4 a. 0.379 at 700. K, 1.28 at 800. K

b. $\Delta H_R^\circ = 56.7 \times 10^3 \text{ J mol}^{-1}$;

$\Delta G_R^\circ (298.15 \text{ K}) = 35.0 \times 10^3 \text{ J mol}^{-1}$

P6.5 c. $4.0 \times 10^3 \text{ bar}$

P6.6 a. 0.141, b. 2.01×10^{-18}

c. 101 kJ mol^{-1}

P6.7 graphite 133 J mol^{-1}, He: 14.3 kJ mol^{-1}, 82.9

P6.8 a. 15.0 kJ mol^{-1}

b. 2.97×10^{-3}

P6.9 b. 0.55, d. 0.72

P6.10 4.76×10^6

P6.11 a. 0.464; $\Delta G_R^\circ = 7.28 \times 10^3 \text{ J mol}^{-1}$

b. $-28.6 \text{ kJ mol}^{-1}$

P6.12 $132.9 \text{ kJ mol}^{-1}$

P6.13 a. 2.10×10^{-2}, b. 0.0969 bar

P6.14 $8.56 \times 10^3 \text{J}$

P6.15 a. 275 kJ mol^{-1}

b. 5.42×10^{-49}

P6.16 $\Delta A = w_{rev} = -0.15 \text{ J}$

$\Delta S = 1.61 \times 10^{-4} \text{ J K}^{-1}$

$\Delta U = -0.10 \text{ J}$

P6.17 $-65.0 \times 10^3 \text{ J mol}^{-1}$

P6.18 -40.99 kJ g^{-1}, -117.6 kJ g^{-1}

P6.19 $1.2 \times 10^3 \text{ K}$, 0.43 Torr

P6.20 $-9.54 \times 10^3 \text{ J}$

P6.21 c. 0.0820, 0.0273, d. -1.49%

P6.22 -17.9×10^3 J, 51.4 J K^{-1}

P6.23 178 g

P6.24 a. $\Delta H_R^\circ = -19.0$ kJ mol^{-1}

$\Delta G_R^\circ(700.^\circ\text{C}) = 3.03$ kJ mol^{-1}

$\Delta S_R^\circ(700.^\circ\text{C}) = -22.6$ J mol^{-1} K^{-1}

b. $x_{CO_2} = 0.408, \quad x_{CO} = 0.592$

P6.27 $9.00, -58.8 \times 10^3$ J mol^{-1}

P6.28 539 K, 1.03×10^4

P6.29 -215.2×10^3 J mol^{-1}, -5.8%

P6.30 0.241

P6.31 -1363 kJ mol^{-1}, -1364 kJ mol^{-1}

P6.32 -257.2×10^3 J mol^{-1} at 298.15 K;
-231.1×10^3 J mol^{-1} at 600. K

P6.33 a. -11.7×10^3 J, -11.7×10^3 J

b. same as a.

P6.34 a. -34.3 kJ; b. -47.3 kJ;

c. -13.0 kJ

P6.35 b. 23.3 Pa;

c. 39.5 Pa

P6.36 c. 9.32×10^{-33} at 600. K and 3.30×10^{-29} at 700. K,

d. 9.32×10^{-33} for 1.00 bar and
2.10×10^{-32} for 2.00 bar

P6.37 28.0 kJ mol^{-1}

P6.38 a. $K_P(700.\ \text{K}) = 3.85$, $K_P(800.\ \text{K}) = 1.56$

b. $\Delta H_R^\circ = -42.0$ kJ mol^{-1}
$\Delta G_R^\circ(700\ \text{K}) = -7.85$ kJ mol^{-1}
$\Delta G_R^\circ(800\ \text{K}) = -2.97$ kJ mol^{-1}
$\Delta S_R^\circ(700\ \text{K}) = -48.7$ J mol^{-1} K^{-1}
$\Delta S_R^\circ(800\ \text{K}) = -48.7$ J mol^{-1} K^{-1}

c. -27.4 kJ mol^{-1}

P6.39 -0.15 J

P6.40 -126 kJ mol^{-1} at 298.15 K, -128 kJ mol^{-1} at 310. K

Chapter 7

P7.1 $b = 3.97 \times 10^{-5}$ m^3 mol^{-1}
$a = 6.43 \times 10^{-2}$ m^6 Pa mol^{-2}

P7.2 a. $R = 0.0833$ L bar K^{-1} mol^{-1}
$a = 5.56$ L^2 bar mol^{-2}, $b = 0.0305$ L mol^{-1}

b. $a = 5.53$ L^2 bar mol^{-2}, $b = 0.0305$ L mol^{-1}

P7.4 310. K ideal gas, 309 K van der Waals

P7.5 670. K, 127 bar

P7.6 $a = 0.692$ L^2 bar mol^{-2}, $\quad b = 0.0537$ L mol^{-1}
$V_m = 24.8$ L

P7.7 $\gamma = 0.414, 0.357, 0.436, 0.755,$ and 1.86
at 150., 250., 350., 450., and 550. bar, respectively.

P7.8 $V_m = 0.0167$ L mol^{-1}, $z = 0.942$

P7.9 ideal gas: 4.65 mol L^{-1}, vdW gas, 4.25 mol L^{-1}

P7.10 ideal gas: -16.9×10^3 J, vdW gas: -17.0×10^3 J, 1%

P7.12 1.62×10^{-10} m

P7.14 0.399 L mol^{-1}, -8.9%

P7.17 $b = 0.04286$ dm^3 mol^{-1}

$a = 3.657$ dm^6 bar mol^{-2}

P7.18 $a = 14.29$ dm^6 bar K$^{\frac{1}{2}}$ mol^{-2}

$b = 0.02010$ dm^3 mol^{-1}

P7.19 ideal gas: 0.521 L, vdW gas: 0.180 L

P7.22 $\rho_{idealgas} = 224$ g L^{-1}, $\rho_{vdW} = 216$ g L^{-1}

P7.24 ideal gas 371 bar

vdW: 408 bar, R-K: 407 bar

P7.25 ideal gas: 0.344 L mol^{-1}

vdW gas: 0.129 L mol^{-1}

R-K gas: 0.115 L mol^{-1}

P7.26 111 K, 426 K, 643 K

Chapter 8

P8.1 32.3 kJ mol^{-1}

P8.2 a. 351 K

b. $\Delta H_{vaporization}(351\ \text{K}) = 40.6$ kJ mol^{-1}
$\Delta H_{vaporization}(298\ \text{K}) = 43.0$ kJ mol^{-1}

P8.4 $\Delta H_{sublimation} = 33.2$ kJ mol^{-1}

$\Delta H_{vaporization} = 30.0$ kJ mol^{-1}

$\Delta H_{fusion} = 3.2$ kJ mol^{-1}

342 K, 6.70×10^4 Pa

P8.7 2.89×10^{-10} J

P8.8 a. 582 bar

b. 2.2×10^2 bar

c. $-1.5\ ^\circ$C

P8.10 1.04×10^4 Pa

P8.11 $\Delta H_{sublimation} = 53.4$ kJ mol^{-1}

$\Delta H_{vaporization} = 31.8$ kJ mol^{-1}

$\Delta H_{fusion} = 21.6$ kJ mol^{-1}

193 K, 18.7 Pa

P8.13 1.95 atm

P8.14 425 Pa

P8.16 51.1 kJ mol^{-1}

P8.17 a. 26.6 Torr, b. 25.2 Torr

P8.19 23.5 kJ mol^{-1}

P8.20 $T_{b,normal} = 342.8$ K, $T_{b,standard} = 343.4$ K

P8.21 a. 0.583 bar, b. 0.553 bar

P8.22 1.76×10^4 Pa

P8.23 6.78×10^4 Pa

P8.24 $\Delta H_{sublimation} = 18.4 \text{ kJ mol}^{-1}$

$\Delta H_{vaporization} = 14.4 \text{ kJ mol}^{-1}, \Delta H_{fusion} = 4.0 \text{ kJ mol}^{-1}$

$T_{triple} = 117 \text{ K}, P_{triple} = 33.5 \text{ Torr}$

P8.25 344 K

P8.26 a. 7.77×10^6 Pa

b. 3.53×10^5 Pa

P8.27 2.33×10^4 Pa, 1.39

P8.28 27.4 kJ mol^{-1}

P8.31 4.40 kJ mol^{-1}

P8.33 -0.717 K at 100. bar and -3.58 K at 500. bar

P8.34 a. $\Delta H_{vaporization} = 32.1 \times 10^3 \text{ J mol}^{-1}$

$\Delta H_{sublimation} = 37.4 \times 10^3 \text{ J mol}^{-1}$

b. $5.4 \times 10^3 \text{ J mol}^{-1}$

c. $349.5 \text{ K}, 91.8 \text{ J mol}^{-1} \text{ K}^{-1}$

d. $T_{tp} = 264 \text{ K}, P_{tp} = 2.84 \times 10^3 \text{ Pa}$

P8.35 $3567 \text{ Pa K}^{-1}, 3564 \text{ Pa K}^{-1}, 0.061\%$

P8.36 $38.4 \text{ J K}^{-1} \text{mol}^{-1}, 16.4 \times 10^3 \text{J mol}^{-1}$

P8.37 $360, \text{ K}, 1.89 \times 10^4 \text{ Pa}$

P8.38 27.1 kJ mol^{-1}

P8.39 1.51×10^3 Pa

P8.40 a. 49.5 J mol^{-1}

b. 267 J mol^{-1}

P8.41 2.62×10^4 Pa

P8.42 a. 56.1 Torr

b. 52.5 Torr

P8.43 $-3.5°C$

P8.44 $142 \text{ K}, 2.94 \times 10^3 \text{ Torr}$

$\Delta H_{sublimation} = 10.1 \times 10^3 \text{ J mol}^{-1}$

$\Delta H_{vaporization} = 9.38 \times 10^3 \text{ J mol}^{-1}$

$\Delta H_{fusion} = 0.69 \times 10^3 \text{ J mol}^{-1}$

P8.45 gas only 868 L, liquid only 0.0200 L

Chapter 9

P9.2 $a_A = 0.489, \gamma_A = 1.58, a_B = 1.00, \gamma_B = 1.45$

P9.4 a. 0.697, b. 0.893

P9.5 -8.0 cm^3

P9.6 $57.9 \text{ cm}^3 \text{ mol}^{-1}$

P9.7 $1.45 \times 10^3 \text{ kg mol}^{-1}$

P9.8 170 Torr

P9.9 0.312

P9.10 7.14×10^{-3}g, 2.67×10^{-3} g

P9.11 a. 25.0 Torr, 0.500

b. $Z_{EB} = (1 - Z_{EC}) = 0.39$

P9.12: a. $P_A = 27.0$ Torr, $P_B = 28.0$ Torr

b. $P_A = 41.4$ Torr, $P_B = 21.0$ Torr

P9.14 $1.86 \text{ K kg mol}^{-1}$

P9.15 a. for ethanol $a_1 = 0.9504, \gamma_1 = 1.055$, for isooctane

$a_2 = 1.411, \gamma_2 = 14.20$

b. 121.8 Torr

P9.16 0.268 mol L^{-1}

P9.17 5.6×10^{-9} M

P9.19 $P_a^* = 0.874$ bar, $P_B^* = 0.608$ bar

P9.20 413 Torr

P9.21 61.9 Torr

P9.22 -0.10 L

P9.24 0.466

P9.25 123 g mol^{-1}

P9.26 $6.15 \text{ m}, 6.01 \times 10^4 \text{ Pa}$

P9.27 0.116 bar

P9.28 $x_{bromo} = 0.769, y_{bromo} = 0.561$

P9.29 0.450

P9.30 a. 2.65×10^3 Torr

b. 0.525

c. $Z_{chloro} = 0.614$

P9.31 $K = 9.5 \times 10^4 \text{ M}^{-1}$

$N = 0.12$

P9.32 $a_{CS_2}^R = 0.872, \gamma_{CS_2}^R = 1.21, a_{CS_2}^H = 0.255, \gamma_{CS_2}^H = 0.353$

P9.33 $M = 41.3 \text{ g mol}^{-1}, \Delta T_f = -1.16 \text{ K}, \dfrac{P_{benzene}}{P_{benzene}^*} = 0.983$

$\pi = 4.93 \times 10^5 \text{ Pa}$

Chapter 10

P10.3 a. 0.91, b. 0.895, c. 0.641

P10.4 a. $0.119 \text{ mol kg}^{-1}$

b. $0.0750 \text{ mol kg}^{-1}$

c. $0.0750 \text{ mol kg}^{-1}$

d. $0.171 \text{ mol kg}^{-1}$

P10.5 a. 48%, b. 60.%, c. 51%

P10.6 0.0547

P10.7 $0.150 \text{ m } 1.13\%, 1.50 \text{ m } 0.372\%,$

Ignoring ionic interactions $0.150 \text{ m } 1.07\%, 1.50 \text{ m } 0.341\%$

P10.8 a. 0.144, b. 0.0663, c. 0.155

P10.9 a. 10.6%, b. 5.92%, c. 4.37%

P10.10 $\Delta H_R^\circ = 17 \text{ kJ mol}^{-1}, \Delta G_R^\circ = 16.5 \text{ kJ mol}^{-1}$

P10.12 $-426 \times 10^3 \text{ J mol}^{-1}$

P10.13 $0.321 \text{ mol kg}^{-1}$

P10.14 $I = 0.072 \text{ mol kg}^{-1}, \gamma_\pm = 0.389, a_\pm = 0.0106$

P10.16 0.41 nm

P10.17 a. $6.28 \times 10^{-5} \text{ mol L}^{-1}$

b. $1.22 \times 10^{-5} \text{ mol kg}^{-1}$

P10.20 $0.239 \text{ mol kg}^{-1}, 0.0539$

P10.21 a. 5.44%, b. 10.9%

P10.22   a.   51.5%,   b. 42.5%

P10.23   1.4 nm

P10.24 $I = 0.215$ mol kg^{-1}

$\gamma_{\pm} = 0.114$

$a_{\pm} = 0.00740$

P10.25 $I = 0.105$ mol kg^{-1}

$\gamma_{\pm} = 0.320$

$a_{\pm} = 0.0128$

P10.26   a.   0.225 mol kg^{-1}

b.   0.0750 mol kg^{-1}

c.   0.300 mol kg^{-1}

d.   0.450 mol kg^{-1}

P10.27   a. 5.5×10^{-3} mol kg^{-1}

b. 8.7×10^{-3} mol kg^{-1}

c. 5.5×10^{-3} mol kg^{-1}

P10.28 $\Delta H_R^\circ = -65.4$ kJ mol^{-1}, $\Delta G_R^\circ = -55.7$ kJ mol^{-1}

P10.29   4.39, 4.63

P10.30   0.791

Chapter 11

P11.1   a. 2.65×10^{6}

b. -36.7 kJ mol^{-1}

P11.2 -103.8 kJ mol^{-1}

P11.3 -131.2 kJ mol^{-1}

P11.4   0.7680 V, 1.38×10^{21}

P11.5   a. 1.216 V,   b. 1.172 V

-3.73%

P11.6   a. 0.337 V, 2.47×10^{11}

P11.7   a. 1.108 V,   b. 1.099 V, -0.804%

P11.8   2.413 V, 7.44×10^{77}

P11.9 -131.1 kJ mol^{-1}

P11.10   1.42 V, 8.55×10^{-5} V K^{-1}

P11.11   1.154 V, 6.61×10^{35}, 204.5 kJ mol^{-1}

P11.12 $\Delta G_R = -33.4$ kJ mol^{-1}, $\Delta S_R = -29.9$ J mol^{-1} K^{-1}

$\Delta H_R = -43.1$ kJ mol^{-1}

P11.13 -210.7 kJ mol^{-1}, 7.21×10^{36}, -210.374 kJ mol^{-1}

P11.15   a. 1.11×10^{8},   b. 6.67×10^{-56}

P11.16   0.769

P11.17   c. -1108 kJ

P11.18   a. 1.100 V,   b. Zn^{2+}: 0.469, Cu^{2+}: 0.410,   c. 1.108 V

P11.19 $\Delta G_R^\circ = -219.0$ kJ mol^{-1}, $\Delta S_R^\circ = -7.91$ J K^{-1}

$\Delta H_R^\circ = -221.4$ kJ mol^{-1}

P11.20 -0.913 V

P11.21 $\Delta G_R = -370.1$ kJ mol^{-1}, -370.6 kJ mol^{-1}

$\Delta S_R = 15.4$ J mol^{-1}K^{-1}, 25.4 J mol^{-1}K^{-1}

$\Delta H_R = -365.8$ kJ mol^{-1}, -362.8 kJ mol^{-1}

P11.22 1.10×10^{-11}

P11.24 -713.2 kJ mol^{-1}, 9.06×10^{124}

P11.25 4.16×10^{-4}

P11.26 8.28×10^{-84}, -1.22869 V

P11.28   a.   9.95,   b.   0.100

P11.29 4.90×10^{-13}

P11.30   a. 6.56×10^{-81},   b. 7.62×10^{-14}

Chapter 12

P12.1 $\Delta v_{H_2} = 1.131$ m s^{-1}, $\dfrac{\Delta v}{v} = 4.55 \times 10^{-4}$

P12.2 $\dfrac{\overline{E}_{osc} - k_B T}{\overline{E}_{osc}} = -0.0219$ for 1000. K. The corresponding

values for 600. K and 200. K are -0.0368 and -0.116.

P12.3 $\widetilde{\nu} = 109{,}737$ cm^{-1}, 27434 cm^{-1}, and 12193 cm^{-1} and

$E_{max} = 2.18 \times 10^{-18}$ J, 5.45×10^{-19} J, and 2.42×10^{-19} J

for the Lyman, Balmer, and Paschen series.

P12.4 7.93×10^{2} m s^{-1}, 3.55×10^{3} m s^{-1}, 7.93×10^{3} m s^{-1}, and

2.51×10^{5} m s^{-1} for 10^4 nm, 500. nm, 100. nm and

0.100 nm; 959 K, 1.92×10^{4} K, 9.59×10^{4} K, and

9.59×10^{7} K for 10^4 nm, 500. nm, 100. nm and 0.100 nm.

P12.5 2.179×10^{-18} J

P12.6 8.18×10^{15} electrons, $E = 2.27 \times 10^{-19}$ J

$v = 7.06 \times 10^{5}$ m s^{-1}

P12.7   at 1100. K, 1.11×10^{-3} J m^{-3}, at 6000. K, 0.981 J m^{-3}

P12.8   0.0467 m s^{-1}

P12.9   0.152 m s^{-1}

P12.10 $\dfrac{E - E_{approx}}{E} = -0.0369, -0.130,$ and -0.933 at 5000. K,

1500. K, and 300. K.

P12.11   25 K for He and 2.6 K for Ar

P12.12 5.82×10^{6} m s^{-1}, 1.54×10^{-17} J

P12.13 1.26×10^{-10} m for H$_2$ at 200. K and 5.93×10^{-11} m for

H$_2$ at 900. K. For Ar, 2.83×10^{-11} m and 1.33×10^{-11} m

at 200. K and 900. K, respectively.

P12.14 4.26×10^{-6} m, 2.50×10^{-6} m and 4.64×10^{-7} m for

675 K, 1150. K and 6200. K

P12.15   4.31 cm

P12.16   16.7 V

P12.17 6.561×10^{-5} m; 3.645×10^{-5} m

P12.18 $\lambda = 0.992$ nm, $n = 2.18 \times 10^{14}$ s^{-1}

P12.19 $h \approx 7.0 \times 10^{-34}$ J s, $\varphi \approx 4.0 \times 10^{-19}$ J or 2.5 eV

P12.20 3.95×10^{26} W

P12.21 $\nu \geq 1.26 \times 10^{15}$ s^{-1}, $v = 5.87 \times 10^{5}$ m s^{-1}

P12.22 5.08×10^{20}

P12.23 1.215×10^{-5} m, 9.118×10^{-6} m

P12.24 6.23×10^{18} s^{-1}

P12.25 1.30×10^{5} J s^{-1}, 0.0475 m

P12.27   a. 3.50×10^{7} J s^{-1},   b. 7.71×10^{17}

Chapter 13

P13.2 $N = \sqrt{\dfrac{2}{d}}$

P13.6 a. $r = \sqrt{11}, \theta = 1.26$ radians; $\varphi = 0.322$ radians

b. $x = -\dfrac{5}{2}, y = \dfrac{5}{2}, z = 5/\sqrt{2}$

P13.9 a. $\sqrt{61} \exp(0.876i)$; b. $2 \exp(-i\,\pi/2)$

c. $4 \exp(-0i)$, d. $\dfrac{\sqrt{26}}{5} \exp(1.125i)$

e. $\sqrt{\dfrac{5}{2}} \exp(-1.249i)$

P13.12 a. no; b. no; c. yes, -1; d. yes $-a^{-2}$; e. no

P13.14 a. yes, -6; b. yes, -1; c. yes, -16

P13.16 $x = 0.796$ m, $t_0 = 5.50 \times 10^{-4}$ s

P13.18 $\dfrac{n_4}{n_1}(125\,\text{K}) = 0.874$, $\dfrac{n_4}{n_1}(750.\,\text{K}) = 3.10$

$\dfrac{n_8}{n_1}(125\,\text{K}) = 0.0135$, $\dfrac{n_8}{n_1}(750.\,\text{K}) = 2.76$

P13.26 for $n_2/n_1 = 0.175$: $T = 127\,\text{K}$

for $n_2/n_1 = 0.750$: $T = 316\,\text{K}$

P13.35 a. $-2i$, b. $(2 + 2i)\sqrt{6}$, c. -1, d. $\dfrac{\sqrt{\dfrac{5}{2}}(1 + i)}{1 + \sqrt{2}}$

Chapter 15

P15.1 a. $\alpha = 7.79 \times 10^{10}$, b. 1.58×10^{-31} J

c. 3.85×10^{-11}

P15.5 c. $3/16, 3/8$, and $7/16$, d. $\langle E \rangle = 43E_1/16$

P15.9 $\sqrt{\dfrac{30}{b}}, \dfrac{b}{2}, \dfrac{2b^2}{7}$

P15.10 $\omega = 5.14 \times 10^{17}$ s^{-1}

$\lambda = 6.67 \times 10^{-11}$ m

P15.12 a. 7.07, b. 24.5, c. 0.223

P15.15 6.34×10^{-6} m

P15.16 1.01×10^{36}

P15.20 a. 8.25×10^{-38} J, b. 1.99×10^{-17}

P15.22 4.92×10^{-21} J

P15.25 a. 2, b. 3

P15.29 3.0×10^{-9} m

P15.34 a. 0.045, b. 0.00041

Chapter 16

P16.1 e. $\left|\dfrac{F}{A}\right|^2 = 0.1$ for $E = 1.5 \times 10^{-19}$ J and 0.02 for

$E = 1.1 \times 10^{-19}$ J, f. 0.2

P16.2 $T_{Si} = 900\,\text{K}, T_C = 4.4 \times 10^3\,\text{K}$

P16.3 b. $\dfrac{\Delta\rho_{total}(x)}{\langle\rho_{total}(x)\rangle} = 0.090$, c. $\dfrac{\Delta\rho_{n=11}(x)}{\langle\Delta\rho_{total}(x)\rangle} = 2.0$, d. 0.76

P16.4 $\lambda = 239$ nm

P16.5 $\lambda = 368$ nm

P16.6 Cu: 2.8×10^6 A/m^2, STM: 1.3×10^9 A/m^2

P16.7 b. 4.76×10^{-20} J, 1.86×10^{-19} J, and 3.95×10^{-19} J

P16.8 reflection 0.016, transmission 0.98

Chapter 17

P17.5 c. for 1.0×10^{-9} s $\Delta\nu = 8.0 \times 10^7$ s^{-1}, 0.00265 cm^{-1}

for 1.0×10^{-11} s, 8.0×10^9 s^{-1} and 0.265 cm^{-1}

P17.10 $p = 2.11 \times 10^{-23}$ kg m s^{-1}, $\dfrac{\lambda}{b} = 0.0031$

P17.12 $E = 0.95$ eV

P17.18 c. $z = \pm 9.42 \times 10^{-3}$ m

P17.23 $\Delta x = 8.5 \times 10^{-34}$ m

Chapter 18

P18.1 $0, 2.59 \times 10^{-22}$ J

P18.3 $1, 7.04, 4.09, 7.95 \times 10^{-2}$

P18.4 $0.420, 0.752, 0.991, 1.20, 1.39$, and 1.57 radians as well as π minus these values

P18.5 vibrational: 1.24×10^{14} s^{-1}, rotational: 1.25×10^{12} s^{-1}

P18.8 $575\,\text{N m}^{-1}, 2.55 \times 10^{-2}$ m

P18.9 $0, 1.23, 4.50, 17.2$

P18.13 $E_0 = 2.78 \times 10^{-32}$ J, $E_0/k_BT = 6.75 \times 10^{-12}$

2.92×10^{-15} m s^{-1}

P18.14 3.15×10^{13} s^{-1}, 9.52×10^{-6} m

P18.16 324 kg s^{-2}; 0.742 kg

P18.18 a. $18.5\,\text{N m}^{-1}$

b. 2.48×10^{-33} J

c. 4.85×10^{-5} J

d. 9.76×10^{28}

P18.19 a. 4.618 pm from F

b. 24.73 pm from D

P18.22 5.56×10^{-21} J, 309 m s^{-1}, $|v|/|v_{rms}| = 0.946$

P18.23 2.29×10^{-20} J, 6.92×10^{13} s^{-1}

P18.24 For I_2: $0.357, 0.127, 891$ K

For H_2: $6.78 \times 10^{-10}, 4.60 \times 10^{-19}, 1.828 \times 10^4$ K,

P18.25 $0, 4.16 \times 10^{-22}$ J

P18.31 for $n = 0, 1$, and 2: $0.0595, 0.103$, and 0.133

P18.34 $E_{rot} = 1.51 \times 10^{-20}$ J, $3.66, E_{vib} = 2.97 \times 10^{-20}$ J,

$7.17, T_{rot} = 1.85 \times 10^{-13}$ s, $T_{vib} = 1.11 \times 10^{-14}$ s,

$\dfrac{T_{rot}}{T_{vib}} = 16.6$

P18.35 a. $0, 9.70 \times 10^{-24}$ J, 2.91×10^{-23} J, 5.82×10^{-23} J,
9.70×10^{-23} J

b. $0, 4.85 \times 10^{-24}$ J, 1.94×10^{-23} J, 4.36×10^{-23} J,
7.76×10^{-23} J

P18.37 8.37×10^{-28} kg, 4.60×10^{-48} kg m^2, 1.49×10^{-34} J s;
4.84×10^{-21} J

Chapter 19

P19.1 $E_0 = 2.94 \times 10^{-20}$ J, $E_1 = 8.65 \times 10^{-20}$ J
$E_2 = 1.41 \times 10^{-19}$ J, $E_3 = 1.93 \times 10^{-19}$ J
$\nu_{0:1} = 8.61 \times 10^{13}$ s^{-1}, $\nu_{0:2} = 1.69 \times 10^{14}$ s^{-1}
$\nu_{0:3} = 2.48 \times 10^{14}$ s^{-1}
Error $(\nu_{0:2}), (\nu_{0:3}) = -2.1\%, -4.4\%$

P19.2 250. N m^{-1}, 3.66×10^{-14} s

P19.3 2.3 cm, 0.54 cm

P19.5 C—O 116.227 pm, O—S 156.014 pm

P19.7 1.09769×10^{-10} m

P19.9 $E_0^{HF} = 4.110 \times 10^{-20}$ J

$E_0^{DF} = 2.980 \times 10^{-20}$ J

$\dfrac{E_0^{HF} - E_0^{DF}}{k_B T} = 2.73$

P19.14 7.523 cm^{-1}, 4.511×10^{11} s^{-1}

P19.16 For F$_2$ at 300. and 1000. K,
$n_1/n_0 = 0.0123$ and 0.267
For F$_2$ at 300. K and 1000. K,
$n_2/n_0 = 1.52 \times 10^{-4}$ and 0.0715
For I$_2$ at 300. K and 1000. K, $n_1/n_0 = 0.357$ and 0.734
For I$_2$ at 300. K and 1000. K, $n_2/n_0 = 0.127$ and 0.539

P19.17 1.4243×10^{-10} m, 1.4880×10^{-10} m

P19.18 8.5×10^5 cm, 3.7×10^2 cm

P19.20 1.5666×10^{-10} m

P19.23 1.277×10^{-20} J

P19.25 0.913

P19.28 1.06×10^{13} s^{-1}, 3.52×10^{-21} J

P19.30 3.89×10^{13} s^{-1}, 2.57×10^{-14} s
1.29×10^{-20} J, 0.0805 eV

P19.31 123.1 pm

P19.33 4.605×10^{-48} kg m^2, 1.3177×10^{-33} J s,
7.943×10^{-23} J

P19.35 1.738×10^{-18} J, 1.717×10^{-18} J/molecule or
1.034×10^3 kJ mol^{-1}

P19.36 11

P19.39 8.6%

P19.40 267.3 pm

P19.41 3.165×10^{-33} J s

P19.42 267.3 pm

Chapter 20

P20.1 most energetic: 109,737, 27434.3 cm^{-1}, 12193.0 cm^{-1}
least energetic: 82302.8, 15241.3, 5334.44 cm^{-1}

P20.4 -4.358×10^{-18} J

P20.5 0.439

P20.6 $r = 4a_0$

P20.7 $1.5a_0$

P20.9 H: 2.179×10^{-18} J
He$^+$: 8.717×10^{-18} J
Li^{2+}: 19.61×10^{-18} J
Be^{3+}: 34.87×10^{-18} J

P20.10 $\langle r \rangle_H = (3/2)a_0$, $\langle r \rangle_{He^+} = (3/4)a_0$,
$\langle r \rangle_{Li^{2+}} = (1/2)a_0$, $\langle r \rangle_{Be^{3+}} = (3/8)a_0$

P20.11 $19, 0.0376$ eV

P20.12 $1.4 \times 10^{-2}, 0.83, 0.999$

P20.13 $2.65\, a_0$

P20.14 145 kg m^{-3}; 4.00×10^{17} kg m^{-3}; 0.016 kg m^{-3};

P20.15 $(3/4)(a_0)^2$

P20.16 $0, a_0^2$

P20.17 $\langle F \rangle_{1s} = -\dfrac{e^2}{2\pi \varepsilon_0 a_0^2}$

$\langle F \rangle_{2pz} = -\dfrac{e^2}{48\pi \varepsilon_0 a_0^2}$

P20.19 $30\, \mu a_0^2$

P20.20 $I_H = 13.60$ eV, $I_{He^+} = 54.42$ eV,
$I_{Li^{2+}} = 122.4$ eV, $I_{Be^{3+}} = 217.7$ eV

P20.21 $5a_0$

P20.27 1.26

P20.29 $(3/2)a_0, a_0$

P20.33 $0.81, 0.12, 6.2 \times 10^{-3}$

Chapter 21

P21.2 $54.7°$ and $125.3°$

P21.6 $2\hbar^2, 2\hbar^2$

P21.7 $2\hbar^2, 0$

P21.8 $\alpha_{optimal} = \dfrac{m_e e^2}{4\pi \varepsilon_0 \hbar^2}$

Chapter 22

P22.7 $\Delta \nu = 1.70 \times 10^9$ s^{-1}

$\dfrac{\Delta \nu}{\nu} = 3.33 \times 10^{-6}$

P22.8 $-\dfrac{16\sqrt{2}}{81}\dfrac{e}{}a_0$

P22.10 b) $3.86 \times 10^{-8}, 8.67 \times 10^{-5}, 4.65 \times 10^{-5}$

P22.11 364

P22.12 Lyman

82258 cm^{-1} $\lambda = 121.6$ nm

97491 cm^{-1} $\lambda = 102.6$ nm

109677 cm^{-1} $\lambda = 91.2$ nm

Balmer

15233 cm^{-1} $\lambda = 656.5$ nm

20565 cm^{-1} $\lambda = 486.3$ nm

23032 cm^{-1} $\lambda = 434.2$ nm

27419 cm^{-1} $\lambda = 364.7$ nm

Paschen

5331.5 cm^{-1} $\lambda = 1876$ nm

7799.3 cm^{-1} $\lambda = 1282$ nm

9139.8 cm^{-1} $\lambda = 1094$ nm

12186.4 cm^{-1} $\lambda = 820.6$ nm

P22.13 $E_{max} = 2.178 \times 10^{-18}$ J, $\nu_{max} = 3.288 \times 10^{15}$ s^{-1}

$\lambda_{max} = 91.18$ nm

$E_{min} = 1.634 \times 10^{-18}$ J, $\nu_{min} = 2.466 \times 10^{15}$ s^{-1}

$\lambda_{min} = 121.6$ nm

P22.18 3.08255×10^{15} s^{-1}, 3.08367×10^{15} s^{-1}

P22.23 $E(3p^2P_{1/2}) = 3.369 \times 10^{-19}$ J $= 2.102$ eV

$E(3p^2P_{3/2}) = 3.373 \times 10^{-19}$ J $= 2.105$ eV

$E(4s^2S_{1/2}) = 5.048 \times 10^{-19}$ J $= 3.150$ eV

$E(5s^2S_{1/2}) = 6.597 \times 10^{-19}$ J $= 4.118$ eV

$E(3d^2D_{3/2}) = 5.797 \times 10^{-19}$ J $= 3.618$ eV

$E(4d^2D_{3/2}) = 6.869 \times 10^{-19}$ J $= 4.287$ eV

P22.28 4, 3, 2, and 1

P22.30 3.39 eV

P22.31 $E(4p^2P) = 6.015 \times 10^{-19}$ J $= 3.754$ eV

$E(5s^2S) = 6.597 \times 10^{-19}$ J $= 4.117$ eV

P22.33 a. 0, 3/2, b. 4, 3/2

c. 1, 1, d. 2, 1/2

P22.35 3.65, 4.70

Chapter 23

P23.1 0.577, 0.875, 0.970

P23.9 For $\varepsilon_2 = -5.93$ eV, $c_{2F} = -0.818$, $c_{2H} = 1.1$

For $\varepsilon_1 = -20.3$ eV, $c_{1F} = 0.76$, $c_{1H} = 0.39$

P23.10 bonding MO: 0.72 on F, 0.28 on H

antibonding MO: 0.28 on F, 0.72 on H

P23.15 8.67×10^{-30} C m $= 2.60$ D

P23.16 $S = 0.15$: $-11.8, -14.9$

$S = 0.30$: $-9.23, -16.0$

$S = 0.45$: $-5.25, -16.8$

P23.21 $S = 0.075$: $-13.4, -18.7$

$S = 0.18$: $-12.4, -19.1$,

$S = 0.40$: $-7.65, -20.1$

P23.22 $S = 0.075$: 0.97

$S = 0.18$: 0.89,

$S = 0.40$: 0.74

Chapter 24

P24.14 a. 3.35×10^{-18}, b. 0.213

Chapter 25

P25.2 1.27 μm, 762 nm

P25.3 1.14, 0.456

P25.5 0, 0.981; 1, 0.0196; 2, 1.96×10^{-4}; 3, 1.31×10^{-6};

4, 6.546×10^{-9}; 5, 2.616×10^{-11}

P25.6 $n = 152$

P25.7 a. $n = 17$, b. 7.34×10^{-19} J, c. $n = 16$, 7.22×10^{-19} J

P25.8 a. $n = 15$, b. 325 cm^{-1}

P25.9 a. 4.72 nm, b. 3.60 nm

P25.10 0.5 nm, 5.5×10^{12} s^{-1}; 1.0 nm, 8.6×10^{10} s^{-1};

2.0 nm, 1.3×10^9 s^{-1}; 3.0 nm, 1.2×10^8 s^{-1};

5.0 nm, 5.5×10^6 s^{-1}

P25.11 2.0 nm, 9.45×10^{10} s^{-1}; 7.0 nm, 5.1×10^7 s^{-1};

12.0 nm, 2.0×10^6 s^{-1}

Chapter 27

P27.3 C_{2h}

P27.4 3 representations, one 2-D and one 1-D

P27.5 a. 30, b. 2, A_1, E, c. A_1 singly degenerate, E four fold degenerate, d. all; e. see c., f. all

P27.6 a. 6, b. 7, A_3, 3E, c. A_2 singly degenerate, E two fold degenerate, d. 12, A_{1g}, E_{1g}, E_{2g}, f. none

P27.10 D_{2d}

P27.12 C_s

P27.13 $\Gamma_{red} = A_1 + A_2 + B_1 + B_2$

P27.15 D_{3h}

P27.18 C_{3v}

P27.21 $\Gamma_{red} = A_1 + A_2 + E$

P27.24 5 representations, four 1-D and one 1-D

Chapter 28

P28.3 5.87 T, 23.3 and 14.5 T

P28.6 -2.04 and 1.12 ppm

Chapter 29

P29.1   a.   4/52
    b.   1/52
    c.   12/52 and 3/52, respectively

P29.2   a.   0.002
    b. 1.52×10^{-6}

P29.3   a.   1/6
    b.   1/9
    c.   21/36

P29.4   a.   8/49
    b.   6/49
    c.   23/49

P29.5   a. 10^7
    b.   160
    c. 1.6×10^9

P29.6   a. 2.56×10^{10}
    b. 5.08×10^9

P29.7   a.   0.372
    b.   0.569

P29.8   a.   0.06
    b.   0.0006

P29.9   a.   720
    b.   360
    c.   1
    d. 3.73×10^{16}

P29.10   120

P29.11   a.   1
    b.   15
    c.   1
    d. 1.03×10^{10}

P29.12   a. 4.57×10^5
    b. 1.76×10^4
    c. 3.59×10^5

P29.13   a. 9.52×10^{-7}
    b. 9.52×10^{-7}
    c. 1.27×10^{-6} and 3.77×10^{-7}

P29.14   a. 1.0×10^{-5}
    b. 2.2×10^{-8}
    c. 3.0×10^{-6}

P29.15   0.004

P29.16   a. 4.52×10^{-8}
    b. 1.04×10^{-6}
    c. 8.66×10^{-9}

P29.17   a.   bosons: 220; fermions: 45
    b.   bosons: 1.72×10^5; fermions: 1.62×10^5

P29.18   a. 7.41×10^{11}
    b. 2.97×10^{10}
    c. 2.04×10^6

P29.19   a. 9.77×10^{-4}
    b.   0.044
    c.   0.246
    d.   0.044

P29.20   a. 1.69×10^{-5}
    b. 3.05×10^{-3}
    c.   0.137
    d.   0.195

P29.21   a. 9.54×10^{-7}
    b.   0.176
    c.   0.015

P29.22   a. $(n)(n-1)$
    b. $(n)(n-1)(n-2)\ldots(n/2+1)/(n+2)!$

P29.24   \$1.91

P29.25   b. $\ln(2)/k$

P29.26   a. $9.6 \times 10^{-6}\,\mathrm{s}$
    b.   0.37

P29.27   c $182\,\mathrm{J\,mol^{-1}}$

P29.28   c.   case 1: 0.245; case 2: 0.618

P29.29   a. $2/a$
    b. $a/2$
    c. $a^2(1/3 - 1/2\pi^2)$
    d. $a^2(1/12 - 1/2\pi^2)$

P29.30   a. $(m/2\pi kT)^{1/2}$
    b.   0
    c. kT/m
    d. kT/m

P29.31 $\langle x \rangle = 0$

P29.32 $1.6 \times 10^4\,\mathrm{m}$

P29.33 $x_{\mathrm{mp}} = (a)^{1/2}$

P29.34   a.   0
    b.   1/3

P29.35   a.   0.175
    b.   0.0653
    c.   0.0563

Chapter 30

P30.1   b. $\exp(693)$
    c. $\exp(673)$

P30.2 10^{694}

P30.4   0.25

P30.5   a. 2.6×10^6
    b.   5148

P30.8 $P_{N_2} = 0.230$ atm; $P_{O_2} = 0.052$ atm

P30.9 a. 250 K

b. 180 K

P30.10 a. 250 K

b. 180 K

P30.11 a. 6.07×10^{-20} J

b. Set C

P30.12 a. 500. cm^{-1}

$W = 1287$

b. 500. cm^{-1}

$W = 2860$

c. 500. cm^{-1}

$W = 858$

P30.13 limiting value is 0.333

P30.14 721 K

P30.15 4152 K

P30.16 1.28, 5.80

P30.17 0.781, 0.00294

P30.18 25,300 K

P30.19 0.999998

P30.20 $a_- = 0.333334$

$a_0 = 0.333333$

$a_+ = 0.333333$

P30.22 432 K

P30.23 1090 K

P30.24 At 300 K, $p = 0.074$ F_2 equivalent at 523 K

At 1000 K, $p = 0.249$ F_2 equivalent at 1740 K

P30.25 1.06×10^{-9}

P30.27 5.85×10^4

P30.29 At 100 K, $p = 0.149$

At 500 K, $p = 0.414$

At 200 K, $p = 0.479$

P30.30 b. 10,300 K

P30.31 1780 K

Chapter 31

P31.1 H_2: 2.77×10^{26}

N_2: 1.43×10^{28}

P31.2 $q_T = 5.66 \times 10^{29}$

$q_T(^{37}Cl_2) = 1.087 \times q_T(^{35}Cl_2)$

P31.3 $q_T = 4.46 \times 10^{24}$

P31.4 $q_T(Ar) = 2.44 \times 10^{29}, T = 590.$ K

P31.5 0.0680 K

P31.6 3.90×10^{17}

P31.7 4.38×10^4

P31.8 1.99×10^5

P31.9 a. 1

b. 2

c. 2

d. 12

e. 2

P31.10 CCl_4: 12

$CFCl_3$: 3

CF_2Cl_2: 2

CF_3Cl: 3

P31.11 rotational: HD; translational: D_2

P31.12 H_2: 1.00; HD: 1.22

P31.13 $q_R = 401$

P31.14 $q_R = 424$

P31.15 a. no

b. no

c. yes

d. yes

P31.16 5840

P31.17 3.77×10^4

P31.18 a. 616 K

b. $J = 5-6$

P31.19 $J = 19-20$ transition

P31.20 4 to 5

P31.21 $q_R = 5.02$; by summation $q_R = 20.6$

P31.22 0.419012 cm^{-1}, $q_R = 494$

P31.23 a.

J	P_J
0	0.041
1	0.113
2	0.160
3	0.175
4	0.167
5	0.132
6	0.095
7	0.062
8	0.037
9	0.019

b.

J	P_J
0	0.043
1	0.117
2	0.116
3	0.179
4	0.163
5	0.131
6	0.093

7	0.059
8	0.034
9	0.018

P31.24 a. $B_A = B_B = B_C = 5.27 \text{ cm}^{-1}$

b. yes

P31.25 0.418 K

P31.26 300 K: $q_V = 1$, $p_0 = 1$

3000 K: $q_V = 1.32$, $p_0 = 0.762$

P31.27 1820

P31.28 IF: 300 K: $q_V = 1.06$, $p_0 = 0.946$, $p_1 = 0.051$, $p_2 = 0.003$

3000 K: $q_V = 3.94$, $p_0 = 0.254$, $p_1 = 0.189$, $p_2 = 0.141$

IBr: 300 K: $q_V = 1.38$, $p_0 = 0.725$, $p_1 = 0.199$, $p_2 = 0.055$

3000 K: $q_V = 8.26$, $p_0 = 0.121$, $p_1 = 0.106$, $p_2 = 0.094$

P31.29 $q_V = 1.67$

P31.30 $q_V = 1.10$

P31.31 $q_V = 1.69$

P31.32 $q_V = 4.09$

P31.33 $p_0 = 1, p_1 = 0$

P31.34 a. 1601 cm^{-1}

P31.35 a. $q_V = 1.00$

b. 1710 K

P31.37 453 cm^{-1}

P31.39 a. $q = L/\Lambda$

b. $q = V/\Lambda^3$

P31.41 a. $q_v = 1$, $q_r = 1.29$

b. $q_v = 1$, $q_r = 1.26$

P31.42 $q_E = 10.2$

P31.43 a. $q_E = 4.77$

b. 2320 K

P31.44 a. $q_E = 3.11$

b. 251 K

P31.45 $q_E = 1.00$

P31.46 $q = 1.71 \times 10^{34}$

P31.47 $q = 1.30 \times 10^{29}$

P31.48 $q_R = 265$

Chapter 32

P32.2 655 K

P32.3 Ensemble b

P32.4 1310 K

P32.5 NkT

P32.6 $U = \dfrac{1}{2} NkT; C_V = \dfrac{1}{2} Nk$

P32.8

Molecule	θ_R (K)	High T?	θ_V (K)	High T?
H^{35}Cl	15.2	No	4150	No
^{12}C^{16}O	2.78	Yes	3120	No
^{39}KI	0.088	Yes	233	No
CsI	0.035	Yes	173	No

P32.9 1.71 kJ mol^{-1}

P32.10 307 J mol^{-1}

P32.12 3.72 kJ mol^{-1}

P32.13 6.20 kJ mol^{-1}

P32.14 U_R are equivalent.

$U_V(\text{BrCl}) - U_V(\text{BrF}) = 434$ J mol^{-1}

P32.16 C_V values (J mol^{-1} K^{-1}) are as follows:

	298 K	500 K	1000 K
2041 cm^{-1}	0.042	0.808	4.24
712 cm^{-1}	3.37	5.93	7.62
3329 cm^{-1}	0.00	0.048	1.56
Total	6.78	12.7	21.0

P32.17 $(0.751)R$

P32.18 $(0.0586)R$

P32.19 $n(1.85 \text{ J mol}^{-1} \text{ K}^{-1})$

P32.20 c. 352 m s^{-1}

P32.21 b. 24.6 J mol^{-1} K^{-1}

P32.22 7.82 J mol^{-1} K^{-1}

P32.25 200 K: 122 J mol^{-1} K^{-1}; 300 K: 128 J mol^{-1} K^{-1}; 500 K: 135 J mol^{-1} K^{-1}

P32.26 1.28×10^{-10} m

P32.27 219 J mol^{-1} K^{-1}

P32.28 256 J mol^{-1} K^{-1}

P32.29 176 J mol^{-1} K^{-1}

P32.30 191 J mol^{-1} K^{-1}, 260 J mol^{-1} K^{-1}

P32.31 186 J mol^{-1} K^{-1}

P32.33 49.1 J mol^{-1} K^{-1}

P32.34 79.5 J mol^{-1} K^{-1}

P32.35 211 J mol^{-1} K^{-1}

P32.38 11.5 J mol^{-1} K^{-1}

P32.39 a. $R \ln 2$ b. $R \ln 4$

c. $R \ln 2$ d. 0

P32.43 Ne: -40.0 kJ mol^{-1}; Kr: -45.3 kJ mol^{-1}

P32.45 -57.2 kJ mol^{-1}

P32.46 $G_{R,m}^{\circ} = -15.4$ kJ mol^{-1}; $G_{V,m}^{\circ} = -0.30$ kJ mol^{-1}

P32.47 2.25×10^{-9}

P32.49 0.824

P32.50 $0.977 \exp(-2720)$

P32.51 a. 143.4 mJ mol^{-1}

b. 28.0

Chapter 33

P33.2

	ν_{mp} (m s^{-1})	ν_{ave} (m s^{-1})	ν_{rms} (m s^{-1})
Ne	495	559	607
Kr	243	274	298
CH$_4$	555	626	680
C$_2$H$_6$	406	458	497
C$_{60}$	82.9	93.6	102

P33.3

	ν_{mp} (m s^{-1})	ν_{ave} (m s^{-1})	ν_{rms} (m s^{-1})
300. K	395	446	484
500. K	510	575	624

$$\nu(H_2) = (3.98)\nu(O_2)$$

P33.4

	H$_2$O	HOD	D$_2$O
ν_{ave} (m s^{-1})	592	576	562

P33.5 ν_{ave} (CCl$_4$) = 202 m s^{-1}; ν_{ave} (O$_2$) = 444 m s^{-1}; $KE_{ave} = 6.17 \times 10^{-21}$ J for both

P33.6 444 m for O$_2$; 274 m for Kr

P33.7 a. 5.66×10^{-4} s
     b. 2.11×10^{-3} s
     c. 0.534

P33.8 $\dfrac{\nu_{ave}}{\nu_{mp}} = \dfrac{2}{\sqrt{\pi}}$; $\dfrac{\nu_{rms}}{\nu_{mp}} = \sqrt{\dfrac{3}{2}}$

P33.9 81.5 K (both cases)

P33.10 1240 K

P33.11 0.843 and 0.157, respectively.

P33.12 a. Ne: 828 m s^{-1}, Kr: 406 m s^{-1}, Ar: 589 m s^{-1}
     b. 2100 K

P33.13 0.392

P33.14 a. 2.10×10^5 K
     b. 3.00×10^4 K

P33.15 298 K: 0.132, 500 K: 0.071

P33.16 b. 557 m s^{-1}

P33.23 1.47×10^{24} collisions s^{-1}

P33.24 a. 4.97 s
     b. 6.03×10^{-4} kg

P33.25 6×10^{11} collisions s^{-1}

P33.26 a. 2.73×10^{23} collisions s^{-1}
     b. 3.60×10^{14} collisions s^{-1}

P33.27 1×10^{-5} m^2

P33.28 a. 7×10^9 collisions s^{-1}
     b. 0.38 atm
     c. 1.3×10^{-7} m

P33.29 a. 2.18×10^{-7} m
     b. 7.78×10^{-5} m

P33.30 a. 8.44×10^{34} m^{-3} s^{-1}
     b. 1380 K

P33.31 a. $z_{11} = 9.35 \times 10^3$ s^{-1}, $\lambda = 0.051$ m
     b. $z_{11} = 9.35 \times 10^{-4}$ s^{-1}, $\lambda = 5.1 \times 10^5$ m

P33.32 a. 1.60×10^{-7} m
     b. 1.60×10^{-5} m
     c. 1.60×10^{-2} m

P33.33 Ne: 2.0×10^{-7} m, Kr: 9.3×10^{-8} m, CH$_4$: 1.0×10^{-7} m

P33.34 4.6×10^{-4} Torr

P33.36 a. 1.00×10^8 s^{-1}
     b. 1.87×10^{28} m^{-3} s^{-1}
     c. 4.40×10^{-6} m

Chapter 34

P34.1 0.318 nm^2

P34.2 1.06×10^{-5} m^2 s^{-1}

P34.3 a. 0.368 nm^2
     b. 0.265 nm^2

P34.4 a. 319 s
     b. 6.13×10^{-10} s

P34.5 a. 1.60×10^{-3} s
     b. 2.40×10^{-3} s

P34.6 b. 2.58×10^{-5} m

P34.7 a. -9.60 J s^{-1}
     b. -33.6 J s^{-1}
     c. -22.8 J s^{-1}

P34.8 No. $\kappa_{Ar} = (2.09)\kappa_{Kr}$

P34.9 -1.80×10^{-4} W cm^{-2}

P34.10 a. 0.0052 J K^{-1} m^{-1} s^{-1}
     b. 0.0025 J K^{-1} m^{-1} s^{-1}
     c. 0.0051 J K^{-1} m^{-1} s^{-1}

P34.11 1.5×10^{-19} m^2

P34.12 1.6×10^{-19} m^2

P34.13 c. 1.5×10^{-19} m^2

P34.14 a. 1.14
     b. 0.659

P34.15 1.33

P34.16 a. 6.22×10^{-19} m^2
     b. 0.00389 J K^{-1} m^{-1} s^{-1}

P34.17 a. 2.05×10^{-3} m^3 s^{-1}
     b. 1.88×10^{-3} m^3 s^{-1}

P34.18 a. 37.3 m s^{-1}
     b. 0.893 m s^{-1}

P34.19 D_2: 119 μP, HD 103 μP

P34.20 1.89 cP

P34.21 22 s

P34.22 b. 1.34×10^{-5} m^2 s^{-1}

P34.23 b. 6.59×10^{-3} J K^{-1} m^{-1} s^{-1}

c. 1.47×10^{-2} J K^{-1} m^{-1} s^{-1}

P34.24 $E = 10.7$ kJ mol^{-1}, A $= 8.26 \times 10^{-3}$

P34.25 4.07 Pa

P34.26 0.265 L s^{-1}

P34.27 a. 1.89 nm

b. 16.8 kg mol^{-1}

P34.28 $f = 7.25 \times 10^{-8}$ g s^{-1}, $r = 3.85$ nm

P34.29 a. catalase: 238 kg mol^{-1}, alcohol dehyd.: 74.2 kg mol^{-1}

b. 3.10×10^4 s

P34.30 a. 1.69×10^{-13} s

b. 1.90 nm

P34.31 77.7 kg mol^{-1}

P34.32 3.75×10^{20} electrons

P34.33 0.0125 S m^2 mol^{-1}

P34.34 Strong electrolyte; $\Lambda_m^\circ = 0.0125$ S m^2 mol^{-1}

P34.35 $\Lambda_m^\circ = 0.00898$ S m^2 mol^{-1}

P34.37 10 steps: 4.39×10^{-2}, 20 steps: 7.39×10^{-2}, 100 steps: 6.66×10^{-2}

P34.38 1.17×10^{-10} m^2 s^{-1}

P34.39 a. 4.4810×10^{-2} S m^{-1} ohm

b. 4.3395×10^{-6} S m^{-1}

P34.40 b. $K = 1.00 \times 10^{-14}$

Chapter 35

P35.2 b. 2.79×10^3 s

c. 425 s

P35.3 First order, $k = 3.79 \times 10^{-5}$ s^{-1}

P35.5 a. First order with respect to (wrt) ClO

First order wrt BrO

Second order overall

Units of k: M^{-1} s^{-1}

b. Second order wrt NO

First order wrt O$_2$

Third order overall

Units of k: M^{-2} s^{-1}

c. Second order wrt HI

First order wrt O$_2$

$-1/2$ order wrt H$^+$

2.5 order overall

Units of k: M$^{-3/2}$ s^{-1}

P35.6 a. Second

b. Third

c. 1.5

P35.7 Second order wrt NO$_2$

First order wrt H$_2$

$k = 1.61 \times 10^{-6}$ kPa^{-2} s^{-1}

P35.8 K $= 1.9 \times 10^{-15}$ molec^{-1} cu^{3}s^{-1}

P35.9 $R = 116$ M^{-1} s^{-1} [Lactose][H$^+$]

P35.11 Second order

$k = 0.317$ M^{-1} s^{-1}

P35.13 a. $k = 2.08 \times 10^{-3}$ s^{-1}

b. $t = 1.11 \times 10^3$ s

P35.14 1.43×10^{24}

P35.15 1.50×10^4 days

P35.16 4.61 decay min^{-1}

P35.17 a. $k = 3.65 \times 10^{-4}$ day^{-1}

b. 1.90×10^3 days

P35.20 20%

P35.23 2.6×10^{-7} s

P35.26 a. $k = 0.03$ M^{-1} s^{-1}

b. 120 s

P35.27 a. 1.08×10^{-12} s

P35.28 $k_1 = 4.76 \times 10^{-10}$ yr^{-1} and $k_2 = 5.70 \times 10^{-11}$ yr^{-1}

P35.29 $\Phi_1 = 0.30$, $\Phi_2 = 0.49$, $\Phi_3 = 0.21$

P35.30 a. Rate $= 2.1 \times 10^{-15}$ M s^{-1}

b. Rate $= 1.6 \times 10^{-15}$ M s^{-1}

c. [Cl] $= 1.1 \times 10^{-18}$ M

[O$_3$] $= 4.2 \times 10^{-11}$ M

P35.32 1800 J mol^{-1}

P35.33 b. 11 kJ mol^{-1}

P35.36 a. 7.0×10^{-11} M^{-1} S^{-1}

b. 3.5×10^{-11} M^{-1} S^{-1}

P35.37 0.15

P35.38 a. $E_a = 1.50 \times 10^5$ J mol^{-1}, A $= 1.02 \times 10^{10}$ M^{-1} s^{-1}

b. $k = 0.0234$ M

P35.39 269 s

P35.40 A $= 7.38 \times 10^{13}$ s^{-1}, $k = 3.86 \times 10^{-5}$ s^{-1}

P35.42 1800 s^{-1} and 1300 M^{-1} s^{-1}

P35.44 3.44×10^{10} M^{-1} s^{-1}

P35.45 $D_{imid} = 1.1 \times 10^{-5}$ cm^2 s^{-1}, $k = 3.6 \times 10^{10}$ M^{-1} s^{-1}

P35.46 No. $k = 6.3 \times 10^{10}$ M^{-1} s^{-1} for diffusion limited

P35.47 a. $E_a = 108$ kJ mol^{-1}, A $= 1.05 \times 10^{10}$ s^{-1}

b. $\Delta S^\ddagger = -62.3$ J mol^{-1} K^{-1}; $\Delta H^\ddagger = 105$ kJ mol^{-1}

P35.48 a. $E_a = 219 \text{ kJ mol}^{-1}$, $A = 7.20 \times 10^{12} \text{ s}^{-1}$

b. $\Delta S^{\ddagger} = -15.3 \text{ J mol}^{-1} \text{ K}^{-1}$; $\Delta H^{\ddagger} = 212 \text{ kJ mol}^{-1}$

P35.49 a. $E_a = 790 \text{ J mol}^{-1}$, $A = 4.88 \times 10^{10} \text{ M}^{-1} \text{ s}^{-1}$

b. $k = 3.17 \times 10^{10} \text{ M}^{-1} \text{ s}^{-1}$

c. $\Delta S^{\ddagger} = -57.9 \text{ J mol}^{-1} \text{ K}^{-1}$; $\Delta H^{\ddagger} = -2.87 \text{ kJ mol}^{-1}$

P35.50 $\Delta S^{\ddagger} = 60.6 \text{ J mol}^{-1} \text{ K}^{-1}$; $\Delta H^{\ddagger} = 270 \text{ kJ mol}^{-1}$

P35.51 $\Delta S^{\ddagger} = 4.57 \text{ J mol}^{-1} \text{ K}^{-1}$; $\Delta H^{\ddagger} = 37.0 \text{ kJ mol}^{-1}$

Chapter 36

P36.8 b. 59 kJ mol^{-1}

c. 434

P36.14 $k_1 = 1.19 \times 10^4 \text{ M}^{-1} \text{ s}^{-1}$; $k_1/k_2 = 1.05 \times 10^3 \text{ M}^{-1}$

P36.16 $2.45 \times 10^{-3} \text{ M s}^{-1}$

P36.17 $R_{max} = 3.75 \times 10^{-2} \text{ M s}^{-1}$

$K_m = 2.63 \times 10^{-2} \text{ M}$

$k_2 = 1.08 \times 10^7 \text{ s}^{-1}$

P36.18 a. $K_m = 2.42 \text{ mM}$, $R_{max} = 0.137 \text{ mM s}^{-1}$

b. $K_m = 73.3 \text{ mM}$, $R_{max} = 0.0298 \text{ mM s}^{-1}$

P36.19 a. $K_M = 2.5 \times 10^{-6} \text{ M}$

$R_M = 6.5 \times 10^{-7} \text{ Ms}^{-1}$

b. $K_M = 1.3 \times 10^{-5} \text{ M}$

$R_M = 7.9 \times 10^{-7} \text{ Ms}^{-1}$

$K_I = 4.8 \times 10^{-5} \text{ M}$

P36.20 0.0431 M

P36.21 $K_m = 6.49 \text{ } \mu\text{M}$, $R_{max} = 4.74 \times 10^{-8} \text{ } \mu\text{M s}^{-1}$,

$K_m^* = 24.9 \text{ } \mu\text{M}$,

$K_i = 70.4 \text{ } \mu\text{M}$

P36.23 $K_m = 0.0392 \text{ M}$, $R_{max} = 0.437 \text{ M s}^{-1}$

P36.28

P (atm)	θ
20.	0.595
50.	0.754
100.	0.853
200.	0.932
300.	0.952

P36.29 $r = 0.674 \text{ mL}$, $s = 4.99 \text{ Torr}^{-1}$

P36.30 $V_m = 1.56 \text{ cm}^3 \text{ g}^{-1}$, $K = 220 \text{ Torr}^{-1}$

P36.32 11 kJ mol^{-1}

P36.36 a. at 298 nm: $7.29 \times 10^{16} \text{ photons cm}^{-2}$;

at 320 nm: $8.05 \times 10^{16} \text{ photons cm}^{-2}$

b. 34.5 s

P36.38 0.95

P36.39 4.55×10^{18} photons, $I = 2.41 \times 10^{-3} \text{ J s}^{-1}$

P36.40 $3.02 \times 10^{19} \text{ photons s}^{-1}$

P36.41 $k_{isc}^S/k_f = 7.33$, $k_P = 3.88 \times 10^{-2} \text{ s}^{-1}$, $k_{isc}^T = 0.260 \text{ s}^{-1}$

P36.42 a. 0.91

b. $7.2 \times 10^{18} \text{ photons cm}^{-2} \text{ s}^{-1}$

c. 1.5 nW

P36.43 b. $A = 0.312$, $B = 0.697$

c. -0.172 atm

P36.44 $k_q = 1.71 \times 10^4 \text{ \%O}_2^{-1} \text{ s}^{-1}$

P36.45 0.39

P36.46 12 Å

P36.48 25 residues

P36.49 a. $3.4 \times 10^{-19} \text{ J}$

b. $1.2 \times 10^{-19} \text{ J}$

P36.50 a. 0.768 Å^{-1}

b. 0.384 Å^{-1}

Credits

Cover

Artwork courtesy of the artist, Kim Kopp; photo credit: Frank Huster.

Chapter 5

Page 114: AP Photo/Frank Franklin II; Imaginechina via AP Images.

Chapter 6

Page 153: Photo courtesy of Professor Dr. Rolf Jurgen Behm/University of Ulm. From T. Diemant, T. Hager, H.E. Hoster, H. Rauscher, and R.J. Behm, *Surf. Sci.* 141(2003) 137, fig. 1a.

Chapter 11

Page 281: Courtesy of Dr. D.M. Kolb, Department of Electrochemistry, University of Ulm.
Page 282: Courtesy of Dr. D.M. Kolb, Department of Electrochemistry, University of Ulm.
Page 283: Courtesy of Dr. D.M. Kolb, Department of Electrochemistry, University of Ulm; Photo courtesy of Professor Dr. Rolf Jurgen Behm/University of Ulm. Magnussen et al. "In-Situ Atomic-Scale Studies of the Mechanisms and Dynamics of Metal STM." *Electrochemica Acta* 46 (2001): 3725–3733, Figure 1.
Page 284: Courtesy of Dr. D.M. Kolb, Department of Electrochemistry, University of Ulm; Photo courtesy of Professor Dr. Rolf Jurgen Behm/University of Ulm. Magnussen et al. "In-Situ Atomic-Scale Studies of the Mechanisms and Dynamics of Metal STM." *Electrochemica Acta* 46 (2001): 3725–3733, Figure 2.
Page 287: Trimmer, A.L., Hudson, J.L., Kock, M., and Schuster, R. *Applied Physics Letters* 82 (2003): 3327. Copyright 2003, American Institute of Physics; Kock, M. et al. *Electrochemica Acta* 48, nos. 20–22 (30 September 2003): Figure 7, p. 3218; Trimmer, A.L., Hudson, J.L., Kock, M., and Schuster, R. *Applied Physics Letters* 82 (2003): 3327. Copyright 2003, American Institute of Physics.

Chapter 16

Page 371: Courtesy of Johnson, Kevin. "The Thermal Decomposition and Desorption Mechanism of Ultra-Thin Oxide on Silicon Studied by Scanning Tunneling Microscopy." PhD thesis, University of Washington, 1991.
Page 372: SEM image of a NANOSENSORS™ AR10 High Aspect Ratio AFM probe, reproduced with permission by NANOSENSORS™.
Page 373: Image reproduced by permission of IBM Research—Zurich. Unauthorized use not permitted.
Page 378: Reproduced with permission from Larson et al. *Science* 300 (30 May 2003). © 2003 American Association for the Advancement of Science.

Chapter 17

Page 388: The Niels Bohr Archive, Copenhagen.

Chapter 22

Page 529: Photo courtesy of Pacific Northwest National Laboratory. © Liang et al. *Proceedings—Electrochemical Society* (2001). Reproduced by permisson of the Electrochemical Society, Inc.

Chapter 24

Chapter 28

Chapter 36

Index

Masses and Natural Abundances for Selected Isotopes			
Nuclide	Symbol	Mass (amu)	Percent Abundance
H	^{1}H	1.0078	99.985
	^{2}H	2.0140	0.015
He	^{3}He	3.0160	0.00013
	^{4}He	4.0026	100
Li	^{6}Li	6.0151	7.42
	^{7}Li	7.0160	92.58
B	^{10}B	10.0129	19.78
	^{11}B	11.0093	80.22
C	^{12}C	12 (exact)	98.89
	^{13}C	13.0034	1.11
N	^{14}N	14.0031	99.63
	^{15}N	15.0001	0.37
O	^{16}O	15.9949	99.76
	^{17}O	16.9991	0.037
	^{18}O	17.9992	0.204
F	^{19}F	18.9984	100
P	^{31}P	30.9738	100
S	^{32}S	31.9721	95.0
	^{33}S	32.9715	0.76
	^{34}S	33.9679	4.22
Cl	^{35}Cl	34.9688	75.53
	^{37}Cl	36.9651	24.4
Br	^{79}Br	79.9183	50.54
	^{81}Br	80.9163	49.46
I	^{127}I	126.9045	100

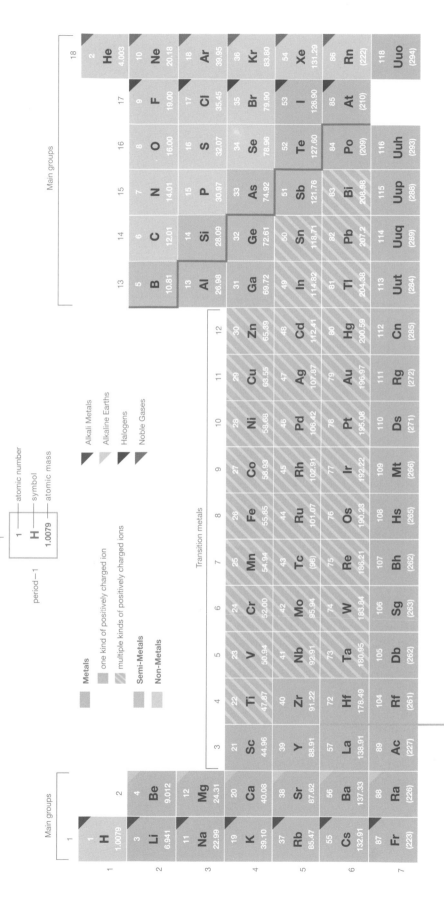